D0160761

GENERAL MICROBIOLOGY

GENERAL MICROBIOLOGY

Robert F. Boyd, PhD

Wirtz, Virginia

With 1172 illustrations and 61 color plates

Original artwork by Sue Solomon Seif, Richmond, Virginia

Times Mirror/Mosby College Publishing

St. Louis / Toronto / Santa Clara 1984

Editor: Diane L. Bowen

Assistant editor: Susan Dust Schapper

Manuscript editors: Teri Merchant, Judith Bange

Design: William Seabright

Production: Margaret B. Bridenbaugh, Jeanne A. Gulledge

A division of The C.V. Mosby Company
11830 Westline Industrial Drive, St. Louis, Missouri 63146

Printed in the United States of America

Cover illustration

Satellite growth of *Haemophilus influenzae* around *Staphylococcus aureus* on blood agar. (×1.)

L.J. Le Beau, University of Illinois Medical Center, Chicago/BPS.

Frontispiece

***Polysiphonia fibrillosa,* a red alga. (×20.)**

Runk/Schoenberger of Grant Heilman Photography.

Library of Congress Cataloging in Publication Data

Main entry under title:

General microbiology.

Bibliography: p
Includes index.
1. Microbiology. I. Boyd, Robert F.
QR41.2.G454 1984 576 83-19306

ISBN 0-8016-0900-3

C/VH/VH 9 8 7 6 5 4 3 2 1 03/A/347

CONTRIBUTORS

MARTIN ALEXANDER, PhD

Department of Agronomy, Cornell University,
Ithaca, New York

GORDON CARTER, PhD

College of Veterinary Medicine, Virginia Polytechnic and State University,
Blacksburg, Virginia

GLENN CHAMBLISS, PhD

Department of Bacteriology, University of Wisconsin—Madison,
Madison, Wisconsin

ROBERT DEIBEL, PhD

Department of Bacteriology, University of Wisconsin—Madison,
Madison, Wisconsin

BRYAN HOERL, PhD

Department of Microbiology, School of Dentistry, Marquette University,
Milwaukee, Wisconsin

GARY HOOPER, PhD

Department of Plant Pathology, Virginia Polytechnic and State University,
Blacksburg, Virginia

F.L. SINGLETON, PhD

Department of Biological Sciences, Old Dominion University,
Norfolk, Virginia

To all scientists,
past and present,
whose endeavors
have made this
book possible

PREFACE

No area of biological science in recent times has become so prominent in the public eye as microbiology. The most exciting discoveries involving microorganisms have important and extensive application. Recombinant DNA technology, for example, has revolutionized microbiology and related areas of science. It behooves all students, whether majors or nonmajors in microbiology, to understand the modern concepts of this science in order to comprehend the many current advances that are affecting and will affect our lives. It is inevitable that microbiology will play an even more important role in the future.

Most microbiology textbooks on the market today are impressive because of their meticulous coverage of specific areas of microbiology (such as ecology, infectious disease, and molecular biology), but few cover in any detail all the major divisions of microbiology. My purpose in writing this book is to provide the student with a glimpse into the major divisions of microbiology. Genetics, immunology, and molecular biology have contributed significantly to our understanding of the science of microbiology. I have tried, therefore, to describe the various aspects of microbiology in more modern terms. The infectious disease section of this book, for example, provides a biochemical rationale for the infectious disease process. The extensive coverage of topics makes this book appropriate for premedical, predental, allied health, agricultural, and animal science students as well as biology and microbiology majors.

DESCRIPTION OF CONTENTS

General Microbiology is divided into eight major parts. The first of six chapters in Part 1, ''Introduction to the Microbial World,'' gives a historical background of microbiology, presenting the most significant past and present discoveries in this science. Chapter 2, ''Biochemical Background for Microbiology,'' presents an overview of biochemistry with particular attention to the types of biochemical reactions and structures that are encountered later in the book. Careful study of this chapter will provide the student with all the biochemical information necessary for comprehension of later material. Chapter 3, ''Introduction to Microscopic Technique,'' is placed in this section so that students may become acquainted with the various types of instruments used to observe microorganisms, because the last two chapters in this section contain numerous photomicrographs. Chapters 4 and 5 discuss the general characteristics of prokaryotes and eukaryotes and are extensively illustrated. The chapter on eukaryotes is more comprehensive than comparable chapters in other textbooks of this type. The current use of eukaryotes in present-day technology necessitates a better understanding of this group of microorganisms. The last chapter of this section, ''Viruses,'' includes the general structure and function of plant, animal, bacterial, and fungal viruses, as well as a discussion of how viruses are cultivated and how they are affected by chemical and physical agents.

Viral replication is reserved for Chapter 16 because it can be considered a ''growth process'' and because replication requires a better understanding of microbial genetics, which is discussed in Chapters 9 through 12.

Part 2, ''Microbial Structure and Function,'' consists of two chapters dealing with the relationship between structure and function in prokaryotes and eukaryotes. Both chapters are abundantly illustrated and give students a chance to apply what they have learned about the general characteristics of eukaryotes and prokaryotes in Chapters 4 and 5.

Part 3, ''Microbial Genetics,'' is divided into four chapters. Three of the chapters discuss the functions of DNA: replication and mutation, synthesis and regulation of proteins, and recombination and gene transfer. These three chapters are presented in a historical sequence because the basics of microbial genetics are still new to the student. The last chapter in this section is devoted to eukaryotic genetics, a subject that receives little attention in other textbooks. The recent discovery of some basic differences in the transcription of DNA in eukaryotes has opened up a whole new field of scientific endeavor that also has applications in prokaryotic systems and in viruses.

Part 4, "Microbial Metabolism and Growth," consists of four chapters. The first chapter, "Energy Production in Microorganisms," deals with all the mechanisms for obtaining energy in microorganisms, including photosynthesis. This is followed by chapters that deal with nutrition (including a discussion of mineral metabolism), microbial growth in eukaryotes as well as prokaryotes, and mechanisms used by microorganisms to control growth at the enzymatic level. The final chapter in this section deals with the replication of viruses and also includes a discussion of the current mechanisms and role of viruses in tumor development.

Part 5, "Environmental and Applied Microbiology," consists of five chapters dealing with the ecology of soil and water microbiology as well as the applied areas of food and industrial microbiology. These chapters are unusual not simply because they are present in the book, but rather because of the nature of the material presented. They introduce the latest concepts in the field of applied microbiology and demonstrate the importance of understanding earlier chapters on structure and function, nutrition, biosynthesis, and metabolic control.

Part 6, "Infection and Immunity," consists of four chapters and is introduced with a chapter on host-parasite interaction. This chapter describes the mechanisms used by microorganisms to cause disease as well as the host's nonspecific mechanisms for warding off infection by virulent microorganisms. The last three chapters deal with immunity and contain an up-to-date discussion of the components of the immune system and a description of the type of tests used to detect infectious agents or their products. The section concludes with a discussion of immune disorders in humans.

Part 7, "Microorganisms and Infectious Disease," consists of eight chapters. Chapter 22, "Host-Parasite Interactions," is an important introduction to this material because it discusses the general mechanisms used by all microorganisms in the infectious disease process. The chapters in this section discuss the general characteristics of infections caused by specific microbial agents (such as bacteria or viruses). The diseases caused by these agents are then separated; based on the organ affected. Only the major diseases are discussed in any detail; the less important ones are outlined in tables. For most of the major diseases discussed I have given some historical background that illustrates a particular aspect of the infectious disease process. I strayed from the usual pattern of organization in Chapters 30 and 33. Sexually transmitted diseases (Chapter 30) are given special place because of the increasing number of microbial agents that are involved in these diseases and the number of cases worldwide. Chapter 33, "Hospital-Associated Diseases," is given special attention because the transmission of these diseases is so different from diseases acquired outside the hospital.

Part 8 is devoted to diseases in animals and plants. The chapter on plant diseases is more extensive than any comparable chapter in other microbiology textbooks. Plants are capable of resisting microbial attack in ways that are not seen in human or animal infection. The fascinating mechanisms associated with the infectious disease process in plants are therefore discussed in some detail.

PEDAGOGICAL FEATURES

This book was written with considerable concern for the students who will be studying it. Each chapter in ***General Microbiology*** begins with an introductory section that provides the student with a chapter purpose. Each chapter is also concluded with a summary of its major concepts. Questions for study and extensive lists of selected readings, separated by books and journals, appear at the ends of chapters to aid the student and to provide for more in-depth understanding. In reading and studying the chapters, students will be greatly helped by the key terms, which appear in ***boldface italic type,*** and the considerable illustrations and line drawings using a second color. A large number of electron micrographs are used judiciously to demonstrate points and are appropriately labeled where needed. The many color plates have been carefully chosen for their instructive value, especially when staining and biochemical reactions are illustrated. An extensive glossary defining the major key terms in each of the chapters is included at the end of the book. Appendixes consist of the classification scheme of *Bergey's Manual,* a basic review of mathematical terms, and a comprehensive review of the techniques used for identification of various groups of microorganisms.

ANCILLARY MATERIALS

General Microbiology Laboratory Manual has been prepared by E. Ronald Wright, PhD, of Case Western Reserve University. Comprehensive and complete, it contains 46 laboratory exercises organized to follow the sequence of topics in ***General Microbiology.*** Each exercise refers to corresponding pages in the text and contains instructions for completing the experiments plus results sheets that may be completed and handed in to the instructor.

General Microbiology Study Guide, also prepared by E. Ronald Wright, is a direct companion to the text and provides students with the opportunity to reinforce their knowledge and develop a further understanding of microbiology. Each chapter includes study outlines, key terms, and carefully designed practice questions with answers.

Instructor's Manual, available on request to all adopters of the text, contains a descriptive rationale for each chapter and an extensive test bank.

ACKNOWLEDGMENTS

No textbook can be completed without the dedicated help of many individuals. The contributors, Martin Alexander, Gordon Carter, Glenn Chambliss, Robert Deibel, Bryan Hoerl, Gary Hooper, and Fred Singleton, require a special thanks for their tireless efforts in the face of their many research and teaching obligations. I would also like to thank Robert Benoit and Fred Rosenberg for their contributions to the development of an outline for this book. I am especially grateful to Sue Seif for the artwork; to the editors of Times Mirror/Mosby College Publishing, Sue Schapper and Diane Bowen; and to the manuscript editors, Teri Merchant and Judi Bange, who shouldered the responsibility for getting this book into print. Reviewers, listed on page x, critiqued the manuscript and provided invaluable suggestions. Last but not least I would like to thank my wife, Beverly, who found time from her busy schedule to type the manuscript from copy that surely must have looked like a foreign language.

Robert F. Boyd

REVIEWERS

I sincerely thank the following people who read and critiqued all or parts of ***General Microbiology*** in manuscript form. Their constructive input was invaluable in carrying out the several revisions, and I hope they will indulge my preferences if I have not always followed their many suggestions. Any errors of fact or judgment that might remain should be assigned to the author, but these people have rescued me from several pitfalls large and small.

Thomas A. Brawner University of Missouri at Columbia
L.E. Casida, Jr. The Pennsylvania State University
Mary Castle Editor, *American Journal of Infection Control*
Norman D. Davis Auburn University
Stephanie Doores The Pennsylvania State University
Christina L. Frazier Southeast Missouri State University
Barbara B. Hemmingsen San Diego State University
Eddie P. Hill Macalester College
Robyn Hillam South Dakota State University
John J. Iandolo Kansas State University
Michael Madigan Southern Illinois University at Carbondale
Charlotte Parker University of Missouri at Columbia
Robert M. Pengra South Dakota State University
Alastair T. Pringle University of California at Los Angeles
William R. Rayburn Washington State University
Norman Reed Montana State University
Gerald D. Schmidt University of Northern Colorado
W.R. Sistrom University of Oregon
Ronald J. Stangeland South Dakota State University
Fred J. Stutzenberger Clemson University
Robert Twarog University of North Carolina at Chapel Hill

CONTENTS

PART 1 INTRODUCTION TO THE MICROBIAL WORLD

PART 2 MICROBIAL STRUCTURE AND FUNCTION

PART 3 MICROBIAL GENETICS

PART 4 MICROBIAL METABOLISM AND GROWTH

PART 5 ENVIRONMENTAL AND APPLIED MICROBIOLOGY

PART 6 INFECTION AND IMMUNITY

PART 7 MICROORGANISMS AND HUMAN INFECTIOUS DISEASE

PART 8 INFECTIOUS DISEASES OF ANIMALS AND PLANTS

APPENDIXES

DETAILED CONTENTS

PART 1 INTRODUCTION TO THE MICROBIAL WORLD

PART 2 MICROBIAL STRUCTURE AND FUNCTION

PART 3 MICROBIAL GENETICS

PART 7 MICROORGANISMS AND HUMAN INFECTIOUS DISEASE

GENERAL MICROBIOLOGY

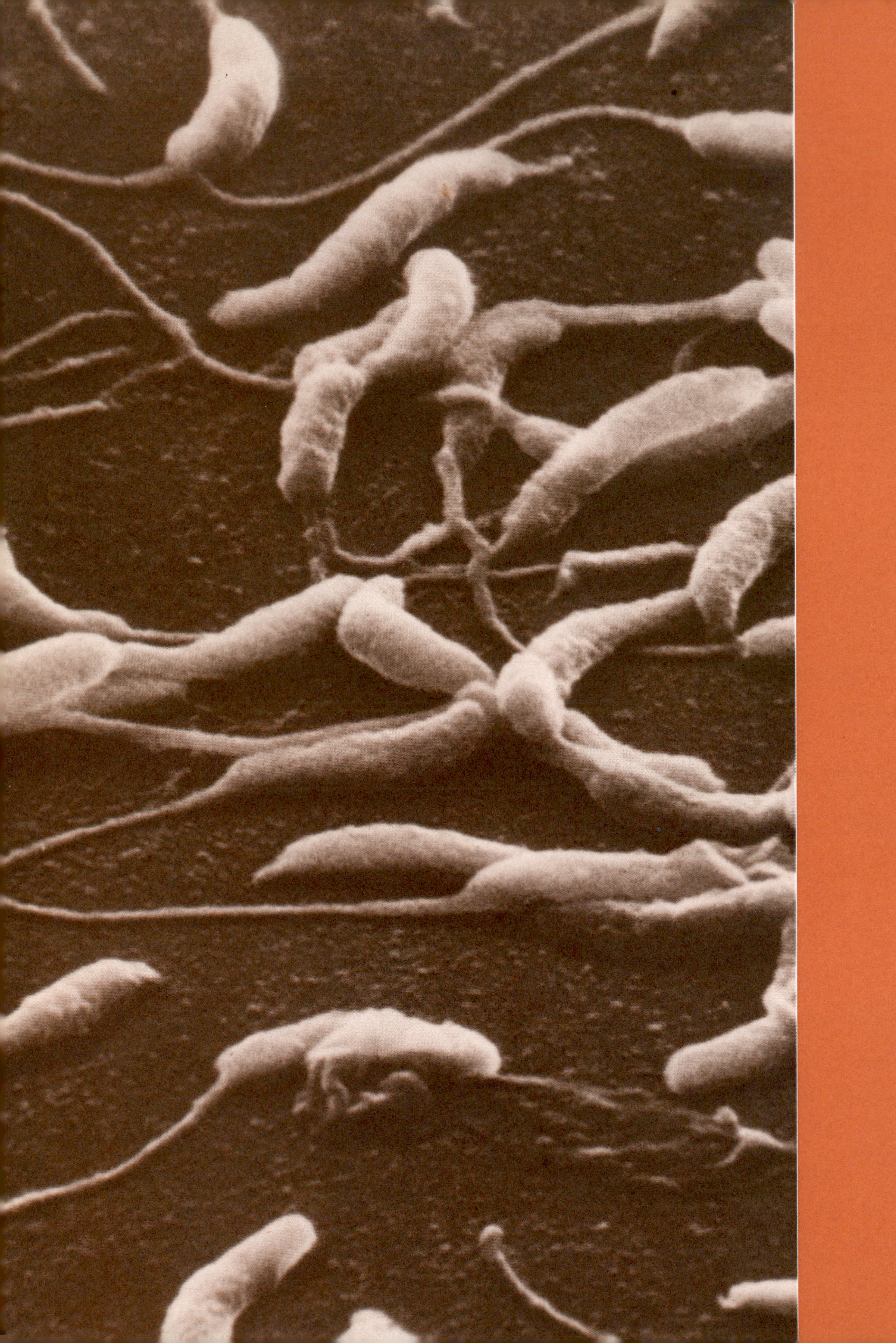

PART 1

INTRODUCTION TO THE MICROBIAL WORLD

Caulobacter sp. (×25.)

Chapter 1 HISTORY OF MICROBIOLOGY

Microbiology is basically a science that is involved with microscopic forms of life. The primary associations of microbiology before the twentieth century were with medicine and hygiene. This was a natural association, since infectious disease had ravaged humankind for centuries. The discovery of the relationship of microorganisms to infectious disease was a monumental event in the late 1800s and led to a search for the causes and cures of the major bacterial diseases of that period: diphtheria, tuberculosis, plague, syphilis, and others. Microbiology has now become a very diverse branch of science that encompasses not only bacteria (bacteriology) but also fungi (mycology), viruses (virology), and animal parasites (parasitology). Theobald Smith, a renowned bacteriologist of the late 1800s and early 1900s, stated very eloquently what microbiology meant to him and should mean to those of us who study it: "Bacteria are not an end in themselves but knowledge concerning them contributes to the solution of higher problems." The following discussion of the history of microbiology endeavors to relate how microbiology has evolved as a science and how it has and will, it is hoped, continue to contribute to the solution of problems that confront civilization.

DISCOVERY OF SMALL "ANIMALCULES"

Leeuwenhoek and primitive microscopes

One of the major obstacles, in early times, to the discovery of microorganisms was an inability to see them. Nearly all microorganisms (bacteria, viruses, protozoa, primitive fungi, and microscopic algae) cannot be seen with the naked eye and require some degree of magnification. Anton van Leeuwenhoek (circa 1685) is thought to be the original discoverer of bacteria. He observed, using primitive lenses (Figure 1-1), various forms of life that he characteristically described as little "animalcules." He observed these microscopic forms in diverse environments such as pond water, scrapings from teeth, and pepper infusions and reported his findings to other members of the scientific community. We can conclude from Leeuwenhoek's drawings that he had observed animal parasites, called protozoa, as well as variously shaped bacteria (Figure 1-2). His findings were confirmed shortly thereafter, but 150 years passed before any serious investigations into the nature of microorganisms were begun. One of the reasons for this was that lenses used before 1820 produced images with spherical and chromatic aberrations (fuzzy images due to the inability of the lens to focus light rays). These aberrations were corrected, and this led to the development of the compound microscope. Later, in 1866 Ernst Abbe developed the technique for using oil immersion in place of water as a medium for the transmission of light rays from the specimen to the lens. Immersion oil prevented loss of light and increased the clarity of the images produced.

Soon after Leeuwenhoek's discovery many investigators attempted to classify microorganisms based on their shape and motility, but poor magnifications made it difficult to distinguish many microorganisms. One interesting concept that prevailed in the nineteenth century was that all microscopic forms of life were merely stages in the development of more complicated forms, such as the fungi. Billroth, for example, in 1876 attempted to prove that spherical and rod-shaped bacteria were stages of a single organism and not individual species. Ferdinand Cohn, one of the

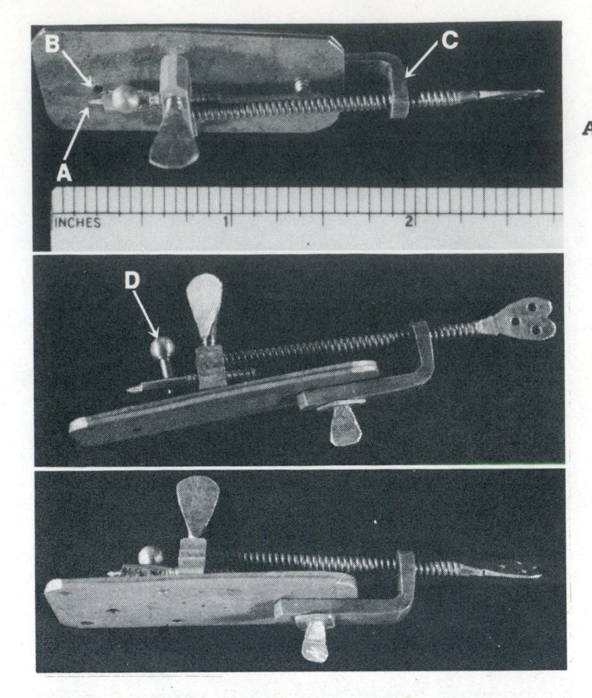

Figure 1-1 Actual-size replica of Leeuwenhoek microscope from Leyden, Holland showing various views of instrument. *A,* Pin on which object is placed for viewing; *B,* lens; *C,* screw for coarse adjustment; *D,* fine adjustment.

From Anderson, D.A., and Sobieski, R.J. Introduction to microbiology *(2nd ed.). The C.V. Mosby Co., St. Louis, 1980.*

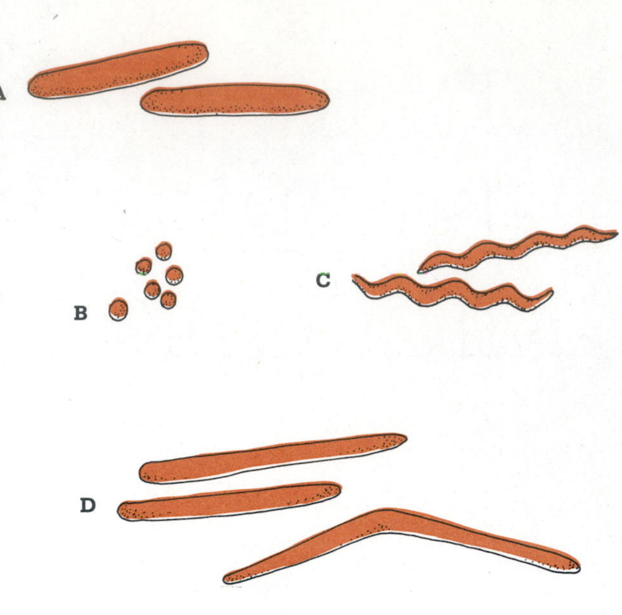

Figure 1-2 Leeuwenhoek's figures of bacteria from human mouth. **A,** Rod-shaped (bacilli). **B,** Spherical (cocci). **C,** Spiral-shaped (spirochetes). **D,** Cigar-shaped rods.

earliest bacteriologists, showed that certain spore-forming bacteria exist in two states: a growing, vegetative state that is sensitive to heat and a dormant, spore state that is resistant to heat. We now know that morphology is but one of many characteristics that serve to distinguish one species of microorganisms from another.

SPONTANEOUS GENERATION

Early experiments

Scientists in the nineteenth century who made observations on microscopic forms of life obtained their small "animalcules" from infusions of various plant and animal products, especially putrefying material. This source of microorganisms represented an important criterion for those who supported the doctrine of spontaneous generation. This early view held that certain forms of life could arise spontaneously from nonliving organic matter. Thus the idea that maggots could arise from decaying meat was accepted as a natural biological process by many scientists of that period.

As unlikely as spontaneous generation may seem to us, biological science in the eighteenth and nineteenth centuries was in its infancy, and little was known about the structure of cells and even less about sterility. John Needham, a Roman Catholic priest, performed experiments in 1749 to prove spontaneous generation. He boiled mutton broth in corked flasks for a short period of time and then observed them periodically thereafter for signs of visible turbidity. Boiling was a technique known to destroy cells, and when Needham's flasks became turbid, he suggested that the organic matter in the flasks was converted to living entities. The results were refuted in 1765 by Lazzaro Spallanzani, who repeated Needham's experiments, except the flasks were hermetically sealed before they were heated. None of the flasks became turbid after a long period of observation, unless air was allowed to enter them. Proponents of spontaneous generation, unswayed by the results of Spallanzani's experiments, maintained that air, which was removed from the flasks, was necessary to support growth. Theodor Schwann and Franz Schulze improved on Spallanzani's experiments by allowing air to enter the flasks only after it had passed through a solution of sulfuric acid and potassium hydroxide. Those who supported spontaneous generation suggested that air was a "vitalizing" power that had been destroyed on passing through the chemicals used by Schwann and Schulze. In 1854

Figure 1-3 Louis Pasteur (1822-1895).
From Historical Picture Service Inc., Chicago.

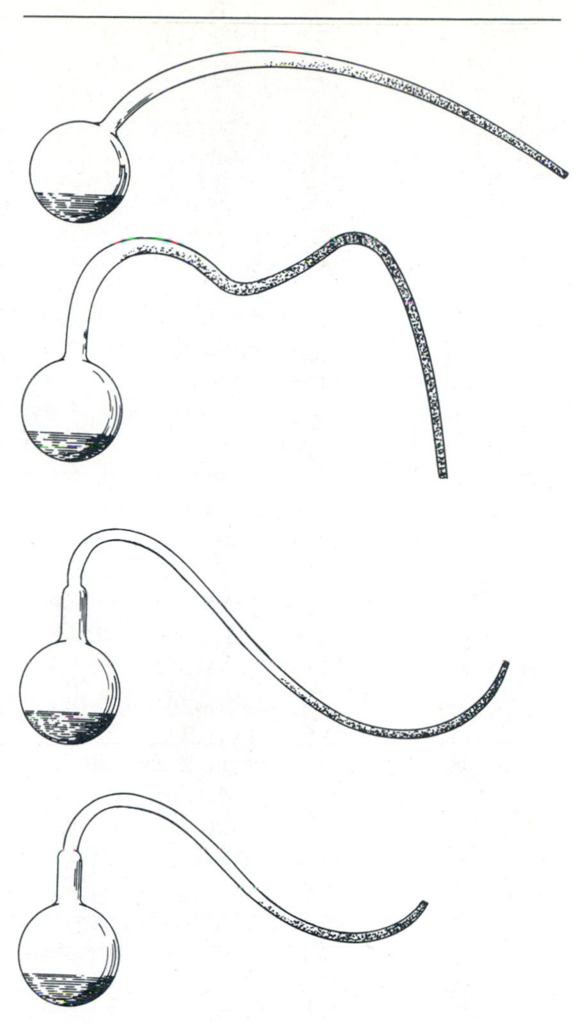

Figure 1-4 Pasteur's swan-necked flasks. Stippling in necks represents microorganisms unable to get into sterile growth medium. Shape of flask permits air to enter, but microorganisms cannot.
From Anderson, D.A., and Sobieski, R.J. Introduction to microbiology (2nd ed.). The C.V. Mosby Co., St. Louis, 1980.

Schröder and von Dusch countered the objections to Schwann and Schulze's experiment by filtering the air before it entered the flasks by passing it through cotton wool. Supporters of spontaneous generation would remain until the experiments of Louis Pasteur.

Pasteur's experiments

Pasteur (Figure 1-3) repeated the experiments of Schröder and von Dusch. The flasks were incubated at 25° to 30° C for several months with no visible growth evident in any of the flasks. If a piece of the cotton plug used to filter air was added to one of the flasks, growth occurred in 24 to 48 hours. Examination of the cotton plug before it had been added to the broth revealed spores and various microbial forms. Later Pasteur used flasks in which the neck of the flask had been drawn out and bent down so that air could enter but microorganisms could not ascend into the flask (Figure 1-4). Again, incubation over several months produced no visible growth. Growth could occur if the neck of the flask were broken or if the flask were tipped and the infusion allowed to make contact with the exposed parts of the neck and then returned to the flask. Pasteur had conclusively shown from these experiments and others that growth in infusions was the result of microbes in the air (Pasteur called them organized corpuscles) and not from organic matter that had been "vitalized." Pasteur concluded his experiments with this statement: "When one incubates the corpuscles and the amorphous debris with which they are associated into liquids which have been subjected to boiling and which would have remained unchanged in previously heated air if this inoculation had not been performed, one observes the appearance in this liquid of exactly the same organisms which they develop in the open air."

Pasteur, whom we consider the Father of Microbiology, made other significant contributions, particularly in the area of fermentation. He demonstrated that specific microbial agents were responsible for fermentation as well as diseases of wines. These studies led to the development of techniques that would directly associate microorganisms with infectious diseases.

ANTISEPSIS

One of the outgrowths of the many attempts to disprove spontaneous generation was the development of sterile techniques. Without sterile techniques the development of microbiology as a science would not have been possible. Sterile technique enables one to destroy unwanted living cells such as microorganisms. Can you imagine surgery with contaminated hands and instruments, or drinking untreated or unfiltered water? It is obvious to us that such procedures can lead to significant disease. Less obvious is the fact that sterility is important in the study of the characteristics of individual species of microorganisms. Sterile technique permits the investigator to use a growth medium devoid of unwanted living entities. Thus if a single cell of a particular microbial species is inoculated into a sterilized medium, a population of cells can be produced that is made up of only one microbial species. Only with a large population of a single species can one effectively study the biochemistry, genetics, physiology, or other characteristics of a single species. These characteristics cannot be studied if the microorganism is growing in a medium contaminated by other microorganisms.

Boiling had been recognized as a technique for destroying cells but had not been used on surgical instruments, since many in the scientific community in the early 1800s had not accepted the concept that microorganisms were the agents of disease. Oliver Wendell Holmes, an American physician, in 1843 published an article on the nature of an infection called puerperal sepsis,* which afflicts mothers following childbirth. Holmes reported that it was safer to have a child at home than in the hospital, where physician handling contributed to disease. Ignaz Semmelweiss (circa 1845), a Hungarian physician, was ridiculed for his insistence that physicians wash their hands before handling pregnant women in the hospital ward. Residents in the hospital frequently handled cadavers in the pathology laboratory before they made their examinations of expectant mothers, usually without washing their hands. Despite the fact that Semmelweiss's experiments showed that handwashing reduced the incidence of disease, physicians refused to admit that they were "unclean" and continued with their septic practices. Joseph Lister, an English physician, in 1867 recognized the importance of Pasteur's earlier experiments demonstrating the relationship between microorganisms and disease. Lister showed that boiling instruments and applying carbolic acid to dressings that covered wounds significantly reduced the incidence of disease following surgical procedures. The experiments of Lister supported the view that microorganisms were the cause of infectious disease.

* *Sepsis* is derived from the Greek, *sepsis,* which means decay. It is used to indicate the presence of pathogenic microorganisms or their toxic products.

GERM THEORY OF DISEASE

Koch's experiments

Pasteur discovered that an infection of silkworms, called pebrine, was associated with a microscopic protozoan, but many scientists still believed that microorganisms were the consequence of disease and not its cause. The direct role of bacteria as agents of infectious disease was demonstrated by Robert Koch (Figure 1-5). Koch, a German physician, began his work on the disease anthrax in 1872. Anthrax is a particularly devastating disease of cattle that is capable of destroying entire herds. The organism causing anthrax, *Bacillus anthracis,* had been microscopically observed by others as early as 1850. Davaine in the 1860s had observed the rod-shaped organism in the blood of anthrax victims, but others could not repeat his experiments and refuted his findings. Koch in 1876 elaborated on the life cycle of the anthrax bacillus and demonstrated how it formed resistant endospores in the soil, thus shedding some light on the epidemiological aspects of the disease. In 1878 Koch published a paper on the role of microorganisms in wound infections. He showed that six different infections in mice were caused by six different bacteria that differed morphologically and biologically from one another. In 1881 he proposed four postulates that would prove whether or not an infectious agent is the cause of a disease:

1. The causative agent of the disease must be present in all cases of the disease and must be absent from healthy animals.
2. The agent of disease can be isolated from the diseased animal and can be grown in pure culture (a population of one species of microorganism).
3. The disease can be reproduced by inoculat-

Figure 1-5 Robert Koch (1843-1910).
From Historical Picture Service Inc., Chicago.

ing a portion of the pure culture into healthy animals.
4. The agent of disease can be reisolated from the infected animal.

Koch's postulates, with few exceptions, can now be demonstrated for all bacterial diseases. They cannot be specifically applied to viruses, since the latter cannot be grown in artificial media. In addition, in some viral diseases the agent may not be directly demonstrable.

Golden Age of Microbiology

The results of Koch's experiments initiated a period of investigation called the Golden Age of Microbiology, in which the causes of many diseases were determined. These are outlined in Table 1-1. This golden age would have been slow to develop if it had not been for specific contributions made by Koch as well as other investigators. Koch's postulates can be proved only if the organism causing disease can be cultivated in the laboratory. In other words, we must have a pure culture. Koch observed that exposure of a potato slice or gelatin to the air resulted in the formation of discrete colonies on its surface and that each colony was made up of a population of cells that had apparently arisen from a single cell. He later demonstrated that if a small population of cells could be thinned out on the surface of a solid nutrient surface, single cells could be deposited that would give rise to individual colonies.

Other important contributions were made that involved the growth medium for bacteria. Potato slices and gelatin are not ideal media for bacterial growth because (1) many microorganisms cannot grow on them because of a lack of specific nutrients, (2) many microorganisms produce enzymes that can liquefy gelatin, and (3) gelatin liquefies at 37° C. The German physician Walter Hesse, who was interested in the bacteriology of air, used gelatin to trap microorganisms from the air, but the gelatin melted at high temperatures. In 1881 Hesse's wife suggested the use of agar-agar, which she had used for many years in preparing fruits and jellies. Agar-agar was more solid at the usual temperatures of incubation for bacterial cultures; it was heat stable; and it was resistant to microbial attack. The discovery was reported to Koch, who immediately adapted the new medium for the cultivation of the agent of tuberculosis. Friedrich Loeffler, who was an associate of Koch, devised a nutrient medium in 1881 that was composed of meat extract and digests of protein. This medium is the basis of a nutrient broth that is still used today. The addition of agar-agar to the nutrient medium would now provide a simple means for isolating and cultivating microorganisms.

Koch also contributed to the more rapid identification of microorganisms through the development of a microbial stain. He developed a technique for staining the bacillus that is the cause of tuberculosis. Koch's pioneering work stimulated other workers to continue the search for new stains. Among these other workers were Paul Ehrlich, Franz Ziehl, Friedrich Neelsen, and Hans Christian Gram, whose names are associated with stains that are routinely used today in microbiological laboratories.

MICROBIOLOGY IN THE TWENTIETH CENTURY

Immunology

The recognition of pathogenic bacteria and their identification were two of the most important advances in the late nineteenth century. Although many diseases were yet to be identified, researchers began to turn their attentions to methods for preventing and treating infectious disease. The study of resistance to infections is one aspect of the discipline called immunology.

Immunology had its beginnings in ancient times when the Chinese inoculated live organisms from smallpox lesions into healthy individ-

Table 1-1 Discoverers of some of the major pathogens during the Golden Age of Microbiology (1879-1900)

DISEASE	CAUSATIVE AGENT	DISCOVERER	DATE
Gonorrhea	*Neisseria gonorrhoeae*	Albert Neisser	1879
Malaria	Species of *Plasmodium*	Charles Laveran; Ronald Ross described life cycle of microbes	1880
Typhoid fever	*Salmonella typhi*	Karl Eberth	1880
Pneumonia	*Streptococcus pneumoniae*	Independent isolation by Louis Pasteur and George Sternberg	1881
Tuberculosis	*Mycobacterium tuberculosis*	Robert Koch	1882
Cholera	*Vibrio cholerae*	Robert Koch	1883
Diphtheria	*Corynebacterium diphtheriae*	Edwin Klebs and Friedrich Loeffler made initial isolation, but Emile Roux and Alexander Yersin established importance of toxin in 1888	1883
Tetanus	*Clostridium tetani*	Arthur Nicolaier	1884
Diarrhea	*Escherichia coli*	Theodor Escherich	1885
Undulant fever	*Brucella* species	David Bruce	1887
Gas gangrene	*Clostridium perfringens*	William Welch and G.H. Nuttall	1892
Plague	*Yersinia pestis*	Shibasaburo Kitasato and Alexander Yersin	1894
Botulism	*Clostridium botulinum*	Emile Ermengem	1896
Dysentery	*Shigella dysenteriae*	Kiyoshi Shiga	1898

Figure 1-6 Edward Jenner (1749-1823).
From Historical Picture Service Inc., Chicago.

uals for the purpose of protection from the disease. Edward Jenner (Figure 1-6) in 1798 noted the pox-free skin of milkmaids and inoculated nonvirulent cowpox to induce immunity against the virulent smallpox (Figure 1-7). These two immunological practices are particularly fascinating, since the agent of disease, a virus, was not discovered until the twenieth century. Pasteur in 1881 recognized that the agent of chicken cholera, when cultured in the laboratory instead of in the animal host, lost some of its virulence (disease-producing qualities). This process of reducing virulence is called attenuation. Pasteur injected attenuated cultures into healthy chickens and noted that they did not die and were resistant to infection when challenged with virulent organisms. Pasteur coined the term *vaccination* (*vacca* means "cow" in Latin) in honor of Jenner's work. Pasteur's most famous immunization technique was that developed against rabies, in which weakened virus used initially to induce resistance was followed by the periodic injection of more potent virus.

These studies by Pasteur suggested that the body possesses specific cells (cellular theory) that are involved in immunity. The Russian zoologist Elie Metchnikoff (circa 1884) (Figure 1-8) discovered certain cells in the body that could eat other cells, and he referred to them as phagocytes (cell eating). There were other workers at this time who believed that immunity was the result of noncellular components found in the blood (humoral theory). Emil von Behring (Figure 1-9) in 1890 demonstrated that immunity to the disease diphtheria is related to the capacity of the cell-free portions of the blood to inactivate diphtheria toxin, which is responsible for the symptoms of disease. In 1903 Almoth Wright showed that immunity is a function of humoral and cellular components. He demonstrated that phagocytosis of bacteria is aided by a noncellular component called opsonins. During this same period Paul Ehrlich (Figure 1-10) made significant contributions to immunology. He proposed the side chain hypothesis of antibody formation, which suggests that a foreign agent (the antigen) when injected into the body induces the formation of a molecule that is specific for it (a lock-and-key relationship). Working in conjunction with von Behring, Ehrlich also standardized antitoxins and toxins that contributed to the treatment of diphtheria.

Immunity from its earliest beginnings was considered a problem in chemistry, and this has become even more evident today. One of the first contributions to the biochemical nature of the immune response came from the laboratory of Karl Landsteiner (1868-1943). Landsteiner demonstrated that small chemical groups could be attached to existing antigens and could impart specificity in the immune response. Landsteiner's work on specificity led to the discovery of the four major blood groups; A, B, AB, and O. His work in cooperation with Alexander Weiner in 1940 led to the discovery of the Rh antigen system.

Modern theories of antibody formation are in large part a result of the studies of Felix Haurowitz and others in the 1930s. They suggested that antigens were held firm by the antibody-forming cells and thus served as a template for the synthesis of antibody. Later, in 1959 Burnet expanded on these ideas and conceived the clonal selection theory. This theory states that cells are endowed with the genetic capacity to produce a specific antibody. The antigen, by combining with a specific cell, causes that cell to divide and to produce significant antibody. Thus antigen, according to Burnet, acts as a trigger for proliferation and not as a template for specificity, as proposed by Haurowitz. Peter Medawar and Burnet in the years 1946-1960 applied the clonal selection theories to the characteristics of immu-

Figure 1-7 Pustules on arm, caused by smallpox virus.
Courtesy S.O. Foster.

Figure 1-8 Elie Metcnnikoff (1845-1916).
From Historical Picture Service Inc., Chicago.

Figure 1-9 Emil van Behring (1854-1917).
From Historical Picture Service Inc., Chicago.

nological tolerance; that is, one does not make antibodies to one's own antigens. Burnet and Medawar shared the Nobel prize in 1960.

Chemotherapy

Diseases such as syphilis and malaria had been treated successfully before the Golden Age of Microbiology with such drugs as mercury and quinine, respectively. Except for these two diseases, treatment of internal infections was provided by the use of immune products such as antitoxins present in the antiserum of animals. One of the major breakthroughs in the field of chemotherapy was provided in 1932 by the laboratory synthesis of a drug called prontosil and the demonstration of its effects on infectious microorganisms. Alexander Fleming in 1929 had noted the antimicrobial activity of a metabolite produced by the mold *Penicillium notatum,* but the implications of this metabolite, which is now called penicillin, was not recognized until 1940 by Abraham and others. Dubos in 1939 furthered the pursuit of antimicrobials when he reported the antimicrobial activity of metabolites isolated from certain soil microorganisms. The discoveries of Fleming and Dubos ushered in a period of extensive research for organisms that produce metabolites with antimicrobial activity. Waksman, who in 1944 suggested that antimicrobial compounds produced by microorganisms be called antibiotics, is best known for his discovery of the antimicrobial called streptomycin. The 1950s were a period of major discoveries of antibiotics that were suitable for the treatment of major bacterial disease.

Virology

Viral diseases have existed since ancient times, but their isolation and cultivation are of recent origin. Virus as a cause of disease was first reported in 1892 by the Russian botanist Iwanowsky. He demonstrated that tobacco mosaic disease could be reproduced in healthy plants by rubbing their leaves with sap from infected plants after the sap had passed through a Chamberlain filter. Chamberlain filters are porcelain and can retain bacteria but permit smaller objects such as viruses to pass through. Much of the scientific community remained skeptical of Iwanowsky's results and did not believe that a new microbial agent had been discovered. Beijerinck, a Dutch microbiologist, in 1898 repeated Iwanowsky's experiments and showed conclusively that the agent passing through the filter was not a bacterium. He named the microbial agent *Contagium vivium fluidum.*

The first animal virus was isolated by Loeffler and Frosch in 1898 from animals with foot-and-mouth disease. Reed in 1902 reported that yellow fever in humans is caused by a filterable "virus" carried by mosquitoes. Rous in 1911 showed that solid tumors of chickens could be transmitted by cell-free filtrates, but his studies remained obscure until 1932, when Shope, Bittner, and others established beyond doubt that certain viruses are capable of causing tumors.

A new class of filterable viruses was discovered in 1915 by Twort, who found that bacteria are also subject to infection. He had noticed that occasionally a single colony of a bacterium called staphylococcus appeared clear or glassy on an agar surface. d'Herelle called the agents infecting the bacteria bacteriophages (bacterial eaters). Methods were soon devised for studying the mechanism of infection and reproduction of these bacterial eaters. The study of bacteriophages thus became the prototype for all viral infection processes because animal viruses had not yet been cultivated in cell culture or chick embryo.

Laboratory animals, such as rabbits and guinea pigs, had been used almost exclusively for animal virus isolation until 1931, when Goodpasture and Woodruff demonstrated the cultivation of virus in chick embryo. They successfully cultivated cowpox and other viruses in the tissues of the developing embryo. In the following 20 years most viruses were cultivated in four main areas of the embryo: the allantoic, amniotic, chorioallantoic, and yolk sacs. Still, several viruses could not be cultivated in chick embryo. In 1948 Dalldorf and Sickler showed that unweaned hamsters, mice, and other rodents were susceptible to infection by previously uncultivated viruses.

One of the most important developments pioneered by microbiology was the cultivation of viruses in cell culture. In order to study the biochemical events involved in the viral replication cycle, pure strains of mammalian cells are required. Virologists initiated the research, but cell biologists such as Puck, Eagle, and Earle perfected the techniques of cultivating mammalian cells. One of the first major breakthroughs in the cultivation of viruses was made by Enders, Weller, and Robbins in 1949. They were able to grow poliovirus in nonneural cells (poliovirus affects neural tissue in humans). Nearly all viruses can now be cultivated in cell culture. The discovery by Enders and others led to an explosion of viral research that led to the cultivation and isolation of previously unknown viruses, the diagnosis of viral disease, and the development of vaccines for the prevention of disease. The importance of cell culture cannot be overstated, since nearly all branches of biology now use it as a basic research material.

Figure 1-10 Paul Ehrlich (1854-1915).
From Historical Picture Service Inc., Chicago.

Research on the properties of viruses began in 1931 when Elford introduced the use of colloidal membrane filters of graded porosities. The filter could retain virus if the latter were larger than the pores of the membrane. Experiments using membranes of different porosities showed that viruses range in size from 20 to 700 nm. Stanley in 1935 isolated and purified the tobacco mosaic virus (TMV) in the form of crystals, a characteristic we associate with molecules and not microorganisms. This seemingly paradoxical situation was resolved later by Fraenkel-Conrat, Gierer, and Schramm, who demonstrated that the viral particle consists of a protein coat that surrounds a core of nucleic acid. They separated the nucleic acid from the protein coat of the virus and demonstrated the former's infectious properties and its ability to initiate replication.

Era of genetics and molecular biology

Microbiology, as well as all of cell biology, made significant advances following the publication of an article by James Watson and Francis Crick in 1953 describing the structure of deoxyribonucleic acid (DNA). Their interpretation of the double helical nature of the DNA molecule, which was later confirmed by experimentation, ushered in the era of molecular genetics. Molecular genetics is a science that investigates the structure and function of nucleic acids using biochemical, biophysical, and genetic techniques. The discovery that bacteria have genetic mechanisms similar to those of higher organisms provided a simple genetic model from which experimental results could be extrapolated to more advanced forms of life.

Microbiology has contributed greatly to our fundamental knowledge of cell biology and promises to make significant contributions to the solution of problems in various areas of scientific endeavor. Two of the most important developments in microbiology that are providing and will provide an impact on humankind are (1) the construction of hybridomas and (2) genetic engineering.

In 1975 Kohler and Milstein were able to fuse a skin cancer cell from a mouse to an antibody-producing white blood cell. This was an important development for two important reasons. First, the antibodies produced by this hybrid are of only one type (monoclonal), which is in sharp contrast to the complex mixtures of antibodies found in normal sera. Second, the hybridoma tissue culture cells can be frozen and used later to inject into animals to produce tumors. Large amounts of homogenous antibody can then be obtained from the affected animal. Monoclonal antibodies have been generated against a great many antigens, but human monoclonal antibodies were not developed until 1980, when Olsson and Kaplan fused a human myeloma cell to immune spleen cells. It is hoped that by using monclonal antibodies, we will now be able to study more precisely the structure and function of proteins produced by viruses during infection. With this knowledge we will be able to construct inhibitors that can be used for therapy against viral diseases. Monoclonal antibodies directed against specific virus components will also aid in their identification in the clinical laboraotry, which today is a time-consuming process. Monoclonal antibodies will probably replace horse and rabbit antibodies that are presently used to treat such diseases as tetanus and transplant rejection. This would prevent the allergic reactions that occur with second injections of animal antibodies. It has also been suggested that immunization with human monoclonal antibodies may become the preferred treatment for drug overdoses and some diseases caused by microorganisms.

Genetic engineering is a technique that theoretically permits the transfer of genes from any source into microorganisms. This technique was first described in 1973 by Chakrabarty, who isolated genes from one organism and inserted them into another for the purpose of degrading oil. This new technique has already resulted in the production of insulin by the bacterium *Escherichia coli,* as well as the production of certain growth hormones. In other words, the genes that code for animal proteins are removed from the animal and inserted into the bacterial DNA. Genetic engineering is a powerful tool that has tremendous potential for the future, such as:

1. Development of safe and effective vaccines to cope with infectious disease. The original foot-and-mouth disease vaccine, for example, was produced from whole virus, which could accidentally cause disease. Genetic engineering is now used to produce a specific viral protein in a bacterium. This single protein can produce immunity but itself does not cause disease.
2. Inexpensive production of high-quality animal protein for use as a diet supplement to help feed the world's population.
3. Manipulation of plant species of agricultural importance to increase their nutritive value or to reduce the importance of synthetic fertilizers for growth. Resistance to plant pests and environmental conditions may also be possible.
4. Development of microorganisms to aid in the degradation of chemicals that contribute to environmental pollution, for example, oil spills.
5. Development of "super" strains of microorganisms used for the production of industrial products such as alcohols, antibiotics, vitamins, enzymes, and others.
6. Insertion of specific genes into microorganisms that normally maintain growth by living on metallic ores. A speeded-up process of oxidation results in the formation of metal oxides that can be physically extracted to recover important industrial metals.

The implications of these developments in microbiology are enormous. I must confess that each day I am amazed and excited at the strides made in microbiology and how it is directly or indirectly affecting our lives. The excitement becomes especially personal when one has some basic knowledge of microbes and their activities. It is hoped that this book will impart some of that knowledge and help you understand the importance of microbiology and maybe even induce you to pursue a career in microbiology or a related discipline.

SUMMARY

Microbiology as a science includes the study of bacteria (bacteriology), viruses (virology), fungi (mycology), and animal parasites (parasitology). Immunology, the study of the immune system, is also considered a branch of microbiology. The history of microbiology had its beginnings with the discovery of microscopic forms of life by Anton van Leeuwenhoek in 1685. The major developments in microbiology did not begin, however, until it was conclusively demonstrated by Louis Pasteur and others in the 1870s that microorganisms are living entities capable of self-replication and are not spontaneously generated from nonliving organic matter. Pasteur is regarded as the Father of Microbiology, not only because of his experiments that disproved spontaneous generation, but also because of his discoveries that microbes can be the agents of disease in humans, plants, and animals, as well as his original ideas, which spanned many branches of science.

Antisepsis was an outgrowth of attempts to disprove spontaneous generation and led to important developments in hygiene as well as basic research. Joseph Lister in 1867 is credited with antiseptic technique because of his experiments demonstrating that boiling instruments and applying carbolic acid to wound dressings reduces the incidence of disease following surgery.

Robert Koch, between the years of 1872 and 1881, showed a direct association between the disease and the infectious agent by using four procedures that we now refer to as Koch's postulates. The period between 1879 and 1900 is referred to as the Golden Age of Microbiology. During this period agents of diseases such as diphtheria, malaria, tuberculosis, and others were isolated and shown to be associated with the infectious disease process.

Microbiology in the twentieth century has developed rapidly, especially in the last 30 years. Developments in immunology, virology, and chemotherapy and associated fields have been due primarily to advances in biochemistry and genetics (molecular genetics). The discovery that initiated the field of molecular genetics was made by Watson and Crick in 1953 when they described the helical structure of deoxyribonucleic acid (DNA). Two of the most important advances in microbiology have occurred since 1973: the construction of hybridomas and techniques in genetic engineering. They are leading to important developments in the prevention and cure of infectious as well as noninfectious diseases and to the solution of many environmental problems that now confront humankind.

Study questions

1. Make a list of the significant contributions made by Robert Koch to the field of microbiology. What do you consider to be his most significant contribution?
2. Why did it take so long for advances in microbiology following the discovery of microscopic forms of life by Anton van Leeuwenhoek?
3. How did Louis Pasteur's experiments disprove spontaneous generation?
4. What did the proponents of spontaneous generation mean by "vitalizing force"?
5. What men made the most significant contributions to the development of hygienic practices during surgery?
6. What was so important about Frau Hesse's suggestion that agar-agar could be used in bacteriological media?
7. Where did the technique of vaccination have its beginnings?
8. What is meant by humoral and cellular immunity? Who proposed the theories of each?
9. What are monoclonal antibodies? Who first suggested the clonal selection theory of antibody formation?
10. What development accelerated viral research in the 1950s? Why?
11. Why were bacterial viruses the model for virus infection? Why not animal viruses?
12. What were the implications of Watson and Crick's description of DNA structure?

Selected readings

Books

Brock, T.D. *Milestones in microbiology.* American Society for Microbiology, Washington, D.C., 1975.

Bulloch W. *The history of bacteriology.* Oxford University Press, London, 1938.

Clark, P.F. *Pioneer microbiologists of America.* University of Wisconsin Press, Madison, Wis., 1961.

DeKruif, P. *Microbe hunters.* Harcourt, Brace & World, New York, 1953.

Dixon, B. *Magnificent microbes,* Atheneum Publishers, New York, 1976.

Dobell, C. (Ed.). *Anton van Leeuwenhoek and his "little animals."* Dover Publications, Inc., New York, 1960.

Dubos, R.J. *Louis Pasteur: free lance of science.* Little, Brown & Co., Boston, 1976.

Lechevalier, H.A., and Solotorovsky, M. *Three centuries of microbiology.* Dover Publications, Inc., New York, 1974.

Reid, R. *Microbes and men.* Saturday Review Press, New York, 1975.

Chapter 2 BIOCHEMICAL BACKGROUND FOR MICROBIOLOGY

Biochemistry is the study of living matter that is composed of both inorganic and organic compounds. The sciences of biochemistry and microbiology have become so intertwined that they are practically inseparable. Biochemistry at one time was a descriptive science that dealt with the properties of the simplest molecules that could be extracted from animal or plant tissue. In the past 30 years biochemistry has become a dynamic science in which the characteristics of growth and metabolism in biological systems are becoming rapidly clarified. Scientists are now probing, using biochemical techniques, even more deeply into the detailed mechanisms that permit certain molecules to behave in specific ways. Much of what we know about biochemical mechanisms has been learned from microbial systems. The microorganism has become a very important tool in the hands of the biochemist, whose knowledge of chemical, physical, and biological techniques has helped the microbiologist to understand the physiological differences between microorganisms. We are now able to more efficiently identify and classify microorganisms, to use microorganisms more effectively in the industrial sector, and to better understand disease processes. The molecular architecture of biological components is also being unraveled through biochemical techniques. Our knowledge of microbial structure and function has increased dramatically with the aid of microscopy and with recently developed techniques for the separation and identification of macromolecules from cellular systems. Any understanding of what makes the microorganism, as well as other biological systems, "tick" cannot be fully appreciated unless some fundamental concepts of biochemistry are understood. In this chapter an attempt is made to concisely outline the basics of biochemistry. It is hoped that this presentation will not only help you understand later chapters that describe biochemical mechanisms, but will also enable you to appreciate biological organization.

CHEMICAL BONDS

The major elements of living matter are carbon, nitrogen, oxygen, phosphorus, hydrogen, and sulfur. Nearly all organic compounds consist of carbon atoms bonded to one another with one, some, or all of the major elements as bonded components. One notices immediately from Table 2-1 that the major elements of organic compounds are among the lightest. The formation of stable molecular bonds is a property associated with the lighter elements.

The atoms of all elements possess shells or orbits that contain up to a maximum, a fixed number of electrons (Table 2-1). The maximum number of electrons for shell 1 is 2 electrons; for shell 2, 8 electrons, and for shell 3, 18 electrons. Electrons possess more energy the farther they are from the nucleus. The electrons found in the last shell are called valence electrons, and the chemical properties of the elements are directly related

Table 2-1 Electronic configuration of the first 18 elements

ELEMENT	ATOMIC NO.	SHELL ELECTRON DISTRIBUTION 1	2	3	FUNCTION IN CELL*
Hydrogen (H)	1	1			Major component of water and organic compounds
Helium (He)	2	2			Not part of normal structure or metabolism of cell
Lithium (Li)	3	2	1		Not part of normal structure or metabolism of cell
Berylium (Be)	4	2	2		Not part of normal structure or metabolism of cell
Boron (B)†	5	2	3		Not part of normal structure or metabolism of cell
Carbon (C)	6	2	4		Major component of organic compounds
Nitrogen (N)	7	2	5		Major component of organic compounds
Oxygen (O)	8	2	6		Major component of organic compounds
Fluorine (F)	9	2	7		Not part of normal structure or metabolism of cell
Neon (Ne)	10	2	8		Not part of normal structure or metabolism of cell
Sodium (Na)‡	11	2	8	1	Not part of normal structure or metabolism of cell
Magnesium (Mg)	12	2	8	2	Required for enzyme activity
Aluminum (Al)	13	2	8	3	Not part of normal structure or metabolism of cell
Silicon (Si)§	14	2	8	4	Not part of normal structure or metabolism of cell
Phosphorus (P)	15	2	8	5	Component of energy compounds
Sulfur (S)	16	2	8	6	Component of certain amino acids; stabilizes proteins
Chlorine (Cl)	17	2	8	7	Not part of normal structure or metabolism of cell
Argon (A)	18	2	8	8	Not part of normal structure or metabolism of cell

*More specific functions are discussed in later chapters.
†Boron is required by certain plants.
‡Sodium is required by marine bacteria.
§Silicon is required by certain algae called diatoms.

to them. Consequently, the electrons in the outer shell are easily excited by the absorption of energy from their surroundings and readily engage in chemical reactions. Various chemical bonds are formed during these chemical reactions that play an important role in terms of their biological function.

The formation of chemical bonds is based on the electronegativity of the atoms involved. Electronegativity is the capacity of atoms to attract electrons of other elements. The major chemical bonds or forces that bind atoms are ***ionic bonds, covalent bonds, hydrogen bonds,*** and ***Van der Waals' forces.*** The amount of energy that exists in these various bonds is described in Table 2-2.

Ionic bonds

Ionic bonds are formed when atoms of high electronegativity combine with atoms of low electronegativity. Atoms such as those containing one, two, or three electrons in their outer shell (atoms

Table 2-2 Bond energies for the various types of chemical bonds

TYPE OF BOND	BOND ENERGY (KILOCALORIES/ MOLE)
Van der Waals	1
Hydrogen	3 to 6
Ionic	4 to 5
Covalent	50 to 100

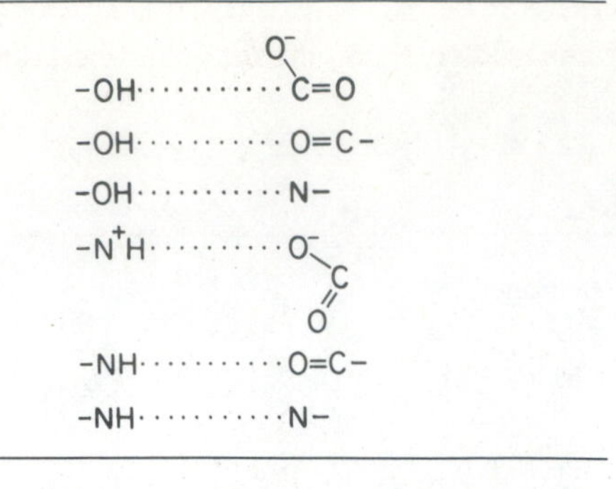

Figure 2-1 Some important hydrogen bond associations in biological systems.

of low electronegativity) form ionic bonds with those atoms that have seven electrons in their outer shell (atoms of high electronegativity). All atoms are neutral, but in the ionic bond there is a transfer of one or more electrons from the weakly electronegative atoms to the strongly electronegative atoms. In the example of sodium and chlorine, sodium is weakly electronegative, and when it combines with chlorine, which is strongly electronegative, an electron is pulled away from sodium and toward the chlorine atom. Sodium chloride (NaCl) therefore represents an ionic type of bond in which sodium is positively charged and chlorine negatively charged.

$$\mathrm{Na\cdot} + \cdot\ddot{\mathrm{Cl}}: \longrightarrow [\mathrm{Na}]^+ [:\ddot{\mathrm{Cl}}:]^-$$

Covalent bonds

Covalent bonds are the strongest bonds and are formed between two atoms that have little difference in their electronegativity. Covalent bond formation is based on the sharing of electrons and occurs between like atoms to produce gases such as hydrogen (H_2), chlorine (Cl_2), and fluorine (F_2). The number of covalent bonds between the atoms is related to the number of electrons shared. Hydrogen, chlorine, and fluorine share two electrons and have one covalent bond, whereas in nitrogen gas (N_2) there are six shared electrons, three from each atom of nitrogen. Nitrogen gas has a total of three covalent bonds and can be represented as $N{\equiv}N$.

Covalent bonds are also formed between different atoms, and although classified as nonionic, they can exhibit ionic properties. The reason for this is that most covalently bonded atoms show an unequal distribution of electrons, which results in the formation of poles. One end of the molecule is positive and the other is negative. Most compounds are to some extent polar, and the degree of polarity determines the solubility of the compound.

Hydrogen bonds

Hydrogen bonds, like ionic bonds, are weak bonds and are 10 times weaker than covalent bonds. The hydrogen atom contains but one proton and one electron. The proton is readily attracted to highly electronegative atoms and bonds with nitrogen and oxygen, which are two of the most electronegative of all the elements. Examples of hydrogen bonding that can take place between organic molecules are illustrated in Figure 2-1. In addition to their role in maintaining macromolecular structure, hydrogen bonds are also important because of their arrangement in the water molecule.

Hydrogen bonds and water. Water is a polar molecule in which the oxygen atom pulls the electrons of hydrogen atoms toward it. This results in a negatively charged oxygen atom and positively charged hydrogen atoms. The dipolar characteristic of water accounts for the hydrogen atoms of one water molecule bonding to the oxygen atoms of adjacent water molecules; consequently, water exists as an aggregate (Figure 2-2) and not as individual molecules. The aggregative nature of water is also responsible for some of its unique properties, one of the most important of which is its behavior as a solvent.

Water is important in biological systems because as a liquid it allows for rapid movement of substances that individual cells and complex living systems require for the maintenance of life. As a solvent, water is important because it is able to neutralize the strong attraction that exists between the positively charged ions of a salt. It can therefore dissociate ions because of its polarity and because it can orient itself close enough to the salt to form a shell of water molecules, a process called hydration. Hydration of the salt potassium chloride (KCl) results in an orientation in which the positively charged hydrogen atoms of water molecules are attracted to the negatively charged chlorine atoms while the negatively charged oxygen atoms of individual water molecules are attracted to the positively charged potassium atoms (Figure 2-3). Water is a good solvent for other polar molecules provided that they are not too large. Organic molecules consisting only of carbon and hydrogen (hydrocarbons), such as benzene, chloroform, and paraffin, are not soluble in water, but replacement of their hydrogen atoms with other groups such as hydroxyl (OH), amino (NH_2), or carboxyl (COOH) makes them polar and soluble in water. The addition of one of these groups is not sufficient to ensure solubility if the hydrocarbon chain is very long.

Most solvents of low molecular weight have low boiling points and cease to function as liquids as the temperature increases, even within ranges normally encountered by living systems. This is not true for water, which, owing to its hydrogen-bonding properties, remains liquid at relatively high temperatures. Water also exhibits other properties that make it an essential chemical material for life. These properties include:

1. Water has a high specific heat. Water is a better insulator than other liquids. A considerable amount of heat must be added or withdrawn in order to raise or lower the water temperature. This is an important property because biological processess occur within a limited temperature range.
2. The density of water decreases between 4° and 0° C and also decreases when it solidifies. These are characteristics not found in other liquids and are important to the maintenance of life. Ice, for example, remains at the surface of a lake, or other body of water, leaving the liquid below at a relatively uniform temperature and capable of supporting life. Cold water would sink to the bottom and freeze if it did not possess its hydrogen-bonding properties.
3. Water has a high heat of vaporization. The vapor pressure of water rises more rapidly than other liquids as the temperature increases, and heat is removed rapidly from systems. The rapidly rising vapor pressure makes for greater cooling efficiency.

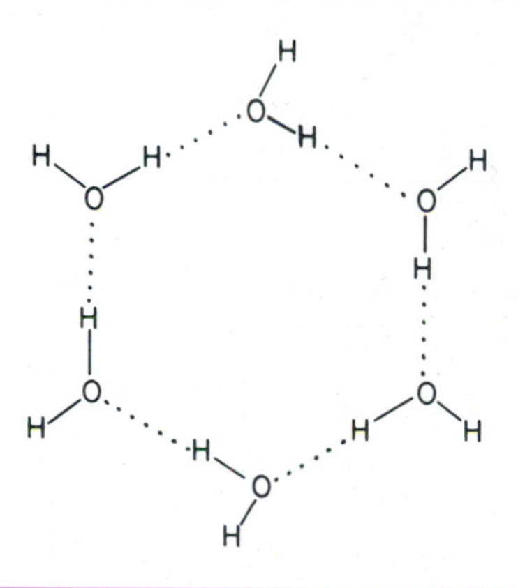

Figure 2-2 Aggregative nature of water due to dipolar characteristics.

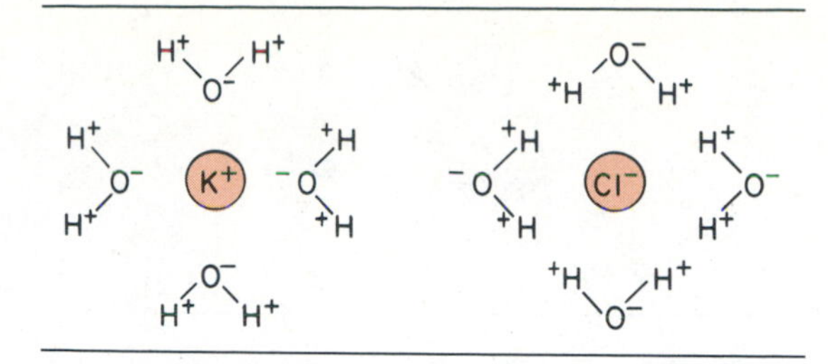

Figure 2-3 Hydration of ions during ionization of potassium chloride (KCl). Negatively charged oxygen of water is attracted to positively charged potassium (K^+), while positively charged hydrogen atoms orient themselves closer to negatively charged chlorine (Cl^-).

Van der Waals' forces

Van der Waals' forces represent a type of bonding between atoms that is based primarily on their proximity to one another (Figure 2-4). This attrac-

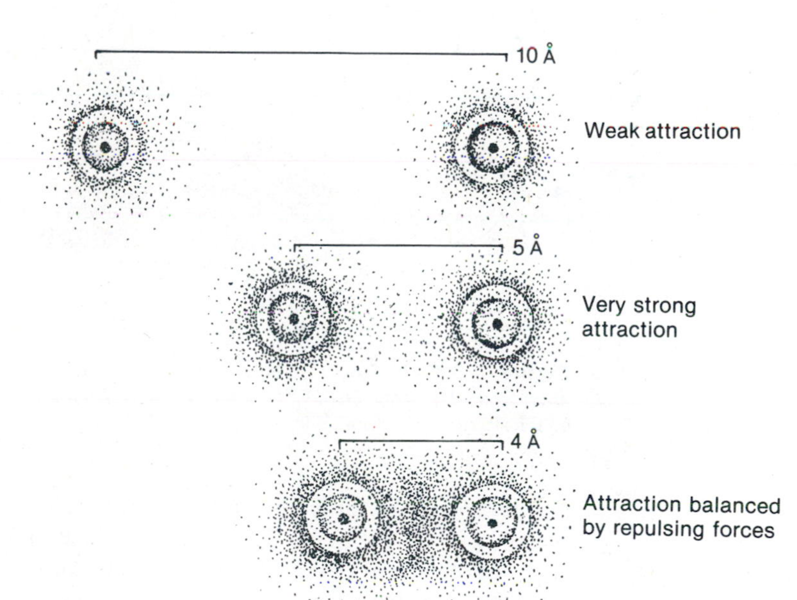

Figure 2-4 Van der Waals' attraction and repulsion forces in relation to electron distribution of monoatomic molecules on inert gas argon.

Redrawn from Pauling, L. General chemistry (3rd ed.). Freeman & Co. Publishers, San Francisco, 1958.

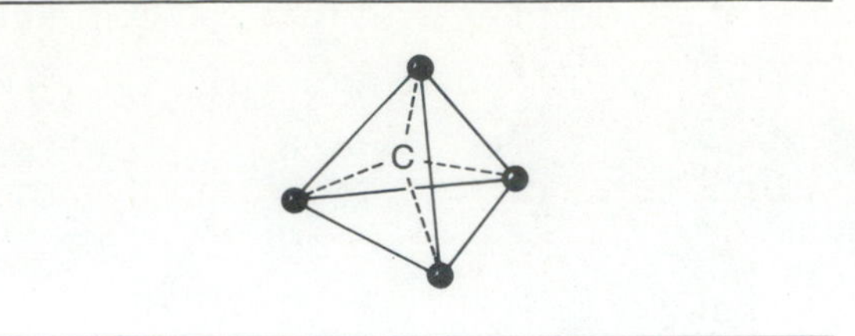

Figure 2-5 Four electrons in outer shell of carbon.

—C—C— Single bond

>C=C< Double bond

—C≡C— Triple bond

Figure 2-6 Carbon-carbon bonds.

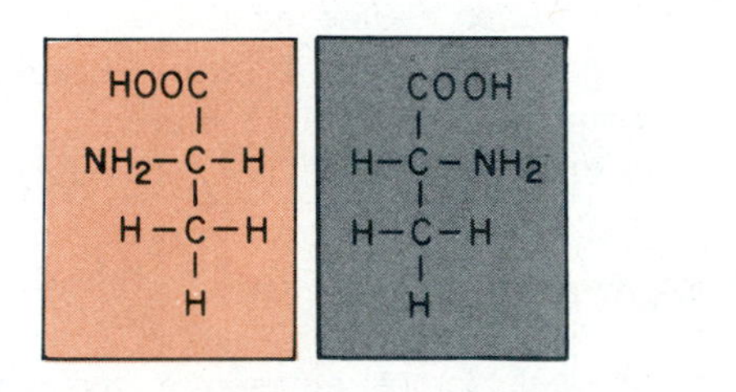

Figure 2-7 Structure of optical isomers. Optical isomer in this figure is amino acid alanine. Carbon atom in color is asymmetric. Isomers are not superimposable.

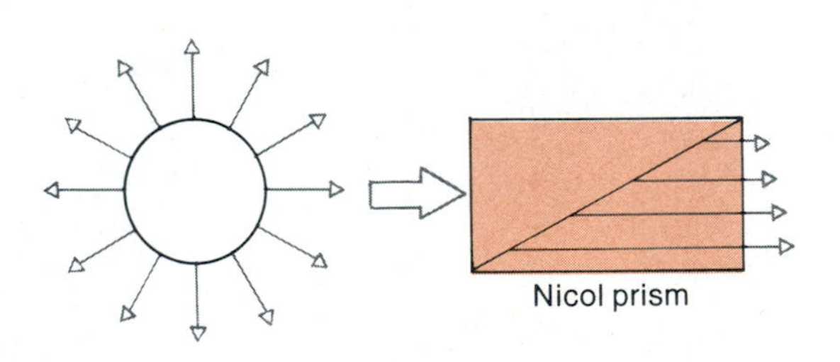

Figure 2-8 Polarization of light. Light vibrates in all directions when unpolarized. Nicol prism polarizes light so that it vibrates in only one direction.

tion between two atoms is strong at certain distances, but as the atoms get closer, the electron shells of the atoms overlap, and a repulsive force balances the attractive force. Van der Waals' forces, although weak interactions, can be important when several atoms of large molecules are closely associated. These forces are more apparent with nonpolar molecules because polar molecules can form other types of bonds.

BUILDING BLOCKS: ELEMENTS AND ORGANIC COMPOUNDS

Carbon

Carbon contains four electrons in its outer shell, which can be represented as being equidistant from each other (Figure 2-5). Carbon therefore contains four binding sites that can be occupied by other elements, in particular, hydrogen, oxygen, and carbon atoms. Most organic molecules are made up primarily of two or more carbon atoms. The maximum number of covalent bonds between two carbon atoms is three (Figure 2-6). Single bonds between carbon atoms permit rotation about the bond axis, and this allows other attached atoms or groups to lie in different planes. Double and triple bonds between carbon atoms reduces the distance between them and does not permit free rotation around the bonds. Hydrogen atoms, for example, attached to these inflexible carbons lie in the same plane. Long-chained molecules, in which there is free rotation about the chemical bond, can assume a variety of different shapes.

A carbon atom (*) with four different atoms or groups of atoms attached to it is said to be ***asymmetric:***

$$
\begin{array}{c}
CH_3 \\
| \\
NH_2—C^*—H \\
| \\
COOH
\end{array}
$$

An asymmetric carbon atom can exist in two ***isomeric*** forms that are mirror images and are not superimposable (Figure 2-7). (Structural isomers have the same molecular formula but different spatial arrangements of their atoms.) Compounds that possess asymmetric atoms can rotate plane-polarized light (Figure 2-8) and are considered optically active. The instrument used to measure optical activity is called a polarimeter. Passing plane-polarized light through a solution containing an optically active compound will result in light being rotated clockwise or counterclockwise as it approaches the viewer's eye. Compounds that rotate the light to the right or clockwise are said to be ***dextrorotatory*** (+), whereas those that rotate the light to the left or counterclockwise are said to be ***levorotatory*** (−). Isomers play a very important role in biological systems because only certain ones are recognized by those enzymes normally found in the cell. The most common groups of molecules found in biological systems that exhibit isomeric properties are amino acids and sugars. An interesting example of how this knowledge is being applied is in the case of commercial sweeteners. Refined sugars are optically active and when ingested are recognized by salivary enzymes that hydrolyze them to smaller metabolic products. Refined sugars are being tested in which the optical activity has been reversed and the sugars are not recognized by human enzymes. The sweet taste is still present, but the sugar cannot be metabolized, an important consideration for diabetics, for those having problems with weight, and for those (many of us) suffering from tooth decay.

Oxygen

Oxygen is a highly reactive molecule that is used by biological systems for the combustion of organic compounds and the release of energy. Molecular oxygen (O_2) is not ordinarily found free in the environment because of its highly reactive nature and must be continually supplied to us through the process of photosynthesis. Oxygen forms more compounds with carbon than does any other element; some of these are discussed later.

Hydrogen

Hydrogen, the lightest element, can share its solitary electron with many other elements. Hydrogen has already been discussed in terms of its ability to form hydrogen bonds, particularly with water molecules. Hydrogen bonds are also important in maintaining the shape and structure of macromolecules such as proteins and nucleic acids (see "Proteins," and "Nucleic Acids").

Organic compounds of carbon, hydrogen, and oxygen

The bonds between carbon atoms, as discussed previously, can be single, double, or triple, leaving three, two, or one electron(s), respectively, available for bonding with other elements. Only the single and double bonds are common in biological systems. The covalent bond between carbon atoms is so strong that carbon chains of great length, or rings of carbon, can be produced to provide an endless number of molecular possibilities.

Hydrocarbons. Chains containing only carbon and hydrogen are referred to as hydrocarbons. Those hydrocarbons in which there are only single bonds between adjacent carbon atoms are referred to as ***saturated*** (Figure 2-9, *A*). ***Unsaturated*** hydrocarbons (Figure 2-9, *B*) are those in which a double bond exists between any of the carbons in the chain. Many hydrocarbons are linear molecules, but some are also cyclic. The arrangement of six carbon atoms into a regular hexagon (unsaturated ring) is characteristic of a group of molecules called ***aromatic*** (having an aroma) compounds. This six-carbon arrangement produces the benzene ring (Figure 2-9, *C*), in which three double bonds contribute to the molecule's unsaturation. The cyclic nature of benzene prevents free rotation about the carbon bonds, and the molecule is planar. The heterocyclic compounds (Figure 2-9, *D*), which contain additional atoms of nitrogen or sulfur, are another important group of hydrocarbons.

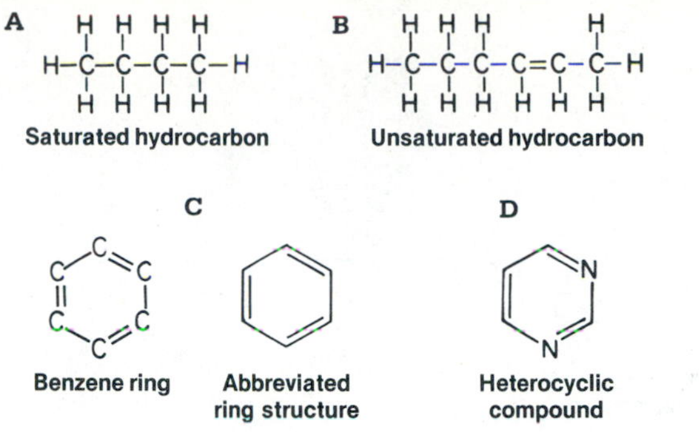

A Saturated hydrocarbon

B Unsaturated hydrocarbon

C Benzene ring; Abbreviated ring structure

D Heterocyclic compound

Figure 2-9 Structure of various hydrocarbons.

Various oxygen-containing groups are found on hydrocarbons, which can be divided into a number of classes, the most important of which are alcohols, aldehydes, ketones, esters, acids, and anhydrides. Each of these classes contains groups that are particularly reactive and enter into many biochemical reactions.

Alcohols. Alcohols are hydroxy derivatives of hydrocarbons and possess a hydroxyl (OH) group. Alcohols such as ethanol, propanol, and butanol (Figure 2-10) are produced as by-products of microbial metabolism. Some of them are of obvious importance in the beverage industry. Other oxygen-containing groups such as aldehydes, ketones, acids, and esters are the result of oxidation of alcohols; discussions follow.

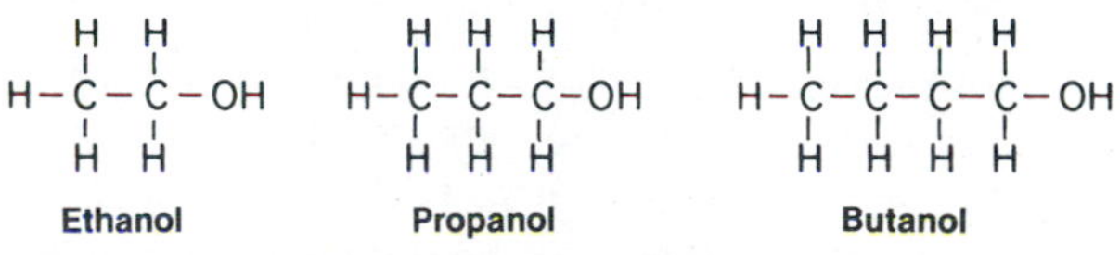

Ethanol; Propanol; Butanol

Figure 2-10 Structure of some important alcohols.

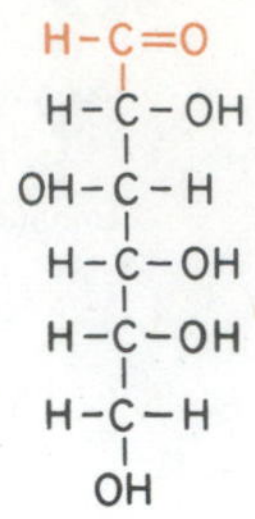

H−C=O
H−C−OH
OH−C−H
H−C−OH
H−C−OH
H−C−H
OH

D-Glucose

Figure 2-11 Structure of D-glucose. Aldehyde group is in color.

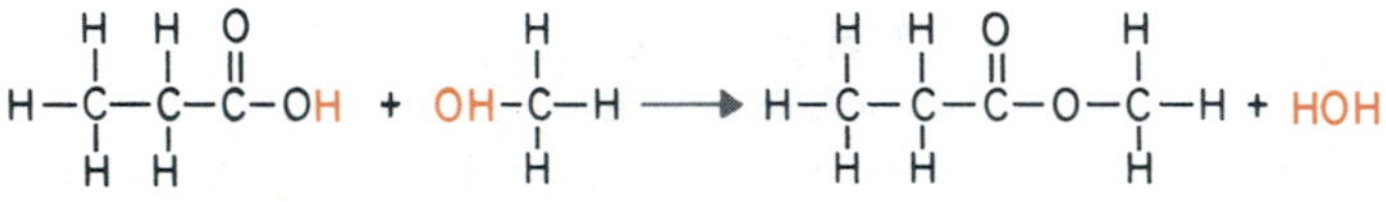

$$CH_3{-}CH_2{-}COOH + HO{-}CH_3 \longrightarrow CH_3{-}CH_2{-}C(=O){-}O{-}CH_3 + HOH$$

Propionic acid — Methyl alcohol — Methyl propionate

Figure 2-12 Formation of simple ester.

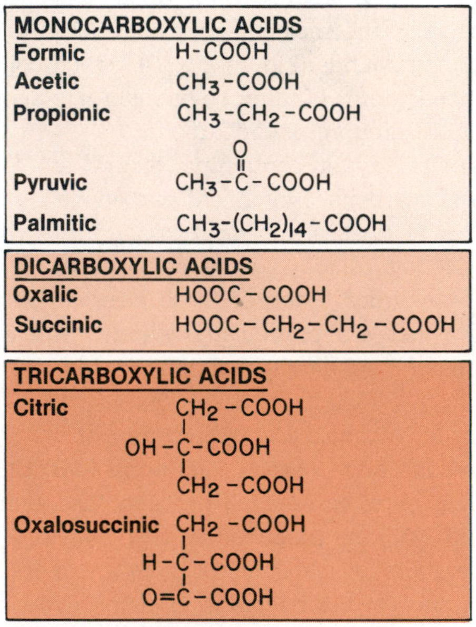

MONOCARBOXYLIC ACIDS	
Formic	H-COOH
Acetic	CH_3-COOH
Propionic	CH_3-CH_2-COOH
Pyruvic	CH_3-C(=O)-COOH
Palmitic	CH_3-$(CH_2)_{14}$-COOH
DICARBOXYLIC ACIDS	
Oxalic	HOOC-COOH
Succinic	HOOC-CH_2-CH_2-COOH
TRICARBOXYLIC ACIDS	
Citric	CH_2-COOH / OH-C-COOH / CH_2-COOH
Oxalosuccinic	CH_2-COOH / H-C-COOH / O=C-COOH

Figure 2-13 Important monocarboxylic, dicarboxylic, and tricarboxylic acids formed in biological systems.

Aldehydes. Aldehydes contain a carbonyl group (C═O) and possess at least one hydrogen atom (H—C═O). Many sugars, such as glucose (Figure 2-11) and ribose, contain an aldehyde group that is very reactive and can bond to other molecules.

Ketones. Ketones are carbonyl groups in which there is no hydrogen atom attached to it (—C—C(=O)—C—). Pyruvic acid, for example, contains a keto group (H_3C—C(=O)—COOH) and is a very important product in the fermentation of sugars. Carbonyl oxygen groups can be replaced by amino groups (NH_2), and many metabolic acids, containing a keto group, can be converted to amino acids in this manner (see "Proteins").

Esters. Esters are produced from the reaction of alcohols with acids, and they contain the following: R—C(=O)—O—R_1 in which R stands for the rest of the molecule. Many lipids, for example, are esters or possess an ester linkage. The formation of a simple ester is described in Figure 2-12. The ester linkage is also important in the formation of nucleotide chains in nucleic acids such as deoxyribonucleic acid (DNA) (see "Nucleic Acids").

Ethers. Ethers have the general formula of R—O—R, in which two groups are attached to oxygen. The R groups may be identical or different. The ethers can be produced in the laboratory by a dehydration reaction involving two molecules of an alcohol:

$$R{-}OH + HO{-}R \xrightarrow{H_2SO_4} R{-}O{-}R + H_2O$$

The ether linkage has some important implications in the bacteria. A group of bacteria recently classified as the archaebacteria possess ether linkages in their membrane lipids instead of the ester linkage found in all other bacteria. This group of bacteria is discussed in Chapter 4, and the importance of the ether linkage is discussed in Chapter 15.

Carboxylic acids. The carboxylic acid group (COOH) on the hydrocarbon consists of a carbonyl group (C═O) and a hydroxyl group (OH). Carboxylic acids are weak acids and may exist as monocarboxylic, dicarboxylic, or tricarboxylic acids (Figure 2-13). Many of these organic acids are produced as end products during metabolism or as intermediates that can be used in the biosynthesis of amino acids, which are the building blocks of proteins, as well as some long-chained fatty acids that are components of lipids.

Anhydrides. The anhydride linkage is formed by the elimination of water from two molecules of acid (Figure 2-14). The acids may be organic or inorganic. One important aspect of the anhydride linkage is that it releases a great deal of energy when the bond is broken. The symbol (~) is used to denote the high-energy bond (see "Phosphorus").

Nitrogen

Nitrogen, which contains five electrons in its outermost shell, can exist in various states of oxidation ranging from −3 to +3. It can substitute very readily for hydrogen on organic compounds to form an amino group ($—NH_2$), which helps to give polarity to the molecule. Nitrogen is available to microorganisms in many inorganic and organic forms. Important inorganic forms are nitrate (NO_3^-), nitrite (NO_2^-), and nitrogen gas (N_2). Organic nitrogen compounds include amino acids and some of the macromolecules from which they are totally or partially derived, such as proteins and nucleic acids. The most important basic organic nitrogen compound is the amino acid, which possesses at least two ionizable groups—an amino group ($—NH_2$) and a carboxyl group (—COOH)—and has the following basic formula:

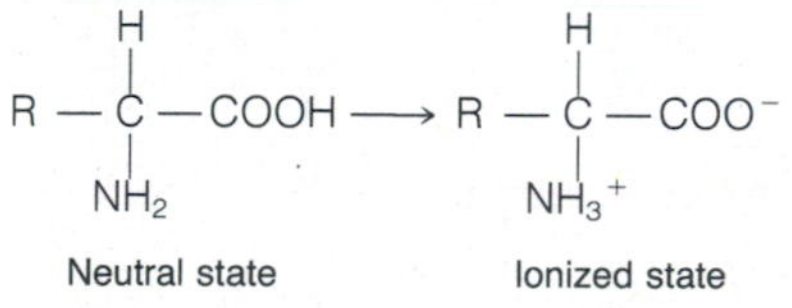

$$R—CH(NH_2)—COOH \longrightarrow R—CH(NH_3^+)—COO^-$$

Neutral state — Ionized state

The R group refers to the different constitutents that may appear attached to the basic amino acid molecule. There are at least 20 different amino acids in biological systems, each with a different R group (Table 2-3). All of the amino acids except glycine contain at least one asymmetric carbon atom and possess two or more isomers.

The amino acids are divided into D- and L-families based on the position of the NH_2 group on the α-asymmetric carbon atom (carbon atom closest to the COOH group). When the COOH group is written at the top and the carbon chain downward, the amino acid belongs to the D-series if the NH_2 group is on the right side (Figure 2-15). Conversely, if the NH_2 group is on the left side, the amino acid belongs to the L-series.

One of the most important reactions involving amino acids is their ability to interact with one another to form an amide linkage, or ***peptide bond*** (Figure 2-16). Peptide bonds are formed by the removal of one molecule of water during the interaction of the carboxyl group of one amino acid with the amino group of the other amino acid. The addition of a third and then fourth amino acid, etc., results in the formation of a tripeptide, tetrapeptide, etc., and eventually polypeptides containing as many as 300 amino acids. The peptide bond (C—N) is a much shorter bond than the normal single covalent bond and therefore behaves as a double bond. This means that there can be no twisting or rotation around the bond; however, the other adjacent bonds have normal single covalent bond lengths and can be twisted. Later, in our discussion of proteins, we will see how twisting of the molecule can produce a particular shape.

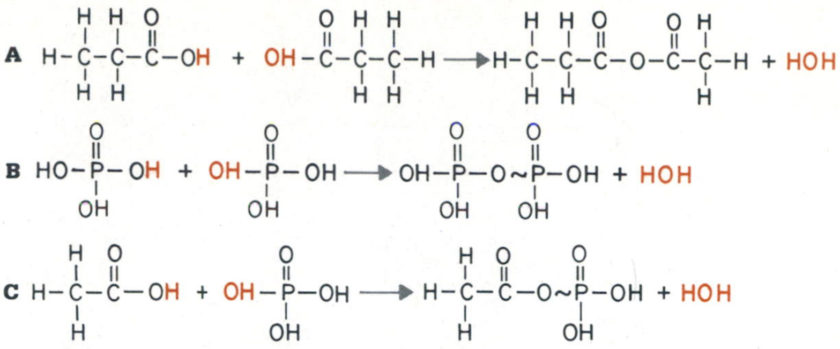

Figure 2-14 Anhydride formation between organic acids, **A;** inorganic acids, **B;** and organic and inorganic acids, **C.**

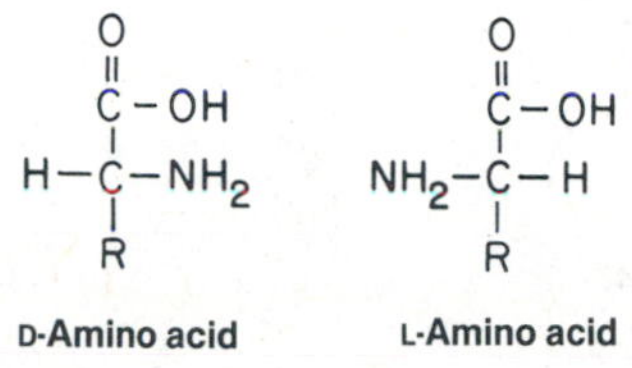

Figure 2-15 Arrangement of amino group in D- and L-amino acids.

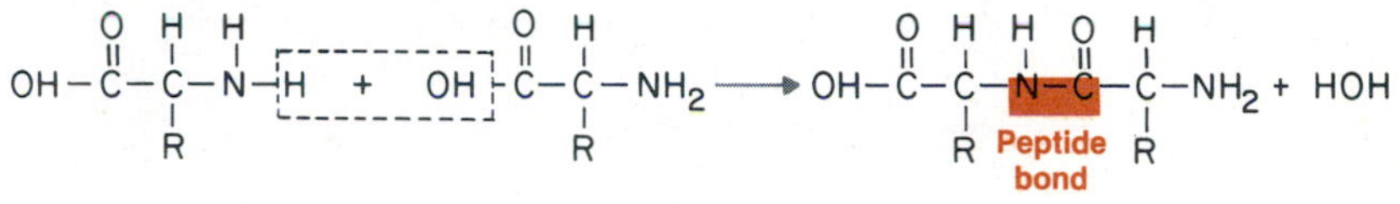

Figure 2-16 Formation of amide (peptide) bond between two amino acids.

Sulfur

Sulfur is similar to oxygen chemically, but only a few organic compounds in biological systems contain sulfur. The most important of these are the amino acids cysteine and methionine (Table

Table 2-3 Twenty amino acids found universally in proteins

SIDE-CHAIN (R GROUP) CHARACTERISTIC	AMINO ACID	CHEMICAL STRUCTURE
Aliphatic, nonpolar	Glycine	$H-CH(NH_2)-COOH$
	Alanine	$CH_3-CH(NH_2)-COOH$
	Valine	$(CH_3)_2CH-CH(NH_2)-COOH$
	Leucine	$(CH_3)_2CH-CH_2-CH(NH_2)-COOH$
	Isoleucine	$CH_3-CH_2-CH(CH_3)-CH(NH_2)-COOH$
Alcoholic, aliphatic, and aromatic	Serine	$HO-CH_2-CH(NH_2)-COOH$
	Threonine	$CH_3-CH(OH)-CH(NH_2)-COOH$
Aromatic	Tyrosine	$HO-C_6H_4-CH_2-CH(NH_2)-COOH$
	Phenylalanine	$C_6H_5-CH_2-CH(NH_2)-COOH$
	Tryptophan	(indole ring: benzene fused with $C=CH-NH$) $-CH_2-CH(NH_2)-COOH$

Table 2-3 Twenty amino acids found universally in proteins

SIDE-CHAIN (R GROUP) CHARACTERISTIC	AMINO ACID	CHEMICAL STRUCTURE
Carboxylic (acidic)	Aspartic	$HO(O)C{-}CH_2{-}CH(NH_2){-}C(O)OH$
	Glutamic	$HO(O)C{-}CH_2{-}CH_2{-}CH(NH_2){-}C(O)OH$
Amine bases (basic)	Lysine	$NH_2{-}CH_2{-}CH_2{-}CH_2{-}CH_2{-}CH(NH_2){-}C(O)OH$
	Arginine	$NH_2{-}C({=}NH){-}NH{-}CH_2{-}CH_2{-}CH_2{-}CH(NH_2){-}C(O)OH$
	Histidine	$HC{=}C(CH_2{-}CH(NH_2){-}C(O)OH){-}NH{-}CH{=}N$ (ring)
Sulfur containing	Cysteine	$HS{-}CH_2{-}CH(NH_2){-}C(O)OH$
	Methionine	$CH_3{-}S{-}CH_2{-}CH_2{-}CH(NH_2){-}C(O)OH$
Amides	Asparagine	$NH_2(O)C{-}CH_2{-}CH(NH_2){-}C(O)OH$
	Glutamine	$NH_2(O)C{-}CH_2{-}CH_2{-}CH(NH_2){-}C(O)OH$
Imino	Proline	$CH_2{-}CH_2{-}CH_2{-}NH{-}CH{-}C(O)OH$ (ring: CH_2—CH_2, CH_2—N(H)—CH)

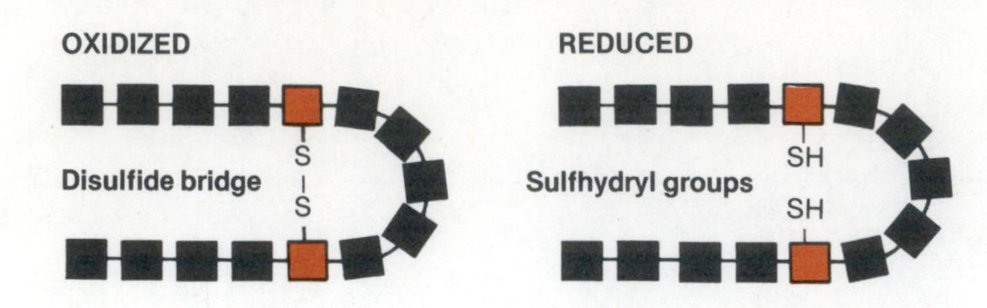

Figure 2-17 Polypeptide containing sulfur amino acids in oxidized state and reduced state. Each box represents an amino acid; colored boxes are cysteine.

$$OH-\overset{\overset{O}{\|}}{\underset{\underset{OH}{|}}{P}}-OH \qquad R_1-O-\overset{\overset{O}{\|}}{\underset{\underset{OH}{|}}{P}}-O-R_2$$

Phosphoric acid — Phosphate ester

Figure 2-18 Structure of phosphoric acid (H_3PO_4). Replacement of two hydrogen atoms with organic groups results in formation of phosphate ester.

2-3), which are found in some proteins. Sulfur in organic compounds is in the form of sulfhydryl groups (—SH). The sulfhydryl group is particularly important in the enzyme molecule because many enzymes maintain their activity only when they possess free sulfhydryl groups. Oxidation of sulfhydryl groups inactivates the enzyme:

$$Enz\begin{matrix}/SH\\ \backslash SH\end{matrix} \underset{2H}{\overset{2H}{\rightleftharpoons}} Enz\begin{matrix}/S\\ \backslash S\end{matrix}$$

Active enzyme — Inactive enzyme

Sulfhydryl groups are also important in the structure of many proteins where cross-linkages within a polypeptide as well as between polypeptides contribute to chain folding. Oxidation of free sulfhydryl groups in the polypeptide strands leads to the formation of a disulfide (S—S) group (Figure 2-17).

Phosphorus

Phosphorus is an important element in biochemistry, since it takes part in many reactions in which energy is either expended or produced. Phosphorus occurs as derivatives of phosphoric acid (H_3PO_4), which is depicted in Figure 2-18. One or two of the hydrogen atoms in the phosphoric acid molecule may be replaced by organic radicals to produce phosphate esters. One important function of the phosphate ester is to act as bridge linking the various nucleoside residues of nucleic acids such as ***deoxyribonucleic acid (DNA)*** and ***ribonucleic acid (RNA)***. Derivatives of phosphoric acid are also capable of interacting to form anhydrides (Figure 2-14), linkages that possess high energy. One of the most important organic molecules possessing the anhydride linkage is adenosine triphosphate (ATP), a molecule that is involved in energy exchange. Hydrolysis of the energy-rich anhydride bond occurs easily because of the repulsion between negative charges on the phosphate groups in the ionized state. The energy released is available for many biological activities, such as movement and metabolism.

CHARACTERISTICS OF THE MAJOR BIOLOGICAL MOLECULES

Carbohydrates

Carbohydrates are so called because their general formula can be represented as $C_n(H_2O)_n$. The most important example is glucose, where $n = 6$, or $C_6H_{12}O_6$. Chemically, carbohydrates may be defined as carbonyl derivatives of polyhydric alcohols. The smallest carbohydrates are called ***monosaccharides,*** which can be linked together to produce more complex molecules called ***polysaccharides.***

Monosaccharides. The most important monosaccharides in biological systems consist of three to seven carbons and can be divided into classes called, respectively, trioses, tetroses, pentoses, hexoses, and heptoses (Table 2-4). Each of these classes can be represented by more than one member, with each member having different chemical properties based on the positions of

Table 2-4 Principal monosaccharides found in microbial systems

NO. OF CARBONS IN MONOSACCHARIDE	NAME OF MONOSACCHARIDE
3	Glyceraldehyde
4	Erythrose
5	Ribose, arabinose, xylose, deoxyribose, xylulose
6	Glucose, mannose, galactose, fructose
7	Sedoheptulose

their hydroxyl groups. Many different hexoses have the general formula $C_6H_{12}O_6$, but the position of their hydroxyl groups differ (Figure 2-19). The carbonyl group can appear either at carbon 1 or carbon 2 of the monosaccharide. The monosaccharide is an aldehyde, if the carbonyl group is on carbon 1, and such monosaccharides are referred to as ***aldoses.*** The monosaccharide is a ketone if the carbonyl group is at carbon 2, and such monosaccharides are called ***ketoses.*** Glucose, for example, is an aldose, but fructose is a ketose (Figure 2-19). The five carbon sugars, which have the general formula $C_5H_{10}O_5$, are especially important as components of the nucleic acids to be described later.

The monosaccharides are often depicted in a linear chain form, but in solution they are usually in the ring form. Ring formation is the result of interaction between the highly reactive carbonyl group and the hydroxyl group of the next to last carbon atom. An example of a six-membered monosaccharide ring structure is depicted in Figure 2-20.

Sugars, like the amino acids, can be further divided into either the D- or L-series, based on the position of the hydroxyl group next to the last carbon atom on the chain. For D-sugars, the hydroxyl group is on the right. In general, the predominant family of sugars in nature are the D-sugars.

D-Glucose D-Galactose D-Mannose D-Fructose

Figure 2-19 Structure of some monosaccharides having general formula $C_6H_{12}O_6$ (hexoses). Numbering system for carbons is indicated in color

Linear form Ring form, all carbons shown Abbreviated ring form

Figure 2-20 Linear and chain forms representing sugar D-glucose.

Disaccharides. Carbohydrates composed of two molecules of hexose-monosaccharides are called ***disaccharides.*** The most frequently encountered disaccharides are lactose, sucrose, and maltose (Figure 2-21), which have the same molecular formula: $C_{12}H_{22}O_{11}$. All are formed by the union of two monosaccharides with the elimination of water. Lactose is composed of glucose and galactose; sucrose, of glucose and fructose; and maltose, of glucose and glucose. These disaccharides are found normally in nature and can be hydrolyzed into their respective monosaccharides by many groups of microorganisms. Note that each disaccharde contains a molecule of glucose, the carbohydrate most readily catabolized by microorganisms.

Polysaccharides. Polysaccharides consist of many units of monosaccharides linked together to produce high molecular weight compounds ranging in size from 10,000 to 4,000,000 daltons.*

* A dalton is an atomic mass unit (amu) and is defined as one twelfth the mass of the most commonly occurring form of carbon, namely carbon 12(^{12}C). Carbon has an atomic mass unit of 12.000 daltons; therefore 1 dalton equals 1 amu.

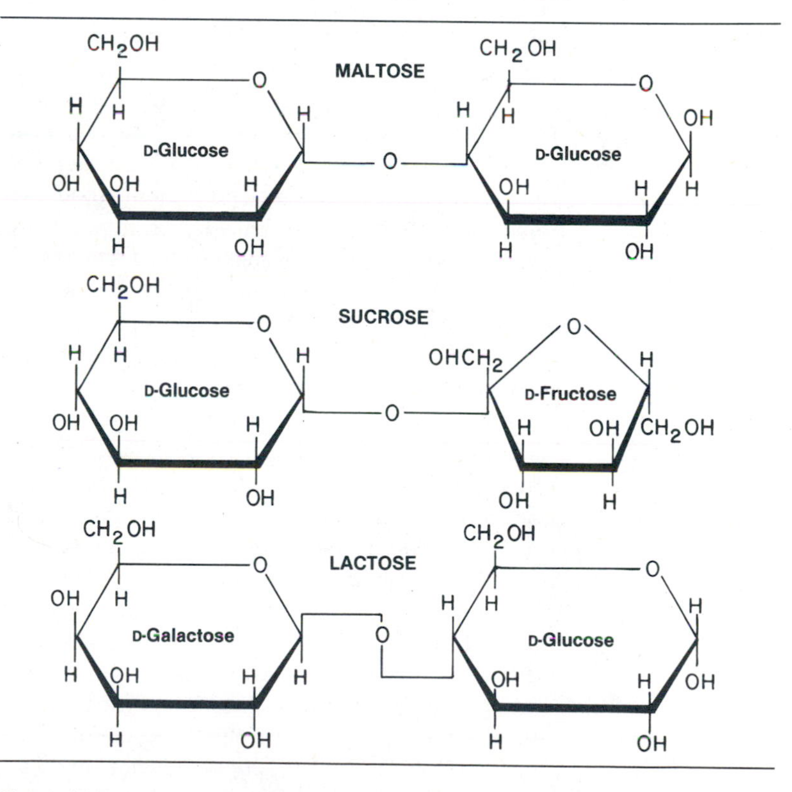

Figure 2-21 Structure of disaccharides: maltose, sucrose, and lactose.

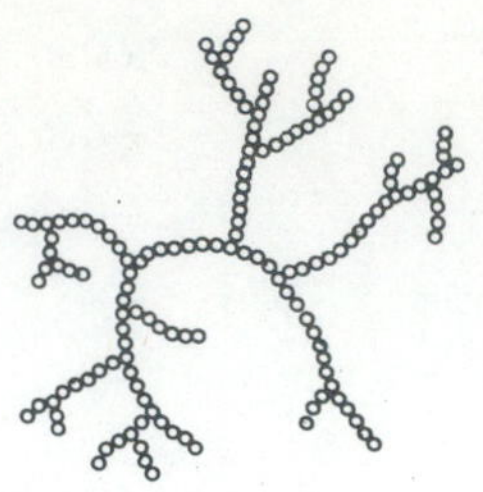

Figure 2-22 Branching in polysaccharides such as glycogen. Each circle represents glucose unit.

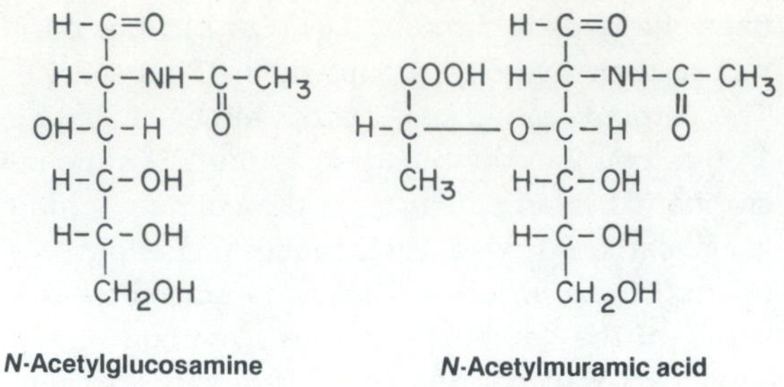

N-Acetylglucosamine *N*-Acetylmuramic acid

Figure 2-23 Structure of amino sugars *N*-acetylglucosamine and *N*-acetylmuramic acid, found in cell wall of bacteria.

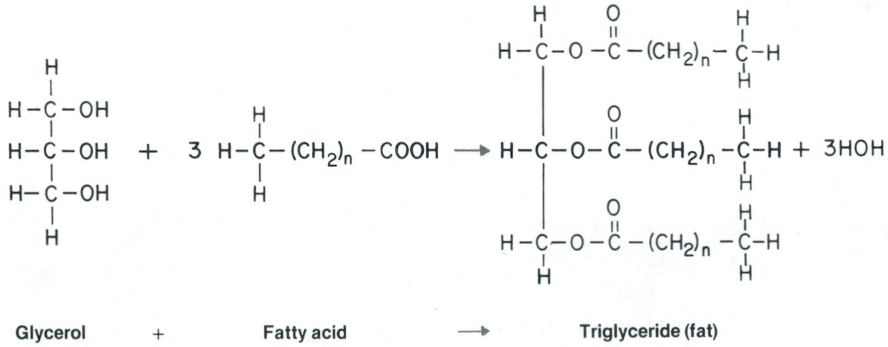

Glycerol + Fatty acid → Triglyceride (fat)

Figure 2-24 Triglyceride formation, or reaction between alcohol and fatty acid.

Some of the major polysaccharides are starch, pectin, cellulose, glycogen, and dextran, each of which consists of glucose units linked in such a way that each can be chemically distinguished from one another. Most linkages are between carbon 1 of one glucose unit and either carbon 4 (1-4 linkage), carbon 6 (1-6 linkage), or, infrequently, carbon 3 of the adjacent glucose unit. These linkages result in extensive branching of the polysaccharide (Figure 2-22). Microorganisms, as well as other organisms, use polysaccharides to make cell walls and as fuel reserves.

Many polysaccharides are composed of monosaccharides and noncarbohydrate subunits, such as proteins and lipids, and have been termed ***heteropolysaccharides.*** The carbohydrate component of many of the heteropolysaccharides is an amino sugar in which an amino group replaces a hydroxyl group at carbon 2. The addition of acetyl and lactic acid groups to one type of amino sugar produces *N*-acetylglucosamine and *N*-acetylmuramic acid (Figure 2-23). They are particularly important because they form the backbone of bacterial cell walls called peptidoglycan and are discussed in Chapters 4 and 7. Chitin is a component of fungal cell walls and is composed of *N*-acetylglucosamine (see Chapters 5 and 8).

Carbohydrates bound to proteins are called glycoproteins and can be found in most organisms. Examples of glycoproteins are blood group antigens, certain hormones, the antiviral substance interferon, and most cell surfaces where the glycoprotein consists primarily of polysaccharide plus small amounts of protein. Certain viruses contain a glycoprotein envelope that contributes to the disease process but that can also be used in the preparation of vaccines to prevent infection.

Lipopolysaccharides are also important macromolecules in biological systems, particularly bacteria, but they are discussed with the lipids.

Lipids

Lipids (fats and fatlike substances) are a very diverse group of organic compounds. They are generally classified as esters or potential esters of fatty acids that are soluble in nonpolar solvents (ether, chlorofrom, benzene, etc.) but insoluble in water. They may be further classified as simple or compound. ***Simple lipids*** are esters of fatty acids and alcohol. The triglycerides are the most important class of simple lipids (the other class is made up of waxes, which are more important in plants and animals) and are esters of glycerol and fatty acids. Glycerol is a derivative of the simplest sugar, glyceraldehyde, and during lipid formation in the cell each hydroxyl group is esterified to a fatty acid (Figure 2-24) that may be 10 to 20 carbons in length. Each fatty acid may be saturated or unsaturated. Simple lipids function primarily as a source of energy because the degradation of large numbers of —CH_2 chains of the fatty acids releases intermediates that lead to the synthesis of ATP.

Compound lipids are also esters of fatty acids with alcohol but contain other groups as well, such as phosphoric acid (Figure 2-25), sulfate, or nitrogen bases. Some compound lipids are also attached to macromolecules such as proteins (lipoproteins) or polysaccharides (lipopolysaccharides). Lipopolysaccharide (LPS) is a component of the cell wall of certain bacteria and contributes to their toxicity during the disease process. Compound lipids function primarily as structural units within cell structures: the cell membrane and cell envelope. The most common compound lipids found in the microbial cell are phospholipids and lipoprotein. These molecules are characterized by the presence of polar as well as nonpolar groups. Polar groups in an aqueous solution have a tendency to orient themselves close to the water molecules and are called ***hydrophilic*** (loving water), whereas nonpolar groups are ***hydrophobic*** (fear water) and tend to huddle together and face away from water. Phospholipids possess a hydrophilic phosphate group and hydrophobic fatty acid chains. The arrangement of these groups and their amount in the membrane control the solubility of molecules that are transported across it (see Chapter 7). Another interesting aspect of the cell membrane is that it is held together by the hydrophobic nature of the R groups of its phospholipid and protein components. Hydrophobic bonds are normally weak bonds, but because of the large number of hydrophobic groups, they lend considerable stability to the cell membrane.

Nucleic acids

Nucleic acids are components of all living cells, as well as acellular viruses. They are large complex molecules that, on the basis of chemical differences, can be divided into two types: DNA and RNA. Each type is made up of two classes of nitrogen bases, called ***purines*** and ***pyrimidines,*** plus sugar phosphates.

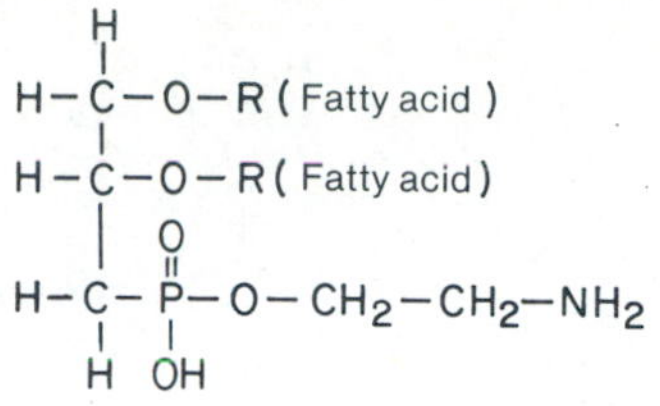

H–C–O–R (Fatty acid)
H–C–O–R (Fatty acid)
H–C–P(=O)(OH)–O–CH_2–CH_2–NH_2

Phospholipid

Figure 2-25 Structure of compound lipid.

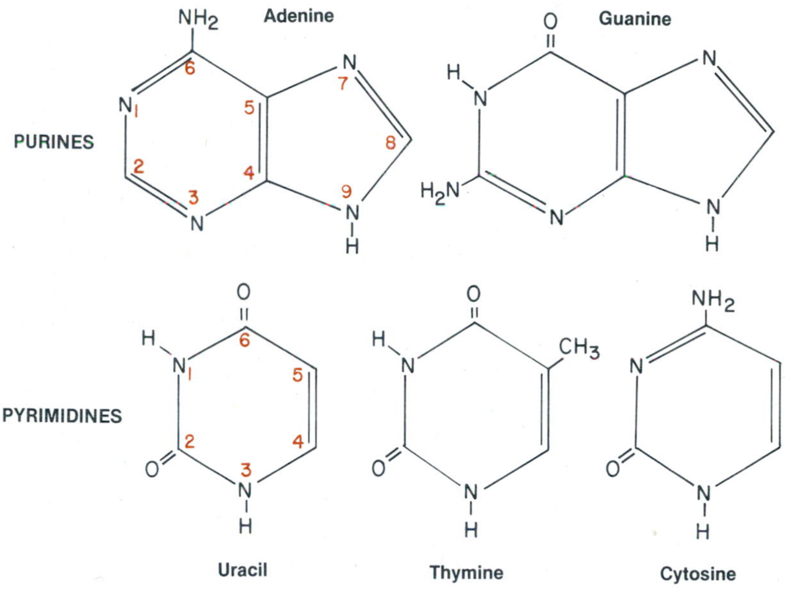

PURINES: Adenine, Guanine

PYRIMIDINES: Uracil, Thymine, Cytosine

Figure 2-26 Structure of purine and pyrmidine bases. Numbering system for atoms is in color.

The structure of the purine and pyrimidine bases is illustrated in Figure 2-26. The purines adenine and guanine are both found in DNA and RNA. The pyrimidines thymine and cytosine appear in DNA, and cytosine and uracil are present in RNA. Both groups of nitrogen bases possess keto and amino groups that, as we will see, can take part in hydrogen bonding. The carbohydrate components of the nucleic acids are D-ribose and D-2-deoxyribose. The latter sugar has an oxygen atom removed from the hydroxyl group on carbon 2 (Figure 2-27). ***Nucleosides*** are compounds that contain the pentose sugar bonded to a purine or pyrimidine base and are referred to as ribonucleosides or deoxyribonucleosides (Figure 2-28). They are more specifically named according to the base attached to them (Table 2-5). Attachment of phosphate either to carbon 5 or to carbon 3 of the ribose or deoxyribose moiety produces what is called nucleoside phosphates (Table 2-5). More

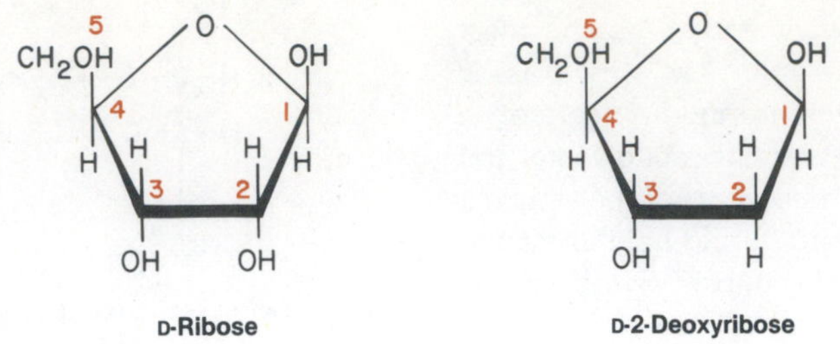

D-Ribose D-2-Deoxyribose

Figure 2-27 Structure of nucleic acid sugars D-ribose and D-2-deoxyribose. Numbering of carbons is in color.

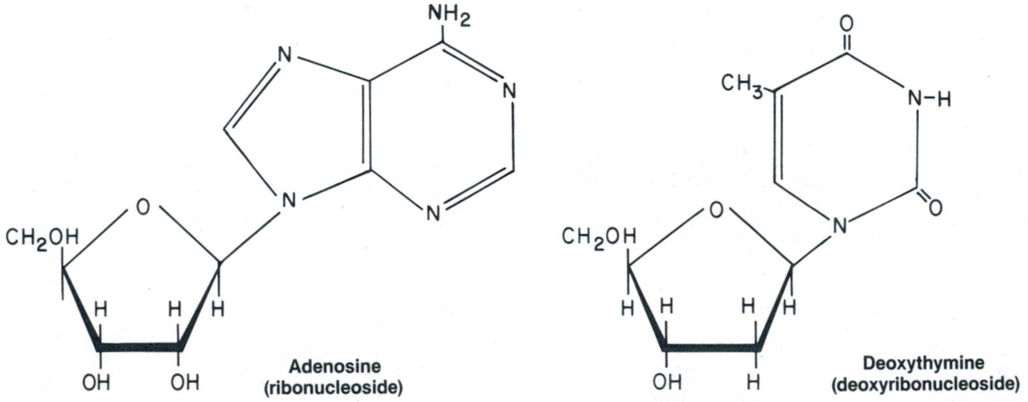

Adenosine (ribonucleoside) Deoxythymine (deoxyribonucleoside)

Figure 2-28 Structure of two nucleosides: adenosine and deoxythymine.

Table 2-5 Nomenclature of some nucleosides and nucleotides

	Nucleoside		Nucleotide	
Base	Ribose	Deoxyribose	Ribose	Deoxyribose
Adenine	Adenosine	Deoxyadenosine	5′-Adenylic acid or adenosine 5′-monophosphate	5′-Deoxyadenylic acid or deoxyadenosine 5′-monophosphate
Cytosine	Cytidine	Deoxycytidine	5′-Cytidylic acid or cytidine 5′-monophosphate	5′-Deoxycytidylic acid or deoxycytidine 5′-monosphosphate
Guanine	Guanosine	Deoxyguanosine	5′-Guanylic acid or guanosine 5′-monophosphate	5′-Deoxyguanylic acid or deoxyguanosine 5′-monophosphate
Uracil	Uridine	Deoxyuridine	5′-Uridylic acid or uridine 5′-monophosphate	5′-Deoxyuridylic acid or deoxyuridine 5′-monophosphate
Thymine	Thymidine	Deoxythymidine	5′-Thymidylic acid or thymidine 5′-monophosphate	5′-Deoxythymidylic acid or deoxythymidine 5′-monophosphate

than one phosphate may be attached to a nucleoside (a nucleoside diphosphate or nucleoside triphosphate), but not if the nucleoside is part of the internal structure of the nucleic acid molecule. The combination of sugar plus phosphate plus base is called a ***nucleotide.*** The arrangement of purines, pyrimidines, and sugar phosphates contributes to their primary as well as secondary structure.

Deoxyribonucleic acid

Primary structure. The DNA molecule is composed to two strands, each of which consists of individual nucleosides linked together via a phosphate ester linkage between the carbon 5′-hydroxyl of one nucleoside and the carbon 3′-hydroxyl of another nucleoside (Figure 2-29). The nucleosides in each strand may number from 1×10^4 to 1×10^8, depending on the species in which they are found. The molecular weight of the DNA molecule may range from 1×10^5 to 2×10^9 daltons.

Secondary structure. Most DNA molecules (except those in some viruses) are double stranded, whereas RNA is single stranded (except in a few viruses that possess double-stranded RNA). X-ray diffraction studies in the 1950s revealed that the two chains of the DNA are twisted around a central axis in the shape of a helix (Figure 2-30). There are 10 bases for each turn of the helix, and each base is separated by 0.34 nm. The bases that project out at right angles to the helix on one chain will hydrogen bond with specific bases on the other chain. The bonding is specific in that one type of purine on one strand pairs with a specific pyrimidine on the other strand, or adenine (A) pairs with thymine (T), and guanine (G) pairs with cytosine (C). There is a maximum of three hydrogen bonds between guanine and cytosine, but only two between adenine and thymine. This is called ***complementary base pairing,*** and because of it the molar ratio of A to T and G to C in the DNA molecule is 1:1. This complementarity exists only when the chains are antiparallel and one chain of the DNA appears to be upside down (Figure 2-29). The base at the top of the chain on the left in Figure 2-29 contains a free, unlinked 5′-phosphate, and the complementary base on the opposite chain contains a free 3′-hydroxyl. The chain on the left therefore runs in the 5′ to 3′ direction, whereas the chain on the right runs from 3′ to 5′, going from top to bottom.

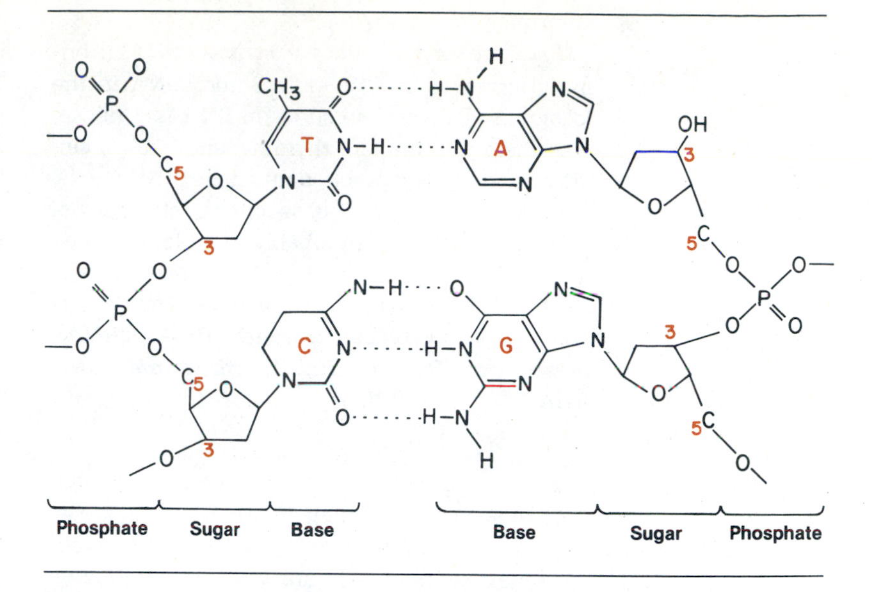

Figure 2-29 Base pair arrangements between adjacent nucleosides on portion of DNA molecule. Note that strands are antiparallel and sugar groups on right chain appear to be upside down.

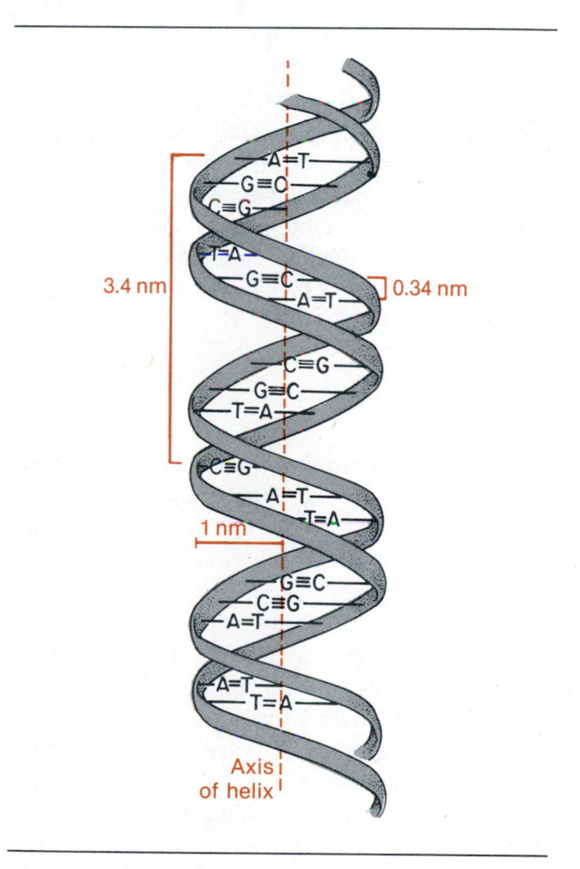

Figure 2-30 DNA helix showing distance between adjacent nucleotides and distance between each turn of helix.

Ribonucleic acid

Primary and secondary structures. The primary structure of RNA differs from DNA in the following respects: the sugar in RNA is D-ribose, uracil is substituted for thymine, and RNA is single stranded (except for some viruses).

Little is known of the secondary structure of RNA except that it is probably a randomly coiled unit.

The RNA found in nature may be of four different functional types: ***transfer RNA (tRNA), messenger RNA (mRNA), ribosomal RNA (rRNA),*** and ***viral RNA.***

1. Transfer RNA is a low molecular weight molecule containing 70 to 80 nucleotides. Its function in the cell is to carry amino acids that ultimately are used in the process of protein synthesis (see Chapter 10).
2. Ribosomal RNA is a component of those structural units in the cell called ribosomes. The ribosome, in addition to its rRNA component, also contains proteins. Ribosomes are the sites of protein synthesis.
3. Messenger RNA is a molecule that is complementary to one of the DNA strands. Its formation is the result of a transcription of the DNA's genetic information. The information in this mRNA will eventually be translated to a protein.
4. Viral RNA is the hereditary material in some viruses and in many instances acts like mRNA.

Proteins

Proteins are high molecular weight polymers consisting of amino acids linked by peptide bonds. Different polypeptides may contain nearly the same number of amino acids, but their activities may vary considerably. It is the number, as well as the sequence, of amino acids in the polypeptide that determines the latter's chemical properties and, hence, function in the cell. The number of totally different polypeptides that could be produced from an average polypeptide containing 200 amino acids would be 20^{200}, assuming that the number of different amino acids is 20. The order in which the amino acids appear in the polypeptide is a function of the hereditary material of the cell. The mechanism by which DNA is translated into specific amino acid sequences is discussed in Chapter 10.

Proteins can be divided into two types based on their composition and solubility. There are the ***simple proteins,*** which are composed only of amino acids, and the ***conjugated proteins,*** which contain a nonprotein or prosthetic group. The prosthetic group may be phosphate (phosphoproteins), carbohydrate (glycoproteins), lipid (lipoproteins), or nucleic acid (nucleoproteins). Proteins may also be divided into groups based on their function, for example, structural, enzymatic, hormonal, or viral. Most of our discussion will be concerned with enzymatic proteins.

Protein structure. The proteins assemble themselves into specific structures based on the interaction of their respective amino acids. This structural organization can be divided into four levels: primary, secondary, tertiary, and quaternary.

Primary structure. The primary structure of a protein is the linear sequence of amino acids. Only a few proteins have been totally sequenced, including insulin, which has 51 amino acid residues, hemoglobin, which has 574 amino acid residues; and the muscle protein myosin, which has 4500 amino acid residues. The peptide bond is the backbone of the polypeptide (Figure 2-31), which means that the backbone is the same for all polypeptides. The only difference in polypeptides is in the R groups projecting from the backbone. Each polypeptide has one free α*-amino group and one free α-carboxyl group, which by convention places the amino group on the left of the polypeptide and the free carboxyl on the right. The R groups projecting from the backbone determine the secondary, tertiary, and quaternary structure of the polypeptide.

Secondary structure. Proteins in solution assume a particular conformation or three-dimensional structure. Secondary structure is a consequence of the hydrogen bonding between the C═O and N—H groups of different peptide bonds. Two types of secondary structure are possible:

1. In the ***pleated sheet*** arrangement there is hydrogen bonding between two polypeptides to produce a sheetlike arrangement (Figure 2-32). Collagen is an example of the pleated sheet arrangement and is characteristic of fibrous proteins, which function as support structures (bone, muscle) in the cell.
2. The ***α-helix,*** which is due to intramolecular hydrogen bonding, is the most stable conformation for the polypeptide. Hydrogen bonding occurs between the C═O of one amino acid and the N—H of a peptide

* *Alpha* refers to the NH_2 and COOH groups that are part of the peptide bond and not part of the R groups.

bond three amino acids down the chain. Once all of the hydrogen bonds between peptide units have been made, the polypeptide assumes the conformation of the α-helix (Figure 2-33). The α-helix gives the appearance of a tightly coiled spring in which the R groups project away from the center of the molecule. Bends can appear in the helix when the α-helical structure is disrupted. Bends appear at those positions occupied by proline or hydroxyproline, the only amino acids in which there is no hydrogen available for bonding once the peptide has been produced.

Tertiary structure. The α-helical coils of the polypeptide can fold into various shapes depending on the interaction of various R groups and the number of proline or hydroxyproline residues. The spatial arrangement of the helical coils is called the tertiary structure. The R group interactions that contribute to the folding are:

1. Disulfide bonds. The disulfide bonds are the only covalent bonds involved in tertiary structure. They can bring together amino acids of the polypeptide that are normally far apart in terms of their position in the polypeptide chain (Figure 2-34).
2. Hydrogen bonds. Hydrogen bonding may occur between various groups: the carboxyl groups of glutamic or aspartic acid and the hydroxyl group of tyrosine.
3. Salt bonds. Ionizable groups such as NH_2 and COOH interact to produce a salt bridge (Figure 2-34), which is an electrostatic interaction.
4. Hydrophobic and hydrophilic interactions. The R groups projecting from the polypeptide chain are either hydrophilic or hydrophobic and can influence tertiary structure. Their interactions serve to arrange the polypeptide into the shape of a sphere, with the hydrophobic groups pushed together in the center of the sphere away from water (Figure 2-34) and the hydrophilic groups projecting away from the sphere and closest to water molecules in solution.

Quaternary structure. Many proteins consist of more than one polypeptide, and the manner in which these polypeptides interact with one another is called the quaternary structure. The R group interactions that contribute to intrachain tertiary structure are also important in interchain quaternary conformation. Disulfide bridges, salt bonds, and ionic bonds all contribute to the qua-

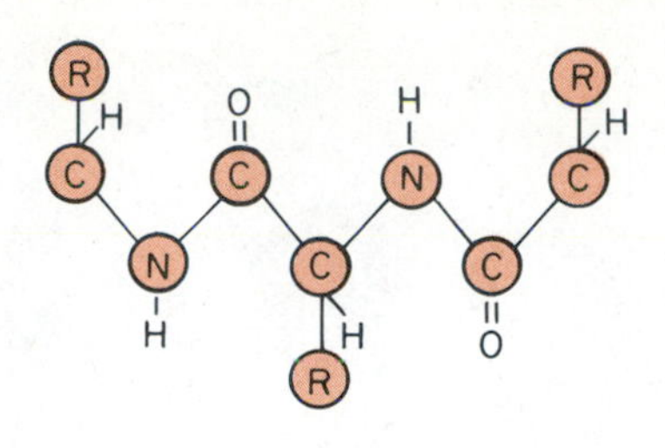

Figure 2-31 Primary structure of polypeptide showing peptide bond backbone and R groups projecting from it.

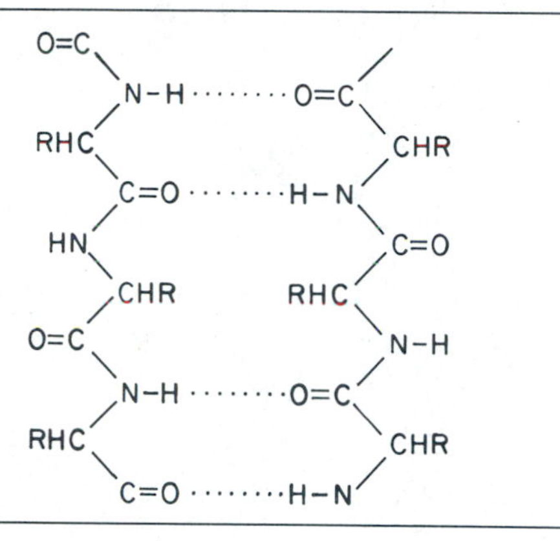

Figure 2-32 Example of pleated sheet secondary structure of certain fibrous proteins. Note that two polypeptides are involved.

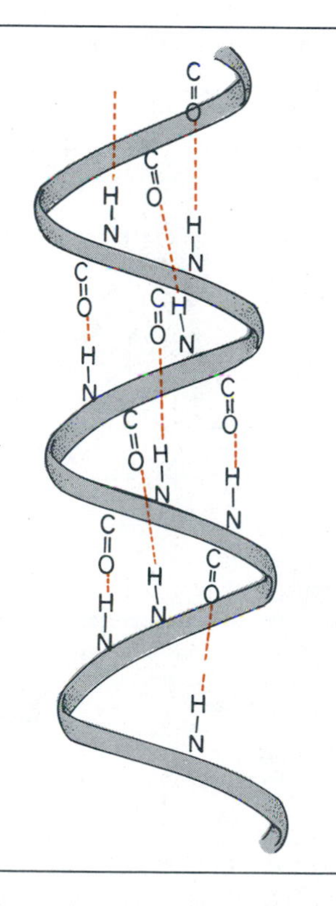

Figure 2-33 Secondary (α-helical) structure of polypeptide. Colored line indicates hydrogen bonding between one peptide group and another three peptides away.

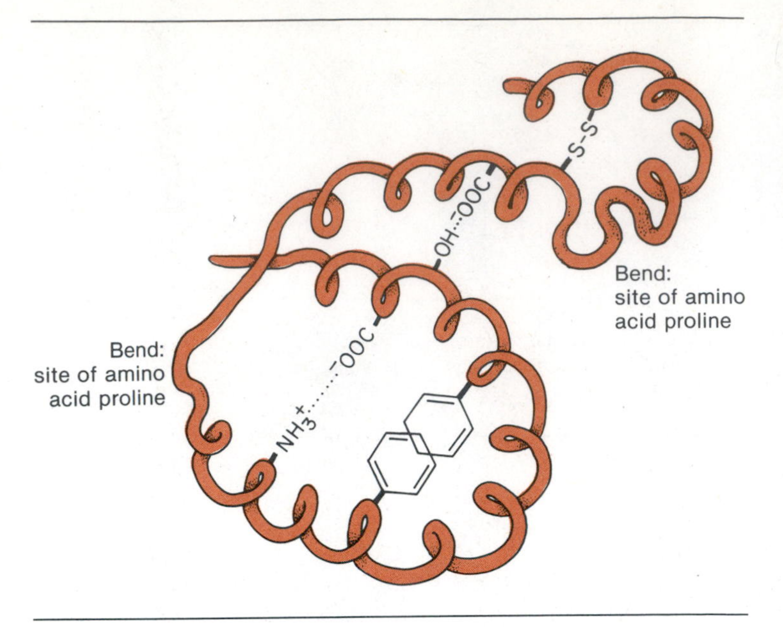

Figure 2-34 Secondary and tertiary structure of protein.

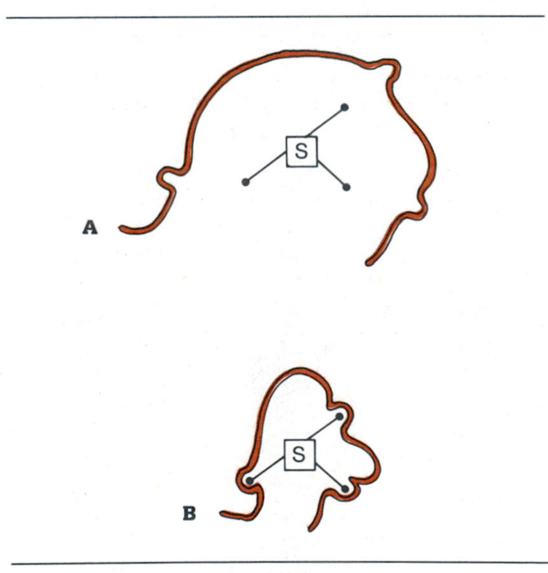

Figure 2-35 Multiple site arrangement for attachment of substrate to active site of enzyme. **A,** Substrate approaches active site of enzyme. **B,** Substrate interacts with amino acids at active site and causes conformational changes in enzyme.

ternary structure. The biological function of proteins, particularly enzymes, is dependent on the interaction of the polypeptide chains.

ENZYMES

Inorganic elements can enter into reactions that result in the formation of compounds, but most require extremes of either temperature, pH, pressure, or radiation. These conditions would be detrimental to living cells such as microorganisms. Chemical reactions occurring in the cell at normal temperatures in the absence of catalysts would not take place or would take place so slowly that before the cell could grow and reproduce, it would die of old age. To speed up these biochemical reactions, nature has provided the cell with enzymes, which act as ***catalysts.*** It is no wonder that most of the information locked up in the hereditary material of all living cells codes for the synthesis of enzymes.

Characteristics

All enzymes are globular proteins that vary in size from 10,000 to over 2,000,000 daltons. Their activity as catalysts is associated with their three-dimensional shape, which creates a site for binding of substrate, called the active site.

Active site. The active site of the enzyme consists of a group of amino acids that are involved in making contact with the substrate. The amino acids at the site may not be contiguous and can be from widely separated areas of the polypeptide. They may even be from different polypeptides. The active site is easily accessible to the substrate, but once substrate is bound, the conformation of the enzyme is believed to change in order to bring the active site–amino acids in closer proximity to the substrate.

Changes in the conformation of an enzyme can lead to alterations at the active site and loss of catalytic activity. These conformational changes, whether caused by high temperature, variations in pH, or other chemical or physical means, is called denaturation. Denaturation is therefore a bond-breaking process (except for the peptide bond) in which the secondary, tertiary, or quaternary structure of the protein has been affected.

Enzyme specificity. The extent of specificity between enzyme and substrate can vary widely among different enzymes. Some enzymes exhibit steric specificity. Recall that D-sugars and L-amino acids are the most predominant types of

Table 2-6 Vitamins and their coenzyme forms

VITAMIN	COENZYME FORM	VITAMIN	COENZYME FORM
Riboflavin (vitamin B_2)	Flavin adenine dinucleotide (FAD) and its reduced form ($FADH_2$); flavin mononucleotide (FMN)	Thiamine (vitamin B_1)	Thiamine pyrophosphate (TPP)
Nicotinic acid (niacin)	Nicotinamide adenine dinucleotide (NAD) and its reduced form ($NADH_2$); nicotinamide adenine dinucleotide phosphate (NADP) and its reduced form ($NADPH_2$)	Pyridoxine (vitamin B_6)	Pyridoxal phosphate; pyridoxamine phosphate
		Folic acid	Tetrahydrofolic acid
		Biotin	Biotin
		Pantothenic acid	Coenzyme A

Table 2-7 Classes of enzymes and examples of the reactions they catalyze*

CLASS OF ENZYME	EXAMPLE AND TRIVIAL NAME	SYSTEMATIC NAME	REACTION CATALYZED
Oxidoreductase	D-Lactate dehydrogenase	D-Lactate: NAD oxidoreductase	D-Lactate + NAD → Pyruvate + $NADH_2$
Transferase	Acetate kinase	ATP: acetate phosphotransferase	ATP + Acetate → ADP + Acetylphosphate
Hydrolase	Lipase	Glycerol ester hydrolase	Triglyceride + H_2O → Diglyceride + Fatty acid
Lyase	Oxalate decarboxylase	Oxalate carboxylyase	Oxalate → Formate + CO_2
Isomerase	Alanine racemase	Alanine racemase	L-Alanine → D-Alanine
Ligase	Methionyl tRNA synthetase	L-Methionine: tRNA ligase (AMP)	ATP + L-Methionine + tRNA → AMP + PP + L-Methionyl tRNA

*Examples are those characteristically found in microbial cells.

stereoisomers in nature. The ability of enzymes to distinguish these differences has resulted in a hypothesis suggesting that there are multiple sites (a maximum of three) for attachment of substrate to enzyme (Figure 2-35). Other enzymes recognize particular groups (amino, carboxyl, ester, peptide, etc.), and any alterations in nonspecific areas of the substrate may have no effect, or minor effects, on enzyme activity. Most enzymes exhibit a specificity that is dependent not only on the catalyzed groups of the substrate but on other regions of the substrate as well. These accessory regions may be required not only for binding to the enzyme but for proper orientation of the substrate at the active site.

Enzyme cofactors. Many enzymes require additional inorganic or organic components, referred to as ***cofactors,*** in order to exhibit catalytic activity. The most important of the cofactors are the organic ***coenzymes,*** without which no catalytic activity occurs (Table 2-6). Most coenzymes are loosely attached to the enzyme, but some organic groups are covalently bound and are called prosthetic groups. Inorganic cofactors include metal ions, which are usually bound by weak bonds (seldom covalent) to specific amino acids at the active site. Some important metal cofactors are magnesium, manganese, cobalt, iron, and molybdenum. When the enzyme protein, called the ***apoenzyme,*** is combined with its cofactor, the active complex is called a ***holoenzyme.***

Classification. Enzymes, for many years, were given names that did not depict the type of reaction in which they were involved. A commission on nomenclature in 1961 declared that biochemical reactions were catalyzed by six general types of enzymes. Various classes, within each type, which represented more specific enzyme activities, were also introduced. The names of all enzymes end in *ase*. Listed below are the six major groups of enzymes (Table 2-7):

1. Oxidoreductases are involved in oxidation-reduction reactions. Some of the enzymes of

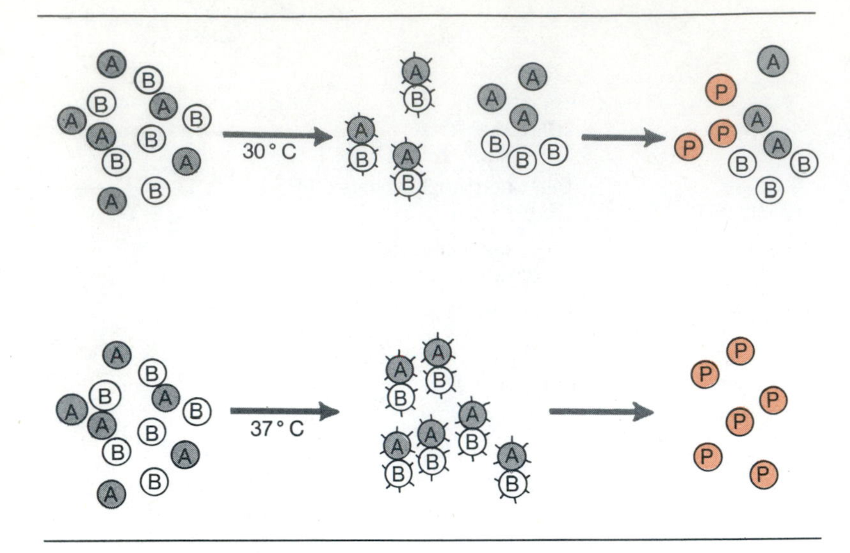

Figure 2-36 Effect of temperature on chemical reaction. Two reactants, *A* and *B*, collide at temperatures of 30° C and 37° C. Only three collisions per unit time occur at 30° C, but this number doubles at 37° C; therefore product *(P)* also doubles.

this group remove hydrogen from a reactant and pass it to another reactant. These are called dehydrogenases. Other enzymes of this group, like the oxidases, incorporate molecular oxygen into one of the reactants.

2. Transferases transfer one group of atoms from one molecule to another. They may transfer an amino group, acetyl group, or phosphate group.
3. Hydrolases bring about the hydrolysis of molecules, such as proteins, carbohydrates, lipids, etc., into their respective monomeric units. Protein hydrolases, for example, cleave peptide bonds and release amino acids.
4. Lyases break covalent bonds and remove groups from a compound other than by hydrolysis. The removal of a carboxyl group or a water molecule from a compound are examples of the activity of lyases.
5. Isomerases bring about the rearrangement of groups of atoms within a molecule. The conversion of a D-amino acid to an L-amino acid is an example of the activity of isomerases.
6. Ligases (sometimes called synthetases) join two molecules together, and during the reaction there is the breakdown of ATP or a related energy molecule. One type of ligase binds an amino acid to a tRNA molecule.

General mechanism of enzyme activity

Some chemical reactions proceed spontaneously with the release of ***free energy.*** This energy, which can be measured in calories, is available to perform work within the cell. Unfortunately, most chemical reactions in biological systems do not proceed spontaneously. All living cells must extract energy from nutrients supplied to them so that it can be used for biosynthesis of molecules, transport of nutrients into the cell, replication, movement, and other microbial functions. In order for a quantity of free energy to be released for useful work from the reaction

$$A + B \longrightarrow C + D + \text{energy}$$

the potential energy locked up in the reactants A and B must be greater than the energy in products C and D. Energy must be put into the system for the reaction to go to completion with the formation of products C and D and energy. If the energy put into the system were in the form of an increase in temperature, some molecules of reactants A and B would acquire enough energy to collide, resulting in the formation of specific amounts of product. The higher the temperature applied to the system, the higher the fraction of activated molecules, the greater the number of collisions, and the greater the number of molecules of product formed (Figure 2-36). The energy used to activate the reactants is called the ***energy of activation.*** Biological systems can survive only within a limited temperature range, and the temperatures required to activate the reactants is so high that it would destroy the cell. Enzymes reduce the activating energy required to promote molecular collisions. Enzymes increase the rate at which a chemical reaction proceeds by acting as a site for the interaction of reactants (we should now call the reactants the ***substrates*** for the enzyme). Enzymes bring the substrates together in the form of an enzyme-substrate complex from which products will be formed.

Specific mechanisms of enzyme activity

The binding of substrate to the active site is due to the specific conformation of the polypeptide and the charges of the R groups of some of its amino acids. In addition, some of the metal ions that are required as cofactors also aid in the binding of substrate to enzyme. The enzyme catalyzed reaction can be represented as:

E	+	S	⟶	ES	⟶	E	+	P
Enzyme		Substrate		Enzyme-substrate complex		Enzyme		Product

Most chemical reactions do not go to completion, and in a reaction in which substrates A and B are converted to products C and D, an equilibrium is established; that is, the reaction is reversible. This can be expressed as:

$$A + B \xrightarrow{k_1} C + D \qquad \text{Forward reaction}$$

$$C + D \xrightarrow{k_{-1}} A + B \qquad \text{Reverse reaction}$$

where k_1 and k_{-1} are the rate constants for the forward and reverse reactions, respectively. When the two reactions are combined, we obtain:

$$A + B \underset{k_{-1}}{\overset{k_1}{\rightleftharpoons}} C + D$$

Applying this equation to that of an enzyme-catalyzed reaction, we obtain:

$$E + S \underset{k_{-1}}{\overset{k_1}{\rightleftharpoons}} ES \underset{k_{-2}}{\overset{k_2}{\rightleftharpoons}} P + E$$

Thus at equilibrium the energy released in the forward reaction equals the energy put into the system to reverse the reaction. Enzymes affect the rate at which equilibrium is established, and more free energy in the cell is realized when the forward reaction is favored. (Free energy is discussed in Chapter 13.)

Factors affecting enzyme activity

Enzyme concentration. In order to determine the effect of enzyme concentration on a reaction, the substrate must be in excess. In other words, the rate of the reaction is independent of the substrate (a zero order reaction or rate = k), and the product produced per unit of time is related directly to the enzyme concentration. Figure 2-37 illustrates that the formation of product is linear with time, and as the time doubles, so does the concentration of product. Depletion of the substrate results in a loss of zero order rates, and there is no longer a direct proportionality between enzyme concentration and enzyme activity. Figure 2-38 illustrates that between points *A* and *B* there is proportionality between enzyme concentration and product formed, but this does not occur between points *B* and *C*. What has happened between *B* and *C* is that only some of the active sites of the enzyme are saturated with substrate; consequently, less product is formed. This type of reaction is called a ***first-order reaction*** because the rate is proportional to the first power of the substrate concentration (rate = k[S]). Enzyme activity is best determined when the enzyme is the only limiting factor.

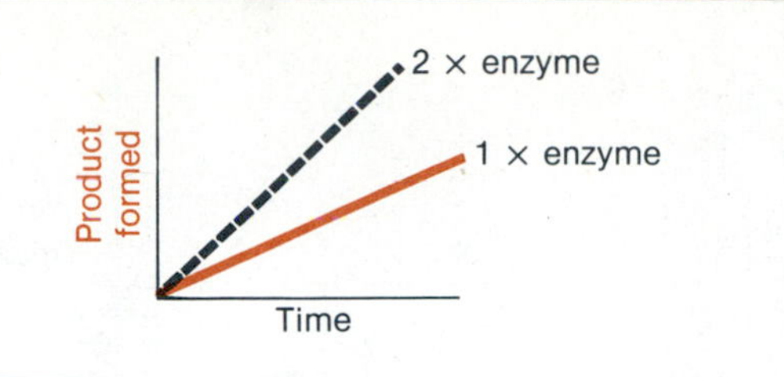

Figure 2-37 Effect of enzyme concentration on activity when substrate is in excess.

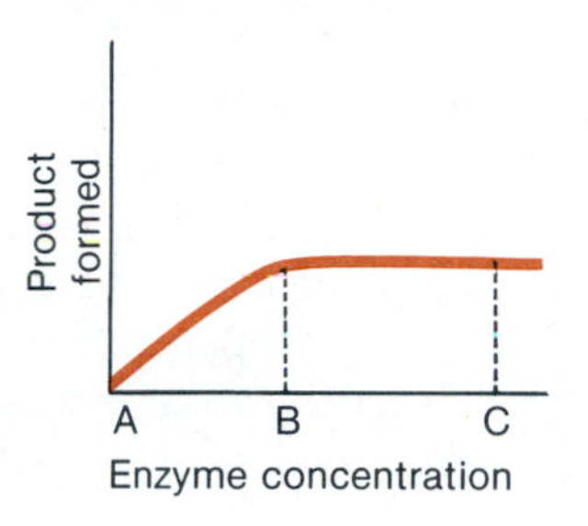

Figure 2-38 Enzyme activity vs. enzyme concentration. Enzyme activity is proportional to enzyme concentration between points *A* and *B*, indicating excess substrate in reaction. Enzyme activity is not proportional to enzyme concentration between points *B* and *C*, because substrate has become limiting.

Substrate concentration. It can be demonstrated that if the enzyme concentration is kept constant and the substrate concentration is increased gradually during a reaction, a point will be reached where the velocity will be maximum. Any further increase in substrate beyond this point has no effect on the velocity of the reaction (Figure 2-39, *A*). Thus at ***maximum velocity*** ***(V_{max})*** all of the enzyme is tied up as an enzyme-substrate complex. Michaelis derived equations to calculate enzyme activity. He proposed the ***Michaelis constant,*** or ***K_m,*** which is defined as the substrate concentration at one-half maximum velocity. K_m can be expressed as:

$$K_m = \frac{k_1 + k_2}{k_{-1}} = \frac{[S]\,V_{max}}{2}$$

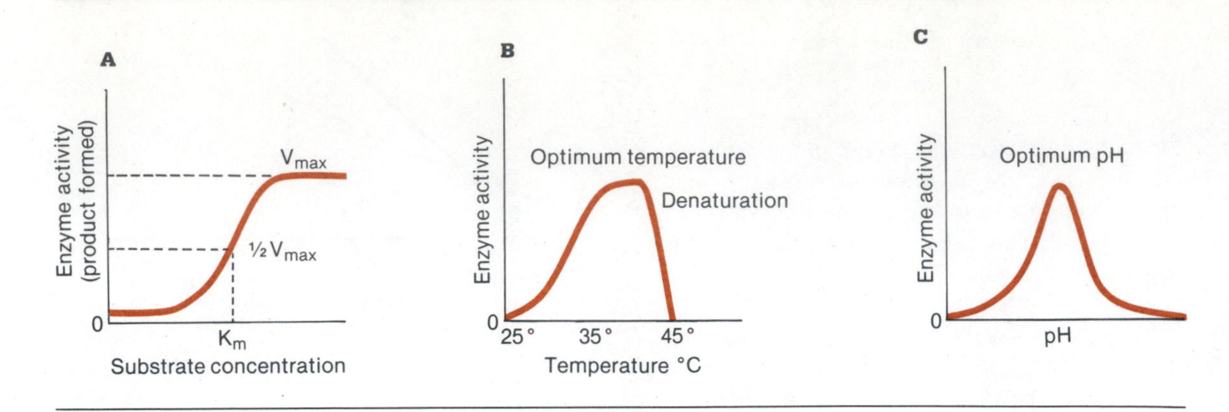

Figure 2-39 Effect of various factors on velocity of enzyme-catalyzed reaction.

The K_m is actually a dissociation constant for the enzyme-substrate complex. The faster the ES complex is converted to E and P, the less chance there is for ES to be converted to E + S in the reverse reaction. The reaction velocity for an enzyme-catalyzed reaction, based on K_m and substrate concentration, can be expressed as:

$$v_t = \frac{V_{max}\,[S]}{K_m + [S]}$$

where v_t equals the velocity of the reaction at any time during the reaction.

Michaelis constants have been determined for most of the commonly used enzymes. They are very descriptive and tell the experimenter some important characteristics of the enzyme. A small K_m says that only small amounts of substrate are necessary to saturate the enzyme, and maximum velocities are obtained at low substrate concentrations. A high K_m indicates that large amounts of substrate are necessary to obtain maximum velocities.

Temperature and pH. An increase in temperature, within certain limits, increases the rate of enzyme-catalyzed reactions. A 10° C rise in temperature will increase the activity of most enzymes by 50% to 100%. The optimum temperature for most enzymes is between 40° and 50° C. Above these temperatures the reaction rate decreases because of denaturation (Figure 2-39, *B*). Most enzyme determinations are carried out at temperatures between 25° and 37° C. If carried out at higher temperatures, the reaction would proceed so quickly that accurate measurements could not be made.

Enzymes are also affected by the hydrogen ion concentration, and the optimum ph may vary considerably from one enzyme to another. The enzyme pepsin, for example, has a pH optimum of 1.5, whereas pancreatic lipase is most active at a pH near 8.0. On either side of the optimum pH, enzyme activity decreases (Figure 2-39, *C*). An extremely high or low pH denatures most enzymes.

Inhibitors. Enzyme inhibitors are substances that alter catalytic activity. They may be of two types: competitive or noncompetitive.

Competitive inhibitors. Competitive inhibitors are compounds that are similar to the normal substrate. They bind to the enzyme and prevent access of normal substrate to the enzyme (Figure 2-40, *B*). The inhibitor is not converted to a product. The velocity of the enzyme-catalyzed reaction will be influenced by the concentration of inhibitor, the concentration of substrate, and the relative affinities of the substrate and inhibitor for the enzyme. Enzyme activity may be abolished if the inhibitor is bound very tightly to the active site. It is possible to reverse the effect of the competitive inhibitor by increasing the substrate concentration, if the inhibitor is not bound tightly to the enzyme. The mechanism of competitive inhibition has been used with certain drugs in the treatment of infectious disease. Sulfa drugs, for example, mimic the normal microbial substrate paraaminobenzoic acid, which is required for folic acid synthesis by disease-causing microorganism (see Chapter 27).

Noncompetitive inhibitors. Noncompetitive inhibition is an irreversible process in which the inhibitor combines with the enzyme at a point other than the active site (Figure 2-40, *C*). This interaction changes the conformation of the enzyme, and the active site is so distorted that it cannot be recognized by its normal substrate. The velocity of this reaction is influenced by the inhibitor concentration. If enough inhibitor is added so that all of the enzyme is complexed with it, there will be no normal enzyme activity, and addition of substrate will have no effect.

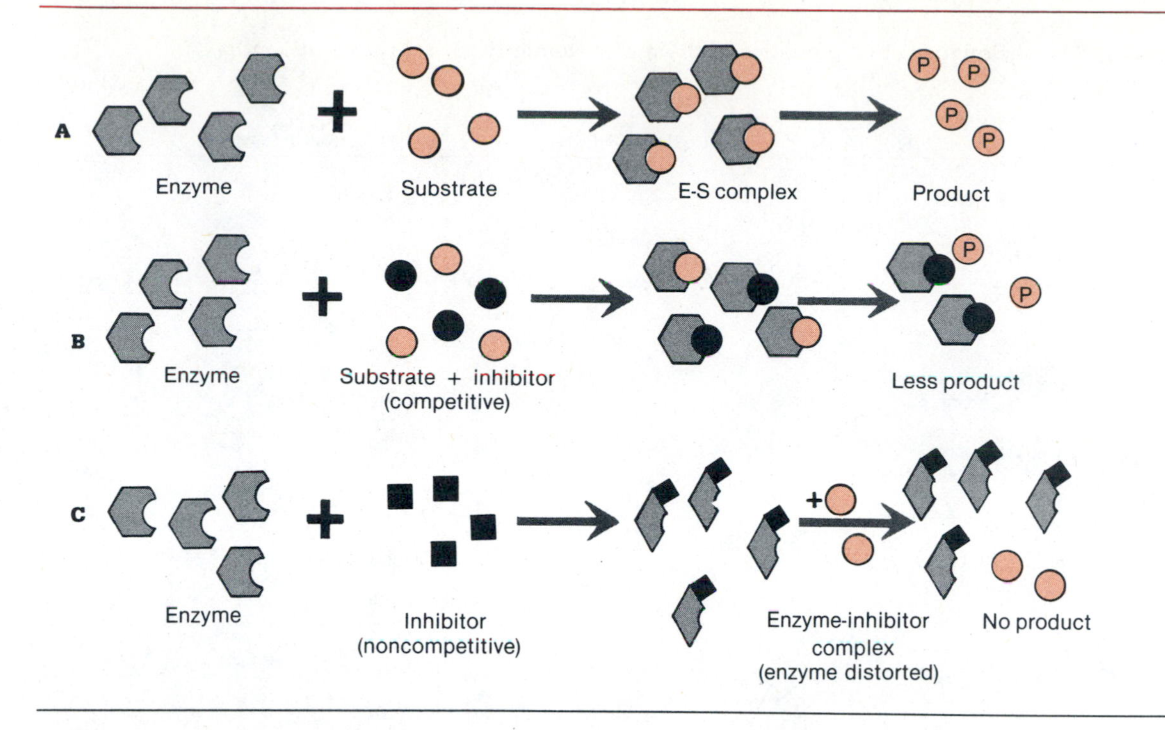

Figure 2-40 Effect of inhibitors on enzyme activity. **A,** No inhibitors. **B,** Competitive inhibitors. **C,** Noncompetitive inhibitors.

SUMMARY

The foundation of all life processes is built around organic compounds whose backbone element is carbon plus other light elements such as nitrogen, sulfur, oxygen, hydrogen, and phosphorus. Because their last orbital shells are incomplete, these elements are capable of sharing electrons and forming stable covalent bonds with other elements in the group. The groups or types of compounds formed by these elements are outlined in Table 2-8.

The major biological molecules of which hydrocarbons are a part are carbohydrates, lipids, proteins, and nucleic acids. Carbohydrates are primary sources of energy to the cell. Lipids are important components of biological membranes and are often combined with proteins or polysaccharides. Most proteins in the cell are enzymes composed of polypeptides whose amino acid sequences vary and give rise to conformational differences and hence differences in activity. These conformational differences are due to amino acid interactions involving disulfide bonds, hydrogen bonds, and salt bonds. Many proteins also form structural components of the cell and can be conjugated to other macromolecules. Nucleic acids are chains of purine and pyrimidine bases held together by sugar phosphates in the form of helical molecules. They are informational molecules, such as DNA and RNA, and are involved primarily in the transcription and translation of genes into enzymatic proteins.

Enzymes are the catalysts for biological reactions in the cell. They exhibit specific activities because of the arrangement of amino acids at the active site. Many factors affect the rate of enzyme-catalyzed reactions, including substrate concentration, cofactors and coenzymes, temperature, pH, and the presence of inhibitors. Enzyme activity can be altered by inhibitors mimicking the substrate (competetive inhibitors) and binding to the active site. Other inhibitors (noncompetitive) bind to sites other than the active site and in so doing alter the conformation of the enzyme and hence reduce its activity.

Table 2-8 Bonding characteristics of the major elements in organic compounds

ELEMENT	ELEMENT COVALENTLY BONDED TO	STRUCTURE OF BOND	TYPE OR CLASS OF COMPOUND
Carbon	Carbon	—C—C— —C═C— —C≡C—	Linear hydrocarbon
	Carbon	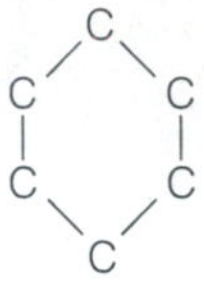 six-membered ring of C atoms (C—C single bonds)	Cyclic hydrocarbon
	Carbon plus nitrogen or sulfur	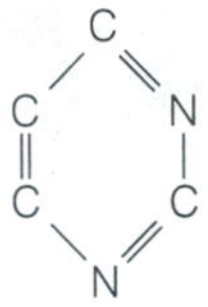 six-membered ring of C, C, N, C, N, C atoms (C═C, C═N and N═C double bonds)	Heterocyclic hydrocarbon
Oxygen	Carbon and hydrogen	C—OH	Alcohol
		H—C═O	Aldehyde
		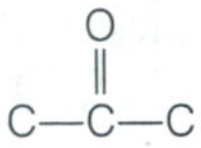 C—C(═O)—C	Ketone
		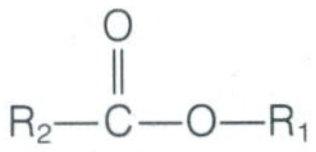 R_2—C(═O)—O—R_1	Ester
		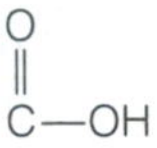 C(═O)—OH	Carboxylic acid
		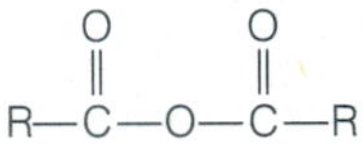 R—C(═O)—O—C(═O)—R	Anhydride
Nitrogen	Carbon and hydrogen	N—C— N═C— N≡C— C—NH_2	Amines (C—NH_2) important in amino acid formation, i.e.: 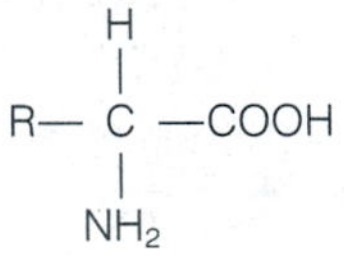 R—C(H)(NH_2)—COOH
Sulfur	Carbon and hydrogen	C—S—H	Sulfhydryl (SH) group found in compounds called thiols
	Sulfur	S—S	Disulfide bond formed between sulfur-containing amino acids
Phosphorus	Oxygen	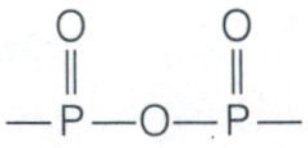 —P(═O)—O—P(═O)—	Anhydride

Study questions

1. What are the primary elements that are essential to all life processes? Are they also found in viruses, which are acellular?
2. What trace elements are important in microbial metabolism? What is their function?
3. Explain why carbon can form so many compounds.
4. Explain the formation of a covalent bond using the example of hydrogen gas. What other kinds of chemical bonds exist between the elements of organic compounds?
5. What is a hydrocarbon? What is meant by saturated hydrocarbons? By unsaturated hydrocarbons?
6. What are compounds called that have the same structural formula but different structural arrangements? What is their significance in biological systems? Can you think of any other advantages?
7. Write out the structural formulas for aldehyde, alcohol, ketone, ester, carboxylic acid, and anhydride. Can you name any biological compounds that contain these groups?
8. Is the hydrogen bond stronger or weaker than the covalent bond? What is the significance of hydrogen bonding in biological systems? Give one or more examples.
9. Write out the basic structural formula for an amino acid. For the D-form. For the L-form.
10. What is the relationship of water in peptide bond formation? In breakage of the peptide bond?
11. What amino acids contain the element sulfur? Does sulfur, of the sulfur-containing amino acids, serve any function in the protein molecule?
12. What is so unique about the anhydride bond, of which phosphorus is involved?
13. In what type of macromolecule is phosphorus most commonly found?
14. How is a D-sugar distinguished from an L-sugar?
15. Write the open-chain formula for the sugars glucose and fructose. What sugar predominates in most polysaccharides? Does this have any significance in terms of microbial metabolism?
16. Define the term *lipid*. What is a simple lipid? Where are they found in the microbial cell? What is their function?
17. In protein organization what is meant by primary, secondary, tertiary, and quaternary structures? What types of bonds are important in maintaining the tertiary structure? In maintaining the quaternary structure?
18. Many macromolecules are intimately associated with other macromolecules. What cell structures possess the following: lipoprotein, lipopolysaccharide, and nucleoprotein?
19. Name the various classes of ribonucleic acids found in the microbial cell. What is their function?
20. What is the function of an enzyme? What is meant by coenzyme? By cofactor? What is the active site of an enzyme?
21. In an enzymatic reaction what is meant by the energy of activation? How can it be increased or decreased?
22. Explain in simple terms or by a flow diagram the sequence of events in an enzyme-catalyzed reaction involving two reactants: A and B.
23. What is the Michaelis constant? How can it be determined graphically? Does it have any significance to the biologist?
24. Explain what happens to the velocity of an enzyme-catalyzed reaction as the temperature is increased, for example, from 37° to 80° C.
25. How do competitive inhibitors affect the active site of any enzyme? How do noncompetitive inhibitors do so?

Selected readings

Books

Christensen, H.N., and Palmer, G.A. *Enzyme kinetics: a learning program for students of biological and medical sciences.* W.B. Saunders Co., Philadelphia, 1967.

Cohen, A. Handbook of cellular chemistry (2nd ed.). The C.V. Mosby Co., St. Louis, 1979.

Conn, E.E., and Stumpf, P.K. *Outline of biochemistry* (4th ed.). John Wiley & Sons, Inc., New York, 1976.

Davies, J., and Littlewood, B.S. *Elementary biochemistry: an introduction to the chemistry of living cells.* Prentice-Hall, Inc., Englewood Cliffs, N.J., 1979.

Debey, H.J. *Introduction to the chemistry of life: biochemistry* (2nd ed.). Addison-Wesley Publishing Co., Inc., Reading, Mass., 1976.

Dickerson, R.E., and Geis, I. *The structure and action of proteins.* Harper & Row, Publishers, Inc., New York, 1969.

Neal, A.L. *Chemistry and biochemistry: a comprehensive introduction.* McGraw-Hill Book Co., New York, 1971.

Speakman, J.C. *Molecules.* McGraw-Hill Book Co., New York, 1966.

Stryer, L. *Biochemistry.* W.H. Freeman & Co., Publishers, San Francisco, 1975.

Suttie, J.W. *Introduction to biochemistry* (2nd ed.). Holt, Rinehart & Winston, New York, 1977.

Wold, F. *Macromolecules: structure and function.* Prentice-Hall, Inc., Englewood Cliffs, N.J., 1971.

Yudkin, M., and Offord, R. *Comprehensible biochemistry.* Longman Group Ltd., London, 1973.

Journal articles

Dickerson, R.E. The structure and history of an ancient protein. *Sci. Am.* **226**(4):58, 1972.

Frieden, E. The chemical elements of life. *Sci. Am.* **227**(1):52, 1972.

Sharon, N. Carbohydrates. *Sci. Am.* **243**(1):90, 1980.

Chapter 3
INTRODUCTION TO MICROSCOPIC TECHNIQUE

Microscopy is the science that deals with the use of microscopes and the interpretation of their magnified images. The development of microscopy as a science parallels that of microbiology. Leeuwenhoek's descriptions of bacteria and protozoa in 1683, using simple glass lenses ground into convex surfaces, represented the first observation of microorganisms that could not be ascertained with the unaided eye. Microscopy has three purposes, all of which are interrelated: (1) to magnify the image, (2) to achieve maximum resolution of the object, and (3) to provide sufficient contrast for observation. It has taken nearly 200 years to develop optical systems that provide these characteristics. Until the 1940s most bacteria observed with our best microscopes looked like spherical or sausage-shaped entities showing little definition. New microscopic systems were developed after the 1940s that provided the microbiologist with structural detail that has given new insight into the relationship between structure and function of the microbial cell. The purpose of this chapter is to briefly describe the basic properties of the various microscopes and to explain how they are used in the numerous branches of microbiology.

BRIGHTFIELD MICROSCOPY

The basic microscopic system used in the microbiological laboratory is the ***compound brightfield microscope*** (Figure 3-1). This system is called *compound* because it possesses at least two lens systems and *brightfield* because visible light is passed directly through the lenses until it reaches the observer's eye. Our discussion will begin with an explanation of the optical properties of brightfield microscopy.

Magnification

Magnification of an image is obtained in the compound microscope by two lens systems: the objective lens, which is closest to the specimen to be examined, and the ocular (eyepiece) lens. The objective lens magnifies the specimen and produces a real image (Figure 3-1, *B*). The real image is projected through the microscope to the ocular lens, which magnifies the real image and produces an image seen by the observer and called the virtual image. Most compound microscopes are equipped with three objective lenses attached to a nosepiece: the low-power lens or 10× magnification, the high-power lens or 45× magnification, and the oil immersion lens or 100× magnification. The total magnification for the lens system is obtained by multiplying the magnification of the objective lens and the magnification of the ocular lens, which is usually 10× magnification. Thus the maximum magnification obtained by using the oil immersion lens is approximately 1000×.

Resolving power

One might think that by increasing the magnification power of the objective lens, a magnification greater than 1000× could be achieved with maximum clarity. Such is not the case, since all lenses have imperfections. The objective lens magnifies the real image, but no lens is capable

A

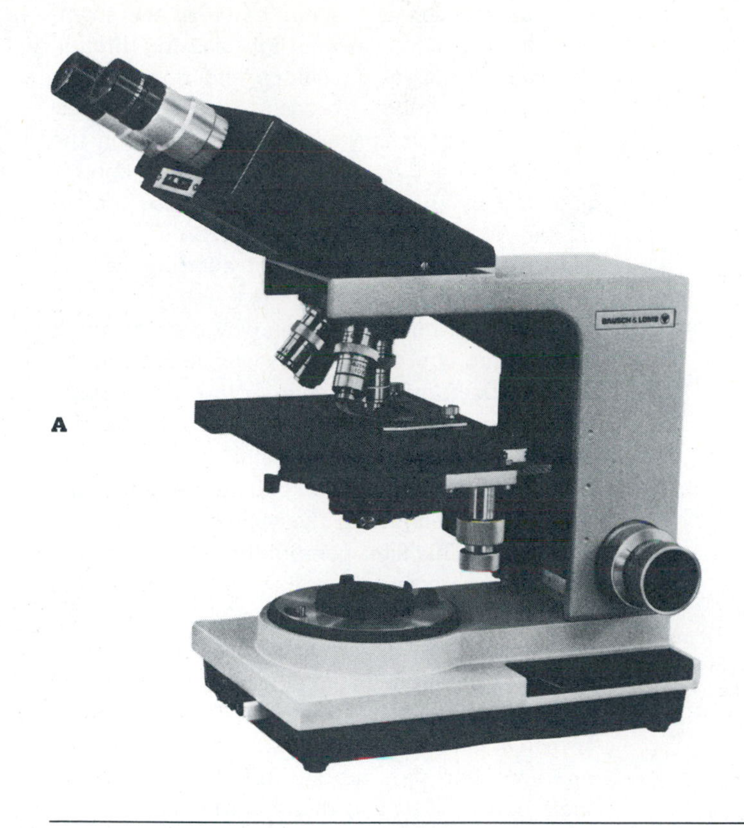

B

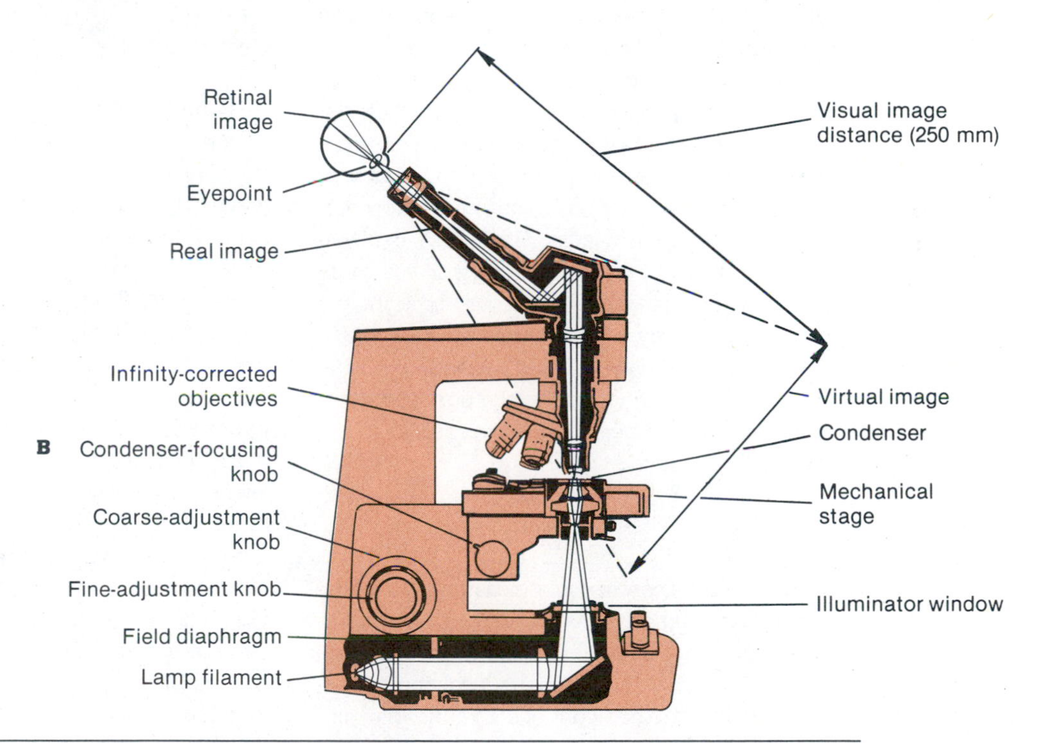

Figure 3-1 **A,** Compound brightfield microscope. **B,** Optical and mechanical features of compound microscope.
A, Courtesy Bausch & Lomb, Rochester, NY; B, Modified from American Optical Corp., Buffalo, NY.

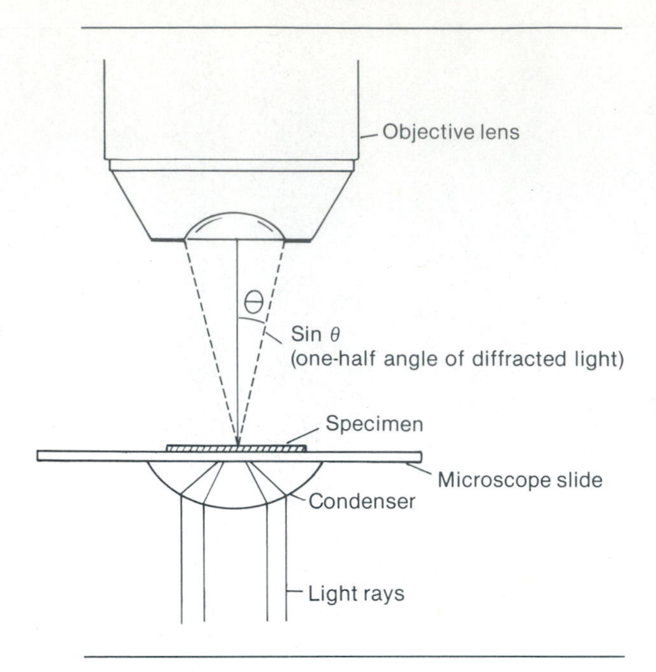

Figure 3-2 Numerical aperture determination of objective lens. N A = n sin θ, where n is refractive index of medium (can be oil or air) and sin θ is one-half angle created by light passing through condenser and specimen and transmitted to objective.

of producing a perfect image, and lens errors (aberrations) are an inherent part of the optical system. Magnification is of little value unless the image also shows clarity. Image clarity is associated with the term ***resolving power,*** which is the ability of the lens to distinguish two objects at a particular distance apart. Resolving power is best explained by assuming that you are microscopically observing two objects. How close can you bring the two objects together and still see them as separate entities? When they are brought too close together, they will be observed as a single object. Resolving power (RP) is dependent on the wavelength (λ) of light used in the optical system and the numerical aperture (NA) according to the following formula:

$$RP = \frac{\lambda}{NA}$$

The ***numerical aperture*** is a measurement of the angle of the maximum cone of light that can enter the objective (Figure 3-2). The greater the numerical aperture, the greater the resolving power. Manipulation of the condenser and iris diaphragm control the amount of light reaching the specimen. The condenser must be at its highest position to ensure that all the light rays focus on the specimen. A reduction in the amount of light reaching the specimen reduces the effective numerical aperture and consequently reduces the resolution. Light from the specimen is collected by the objective and transmitted to the ocular lens. Since there is a space between the specimen and the objective lens, light passing through this space is lost as a result of refraction because the refractive indices* of glass and air are different. When immersion oil is placed between the specimen and the objective lens, more light is gathered by the objective lens because the refractive indices of glass and oil are comparable. Thus the maximum numerical aperture can be achieved.

Compound microscopes employ a tungsten filament lamp, which generates wavelengths in the visible spectrum. The shorter the wavelength, the greater the resolving power according to the preceding formula. A blue filter is inserted between the light source and the slide, since the shorter wavelengths in the visible spectrum are blue-violet. This filter permits the transmission of blue-violet wavelengths and the absorption of the longer (yellow, green, and red) wavelengths. The resolving power of the compound microscope is about 0.2 μm when oil immersion is used and the numerical aperture is maximum. This means that two objects can be distinguished as two separate entities when they are separated by 0.2 μm or more but not if by less than 0.2 μm.

Contrast

Unstained microorganisms, such as bacteria, are transparent when observed with the compound microscope, but this problem can be partially corrected by reducing the amount of light used to illuminate the specimen. Compound microscopes have an iris diaphragm, located between the condenser and the light source, which can be adjusted to reduce illumination. Even this procedure does not always produce sufficient differentiation between the background and the microbial cell. Staining the microbial cell can make it readily visible even at magnifications lower than 1000×. Microbiological stains are called dyes. A dye is an organic compound containing a ***chromophore*** group, which imparts color to the compound, and an ***auxochrome,*** which is a dissociable group that binds to the substrate and intensifies the color. The bacterial cell contains many negatively charged groups (carboxyl, for example), and only basic dyes, which contain positively charged ionizable groups, will bind to them. The more commonly used basic dyes are methylene blue, crystal violet, safranin, and malachite green. These stains

* ***Refractive index*** is related to the change in direction of light as it passes from air into a more dense medium such as a liquid.

may be used individually to stain the cell, and the staining procedure is called ***simple staining.*** There are some procedures that use more than one stain in order to distinguish differences in the chemical composition of the bacterial cell. These are called differential staining procedures. The most important differential staining techniques for bacteria are the ***Gram stain*** and the ***acid-fast stain.*** The steps involved in these staining techniques are described in Appendix B. In addition to staining the entire cell, individual components of the cell such as flagella and endospores may also be stained. There is a technique called ***negative staining*** in which the background is filled with particles of dye, making it appear dark while the cells remain clear and transparent. India ink particles or nigrosin dyes are frequently used to "stain" the background around cells that produce extracellular polymeric substances called capsules (Figure 3-3). Nigrosin is an acidic dye and does not bind to the negatively charged bacterial cell.

One of the advantages of brightfield microscopy is that living cells can be examined.

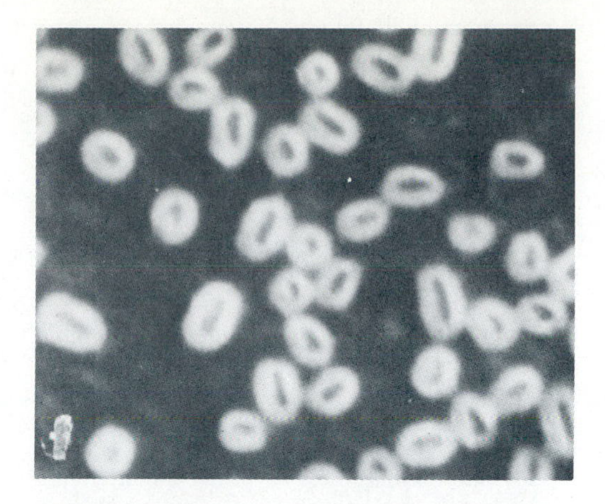

Figure 3-3 Negative staining. India ink preparation of bacterial cells observed by brightfield microscopy. Bacterial cells are rod shaped and appear to be surrounded by white halo. White halo is extracellular component, called capsule, that does not stain.
From Babb, J.L. Infect. Immun. ***19:****1088, 1978.*

DARKFIELD MICROSCOPY

Some bacteria, such as certain spirochetes, have a diameter (0.1 to 0.15 μm) that is less than the resolving power of the brightfield microscope and are difficult to observe. Transparent objects, like bacteria, can be more easily observed when the background is dark (Plate 1). The compound microscope can be fitted with a special condenser called a darkfield condenser, which has a numerical aperture greater than the objective and blocks all direct light from reaching the objective. The condenser causes only diffracted light, passing through the specimen, to reach the objective lens. The cells in the specimen appear light against a dark background because of light scattering. There is no increase in resolution, since the wavelengths and numerical aperture used are the same as those in brightfield microscopy.

PHASE-CONTRAST MICROSCOPY

As we have already noted, the transparency of whole cells makes them difficult to distinguish from the background. The same difficulty is encountered in trying to differentiate organelles or other dense components within a single cell. The phase-contrast microscope consists of special condensers and objectives that enable one to increase the contrast between the transparent

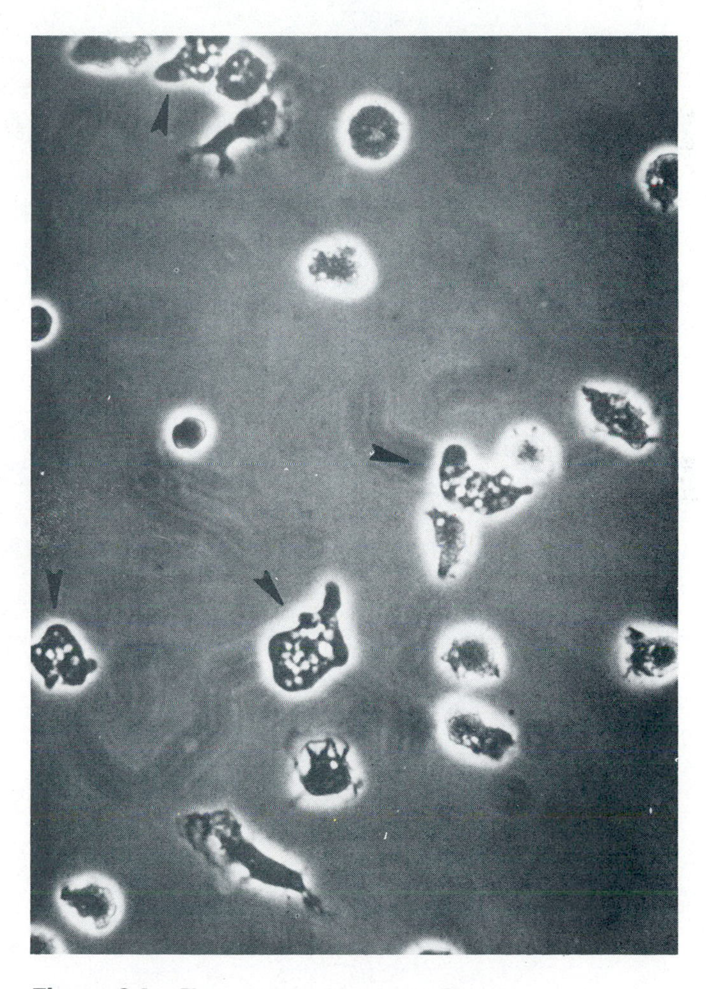

Figure 3-4 Phase-contrast microscopy. Phase-contrast wet smear of cerebrospinal fluid of patient. (×2500.) All of cellular material appears to have halo. Arrow points to parasitic amebas, which contain many vacuoles and are distinguished from more spherical red blood cells.
From Stevens, A.R., et. al. J. Infect. Dis. ***143:****193, 1981.*

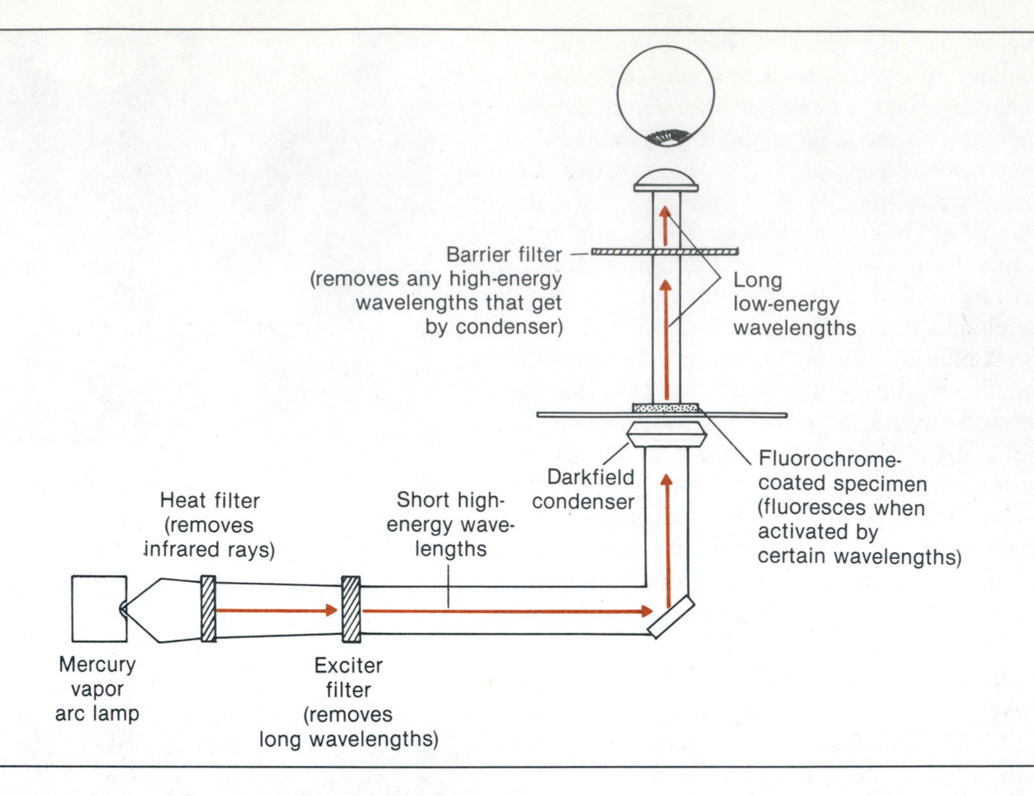

Figure 3-5 How fluorescence is achieved in fluorescent microscope.

components in the cell by exploiting differences in their densities. When light strikes the transparent cell, some of the light rays, called direct rays, pass through and are unchanged in terms of amplitude and phase. Other light rays that strike components in the cell have different densities and emerge with their wavelength retarded by one fourth. The amplitude of direct and indirect rays when brought into phase is the sum of the two waves, and this gives the specimen the appearance of being bright against a dark background. If the direct and diffracted rays are made to be out of phase (one-half wavelength off), their amplitudes cancel each other out and the object appears dark against a bright background. Phase plates are present in the objective lens that will either retard (positive-phase plate) or increase (negative-phase plate) the phase of the diffracted light relative to the direct light. The technique is called bright-phase microscopy when the image appears brighter than the background and dark-phase microscopy when the image is darker than the background. An example of phase-contrast microscopy can be seen in Figure 3-4.

FLUORESCENT MICROSCOPY

Certain molecules when struck by light waves emit light of a longer wavelength. The electromagnetic wave striking the molecule imparts energy and causes the electrons of the molecule to oscillate. The energy in these oscillations is released by the molecule in the form of heat or light and can be used to initiate chemical reactions. If the energized molecule releases its energy in the form of light, whose wavelength is greater than the exciting wave, the compound is called a ***fluorochrome.*** Fluorochromes, therefore, exhibit the property of ***fluorescence.*** Some of the important fluorochromes are auramine O, acridine orange, fluorescein, and rhodamine. Microbiological material can be examined by coating it with fluorochromes and using the fluorescent microscope, which differs considerably from the brightfield microscope.

The fluorescent microscope uses a mercury vapor lamp instead of the incandescent lamp used in the brightfield microscope. The mercury lamp produces shorter wavelengths (ultraviolet, violet, blue), which are needed to impart a sufficient amount of energy to the fluorochromes. Two filters are placed between the mercury lamp and the fluorochrome-coated specimen (Figure 3-5). They serve to remove the longer wavelengths generated by the mercury lamp and transmit the shorter wavelengths. A darkfield condenser is used to produce a dark background, which provides the best contrast. The condenser also deflects most of the shorter wavelengths, such as ultraviolet, which could be detrimental to the observer's eyes. After the fluorochrome has been

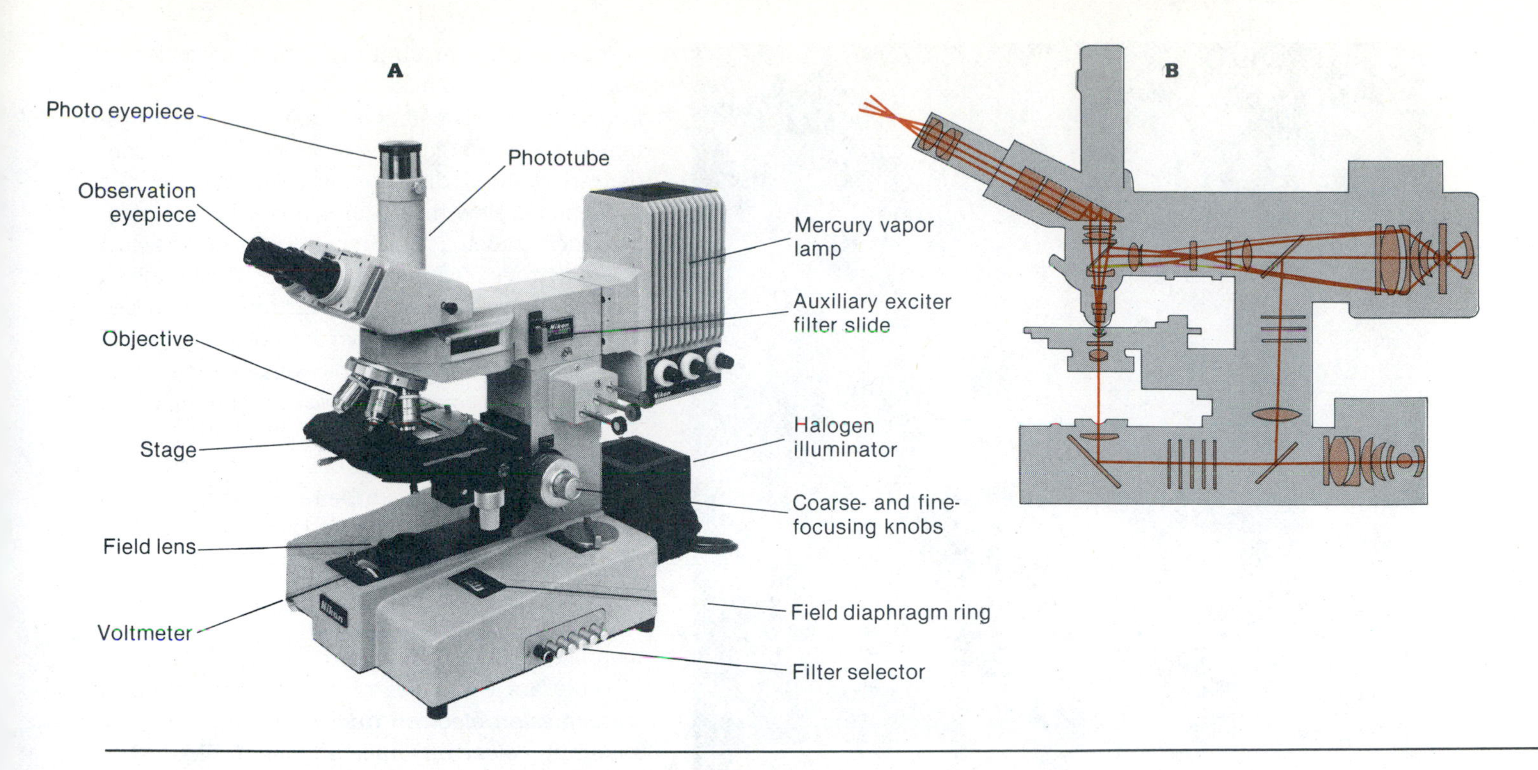

Figure 3-6 Mechanical and optical features of fluorescent microscope. **A,** Mechanical features of Nikon fluorescent microscope. **B,** Optical properties of Nikon fluorescent microscope.
Courtesy Nikon, Inc., Instrument Division, Garden City, N.Y.

excited, another filter is placed between the specimen and the observer, which removes the shorter wavelengths and retains the longer wavelengths such as the fluorescing wavelengths. The optical system of the fluorescent microscope is outlined in Figure 3-6, *B*.

Fluorescent microscopes are expensive, but they have played an extremely important role in the identification of infectious agents or their products in clinical laboratories. The most widely used fluorescent technique is that of immunofluorescence, in which an antibody is conjugated (coated or tagged) with a fluorochrome (Figure 3-7). For example, when an animal is injected with an antigen such as the rabies virus, antibodies are produced that are specific for that virus. The virus-specific antibodies can be recovered from the animal and conjugated with a fluorochrome. Later, it may be necessary to determine if an animal has been infected by the rabies virus. An investigator removes a specimen of nervous tissue from the rabid animal, or animal suspected of being rabid, and applies fluorochrome-coated rabies antibodies. Labeled antibodies will bind and coat the virus if the specimen does contain the rabies virus, and when observed microscopically, fluorescent particles will be apparent. It is also possible to label the antigen with a fluorochrome and use it to detect the presence of antibodies

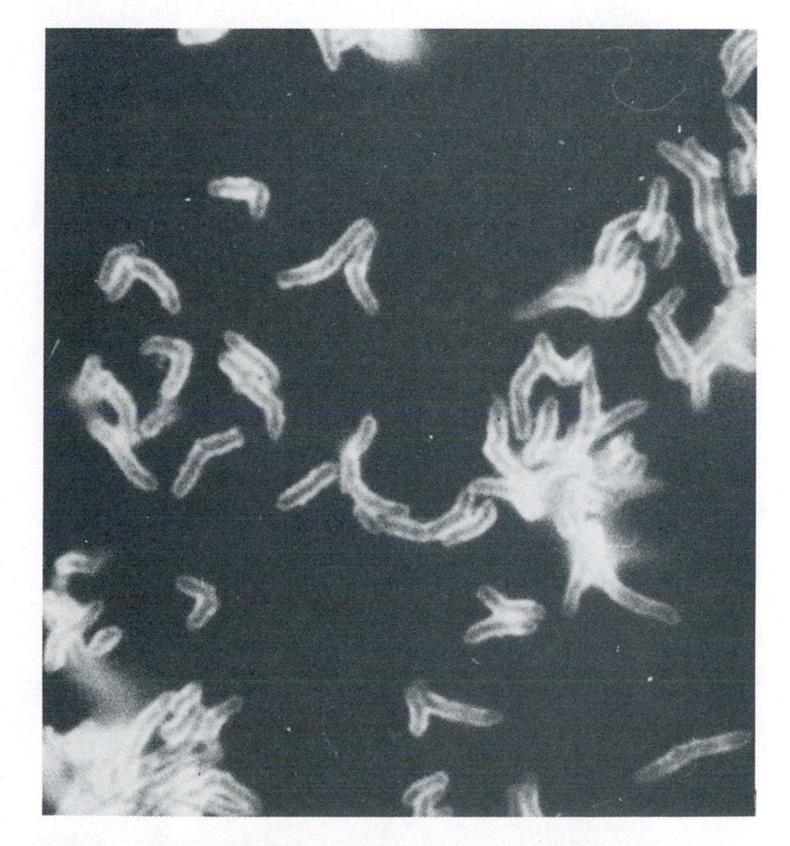

Figure 3-7 Fluorescent microscopy. Fluorescent antibody is used to detect a filamentous bacterium, *Actinomyces naeslundii.*
From Lai, C.H. Infect. Immun. ***25:****1016, 1979.*

Figure 3-8 Transmission electron microscope.
Courtesy Carl Zeiss, Inc., New York.

specific for the antigen. Some other important immunofluorescence tests include the determination of spyhilis infection (see Appendix B) and the mechanism of cell wall growth (see Chapter 15).

ELECTRON MICROSCOPY

The electron microscope (Figure 3-8) revolutionized the science of microscopy because it improved resolution obtainable from biological specimens from 0.2 to 0.0005 μm. Its role in the elucidation of bacterial structure and function cannot be overemphasized. The only drawback to electron microscopy is that living cells cannot be examined.

Electrons are substituted for light waves in the electron microscope. The wavelengths of electrons are 100,000 times shorter than light waves, and, as you have already learned, resolving power is related to the wavelength. A tungsten filament is placed in a vacuum and heated to emit electrons that are accelerated and passed through an evacuated cylinder until they reach a condenser system. The condenser system consists of electromagnetic coils and is analgous to the lens system of the brightfield microscope. The first condenser system directs the electrons to the specimen. Below the specimen is a second magnetic coil, which acts as an objective and collects the electrons that have passed through and around the specimen. The electrons are passed to a third magnetic coil, which enlarges the image produced by the passage of electrons through the specimen and focuses the image on a fluorescent screen that can be flipped up to take a photograph. A final coil system acts as an eyepiece. The image can be recorded either on a plate or a 35 mm roll of film. Direct screen magnifications of 1,000,000× are possible. Most large electron microscopes go up to magnifications of 500,000×, whereas most small ones go up to magnifications of 100,000×.

There are two types of electron microscopes: ***transmission electron microscopes*** (TEMs) and ***scanning electron microscopes*** (SEMs). The electrons generated in the tungsten filament in TEMs pass through the specimen and onto a photographic plate. Using the highest accelerated voltages, specimens up to 2 μm in thickness can be used. The specimens to be examined by electron microscopy are prepared by techniques that include negative staining, thin sectioning, and ***freeze-etching.*** The contrast required to differentiate cellular components in electron microscopy is directly related to the diffraction of electrons by the specimen. The greater the mass of the atoms in the cellular organelle or component, the greater the scattering of electrons. The elements in biological materials have small masses, and thus the contrast is not great when electrons are diffracted. The contrast can be enhanced by staining the background (negative staining) with heavy metals (uranium salts or phosphotungstic acid). Negative staining is used especially for the study of macromolecules, viruses, microbial organelles, and other components in a cell-free system.

The observation of whole cells by electron microscopy in order to examine the relationship of their surface and internal components requires that the cells be thin sectioned. Specimens are fixed with chemicals such as osmium tetroxide or glutaraldehyde, then dehydrated in alcohol or acetone, and finally impregnated with polymers such as epoxy resins. The polymer-impregnated specimen can be sliced to produce sections as thin as 0.03 μm. This procedure, commonly called thin sectioning, is performed with an ultramicrotome. The specimen can be prestained with a heavy metal such as uranyl acetate, lead citrate, or lead hydroxide to increase contrast. One of the major problems in interpretation of fixed preparations is that chemical and physical procedures create artifacts by alteration of the cellular components. (This is in addition to the usual drying procedures for any electron microscopic observation.) Such interpretations must therefore be corroborated with other studies that may be of a microscopic or biochemical nature.

Chemical fixation can be bypassed by using the technique of freeze-etching, which involves rapid freezing and fracturing of the specimen. Different surfaces of the cell are exposed (Figure 3-9) by fracturing with special knives. The fractured specimen surface is shadowed with a heavy metal such as platinum, and then a layer of carbon is evaporated over this surface. Chemical treatment destroys the shadow, and the platinum-carbon replica of the specimen is examined. Freeze-etching has proved to be invaluable in the examination of microbial membranes and cell walls.

The principal advantage of scanning electron microscopy is the depth of the field that can be observed. SEMs provide a three-dimensional image as opposed to the two-dimensional TEM image. The specimen, for examination by scanning electron microscopy, is coated with a heavy metal and then scanned with a focused beam of electrons. When the beam of electrons strikes the specimen, secondary electrons are released, collected, and used to produce an image on a cathode ray tube. Magnifications of up to 100,000× are possible with a resolution of 0.0025 μm. An example of the three-dimensional effect of scanning electron microscopy is illustrated in Figure 3-10.

The student might consider microscopy to be more appropriate for the laboratory section of a microbiology course. It is true that microscopic identification of microorganisms is often necessary in determining the cause of infectious diseases or the source of pollution. Keep in mind, however, that our understanding of the biology of microorganisms is directly associated with the correlations that have been made between biochemical properties and structure, as revealed by microscopy. This correlation becomes evident in the next chapters, which describe the relationship between structure and function in the microbial cell.

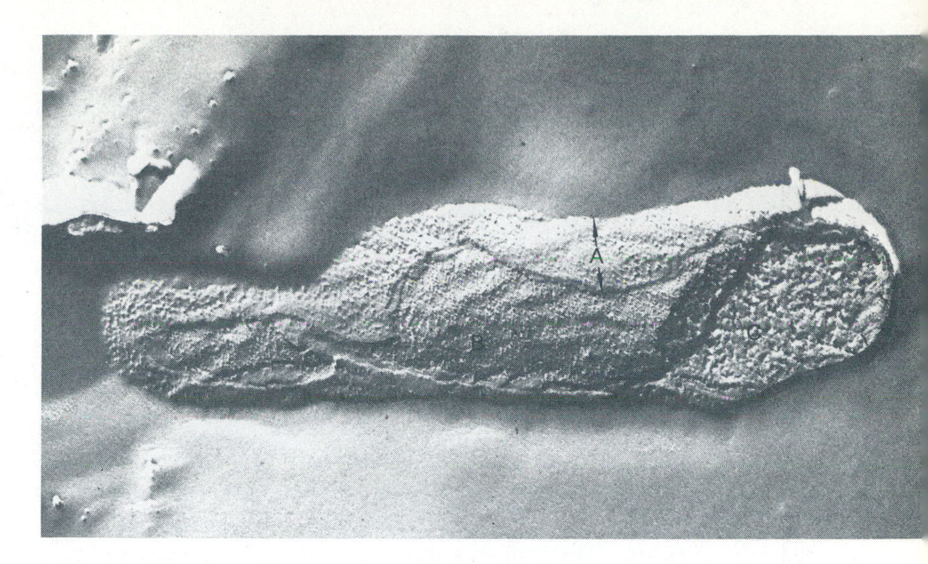

Figure 3-9 Transmission electron micrograph of replica of freeze-etched preparation of *Escherichia coli. A,* Cell wall; *B,* cytoplasmic membrane; *C,* cytoplasm.

From Costerton, J.W., Annu. Rev. Microbiol. ***33:****459, 1979.*

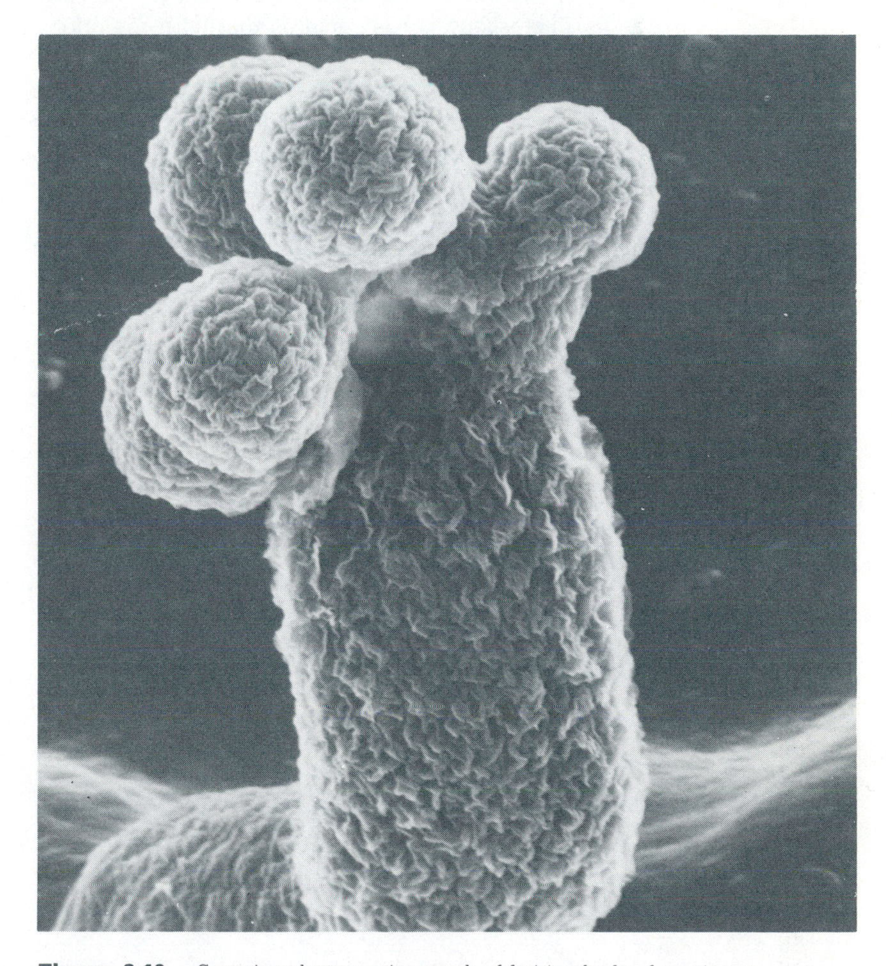

Figure 3-10 Scanning electron micrograph of fruiting body of species of *Myxobacteria.* Note detail as well as three-dimensional aspects. Bar = 20 μm.

From Qualls, G.T., et. al. Science ***201:****444, 1978. Copyright 1978 by the American Association for the Advancement of Science.*

SUMMARY

The basic laboratory microscope is the compound brightfield microscope. It is equipped with objective lenses and can obtain magnifications of 50× to 1000×, and its resolving power reaches 0.2 μm. The resolving power, or image clarity, is defined as:

$$\text{Resolving power (RP)} = \frac{\text{Wavelength of light } (\lambda)}{\text{Numerical aperture (NA)}}$$

Because bacteria are transparent, the contrast of the unstained specimen is reduced when the specimen is illuminated. Contrast can be obtained by reducing the amount of light reaching the specimen using the iris diaphragm. When the specimen is stained for microscopic visualization, the technique is called ***positive staining.*** If only the background is stained, the technique is called negative staining.

The darkfield microscope is equipped with lenses that diffract light in such a way that the specimen appears bright against a dark background. The resolution obtained with darkfield microscopy is the same as that with the brightfield microscope.

In phase-contrast microscopy differences in the density between bacterial components and the cytoplasm can be exploited through the use of special lenses. Certain intracellular components can be resolved using phase contrast, however, the resolution is the same as in brightfield microscopy.

In fluorescent microscopy the biological material is coated with a fluorescing compound called a fluorochrome. A mercury lamp is used to generate shorter wavelengths in order to impart enough energy to the fluorochrome to make it fluoresce. Immunofluorescence, in which an antibody is tagged with a fluorochrome, is the most widely used fluorescent technique. The fluorescing antibody will bind to specific antigens. Antigens can also be tagged if one wishes to detect specific antibodies in a specimen.

The electron microscope uses electrons instead of light waves, and electromagnetic coils instead of lenses. Magnifications of close to 1,000,000× can be obtained with a resolving power of 0.0005 μm. Two types of electron microscopes are available: transmission electron microscopes (TEMs) and scanning electron microscopes (SEMs). In transmission electron microscopy electrons pass through the specimen and are diffracted to different degrees depending on the mass of the atoms in the cellular component. Specimens for transmission electron microscopy are nonliving and must be prepared by staining or freeze-etching. Scanning electron microscopy provides a three-dimensional image in which the specimen is coated with a heavy metal and is then scanned with a beam of electrons. Magnifications for SEMs approximate 100,000×.

Study questions

1. What is a compound microscope?
2. How does the background appear around an object being observed by brightfield microscopy? By darkfield microscopy?
3. Will it be possible someday to have unlimited magnification in microscopic systems? Explain why or why not.
4. When the brightfield microscope is being used, which of the following wavelengths will give the best resolution in the visible spectrum: blue, red, green, or yellow? What accommodations are made on the microscope to obtain this maximum resolution?
5. What is the resolving power of the brightfield microscope? Of the electron microscope? Define resolving power in terms of these values.
6. What is the value of negative staining in microscopy?
7. What is meant by phase-contrast microscopy? What is the mechanism involved?
8. What are the differences between fluorescent and brightfield microscopy? Is there any difference in resolving power between the two? Under what circumstances is fluorescent microscopy generally used?
9. Why are greater resolving powers obtained in electron microscopy? How does scanning electron microscopy differ from transmission electron microscopy?
10. List the advantages and disadvantages of the various types of microscopes. Explain the conditions under which they are best used.

Selected readings

Books

Barer, R. Microscopy. In Bourne, G.H. (Ed.). *Cytology and cell physiology.* Academic Press, Inc., New York, 1964.

Glauert, A.M. (Ed.). *Practical methods in electron microscopy.* North-Holland Publishing Co., Amsterdam, 1975.

James, J. *Light microscopic techniques in biology and medicine.* Martinus Nijhoff Publishers, Medical Division, Amsterdam, 1976.

Lickfeld, K.G. Transmission electron microscopy of bacteria. In Norris, J.R. (Ed.). *Methods in microbiology* (Vol. 9). Academic Press, Inc., Ltd., London, 1976.

Quesnel, L.B. Microscopy and microtomy. In Norris, J.R., and Ribbons, D.W. (Eds.). *Methods in microbiology* (Vol. 5a). Academic Press, Inc., Ltd., London, 1971.

Slayter, E.M. *Optical methods in biology.* Wiley-Interscience, New York, 1970.

Wilson, M.B. *The science and art of basic microscopy.* American Society for Medical Technology, Bellaire, Tex., 1976.

Journal articles

Costerton, J.W. The role of the electron microscope in the elucidation of bacterial structure and function. *Annu. Rev. Microbiol.* **33:**459, 1979.

Chapter 4 CHARACTERISTICS OF THE MICROBIAL WORLD Prokaryotes

The term *microorganism* does not signify any particular group but merely describes a characteristic of certain organisms: they are small. Until the discovery of the compound light microscope, organisms were placed into two kingdoms: Animal and Plant. This division sufficed until microscopic observations revealed that some organisms possessed traits that could not be attributed to either plants or animals. Scientists, in the late 1800s, proposed that a third kingdom be created to encompass this group of organisms: the Protista. The Protista were to include those organisms that do not possess any tissue systems, that is, different cell types performing different functions in a given organism. The Protista therefore included the newly discovered microscopic organisms—the microorganisms—and excluded the microscopic animals such as rotifers and nematodes.

CURRENT CLASSIFICATION OF MICROORGANISMS (PROKARYOTES vs. EUKARYOTES)

The Protista were later divided into two major groups—the prokaryotes and the eukaryotes—based primarily on the presence or absence of a nuclear membrane. Eukaryotes (Gr. *eu*, true; *karyon*, nucleus) possess a nuclear membrane, but prokaryotes (Gr. *protos*, primitive; *karyon*, nucleus) do not possess a nuclear membrane. There are also other fundamental differences between these two groups, which are outlined in Table 4-1.

The many differences as well as similarities that exist between eukaryotic and prokaryotic microorganisms have resulted in various classification schemes. One recent (1978) scheme proposed by Whittaker and Margulis still divides the living world into prokaryotes and eukaryotes but also includes five kingdoms to encompass all biological groups. The scheme is as follows:

- Superkingdom: Prokaryotae
  - Kingdom: Monera (Bacteria and Cyanobacteria)*
- Superkingdom: Eukaryotae
  - Kingdom: Protista
    - Branch Protophyta (plantlike, for example, primitive algae)
    - Branch Protomycota (funguslike, for example, slime molds)
    - Branch Protozoa (animal-like, for example, protozoans)
  - Kingdom: Fungi (unicellular yeasts and multicellular molds)
  - Kingdom: Plantae (complex algae and plants such as mosses and ferns)
  - Kingdom: Animalia (multicellular vertebrate and invertebrate animals)

* *Bergey's Manual of Determinative Bacteriology* (8th ed.), which is the most extensive work for classifying bacteria, classifies Bacteria and Cyanobacteria as divisions under the kingdom Prokaryotae.

Title: BOYD GENERAL MICRO

P18 012684 00900 000230009908

DR JOHN A CARPENTER
UNIVERSITY OF GEORGIA
DEPT OF FOOD SCIENCE
570 ADERHOLD
ATHENS GA 30602

1. Is this book suitable for your course(s)?

   Yes ______

   No ______

2. If yes, do you plan to adopt this book?

   Yes ______ How many? ______

   No ______

3. Please identify some of the features that lead you to select this text for your course(s).

   a.) ______

   b.) ______

   c.) ______

4. If you have chosen not to adopt this text, please explain any deficiencies you may have encountered:

   Content ______ Comprehension Level/Too High ______

   Presentation ______ Comprehension Level/Too Low ______

   Comments: ______

5. What book(s) are you now using in your course?

   ______

   Why did you choose this book?

   ______

6. If you have not yet adopted a textbook for your course, what is your decision date? ______

Your comments are welcome. We rely on you to help us build better Mosby Books!

Please write: Marketing Services
C.V. Mosby
11830 Westline Industrial Drive
St. Louis, MO 63146

Or Call: (800) 325-4177 ext. 588
In Missouri call collect:
(314) 872-8370 ext. 588

Would you be willing to discuss this questionnaire with us? If so, please indicate your phone number. ______

Fold, moisten and mail.

HELP US BUILD BETTER BOOKS

Meeting your needs is our business. You can help us meet these needs by sharing your opinions with us. This MOSBY text has been sent to you with our compliments. We hope you'll share in our enthusiasm over this excellent text. Please share your opinions with us . . . We need you to help us build better books!

BUSINESS REPLY MAIL
FIRST CLASS PERMIT NO. 135 St. Louis, MO.

POSTAGE WILL BE PAID BY

The C. V. MOSBY Company
Attn: Marketing Services
11830 Westline Industrial Drive
St. Louis, Missouri 63146

Table 4-1 Characteristics distinguishing prokaryotes from eukaryotes*

CHARACTERISTIC OR STRUCTURE	EUKARYOTE	PROKARYOTE
Nucleus	Present; DNA is in form of chromosomes enclosed in nucleus with nuclear membrane	Absent; DNA dispersed in cell into nuclear areas (nucleoids) but no nuclear membrane
Nucleolus	Present	Absent
DNA	Several DNA molecules make up a chromosome; more than one chromosome per cell; DNA complexes with proteins called histones; organelles, such as chloroplasts and mitochondria, contain their own DNA, which resembles prokaryotic DNA	Single chromosome made up of only one molecule of DNA; histones absent; plasmid DNA common extrachromosomal component
Cell wall	Present in algae and fungi; absent in protozoa; absence of peptidoglycan	Present in all groups except mycoplasmas; peptidoglycan present except in mycoplasmas and archaebacteria
Cytoplasmic membrane	Contains sterols	Sterols generally absent
Cytoplasmic streaming	Present	Absent
Mitosis and meiosis	Present	Absent
Ribosomes	Larger than prokaryotes; sedimentation coefficient of 80S; attached to membranous system called endoplasmic reticulum; ribosomes similar to prokaryotes found in mitochondria and chloroplasts	Sedimentation coefficient of 70S; exist independently in cytoplasm
Respiration	Associated with organelle called a mitochondrion	Associated with cytoplasmic membrane
Photosynthesis	Present; associated with membranes called thylakoids that are enclosed within an organelle called a chloroplast	Present; some associated with thylakoids, others not; none associated with organelle
Motility	Flagella (cilia) or ameboid movement; flagellum more complex than prokaryote; some gliders	Flagella, axial filament, and some gliders; flagellum has simple construction
Reproduction	Sexual or asexual; conjugation part of reproductive process	Asexual (binary fission); conjugation rare and not part of reproduction

*Differences in metabolism and genetics are discussed in later chapters.

A group of prokaryotes has recently been shown to differ biochemically and in the structure of certain large molecules from other prokaryotes. This group has been called Archaebacteria. Many genetic, metabolic, and physiological differences exist among the Archaebacteria, Prokaryotae, and Eukaryotae, which are dealt with later in this chapter as well as in other chapters. The purpose of the remainder of this chapter is to briefly describe some of the more obvious properties of the prokaryotic bacteria and cyanobacteria. More detailed discussions of structure and function in prokaryotes are presented in Chapter 7.

GENERAL CHARACTERISTICS OF BACTERIA AS PROKARYOTIC CELLS

The following discussion is a cursory examination of some of the characteristics of bacteria as prokaryotic cells. Many of these characteristics, which are morphological, are more easily represented in the illustration of a typical bacteria cell presented in Figure 4-1. Refer to this illustration when reading the following paragraph. It will help to clarify the discussion that follows.

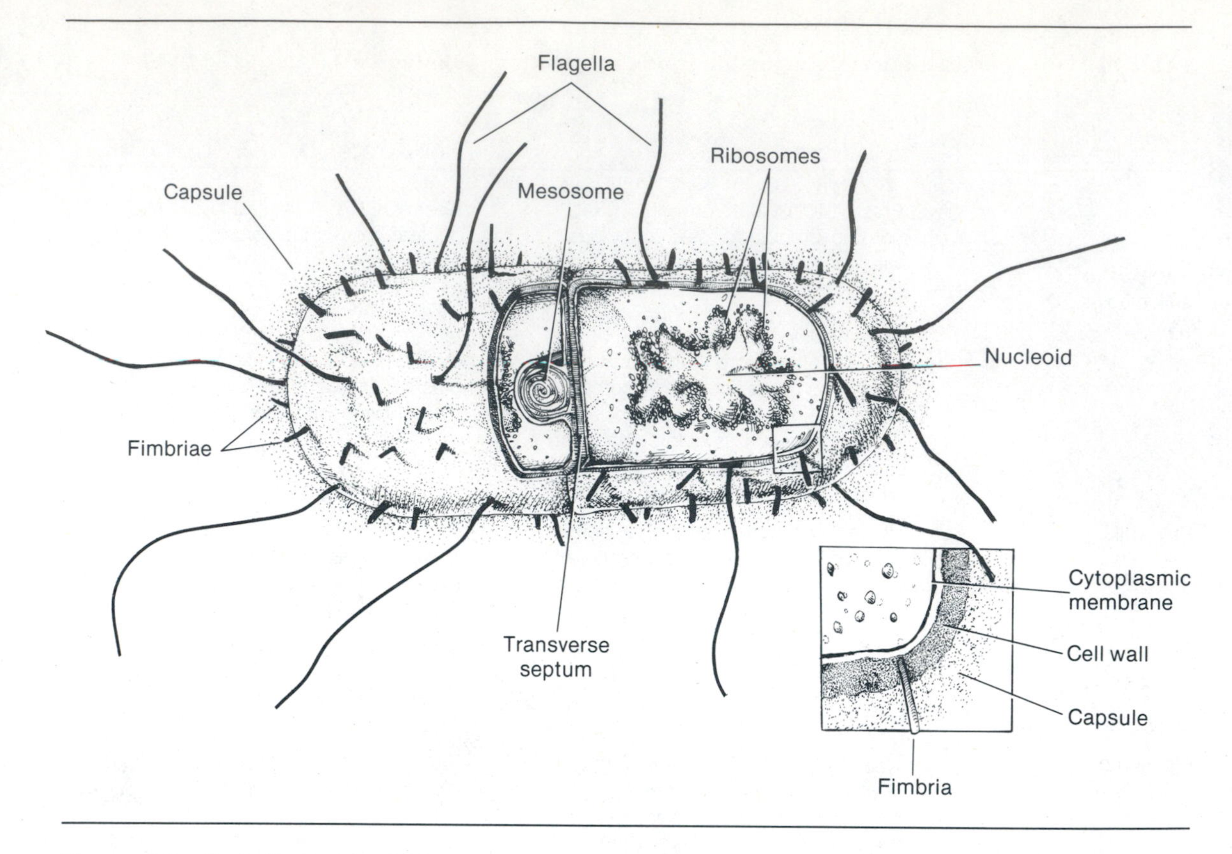

Figure 4-1 Typical bacterial cell. Not all morphological features presented here are necessarily present in a single cell type.

Size and morphology

Bacteria are characteristically smaller than the more advanced microscopic eukaryotes and range in size from 0.2 μm, in width, which is barely visible with the compound microscope, to some spiral-shaped species that reach 400 to 500 μm in length. Generally speaking, most bacteria are 1 to 6 μm in length with diameters between 0.2 and 1.5 μm. The considerable size difference between bacteria and the more advanced eukaryotes is significant. The rate of metabolism increases with an increase in the surface/volume ratio, and in prokaryotes this ratio is much greater than in eukaryotes. The surface/volume ratio for prokaryotes, such as spherical bacteria with a diameter of 0.5 μm, is approximately 100,000, whereas in a hen's egg, whose diameter may be 1.5 inches, the surface/volume ratio is between 1 and 2. The bacterium, therefore, with its proportionately increased surface area, can effect a rapid exchange of nutrients and other materials with the surrounding environment. A bacterium such as *Escherichia coli,* a common inhabitant of the intestinal tract, can degrade over 1000 times its own weight of glucose within an hour at 37° C. It would take a human one half a lifetime to degrade glucose that amounted to 1000 times his own weight. The prokaryotic cell not only experiences a faster rate of metabolism but also divides more rapidly than the eukaryotic cell. These properties are complemented by the vast metabolic potential of prokaryotes as a whole, a characteristic not evident in most eukaryotes.

The bacteria can be separated into the following shapes: spherical, rod-shaped, curved or spiral, and square, with variations in most types (Figure 4-2).

Spherical bacteria. Spherical bacteria are referred to as ***cocci*** (sing. coccus) which, depending on the species, exist in different arrangements depending on the plane of division during reproduction (Figure 4-3). Division in one plane results in the formation of either a pair of cells, called ***diplococci,*** or a chain of cells, called ***streptococci.*** Division in two planes results in the formation of tetrads. Division in three planes produces cubical masses that appear as groups of eight called ***sarcinae*** or as clusters of cells called ***staphylococci.***

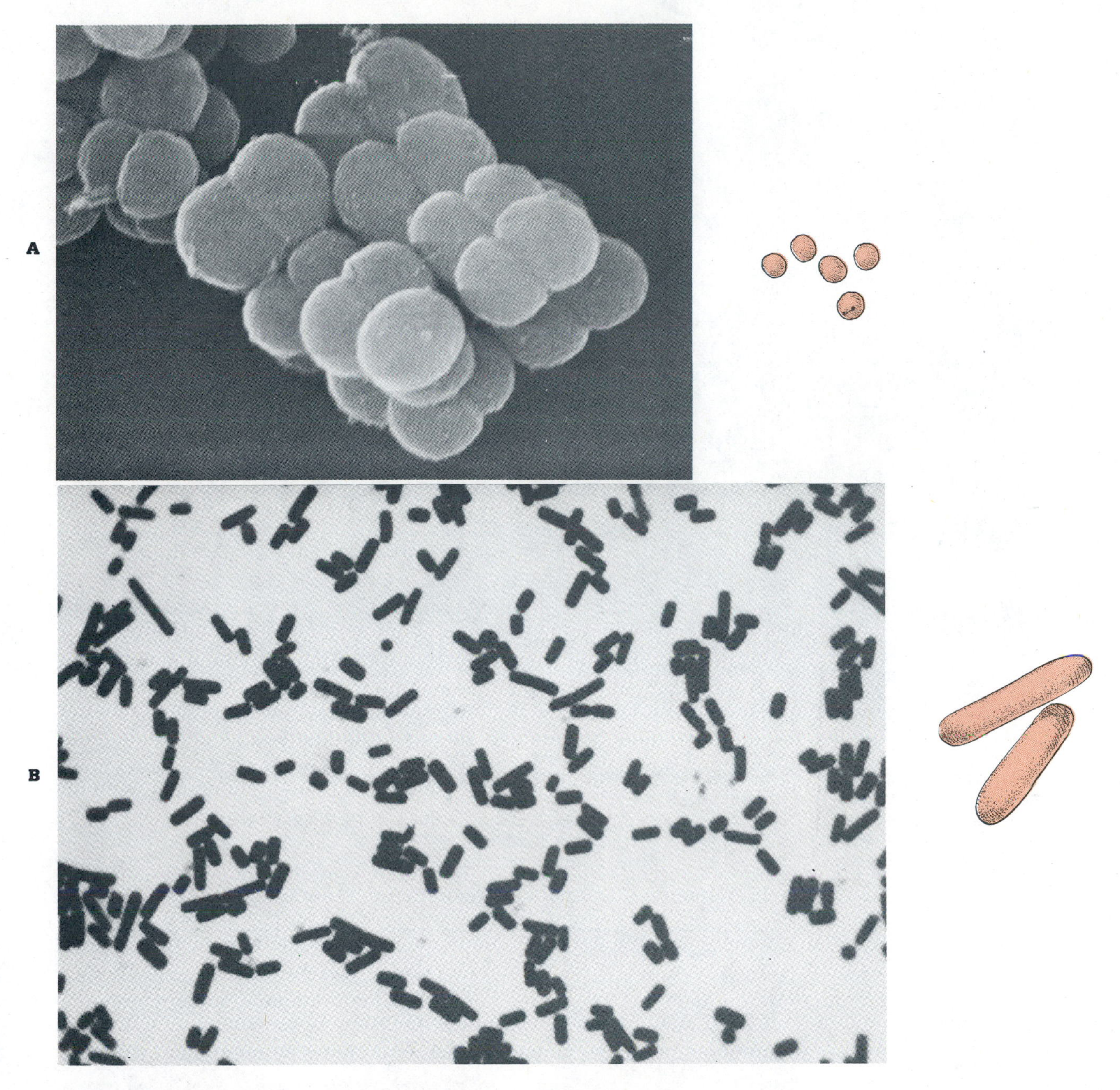

Figure 4-2 Four morphological types of bacteria. **A,** Spherical (coccus). Scanning electron micrograph is of *Staphylococcus aureus.* **B,** Rod (bacillus). Light micrograph is of *Clostridium perfringens.* (×2560.) *Continued.*

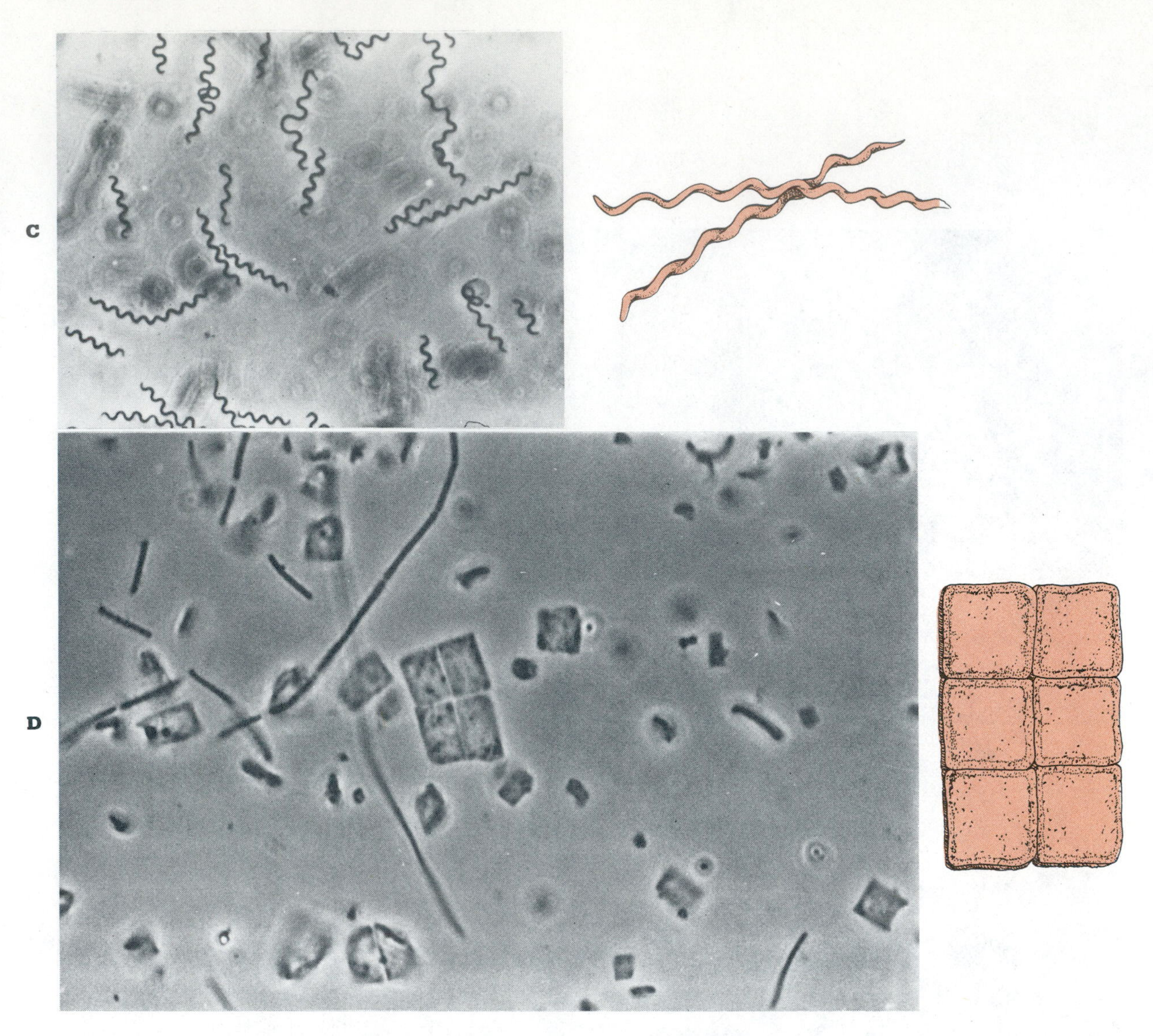

Figure 4-2 cont'd **C,** Spiral. Phase-contrast micrograph is of *Treponema* species. (×2000.) **D,** Square. Micrograph is phase contrast. (×1600.)

A, Micrograph from Kayama, T., et al. J. Bacteriol. ***129**:1518, 1977. **B,** Micrograph courtesy P.D. Walker. **C,** Micrograph from Ziolecki, A.* Appl. Environ. Microbiol. ***37**:131, 1979. **D,** Micrograph from Stoeckenius, D.* J. Bacteriol. ***148**:352, 1981.*

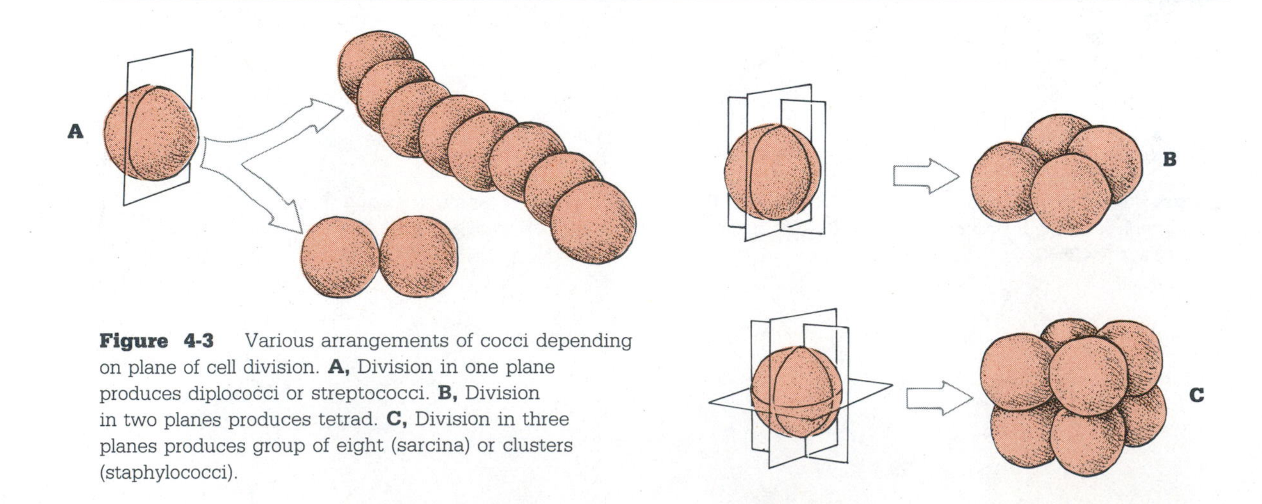

Figure 4-3 Various arrangements of cocci depending on plane of cell division. **A,** Division in one plane produces diplococci or streptococci. **B,** Division in two planes produces tetrad. **C,** Division in three planes produces group of eight (sarcina) or clusters (staphylococci).

Rod-shaped bacteria. Rod-shaped organisms are called ***bacilli*** (sing. bacillus). The term *bacillus* is used here as a morphological characteristic, but it can also be used to indicate a taxonomic group; the genus *Bacillus*. Rod-shaped bacteria exhibit considerable variation in length and diameter. The length and diameter of some rods is so similar that they are difficult to distinguish from cocci. The ends of rods also vary in shape and may be flat, rounded, bifurcated, or cigar shaped (Figure 4-4). Cell division in most rod-shaped bacteria results in the formation of two similar daughter cells, but this may be subject to variations. The actinomycetes actually form branches and develop into filaments, called ***hyphae*** (sing. hypha). The filaments may become so extensive that large mats of filaments called ***mycelia*** (sing. mycelium) are formed. This type of growth is also characteristic of the fungi that the actinomycetes resemble in many other ways.

Spiral-shaped bacteria. Some rod-shaped organisms are curved or exist in the shape of a helix (Figure 4-2, *C*). The rods of the spiral bacteria may be flexible or rigid, depending on the species. Motility is typical of most members of this group and is characterized by a corkscrew movement.

Square bacteria. One of the oddities of the microbial world is the square bacteria. These organisms were originally discovered in 1981 by Walsby on the shores of the Red Sea (Figure 4-2, *D*). The sides of these halophili (salt-loving) bacteria range from 2 to 4 μm and are 0.25 μm thick. They possess a cell wall, many gas vacuoles, and apparently a pigment similar to bacteriorhodopsin. These characteristics indicate that the square bacteria may be members of the archaebacteria.

Pleomorphism. Pleomorphism can be defined as the existence of an organism in different morphological forms. Environmental factors influence the size and sometimes the shape of bacteria. Size variations, which are more readily detected in rod-shaped microorganisms than in spherical cells, are associated with changes in the biosynthesis and growth of the cell wall. Rod-shaped cells, cultured in media in which the cells are dividing at a rapid rate, are two to three times larger than cells that are dividing at lower rates. This variation in size is related to the growth-supporting ability of the medium as well as the ambient temperature for growth.

Microorganisms that are actively dividing and then deprived of nutrients produce ***autolytic enzymes*** that influence the shape of the cell by their activity on the cell wall. Many cells are destroyed by these autolytic enzymes, but others can be observed as abnormal or aberrant forms. Large filaments, for example, are especially noticeable in rod-shaped organisms when the growth medium contains penicillin or similar inhibitors of cell wall synthesis. Penicillin in low concentrations permits the cell to grow, but the cell wall does not maintain its normal shape, and filamentous extensions are produced (Figure 4-5).

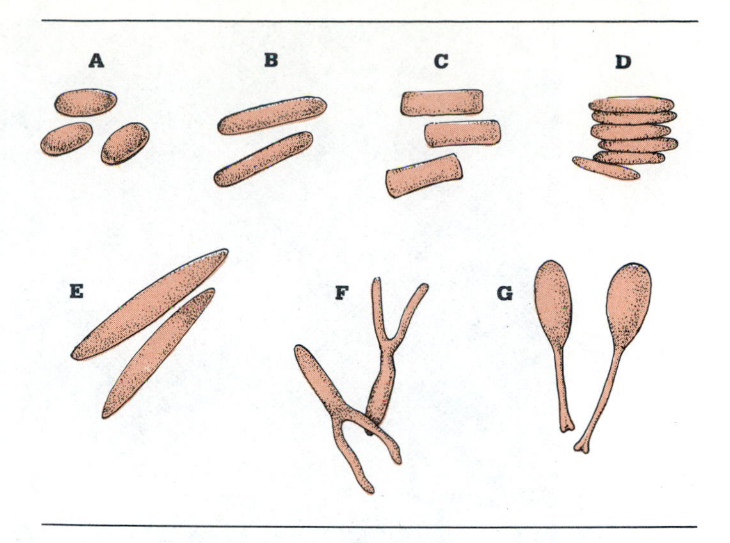

Figure 4-4 Variations in morphology and arrangement of rod-shaped bacteria. **A,** Coccobacillus. **B,** Rods with rounded ends. **C,** Rods with flat ends. **D,** Rods stacked in palisade arrangement. **E,** Cigar-shaped rods. **F,** Bifurcated rods. **G,** Rods with stalks.

Cell wall and cytoplasmic membrane

All bacteria except the mycoplasmas possess cell walls that protect the cell and determine to a great extent the cell's shape. The cell walls of nearly all bacteria, except the mycoplasmas, and archaebacteria possess a cell wall component called peptidoglycan. The bacteria have been divided into two groups based on the chemical components that lie outside this peptidoglycan layer: ***gram-positive*** and ***gram-negative*** bacteria. This division is based on the staining characteristics of the cell wall components. Hans Christian Gram discovered that some bacteria resist decolorization with alcohol following the application of a dye called crystal violet and iodine. Those microorganisms that resist decolorization and retain the crystal violet–iodine complex are called gram-positive, but those that are decolorized can be counterstained with a red dye, called safranin, and are referred to as gram-negative

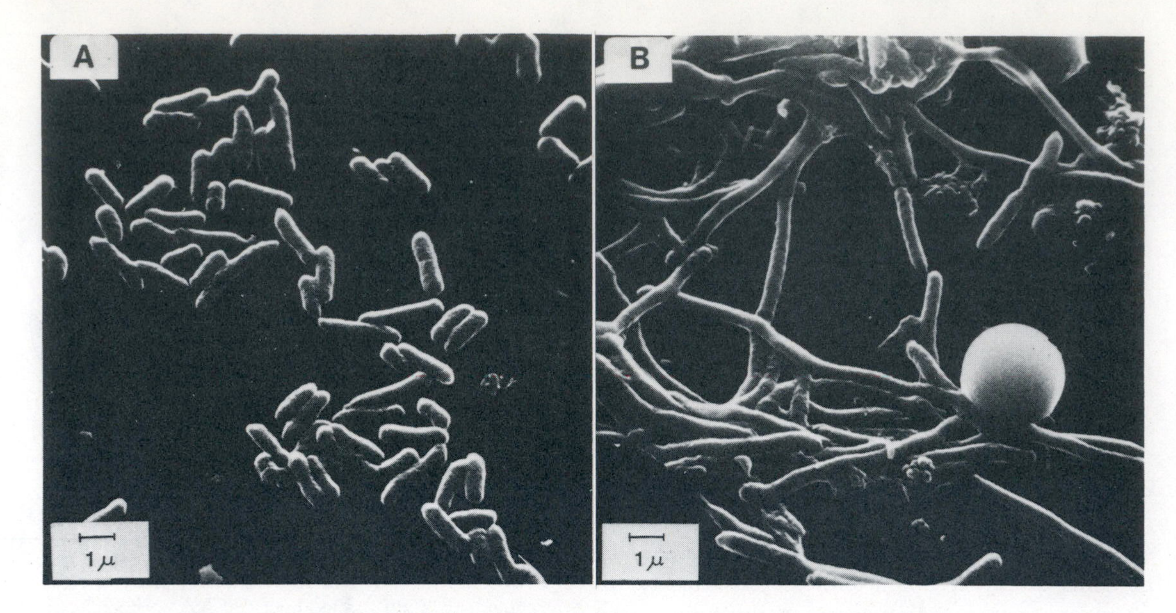

Figure 4-5 Effect of penicillin derivative (ampicillin) on growth and morphology of rod-shaped bacterium *Escherichia coli.* **A,** Untreated control. **B,** Treatment with ampicillin. Note formation of long filaments.

From Zak, O., and Kradolfer, F. Rev. Infect. Dis. *1(5):862, 1979.*

(Plates 2 and 3). Wall porosity is believed to be an important factor in determining the ability of cell wall components to retain or not retain the crystal violet–iodine complex, but other factors may also be important (see Chapter 7). Porosity is influenced by the cell wall lipid concentration, which is considerably higher in gram-negative bacteria than in gram-positive bacteria. Lipids are disrupted and extracted by alcohol, which alters membrane permeability. (See Appendix B for details of the Gram stain.)

The cytoplasmic membrane of bacteria, as well as other organisms, functions primarily in the transport of nutrients into the cell and wastes out of the cell. The cytoplasmic membrane of bacteria also serves as an anchor for the attachment of DNA during replication and possesses enzymes that function in energy production as well as cell wall synthesis. Those bacteria that do not possess cell walls have cytoplasmic membranes that are rendered structurally more stable by the presence of ***sterols,*** which are absent from wall-containing bacteria.

Reproduction

Sexual reproduction, a characteristic of more advanced forms of life, is absent in bacteria. ***Binary fission,*** an asexual process, is the most common method of bacterial reproduction and is characterized by an approximately equal separation of the cell into two daughter cells. Division is perpendicular to the long axis of the cell in rods and spherical cells (Figure 4-6) and is preceded by intercalary growth in which only a small portion of new cell wall is synthesized. The daughter cells, especially of spherical bacteria, do not always separate, and depending on the plane of division, different morphological arrangements are produced (see Figure 4-3).

Budding is another method of asexual reproduction but is relatively infrequent in bacteria, except for a few species. Budding, which is a more common type of reproduction in the eukaryotic yeasts, is usually polar, with one end of the parent cell bulging and producing a spherical daughter body that separates from the parent (Figure 4-7). Budding may appear to be similar to binary fission, but there is a difference. Budding is considered a reproductive process only if the daughter cell is smaller than the parent cell and all or most of the bud-cell wall and cytoplasmic material are newly synthesized. Variations of these reproductive processes are discussed later in the special groups with which they are associated.

Motility

Flagella. The ***flagellum*** (pl. flagella) is the principal organelle associated with movement in bacteria. It is a long, helical, unbranched filament ranging in length from 10 to 20 μm, which makes it several times longer than the average bacterial cell. It is so thin (0.2 μm), however, that it cannot be observed unstained with the compound mi-

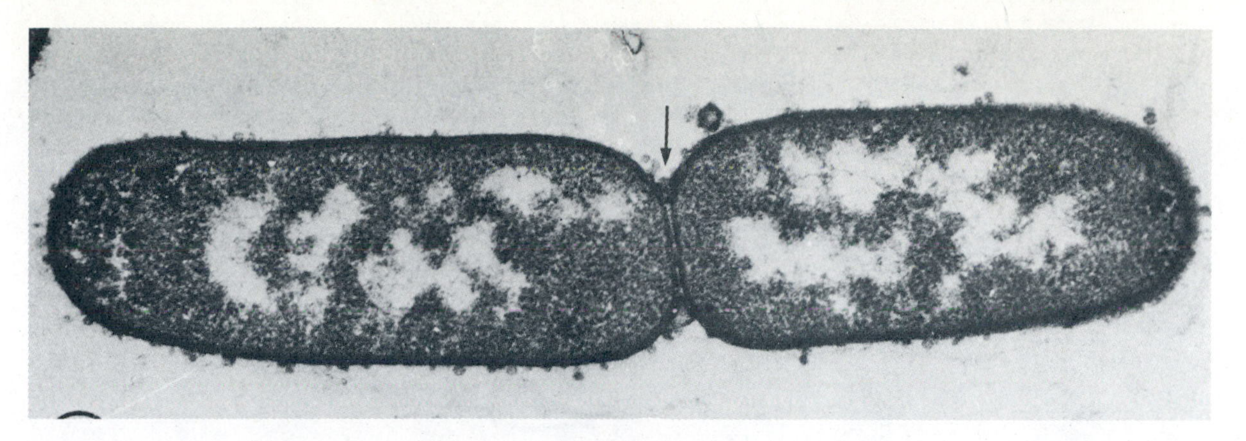

Figure 4-6 Division in rod-shaped bacterial cell. Division is perpendicular to long axis of cell *(arrow)*. Point of indentation where cell divides is called septum
From Burdett, I.D.J., and Murray, R.G.E. J. Bacteriol. ***119:****1039, 1974.*

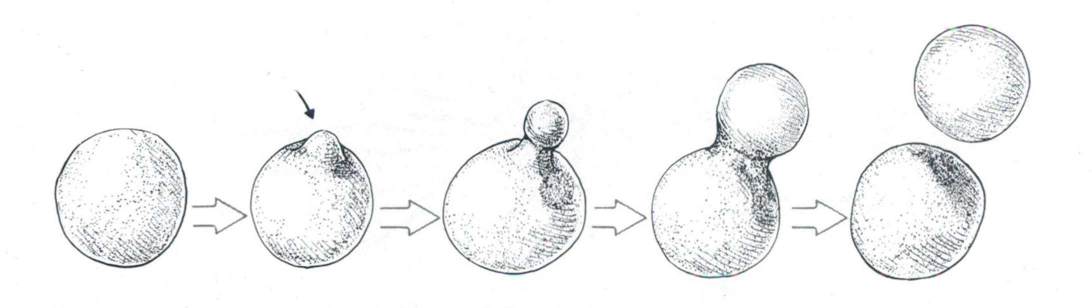

Figure 4-7 Budding as division process in bacteria. Daughter cell *(arrow)* arises as polar outgrowth of mother cell.

croscope. Flagella may be arranged on the cell according to four basic types (Figure 4-8): ***peritrichous*** (flagella cover the entire cell surface), ***monotrichous*** (a single polar flagellum), ***lophotrichous*** (two or more flagella at one pole of the cell), and ***amphitrichous*** (tufts of flagella at both poles of the cell). The arrangement of flagella on an organism is constant, and this is of some taxonomic value. Recent studies have shown that nutritional alterations can cause variations in the flagellar arrangement.

Flagella are better suited for movement through relatively nonviscous liquids. When the viscosity is increased, flagellar locomotion is severely impeded, probably because the flagella are directly exposed to the environment. Some flagellated species such as *Pseudomonas aeruginosa,* which is 3 to 4 μm in length, can move 37 cell lengths/sec. This is a remarkable rate, since horses move about 6 lengths per second.

Flagella-like filaments called ***axial filaments,*** which do not project from the surface of the cell but are wrapped about the body of the cell, are characteristic of the spirally shaped bacteria called ***spirochetes*** (Figure 4-9). Spirochete locomotion is not hampered when the viscosity of the medium is increased, and in many instances the velocity actually increases. This is probably due to the location of the axial fibers, which are beneath the outer sheath of the cell. The axial fibers are therefore not in direct contact with the extracellular environment. The velocity of the spirochetes depends on the viscosity of the medium and is somewhere between 20 and 30 μm/sec.

Gliding motility. Some prokaryotes, especially cyanobacteria and myxobacteria, show a gliding type of motility over solid or semisolid surfaces in which no apparent organelle has been identified. The only characteristic common to all of the gliders is the secretion of an extracellular slime layer outside the cell wall. Gliding may be associated with single cells or a filament of cells (trichome). The velocity of movement is affected by moisture levels, temperature, and the nature of the substratum. Some single-cell gliders move only 1 to 2 μm/min, whereas others may move as fast as 150 μm/min. Trichomes of some cyanobacteria move as rapidly as 600 μm/min. None of these velocities compares with the rapid movements of flagellated bacteria. The exact mechanism(s) for gliding movement is as yet unknown.

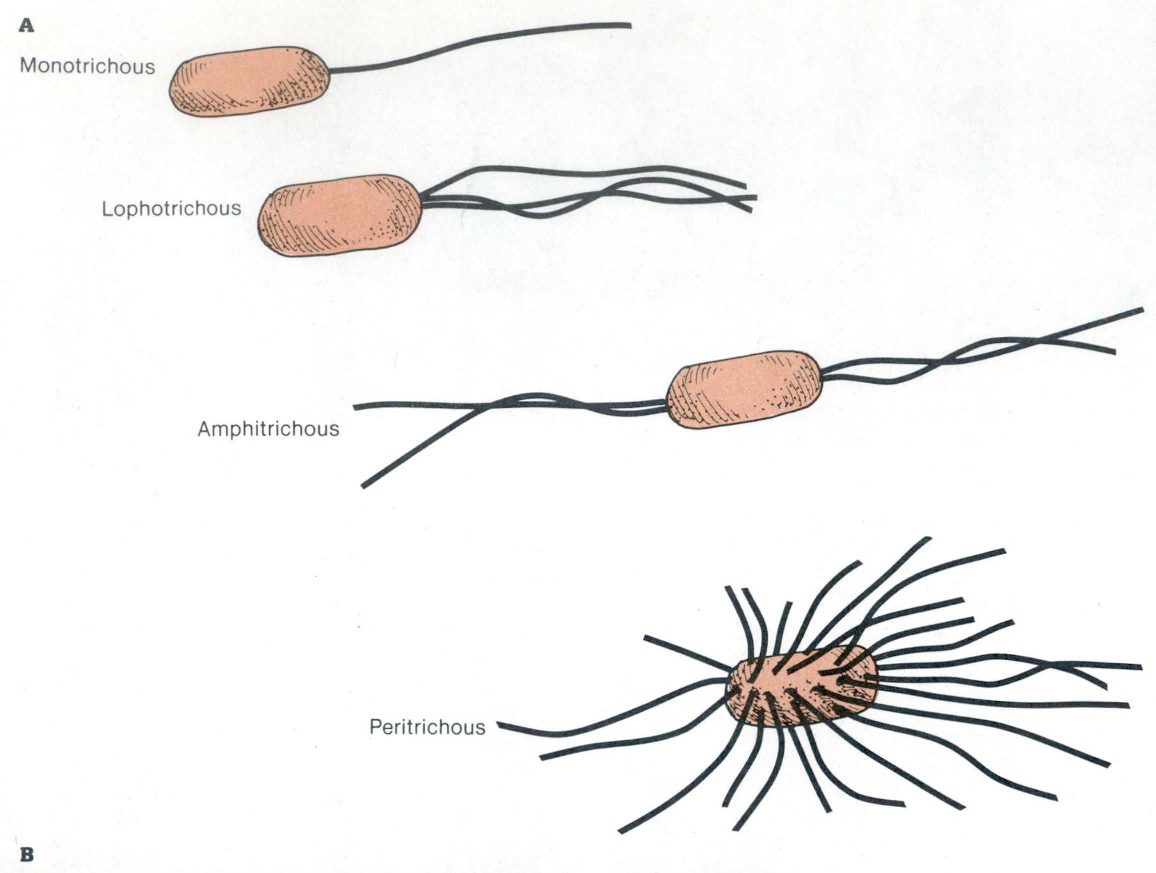

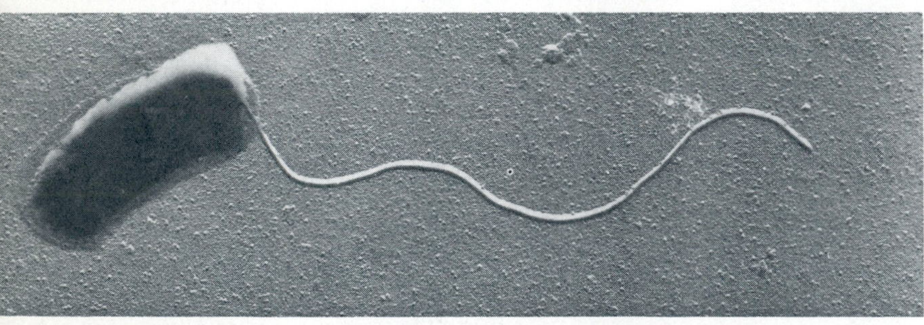

Figure 4-8 **A,** Flagellar types. **B** to **D,** Electron micrographs of some representative flagellar types. **B,** Monotrichous type, species of *Vibrio*. (×24,000.)
***B** and **C,** Courtesy W. Hodgkiss. **D,** From Hoeniger, J. F. M. J. Gen. Microbiol. **40:**29, 1965).*

Functions of motility. Many microorganisms exhibit some type of motility, yet there are many that do not. What is the advantage of motility to a microorganism? Is locomotion necessary for the survival of the cell? Independent movement in bacteria is either directed to or away from some stimulus in the environment. Some microorganisms, because of their particular environmental niche, are directed to sources of nutrients and away from toxic factors by different locomotor responses. Movement to or from a chemical in the environment is called ***chemotaxis*** (see Chapter 7). Motility is usually associated with single cells, but this need not be the case. The gliding bacteria appear to use motility to form aggregates of cells that swarm as a single unit. ***Swarming*** may be a mechanism for feeding or for forming a fruiting body, which may be of survival value to the species.

Cellular interactions

Prokaryotes, for the most part, do not exhibit cell-to-cell interactions. Many bacteria form ***colonies*** when they reproduce on or in a substratum (Figure 4-10); these represent random accumulations of cells that do not enter into a cooperative state. Bacteria within the colonies are also competing for existing nutrients and do not differentiate into groups with specific activities, which is characteristic of multicellular organisms. Eukaryotic microorganisms produce more complex colonies in which there is coordination of activities, and it is this colonial coordination of cells that has permitted eukaryotes to evolve into larger multicellular life forms. There are some prokaryotes, however, that produce colonies in which some of the cells exhibit specific roles. The rod-shaped cells of the actinomycetes, for example, develop into long filaments, some of which penetrate the substratum to seek nutrients, whereas other filaments rise above the substratum and produce aerial spores,

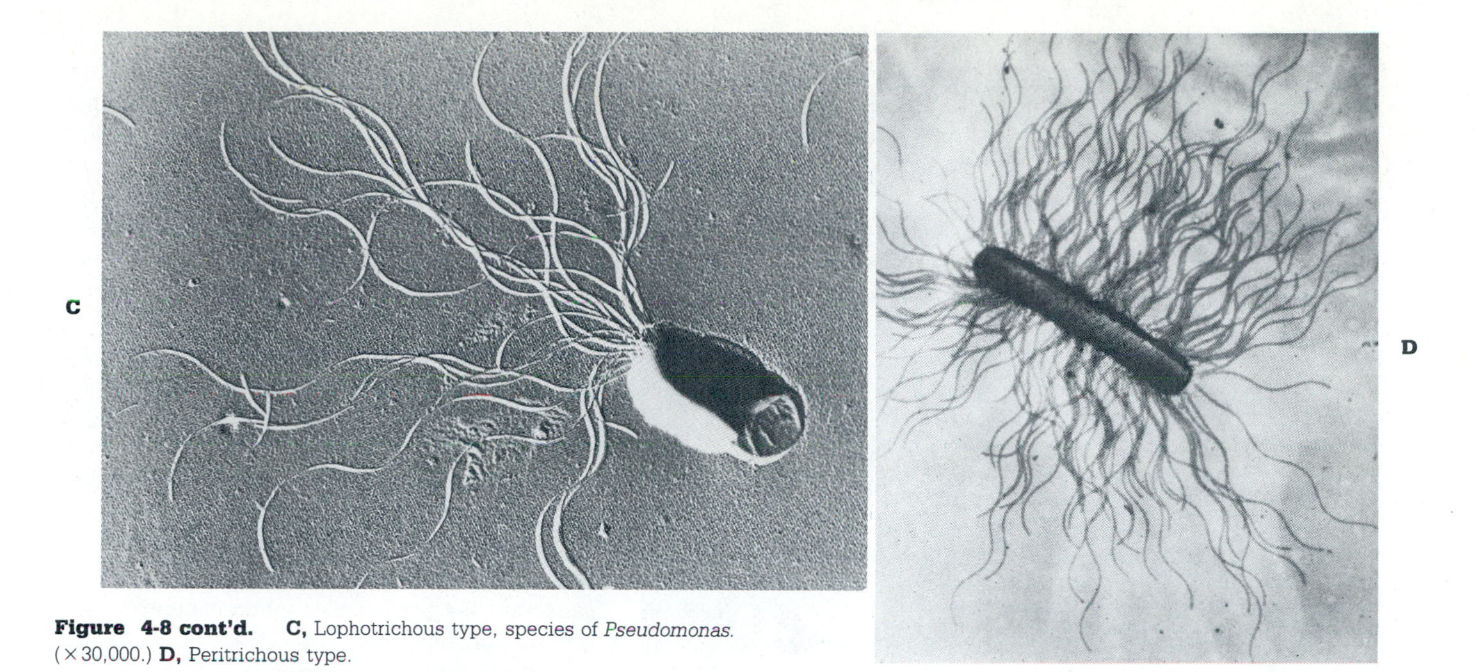

Figure 4-8 cont'd. **C,** Lophotrichous type, species of *Pseudomonas.* (×30,000.) **D,** Peritrichous type.

a cell type that can be released to the surface to produce individual colonies.

Symbiosi Most environmental niches contain various groups of organisms that interact to create a variety of relationships called ***symbioses.*** Microorganisms can interact with plants and animals as well as with other microorganisms to create ***endosymbiotic*** or ***ectosymbiotic*** relationships. Microorganisms that exhibit endosymbiosis grow within the host cell, whereas those microorganisms that remain attached but outside the host cell demonstrate ectosymbiosis. Ectosymbiotic or endosymbiotic relationships can be categorized as ***mutualistic*** if both partners benefit from the association; ***commensalistic*** if one partner is benefited and one is unaffected; and ***parasitic*** if one organism is harmed and the other is benefited. The most widely studied and clearly defined types of symbioses have been those existing between microorganisms and plants or animals. Bacteria can interact with other microorganisms, but this type of relationship has been difficult to evaluate in their natural environment, and most of what we know comes from laboratory studies. Many bacteria, because of their small size, are the prey for eukaryotic microorganisms, and laboratory studies have demonstrated that the ingested prokaryote occasionally survives in its host and becomes an endosymbiont. The eukaryotic microorganism is always the host when endosymbiosis occurs, and this is believed to be due to the greater flexibility

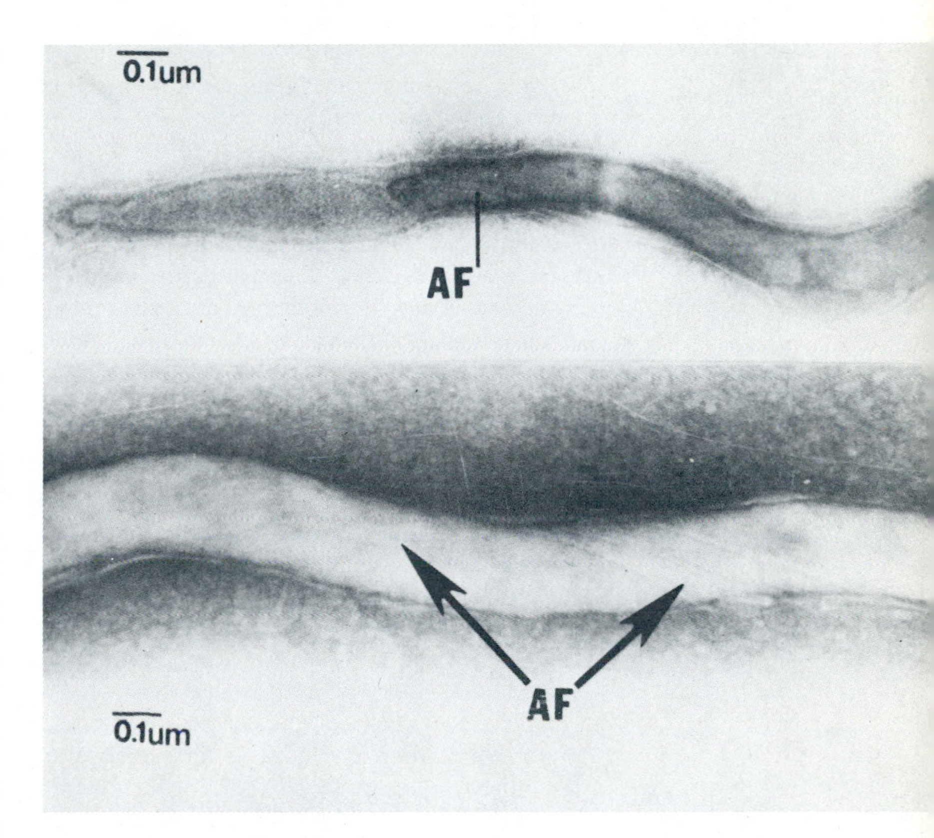

Figure 4-9 Flagella-like filaments called axial filaments *(AF)* of spirochetes. Axial fibers *(AF)* are seen to wrap about body of cell.

*From Holt, S.C. Microbiol. Rev. **42:**114, 1978.*

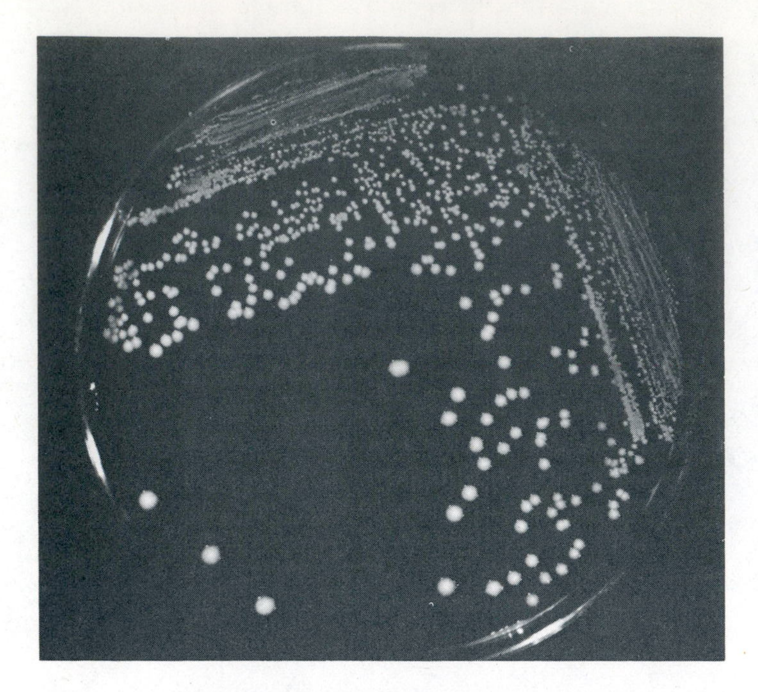

Figure 4-10 Bacterial colonies on agar surface. Colonies appear as they would to unaided eye.

of its cytoplasmic membrane. The cytoplasmic membrane of eukaryotic microorganisms can surround and engulf the prokaryote, but the prokaryotic cytoplasmic membrane is inflexible, and large particles, such as microorganisms, cannot penetrate into the cytoplasm. Prokaryotes can become hosts only when the other partner engages in ectosymbiosis. Symbiotic relationships have great ecological significance and are discussed in more detail in later chapters.

Classification

Taxonomy is a means of classifying organisms based on mutually similar characteristics. The basic classification unit is the ***species.*** A bacterial species can be defined as a distinct member of a group of microorganisms that possesses similar characteristics yet differs significantly from related groups of bacteria in other independent characteristics. In certain instances a member of this distinct group may not be totally identical to the other members of the species and is referred to as a ***strain.*** *Escherichia coli,* for example, could be manipulated in the laboratory to produce an organism with an altered response to a bacterial virus. We would still call the organism *E. coli,* but with an arbitrary designation, such as K10. Any letters, numbers, or names following the species name indicate the strain. A species may therefore be defined as a group of related strains. Taxonomists often disagree as to which characteristics are considered to be most crucial in separating and delineating the bacterial species. Consequently, many changes in classification occur as new microbial features are elucidated. The characteristics that have been used to determine the degree of similarity between organisms include morphological, biochemical, ecological, and serological. The importance of each characteristic is and will always be a matter of debate. Ideally one strives to have a natural classification that is based on evolutionary ***(phylogenetic)*** trends as well as biological relationships. There is very little that the taxonomists can do about evolutionary trends because bacterial microfossils discovered in Australia and elsewhere have told us the age of bacteria (3.5 billion years) and general shape but little else. The history of the bacterial cell is also recorded in the sequence of amino acids in proteins and nucleotide sequences of DNA and RNA, and their analysis has provided some important discoveries to be discussed later.

Nomenclature (naming of species). The method for naming an organism is based on the ***binomial*** system of nomenclature, first used by Linnaeus in the eighteenth century. It employs two words, which are given Latin endings, to describe a species. The first word indicates the ***genus*** (pl. genera), with the first letter of the genus capitalized. The second word is not capitalized and indicates the ***species*** of the genus. Both genus and species are italicized, as in *Bacillus anthracis.* The genus and species names are very descriptive and designate certain characteristics to the reader that may be based on the organism's metabolism, shape, or colony color. Genera such as *Bacillus* (rod), *Streptococcus* (chain of spherical cells), and *Spirillum* (spiral) indicate shape, whereas species names such as *aureus* (gold) and *albus* (white) indicate the color that develops when the organism forms a colony. Not all genus-species nomenclature is so descriptive, and some organisms are named according to the individual who isolated them or the geographic site from which they were isolated. Theodor Escherich, a German microbiologist, for example, first described the organism we now call *Escherichia coli.*

The classification of bacteria began early in the twentieth century and was based on specific physiological characteristics, but since that time classification has undergone constant revision. Currently, the most extensive classification scheme for the bacteria is *Bergey's Manual of Determinative Bacteriology,* which first appeared in 1923. The classification originally included not only genera and species but other related divisions as well. Related genera were grouped into families, families into orders, orders into classes, classes into phyla, and phyla into a kingdom. The taxonomic groups have undergone many changes, and in the current eighth edition the kingdom is called ***Procaryotae,*** with two divisions: **division I, *Cyanobacteria,*** and ***division II, Bacteria.*** The bacteria are divided into ***19 parts,*** each with a specific categorical name. The separation into parts is based primarily on the Gram stain, but other parameters are also used, including morphology, oxygen requirements, physiological responses, and spore formation. Each part includes a number of genera, and some genera are grouped into families and orders. These divisions are outlined in Appendix C. The characteristics of all 19 parts are discussed shortly.

$$\%S = \frac{Nsp + Nsn}{Nsp + Nd + Nsn}$$

Figure 4-11 Formula for determining similarity coefficient for two bacterial strains. *Nsp,* Number of similar positive matches; *Nsn,* number of similar negative matches for both organisms; *Nd,* number of dissimilar matches, that is, positive for one organism but negative for other organisms; *%S,* percent similarity.

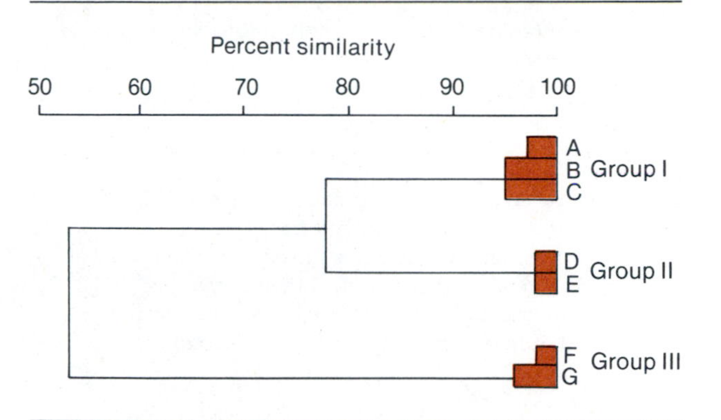

Figure 4-12 Hypothetical dendrogram showing similarity profile of seven bacterial strains A through G. Strains may be members of same or different species. Groups I, II, and III represent separate species. Groups I and II could contain members of same genus, since their similarity profile is about 80%, but Group III's similarity to Groups I and II is only 50%; therefore they probably belong to another genus.

Approaches to taxonomy

Numerical taxonomy. Phylogenetic relationships based on sequences of amino acids in proteins and nucleotides in RNA and DNA are still in their infancy. The taxonomist must therefore rely on other techniques for demonstrating relatedness among microorganisms. One approach is ***numerical taxonomy,*** in which characteristics such as morphology, staining, and biochemical and physiological traits are compared. Numerical taxonomy was first suggested by Adanson (adansonian taxonomy), who showed that the degree of similarity between two species is best determined by comparing as many characteristics as possible. Each characteristic is given some value, often of equal weight (although this need not always be true), and from such data a similarity profile or dendrogram can be established. Similarity between pairs of organisms can be determined using a formula that compares similarities only, or a formula that compares positive as well as negative matches (Figure 4-11). A computer is useful in matching characteristics. After the pairs of organisms have been compared, the dendrogram allows one to place individual microorganisms into specific groupings and also to join groups based on similarities (Figure 4-12). If two organisms show a 90% or more similarity in characteristics, they are usually considered as belonging to a single species. The number of identifiable characteristics increases as new techniques or tests become available, and this is one reason that classification schemes change so rapidly.

Genetic analysis. Since the advent of molecular biology, genetic analysis has revealed much about the relatedness of microorganisms. A technique for sequencing bases in specific genes, for all microorganisms, would be the ideal tool for classification. This technique is not yet available, but there are several genetic approaches that have been used to determine DNA relatedness between microorganisms: ***genome size, guanine plus cytosine content,*** and ***hybridization techniques.***

The genome size of bacteria varies from 1×10^9 to 1×10^{10} daltons, a difference that is not large enough to be used as a parameter for separation. Many groups of microorganisms, for example, that differ considerably in other characteristics may have a similar genome size.

The ratio of G:C and A:T in the DNA molecule is one, since guanine pairs with cytosine and adenine with thymine. The molar ratio of G:C base pairs to A:T base pairs, however, will vary depending on the species of microorganism examined. The ratio will remain the same for strains from the same species. The G:C content ranges

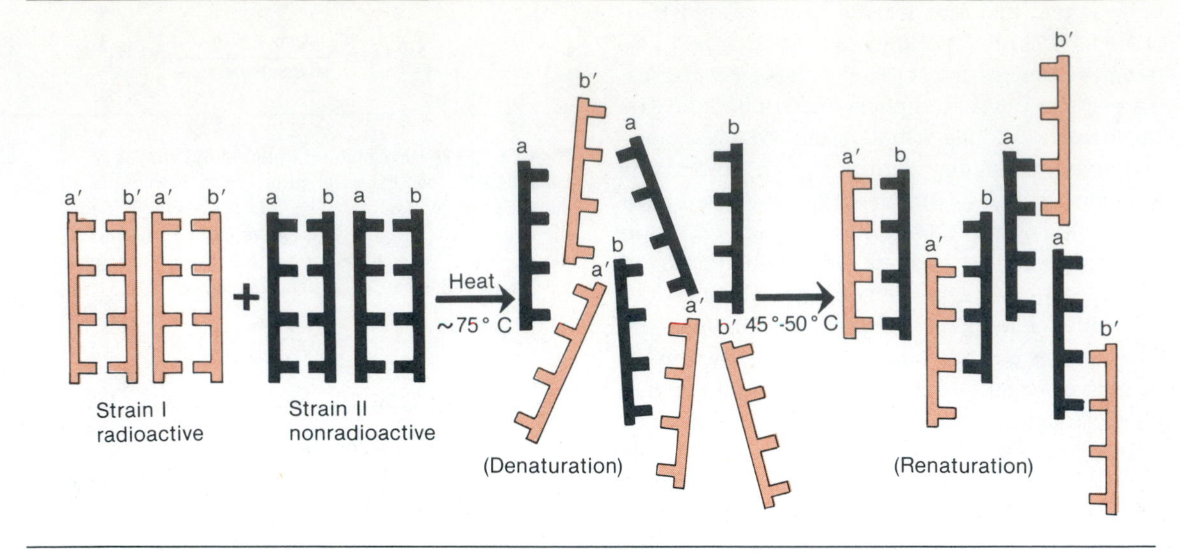

Figure 4-13 Hybridization experiment. DNA is isolated from strain I, which was grown in radioactive isotope (color) and from strain II, which was grown in nonradioactive medium (black). DNA from the two strains is mixed and then heated at 75° C to denature molecules and produce single strands. Temperature is then reduced 25° to 30° C below denaturation temperature, permitting homologous areas from different strands to bind, for example, a′ from strain I binding to b from strain II and b′ from strain I binding to a from strain II.

from 23% to 75%* in bacteria and is determined by finding the temperature at which the DNA melts, that is, the temperature at which the double-stranded DNA unwinds and becomes denatured. There is a gradual increase in absorbency, at wavelengths of 260 nm, as the DNA slowly unwinds. The denaturation process, or breaking of hydrogen bonds, is directly related to the G:C content because of the extra hydrogen bond existing between these nucleotide pairs. Two microorganisms having the same G:C content do not necessarily belong to the same species, but a wide variation in G:C content indicates that the two microorganisms should be placed in different species.

DNA melting results in the separation of single strands; therefore DNA strands from two different microorganisms can be mixed to determine the degree of homology or reassociation between them. ***Reassociation*** refers to the binding (annealing) of complementary DNA strands. Ordinarily the rapid cooling of melted DNA does not permit the binding of complementary DNA strands, but reassociation can occur if the temperature is held 10° to 30° C below the melting temperature. Single strands from one strain can therefore reassociate with single strands from a second strain to form ***hybrids.*** Related microorganisms show a greater degree of reassociation between their DNA strands than do unrelated species. A typical experiment involves cultivating strain I in a medium in which a specific radioactive isotope is incorporated into the DNA while strain II is grown in a medium that contains no isotopes. The DNA is then extracted from both strains. Strain I contains radioactive DNA, whereas the DNA from strain II is nonradioactive. The two DNA samples are mixed, denatured, and then reassociated at a temperature below the melting temperature. If the two strains are totally unrelated, then the pairing that takes place is primarily between the complementary single strands of the DNA from the same organism; that is, there are few if any hybrids (Figure 4-13). There is usually some hybridization, the degree of which depends on the relatedness of the two microorganisms. The hybridization specificity can be determined by performing a reassociation experiment at two different temperatures. The greater the degree of homology between DNA strains, the lesser the effect high temperature has on the binding. If the degree of homology is low (fewer specific base pairings between strands), a higher temperature will reduce the reassociation (Table 4-2). Strains from the same species typically show 60% to 100% relatedness at the higher temperatures, but strains from different species show less than 50% relatedness.

* Percent G + C = $\frac{G + C}{A + T + G + C} \times 100\%$

Table 4-2 Effect of temperature on the hybridization of DNA from different strains*

SPECIES	PERCENT OF HYBRIDIZATION AT 60° C†	75° C‡
A/A	100	100
A/B	75	74
A/C	72	40
A/D	25	40

*This hypothetical experiment was designed to compare species A with three other organisms: B, C, and D. A/A is a control in which two strains of A have been normalized to 100%. From the data it appears that at an incubation temperature of 60° C, organisms B and C are related, but when the values are compared at higher temperatures there is considerably less homology between them. Organisms A and B, although not the same species, show a degree of relatedness. Organism D is unrelated to A, B, or C.
†This value is 30° C below the melting temperature for DNA.
‡This value is 15° C below the melting temperature for DNA.

Sequence analysis. Macromolecules act as documents of evolutionary history. Sequence analysis of informational molecules such as nucleic acids and proteins provides a historical clue to the relatedness between microorganisms. Proteins such as ***cytochrome*** c and ***ferredoxin,*** which are involved in electron transport, have been sequenced in various organisms. This technique has been invaluable in speciating some groups of bacteria, such as the purple nonsulfur bacteria. The same proteins are not always present in the different organisms examined and, even when they are, are not always similar in terms of structure. Protein analysis therefore has its limitations. The ribosome, which is involved in protein synthesis and is present in all organisms, contains a ribonucleic acid component ***(16S rRNA),*** which is being used to show the relatedness or unrelatedness of various species. The 16S component possesses two regions, one of which has shown variability during the passage of time while the other has remained fairly stable. Since the molecule possesses a large number of bases in both regions, a statistical analysis of recent as well as distant relationships can be established.

The 16S rRNA of the organisms to be compared is digested with a viral nuclease. The digestion products consist of a variety of oligonucleotides (pentamers [5], hexamers [6], heptamers [7], etc.) Comparisons are made between different organisms, reflecting the number of oligonucleotide sequences common to them. The process is expensive and time consuming and is used only as an adjunct to other studies. Several

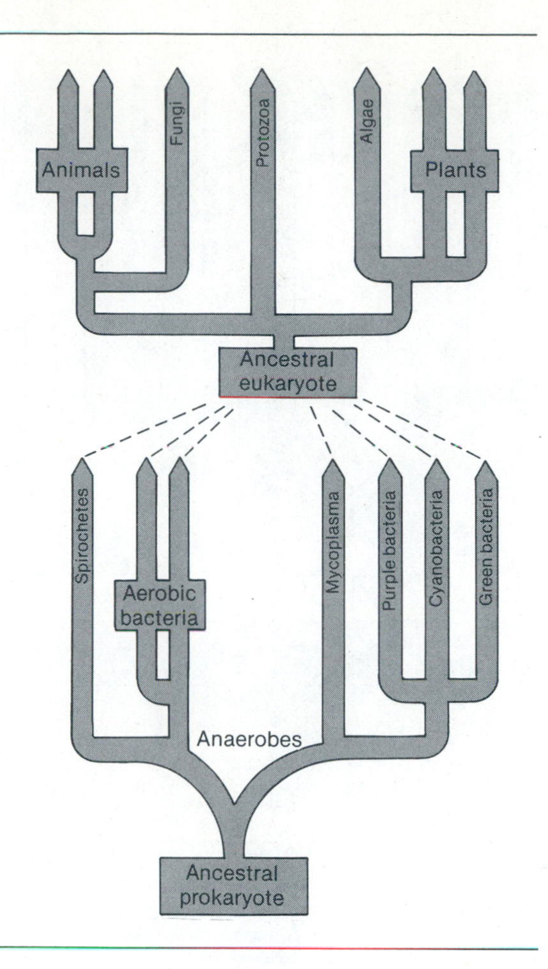

Figure 4-14 Phylogenetic tree based on conventional taxonomic schemes.
*Modified from Woese, C.R., Sci. Am. **244**(6):98, 1981.*

species of *Bacillus* in one study, for example, had been placed in three different groups based on spore shape and swelling of sporangium, and this rather arbitrary separation was actually confirmed by using 16S rRNA analysis. Analyses of this type have also demonstrated that the chloroplast 16S rRNA of red algae is not related to its own cytoplasmic 18S rRNA but is more closely related to bacterial 16S rRNA. These studies have suggested that the chloroplast may have evolved from endosymbiotic bacteria.

Over 170 individual species have been characterized using 16S rRNA sequencing. The results of these studies have demonstrated that the older traditional methods used to define taxonomic levels may not always be reliable. Before these studies it was believed that prokaryotes derived from an ancestral prototype and that eukaryotes in turn developed from prokaryotes (Figure 4-14). The first line of prokaryotes was believed to have been anaerobic cells that derived their energy from fermentation. Later, when oxygen appeared in the atmosphere, certain anaerobes, that had

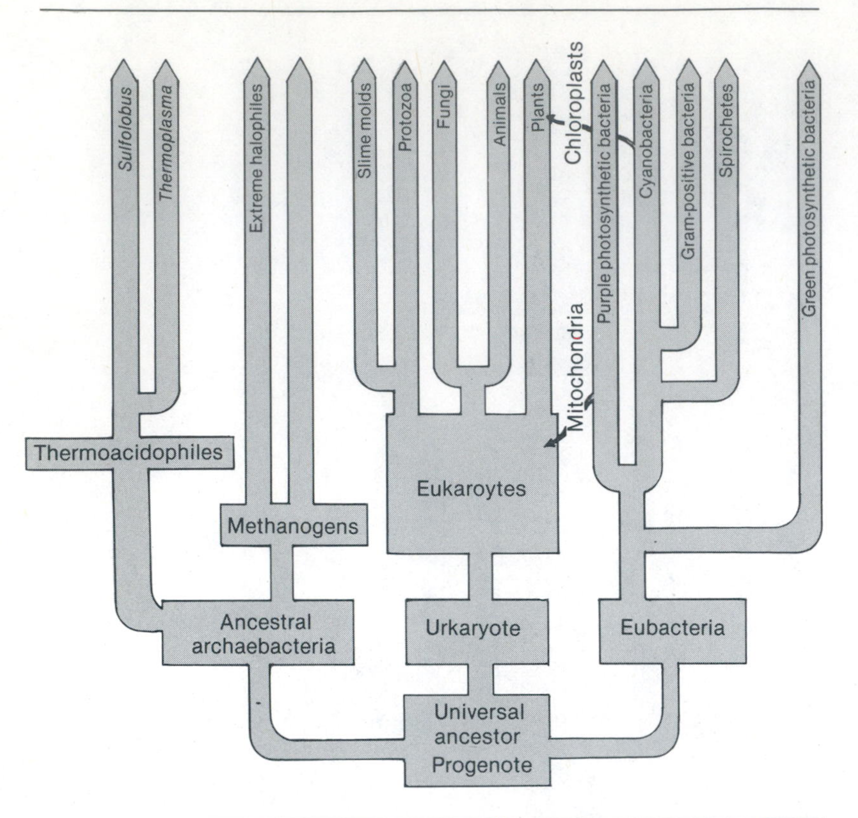

Figure 4-15 Major lines of prokaryotic descent derived in part from 16S rRNA sequence analysis.
*Redrawn from Fox, G.E., et. al. Science **209:**457, 1980. Copyright 1980 by the American Association for the Advancement of Science.*

lost their cell wall (the mycoplasmas) ingested smaller bacteria, with whom an endosymbiotic relationship developed. This cell type is believed to have evoled into the mitochondrion, the organelle of all eukaryotes. A photosynthesizing cyanobacterium could also have developed into the chloroplast, a structure found in all plant cells.

The 16S rRNA studies have led several investigators to suggest that there are three lines of descent from a common ancestor called the ***progenote*** (Figure 4-15): the ***eubacteria*** or true bacteria, which represent the majority of bacterial species; the ***archaebacteria,*** which contain the methanogens, extreme halophiles, and certain thermoacidophiles; and the ***urkaryote,*** or original eukaryotic cell. Eukaryotes therefore evolved only when the urkaryote became a host for bacterial endosymbionts and, according to the scheme, cannot be considered as a descendant comparable to the two bacterial lines. The three groups of archaebacteria are found in extreme environments. The methanogens are found in bogs, lake sediments, sewage treatment plants, and rumen of ruminants, where oxygen is absent and hydrogen is present. The extreme halophiles live in areas of high salinity. They have a simple photosynthetic system based on membrane-bound bacterial rhodopsin and not chlorophyll. The thermoacidophiles are found in hot acid environments such as sulfur springs and acidic coal piles. Archaebacteria and eubacteria are both prokaryotic cells but are considered separate because of some of the following differences:

1. Muramic acid, a component of all cell walls in eubacteria, is absent in archaebacteria.
2. Archaebacteria have membranes whose major component is a branched-chain, ether-linked lipid. Eubacteria have ester-linked, straight-chain lipids.
3. Transfer RNA in archaebacteria is devoid of ribothymidine in the thymidine-pseudouracil-cytosine loop, which is present in eubacteria.
4. Archaebacteria have a distinctive RNA polymerase subunit structure not found in eubacteria.

CHARACTERISTICS OF THE MAJOR GROUPS OF PROKARYOTES

Prokaryotes can be found in any type of habitat that is capable of supporting the growth of living organisms. They are found in the Arctic, in the depths of oceans and lakes, in hot springs, in highly acid or alkaline environments, and as parasites of living organisms. Many prokaryotes, like the fungi that are found in the soil, play an important role in the decomposition of organic materials and in recycling certain elements, such as sulfur, in the natural environment. They also inhabit animals, including humans—a relationship that may be beneficial, harmful, or neutral, depending on the species and the state of the host.

It is not possible to describe in any detail the many characteristics of all the prokaryotes classified in *Bergey's Manual* (see Appendix C). More detailed discussions of the important aspects of genetics, metabolism, ecology, or pathogenesis of these groups are presented in later chapters.

Cyanobacteria

The cyanobacteria belong to a group of prokaryotes commonly called phototrophs, which also includes members of division II, Phototrophic Bacteria. The term *phototroph* means that the organism obtains energy from light. The cyanobacteria were originally called blue-green algae. The older terminology was not without justification, since the cells of this group are considerably larger than bacteria and approach algae in size. The cyanobacteria possess the photosynthetic pigment chlorophyll a, which is involved in a process that results in the evolution of oxygen, a characteristic associated with eukaryotic algae. Chlorophyll a and the pigment phycocyanin are responsible for the blue-green color. Investigations into the cellular organization of the cyanobacteria have revealed that they are truly prokaryotes; they possess a gram-negative cell wall that contains peptidoglycan, there is no nuclear membrane, and there are no organelles, such as mitochondria or chloroplasts. In evolutionary terms it is believed that this group arose in adaptation to the depletion of organic nutrients from the primordial environment and evolved means of making their own organic materials. Once the cyanobacteria became members of the microbial community, they became the major source of oxygen in the environment. The cyanobacteria may represent a special link between prokaryotes and eukaryotes, especially photosynthetic eukaryotes. The nucleotide sequences of chloroplast ribosomal RNA is specifically related to ribosomal RNA sequences in cyanobacteria.

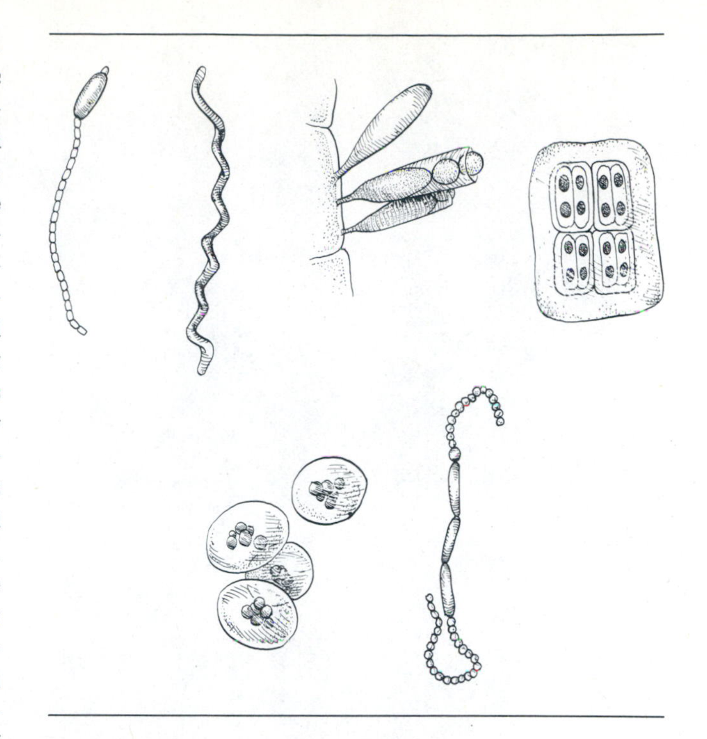

Figure 4-16 Morphological types of cyanobacteria.

General characteristics. Various groups of cynobacteria exist in marine and freshwater environments; some inhabit hot sulfur springs; and many are common to all types of soil. They range in size from the smaller unicellular rods and cocci to the filamentous types (Figure 4-16). The unicellular types divide either by binary fission or by budding. The filamentous types (a filament is called a ***trichome***) multiply by breaking into smaller units of cells called ***hormogonia.*** The trichomes of some groups contain, in addition to their chain of vegetative cells, differentiated cell types called ***heterocysts*** and ***akinetes,*** which are discussed later. Some but not all species exhibit a gliding type of motility that is associated with an extracellularly produced slime layer. Cyanobacteria produce other extracellular products, one of which is called anatoxin a, a compound similar in structure to cocaine, which acts as a neuromuscular blocking agent. A compound called ***geosmin*** is also produced by cyanobacteria and is responsible for the earthy odor associated with drinking water contaminated by these organisms. Geosmin is also produced by soil bacteria, the actinomycetes (p. 84).

The photosynthetic process is discussed in more detail in Chapter 13, but a fundamental aspect of this metabolic process is discussed here in order to give some understanding of one of the differences between phototrophic microorganisms. Photosynthesis, like other metabolic processes, uses electron donors. The electron donor in the cyanobacteria is water whose oxidation results in the liberation of oxygen ***(oxygenic photosynthesis)**** and reducing power in the form of hydrogen. Reducing power is used to fix carbon dioxide or, more specifically, to convert it to cellular organic material. Other photosynthetic microorganisms, such as the green and purple bacteria, use different electron donors, and oxygen is not produced ***(anoxygenic photosynthesis).*** These differences will become more evident when the green and purple bacteria are discussed.

Cell wall and cytoplasmic membrane. The cell wall of cyanobacteria is a gram-negative envelope that is sensitive to penicillin and lysozyme, as are other prokaryotes that possess a cell wall. Pili project from its surface, as in other gram-negative bacteria, and several functions have been suggested for their presence, including adhesion to surfaces on which the cell glides. Outside the wall is a mucilagenous layer of slime (Figure 4-17).

The cytoplasmic membrane (also called plasmalemma), which abuts against the cell wall, is also the source of the photosynthetic membranes called ***thylakoids*** (Figure 4-17), which traverse

* There are certain nutritional and environmental conditions in which cyanobacteria can carry out anoxygenic (no oxygen is liberated) photosynthesis.

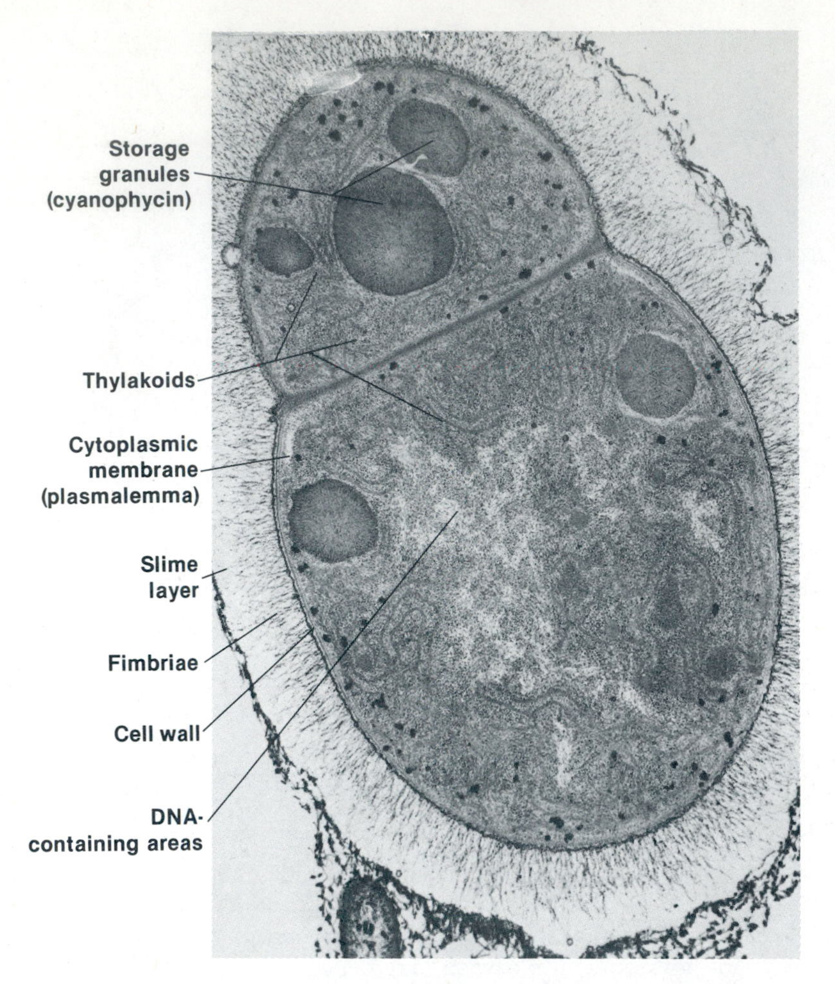

Figure 4-17 Electron micrograph of typical cyanobacterial cell.
Courtesy Lee V. Leak.

the cytoplasm of the cell. Most prokaryotes are unable to synthesize polyunsaturated fatty acids, and membrane lipids are usually esterified with saturated and monounsaturated fatty acids. Some cyanobacteria conform to this rule; however, others have fatty acids that resemble the eukaryotic chloroplast, that is, they are polyunsaturated fatty acids. This characteristic is helpful in taxonomic studies. Photosynthetic bacteria always use bacteriochlorophyll and carotenoids to capture light; however, the cyanobacteria use ***chlorophyll a, carotenoids,*** and ***phycobiliproteins,*** which are attached to the outer surface of the thylakoids in aggregates called ***phycobilisomes.***

Cellular inclusions. The cyanobacteria contain several types of inclusions, some of which are membrane bound and others of which are not. The nonmembrane-bound inclusions are the previously mentioned phycobilisomes and glycogen and cyanophycin granules (Figure 4-17), which are reserves of carbon and nitrogen, respectively. Membrane-bound inclusions include carboxysomes, which contain enzymes involved in carbon dioxide fixation and gas vacuoles, which provide buoyancy. These and other membrane-bound inclusions are discussed at more length in Chapter 7.

Differentiated cell types

Heterocysts. Two of the unusual characteristics of the cyanobacteria are their ability to carry out photosynthesis with the evolution of oxygen, and nitrogen fixation, which takes place under anaerobic conditions. These two processes can be carried out simultaneously in cyanobacteria by means of a cellular differentiation process in which some of the vegetative* cells of the filament give rise to specialized cells called heterocysts. Heterocysts are often spaced at regular intervals along the filament (Figure 4-18). After differentiation takes place, the vegetative cells of the filament are involved in photosynthetic reactions leading to the evolution of oxygen, whereas the heterocysts engage only in the fixation of nitrogen and photosynthetic adenosine triphosphate (ATP) production. The mechanism by which the cyanobacteria can accomplish two apparently mutually exclusive functions is discussed in Chapter 14.

Akinetes. Akinetes are spores found in the multicellular filaments of cyanobacteria. They are similar in appearance to vegetative cells but are considerably larger (Figure 4-19). In most of their physiological processes akinetes are similar to vegetative cells. Akinetes, although more resistant to environmental factors than vegetative cells, do not approach the degree of resistance of bacterial endospores. Akinetes do not fix nitrogen but do carry out photosynthetic processes like the vegetative cells. Akinetes often develop from vegetative cells that are adjacent to heterocysts, although akinetes are not always adjacent to them.

Phototrophic bacteria (part 1—Bergey)

General characteristics. The phototrophic bacteria have been given the common name of ***green*** and ***purple bacteria.*** Morphologically they may be single spherical or helical cells, whereas others may be filamentous or appear as cubical packets. Differences between the green and pur-

* *Vegetative* refers to cells that are actively metabolizing and reproducing and is in contrast to the dormant, nonreproductive spore state.

ple bacteria and the cyanobacteria are based on the type of photosynthesis (oxygenic or anoxygenic) and the nature of the photosynthetic pigments. Photosynthesis may be a common link between the green and purple bacteria, but there are differences between these two groups (Table 4-3). Unlike the cyanobacteria, which produce oxygen as a product of photosynthesis, the green and purple bacteria do not produce oxygen.

Sulfur metabolism is an especially important aspect of growth in the green and purple bacteria. The electron donor in anoxygenic photosynthesis is very often a sulfur compound more reduced than sulfate (other electron donors such as molecular hydrogen or thiosulfate are occasionally used). Hydrogen sulfide (H_2S) is the most frequently used electron donor among this group of bacteria, and its oxidation is used to fix carbon dioxide and to produce elemental sulfur. Elemental sulfur is deposited as sulfur granules or globules within the cell or outside the cell, depending on the species, according to the following reaction:

$$2H_2S + CO_2 \xrightarrow{\text{light}} 2\ S^\circ + H_2O + \underset{\text{(carbohydrate)}}{(CH_2O)}$$

The oxidation of hydrogen sulfide to elemental sulfur (S^o) takes place rapidly in the cell, but then the sulfur is slowly oxidized to sulfate according to the following reaction:

$$2S^\circ + 3\ CO_2 + 5H_2O \xrightarrow{\text{light}} 2H_2SO_4 + 3(CH_2O)$$

The overall conversion of sulfide to sulfate may be expressed as:

$$2H_2S + 4\ CO_2 + 4H_2O \xrightarrow{\text{light}} 2H_2SO_4 + 4(CH_2O)$$

The purple and green bacteria are characteristically found in the anaerobic layer of acquatic environments where H_2S, H_2, CO_2 and small amounts of organic substances are available. Hydrogen sulfide which is produced in the water by anaerobic microorganisms that reduce sulfate, is a toxic component but, as can be seen in the equations above, is photooxidized by the phototrophic bacteria to harmless sulfate. Thus a cycle driven by light energy is maintained between the anaerobic microflora and the phototrophic bacteria.

The phototrophic bacteria consist of four families, but only three are part of the Bergey classification. The three families are Rhodospirillaceae, Chromatiaceae, and Chlorobiaceae. The fourth family, Chloroflexaceae, include the filamentous gliding green bacteria, which could have been classified with the gliding bacteria, but because

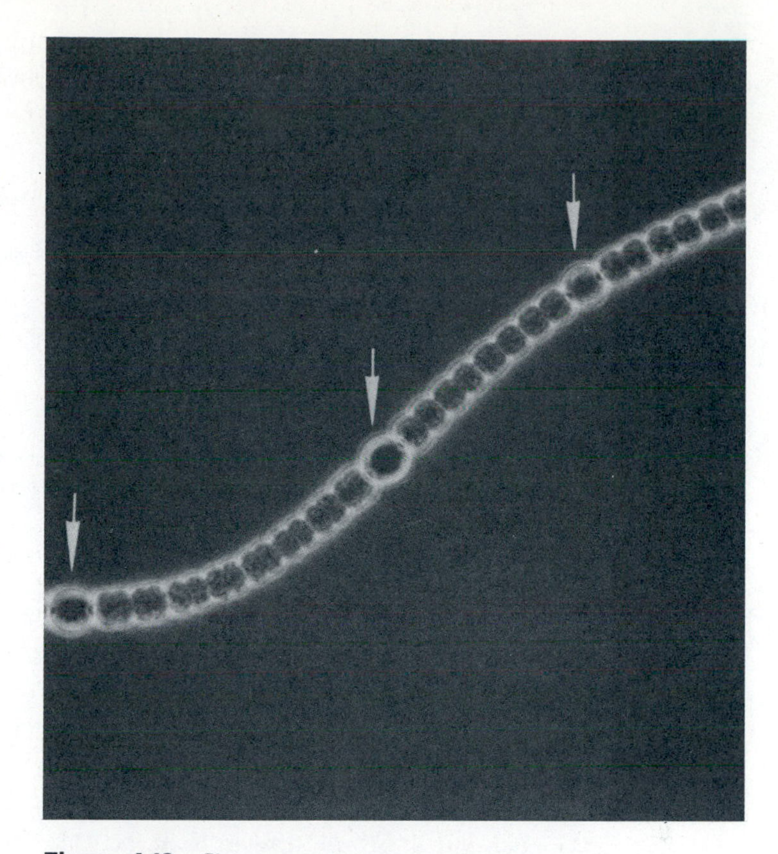

Figure 4-18 Phase-contrast photomicrograph of heterocysts *(arrow)* in cyanobacteria.
From Stacey, G., Bottomley P., Van Baalen, C., and Tabita, F.R. J. Bacteriol. ***137****:321, 1979.*

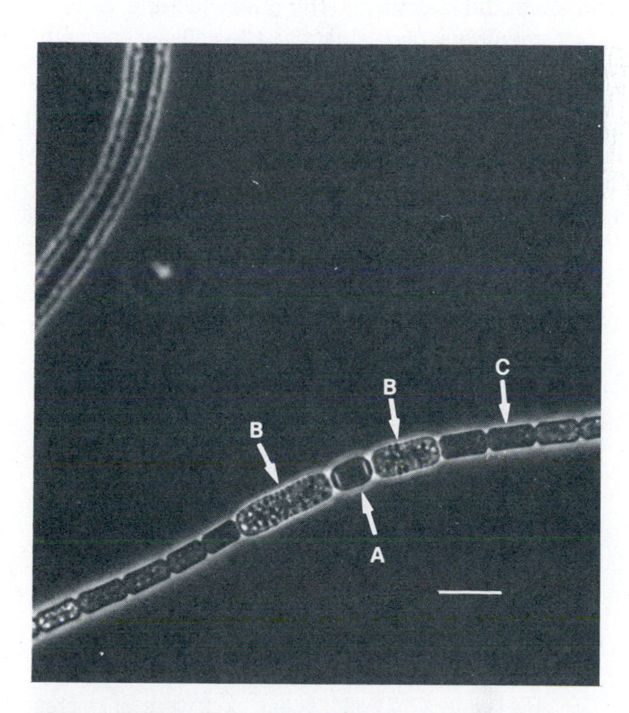

Figure 4-19 Phase-contrast photomicrograph of differentiated cells within filamentous cyanobacteria. Bar = 10 μm. *A*, Heterocyst; *B*, akinete; *C*, vegetative cell.
Courtesy N.G. Carr.

Table 4-3 Distinctions between the various photosynthetic prokaryotes, the cyanobacteria, and the green and purple bacteria

MICROBIAL GROUP	BERGEY CLASSIFICATION	CHLOROPHYLL	PHOTOSYNTHETIC MEMBRANE	MOTILITY
Cyanobacteria	Division I	Chlorophyll a	Photosynthetic membrane often multilayered and complex	Some species exhibit gliding motility
Purple bacteria	Division II (part 1) Family: Rhodospirillaceae Chromatiaceae	Bacteriochlorophyll a and b	Photosynthetic membrane apparatus continuous with cytoplasmic membrane	Most species flagellated
Green bacteria	Division II (part 1) Family: Chlorobiaceae	Bacteriochlorophyll c, d, and e	Vesicles called chlorobium vesicles attached to cytoplasmic membrane but not continuous with it.	Immotile

of their photosynthetic properties remain as phototrophic bacteria.

Characteristics of the green bacteria. The green bacteria *(Chlorobiaceae)* also have deep brown counterparts. The color difference is due to the possession of carotenoids by the brown species, which have a maximum absorption of light between 480 and 550 nm (the blue-green region of the spectrum). The green bacteria, except for members of the family Chloroflexaceae, are nonmotile. They carry their photosynthetic pigments in vesicles, called ***chlorobium vesicles,*** which are attached to the cytoplasmic membrane but are not continuous with it (see Chapter 7). All species are strict anaerobes. Those species that do not possess gas vacuoles are found in the sulfide-rich black muds at the bottoms of freshwater lakes and ponds. The gas vacuole–containing species form green to brownish colored blooms in the sulfide-rich anaerobic water layer. Lakes and ponds have vertical gradients of light from above and hydrogen sulfide from below. The green bacteria are restricted to certain levels in the lake because of hydrogen sulfide concentration and light intensity. This is to be contrasted with the purple bacteria, which are flagellated and can move through the water to position themselves according to light intensity and sulfide concentration.

The filamentous gliding green bacteria belong to the family Chloroflexaceae and were only first described in 1971. They are primarily photoheterotrophs (they can use organic compounds as sources of carbon instead of carbon dioxide), even though they can carry out some photoautotrophic growth using sulfide as an electron donor. The habitat of these organisms are hot springs, where they can form orange-colored mats of growth when the sulfide concentration is very low and green- to orange-colored mats when sulfide is at a high level.

Characteristics of the purple bacteria. Most of the purple bacteria are flagellated and can be attracted or repelled by light ***(phototaxis)*** or chemicals ***(chemotaxis)*** in their aquatic environment. The photosynthetic membrane system of the purple bacteria is continuous with the cytoplasmic membrane, a characteristic that distinguishes it from the green bacteria. The membrane system may appear as lamellar sheets, tubules, or vesicles (Figure 4-20). The purple bacteria have more metabolic potential than the green bacteria because of their ability to use organic compounds as sources of electrons. In addition, except for one species, sulfur is deposited intracellularly and can be used as an electron donor for carbon dioxide fixation. The purple bacteria are made up of two families: the Rhodospirillaceae (purple and brown nonsulfur bacteria) and the Chromatiaceae (purple sulfur bacteria).

The Rhodospirillaceae are referred to as nonsulfur bacteria because of their inability to oxidize sulfide; however, it has been discovered that there can be some oxidation if the concentration of sulfide in the environment is very low. When sulfide is used as an electron donor, sulfate is formed during the oxidation but elemental sulfur is not formed, as in other phototrophs. Members of this family can exist not only as anaerobic phototrophs but can also thrive in the presence of oxygen when using organic compounds as carbon sources. This is a characteristic that distinguishes them from other phototrophic bacteria.

The Chromatiaceae, which are found in sulfur springs and anaerobic zones of lakes, can oxidize sulfide and elemental sulfur to sulfate. The sulfur globules may be located inside or outside the cell. During periods of environmental stress due to the presence of air, high sulfide concentrations, or high light intensities, single cells lose their motility and form aggregates surrounded by a slime layer. The slime layer appears to be a protective device, but when the environmental stress has been removed, the aggregate disperses into individual motile cells.

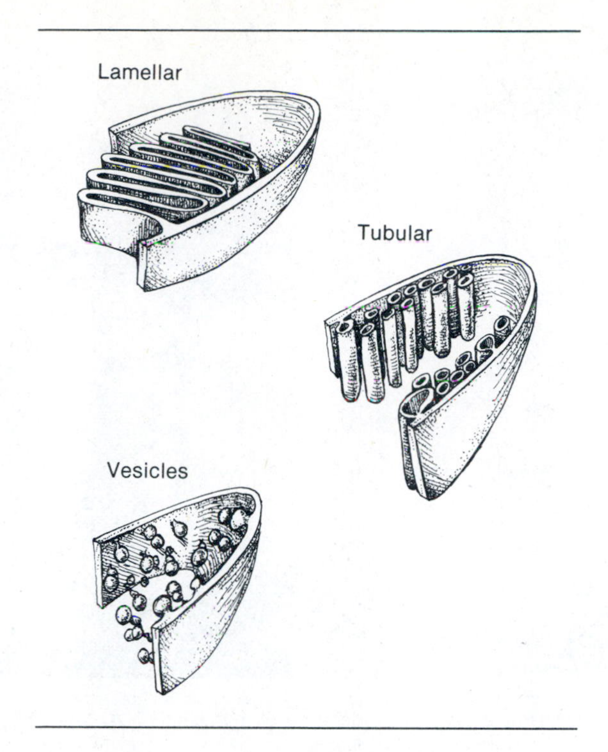

Figure 4-20 Photosynthetic membrane systems in purple bacteria.

Gliding bacteria (part 2—Bergey)

The gliding bacteria* are gram-negative microorganisms that move over surfaces in a gliding motion but do not swim like other motile bacteria. They glide on solid surfaces and leave behind a trail of polysaccharide slime, but no visible structures of locomotion have been detected. Several groups are included among the gliding bacteria, but the most thoroughly studied are the myxobacteria.

The myxobacteria are an interesting group, not only because of their gliding ability but also because of their differentiation process. During the developmental cycle individual rod-shaped cells aggregate and form multicellular ***fruiting bodies*** (Figure 4-21), which in some species are very simple. Other species produce a stalk, which arises from an aggregated mass of cells, with an apical bulb (Figure 4-22). Individual vegetative cells of myxobacteria remain dispersed in a nutrient-sufficient medium, but when nutrients become depleted, the dispersed cells begin to aggregate and switch from vegetative growth to fruiting body formation. In response to starvation, individual cells in the fruiting bodies of some species enter into a dormant state and are called ***myxospores.*** The vegetative cells in other species become encased in a protective structure called a cyst. Under proper environmental and nutritional conditions either the myxospores are released to germinate or encysted vegetative cells are released to reproduce immediately. Vegetative cells, produced from the fruiting body, form swarms, which enter into a feeding process.

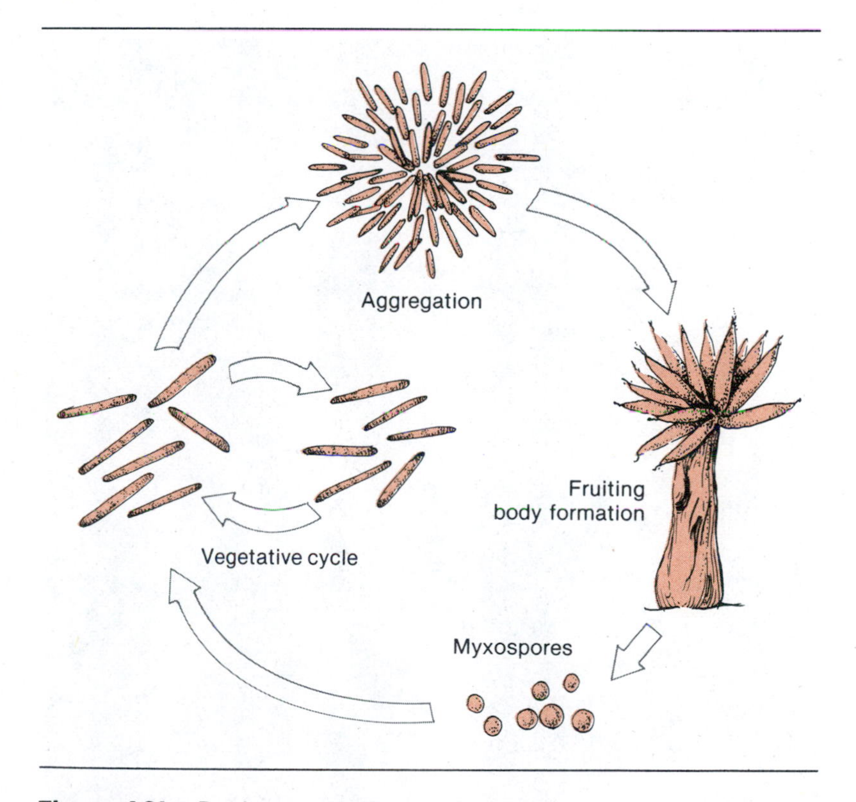

Figure 4-21 Developmental cycle of myxobacteria. Individual vegetative cells under adverse environmental conditions, such as nutrient depletion, aggregate to form mass of cells that develop into fruiting body that has stalk and simple or complex bulb. Many of these fruiting body cells develop spores that under better environmental conditions will germinate to produce vegetative cells.

* Even though the gliding bacteria are given taxonomic status in *Bergey's Manual,* this type of motility is also found in other groups such as the cyanobacteria.

A B C

Figure 4-22 Multicellular fruiting bodies of members of myxobacteria **A,** *Myxococcus.* **B,** *Stigmatella aurantiaca.* **C,** *Chondromyces apiculatus.*

From Dworkin, M., and Reichenbach, H. Symp. Soc. Gen. Microbiol. ***23:****125, 1973. Cambridge University Press.*

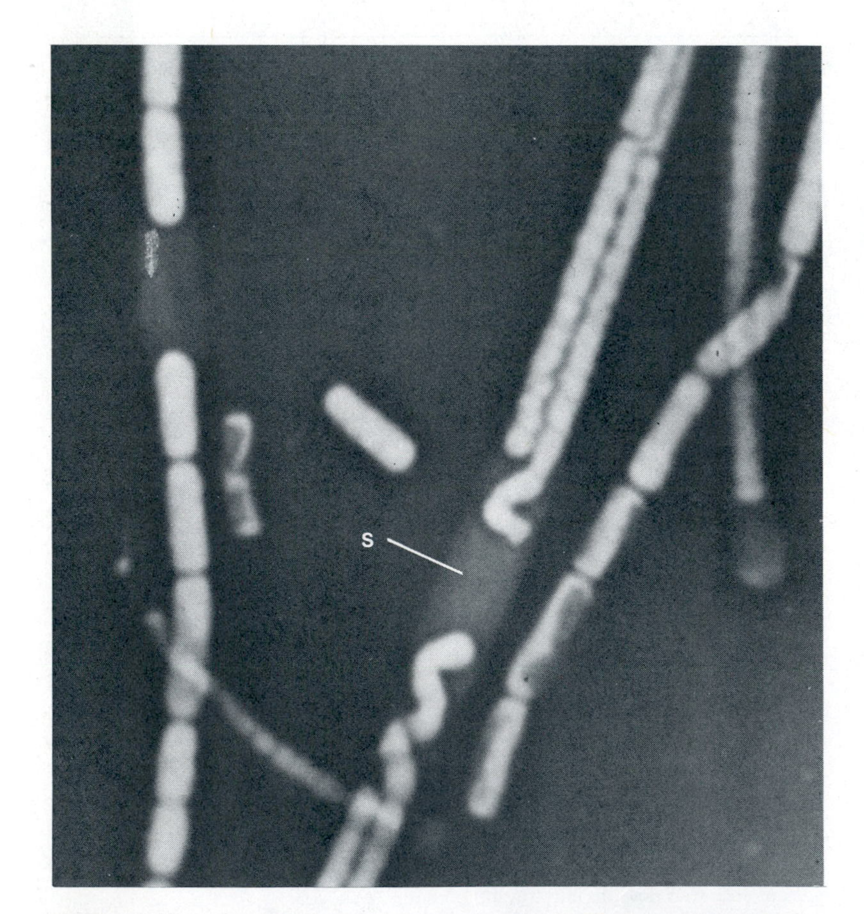

Figure 4-23 Sheathed bacteria. Negatively stained preparation of species of *Leptothrix.* (× 4500.) Sheath *(S)* is shown to enclose two strands of rod-shaped cells.

From Stokes, J.L. J. Bacteriol. ***67:****278, 1954.*

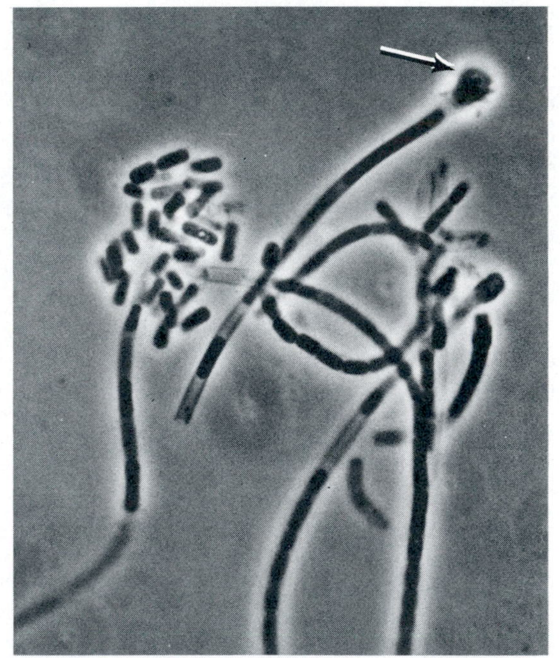

Figure 4-24 Holdfast *(arrow)* of sheathed bacterium *Sphaerotilus natans.*

From van Veen, W.L., Mulder, E.G., and Deinema, M. Microbiol. Rev. ***42:****329, 1978.*

The differentiation process, which involves single cells as well as fruiting bodies, suggests that a chemical signal (chemotactic response) initiates cell-to-cell interaction. This process is believed to be similar to one that occurs in the slime molds (see Chapter 5).

The myxobacteria are found in soil, in decaying vegetation, on the bark of trees, and in animal dung. Gliding allows movement within dense substrates such as soil or rotting wood. Many species appear as brightly colored orange, red, or yellow colonies. They degrade and feed on macromolecules by releasing hydrolytic enzymes and have the capacity to degrade viable as well as dead bacterial cells. This bacteriolytic activity is possible because the extracellular enzymes released from the myxobacteria degrade the cell wall. Other extracellular enzymes such as proteases, nucleases, and lipases are also important for the hydrolysis of proteins, nucleic acids, and lipids, respectively, which are released by lysed bacterial cells.

Sheathed bacteria (part 3—Bergey)

The common characteristic of the sheathed bacteria is the presence of a gelatinous sheath that surrounds the cell or chain of cells. This sheath of material is morphologically distinct from the gelatinous material that surrounds the cell envelope of several other bacterial groups. *Leptothrix* and *Sphaerotilus,* which are the two major genera of sheathed bacteria, are found in nutrient-poor and polluted waters, as well as in activated sludge in sewage treatment facilities. The organisms are regularly found as gray or brownish masses on stones or plants in polluted streams.

The gelatinous sheath (Figure 4-23) enables the members of this group to attach to solid substrates, which is an important device for obtaining nutrients in nutrient-deficient waters. The sheath contains several cells, some of which can move out of the sheath and subsequently form new sheaths. The sheath is also a protective device against parasites and predators.

Like the budding bacteria, sheathed bacteria produce ***holdfasts*** (Figure 4-24) as a means of attachment to solid surfaces. Many sheathed bacteria in nutrient-deficient environments appear to radiate from the holdfast (Figure 4-25).

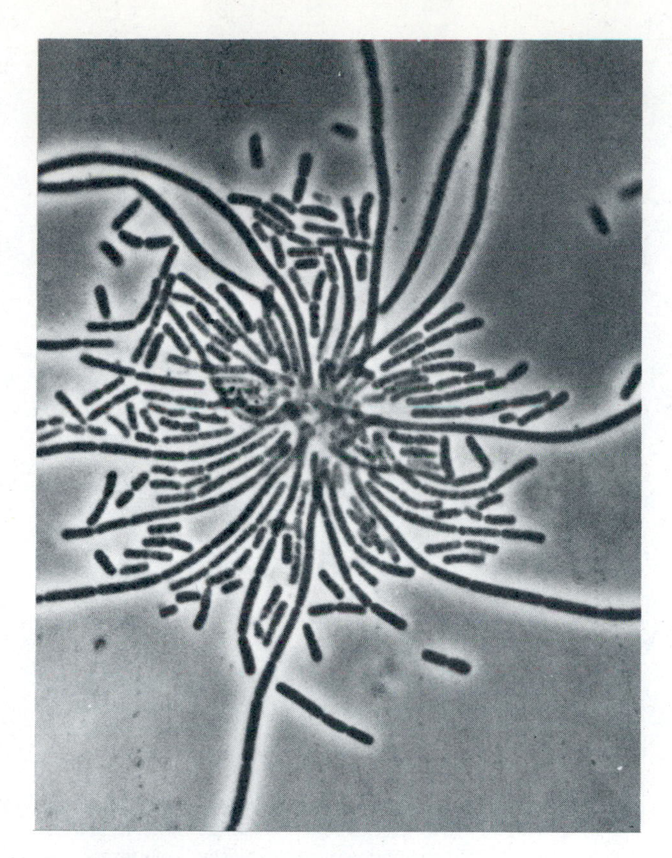

Figure 4-25 Photomicrograph of species of *Leptothrix* in which many sheaths (trichomes) radiate from common holdfast. (×1625.)

From van Veen, W.E., Mulder, E.G., and Deinema, M. Microbiol. Rev. ***42:****329, 1978.*

Budding and appendaged bacteria (part 4—Bergey)

The budding and appendaged bacteria are an unusual and diverse group of microorganisms that produce stalks, hyphae, and various appendages during the growth cycle. Protrusions of the cell that contain cytoplasm and are derived totally or in part from the cell envelope are termed ***prosthecae.*** The cell cycle of these organisms can be followed morphologically, and for this reason their differentiation process has been extensively studied. Two of the most widely studied genera are *Caulobacter* and *Hyphomicrobium,* both of which demonstrate unequal cell division and the formation of dissimilar daughter cells by budding.

The cell cycle in *Caulobacter* begins with a motile ***swarm cell*** containing a polar flagellum (Figure 4-26). The swarm cell during the maturation process loses its flagellum, and an outrowth develops at the previous flagellar site, which is referred to as the stalk. The stalk–mother cell develops a holdfast that is used to anchor to a substrate. Cell division begins in the mother cell, and the future daughter cell develops a flagellum (Figure 4-27). Completion of cell division results in the release of the smaller flagellated daughter cell. The released daughter cell (now referred to

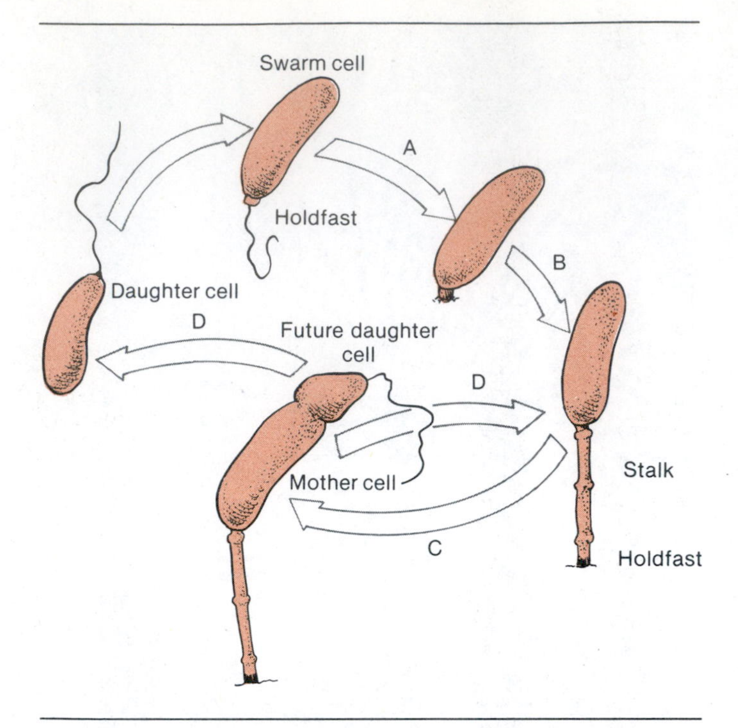

Figure 4-26 *Caulobacter* cell cycle. *A,* Swarm cell with flagellum goes through maturation process in which flagellum is shed. *B,* Cell retains holdfast and synthesizes stalk. *C,* Stalk cell undergoes cell division with future daughter cell portion synthesizing flagellum. Mother cell retains stalk. *D.* Cell division releases mother cell, which can bypass swarm cell development, while daughter cell develops into swarm cell.

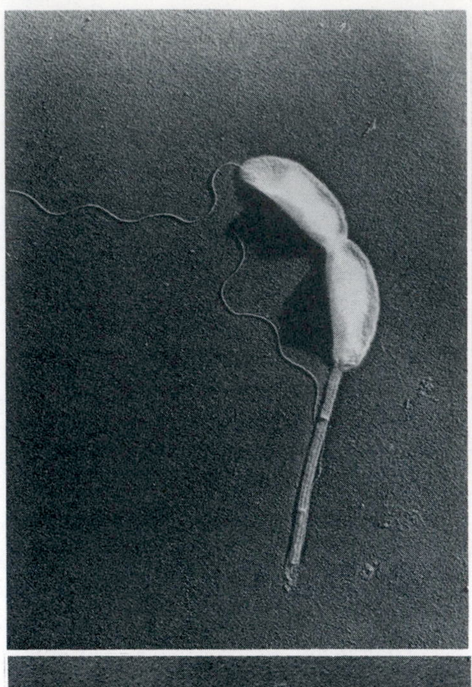

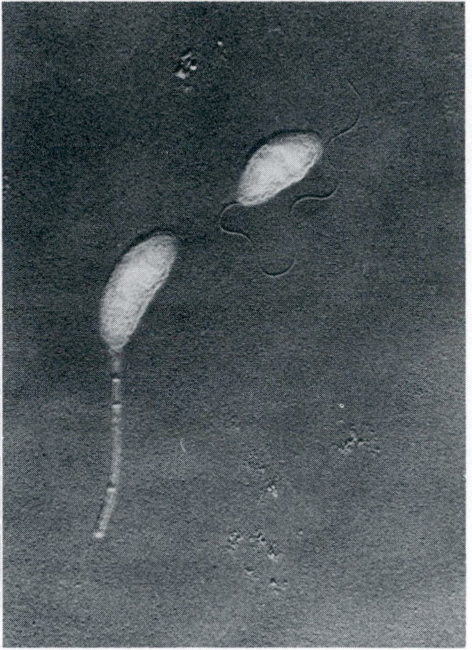

Figure 4-27 Electron micrographs demonstrating *Caulobacter* cell division. (×9000.) **A,** Stalked mother cell is in process of division. Future daughter cell possesses flagellum. **B,** Cell division releases flagellated daughter cell and stalked mother cell.

Courtesy Jeanne S. Poindexter.

as a swarm cell) can repeat the entire differentiation process while the mother cell bypasses the maturation steps and can immediately initiate new daughter cell formation.

Hyphomicrobium also has a life cycle somewhat similar to *Caulobacter.* Swarm cells usually possess a polar flagellum plus a short remnant of the hypha from which the cell was originally produced and that appears as a slight bulge at one of the poles. Flagella appear to be important not only for movement but for attachment to substrates. The flagellum is shed as the swarm cell undergoes maturation and the hypha begins to elongate. The hyphae are usually single and develop at one of the poles, but branching may also occur. The length of the hypha appears to be related to the concentration of substrate in the medium. One theory is that hyphae increase in length in nutrient-deficient media in order to increase the efficiency of nutrient uptake. Daughter cell formation is observed as a swelling at the tip of the hypha (Figure 4-28). The bud increases in size, acquires a flagellum, and then separates from the mother cell. The number of daughter cells produced from a single mother cell is eight in species of *Hyphomicrobium.* The limitation to eight is because the components of the mother cell are not renewed, as in ordinary binary fission, and consequently the mother cell goes through an aging process.

Species of *Hyphomicrobium* and *Caulobacter* are found in practically any aquatic habitat, particularly in those niches that are nutrient deficient. They have been found in seawater, fresh

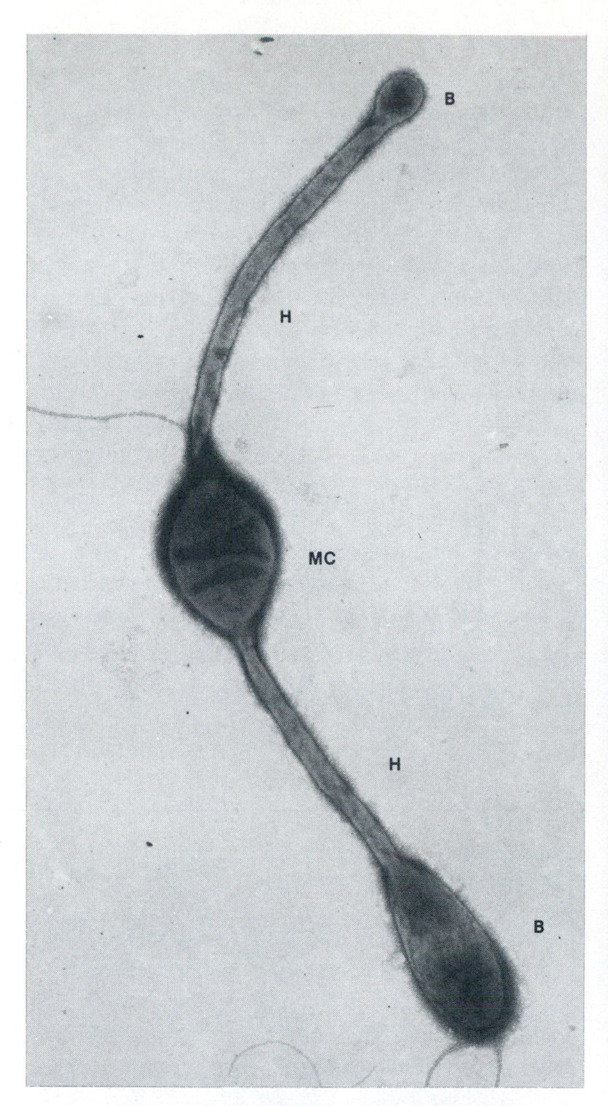

Figure 4-28 Hyphomicrobium. Electron micrograph of species of *Hyphomicrobium*. Mother cell *(MC)* has two hyphae *(H)*, at tips of which are two budding daughter cells *(B)* in different stages of development.

From Hirsch P. In Buchanan, R.E., and Gibbons, N.E. (Eds.). Bergey's manual of determinative bacteriology *(8th ed.).* © *1974, The Williams & Wilkins Co., Baltimore.*

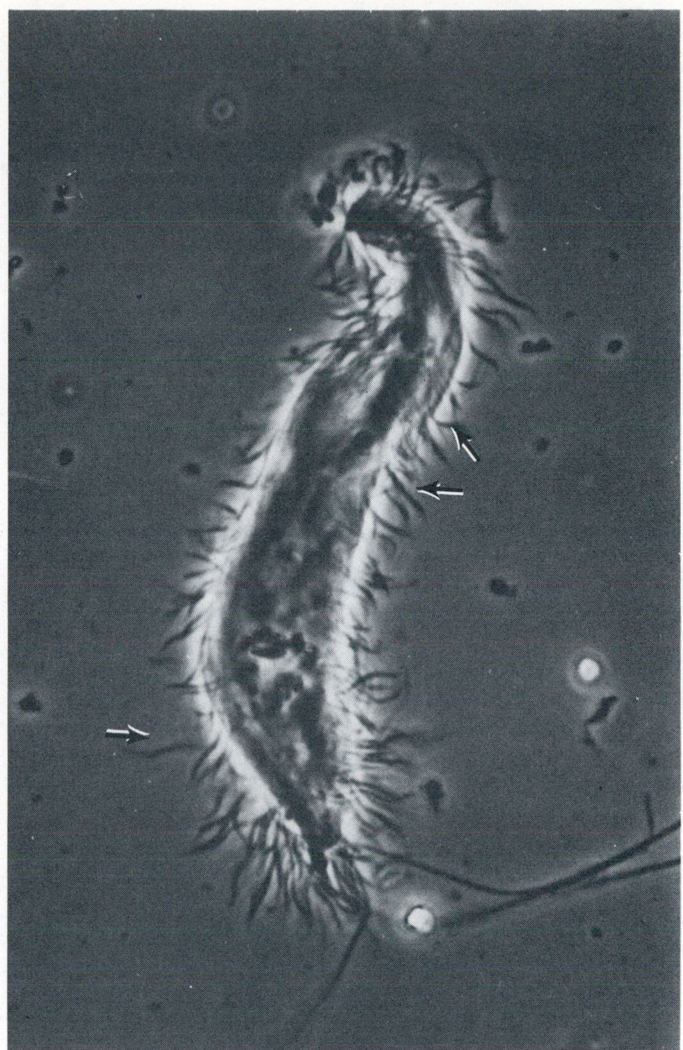

Figure 4-29 Spirochete attachment to surface of protozoan. Large eukaryotic protozoa is covered with many spirochetes *(arrows)*, which help propel larger cell.

From Arnott, H.J. Trans. Am. Microsc. Soc. ***93:****180, 1974.*

water, manganese deposits, water pipes, laboratory water baths, and acid mines. Some species can become permanently attached to substrates and in nature have been found attached to algae and fungi. *Caulobacter* is often observed in the form of a rosette in which several cells are attached by their stalks. The stalks of *Caulobacter* and hyphae of *Hyphomicrobium* appear to be adapted for nutrient absorption, but this may not always be true, since some appendaged bacteria are also found in nutrient-rich environments. *Hyphomicrobium, Caulobacter,* and other related groups oxidize minerals that become encrusted on the older portions of the cell. Some investigators have suggested that hyphae production ensures daughter cell formation at a distance from the encrusted mother cell. Others believe the hyphae offer the cell an alternate site of division for cells in which complex internal structures are not evenly divided between daughter cells during binary fission. One certain function, however, is the increase in surface-to-volume ratio, which helps in the absorption of nutrients at low concentrations in the environment.

Spirochetes (part 5—Bergey)

The spirochetes are helically shaped gram-negative bacteria that possess a multilayered outer envelope that lies outside the peptidoglycan layer of the cell wall. The one characteristic that distinguishes this group from other prokaryotes is the location of the filaments for locomotion, called axial filaments or axial fibers (see Figure 4-9).

Spirochetes, which can be found in many environments, can exist as free-living organisms and have been found in fresh water as well as marine waters. Some species require a living host; some engage in a nonpathogenic relationship; and others are parasites and act as pathogens. An unusual nonpathogenic relationship has been observed between a spirochete and a protozoan found in the gut of termites. The spirochete attaches itself to the surface of the protozoan and helps propel the protozoan by undulations of its cell surface (Figure 4-29). Other nonpathogenic spirochetes are common inhabitants of the large intestine of animals, especially

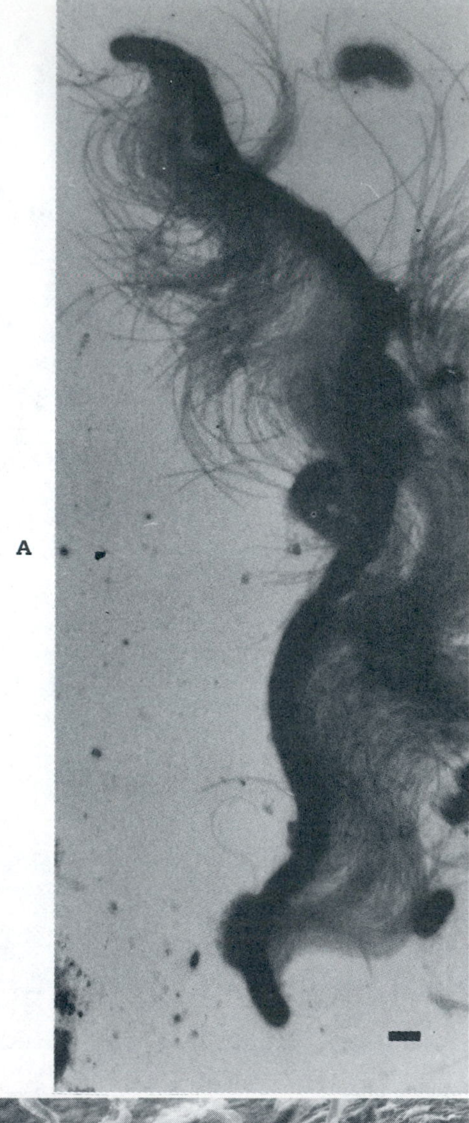

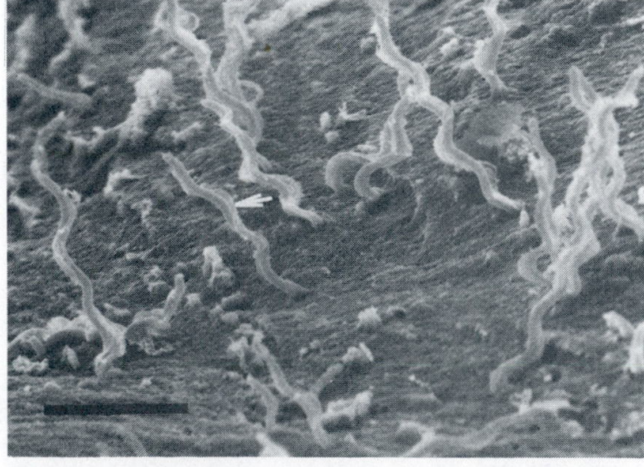

Figure 4-30 Electron micrographs of *Cristispira* species. **A,** Transmission electron micrograph of negatively stained specimen. Note numerous axial filaments that are packed into bundle called crista. Bar = 1.0 μm **B,** Scanning electron micrograph of species attached to epithelial surface in oysters. Arrow points to crista. Bar = 10 μm.
From Tall, B.D., and Nauman, R.K. Appl. Environ. Microbiol. ***42:****336, 1981*

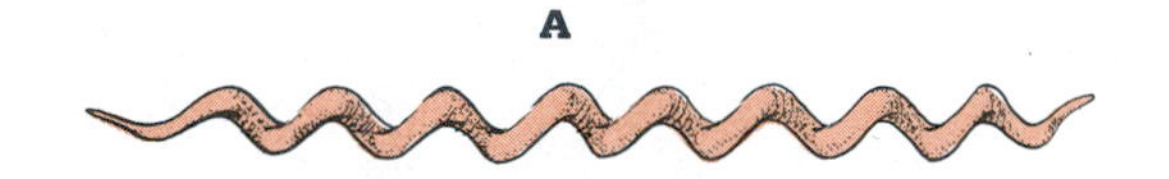

Figure 4-31 Morphology of principal pathogenic spirochetes. **A,** Treponema.

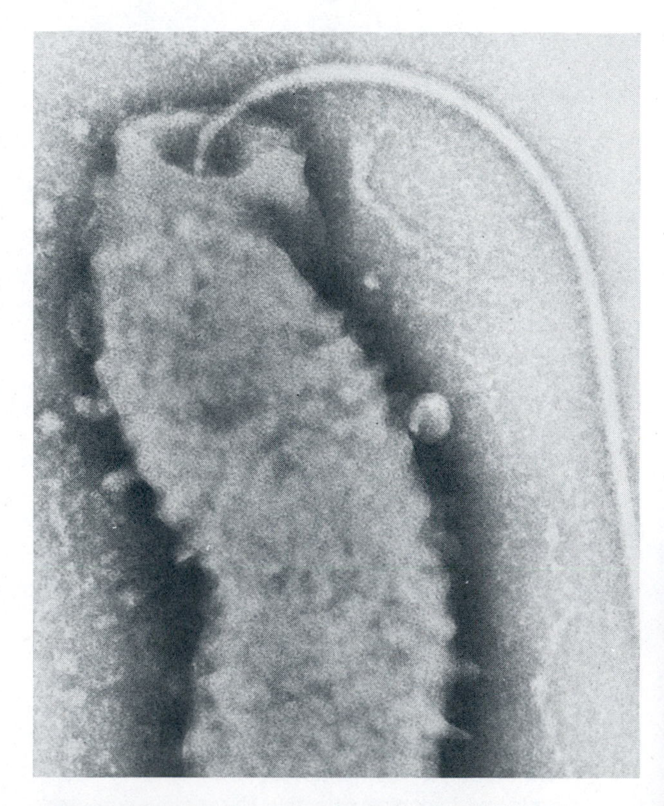

Figure 4-32 Electron micrograph of *Campylobacter* species showing the polar flagellum. (×126,000.)
From Pead, P.J. J. Med. Microbiol. ***12:****383, 1979.*

ruminants such as cattle, and several species can be found in the oral cavity of humans. One of the largest spirochetes (species of *Cristispira*) are found in the digestive tract of freshwater and marine mollusks (Figure 4-30). The principal pathogenic species belong to three genera: *Treponema, Borrelia,* and *Leptospira.* They are distinguished by their size and morphology (Figure 4-31). The diseases they cause in humans are syphilis *(Treponema pallidum),* relapsing fever *(Borrelia recurrentis),* and leptospirosis *(Leptospira interrogans).*

Spiral and curved bacteria (part 6—Bergey)

Spiral and curved bacteria are gram-negative rigid helical cells whose movement is associated with flagella. Unlike the spirochetes, the flagella of the spiral and curved bacteria emanate externally from the poles and may appear singly (Figure 4-32) or in large tufts (Figure 4-33). Some species are free living in fresh or marine waters, whereas others are saprophytic or parasitic. Few are pathogenic. *Bergey's Manual* described only two genera in the family Spirillaceae: *Spirillum* and *Campylobacter;* however, two other genera—*Aquaspirillum* and *Oceanospirillum*—are generally recognized by other taxonomists. All of the genera oxidize organic substrates for energy, especially the salts of organic acids. They are not fermentative organisms, and maximal growth is obtained under conditions in which the oxygen concentration is less than 1 atmosphere. The general characteristics of these genera are outlined in Table 4-4.

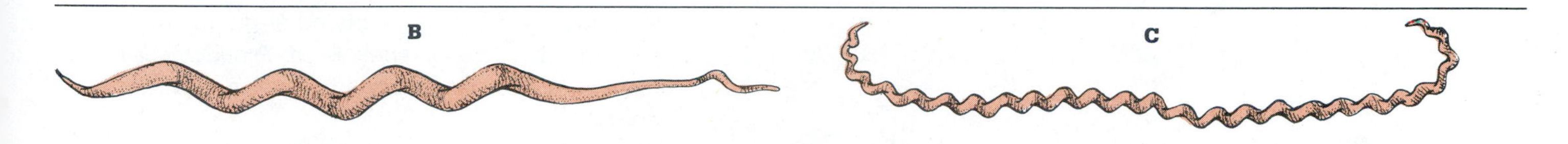

Figure 4-31, cont'd **B,** Borrelia. **C,** Leptospira.

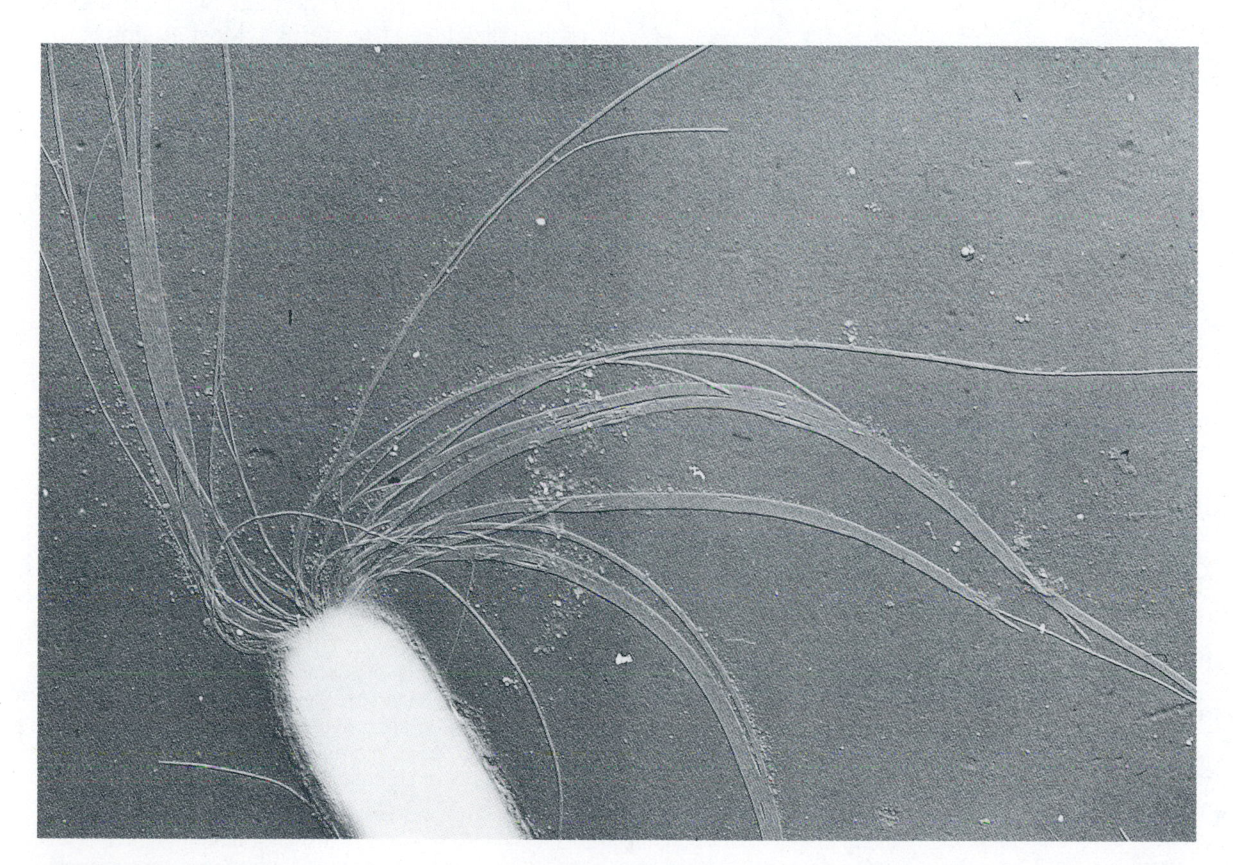

Figure 4-33 Electron micrograph of *Spirillum volutans* demonstrating tuft of flagella at one of the poles. ($\times$16,000.)
Courtesy Noel Krieg.

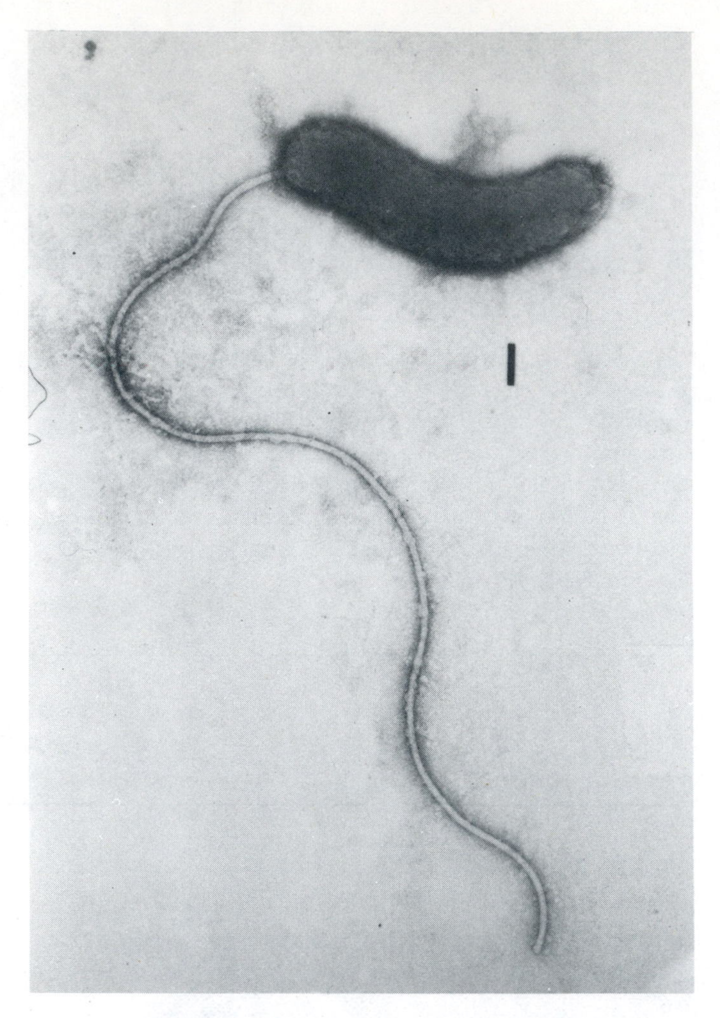

Figure 4-34 Electron micrograph (bar = 0.5 μm) of *Bdellovibrio* species with its polar flagellum.
From Burnham, G.F., et. al. J. Bacteriol. ***96****:1366, 1968*

Gram-negative aerobic rods and cocci (part 7—Bergey)

The gram-negative aerobic rods and cocci obtain their energy from the aerobic oxidation of organic compounds. They do not demonstrate anaerobic or fermentative metabolism. All species are motile by polar flagella, which may appear singly or in tufts. There are five families, whose characteristics are described below. Many of their special properties are discussed at more length in other chapters.

Pseudomonadaceae. *Pseudomonas* is the most important genus in the family Pseudomonadaceae, and its species are often found as the predominant organism in many fresh and marine waters, as well as on the surface of plants. One of the most significant characteristics of this group is their ability to catabolize a wider range of organic substrates than any other group of microorganisms. Several species of *Pseudomonas* produce fluorescent pigments, and a few are pathogenic.

Pseudomonas aeruginosa is a common inhabitant of soil, water, sewage, the mammalian gut, and plant surfaces. It can be a particularly devastating pathogen for humans who have been hospitalized (see Chapter 33). Most strains of this species produce a bluish green pigment called ***pyocyanin,*** which is not produced by any other microorganism. Several other pigments are also produced by *P. aeruginosa,* as well as other species, and they may be important as antimicrobial

Table 4-4 Characteristics of the major genera of the spiral and curved bacteria (part 6—Bergey)

GENUS	CHARACTERISTICS
Spirillum	*S. volutans* only recognized species; large cells with diameter of 1.4 to 1.7 μm and length of up to 60 μm; large tuft of flagella at both poles; found in fresh water
Aquaspirillum	Contains several species that are found in fresh water; most species possess bipolar flagellar tufts; cell diameter is from 0.2 to 1.4 μm, and length seldom exceeds 10 μm
Oceanospirillum	All are marine species, except for one; cell diameter is from 0.3 to 1.2 μm, and length is from 2 to 10 μm; none can ferment or oxidize carbohydrates
Campylobacter	Are curved spiral rods that have single polar flagellum; cell diameter is from 0.3 to 1.3 μm, and cell length varies from 1.5 to 3.5 μm; do not grow well anaerobically; some are found as normal inhabitants of reproductive organs, intestinal tract, and oral cavity of humans and animals; some species are pathogenic; *C. fetus (ssp. fetus)* is associated with abortion and reproductive disease in cattle; *C. fetus (ssp. jejuni)* is associated with disease in humans (see Chapter 28)
Bdellovibrio	Small curved rod with cell diameter of 0.25 to 0.4 μm and length of from 0.5 to 0.8 μm (Figure 4-34); found in soil as well as in fresh and marine waters; is a parasite for other bacteria, especially gram-negative bacteria (Figure 4-35); requires oxygen at 1 atmosphere but cannot oxidize carbohydrates; energy is obtained primarily from oxidation of amino acids

agents that protect the microorganisms from others in the environment. Other species that may infect humans are *P. cepacia, P. maltophila, P. pseudomallei,* and *P. mallei. P. cepacia* is a frequent contaminant of water supplies and has been implicated in hospital-associated infections. This organism, like most pathogenic species of *Pseudomonas,* is difficult to eradicate because of its resistance to many of the prescribed antibiotics. *P. pseudomallei* is found in tropical soils and water in the Far East and Australia. It causes a chronic ulcerating disease in animals and is occasionally transmitted to humans. *P. mallei* causes glanders in horses, a disease characterized by nasal, cutaneous, or pulmonary ulcerations. Occasionally humans can also be infected by *P. mallei.* Over 80 species of *Pseudomonas* cause disease in plants, and among the more important are *P. solanacearum, P. syringae,* and *P. cepacia.*

Rhizobiaceae. Two genera of soil microorganisms, *Rhizobium* and *Agrobacterium,* make up the family Rhizobiaceae. *Rhizobium* carries out gaseous nitrogen fixation but, unlike *Azotobacter,* only in association with leguminous plants (soybeans, peanuts, etc.) through the formation of root nodules. A discussion of this process appears in Chapter 14.

Agrobacterium, which lives in the soil, is capable of infecting the roots and stems of plants to produce tumors called galls. The species associated with gall formation is *A. tumefaciens,* which has a unique technique for infecting plant tissue (see Chapters 9 and 35).

Methylomonadaceae. The family Methylomonadaceae is made up of two genera—*Methylomonas* and *Methylococcus*—that comprise a large group of organisms that are metabolically categorized as methane-oxidizing bacteria. Methane is produced anaerobically in the environment by a number of microorganisms, but it is also used as an energy source by the methane-oxidizing bacteria. Methane-oxidizing bacteria are found primarily in the anaerobic portions of lakes, in the intestinal tract, and in sewage disposal tanks. The chemistry of methane oxidation is described in Chapter 13.

Halobacteriaceae. The genera *Halobacterium* and *Halococcus* make up the family Halobacteriaceae. These organisms derive their name from their requirement for salt (*halophilic,* salt loving). They are found in salt lakes such as the Great Salt Lake of Utah, as well as the Dead Sea and the Caspian Sea. The sodium found in salt is re-

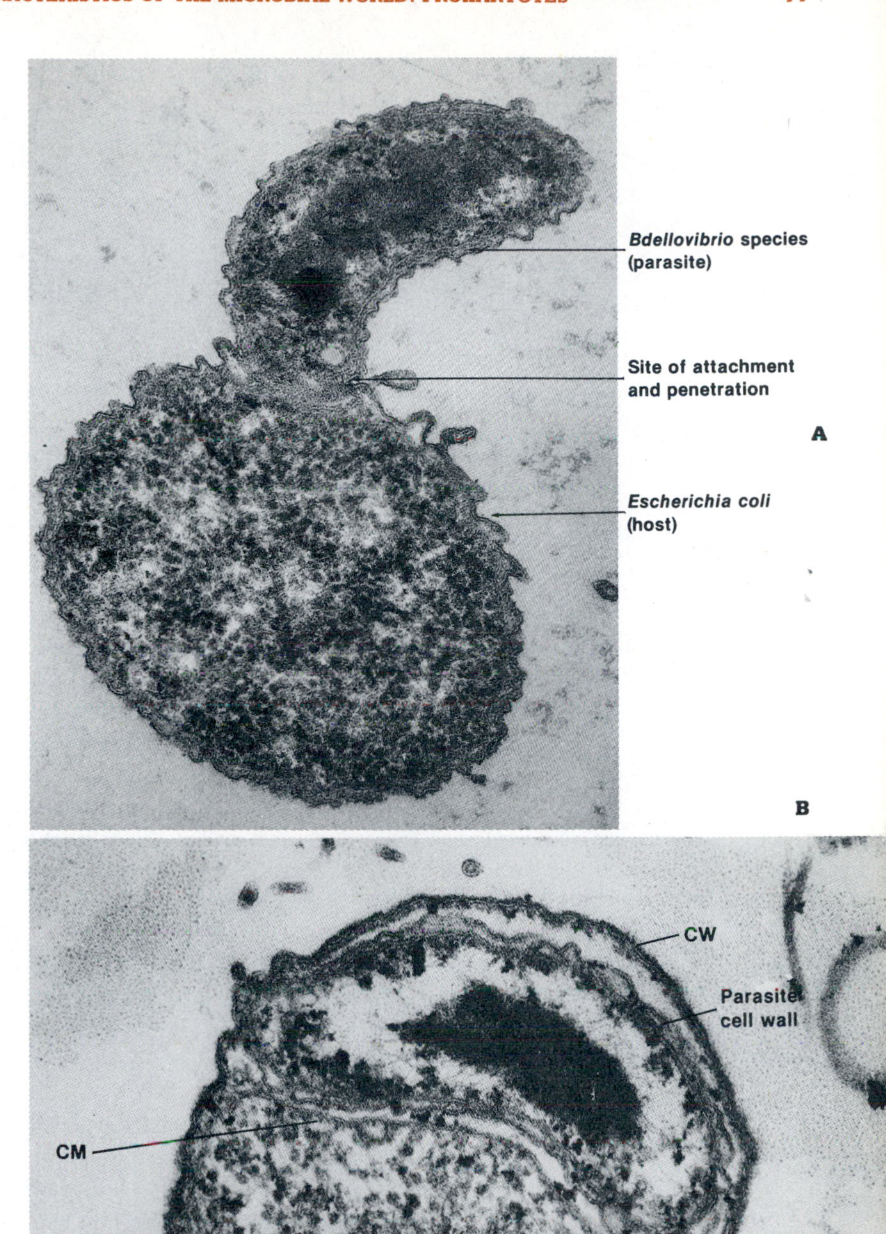

Figure 4-35 Parasitization of *Escherichia coli* by *Bdellovibrio* species. **A,** Electron micrograph demonstrating attachment of parasite to host bacterium. (× 154,000.) **B,** Parasite has not penetrated cytoplasmic membrane *(CM)* of host but occupies space between it and cell wall *(CW).*
From Burnham, G.F., et al. J. Bacteriol. **96:**1366, 1968

quired to stabilize the cell wall of these organisms, which lack muramic acid. The absence of muramic acid is one of the characteristics that place these microorganisms in the group called Archaebacteria. The effects of salt on these organisms and the mechanisms for stabilization in the high-salt environment are discussed in Chapter 15. *Halobacterium* is also unique because light-associated ATP synthesis does not involve the photosynthetic pigment chlorophyll. Instead, a pigment called bacteriorhodopsin appears as purple patches in the cytoplasmic membrane (see Chapter 13 for a discussion of ATP synthesis in the halobacteria).

Gram-negative facultatively anaerobic rods (part 8—Bergey)

The gram-negative facultatively anaerobic rods are an extremely large group of bacteria, consisting of 26 genera, many of which are medically important. Many of its members are motile by flagella that may be arranged peritrichously or polarly on the cell. The term *facultative anaerobe* implies that the organism can live either in the presence or in the absence of oxygen. There are two families in this group of bacteria: the Enterobacteriaceae and the Vibrionaceae.

Enterobacteriaceae. Much discussion about this family of microorganisms is presented here for the following reasons:

1. Some of its members, notably *Escherichia coli,* are important tools of the microbiologist. These organisms are easily cultivated in the laboratory and can be grown on synthetic media containing only a carbon source such as glucose plus inorganic minerals. In organic media they can divide as rapidly as every 20 minutes, and large quantities of microorganisms can be obtained for experimental purposes in just a few hours.
2. Some of its members, particularly *E. coli* and species of *Salmonella,* readily engage in conjugation and other genetic processes from which a considerable amount of information on genetic and microbial metabolism has been learned. This information has been successfully applied to an understanding of metabolic pathways and the nature of the gene in mammalian systems, including humans.
3. Many species that are inhabitants of the intestinal tract, such as *E. coli* and species of *Proteus, Klebsiella,* and *Citrobacter,* as well as species found in the soil and water, such as *Serratia* and *Enterobacter,* are the major cause of hospital-associated infections. These organisms are referred to as ***opportunistic pathogens*** because they cause no disease in the healthy person, but in the patient debilitated by some underlying illness or the trauma of surgery or some other injury, they are a frequent cause of infection. Other members such as species of *Shigella* and *Salmonella* are primary pathogens of humans and cause severe forms of gastroenteritis, and *Yersinia pestis* is the causative agent of plague.

   Diseases of the intestinal tract (enteric diseases) are transmitted primarily by fecal contamination. *E. coli,* which is indigenous to the mammalian intestinal tract, is an indicator of fecal contamination when recovered from water samples. The methods of sanitary analysis are discussed in Chapter 18.
4. Species of *Erwinia* are important plant pathogens. Members of this genus, unlike other Enterobacteriaceae, are capable of hydrolyzing the polysaccharide pectin, which is an important component of plant tissue. Because of their pectolytic activity, some species are causes of tissue-macerating plant diseases and storage rots (see Chapter 35). One species, *E. amylovora,* causes wilt diseases of plants, and *E. herbicola* (also called *E. agglomerans*) produces yellow pigments (carotenoids) and has been associated with opportunistic infections in humans.

Vibrionaceae. The family Vibrionaceae contains both freshwater and marine species. It is characterized by straight or curved rods in which motility is due primarily to polar flagellation. The principal genera are *Vibrio, Aeromonas, Photobacterium, Lucibacterium* and *Chromobacterium.*

The genus *Vibrio* contains two species pathogenic for humans: *V. cholerae,* the causative agent of cholera, and *V. parahemolyticus* (formerly called *Beneckea parahemolyticus*), the causative agent of gastroenteritis contracted by eating contaminated shellfish.

Two genera of marine bacteria are important because they emit light: *Lucibacterium* (includes many species originally classified as *Beneckea*) and *Photobacterium.* They are widely distributed in the marine environment and may be free living, saprobes, commensalistic, or parasitic. The largest number are found as commensals in the intestinal tract of marine animals. These species hydrolyse ***chitin,*** a major component of the shells of marine invertebrates. The marine luminous bacteria differ from their terrestrial counterparts in that the former have a specific and relatively high requirement for sodium. Sodium is required for certain enzymatic reactions as well as for maintenance of the integrity of the cell wall. Many species of fish produce their own light, but some have specialized organs in which the luminous bacteria reside and provide light. This relationship benefits the bacteria, which receive nutrients from the host while the fish is supplied light, which may be used for attracting prey, as a means of communication, or as a mechanism for avoiding predation. The light-emitting reaction of luminous bacteria is catalyzed by the enzyme ***luciferase*** (Plates 4 to 7). Molecular oxygen is used to oxidize reduced flavin mononucleotide (FMN) and an aliphatic aldehyde according to the following reaction:

$$FMNH_2 + O_2 + RCHO \xrightarrow{\text{luciferase}} FMN + H_2O + RCOOH + \text{light}$$

The light emitter is believed to be a luciferase-flavin complex in which the flavin is brought to an excited state and a portion of the excitation energy is released as a photon of light. The function of the reaction for the bacterial cell is not clear, but it may be an alternate route for electron transport and the formation of energy during periods of low oxygen tension, since luciferase is extremely sensitive to even small amounts of oxygen.

Terrestrial luminous bacteria have been observed for over a 100 years. In the 1800s wounds incurred during war were reported to contain harmless luminous bacteria. The presence of these organisms was supposed to indicate that the wound would heal uneventfully. The organisms were never identified, but the recently discovered genus *Xenorhabdus* is a terrestrial luminous species that also produces antibiotics. This could explain the reason why wound luminous bacteria were prognosticators of wound healing, if in fact *Xenorhabdus* was the originally observed microorganism.

Gram-negative anaerobic bacteria (part 9—Bergey)

Bergey's Manual lists one family among the gram-negative anaerobic bacteria—the Bacteroidaceae—which is composed of three genera: *Bacteroides, Fusobacterium,* and *Leptotrichia.* These genera are found primarily in the anaerobic confines of the intestinal tract, upper respiratory tract, and genitourinary tract. Nearly 90% of the intestinal flora is composed of these species. One gram of feces, for example, may contain up to 1×10^{10} of species of *Bacteroides.* In the oral cavity these species are frequently found in the gingival sulcus (the area between the gums and the tooth surface).

The gram-negative anaerobic bacteria are not ordinarily involved in infection, but they can be pathogenic when the animal or human host is compromised by some underlying condition (surgery, a deep wound) or disease (cancer).

The gram-negative anaerobic bacteria play an especially important role in the ruminants (those animals such as cows, sheep, and goats that possess a special digestive organ called the rumen). The rumen is a type of stomach in which cellulose (the primary component of grasses) digestion occurs and where a complex of microorganisms engage in a variety of metabolic reactions. Some of the products of microbial metabolism are used as energy sources by the ruminants, and some products are used by other microorganisms as energy sources. Many of the ruminant organisms are classified in *Bergey's Manual* as genera of uncertain affiliation and include *Butyrvibrio, Selenomonas, Succinimonas, Lachnospira,* and *Succinivibrio.* Rumen metabolism is discussed at greater length in Chapter 13. One genus of uncertain affiliation, *Desulfovibrio,* is found in aquatic habitats. This organism belongs to a unique group in which sulfate is used as an electron acceptor during energy production. Sulfate metabolism is discussed in Chapter 13.

Gram-negative cocci and coccobacilli: aerobes (part 10—Bergey)

One family, Neisseriaceae, is classified among the gram-negative aerobic cocci and coccobacilli. It contains the genera *Neisseria, Moraxella, Branhamella,* and *Acinetobacter.* These organisms exist as coccal or coccobacillary forms and are arranged in pairs. The cocci measure 0.6 to 0.8 μm in diameter. Some species exhibit a type of movement referred to as "twitching," which appears to be associated with polar fimbriation. *Moraxella, Neisseria,* and *Branhamella* show greater relatedness to one another than they do to *Acinetobacter. Moraxella, Branhamella,* and *Neisseria* all produce the enzyme cytochrome oxidase, which is absent in *Acinetobacter. Neisseria, Branhamella,* and *Moraxella* are biochemically inactive, and few sugars are metabolized. *Acinetobacter,* however, can use a wide range of organic compounds as carbon and energy sources.

Neisseria, Moraxella, and *Branhamella* have been isolated only from the mucous membranes of warm-blooded animals, whereas *Acinetobacter* is commonly found in the soil and water. *Neisseria* species such as *N. gonorrhoeae* and *N. meningitidis* are the causative agents of the venereal disease gonorrhoea and the neural condition called meningitis, respectively. These diseases are discussed in Chapters 28 and 30. *Branhamella* and *Moraxella* are not considered as primary pathogens, but they can cause disease in patients undergoing long-term chemotherapy or those taking immunosuppressive drugs.

Gram-negative cocci: anaerobes (part 11—Bergey)

One family, Veillonellaceae, is classified among the gram-negative anaerobic cocci. This family is composed of three genera: *Veillonella, Acidaminococcus,* and *Megasphaera.* These organisms appear in pairs or short chains and are slightly smaller than *Neisseria.* They are found primarily in the intestinal tract of animals, but *Veillonella* species are found in considerable numbers in the oral cavity. None of the genera are considered pathogenic.

Gram-negative chemolithotrophic bacteria (part 12—Bergey)

The term *chemolithotrophic* refers to the ability of microorganisms to obtain their energy from the oxidation of inorganic compounds. They have been divided into three groups: organisms oxidizing ammonia and nitrite, organisms metabolizing sulfur, and organisms depositing iron or manganese oxide. All these organisms play an important role in the ecology of soil and water and are discussed at greater length in subsequent chapters, particularly Chapter 13.

Methane-producing bacteria (part 13—Bergey)

A discussion of a group of microorganisms that produce a natural gas seems appropriate in this era of energy conservation. Biological methane production is associated with anaerobic degradation of organic matter. Methane producers ***(methanogens)*** use hydrogen gas and carbon dioxide, or methanol or acetate produced by other anaerobic bacteria and convert it to methane. Such processes occur in the rumen and intestinal tract of animals, in sewage digesters, and in sediments and muds of aquatic habitats. The biology of methane-producing bacteria has only recently been studied, and there have been some unusual findings.

Methanogens have been organized into one group based solely on a physiological response, methane production. Morphologically, they are a most diverse group of microorganisms, including spiral-, rod-, and coccus-shaped types with each demonstrating variations in organization as well as shape. Investigations of the cell wall structure and physiology have revealed that there appear to be gram-positive and gram-negative species within the group. The most unique characteristic of the cell wall is the absence of muramic acid in the peptidoglycan layer and the presence of components not found in other bacterial species, one of which is ***N-acetyltalosaminuronic*** acid. These results have suggested that the Gram-staining reaction may depend more on structural organization than on chemical differences such as lipid concentration. In addition, sequence analysis of the ribosomal RNA indicates that the methanogens, like extreme halophiles and thermoacidophiles, at some time early in evolution branched off as an independent group maintaining their unique characteristics (see Figure 4-15). They are currently considered members of the archaebacteria.

Gram-positive cocci (part 14—Bergey)

The gram-positive cocci consist of a very diverse physiological group of microorganisms, among which are three important genera: *Micrococcus, Streptococcus,* and *Staphylococcus.* The streptococci are often grouped with the lactobacilli because of their physiological similarities (ability to produce excess lactic acid during growth). *Bergey's Manual,* however, retains the streptococci with other gram-positive cocci, and we shall do the same.

Micrococcus. The micrococci are obligate aerobes that are found in the soil and water. Samples of water or soil incubated on agar plates will often reveal them as pigmented types. Microscopic examination shows that the micrococci are morphologically similar to the staphylococci and appear in clusters. They differ from the streptococci in that acid is produced by aerobic oxidation and not by fermentation.

Staphylococcus. The staphylococci (see Figure 4-2) can grow either in the presence of air (aerobic) or in the absence of air (anaerobic). They are found primarily on mammalian skin, including the anterior nares of humans, but can also be found, although infrequently, on other body sites such as the mouth, throat, mammary glands, and intestinal tract. Several species have now been identified, but the most important in terms of the human condition are *S. aureus* and *S. epidermidis*. *S. epidermidis* is the primary inhabitant of the skin, and 90% to 100% of the species isolated from the anterior nares are *S. epidermidis*. Ten to forty percent of individuals in the community (excludes persons confined to hospitals as well as hospital personnel) carry *S. aureus* in the anterior nares to the exclusion of *S. epidermidis*. Most of us at one time or another have suffered from *S. aureus* infections (boils, carbuncles—see Chapter 28). Hospitalized patients (see Chapter 33) whose resistance to infection is already impaired are very susceptible to infection by *S. aureus*. Carriage of *S. aureus* by humans may persist for variable periods of time (weeks to years). Human carriers and animals are the primary source of strains of *S. aureus* that produce an enterotoxin and are the cause of a type of food poisoning (see Chapters 20 and 28).

Staphylococci are unusually resistant to drying and to high concentrations of salt. This latter property is used by the microbiologist in isolation procedures because most bacteria (particularly those in clinical material) will not grow at elevated concentrations of salt. *S. aureus* produces coagulase, an enzyme that clots plasma, and this is a property used to distinguish it from *S. epidermidis* (Table 4-5).

Table 4-5 Some of the characteristics used to differentiate *Staphylococcus aureus* from *S. epidermidis* and the micrococci

PROPERTY	*S. AUREUS*	*S. EPIDERMIDIS*	MICROCOCCI
Coagulase production	+	−	−
Mannitol utilization	(+)*	(−)†	−
β-Hemolysis‡	+	−	−
Phosphatase production	+	−	−
Gelatinase	(+)	−	−

*(+), Most are positive.
†(−), Most are negative.
‡β-*Hemolysis* refers to the complete lysis of red blood cells on blood agar and the formation of a clear zone around the bacterial colony. (See Appendix B for a discussion of the various types of hemolysis)

Streptococcus. The streptococci belong to the family Streptococcaceae, which also includes the genera *Aerococcus, Gemella, Leuconostoc,* and *Pediococcus*. They appear in chains of varying length and grow in the presence or absence of air. The streptococci are widely distributed in nature, and several species are indigenous to animals and humans. They can be found in humans on the skin and in the mucous membranes of the upper respiratory tract, digestive tract, and urogenital tract. Others have been found on plants and dairy products. Many of the streptococci have been classified into Lancefield (Rebecca Lancefield was the first to initiate research in this area) groups A through O, based on a carbohydrate called the ***C substance*** found in the cell envelope. The C substance is a branched polymer of L-rhamnose and *N*-acetyl D-glucosamine, with the latter being the important determinant. Many of the Lancefield groups contain universally accepted species, whereas others do not. Lancefield group A, for example, is called *S. pyogenes*, and this group is the major cause of infections in humans (pharyngitis, impetigo). Lancefield group A can be further subdivided into over 50 different types based on differences in a protein found in the cell envelope called ***protein A.*** The streptococci found on plants and dairy products (such as *S. lactis*) belong to group N. Some of these streptococci are used as starter cultures in the production of some dairy foods, such as cottage cheese (see Chapter 20). The lysis of red blood cells on blood agar is one of the most useful characteristics for identifying streptococci (see Appendix B). For example, the most pathogenic streptococci are β-hemolytic and belong to groups A, B, C, F, and G (Table 4-6).

Other important streptococci in which the C substance has not been found or is heterogenous include the viridans streptococci, *S. pneumoniae*, and the anaerobic streptococci. The viridans streptococci are α-hemolytic (produce greening around the bacterial colony on blood agar) and include species found normally in the oral cavity, for example, *S. salivarius., S. sanguis.,* and *S. mutans*. *S. mutans* is the etiological agent of tooth decay (caries) and is discussed in Chapter 22.

S. pneumoniae appears as a diplococcus or in short chains surrounded by a polysaccharide capsule. This organism is responsible for most of the cases of bacterial pneumonia in the United States (see Chapter 28).

Anaerobic streptococci include four genera: *Peptococcus, Peptostreptococcus, Ruminococcus,* and *Sarcina*. *Peptococcus* and *Peptostreptococcus* are found as indigenous members of the oral cavity, gastrointestinal tract, and genitourinary tract. Some species have been causes of infec-

Table 4-6 Characteristics of the Lancefield groups A through G of the streptococci.

GROUP	SPECIES NAME(S)	PATHOGENIC POTENTIAL
A	*Streptococcus pyogenes*	Major cause of disease in humans
B	*S. agalactiae*	Infections in newborn infants; cause of bovine mastitis
C	*S. zooepidemicus*	Infections in animals
	S. equi	Infections in horses
	S. equisimilis	Infections primarily in animals
	S. dysgalactiae	Bovine mastitis
D	*S. faecalis* (enterococcus)	Infections of urinary tract in animals; infrequent infections in humans
	S. suis	Infections in pigs
E	?	Infections primarily in swine
F	*S. anginosus*	Occasional infections in humans
G	*S. canis* (not universally accepted)	Infections in dogs, cattle, and cats; occasional infections in humans

tion, especially in compromised patients (those undergoing surgery involving the gastrointestinal tract and genitourinary tract). Species of *Peptostreptococcus* and *Ruminococcus* are also found in the rumen (first stomach of animals such as cows) and are engaged in the decomposition of plant polysaccharides such as cellulose and starch (see Chapter 13).

Endospore-forming rods and cocci (part 15–Bergey)

A differentiation process takes place in some bacteria in which the vegetative cell, responding to harsh environmental factors, such as nutrient depletion, engages in the formation of a dormant cell type called the ***endospore*** (Figure 4-36). Most endospore-forming bacteria are gram-positive rods such as members of the genera *Bacillus* and *Clostridium,* but one group is coccoid. The changes that take place in the cell during endospore formation (sporulation) are discussed in Chapter 7.

The genus *Bacillus* includes aerobic or facultative anaerobic organisms, whereas the clostridia are strict anaerobes. Both genera are inhabitants of the soil, where they exist primarily in the dormant spore state. Spores can infect humans or sometimes animals, and they can germinate into vegetative cells and produce some of the most potent poisons (toxins) known to humans. For example, one of many different food poisonings, ***botulism,*** results from the ingestion of toxin released in foods by vegetative cells of *Clostridium botulinum.* Gas gangrene and tetanus are caused by the contamination of wounds with spores of *C. perfringens* and *C. tetani,* respectively, which germinate at the site of the wound and produce toxins that enter the bloodstream. Anthrax, a devastating infection of cattle, results from the ingestion of spores of *Bacillus anthracis,* which are found on vegetation that cattle feed on. Anthrax in humans usually results from contact with animal products (hides, wool, etc.) that have been contaminated by anthrax spores.

The clostridia and bacilli have significance in medicine and industry. The bacilli are important because many species produce antibiotics such as polymyxin, bacitracin, and gramicidin. Two species of *Bacillus*—*B. popilliae* and *B. thuringiensis*—are used as insecticides. *Clostridium acetobutylicum* is used in the industrial production of acetone and butanol (see Chapter 21).

Gram-positive nonsporing rod-shaped bacteria (part 16–Bergey)

There is one family among the gram-positive nonsporing rod-shaped bacteria, the Lactobacillaceae, which is made up of the single genus *Lactobacillus.* The lactobacilli are rod-shaped immotile microorganisms that are ubiquitous in nature and can be found in humans, animals, various dairy products, fermented beverages, and plants.

The lactobacilli are members of a group given the common name of the lactic acid bacteria. The other members are gram-positive cocci, such as

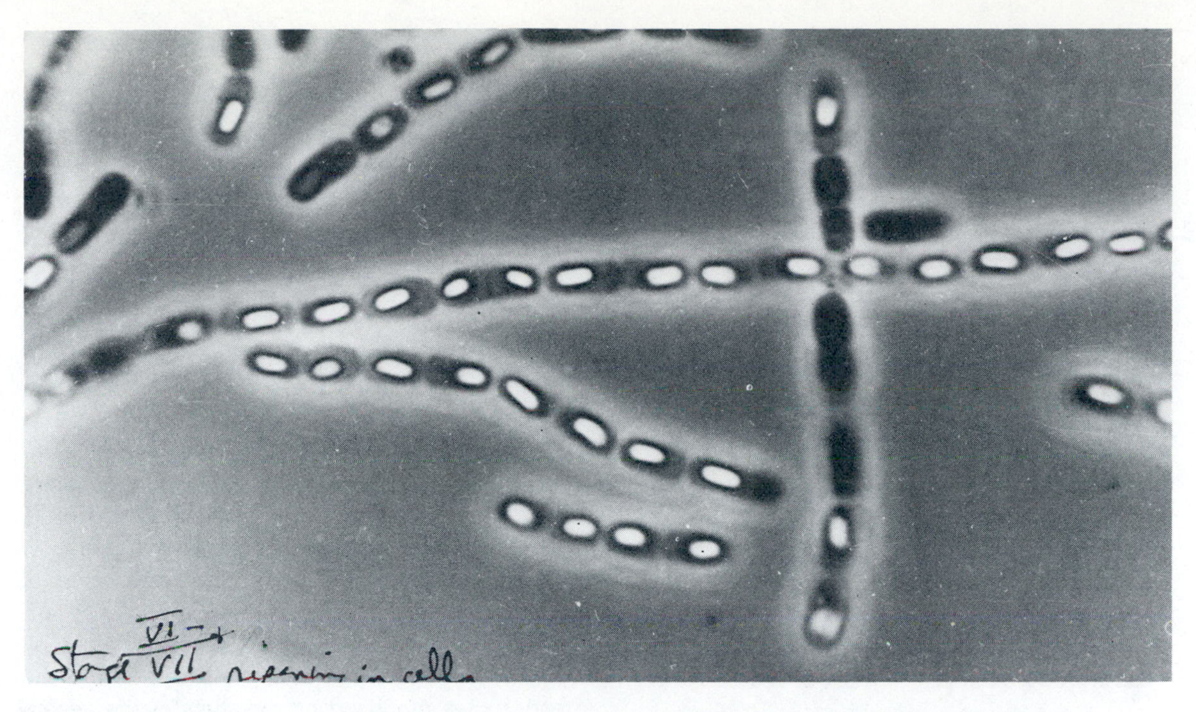

Figure 4-36 Endospores. Photomicrograph of species of *Clostridium.* Spore is refractile gray-white body within vegetative cell.
Courtesy P.C. Fitz-James.

the streptococci, pediococci, and *Leuconostoc,* which belong to part 14 of Bergey's classification. The streptococci are primarily associated with animals, birds, and humans, and among the lactic acid group they are more capable of causing infections in humans (''strep'' throat, rheumatic fever, impetigo, and scarlet fever). The pediococci and *Leuconostoc* are usually associated with plants or plant products. The lactic acid bacteria are characterized by their complex nutritional requirements and the absence of a functional electron transport system for the aerobic production of energy in the form of ATP. They grow in the presence or absence of air, but energy is obtained only from the fermentation process, that is, the degradation of organic matter in the absence of oxygen. The lactic acid bacteria can be divided into two physiological groups based on their fermentation end products. The ***homofermentative*** species produce primarily lactic acid (at least 85%) as the fermentation end product, whereas the ***heterofermentative*** species produce lactic acid (less than 85%) and significant quantities of other products such as carbon dioxide and ethanol.

The lactobacilli not only produce high quantities of lactic acid but are also more tolerant of acid than any other group of microorganisms. Consequently, in environments rich in organic materials where microbial metabolism has produced significant amounts of acid, the lactobacilli can multiply, whereas other microorganisms are incapable of growth. These physiological properties have been exploited for industrial purposes. Lactobacilli have been used in the following processes: (1) the fermentation of cabbage to produce sauerkraut *(Lactobacillus plantarum),* (2) ripening of cheddar cheese from raw milk *(L. casei),* and (3) in brewer's mash to lower the pH and prevent other microbial growth before inoculation with yeast *(L. delbrueckii).* A more detailed discussion of the industrial importance of the lactobacilli as well as other lactic acid bacteria appears in Chapter 21.

Lactobacilli are found in the oral cavity, intestinal tract, and vagina of humans. It is postulated that species of lactobacilli are instrumental in keeping the population of other microbial species under control because of the acidic nature of their fermentation products. They are not considered to be highly infectious for humans. Some studies have demonstrated, however, that various species in the oral cavity have been involved in serious infection, particularly in those individuals undergoing dental surgery who also have severe heart disease. Infections are contracted when the lactobacilli enter the bloodstream via an abrasion in the oral cavity produced by instrument manipulation during oral surgery. The organisms find their way to damaged heart tissue, where they adhere, divide, and cause further damage to the organ.

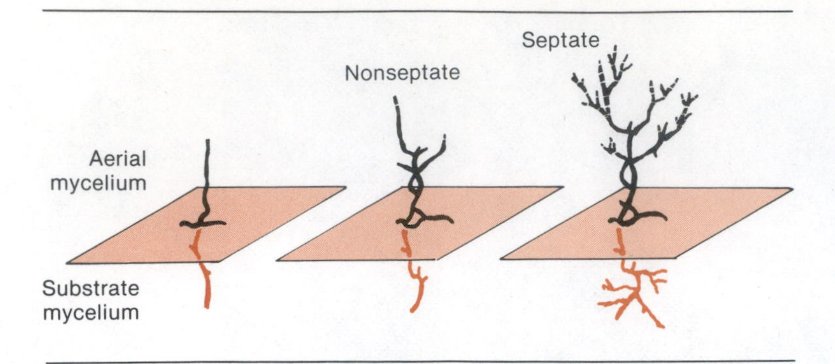

Figure 4-37 Hyphal development in *Actinomycetales*. Septae that develop in some filaments may be complete or incomplete, depending on species. Hyphal mat that develops may remain attached to substrate (substrate mycelium) or may give rise to aerial hyphae from which spores can develop.

Actinomycetes and related microorganisms (part 17—Bergey)

The actinomycetes and related microorganisms probably represent the most diverse and largest group of bacteria. They are composed primarily of gram-positive, rod-shaped, unicellular, and branching microorganisms. The branching species resemble in many ways the filamentous fungi. The actinomycetes comprise the Coryneform group and an order of microorganisms called the Actinomycetales.

The Coryneform group seldom produces filaments but does possess cell wall components that are similar to those of the members of Actinomycetales. They are gram-positive rods that may appear club shaped, and following cell division they arrange themselves into ***palisades,*** or "Chinese letter" configurations. One of the principal genera is *Corynebacterium,* whose species are pathogens of humans, animals, and plants. Diphtheria is an important health hazard for humans and is caused by *Corynebacterium diphtheriae.* Corynebacteria that infect plants are a frequent cause of wilt in wheat and other grains, as well as in many vegetables. The genus *Arthrobacter* and other coryneforms make up a large percentage of soil microorganisms, and many are capable of slowly degrading recalcitrant (resistant to biological degradation but disappear at a slow rate) substrates such as petroleum hydrocarbons and insecticides. Some species of *Arthrobacter* can emulsify oil when growing on hydrocarbons by producing compounds such as fatty acids. These emulsifying agents help in making the recalcitrant molecules more susceptible to biological and abiological degradative processes.

The order Actinomycetales is composed of 10 families, nearly all of which produce filaments that show extensive branching and the formation of mycelia. A mycelium may remain attached to the surface ***(substrate mycelium),*** or ***aerial filaments*** may arise directly from the substrate mycelium. The individual filaments or hyphae of the mycelium are subdivided into units in which there is an inward growth of the cell wall at regular intervals along the filament (a process called septation) (Figure 4-37). Each septated unit in the filament contains at least one DNA molecule. Some of the different types of filaments produced by Actinomycetales are illustrated in Figure 4-38. One of the unusual characteristics of this group is that reproduction often leads to the formation of spores that are produced on specialized hyphal branches, many of which develop on aerial filaments and may appear singly or in chains. The spores produced by most species are not heat resistant, as are the endospores produced by members of the genera *Bacillus* or *Clostridium.* Only one family of Actinomycetales, Thermoactinomycetaceae, produces true spores that are heat resistant and similar in chemical composition to endospores of *Bacillus* and *Clostridium.*

Actinomycetes make up approximately 20% to 60% of the total microbial population of the soil. The so-called earth smell of soil is due to their metabolic activity and the production of pigments, perpenoids (geosmins), and extracellular enzymes. Their morphological diversity is suited to their environment, the solid surfaces of soil particles, and they are similar to the eukaryotic fungi, which are also primary inhabitants of the soil. The actinomycetes, for example, produce a multitude of extracellular enzymes that are capable of degrading plant and animal organic matter. The family Micromonosporaceae is abundant in the mud bottoms of freshwater lakes and rivers and may constitute up to 50% of the microbial population. Only 3 of the 10 families of Actinomycetales are discussed here.

Actinomycetaceae. The family Actinomycetaceae is composed of genera that are primarily inhabitants of the soil. The genus *Actinomyces,* which is but one of six genera, is composed of species that are indigenous to humans and lower animals. It is a non-spore-forming bacillus that branches and can be found in the oral cavity and intestinal tract of mammals. The infections they cause are referred to as ***actinomycoses*** (sing. actinomycosis). The major pathogenic species is *Actinomyces israelii.*

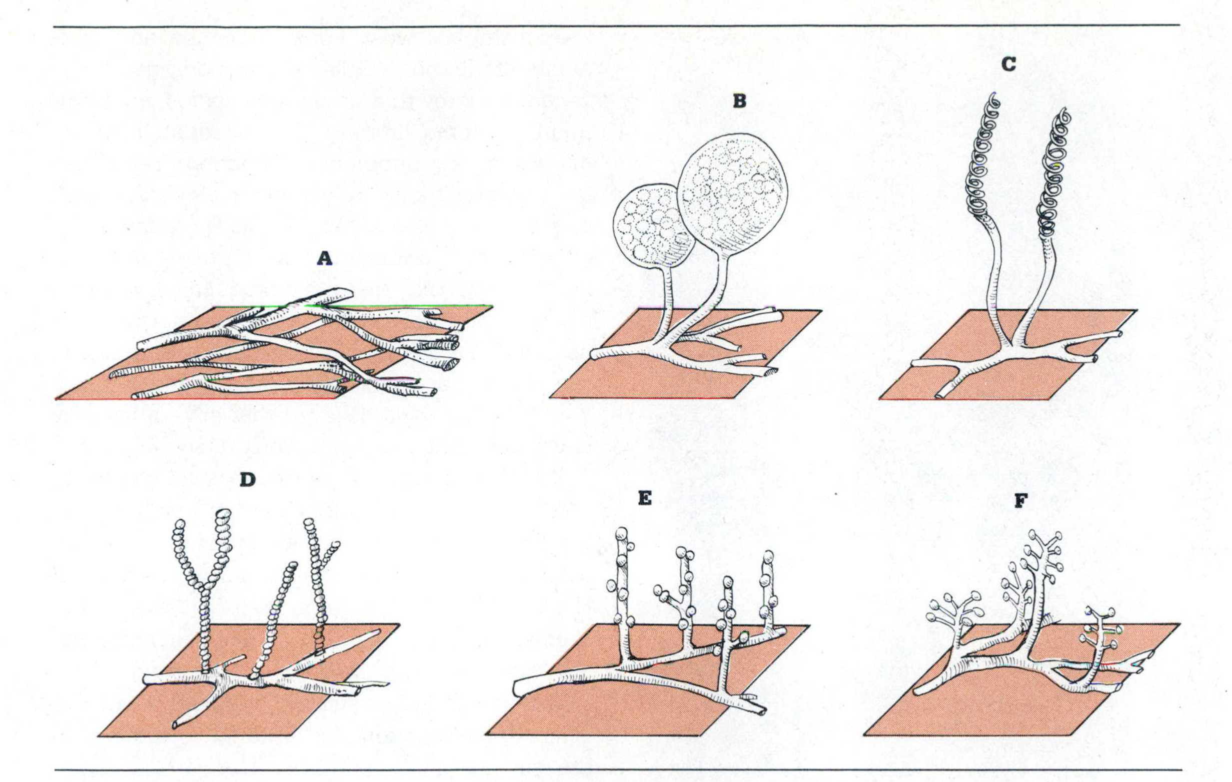

Figure 4-38 Filament morphology in Actinomycetales demonstrating substrate and aerial hyphae. **A,** Substrate mycelium, no aerial branches. **B,** Substrate mycelium with aerial sacs (sporangia) that contain spores. **C,** Substrate mycelium with spiral spore chains. **D,** Substrate mycelium with straight-branching aerial spores. **E,** Substrate mycelium with aerial hyphae and spores. **F,** Small branching mycelia containing spores.

Mycobacteriaceae. Members of the family Mycobacteriaceae, in which the principal genus is *Mycobacterium,* are non–spore forming and do not usually form mycelia. Filaments may form that are extremely fragile and fragment on the slightest disturbance. One of their unusual characteristics is a slow growth rate. The development of a visible colony on laboratory media may take from 2 days to 3 weeks, depending on the species. This is to be compared with the 18 to 24 hours required for colony formation by most other bacteria. Many mycobacteria are found in the soil and water, but some are obligate parasites in warm- and cold-blooded animals. The two most important human pathogens are *Mycobacterium tuberculosis* and *M. leprae,* the causative agents of tuberculosis and leprosy, respectively.

Most of the bacterial world is divided into two cell wall types: gram-positive and gram-negative, based on the Gram stain. The mycobacteria, however, do not lend themselves to the usual staining procedures because of the nature of their cell envelope. They possess a basal layer containing peptidoglycan, as do other bacteria, and would stain gram-positive if only the basal layer were involved. The peptidoglycan, however, is bound to an outside layer that contains sugars and fatty acids called ***mycolic acids.*** Adjacent to this component are layers of lipids, which along with mycolic acids, make Gram-staining impossible. As much as 60% of the wall of mycobacteria is composed of lipids, as compared with 20% for gram-negative bacteria and 1% to 5% for gram-positive bacteria. Large accumulations of lipid in the cytoplasm are characteristic of the mycobacteria during certain stages of the growth cycle. The mycobacteria are stained by a technique called the ***acid-fast stain*** (also called Ziehl-Neelsen stain). See Appendix B for further details on this staining procedure.

Streptomycetaceae. The principal genus in the family Streptomycetaceae is *Streptomyces.* Within this genus are several hundred species that are primary inhabitants of the soil. Like many of their soil relatives, *Streptomyces* species produce a number of degradative enzymes that enable them to break down organic molecules that are resistant to degradation by most other bacteria. *Streptomyces* species produce special-

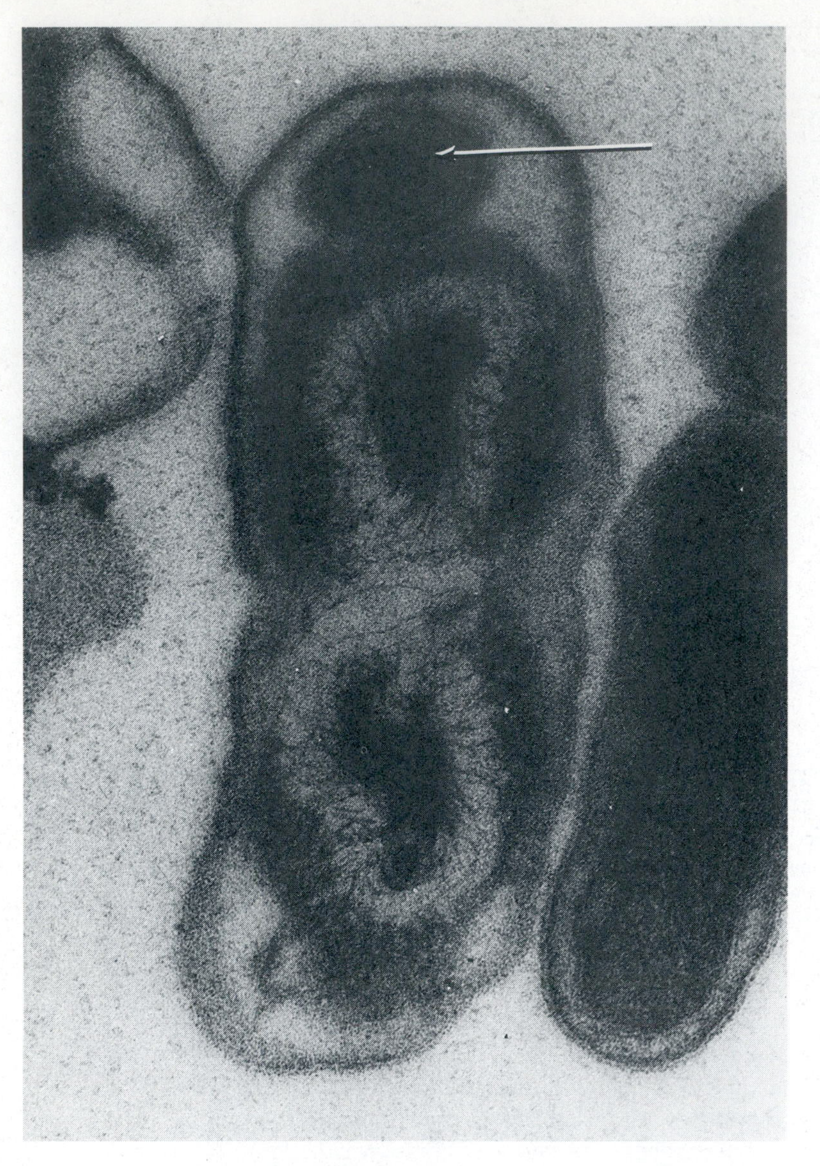

Figure 4-39 Electron micrograph of *Coxiella burnetii* demonstrating endospore-like body *(arrow)*.
From McCaul, T.F., and Williams, J.C. J. Bacteriol. ***147:****1063, 1981.*

ized hyphae from which ***spores (conidia)*** are formed; however, they are more densely packed in the spore than in the vegetative cell, which probably accounts for their resistance to dessication. Many secondary metabolites (by-products produced during the formation of essential primary metabolites) are synthesized by *Streptomyces* and have industrial and commercial value. One group of secondary metabolites is the ***antibiotics*** (compounds produced by microorganisms that inhibit the activity of other, usually unrelated microorganisms). Nearly 90% of all the known antibiotics are produced by species of *Streptomyces,* for example, streptomycin, kanamycin, erythromycin, and amphotericin B, to name just a few. There appears to be a very close relationship between spore formation and antibiotic production. Cells of *Streptomyces* cease vegetative growth and produce spores when nutrients become limiting. Spore formation is accompanied by antibiotic production. Yet it has been demonstrated by genetic analysis that species of *Streptomyces* blocked in spore formation still produce antibiotics. In vitro experiments clearly show that these soil-produced antibiotics can inhibit the growth of microorganisms that normally compete with *Streptomyces,* but because only very small amounts are actually produced in situ, it is believed that antibiotics may have more important functions. One suggestion is that antibiotics are packaged in the spore to prevent germination until conditions are more favorable. For example, one antibiotic produced by a species of *Streptomyces* has been isolated and shown to inhibit the germination process. Other species of *Streptomyces* produce an antibiotic that acts like an alkali-metal transporting agent. Much research must still be performed in order to evaluate the function(s) of antibiotics produced in the soil.

Streptomyces, as well as other antibiotic-producing microorganisms, are found in soils throughout the world, and investigators are continually trying to isolate new species. The steps involved in the isolation of antibiotics used in the treatment of human disease are discussed in Chapter 27, and the industrial production of antibiotics are elaborated on in Chapter 21.

Rickettsias (part 18—Bergey)

The rickettsias are gram-negative bacteria that are obligate (restricted to one condition) intracellular parasites that (except for one species, *Rochalimaea quintana*) cannot be cultivated on ordinary laboratory media. They are deficient in one or another metabolic functions that may include the biosynthesis of amino acids, nucleotides, or ATP and therefore require a living host in order to replicate. Within this group are three important genera: *Rickettsia, Chlamydia,* and *Coxiella.*

Rickettsia. *Rickettsia* species are small rod-shaped microorganisms that are seldom longer than 2 to 3 μm and, like most other bacteria, divide by binary fission. Many species of *Rickettsia* are pathogenic for humans and are transmitted to humans by arthropod vectors such as ticks and lice. Outside of the host *Rickettsia* species are extremely sensitive to chemical and physical agents, which may explain their transmission from vertebrate host to vertebrate host by arthropod vectors. Inside the host the microorganism is initially enveloped by the cytoplasmic membrane of the eukaryotic cell. The membrane surrounding the enveloped organism must be digested by a phospholipase in order to be released into the cytoplasm, where it multiplies. Diseases caused by *Rickettsia* include ***Rocky Mountain spotted fever*** and ***typhus.***

Rickettsia species were believed to have leaky membranes because ATP, which will not penetrate other bacteria, restored or prevented the loss of metabolic activity when added to rickettsial cells. This is untrue, and *Rickettsia* species have been shown to possess very tight membranes and to have a highly specific transport system for the uptake of ATP, which is similar to that of the mitochondrion.

Coxiella. *Coxiella burnetii* differs from other rickettsias in that it has a variable gram reaction and shows a high degree of resistance to chemical and physical agents. The organism causes human infection and is transmitted by ingestion of contaminated milk or by inhalation of contaminated dust. It appears that there are two morphological cell types of *C. burnetii,* which accounts for its differences from other rickettsias. A smaller morphological cell type is the resistant form of the organism and is believed to form an endospore (Figure 4-39). The resistance of the smaller cell type is considerable, since in milk it can withstand a temperature of 63° C for 30 minutes and at −20° C may remain viable for up to 2 years. The organism is the agent of ***Q fever,*** which is ordinarily transmitted to humans by inhalation of airborne dust or contact with infected animal tissues. Contamination of the environment from infected ewe or cow placentas is perhaps the most important source of infection for humans. Pasteurization is required to kill the organism.

Chlamydia. *Chlamydia* species are ubiquitous obligate parasites found in the animal kingdom, but some can parasitize humans. They cause a variety of diseases, including those that affect the eye ***(trachoma)*** and the urogenital tract ***(urethritis*** and ***proctitis***) of humans. The chlamydiae have become very important agents of sexually transmitted diseases. The number of sexually transmitted chlamydial infections is slowly approaching that of gonorrhea (see Chapter 30).

Chlamydiae seldom measure more than 0.2 to 0.4 μm in length (Figure 4-40). They reproduce by binary fission but carry out an unusual developmental cycle in the host cell. The microorganism exists in two states: one as an extracellular unit called the ***elementary body*** and one as an intracellular unit called the ***reticulate body.*** Both bodies contain DNA and ribosomes and are limited by a cytoplasmic membrane and cell wall. The developmental cycle can be broken down into the following states (Figure 4-41).

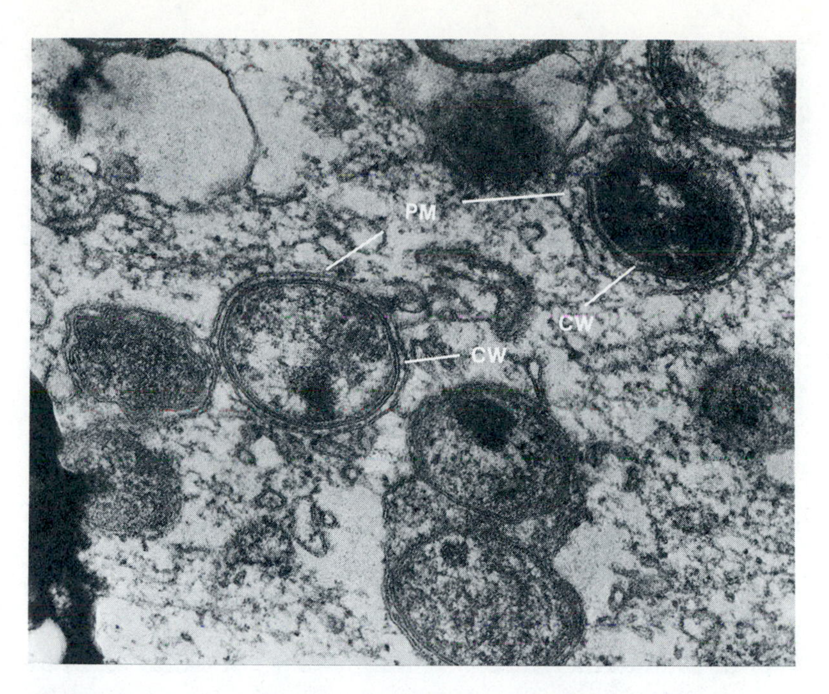

Figure 4-40 Electron micrograph of infected mammalian cell containing cells of species of *Chlamydia.* Chlamydial cell can be seen with its cell wall *(CW)* surrounded by cytoplasmic membrane unit called phagosome membrane *(PM).*
From Wyrick, P., and Brownridge, E.A. Infect. Immun. ***19:****1054, 1978.*

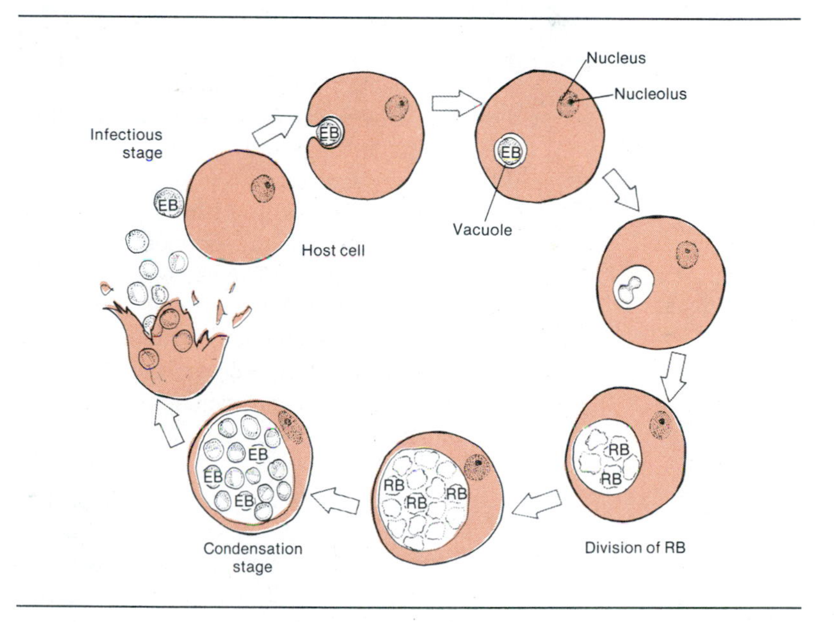

Figure 4-41 Developmental cycle of *Chlamydia.* Elementary body *(EB),* which is infectious stage of organism, attaches to host cell. Cytoplasmic membrane of host cell forms vacuole, within which elementary body develops into reticulate body *(RB).* Reticulate bodies divide and give rise to other reticulate bodies that later condense into elementary bodies and are released from cell.

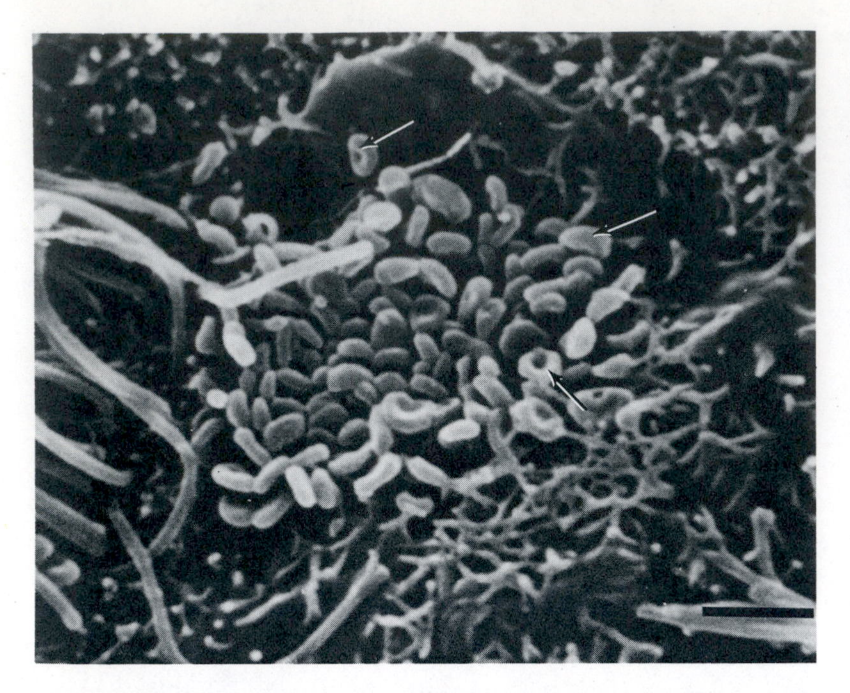

Figure 4-42 Electron micrograph of colony of *Mycoplasma pneumoniae* attached to epithelial surface. Mycoplasmas are seen as doughnut-shaped *(arrows)* objects.
From Kohn, D.F., and Chinookoswong, N. Infect. Immun. ***34**:292, 1981*

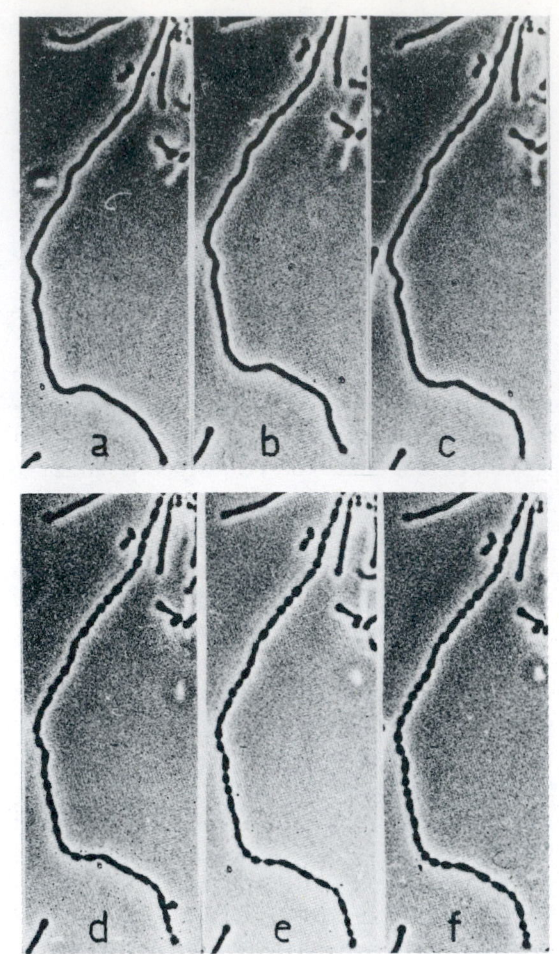

Figure 4-43 Transformation of mycoplasmal filament into chain of coccilike cells.
From Bredt, W. J. Bacteriol. ***113**:1233, 1973.*

1. The elementary body, lying outside the host cell, is in a dormant state exhibiting little metabolic activity and therefore resembles prokaryotic endospores.
2. The elementary body adsorbs to the host cell membrane, which responds by forming a vacuole around it. The vacuole and developing elementary body are called an inclusion body. The elementary body engages in the synthesis of ribosomes, proteins, and DNA and in so doing increases in size, as does the vacuole surrounding it. The elementary body eventually becomes an enlarged reticulate body that divides by binary fission, giving rise to other reticulate bodies. DNA synthesis in the reticulate bodies is followed by binary fission and the formation of elementary bodies. The enlarged vacuole is finally disrupted, and mature elementary bodies are released as the host cell dies.

Mycoplasmas (part 19—Bergey)

The term ***mycoplasmas*** is a trivial name for all members of the class Mollicutes, which contains one order: Mycoplasmatales. *Bergey's Manual* lists two families—Mycoplasmataceae and Acholeplasmataceae—but a third family—Spiroplasmataceae—has since been added.

All mycoplasmas lack a cell wall (Figure 4-42) and are among the smallest prokaryotic cells (0.2 to 0.3 μm). They are found in plants, animals, and insects, and one species, *Thermoplasma acidophilum,* is found in highly acidic coal refuse piles and is a member of the archaebacteria. The mycoplasmas have complex nutritional requirements and are the only prokaryotes that cannot synthesize fatty acids. Sterols, such as ***cholesterol,*** are present in the cell membrane of the genera *Mycoplasma* and *Spiroplasma* but are absent from *Acholeplasma* and *Thermoplasma.* Sterols enable the cell membrane to be relatively rigid in the absence of a cell wall. Even so, the mycoplasmas are osmotically sensitive, and cultivation in laboratory media has been accomplished with only a few species. Morphologically, the mycoplasmas are pleomorphic. They are usually observed as spherical cells, but under optimum growth conditions species of *Mycoplasma* produce filaments. A decrease in the growth rate causes the filaments to be transformed into a

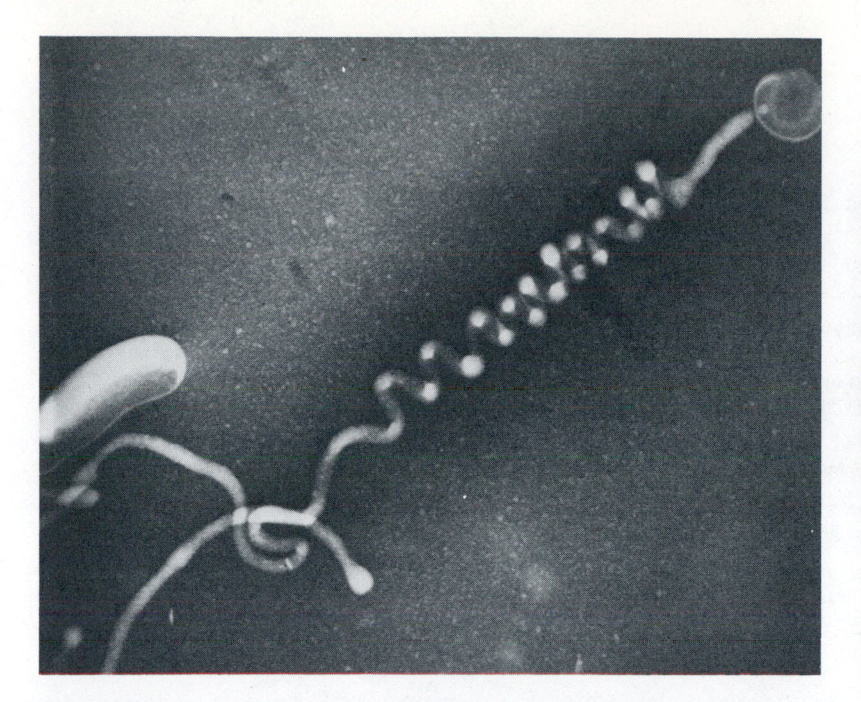

Figure 4-44 Electron micrograph of *Spiroplasma citri,* agent of plant disease. (×34,500.) Note spiral nature of this bacterium.
Courtesy Roger M. Cole.

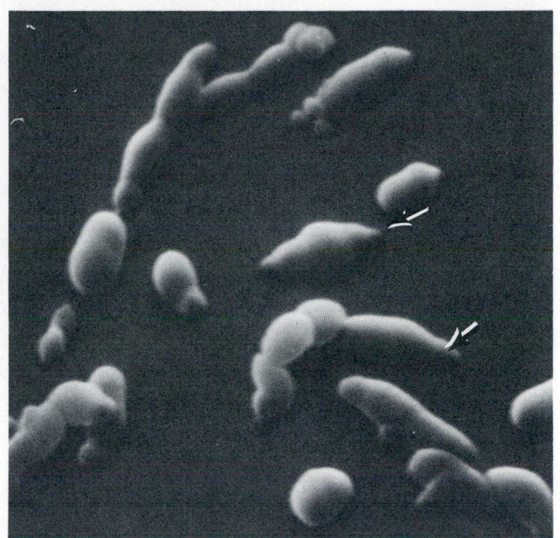

Figure 4-45 Electron micrograph of *Mycoplasma gallisepticum* showing blebs *(arrows)* that may be involved in motility.
From Razin, S., et. al. Infect. Immun. ***30:****538, 1980.*

chain of cocci by constriction of the cell membrane (Figure 4-43). ***Spiroplasma*** organisms are also pleomorphic but produce primarily short helical filaments (Figure 4-44). Some form small round cells when the rate of growth decreases.

Motility among the mycoplasmas, except for *Spiroplasma,* appears to be of a gliding nature and is believed to be associated with specialized polar structures such as stalks, blebs, or tapered tips (Figure 4-45). These structures apparently help the organism to attach to surfaces and to push forward in a gliding movement. *Spiroplasma* organisms appear to exhibit the same type of motility as the spirochetes. *Spiroplasma* organisms, however, lack flagella and have no cell wall or axial filaments. Membrane-associated fibrils similar to the actin proteins of muscle have been detected in *Spiroplasma* and may be involved in the motility of this group.

Very few species of *Mycoplasma* are agents of disease in humans. *Mycoplasma pneumoniae* is a cause of a minor pneumonia, but most mycoplasmas are more frequently found as agents of disease in cattle and plants. Some mycoplasmas are commensals in the upper respiratory tract of animals and are frequent causes of respiratory disease, particularly in many avian species. *M. mycoides* is the cause of contagious bovine pleuropneumonia, a major disease of cattle in parts of Africa and Asia. Mycoplasmas inhabit the phloem of plants and are transmitted by leaf hoppers, who naturally feed on phloem. Many of the plant mycoplasmas have not been completely characterized because of difficulties in cultivation and have been referred to as mycoplasma-like organisms (MLOs). MLOs are associated with the yellow diseases of a large variety of plants (see Chapter 35), whereas *Spiroplasma* organisms are known to cause stubborn disease of citrus and corn stunt disease.

Mycoplasmas are important research tools because of their limited genetic information, minimal biochemical activity, and reduced size. They possess a minimum set of structures that are necessary for self-replication, a cell membrane, DNA, and ribosomes. It has not been determined if on the evolutionary scale they are at the bottom of the prokaryotic ladder or represent a more developed prokaryotic cell that during evolution lost some genetic information.

Table 4-7 General characteristics of prokaryotes and eukaryotes

CHARACTERISTIC	COMPARISON BETWEEN PROKARYOTES AND EUKARYOTES
Size and metabolism	Prokaryotes usually smaller than eukaryotes and have greater ratio of surface area to volume; prokaryotes have faster rate of metabolism and cell division
Cellular interactions	Most prokaryotes produce colonies with no cooperation between cells; eukaryotes have complex cell aggregations with coordinated activities
Cell wall and membrane	Most prokaryotes have cell wall (except *Mycoplasma*); eukaryotes such as fungi and algae have cell wall, whereas protozoa have cell membrane with no cell wall; eukaryotes have specialized membrane system: endoplasmic reticulum; prokaryotic cell membrane carries out function of respiration, whereas mitochondria carry out this function in eukaryotes
Nuclear organization	Prokaryotes have haploid chromosome; eukaryotes have more than one chromosome and are diploid; nuclear membrane present in eukaryotes and absent in prokaryotes
Motility	Both possess flagella, but it is more complex in structure in eukaryotes; cilia appear only in eukaryotes; ameboid movement by pseudopodia occurs in eukaryotes such as protozoa
Endosymbiosis	Prokaryotes are never the host in endosymbiotic relationships; eukaryotes are nearly always the host in such relationships

SUMMARY

Microorganisms belong to the Protista, a group that does not possess any tissue systems. The higher protists, such as the protozoa, algae, and fungi, possess a nuclear membrane and are referred to as eukaryotes, and the lower protists, such as the cyanobacteria and bacteria, possess no nuclear membrane and are called prokaryotes. A group of prokaryotes with characteristics different from other prokaryotes are now referred to as archaebacteria. Some of the general characteristics of eukaryotes and prokaryotes are outlined in Table 4-7, whereas the more fundamental differences used to differentiate them are presented in Table 4-1.

Bacterial classification relies heavily on characteristics of the genus and species. Techniques used to differentiate and classify bacterial species include numerical taxonomy, in which many characteristics are weighed and evaluated; genetic analysis, in which either the guanine and cytosine content is determined or the extent of DNA hybridization between two microorganisms is evaluated; and sequence analysis, in which the sequence of certain amino acids, for example, in comparable proteins, is determined.

Morphologically, bacteria exist either as spheres, straight rods, square, or curved or spiral rods. They reproduce primarily by binary fission. Budding occurs infrequently and primarily in one group (budding bacteria).

In *Bergey's Manual of Determinative Bacteriology* (eighth edition) the bacteria have been divided into 19 parts. These parts have been selected according to special characteristics that include motility, morphology, staining properties, spore formation and oxygen tension.

Cyanobacteria are now considered to be prokaryotes, even though they resemble algae in many respects, such as color, size, and photosynthetic properties. Their prokaryotic characteristics include a gram-negative cell wall, absence of organelles, and absence of a nuclear membrane. Filamentous types are characterized by the ability to carry out aerobic and anaerobic processes simultaneously with the aid of a specialized cell called the heterocyst.

Study questions

1. What is the fundamental difference between the eukaryotic and the prokaryotic cell? How does this difference influence other properties of the cell? Explain.
2. Explain very briefly how prokaryotes differ from eukaryotes in terms of size, motility, cell envelope characteristics, nuclear organization, and endosymbiosis.
3. What characteristics of a specific plant or animal allow for its inclusion in distinct species? Can this distinction be applied to bacteria?
4. What parameters are used to classify bacteria into various groups as well as genera and species? What would be the ultimate test for differentiating microbial species, or is there no single test that would be appropriate?

5. Explain the techniques of DNA hybridization. What is the role of temperature in this process?
6. Why cannot size or shape be used in speciating bacteria? Could it be used for other groups of microorganisms?
7. Show diagrammatically how the various arrangements of cocci can be produced.
8. If bacteria are haploid, why can four nuclear areas be observed in some bacteria?
9. List and explain the ways in which bacteria reproduce.
10. What is the purpose of fruiting body formation in the myxobacteria? Do the myxobacteria perform any important function in their ecological niche?
11. What two groups of bacteria produce holdfasts? What is their function?
12. What is so unique about the flagella found in spirochetes in terms of location? What three genera are important in human infections, and how are they distinguished from one another?
13. What group of bacteria cannot be cultivated on ordinary laboratory media?
14. What characteristic of the bacilli and clostridia permits their separation into a specific classification group? Could this characteristic have any effect on human endeavors?
15. What group of bacteria is without a cell wall? What limitations does this impose on their choice of habitats?
16. In what ecological niche(s) are the methane-producing bacteria found? What unique characteristic of the cell wall distinguishes them from other bacteria?
17. What group of bacteria most resembles the fungi in morphology and reproduction?
18. Why cannot the Gram stain be used in the identification of the mycobacteria? What important diseases are caused by species of *Mycobacterium?*
19. What family of bacteria of the actinomycetes produces the most antibiotics isolated by humans?
20. What characteristics of cyanobacteria make them appear as prokaryotes? As eukaryotes?
21. How are the filamentous cyanobacteria able to carry out anaerobic processes when they also produce oxygen?
22. What role do the cyanobacteria play as endosymbionts?

Selected readings

Books

Buchanan, R.E., and Gibbons, N.E. (Eds.). *Bergey's manual of determinative bacteriology* (8th ed.). The Williams & Wilkins Co., Baltimore, 1974.

Brock, T. *Biology of microorganisms* (3rd ed.). Prentice-Hall, Englewood Cliffs, N.J., 1979.

Carlile, M.J. (Ed.). *Primitive sensory and communication systems: the taxes and tropisms of microorganisms and cells.* Academic Press, Inc., New York, 1975.

Carlile, M.J., Collins, J.F., and Moseley, B.E.B. (Eds.). *Molecular and cellular aspects of microbial evolution: thirty-second symposium of the Society for General Microbiolgy.* Cambridge University Press, New York, 1981.

Davis, B., et al.: *Microbiology* (3rd ed.). Harper & Row, Publishers, Inc., 1980.

Gibbons, N.E., Pattee, K.B., and Holt, J.G. (Eds.). *Supplement to index bergeyana.* The Williams & Wilkins Co., Baltimore, 1981.

Goldman, R. (Ed.). *Cell motility.* Cold Springs Harbor Lab., Cold Springs Harbor, N.Y., 1976.

Halvorson, H.O., and Van Holde, K.E. (Eds.). *The origins of life and evolution.* Alan R. Liss, Inc., New York, 1980.

Hazelbauer, G.L. (Ed.). *Taxis and behavior.* Chapman & Hall, New York, 1978.

Stanier, R. Y., Adelberg, E.A., and Ingraham, J. *The microbial world* (4th ed.). Prentice-Hall, Inc., Englewood Cliffs, N.J., 1976.

Starr, M.P., et al. (Eds.). *The prokaryotes.* Springer-Verlag, New York, Inc., New York, 1981.

Journal articles

Balch, W.E., et al. Methanogens: reevaluation of a unique biological group. *Microbiol. Rev.* **43:**260, 1979.

Barksdale, L., and Kim, K.S. Mycobacterium. *Bacteriol. Rev.* **41:**217, 1977.

Burchard, R.P. Gliding motility of prokaryotes: ultrastructure, physiology and genetics. *Annu. Rev. Microbiol.* **35:**497, 1981.

Canale-Parola, E. Motility and chemotaxis of spirochetes. *Annu. Rev. Microbiol.* **32:**69, 1977.

Fox, G.E., et al. The phylogeny of procaryotes. *Science* **209:**457, 1980.

Johnson, R.C. The spirochetes. *Annu. Rev. Microbiol.* **31:**89, 1977.

Kaiser, D., Manoil, C. and Dworkin, M. Myxobacteria: cell interactions, genetics and development. *Annu. Rev. Microbiol.* **33:**595, 1979.

Moore, R.L. The biology of *Hyphomicrobium* and other prosthecate budding bacteria. *Annu. Rev. Microbiol.* **35:**567, 1981.

Poindexter, J.S. The caulobacters: ubiquitous unusual bacteria. *Microbiol. Rev.* **45:**123, 1981.

Razin, S. The mycoplasmas. *Microbiol. Rev.* **42:**414, 1978.

Schachter, J., and Caldwell, H.D. Chlamydiae. *Annu. Rev. Microbiol.* **34:**285, 1980.

Stackebrandt, E., and Woese, C.R. The evolution of prokaryotes. *Symp. Soc. Gen. Microbiol.* **32:**1, 1981.

Stanier, R.Y., and Cohen-Bazire, G. Phototrophic procaryotes: the cyanobacteria. *Annu. Rev. Microbiol.* **31:**225, 1977.

Walsby, A.E. The gas vacuoles of blue-green algae. *Sci. Am.* **237:**90, 1977.

Woese, C.R. Archaebacteria, *Sci. Am.* **244:**98, 1981.

Chapter 5 CHARACTERISTICS OF THE MICROBIAL WORLD Eukaryotes

The eukaryotic microorganisms consist of three groups: the Fungi, the Protozoa, and the Algae. Eukaryotes carry out the same biological processes as prokaryotes, such as energy metabolism, nutrient transport, motility, and translation of genetic information, but they carry out these processes in different ways and often with more complex structures than do prokaryotes. Those characteristics that separate eukaryotes from prokaryotes are elaborated on in Chapter 4, and there is no need for reiteration here. Many areas of eukaryotic research have not reached the levels of those for prokaryotes; this is understandable, since most eukaryotes require a longer time for cell division and are much more difficult to cultivate in the laboratory. Studies on eukaryotic morphology, genetics, metabolism, and physiology have important implications for humans, because we too are eukaryotes. The discovery, for example, of the segmented nature of genetic information in eukaryotes was an unexpected and unusual finding. This discovery will undoubtedly help us in the future to understand more fully the structure and function of the eukaryotic genome. Bacteria have been an important tool in the research laboratory, but priorities change, and it is without question that a significantly greater amount of research in the future will be directed toward the role of eukaryotic microorganisms in the fields of agriculture, geochemical activities, and pollution control. Such studies may be necessary for the survival of mankind.

The purpose of this chapter is twofold: first, to describe in general terms some of the organizational characteristics of eukaryotic cells; second, to summarize some of the characteristics unique to each group of eukaryotic microorganisms. Figure 5-1 illustrates a typical eukaryotic cell.

CHARACTERISTICS OF A EUKARYOTIC CELL

The cytoplasmic contents of a eukaryotic cell are protected in much the same way as prokaryotic cells: a cell wall or a specifically adapted cell membrane. Cell walls are present in fungi and algae, but their chemical composition is radically different from those of prokaryotic cells. A cell wall is absent in the protozoa and some algae, but this is often compensated for by modifications of the cytoplasmic membrane. The flexibility of the cytoplasmic membrane accounts in part for the phenomena of intracellular symbioses in which eukaryotic and prokaryotic microorganisms are ingested intact into the host cell. The cytoplasmic membrane of eukaryotes is chemically similar to that of prokaryotes; the major distinction is the presence of sterols in eukaryotes and their absence in prokaryotes (except mycoplasmas and some other prokaryotes). The eukaryotic membrane system is organizationally different from prokaryotes and is characterized by the presence of an endoplasmic reticulum—a membrane system that traverses the cytoplasm of the cell. Some portions of the endoplasmic reticulum contain ribosomes, whereas other regions are

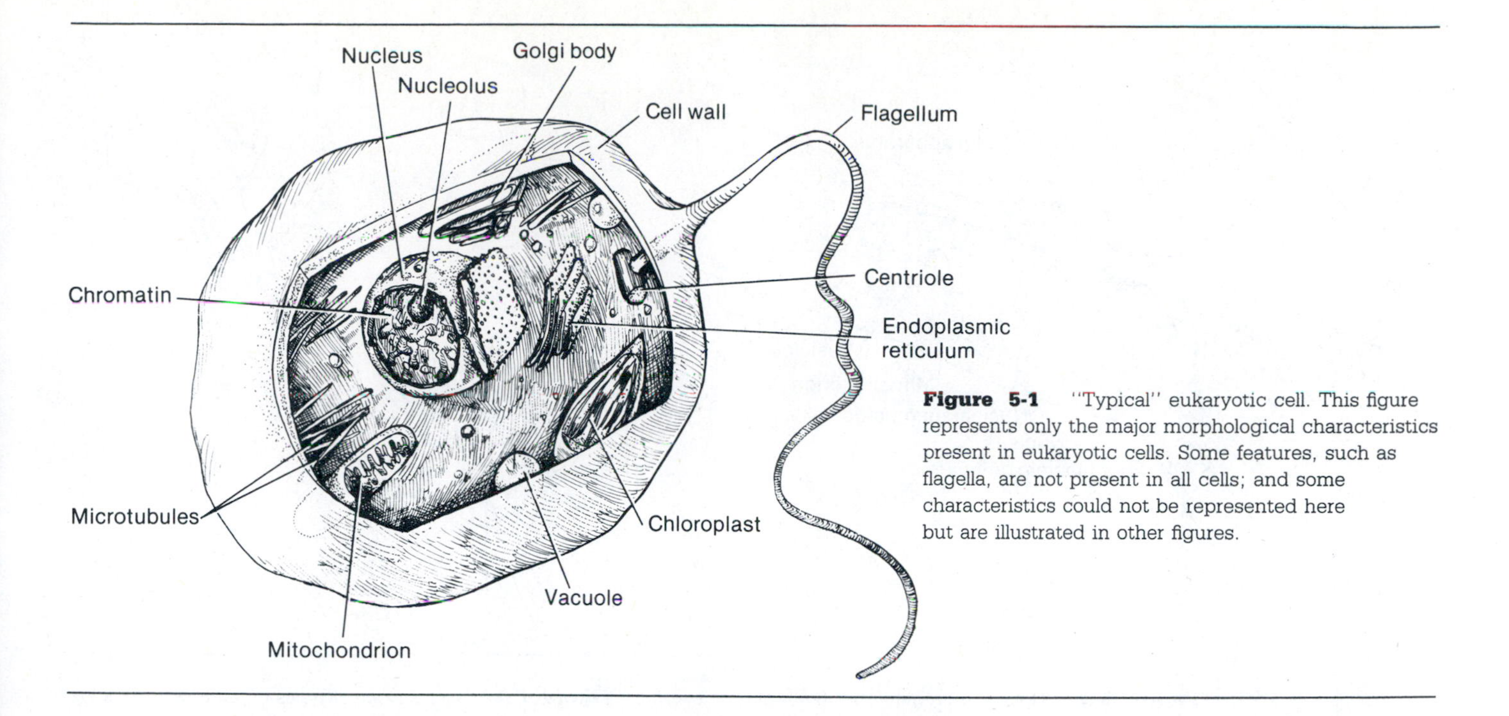

Figure 5-1 "Typical" eukaryotic cell. This figure represents only the major morphological characteristics present in eukaryotic cells. Some features, such as flagella, are not present in all cells; and some characteristics could not be represented here but are illustrated in other figures.

continuous with the membrane surrounding the nucleus.

Located within the cytoplasm of the eukaryotic cell are organelles, such as mitochondria, chloroplasts, and the Golgi appartuses, each of which has its own characteristic membrane system. The mitochondrial membrane system, for example, contains enzymes for energy production—a characteristic associated with the cytoplasmic membrane in prokaryotes.

Eukaryotes, like prokaryotes, can use flagella as structures of locomotion. Eukaryotic flagella are considerably longer and wider than those in prokaryotes and range from 20 to 250 μm long. Shorter versions of flagella are ***cilia,*** which are chemically and structurally the same as flagella but are found on the eukaryotic cell in larger numbers (several thousand may appear on a single cell). Another type of motility is characteristic of some protozoa and is called ameboid movement—so named because of its discovery in the protozoan called *Amoeba.* The cytoplasm flows freely in eukaryotes (called ***cytoplasmic streaming***), and in amebas where there is a flexible cytoplasmic membrane, the cell can change its shape and form projections called ***pseudopodia*** (Figure 5-2). This type of movement allows the ameba to gradually translocate itself over a solid surface, but the movement is less rapid than in cells whose locomotion is by flagella. Most spore-forming parasitic protozoa (sporozoans) exhibit a gliding motility through fluids, but the mechanism is unknown. Electron microscope studies have revealed the formation of waves in the cytoplasmic membrane that move from the anterior end to the posterior end of the parasite.

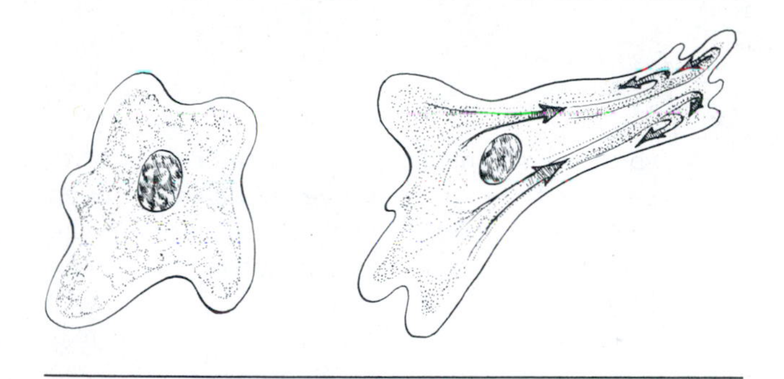

Figure 5-2 . Direction of cytoplasmic streaming one-celled protozoa, ameba.

Sexual processes are characteristic of eukaryotic systems, and cells are produced that possess two complete sets of genetic information (the diploid state). The diploid state is a result of the fusion of male and female gametes, each of which possesses only one set (haploid state) of genetic information. The diploid state appears fleetingly in some eukaryotic microorganisms; most of their life cycle is spent in the haploid state. This is contrasted with more complex eukaryotic microorganisms as well as higher eukaryotes such as plants and animals, including humans, in which the diploid state predominates throughout the life cycle. The extra set of genetic information in eukaryotes provides the cell with a greater variety of potential responses to an unfa-

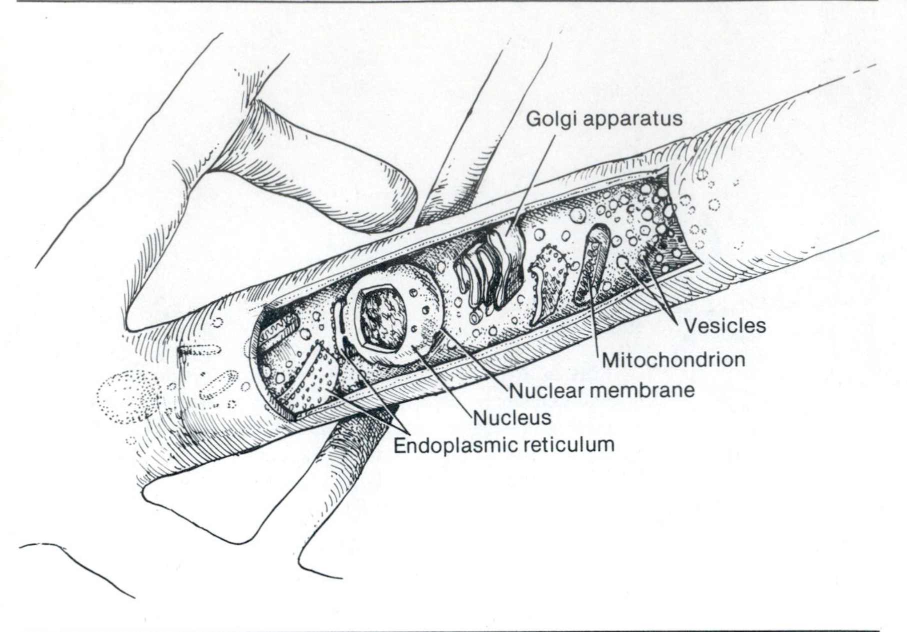

Figure 5-3 Thallus, a fungal vegetative unit.

Figure 5-4 Morphological types of fungi. **A,** Spherical yeast cells. **B,** Filamentous mold with cross walls (septa). **C,** Filamentous mold without septa.

vorable environment. In addition, diploidy has permitted the eukaryotic cell to achieve, during evolution, a greater structural complexity—more than could be achieved by primitive haploid prokaryotes.

FUNGI

Classification

Fungi are a diverse group of organisms ranging in size from some unicellular microscopic types called yeasts to some large multicellular branching forms such as mushrooms. As in the case of bacteria, no fossil records are available to establish an evolutionary history. Morphological and physiological characteristics are the major means of classification, but environmental factors play such a major role in determining fungal form and physiology that classification has been difficult and frustrating. Fungi were originally classified as plants, but because of their variability, diversity, and absorptive nutrition have been placed in a separate kingdom: the Fungi, or Myceteae. Most mycologists (those who study fungi) have separated the kingdom into two divisions: the Myxomycota, which are slime molds, and the Eumycota, which are considered true fungi. The Eumycota are separated into five subdivisions: Zygomycotina, Mastigomycotina, Ascomycotina, Basidiomycotina, and Deuteromycotina.

Morphology

The basic vegetative unit of fungi, called the ***thallus,*** is bounded by a cell wall made up to chitin, cellulose, or other glucans; proteins, lipids, and other constituents are also present. The cytoplasm contains a membrane-bound nucleus as well as the usual organelles (Figure 5-3). Fungi that are filamentous and exhibit extensive branching are called ***molds,*** whereas unicellular, spherical fungi are called ***yeasts*** (Figure 5-4). The terms *yeast* and *mold* have no taxonomic value because some species of fungi may exist in either state, depending on temperature or other environmental conditions. This potential of a single species to exist in either state is called ***dimorphism*** and is characteristic of certain fungi pathogenic to humans. Yeasts and molds are also distinguished by the fact that yeasts have a single nucleus, whereas filamentous forms are usually multinucleated. Like the prokaryotic actinomyces, the filamentous extensions of the mold are called ***hyphae,*** and a mat of hyphae is called a ***mycelium.*** Mycelial filaments that penetrate the substrate, such as organic debris in the soil, are called ***rhizoids.*** Rhizoids (Figure 5-5) are

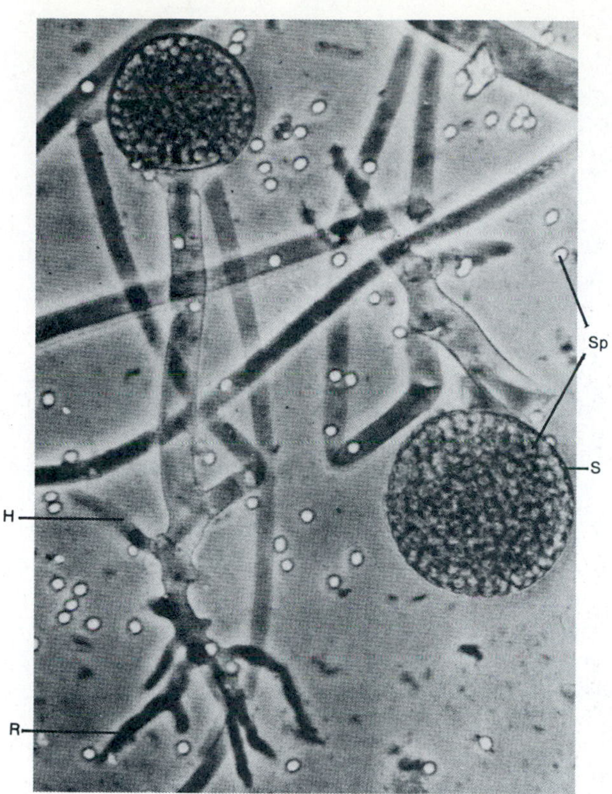

Figure 5-5 Photomicrograph of a species of *Rhizopus* demonstrating rhizoids. Note that hyphae have no septa (coenocytic), and rhizoids have finely tapered ends. *S,* Sporangium; *Sp,* spores; *H,* substrate hyphae; *R,* rhizoids.
From Bottone, E.J. J. Clin. Microbiol. ***9:****530, 1979.*

more finely tapered than hyphae and are usually free of nuclei; that is, they are sterile cells. Some fungal species parasitize plants or animals,* and the mycelia that penetrate the host to obtain nutrients are called ***haustoria*** (Figure 5-6). The mycelia in many species give rise to aerial structures that form airborne spores and thus help disperse the species. The filaments of the mycelium may be continuous ***(coenocytic),*** in which case all components of the mycelium may mingle (Figure 5-7); or the filaments may be subdivided by cross walls called ***septa*** (sing. septum). The presence of septa does not ensure that the cell cytoplasms are separated, because some septa are perforated and permit the intermingling of nucleic and other cytoplasmic contents.

Nutrition

Fungi are heterotrophic (they require organic compounds as a source of carbon) and use organic molecules as carbon and energy sources. This heterotrophism is greatly facilitated by the production of extracellular enzymes that degrade

* The ability of fungi to parasitize plants has been used commercially as a mechanism for biological control. Fungal plant pathogens, termed mycoherbicides, are being used to kill or inhibit the growth of weeds such as jointvetch in the rice fields of Arkansas, Mississippi, and Louisiana and water hyacinth in many southern states.

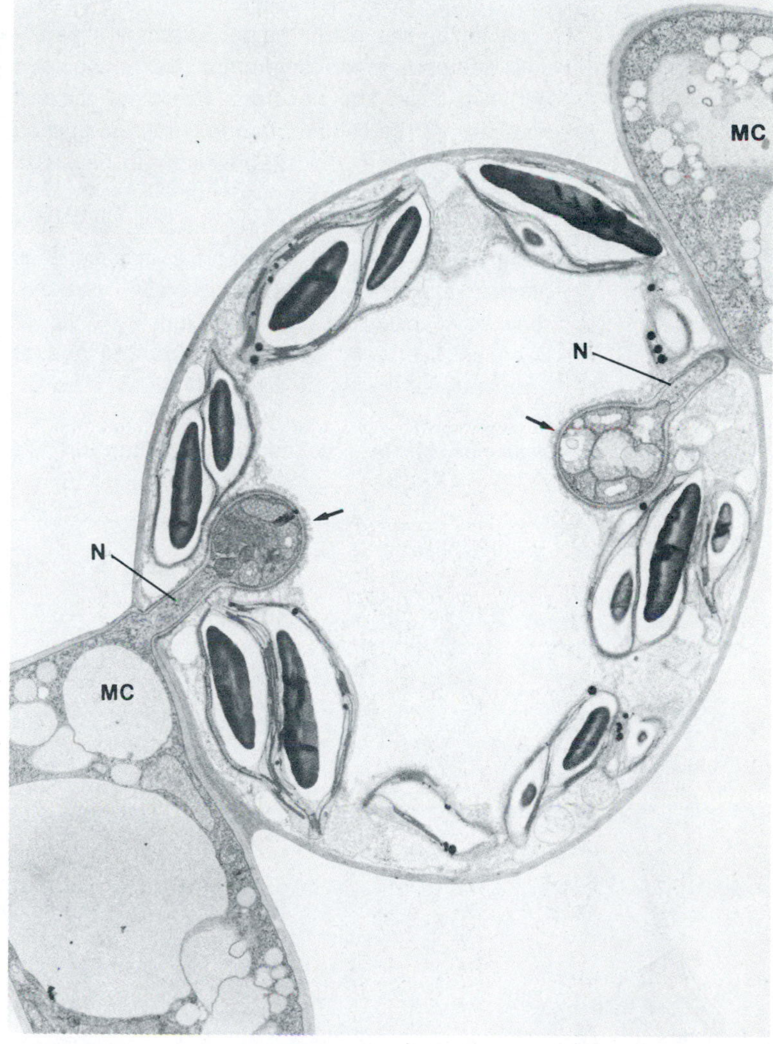

Figure 5-6 Haustoria. Section through plant cell showing presence of two haustorial bodies *(arrows)* connected by neck *(N)* to their respective mother cells *(MC).*
From Coffey, M. Can. J. Botany ***53:****1285, 1975.*

the large macromolecules that are found in the environment: soil, plants, animals, or fabrics. Many fungi exist only as ***saprobes*** (which live on dead organic material), but a few are parasitic and obtain their nutrients from living hosts. Many species can be either saprobes or parasites; for example, many of the fungi that infect humans are saprobic filamentous forms whose normal habitat is the soil. This filamentous arrangement provides an increased surface area that is adapted for obtaining nutrients from the soil, whereas inside the human host this same species may exist in a spherical state. The temperature inside the host (37° C) is higher than the normal soil habitat of the fungi, and the species causing infections are confronted with a hostile environment. It is suspected that the pathogenic yeast form is an induced response to this harsh environment. Fungi that parasitize plants do not show dimorphism. The hyphae penetrate the tough outer coat of the plant (Figure 5-6), and once inside the plant, fungal enzymes separate cells and permit extensive branching of haustoria between cells. The haustoria penetrate the cell wall but not the cell membrane, which is pushed aside to produce a pocket where the haustoria can develop and absorb nutrients.

Most fungi are easily cultivated in the laboratory, requiring only a carbon source such as glucose and mineral salts. Generally speaking, they grow best in acid conditions (pH values between 5 and 6) with high humidity and at temperatures between 20° and 25° C. The dimorphic fungi can be differentiated by the appearance of their colonies, depending on the temperature of incubation and certain nutritional factors. The mycelial phase, which develops at room temperature (20° C), is usually represented as a fuzzy colony that produces aerial spores, whereas the yeast phase, which can be detected at 37° C, is granular or mucoid and produces no aerial spores (Plates 8 and 9).

Reproduction

Reproduction in fungi may occur by sexual or asexual means. Asexual reproduction by budding (Figure 5-8) or by fission occurs in a very few species, such as the yeasts. The reproductive unit in most fungi, however, is the ***spore,*** which may be produced asexually or sexually. Most spores are produced asexually and may arise directly from the mycelium, of which there are different types (Figure 5-9), or they may be produced on or within specialized structures. For example, among the more primitive fungi, spores called ***sporangiospores*** are produced within a structure called a ***sporangium,*** and spores called ***conidia (conidiospores)*** are usually borne externally on a modified hyphal tip called a ***conidiophore*** (Figure 5-10). Conidia may be small and single celled (microconidia) or large and multicelled (macroconidia) (Figure 5-11). The size, shape, color, and number of conidia and conidiophores are useful in the identification of fungal species. In more complex fungi sexual spores are produced on or in structures called ***fruiting bodies.*** These are discussed later.

Spores produced sexually (sexual reproduction is called the perfect state) are of four types: ***oo-***

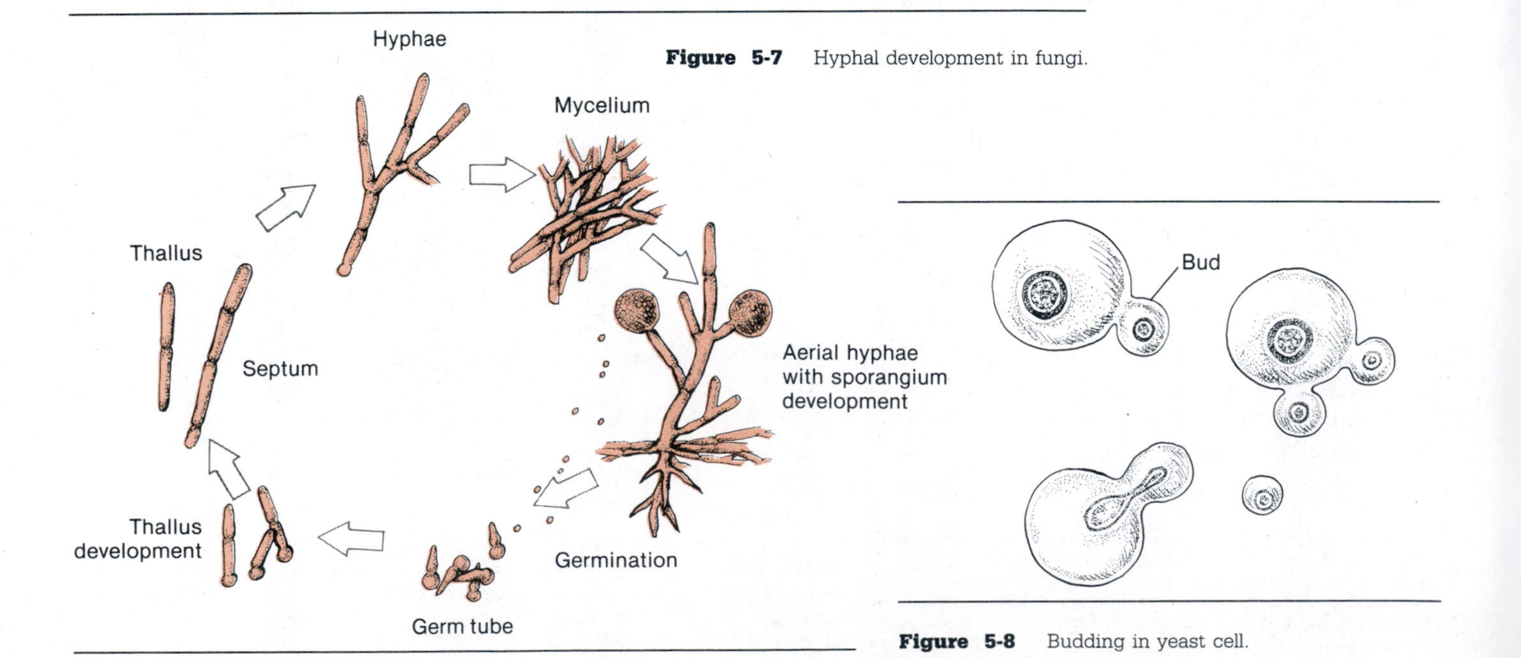

Figure 5-7 Hyphal development in fungi.

Figure 5-8 Budding in yeast cell.

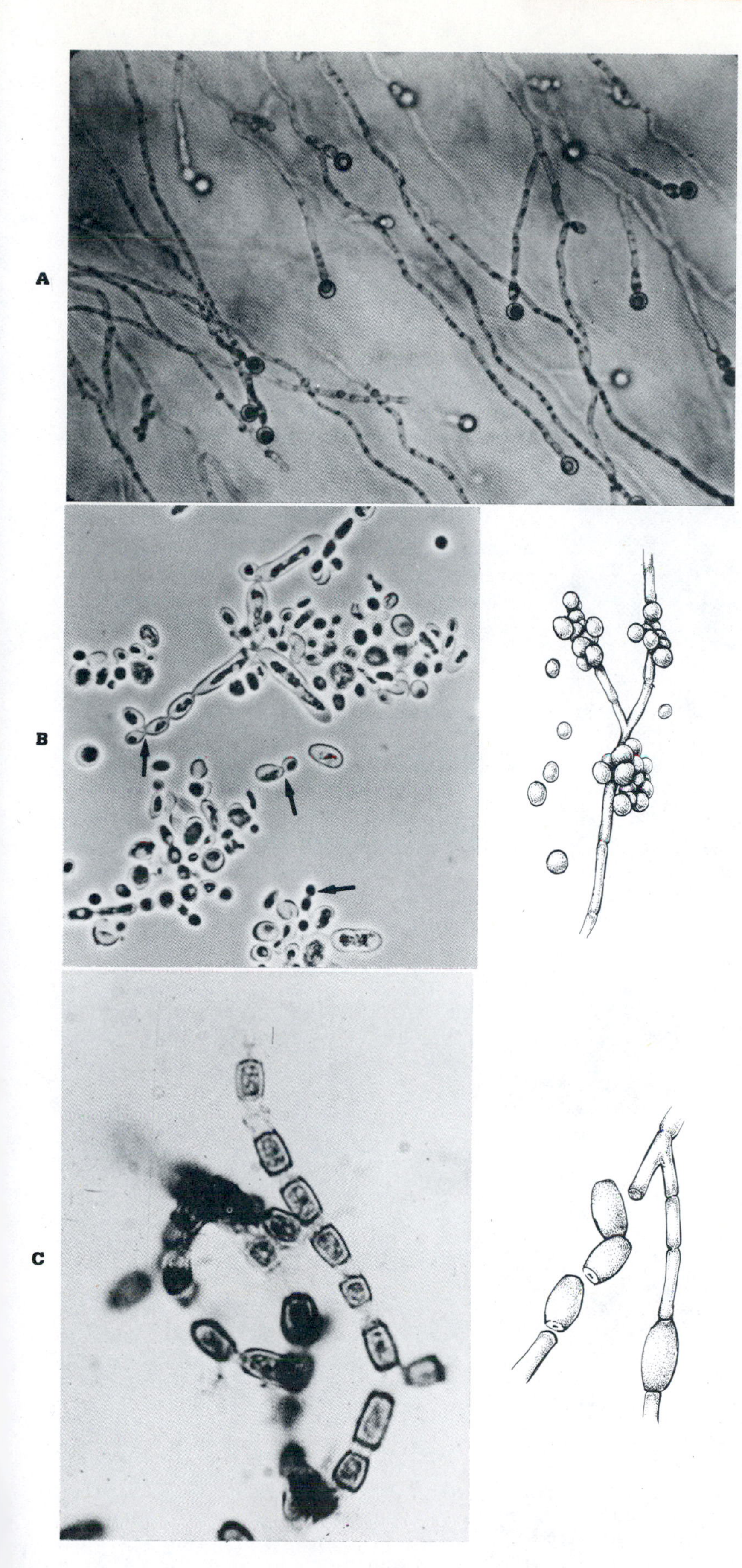

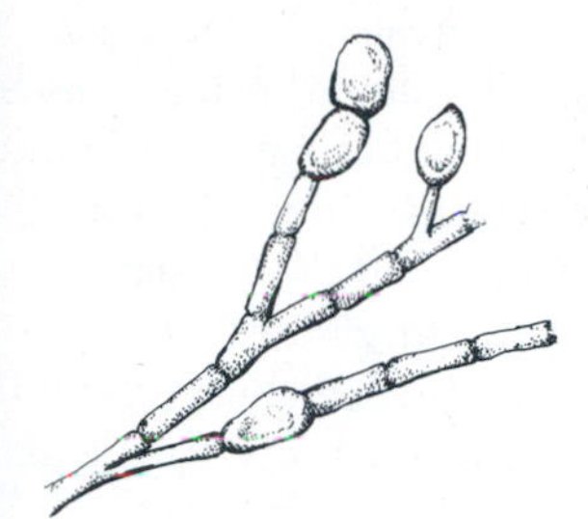

Figure 5-9 Asexual fungal spores that arise directly from mycelium. **A,** Chlamydospores arise by rounding and thickening of vegetative (thallus) cell wall. They may appear terminally as shown or within hypha. **B,** Blastospores *(arrows)* are buds that arise by separation from vegetative mother cell. **C,** Arthrospores arise by fragmentation of hypha.

***A,** Courtesy Glenn Roberts. **B,** From Rogers, A.L. In* Manual of clinical microbiology *(3rd ed.). American Society for Microbiology, Washington, D.C., 1980. **C,** From Larsh, H.W., and Goodman, N.L. In* Manual of clinical microbiology *(3rd ed.). American Society for Clinical Microbiology, Washington, D.C., 1980.*

spores, basidiospores, zygospores, and ***ascospores.*** Zygospores and ascospores are the most frequently encountered. Sexual reproduction may involve the union of either motile gametes (sex cells), specialized sex organs, or vegetative hyphae. The union in most instances involves two sex organs: the male antheridium and the female ascogonium. Regardless of the sexual method, the final result is the mingling of cell cytoplasms ***(plasmogamy),*** followed by fusion of their respective nuclei ***(karyogamy),*** and meiosis, in which genetic units are exchanged and four haploid spores are produced.

Fungal sporulation, as in bacteria, is usually triggered by a depletion of nutrients. Other factors such as light or chemical products in the environment can also induce spore formation; however, little is known of the specific mechanisms involved. Many spores are produced during sporulation and, if left to germinate at one site on a substrate, they would compete for nutrients. It is to the advantage of the species that spores are dispersed at some distance from the parent to reduce such competition. The spores, or in some species, the spore-bearing structures (sporangia), are violently released from the mycelium when a pressure change occurs within the spore sac. The triggering condition may be light, temperature, or a physical jolt, and the released spores may travel from a few centimeters to 1 or 2 m. Some spores can be disseminated by air currents, animals, or water vectors.* Germination of the spores will occur if environmental conditions of temperature, humidity, and nutrients are ade-

*The importance of water in spore dispersal is especially recognized by rose growers. A fungus disease called black spot attacks the rose leaves and causes them to turn yellow and fall off. The infection process will continue the following spring if the dead leaves are not removed from the ground surrounding the plant because rain drops splashing on the dead leaves cause the dispersal of spores, which can attack healthy leaves.

A

B

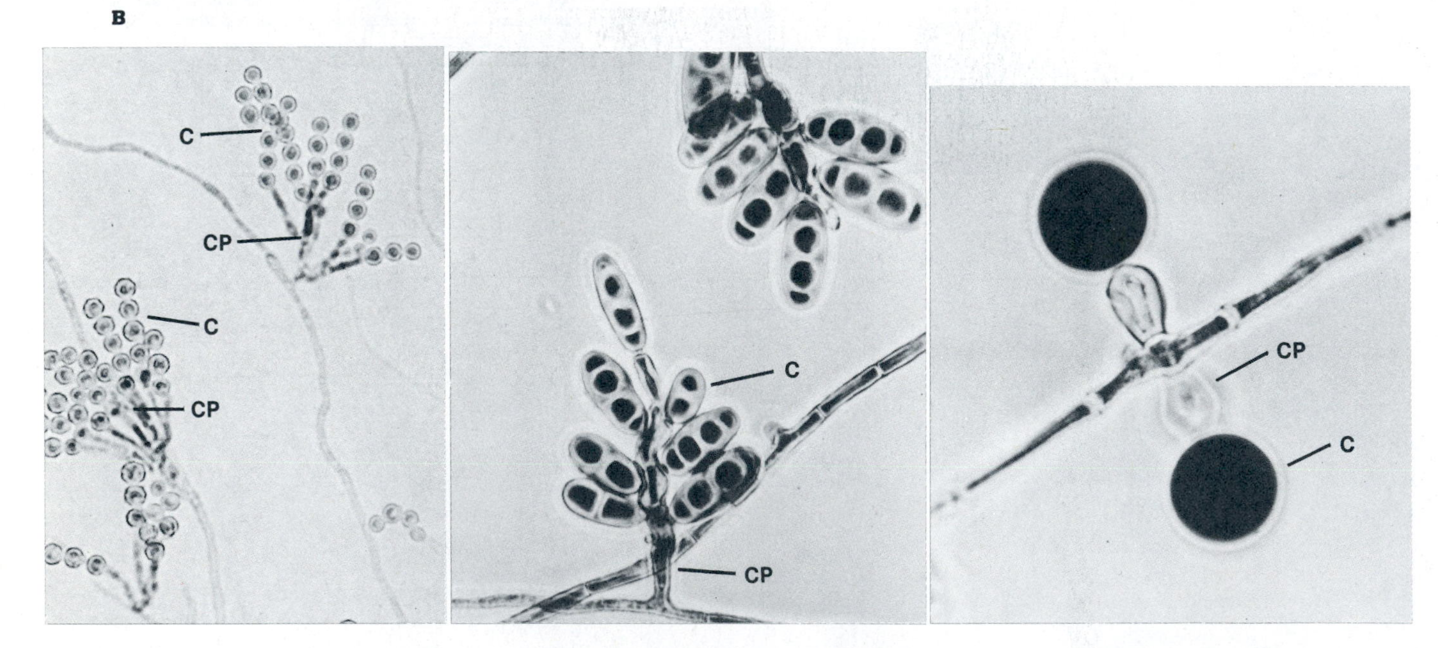

Figure 5-10 Specialized fungal structures that bear spores. **A,** Sporangium and sporangiophore. Sporangium *(S)*, or spore sac, is borne on modified hyphal filament, sporangiophore *(Sp)*. One sporangium *(arrow)* has released spores. **B,** Conidiophore *(Cp)* and arrangement of conidia *(C)* that they bear.

From Rogers, A.L. In Manual of clinical microbiology *(3rd ed.). American Society for Microbiology, Washington, D.C., 1980.*

quate. The germination of a spore results in the formation of a protuberance called a ***germ tube*** (Figure 5-7), which elongates and forms a hyphal filament. The hypha in turn grows and produces a mycelium from which new spore-bearing structures may be produced, and the cycle is repeated.

Characteristics of the major subdivisions of fungi

Zygomycotina. The principal class of fungi in the subdivision Zygomycotina is the Zygomycetes, which is composed of three orders: the Mucorales, the Entomophthorales, and the Zoopagales.

The Mucorales are terrestrial fungi that can be isolated from soil, dung, and other decaying matter. Other species can cause severe infections in plants and animals. Asexual spore production is usually associated with the formation of a sporangium (Plate 10), whereas sexual reproduction is initiated by the formation of sex organs that fuse and form a ***zygospore*** in which nuclear fusion and meiosis occur (Figure 5-12). Species of *mucor* and *Rhizopus* are involved in fatal infections, particularly in diabetic persons. *Rhizopus nigricans,* the common bread mold, is used in industry for the transformation of steroids. Some species of *Rhizopus* are used in the production of Oriental foods such as sufu and tempeh that are made from soybeans.

The Entomophthorales have a sexual cycle similar to the Mucorales. Asexual reproduction is associated with the formation of a conidium produced at the tip of a swollen conidiophore. Many species in this order are parasites of insects. The fungus grows in the abdominal cavity of a fly, for example, until all organs are digested and the mycelium occupies the entire abdominal cavity. Later the conidiophores penetrate between the abdominal segments to the exterior and release their spores, which cover the surface of the fly (Figure 5-13).

The majority of the Zoopagales are microscopic fungi that parasitize microscopic organisms such as protozoans, rotifers, and nematodes. Those species that capture small animals and consume their body contents produce a sticky substance that coats the mycelium. Once the microscopic animal comes in contact with the fungus, animal and fungus stick together. The hyphae penetrate the animal and digest the internal contents (Figure 5-14).

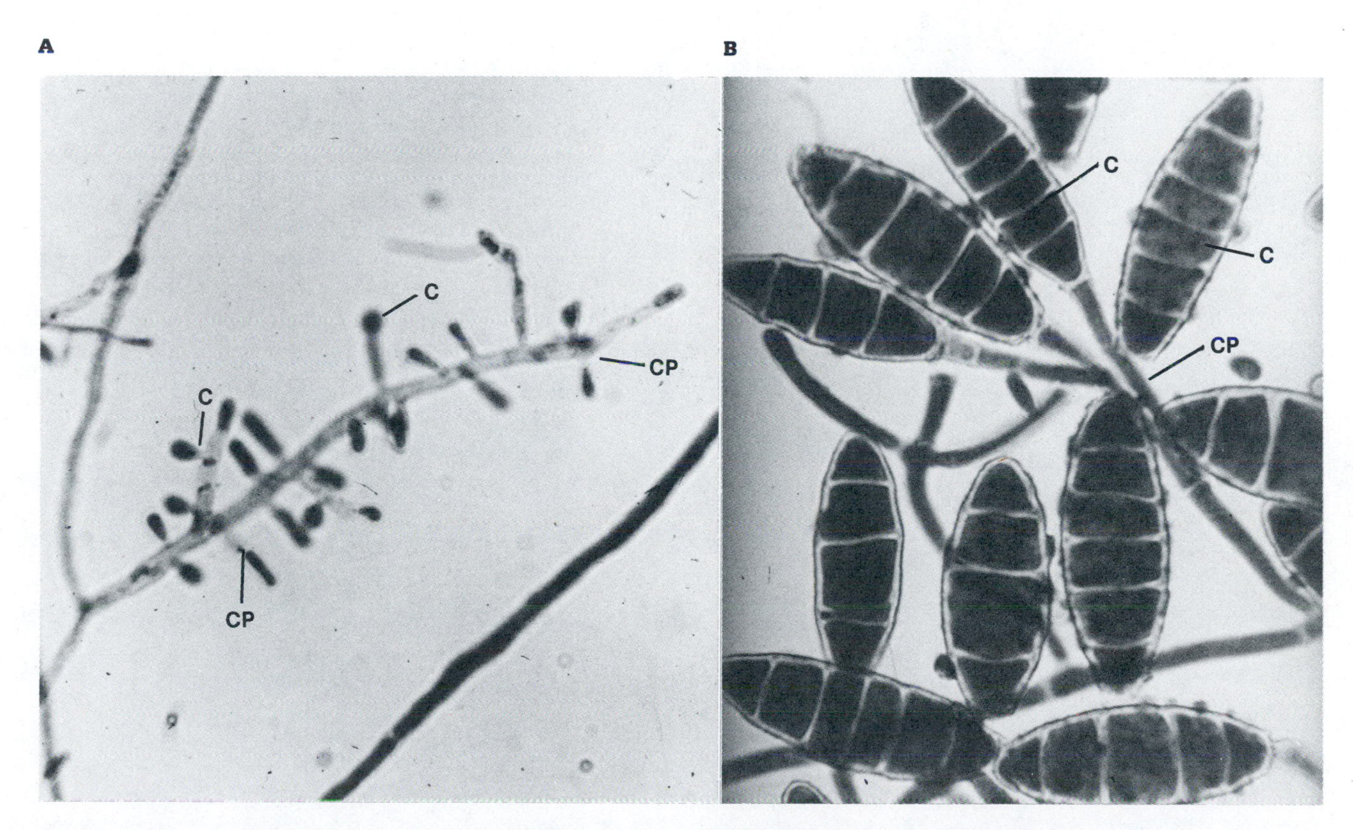

Figure 5-11 Conidia. **A,** Microconidia. (×1070.) **B.** Macroconidia. (×487.) *C,* Conidia; *CP,* conidiophore.
From Ajello, L. In Manual of clinical microbiology *(3rd ed.). American Society for Microbiology, Washington, D.C., 1980.*

A

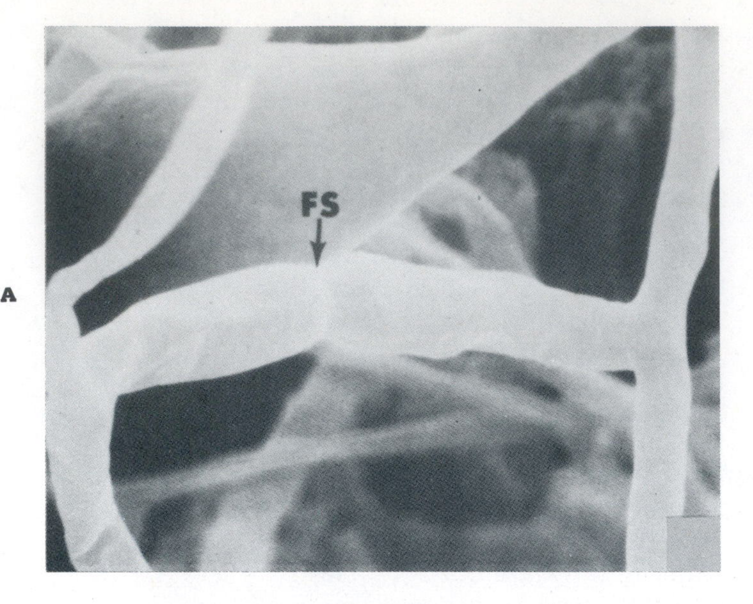

Mastigomycotina. The Mastigomycotina contain one class called Chytridomycetes, of which there are three orders: Chytridiales, Blastocladiales, and Monoblepharidales. This subdivision includes members that are aquatic as well as terrestrial. All produce motile cells that possess a single flagellum—a characteristic that separates them from nearly all other fungi.

The Chytridiales (the most primitive of the group) do not produce a mycelium, although a rhizoid system is present in some species. They are parasites of algae, water molds, and vascular plants. This order can be divided into two groups based on the morphology of the thallus. One group consists of a single thallus that has no appendages (rhizoids), and during its life cycle is transformed directly into a reproductive structure. These are called ***holocarpic*** species. The second group, called ***eucarpic,*** consists of species in which part of the thallus is rhizoidal. These thallic forms can also be found in other orders of this subdivision. Holocarpic species cannot exist as saprobes, living on dead organic material, because of the absence of rhizoids; therefore, nearly all of them are found within living hosts, such as algal cells. The life cycle of a holocarpic species is illustrated in Figure 5-15.

B

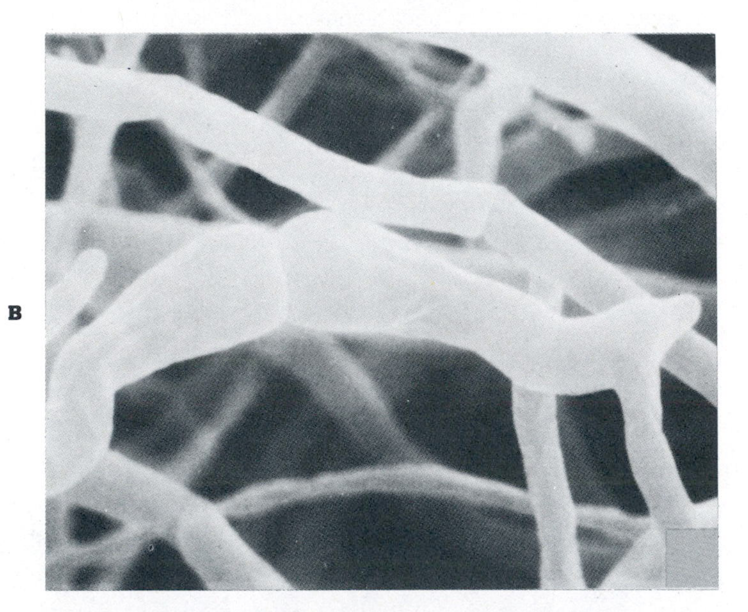

Figure 5-12 Zygospore formation. **A,** Contact of hyphal tips (gametes) and formation of fusion septum *(FS)*. **B,** Enlargement of hyphal contact point due to flowing of protoplasm and nuclei between hyphae. **C,** Rupture of primary wall at point where zygospore is developing. **D,** Zygospore expanding *(arrow)*. **E,** Maturing zygospore showing warty surface.
From O'Donnel, K.L., Ellis, J.J., and Hesseltine, C.W. Can. J. Bot. ***55:****662, 1977.*

C

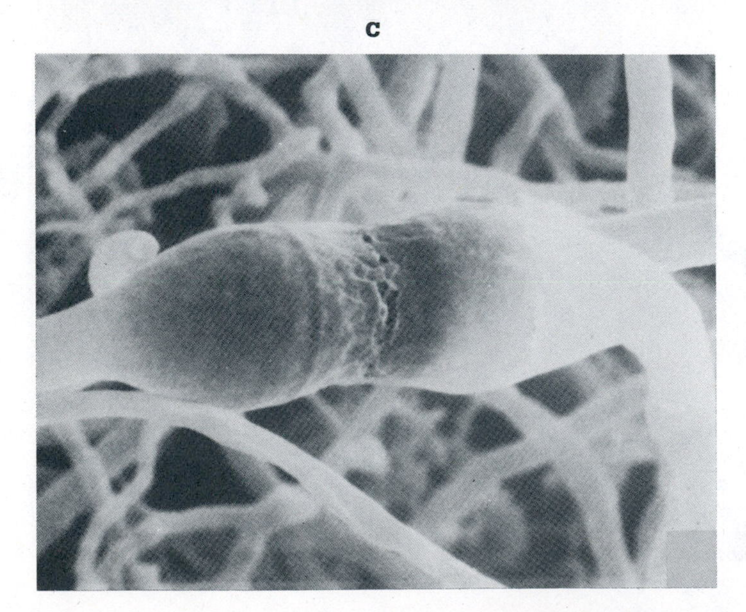

D

Eucarpic species can also live within a host, but they use rhizoids to penetrate the host while the remainder of the mycelium lies outside. Some aquatic eucarpic species float free in the water with only the distal ends of the rhizoids attached to nutrient substrates. One member of this order, *Synchytrium endobioticum,* is a cause of potato wart. Worldwide in distribution, the disease is characterized by masses of warty tissue covering the tuber.

The Blastocladiales have a well-defined mycelium. Most of the members of this order are aquatic, but some are obligate parasites on insects such as mosquito larvae. The motile spores of this group all possess a ***nuclear cap*** (Figure 5-16), which is composed of densely packed ribosomes. The reason for this is not clear. Another organelle found almost exclusively in the zoospore is the ***side body complex*** (Figure 5-17), which consists of lipid droplets, mitochondria, microbodies, and membranes. The third organelle present in the motile spores is the ***gamma particle,*** or ***endocytosome,*** which contains DNA, RNA, and the enzyme chitin synthetase. The function of the gamma particle is believed to be in controlling the encystment of the zoospore through chitin metabolism.

Figure 5-13 Conidia covering housefly following penetration of conidiophores through abdominal cavity.
From Ross, I.K. Biology of the fungi, *McGraw-Hill Book Co., New York, 1979.*

E

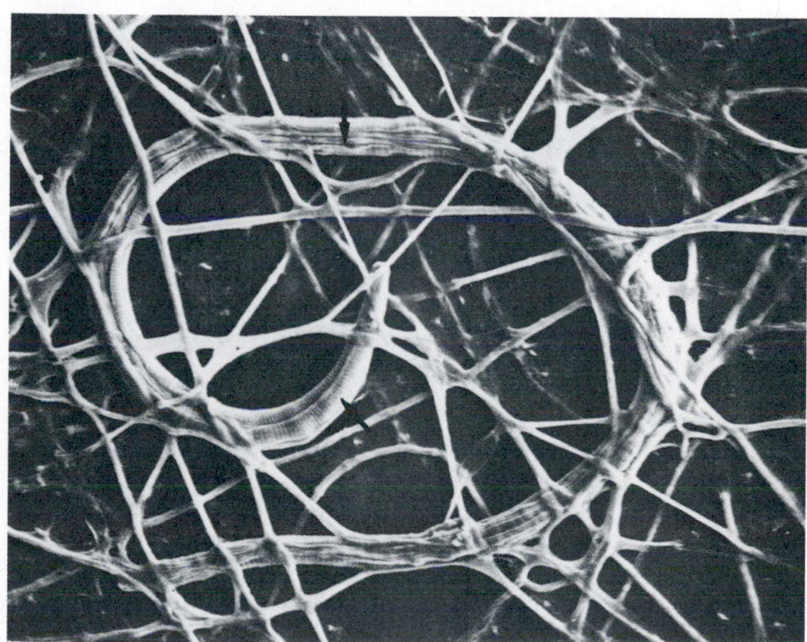

Figure 5-14 Scanning electron micrograph of nematode *(arrows)* entrapment by adhesive network of filaments produced by a member of Zoopagales. (×1900).
Reprinted by permission from Mycologia ***71****:379. Copyright 1979, Dowsett, J.A., and Reid, J.J. and The New York Botanical Garden.*

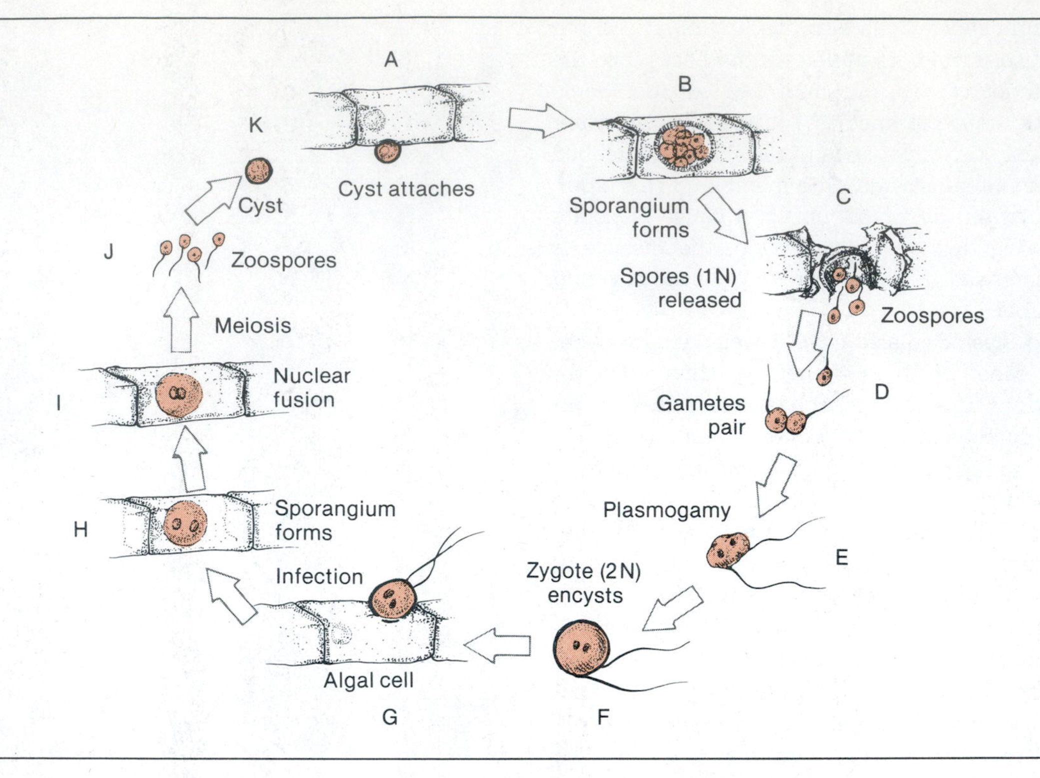

Figure 5 15 Life cycle of holocarpic species of order Chytridiales. *A,* Nonmotile cyst attaches to algal cell wall and dissolves hole in wall. Cyst envelope releases fungal protoplast (wall-less) that penetrates algal cell wall *B,* Envelope forms in algal cell that surrounds protoplast; this becomes sporangium where fungal nucleus divides repeatedly and produces spores. *C,* Lysis of cell wall and release of motile haploid spores called zoospores by sporangium. *D,* Spores acting as gametes pair. Gametes are usually derived from different sporangia, but sister spores may also pair. *E,* Plasmogamy (gamete fusion) occurs and produces flagellated binucleate cell *F,* Zygote encysts *G,* Cyst infects algal cell just as in *A. H,* Sporangium is produced in algal cell, but this is resting sporangium that is maintained during winter and will not develop until more favorable environmental conditions arise. *I,* Nuclei in sporangium fuse. *J,* Meiosis results in flagellated zoospores. *K,* Zoospores encyst.

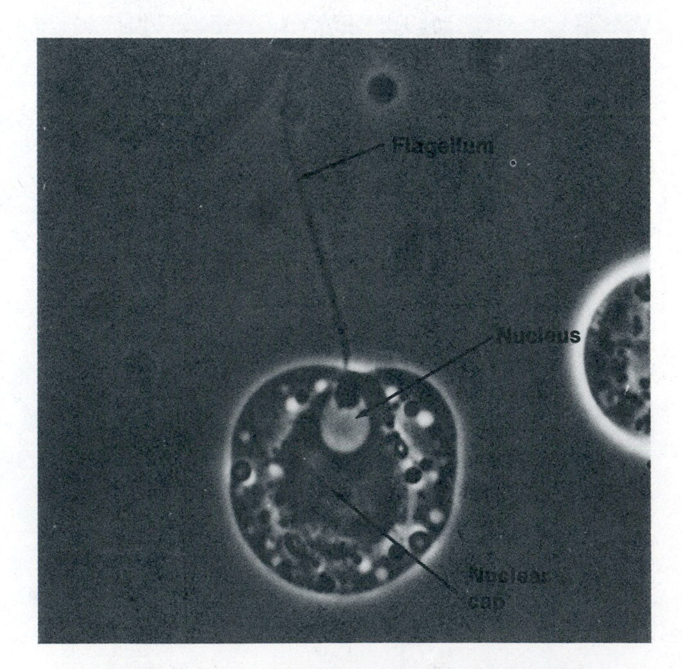

Figure 5-16 Electron micrograph of nuclear cap in motile spores of order Blastocladiales. (×2800.)

Reprinted bv permission from Mycologia *69:1. Copyright 1977, Fuller, M. and The New York Botanical Garden.*

The Monoblepharidiales are aquatic fungi that have a unique sexual reproduction scheme (Figure 5-18), which is found nowhere else among the fungi. Sexual fusion involves a small motile male gamete and a large nonmotile female gamete—a mechanism called ***oogamy.*** The male antheridium (organ that produces male sex cells) is formed on the female sex organ (called the oogonium). The male antheridium, when mature, releases several gametes that swim to the female oogonium, in which only one egg is produced. Fertilization results in the formation of a zygote, which develops into an oospore. The oospore germinates by splitting its wall, from which emerges a germ tube.

Ascomycotina. The members of this subdivision are referred to as ascomycetes and include both unicellular and filamentous forms. They possess complex structures that exhibit great variation in shape and subcellular organization. The one characteristic that distinguishes this group from

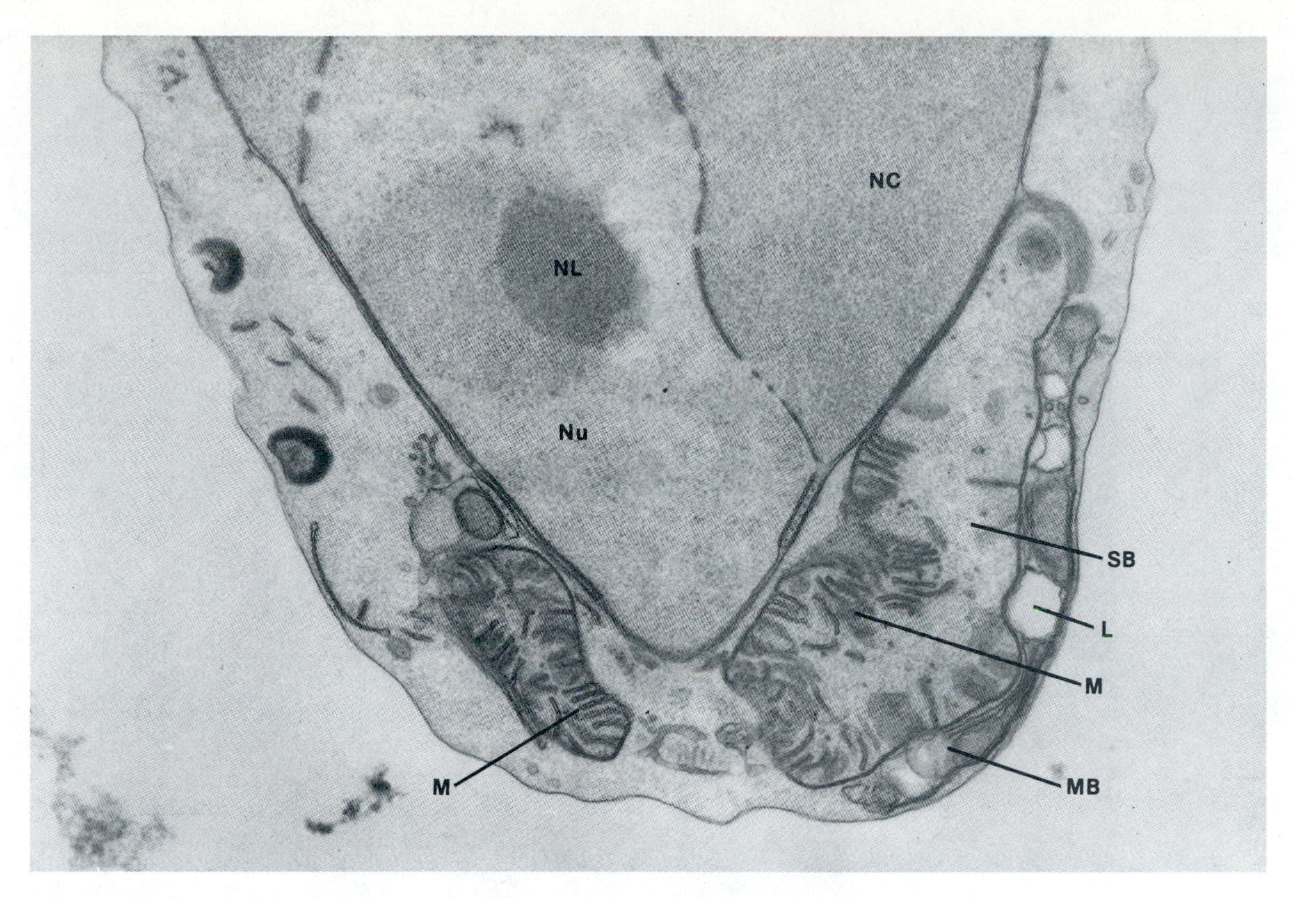

Figure 5-17 Electron micrograph of posterior half of zoospore demonstrating "side body complex." (×64,000.) *NL*, Nucleolus; *Nu*, nucleus; *M*, mitochondrion; *SB*, side body complex; *L*, lipid; *MB*, microbody.

From Fuller, M. Am. J. Bot. ***54:****81, 1967.*

all others is the ***ascus***—a sac that holds the ***ascospores,*** the products of meiosis. There may be four ascospores produced if only meiosis takes place, or eight if meiosis and mitosis occur (Figure 5-19).

Many of the yeastlike species belong to the order called Endomycetales. Yeasts reproduce asexually by budding or by fission. During the budding process chains of cells appear that do not dislodge from the parent. These chains are referred to as ***pseudomycelia*** or ***pseudohyphae*** (Figure 5-20). (They are not true filaments.) Sexual reproduction in yeasts is characterized by union of vegetative cells, the product of which develops into an ascus.

Asexual reproduction in the filamentous forms is primarily by the formation of conidia. Sexual reproduction in the filamentous ascomycetes is characterized by meiosis in the ascus. The asci produced are not free on the vegetative cell but are enclosed in a fruiting body called an ***ascocarp.*** The ascocarp is a sterile tissue that is com-

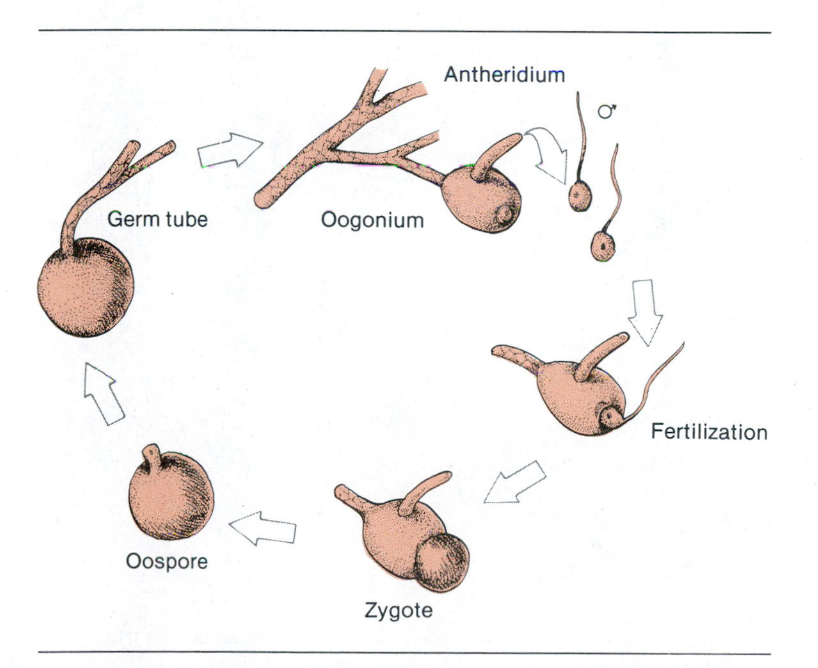

Figure 5-18 Sexual cycle in a species of Monoblepharidales.

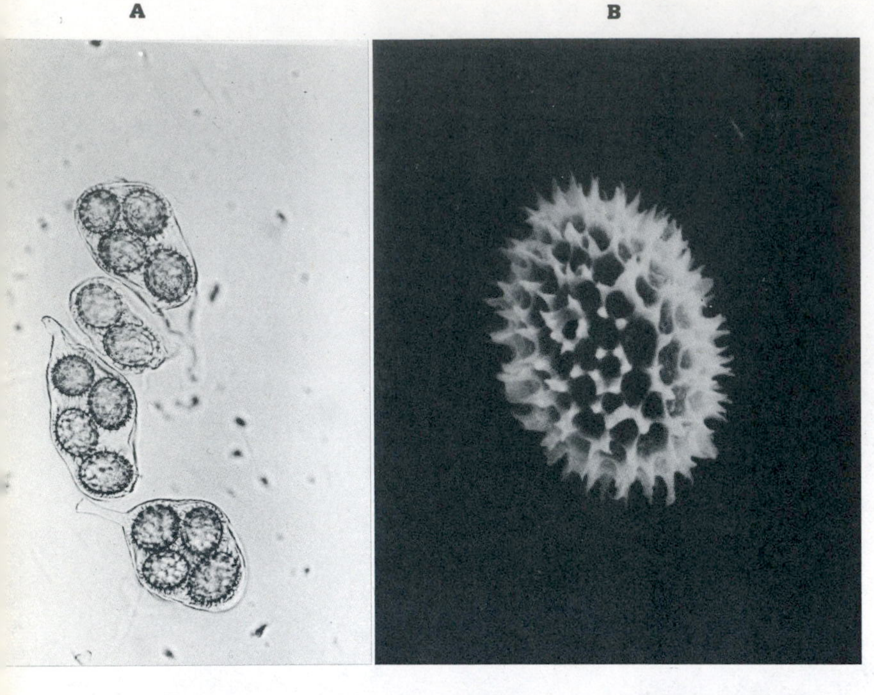

Figure 5-19 Ascospores. **A.** Photomicrograph of asci containing ascospores. (×360.) **B.** Scanning electron micrograph of ascospore.
Reprinted with permission from Mycologia *69:628 Copyright 1977, Uecker, F.A., and Burdsall, H.H., Jr. and The New York Botanical Garden.*

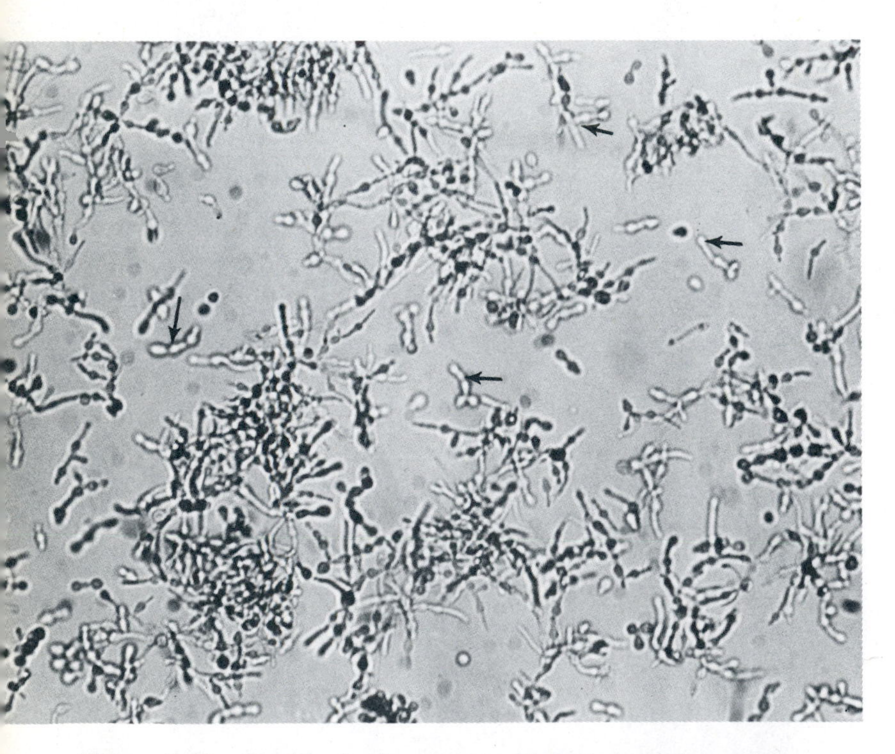

Figure 5-20 Pseudohyphae in yeast cells. Filaments *(arrows)* are not true hyphae. (True hyphae have parallel walls with no invaginations at septa.)
From Bulmer, G.S., et al. Infect. Immun. *20:262, 1978.*

posed of several different cell types that have arisen from modified hyphae. The organization and structure of these fruiting bodies (Figure 5-21) are important for the separation of various groups.

Many members of the ascomycetes are particularly important to scientists and the general public (see also Chapters 20 and 21) for the following reasons.

1. *Saccharomyces cerevesiae* because of its fermentative properties is used in the baking and brewery industries, producing carbon dioxide for the raising of dough and alcohol in breweries.
2. *Penicillium roquefortii* is used in the production of roquefort cheese.
3. *Penicillium chrysogenum* is the source of the antibiotic penicillin.
4. *Taphrina deformans* is a parasite of vascular plants that causes leaf curl and witches'-broom (characteristic appearance of the foilage) in such plants as peach and almond trees (Plate 11).
5. *Histoplasma capsulatum* can cause a serious, sometimes fatal disease called histoplasmosis.
6. *Trichophyton* species (Plate 12) are principal causes of ringworms that appear on various parts of the body surface (athletes' foot, for example).
7. *Aspergillus flavus* is responsible for the formation of a group of toxins called aflatoxins that have been implicated in liver cancer, particularly among Asians.
8. *Claviceps purpurea* parasitizes the rye plant and is a cause of ***ergot.*** Cattle grazing on grasses contaminated by this fungus are prone to disease, and cows may abort their calves. The improper cleaning of grain in earlier times often resulted in the contamination of flour and was the cause of disease symptoms characterized by convulsions and often death. The compounds causing ergot are alkaloids. Semisynthetic ergot alkaloids have been used for a variety of purposes: induction of labor, prevention of postpartum hemorrhage during childbirth, and suppression of postpartum lactation. The work on LSD is an outgrowth of studies on ergot alkaloids.
9. Truffles are underground fruiting bodies of certain ascomycetes. They are not only extremely delicious but expensive (approximately $200 an ounce). Pigs or dogs are used to snuff them out. Truffles are usually

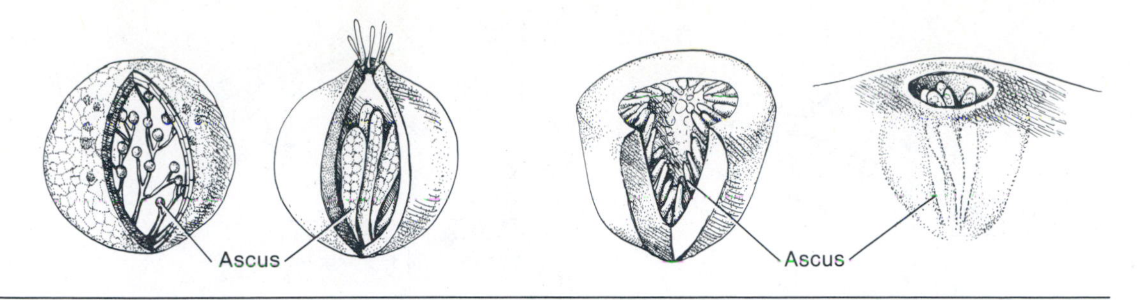

Figure 5-21 Organization and structure of some ascocarps. Asci of different types are seen within ascocarp.

found associated with the roots of oak and beech trees.

10. Morels, which belong to the genus *Morchella,* are also edible; unlike truffles, the fruiting body is above ground (Figure 5-22).

Figure 5-22 Morchella.
Courtesy Carolina Biological Supply Co.

Deuteromycotina (fungi imperfecti). The Deuteromycotina are a heterogenous assembly of fungi that have been considered as conidial stages of ascomycetes and sometimes basidiomycetes whose sexual stages have not been discovered. They have been classified primarily on morphological grounds and by differences in how conidia are produced. An artificial means of classification has been used to separate the Deuteromycotina. Two groups in this subdivision are considered here: the Hyphomycetes, in which conidia are borne free on the mycelium or at the ends of the conidiophore, and the Coelomycetes, which produce conidia inside a cavity on the substrate or a special fruiting body. Most of the fungi pathogenic to animals and humans are included in this group, among which are species of *Candida, Coccidioides, Cryptococcus, Sporotrichum, Geothrichum,* and *Paracoccidioides.* The pathogenic fungi are discussed in Chapter 31.

Basidiomycotina. The Basidiomycotina are the most familiar to humans because among its members are the edible and inedible mushrooms (Plates 13 and 14), puffballs, stinkhorns, rusts and smuts, and many species that attack lumber. This group differs from all others because of the basidium, a swollen shaped cell in which meiosis occurs and spores (basidiospores) are produced.

The mycelium, which can be observed on wet dead leaves, bark, or other organic matter, penetrates its host to obtain nutrients. The sexual reproductive cycle in mycelial development is relatively simple in most Basidiomycetes (Figure 5-23). Genetically different thalli (called ***monokaryons*** because they contain only one kind of genetic nucleus) fuse and produce a ***dikaryon.*** The nuclei derived from the two thalli do not fuse, and the characteristics of the dikaryon are determined by the two sets of haploid genes. The dikaryotic hyphae grow and form a mycelium that attaches to their normal substrate: bark, grain, soil, etc. An unusual characteristic of mycelial growth is the formation of structures called ***clamps*** (Figure 5-24). Clamps represent a mechanism for ensuring that each cell produced from division remains a dikaryon and does not receive two identical nuclei. A clamp is produced near one of the nuclei when the tip of a hyphal branch

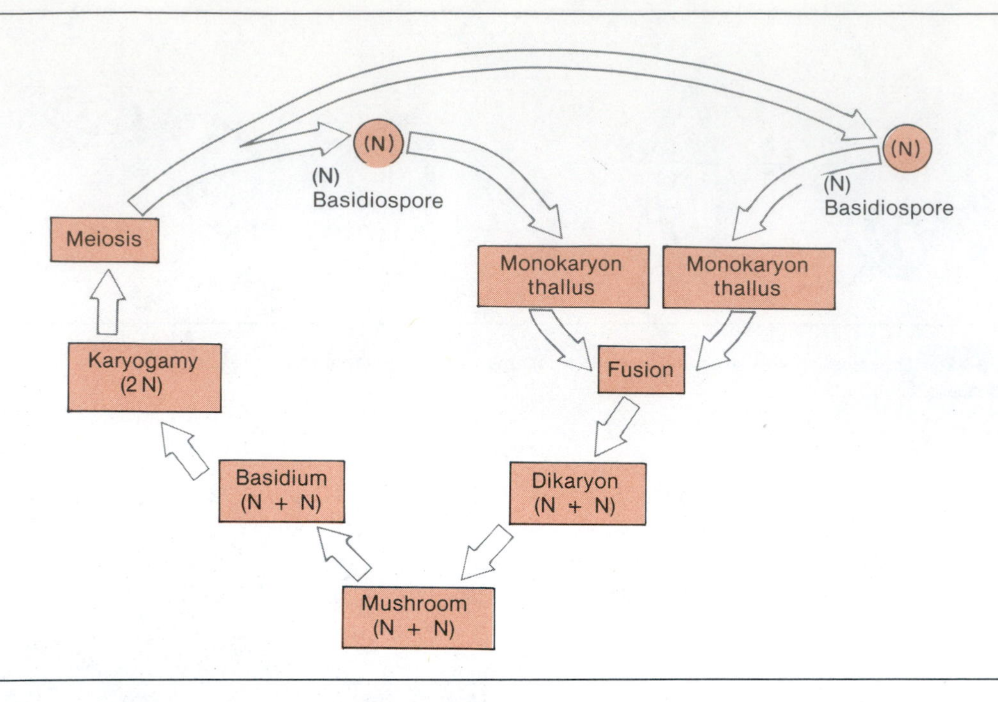

Figure 5-23 Sexual reproductive cycle of Basidiomycotina. *N,* Haploid genetic state; *2N,* diploid genetic state.

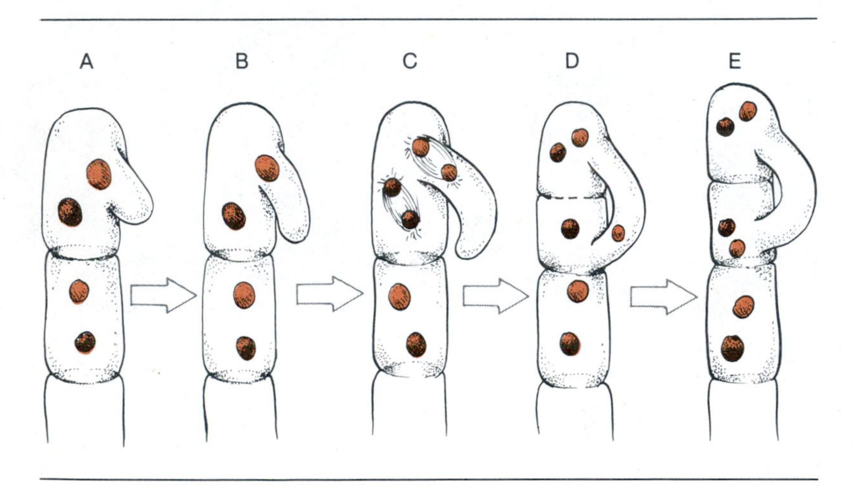

Figure 5-24 Clamp connection in Basidiomycotina. *A,* Clamp begins to form next to nucleus. *B* and *C,* Mitosis of both nuclei followed by separation of nuclei. Movement of one nucleus into apical portion of clamp. *D,* Clamp connection made with subapical portion of hypha. Septum *(dotted line)* begins to form. *E,* Nucleus in clamp migrates to subapical portion. Septum is now well defined.

has finished its growth and a septum is to be formed. The nuclei divide, and a septum is formed, resulting in four nuclei: one is in the clamp, one is in the subapical region, and two are in the apical portion of the hypha. The clamp grows backward and fuses with the subapical region to which the clamp nucleus has now migrated.

A ***basidiocarp*** and ***basidium*** or a basidium only will develop from the mycelium. The basidiocarp, the fruiting body in the Basidiomycetes, may vary in size and shape (Figure 5-25). Mushrooms, puffballs, and others are examples of basidiocarps of the fungi producing them. The basidiocarp bears basidia (sing. basidium), which are arranged in layers called ***hymenia*** (Figure 5-26) and may cover the entire surface or only a portion of the basidiocarp. The arrangement and microscopic features of the hymenia are used to classify the various Basidiomycetes. The basidia arise initially from a binucleate hypha and then enlarge (Figure 5-27). During their enlargement the nuclei fuse and undergo meiosis to produce four haploid nuclei followed either immediately or thereafter by mitosis. Four outgrowths appear on the top of the basidium called sterigmas. One nucleus moves into each of the sterigmas, which are now called ***basidiospores.**** Movements of air may stimulate the discharge of basidiospores (for example, from the gills of mushrooms). Stinkhorns initially develop beneath the ground in thick, leathery sacs from which emerges the basidiocarp. The stalk ***(stipe)*** of the basidiocarp penetrates the underground sac as the fungus approaches maturity and emerges above the ground with further tissue development. The hy-

* The number of basidiospores as well as the sequence of events in basidiospore formation may vary.

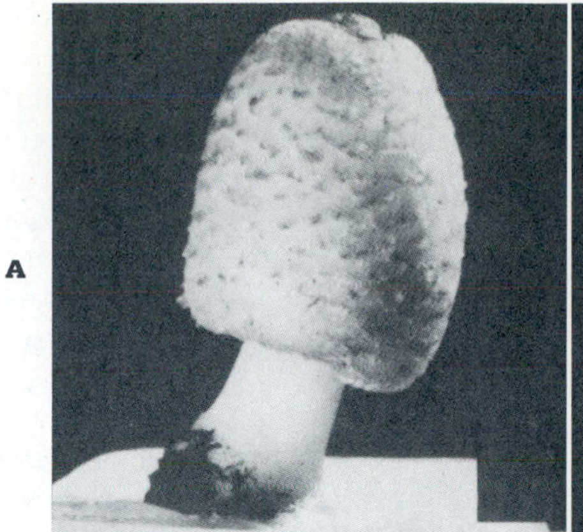

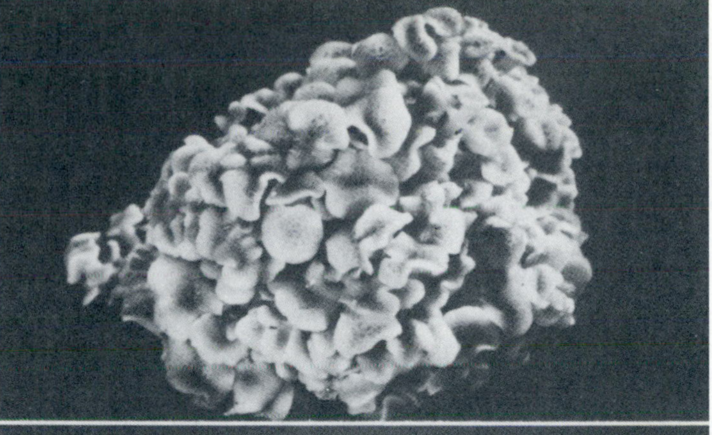

Figure 5-25 Basidiocarps (fruiting bodies) of some mushrooms. **A,** Species of *Coprinus.* **B,** Compound basidiocarp of species of *Polyporus.* **C,** Species of *Coprinus.*

Reprinted by permission from Mycologia ***71**:445. Copyright 1979, Perry, J.W., Fisher, D.G., and Kuter, G.A. and The New York Botanical Garden.*

menium (also called ***gleba***) dissolves into a sticky, foul-smelling mass of spores that attracts flies and other insects. Insects aid in the dissemination of spores.

Rusts and ***smuts*** are parasites for plants. The ravages of rust (the color of the fungus) were known by the Romans. One of their gods, Robigus, was believed to be the cause of rust epidemics. A festival to appease Robigus was held in which red wine was poured over an altar and a red dog sacrificied. The red dog was apparently a substitute for the red rust and thus allowed the Romans to have an abundant harvest. The mycelial hyphae of rusts and smuts grow either intercellularly or intracellularly on host tissue. The entire life cycle includes two distinct and unrelated hosts and the formation of up to five different types of spores. The basidiospores of species of wheat rust infect only the barberry leaf during the spring months and are coated with a nectar that attracts insects and aids in their dissemination. These spores can infect only wheat and form a spreading mycelium from which spores are produced throughout the summer. The spores are wind disseminated and infect other wheat plants; later during the summer another type of spore is produced that is capable of surviving harsh winters. These cold-resistant spores fuse during the spring, mature, and later following meiosis, produce the basidiospores, completing the life cycle (Figure 5-28).

Slime molds

Slime molds represent a large heterogenous group of microorganisms. The taxonomic relationship of the various slime molds has not been satisfactorily reconciled because they exhibit protozoan as well as fungal characteristics. We will consider them fungi and dispense with any for-

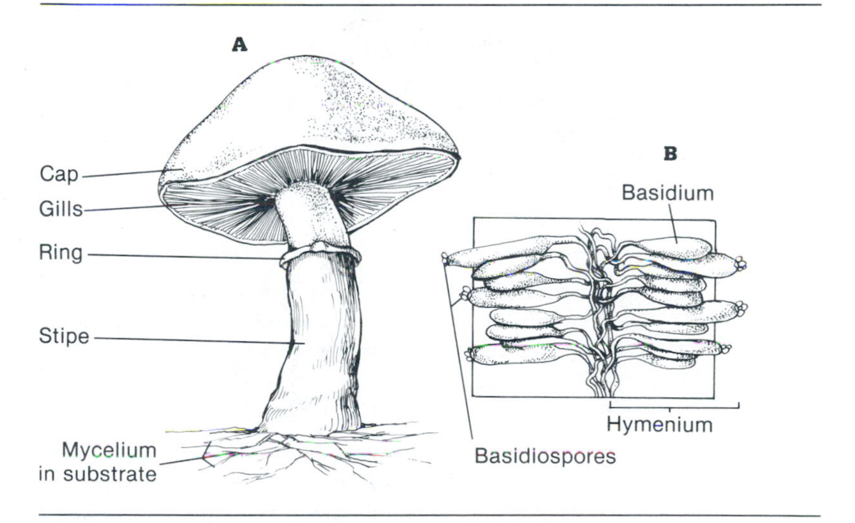

Figure 5-26 Morphological characteristics of basidiocarp of mushroom. **A,** Mature basidiocarp. **B,** Vertical section through gill demonstrating structure of hymenium with its basidia and basidiospores.

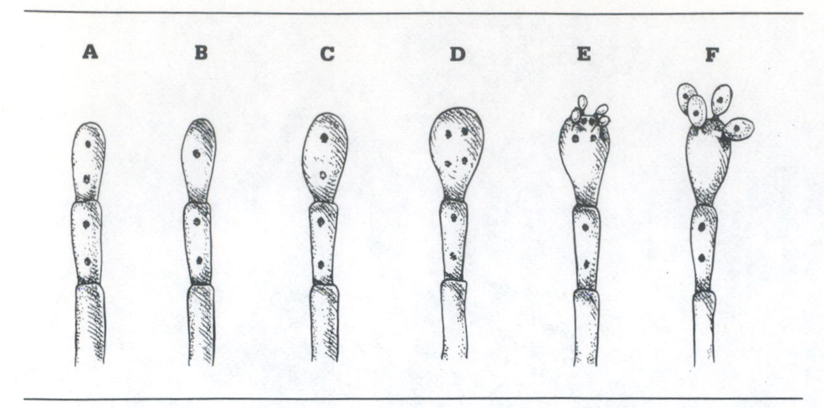

Figure 5-27 Stages in development of basidium and formation of basidiospores. **A,** Binucleate hyphal tip. **B,** Fusion of nuclei. **C,** First meiotic division (two nuclei). **D,** Second meiotic division (four nuclei). **E,** Basidiospores developing on filamentous projections (sterigmas), but nuclei have not migrated into developing spores. **F,** Mature basidium with four uninucleated basidiospores.

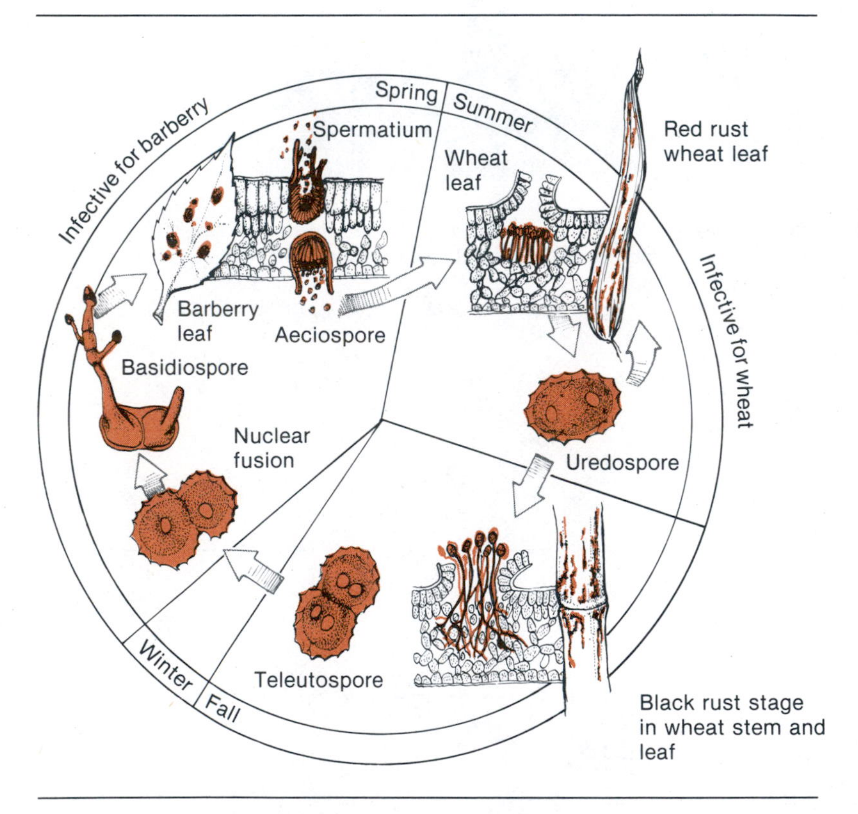

Figure 5-28 Life history of wheat rust fungus.
Redrawn from Poindexter, J.S. Microbiology, Macmillan, Inc., New York, 1971.

mal classification. All of them at some stage in their life cycle consist of single cells that are motile. They resemble protozoans in that the vegetative unit consists of a protoplast (no cell wall) that can show ameboid movement; they resemble fungi in that fruiting bodies are produced. Whenever mycologists isolate true fungi from the environment, slime molds are also found in the same ecological niche. The fruiting body in many groups of slime molds is the direct result of the aggregation of individual cells. This is a characteristic that has already been observed in the prokaryotic myxobacteria. There are several taxonomic divisions of the slime molds, but this discussion centers only on those groups we consider to be the most interesting.

Net slime molds. Net slime molds are aquatic, and some are known to parasitize algae and plants. The basic cellular unit is the ***spindle cell,*** which is nonflagellated and possesses indentations on the periphery. These indentations correspond to special organelles called ***saegenogens,*** from which extrude membranous sacs that eventually coalesce to form a membrane ***(ectoplasmic sheath)*** around the spindle cells (Figure 5-29). The sheath serves not only to capture prey such as bacteria and yeasts but also to aid the spindle cells in locomotion, which is characterized by a gliding movement. The spindle cells migrate to the food source when prey is captured, but if no food is available, the net begins to spread in all directions as if hunting. Under adverse conditions the vegetative state is followed by cell aggregation and sporangium formation. Flagellated zoospores, which develop from the sporangium, become spindle cells.

Cellular slime molds. Cellular slime molds are a more complex group than other slime mold divisions. One species, *Dictyostelium discoideum,* has been extensively studied because of the nature of its differentiation process, which involves the formation of a fruiting body. The principal vegetative unit is an ameba that feeds on bacteria and divides by mitosis, forming a large population of cells. Each cell behaves independently of other cells if the food supply is adequate, but under conditions of nutrient depletion the amebas begin to swarm and form individual aggregates that move to a central point, forming a mass of cells that develops into a ***slug*** (Figure 5-30). The slug can move over the substrate and responds to light (phototropism) as well as temperature (thermotropism). Some of the amebas in the slug are responsible for the formation of a stalk, whereas other cells secrete a substance that gives the stalk a netlike appearance. At the tip of the stalk is a ***sorocarp,*** where individual amebas are transformed into spores. The mechanism of aggregation in these slime molds is discussed at more length in Chapter 8. Little is known of the sexual cycle in this group.

True slime molds. The propagative unit of the true slime molds is a dormant, spherical, thick-walled spore that is often covered with spines. The first stage in the life cycle of this group is characterized by spore germination and the release of a uninucleate ameboid cell, which becomes flagellated (Figure 5-31, *A*). The second stage involves the formation of a plasmodium, a multinucleated, wall-less structure arising from one ameba or from the fusion of ameboid units (Figure 5-31, *B*). The plasmodium is the primary vegetative phase and may vary in size from a few hundred micrometers to several square meters. The plasmodium feeds by ingesting microscopic organisms, such as bacteria and protozoa, but later ceases to feed and forms fruiting bodies. The fruiting bodies (sporangia) are usually one of two types: a sporangium in which the stalk is hollow or one in which the interior of the stalk is fibrous. The life cycle is completed by cleavage of the multinucleated protoplasm into uninucleate segments, which acquire a thick wall and become spores (Plates 15 to 17).

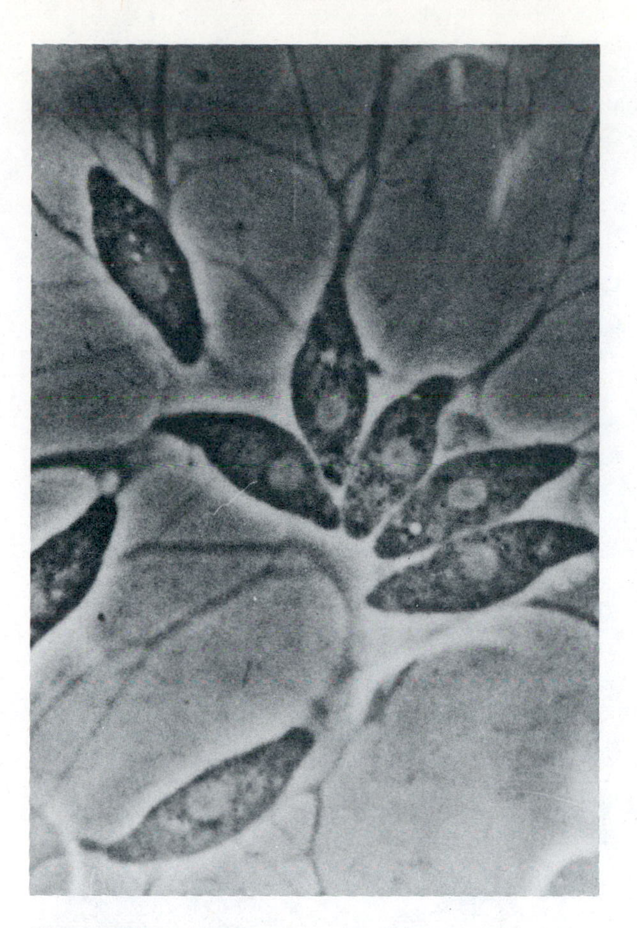

Figure 5-29 Ectoplasm sheath and spindle cells of net slime mold.
From Ross, I.K. Biology of the fungi, *McGraw-Hill Book Co., New York, 1979.*

Fungal symbioses

We have discovered that the fungi act as predators by trapping and digesting various microscopic animals such as rotifers, nematodes, protozoa, and bacteria. They can also parasitize insects, higher animals, and humans, Many of these parasitic associations are discussed in later chapters. Fungi can also exist in a state of mutualism with other organisms whereby both partners benefit from the association. One of the most widely recognized mutualistic associations is that found in ***lichens,*** or those plants composed of algae and fungi. The fungus and alga are intertwined to form a single thallus. The relationship between the two is delicate, and any condition that favors one may result in dissociation or destruction of one of the partners. The fungal partner is usually a member of the ascomycetes, whereas the algal partner is usually the green alga *Trebouxia* or a member of the cyanobacteria or a combination of green algae and cyanobacteria. The fungus absorbs water and inorganic nutrients that are transported to the fungus. If the lichen contains a cyanobacterium as one of the partners, fixed nitrogen is also supplied in large quantities. The fungus apparently protects the alga from high light intensities and dessication. Lichens are ubiquitous and can be found in habitats such as the Arctic and Antarctic and solidifed laval flows in Hawaii. Many are found on the bark of trees or upon rock. The reindeer moss is a lichen and provides some winter food for car-

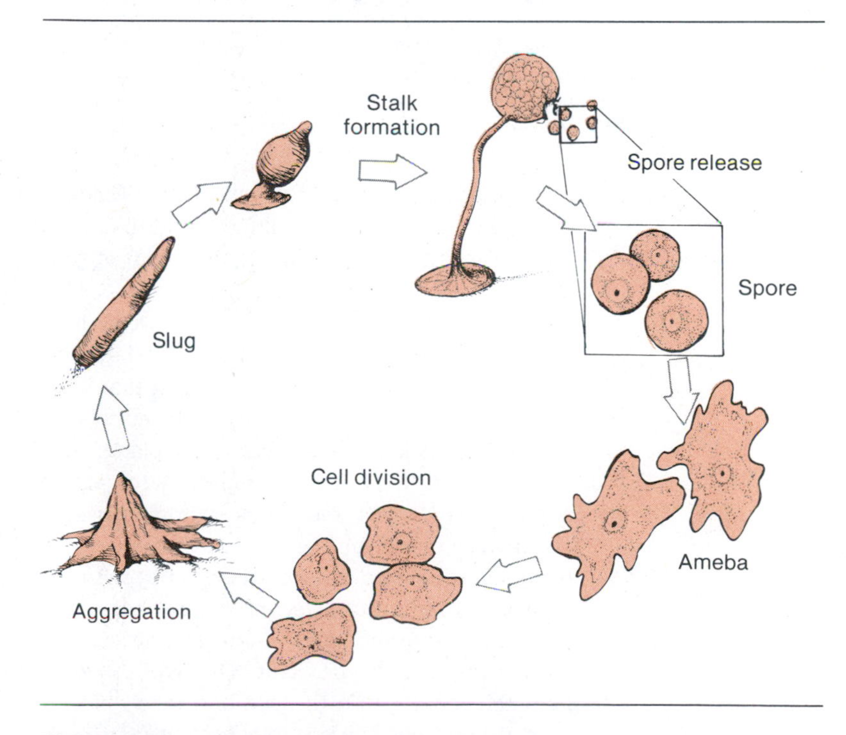

Figure 5-30 Morphological changes in life cycle of slime mold *Dictyostelium discoideum.*

A

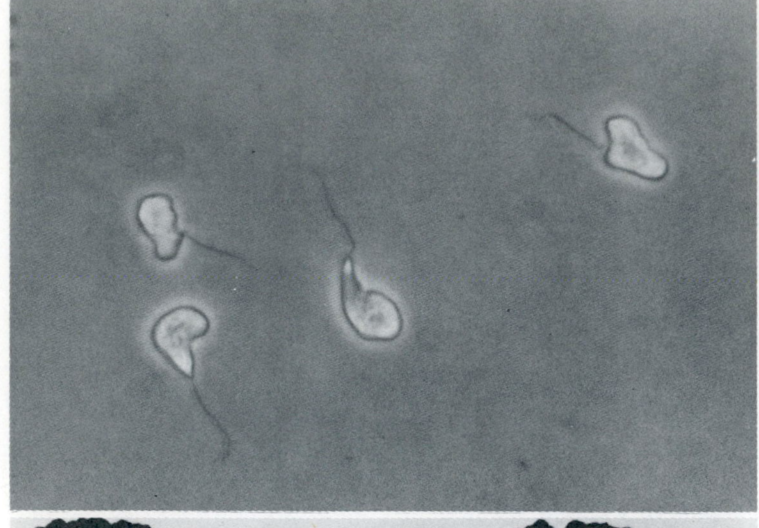

B

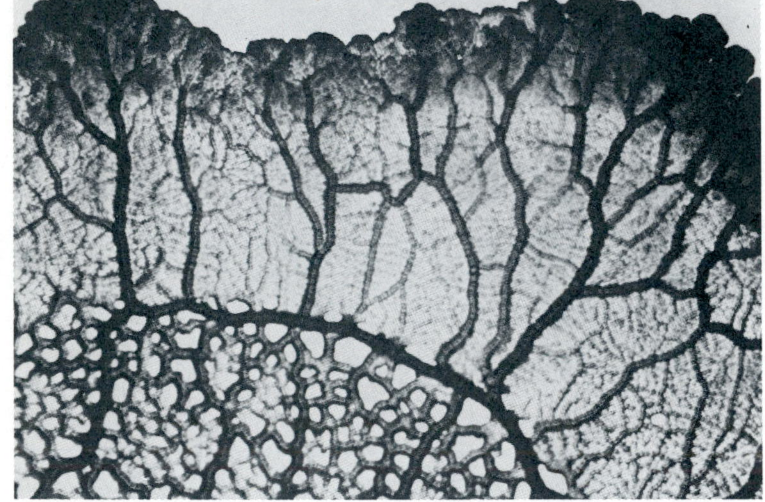

Figure 5-31 True slime molds. **A,** Flagellated ameboid cell of genus *Physarum* **B,** Plasmodium of genus *Physarum.*
Courtesy Carolina Biological Supply Co.

ibou and reindeer in Europe and North America. Lichens are sensitive to pollution, and their numbers, or lack thereof, in the environment is some measure of the degree of pollution.

Mycorrhizae are associations between fungi and plant roots in which the fungus is beneficial to the plant and does not cause disease. It appears that most plants form mycorrhizae to some degree, and it is believed that many plants could not survive in their habitat without such associations. There are three types: ectomycorrhizae, endomycorrhizae, and ectendomycorrhizae. ***Ectomycorrhizae*** are characterized by hyphae that form a mantle around feeder roots. These fungi enter the roots and grow intercellularly between the cortical cells of the plant. The hyphae extend from the mantle into the soil and take nutrients from it and pass them into the plant cortex. Ectomycorrhizae are formed primarily on forest trees by members of the basidiomycetes. ***Endomycorrhizae*** appear similar to ectomycorrhizae except that the fungus grows internally within the cortical cells of the plant. This type of association is obligatory in plants such as orchids, where dependence appears to be due to a lack of plant enzymes required for metabolism of storage products used for orchid growth. Endomycorrhizae are found on most cultivated plants and some trees. The ***ectendomycorrhizae*** are characterized by hyphae that penetrate into and between cortical cells of the plant, and they may or may not have a fungal mantle covering the roots. Mycorrhizae associations benefit the fungus, which receives carbohydrates from the plant. Benefits to the plant include the following.

1. Possible compensation for loss of roots caused by disease.
2. Increase in the availability of mineral elements in infertile soils. Phosphorus, for example, appears to be concentrated by the fungus and then slowly released to the plant. Mycorrhizae also enhance the uptake of elements such as nitrogen, iron, copper, boron, and aluminum.
3. Deterrence of the infection of feeder roots by soil-borne pathogens by reduction of the level of free carbohydrates and other nutrients in root exudate that normally stimulate pathogen activity. This may be accomplished by establishing a physical barrier to penetration or by releasing inhibitory compounds. Some ectomycorrhizae provide protection by producing antibiotics. Certain endomycorrhizae increase the arginine content in the roots of tobacco plants, which suppresses chlamydospore formation by some plant pathogens.
4. Reduction of the amount of susceptible plant tissue to infection.

Fungi are also associated with viruses; such associations are discussed in Chapter 6.

PROTOZOA

Classification

The term *protozoa* signifies a unicellular microscopic animal, a term that probably does not mean much to the average reader until he or she examines a drop of pond water with a microscope. That microscopic world is filled with small unicellular animals whose structural clairty and movements through the watery medium remind one of a ballet. It is a mesmerizing experience that one never forgets and is always willing to repeat. Perhaps that is one of the reasons some scientists become protozoologists.

Protozoa are important organisms for study for a number of reasons. They are parasitic to humans and cause such diseases as malaria, dysentery, sleeping sickness, and leishmaniasis. These diseases are often associated with tropical areas or countries of the Third World, but with the increase in travel between continents and the increase in immigrants from these areas, some of these diseases are surfacing in areas where they had been previously absent. Protozoa can parasitize animals, and some are fatal to cattle and sheep as well as poultry. Protozoa also play some role in the degradation of sewage through their oxidative metabolic activities and ability to remove suspended bacteria. One group, called the ***foraminifera,*** have flourished within different geological eras and have deposited their calcium carbonate shells, which serve as markers of their existence. It has been possible to locate oil deposits by looking for specific foraminifera deposits. Protozoa are also excellent tools for research because they are easily cultivated in the laboratory. They reproduce asexually, and clones are easily generated with the same genotype as any desired "parent." The effects of environmental factors, such as radiation, are easily analyzed in these organisms.

The protozoa are classified in the Kingdom Protista, but they are difficult in many ways to distinguish from other eukaryotes. For our purposes we are only concerned with four groups, based on their mechanism of locomotion.

Sarcomastigophora—locomotion by flagella or ameboid movement
Sporozoa—spore-forming protozoa, no locomotor organelles, but movement by gliding in some species
Myxospora—spores that contain a filament, vegetative cells that use ameboid movement
Ciliophora—locomotion by cilia

The subdivisions presented above are an artificial and simplified scheme; a more recent and detailed classification scheme is presented in Table 5-1.

Table 5-1 Current classification scheme (abbreviated) for the Protozoa

CLASSIFICATION	CHARACTERISTICS	EXAMPLES
Phylum: Sarcomastigophora	Flagella, pseudopodia, or both types of locomotor organelles	
Subphylum: Mastigophora	One or more flagella present	
Class: Phytomastigophora*	Chloroplasts present; mostly free living	*Euglena*, *Peranema*, and dinoflagellates such as *Gymnodinium*
Class: Zoomastigophora	Chloroplasts absent; one to many flagella	Parasitic, such as *Leishmania* and *Trypanosoma*
Subphylum: Opalinata	Numerous cilia in oblique rows over entire body; primarily parasitic	
Subphylum: Sarcodina	Primarily pseudopodia; shell (tests) often present	
Superclass: Rhizopoda	Locomotion by lobopodia, filopodia, or reticulopodia but no discrete pseudopodia; some contain tests such as the foraminifera, whereas others such as the slime molds produce fruiting bodies	*Amoeba* species
Superclass: Actinopoda	Spherical, usually planktonic; produce slender radiating pseudopodia (axopodia); skeleton when present often composed of silica or strontium	*Aulacantha* species
Phylum: Labyrinthomorpha	Ectoplasmic network with spindle-shaped or spherical nonameboid cells; some ameboid cells move within a network by gliding; many parasitic to algae	
Phylum: Apicomplexa	All species parasitic; cysts often present; many contain spores with sporozoites	

From *J. Protozool.* **27:**37, 1980.
*We have opted to include this group with the algae.

Continued.

Table 5-1 Current classification scheme (abbreviated) for the Protozoa—cont'd

CLASSIFICATION	CHARACTERISTICS	EXAMPLES
Class: Perkinsea	Flagellated zoospores; parasitic to oysters	
Class: Sporozoa	All species parasitic for vertebrates and invertebrates; spores present with no polar filament; many transmitted by blood-sucking insects	*Toxoplasma, Babesia,* and *Plasmodium*
Phylum: Microspora	Unicellular spores with spiroplasm containing simple or complex polar filaments; obligatory intracellular parasites in nearly all animal groups	
Class: Rudimicrosporea	Simple polar filament; parasites of annelids	
Class: Microsporea	Complex polar filament	
Phylum: Ascetospora	Spore with one or more spiroplasms; no polar capsules or polar filaments; all parasitic	
Class: Stellatosporea	Spore with one or more spiroplasms	
Class: Paramyxea	Spore with one spiroplasm	
Phylum: Myxozoa	Spores of multicellular origin with one or more polar capsules; all parasitic	
Class: Myxosporea	Spore with one or two sporoplasms; parasitic for cold-blooded vertebrates	
Class: Actinosporea	Spores with three polar capsules each with polar filament; parasitic primarily for annelids	
Phylum: Ciliophora	Simple cilia or compound ciliary organelles in at least one stage of life cycle; contractile vacuole present; most species free-living but some commensals, parasites, and many symbionts in a variety of hosts	
Class: Kinetofragminophorea	Compound ciliature typically absent	
Subclass: Gymnostomatia	Toxicysts common, somatic ciliation uniform; simple oral groove	
Subclass: Vestibuliferia	Free-living or parasitic, particularly in digestive tract of vertebrates and invertebrates	
Subclass: Hypostomatia	Cytosome on ventral surface; body often cylindrical or flattened dorsoventrally; free-living ectocommensals or endocommensals principally of invertebrates	
Subclass: Suctoria	Suctorial tentacles; adult body sessile and sedentary; reproduction by budding	
Class: Oligohymenophorea	Generally well-defined oral apparatus in most but absent in one; colony formation common in some members	
Subclass: Hymenostomatia	Body ciliation uniform and heavy; predominantly fresh water; oral area on ventral surface	*Tetrahymena, Paramecium*
Subclass: Peritrichia	Oral cilia prominent; mucocysts universal; conjugation total and involves fusion of microconjugants and macroconjugants	*Vorticella*
Class: Polyhymenophorea	Well-developed buccal ciliary features; cirri present in one group; cystosome at bottom of buccal cavity; cysts common in some members; large and commonly free living	*Stentor*
Subclass: Spirotrichia	Characteristics of the class	

Morphology

Protozoa are extremely variable in size. Some free-living giant amebas are 1 to 5 mm long (Figure 5-32), whereas parasitic amebas are only 4 to 5 μm long. There is also a great variability in shape: some are spherical, some are wormlike, and others, because of the flexibility of their cytoplasmic membrane or because of secreted surface structures, can assume a variety of shapes (Figure 5-33).

The protozoa possess the same basic structure and organization that appear in other eukaryotic microorganisms, but more than any other group of eukaryotes they show many subcellular specializations and modifications (Table 5-2). Diagrammatic illustrations of the location and general morphology of some of these structures are shown in Figure 5-34.

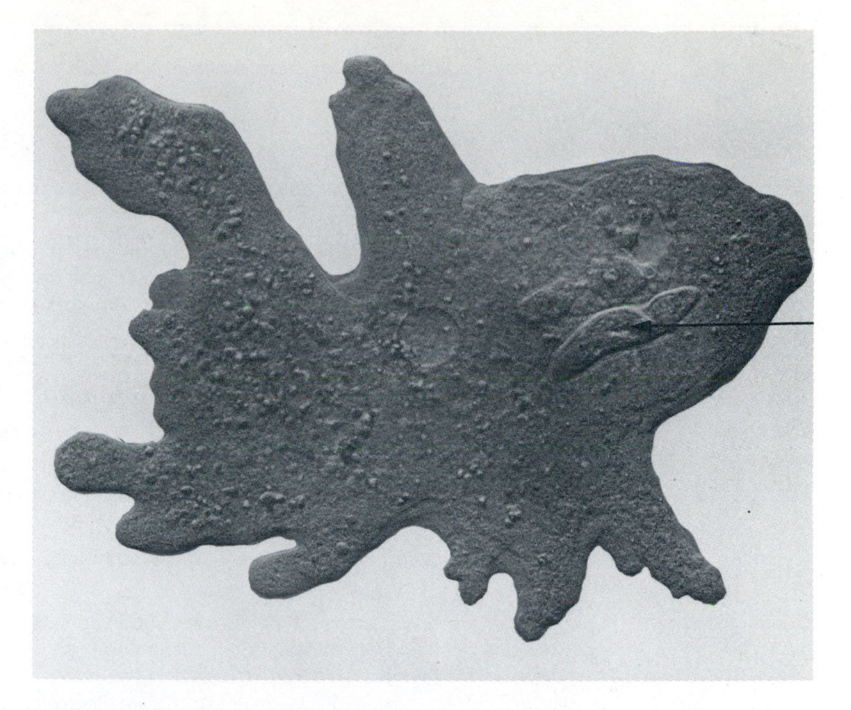

Figure 5-32 Photomicrograph of giant ameba. Note that arrow points to another protozoan, called a paramecium, that has been ingested by ameba.
Courtesy Carolina Biological Supply Co.

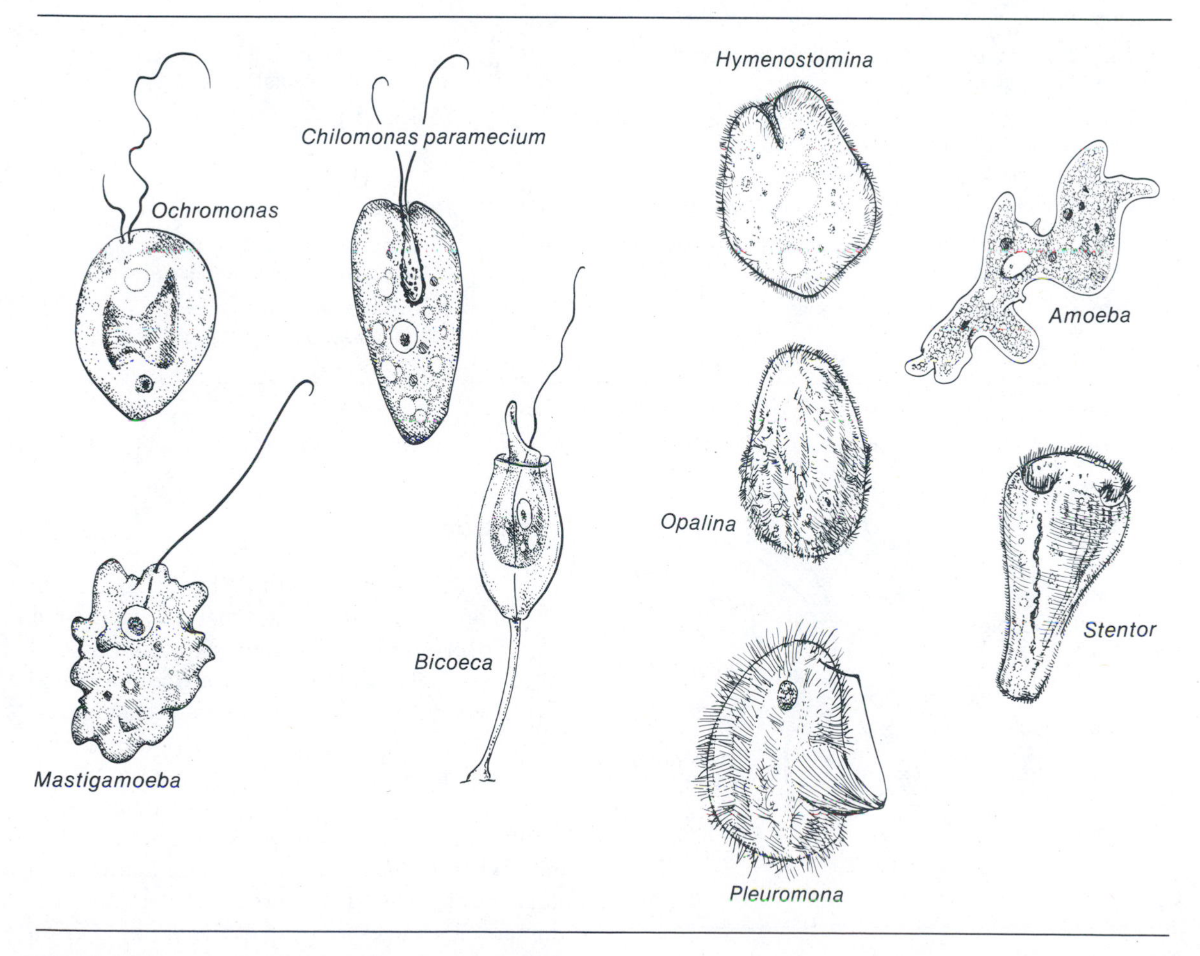

Figure 5-33 Morphological varieties among protozoa.

Table 5-2 Modification of some surface structures in the Protozoa

SURFACE FEATURE	CHARACTERISTICS	FUNCTION
Pellicle	A thick covering over the cell membrane and believed to be derived from it	Protection or support
Shell (test)	A capsular-like covering secreted by the organism—some are like skeletons and contain calcium carbonate; others secrete a gluelike material that can be mixed with sand to produce a matrix	Protection; sometimes used to attach to a substrate.
Mucocyst	Sacs that lie beneath but pressed against pellicle; discharge mucus to the exterior	Protective device by aiding in cyst development; aids phagocytosis of particles
Cytosome	An opening through which food is ingested—may be round or slitlike	Feeding
Oral groove	Indentation of pellicle of ciliates; the cilia around the oral groove appear to fuse and give appearance of undulating membrane	Mechanism for concentrating food.
Cytopharynx	Region associated with cytosome in which food passes before being pinched off in a vacuole	Mechanism for concentrating digested food
Contractile vacuole	Fixed or variably positioned membrane-bound vacuole; cytoplasmic fluid fills vacuole; When pushed against the cytoplasmic membrane, fluid is expelled	Osmoregulatory device; may also excrete products such as ammonia and urea
Food vacuoles	Membrane-bound vesicles that contain food ingested through cytosome or food acquired at cell surface by pinocytosis; vacuole will acquire enzymes to digest food and then move through cytoplasm	Food digestion
Cytopyge	Fixed opening through which digested food is eliminated	Waste elimination
Extrusomes	Filamentous structures whose function varies; there are four types:	
	TRICHOCYSTS: occur in ciliates and flagellates; located beneath pellicle, they can eject filament	Function in doubt but may be protective response to irritants
	HAPTOCYSTS: small sacs located at the tips of tentacles of aquatic sessile species; when tentacle touches a prey, the haptocyst penetrates pellicle and binds prey	To capture prey
	TOXICYSTS: fibers concentrated around the cytosome	To capture and paralyze prey
	EJECTOSOMES: coiled refractile bodies located beneath the pellicle of one class of protozoa	When activated, believed to propel organism away from danger
Uroid	Posterior protrusion of certain amebas	Believed to be associated with pseudopodial activity or movement

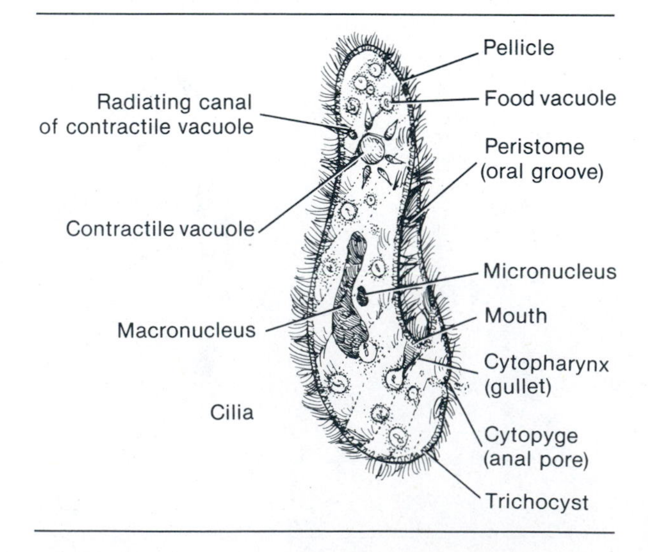

Figure 5-34 Structural features of protozoan.

Nutrition

Protozoa can be found as free-living species in the soil and in both fresh and marine waters, but some species are parasitic to invertebrates and vertebrates. Free-living protozoa require some level of water for survival, such as ditches, bogs, moist soil, lakes, or oceans. Many protozoa form ***cysts*** that can be dispersed by air currents, insects, and birds, if water becomes unavailable. Bacteria are an important protozoan food source, and their presence often determines the relative numbers of protozoa that are present in an environment. Most protozoa live at the edges of or on the bottom of shallow bodies of water, where decaying vegetation, temperature, and oxygen levels are adequate for their survival.

Unlike bacteria and fungi, protozoa are not primary decomposers. They are heterotrophic and can ingest organic debris or feed on other organisms, such as bacteria, algae, and nematodes, as well as other protozoa. Food may be taken into the cell by (1) passing through permanent structures such as the mouth or cytosome of some flagellates and ciliates, (2) dissimilation either by pinocytosis at various sites on the body surface as in ameboid protozoa, or (3) diffusion through the cell membrane. Food ingested by amebas is surrounded by a membrane to produce a ***food vacuole,*** in which digestive enzymes break down the food. The food vacuole circulates freely in the cytoplasm. The ciliates and some flagellates have a permanent indentation at the cell surface called the ***oral groove*** that leads to the ***cytosome,*** an area for digestion of ingested food. Food taken into the oral groove passes through the ***cytopharynx*** before it is pinched off into a food vacuole in the cytosome. Undigested material left in food vacuoles is eliminated at a site at the cell surface called the ***cytopyge.*** Freshwater protozoa, because they live in an environment that is hypotonic, continually take water into the cell to maintain an osmotic balance. The cell might burst if it were not for the ***contractile vacuole,*** which maintains water balance by pumping out water after the cell has reached a critical size.

Most free-living protozoa are aerobic. Anaerobic forms are found primarily as parasites, such as those species that inhabit the intestinal tracts of vertebrates and invertebrates.

Reproduction

Asexual reproduction. Asexual reproduction occurs in all protozoa and is the most frequent means of propogation. Binary fission is the most common method of asexual reproduction, but other mechanisms including budding, multiple fission, and plasmotomy are available to some species.

Binary fission. Binary fission in flagellates is longitudinal (Figure 5-35, *A*). The kinetosome and the nucleus divide during binary fission, and when the division is complete, each daughter cell receives a nucleus as well as a flagellum. Binary fission in ciliates is transverse and occurs at right angles to the long axis of the cell (Figure 5-35, *C*)

Budding. Budding occurs more frequently in the protozoa that are sessile and anchored to stones, algae, or detritus (disintegrated materials). A portion of the adult constricts and leaves the parent cell during the budding process.

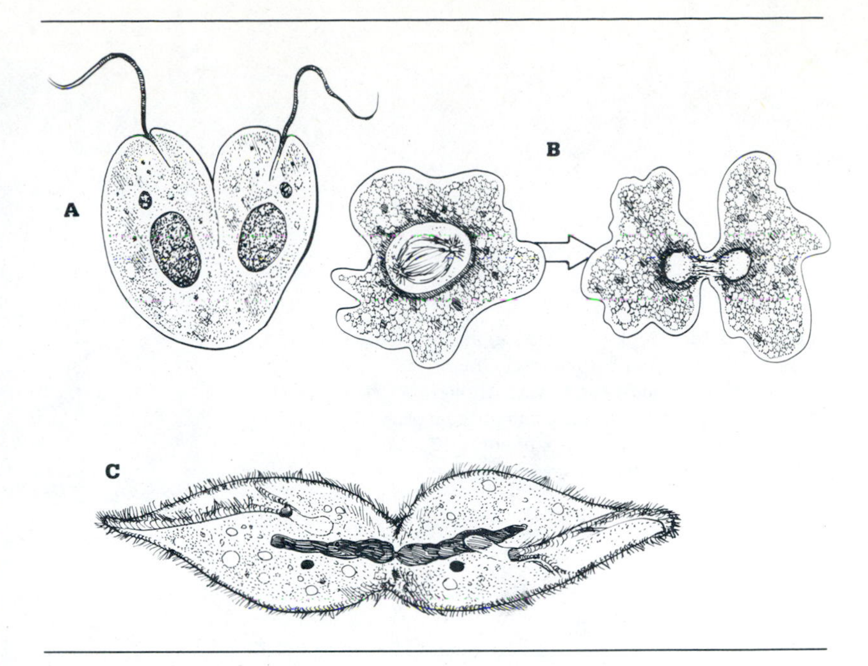

Figure 5-35 Binary fission in protozoa. **A,** Longitudinal fission in flagellate. **B,** Binary fission in ameba. **C,** Transverse fission in ciliate.

Multiple fission (schizogony). Multiple fission is a process in which the nucleus within a cell divides repeatedly to produce as many as several thousand nuclei. Each nucleus becomes surrounded by a small portion of cytoplasm and breaks away from the multinucleated mass as an individual uninucleated cell. This type of process is characteristic of the sporozoa that are parasites of animals, including humans (Figure 5-36). The high fever and concomitant chills that appear periodically in malaria are a result of multiple fission in the red blood cells. Thousands of these uninucleated forms are released when the infected red blood cell lyses and the body responds to them by an increase in temperature.

Plasmotomy. Plasmotomy is an asexual process that occurs in multinucleated protozoans. The cytoplasm divides independently of the nuclei, and consequently reproduction results in the formation of individuals with two or more nuclei.

Sexual reproduction. Sexual reproduction appears during the life cycle of most protozoa except in those species that are parasitic; however, sexual reproduction is influenced primarily by environmental conditions. Sexual reproductive patterns vary in the protozoa as they do in the fungi. The vegetative cell of some species spends the majority of its life cycle in the haploid state; in others, it is the diploid cell that predominates.

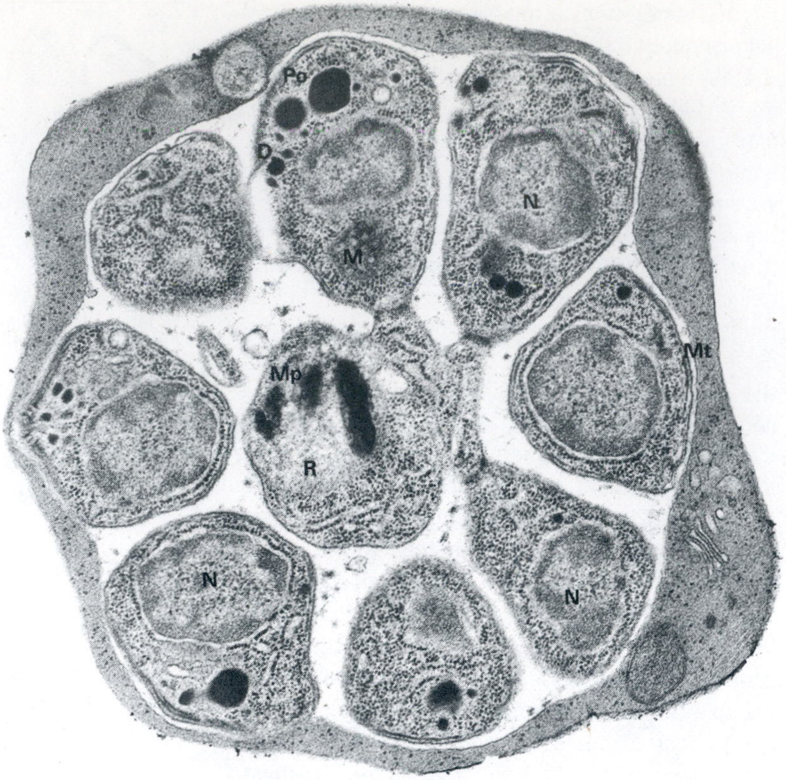

Figure 5-36 Multiple fission in avian malaria parasite *Plasmodium cathemerium*. Eight merozoites surround central residual mass *(R)*. *N*, nuclei; *Mt*, microtubules; *Po*, paired organelles; *D*, dense bodies; *Mp*, malarial pigment. (×30,000.)
From Aikawa, M. Am. J. Trop. Med. Hyg. ***15**:449, 1966.*

Finally, there are those species that show an equivalent alternation between the haploid and diploid states.

Basically two types of sexual reproduction occur in protozoa: ***syngamy*** and ***conjugation.*** Syngamy involves the fusion of free-swimming gametes that vary in size and morphology. Conjugation occurs only in ciliates, and during the process two parent cells, called ***conjugants,*** initially bind to each other at their anterior ends and then finally in the area of the oral groove (Figure 5-37). A bridge is produced after the binding process that permits cytoplasmic flow between the conjugants.

Cysts

Many protozoa, under appropriate conditions, can reduce their size, eliminate wastes and water, store food reserves, and develop a thick cuticle, all of which results in the formation of a ***cyst.*** The cyst is usually a protective device for the cell, but it can also function in a reproductive manner, or as an agent of transmission. Cyst formation (encystment) can be stimulated by a number of environmental conditions including evaporation, pH changes, nutrient depletion, and overcrowding. These factors initiate encystment for the purpose of protection of the genetic material in the cell. The nucleus within the cyst divides into four and sometimes eight nuclei in some species—a typical reproductive function. Some species are parasitic, however, and the cyst also becomes a unit of transmission. The amebas causing dysentery in humans, for example, will encyst while in the host and will become infective only when eliminated in the feces and ingested by a suitable host. Emergence from the encysted state (excystment) is usually stimulated by conditions that favor vegetative growth. Excystment results in the digestion of the cyst by enzymes present in the gut of the host. Some cysts have a preformed hole or plug that is closed during encystment but can be dissolved during excystment, releasing a vegetative cell.

Protozoan symbioses

Bacteria and algae, which are frequent food sources for many protozoa, sometimes survive digestion and become endosymbionts. Certain heterotrophic ciliates possess an algal endosymbiont, and when the host's food supplies become exhausted, growth is not diminished. During pe-

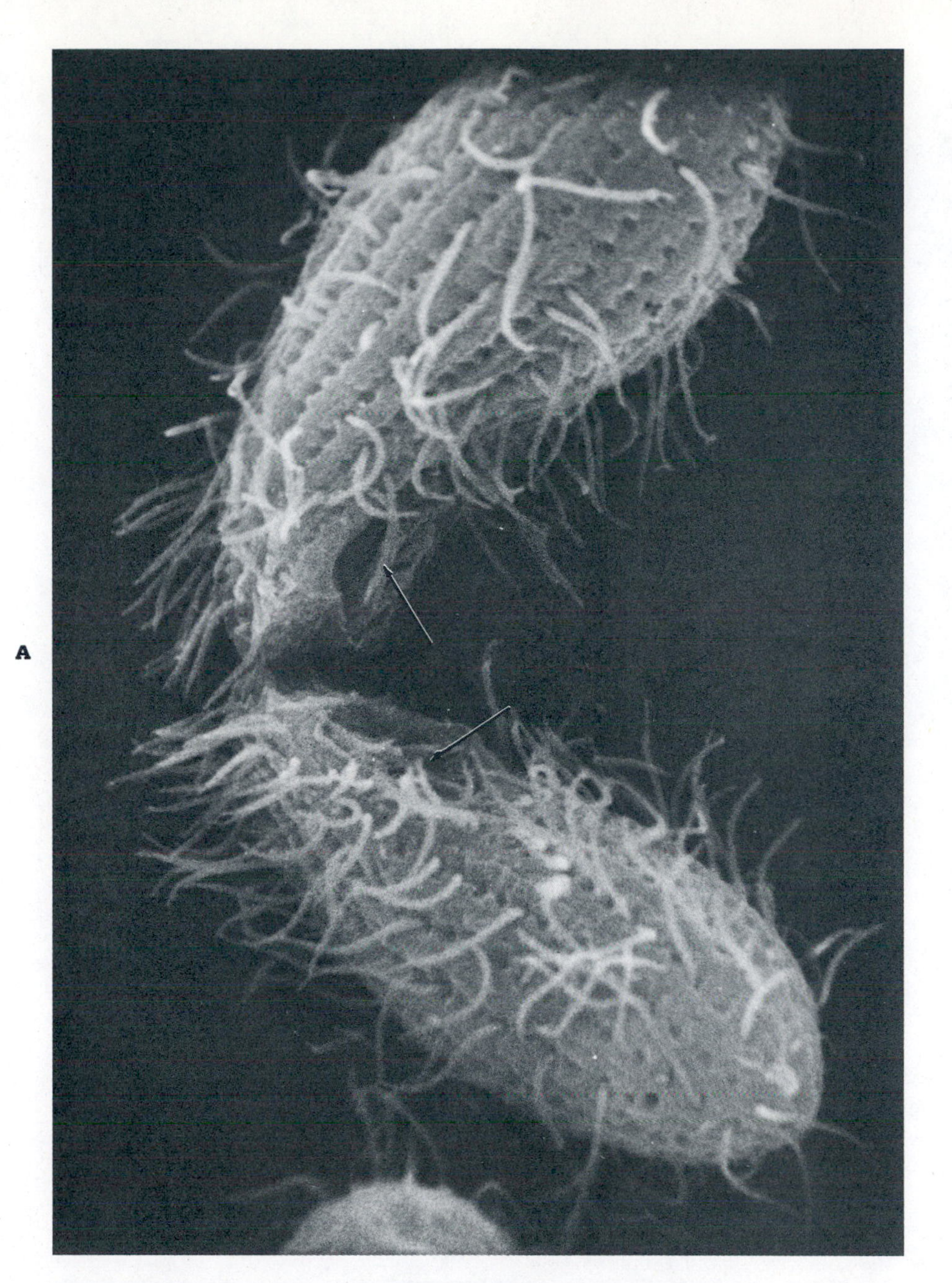

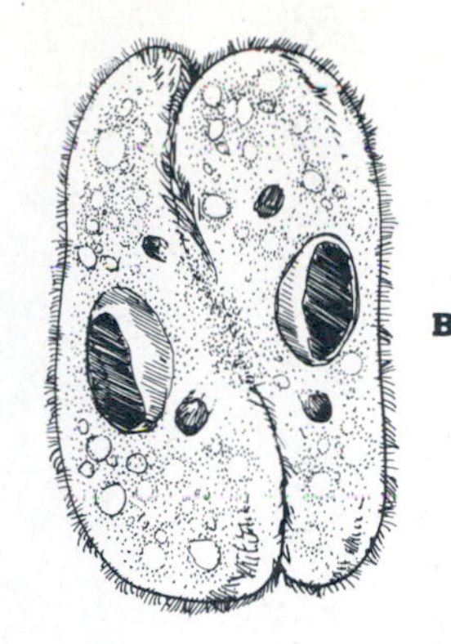

Figure 5-37 Conjugation in ciliates. **A,** Scanning electron micrograph of two ciliates. Attachment occurs at anterior ends of each ciliate near oral grooves *(arrows)*. **B,** Conjugation.
A, From Wolfe, J., and Grimes, G.W. J. Protozool. ***26:****82, 1979.*

riods of abundant food supplies, especially in the dark, the algal endosymbiont is supplied nutrients provided by the protozoa. The algal endosymbiont reciprocates by providing the protozoa with nutrients produced during photosynthetic processes that occur in the light.

Some ciliates contain endosymbionts that confer the ***killer*** trait (referred to as kappa, sigma, gamma, etc., depending on the species involved). The endosymbionts are bacterial (Figure 5-38) and produce a toxic substance that, when released, is lethal to strains of protozoa that do not possess the killer factor. The bacterial endosymbionts involved have been differentiated by electron microscopy. They vary in shape from flagellated rodlike particles to coccal forms.

Characteristics of the major groups of protozoa

Sarcomastigophora. Members of this group include not only flagellates and ameboid forms but species that can utilize both methods of movement. Flagella, when present, may be single or

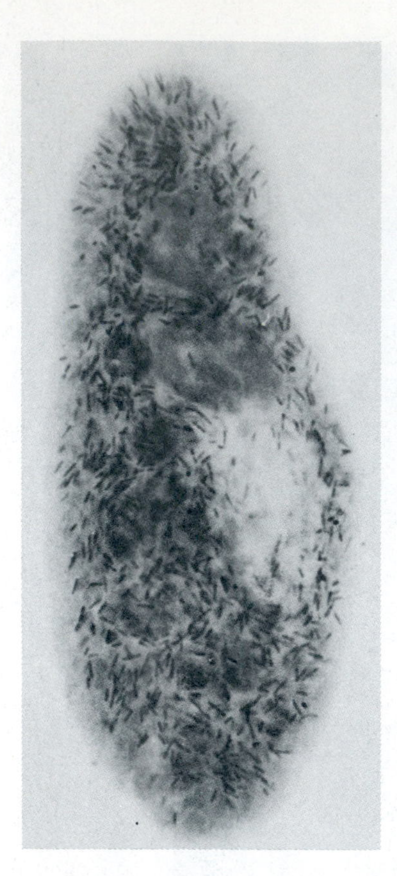

Figure 5-38 Phase contrast photomicrograph of protozoa (species of *Paramecium*). (×500.) Protozoa contains many endosymbionts (bacteria) that appear as dark-stained rods.

From Preer, J.R. et al. Microbiol. Rev. ***38:****113, 1974.*

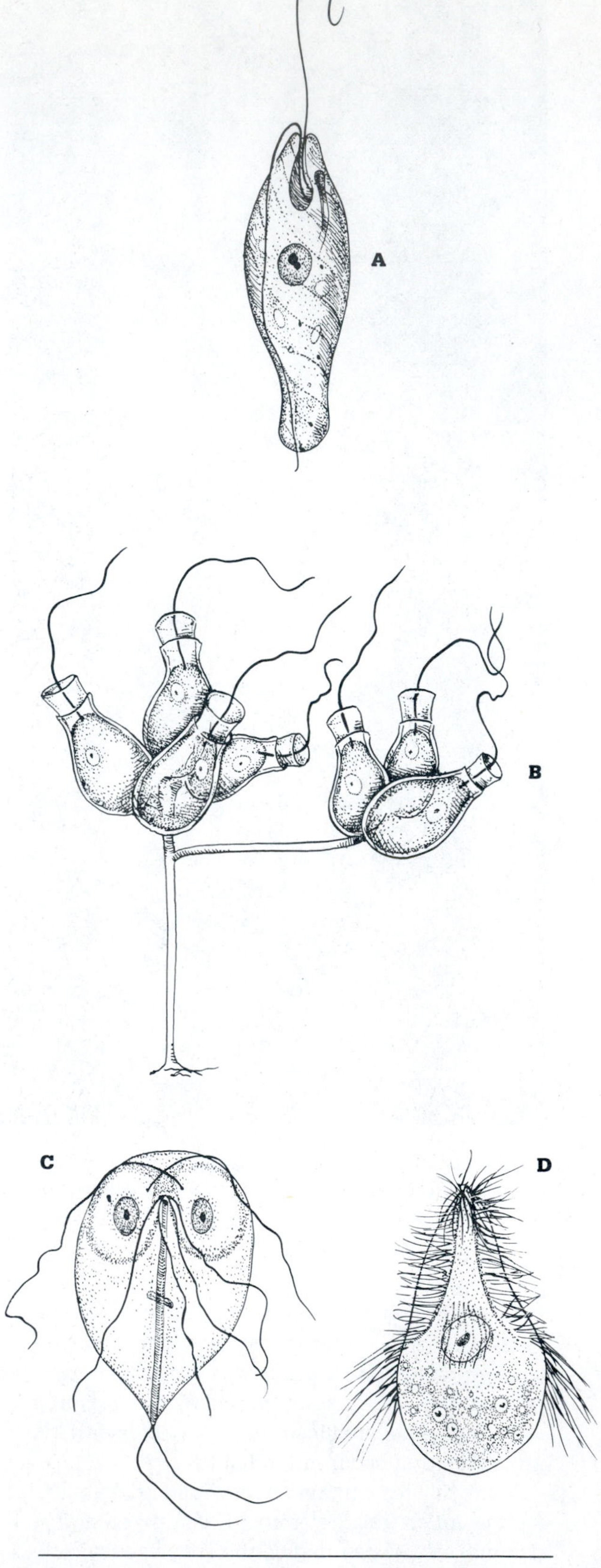

Figure 5-39 Flagellated protozoa. **A,** *Peranema* species. **B,** *Codosiga* species. **C,** *Giardia* species. **D,** *Trichonympha campanula.*

many and can be used not only for movement but for feeding, as a sensory organ, and for surface cleaning, attachment, and gamete locomotion. Some examples of flagellated species are illustrated in Figure 5-39.

Flagellates are heterotrophic in their nutrition. Many species possess one to several flagella, but they can also produce pseudopodia. Some species are found in water as solitary or colonial forms, whereas others are commensals or parasites of vertebrates and invertebrate hosts such as snails, insects, mollusks, nematodes, and humans. Some diseases of humans are caused by protozoans transmitted by insect vectors. For example, leishmaniasis is caused by a protozoan transmitted to humans by sandflies. The sandfly acquires the disease agent by feeding on an infected human and transmits it to other humans via biting. The genera of flagellates that are parasitic to humans are *Giardia, Trichomonas, Leishmania,* and *Trypanosoma.* Certain species of *Trichomonas* can also infect poultry and wild birds, causing interference with feeding and sometimes death.

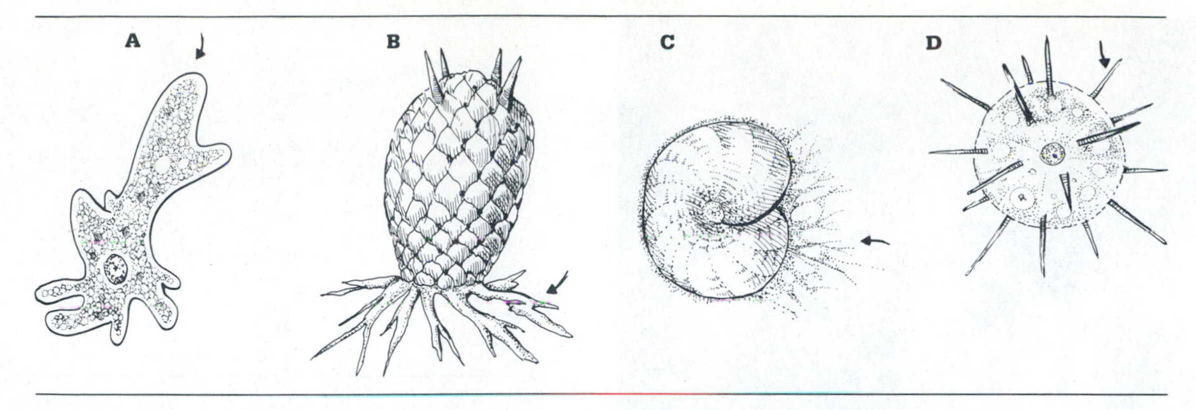

Figure 5-40 Pseudopodia produced by ameboid forms of protozoa. **A,** Lobopodia—blunt and rounded as in amebas. **B,** Filopodia—slender and tapered to a point **C,** Myxopodia—slender pseudopodia anastamose to form mesh. **D,** Axopodia—axial filaments about which cytoplasm flows.

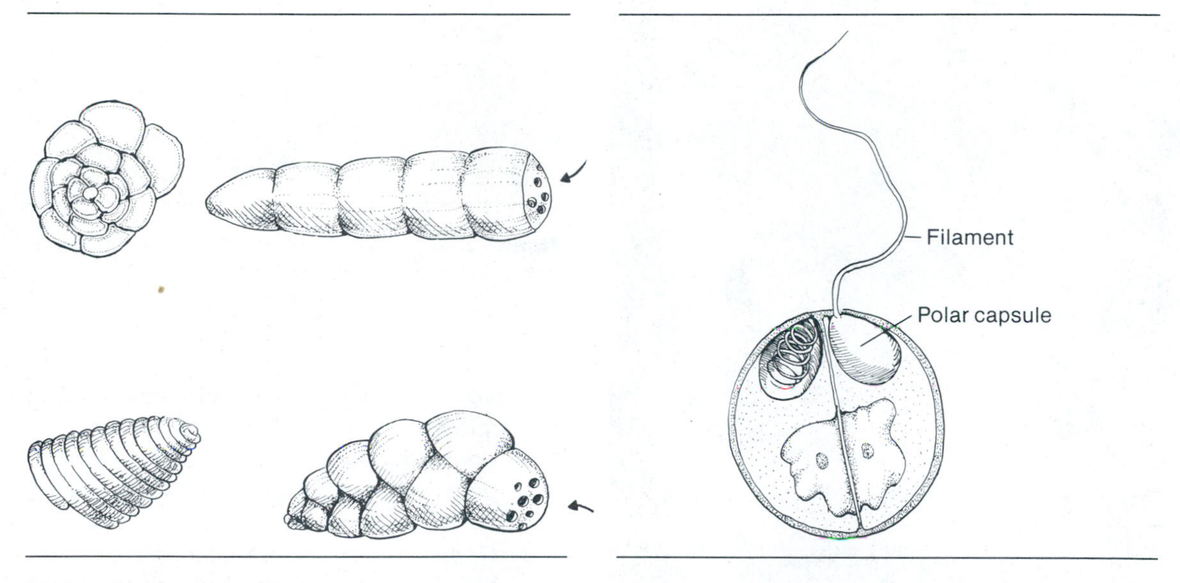

Figure 5-41 Shells (tests) of some foraminifera. The shells contain pores at tips *(arrows)* through which cytoplasmic extensions flow.

Figure 5-42 Polar capsules found in spore of *Myxospora* species. Note one filament is extruded from polar capsule.

The ameboid members of the Sarcomastigophora utilize the pseudopodial movements of the cytoplasm not only for movement but to capture food. Amebas feed on bacteria, algae diatoms, and ciliated and flagellated protozoa. They are found in both fresh and marine waters, and some parasitic forms can be found in the intestinal tract of both vertebrates and invertebrates. Some species are also commensalistic in the oral cavity of humans and the intestinal tract of animals and humans. The pseudopodia produced by ameboid forms have characteristics that are useful in classification (Figure 5-40). Some species secrete a ***shell (test),*** or skeleton, that is also used for identification purposes (Figure 5-41). The foraminifera, which produce calcium carbonate shells that are useful in acquiring geological data, are almost exclusively marine organisms that move slowly along the ocean floor or float in the water. They produce shells of various shapes and exhibit colors whose hue depends on the mineral content of the ocean environment. Some of the larger species contain algal symbionts.

Sporozoa. Sporozoa do not have flagella (except for some gametes), and movement is usually accomplished by flexing the body, which creates an undulating movement. The sporozoa are all parasites and infect nearly all species of the animal kingdom. They are responsible for large economic losses in cattle and poultry, as well as serious infections of wildlife. Human diseases, such as malaria and toxoplasmosis, are also caused by members of this group. Nearly all members exhibit a

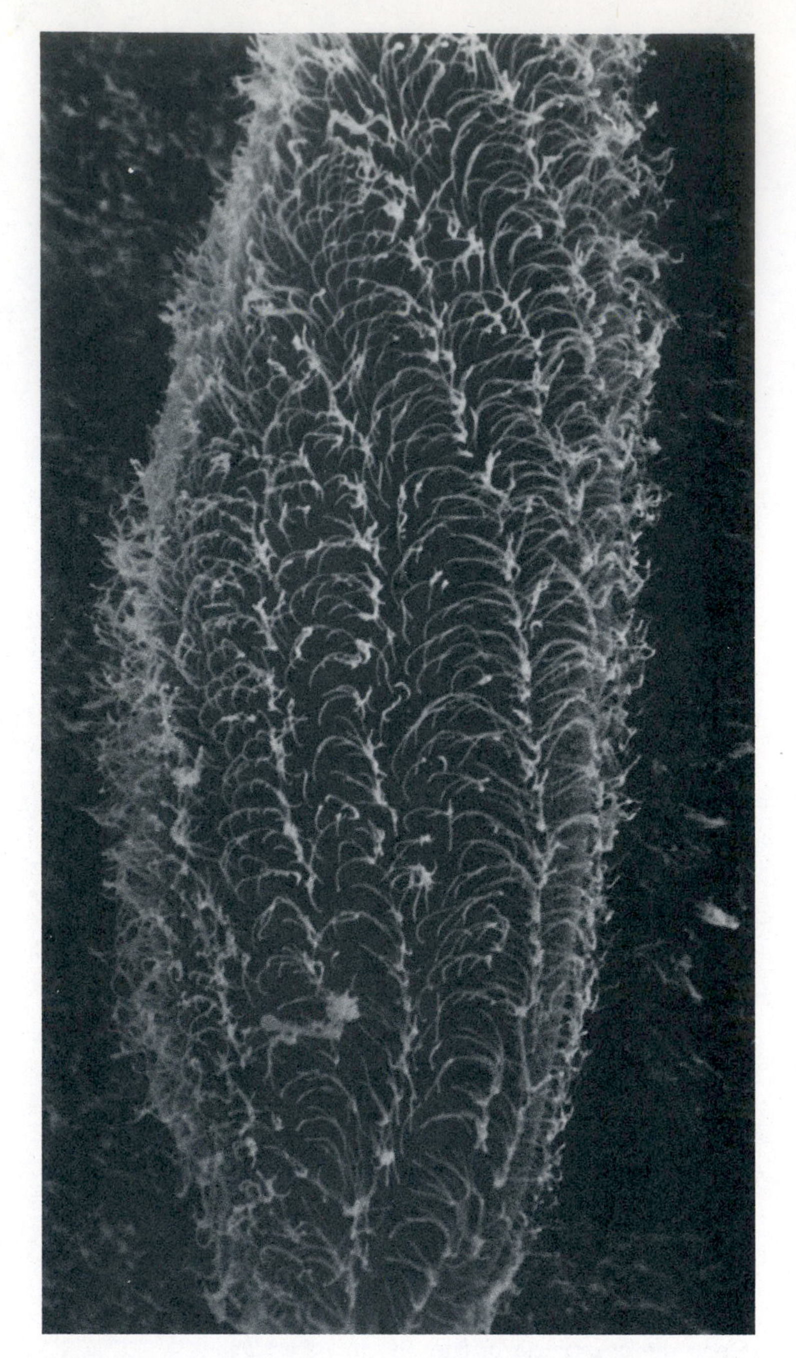

Figure 5-43 Scanning electron micrograph of ciliate *Paramecium caudatum.* Organism has been preserved in state of movement in which cilia are in reverse wave.
From Barlow, D.I., and Sleigh, M.A. J. Microscopy ***115:****81, 1979.*

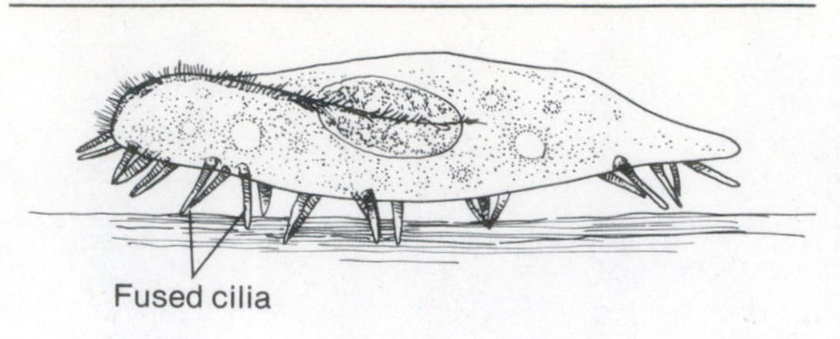

Figure 5-44 Ciliate using fused cilia to move along surface.

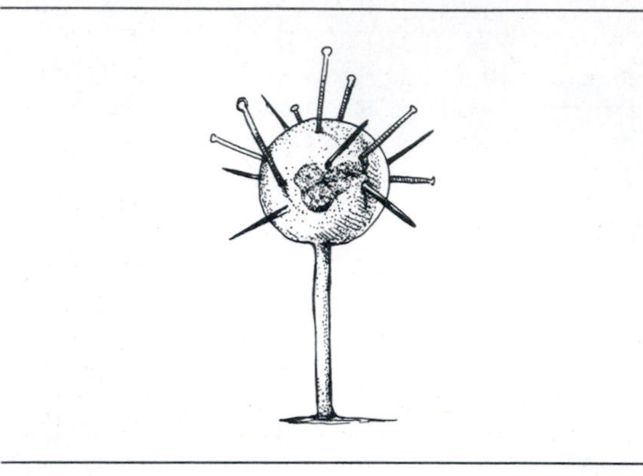

Figure 5-45 Ciliate (suctorian) in which tentacles are used in adult to capture food.

spore-forming stage called an oocyst. Sporozoa life cycles usually involve three phases.

1. The infective stage of sporozoa is called a ***sporozoite,*** which invades the host. Some sporozoites, which develop in the mosquito and are transmitted to humans via a bite, undergo multiple fission in the human host, producing uninucleate cells called ***merozoites.***
2. Merozoites undergo differentiation into gametes (male microgametocyte and female macrogametocyte).
3. ***Gametocytes*** fuse to produce a zygote. Meiosis in the zygote results in the formation of oocysts, which engage in multiple fissions and the formation of many uninucleate sporozoites (a repeat of the cycle).

Myxospora. Myxospora are parasites of invertebrates and lower vertebrates, especially amphibians, fish, mollusks, and insects. In some instances there is no cure for infection, and the animals must be destroyed. Spores are produced within cells that are lodged in various tissues and organs of the host. The spores possess unique structures called ***polar capsules*** (Figure 5-42), which contain filaments that are either coiled within the capsule or extruding from the cell. After the host ingests these spores, the polar filaments are extruded and used to attach to the host tissue, such as the epithelium of the intestinal tract. During this period of attachment the ameboid unit within the spore, called a sporoplasm, is released, and it migrates to specific organs where it develops into a multinucleated

Table 5-3 Major divisions of the algae and their characteristics

DIVISION	COMMON NAME	MORPHOLOGY	CHLOROPHYLLS	OTHER PIGMENTS	CELL COVERING	STORAGE PRODUCTS	OTHER CHARACTERISTICS
Euglenophyta	Euglenoids	Most unicellular	a and b, but most lack chlorophyll	Carotenoids, xanthophylls	Pellicle	Paramylon	Colorless species common, freshwater
Chlorophyta	Green algae	Unicellular, colonial, multicellular, and filamentous	a and b	Carotenoids, xanthophylls	Cellulose when wall present	Starch in pyrenoid	90% freshwater species
Bacillariophyta	Diatoms	Most unicellular	a and c	Fucoxanthin	Silicon dioxide, protein, and carbohydrate	Chrysolaminarin	Half marine and other half fresh water; important producers of oxygen
Pyrrophyta	Dinoflagellates	Unicellular	a and c	Carotenoids, peridinin	None, or cellulose thecal plate under cytoplasmic membrane	Starch	Primarily plankton in oceans; some parasites of animals, symbionts of marine animals (corals); some cause blooms as "red tide"
Phaeophyta	Brown algae	Multicellular	a and c	Carotenoids, fucoxanthin	Cellulose and alginic acid	Laminarin	Most marine; commercial source of alginates used in foods
Rhodophyta	Red algae	Most multicellular	a	Phycobilins, carotenoids, xanthophylls	Cellulose, or mannan microfibrils embedded in mucilage such as agars or carrageenans, both composed of galactose; some deposit $CaCO_3$ in walls	Floridean starch	Most species marine; commercial source of agar for microbiology laboratories and for foods; carrageenans used as gels in ice cream and puddings

unit. One preferred site is the cartilage of the head and spine of fish. Infection results in an interference with coordinated swimming, and the fish often falls victim to predators. If the fish survives, deformities appear in the head.

Ciliophora. Ciliophora possess both simple and complex ciliary organelles whose distribution and organization on the cell have been useful for taxonomic purposes. Cilia completely cover some species and function together in a concerted manner for movement (Figure 5-43). A wave of movement takes place over the organism that is believed to result from the stimulation of one cilium a fraction behind the cilium in front of it. Cilia in some species may be more concentrated near the oral groove where their movement aids in the feeding process. Some cilia fuse to form an undulating membrane that is used to generate a current of water through the oral cavity, bringing particles near the cytosome. Other cilia are fused into bundles called ***cirri,*** which increase the force of ciliary beating and also help the organism stand or move along a solid surface (Figure 5-44). There are also members of this group in which cilia are present only in the larval stages, and feeding in the adult is by tentacles (Figure 5-45). Examples of other ciliated species are illustrated in Figure 5-33.

Ciliophora are unique in that they possess two

different nuclei: a ***macronucleus*** and a ***micronucleus.*** Micronuclei, which are small and spherical, can give rise to macronuclei. Micronuclei are diploid and contain genetic information that can be exchanged during sexual conjugation. Macronuclei, which are large and variable in shape, are polyploid (containing several sets of chromosomes) and function in the control of metabolic processes in the cell. The micronucleus undergoes mitotic division when ciliates divide by binary fission, but the macronucleus merely elongates and is cut in half as the cell divides.

The ciliates can be found in a variety of watery habitats including bogs, fresh and marine waters, and polluted environments. Those ciliates that form cysts are found primarily in soil environments. Ciliophora swim freely in their environment, but a few are sessile and remain attached to substrates. Only a few species are parasitic to marine or freshwater animals, and only one species *(Balantidium coli)* is parasitic to humans. Humans who are in contact with swine are more likely to acquire disease by this organism. Humans are infected by ingestion of cysts released by infected animals. The ciliate excysts in the cecum or colon and begins to feed on the intestinal epithelium. An ulcer develops that can lead to hemorrhage and perforation of the intestine.

ALGAE

Classification

The study of algae is called ***phycology*** (Gr. *phykos,* seaweed). The algae are photosynthetic eukaryotes found in aquatic environments including fresh, marine, and brackish water and soil environments. Most are free living but some exhibit symbioses with plants, animals, protozoa, and fungi. Algae are rarely parasites. They range in size from microscopic unicellular types, such as diatoms, to some of the largest plants known to humans, the giant kelp. Algae play an important role in nature. The diatoms, for example, are the major primary producers in marine water, serving as a food for a variety of marine organisms and are probably the most important fixers of carbon dioxide. Kelp, which is found in marine waters, possesses holdfasts that anchor this multicellular algae to the ocean floor, sometimes as deep as 200 meters. Giant kelp beds perform many functions: (1) they are a food source for marine animals; (2) they provide shelter for spawning fishes; (3) they can be used by humans for food, fertilizer, and as a source of algin (a hydrophilic colloid that is used as a stabilizer and emulsifier for such products as ice cream, frozen custards, salad dressings, and other foods).

A

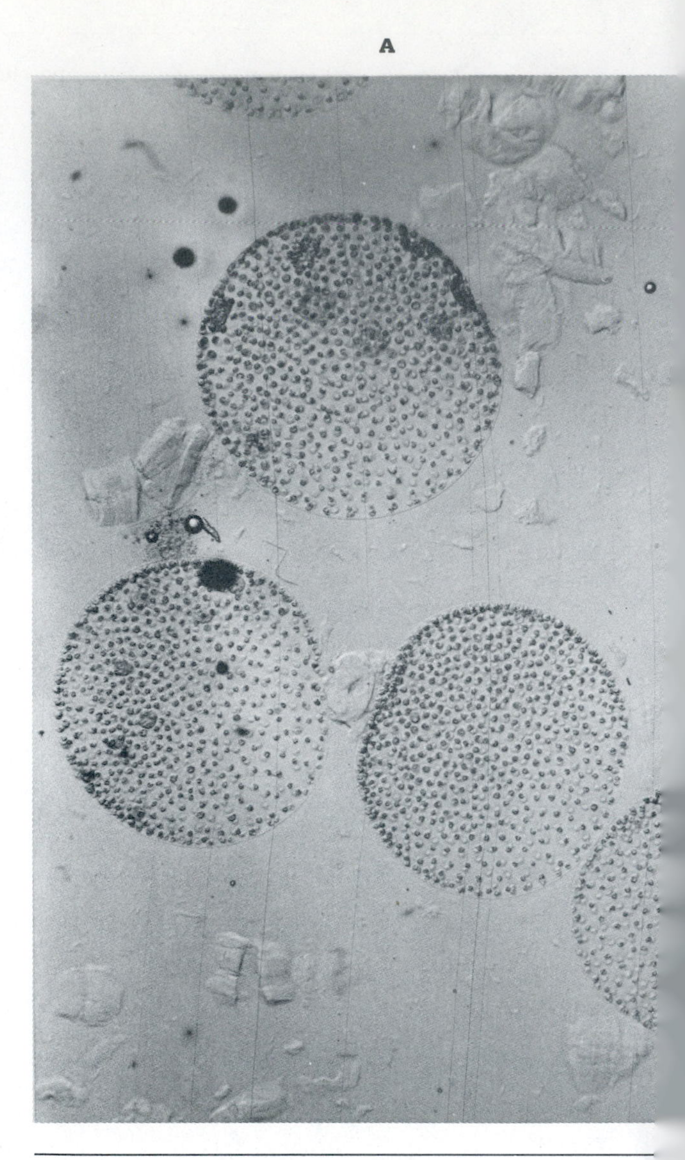

Figure 5-46 Morphological types of microscopic algae. **A.** Colonial *(Volvox).* (×33.)
Courtesy Fred Rosenberg.

The algae are classified primarily according to the pigments of the organism. They all contain chlorophyll, but other pigments may mask the color in some species. A classification scheme for all the algae is provided in Table 5-3, but our attention is devoted exclusively to the unicellular algae: Euglenophyta, Chlorophyta, Bacillariophyta, and Pyrrophyta.

Morphology

Cell envelope. Algae may appear as single cells or as multicellular organisms in which the indi-

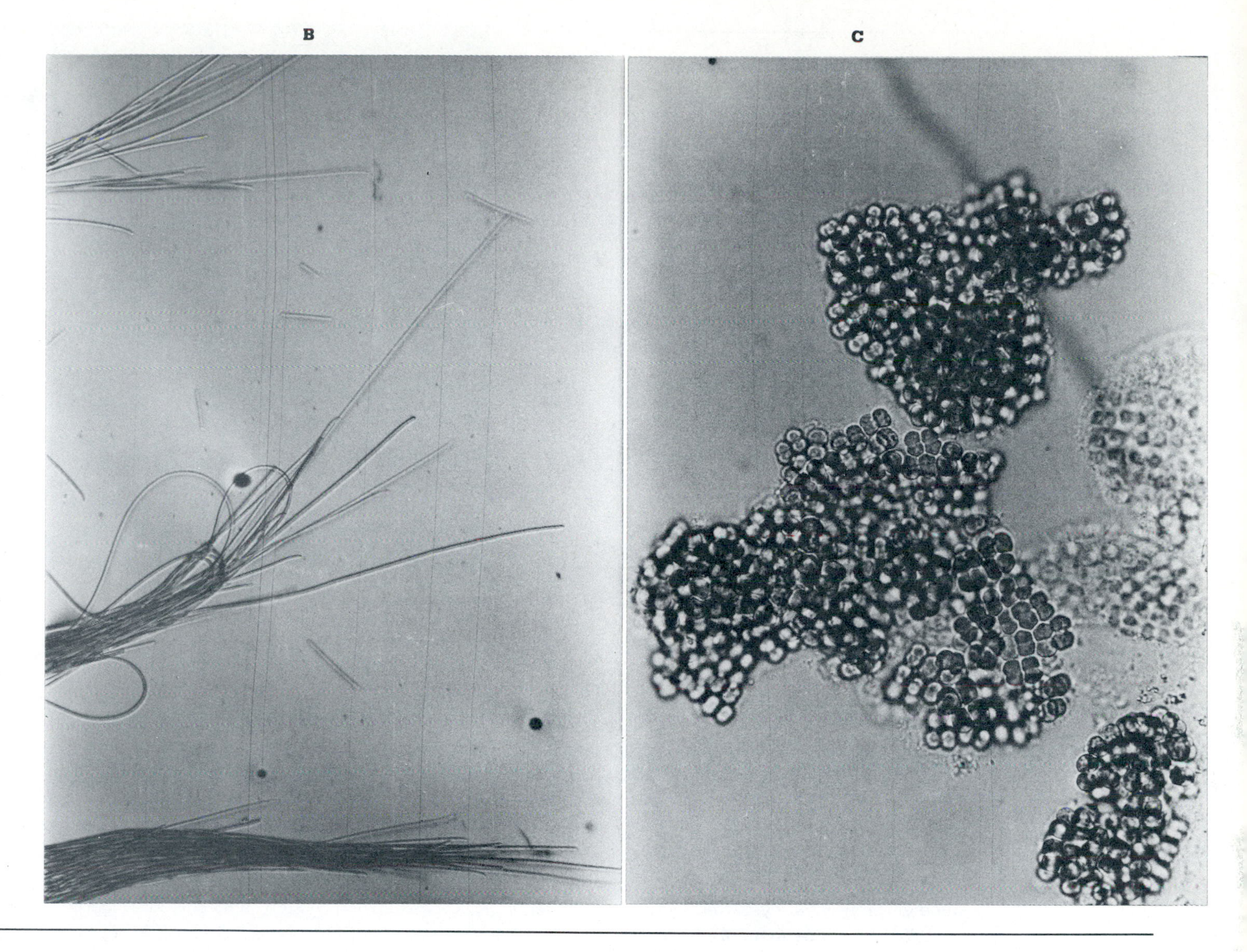

B, Filamentous *(Oscillatoria).* (×33.)

C, Sheets *(Merismopedia).* (×33.)

vidual cells are arranged in colonies (Figure 5-46, *A*), filaments (Figure 5-46, *B*), or sheets (Figure 5-46, *C*). There is some variation in the structure of cell walls among the unicellular algae. The cell walls of most green algae (Chlorophyta) are composed of an inner layer of cellulose (β-1,4-polymer of glucose) and an outer layer of pectin (a polysaccharide containing galactose, arabinose, and galacturonic acid). Some green algae have cell walls composed of a polymer of mannose. The cell walls of diatoms (Bacillariophyta) are composed of a protein (and probably lipid) matrix on which silicon dioxide (SiO_2) is deposited. Many unicellular algae have no cell wall but only a plasma membrane that may be naked, thickened, covered with scales, or surrounded by a gelatinous coat. Various minerals and other substances may be impregnated in these plasma membrane coverings, including silicon, calcium carbonate, iron, cellulose, and protein. A flexible membrane called the ***pellicle*** is located beneath the plasma membrane in the Euglenophyta and is composed primarily of protein (Figure 5-47). The pellicle is flexible enough to permit the organism to change shapes and to squeeze through small openings.

Muciferous bodies, which are located beneath the pellicle in Euglenophyta (and also in the Pyrrophyta) produce a gelatinous material that can coat the outer surface of the cell. The mucilage is produced in such quantities that it can form a

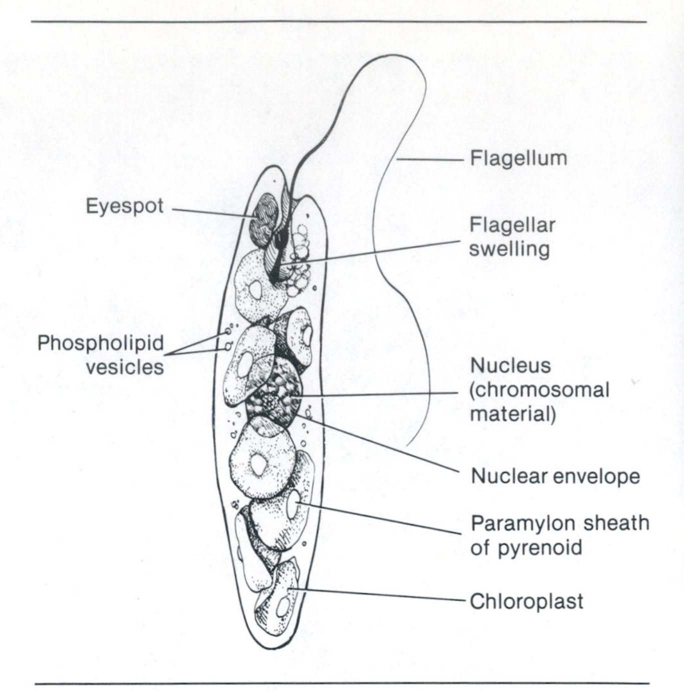

Figure 5-47 Various structures in *Euglena,* a member of the Euglenophyta.

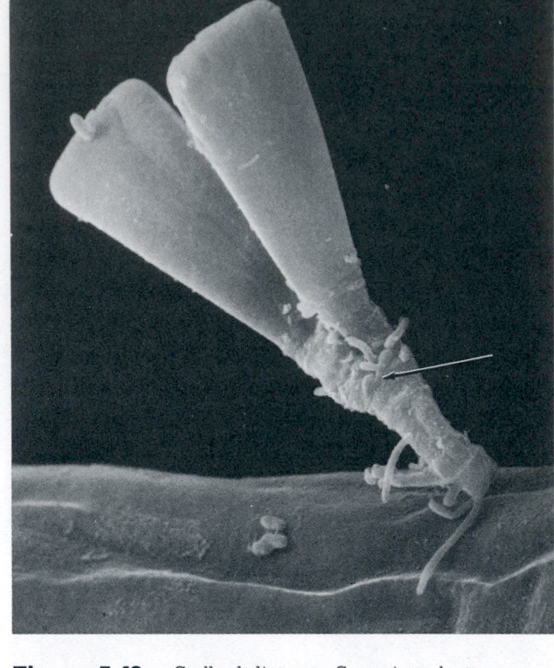

Figure 5-48 Stalked diatoms. Scanning electron micrograph of diatoms attached by polysaccharide stalk *(arrow)* to green filamentous algae. (×5800.) Note bacteria attached to stalk.
From Lowe, R.L., Rosen, R., and Kingston, J.C. J. Great Lakes Res. ***8:**164, 1982.*

stalk for sessile types. Some sessile types that do not have muciferous bodies are equipped with a polysaccharide stalk suitable for attachment to a substrate (Figure 5-48). The muciferous bodies of the Euglenophyta also secrete a layer of mucilage around the cell under adverse conditions and produce a protective cyst that germinates when conditions become more favorable for growth.

Cytoplasmic organelles and inclusions. Algae possess the usual eukaryotic structures: Golgi apparatuses, mitochondria, an endoplasmic reticulum, and a nucleus. Their structure and function are typical of other eukaryotes; however, the chloroplast forms a distinctive structure within the cell. Chloroplasts are found free in the cytoplasm of most algae, but their thylakoid arrangement differs among the various groups. A double membrane surrounds the chloroplast, but in some species an additional outer membrane, continuous with the endoplasmic reticulum, is present. The membrane of the thylakoids contains various pigments, depending on the species—chlorophyll a, b, or c, various carotenoids (yellow orange such as carotenes), and xanthophylls (orange red). Fucoxanthin, the principal xanthophyll in golden brown algae, masks the green chlorophyll and gives the cells their golden brown appearance. Peridinin is a carotenoid found in the Pyrrophyta. Located within the chloroplast of many algae is a protein structure called the ***pyrenoid,*** which apparently functions in the synthesis of the storage product of the cell. These storage products vary from one group to another and include paramylon, starch, and chrysolaminarin—all polymers of glucose (Table 5-3). An ***eyespot*** (Figure 5-47) is found associated with the chloroplast in many motile algae and contains carotene-lipid complexes that are believed to shade the photoreceptor located near the flagellum. They are either red or orange.

Nutrition

Most algae are characterized by their ability to use carbon dioxide as a carbon source (autotrophs) and light as an energy source (phototrophs). There are species of algae that, although photosynthetic and autotrophic, still require certain preformed organic compounds such as amino acids, vitamins, and other nutrients in order to survive. There are even some colorless species that rely entirely on preformed organic molecules for energy (heterotrophic). Some species of *Euglena,* although phototrophic, lose their chloroplasts when grown in the dark for long periods of time and are capable of a heterotrophic existence. Chloroplast formation is possible when the organisms are placed again in the light. Finally, some heterotrophic species behave like protozoa and ingest other living organisms (phagotrophs), especially other algae.

Motility

Flagellar motility is especially common among the unicellular algae and occurs in a spiraling manner. Anywhere from one to four flagella may be present on the cell. Those species that do not have motile adult stages usually produce motile gametes. Fusion of motile gametes in sessile forms is characterized by resorption of the flagella, but later, when the adult produces gametes, the flagella become reassembled. Structurally, the flagella of algae are typical of other eukaryotic flagella (see Chapter 8).

Diatoms (Bacillariophyta) are without flagella (except for male gametes of certain marine species), but gliding movement is still possible in some species because of mucus secretions. They have been observed to move at a rate of 2 to 14 μm/sec. This type of movement is similar to that found in certain bacteria, but the mechanism is controversial (see Chapter 7).

Reproduction

Both sexual and asexual reproduction occur in the algae, but some groups have no apparent sexual mechanisms. Asexual reproduction occurs by mitosis, and in multicellular forms this is often in the form of fragmentation. Mitosis in unicellular forms is usually by binary fission, in which nuclear division is followed by longitudinal division of the cytoplasm. The vegetative cell in many nonmotile algae forms a motile zoospore,* which can later develop into a vegetative cell following germination.

In most instances during sexual reproduction the male and female gametes are motile, and sexual union occurs by random collision. Gamete collision in one species results in an agglutination reaction between their respective flagella, and this is soon followed by nuclear fusion. In other groups only the male gamete is motile, and it is attracted to the nonmotile female gamete by chemotactic mechanisms. For example, in the colonial algae, called *Volvox,* a chemical factor is produced in the male colony that causes asexual colonies to produce male and female gametes. The chemotactic factor is produced in response to nutrient depletion and decreased light intensity. Some algae produce colonies that possess distinct arrangements and numbers of cells. The entire colony, called a ***coenobium,*** can reproduce asexually to produce a daughter colony with the same arrangement and number of cells as the parent colony.

* Some zoospores are produced in specialized cells called sporangia.

Algal symbioses

One of the clearest examples of endosymbioses in the algae is the lichen, in which an association between an alga and a fungus takes place. A number of invertebrate marine animals such as sea anemones, coral polyps, and hydra also possess endosymbiotic algae. Symbioses between dinoflagellates and reef corals, for example, are common. The growth of the corals occurs more rapidly in light than in darkness and suggests an important function for the dinoflagellates. Three mechanisms have been proposed to explain this phenomenon: (1) excreted organic products of photosynthesis permit the increased calcification of coral; (2) removal of carbon dioxide in photosynthesis favors precipitation of calcium carbonate; and (3) algae remove phosphate that may inhibit calcification, which thus favors crystallization of calcium carbonate. It is not known if the algae derive any benefit from these associations.

Characteristics of the unicellular algal groups

Euglenophyta. Members of the division Euglenophyta are called ***euglenoids.*** They are an unusual group because they exhibit both animal and plant characteristics. They are unicellular and lack a cell wall but possess a tough flexible pellicle located underneath the cytoplasmic membrane. They contain chlorophyll a and b but none are totally photoautotrophic (capable of growth in a medium devoid of all organic compounds with carbon dioxide as a carbon source, inorganic nitrogen as a nitrogen source, and light as a source of energy). Nearly all species can use preformed organic nutrients, either as a complete source of carbon and energy or as a partial source of carbon in the case of some photosynthetic types. Most algal species require at least one vitamin, and for this reason some species have been used to quantitate vitamins in solutions. A few species are phagotrophs and obtain their nutrients by ingesting other organisms, usually algae. The euglenoids are found primarily in fresh water where high concentration of organic material are found

Figure 5-49 Morphological types of diatoms

Courtesy Barry Rosen, The Ohio State University.

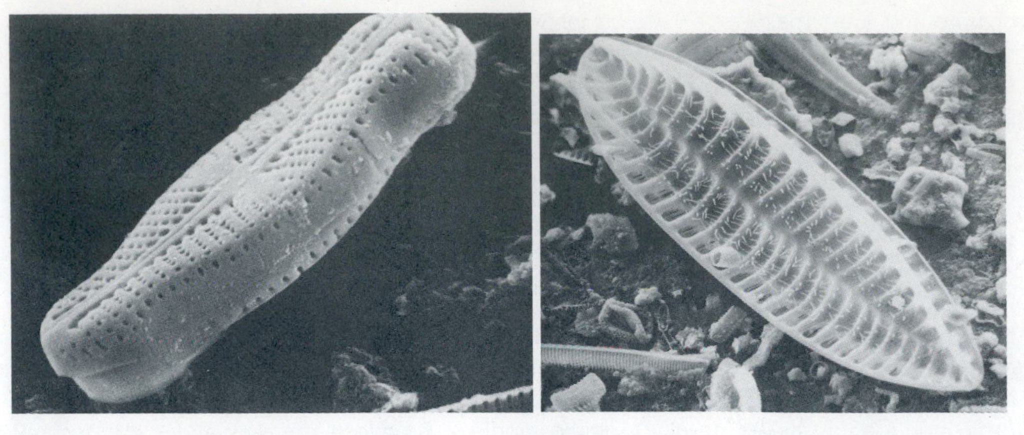

A, *Acnnanthes* species. (×10,000.)

B, *Surirella* species. (×500.)

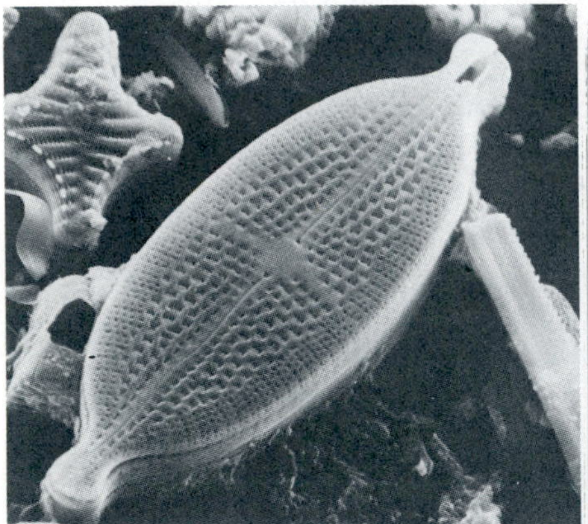

C, *Navicula tuscula.* (×2750.)

D, *Cyclotella* species. (×2800.)

E, *Campylodiscus hibernicus.* (×700.)

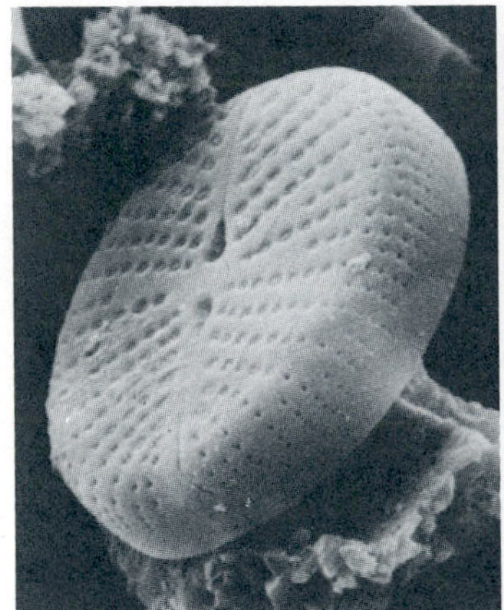

F, *Navicula scutiloides.* (×2200.)

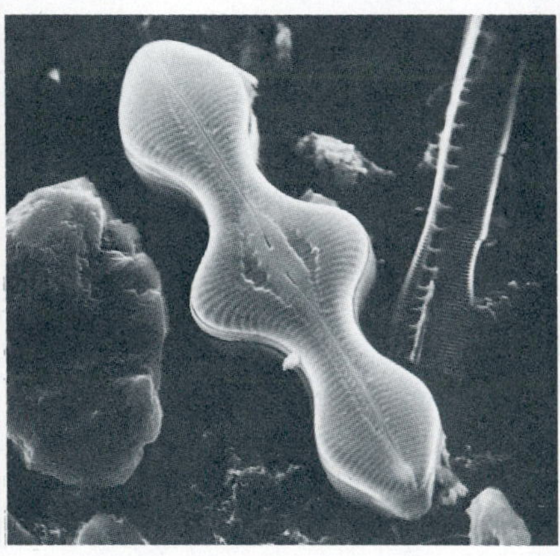

G, *Coloneis lewissii.* (×1200.)

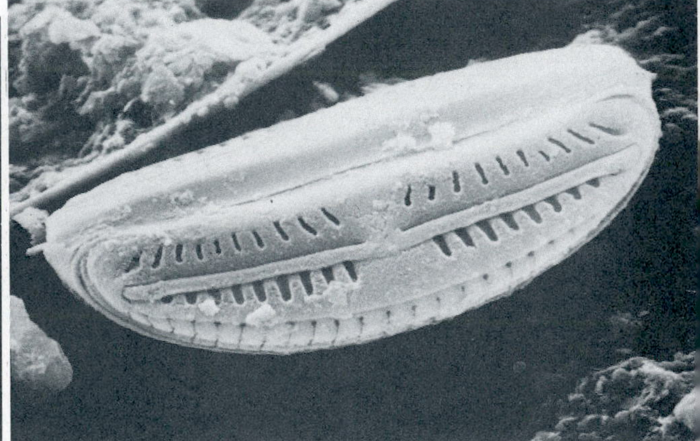

H, *Ampnora* species. (×5100.)

such as in farm ponds, but some species are marine.

Encystment is characteristic of the euglenoids and is the result of mucilage production by muciferous glands. A wall-like material is also produced during encystment that surrounds the cyst and protects the internal protoplasm. The germination of cysts in ditches and ponds is responsible for the unmistakable green and red blooms observed after heavy rains.

The eyespot (Figure 5-47) appears to act as a shading device for the flagellar swelling and controls the direction in which the organism travels. Euglenoids move toward light (positive phototaxis) of low intensity and away from light (negative phototaxis) of high intensity. These photo-

Figure 5-50 Scanning electron micrograph of centric diatom *(Cyclotella meneghiniana)* demonstrating frustules *(F)*. Note marginal spines *(arrows)* and ornamentation of mantle *(M)*.
From Hoops, H.J., and Floyd, G.L. Phycologia ***18:****424, 1979.*

tactic responses are particularly evident in those euglenoids found in tidal mud flats. The organisms respond negatively to light at high tide and burrow into the mud, whereas at low tide they respond positively and migrate to the surface.

One species of *Euglena (Euglena sanguinea)* produces a red pigment, astaxanthin, which colors the entire organism when the light intensity is high. Large populations of these organisms are present in water high in organic nutrients and are responsible for the appearance of its red color. It has been suspected that the biblical plague in which the Nile River turned blood red may have been caused by this organism.

One of the heterotrophic euglenoids lives in the gut of dragonfly nymphs in the winter, causing their posterior to appear green. In the spring the algae leave the dragonfly rectum as free-swimming cells.

Chlorophyta. The Chlorophyta, or green algae, are for the most part unicellular organisms, but colonial forms, such as *Volvox* (Figure 5-46), multicellular forms such as *Ulva,* and filamentous types such as *Oedogonium* are also part of this group. They are more frequently found in shallow waters, on tree trunks and sheltered sides of buildings, and in snow, ice, and soil. Some species also appear in marine environments. In other words, any environmental niche that has inorganic nutrients, is moist, and receives sufficient light will support their existence. They are not usually found in deep water because their pigments, chlorophyll a and b, absorb the red (650

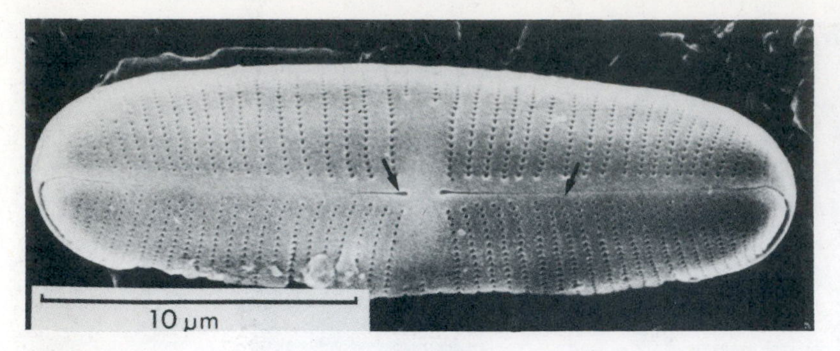

Figure 5-51 Scanning electron micrograph of motile diatom demonstrating furrow called a raphe *(arrows)*, which extends between two poles.
From Hanic, L. J. Phycol. ***15**:174, 1979.*

nm) and blue (450 nm) wavelengths that do not penetrate deeply into waters that are dense. In clearer tropical waters light can penetrate more deeply, and some green algae are found at depths of up to 100 to 200 m.

The unicellular green algae are the predominant phytoplankton found in fresh waters. The dense mats of algae found in confined bodies of water such as ponds and fish tanks are green algae. Some species of *Chlamydomonas* are found in the snows of some mountains in North America; one species gives a red color to the snowfields.

Some unicellular species such as species of *Chlorella* and *Chlamydomonas* are used in sewage treatment facilities. The algae help bacterial decomposition by providing oxygen, and they also recover mineral nutrients from the sewage that would otherwise be lost in the effluent. The excess algae can be used to feed livestock. The Japanese and Chinese use species of *Chlorella, Ulva,* and others in soups and for flavoring meat and other commercial products. The green algae are of little medical interest, although a few species parasitize the leaves of certain plants such as *Magnolia* and *Rhodedendron* and *Thea* (tea leaves).

The green algae have been the most widely studied because of their size and ease of cultivation in the laboratory. They have been studied because of their photosynthetic and nutritional characteristics and because they are believed to be the direct ancestors of green land plants. The characteristics they have in common with the higher plants are (1) cellulose in the cell wall, (2) chlorophyll a and b, (3) starch as a storage product, and (4) similar chloroplast structure.

Bacillariophyta. The Bacillariophyta are primarily unicellular and are characterized by the presence of large quantities of the brown pigment fucoxanthin, a xanthophyll. The ***diatoms*** are perhaps the most important class in this division and appear in a variety of striking shapes (Figure 5-49). They are the most important of the primary producers in marine environments. The siliceous cell walls of diatoms are composed of two overlapping halves called ***frustules,*** which give the diatom the appearance of a pill box or Petri dish (Figure 5-50). When the internal protoplast dies, the siliceous cell walls that remain are deposited (creating ***diatomaceous earth***). Deposits of diatom frustules have been used as geological markers in an attempt to locate oil deposits. The diatoms are divided into two groups based on the symmetry of their cell walls: radial symmetry and bilateral symmetry. In motile diatoms with bilateral symmetry a furrow called a ***raphe*** extends between the poles of the cell (Figure 5-51). The secretion of a mucilagenous substance through the raphe is believed to be involved in a gliding movement. The sticky substance attaches to the substrate and during locomotion leaves behind a slime trail.

The peculiar nature of the cell wall of diatoms provides the cell with an unusual division process. The overlapping valves during mitosis come apart, and each produces a new bottom. Thus, the bottom part of the parent cell becomes the top part of the daughter cell. Successive division results in a population of cells, many of which are considerably smaller than the original parent cell. It appears that when a critical minimal size is reached, a specialized growth spore is formed from the union of two gametes in a sexual process. The growth spore is called an ***auxospore,*** which, following germination, produces a diatom whose size is the same as the original parent.

Diatoms are the major component of fresh and marine water phytoplankton. As primary producers they must remain near the surface in a zone that receives sufficient light (photic zone). Diatoms have a cytoplasmic density of 1.03 to 1.10 g/cc, which makes them more dense than sea water, yet they are still capable of buoyancy. Buoyancy appears to be provided by a large vacuole that excludes heavier ions and thus its contents are a lower density than sea water. This type of mechanism does not work in fresh water, but some colonial diatoms are capable of flotation because of a gelatinous sac that surrounds the aggregate.

Diatoms vary considerably in their response to environmental conditions and because of this have been used to determine the quality of water in respect to such factors as alkalinity, acidity, temperature, and pollution. Certain species, for example, are present only at certain alkalinities, and by finding which species is present, an estimate of the alkalinity of water can be obtained.

Pyrrophyta. The Pyrrophyta, or dinoflagellates, have both plant and animal characteristics. They are usually motile organisms found in both fresh and marine waters and are best known to the layman as the organisms involved in the ***red tide.*** The cells are usually flattened and have a transverse constriction around the equator of the cell called the ***girdle*** (Figure 5-52). Flagella are inserted in the girdle; one circles the cell and is flat, and the other trails the cell.

The cell covering ***(thecal plate)*** of the dinoflagellates is a cytoplasmic membrane beneath which lies a layer of flattened vesicles. The vesicles contain plates of cellulose that give the cell covering its characteristic structure. The plates may be unornamental, or they may contain a series of cellulose plates that overlap and are held together at their joints. The plates appear as horns or wings. Few species have a true cell wall.

Blooms of some species of dinoflagellates (*Gonyaulax* and *Gymnodinium*) in the marine environment are responsible for what has been referred to as the red tide. Several factors are believed to control blooms, including temperature, salinity, pH, illumination, and trace elements in the water. Initially, the bloom gives the water a yellow color, which is followed by brown and then red coloration. These color changes are caused by changes in pigment concentration within the cells. As many as 10^8 cells can be found per liter of water with blooms extending over several kilometers. A toxin produced by some of the species is often released as the cells break up during passage through the gills of fish. The toxin adversely affects the fish, and as many as 50 million fish have been killed from one bloom in the Gulf of Mexico. Paralytic shellfish poisoning in humans occurs when shellfish are eaten that have filtered out of the water a toxic species of dinoflagellate. The toxin concentrates in the tissues of the shellfish. The toxin blocks nerve transmission, which can result in death resulting from respiratory paralysis. Over 220 human fatalities have been recorded from shellfish poisoning. The toxin in some dinoflagellates is heat stable and can remain within the shellfish for up to a year after the algal bloom has disappeared.

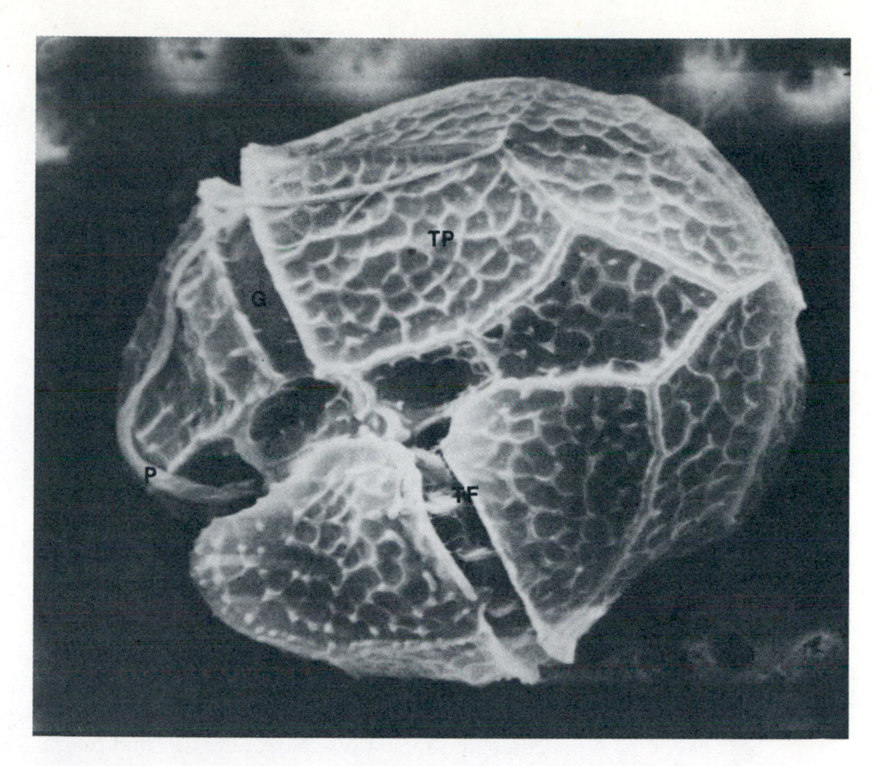

Figure 5-52 Scanning electron micrograph of dinoflagellate. Girdle *(G)* surrounds cell, which is covered by thecal plates *(TP).* Transverse flagellum *(TF)* is inserted in girdle; posterior flagellum (P) is partially seen.
From Berdach, J. J. Phycol. ***13:****243, 1977.*

SUMMARY

Fungi

The thallus is the basic vegetative unit of the fungi. If the thallus is filamentous, it is called a mold. If the thallus is spherical, it is called a yeast. Individual filaments in the mold are called hyphae, which when organized into a dense mat are called a mycelium. Hyphae show variation in structure and function depending on the species involved. Fungi are found primarily in the soil. They produce extracellular enzymes that digest organic debris found in the environment and are therefore important in the recycling of elements and in maintaining the fertility of the soil. Only a few species are infectious to animals or humans. Some of these infectious species are dimorphic and can exist in either a mold or yeastlike state, depending on environmental conditions.

Fungal reproduction may be sexual or asexual. Budding or binary fission occurs in some yeasts, but the spore is the major reproductive unit, and it can be produced sexually or asexually. Asexual spores may arise directly from hyphae, or they may be produced on or in specialized structures (e.g., conidiophores) arising from the hyphae.

The structure of the spore-bearing structures and the morphology of the spores are important criteria in the identification and classification of fungi.

The Fungi are separated into two divisions: the slime molds (Myxomycota) and the true fungi (Eumycota). The Eumycota are further separated into five subdivisions. The Zygomycotina are terrestrial species, many of which parasitize plants, animals, and humans. The Mastigomycotina, which includes aquatic as well as terrestrial species, possess a flagellum that distinguishes them from other fungi. The Ascomycotina are morphologically a more complex group and are characterized by a spore-bearing sac called the ascus. The Deuteromycotina have no apparent sexual stage. Most of the fungi pathogenic to man are included in this group. The Basidiomycotina are a structurally complex group characterized by a spore-bearing structure called the basidium, which in some species is enclosed by a fruiting body, the basidiocarp.

The slime molds possess protozoan and fungal qualities, and their taxonomic status has not been resolved. In the environment they are associated with true fungi and contribute to degradation of organic cellular debris.

Protozoa

Protozoa are unicellular microorganisms that are important primarily because of their parasitic capabilities in animals and humans. They are not decomposers like the fungi and bacteria. Protozoa can be found in fresh and marine waters. They have developed specialized structures within and on the surface of the cell to carry out biological processes. For example, they have digestive systems that are designed to concentrate, capture, and digest their prey. The ingested material is degraded within special vacuoles.

Asexual reproduction is the most common type of propagation, with binary fission the most frequently used form. Sexual reproduction occurs in most protozoa but is influenced by environmental conditions. Two types of sexual reproduction occur: conjugation, which is characteristic of ciliated protozoa and is carried out by the union of two parental cells, and syngamy, which is the fusion of free-swimming gametes. Cysts are part of the sexual and asexual reproductive cycle, and their formation is stimulated by environmental conditions. The cyst can serve in a reproductive capacity, or it can be used as a unit of transmission in disease.

The major subdivisions of protozoa include the Sarcomastigophora, which are either ameboid or flagellated. The group includes parasitic as well as commensalistic amebas, and the foraminifera, whose shell deposits are indicators of oil pockets in the earth. The Sporozoa are all parasites and move by undulating motions. The infective stage is called a sporozoite. They are responsible for considerable loss of animal and human life. The Myxospora are parasites of fish, amphibians, and insects. Spores are produced in the tissue of certain hosts, which if ingested by humans, become activated and release an ameboid unit that migrates to various organs of the body. The Ciliophora possess cilia that cover all or just parts of the organism. They possess two nuclei: a macronucleus that controls metabolism and a micronucleus that is used for propagation purposes.

Many protozoa carry bacteria and algae as endosymbionts. Some are helpful in supplying nutrients to the protozoa, whereas others produce toxic substances that help the protozoa compete with other protozoa in the environment.

Algae

The algae are photosynthetic eukaryotes that appear in marine, fresh, and brackish waters. They are composed of unicellular and multicellular forms, each important to humans and their environment. Some species are the primary producers in water and serve as the first link in the food chain.

Like other eukaryotes, algae possess the usual cytoplasmic organelles, but there are some important differences. In many of the unicellular species a cell wall is absent and the cytoplasmic membrane is modified to protect the cytoplasmic contents. The chloroplast is functionally the same in all algae, but the thylakoid arrangement differs from group to group. Within the chloroplast is a structure called a pyrenoid that serves in the synthesis of storage compounds.

Most algae are photoautotrophs (they use light as a source of energy and carbon dioxide as a source of carbon), but many also have requirements for preformed organic compounds such as vitamins.

Motility is common in the adult stages of most unicellular algae. If motility is absent in the adult, it is present in the gametes. Some diatoms have no flagella but exhibit motility by secreting a slime layer.

Both asexual and sexual reproduction occurs in the algae. In the unicellular forms asexual reproduction occurs by mitosis and is characterized by

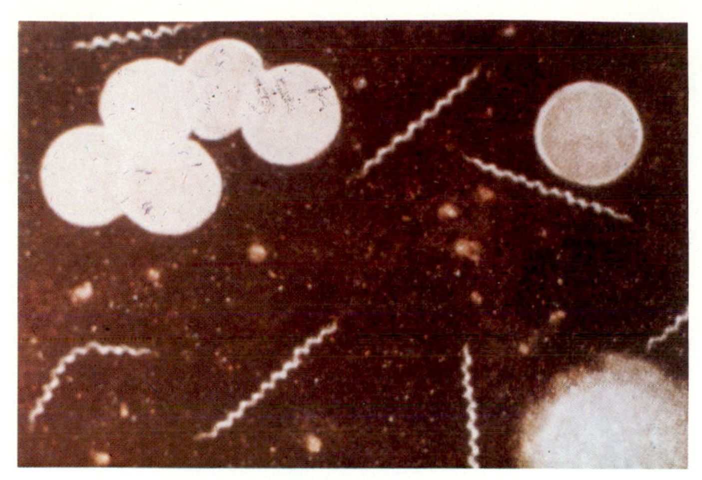

Plate 1 Darkfield microscopy of spirochete, *Treponema pallidum*.
Courtesy The Centers for Disease Control, Atlanta.

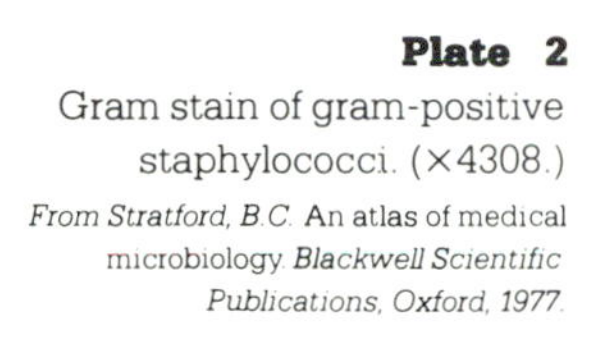

Plate 2
Gram stain of gram-positive staphylococci. (×4308.)
From Stratford, B.C. An atlas of medical microbiology. *Blackwell Scientific Publications, Oxford, 1977.*

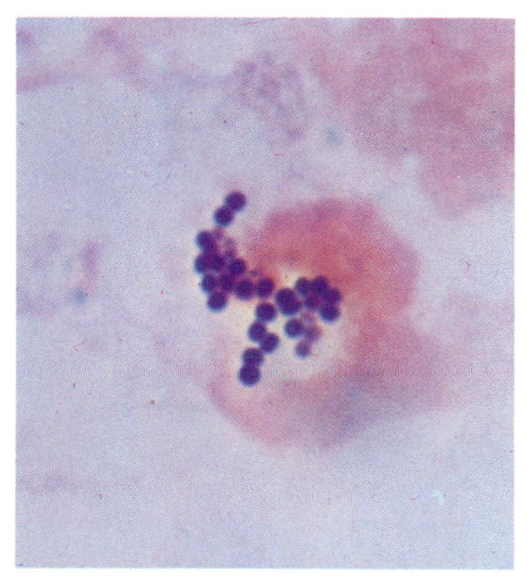

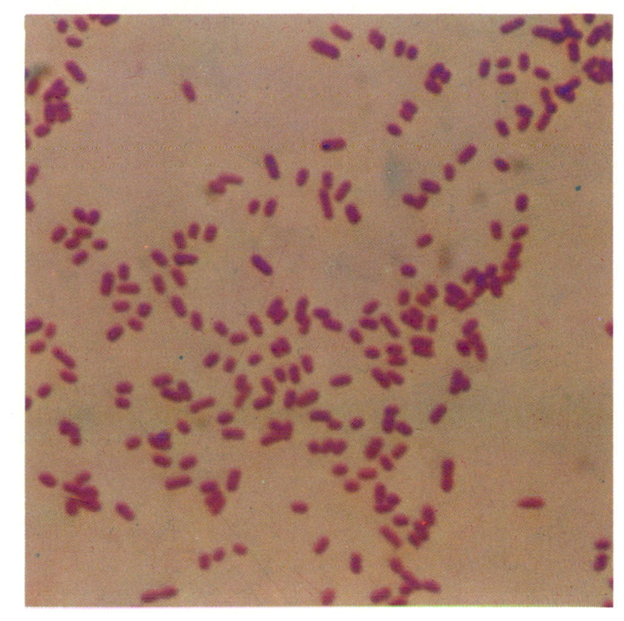

Plate 3 Gram stain of gram-negative rod *Escherichia coli*. (×4342.)
From Stratford, B.C. An atlas of medical microbiology. *Blackwell Scientific Publications, Oxford, 1977.*

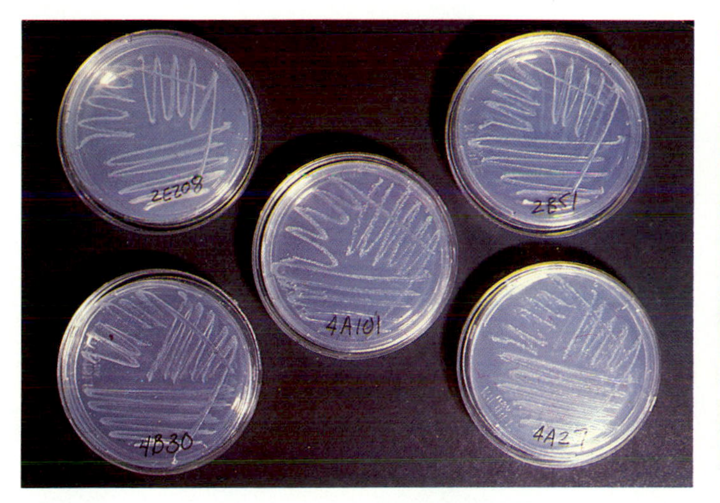

Plate 4 Luminous bacteria in agar culture in room light.
Courtesy F.A. Singleton.

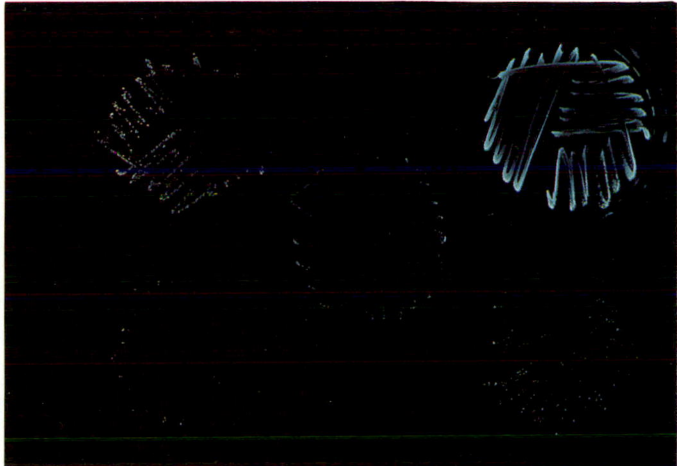

Plate 5 Luminous bacteria in agar culture in its own light.
Courtesy F.A. Singleton.

Plate 6 Luminous bacteria in broth in room light.
Courtesy J.W. Hastings, Harvard University.

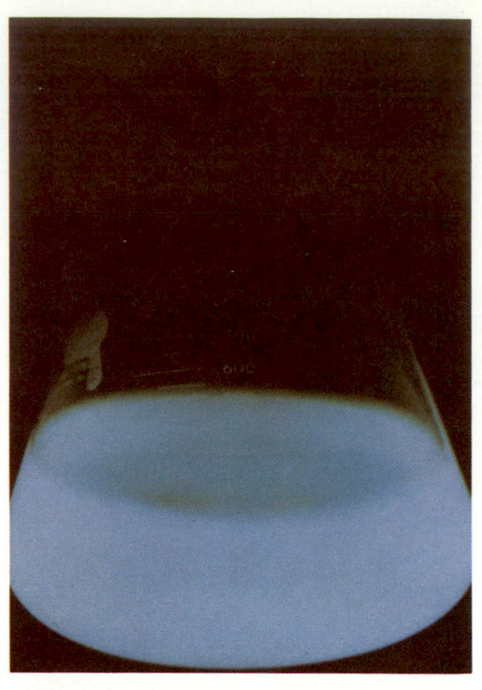

Plate 7 Luminous bacteria in broth in its own light.
Courtesy J.W. Hastings, Harvard University.

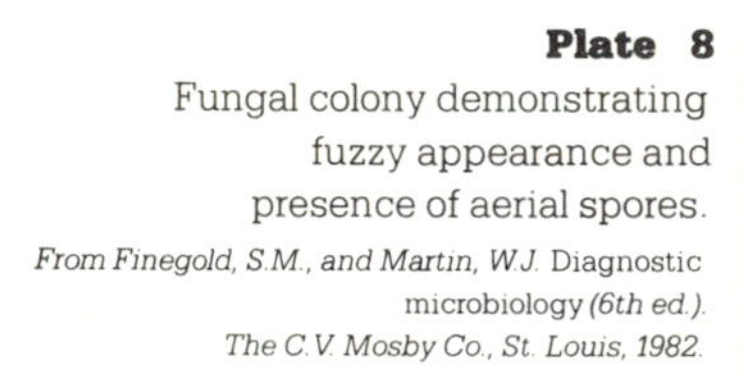

Plate 8 Fungal colony demonstrating fuzzy appearance and presence of aerial spores.
From Finegold, S.M., and Martin, W.J. Diagnostic microbiology (6th ed.). The C.V. Mosby Co., St. Louis, 1982.

Plate 9 Fungal colonies on agar demonstrating mucoid appearance and absence of aerial spores.
From Finegold, S.M., and Martin, W.J. Diagnostic microbiology (6th ed.). The C.V. Mosby Co., St. Louis, 1982.

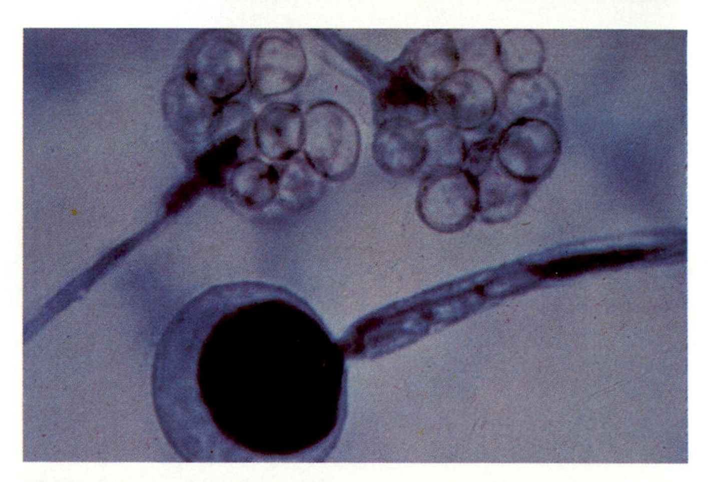

Plate 10 Species of *Mucor* in which spores can be observed inside sporangium.
From Finegold, S.M., and Martin, W.J. Diagnostic microbiology (6th ed.). The C.V. Mosby Co., St. Louis, 1982.

Plate 11 Peach leaf curl in apricot leaf caused by species of *Taphrina*.

Courtesy Robert Wick, Virginia Polytechnic Institute and State University.

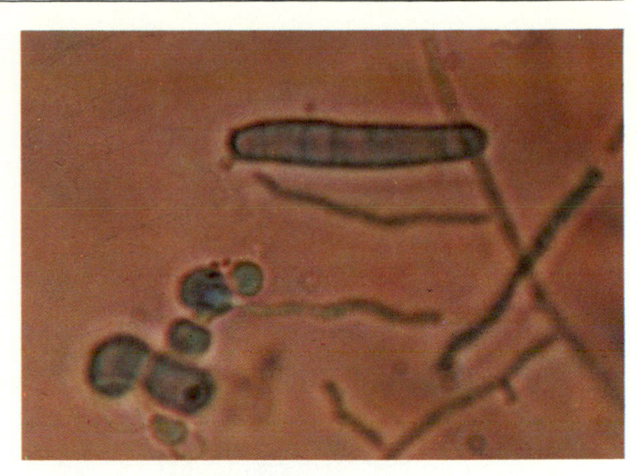

Plate 12 Preparation of *Trichophyton rubrum*, a cause of ringworm. Microconidia appear tear shaped, whereas macroconidia are long and slender. (×688.)

From Stratford, B.C. An atlas of medical microbiology. Blackwell Scientific Publications, Oxford, 1977.

Plate 13 Poisonous mushroom, *Amanita muscaria*.

Courtesy Stanley Flegler, Michigan State University.

Plate 14 Poisonous mushroom, *Amanita verna*, also called Angel of Death.

Courtesy Stanley Flegler, Michigan State University.

Plate 15 Slime mold *Physarum polycephalum*; plasmodium is yellow.

Courtesy Stanley Flegler, Michigan State University.

Plate 16 Slime mold *Ceratiomyxa fruticulosa* showing white fruiting body.
Courtesy Stanley Flegler, Michigan State University.

Plate 17 Slime mold *Hemitrichia stipitata* showing stalked red fruiting bodies on surface of log.
Courtesy Stanley Flegler, Michigan State University.

Plate 18 Capsule stain. *Pasteurella haemolytica* stained with fluorescein isothiocyanate–conjugated antiserum. (×500.) Wide capsules can be observed surrounding cell.
From Corstvet, R.E., et al. J. Clin. Microbiol. **16:***1123, 1982.*

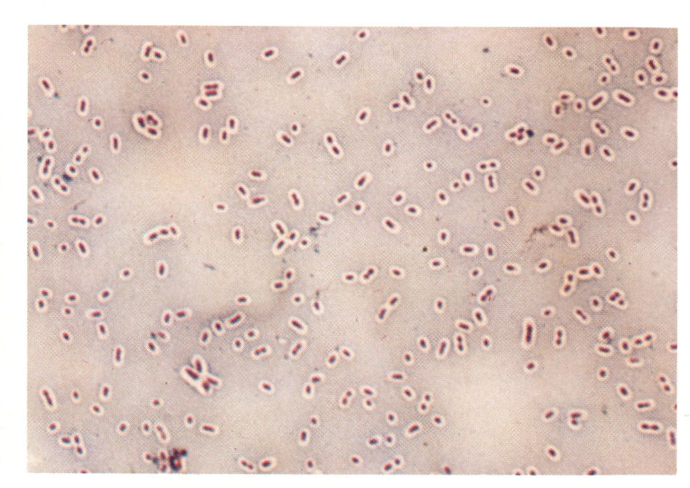

Plate 19 *Pasteurella haemolytica* stained with Congo red and counterstained with Maneval stain. Cells appear red surrounded by clear capsular zone.
From Corstvet, R.E., et al. J. Clin. Microbiol. **16:***1123, 1982.*

binary fission. Sexual reproduction occurs primarily by random collision and fertilization of male and female gametes.

Algae enter into endosymbiotic relationships with other organisms such as fungi (lichen) and various marine animals. They supply nutrients to the host, but their benefit from the associations, particularly in marine animals, is not known.

The principal unicellular algae belong to the following groups (see also Table 5-3).

1. The Euglenophyta (euglenoids) lack a cell wall and are primarily freshwater organisms. They give color to algal blooms, particularly in ditches and small ponds.
2. The Chlorophyta (green algae) are primarily freshwater species. They are predominant phytoplankton in fresh water. They are believed to be direct ancestors of the green land plants.
3. The Bacillariophyta (diatoms) are the primary producers in marine environments. Their cell wall is composed almost entirely of silicon dioxide. Their fossil records have been important to geologists and environmentalists alike.
4. The Pyrrophyta (dinoflagellates) are characterized by the modifications that occur in their cytoplasmic membrane. They are important as the cause of the red tide and shellfish poisoning in humans.

Study questions

1. What role(s) do the protozoa play in the environment?
2. How does the nutrition of protozoa differ from that of prokaryotes? Are there any special structures that are associated with protozoan nutrition that are not found in prokaryotes?
3. In what ways are the reproductive processes of protozoa similar to bacteria? Dissimilar?
4. What is the function of the micronucleus in ciliates? The macronucleus?
5. Does the process of schizony offer any advantage to protozoa?
6. How does the protozoan cyst differ from the bacterial spore? Are they similar?
7. Name the major groups of protozoa, listing at least one important characteristic.
8. How does one distinguish between a yeast and a mold?
9. What are dimorphic fungi? Does this characteristic have any diagnostic significance?
10. What, if any, are the structural and functional differences in the filaments or hyphae of the fungi?
11. Compare and contrast the nutrition of fungi, bacteria, and protozoa.
12. What fungal characteristic is most useful in identification and classification?
13. Define the following terms: holocarp, eucarp, germ tube, conidia, conidiophore, ascocarp, dikaryon, thallus, rhizoids, haustoria.
14. Why is the classification of slime molds, as to type of microorganisms, so difficult?
15. What beneficial function do the slime molds serve in the environment?
16. What is meant by mycorrhizae? Are they of any economic value?
17. How does the cell wall of algae differ from the cell wall of fungi? Of bacteria? What modifications in the cell envelope compensate for this deficiency?
18. What is the function of the eyespot? What counterparts, if any, exist in other microorganisms?
19. What is meant by phytoplankton? What do they all have in common?
20. Describe the relationship (symbiosis) that exists in lichens.
21. Why are the green algae considered direct ancestors of green land plants?
22. Which groups of algae are considered the predominant phytoplankton in fresh water?
23. What is meant by the red tide? What is its importance to humans?
24. What algae are considered the primary producers in marine environments?
25. What are diatoms so important to geologists while other algae are not?

Selected readings

Books

Alexopolous, C.J., and Mims, C.W. *Introductory mycology* (3rd ed.). John Wiley & Sons, Inc., New York, 1980.

Ashworth, J.M., and Dee, J. *The biology of slime molds.* Edward Arnold (Publishers) Ltd., London, 1975.

Deacon, J.W. *Introduction to modern mycology.* John Wiley & Sons, Inc., New York, 1980.

Farmer, J.N. *The protozoa.* The C.V. Mosby Co., St. Louis, 1980.

Goodaa, G.W., Lloyd, D., and Trinici, A.P.J. (Eds.). *The eukaryotic microbial cell.* In the 30th Symposium of the Society of General Microbiology, Cambridge University Press, New York, 1980.

Krier, J.P. *Parasitic protozoa.* Academic Press Inc., New York, 1978.

Kumar, H.D., and Singh, H.N. *A textbook on algae.* Macmillan Publishing Co., Inc., New York, 1979.

Ross, I.K. *Biology of the fungi.* McGraw-Hill Book Co., New York, 1979.

Round, F.E. *The biology of the algae* (2nd ed.). Edward Arnold (Publishers) Ltd., London, 1973.

Schmidt, G.D., and Roberts, L.S. *Foundations of parasitology* (2nd ed.). The C.V. Mosby Co., St. Louis, 1981.

Sleigh, M.A. *The biology of the protozoa.* American Elsevier, New York, 1973.

Trainor, F.R. *Introductory phycology.* John Wiley & Sons, Inc., New York, 1978.

Webster, J. *Introduction to fungi.* Cambridge University Press, Cambridge, London, 1977.

Chapter 6 VIRUSES

Viruses have plagued humans since their existence, but an understanding of their nature is only of recent origin. They were probably observed microscopically in 1886 in a preparation derived from the lymph gland of a patient with smallpox. The spherically stained particles were thought to be spores of the bacterial genus *Micrococcus*. Coincidentally, the smallpox virus happens to be the largest virus and the only one observable with the compound microscope. Iwanowsky and Beijerinck in the 1890s demonstrated that material from tobacco plants afflicted with tobacco mosaic disease passed through a bacterial filter and was capable of causing the same disease when applied to healthy plants. Filterable agents were later found to be responsible for infections of many other plants, as well as animals. The filterable agents eventually were referred to as ***virus***, the Latin word for poison.

In the interval between 1935 and 1960 the true nature of the virus was unveiled and finally appreciated. During this period the following events contributed to virus elucidation:

1. The tobacco mosaic virus (TMV) was isolated in the form of crystals from infected tobacco plants by Stanley in 1935. The virus had the ability to reproduce and to cause disease before and after repeated crystallization in the laboratory.
2. The study of bacterial viruses ***(bacteriophages)*** revealed that only the viral nucleic acid enters the host cell (this is not true for animal viruses, in which the entire virus particle enters the cell). This discovery demonstrated the importance of the nucleic acid in determining the characteristics of the virus.
3. The development of tissue culture provided a means of growing animal virus in large quantities.
4. The development of the electron microscope permitted scientists to visualize the virus and to morphologically distinguish one type from another.

The purpose of this chapter is to characterize the basic morphology of the major groups of viruses and to describe the effects of various chemical and physical agents on viral activity. Viral replication in host cells is discussed in Chapter 16.

GENERAL CHARACTERISTICS

Viruses are given independent status because they are the smallest of microbial agents and because they are obligate intracellular parasites; that is, the virus is totally dependent on a living host for its survival. Viruses are infectious agents that can parasitize all forms of life. The absolute necessity of a virus for its host is emphasized by the fact that viruses cannot make their own ATP; they lack the enzymatic capacity to totally replicate themselves, and they are unable to demonstrate any significant metabolic activity. The relationship between the host and the infecting virus is a specific one, and any given virus may have a restricted host range. Certain viruses have been shown to have a broad range and may be able to infect humans, hogs, horses, some birds, and insects. At the other extreme, a virus may be able to infect only one strain of a species; for example, there is one bacterial virus that can infect only one strain of *Escherichia coli*. Since many viruses infect only one strain of bacteria, they are sometimes used to type organisms.

Size

All viruses, except the smallpox virus, are beyond the limits of resolution with the compound microscope. Since organisms smaller than 0.2 μm cannot be observed with the compound microscope, it follows that most viruses must be less than 0.2

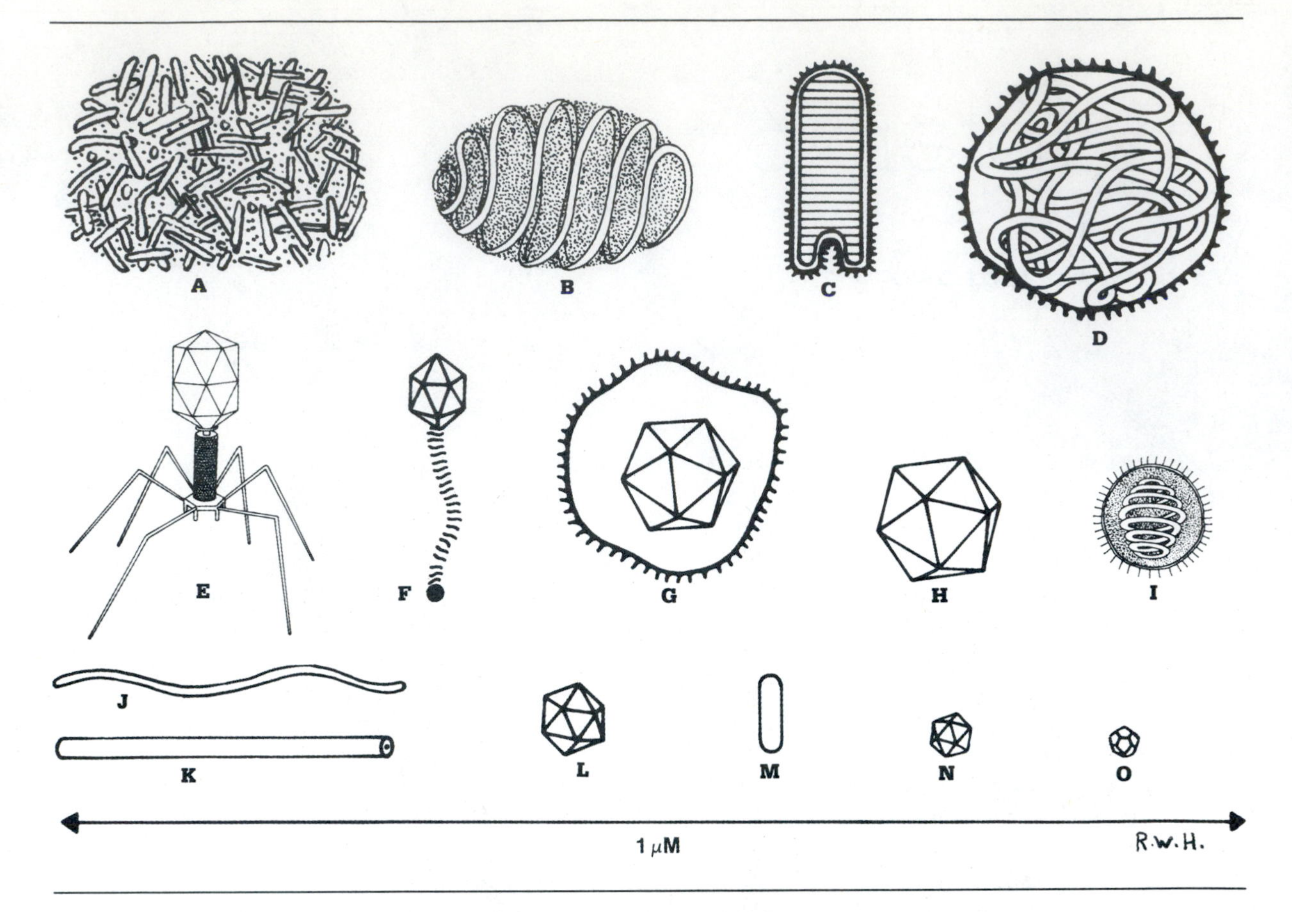

Figure 6-1 Comparative size and shape of various groups of viruses. **A,** Poxviruses. **B,** Orf virus (poxvirus). **C,** Rhabdovirus. **D,** Parainfluenza viruses (mumps). **E,** T-even bacteriophage. **F,** Flexuous-tailed bacteriophage. **G,** Herpesvirus. **H,** Adenovirus. **I,** Influenza viruses. **J.** Filamentous flexuous virus. **K,** Rodlike virus. **L,** Polyoma/papilloma virus. **M,** Alfalfa mosaic virus. **N,** Picornavirus (poliovirus and plant viruses). **O,** Bacteriophage ϕx174.

From Horne, R.W. The structure and function of viruses.

μm. The smallpox virus is 0.250 μm (250 nm) in diameter, and the poliovirus, one of the smallest viruses, is approximately 18 nm in diameter. Even though microbiologists in the early twentieth century were unable to see viral particles, they were still able to isolate and experiment with them, free from contaminating material, by passing virally contaminated culture or tissue extracts through filters that other microbial agents could not penetrate. The comparative size and shape of several groups of viruses are illustrated in Figure 6-1. The structural components found in a virus particle are demonstrated in Figure 6-2.

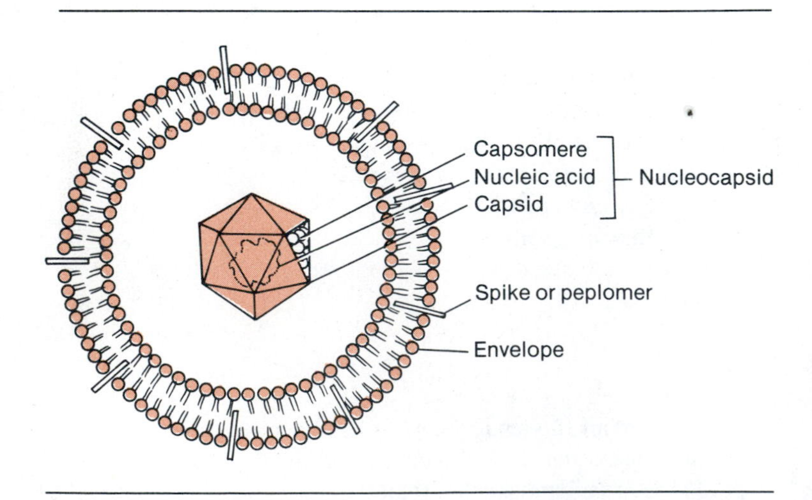

Figure 6-2 Basic structure of virus using animal virus as model.

Chemical composition

Nucleic acid. Viruses, with only minor modifications, are composed of a central core of nucleic acid surrounded by a protein coat called the ***capsid.*** A completely assembled and infective virus particle is called a ***virion.**** The virus is unique because the genetic information is encoded in either DNA or RNA, but not both, and the nucleic acid may be either single stranded or double stranded. The nucleic acid when isolated from the virus may be a linear or circular molecule, and in certain groups of viruses the nucleic acid may even appear in several distinct segments.

Unlike other microorganisms in which a variety of metabolic enzymes and organelles, such as ribosomes and mitochondria, are present in the cell cytoplasm, the virus contains practically nothing, save its nucleic acid. There are exceptions, of course; for example, a few viruses contain either a DNA or an RNA polymerase associated with the nucleic acid core. These polymerases are used by the virus for replication during the infection process. One group of viruses, called oncogenic (cancer-causing) RNA viruses, possesses a unique enzyme, reverse transcriptase, which is a DNA polymerase that requires an RNA template. The significance of these enzymes is discussed in Chapter 16.

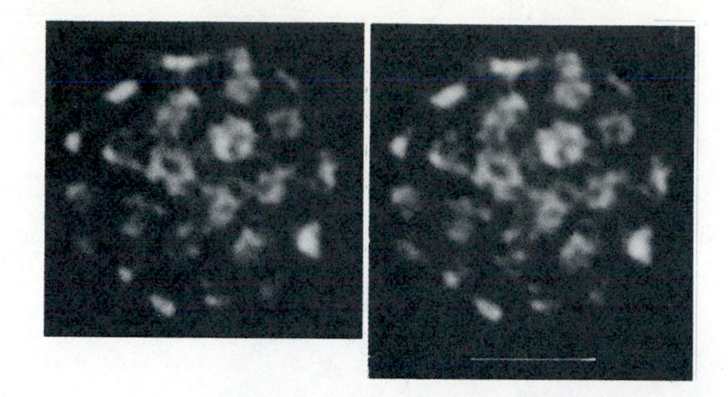

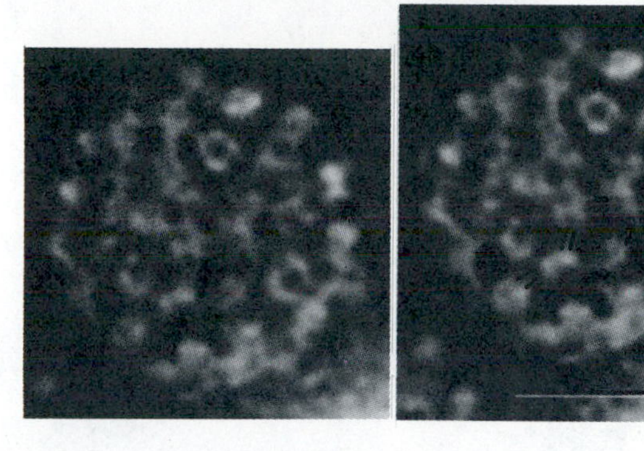

Figure 6-3 Electron micrograph of papilloma virus demonstrating individual capsomeres that make up capsid. Photographs at right indicate bridges that appear to connect capsomeres in upper micrographs. Scale lines = 250 Å.

From Yabe, Y., et.al. Virology ***96:****547, 1979.*

Protein coat. The protein coat surrounding the nucleic acid is called the capsid and is made up of individual protein subunits called ***capsomeres*** (Figure 6-3). The capsid subunits are arranged in different types of symmetry, which are discussed later. One genus of virus, *Reovirus,* possesses two capsids, both containing capsomeres (Figure 6-4), but this is the exception and not the rule. The capsid with its enclosed nucleic acid is called a ***nucleocapsid.*** The principal function of the capsid is to protect the nucleic acid from any environmental conditions that might affect it. This protection is due to the manner in which the amino acids of the polypeptide are arranged on the external surface of the capsid. It appears that the amino acid R groups on the external capsid surface are made up of nonpolar groups that interact (hydrophobic bonding) to exclude water, thus making the protein unusually resistant to pH, temperature, and radiation. This technique of water exclusion is similar to that found in the spore coats of *Bacillus* and *Clostridium.*

Figure 6-4 Electron micrograph of *Rotavirus* showing apparent double-capsid symmetry. *IC,* Inner capsid; *OC,* outer capsid.

From Roseto, A., et.al. Virology ***98:****471, 1979.*

Lipid envelope. Some viruses contain a membrane or ***envelope*** that surrounds the nucleocapsid (see Figure 6-2). The membrane is acquired by the virus, after replication, as it passes through the host's cytoplasmic membrane or nuclear membrane and is composed of lipids and proteins. Glycoproteins make up the outer portion

* There are instances where nucleic acid alone may be infective—a process called ***transfection.***

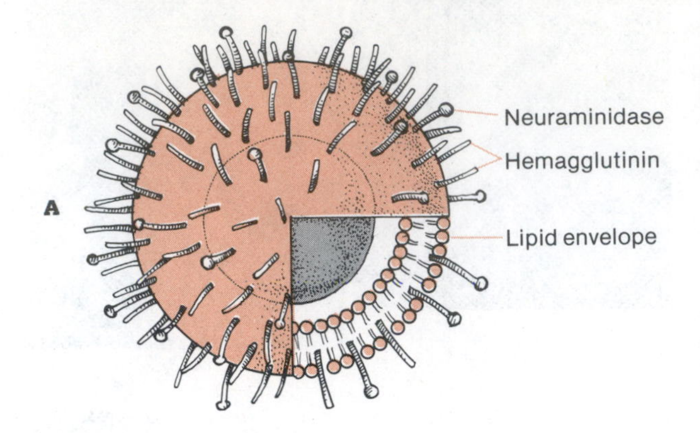

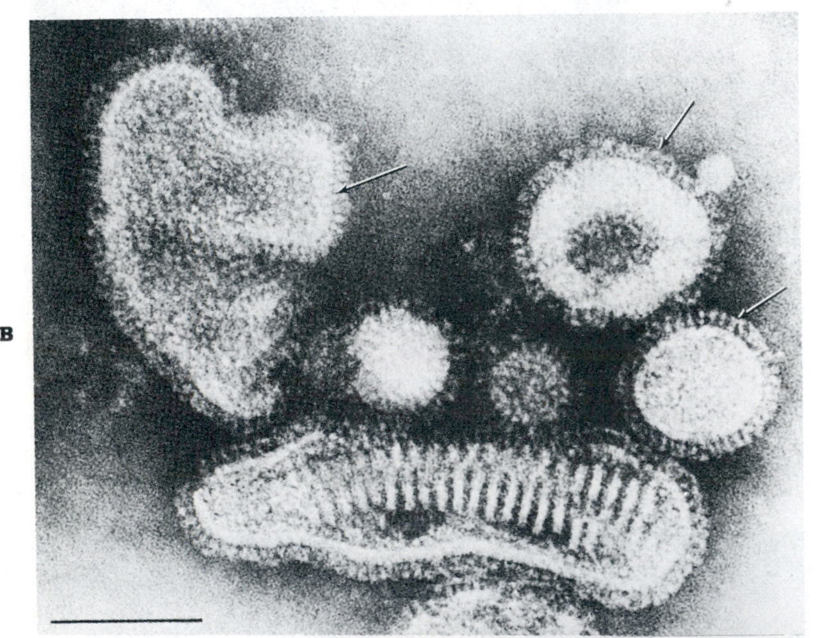

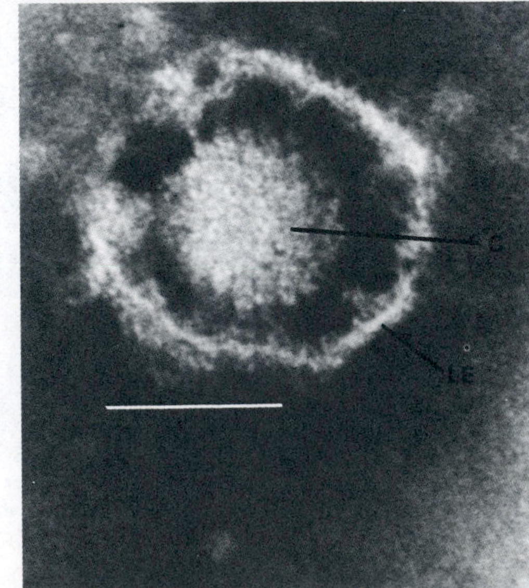

Figure 6-5 Lipid envelope of viruses. **A,** Influenza virus containing two types of spikes in its lipid envelope: neuraminidase and hemagglutinin. **B,** Electron micrograph of influenza virus. Particles are pleomorphic (some appear to be spherical, and one is rod shaped). Spikes *(arrows)* are easily revealed. Rodlike particle has its envelope penetrated by phosphotungstate stain revealing coiled internal ribonucleoprotein component. Bar = 100 mm. **C,** Electron micrograph of equine herpesvirus demonstrating capsid *(C)* and lipid envelope *(LE)*, which contains no spikes. Bar = 100 nm.

B, Courtesy Mrs. J.D. Almeida and F. Fenner. C, From O'Callaghan, D. Prog. Med. Virol. 22:152, 1976. S. Karger AG, Basel.

of the envelope, and they often appear as ***spikes*** (also called ***peplomers***) in the electron microscope. The protein components of the glycoprotein are viral in origin, but the carbohydrate components, as well as the lipid moieties, are derived from the host cell. The protein components of the envelope can have a variety of functions. The spikes on the influenza virus, for example, are of two types (Figure 6-5, *A* and *B*): one is an enzyme called ***neuraminidase,*** and the other is a ***hemagglutinin.*** Neuraminidase removes *N*-acetylneuraminic acid from the polysaccharide found on the surface of mammalian cells. The enzyme functions not only for attachment and spread of virus on epithelial surfaces but also as a means of intracellular release from already-infected cells. Hemagglutinins bind to red blood cells and cause them to form aggregates; that is, they become agglutinated. Not all enveloped viruses possess spikes (Figure 6-5, *C*), but those that do appear to have a built-in mechanism for attachment to host tissue. Enveloped viruses lose their infectivity once the envelope is destroyed or removed by agents such as bile salts, which are present in the alimentary tract. It is not surprising that enveloped viruses do not use the alimentary tract as a means of infecting the host.

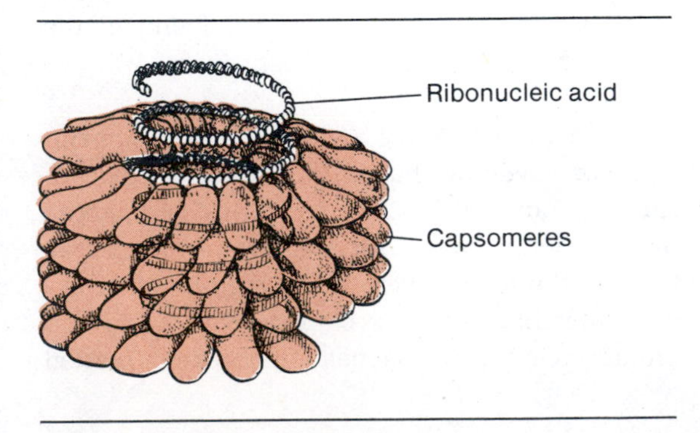

Figure 6-6 Tobacco mosaic virus.

Shape (capsid symmetry)

The shape as well as the size of the virus is related to the size of the viral genome and to the nature of the capsid proteins. The viral genome is so small that it can code for only a few structural proteins. Viruses that exhibit helical capsid symmetry have only one type of capsid protein, whereas viruses with other types of symmetry may have several types of capsid proteins. The interactions of these protein subunits (the bonding between their amino acid R groups) determines the specific geometric patterns of the virus capsid, which may be represented as helical, icosahedral, or complex.

Helical capsid symmetry. The identical protein subunits of helical capsids are arranged end to end in the shape of a rod, with each turn of the rod possessing the same number of subunits. Our understanding of this arrangement has come from x-ray diffraction studies of the TMV. A model of the virus, illustrated in Figure 6-6, demonstrates that the protein subunits form a spiral-shaped structure. Noncovalent bonds are formed between amino acids on adjacent subunits on the same turn of the spiral, as well as between adjacent turns of the spiral. The vast number of noncovalent bonds gives the capsid tremendous structural stability, similar to the hydrogen bonding between strands of the DNA molecule. A chain of RNA follows the protein helix and is located between the protein subunits. This produces an empty hole extending along the axis of the virus, which can be revealed by electron microscopic examination (Figure 6-7). The TMV is a naked virus, and its helical nucleocapsid gives it the appearance of a rod. Many helical animal viruses, such as the influenza virus, possess envelopes that gives the virus a nearly spherical shape. The nucleocapsid of the influenza virus exhibits helical symmetry, but its rod-shaped spring is flexible (this means that there are fewer forces holding the neighboring coils of the spring together). The nucleocapsid therefore is coiled within the envelope (see Figure 6-5). The flexibility of the spring could be compared to the bending of α-helical proteins due to the positioning of proline or hydroxyproline residues (see Chapter 2).

Icosahedral capsid symmetry. Spherically shaped nucleocapsids are constructed according to an icosahedral (20-sided) symmetry (Figure 6-8). Capsomeres that are connected to five neighboring capsomeres are called ***pentamers.*** Capsomeres that are connected to six neighboring

Figure 6-7 Electron micrograph of TMV particles. Stain has penetrated into axial hole *(arrows),* which has 4 nm diameter.

Courtesy R.W. Horne.

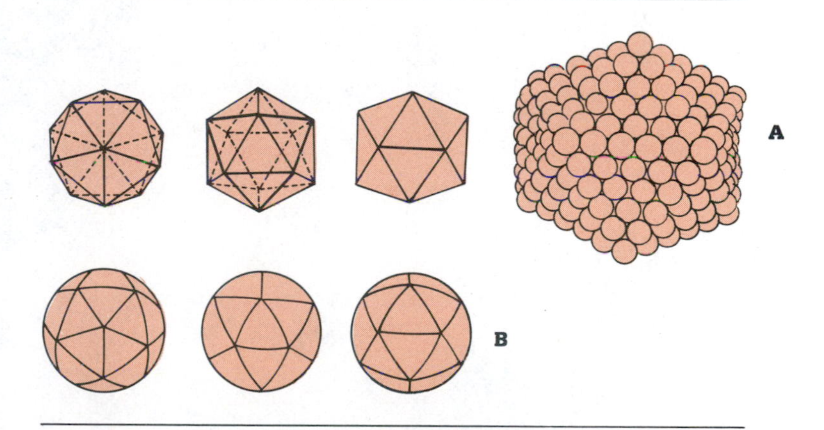

Figure 6-8 Icosahedron viewed in three positions. **A,** Icosahedra *(upper left to right)* seen in fivefold, threefold, and twofold axis of rotational symmetry. On far right is model showing fivefold axis *(arrow),* using 252 table tennis balls, of rotational symmetry and capsomere position. **B,** Spherical models with icosahedral symmetry with same axes as in **A.** Lower surface not shown in any view.

From Mattern, C. Symmetry in virions. In Nayak, D. (Ed.). Molecular biology of animal viruses. *Marcel Dekker, Inc., New York, 1979. Reprinted by Courtesy of Marcel Dekker, Inc.*

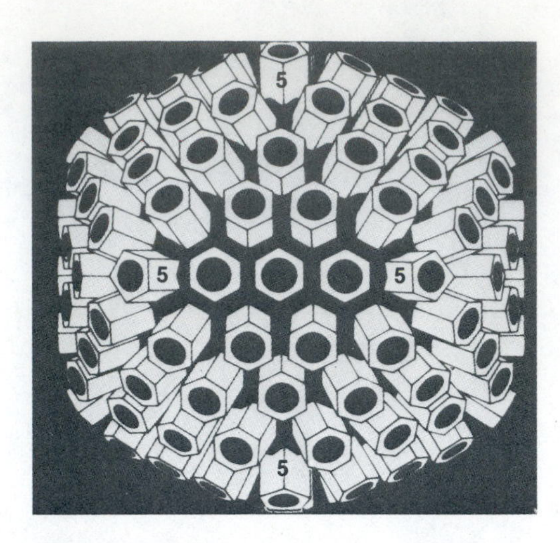

Figure 6-9 Pentamer location in icosahedron. Only four pentamers from this plane are labeled. Remaining capsomeres are hexamers (six sided).
From Horne, R.W. Virology ***15****:348, 1968.*

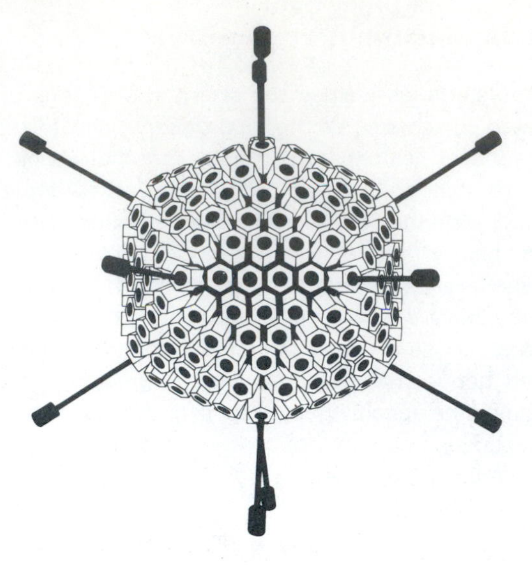

Figure 6-10 Adenovirus. There are 240 hexamers and 12 pentomers. Antenna-like fibers are called pentons and are attached to 12 pentamers.
Courtesy R.W. Horne.

A **B**

Figure 6-11 Electron micrographs of two viruses that possess icosahedral symmetry. **A,** *Alphavirus.* **B,** Adenovirus.
A, *From Enzmann, P.J. Virology* ***95****:501, 1979.* ***B,*** *Courtesy H.D. Mayor.*

capsomeres are called ***hexamers.*** A model demonstrating this arrangement in spherical viruses is illustrated in Figure 6-9. The most basic capsid is composed of 12 capsomeres. The number of capsomeres increases with larger viruses, but the basic symmetry is generally maintained. To date, the maximum number of virus capsomeres is 252, which can be found in the adenoviruses (Figure 6-10). The arrangement of the nucleic acid in the core of icosahedral viruses is not completely understood. It appears to be one tightly coiled unit in some viruses, whereas in others the nucleic acid is in coiled fragments. Electron micrographs of some of the viruses that possess icosahedral symmetry are presented in Figure 6-11.

Complex or combined symmetries. Some viruses have a symmetric construction that is complex or a combination of helical and icosahedral symmetry. Several viruses fall into this group, including poxviruses, and orf virus. The poxvirus in electron micrographs looks to be brick shaped, with filaments or tubules appearing to be wrapped about the virus particles (Figure 6-12). The best examples demonstrating combined symmetries are the T-even bacterial viruses (bacteriophages), which infect the bacterium *Escherichia coli* and are commonly referred to as ***coliphages.*** Nearly all bacterial viruses possess icosahedral symmetry, but T-even phage possesses additional components. The basic features of this virus are illustrated in Figure 6-13. The head of the virus exhibits the typical icosahedral symmetry, with the nucleic acid tightly packed into the capsid. A ***collar*** structure is located at the base of the head and connects to a ***tail*** assembly that is in the shape of a helical sheath surrounding an inner hollow core. It is through

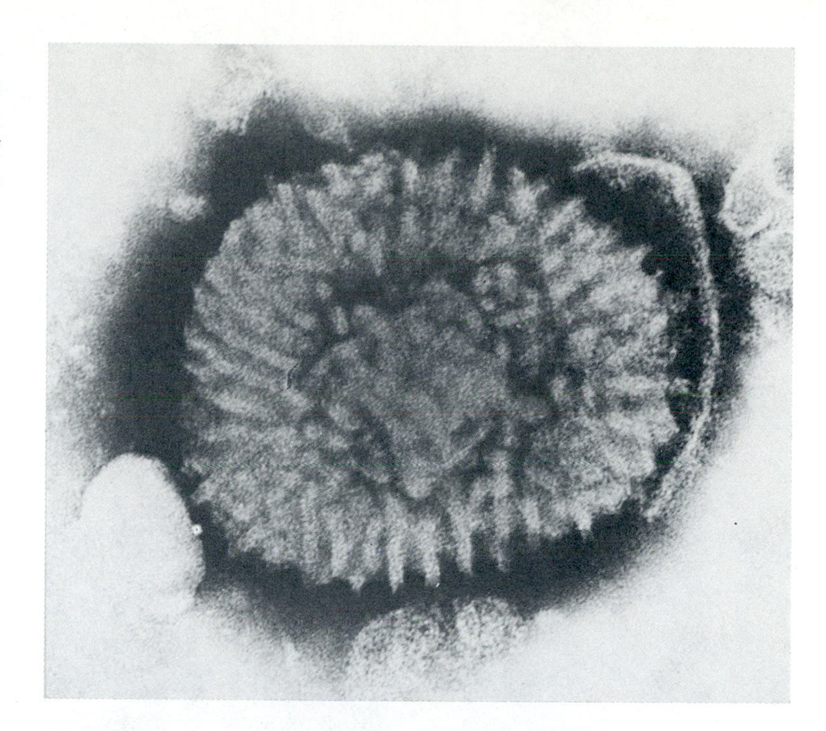

Figure 6-12 Electron micrograph of poxvirus (vaccinia) demonstrating central core and lateral bodies surrounded by outer membrane.

From Dales, S. Am. J. Med ***38**:699, 1965.*

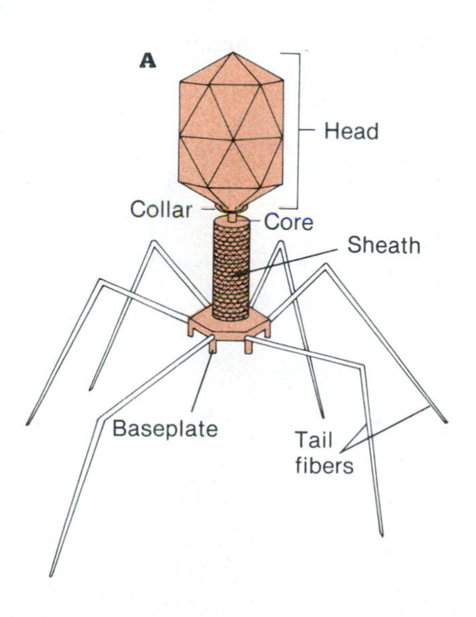

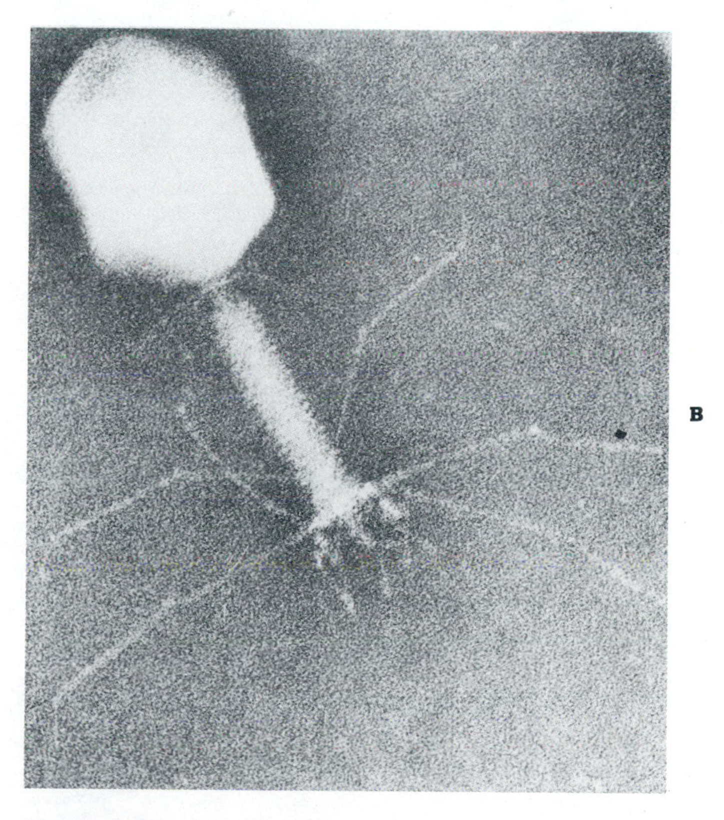

Figure 6-13 **A,** T_4 bacteriophage model. **B,** Electron micrograph of T_4 bacteriophage.

B, *From Williams, R.C., and Fisher, H.W.* An electron micrograph atlas of viruses. *Charles C Thomas, Publisher, Springfield, Ill., 1974.*

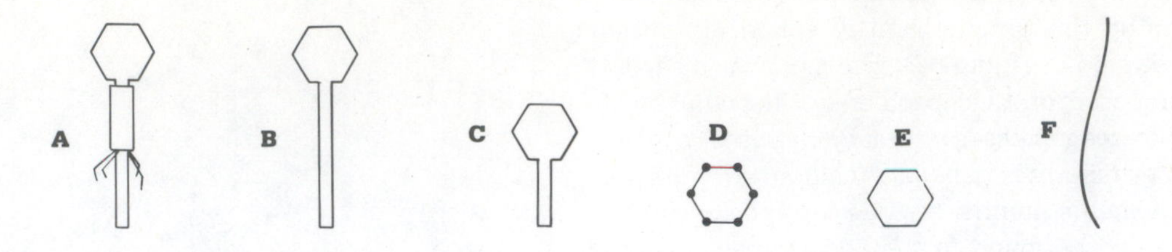

Figure 6-14 Characteristics of various morphological forms of bacteriophages. Examples are those that infect *Escherichia coli.* **A,** Contractile tail, double-stranded DNA: T_2 and T_4 bacteriophages. **B,** Long noncontractile tail, double-stranded DNA: T_1 bacteriophage and lambda (λ). **C,** Short noncontractile tail, double-stranded DNA: T_3 bacteriophage. **D,** No tail, large capsomeres, single-stranded DNA: bacteriophage $\phi\chi 174$. **E,** No tail, small capsomeres, single-stranded RNA: MS2 bacteriophage. **F,** Filamentous, no head, single-stranded DNA: M13 bacteriophage.

Redrawn from Bradley, D.E. Microbiol. Rev. ***31**:230, 1967.*

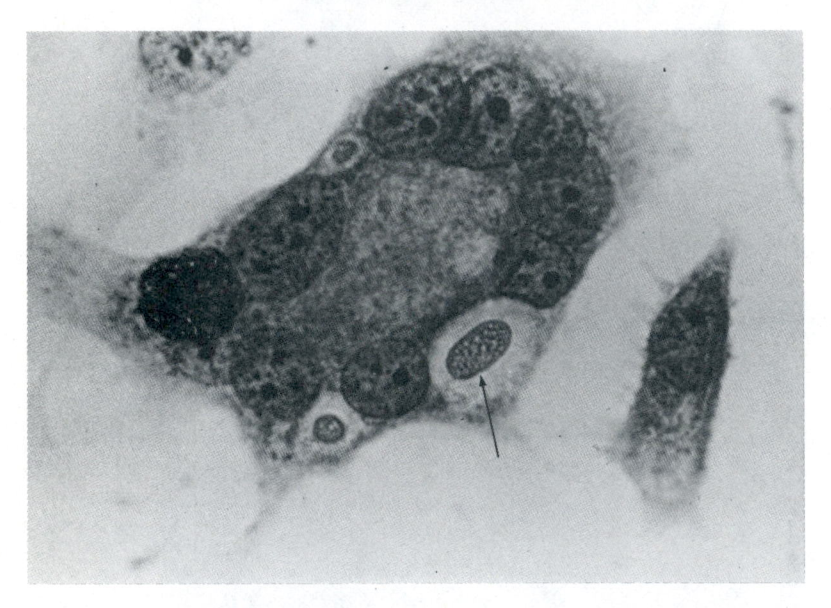

Figure 6-15 Cytopathic effects of viral infection. Multinucleated giant cells with intranuclear inclusion *(arrows).*

From Yamashiroya, H.M. In Boyd, R.F., and Marr, J. (eds.). Medical microbiology. *Little, Brown & Co., Boston, 1980.*

this core that DNA, located in the head, can be injected into the bacterium during the infection process. See Chapter 16 for details of this process. A hexagonal *baseplate* and ***tail fibers,*** which function in the adsorption of virus to the bacterial cell wall, are located at the proximal end of the sheath. The basic shapes of other bacterial viruses are illustrated in Figure 6-14.

CULTIVATION

Viruses cannot be cultivated in nutrient media, but they can be cultivated in their appropriate living hosts, such as bacteria, fungi, plants, and animals. For those viruses that infect higher eukaryotes the simplest technique is to cultivate them in tissue or cell culture. A variety of animal cells may be infected by one virus, or the virus may be restricted to growth in one cell type. Hu-

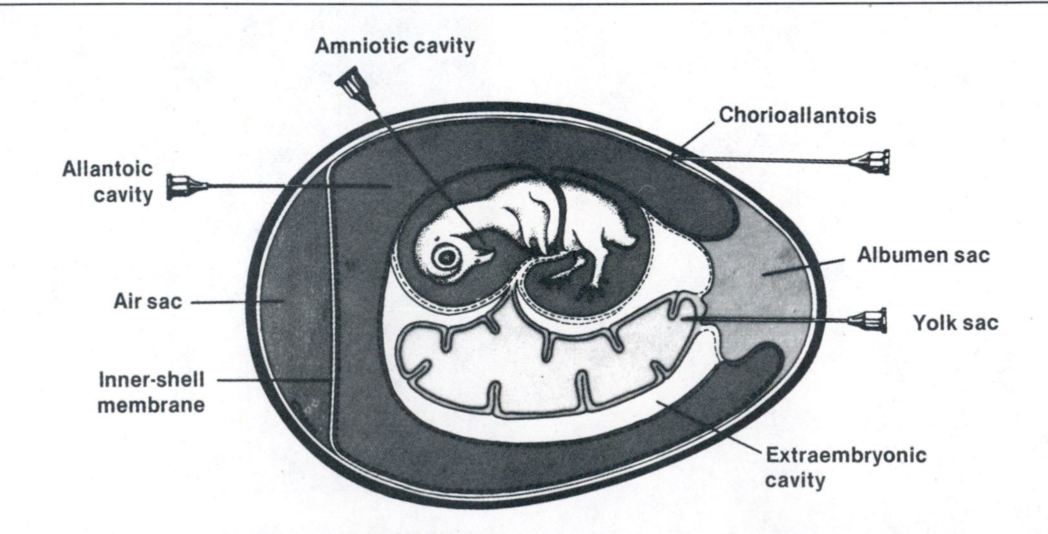

Figure 6-16 Possible sites of viral inoculation into embryonated egg.

Courtesy Audiovisual Service, Department of Pathology, University of Illinois Medical Center, Chicago.

man and monkey kidney cells, for example, support the growth of many animal viruses. Cultivation can be used to increase the number of viral particles so that they can be used for biochemical or biophysical studies and in the preparation of vaccines. Growth in tissue or cell culture also permits the isolation of virus from infected patients and thus serves as a diagnostic tool. Cells are first grown in a flask or dish containing the appropriate medium, and as they divide they form a one-cell-thick layer (monolayer) over the glass or plastic surface. Virus is then added to this culture, and viral growth is monitored by some change in the host cells called ***cytopathic effect.*** Cytopathic effects are observable with the compound microscope and sometimes with the naked eye. They may take the form of cell rounding, fusion of cells into giant cells, or the formation of ***inclusion bodies*** (Figure 6-15). Inclusion bodies, which are usually observed following some staining procedure, are actually viral proteins, nucleic acids, or some unusual cell product that accumulates in the cell during infection. Some inclusion bodies are so distinctive, for example, in the case of rabies virus infection, that they are of diagnostic significance. A few viruses that are difficult to grow in cell culture can be cultivated in an embryonated egg. Virus can be introduced into the chorioallantoic membrane, amniotic sac, allantoic sac, or yolk sac (Figure 6-16). Viral multiplication is indicated by death of the embryo, the formation of lesions called ***pocks*** on the various membranes (Figure 6-17), or by assaying the fluid from infected eggs.

Virus particles can be separated from the host cells and concentrated following the infection process. Separation of virus from extraneous host material can be accomplished using such techniques as chromatography, centrifugation, and electrophoresis. Once the viral particles have been concentrated, it may be necessary to determine those particles that are infectious and those that are not; that is, one must determine the titer (quantity) of the virus suspension. Infectious virus can be treated with inactivating agents, such as phenol or formaldehyde, if it is to be used in certain vaccines.

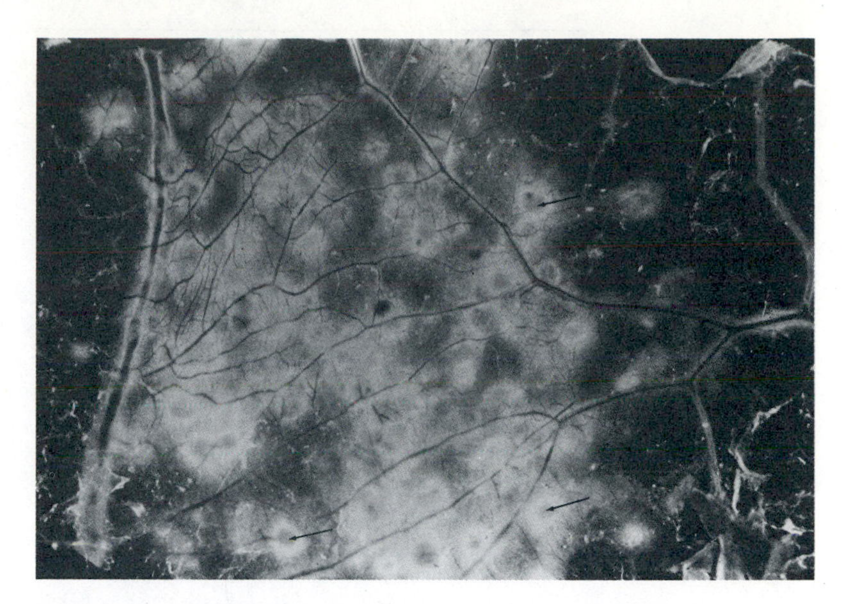

Figure 6-17 Numerous pocks *(arrows)* produced on chorioallantoic membrane of embryonated hen's egg after 5 days of incubation at 35.5° C.

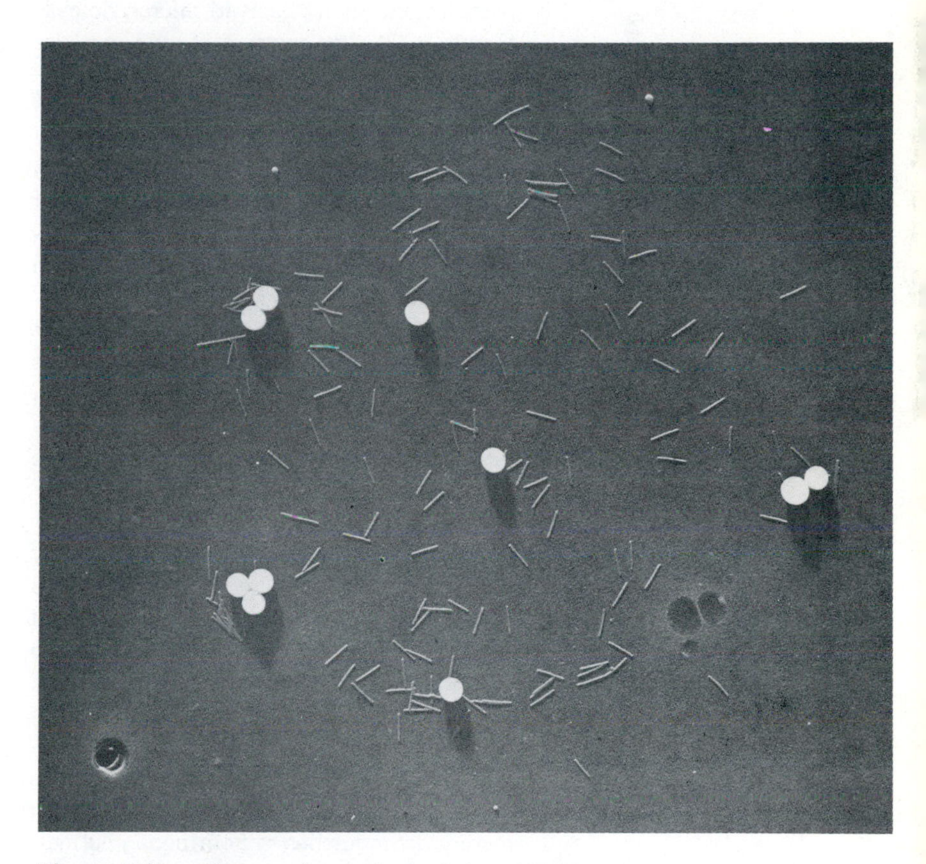

Figure 6-18 Electron micrograph of rod-shaped TMV particle suspension mixed with spherical latex particles.
From Symington, J., Commoner, B., and Yamada, M. Natl. Acad. Sci. USA ***48:****1675, 1962.*

ASSAY OF VIRUSES

Determining the number of viable or nonviable particles in viral suspensions is not as easy as the colony-counting techniques or other measurements used in the enumeration of bacteria. Three of the most useful techniques for counting viruses are:

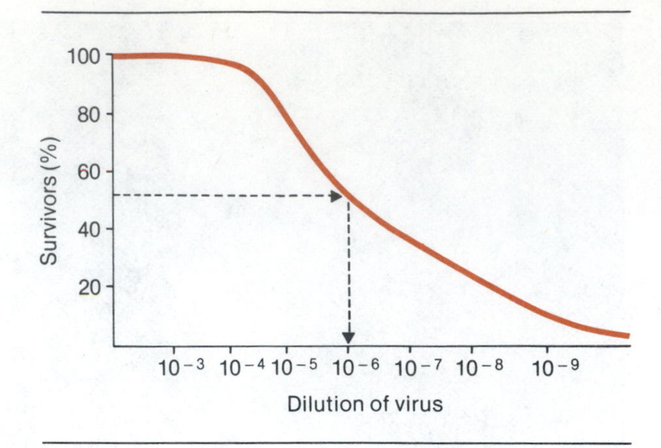

Figure 6-19 Hypothetical plot determining lethal dose (LD_{50}) for 50% of animals inoculated with dilution of virus. Extrapolation *(dashed lines)* of curve shows that at 10^{-6} dilution of virus 50% of animals were killed.

1. Counting the total number of particles that look like virus with the aid of the electron microscope. Virus particles can be mixed with a known diluted suspension of latex particles and then observed microscopically (Figure 6-18). The number of viral particles observed times the dilution factor for the latex particles provides a measure of the concentration of virus. For example, if there were 10^9 latex spheres per milliliter of suspension and the electron microscope count revealed 20 virus particles for each latex sphere, then the number of viral particles in the original suspension would be 20×10^9/ml or 2×10^{10}/ml. This technique does not differentiate infectious from noninfectious virus.
2. Agglutination of red blood cells (hemagglutination) is a characteristic property of the proteins found on some viruses. Hemagglutination is usually due to the intact virus but may also result from proteins released by denaturation of virus or by viral proteins bound to the membrane of virus-infected cells. The hemagglutinin acts to join two red blood cells by forming a bridge between them. Large aggregates of red blood cells will form, if sufficient hemagglutinin is present, and fall to the bottom of the tube. The characteristic pattern of hemagglutination readily distinguishes it from nonagglutinated red blood cells. In an actual test virus-containing fluid is first prepared at various dilutions. The highest dilution of virus that can still cause agglutination of red blood cells is called the hemagglutination titer. Both infectious and noninfectious particles as well as fragments of infected cells may cause agglutination of red blood cells, but differentiation between the former two is not possible with this technique.
3. Biological effects such as animal death, cytopathic effects, the number of animals infected, or the number of pocks produced can be quantified. The virus titer is expressed as the ***infectious dose (ID_{50})*** or ***lethal dose (LD_{50})*** that causes infection or death, respectively, in 50% of the animals inoculated. For example, suppose we have prepared a virus suspension of unknown concentration. We can prepare dilutions of the virus ranging from 10^{-3} to 10^{-8}; that is, we have diluted the suspension from 1000 to 100,000,000 times. Next, an aliquot of each dilution is used to infect 100 animals. A period of time is allowed for the virus to cause a positive response in the animals. The number of animals showing a positive response at each dilution is plotted (Figure 6-19), and the dilution at which 50% of the animals show a positive response is extrapolated.

The plaque assay is one of the most widely used assays for animal virus. Monolayers of cells are inoculated with various dilutions of virus. After several days the cytopathic effects can be macroscopically distinguished from uninfected cells (see Figure 29-1). The sites of infection are referred to as plaque-forming units (PFUs). There is a direct proportionality between the number of virus particles and the number of PFUs.

Bacterial viruses also produce plaques on susceptible bacteria. The plaques appear as areas of clearing on a lawn of bacteria (see Figure 16-5). A more detailed discussion of the bacterial replication process is presented in Chapter 16.

EFFECT OF CHEMICAL AND PHYSICAL AGENTS ON VIRUSES

Because of the chemical simplicity of viruses, only a few components are susceptible to the effects of chemical and physical agents, such as the nucleic acid, protein coat, or lipid membrane of enveloped viruses. Some of the chemical and physical agents are of practical value in terms of their ability to inactivate virus and prevent the transmission of virus, as well as their use in the preparation of vaccines. Studies in which the structure and arrangement of various viral components are being explored also require the use of some of these agents. Antiviral agents used in the treatment of human disease are discussed in Chapter 27.

Chemical agents

Many chemical agents can inactivate viruses, but most cannot be used in the treatment of viral infections because of their deleterious effects on host tissue. Some of the more commonly used agents to inactivate viruses are:

1. Lipid solvents. Agents such as ether, chloroform, and anionic detergents solubilize lipid, a component of the membrane of enveloped viruses. Nearly all enveloped viruses are inactivated by lipid solvents.
2. Phenol. Phenol is commonly used to denature proteins and is active against the capsid proteins and protein components of enveloped viruses. The peplomers of the influenza virus, for example, are destroyed by phenol, which is a component of several disinfectants (see Chapter 27).
3. Formaldehyde. Formaldehyde reacts with the amino groups of nucleic acids and proteins. Formaldehyde does not drastically affect the antigenic properties of viruses and for this reason has been used in the preparation of vaccines.
4. Other agents. Other disinfecting agents that can inactivate viruses include 70% to 90% isopropyl or ethyl alcohol, iodophors, sodium hypochlorite, and 2% glutaraldehyde. These agents are also discussed in Chapter 27.

Photoinactivation. Certain dyes such as neutral red, proflavine, and toluidine blue can penetrate the capsid of some viruses and bind to the nucleic acid. A ***photooxidation*** process, which renders the nucleic acid inactive, takes place if the virus is subjected to visible light. Neutral red has been used to treat localized herpesvirus (coldsores) infections in humans. Such treatment, however, has been discontinued because of the dye's allergenicity and oncogenic potential when used in animal studies.

Physical agents

Temperature. Viruses are usually preserved by storage at subfreezing temperatures (−70° C in a mechanical freezer or −196° C in liquid nitrogen) but are inactivated by temperatures of 50° to 70° C for 1 hour. Heat alters the conformation of proteins and nucleic acids by causing them to unfold and become denatured. Sterilization temperatures, such as autoclaving at 121° C for 15 minutes at 15 lb/in^2 pressure or dry air sterilization at 160° to 180° C for 2 to 3 hours will destroy all viruses.

Radiation. X-rays, ultraviolet (UV) light, and some ionizing particles can inactivate viruses. X-rays impart energy to the medium surrounding the nucleic acid of the virus, which causes the stripping of electrons from some elements and their capture by others. These interactions result in the formation of ions and various radicals that can affect sensitive components of the nucleic acid. In addition to these indirect effects, it is believed that x-rays may also directly affect the nucleic acid by causing breaks in the sugar-phosphate chains. The DNA molecule remains intact if a break occurs in only one strand, but a second break directly opposite the first break will result in breakage of the entire molecule. UV light also affects nucleic acids and does so directly. The bases on the nucleic acid have a maximum absorption in the UV spectrum at 2600 Å. One of the effects of UV absorption is the formation of dimers, which result from the abnormal bonding between adjacent thymine residues. Dimerization causes an inflexibility within the molecule that twists the molecule out of shape, making replication difficult. See Chapter 9 for a discussion of dimerization and mutant formation.

PLANT VIRUSES

There are records of plant diseases as early as the sixteenth century—long before viruses were recognized as infectious agents. Variegation in the color of tulips was observed in 1570 and in 1670, and it was suggested that this unusual characteristic was due to a transmissible disease. It was not until the twentieth century that tulip variegation was discovered to be due to a virus transmitted by aphids. Despite their early discovery, plant viruses (other than TMV) have not received as much attention as animal or bacterial viruses. They are more difficult to isolate and cultivate than animal and bacterial viruses and, except for a few types, much less is known concerning their morphology and ultrastructure. Viral diseases in plants are discussed in Chapter 35.

Characteristics

Size and shape. Plant viruses, like other viruses, are composed of a protein coat and an inner core of nucleic acid. They are of two types: helical, or rod-shaped, particles and icosahedral particles. The size of rod-shaped particles ranges from 15 nm in diameter by 300 nm in length for the TMV to 66 nm in diameter by 227 nm in length for the larger lettuce necrotic virus. Much more is known about the morphology of rod-shaped particles

A

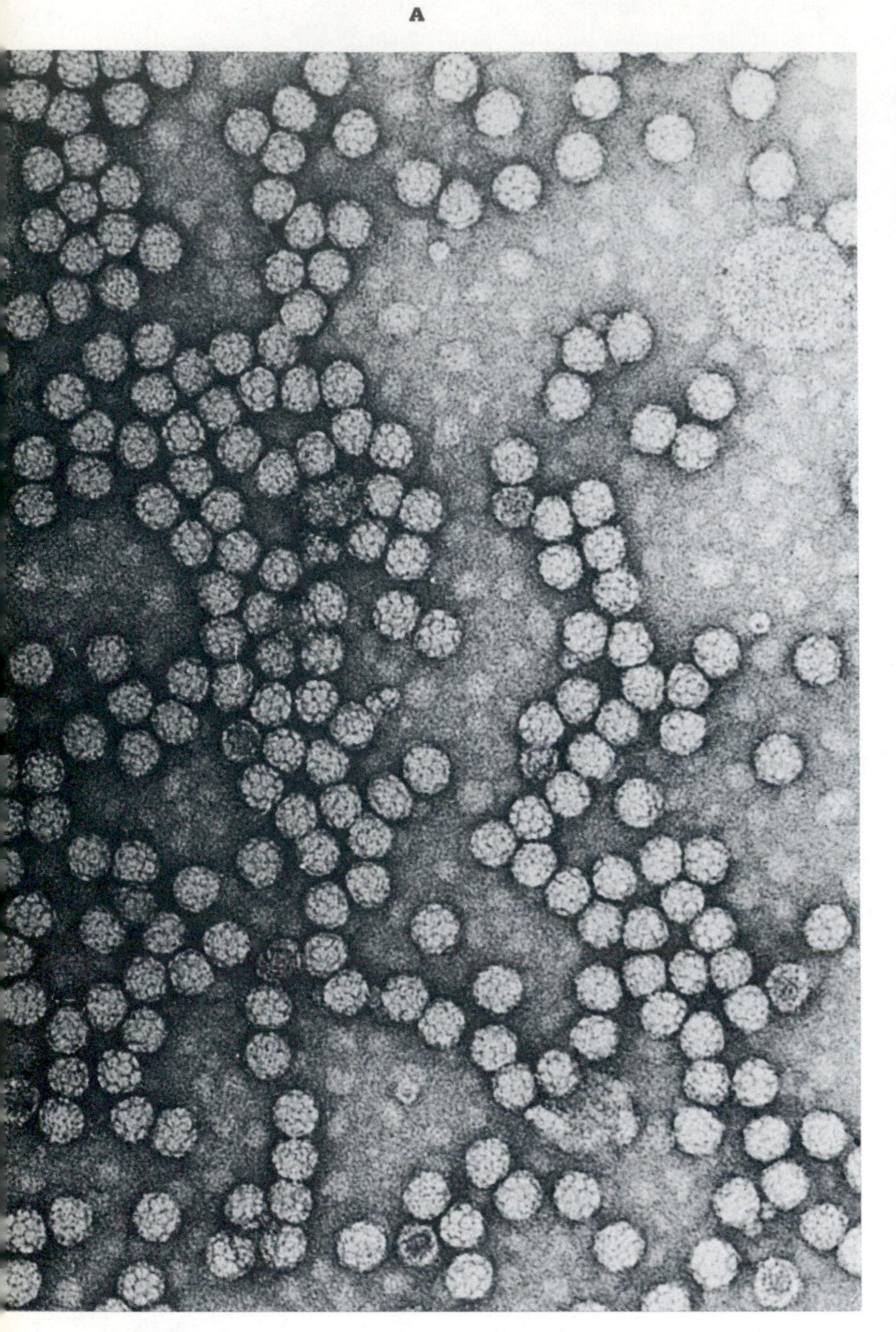

B

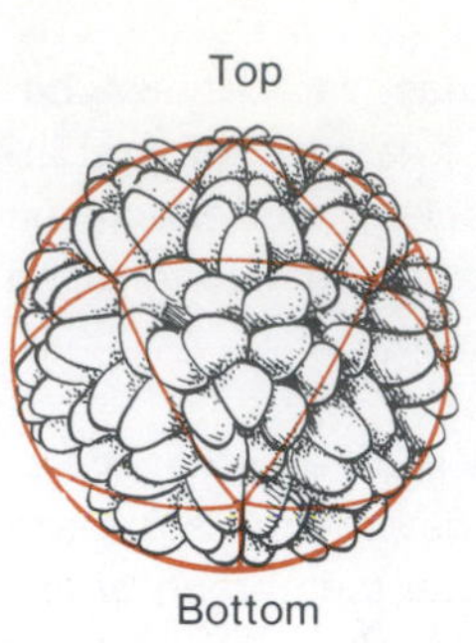

Figure 6-20 Turnip yellow mosaic virus (TYMV). **A,** Electron micrograph of TYMV. **B,** Model of how outer surface of TYMV is believed to appear.

A, Courtesy L. Amos. B, Based on model by Finch, J.T., and Klug, A. J. Mol. Biol. ***15:****315, 1966. Copyright: Academic Press, Inc. (London), Ltd.*

than about the morphology of icosahedral viruses, which range from 30 to 60 nm in diameter and are difficult to distinguish from cellular structures. The basic morphology of rod-shaped viruses is discussed earlier in this chapter under "Helical Capsid Symmetry." Although, unfortunately, less information is available concerning the morphology of icosahedral plant viruses, one of the more widely studied icosahedral viruses is the turnip yellow mosaic virus (Figure 6-20).

Nucleic acid. There are over 300 different plant viruses belonging to approxmately 20 taxonomic groups, and 18 of these groups possess an RNA genome that is single stranded. The two groups containing DNA viruses are the caulimoviruses (double-stranded DNA) and the geminiviruses (single-stranded DNA). The manner in which RNA is associated with the capsid protein is known only for some of the rod-shaped viruses such as the TMV (see Fig 6-6).

It was discovered in 1956 that, unlike many other viruses, the nucleic acid of plant viruses was by itself infectious. Purification of nucleic acid and storage at subfreezing temperatures for over 2 years did not reduce its infectivity. Experiments comparing the infectivity of naked viral nucleic acid and intact virions demonstrated that the lesions, characteristic of infection, appeared more rapidly when naked nucleic acid was used. It is assumed that the time required to uncoat intact viral particles is responsible for the delay in the appearance of cytopathic effects.

Cultivation. Plant viruses can be cultivated by both tissue culture and cell culture. Growth of virus in tissue culture is best obtained if the tissue, for example, root or stem tips, is removed from infected plants and cultivated in nutrient media in the laboratory. Infection of healthy tissue in vitro can be brought about by pricking an inoculum of virus into the tissue by means of pins or through the use of an abrasive material. It is also possible to infect the tissue culture by using the natural vector of the virus. Zoospores of certain fungi, for example, are known to infect plants and transmit virus. Insect vectors can also be used to infect cultures in vitro.

The use of cell culture for cultivation of plant viruses has not reached the state of the art that it has for animal virus cultivation. Isolation of single plant cells for cultivation in the laboratory is more difficult than for animal cells. One technique that has proved valuable is the use of monolayers of insect cells isolated from such vectors as leaf hoppers and aphids.

Testing for virus. A rapid technique for determining whether the sap of a plant is infected by virus is an important asset in horticulture and has many economic implications. The sap of infected plants often contains only a few viral particles, and microscopic detection, or use of cell or tissue cultures, is futile.* ***Indicator plants,*** such as tobacco (Figure 6-21) or spinach, have been used successfully to indicate infection by viruses that infect the same or different species of a plant. Indicator plants can be used to determine what percentage of a crop is virus infected; this is especially important for those viruses that cause ***latent*** infection (they do not initially elicit symptoms in their natural host and require inducement). Often when a latent virus is transferred from its natural host to another plant species, such as an indicator plant, symptoms of infection appear within a few days. This technique has been especially important when grafting procedures are being considered and also in testing potatoes and stone fruits (pears, plums, peaches, etc.) for latent virus. The indicator plant should have large leaves and demonstrate discrete localized lesions following infection. In some instances virus can actually be assayed on the indicator plant if the lesions are discrete and do not coalesce.

Figure 6-21 Tobacco leaf used as indicator plant for viral infection. **A,** Uninfected leaf. **B,** Infection by lettuce speckle mottle virus. **C,** Infection by beet western yellow virus. **D,** Infection by both viruses.
From Falk, B.W. Phytopathology ***69:****612, 1979.*

Viroids. Viroids are viruslike agents that are smaller than ordinary viruses and do not possess a protein coat. The discovery of viroids, like so many other scientific discoveries, reads like a detective story. Spindle tuber disease in potatoes had been recognized by pathologists since the early 1920s, but no infectious agent could be recovered from diseased tissue. The disease could be transmitted to healthy tissue, even tomatoes, by using extracts of infected tissue. A virus was suspected of causing the disease, and extracts of infected potato tissue were centrifuged, first at low speeds to remove cellular debris, and then at higher speeds to pellet the infectious agent. Infectious activity remained, however, in the supernatant and not in the pellet. The reason for this would become clear when later studies revealed that the agent had a molecular weight of 130,000 (most viruses have a molecular weight of 1 million). The infectious supernatant material was insensitive to the chemical agents that solubilize lipid and was also insensitive to phenol. These results indicated that the agent did not possess an envelope or, for that matter, a protein coat. The infectious material was sensitive to RNase but not DNase. Chemical and electron microscopic studies have clearly shown that the viruslike agent is a naked RNA molecule devoid of a protein coat. It is a closed, circular, single-stranded RNA molecule in which short regions show ''in-strand'' base pairing while the other regions remain single stranded. No other agent in nature has been found to possess these characteristics. Viroids are known to cause disease in six other plants, including cucumbers and chry-

* The number of virus particles required to infect plant cells is often very large and ranges from 10^5 to 10^9 particles.

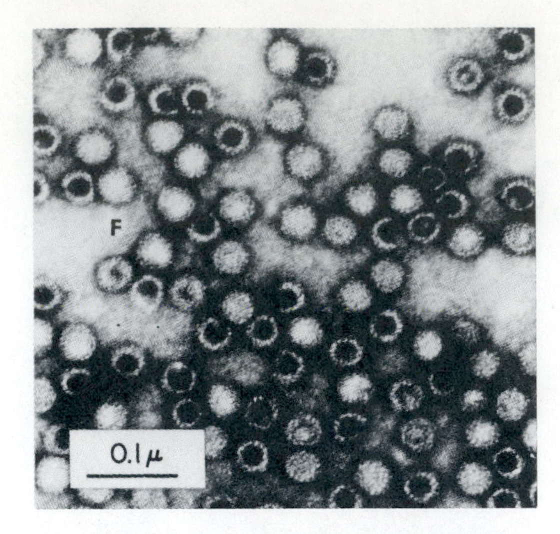

Figure 6-22 Electron micrograph of virus isolated from fungus *Penicillium stoloniferum*.
From Bozarth, R.F., et.al. Virology **45:**516, 1971.

santhemums. How viroids interfere with the growth process in plants is still a mystery. Viroids have not been found in animal cells but it has been suggested that they may be responsible for degenerative diseases such as Kuru and Creutzfeldt-Jakob disease in humans.

OTHER VIRUS-HOST ASSOCIATIONS

There is much less information available on the structure, composition, and physiology of infection by viruses that infect or are associated with other organisms such as fungi, protozoa, and some insects. The information that is available is most interesting because it points out some unusual relationships that exist between virus and host.

Fungal viruses

Until 1960 the existence of fungal viruses (mycoviruses) was suspected, but none had been visualized. During the search for an antiviral agent that would combat such diseases as paralytic polio, an unusual chemical was found in the mold *Penicillium*. The chemical was capable of stimulating an antiviral compound called ***interferon*** in animal cells. Investigators later discovered that the interferon inducer was a double-stranded RNA particle. All fungi examined so far contain double-stranded RNA viruses, most of which are spherical with diameters of 25 to 48 nm (Figure 6-22). The free virus particles isolated from fungi are not infectious, nor are they considered to be defective virus particles. The virus probably exists as a latent or persistent virus in the fungal cell and does not replicate until induced by some environmental factor. The genome of the virus is segmented, and more than one double-stranded RNA species is distributed among the viral particles in the cytoplasm. Genome segmentation is characteristic of all viruses containing double-stranded RNA.

Fungal viruses are usually found in aggregates in older hyphal cells and during vegetative growth are apparently transmitted by cytoplasmic exchange (plasmogamy) with adjacent cells. This type of transmission is an alternative to the usual type of transmission in which attachment to a cell surface is required. Virus can also be found in fungal spores, which represents a means of transmission during the nonvegetative phases of the fungal life cycle.

Fungal viruses are usually considered avirulent, but their presence in tissue causes structural modification and reduced growth of mushrooms. Another example of virus alteration of host phenotype is exemplified in the ***"killer"*** system of the yeast *Saccharomyces cerevisiae*. Killer strains of yeast secrete a toxin that is lethal to sensitive strains of yeast but to which the killer is immune (Figure 6-23). The killer factor is a double-stranded RNA, probably a fungal virus, and is believed to be controlled by certain chromosomal genes of the host. One set of chromosomal genes maintains the killer characteristic, and a mutation in this set of genes can result in loss of the killer virus, or there can be an inability of the yeast to grow. Another set of genes controls the ability to secrete the toxin as well as maintain resistance to toxin.

Insect viruses

There are over 300 insect viruses. They are either rod shaped or have icosahedral symmetry. Some have caused epizooitics (epidemics among animals) that have ravaged large populations. The majority of insect viruses differ from animal and plant viruses in that individual virus particles are encased either singly or in large numbers in protein crystals that are insoluble in water. The protein-encased viral particles are called ***polyhedral inclusion bodies*** (Figure 6-24). Polyhedroses may form in the nucleus or cytoplasm of the infected insect. Nuclear polyhedroses multiply primarily in skin, blood, and tracheal tissue, and cytoplasmic polyhedroses multiply in epithelial cells of the insect gut. These protein crystals are produced in large numbers in infected insect larvae and are released when the insect dies and disintegrates. The protein crystals can remain infectious for years in the environment because they are insoluble in water when released from the host.

Viruses are being used to control certain insect pests, such as the sawfly in Canada. This pest was accidentally introduced into Canada from Europe, and, without any natural enemies, it decimated forests. Viruses are now being used to control the sawfly. One viral preparation (Virion H) is being used in the United States to control the cotton bollworm and the tobacco budworm. The techniques for propagation of virus for biological control are infection of insect larvae and infection of stable insect tissue culture lines. Once the virus is purified, it can be administered with a spraying machine for fields, or it can be used for individual treatment of trees.

Some insects, such as bloodsucking arthropods (mites, ticks, or mosquitoes, for example), acquire virus by feeding on vertebrate animals. The virus usually multiplies in the salivary glands but does no obvious damage to the arthropod. The virus can be transmitted to vertebrates, including humans. The viruses causing such infections belong to several taxonomic groups, the majority of which are togaviruses.

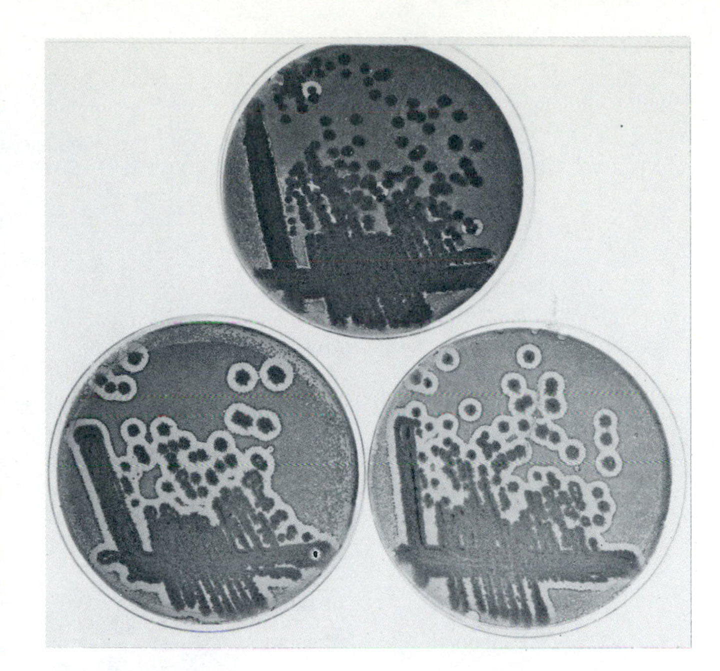

Figure 6-23 Killer system in yeast. Yeast strain sensitive to "killer" factor is seeded into agar over which is streaked test strains of yeast. Upper photograph shows that killer-sensitive yeast strain, which covers surface of plate, grows up to colonies of test strain. Test strain therefore produces no diffusible "killer" factor. Lower photograph shows clear areas of no growth around test strain, indicating that toxic diffusible factors had been produced.

Courtesy J.R. Preer, Jr.

CLASSIFICATION

Living organisms such as bacteria and fungi are classified based on their many biological properties, with size and shape playing minor roles. Biological properties cannot be used in viral classification schemes, since the virus relies on the host's biological properties for its own development. Logically, size, shape, and structure of viral components would seem to be the most useful characteristics for classification. Advances in microscopy and biochemical techniques have provided a more extensive classification scheme, but they are of recent origin, and many viruses have eluded detailed microscopic or biochemical characterization. The principal viral characteristics that are being used today for classification include type of nucleic acid, capsid symmetry, number of capsomeres, diameter of the virus, molecular weight of nucleic acid, and number of genes. Table 6-1 illustrates a typical scheme for the classification of certain plant, animal, and bacterial viruses. A detailed scheme for the classification of viruses that infect humans is presented in Chapter 29.

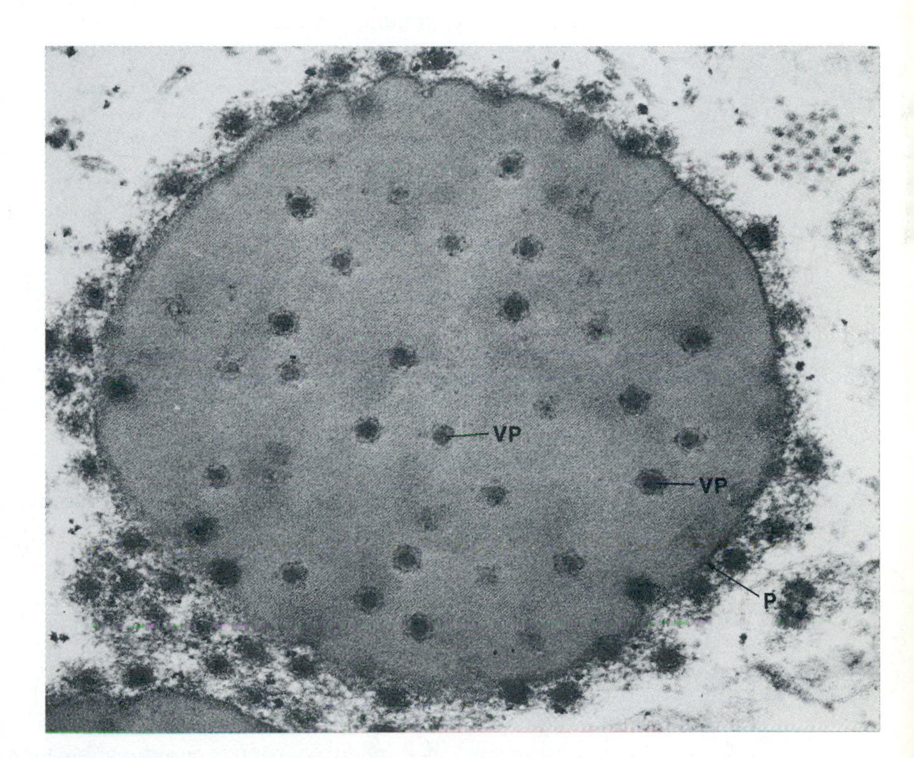

Figure 6-24 Cytoplasmic polyhedroses (insect virus). Electron micrograph of midgut cell in early stage of infection. (×150,000.) Polyhedroses *(P)* contain dense bodies, which are viral particles *(VP)*.

From Arnott, H.J. J. Ultrastruct. Res. ***24**:479, 1968.*

Table 6-1 Scheme for the classification of some bacterial, plant, and animal viruses

NUCLEIC ACID	MOL. WT. × 10^6	CAPSID SYMMETRY	ENVELOPE	NO. OF CAPSOMERES	PHYSICAL TYPE OF NUCLEIC ACID	NO. OF GENES	EXAMPLE	DISEASE OR COMMON NAME
DNA	200	Helical	Present	—	DS	400	Poxvirus	Smallpox (animal virus)
	1.6	Icosahedral	Absent	12	SS	7	$\phi\chi$174	Bacterial virus (bacteriophage)
	30	Complex (icosahedral head, helical tail)	Absent	32	DS	—	T phage	Bacterial virus (bacteriophage)
	23	Icosahedral	Absent	252	DS	50	Adenovirus	Respiratory disease
RNA	2	Helical	Absent	—	SS	10	Tobacco mosaic virus (TMV)	TMV disease
	4	Helical	Present	—	SS	15	Influenza virus	Influenza
	15	Icosahedral	Absent	—	DS	40	*Reovirus*	Gastroenteritis
	2	Icosahedral	Absent	—	SS	—	Napovirus	Turnip yellow mosaic virus (TYMV)

DS, Double stranded; *SS*, single stranded; (–), unknown.

SUMMARY

Viruses are submicroscopic, obligate, intracellular parasites that rely on their host for energy and nutrients. They range in size from 18 nm to over 250 nm. Chemically, viruses consist of a core of nucleic acid, either RNA or DNA, surrounded by a protein coat. The coat protein, or capsid, is composed of individual protein units called capsomeres. Some viruses also contain a lipid envelope, which surrounds the viral capsid. Some lipid envelopes contain spikes or protein called peplomers. The complete infective viral particle is called a virion.

The shape of the viral capsid conforms to certain geometric patterns: helical, icosahedral, or complex. Helical capsids are usually associated with rod-shaped viruses, whereas icosahedral capsids are found in spherical viruses. Complex symmetry is usually a combination of helical and icosahedral capsid symmetries.

Viruses can be grown in tissue culture or embryonated eggs. Growth often results in cellular degeneration, which can be detected as a cytopathic effect. Cytopathic effects are visible with the naked eye or with a compound microscope. The number of viruses in a suspension can be determined by applying certain techniques: electron microscopy, agglutination of red blood cells, or measurement of cytopathic effects.

Viruses can be inactivated by various chemical and physical agents. Chemicals such as formaldehyde, phenol, and lipid solvents can inactivate viruses. Photoinactivation following staining of viruses can also be used. Radiation can inactivate viruses directly via interaction with the viral nucleic acid or indirectly by generating oxidizing agents as it interacts with the cytoplasm of the cell.

Plant viruses are rod shaped or spherical. Rod-shaped viruses can be as large as 227 nm. Most plant viruses contain RNA, and the nucleic acid by itself is infectious, a characteristic that distinguishes it from nearly all other viruses. Virus can be assayed by determining cytopathic effects on certain indicator plant leaves. Certain viruslike particles are without a capsid but are infectious for certain plants and are called ***viroids.***

Fungal viruses all contain double-stranded RNA and are spherical. Nearly all appear to be avirulent. It is believed that they are probably defective and exist in a latent state in the cell.

Insect viruses are rod shaped or icosahedral. Viral particles are encased in protein and are referred to as polyhedral inclusion bodies. They may be found in the nucleus or cytoplasm of infected insect cells. Some insect viruses are used to control insect populations that devastate crops or other plants.

Viruses are classified according to certain parameters: type of nucleic acid, capsid symmetry, capsomere number, size of virus, and molecular weight of nucleic acid.

Study questions

1. What biological and nonbiological properties are characteristic of viruses?
2. Compare the size of viruses with bacteria, fungi, algae, and protozoa.
3. What is the relationship of capsids, capsomeres, and nucleocapsids in viral structure?
4. What is the difference, if any, between a virus and a virion?
5. In addition to its protein coat and nucleic acid core, what other structural or biochemical components may be present in some animal viruses? In bacterial viruses?
6. Diagrammatically show how the nucleocapsid of a helical virus could be in the shape of a filament with the virus appearing spherical.
7. How can you know if a virus is present in a cell culture without microscopic aids?
8. What methods are used to assay large numbers of viral particles?
9. Name four chemical agents that can be used in disinfecting procedures to destroy virus?
10. What type of nucleic acid is present in animal viruses, bacterial viruses, plant viruses, and fungal viruses?
11. How do viroids differ from "normal" viruses?
12. In terms of parasitism, how do fungal viruses differ from all other viruses?
13. What single characteristic distinguishes insect viruses from plant and animal viruses? Of what value is this to the virus?
14. What is a polyhedral inclusion body?
15. What properties are presently used in the differentiation and classification of viruses?

Selected readings

Books

Diener, T.O. *Viroids and viroid diseases.* John Wiley & Sons, Inc., New York, 1979.

Dulbecco, R., and Ginsberg, H.S. *Virology.* Harper & Row, Publishers, Inc., New York, 1980.

Fenner, F., and White, D.O. *Animal virology.* Academic Press, Inc., New York, 1976.

Fraenkel-Conrat, H., and Wagner, R.R. *Reproduction, bacterial DNA viruses.* Plenum Press, New York, 1977.

Fraenkel-Conrat, H., and Wagner R.R. (Eds.). *Newly characterized vertebrate viruses.* Plenum Press, New York, 1979.

Horne, R.W. *Virus structure.* Academic Press, Inc., New York, 1974.

Joklik, W.K. (Ed.). *Principles of animal virology.* Appleton-Century-Crofts, New York, 1980.

Knight, C.A. *Chemistry of viruses* (2nd ed.). Springer-Verlag, New York, Inc., New York, 1975.

Levy, H.B. *Biochemistry of viruses.* Marcel Dekker, Inc., New York, 1969.

Luria, S.E., Darnell, J.E., Jr., and Campbell, A. *General virology* (3rd. ed). John Wiley & Sons, Inc., New York, 1978.

Mathews, C.K. *Bacteriophage biochemistry.* Van Nostrand Reinhold Co., Inc., New York, 1971.

Smith, K.M. *Plant viruses* (6th ed.). Chapman & Hall, Ltd., London, 1977.

Smith, K.M. *Introduction to virology.* John Wiley & Sons, Inc., New York, 1980.

Stent, G.S. *Molecular biology of bacterial viruses.* W.H. Freeman & Co., Publishers, San Francisco, 1963.

Viruses and plasmids in fungi and protozoa. In Schlessinger, D. (Ed). *Microbiology—1977.* American Society for Microbiology, Washington, D.C., 1977.

Journal article

Durham, A.C.H., and Finch, J.T. Structure and roles of polymorphic forms of tobacco mosaic virus. *J. Mol. Biol.* **67:**307, 1972.

PART 2

MICROBIAL STRUCTURE AND FUNCTION

Freeze-etch of *Trichomonas vaginalis.* (Approx. ×33,000.)

S.C. Holt, University of Massachusetts/BPS.

Chapter 7 PROKARYOTIC ULTRASTRUCTURE AND FUNCTION

For many years microbiology and many other areas of biology were merely descriptive sciences in which the morphological properties of the cell were observed either macroscopically or at the very most with the compound microscope. The advent of the electron microscope, along with advances made in molecular biology and biochemistry, have permitted scientists to probe more deeply into the molecular structure of microorganisms. Ultrastructural studies have revealed morphological characteristics that were either invisible with the light microscope or only speculated on from biochemical studies. The structure of the cell wall and that of the cytoplasmic membrane, for example, which were hypothesized previously from biochemical studies, have been detailed from ultrastructural investigations.

The purpose of this chapter is to describe the ultrastructural properties of prokaryotic cells and to demonstrate the relationship between structure and function. The principal structures found in the prokaryotic cell are illustrated in Fig. 4-1. An electron micrograph revealing many of the major components to be discussed in this chapter is demonstrated in Figure 7-1.

CELL WALL

All prokaryotes, except the mycoplasmas and L-forms, possess cell walls. The cell wall lies outside the cytoplasmic membrane and lends structural support and rigidity to the cell. The wall represents the first structural component of the bacterial cell that reacts with the external environment. The fact that bacteria can be found in extremely harsh environments, such as the high temperatures and acidity of hot sulfur springs, the high alkalinity of areas such as the Dead Sea, and the arid areas of deserts, suggests that the wall is a formidable structure. The cell wall has a basic unit of design that has apparently evolved various modifications to accommodate the harsh environments to which some bacteria are exposed. Some of these modifications have already been characterized in the thermoacidophiles, extreme halophiles, and methanogens, which are now designated archaebacteria. We will examine the basic design and some of the modifications of the two major groups of bacteria that have been separated based on the staining characteristics of the cell wall: the gram-positive and gram-negative bacteria.

Gram-positive cell types

Peptidoglycan. The cell wall of gram-positive bacteria is of simple design compared with the gram-negative cell wall, but both possess the basic unit called the ***peptidoglycan*** (also called murein) layer (Figure 7-2). The peptidoglycan forms a sac, or meshwork, that covers the surface of the cell and is sometimes referred to as the ***murein sacculus.*** The glycan portion of the peptidoglycan is a network of linear strands of disaccharides (may be 20 to 100 disaccharides long)

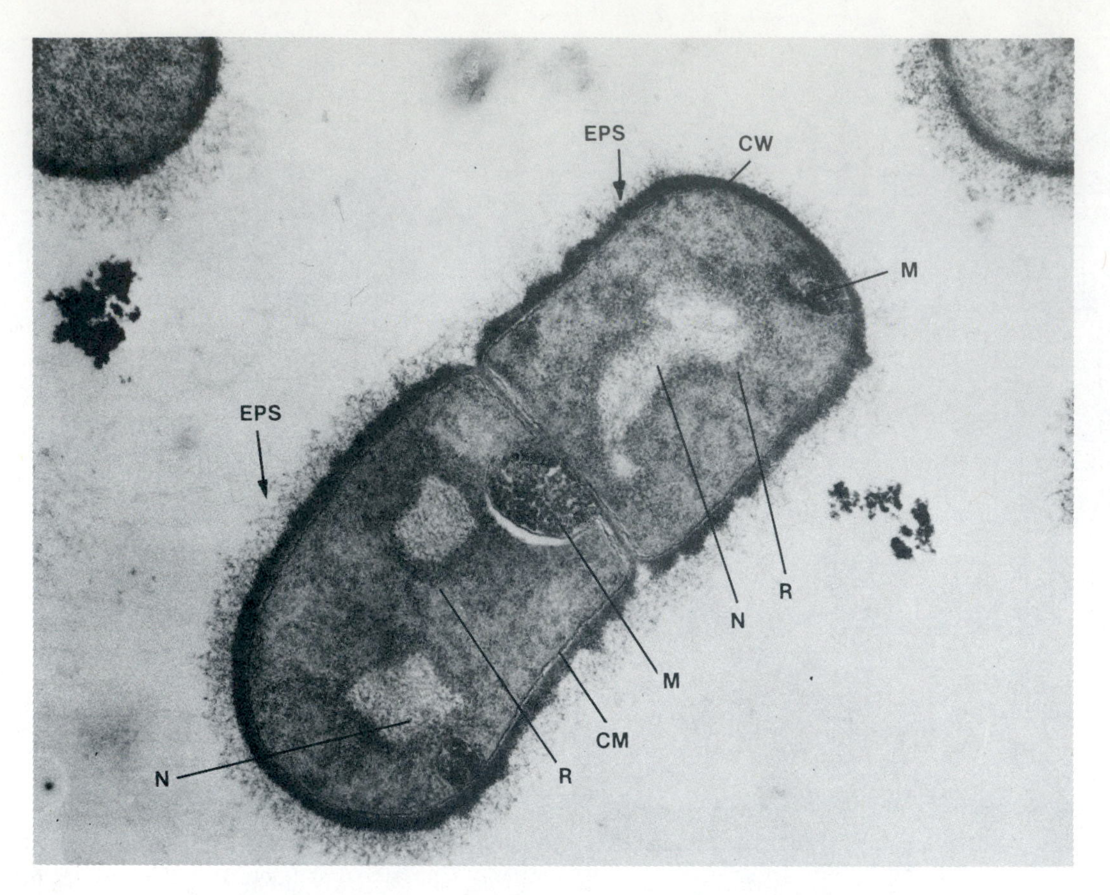

Figure 7-1 Electron micrograph of bacterial cell demonstrating some principal structures. *EPS*, Extracellular polymeric substance; *CM*, cytoplasmic membrane; *CW*, cell wall; *M*, mesosome; *N*, nucleoid; *R*, ribosomes.

From Ellar, D.J., Lundgren, D.G., and Slepecky, R.A. J. Bacteriol. ***94:****1189, 1967.*

that consist of the repeating units ***N-acetylglucosamine (NAG)*** and ***N-acetylmuramic acid (NAM)*** (see Figure 2-23). The strands are linked randomly to adjacent strands by tetrapeptide branches that are covalently attached to the NAM residues. The peptide branches are linked by interpeptide bridges that may be direct and involve the amino acids of the branches, or the bridge may consist of one or several amino acids. Pentaglycine, for example, is the interpeptide bridge in *Staphylococcus aureus* (Figure 7-3, *B*). Not all of the tetrapeptide branches are linked by an interpeptide bridge, and as much as 50% are free in some bacterial species. The greater the degree of cross-linking, the more rigid the glycan strands and their peptide branches. Secondary polymers, such as ***teichoic acid*** and ***teichuronic acid,*** may also be covalently linked to NAM, but these are discussed later. The peptidoglycan may account for 15% to 50% of the dry weight of the gram-positive cell wall, depending on the species as well as the environmental conditions, such as availability of nutrients.

Teichoic acid. Teichoic acids are acidic polysaccharides consisting of repeating units of glycerol or ribotol that are linked by phosphate esters and have other sugars and D-alanine attached (Figure 7-4). The relationship of teichoic acids to the peptidoglycan layer has not been entirely clear in many gram-positive species. Teichoic acid in *Bacillus subtilis* is 20 to 30 residues in length and makes up as much as 50% of the dry weight of the cell wall. One half of the teichoic acid is intertwined within the peptidoglycan matrix and is covalently bonded to the 6′-hydroxyl group of NAM. Some of the teichoic acid is also exposed on the surface and is associated with the outer face of the peptidoglycan. Surface teichoic acid forms part of the receptor for various bacteriophages. Teichuronic acid (*N*-acetylgalactosamine plus glucuronic acid linked as a disaccharide) contains no phosphorus and is a secondary polymer found in species of *Bacillus*. Teichuronic acid can replace teichoic acid when the phosphate level in the medium is very low.

Lipoteichoic acid may also be found in the

A

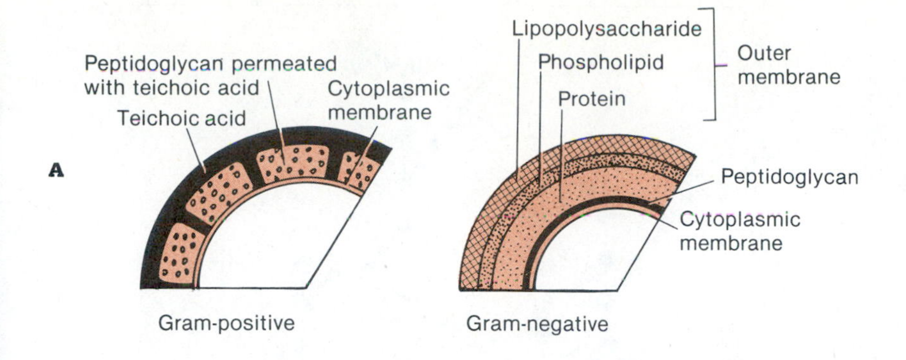

Figure 7-2 Bacterial cell wall. **A,** Chemical composition and structural relationships of gram-negative and gram-positive cell wall. Note that layers of gram-negative cell wall are more indicative of chemical composition than structural relationships. See Figure 7-5 for more representative picture of structural relationships. **B** and **C,** Electron micrographs demonstrating cell wall of gram-positive (×124,000), **B,** and gram-negative (×60,000), **C,** bacteria. *EPS,* Extracellular polymeric substance (not part of cell wall); *OM,* outer membrane; *pg,* peptidoglycan; CM, cytoplasmic membrane.

B, Courtesy Michael Higgins. C, From Mansheim, B.J., Rev. Infect. Dis. *1(2):263, 1979.*

B

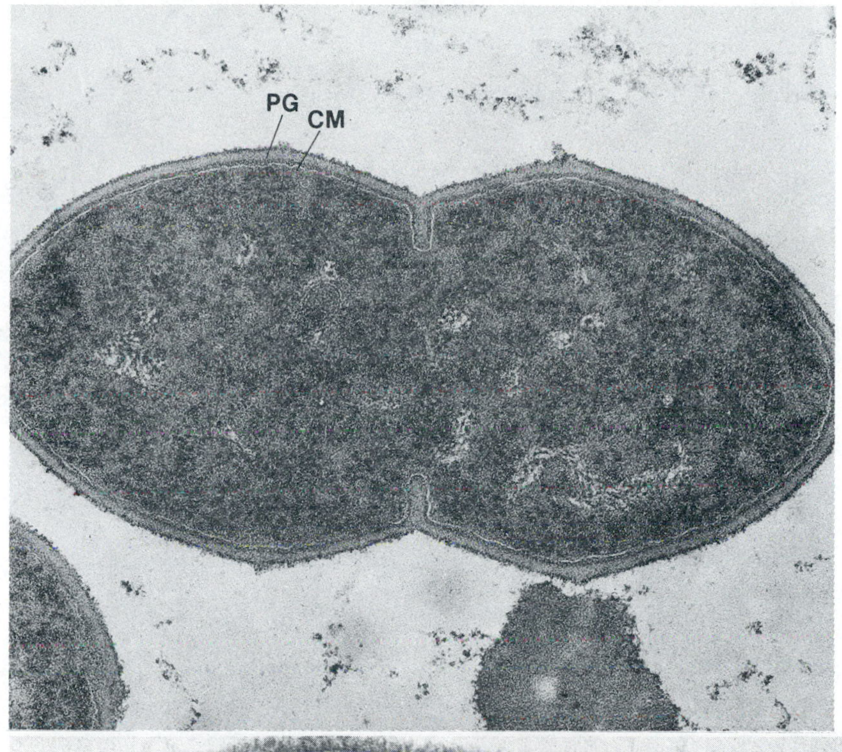

C

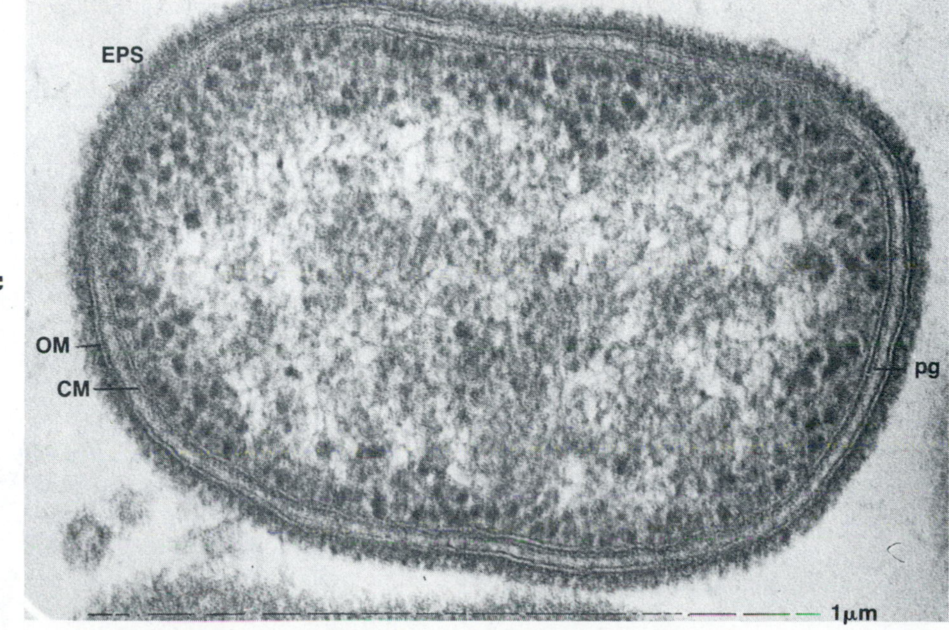

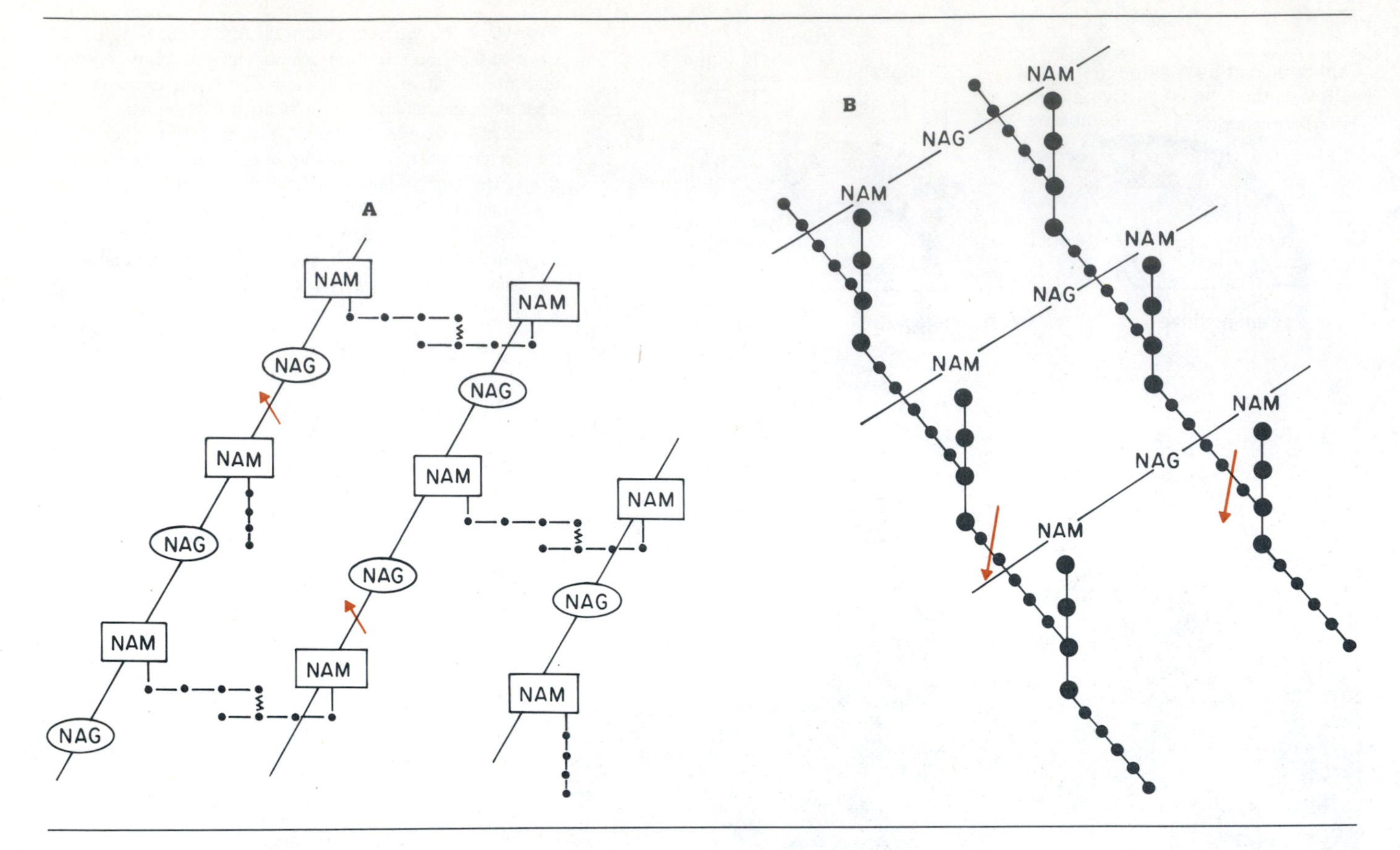

Figure 7-3 Structure of peptidoglycan layer in gram-negative and gram-positive cell walls. **A,** Gram-negative cell wall as in *Escherichia coli.* Colored arrow points to site of breakage by enzyme lysozyme. Tetrapeptides (●) are attached to muramic acid (NAM) residues. Tetrapeptide on one chain can be connected to tetrapeptide on adjacent chain by peptide bond (〰). Note that not all of tetrapeptides produce interchain bridges, and gram-negative cell wall is structurally less stable than gram-positive cell wall but is more flexible than gram-positive cell wall. **B,** Gram-positive cell wall as in *Staphylococcus aureus.* Tetrapeptides (●●●●) attached to muramic acid residues are covalently bridged to muramic acid residues on adjacent chains by pentaglycine (●●●●●). Colored arrow points to site of action of penicillin. *NAM, N*-acetylmuramic acid; *NAG, N*-acetylglucosamine.

A

----O—CH2 O O—CH2 O O—CH2 O
RO—CH P alaO—CH P O—CH P
OH OH OH
CH2O CH2O CH2O

B

O—CH2 O—CH2 O—CH2
H—C—O—ala O H—C—O—ala O
H—C—OH P H—C—OH P
H—C—OR OH H—C—OR OH
H—C—O H—C—O
H H

Figure 7-4 Two types of teichoic acid found in gram-positive bacterial cell wall. **A,** Glycerol teichoic acid. **B,** Ribitol teichoic acid. *Ala,* alanine; *R,* different organic groups.

cell walls of some species of *Lactobacillus.* Lipoteichoic acid has its polyglycerol phosphate chains extending into the wall, with some reaching the cell surface while the glycolipid is inserted into the cytoplasmic membrane.

Function of the gram-positive wall components. The peptidoglycan layer, with its cross-linking and interpeptide bridges, gives structural rigidity to the wall, but other components also play a role. Both teichoic acid and teichuronic acid contribute to the negative charge associated with the surface of the gram-positive cell wall. Both polymers bind divalent cations, such as magnesium, which allows the wall to become more compact (less electronegative repulsion) but does not alter the porosity of the wall. When mutants of *B. subtilis* are produced that are deficient in teichoic acid or teichuronic acid, the cells exhibit coccoid shapes but return to their normal

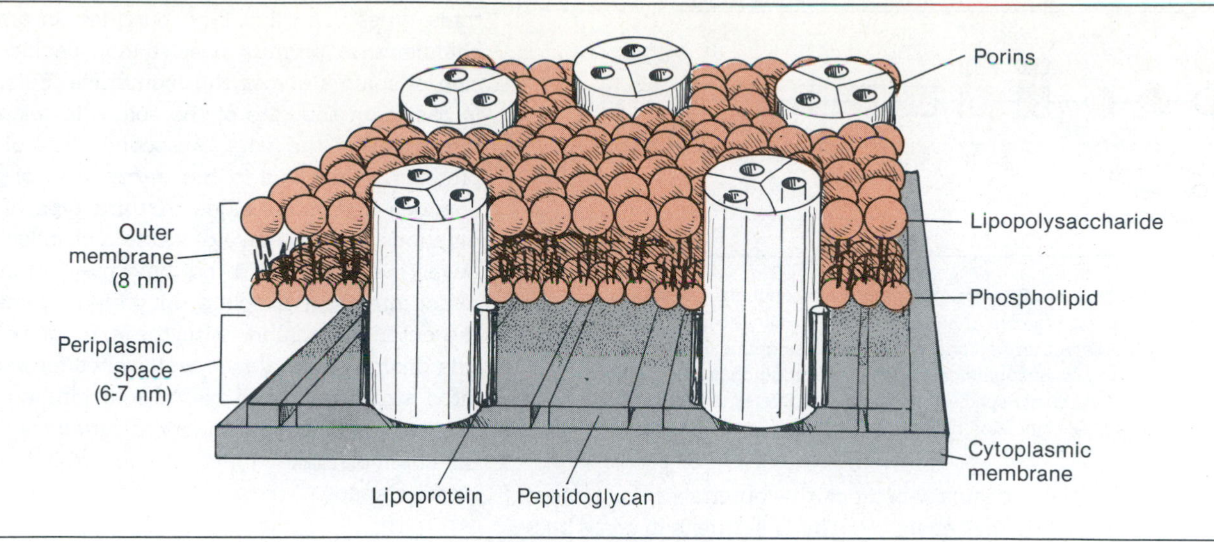

Figure 7-5 Tentative structure of cell wall of gram-negative bacterial cell. Outer membrane consists of sheet of LPS molecules in which are embedded proteins, called porins, which are believed to form channels for transport of some molecules. Periplasmic space, or intermediate zone, is believed to possess enzymes involved in outer membrane synthesis and transport of molecules.
*Redrawn from Nikaido, H., and Nakae, T. Adv. Microb. Physiol. **20**:163, 1979. Copyright: Academic Press, Inc. (London), Ltd.*

rod shape under conditions that permit insertion of these polymers. This suggests that teichoic acid and teichuronic acid lend structural support to the peptidoglycan. Teichoic acid and teichuronic acid are also involved in the control of ***autolysins,*** enzymes that hydrolyze the cell wall. Controlled hydrolysis of cell wall components is necessary in the normal growth cycle because old cell wall must be cleaved to allow for the insertion of new cell wall components. Autolysis may also be totally destructive, and some cells lyse when vegetative growth has ceased as a result of complete hydrolysis of the cell wall. Teichoic acid and teichuronic acid appear to control this lytic system, which is discussed at more length in Chapter 15.

Gram-negative cell types

The cell wall of gram-negative bacteria is a complex unit consisting of an ***outer membrane*** and an underlying peptidoglycan, which may be loosely or strongly adherent to the cytoplasmic membrane. The peptidoglycan is also part of a space called the periplasmic space, which is adjacent to the cytoplasmic membrane. The structural arrangement of the gram-negative cell wall is illustrated in Figure 7-5.

Outer membrane. The outer membrane consists of 20% to 25% ***phospholipid,*** 30% ***lipopolysaccharide (LPS),*** and 45% to 50% protein. The outer face of this membrane is composed of LPS and proteins, and the inner portion is primarily composed of phospholipid. This arrangement creates a lipid bilayer in which the fatty acid chains are arranged at right angles to the plane of the membrane, thus forming a hydrophobic (water-repelling) interior. The polysaccharide portion of the LPS produces a hydrophilic (water-loving) exterior. Proteins are located in the outer membrane and include the matrix proteins and lipoproteins. The matrix protein forms a lattice that is bound to the peptidoglycan by ionic bonds. Some of the matrix proteins are arranged in such a way as to form channels or pores ***(porins)*** for the passive transfer of solutes (Figure 7-5). Lipoproteins are smaller than matrix proteins and are the most abundant proteins (7.2×10^5 per cell). One third of the lipoprotein is covalently bonded by an amino group to every tenth or twelfth meso-diaminopimelic acid of the peptidoglycan, whereas two thirds of the lipoprotein remains free. The covalently bound lipoprotein provides a bridge holding the outer membrane to the peptidoglycan.

Functions. The polysaccharide portion of the LPS is composed of a core of sugars that is common to one group of bacteria. There are sugar side chains, however, that vary from one species to another within the bacterial groups (Figure 7-6). These side chains are antigenic, since they stimulate the formation of immunoglobulins (antibodies) when injected into an animal, and are called ***O antigens.*** Unless an extracellular polymeric substance is present on the cell, the sugar

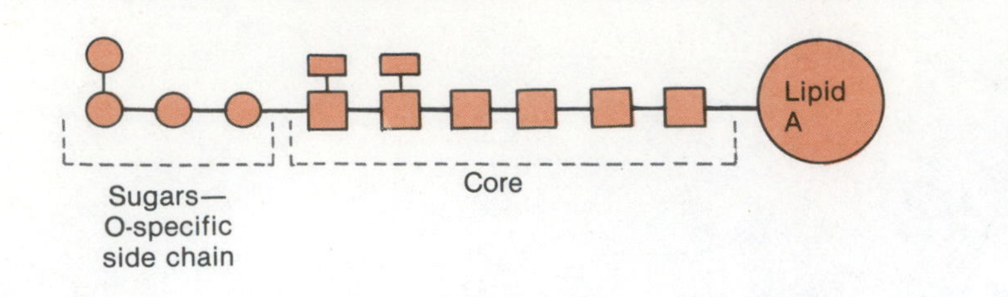

Figure 7-6 Structure of LPS in gram-negative bacterial cell walls. Lipid A, toxic component, is bound to core of sugars (□—□—□—□) whose composition remains relatively stable within bacterial group, such as genus *Salmonella*. Core is bound to variable region of sugars called O-specific side chain. Serological differences between *Salmonella* species are related to differences in sugar composition of these O-specific side chains.

side chains represent the outermost component of the cell envelope. The O antigens in some bacterial species protect the cell from digestion (phagocytosis) by white blood cells in the vertebrate host. The resistance is due to the composition of the sugar side chains. Outside of the vertebrate host, these same bacteria are apparently protected from digestion by free-living amebas, which also employ the phagocytic process.

The LPS is often referred to as ***endotoxin,*** and some types are capable of inducing a physiological response such as fever in animal hosts, including humans. The toxic component of LPS is the lipid portion, which is called ***lipid A.*** Lipid A is covalently bound to the core of the polysaccharide (Figure 7-6).

The outer membrane acts as a barrier to the penetration of certain molecules by the very nature of its construction and chemical composition. Certain dyes such as crystal violet, antibiotics such as erythromycin and actinomycin, heavy metals, and bile salts do not readily penetrate the gram-negative cell wall but do penetrate the gram-positive cell. The mechanism of transport of molecules across the cell wall remains highly speculative. The tightly clustered hydrophobic fatty acid side chains of the LPS are believed to slow down the penetration process, particularly of hydrophobic molecules. Once hydrophobic molecules penetrate the LPS, they easily pass through the phospholipid bilayer. This relative impermeability to hydrophobic compounds may explain the large number of gram-negative bacteria present in the intestinal tract, where bile salts are continuously present. The outer membrane thus offers the gram-negative cell a selective advantage. Hydrophilic molecules that are small enough are thought to pass through the outer membrane by diffusion through the porins. The porins presumably exclude hydrophobic compounds. The porin proteins appear to be of three types. The first type provides an aqueous channel that permits passive nonspecific diffusion of solutes across the membrane. Selectivity is based on the size of the solute in relation to the radius of the pore. A second class of porin proteins is involved in the transport of oligosaccharides and nucleosides. A third type of porin functions in the uptake of solute molecules of relatively large size, and this type may actually interact directly with solute. All porin proteins span the outer membrane, with the exterior polypeptide chains serving as receptor sites for bacterial virus attachment and bacteriocin binding.* The polypeptide chains on the inner surface of the porin channel make direct contact with the periplasmic space.

Peptidoglycan. The peptidoglycan of gram-negative bacteria performs the same function as the peptidoglycan of gram-positive bacteria. The gram-negative peptidoglycan, however, differs in amount and location from the gram-positive component. The peptidoglycan accounts for 5% to 10% of the dry weight of the cell wall. The interpeptide linkage of most gram-negative types are of the direct type, such as appears in *Escherichia coli* (see Figure 7-3, *A*). The attachment of the peptidoglycan to the outer membrane has already been discussed.

Periplasmic space. The periplasmic space separates the outer membrane from the cytoplasmic membrane (see Figure 7-5). It contains all of the peripherally attached proteins of the inner face of the outer membrane and outer face of the cytoplasmic membrane. The periplasm contains several enzymes whose functions are discussed below.

Functions. The periplasm is in equilibrium with the cytoplasm of the cell; that is, it is isosmotic. If the osmotic pressure outside the cell is increased, the periplasmic space also enlarges and thus acts as an osmotic buffer zone between the cytoplasm and its external environment.

Many of the solutes transported across the outer membrane are processed in the periplasmic space. Some are broken down by various degradative enzymes (proteases, lipases, etc.) into smaller units before their final transport across the cytoplasmic membrane. Specific binding proteins are also located within the periplasmic space and help to transfer some solutes to specific carrier proteins attached to the cytoplasmic membrane. Experiments with membrane vesicles

* Bacteriocins are chemical compounds produced by bacteria that inhibit or kill closely related species.

(closed cytoplasmic membranes prepared by cold osmotic shock) have shown that some solute transport across the cytoplasmic membrane does not take place in the absence of the periplasmic space and its binding proteins.

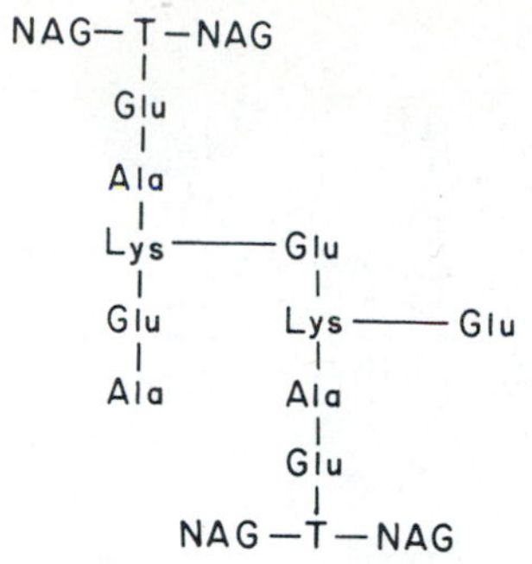

Figure 7-7 Arrangement of peptidoglycan in *Methanobacterium thermoautotrophicum*, member of archaebacteria. *NAG*, *N*-acetylglucosamine; *T*, *N*-acetyltalosaminuronic acid; *Glu*, glutamic acid; *Ala*, alanine; *Lys*, lysine.

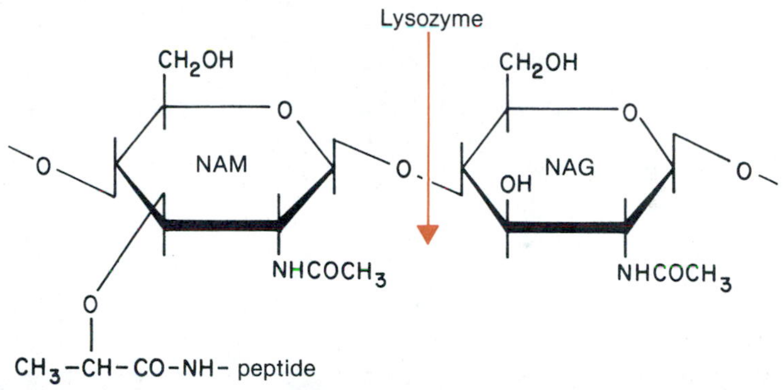

Figure 7-8 Site of action of enzyme lysozyme on bacterial cell wall. *NAM*, *N*-acetylmuramic acid; *NAG*, *N*-acetylglucosamine.

Cell walls of the archaebacteria. The archaebacteria have atypical peptidoglycans or related cell wall structures and provide an interesting insight into wall modifications. *Methanobacterium thermoautrophicum* contains an aminohexuronic acid called *N*-acetyltalosaminuronic acid, which substitutes for NAM (Figure 7-7). *Halococcus morrhuae* contains *N*-glycylglucosamine, an example of the amino acid glycine being substituted for an amino group on NAG. The glycine residues are connected by peptide bonds to sugars in the cell wall. *Methanococcus vannielii* contains neither a rigid cell wall nor sheath but exhibits an outer protein layer consisting of subunits that disintegrate very easily. One interesting observation from these studies is that many of the species show a gram-negative reaction that is based not on chemical characteristics but on the thickness of the cell wall. Those with thick cell walls are gram-positive, whereas those with thin cell walls are gram-negative.

Cell wall and osmosis. In addition to its function in maintaining the shape of the cell, the cell wall is a dynamic structure that responds to its environment. The concentration of solute outside the cell for many bacteria in their natural environment is hypotonic (the concentration of salts is lower outside the cell than inside the cell). Under such circumstances water tends to flow into the cell to compensate for this concentration difference, a process called ***osmosis.*** In a hypotonic medium water rushes into the cell and causes the cytoplasmic membrane to expand. The cell wall also expands but does not burst. It acts as a cushion for the expanding cytoplasmic membrane. The protective effect of the cell wall can be demonstrated if a bacterial culture is placed in a hypotonic medium containing lysozyme, an enzyme that can affect the removal of the cell wall. Lysozyme is found naturally in tears, saliva, and other body fluids. It can cause the breakdown of the cell wall by cleaving the covalent linkage between NAM and NAG (Figure 7-8). As the cell wall is being removed or weakened in a hypotonic medium, water rushes into the cell, causing the unprotected cytoplasmic membrane to expand and eventually burst; that is, the cell lyses. The removal of the bacterial cell wall need not result in cell lysis if the medium contains solutes in such a concentration that osmotic forces outside the cell balance osmotic forces inside the cell. Under such circumstances a spherically shaped wall-less cell, called a ***protoplast,*** is formed (Figure 7-9). Protoplasts are spherical whether the original cell was spherical or rod shaped. If the solute concentration outside the protoplast is higher than that inside the cell, water will flow out of the cell (dehydration), and it will collapse, a process called ***plasmolysis.*** Many bacteria are inhibited by high concentrations of solutes such as salt and sugars, and this property is used in the curing of various meats and in the preparation of jellies, respectively (see Chapter 20).

Cell wall synthesis is inhibited by several antibiotics such as the penicillins and cephalosporins. These agents affect the ***transpeptidation*** reaction that occurs between the adjacent polysaccharide strands in the cell wall (see Figure 7-3). The effects of antibiotics on the cell wall are discussed in detail in Chapter 27.

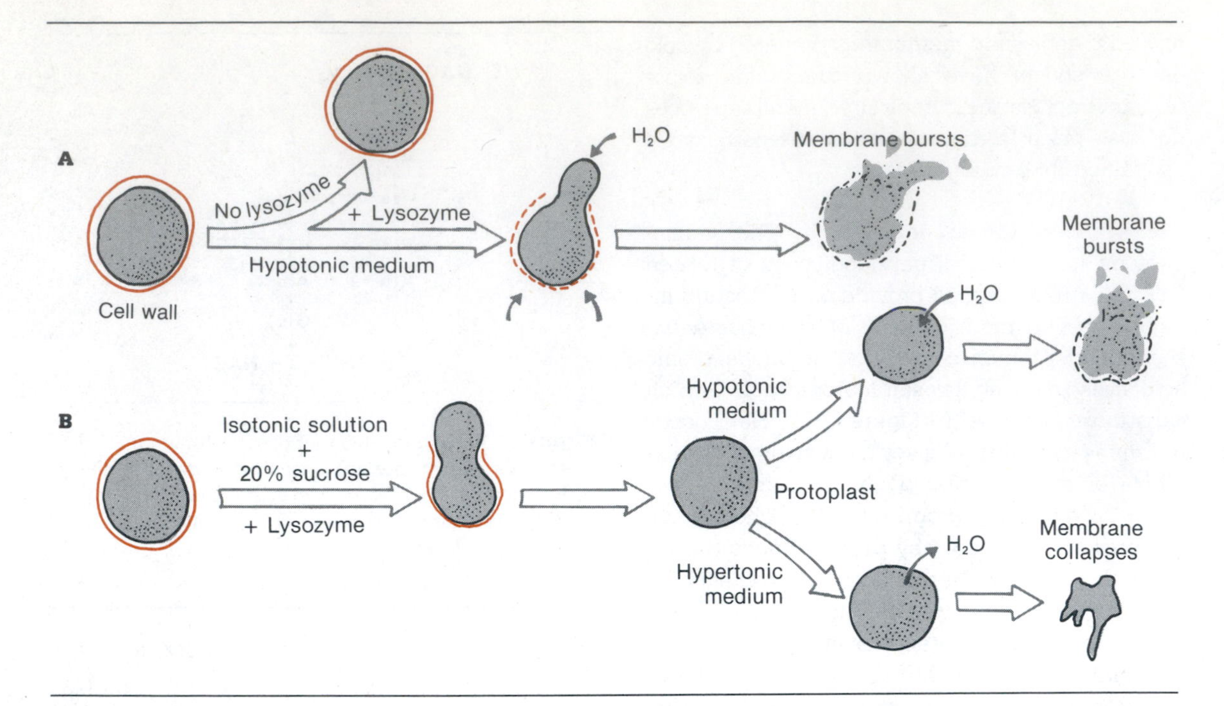

Figure 7-9 Effect of wall-lysing agents on bacterial cell in media of different osmotic pressure. **A,** Hypotonic media in presence and absence of lysozyme. **B,** Isotonic medium containing 20% sucrose plus lysozyme.

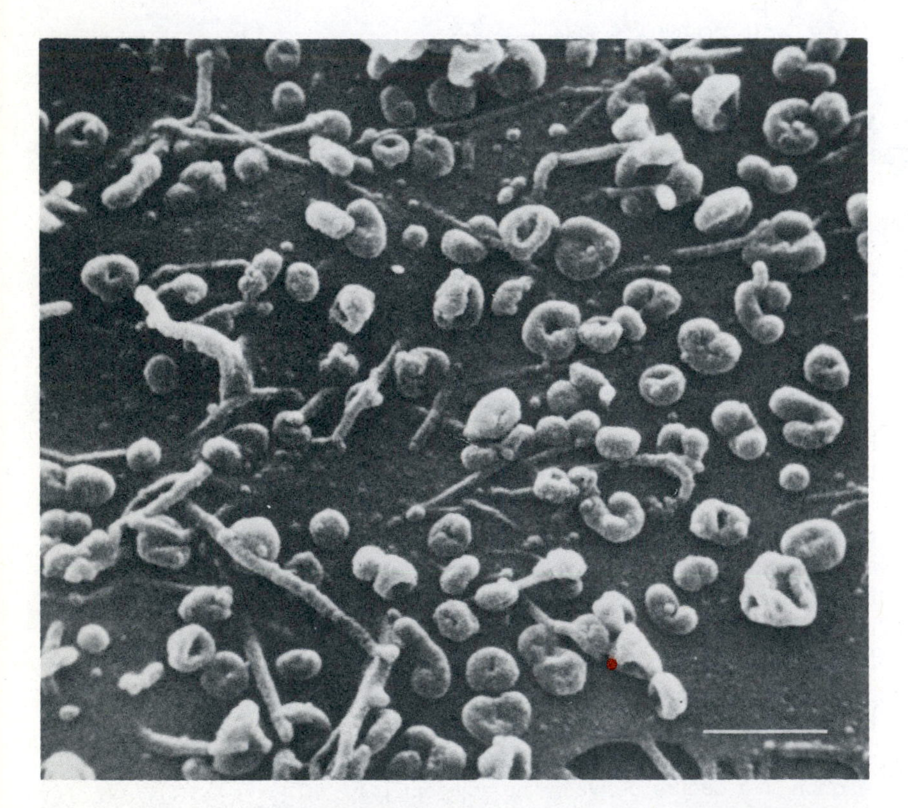

Figure 7-10 Electron micrograph of a species of *Mycoplasma.*
Courtesy E. Boatman.

Wall-less bacteria

Wall-less bacteria can be divided into two groups: naturally occurring wall-less bacteria called ***mycoplasmas,*** and ***L-forms,*** which are derived from gram-negative and gram-positive bacteria. Mycoplasmas are the smallest free-living organisms capable of self-replication (Figure 7-10). They possess no peptidoglycan layer, and the cell cytoplasm is bounded only by a phospholipid bilayer with associated proteins. The stability of the mycoplasmas is related to their small size and the high concentration of cholesterol or other sterols that are incorporated into the cytoplasmic membrane. The mycoplasmas are intracellular organisms and do not multiply outside of the animal cell except on special laboratory media that are hypertonic and contain cholesterol. Mycoplasmas can infect humans, plants, and animals.

L-forms were originally described at the Lister Institute in London, hence their name. L-forms occur naturally, but they can be produced from any gram-positive or gram-negative bacterium using cell wall–degrading agents, such as lysozyme, or an inhibitor of cell wall activity, penicillin. Once isolated, the L-form may be stable or unstable. Stable L-forms maintain their wall-less state in the absence of wall-lysing agents. Naturally occurring L-forms are believed to represent a technique for survival. Under unfavorable conditions bacteria may lose their cell wall by the induction of autolytic enzymes, and only when conditions are favorable for normal growth and reproduction will the cell revert back to its natural state. Bacteria that infect a host, for example, may be exposed to antibiotics such as penicillin or to immune responses by the host that result in destruction of the cell wall. If left exposed to these agents, the bacterial population would cease to exist, but in the absence of a cell wall these agents have no effect and the survival of the microbial cell is ensured.

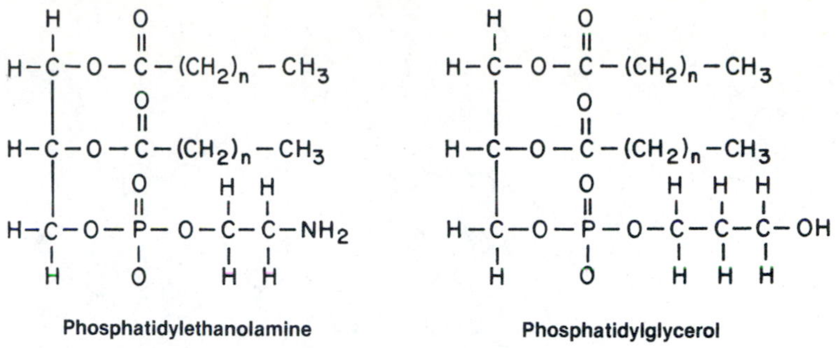

Phosphatidylethanolamine

Phosphatidylglycerol

Figure 7-11 Two common phospholipids found in cytoplasmic membrane of prokaryotic cells. Hydrocarbon chains (saturated or unsaturated) are referred to as $(CH_2)_n$.

CYTOPLASMIC MEMBRANE

The cytoplasmic membrane (also called plasma membrane) separates the cytoplasm of the cell from its external environment. Damage to the membrane usually results in the loss of important cytoplasmic constituents, and the cell dies. The cytoplasmic membrane of the bacterial cell is a prototype for other biological membranes. This universal membrane structure is referred to as a ***unit membrane,*** which distinguishes it from complex membranes. The currently devised evolutionary scheme proposes that prokaryotes preceded eukaryotes, and it is not surprising to find that the cytoplasmic membrane of prokaryotes carries out many of the functional roles of the membranes surrounding the organelles of eukaryotic cells.

Structure

The cytoplasmic membrane is chemically composed of 40% lipid and 60% protein. Phospholipids represent the major lipid components, and two of the principal ones found in bacteria are illustrated in Figure 7-11. The fatty acid side chains of microbial lipids vary in length, but most are between 16 to 18 carbon units long. The lipid composition of the prokaryotic membrane may vary considerably from that of the eukaryotic membrane. The chemical and physical properties of these membrane lipids, particularly their fatty acid side chains, permit some prokaryotes to survive in environments where there are chemical and/or physical extremes. Prokaryotes are therefore found in nearly all ecological niches, but eukaryotes are not. The survival value of these lipids is discussed in Chapter 15. The phospholipids consist of two long fatty acid chains that are insoluble in water and impart a hydrophobic character to the molecule. The charged phosphate groups in the phospholipid are hydrophilic; thus one end of the phospholipid avoids water and the other end is attracted to it. In their normal aqueous environment the hydrophilic regions of the membrane face outward toward water, and the hydrophobic ends face inward and away from the aqueous environment. The molecular arrangement that is currently believed to satisfy this property is the ***phospholipid bilayer*** (Figure 7-12, *A*). The positioning of the proteins in the phospholipid bilayer is dependent on their hydrophilic and hydrophobic components. The hydrophilic parts of the protein appear oriented adjacent to the membrane, whereas the hydrophobic portions are buried in the lipid bilayer (Figure 7-12, *B*). Some proteins are large enough to span the bilayer, whereas others are distributed at various levels. There appears to be no covalent bonding between the proteins and lipid; thus the membrane demonstrates fluidity and some freedom of movement. The proteins embedded in the phospholipid bilayer lend structural integrity and may serve a functional role in metabolism and/or solute transport. Structural stability may also be achieved by the binding of divalent cations such as magnesium and calcium to the anionic phosphates of the phospholipid.

Teichoic acids are also present in the cytoplasmic membrane of some bacteria and are referred to as lipoteichoic acids because of their structure. They consist of a teichoic acid chain attached to a lipid moiety. The hydrophobic lipid component anchors the molecule in the cytoplasmic membrane, and the hydrophilic teichoic acid projects from the membrane surface. In certain bacteria, as discussed previously, the teichoic acid may penetrate the cell wall and even reach the surface of the wall. Lipoteichoic acids therefore provide a physical bridge binding the cell wall to the cytoplasmic membrane in some bacteria.

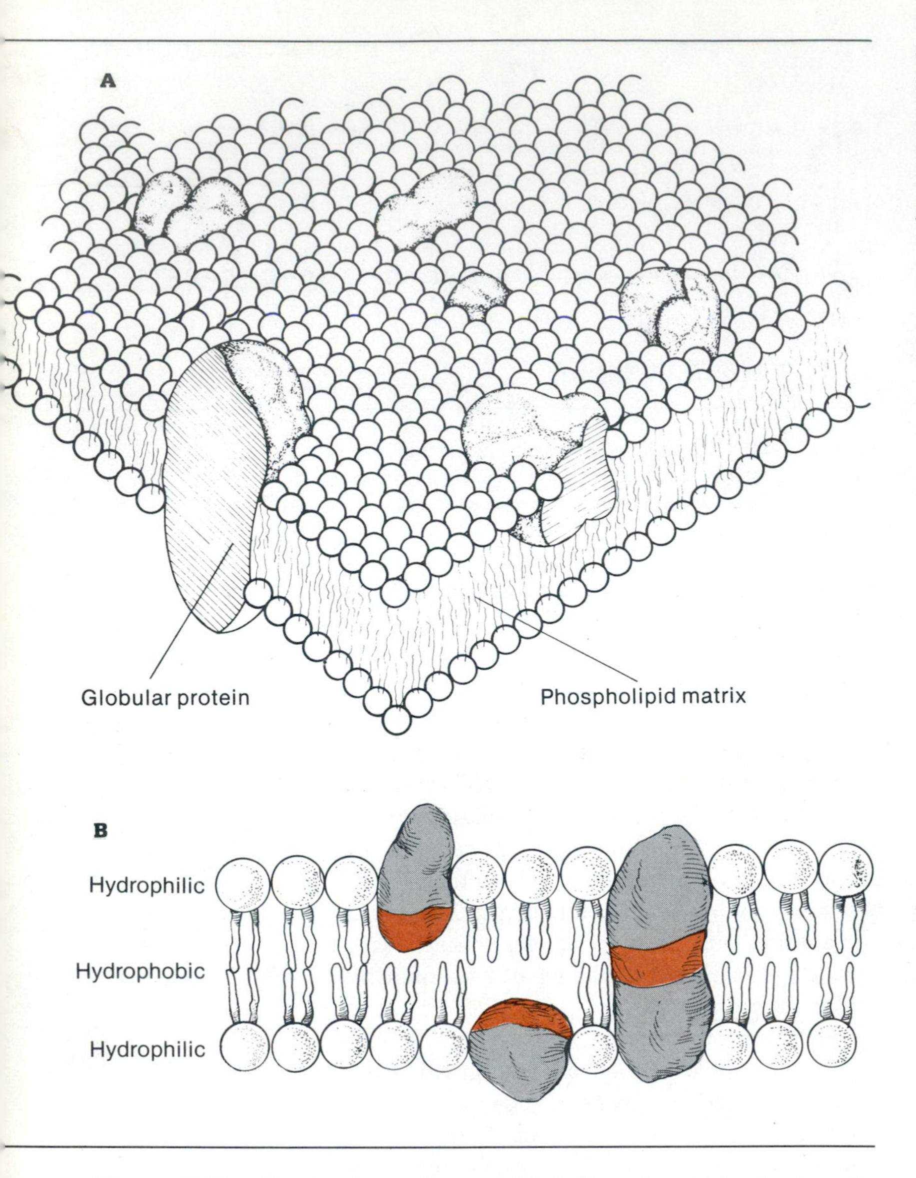

Figure 7-12 Cytoplasmic membrane model. **A,** Current model for structure of cytoplasmic membrane. Globular proteins are embedded in continuous phospholipid layer (fluid mosaic model). **B,** Relationship of hydrophobic and hydrophilic groups in cell membrane based on fluid mosaic model.

***A,** From Singer, S.J., and Nicolson, G.I. Science **175**:720, 1972.*

Functions

The functions of the cytoplasmic membrane can be divided into those activities associated with solute transport, energy formation, biosynthesis, and secretion.

Solute transport. The cytoplasmic membrane does not appear to have pores, like the outer membrane of gram-negative bacteria, that permit passage of solutes into the cytoplasm. The membrane consists of highly charged phosphate groups and hydrophobic fatty acid chains that are highly selective to molecular transport. This selectivity is based on size as well as electrostatic charge of the solute. The mechanisms that operate to select molecules for transport include passive and active transport systems. ***Passive transport*** permits solutes to pass across the membrane in the direction of decreasing concentration, that is, from high solute concentration to low solute concentrations with no expenditure of energy. The molecular size, shape, charge, solubility, and concentration of the solute on either side of the membrane are limiting factors. Passive diffusion is nonspecific and usually very slow. Under most environmental conditions the concentration of solute is less than that inside the cell. The survival of the microbial cell depends on moving these solutes against a concentration gradient, that is, from a point of low concentration to a point of high concentration. The ability to concentrate nutrients when they are available gives the cell a better chance of survival when such nutrients are temporarily absent from the environment. Moving solutes against a concentration gradient requires an expenditure of energy called ***active transport.*** Active transport requires the use of carrier proteins called ***permeases*** located within the cytoplasmic membrane (Figure 7-13). Carriers show specificity and permit a more rapid rate of solute transfer across the cytoplasmic membrane. Molecules can also be transported on carriers without the expenditure of energy in a process called ***facilitated diffusion.*** Facilitated diffusion does not concentrate solutes in the cell like active transport but does aid in the rapid equilibration of solutes across the cytoplasmic membrane. The only carbohydrate transferred by facilitated diffusion in the systems so far examined is glycerol. ***Group translocation*** is also a mechanism using carriers. This mechanism is unique in that the solute is altered chemically during transport on the carrier. The best-characterized group translocation system is the phosphoenolpyruvate-dependent sugar phosphotransferase system, in which the solute becomes phosphorylated during its transfer across the cytoplasmic membrane (Figure 7-14). Several sugars, including fructose, mannitol, and rhamnose, are transported by this mechanism.

Energy production. Most energy in eukaryotes is produced in organelles called mitochondria. There are no mitochondria in prokaryotes, and some energy production is carried out in the cytoplasmic membrane that contains the appropriate enzymes (see Chapter 13).

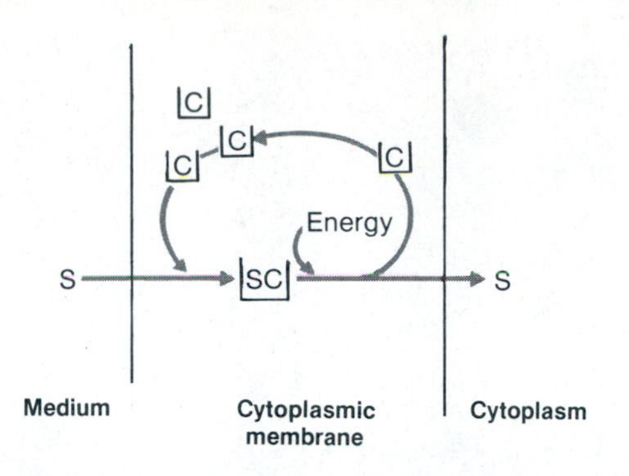

Figure 7-13 Mechanism of selective transport. Solutes *(S)* outside cell reach cytoplasmic membrane, where they interact with carrier *(C)* proteins called permeases. Many permease transport systems require energy (active transport) to transfer solutes across cytoplasmic membrane into cytoplasm. Carrier is released before solute enters cytoplasm and can pick up another solute molecule.

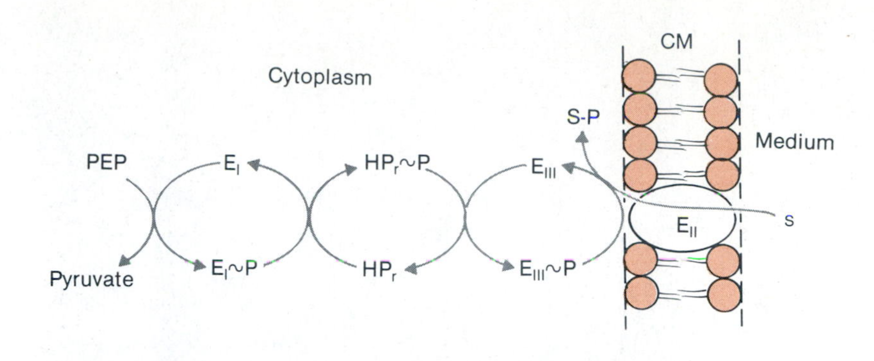

Figure 7-14 Group translocation (phosphoenolpyruvate-dependent sugar phosphotransferase system). Phosphoenolpyruvate transfers its high-energy phosphate to enzyme I (E_I) and then to heat-stable protein (HP_r). High-energy phosphate on HP_r is transferred to enzyme III (E_{III}), which in turn interacts with sugar and enzyme II (E_{II}) to facilitate transfer of solute *(S)* across cytoplasmic membrane into cytoplasm of cell as phosphorylated derivative (S-T).

Biosynthesis. Many enzymes and enzyme systems attached to the cytoplasmic membrane are involved in biosynthetic activities, including the following:

1. Biosynthesis of cell membrane lipid components
2. Biosynthesis of cell wall polymers such as peptidoglycan and teichoic acids
3. Biosynthesis of polysaccharides employed as extracellular polymeric substances
4. Biosynthesis of the LPSs that make up part of the outer membrane of gram-negative bacteria

Secretion The cytoplasmic membrane is involved in the release of extracellular proteins that include various toxins, bacteriocins, and enzymes produced in the cytoplasm of the cell. These products are important to the survival of the microbial species. Many of the enzymes, for example, are used to hydrolyze nontransportable macromolecules such as proteins, polysaccharides and lipids, into lower molecular weight units for transport from the environment into the cytoplasm of the cell. Bacteriocins released by the bacterial cell are able to inhibit the growth of organisms that compete with them for nutrients. Bacteriocins are specific in that they inhibit only related species of bacteria (see Chapter 9).

During infection of mammalian hosts some bacteria produce toxins that, when released, inhibit the defense mechanisms of the host (see Chapter 22). Most of the secreted molecules appear to be too large or hydrophilic to pass through the cytoplasmic membrane, and their ability to be transported, except in a few isolated cases, is not understood. The enzyme penicillinase, which degrades and inactivates the penicillin molecule, is secreted by a few species of bacteria. The penicillinase molecule is very hydrophilic but is made hydrophobic by being covalently attached to a ***phospholipid peptide (PLP)*** in the cytoplasmic membrane. The PLP-penicillinase molecule passes through the cytoplasmic membrane until it reaches the junction of the cell wall–cytoplasmic membrane where the PLP moiety is cleaved by a protease releasing the active penicillinase molecule.

There are apparent variations on the mechanism of secretion of proteins from the cell. The lipoprotein, for example, which is a component of the outer membrane of gram-negative bacteria, passes through the cytoplasmic membrane during cell wall formation using a different mechanism. A new form of cell wall lipoprotein has recently been isolated that has an extra sequence of hydrophobic amino acids at the amino end of the peptide. Experiments performed in mammalian cells have resulted in the formulation of a model to explain the transport of secretory proteins and molecules such as the cell wall lipoprotein through the cytoplasmic membrane. It is called the ***linear model*** or ***signal hypothesis*** (Figure 7-15). This model proposes that near the amino end of the polypeptide is a short sequence of amino acids (such as the hydrophobic amino acids on the lipoprotein) that acts as a signal for the secretion of the protein. The signal sequence is recognized by cytoplasmic proteins, which then form a tunnel for linear translocation of the precursor protein from inside the cell through the

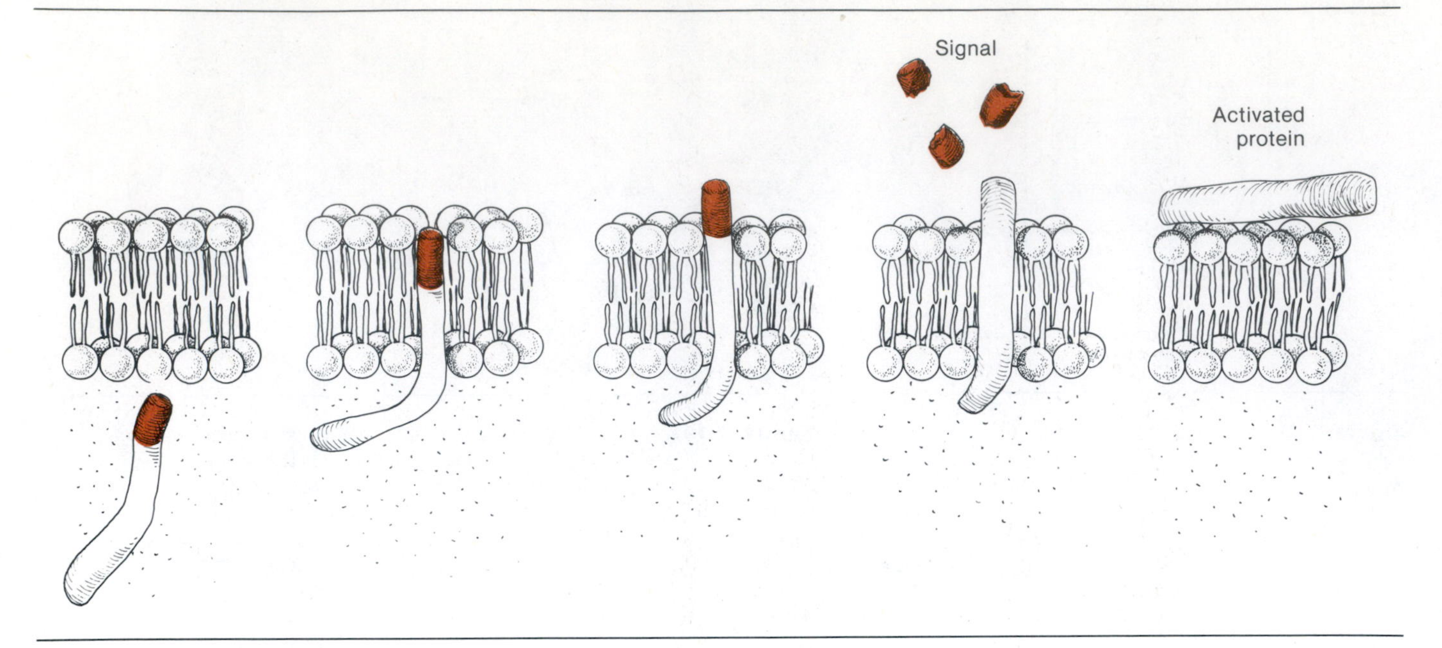

Figure 7-15 Linear model (signal hypothesis) for transport of secretory proteins through cytoplasmic membrane. "Signal" end or extension peptide of secretory protein is represented in color. Secretory protein moves through channel in membrane until "signal" portion is outside membrane, where it is cleaved, leaving activated protein that can be transported through cell wall.

cytoplasmic membrane to the exterior of the cell. Once outside the cell, the signal is cleaved. This model assumes that secretory proteins are synthesized initially as inactive precursor molecules. When the signal portion of the molecule is cleaved following transport through the cytoplasmic membrane, the secretory protein becomes activated. Studies with microbial mutants indicate that factors other than signal recognition are important in the vectorial transport process, and size and protein structure may also play a critical role. An alternative to the linear signal hypothesis is the ***loop model,*** which suggests that the signal remains in the cytoplasmic membrane, where it is cleaved by proteolytic enzymes.

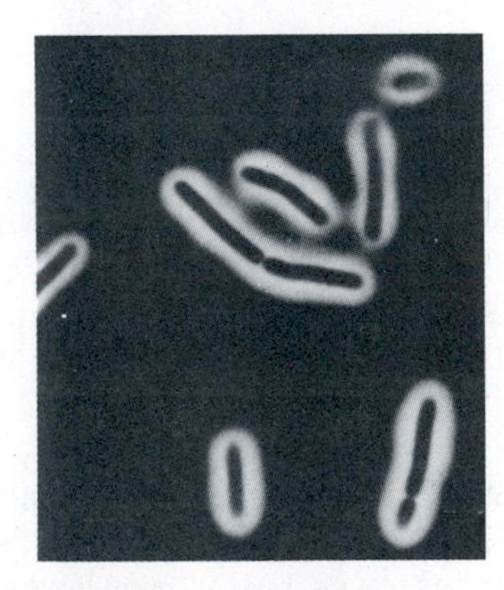

Figure 7-16 India ink preparation demonstrating bacterial capsule. Capsule appears as refractile halo around bacillus.
From Babb, J.L. Infect. Immun. ***19:****1088, 1978.*

Mesosomes

Electron micrographs of many bacteria reveal the cytoplasmic membrane as an invaginating structure that permeates the cytoplasm of the cell, forming organelle-like structures called ***mesosomes*** (see Figure 7-1). No function has been found for these structures, even though they are believed by some investigators to be the sites for DNA attachment and ATP production during replication. Many investigators believe that they are artifacts generated during processing of the cell for electron microscopy. Some groups of bacteria have intracytoplasmic membranes that are natural invaginations of the cytoplasmic membrane. These invaginations have a variety of arrangements (see Figure 4-20) in both photosynthetic and nonphotosynthetic bacteria.

EXTRACELLULAR POLYMERIC SUBSTANCES

Considerable disagreement has arisen concerning the term to be used for those polymeric substances that surround a single cell or colony of cells. I prefer the term ***extracellular polymeric substances (EPSs).*** The EPS may assume a form that has structural integrity and is firmly adherent to the cell envelope, in which case it is called a ***capsule*** (Figure 7-16). If the EPS is released by the cell and exists as an amorphous mass surrounding the cell, it is called ***slime.*** It appears that EPSs of various dimensions may be associated with more bacterial species than was once suspected. Earlier studies were based on transmission electron microscopy, which used dehydrated specimens, and thus the extent and distribution of many extracellular components went undetected or were barely recognized. Recent ultrastructural studies employing ruthenium red, which stains the EPS, and stabilization of the extracellular polymer with specific antibodies have given us a better understanding of the organization of EPSs in some bacterial species.

Structure

Most EPSs are either homopolymers or heteropolymers of a wide variety of monosaccharides, such as hexoses, uronic acid, and amino sugars, as well as other constitutents. The chemical composition of the EPSs of some bacterial species is described in Table 7-1. Some strains of a species will produce the same extracellular polymer, whereas strains of some species, such as *Streptococcus pneumoniae,* may produce different types. The consequences of such a characteristic are discussed later. Some EPSs are made of protein; for example, *Bacillus anthracis* produces a polyglutamic acid EPS. The EPS is a very hydrated and sticky substance and therefore adheres to various substrates or is adhered to by the same or other species in its particular habitat.

Table 7-1 Chemical composition of the extracellular polymeric substances of some bacterial species

BACTERIUM	CHEMICAL COMPOSITION
Acinetobacter calcoaceticus	Glucose plus glucouronic acid
Azotobacter vinelandii	Alginate
Bacillus anthracis	Polyglutamic acid
Erwinia amylovora	Polysaccharide
Leuconostoc mesenteroides	Dextran-dextran-sucrose
Rhizobium meliloti	Glucose plus galactopyruvate
Ruminococcus albus	Glycoprotein
Streptococcus mutans	Dextran plus other glucan
Streptococcus (group B)	Galactose plus *N*-acetylglucosamine or sialic acid plus galactose plus heptose plus glucose plus glucosamine plus mannose
Streptococcus pneumoniae	Various oligosaccharides
Yersinia pestis	Protein plus glycoprotein containing oligomeric galactan

Functions and properties

Adherence. Some groups of microorganisms, such as the myxobacteria, use the EPS for aggregation when nutrients become reduced in the environment. Many species in aquatic habitats use the EPS to form aggregates of cells of the same or different species. The EPS may afford the cell protection and permit the cell(s) to attach to a specific substrate in order to remain sessile in fast-moving streams and/or obtain nutrients from the substrate. Once the EPS is formed, bacteria can divide and form a microcolony of many cells (Figure 7-17). The EPS acts like a resin exchange column that can retain some ions and molecules and let other molecules pass through. This permits the microcolony to be protected from heavy metals, adsorption by bacterial viruses, and the action of antibiotics and bacteriocins produced by other microorganisms.

The EPS is used as a recognition site for the attachment of nitrogen-fixing bacteria, such as species of *Rhizobium,* to leguminous plants (Figure 7-18). Plant surface proteins or glycoproteins called ***lectins*** act as the binding site for the encapsulated bacteria.

The bacterium *Streptococcus mutans,* known to cause tooth decay (caries), produces an extracellular slime layer composed of a water-soluble glucose polymer called ***dextran*** (mutan) (Figure 7-19). The polymer enables the organism to bind to the enamel surface and also serves as the site for accumulation of other microbial species. As the number of *S. mutans* cells increases in the microcolony, and when sweets containing sucrose are ingested, fermentation results in excessive acid production, which reduces the pH level to a point where demineralization of enamel takes place.

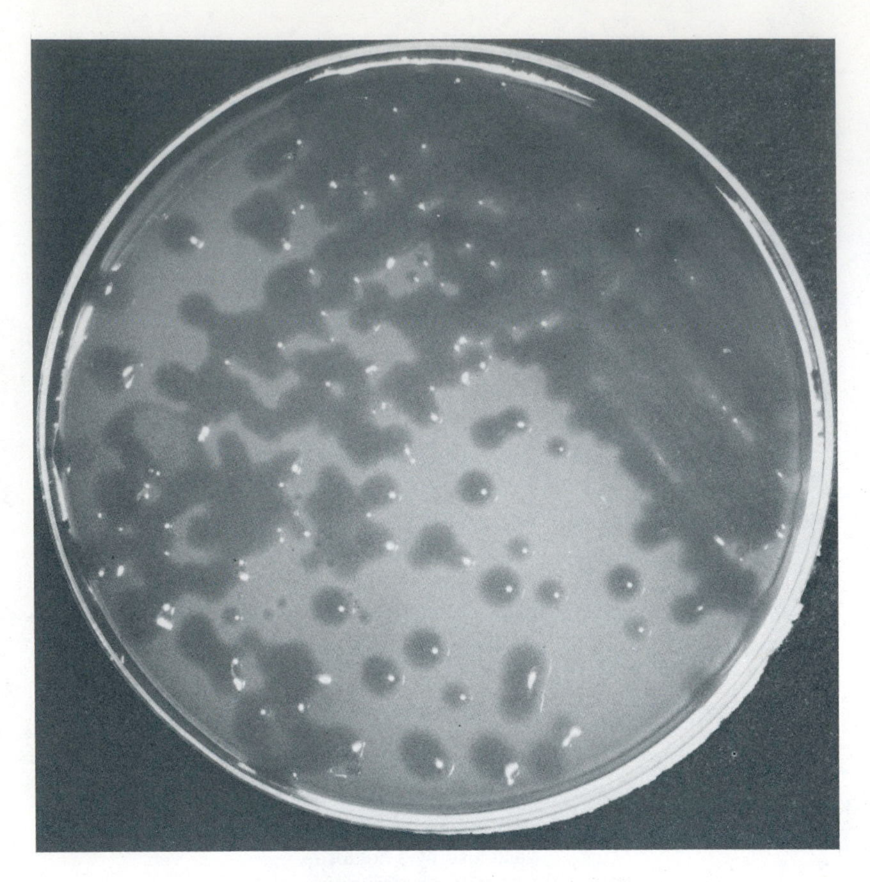

Figure 7-17 Mucoid appearance of colonies on agar surface produced by bacterial cells that release extracellular polymeric substances (EPSs).

Virulence. The disease-producing potential (virulence) of several microorganisms is related to the production of capsules or slime (Plates 18 and 19). Some EPSs protect the cell from phagocytosis (digestion by phagocytes) during the infection process. The EPS can be a factor in the spread of the infectious agent because of the latter's ability to multiply within the phagocyte. *Streptococcus pneumoniae,* the most common cause of pneumonia, is virulent only when it produces a capsule, and over 100 capsular types have been described. Mutants of *S. pneumoniae* unable to produce capsules are readily digested by phagocytes and are no longer virulent. The EPSs produced by certain species of human pathogens can also inhibit the transport of certain molecules into the cell. Species of *Pseudomonas,* for example, that are normally sensitive to certain antibiotics become resistant when, through genetic changes, they develop the capacity to produce EPSs.

Vaccines can be produced from some EPS-producing bacteria, such as *S. pneumoniae,* because many capsular components are strongly antigenic. Vaccines stimulate the formation of antibodies but do not cause disease and thus make the host resistant to future infections by the same species. Many capsular types of *S. pneumoniae* cause disease; therefore, the vaccine used must be composed of more than one type of capsule (that is, be polyvalent).

FLAGELLA

Structure

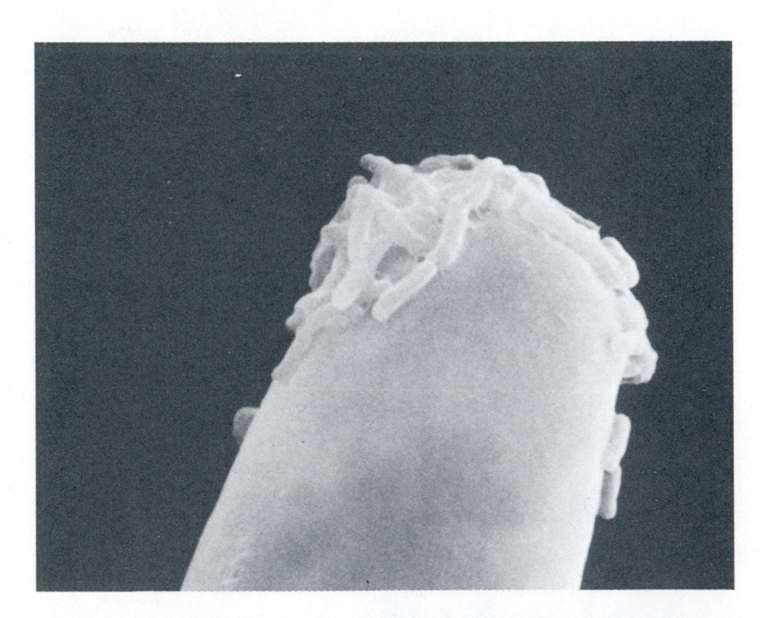

Figure 7-18 Adherence of nitrogen-fixing bacteria (species of *Rhizobium*) to clover root hair.
From Dazzo, F.B., and Brill, W.J. J. Bacteriol. ***137:****1362, 1969.*

The flagellum in most bacteria is composed of a filament, hook, and basal structure (Figure 7-20). The ***filament*** is made up of a single protein subunit called ***flagellin,*** which is constructed by the addition of flagellin subunits to its distal end. The subunits are added in a helical fashion around a central hollow core. Most filaments are naked, but in some genera (*Bdellovibrio, Vibrio, Pseudomonas,* and one *Rhizobium* species) a sheath of protein covers the filament. The sheath protein differs from the core protein and appears to be an extension of the outer membrane of the gram-negative bacterial cell. The ***hook*** structure is located at the base of the flagellum and is also composed of a single polypeptide that is different from that of the filament. It has been suggested that the hook acts as a universal joint that transmits rotational motion to the rest of the flagellum. The ***basal structure*** is the more complex component and is composed of several different polypeptides. There are four ringlike structures in gram-negative bacteria attached to the rod of the

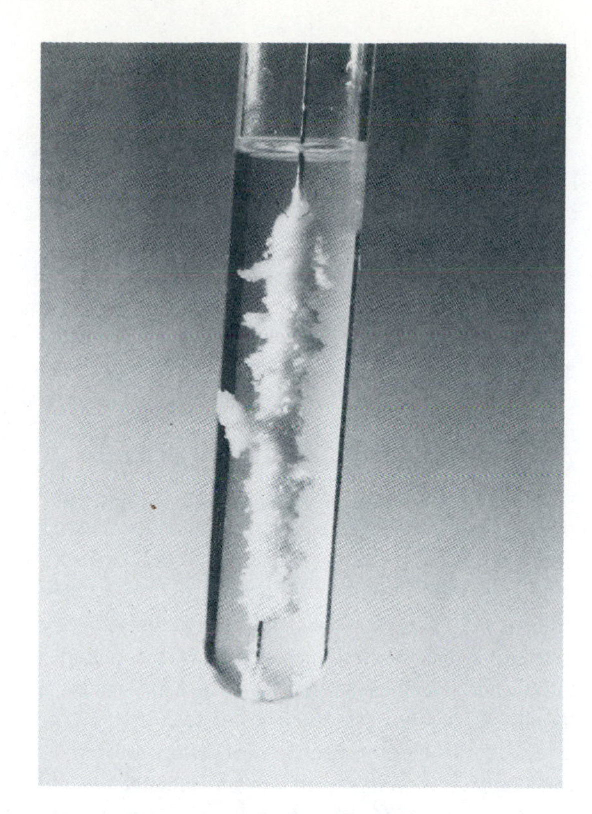

Figure 7-19 *Streptococcus mutans* grown in culture tube containing synthetic medium plus sucrose. Wire in tube is covered with sticky white polysaccharide (dextran) produced by growing microorganisms.

basal structure (Figure 7-20, *A*). The outer rings, called L and P, are attached to the LPS and peptidoglycan layers, respectively, of the cell. The two inner rings are called S and M. The S ring lies between the cytoplasmic membrane and peptidoglycan, and the M ring is attached to the cytoplasmic membrane. Only two rings exist in gram-positive bacteria, one attached to the cytoplasmic membrane and one presumably to the teichoic acid.

The spirochetes have ***axial fibers*** that arise from each pole of the cell and appear between the protoplasmic cylinder and the outer sheath (Figure 7-21). One end of each fiber is inserted into areas called insertion pores located near the poles (Figure 7-22). There may be 2 to 100 axial fibers per cell, with each consisting of a filament, hook, and basal structure similar to bacterial flagella.

Functions

There is considerable difference in flagellar movement between prokaryotes and eukaryotes. There is no contractile mechanism in the prokaryotic flagellum, nor can waves of motion be transmit-

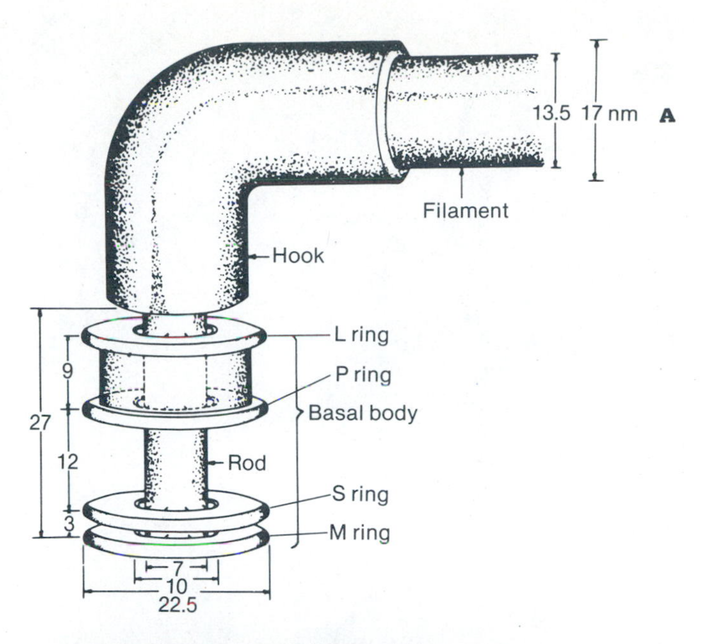

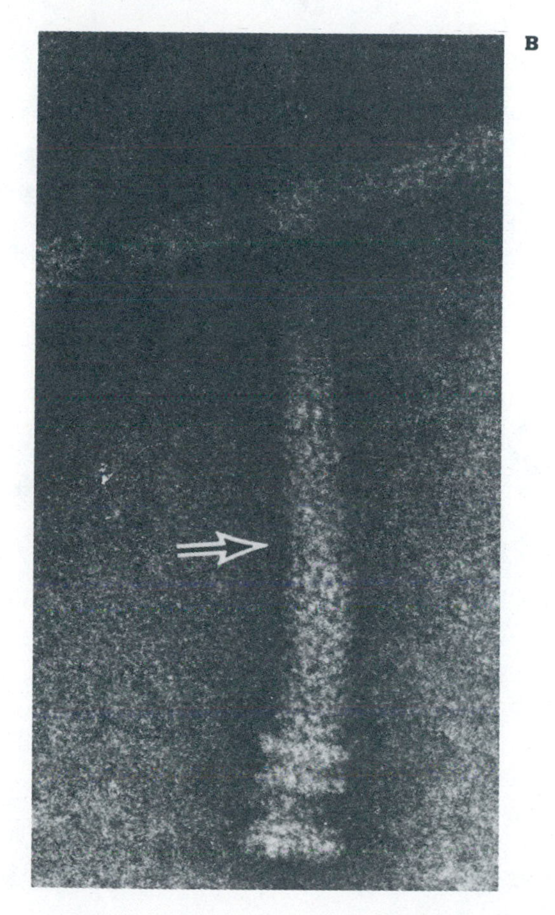

Figure 7-20 Structure of prokaryotic flagellum. **A,** Structure and size of flagellum. L and P rings in gram-negative bacteria are anchored to outer membrane of cell envelope, and S and M rings are attached to inner membrane. Gram-positive cells possess flagella in which only S and M rings are present. Dimensions are in nanometers. **B,** Electron micrograph of flagellum from gram-negative cell. Arrow points to junction between hook and filament.

A, From DePamphilus, M.L., and Adler, J. J. Bacteriol. ***105****:384, 1971.* ***B,*** *From DePamphilus, M.L., and Adler, J. J. Bacteriol.* ***105****:396, 1971.*

A

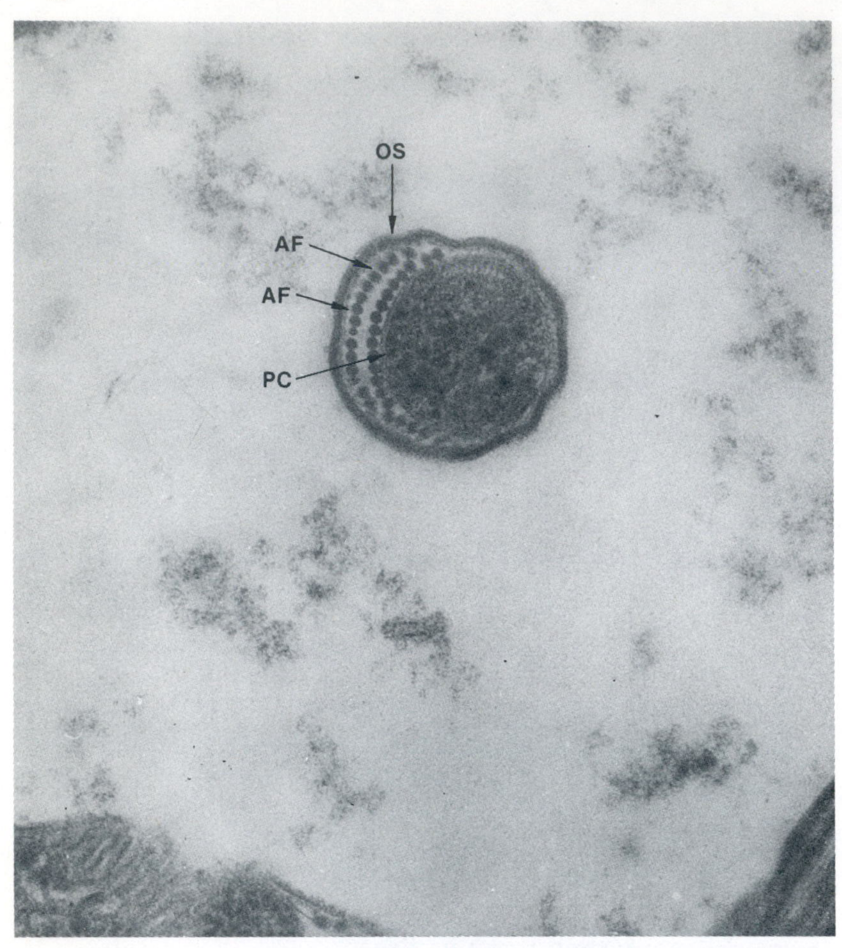

B

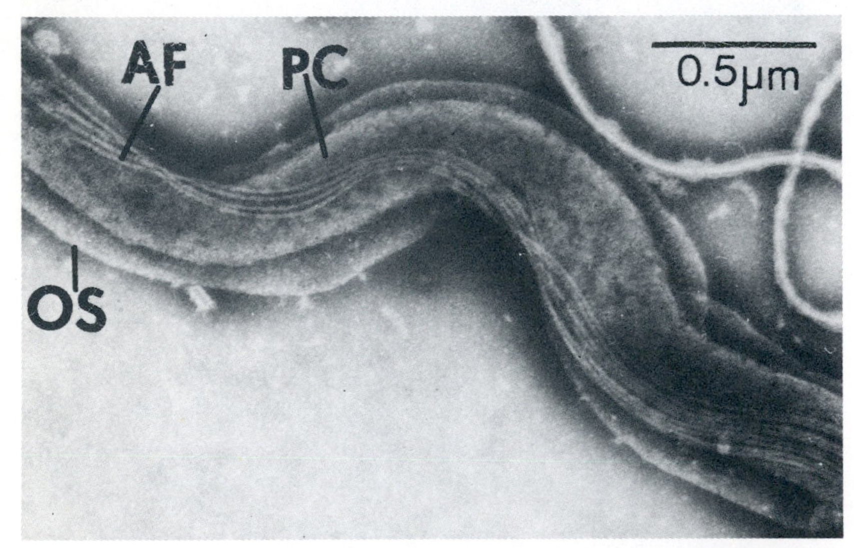

Figure 7-21 Axial fibers of spirochetes. **A,** Electron micrograph showing cross-section of spirochete. (×183,000.) Axial fibers *(AF)* are many and lie between outer sheath *(OS)* and protoplasmic cylinder *(PC)*. Outer part of protoplasmic cylinder is cell wall. **B,** Electron micrograph showing relationship of axial fibers to other components of spirochete. *OS,* Outer sheath; *PC,* protoplasmic cylinder; *AF,* axial fibers.

A, *From Lisgarten, M.A., and Socransky, S.S.* J. Bacteriol. ***88:****1093, 1964.* ***B,*** *From Ziolecki, A.* Appl. Environ. Microbiol. ***39:****919, 1980.*

ted along it. Instead, motility results from rotation of the flagellum with the help of a basal body that acts as a rotary motor. Mechanically, the M ring is considered the motor, the S ring a stator, and P and L rings bushings. Torque is believed to be generated between the S and M rings, with the M ring rotating in the cytoplasmic membrane. It has been observed that when latex beads are attached to the flagellum of a bacterial cell, the beads can be seen to rotate about the axis of the flagellum during movement while the body of the cell rotates in a different direction. Motile bacteria, such as the peritrichously flagellated *E. coli,* swim in almost a smooth straight-line fashion, with each run being interrupted by jiggling motions called ***twiddles*** or ***tumbles.*** Following the short tumbling period, the bacterial cell begins a new run, but in a random direction. These changes in direction are also accompanied by a change in the direction of rotation of the flagellum. The direction of twist of the rigid bacterial flagellum is left handed, and smooth swimming is characterized by counterclockwise rotation of the flagella, whereas tumbles are clockwise. The flagella in *E. coli* form a coordinated bundle during smooth swimming (Figure 7-23, *A*) that enables the cell to move and rotate in one direction. A change of direction or tumbling results in clockwise rotation, and the flagellar bundle flies apart (Figure 7-23, *B*). Not all bacteria tumble in order to change direction. Some monoflagellated bacteria "back up" after a period of smooth swimming and then change direction.

The mechanism by which spirochetes move is not completely understood. They move in a screwlike motion when observed in vivo, and many investigators believe they rotate about their longitudinal axes. One mechanism that has been proposed suggests that the rotation of axial fibers in one direction causes the rotation of the protoplasmic cylinder in the opposite direction. The outer sheath, which is believed to be flexible, rotates in a direction opposite that of the protoplasmic cylinder.

Taxes and tropisms. Microorganisms have sensory receptors that enable them to respond to stimuli of a chemical or physical nature. These responses may be in the form of taxes and tropisms. ***Tactic*** responses are movements by cells that are motile, and ***tropisms*** are growth orientations of organisms that are normally attached to a surface and are unable to demonstrate locomotion. Tropisms are more characteristic of eukaryotic microorganisms. The tactic response in prokaryotes may be due to stimuli such as light, heat, electricity, and chemical agents. The most widely studied responses are those resulting from the presence of chemical agents, that is, chemotactic responses.

Chemotaxis. Chemotaxis is a biological phenomenon that is common to the microbial world as well as higher organisms. It is a property that has been demonstrated to be important in controlling many behavioral responses, including sexual activity, feeding, predatory-prey relationships, and aggregation. Bacteria may be attracted to chemicals ***(positive chemotaxis)*** or repelled by chemicals ***(negative chemotaxis).*** Chemotaxis can be demonstrated by creating a concentration gradient, and one technique is to use a capillary tube containing a specific chemical. If the capillary tube contains an attractant and is placed in a suspension of bacteria, the bacteria will soon be found in high concentrations around the capillary opening and within the capillary tube (Figure 7-24). By removing the capillary tube at intervals of time, it can be seen that the number of bacteria in the capillary tube increases with exposure time. If a repellent is added to the bacterial suspension, there will be a rapid increase in the number of bacteria within the capillary tube where the concentration of repellent is least. Since chemotaxis is associated with flagellar movement, many factors that affect flagellar activity will also affect chemotaxis. Such factors include pH, temperature, and heavy metal concentrations.

How does the cell respond in its movement toward an attractant or away from a repellent? A cell, in the absence of a chemical gradient, will move randomly in one direction and then in another with little progress in any one direction. Once a chemical is added to the medium, the cell responds in one of two ways. If the chemical is an attractant, the microbial cell moves to the area where the concentration of attractant is highest. This movement is characterized by less tumbling and longer periods of smooth swimming. The cell, in the presence of a chemical that acts as a repellent, swims away from the highest concentration of repellent to the area of lowest concentration of repellent. Again the tumbling is decreased. In both of the above instances there is specific orientation and progress to or away from the stimulus.

Microorganisms may be attracted to amino acids, sugars, oxygen, vitamins, and inorganic ions; however, it is not necessary that these attractants be actually metabolized by the cell. It has been observed in some instances that the attractant can become a repellent as the concentration of the chemical increases or if other activities in the cell change. Aerobic microorganisms, for ex-

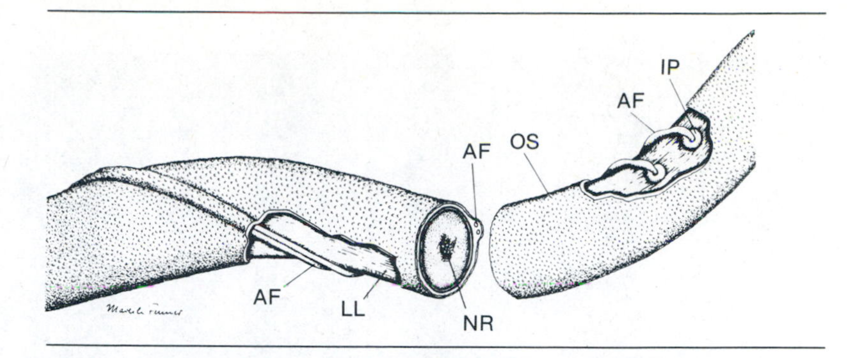

Figure 7-22 Relationship of flagella-like fibers (axial fibers) in spirochetes to other cell components. Axial fibers *(AF)* are located between outer sheath *(OS)* and lipoprotein layer *(LL)* of cell wall. Axial fibers are inserted into protoplasmic cylinder by insertion pores *(IP)*. *NR*, Nuclear region.
From Holt, S. Microbiol. Rev. ***42:****114, 1978.*

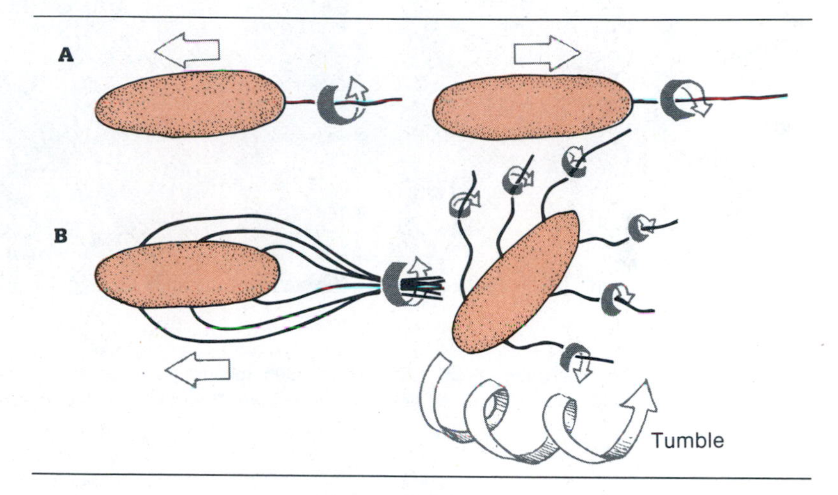

Figure 7-23 Flagellar rotation and movement. **A,** Response of cell with polar flagellum. On left is normal body rotation in which single flagellum rotates counterclockwise and cell moves forward (period of smooth swimming) with flagellum trailing cell. On right, single flagellum shows clockwise rotation. Direction of movement (period of tumbling) is with flagellum ahead of cell. **B,** Response of cell with peritrichous flagellation. On left, counterclockwise rotation (period of smooth swimming) causes flagella to form bundle behind cell. On right, clockwise rotation (period of tumbling) causes individual flagella to fly apart.

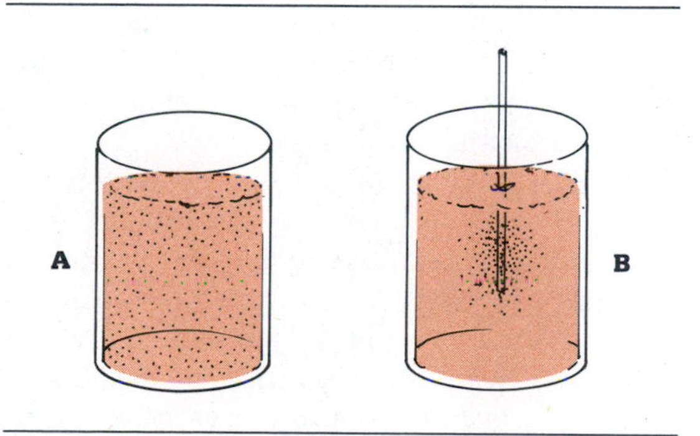

Figure 7-24 Demonstration of bacterial chemotaxis. **A,** Bacteria evenly dispersed in flask. **B,** Capillary tube with chemical attractant is added to culture medium. Bacteria are seen to aggregate around and within capillary tube.

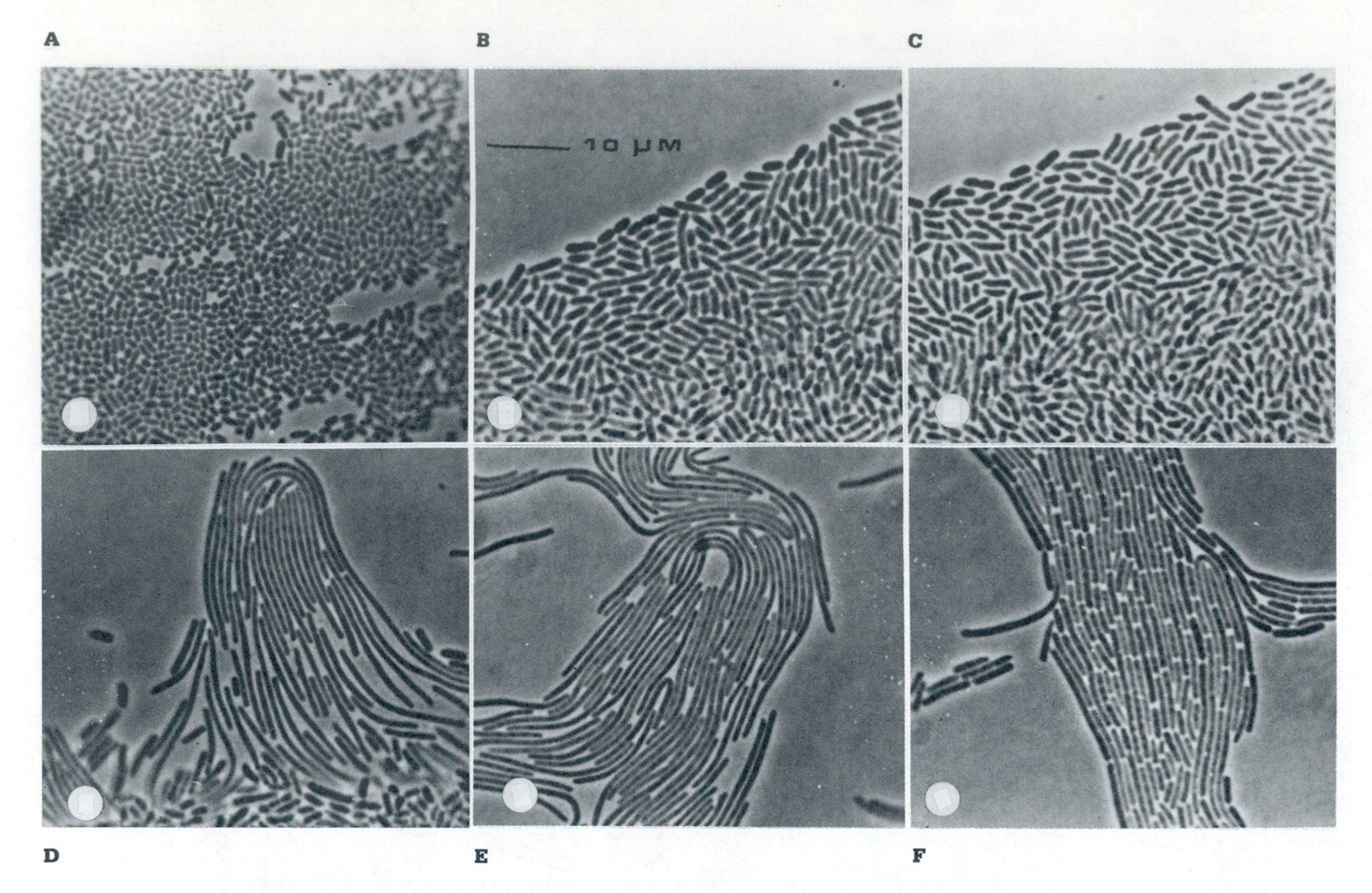

Figure 7-25 Swarming phenomenon. Phase-contrast micrographs of *Proteus mirabilis* within colony. **A,** Cells are initially coccoid. **B** and **C,** Cells at edge of colony begin to enlarge. **D** and **E,** Cells at edge of colony move out into agar in large groups of long swarm cells. **F,** Swarm cells begin to divide and form short rods as they consolidate.

From Williams, F.D., and Schwarzhoff, R.H. Reproduced, with permission, from the Annual Review of Microbiology, *Volume 32.* © *1978 by Annual Reviews, Inc.*

ample, can be attracted to a high concentration of oxygen, but when respiration slows down because of a lack of nutrients in the medium, oxygen acts as a repellent. Negative chemotaxis is most frequently observed with chemicals that produce extremes of pH, that is, pH values between 2 and 4, and 10 and 12. Such responses to stimuli are of obvious selective advantage to the microorganism because they provide the cell with a mechanism for avoiding lethal or deleterious regions of its environment.

Mechanism of chemotaxis. The sensory response of bacteria to various chemicals is associated with ***chemoreceptor proteins*** located in the cell. Two types of receptors have been detected: those located in the periplasm and those associated with the cytoplasmic membrane. Three ***periplasmic receptors*** have been identified: the galactose receptor, the ribose receptor, and the maltose receptor. Each of these receptors is also involved in the transport of these sugars from the periplasm into the cytoplasm of the cell.

The first step in microbial chemotactic response is initiated by the binding of the chemoeffector to the periplasmic receptor. The binding induces a conformational change that permits the periplasmic receptor to bind to the cytoplasmic membrane receptor. This conformational change results in the transmission of the signal to the cytoplasm of the cell. The cytoplasmic membrane protein also undergoes a ***methylation-demethylation*** step that is intimately tied to the sensory process. Methylation is apparently tied to smooth swimming of the organism, since the presence of an attractant increases methylation and decreases demethylation, and demethylation is associated with tumbling and the presence of a repellent. Once inside the cytoplasm, the signal is apparently processed and finally translated by the rotary motor of the flagellum.

Swarming. Motile species of *Proteus* exhibit a response on the surface of agar called ***swarming.*** The response is observed when cells are inoculated near the center of the plate. The cells undergo division and form colonies, but later the cells on the perimeter of the colony change mor-

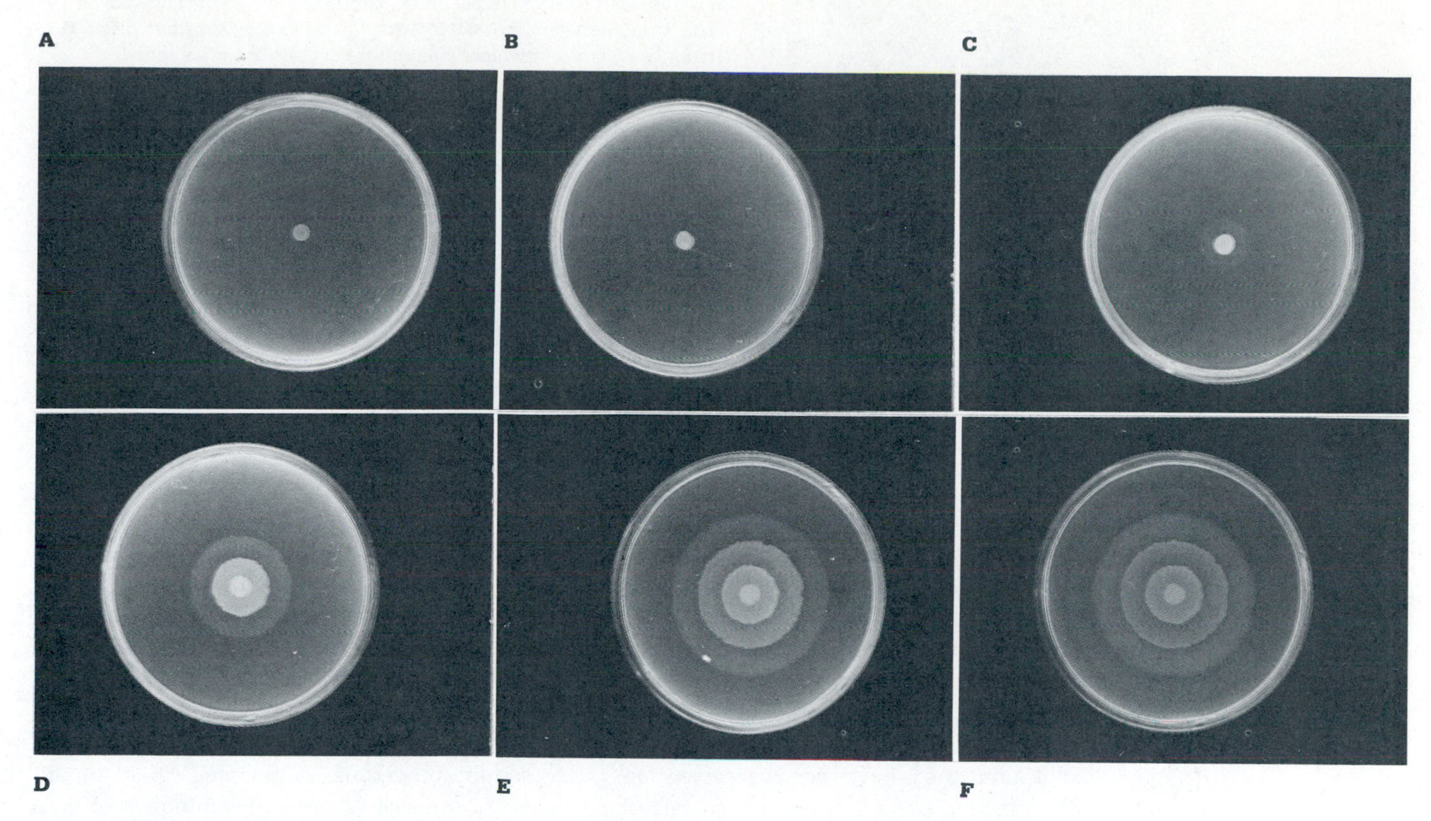

Figure 7-26 Appearance of agar plate demonstrating swarming of *Proteus mirabilis* after 2 hours, **A;** 3 hours, **B;** 4 hours, **C;** 8 hours, **D;** 12 hours, **E;** and 16 hours, **F** (see text for details).

From Williams, F.D., and Schwarzhoff, R.H. Reproduced, with permission, from the Annual Review of Microbiology, *Volume 32.* © *1978 by Annual Reviews, Inc.*

phologically and swarm to fresh areas. The process will continue until the entire surface is covered with microbial growth. The mechanism controlling this response is not understood, but it is not believed to be a chemotactic response. The most interesting aspect of this phenomenon is the morphological changes that occur (Figure 7-25). First, as the colony develops, cells at the edge of the colony elongate and the number of flagella per cell increases. The elongated (swarm) cells then move out from the colony in larger and larger groups, and later the swarm cells consolidate and undergo normal growth and division to produce short cells. As the colony enlarges, short cells at the edge of the colony again elongate, and the whole process is repeated until the plate is covered with zones of light and heavy growth (Figure 7-26).

Gliding motility

Evidence is accumulating that suggests that gliding motility may be associated with cellular structures. ***Rotary assemblies*** have been located in the cell envelope of some species of *Cyanobacteria*. The rotary particles are composed of protein and resemble the basal structures found in flagellated bacterial species (Figure 7-27). Latex spheres added to a cell culture of certain gliding bacteria adhere to the surface of the cell. Microscopic examination of the culture shows the latex particles moving over the surface of the cell. Sometimes the latex spheres bind to one of the poles of the cell and appear to spin in place. It has been suggested that the rotary assemblies can spin in one direction or another and that somehow they are associated with gliding movement.

PILI AND FIMBRIAE

Pili (Latin: hair or threadlike) and *fimbriae* (Latin: thread or fiber) are terms that have been used interchangeably to designate proteinaceous nonflagellar appendages. *Fimbriae* is currently used

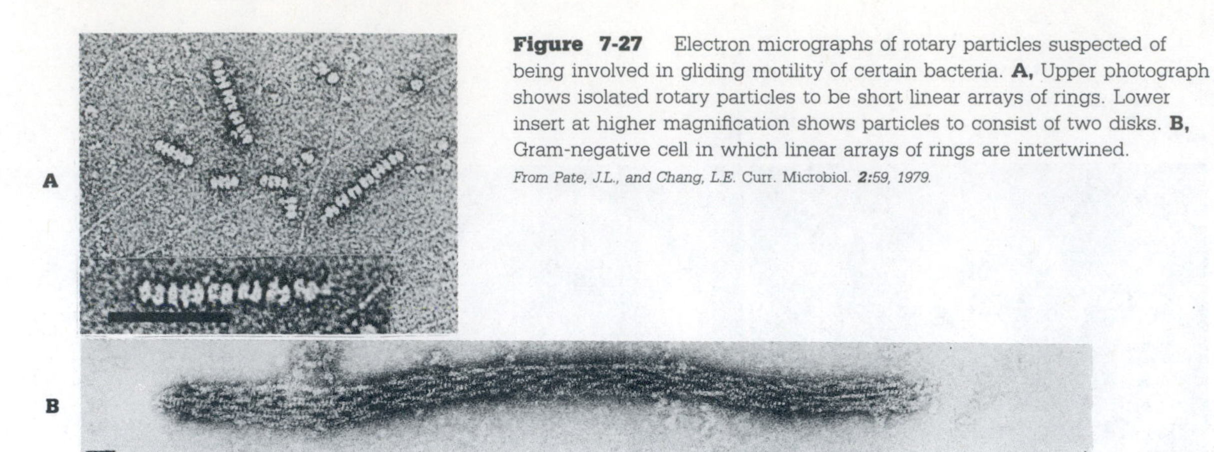

Figure 7-27 Electron micrographs of rotary particles suspected of being involved in gliding motility of certain bacteria. **A,** Upper photograph shows isolated rotary particles to be short linear arrays of rings. Lower insert at higher magnification shows particles to consist of two disks. **B,** Gram-negative cell in which linear arrays of rings are intertwined.
*From Pate, J.L., and Chang, L.E. Curr. Microbiol. **2**:59, 1979.*

to denote nonflagellar appendages other than those involved in sexual conjugation. *Pili* refers to appendages that are involved in the transfer of nucleic acid during a sexual process between bacteria called conjugation. (See Chapter 11 for details of the conjugation process.) Fimbriae and sex pili are for the most part found on gram-negative bacteria, particularly members of the family Enterobacteriaceae (Figure 7-28). Fimbriae or fimbriae-like appendages have been observed on some gram-positive bacteria, such as species of *Streptococcus* (Figure 7-29) and *Corynebacterium,* but sex pili have not been found in gram-positive bacteria. Neither fimbriae nor sex pili are absolutely essential to the survival of the cell, but they do provide the cell with certain advantages over those cells that do not carry them.

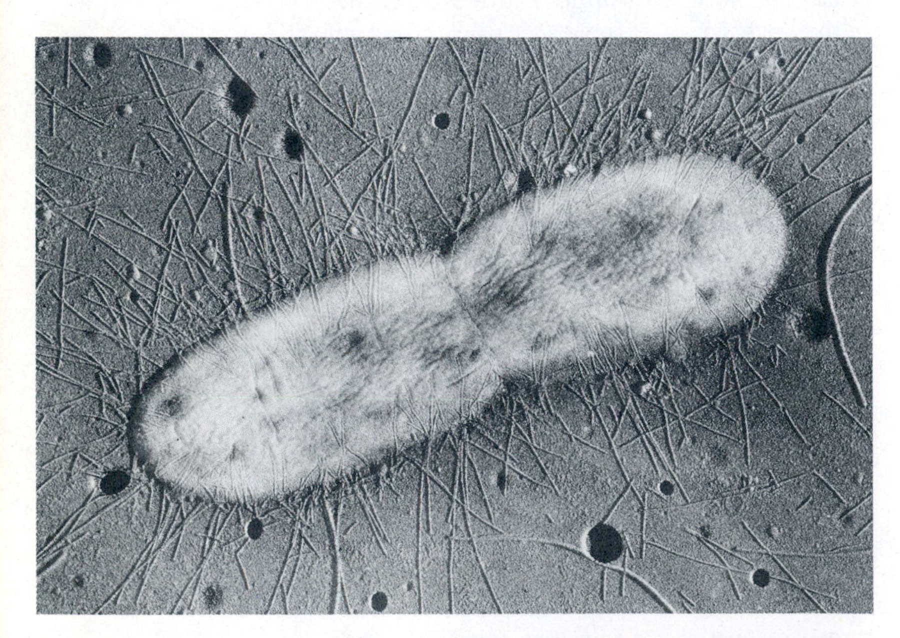

Figure 7-28 Electron micrograph of fimbriated gram-negative *(E. coli)* bacterium.
Courtesy A. Ryter, Institute Pasteur.

Fimbriae arise from the cytoplasm of the cell, and the entire surface may be covered by them. A cell may carry from one to 100 fimbriae, which can vary in length from 0.2 to 20 μm, but a few species have fimbriae located only at the poles. Fimbriae production is controlled by genes on the chromosome, in contrast to pili, which are governed by genes located on extrachromosomal DNA called plasmids. Although fimbriae do confer certain properties on the cell, their significance in terms of selective advantage over cells without them has been difficult to evaluate. Some of the suggested properties are:

1. ***Antigenicity.*** Pili and fimbriae are antigenic, and their presence on the bacterial cell surface makes them readily accessible to those immune products of mammalian systems that engage in antibody formation. Fimbriae formation by the causative agent of gonorrhoea *(Neisseria gonorrhoeae)* has prompted an attempt at producing a vaccine by using purified fimbriae-protein.
2. ***Adherence.*** Fimbriae confer adhesive properties to the cell and enable it to bind to epithelial surfaces such as those found on the respiratory, intestinal, and genitourinary tracts of animals and humans (Figure 7-30). Attachment permits microbial cells to replicate and increase to higher numbers than cells that do not have mechanisms for attachment.
3. ***Aggregation.*** Fimbriae cause certain bacteria to aggregate and form pellicles or thin layers of cells packed together on a surface of a liquid medium. This characteristic may facilitate oxygen delivery to the cell in oxygen-poor or oxygen-depleted areas.

Sex pili are nonflagellar appendages on the cell surface that are involved in the transfer of nucleic acid from one cell to another (Figure 7-31). The sexual preference may be between species of the same genus or between species from different but related gram-negative genera. The DNA transferred may be totally chromosomal, viral, or plasmid (extrachromosomal DNA). The information for the production of sex pili (sing., pilus) is referred to as the fertility factor or F^+ factor. The cell producing the sex pilus is called the male or donor cell. Sex pili are easily distinguished from fimbriae by their greater length (usually 10 to 20 μm) and larger diameter. Only 1 to 10 sex pili can be found on any one cell. Electron micrographs indicate they have an axial hole and knobs, which might indicate a membrane base. In addition to their role in DNA transfer during the conjugation process, sex pili are also antigenic and can be distinguished by immunological techniques. They also act as organs of attachment for certain RNA bacteriophages and filamentous DNA phages that invade and lyse specific bacterial species.

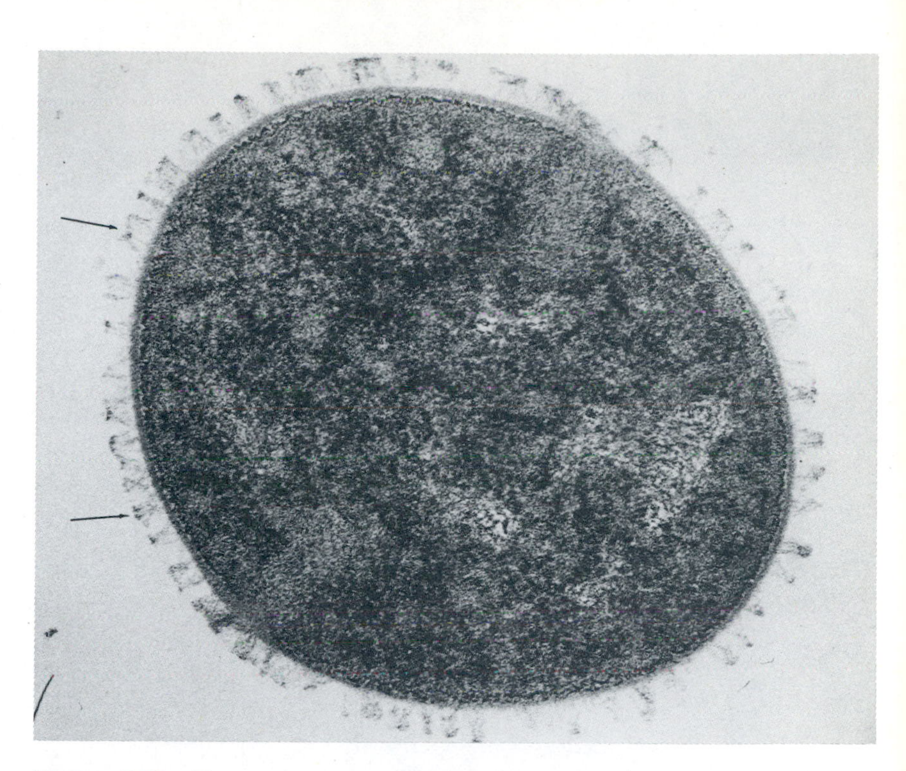

Figure 7-29 Electron micrograph of fimbriae *(arrows)* on surface of gram-positive bacterium *(Streptococcus pyrogenes)*. (×140,000.)
Courtesy Roger M. Cole.

CYTOPLASMIC INCLUSIONS

Cytoplasmic inclusions are insoluble structures or components in the cytoplasm that perform specific functions in the cell. They include ribosomes, nucleoids, gas vacuoles, carboxysomes, chlorosomes, magnetosomes, and areas of accumulated polymers.

Ribosomes

Ribosomes are cytoplasmic particles that engage in protein synthesis. They can be seen with the electron microscope when the cell is carefully prepared and stained (Figure 7-32). The prokaryotic ribosome, which is composed of 60% RNA and 40% protein, has a sedimentation constant* of ***70S*** and is composed of two subunits: ***30S*** and ***50S***. The 70S ribosomes are stable in solutions with relatively high magnesium concentrations, but when the magnesium concentration is reduced, the two subunits can be disassociated. As the growth rate of a cell increases, there is also an increase in the number of ribosomes. The structure and biosynthesis of the ribosome differs considerably between prokaryotes and eukar-

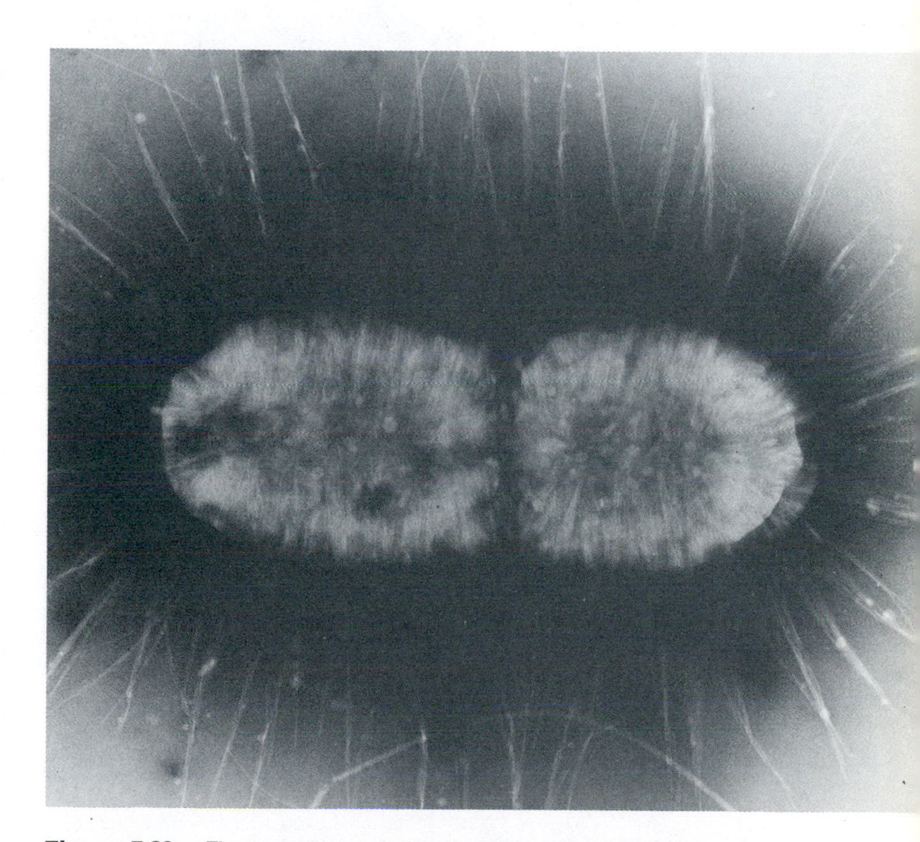

Figure 7-30 Electron micrograph demonstrating fimbriae of *E. coli* used to attach to intestinal epithelium. (×60,000.) This is strain that can cause diarrhea.
From Deneke, C.F., et. al. Infect. Immun. ***26:****362, 1979.*

* *S* stands for Svedberg units. They represent the relative sedimentation constant of a component being centrifuged at very high speeds. The greater the molecular weight of a component, the faster it sediments in a high-viscosity suspending medium, and the larger the S value.

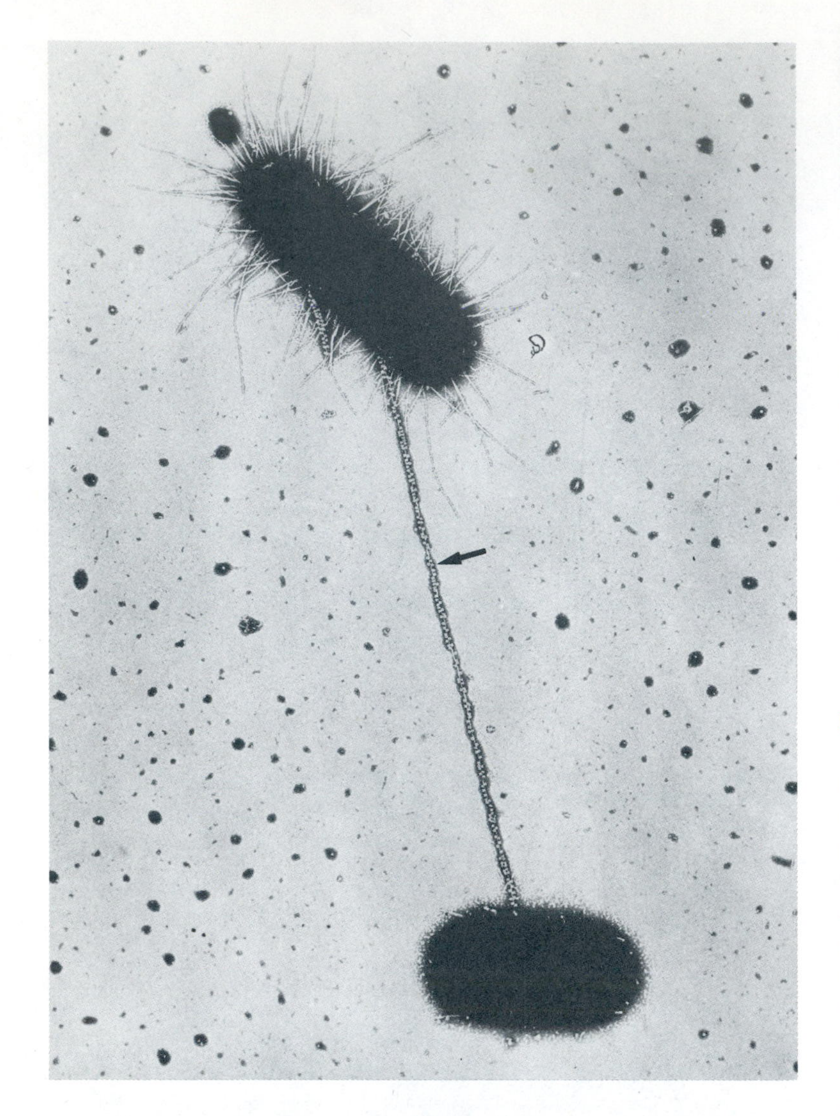

Figure 7-31 Electron micrograph demonstrating sex pilus in *E. coli*. Upper bacterial cell is covered with fimbriae, but there is only one sex pilus *(arrow)*, which in this photo is covered with special viral particles. Fimbriated cell with sex pilus is male, or donor, cell, and cell in lower portion of photograph is female, or recipient.
Courtesy J. Brinton.

yotes, even though ribosomal function and the overall mechanics of protein synthesis are the same. This disparity is revealed indirectly when their sensitivity to certain antibiotics are compared. Streptomycin, for example, binds to prokaryotic ribosomes and inhibits protein synthesis but has no effect on eukaryotic ribosomes. This difference indicates a possible independent evolutionary origin of the ribosome or ribosomal proteins for the two groups. The mechanisms of protein synthesis are discussed in Chapter 10.

Nucleoids

One of the most distinctive differences between eukaryotic and prokaryotic cells is the absence of a nuclear membrane in prokaryotic cells. The nuclear material, or DNA, occupies a position in the cell called the ***nucleoid*** region. Under optimal growth conditions as many as four macromolecules of DNA may be present in the cell, each representing a nucleoid region. Nucleoid regions are easily observed in electron micrographs (Figure 7-32) as well as in phase-contrast microscopy (Figure 7-33).

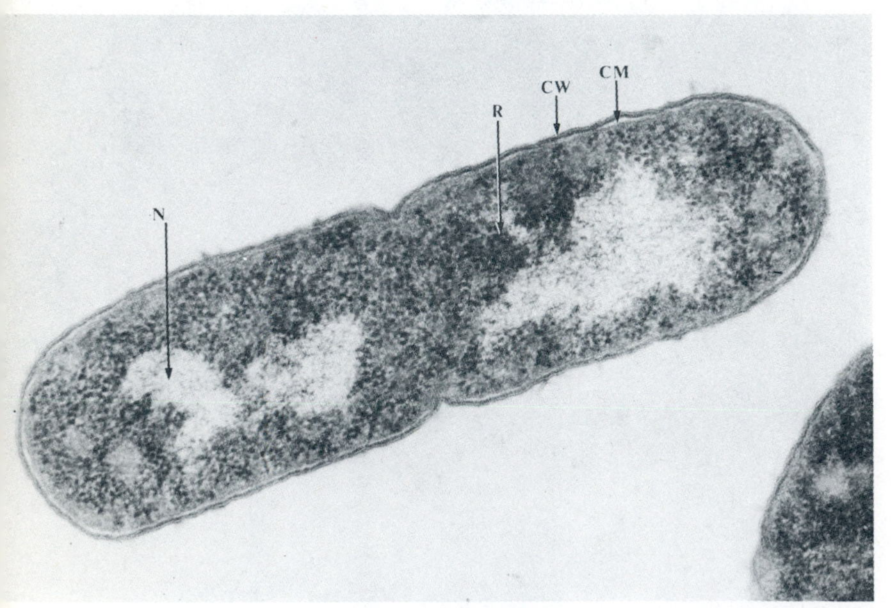

Figure 7-32 Ribosomes. Electron micrograph of *E. coli* demonstrating ribosomes. (×60,000.) *CW*, Cell wall; *CM*, cytoplasmic membrane; *N*, nucleoid; *R*, ribosomes.
Courtesy A. Ryter, Institute Pasteur.

Gas vacuoles

Gas vacuoles are hollow cavities made up of protein ***vesicles*** and are found (Figure 7-34) in many aquatic prokaryotes, such as the cyanobacteria, photosynthetic bacteria, and halobacteria. The protein subunits are assembled in the shape of a cylinder, producing a hollow space within the protein clusters. Gas enters the vesicles by passive diffusion through pores in the protein, but water cannot penetrate. The inner surface of the vesicle appears to be hydrophobic, and this prevents the entrance of water. Gas vesicles, which have no counterpart in eukaryotes, are not bounded by a unit membrane, which is characteristic of other organelles. The function of the vesicles is to provide buoyancy, and this keeps the organism at a depth where nutrients, oxygen, light, etc., are most suitable for microbial growth. Once the vesicles collapse, they cannot be reinflated, and new vesicles must be produced to replace them.

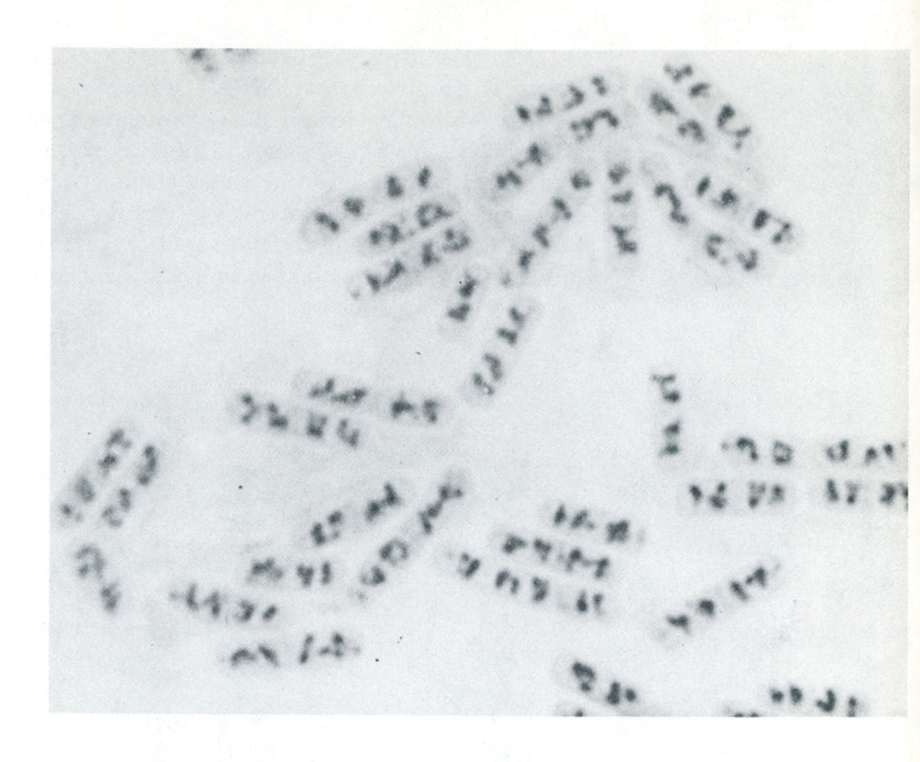

Figure 7-33 Nucleoids of *Bacillus cereus*. Nucleoids have been fixed in osmium tetroxide and stained with Giemsa.
Courtesy C.F. Robinow.

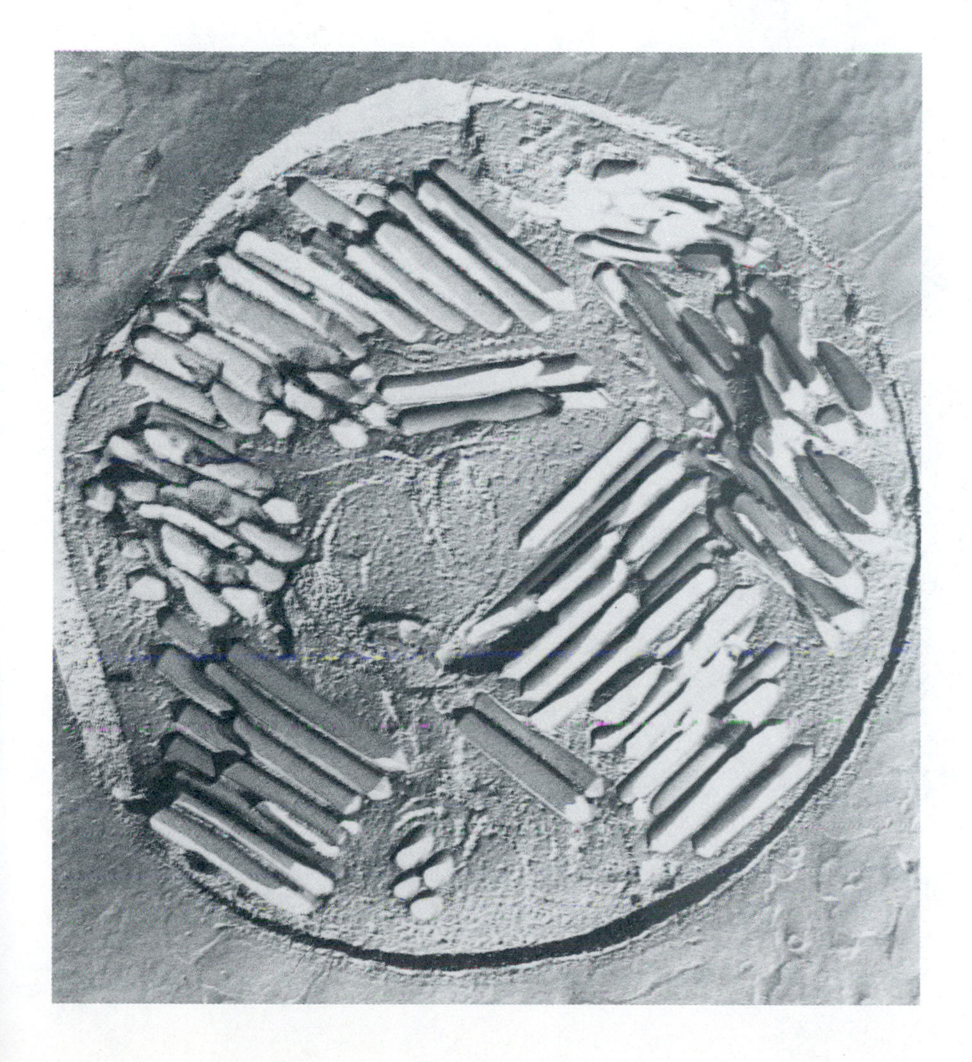

Figure 7-34 Gas vacuoles. Freeze-etched electron micrograph showing cross section of gas vacuoles in cyanobacterium.

Nearly all gas vesicle–containing bacteria are aquatic, and few possess flagella. Aquatic organisms that do not possess gas vacuoles are flagellated. Thus the ability of aquatic microorganisms to position themselves in water where various chemical gradients exist may be accomplished by either motility and chemotaxis or by the property of buoyancy associated with gas vacuoles.

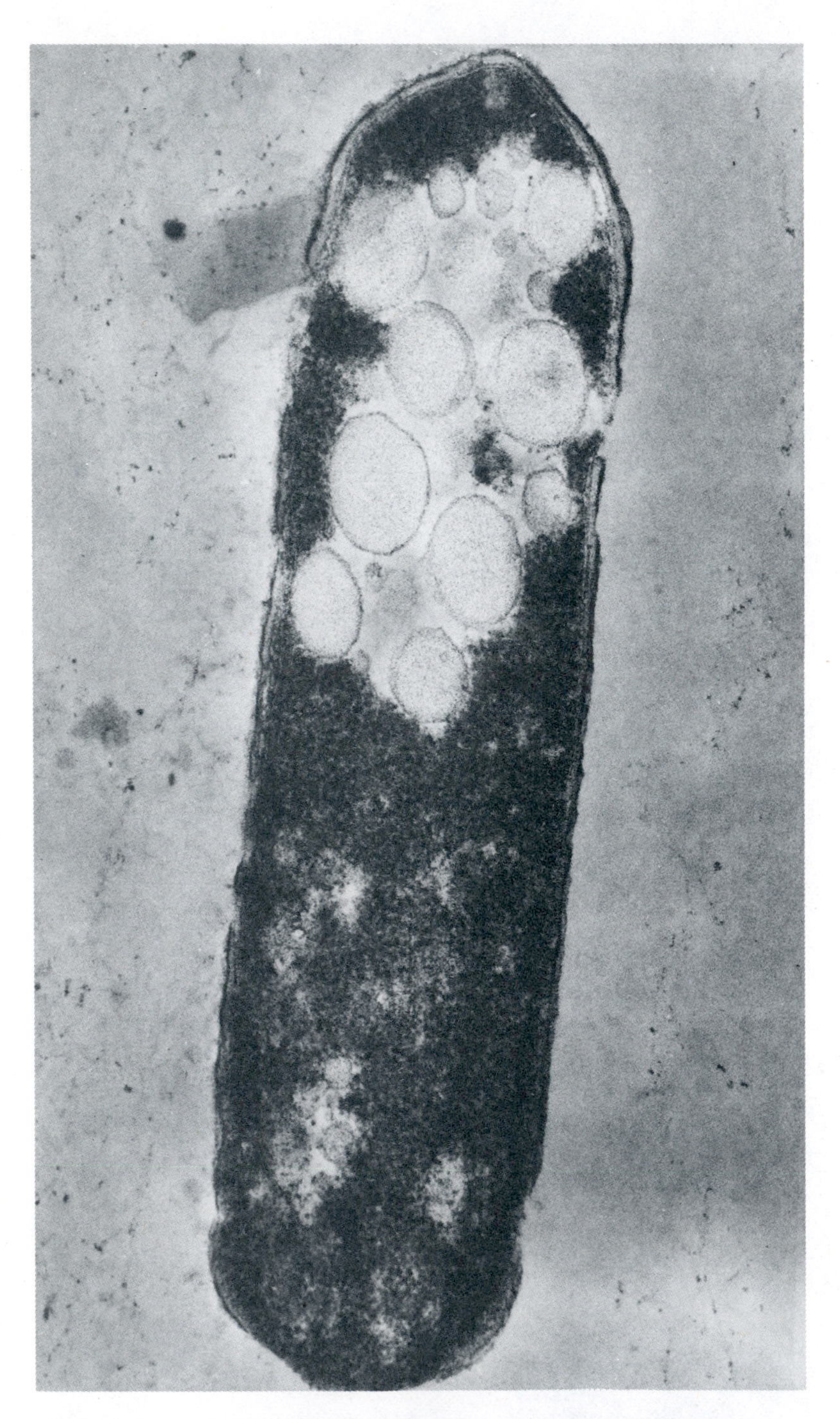

Figure 7-35 Electron micrograph of glycogen granules from *Clostridium pasteurianum*. Cell is damaged, revealing membranous coats surrounding granules. *From Laishley, E.J., et al. Can. J. Microbiol.* **19:**991, 1973.

Glycogen granules

Glucose is stored by many bacterial cells, in the form of a polymer called glycogen, under conditions of carbon excess and certain nutrient limitations such as nitrogen depletion. The accumulated glycogen (Figure 7-35) forms inclusions that in most bacteria is surrounded by a membrane. The glycogen reserve is later hydrolyzed to glucose when nutrients become available.

Polyphosphate granules

Inorganic phosphate accumulates in some bacterial cells during maximum growth. The phosphates are converted to polyphosphates, which form inclusions in the cell called ***metachromatic granules*** (also called volutin or Babés-Ernst bodies). Their only function is as a reserve of phosphate. They are useful in the identification of corynebacteria.

Poly β-hydroxybutyrate granules

Poly β-hydroxybutyrate (PHB) granules are lipid storage components accumulated by many bacteria, including the cyanobacteria. They have also been observed to accumulate in sporulating cells of *Clostridium botulinum* type E, the causative agent of a type of food poisoning called botulism. PHB granules are cellular reserves of carbon and energy or energy alone, and when excess carbon and energy sources are present in the medium, they may account for up to 50% of the dry weight of the cell. The granules are surrounded by a nonunit membrane that is believed to contain the enzymes for PHB synthesis (Figure 7-36).

Sulfur granules

Certain bacteria contain inorganic sulfur in the form of inclusions (Figure 7-37). Sulfur granules are surrounded by a nonunit membrane that is believed to be proteinaceous. Sulfur granules are produced by the nonphotosynthetic *Beggiatoa* during the oxidation of hydrogen sulfide, the organism's energy source. (See Chapter 13 for a discussion of the sulfur-oxidizing bacteria.) Intracellular sulfur can be oxidized to sulfate to produce energy when hydrogen sulfide becomes depleted as an energy source.

Figure 7-36 Electron micrograph of thin section of *Ferrobacillus ferrooxidans* showing poly β-hydroxybutyrate (PHB) granules.
From Wang, W.S., and Lundgren, D.G. J. Bacteriol. ***97:**947, 1969.*

Magnetosomes

Bacteria collected from marine marsh muds in Cape Cod, Massachusetts, in 1975 were observed to be magnetotactic. When observed microscopically (without the involvement of light), the microorganisms moved to one side of a drop of mud and changed direction only when magnets were moved in their vicinity. Ultrastructural studies of these organisms revealed the presence of a chain of electron-dense particles (Figure 7-38), each of which was surrounded by a nonunit protein membrane. These particles are rich in cellular iron resembling magnetite and are now referred to as ***magnetosomes.*** Cell motility is required for magnetotaxis and results in the orientation of the organism in its environment. The organisms possessing magnetosomes are microaerophilic or strict anaerobes, and it has been suggested that magnetotaxis serves to direct organisms downward toward sediments and anaerobic areas favorable to their growth.

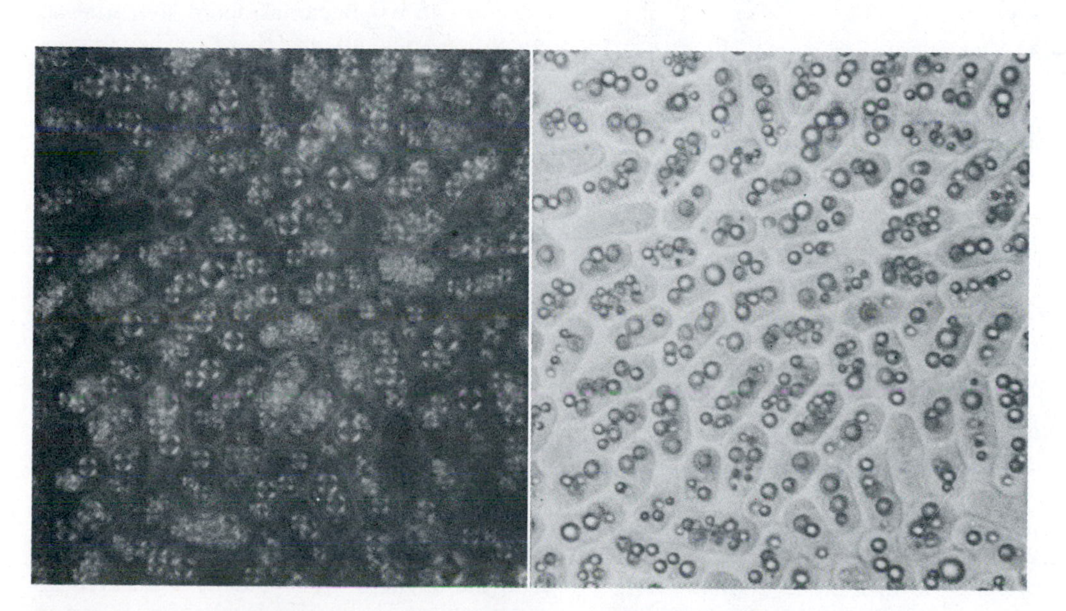

Figure 7-37 Sulfur inclusions. Photomicrograph of sulfur granules in species of *Chromatium.*
From Hageage, G., Jr., Eanes E., and Gherna, R.L. J. Bacteriol. ***101:**464, 1970.*

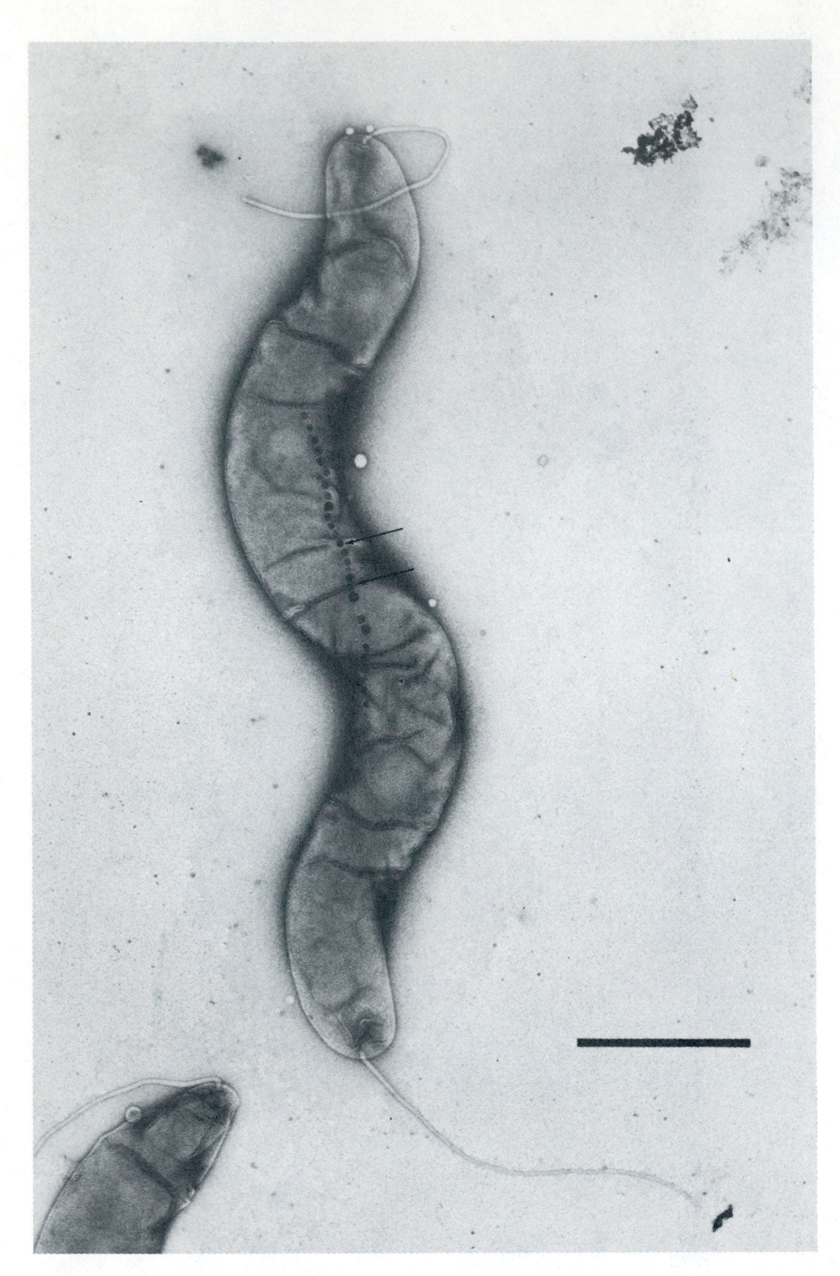

Figure 7-38 Magnetosomes. Electron micrograph of magnetotactic spirillum showing chain of magnetosomes *(arrows)*. Bar = 1 ™m.
From Balkwill, D.L., Maratea, D, and Blakemore, R.P. J. Bacteriol. ***141:****1399, 1980.*

Chlorobium vesicles (chlorosomes)

Chlorobium vesicles are light-harvesting structures found in the photosynthetic green bacteria *(Chlorobiaceae* and *Chloroflexaceae)* that contain bacteriochlorophylls c, d, and e. The chlorobium vesicles serve to distinguish the green bacteria from other photosynthetic bacteria because they are vesicular structures that lie immediately beneath the cytoplasmic membrane (Figure 7-39). Each vesicle is surrounded by an electron-dense nonunit membrane.

Carboxysomes

Carboxysomes are polyhedral or hexagonal bodies that are found in a wide variety of prokaryotes, particularly the cyanobacteria (see Chapter 4), the nitrifying bacteria, and the thiobacilli (Figure 7-40). Carboxysomes, which are surrounded by a nonunit membrane, contain a mass of particles that have been identified as the enzyme ***ribulose 1, 5-diphosphate carboxylase,*** hence the name *carboxysomes.* This enzyme is used for carbon dioxide fixation (the organisms possessing it use carbon dioxide as their sole carbon source). It has not been determined if the carboxysomes are active in carbon dioxide fixation or if they are merely a storage area for the enzyme.

BACTERIAL ENDOSPORES (DIFFERENTIATION PROCESS)

Differentiation in the biological world implies that during an organism's cycle of development or growth some of the cells take on specialized functions; that is, they become differentiated. Differentiation is a property generally associated with eukaryotic organisms or those possessing more than one cell type. One type of differentiation process in prokaryotes is spore formation, which is a response to detrimental environmental conditions. Spore formation does not take place in all bacterial species. Most species that are indigenous to animals and humans do not form spores, since they inhabit areas of the host that are most suitable for their growth and development and where dramatic changes in their environment are unlikely to occur. Bacterial spore formation is more often associated with those species that inhabit the soil, such as species of *Bacillus, Clostridium, Sporosarcina, Thermoactinomyces, Actinobifida,* and *Streptomyces.* All of the above except *Streptomyces* produce endospores. *Streptomyces* species produce spores in specialized multinucleated hyphal filaments that are resistant to dessication but in no way compare with the resistance of endospores to both chemical and physical agents. Spores are a mode of reproduction in *Streptomyces.*

Endospore formation (sporulation process)

Endospore formation occurs most frequently when there is a depletion of an essential metabolite in the environment. Nutrient depletion triggers biochemical as well as morphological changes in the cell, during which time vegetative

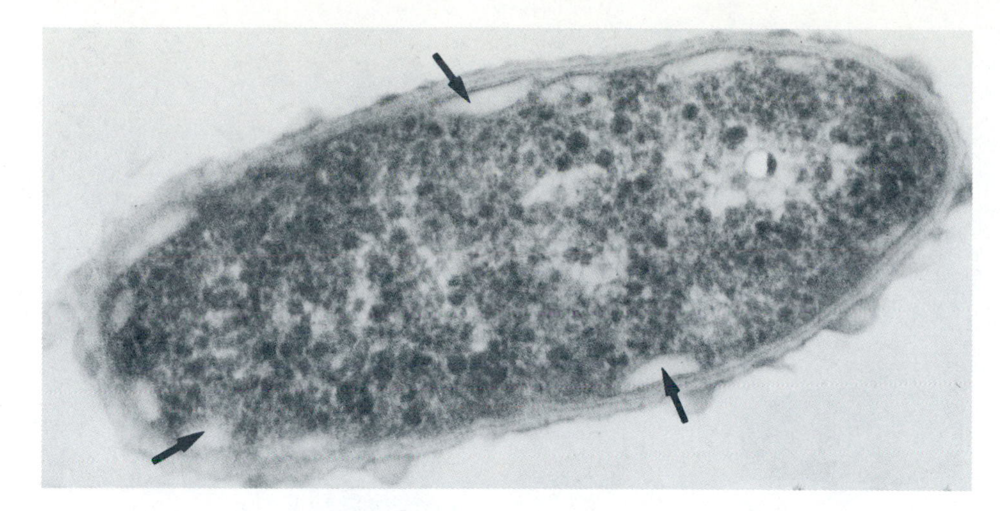

Figure 7-39 Chlorosomes. Electron micrograph of chlorosomes *(arrows)* in species of green bacteria (×90,000.)
From Staehelin, L. Biochem. Biophys. Acta. ***589:****30, 1980.*

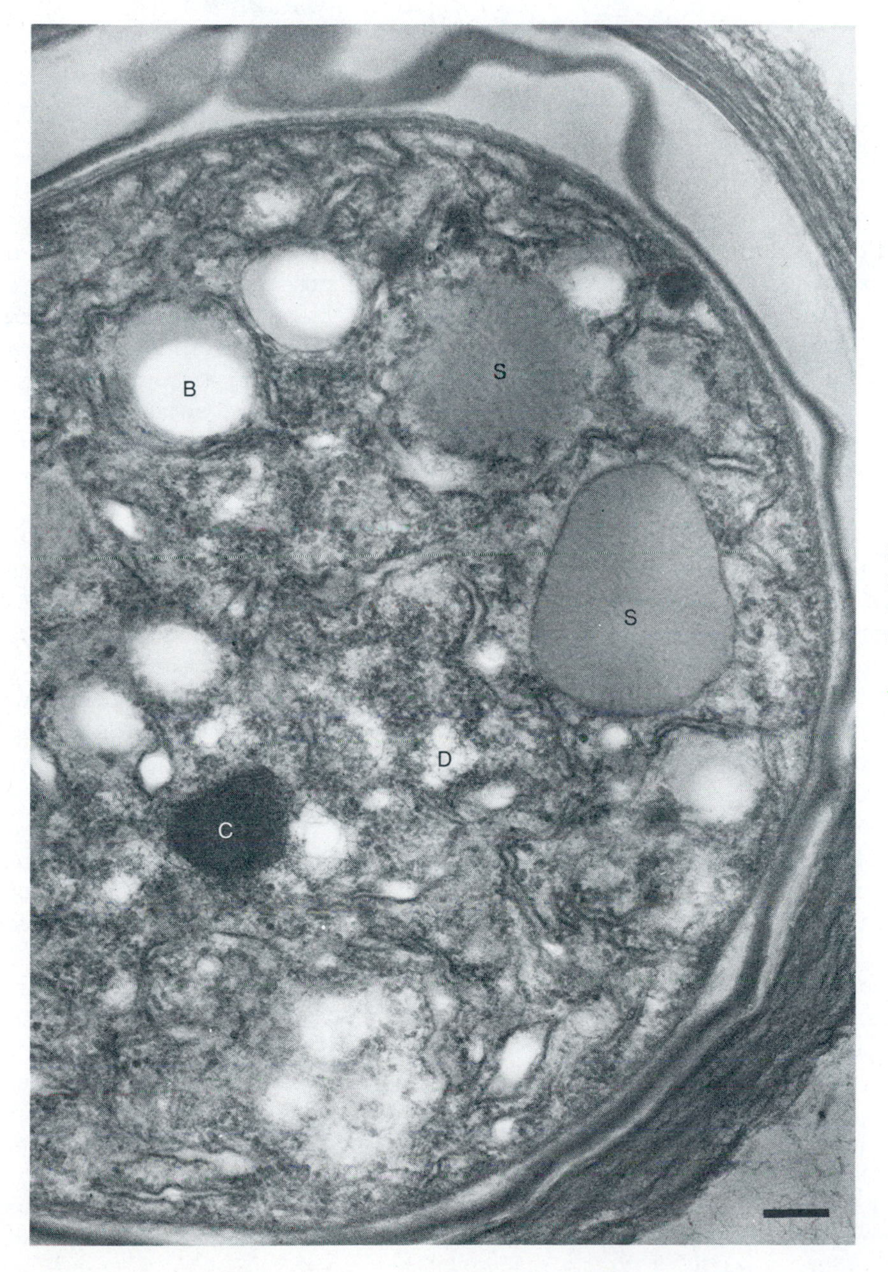

Figure 7-40 Carboxysomes. Electron micrograph of cyanobacterial cell illustrating carboxysome. *C,* Carboxysome; *S,* structural granules; *B,* poly β-hydroxybutyrate granule; *T,* thylakoid; *R,* ribosomes; *D,* DNA.
From Jensen, T.E., and Sicko-Goad, L. J. Bacteriol. ***106:****683, 1971.*

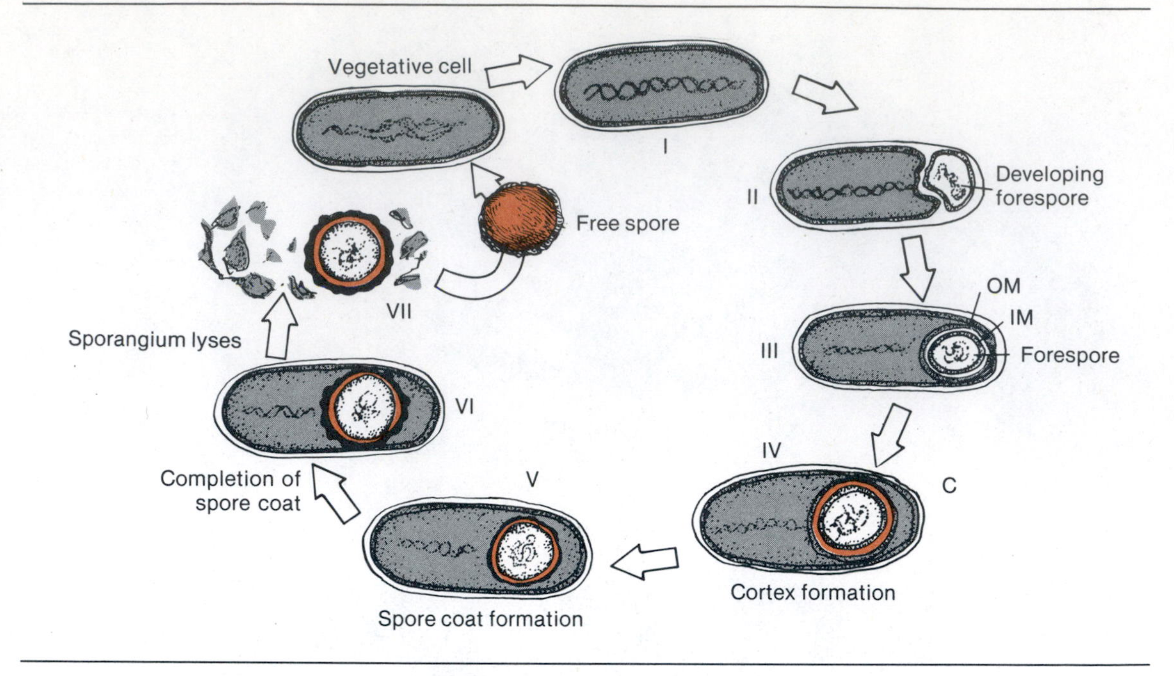

Figure 7-41 Stages in formation of endospore (see text for details).

growth and cell division cease and a spore (endospore) develops in the cytoplasm of the cell (the vegetative part of the cell is called the sporangium). When the mature spore is released from the sporangium, it does not exhibit any metabolic processes, and it is able to resist heat, radiation, and bactericidal agents. The endospore is therefore the dormant stage of the vegetative cell. Once environmental conditions are again suitable for vegetative growth, a previously silent mechanism is triggered, which results in germination and the establishment of vegetative growth. The biochemical and morphological events that occur during sporulation are described below. They occur in stages and coincide with those of Figure 7-41.

Stage I. The bacterial chromosome changes from a compact coiled unit to a filamentous one.

Stage II. There is invagination of the cytoplasmic membrane and formation of a double-membrane septum. The cell now appears to possess two compartments: one small and one large. The smaller unit will be the site of the developing endospore. Just before septum formation the DNA divides, ensuring a DNA molecule in each compartment.

Stage III. The smaller compartment of the cell invaginates into the larger one. As the engulfment continues, the smaller compartment is seen to be composed of a complete double membrane, which is called the ***forespore.***

Stage IV. Peptidoglycan is laid down between the inner and outer forespore membrane and is called the ***cortex.*** During this period calcium is taken up by the forespore from the sporangium and ***dipicolinic acid*** (dipicolinic acid is not found during vegetative growth) is synthesized in the forespore. There is a 1:1 ratio of dipicolinic acid to calcium. Calcium is believed to pass into the forespore by facilitated diffusion, where it binds (chelates) to dipicolinic acid to form calcium dipicolinate, which accounts for up to 15% of the dry weight of the mature spore.

Stage V. Protein coats high in the amino acid cysteine are deposited outside the cortex. These spore coats are electron dense when viewed by the electron microscope and give the appearance of a white refractile body (Figure 7-42).

Stage VI. The forespore structures are completed, and the endospore at this stage demonstrates resistance to heat.

Stage VII. The sporangium lyses, releasing the mature spore into the environment.

Although the sporulation process is described in stages, there is no rigid separation, because many of the activities overlap into other stages. For example, during the entire process water is slowly removed from the spore.

Mechanism of spore resistance and its implications. An organism's sensitivity to chemicals is usually the result of destruction of important proteins (enzymes) or nucleic acids. These chemicals require the presence of free water for chemical reactions to occur. The spore is dehydrated and is resistant to many chemicals that would destroy vegetative cells. The mechanism of dehydration has not been fully explained. It has been suggested that heat resistance may be due to the immobilization of highly labile molecules such as nucleic acids. Calcium dipicolinate, for example, has been shown to be capable of intercalation into the DNA molecule.

The mechanism of resistance to ultraviolet (UV) radiation has not been determined. It has been suggested that a special class of proteins are synthesized in the developing spore that interact with the nucleic acids, changing their conformation and making them resistant to UV light. During the germination process these proteins are degraded, making the DNA susceptible to UV light.

In addition to their survival value in harsh environments, spores also present problems in the medical community, especially spores of the genera *Bacillus* and *Clostridium*. For example, *Bacillus anthracis* is responsible for a highly fatal disease in cattle and other herbivores called anthrax; *Clostridium botulinum* is the etiological agent of one type of food poisoning called botulism; *Clostridium tetani* is the etiological agent of tetanus; and *Clostridium perfringens* is one of the etiological agents of gas gangrene and food poisoning. These organisms in their spore state exhibit resistance to boiling, antibiotics, and many antiseptics. Vegetative cells of bacteria and yeasts, for example, are killed in 5 to 10 minutes at 80° C. Studies have shown that it takes 330 minutes of boiling to kill botulinum spores, 90 minutes to kill tetanus spores, and 30 minutes to kill the spores of the gas gangrene organisms. They all can be destroyed by autoclaving, that is, steam at 121° C under 15 pounds pressure per square inch for 15 minutes. Wounds or abrasions, as well as food contaminated by some of these soil-residing spores, offer environments most suitable for rapid germination and growth. They produce deadly toxins in their vegetative states, but in one species, *C. perfringens*, the toxin responsible for food poisoning is released when sporulation occurs in the intestinal tract (see Chapters 20 and 28).

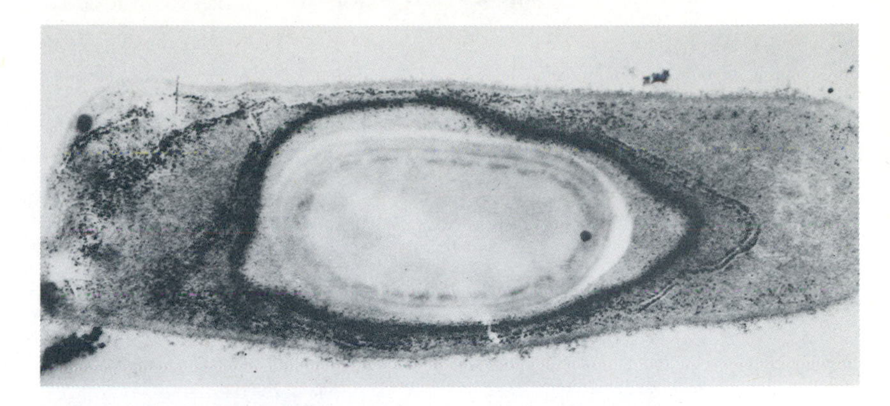

Figure 7-42 Electron micrograph of ultrathin section of *Bacillus cereus*. (×76,250.) Endospore is white refractile body in center of vegetative cell.
Courtesy P.D. Walker.

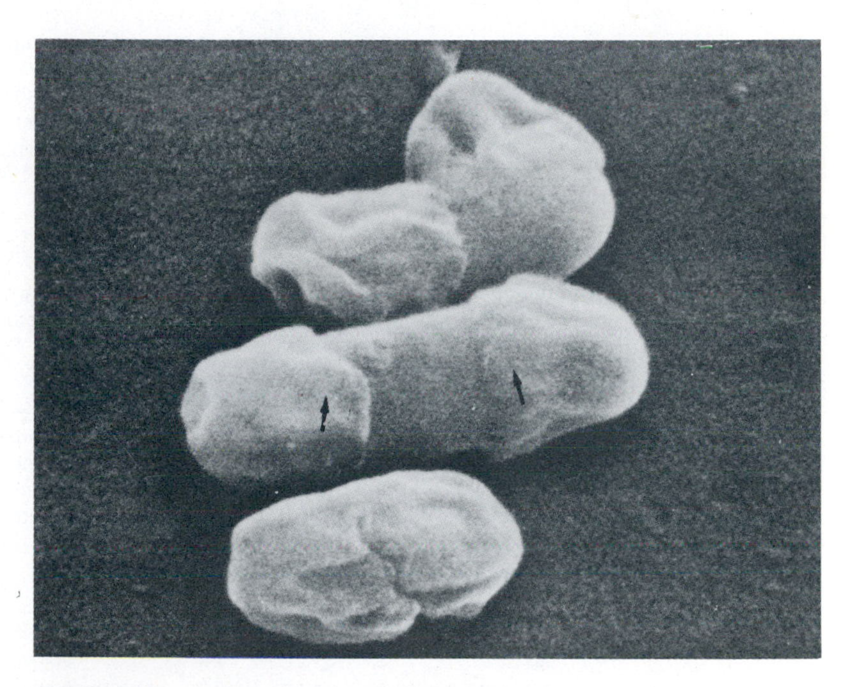

Figure 7-43 Scanning electron micrograph of spores in process of germination and breakage of spore coat *(arrows)*. (×30,000.)
Courtesy of Z. Yoshii.

Spore-forming species, particularly those of *Bacillus* and *Streptomyces*, are responsible for the synthesis of most of our antibiotics. They are produced just before and during the sporulation process. They are discussed in Chapter 4 and are discussed further in Chapter 21.

Germination

Germination, or the conversion of the spore to the vegetative state, is preceded by a stage of activation. This preparatory stage depends on the

Table 7-2 Characteristics of prokaryotic structures and inclusions

STRUCTURE OR INCLUSION	CHEMICAL COMPOSITION	FUNCTION	COMMENTS
Cell wall	Peptidoglycan is basic structural unit, making up 15% to 50% of dry weight of gram-positive cell wall but only 5% to 15% of gram-negative cell wall; gram-negative cell has outer membrane made up of lipopolysaccharide (LPS), protein, and phospholipid; gram-positive cell has teichoic acid	Structural support to cell; also acts as osmotic barrier to certain molecules	There are natural wall-less bacteria called mycoplasmas and artificially induced wall-less types called L-forms; archaebacteria have modified walls
Cytoplasmic membrane	Consists of phospholipid bilayer with proteins embedded in matrix	Abuts against cell wall but may invaginate in some bacteria; associated with activities such as solute transport, respiratory activity, secretion of molecules, and biosynthesis of cell membrane and cell wall components	Mesosomes may be cytoplasmic extensions or artifacts
Extracellular polymeric substances (EPS)	Usually polysaccharide with occasional amino sugars; one species of *Bacillus* produces polypeptide EPS; EPS may be adherent to cell (capsule) or an amorphous mass (slime)	Surrounds cell envelope; function is uncertain, but they protect some cells from phagocytosis; may be used for adherence to substrates	Vaccines are produced from some encapsulated species such as *Streptococcus pneumoniae*
Flagella	Protein	These are appendages extending from surface of cell; are involved in locomotion toward or away from chemical or physical stimuli	Spirochetes have flagella-like filaments called axial fibers that wrap around body of cell and beneath a sheath
Pili and fimbriae	Protein	Hairlike appendages found on surface of cell; pili are engaged in sexual conjugation; fimbriae may be useful in adherence to substrates or for aggregation	Found primarily on gram-negative bacteria
Gas vacuole	Protein	Provides buoyancy to organism	Found primarily in aquatic bacteria
Nucleoid	DNA	Contains hereditary information	Nucleoids increase with increase in growth rate; as many as four nucleoids may appear in cytoplasm
Ribosomes	RNA and protein	Protein synthesis	Sedimentation constant for prokaryotic ribosome is 70S; that is, it contains 50S and 30S subunits
Magnetosomes	Iron (magnetite?)	Orientation in environment	Found in marine marsh muds
Chlorobium vesicles	Bacteriochlorophylls c, d, and e	Light harvesting	—
Carboxysomes	Ribulose 1,5-diphosphate carboxylase	Carbon dioxide fixation or merely storage area for enzyme	—
Granules	Polyphosphate, poly-β hydroxybutyrate (PHB), sulfur, and glycogen	Storage products used primarily as energy, or energy and carbon sources	Type of storage product depends on species as well as nature of medium on which they are growing

presence of water and specific temperatures. When placed in water at 60° C for 1 hour, spores of many species of *Bacillus* can be activated, a process that results in swelling, elongation, and bursting of the spore coats (Figure 7-43). The presence of suitable nutrients in the environment activates RNA, DNA, and protein synthesis, which results in the formation of vegetative components. The period ends with division of the vegetative cell.

SUMMARY

The prokaryotic cell, that is, bacterial cell, possesses a variety of structures that have chemical and functional characteristics that distinguish them from eukaryotic cells. Most species of bacteria have been divided into two groups based on the staining characteristics of their cell wall: gram-positive and gram-negative. The characteristics of the major structures found in the prokaryotic cell are outlined in Table 7-2.

Some bacterial species, primarily members of the genera *Bacillus* and *Clostridium,* produce endospores. The formation of a spore is not a reproductive process, as in *Streptomyces* and some eukaryotes, but merely a mechanism used by the vegetative cell to survive stringent environmental conditions. During sporulation there are biochemical and morphological changes that distinguish it from the vegetative cell. During sporulation there is a depletion of water and the formation of a dense wall-like material around a DNA molecule. The spore contains high concentrations of calcium and dipicolinic acid, which are not found in the vegetative cell. The spore is unusually resistant to heat and radiation as well as various chemical and physical conditions that might be detrimental to the vegetative cell.

Study questions

1. Diagrammatically illustrate the structures found in a typical prokaryotic (bacterial) cell.
2. Compare and contrast the chemical composition of the cell wall of gram-positive and gram-negative bacteria. Do these chemical differences provide any advantages or disadvantages to the cell in question?
3. What is the significance of the outer membrane in the gram-negative cell wall?
4. What are the functions of the cell wall? In wall-less bacteria are these functions compensated for? How?
5. How do naturally occurring wall-less bacteria differ from L-forms or artificially induced wall-less bacteria?
6. What biological activities are associated with the bacterial cytoplasmic membrane?
7. Briefly describe the different ways in which molecules pass through the cytoplasmic membrane into the cytoplasm of the cell.
8. Describe one model that demonstrates how a macromolecule passes from inside the cell to the exterior of the cell.
9. What is the usual chemical composition of capsules? What is the function of the capsule?
10. What are the advantages of flagella on bacterial cells? What are the names of the three major components of the flagellum?
11. What are the differences in flagellar movement when a bacterial cell moves toward an attractant? Away from a repellent?
12. What is the difference between tactic responses and tropisms?
13. Discuss the role that pili or fimbriae play in microbial activities.
14. What is the function of gas vacuoles in microorganisms? In what type of environment are these microorganisms found?
15. What is the structure of the prokaryotic ribosome? What is meant by a Svedberg unit?
16. What is the chemical composition of some of the granules stored by the microbial cell? What is their function?
17. Why does a bacterial cell form a spore? How does bacterial spore formation differ from vegetative growth?
18. What components in the spore are absent or at the very least at very low concentrations in the vegetative cell?
19. Why is the endospore resistant to heat, chemicals, and radiation?
20. Does sporulation have any significance to the bacteriologist? Explain.
21. Define the following: permease, mesosome, teichoic acid, swarming, peptidoglycan, porin, lysozyme, O antigen, active transport, passive transport, bacteriocins, chemotaxis, axial filament, nuclear area, and germination.

Selected readings

Books

Brock, T. *Biology of microorganisms* (3rd ed.). Prentice-Hall, Inc., Englewood Cliffs, N.J., 1979.

Carlile, M.J. (Ed.). *Primitive sensory and communication systems: the taxes and tropisms of microorganisms and cells.* Academic Press, Inc., New York, 1975.

Goldman, R. (Ed.). *Cell motility,* Cold Spring Harbor Laboratory, Cold Spring Harbor, N.Y., 1976.

Hazelbauer, G.L. (Ed.). *Taxis and behavior,* Chapman & Hall, Ltd., London, 1978.

Mandelstam, J., and McQuillen, K. *Biochemistry of bacterial growth* (2nd ed.). John Wiley & Sons, Inc. New York, 1973.

Parish, J.H. *Developmental biology of prokaryotes.* University of California Press, Berkeley, 1979.

Peberdy, J.F. *Developmental microbiology.* Blackie & Son, Ltd., Glasgow, Scotland, 1980.

Reiss, J.L. (Ed.). *Microbial interactions.* Chapman and Hall, Ltd., London, 1977.

Rose, A.H. *Chemical microbiology* (3rd ed.). Plenum Publishing Corp., New York, 1976.

Rosen, B.P. (Ed.). *Bacterial transport.* Marcel Dekker, Inc., New York, 1978.

Stanier, R.Y., Adelberg, E.A., and Ingraham, J. *The microbial world* (4th ed). Prentice-Hall, Inc. Englewoods Cliffs, N.J., 1976.

Stanier, R.Y., Rogers, H.J., and Ward, J.B. (Eds.). *Relations between structure and function in the prokaryotic cell: twenty-eighth symposium of the Society for General Microbiology,* Cambridge University Press, New York, 1978.

Journal articles

Adler, J. The sensing of chemicals by bacteria. *Sci. Am.* **234:**40, 1976.

Beveridge, T.J. Ultrastructure, chemistry, and function of the bacterial cell wall. *Int. Rev. Cytol.* **72:**229, 1981.

Burchard, R.P. Gliding motility of prokaryotes: ultrastructure, physiology, and genetics. *Annu. Rev. Microbiol.* **35:**497, 1981.

Canale-Parola, E. Motility and chemotaxis of spirochetes. *Annu. Rev. Microbiol.* **32:**69, 1977.

Costerton, J.W., Ingram, J.M., and Cheng, K.J. Structure and function of the cell envelope of gram-negative bacteria. *Microbiol. Rev.* **38:**87, 1974.

Dawes, E.A., and Senior, P.J. The role and regulation of energy reserve polymers in microorganisms. *Adv. Microb. Physiol.* **10:**135, 1973.

Dills, S.S., et al. Carbohydrate transport in bacteria. *Microbiol. Rev.* **44:**385, 1980.

Gottow, J.C.G. Ecology, physiology and genetics of fimbriae and pili. *Annu. Rev. Microbiol.* **29:**79, 1975.

Gould, C.W., and Dring, G.J. Heat resistance of bacterial endospores and concept of expanded osmoregulatory cortex. *Nature* **258:**402, 1975.

Keynan, A. The transformation of bacterial endospores into vegetative cells. *Symp. Soc. Gen. Microbiol.* **23:**85, 1973.

Kurland, C.G. Structure and function of the bacterial ribosome. *Annu. Rev. Microbiol.* **46:**173, 1977.

Lodish, H., and Rothman, J.E. The assembly of cell membranes. *Sci. Am.* **240:**48, 1979.

Marahiel, M.A., et al. Biological role of gramicidin S in spore functions. *Eur. J. Biochem.* **99:**49, 1979.

Salton, M.R.J., and Owen, P. Bacterial membrane structure. *Annu. Rev. Microbiol.* **30:**451, 1975.

Silverman, M., and Simon, M.I. Bacterial flagella. *Annu. Rev. Microbiol.* **31:**397, 1977.

Ward, J.B. Teichoic and teichuronic acids. *Microbiol. Rev.* **45:**211, 1981.

Chapter 8 EUKARYOTIC ULTRASTRUCTURE AND FUNCTION

Eukaryotic microorganisms are more recent inhabitants of the earth than prokaryotes and are believed to have evolved from prokaryotes somewhere between 1.2 and 2.1 billion years ago. Their formation is believed to have coincided with the evolution of aerobic ecosystems (see Chapter 4). Structurally diverse eukaryotic species dominate many aerobic ecosystems, whereas prokaryotes dominate anaerobic as well as aerobic environments. There are several characteristics that distinguish eukaryotes from prokaryotes. Some of the more obvious differences are outlined in Table 4-1. The structural changes or modifications that occurred during the evolution of the eukaryotic cell were often accompanied by variations in function that this chapter will try to elucidate. The primary purposes of this chapter are twofold: first, to describe chemically those structures that are unique to eukaryotes or have similar counterparts in prokaryotic cells, and second, to assess the functional significance of these structural changes or modifications. Figures 4-1 and 5-1 give some morphological perspective of the similarities and differences between eukaryotic and prokaryotic cells. The major emphasis in this chapter is on the fungi and protozoa.

CELL WALL

The cell wall in eukaryotic microorganisms is a feature of the fungi and the algae. Chemically the cell wall of fungi differs considerably from the cell wall of bacteria although both function in the same way. The hypha, which represents the unit of fungal growth, has a cell wall whose appearance varies with age (Figure 8-1). The growing point or tip of the hypha reveals a primary cell wall that is a smooth-structured unit showing little demarcation. Older areas of the hypha contain not only a primary wall but a secondary layer of structureless material. The primary wall, which is adjacent to the cytoplasmic membrane, lends structural support to the cell. Polysaccharides make up from 70% to 80% of the dry weight of the fungal cell wall, with protein making up most of the remainder. Little lipid is present. The cell wall in most fungi is composed of such polysaccharides as chitin (a polymer of *N*-acetylglucosamine) or cellulose (a polymer of glucose) both of which are β-glucans. They appear as crystalline ***microfibrils*** (Figure 8-2) that are believed to have special orientation depending on their location in the wall. A thin layer of protein overlaps this polysaccharide and is believed to be covalently attached to it. The secondary wall, which increases in thickness with distance from the tip, is composed of layers of glycoprotein and mixtures of glucans. The lack of peptidoglycan and *N*-acetylmuramic acid in the fungal cell wall accounts for the inability of bacterial cell wall inhibitors such as penicillin to have any effect on fungal growth. Large molecules can pass into and out of the fungal cell wall, indicating less compactness than in prokaryotic cells. It is not known whether per-

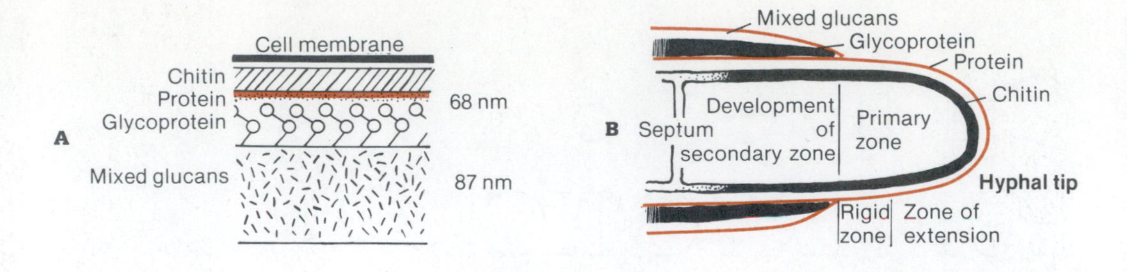

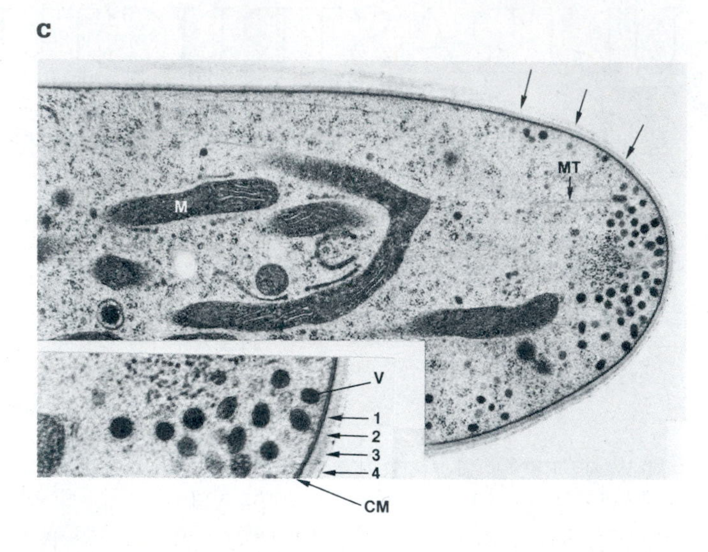

Figure 8-1 Composition and structure of fungal cell wall. **A,** Chemical composition and relative position of fungal cell wall components. **B,** Hyphal cell wall tip. Wall increases in thickness farther away from tip of hypha. Tip grows in length only in zone of extension or primary zone. Hypha increases in thickness in secondary zone. **C,** Electron micrograph of hyphal tip cell. Lower photograph is enlargement of boxed area in larger photograph. *MT,* Microtubule; *M,* mitochondrion; *V,* vesicles; *CM,* cytoplasmic membrane. Arrows in upper photograph point to outer cell wall layer. Note four layers in cell wall of insert.

C, *From Howard, R.J., and Aist, J.R.* J. Ultrastruct. Res. ***66:****224, 1979.*

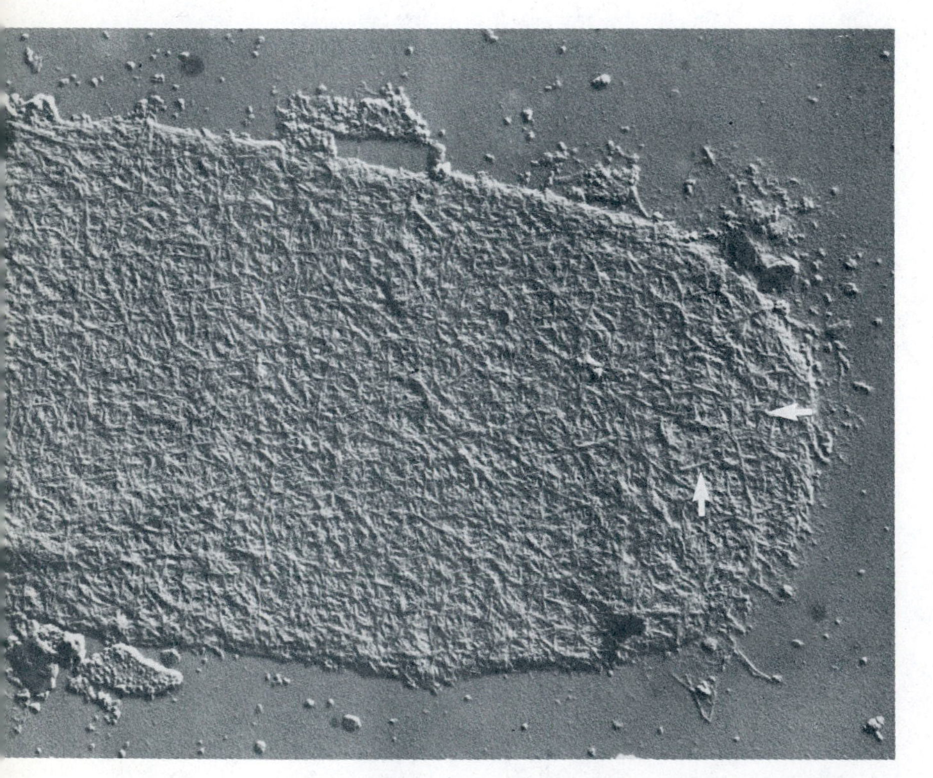

Figure 8-2 Electron micrograph of tip of hypha demonstrating thin microfibrils *(arrow).*

From Hobot, J.A., and Gull, K. Symp. Soc. Gen. Microbiol. ***30:****207, 1980. Cambridge University Press.*

manent pores exist in the cell wall as avenues for transport or whether pores are formed only when contact is made between specific cell wall components and the penetrating molecule.

CYTOPLASMIC MEMBRANE

There are some chemical differences between prokaryotic and eukaryotic cytoplasmic membranes even though both are unit membranes. One of the major differences between them is the presence of ***sterols*** in the membrane of eukaryotes and the absence of sterols in prokaryotes (except mycoplasmas and some others). Sterols stabilize the membrane, a function that is readily noted when one compares the instability of laboratory-induced wall-less bacteria (L-forms) with the stability of mycoplasmas in isotonic media. The cytoplasmic membrane, in nearly all prokaryotes, is the only unit membrane system. The exception is the cyanobacterium, in which there is also present the photosynthetic membrane apparatus called thylakoids. A network of unit mem-

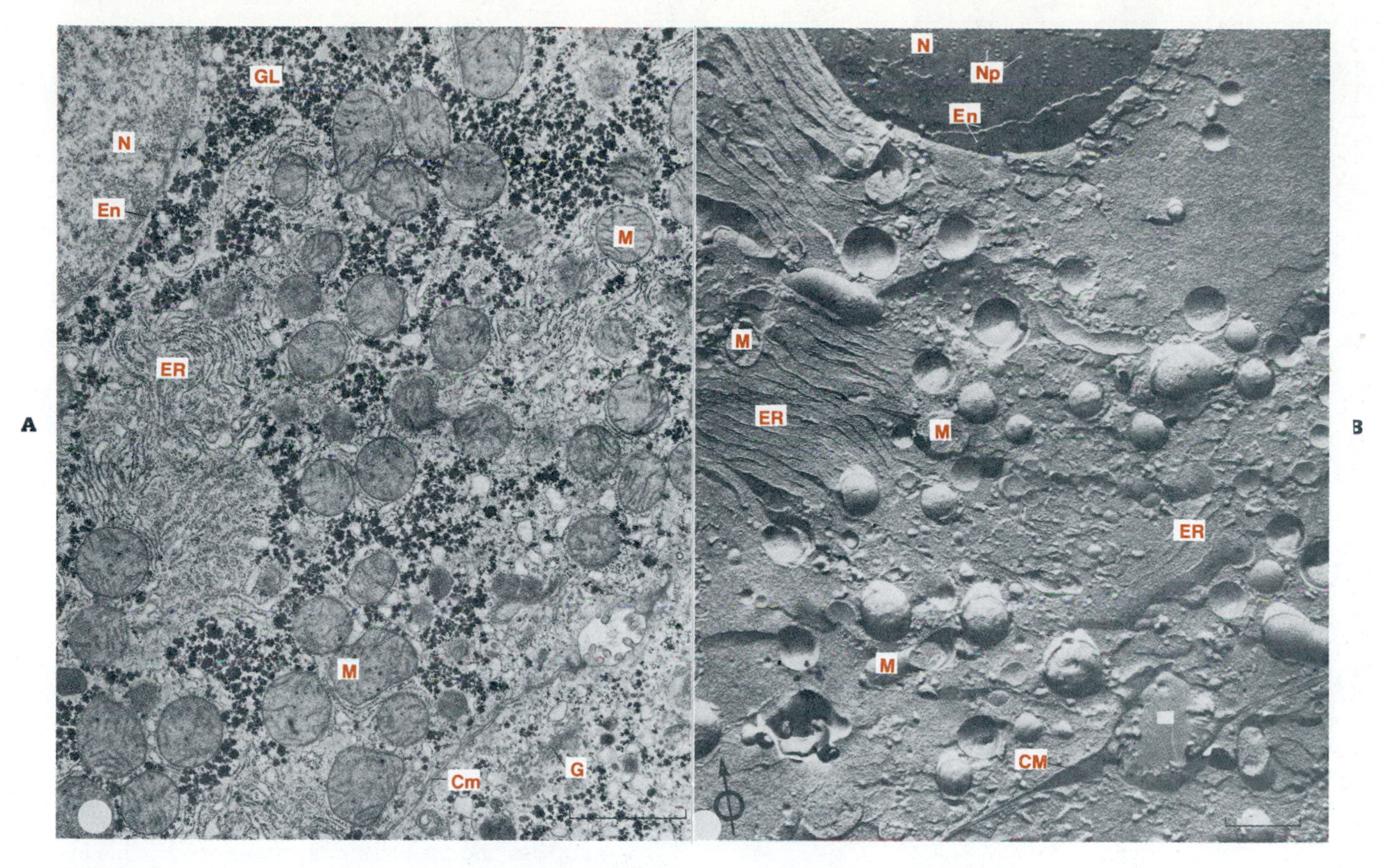

Figure 8-3 Electron micrograph of cytoplasmic components of eukaryotic cell **A,** Tissue was fixed chemically, embedded in plastic, sectioned, and stained. (×36,000.) *G,* Golgi apparatus; *GL,* glycogen. **B,** Freeze fractured tissue. (×23,000.) *M,* Mitochondrion; *CM,* cytoplasmic membrane; *EN,* nuclear envelope; *N,* nucleus; *Np,* nuclear pore; *ER,* endoplasmic reticulum.
From Roland, J., Szollos, A., and Szollos, D. Atlas of cell biology, *Little, Brown & Co., Boston, 1977.*

branes in eukaryotes that form channels continuous with the cytoplasmic membrane and nuclear membrane is called the endoplasmic reticulum (Figure 8-3). The ***endoplasmic reticulum*** (ER), which is expanded around the nucleus as a nuclear membrane, provides a channel between the cytoplasm and the nucleus (Figure 8-3, *B*). The ER network compartmentalizes the cytoplasm and maintains the various organelles in a relatively constant position. The channels formed by the endoplasmic reticulum also aid in the diffusion of small molecules within the cytoplasm. The major portion of the endoplasmic reticulum has ribosomes attached to it (ribosomes are also free in the cytoplasm) and is referred to as the ***rough endoplasmic reticulum.*** Portions that do not contain ribosomes are called ***smooth endoplasmic reticulum.*** The proteins synthesized on the rough ER are secretory proteins and membrane proteins as compared with the enzymatic proteins produced primarily by free ribosomes. The proteins synthesized on the rough ER are contaminated by ER membrane proteins, and they must be modified before they reach their final destination. This modification step appears to take place in the Golgi apparatus, discussed later. The smooth ER can provide membrane components to be used for other organelles of the cell as well as the cytoplasmic membrane. The smooth ER also possesses enzymes that are involved in the oxidation of drugs, a process called ***biotransformation.*** Eukaryotic cells contain several organelles and structures that possess membranes—Golgi apparatuses, chloroplasts, mitochondria, lysosomes, and vacuoles.

Two of the principal functions of the cytoplasmic membrane in both eukaryotes and prokaryotes is transport of nutrients into the cell and excretion of wastes or other products from the cell into the environment. These functions are shared by both, but the mechanism differs in those eukaryotic cells that are wall-less—for example, the protozoa. The cytoplasmic membrane in prokaryotic cells is a much more selective barrier to the passage of nutrients into the cell. The cytoplasmic membrane in those eukaryotes without walls is relatively pliable and provides the cell with unique methods for nutrient and waste

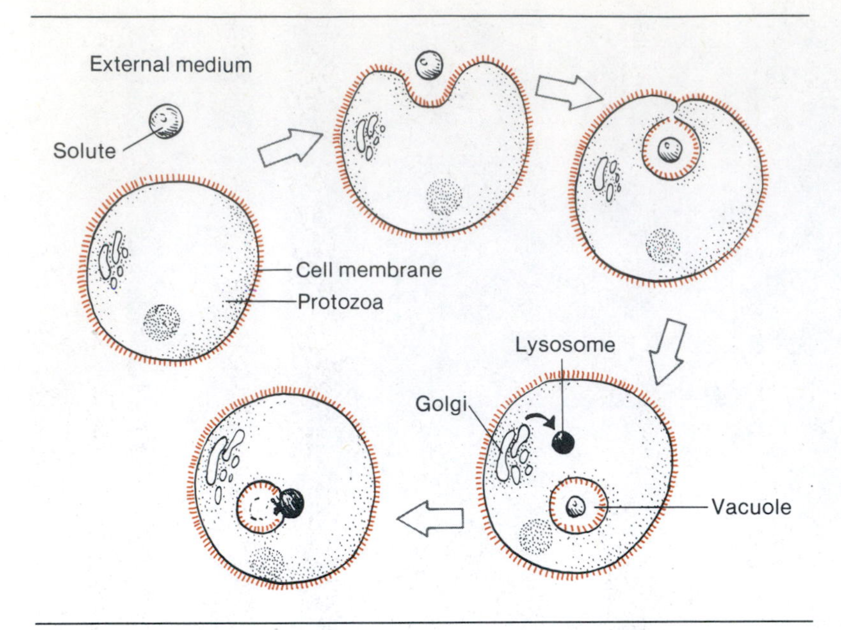

Figure 8-4 Mechanism of pinocytosis as exemplified in protozoa. Solute particle is engulfed by cytoplasmic membrane. Vacuole, consisting of detached cytoplasmic membrane plus solute moves into cytoplasm. Lysosome, produced by Golgi apparatus and containing hydrolytic enzymes, moves into cytoplasm to engage vacuole. Lysosome fuses with vacuole (see Figure 8-5).

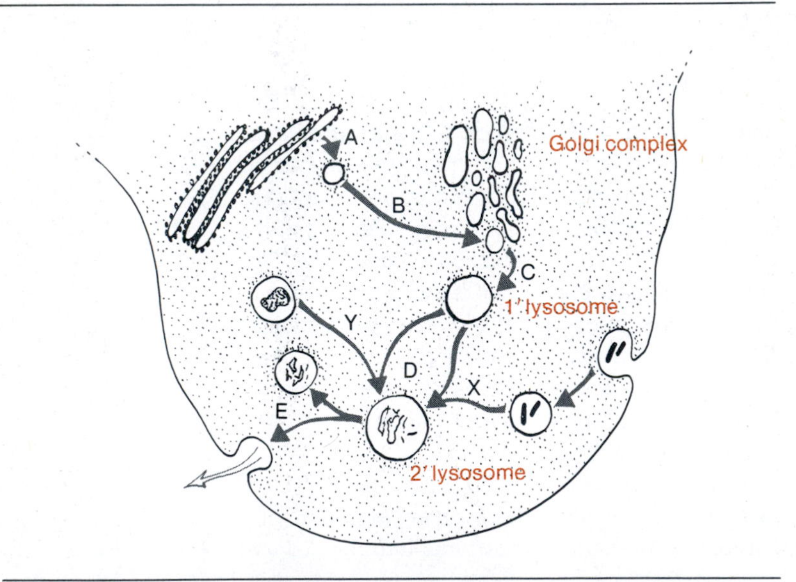

Figure 8-5 Relationship of lysosomes to phagocytic or pinocytic capture of food particles. *A*, Vesicle produced by endoplasmic reticulum contains hydrolytic enzymes. *B* and *C*, Vesicle interacts with Golgi and enzymes become concentrated to produce primary *(1')* lysosome. *D*, Lysosome fuses with vacuole or vesicle-containing food particles *(X)* or cellular debris *(Y)* to produce secondary *(2')* lysosome. *E*, Particles or debris in secondary lysosome may be expelled at surface or some debris, such as cellular components, may remain in secondary lysosomes.

transport: pinocytosis, phagocytosis, and exocytosis. ***Pinocytosis*** is a transport process in which fluids or suspensions of material from the external environment are delivered across the cytoplasmic membrane in a closed membrane produced at the cell surface (Figure 8-4). ***Phagocytosis*** is a transport process in which the material being transported is particulate and not in solution. Pinocytosis and phagocytosis are referred to as ***endocytoses*** (sing. endocytosis). A more lengthy discussion of phagocytosis, as carried out by cells of the immune system in humans, is presented in Chapter 22.

Endocytosis does not entail the indiscriminate uptake of molecules from the environment. It is a process that is induced by the chemical nature of the solute or solution in the cell's environment. There is also a limit to the number of endocytic activities that can take place within any period of time. Since the cytoplasmic membrane of the cell is used to produce a closed structure around the solute during endocytosis, new cytoplasmic membrane must of necessity be regenerated. It is believed that the number of endocytoses is controlled by the amount of cytoplasmic membrane precursor available in the cell. Endocytosis provides the wall-less eukaryotes with a means of capturing their "meals" in a concentrated particulate form. This is a selective advantage for the eukaryotic cell over the prokaryotic cell because eukaryotes do not have to produce extracellular enzymes to break down large molecules or food particles that cannot penetrate the cytoplasmic membrane, a necessity for most prokaryotic cells.

Endocytosis also explains the presence of endosymbionts. Prokaryotes are often the prey for eukaryotes; for example, protozoa phagocytose bacteria as a source of food. Sometimes the engulfed prokaryote survives in the host, and the host may even benefit from the association. These relationships are discussed in Chapter 4.

The extrusion of materials from within the cell to the environment also takes place in eukaryotes in a process that is the reverse of endocytosis called ***excytosis.*** Products in the cell are extruded in vacuoles that have been produced by a sequestering of membrane from the smooth endoplasmic reticulum or Golgi apparatus, if present. The vacuole moves to the surface of the cell where the vacuolar membrane fuses with the cytoplasmic membrane and the products are excreted (Figure 8-5).

VACUOLES, VESICLES, AND LYSOSOMES

The Golgi apparatus and endplasmic reticulum are involved in the formation of membranous structures called vacuoles, vesicles, and lysosomes. These three membranous structures can be defined as separate entities, but their functions in the cell are interrelated. For example, a cell that captures its food by endocytosis has the food particles surrounded by a closed membranous structure. If phagocytosis is the mechanism of capture, the membranous structure is called a ***vacuole;*** if pinocytosis is the mechanism of capture, the membranous structure is called a ***vesicle.*** Vesicles may also contain endogenous cellular components such as mitochondria and ribosomes that undergo self-digestion. The digestion of exogenous or endogenous material occurs in a series of events initiated first by the formation of hydrolytic enzymes on the endoplasmic reticulum. The enzymes are then packaged within a membranous unit, produced by the Golgi apparatus, called a ***lysosome.*** Cellular constituents would be hydrolyzed by the lysosomal enzymes if they were not packaged in these individual membranous sacs. Fusion of the lysosomal membrane with that of the particle-containing vacuole or vesicle results in the formation of a single membranous structure called the ***secondary lysosome,*** where digestion of the captured particles occurs. Digested material may pass directly through the lysosome membrane and back into the cytoplasm, whereas undigested material or particles may remain within the cell or be expelled at the cell surface (Figure 8-5). A lysosome can, therefore, be considered a specialized vesicle.

A specialized type of vacuole is produced in many protozoa and algae called the ***contractile vacuole.*** It functions as a receptacle for water that accumulates in the cell and is discussed in Chapter 5.

Vesicle formation at the surface of the cell is also a mechanism for transport of viruses into susceptible animal cells. Once the virus makes contact with the surface of the host cell it becomes enclosed in the vesicle and enters the cytoplasm where the vesicle membrane is later dissolved (Figure 8-6). Vesicles may also be involved in certain biosynthetic reactions within the cell—for example, during fungal growth in which the hyphal tips are actively engaged in cell wall synthesis and tip elongation. It is believed that vesicles, formed in the cytoplasm, carry various enzymes and polysaccharides that are involved in cell wall synthesis. The vesicles fuse with the cytoplasmic membrane and release their contents at the hyphal tip.

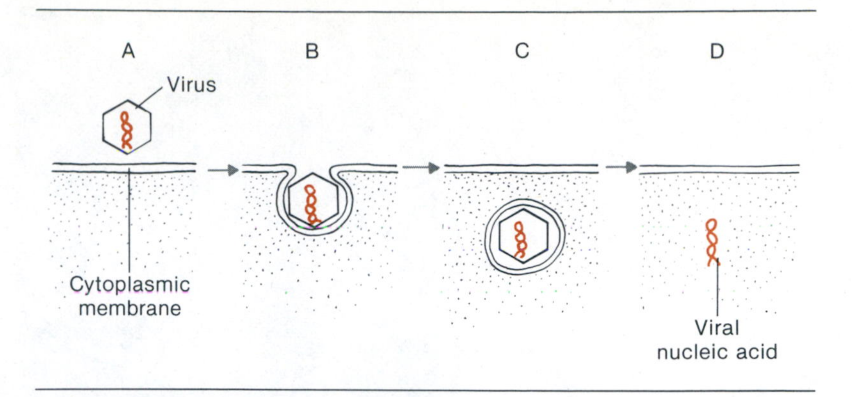

Figure 8-6 Virus penetration into cell by vesicle formation. *A,* Binding of virus to cytoplasmic membrane *B,* Invagination of cytoplasmic membrane about virus particle. *C,* Vesicle, which contains virus, is pinched off from cytoplasmic membrane. *D,* Vesicle membrane disappears and viral coat is removed, releasing viral nucleic acid into cytoplasm.

NUCLEUS AND DNA

The genetic material in eukaryotes is enclosed in a membranous structure called the ***nucleus.*** The nuclear membrane is composed of two layers between which is a perinuclear space (Figure 8-7). The nuclear membrane is perforated by pores that may provide an avenue for transport of macromolecules into and out of the cytoplasm. At some locations on the nuclear envelope the membrane is continuous with the endoplasmic reticulum. Within the nucleus is a densely stained region called the ***nucleolus,*** which is the site of ribosomal RNA synthesis and assembly of ribosomes. These ribosomes will eventually make their way through the nuclear envelope into the cytoplasm to engage in protein synthesis. During cell division various structures appear in the nucleus; they are discussed in Chapter 12.

Eukaryotic DNA organization and structure

The DNA of eukaryotic cells, which is a linear molecule and not circular like that of prokaryotes, is coiled into compact units called ***chromosomes.*** Eukaryotes, unlike prokaryotes, possess more than one chromosome, and in each chromosome there is one DNA molecule. The DNA in the chromosome is much longer than the chromosome length because it is folded many times

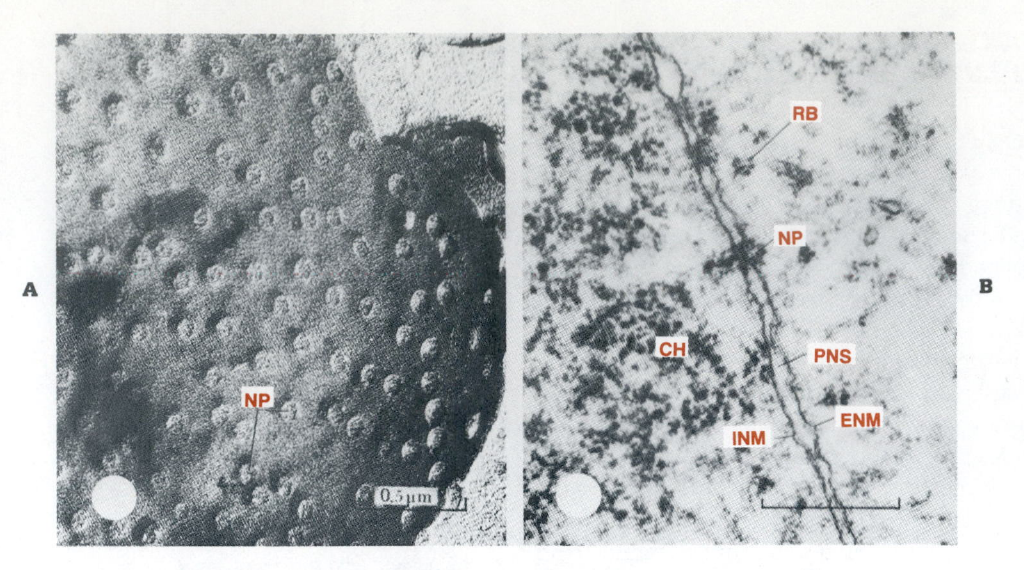

Figure 8-7 Nucleus. **A,** Electron micrograph of freeze-etched preparation of plant cell showing nucleus and nuclear pores *(NP)*. (×50,000). **B,** Electron micrograph of nucleus and nuclear envelope. (×65,000.) *NP,* Nuclear pores; *ENM,* external nuclear membrane; *INM,* internal nuclear membrane; *RB,* ribosomes; *PNS,* perinuclear space; *CH,* chromatin.

From Roland, J., Szollos, A., and Szollos, D., Atlas of cell biology, *Little, Brown & Co., Boston, 1977.*

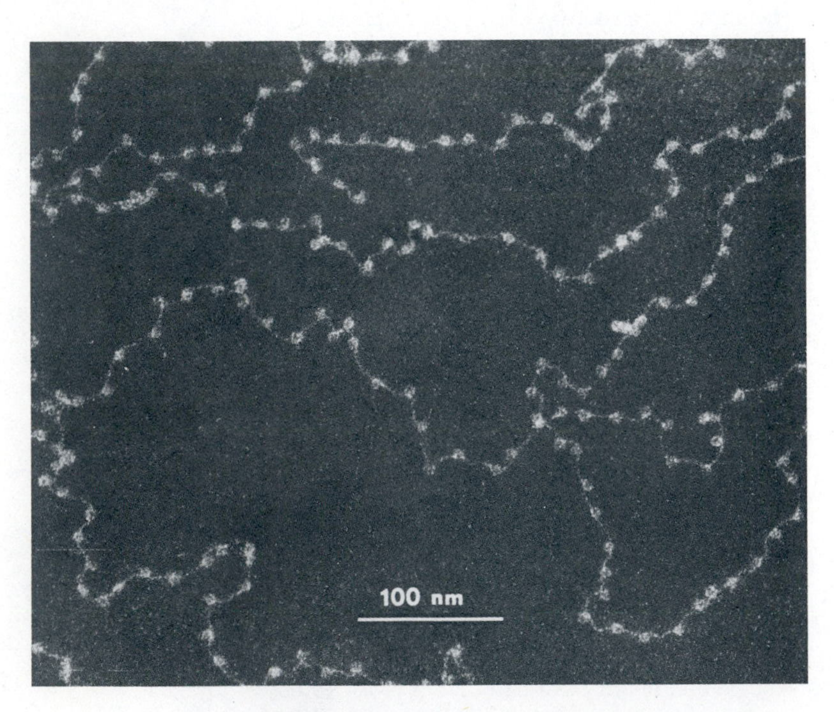

Figure 8-8 Nucleosomes. Darkfield electron micrograph of edge of chicken erythrocyte nucleus demonstrating beaded nucleosomes associated with eukaryotic chromosome.

From Olins, A.L., and Olins, D.E. Am. Sci. ***66:****704, 1977. Reprinted by permission,* American Scientist, *Journal of Sigma Xi, The Scientific Research Society, Inc.*

into a compact unit. Intimately associated with the chromosome are proteins called ***histones*** that are rich in basic amino acids such as lysine. Basic amino acids are important in neutralizing the negatively charged DNA molecules. The combination of histone and DNA is referred to as ***chromatin.*** Histones are arranged on eukaryotic DNA in such a way as to give the appearance of a repeating pattern of nucleoprotein particles or beads (Figure 8-8). These repeating nucleoprotein particles are called ***nucleosomes*** (also called chromatin subunits). The nucleosome is believed to be composed of a core containing 140 base pairs of DNA wrapped around four histone proteins, plus a spacer containing 40 to 60 base pairs of DNA surrounding a single histone protein. Histone synthesis is synchronized with DNA replication; if histone synthesis is inhibited, DNA replication is also depressed. Nucleosomal organization appears to differ between cells of the same species as well as between higher and lower eukaryotes. Modification of histones by methylation, acetylation, and phosphorylation has been observed to occur during DNA replication in higher eukaryotes (plants and vertebrates). These characteristics and activities suggest that histones may somehow control gene activity.

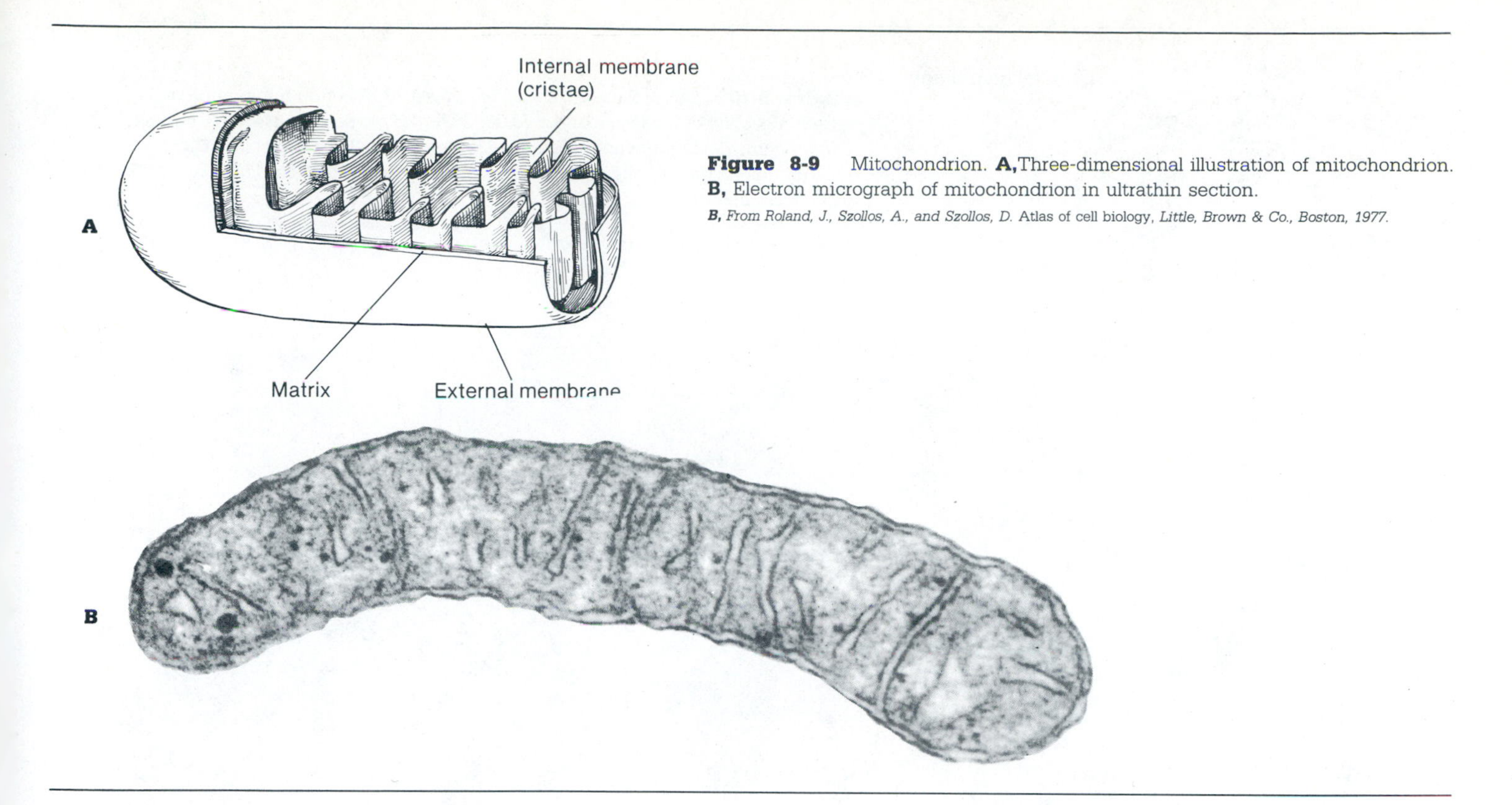

Figure 8-9 Mitochondrion. **A,** Three-dimensional illustration of mitochondrion. **B,** Electron micrograph of mitochondrion in ultrathin section.
B, From Roland, J., Szollos, A., and Szollos, D. Atlas of cell biology, *Little, Brown & Co., Boston, 1977.*

MITOCHONDRIA

We have indicated so far that the eukaryotic cell is a compartmentalized unit in which certain cell functions are restricted to specific areas of the cell. Thus, the substrates and products of metabolic reactions are confined within membrane-enclosed structures, or organelles. One of these organelles, the ***mitochondrion,*** is found in all eukaryotic cells except anaerobic protozoa. Mitochondria are themselves compartmentalized, and it is this compartmentalization that is associated with specific biochemical functions. The mitochondrion is a complex organization of enzymes whose activity is associated with aerobic respiration, that is, aerobic energy production. Structurally, the mitochondrion is a double-membrane organelle (Figure 8-9). The smooth outer membrane is the limiting membrane of this organelle and exhibits no apparent enzyme activity, whereas the inner membrane is folded into units called ***cristae.*** The inner membrane delineates the viscous inner chamber of the mitochondrion called the ***matrix.*** The inner membrane is composed of knob-like units that are made of a spherical headpiece, a stalk, and a base. The cristae engage in electron transport and ATP production, whereas the matrix is made up of enzymes that take part in the tricarboxylic acid (TCA) cycle.

The mitochondrion can multiply by division and possesses its own genetic system—that is, DNA, ribosomes, and other machinery associated with DNA replication and translation. Not all of the genes necessary for replication and other functions are found in mitochondrial DNA. Mitochondrial DNA is much smaller than nuclear DNA and ranges in size from 9×10^6 to 7×10^7 daltons, depending on the microorganism. Mitochondrial DNA is also naked and not associated with histones as is nuclear DNA. The ribosomes in the mitochondrion are more closely related to prokaryotic ribosomes than the ribosomes in the eukaryotic cell cytoplasm. These latter findings have suggested that organelles such as mitochondria that contain their own genetic systems may have evolved from endosymbiotic bacteria. Thus, it is possible that bacteria parasitizing eukaryotes lost their cell wall and evolved as membranous organelles with their own genetic system but incapable of synthesizing cell wall. During evolution the eukaryotes became dependent on the endosymbiont for survival and vice versa.

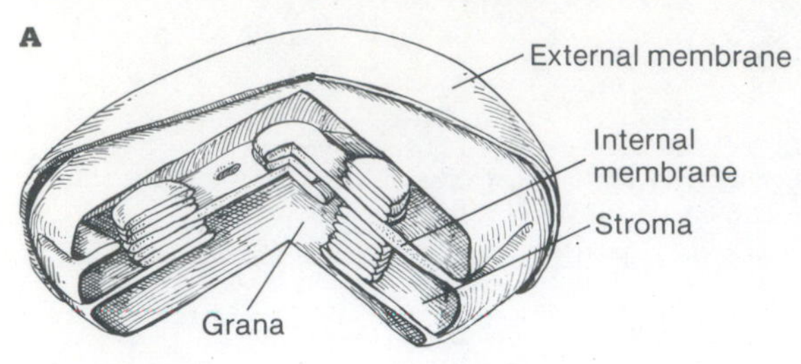

Figure 8-10 Chloroplast. **A,** Plant chloroplast. **B,** Electron micrograph of ultrathin section of tobacco leaf. (×72,000.) *CE,* Chloroplast envelope; *S,* stroma; *GR,* granum; *A,* starch-bearing vesicle; *Th,* thylakoids; *L,* lipid globule. **C,** Electron micrograph of chloroplast from algal cell. (×20,500.) *Th,* Thylakoids; *CE,* chloroplast membrane; *L,* lipid droplets.

A* and *B*,** *From Roland, J., Szollos, A., and Szollos, D. Atlas of cell biology, Little, Brown & Co., Boston, 1977.* ***C*,** *From Heywood, P. J. Phycol.* ***13*:68, 1977.*

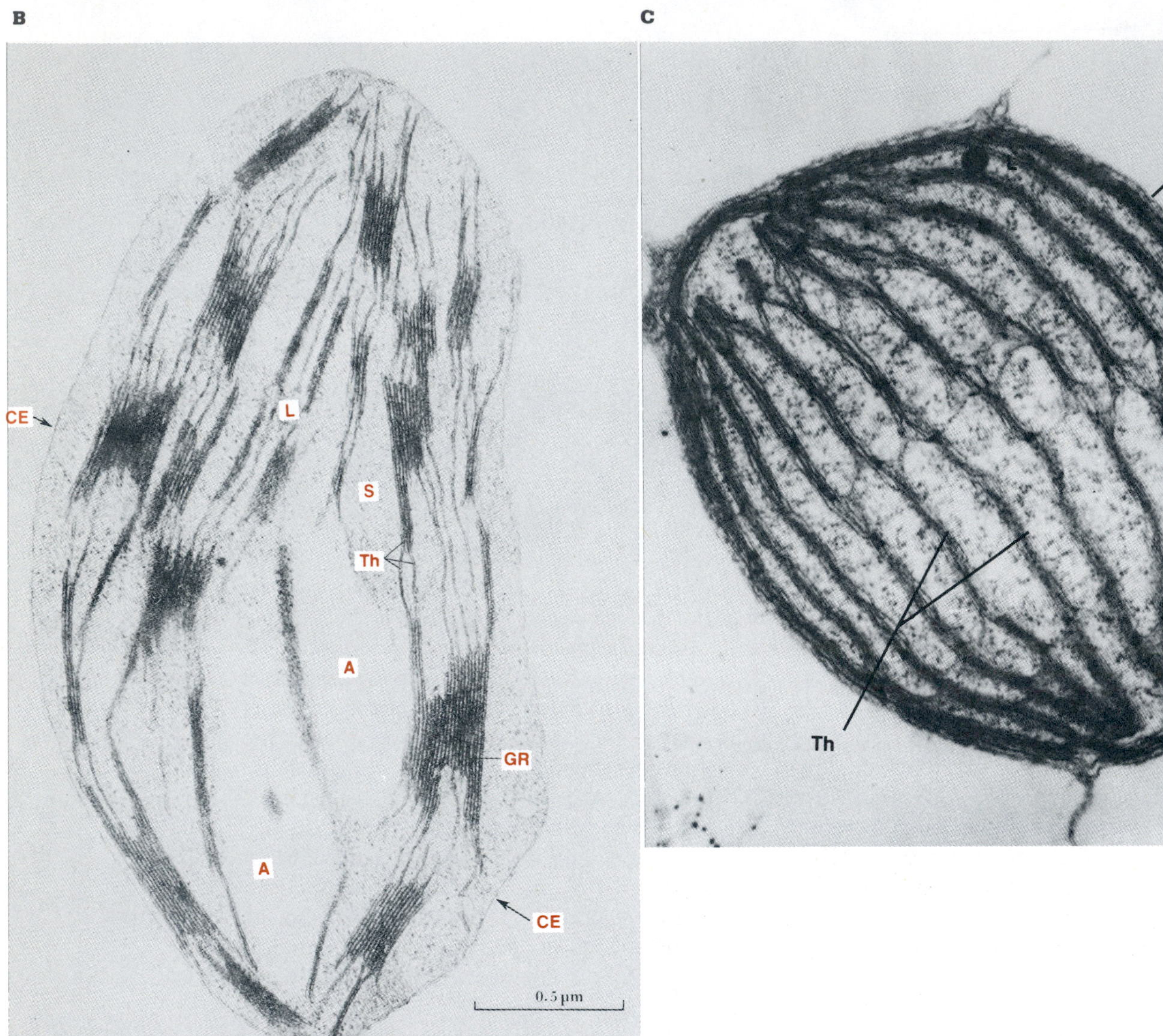

CHLOROPLASTS

Chloroplasts are organelles found in those eukaryotes that carry out photosynthetic reactions and result in the production of oxygen. Photosynthesis can also occur in prokaryotes, but the chloroplast as an organelle is found only in eukaryotic cells. The generalized structure of a chloroplast (Figure 8-10, *A*), like a mitochondrion, possesses an external envelope and internal membrane system. The outer envelope consists of an outer and inner membrane. The outer membrane of the envelope is permeable to molecules of molecular weight less than 1000, whereas the inner membrane is passively permeable only to small molecules such as water, ammonia, and carbon dioxide. Only in chloroplasts does a rough endoplasmic reticulum completely envelop the organelle and isolate it from the cell's cytoplasm. The chloroplast is usually pressed against the nucleus, and the outer membrane of the nuclear envelope is continuous with the sheath of endoplasmic reticulum surrounding the chloroplast. The internal homogenous matrix of the chloroplast is called the ***stroma*** in which are embedded the system of internal membranes called the ***thylakoids*** (Figure 8-10, *B* and *C*). Thylakoids may be free within the stroma of the chloroplast, or they may be stacked into units called ***grana*** (Figure 8-10, *B*) such as those in the Chlorophyta. The thylakoids are made up of chlorophyll and other pigments associated with photosynthesis, as well as glycolipids, phospholipids, and proteins. The conversion of light energy to ATP production occurs within the thylakoid membrane. The stroma is the site of reduction of carbon into organic materials (see Chapter 13 for a discussion of photosynthesis). Chloroplasts, like mitochondria, contain DNA and ribosomes and other protein-synthesizing components. The DNA and ribosomes of the chloroplast show a greater relatedness to prokaryotic DNA and ribosomes than they do to eukaryotic nuclear DNA and cytoplasmic ribosomes. This kind of similarity has also lent some credence to the hypothesis that chloroplasts, like mitochondria, arose initially as endogenous endosymbionts. Another hypothesis has also been suggested based on the following: photosynthesis in prokaryoates is carried out on thylakoids, but they are not part of an organelle. The thylakoids, for example, are in intimate contact with the cytoplasmic membrane, and it has been suggested that they pinched off from the cytoplasmic membrane to form discrete photosynthetic units in the cell and thus became precursors to chloroplasts.

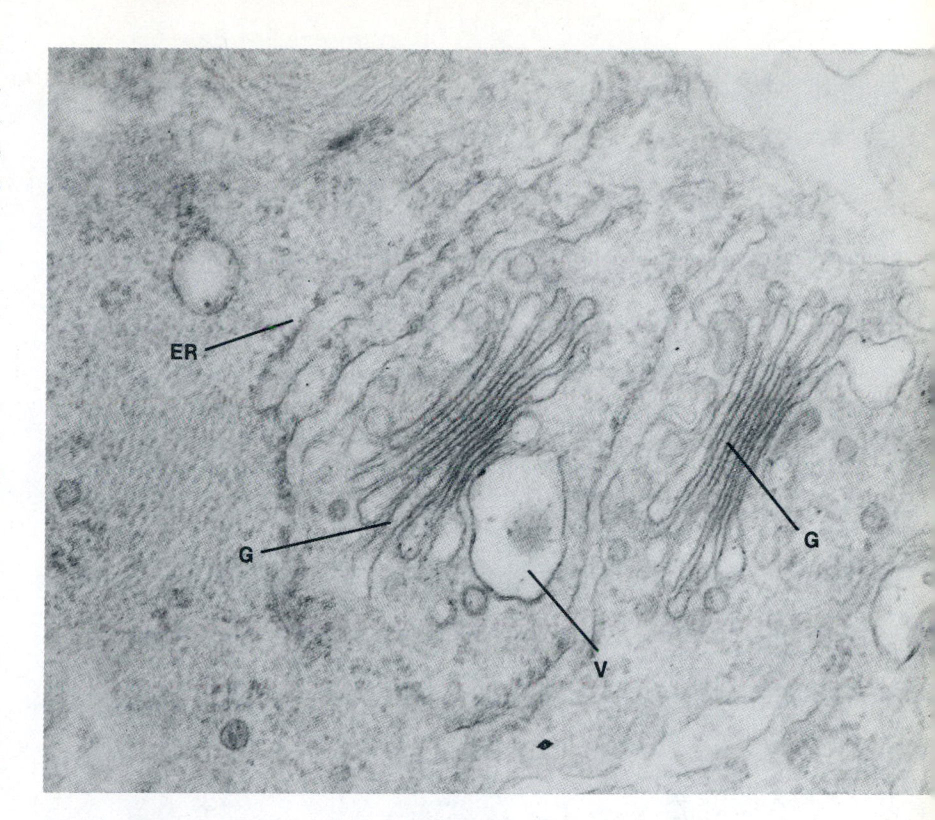

Figure 8-11 Golgi complex. Electron micrograph of algal cell showing Golgi complex. *G*, Golgi complex; *V*, vesicle containing granular material; *ER*, endoplasmic reticulum.

From Browitzka, M.A. J. Physol. ***13****:6, 1977.*

GOLGI APPARATUS

The Golgi apparatus is a complex membranous network found exclusively in eukaryotic cells, but not all eukaryotes possess them. The Golgi apparatus packages several products such as proteins and lipids that are transferred to selected areas of the cell. The formation of lysosomes by the Golgi complex has already been discussed. The Golgi complex consists of a stack of six or more flattened membrane-bound sacs. Small vesicles can often be seen to bud from the ends of the sacs. The surface of the Golgi that is closest to the nuclear membrane (the ***forming face***) is composed of sacs that are flatter than those at the opposite end ***(maturation face)*** of the stack (Figure 8-11). The Golgi surface closest to the nuclear membrane is also adjacent to the endoplasmic reticulum, which contains ribosomes. One of the perplexing aspects of Golgi function has been the mechanism by which proteins are processed after their synthesis on the endoplasmic reticulum. Experimental studies by Rothman and co-workers suggest that the Golgi complex acts like a distilling apparatus that takes crude unfinished proteins and refines them to a product ready for delivery to various cellular compartments. The process works in the following way. Many proteins synthesized on the endoplasmic reticulum are used for surface membranes, secretion granules, and lysosomes. Not only are these proteins contaminated with endoplasmic reticulum proteins when they leave the endoplasmic reticulum, but they must also be modified before delivery. For example, many membrane proteins and secretion proteins have carbohydrates attached to them. The Golgi apparatus is believed to remove contaminating endoplasmic reticulum protein and to perform many of the modification steps, which include acetylation, sulfation, and adding and removing various sugars. Once the membrane proteins leave the endoplasmic reticulum, they are refined in the Golgi sacs closest to the nuclear membrane. As the protein passes through the Golgi complex, the contaminating endoplasmic reticulum proteins are retrieved and sent back to the endo-

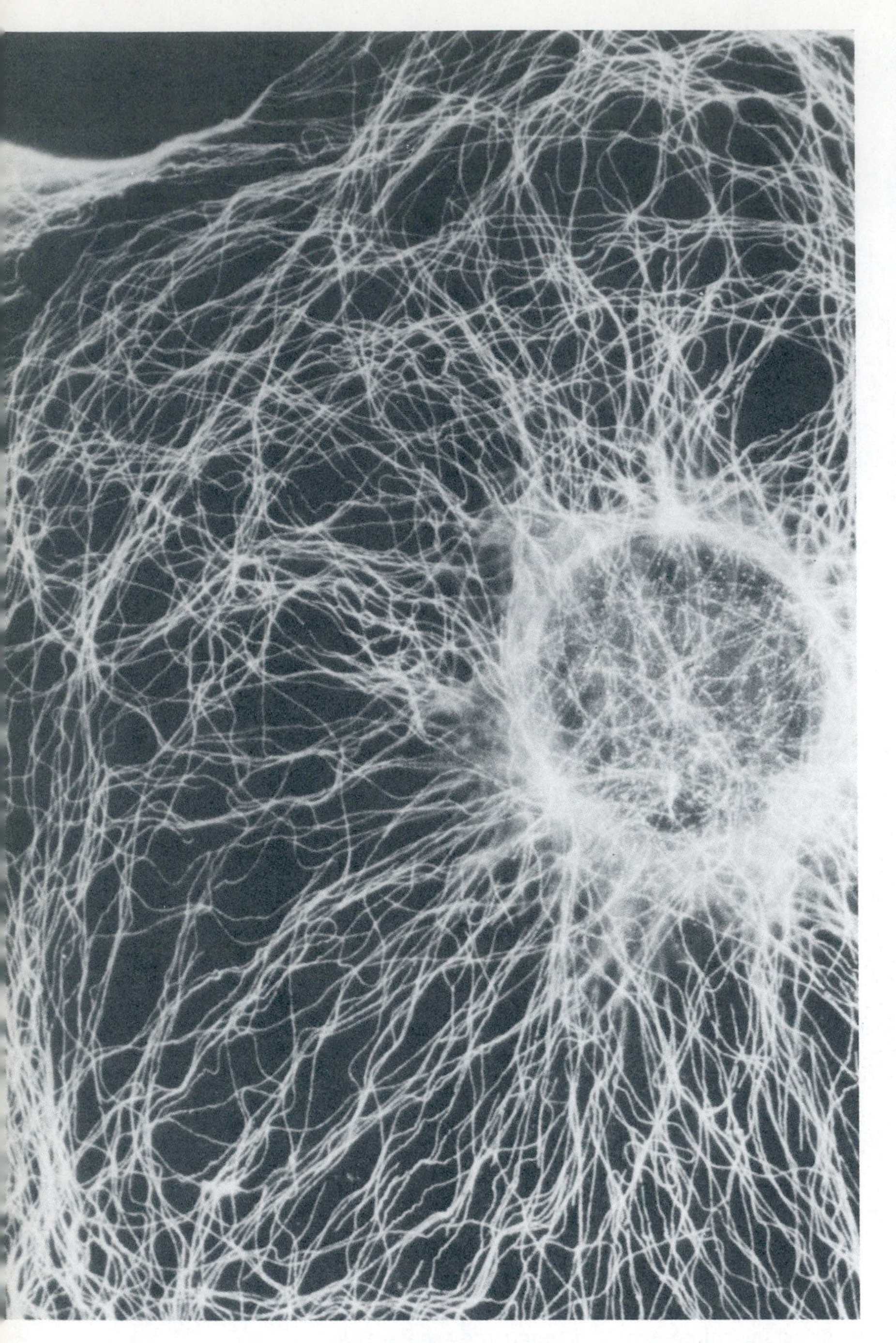

Figure 8-12 Photomicrograph of tubulin fibers surrounding nucleus of mouse cell.
From Weber, K. Proc. Natl. Acad. Sci. *(US)* ***75:****1820, 1978.*

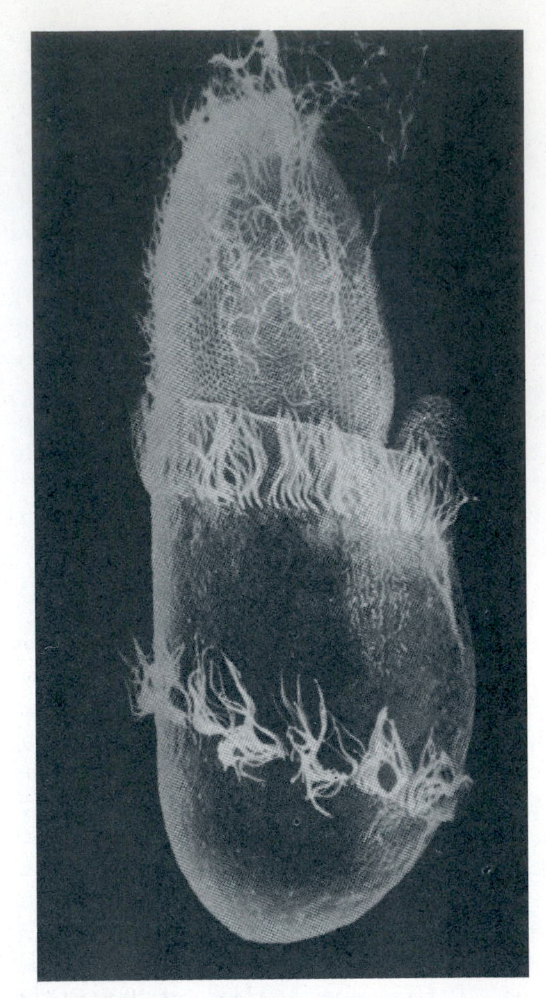

Figure 8-13 Partially engulfed Paramecium species.
From Wessenberg, H., and Antipa, G.J. Protozool. ***17:****250, 1970.*

plasmic reticulum. The refined protein is then processed further as it passes through the other Golgi sacs until a pure product is released.

MICROTUBULES

Microtubules are rigid hollow filaments found in all eukaryotic cells. They are usually long and between 150 and 250 Å wide. Most of the microtubules so far isolated and chemically analyzed are composed of protein subunits called ***tubulin.*** Once tubulin is disaggregated, the protein subunits are capable of self-assembly, a process that is also characteristic of the protein subunits of flagella and viruses. Microtubules are known to be associated with motility because they are components of flagella. Microtubules are also involved in nuclear division since they form part of the mitotic spindle (Figure 8-12). Because of their location in the cytoplasm and their apparent mobility, it has been suggested that microtubules may affect cell shape in wall-less eukaryotes such as the amebas and that they may be involved in the transport of material for biosynthesis. During fungal cell wall growth, for example, vesicles carrying wall-lysing enzymes are believed to be transported to the hyphal tip by microtubules. The association of microtubules with motility is discussed later in the chapter.

FLAGELLA, CILIA, AND CELLULAR MOVEMENT

Movement of many eukaryotic microorganisms is brought about by the action of either flagella or cilia. Structurally and functionally both flagella and cilia are the same, but cilia are more numerous and do not reach the lengths of flagella. Flagella may reach lengths of 250 μm, whereas cilia are usually no larger than 5 μm. Smaller eukaryotes such as microscopic algae possess one or more flagella, whereas larger eukaryotes such as certain protozoa (Figure 8-13) often employ several hundred cilia. Their ultrastructure is illustrated in Figure 8-14, *A* and reveals two central microtubules surrounded by nine doublet microtubules enclosed in a sheath. Each microtubule is made up of protein subunits (tubulin) that extend from one end of the cilium to the other. Attached to one microtubule of each doublet are protein linkages called ***dynein arms.*** They possess ATPase activity and are involved in the mechanism of movement. The nine doublet microtubules appear to be linked to each other by structural units called ***nexin links.*** Spokes are present that appear to bond each doublet microtubule to the central microtubules. Surrounding the entire microtubular system is a membrane that is an extension of the cytoplasmic membrane of the cell (a membrane is not found in most prokaryotic flagella). The cytoplasmic cell structure to which the cilium or flagellum is anchored is called a ***centriole*** (also called ***basal body*** or ***kinetosome***). If a cilium or flagellum is broken off at the cell surface, the centriole serves as a center for synthesis of new ciliary or flagellar material.

Ciliary or flagellar movement may be oarlike or undulating. The oarlike motion results from a bending at the base of the cilium, causing the tip to produce a circular arc (Figure 8-15). Undulating movements are generally observed when a cell possesses a single flagellum, whereas the oarlike movements are observed when many cilia are present on the cell. Like prokaryotes, motile eukaryotes try to avoid inhospitable environments and thus can change their direction of movement. The direction of the wave in avoiding a stimulus may be the same or different from the wave direction that involves attraction to a stimulus. The wave motion in some protozoa, for example, originates at the tip of the cilium or flagellum and proceeds to the base, but movement away from a stimulus initiates a wave at the base of the cilium or flagellum that travels to the tip (Figure 8-16).

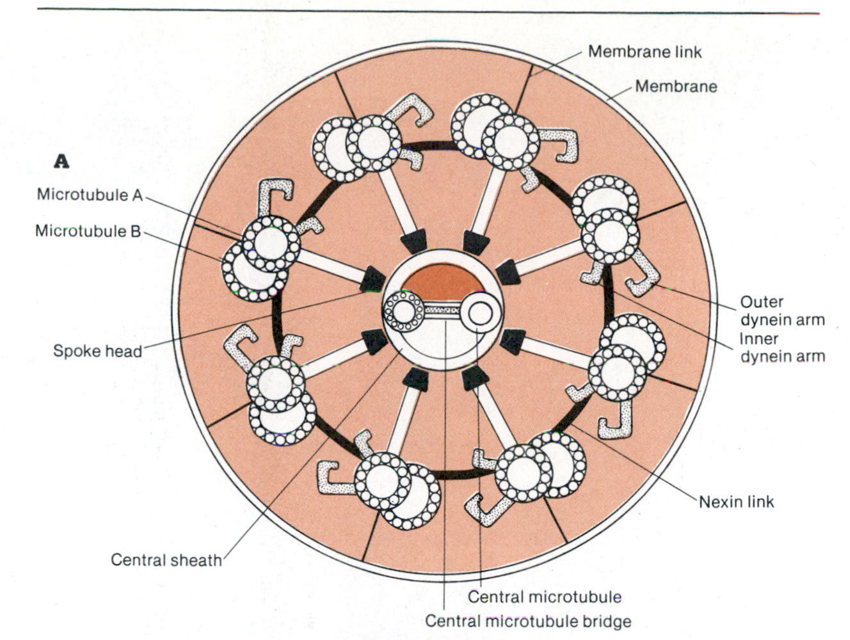

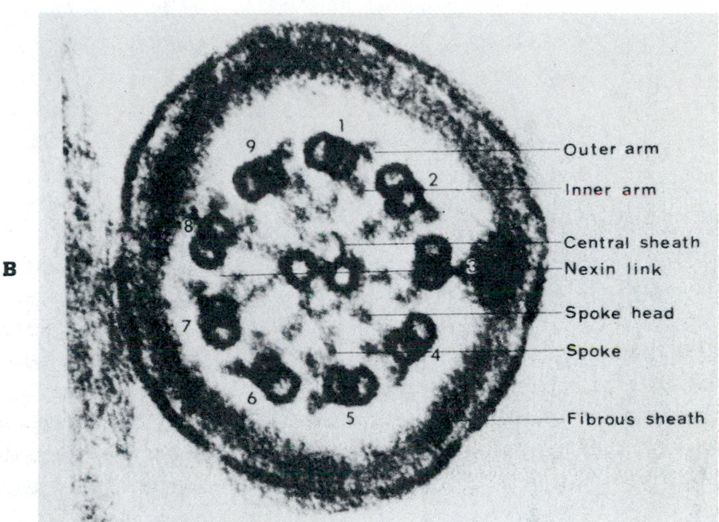

Figure 8-14 Cilium ultrastructure. **A,** Cross section of typical cilium. Two central microtubules are surrounded by nine doublet microtubules. **B,** Electron micrograph of cross section of cilium. Numbers represent doubled tubules.

***B,** From Afzelius, B.A. J. Ultrastruct. Res. **69:**43, 1978.*

The mechanism of movement of cilia or flagella is not completely understood but is believed to be caused by a sliding of the outer doublet microtubules with respect to each other. These sliding forces cause a bending of the cilium that is needed for ciliary movement. The force provided for the sliding is believed to be caused by the interaction of dynein and tubulin and the hydrolysis of ATP by ATPase contained in the dynein. The ability to control ciliary movement has been studied, and results indicate that calcium plays a major role. It has been suggested that the cell responds to mechanical or

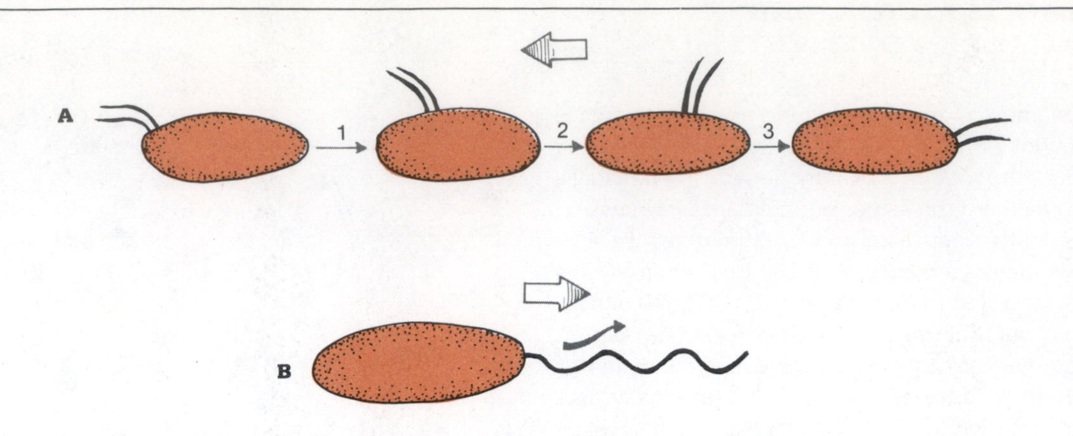

Figure 8-15 Wave forms generated in eukaryotes during movement by cilia, **A.** Only two cilia per cell are shown. Sequence 1, 2, and 3 shows direction of ciliary movement, which is oarlike. Large arrow indicates direction of cell movement. **B,** Wave forms generated in eukaryotes during movement by flagellum. Undulating movement of flagellum with wave *(arrow)* traveling to distal tip. Larger arrow indicates direction of cell movement.

chemical stimuli by inducing an electrical response in the membrane that in some way alters the calcium concentration of binding sites within the cell. Calcium apparently acts on specialized sites at the base of the cilium that control its behavioral responses.

The flagella and cilia of eukaryotes often perform functions in addition to movement to or away from chemical stimuli. Many eukaryotes use motility to aggregate with members of the same species. The aggregation process often culminates in the formation of a colony of cells that undergoes differentiation, with different cell types performing specific functions. The colonial state is of survival value to the species. Some eukaryotes show no motility in the adult stage, but their gametes (sex cells) are motile, particularly in the presence of chemical attractants. This ensures that there will be rapid zygote formation before the cells succumb to environmental influences. Flagella and cilia in the protozoa play many roles in addition to movement; these are discussed in Chapter 5.

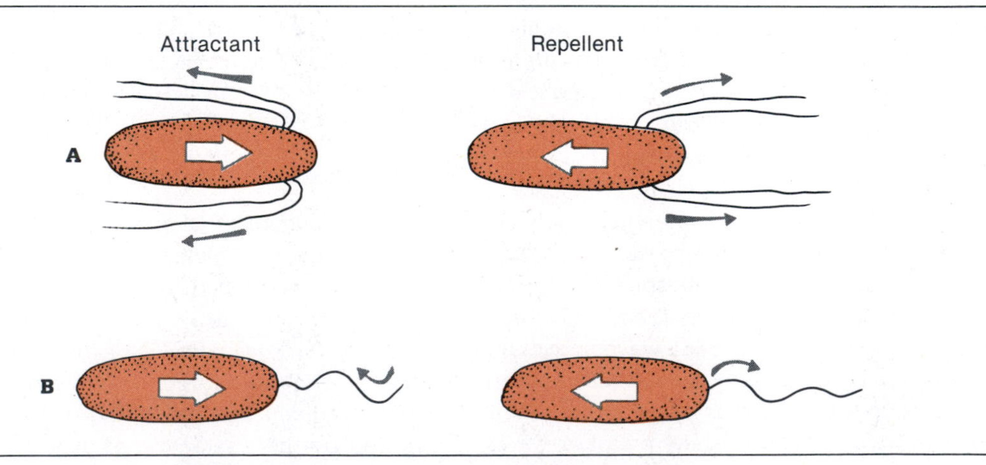

Figure 8-16 Response of motile cells to stimulus. Cells on left are reacting to attractant, whereas cells on right are responding to repellent. Arrows inside cell indicate cell movement; arrows outside cell indicate direction of cilia or flagella. **A,** Attraction to stimulus results in cilia pointing posteriorly, whereas in avoiding reaction they are pointed anteriorly. **B,** In attraction to stimulus, flagellar wave moves from tip to base; in avoiding stimulus, wave movement is initiated at base and moves to tip.

Cytoplasmic streaming

Observations of the cytoplasm of some eukaryotes reveals a characteristic movement referred to as cytoplasmic streaming. Cytoplasmic streaming is associated with cellular fluids that exhibit low viscosity, a property not found in prokaryotic cells. Protoplasmic streaming plays an important role in eukaryotes. Microorganisms such as the protozoa phagocytose nutrients into the cytoplasm where they are digested and released from vacuoles. The released nutrients must reach all parts of the cell. The large size of the eukaryotes makes simple diffusion of nutrients an inefficient process. Even with the help of protoplasmic streaming, which may reach 1 mm/sec, nutrient distribution is not as efficient as in prokaryotes. Eukaryotes have compensated for this inefficiency by exhibiting reduced metabolic rates. The fastest growing eukaryote may have a generation time (time for one division) of 1 hour, whereas the fastest prokaryote *(Vibrio natriegens)* may divide six times in 1 hour.

Cytoplasmic streaming also accounts for the movement over solid surfaces by such eukaryotes as slime molds and amebas. Located within the cytoplasm are channels in which the cytoplasm flows. Cytoplasmic contractibility is the source of mechanical work that brings about cytoplasmic streaming and locomotion. Rhythmic contractions take place at different times and places in the cell. If, for example, the volume of protoplasm projected to the forward parts of the cell is greater than that in the posterior sections of the cell, then the loose cytoplasmic membrane projects forward to produce an elongation called a ***pseudopodium.*** The direction of movement of the protoplasm depends on cytoplasmic contractions as well as the resistance offered by the environment outside the cell.

Chemotaxis in eukaryotes

Chemotaxis is a property that is found in eukaryotic microorganisms as well as prokaryotes. The chemotactic response in several eukaryotic groups is involved in the aggregation of single cells into multicellular bodies that undergo differentiation processes. One of the more widely studied eukaryotic chemotactic responses is in the cellular slime mold *Dictyostelium discoideum,* discussed briefly in Chapter 5. Recall that nutrient depletion triggers a differentiation process in which individual amebas aggregate to produce a multicellular body that may exist as a motile slug or as a stationary fruiting body.

Differentiation process in *Dictyostelium discoideum*

Laboratory experiments have demonstrated that after a few hours of starvation randomly dispersed cells of *D. discoideum* elongate and make contact with other cells. This is followed by the formation of long streams of cells that converge at a central point. On an agar surface several aggregates can be seen, each one containing from 10^3 to 10^5 cells (Figure 8-17). Aggregation occurs in a series of steps in which an attractant, called ***cyclic adenosine monophosphate*** (cAMP), is released from the center of the aggregate in a series of pulses. Cells at a distance from the center are activated by the cAMP and they in turn secrete cAMP and enter a refractory period that usually lasts from 5 to 10 minutes. The refractory period is an interval that permits activated cells to increase the production of cAMP to a level that will influence amebic movement. A second pulse of activity ensues when the concentration is high enough. Thus, a concentration zone of cAMP moves toward the periphery and away from the center of the aggregate. Therefore, for each outward movement of cAMP there is an inward movement of amebas (Figure 8-17, *D*).

Once aggregation has taken place, the mass of cells begins to move, and in the process a layer of slime is deposited. The migration mass of cells is referred to as a ***slug*** (see Figure 5-30). The next stage in development involves a cessation of movement and the differentiation of the slug into a ***fruiting body.*** The cells at the front of the slug during migration do not move, but the cells in the rear do move. Consequently, the cells in the rear push the stationary cells in the front of the slug into a vertical mass. The cells in the center of the tip of the vertical mass are called ***prestalk*** cells. They eventually push downward toward the base of the slug. The prestalk cells synthesize cellulose and produce a rigid structure called the ***stalk,*** and the most posterior stalk cells develop into a basal disk. The cells that initially were outside the prestalk cells at the tip of the vertical mass develop into spores. The spore cells synthesize cellulose, which they use to surround themselves and protect them until conditions are more favorable for germination. The remaining cells of the fruiting body eventually disintegrate and die. The biochemical mechanisms that are involved in this developmental process are currently under investigation. Studies indicate that cAMP probably plays an important role in the formation of the differentiated cell types: stalk, basal disk, and spore.

Figure 8-17 Aggregation in cellular slime mold *Dictyostelium discoideum*. **A,** Starved ameboid cells begin to elongate. (×200.) **B,** Streams of elongated cells merge toward center. (×200.) **C,** Lower magnification of aggregating amebas forming discrete colonies. (×10.)

A, B, *and* ***C,*** *Courtesy P. Brachet.*

D

D, Concentric wave pattern of ameba observed during aggregation using darkfield optics.

D, *From Alcantara, F. J. Gen. Microbiol.* ***85:****321, 1974.*

SUMMARY

Eukaryotes are structurally and functionally more complex than prokaryotes. Some of these differences are outlined in Table 4-1. Listed in Table 8-1 are the characteristics of the basic structures of eukaryotes. Additional structures are unique to a particular eukaryotic group, but these are discussed at more length in Chapter 5.

Table 8-1 Characteristics of eukaryotic structures and inclusions

STRUCTURE OR INCLUSION	CHEMICAL STRUCTURE	FUNCTION	COMMENTS
Cell wall	In green algae primarily cellulose; in other algae, silica, $CaCO_3$ and other polysaccharides in addition to cellulose	Structural support to the cell	Protozoa have no cell wall; wall is more porous than in prokaryotes
Cytoplasmic membrane	A unit membrane, but in addition contains sterols	Nutrient transport; in protozoa the membrane can capture food (endocytosis)	
Endoplasmic reticulum (ER)	A unit membrane with a rough portion containing ribosomes and a smooth portion that does not contain ribosomes	Permeation of cytoplasm; smooth portion functions to produce membranous structures such as vacuoles; rough ER engages in protein synthesis	Proteins synthesized are primarily secretory or membrane
Mitochondria	Membranous structure containing many enzymes, cytochromes, quinones; also contains its own DNA	Fatty acid oxidation, electron transport, and ATP production	Independent unit capable of self-duplication inside cell only
Chloroplast	Membranous system with an inner membrane organized into thylakoids containing glycolipid, phospholipid, and protein; also contains pigments such as chlorophyll and its own DNA	Conversion of light energy into ATP and reduction of carbon dioxide to organic molecules	Independent unit capable of self-duplication inside cell only
Golgi apparatus	Membranous organization of sacs	Synthesis and formation of lysosomes; may be a site for refinement of protein synthesized on ER	Probably modifies proteins by acetylation, sulfation, and adding or removing sugars
Microtubules	Protein units called tubulin	Movement and reproduction; make up structural components of flagella and mitotic spindle	May also be involved in intracellular transport and in maintaining shape of wall-less protozoa; movement of chromosomes
Flagella or cilia	Both made up of protein microtubules; also contain ATPase	Movement associated with feeding, sexual activity, and aggregation; cilia shorter and more numerous on cell than flagella	Flagella or cilia more complex structurally than prokaryotic flagella
Nucleus	A porous unit membrane structure containing DNA; membrane continuous with endoplasmic reticulum	Contains hereditary information for cell; nucleolus involved in production of ribosomal RNA	DNA in nucleus associated with proteins called histones; organized into unit called chromosomes

Study questions

1. What are the similarities and dissimilarities between the bacterial and fungal cell wall (in terms of chemical and structural organization)?
2. What is the function of the microfibrils found in the fungal cell wall?
3. What chemical component found in the fungal cell membrane is absent from the bacterial cell membrane?
4. Does the cytoplasmic membrane of eukaryotes differ in a functional way from prokaryotic cell membranes?
5. List those structural components found in the eukaryotic cell that are not found in the prokaryotic cell. What is their function?
6. How does pinocytosis differ from phagocytosis?
7. What group of eukaryotes uses phagocytosis as a means of nutrient transport?
8. What is the relationship of vacuoles to vesicles and lysosomes? What special functions do vesicles serve in the microbial cell?
9. Besides chemical structure and function, what makes mitochondria and chloroplasts different from other organelles in the eukaryotic cell?
10. How do the Golgi apparatus and endoplasmic reticulum function together in the cell?
11. What are microtubules? Where are they found in the cell and what are their functions?
12. List the chemical and structural differences between eukaryotic and prokaryotic flagella. How do cilia and flagella differ? How do the movements of cilia differ from those of flagella?
13. What is the purpose of protoplasmic streaming in eukaryotes? What might be the reason for its absence in prokaryotes?
14. What are some of the additional functions of chemotaxis in eukaryotes that are not required in prokaryotes?
15. What is the role of cyclic AMP in the aggregating of cellular slime molds? What function does aggregation serve in these microorganisms?
16. How do DNA organization and structure in eukaryotes differ from those in prokaryotes?
17. Define the following: thylaklids, hypha, endoplasmic reticulum, endocytosis, tubulin, pseudopodium, chromatin, histones, nucleolus, and nucleosomes.

Selected readings

Books

Brock, T. *Biology of microorganisms* (3rd ed.). Prentice-Hall, Inc., Englewood Cliffs, N.J., 1979.

Carlile, M.J. (Ed.). *Primitive sensory and communication systems: the taxes and tropisms of microorganisms and cells.* Academic Press, Inc., New York, 1975.

Cole, G.T. Architecture and chemistry of the cell walls of higher fungi. In Schlessinger, D. (Ed.). *Microbiology 1981.* American Society for Microbiology, Washington, D.C.

Dyson, R.D. *Essentials of cell biology* (2nd ed.). Allyn & Bacon, Boston, 1978.

Goldman, R. (Ed.). *Cell motility.* Cold Spring Harbor Laboratory, Cold Spring Harbor, N.Y., 1976.

Goodaa, G.W., Lloyd, D., and Trinci, A.P.J. (Eds.). *The eukaryotic microbial cell: thirtieth symposium of the Society for General Microbiology,* Cambridge University Press, New York, 1980.

Hawker, L.E., and Linton, A. (Eds.). *Microorganisms: function, form, and environment* (2nd ed.). University Park Press, Baltimore, 1979.

Hazelbauer, G.L. (Ed.). *Taxis and behavior.* Chapman and Hall, Ltd., London, 1978.

Krier, J.P. *Parasitic protozoa.* Academic Press, Inc., New York, 1978.

Peberdy, J.F. *Developmental microbiology.* Blackie & Son, Ltd., Glasgow, Scotland, 1980.

Reissig, J.L. (Ed.). *Microbial interactions.* Chapman and Hall, Ltd., London, 1977.

Round, F.E. *The biology of the algae* (2nd ed.). Edward Arnold (Publishers), Ltd., London, 1973.

Sleigh, M.A. *The biology of protozoa.* American Elsevier Publishing, New York, 1973.

Stanier, R.Y., Adelberg, E.A., and Ingraham, J. *The microbial world* (4th ed.). Prentice-Hall, Inc. Englewood Cliffs, N.J., 1976.

Journal articles

Calvalier-Smith, T. The origin and early evolution of the eukaryotic cell. *Symp. Soc. Gen. Microbiol.* **32:**33, 1981.

Farkas, V. Biosynthesis of cell walls of fungi. *Microbiol. Rev.* **43:**117, 1979.

Rothman, J.E. The Golgi apparatus. *Science* **213:**1212, 1981.

Schwartz, R.M., and Dayhoff, M.O. Origins of prokaryotes, eukaryotes, mitochondria and chloroplasts. *Science* **199:**395, 1978.

Wood, I.G. The structure and function of eukaryotic ribosomes. *Annu. Rev. Biochem.* **48:**719, 1979.

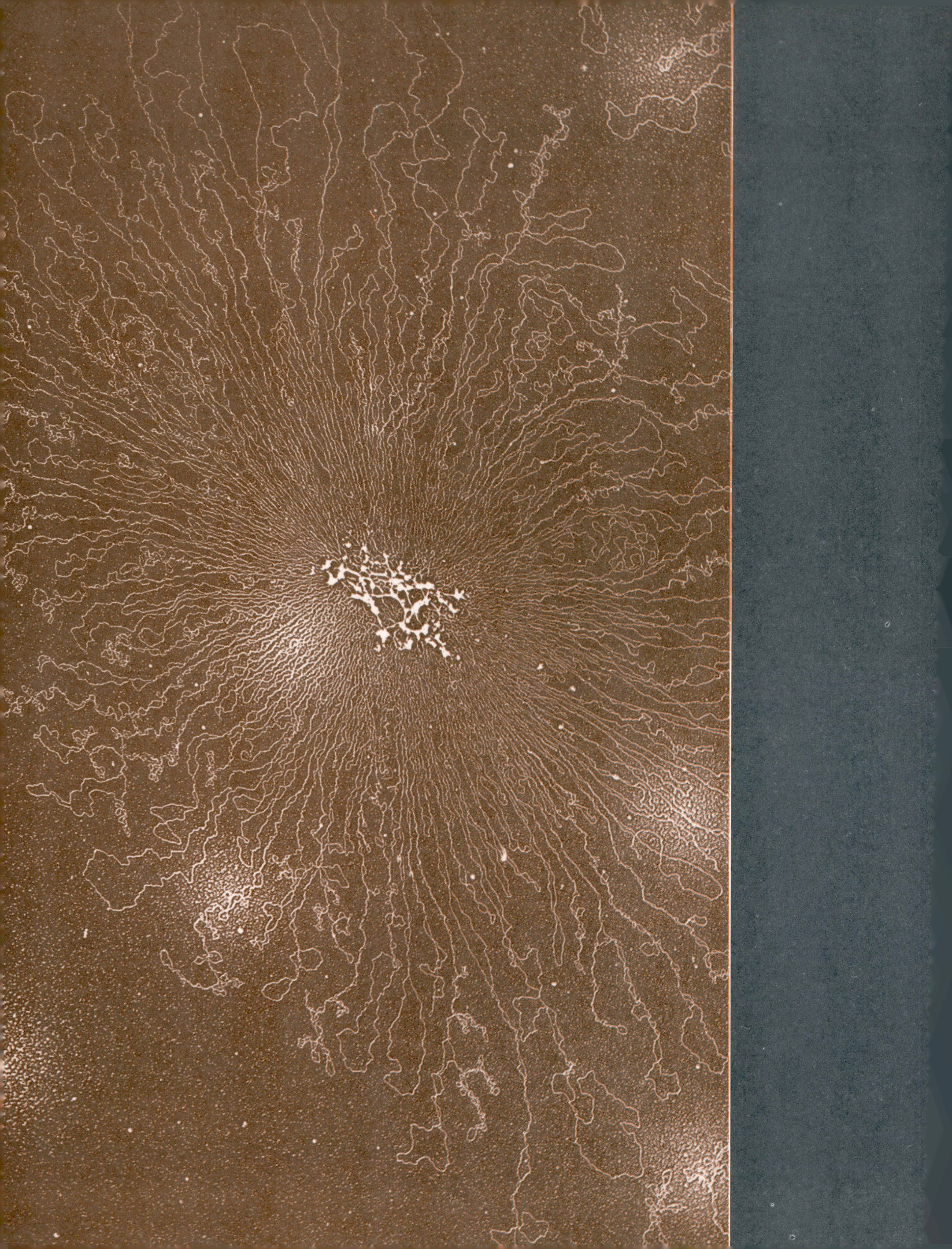

PART 3

MICROBIAL GENETICS

Folded chromosome from *Escherichia coli* spread out on salt solution.

R. Kavenoff and B.C. Bowen, University of California, San Diego/BPS.

Chapter 9
DNA REPLICATION AND MUTATION

Genetics is the science that deals with the mechanisms by which organisms pass traits from one generation to another. There are considerable differences in the traits expressed by various groups of organisms, but the nature of the hereditary material and the chemical and physical factors that affect it are the same. Microorganisms, especially bacteria, are being used extensively to study genetic mechanisms and along with fruit flies, mice, and corn are an important genetic tool. Bacteria are used because they multiply rapidly, are easily cultivated, are not burdened by complex cellular interactions as are multicellular organisms, are haploid, and have genetic material that is not complexed with other organic molecules or surrounded by a nuclear membrane. Genetic studies with prokaryotic organisms have made it possible to analyze genetic mechanisms in higher organisms.

One of the purposes of this chapter is to describe those mechanisms that we now believe to be responsible for the replication of DNA in microorganisms. In addition, the chemical and physical factors are described that affect the DNA molecule and that subsequently alter the hereditary traits of the microorganism.

DNA REPLICATION

Nature of the hereditary material

The nature of the hereditary material remained a mystery until a series of experiments, beginning in 1928, revealed some important genetic mechanisms. Each experiment, in its own way, provided vital information that would eventually help us understand the replication process.

Griffith in 1928 was able to show that the virulence (disease-producing ability) of *Streptococcus pneumoniae* was due to the formation of a capsule. Encapsulated strains produce a smooth colony on laboratory media and are referred to as *S* strains. Strains of *S. pneumoniae* that do not produce a capsule form rough colonies and are referred to as *R* strains. Laboratory mice die of septicemia when injected with a smooth strain but are unaffected by injected R strains (Figure 9-1). In later experiments Griffith used a combination of live R bacteria and heat-killed S bacteria to inject into mice. He noted that not only did the mice die, but smooth strains could be isolated from their blood. Griffith reasoned that something from the dead S strains had converted the R strains into smooth strains—a process called ***transformation.***

Avery and others in 1944 more clearly established that the ***transforming principle*** was DNA. They isolated DNA from S strains of *S. pneumoniae* and demonstrated in vitro that the nucleic acid could transform an R strain into an S strain (Figure 9-2). Furthermore, the isolated DNA fraction, when incubated in the presence of a DNA-hydrolyzing enzyme called deoxyribonuclease (DNase), lost its transforming properties.

Chemical analysis of the nucleic acid by Chagraff and others in the 1950s revealed it to be composed of phosphate, sugar, and purine and pyrimidine bases. They also established that the total amount of purines in the DNA was equal to the total amount of pyrimidines. Shortly thereaf-

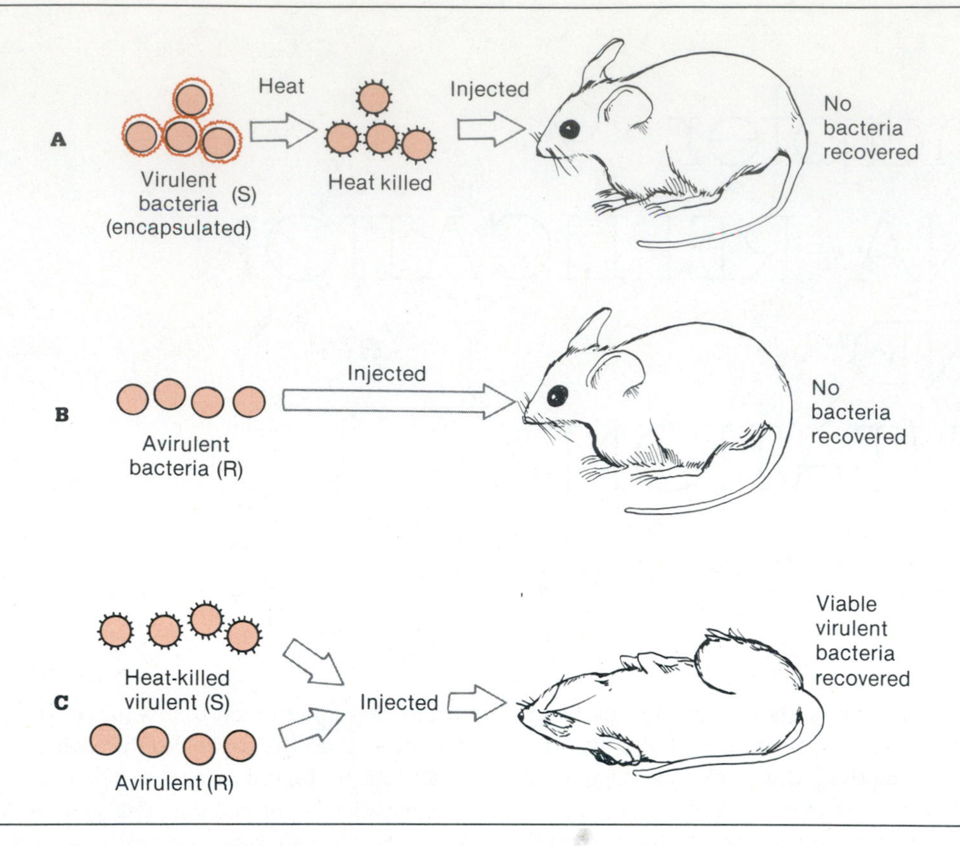

Figure 9-1 Transformation experiment. **A,** Virulent encapsulated *Streptococcus pneumoniae* cells (S type) are heat killed and injected into mouse. Mouse survives, and no bacteria can be recovered. **B,** Avirulent *S. pneumoniae* (R type) possessing no capsules are injected into mouse. Mouse survives, and no bacteria can be recovered. **C,** Experiments **A** and **B** are combined by injecting heat-killed virulent plus live avirulent bacteria into mouse. Mouse dies, and viable virulent bacteria are recovered, indicating that avirulent cells were somehow transformed into virulent types.

ter, Wilkins and co-workers, using x-ray diffraction techniques, noted that DNA was in the form of a helix. This discovery prompted Watson and Crick to propose a model for the DNA molecule. The chemical and physical properties of this molecule are discussed in Chapter 2.

Hershey and Chase in 1952 demonstrated that the genetic material could be transferred to cells by means other than transformation. They studied the viral infection process in the bacterium *Escherichia coli* by labeling the viral protein and nucleic acid with radioactive isotopes. Viral particles can be labeled by allowing them to replicate in bacteria in which the bacterial growth medium contains radioactive sulfur (^{35}S) and phosphorus (^{32}P). Radioactive sulfur is incorporated preferentially into viral protein, and radioactive phosphorus is incorporated into DNA (Figure 9-3). The actual experiment was carried out by taking unlabeled bacteria and infecting them with doubly labeled virus. A few minutes after the adsorption process, the suspension of infected bacteria was agitated in a blender. The agitation procedure removed the radioactive protein coat from the bacterial surface but did not interfere with the production of new viral particles, because viral DNA had penetrated the cell. DNA was therefore considered the hereditary material and was thought to be responsible in some way for the production of new viral particles.

The hereditary material is usually a double-stranded DNA molecule, although there are variations. For example, in some viruses the hereditary material is RNA, existing either as a single strand or double-stranded molecule. A few viruses possess single-stranded DNA instead of the double-stranded variety. The chemical and physical properties of the hereditary material are basically the same despite these variations, and they all share the same properties or functions in the cell which include:

1. The ability to replicate and transfer hereditary molecules to progeny during cell division

2. The ability to mutate
3. The ability to act as a blueprint for the expression of nucleic acid information into functional proteins (enzymes) or molecules that regulate metabolic processes
4. The ability to be transferred wholly or in part to other cell types (outside of cell division) and to engage in recombination with other nucleic acid molecules

Semiconservative replication

What sorts of mechanisms in the cell ensure that genetic determinants will be passed faithfully from one generation to another? Watson and Crick suggested that during replication the chains of the DNA molecule separate and that each acts as a template from which a complementary chain can be produced. The evidence that strand separation did occur came from the laboratories of Meselsohn and Stahl in 1958. They grew *E. coli* in a medium containing heavy nitrogen (^{15}N) for several generations until all of the DNA contained the heavy isotope. The cells were collected and then transferred to a medium containing light nitrogen (^{14}N). Some cells were allowed to divide just once and were then removed from the medium, and others were removed after two or more generations. DNA was extracted from the cells and centrifuged in cesium chloride. DNA containing the heavier isotope had a greater density and sedimented faster, DNA containing light nitrogen did not sediment as rapidly and remained near the top of the centrifuge tube. DNA isolated after the cell population had doubled was shown to be intermediate in density between DNA containing all light nitrogen and DNA containing all heavy nitrogen. This result demonstrated that the parent molecule does indeed separate and act as a template for the synthesis of a complementary strand (Figure 9-4). Their conclusion was that replication of DNA is semiconservative; that is, one of the parent DNA strands is conserved during the replication process, whereas its complementary strand is newly synthesized.

DNA polymerase and the replication process

The work of Meselsohn and Stahl confirmed the theory of strand separation postulated by Watson and Crick but gave no idea as to the mechanism(s) involved or the enzymes required. Much of the enzymatic details of replication were discovered by Kornberg and his colleagues in the

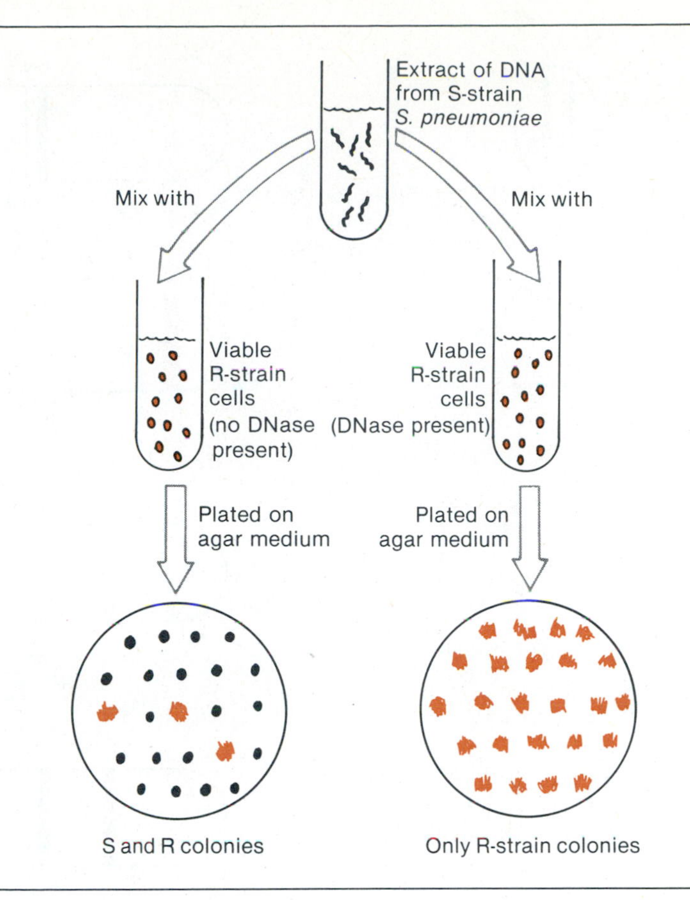

Figure 9-2 In vitro transformation process. DNA is extracted from smooth *(S)* strain (virulent) of *S. pneumoniae* and is mixed with viable rough *(R)* strains (avirulent) with and without DNase. In absence of DNase, some R strains are converted to S strains, but in presence of DNase, only R strains appear following cultivation. Transforming factor was hydrolyzed in presence of DNase.

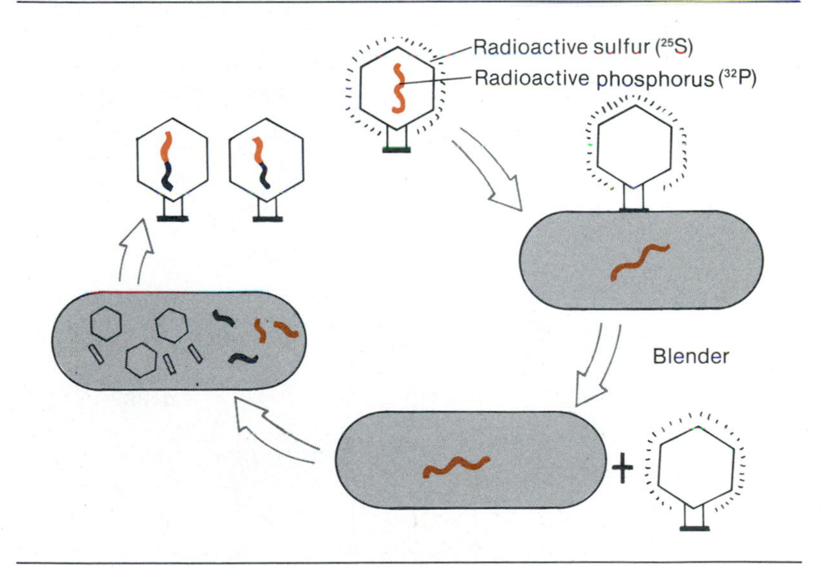

Figure 9-3 Experiment used to prove that only DNA and not protein of virus enter bacterial cell following infection. Agitation in blender removes radioactive sulfur from cell, but radioactive phosphorus, which is part of viral DNA (Color), remains inside cell. Newly produced viral particles contain radioactive phosphorus but not radioactive sulfur. See text for details.

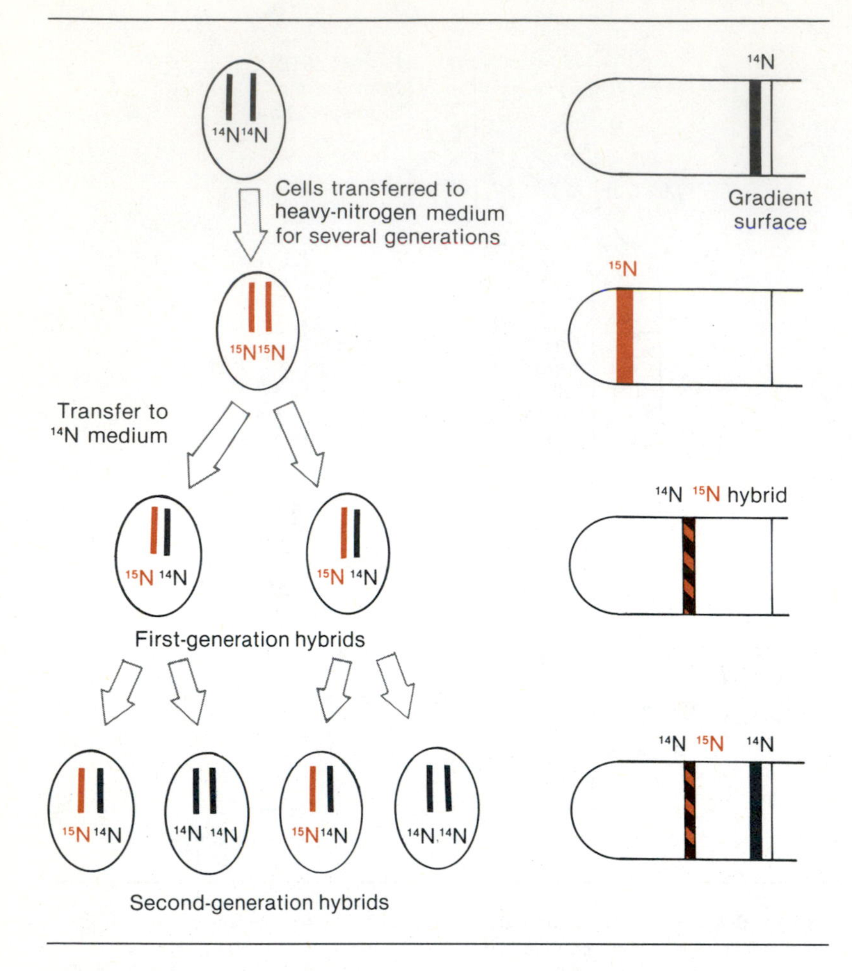

Figure 9-4 Schematic representation of Meselsohn-Stahl experiment demonstrating semiconservative replication. *E. coli* cells containing light nitrogen (^{14}N) are grown for several generations in media containing heavy nitrogen (^{15}N). Heavy cells are transferred to medium containing only light nitrogen and are then harvested after one generation (one doubling) and then harvested after subsequent generations. All cells of first generation are hybrids in which DNA contains one strand of ^{14}N and one strand ^{15}N. Population of cells harvested after two generations is of two types. One half of population are hybrid, and one half contain totally light DNA. Positions of various cell populations following density gradient centrifugation are shown on right.

1950s. They found that in vitro DNA synthesis required the following components:

1. A mixture of all the deoxyribonucleoside 5′-triphosphates: ATP, GTP, CTP, and TTP (Omission of any nucleoside resulted in no replication.)
2. Magnesium ions (Mg^{++}) (Magnesium stabilizes the DNA by binding to negatively charged phosphate groups.)
3. High molecular weight DNA (The DNA acts as a ***primer,*** or growing point, for the addition of nucleotides as well as a ***template;*** that is, the product formed is complementary to the original DNA molecule.)

The DNA polymerase enzyme isolated by Kornberg required a free 3′ hydroxyl on the DNA strands, which meant that the direction of DNA synthesis was only in the 5′ to 3′ direction (Figure 9-5). The DNA used as template could be from sources other than *E. coli.* Every experiment of Kornberg's using *E. coli* DNA or other DNA showed that the ratios of each nucleotide in the template were equivalent to those in the product. For example, if the template contained adenine, thymine, guanine, and cytosine, nucleotides in the ratio 0.64 to 0.66 to 1.34 to 1.35, the same ratio would appear in the product. Thus the Kornberg enzyme could catalyze the self-copying of DNA. Experiments by other investigators would later reveal Kornberg's DNA polymerase to be actually a DNA repair enzyme called DNA polymerase I and that other enzymes are involved in the replication process.* These enzymes are discussed later.

* One of these later experiments involved a mutant strain of *E. coli* that could replicate but from which no Kornberg enzyme could be recovered.

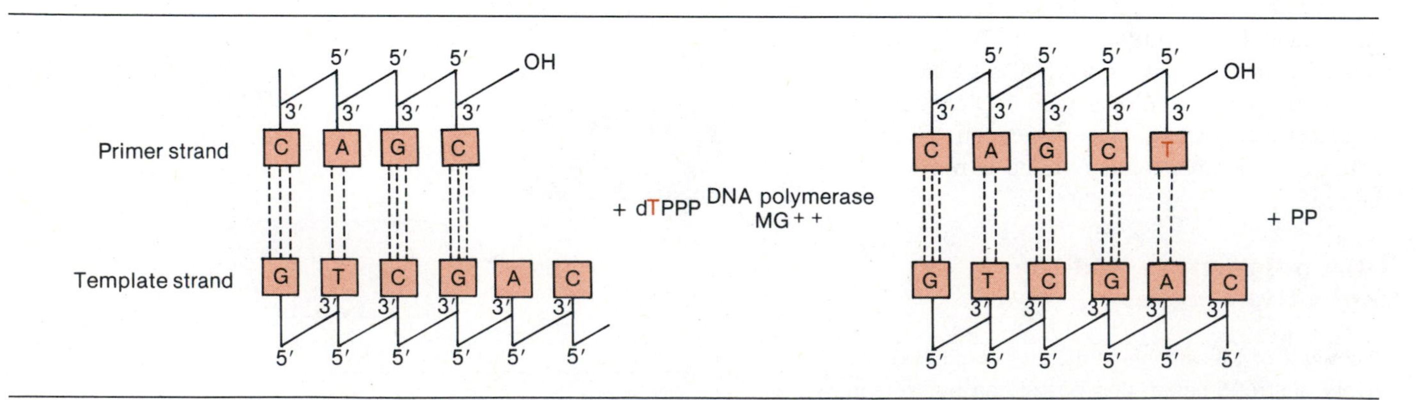

Figure 9-5 Requirements for in vitro DNA synthesis. Double-stranded DNA required for synthesis, with one strand containing free 3′ hydroxyl *(OH)* group for continued synthesis. Through action of DNA polymerase, 3′ hydroxyl *(OH)* of cytosine on primer cleaves pyrophosphate *(PP)* from deoxythymidine triphosphate *(dTPPP).*

Strand separation (replicating fork)

Bacterial DNA can exist as a rod or circle, and both forms are interchangeable in the laboratory when appropriate enzymes are available (Figure 9-6). DNA is usually found attached at some point to the cell membrane, which serves as an anchoring site for untwisting of the DNA during replication. When DNA is carefully isolated from the bacterial cell, electron microscopy reveals a circular molecule folded into loops and supercoils. How does replication occur in a circular molecule that exists in the cell in a tightly coiled state? Do the strands separate completely before DNA synthesis begins, or does DNA synthesis occur as the strands unwind? Cairns in 1962 used autoradiography to show how DNA synthesis occurs. In the experiment originated by Cairns, bacteria are grown in a medium containing tritiated (^{3}H) thymidine. The DNA, which is labeled, is gently isolated from the bacteria and covered with a photographic emulsion on a microscopic slide. Development of the slide and observation with a light microscope reveal a pattern of silver grains that outlines the structure of the chromosome. Interpretation of the autoradiograph (Figure 9-7) is based on the varying densities of the lines of grains. The experiment demonstrates that DNA synthesis in vivo occurs on both strands and begins at one site, the ***origin.*** The point at which the strands unwind creates a replicating fork, and as DNA synthesis continues, the replicating fork travels around the circular DNA molecule. This experiment of Cairns, although resolving a fundamental question, raised two others. First, how can DNA synthesis take place on both strands if they are antiparallel and if synthesis can proceed only in a 5′ to 3′ direction? Second, if the strands unwind at one end, the opposite end must rotate continuously. A closed circle would seem to prevent this necessary type of rotation.

Bidirectional DNA synthesis

Okazaki and others in the early 1970s sought to resolve the problem of how DNA synthesis could proceed on both strands of the DNA from a fixed point. They added radioactive DNA precursors to cultures of *E. coli.* DNA was isolated from the cultures at various times before one round of replication had taken place. The DNA was extracted and centrifuged in a sucrose gradient and analyzed for radioactivity. Their results demonstrated that at early times during the replication process most of the radioactivity was associated with low molecular weight DNA, whereas at later times in the replication process the radioactivity was predominantly part of high molecular weight DNA. Their conclusion was that short pieces of DNA are synthesized early (the short pieces are now referred to as ***Okazaki fragments***) and later are covalently bonded to produce high molecular weight DNA. The short pieces were thought to represent a discontinuous synthesis on the DNA molecule. Let us see how the actual process of discontinuous DNA replication is now believed to take place in the bacterial cell (Figure 9-8):

1. A multienzyme complex initiates DNA synthesis at a site called the origin. One strand of the tightly coiled DNA molecule is nicked by an endonuclease near the origin, which permits eventual unwinding of the DNA molecule. The unwinding is accomplished with the aid of helix-destabilizing proteins

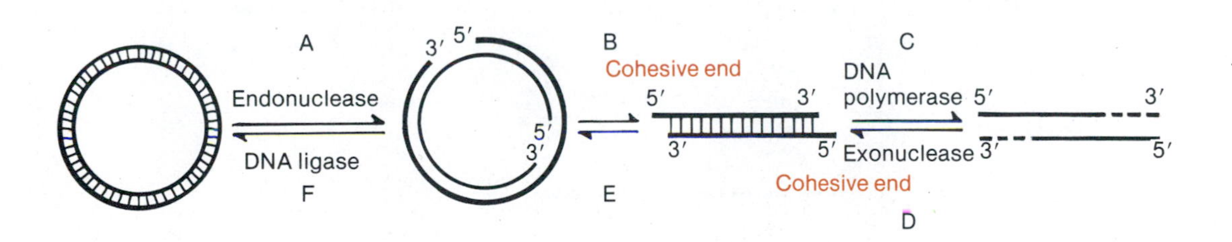

Figure 9-6 Interconversion of circular and linear forms of DNA molecule. *A,* Endonuclease cleaves circular DNA molecule at two specific sites. *B,* Circular DNA becomes linear, and each 3′ end of DNA strand stops short of its complementary strand. *C,* Complementary bases to cohesive ends can be sealed with nucleotides through action of DNA polymerase. *D,* Linear molecule of DNA can be cleaved by exonuclease to produce cohesive ends. *E,* Linear molecule circularizes. *F,* Nucleotide gaps in circular molecule are filled through action of DNA ligase.

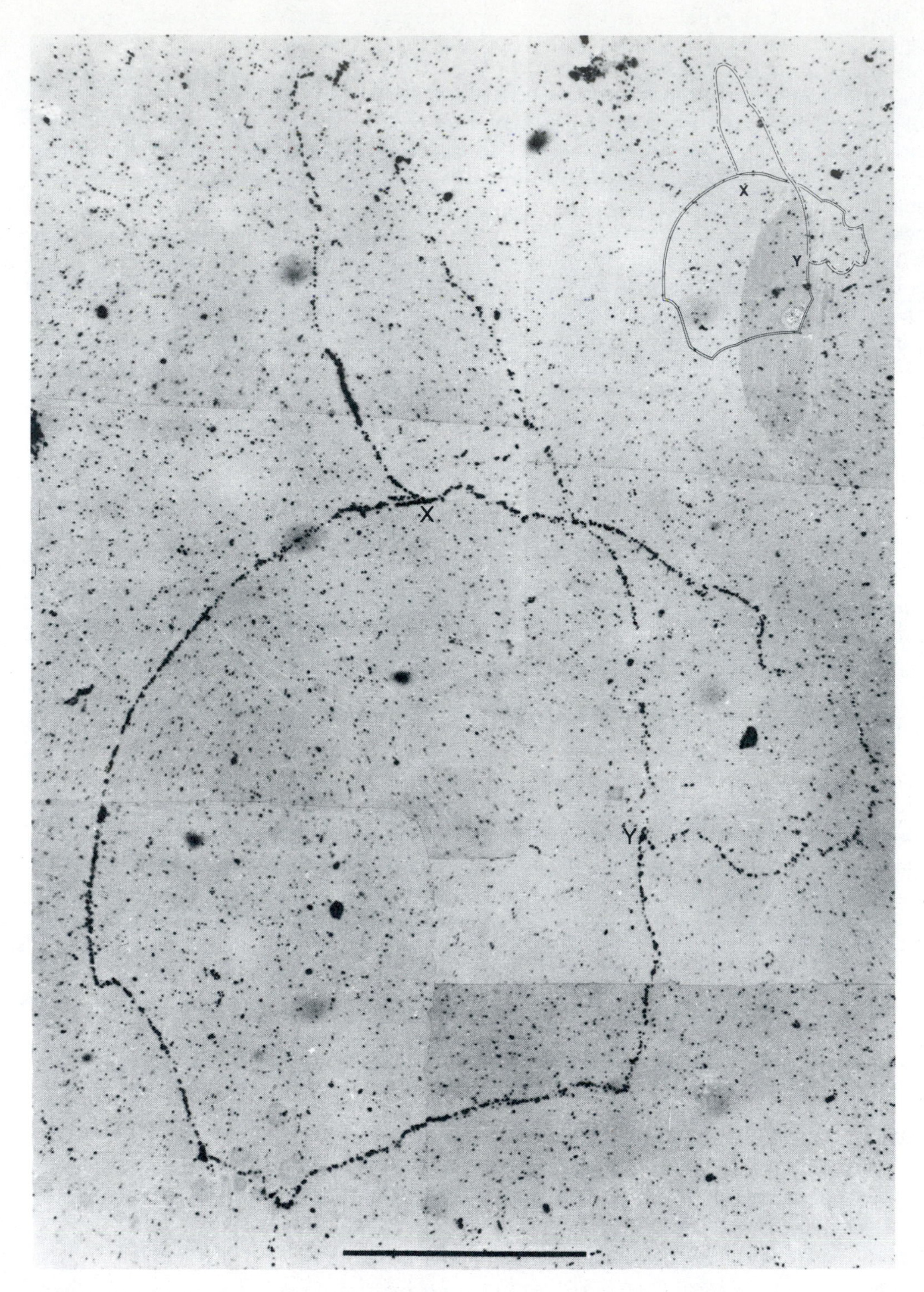

Figure 9-7 Autoradiograph technique demonstrating duplication of *E. coli* chromosome. Tritiated thymidine is revealed as line of dark grains in photographic emulsion. Interpretation of autoradiograph is based on varying densities of grains. Insert shows doubly labeled DNA as two solid lines; singly labeled DNA is solid line and dashed line. Duplication began at *X* with newly labeled strands extending as far as *B*.

From Cairns, J. Cold Spring Harbor Symp. Quant. Biol. ***28:**43, 1963.*

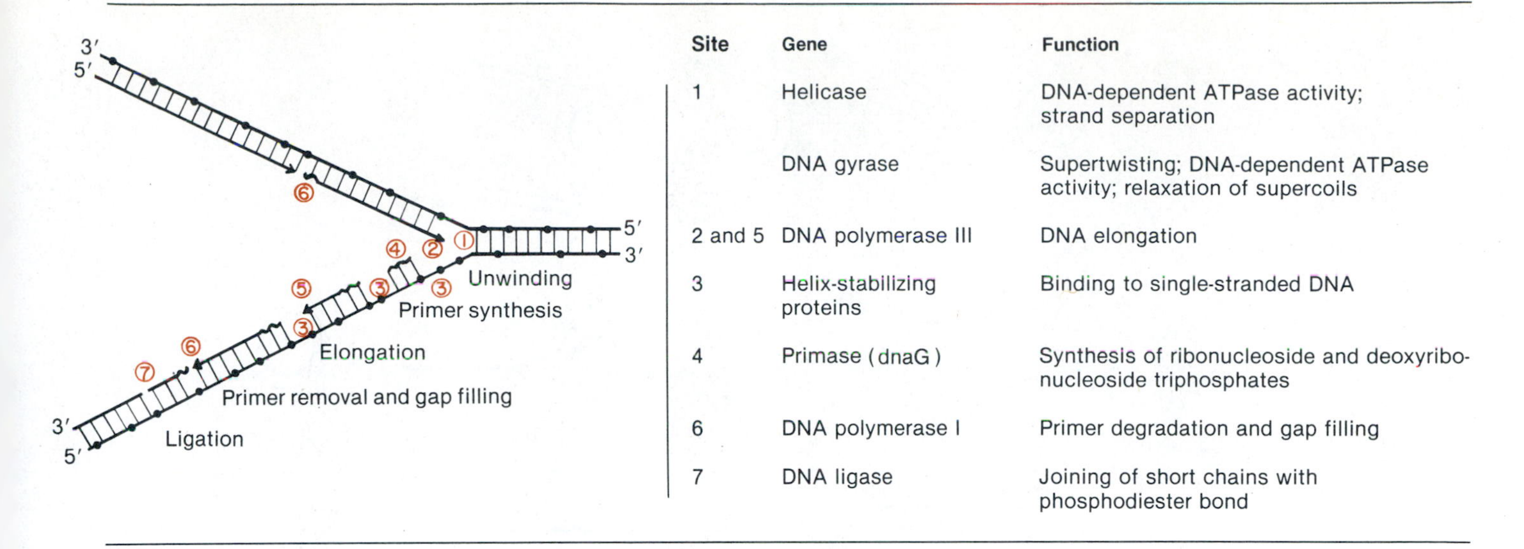

Site	Gene	Function
1	Helicase	DNA-dependent ATPase activity; strand separation
	DNA gyrase	Supertwisting; DNA-dependent ATPase activity; relaxation of supercoils
2 and 5	DNA polymerase III	DNA elongation
3	Helix-stabilizing proteins	Binding to single-stranded DNA
4	Primase (dnaG)	Synthesis of ribonucleoside and deoxyribonucleoside triphosphates
6	DNA polymerase I	Primer degradation and gap filling
7	DNA ligase	Joining of short chains with phosphodiester bond

Figure 9-8 Scheme for discontinuous replication of DNA. (See text for details.) Closed circles in parental strand represent potential initiation sites for primer synthesis. Sites of action of replication proteins are indicated by circled numbers. Description of replicating proteins are to right of figure.
From Ogawa, T., and Okazaki, T. Reproduced, with permission, from the Annual Review of Biochemistry, *Volume 49.*

(DNA-binding proteins), a helicase, and DNA gyrase.* The helix-destabilizing proteins bind to single-stranded DNA, but they have no enzyme activity. The helicase and DNA gyrase are involved in relaxation of the supercoils and in keeping the strands separated. They have DNA-dependent ATPase activity; thus energy for these functions is derived from ATP hydrolysis. By keeping the strands separated, a bubble is created in the DNA molecule, in which two growing points are produced.

2. Unlike the DNA used by Kornberg in his in vitro experiments, in vivo DNA does not have a primer. How, then, are new DNA chains synthesized? The initiation of DNA synthesis appears to require the synthesis of a short RNA primer catalyzed by a primase (dnaG protein). Primase polymerizes both ribonucleoside and deoxyribonucleoside triphosphates to variable chain lengths.
3. The growth of the DNA chain occurs in the 5′ to 3′ direction from the 3′ hydroxyl terminus of the primer. The activity is catalyzed by DNA polymerase III.
4. The short RNA primers are removed by DNA polymerase I, which has exonuclease activity. The gaps left by removal of the RNA primer are filled with deoxyribonucleotides, and this activity is also catalyzed by DNA polymerase I. The gaps are filled until a single phosphodiester bond remains between the chains. These last gaps are sealed by the enzyme DNA ligase. It is still in doubt whether both chains are synthesized discontinuously using Okazaki fragments, since one of the chains normally has a reactive 3′ hydroxyl for continuous DNA synthesis.
5. The replication process continues in both directions from the replicating origin until the two growing points meet and two double-stranded molecules of DNA have been produced. An enzyme is believed to separate the two daughter molecules at the termination site (Figure 9-9), and one DNA molecule is distributed to each of the

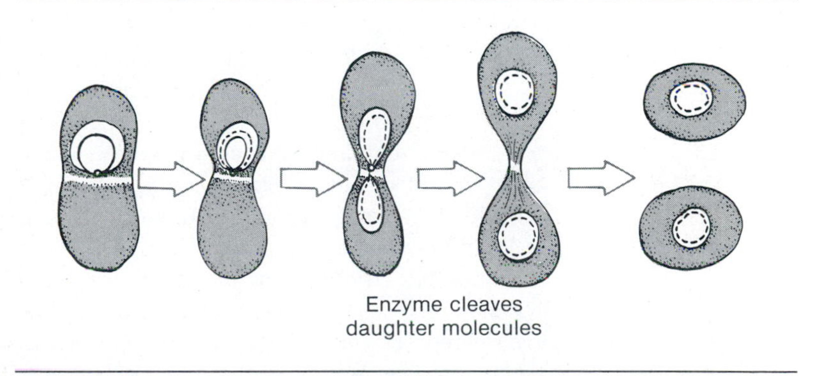

Figure 9-9 Model to explain separation of daughter DNA molecules produced during cell division. Dashed line represents newly synthesized DNA.

* Gyrases belong to a group of enzymes called topoisomerases. They can break and rejoin the DNA backbone and thus allow the interconversion of the superhelical and relaxed forms of DNA.

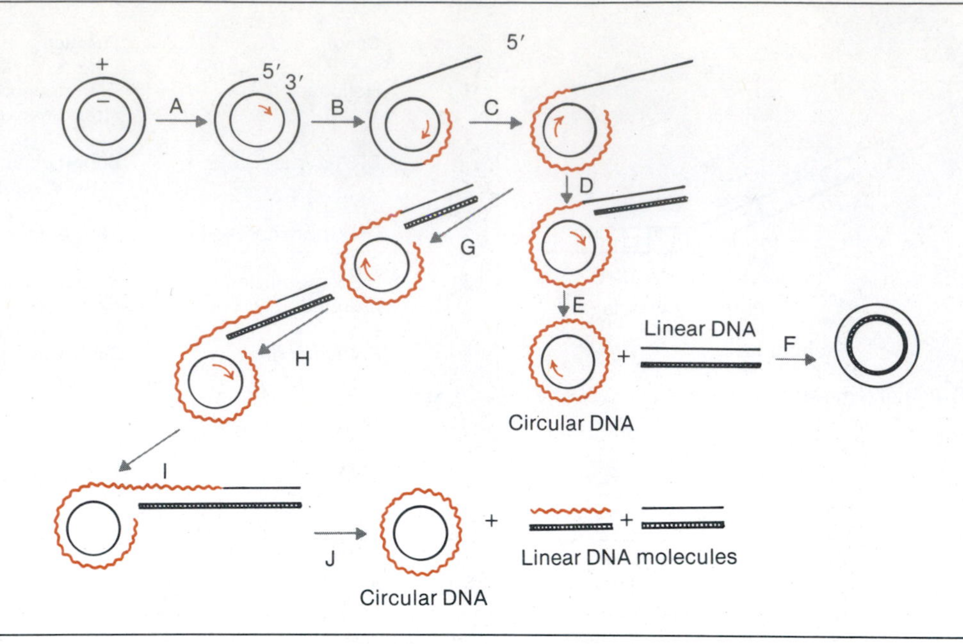

Figure 9-10 Rolling circle mechanism of DNA synthesis. *A*, Double-stranded DNA with positive (+) and negative (−) strand. Positive strand is nicked by endonuclease. *B*, Intact negative strand begins to roll, and 5′ end of positive strand is displaced (attached to membrane of cell). DNA synthesis on positive strand is begun as negative strand rolls with nucleotides added to 3′ end of positive strand (color). *C*, There is further rolling of negative strand, and more DNA is added to positive strand. *D*, Displaced 5′ end of positive strand now acts as template, and complementary strand (▬) synthesis is begun. *E*, Negative strand has completed 360-degree revolution. Displaced positive strand is cleaved by enzyme and releases circular and linear molecule of DNA. *F*, Linear molecule circularizes. *G* and *H*, Rolling circle can continue revolving another 360 degrees, resulting in production of long tail containing concatameric (multiple copies) DNA. *I* and *J*, Enzyme cleaves concatameric DNA, leaving behind circular DNA molecule. Linear DNA molecules can circularize. Many multiple copies of DNA could be made before cleavage of linear molecules.

daughter cells. See Chapter 15 for a discussion of the relationship of DNA replication to cell growth.

Rolling circle mechanism of replication

A second type of circular replicating DNA believed to occur in some viruses and during the sexual mating process of bacteria is called the rolling circle (Figure 9-10). This is an example of unidirectional replication. One of the strands of the DNA molecule is cut by an endonuclease to produce a linear strand with 5′ and 3′ ends. The other strand remains as a closed intact circle. The 5′ end of the linear strand is displaced and becomes attached to a membrane site, whereas the 3′ end acts as a primer for chain elongation, using the closed intact strand as a template. DNA synthesis is completed when the intact single-stranded circle has revolved 360 degrees and all DNA complementary to the intact strand has been synthesized. The displaced portion of the replicating strand serves as a template for the synthesis of its complement. One of the advantages of this technique is that many copies of DNA can be produced (the term *concatameric* is used to describe copies of DNA that are joined together). The linear copies can be separated by the action of a cleaving enzyme, and the DNA molecules can then circularize.

EXTRACHROMOSOMAL DNA (PLASMIDS)

Characteristics

Some of the genetic material in the microbial cell is extrachromosomal; that is, it is not part of the genome of the cell. These independent pieces of DNA are called ***plasmids.*** They are closed circles of double-stranded DNA having molecular weights of 10^6 to 10^8 daltons (for comparison, the genome of *E. coli* has a molecular weight of 2×10^9 daltons) and are passed to daughter cells following cell division. Plasmids can also be "cured" from the cell when the latter is subjected to certain environmental stresses such as changes in temperature, the presence of certain dyes, or nutrient depletion. Plasmids are not indispensable to the cell, but they may confer on the cell certain selective advantages over other organisms. Some plasmids, for example, carry information for the degradation of certain substrates, and antibiotic resistance plasmids confer resistance to specific antibiotics that may be present in the microorganism's environment. Other phenotypic properties carried by plasmids are described later. The first plasmid to be discovered was one that conferred the ability to transfer itself to a recipient cell in a process called conjugation. (See Chapter 11 for details of this process.) It was called the sex factor, fertility factor, or F^+ factor.

Plasmid replication can occur at two different times: (1) when the bacterial cell divides (plasmid DNA also divides, ensuring that both daughter cells receive a plasmid) and (2) during the conjugation process, when a replicated molecule can enter the recipient cell. It appears that during plasmid replication the DNA attaches to the cytoplasmic membrane and uses the same enzymes and machinery that are used in the replication of chromosomal DNA. Replication may be unidirectional (one replicating fork) or bidirectional (two replicating forks), depending on the type of plasmid.

Many plasmids confer the property of conjugation, whereas others are nonconjugative and cannot effect their own transfer. Antibiotic-resistant plasmids in gram-positive bacteria such as the staphylococci cannot be transferred by a conjugation process, because conjugation does not exist or at least has not been discovered in most gram-positive bacteria.* The plasmids in the gram-positive staphylococci can only be transferred during a process called transduction, which involves a viral agent. (See Chapter 11 for details of this process.) Nonconjugative plasmids in gram-negative bacteria can be transferred only if the cell also contains conjugative plasmids. The transfer factor on one plasmid can therefore effect the transfer of other plasmids, a process called ***plasmid mobilization.***

Bacteria may contain a number of plasmids. The Enterobacteriaceae, for example, possess one to two conjugative plasmids per chromosome but can also carry 10 to 15 nonconjugative plasmids per chromosome. When two or more different plasmids are stably inherited, they are said to be ***compatible.*** Other plasmids are considered ***incompatible*** when after a few cell divisions one of the plasmid types is lost. The molecular mechanisms controlling this incompatibility are not known but may be due to genetic factors controlling plasmid replication and segregation at cell division.

* Conjugation occurs in *Bacillus subtilis* and in some species of *Streptococcus* and *Streptomyces.* Conjugation plasmids coding for antibiotic resistance have been found in *Streptococcus faecalis.*

Classes

The characteristics conferred on the host by plasmids are extensive, and some of them are indicated below.

Sex factor plasmids. As previously discussed, the sex factor is important for plasmid transfer to recipient cells. In addition, the sex factor can be integrated into the host chromosome and in this state is called an ***episome.*** An integrated sex factor enables the cell to mobilize the transfer of the bacterial chromosome during the conjugation process (see Chapter 11).

R plasmids. Antibiotic resistance in many microorganisms is due to the presence of plasmids that contain specific information for the synthesis of enzymes that inactivate specific antibiotics. These are called resistance or ***R factor*** plasmids. R factors were first discovered in Japan during an epidemic of bacillary dysentery. It was discovered that a strain of *Shigella dysenteriae* had acquired resistance to four different antibiotics and that these determinants could spread rapidly in a population of related gram-negative species that were initially sensitive to the same antibiotics. R factors are composed of two components: the ***resistance (r) determinant*** and the ***resistance transfer factor (RTF),*** which can be transferred independently of the resistance determinants or transferred with it. The transfer of the r determinants to recipient cells during the conjugation process is controlled by the RTF. Both r and RTF are capable of autonomous replication in their independent states and can integrate themselves into other extrachromosomal elements as well as chromosomal units.

Conjugative R plasmids conferring resistance to antibiotics such as kanamycin, chloramphenicol, tetracycline, penicillin, and ampicillin, as well as others, have been discovered in several virulent strains of bacteria (see Chapter 27).

Col plasmids. Many bacteria produce proteins, called ***bacteriocins,*** that are released into the environment and are lethal to related species of bacteria. Most are produced in gram-negative bacteria such as *E. coli,* in which the bacteriocin is called ***colicin*** and the plasmid encoding them

Figure 9-11 Characteristic crown gall observed on stem of willow tree.
Courtesy R.C. Lambe, Virginia Polytechnic Institute and State University.

is called ***Col plasmid.*** Col plasmids are divided into two groups: those that are conjugative and those that are nonconjugative. Some of the conjugative Col plasmids may also code for drug resistance or some virulence factor. Some colicins are a complex of two proteins: one that is active in killing sensitive cells and the other, which confers immunity. Every Col plasmid confers immunity to the colicin for which it encodes. The bacterial cell, however, may be susceptible to other types of colicins. All colicins bind to specific receptors on the cell envelope of the target cell. The mechanism by which colicins act to kill sensitive cells has not been clarified. Some may act on specific intracellular sites such as the DNA or ribosomes. Others appear to attack the cytoplasmic membrane, affecting its permeability and causing the loss of important cytoplasmic cations such as potassium.

The effectiveness of colicins in offering a selective advantage to cells in their natural environment is unclear. It is not known, for example, if colicin is produced in significant quantities in the intestinal tract, where *E. coli* normally resides, to control the population of other gram-negative species.

Only a small percentage of gram-positive bacteria, such as the staphylococci (staphylococcin), streptococci (streptococcin), and *Clostridium perfringens* (perfringocin), are known to produce bacteriocins.

Mercury and other heavy metal ion–resistant plasmids. Some bacterial species such as *Staphylococcus aureus* and species of *Pseudomonas* are resistant to certain heavy metals such as mercury, arsenic, cadmium, and lead. Mercury resistance has received much scientific and public attention. Inorganic mercury compounds, which are nontoxic, can be converted by anaerobic bacteria into methylated organic compounds that possess considerable neurotoxic activity for humans. Inorganic mercury resistance in microorganisms is due to the enzyme mercuric reductase, which reduces inorganic mercury according to the following reaction:

$$Hg^{2+} \xrightarrow{\text{Mercuric reductase}} Hg^{\circ}$$

Organic mercury resistance is due to hydrolases that separate mercury from the organic moiety.

Plasmids that confer resistance to metal ions are selected in those environments containing high levels of the metal. Plasmid-associated resistance to silver, for example, has been found in those bacteria isolated from sludge produced by industrial processing of used photographic film.

Plasmids and catabolic activity. The ability of certain bacterial species, particularly species of *Pseudomonas,* to extract carbon and energy from unusual organic compounds is determined by plasmids. For example, some species can catabolize over 100 different compounds, and the enzymes responsible for the breakdown of some of them are encoded in plasmids. Plasmid-encoded enzymes can break down such products as naphthalene, xylene, camphor, and octane into metabolites such as pyruvate and acetate.

Plasmids are used in the field of genetic engineering for the creation of microorganisms with special properties. Chakrabarty, for example, constructed an organism that could degrade 75% of the components making up crude petroleum. This constructed organism has been patented and may someday play a role in the clearing of oil spills. How plasmids are used in genetic engineering is discussed in Chapter 11.

Plasmids and plant hyperplasias. Crown gall is one type of cancerous disease of plants caused by a bacterium, *Agrobacterium tumefaciens* (Figure 9-11). The information for tumor (hyperplasia) development is carried on the bacterial plasmid. The evidence suggesting that the bacterium can transfer genetic information to the plant has come from in vitro studies. The excised cancerous tissue, which can be subcultured in the laboratory without the addition of plant hormones, exhibits tumor-specific properties. (More detailed information on crown gall appears in Chapter 35.) Gall formation also occurs on oleander plant stems infected with the bacterium *Pseudomonas savastanoi.* The organism produces two enzymes involved in the synthesis of the plant growth compound, indole acetic acid (IAA). The genes coding for the two enzymes are located on a plasmid carried by the bacterium. Mutants of *P. savastanoi* lacking IAA synthetic capabilities fail to induce gall formation, and those with increased capacity for IAA synthesis induce more severe symptoms in the plant.

Virulence plasmids for mammals. Plasmids in various bacteria carry information that contributes to their virulence during infection of mammals, including man. Some of this plasmid information codes for (1) the production of an enzyme (hemolysin) that hemolyzes red blood cells, (2) toxin production, and (3) the production of cell surface antigens that permit certain bacteria to adhere to epithelial surfaces. Some of these are discussed in Chapters 22 and 34.

MUTATION

One question a student might ask after reading the discussion of DNA replication is, "Do errors occur during the replication process?" The answer is yes. Errors are sometimes made that result in alterations in the coded information, that is, the ***genotype,*** which may or may not cause changes in the expressed traits, that is, the ***phenotype,*** of the cell. The genetic material is subject to internal as well as environmental influences of both a chemical and physical nature. Hereditary changes in the genotype of a cell are called ***mutations,*** and the physical and chemical agents causing the mutations are called ***mutagens*** (Table 9-1). Mutations may be of two general types: spontaneous and induced. Spontaneous mutations occur in the absence of human intervention. Induced mutations are alterations caused by specific chemical or physical agents. The primary purposes of this part of the chapter are to describe (1) the molecular mechanisms of spontaneous and induced mutations, (2) how microorganisms can repair some mutations, (3) the significance of mutation in microbial evolution, and (4) some practical applications of microbial mutants.

The effect that mutation has on coded information in the cell and its translation are discussed in Chapter 10.

Characteristics of the mutation process

Mutations are rare events, and the rate at which they appear in a bacterial cell ranges from 1×10^{-4} to 1×10^{-10} per generation. In other words, one bacterial cell in 10,000 or one in 10 billion may show a mutation. Mutation rates are generally defined as the probability of mutation per cell per generation. The mutation rate varies for each characteristic examined and also varies for the same characteristic between species or even between strains of the same species. Some areas of the DNA are more susceptible to mutation, and these are called "hot spots." Thus when mutants of a bacterial species are isolated, genetic analysis may reveal that many of them occur at a single site on the chromosome. The mutation rate for a particular characteristic will also vary depending on the type of mutagen used. The spontaneous mutation rate for penicillin sensitivity (Pen^s) might be 1×10^{-10}, but in the presence of a chemical or physical mutagen the rate could change to 1×10^{-4}.

In biology the organism isolated from nature that has not been exposed to a mutagen in the laboratory is called the ***wild type.*** The progeny of the wild type in which the mutation has been expressed are ***mutants.*** The mutant is referred to as an ***auxotroph*** if the mutation involves the requirement of a growth factor such as an amino acid, vitamin, or purine or pyrimidine base. An auxotroph, for example, unable to produce histidine is designated His^-. The parent of the auxotroph that does not require the growth factor is called the ***prototroph*** and in the case of histidine synthesis is designated His^+.

Some mutations remain stable and are transferred to future generations in the altered state. A mutation from the wild type, such as His^+, to the auxotroph, such as his^-, is called a forward mutation, but a mutation could reverse the direction and go from His^- to His^+ and thus restore the function of histidine synthesis. The chances for the forward mutation from His^+ to His^- and a loss of function are many, since there are nine enzymes involved in the synthesis of histidine, and a mutation in any one could result in a loss of function. In order to restore function (His^- to

Table 9-1 Characteristics of the common microbial mutagens

	MUTAGEN	MECHANISM OF ACTION	EFFECT OF MUTATION
Physical	Ultraviolet light	Causes dimerization between pyrimidine residues on DNA	Inhibits replication; repair may lead to deletion of base
	X-rays	Causes formation of free radicals such as hydroxyl, as well as superoxide ion	Oxidation and breakage of DNA chains
Chemical	5-Bromouracil	Analogue of thymine that in its rare "enol" form pairs with guanine	Causes transitions from A:T base pairs to G:C bases pairs, and vice versa
	2-Aminopurine	Analogue of adenine that in its rare imino form bonds to cytosine; in natural state can also bind to cytosine	Causes transitions from A:T base pairs to G:C bases pairs, and vice versa
	Nitrous acid	Oxidatively deaminates bases such as adenine, cytosine, and guanine	Causes mispairing and transitions from A:T base pairs to G:C base pairs, and vice versa
	Hydroxylamine	Causes deamination of cytosine	Transitions from G:C base pairs to A:T base pairs only
	Nitrogen mustard (Figure 9-12)	Alkylating agent that can act as binding component between purines on opposite strands of DNA	Causes removal of bases that can be later replaced by any of the four bases; thus transitions and transversions are possible
	Nitrosoguanidine	Like nitrogen mustard causes interstrand linkage of purine bases	Similar to nitrogen mustard
	Ethylethanesulfonate	Alkylating agent like nitrogen mustard	Causes removal of purines, especially guanine
	Acridines	Insert themselves between bases on DNA strand	Depending on whether DNA strand is parental or is being synthesized, addition or deletion, respectively, will occur
Biological	Insertion sequences	Transposable from site to site on chromosome	Alter expression of genes into which they insert themselves as well as genes near insertion site
	Transposons	Same as insertion sequence	Same as insertion sequences except that transposons carry a specific trait that is linked to genes where they insert on chromosome

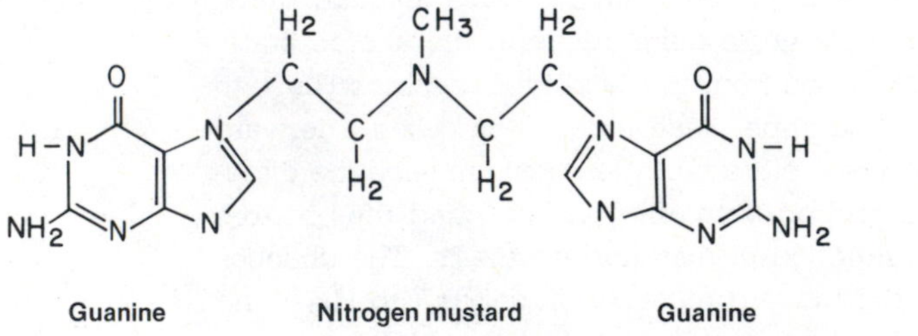

Figure 9-12 Nitrogen mustard alkylating adjacent guanines.

His^{+}), the mutation must be at the one genetic site involved in the alteration. These reversed mutations are called ***revertants.*** The reversion process may actually involve changes at the original mutation site, in which case the mutation is called a ***true back mutation.*** A reversion that occurs by alteration of the genetic material at a site distant from the original mutation is called a ***suppressor mutation.*** Suppressor mutations are discussed later.

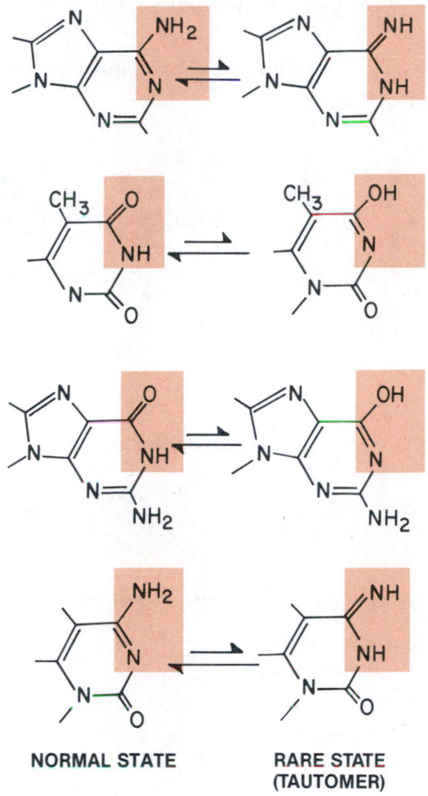

Figure 9-13 Tautomeric shifts of purine and pyrimidine bases to rare molecular state.

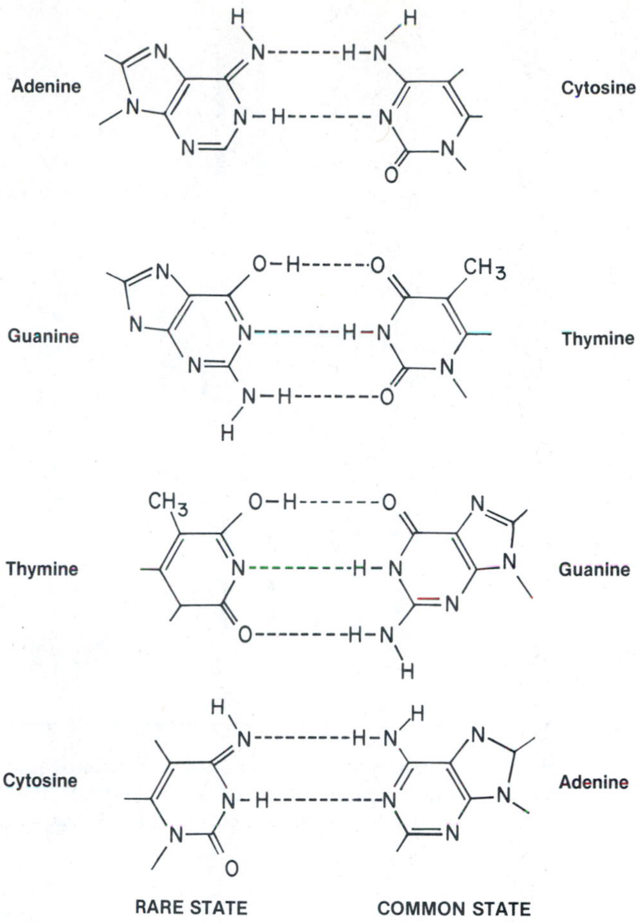

Figure 9-14 Abnormal base pairing resulting from formation of tautomers. All of "forbidden" base pairings in illustration are called transitions.

Types

Base pair substitutions

Spontaneous base pair substitutions. Watson and Crick had predicted that if the two strands of DNA unwound and acted as a template in the replication process, errors could be made as the complementary strands were being synthesized. Such errors can take place because "abnormal" base pairing is possible between bases on the DNA molecule. Abnormal base pairing is a consequence of rearrangements in the distribution of electrons on the purine and pyrimidine bases. These arrangements, called tautomeric shifts, result in the formation of tautomers. Tautomers are compounds that exist in equilibrium but differ in the arrangement of their atoms (Figure 9-13). The normal hydrogen bonding between purine and pyrimidine bases is associated with those forms that possess amino (NH_2) and keto (C=O) groups. Tautomers are formed on rare occasions where the hydrogen bonding involves imino (—NH) and enol (—OH) groups (Figure 9-14) and normal purine-to-pyrimidine pairing is not possible. The replacement of a purine or pyrimidine on a DNA strand by a different purine or pyrimidine, respectively, is called a ***transition.*** For example, adenine, which normally pairs with thymine, can pair with cytosine when adenine is in its rare imino state. The primary consequence of the initial pairing mistake is the formation of a DNA molecule, following replication, that contains a "forbidden" base pair; that is, there is a transition from A:T to G:C or G:C to A:T (Figure 9-15).

Induced base pair substitutions. Three of the more widely used chemical mutagens that cause transitions are 5-bromouracil, nitrous acid, and alkylating agents such as nitrogen mustard (Figure 9-12) and ethylmethane sulfonate. ***5-Bromouracil (5-Bu)*** is an analogue of thymine and has a tendency to ionize more frequently than thymine. 5-Bu in its normal keto state pairs with ad-

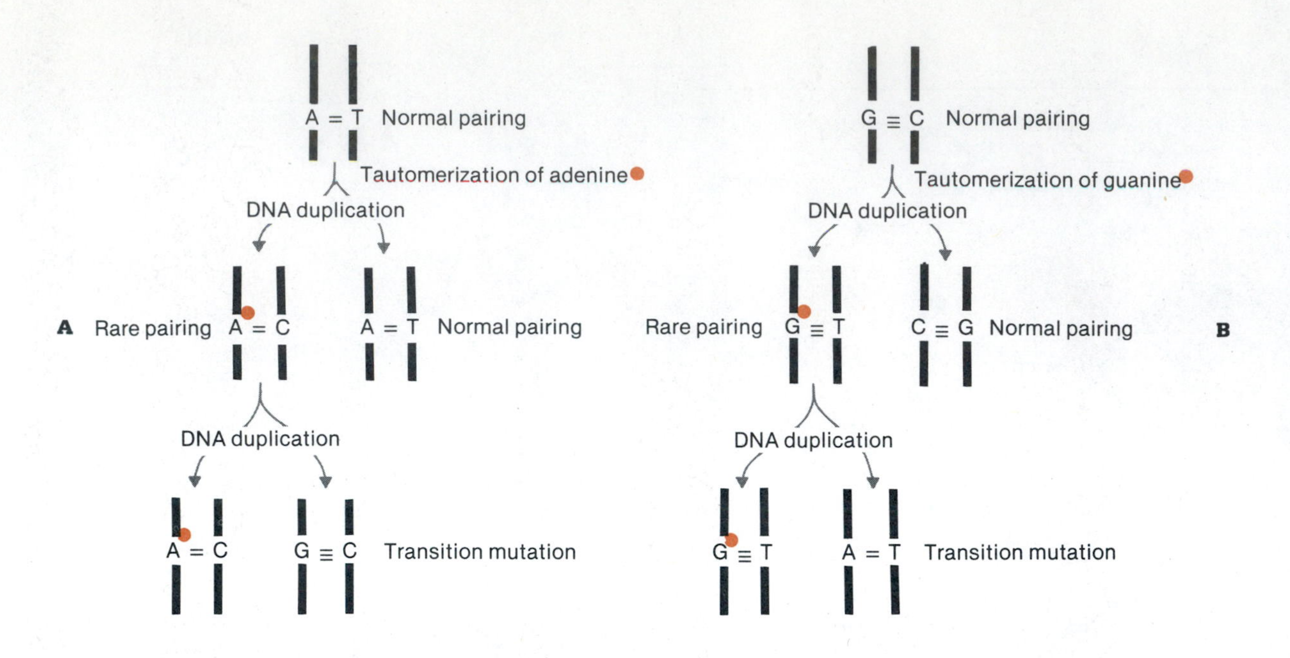

Figure 9-15 Consequences of tautomerization involving bases adenine and guanine. **A,** Tautomerization of adenine and transition from A:T to G:C base pairing. **B,** Tautomerization of guanine and transition from G:C to A:T base pairing.

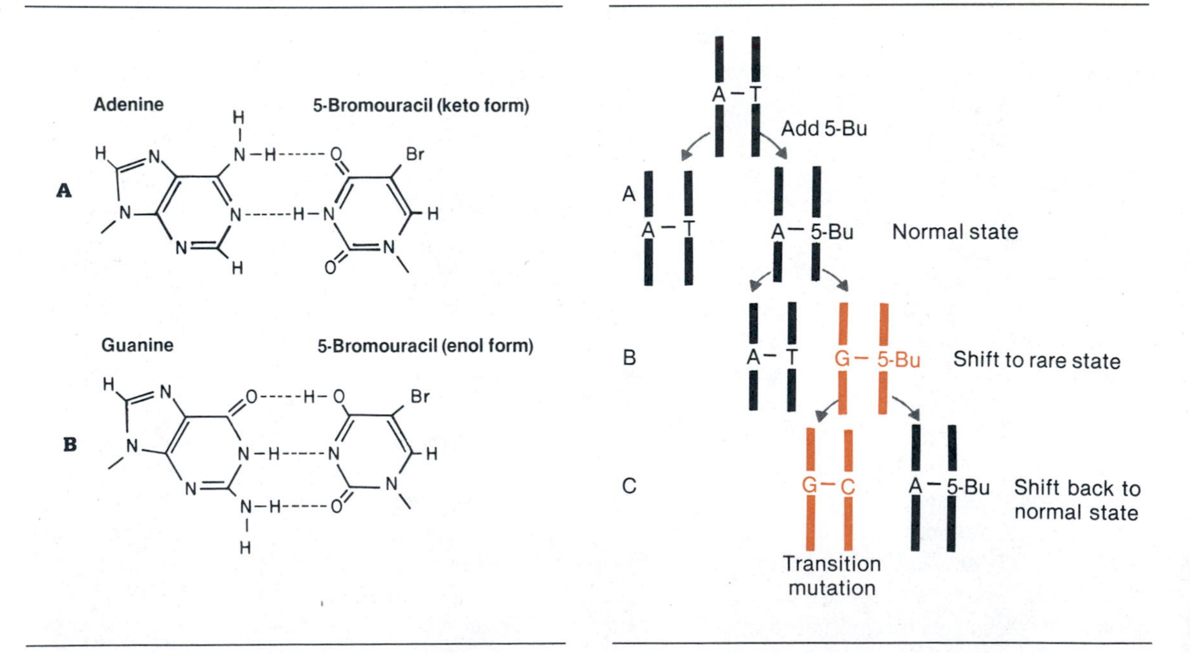

Figure 9-16 Pairing properties of 5-bromouracil (5-Bu) in its normal keto state, **A,** and its abnormal enol state, **B.**

Figure 9-17 How 5-Bu causes transitions from A:T to G:C. *A,* 5-Bu acts like thymine in its normal state and is base paired with adenine. *B,* During duplication of DNA, 5-Bu shifts to rare enol form and behaves like cytosine and therefore pairs with guanine. *C,* During next round of replication, guanine on template strand pairs with cytosine of complementary strand.

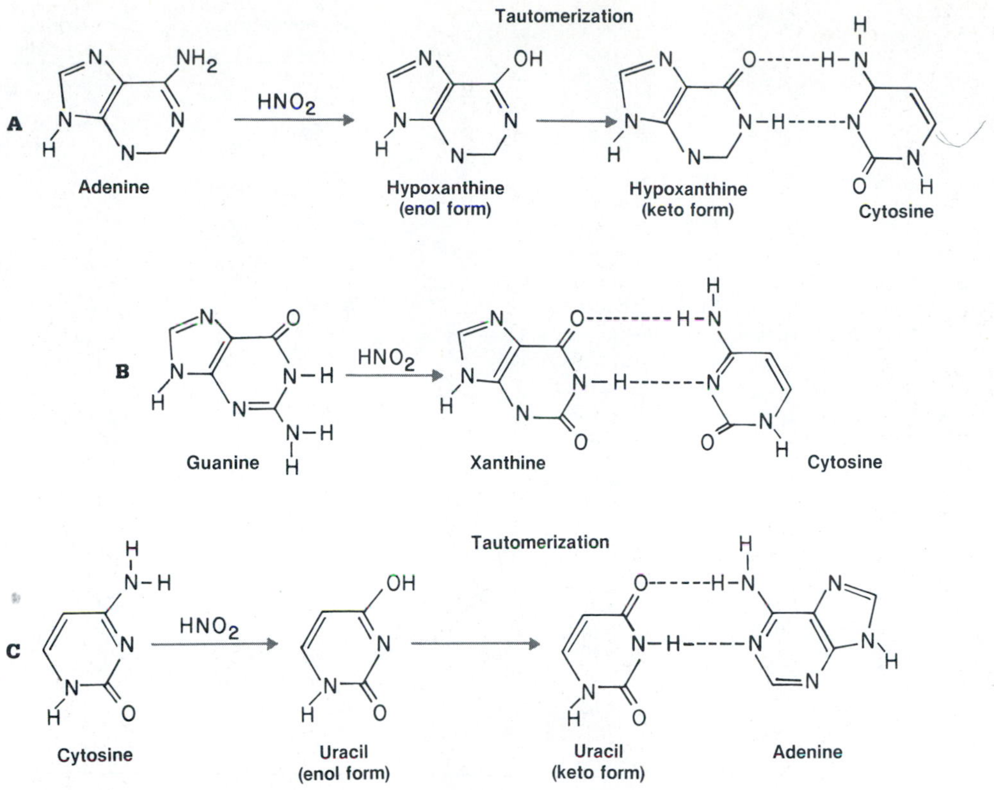

A
Adenine → HNO_2 → Hypoxanthine (enol form) → Tautomerization → Hypoxanthine (keto form) — Cytosine

B
Guanine → HNO_2 → Xanthine — Cytosine

C
Cytosine → HNO_2 → Uracil (enol form) → Tautomerization → Uracil (keto form) — Adenine

Figure 9-18 Effects of nitrous acid on purine and pyrimidine bases. **A,** Nitrous acid deaminates adenine to hypoxanthine, which can pair with cytosine, thus causing transition mutation. **B,** Deamination of guanine with nitrous acid produces xanthine, which pairs with cytosine, and no base pair substitution occurs. **C,** Deamination of cytosine produces uracil, which pairs with adenine, therefore causing transition mutation.

enine, just as its analogue thymine does, but in its rare enol state it pairs with guanine (Figure 9-16). The transition from A:T to G:C or G:C to A:T, following incorporation of 5-Bu into one of the DNA strands, appears after two duplications (Figure 9-17).

Nitrous acid (HNO_2), unlike 5-Bu, exerts its mutagenic effects on nonreplicating DNA. Nitrous acid deaminates amino groups from purines and pyrimidines and causes the substitution of hydroxyl (OH) groups for amino groups. Cytosine and adenine appear to be the most sensitive to the action of nitrous acid. Deamination of cytosine leads to the formation of uracil, which has a hydrogen binding specificity for thymine, and deamination of adenine leads to the formation of hypoxanthine, which has hydrogen-bonding specificities for cytosine (Figure 9-18). One of the characteristics of nitrous acid is its ability to cause transitions in both directions:

$$A:T \rightleftharpoons G:C$$

Alkylating agents, such as ethylethane sulfonate ($CH_3CH_2SO_3CH_2CH_3$) and ethylmethane sulfonate ($CH_3CH_2SO_3CH_3$), possess ethyl groups that can be donated to nitrogen in the 7-position of purine bases, particularly guanine. Alkylation causes hydrolysis of the base-sugar linkage and results in the release of the base from the DNA molecule. This creates a gap in the DNA chain where the base had formerly been present. Any base can be inserted into the gap (Figure 9-19) during the replication process. The inserted bases can be the same as the normal one, or they can be a different base. A purine or pyrimidine replacement may be in the nature of a transition, but the replacement of purine for a pyrimidine or a pyrimidine for a purine can also take place. Such replacements are called ***transversions.*** Nitrogen mustard is also an important alkylating agent.

Deletions and insertions. ***Acridines,*** such as proflavine (Figure 9-20), are dyes that resemble purine bases. They can insert (intercalate) themselves and widen the gap between bases on the DNA molecule, but acridines do not become part of the DNA molecule. Intercalation may result in

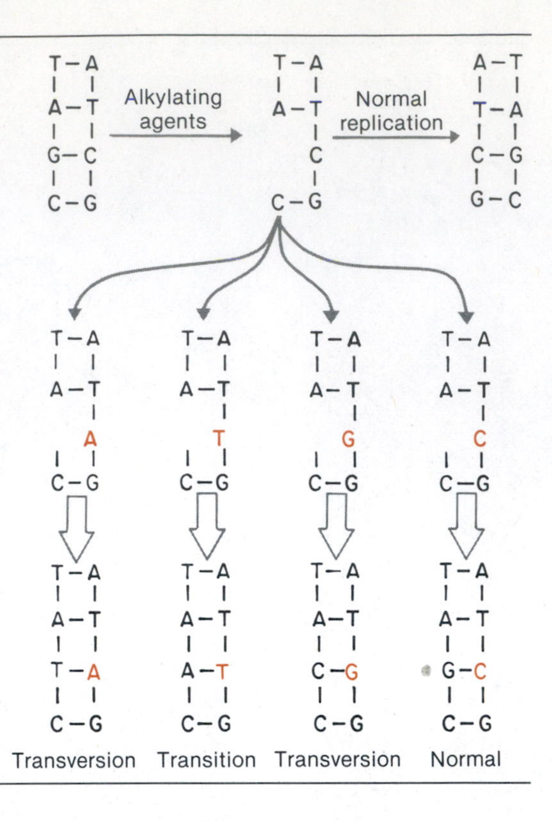

Figure 9-19 Consequences of alkylation of guanine. Removal of guanine by alkylation creates gap in one strand of DNA. At next replication unaffected strand gives rise to normal double-stranded DNA molecules. Strand with gap can give rise to complementary strand with one of four possible bases. Only one out of four DNA molecules will be normal; the rest will be mutations.

Redrawn from William Hexter/Henry T. Yost, Jr. The science of genetics, © 1976, p. 426. Reprinted by permission of Prentice-Hall, Inc., Englewood Cliffs, N.J.

H_2N — N — NH_2

Figure 9-20 Structure of proflavine.

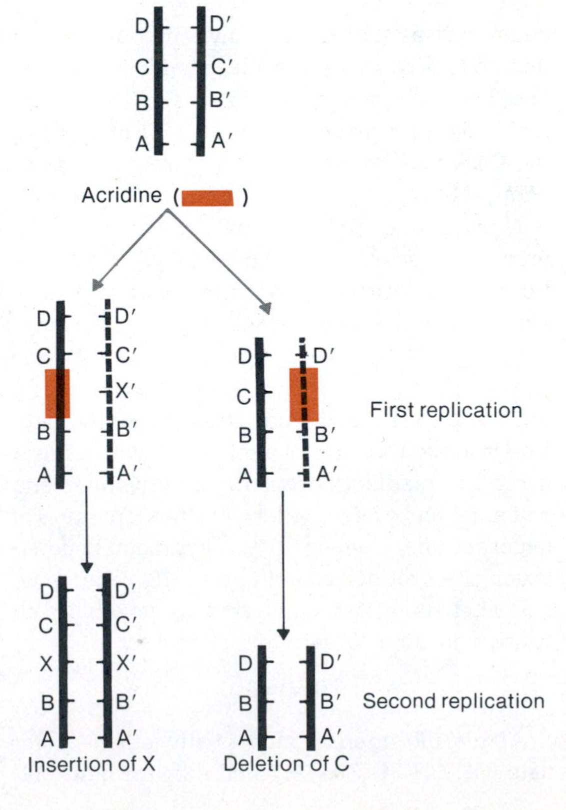

Figure 9-21 Effect of acridines on DNA. DNA molecule on left has acridine inserted into template strand, separating nucleotides *C* and *B*. During next duplication, base is inserted into complementary strand. On right, acridine is inserted into daughter strand, masking normal base so that at next round of replication a base is deleted.

Redrawn from Hayes, W. The genetics of bacteria and their viruses (2nd ed.). Blackwell Scientific Publications, Ltd., Oxford, 1968.

the deletion or insertion of a base, depending on whether the DNA molecule is in the process of replication or recombination (Figure 9-21). Acridines can be intercalated into the template strand, and when the DNA polymerase encounters it during complementary strand formation, a new base is inserted to pair with it. The acridine can also be intercalated into the complementary strand as it is being synthesized. This means that the acridine will mask the base in the template strand and no base will pair with it. A ***deletion*** is produced during the next round of DNA replication. The effects of acridines on gene activity are discussed in Chapter 10.

X-ray–induced mutations. DNA is the most radiosensitive component of cells. X-rays can cause mutations by directly affecting the DNA molecule through the absorption of energy. X-ray photons can cause deletions in the DNA backbone by breaking both chains at opposing sites. X-rays can also indirectly affect the DNA molecule when their energies are absorbed by small molecules in the surrounding medium. Various free radicals such as HO_2 as well as organic peroxides can react with the DNA and chemically alter it (see Chapter 27).

Ultraviolet light–induced mutations. Ultraviolet (UV) light causes mutations through the dimerization of pyrimidines, particularly thymine. Dimerization is the covalent bonding of adjacent residues, which distorts the DNA (Figure 9-22) and inhibits pairing with purines of the opposite strand. The effects of UV light can be reversed by special mechanisms, the most important of which are photoreactivation and excision repair.

Photoreactivation repair. Photoreactivation repair is accomplished through the activity of a photoreactivating enzyme (PRE) produced by the microorganism. PRE, which binds to the dimer in the dark, becomes activated and splits the dimer (Figure 9-23, *A*) when the organism is exposed to certain wavelengths in the visible spectrum (310 to 400 nm).

Excision repair. Excision repair mechanisms can take place in the dark when special enzymes are induced. One of the enzymes makes a cut in the DNA on the 5′ side of the dimer (Figure 9-23, *B*), and a second enzyme excises not only the dimer from the DNA but adjacent bases as well. This leaves a small gap in the one strand of DNA, but this is easily repaired because the complementary bases on the other strand are intact and a DNA polymerase catalyzes the resynthesis of the daughter strand. In the final enzymatic step the sugar-phosphate linkage is sealed by a DNA ligase. Recall that in our previous discussion of DNA replication a mutant was found that could replicate in the absence of the Kornberg DNA polymerase. The mutant was subsequently found to be sensitive to UV light and could not repair the damage to its DNA. This led to the realization that the Kornberg enzyme is actually a repair enzyme, or DNA polymerase I.

How does UV light cause mutations when such repair mechanisms are available to the cell? It has been suggested that some kinds of UV damage induce an "error-prone" DNA polymerase repair system. One such postulated mechanism is illustrated in Figure 9-24.

Figure 9-22 Thymine dimer formation. Adjacent thymine residues (can be interchain or intrachain), when exposed to UV light, produce double bonds between carbons 4 and 5.

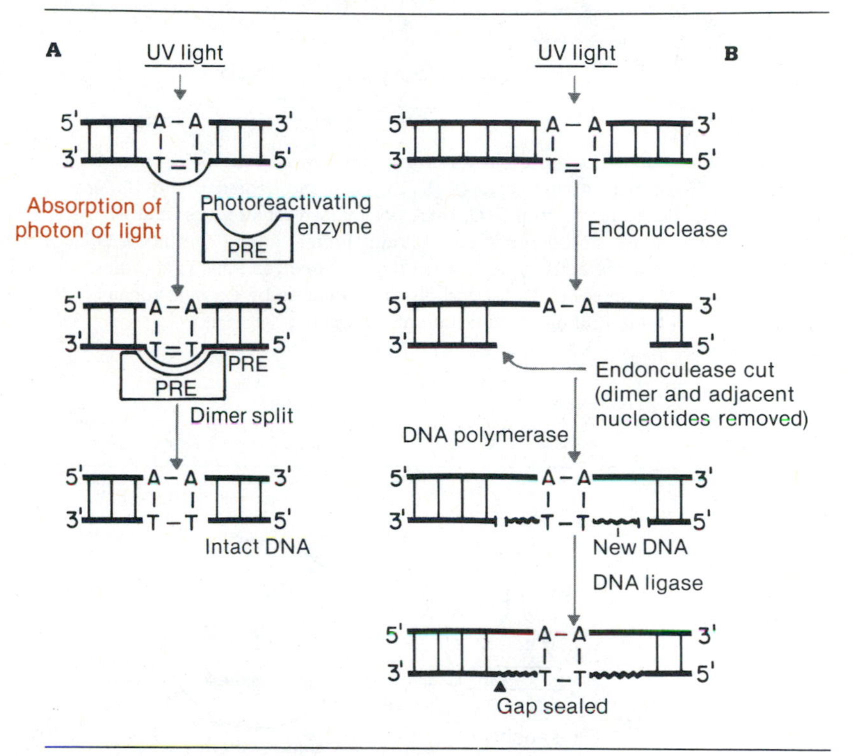

Figure 9-23 Principal mechanisms of DNA repair following damage and thymine dimer formation. **A,** Photoreactivation. **B,** Excision repair.
Redrawn from Witkin, E.M. Microbiol. Rev. ***40****:869, 1976.*

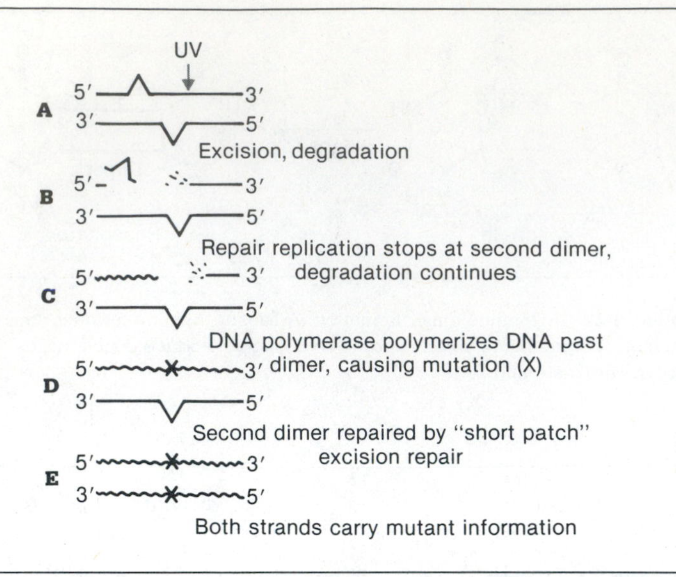

Figure 9-24 Model for mutation via error-prone excision repair mechanism. **A,** Dimers ( ) on both strands of DNA. **B,** Excision and degradation of dimer and adjacent nucleotides on one strand. **C,** DNA polymerase repair stops at site opposite second dimer, and degradation of original dimer strand continues. **D,** Special DNA polymerase activity is induced that polymerizes DNA past dimer, thereby causing mutation *(X)*. **E,** Second dimer is repaired by excision repair mechanism, and both strands contain mutant information

Redrawn from Witkin, E.M. Microbiol. Rev. ***40**:869, 1976.*

Spontaneous mutation

Mutant isolation. At one time a controversy existed as to the nature of spontaneous mutation. Many believed that a specific mutation, for example, from bacteriophage sensitivity to bacteriophage resistance was due to prior contact of the bacterial cells with the bacterial virus. It was demonstrated in the early 1940s that resistance could occur in the absence of the bacterial virus and that the only function of the virus was to select those mutants that arose in its absence. These experiments led to the use of indirect selection techniques to detect mutants in the absence of those conditions that favor their establishment. One of these indirect techniques is called ***replica plating,*** which is still used today to detect mutants. The following is a description of the technique using penicillin resistance as the mutational event (Figure 9-25):

1. A large population of penicillin-sensitive microorganisms such as *E. coli* are streaked on a drug-free agar surface to produce isolated colonies. This is the master plate on which both penicillin-sensitive and penicillin-resistant cells grow.

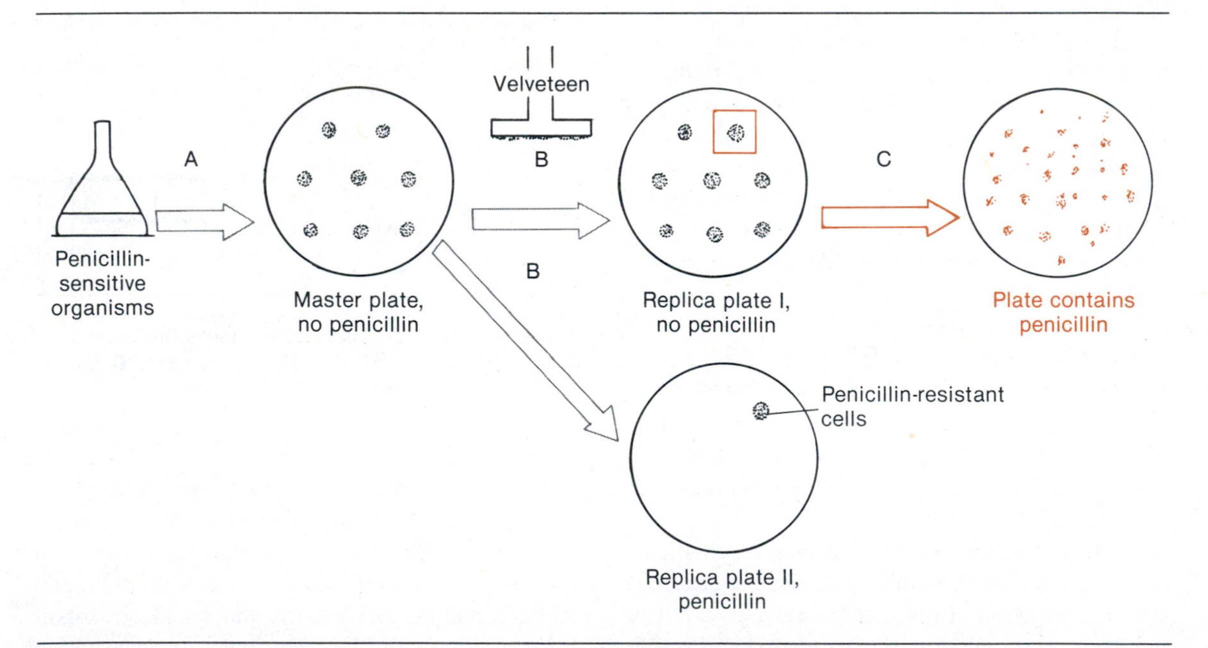

Figure 9-25 Replica-plating technique for selection of penicillin-resistant mutants. *A,* Penicillin-sensitive organisms are streaked on agar plate with no penicillin present. Colonies develop on plate after suitable incubation. *B,* Velveteen pad is placed over master plate, and parts of colonies are picked up and can be transferred to plate with penicillin (replica plate II) and plate with no penicillin (replica plate I). Colony on replica plate II contains penicillin-resistant cells. *C,* Corresponding colony on replica plate I is dispersed in saline, and aliquot is streaked on agar plate containing penicillin. If penicillin-resistant cells arise spontaneously from corresponding colony on master plate, then most cells of that colony will be penicillin resistant.

2. The master plate is pressed over a pad of sterile velveteen or velvet. The fibers of the pad pick up samples of the colonies.
3. The impregnated velveteen or velvet pad is used like a stamp and is first pressed on the surface of a drug-free agar plate and then is pressed on a second plate that contains penicillin.
4. Growth will occur on the penicillin-containing plate only when a drug-resistant mutant is present.
5. The drug-resistant colonies are compared with the corresponding colonies on the drug-free plates. The suspected drug-resistant colonies from the drug-free plate are then tested for their resistance to the drug. Colonies should be rich in resistant mutants if a mutation arose in the absence of the drug, as compared with those colonies that were not mutants.

The replica-plating technique is useful for the detection of biochemical mutants as well as for determining the degree of antibiotic sensitivity in microorganisms. Certain procedures can also use antibiotic sensitivity to isolate biochemical mutants. For example, if we look for auxotrophs in a mutagenized population of *E. coli*, a medium can be used in which the auxotrophic requirement is missing. In other words, the auxotroph will not be able to grow on the deficient medium but the prototrophs will. A medium containing penicillin to inhibit growing cells will be lethal to prototrophs but will not affect auxotrophs (Figure 9-26).

Insertion sequences and spontaneous mutation. Spontaneous mutations were thought to arise primarily from products formed during metabolism, such as formaldehyde, peroxides, and base analogues. Recently the discovery of certain DNA sequences, called ***insertion sequences,*** has suggested a biological cause of spontaneous mutation. DNA insertion sequences consist of 700 to 1400 contiguous base pairs that are found at one or more sites on the bacterial chromosome. They are natural components of the chromosome or plasmid but do not contain specific genetic information. They can insert themselves from their natural position on the chromosome into functional genes, resulting in a mutant phenotype. Genes distal to the insertion sequences can be depressed in their activity, particularly those that are part of a cluster of genes. These distal muta-

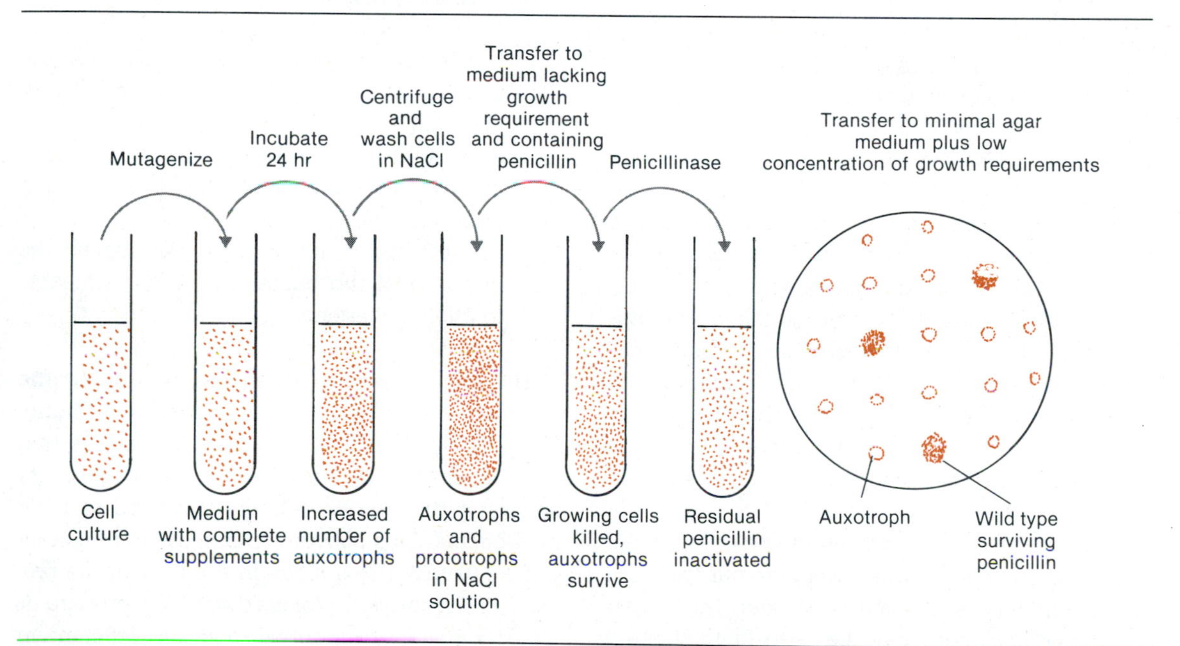

Figure 9-26 Isolation of auxotrophic mutant. Large population of cells is mutagenized and then grown in nutritionally supplemented medium. Culture is incubated for 24 hours to increase number of auxotrophs. Culture is centrifuged and washed with sodium chloride solution to remove any traces of medium. Washed cells, containing auxotrophs and prototrophs, are transferred to medium that is deficient in auxotrophic requirement and contains penicillin. Growing cells are killed by penicillin, but auxotrophs survive. Penicillinase is later added to destroy any residual penicillin. Culture is spread on agar surface containing very minimal amounts of auxotrophic requirement. Mutants will produce very small colonies on medium, whereas any wild-type cells that survived penicillin treatment will be much larger.

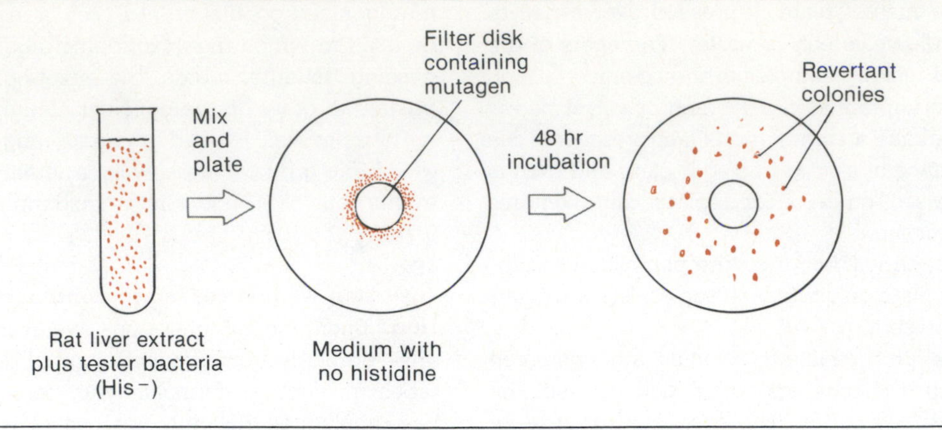

Figure 9-27 Ames test for identifying potential carcinogens. Tester bacteria containing histidine mutation are mixed with rat liver extract. Mixture is spread on agar medium deficient in histidine so that confluent layer of bacteria covers surface. Specific amount of potential chemical mutagen is soaked into filter disk and dried and then placed on surface of plate. During 48-hour incubation period, chemical in filter disk will diffuse outward with highest concentration nearest disk. If chemical gives rise to mutations, histidine mutants will revert to wild type and give rise to individual colonies. Test can also be performed by mixing bacteria, chemical mutagen, and liver extract and then plating.

tions are called ***polar mutations*** and are discussed in Chapter 10. It is believed that insertion sequences may be the principal cause of spontaneous mutations. Insertion sequences are also discussed in Chapter 11.

Practical applications of microbial mutants

Microorganisms have been used for many years in industrial processes. Production of antibiotics, amino acids, enzymes, ethanol, and single-cell protein are some of the major industrial uses of microorganisms. Strain improvement, in each microbial-related process to increase the yield of microbial products, is usually obtained through mutation. The uses of microbial mutants are discussed in more detail in other chapters. Let us look at two additional uses for microbial mutants that have some very practical applications.

Indicators of carcinogenic mutagens. Industrial chemicals have found their way into our food and water supplies, and some have been implicated as mutagens that may be also carcinogenic (causing malignant tumor formation). One technique used to detect the mutagenic potential of these chemicals is to test them on preformed mutants of *Salmonella typhimurium* that have an absolute requirement for the amino acid histidine. The histidine auxotrophs have a characteristic low frequency of reversion that can be increased when another mutation is introduced, for example, by a chemical compound. This is the theoretical basis of a test called the Ames mutagenicity test. The tester strain (the histidine auxotroph) has three other modifications that make it more sensitive and suitable for the test:

1. It is made more sensitive to DNA damage by eliminating its ability to perform DNA repair following mutation.
2. A plasmid is introduced into the strain that makes DNA more susceptible to replicating errors.
3. A mutation is introduced that makes the envelope of the bacterium more permeable to chemical agents.

The Ames test is carried out by adding the chemical to be tested to a rat liver extract (Figure 9-27). The reason for the rat liver extract is that many chemicals do not become carcinogenic for animals until they have been converted to other metabolites. This conversion process in the intact animal usually takes place in the liver in the process called ***biotransformation.*** The mixture is plated on a solid nutrient medium deficient in histidine. After 48 hours the plates are examined for any revertants (only organisms that can synthesize histidine will be able to grow). The frequency of reversion to the wild type in the presence of the mutagenic agent is compared with the reversion frequency in the absence of the mutagen. The number of colonies per mole of mutagen is a quantitative estimate of the mutagenic potential of the chemical. False-negative results can arise because the mutagen may not restore the specific histidine function and may thus go undetected. This type of problem can be prevented by using other tester bacteria. A very high reversion frequency in the presence of the mutagen indicates that the chemical agent may be carcinogenic, and further tests are initiated using laboratory animals. It may be necessary to consider removal of the chemical from public or industrial use if the animal tests are positive.

Vaccine production. Vaccines are made up of either living or dead microorganisms. They induce immunity to infection by the same microorganism but do not cause disease. Mutants are used to reduce the disease-producing qualities of some living microorganisms that are to be used as vaccines. A very recent type of vaccine being experimented on is one in which temperature-sensitive organisms are used; that is, the mutants are able to grow at one temperature but not another. For example, a vaccine to prevent respiratory tract damage following vaccination has been developed. Respiratory tract damage can occur in those areas where host defense mechanisms are less effective. The temperature-sensitive strain to be used as the vaccinating agent will not grow at 37° C, the temperature of the lower respiratory tract where tissue damage could occur, but will grow in the upper respiratory tract, where the temperature is cooler and respiratory tissue damage will not occur.

Mutation and microbial evolution

Sexual mechanisms of reproduction in eukaryotic systems permit variations in genotypes to be produced. Genetic variation in prokaryotes is believed to occur primarily through spontaneous mutation and selection of those mutants with survival advantages. The microbial world is supplied with an almost inexhaustible supply of genetic possibilities because of the short generation time and randomness of mutation. Spontaneous mutations in the laboratory, for example, have been detected for practically every gene locus. The mutation rate depends on the gene locus involved. It is speculated that during evolution the first microorganisms possessed very few genes and that the products of gene expression—the enzymes—demonstrated a very broad substrate specificity, which enabled them to act as catalysts on a wide range of related substrates. Greater enzyme specificity or specialization evolved later. For example, a gene could be duplicated with alterations occurring in only one

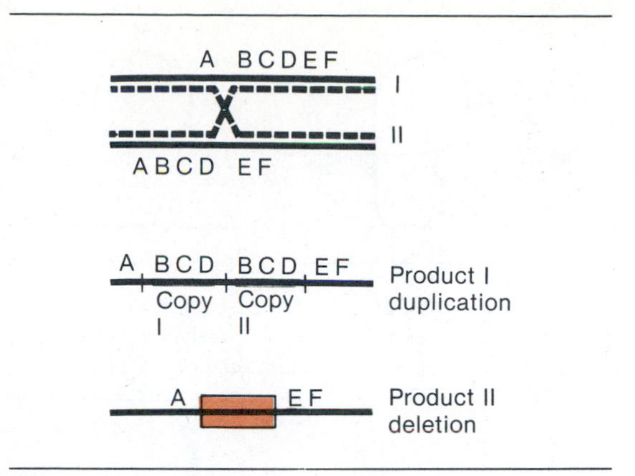

Figure 9-28 Genetic exchange between two DNA molecules resulting in tandem duplication of a genetic sequence B, C, and D, Solid lines indicate double-stranded DNA. Dashed lines represent reciprocal recombinational event.

copy. Several naturally occurring duplications are known in *E. coli,* in which there are at least seven copies of the genes that code for rRNA. The duplicated genes may be in tandem, or they may appear at different sites on the chromosome. A mechanism for tandem duplication could involve a reciprocal recombinational event between identical or nearly identical genetic sequences located at different sites on the chromosome (Figure 9-28). The mutated copy could produce an inactive product while the parent gene retained its original product specificity. The inactive gene, through other mutations or reversions, could express a product with the same specificities as the original unmutated gene, or it could possess new substrate specificities. The new gene would enable the cell to have a selective advantage over other cells, provided that the substrate for the new enzyme was in the environment.

Another mechanism for acquiring new cellular functions may be associated with the activities of microbial communities. Microbial communities are better equipped to handle novel substrates that are introduced into the environment. It is believed that one of the microbial members possesses an enzyme that recognizes the new substrate and at the very least modifies it. The modified product may be recognized by other enzymes produced by other members of the community. The gene coding for the novel enzyme could be transferred to other members of the community. This new gene could be maintained as long as the novel substrate stayed in the environment and offered a selective advantage to the microorganisms. We already know that the maintenance of certain antibiotic-resistant plasmids in humans is associated with the continued

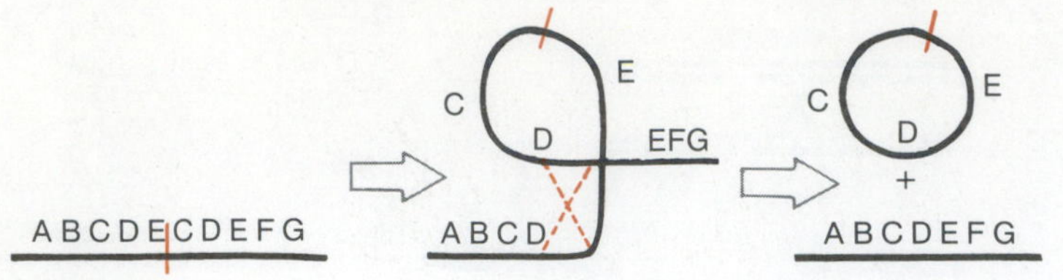

Figure 9-29 Hypothetical mechanism for formation of plasmid from chromosome containing tandem sequence of genes. Tandem sequence is CDE. Single crossover event could produce unduplicated chromosome (ABCDEFG) and DNA circle containing genes CDE. Colored line indicates point on chromosome where tandem genes are fused.

presence of the antibiotic. Once the antibiotic is no longer taken by the patient, the antibiotic-resistant strains are supplanted by the antibiotic-sensitive strains. Experiments in the laboratory have also demonstrated that when a community of different microorganisms is permitted to grow in a medium containing degradable carbon and nitrogen sources plus a nondegradable carbon source, for example, a herbicide, one of the organisms in the community through mutation may acquire the ability to convert the herbicide into a product that can be degraded by other microorganisms.

It can be expected that we will continue to challenge microorganisms with new chemicals that make their way into the environment. It is hoped and expected that this challenge will be met because of the ability of microorganisms to produce a multitude of genetic changes during evolution.

The role of spontaneous mutation in evolution has been questioned more recently because of the discovery of plasmids, as well as the knowledge we have gained from a better understanding of the functional role of chromosome organization. The F factor plasmid, which is responsible for sexual contact and genetic transfer between related species, is believed to play a minor role in evolution via the recombination process between chromosomes. Investigations into other types of plasmids has revealed that factors other than chromosomal recombination may play a vital role in evolution. Insertion sequences have been discovered in several locations outside of structural genes on both chromosomes and plasmids. What this means is that genetic exchange between insertion sequence (IS) elements can occur when these elements are present on plasmids and chromosomes and that structural genes may be carried with them. This type of genetic exchange can occur within the cell or between cells if the mating process is involved. In addition, a method for the formation of plasmids has been suggested. We already know that gene duplication occurs in prokaryotes as well as eukaryotes. Duplicated genes located in tandem on the chromosome are also unstable and create an avenue for genetic exchange, that is, reciprocal crossover between the homologous genes. This can lead to the formation of plasmids (Figure 9-29), and genes on these plasmids may contain insertion sequences as well as transfer potential. Plasmids can be transferred to other species and inserted into any replicon: viral, plasmid, or chromosomal DNA. Proponents of this theory believe that plasmids play a major and not a minor role in evolution among bacteria.

SUMMARY

The DNA of bacterial cells is capable of engaging in a number of processes. These processes or characteristics include engaging in duplication and transfer of the duplicated material to daughter cells, mutating, being translated into proteins, and being transferred totally or in part to other cell types with the possibility of recombination.

In the bacterial cell DNA appears in a circular form. DNA duplication occurs in a semiconservative manner, resulting in the formation of two DNA molecules, each containing one parental strand and one strand of newly synthesized DNA. In vitro studies have revealed that for DNA synthesis to occur, a double-stranded DNA molecule is required, with one strand acting as a template and the other as a primer strand. Also required is DNA polymerase plus all the deoxyribonucleotides.

DNA replication begins at a site called the origin and proceeds in both directions on the molecule. On one strand DNA synthesis is discontinuous, but it is probably continuous on the opposite strand. During DNA replication the circular strand is nicked with an enzyme and is then unwound through the aid of unwinding proteins. As the DNA is unwound, new DNA is synthesized on both strands until the termination site on the DNA is reached, and a duplicate DNA is produced. In some viruses DNA is replicated by a rolling circle mechanism in which one strand is cut with an enzyme, making it linear, while the other strand remains intact. When the initial single strand revolves 360 degrees around the circular strand, DNA synthesis is complete and a duplicate DNA molecule is produced.

Plasmids (extrachromosomal DNA) appear in all bacterial cells. Extrachromosomal DNA is similar to chromosomal DNA in structure and function but is smaller and is subject to loss from the cell by environmental manipulation. Plasmids may be conjugative or nonconjugative, depending on the presence or absence, respectively, of a sex (F^+) factor. Plasmids are not essential to the survival of the cell but offer properties that make it more competitive in its environment. Some of these properties include the ability to conjugate with related species, resistance to antimicrobials, and increased catabolic activity. In addition, plasmids also carry virulence factors that cause disease in plants and animals.

The word *mutation* in genetics means a change or alteration in the chemical nature of the hereditary material that may or may not result in observable changes in the phenotype of the cell. Mutations are caused by chemical and physical agents that are collectively called mutagens.

Mutation may be produced in a number of different ways: (1) by base pair substitutions, either induced or spontaneous, which result in abnormal base pairing between DNA strands, (2) by deletions or insertions caused by dyes, resembling purine bases, which are inserted between bases on the DNA molecule, and (3) by radiation effects in which x-rays may cause breakage of the DNA backbone or UV light may cause covalent bonding between bases and prevent unwinding of the DNA. Some microorganisms possess the enzymatic potential to repair UV damage by such processes as photoreactivation repair or excision repair. These processes basically result in the excision of the damaged base(s) and the resynthesis of nucleotides.

Spontaneous mutation may be due to rearrangements that occur naturally in the DNA molecules because of special DNA sequences called insertion sequences. Spontaneous mutations can be detected using replica-plating techniques, which select mutants in the absence of any inducing agent.

Microbial mutants are used in industrial processes involving the production of antibiotics, vaccines, amino acids, and enzymes. They have also been used to detect potential carcinogens that are part of our environment.

Mutation has played an important part in the evolution of microorganisms, particularly in respect to their ability to metabolize a wide variety of organic substrates. It has been suggested that some novel metabolic activities could be handled in a sequential manner by two or more members of the microbial community. The metabolic potential of one member could be transferred to other members of the community, either in the form of chromosomal or extrachromosomal DNA (plasmids).

Study questions

1. What is necessary to carry out the transformation of a bacterial cell in an in vitro system?
2. Name the functions or properties of all nucleic acids.
3. The text describes semiconservative replication. Can you illustrate conservative replication? Can you speculate as to the advantages or disadvantages of each mechanism?
4. If you were performing an in vitro DNA synthesis experiment, what specifically would be required?
5. What is meant by bidirectional DNA synthesis? Does the mechanism of synthesis of one strand differ from that of its complementary strand? Explain.
6. Why are unwinding proteins required in DNA replication?
7. What are the properties of DNA polymerase I? DNA polymerase III?
8. How does the rolling circle mechanism of replication differ from that of the bidirectional model? What are its advantages? In what microbial system would such a mechanism be extremely advantageous? Explain.
9. Distinguish between the terms *plasmid* and *episome*.
10. What are the advantages of plasmids to the microbial cell? Give examples.
11. What is the basic difference between a conjugative and nonconjugative plasmid?
12. What is meant by mutation rate and how can it be altered in a microbial species?
13. What type of change in the DNA molecule would most likely result in a transversion?
14. Describe at least one mechanism used by bacterial systems to repair UV damage.
15. Describe some ways in which microbial mutants have been used successfully in industry.
16. What type of genetic change do you think has been most responsible for evolution in microorganisms?
17. What are insertion sequences? Where are they found? How are they believed to be involved in microbial evolution?

18. What is the purpose of replica plating?
19. How does one determine whether acridines will cause deletions or insertions in the DNA?
20. Define the following: prototroph, auxotroph, mutagen, transition, transversion, dimerization, and carcinogen.

Selected readings

Books

Bainbridge, B.W. *The genetics of microbes.* Blackie & Son, Ltd., Glasgow, Scotland, 1980.

Birge, E.A. *Bacterial and bacteriophage genetics.* Springer-Verlag, New York, Inc., 1981.

Herskowitz, I.H. *The elements of genetics.* Macmillan Publishing Co., Inc., New York, 1979.

Hexter, W., and Yost, H.T. *The science of genetics.* Prentice-Hall, Inc., Englewoods Cliffs, N.J., 1976.

Jacob, F., and Wollman, E.L. *Sexuality and the genetics of bacteria.* Academic Press, Inc., New York, 1961.

Kornberg, A. *DNA synthesis.* W.H. Freeman & Co., Publishers, Inc., San Francisco, 1974.

Russel, P.J. *Lecture notes on genetics.* Blackwell Scientific Publications, Ltd., Oxford, 1980.

Watson, J. *Molecular biology of the gene* (3rd ed.). Benjamin/Cummings Publishing Co., Inc., Menlo Park, Calif., 1976.

Journal articles

Bauer, W.R., Crick, F.H.C., and White, J.H. Supercoiled DNA. *Sci. Am.* **243:**118, 1980.

Beadle, G.W., and Tatum, E.L., Neurospora II: methods of producing and detecting mutations concerned with nutritional requirements. *Am. J. Bot.* **32:**678, 1945.

Benzer, S. On the topography of the genetic fine structure. *Proc. Natl. Acad. Sci.* **47:**403, 1961.

Campbell, A. Evolutionary significance of accessory DNA elements in bacteria. *Annu. Rev. Microbiol.* **35:**55, 1981.

Devoret, R. Bacterial tests for potential carcinogens. *Sci. Am.* **241:**40, 1979.

Drake, J.W., and Baltz, R.H. The biochemistry of mutagenesis. *Annu. Rev. Biochem.* **45:**11, 1976.

Eigen, M., et. al. The origin of genetic information. *Sci. Am.* **244:**88, 1981.

Gefter, M.L. DNA replication. *Annu. Rev. Biochem.* **44:**45, 1975.

Hanawalt, P.C., et. al. DNA repair in bacteria and mammalian cells. *Annu. Rev. Biochem.* **48:**783, 1979.

Lederberg, J., and Lederberg, E.M. Replica plating and indirect selection of bacterial mutants. *J. Bacteriol.* **63:**399, 1952.

Meselson, M., and Stahl, F.W. The replication of DNA in *Escherichia coli. Proc. Natl. Acad. Sci.* USA **44:**671, 1958.

Novick, R.P. Plasmids. *Sci. Am.* **243:**102, 1980.

Ogawa, T., and Okazaki, T. Discontinuous DNA replication. *Annu. Rev. Biochem.* **49:**421, 1980.

Riley, M., and Anilonis, A. Evolution of the bacterial genome. *Annu. Rev. Microbiol.* **32:**519, 1978.

Witkin, E.W. Ultraviolet mutagenesis and inducible DNA repair in *Escherichia coli. Microbiol. Rev.* **40:**869, 1976.

Chapter 10 GENETIC CODE AND PROTEIN SYNTHESIS

GENETIC CODE

Previous topics have dealt with DNA as a molecule of replication and mutation. We have not discussed, however, how DNA carries out its role as an informational molecule. It was realized early in the twentieth century that the primary function of DNA was to control the activities of specific proteins, particularly enzymes. Beadle and Tatum in the 1940s, using biochemical mutants, demonstrated that genes control sequentially related metabolic reactions. Their work led others to verify that mutations do result in a loss of specific enzyme activity and that enzyme function is under the control of specific genes. Investigations into the mechanism of protein synthesis had revealed that proteins were made on structural units called ribosomes and that a highly unstable RNA called messenger RNA (mRNA) carried the information for protein synthesis. The flow of information in the cell had thus been established:

$$\text{DNA} \longrightarrow \text{RNA} \longrightarrow \text{Protein.}$$

The major questions left unanswered were how the information on the gene is converted to a polypeptide and how the gene is able to exert its control over the vast number of proteins in the cell. The primary purpose of this chapter is to discuss how those questions were answered and how the gene carries out its role as an informational molecule by translating a sequence of nucleotides into completely different chemical units called amino acids on the polypeptide. Later in the chapter the mechanics of protein synthesis are discussed, as well as how the gene controls the level of protein in the cell.

Characteristics

Triplet code. It was apparent that the conversion of nucleotide information into amino acids could take place only if the DNA contained coded information. One of the first considerations in deciphering a code is to evaluate the components of the system. The number of different nucleotides in the DNA is 4, and the number of amino acids to be coded is 20. The simplest code is one in which a combination of nucleotides stands for one amino acid. A two-letter code gives only 16 possibilities, that is, 4^2, and leaves us 4 short of the 20 code words required. A three-letter code gives 64 possibilities, that is, 4^3, which is far in excess of what is needed. Experiments to be described later confirmed this three-letter code.

Nonoverlapping code. The nucleotide sequences on the DNA molecule are obviously very long. The question that had to be answered was whether the code words should be read in an overlapping or a nonoverlapping manner (Figure 10-1). A nonoverlapping code had been suggested from earlier studies with the hemoglobin molecule. Hemoglobin from normal adults was compared with hemoglobin from those with a type of sickle cell anemia. The hemoglobin in patients with sickle cell anemia is altered in such a way that it results in an increase in the rate of destruction of erythrocytes, which leads to hemolytic anemia. Analysis of the polypeptides from these two groups showed that the only difference between them is a single amino acid on one of the polypeptides; for example, in one type of sickle cell anemia, valine, an uncharged amino

```
A C C|T A C|A T T   DNA
A C C|T A C|A T T   nonoverlapping
A C C                overlapping
  C C|T
    C|T A
     |T A C
         A C|A
           C|A T
            |A T T
```

Figure 10-1 Effect of mutational event involving single base (color-coded base) in genetic systems that possess either overlapping or nonoverlapping code. Note that in overlapping code, mutation in one base will cause changes in three codons.

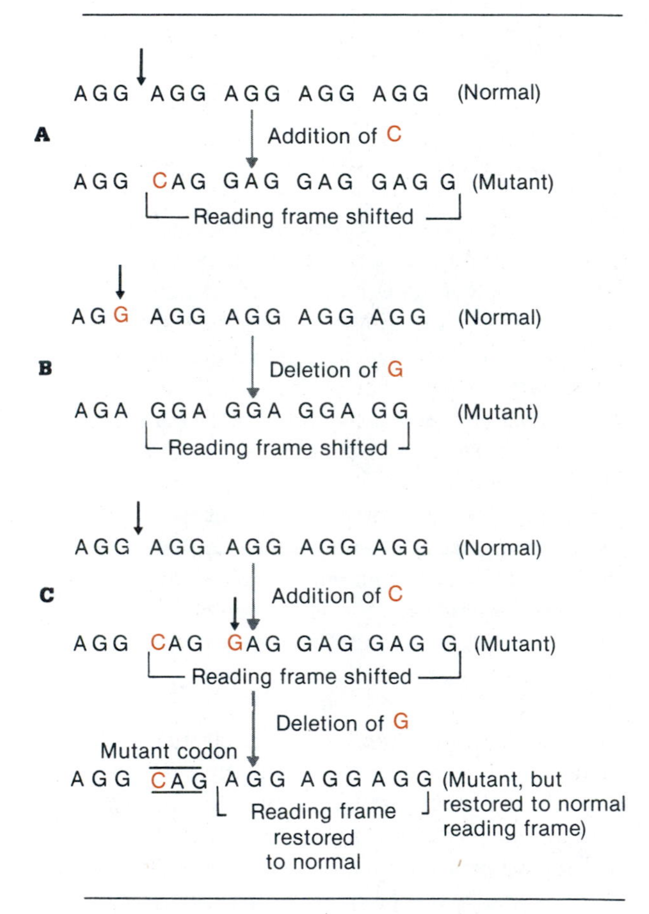

Figure 10-2 Effect of deletion and/or insertion of bases in DNA. **A,** Addition of base. **B,** Deletion of base. **C,** Addition of base followed by deletion of base to restore reading frame; arrows point to sites for addition or deletion of base.

acid (nonpolar) is found in place of the normal glutamic acid, a charged (polar) amino acid. This single amino acid substitution is enough to change the conformation of the hemoglobin molecule and make it function abnormally. It was reasoned that a mutational event in an overlapping code would produce three new code words and hence three amino acid changes in the polypeptide, but a mutation in a nonoverlapping code would produce only a single amino acid change.

Reading of the code. Can the code be read from any point, or is it read from one end to the other? Experiments were performed on T_4 bacteriophage to determine the manner in which the code should be read. The T_4 virus was treated with proflavine (a mutagen that can produce deletions and insertions) to produce mutants referred to as rII, which cannot grow on a particular strain of *Escherichia coli*. The rII mutants were treated with proflavine, which resulted in the formation of revertants that could grow in the *E. coli* strain. The ability of the revertants to grow on *E. coli* is best understood by examining Figure 10-2. The deletion or addition of a base causes a change in the reading frame of the code when we assume a triplet nonoverlapping code. The addition or deletion of a base changes the reading frame of the message from the point of the addition or deletion. These code changes result in changes in the amino acids of the polypeptide for which the gene codes. The rII revertants therefore represent double mutants in which the addition or deletion of a base restores part or all of the reading frame, depending on its location (Figure 10-2, *C*) in the message. For example, if the second mutation is close to the first mutation, only a few amino acids will be changed in the polypeptide because most of the original reading frame has been restored. A second mutation located some distance from the first also restores the reading frame but will leave a large part of the altered message unchanged. The polypeptide produced under the latter circumstance will probably be nonfunctional. Experiments showed that if the number of nucleotides deleted or added to the gene equaled the codon number (or multiple thereof), some activity could be restored in the polypeptide. These results demonstrated that the code contained no punctuation, such as commas, and that the nucleotides were used in successive nonoverlapping groups of probably three.

Amino acid code. The above-mentioned experiments provided valuable information as to the nature of the code but told us nothing about the specific code words. Remember, there are 64 code words for 20 amino acids. Scientists had proposed that the code was actually degenerate; that is, there might be more than one code word for a single amino acid. This means that there could exist a maximum of three code words for each amino acid, with four left over. The solution to this coding problem was made possible when an in vitro protein-synthesizing system was developed by Nirenberg in 1961. The cell-free system consisted of ribosomes, transfer RNAs (tRNAs),* amino acids, energy sources, protein-synthesizing enzymes (aminoacyl tRNA synthetases),* and cofactors. Protein could be synthesized in this system only when mRNA was added. A series of experiments were set up in which reaction mixtures contained a different radioactive amino acid. The first experiment used synthetic messages containing one base, for example, polyuridylic acid (poly U) or polyadenylic acid (poly A). The polypeptides produced contained polyphenylalanine or polylysine, respectively; that is, the code word for phenylalanine was UUU, and the code word for lysine, AAA.† These results prompted further experiments using synthetic mRNAs in which different triplets could be produced. Difficulties were encountered in distinguishing the arrangement of codons on a large synthetic message. These difficulties were solved by the development of the tRNA-binding technique. The basis of the technique is that ribosomes and mRNA bind to cellulose nitrate filters. More important, tRNAs, with their attached amino acids, will also bind to the ribosome mRNA complex as long as they are specific for a particular triplet on the message. A filter with bound poly (U) mRNA and ribosomes will bind only those tRNAs carrying the radioactive amino acid phenylalanine, whereas all other tRNAs will pass through the filter. Experiments were designed so that trinucleotide mRNAs such as UAC, CCA, and AGA could be added to a protein-synthesizing system to determine which aminoacyl tRNA bound to the ribosome-mRNA complex. These studies led to the complete deciphering of the code (Table 10-1). Examination of the code words reveals that there are 61 code words for the 20 amino acids ***(degeneracy).*** Three of the codons—UAA, UGA, and UAG—represent no amino acids and are called ***nonsense.*** These are discussed later.

Table 10-1 The genetic code

FIRST LETTER	SECOND LETTER U	C	A	G	THIRD LETTER
U	Phe	Ser	Tyr	Cys	U
	Phe	Ser	Tyr	Cys	C
	Leu	Ser	Nonsense	Nonsense	A
	Leu	Ser	Nonsense	Trp	G
C	Leu	Pro	His	Arg	U
	Leu	Pro	His	Arg	C
	Leu	Pro	Gln	Arg	A
	Leu	Pro	Gln	Arg	G
A	Ileu	Thr	Asn	Ser	U
	Ileu	Thr	Asn	Ser	C
	Ileu	Thr	Lys	Arg	A
	Met	Thr	Lys	Arg	G
G	Val	Ala	Asp	Gly	U
	Val	Ala	Asp	Gly	C
	Val	Ala	Glu	Gly	A
	Val	Ala	Glu	Gly	G

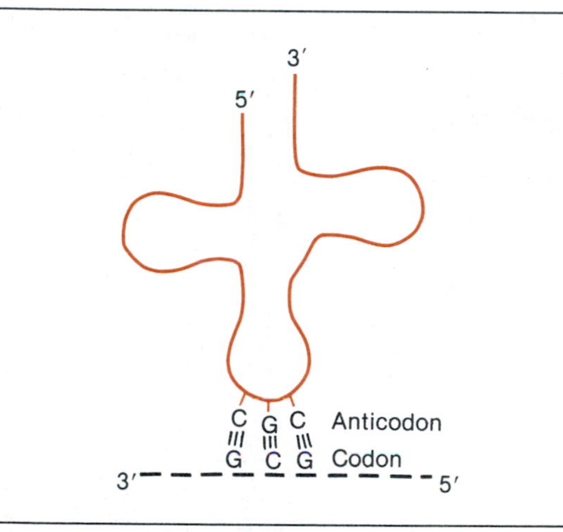

Figure 10-3 Binding between mRNA codon and tRNA anticodon.

Wobble hypothesis

Studies on tRNA, at the time of the decoding experiments, indicated that an amino acid may be bound to more than one tRNA species. Does that mean that there are 61 different tRNA molecules in the cell, each one recognizing a different codon? Each tRNA has a triplet, called the ***anticodon,*** that is complementary to the ***codon*** on the mRNA (Figure 10-3). Nirenberg's experiments with triplet messages demonstrated that several tRNA species can carry the same amino acid and that some tRNA species recognize the same codon. One proposal formulated to explain the range of binding activities for tRNA is called the wobble hypothesis. This proposal envisages that only the first two letters of the code, except the code words for serine, arginine, and leucine,

* Transfer RNAs carry amino acids. The charging of the amino acid to a tRNA molecule is catalyzed by aminoacyl tRNA synthetases. Transfer RNAs are discussed at more length on p. 237

† Artificial mRNA polymers are poor stimulators of polypeptide synthesis unless the first codon is AUG. The AUG codon is called the initiating codon and is important for ribosome binding on the mRNA (see p. 237). Nirenberg's experiments worked only because artificial mRNAs in the presence of high concentrations of magnesium ions are not dependent on the initiating codon.

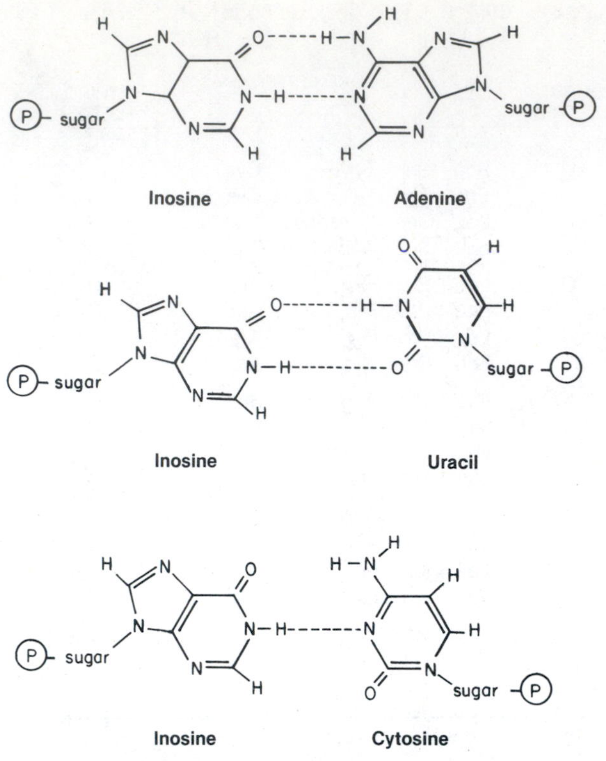

Figure 10-4 Pairing properties of base inosine, which is found in tRNA molecule. Inosine can pair with three bases because it can rotate slightly on its sugar-phosphate linkage, thus producing "wobble."

Anticodon (5′ end)	Codon (3′ end)
Inosine	U C A
Guanine	U C
Adenine	U
Cytosine	G
Uracil	G A

Figure 10-5 Base pair possibilities for third (5′) base of anticodon with its codon.

specify the amino acid. For example, CCU, CCA, CCG, and CCC code for proline. According to the wobble hypothesis, wobble can make possible the formation of more than one type of codon-anticodon interaction at the third codon position. This flexibility in hydrogen bonding is believed to be due to the minimal steric constraints at the 5′ end of the anticodon, where there is a turn in the molecule (see Figure 10-11), and the appearance of minor bases such as inosine at the 5′ end of the anticodon. Inosine, for example, can pair with adenine, uracil, and cytosine (Figure 10-4). The range of base interactions at the third codon position are illustrated in Figure 10-5.

Mutation and the genetic code

Four types of mutations alter the activity of polypeptides: missense, nonsense, reading frame, and suppressor mutations. Their ability to affect the phenotype of the cell depends on their location in the gene.

Missense mutations. Mutations that involve a single base pair substitution in the DNA (see Chapter 9) may result in the substitution of one amino acid for another in the polypeptide. Occasionally the codon change is such that the replacement amino acid is the same as the original amino acid in the polypeptide. This is referred to as a ***sense codon. Missense codons*** often alter the activity of the polypeptide because of the type of substitution. Substitution of polar for nonpolar amino acids or vice versa and replacement of cysteine by another amino acid are important causes of altered activity. Such amino acid substitutions are missense and result in changes in the conformation of the polypeptide and hence its activity (Figure 10-6, *A*).

Nonsense mutation. Nonsense codons do not specify for any amino acids. They are recognized by ribosomes as termination signals. Ribosomes, when they meet the termination codon, fall off the message, and this results in the formation of a polypeptide that is shorter than normal (Figure 10-6, *B*). ***Nonsense mutations*** located near the end of the message may not result in a loss of polypeptide activity, but the closer the mutation is to the 5′ end (the origin of the message), the greater the loss of polypeptide activity.

Reading frame mutations. The addition or deletion of a base produces reading frame mutations, resulting in changes in the amino acids in the polypeptide from the mutation site to the carboxyl end of the polypeptide. Reading frame or frameshift mutations are discussed in Chapter 9.

Suppressor mutations. Some mutations can be suppressed by the introduction of a second mutation. These mutations, or ***suppressor mutations,*** may be ***intragenic;*** that is, the second mutation occurs in the gene near the first mutation. There are two types of intragenic suppression: (1) the addition or deletion of a base suppresses the deletion or addition, respectively, of the first mutation, and (2) an addition or deletion of a base alters the reading of a nonsense triplet in such a way that it makes sense. Each type of suppression results in a restoration of some polypeptide activity (also called the ***pseudowild type***) but seldom in a restoration of full activity.

Mutations at one genetic site that suppress a mutation at a different site (for example, on another gene) are called ***intergenic*** suppressor mutations. Some of the most widely studied intergenic suppressor mutations were discovered in T_4 bacteriophage, which infects the bacterium *E. coli*. These suppressor mutations are called amber* mutations. The virus possessing the amber mutation can grow, albeit poorly, in one strain of *E. coli* (the permissive strain) but not in another (the nonpermissive strain). What is different about the strain of *E. coli* that permits it to support viral growth? Amber mutations are nonsense mutations, or chain-terminating codons—UAA, UAG, or UGA—that are not read as nonsense in the permissive strain. The simplest interpretation is that some of the tRNAs in the permissive strain do not recognize the amber mutations as nonsense but as a codon for some amino acid. The amino acids recognized in the permissive strain were discovered to be tyrosine, glutamine, and serine. What this means is that the suppressor mutation causes tyrosine, serine, or glutamine to be inserted in the position that would normally be chain terminating. The reason that tyrosine, serine, or glutamine is inserted is that only one base change is required (Figure 10-7) to change from nonsense to sense. The interpretation for this type of suppression is that the mutation involves a genetic site that controls tRNA synthesis and that the suppressor mutation can suppress a limited number of mutational events to restore partial activity.

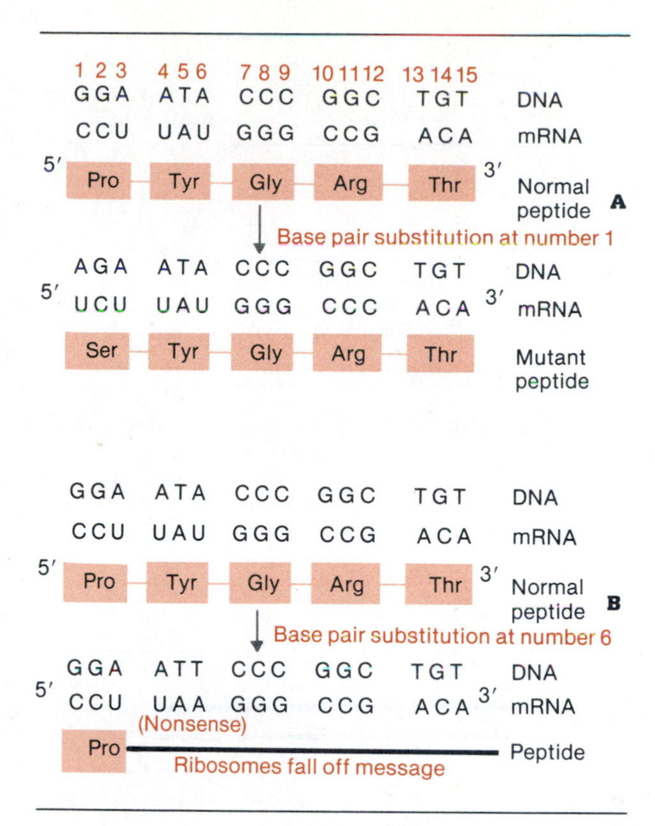

Figure 10-6 Effect of missense and nonsense mutations on polypeptide synthesis. **A,** Missense mutation: substitution of adenine for guanine at position 1. **B,** Nonsense mutation: base pair substitution at position 6.

	Normal codon	Nonsense codon
Serine	UCA	UAA
Tyrosine	UAU	UAA
	UAC	UAG
Glutamic acid	GAA	UAA
	GAG	UAG

Figure 10-7 Base changes in serine, tyrosine, and glutamic acid codons that are read as nonsense.

PROTEIN SYNTHESIS

The preceding discussion of the genetic code has given us some concept as to the nature of the gene. A ***gene*** can be defined as a segment of DNA (in some viruses the hereditary material is RNA) in which a sequence of nucleotides is transcribed into an RNA molecule (mRNA), which in turn is translated into a sequence of amino acids in a polypeptide chain. Not all DNA sequences are transcribed into mRNA; some are transcribed into ribosomal RNA (rRNA) and tRNA. In the following paragraphs we will discuss the manner in which genetic information is transcribed and translated in the cytoplasm of the prokaryotic cell.

Transcription process

Transcription of messenger RNA. Information on the DNA molecule is first ***transcribed*** into an

* The amber connotation is actually a translation of Bernstein, the name of the original discoverer, into English.

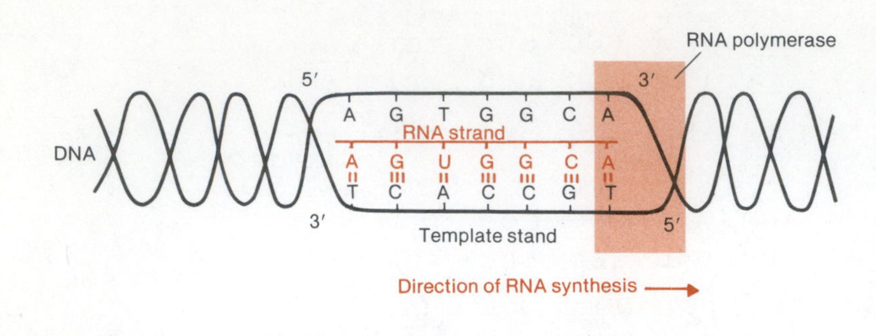

Figure 10-8 Transcription of RNA from DNA double helix. DNA unwinds, exposing both strands, but only one (template strand) is transcribed in 5′ to 3′ direction through activity of RNA polymerase. (See Figure 10-9 for mechanism of action of RNA polymerase.)

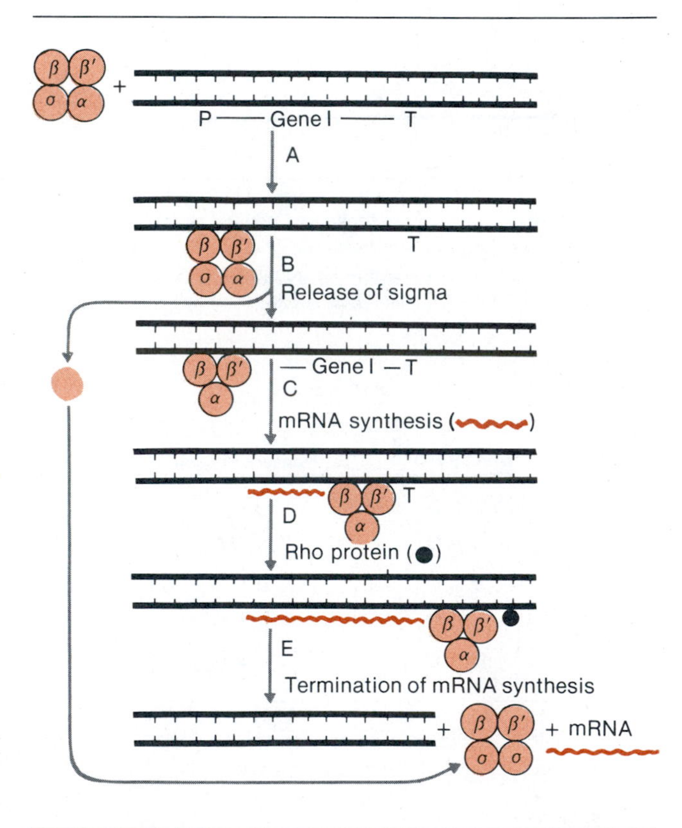

Figure 10-9 Mechanism of action of RNA polymerase in mRNA synthesis. *P*, Promoter site; *T*, termination site. *A*, RNA polymerase binds to promoter region of template strand of gene I. *B*, Sigma subunit of RNA polymerase is released. *C*, RNA polymerase (core enzyme) catalyzes transcription of gene I into mRNA molecule. *D*, Rho protein(●) causes termination of mRNA synthesis. *E*, Core enzyme is released from DNA strand and combines with recycled sigma factor, and mRNA molecule is completely synthesized.

RNA molecule called ***messenger RNA (mRNA).*** Recall that the nucleotides in DNA and RNA are the same except that uracil substitutes for thymine in the RNA molecule.

A strand of DNA in the gene serves as a template for the synthesis of mRNA. The enzyme that directs the synthesis of mRNA from the DNA template is a DNA-dependent RNA polymerase (Figure 10-8). Unlike the eukaryotic cell, the prokaryotic cell uses this same enzyme to produce all of its RNAs. The transcription process is an exact one because of the manner in which hydrogen bonding occurs between complementary bases. The complementarity between the mRNA and its DNA template has been tested by infecting *E. coli* with bacteriophage. Viral mRNA isolated from infected cells binds only to viral DNA and not to bacterial DNA. The size of the mRNA is a function of the size of the polypeptides for which it codes. The mRNAs of prokaryotes are with few exceptions very unstable molecules and are degraded by ribonucleases in 2 minutes or less following their transcription. Messenger RNA instability is actually an invaluable property to the cell. The enzymatic needs of the cell vary considerably, depending on nutritional sources and environmental conditions. Genes are turned on and off to meet these needs. A stable mRNA, producing more protein, after the gene controlling its transcription has been turned off, causes a waste of energy, but an unstable mRNA allows the cell to respond rapidly to environmental changes.

Messenger RNA that is synthesized on the DNA template in the 5′ to 3′ direction (Figure 10-8) requires an initiation as well as a termination signal. The signals are associated with the RNA polymerase, which is composed of subunits designated beta (β), beta prime (β′), alpha (α), and sigma (σ). The sigma factor is not bound as tightly as the other subunits and readily dissociates from them, leaving behind a β, β′, α enzyme, referred to as the ***core enzyme***. Messenger RNA synthesis is initiated when the core enzyme binds to a sequence of nucleotides on the DNA called the ***promoter*** (Figure 10-9). The sigma subunit is especially important to core enzyme specificity because in its absence the enzyme binds to random sites on the DNA. Sigma is a recycled element whose only function is initiation of mRNA synthesis. Pribnow discovered an unusual characteristic of nucleotide sequences on the promoter. These sequences consisted of seven nucleotides having the general form of 3′···A-T-A-pyrimidine-T-A-pyrimidine···5′ and are now referred to as ***Pribnow sequences.*** These sequences are believed to be important for the binding of the sigma subunit. The promoter nucleotides are not transcribed by RNA polymerase. Instead, RNA polymerase moves to the end of the promoter and begins transcription only when the first two nucleotides of the adjacent gene have been encountered. Termination of transcription is also associated with a recognition factor that is a protein called the ***rho factor.*** Rho binds to the termination sequences on the DNA. It has been suggested that the termination sites on the DNA consist of the sequences AAAAAA, which when transcribed are UUUUUU, a termination signal for the RNA polymerase.

One of the questions you might ask at this time is, "How does the RNA polymerase enzyme distinguish the strands to be transcribed?" The answer is not quite so simple. It might be reasoned that since the DNA strands have opposite polarity, the enzyme can distinguish this difference. Unfortunately, experiments have shown that sometimes both strands can be read. For example, the area coding for the enzymes involved in histidine biosynthesis can be read on one strand, and the area coding for enzymes involved in lactose catabolism can be read from another strand. Under no circumstances, however, is a single region transcribed from both strands.

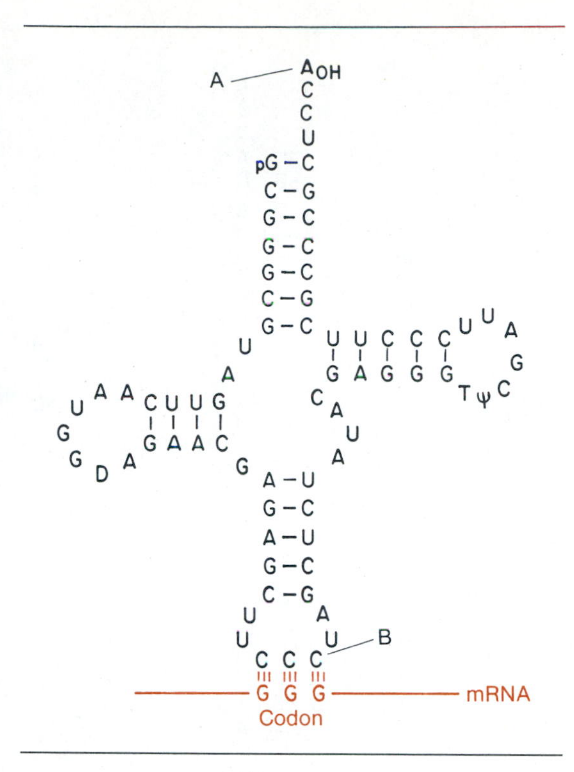

Figure 10-10 Cloverleaf structure of glycyl tRNA. Site A is part of CCA end of molecule in which glycine is covalently attached to adenine base. Site B is anticodon site for attachment of tRNA to ribosome. *D,* 5,6-Dihydrouracil; ψ, pseudouracil.

Transcription of transfer RNA and ribosomal RNA. The smallest RNA molecules in the cell, ***transfer RNAs (tRNAs),*** are single stranded and contain from 70 to 80 nucleotides. Each carries a specific amino acid, and all are involved in protein synthesis. There is at least one specific tRNA molecule for each of the 20 amino acids found in the cell. The tRNA molecule contains several functional areas that are associated with protein synthesis. One end of the molecule always contains the sequence CCA, whereas G is at the other end of the molecule. The specific amino acid carried by the tRNA is attached to the adenine of the CCA end of the molecule. A second site consists of three bases called the anticodon, which is complementary to a triplet sequence on the mRNA (this specificity is discussed later). A third site is associated with binding to the ribosome during protein synthesis. A fourth site engages the enzyme (aminoacyl tRNA synthetase), which covalently attaches the amino acid to the CCA end of the molecule. The tRNA exhibits a helical structure, even though it is single stranded. Several tRNAs have been subjected to nucleotide analysis, and they are believed to exhibit a secondary structure similar to that of a "cloverleaf" (Figure 10-10). Other hydrogen bonding is believed to bend the cloverleaf into a tertiary structure that has a more stable conformation (Figure 10-11). Transfer RNAs contain several "abnormal" bases, such as pseudouracil, 5,6-dihydrouarcil, and dimethylguanine. The normal bases are produced on the tRNA during transciption from the DNA, and the abnormal bases are produced (for example, methylation of guanine) after transcription. The modification of certain bases on the tRNA may protect the molecule from degradation by RNA nucleases in the cell. Transfer RNA molecules, unlike mRNA, are used repeatedly during protein synthesis and may require this modification process to resist digestion.

Ribosomal RNA (rRNA) and proteins are components of ribosomes. The rRNA found in prokaryotes consists of 23S, 16S, and 5S molecules. The 23S and 5S components combine with a large number of proteins to produce a 50S ribosomal subunit, and the 16S rRNA combines with other proteins to produce a 30S ribosomal subunit. The 50S and 30S subunits combine to produce the typical 70S prokaryotic ribosome. Similar rRNA subunits in eukaryotic cells combine with proteins to produce 60S and 40S ribosomal subunits, which combine to form the typical 80S ribosome (Figure 10-12). There are many copies of rRNA genes on the prokaryotic chro-

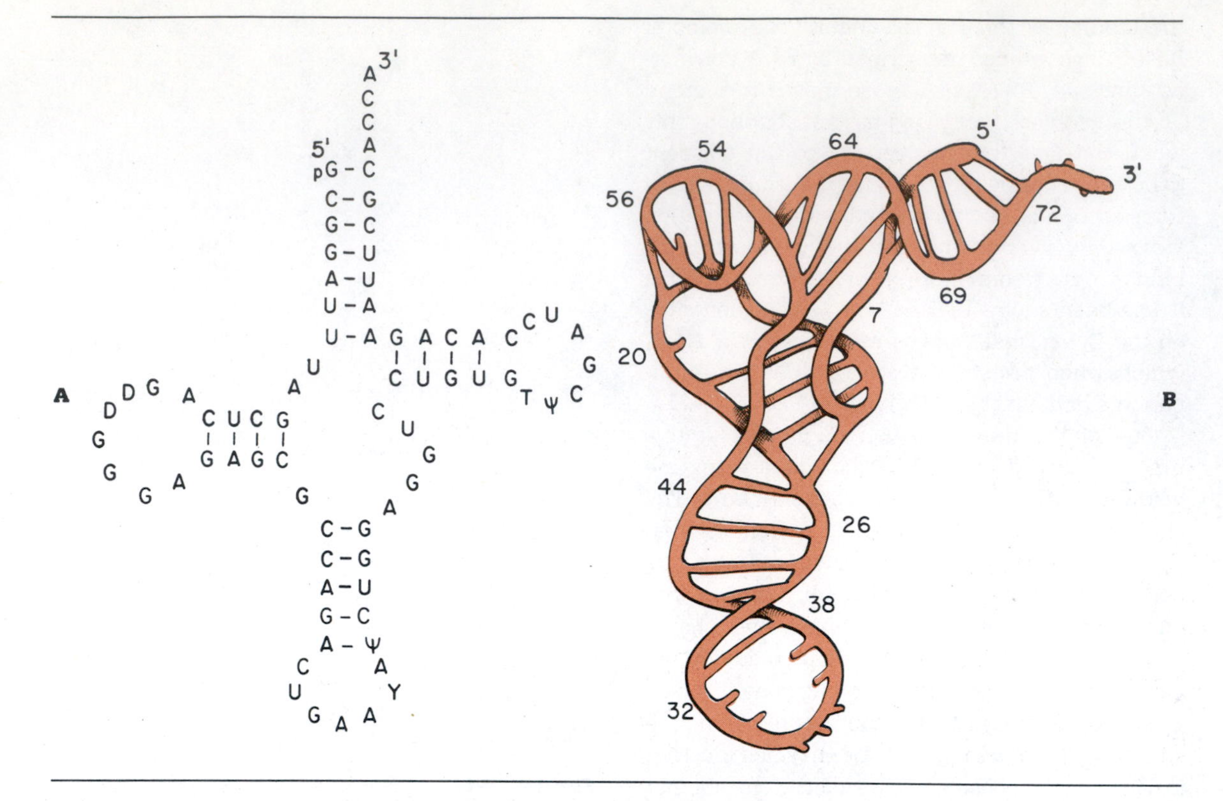

Figure 10-11 Yeast phenylalanyl tRNA. **A,** Primary structure of phenylalaninyl tRNA demonstrating nucleotide positions. **B,** Tertiary structure of phenylalanyl tRNA. Ribose phosphate backbone is drawn as continuous cylinder with various bars to show tertiary interactions. Numbers indicate nucleotide positions starting at 5′ end (compare with **A**).

Redrawn from Kim, S.H., et al. Science ***185****:435, 1974. Copyright 1974 by the American Association for the Advancement of Science.*

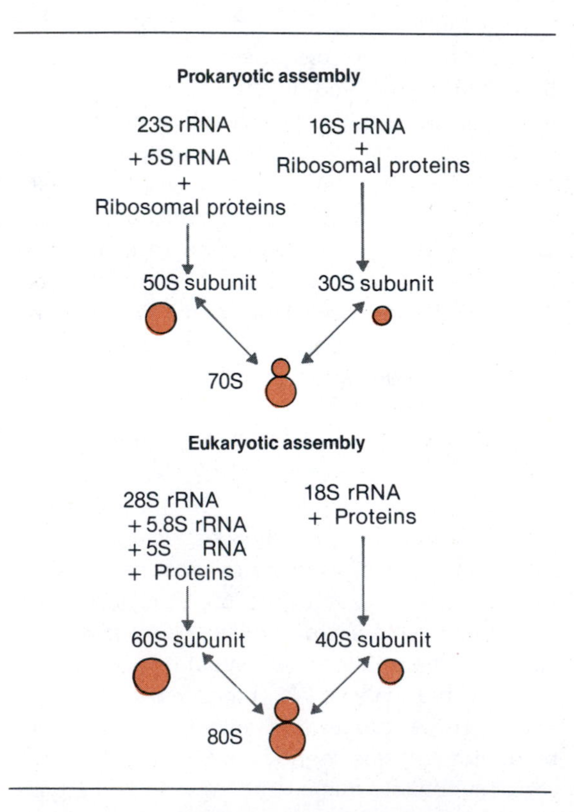

Figure 10-12 Subunit association scheme for prokaryotic and eukaryotic ribosomes.

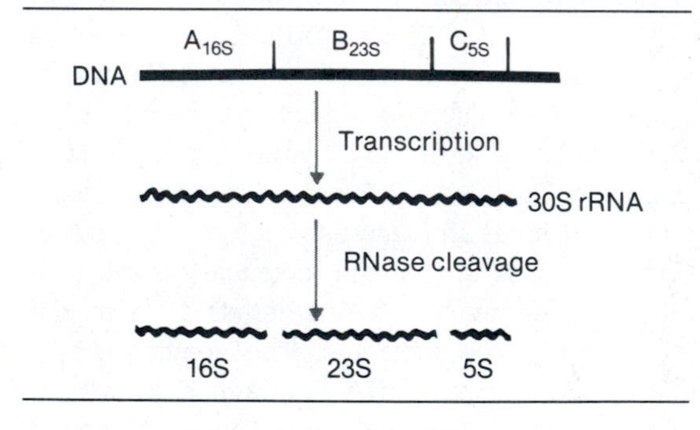

Figure 10-13 Proposed scheme for transcription of ribosomal genes. Genes for rRNAs indicated in illustration are 16S *(A)*, 23S *(B)*, and 5S *(C)*, and they are linked together at various sites on chromosome. The three genes are transcribed into single 30S rRNA precursor, which is later cleaved by RNase into 16S, 23S, and 5S rRNA.

mosome, and these are arranged as tandem repeats for the 16S, 23S, and 5S rRNA sequences. Each repeat unit is initially transcribed as a precursor 30S molecule that is later cleaved by a ribonuclease (RNase) into 16S, 23S, and 5S subunits (Figure 10-13). Once the ribosomal proteins are assembled, the rRNAs interact with them in such a way that they become buried and protected from nuclease activity.

Translation process (mechanics of protein synthesis)

Once the sequences of bases on the mRNA have been transcribed, the molecule is ready for translation into a specific sequence of amino acids that make up the polypeptide chain. A sequence of three nucleotides on the mRNA, called the codon, codes for a single amino acid in the polypeptide chain. (The characteristics of the genetic code are discussed earlier). This conversion process is called ***protein synthesis.***

Charging of amino acids in preparation for protein synthesis. Amino acids do not attach directly to the mRNA, since they have no specific chemical or physical affinity for nucleic acids. Amino acids are first charged and attached to small carrier molecules called tRNA. The charged amino acids float free in the cytoplasm and are waiting to be engaged in protein synthesis. The tRNA molecules in a starving cell are devoid of their passenger amino acids, but in rapidly growing cells, where protein synthesis is at a maximum, they are fully charged.

Each amino acid is carried by a specific tRNA molecule. This specificity is maintained because of enzymes that recognize only certain amino acids and specific tRNA molecules. These enzymes are called aminoacyl tRNA synthetases, and each possesses a site for attachment of the amino acids and a site for the specific tRNA. The activation of amino acids and tRNA binding occur in the following manner:

$$\text{Amino acid} + \text{ATP} \xrightleftharpoons[\text{synthetase}]{\text{Aminoacyl tRNA}} \text{Aminoacyl AMP} + \text{PP}$$

$$\text{Aminoacyl AMP} + \text{tRNA} \xrightleftharpoons[\text{synthetase}]{\text{Aminoacyl tRNA}} \text{Aminoacyl tRNA} + \text{AMP}$$

On each tRNA molecule, as mentioned earlier, is a triplet sequence that is complementary to the triplet code on the mRNA. The triplet code on the mRNA is called a codon, whereas its complementary triplet sequence on the tRNA is called an anticodon.

Translation of messenger RNA. The translation of mRNA occurs on the ribosomes that contain sites for the binding of mRNA, aminoacyl tRNA, and the growing polypeptide chain. Just as each sentence has a beginning and an end, so does the polypeptide and the mRNA for which it codes. The polypeptide is synthesized in a stepwise manner, one amino acid at a time, beginning from the amino end (the first amino acid has a free amino group) and terminating with an amino acid containing a free carboxyl end. The synthesis of the polypeptide can be divided into the following steps.

```
    H       H   H   H
    |       |   |   |
H - C - S - C - C - C - COOH
    |       |   |   |
    H       H   H   N - C - H
                    |   ||
                    H   O
```

Figure 10-14 Structure of *N*-formylmethionine. *N*-formyl group is shaded.

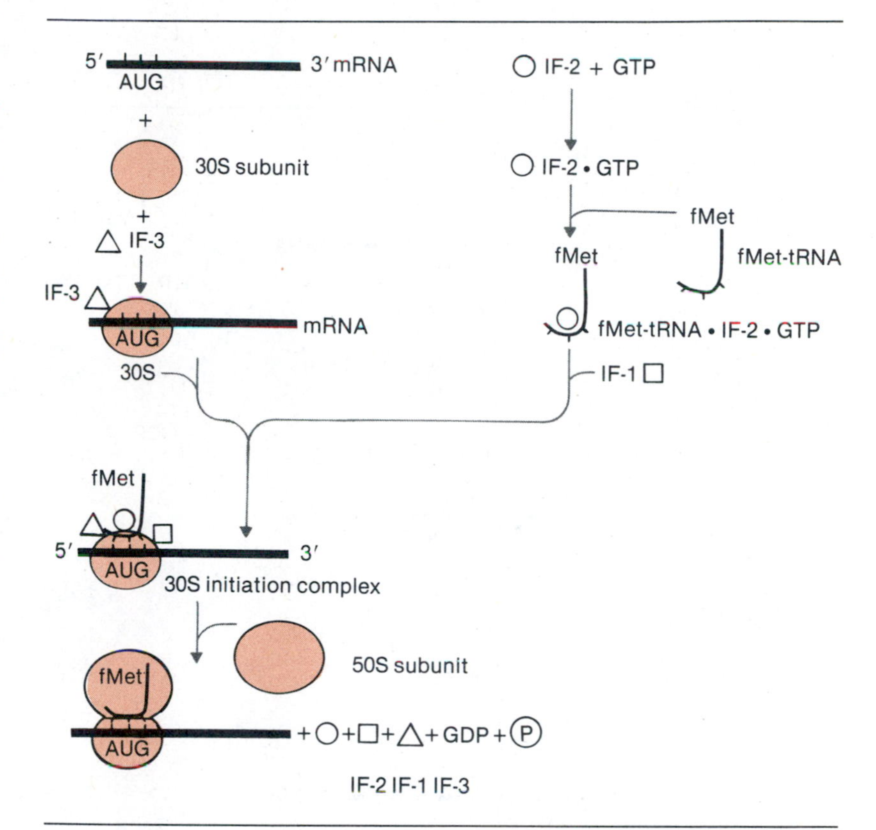

Figure 10-15 Summary of steps involved in initiation of protein synthesis. *IF,* Initiation factor; *fMet,* formylmethionine.

Modified from Watson, J.D. Reprinted by permission from Watson, James D. Molecular biology of the gene, *third edition, Menlo Park, California, The Benjamin/Cummings Publishing Company, 1977, Figure 12-23, p. 331.*

Initiation. The initiation of polypeptide synthesis in all prokaryotes begins with the binding of a tRNA that carries *N*-formylmethionine (tRNA·fMet). The tRNA·fMet carries an anticodon that recognizes the initial mRNA codon, AUG or GUG. The formyl unit attached to the amino group of methionine (Figure 10-14) prevents the amino acid from engaging in peptide bond formation and therefore will never appear in the interior of the polypeptide. The formyl group is subsequently removed by an enzyme, leaving

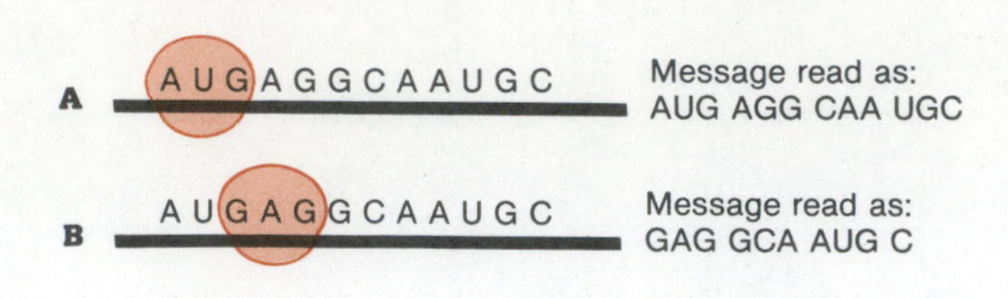

Figure 10-16 Effect of incorrect orientation of ribosome on mRNA. **A,** Correct orientation. **B,** Incorrect orientation.

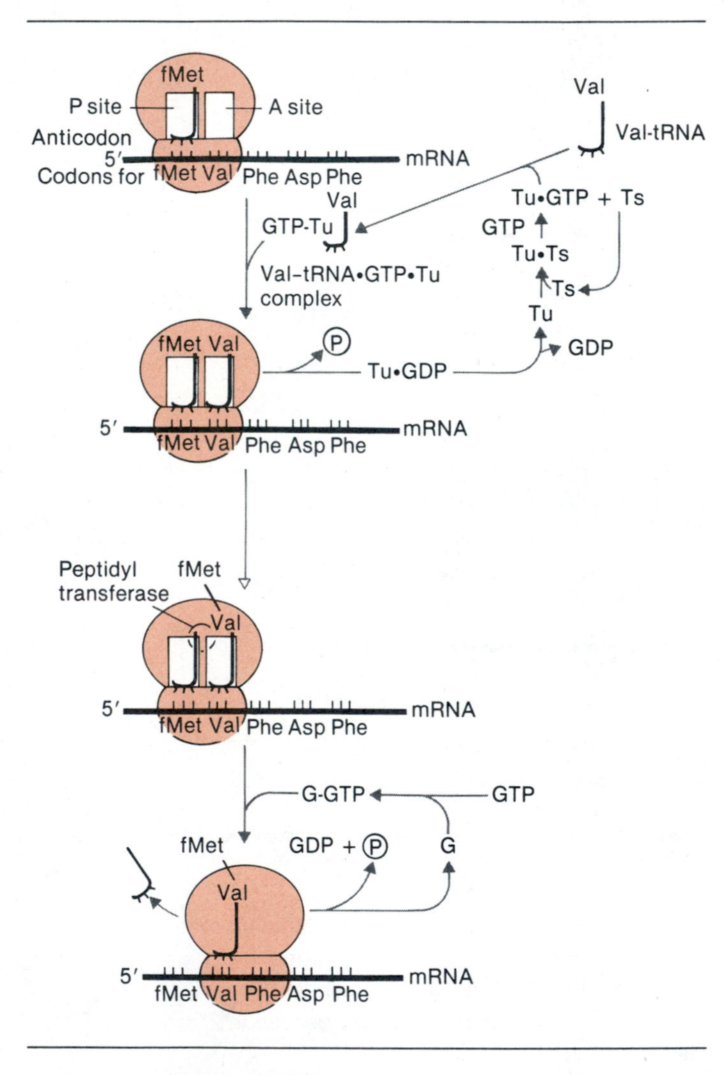

Figure 10-17 Summary of elongation and translocation steps in protein synthesis. *Tu* and *Ts* are subunits of elongation factor *T* (EF-T); *G* is elongation factor EF-G; Ⓟ is phosphate group. See text for discussion.

Redrawn from Watson, J. D. Reprinted by permission from Watson, James D. Molecular biology of the gene, *third edition, Menlo Park, California, The Benjamin/Cummings Publishing Company, 1977, Figure 12-24, p. 333.*

a polypeptide that begins with methionine. Sometimes the entire formylmethionine is cleaved from the polypeptide, and an amino acid other than methionine is located at the amino end.

The initiation step also involves binding to the ribosome (Figure 10-15), and a complex is formed between tRNA·fMet, mRNA, and the 30S ribosomal subunit. Prokaryotic mRNA may contain information for more than one polypeptide; that is, the mRNA is ***polycistronic.**** Each unit of the mRNA coding for a polypeptide begins with a sequence that must be correctly oriented with the ribosomes so that the message is read completely and in the correct reading frame (Figure 10-16). These initiating sequnces on the mRNA are called ***ribosome-binding sites.*** Three proteins are required for the initiation process and are called initiation factors: IF-1, IF-2, and IF-3. Simply stated, these initiation factors help to stabilize the tRNA·fMet-mRNA-30S complex. Once the complex is formed, the 50S ribosome subunit binds to form the 70S initiation complex. The complex is now ready for the next step, called elongation, or the addition of amino acids and the formation of peptide bonds.

Elongation. Two sites exist on the 70S ribosome for binding of aminoacyl tRNA: the A site (refers to acceptor site) and the P site (refers to peptidyl site). *N*-formylmethionyl tRNA binds to the P site (Figure 10-17). The adjacent A site binds the second aminoacyl tRNA because the second codon on the mRNA is exposed. The binding of the second amino acid involves a protein called elongation factor T (EF-T), which is made up of two subunits: Tu and Ts. EF-T and GTP form a complex with the second aminoacyl tRNA, and once bound to the ribosome, GTP is hydrolyzed, releasing an EF-TuGDP complex. The significance of EF-T is described in Figure 10-18. An amino acid polymerase, which is part of the 50S subunit, catalyzes the peptide bond formation between the carbonyl group of tRNA·fMet and the free amino group of the second amino acid. This process releases the first tRNA, which is now free to pick up another formylmethionine in the cytoplasm. The dipeptide that has been formed is now located on the A site, but before a third aminoacyl tRNA can enter the reaction, the dipeptide must be translocated to the P site (note that the P site is free because of the release of tRNA·fMet).

Translocation results in the movement of the mRNA relative to the ribosomes so that the third

* A ***cistron,*** for all practical purposes, is the same as a gene—a functional unit that has been recognized through mutant testing.

1. EF-T (Tu · Ts) + GTP ⇌ EF-T (Tu · GTP) + Ts
2. EF-T (Tu · GTP) + Aminoacyl tRNA → Aminoacyl tRNA · EF-T (Tu · GTP)
3. Aminoacyl tRNA · EF-T (Tu · GTP) + 70S ribosome → Aminoacyl tRNA · 70S (Binds to A site) + EF-T (Tu · GDP) (Released from ribosome) + P_i
4. EF-T (Tu · GDP) + Ts → EF-T (Tu · Ts) (Can reenter reaction 1.) + GDP

Figure 10-18 Summary of activity of elongation factor T (EF-T) and its subunits Tu and Ts.

codon on the mRNA becomes exposed and the A site is free to accept the third aminoacyl tRNA. Each translocation requires elongation factor G (EF-G) and hydrolysis of GTP. Translocations are required for each additional amino acid that is added to the growing polypeptide chain.

Termination. Each mRNA has a termination signal that does not recognize any aminoacyl tRNA anticodon. Termination codons are UAA, UAG, and UGA. The polypeptide is released from the last bound tRNA through the activity of the enzyme peptidyl transferase. The ribosome will continue to move along the mRNA until a new initiation sequence is reached (for example, in a polycistronic message), or the ribosomes will dissociate from the mRNA. The synthesized polypeptide begins to assume a three-dimensional structure even before it is released from the ribosome.

Other aspects of prokaryotic protein synthesis. The mRNA of prokaryotic cells will have several ribosomes bound to it during protein synthesis, and this complex is called a ***polyribosome*** (polysome). This means that more than one polypeptide can be made from a single mRNA before the latter is degraded by nucleases. At any one time on the mRNA the polypeptide appearing on each ribosome will vary in length, depending on the distance of the ribosome from the 5′ end (initiation site) of the mRNA (Figure 10-19).

It appears that in prokaryotic cells ribosomes bind to the mRNA and engage in protein synthesis as it is being transcribed from the DNA (Figure 10-20). This type of arrangement would be impossible in eukaryotic cells, where a nuclear membrane separates the transcription process from the translation process (see Chapter 12).

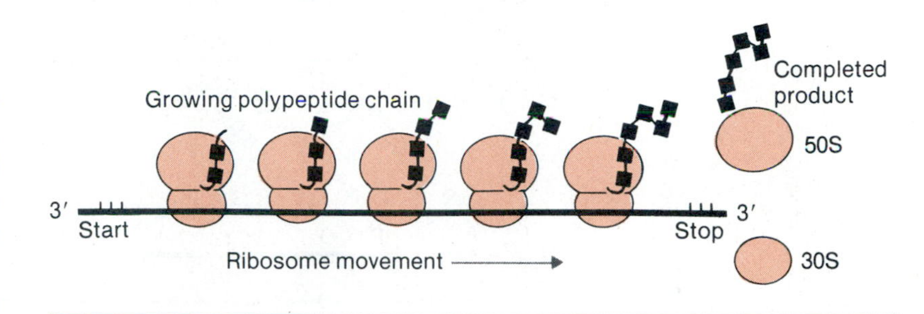

Figure 10-19 Polysome engaged in protein synthesis.

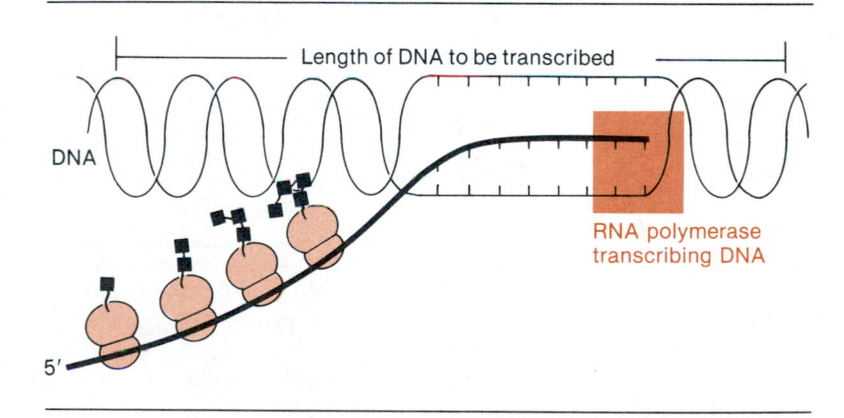

Figure 10-20 Translation of mRNA in prokaryotic cells. Translation apparently begins as soon as mRNA is being transcribed on chromosome.

REGULATION OF PROTEIN SYNTHESIS

The availability of nutrients, as well as many other environmental factors, determines what enzymes are required by the cell for adequate growth. The synthesis of all the potentially usable enzymes would be a waste of energy; therefore regulatory mechanisms have evolved in biological systems. This regulation is achieved by control

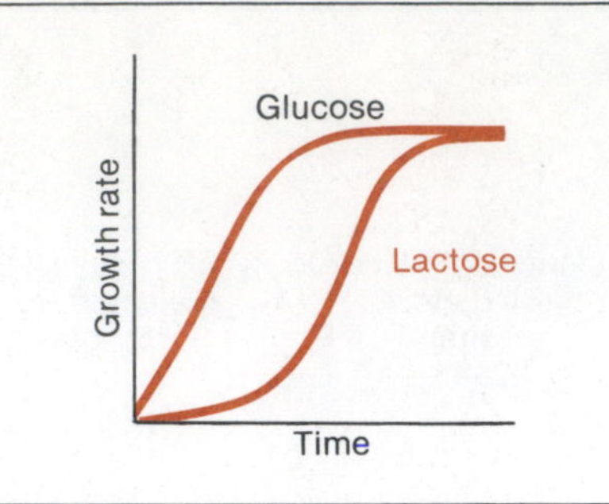

Figure 10-21 Graphic description of enzyme induction. Bacteria *(E. coli)* grown on glucose show no lag in growth, but cells grown on lactose do show growth lag. Once enzymes for lactose catabolism have been synthesized during lag period, lactose-grown cells have same rate of growth as glucose-grown cells.

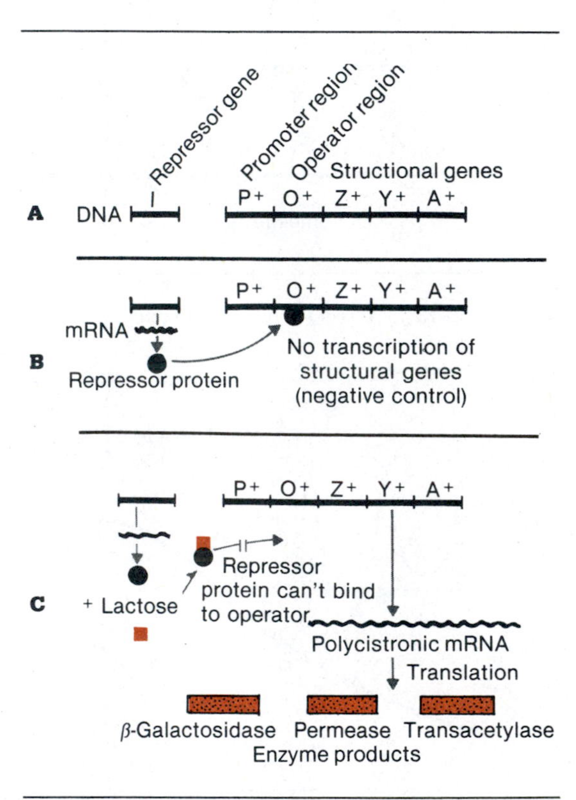

Figure 10-22 Features of lactose operon. **A,** Organizational features of lactose operon. **B,** Condition of operon in absence of inducer, lactose. Operon operates negatively, and genes for lactose catabolism are repressed. **C,** Condition of operon in presence of lactose. Cell requires lactose catabolic enzymes, and inducer prevents repressor from shutting down translation of lactose structural genes.

mechanisms that act at the level of the gene and control the synthesis of enzymes, or that act at the level of the enzyme and control the activity of already-synthesized enzymes. Control of enzyme activity is discussed in Chapters 2 and 14. The following discussion is concerned with the genetic control of enzyme synthesis.

Induction and repression (lactose operon)

The phenomenon of enzyme ***induction*** was recognized long before the mechanism of protein synthesis had been elucidated. Investigators in the 1940s noted that microbial cells growing on a carbon source such as glucose could adapt to growth on other carbon sources, such as galactose, when glucose became exhausted in the medium. An initial lag in growth was observed when galactose was added to the glucose-exhausted medium, but later the rate of growth was equal to that of cells grown with only glucose present in the medium (Figure 10-21). The lag period was recognized as the period for synthesis of enzymes that had been absent during growth on glucose; consequently, the enzymes for galactose catabolism (degradation) were inducible. Another interesting phenomenon was observed when the cells growing on galactose had exhausted their carbon source and were transferred to media containing glucose. There was no lag in the growth rate following this transfer. This meant that the enzymes for glucose catabolism are present in the cell whether or not glucose is present in the medium; that is, the enzymes for glucose catabolism are ***constitutive,*** whereas ***inducible*** enzymes are produced only in the presence of their substrate. Later investigations revealed that one of the enzymes induced by the substrate galactose is a ***permease.*** Permeases are responsible for the transport of various nutrient molecules across the cytoplasmic membrane and play a role in microbial metabolism. The lag period observed in enzyme induction experiments, for example, was found to be due to a lag in production of permeases and not to a lag in production of enzymes that catabolize galactose. How does the inducer molecule control the synthesis of enzymes involved in its own metabolism? The answer to this question came from genetic studies on *E. coli* and the brilliant work of two French investigators: Jacob and Monod, who studied the induction of enzymes involved in the catabolism of the disaccharide lactose. Lactose as a disaccharide is not metabolized and must first be broken down into its respective monosaccharide units: glucose and galactose. Jacob and Monod were able to show that enzyme induction is under genetic control because the enzymes for lactose catabolism can be produced constitutively through mutation.

The experiments of Jacob and Monod resulted in the formulation of the operon theory of control. An ***operon*** is a contiguous cluster of genes whose expression involves a particular trait (for example, the enzymes involved in the biosynthesis of histidine or the enzymes involved in the catabolism of lactose) and is in turn under the control of an operator and a repressor. Organizationally, the operon consists of ***structural genes,*** an ***operator,*** and a ***promoter*** region. An associated gene, which is located outside the operon, is the ***repressor gene.*** The characteristics of these genetic units as they apply to the lactose operon are discussed below. The organizational features of the lactose operon are illustrated in Figure 10-22, *A*.

Structural genes. The lactose operon possesses three structural genes: the gene for β-galactosidase (Z), which cleaves lactose into glucose and galactose; the permease gene (Y), which transports lactose across the cytoplasmic membrane; and the transacetylase gene (A), whose function is as yet not fully understood.

The structural genes are transcribed as a single polycistronic mRNA (Figure 10-22, *C*). Ribosomes translate the message from the 5′ end of the gene and stop at the termination site of the Z gene and then continue to translate the subsequent genes in the same manner. Ribosomes cannot initiate translation at the start regions of the Y or A genes, because only the 5′ end of the mRNA contains the "correct" initiation sequence.

Missense mutations in a structural gene may or may not result in a loss of enzyme activity for which the gene codes, but there is no loss of activity to adjacent structural genes. Nonsense mutations, particularly those nearest the promoter end of the gene, such as the Z gene, may cause a loss of activity not only for that gene but also for those genes distal to it (the Y and A genes). This reduction in activity of genes distal to the mutation is called a ***polar effect.*** The polar effect is a result of the relative inability of some ribosomes to "jump" from the mutation site, which creates a premature stop signal, to the start signal of the next gene (Figure 10-23). The greater the distance the ribosomes must "jump," the greater the polar effect.

Operator. The operator is only a sequence of nucleotides adjoining the structural genes and is not transcribed into an mRNA or other functional nucleic acid molecule. The function of the operator is to control the expression of the structural genes by binding the repressor in the absence of the inducer lactose* and preventing transcription of the structural genes. Mutations in the operator

* Lactose is converted to allolactose, which is the actual inducer of the lactose operon.

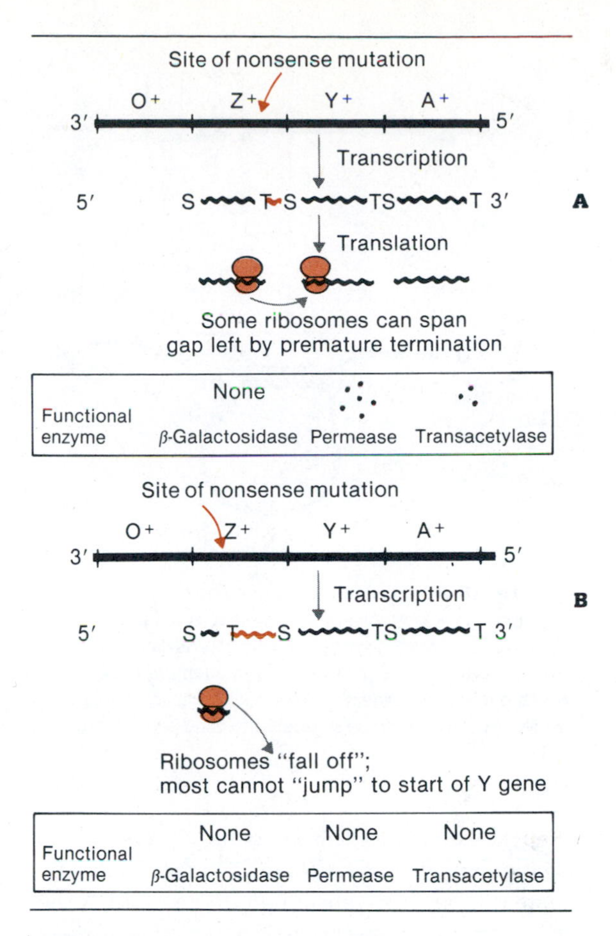

Figure 10-23 Polar effect of mutation on enzymes coded by lactose operon. *S,* Start signal for initiation of mRNA translation; *T,* termination site of mRNA. **A,** Effect of nonsense mutation at distal end of Z gene. Gap (colored wave) left by mutation is small enough that some ribosomes can "jump" gap and initiate translation of genes Y and A. **B,** Nonsense mutation is very close to initiating site of Z gene. Gap (colored wave) produced by mutation is so large that ribosomes cannot bridge distance to Y gene, and they "fall off." Thus no enzymes are produced.

can result in the inability of the repressor to bind to the operator, and a permanent constitutive state can be obtained. Any condition or mutation that results in either a faulty operator or a faulty repressor creates the constitutive state.

Promoter. The promoter region lies on the 5′ side of the operator region and like the operator does not produce an mRNA or functional nucleic acid. The promoter is the binding site for RNA polymerase (see p. 234), whose function is to initiate the transcription of the structural genes. Mutations involving the promoter region prevent transcription of structural genes and produce a condition of repression without the help of a repressor molecule. The promoter also binds a protein called catabolite activator protein (CAP), which we will discuss shortly.

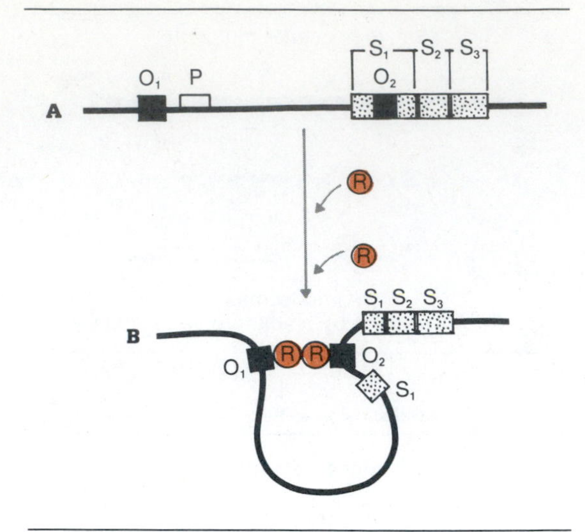

Figure 10-24 Galactose operon showing two operators. **A,** Operator *(O_1)* adjacent to promoter *(P)* and operator *(O_2)* located within first structural gene *(S_1)*. **B,** Repressor molecules *(R)* in form of monomers bind to both operators, causing them to loop out. Repressor forms dimer. RNA polymerase is unable to bind to promoter.

Repressor. The repressor gene (i gene) is not contiguous with the operon that it controls. The repressor gene is transcribed into an mRNA that codes for the repressor protein, which is made up of two subunits (dimer) and is subject to conformational changes. The repressor molecule, in the absence of an inducer, exists in a conformational state that binds to the operator region. This complex prevents the binding of RNA polymerase to the promoter region, and structural gene transcription cannot take place. Mutation in the repressor gene results in the formation of a repressor molecule with binding properties ranging from a state of inability to bind to the operator to a state of increased binding. The various binding properties are the result of changes in the conformation of the molecule.

The induction-repression mechanism for the lactose operon works in the following way.

1. The repressor molecule is synthesized in the absence of an inducer (lactose) and is free to bind to the operator. The transcription of structural genes is prevented, and no lactose-metabolizing enzymes are synthesized. Thus the lactose operon is under negative control.
2. Lactose, when present in the medium, binds to the repressor molecule, preventing its attachment to the operator. The operon is free to bind RNA polymerase, and the structural genes are transcribed into mRNA (Figure 10-22, *C*). The mRNA is translated into the enzymes involved in lactose metabolism.

Multiple operators and repression. Examination of repression in the galactose operon has revealed that more than one operator is involved in control. The typical operator is located adjacent to the structural genes, but a second operator has also been discovered within the first gene of the galactose operon (Figure 10-24). What seems to happen is that repressor binds to both operators, and each repressor molecule binds to each other to form a dimer. The binding of the two repressor molecules causes a physical looping out of the promoter region. This results in the inability of RNA polymerase to bind to the promoter, and the transcription of galactose structural genes is prevented. Other operator sites had been discovered in the lactose structural genes; however, their capacity for binding repressor was often 100 times less than that of the primary operator. Such studies indicate that our knowledge of control mechanisms is still not completely understood.

Catabolite repression (positive control in the lactose operon)

We have already noted that the lactose operon is under negative control via the action of a repressor molecule that binds to the operator region. A positive control mechanism also exists in the lactose operon, as well as in other inducible operons. The term *positive* refers to the enhanced activity of the lactose promoter in binding RNA polymerase and thus initiating transcription of structural genes. A phenomenon called ***catabolite repression*** demonstrates how a positive control mechanism operates in the lactose operon. Bacterial cells, such as *E. coli,* growing in the presence of two carbon sources—glucose and lactose—preferentially metabolize glucose, and the lactose operon is repressed. Repression of the lactose operon is the result of a metabolic product of glucose catabolism, hence the name *catabolite repression,* or the ***glucose effect.*** Repression of the lactose operon is brought about by the ability of the metabolite produced during glucose metabolism to lower the concentration of cyclic AMP (cAMP) in the cytoplasm of the cell. When cyclic AMP is in sufficient concentration in the cell, it combines with a catabolite gene-activator protein (CAP) to produce a complex that binds to the promoter and initiates RNA polymerase binding and activity. These factors promote the transcription of the structural genes (Figure 10-25). The presence of glucose prevents the binding of CAP to its site on the promoter, and transcription of lactose genes is prevented. Catabolite repression is an important energy-saving device for cells growing in the presence of two carbon sources such as glucose and lactose. Glucose is much more easily metabolized than lactose, and with the lactose operon repressed, the enzymes for which it codes are not produced and energy is saved. Many catabolic enzymes in *E. coli* and other enteric bacteria are sensitive to glucose repression. In addition, flagellum biosynthesis is also repressed by glucose.

Studies have indicated that naturally occurring metabolites other than cAMP can cause the induction of the L-arabinose operon. Mutants of *E. coli* with defects in the gene coding for adenyl cyclase (catalyzes the formation of cAMP) can be induced to produce L-arabinose by imidazole metabolites such as histidine, histamine, and urocanic acid. None of these naturally occurring metabolites is used as a carbon source by *E. coli,* but each does have a ring structure similar to cAMP. It has been hypothesized that the imidazole metabolites form a complex with the promoter region on the genome and create a site having a high affinity for RNA polymerase. RNA polymerase binds to the promoter region and initiates transcription. Neither the lactose nor the maltose operon is affected by the imidazole metabolites, which suggests that variations exist in the various promoter regions.

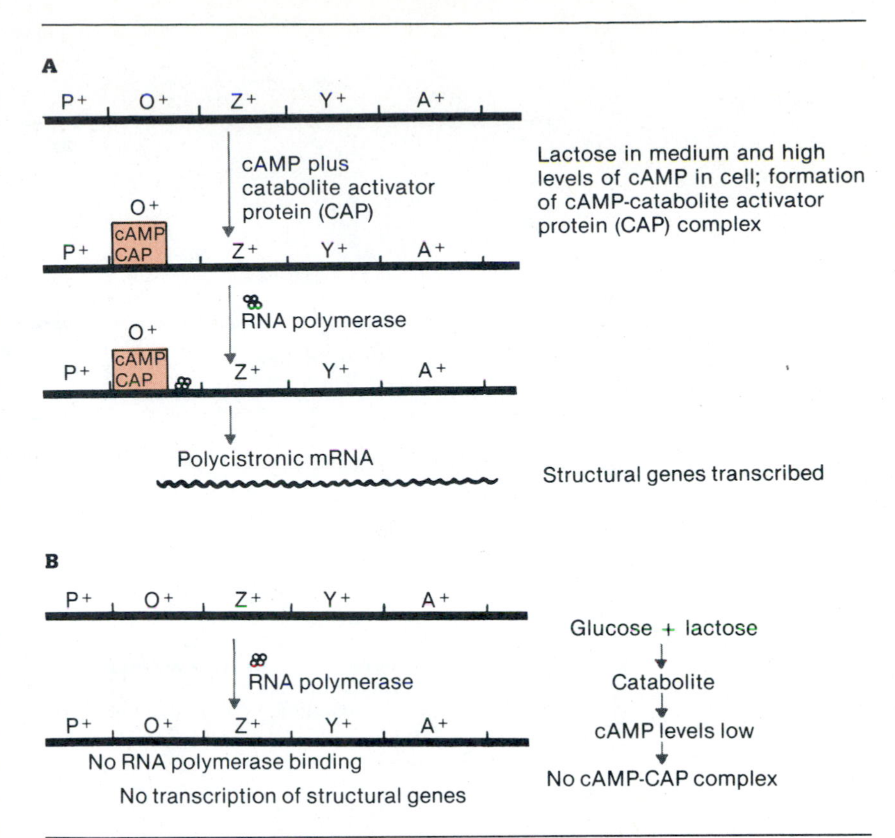

Figure 10-25 Catabolite repression in lactose operon. **A,** Lactose is present in medium. No catabolite repression occurs, because there are high levels of cAMP, which forms cAMP-catabolite activator protein (CAP) complex, which is required for RNA polymerase attachment. **B,** Lactose and glucose are present in medium. Product of glucose catabolism causes decrease in cAMP, which prevents binding of cAMP-CAP complex and in turn prevents binding of RNA polymerase.

Repressible operons

So far, we have talked about the lactose operon as a type of inducible system in which the operon is turned off in the absence of lactose. The lactose operon, as well as other operons that control catabolic reactions, exhibits this inducible characteristic. There are negative control systems in which the operon is turned on and the structural genes are transcribed in the absence of the product of structural gene activity. These operons control reactions in the cell that are anabolic, for example, reactions involved in the synthesis of products such as amino acids. They are termed ***repressible operons*** because the product of structural gene activity, called the ***corepressor,*** facilitates the binding of an inactive repressor protein, called the ***aporepressor,**** to the operator, and the operon is turned off (Figure 10-26) The corepressor may be an amino acid or aminoacyl tRNA. The tryptophan operon in *E. coli* is one of the more widely studied repressible systems. A tryptophan-activated repressor controls the binding of RNA polymerase to the promoter.

Apparently, repression is not the only way that the microbial cell can regulate the tryptophan operon. Mutant analysis has shown that even if a functional repressor is lacking, the cell can still respond to tryptophan starvation by increasing the rate of tryptophan mRNA synthesis. In addition, in this analysis some mutants with deletions in the transcribed region of the operon had a sixfold increase in expression of the remaining genes of the operon. Repressor control was not affected by these deletions, suggesting that the deleted region was distinct from the operator but still was involved in control of the operon. The tryptophan operon possesses some divisional elements not found in the lactose operon (Figure 10-27). A leader sequence segment (Trp L) of the operon was found between the operator and the structural genes. Within this leader sequence is a transcription termination site, which is part of a sequence of bases called the ***attenuator. Attenuation*** is a term now used to denote regulation by transcription termination. The leader se-

* Not all operons coding for biosynthetic enzymes are controlled by a regulator gene product; for example, histidine, leucine, and isoleucine operons are not.

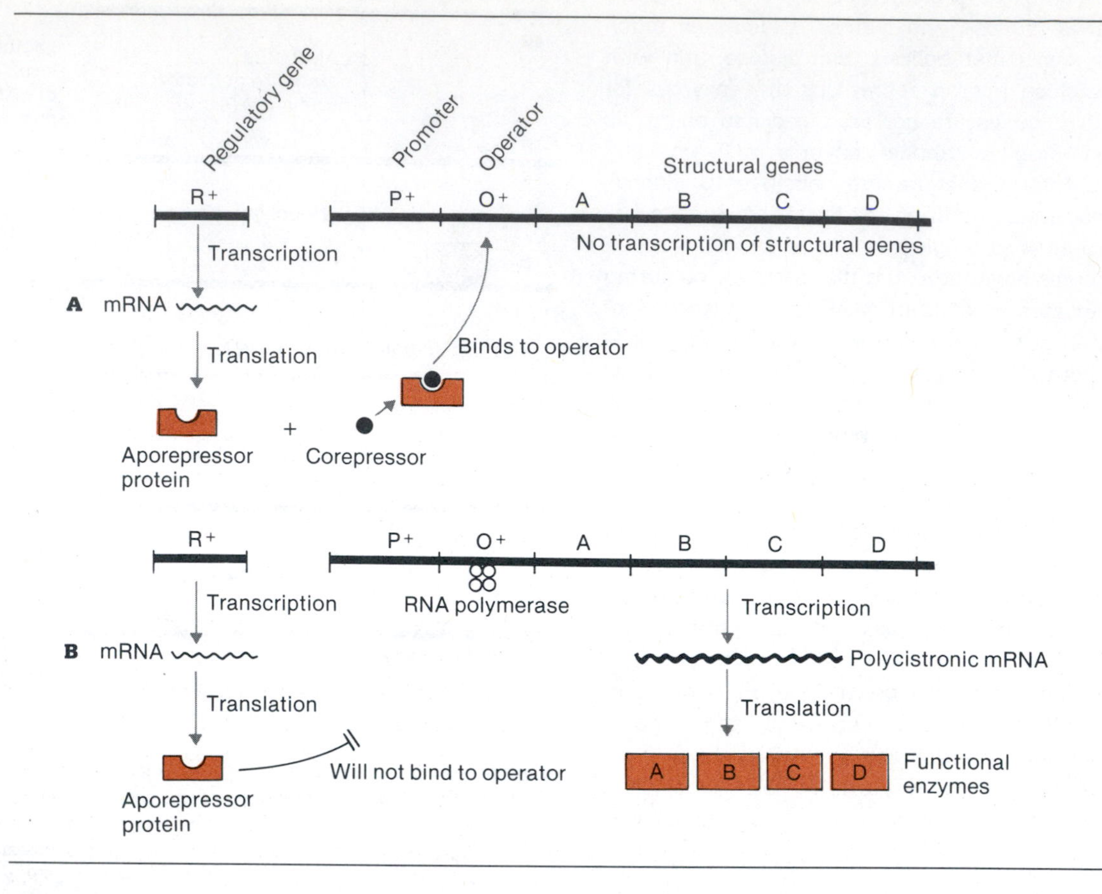

Figure 10-26 Characteristics of repressible operon. **A,** Presence of corepressor. **B,** Absence of corepressor.

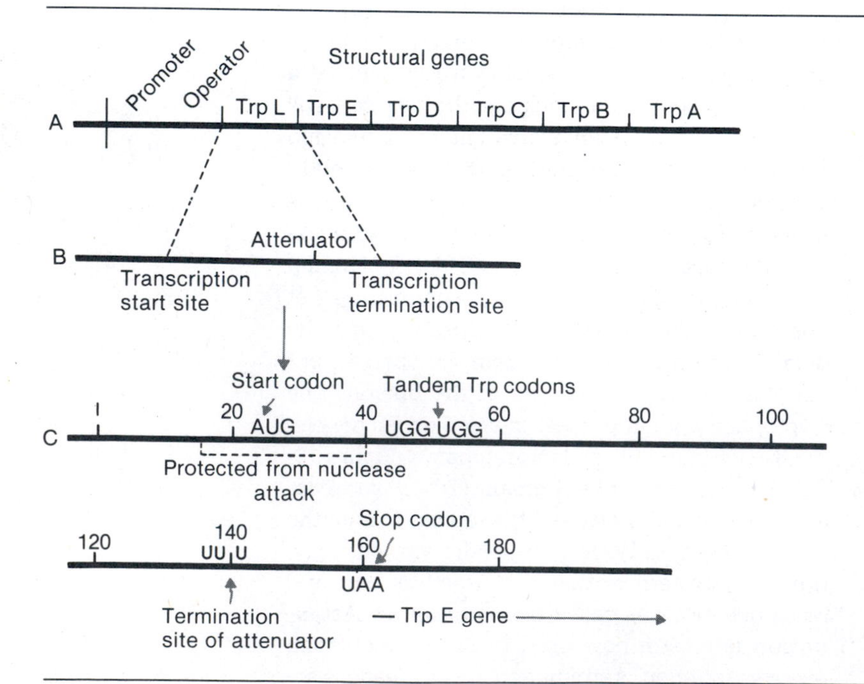

Figure 10-27 Organizational features of tryptophan operon, with special reference to attenuator. *A,* Tryptophan operon. To right of operator is nucleotide sequence called Trp L or leader sequence. Leader sequence precedes structural genes Trp E, D, C, B, and A. *B,* Expanded leader sequence showing transcription start site and transcription termination site, which is involved in attenuation of tryptophan structural genes. *C,* Leader sequence is divided into nucleotide regions and consists of 160 nucleotides. Small region (nucleotides 10 to 40) is protected from nuclease attack, and within region at approximately nucleotide 23 is start codon. At approximately nucleotides 59 and 60 are located two tryptophan codons, UGG. At nucleotide 140 is termination site for attenuator, and stop codon for leader sequence is at nucleotide 160.

quence, when fully transcribed, contains approximately 160 nucleotides (Figure 10-27, *C*). A 20-base segment in the early portion of the leader sequence is protected by ribosomes from nuclease digestion, and within this segment is a potential AUG start codon. The peptide translated from the leader sequence contains two tryptophan residues in tandem.* It is believed that the ribosome translating the leader sequence will either stall over the tryptophan codons or continue to read through. Stalling apparently permits the suceeding segment of the transcript to form a secondary structure that permits RNA polymerase molecules to continue with transcription. Thus the RNA polymerase either terminates transcription or continues to read through into the structural genes of the operon. Unlike repression, which is controlled by the level of tryptophan in the cell, the attenuator is apparently controlled by the level of tryptophan tRNA. Under conditions in which the cell requires tryptophan, the concentration of charged tryptophan tRNA is very low. This creates a situation for stalling of the translation process. Thus there will be little or no termination at the attenuator, and the tryptophan structural genes will be expressed. When the level of charged tryptophan tRNA is very high in the cell (indicating little need for more tryptophan), there is maximal termination at the attenuator. Under this set of conditions the tryptophan genes will be expressed very little or not at all.

Control at the level of translation

Our discussion of protein synthesis regulation up to now has centered on mechanisms that affect transcription of the gene. There are ways in which the cell might control protein synthesis at the level of translation. The products of some polycistronic mRNAs are produced in stoichiometrically equal amounts. The lactose operon contains three different genes whose coded products are not produced in equal amounts. The enzymes coded by the Z, Y, and A genes occur in the relative proportions of 10:5:2. Two possible mechanisms may account for the unequal production of enzymes coded by a polycistronic mRNA:

1. Nucleotide sequences (intercistronic) exist between the termination signal of the Z gene and the initiation signal of the Y gene, as well as between the termination signal of the Y gene and the initiation signal of the A gene. Some ribosomes translating the Z gene could fall off the mRNA before reaching the Y gene. Proportionately more ribosomes would drop off during their intercistronic movement from the Y gene to the A gene.
2. A second mechanism for translation control might involve the use of different codons for the same amino acid. For example, the codon for leucine in gene Z might be different from that in gene Y. Disproportionate amounts of enzyme could be produced if the concentration of one charged tRNA · Leu was different from another.

* The peptides of other operons will be rich in the amino acid that their operons control.

SUMMARY

The hereditary information of the cell is present in the DNA in the form of genes. This DNA information (genotype) is translated primarily into enzymatic proteins, which impart to the cell certain characteristics collectively called the phenotype of the cell. In addition, some areas of the DNA can act as a template for the formation of transfer RNA (tRNA) and ribosomal RNA (rRNA).

The genetic code was found to be a triplet code in which most triplets code for an amino acid. There are 64 code words for the 20 biologically available amino acids; therefore an amino acid may be coded for by more than one triplet (the code is degenerate). The code, which is nonoverlapping and contains no punctuation, is read from a fixed point on the message. The addition or deletion of a base in the message produces a shift in the reading frame, which in turn results in changes in the amino acids of the polypeptide. Four types of mutation alter the activity of polypeptides: missense, nonsense reading frame (see Chapter 9), and suppressor mutations. Missense mutations involve a single base pair substitution, and nonsense mutations result in the formation of codons that specify for no amino acid. Mutations that suppress a primary mutation are called suppressor mutations and may be within the same gene (intragenic) or in other genes (intergenic).

There are 20 molecular species of tRNA, but several can carry more than one amino acid. The ***wobble hypothesis*** explains this characteristic and assumes that a rigid specificity exists in the first two letters of the codon-anticodon interaction, whereas the third-letter interactions show more flexibility (wobble).

The information in the DNA molecule during protein synthesis is first transcribed into messenger RNA (mRNA), following the binding of RNA polymerase to the promoter site on the DNA. The mRNA is translated into protein in a series of events: first, there is activation of the amino acids and their attachment to specific tRNA molecules; second, the mRNA is translated into polypeptides in which four processes take place: initiation, elongation, translocation, and termination.

The cell must regulate the synthesis of certain proteins not always required by the cell under certain environmental conditions. The synthesis of these proteins is controlled by two types of molecules: inducers and repressors. Under most conditions enzymes are negatively controlled; that is, they are in a repressed state. Repression takes place on the DNA molecule in special areas called operons, which are contiguous units of DNA nucleotides composed of structural genes, an operator, and a promoter. Outside many operons there is a repressor gene whose product inhibits the operation of the operon and hence the synthesis of specific proteins. The operon is inhibited by the repressor until a specific inducer is present to inactivate the repressor. There are operons in the cell in which the repressor exerts a positive as well as a negative control, but these are few.

Catabolite repression is a special mechanism used by cells growing on glucose plus another sugar (one that is inducible). Glucose is metabolized to a product that represses the induction of enzymes that would catabolize the second sugar. Some repressible operons also operate in the microbial cell, in which a product of the structural gene facilitates the binding of the repressor to the operator and the operator is turned off.

Study questions

1. Define the term *gene* in terms of function as well as structure.
2. Describe five characteristics of the genetic code.
3. What would be the effect on translation if there were punctuation in the code? Would there be any effects on the evolutionary process?
4. Distinguish between missense and nonsense mutations.
5. What are the termination signals on the messenger RNA molecule?
6. Describe the sequence of events in which an amino acid becomes activated before its involvement in protein synthesis.
7. What is the importance of *N*-formylmethionine in bacterial protein synthesis?
8. In genetic terms define induction and repression. Why aren't all enzymes in the cell inducible and repressible? What is the advantage of the induction-repression system?
9. Diagrammatically illustrate the components of an operon. Is the entire DNA molecule made up of operons? If not, what other genetic units are also present?
10. What is the role of cyclic AMP in catabolite repression?
11. Define the following: catabolite activator protein, promoter, phenotype, constitutive enzyme, anticodon, rho factor, and nonsense codon.

Selected readings

Books

Bainbridge, B.W. *The genetics of microbes.* Blackie & Son, Ltd., Glasgow, Scotland, 1980.

Beckwith, J.R., and Zipser, D. *The lactose operon.* Cold Spring Harbor Laboratory, Cold Spring Harbor, New York, 1970.

Birge, E.A. *Bacterial and bacteriophage genetics.* Springer-Verlag, New York, Inc., New York, 1981.

Genetic code: Cold Spring Harbor Symposium for Quantitative Biology (Vol. 31). Cold Spring Harbor Laboratory, Cold Spring Harbor, New York, 1966.

Herskowitz, I.H. *The elements of genetics.* Macmillan Publishing Co., Inc., New York, 1979.

Hexter, W., and Yost, H.T. *The science of genetics.* Prentice-Hall, Inc., Englewood Cliffs, N.J., 1976.

Miller, J.H., and Reznikoff, W.S. *The operon.* Cold Spring Harbor Laboratory, Cold Spring Harbor, N.Y., 1978.

Russel, P.J. *Lecture notes on genetics.* Blackwell Scientific Publications, Ltd., Oxford, 1980.

Stanier, R.Y., Adelberg, E.A., and Ingraham, J. *The microbial world* (4th ed.). Prentice-Hall, Inc., Englewood Cliffs, N.J., 1976.

Stewart, P.R., and Letham, D.S. (Eds.). *The ribonucleic acids* (2nd ed.). Springer-Verlag, New York, Inc., New York, 1977.

Watson, J. *Molecular biology of the gene* (3rd ed.). Benjamin/Cummings Publishing Co., Inc., Menlo Park, Calif., 1976.

Weissbach, H., and Pestka, S. (Eds.). *Molecular mechanisms of protein synthesis.* Academic Press, Inc., New York, 1977.

Journal articles

Bertrand, L., et. al. New features of the structure, function, and regulation of the tryptophan operon. *Science* **189:**22, 1975.

Brenner, S., Jacob, F., and Meselson, M. An unstable intermediate carrying information from genes to ribosomes for protein synthesis. *Nature* **190:**576, 1961.

Calvo, J.M., and Fink, G.R. Regulation of biosynthetic pathways in bacteria and fungi. *Annu. Rev. Biochem.* **40:**943, 1971.

Clemens, M.J. (Eds.). *Gene expression.* CRC Press, Inc., Boca Raton, Fla., 1980.

Crick, F.H.C. Codon-anticodon pairing: the wobble hypothesis. *J. Mol. Biol.* **19:**548, 1966.

Davidson, E.H., and Britten, R.J. Regulation of gene expression: possible roles of repetitive sequences. *Science* **204:**1052, 1979.

Dickson, R.C., et. al. Genetic regulation: the lac control region. *Science* **187:**27, 1975.

Englesberg, E., and Wilcox, G. Regulation: positive control. *Annu. Rev. Genet.* **8:**219, 1974.

Garen, A. Sense and nonsense in the genetic code. *Science* **160:** 149, 1968.

Holley, R.W., et. al. Structure of a ribonucleic acid. *Science* **147:**1462, 1965.

Jacob, F., and Monod, J. Genetic regulatory mechanisms in the synthesis of proteins. *J. Mol. Biol.* **3:**318, 1961.

Khorana, H.G. Polynucleotide synthesis and the genetic code. *Harvey Lect.* **62:**79, 1976-1977.

Kurland, C.G. Structure and function of bacterial ribosomes. *Annu. Rev. Biochem.* **46:**173, 1977.

Maniatis, T., and Ptashne, M. DNA-operator repressor systems. *Sci. Am.* **234**(1):64, 1976.

Nirenberg, M., and Leder, P. RNA code words and protein synthesis. *Science* **145:**1399,1964.

Nomura, M. Assembly of bacterial ribosomes. *Science* **179:**864, 1973.

Ptashne, M., and Gilbert, W. Genetic repressors. *Sci. Am.* **222:**36, 1970.

Streisenger, G., et al. Frameshift mutations and the genetic code. *Cold Spring Harbor Symp. Soc. Quant. Biol.* **31:**77, 1966.

Umbarger, H.E. Feedback control by end-product inhibition. *Cold Spring Harbor Symp. Soc. Quant. Biol.* **26:**301, 1961.

Chapter 11 RECOMBINATION AND GENE TRANSFER

Any understanding of evolution invariably must include a knowledge of genetic systems and the processes of mutation and recombination. Mutation ensures that variability will occur in the gene, but genetic recombination guarantees that a number of different gene combinations will be produced. Any significant complexity in organisms could not have arisen without these processes. Recombination in bacteria, in which some of the genetic traits of two separate cells have been exchanged, occurs as the result of processes referred to as transformation, transduction, and conjugation. The purpose of this chapter is to discuss the mechanics of the recombination process and the ways in which genes can be transferred to a bacterial cell for participation in this process.

RECOMBINATION

The term ***recombination*** in bacterial genetics has undergone some changes in the past 25 years. Originally the term was used to denote the transfer of DNA from one cell to another and the establishment of some of this genetic information in the recipient cell. Today, recombination is thought of as a reassortment of nucleotide sequences within the DNA molecule and may take two forms. First, the reassortment may result from the rearrangement of pieces of DNA derived from two parent molecules. The recombination could involve chromosomal DNA from a donor cell and a recipient cell, plasmid DNA and plasmid DNA, viral DNA and chromosomal DNA, and plasmid DNA and chromosomal DNA. The interactions of these molecules may produce a single molecule derived from each parent molecule or two molecules derived in part from each parent molecule. Sometimes there is no interaction between two DNA molecules and the rearrangement of nucleotides may occur on a single DNA molecule. Deletions, inversions, duplications, or transpositions may lead to such rearrangements, and these are discussed later. The modes of genetic recombination may be divided into the following types: general, site specific, and illegitimate.

General recombination

General recombination occurs more frequently than other types of recombination because of the homology that exists between the nucleotide sequences of the two DNA molecules. General recombination is common in processes such as transformation, transduction, and conjugation where the DNA molecules are from the same or similar species. Binding of homologous areas between two DNA molecules usually affects a rapid recombinational event but requires the products of specific genes called ***Rec genes.*** The enzymes coded by these genes are apparently involved in unwinding of the DNA molecule as well as strand cleavage (exonuclease and endonuclease) and synthesis (polymerases).

Model for general recombination. The Holliday model is the most widely accepted model that explains the outcome of general recombination in which there is a reciprocal genetic exchange (Figure 11-1). The model predicts that two double-stranded DNA molecules are held together at homologous sequences. The single chains of the homologous sequences are broken at corresponding sites and then exchanged. Finally, the exchanged segments are joined to the resident chains. DNA molecules heterozygous (different alleles of a gene) for a trait may be produced if one of the DNA sequences involved in the exchange is mutant.

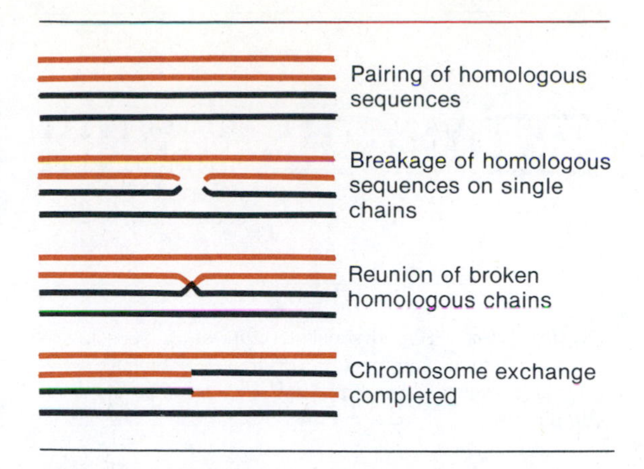

Figure 11-1 Holliday model for general recombination.

Site-specific recombination

There is very little, if any, homology between DNA molecules that engage in site-specific recombination, but there is a special crossover position on one or both of the DNA molecules where recombination takes place. This type of recombination may or may not be frequent, depending on the system. One example of site-specific recombination is that observed during the integration* and excision of bacteriophage lambda (Figure 11-2). Lambda (λ) can integrate into the bacterial chromosome because of a recombinational event between a specific phage attachment site on the virus, called ***att P,*** and a bacterial attachment site on the bacterial chromosome, called ***att B*** (Figure 11-3). The attachment sites are several hundred base pairs in length, but there is homology only between core sections that are 15 base pairs in length. Crossover occurs only within the core area of the attachment sites, because the remainder of the attachment site bases are nonhomologous. Recombination results in the integration of viral DNA into the host chromosome with hybrid (part bacterial, part viral) attachment sites flanking the viral DNA (see Figure 11-2). The integration process requires two host gene products and a viral-coded enzyme called ***integrase,*** which recognizes the core sequence. The excision of virus from the bacterial chromosome can also take place and is a reversal of the integration process. There is a reciprocal crossover between the attachment sites that flank the integrated viral genome, and the products of this event are a fully circular viral chromosome and an intact bacterial chromosome.

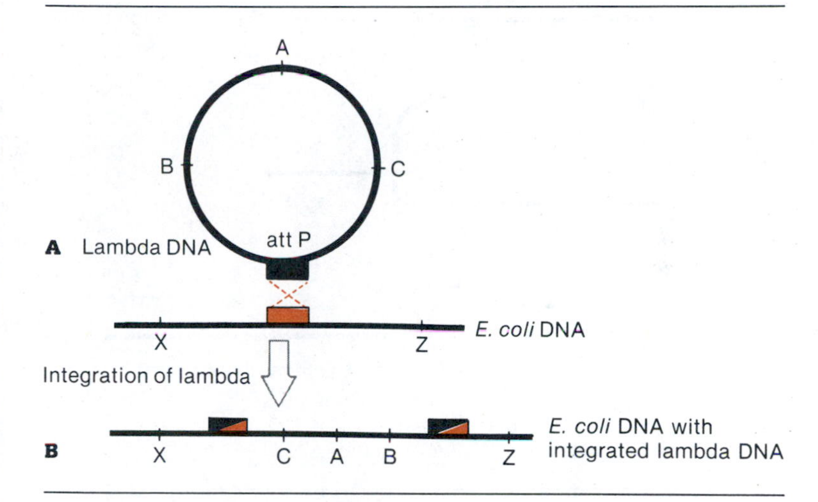

Figure 11-2 Site-specific recombination. Integration of bacteriophage lambda (λ) into chromosome of *E. coli,* example of site-specific recombination. **A,** Lambda DNA is represented as circular molecule and bacterial DNA as linear molecule only for ease of description. Letters *A, B,* and *C,* and *X* and *Z* represent genes on viral and bacterial chromosomes, respectively. Single reciprocal crossover occurs between viral attachment site (black) and bacterial attachment site (color). **B,** *Escherichia coli* chromosome contains integrated lambda genome flanked by hybrid attachment sites.

Transposable genetic elements. The ability of certain DNA sequences to promote the movement of certain functional genes from one DNA molecule to another or within the same DNA molecule is called ***transposition.*** Transposable genetic elements include ***transposons, insertion***

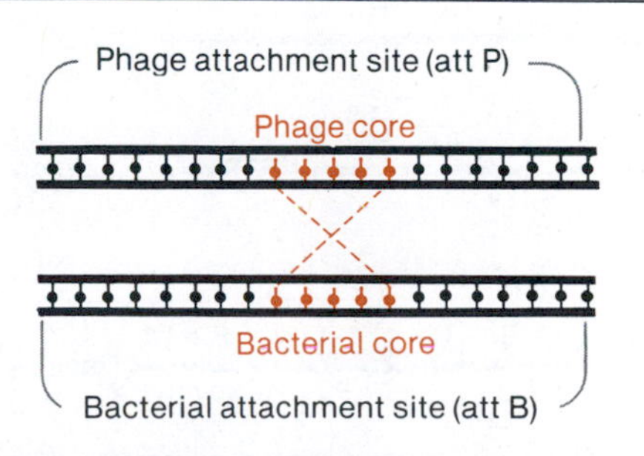

Figure 11-3 Attachment sites of lambda (λ) and bacterial DNA. Reciprocal crossovers (X) occur between core regions of phage attachment sites.

* Viruses that are integrated into the host genome exist in what is referred to as the ***provirus*** state. In the case of bacterial viruses it is called the ***prophage*** state.

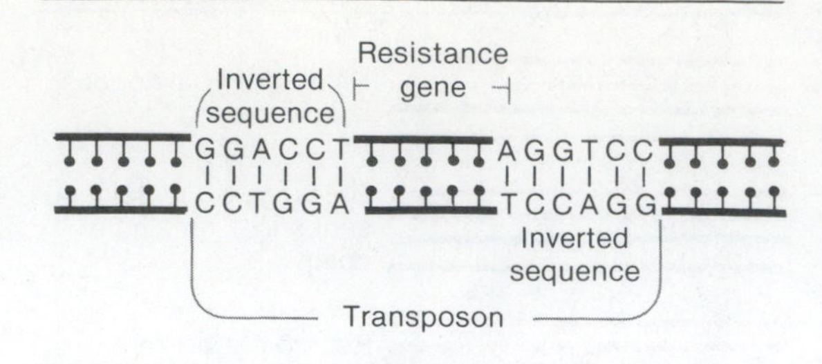

Figure 11-4 Hypothetical structure of transposon. Inverted sequences appear on either side of specific gene (in this instance, gene for resistance to drugs or other compounds).

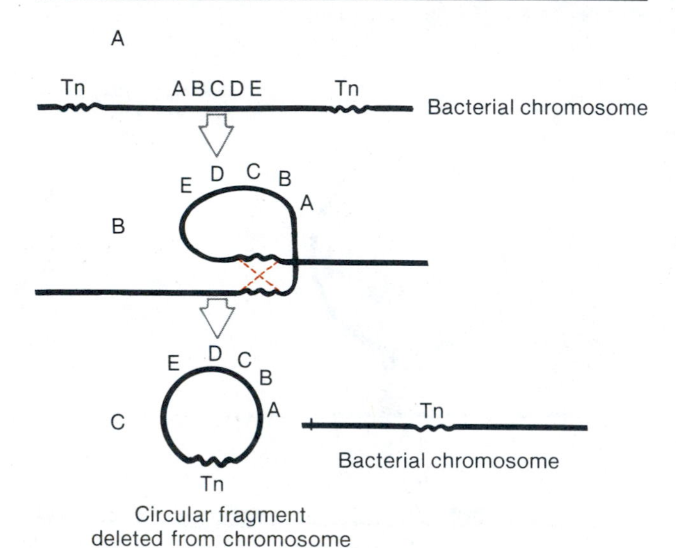

Figure 11-5 Recombination between transposon sequences that result in deletion of gene. *A,* Genes A, B, C, D, and E on bacterial chromosome are flanked by two similar transposons *(Tn). B,* Transposons engage in crossover event probably between their inverted sequences. *C,* recombination produces bacterial chromosome from which genes A, B, C, D, and E have been deleted plus circular chromosome fragment containing genes A, B, C, D, and E, as well as transposon.

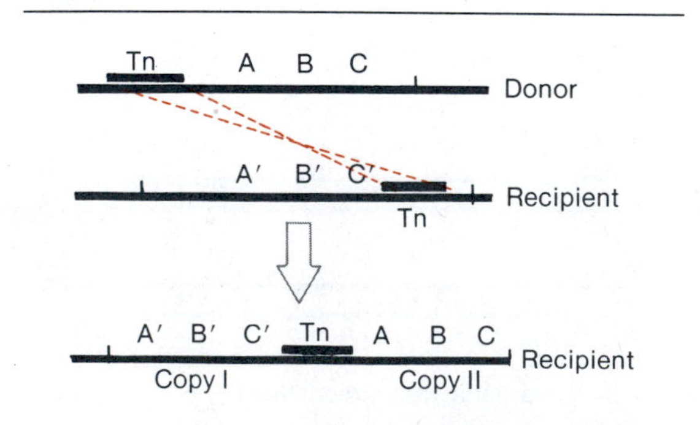

Figure 11-6 Recombination between transposon sequences that leads to duplication of genes A, B, and C. Recombination is uneven.

sequences, and the ***bacteriophage Mu.*** These elements had earlier been recognized as DNA molecules that showed no apparent homology with the DNA molecule with which they interacted, but recent discoveries indicate that they also exhibit site-specific recombination. Transposable genetic elements have become important tools in genetic analysis because of their ability to cause mutations and the fact that they carry selectable genetic traits. The organization of operons can be studied using these transposable genetic elements.

Transposons. Transposons, unlike plasmids, cannot exist independently in the cell. They contain terminal repeat sequences called palindromic sequences (Figure 11-4) and carry genes, many of which confer antibiotic resistance when expressed by the host bacterium. Transposons are believed to also contain information for transposition, such as a transposase, as well as a repressor that regulates transposase synthesis. Transposition occurs because of a specificity between the inverted sequences at the end of the transposon and the site of insertion. The specificity may be such that several sites on the chromosome recognize the inverted sequences of the transposons. Consequently, many areas of the chromosome can have new genetic elements inserted within or between genes.

The insertion of transposons causes mutations that are polar in terms of the expression of distal genes in the operon. Polarity is the result of inserted transposons containing sequences that terminate transcription, and any gene distal to the transposon is not transcribed. Transposons, as well as other transposable elements, can also cause deletions (Figure 11-5) or duplications (Figure 11-6) in the chromosome because of recombination between inverted sequences. The insertion of any gene between transposable genetic elements makes possible the transfer of a gene to a structurally unrelated DNA molecule. This type of activity is particularly important with respect to microbial antibiotic resistance, because transposons can be shuffled into a variety of plasmids, which in turn can be transferred to other cells during the conjugation process (see "Conjugation," p. 255). Drug-resistant plasmids appear to have evolved as collections of transposons, since plasmids isolated in different parts of the world show extensive nucleotide homology in certain segments of the molecule and absolutely no homology in others. The discovery of transposons and other transposable genetic elements has clarified to some degree the great biological diversity that exists in many systems and has provided an alternative theory for spontaneous mutation.

Insertion sequences. Insertion sequences are considerably smaller than transposons and contain only 200 to 1500 base pairs. Insertion sequence (IS) elements are closely related to transposons. The essential difference between them is that IS elements code for no observable phenotype, whereas transposons do (such as antibiotic resistance). Some transposons actually contain insertion sequences flanking the transposon gene. Several copies of IS elements have been found within individual plasmid, viral, and bacterial DNA molecules. They have inverted terminal repeat sequences that show chromosomal specificity, but they are not always perfect. Many IS elements possess sequences that can behave as promoters as well as terminators of transcription. Their insertion into the chromosome can result in the turning off or turning on of gene transcription. IS elements inserted into the bacterial chromosome also contain sequences that are recognized by the sex factor plasmid and are important for the latter's integration or excision during the conjugation process (see p. 255).

The transposition of an IS element is not simply removal of the element from its original site and insertion into a new site. A copy of the IS element is first made and then transferred to the site, leaving the original IS element intact. Once the IS element has been inserted, there is a duplication of some (3 to 12 base pairs) of the nucleotide sequences at the insertion site. In other words, the newly inserted IS element is flanked by the same nucleotide sequences.

Bacteriophage Mu. Bacteriophage Mu is a temperate virus that can integrate into the bacterial chromosome as a prophage. The term *Mu* was enlisted because of the propensity of the virus to cause mutations. Mu can integrate into any place on the bacterial chromosome to form a prophage and when it is released carries part of the bacterial chromosome. For this reason, the ends of Mu are found to be of bacterial origin. There is a particularly interesting aspect of the Mu integration process as compared with other transposable genetic elements: the host sequences on the ends of Mu are different for each virus strain, and none are integrated with phage DNA into the host chromosome during the insertion process. It appears that there are specific Mu sequences adjacent to the host sequences that recombine with host DNA during the insertion process. The host sequences that are attached to the infecting Mu DNA are lost during the integration process (Figure 11-7). Bacterial genes at each end of the inserted Mu are released with the virus during excision. Like transposons and insertion sequences, Mu, when inserted into the bacterial chromosome, can cause polar mutations.

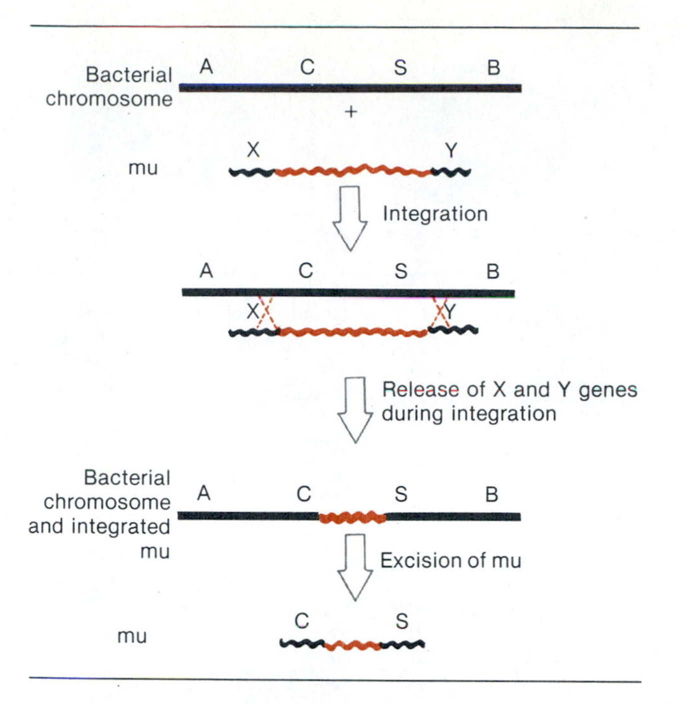

Figure 11-7 Integration and excision of bacteriophage Mu, demonstrating loss of Mu genes during integration process. Bacterial chromosome binds to Mu (color), which contains bacterial genes X and Y. Recombination that occurs between bacterial chromosome and sequences adjacent to X and Y genes on Mu molecule results in release of X and Y genes. Excision of Mu results in release of bacterial genes C and S. Mu now contains new set of terminal bacterial genes.

Illegitimate recombination

Recombination can take place at very low frequencies at random sites on the chromosome with no apparent homology between the nucleotide sequences and no apparent specific sites for recombination. This type of recombination is referred to as ***illegitimate.*** Duplications and deletions (other than those caused by transposons or insertion sequences) are caused by this type of recombination; however, little is known of the mechanisms involved.

GENE TRANSFER

Recombination in bacteria is facilitated by mechanisms that transfer DNA from various donor sources into a recipient cell. The fate of the transferred DNA depends on its capacity to be taken up by the host cell, its stability in the host cell, as well as its homology with the host chromosome. There are three mechanisms for gene

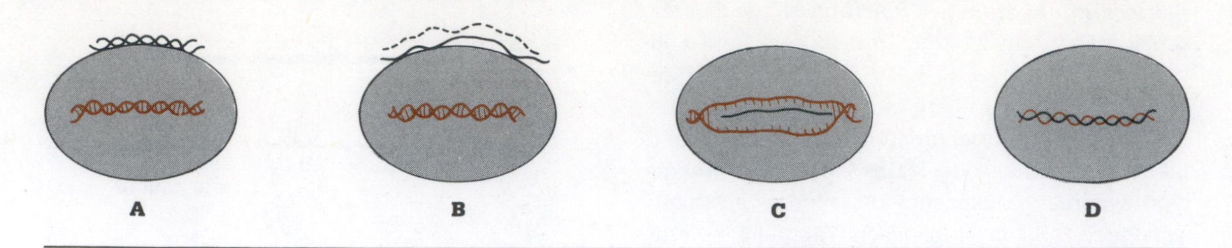

Figure 11-8 Transformation process. **A,** Binding of donor DNA to recipient cell. **B,** Endonuclease attack of donor DNA, releasing single-stranded DNA. **C,** Localized denaturation of recipient DNA and binding of homologous donor DNA strand. **D,** Endonuclease digestion of one strand of recipient DNA and integration of donor strand.

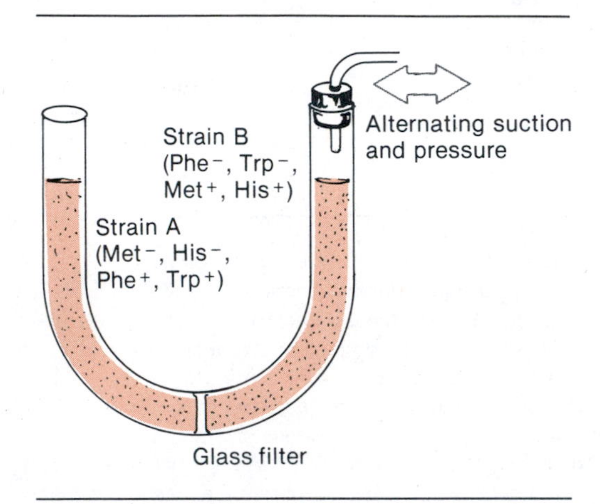

Figure 11-9 U tube used to demonstrate genetic transfer by virus of bacterial genes (transduction). Bacterial strain on left cannot synthesize methionine *(Met⁻)* or histidine *(His⁻)*, and bacterial strain on right cannot synthesize phenylalanine *(Phe⁻)* or tryptophan *(Trp⁻)*. Glass filter prevents bacterial transfer but not viral transfer. Alternating suction and pressure is used to transfer media between arms of U tube. See text for details of experiment.

transfer: transformation, transduction, and conjugation.

Transformation

Transformation is discussed earlier as a microbial mechanism that implicated DNA as the genetic material (see Chapter 9). In the following discussion the mechanics of the transformation process are explored, as well as its significance in recombination. Transformation is defined as a process in which free DNA is taken up by a cell, resulting in a genotypic change in the recipient. Not all cells can be transformed, but those that can are referred to as ***competent.*** The composition of the medium in which the transformation is to take place, as well as the phase of the bacterial cell cycle, influences the transformation process. Most transformation experiments have been performed with gram-positive species of *Streptococcus* and *Bacillus,* both of which show similarities in the process. Experiments with streptococci have shown that the entire culture becomes competent when the cell concentration is from 10^7 to 10^8 cells/ml. This response is due to the release of a small activating protein called the ***competence factor.*** The competence factor, which increases as the culture density increases, has been isolated and shown to induce competence in noncompetent cells. The mechanism of action of this protein is unknown.

Double-stranded DNA ranging in size from 3×10^5 to 1×10^7 daltons or greater is required for effective transformation. The donor DNA, which is added to a bacterial culture, binds to the surface of the cell, where an endonuclease cuts the molecules, releasing single-stranded (Figure 11-8) molecules of varying lengths. The DNA must be single stranded and, depending on the system, must be approximately 7 to 30 thousand bases in length before it can enter the cell cytoplasm. The transforming DNA, once inside the cell, enters an eclipse period before it becomes integrated and during this period is protected from DNase digestion. The frequency of integration (proportion of recipients that are transformed to the wild type) will depend in large part on the degree of homology between the single-stranded transforming DNA and its counterpart on the host DNA. Transformation can be used to map genes on the bacterial chromosome, but more appropriate techniques are used, and these are discussed later in the chapter. Transformation in nature is apparently a very infrequent process, and its contribution to evolution of the bacterial genome is small.

Transformation can be artificially induced in some bacteria, such as gram-negative *E. coli,* by treating the culture with calcium chloride. The artificially treated cell is capable not only of taking up chromosomal DNA but will also incorporate intact plasmid DNA or intact viral DNA. The process of viral nucleic acid uptake, which results in infection of the cell and the release of viral particles, is called ***transfection.*** (This is an abnormal infection process because of the absence of the viral protein coat.) Artificially induced transformation has important implications in recombination DNA research and is discussed later.

Transduction

Transduction is a process in which bacterial DNA is transferred from one cell to another with the help of a virus. It was discovered in 1952 by Zinder and Lederberg, whose experiments were designed to demonstrate that transfer of genetic material between strains of *Salmonella* does not require cell-to-cell contact (up until this time only conjugation and transformation were known as mechanisms of gene transfer). They used a U tube (Figure 11-9) in which a glass filter separated the two arms of the tube. The glass filter was porous, but the pores were too small for bacteria to pass through them. A minimal medium, which could not support the growth of biosynthetic mutants, was placed in the tube. One strain of a double amino acid auxotroph was placed in one arm of the tube, and another auxotroph requiring two different amino acids was placed in the other arm. Wild-type bacteria appeared on only one side of the glass filter, suggesting that a filterable agent* had passed from one arm into the other arm.

The explanation for this discovery was that one strain harbored a virus that could transfer bacterial genes into the other arm of the U tube. The process has been called transduction and is believed to occur in the following way. The strain on the right in Figure 11-9 carries a temperate virus, which integrates into the bacterial chromosome (the bacterium carrying the temperate virus is said to be ***lysogenic,*** and the integrated virus is called a prophage). Occasionally the temperate virus is excised from the bacterial chromosome and begins to replicate (vegetative virus), producing virulent viral particles that will lyse the cell. The released phage particles move across the glass filter and infect and lyse the bacterial cells on the left, which are not lysogenic. Infrequently, some of the phage particles incorporate bacterial DNA instead of viral DNA. The phage particles move across the glass filter and lysogenize the bacterial cells on the right. Wild-

* The agent was not naked DNA or RNA, because neither DNase nor RNase treatment interfered with the transduction. Thus transformation was eliminated as a possibility.

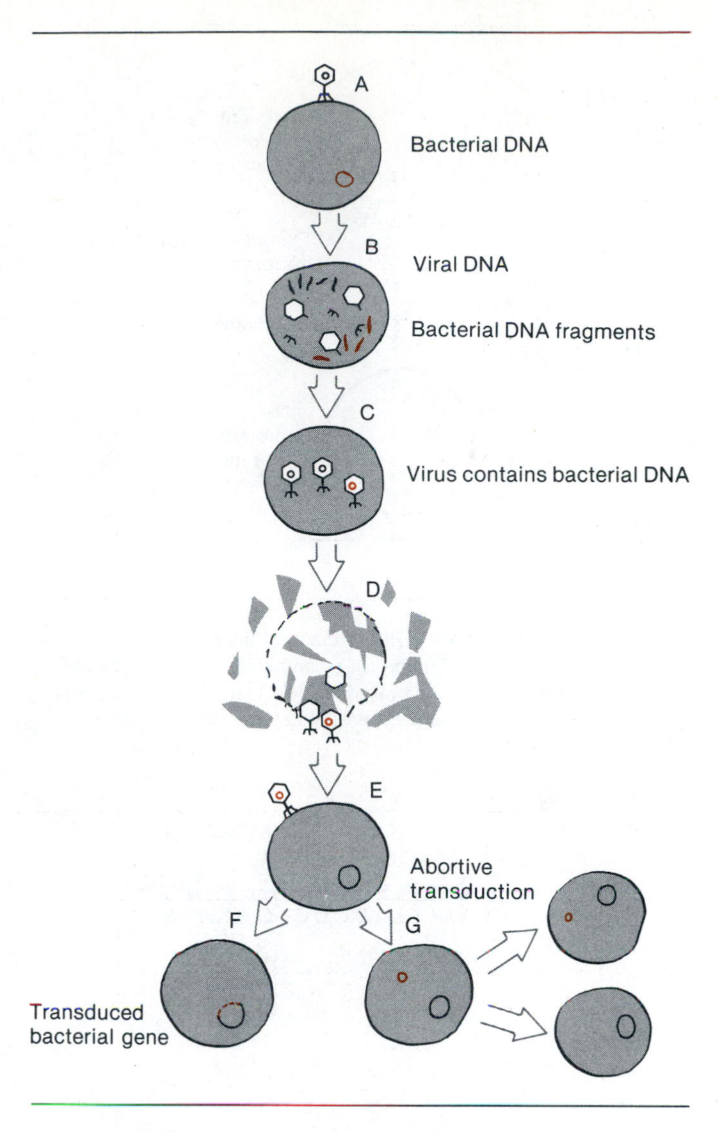

Figure 11-10 Sequence of events in generalized transduction. *A*, Bacteriophage infects bacterial cell. *B*, Viral DNA and protein are synthesized. Bacterial DNA is degraded. *C*, Some viral particles have bacterial DNA packaged in head. *D*, Bacterial cell lyses, releasing virus, including eventual transducing virus. *E* to *F*, Transducing virus infects new host, but no virus is produced, because it contains only bacterial DNA. *E* to *G*, Sometimes transduced fragment does not become incorporated into host genome, resulting in abortive transduction.

type bacterial recombinants appear on the right side if the virus lysogenizing them carries the necessary wild-type genes. Wild-type genes cannot appear on the left, because the cells are nonlysogenic. Lysogeny is discussed in more detail in Chapter 16.

There are two types of transduction processes: generalized and specialized.

Generalized transduction. The virus that engages in generalized transduction carries primar-

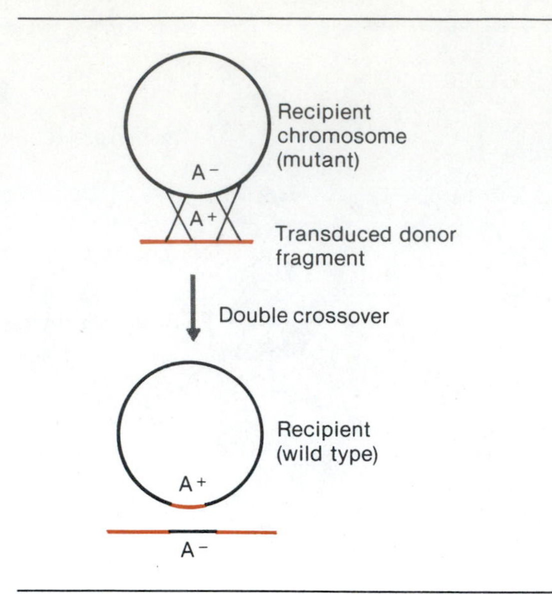

Figure 11-11 Recombination during generalized transduction. Reciprocal recombinational event (double crossover) results in exchange of homologous genes and formation of recipient cell containing wild-type gene.

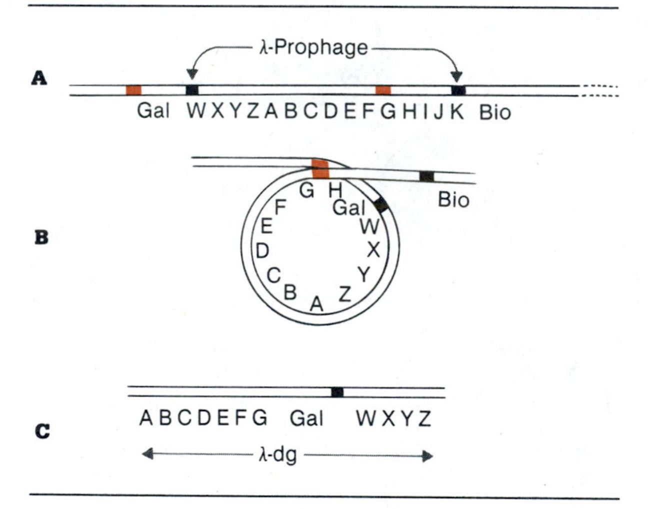

Figure 11-12 Formation of defective transducing phage (λ-dg). **A,** λ-Prophage as it appears during its integrated state within bacterial chromosome. Lambda is flanked by bacterial galactose *(Gal)* and biotin *(Bio)* genes. Colored blocks indicate areas of homology. **B,** Bacterial chromosome loops out, and homologous areas undergo crossover event. **C,** Deintegrated λ-dg contains bacterial galactose gene.

ily bacterial DNA. This was the type of transduction observed by Zinder and Lederberg in 1952. The process is carried out by preparing a high concentration (titer) of phage obtained by lytic infection of a bacterial strain (Figure 11-10). Some of the viral particles produced in the cell package bacterial DNA instead of viral DNA, and these represent the transducing particles. Once they are released from the bacterial cell, the transducing particles can infect susceptible bacterial cells but cannot lyse them, because they do not contain viral information. A recombinational event can occur between transduced and recipient DNA. The recombination (Figure 11-11) involves the incorporation of a double-stranded segment of transducing DNA in exchange for recipient bacterial genes (this differs from transformation in which a single-stranded DNA is incorporated into the bacterial chromosome). Abortive transduction can occur if there is no integration of transduced DNA into the recipient DNA. As the bacterial population multiplies, only one cell carries the transduced DNA (see Figure 11-10, *D*).

Specialized transduction. Specialized transduction involves the transfer of specific bacterial genes that are located near the viral integration site. A variety of temperate viruses can take part in specialized transduction, including the virus lambda, which integrates into the bacterial chromosome near the genes that control galactose and biotin metabolism. Lambda, which has been the most widely studied, rarely is excised from the bacterial chromosome but can be induced to deintegrate by ultraviolet (UV) light. Abnormal excision, involving crossover events between lambda and the bacterial chromosome, releases a transducing particle that carries the galactose or biotin genes (Figure 11-12). The excised transducing particle also lacks many viral genes because they are left behind during the excision and crossover events. The transducing particles are usually defective and cannot become lytic, but some are infectious. Defective lambda particles carrying the galactose or biotin genes are denoted λ-d Gal or λ-d Bio. Different temperate phages, because of their ability to integrate at different sites on the bacterial chromosome, can transduce specific bacterial genes and are used to map the bacterial chromosome.

Transduction and gene mapping. Transduction is a suitable method for determining the order of genes, as well as the relative distances between genes, when two or more genes are incorporated into the transducing viral particle. Phage P_1 (a transducing phage of *E. coli*) can be used in a typical experiment to transduce genes from *E. coli* to recipient strains of this species. The phage progeny carry various segments of bacterial DNA from the donor some of which carry the selected genetic markers. The phages are used to transduce the recipient bacterium, which is mutant for the selected markers. The order of transduced genes can best be determined by examining

Table 11-1 Cotransduction frequencies for markers in P_1 phage involving the cross Ara^+ Thr^+ Leu^+ with Ara^- Thr^- Leu^-

SELECTED MARKER	APPEARANCE OF UNSELECTED MARKER
Ara^+	75% = Leu^+
	0% = Thr^+
Thr^+ Leu^+	85% = Ara^+

three genes, two of which may or may not be cotransduced with a third gene (the frequency of cotransduction decreases as the distance between markers increases). For example, if we used the genetic markers for threonine synthesis (Thr^+), leucine synthesis (Leu^+), and the ability to ferment arabinose (Ara^+), the arrangement of genes on the chromosome could be determined (Table 11-1). Results of such an experiment might show that when arabinose utilization is selected in the transductants, 75% also carry the leucine gene, but the threonine gene does not appear. The arabinose gene is therefore linked more closely to the leucine gene than to the threonine gene. Consequently, the gene order could be Ara-Leu-Thr or Leu-Ara-Thr. When the threonine-leucine transductants are sought, 85% of the time the arabinose gene is also cotransduced, suggesting that the gene order is Leu-Ara-Thr. Wild-type leucine and threonine transductants would transduce very infrequently if the arabinose gene were between leucine and threonine.

More accurate measurements of the linkage order between two genes can be obtained by performing reciprocal crosses in which each parent is used alternatively as donor and recipient. Such crosses are set up by growing the transducing phage on each mutant strain of bacteria. Let us assume that we are examining three genes (A, B, and C) and we do not know if the gene order is CBA or BCA (Figure 11-13). Various crossovers between the mutant donor and recipient DNA molecules must occur in order to obtain the wild type in the recipient cell.

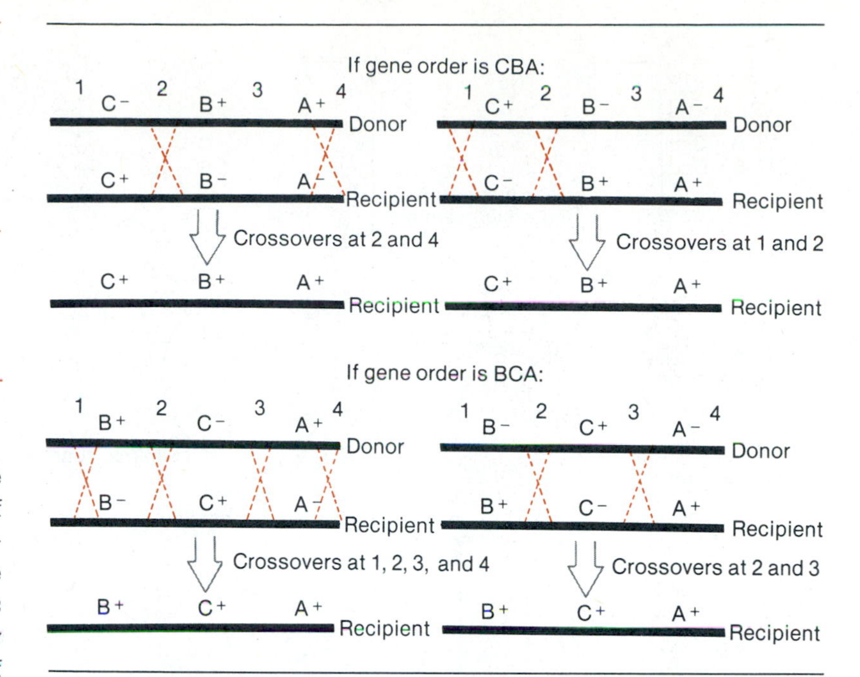

Figure 11-13 Determination of gene order through transduction experiments by using reciprocal crosses. Postulated gene order is CBA or BCA. Number of crossovers required to obtain prototrophs (C^+, B^+, A^+) for each postulated gene order is indicated. If gene order were BCA, four-crossover events would be required in one of reciprocal crosses, which is very infrequent. If gene order were CBA, ratio of recombinants in upper figure would be approximately one. Ratio of recombinants for reciprocal crosses in lower figure would be much less than one because of infrequency with which a quadruple crossover can occur.

Conjugation

Conjugation in bacteria is a mechanism for gene transfer that requires cell-to-cell contact. The original experiments used to demonstrate conjugation were designed by Lederberg and Tatum in 1947. Their discovery was based on the formation of prototrophs of *E. coli* from a mixture of cells that consisted of two kinds of auxotrophs. The auxotrophs can be represented as $A^+B^+C^+D^-E^-F^-$ and/or $A^-B^-C^-D^+E^+F^+$ (Figure 11-14). The auxotrophs were cultivated together overnight in a complete medium (supporting the growth of mutants). The cells were centrifuged, washed, and then plated on a minimal medium (supporting only the growth of prototrophs). One prototroph for every 1×10^7 cells plated appeared on the minimal agar. (No prototrophs appeared when each of the auxotrophs was plated on a minimal medium.*) A U tube similar to the one used in transduction experiments (Figure 11-9) was used to show that mutant strains placed on either side of the glass filter produced no prototrophs unless the filter was removed and cell contact took place. Lederberg and Tatum originally believed that the two bacterial cells entering the conjugation process played equal roles, but this was discredited by Hayes' discovery of the sex factor.

* The reason that there were no reversions is that the probability of a reversion to the wild type for a cell containing three mutations is the product of the individual probabilities. For example, if the reversion frequency for each of three mutations is 1×10^{-6}, then the reversion frequency for a cell containing three mutations will be 1×10^{-18} (1 cell in 1×10^{18} will be a prototroph).

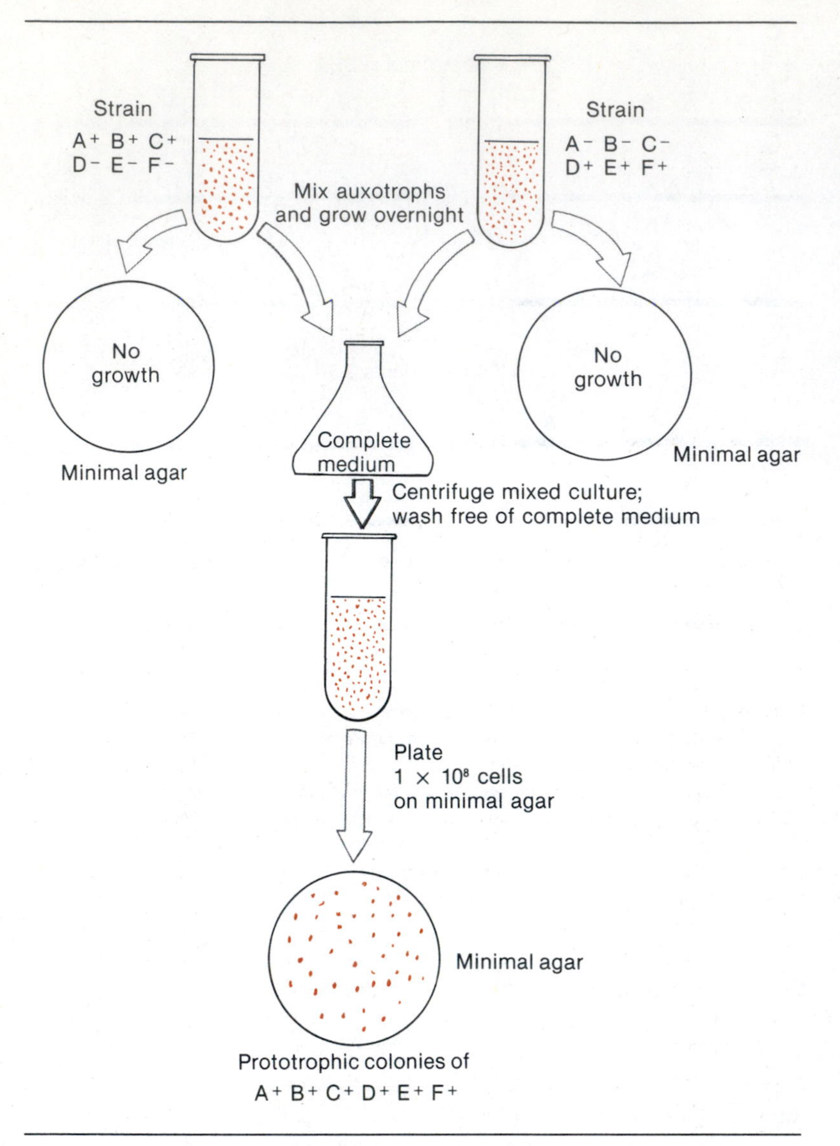

Figure 11-14 Lederberg and Tatum experiment demonstrating conjugation between two auxotrophic strains of *E. coli* K12.

Sex factor and conjugation. Hayes in the early 1950s discovered that treatment of one of the Lederberg and Tatum auxotrophs with streptomycin (strain A), before the conjugation process took place, had no effect on the number of prototrophs produced. Similar treatment of the other auxotrophic strain (strain B), however, prevented recombination. This experiment suggested that recombination takes place in only one direction and that one cell (strain A) acts as a donor while the other (strain B) acts as a recipient. In other words, in the experiment the donor cell could be inhibited in its growth and still enter into the recombination process, but the recipient cell, which was sensitive to streptomycin, could not grow and hence was unable to produce recombinant colonies on an agar surface. Further experiments involving strain A and strain B cells showed that some failed to produce recombinants. Crossing one of these strains with the corresponding partner of known fertility showed that strain A had become defective and was no longer acting as a donor. The defective donor could be made fertile if it were crossed with a normal donor strain. Hayes proposed that the donor strain contained an infectious particle, or ***sex factor***, that could be transferred during the conjugation process. Cells containing the sex factor could therefore donate genes to a recipient. Cells that possess the sex factor are called F^+, and those that do not contain the sex factor are called F^- This means that $F^+ \times F^-$ crosses are fertile, whereas $F^- \times F^-$ crosses are infertile.

Extraction of DNA from cells containing the sex factor reveals a satellite band of DNA in addition to chromosomal DNA. The F factor is a circular double-stranded DNA molecule whose molecular weight is approximately 5×10^7 daltons. It can exist independently of the host chromosome and is referred to as a ***plasmid.***

The F factor codes for about 22 genes, most of which are concerned with the conjugation process.* The most important function of these genes is associated with the formation of a conjugation bridge between donor and recipient. This bridge is called a sex pilus (Figure 7-31), and once it has been formed, the F factor, which is attached to a special membrane site in the cell, replicates independently of the host chromosome (Figure 11-15). A single strand of the F factor, beginning with the 5′ end, passes through the conjugation bridge into the recipient, where the complementary strand is synthesized (the untransferred donor strand acts as a template for synthesis of a complementary strand). A circular F factor is produced in the recipient, and the conversion of F^- to F^+ is complete). The F factor replication mechanism is patterned after the rolling circle model described in Chapter 9. The transfer process takes only a few minutes, so that in a large population of F^- cells and only a few F^+ cells the entire population can become F^+ after only a few cell divisions. Only one recipient in a million that has been converted to F^+ shows recombination of chromosomal genes.

Sex factor and the integrated state. Hayes and others discovered a strain of *E. coli* in which recombinants appeared 100 to 1000 times more frequently than the Lederberg and Tatum strains (recombination appeared once for every 10 to 20

* Approximately 9 genes of a 12-gene operon are associated with conjugation. Mutation in any of the 9 results in lack of pilus formation and inability to adsorb bacteriophage. Mutations in the other 3 genes of the operon do not affect pili production, but the donor capacity of the cell is hindered.

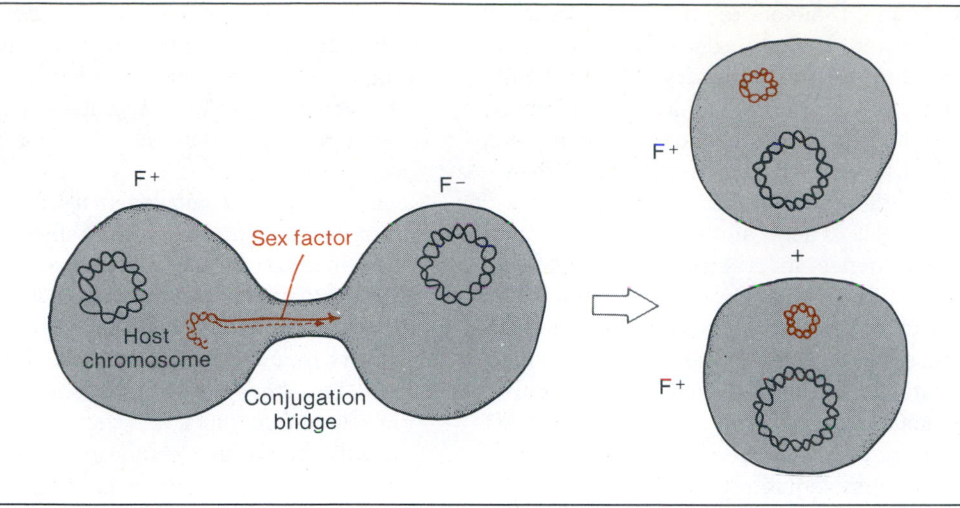

Figure 11-15 Conjugation between F^+ and F^- cells. Sex factor (color) replicates independently of host chromosome (black) and is transferred from F^+ cell to F^- cell through conjugation bridge.

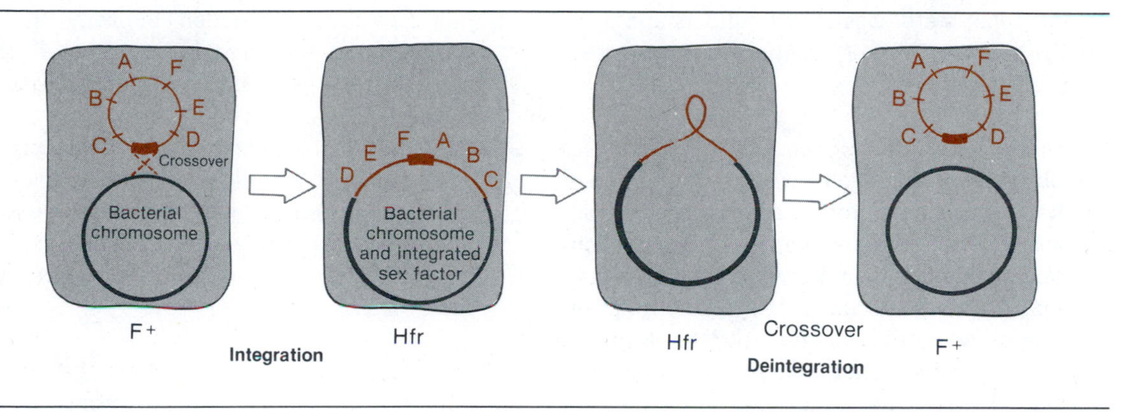

Figure 11-16 Integration and deintegration of sex factor. Single crossover between sex factor and bacterial chromosome brings about integration of sex factor and conversion of bacterial cell from F^+ to Hfr. Looping-out process in Hfr chromosome brings about deintegration of sex factor and conversion of cell from Hfr to F^+.

thousand cells examined). This "super male" strain was referred to as ***Hfr*** (high frequency of recombination). The sex factor in Hfr bacteria is integrated into the chromosome and behaves like an episome. The integration is a rare recombinational event between the F^+ DNA molecule and the bacterial chromosome (Figure 11-16). The recombination involves a single crossover and can take place at different locations on the chromosome depending on the strain (there is an apparent insertion sequence recognition between the sex factor and specific chromosomal sites). The F factor can also be deintegrated, converting the Hfr to the F^+ state. The mating process between an Hfr and an F^- is the same as in $F^+ \times F^-$ matings. The major difference is that chromosomal DNA is transferred from the donor to the recipient (Figure 11-17) in an Hfr $\times$ F^- mating. The transfer can be interrupted by breakage of

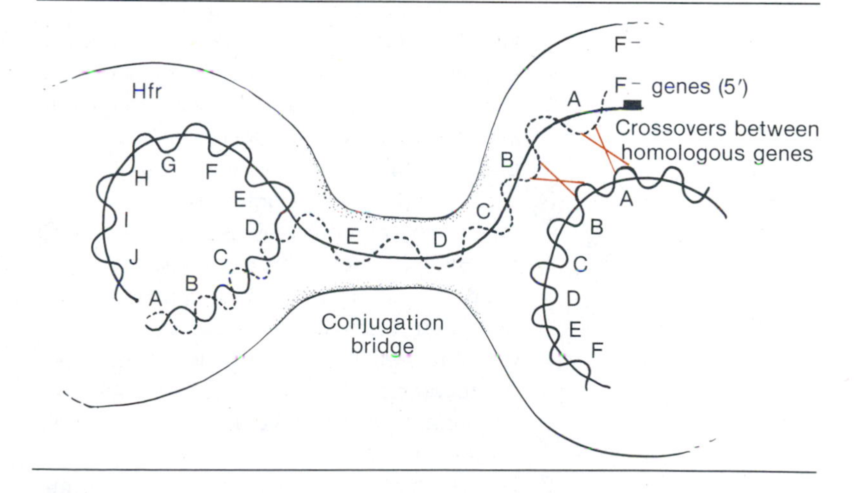

Figure 11-17 Transfer of genes from Hfr to F^- during conjugation process. Dashed lines represent newly synthesized DNA strands. 5′ Terminus of Hfr strand is first to enter F^- cell through conjugation bridge, and first genes to enter are sex factor genes. Complementary DNA is synthesized in Hfr and F^-. Recombination occurs between homologous genes in recipient (F^-) cell.

the chromosome; therefore the number of recombinants detected will be dependent on the amount of chromosome transferred. All bacteria of one Hfr strain transfer their chromosomal genes in a particular sequence, such as ABCDEFG. Different strains may transfer their genes in a different sequence, for example, CDEFG or FGHI. There are some strains that transfer their genes in reverse order, such as HGFEDC or ZYXW. The relationship of the various genes to one another, however, remains the same. Consequently, there is an origin and termination site on the bacterial chromosome during the conjugation process. The origin, as we have noted, may vary from one Hfr strain to another, but the first segment of genes attached to the origin, which enters the recipient, are some F factor genes that do not control pilus formation or the donor state. The last chromosomal segment to enter are also F factor genes that do control the donor state. It is easy to understand why few F^- cells in Hfr × F^- matings are converted to the Hfr state.

Mapping of the chromosome by conjugation. The observation that each Hfr strain tested transfers its chromosomes from a specific site offered an opportunity for gene mapping. Conjugation mapping is different from transformation or transduction mapping because the relationship of genes is based on time of entry into the recipient and not on recombination frequencies. Jacob and Wollman devised the interrupted mating technique as a method for mapping genes by conjugation. Their experiment was set up in the following way:

1. The genetic cross involved a prototrophic Hfr strain and an F^- strain that was auxotrophic for the biosynthesis of threonine (Thr^-) and leucine (Leu^-), was sensitive to sodium azide (Azi^s), was sensitive to T_1 phage (Ton^s), and was unable to ferment galactose (Gal^-) and lactose (Lac^-). In order to ensure that only F^- recombinants and not Hfr cells would be recovered after conjugation, the Hrf strain selected was sensitive to streptomycin, whereas the F^- strain was resistant to the antibiotic. Thus when the recombinants were plated on a medium containing streptomycin, only F^- cells would survive and produce colonies.
2. The Hfr and F^- cells were mixed in a nutrient medium for about 5 minutes to promote conjugation and then were diluted to prevent the formation of new mating pairs. Samples of the mating pairs, usually taken at 1-minute intervals, were agitated in a blender in order to break the conjugation bridge and interrupt chromosome transfer. A large number of recipient cells therefore contain various lengths of donor chromosome.
3. The experiment showed that the threonine and leucine genes were transferred first. The medium on which these recombinants were produced did not contain threonine or leucine but did contain all the other nutrients required for growth, plus streptomycin. The other genetic markers represented unselected markers, and their distance from the threonine and leucine genes could also be determined. The results of these experiments are illustrated in Figure 11-18. The marker closest to the threonine and leucine genes was the gene coding for resistance or sensitivity to sodium azide and appears 8 minutes after conjugation. Similarly, the T_1 phage gene appeared at 10 minutes, the lactose gene at 17 minutes, and the galactose gene at 23 minutes. Selection of these markers was obtained by plating recombinants on different selective media (Table 11-2). A small genetic map of *E. coli* was constructed from the results of these experiments which read: origin-Thr-Leu-Azi-Ton-Lac-Gal.

The more distant the gene is from the origin, the more difficult it is to obtain recombinants because of chromosome breakage. This problem was rectified by using different Hfr strains whose origins are all different. The results of these experiments has permitted a rather detailed genetic map of *E. coli* and some other species. The *E. coli* map is divided into 90 one-minute divisions (90 minutes is the time it takes for transfer of the entire chromosome when the conjugation process takes place at 37° C), with the threonine gene indicating the approximate site of the origin. A modified version of this map is presented in Figure 11-19.

Conjugation in gram-positive bacteria. Only a few species of gram-positive bacteria are known to be capable of conjugation. Most of the studies to date have involved species of *Streptococcus.* Conjugation and chromosomal transfer apparently do occur in gram-positive bacteria, but not at a very high frequency. Plasmid transfer by conjugation, however, does occur at a high frequency, and the mechanism used is unique for bacteria. *Streptococcus faecalis* can transfer some of its plasmids to recipient cells at a very

Table 11-2 Medium used to select the genetic markers in a cross involving prototrophic Hfr and auxotrophic F^-

MARKER*	MEDIUM CONTAINS	MEDIUM LACKS
Thr^+	Leucine	Threonine
Leu^+	—	Leucine and threonine
azi^R	Sodium azide	Leucine and threonine
T_1^R	Sodium azide and T_1 phage	Leucine and threonine
Gal^+	Galactose and sodium azide	Leucine and threonine
Lac^+	Lactose and sodium azide	Leucine and threonine

*The genetic markers sought are the ability to synthesize threonine (Thr^+) and leucine (Leu^+), resistance to sodium azide (Azi^R) and T_1 phage infection (Ton^R), and the ability to ferment galactose (Gal^+) and lactose (Lac^+).

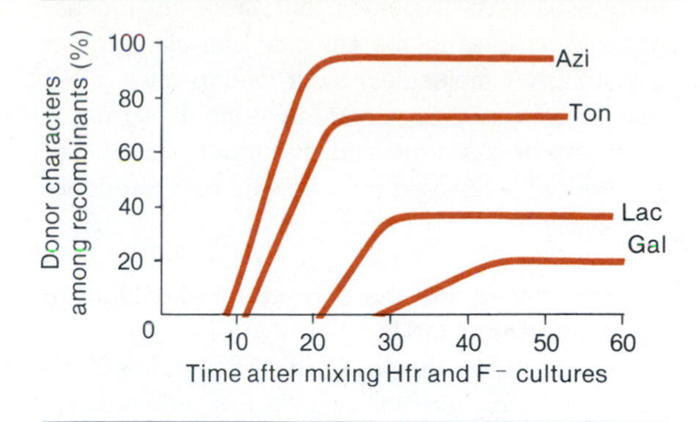

Figure 11-18 Kinetics of inheritance of unselected markers after conjugation between prototrophic Hfr strain (Thr^+, Leu^+, Azi^+, Ton^+, Lac^+, and Gal^+) and F^- cell (Thr^-, Leu^-, Azi^-, Ton^-, Lac^-, and Gal^-). Conjugation is interrupted by agitation in blender. Selected markers among recombinants are Thr^+ and Leu^+, which appear 8 to 8½ minutes after conjugation. Appearance of unselected markers is indicated in graph.

Redrawn from Jacob, F., and Wollman, E.L. Comptes Rendus Acad. Sci. ***240:****2450, 1955.)*

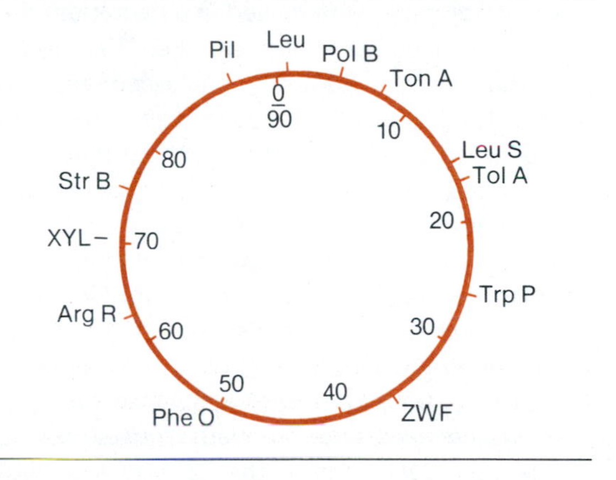

Figure 11-19 Circular map of *E. coli* showing only a few of the genetic markers obtained from conjugation experiments involving several Hfr strains. *Leu,* Leucine; *Pol B,* DNA polymerase II; *Ton A,* T_1 bacteriophage; *Leu S,* leucyl tRNA synthetase; *Tol A,* tolerance to colicins; *Trp P,* tryptophan permease; *ZWF,* glucose 6-phosphate dehydrogenase; *Phe O,* operator locus for phenylalanine; *Arg R,* arginine regulatory gene; *XYL,* use of D-xylose; *Str B,* streptomycin resistance; *Pil,* pili or fimbriae gene.

Redrawn from Taylor, A.L., and Trotter, C.L. Microbiol. Rev. ***36:****504, 1972).*

high frequency (as great as 10^{-1} per donor), but others are transferred poorly (less than 10^{-6} per donor). In those systems in which there is efficient plasmid transfer, the recipient cells release small peptides, or ***sex pheromones.*** The use of sex pheromones by higher organisms (fungi and higher eukaryotes) had been recognized earlier, but they had not previously been found in bacteria. When these peptides make contact with the donor cell, the latter synthesizes surface proteins that facilitate the formation of aggregates of donor-recipient cells. The pheromones also induce the transfer of the plasmid to the recipient cell. The mechanism of plasmid transfer is not known.

Recombinant DNA and restriction endonucleases (genetic engineering)

Our discussion of recombination to this point has involved the rearrangement or reciprocal exchange of genetic material from the same or related species of bacteria. Recently, experimental procedures have been developed that permit scientists not only to fuse genes but also to insert genes from any eukaryotic or prokaryotic source into a bacterial replicon using phage or plasmid DNA. These developments have led to new research interests referred to as ***recombinant DNA*** technology or genetic engineering. One of the major discoveries that permitted this development was the isolation of enzymes called ***restriction endonucleases.***

Restriction endonucleases catalyze the cleavage of specific nucleotide base pairs in the DNA molecule. They produce double-stranded breaks on the phosphodiester bonds at approximately every 1000 or more nucleotide pairs. Bacteria can protect themselves from invasion by viruses or other foreign DNA molecules by using restriction endonucleases. The ability of the endonucleases to cleave the DNA molecule at specific sites leaves the foreign DNA with no functional or structural integrity. The fact that DNA of bacterial cells is left intact during growth and that many viruses are still capable of successfully infecting bacteria indicates that there are mechanisms operating in the cell that can also protect certain DNA molecules, even foreign ones. DNA can be protected from endonuclease digestion by methylation of purine and pyrimidine bases that are normally selected for cleavage by the endonuclease.

Requirements for the formation and cloning of recombinant DNA

Source of the foreign gene. The source of the gene to be cloned will dictate the methodology used in its isolation. Genes in eukaryotic organisms are often quite complex because they contain base sequences, called intervening ***(introns)*** sequences, that are not involved in determining the amino acid sequence of the protein encoded by the gene. When the gene is transcribed, the resultant RNA molecule is processed by special nucleases that delete specific portions of the RNA molecule so that the final mRNA is shorter than the original gene from which it was transcribed. If the entire gene were cloned in a bacterium that lacked the special nucleases, then the mRNA from that gene would not be processed, and consequently the protein encoded by the unprocessed mRNA would be different from that found in the eukaryotic cell and probably would not be functional. It is therefore, best to start with the mRNA for the eukaryotic protein of interest. A copy of DNA can be made from the mRNA by the enzyme reverse transcriptase, which is isolated from avian myeloblastosis virus. The DNA of the hybrid RNA-DNA produced by reverse transcriptase is later used as a template to make a complementary DNA strand (the RNA of the RNA-DNA hybrid can be removed by RNase). The double-stranded DNA product contains the exact eukaryotic coding information; hence mRNA transcribed from it will contain only protein-encoded information.

Bacterial genes do not contain introns and can be cloned directly without the necessity of first isolating mRNA and making DNA from it.

Need for a vector. A gene or specific genetic sequence that is to be transferred to another cell must first be inserted into a molecule (vector) that will permit its transfer. The vector must be capable of entering the host cell, and it must be able to replicate. Two vectors that are small enough for this task are plasmids and bacteriophages. The plasmids used at the present time are from gram-negative bacterial species. They have sex factor genes, or at the very least they can be mobilized by other transfer factors. The bacteriophage used most frequently is phage lambda. Plasmids have a narrow or wide bacterial

Endonuclease	Cleavage sites
Bam III	5′ G↓G A T C G 3′ 3′ C C T A G↑G 5′
Eco RI	5′ G↓A A T T C 3′ 3′ C T T A A↑G 5′
Hind III	5′ A↓A G C T T 3′ 3′ T T C G A↑A 5′
Hae III	5′ G G↓C C 3′ 3′ C C↑G G 5′
Sma I	5′ C C C↓G G G 3′ 3′ G G G↑C C C 5′

Figure 11-20 Recognition sites *(arrows)* for some restriction endonucleases. *Bam, Bacillus amyloliquefaciens; Eco R1, E. coli* RY13; *Hind, Haemophilus influenzae, Hae, Haemophilus aegyptius; Sma, Serratia marcescens.*

host range, depending on the plasmid used, but lambda is limited to a few strains of *E. coli.* Vectors for eukaryotic cells have also been developed, such as a hybrid bacterial-yeast plasmid that can transfer its DNA to either bacteria or yeast. The DNA from some tumor viruses, such as SV40, can also transduce recombinant DNA to a number of different eukaryotic cell types. Vectors should also have at least one site where the integration of the foreign DNA will not inhibit a function such as the normal replication cycle of the vector.

Method for inserting genes into the vector. Restriction endonucleases and other enzymes provide the means for isolation as well as insertion of genes into the vector. ***Restriction endonucleases*** cleave specific sites on the DNA that are present on both strands of the DNA molecule. Some of these specificities are illustrated in Figure 11-20. Most of the sequences recognized by endonucleases are palindromic (inverted repeat sequences) and contain 3 to 6 base pairs. Sometimes the endonuclease cuts are made at some distance from the recognition site and are not palindromic, but others are made at the recognition site. Staggered cuts in the two DNA strands made by endonucleases will produce fragments that have identical base pair sequences when they are read from the 5′ ends of the strands. For example, the Eco R1 restriction endonuclease (Figure 11-20) produces ends that are:

```
       3′           5′
 —G—            —AATT C
  |        5′         |
  C TTAA—             G—3′
```

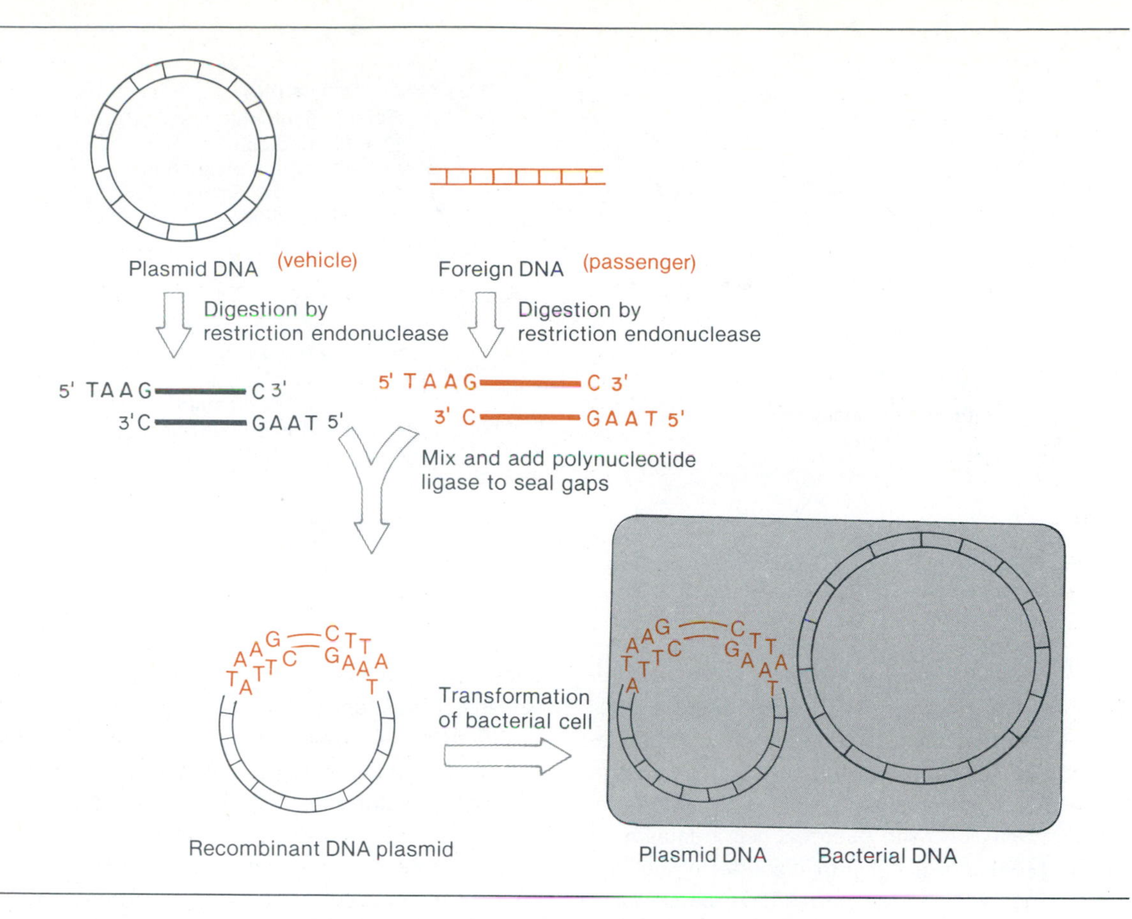

Figure 11-21 Recombinant DNA experiment. Endonuclease digestion of plasmid DNA and foreign DNA produces fragments with complementary "sticky" ends. Foreign gene with sticky ends complementary to plasmid DNA will be incorporated into plasmid. Gaps in plasmid are sealed by using polynucleotide ligase, and circular molecule is produced. Recombinant DNA plasmid is then used to transform bacterial cell. Transformed cell can be cloned, and then foreign DNA information can be translated.

The complementary ends of the fragment are called ***cohesive*** or ***sticky ends*** because they can hydrogen bond with any other fragmented ends produced by the same restriction endonuclease.

Restriction endonucleases can be used to cleave DNA from eukaryotic as well as prokaryotic sources. In other words, we could hypothetically take a piece of DNA from a pig, insert it into a plasmid that serves as the vector, and then transfer the foreign DNA vector into a bacterial cell. The experiment (Figure 11-21) would proceed in the following way. The foreign DNA and plasmid DNA are treated with the same restriction endonuclease and are then mixed together. Some of the foreign DNA will hydrogen bond with the plasmid DNA. The enzyme DNA ligase is used to zipper up (produce covalent bonds between the phosphate and sugar groups) the plasmid and its foreign DNA, thus producing a recombinant DNA molecule. The plasmid, because it is derived from more than one source, is called a ***chimeric plasmid*** (from the mythological creature Chimera, which was part goat, lion, and serpent). At this point in the experiment we have produced a number of plasmids containing different foreign genes.

There is a problem when double-stranded DNA to be cloned is synthesized from mRNA. This DNA possesses ends that are not complementary to those of the cloning vector and cannot be incorporated as such into the vector. This problem has been circumvented by the chemical synthesis of short pieces of double-stranded DNA that contain the cleavage site for a restriction endonuclease. These are called ***synthetic linkers,*** and they can be chemically attached to each end of the piece of DNA that is to be cloned. Treatment of DNA that has synthetic linkers with specific endonuclease generates cohesive ends as the enzyme cuts at its cleavage site contained in the linkers.

Once the chimeric plasmid has been constructed, the trick is to put it into a viable host so that this new information can be expressed. It is at this point that our next requirement must be discussed.

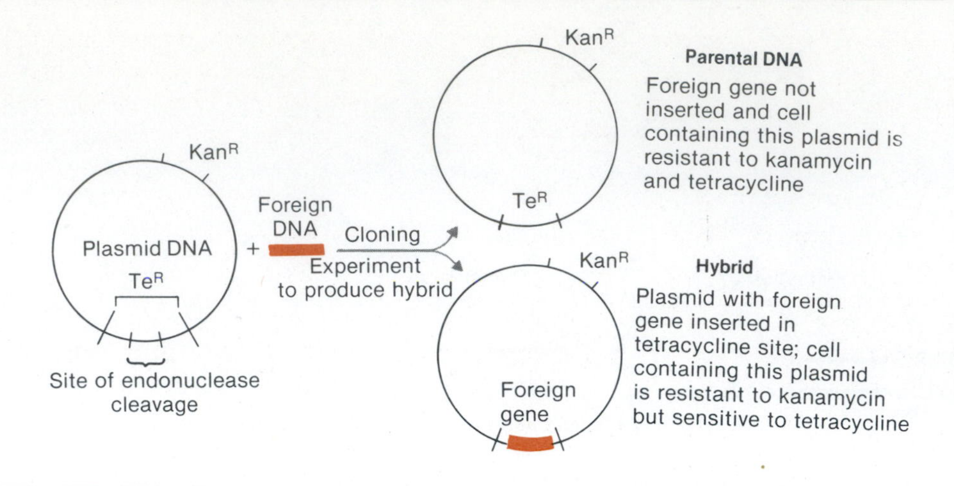

Figure 11-22 Technique for selection of recombinant gene using antibiotic-resistant markers. *Kan*^{R}, Kanamycin resistance; *Te*^{R}, tetracycline resistance.

Insertion of vector foreign recombinant DNA into the host cell. The chimeric plasmids can be inserted into bacteria (although other recipients could theoretically be used) in a transformation process (Figure 11-21). The uptake of plasmid DNA into the bacterial cell is facilitated by the presence of calcium chloride in the medium. Once the foreign genes have been transferred to the host bacterial cell, they are plated and cloned on an agar medium. We now have a large number of bacterial cells containing recombinant DNA plasmids, and this presents us with a problem: selection of the specific recombinant.

Selection of recombinants. The population of cells containing recombinant DNA will be quite numerous, and some of these cells will have nonrecombinant plasmids that circularized spontaneously following endonuclease digestion. Others will be recombinants that the experimenter will not want. How does one select a clone in which cells carry the specific gene that the experimenter wishes to study. First, one must select the recombinants. One technique is to initially use a plasmid that contains two drug-resistant markers, one of which has a specific restriction endonuclease site located within it. Three types of cells will be produced following the formation of recombinant DNA and the transformation process if we use the antibiotic-resistant markers: (1) those cells in which no plasmids appear and that are sensitive to both antibiotics; (2) those cells in which the foreign gene is inserted into the site originally occupied by the antibiotic-resistant marker (these recombinant cells will be sensitive to one of the antibiotics [Figure 11-22] and resistant to the other); and (3) those cells containing the plasmid but no recombinant DNA (these are resistant to both antibiotics).

Selection of cloned genes. The use of several restriction endonucleases can cut up an entire genome, resulting in the formation of a large number of recombinant DNA clones. If the gene being sought codes for an enzyme, colonies can be assayed for its presence, but it is more difficult if the gene codes for a nonenzymatic protein. One technique uses immunological principles in which antibodies to the protein are used to detect the presence of the protein in cloned cells. A second technique employs a radioactive probe. The experiment is carried out by replica plating all the clones obtained from the agar plate onto a nitrocellulose filter. The cells are exposed to a chemical that lyses the cells and denatures DNA. The denatured DNA remains attached to the filter. The filter is placed in a solution containing a radioactive probe that consists of either ^{32}P-labeled mRNA specific for the gene or a complementary DNA, either of which will hybridize with the denatured DNA. The filter is washed to remove unhybridized material, and a photographic emulsion is placed over the filter. Complementary fragments are detected by autoradiography.

Significance of recombinant DNA research

Recombinant DNA research has many applications, the most important of which is genetic analysis. Eukaryotic and prokaryotic genomes alike have been analyzed to determine the function of certain genes as well as to map entire genomes such as the SV40 virus and bacteriophage $\phi\chi$174. One of the fundamental advantages of examining eukaryotic genes in a prokaryotic environment is that eukaryotic genes are free of their eukaryotic regulatory elements. Thus gene function is more easily ascertained.

Recombinant DNA research also has great potential for mankind. It has been used for the following purposes, some of which are referred to in later chapters.

1. Formation of a "superbug" that can metabolize some of the major hydrocarbons of petroleum. This organism carries the fragments of four plasmids that code for enzymes that digest four different hydrocarbons. It is hoped that this organism, or one similar to it, can be used to clean oil spills. This is the first organism for which a patent has been upheld by the Supreme Court of the United States.
2. Formation of proteins not normally synthesized by microorganisms. Mammalian genes coding for insulin, somatostatin (a hypothalmic hormone that controls the release of several hormones from the pituitary), and human growth hormone has been inserted as recombinant plasmids in *E. coli.* Interferon is also a protein, produced normally by the body in response to viral infections, that has been shown to restrict the growth of certain tumors. Interferon is produced in very small amounts in cell cultures but is produced in greater quantities in bacterial cells carrying the gene for interferon synthesis.
3. Formation of an organism that fixes nitrogen more efficiently than the normal cell. Some bacteria, such as species of *Rhizobium,* live symbiotically in the roots of leguminous plants. The bacteria supply the plant with fixed nitrogen, necessary for photosynthesis, while the plant supplies the bacteria with nutrients. The normal nitrogen fixation process wastes energy by releasing hydrogen gas. Insertion of a gene coding for the enzyme hydrogenase into species of *Rhizobium* results in the breakdown of hydrogen into free protons (H^+) and electrons (e^-) that drive the reactions of photosynthesis and conserve energy.
4. Technique for producing site-specific mutagenesis. Ordinarily mutations occur randomly in the bacterial genome or within a particular gene, and it is difficult to produce specific mutations in the laboratory. Isolation of genes from clones offers an opportunity to produce point mutations (mutations involving a single base pair). The isolated gene can be treated in the laboratory with specific agents and then reinserted into the host cell and exchanged for the wild-type gene. This type of mutagenesis has tremendous implications in industrial microbiology, where an increase in production of special compounds can be controlled by genetic means other than random selection of mutants.

The possible benefits of recombinant DNA research are innumerable, but it is possible to produce organisms with less than desirable attributes. Resistance to all known antibiotics and toxin production in normally nonpathogenic microorganisms that are indigenous to humans are just two examples of potential products of this research. Considerable controversy surrounded recombinant DNA research, and strict guidelines were initiated but have subsequently been relaxed. The strains used in such experiments, for example, *E. coli,* have been selected with many mutational defects that ensure their own destruction outside the laboratory. *E. coli* K12, the strain being currently used, has several mutations that prevent it from being a biohazard: (1) it cannot synthesize the peptidoglycan of its cell wall; (2) it cannot synthesize colanic acid, which, if produced, could permit *E. coli* to survive without a peptidoglycan; (3) it is sensitive to bile salts and detergents; (4) it degrades its own DNA in environments devoid of thymine or thymidine; (5) it is resistant to a number of *E. coli*–transducing phages; (6) it is unable to act as a recipient in matings with donors possessing conjugative plasmids commonly found in enteric microorganisms; (7) it is very sensitive to UV irradiation; and (8) it is defective as a recipient in matings with donors possessing other conjugative types.

Even though the fears of using *E. coli* in recombinant DNA research are no longer prevalent, genetic engineers are searching for other organisms to manipulate. *Bacillus subtilis,* for example, a free-living soil bacterium, is considered to be a safer organism. This organism, unlike *E. coli,* ordinarily secretes its products, has no surface toxins, and is not infectious. Considerable research needs to be performed, elucidating the genetic control mechanisms of *B. subtilis,* before it can be substituted for *E. coli.* Yeast cells are also being actively explored for recombinant DNA research. They have the advantage of being 50 times larger than bacteria, easy to grow, and easy to separate from the medium in which they are cultivated. They also have a relatively thick cell wall that is resistant to relatively high concentrations of byproducts such as alcohol. Yeast is being actively pursued for the enzymatic conversion of cellulose to alcohol (see Chapter 21).

SUMMARY

Nucleotide sequences in microbial systems can be reassorted, and this often (but not always) involves an exchange or recombinational event between two DNA molecules. Recombination can take place by one of three methods: general, site specific, and illegitimate. The differences between them are related primarily to the homology (relatedness) of the interacting DNA molecules.

General recombination is the most common type of recombination and involves considerable relatedness between two DNA molecules. In site-specific recombination there is homology at sites on the interacting DNA molecules. Site-specific recombination has received the most attention because it may involve insertion sequences. Insertion sequences, which are specific nucleotide sequences without genetic information, are attached to genes and are transposable. Other transposable genetic elements include the bacteriophage Mu and transposons. Some genes are transferred or inserted into other strands of DNA, where homology exists primarily between the insertion sequences attached to the genes. It is known that insertion sequence (IS) elements exist as multiple copies on a single DNA molecule and that recombination can take place within a single DNA molecule. Illegitimate recombination involves no apparent homology between DNA molecules, and little is known of this type of event.

Recombination is an event that can occur within a single DNA but is usually between DNA molecules from different cells. Recombination can occur when genetic information is transferred from one cell to another in one of three processes: transformation, transduction, or conjugation. In transformation naked DNA (no viable cell is required) is taken up by a cell and part of the DNA is incorporated into the recipient DNA. In transduction viral DNA is used to transfer bacterial DNA from one cell to another, resulting in an exchange of bacterial DNAs in the recipient cell. Sometimes the DNA transferred is mostly viral (specialized transduction), but usually it is bacterial) (generalized transduction). A cell-to-cell interaction occurs in conjugation with the transfer of DNA from a male donor (F^+) to a female recipient (F^-). The sex factor may be the only information transferred if it exists as a plasmid outside the chromosomal DNA. If the sex factor is part of the chromosome (an episome), all or part of the chromosomal DNA can be transferred to engage in recombination.

Several discoveries have led to the new science of genetic engineering in which genes from a variety of sources can be cloned in a host cell (bacterium). The genes are carried on vectors such as bacteriophages and are removed from their sources by restriction enzymes. The restriction enzymes are also used to insert the genes into the vector. The vector and its new genes are then taken up by a bacterial cell in the process of transformation. The transferred genes within the new host act like plasmids and can be expressed and transferred to other bacterial progeny.

Study questions

1. In what specific way does general recombination differ from site-specific and illegitimate recombination?
2. How has the discovery of insertion sequences changed our thinking about recombination between DNA molecules?
3. In transformation what is meant by competence?
4. How can transformation be induced in cells that normally are incapable of being transformed?
5. How can one clone a gene? Has cloning offered any oppotunities to mankind? Explain. Any disadvantages?
6. During the transduction process why doesn't the transduced cell lyse?
7. How do generalized and specialized transduction differ?
8. On what types of DNA can the sex factor (F factor) be found? Why doesn't it appear in viral DNA?
9. How can one distinguish between a cell carrying a free F factor and one carrying an integrated one?
10. Of what value is the possession of restriction endonucleases to bacterial cells? Of what value has it been to microbiologists? To mankind in general?
11. In genetic engineering experiments what are vectors? Name the ones being currently used.
12. What is the composition of a chimeric plasmid?
13. Define the following: prophage, palindromic, transposon, transduction, lambda, and genetic engineering.

Selected readings

Books

Archer, L.J. *Bacterial transformation.* Academic Press, Inc., New York, 1973.

Bainbridge, B.W. *The genetic of microbes.* Blackie & Son, Ltd., Glasgow, Scotland, 1980.

Birge, E.A. *Bacterial and bacteriophage genetics.* Springer-Verlag, New York, Inc., New York, 1981.

Bukhari, A.I., and Hicks, J.B., (Eds.). *Movable genetic elements.* Cold Spring Harbor Laboratory, Cold Spring Harbor, N.Y., 1981.

Herskowitz, I.H. *The elements of genetics.* Macmillan Publishing Co., Inc., New York, 1979.

Hexter, W., and Yost, H.T. *The science of genetics.* Prentice-Hall, Inc., Englewoods Cliffs, N.J., 1976.

Russell, P.J. *Lecture notes on genetics.* Blackwell Scientific Publications, Ltd., Oxford, 1980.

Stanier, R.Y., Adelberg, E.A., and Ingraham, J. *The microbial world* (4th ed.). Prentice-Hall, Inc., Englewood Cliffs, N.J., 1976.

Watson, J. *The molecular biology of the gene* (3rd ed.). Benjamin/Cummings Publishing Co., Inc., Menlo Park, Calif., 1976.

Journal articles

Bachman, B.J., Low, K.B., and Taylor, A.L. Recalibrated linkage map of *Escherichia coli* K-12. *Microbiol. Rev.* **40:**116, 1976.

Bukhari, A.I. Bacteriophage Mu as a transposition element. *Annu. Rev. Genet.* **10:**389, 1976.

Calos, M., and Miller, J.H. Transposable elements. *Cell* **20:**579, 1980.

Cohen, S.N. The manipulation of genes. *Sci. Am.* **235:**25, 1975.

Cohen S.N., and Shapiro, J.A. Transposable genetic elements. *Sci. Am.* **242:**40, 1980.

Gilbert, W., and Villa-Komaroff, L. Useful proteins from recombinant bacteria. *Sci. Am.* **242**(4):74, 1980.

Grobstein, C. The recombinant DNA debate. *Sci. Am.* **237**(1):22, 1977.

Malik, V.S. Recombinant DNA technology. *Adv. Appl. Microbiol.* **27:**1, 1980.

Morse, M.L., Lederberg, E.M., and Lederberg, J. Transduction in *Escherichia coli* K-12. *Genetics* **41:**142, 1956.

Nash, H.A. Integration and excision of bacteriophage lambda. *Curr. Top. Microbiol. Immunol.* **78:**171, 1977.

Nathans, D., and Smith, H.O. Restriction endonucleases in the analysis and restructuring of DNA molecules. *Annu. Rev. Biochem.* **44:**273, 1975.

Nevers, P., and Saedler, H. Transposable genetic elements as agents of gene instability and chromosomal rearrangements. *Nature* **268:**109, 1977.

Shapiro, J.A. Molecular model for the transposition and replication of bacteriophage Mu and other transposable elements. *Proc. Natl. Acad. Sci.* USA **76:**1933, 1979.

Smith, H.O., and Danner, D.B., Genetic transformation. *Annu. Rev. Biochem.* **50:**41, 1981.

Starlinger, P. IS elements and transposons. *Plasmids* **3:**241, 1980.

Starlinger, P., and Saedler, H. IS-elements in microorganisms. *Curr. Top. Microbiol. Immunol.* **75:**111, 1976.

Tartof, K.D. Redundant genes. *Annu. Rev. Genet.* **9:**355, 1975.

Wetzel, R. Applications of recombinant DNA technology. *Am. Sci.* **68:**644, 1980.

Zinder, N.D., and Lederberg, J.L. Genetic exchange in *Salmonella. J. Bacteriol.* **64:**679, 1952.

Chapter 12 EUKARYOTIC GENETICS

One of the most important differences between eukaryotes and prokaryotes, as mentioned earlier, is the presence of a nuclear membrane in eukaryotes and its absence in prokaryotes. DNA synthesis and transcription, as well as messenger RNA (mRNA) translation, are processes common to both groups, but the presence of a nuclear membrane has enabled eukaryotes to modify these processes. These modifications coincidentally also occur for viruses that replicate in the nucleus of infected eukaryotic cells. Differences in the features of genetic organization, transcription, and translation, which serve to distinguish eukaryotes from prokaryotes, are outlined in Table 12-1. These features are discussed in this chapter, although some of them are alluded to in earlier chapters. An additional distinction between eukaryotes and prokaryotes is the absence of a sexual process in prokaryotes and its presence in eukaryotes. Mutation, in asexual organisms as well as in viruses, is the only major mechanism for obtaining variability within the population. Meiosis in sexual organisms leads to the segregation of ***alleles*** (one or more forms of a gene) and along with mutation enhances the probability of variation in the cell. This chapter includes a discussion of this meiotic process, as well as a demonstration of how changes can be made in the genotype. The student should first review those chapters that deal with prokaryotic DNA and RNA synthesis in order to understand eukaryotic modifications.

DNA ORGANIZATION AND STRUCTURE

Most eukaryotic DNA is found as chromosomes in the nucleus, but some DNA is also found in mitochondria and chloroplasts. Each chromosome contains a double-stranded molecule of DNA, which is associated with basic proteins called ***histones,*** but many nonhistone acidic proteins are also associated with the DNA molecule. They include such enzymes as DNA and RNA polymerases and other enzymes involved in the control of DNA and RNA synthesis. Chromosomes or segments of chromosomes that appear tightly coiled and compact stain intensely and are referred to as ***heterochromatin*** material. Chromosomes or segments of chromosomes that appear diffuse stain poorly and are referred to as ***euchromatin.***

It is reasonable to assume that more genetic information is required for eukaryotes because of their diversity in structure and function (Figure 12-1). What has been surprising is that eukaryotes possess more DNA than appears to be needed to code for the proteins required by the cell. Experiments have demonstrated that the genome of eukaryotic microorganisms, as well as those of viruses that replicate in the nucleus of animal cells, is not expressed as a continuous unit and some sequences within the gene do not appear to code for any information (see "Transcription" and "Translation," pp. 272 to 274).

Eukaryotes contain repetitive DNA sequences in which genetic units of up to 300 nucleotides in length may be represented by as many as 100,000 copies in the cell. Multiple copies of the genes for ribosomal and transfer RNA (rRNA and tRNA) are known to exist in eukaryotic systems, but some repetitive sequences are not transcribed, and their function is still a mystery. The major portion of eukaryotic DNA is nonrepetitive and codes for cellular proteins. There appears to be more economical use of the genome in prokaryotes, and especially in viruses, where less unused DNA is present.

The functional unit of prokaryotic DNA, called the operon, is not present in eukaryotic genetic systems. Genes with related functions, such as those associated with the synthesis of a single amino acid, are usually unlinked and scattered within and between chromosomes in eukaryotic

Table 12-1 Comparative features of the genetic systems of prokaryotes and eukaryotes

	EUKARYOTES	PROKARYOTES
DNA organization	More than one chromosome; DNA associated with proteins as chromatin; chromatin organized into nucleosomes; DNA in excess of that needed for coding of proteins, tRNA, and rRNA; no operons—genes concerned with metabolic pathways are not clustered; noncoding sequences (introns) within nuclear genes; repetitive sequences common—many copies (100 for rRNA)	One chromosome; DNA not associated with protein; DNA not in excess of that needed for coding of proteins, tRNA, or rRNA; operons present and contain clusters of genes that associate with single metabolic pathway; noncoding sequences not present in genes; copies of single gene seldom in excess of two or three
Transcription	Separate RNA polymerases for mRNA, tRNA, and rRNA; primary mRNA transcript often much larger than final functional mRNA; mRNA codes for only one polypeptide (monocistronic); capping of 5′ end of mRNA with methylated structure; addition of approximately 100 to 200 nucleotides to 3′ end of mRNA; excision of sequences specified by introns and splicing to give tRNA, rRNA, and mRNA	One RNA polymerase for mRNA, tRNA, and rRNA; no bases are removed from primary mRNA transcript; mRNA is polycistronic; that is, it contains information for several gene products; mRNA is relatively unstable, and no modifications are apparent
Translation	Ribosomes larger (80S) and more complex than in prokaryotes; species differ considerably in ribosome structure; initiation of translation limited to single site on mRNA	Ribosomes smaller (70S) than eukaryotes and not as complex; initiation of translation can occur at more than one site on mRNA

DNA. Clustered eukaryotic genes of related function are controlled coordinately like those in prokaryotic cells.

DNA REPLICATION AND THE CELL CYCLE

Cell division and the events leading up to it are more complicated in eukaryotes than in prokaryotes. DNA synthesis in eukaryotes occurs only periodically and does not occupy the entire cell cycle as in prokaryotes. There are four stages (Figure 12-2), or phases, associated with the cell cycle in eukaryotes: G_1 (gap 1), S (synthesis), G_2 (gap 2), and M (mitosis). G_1, S, and G_2 are referred to as the interphase. These stages are followed by the separation of the cell into daughter cells.

G_1 phase

The G_1 phase is characterized by a number of biochemical events. Once these biochemical events have been completed, the cell is committed to the initiation of DNA synthesis. The chromosomes during this period appear to be very dispersive and are not tightly coiled as in later stages. This dispersive characteristic probably al-

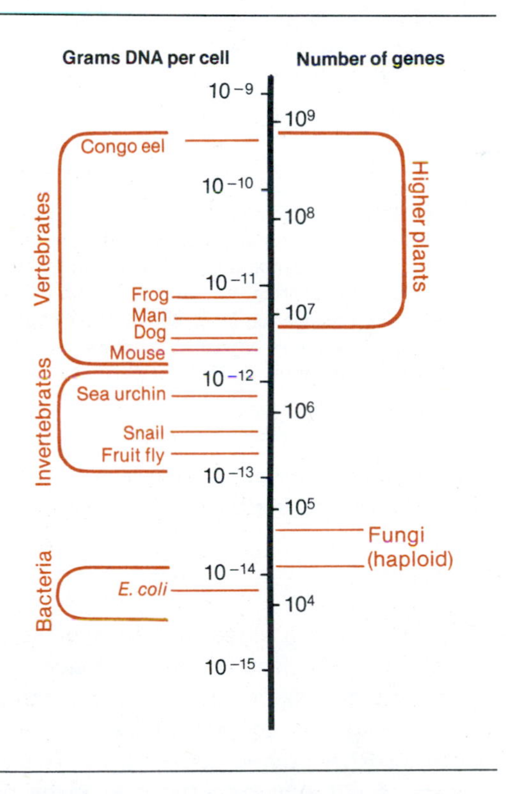

Figure 12-1 DNA content and gene load per cell in various eukaryotes as compared with *E. coli.*

Modified from Holliday, R. Symp. Soc. Gen. Microbiol. ***20:****359, 1970. Cambridge University Press.*

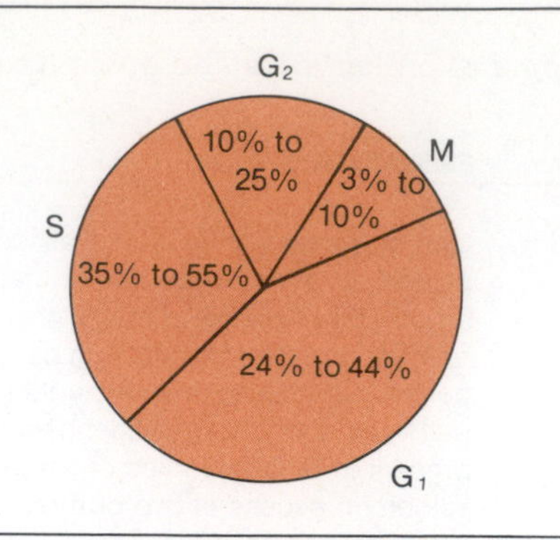

Figure 12-2 Phases of eukaryotic cell cycle. Percentages represent variations in time that different systems may occupy for that particular phase.

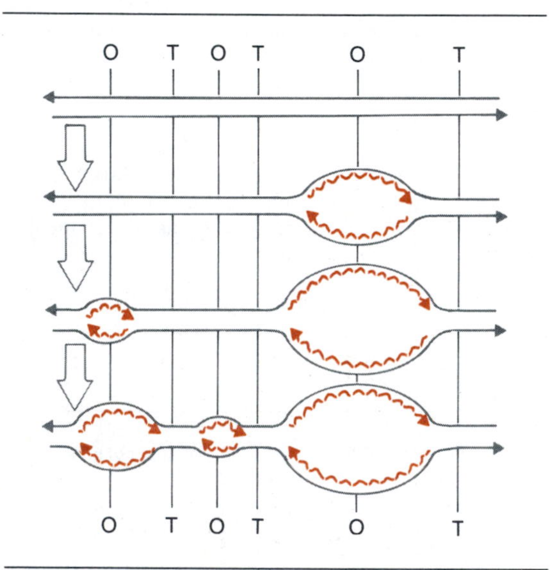

Figure 12-3 Replication of eukaryotic DNA. DNA synthesis begins at various replicating units, with each replicating unit having an origin *(O)* and two termination sites *(T)*. Each replicating unit (represented as bubble in figure) has two replicating forks, and as DNA synthesis continues, new bubbles appear in chromosome where DNA synthesis is initiated. DNA synthesis is discontinuous.

lows strand separation to take place, which is required for the transcription process.

S phase

The S period is a stage of DNA synthesis and chromosome duplication. The evidence to date suggests that DNA synthesis is a discontinuous process, like that in prokaryotes, but the evidence for RNA primers initiating the replication process is still unsubstantiated in many eukaryotic systems. Okazaki fragments in some eukaryotes are much smaller (100 to 150 nucleotides) than in prokaryotes (1000 to 2000 nucleotides).

Replication of eukaryotic DNA, like the replication of prokaryotic DNA, occurs in a semiconservative manner. DNA replication is initiated on a single chromosome at multiple origins*; therefore, there are several replicating units, or replicons. The number of replicating units varies from organism to organism. Figure 12-3 illustrates the arrangement of these replicating units, each of which corresponds to the replicating forks produced at the single origin site of replicating DNA in the bacterium *Escherichia coli.*

G_2 phase

G_2 is the shortest of the interphase stages. It immediately precedes mitosis and is characterized by chromosome condensation, and reduced RNA and protein synthesis.

M phase

The chromosomes in the M phase, or mitosis, engage in a cycle of events that ensures their equal segregation between two daughter cells. The large number of chromosomes in eukaryotic cells are enclosed in a nuclear membrane that is also involved in the division process. Spindle fibers are present in the nucleus and perform a function somewhat similar to that of the prokaryotic cytoplasmic membrane, which serves as an anchoring site for chromosome replication. The relationship of the spindle to the mitotic and meiotic processes is discussed in the following section.

MITOSIS AND MEIOSIS

Mitosis

Mitosis is a mechanism used by the cell to maintain a constant chromosome number during successive cell divisions. Eukaryotes possess chromosomes in pairs; for example, there are 23 pairs in humans, whereas yeast cells have 2 to 4 pairs, depending on the species. Somatic cells of most eukaryotic organisms, which contain the full complement of chromosomes, are said to be diploid (symbolized as 2N). The gametes or sex cells of eukaryotes contain one half the chromosome complement and are referred to as haploid (symbolized as N). Mitosis, as discussed earlier, is associated with the appearance in the nucleus of a microtubular assembly called the ***mitotic spindle*** (Figure 12-4). The mitotic spindle provides the cell with a mechanism for equal partitioning of the duplicated chromosomal material. The spin-

* Multiple origins are presumably required because the length of eukaryotic DNA is so much longer than in prokaryotes.

dle is composed of two types of microtubular fibers: chromosomal fibers, which link the spindle to special chromosomal areas called ***kinetochores,*** and polar fibers, which connect the two poles of the mitotic spindle. The kinetochore is believed to be the organization center for the microtubules of the chromosomal fibers. The sequence of events that characterize the mitotic cycle (Figure 12-5) are summarized below.

Prophase. The first phase of mitosis, prophase, is marked by the appearance of highly coiled and compact chromosomes. Faithful separation of chromosomes during mitosis would be extremely difficult if the chromosomes remained in the diffuse state exhibited during interphase. A compact rodlike structure reduces the amount of space required for assembly and eventual separation. The compactness also reduces the ability of chromosomes to engage in activities such as transcription; therefore chromosomes show no functions during mitosis. Each chromosome is observed to be composed of two ***chromatids,*** or threads, coiled about each other. The chromatids are exact copies of one another (homologous chromosomes) and are held together at the kinetochore. The chromatids will not separate until later, when the kinetochore has duplicated.

The ***nuclear membrane*** in higher eukaryotes disappears during the prophase or just before it in the interphase (the period between successive mitoses), but the nuclear membrane in eukaryotic microorganisms remains intact during mitosis.

Metaphase. The chromosomes during metaphase are aligned in the middle of the spindle. Metaphase prepares the chromosomes for eventual separation because during this period the kinetochore divides. Now the spindle fibers are capable of pulling or separating the two chromatids of each chromosome and moving them toward the poles of the cell.

Anaphase. Contraction of the spindle fibers during anaphase separates the chromosomes. The force responsible for their movement has not been determined, but a mechanism similar to the role played by actomyosin in muscle contraction has been suggested. Another possibility is the disassembly of microtubules, which would cause them to become increasingly shorter. Chromosome movement through the cytoplasm is in the form of a V shape with the arms of the chromosome trailing behind the kinetochore.

Telophase. The chromosomes have reached the poles of the spindle and begin to elongate. The

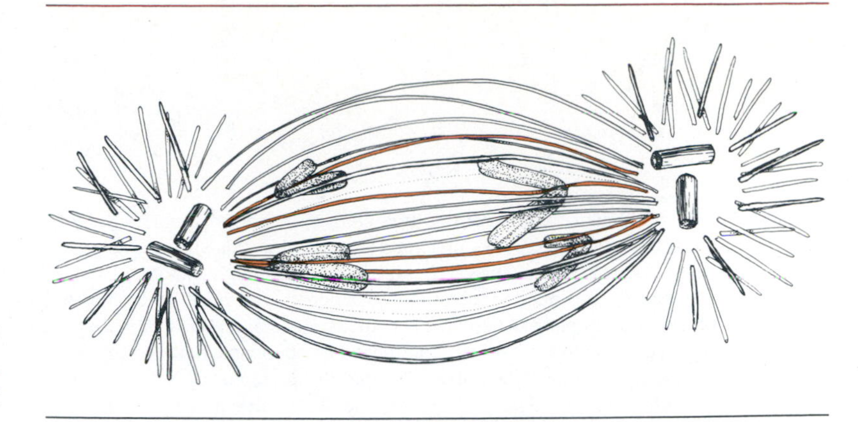

Figure 12-4 Mitotic spindle. Chromosomal microtubular fibers, which link spindle to chromosomal areas called kinetochores, are in color; remaining fibers are polar fibers and connect poles.

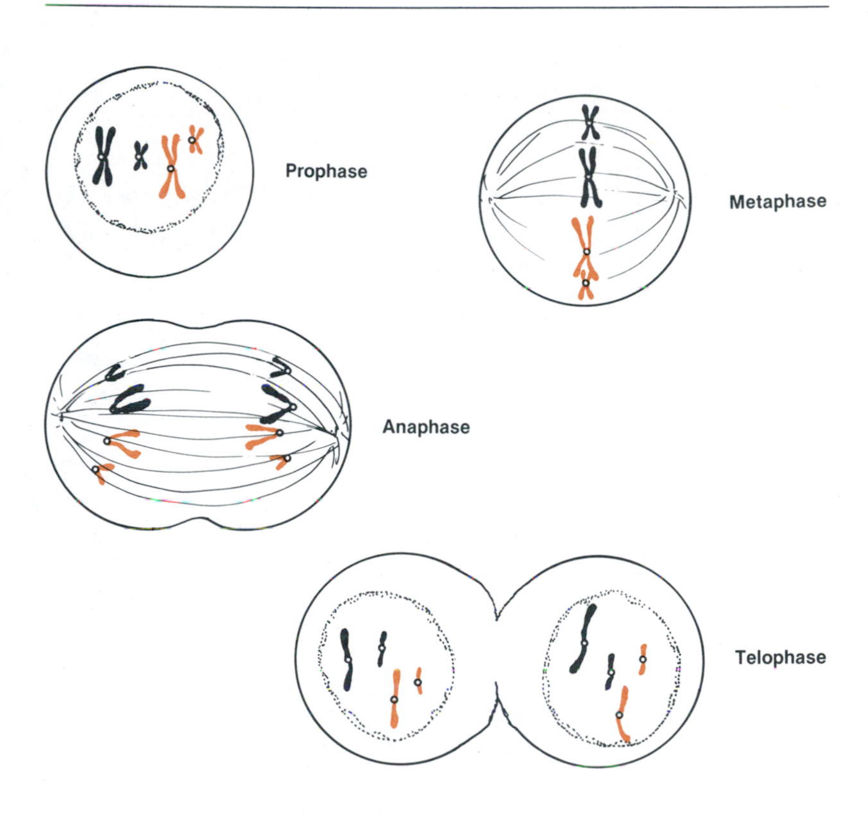

Figure 12-5 Mitosis in eukaryotic cell with haploid chromosome number of 2. Prophase: chromosomes become coiled and compact; each chromosome consists of two chromatids; nuclear membrane disappears at end of prophase. Metaphase: spindle fibers radiate from poles and align chromosomes at equatorial metaphase plate. Anaphase: centromeres divide, and sister chromatids separate; centromeres migrate to poles; cell begins to divide (cytokinesis). Telophase: nuclear membrane begins to form around chromosomes; spindle fibers disappear; nucleolus reforms; chromosomes will begin to become uncoiled and diffuse; complete cell division follows, and each nucleus enters interphase.

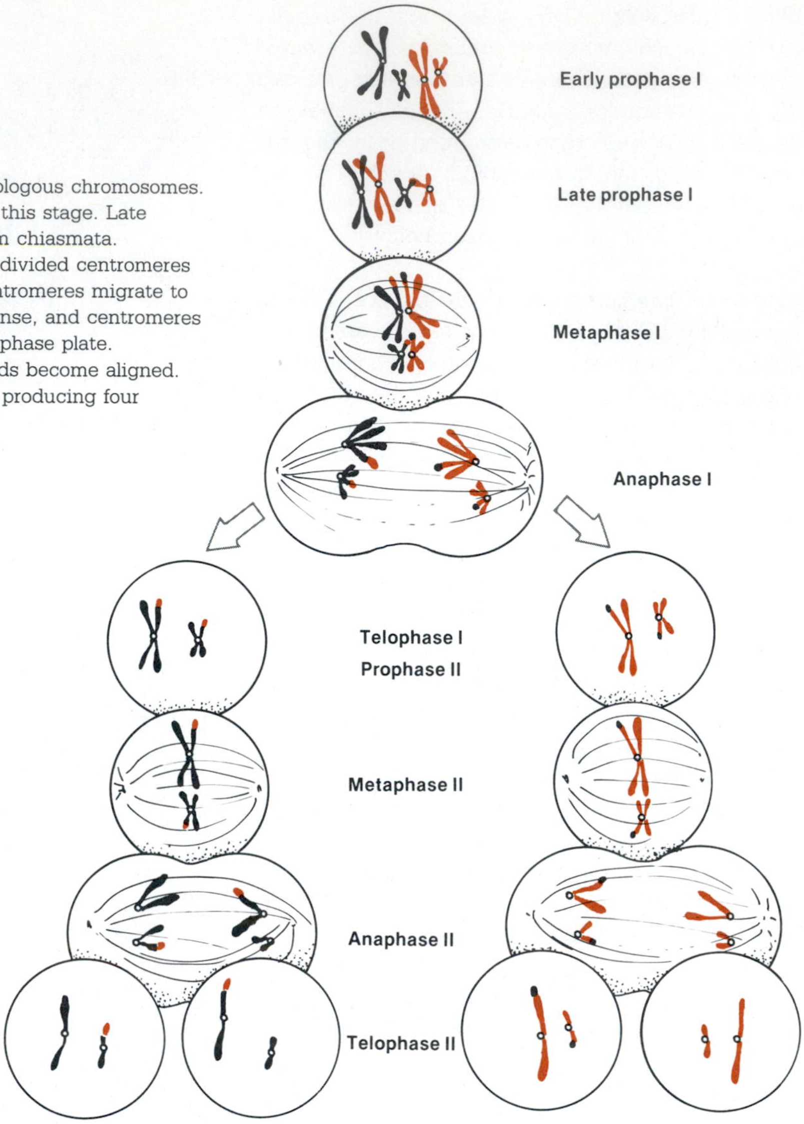

Figure 12-6 Meiosis in cell containing two pairs of homologous chromosomes. Early prophase I: homologous chromosomes are separated at this stage. Late prophase I: homologous chromosomes pair (synapse) and form chiasmata. Metaphase I: nucleolus and nuclear membrane disappear; undivided centromeres are aligned on metaphase plate. Anaphase I: homologous centromeres migrate to opposite poles. Telophase I, prophase II: chromosomes condense, and centromeres divide. Metaphase II: chromosomes become oriented at metaphase plate. Anaphase II: centromeres go to opposite poles, and chromatids become aligned. Telophase II: both cells produced during first division divide, producing four haploid cells.

nuclear membrane, when present, splits into two daughter nuclei, and the spindle disintegrates. Daughter nuclei with or without a nuclear membrane form around each set of chromosomes. Cytoplasmic division ***(cytokinesis)*** separates the nuclei and their chromosome complement. Once division is completed and the cells are totally separated, the chromosomes become uncoiled and engage in the resynthesis of an exact copy in preparation for the next round of division. The diploid state is therefore maintained.

Meiosis

Meiosis is a reductive process used by organisms, reproducing sexually, to produce haploid gametes from the diploid state (if each parent produced diploid sex cells, the chromosome number would double each generation). The fusion of male and female haploid gametes during sexual reproduction results in the formation of a zygote containing the diploid chromosome number. The vegetative cells of lower eukaryotes, such as the more primitive fungi, are haploid throughout most of their life cycle, whereas the diploid zygote state is very transient. Zygote formation in lower eukaryotes is followed almost immediately by chromosome reduction, or meiosis, and the formation of vegetative cells. Vegetative cells in higher eukaryotic microorganisms, such as the algae and more complex fungi, are usually diploid during most of the stages of their life cycle. The terminology applied to the meiotic process is extensive, and entire books are devoted to it. Meiosis and mitosis are very similar in terms of the mechanics of division, such as spindle formation and nuclear division, and our discussion will be brief and limited to those important aspects of meiosis that distinguish it from mitosis. Meiosis is divided into two parts: meiosis I and meiosis II (Figure 12-6).

Meiosis I. The most significant stage in the meiotic I process occurs during the first prophase. First, however, we should mention that before prophase I, during interphase, the diploid chromosomes replicate in the zygote. This means that there are four copies of the genome (or four haploid sets of information or four alleles for every gene). Initially during prophase the chromosomes appear as threads, but soon they thicken and become more compact. The chromosomes begin to pair ***(synapse)*** as they thicken. The pairing between homologous chromosomes occurs along their entire length. Synapsed homologous chromosomes are called ***bivalents.*** Actually, each bivalent is made up of four chromatids (also called a tetrad), two from each parent. The coiling and condensing of chromosomes later during prophase I becomes more pronounced, and visible regions of attachment, called ***chiasma*** (pl. chiasmata), give the chromosome a crosslike or X appearance. Each chiasma represents a region where a physical exchange of parts of the nonsister chromatids (chromatids of a different pair of homologous chromosomes) takes place (Figure 12-7). In other words, a crossing over (recombination) occurs between homologous chromosomes, and new combinations of alleles may arise on the same chromosome. This crossover process is believed to be similar to that which occurs in bacteria and viruses. Each crossover event involves only two chromatids. The chiasmata are pushed to the ends of the paired chromatids by the end of prophase, and the chromosome pairs appear to be joined at their ends. The remaining stages of meiosis are like those of mitosis. One of the major differences that distinguishes meiosis I from mitosis is that there is no division of the centromeres during anaphase. Telophase I therefore consists of two nuclei with the haploid number of chromosomes, but each still contains double the amount of DNA.

Meiosis II. The interphase following meiosis I does not involve DNA duplication. Meiosis II is also mechanically similar to mitosis, and there is spindle formation with centromere division and separation at anaphase II. A nuclear membrane is formed around each group of chromosomes in telophase, resulting in the formation of four nuclei, each with a chromosome consisting of a single chromatid.

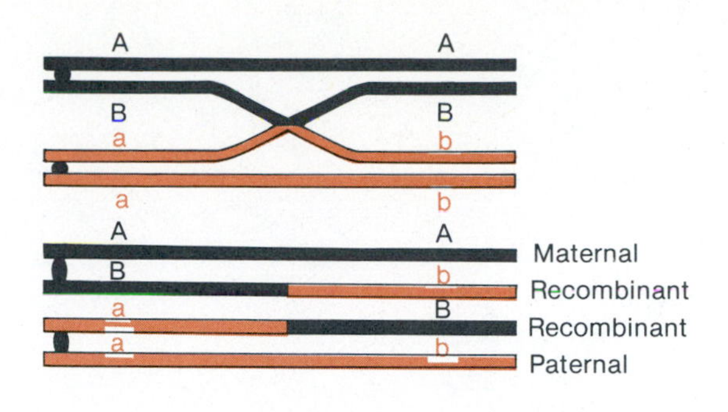

Figure 12-7 Chiasmata formation between maternal (black) and paternal (colored) homologous chromosomes (tetrad). Maternal chromosomes contain wild-type alleles A and B, whereas paternal chromosomes contain mutant alleles a and b.

RECOMBINATION IN EUKARYOTES

The meiotic process can produce gametes in which different combinations of chromosomes may appear by a process of crossing over, or recombination, between homologous chromosomes. An example of the results of a crossover event between the maternal and paternal homologous chromosomes is illustrated in Figure 12-7. A single chromatid exchange is seen to occur at the site of chiasmata formation between nonsister chromatids. Four gametes are produced, in which one chromosome is entirely paternal, one chromosome is entirely maternal, and two are recombinants. Chiasmata appear at several sites on the tetrad unless one of the chromosome arms (the appearance of the chromosomes due to centromere location) is short, and recombination can take place at a number of them. Usually each of the four chromatids becomes involved in a recombinational event.

The order of arrangement of linked genes that can be separated on the chromosome and the relative distances between genes can be determined by analysis of recombination data. A genetic map can also be constructed from this data. Let us look at a hypothetical example involving three linked genes (A, B, and C) and their respective mutant alleles (a, b, and c) (Table 12-2). The number of different phenotypes that can be obtained following meiosis equals eight. The number of offspring in which there is a crossover between the genes A and B equals 36.

$$\left(\frac{293}{800} \times 100 = \sim 36\%\right).$$

Table 12-2 Results of a cross of three mutants (a, b, and c) with the wild type (A, B, and C)

PHENOTYPES PRODUCED	TOTAL OFFSPRING FROM TEST CROSS	RECOMBINANTS RESULTING FROM GENETIC EXCHANGE BETWEEN GENES		
		A AND B	B AND C	A AND C
ABC	205	—	—	—
abc	226	—	—	—
Abc	139	139	—	139
aBC	145	145	—	145
ABc	35	—	35	35
abC	41	—	41	41
AbC	5	5	5	—
ABc	4	4	4	—
TOTALS	800	293	85	360

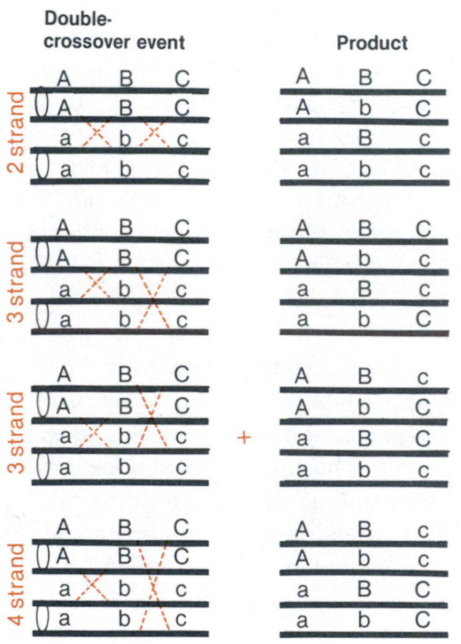

Figure 12-8 Four possible types of double crossovers between wild-type genes A, B, and C and mutant genes a, b, and c.

The frequency of recombination* is 36%, or the distance between A and B is 36 map units (one map unit equals 1% of recombination). The recombination frequency between B and C is 10%, or 10 map units.

$$\left(\frac{85}{800} \times 100 = \sim 10\%\right)$$

* The frequency of recombination $= \frac{\text{No. of recombinants}}{\text{Total progeny}} \times 100$

Since the distance between A and C is calculated to be approximately 45 map units, C must be to the right of B. The genetic map is therefore:

36 10
A B C or C B A
— 45 —

The values indicating distance between genes can be misleading if double crossovers occur between chromatids. Double crossovers seldom occur between genes located very close to each other, but they increase in frequency as the distance between genetic loci becomes greater. The four possible types of double crossovers are illustrated in Figure 12-8. Ordinarily, an accurate linkage map can be constructed by performing a series of crosses where each pair of genes is closely linked, for example, when the distance between genes is under 5 map units.

TRANSCRIPTION

The transcription of eukaryotic DNA, like that of prokaryotic DNA, requires RNA polymerases. Eukaryotic systems, however, possess three types of RNA polymerase. RNA polymerase I is found in the nucleolus and is involved in the transcription of the genes for 28S, 5.8S, and 18S rRNA. RNA polymerase II is associated with mRNA synthesis, and RNA polymerase III is involved in the transcription of tRNA and 5S rRNA.

A region similar to the promoter in prokaryotes is found in eukaryotes and is called the ***5′-flanking region.*** This region contains two areas of interest. One is called the ***Goldberg-Hogness box*** or ***TATA box*** and consists of the sequences TATXAX, where X is either T or A. The TATA box is found about 30 bases upstream from the point of initial gene transcription. Mutations in this area result in the inability of RNA polymerase II to initiate transcription. A second region appears anywhere from 50 to 500 nucleotides upstream and is also important in the transcription process. Deletions in this area severely curtail mRNA transcription on adjacent genes.

The size of the mRNA produced in eukaryotic systems does not correlate with genome size. The eukaryotic genome, as stated earlier, contains sequences that apparently do not code for any information. Experiments have already confirmed that mRNA is spliced during the transcription process and that some of its sequences are removed. The enzymes involved in gene splicing have not been characterized, but the process is believed to occur in the following way (Figure 12-9). A gene in the DNA is transcribed into a complete mRNA unit. Later, some sequences on the mRNA, which are not present in the cytoplasmic mRNA, are removed, and the functional sequences are spliced together. The consequence of this is the formation of an mRNA that is much shorter that the DNA from which it was originally transcribed. The functional sequences on the DNA are called ***exons*** because their transcription product will go out into the cytoplasm, whereas the nonfunctional coding sequences are called ***introns,**** since their transcription product will remain within the nucleus, possibly to be degraded. The 5′ end of the mRNA, called the ***leader sequence,*** is spliced to sequences (exons) at various distances from the 3′ end during the splicing process. The number of different exon segments in some animal cells that can be spliced together is as high as 30 (the genes for collagen or serum albumin). The size of the introns range from 10 to 600 base pairs, whereas exons appear to have approximately 300 base pairs. It has been estimated that the amount of intron material in the DNA may exceed that of exon material by 10 to 100 times. This could explain the excess DNA found in eukaryotes, however, the function of introns is still unexplained. Splicing in viruses offers a mechanism for obtaining different mRNAs that code for different polypeptides from the same DNA transcript, a mechanism for genetic conservation.

Eukaryotic mRNA is more stable than prokaryotic mRNA. This stability is believed to be due to modification steps immediately following transcription. One modification system is called ***capping.*** The 5′ ends of nearly all eukaryotic mRNAs are capped with a guanosine that protects the next to last (penultimate) base (Figure 12-10). The 5′ cap is unusual because it is methylated and is attached to the penultimate base by a unique triphosphate linkage. The penultimate base is also methylated. Methylation renders the bases polar and thus capable of interaction with charged molecules. Capping, in addition to lending stability to the molecule, also provides a means of ri-

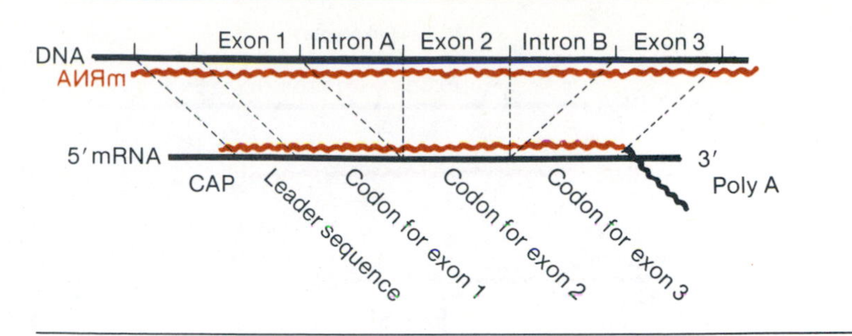

Figure 12-9 Proposed mechanism of gene splicing during DNA transcription in eukaryotic cells. DNA is transcribed into complementary mRNA that is cleaved by RNA-splicing enzymes and then assembled into shorter mRNA containing leader sequence of nucleotides plus codons for exons I, II, and III. At 5′ and 3′ ends of mRNA are cap and poly A sequences, respectively (see Figure 12-10). Shortened mRNA codes for one polypeptide.

Figure 12-10 Structure of 5′ cap present on eukaryotic mRNAs. If base is adenine, then nitrogen 6 of adenine is methylated in addition to methylated ribose sugar.

* Recent results have shown that mitochondrial introns in yeast code for proteins called maturases. These proteins are apparently used for the splicing of introns.

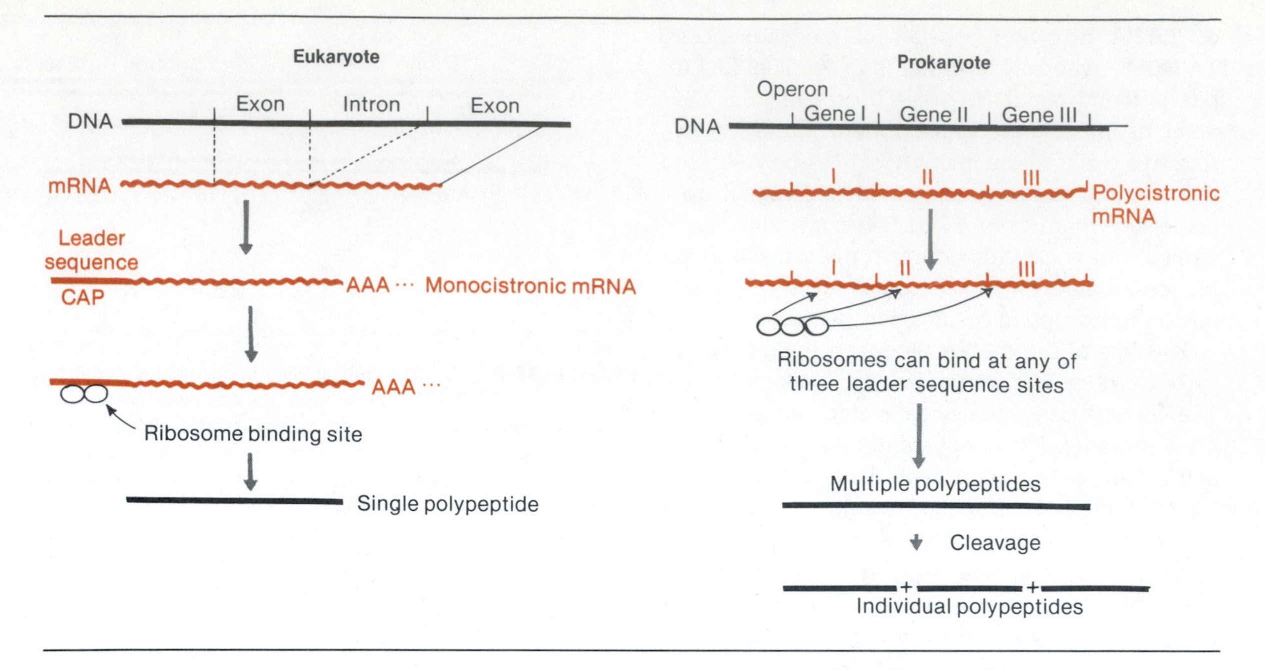

Figure 12-11 Comparative aspects of DNA transcription and translation in eukaryotic and prokaryotic cells (see also Table 12-1).

bosome recognition during translation of mRNA. Eukaryotic mRNA is also modified by the addition of poly A sequences to the 3′ end of the molecule. The length of the poly A sequences may range from 30 to 200 bases. Their function is not understood at this time, but they may be a recognition signal during translation, or they may be a binding site for some cytoplasmic component.

TRANSLATION

Eukaryotic mRNA must migrate from the nucleus into the cytoplasm before translation can be initiated. The 5′ capping unit on the mRNA is apparently required for efficient binding to the ribosome. The initiation codon is AUG, which is recognized by methionyl tRNA, but the molecule is not formylated as in prokaryotic systems.

A monocistronic mRNA is formed in eukaryotic systems, where splicing connects the 5′ leader sequence to noncontiguous areas of the mRNA. Ribosomes bind only at the 5′ site, in contrast with prokaryotic cells, where binding can occur at other sites on the polycistronic mRNA (Figure 12-11).

Eukaryotic ribosomes, as discussed in Chapter 8, are larger than prokaryotic ribosomes and consist of 60S and 40S subunits. A precursor rRNA is transcribed from the rRNA genes and is later cleaved and modified into mature 28S, 18S, and 5.8S rRNAs (there are some variations in the size of these rRNAs, depending on the eukaryotic species). Ribosomal proteins are assembled with rRNAs in the nucleolus to produce the 60S and 40S subunits. Polypeptide synthesis cannot begin until the subunits are released through the nuclear membrane into the cytoplasm and have bound to mRNA. Recall that mRNA translation in prokaryotic cells begins even before the mRNA is completely transcribed from the DNA. This difference in translation may be associated with the differences in mRNA stability between eukaryotes and prokaryotes. Finally, the elongation process and chain termination are basically the same for eukaryotes and prokaryotes. The protein elongation factors in eukaryotes are equivalent to those in prokaryotes, even though they are not interchangeable, and some termination codons operate in both groups.

EXTRANUCLEAR INHERITANCE

Eukaryotes contain mitochondria and chloroplasts that have their own genetic systems. The genes found in these organelles have been shown to influence the phenotype of the cell. Much of our knowledge of extranuclear inheritance comes from studies involving mitochondria. Mitochondrial genes play a very important role in cellular metabolism because they code for some of the cytochrome-containing enzymes that are involved in respiratory activity. Mitochondrial DNA also contains the genes for rRNA, some tRNAs, and a few ribosomal proteins, but most of the ribosomal proteins and cytochrome-containing enzymes are coded for by nuclear genes. The genetic and protein-synthesizing systems of mitochondria are similar to those found in bacteria. These similarities include the following:

1. Circular, naked (no association with proteins), and double-stranded DNA
2. Similarity of ribosomes in size and in sensitivity to certain antibiotics
3. Protein synthesis requirement of the initiator codon AUG and a specific formylmethionyl tRNA

Scientists have used these findings to hypothesize that mitochondria evolved from prokaryotic endosymbionts. The main objection to this hypothesis is that mitochondrial mRNA is spliced like nuclear eukaryotic mRNA.

Extranuclear inheritance does not conform to the usual patterns of mendelian inheritance. For example, according to mendelian law, the alleles of a particular trait segregate independently at meiosis. Examination of a reciprocal cross between wild type and mutant for an extranuclear gene reveals that there is no 2:2 segregation of the gene in the four meiotic products. Instead, all of the progeny show the phenotype of the female parent. The reason for this is that in most plants and animals the female gamete supplies most of the cytoplasm, whereas the male contributes only a nucleus. Note that in the fusion of yeast cells it is not possible to determine which is male and which is female; however, the segregation pattern of the particular allele is a clue to extranuclear inheritance. Let us look at the genetics of the yeast *Saccharomyces cerevisiae*, in which respiratory deficiencies may be due to nuclear as well as mitochondrial mutation. Some cells of *S. cerevisiae*, when grown on an agar medium, produce extremely small (called ***petite***) colonies in comparison with the normal colony. Petite mutants carry mutations that affect respiration and hence influence cell growth. Figure 12-12 illustrates the phenotypic pattern obtained when nuclear or extranuclear inheritance is involved. A cross between a wild type and a petite mutant in which nuclear genes are involved produces a zygote that exhibits normal growth. Sporulation (meiosis) produces four haploid spores, two of which are normal and two of which are petite. A similar cross in which the petite mutant is due to

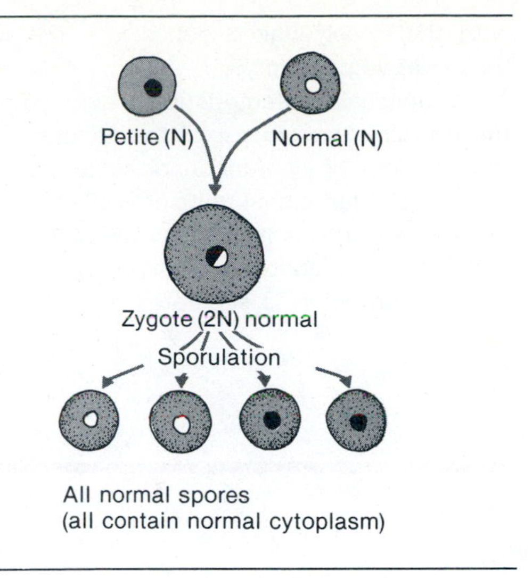

Figure 12-12 Results of cross involving haploid petite mutant and haploid normal gamete of *Saccharomyces cerevisiae*.

extranuclear inheritance produces a phenotypically normal zygote, but all the meiotic products are normal, with no petite spores. All spores are normal because the cytoplasm of those cells that carry the petite mutant have been inherited from a normal cell.

Viruses and plasmids also represent extranuclear inheritance, but they are dispensible to the cell. For example, the "killer" factor in yeast is associated with a virus (see Chapter 6).

SUMMARY

DNA organization and structure in eukaryotes differs considerably from that of prokaryotes. These differences are outlined in Table 12-1. One of the most basic differences is that only in eukaryotes does the nucleus undergo a series of events called mitosis, in which chromosomes are faithfully duplicated and partitioned between the dividing nuclei. Also, in those eukaryotes with a sexual cycle the fusion of sex gametes results in a process of chromosome reduction called meiosis, in which haploid cells are produced. During the meiotic period pairing between homologous chromosomes results in recombinational events that can lead to the formation of new combinations of alleles. Genetic maps for eukaryotes can be constructed through recombination studies performing a series of crosses in which one pair of genes is closely linked. The frequency with which other genes are recombined with the closely linked genes is a measure of their relative proximity.

An unusual characteristic of eukaryotes and the transcription process is the apparent excessive amount of nonfunctional sequences called introns. The functional units are called exons. An mRNA transcript is made that contains the functional and nonfunctional sequences. The exon-coded information is separated from the intron-coded information and spliced into a smaller but totally functional mRNA.

Eukaryotic organelles, such as mitochondria and chloroplasts, carry their own genetic information. This extranuclear genetic information is important to the survival of the cell. extranuclear traits cannot be mapped on the nuclear genome, and their pattern of inheritance differs from those genes of chromosomal origin.

Study questions

1. What might be the advantage of having multiple copies of a gene? In what type of microorganisms are multiple copies found? Does this multiplicity have any evolutionary significance?
2. How do microtubules function in eukaryotic genetics?
3. What is the major mechanism for genetic diversity in eukaryotic cells? Explain the process.
4. Why is the mRNA molecule initially transcribed from eukaryotic DNA so much larger than the actual mRNA to be translated? Could this provide the microorganism with any advantages?
5. How does translation of mRNA in eukaryotes differ from that in prokaryotes?
6. How is mRNA in eukaryotes modified in order to provide stability to the molecule?
7. What types of activity in eukaryotic cells are controlled by extranuclear genes (nonchromosomal)?
8. What is the principal distinction between mitosis and meiosis in terms of function to the cell?
9. What arguments suggest that mitochondria are prokaryotic endosymbionts? What argument suggests that they have not evolved from prokaryotes?
10. Define the following: intron, exon, kinetochore, chromatid, bivalent, polycistronic, and monocistronic.

Selected readings

Books

Ayala, F.J., and Kiger, J. *Modern genetics.* Benjamin/Cummings Publishing Co., Inc., Menlo Park, Calif., 1980.

Brachet, J., and Mirsky, A.E. (Eds). *The cell, Vol. 3, Meiosis and mitosis.* Academic Press, Inc., New York, 1961.

Bradbury, E.M., and Javaherian, K. (Eds.) *The organization and expression of the eukaryotic chromosome.* Academic Press, Inc., New York, 1977.

Burns, G.W., *The science of genetics* (4th ed.). Macmillan Publishing Co., Inc., New York, 1980.

Chromatin: Cold Spring Harbor symposia for quantitative biology (Vol. 42). Cold Spring Harbor Laboratory, Cold Spring Harbor, N.Y., 1977.

Goodenough, U. *Genetics* (2nd ed.). Holt, Rinehart & Winston, New York, 1978.

Herskowitz, I.H. *The elements of genetics.* Macmillan Publishing Co., Inc., New York, 1979.

Hexter, W., and Yost, H.T. *The science of genetics.* Prentice-Hall, Inc., Englewood Cliffs, N.J., 1976.

Russel, P.J. *Lecture notes on genetics.* Blackwell Scientific Publications, Ltd., Oxford, 1980.

Wagner, R.P., et. al. *Introduction to modern genetics.* John Wiley & Sons, Inc., New York, 1980.

Journal articles

Anderson, W.F., and Diacumakos, E.G. Genetic engineering in mammalian cells. *Sci. Am.* **245:**106, 1981.

Banerjee, A.K. 5′ terminal cap structure in eukaryotic messenger ribonucleic acids. *Microbiol. Rev.* **44:**175, 1980.

Brauerman, G. Eukaryotic messenger RNA. *Annu. Rev. Biochem.* **43:**621, 1974.

Chambon, P. Eukaryotic nuclear RNA polymerases. *Annu. Rev. Biochem.* **44:**613, 1975.

Chambon, P. Split genes. *Sci. Am* **244:**60, 1981.

Crick, F. Split genes and RNA splicing. *Sci. Am.* **204:**264, 1979.

Cullum, J., and Saedler, H. DNA rearrangements and evolution. *Symp. Soc. Gen. Microbiol.* **32:**131, 1981.

Darnell, J.E. Implications of RNA: RNA splicing in evolution of eukaryotic cells. *Science* **202:**1257, 1978.

DePamphilis, M.L., and Wassarman, P.M. Replication of eukaryotic chromosomes: a close-up of the replication fork. *Annu. Rev. Biochem.* **49:**627, 1980.

Edenberg, H.J., and Huberman, J.A. Eukaryotic chromosome replication. *Annu. Rev. Genet.* **9:**245, 1975.

Elgin, S.C.R., and Weintraub, H. Chromosomal proteins and chromatin structure. *Annu. Rev. Biochem.* **44:**725, 1975.

Hartwell, L.H. *Saccharomyces cerevisiae* cell cycle. *Microbiol. Rev.* **38:**164, 1974.

Kornberg, R.D. Structure of chromatin. *Annu. Rev. Biochem.* **46:**931, 1977.

Kornberg, R.D., and Klug, A. The nucleosome. *Sci. Am.* **244**(2):52, 1981.

Mazia, D. The cell cycle. *Sci. Am.* **230**(1):54, 1974.

O'Malley, B.W., et al. Regulation of gene expression in eukaryotes. *Annu. Rev. Genet.* **11:**239, 1977.

Revel, M., and Groner, Y. Post transcriptional and translational controls of gene expression in eukaryotes. *Annu. Rev. Biochem.* **47:**1079, 1978.

Sheinin, R. Humbert, J., and Pearlman, R.E. Some aspects of eukaryotic DNA replication. *Annu. Rev. Biochem.* **47:**277, 1978.

Weissbach, A. Eukaryotic DNA polymerases. *Annu. Rev. Biochem.* **46:**25, 1977.

Winicov, I., and Perry, R.P. Synthesis, methylation and capping of nuclear RNA by a subcellular system. *Biochemistry* **15:**5039, 1976.

Worcel, A. Molecular architecture of the chromosome fiber. *Cold Spring Harbor Symp. Quant. Biol.* **42:**313, 1977.

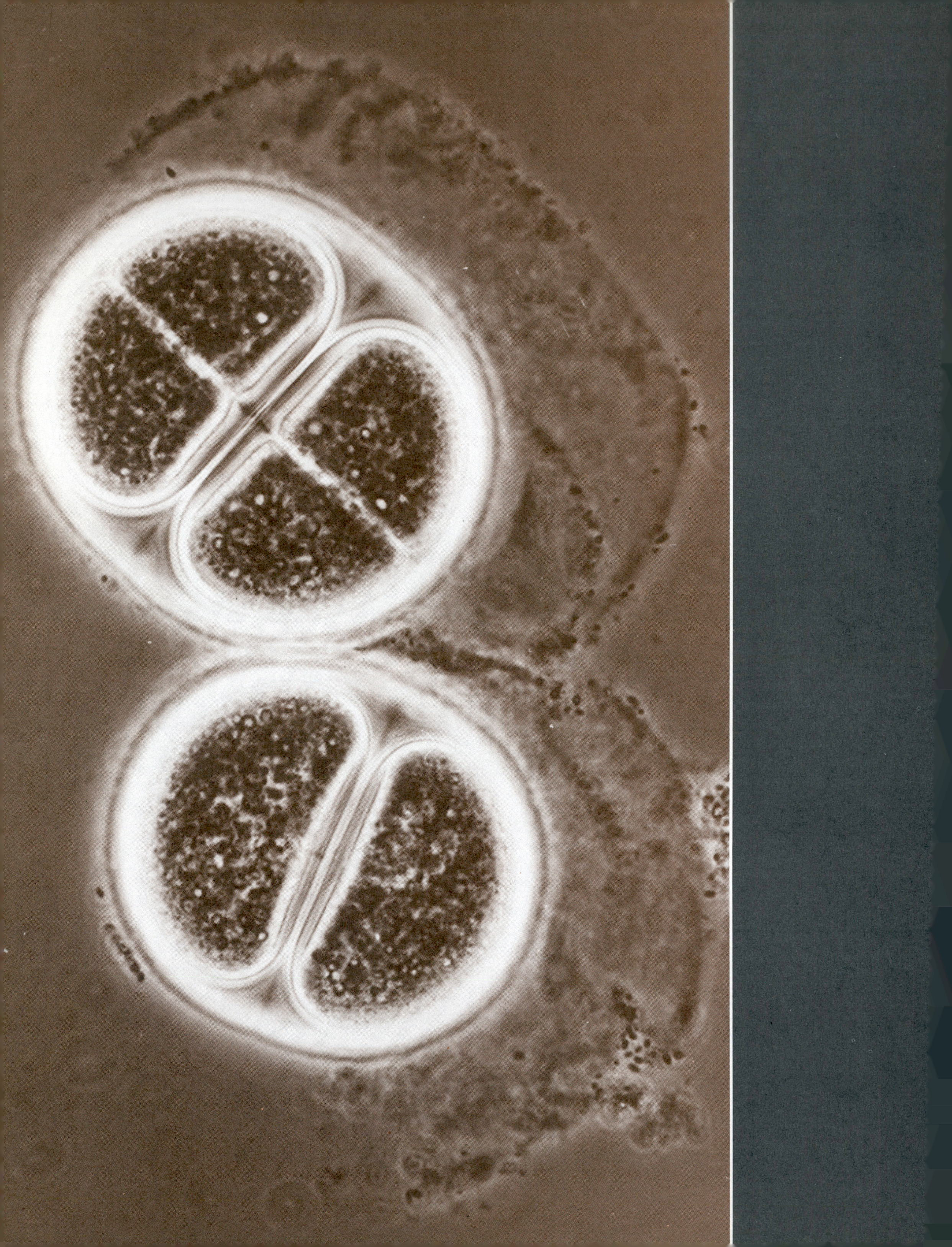

PART 4

MICROBIAL METABOLISM AND GROWTH

Chroococcus turgidus, a blue-green alga. (×1280.)

J. R. Waaland, University of Washington/BPS

Chapter 13 ENERGY PRODUCTION IN MICROORGANISMS

Energy, which can be defined as the capacity to do work, may exist in various states, such as potential, radiant, thermal, and electrical. Survival in the biological world requires energy for certain cellular functions—transport of solutes across the cytoplasmic membrane, movement, and biosynthesis. The energy to perform these functions can be obtained by the cell in two ways. First, organisms called ***phototrophs*** convert the radiant energy from sunlight into chemical energy. Second, some organisms called ***chemotrophs*** transfer energy from preformed organic or inorganic molecules into specific energy molecules. The purpose of this chapter is to discuss these various energy-transformation processes as they occur in two nutritional types of microorganisms: those that obtain their carbon for biosynthesis from organic compounds ***(heterotrophs)*** and those that obtain their carbon for biosynthesis from carbon dioxide ***(autotrophs).***

LAWS OF ENERGY TRANSFORMATION (THERMODYNAMICS)

Energy transformations, which involve the production or expenditure of energy, are the basis of the science of thermodynamics. Every chemical or physical event that involves an exchange of energy between the system (such as a chemical reaction) and the environment outside the system is governed by two laws of thermodynamics. First, the total energy of the universe remains constant, or the energy gained by one reaction is lost by another, and vice versa. Second, only part of the energy released in a reaction can be used for work. The usable energy obtained from a chemical reaction is called the ***free energy,*** and as it declines there is an increase in the amount of nonusable energy, called ***entropy.*** Reactions in which two compounds, A and B, are converted to products C and D can be described thermodynamically since there is a change in their energy contents. Gibbs explained the changes in the energy content of such a system with the following formula:

$$\Delta G = \Delta H - T\Delta S$$

Let us examine each of the components of the Gibbs formula:

ΔG refers to the change (Δ) in free energy of a system or the energy that can be obtained from a chemical reaction to do work. ΔG is the difference in free energy between products and reactants, or

$$\Delta G = (G_C + G_D) - (G_A + G_B)$$

ΔH refers to the heat that is transferred between the system and its surroundings, or the ***enthalpy*** change. The reactants and products in a chemical reaction each have a characteristic ***heat of combustion*** in which a specific amount of heat (in calories/gram*) is released when each compound is burned in oxygen (Table 13-1). The ΔH is, therefore, equal to the difference between the sum of the ΔH of the products and the sum of the ΔH of the reactants or ΔH (products) $-$ ΔH (reactants). In most instances the ΔH is nearly equal to the ΔG, and the two are often used in-

* A ***calorie*** is the amount of energy required to raise the temperature of one gram of water from 14.5° to 15.5° C.

Table 13-1 Heats of combustion of some cellular fuels

FUEL	MOLECULAR WEIGHT (DALTONS)	ΔH (CAL/MOLE)	CALORIC VALUE (KCAL/GM)
Carbohydrate: Glucose (ring structure with CH_2OH, H, OH, O)	180	−673,000	3.74
$CH_3—CH(OH)—COOH$ Lactic acid	90	−326,000	3.62
Lipids: $CH_3(CH_2)_{14}COOH$ Palmitic acid	256	−2,380,000	9.30
Amino acid: $H—CH(NH_2)—COOH$ Glycine	75	−234,000	3.12

Adapted from Lehninger, A.L. *Bioenergetics*. Menlo Park, California, W.A. Benjamin Inc., 1965.

terchangeably. Let us use the reversible reaction $A + B \rightleftharpoons C + D$ and assume three different conditions in which the ΔH is different for products and reactants:

A + B	**⇌**	**C + D**	**ΔH for reaction**
Reaction 1	ΔH = 1000 calories/mole	ΔH = 900 cal/mole	+100 cal/mole
Reaction 2	ΔH = 900 cal/mole	ΔH = 1000 cal/mole	−100 cal/mole
Reaction 3	ΔH = 900 cal/mole	ΔH = 900 cal/mole	0 cal/mole

Reaction 1, in which the ΔH is positive, is referred to as an ***endergonic*** reaction, which means that the forward reaction cannot proceed unless energy is put into the system. The reverse reaction, however, will proceed spontaneously. Reaction 2, in which the ΔH is negative, is referred to as an ***exergonic*** reaction and tends to proceed spontaneously in the forward direction with the release of free energy. Reaction 3, in which the ΔH is 0, is in a state of equilibrium, and no chemical change takes place.

In TΔS, T represents the temperature of the reaction, and ΔS is the change in entropy. Entropy is a measure of the randomness of a system or simply the energy in a system that tends to go downhill and become randomized. TΔS represents that part of the total energy change that is lost to the system in the form of random molecular motions and is therefore not available to do work. In most biochemical reactions in a cell the TΔS is usually small in comparison with the quantities for ΔG and ΔH. The oxidation of glucose, for example, by molecular oxygen can be represented in the following way:

$$\text{Glucose} + 6\ O_2 \longrightarrow 6\ CO_2 + 6\ H_2O$$

where

$\Delta G^\circ = -686{,}000$ cal/mole (−686 Kcal/mole)

$\Delta H = -673{,}000$ cal/mole (−673 Kcal/mole)

$\Delta S = +43.6$ cal/mole-degree

THERMODYNAMICS AND METABOLISM

Metabolism is the sum total of all the chemical reactions that occur in the cell and, like other chemical reactions, is governed by laws of ther-

Table 13-2 Energy rich compounds found in microbial cells

COMPOUND	GENERAL FORMULA FOR ENERGY LINKAGE	ΔG° (CAL/MOLE)
Phosphoenolpyruvate	$R{-}C(=CH_2){-}O{\sim}P$	−12,800
Adenosine triphosphate (ATP)	$R{-}O{-}P(=O)(OH){-}O{-}P(=O)(OH){-}O{\sim}P$	−7000
Acetyl CoA	$R{-}C(=O){\sim}SH$	−10,500
Acetyl phosphate	$R{-}C(=O){-}O{\sim}P$	−10,100

modynamics. Thousands of metabolic reactions occur in the cell; some are exergonic and others endergonic. Exergonic reactions are those processes that release free energy such as the oxidations involving carbohydrates, fats, and proteins, as well as the hydrolytic reactions involving energy-rich molecules such as ATP. Endergonic reactions are processes that require the addition of energy, for example, flagellar movement, nutrient transport, and reactions that are involved in the synthesis of amino acids, lipids, and carbohydrates.

A chemical reaction will reach a final equilibrium point, that is, a point at which no further chemical change takes place in the reaction, and the rate of conversion of reactants A and B to products C and D will be balanced by the rate of conversion of C and D to A and B. This can be expressed as an equilibrium constant, K_{eq} or:

$$K_{eq} = \frac{\text{Product of concentrations of products}}{\text{Product of concentrations of reactants}} = \frac{[C][D]}{[A][B]}$$

K_{eq} is a function of the free energy change of the components of the reaction and is related to it in the following equation:

$$\Delta G^\circ = -RT \ln K_{eq}$$

where ΔG° is the standard free energy in which 1 mole of reactant is converted to 1 mole of product under standard conditions of temperature and pressure. R is the gas constant (1.987 cal/mole-degree), T is the absolute temperature, and $\ln K_{eq}$ is the natural log of the equilibrium constant. If, for example, the equilibrium constant is high (1000) it means that more product has been released, the reaction tends to go to completion, and the free energy change is negative. If the equilibrium constant is low (0.001) it means that little product is formed, the reaction does not go far to the right, and the free energy change is positive. If the equilibrium constant is low, converting 1 mole of reactant to 1 mole of product requires that energy be put into the system. High-energy compounds, such as ATP, are responsible for supplying most of the energy required for endergonic reactions. Hydrolysis of the terminal phosphate of ATP releases approximately −7000 cal/mole of free energy. Other energy-rich compounds that may be found in microbial cells are described in Table 13-2.

The ΔG of metabolic reactions refers only to the potential of the reaction occurring. Even though a reaction has a high negative free energy and proceeds spontaneously, it may take years for the reaction to reach its final equilibrium point. The rate at which the reactants are converted to products is influenced by enzymes discussed in Chapter 2.

ENERGY PRODUCTION IN HETEROTROPHS

Heterotrophs use organic compounds for their source of carbon, but energy sources may vary depending on the species. Some heterotrophs

Table 13-3 Nutritional classification of microorganisms based on carbon and energy sources

NUTRITIONAL TYPE	CARBON SOURCE	ENERGY SOURCE	EXAMPLES
Chemoautotroph*	Carbon dioxide	Chemicals	Hydrogen bacteria, colorless sulfur bacteria, iron bacteria, nitrifying bacteria
Chemoheterotroph	Organic compounds	Chemicals	Animals (including humans), most bacteria, fungi, and protozoa
Photoautotroph	Carbon dioxide	Light	Green plants, most algae, purple and green bacteria
Photoheterotroph	Organic compounds	Light	Some cyanobacteria, some algae, some purple and green bacteria

*Also called chemolithotrophs. All chemoautotrophs are chemolithotrophs but not all lithotrophs are autotrophic since some can use organic carbon. See text for distinctions between these two groups.

use sunlight as a source of energy and are called ***photoheterotrophs;*** others use organic compounds and are called ***chemoheterotrophs.*** Most of the chemoheterotrophs can use the same organic nutrient, such as glucose, as a source of carbon as well as energy. Chemoheterotrophy is characteristic of many groups of microorganisms, as well as plants and animals, including humans. Table 13-3 characterizes the various nutritional types.

Modes of energy (ATP) production

Energy production in the cell is based on oxidation-reduction reactions in which some molecule donates electrons (the compound to be oxidized) and another molecule accepts electrons (the compound to be reduced). The energy required for removal of electrons is provided by the accompanying reduction. Expressed another way, electrons will flow from a more negative compound to a more positive one with a decline in free energy ($-\Delta G$). Some of this energy is lost as heat, but some can be trapped by the cell in phosphorylated compounds such as the nucleoside triphosphates, ATP and GTP. Most of the energy supplied to endergonic (reactions that require energy) reactions is provided by ATP, which constitutes the "universal currency" of free energy. Heterotrophs, depending on the kind of electron acceptor, can produce ATP using different or modified metabolic pathways, including the following:

1. Fermentation. Fermentation is an oxidation-reduction reaction in which the donor molecule is organic and the acceptor is organic, but oxygen is not involved. It is an incomplete oxidation process in which the end products formed possess some extractable energy. These compounds are usually various types of acids (such as lactic, acetic, propionic) or alcohols that are released by the cell into the environment.
2. Respiration. ***Respiration*** is a process in which either an organic (could be the same compound used in fermentation) or inorganic electron donor is oxidized and the final electron acceptor is oxygen. This is a complete oxidation process resulting in the extraction of all available energy from the donor molecule. Complete oxidations of organic molecules are characterized by the formation of carbon dioxide and water.
3. Anaerobic respiration. The donor of electrons in anaerobic respiration is an organic or inorganic compound. Inorganic molecules other than oxygen, such as nitrate, sulfate, or carbon dioxide, can be the final electron acceptor.

Fermentation and glycolysis. The major gases in the atmosphere at the time of the appearance of the first microorganisms are believed to be carbon dioxide, methane, argon, nitrogen, and water vapor—but no oxygen was present. Organisms that can live in the absence of oxygen are called *anaerobes.* The wide variety of organic materials in the primordial environment, as well as the absence of oxygen, were selective pressures that favored the development of microorganisms whose method of energy production was fermentation. There is a wide variety of substrates that can be fermented: carbohydrates (sugars), amino acids, and pyrimidine and purine bases. Some microorganisms can ferment many different organic substrates, whereas others are limited to one. The most widespread type of fermentation among microorganisms uses the carbohydrate glucose as a source of fuel. The anaerobic breakdown of glucose is called ***glycolysis,*** or the ***Embden-Meyerhoff-Parnas (EMP) pathway*** (after the men who made the most significant contributions to its discovery). The final electron acceptors in glycolysis are two three-carbon units, called ***pyruvate,*** (Figure 13-1), which can be reduced to lactate and other products that are discussed later. Electrons are shifted or transferred during glycolysis, and in the process free energy is released. Some of the energy (25% to 27%) is conserved through the formation of ATP. There are two reactions in glycolysis in which a high-energy phosphoryl group is produced on a substrate and then later transferred to form ATP; they are outlined in Figure 13-2. This type of energy production is called ***substrate phosphorylation*** because the energy is tied up initially in an organic substrate molecule. Overall, the process of glycolysis results in the incomplete oxidation of glucose to two molecules of pyruvate, the formation of NADH, and the formation of ATP. A total of four ATPs is produced during glycolysis, but since two ATPs are used in the pathway, the net gain is only two ATPs per molecule of glucose.

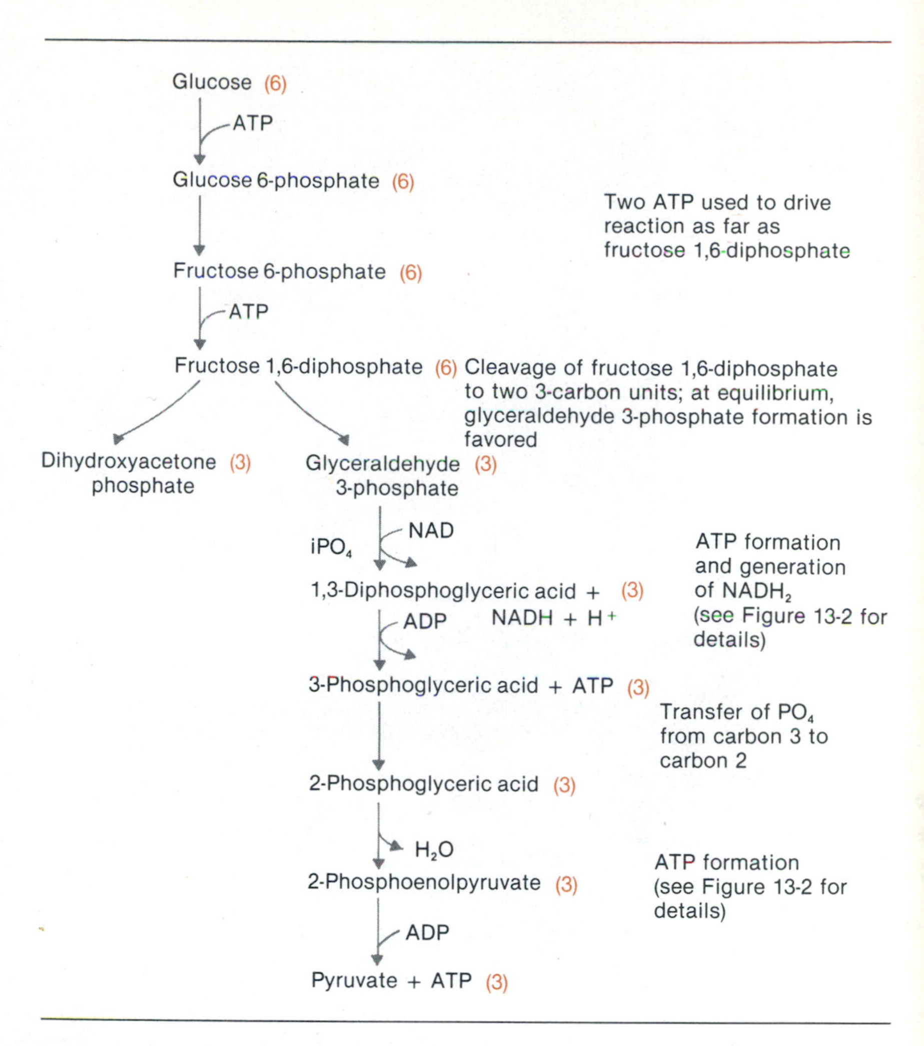

Figure 13-1 Abbreviated scheme for glycolysis (glucose fermentation). Number of carbons for each molecule is indicated in parentheses. Note that final number of carbons equals original number and that ATP and hydrogen have been released.

Products of pyruvate metabolism. Pyruvate, which is the final electron acceptor in glycolysis, can be reduced to lactate—a process that regenerates NAD for reuse. Species of microorganisms differ considerably in their fermentation products. When lactic acid is the sole end product of glycolysis, the fermentation is called ***homolactic,*** but when microbes produce lactic acid plus large amounts of other products, the fermentation is called ***heterolactic.*** Microorganisms have evolved different enzyme systems for the metabolism of pyruvate. Some do not produce any appreciable amounts of lactic acid but do accumulate other products, such as acetic acid, propionic acid, butyric acid, and formic acid, as well as gases, such as hydrogen and carbon dioxide. How these products are derived from pyruvate is illustrated in Figure 13-3. The end products of pyruvate metabolism often give off distinctive odors that are readily detected following growth on media containing glucose. Yeast fermentation, for example, results in the production of alcohols, whereas species of the genus *Clostridium* ferment glucose with the production of butyric acid, which has a distinctive odor. Some microbes that infect the human body ferment tissue components such as the amino acids of proteins, resulting in the production of a foul and distinctive odor. The compounds associated with this odor are cadaverine and putrescine.

In anaerobic environments such as the gastrointestinal tract of animals (for example, the rumen of ruminants), sludge digestors, black muds, and marshes, the breakdown of organic compounds leads to the formation of methane and carbon dioxide. Methane formation (fermentation) is the result of specific bacteria called ***methanogens,*** which are the terminal organisms in the

Glyceraldehyde 3-PO_4 + $H{-}O{-}PO_3H_2$ $\xrightarrow{NAD}$ 1,3-Diphosphoglycerate + $NADH + H^+$

1,3-Diphosphoglycerate $\xrightarrow{ADP}$ 3-Phosphoglycerate + ATP

2-Phosphoglycerate $\xrightarrow{-H_2O}$ 2-Phosphoenolpyruvate + H_2O

2-Phosphoenolpyruvate $\xrightarrow{ADP}$ Pyruvate + ATP

Figure 13-2 Reactions in glycolysis in which high-energy phosphate ($\sim PO_4$) is produced.

microbial food chain. Methanogens rely on the fermentation products of other microorganisms in order to produce energy. One of the most available fermentation products for methanogenic metabolism is acetate. Acetate conversion to methane also requires hydrogen gas, which can be supplied by the metabolism of nonmethanogenic microorganisms or from geochemical processes (volcanic activity). The equation for energy metabolism by methanogenic bacteria using acetic acid as a substrate follows.

$$CH_3COOH + H_2 \longrightarrow CH_4 + CO_2 \quad \Delta G = 8600 \text{ cal/mole}$$

Approximately one third of the human population consistently forms methane from intestinal fermentation by microorganisms. A familial correlation was found among those who produced

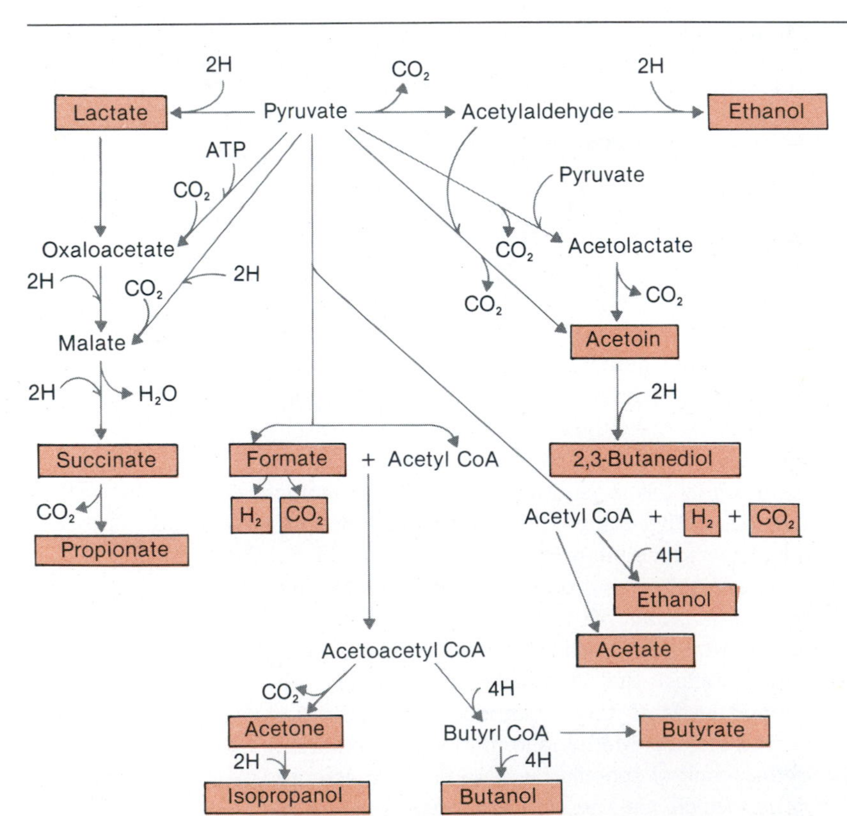

Figure 13-3 Products of pyruvate metabolism. Products enclosed in boxes are those that accumulate in various microbial species.

methane, indicating genetic or environmental influences. Apparently the rate at which food passes through the digestive tract influences methanogenic colonization. If food is digested rapidly and quickly eliminated, the products of fermentation by nonmethanogenic microorganisms are not available in sufficient quantity to support the growth or colonization of methanogens, which divide slowly (note that the amount of energy produced from acetate metabolism is slightly more than the energy derived from the hydrolysis of ATP). Methane is also the most abundant gas evolved from marshes and lake sediments, but some of this methane can be oxidized by certain groups of microorganisms called ***methyltrophs*** (see p. 302).

Methane production by microorganisms is common in ruminants, such as cattle, deer, and camels. Ruminants have a four-chambered stomach, and as much as 100 liters of material can occupy the first two chambers. These chambers are analagous to a large fermentation vat whose anaerobic microorganisms (bacteria and protozoa) break down the major components of grasses, such as cellulose, starch, and pectin, to sugars, which are then decomposed to a variety of fermentation products, such as succinate, acetate, formate, butyrate, propionate, lactate, carbon dioxide, and hydrogen. The various acids are removed by absorption through the rumen wall to be metabolized by the host while carbon dioxide and hydrogen are used by the methanogenic bacteria and converted to methane. Large populations of the rumen bacteria including the methane producers are eventually passed into the gastrointestinal tract where they are digested by intestinal enzymes and provide the ruminant with a source of vitamins, amino acids, and other nutrients required for growth. The ecology of the rumen is discussed later in this chapter.

It is no accident that there are variations in the fermentation end products released by microorganisms. Some of them have been shown to inhibit growth of other species capable of inhabiting the same ecological niche. Anaerobes in the intestinal tract produce fatty acids, such as butyric and propionic acids, that prevent the colonization of pathogenic species such as *Salmonella* and *Shigella,* which can cause gastroenteritis, dysentery, and typhoid fever. There are organisms in the oral cavity, such as *Veillonella,* that do not possess enzymes for the fermentation of glucose but can metabolize the products of glycolysis. These products include such compounds as lactate and acetate, both of which are readily supplied via the fermentation of glucose by other microorganisms. There are also organisms whose growth is severely restricted because of the products of their own metabolism. This problem can also be handled in the microbial world by interactions between microbial species such as the methanogenic bacteria. In anaerobic environments fermentation can result in the production of ethanol, which can be metabolized by species of *Methanobacillus* (organism I) to acetate and hydrogen gas according to the following reaction:

$$\text{Organism I} \quad 2\ CH_3CH_2OH + H_2O \longrightarrow 2\ CH_3COOH + 4\ H_2$$

The evolution of H_2 from NADH (NADH + H^+ → NAD + H_2) is a thermodynamically unfavorable reaction in which the standard free energy is +4.3 kcal/mole under standard conditions (one atmosphere of hydrogen, 25° C, and pH 7.0). The accumulation of hydrogen gas results in a reduction in the amount of substrate that can be metabolized and consequently causes a reduction in growth. If hydrogen gas is removed rapidly from the environment, the free energy change becomes negative and hydrogen can be evolved from NADH. A species of *Methanobacterium* (organism II) supplies this function by removing hydrogen gas and producing methane according to the following reaction:

$$\text{Organism II} \quad 4\ H_2 + CO_2 \longrightarrow CH_4 + 2\ H_2O$$

Thus organism II is necessary for the successful development of organism I in the environment.

Entry of sugars into the glycolytic pathway. It might appear that for an organism to use the glycolytic pathway glucose must be present. Most microorganisms in the environment do not have the luxury of glucose in their surroundings. Polysaccharides, disaccharides, trisaccharides, and other complex carbohydrates are sometimes present, and most are characterized by the presence of glucose as one of their covalently bonded constituents. Some polysaccharides, such as starch and cellulose, have molecular weights of several hundred thousand and are composed of glucose units linked to each other in specific arrangements. The disaccharides maltose, sucrose, and lactose, as well as certain trisaccharides also contain glucose as one of their components. Complex polysaccharides cannot penetrate the cytoplasmic membrane, but many microorganisms produce inducible extracellular enzymes that can hydrolyze the polysaccharide and release the individual glucose or other monosaccharide units so they can be transported into the cell. Some microbes also have the enzymatic potential to hydrolyze disaccharides and trisaccharides found in the environment and release the individual monosaccharides. Monosaccharides such as

H–C=O / OH–C–H / OH–C–H / H–C–OH / H–C–OH / CH_2OH (D-Mannose) + ATP —Kinase→ H–C=O / OH–C–H / OH–C–H / H–C–OH / H–C–OH / $CH_2–O–PO_3H_2$ (D-Mannose 6-phosphate) —Isomerase→ CH_2OH / C=O / OH–C–H / H–C–OH / H–C–OH / $CH_2–O–PO_3H_2$ (Fructose 6-phosphate)

D-Mannose | **D-Mannose 6-phosphate** | **Fructose 6-phosphate**

H–C=O / H–C–OH / OH–C–H / OH–C–H / H–C–OH / CH_2OH (D-Galactose) + ATP —Kinase→ $H_2C–O–PO_3H_2$ / H–C–OH / OH–C–H / OH–C–H / H–C–OH / CH_2OH (D-Galactose 1-phosphate) —UDP-glucose→ $H_2C–O–PO_3H_2$ / H–C–OH / OH–C–H / H–C–OH / H–C–OH / CH_2OH (D-Glucose 1-phosphate) —Mutase→ H–C=O / H–C–OH / OH–C–H / H–C–OH / H–C–OH / $CH–O–PO_3H_2$ (D-Glucose 6-phosphate)

D-Galactose | **D-Galactose 1-phosphate** | **D-Glucose 1-phosphate** | **D-Glucose 6-phosphate**

$H_2C–OH$ / C=O / OH–C–H / H–C–OH / H–C–OH / CH_2OH (D-Fructose) + ATP —Kinase→ $H_2C–OH$ / C=O / OH–C–H / H–C–OH / H–C–OH / $CH_2–O–PO_3H_2$ (D-Fructose 6-phosphate)

D-Fructose | **D-Fructose 6-phosphate**

Figure 13-4 Reactions permitting entry of carbohydrates other than glucose into glycolytic pathway. Boxed areas represent groups that are enzymatically rearranged.

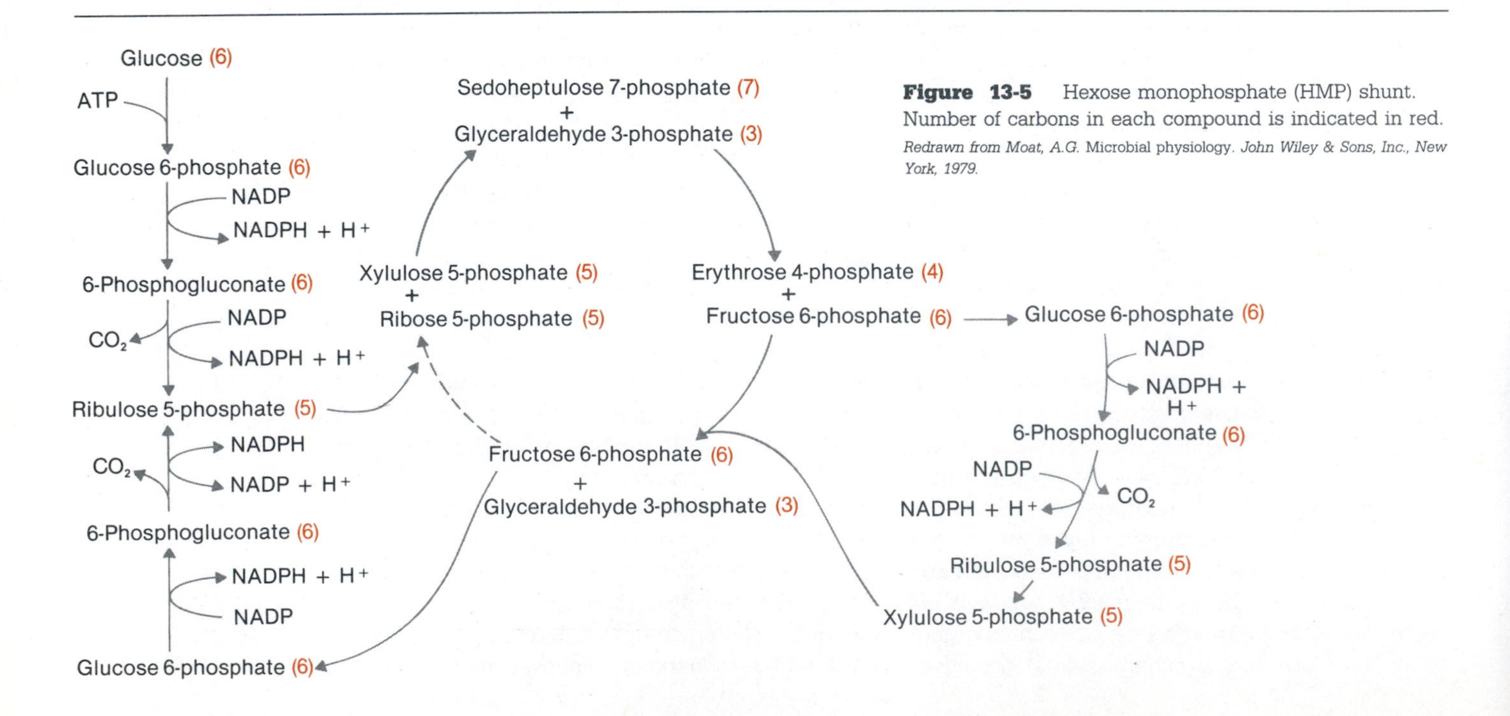

Figure 13-5 Hexose monophosphate (HMP) shunt. Number of carbons in each compound is indicated in red.
Redrawn from Moat, A.G. Microbial physiology. John Wiley & Sons, Inc., New York, 1979.

galactose (a component of lactose), mannose, and fructose (a component of sucrose) can be converted into glycolytic intermediates by phosphorylation and by rearrangement of their hydroxyl groups (Figure 13-4).

Other pathways of glucose catabolism. There are variations of glycolysis in which different intermediates are produced, and pyruvate is not always the final electron acceptor. These pathways represent alternative routes for glucose catabolism, but in other species they may be the sole pathway of fermentation. These pathways include the following.

Hexose monophosphate shunt and phosphoketolase pathways. The hexose monophosphate (HMP) shunt (also called the pentose phosphate pathway) is a cyclic pathway that in its entirety is seldom used by anaerobic microorganisms. The complete pathway is used by aerobic microorganisms and provides ribose 5-phosphate (Figure 13-5), which is important in the biosynthesis of nucleic acids and also provides reduced NADP, which is used in biosynthesis (see "Respiration"). Anaerobes use the HMP pathway only as far as the formation of pentose. The anaerobic oxidation of pentose takes place in another series of reactions called the ***phosphoketolase pathway,*** in which xylulose 5-phosphate is cleaved to glyceraldehyde 3-phosphate and acetylphosphate (Figure 13-6). The latter products are subsequently converted to lactate and ethanol, respectively, with a net gain of one ATP.

Entner Doudoroff (ED) pathway. Glucose is converted in the ED pathway to a six-carbon unit called 2-keto-3-deoxy-6-phosphogluconate, which is later cleaved by an enzyme called aldolase into pyruvate and glyceraldehyde 3-phosphate. The final electron acceptor in the ED pathway (Figure 13-7) is acetaldehyde (in some cases pyruvate may also be reduced to lactate), and only one ATP is gained in this series of reactions. The ED pathway is used frequently by aerobic microorganisms but not anaerobes, who can obtain more energy from the EMP pathway.

Respiration. It is speculated that oxygen accumulated in the primordial atmosphere only after the appearance of oxygenic photosynthetic microorganisms and that heterotrophs later evolved the machinery for using oxygen as a final electron acceptor. The potential for energy release during respiration is greater than in fermentation because the organic electron donor is completely oxidized to carbon dioxide and water with the release of many electron pairs:

$$\text{Hydrogenated substrate} \xrightarrow{\text{Dehydrogenation}} 2\,H^+ + 2\,e^- + (CO_2 \text{ and } H_2O \text{ release})$$

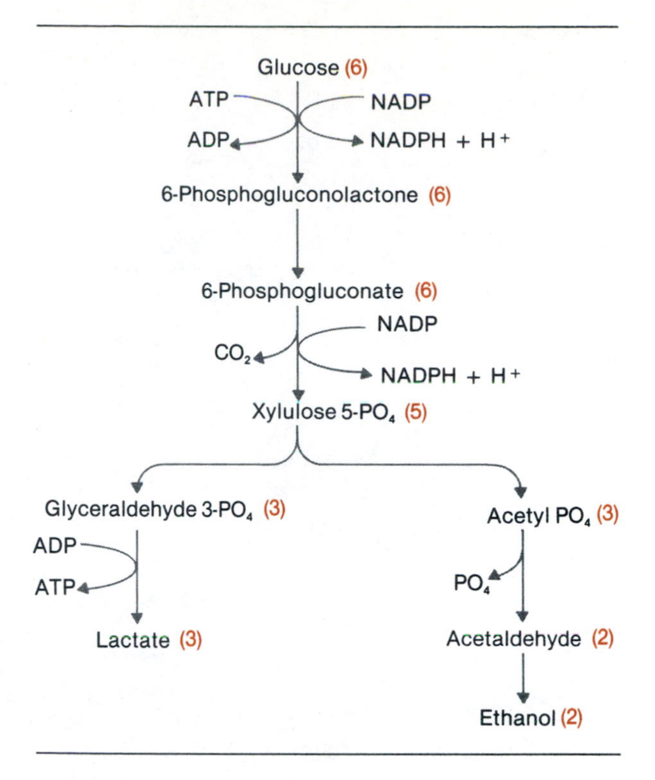

Figure 13-6 Phosphoketolase pathway. Number of carbons in each intermediate is indicated in color.

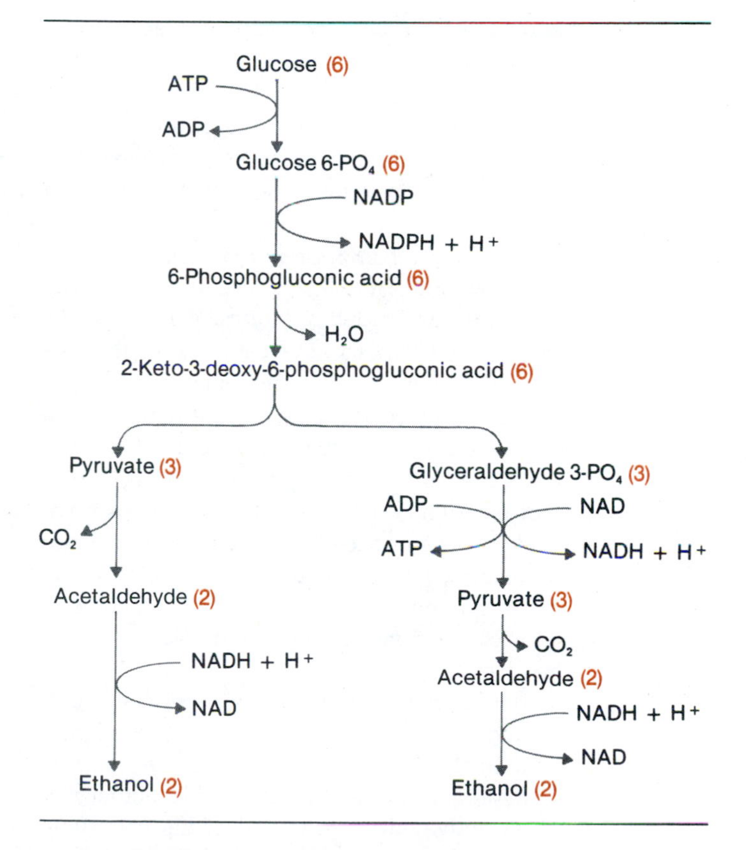

Figure 13-7 Entner-Doudoroff pathway. Number of carbons in each intermediate is indicated in color.

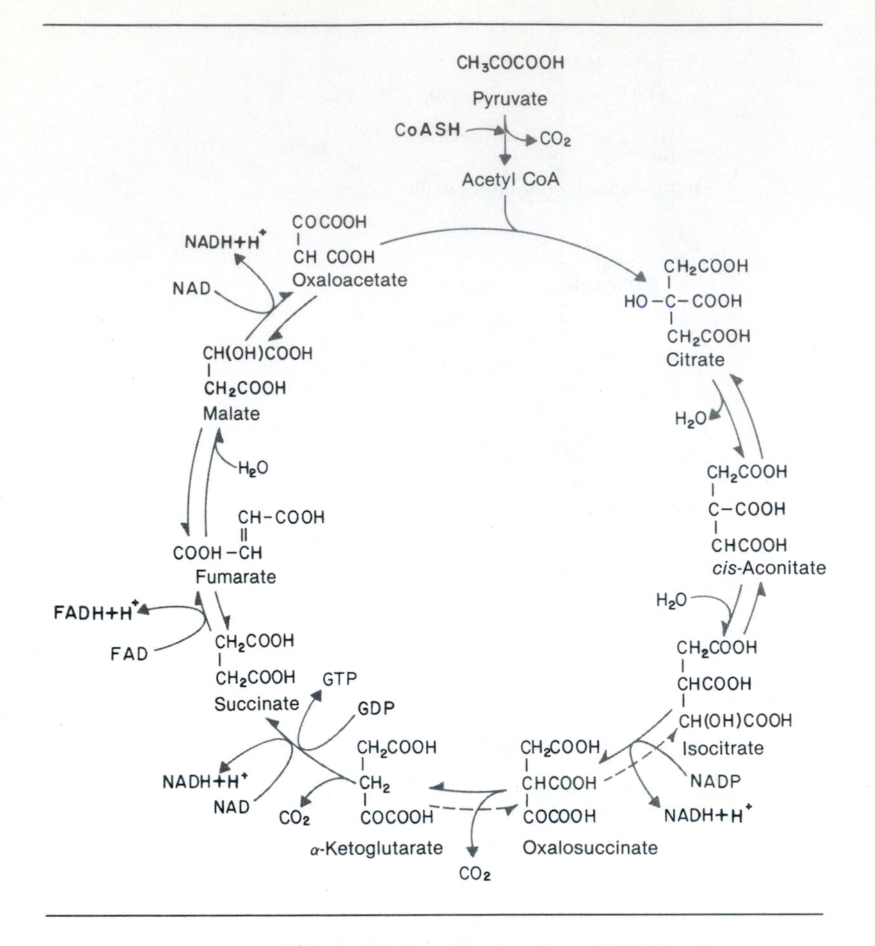

Figure 13-8 Tricarboxylic acid (TCA) cycle.

The first thing to consider in respiration is how the organic compound becomes completely oxidized; the second is how energy can be realized in the process.

How organic compounds are completely oxidized (tricarboxylic acid [TCA] cycle). The TCA cycle, also called the ***citric acid*** or ***Krebs cycle,*** can be viewed as a pathway for the oxidation of acetate derived from the fermentation of glucose, but acetate can also be derived from the catabolism of proteins and lipids. The TCA cycle is responsible for the release of a large number of electrons and protons and produces compounds that can be used as intermediates for biosynthesis (see Chapter 14). The complete TCA cycle is summarized in Figure 13-8, and some of the major reactions are discussed below.

1. Before initiation of the cycle, pyruvate, produced in glycolysis, is oxidized to acetate with the release of carbon dioxide and two hydrogen atoms. The acetate is actually in an activated state as part of the coenzyme A (CoA) molecule (Figure 13-9). CoA is a carrier of acetyl groups, and the bond between the acetate and the sulfhydryl (SH) group of CoA is a high-energy linkage. Hydrolysis of the SH-acetate linkage results in the release of free energy higher than that obtained from the hydrolysis of the terminal phosphate of ATP (Table 13-2). The hydrogen atoms and their electrons are picked up by specific carriers such as NAD (nicotinamide adenine dinucleotide) to be transported for the purpose of ATP formation (see "Electron Transport").
2. The acetyl group of acetyl CoA is transferred enzymatically to oxaloacetate (OAA), a four-carbon unit, to produce citrate, a six-carbon compound. It is this reaction that initiates the TCA cycle and regenerates CoA.
3. Citrate, in a series of reactions, loses a carbon atom as carbon dioxide and two hydrogen atoms to produce a five-carbon intermediate called α-ketoglutarate (α kg). NAD again is the carrier of the hydrogen atoms and electrons.
4. Alpha ketoglutarate loses a carbon atom as carbon dioxide to form succinate, a four-carbon intermediate, and two electrons and a hydrogen atom are released and trapped by NAD.
5. Succinate is oxidized to fumarate, a four-carbon intermediate, with the release of hydrogen, which is captured by flavine adenine dinucleotide (FAD) Fumarate is hydrated to become malate.
6. Malate, a four-carbon compound, is oxidized to the four-carbon oxaloacetate with the release of hydrogen, which is captured by NAD.

The TCA cycle can be said to have resulted in the following: (1) pyruvate generated during glycolysis loses three carbon atoms as carbon dioxide in three reactions; (2) electrons and hydrogen atoms are generated in five reactions and captured by electron carriers, such as NAD and FAD (their fate is discussed shortly); and (3) intermediates, such as oxaloacetate, α-kg, and malate, can be used in the biosynthesis of amino acids as well as purines and pyrimidines

Electron transport. We have already noted that during the oxidation of certain cellular fuels, electrons are released and captured by specific electron acceptors such as NAD, NADP (nicotinamide adenine dinucleotide phosphate), and FAD. These electron acceptors are coenzymes that function in coordination with dehydrogenases. They transfer their electrons to a respiratory chain of proteins and nonproteins located in the cytoplasmic membrane of prokaryotes and the

Figure 13-9 Formula for acetyl coenzyme A.

Acetate unit
Cysteine unit
Pantothenate unit
ATP unit
Coenzyme A unit

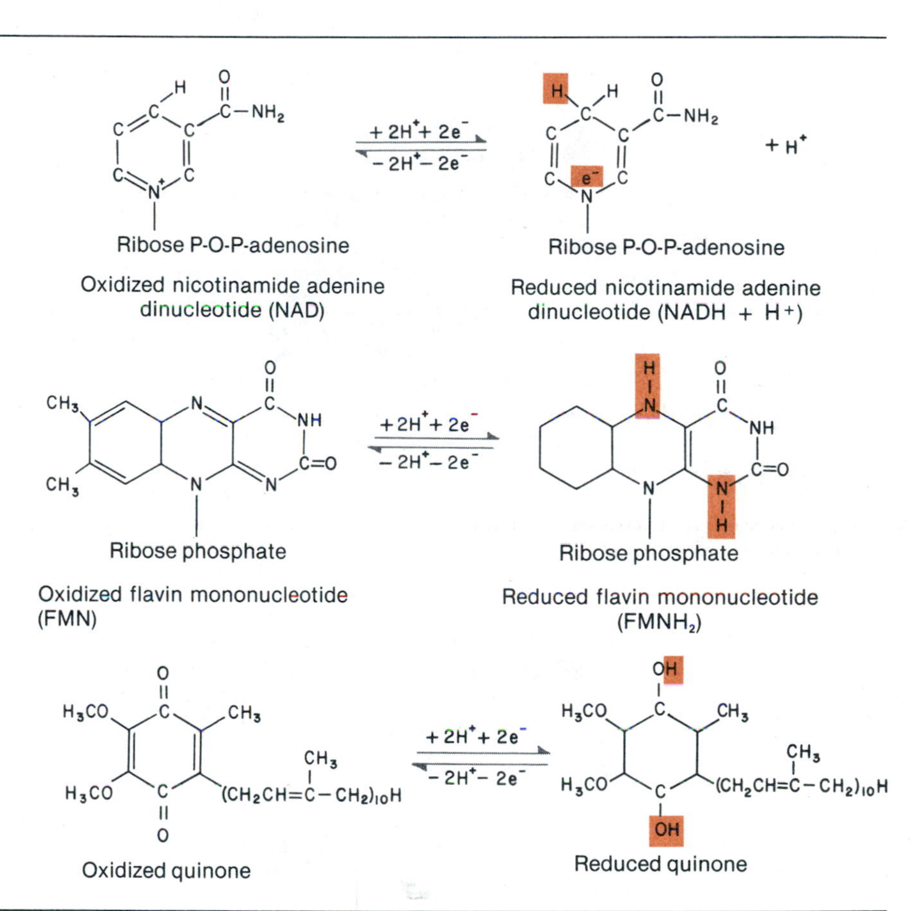

Figure 13-10 Modified chemical structures of some common carriers in electron transport system. Nicotinamide adenine dinucleotide (NAD), flavin mononucleotide (FMN), and quinone. Nicotinamide adenine dinucleotide phosphate (NADP) is formed by adding phosphate to carbon 2 of adenosine. Flavin adenine dinucleotide (FAD) is formed by adding adenosine 5-phosphate to ribose phosphate of FMN.

mitochondria of eukaryotes. It is during the passage of electrons that energy is released and captured in the form of ATP.

The respiratory chain is made up of different classes of electron carriers, which have minor modifications in certain species. Basically, the carriers include flavoproteins and cytochromes, which are proteins, and the quinones, which are nonproteins. The first electron carrier found in the respiratory chain of most microbial systems is a ***flavoprotein,*** which contains a prosthetic group of either FMN (flavine adenine mononucleotide) or FAD (Figure 13-10). The flavoproteins accept hydrogen atoms (protons) and electrons from reduced NAD generated during the oxidation of fuels and transfer them to the ***quinones.*** The quinones release the protons into the cytoplasm, but the electrons are transferred to a series of en-

Table 13-4 The oxidation-reduction potential (E°, expressed in volts) of some redox pairs in microbial systems

REDOX PAIRS	OXIDATION-REDUCTION POTENTIAL (VOLTS)
CO_2/formate	−0.432
H^+/H_2	−0.420
NAD/NADH	−0.320
S°/HS^-	−0.280
FAD/FADH	−0.22
FMN/FMNH (flavoprotein)	−0.19
$SO_3^=/S$	−0.11
Menaquinone ox/red	−0.075
Cytochrome b_{ox}/cytochrome b_{red}	+0.030
Ubiquinone ox/red	+0.11
Cytochrome c_1 ox/cytochrome c_1 red	+0.23
Cytochrome a_3 ox/cytochrome a_3 red	+0.385
NO_3^-/NO_2^-	+0.43
Fe^{+++}/Fe^{++}	+0.770
O_2/H_2O	+0.82

The numerical values for the oxidation-reduction potentials reflect the reduction potential relative to the $H^+ + e^- \longrightarrow \frac{1}{2} H_2$ half reaction, which is −0.42 volts at pH 7.0. A complete oxidation-reduction reaction is composed of two half reactions, one in which a substance gains electrons and one in which a substance loses electrons. The E°, therefore, represents the voltage required to remove an electron from a substance. One can calculate the free energy change that occurs during the transfer of a pair of electrons between two redox pairs when the E° values are known. The equation representing this change is $\Delta G^{o'} = nF\ \Delta E^{o'}$, where $\Delta G^{o'}$ is the standard free energy at pH 7.0; n is the number of electrons transferred; F is a faraday, a constant equal to 23,000 cal/volt; and $\Delta E^{o'}$ is the difference in volts between the two redox pairs. For example, the free energy potential between NAD/NADH and O_2/H_2O during the transfer of electrons from NAD to oxygen follows.

$$\begin{aligned} \Delta G^{o'} &= -2 \times 23{,}000 \text{ cal/volt} \times (-0.320) - (+0.82) \\ &= 46{,}000 \text{ cal/volt} \times 1.40 \text{ volts} \\ &= -52{,}440 \text{ cal/mole} \end{aligned}$$

zymes called the ***cytochromes,*** which are designated by letters such as cytochrome a (cyt a), cytochrome b (cyt b) and so on. The reversible oxidation of the cytochrome takes place on an iron atom that is bound to a porphyrin prosthetic group on the molecule.

$$Fe^{2+}_{ox} + 2\,e^- \longrightarrow Fe^{3+}_{red} + e^-$$ (Cytochromes can accept and transfer only one electron at a time.)

This series of oxidations and reduction is terminated by reduction of the final electron acceptor, which in aerobic respiration is oxygen. Reduction of oxygen is catalyzed by cytochrome a*, also called cytochrome oxidase. The overall reaction from the capture of two electrons by NAD to the reduction of oxygen follows.

* Other terminal cytochromes may be part of the respiratory chain, such as cytochrome o, but they still have the same function.

$$NADH + H^+ + \frac{1}{2} O_2 \longrightarrow NAD + H_2O$$

Cytochrome c is missing in some bacterial species, and this characteristic is the basis of a laboratory test called the oxidase test. The oxidase test is used for the presumptive identification of neisserial species (*Neisseria meningitidis* and *N. gonorrhoeae* are the causative agents of a type of meningitis and gonorrhea, respectively). The test is usually performed by removing a portion of a bacterial colony from an agar plate and rubbing it on a strip of filter paper that has been impregnated with oxidase reagent. The area rubbed with bacteria will turn dark purple in 8 to 10 seconds. The oxidase reagent is tetramethyl p-phenylenediamine-dihydrochloride, a dye that is oxidized by cellular cytochrome oxidase in the presence of oxygen to the colored compound indophenol. There is no color change (oxidase negative) if the bacterial cells contain no cytochrome c.

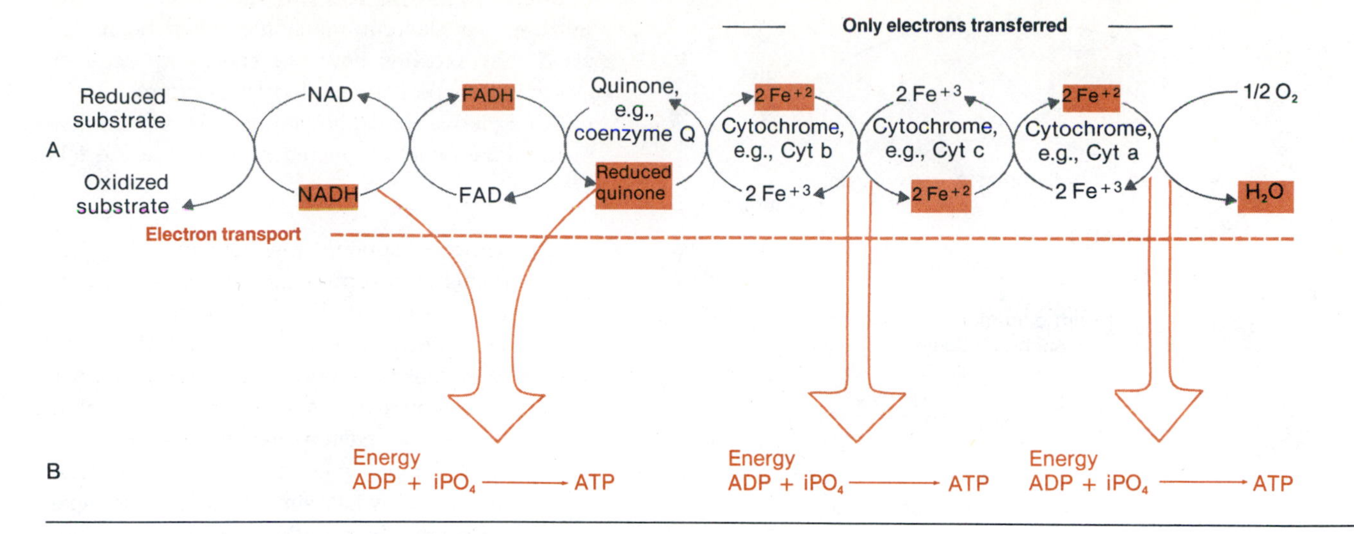

Figure 13-11 Coupling of electron transport *(A)* to oxidative phosphorylation *(B)*. During oxidation of substrate, hydrogens and concomitant electrons are removed and transferred along a series of transport molecules. Hydrogens are carried only as far as quinone, whereas electrons are carried along entire respiratory chain. Sufficient energy (oxidation-reduction potential) is generated at three points along chain, which is then coupled to oxidative phosphorylation and formation of ATP. In final step, deenergized electrons and hydrogens are combined with oxygen to produce water. Reduced respiratory components are demonstrated in shaded boxes. Some reduced substrates such as succinate may transfer their hydrogen atoms to flavoprotein component of chain, in which cases only two ATP will be produced. There is variation in cytochromes of different bacterial species, and in many, alternative cytochrome pathways are also present.

Oxidative phosphorylation (ATP formation). We discussed previously that during oxidation-reduction processes free energy can be realized. The burning of glucose by molecular oxygen, for example, releases considerable free energy (−686,000 cal/mole). If the cell were to burn fuel in this manner, it would literally burn itself up. The cell oxidizes glucose and other fuels by ***dehydrogenations,*** which release hydrogen (protons) and electrons. The potential energy in the electrons is captured by electron-deficient molecules such as NAD. Reduced NAD is a high-level energy compound, and if it were used by the cell to reduce oxygen directly, the amount of free energy realized would be:

$$\text{NAD} + \text{H}^+ + \tfrac{1}{2}\,\text{O}_2 \longrightarrow \text{NAD}^+ + \text{H}_2\text{O}$$
$$\Delta G = -52{,}440 \text{ cal/mole}$$

The 52,440 calories could not be captured all at once by a single ATP molecule, and much of the energy would be lost as heat. The cell has evolved a chain of electron transport components (the respiratory chain) so that electrons are transferred from atom to atom in molecules of lower and lower energy. Consequently, small bursts of energy are released in this downhill transfer of electrons until the final electron acceptor, oxygen is reached.

Each of the electron donors in the respiratory chain has a characteristic standard oxidation-reduction potential (E°, expressed in volts), which represents their relative electronegativity (Table 13-4). Electrons tend to travel from a more negative carrier to a more positive one, and the decline in free energy between adjacent respiratory components is small enough that energy is released in small increments. The amount of energy released is related to the magnitude of difference in the oxidation-reduction potentials of the adjacent respiratory-chain components.

The transfer of a pair of electrons down the respiratory chain results in the formation of three molecules of ATP. There are three points in the chain (Figure 13-11) where there is a sufficient drop in free energy that a high-energy phosphate can be donated to ADP to form ATP—a process called ***oxidative phosphorylation.*** If the electrons removed from organic substrates during oxidation are first captured by FAD, then the NAD step is bypassed and only two ATPs are produced (Figure 13-11). Since the free energy of hydrolysis of a single molecule of ATP is −7000 cal/mole, the efficiency of the respiratory chain can be calculated as follows.

$$3(-7000 \text{ cal/mole})/-52{,}440 \text{ cal/mole or}$$
$$\frac{-21{,}000 \text{ cal/mole}}{-52{,}440 \text{ cal/mole}} = 40\%$$

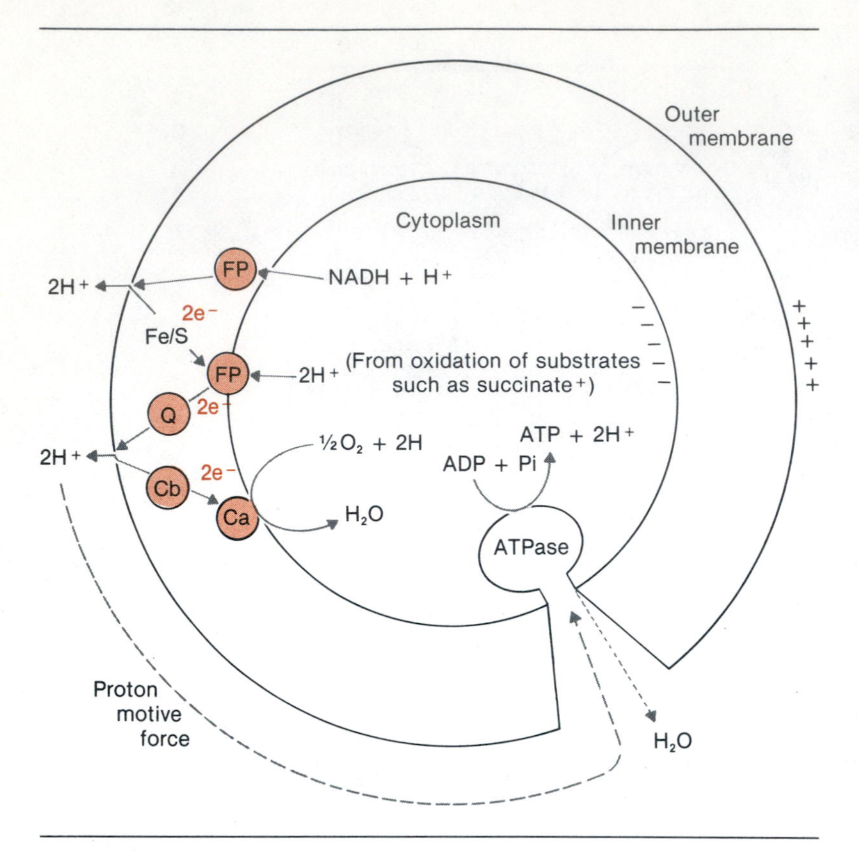

Figure 13-12 Schematic representation of proton translocation and ATP synthesis according to chemiosmotic theory. Protons generated by metabolism and transported by carriers in respiratory chain produce proton motive force whose energy is used to drive ATPase complex in direction of ATP synthesis. *FP*, Flavoprotein; *Fe/S*, iron-sulfur protein; *Q*, quinone; *Cb*, cytochrome b; *Ca*, cytochrome a or final electron acceptor. Some electron carriers in this scheme may vary depending on microbial species.

The total number of ATPs gained from the complete oxidation of glucose follows.

Glucose oxidized to pyruvate (substrate phosphorylation)		2 ATP
Glyceraldehyde 3-phosphate⇌1,3-diphosphoglycerate + 2 NADH (to electron transport system)		6 ATP
Pyruvate → Acetyl CoA + 2NADH (to electron transport system		6 ATP
TCA cycle	6 NADH	18 ATP
TCA cycle	2 FADH	4 ATP
2 succinyl~CoA → 2 succinate + 2ATP*		2 ATP
	TOTAL	38 ATP

The 38 ATPS gained through the complete oxidation of glucose to carbon dioxide and water indicates the relative efficiency of the respiratory chain in extracting energy compared with the two ATPs gained by the incomplete oxidation of glucose under anaerobic conditions.

* The overall reaction involves GTP:

$$\text{Succinyl}{\sim}\text{CoA} + \text{GDP} \xrightarrow{\text{iPO}_4} \text{Succinate} + \text{GTP}$$
$$\text{GTP} + \text{ADP} \longrightarrow \text{GDP} + \text{ATP}$$

Coupling of electron transport and ATP formation. No clearcut mechanism has been devised that explains how the energy released in electron transport is coupled to the formation of ATP (oxidative phosphorylation). Three mechanisms have been suggested and include the following:

1. Chemical coupling mechanism. This hypothesis is based on the availability of carrier proteins at the ATP-forming sites. These intermediates trap energy generated by electron transport and transfer it to other high-energy systems such as ATP. No such intermediates have as yet been isolated.
2. Conformational mechanism. This mechanism is somewhat similar to the chemical mechanism except that the intermediate carrier can exist in different conformations. The energy released during electron transport is trapped in the intermediate in the form of a conformational change. The noncovalent forces that hold this energy within the intermediate can be broken to release the energy for the phosphorylation of ADP to ATP. The intermediate, following energy release, assumes a low-energy conformation state.
3. Chemiosmotic hypothesis. Mitchell in 1961 proposed a mechanism for coupling electron transport to ATP formation based on the impermeability of the mitochondrial membrane to hydrogen ions (protons), as well as the arrangement of the respiratory carriers across the membrane. This mechanism, called chemiosmosis, is the most widely accepted hypothesis to date. Respiratory enzymes, as well as other energy-generating systems, are arranged in loops across or within the cytoplasmic membrane of prokaryotes and mitochondrial membrane of eukaryotes (Figure 13-12). Protons and electrons, generated via oxidation of fuels and accepted by NADH, are transferred to flavoproteins located on the inner portion of the membrane. The two protons are extruded outside the membrane, but the electrons are returned by iron-sulfur proteins to the inner surface of the membrane. These electrons, together with protons from the cytoplasm (also generated from the oxidation of fuels) are transferred to quinones, which also extrude protons to the outside of the membrane. The electrons are shuttled to the cytochromes and finally to the terminal cytochrome where oxygen is reduced to water. The extrusion of protons creates

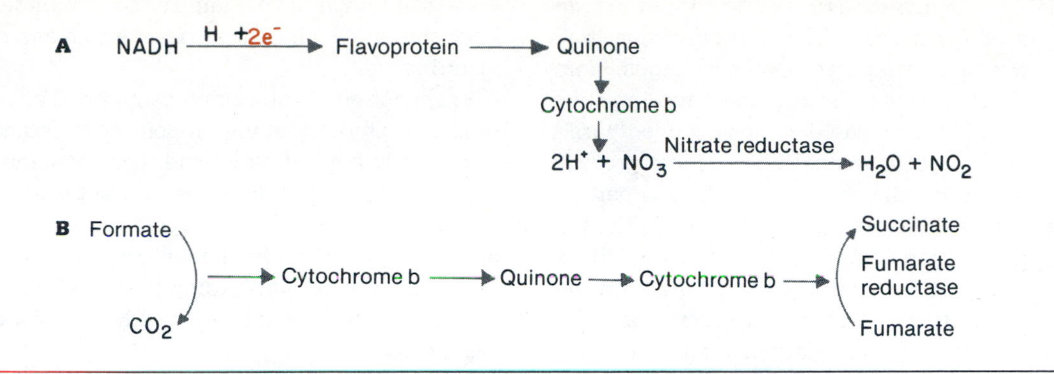

Figure 13-13 Anaerobic respiration in heterotroph in which nitrate (NO_3), **A,** is final electron acceptor and fumarate, **B,** is final electron acceptor. NADH and formate are the electron donors for these reactions; however, other donors are possible.

an electrochemical gradient across the membrane in which the outside becomes positively charged and acidic while the inside becomes negatively charged and alkaline. The gradient generated by proton movement is referred to as the ***proton motive force (PMF).*** The PMF provides energy that can be used to rotate flagella, provide for the active transport of certain ions, sugars, and amino acids, and, as we shall see later, for the phosphorylation of ADP to ATP.

The synthesis of ATP is brought about by the translocation of protons from outside the membrane into the cytoplasm of the cell. Protons transported in this way exhibit a drop in potential, and the free energy that is generated is used to phosphorylate ADP to ATP. The membrane is ordinarily impermeable to hydrogen ions, but a channel is provided by an enzyme complex called ATPase, which consists of two proteins: a headpiece and a basepiece. The basepiece is arranged across the membrane and serves as the channel for proton translocation. ATPase catalyzes a reversible reaction in which ATP can be produced or hydrolyzed.

$$ADP + PO_4 \xrightleftharpoons{ATPase} ATP$$

The hydrolysis of ATP is associated with the translocation of hydrogen ions from the cytoplasm of the cell to the outside portion of the membrane. The free energy from electron transport, however, is used to reverse the hydrolyzing activity of ATPase, and ATP is produced.

Anaerobic respiration. The mechanism of electron transport and oxidative phosphorylation is basically the same in anaerobes as in aerobes with minor variations.

1. The final electron acceptor is not oxygen but is some other inorganic molecule, such as nitrate (NO_3), sulfate (SO_4), nitrite (NO_2), or an organic molecule, such as fumarate.
2. The terminal enzyme component in electron transport in anaerobes is not cytochrome oxidase but enzymes such as reductases that catalyze the reduction of nitrate to nitrite, sulfate to hydrogen sulfide, and nitrite to molecular nitrogen. One of the most frequently used inorganic electron acceptors is nitrate, whereas fumarate is a more common organic electron acceptor (Figure 13-13). The synthesis of terminal reductases is repressed and their activities inhibited by oxygen. This mechanism is especially important in facultative anaerobes that switch from anaerobic to aerobic growth when oxygen becomes available.
3. Less energy is realized during electron transport under anaerobic conditions. The difference in the oxidation-reduction potential between the donor and acceptor (for example, between NADH and nitrate) (see Table 13-3) is not sufficient to generate three molecules of ATP. Depending on the donor and acceptor, the number of ATP molecules generated during electron transport under anaerobic conditions is usually between one and two. The rate of growth of anaerobes is, therefore, considerably slower than that of aerobes. A facultative anaerobe, for example, under anaerobic conditions might require 48 hours of optimal growth conditions to produce a population of cells that under aerobic conditions could be established in 16 hours or less.

Anaerobe metabolism in the rumen (an example of symbiosis). The gastrointestinal tract of mammals is composed of an esophagus, stomach, small intestine, cecum, and large intestine. In many mammals, such as humans, food passes into the stomach where high acidity prevents colonization by microorganisms. Food is partially digested in the stomach through the activity of host enzymes and acids. Further chemical digestion occurs in the small intestine where the nutrients produced are absorbed into the animal's bloodstream. Only in the large intestine do microorganisms exist in large numbers, and there they ferment undigested residues. In the ruminants (sheep, cattle, goats, giraffes) plant material is indigestible by mammalian enzymes, and microorganisms are the primary digesters of food. The rumen stomach, unlike the nonruminant stomach, is divided into compartments, the first of which acts like a large fermentation vessel that in cattle may hold up to 100 liters. This first chamber is kept at an alkaline pH and a temperature of 39° C, which is ideal for microbial growth. The number of microorganisms present in the rumen equals that present in the large intestine of nonruminant mammals. The oxidation-reduction potential in the rumen is approximately −30 mV; therefore all the indigenous microorganisms engage in anaerobic metabolism. The principal microorganisms found in the rumen are bacteria, but some protozoa are also present. Most of the bacteria are gram-negative rods, but some cocci are also present.

The diet of rumens consists primarily of plant material that is made up of cellulose, hemicellulose, and pectin but little protein, fats, or digestible carbohydrates. Fibrolytic microorganisms, which make up only 5% of the rumen population, produce extracellular enzymes that break down plant polysaccharides into soluble sugars such as glucose and cellobiose. Some of these decomposers include species of *Bacteroides, Ruminococcus, Clostridium,* and *Streptococcus.* A large percentage of the soluble sugars are utilized by nonfibrolytic microorganisms as sources of energy, and thus their population numbers are maintained in the rumen. The products of rumen fermentation include gases such as hydrogen (H_2), methane (CH_4), and carbon dioxide (CO_2) plus a variety of organic acids such as acetic, propionic, formic, succinic, lactic, and valeric. Gases produced in the rumen are eliminated by constant belching by the animal. The organic acids pass through the rumen epithelium and are used as energy sources by the animal. This represents a drastic departure from most mammals, in whom sugars are the principal energy source. Thus, the rumen is an example of a symbiotic relationship in which both microorganism and ruminant benefit.

Examination of the rumen contents at any one time reveals a constant proportion of propionic acid, acetic acid, butyric acid, and methane but little hydrogen gas, formic acid, lactic acid, or succinic acids. Studies using radioactive isotopes of these compounds have indicated that they are rapidly metabolized according to the following reactions by some of the indicated microbial species.

1. $HCOOH$ (formic acid) $\longrightarrow H_2 + CO_2$
   (Ruminococcus albus)
2. $4\,H_2 + CO_2 \longrightarrow CH_4 + 2\,H_2O$
   (Methanobacterium ruminantium)
3. $CH_3CHOHCOOH$ (Lactic acid) $\longrightarrow CH_3COOH$ (Acetic acid) or CH_3CH_2COOH (Propionic acid) or $CH_3CH_2CH_2COOH$ (Butyric acid)
4. $COOHCH_2CH_2COOH$ (Succinic acid) $\longrightarrow$ CH_3CH_2COOH (Propionic acid) + CO_2

Experiments with pure cultures of organisms that produce methane (methanogens) and nonmethanogenic bacteria indicate that methanogens play a most important role in rumen metabolism. In the absence of methanogens the products of metabolism in the rumen are hydrogen, carbon dioxide, formic acid, and succinic acid. In the presence of methanogens the products of metabolism in the rumen are primarily carbon dioxide, methane, and unusually low levels of hydrogen. Methanogens perform an important role by capturing electrons produced during fermentation and converting the copious amounts of hydrogen produced by nonmethanogenic bacteria to methane. By the use of electrons for the formation of hydrogen and methane, pyruvate is converted to acetyl CoA, and energy is generated in the form of ATP. If the electrons generated during fermentation were not used to produce hydrogen and methane, pyruvate would be reduced to ethanol and lactate, and considerable potential energy would be lost.

Most of the rumen bacteria can use ammonium (NH_4) as a source of nitrogen; thus inorganic nitrogen is converted to amino acids and thence to protein in the microbial cell. Microbial cells pass out of the rumen together with undigested plant material, such as lignin, into other stomach compartments. These microorganisms are digested by host enzymes such as proteases; thus the microorganism serves as a source of nitrogen for the ruminant and not protein from ingested food. Since ammonium is a source of nitrogen for the ruminant bacteria, urea (NH_2-CO-NH_2), which is normally a waste product, can be used as a supplemental nitrogen source for the ruminant.

The rumen is a delicately balanced type of chemostat in which soluble sugars are at a relatively low concentration because of slow digestion of plant material. If excess soluble sugars are added to the diet, fermentation occurs rapidly, producing excess acid and dropping the rumen pH to as low as 4.0. Such a change in the rumen causes loss of rumen motility, and the animal may die within 24 hours.

The role of protozoa in rumen metabolism is not completely understood. They can ferment various sugars and acids produced by bacterial catabolism, but they also feed on bacteria. It has been suggested that protozoa through predation control the level of bacteria in the rumen.

Effect of oxygen on energy production

Microorganisms can be divided into four groups based on the role that oxygen plays in their nutrition.

1. ***Obligate aerobes*** have an absolute requirement for oxygen usually at partial pressures of 0.2 atm.
2. ***Microaerophiles*** require oxygen but usually at partial pressures less than that present in air.
3. ***Obligate anaerobes*** cannot grow in the presence of oxygen.
4. ***Facultative anaerobes*** can use aerobic or anaerobic respiration or anaerobic fermentation systems, depending on the availability of oxygen.

Facultative anaerobes and microaerophiles can use oxygen as a terminal electron acceptor for energy production, but the growth of microaerophiles and even some aerobes can be inhibited, depending on the concentration of oxygen. Oxygen is lethal to some obligate anaerobes, but in others oxygen may be tolerated but no growth occurs. Aerotolerant anaerobes such as lactic acid bacteria, however, can grow in the presence of oxygen.

Oxygen and aerobic bacteria. Cells that use oxygen as a terminal electron acceptor possess cytochrome oxidase or a comparable oxidase to catalyze the final transfer of electrons and the subsequent formation of water. Nearly all aerobes possess additional enzymes that use oxygen as an electron acceptor, but products other than water are formed, and no energy is produced. Two of these products, hydrogen peroxide and superoxide, are toxic to the cell. Most cells synthesize flavoprotein oxidases that catalyze the divalent (a two-electron transfer) reduction of oxygen with the formation of hydrogen peroxide (H_2O_2).

$$H_2 + O_2 \xrightarrow[\text{oxidase}]{\text{Flavoprotein}} H_2O_2$$

Microorganisms produce two kinds of enzymes that can destroy hydrogen peroxide: ***catalase*** and ***peroxidase.*** Catalase catalyzes the following reaction.

$$2\ H_2O_2 \xrightarrow{\text{Catalase}} 2\ H_2O + O_2$$

Peroxidases catalyze the reduction of hydrogen peroxide to water through the oxidation of organic compounds.

Oxygen can also accept a single electron (univalent reduction) from reduced flavines and be converted to the superoxide anion, O_2^-, as well as other intermediates such as the hydroxyl radical. The sequence of univalent reductions of oxygen can be represented in the following way:

$$O_2 \xrightarrow{e^-} \underset{\text{Superoxide anion}}{O_2^-} \xrightarrow{e^- + 2H^+} \underset{\text{Hydrogen peroxide}}{H_2O_2} \xrightarrow{e^- + H^+} \underset{\text{Hydroxyl radical}}{OH^-} \xrightarrow{e^- + H^+} H_2O$$

The superoxide radical is more stable than the hydroxyl radical, but the latter is believed to be the intermediate that is most toxic to the cell. Where superoxide and hydrogen peroxide are generated simultaneously, the possibility of hydroxyl radical production exists. The microbial cell prevents the buildup of superoxide and other intermediates through the synthesis of the enzyme superoxide dismutase, which removes the toxic anion according to the following reaction:

$$O_2^- + O_2^- \xrightarrow[\text{dismutase}]{\text{Superoxide}} H_2O_2 + O_2$$

Hydrogen peroxide is not as deleterious as once suspected, and in most aerobes catalase or peroxidase can handle any excess of this compound. Nearly all aerobic cells possess superoxide dismutase, but in some microbial species excess oxygen may inhibit the enzyme. Key enzymes important in energy formation may be inhibited by elevated concentrations of superoxide, and this may be the reason that microaerophiles cannot handle oxygen at partial pressures greater than 0.2 atmospheres.

The formation of superoxide dismutase indicates that in addition to the divalent reduction of oxygen via the usual respiratory enzymes such as cytochrome oxidase some univalent reductions do occur. A cell, for example, might reduce 95% of the available oxygen to water, but 5% could be converted to the superoxide radical.

Facultative anaerobes have the luxury of growing in the presence or absence of oxygen. Respiratory enzymes in facultative anaerobes are induced by oxygen when the gas is present in the environment, but anaerobic energy-forming systems are inhibited. The enzyme lactate dehydrogenase, for example, which catalyzes the conversion of pyruvate to lactate during fermentation, is not only inhibited by oxygen, but its synthesis is also repressed. Yet in aerotolerant anaerobes such as lactic acid bacteria oxygen does not inhibit the enzyme.

Oxygen and anaerobic bacteria. Obligate anaerobes are not able to grow in the presence of air, but they do show different sensitivities to oxygen. Some anaerobic bacterial populations are almost completely killed when exposed to oxygen for 3 minutes, whereas others remain viable even though they cannot grow. Indifference to oxygen by some anaerobes is the result of their inability to reduce oxygen; therefore toxic intermediates are not produced. In other anaerobes the concentration of oxygen is the critical factor. Obligate anaerobes are often found associated with facultative anaerobes in areas where oxygen is present. Facultative anaerobes can remove oxygen from the environment by using it in aerobic respiration, thus making the environment suitable for the growth of obligate anaerobes. Studies with oxygen-tolerant anaerobes have demonstrated that a positive correlation exists between oxygen tolerance and the content of superoxide dismutase in the cell. It is now clear that microbial cells whether aerobes or anaerobes rely on scavenger systems such as catalase, peroxidase, and superoxide dismutase to remove toxic intermediates produced by the reduction of oxygen.

Energy production from organic substrates other than carbohydrates. One might get the impression after reading this chapter that heterotrophic microorganisms can use only carbohydrates as organic energy sources. It is true that the principal pathway involved in energy production is the Embden-Meyerhoff-Parnas pathway, where glucose and other sugar intermediates are oxidized. Glucose or other simple sugars are not primary sources of energy for many microorganisms in their environment. Instead, they must extract energy from other organic compounds that might be present, such as amino acids, purines or pyrimidines, fatty acids, and a variety of hydrocarbons. These compounds are either channeled directly into energy-producing pathways or oxidized to products that can be channeled. Some of the reactions that are used by the cell to derive energy from organic compounds other than carbohydrates follow.

Amino acid oxidation or hydrolysis. Amino acids may be free in the environment, or they may be the products of protein hydrolysis by proteases excreted by microorganisms. Amino acids can be oxidatively deaminated to keto acids, which can directly enter the TCA cycle. Other amino acids may be reductively deaminated to produce saturated fatty acids, for example:

$$R{-}CH(NH_2){-}COOH \longrightarrow R{-}CH_2{-}COOH + NH_3$$

A very few microbes can obtain their energy directly from the hydrolysis of an amino acid such as arginine according to the following reactions:

$$\text{Arginine} \longrightarrow \text{Citrulline} + NH_3$$
$$\text{Citrulline} + \text{Pi} + \text{ADP} \longrightarrow \text{Ornithine} + CO_2 + NH_3 + \text{ATP}$$

The hydrolysis of arginine is a significant source of energy for some fermentative bacteria but plays a minor role in aerobic energy production.

One of the characteristics of the anaerobic clostridia and some other anaerobes is that they obtain energy from the fermentation of amino acids. Some clostridia are unable to ferment sugars and therefore rely upon amino acid fementation as a means of extracting energy. Many of the clostridia that ferment amino acids are also proteolytic, for example, *Clostridium perfringens* and *C. histolyticum,* which are causative agents of gas gangrene in animals. Proteolysis of muscle by these organisms and subsequent amino acid fermentation results in the formation of foul-smelling products characteristic of disease. One of the most important discoveries associated with microbial fermentation was recorded in 1934 by Stickland, who observed that some clostridia ferment single amino acids poorly but readily degrade certain pairs of amino acids. The amino acid pairs are coupled in oxidation-reduction reactions that are referred to as ***Stickland reactions.*** The oxidation reactions are similar to those catalyzed by aerobic organisms except that oxygen is absent. The oxidations include oxidative deaminations or transaminations, but the reduction reactions are distinctive because a suitable electron acceptor must be generated from the amino acids that the anaerobe metabolizes. The major electron acceptors are various amino acids, keto acids, protons, or unsaturated acids or their CoA thiolesters. The simplest example of a Stickland reaction involves two amino acids, glycine and alanine, fermented by *Clostridium sporogenes*. The overall reaction is

$$CH_3CHNH_2COOH + 2\ CH_2NH_2COOH + ADP + P_i \longrightarrow 3\ NH_3 + 3\ CH_3COOH + CO_2 + ATP$$

Alanine Glycine

in which glycine is the electron acceptor and alanine the electron donor. The various reactions in this fermentation process are shown at right.

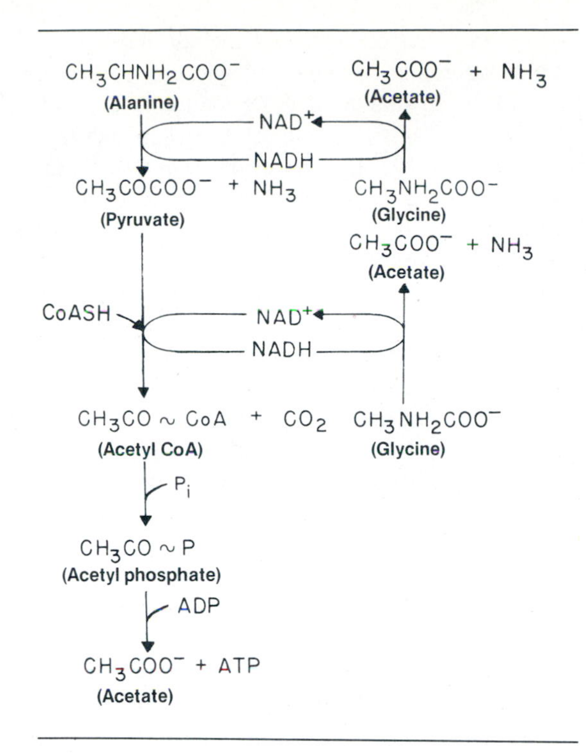

Purine and pyrimidine oxidation. The hydrolysis of nucleic acids results in the release of nucleosides and nucleotides. The purines and pyrimidines can be converted to a variety of products, including glycine, formate, aspartate, alanine, propionic acid, and isobutyric acid. Many of these compounds can be used as precursors in the biosynthesis of new purine and pyrimidine bases (see p. 326) and can also be channeled into TCA intermediates.

Lipids and fatty acid oxidation. Lipid degradation is not a common mechanism among microorganisms for obtaining their energy, because lipids are not always a frequent commodity in the environment. Lipid degradation is often associated with microorganisms that are parasites of animals. Staphylococci, for example, inhabit the skin of humans and break down surface lipids to fatty acids, which may be assimilated directly by the cell into lipids or they may be further oxidized to acetyl CoA units in a process called ***beta-oxidation.*** Fatty acid catabolism usually takes place when the chain length is more than eight carbons. The fatty acid is first converted to a CoA ester by acyl CoA-synthetase. The CoA ester is then oxidized in the beta position:

$$RCH_2CH_2COCoA \xrightarrow{\text{Oxidation}} RCH{=}CHCOCoA$$

Beta

Subsequently, acetyl CoA is released, leaving behind a CoA ester of the fatty acid shortened by two carbons. Even-numbered fatty acids yield only acetyl CoA units, whereas odd-numbered fatty acids yield acetyl CoA units plus propionyl CoA. Reduced NAD and FAD are generated during fatty acid oxidation and can be transported to the respiratory chain to generate energy. Acetyl CoA can also enter the glycoxylate cycle, which functions in the conservation of carbon (see Chapter 14).

Hydrocarbon oxidation. The ability to hydrolyze hydrocarbons, that is, alkanes and alkenes, is found among several species of microorganisms. This type of reaction has recently drawn considerable attention. Microorganisms have been genetically manipulated to produce enzymes that rapidly degrade petroleum products and related hydrocarbons. They are being considered for use in oil spills so that degradation processes can be speeded up and environmental recovery enhanced. The oxidation of alkanes by microorganisms takes place according to the following reactions:

$$R{-}(CH_2)_nCH_3 \xrightarrow{NAD} R{-}(CH_2)_nCH_2OH \xrightarrow{NAD} R{-}(CH_2)_nCHO \longrightarrow R{-}(CH_2)_nCOOH$$

Alkane Oxygen Alcohol Aldehyde Fatty acid

One of the major problems for microorganisms that degrade hydrocarbons and use them for carbon and energy sources is that hydrocarbons are deficient in nitrogen, phosphorus, and oxygen. Since most natural aquatic environments are deficient in usable nitrogen and phosphorus, the latter nutrients must be added to the system. One technique that is being considered to eliminate the need for additions of nitrogen is to manipulate the microorganism so that it is genetically capable of fixing nitrogen.

The manufacture of many toxic synthetic hydrocarbons and their subsequent release into the environment are causing considerable ecological problems. Bacteria and fungi can metabolize many diverse hydrocarbons, but a few, such as pesticides and herbicides, are unusually recalcitrant. The chlorinated insecticides, such as DDT, aldrin, and chlordane, are among the most resistant to microbial degradation* and can persist in the soil in an unaltered state for up to 5 years. The metabolism of some of these recalcitrant molecules can be speeded up if an additional organic energy source, which can be readily degraded, is also present. This phenomenon is called ***cometabolism.***

Organisms that cannot generate their own energy. Some microorganisms are obligate intracellular parasites unable to survive outside of a living host. Many of these parasites require a variety of different nutrients or enzymes that can be supplied only by the host cell. The parasite, in some instances, is unable to produce any energy, and two groups of parasites fall into this category: chlamydia, which are gram-negative bacteria, and viruses. Chlamydia possess RNA, DNA, and ribosomes but lack certain key enzymes that lead to the synthesis of certain amino acids and nucleosides. In addition, they also require the mitochondria of the host cell to supply energy. These pathogens are permeable to ATP that other pathogens are not. Viruses contain substantially less genetic information than eukaryotes or prokaryotes, and the enzymes necessary for generating energy are missing. Viruses can produce new virus particles only as long as the host's energy-generating systems are intact. Once the energy-producing components of the host have been destroyed, no new viral particles can be synthesized.

ENERGY PRODUCTION IN AUTOTROPHS

Autotrophs obtain their carbon for biosynthesis from carbon dioxide. They are divided into two groups based on their method of extracting energy. Chemoautotrophs obtain their energy from the oxidation of inorganic compounds. The term ***chemolithotroph*** is sometimes used in place of chemoautotroph. Lithotroph means "rock eating" and implies an existence on minerals. The photoautotrophs obtain their energy from light and are referred to as photosynthetic.

Chemoautotrophs (chemolithotrophs)

Chemoautotrophs, according to present-day view, represent a more complex group of microorganisms than was once supposed. Scientists had previously considered them to be among the most primitive prokaryotes. This was an understandable conclusion since they have an unusually slow rate of growth and extract their energy and carbon from inorganic sources. Reduced organ compounds were believed to be abundant during the anaerobic period of the earth's history. The fermentative bacteria that were present during this period flourished abundantly, and when the organic material diminished, there began a competition for energy sources. Reduced inorganic compounds, which were undoubtedly present in large quantities during this period, represented alternative energy sources, and therefore autotrophic microorganisms had the best chance of survival.

Most chemoautotrophs can survive in an entirely inorganic environment as long as carbon dioxide and oxygen are available. The growth of many species is hindered by the presence of organic compounds, but some species can use organic substrates, in a limited way, as carbon and energy sources. Chemoautotrophs are found in many diverse habitats in the biosphere—soil and fresh and marine waters—and are engaged in the cycling of elements in the biosphere (these activities are discussed in Chapters 17 to 19). The principal inorganic compounds used as energy sources by this group of microorganisms are listed in Table 13-5.

Inorganic substrates oxidized by chemoautotrophs

Hydrogen. Several species of bacteria in several genera use the oxidation of hydrogen to generate energy as ATP and to couple this energy to the fixation of carbon dioxide (see page 308). Hydrogen is used by the hydrogen bacteria in two types of reactions: (1) a reaction that ultimately ends in the generation of ATP through electron transport and oxidative phosphorylation:

$$H_2 + \frac{1}{2} O_2 \longrightarrow H_2O + \text{Energy from electron transport}$$

and (2) a reductive process to generate NADH:

$$H_2 + NAD \longrightarrow NADH + H^+$$

The NADH is used to reduce NADP, which in turn is used in biosynthetic reactions and in chemoautotrophs to fix carbon dioxide and produce organic compounds.

Iron. Only in the organism *Thiobacillus ferrooxidans* has iron oxidation been studied in any detail. The oxidation reaction that ultimately results in the production of energy follows:

$$4\,FeSO_4 + O_2 + 2\,H_2SO_4 \longrightarrow 2\,Fe_2(SO_4)_3 + 2\,H_2O$$

* Recent genetic engineering experiments have demonstrated that a gene coding for the enzymatic removal of chlorine can be cloned in bacteria. The dechlorinated compounds are more readily metabolized by microorganisms normally present in the environment.

Table 13-5 Principal inorganic energy sources used by chemoautotrophs (chemolithotrophs)

ENERGY SOURCE (ELECTRON DONOR)	ELECTRON ACCEPTOR	METABOLIC PRODUCTS	PHYSIOLOGICAL GROUP	BACTERIAL SPECIES
H_2	O_2	H_2O	Hydrogen bacteria	*Hydrogenomonas* sp.
S	O_2	H_2SO_4	Sulfur-oxidizing bacteria	*Thiobacillus* sp.
H_2S	O_2	H_2SO_4	Sulfur-oxidizing bacteria	*Thiobacillus* sp.
S	NO_3	SO_4, N_2	Sulfur-oxidizing bacteria	*Thiobacillus denitrificans*
Fe^{++}	O_2	Fe^{+++}	Iron-oxidizing bacteria	*Thiobacillus ferrooxidans*
H_2	SO_4	S^{--}, SO_3^{--}, H_2O	Dissimilatory sulfate reducers	*Desulfovibrio* sp.
NO_2	O_2	NO_3^-, H_2O	Nitrifying bacteria	*Nitrobacter* sp.
NH_4^+	O_2	NO_2^-, H_2O	Nitrifying bacteria	*Nitrosomonas* sp.

The sulfuric acid required in the oxidation of $FeSO_4$ is generated by sulfur-oxidizing bacteria. Many of the iron oxidizers can also obtain their energy from the oxidation of sulfide in such compounds as the sulfides of copper, lead, nickel, zinc, and cobalt.

Ammonia and nitrite. The nitrifying bacteria oxidize but two inorganic nitrogen compounds, either ammonium (NH_4^+) or nitrite (NO_2^-). Ammonium is produced in the environment following mineralization of organic nitrogen—a process called ammonification. Nitrite is found in the environment often as the result of the microbial oxidation of ammonia. In anaerobic environments nitrite occurs as a result of anaerobic respiration. Ammonium and nitrite oxidation results in the production of energy according to the following reactions:

$$NH_4^+ + 1^1/_2\,O_2 \longrightarrow NO_2^- + H_2O + 2\,H^+$$

$$NO_2^- + {}^1/_2\,O_2 \longrightarrow NO_3^-$$

Nitrification in soils and water plays an important role in agriculture and water quality; these aspects are discussed in Chapters 17, 18, and 19.

Sulfur. Several types of reduced sulfur compounds found in the environment may be oxidized by various species of bacteria. Species of *Thiobacillus,* for example, can oxidize the following sulfur compounds:

$$S_8 + 12\,O_2 + 8\,H_2O \longrightarrow 8\,H_2SO_4$$

$$H_2S + 2\,O_2 \longrightarrow H_2SO_4$$

$$Na_2S_2O_3 + 2\,O_2 + H_2O \longrightarrow Na_2SO_4 + H_2SO_4$$

Sulfur is polymerized in its elemental form and is represented as S_8 (Figure 13-14). The sulfide and thiosulfate states of sulfur are represented as H_2S

Figure 13-14 Molecular form of elemental sulfur.

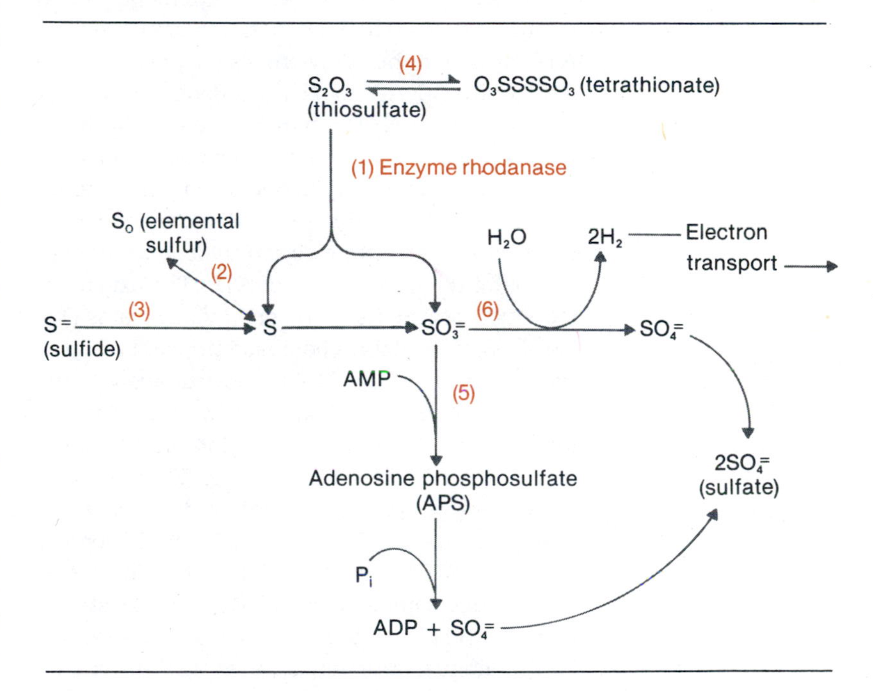

Figure 13-15 Steps in oxidation of different sulfur compounds. Thiosulfate, sulfide, and elemental sulfur are oxidized in reactions 1, 2, and 3, respectively. Thiosulfate is cleaved to S and SO_3 by enzyme rhodanase. Elemental sulfur and sulfide are oxidized to S. Reaction 4 indicates that under certain conditions of growth tetrathionate can also be produced. Tetrathionate is excreted into medium and can be further oxidized. Sulfite in reaction 5 is oxidized by adenosine phosphosulfate reductase, which ultimately results in formation of ADP and finally ATP production in electron transport system. Sulfite in reaction 6 can be oxidized to sulfate by sulfite-cytochrome c-oxidoreductase.

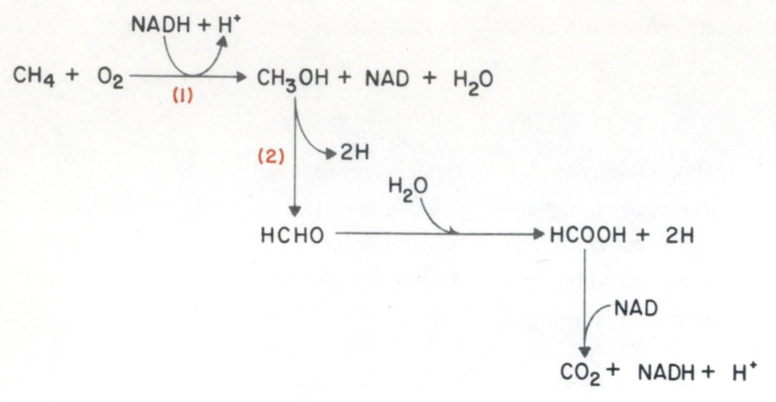

NADH + H^+

CH_4 + O_2 (1) → CH_3OH + NAD + H_2O

(2) 2H

H_2O

HCHO → HCOOH + 2H

NAD

CO_2 + NADH + H^+

Figure 13-16 Pathway for oxidation of methane (CH_4) and methanol (CH_3OH) as well as other one-carbon substrates.

and S_2O_3, respectively. The interrelationships of these three sulfur states during oxidation and energy formation are outlined in Figure 13-15.

Mechanism of energy production in chemoautotrophs. Chemoautotrophs use the conventional methods of electron transport and oxidative phosphorylation and, therefore, possess flavoproteins, quinones, and cytochromes. Electrons generated during the oxidation of fuels are important for ATP formation, but they are also important in the reduction of NAD to NADH. Reduced NAD is used to reduce NADP, which is important in the biosynthesis of the cell's metabolites. The hydrogen bacteria couple hydrogen oxidation to the reduction of NAD, and the entire electron transport chain is available for electron transport and generation of three molecules of ATP. The oxidation-reduction potentials of the energy substrates (Table 13-3) for other chemoautotrophs is higher (more positive) than NAD, and oxidation of the energy source cannot be directly coupled to the reduction of NAD. The oxidation-reduction potential of the Fe^{++}/Fe^{+++} couple ($E° = +0.770$ volts), for example, is too high to drive the reaction in which NAD is reduced (the E° for the NAD^+/NADH couple is -0.320 volts). The chemolithotrophs compensate for this characteristic by using "uphill" reactions that are just the reverse of the normal electron flow in the respiratory chain. Oxidation of iron, for example, releases electrons that probably enter the respiratory chain at cytochrome c. The energy gained in their oxidation is used to drive the uphill reaction that terminates in the reduction of NAD or NADP. Thus most chemolithotrophs, because of reverse electron flow, do not produce large amounts of energy for growth. They all grow slowly, and doubling rates of up to 6 hours are obtained and only then when their energy source is in excess.

Methylotrophs. Methylotrophs are microorganisms that gain their energy from the oxidation of one-carbon (C1) compounds other than carbon dioxide. These carbon compounds include formaldehyde (HCHO), methanol (CH_3OH), methane (CH_4), and other related compounds. The methylotrophs, even though they oxidize organic compounds, exhibit properties similar to the autotrophs and have, therefore, been included here. The oxidation of one-carbon compounds results in the formation of carbon dioxide or formaldehyde. The methylotrophs, unlike the autotrophs, do not reduce atmospheric carbon dioxide to cell material*—methylotrophs require a compound more reduced than carbon dioxide, for example, formaldehyde or a mixture of formaldehyde and carbon dioxide. The overall scheme for the oxidation of carbon compounds is presented in Figure 13-16. The oxidation reactions for the principal carbon substrates are outlined below.

Oxidations involving single-carbon substrates. Methane is found in coal and oil deposits and is associated with the anaerobic decomposition of organic matter, sewage sludge digestors, sediments and muds of various aquatic environments, and the intestinal tract of animals and humans. Methanol is a degradation product of several naturally occurring substances, such as pectins (polymer of galacturonic acid), which are found especially in fruits. The oxidation of methane and methanol proceeds via reactions outlined in Figure 13-16. Their oxidation results in the formation of formaldehyde, which is used to produce cellular organic material.

Formaldehyde as an energy source may be oxidized by two different routes. The most direct route involves two reactions catalyzed by separate enzymes. Formate is the product of the first reaction and is later cleaved to carbon dioxide according to the following reactions:

$$HCHO + X + H_2O \longrightarrow HCOO^- + XH_2$$

$$HCOO^- + NAD \longrightarrow CO_2 + NADH + H^+$$

Methylamines are substrates for many bacterial species. These compounds are found as degradation products of various plants and animal tissue. Oxidation of methylamines occurs in a series of reactions in which successive carbon-nitrogen bonds are cleaved and methyl groups re-

* Other microorganisms using one-carbon compounds convert carbon dioxide to usable organic compounds in the cell via the Calvin Cycle (see p. 308).

$$(CH_3)_3NH^+ + \text{Flavoprotein} + H_2O \longrightarrow (CH_3)_2NH_2^+ + HCHO + \text{Reduced flavoprotein}$$

Figure 13-17 Oxidation of methylamine.

Table 13-6 Characteristics of the photosynthetic structures and mechanisms found in eukaryotes and prokaryotes.

		PROKARYOTES		
	EUKARYOTES	CYANOBACTERIA	PURPLE BACTERIA	GREEN BACTERIA
Electron donors	H_2O	H_2O; some can use H_2S	H_2S, S_0, H_2, S_2O_3 organic compounds	H_2S, S_0, H_2, S_2O_3 organic compounds
Site of photosynthesis	Thylakoids	Thylakoids	Cell membrane	Chlorosomes
Oxygen production	Yes	Yes	No	No
Chlorophyll type	Chlorophyll a	Chlorophyll a	Bacteriochlorophyll a or b	Bacteriochlorophyll a and c, d or e
Photosystem I	Present	Present	Present	Present
Photosystem II	Present	Present	Absent	Absent

leased as formaldehyde. The oxidation of a typical methylamine is outlined in Figure 13-17.

Photoautotrophs (photosynthesis)

Photosynthesis is a biological process in which light energy is converted to chemical energy. The chemical energy is used to convert carbon dioxide into organic material; therefore, photosynthesis provides a mechanism for supplying the energy that will be the driving force for all life processes. The organic material produced by plants, for example, is used as fuel for animals, who in turn oxidize it and release carbon dioxide to the atmosphere. The released carbon dioxide is used again by plants and other photosynthetic organisms to produce more organic material. Oxygen, which is a byproduct of most photosynthetic organisms, is released into the atmosphere and thus sustains those animals and other organisms that require oxygen for respiration. In 1779 Jan Ingen-Housz demonstrated that oxygen production occurs only in the presence of sunlight, and the plant, when left in the dark, carried out respiration just like members of the animal kingdom.

Photosynthesis occurs in both eukaryotic and prokaryotic microorganisms. Photosynthesis in prokaryotes is characteristic of cyanobacteria, purple bacteria, and green bacteria, whereas in eukaryotic microorganisms photosynthesis occurs in algae. Oxygen production ***(oxygenic)*** is a part of the photosynthetic process in eukaryotic microorganisms as well as higher plants, but in prokaryotes oxygen may or may not ***(anoxygenic)*** be produced. Among the prokaryotes only cyanobacteria possess a photosynthetic mechanism that is oxygenic. The evolutionary significance of this similarity between prokaryotic cyanobacteria and eukaryotes is discussed in Chapter 4. Some of the characteristics of the photosynthetic mechanisms and structures associated with photosynthesis in eukaryotes and prokaryotes are outlined in Table 13-6.

Photosynthetic process. Before some of the more detailed aspects of photosynthesis are discussed, a diagrammatic explanation of the overall process of photosynthesis will be useful (Figure 13-18). The first phase of photosynthesis is called the ***light phase*** and takes place in a pigmented membrane system. Light is captured and used in oxygenic systems to break down water ***(photolysis)*** into protons, electrons, and oxygen. The oxygen is released into the environment (in oxygenic systems). Electron transport and photophosphorylation result in the formation of NADPH to be used for reduction of carbon dioxide, and ATP, to be used as the energy source for biosynthesis. The second phase is the ***dark phase*** and

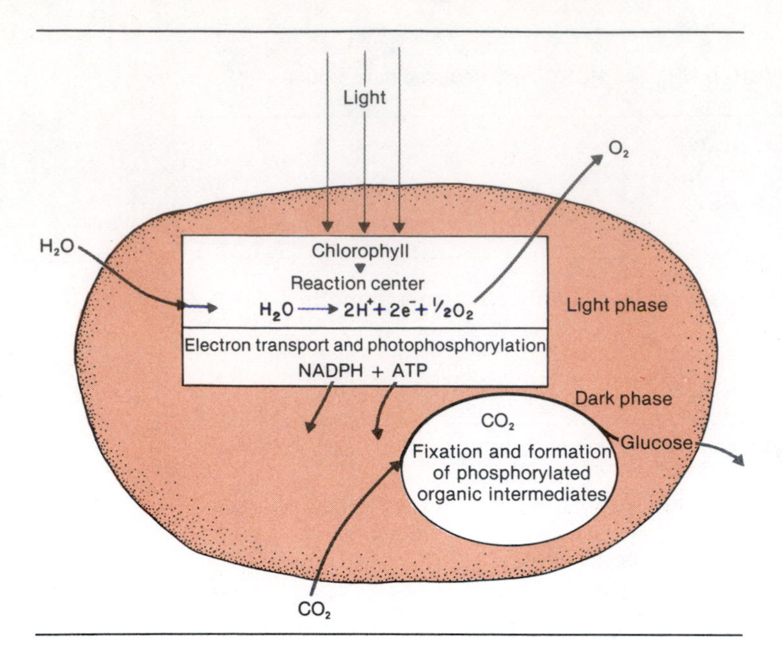

Figure 13-18 Summary of photosynthetic process as it would occur in chloroplast. Light striking chlorophyll molecule is transmitted to reaction center where there is photolysis of water into electrons, hydrogen, and oxygen. Electron transport system and photophosphorylation systems produce NADPH and ATP, which pass into stroma and are used to fix (reduce) carbon dioxide in order to produce organic intermediates, which culminate in glucose formation.

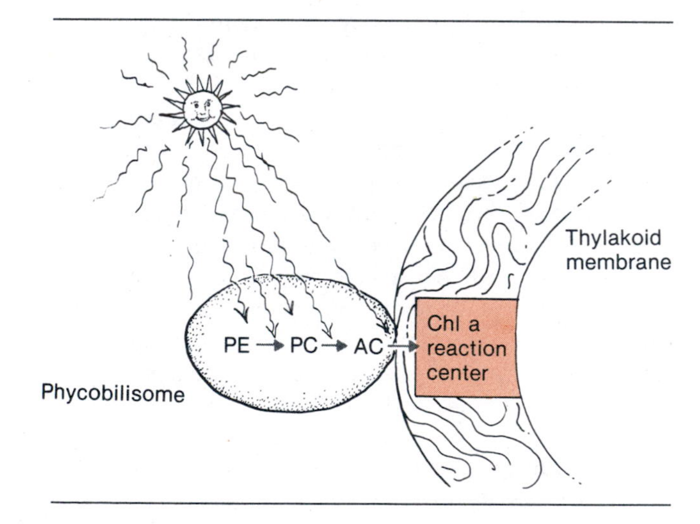

Figure 13-19 Association of light harvesting pigments of cyanobacteria with thylakoids. Light waves strike pigments in phycobilisomes, which lie outside thylakoid membrane system. Light energy is transferred to chlorophyll a in reaction center. *Chl a,* Chlorophyll a; *PE,* phycoerythrin; *PC,* phycocyanin; *AC,* allophycocyanin.

occurs in the stroma of the chloroplast in eukaryotes or the cytoplasm of the cell in prokaryotes. NADH and ATP generated in the light phase are used to reduce carbon dioxide and provide organic materials during the dark phase.

Photosynthetic structures. The structures in plants and algae for harvesting light energy are called chloroplasts, and their systems of membranes are called thylakoids. Chlorophyll pigments are located within the thylakoid membrane and are organized into stacks because of the folding of the membrane. Packets of light, called ***quanta,*** are absorbed by pigment molecules, such as chlorophyll or some other molecule during photosynthesis. The absorption of light causes the pigment molecules to become excited with the release of an electron. This excitation travels easily from one pigment molecule to another because of the stacking arrangement in the membrane. Somewhere between 200 and 300 pigment molecules harvest the light and then transfer this energy to a special trap, or reaction center, composed of chlorophyll a (Chl_a), which converts the light energy into chemical energy.

Light-capturing pigments in the prokaryotes are organized into membranous structures related to chloroplasts, yet significantly different. The membranes (thylakoids), for example, in the cyanobacteria traverse the cytoplasm of the cell (Figure 13-19). The pigments that harvest light in cyanobacteria are called ***phycobilisomes,*** which are stacked in the outer thylakoid membrane. Chlorophyll a also acts as a reaction center, and it too is membrane bound. Bacteriochlorophylls c, d, and e in the green bacteria are the principal light-harvesting pigments and are located in precisely structured vesicles called ***chlorosomes.*** Bacteriochlorophylls a and b in purple bacteria are the light-harvesting pigments, and they are found in the cytoplasmic membrane.

Pigments and energy capture. The pigments found in the different photosynthetic systems absorb light in all bands of the visible spectrum (Figure 13-20). Some of the pigments have two major absorption bands, whereas others have only one. All light-harvesting pigments are related to two structural types: tetrapyrroles and isoprenoids (Figures 13-21 and 13-22). Tetrapyrroles are composed of two classes: the green pigmented chlorophylls a, b, c, d, and e, all of which are closed-chain molecules that absorb primarily in the red and blue region of the visible spectrum (660 to 800 mm and 400 to 460 mm, respectively.) Chlorophyll, a, for example, is a cyclic tetrapyrrole found in all photosynthetic organisms that release oxygen. Chlorophylls b, c, d, and e are termed bacteriochlorophylls (bchl). They are synthesized through the same pathways as chlorophyll a but have variable substituents (see Figure 13-21 and Table 13-7). The distribution of bacteriochlorophylls in prokaryotes is

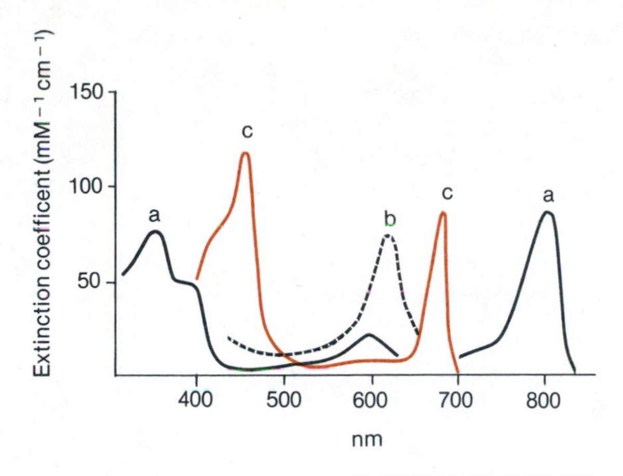

Figure 13-20 Absorption spectra of some photosynthetic pigments. Letters refer to type of chlorophyll.

Figure 13-21 Tetrapyrroles as components of light harvesting pigments. Basic structure of chlorophyll molecule. Nature of R groups is basis of classification of bacteriochlorophylls (see Table 13-5). Tetrapyrrole rings (A, B, C, and D) are connected in form of heterocyclic compound.

outlined in Table 13-7. Bacteriochlorophylls a and b are the only bacteriochlorophylls that possess a saturated double bond in their substituent groups. This alters the absorption spectra of bchl a and bchl b relative to the other chlorophylls (Table 13-8) and help distinguish the green from the purple bacteria.

Another class of tetrapyrroles are bile pigments, which are open chained and covalently attached to proteins such as ***phycobiliproteins*** found in cyanobacteria. Phycobiliproteins are blue or red pigments that absorb primarily in the 500- to 600-nm range. Species harboring these pigments are found in aquatic environments where available light is in the green region of the spectrum.

The ***isoprenoids*** (Figure 13-22) are open-chained molecules, such as beta carotene, that do not appear to have a major role in harvesting light. They may function to quench excessive excitation of chlorophyll and, thereby, prevent photooxidation of the photosynthetic system.

Electron flow and the photosystems. Energy transfer occurs between the various pigments, and during photosynthesis the most efficient energy transfer occurs when there is flow toward the pigment with the longest absorption band. Energy transfer and electron flow in green plants take place in two systems: photosystem I and photosystem II. Each system consists of different types of chlorophyll and different electron acceptors.

PHOTOSYSTEM I (Figure 13-23). The reaction center in photosystem I contains chlorophyll a which absorbs at the longer wavelength of 700 nm and is called chlorophyll P_{700} (P stands for pigment). When wavelengths of light at 700 nm and above strike P_{700}, the dislodged electron is accepted by an unidentified electron acceptor called X. The reduced electron acceptor X transfers its electron to ferredoxin, a nonheme protein, which in turn reduces NADP to NADPH with the aid of hydrogen produced from the photolysis of water in photosystem II.

Figure 13-22 Isoprenoids. **A,** Basic isoprene unit. **B,** Structure of carotenoid in which two rings (1 and 2) are connected by chain of repeating isoprene linkages.

ATP synthesis in photosystem I is believed to be a cyclic process and occurs when reducing power in the form of NADH is not required. The electron acceptor X transfers its electrons to cytochrome b_6 and cytochrome f. The reduction of

Table 13-7 Chemical differences between chlorophyll a and bacteriochlorophylls (Bchl)*

CHLOROPHYLL	R_1	R_2	R_3	R_4	R_5	R_6	R_7
a	$-CH-CH_2$	$-CH_3$	$-C_2H_5$	$-CH_3$	$-C(=O)-OCH_3$	Phytyl†	$-H$
Bchl a	$-C(=O)-CH_3$	$-CH_3$‡	$-C_2H_5$‡	$-CH_3$	$-C(=O)-O-CH_3$	Phytyl or geranylgeraniol§	$-H$
Bchl b	$C(=O)-CH_3$	$-CH_3$‖	$-CH-CH_3$‖	$-CH_3$	$C(=O)-O-CH_3$	Phytyl	$-H$
Bchl c	$-C(H)(OH)-CH_3$	$-CH_3$	$-C_2H_5$	$-C_2H_5$	$-H$	Farnesyl¶	$-CH_3$
Bchl d	$-C(H)(OH)-CH_3$	$-CH_3$	$-C_2H_5$	$-C_2H_5$	$-H$	Farnesyl	$-H$
Bchl e	$-C(H)(OH)-CH_3$	$-C(H)-O$	$-C_2H_5$	$-C_2H_5$	$-H$	Farnesyl	$-CH_3$

*See Figure 13-21 for structure of chlorophyll molecule
†Phytyl ester ($C_{20}H_{39}O-$)
‡No double bond between C—3 and C—4; additional —H atoms at C—3 and C—4.
§Geranylgeraniol ester ($C_{10}H_{17}O-$)
‖No double bond between C—3 and C—4; additional —H atom at C—3.
¶Farnesyl ester ($C_{15}H_{25}O-$)

Table 13-8 Spectroscopic properties of purified chlorophylls in ether solution

ABSORPTION MAXIMA (WAVELENGTHS IN NM; RELATIVE PEAK HEIGHTS IN PARENTHESES)

Chlorophyll a			430 (100)		615 (13)	662 (77)	
Bchl c			428 (100)		622 (29)	660 (63)	
Bchl d			424 (100)		608 (17)	654 (61)	
Bchl e			458 (100)		593 (12)	647 (32)	
Bchl a	358 (100)	391 (68)		577 (29)			773 (126)
Bchl b	368 (100)	407 (87)		582 (30)			795 (96)

cytochrome is coupled to the phosphorylation of ADP to ATP. The cytochromes return the electron to chlorophyll, and hence no external electron donor or acceptor is required.

Photosystem I is also present in the green and purple bacteria, but photosystem II is absent (Figure 13-24). The electron donor in these two groups of bacteria is not water but other compounds, such as sulfur, hydrogen, or organic compounds. The reaction center pigment is a bacteriochlorophyll molecule whose wavelength of maximum absorption is 870 nm, which is in the infrared range. The emitted electron is accepted by an unidentified component also called X. Electron flow from X involves ubiquinones and cytochromes in a cyclic process that results in

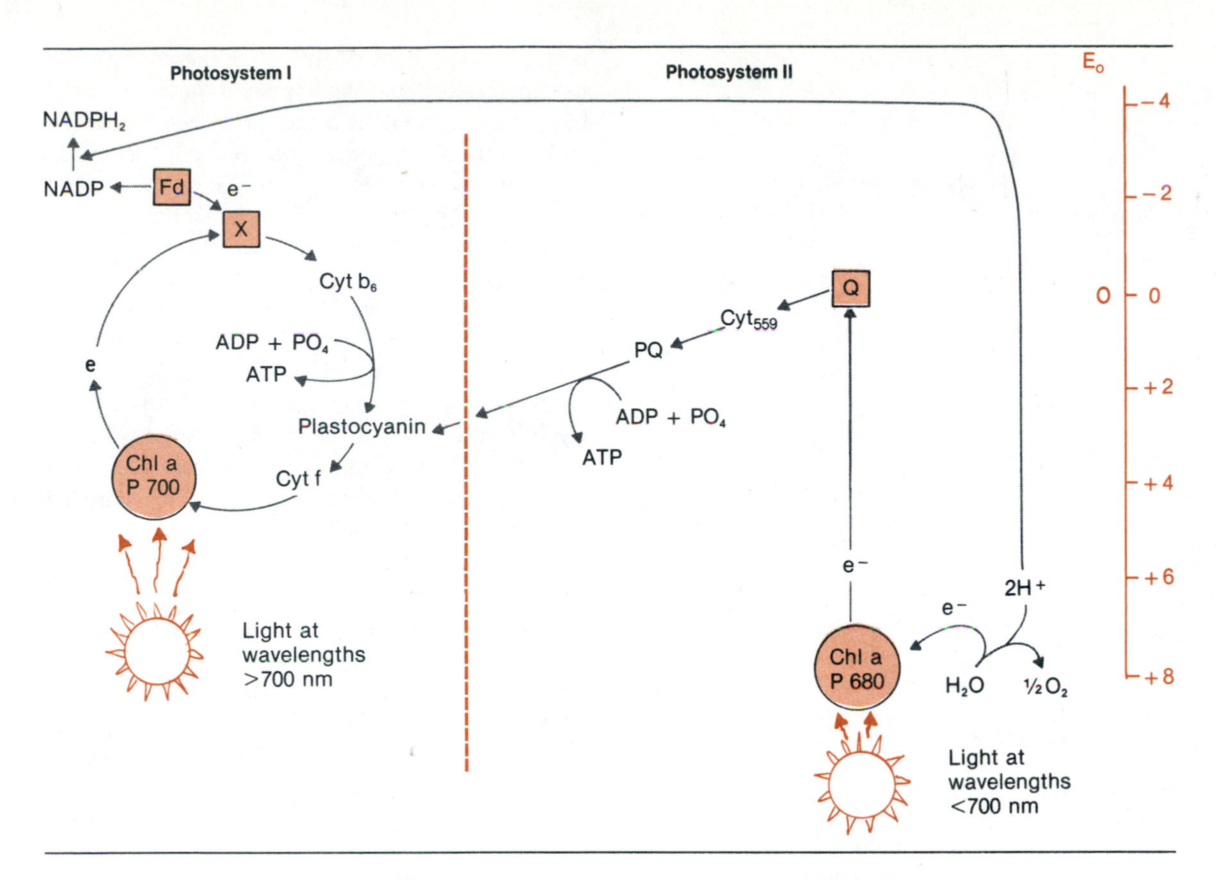

Figure 13-23 Postulated mechanism of electron flow in chloroplasts containing photosystems I and II. On right are oxidation-reduction potentials (E_0), which are aligned with specific components of photosystems. *X*, Electron acceptor of photosystem I; *Q*, electron acceptor of photosystem II; *PQ*, plastoquinone; Fd, ferredoxin; Chl a, chlorophyll a. See text for details of this system.

ATP synthesis. The path leading to the formation of reduced NADP is still in doubt. It is believed to be formed by an energy-requiring reaction rather than by direct photooxidation of an electron acceptor as occurs in those photosynthetic organisms that possess two photosystems. The evidence for this postulation is derived from experiments in which ATP was able to replace light in the reduction of NADP to NADPH.

PHOTOSYSTEM II. Photosystem II (Figure 13-23) contains chlorophyll a type P_{680}, which absorbs light at wavelengths shorter than 700 nm. The electron dislodged from P_{680} is transferred to the primary electron acceptor Q (Q stands for quencher). The electron is then transferred to $cytochrome_{559}$, plastoquinone, and plastocyanin. ATP synthesis is believed to occur during electron transport between plastoquinone and plastocyanin. ATP synthesis is noncyclic since electrons are removed from the system. The photolysis of water is also associated with photosystem II, and this results in the evolution of oxygen, and production of electrons and hydrogen atoms (protons). A special catalytic system using manganese is required in this reaction. The

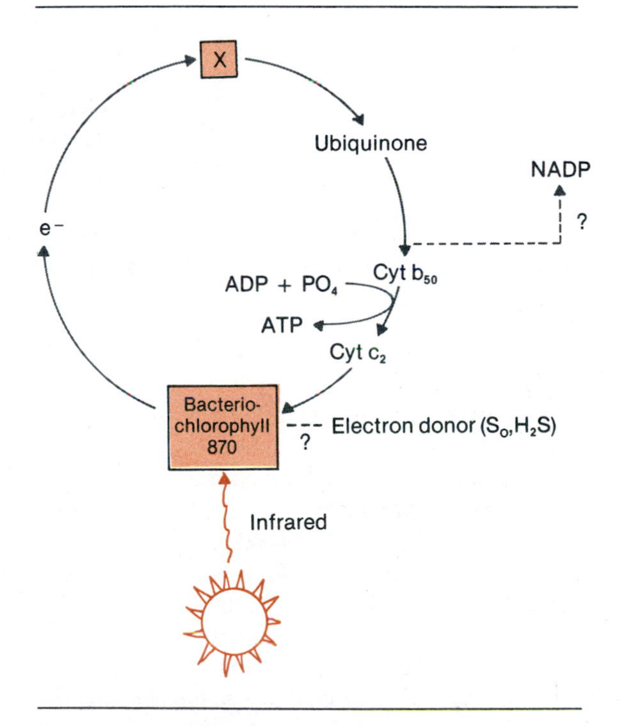

Figure 13-24 Single photosystem in green and purple bacteria. Association of reactions involving electron donor and reduced NADP formation in this photosystem have not been elucidated.

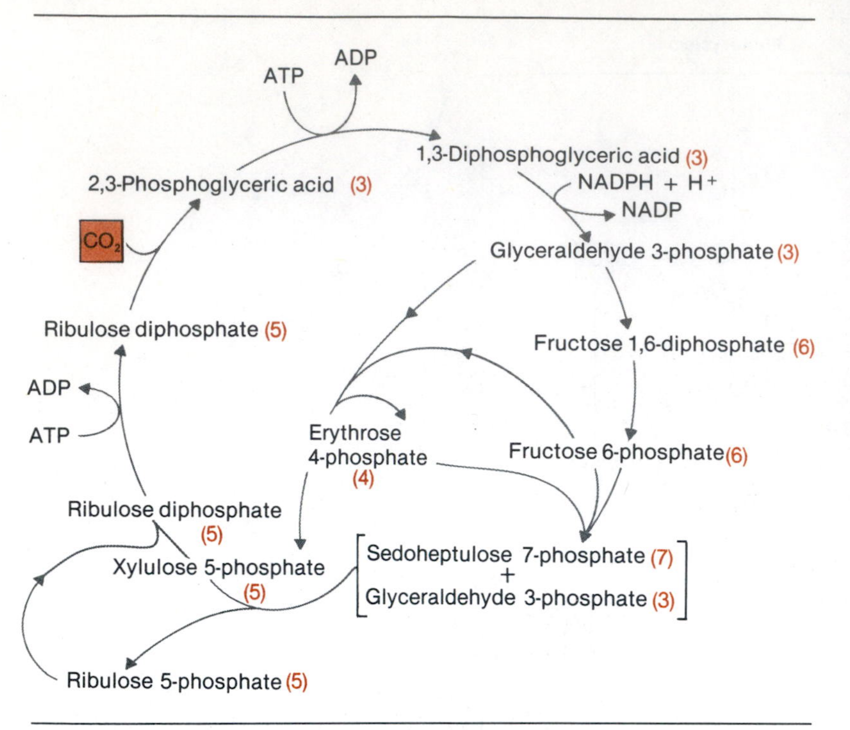

Figure 13-25 Calvin-Benson cycle. Numbers in parentheses indicate number of carbons in intermediate. Fructose 6-phosphate is primary product and can be converted to glucose to be used in glycolysis or be stored as polysaccharide in the cell.

electron dislodged from water during photolysis is used to neutralize the positively charged reaction center of photosystem II.

Photosystem II is not as stable as photosystem I and can be affected by detergents as well as specific inhibitors such as 3-(3,4,-dichlorophenyl 1)-1,1-dimethylurea (DCMU). DCMU is a herbicide that blocks photosystem II by interrupting the electron carrier chain between photosystems I and II, and thus one can examine the specific properties of photosystem I without interference from photosystem II.

• • •

Both photosystems I and II are required to take electrons expelled from the chlorophylls and raise them from positive oxidation-reduction potentials to the negative oxidation-reduction potential of NADPH—from the + 0.820 volts of the O_2/H_2O couple to the −0.320 volts of the NADP/NADPH couple. The electrons are capable of yielding considerable energy when transferred to electron acceptors in the system.

Electron flow and ATP synthesis. Several hypotheses have been proposed to explain the coupling of electron flow to ATP synthesis (see p. 294). Currently the chemiosmotic hypothesis is favored in the light of the experiments with the halobacteria. Some halobacteria have a membrane that appears as a mosaic of patches of red and purple. The purple membrane consists almost entirely of a single protein called ***bacteriorhodopsin,*** which is similar to the protein found in the retina of the human eye, plus some lipid. The color of the membrane is the result of the chromophore, ***retinal,*** which is complexed with bacteriorhodopsin. The retinal-protein complex absorbs photons of light, which establishes an electrochemical gradient resulting in the pumping of protons across the membrane from inside to outside. The resulting proton-motive force is coupled to ATP synthesis by an ATPase complex located in the red membrane. The purple membrane is, therefore, an alternate system to the respiratory chain used by other prokaryotes to generate energy. Carotenoids such as bacteriorubrins are also located in the red membrane and protect the cell from ultraviolet light and photodynamic oxidation.

Carbon dioxide fixation in photosynthetic systems. Reduced NADP produced in photosystem I is used by eukaryotes to reduce carbon dioxide to organic material. This reduction takes place in a series of reactions called the Calvin-Benson cycle. The cycle occurs in the dark but requires energy in the form of ATP and reducing equivalents in the form of NADPH to be supplied from the light phases of photosynthesis. The cycle involves several enzymes and is relatively complex. The enzymes for carbon dioxide fixation in prokaryotes are located in ***carboxysomes,*** whereas in eukaryotes they appear within the stroma of the chloroplast. The cycle is most respects resembles the pentose cycle discussed on p. 289, and its intermediates serve similar functions. Instead of discussing each reaction in the cycle, I will point out only the most significant ones. The complete cycle is outlined in Figure 13-25. The major reactions are shown on p. 309.

Overall the cycle consists principally of carbon dioxide fixation (carboxylation), a reduction, and two reactions requiring ATP. All of the other reactions are rearrangements that eventually result in the formation of ribulose monophosphate. Six turns of the cycle are required to produce a hexose.

$$\text{6 ribulose 1,5-diphosphate} + 6\,CO_2 + 18\,ATP + 12\,NADPH + 12\,H^+ \longrightarrow \text{6 ribulose 1,5-diphosphate} + \text{hexose} + 18\,P_i + 18\,ADP + 12\,NADP$$

The hexoses produced in the cycle are stored in the cell as starch or glycogen in eukaryotes and cyanobacteria but as β-polyhydroxybutyric acid in the purple and green bacteria.

1. Formation of ribulose diphosphate. Ribulose monophosphate, a pentose sugar, is elevated to a higher energy level via phosphorylation and the formation of ribulose diphosphate. This reaction is catalyzed by phosphoribulokinase, which is found only in cells that carry out the Calvin-Benson cycle (Figure 13-25).

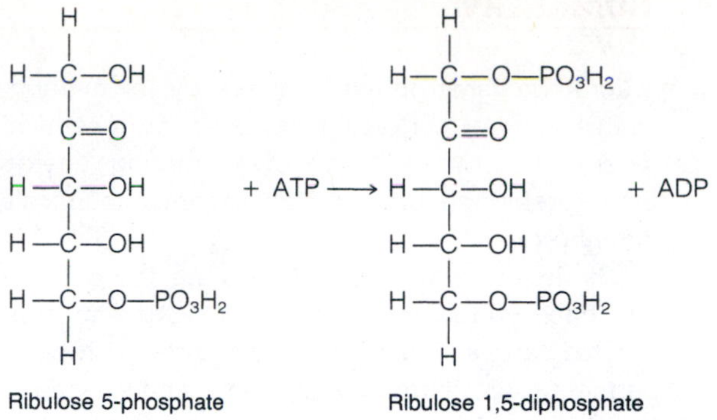

```
    H                                    H
    |                                    |
H—C—OH                            H—C—O—PO3H2
    |                                    |
    C=O                                  C=O
    |                                    |
H—C—OH            + ATP ⟶        H—C—OH              + ADP
    |                                    |
H—C—OH                            H—C—OH
    |                                    |
H—C—O—PO3H2                       H—C—O—PO3H2
    |                                    |
    H                                    H
```

Ribulose 5-phosphate — Ribulose 1,5-diphosphate

2. Carbon dioxide fixation. Carbon dioxide fixation is a carboxylation reaction involving ribulose diphosphate. This reaction is immediately followed by the cleavage of the unstable six-carbon intermediate to two molecules of phosphoglyceric acid. The enzyme catalyzing this reaction is ribulose diphosphate carboxylase.

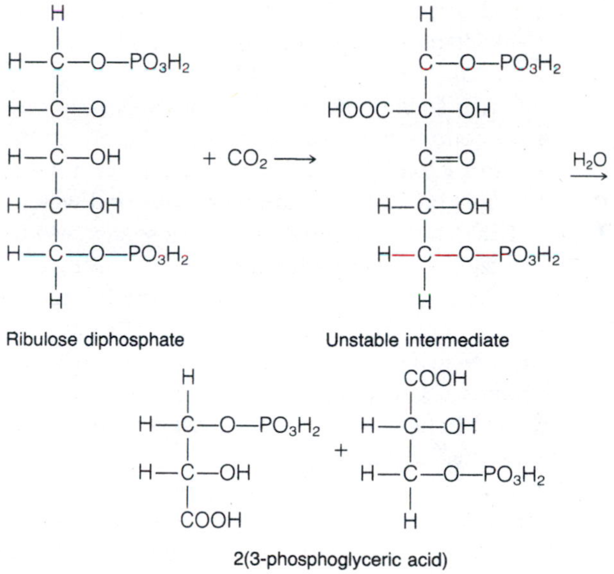

```
    H                                     H
    |                                     |
H—C—O—PO3H2                           C—O—PO3H2
    |                                     |
H—C=O                          HOOC—C—OH
    |                                     |
H—C—OH          + CO2 ⟶                 C=O            H2O
    |                                     |              ⟶
H—C—OH                              H—C—OH
    |                                     |
H—C—O—PO3H2                         H—C—O—PO3H2
    |                                     |
    H                                     H
```

Ribulose diphosphate — Unstable intermediate

```
        H                     COOH
        |                      |
    H—C—O—PO3H2          H—C—OH
        |           +          |
    H—C—OH               H—C—O—PO3H2
        |                      |
      COOH                     H
```

2(3-phosphoglyceric acid)

3. 3-Phosphoglyceraldehyde formation. The formation of 3-phosphoglyceraldehyde takes place in two steps. Note the requirement of NADPH and ATP supplied by the light phase of photosynthesis in these reactions. These steps are catalyzed by the same enzymes found in glycolysis.

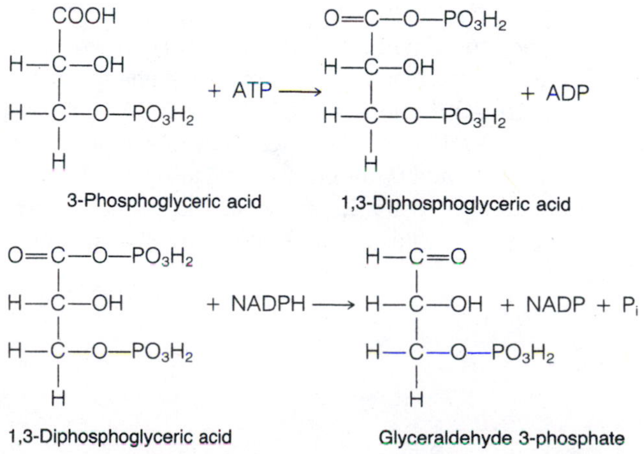

```
    COOH                          O=C—O—PO3H2
    |                                 |
H—C—OH                            H—C—OH
    |           + ATP ⟶               |            + ADP
H—C—O—PO3H2                       H—C—O—PO3H2
    |                                 |
    H                                 H
```

3-Phosphoglyceric acid — 1,3-Diphosphoglyceric acid

```
O=C—O—PO3H2                          H—C=O
    |                                    |
H—C—OH          + NADPH ⟶           H—C—OH  + NADP + Pi
    |                                    |
H—C—O—PO3H2                          H—C—O—PO3H2
    |                                    |
    H                                    H
```

1,3-Diphosphoglyceric acid — Glyceraldehyde 3-phosphate

SUMMARY

Microorganisms obtain their energy for metabolism by either converting sunlight into chemical energy (phototrophs), or by extracting energy from preformed organic or inorganic chemicals (chemotrophs).

Oxidation-reduction reactions are the basis of all energy-producing reactions. Organic substrates can be oxidized by one of three pathways: fermentation, aerobic respiration, and anaerobic respiration. Fermentation uses an organic electron donor, such as glucose, and an organic electron acceptor, usually derived from the donor, such as pyruvate. The fermentation pathway using glucose is called the Embden-Meyerhoff-Parnas pathway. Aerobic respiration uses an organic electron donor and oxygen as the final electron acceptor. Anaerobic respiration also uses an organic electron donor, but the final electron acceptor is an inorganic molecule other than oxygen, such as sulfate and nitrate. More energy is realized from aerobic respiration than the others because the oxidation process is complete; that is, the final products are carbon dioxide and water.

Besides the EMP pathway glucose can also be oxidized in other pathways such as the Entner-Doudoroff pathway, the hexose monophosphate shunt, and the phosphoketolase pathway. The hexose monophosphate shunt is especially important because it provides reduced NADPH for biosynthetic reactions and ribose sugars for nucleic acid biosynthesis.

Pyruvate, one of the major products of glucose fermentation, can also be converted to a variety of products including many different organic acids, alcohols, and gases. Microbial species show a variation in the types of products derived from pyruvate, and these products are often a clue to their identity.

During aerobic respiration organic substrates such as glucose become completely oxidized in the tricarboxylic acid cycle (TCA). The electrons generated in the TCA cycle are carried to the electron transport system (ETS) where their movement results in the release of energy that is captured in the form of ATP. The TCA cycle also provides intermediates for biosynthesis.

The coupling of electron oxidation to ATP formation is believed to occur via a mechanism called chemiosmosis. Protons (H^+) transported across the cell membrane, where the respiratory enzymes reside, create a proton motive force that is coupled to ATP synthesis. In addition, an ATPase enzyme complex in the membrane accounts for the hydrolysis of ATP and release of free energy.

Oxygen not only plays an important role as a final electron acceptor in energy production in aerobic microorganisms but can also influence cells that do not use it for energy production. Oxygen is often reduced to intermediates such as the superoxide radical that are toxic to anaerobes. Superoxide produced in aerobes is removed by the enzyme superoxide dismutase, which is generally absent in anaerobes.

Although carbohydrates are the most frequently metabolized organic compounds, other molecules such as amino acids, purines and pyrimidines, lipids and fatty acids, and hydrocarbons can also be used as energy sources.

Energy from the oxidation of inorganic compounds is characteristic of the group called the chemolithotrophs. This group extracts both carbon and energy from inorganic sources. The energy sources include hydrogen, iron, ammonia, and nitrite. These bacteria use the electron transport chain, but growth is usually slow, and the number of ATP produced per electron pair is less than that obtained by organisms that use organic substrates.

One-carbon compounds such as methane and methanol can also be metabolized to produce energy. During the oxidation process formaldehyde is formed, which is used to produce organic material in the cell.

A second group of autotrophs obtain their energy from the sun and convert it into chemical energy in the form of ATP—photoautotrophs. Various light-harvesting pigments are found in photosynthetic organisms. The illuminated pigments release an electron that is transferred to various electron acceptors for the purpose of producing ATP and, in oxygenic photosynthetic organisms, reduced NADP. Both of these molecules are used in a series of dark reactions called the Calvin-Benson cycle. In this cycle carbon dioxide from the atmosphere is fixed to an organic molecule and in a series of reactions is converted to fructose 6-phosphate. This hexose can be converted to glucose and stored by the cell in the form of polysaccharides.

Study questions

1. What are the terms used to express the different carbon sources of an organism? Their energy sources?
2. Into what two divisions are metabolic reactions divided?
3. What are the energy-producing reactions called that involve an oxygen donor and an organic acceptor? What is the relationship of oxygen tension to these reactions?
4. What is the purpose of pyruvate reduction in glycolysis?
5. Name some products that are the result of pyruvate metabolism by microorganisms. Do these products have any significance to a microbiologist?
6. What is the basic difference between substrate phosphorylation and oxidative phosphorylation?
7. Write out the two reactions in which ATP is produced in glycolysis.
8. What is the function of the phosphoketolase pathway? The hexose monophosphate shunt? What reactions are common to both pathways?
9. Can sugars other than glucose enter the EMP pathway? How? How does a cell metabolically respond to extracellular polysaccharides found in its environment?
10. What is the energy production resulting from the oxidation of an organic donor and inorganic acceptor called? What is the relationship of oxygen tension?
11. What are the functions of the TCA cycle? Write out those reactions in which carbon is lost as carbon dioxide. Write out those reactions in which reducing equivalents are generated.
12. In the electron transport system how is the oxidative process coupled to energy formation? What is meant by oxidation-reduction potential?
13. At what stages in the electron transport system are the hydrogens released? Do they reenter the ETS again? What happens to the hydrogens?
14. What metabolic oxidation process in the cell produces the most energy? The least?
15. What is meant by proton motive force? In just two or three sentences explain how these forces can be used by cells to release energy through the hydrolysis of ATP and to produce it. What is the role of ATPase in this system?

Selected readings

Books

Bothe, H., and Trebst, A. *Biology of inorganic nitrogen and sulfur.* Springer-Verlag New York, Inc., New York, 1981.

Doelle, H.W. *Bacterial metabolism* (2nd ed.). Academic Press, Inc., New York, 1975.

Gottschalk, G. *Bacterial metabolism.* Springer-Verlag New York, Inc., New York, 1979.

Knowles, C.J. *Diversity of bacterial respiratory systems.* CRC Press, Inc., Boca Raton, Fla., 1980.

Mandelstam, J., and McQuillen, K. *Biochemistry of bacterial growth* (2nd ed.). John Wiley & Sons, Inc., New York, 1973.

Moat, A.G. *Microbial physiology.* John Wiley & Sons, New York, 1979.

Rose, A.H. *Chemical microbiology* (3rd ed.). Plenum Publishing Corp., New York, 1976.

Weinberg, E.D. (Ed.). *Microorganisms and minerals.* Marcel Dekker, Inc., New York, 1977.

Journal articles

Biebl, H., and Pfennig, N. Growth of sulfate-reducing bacteria with sulfur as electron acceptor. *Arch. Microbiol.* **112:**115, 1977.

Colby, J., Dalton, H., and Whittenbury, R. Biological and biochemical aspects of microbial growth on C_1 compounds. *Annu. Rev. Microbiol.* **33:**481, 1979.

Cross, R.L. The mechanism and regulation of ATP synthesis by F_1-ATPases, *Annu. Rev. Biochem.* **50:**681, 1981.

Evans, M.C. The mechanism of energy conversion in photosynthesis. *Science Prog.* **62:**543, 1975.

Fillingame, R.H. The proton-translocating pumps of oxidative phosphorylation. *Annu. Rev. Biochem.* **49:**1079, 1980.

Haddock, B.A., and Jones, C.W. Bacterial respiration, *Microbiol. Rev.* **41:**47, 1977.

Harold, F.M. Ion currents and physiological functions in microbiology. *Annu. Rev. Microbiol.* **31:**181, 1977.

Higgins, I.J., et al. Methane-oxidizing microorganisms. *Microbiol. Rev.* **45:**556, 1981.

Hinkle, P.C., and McCarty, R.E. How cells make ATP. *Sci. Am.* **238:**104, 1978.

Mah, R.A., et al. Biogenesis of methane. *Annu. Rev. Microbiol.* **31:**309, 1977.

Morris, J.G. The physiology of obligate anaerobiosis. *Adv. Microb. Physiol.* **12:**169, 1975.

Parson, W.W. Bacterial photosynthesis. *Annu. Rev. Microbiol.* **28:**41, 1974.

Smith, A.J., Hoare, D.S. Specialist phototrophs, lithotrophs, and methyltrophs: a unity among a diversity of procaryotes. *Microbiol. Rev.* **41:**419, 1977.

Thauer, R.K., Jungermann, K., and Decker, K. Energy conservation in chemotrophic anaerobic bacteria. *Microbiol. Rev.* **41:**100, 1977.

Truper, H.G. The enzymology of sulfur metabolism in phototrophic bacteria: a review. *Plant Soil* **43:**29, 1975.

Zelitch, I. Pathways of carbon fixation in green plants. *Annu. Rev. Biochem.* **44:**123, 1975.

Chapter 14 NUTRITION, BIOSYNTHESIS, AND METABOLIC CONTROL

The microbial cell can replicate only if it is capable of producing its own energy or, if it is a parasite, by using the energy compounds of the host. The energy-requiring growth process is characterized by the continual breakdown of cellular components, as well as by the biosynthesis of new cellular building blocks. The materials that are used to construct new cellular components are called ***nutrients.*** As you have already learned, the major elements of organic material are carbon, hydrogen, nitrogen, sulfur, phosphorus, and oxygen. The nutrients required for growth are those inorganic or organic compounds that supply these major elements to the cell, as well as micronutrients, which are not major components of building block materials but are still required for growth. ***Micronutrients*** are minerals, such as magnesium, iron, and manganese, that are frequently cofactors for enzymes or, in some cases, are energy sources. Other nutrients may be organic compounds, such as vitamins, that can be directly incorporated into building block materials or used as energy sources or enzyme cofactors.

The purpose of this chapter is, first, to discuss those inorganic and organic compounds that are required for growth and, second, to outline the basic biosynthetic pathways used by the cell to convert nutrients into the monomers, such as amino acids, purines and pyrimidines, and fatty acids, that are incorporated into macromolecules. The last part of the chapter is devoted to those mechanisms used by the cell to control metabolic reactions.

NUTRITION

Inorganic nutrients and metabolism

Carbon. Carbon exists in many different chemical forms in nature. It may appear as inorganic carbon, such as carbonate or gaseous carbon dioxide, or as a component of simple or complex organic compounds, such as the various hydrocarbons that are abundant in the environment. Some microorganisms can use carbon dioxide from the atmosphere to synthesize more complex organic compounds. Such microbes are called autotrophs and include such organisms as plants, algae, and the photosynthetic bacteria discussed in Chapter 13.

Those microorganisms that require an organic source of carbon are called heterotrophs. Heterotrophism is characteristic of most eukaryotes, including humans. The ability to use one or more organic carbon sources is determined by the enzymatic machinery of the microorganism. Some microorganisms are limited to one or a few carbon sources, whereas others, such as species of *Pseudomonas,* have the potential to catabolize up to 100 different carbon sources. Each carbon

Table 14-1 Principal carbohydrate polymers found in nature and degradable by microorganisms

POLYMER	CHEMICAL COMPOSITION	SOURCE IN NATURE	DEGRADING ENZYME	MICROORGANISM DEGRADING THE POLYMER
Starch	Polymer of D-glucose (α-1,4 linkage)	Plants	α-Amylase, β-amylase	Many bacteria and fungi
Cellulose	Polymer of D-glucose (β-1,4 linkage)	Plants	Cellulase	Several groups of microorganisms, particularly those in the stomach of ruminants such as cows
Pectin	Polymer of methyl D-galacturonic acid	Plants	Pectinase	Many microorganisms especially members of the *Enterobacteriaceae*
Agar	Polymer of D- and L-galactose	Marine algae	Agarase	Degradable by only a few microorganisms
Glycogen	Polymer of D-glucose (α-1,4, linkage)	Animal storage product	Amylase	Many microorganisms
Chitin	Polymer of *N*-acetyl glucosamine	Shells of invertebrates such as lobsters and exoskeletons of insects	Chitinase	Many bacteria and fungi; species of *Streptomyces* are most active decomposers
Hyaluronic acid	Polymer of N-acetyl glucosamine and glucuronic acid	Connective tissue of animals	Hyaluronidase	Primarily species of *Clostridia*
Paramylon	Polymer of glucose (β-1,3 linkage)	Energy storage product of certain algae	β-1,3-Glucanase	Many microorganisms
Laminarin	Polymer of glucose (β-1,3 linkage)	Energy storage product of certain algae	β-1,3-Glucanase	Many microorganisms

source must be transported across the cytoplasmic membrane and enzymatically metabolized in the cytoplasm. Some carbon sources, such as the polymers of glucose (cellulose, starch, etc.) have molecular weights of 1×10^6 daltons or more and cannot be transported into the cell as such but are hydrolyzed outside the cell (Table 14-1). Microbial cells can produce extracellular enzymes that degrade these polymers into smaller units that are transportable across the cytoplasmic membrane. Some eukaryotes, such as the protozoa, can use the mechanism of phagocytosis to transport polymers into the cell. Most carbohydrate polymers, except for those such as agar and cellulose, are more easily hydrolyzed by microorganisms. Other polymers such as lipids, proteins, and nucleic acids are also potential sources of carbon skeletons and can be hydrolyzed into monomers by lipases, proteases, and nucleases, respectively. Hydrolytic enzymes play a special role in the breakdown of organic material by fungi and bacteria in the soil and water and thus have ecological significance.

Nitrogen. Nitrogen can exist in several valence states and as a consequence can appear in reduced (NH_3) or oxidized (NO_3^-, NO_2^-) forms. Nitrogen also exists in an elemental form (N_2) in the atmosphere. Inorganic nitrogen can be converted by the cell into organic nitrogen compounds such as amino acids, nucleotides, and other nitrogen-containing molecules (Figure 14-1).

Nitrogen gas. Nitrogen gas has a valence of 0 and must be reduced to NH_3 before it can be incorporated directly into cellular organic compounds (the amino group of amino acids). Nitrogen gas, which makes up 78% of the atmosphere, can be used by several prokaryotic cells in a process called ***nitrogen fixation.*** Nitrogen fixation can occur in the atmosphere (lightning discharges, for example), but most occurs in biological processes. Biological nitrogen fixation is as-

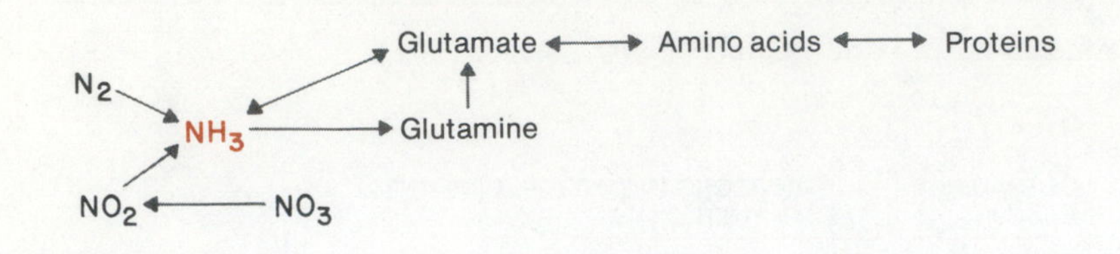

Figure 14-1 Pathways of nitrogen metabolism. Ammonia is basic nitrogen compound from which organic nitrogen compounds will be synthesized.

sociated primarily with agricultural plants in which bacteria in a symbiotic or free-living state (Table 14-2) fix atmospheric nitrogen and produce NH_3 for use by the plant. The fixation of nitrogen is a complex process that is catalyzed by the enzyme ***nitrogenase.*** The production of active cell-free extracts and the use of acetylene reduction for determining nitrogenase activity have contributed greatly to our understanding of this very important biological process. Nitrogenase is an iron-sulfur protein made up of two units referred to as ***component I*** and ***component II.*** The nitrogenases from different organisms are similar, and component I from one genus can interact with component II of another genus to produce enzyme activity. The characteristics of these components are as follows:

1. Component I. Component I is the larger of the two components and consists of two molybdenum atoms and 28 to 34 nonheme iron atoms. It functions in the binding as well as reduction of substrate (N_2). Note that substrates other than N_2 can bind to component I and be reduced, for example, acetylene (C_2H_2), protons, and cyanide (CN).
2. Component II. Component II contains no molybdenum but does possess four nonheme iron atoms. Its function is to supply electrons, one at a time, to component I. It also binds ATP and magnesium.

Six electrons are required for the reduction of nitrogen, but the sequence of the steps has not been determined. The overall reaction is:

$$6\,H + 6\,e^- + N{\equiv}N \longrightarrow 2\,NH_3$$

The reduction of the triple-bonded nitrogen (or C≡C in acetylene or C≡N in cyanide) requires energy that is released from the hydrolysis of ATP. The total ATP requirement in vivo varies from one genus of bacteria to another. Only 4 to 5 ATPs are required for species of *Azotobacter,* but as many as 20 ATPs are required for nitrogen fixation by *Clostridium pasteurianum.* ATP binds to component II and is not hydrolyzed until component II transfers an electron to component I.

Table 14-2 Examples of nitrogen-fixing microorganisms*

	ORGANISMS
Free living (nonsymbiotic)	
Aerobic	*Azotobacter*
	Beijerinckia
	Azomonas
	Aquaspirillum peregrinum
	Cyanobacteria (heterocystous species)
	Gloeocapsa (coccoid cyanobacterium)
Microaerophilic	*Xanthobacter autotrophicum*
	Azospirillum lipoferum
	Thiobacillus ferrooxidans
	Alcaligenes latus
	Many nonheterocystous cyanobacteria
	Rhizobium japonicum
	Arthrobacter species
Facultative	*Klebsiella pneumoniae*
	Klebsiella aerogenes
	Klebsiella oxytoca
	Enterobacter species
	Citrobacter freundii
	Bacillus polymyxa
	Bacillus macerans
Anaerobic	*Clostridium* species
	Photosynthetic species of *Chromatium, Chlorobium,* and *Rhodospirillum*
Symbiotic	*Rhizobium* species (nodulate leguminous plants such as peas, soybeans, and clover)
	Certain actinomycetes (certain nodulate plants such as *Comptonia* and alder)
	Anabaena azollae (fixes nitrogen within leaf pores of the water fern *Azolla*)

*No eukaryotes are known to fix nitrogen.

Electrons for nitrogen fixation may be derived from photosynthetic reactions, such as occur in the cyanobacteria, or from carbon substrates such as pyruvate, as in *Clostridium pasteu-*

rianum. The carriers appear to be ferredoxin (Figure 14-2) or NADPH.

Both components I and II of the nitrogenase enzyme are inhibited by oxygen. This presents no special problem for the strict anaerobes, such as *Clostridium pasteurianum,* which grows only in the absence of oxygen. Facultative anaerobes, such as species of *Klebsiella,* fix nitrogen only under anaerobic conditions. Some organisms that carry out growth under conditions in which oxygen is present have evolved systems that still permit nitrogen fixation. *Azotobacter* species, for example, are strict aerobes, but their respiratory activity is so high that oxygen is quickly reduced to water. The rapid reduction of oxygen occurs as long as carbon substrates are plentiful and protons can be generated. When carbon substrates become limited, the oxygen produced during respiration is not quickly reduced and could potentially inhibit nitrogenase activity. *Azotobacter,* however, can cope with this condition because it produces another iron-sulfur protein that complexes with nitrogenase and protects it from inactivation. This new protein prevents nitrogenase from fixing nitrogen, but as soon as the respiratory activity of the organism increases, it dissociates from the nitrogenase and activity is restored. Oxygen lability is also a potential problem for some of the photosynthetic cyanobacteria that can fix nitrogen. They exist in an environment of oxygen and also produce it from photosynthetic reactions.

Some species can fix nitrogen under microaerophilic or anaerobic conditions, but others produce ***heterocysts*** that enable them to have active nitrogenase under aerobic as well as anerobic conditions. Heterocysts are thick-walled refractive structures interspersed among the vegetative cells of the filamentous species (see Chapter 4). Vegetative cells are produced only when the concentration of nitrogenous compounds such as NH_3 and NO_3 are at sufficient levels in the environment. When these compounds fall below a critical level, nitrogen fixation is the only mechanism for obtaining combined nitrogen, and the cell responds by producing heterocysts. The vegetative cells, which lie on either side of the heterocyst, supply nutrients to the heterocyst. The septum separating the vegetative cells from the heterocysts become very small, and only small pores called ***microplasmadesmatas*** connect their cytoplasms. It is through these pores that nutrients are carried from the vegetative cell to the heterocyst (Figure 14-3). The glycolipid envelope of the heterocyst provides a barrier to the entrance of oxygen. Studies with mutants have shown that cells deficient in this glycoplipid envelope are more sensitive to oxygen than normal cells. The heterocysts possess photosystem I activity and generate large amounts of the ATP required for nitrogen fixation, but they do not contain photosystem II activity, which generates oxygen.

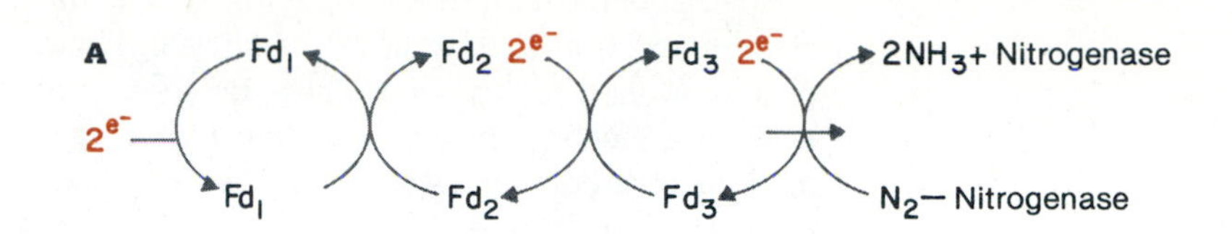

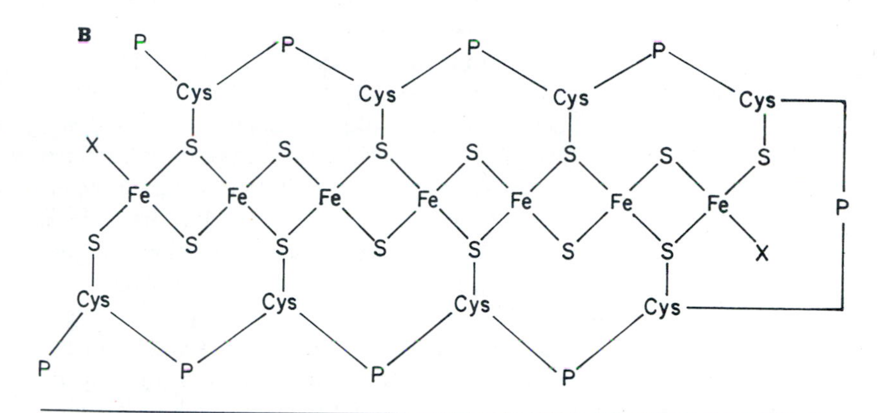

Figure 14-2 Nitrogen fixation. **A,** Nitrogen fixation in bacteria. Two electrons *($2e^-$)* generated by electron source are picked up and transferred by three different ferredoxin components *(Fd_1, Fd_2* and *Fd_3)* of nitrogenase enzyme system. Hydrogen atoms (not shown) are also believed to be carried on intermediates bound to enzyme system. Energy for fixation of nitrogen comes from ATP hydrolysis. **B,** Partial structure of ferredoxin molecule. *Fe,* Iron; *S,* sulfur; *Cys,* cysteine; *P,* segment of protein chain.

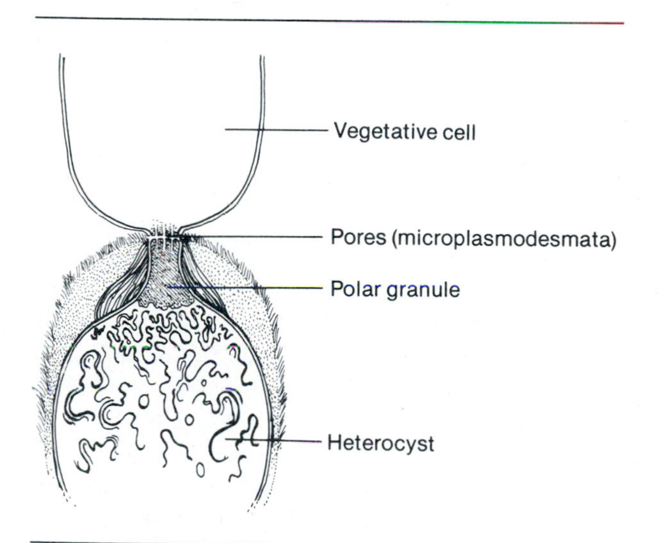

Figure 14-3 Cytoplasmic association of heterocyst and adjacent vegetative cell. Cytoplasm of heterocyst and vegetative cell are connected by small pores (microplasmadesmatas).

Species of bacteria belonging to the genus *Rhizobium* exist in a symbiotic relationship with plants and also fix nitrogen. These bacteria have evolved a mechanism for protecting nitrogenase from inactivation by oxygen. Species of *Rhizobium* are able to infect the root hairs of certain leguminous plants (peas, alfalfa, clover, peanuts, and beans). Within the epidermal cells of the plant the bacteria become embedded in a sheath called an ***infection thread.*** The infection thread and its enclosed bacteria penetrate into the cortical tissue of the root. Bacteria are released within the root cells, and the latter are induced to increase their rate of cell division and to differentiate into nodules. Bacteria within the nodules are surrounded by a membrane envelope and are called ***bacteroids.*** Surrounding the bacteroids are large quantities of ***leghemoglobin,*** whose function is to bind oxygen and facilitate its diffusion to rhizobia while preventing free oxygen from inactivating nitrogenase. The heme portion of leghemoglobin is contributed by the bacterium, and the globin portion is coded for by the plant. Bacteria in the nodules fix nitrogen, and the ammonia produced is used by the plant to synthesize amino acids. Thus soil deficient in fixed nitrogen can support the growth of leguminous plants because the bacterial symbiont takes nitrogen from the atmosphere and not from the soil. Leguminous plants can be plowed under to replace nitrogen lost when the soil has been depleted of nitrogen by nonleguminous plants.

The industrial conversion of gaseous nitrogen to ammonia for use as a fertilizer is expensive and difficult because of the need for extremes of temperature and pressure. A great deal of research has been devoted to understanding the enzymatic mechanism of nitrogen fixation in the microbial cell. The development of an industrial catalyst similar in activity to biological systems could be an important energy-saving device in the production of commercial fertilizers. The genes for nitrogen-fixing microorganisms are transferable by conjugation and can be manipulated on plasmids. The transfer of nitrogen-fixing genes through recombinant DNA technology to other microorganisms or to plants may someday increase crop yields and help to rejuvenate grazing lands. See Chapter 19 for further discussion of nitrogen fixation.

Nitrate. The nitrogen in nitrate has a valence of +5 and to be assimilated is reduced to +3, the valence of nitrogen in NH_3. The assimilatory process is accomplished through the activity of two enzymes—nitrate reductase and nitrite reductase—with electrons supplied by $NADPH_2$ and is not coupled to ATP formation.

$$NO_3 \xrightarrow[\text{reductase}]{\text{Nitrate}} NO_2 \xrightarrow[\text{reductase}]{\text{Nitrite}} NH_3$$

Nitrite, which is toxic, is seldom excreted as such and is rapidly reduced to ammonia. Nitrites have been used for many years as preservatives for certain meats and meat products such as bacon and frankfurters. Concern over their use has arisen because nitrites are converted in the body to nitrosamines, which have been described as carcinogenic in animals (see Chapter 20).

Nitrate can also be used by some microorganisms under anaerobic conditions as a terminal electron acceptor in ATP formation—a dissimilatory reduction process. This function is discussed in Chapter 13.

Ammonia. Ammonia (NH_3) is the most efficiently used nitrogen source for microorganisms because it can be incorporated directly into carbon skeletons to produce amino acids. The concentration of ammonia in the environment, as well as the level of energy sources, determines its fate. The scheme of the bacterial cell is to use ammonia immediately, if biosynthesis of amino acids and other nitrogen-containing material is required, or to store it for future use when biosynthesis is repressed. A ***glutamine synthetase/glutamate synthase*** enzyme system (Figure 14-4) exists in the bacterial cell to control the assimilation of nitrogen. When adequate concentrations of ammonia are available to the cell, glutamate dehydrogenase catalyzes the assimilation of nitrogen into glutamate, and some of the glutamate is converted into glutamine. Under limited ammonia availability all of the nitrogen is incorporated into glutamine via the glutamine synthetase reaction. Basically, the nitrogen in the ammonia molecule under conditions of biosynthesis will become part of the glutamate molecule according to the following reaction:

$$\underset{\alpha\text{-Ketoglutarate}}{\begin{matrix} COO^- \\ | \\ C{=}O \\ | \\ H{-}C{-}H \\ | \\ H{-}C{-}H \\ | \\ COO^- \end{matrix}} + \underset{\text{Ammonia}}{NH_3} \xrightarrow{NADPH_2} \underset{\text{Glutamate}}{\begin{matrix} COO^- \\ | \\ H{-}C{-}NH_2 \\ | \\ H{-}C{-}H \\ | \\ H{-}C{-}H \\ | \\ COO^- \end{matrix}} + H_2O + NADP$$

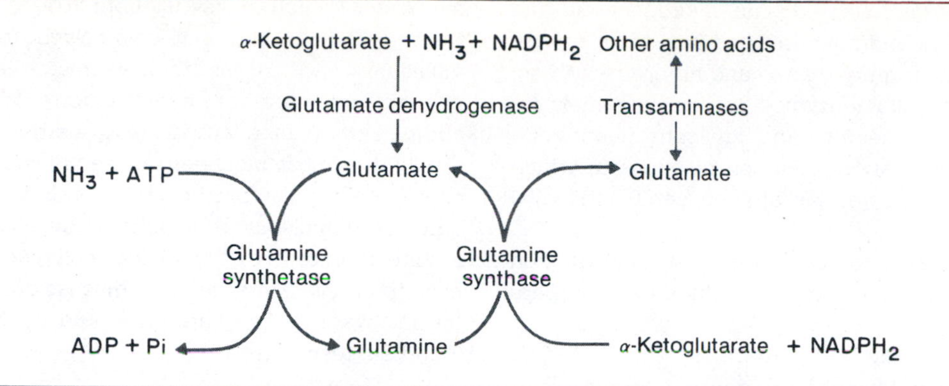

Figure 14-4 Mechanism used by cell to control assimilation of nitrogen: glutamine synthetase/ glutamate synthase enzyme system. See text for discussion.

Glutamate provides the amino (NH_2) group for most other amino acids by participating in reactions called ***transaminations,*** such as:

$$\underset{\text{Pyruvate}}{CH_3-\overset{\overset{O}{\|}}{C}-COO^-} + \underset{\text{Glutamate}}{\begin{array}{c} COO^- \\ | \\ H-C-NH_2 \\ | \\ H-C-H \\ | \\ H-C-H \\ | \\ COO^- \end{array}} \xrightarrow{\text{Transaminase}} \underset{\text{Alanine}}{\begin{array}{c} CH_3 \\ | \\ H-C-NH_2 \\ | \\ COOH \end{array}} + \underset{\alpha\text{-Ketoglutarate}}{\begin{array}{c} COO^- \\ | \\ C=O \\ | \\ H-C-H \\ | \\ H-C-H \\ | \\ COO^- \end{array}}$$

The glutamine produced in the cell supplies nitrogen to a number of metabolites such as the amino sugars and nucleotides, and the amino acids histidine and tryptophan.

Sulfur. Sulfur appears in the environment primarily as inorganic sulfate ($SO_4^=$) which can be assimilated by microorganisms and reduced, in a series of reactions, to hydrogen sulfide (H_2S). Inorganic sulfur can become part of organic molecules in the following reaction:

$$\underset{\text{Hydrogen sulfide}}{H_2S} + \underset{\text{Serine}}{CH_2OH-\overset{\overset{NH_2}{|}}{\underset{\underset{H}{|}}{C}}-COOH} \longrightarrow \underset{\text{Cysteine}}{\overset{\overset{SH}{|}}{CH_2}-\overset{\overset{NH_2}{|}}{\underset{\underset{H}{|}}{C}}-COOH} + \underset{\text{Water}}{H_2O}$$

Cysteine helps maintain the tertiary structure of proteins through the formation of sulfur-sulfur bonds.

The formation of hydrogen sulfide occurs in the following steps:

$$\text{ATP} + \text{Sulfate} \xrightarrow{\text{ATP sulfurylase}} \text{Adenosine 5'-phosphosulfate (APS)} + PP_i$$

$$\text{ATP} + \text{APS} \xrightarrow{\text{APS phosphokinase}} \text{Adenosine 3'Phosphate 5'-phosphosulfate (PAPS)} + \text{ADP}$$

$$\text{PAPS} + NADPH_2 \xrightarrow{\text{PAPS reductase}} SO_3^- + \text{AMP 3'-phosphate} + \text{NADP}$$

$$SO_3^- + NADPH_2 \xrightarrow{\text{Sulfite reductase}} H_2S + \text{NADP}$$

Inorganic sulfur is also an important component of certain proteins such as ferredoxin and of oxidation-reduction enzymes, where it behaves as a cofactor. The inorganic sulfur in these proteins is usually associated with other cofactors:

iron-sulfur, as in ferredoxins, hydrogenases, and succinate dehydrogenase; and iron-sulfur molybdenum, as in nitrogenases and nitrate reductase. Many iron-sulfur enzymes possess extremely reduced redox centers and are consequently involved in oxidation-reduction reactions in which the reduction potentials are between 0 and −550 mV.

Sulfate can also be used as a terminal electron acceptor, and this function is discussed in Chapter 13.

Phosphorus. Phosphorus appears in organic molecules primarily as a component of nucleotides, such as ATP, which is important as a carrier of energy and phosphate and as a constituent of nucleic acids. Some microorganisms transport phosphate into the cell and accumulate it in the form of granules called ***metachromatic granules*** or ***volutin.*** The microbial cell assimilates inorganic phosphate into organic compounds through the following reaction:

$$\begin{array}{c} COO^- \\ | \\ H-C-OH \\ | \\ H-C-O-PO_3H_2 \\ | \\ H \end{array} + \underset{\text{Inorganic phosphate}}{iPO_4} \longrightarrow \begin{array}{c} CO-PO_3H_2 \\ | \\ H-C-OH \\ | \\ H-C-O-PO_3H_2 \\ | \\ H \end{array}$$

3-Phosphoglycerate — 1,3-Diphosphoglycerate

Media in which microorganisms, especially bacteria, are cultivated is prepared with salts of phosphate, such as ammonium phosphate and potassium phosphate. The phosphates serve not only as a source of phosphorus but also as a buffering agent for the medium, since they are capable of binding one, two, or three hydrogen atoms:

$$Na_3PO_4 \xrightarrow{H^+} Na_2HPO_4 \xrightarrow{H^+} NaH_2PO_4 \xrightarrow{H^+} H_3PO_4$$

Phosphorylation of proteins is also an important process in the cell, but its exact function(s) is not completely understood. Possible functions include protein inactivation, protein activation, a regulatory device, and a protective device against protease digestion (see p. 330).

Oxygen. Oxygen is the final electron acceptor during aerobic respiration and is converted to nontoxic as well as toxic intermediates. These aspects of oxygen metabolism are discussed in Chapter 13.

Water. The principal component of living matter, at least quantitatively, in the microbial cell is water. Water, which makes up from 70% to 80% of the weight of the cell, acts as a solvent for many molecules and carries them across the cytoplasmic membrane in a hydrated form. Water, in addition to its inert activity as a carrier, is also involved in most biochemical reactions and, because of its polar nature and capacity to bind other polar molecules, influences the shape and architecture of many biological polymers, especially proteins. More detailed discussions of water and water activity are presented in Chapters 2 and 20, respectively.

Mineral metabolism

The crust of the earth is composed of many minerals that are important in the life of the microbial cell. These elements include magnesium, calcium, manganese, zinc, iron, selenium, potassium, sodium, cobalt, copper, and molybdenum. Some of these minerals are major nutrients and are required in high concentration in all cells, whereas others are required only by certain species. Minor nutrients are those that are used by cells at low concentrations, and laboratory media will often contain sufficient impurities to satisfy the cell's needs.

Magnesium. Magnesium is a divalent cation that plays an important role in metabolism. All cells require magnesium and actively accumulate it by specific energy-dependent transport systems. Most of the magnesium in the cell is associated with the nucleic acids, especially ribosomes; the magnesium neutralizes the negatively charged phosphate groups and thus acts to stabilize ribosomes. Magnesium deficiency can lead to filament formation in ordinarily unicellular forms, to ribosome degradation, and to a decrease in nucleic acid synthesis. Magnesium also binds to the negatively charged teichoic acids, which are believed to act as a respository for magnesium ions in gram-positive cells. Magnesium is a cofactor for many enzymes and is also a component of the chlorophyll molecule.

Calcium. The calcium level in microbial cells is very low compared with its concentration in the external environment. Calcium represents a special case because it is secreted by metabolically active processes and is not actively accumulated like nearly all other minerals. Lowering the temperature or using inhibitors of energy formation causes microbial cells to accumulate calcium, thus indicating a carrier-mediated transport system.

The reverse orientation of calcium transport

can be demonstrated by a very unusual technique of preparing microbial cell membranes with different orientations. In other words, the cell membrane can be prepared in a normal way in which the outer part of the membrane remains outside (right side out), or it can be prepared abnormally, in which the membrane is turned inside out (Figure 14-5). The right-side-out orientation shows that calcium is secreted, but in the inside-out orientation calcium is accumulated.

The only instance where calcium has been observed to be accumulated by bacterial cells is during the sporulation process by members of the genus *Bacillus.* A carrier is believed to be involved, but it is not known if it is the same one used for secretion of calcium by nonsporulating cells.

Calcium appears to play some specific roles in eukaryotic cells. It may be important for the aggregation associated with the differentiation process in the slime mold *Dictyostelium discoideum.* Calcium increases dramatically in cells during slug migration and stalk and spore formation. Calcium is also associated with ciliary movement in protozoa such as *Paramecium.* Changes in calcium concentration appear to be associated with changes in direction of movement of the protozoa. When the organism meets an obstacle and must swim backward, the calcium concentration increases in the cell; for the organism to swim forward, calcium is pumped out of the cell.

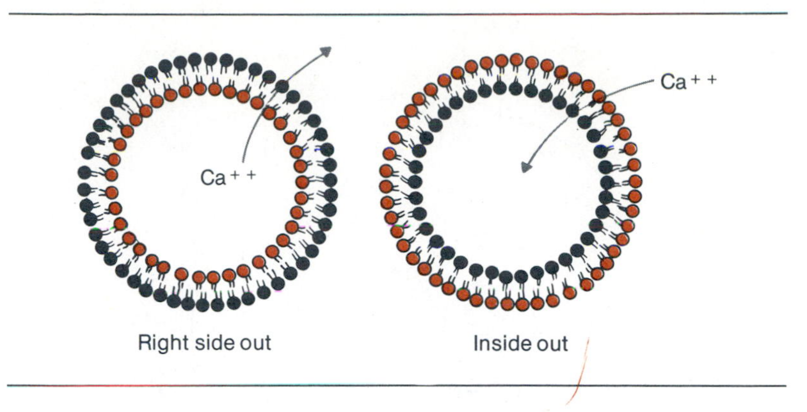

Figure 14-5 Direction of movement of calcium in membranes prepared right side out and inside out.

Figure 14-6 Scanning electron micrograph of microorganisms encrusted with metal deposits on ocean floor. Microorganisms are attached to mussel beds in vicinity of thermal submarine vents at Galapagos rift at depth of over 7600 feet. Spherical cells are encrusted with manganese and/or iron and are interspersed with clusters of filamentous forms. Hydrogen sulfide is source of energy for these bacteria, which are sole source of food for animal populations at these extreme depths. Bar = 10 μm.

From Jannasch, H.W., and Wirsen, C.O. Appl. Environ. Microbiol. ***41****:528, 1981.*

Manganese. Manganese is required by microbial cells in trace amounts, and its functions are largely unknown. It can serve as an alternate cofactor for magnesium during magnesium deficiency, particularly in magnesium-requiring enzymes.

Microbial oxidation of manganese occurs among many soil and marine species. Nodules of ferromanganese (Figure 14-6) are produced on the ocean floor, presumably through the activity of microbial enzymes, but the exact mechanism is not known. Soil bacteria and fungi reduce insoluble salts of manganese (Mn^{4+}) to soluble manganese (Mn^{2+}) compounds. This reduction process, which is characteristic of some bacteria that inhabit the roots of some plants, makes manganese available in a soluble form. The sheathed bacteria, such as *Leptothrix* and *Sphaerotilus,* oxidize manganese (Mn^{2+}) and accumulate it within the sheath as deposits of manganese oxides.

Manganese is required for the synthesis of secondary metabolites such as antibiotics. Peptide antibiotics produced industrially by species of *Bacillus,* for example, cannot be synthesized in the absence of manganese. Peptide antibiotics, which are produced during sporulation, are believed to act as carriers of calcium and thus may help the cell to accumulate this divalent cation.

Zinc. Zinc is required only in trace amounts (0.05 to 0.1 μM) by the microbial cell. It is essential for the activity of several enzymes, including DNA and RNA polymerase. Zinc is also essential for the synthesis of many secondary metabolites pro-

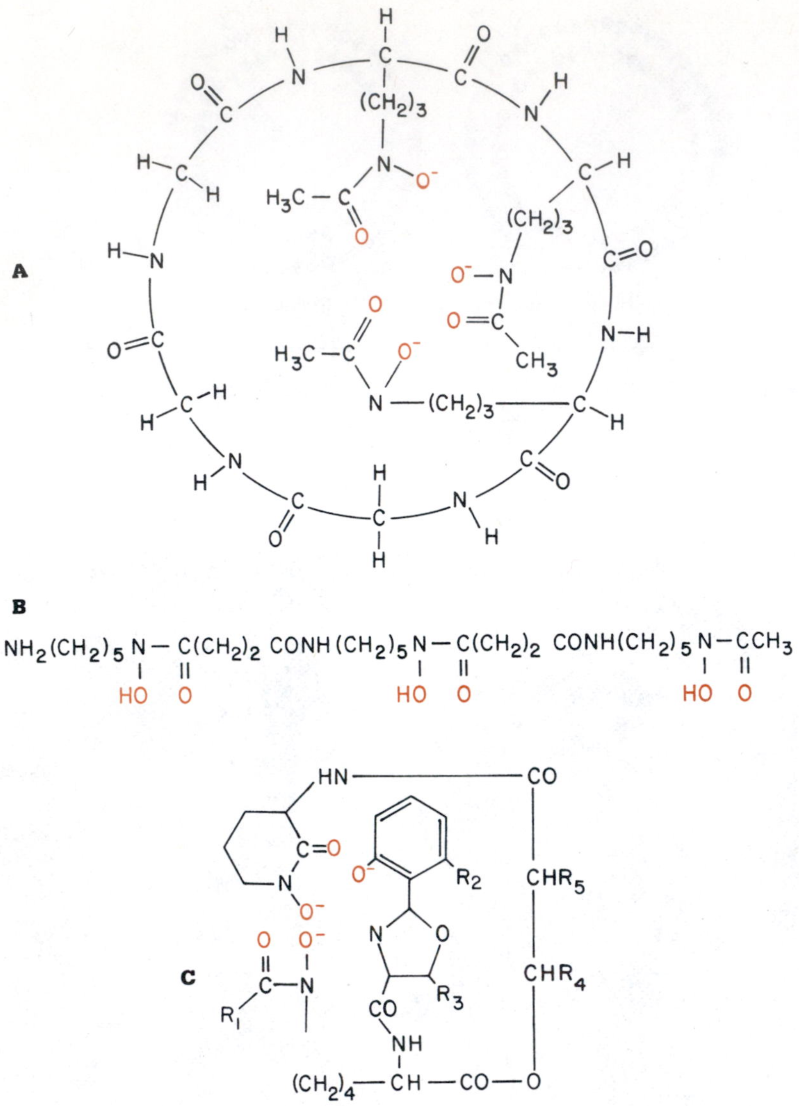

Figure 14-7 Structure of three hydroxamate derivatives. Colored atoms are sites of iron chelation. **A,** Ferrichrome. **B,** Ferrioxamine B. **C,** Mycobactin.

duced by microorganisms that are used industrially and medically.

Iron. The earth was originally enveloped in an anaerobic atmosphere, and one of its primary elements, iron, existed in the ferrous state (Fe^{++}). The appearance of oxygen from photosynthetic processes resulted in the oxidation of much of the ferrous iron to the ferric (Fe^{+++}) state. Oxidation produces iron complexes that make the mineral unavailable for biological transport in the ionic form. Microbial cells have evolved techniques to compensate for this transport deficiency. Ferric iron is made available to the cell through the production of metal-binding (chelator) molecules called ***siderophores.*** The iron is bound to these transport molecules via an oxygen, sulfur, or nitrogen atom. There are three classes of natural iron transport molecules: hydroxamic acids, catechols (derivatives of 2,3-dihydroxybenzoic acid), and citrate. Three identifiable hydroxamates are ferrichrome, ferrioxamine B, and mycobactin (Figure 14-7). Mycobactin is found in the mycobacteria, which have a high lipid content in their envelope. Mycobactin, interestingly enough, has the greatest lipid solubility among the siderophores. An important catechol derivative is enterobactin (Figure 14-8), which is produced by members of the *Enterobacteriaceae.* Specific siderochromes are produced by different genera of bacteria, but some microorganisms can use the chelating agents produced by microorganisms from other genera.

Special receptors and transport components are required for the uptake of the chelator-iron complex into the cytoplasm of the cell. The metal is withdrawn from the chelator by reduction of iron to the ferrous state or by enzymatic hydrolysis of the ligands (the atoms on the chelator that bind the iron). The importance of iron to the survival of the microbial cell is emphasized by the fact that *Escherichia coli* has at least 13 genes involved in siderophore synthesis and uptake—a system that undoubtedly requires a great expenditure of energy. Siderophore production is depressed when concentrations of iron are high, but siderophore synthesis is increased at low concentrations of iron.

Iron is a metal required by all bacteria, and its major role is in electron/hydrogen movement. Iron is a component of the hemin molecule that makes up part of the cytochromes and thus functions in the transport of electrons by being alternatively reduced and oxidized. Other heme proteins in which iron is a component are catalase and peroxidase, two enzymes that protect the cell from the deleterious effects of hydrogen peroxide generated during aerobic respiration. Nonheme iron proteins involved in electron transport are hydrogenase, flavoproteins, and ferredoxin. Iron is also believed to play an important role in the host's response to invading microbial pathogens. Nearly all vertebrates respond to the infection process by altering their metabolic processes in such a way as to deprive the invading pathogen of iron. One vertebrate chelator of iron is ***transferrin,*** which is found in other secretions such as milk, tears, saliva, hepatic bile, synovial fluid, and seminal fluid. Iron-binding proteins such as transferrin increase in concentration at the site of infection and are believed to behave as bacteriostatic agents by siphoning away iron that could be

Table 14-3 Vitamins and their function in microbial metabolism

VITAMIN	FUNCTION
Thiamine	As thiamine pyrophosphate, a coenzyme involved in decarboxylation reactions
Riboflavin	In combined form such as flavin adenine dinucleotide (FAD), a coenzyme important in oxidation-reduction reactions
Niacin	In combined form such as nicotinamide adenine dinucleotide (NAD) or its phosphorylated derivative (NADP), a coenzyme in oxidation-reduction reactions
Pantothenic acid	Part of the coenzyme A (CoA) molecule; functions as a carrier of acyl groups; precursor of prosthetic group of acyl carrier protein (ACP) involved in lipid metabolism
Pyridoxal	In phosphorylated form, functions as a coenzyme; important in amino acid metabolism, particularly group transfer reactions involving the amino group
Biotin	Coenzyme important as a carrier of carbon dioxide
Folic acid	In reduced form (tetrahydrofolic acid), acts as a carrier of formyl groups; especially important in purine biosyntheis
Vitamin B_{12}	Coenzyme functioning as a carrier of alkyl groups
Vitamin K	Precursor of menaquinone, which is involved in electron transport

used for growth of the invading parasite.

Toxin production in *Corynebacterium diphtheriae,* the causative agent of diphtheria, is controlled by the concentration of iron, and this aspect of the host-parasite relationship is discussed in Chapter 22.

Potassium. Potassium is required by all microbial species tested to date and is especially important in ribosome function. Cells deprived of potassium or mutants unable to transport potassium into the cell are unable to carry out protein synthesis.

Other minerals. Molybdenum, cobalt, and copper are minor nutrients that play very specific roles in the function of some microorganisms. Molybdenum plays a role in nitrogen metabolism, since it is an essential component of ferredoxins, nitrogenase, and nitrate reductase. Cobalt is part of the vitamin B_{12} molecule (Table 14-3), and copper is a component of cytochrome oxidase and is involved in electron transport during aerobic respiration.

Specific organic nutrients (growth factors)

Some microorganisms are unable to take inorganic or organic nutrients from the environment and convert them to organic metabolites. Consequently, these organisms require supplements of specific organic products, called growth factors. Growth factors include vitamins, amino acids, and nucleotides. Vitamins are especially important because of their role in enzymatic reactions. They form the coenzyme portion of many enzymes and have specific functions, which are described in Table 14-3. Many microorganisms isolated from their natural environment and cultivated in the laboratory require growth factors. Microorganisms in their natural environment may have these requirements supplied by other organisms that inhabit the same ecological niche. Microorganisms that normally colonize an-

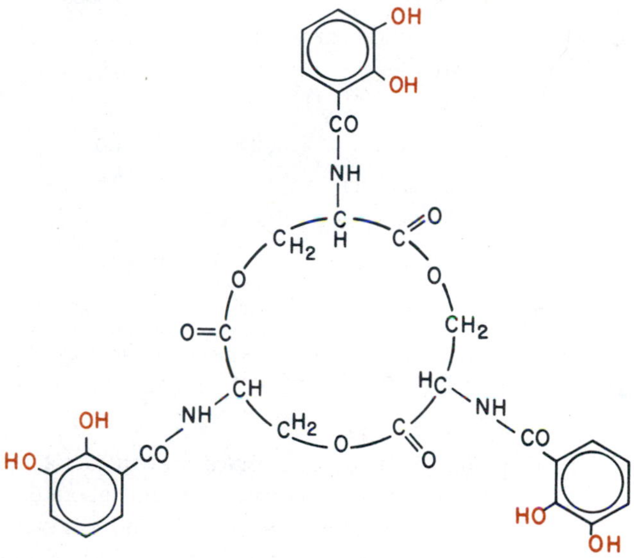

Figure 14-8 Structure of siderophore enterobactin. Atoms in color represent sites for chelation of iron.

Table 14-4 Chemical composition of two kinds of synthetic media*

MEDIUM	MG/L
Medium A	
$CaCl_2$	15
$MgSO_4$	120
$(NH_4)_2SO_4$	1200
Na_2HPO_4	7000
NaH_2PO_4	200
Glucose	10,000
Medium B	
Cysteine	15
Sodium glutamate	500
$(NH_4)_2SO_4$	1200
Sodium acetate	10,000
Na_2HPO_4	7000
NaH_2PO_4	1200
Folic acid	0.005
Biotin	0.0025
Para-aminobenzoic acid	0.1
Thiamine	0.5
Riboflavin	0.5
Pyridoxal	1.0
Pantothenate	0.5
Nicotinic acid	1.0
Glucose	10,000

*Medium A is a basal salts minimal medium used to cultivate such enteric organisms as *E. coli*. Medium B is a more complex medium used to cultivate some species of oral streptococci.

imals and humans often have growth factor requirements when cultivated in the laboratory. Scientists have suggested that during evolution these microorganisms could produce all of the necessary metabolites from simple carbon sources; however, once they became permanent colonizers of mammals, who could supply preformed metabolites, biosynthetically deficient mutants were selected. One especially important exception is *E. coli,* which inhabits the intestinal tract of animals. Most strains of this species are capable of growth on a simple defined medium (Table 14-4) and can synthesize all of their organic requirements from inorganic salts and a single organic carbon source such as glucose. Compare this with the requirement of a species of *Streptococcus* isolated from the oral cavity (Table 14-4). The ability of *E. coli* to synthesize all of its organic needs is one of the reasons why this organism has been used extensively in genetic and metabolic studies.

BIOSYNTHESIS

How energy is extracted from inorganic and organic compounds and how this energy is used for movement and nutrient transport is demonstrated in earlier chapters. Energy is also used to convert the basic nutrients, discussed earlier in this chapter, into organic compounds that make up the structural and enzymatic components of the cell. What organic compounds can be synthesized by each cell is determined by the genetic content of the cell and by what enzymes are potentially inducible. This biosynthetic capacity may be the result of natural evolutionary processes, or it may be due to unnatural laboratory manipulation, such as genetic engineering, in which foreign genes are introduced into the cell.

Biosynthetic capacity

The biosynthetic potential of microorganisms is as variable as any other property that can be measured. We have already discussed the fact that some microorganisms are able to take inorganic nutrients and assimilate them into organic molecules whereas other organisms require external supplies of many organic compounds in order to grow. Biosynthesis is a continual process in cells, and whether the cell has a short (20-minute) or long (24-hour) generation time, biosynthesis is required to construct new cell material and to replenish certain molecules in the cell that have been degraded. Macromolecules, such as proteins and nucleic acids, which are continually being degraded during the cell cycle, may have a short or long half-life. Some messenger RNA (mRNA) molecules in bacteria have a half-life of 60 to 120 seconds, whereas in eukaryotic cells the mRNA half-life may be up to 20 minutes. Biosynthesis occurs at a faster rate in prokaryotic cells (mycobacteria are exceptions) than in eukaryotic cells—a property that is especially important when antimicrobials are administered to humans suffering from infectious diseases.

One of the most significant properties of both eukaryotic and prokaryotic cells is that the pathways leading to the biosynthesis of a central metabolite, such as an amino acid, are invariably the same or similar. This indicates that little change has occurred after the evolutionary divergence of these cellular types. This is one of the reasons why research of biosynthetic processes in micro-

Table 14-5 Metabolic intermediates serving as precursors in the biosynthesis of amino acids

PRECURSOR(S)	AMINO ACID END PRODUCT(S)
Pyruvic acid	Alanine, valine, leucine
Oxaloacetic acid	Aspartic acid, threonine, isoleucine, lysine, methionine
α-Ketoglutaric acid	Lysine, glutamic acid, proline, arginine
3-Phosphoglyceric acid	Serine, glycine, cysteine, cystine
Erythrose phosphate and phosphoenolpyruvate	Tyrosine, phenylalanine, tryptophan
Pentose phosphate	Histidine

$$\alpha\text{-Ketoglutarate} + NH_4^+ + NADPH_2 \longrightarrow \text{L-Glutamate} + NADP + H_2O + H^+$$
$$\text{Pyruvate} + NADH_2 + NH_4^+ \longrightarrow \text{L-Alanine} + NAD + H_2O + H^+$$
$$\text{Fumarate} + NH_4^+ \longrightarrow \text{L-Aspartate} + H^+$$

Figure 14-9 Three reactions in microorganisms in which ammonia is assimilated directly into amino acids.

organisms has been so helpful in understanding human biosynthetic processes.

Biosynthetic pathways

Biosynthesis of amino acids. We discussed earlier how nitrogen in all of its inorganic forms is converted to ammonia, which in turn is incorporated into organic compounds to produce amino acids. Four different mechanisms are available to microorganisms for the assimilation of ammonia, but the reactions are basically the same—the amination of a keto acid. Three of the reactions are outlined in Figure 14-9. The fourth reaction involves a system devised to aid the cell in ammonia assimilation when high or low concentrations of the ion are present in the medium, as discussed earlier in the chapter.

One of the principal functions of the TCA cycle is to produce intermediates that act as precursors to the amino acids. The amino acids have been divided into families based on their precursor types (Table 14-5). Only L-lysine, of all the amino acids, is synthesized via two different pathways (Figure 14-10). This is an example of the type of divergence that is believed to have accompanied the evolutionary split between prokaryotes and eukaryotes. L-Lysine in prokaryotes is synthesized from the amino acid L-aspartate, whereas in eukaryotic fungi and most algae L-glutamate is the primary metabolite from which L-lysine is produced. The aspartate pathway provides two essential intermediates that are an important part of the bacterial life cycle: diaminopimelic acid (DAP) and dipicolinic acid (Figure 14-11). DAP is

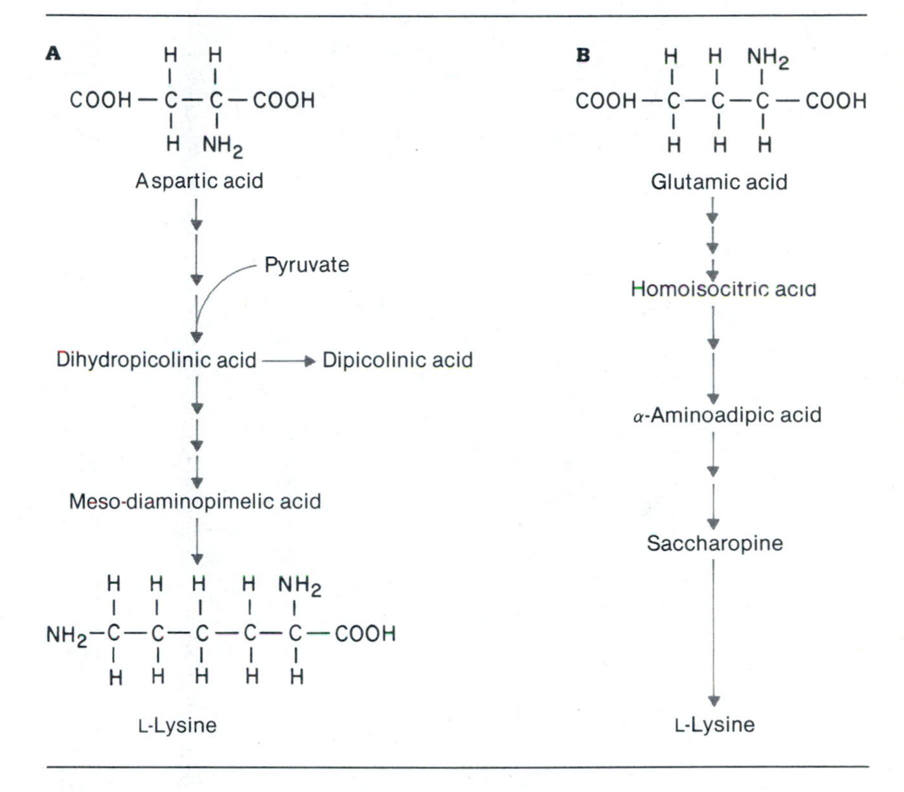

Figure 14-10 Alternative pathways of lysine biosynthesis (not all intermediates are indicated). **A,** Pathway used in bacteria. **B,** Pathway used by fungi and most algae.

a component in the peptidoglycan layer of some bacterial cell walls, and dipicolinic acid is a component of the cortex of the bacterial endospore. Neither of these intermediates are used by eukaryotic cells and consequently are not produced.

Some microorganisms cannot synthesize all of the necessary amino acids, and they must be supplied preformed in the medium, either as free amino acids or in the form of proteins. Many microbial species are genetically equipped to produce extracellular proteases that hydrolyze proteins into amino acid residues. The clostridia, which are capable of causing gas gangrene in humans, produce proteases that degrade muscle protein during the infection process. The amino acids released by hydrolysis are transported across the cytoplasmic membrane to be fermented and to engage in protein synthesis. The

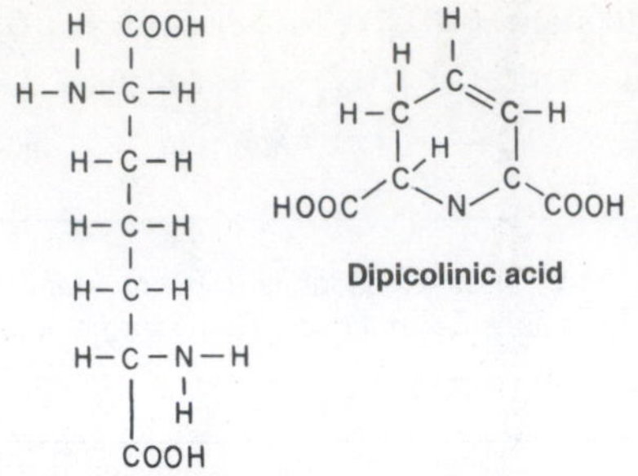

Diaminopimelic acid

Dipicolinic acid

Figure 14-11 Structure of diaminopimelic acid (DAP) and dipicolinic acid.

virulence of pathogenic species of *Neisseria* and *Streptococcus* appears to be related to the production of specific proteases. *Neisseria gonorrhoeae* (the causative agent of gonorrhoea) owes part of its virulence to its ability to adhere to epithelial cells of the urogenital tract. Epithelial secretions in the urogenital tract (or other secretory surfaces) contain secretory antibodies, called secretory IgA (sIgA), that can prevent the attachment of pathogenic species to the epithelial surface. *N. gonorrhoeae* however, produces a protease that cleaves specific sIgA molecules and inactivates them—a process that permits attachment of the virulent organisms to the host. Nonpathogenic species of *Neisseria* apparently do not produce the protease.

Even if a microbial cell cannot synthesize all of its required amino acids, it is not necessary to have all of them supplied in the medium. Certain amino acids can donate their amino groups to keto acids to produce other amino acids in a process described earlier as transamination. The basic reaction is:

$$\text{Amino acid}_1 + \text{Keto acid}_2 \rightleftharpoons \text{Keto acid}_1 + \text{Amino acid}_2$$

A specific example of this type of reaction is:

$$\text{L-Aspartate} + \alpha\text{-Ketoglutarate} \longrightarrow \text{Oxaloacetate} + \text{L-Glutamate}$$

Most of the amino acids in nature are L-isomers, and cellular enzymes recognize only that form. There are some instances, for example, in the bacterial cell wall, where the D-isomer of an amino acid exists (see Chapter 27). Conversion of the L-isomer to the D-isomer is catalyzed by enzymes called ***racemases.*** Without racemases a microorganism would have to be supplied preformed D-amino acids, which are minor forms in nature.

Biosynthesis of fatty acids and lipids. Lipids play a very important role in the construction of microbial membranes. We have already discussed in previous chapters the importance of cytoplasmic membrane lipids in terms of solute transport, excretion of proteins, and stability to environmental extremes. The membrane lipids also provide a protective device surrounding the capsid of some viruses that obtain their membrane when released from the infected cell (see Chapters 6 and 16).

Microbial lipids. The principal lipids found in the microbial membrane are glycerophospholipids. Their basic structure is illustrated in Figure 14-12. The fatty acid components of microbial lipids are most frequently chains containing 15 to 19 carbons. The chains are usually straight, but occasionally they are cyclic and saturated. Any unsaturation of the molecule is usually at carbon 2 of the fatty acid. Unsaturated fatty acids are associated with psychrophiles, in whom fluidity of the membrane can be maintained at low temperatures because of the lower melting point of unsaturated fatty acids (see Chapter 15). Glycolipids, which are also present in bacteria, contain hexoses as the sugar moiety. Glycolipids make up part of the cell wall–outer membrane of gram-negative bacteria and the cytoplasmic membrane of gram-positive bacteria. Glycoplipids are also important in the cell envelope of the mycobacteria (some species cause tuberculosis and leprosy), in which the fatty acid side chains are mycolic acid. These glycolipids are believed to provide the mycobacteria with their disease-producing potential. The basic structure of some glycolipids is illustrated in Figure 14-12.

Biosynthesis of fatty acids. Two components in the microbial cell are especially important during the formation of saturated fatty acids: acetyl CoA and acyl carrier protein (ACP). Very simply, fatty acid biosynthesis involves the repeated addition of two carbons in the form of acetyl groups to a growing hydrocarbon chain. During the addition of these two carbon units the acyl carrier protein binds to the growing fatty acid moiety. Seven steps are required to extend the length of the fatty acid chain, and these are repeated until the final chain length is attained. The synthesis of long-chain, even-numbered fatty acids is referred to as the malonyl CoA pathway, in which acetyl CoA acts as a primer molecule to which are added two carbon units at a time from malonyl CoA (COOH-CH_2-CO-CoA). The seven steps involved in the synthesis of even-numbered fatty acids are illustrated in Figure 14-13. The overall reaction for the biosynthesis of a fatty acid containing 18 carbons (stearic acid) would be:

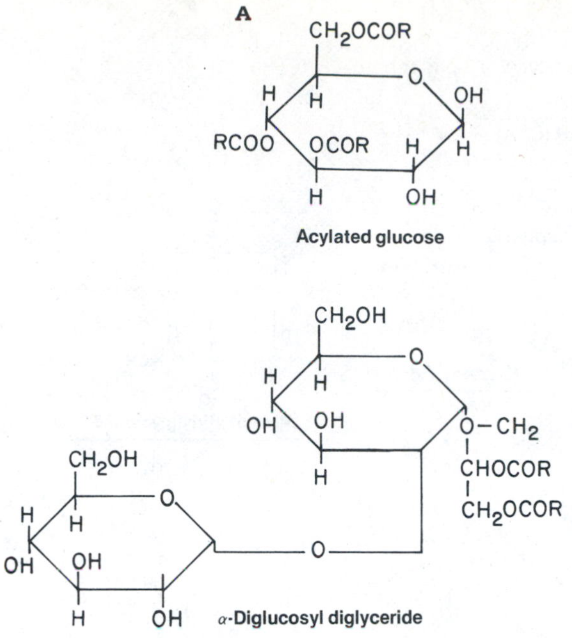

A

Acylated glucose

α-Diglucosyl diglyceride

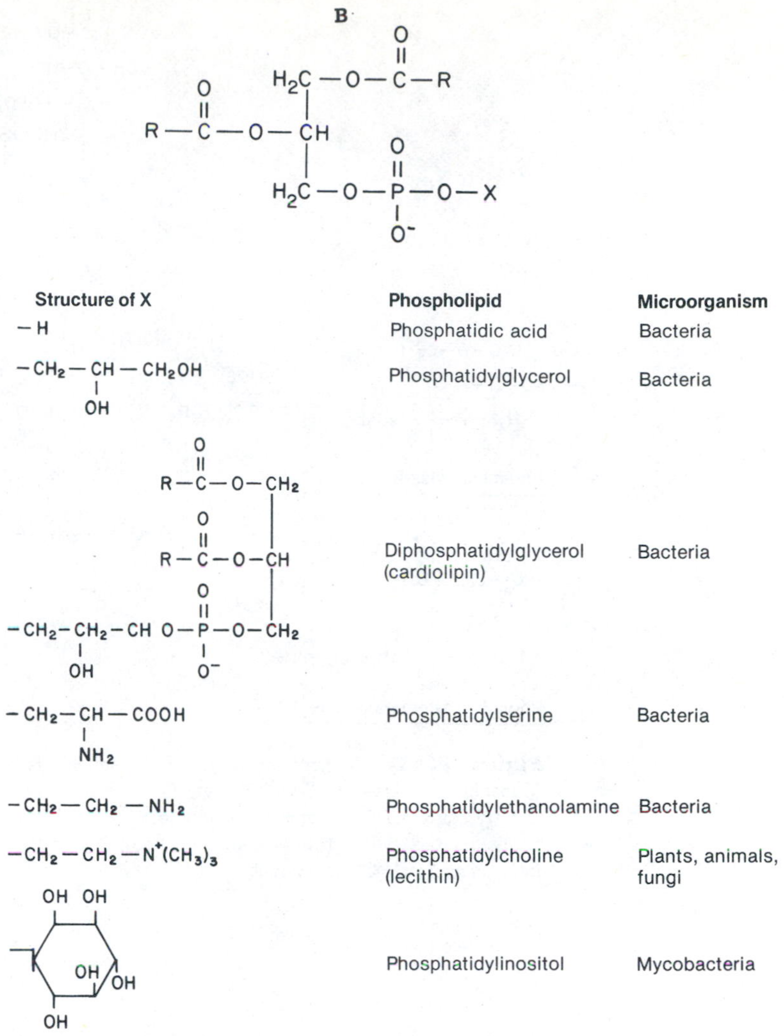

B

Structure of X	Phospholipid	Microorganism
$-H$	Phosphatidic acid	Bacteria
$-CH_2-CH(OH)-CH_2OH$	Phosphatidylglycerol	Bacteria
$-CH_2-CH_2-CH(OH)-O-P(O)(O^-)-O-CH_2-CH(OCOR)-CH_2OCOR$	Diphosphatidylglycerol (cardiolipin)	Bacteria
$-CH_2-CH(NH_2)-COOH$	Phosphatidylserine	Bacteria
$-CH_2-CH_2-NH_2$	Phosphatidylethanolamine	Bacteria
$-CH_2-CH_2-N^+(CH_3)_3$	Phosphatidylcholine (lecithin)	Plants, animals, fungi
Inositol	Phosphatidylinositol	Mycobacteria

Figure 14-12 Microbial lipids. *R* represents fatty acid constituents. **A,** Two types of glycolipids found in bacteria. **B,** Basic structure of microbial glycerophospholipids.

1. $ACP-SH + CH_3CO-S-CoA \rightleftharpoons CH_3CO-S-ACP + CoASH$
2. $COOH-CH_2CO-S-CoA + ACP-SH \rightleftharpoons COOH-CH_2CO-SACP + CoASH$
3. $CH_3-CO-S-ACP + COOH-CH_2COS-ACP \rightleftharpoons CH_3-CO-CH_2CO-S-ACP + HS-ACP + CO_2$
4. $CH_3-CO-CH_2CO-S-ACP + NADPH + H^+ \rightleftharpoons CH_3-CHOH-CH_2CO-S-ACP + NADP^+$
5. $CH_3-CHOH-CH_2CO-S-ACP \rightleftharpoons CH_3-CH=CH-CO-S-ACP + H_2O$
6. $CH_3-CH=CH-CO-S-ACP + NADPH + H^+ \rightleftharpoons CH_3-CH_2CO-S-ACP + NADP^+$
7. $CH_3CH_2-CH_2CO-S-ACP + CoASH \rightleftharpoons CH_3-CH_2-CH_2CO-S-CoA + ACP-SH$

Figure 14-13 Seven steps required in biosynthesis of fatty acids. Reactions 1 and 2 are involved in formation of acetyl-ACP and malonyl-ACP, respectively. In reaction 3, acetyl-ACP provides first two carbons of fatty acid. Remaining reactions involve reduction processes to produce butyryl-ACP.

$$(CH_3\text{-}CO\text{-}CoA)_{16} + COOH\text{-}CH_2\text{-}CO\text{-}S\text{-}CoA + 16\ NADPH_2 \longrightarrow CH_3(CH_2)_{16}CO\text{-}SCoA + 8\ CoASH + 16\ NADP + 16\ H_2O$$

The synthesis of odd-numbered fatty acid chains in some bacteria can take place by using valeral CoA (a five-carbon compound). Unsaturated fatty acids such as those containing one, two, or three double bonds are also components of microbial lipids. Their synthesis occurs by one of two mechanisms: one that requires aerobic conditions and one that requires anaerobic conditions. The aerobic system is characterized by double-bond insertion between carbon 9 and carbon 10, whereas in the anaerobic system, which

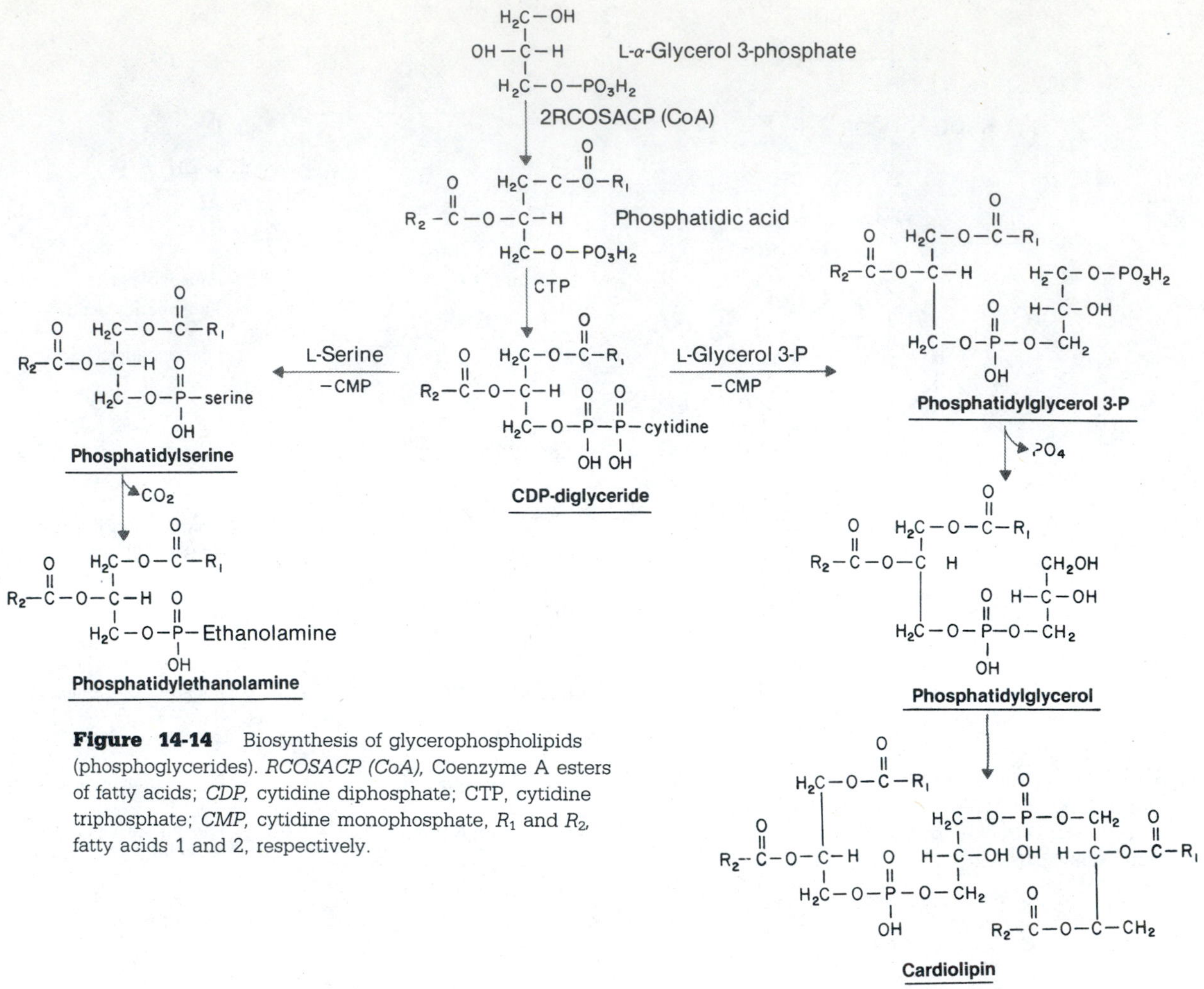

L-α-Glycerol 3-phosphate

2RCOSACP (CoA)

Phosphatidic acid

CTP

CDP-diglyceride

L-Serine

−CMP

Phosphatidylserine

CO_2

Phosphatidylethanolamine

L-Glycerol 3-P

−CMP

Phosphatidylglycerol 3-P

PO_4

Phosphatidylglycerol

Cardiolipin

Figure 14-14 Biosynthesis of glycerophospholipids (phosphoglycerides). *RCOSACP (CoA),* Coenzyme A esters of fatty acids; *CDP,* cytidine diphosphate; CTP, cytidine triphosphate; *CMP,* cytidine monophosphate, R_1 and R_2, fatty acids 1 and 2, respectively.

occurs in most bacteria, dehydration and double-bond formation occur at carbon 10.

Biosynthesis of glycerophospholipids. The basic components of glycerophospholipids are glycerol phosphate, which is an intermediate in the glycolytic pathway, and fatty acids. Carbons 1 and 2 of glycerol phosphate are acetylated with the coenzyme A esters of fatty acids to produce phosphatidic acid. The sequence of reactions by which phosphatidic acid is converted to other phospholipids is illustrated in Figure 14-14.

Biosynthesis of purines and pyrimidines. The structure of purine and pyrimidine bases is illustrated in Chapter 2. These cyclic compounds are made up of elements derived from various metabolic intermediates. The structure and numbering of the atoms in the purine and pyrimidine molecules are illustrated in Figure 2-26. The origin of these atoms is described in Table 14-6.

Biosynthesis of purines. The basic purine ribonucleotide molecule, inosine 5-phosphate (ino-

Table 14-6 Origin of the atoms of purine and pyrimidine bases*

ATOM AND POSITION	PURINE (MOLECULE DERIVED FROM)	PYRIMIDINE (MOLECULE DERIVED FROM)
N-1	Aspartate	Aspartate
C-2	Formate	Carbon dioxide
N-3	Glutamine	Ammonia
C-4	Glycine	Aspartate
C-5	Glycine	Aspartate
C-6	Carbon dioxide	Aspartate
N-7	Glycine	—
C-8	Formate	—
N-9	Glutamine	—

* The atom numbers and positions are illustrated in Figure 2-26.

sinic acid), is synthesized in a series of reactions outlined in Figure 14-15. Inosinic acid serves as precursor to two of the major purines: guanylic acid (GMP) and adenylic acid (AMP). The pathways leading to the synthesis of purines are outlined in Figure 14-16.

Biosynthesis of pyrimidines. The pyrimidines—thymine, cytosine, and uracil—are synthesized in a series of reactions in which the initial precursor molecules are aspartate and carbamoyl phosphate (Figure 14-17).

Most cells can synthesize purines and pyrimidines de novo (from basic nutrients in the cell), but preformed bases supplied in the growth medium can be converted directly into ribonucleotides, or they may be converted to other bases if the organism lacks an enzyme in the de novo pathway. Some microorganisms can convert free bases such as adenine to a nucleotide by reacting them with phosphoribosylpyrophosphate (PRPP):

$$\text{Adenine} + \text{PRPP} \longrightarrow \text{AMP} + \text{PP}_i$$

Biosynthesis of carbohydrates and polysaccharides. Carbohydrates, in addition to their role in energy production, are also important as structural components of the cell. Capsules, the cell wall, and the cytoplasmic membrane are composed of carbohydrates that appear either as oligosaccharides or as polysaccharides. Monosaccharides can be activated by nucleoside triphosphates such as CTP, UTP, and ATP to produce sugar nucleotides, which are capable of entering into several biosynthetic processes. Glucose, for example, can become activated and ready for participation in biosynthetic reactions in the following type of reaction:

$$\text{Glucose 1-phosphate} + \text{UTP} \longrightarrow \text{UDP-glucose} + \text{PP}_i$$

The major biosynthetic reactions in which sugar nucleotides participate are the formation of monosaccharides other than glucose and polysaccharide synthesis.

Biosynthesis of monosaccharides other than glucose. Glucose can serve as a precursor to other monosaccharides. Galactose, glucuronic acid, *N*-acetylglucosamine, arabinose, and other sugar derivatives can be produced from sugar nucleotides such as UDP-glucose. Many of these derivatives are important in cell wall peptidoglycan formation. (See Chapter 27 for discussion of cell wall synthesis.)

Biosynthesis of polysaccharides. A variety of polysaccharides, either linear or branched, are synthesized by microbial cells. The most common one is a high molecular weight polymer called

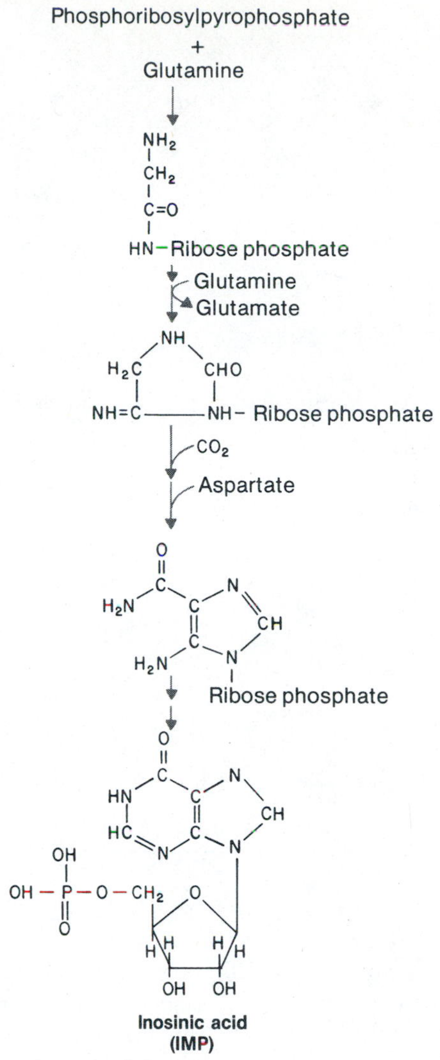

Phosphoribosylpyrophosphate
+
Glutamine
HN–Ribose phosphate
Glutamine
Glutamate
NH– Ribose phosphate
CO_2
Aspartate
Ribose phosphate
Inosinic acid
(IMP)

Figure 14-15 Biosynthesis of inosinic acid. Sequence is not complete, and several intermediates have been omitted. Note importance of amino acids in pathway.

glycogen, which acts as an energy storage form of glucose. Glycogen synthesis occurs under conditions where there is excess glucose and a deficiency in a specific nutrient required for biosynthesis. The cell, instead of hydrolyzing glucose to produce more energy as ATP, will convert the glucose into glycogen. The conversion of glucose to glycogen includes the following reactions:

$$\text{Glucose 6-phosphate} \longrightarrow \text{Glucose 1-phosphate}$$
$$\text{Glucose 1-phosphate} + \text{ATP} \longrightarrow \text{ADP-glucose} + \text{PP}_i$$
$$\text{ADP-glucose} + \text{Glucose primer (n)} \longrightarrow (\text{Glucose})_{n+1} + \text{ADP}$$

During the synthesis of some polysaccharides (such as in cell wall formation) the sugar nucleotide is not transferred directly to the growing polymer but is first linked to a lipid. The lipid has

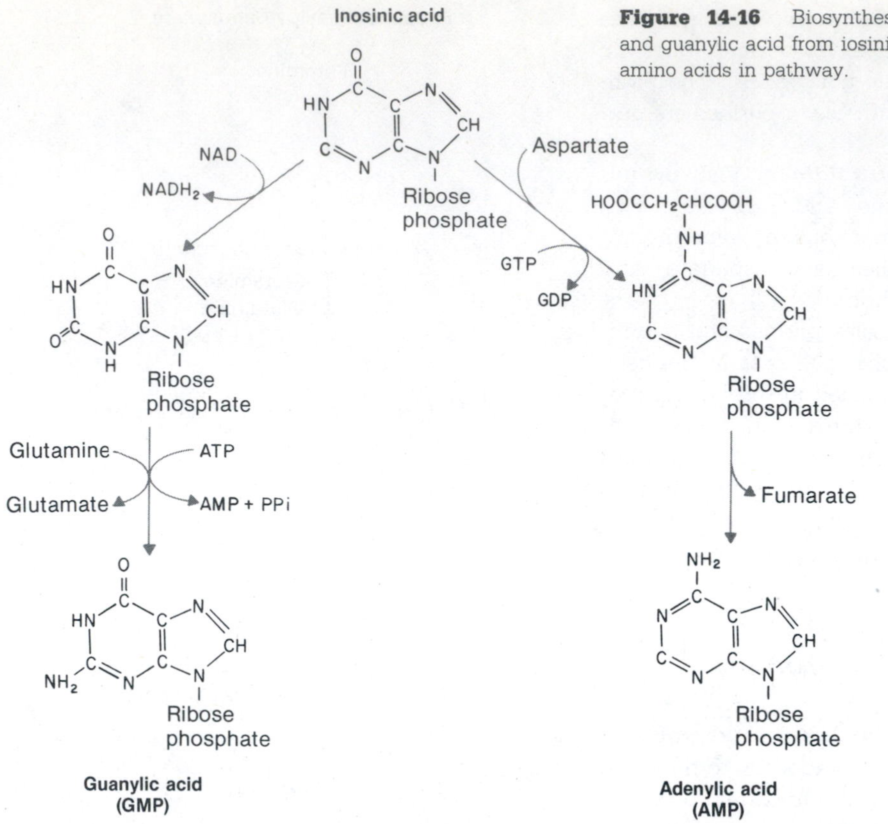

Figure 14-16 Biosynthesis of purines adenylic acid and guanylic acid from iosinic acid. Note importance of amino acids in pathway.

been identified as a C_{55} polyisoprenoid (Figure 14-18), which must be phosphorylated before it accepts a glycosyl unit. The C_{55} lipid carrier is involved in the biosynthesis of polysaccharides, such as the lipopolysaccharide of gram-negative bacteria, and the teichoic acids of the gram-positive bacteria, and the peptidoglycan of the bacterial cell wall. The reactions in cell wall biosynthesis and the antimicrobial drugs that inhibit them are discussed in Chapter 27.

Gluconeogenesis. Glucose can be synthesized by the reversal of most of the reactions of glycolysis—a process called ***gluconeogenesis.*** Many microorganisms in their natural environment seldom encounter free extracellular glucose and must therefore synthesize it from other metabolites such as those in the TCA cycle. This means that a TCA intermediate can be converted to glucose only by the reversal of glycolysis. During the reversal process, however, the conversion of pyruvate to phosphoenolpyruvate is irreversible, and the cell bypasses the reaction by converting the TCA intermediate to phosphoenolpyruvate in the following reaction:

$$\text{Oxaloacetate} + \text{ATP} \longrightarrow \text{Phosphoenolpyruvate} + CO_2 + \text{ADP}$$

Phosphoenolpyruvate is easily converted to glucose by the reversal of glycolysis.

METABOLIC CONTROL

Up to this point we have observed that the primary function of glycolysis is to provide energy, which can be used in the biosynthesis of metabolites whose precursors are formed in the TCA cycle. The microbial cell is often confronted with metabolic conditions where either energy is not available and biosynthesis must be discontinued, or excess sources of energy are available but certain nutrients are unavailable for biosynthesis. Control mechanisms are available to the cell that enable it to conserve energy sources when they are available to prevent the biosynthesis of unneeded metabolites.

Control mechanisms

We learned in Chapter 10 that one of the mechanisms used by the cell to control metabolic reactions is of genetic origin and that enzymes controlling certain reactions in the cell are sub-

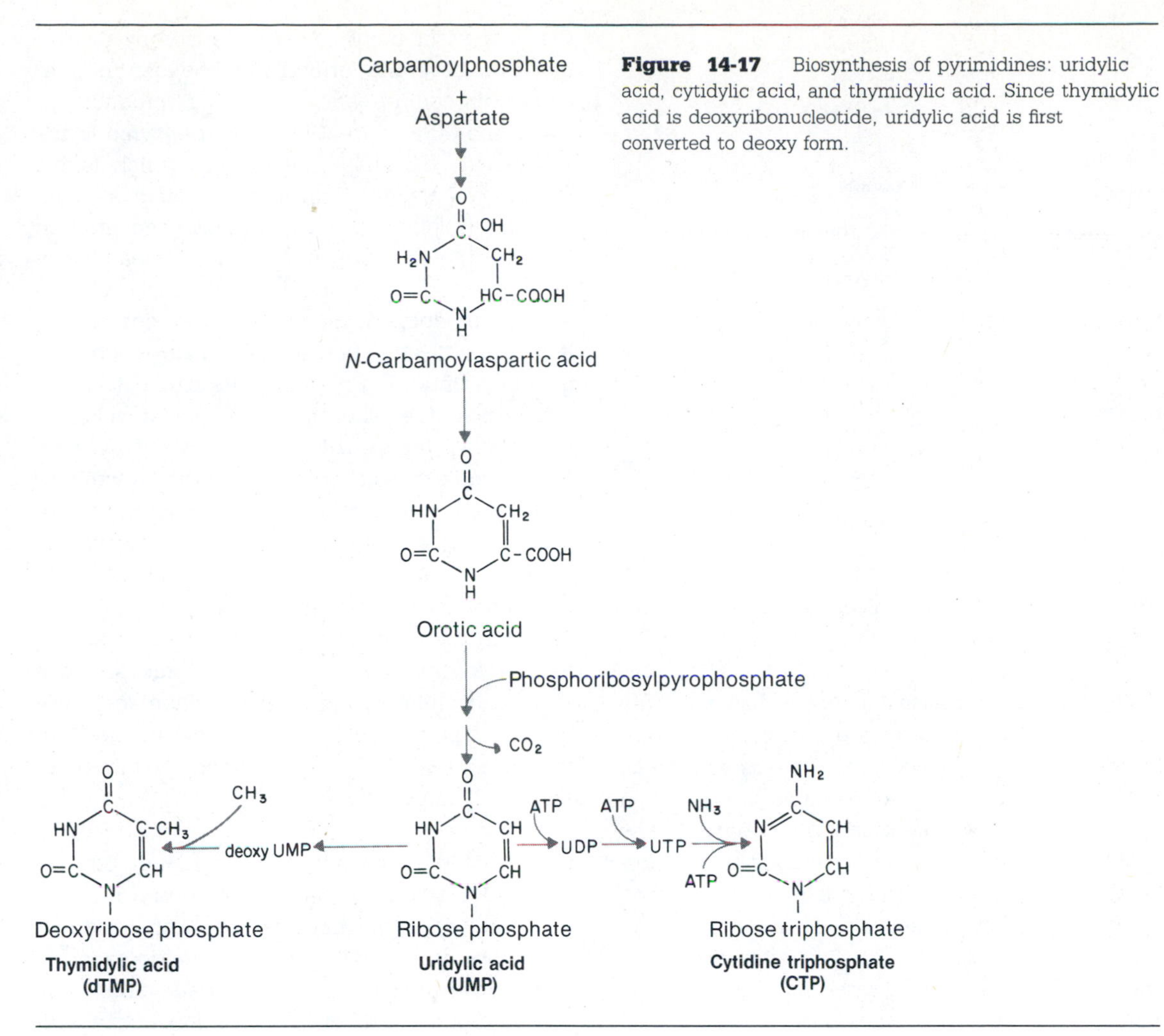

Figure 14-17 Biosynthesis of pyrimidines: uridylic acid, cytidylic acid, and thymidylic acid. Since thymidylic acid is deoxyribonucleotide, uridylic acid is first converted to deoxy form.

ject to induction or repression. This genetic control affects the synthesis of enzymes but has no effect on enzymes that are already present in the cell. A second major control system has evolved in organisms that controls the activity of preformed enzymes.* Every chemical reaction in the cell is catalyzed by an enzyme, but at strategic places in many metabolic pathways there are special enzymes that control the amount of product produced in the pathway. These enzymes are called ***allosteric*** (means other site) or ***regulatory enzymes.*** Regulatory enzymes consist of subunits (usually four) on which both substrate as well as effector molecules may bind. Effectors modulate the activity of regulatory enzymes by binding at a site or sites separate from the substrate binding site. The activity curves of regulatory enzymes are different from those of nonregulatory enzymes (Figure 14-19). In reactions catalyzed by nonregulatory enzymes the velocity

```
CH3H  H          CH3              H  CH3      H        OH
 |  |  |          |               |  |        |        |
CH3-C=C-C-(CH2-C=C-CH2)9-C-C=C-C-O-P-OH
       |              |           |     |  |       ||
       H              H           H     H  H       O
```

Figure 14-18 Structure of lipid carrier polyisoprenoid (phosphorylated derivative).

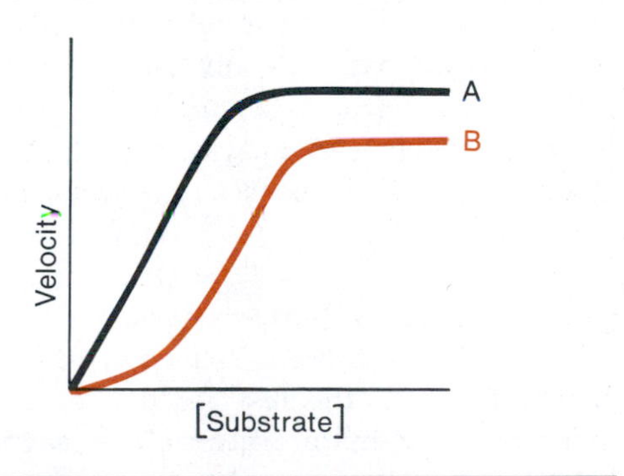

Figure 14-19 Comparison of velocity curves for nonregulatory *(A)* and regulatory *(B)* enzymes.

* A third but less-understood mechanism of control is called enzyme inactivation and is discussed on p. 330.

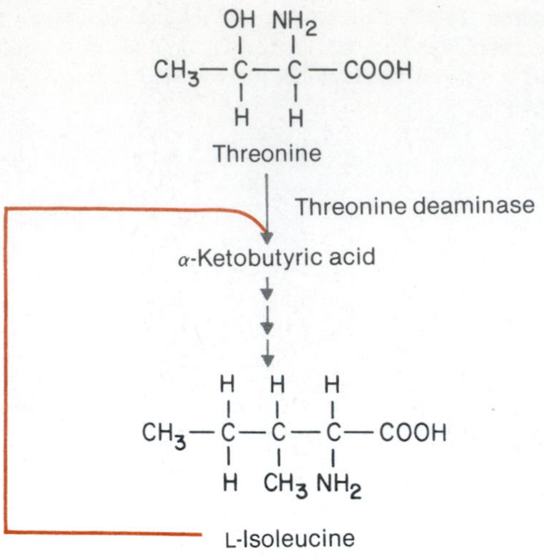

Figure 14-20 Feedback inhibition in biosynthesis of L-isoleucine.

of the reaction increases linearly at low substrate concentrations, but in reactions catalyzed by regulatory enzymes the velocity curve is sigmoidal. The reason for the sigmoidal curve is that when one substrate molecule binds to a subunit at low substrate concentrations, it changes the conformation of the enzyme and enhances the binding of other substrate molecules. As more substrate becomes available, the binding sites are made more accessible to substrate and the curve assumes a linear (at low substrate concentrations) character and then a hyperbolic (at high substrate concentrations) character. The effector molecules are usually low molecular weight metabolites and include such molecules as amino acids, TCA or glycolytic intermediates, ATP, and ADP. The effector molecules usually exhibit a negative effect by shutting down the activity of the enzyme, but there are some enzymes in which the effector actually stimulates the activity of the enzyme. The negative effector molecule binds to a subunit and changes the conformation of the enzyme so that the substrate-binding sites become inaccessible to substrate. As more negative effector becomes bound, the activity of the enzyme is decreased. Enzyme activity can be restored only when excess substrate becomes available.

Control in biosynthetic pathways

Feedback inhibition. Regulatory enzymes are usually found at the first reaction of a biosynthetic pathway, but not always. The reactions involved in the synthesis of the amino acid L-isoleucine is a typical example of a pathway controlled by feedback inhibition. The biosynthesis of L-isoleucine is characterized by the formation of several intermediates between the precursor threonine and the final product, L-isoleucine (Figure 14-20). The first enzyme in the pathway is threonine deaminase, which is the regulatory enzyme. The allosteric effector molecule is L-isoleucine, which in high concentration inhibits the activity of threonine deaminase. Thus the product of threonine deaminase, as well as other products in the pathway, are not formed. If a bacterial cell such as *E. coli* is grown in a medium containing mineral salts plus glucose as a carbon source, all of the amino acids, including L-isoleucine, are synthesized de novo. If we were now to add excess L-isoleucine to the growth medium, the cell would have no need to synthesize L-isoleucine, and threonine deaminase would be inhibited. This type of inhibition, in which the end product inhibits the activity of earlier enzymes in the pathway, is called (as noted above) ***feedback inhibition.*** There are several variations of this mechanism, particularly when there are branched pathways leading to the synthesis of more than one product and where certain reactions are shared.

Inactivation of enzymes. A third but not entirely understood mechanism of control involves the inactivation of enzymes. The term inactivation does not mean that enzymes are degraded to their constituent amino acids. Inactivation may be due to a physical or conformational change in the enzyme resulting from the binding of low molecular weight molecules such as NH_4^+, a chemical modification of the enzyme, or a cleavage of a peptide bond in the enzyme. One type of modification, such as the phosphorylation of proteins, is mentioned earlier in the chapter. Other types of modification may include glucosylation, acetylation or deactylation, and methylation. You may recall that glucosylation and/or methylation of DNA is a mechanism for protecting bacteriophage DNA from host endonuclease attack. Most examples of enzyme inactivation occur during shifts in carbon or nitrogen metabolism. An organism on one substrate, for example, may exhaust its supply and have to use other substrates. One of the most extensively studied inactivation systems occurs in the organism *Rhodopseudomonas gelatinosa* during its growth on citrate under anaerobic conditions. Growth on citrate results in the catabolism of the molecule to acetate, but when citrate is exhausted, the organism must shift to citrate synthesis. Citrate-catabolizing enzymes that are already present in the cell are prevented from engaging in citrate synthesis because the first enzyme in the catabolic pathway, citrate lyase, is inactivated by another enzyme, *S*-acetylcitrate lyase deactylase. The inactivating enzyme removes an acetyl group from a sulfhydryl group on citrate lyase. Should citrate later accumulate in the cell, the inactivated citrate lyase can be reactivated by an acetylation process.

Control in catabolic pathways. Regulatory enzymes are also a feature of catabolic systems. Since catabolism is involved in energy formation, it should not seem unusual that molecules such as ATP and AMP are effector molecules. One of the principal control sites in glycolysis is in the reaction:

$$\text{Fructose 6-phosphate} + \text{ATP} \rightleftharpoons \text{Fructose 1,6-diphosphate} + \text{ADP}$$

The forward reaction is catalyzed by an enzyme that is inhibited by high concentrations of ATP but stimulated by ADP or AMP. In other words, if the cell has excess ATP (the ATP/AMP + ADP ratio is high), then the forward reaction, in which more ATP will eventually be produced, is unnecessary. High concentrations of AMP or ADP are signals to the cell that ATP has been used up in biosynthetic reactions and more is needed; hence the forward reaction is stimulated. The reverse reaction is stimulated by ATP and inhibited by AMP or ADP. Stimulating the reverse reaction prevents the forward reaction from proceeding and vice versa. Inhibition of enzyme activity is the only means of control in glycolysis, since all of the enzymes are synthesized constitutively and are not subject to genetic control.

Anaplerotic pathways

A third class of reactions occurring in the cell, which are neither catabolic or biosynthetic, are called ***anaplerotic.*** Anaplerotic pathways function to replenish intermediates that have been drained from cyclic processes such as acetylation of oxaloacetate to citrate in the TCA cycle. At maximum activity the TCA cycle acts primarily as a source of intermediates for biosynthesis and secondarily as a supplier of reducing equivalents for energy production. Replenishing TCA intermediates can take place in two different ways, depending on the organism's carbon source:

1. If the organism is growing on glucose, intermediates of glucose can be carboxylated:

$$\text{Pyruvate} + CO_2 + \text{ATP} \longrightarrow \text{Oxaloacetate} + \text{ADP}$$

or

$$\text{Phosphoenolpyruvate} + CO_2 \longrightarrow \text{Oxaloacetate} + \text{iP}$$

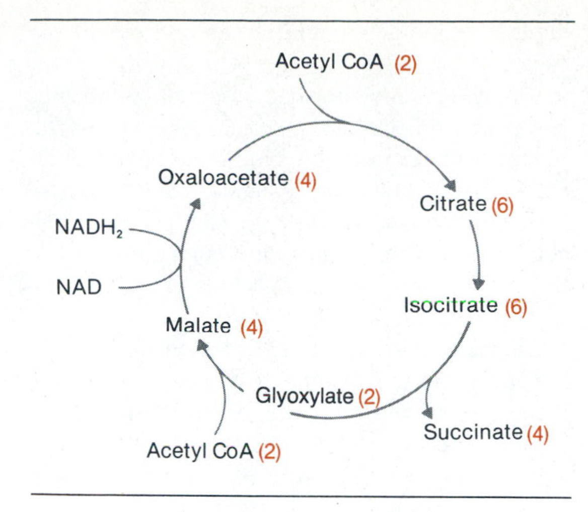

Figure 14-21 Glyoxylate cycle. Number of carbons in each intermediate is indicated in color. Compare with TCA cycle.

2. If the organism is growing on a two-carbon source such as acetate, the glyoxylate cycle provides a means of replenishment (Figure 14-21). The glyoxylate cycle is an abbreviated version of the TCA cycle, but unlike the TCA cycle possesses enzymes that conserve carbon and do not lose it as carbon dioxide.

SUMMARY

Nutrients are those elements or compounds that must be supplied to the cell for its normal growth and development. Inorganic or organic forms of carbon, nitrogen, sulfur, and phosphorus represent the major nutrient sources, whereas various minerals, such as magnesium, manganese, iron, and others, are considered minor nutrients. Most microorganisms, except for photosynthetic species, require carbon in its organic form, such as carbohydrates. Nitrogen may appear in many forms, including amino acids, proteins, and purine and pyrimidine bases, but inorganic nitrogen as ammonia or nitrate is the principal source of nitrogen for conversion into organic material in the cell. Sulfur and phosphorus as components of organic molecules are most readily available as nutrients to the cell in the form of inorganic sulfate and phosphate, respectively. The minor nutrients are usually cofactors for enzymes or are involved in energy reactions. Growth factors are cellular metabolites that cannot be synthesized from nutrients supplied in the environment and must be added to the medium. They include vitamins, amino acids, and purine and pyrimidine bases.

The ability to synthesize essential metabolites varies from species to species. Generally, saprophytes from the soil can synthesize nearly all organic metabolites from a single carbon source. Parasites, however, have lost much of their biosynthetic potential and when cultivated outside the host often require a multitude of organic metabolites.

Amino acid synthesis is usually dependent on the incorporation of ammonia into a keto acid. Precursors to the amino acids are therefore derived primarily from the TCA cycle. The pathways leading to the synthesis of amino acids in both eukaryotes and prokaryotes are similar except for L-lysine. The intermediates produced during L-lysine synthesis in prokaryotes are important for the construction of the cell wall.

The principal lipids synthesized in the microbial cell are glycerophospholipids and glycolipids. The fatty acid components of lipids are synthesized through the repeated addition of two carbons, in the form of an acetyl group, to a growing hydrocarbon chain. The two carbon units are transferred by a carrier called the acyl carrier protein (ACP).

Purines and pyrimidines are synthesized from precursor molecules, particularly amino acids such as aspartic acid, glutamine, and glycine. Inosinic acid is the intermediate in the formation of the purines guanylic and adenylic acid. In pyrimidine biosynthesis uridylic acid is produced, which then serves as an intermediate in the formation of thymidylic acid and cytidylic acid.

Carbohydrate or polysaccharide biosynthesis requires activation of monomeric sugars into sugar nucleotides, using CTP, UTP, or ATP. During some polysaccharide syntheses the sugar nucleotides are transferred directly to a lipid carrier such as C_{55} isoprenoid before incorporation into the growing polysaccharide chain. When glucose or another carbohydrate is not available to the cell for incorporation into cell components, glucose can be synthesized by the reversal of glycolysis.

Control of metabolism at the level of the enzyme is possible through the action of enzymes that have regulatory subunits, that is, allosteric enzymes. There are sites on the allosteric enzymes for substrate as well as activators and inhibitors. Regulatory enzymes can be found in biosynthetic pathways such as those leading to the synthesis of amino acids. The amino acid as an end product can provide feedback to the first enzyme in the pathway, which is the allosteric enzyme. Allosteric enzymes are also present in catabolic pathways where energy compounds such as ADP, AMP, and ATP act as inhibitors and activators. A less understood mechanism of enzymatic control is called enzyme inactivation. Anaplerotic reactions are neither catabolic nor anabolic. They function to replenish intermediates that have been drained from metabolic cycles such as the TCA cycle. The glycoxylate cycle is an example of an anaplerotic pathway.

Study questions

1. What is a nutrient? Are they compounds or elements, or both?
2. What types of carbon compounds are most frequently used by microorganisms as an energy source and as a carbon source? How is carbon dioxide used by the cell?
3. Inorganic nitrogen is used by the cell for conversion to what important organic molecules? Can any inorganic nitrogen compounds serve other functions in the cell other than as part of an organic molecule?
4. What is the name of the electron transport carrier involved in nitrogen fixation?
5. What is the name of the reaction responsible for the transfer of ammonia (NH_3) into an organic molecule producing an amino acid?
6. How is inorganic sulfur integrated into organic molecules?
7. What function does phosphorus serve in the cell?
8. How does a growth factor differ from a nutrient?
9. Why are minerals, such as magnesium, important to the function of the cell?
10. In what specific reactions in the TCA cycle and EMP pathway are biosynthetic intermediates provided to the cell?
11. If a microorganism cannot synthesize some amino acids, how can it survive?
12. What effect does a high lipid content have on the Gram-staining reaction? In what group of microorganisms is this evident?
13. Briefly explain the importance of the acyl carrier protein (ACP) in fatty acid biosynthesis.
14. What carrier molecule is important in the synthesis of polysaccharides?
15. How can a cell synthesize glucose or its intermediates if its carbon source is a nonsugar, such as an amino acid?
16. Could a cell survive without regulatory enzymes? What is their significance to the cell?
17. How does the conformation of a regulatory enzyme change when exposed to a positive effector molecule? To a negative effector molecule?
18. What are the advantages and disadvantages of feedback inhibition to genetic control?
19. What are anaplerotic pathways? Can a cell survive without them?
20. What is the relationship of iron to bacterial virulence?

Selected readings

Books

Bothe, H., and Trebst, A. *Biology of inorganic nitrogen and sulfur.* Springer-Verlag, New York, Inc., New York, 1981.

Doelle, H.W. *Bacterial metabolism* (2nd ed.). Academic Press, Inc., New York, 1975.

Fenchel, T. *Bacteria and mineral cycling.* Academic Press, Inc., New York, 1979.

Gottschalk, G. *Bacterial metabolism.* Springer-Verlag, New York, Inc., New York, 1979.

Mandelstam, J., and McQuillen, K. *Biochemistry of bacterial growth* (2nd ed.). John Wiley & Sons, Inc., New York, 1973.

Moat, A.G. *Microbial physiology.* John Wiley & Sons, Inc., New York, 1979.

Weinberg, E.D. (Ed.). *Microorganisms and minerals.* Marcel Dekker, Inc., New York, 1977.

Journal articles

Bailey, S.T., and Morton, R.A. Recent developments on the molecular biology of extremely halophilic bacteria. *CRC Crit. Rev. Microbiol.* **6:**151, 1978.

Biebl, H., and Pfennig, N. Growth of sulfate-reducing bacteria with sulfur as electron acceptor. *Arch. Microbiol.* **112:**115, 1977.

Bullen, J.J. The significance of iron in infection. *Rev. Infect. Dis.* **3**(6):1127, 1981.

Calvo, J.M., and Fink, G.R. Regulation of biosynthetic pathways in bacteria and fungi. *Annu. Rev. Biochem.* **40:**943, 1971.

Dills, S.S., et al. Carbohydrate transport in bacteria. *Microbiol. Rev.* **44:**385, 1980.

Haddock, B.A., and Jones, C.W. Bacterial respiration. *Microbiol. Rev.* **41:**47, 1977.

Knowles, C.J. Microbial metabolic regulation by adenine nucleotide pools. *Symp. Soc. Gen. Microbiol.* **27:**241, 1977.

Monod, J. The growth of bacterial cultures. *Annu. Rev. Microbiol.* **3:**371, 1949.

Monod, J., Changeux, J.P., and Jacob, F. Allosteric proteins and cellular control systems. *J. Mol. Biol.* **6:**306, 1963.

Morris, J.G. The physiology of obligate anaerobiosis. *Adv. Microb. Physiol.* **12:**169, 1975.

Neilands, J.B. Microbial iron compounds. *Annu. Rev. Biochem.* **50:**715, 1981.

Nierlich, D.P. Regulation of bacterial growth, RNA and protein synthesis. *Annu. Rev. Microbiol.* **32:**393, 1978.

Nierlich, D.P., and Magasanik, B. Regulation of purine ribonucleotide synthesis by end product inhibition. *J. Biol. Chem.* **240:**358, 1965.

Pastan, I., and Perlman, R. Cyclic AMP in *Escherichia coli. Microbiol. Rev.* **40:**527, 1976.

Sanwal, B.D. Allosteric controls of amphibolic pathways in bacteria. *Microbiol. Rev.* **34:**20, 1970.

Switzer, R.L. The inactivation of microbial enzymes in vivo. *Annu. Rev. Microbiol.* **31:**135, 1977.

Umbarger, H.E. Regulation of amino acid metabolism. *Annu. Rev. Biochem.* **38:**323, 1969.

Umbarger, H.E. Amino acid biosynthesis and its regulation. *Annu. Rev. Biochem.* **47:**533, 1978.

Chapter 15 MICROBIAL GROWTH AND FACTORS AFFECTING GROWTH

Growth, in physical terms, may be defined as the increase in size of an individual or the increase in the number of individuals within a population. The physical processes of growth are only a culmination of the biochemical reactions that have occurred within the cell. When both chemical and physical processes are coordinated, the cell will divide and give rise to two daughter cells, both of which will receive a copy of replicated DNA. Many factors, obviously, determine whether a cell will grow and divide: the availability of carbon and energy sources, the ability of DNA to replicate, the size and mass of the cell, and the effect of chemical and physical agents (such as temperature and pressure) in the environment. Many of these factors are discussed in other chapters. The purposes of this chapter are, first, to describe those internal factors that control the growth of cell populations; second, to illustrate how growth is measured; and, third, to discuss those environmental factors affecting growth.

GROWTH IN PROKARYOTES

Growth of individual cells

The physical growth of a cell, such as its increase in size and its subsequent division, can be visualized by direct microscopic observation. Unfortunately, we do not have devices that can detect biochemical changes within a single cell; for this reason, large populations of cells are used to analyze biochemical data. Ordinarily, when one cultures a population of bacteria in the laboratory, the cells will be in various stages of growth. Some cells will be dividing, others dying, and others in various stages of the growth cycle. One could not accurately determine, for example, the rate of RNA synthesis in a cell if in the population of cells some were synthesizing RNA and others were not. For this reason, the population of cells must be synchronized so that most cells are in the same stage of the growth process.

Synchronization of cells. Synchronized cultures may be obtained by an induction process in which the population is subjected to environmental manipulation or by a selection process that is based on the physical state of the organism. The whole purpose is to obtain a culture in which all of the cells divide at the same time.

Environmental manipulations. Several chemical or physical techniques are used to obtain a synchronous culture. The technique to be employed is often related to the type of organism used in the experiment. One method uses temperature variations during growth; for example, a culture may be grown at 42° C for a period of time and then at 28° C. Repetition of these temperature changes in some way coordinates the population as long as the length of time at each temperature closely follows the generation time

(time required for a cell to divide) of the organism or the population of cells. Photosynthetic microorganisms can be synchronized by subjecting a culture to alternating cycles of light and dark exposure. If one has a species of bacteria that produces spores, germination can provide some synchrony for a short period of time. A frequently used technique is to starve cells by placing them in a medium in which one of the nutrients is in short supply. In other words, the concentration of nutrient in the medium will be enough to support the growth of only a specific number of cells. Once that population number has been reached, there is not enough nutrient for future generations. Presumably, during the starvation period the cell population approaches a uniform state in the cell cycle. Later the cells are placed in a nutritionally complete medium where they will initiate growth at approximately the same stage in the cell cycle.

Selection techniques. Selection techniques are of two types, and both are related to the physical aspects of cell division. Immediately following cell division bacterial cells are smallest. In a nutritionally complete medium the cell will enlarge as it grows (note that this enlargement is barely visible in spherical bacterial cells). One selection method uses a membrane filter (Figure 15-1, *A*). The asynchronous culture is poured over a filter on which the cells adhere but whose porosity prevents the bacterial cells from passing through. The filter is inverted, and fresh medium is added to the filter. The cells on the filter will divide, with one of the daughter cells adhering to the filter and the other daughter cell being eluted. The eluted cells represent newly divided cells. The second synchronization technique uses density centrifugation. An asynchronous culture is placed in a centrifuge tube containing a nonmetabolizable solute such as sucrose (30% solution) and is centrifuged for a specific period of time. A density gradient of sucrose is produced during centrifugation that increases with the distance from the axis of rotation; that is, the density is greatest at the bottom of the centrifuge tube (Figure 15-1, *B*). Bacteria will also form a gradient when centrifuged in this medium, with the larger and heavier population of cells at the bottom of the centrifuge tube and the smaller (recently divided) cells near the top. This same technique is also used to separate viruses, macromolecules, and other particulate matter.

Once the microbial population has been synchronized, we are ready to investigate the cycle of events that is associated with the growth and division of the microbial cell.

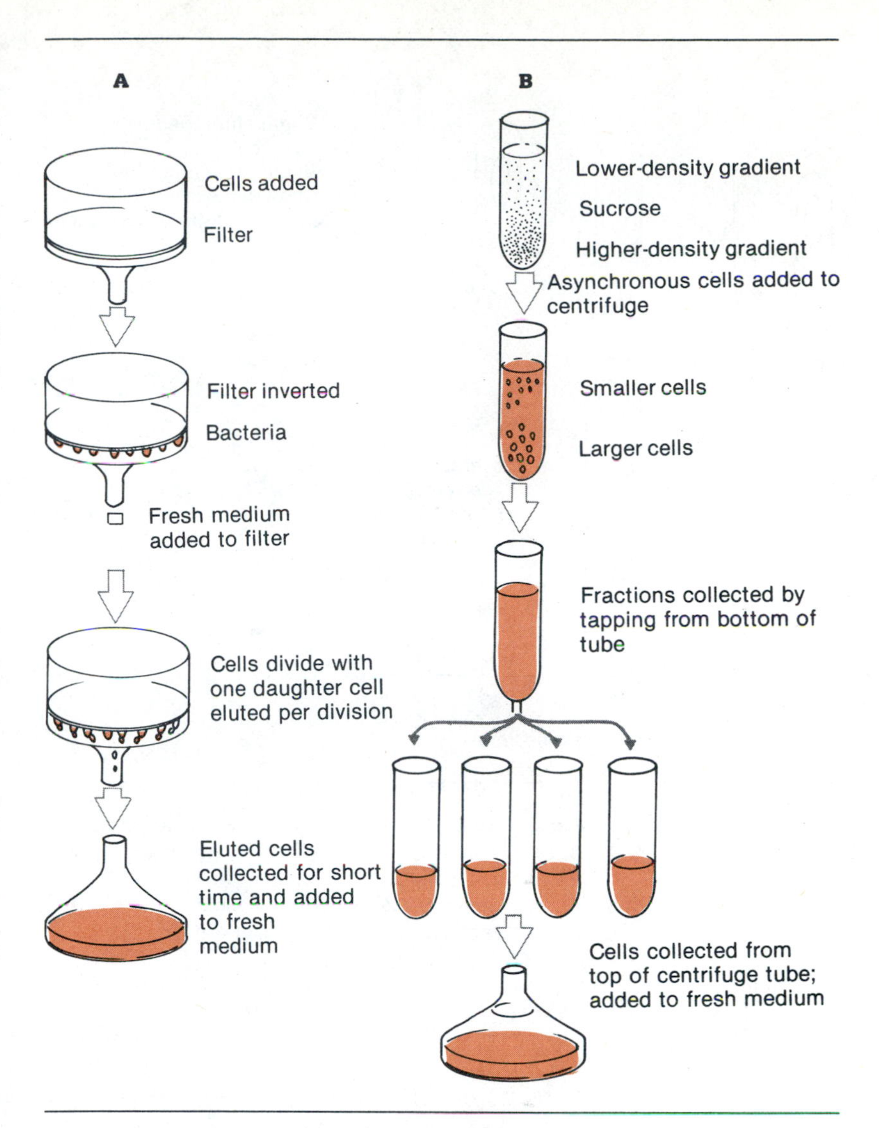

Figure 15-1 Selection techniques for obtaining synchronized cultures. **A,** Membrane filter technique. **B,** Centrifugation technique.

Bacterial cell cycle

A number of biochemical and physical events occur in the bacterial cell from the time it is derived from a parent cell until it divides into two daughter cells. This period of time, called the bacterial cell cycle, is characterized by a number of events. These events include (1) growth, or the coordinated synthesis of macromolecules as well as cellular components in the cell; (2) septum formation; and (3) division, or the process in which each daughter cell receives a copy of DNA. Many individual aspects of control have been investigated, but as yet they have not been pieced together into an integrated cycle of events. The most significant results have come from studies that have attempted to correlate DNA synthesis and cell envelope synthesis with cell division.

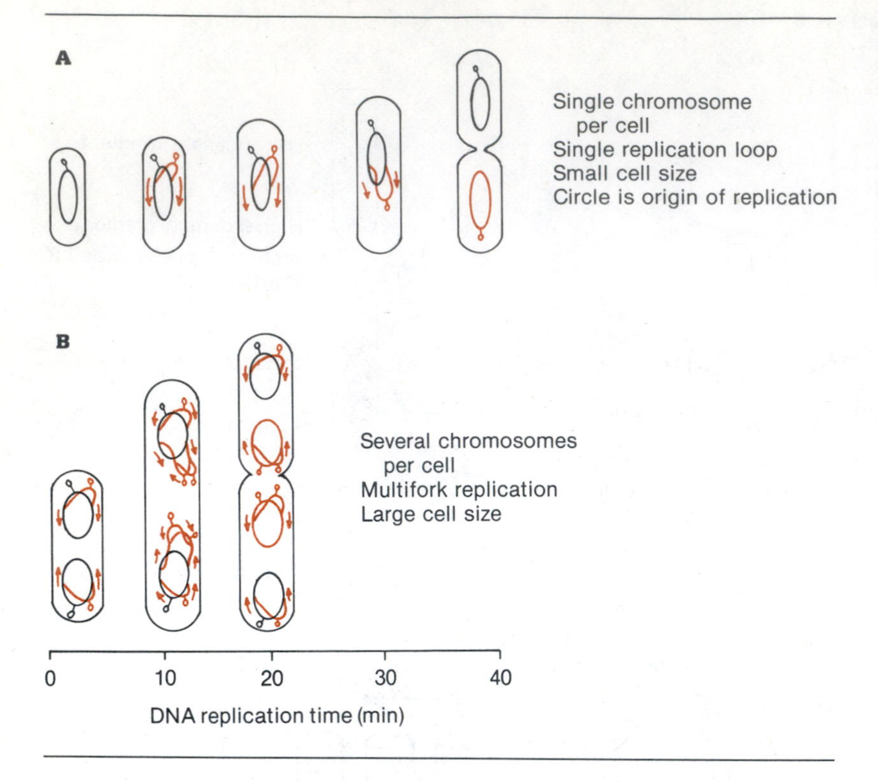

Figure 15-2 Replication of DNA in bacterium *E. coli.* **A,** At low growth rates there is single chromosome per cell and no reinitiation at origin before completion of round of replication. **B,** There are multiple initiations of replication at origin and more than one chromosome per cell during higher growth rates.

Redrawn from Dawes, I.W., and Sutherland, I.W., Microbial physiology. *Blackwell Scientific Publications, Ltd., Oxford, 1976.*

DNA Replication. DNA synthesis occupies the entire bacterial cell cycle (in eukaryotes DNA synthesis occupies only a portion of the cell cycle) and has been shown to be involved in control. Studies have demonstrated that if DNA synthesis is interrupted, cell division is also interrupted; however, if cell division is interrupted by an antimicrobial, such as penicillin, DNA synthesis still continues. This indicates that DNA replication is somehow involved in the cell division process.

The time it takes for a bacterial cell such as *Escherichia coli* to divide can be controlled by adjusting those factors that affect growth, such as nutrients, pH, and oxygen tension. *E. coli* grown under ideal conditions in a medium rich in nutrients at a temperature of 37° C will divide every 20 to 25 minutes or less. The generation time is 60 minutes if the same medium is used, but the temperature of incubation is 30° C. The reason for mentioning this is that in *E. coli,* where the cell cycle has been most extensively studied, DNA replicates once every 40 minutes (provided that the cell divides in a time period between 25 and 70 minutes). The generation time for most bacteria is between 40 and 60 minutes. The question that probably comes to mind is, how can a cell that is dividing every 20 to 30 minutes synthesize a complete molecule of DNA if DNA replicates every 40 minutes? The problem was resolved when it was discovered that multiple replication forks are produced while the DNA is being replicated. For example, prior to cell division a new round of DNA replication occurs before the initial one has been completed (Figure 15-2). One of the consequences of multiple replication sites is that in a rich medium in which several generations have elapsed, cells can be observed to have 2 to 4 nucleoids.

A model to demonstrate how DNA replication is coordinated with cell division is presented in Figure 15-3. This model serves to illustrate the following:

1. A new round of DNA replication occurs when the initial mass (M_i) of the cell has doubled. In cells that divide every 25 minutes, a new round of DNA replication begins at the origin every 25 minutes.
2. At each doubling, two processes are initiated: first, DNA synthesis begins and continues until a termination protein is synthesized. This process takes 40 to 45 minutes to complete. Second, protein synthesis occurs during this period and is followed by what is believed to be the formation of a septum precursor.
3. When DNA replication is finished, the synthesized termination protein is believed to interact with a septum precursor molecule that initiates septum formation and cell division.

Cell envelope and cell division in cocci. The cell envelope, which is composed of the cell wall and cytoplasmic membrane, is intimately associated with growth and cell division. Studies of only the cell wall, however, have yielded sufficient data to obtain a clear picture of how cell surfaces respond during the cell cycle.

Cytoplasmic membrane. The cytoplasmic membrane is believed to play a role in the regulation of DNA initiation as well as in the segregation of replicated DNA. The ***replicon hypothesis*** proposed by Jacob and others assumed that at an early stage in the cell cycle the DNA is attached to a special membrane site, which signals the start of DNA replication. The membrane attachment site duplicates, and as the cell membrane becomes synthesized, the replicating genomes also become segregated (see Figure 9-10). Most of the evidence for this type of mechanism

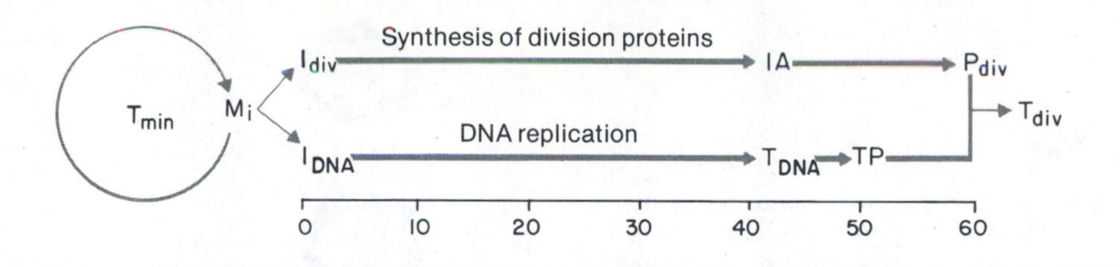

Figure 15-3 Model of cell cycle in *E. coli.* There is coordination between synthesis of division proteins (I_{div}) and initiation of DNA replication (I_{DNA}) as initiation mass, M_I, of cell doubles. DNA replication time (T_{DNA}) is 40 minutes, at which time synthesis of termination protein *(TP)* and initiation assembly *(IA)* of some septum precursor begins. Termination protein and septum precursor or division protein (P_{div}) interact at 60 minutes to initiate division (T_{div}).

Redrawn from Donachie, W.D., Jones, N.C., and Teather, R. Symp. Soc. Gen. Microbiol ***23:****9, 1973. Cambridge University Press.*

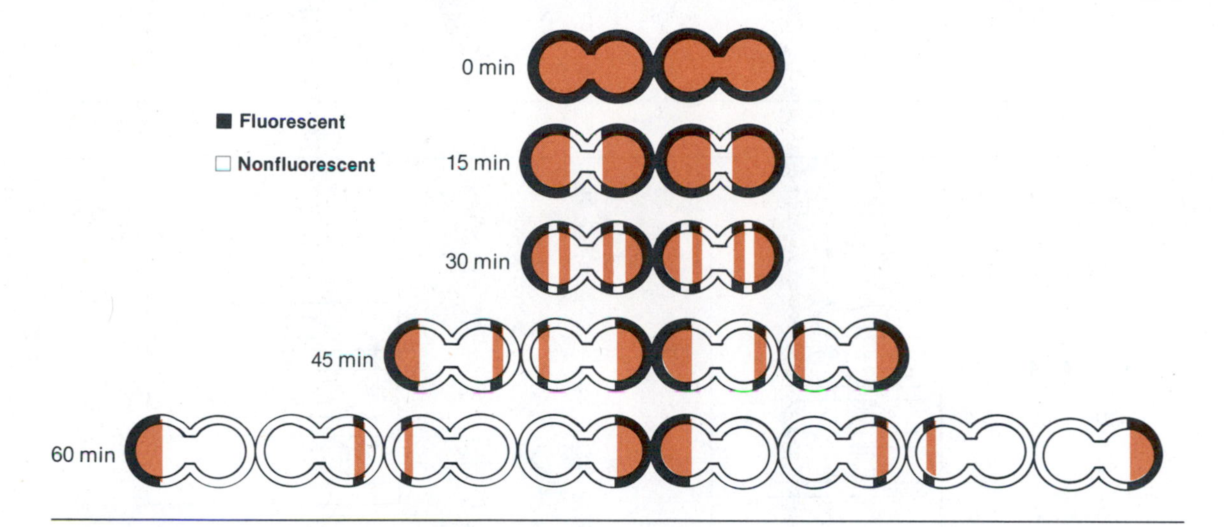

Figure 15-4 Cell wall growth in *Streptococcus pyogenes.* Bacteria are initially grown in medium containing fluorescent antibody that attaches to cell envelope. Fluorescent cells are then transferred to medium containing no fluorescent antibody. Times indicate period of incubation in nonfluorescent medium.

Redrawn from Cole, R.M. Bacteriol. Rev. ***29:****326, 1965.*

comes from biochemical studies. Whether or not the membrane attachment sites coincide with the mesosome is still a matter of speculation, since many scientists consider mesosomes to be artifacts.

Cell wall. Experiments have shown that cell wall synthesis does not occur in the same way in all species of bacteria. The clearest picture to date of cell wall synthesis during the cell cycle has been obtained with those organisms that divide in one plane, such as the streptococci. Streptococci can be grown in a medium that contains an antibody capable of reacting with cell wall antigen. The antibody is first labeled with a fluorescent molecule, called a fluorochrome, thus making it observable microscopically. Once the streptococci have been labeled and the entire surface appears to fluoresce, the microorganisms are removed from the medium and inoculated into a medium containing nonfluorescent antibody. As the population of streptococci grows, only newly synthesized cell wall will react with nonfluorescent antibody. The results of such an experiment are illustrated in Figure 15-4. It can be seen from this figure that new cell wall is laid down centripetally, so that the new hemispheres for each of the eventual daughter cells are back to back. Figure 15-5 is a model demonstrating how cell wall synthesis is organized. The model illustrates the following:

1. New cell wall synthesis begins at areas of the cell envelope, called ***wall bands,*** that have been produced during the previous cell cycle (Figure 15-5). A membrane (mesosome) forms beneath each wall band, which represents the membrane attachment site.

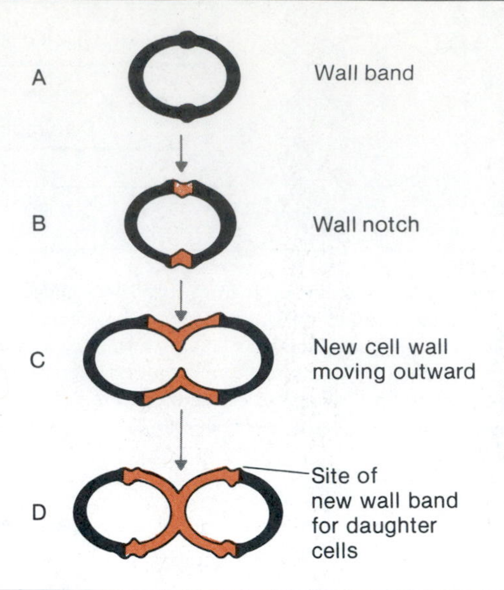

Figure 15-5 Cell wall growth in *Streptococcus faecalis*. *A*, Raised area on cell envelope, called wall band, appears. *B*, Wall band invaginates and forms wall notch, which becomes septum. In *B* through *D*, new cell wall is synthesized and moves outward from wall notch. *D*, New cell wall bands are being initiated at center of each daughter coccus even before cell divides.

Based on Higgins, M.L., and Shockman, G.D. CRC Crit. Rev. Microbiol. ***1**:29, 1971. Copyright the Chemical Rubber Co., CRC Press Inc.*

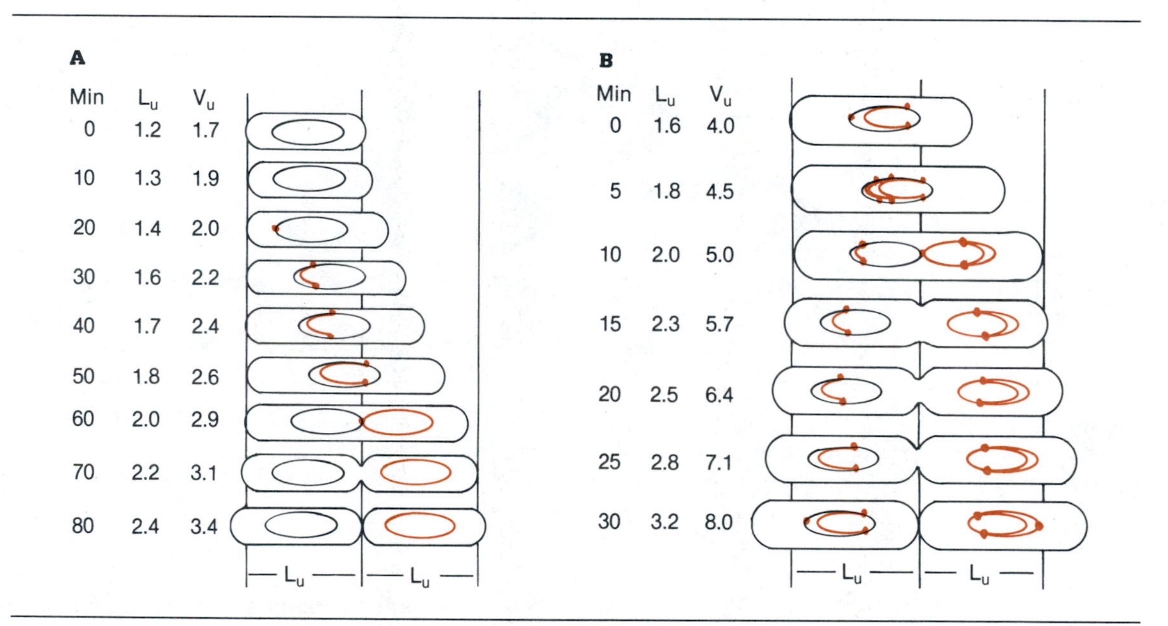

Figure 15-6 Growth in rod-shaped organisms such as *E. coli.* **A,** Generation time of 80 minutes. **B,** Generation time of 30 minutes. L_u, Minimum unit cell length; V_u, minimum unit cell volume. See text for details.

Redrawn from Donachie, W. In John, P.C.L. (Ed.). The cell cycle. *Oxford University Press, Oxford, 1981.*

2. Cell wall precursors are added below the wall band, and as they are added, the membrane invaginates and forms a ***wall notch.*** The wall notch will eventually become a ***septum,*** or point of cell division. The addition of new cell wall precursors requires certain cell wall lytic enzymes to breakdown preexisting peptidoglycan.
3. The cytoplasm becomes completely divided as new cell wall precursors are added centripetally, and the cell is cleaved into two daughter cells. Each daughter cell possesses a wall band that represents cell wall material produced during two different cell generations.

Cell envelope and cell division in rod-shaped bacteria. Experiments to identify cell wall growing points in rod-shaped bacteria have been difficult to evaluate because of conflicting results. The model that seems to best integrate the data obtained from cocci and bacilli alike is presented in Figure 15-6. This model, which is based on the concept of minimum unit cell volume (V_u) and cell length (L_u) is best explained by comparing the growth of *E. coli* in media in which the generation time is 80 minutes with growth in media in which the generation time is 30 minutes. According to the model, cell wall growth and the addition of cell wall precursors occur in only one direction and toward one pole until the minimum cell length is doubled, or $2 \cdot L_u$. When $2 \cdot L_u$ is reached, a septum begins to form at a point midway between the two cell ends, and cell wall growth becomes bipolar. Septum formation takes approximately 20 minutes to complete, and then the cell divides. Rounds of chromosome replication are initiated at each doubling of the unit volume, or $2 \cdot V_u$. The chromosome is completely replicated after 40 minutes, and termination of rounds of replication always coincide with the time at which the cell length reaches $2 \cdot L_u$. If the generation time is less than 40 minutes (Figure 15-6, *B*), then new rounds of replication are initiated before preceding rounds are completed.

Population growth

Large populations of cells are routinely used in laboratory experiments in which microbial growth and related processes are measured. A small inoculum of cells is usually introduced into a culture medium, and after several hours of growth the cells are harvested. Microbial populations may be prepared in one of two ways: batch culture or continuous culture.

Batch culture. The rate of growth of a cell population will be seen to vary with time following the inoculation of a small number of cells into the culture medium. In other words, from the time of inoculation of cells into the medium until growth ceases, the cell population will go through a number of phases that can be expressed as a growth curve. The growth curve can be separated into the following stages (Figure 15-7).

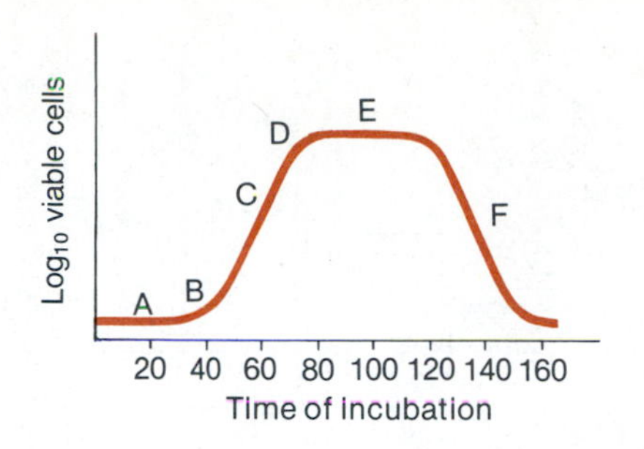

Figure 15-7 Growth curve. *A,* Lag phase; *B,* acceleration phase; *C,* logarithmic phase; *D,* deceleration phase; *E,* stationary phase; *F,* death or decline phase.

Lag phase. Following inoculation into a medium, microorganisms require a period of time to adjust to their new environment. This lag time refers only to a lag in cell division and not to other aspects of growth. The cells, for example, increase in size, and there is synthesis of enzymes to ensure that nutrients in the medium are metabolized. There is an increase in the synthesis of macromolecules such as RNA and protein, but DNA synthesis remains constant. The length of the lag phase is related to the physiological and genetic state of the species at the time of inoculation into the culture medium. Cells that are nutritionally starved before their inoculation into a medium demonstrate a long lag phase, but if the original inoculum of cells contains an abundance of enzymes from previous growth in the same type of medium, the lag phase will be short. Organisms that do not have the necessary inducible enzymes exhibit an indefinite lag phase until the needed nutrients or other factors have been supplied to the culture. At the end of the lag phase a very short period of acceleration occurs, in which some of the cells are entering a period of cell division but not yet at a constant rate.

Logarithmic phase. The logarithmic phase (also called log or exponential phase) represents that period of time in which cell division is maintained at a constant and maximum rate. *E. coli,* for example, in one type of medium might divide every 30 minutes in the log phase, but if it is grown in a less-rich medium or at a different temperature, the division time could be 60 minutes or longer. The growth rate, or doubling time, remains constant for any organism during the log phase regardless of the medium. The culture in the log phase is said to be in ***balanced growth.*** Balanced growth also means that if the cell population doubles, so do all of the components of the cell, such as DNA, RNA, and protein. Cultures in a closed system, in which no nutrients enter the culture flask and no waste products are removed, reach a point where environmental factors slow down the rate of growth. These factors might include overcrowding, nutrient depletion, and pH changes due to acid formation. This slowing-down process is referred to as a deceleration phase and precedes the next stage, called the stationary phase.

Stationary phase. The number of microorganisms in the stationary phase remains constant. The conditions that initiate the deceleration phase are accentuated in the stationary phase. This is a period of ***unbalanced growth,*** and cellular components and macromolecules are synthesized at different rates. Many cells are dividing, but just as many are dying. Those that are

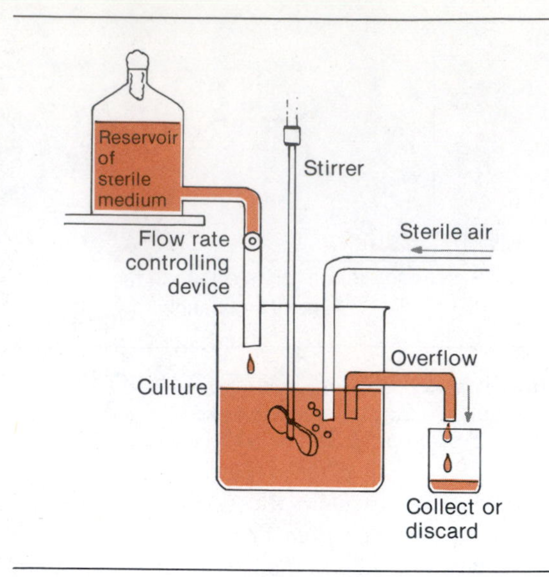

Figure 15-8 Chemostat.

alive are producing more acids and waste products, which further inhibit cell division. Some dying cells lyse and release autolytic enzymes, such as proteases, nucleases, and lipases, that break down macromolecules into their respective building blocks, which can be used by those cells that have survived. The stationary phase can be short or very long, depending on the bacterial species.

Death or decline phase. The rate at which a population dies varies with the microbial species involved. The rapid lysis of some microbial species, such as *Neisseria* species, during this period is brought about by a self-destruct mechanism that is characterized by the induction of autolytic enzymes that degrade the cell envelope. These autolytic enzymes are the same enzymes that are necessary for growth, because they allow the insertion of newly formed peptidoglycan into the cell wall. Obviously, a balance must be maintained in the cell between cell wall lytic action and cell wall synthesis. Studies with *Streptococcus pneumoniae* indicate that the nature of the lipoteichoic acid in the cell wall is an important factor in control. Apparently, when the lipoteichoic acid is in an acylated state, the cell wall hydrolytic enzyme (β-1,4-*N*-acetylmuramide glycan hydrolase) is inhibited. The deacylated form does not inhibit the enzyme, suggesting that a deacylase enzyme may be the controlling factor. Other mechanisms must also exist for cells without lipoteichoic acid in the cell envelope.

The rate of death can be exponential, but sometimes it is not. It is believed that if the cell requires only one target site to be affected in order to cause death, then the death curve will be exponential. If two or more sites are required, then a nonexponential rate of death takes place in the population.

Continuous culture. One of the disadvantages of batch culture is that one cannot independently control those factors that decrease the growth rate, such as nutrient supply and waste removal, although air supply and pH can be controlled. A technique for overcoming this problem involves the use of an open-culture system called ***continuous culture.*** The primary purpose of continuous culture is to maintain a population of cells at a constant growth rate in the exponential phase of growth for an indefinite period of time. The devices used to produce continuous culture are of two types: ***chemostats*** (also called bactogens) and ***turbidostats.***

Chemostats. The simple chemostat (Figure 15-8) consists of a vessel in which the culture is grown. Devices are connected to the growth vessel that control the amount of air and fresh medium to be added to the culture, as well the overflow of medium containing cells and waste products. The culture will reach a constant volume and density when an equilibrium between inflow of fresh medium and outflow of medium is reached. All of the constituents of the medium in a chemostat except one are in excess. The deficient nutrient will support the growth of only a limited number of microorganisms; therefore the density of the culture in a chemostat is controlled by nutrient limitation. The carbon source, for example, glucose, may be used as the limiting nutrient, but other nutrients such as nitrogen, magnesium, sulfate, or phosphorus may also be used.

By controlling the rate at which nutrients are added to the chemostat, the rate of bacterial growth (k) may also be controlled. The rate of nutrient addition is referred to as the dilution rate (D). The dilution rate is the volume of nutrient added per hour and is expressed as a fraction of the volume of the vessel. The dilution rate (D) can be expressed as:

$$Dx = \frac{dx}{dt}$$

where $\frac{dx}{dt}$ is the change in numbers of bacteria between two time periods. If the growth rate constant (k) is also expressed as:

$$kx = \frac{dx}{dt}$$

then the dilution rate equals the growth rate constant. Any changes in cell concentration can be defined as:

$$\frac{dx}{dt} = kx - Dx$$

or a change in cell concentration equals growth of bacteria minus dilution of bacteria. At low dilution rates (D_o) the concentration of cells in a continuous culture system is at its highest (Figure 15-9). A very gradual decline in bacterial concentration and a gradual increase in substrate concentration occur when the dilution rate begins to increase. A point is reached during this increase in dilution rate (D_c) where the rate at which bacteria are diluted out exceeds the maximum growth of the culture and there is a rapid decline in the concentration of bacteria. In other words, more bacteria are washed out of the culture vessel than are produced by cell division.

Turbidostats. Cell density in turbidostats is controlled by a device that increases or decreases the flow of culture medium into the growth vessel. There is no limiting nutrient. A photoelectric device is attached to the turbidostat that measures cell density. Any deviations in culture density are passed on as a signal to a pump that increases or decreases the flow of growth medium into the growth vessel.

Continuous cultures have played a very important role in studies of microbial physiology and have been used extensively in industry. The effects of nutrient limitation have revealed many important features of cell composition, enzyme production, and growth, that could not have been obtained with batch cultures. Continuous cultures have been used to analyze mutation rates of organisms, and many of the regulatory devices in microbial systems have been amenable to study through this technique. The results of these studies have led to a greater understanding of control mechanisms in animal systems. The technique of continuous culture has also been applied in industry to the production of single-celled proteins. Microbes are being considered as a food source because of their ability to produce a high yield of protein when fed certain carbon compounds. Large-scale single-celled protein processes use carbon sources as a limiting nutrient in continuous culture (see Chapter 21).

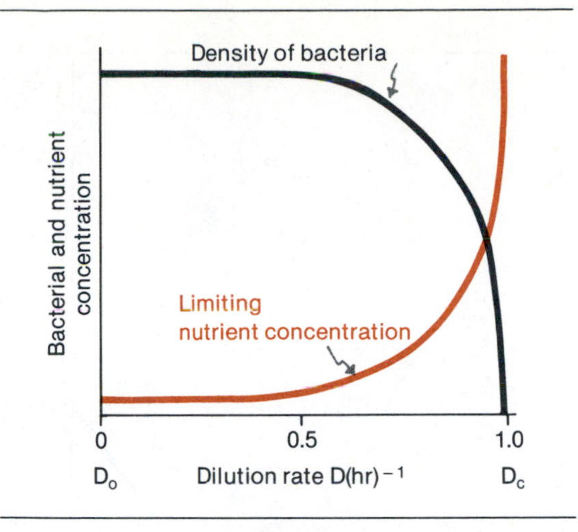

Figure 15-9 Relationship between cell density, limited nutrient concentration, and dilution rate in continuous culture system. See text for details.

Mathematics of bacterial growth

Bacterial cells divide in geometric or exponential fashion, that is, 1-2-4-8-16-32, etc. The population therefore increases as an exponent of 2 or 2^n. For example, if we were to start with 1 cell, the number of cells after 3 generations or doublings would be 2^3 or $2 \times 2 \times 2$ or 8. The actual number of cells measured under laboratory conditions is much larger, often between 1×10^5 to 1×10^9 cells per milliliter of suspension. Suppose, for example, we have two sample cultures that were collected at two different times during the growth cycle. Further suppose that we have a method (to be discussed later) for determining the number of cells in the sample. First, we could either plot the number of microorganisms as a function of time or plot the logarithm of the number of cells as a function of time (Figure 15-10). The disadvantage of plotting the arithmetic number of cells versus time is that at higher cell concentrations the curve is nearly impossible to evaluate. The logarithm of the number of organisms is usually plotted as the logarithm to the base 10 (obtained by multiplying the logarithm to the base 2 of a number by 0.3010). The number of generations (n) can be determined by using the following formula:

$$n = \frac{\log_{10}N_2 - \log_{10}N_1}{\log_{10}2}$$

where N_1 and N_2 are the number of organisms at times T_1 and T_2, respectively. If in our example N_1 equals 100 organisms and N_2 equals 800 organisms, then the equation reads:

$$n = \frac{\log_{10}800 - \log_{10}100}{\log_{10}2} = \frac{2.903 - 2.0}{0.3010} = 3 \text{ generations}$$

Since the generation time (gt) is equal to the time interval between T_1 and T_2 (assume it to be 90 minutes) divided by the number of generations, then:

$$gt = \frac{T_2 - T_1}{n} = \frac{90 \text{ min}}{3} = 30 \text{ min}$$

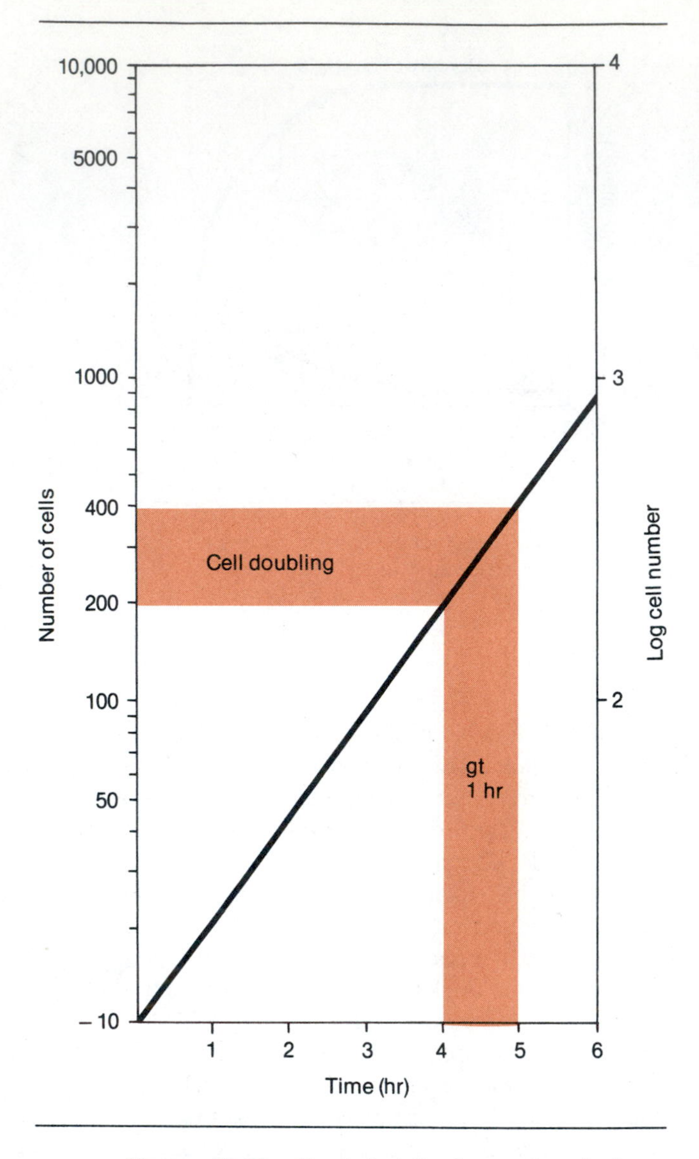

Figure 15-10 Graph depicting log number of cells as function of time when cultivated in laboratory growth medium.

From the previous equations one can also determine the specific growth rate constant (k) for any culture during unrestricted growth, or:

$$k = \frac{\ln N_2 - N_1}{T_2 - T_1}$$

where N may be the cell number or cell mass or other measurable parameter of the cell. The growth rate constant (k) is therefore equal to the natural logarithm of the increase in a property of the culture divided by the time it takes for that increase to occur. In those instances where the culture has doubled, then:

$$k = \frac{\ln_2}{gt} = \frac{0.693}{gt(hr)} \quad \text{(0.693 is the natural logarithm of 2)}$$

If the time it takes for the culture to double is known, then the growth rate constant is easily calculated. The growth rate constant will be the same during exponential growth regardless of the component that is measured (cell mass, cell number, protein, carbon, etc.), but it can be influenced by a variety of environmental conditions, including oxygen tension, pH, temperature, and type of growth medium. The growth rate constant provides the researcher with a valuable tool for comparison between different microbial species provided that standard environmental conditions are maintained.

Laboratory methods for determination of growth

There are many instances in research, industry, and medicine where the number of microorganisms in a sample must be determined. Determining the potability of water requires that specific bacterial counts be performed daily in water treatment facilities. Many food products are subjected to microbial counts to determine their suitability for human consumption. Bacterial counts are sometimes required by the physician in order to evaluate the necessity for antimicrobial therapy. Whatever the measurement of growth that is used, it is important to distinguish between total counts and viable counts. Total-count measurements include dead as well as viable cells, whereas viable counts include only those cells that are alive and capable of growth. Both physical and chemical methods are used to determine bacterial growth and to distinguish between total and viable counts.

Cell count

Microscopy. Microorganisms can be counted directly by using a specially etched microscopic slide. A grid is engraved on a microscopic slide containing a pattern of squares (Figure 15-11). Each small square on the grid is 50 μm on each side. On either side of the grid are raised edges so that when a drop of fluid is placed on the grid and covered with a coverslip, the underside of the coverslip is exactly 20 μm above the grid. The bacteria are counted in the squares when observed microscopically, and this figure is multiplied by a volume figure (derived from the size of the grid squares and the height above them). The cell count will have to be multiplied by a dilution factor if the sample was originally diluted. The engraved grid used for this technique is called the ***Petroff-Hauser counter.*** One cannot, however, differentiate living from nonviable cells using this technique.

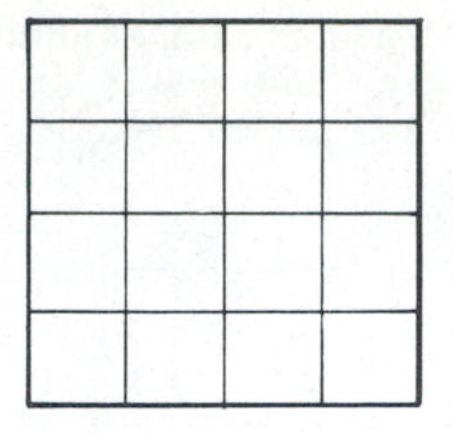

Figure 15-11 Representation of etched grid found on slide chamber of Petroff-Hauser counting chamber. Dimensions of each small square are 0.05 mm × 0.05 mm × 20μm, or volume of 50,000 cubic μm. Bacteria are counted microscopically in large square, and this figure is multiplied by volume figure times original dilution factor if original sample was diluted.

Colony count. A colony count is the most effective way to distinguish between viable and nonviable cells in a bacterial suspension. This is an indirect technique in which a sample of bacteria is diluted to an appropriate number of cells that can be spread on the surface of agar in a Petri dish. Each cell plated on the agar surface, following suitable incubation, will give rise to discrete aggregates of cells called colonies. The number of colonies that can be counted and that is of statistical significance is between 30 and 300. A bacterial suspension that is faintly turbid to the eye will contain approximately 1×10^7 bacteria per milliliter and will require dilution in order to obtain a measurable number of colonies (Figure 15-12). Once diluted, an aliquot of the suspension is spread on the surface of an agar plate and often incubated for 18 to 24 hours between 35° and 37° C. (Some bacterial species require longer times and different temperatures of incubation.) Most colonies will be visible after this period of time. An alternative method, called the pour tube technique, can also be used. The original diluted bacterial suspension is placed in molten (42° C) agar and then poured into the bottom of a Petri dish. Once the agar has hardened and after suitable incubation, colonies will be produced not only on the surface but also within the agar.

Electrical resistance. Microbial cells are poor conductors of electricity and offer resistance to current flow. The ***Coulter counter*** is a device in which a suspension of cells is made to pass through an opening on either side of which are electrodes. A microbial cell appearing between the electrodes causes the current to be impeded, and there is a drop in voltage. This voltage drop is recorded as a pulse, the height of which is proportional to the volume of the cell. Cell number and cell size can therefore be determined using this technique. No distinctions, however, can be made between viable and nonviable cells, and contaminating dust particles or fibers in the bacterial suspension can be a source of error.

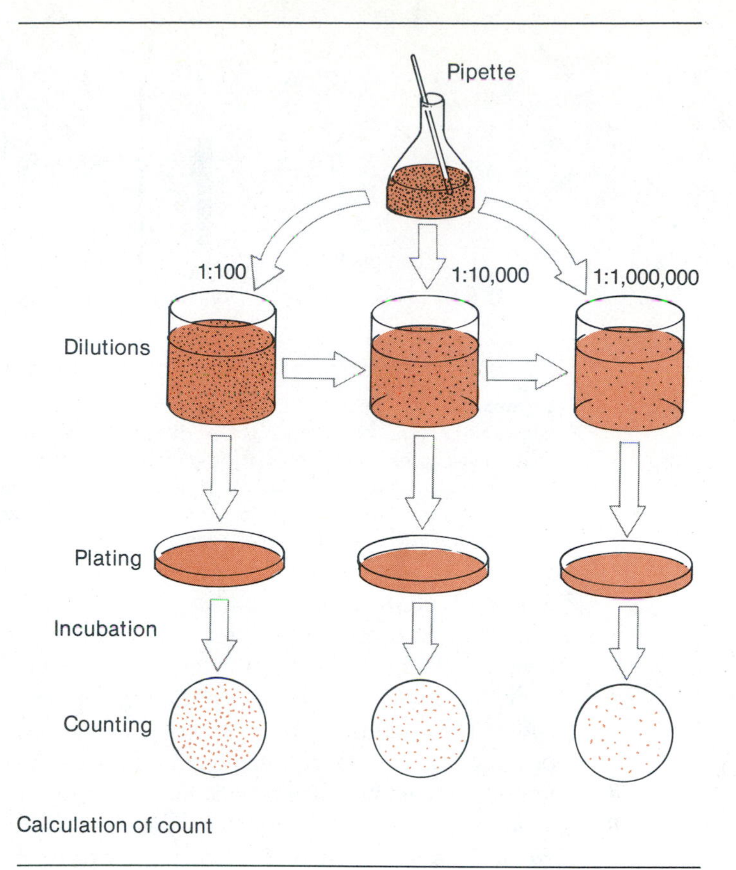

Figure 15-12 Dilution technique for determining number of viable microorganisms (bacteria) in culture.

Cell density. Photometric devices are the most widely used methods for determination of microbial growth. ***Photometry*** is a technique that employs a beam of light to estimate bacterial populations. A beam of light focused on the bacterial suspension (Figure 15-13) is scattered once it strikes the cell wall. The undeviated light can be measured by a technique called ***turbidometry,*** and scattered light is measured by a technique called ***nephelometry.*** Turbidometry is used almost exclusively over nephelometry for growth measurements.

The instruments used to measure light absorbed by a microbial population are called photometers or spectrophotometers. The theory behind the use of these instruments is related to the Beer-Lambert law for solutes, which states that there is a straight-line relationship between absorbency (also called optical density or OD) of

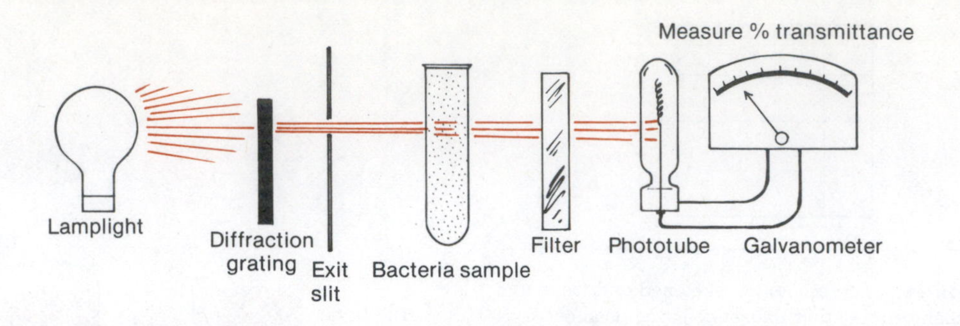

Figure 15-13 Spectrophotometer used to determine turbidity of bacterial suspension. White light generated by tungsten filament strikes diffracting grating and produces spectrum of wavelengths. Only wavelengths that fall on exit slit go through to strike sample (wavelengths are adjustable). Light that passes through sample strikes phototube, which activates galvanometer. The more light transmitted, the less concentrated the sample, whereas the less light transmitted, the greater the turbidity of concentration of cells in suspension.

a solution and the concentration of the solute).* A bacterial suspension can be considered the solute, and therefore a linear relationship exists between its optical density and cell number per milliliter. Measurements are carried out by first placing the bacterial suspension in a tube of uniform-bore diameter. The tube is inserted into the spectrophotometer, in which a beam of light strikes the suspension, and the light passing through the suspension is measured. The intensity of the light passing through the suspension will be governed by the concentration of bacteria in the suspension, the intensity of light, the wavelength of light, and the diameter of the tube containing the bacterial suspension. A photodetector measures the intensity of light passing through the bacterial suspension and translates it on a meter or chart as percent transmittance (% T) and/or absorbency (A).

The values of absorbency and/or percent transmittance can be correlated with the cell number by using additional procedures, provided that one uses the same species of bacteria for experiments over a long period of time. This can be accomplished by doing a growth curve for the bacterial species in question. The samples taken during growth should be subjected to cell counts as well as optical density measurements. This type of correlation will be of considerable value in future experiments because any optical density measurement can be extrapolated on the growth curve. Cell size influences the results obtained by turbidometry; therefore the same organism must be used, and samples should be taken during the exponential phase of growth in order for the data to be of value.

* The optical density is a logarithmic function of the original intensity of the light beam (I_o) and the intensity of the beam (I) passing through the solution, or:

$$\text{Optical density} = \frac{\log I_o}{I}$$

Cell mass. As pointed out previously, one can determine the rate of growth during the exponential phase by measuring other parameters such as RNA, DNA, and protein content, as well as cell number. One can also measure cell mass by determining the dry or wet weight of a population of bacterial cells. In order to determine the dry weight, the cell suspension (the sample should contain approximately 1×10^{11} cells) is centrifuged and washed several times with distilled water and recentrifuged. The sample is placed in a dish and dried in an oven at 100° C or in a vacuum at 80° C. Wet-weight determinations are made by collecting the bacterial cells on a filter that has been previously dried to a constant weight. Extracellular water is removed when the membrane and bacteria are dried in a vacuum at 40° C.

The most consistent relationship between turbidity and cell density is obtained when correlations are made between optical density and cell mass and not between optical density and cell number, since cell size influences optical density. Determining cell mass is a time-consuming process and is used only under special research conditions, such as those in which the culture forms filaments or the culture clumps excessively.

Other methods

Chemical methods. Several chemical components of the cell can be analyzed by colorimetric means. One can measure RNA, DNA, or protein by treating a washed suspension of cells with agents that release these molecules from the cell and hydrolyze them. The addition of a specific reagent produces a color complex with one of the

Table 15-1 Characteristics of the stages in the eukaryotic cell cycle

STAGE	PERCENTAGE OF CYCLE*	CHARACTERISTICS
G_1	Up to 50%	Most RNA and protein synthesis occurs in this period; dry mass increase is greatest in G_1
S	From 20% to 25%	DNA duplication; DNA content doubles
G_2	Up to 25%	Very little synthetic activity except for precursors for mitotic spindle and precursors to components involved in cytoplasmic division (cytokinesis)
M	Approximately 5%	Period of mitosis; practically no synthesis; spindle is visible; period terminates with nuclear division and separation of daughter cell from parent

*Applies only to the simplest of eukaryotic microorganisms, such as the single-celled fungi.

macromolecules, whose intensity can be measured with a spectrophotometer. Although seldom used as a technique for determining growth, colorimetric methods have certain advantages in that fewer cells are needed and the test is not as time consuming as some other techniques.

Estimations of the efficiency of cellular growth can also be made by measuring the total ATP produced. The ability of an organism to convert the carbons of a fermentable carbohydrate, such as glucose, to cellular carbon is a function of the organism's efficiency at producing ATP. The total ATP obtained from the fermentation of carbohydrates is known for many microorganisms and is relatively constant when total cell mass is measured. When the ATP yield is compared with the amount of carbohydrate fermented, this equivalency disappears. This difference is related to the different pathways that some microorganisms use to catabolize a carbon source (see Chapter 13). Unknown catabolic pathways can be evaluated from known ATP yields.

Metabolic uptake. When environmental conditions are standardized, growth can be related to the oxygen consumed by an organism or to the excretion or uptake of certain metabolites, such as fermentable sugars. One can also measure the amount of acid produced from the fermentation of sugars, and this can be related to the growth yield, defined as the milligrams of dry weight per mole of acid produced.

GROWTH IN EUKARYOTES

Eukaryotic cell cycle

One of the major differences between eukaryotic and prokaryotic cell cycles is the pattern of DNA synthesis. DNA synthesis in eukaryotes occurs only periodically and does not occupy the entire cell cycle, as it does in prokaryotes. In addition, the mitochondria and chloroplasts of eukaryotes also contain DNA with functional genes. There are more morphological events associated with eukaryotic growth, and this tends to make the eukaryotic cell cycle more complex than the cycle of events in prokaryotes.

Stages in the eukaryotic cell cycle. The eukaryotic cell cycle can be divided into four stages: G_1, S, G_2, and M, which are described in Chapter 12. The duration of each cycle, which varies from species to species, may take from 2 hours to several days and is controlled by the growth conditions, particularly nutrient supply. Table 15-1 reviews the stages in the cell cycle and describes some of the cellular events that occur in them.

Cell cycle in a single-celled eukaryote. Our understanding of the eukaryotic cell cycle events, particularly their control, is very meager. Most of what we know about cell cycle events in eukaryotes has been revealed from studies with single-celled microorganisms, such as *Saccharomyces cerevisiae,* a yeast that divides by budding.

The cell cycle in *S. cerevisiae* is best explained by starting with an unbudded cell in the G_1 period (Figure 15-14) and following the events described below:

1. The G_1 period prepares the cell for DNA synthesis. Several biochemical events occur during this period, and a point is reached where the cell is irreversibly committed to initiate DNA synthesis and complete cell division. The period is terminated by the initiation of DNA synthesis, the duplication of the ***spindle plaque*** (centrioles), and the

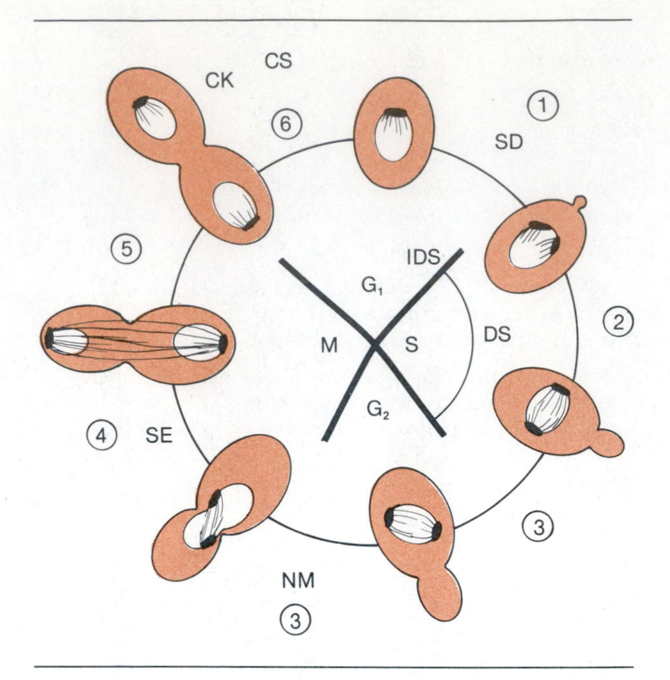

Figure 15-14 Cell division cycle in yeast *Saccharomyces cerevisiae.* *1,* Mitotic spindle duplicates, and bud emerges. *2,* Spindle separates, and bud enlarges. *3,* Nuclear migration. *4,* Spindle elongates, and nuclear division occurs. *5,* Nuclear division; bud approaches size of parent. *6,* Cytokinesis and cell separation. *SD,* Spindle duplication; *IDS,* initiation of DNA synthesis; *SE,* spindle elongation; *CK,* cytokinesis; *CS,* cell separation
Redrawn from Hartwell, L.H. Bacteriol. Rev. **38:**164, 1974.

emergence of the bud. The spindle plaque is comparable to the centrioles of higher eukaryotes and is concerned with microtubule and spindle formation. The DNA division cycle is initiated only when the cell has reached a critical size. The faster the growth rate, the shorter the time required to attain this critical size.

2. The S period occupies about 25% of the cell cycle and is characterized by the duplication of the chromosomes. The spindle plaques produced in G_1 separate to produce a complete spindle, and the bud that was formed in G_1 continues to enlarge.
3. The G_2 period is characterized by migration of the nucleus to the neck. The nucleus undergoes its first stage of nuclear division and spindle elongation.
4. The M period is represented by further spindle elongation as well as the completion of the remaining phases of nuclear division. At the end of the M period there is cell membrane separation (cytokinesis) and cell wall separation. Finally, the daughter cells separate, and two unbudded cells are produced. The division of a budding yeast cell is asymmetric, yielding two cells, one of which is the older parent and one of which is the new daughter cell. The daughter cell is always smaller than the parent. A ring of chitin builds up at the bud isthmus when the yeast cell buds. This chitin remains on the parent after the bud has separated and is called a ***bud scar,*** which can be visualized microscopically. It is therefore possible to determine the number of division cycles that a cell has passed through by determining the number of bud scars (Figure 15-15). The parent cell after cell division can immediately reenter the division cycle, but the daughter cell must wait until a critical size has been reached.

Cell wall growth during the cell cycle

The mechanism of cell wall growth in fungi is even less understood than it is in bacterial cells. One of the basic premises of fungal metabolism is that growth in the fungal hypha (comparable to the rod-shaped bacterial cell) occurs only at the wall apex. How then, does the cell wall grow in a spherical cell such as *S. cerevisiae?* We are really not sure; however, there are some aspects of cell wall growth that seem to apply to both spherical as well as hyphal cell types. The student should reexamine the structure of the fungal cell first described in Chapter 5. Listed below are some of the pertinent aspects of fungal wall growth:

1. Preexisting cell wall near the apical region of the hypha is hydrolyzed by hydrolases. This activity loosens the cell wall and produces a region that is involved either in growth by elongation or the formation of a tiny bud. The major cell wall component in yeast is a 1,3-β-glucan. A 1,3-β-glucanase has been isolated from yeast, and it has been postulated to be the autolytic enzyme involved in modification and limited breakdown of the cell wall during the life cycle.
2. Protein polysaccharide complexes are polymerized by enzymes found in the smooth endoplasmic reticulum. The polymerized units are packed into vesicles and then transported to sites on the plasmalemma (cytoplasmic membrane) adjacent to the growing sites of the wall. The contents of the vesicles are then discharged into the periplasmic space of the cell wall (space between the glucan and protein-polysaccharide complex).
3. Skeletal polysaccharides are synthesized by

synthases located within the plasmalemma.

4. The wall grows by the addition of materials brought by vesicles. The growth of the wall in hyphal cells is restricted to the apical region, but in spherical cells, cell wall precursors are incorporated over the entire surface.

FACTORS AFFECTING GROWTH

In addition to their chemical needs, microorganisms are affected by the physical nature of their environment. Environmental factors, such as temperature, pH, osmotic pressure, and radiation, influence what type of microorganism can survive in a particular ecological niche. Habitats that are exposed to extremes of temperature, pH, etc., are colonized by microorganisms that exhibit unusual structural or chemical characteristics.

Temperature

Growth is the culmination of a series of enzymatic reactions, and because of this, temperature plays a very important role in the cell cycle. The optimum* temperature for enzyme activity in microorganisms is dependent on the ecological site of the species under consideration and its location (extracellular or intracellular). Certain temperatures are lethal to some microorganisms, but other microorganisms not only survive at these temperatures but are actually capable of growth. Microorganisms can grow over a temperature range of 30° C or more, but the ***optimum growth temperature*** is within a very narrow range of temperatures (Figure 15-16). Above the optimum range enzymes become denatured, and the rate of growth drops precipitously, producing what is called the ***maximum growth temperature.*** Below the optimum temperature range the rate of growth decreases slowly until a point is reached, called the ***minimum growth temperature,*** where growth ceases. Microorganisms have been divided into three groups based on their optimum growth temperature: ***psychrophiles*** (low temperatures), ***mesophiles*** (moderate temperatures), and ***thermophiles*** (high temperatures). How each group has evolved to grow within particular temperature ranges and how each responds to temperatures outside its range is the subject of the following discussion.

Psychrophiles. Generally speaking, psychrophiles exhibit an optimum range of growth between 0° and 20° C. This is not to say that psychrophiles cannot grow above these temperatures. Some can, albeit more slowly than at psychrophilic temperatures. Psychrophiles, as one might expect, are found in great numbers in the Antarctic, where temperatures fall below 0°, but they are also found in cold soil, streams, ocean water below the thermocline, rivers, and lake muds of temperate climates. Consequently, psychrophiles can play a role in biodegradation in cold-water lakes and in refrigerated foods.

One of the most interesting questions confronting scientists concerns the physiological

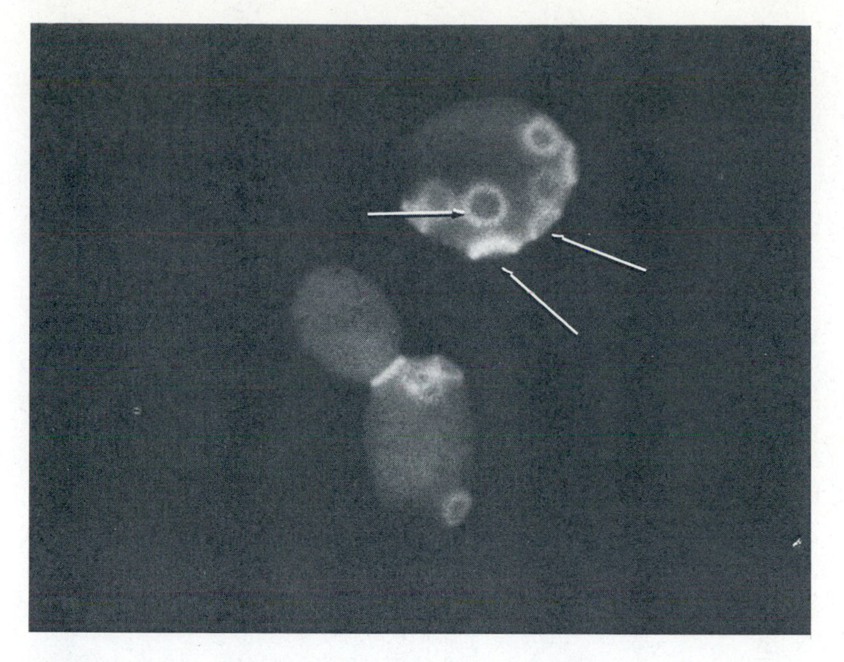

Figure 15-15 Bud scars *(arrows)* on yeast *Saccharomyces cerevisiae.*

With permission from Beran, K., and Streibloua, E. Adv. Microb. Physiol. ***2****:143, 1968. Copyright Academic Press, Inc., (London). Ltd.*

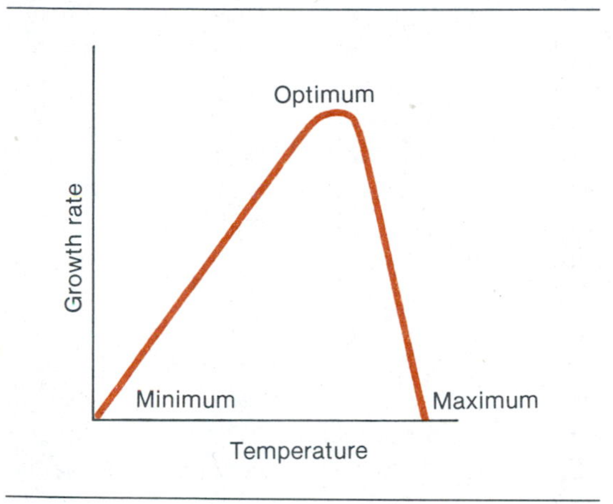

Figure 15-16 Effect of temperature on growth rate.

* *Optimum* refers to the temperature at which the growth is maximal or enzymes work the fastest.

bases of psychrophilic growth. Experiments have demonstrated that temperature may affect a wide variety of different microbial processes. The ribosomes in some psychrophiles become more unstable as the temperature increases above 20° C, and this results in a reduction in the rate of protein synthesis. The ribosomes of some psychrophiles possess a protein factor that enables protein synthesis to occur at 0° C. Removal of the factor causes protein synthesis at 0° C to cease, but at higher temperatures (25° C) protein synthesis is not affected. Many organisms appear to protect themselves from colder temperatures by increasing the amount of unsaturated fatty acids in membranes as the temperature decreases and approaches 0° C. Comparison of mesophilic and psychrophilic strains of the genus *Bacillus* has revealed that the saturated fatty acid content is considerably higher in the psychrophilic strains than in the mesophilic strains. Elevated temperatures, such as those in the mesophilic range, affect the cell membrane of psychrophiles and cause the leakage of intracellular contents, which may lead to lysis of the cell. The inability of psychrophiles to grow at elevated temperatures may also be due to heat inactivation of certain enzymes. There are many key enzymes in metabolism, and inactivation of any one of them could inhibit growth—for example, an enzyme in glycolysis or an enzyme required in protein synthesis. The increase in temperature above the psychrophilic range is not always lethal, and a return to psychrophilic temperatures restores the growth of the microorganism.

Thermophiles. The temperature span for optimum growth of different thermophiles is 40° to 80° C, but few thermophiles can live at temperatures of 90° to 100° C. Thermophiles are found among many groups of microorganisms, such as the cyanobacteria, the photosynthetic bacteria, gram-positive and gram-negative bacilli, and the protozoa. Most species are found in hot springs in many areas of the world. Thermophiles have ribosomes, enzymes, enzyme systems, metabolic pathways, and regulatory devices that are functionally the same as those found in mesophiles. The ability of thermophiles (but not mesophiles) to survive at elevated temperatures is a characteristic that has not been fully explained. Many theories have been proposed, but it may be that the molecular mechanisms vary from one group to another.

One mechanism that has received considerable support is that macromolecules are inherently stabilized. Stability could be the result of the substitution of polar amino acids for nonpolar ones, resulting in an increase in ionic and hydrogen bonding, particularly in the interior of the molecule. This would alter the conformational properties of the macromolecule and make it resistant to denaturation by heat. A comparison of the lactic dehydrogenases of thermophilic and mesophilic species of bacteria has revealed that this type of change does take place. Stability could also be obtained by a decrease in size of macromolecules.

A second mechanism for stability may be the result of biochemical modification. The transfer RNA of an extreme thermophile (*Thermus thermophilus*) shows modifications not evident in moderately thermophilic microorganisms. The modification involves the replacement of ribothymidine by 5-methyl 2-thiouridine, which apparently stabilizes the molecule at high temperatures. The replacement reaction is activated when the organism is grown at higher temperatures, and the mole percent of the replacement component increases with the increase in temperature.

A third mechanism that may be used for stability involves the interaction of macromolecules with cofactors such as divalent cations (Ca^{++}, Mg^{++}, etc.) and polyamines. Thermophiles contain several types of polyamines, some of which are unique. They appear to stabilize the ribosome-tRNA-mRNA complex and ensure its activity. The complex is inactive at higher temperatures in mutants that are unable to synthesize polyamines. Most of the polyamines are related to spermidine, which is found abundantly in nature and has the structure of:

```
 H    H   H   H    H    H   H   H   H    H
 |    |   |   |    |    |   |   |   |    |
H—N⁺—C—C—C—N⁺—C—C—C—C—N⁺—H
 |    |   |   |    |    |   |   |   |    |
 H    H   H   H    H    H   H   H   H    H
```

The cell membrane of microorganisms also changes with variations in temperature, and these changes are believed to play a minor role in cellular stability. The melting point of the membrane fatty acids also increases with increases in temperature; thus the fatty acids synthesized at elevated temperatures become more branched, longer, and more saturated.

It has been proposed that most thermophilic groups of microorganisms evolved from mesophily to thermophily, since the prokaryotes are believed to have evolved in a sea that was cool to warm but not hot. The evolution to thermophily may have been one of selective advantage in order to escape predation and eliminate other competitors for nutrients. It has been assumed that this evolutionary process occurred in a stepwise manner through a series of single mutations, with each mutation resulting in an increase in heat stability for only a single protein. Others have suggested, however, that the potentially thermostable components of the cell are controlled by the products of a few genes. The genes could conceivably be plasmid associated and easily transferable to other related species.

Resistance to heat is a property associated with the endospores of such bacteria as the clostridia and is a constant problem in home canning because of the organism's ability to release potent toxins (*Clostridium botulinum* toxin). Some spore-forming clostridia are also thermophilic, and the vegetative cells can grow at high temperatures. Thermophilic clostridia can ferment a wide variety of organic substrates, including the high molecular weight cellulose. They break down cellulose into a number of products, including acetic acid, ethanol, hydrogen gas, carbon dioxide, lactic acid, and butyric acid. These products, particularly the organic acids and ethanol, are presently produced from petroleum and natural gas. The clostridia have been considered for production of these chemicals to conserve petroleum. Thermophilic clostridia that ferment cellulose can be grown in mixed cultures with methane-producing bacteria. The products of clostridial fermentation (acetic acid and hydrogen gas) are used by the methane-producing bacteria to generate methane gas. Methane gas is also a source of fuel for certain industrial operations:

$$\underset{\text{Acetic acid}}{CH_3COOH} + \underset{\text{Hydrogen gas}}{2H_2} \longrightarrow \underset{\text{Methane}}{2CH_4}$$

Mesophiles. Mesophilic microorganisms make up the majority of the species known to humans. Some species are indigenous as well as pathogenic to humans and other warm-blooded animals. This explains why indigenous microflora in and on the surface of humans grows best at 37° C (body temperature). The optimum temperature for growth of mesophiles is between 25° and 40° C, but they have minimum and maximum temperatures of 10° and 50° C, respectively. The ability of mesophiles to survive at cold as well as hot temperatures appears to be related to the composition of the fatty acids in the cell membrane. When the membrane fatty acids of mesophiles are compared at psychrophilic and thermophilic temperatures, there is observed to be an increase in the melting point of the fatty acids as the temperature increases. Fatty acid changes in the membrane alter the viscosity, which in turn affects nutrient transport and ultimately growth.

Temperature and mutation. Mutations in the microbial genome can produce temperature-sensitive mutants unable to grow at temperatures that normally support growth, such as 37° C, but able to grow at lower temperatures, such as 25° C. Temperature-sensitive mutants are the result of genetic alterations that affect enzyme structure at the tertiary level. In other words, at the nonpermissive temperature the configuration of the enzyme is distorted and normal tertiary bonding is absent, whereas at a lower, permissive temperature the tertiary structure is partially or completely restored. Temperature-sensitive mutants have been important tools for the researcher wishing to study cellular processes. Mutations involving certain cellular activities can be assayed because the mutant can express itself at one temperature but not at another. Colonies of cells can be detected at the permissive temperature and then switched to the nonpermissive temperature to assay for the defect.

Temperature-sensitive mutants of a large number of viruses have also been studied in animal models in order to characterize the biochemical, replicative, and genetic events that are characteristic of the virus-infection process. Temperature-sensitive mutants of virulent viruses, such as the influenza virus, cytomegalovirus, and respiratory syncytial virus, are being evaluated for use as vaccines. The temperature-sensitive viruses are not able to replicate at the nonpermissive temperature of specific host tissue (for example, the lower respiratory tract) but can still stimulate the formation of antibodies. One of the major difficulties with the use of temperature-sensitive mutants in vaccines is their potential for reversion to the wild type and the establishment of a persistent and possibly fatal infection.

Temperature has been used as a technique to cure microorganisms of plasmid activity. A microorganism that is virulent because of plasmids that control the synthesis of toxins can be made avirulent by subculturing at higher temperatures. In some instances the avirulent species appears to carry a lower molecular weight plasmid than the virulent species, which suggests that a deletion has taken place.

Osmotic pressure

Microorganisms are often in environments in which the solute concentration is lower (hypotonic) than the cell cytoplasm, but the cell does not lyse because of the nature of the rigid cell

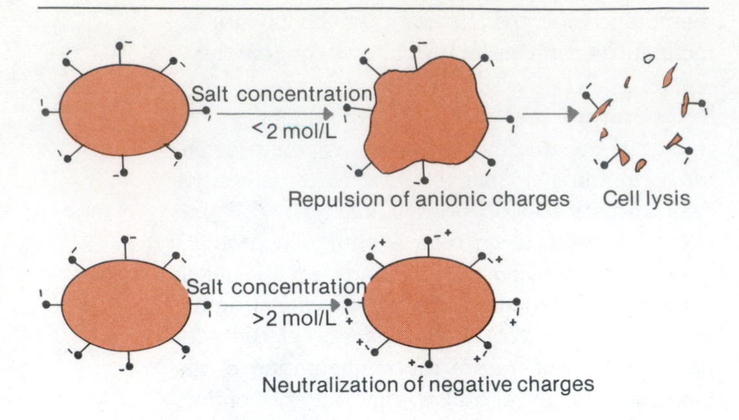

Figure 15-17 Effect of high and low salt concentration on halophiles.

wall. (See Chapter 7 for a discussion of osmosis.) There are a few ecological sites, however, in which the environment outside the cell is extremely hypertonic, such as the Dead Sea and inland lakes where evaporation has created a saline environment. The organisms living in these areas not only have evolved mechanisms that enable them to tolerate a saline environment and prevent dehydration, but some are unable to survive unless the concentration of salt is between 3 and 6 mol/L. These organisms are called ***obligate*** or ***extreme halophiles*** and belong to the genera of bacteria called *Halobacterium* and *Halococcus.* The structure of the cytoplasmic membrane is one factor that enables these organisms to exist in an environment that will not support any other cell type (Figure 15-17). The lipids of halophilic bacteria are highly acidic (negatively charged), and when the organisms are placed in a medium in which the salt concentration is below 2 mol/L, the outer portion of the envelope disintegrates. Presumably, there is a repulsion between the anions of the lipid, which makes the envelope unstable, but at elevated salt concentrations this ef fect is neutralized by the abundant cations.

Proteins are also affected by high concentrations of salt. The nonpolar hydrophobic groups that arrange themselves on the inside of protein molecules are responsible for the conformational properties of proteins and hence their activity. Hydrophobic bonds are weak, and high concentrations of salt increase the strength of these hydrophobic interactions, making the molecule more rigid and less susceptible to unfolding and denaturation. The ability of the halophiles to survive and prevent dehydration, an osmotic problem, is related to their ability to concentrate certain solutes, for example, polyhydric alcohols such as glycerol by xerotolerant (sugar-tolerant) yeasts and potassium by salt-tolerant *Halobacterium* and *Halococcus.* In the case of *Halobacterium* and *Halococcus* in vitro experiments have shown that their enzymes have optimum activity when the sodium chloride concentration is between 0.5 and 1.0 mol/L and that concentrations between 2 and 6 mol/L are inhibitory. Potassium, however, at concentrations between 2 and 6 mol/L is a relatively poor inhibitor of enzyme activity; thus halophilic enzymes can maintain maximum activity at high intracellular levels of salt by concentrating potassium.

Halophiles have generated considerable interest and research in the National Aeronautics and Space Administration (NASA). The search for extraterrestrial life on planets where there is low water activity makes halophilic microorganisms of particular interest.

One of the reasons for mixing high concentrations of salt or sugar with foods is to prevent growth of microorganisms through dehydration. The addition of 12% sodium chloride is a favorite technique for reducing microbial growth, but this technique has no effect on certain halophiles that can spoil many types of dried foods. Halophiles have been implicated in the spoilage of brines used to pickle olives (see Chapter 20).

Hydrostatic pressure

The pressure exerted by a column of water on the bottom of the column is referred to as ***hydrostatic pressure.*** Hydrostatic pressure is a physical characteristic that affects all life but especially life in the ocean's depths. The ocean has an average depth of 3,500 m, but in areas such as the bottom of the Mariana Trench, located near the Philippines, the depth is approximately 10,800 m. Pressure increases 1 atm for every 10 m, and at the bottom of the Mariana Trench there is 1160 atm of pressure. The study of effects of hydrostatic pressure on organisms is called ***barobiology.*** Microorganisms that tolerate elevated hydrostatic pressures are called ***barotolerant,*** and those that grow better at elevated hydrostatic pressures than at normal atmospheric pressure are called ***barophilic.***

Microorganisms show a wide range of sensitivity to pressure; the growth of some microorganisms is inhibited by 100 atm of pressure, whereas others are able to grow under pressures of 1000 atm. Those microorganisms possessing gas vacuoles are extremely sensitive to changes in hydrostatic pressure because an increase in only a few atmospheres will cause the vacuoles to collapse. Most microorganisms do not contain gas vacuoles, and increases in hydrostatic pressure affect them in different ways. In prokaryotic cells increased hydrostatic pressures affect biochemical processes, whereas in eukaryotic cells microtubular structures are disrupted. Most of the microorganisms isolated from the ocean's depths grow better at ambient pressures than in their normal habitat and are referred to as ***facultative barophiles.*** Only recently has an obligate barophilic bacterium been isolated from the ocean. Yayanos and others in 1981 isolated an obligate barophilic bacterium from the Mariana Trench. This bacterium, which was recovered from a decomposing crustacean, is unable to grow at 2° C at pressures less than 400 atm but can grow at pressures as high as 1035 atm. Growth of the obligate barophile is extremely slow (30-hour generation time), and this is due in part to the extremely low temperature (approximately 1° to 2° C) that it encounters in its environment.

Microorganisms that are normally found at ambient temperatures have a wide range of response to increases in pressure in accordance with their physiological state. Sensitivity to atmospheric pressures has been studied in the bacterium *Streptococcus faecalis. S. faecalis* is aciduric and can tolerate high concentrations of acid produced during growth. Acid toleration is a result of the cell's ability to concentrate potassium ions during the transport of protons out of the cell through the cytoplasmic membrane—ATPase. Protons released into the medium are therefore prevented from reentry into the cell. When *S. faecalis* is grown at high hydrostatic pressure, ATPase is severely affected, and growth declines precipitously because of the reentry of protons into the cytoplasm of the cell. The growth of many other bacterial species also appears to be inhibited by high pressures due to the accumulation of acids in the cell. *S. faecalis* barotolerance can be shown to be associated with the type of carbohydrate catabolized. When pyruvate is the carbohydrate source, the organism can tolerate only 200 atm of pressure. When ribose is the carbohydrate source, the organism can tolerate 450 atm of pressure, and when the carbohydrate is a hexose such as glucose or galactose, the organism can grow at hydrostatic pressures of 550 atm. Neutralization of acid by various ions is important for barotolerance; for example, ammonia production from the hydrolysis of arginine during glycolysis or the presence of calcium or magnesium in the growth medium enables the microorganism to grow at elevated hydrostatic pressures. It also appears that some microorganisms may reduce their size to become more resistant to pressure. A psychrophilic bacterium isolated from the Antarctic in 1976 responded to starvation by decreasing its cell size and in this state was also found to be more resistant to increases in pressure.

Hydrogen ion concentration (pH)

When cultivated in closed systems, such as a laboratory flask, microorganisms excrete products of metabolism, such as acids, into the surrounding medium. Eventually the pH of the medium drops, and growth is inhibited because enzyme activity is affected by the hydrogen ion concentration. Buffer systems are normally added to a culture medium to prevent rapid drops in pH. One of the most effective buffering systems uses a mixture of K_2HPO_4 and KH_2PO_4, which, when added to nutrient media, enables one to produce a large population of microorganisms in a relatively short period of time.* Most eukaryotic and prokaryotic microorganisms carry out metabolic activities at pHs between 5.0 and 8.0, the most effective range for phosphate-buffering capacity. There are environments outside the laboratory in which the pH is as low as 2.0 and where microorganisms have not only adjusted to such conditions but actually require them for survival. These organisms disintegrate if the pH increases to 5.0.

Two of the more thoroughly studied acid-loving ***(acidophilic)*** bacteria—*Sulfolobus acidocaldarius* and *Thermoplasma acidophilum*—also exist as thermophiles. *S. acidocaldarius* is found throughout the world in thermal acidic soils and acidic hot springs, where it oxidizes elemental sulfur to sulfuric acid, creating an environment in which the pH is less than 3.0. The cell wall of this species is unique in that no peptidoglycan is present; however, the cell membrane is covered with a regular polyhexagonal array of protein subunits that provide stability. *T. Acidophilum* belongs to the mycoplasmas, which do not possess cell walls. Its habitat is restricted to hot acidic environments associated with heating coal refuse piles, where optimum growth is obtained at pH 2.0 and 59° C. The internal pH in both bacteria is approximately 6.0. The mechanism of survival when the external pH is 3.0 or below has been a subject for considerable speculation. Again, the cell membrane appears to play a vital role. Both acidophilic species possess membrane lipids that contribute to thermophilic survival at high temperatures. The membrane lipids possess ether linkages instead of the usual ester linkages seen in mesophilic fatty acids. Some diethers consist of two glycerol moieties joined together by two

* Complex media at pHs between 6.0 and 8.0 have some buffering capacity because of the presence of proteins and amino acids.

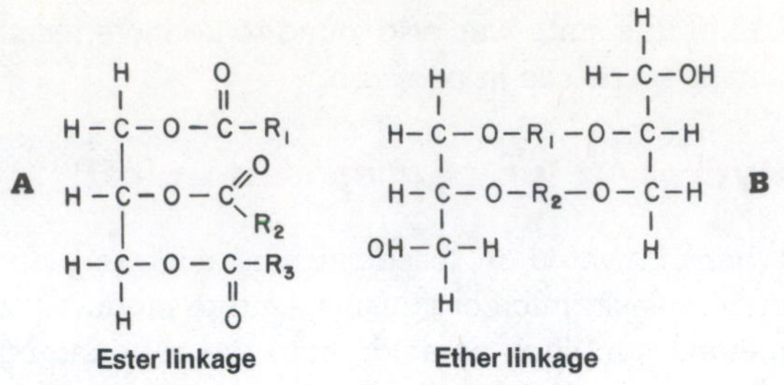

A Ester linkage

Ether linkage B

Figure 15-18 Lipid structure in mesophiles and thermophiles. **A,** Ester linkage in mesophiles. **B,** Ether linkage in thermophiles. *R* represents fatty acid chains.

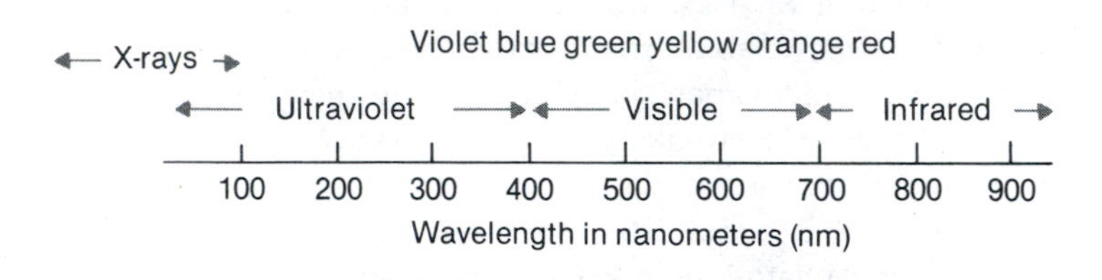

Figure 15-19 Spectrum of ultraviolet, visible, and infrared regions.

long-chain alkyl groups (Figure 15-18). These ether linkages are believed to increase the stability of the membrane in acidic environments. An additional characteristic of *T. acidophilum* has been examined that appears to enable this organism to survive acidic pH. The proteins of the membrane are similar to those of mesophiles except that in *T. acidophilum* the number of free carboxyl groups and amino groups is reduced by one half. This is believed to improve stabilization of the membrane over wider pH ranges. If the mesophilic content of carboxyl groups were present in the acidophile, then ionization at low pHs would occur, resulting in charge repulsion on the membrane and subsequent disintegration.

There is a third type of thermoacidophile, called *Bacillus acidocaldarius,* which differs from the previously mentioned types. This organism is a typical prokaryote having a peptidoglycan-containing cell wall, but it does not contain the tetraether-type fatty acids. It does possess a high content of cyclic fatty acids, which are believed to create a tighter and more stable membrane.

Electromagnetic radiation

Electromagnetic radiation is the form in which energy can be transferred from one place to another through space. It is transmitted with the speed of light through space as ***photons*** or particles of radiation. There is a spectrum of electromagnetic radiation that is characterized by differences in wavelength (Figure 15-19); the shorter the wavelength, the greater the energy transferred. The visible portion of the spectrum, which we call light, is required for the growth of photosynthetic organisms because it can be converted to a chemical form of energy in the cell (see Chapter 13). Other radiations that are invisible, including x-rays, gamma rays, and ultraviolet light, are harmful to nearly all forms of microbial life (see Chapter 9).

The photosynthetic microorganisms requiring light energy include cyanobacteria and purple and green bacteria. Each possesses pigments, such as chlorophylls, carotenoids, and phycobiliproteins, that absorb light at various wavelengths. These pigments transfer their energy to chlorophyll molecules. The chlorophyll in its excited state (electron at a higher energy level) can react with unactivated oxygen ($^3O_2$), converting it to a very reactive singlet oxygen ($^1O_2$) that is capable of photooxidation. The photooxidation of cellular components is lethal to the cell, but phototrophic microorganisms that are anaerobic are not confronted with this problem. Phototrophic microorganisms that do require oxygen can be protected from photooxidation by carotenoid pigments (see Figure 13-22) possessing a large number of conjugated double bonds that act as quenching agents of excited oxygen and hence remove it from solution. Sometimes the carotenoids are overwhelmed and unable to protect the cell. Cyanobacteria, which are aerobes, form blooms on lakes and are also protected by carotenoids. They can form dense populations under certain conditions at the surface, but in bright sunlight oxygen becomes overabundant and saturates the water, resulting in photooxidative processes and destruction of the bacterial population. Oxygen toxicity from this source is found not only in organisms that use light as an energy source but also in those that can use only preformed chemicals as energy sources, such as the chemotrophs discussed in Chapter 13.

SUMMARY

Because of its size, a single prokaryotic cell cannot be properly evaluated, but if a cell culture is synchronized so that all of the cells are in the same stage of development, the growth data can be extrapolated to single cells. Synchronization of cultures can be obtained by manipulating environmental factors such as temperature, light, and nutrition or by using selection techniques. Two selection techniques include filtration and density centrifugation, in which cell size is the parameter used to separate synchronized from nonsynchronized cells.

The bacterial cell cycle, which includes growth, septum formation, and cell division, is controlled by certain factors, only two of which have been amenable to investigation: DNA replication and cell envelope synthesis. DNA synthesis continues throughout the entire cell cycle and is coordinated with cell envelope synthesis and septum formation. This coordination terminates with each cell receiving a complete molecule of DNA.

The growth of a population of cells is easily measured in the laboratory using either ***batch*** or ***continuous*** cultures. In batch cultures the bacterial population goes through a number of different growth phases, which include the lag, log, stationary, and death phases. Continuous cultures are prepared in devices such as the chemostat, in which the bacterial population is maintained at a particular density and in a state of logarithmic growth.

The actual measurements of growth, that is, the number of cells present or the rate of growth of a culture sample, can be determined by employing a number of techniques, which include measurements involving direct microscopic count using special engraved slides, colony count, electrical resistance of cells, cell density, and cell mass. The most frequently used technique measures cell density by employing a spectrophotometric device that measures the light scattered by a bacterial population.

The growth of eukaryotic cells is more complex than that of prokaryotes. In addition, DNA synthesis in eukaryotes occupies only a part of the cell cycle and not the entire cycle as in prokaryotes. The cell cycle is divided into four phases: G_1, S, G_2, and M, which vary in time from species to species. The eukaryotic cell cycle has been studied extensively in only a few species, such as the yeast *Saccharomyces cirevesiae.* In this species there occurs the following: (1) the G_1 period, in which there is microtubule formation as well as the initiation of DNA synthesis and the formation of a bud; (2) the S period, in which the complete mitotic spindle is formed; (3) the G_2 period, in which nuclear division and spindle elongation occur and the nucleus migrates into the neck of the bud; and (4) the M period, in which there is completion of nuclear division and separation of the daughter cells.

The growth of microorganisms is affected favorably or adversely by environmental factors. One of the most widely studied factors is temperature. Microorganisms, because of their wide variations in temperature requirements, have been divided into three groups: psychrophiles, mesophiles, and thermophiles. This temperature variation is related to the chemical composition as well as the molecular architecture of the bacterial cell. Other factors that influence the cell are pH, osmotic pressure, hydrostatic pressure, and electromagnetic radiation.

Study questions

1. Explain the ways in which a cell culture can be synchronized to divide at approximately the same time.
2. What type of experiment demonstrates that DNA replication is involved in the cell division process?
3. What sequence of events occurs in the prokaryotic cell that ultimately leads to cell division?
4. What is the role of the cytoplasmic membrane in DNA replication and cell division?
5. What is meant by the "unit cell" concept in cell division?
6. Graph a typical growth curve, labeling each of the stages. Define each stage of growth.
7. How could one mathematically express the log phase of growth?
8. What is the state of the ribosomes in the lag, log, and stationary phases of growth?
9. What is the distinction between a closed and open growth system?
10. What is the function of a chemostat? Does it serve any useful purpose for the research microbiologist?
11. What do you think is the easiest and most reliable method for determining the rate of bacterial growth?
12. List all the methods that are used to measure growth in the laboratory.
13. What is the difference between turbidometric and nephelometric devices?
14. What property distinguishes the eukaryotic cell cycle from the prokaryotic cell cycle?
15. What is the relationship of the S and M stages in the eukaryotic cell cycle?
16. Describe the growth of the cell wall in the hyphae of fungal cells. Compare it with the growth of a bacterial cell wall.
17. Name the three groups of microorganisms based on their optimum temperature for growth. What are the temperature ranges for each group? Can one group survive in the temperature ranges of the other groups?
18. What physiological and/or structural changes occur in the bacterial cell that enable it to survive over a wide range of temperatures?
19. Is spore formation a response to sudden changes in temperature? Are the vegetative cells of all spore formers resistant to high temperatures?
20. How do halobacteria survive in high salt (sodium chloride) concentrations when their enzyme systems are so sensitive to sodium?
21. What characterisitcs of halobacteria make them more closely related to eukaryotes than to prokaryotes?
22. What is so unusual about the acidophilic bacteria? What structural or physiological changes enable them to survive at extremely low pHs?
23. What molecules in the cell are used to neutralize oxygen that has been excited by chlorophyll during the process of photosynthesis? How is this related to algal blooms on lakes?

Selected readings

Books

Brock, T.D. *Thermophilic microorganisms and life at high temperatures.* Springer Verlag, New York, Inc., New York, 1978.

Dawes, I.W., and Sutherland, I.W. *Microbial physiology.* Halsted Press, New York, 1976.

Doelle, H.W. *Bacterial metabolism* (2nd ed.). Academic Press, Inc., New York, 1975.

Kushner, D. (Ed.). *Microbial life in extreme environments.* Academic Press, Inc., New York, 1978.

Mandelstam, J., and McQuillen, K. (Eds.). *Biochemistry of bacterial growth* (2nd ed.). John Wiley & Sons, Inc., New York, 1973.

Moat, A.G. *Microbial physiology.* John Wiley & Sons, Inc., New York, 1979.

Shilo, M. (Ed.). *Strategies of microbial life in extreme environments.* Verlag Chemie International, Inc., New York, 1979.

Tempest, D.W. The continuous cultivation of microorganisms, Part 1, Theory of the chemostat. In Norris, J.R., and Ribbons, D.W. (Ed.). *Methods in microbiology* (Vol 2). Academic Press, Inc., New York, 1976.

Journal articles

Bayley, S.T., and Morton, R.A., Recent developments on the molecular biology of extremely halophilic bacteria. *CRC Crit. Rev. Microbiol.* **6:**151, 1978.

Brown, A.D. Microbial water stress. *Microbiol. Rev.* **40:**803, 1976.

Cole, R.M. Symposium on the fine structure and replication of bacteria and their parts. III. Bacterial cell-wall replication followed by immunofluorescence. *Microbiol. Rev.* **29:**326, 1963.

Donachie, W.D., Jones, N.C., and Teather, R. The bacterial cell cycle. *Symp. Soc. Gen. Microbiol.* **23:**9, 1973.

Higgins, M.L., and Shockman, G.R. Prokaryotic cell division with respect to wall and membranes. *CRC Crit. Rev. Microbiol.* **1:**29, 1971.

Marquis, R.E. Microbial barobiology, *Bioscience* **32**(4):267, 1982.

Matsushita, T., and Kubitschek, H.E. DNA replication in bacteria. *Adv. Microb. Physiol.* **12:**247, 1973.

Morita, R.Y. Psychrophiles bacteria. *Microbiol. Rev.* **39:**144, 1975.

Rogers, H.J. Bacterial growth and the cell envelope. *Microbiol. Rev.* **36:**347, 1972.

Slater, M., and Schaechter, M. Control of cell division in bacteria. *Microbiol. Rev.* **38:**199, 1974.

Yayanos, A.A., Dietz, A.S., and van Boxtel, R. Obligately barophilic bacterium from the Mariana Trench. *Proc. Natl. Acad. Sci. USA* **78:**5212, 1981.

Chapter 16 VIRAL REPLICATION AND TUMOR DEVELOPMENT

Viral replication occurs only in hosts in which metabolites and metabolic machinery can be subverted for viral needs. There is a wide range of organisms infected by viruses, but the most widely studied hosts are the bacteria. This chapter is devoted to a discussion of the scheme of replication of the bacterial virus (bacteriophage) and the barriers it encounters during infection of the cell. Unless otherwise stated, the T-even series of bacteriophage, specifically T_4 (*T* stands for type), is the principal bacterial virus discussed. (See Chapter 6 for a discussion of T_4 structure.) This is followed by discussions of replication of animal and plant viruses and, finally, of the role of viruses in cancer and tumor development.

INFECTION BY BACTERIAL VIRUSES (BACTERIOPHAGES)

At first glance the reproduction of a virus within a host might appear to be a relatively simple task, since the virus is composed of only a core of nucleic acid surrounded by a protein coat (plus a lipid envelope in some animal viruses). There are several barriers that the virus must overcome before it can faithfully reproduce itself in the host cytoplasm. The virus must first attach to the surface of the bacterial cell, and this requires some receptor specificity. Second, once the viral nucleic acid is present in the cytoplasm of the cell, it must resist host defense mechanisms in order to replicate. Third, viral protein synthesis and viral maturation must compete with host metabolic activities.

Adsorption to the bacterial cell

Bacterial viruses, like other viruses, show a high degree of specificity in terms of the bacterial species that they can infect. The interaction between specific adsorption sites on the virus and on the host cell is genetically determined, since mutations in the bacterial or viral genome, which alter surface structures, greatly influence adsorption. The T series of bacteriophages that infect gram-negative bacteria interact with bacterial receptor sites that may include (1) the lipopolysaccharide of the outer membrane of the cell wall, (2) the lipoprotein of the outer membrane of the cell wall, and (3) appendages such as pili and flagella. The adsorption to pili occurs most frequently among the RNA viruses and the filamentous DNA viruses. Adsorption to pili occurs only if the bacterial cell is "male," and if the pili are mechanically removed, no phage adsorb. The viral receptor sites also vary from one group of viruses to another. The T series possess ***tail fibers*** (Figure 16-1) that aid in attachment, but not all bacterial viruses possess such appendages. T_4 is believed to bind to a lipopolysaccharide receptor (such as a sugar residue of the O antigen) by a single tail fiber. Once the virus is attached to the bacterial surface, interaction between other tail fibers and other lipopolysaccharide receptor sites prevent it from leaving the bacterial surface. Irreversible anchorage occurs by interaction of baseplate pins and the bacterial surface. The specific site-interaction between host and virus in some systems is influenced by certain cations. The addition of specific ions, such as calcium, promotes adsorption, and this may be related to the surfaces of

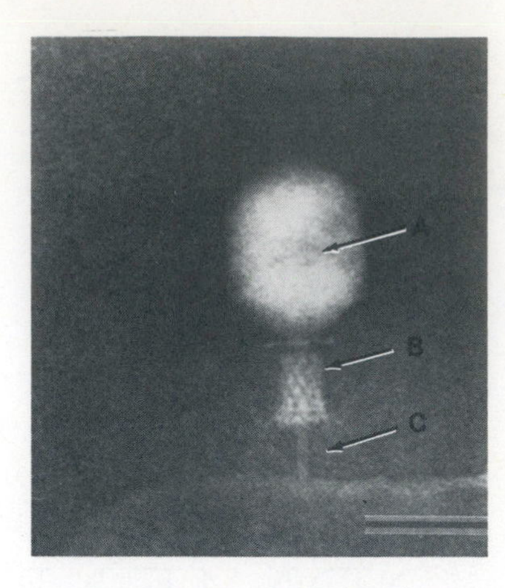

Figure 16-1 Electron micrograph demonstrating attachment of T_4 bacteriophage to cell wall. *A.* Head; *B,* sheath; *C,* tail fibers.
From Mizushima, S. J. Bacteriol. ***140**:1071, 1979.*

the virus and host, which are naturally electronegative and tend to repel one another. The specificity between host cell wall receptors and viral surface receptors is best demonstrated by the fact that the nucleic acid of some viruses is capable of replication in the cytoplasm of a wide range of different host cells, whereas these same viruses with intact protein coats and/or envelopes can infect only one cell type.

Penetration of the cell wall

One of the major differences between bacterial and animal virus infection is that only the nucleic acid of bacterial viruses enters the cytoplasm of the cell. Once the baseplate pins are attached to the bacterial surface, there is contraction of the tail sheath. This forces the phage head and collar to be pulled closer to the bacterial surface and also causes a portion of the tail to be forced through the cell wall. DNA is released by the capsid and injected through a membrane pore into the cytoplasm (Figure 16-2). The cytoplasmic membrane becomes temporarily leaky following penetration, and if many viral particles were to adsorb and penetrate the cell wall, the exposed membrane would disintegrate, resulting in lysis of the cell. The latter process is called "lysis from without"; that is, the cells lyse before the virus has had a chance to replicate in the bacterial cytoplasm and cause "lysis from within."

Viral growth and maturation

Protection from intracellular enzymes. The viral nucleic acid is potentially susceptible to hydrolytic enzymes (nucleases) within the cell. Bacterial cells and other organisms protect themselves from invasion by foreign DNA by using ***endonucleases*** that cleave specific internal areas of the foreign DNA molecule. There are also bacterial enzymes, called ***modifying enzymes,*** that methylate the viral DNA, thus rendering it resistant to endonuclease activity. The DNA sites to which the endonuclease and methylase are specific varies from species to species of bacteria. Their enzyme specificities are discussed in Chapter 11. Sometimes the nucleic acid of the invading virus already possesses unusual nucleotides such as hydroxymethylcytosine or glucosylated hydroxymethylcytosine, which confer some resistance to cell-mediated nucleases.

Breakdown or modification of host metabolism. Replication of the viral genome and synthesis of viral proteins require the energy-generating systems of the host. The bacterial ribosomes as well as many of the bacterial enzymes required for protein and nucleic acid synthesis are also used by the virus. Energy-producing systems in the bacterial cell are not significantly altered following viral infection, but the virus will either shut down or modify other host metabolic systems. Some viruses, for example, increase the activity of already-existing enzymes to ensure maximum activity during replication. There are several potential sites for shutting down or modifying host metabolism. Degradation of host DNA is one mechanism used to prevent bacterial DNA synthesis, but this activity is probably more important in supplying nucleotides for viral nucleic acid synthesis in some bacterial virus infections. In other bacterial virus infections degraded DNA supplies only 20% of the nucleotide precursors. Shutdown of host DNA synthesis to prevent depletion of nucleotide precursors appears to be a more important mechanism than DNA degradation in many viral infections. Another mechanism for shutting down host DNA synthesis is nuclear disruption, in which nuclear areas (nucleoids), located centrally before infection, move toward the cell membrane after infection. Studies have shown that it takes much longer for host DNA synthesis to be shut off in mutants in which this morphological change does not take place.

The shutdown of host RNA and protein synthesis is believed to involve modification techniques. The transcription of host DNA to messenger RNA (mRNA) can be prevented by modification of the protein subunits on the RNA polymerase.* Such changes can prevent the RNA polymerase from recognizing host DNA promoter sites. It has been speculated that the translation of host mRNA can also be shut down by modification of host transfer RNA (tRNA) or by alteration of the ribosomes. Phage-induced alteration of ribosomes appears to also be a major mechanism of modification, even though phage-induced modification of tRNA has been observed in some infected bacterial cells. Protein-modified ribosomes are unable to bind to host mRNA, thus preventing the latter's translation.

Viral reproduction. The replication scheme by bacterial viruses is dependent on whether the nucleic acid is RNA or DNA and whether the nucleic acid is single or double stranded. The T-even phages, such as T_4, possess double-stranded DNA. T_4 DNA is a very long linear molecule that is replicated bidirectionally. One of the major differences between viral DNA synthesis and cellular DNA synthesis is that repeated initiations (there are several origins on the DNA) can take place on the viral chromosome but not on the cellular chromosome. Multiple initiations lead to the formation of a large number of DNA copies. Recombination between these DNA molecules results in the formation of molecules of a length greater than that found in the mature phage (Figure 16-3). These molecules are called ***concatemers*** (concatemer formation by the rolling-circle mechanism of replication is discussed in Chapter 9). T_4 concatemers are cut by an endonuclease, producing mature DNA of constant length and containing ***repetitious ends*** and ***permuted sequences*** (Figure 16-4). Permuted refers to the different nucleotide sequences at the ends of the different DNA molecules.

Single-stranded DNA is characteristic of the bacterial virus $\phi\chi$174. This virus uses the single strand as a template to produce a double-stranded molecule called the ***replicative*** form, but before new viral particles are assembled, one of the strands is discarded. The other types of replicative schemes used by viruses that have single- or double-stranded RNA are discussed later in this chapter under "Infection by Animal Viruses."

Regardless of the type of nucleic acid, viral replication is dependent on producing viral proteins from mRNA molecules, which are usually polycistronic. The viral protein synthesized in the host can be divided into early proteins or late proteins. Early proteins are synthesized during a period called the ***eclipse,*** in which no mature viral particles are produced (Figure 16-5). Some of the early viral proteins include enzymes used for viral nucleic acid production, such as DNA polymerase and enzymes that aid in the synthesis of nucleotides. Other viral gene products shut off host metabolism. Later during the eclipse period several hundred molecules of viral nucleic acid are synthesized. The transcription of T_4 into mRNA uses a modified host RNA polymerase. This is to be contrasted with T_7 bacteriophage, which initially uses host RNA polymerase for synthesis of

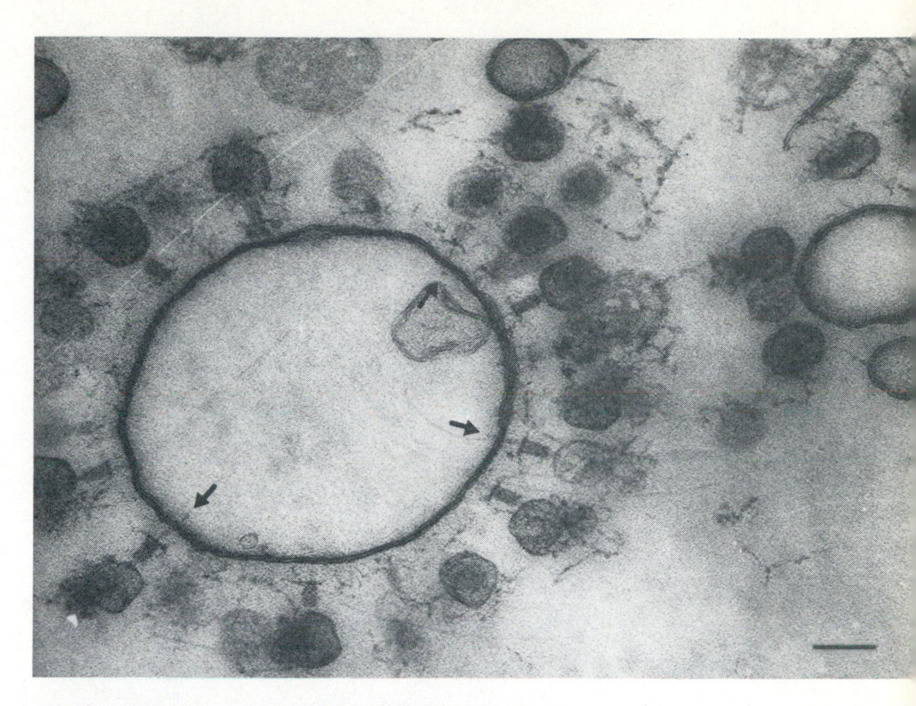

Figure 16-2 Electron micrograph demonstrating tail tube penetration of cell wall. Several T_4 bacteriophage particles can be seen to be adsorbed to reconstituted cell wall. Arrows point to tail tube of virus, which has punctured cell wall.

*From Mizushima, S. J. Bacteriol. **140:1071, 1979.***

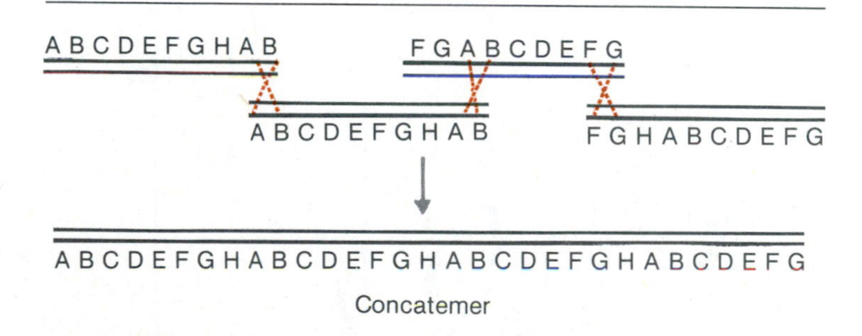

Figure 16-3 Concatemer formation by recombination.

* About 30 seconds after infection one of the α-subunits is modified (ADP ribosylation) by a T_4 protein, which is a component of mature phage. A second ADP ribosylation occurs about 2 minutes after infection and involves both α-subunits.

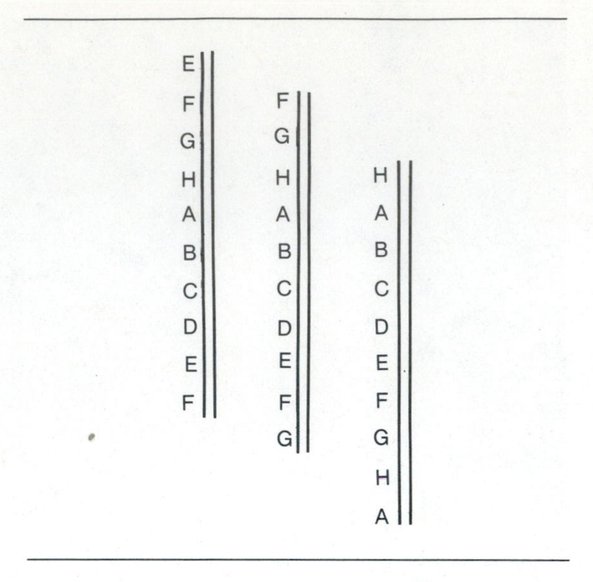

Figure 16-4 T_4 DNA molecules showing repetitious ends as well as different ends for each molecule.

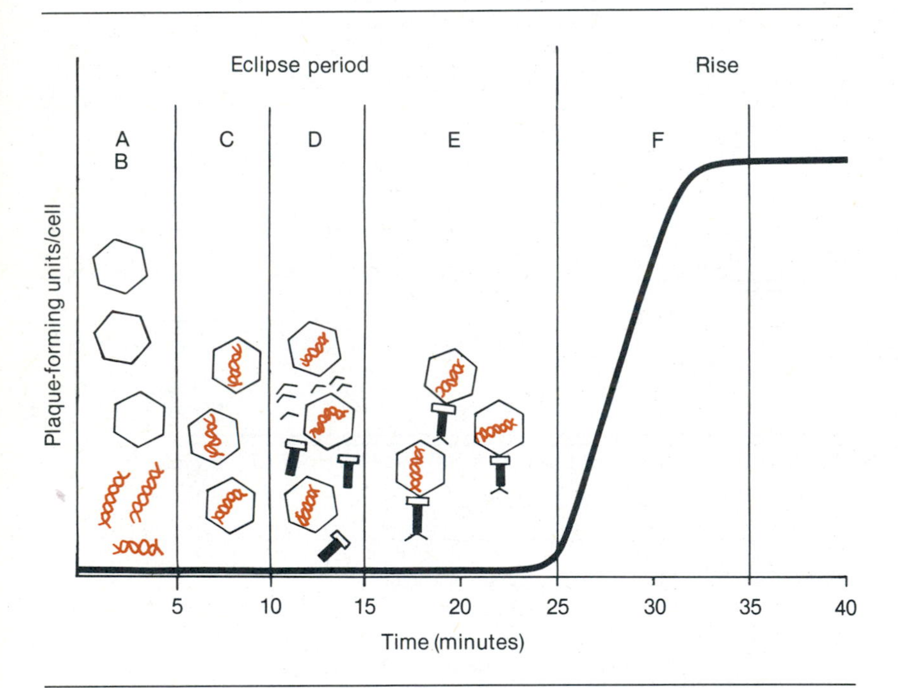

Figure 16-5 Growth curve of T_4 bacteriophage. Intervals *A* through *F* depict approximate time that various viral components are produced and assembled in infected bacterial cell. Mature viral particles are not assembled until 20 to 25 minutes following infection. Sudden rise in detectable viral particles is due to lysis of bacterial cell(s) and immediate release of all viral particles.

early proteins but then produces an mRNA that codes for a virus-specific RNA polymerase for synthesis of late proteins. Late viral proteins are also synthesized during the eclipse period and include the structural proteins of the phage, such as head and tail fibers, as well as the enzyme lysozyme.

The switch from synthesis of early proteins to late proteins in T_4 is related to a control mechanism involving a modification of host RNA polymerase. An early T_4 protein apparently binds to RNA polymerase, and as the concentration of this protein increases, the initiation of promoters in the early phage genes is prevented. The modified RNA polymerase molecules are now able to recognize the promoters of late genes. An interesting aspect of this control mechanism is that the late genes are transcribed on the opposite DNA strand, with RNA polymerase molecules moving in a direction opposite that for early mRNA synthesis.

Viral maturation and release from the cell. The first step in viral maturation is the condensation of phage DNA into a tightly coiled unit about which the head protein forms a shell. The packaging of DNA into the head also explains the occasional incorporation of fragments of bacterial DNA during the process of transduction by bacterial viruses (transduction is discussed in Chapter 11). The tail with its tail plate binds to the head protein, and, finally, the tail fibers attach to the tail plate. The eclipse period ends with assembly of the first mature viral particle (Figure 16-6). Most viral infections of bacteria result in lysis of the host cell. Newly synthesized viruses are released by the action of virally induced lysozyme on the bacterial cell wall. The eclipse period for bacterial viruses is short compared with that for animal viruses. Lysis of the bacterial cell often occurs between 30 and 60 minutes following infection and can be detected by the formation of areas of infection called viral ***plaques.*** The number of free bacteriophage particles in a suspension can be easily determined in the laboratory by using the following procedure:

1. Serial dilutions of the unknown phage suspension are made, for example, 10^2- to 10^8-fold dilutions.
2. An overnight culture of indicator bacteria (the bacteria to be infected) is prepared.
3. To a culture tube containing 2.5 ml of melted agar (called ***top agar***) is added a drop of the indicator bacteria. This is followed by the addition of 0.1 ml of a phage dilution, and the tube is agitated to distrib-

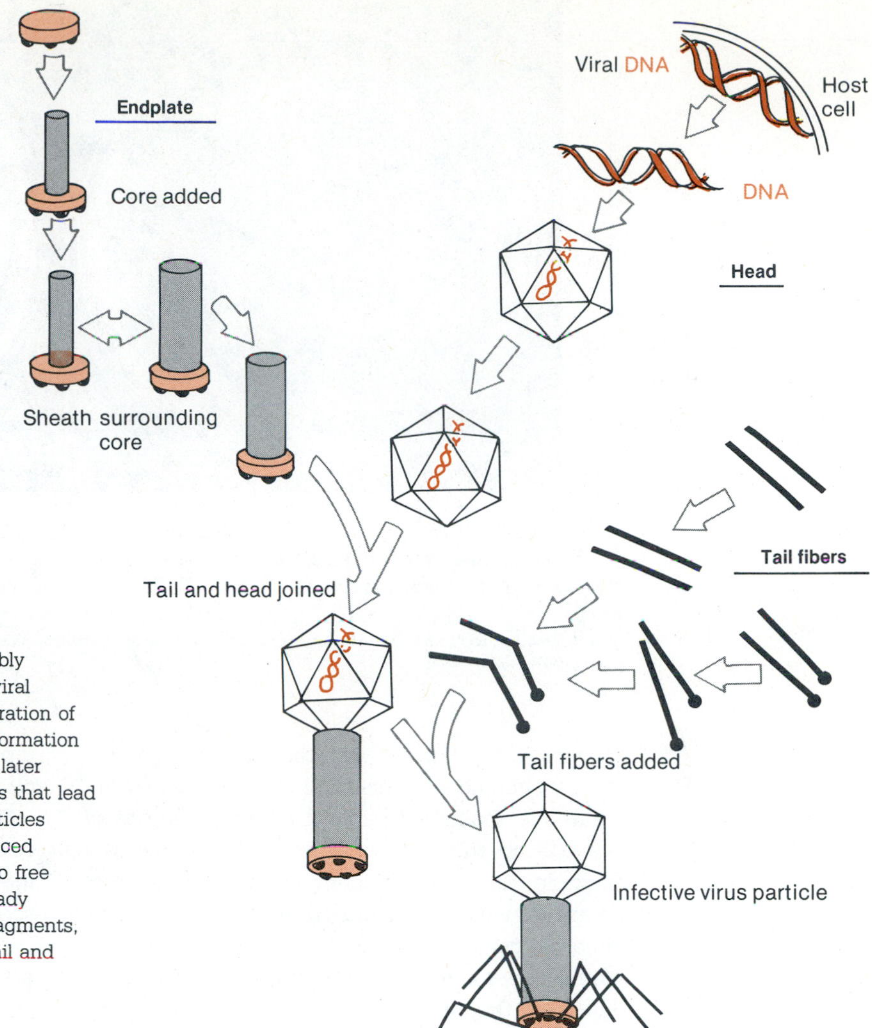

Figure 16-6 Abbreviated scheme showing assembly pathway of bacteriophage T_4. Formation of complete viral particle in T_4 bacteriophage is brought about by integration of three independent pathways. Each pathway leads to formation of structural unit: head, tail, and tail fibers, which are later joined to produce infectious particle. If pathway events that lead to formation of head are out of sequence, abortive particles are produced. Prehead particle is believed to be produced and is later filled with DNA. Tail fibers do not attach to free base plates or to fully assembled tail unless tail is already attached to head. Tail has to combine with specific fragments, and head must interact with other fragments before tail and head join firmly.

ute the bacteria. The procedure is repeated for all of the dilutions to be used.

4. The melted top agar suspension is then poured over the surface of a Petri dish containing hardened agar (called ***bottom agar***). The top agar is allowed to cool and harden, and the Petri plate is then incubated overnight at 37° C.
5. The titer of the original unknown suspension is determined by examining the incubated plates. The plates will be seen to possess a lawn of confluent bacterial growth, within which appear small, clear, and nearly spherical areas called ***plaques*** (Figure 16-7).

Plaques represent the sites where phage-infected cells are deposited. The plaques are not produced by the initially infected bacterial cells but by infection of many thousands of adjacent cells. Each initially infected cell releases approximately 75 to 100 phage particles. These phage particles in turn infect adjacent bacterial cells, which become lysed and produce the typical plaque.

Lysogeny

Some infections by potentially lytic bacterial viruses do not always result in lysis of the bacterial cell because of repression of many viral genes. The bacterial cell carrying the nonlytic virus is called a ***lysogen.*** Nonlytic viruses may exist in two different states within the bacterial cell. Some act as ***prophages*** and integrate into the bacterial chromosome, creating a condition similar to transformation in virally infected animal cells (see p. 373). Viruses that integrate into the bacterial genome are also referred to as ***temperate viruses.*** The virus lambda (λ), for example, which is one of the most widely studied temper-

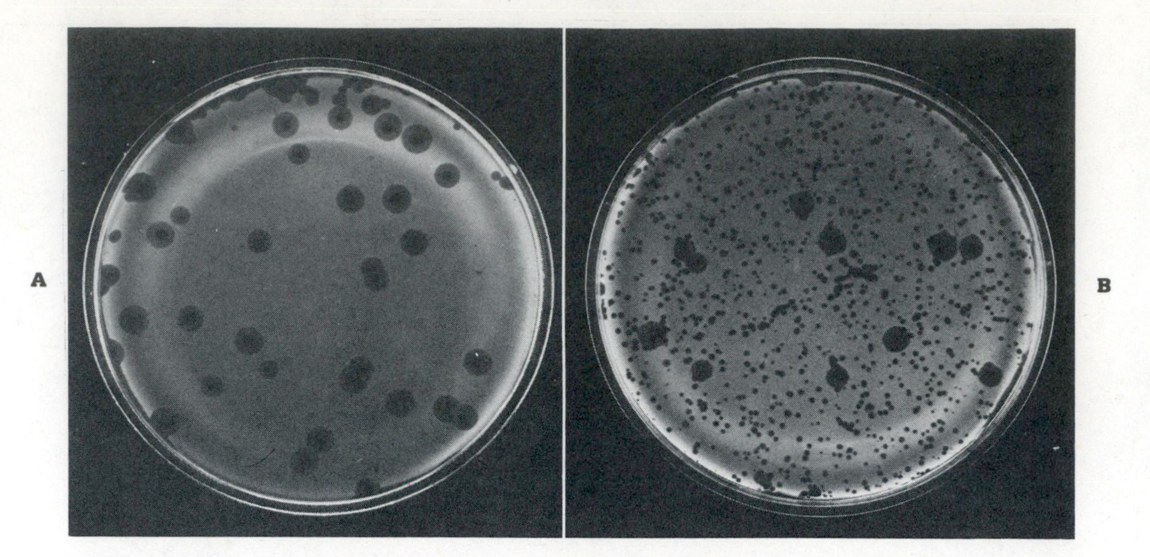

Figure 16-7 Bacterial plaques. Clear areas represent sites on lawn of bacteria where viral particles have been deposited and have reproduced, resulting in lysis of bacterial cells. **A,** Uniform plaques produced from single strain of phage. **B,** Plaques of varying size due to mixture of phage strains.

From Hawker, L.E., and Linton, A.H. Microorganisms: function, form, and environment *(2nd ed.). Edward Arnold (Publishers), Ltd., London, 1979.*

ate viruses, integrates into the bacterial genome at very specific sites (see Chapter 11). Some temperate viruses may be able to attach at more than one site on the bacterial genome. Phage P_1, for example, has nine attachment sites. Some temperate viruses do not integrate into the bacterial genome but merely bind to the cytoplasmic membrane, expressing their information independently of the host chromosome. Only a limited amount of genetic information is expressed, usually one or two genes.

Lysogeny has played a very important role in human disease. The rash associated with scarlet fever is due to infection by streptococci that have been lysogenized. The information for production of a toxin that causes a rash is carried in the viral nucleic acid. The toxin that causes the symptoms of diphtheria is also produced from genetic information carried on the virus that lysogenizes the bacterium *Corynebacterium diphtheriae* (see Chapter 22). Nonlysogenic *C. diphtheriae* does not cause severe disease in humans. In the two examples cited, the bacterial cell has acquired new characteristics due to lysogeny by a bacteriophage, a process called ***lysogenic conversion.***

Lysogeny vs. the lytic state. A lysogenized cell cannot be lysed at some later time by strains of the same virus. The lysogenized cell can, however, produce a vegetative phage that will lyse the cell under a complex set of circumstances (Figure 16-8). In the case of lambda two types of repressor molecules, both of which are coded for by the lambda genome, are involved.

The lambda repressor gene ***(cI)*** ordinarily codes for a product that maintains lambda in the integrated state. The repressor prevents the transcription of viral genes except the one that codes for its own synthesis. In other words, cI exerts a positive as well as negative control (the term *repressor* in this instance appears to be a misnomer; however, its positive effects were not realized until recently). The second repressor gene, called ***cro,*** is ordinarily repressed during lysogeny, but under certain circumstances it escapes repression and is transcribed. These special circumstances may include a mutation that occurs in the infecting genome or the use of ultraviolet (UV) light, increased temperature (44° C), or certain chemicals. Apparently, the viral DNA is damaged and is released from its integrated state. This results in viral replication and lysis of the bacterial cell. During the lytic state the cro repressor protein blocks the transcription of the cI repressor gene, and consequently all other viral genes are released from repression.

Mechanism of action of cI and cro repressors. Both cI and cro repressor proteins, in the form of dimers, bind to the same regions of lambda DNA, called the right operator (O_R). O_R is 80 base pairs long and consists of three sites: O_R1, O_R2, and O_R3. Two promoters are located between cro and cI; one directs RNA polymerase to move leftward and transcribe the cI gene, and the other directs

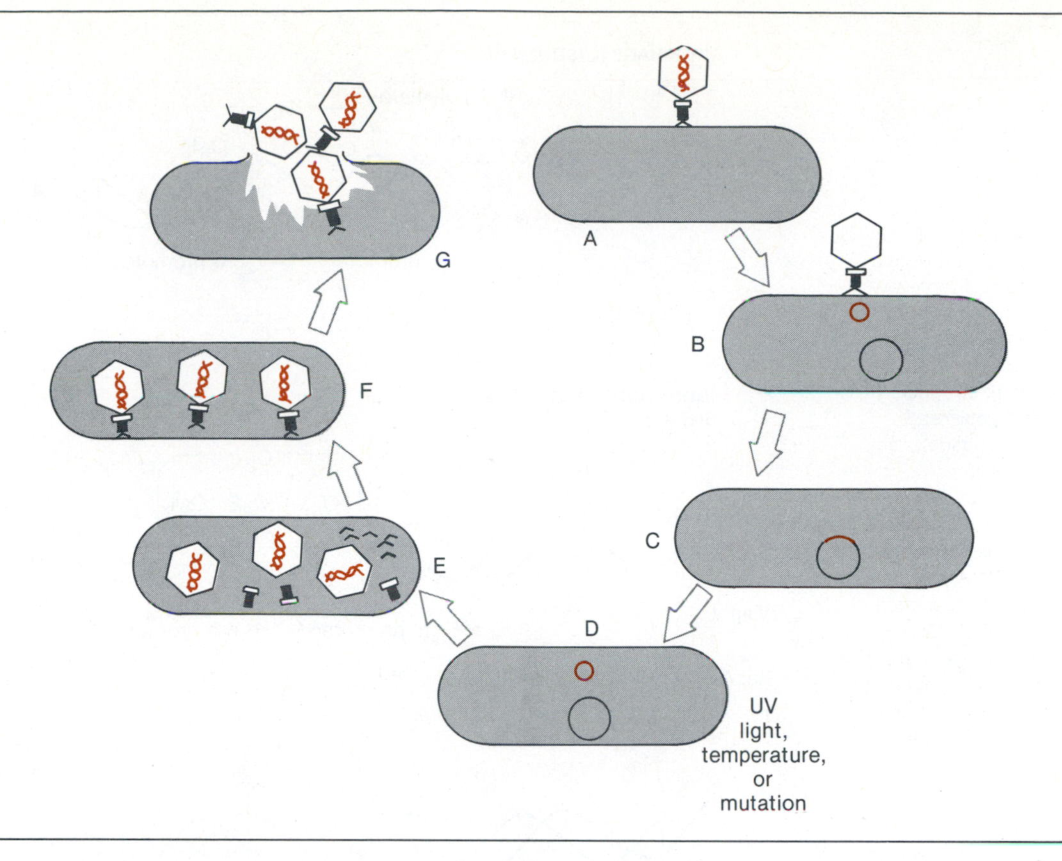

Figure 16-8 Lysogenic process. *A* through *C* represent integration of virus into bacterial DNA to produce prophage state. This state can be obtained by infection of lysogenic bacterial cell with temperate (nonlytic) virus such as lambda. *D* through *G* represent deintegration of prophage from bacterial chromosome by mutation, chemical, temperature, or UV light. Virus is free to replicate and produce viral particles that can ultimately lyse cell.

RNA polymerase to move rightward and transcribe the cro gene (Figure 16-9). Transcription to the right or left is blocked by a regulatory molecule that binds to a specific site on the operator.

Lysogeny. cI and cro repressor proteins have different affinities for the three sites on O_R. During lysogeny cI dimers bind to O_R1 and O_R2, but not O_R3 (unless the concentration of the cI repressor is very high). The bound repressor molecules mask the sequences that permit RNA polymerase transcription to the right; that is, the transcription of cro is prevented. RNA polymerase binds to the promoter of the cI gene, and leftward transcription is initiated; that is, the cI repressor gene is transcribed (Figure 16-10, *A*), and lambda is maintained as a prophage.

Lytic state. Damage to the lambda genome from UV light or other conditions causes a protein called RecA to behave as a protease. Recall that RecA is normally involved in genetic recombination. Protease cleaves the cI repressor dimers, and they subsequently fall off O_R1 and O_R2, leaving these sites vacant (Figure 16-10, *B*). The re-

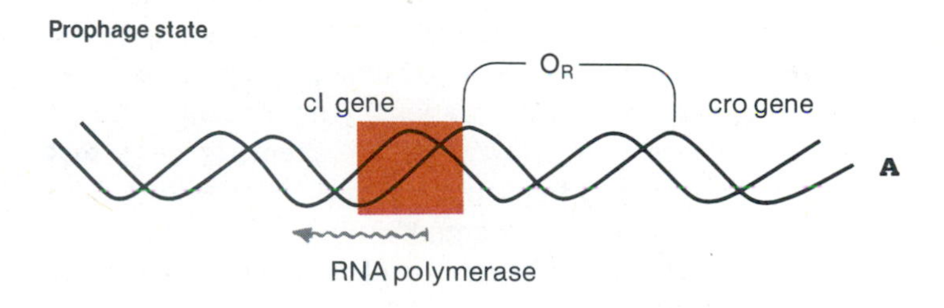

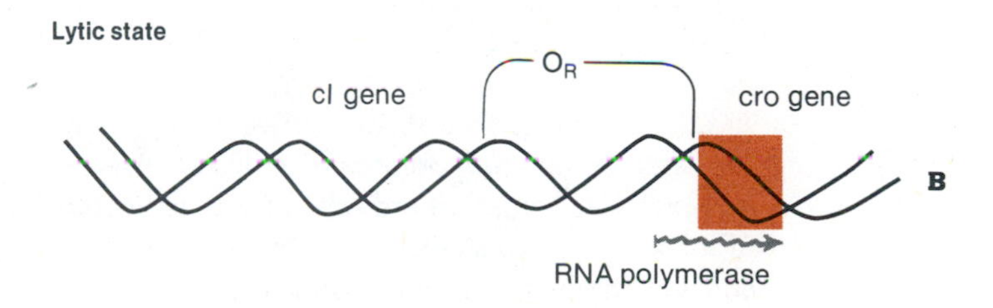

Figure 16-9 Direction of RNA polymerase transcription during prophage, **A,** and lytic, **B,** state. *cI, cro,* lambda repressor genes; O_R, right operator.

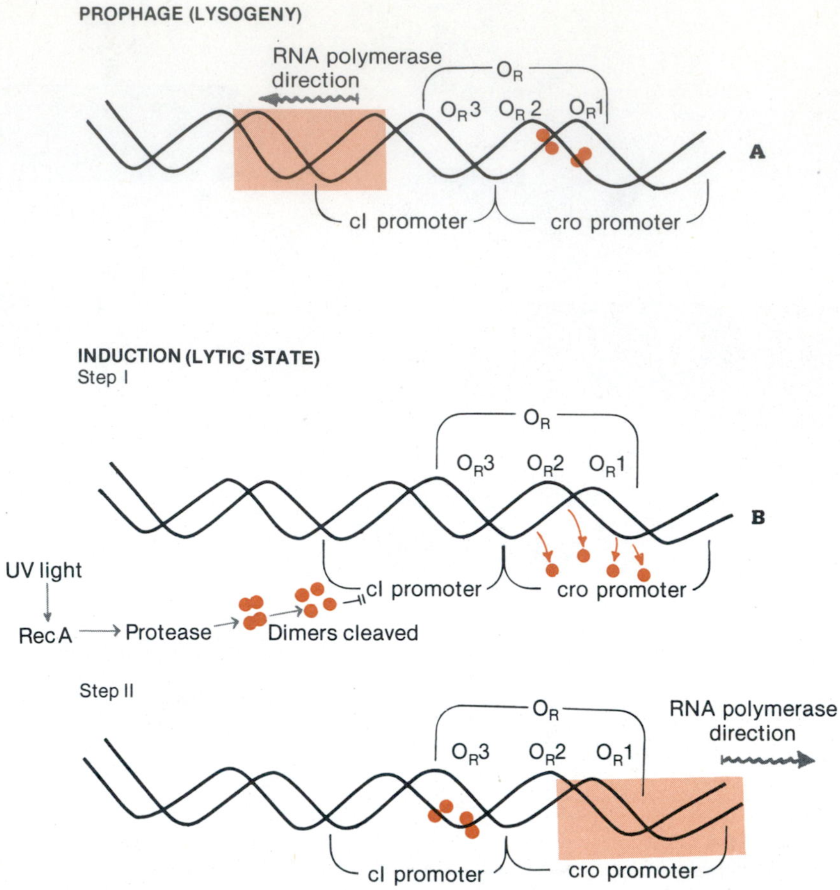

Figure 16-10 Mechanism of action of cro and cI repressors during prophage (lysogeny) and during induction of lytic state. **A,** Lysogeny. cI repressor-protein molecules in form of dimers (colored circles) bind to sites O_R2 and O_R1 of right operator (O_R). Repressor molecules mask DNA sequences on cro promoter and prevent transcription of cro gene. RNA polymerase transcribes cI repressor gene, and virus is maintained in integrated (prophage) state. **B,** Induction. In first step UV light induces conversion of RecA protein (normally involved in recombination process) to protease that cleaves cro protein dimers and causes their release from cro promoter. In second induction step RNA polymerase binds to cro promoter, which results in formation of cro repressor molecules that bind to O_R3. Cro repressor dimers mask cI promoter and prevent transcription of cI repressor gene. Transcription of cro by RNA polymerase is followed by transcription of genes involved in early lytic growth of lambda.

lease of the repressor molecules from the operator sites also leaves the promoter for rightward transcription free to bind RNA polymerase and to transcribe cro and other genes required for lytic growth. Cro repressor protein binds tightly to O_R3 but not to O_R1 and O_R2 (unless the repressor is at very high concentrations). The binding of cro repressor to O_R3 masks the promoter gene that controls cI transcription; thus cI repressor protein synthesis is prevented.

As the concentration of cro repressor increases during lytic growth, O_R1 and O_R2 sites are filled. This turns off the transcription of all the early genes and permits the transcription of late lytic genes.

Lambda and recombinant DNA technology. Variants of lambda defective in lysogenization and plasmid formation are being used as vectors in recombinant DNA experiments. There are several advantages for using variants of lambda as vectors:

1. The bacterial strain infected by lambda is *Escherichia coli* K12, one of the principal organisms employed in the host-vector systems in recombinant DNA research. Hybrid DNA can be prepared in great quantities, since each bacterial cell can produce several hundred copies of variant lambda DNA.
2. Lambda variants have been produced that have only one or two sites susceptible to the restriction endonuclease (Eco RI) used in the system. This characteristic facilitates the isolation of bacterial genes.
3. Middle segments of lambda DNA are not required for lytic growth. Therefore it is possible to insert foreign DNA in place of the middle segments. In fact, the insertion of foreign DNA can be made to be essential for plaque formation.

Let us consider the sequence of events that led to the use of lambda as a vector. In order for lambda to be produced in great quantities in the cell and to produce plaques, it cannot lysogenize *E. coli* K12. A variant of lambda was produced that was defective in the integration process. It was also learned during mutant formation that a deletion of the middle region of the lambda genome did not destroy any necessary lambda genes. The middle region is not only the integra-

Table 16-1 Enveloped animal viruses

VIRAL GROUPS	TYPE OF NUCLEIC ACID	REPRESENTATIVE TYPES	ASSEMBLY ON CELLULAR MEMBRANE
Arena	RNA; ss	Lassa fever, lymphocytic choriomeningitis	Yes
Bunya	RNA; ss	LaCrosse, snowshoe hare	Yes
Corona	RNA; ss	Murine hepatitis, human respiratory tract infections	Yes
Herpes	DNA; ds	Herpes simplex, Epstein-Barr, pseudorabies	Yes
Irido	DNA; ds	African swine fever, Tipula iridescent	No*
Myxo	RNA; ss	Influenza	Yes
Paramyxo	RNA; ss	Parainfluenza, Newcastle disease; measles; mumps; canine distemper	Yes
Pox	DNA; ds	Vaccinia, smallpox, fowlpox	No
Retro	RNA; ss	Avian leukemia, avian sarcoma, murine mammary carcinoma, feline leukemia, primate leukemia, murine leukemia	Yes
Rhabdo	RNA; ss	Vesicular stomatitis, rabies	Yes
Toga	RNA; ss	Semliki Forest, Sindbis, rubella	Yes

ss, Single strand; *ds*, double strand.
*The envelopes of these viruses are not derived from the preexisting cytoplasmic membrane but are made de novo in the cytoplasm of the cell.

tion site associated with the lysogenic process, but two of the restriction enzyme targets are also within it. It would seem that elimination of the middle region solved two problems, but it did not. Although no necessary lambda genes are in the central region, the remaining lambda DNA is not of sufficient length for packaging into a viable phage particle. The exciting discovery was that any DNA, including eukaryotic DNA, of sufficient length could be inserted into the middle region and make a phage particle. Thus we have a system in which the foreign DNA is absolutely essential for plaque formation.

Lambda, as an infectious particle, could conceivably escape from the laboratory. If the hybrid DNA contained virulent genes, then lambda could be a potential laboratory hazard, but several characteristics of lambda and its variants make this unlikely. *E. coli* strains sensitive to lambda are infrequent. *E. coli* cells in the gut of animals, including humans, that are sensitive to lambda have not been found. Lambda is also very sensitive to acids in the stomach as well as being sensitive to dessication. Phage lysates are routinely treated with chloroform in the laboratory, and this eliminates any surviving bacteria, including any that may have been lysogenized or any in which lambda may have been converted to the plasmid state. In addition, mutations are inserted into the lambda genome that make its survival possible only under laboratory conditions.

INFECTION BY ANIMAL VIRUSES

Adsorption to the animal cell

Each animal virus has a specific range of animal hosts that it can infect. Infection is initiated by contact of the virus with the host cytoplasmic membrane (Figure 16-11). The receptor sites on the animal virus vary from one virus to another. Some viruses are naked and possess only a capsid protein coat consisting of protein subunits. Some viruses, such as the adenoviruses, possess appendages called penton fibers (see Figure 22-15) projecting from the capsid, which are believed to serve as adsorption sites. Several viruses such as the agents of herpes, influenza, smallpox, rabies, and yellow fever are enveloped (Table 16-1) and possess ***glycoproteins*** that serve as receptors. Most of our knowledge of receptor interaction between host and virus has come from studies of viral glycoproteins. Recall that the glycoproteins (see Chapter 6) project from the external lipid bilayer of the viral envelope, and it is understandable that they would be involved in adsorption. Some viruses, such as the agents of measles and mumps, possess a single glycoprotein that is involved in adsorption, but in the influenza virus there are two major glycoproteins: the ***hemagglutinin spike*** and the ***neuraminidase spike.*** The hemagglutinin glycoprotein is associated with adsorption to host cells or erythrocytes by interaction with neuraminic acid.

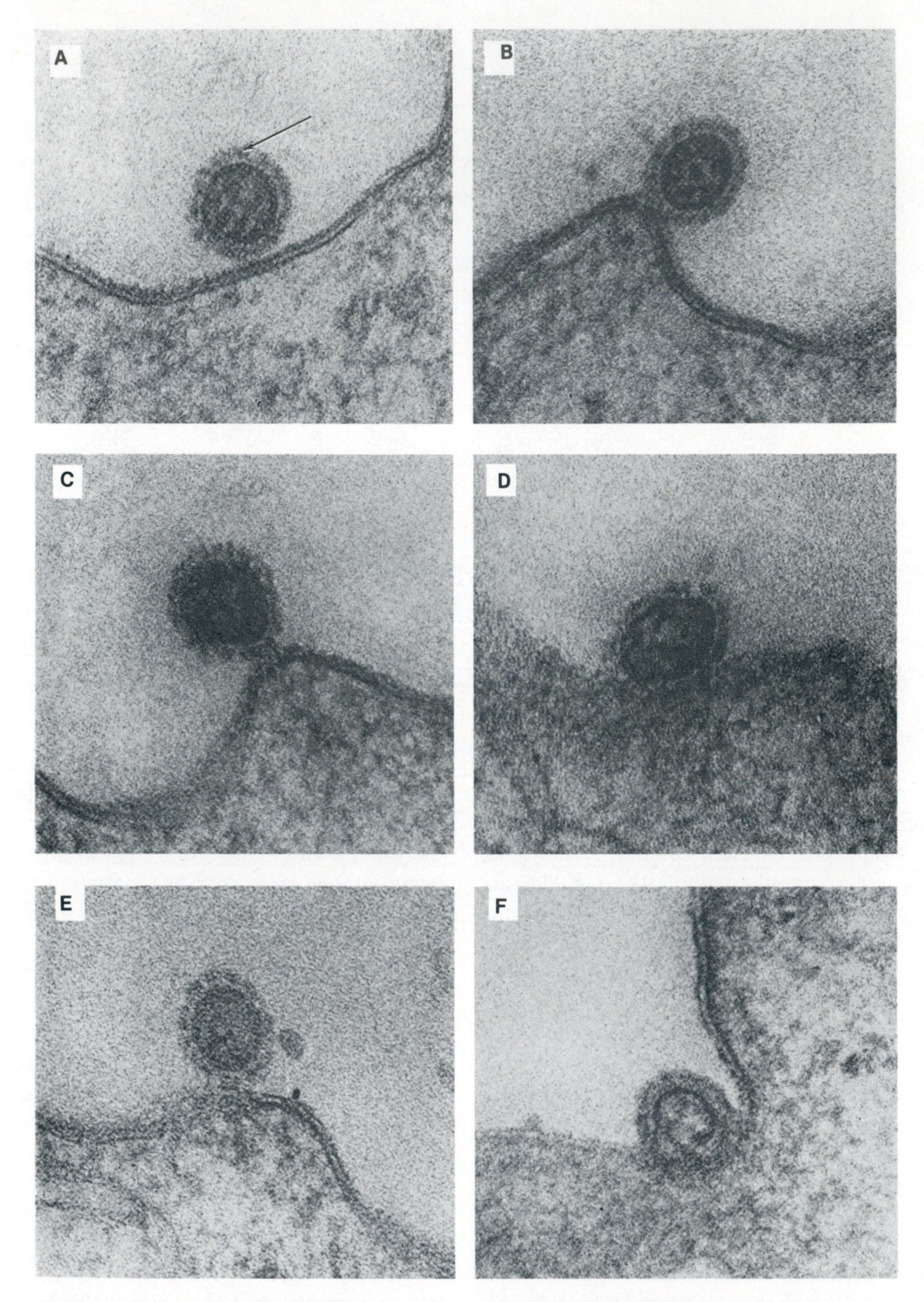

Figure 16-11 Electron micrograph demonstrating animal virus adsorption to cytoplasmic membrane. (×200,000.) **A** and **B,** Virus has envelope *(arrow)* that makes contact with cell membrane. **C,** Bridge appears to connect virus with cell membrane. **D** and **E,** Cell membrane appears indistinct. **F,** Virus has merged with cell, and envelope surrounding core of virus appears to be disrupted.
Courtesy Walter Brandt.

Neuraminic acid is found on the surface of most vertebrate cells, and this fact suggests that the virus could adsorb to most vertebrate cells, which it does. The viral neuraminidase spike cleaves neuraminic acid residues from cell membrane glycoprotein, and it is this activity that is believed to facilitate the spread of virus from one host cell to another. In addition to their role in adsorption and penetration, the viral glycoproteins also induce antibody formation. Antibodies to the hemagglutinin glycoprotein are capable of neutralizing the virus and preventing infection, and antibodies to the neuraminidase glycoprotein reduce the rate of spread of virus.

Penetration

Nonenveloped viruses can penetrate into the cytoplasm of the cell by a process similar to phagocytosis; that is, the virus is engulfed by the cell membrane and carried into the cell cytoplasm in ***phagocytic vesicles*** (Figure 16-12). It is not entirely clear how many enveloped viruses penetrate the cell, but some appear to fuse with the host's cytoplasmic membrane, which is followed by passage into the cytoplasm. There does appear to be a host-specific activity that is required for penetration of the enveloped viruses. The host produces a ***protease*** that cleaves the viral glycoprotein, a process that activates the glycoprotein and promotes penetration and fusion of the viral envelope with the cytoplasmic membrane. Experiments have demonstrated that in cells in which the protease is inactivated the virus adsorbs to the cell, but penetration is inhibited, and thus virulence is lost. The protease is not specified by the virus, and its biological role in uninfected cells is not known.

Uncoating

The viral protein coat of the nucleocapsid is removed by proteolytic enzymes whether the virus is in the phagocytic vesicle or left free in the cytoplasm. The nucleocapsid in some viruses such as the herpesvirus is carried first into the nucleus of the cell before uncoating begins. Most nucleocapsids are completely uncoated, leaving the nucleic acid naked in the cytoplasm or nucleoplasm, but a few viruses are never completely uncoated.

Replication

The prime function of the animal virus, as with other viruses, is to replicate and make more viral particles. The length of the eclipse period (see "Infection by Bacterial Viruses [Bacteriophages]" for a discussion of the eclipse) in animal virus infection can vary from a few hours to several days, depending on the virus. Early during the eclipse viral nucleic acid is replicated to produce more viral genomes, and some viral proteins are synthesized that totally or partially inhibit host DNA, RNA, and protein synthesis. The DNA of animal viruses may be double-stranded linear (adenoviruses, herpesviruses, and poxviruses), double-stranded circular (papovaviruses such as SV40), or single-stranded linear (parvovirus). The pattern of replication is similar to that for bacteriophages. The mechanisms of transcription of the viral genome into mRNA is dependent on the type of nucleic acid present.

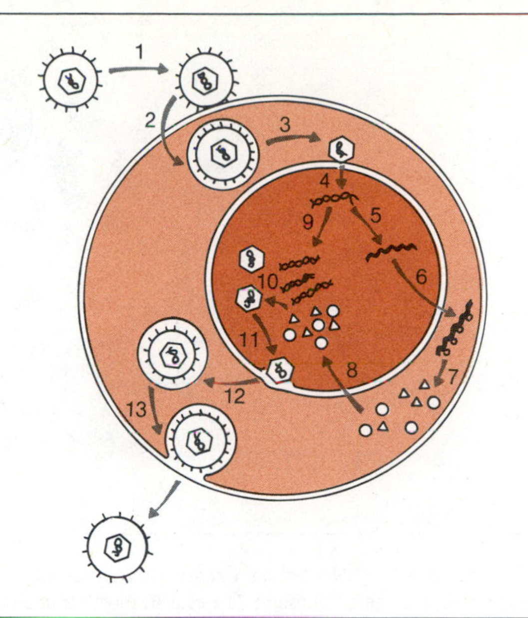

Figure 16-12 Mechanism of infection by enveloped virus such as herpesvirus. *1,* Virus adsorbs to cytoplasmic membrane. *2,* Virus is enclosed in vesicle within cytoplasm. *3,* Virus is released from vesicle. *4,* Viral nucleic acid enters nucleus. *5,* Viral mRNA is synthesized in nucleus. *6,* Viral mRNA is transported to cytoplasm. *7,* Viral mRNA and attached ribosomes engage in synthesis of viral proteins (○△). *8,* Viral proteins enter nucleus. *9,* Replication of viral DNA in nucleus. *10,* Assembly of viral particles in nucleus. *11,* Virus fuses with nuclear membrane. *12,* Complete viral particle with nuclear envelope and enclosed in vesicle. *13,* Virus-containing vesicle fuses with cytoplasmic membrane and is released from cell.

DNA virus transcription. The DNA viruses, except for one type of animal virus, possess double-stranded DNA. Viral DNA replication and mRNA synthesis in mammalian cells are essentially the same as cellular DNA replication and transcription. The transcription process, except for poxvirus, takes place in the nucleus and is mediated by a virus-specific polymerase. The mRNA that is

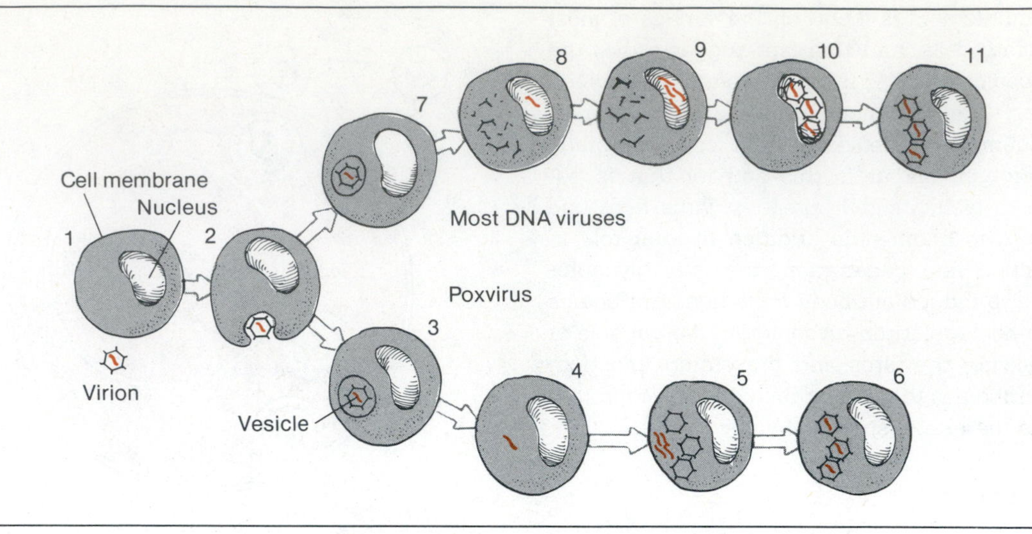

Figure 16-13 DNA animal virus transcription. Sequences *1* through *6* occur in poxviruses, and sequences *1, 2,* and *7* through *11* occur in most other DNA viruses. *1,* Interaction of virus with animal cell. *2,* Virus is engulfed by cytoplasmic membrane. *3,* Virus penetrates into cytoplasm within phagocytic vesicle. *4,* Phagocytic membrane is dissolved, and protein coat of virus is removed, leaving free viral nucleic acid in cytoplasm. *5,* Viral genome is replicated and coat protein synthesized. *6,* Viral particles are assembled in cytoplasm. *7,* Virus penetrates nucleus of cell. *8,* Viral phagocytic vesicle is dissolved and protein coat of virus removed. *9,* Viral nucleic acid replicates in nucleus. Viral mRNA leaves nucleus and enters cytoplasm, where ribosomes are present. Viral protein is synthesized in cytoplasm. *10,* Viral protein coats enter nucleus where viral particles are assembled. *11,* Complete viral particles leave nucleus and enter cytoplasm.

produced leaves the nucleus and enters the cytoplasm to bind to ribosomes, where structural and enzymatic proteins are synthesized (Figure 16-13). Only part of the animal viral genome, as in bacteriophage, is transcribed into early mRNA. Most of the late mRNA is synthesized after viral DNA synthesis. DNA replication in the smaller viruses depends primarily on the use of host cell DNA–replicating enzymes such as DNA polymerase, but in the larger viruses, such as the herpesviruses and poxviruses, some virus-specific enzymes such as DNA polymerase are available.

RNA virus transcription. Transcription of most RNA viruses may be accomplished by an RNA-dependent RNA polymerase, but in those RNA viruses that form tumors (***oncogenic*** viruses) an RNA-dependent DNA polymerase is used. The RNA-dependent RNA polymerase ***(replicase)*** is virus induced and replicates the single strand of viral RNA. In the poliovirus, for example, the RNA carried by the viral particle is called a positive strand because it functions as mRNA and can direct the synthesis of viral protein. On infection, the positive strand in the poliovirus first acts as mRNA and is translated into RNA polymerase. It then acts as a template for the synthesis of a complementary negative strand of RNA. The negative strand in turn acts as a template for the synthesis of many positive strands that can be used as mRNA or incorporated into new viral particles (Figure 16-14, *A*). The positive-stranded poliovirus mRNA is monocistronic, and the protein product must be cleaved to form the functional proteins.

The viral genome can be considered negative stranded when it will not function as mRNA. The negative strand must first be transcribed into a positive strand, which can then be used as mRNA (the message is not monocistronic) or as a template for the synthesis of negative strands. The negative strands are available as genomes for future viral particles (Figure 16-14, *B*). The reoviruses are different from all other animal RNA viruses because their genomes are double stranded and exist in segments (10 in reoviruses). Each double-stranded segment has a positive strand and a negative strand. Only the negative strand of the viral genome is used, and from it positive-stranded mRNAs are produced. The newly synthesized positive-stranded RNAs serve as mRNAs and as templates for the synthesis of many negative strands, which pair with the positive strands to produce double-stranded genome segments (Figure 16-14, *C*)

The RNA-dependent DNA polymerase used by tumor-forming viruses is called reverse transcriptase because it catalyzes the formation of a DNA

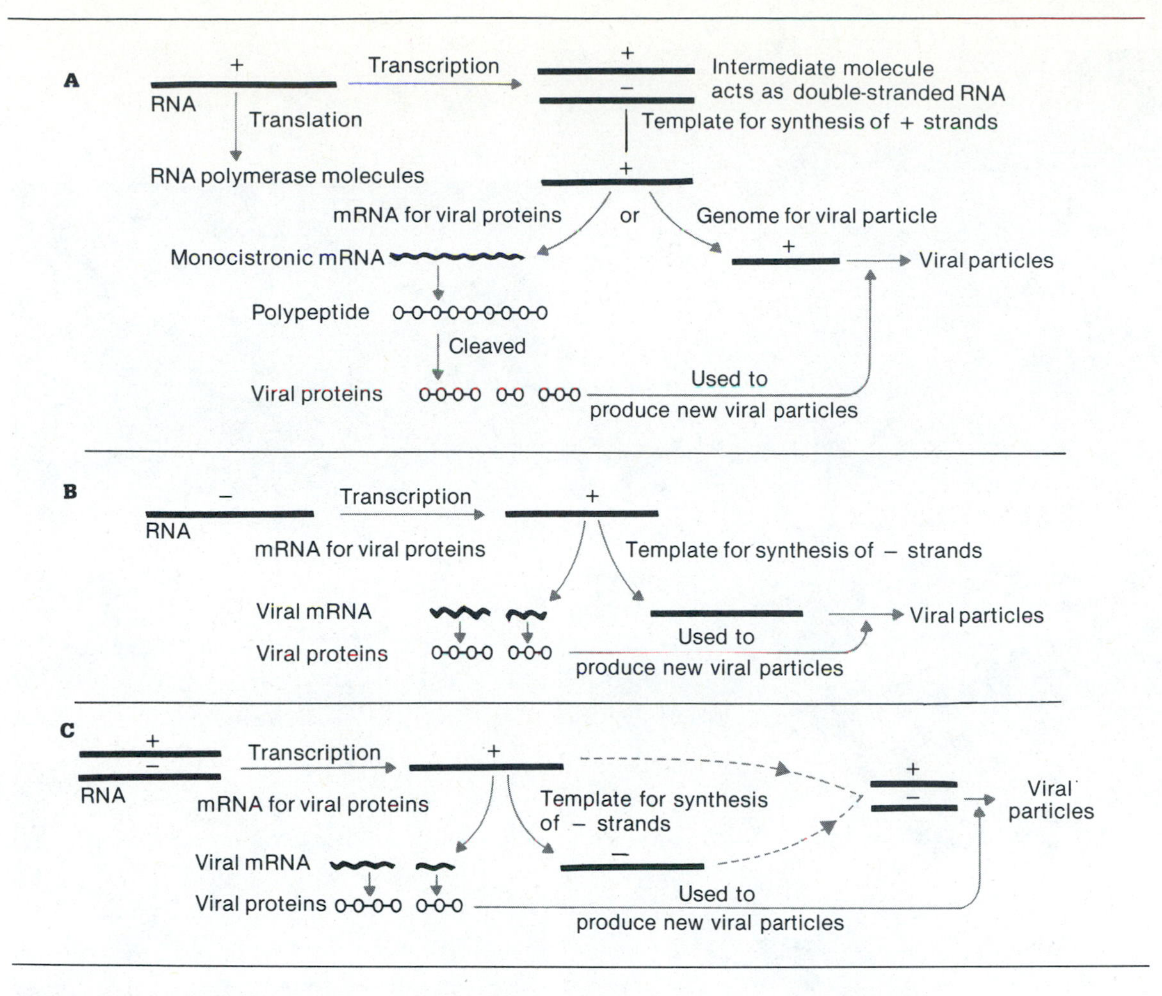

Figure 16-14 Replication scheme for RNA viruses using RNA-dependent RNA polymerase. **A,** Positive-stranded RNA. **B,** Negative-stranded RNA. **C,** Double-stranded RNA.

molecule from an RNA template. How this enzyme is used by tumor-forming viruses is discussed later in the chapter.

Viral maturation

Shortly after viral nucleic acid replication has begun, the synthesis of structural proteins, such as capsids, is initiated in the infected cell. Viral maturation involves the formation of ***nucleocapsids,*** a process that takes place in the nucleus or cytoplasm of the cell, depending on the viral type (Figure 16-15). Additional maturation steps are required for the enveloped viruses, which acquire their envelope by passing through the cytoplasmic membrane or nuclear membrane of the cell in a process called ***budding*** (Figure 16-16). The sequence of events involved in the maturation of a typical enveloped RNA virus include the following:

1. The viral mRNA for glycoprotein is translated on the ribosomes. The nascent polypeptide that is synthesized has carbohydrate components added and is then inserted into the endoplasmic reticulum.
2. The glycoprotein, like the secretory proteins (see Chapter 7), has a signal sequence that is removed while it is attached to the endoplasmic reticulum. This activation step is part of the maturation process of the glycoprotein.
3. The glycoprotein is transferred from the endoplasmic reticulum to the smooth membrane and then to the Golgi apparatus and finally to the cytoplasmic membrane, where it is inserted.
4. The nucleocapsid of the virus is transported from the cytoplasm to the cytoplasmic membrane, where it associates with the modified membrane and acquires the glycoprotein envelope by budding. The lipid of the viral envelope, as mentioned previously, is also derived from the host cell membrane but it is not virus specific.

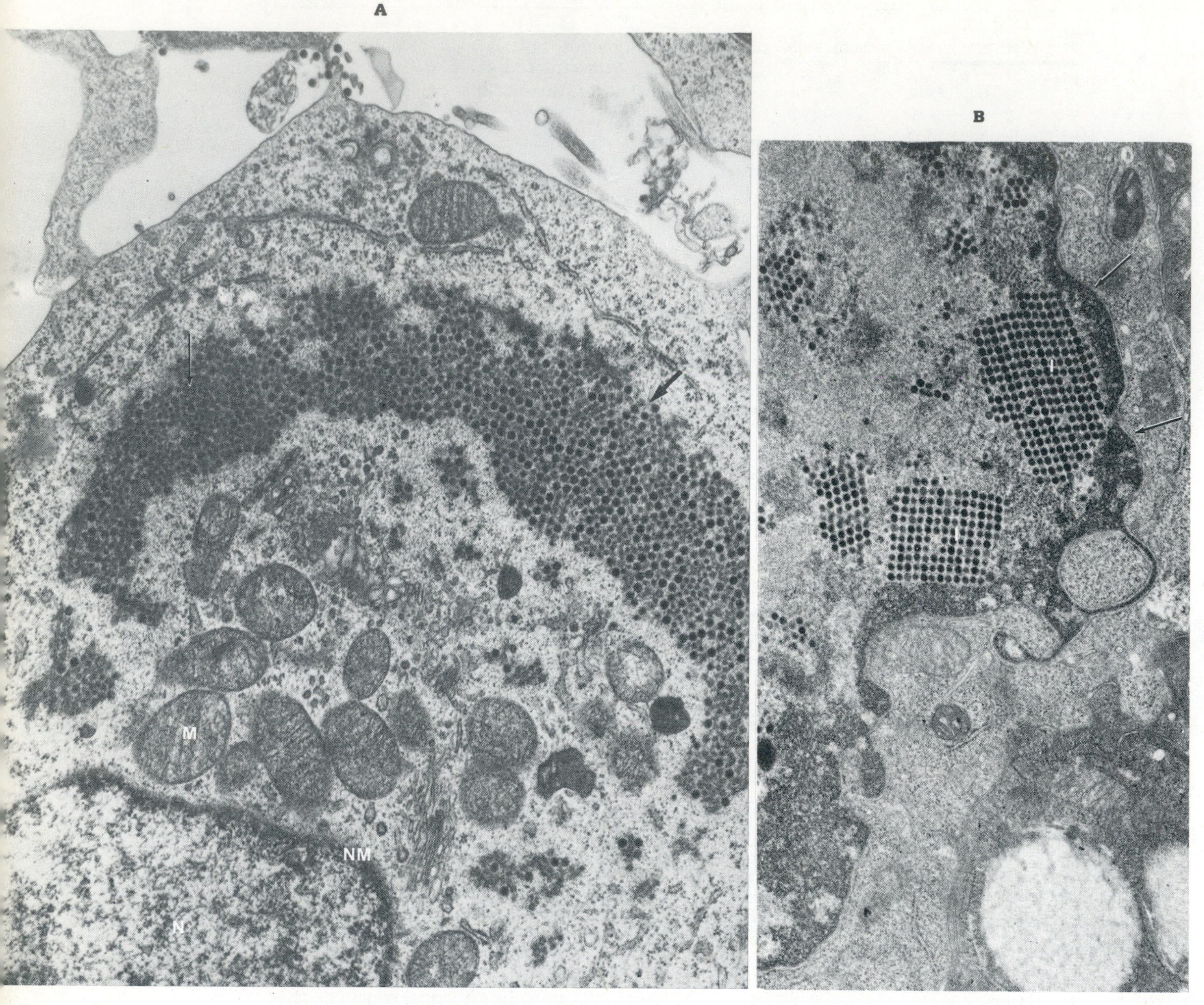

Figure 16-15 Electron micrographs demonstrating assembled viral particles in cytoplasm or nucleus of cell. **A,** Cytoplasmic inclusions (reovirus). Arrows point to viral inclusions. (×50,000.) *N,* Nucleus; *NM,* nuclear membrane; *M,* mitochondria. **B,** Nuclear inclusions (adenovirus). (×19,000.) Viral inclusions *(I)* are surrounded by nuclear membrane *(arrows).*

A, From Karpas, A., et al. J. Gen. Virol. ***35:191,*** *1977. B, From Takeuchi, A., and Hashimoto, K.* Infect. Immun. ***13:569,*** *1976.*

Effect of viral infection on the cell

Animal virus infection can result in one of basically two types of cellular response: lysis of the cell or persistence of the virus in the host.*

Cell lysis. Lysis of the cell is the most common response to viral infection. Those viruses that do not possess an envelope rupture through the cytoplasmic membrane and are released as free virions. The enveloped viruses are released slowly over a long period of time without lysing host cells.

Persistence. Viruses that persist in the host cell or tissue can cause a number of different cellular responses, depending on the virus and the initial response of the host to infection. Included among these responses are the following.

Immunopathological cell damage (immune complex disease). During infection antibodies to

* Abortive infections are known to occur in which no infectious virus is produced. This may be due to a defect in the viral replication process or to the type of cell being infected.

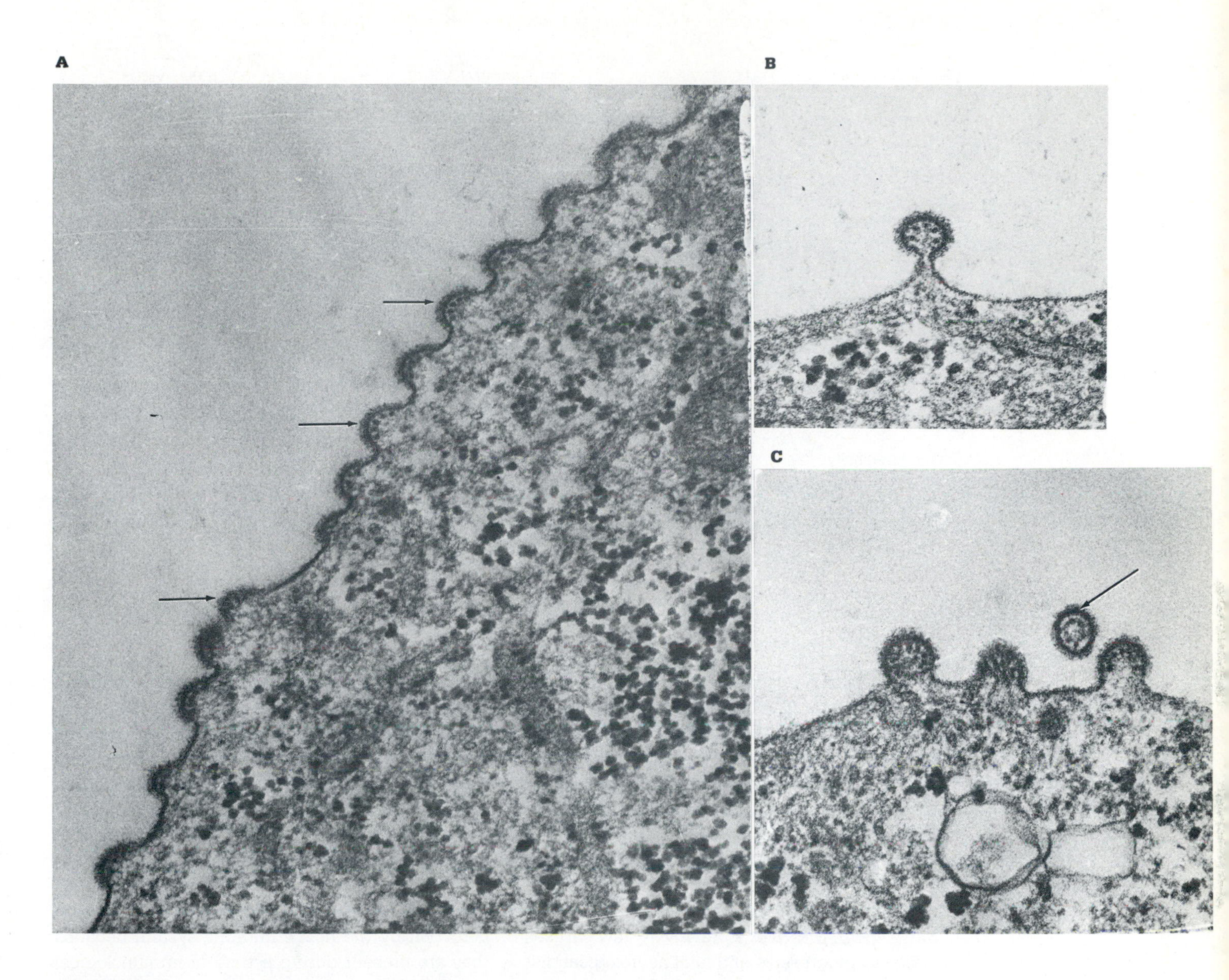

Figure 16-16 Electron micrographs demonstrating process of budding in animal virus (respiratory syncytial virus). **A,** Viral particles *(arrows)* appear to bulge at cell membrane. (×75,000.) **B,** Virion just before pinching off from cytoplasmic membrane. (×86,000.) **C,** One virion has budded *(arrow),* and three others are about to bud. (×76,000.)

From Norby, E., et al. J. Virol. *6:237, 1970.*

certain viruses bind viral antigens that are either circulating free in the host or are bound to cell surfaces. These antigen-antibody complexes may circulate freely and become deposited on tissues, where they elicit further immunological responses that damage the affected tissue. The antigen-antibody complexes produced during hepatitis B virus infection deposit on the glomeruli of the kidneys as well as vascular tissue, causing inflammatory diseases such as glomerulonephritis and vasculitis, respectively.

Antigenic modulation. The viral surface is altered during some viral infections. This activity hinders the host's antiviral immune mechanisms and results in other cellular responses to infection. Two very important examples of this are subacute sclerosing panencephalitis (SSPE) and recurrent herpes infection of the facial area. In the cases of SSPE, a condition resulting from chronic measles infection, the viral glycoproteins are apparently removed from the cell surface, which ultimately results in the loss of measles virus antigens. SSPE is a progressive degenerative disease of the central nervous system in which

Table 16-2 Comparison of plant viruses with bacterial and animal viruses

CHARACTERISTIC	PLANT VIRUS	ANIMAL VIRUS	BACTERIAL VIRUS
Nucleic acid present	RNA, except for cauliflower mosaic virus	RNA and DNA	RNA and DNA
Host range	Limited almost exclusively to plants	Animals, but some can replicate in arthropods	Limited to bacteria
Method of spread in nature	Requires vectors (insects, fungi, animal parasites)	Most do not require vectors; direct or indirect contact is most predominant method	Does not require vectors
Mechanism of infecting host tissue	Requires abraded area on plant surface	Tissue need not be damaged for infection to occur	Infection does not require damage to bacterial cell envelope
Survival in host	Indefinite, since host lacks antibodies	Host can inactivate virus because of antibodies, but some can remain latent for many years	Except for lysogeny, infection results in lysis of host cell
Tissue specificity	Nearly all plant tissue of the infected species (except for certain meristematic tissue) is infected	Most animal viruses are limited to specific tissue	

the virus persists despite high levels of antibody to measles virus. The herpes simplex virus can persist in nervous tissue for long periods of time, causing no apparent harm, but when activated migrates to the surface tissue and causes the characteristic tissue damage known as coldsores. It has been hypothesized that a specific type of antibody (IgG) binds to the surface of the virus-containing cell, inhibiting viral replication and allowing the virus to persist in the tissue.

Slow viral disease. A relatively few viruses give rise to progressive and fatal neurological disease after many years of incubation. Kuru and Creutzfeld-Jacob disease are examples of slow viral infections in humans, and scrapie is a naturally occurring slow viral disease of sheep (see Chapter 26). There is some evidence to suggest that some of the viruses causing slow viral disease possess incomplete genomes. These incomplete particles are called ***defective interfering particles*** because they interfere with the normal viral maturation cycle and prevent cell lysis but contain sufficient viral information to cause serious disease.

Transforming virus. ***Transformation**** is a process associated with tumor-forming viruses. These viruses may persist in tissue for years without damaging tissue but then suddenly cause malignancies (tumors) through some activating factor. The principal consequence of transformation is the alteration of the cytoplasmic membrane of the cell and the latter's acquisition of new antigens. These changes affect regulatory processes in the cell and are discussed at more length later in this chapter.

Immunological escape. Some viruses, such as the rubella virus (German measles), may escape detection by the host's immune system because they are present during a time of immunological immaturity, that is, during fetal development. The rubella virus, which may be acquired during pregnancy, can persist in the newborn for 12 to 18 months after birth. The virus reduces the rate of growth of host cells and is a cause of severe birth defects.

* This is not the same as the genetic transformation experiments demonstrating recombination in bacteria.

INFECTION BY PLANT VIRUSES

Before discussing the growth of plant viruses it would be informative for the student to first note some of the characteristic similarities and differences between plant, bacterial, and animal viruses. These characteristics are elaborated on in Table 16-2. A more detailed discussion of plant diseases is presented in Chapter 35.

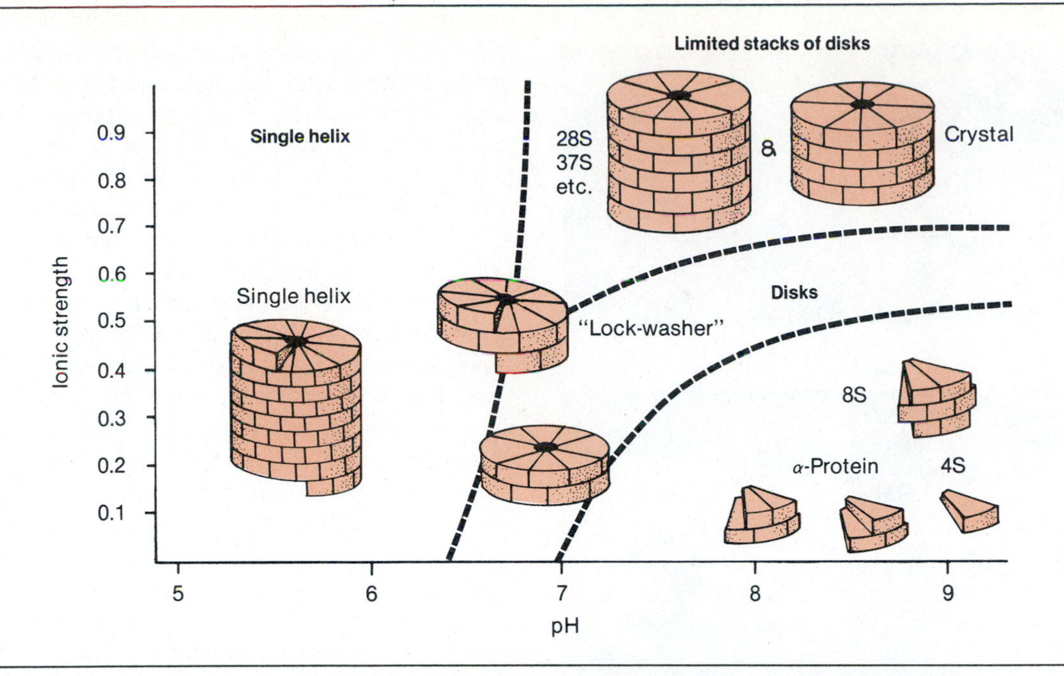

Figure 16-17 Ranges of pH and ionic strength over which specific reproducible units of tobacco mosaic virus (TMV) protein engage in formation of helical unit. Aggregates (8S and 4S) are formed initially at alkaline pH and low ionic strength. These aggregates form disks (20S and 28S) that polymerize at acidic pH to produce helical structure.

Redrawn from Durham, A.C.H., Finch, J.T., and Klug, A. Nature *229:47, 1971.*

Virus-cell interaction

The usual adsorption step observed in bacterial and animal virus infection processes is not part of the plant virus infection process. The cell wall of plants is much thicker than that of bacteria and fungi and cannot be penetrated by viruses. Plants can be infected when the surface tissue is abraded, for example, when sap from an infected plant comes in contact with an abraded area of a healthy plant. Most plant virus infections, however, are the result of the injection of virus into tissue via plant-sucking and chewing insects (mites, aphids, beetles, and leaf hoppers). Occasionally some soil-inhabiting vectors such as fungi and nematodes (roundworms) can damage plant tissue and transmit virus. A large inoculum of virus (up to 1×10^5 viral particles) is required to infect plant tissue (at least in the laboratory). Once the plant cells have been infected, viral particles are disassembled and viral nucleic acid spreads to surrounding cells and replicates in them. Viral titers may continue to increase and remain at high levels if the plant tissue does not die.

Assembly and maturation

Very little is known concerning the biochemistry of replication of plant viruses. The ***tobacco mosaic virus*** (TMV) is the most widely studied, but many of the subtle viral mechanisms that are known for bacterial viruses remain an enigma for plant virus infections. Nearly all plant viruses have an RNA genome, either single or double stranded. Only single-stranded RNA is potentially capable of being infective in the absence of a protein coat. Single-stranded RNA acts as mRNA and does not require transcription. The host ribosomes are used by plant viruses just as in bacterial and animal virus infections. Although plant organelles such as chloroplasts and mitochondria possess ribosomes, it is the cytoplasmic ribosomes that are used by the plant viruses. Few proteins other than the coat protein have been identified in infections by plant viruses or in plant viruses.

It appears that different viruses use different cellular sites for nucleic acid replication. Virus-associated material in TMV, for example, has been found to pass from the nucleolus through the nuclear membrane and into the cytoplasm. RNA synthesis in the turnip yellow mosaic virus occurs in the chloroplasts.

All plant viruses are assembled in the cytoplasm. The TMV consists of 2200 identical protein subunits that are arranged in the shape of a helix, with 16⅓ subunits per turn of the helix. RNA is embedded between successive turns of the helix (Figure 6-6). Laboratory studies have

A

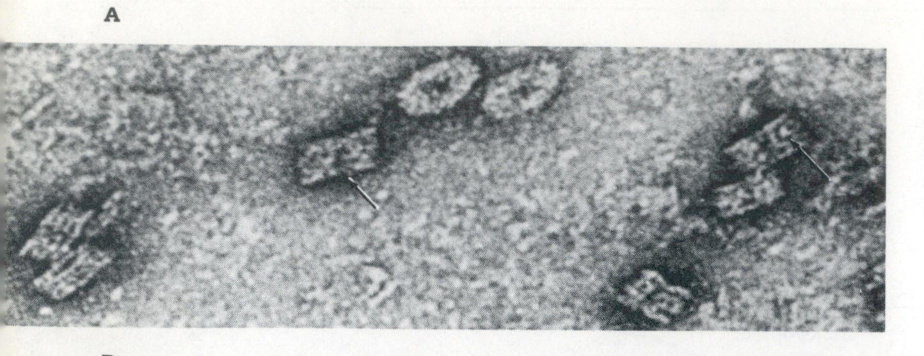

B

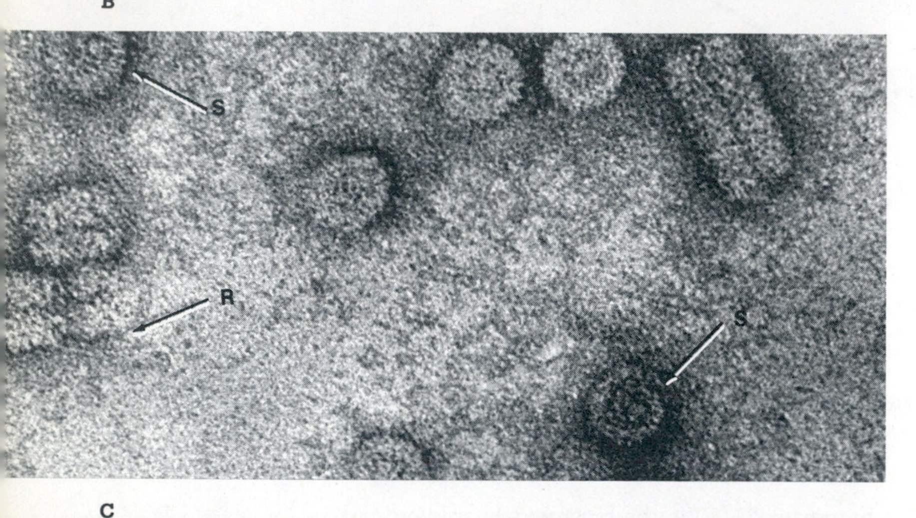

C

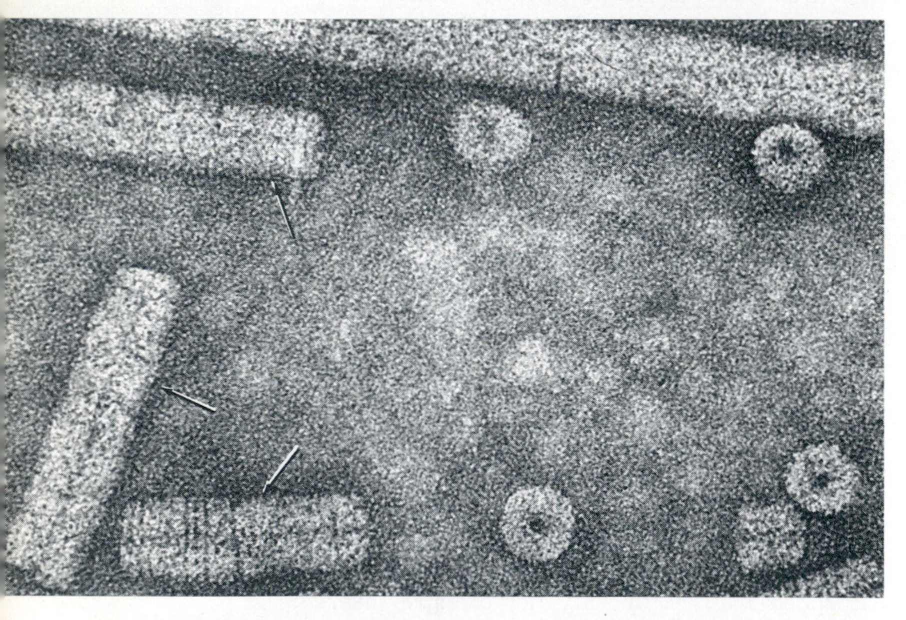

Figure 16-18 Electron micrographs demonstrating effect of pH on TMV protein aggregation and formation of helical structure. **A,** Starting material is 28S aggregate formed at pH 8.0. Most of protein is in form of stacks of 4 rings or 2 disks *(arrows)*. **B,** Protein has been placed in acetate buffer pH 5.0 for 10 minutes. Note that some short rods *(R)* are visible, but most protein aggregates are short, and some have their centers filled and show spiral *(S)* appearance or transformation to helix. **C,** Protein has been in acetate buffer pH 5.0 for 3 hours. Although there are some perfect single helical rods *(arrows)*, there are some short stacks of disks present, and when seen end on, they do not have filled centers or spiral appearance. (×500,000.)

From Durham, A.C.H., et al. With permission from J. Mol. Biol. ***67****:307, 1972. Copyright Academic Press, Inc. (London) Ltd.*

demonstrated that the protein subunits form aggregates of various sizes that can polymerize, depending on the ionic strength and pH of the solution (Figure 16-17). Subunit aggregates are produced at alkaline pH that result in the formation of disks or cylinders. Larger aggregates or cylinders are formed as the pH is decreased, and at acidic pH the aggregates are converted into helical structures (Figure 16-18). Intact disks may then be added to the growing helix. The assembly of the intact TMV particles is believed to occur in the same manner as protein assembly in vitro. RNA and protein subunits interact to produce one or two turns of the helix. This then acts as a nucleus for further polymerization and elongation of the helix. The initiation of polymerization is believed to occur at a unique sequence of nucleotides in the TMV RNA.

Mature viral particles can spread from one cell to another through the plasmodesmata or through the vascular tissues (phloem) until the entire plant is infected.

Satellite virus

In satellitism one form of a virus supplies a function lacking in a second form of the same virus. One of the functions could be in replication or assembly. The form unable to perform the function is called a satellite virus. The tobacco rattle virus, for example, occurs in two forms in a plant: long rods and short rods. Both particles possess an RNA genome and coat protein, but apparently a gene coding for a protein required by the long-rod particle is missing, and stable long-rod particles are not produced unless short-rod RNA is also present. The short-rod particle RNA codes for the protein required by the long-rod particle form for complete maturation.

Viroids

Certain plant diseases, particularly potato spindle tuber disease, is caused by a viral agent that exists as a free nucleic acid. The unconventional viral agent, which does not exist as a complete viral particle, is called a viroid. Viroids are discussed in Chapter 6.

VIRUSES AND CANCER

We have been able to define certain cancer-causing agents and have been able to control certain types of cancer, but we are still unable to determine the mechanisms of cancer formation. ***Cancer*** may be defined as a malignant growth or tumor resulting from the transformation of cells that demonstrate unrestricted growth. The tumor may remain localized, or it may spread (metastasize) to other sites. Viruses that have been suggested as possible causes of some cancer are referred to as ***oncogenic.*** Virally transformed cells are characterized by the integration of the viral genome into the host genome. This integrated state is similar to the process of lysogeny discussed earlier. This occurs when temperate bacterial viruses infect the cell and the viral genome becomes integrated or intimately associated with the bacterial genome, thereby conferring additional properties on the bacterial cell.

Integration of the viral genome

Viruses that transform cells may possess either an RNA or a DNA genome. Integration of viral DNA into the cellular genome presents no major problem, since both genomes are chemically similar. The RNA virus presents a different problem, because integration would mean the incorporation of an RNA molecule into a DNA molecule. Temin, Baltimore, and Mizutan in 1970 discovered that oncogenic RNA viruses (retroviruses) carry with them an enzyme called ***reverse transcriptase,*** which is an RNA-dependent DNA polymerase (as contrasted to the normal DNA-dependent DNA polymerase). This enzyme is capable of producing, from an RNA genome, a DNA product that can be incorporated into the host DNA (Figure 16-19). The sequence of events that follow infection by an oncogenic RNA virus can be described in the following way. Reverse transcriptase uses the viral RNA (called positive strand) as a template for the synthesis of a complementary DNA strand (called negative strand). This results in the formation of a DNA:RNA hybrid. The negative-stranded DNA in turn acts as a template for the synthesis of a complementary DNA strand, thus producing a double-stranded DNA molecule. The DNA is integrated into the host genome and in its integrated state is called a ***provirus.*** The provirus codes for information (transformation protein) that can transform the host cell, resulting in changes in metabolism, growth, and physiology. The provirus DNA can be transcribed into RNA, some of which acts as mRNA coding for viral proteins, and the remainder becomes RNA genomes. In some cells infected by oncogenic viruses only one gene is transcribed, and viral replication does not occur. An oncogenic virus that is replicated can be packaged in the cell and migrate to the cytoplasmic membrane, where it buds and is released by the cell. Ordinarily, if too many viral particles are produced in the cell, the cell will lyse. Stable transformation of the cell requires that very little viral multiplication occur.

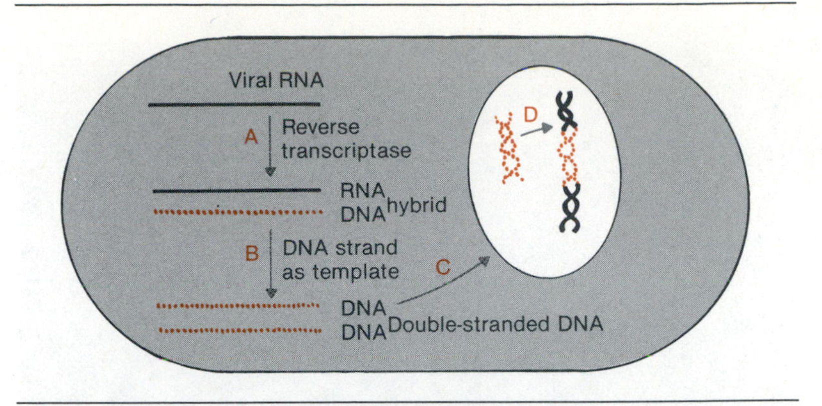

Figure 16-19 Replication scheme for RNA viruses that use RNA-dependent DNA polymerase (reverse transcriptase). *A,* Single-stranded RNA is transcribed by reverse transcriptase into RNA:DNA hybrid. *B,* Only DNA strand of hybrid acts as template for synthesis of complementary DNA strand and results in formation of double-stranded DNA molecule. *C* and *D,* Double-stranded DNA enters nucleus of cell and integrates with host DNA, thus becoming provirus.

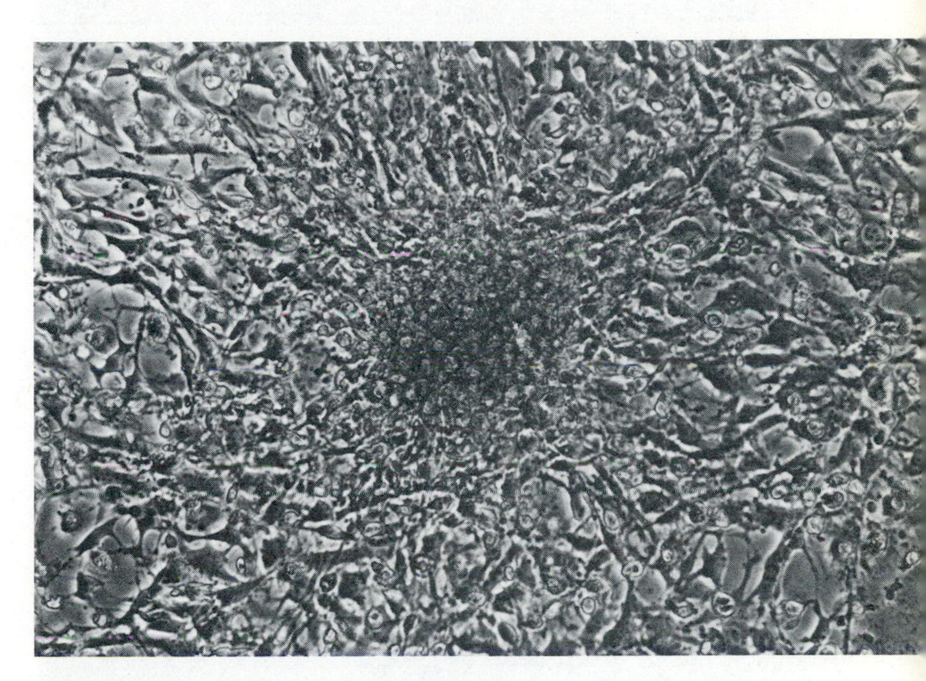

Figure 16-20 Microtumors produced in cell culture resulting from viral infection and transformation. Central aggregated cells represent focus of transformed cells produced by oncogenic virus.
From Hamada, K., et al. J. Virol. ***38**:327, 1981.*

Characteristics of the transformed cell

Transformed cells exhibit certain characteristics not observed in uninfected cells. Uninfected cells that grow on glass or plastic surfaces in tissue culture stop division once a monolayer has

A

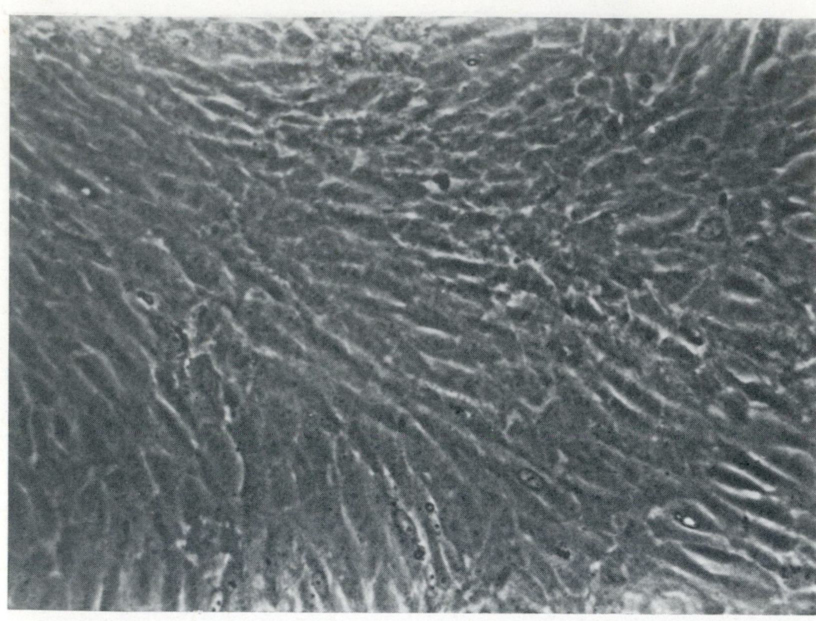

B

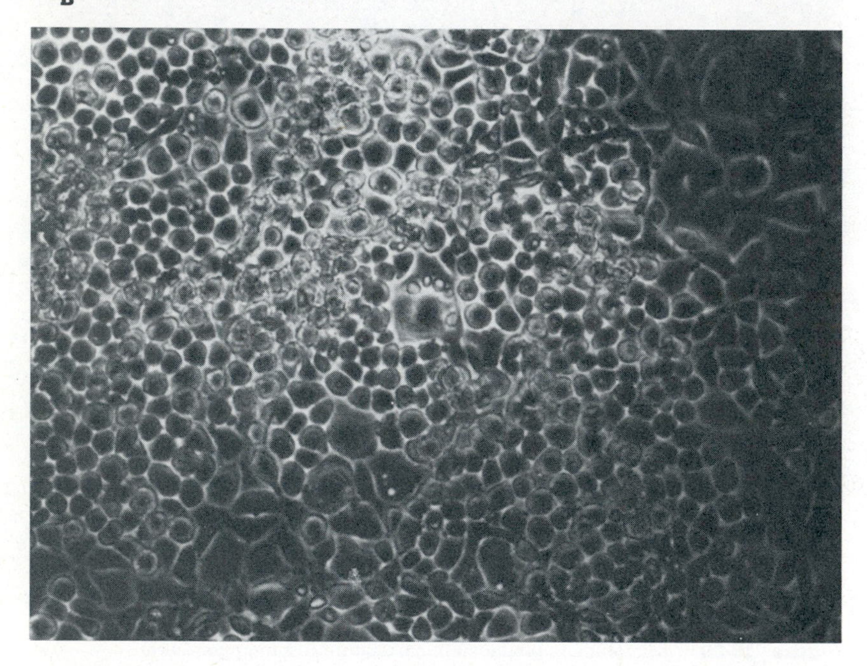

Figure 16-21 Characteristic of transformed cells. Phase-contrast micrographs of rat kidney cells (fibroblasts). (×100) **A,** Uninfected. **B,** Infected with avian sarcoma virus. Note rounding up of cells.

From Turek, L., and Oppermann, H. J. Virol. ***35:****466, 1980.*

formed. Some of the cells lose the property of contact inhibition if certain oncogenic viruses are allowed to infect this monolayer of cells. The infected cells begin to divide, and the progeny cells accumulate at the site of infection, producing microtumors (Figure 16-20). The site of the tumor is referred to as a ***focus,*** and the assay used to quantitate a tumor is called focus assay. Transformed cells also exhibit changes in their cytoplasmic membrane, and during infection virus-specific antigens are incorporated into the membrane. These antigens can cause changes in the immunological response of the host to the transformed cell and cause changes in the metabolism and growth of the cell.

If injected into susceptible animal hosts, transformed cells can cause tumors, but they cannot if they are injected into animals of the same species. Transformed cells also exhibit different morphological properties compared with the uninfected cell. The transformed cells often round up and exhibit irregular patterns of alignment (Figure 16-21).

Suspected human oncogenic viruses

The idea of viruses being causative agents of human cancer was hypothesized over 30 years ago when it was discovered that animal leukemias caused by animal viruses were similar to leukemias in humans. Many human viruses can transform certain strains of animal cells, but no virus to date has been demonstrated to be a cause of human cancer.

Indirect methods must be used in human studies to document viral agents as causes of cancer, because one cannot isolate the suspected virus and inoculate it into human subjects, as might be done with animals. These indirect methods include electron microscopy, serology, and the culture and inoculation of experimental animals. Electron microscopy is handicapped by the fact that few complete viral particles are present in the infected tissue. Serology is hindered because some tumor-forming viruses in animals do not elicit an antibody response in the host. The absence of antibodies to the virus therefore is not indicative of lack of infection. The use of cell culture and inoculation of experimental animals often represents an artificially manipulated condition. Usually the cell line or animal used is from another species, or the virus and/or culture must be preconditioned in order to produce positive results.

Oncogenic DNA viruses. The herpesvirus, a DNA virus, is the most likely candidate as an agent of human cancer. A member of the herpesvirus group, the Epstein-Barr virus (EBV), is the cause of ***infectious mononucleosis*** and has also been associated with ***Burkitt's lymphoma.*** Burkitt's lymphoma is endemic in Africa and is an infection of the lymphatic tissue that localizes in the face and neck. The lymphocytes affected are end stage under normal conditions and cannot multiply; however, when infected by EBV, they produce permanently growing cell lines. The EBV genome has been detected in the lymphocytes of those afflicted with Burkitt's lymphoma. In addition, the infected lymphocytes, when injected into certain animals such as marmosets, give rise to fatal malignancies. A second member of the herpesvirus group, the ***Herpes simplex virus (HSV)*** type 2, can be transmitted venereally (sexually).* The virus has been implicated as a potential cause of cancer of the cervix. The evidence is only epidemiological, but women who possess high levels of antibody to the virus are at a greater risk of acquiring cancer of the cervix than those who do not possess antibodies to HSV.

Oncogenic RNA viruses. RNA-containing tumor viruses belong to the family Retroviridae. ***Type C oncovirus,*** a member of this family, has been frequently implicated as a possible cause of human cancer. Type C oncoviruses consist of two types: (1) leukosis viruses that cause lymphomas or leukemia in animals and are nondefective (can replicate on their own in the host cell) and (2) leukosis or sarcoma viruses that often are defective (require a helper virus to supply important genes for replication). Mammary tumors in mice as well as leukemias and lymphomas in cats are caused by type C oncoviruses. Certain species of the virus have been isolated from cat tumors and injected into healthy cats to produce leukemias or lymphomas and have then been reisolated, thus fulfilling Kock's postulates. Morphologically similar RNA viruses have been sought in humans with carcinomas, but the results have been inconclusive. Reverse transcriptase and high molecular weight RNAs that are related to some viruses that cause animal cancers have been recovered from humans with different neoplasias. Conclusive proof that viruses cause cancer would be possible if integrated provirus could be detected and shown to be active in transformed human cells.

The transforming gene of the ***Rous sarcoma virus (RSV),*** which causes sarcomas in chickens, has been isolated. The RSV oncogene is called the ***src*** gene and codes for a protein called pp60v-src (*pp* represents phosphoprotein; *60* represents 60,000 daltons, the molecular weight of the protein; and *v-src* represents virus sarcoma). It is believed that pp60v-src phosphorylates the amino acid residues of proteins, especially tyrosine of proteins, in the cytoplasmic membrane of the cell. It appears that during viral infection by RSV the src gene codes for a protein that causes the cytoplasmic membrane to lose its adhesive properties and hence to pile up and metastasize, one of the properties of cancer cells. One of the most interesting results of these studies is the relatedness of the src gene and other oncogenes to normal vertebrate genes. There are 17 known retrovirus oncogenes, and, using DNA hybridization studies, 16 are known to be closely related to the normal genes of vertebrate species. Thus vertebrates have a cellular src gene, called ***c-src,*** which produces a protein pp60c-src that is indistinguishable in regard to its function from pp60v-src. These results support the view that oncogenes are cellular genes that have gone awry.

* Recent evidence indicates that HSV types 1 and 2 can cause both oral and genital herpes.

Theories of viral oncogenesis

Viruses may be transmitted horizontally or vertically. ***Horizontal transmission*** results from direct or indirect contact with contaminated secretions. For example, the viral agent could be transferred from parent to offspring in the mother's milk. ***Vertical transmission*** is characterized by transmission of the viral genome from parent to progeny during cell division or through the germ cells. Vertically transmitted viruses are said to be endogenous. Three theories have been proposed to explain the presence of oncogenes: the provirus theory, the oncogene-virogene, theory, and the protovirus theory.

Provirus theory. The provirus theory proposes that viral infection occurs horizontally. The viral reverse transcriptase is used to synthesize a DNA molecule that becomes integrated into the host genome. The cell is not transformed at this point and cannot be differentiated from uninfected cells. Viral genes can be activated by chemical or physical agents, causing the induction of a gene coding for a transformation protein. Virus could also be produced that would allow infection of adjacent cells.

Oncogene-virogene theory. The oncogene-virogene theory, proposed by Huebner and Todaro, is similar to the provirus theory but is based on the supposition that viral information (virogene) is part of all cells (endogenous) rather than exogenous and that it is transmitted vertically rather than horizontally. It has been further theorized that some of the virogene information is oncogenic (oncogenes). An explanation of this theory based on the evidence obtained from the RSV studies would be as follows. Retrovirus genes are probably copies of cellular genes. The viral oncogenes differ from their cellular counterparts because of mutations that were introduced when the cellular genes were copied into the retrovirus genome. The virogene-oncogene information may or may not be expressed during the lifetime of the cell. Viral information is normally in a repressed state within the genome of the cell. Carcinogenic agents (chemical radiation, nononcogenic viruses) could release the viral genes from cellular repression, and in a derepressed state either virogene or oncogene information could be expressed. The mutated cellular gene probably transforms the cell because the phosphorylating protein is either overproduced or has a different target in the cell. Three cellular oncogenes isolated from mice and rats have been shown to induce cancerous growth in cultured cells.

Protovirus theory. The protovirus theory assumes that certain normal genes (the ***protoviruses***) are transcribed into RNA and back into DNA by reverse transcriptase.* The duplicated DNA is integrated at different sites on the genome and produces new genetic sequences. This misplaced genetic information results in new cellular functions that may result in malignancy. The transcribed genetic information may be incorporated into the genome of the same or adjacent cells, thereby transforming them.

SUMMARY

Infection by bacterial viruses follows a pattern of adsorption to the cell surface, penetration of the nucleic acid into the cytoplasm, nucleic acid replication, viral assembly, and finally release from the cell. The bacterial virus subverts or breaks down the metabolism of the host, and in so doing, the virus is supplied the necessary nucleotides required for its own nucleic acid synthesis. In addition, modification of the protein-synthesizing system provides a mechanism for recognition of viral mRNA but not bacterial mRNA.

The viral reproduction cycle in the cell is initiated by the formation of early and late proteins, but no mature viral particles are yet produced. This period is called the eclipse. The viral nucleic acid is then packaged into the phage head protein, and this is followed by the addition of tail and tail fiber proteins. Virally induced lysozyme breaks down the cell wall and causes the release of the mature virus. Some viral infections do not result in lysis, and the virus remains within the host, either integrated into the host's nucleic acid or attached to the cell membrane. Under such circumstances the bacterial cell is said to be lysogenized.

During animal virus infection the entire virus, not just the nucleic acid, is taken into the cell in a process similar to phagocytosis. An eclipse period is also present, during which many viral components are synthesized but no mature virus is formed. Following assembly into mature particles, the virus may lyse the cell (most common response), transform the cell, or cause persistent infection.

The mechanisms of plant virus replication are not well understood. Plant viruses usually enter the cell via abrasions or by injection via plant-sucking insects. The viral nucleic acid is usually RNA, and replication may occur at different sites within the cell. Following assembly into mature virus, viral particles can spread from one cell to another via the plasmodesmata or through the vascular phloem.

Transformation of animal cells by virus can result in tumor formation. These tumor-forming viruses are called oncogenic. Transformation requires incorporation of the viral nucleic acid into the host genome. If the viral nucleic acid is RNA, a special enzyme called reverse transcriptase is carried by the virus to convert RNA into DNA before incorporation into the DNA of the host. The viral genome codes for information that transforms cells; that is, the cells pile up, cell membrane antigens are changed, and cell metabolism is altered. Cancer-causing viruses are known for some animals. There is some evidence that viruses can also cause cancer in humans. For example, one DNA virus, the Epstein-Barr virus, has been implicated in lymphomas of African children.

* Note that reverse transcriptase can also catalyze the normal DNA to RNA flow of information, and its presence in cells would not be that unusual.

Study questions

1. What is the sequence of events in the infection of a bacterium by a virus? What are the similarities and dissimilarities to animal virus infection?
2. What kind of enzymatic interplay occurs in the cell that causes a bacterium to protect itself from infection or to be destroyed by infection?
3. What are the sources of nucleotides for viral nucleic acid synthesis?
4. Describe some of the modification processes that occur during viral infection of bacteria. What is their purpose?
5. What are viral plaques? Can they be measured with all viruses?
6. What is the special association of lambda with the bacterial cell? How can this association be altered?
7. Does lysogeny have any significance to human disease?
8. How does animal virus infection compare with bacterial virus infection in terms of adsorption, penetration, and replication?
9. What is the distinction between positive- and negative-stranded RNA during replication of RNA viruses?
10. Are all animal cells lysed by infecting viruses? What is the significance of lysogeny?
11. Why are some viral infections persistent?
12. Why is the infection process of plant viruses different from that of other viral infections?
13. What is the function of satellite viruses?
14. What is the relationship of reverse transcriptase and oncogenic viruses?
15. What changes occur in cells that have been transformed by viruses?
16. What is the distinction between a provirus and a virion?
17. What viruses are believed to cause cancer in humans? In animals?
18. Distinguish between vertical and horizontal transmission of virus.
19. What are the characteristics of oncogenes in terms of three theories used to explain their existence?
20. What techniques can be used to determine the carcinogenic potential of a chemical? (See Chapter 9.)

Selected readings

Books

Davis, B. et al. *Microbiology* (3rd ed.). Harper & Row, Publishers, Inc., New York, 1980.

Dulbecco, R., and Ginsberg, H.S. *Virology.* Harper & Row, Publishers, Inc., New York, 1980.

Fenner, F., and White, D.O. *Animal virology.* Academic Press, Inc., New York, 1976.

Joklik, W.K. (Ed.). *Principles of animal virology.* Appleton-Century-Crofts, New York, 1980.

Knight, C.A. *Chemistry of viruses* (2nd ed.). Springer-Vertag, New York, Inc., New York, 1975.

Levy, H.B. *Biochemistry of viruses.* Marcel Dekker, Inc., New York, 1969.

Luria, S.E., Darnell, J.E., Jr., and Campbell, A. *General virology* (3rd ed.). John Wiley & Sons, Inc., New York, 1978.

Mathews, C.K. *Bacteriophage biochemistry.* Van Nostrand Reinhold Co., New York, 1971.

Smith, K.M. *Plant viruses* (6th ed.). Chapman & Hall, Ltd., London, 1977.

Journal articles

Baltimore, D. Retroviruses and cancer. *Hosp. Pract.* **13:**49, 1979.

Baltimore, D. Viral RNA–dependent DNA polymerase. *Nature* **226:**1209, 1970.

Bishop, J.M. Retroviruses. *Annu. Rev. Biochem.* **47:**35, 1978.

Butler, P.J., and Klug, A. The assembly of a virus. *Sci. Am.* **239:**62, 1978.

Casjens, S., and King, J. Virus assembly. *Annu. Rev. Biochem.* **44:**555, 1975.

Delbruck, M. The growth of bacteriophage and lysis of the host. *J. Gen. Physiol.* **23:**643, 1940.

Durham, A.C.H., and Finch, J.T. Structure and roles of polymorphic forms of tobacco mosiac virus. *J. Mol. Biol.* **67:**307, 1972.

Nash, H. Integration and excision of bacteriophage λ. *Curr. Top. Microbiol. Immunol.* **78:**171, 1977.

Temin, H.M. The DNA provirus hypothesis. *Science* **192:**1075, 1976.

PART 5

ENVIRONMENTAL AND APPLIED MICROBIOLOGY

Zygnema insigne, a green alga. (×50.)

Runk/Schoenberger of Grant Heilman Photography.

Chapter 17 CONCEPTS OF MICROBIAL ECOLOGY

Martin Alexander

Every microbial species has one or more habitats in which it lives. This habitat may be the sea, fresh waters, soils, decaying organic matter, plant or animal surfaces, or the tissues or body fluids of humans. The habitat may be restricted to the surface of a clear lake, or it may encompass much of the sea. Often for the microbiologist, the ***microhabitat*** is more important from a conceptual viewpoint because of the size of individual microbial cells or filaments. Microbial ecology is concerned with the interrelationships between the microorganisms, be it a unicellular or filamentous microorganism, and the environment or habitat in which that organism lives.

Recent years have witnessed an enormous increase in research on microbial ecology. Part of this activity comes from microbiologists, and part from ecologists and other environmental scientists. Microorganisms are unique tools for ecological studies, and the ability to use microbial mutants deficient in traits of presumed ecological importance and to perform rapidly a variety of studies assessing biochemical interrelationships have made microorganisms especially valuable for defining ecological principles. Microbiologists have always had considerable interest in defining the interrelationships between the species of concern to them and the environment in which they reside. In addition, many applied problems have appeared in recent years for which information on microbial ecology is of practical importance. Thus, the increase in fertility of fresh waters resulting from human activities has led to unwanted and often obnoxious blooms of algae and cyanobacteria, and attempts to control these blooms have fostered interest in the nutrition of the responsible microorganisms. Research on chemical pollution of waters and soils has established that microorganisms are the sole or major agents of ***biodegradation,*** and this knowledge has prompted many studies of microbial transformations of polluting chemicals. Moreover, microorganisms have recently been found to produce a variety of hazardous products in waters and soils, and in this way, microorganisms themselves are responsible for chemical pollution. Furthermore, the growing awareness of the limited food supply for the human population of the globe has focused interest on the crucial role of microorganisms in food production on land and in natural waters.

CONCEPTS AND TERMINOLOGY

From the ecological viewpoint, it is important to understand certain terms in what is sometimes called the ecological hierachy. At the lowest level of that hierarchy is the individual microorganism. This may be the single cell of the unicellular bacterium or protozoan, the resting body of a fungus, or a hyphal strand. All cells or hyphae of a single type that reside in a single habitat, or microhabitat, represent a distinct ***population.*** Those populations that represent a strain or species of economic or health importance have achieved special prominence. The various populations inhabitating a given site represent the ***community.*** Considerable information exists on community composition when the environment is the body of the human or higher animal, but less information exists on the communities of marine, freshwater, or terrestrial environments. The var-

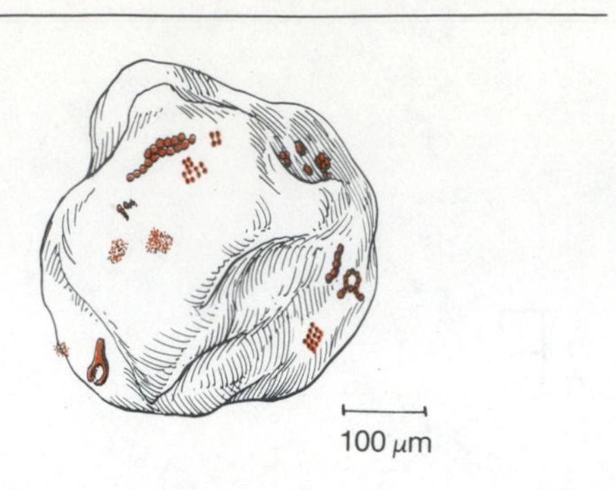

Figure 17-1 Occurrence of microorganisms on surface of marine sand particle.
Redrawn from Meadows, P.S., and Anderson, J.G. J. Marine Biol. Assoc. U.K. ***48:****161, 1968.*

ious communities in a particular environment together with the nonliving, or ***abiotic,*** surroundings constitute the ***ecosystem.*** Each ecosystem has a unique collection of organisms and inorganic and organic components.

Many misconceptions about microbial ecology exist because of the lack of understanding of the microenvironment. At the microscopic level, which is appropriate for considering microorganisms in nature because of their minute size, there may be major gradations in ecologically important variables. In a zone that may have a diameter of less than 0.1 mm, enormous differences in the amounts of organic substrates, inorganic nutrients, inhibitory substances, and oxygen may prevail. At one site, the level of one or more of these factors may be high, and at a site less than 50 μm away, the concentrations may be low. Thus, two distinctively different populations may reside in what appears, to the macroscopic viewer at least, to be the same site, but the organisms in fact are living and functioning in different microenvironments (Figure 17-1). Analyses of microhabitats are difficult because most of the currently available physical, chemical, and often biological procedures are not useful at the microscopic scale, and hence a conclusion based on a property of the macroenvironment and analyses performed by techniques not suitable for the microenvironment frequently is at odds with the presumed properties of the organisms based on tests carried out in laboratory culture. It is critical to bear in mind that the microenvironment is the functional habitat of the microorganism, whether that microorganism be performing its various metabolic activities and reproducing in soils, waters, plant tissues, or different sites within the animal or human body.

The obviously marked differences in species composition of microbial communities are at odds with the oft-stated view that microorganisms are potentially everywhere. This view is derived from observations that microorganisms are easily transported with currents of air and water or on animal and plant surfaces, and they are ingested or otherwise penetrate animal bodies and plant tissues. Furthermore, many microbial species grow rapidly in the laboratory as well as presumably in nature. Given that microorganisms are so easily disseminated and grow so rapidly, how can one explain the fact that individual ecosystems have characteristic microbial communities? These characteristic communities are evident in oceans, lakes, soils, plant surfaces, the human mouth, or in stool samples.

The reason is simply that, although microorganisms indeed are potentially everywhere and some can grow rapidly, the environment selects those species from among the invaders and permits only a few to become established. Each ecosystem has certain physical, chemical, and biological factors that govern the community composition. These factors may be types and concentrations of organic nutrients, toxins, surfaces receptive to microbial development, temperature, and acidity. These crucial factors determine whether a species that has reached that environment will or will not grow or even survive. The environment sorts out the arrivals with those physiological properties that are attuned to the particular local circumstances. Populations not meeting these environmental stresses are eliminated early in the development of the community. The environment thus selects the fit microorganisms and rejects those that are unfit. The former will survive and multiply, and the latter are eliminated.

NATURAL SELECTION AND FITNESS

Each microbial species has unique biochemical or morphological properties. These properties permit the development of that organism in some environments but not in others, and it is these traits that determine the geographical patterns of microorganisms or the microenvironments in which they reside. It is on these traits that natural selection operates, and the interactions between natural selection and these properties determine the distribution of the species. Based on these traits, some species have a selective advantage in one environment, and others are favored in dissimilar environments. The outcome of the selection is survival or, at some time, active metabolism. To survive the selective processes, the cell or filament must win in the struggle for existence. The unsuited species, in contrast, has a small chance for survival at that site, and is usually eliminated.

The identities of most of these ecologically important attributes are difficult to establish because most microbial communities contain many different species, the organisms are extremely small, and the interactions among the populations are numerous. A few of the attributes are simple to establish. Thus, the fit species must be able to tolerate the extremes of temperature, salinity, or osmotic pressure to which the environment is exposed. It may need to endure periods of drying. If it is in the body, it must be able to endure attack by phagocytes. These are all necessary conditions for an organism to exist, but they are not sufficient conditions. Usually, there are many more covert than overt traits of ecological importance. The lack of a single trait may result in the elimination of a species. However, only the most obvious of these traits are known. Let us examine a few of those properties of microorganisms that determine fitness.

Factors that contribute to the selection process

The obvious factors contributing to selection can be dispensed with quickly: the inhabitant of any environment must be able to tolerate the ambient moisture levels, pH, hydrostatic pressure if it exists in the water column of the ocean, oxygen levels, salinities, and temperature of that locality. Because many microbial communities are exposed to intense sunlight, the ability of an organism to maintain its viability in the presence of lethal light intensities is of crucial importance. Sunlight kills many populations. To cope with the potential harm from sunlight, some species have carotenoid or melanin pigments that are chief fitness traits in sun-drenched environments. Populations possessing these pigments are common on leaf and water surfaces and in salt ponds exposed to high light intensities, and they dominate among the species that are transmitted for long distances through the atmosphere. Many populations are retained on surfaces in regions with moving liquids, for example, microorganisms growing on rocks in streams, bacteria on pipes containing flowing fresh water, protozoa in the intestine, or bacteria on the epithelial surfaces of the mouth. These organisms have a characteristic fitness trait: an attachment devise or other means to allow the organism to be retained on the surface. Because the cell constituents of many microorganisms have high specific gravities, they would usually settle to the bottom of lakes and the ocean, but the organism that is native to aquatic communities must have some special buoyancy property that allows it to be maintained in such environments. These populations commonly have cell constituents that compensate for the otherwise high specific gravity of the cell or filament. For some protozoa, small size is an ecologically critical trait because without it, they would be unable to reside in soil or sand, which have very small pores between the soil particles or sand grains. These pores would prevent the active movement and feeding of cells having large sizes. In contrast, large size is important in some aquatic environments because big cells are not easily consumed by small predators, which often avidly feed on minute microorganisms.

Natural selection is also obvious among microorganisms that are associated with human or animal disease. In these instances, the selective force is often the host. The successful parasite or disease agent makes use of the many nutrients within the host, and it copes with host defenses. In contrast, a nonparasitic species often owes its failings to its inability to become established in the environment, which in this case is the body of the potential host. This lack of establishment may be related to the absence of enzymes, allowing the population to make use of host constituents or of structural properties that allow it to ward off host defenses. In these instances, the enzymes or morphological properties are essential fitness traits.

Natural selection is also evident in human medicine and in the gain in properties by an invasive microorganism that converts the invader from an avirulent to a virulent strain. Antibiotics are of enormous importance in human medicine, but associated with their use is the appearance of drug-resistant bacteria. Their appearance is a reflection of the forces of natural selection because the drug-resistant variant form of the pathogen has a unique fitness trait that is absent from the original species: it possesses physiological or structural properties that allow the variant to grow in the presence of a stress not tolerated by the original population. The latter did not have these special traits. This is evident in the increase in resistance among many bacteria to penicillin, an outcome of the widespread use of the antibiotic in chemotherapy (Figure 17-2). Similarly, the development of many microorganisms in human, animal, and plant hosts results in a shift in the population from one dominated by avirulent individuals to one dominated by virulent cells. The host is the selective factor for the vir-

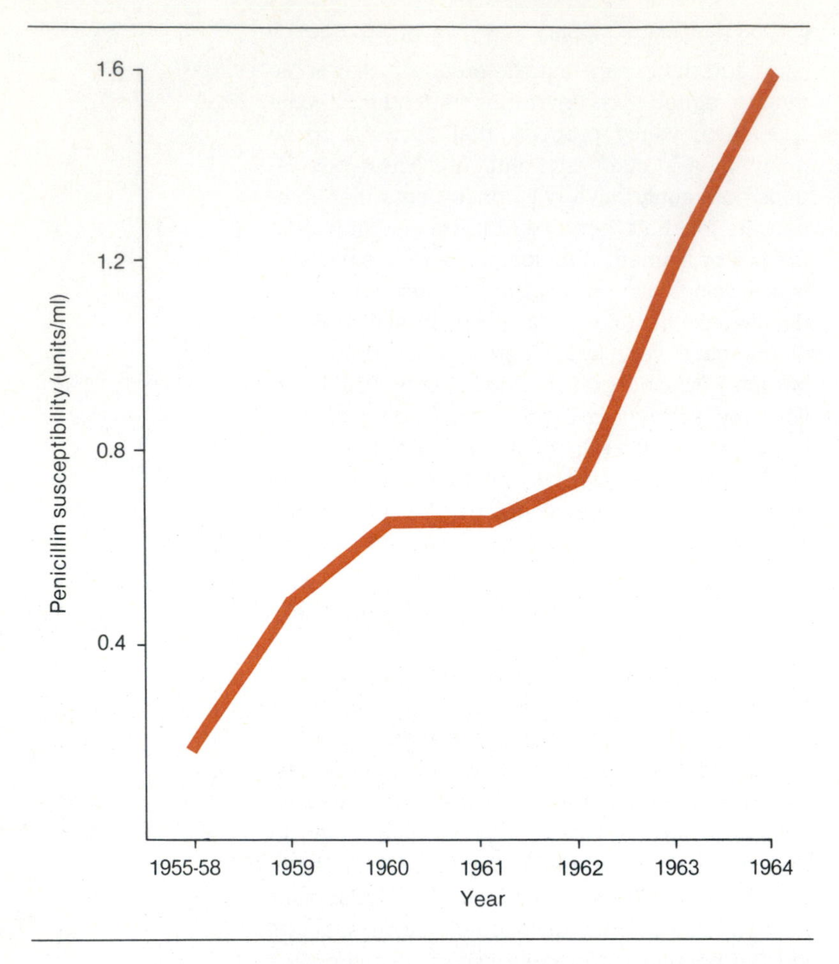

Figure 17-2 Rise in resistance to penicillin of isolates of *Neisseria gonorrhoeae* not eliminated by concentrations of antibiotic commonly used in therapy.

Modified from Thayer, J.D., et al. In Antimicrobial agents and chemotherapy—1964. Am. Soc. Microbiol. *Ann Arbor, Mich., p 433, 1964.*

ulent pathogens. The avirulent cells, merely because they do not have the requisite fitness traits, do not win the struggle for existence in the environment. Here, the host is the agent of natural selection.

SPECIES DIVERSITY

The number of species in a given habitat or microhabitat is often large, and soils and sewage harbor a multitude of species in a small area. In contrast, some communities have few species, and some may have only one. The reasons for these differences in species diversity and numbers are usually unknown. Few species are present in environments having a high intensity of some deleterious factor, as in localities with high temperatures, where drought is common, or with high levels of salts or, as in many foods, sugars. Conversely, the presence of many different populations at a particular site is linked with nutrient deficiencies; many algae are present in nutrient-poor waters, and many bacteria are present in soils having little available organic matter. Why so many species coexist at a particular site has long intrigued ecologists. Possibly, the coexistence is attributable to the dissimilar functions of the various populations, and species possessing the diverse physiological properties are able to live together. In many environments, a species that might be expected to dominate is suppressed by parasites or predators, and thus overexploitation by a single species is rare. Some environments have mechanical barriers or gradients of stresses or nutrients, so that one species exists at only one level of the gradient or is protected from harmful neighbors by the barriers.

Food chains

Food chains also permit the coexistence at a particular site of many dissimilar species. Growth of the indigenous populations requires sources of energy, carbon, other elements, and growth factors. The energy source for higher plants, algae, cyanobacteria, or photosynthetic bacteria is light; for heterotrophs it is an organic molecule; and for chemoautotrophs it is an inorganic ion or molecule (Figure 17-3). These organisms, which are at the base of food chains (first ***trophic level***), proliferate at the expense of their energy source and excrete organic compounds. These excretions, or products released when the cells die, sustain other populations. Moreover, the cells of the populations using the first energy sources serve as nutrients for predators such as protozoa and for a variety of microbial parasites. The organisms at the second trophic level live using products or cells of those species using the energy sources that enter the environment from the outside. In turn, these secondary feeders will themselves be nutrients or will excrete nutrients for new feeders (the third trophic level), and the latter in their turn provide nutrients for still other organisms. Complex food chains of these sorts are common in soils, natural waters, and sewage. Hence, even in an environment in which the only energy source is sunlight, a variety of species will coexist.

NUTRITION AND ECOLOGY

The habitat of a species is the source of all of its nutrients, inorganic or organic. From its surroundings, each species must obtain a variety of inorganic substances, an energy and a carbon source, as well as any ancillary compounds that

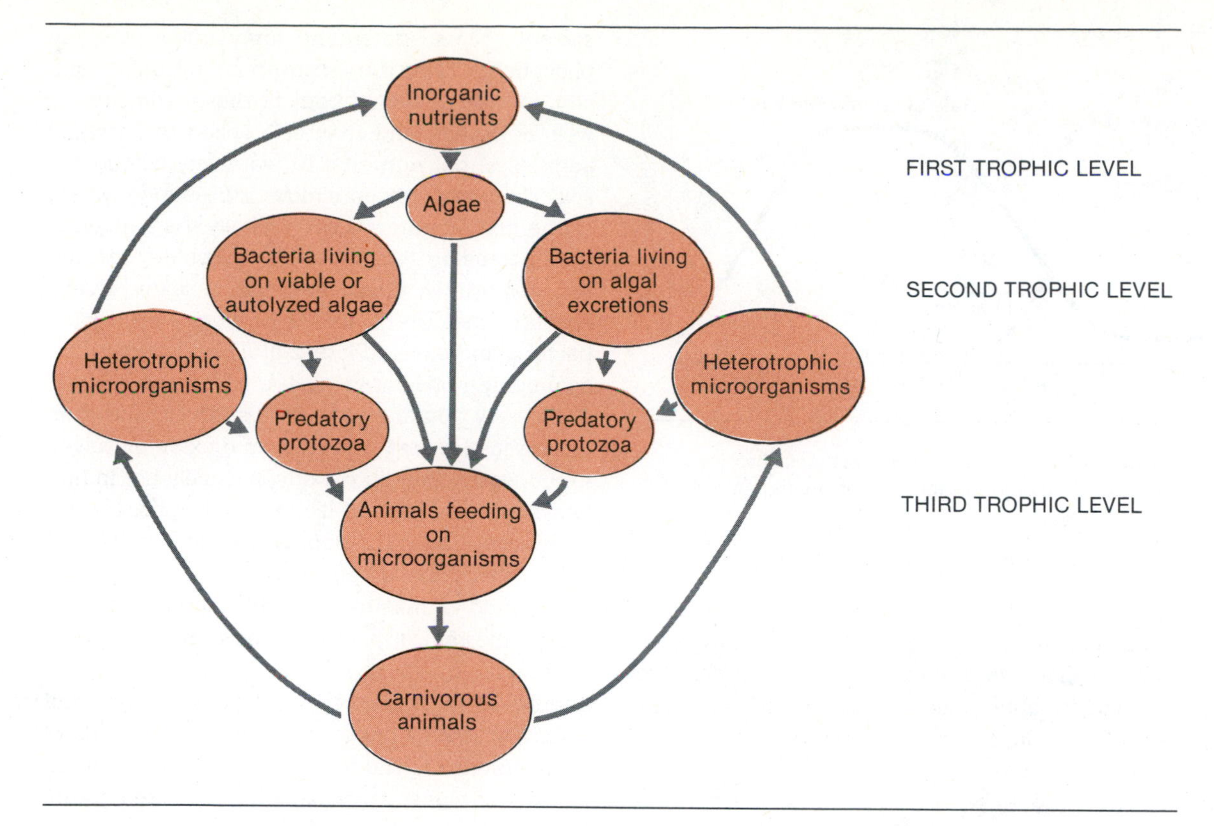

Figure 17-3 Typical food chain.
Modified from Alexander, M.A. Microbial ecology, *John Wiley & Sons, Inc., New York, 1971.*

are required for multiplication. The absence of a species from an environment could result from the lack of one or more of its essential nutrients. Conversely, if a population grows, all of its needed nutrients must be provided. Nevertheless, chemical analysis of that environment may show none of the nutrient to be present, but this results from the entire supply being assimilated by the proliferating populations.

Not all nutrients are ecologically significant. Should the supply of a particular needed compound or ion be greater than the demand of the entire community, that substance probably is not significant in determining the activity or the occurrence of individual species: all inhabitants get what they require. In contrast, should the supply be less than the demand, the species will interact for the limited supply, and an organism that is unable to obtain what it requires will not be established.

Growth factors

Particular attention has been given to growth factors because many bacteria, protozoa, fungi, and even aquatic algae require them for development. Each of these exacting species is unable to multiply unless its demand for these essential organic compounds is met. The microorganism that requires several growth factors is at a disadvantage in many ecosystems because these factors may not always be present. Nevertheless, microorganisms that need growth factors are widespread, meaning that the species at some time and place must find a source of their essential nutrients. Indeed, many environments that one might not consider to contain growth factors do in fact contain them. For example, a high percentage of the bacteria in soil and marine sediments and algae in the oceans requires these organic compounds, and a supply must thus be present. Moreover, if the gain of a requirement for growth factors makes a microorganism more versatile physiologically, then a fastidious species may be favored provided its nutritional demands are satisfied. Growth factor–requiring species are often highly active metabolically, so that natural selection may favor them over fastidious organisms in environments that have an adequate supply of the needed substances.

For a species to cause disease, it must find a host, invade it, and multiply. In many instances, the potential parasite is unable to proliferate because of defense mechanisms of its host. In some instances, the lack of proliferation is the result of the absence from the site in the host body of a

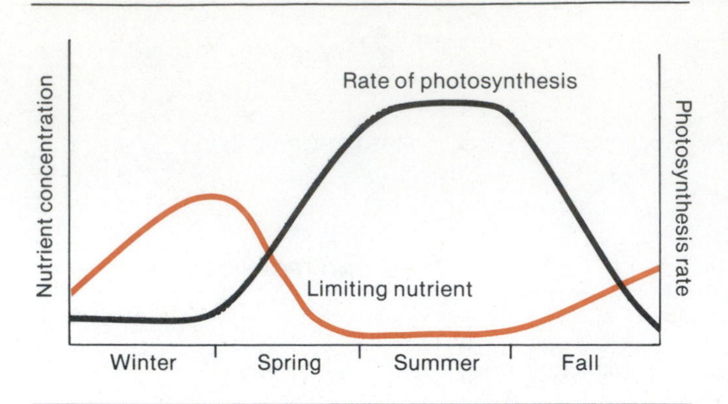

Figure 17-4 Relation between rate of algal photosynthesis and concentration of limiting nutrients in lake water.

full complement of microbial nutrients. A number of bacteria requiring amino acids for growth fail to cause harm to animal or plant hosts because they cannot obtain needed amino acids and thus cannot reproduce and cause disease.

Nutrient levels in the biosphere. Most of the inhabitable sites of the biosphere contain extremely low levels of available organic matter. About 99% of the volume of biosphere is made of up of oceans, and in oceans and lakes the total amount of organic carbon is usually 1 to 25 μg/ml. If this is the total amount of organic matter, the supply of available organic carbon must be much lower. Nevertheless, many microbiologists use extremely rich media in the laboratory, and hence the organisms that they isolate may not represent those that function in the natural nutrient-poor habitat. Based on the concentration of nutrients they need, microorganisms may be divided in two categories: ***eutrophs*** that grow at the high levels of nutrients common in laboratory media but only in certain environments, and ***oligotrophs*** that grow at low concentrations of nutrients, especially organic carbon. It is still not clear why the latter do not grow at high nutrient concentrations, and the former do not multiply at low nutrient levels. Oligotrophs are probably important in natural waters because of the low levels of organic matter. Moreover, the widespread concern with chemical pollution of waters and soils is directed at compounds whose concentrations are usually no greater than 100 parts per billion. At these concentrations, only oligotrophs are presumed to grow.

Lakes, rivers, and streams have been receiving an increasing supply of inorganic nutrients. These nutrients come in part from natural sources. In part, they come from the activities of society. Thus, detergent formulations contain phosphorus, fertilizers contribute nitrogen, and city sewage provides both of these elements as well as others, and manure applied to farmland adds essential nutrients to water underlying the soil. The latter nutrients move into surface waters as the groundwaters move laterally. As a result of the increasing amounts of phosphate, nitrate, and sometimes other nutrients in lakes, rivers, and streams, the ***biomass*** of algae and cyanobacteria increases (Plate 20). These species are commonly not limited by the energy source because they obtain energy from sunlight, and their carbon is obtained from carbon dioxide dissolved in the water. Carbon dioxide is usually not in limited supply in waters. If nitrogen or phosphorus, for example, limit the photosynthetic populations, the introduction of phosphate or nitrate into a lake would stimulate these populations and increase the rate of photosynthesis. In the fall, the organisms die and release the element again (Figure 17-4). The excessive growth of algae and cyanobacteria is one of the obvious problems of ***eutrophication,*** the increasing fertility of waters. Growth of algae, cyanobacteria, or rooted plants in eutrophic waters may be so massive that the movement of boats through waterways is reduced (Plate 21). More often, however, the practical concern with eutrophication arises from dying of the algae and cyanobacteria. During the decline of these populations, bacteria proliferate and produce offensive tastes and odors. The bacterial activity may be so pronounced that oxygen is totally removed from the water. This leads to the death of fish, which are aerobes. In addition, toxins may appear in the water, and these toxins may affect fish as well as human consumers of what was once a supply of potable water. The practical consequences of eutrophication have led to governmental action and also a renewed interest in the nutrition of aquatic populations.

COLONIZATION AND SUCCESSION

Many environments initially have a small biomass and few populations. This is true of lava that has cooled after a volcanic eruption and the surfaces of rocks that become exposed by a breakage of the rock. Tissues of plants and animals and body fluids are also usually free of microbial life. The newborn infant and the seedling emerging from the seed similarly do not initially contain a microbial community. Fresh fruits that become bruised or canned foods on exposure to the air also are initially free of any forms of life. These materials

Table 17-1 Succession in aquatic anaerobic sediments governed by energy-yielding oxidation-reduction reactions

METABOLIC PROCESS	BIOGEOCHEMICAL ZONE	PROMINENT BACTERIA	CHEMICAL EQUATION	ΔG°*
Aerobic respiration	Aerobic zone	Aerobic heterotrophic	$CH_2O + O_2 \longrightarrow CO_2 + H_2O$	−686
Anaerobic respiration	Redoxocline	Nitrate-reducing and denitrifying	$5\ CH_2O + 4\ NO_3^- + 4\ H^+ \longrightarrow 2\ N_2 + 5\ CO_2 + 7\ H_2O$	−579
Anaerobic respiration	Sulfate-reducing zone	*Desulfovibrio, Desulfotomaculum*	$2\ CH_2O + SO_4^{2-} \longrightarrow S^{2-} + 2\ CO_2 + 2\ H_2O$	−220
Anaerobic respiration	Carbonate-reducing zone	Methanogenic	$2\ CH_2O \longrightarrow CH_4 + CO_2$	−99

From Jones, G.E. Biogeochemical succession of bacterial activities in aquatic sediments. In *Microbiology 1980*. American Society for Microbiology, Washington, D.C., 1980.
*Expressed in kilocalories per mole of glucose equivalent oxidized; ΔG°, change in Gibbs free energy.

rapidly become invaded, and microbial colonization is initiated. ***Succession*** is the replacement of populations and community types that occurs with time. The initial populations in the lifeless regions are the pioneer species, but a community dominated by pioneer species is soon replaced through a series of successional events. These events are rarely abrupt, and a slow progression of species changes is more characteristic of sites that are subject to microbial colonization.

The identity of the pioneer is determined by the physical and chemical properties of the environment undergoing colonization. This is evident in the differences in pioneer communities of different types of food products and in the dissimilarities in the bacterial communities of the gastrointestinal tracts of newborn children fed mother's and cow's milk.

In an environment with little organic carbon, as in waters or on the surfaces of rocks, the pioneers are usually algae, cyanobacteria, or lichens. The lichen is a symbiosis of a fungus with an alga or a cyanobacterium (Plates 22 and 23). In aquatic sediments where there is considerable organic matter, the pioneers are different. The succession of bacteria in these environments, where the rate of organic deposition exceeds the supply of dissolved oxygen, is related to depth of sediment and availability of electron acceptors. A thin aerobic zone appears at the surface that is dominated by organisms using aerobic respiration (Table 17-1). The zones below the aerobic zone are anaerobic, and the organisms use terminal electron acceptors other than oxygen, that is, nitrate, sulfate, and carbonate, as well as organic molecules. The metabolically efficient microorganisms appear near the surface, whereas the less efficient ones appear at greater depths in the sediment. In contrast, initially sterile sites that are rich in organic materials have quite different pioneers. In this regard, considerable attention has been given to the colonization of the newborn baby because such colonization is of considerable practical as well as ecological interest, especially if the potential colonist is a pathogen.

Table 17-2 Microorganisms found in the saliva of humans

ORGANISM OR GROUP	APPROXIMATE PERCENTAGE
Streptococci	40 to 50
S. mitis	
S. salivarius	
S. mutans	
S. sanguis	
Enterococci	
Diphtheroids	5
Bacteroides	5
Fusobacterium	2 to 3
Corynebacterium	10 to 12
Veillonella	14 to 15
Peptostreptococcus	10 to 12
Neisseria	1
Actinomyces	1 to 2
Vibrio	1

Specificity of colonization

Microorganisms are specific for the sites that they colonize. This specificity, in the newborn infant, for example, may be associated with development of certain species only in the mouth

Table 17-3 Barriers to colonization by microorganisms

ENVIRONMENT	BARRIER POSTULATED OR DEMONSTRATED
Skin	Fatty acids
Lungs, stomach	Mucosa
Nose	Lysozyme
Blood	Phagocytes, antibodies
Intestine	Mucosa
Fish tissue	Protamines
Milk	Peroxidase, agglutinins
Roots	Cork layer
Fruits	Cuticle layer
Trees	Gums, resins, tannins
Plant tissue	Phenolic compounds, glycosides
Virus-infected cells	Interferon

From Alexander, M. *Microbial ecology*, John Wiley & Sons, Inc., New York, 1971.

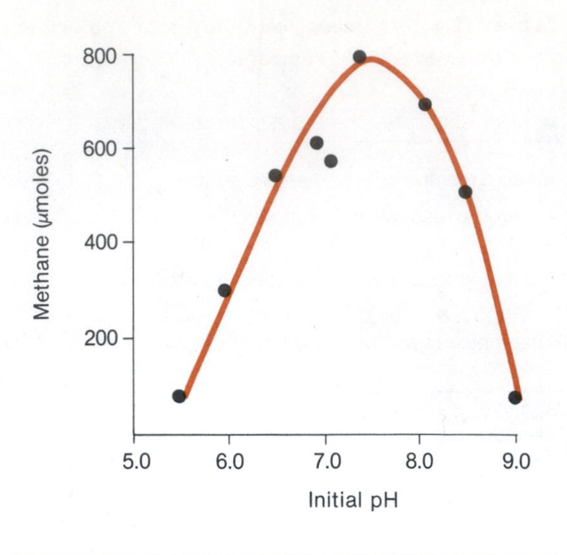

Figure 17-5 Effect of pH on methane production in thermophilic methanogen cultivated in laboratory. *Redrawn from Ferguson, T.J., and Mah, R.A.* Appl. Environ. Microbiol. ***45**:265, 1983.*

(Table 17-2). The successful colonist of this environment must be able to adhere to the epithelial cells of the cheek or tongue and grow once retained on the surface. In the absence of adherence, the organism would be washed out and into the stomach. Colonization of specific sites is also evident among the algae that are uniquely adapted to grow on rocks at the seashore of coastlines that are buffeted by intense wave action. Toxin-producing bacteria associated with diarrhea are also specific for their sites of colonization, and these organisms are specialized for the colonization of surfaces of the intestine. Often, the specficity of microoganisms for individual sites is related to their having a peculiar surface constituent. In turn, the habitat may have specific receptors or other structural features that allow for colonization.

Barriers to colonization

No species is native to every site to which it is transported. The reason is that properties of the environment exclude many arrivals. These properties are often called ***barriers to colonization,*** that is, conditions restricting population size or activity or preventing the establishment of an arriving species. The barriers are frequently mechanical. Mechanical barriers include the skin, mucous membranes, the cuticle of fruits, the bark of trees, and the gum that is exuded by certain plants when they are wounded (Table 17-3). Breaching of these barriers permits colonization, as when the human skin is damaged or plants are bruised.

A particular species may fail to become established because of a chemical barrier in a given environment. Acidity is a common chemical barrier (Figure 17-5) and probably explains the failure of many species to become established in the stomach or in acid foods, such as citrus and tomato products. Organic acids themselves, in addition to the low pH associated with them, are also major barriers to the establishment of microorganisms; witness the long life of vinegar that contains acetic acid or of pickles and sauerkraut, products that contain lactic acid. The inability of many bacteria to colonize the surface of the eyes or the mucosa of the nose has been related to the presence of an enzyme, lysozyme, that lyses many of the potential colonists. The defense mechanisms of humans, animals, and plants also are often chemical barriers, as in the cases of antibodies and the antimicrobial agents produced by plants.

Composition of the community

As the pioneers are displaced and new species arise in succession, the community is altered. Some organisms are totally eliminated, and others become relatively rare. At the same time, new organisms come to be dominant. The changes in community composition as a result of succession may arise because the early arrivals produce compounds that inhibit themselves, they produce nutrients useful to the late arrivals, or they expose organic materials that were not previously accessible. The composition of the community may also be altered because of modifications not brought about by the indigenous populations; for example, changes in light intensity and temperature with the seasons of the year or the feeding on microorganisms in waters by small animals. In some cases, the pioneers or other species early in succession reduce the supply of inorganic or organic nutrients, and the new arrival is better able to grow at the new nutrient concentrations than are the earlier populations.

Regardless of whether a species is evident during initial colonization, late in succession, or in the finally established community, it has certain features that dictate at which stage it appears. For some organisms, the unique trait in colonization, succession, or community establishment is their rapid growth. For other species, it is their capacity to tolerate nonbiological stresses or, for pathogens, to overcome barriers erected by the animal or plant host. For a population growing on rocks in a stream, its particular position in the colonization-succession sequence may result from its capacity to produce specialized structures allowing it to adhere to the surfaces of the rocks. However, only a few of the distinctive advantages possessed by individual species are presently known.

In most habitats, wave after wave of populations appear and then recede (Figure 17-6). These waves may succeed one another rapidly, or they may proceed slowly. Ultimately, a community is established that is relatively stable. In this community, known as the ***climax community,*** the dominant organisms and the relative abundance of the species are reasonably constant. The species are in physiological and nutritional equilibrium with their environment, and these are the species that have been selected out among the numerous arrivals for their ability to grow and tolerate all of the biological and nonbiological stresses to which that environment is exposed. To maintain the climax community, there must be a continual inflow of energy. That energy may come from the organic carbon formed by algae in photosynthesis, the excretions of plant roots, food ingested by animals, or the organic matter introduced into streams and rivers. It is because of the known composition of many climax communities that microbiologists know where to look for a particular organism they wish to isolate.

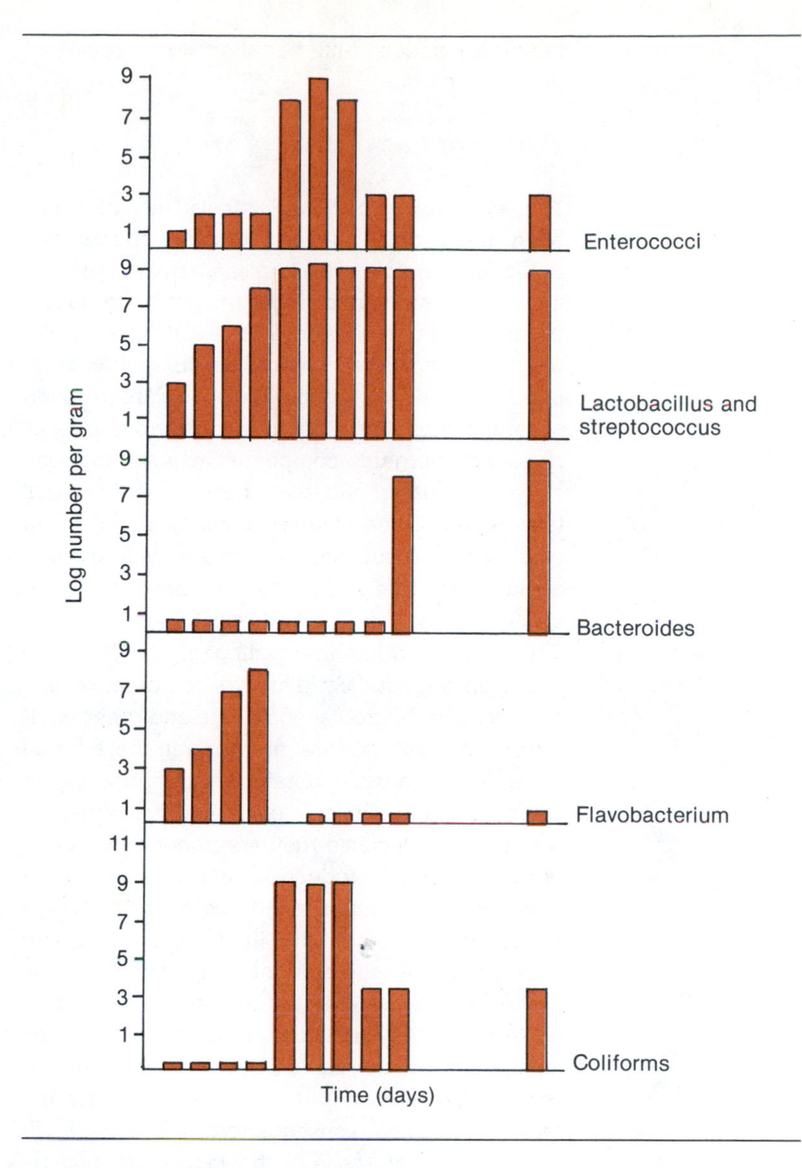

Figure 17-6 Succession of microbial populations in large intestine of mice following birth. No samples collected after 18 days. Initial colonizers are lactobacilli, enterococci, and flavobacterium. This is followed by facultative anaerobes such as coliforms. Finally, the anaerobic colonizers, bacteroides, predominate.
Redrawn from Schaedler, R.W., et al. J. Exp. Med. ***122:****59, 1965.*

MICROORGANISMS AND BIOGEOCHEMISTRY

Biogeochemistry is concerned with biological activities that affect the chemistry of the earth. Microoranisms are extremely important in this regard because processes they bring about alter the chemistry of the atmosphere, oceanic and inland waters, marine and freshwater sediments, and soils. In some of the geochemical processes in which they are implicated, microorganisms bring out changes that are essential to life on earth. Other of these reactions are not critical for the maintenance of life but alter the chemistry of the earth.

Microbial activities

Biogeochemical cycles. Particular attention has been given to the role of microbial communities in biogeochemical cycles. In some cycles, the elements are circulated from inorganic to protoplasmic and thus organic forms, and then the elements are converted back to the inorganic state. In other cycles, no organic molecules are involved or are significant, but rather microorganisms transform inorganic compounds or ions from one state to another and then back to the original form again. Some of these processes involve an oxidation, and subsequently in the cycle there is a reduction; this is true, for example, of the cycles of nitrogen, sulfur, and iron.

Interest in microbial cycling of the elements has risen dramatically. This has occurred for several reasons. Microbial ecologists and other environmental scientists are interested in these transformations in order to obtain new information. On the other hand, there is enormous concern with these cycles because they are important, indeed critical, for food production and, hence, for human life. For example, the rates of certain steps in the nitrogen cycle often limit plant growth and thus, indirectly, the amount of meat that can be produced. In waters, the microbial cycling of phosphorus may limit the development of algae and hence control fish production. Another reason for the interest is to be able to predict the regional or global consequences of new technologies. Some of these technologies are not restricted to small regions and could affect the global cycling of an element that is essential for human health or well being. It has been suggested that nitrogen-containing pollutants discharged from cities and agriculture or sulfur emitted in the burning of coal and the operation of mines may have catastrophic consequences. This had led many scientists to try to find out whether these pollutants will seriously affect global or regional cycles essential for life.

Another reason to examine the role of microorganisms in biogeochemical cycles is to explain anomalies in the amounts of certain pollutants in the atmosphere. Studies designed to explain these anomalies have disclosed that microbial communities are probably major sources of some air pollutants and are active in preventing the accumulation of other pollutants in the atmosphere. This they do by activities not in the air but rather in the underlying soils and oceans. These findings have been made as part of assessments of the cycling of sulfur, carbon monoxide, and hydrocarbons in the atmosphere.

For elements for which the cycle is complete, there is no long-term upset in the earth's chemistry as a result of the cycling. Thus, many of the steps in the nitrogen cycle result in major chemical changes, but these are balanced by reactions that go in the reverse direction. In contrast, certain elements undergo noncyclic modifications, and these lead to lasting modifications in the distribution of the elements or compounds wherein the element is found. Such modifications take place either on or below the earth's surface.

Some of the elements that are transformed microbiologically follow.

- Important in metabolic reactions or cells of microorganisms
  - Major: C, H, O, N, P, S, K
  - Minor: Fe, Mg, Cu?, Zn?, Co?
  - Needed only by some microorganisms: Si, Mn, Ca, Na, Mo, B, V, Cl, Se, Ni
- Not biologically important (in cell structure)
  - Field evidence for transformation: Al, As, Hg, Pb
  - Only laboratory evidence: U, Bi, Te, Os, Cd, Ag, Au, Sb, Ba

Some of the elements are important in cell metabolism. Of these, elements such as carbon, oxygen, nitrogen, hydrogen, and phosphorus are major cell constituents. Others are only minor cell components, and thus the microbial biomass in waters and soils does not contain much of the element. In contrast, a variety of elements are transformed microbiologically, although they are not important to the species bringing about the transformations. Microbial communties oxidize or reduce these elements or form complexes with them that are of geochemical significance.

Mineralization. One of the major microbial processes in geochemistry is mineralization. Mineralization is the conversion of the organic form of an element to the inorganic state. If the element in inorganic form is important as a nutrient (such as nitrate, phosphate, and sulfate), mineralization, by converting the organic complexes to inorganic products, regenerates the nutrients for aquatic or terrestrial plants. Carbon, nitrogen, phosphorus, and sulfur, because they exist in organic complexes in plants, animals, and microorganisms, are thus subject to mineralization. In geochemical terms, mineralization decreases the biochemical complexity of the ecosystem, and it also prevents the accumulation of organic materials in waters and soils. At the same time, the mineralization of carbon provides energy to heterotrophs, and thus carbon mineralization is a driving force for many biogeochemical cycles. Nitrogen mineralization is critical in many soils because it is inorganic nitrogen that is assimilated and used by higher plants. The rate of formation of inorganic nitrogen by microbial mineralization of organic compounds thus ultimately regulates plant growth. Similarly, phosphorus mineralization is important in many waters because aquatic algae use phosphorus in the inorganic state, and the phosphorus supply is often less than the algal demand.

Oxidation and reduction. Microbial oxidations are also prominent in waters and soils. The elements subject to oxidation include carbon, nitrogen, sulfur, hydrogen, and iron. The energy released in the oxidations may be captured by individual populations. It is, by definition, the heterotroph that is able to capture the energy released by oxidation of carbon compounds. Also, by definition, it is the autotrophs that capture energy released by oxidation of other elements. However, the energy released in oxidations of some elements may not be captured. In these cases, the energy is lost as heat. in many natural environments, the sulfur and nitrogen in organic or inorganic compounds may be oxidized by microbial activities, yet the responsible populations acquire little or no energy thereby. This contrasts with autotrophs that oxidize inorganic nitrogen or sulfur compounds or ions, getting benefit from the reactions they carry out. With some elements, the oxidation is not the result of an enzymatic process but rather results from a product excreted by a microorganism; this is one way that microorganisms alter iron or manganese ions.

Reductions are also characteristic of microbial activities in natural environments. The reduction may be of an organic substrate, in which case reduced carbon-containing products accumulate. In habitats having oxygen, a consequence of the reducing activity of microbial communities is to convert oxygen to water. This is, of course, the key reaction of aerobic populations. When oxygen is not present but oxidized forms of iron, manganese, sulfur, or nitrate or carbonate ions are, these ions may be reduced. Such reductions are believed to account for significant accumulations of reduced forms of iron and manganese, the appearance of sulfides, the production of dinitrogen and nitrous oxide from soils, and the generation of methane from anaerobic habitats.

Methylation. Microbial methylation may occasionally be ecologically important. In methylation, a methyl group is attached to an element to give methyl derivatives. Depending on the element, one, two, three, or four methyl groups may be attached to mercury, sulfur, tellurium, arsenic, or lead. A few of these methylations may be geochemically prominent; methyl derivatives of sulfur, for example, are important means for the emission from soil of volatile sulfur. Such compounds may be major contributors to the sulfur content of the atmosphere. Great attention has been devoted to methylations that convert an element into a highly toxic form. By adding methyl groups to mercury in sediments, for example, microorganisms create major contamination problems in fresh waters. Methyl forms of arsenic are also highly toxic and are readily produced by microbial action (Figure 17-7).

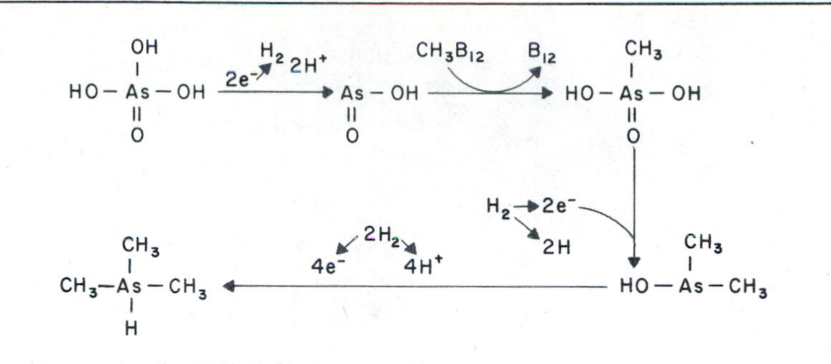

Figure 17-7 Methylation of arsenic. CH_3B_{12}, methylated form of vitamin B_{12}.
Redrawn from Wood, J.M. Environmental pollution by mercury. Adv. Env. Sci. Tech. **2***:39, 1971. Copyright John Wiley & Sons.*

Carbon cycle

Microbial communities are major participants in the carbon cycle. In the oceans, algae dominate photosynthetic activities. The algae concentrate carbon from the atmosphere (where, as carbon dioxide, it is extremely dilute) and convert it to products that can be used by other species. The light energy is thus made available to heterotrophs. In geological terms, the rate of photosynthesis is rapid relative to the amount of carbon available in the atmosphere. Thus, possibly 5% to 10% of the carbon dioxide in the air is converted to organic carbon in algal cells inhabiting marine waters and in higher plants on land each year. To prevent depletion of the atmospheric supply of carbon dioxide, the algal and plant carbon must be returned rapidly to the atmosphere. To explain the almost constant level of carbon dioxide, it is generally assumed that the rate of photosynthesis equals the rate of carbon dioxide formation. The latter, of course, is carbon mineralization. It is widely believed that most of the carbon dioxide that is returned to the atmosphere comes from microbial activity in soils and in waters, although

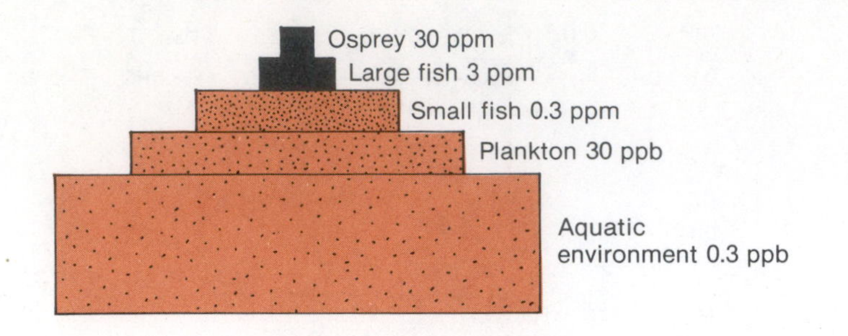

Figure 17-8 Biomagnification of DDT: Dissolved DDT in aquatic environment moves through plankton and other trophic levels where it is finally concentrated in osprey tissue. DDT concentrations are idealized and, although typical, they do not represent actual analytical data.

Redrawn from Atlas, R.M., and Bartha, R. Microbial ecology: fundamentals and applications, *© 1981, Addison-Wesley Publishing Co, Inc., Reading, Mass., p. 417, figure 13.2. Reprinted with permission.*

Table 17-4 Recalcitrant molecules

PESTICIDES	POLYMERS	OTHER SYNTHETICS
DDT	Nylon	TNT
Chlordane	PVC	Dioxin
Heptachlor	Polyethylene	Chlorinated alkane
Aldrin	Polypropylene	ABS surfactants

respiration of plants makes a significant contribution to the regeneration of carbon dioxide.

The populations responsible for mineralization are unknown. It is generally assumed that bacteria are particularly important, especially in waters, but that fungi and bacteria dominate in soil. Although the responsible populations have yet to be identified, no doubt exists that the number of different organic compounds that can be mineralized is enormous. Any compound that is biologically formed is mineralizable in environments where oxygen is available. Thus, all of the polysaccharides, lignins, proteins, nucleic acids, and lipids of plants, microorganisms, and animals are transformed, either rapidly or slowly, to the inorganic form of carbon that enters the atmosphere. Were it not true that every biologically formed organic compound is mineralized, then those resistant compounds would accumulate in nature. Given that the biosphere is more than 1 billion years old, that accumulation would be massive. The only places at the surface of the earth where such accumulations do in fact occur are in the deposits of peat and related organic materials. These in turn may be subjected to nonbiological forces to give rise to coal. The fact that such massive deposits of organic matter exist points to a deficiency of microorganisms, namely, that they do not mineralize rapidly or completely in natural habitats that have little or no oxygen. It is in these anaerobic environments where such accumulations do arise.

Although it is generally accepted that every biologically produced organic compound is mineralizable when oxygen is present, the same is not true for many synthetic organic compounds. Some synthetic organic molecules are resistant to microbial attack, either partially or completely, and thus their release into waters and soils is not a prelude to mineralization or biodegradation. If these resistant organic molecules happen to be toxic to humans, animals, or plants, they are deemed to be pollutants. Such is the case for some of the early insecticides, such as DDT, and it is also true of a variety of pollutants from the chemical industry, such as polychlorinated biphenyls (PCBs). Although the levels of these pollutants may be relatively small in the aquatic or soil environment, they accumulate in both eukaryotic and prokaryotic cells. By the time the pollutant has reached the higher trophic layers in the food chain, it has increased severalfold, a process called ***biomagnification*** (Figure 17-8).

Let us use one example in the carbon cycle to show how basic studies of microbial biogeochemistry point to potential problems. The atmosphere contains about 10^{12} kg of CH_4 (methane), but the actual concentration in the air is small. In the past, no one has worried about methane in the atmosphere because it is relatively nonreactive. Geochemists estimate that 80% of the methane that is released to the atmosphere is produced by microorganisms in anaerobic environments. These environments may be marshes, swampy regions, or the paddy fields of Asia, but significant amounts also come from fermentation within the gastrointestinal tract of animals. The quantity of methane thus formed is large, and if there were no means for destryoing the methane in the atmosphere, the concentration should increase quite rapidly. Yet, the concentration is essentially constant. This indicates that there is a major means for destroying the methane in the atmosphere. It is currently believed that one of the chief mechanisms by which this methane is destroyed is by chemical oxidation in the upper atmosphere. However, this gives rise to carbon monoxide (CO), and carbon monoxide in the atmosphere, even at a very low concentration, is a significant pollutant. Such extrapolations lead to a surprising hypothesis: by

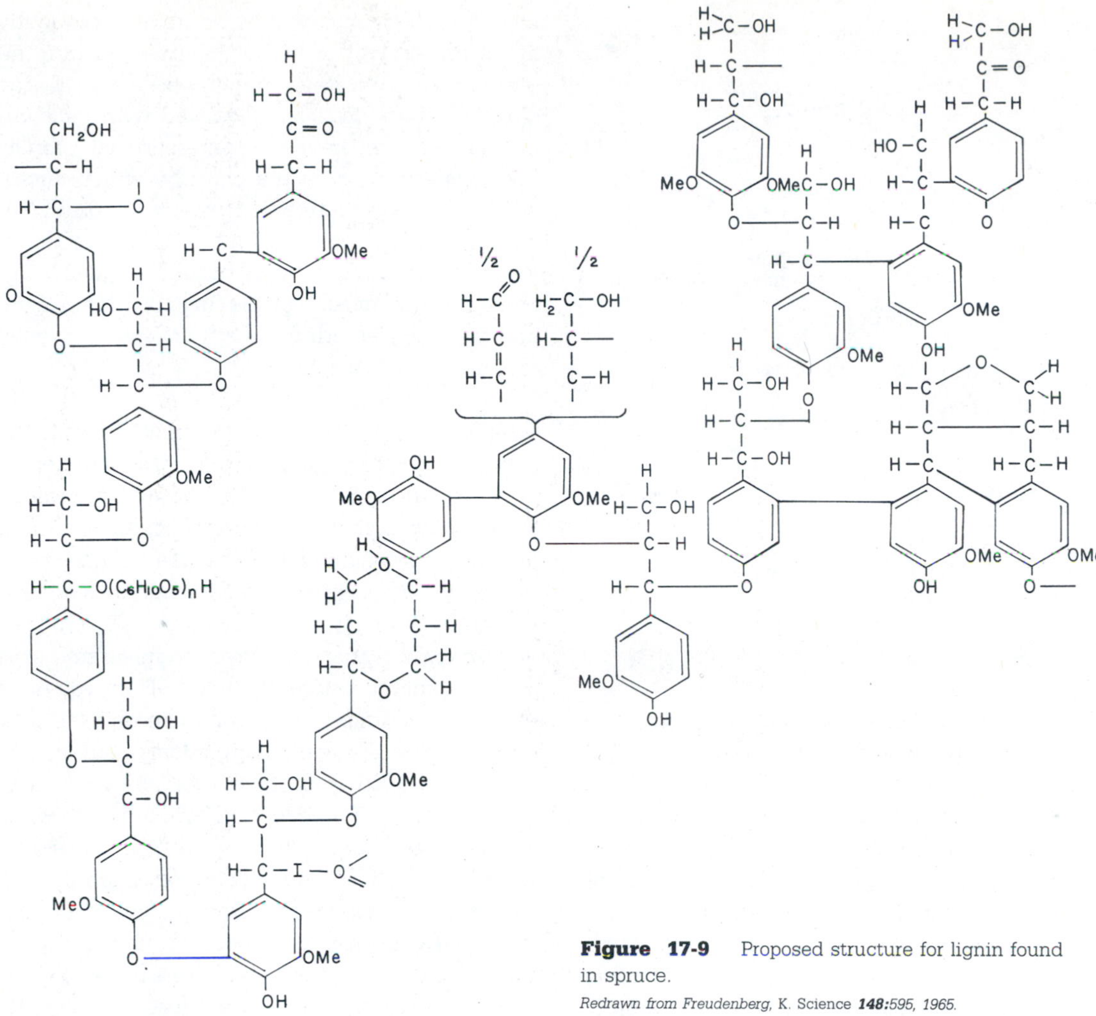

Figure 17-9 Proposed structure for lignin found in spruce.
Redrawn from Freudenberg, K. Science ***148****:595, 1965.*

producing methane, which is then converted in the atmosphere to carbon monoxide, microorganisms in marshes, flooded soils, and the alimentary tract of animals are major sources of a significant air pollutant, indirect though their contribution is.

Recalcitrant molecules. Microbiologists whose focus of interest is solely the laboratory are proud of their successes in obtaining isolates that are able to grow on, and therefore usually mineralize, a variety of organic compounds. As pointed out previously, microbial communities in natural habitats are extremely versatile, and individual species are able to metabolize a wide variety of organic compounds of biosynthetic origin. Summing up the versatilities of all of the representative populations within the community results in a community of impressive mineralizing potential. Based on this versatility, the assumption gained popularity that microbial communities were omnipotent and that they could degrade not only the compounds of natural origin but any organic compound, be it of synthetic or biological origin. Nevertheless, environmental scientists know that this assumption is not true. Many biologically formed products are resistant to microbial attack under certain circumstances, and several synthetic chemicals are resistant under all circumstances to mineralization. These chemicals are known as ***recalcitrant molecules*** (Table 17-4).

Among the chemicals or substances that are recalcitrant and persist for long periods in nature are lignin, which is a major constituent of higher plants, and the humus of soils (Figure 17-9). In addition, because the spores of bacteria and resting structures of various fungi endure in well-aerated soils, which contain highly diverse and active communities, one must conclude that the organic compounds at the surfaces of these structures are recalcitrant. Were they not resis-

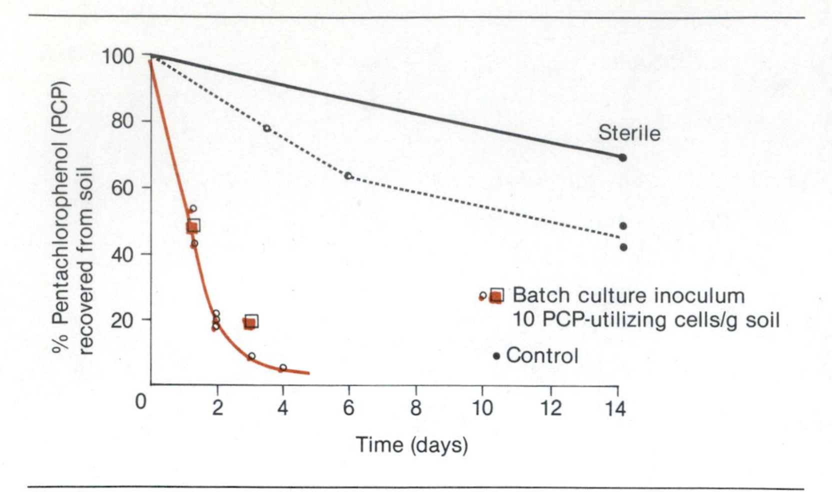

Figure 17-10 Degradation of pentachlorophenol in soil by indigenous and inoculated bacteria under laboratory conditions at 30° C.

Redrawn from Edgehill, R.U., and Finn, R.K. Appl. Environ. Microbiol. ***45:****1122, 1983.*

tant to microbial mineralization, the spores and resting structures would be destroyed.

Recalcitrant synthetic compounds have achieved prominence because they are sometimes significant pollutants. In this category are many pesticides, a large number of organic compounds in industrial wastes, synthetic polymers used for fabrics or packing materials, and products formed when organic compounds that are in natural waters are chlorinated as a step in municipal water treatment. If, in fact, carbon mineralization in nature is a result largely or entirely of microbial activities, the persistence of synthetic organic compounds illustrates that microbial communties are not infallible. Such a metabolic failing is revealed when pollutants persist in waters and soils. Establishing the reasons why organic compounds resist mineralization and determining the physiological bases for the failings of microorganisms are major areas of current research.

Inoculation of soils with microorganisms having special degradative properties is being used to clean up areas where there have been accidental spills of recalcitrant products or where toxic chemicals are stored and leak into the environment (Figure 17-10). Recombinant DNA technology is also being used to construct microorganisms with special degradative properties.

Nitrogen cycle

Many reasons exist for microbiologists to focus on the nitrogen cycle. This element is essential for plants and animals. On land and in certain inland and coastal waters, nitrogen frequently limits the growth of higher plants or aquatic algae. More often than any other element, the global food shortage is associated with deficiencies of nitrogen; hence, an international effort is underway to increase the ways by which microorganisms make nitrogen available to plants. In addition, excess nitrogen in fresh waters represents a significant form of pollution. Finally, microorganisms make volatile oxides of nitrogen, and these, on entering the air, contribute greatly to atmospheric pollution.

Mineralization. One of the critical steps in the nitrogen cycle is mineralization. As nitrogen is mineralized, it is converted from the organic form, which higher plants and most algae cannot use, to the inorganic forms that are readily assimilated by photosynthetic organisms on land and in waters. Nitrogen, as ammonium (NH_4^+), is also oxidized and is thereby converted in aerated soils and oxygenated waters to nitrate. This process is known as ***nitrification.*** Nitrate can be used as a nutrient source by microorganisms, but it also may be reduced to gaseous products when oxygen is absent. This reduction, which is known as ***dentrification,*** removes nitrate from soils and waters and converts it into dinitrogen, nitrous oxide, and other oxides of nitrogen. Dinitrogen (N_2) is used by a variety of heterotrophic bacteria, cyanobacteria, and a few actinomycetes; this process of ***nitrogen fixation*** (see Chapter 13) returns nitrogen from the gaseous form to compounds that other microorganisms, plants, and animals can assimilate. These processes occur in soils, the oceans, and lakes.

The substrates for mineralization are plants on land, the plankton in waters, the organic materials native to soils and sediments, and other materials that may be introduced above or below the surface of terrestrial and aquatic ecosystems. The microorganisms involved in mineralization are largely unknown, but many bacteria and fungi are able to use organic complexes of this element and convert them to ammonium. However, the product of the reaction is often at indetectable or low levels because algae or higher plants occupying the habitat assimilate the ammonium, or it is used by microorganisms either as a nutrient or as a substrate for nitrification.

Nitrification. In nitrification, ammonium is converted to nitrate (see Chapter 14). Other compounds or ions of nitrogen may also be substrates for the nitrifying microorganisms, and these substances are converted by the responsible populations to nitrate. Often, nitrite is an intermediate in the transformation. In most environments, nearly all of the nitrate found is formed microbiologically, and only a small amount is returned to waters and soils in the form of precipitation. The nitrate that is generated by the indigenous community is a nutrient for plants or microorganisms, but it can also be utilized by denitrifying bacteria. The nitrate that is formed in soil is important, moreover, not only because it is a nutrient for plants, but it is readily leached out of soil and thus enters underground waters, from which it may enter supplies of water used for drinking purposes. High levels of nitrate are undesirable in drinking water because of deleterious effects on infants.

It is generally believed that nitrification is carried out by autotrophic bacteria. These bacteria are represented by such genera as *Nitrosomonas* and *Nitrobacter*. The former oxidizes ammonium to nitrite, the latter oxidizes the resulting nitrite to nitrate. Several related genera of bacteria common in many soils and waters carry out the same reactions. However, indirect evidence exists for the functioning, at least at terrestrial environments, of heterotrophs that oxidize nitrogen in organic compounds, but they get little or no energy from the transformation. These organisms require a carbon source to provide energy for biosynthesis. The identity of these heterotrophs has yet to be determined.

The process of nitrification is extremely sensitive to environmental stress. Indeed, it is one of the most sensitive microbial processes in aquatic and terrestrial environments. For this reason, it is often used as a measure of stresses, and a pollutant that does not inhibit nitrification probably will not reduce the rate of any other process brought about by nonphotosynthetic microorganisms. Conversely, an inhibition of the nitrifying populations suggests that other species may be subject to environmental or chemical stress.

Denitrification. Microorganisms that carry out denitrification have been known for a long time. Only bacteria are responsible for this process. These bacteria use nitrate when oxygen is no longer present in the environment. They then reduce the nitrate to one of a series of products. Which product accumulates in the habitat of the organism is of great importance. It was once widely believed that the product in nature was solely dinitrogen. However, recent research has shown that nitrous oxide (N_2O) and nitric oxide (NO) are also released during denitrification, and these gases may be emitted from soil to the atmosphere. The latter gas reacts in the air with oxygen to yield nitrogen dioxide (NO_2). These three oxides are significant pollutants. Nitrous oxide diffuses from soils to the lower atmosphere, and from there to the stratosphere. In the stratosphere, it reacts with ozone (O_3). The ozone is then partially destroyed, and this destruction results in an increased incidence of ultraviolet light reaching the earth. Ultraviolet light is hazardous to many animals and plants and is a cause of skin cancer of humans. Most or nearly all of the nitrous oxide is released as a result of microbial activity, especially in soils, and it has thus been suggested that microorganisms are, indirectly to be sure, causes of skin cancer. The other two oxides of nitrogen, nitric oxide and nitrogen dioxide, are themselves toxic to humans, and the calculations of atmospheric scientists suggest that most of these two pollutants are generated not by automobiles and other industrial processes, as is widely believed, but rather result from microbial activity in nature, especially from soils.

Nitrogen fixation. The sole organisms that can bring about nitrogen fixation are microorganisms. The responsible species either live free in soils and waters or form symbiotic associations with higher plants. Some (the cyanobacteria) are symbionts with fungi in lichens. The symbiotic associations with higher plants are usually localized in structures on the roots, called ***nodules*** (Figure 17-11). Especially prominent are the nodules on legumes, which are induced by members of the bacterial genus, *Rhizobium*. In addition, members of the actinomycete genus, *Frankia*, also bring about nitrogen fixation in the nodules of a few nonlegumes (Table 17-5).

The processes involved in nitrogen fixation make the element available to the responsible microorganisms. If the association involves a higher plant, that nitrogen is incorporated into plant tissue. In the free-living bacteria and cyanobacteria, the nitrogen is released as a result of (1) excretion by the active populations, (2) their decomposition by other organisms, or (3) the feeding on these populations by predators. In this way, the nitrogen, which orginally was in an unavailable form (dinitrogen); is provided to other organisms inhabiting the same ecosystem. Because nitrogen fixation serves much the same function as chemical fertilizers in agriculture and fertilizers are expensive in developed countries and unavailable in many of the poor countries, considerable research is underway to enhance nitrogen fixation. Particular attention is being given to increasing the activity of the symbiosis between *Rhizobium* and leguminous plants. In addition, attempts are

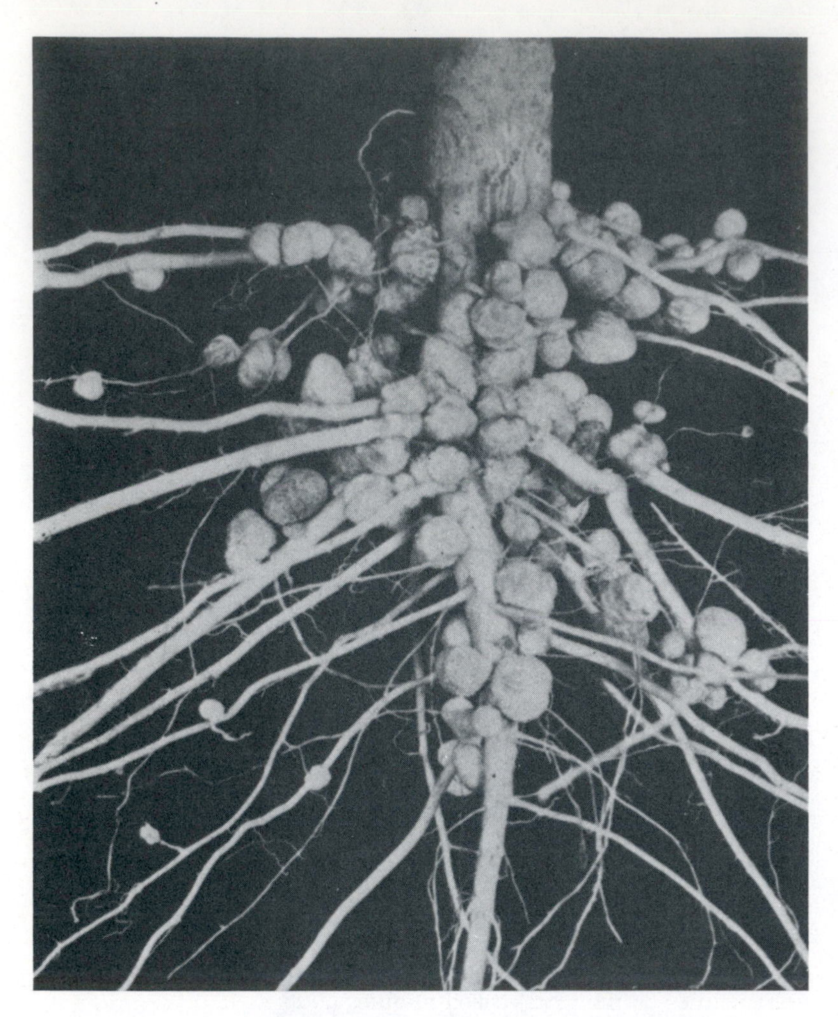

Figure 17-11 Nodulation on root of soybean plant.
Courtesy Raymond C. Valentine, University of California at Davis.

being made to determine how to increase the nitrogen-fixing activity of bacteria around the roots of cereals and of cyanobacteria that inhabit paddy fields, which are important for rice production in much of Asia.

Sulfur cycle

Microbial populations at or below the soil surface, in marine and freshwater sediments, and in natural waters are also important in the cycling of sulfur (Figure 17-12). Sulfur is present in ores and rocks as sulfides, and this abundant form of sulfur is oxidized when oxygen is introduced into the environment. The responsible populations are members of the genus *Thiobacillus.* These autotrophic bacteria oxidize the sulfides and use them as energy sources for growth. As a result, sulfate is produced, and the acidity of the surroundings increases. The acidity increases because the reaction gives rise not simply to sulfate but rather to sulfuric acid. Oxidations that are carried out by heterotrophs occur in soils and in waters, where the energy sources are organic compounds of sulfur. The sulfur in these compounds is unavailable for higher plants and for many microorganisms. However, the responsible populations mineralize the sulfur and also convert it to sulfate, but in these instances, the populations are heterotrophic. This sulfate is then assimilated by the indigenous populations of plants and microorganisms.

In lakes, photosynthetic bacteria may be quite important. These organisms use light as their energy source, but at the same time, some oxidize hydrogen sulfide. The hydrogen sulfide that they use is generated not in the lighted zone of the water, where the photosynthetic species grow, but rather in the underlying anaerobic region.

In marshes, sediments, poorly drained soils, and other anaerobic environments, the sulfate is reduced by bacteria that do not use oxygen in their metabolism. These bacteria, a representative of which is *Desulfovibrio,* proliferate and convert sulfate to hydrogen sulfide. For the process, these anaerobes require an energy source, which is often organic, but it may also be hydrogen gas. The hydrogen sulfide that these bacteria excrete may reach high levels and become toxic to fish in waters or to higher plants growing in poorly aerated soils.

Geological microbiologists have been especially interested in the sulfur cycle. They have given particular attention to sulfuric acid formation from ores because of the acidity that is generated (Figure 17-13). This acid may solubilize many elements, and the elements thus solubilized are mobile and move from place to place. In some instances, the soluble form of the element is toxic. Geological microbiologists have also been interested in sulfate reduction, which, by generating hydrogen sulfide, causes the precipitation of metallic sulfides. Geological deposits of sulfides may be major sources of the metal associated with the sulfur.

MICROORGANISMS AND POLLUTION

Pollution has a major impact on microbial processes, and microorganisms in turn may have major impacts on pollutants. Classically, the microbial role in pollution was largely assigned to pathogenic species, which themselves are pollutants. However, within the last two decades, it has become evident that microbial processes are affected by chemical pollutants introduced into waters and soils, and that the indigenous com-

Table 17-5 Principal economically important nitrogen-fixing systems

PLANT	SYMBIONT	PRINCIPAL USES
Food and grain legumes (peas, beans, soybeans, chick peas, peanuts)	*Rhizobium* species	Human and animal food and food products (e.g., soya protein)
Forage legumes (alfalfa, clover)	*Rhizobium* species	Livestock fodder, green manure
Sward legumes (clover, subclover, vetch)	*Rhizobium* species	Upgrading grazing land
Leguminous shrubs and flowering plants (lupines, sweet peas)	*Rhizobium* species	Horticulture, landscaping
Azolla	*Anabaena*	Green manure
Alder	*Frankia*	Reforestation, interplanting

Postgate, J.R., and Hill, S. Economic microbial ecology. in Lynch, J.M., and Poole, N.J. (Eds.). *Microbial ecology: a concept approach.* Blackwell Scientific Publications, Oxford, 1979.

munities of these environments are major means both for destroying and for generating chemical pollutants.

A chemical that enters water or soil as a result of human activities may reduce the rate of a microbial process in a fashion that is deleterious to other species residing in the same environment. In waters, microbial photosynthesis is extremely important. The photosynthetic algae and cyanobacteria are at the base of food chains on which fish and other aquatic animals rely. An inhibition of the rate of microbial photosynthesis in these habitats could result in a diminution in the food supply for, and hence in the yield of, fish or other animals. Some chemicals, it now appears, do cause such inhibitions. A reduction in the rate of the microbial mineralization of an element that is in insufficient supply in a terrestrial or aquatic ecosystem likewise will have a major effect on the function of that ecosystem. For example, should a pesticide or industrial chemical introduced into soil reduce the rate of microbial conversion of organic nitrogen to the inorganic forms used by higher plants, this suppression of microbial activities will be reflected in a diminished rate or extent of plant growth. Similarly, because nitrogen fixation by legumes is important in many agricultural lands as well as in natural ecosystems, an inhibition of *Rhizobium* will reduce the amount of nitrogen that is returned to the environment as a result of the legume-*Rhizobium* symbiosis.

The microbial modification of organic and inorganic compounds or ions may likewise be of profound importance to environmental quality. Thus, microbial populations mineralize many industrial pollutants and pesticides, and these min-

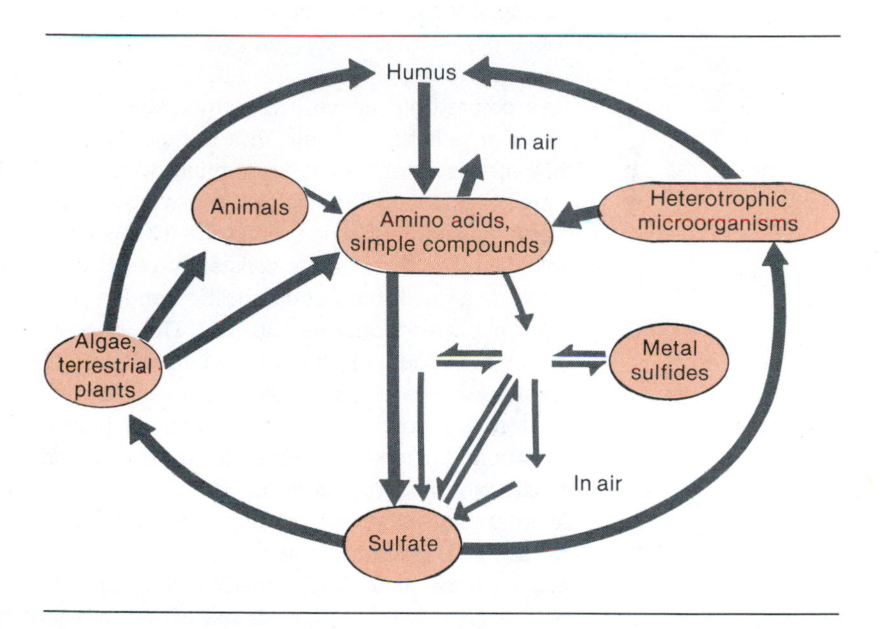

Figure 17-12 Sulfur cycle.

$$FeS_2 + 3\tfrac{1}{2}O_2 + H_2 \xrightarrow{\text{Abiotic}} FeSO_4 + H_2SO_4$$

$$2FeSO_4 + \tfrac{1}{2}O_2 + H_2SO_4 \xrightarrow{\text{Bacteria}} Fe_2(SO_4)_3 + H_2O$$

$$FeS_2 + Fe_2(SO_4)_3 \xrightarrow{\text{Abiotic}} 3FeSO_4 + 2S$$

$$2S + 3O_2 + 2H_2O \xrightarrow{\text{Bacteria}} 2H_2SO_4$$

Figure 17-13 Formation of sulfuric acid (H_2SO_4) by microbial oxidation of pyrite (FeS_2).

eralization reactions, by converting the organic chemical to inorganic products, destroy the toxic substances and convert them to harmless substances. In nature, except in areas of high light intensity, the only means to destroy organic pollutants completely and to convert them to inorganic products is by a microbial transformation. These conversions are known as ***detoxications*** because the processes destroy the inherent toxicity of the molecule. On the other hand, microorganisms also bring about ***activation,*** a reaction in which a compound of little or no toxicity is converted to a highly toxic molecule. Microorganisms are now recognized as being involved in several activation reactions and in thus generating toxic chemicals. In this way, microorganisms are not solely agents of destroying pollutants, but they also create them. Many examples of microbial detoxication are known, but only a few instances of the formation of toxic chemicals by microbial populations in soils and waters have been described to date.

As pointed out above, microorganisms are involved in polluting the air with nitrogen oxides. They are also responsible for forming other types of air pollutants as well as for destroying certain volatile pollutants. Carbon monoxide (CO) is a notable example. This gas is constantly being generated as a result both of combustion and of photochemical reactions in the air. The level of carbon monoxide in the atmosphere is not increasing, however, because microbial activity, apparently in soil, converts the carbon monoxide to carbon dioxide. Although this oxidation and detoxication occurs readily in soil, the responsible organisms—bacteria or fungi or even the genera involved—are not known as yet. Microorganisms are also able to form carbon monoxide in natural environments, but the net effect appears to be one of destruction rather than emission of this highly toxic gas.

Volatile hydrocarbons are also produced by microorganisms. Although the concentration of hydrocarbons in the atmosphere is small, natural communities are constantly generating a number of these volatile compounds. Conversely, microorganisms are active in destroying volatile hydrocarbons. Some of this destruction may take place in aquatic environments, but probably most occurs in soil.

Sulfur and other elements are also converted to volatile products. Some of these are of interest to toxicologists, and others are purely of interest to geochemists and atmospheric scientists. The volatilization is often the consequence of microbial addition of methyl groups to an ionic form of the element. In addition, microorganisms generate other volatile products, for example, carbon disulfide.

This recently recognized role of microorganisms in destroying many synthetic pollutants and in creating new pollutants is serving as impetus for much research. Nevertheless, it still is not clear to what extent microorganisms are major contributors to pollution, but they are unquestionably major agents in preventing the deterioration of environmental quality.

SUMMARY

The composition of the microbial community is determined by the available nutrient supply, certain critical physical and chemical factors, and interactions among these species that are able to tolerate the abiotic stresses. Natural selection operates on the species that enter an environment, and the species that survive and grow have fitness traits allowing them to persist and multiply in that habitat. The communities within the ecosystem may have a low or a high species diversity. These species interact in food chains that permit the coexistence of populations with different nutritional patterns. Some organisms are characteristically pioneers and appear early, and others appear late in successions, but some microorganisms are typically members only of climax communities. Only a few of the physiological or morphological traits that determine the position of a species in succession are known. Microbial transformations in soils and natural waters affect the chemistry of the earth and of the atmosphere, and these processes may also have immediate, practical effects. The reactions alter the chemistry of soils and waters because of mineralizations, oxidations, and reductions. Some of these microbial processes may bring about the cycling in nature of carbon, nitrogen, sulfur, and other elements. Some lead to the destruction or formation of polluting compounds in natural environments.

Study questions

1. Distinguish between population, community, and ecosystem.
2. Why is it difficult to establish the ecologically significant factors in the microenvironment?
3. Explain how natural selection operates in microbial communities.
4. What abiotic factors are important to the ecology of microorganisms?
5. Why do many species coexist in an environment receiving carbon sources that only a few of them can metabolize?
6. Distinguish between first and second trophic levels.
7. Distinguish between essential nutrients that are ecologically significant and those that are not.
8. Why do some species appear early and some late in a succession?
9. Why is their interest from the practical viewpoint in the role of microorganisms in biogeochemistry?
10. What is the evidence that all biologically formed organic molecules are mineralized aerobically?
11. How may microorganisms make compounds important in chemical pollution?
12. Why is there special interest in the microbial role in the nitrogen cycle?
13. What microbial process is a sensitive indicator of deleterious environmental changes?

Selected readings

Books

Alexander, M. *Microbial ecology.* John Wiley & Sons, New York, 1971.

Atlas, R.M., and Bartha, R. *Microbial ecology: fundamentals and applications.* Addison-Wesley Publishing Co., Inc., Reading, Mass., 1981.

Clark, F.E., and Rosswall, T. (Eds.). *Terrestrial nitrogen cycles.* Ecol. Bull. 33. Swedish National Research Council, Stockholm, 1981.

Ehrlich, H.L. *Geomicrobiology.* Marcel Dekker, Inc., New York, 1981.

Elliwood, D.C., et al. (Eds.). *Contemporary microbial ecology,* Academic Press, Inc., New York, 1980.

Fenchel, T., and Blackburn, T.H. *Bacteria and mineral cycling.* Academic Press, Inc., New York, 1977.

Lynch, J.M., and Poole, N.J. (Eds.). *Microbial ecology: a conceptual approach,* Blackwell Scientific Publications, Oxford, 1979.

Marples, M.J. *The ecology of the human skin,* Charles C Thomas, Publisher, Springfield, Ill., 1965.

Rosebury, T. *Microorganisms indigenous to man.* McGraw-Hill Book Co., New York, 1962.

Skinner, F.A., and Carr, J.G. (Eds.). *The normal microbial flora of man.* Academic Press, Inc. New York 1974.

Stevenson, L.H., and Colwell, R.R. (Eds.). *Estuarine microbial ecology.* University of South Carolina Press, Columbia, S.C., 1973.

Journal articles

Alexander, M. Nonbiodegradable and other recalcitrant molecules. *Biotechnol. Bioeng.* **15:**611, 1973.

Alexander, M. Biodegradation of chemicals of environmental concern. *Science* **211:**132, 1981.

Bowden, G.H.W., Ellwood, D.C., and Hamilton, I.R. Microbial ecology of the oral cavity. *Adv. Microb. Ecol.* **3:**135, 1978.

Bremner, J.M., and Steele, C.G. Role of microorganisms in the atmospheric sulfur cycle. *Adv. Microb. Ecol.* **2:**155, 1979.

Cosgrove, D.J. Microbial transformation in the phosphorus cycle. *Adv. Microb. Ecol.* **1:**95, 1976.

Delwiche, C.C. The nitrogen cycle. *Sci. Am.* **223**(5):137, 1970.

Hurst, A., and Collins-Thompson, D.L. Food as a microbial habitat. *Adv. Microb. Ecol.* **3:**79, 1978.

Poindexter, J.S. Oligotrophy: feast and famine existence. *Adv. Microb. Ecol.* **5:**63, 1981.

Stout, J.D. The role of protozoa in nutrient cycling and energy flow. *Adv. Microb. Ecol.* **4:**1, 1980.

Chapter 18 AQUATIC MICROBIOLOGY

F.L. Singleton

The ***hydrosphere,*** that is, the earth's aquatic environment, can be divided into two categories on the basis of location: surface waters and subsurface waters (Figure 18-1). Of the surface waters, two types are common. Marine waters, characterized by high salt concentration (high salinity), cover the majority of the surface of the earth. Freshwater environments, located inland, contain low concentrations of salts. A third type of aquatic environment shares characteristics of both marine and freshwater environments. These systems, ***estuaries,*** are located in coastal regions, where the land's topography permits the entrance of fresh water into a basin that has a free exchange of waters with the ocean (Figure 18-1). Estuaries are therefore transition areas between freshwater systems and the ocean.

In the past eukaryotic microorganisms (algae and invertebrate animals) were the subject of the majority of studies on the microbial community of aquatic systems. This was due in large part to the availability of techniques to study populations and communities of these organisms. In recent years techniques have become available for the qualitative and quantitative study of aquatic prokaryotic communities. This chapter deals primarily with the role of prokaryotic populations and communities in aquatic ecosystems. Refer to the selected readings at the end of this chapter for references devoted to eukaryotic organisms in the aquatic environment.

In the past the bacterial component of the microbial community of aquatic systems was considered to be important only in the role of decomposers. Although many types of aquatic bacteria are involved in the decomposition of organic materials, they contribute other processes that are vital to the functioning of an ecosystem. Such processes include nutrient cycling and regeneration, serving as a food source for higher-trophic-level organisms (filter-feeding invertebrates), and contributing to the overall metabolic activities involved in maintaining homeostasis of an ecosystem.

METHODS OF STUDYING AQUATIC MICROORGANISMS

The methods employed to study aquatic microorganisms include those routinely used in other disciplines of microbiology as well as some that have been developed for use with aquatic microorganisms. Such methods include procedures to enumerate the organisms and to study their activity.

Direct techniques

The size of aquatic microbial populations can be estimated by direct or indirect techniques. The direct technique that has been employed for decades to enumerate aquatic bacteria employs solid culture media and is referred to as the ***spread plate technique.*** An appropriate dilution of a water sample is spread over the surface of the medium and after incubation the colonies are counted. If it is assumed that one colony represents one cell in the dilution, the number of viable organisms can be determined. This technique yields only an estimate of the bacterial community, since more than one cell can give rise to a single colony. More important, no culture medium or laboratory condition provides the environmental conditions necessary for the growth of all the viable organisms in a natural mixed bacterial community, which may include populations of heterotrophs, autotrophs, mesophiles, psychrophiles, and others. The spread plate technique is superior to the ***pour plate technique,*** since the former does not expose the organisms to temperatures required to keep agar molten (>42° C). In the pour plate procedure a water sample is mixed with molten agar, and the agar is poured into a sterile petri dish. Bacterial colonies develop on and within the agar.

Often as few as 0.01% (or less) of the total community will grow on a culture medium. Although they constitute only a fraction of the total community, the culturable bacteria have been employed to study the bacterial community of aquatic systems. Although the spread plate technique provides a means for enumerating only a small fraction of a bacterial community, it is used today not only as a means of counting bacteria but also for obtaining data that can be compared with previous studies.

Other techniques employed to enumerate aquatic bacteria or specific bacterial populations include extinction dilution, most probable number (MPN), and membrane filters. All such techniques require the organisms to grow on or in a culture medium. For example, the ***MPN technique*** provides a statistical estimate of the number of organisms in a water sample when multiple samples of a series of dilutions are inoculated into a culture medium. Membrane filters are employed to cultivate and enumerate aquatic bacteria by filtering a known volume of water and placing the filter onto an appropriate culture medium. After a suitable incubation period, colonies are counted.

The most direct method to determine the number of individuals in a microbial population or community is by direct microscopic observation. Numbers of planktonic organisms are usually determined in this manner. The planktonic community can be divided into three major groups: (1) ***phytoplankton*** (free-floating algae), (2) ***zooplankton*** (free-floating animals), and (3) bacteria, which are often referred to as ***bacterioplankton.*** However, when one is dealing with bacteria, microscopic observation of a water sample is of limited value, since bacteria are transparent and the number of bacteria in most waters (in the range of 1×10^6 per milliliter) is too low to allow for accurate enumeration. Therefore the organisms must either be stained to be visualized or observed by a microscopic technique other than one that employs ordinary light microscopy. In either case, when one is working with most natural water sources, the bacteria must be concentrated. This is usually accomplished by filtering a water sample through a membrane filter with a pore size of 0.2 or 0.45 μm.

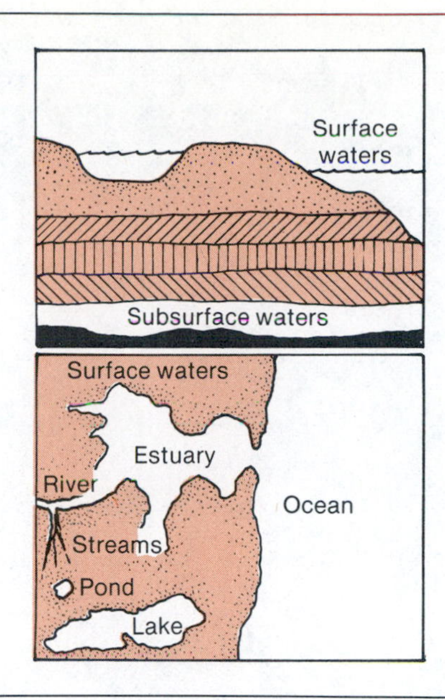

Figure 18-1 Major types of aquatic systems.

A combination technique of membrane filtration and staining the cells can be of limited value, since it is not always possible to differentiate between cells and small particulate materials. Therefore improvements in methods of microscopic counting of aquatic bacteria include the use of fluorescent dyes (such as acridine orange) that have an affinity for binding to specific cellular components. When used in conjunction with fluorescent microscopy (see Chapter 3), such stains allow the investigator to differentiate cells from particulate materials (Figure 18-2).

Another application of fluorescent microscopy to aquatic microbiology includes the use of fluorescent antibodies to enumerate specific types of aquatic microorganisms. When properly applied to a water sample containing cells with which the antibody is reactive and observed with a fluorescent microscope, only those cells with the antibody attached will be fluorescent.

One of the major drawbacks of microscopic techniques to enumerate bacteria is the inabiliity to differentiate living and dead cells. Microscopic counting of aquatic bacterial populations will yield counts that overestimate the metabolic potential of these populations in the system, since both living and dead cells are counted.

Indirect techniques

Techniques have been developed that permit estimating the size of the microbial community indirectly. These are based on quantifying a specific cellular component (Table 18-1). In the past

Figure 18-2 Bacterial cells (light areas) stained with fluorescent dye and observed with epifluorescent miscroscope.
Courtesy R.M. Baker and M.A. Hood.

Table 18-1 Components of microbial cells that have been employed as a means of determining the microbial biomass in aquatic systems

COMPONENT	COMMENT
Total organic carbon	Assay does not distinguish between living and nonliving
Total organic nitrogen	Assay does not distinguish between living and nonliving
Adenosine triphosphate	Assay measures only living organisms; does not differentiate types of organisms
Lipopolysaccharide	Assay is specific for gram-negative bacteria
Poly-β-hydroxybutyrate	Assay is specific for this bacterial storage product
Lipids Phospholipids Glycolipids	Assays measure total quantity of indicated lipids, both living and nonliving

determinations of cellular carbon, nitrogen, phosphorus, or other chemicals were employed as an index of the biomass, that is, the quantity of organisms, in a pure culture. However, such techniques are of little value for natural water samples, since the water may contain noncellular quantities of such materials. Therefore specific cellular components such as ATP, muramic acid, poly-β-hydroxybutyrate, and lipids have been successfully employed to estimate the microbial biomass in water samples or aquatic sediments. In addition, by differentially filtering water through filters of selected pore sizes, it is possible to separate the various components of the community, that is, separate free-living bacteria from phytoplankton and zooplankton.

The heterotrophic activity of aquatic microbial communities is usually measured with radio-labeled substrates such as ^{14}C glucose, amino acids, or other organic substrates. In heterotrophic activity assays, a water sample is placed in a clean, sterile container and a low concentration of the substrate is added. After appropriate incubation periods the organisms are killed, and acid is added to the samples to drive off any ^{14}C carbon dioxide formed during metabolism of the substrate. The cells and the ^{14}C carbon dioxide are collected, and the amount of radioactivity is measured with a scintillation counter. From this information the rate of uptake and mineralization of the substrate is calculated. These parameters are indicative of the potential activity of the microbial community.

ADAPTATION TO THE AQUATIC ENVIRONMENT

Organisms inhabiting an aquatic environment are subjected to spatial and temporal fluctuations in physicochemical parameters. For example, in a deep lake the difference in temperature between surface and bottom waters is several degrees. Likewise, the temperature variation between seasons can be as much as 20° to 30° C. Similar variations in dissolved oxygen concentrations, pH, light intensity, and other parameters occur.

Aquatic microorganisms exist in an environment in which the concentration of nutrients is usually low. For example, the concentration of dissolved organic carbon in open ocean waters is in the milligrams-per-liter range. Bacteria, that grow and survive on these low concentrations of nutrients must possess efficient uptake mechanisms as well as efficiently use nutrients that become available. Such bacteria are referred to as ***oligotrophic.*** Not only are oligotrophic bacteria

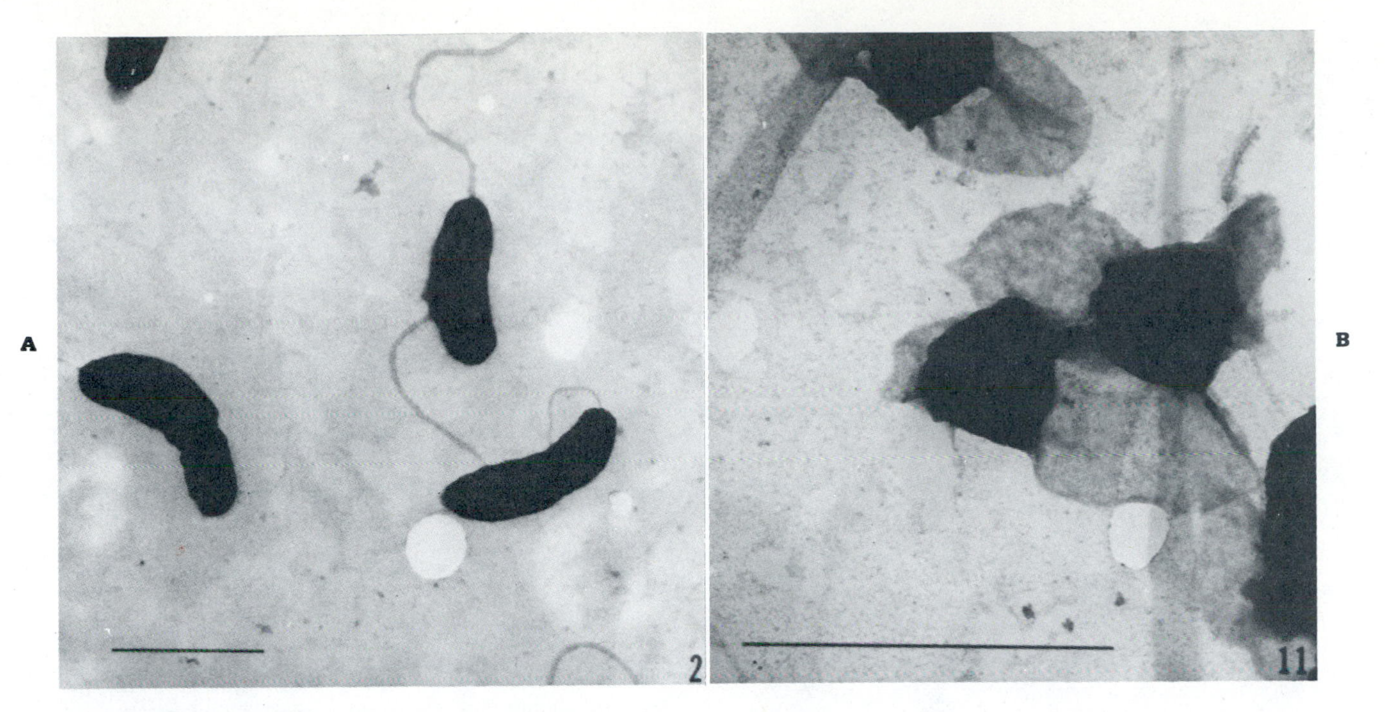

Figure 18-3 One effect of starvation on an aquatic bacterial species. Transmission electron micrographs of normal cells, **A,** and starved cells, **B.** Starved cells were maintained for 5 days in nutrient-depleted liquid medium.
Courtesy R.M. Baker and M.A. Hood.

adapted for growth on low concentrations of nutrients, but many are inhibited by high nutrient concentrations that are characteristic of common laboratory culture media, that is, in the range of grams per liter.

Not all aquatic bacteria are oligotrophic. Indeed, in many nutrient-rich waters a large proportion of the culturable bacteria are ***eutrophic (copiotrophic)*** and are not inhibited by high nutrient concentrations. These bacteria are sometimes referred to as heterotrophic. However, this term is not preferred by many microbiologists when used in this regard, since it can be confused with the term defining the type of metabolism employed by an organism to obtain carbon and energy.

The occurrence and distribution of oligotrophic and eutrophic bacteria in an aquatic environment is dependent on the concentration of nutrients. Although both nutritional types occur in most aquatic environments, some investigators consider the oligotrophs to be "true" aquatic bacteria, since they posess mechanisms for survival in a nutrient-poor environment. For example, in response to extended periods of limited nutrient supply, some aquatic bacteria decrease their biomass and metabolic activities until nutrients become available (Figure 18-3). These cells of reduced size and activity survive for periods of weeks and months and, when nutrients become available, resume normal metabolic activities. Other aquatic bacteria overcome nutrient scarcity by becoming located in regions of increased nutrient concentrations, such as in association with other organisms or on the surface of particulate materials. When associated with another organism or a particle, the bacteria are provided with a favorable microenvironment. Not only do these associations provide a favorable microenvironment for the bacteria, but this is another ecological strategy for several species to coexist within a common environment.

The distribution of algae in aquatic environments is also influenced by the concentration of nutrients in the water. For example, diatoms are more abundant in waters of low nutrient concentration. Since phytoplankton are photosynthetic, their distribution is influenced by the availability of nutrients required for metabolic processes involved in photosynthesis.

Whereas the distribution of bacteria and phytoplankton are directly influenced by the availability of nutrients, zooplankton are seemingly indirectly influenced by this parameter. The major food sources of zooplankton are bacteria and algae. Therefore the availability of a food source determines the ability of a predator species to survive within a system.

Table 18-2 Characteristics of the major groups of algae

DIVISION (COMMON NAME) EXAMPLE	WHERE COMMONLY FOUND	DISTINGUISHING CHARACTERISTIC
Chlorophyta (Green algae) *Ulva, Enteromorpha Cholorella, Volvox*	Common to marine and freshwater systems	Major storage material is starch; cellulose in cell walls
Chrysophyta (Yellow-green, golden-brown algae and diatoms) *Asterionella, Tabellaria*	Common to marine and freshwater systems	Silica in cell walls; major storage material is carbohydrate or oil
Phaeophyta (Brown algae—seaweed) *Sargassum*	Many marine species	Primary storage material is carbohydrate (mannitol and laminarin)
Rhodophyta (Red algae)	Majority are marine	Primary storage material is starch*

*Starch produced by Rhodophyta is referred to as floridian starch. It does not react with iodine.

MICROBIAL ASSOCIATIONS IN THE AQUATIC ENVIRONMENT

Bacteria and animals

Bacteria that exist in association with zooplankton (Figure 18-4) or other aquatic animals are referred to as ***epizooic.*** There is doubt as to whether specific associations exist between aquatic animals and specific bacterial inhabitants in the sense that a given type of bacteria will associate with a particular type of organism. Rather, associations such as this appear to depend on which bacteria were first to colonize the host. In such an association the bacteria are provided with a convenient source of nutrients since many bacteria colonize the feeding appendages and the mouth parts (Figure 18-5). It is believed by many that in an association between bacteria and zooplankton the host derives the benefit of a convenient source of vitamins and other growth factors.

Bacteria and plants (algae)

Bacteria that exist in association with plants or algae are referred to as ***epiphytic.*** Epiphytic bacteria have been the subject of numerous studies designed to determine the exact relationships between these organisms. It is to the benefit of both the algae and the bacteria that such associations occur. Within an aquatic system most of the photosynthesis is carried out by the algal community (see Table 18-2). However, of the organic carbon produced during photosynthesis, a large fraction (30% or more) can be lost from some algal cells to the environment. Loss of organic carbon usually is in the form of glycolic acid. Bacteria in the slime layer surrounding the algal cells, as well as planktonic bacteria, use glycolic acid and similar organic compounds as a growth substrate. In turn, the bacteria provide the algae with some vitamins and other growth factors.

Evidence for the association between algae and bacteria has been obtained from studies carried out in many many aquatic environments; these studies have demonstrated the distribution patterns of these organisms to be parallel. Also, examination of the slime layer of many algal cells reveals a very heavy bacterial load. However, it is important to note that not all members of the phytoplankton are covered with bacteria. Diatoms are known to be free of bacteria. This is important for the survival of the diatoms, since they exist within delicate exoskeletons containing pores to allow the entrance and exit of water (and nutrients). A heavy bacterial load on the surface of these organisms would interfere with the exchange of water. Diatoms may prevent bacteria from attaching to the exoskeleton by releasing various organic acids that are inhibitory to the bacteria.

Organisms and inanimate objects (surfaces)

Aquatic microorganisms are associated with inanimate particulate materials in the water column as well as on the surface of the water. Such organisms are referred to as ***epilithic.*** This type of association is apparently more beneficial to bacteria than to other microbes. These bacteria are present at the interface between a particle and the water. The other major interface in the

Figure 18-4 Bacteria attached to mouth parts of copepod.
Courtesy A. Huq and R.R. Colwell.

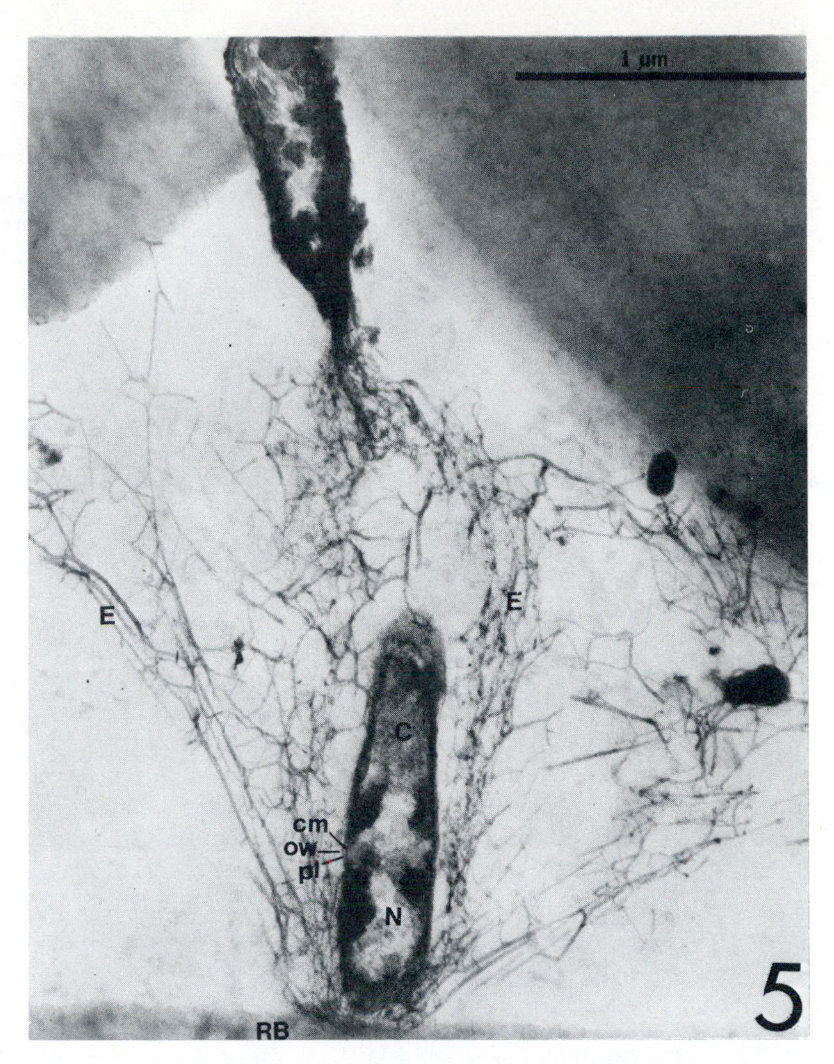

Figure 18-5 Transmission electron micrograph of thin section of bacteria attached to surface of resin block *(RB)*. Exopolymer *(E)* appears as thin fibrils attaching cells to surface. Note internal structure of attached bacterium. *C,* Cytoplasm; *N,* nuclear region; *cm,* cytoplasmic membrane; *pl,* peptidoglycan layer; *ow,* outer wall.
Courtesy C.A. Liebert and M.A. Hood.

aquatic environment is between the water surface and the atmosphere. Organisms located at interfaces have a greater chance of survival, since the nutrient concentration is many times that found in the water column. Nutrient concentrations at interfaces can support the vigorous growth of bacteria. Therefore the number of bacteria present at an interface is much larger than that in the open water column.

At solid-liquid interfaces proteins, various other organic molecules, and inorganic molecules are concentrated as a result of electrostatic attraction and other physical forces. At the liquid-gas interface on the water's surface hydrophobic molecules such as lipids and some types of proteins are found in increased concentrations. Regardless of the type of interface, many of the materials present can be used by bacteria as a growth substrate.

Microorganisms become located at interfaces by chemotactic response to the region of increased nutrients (see Chapter 7) or, they are transported there by currents. Also, some bacteria become localized at the surface interface by increasing the hydrophobicity of the cells. Cells located at the liquid-air interface remain there by hydrophobic forces. However, those at a liquid-solid interface are initially maintained in position by electrostatic attraction. Many bacteria become fixed in place by producing an extracellular material (usually polysaccharide or protein in composition) that serves to glue the cells to the surface of the particle (Figure 18-5).

AQUATIC MICROBIAL COMMUNITIES

The presence of free-floating bacteria in virtually all aquatic systems has been observed. However, the numbers and types of planktonic bacteria depend in large part on the characteristics of a system. The existence of a true planktonic community in streams has been questioned, since any planktonic organisms would be moved downstream. Therefore there would have to be a continual input of organisms in the headwaters of those systems. However, in open-water sources such as lakes and the ocean, planktonic bacteria are known to exist.

Marine environment

Open ocean surface waters. The marine environment covers approximately 71% of the earth's surface and has an average depth of approximately 3500 m. Although the majority of the solar radiation reaching the earth enters the marine environment, the average temperature of ocean waters is less than 4° C. Another characteristic of marine waters is a high salt content. The salinity of open ocean waters ranges from 34.0 to 35.8 parts per thousand (ppt), depending on the latitude. Thus marine waters have a relatively constant chemical composition. The typical salt content of an open ocean water sample is given in Table 18-3. Note that the nine ions listed constitute more than 99% of the total amount of salt in the sample. Other elements are present in ocean waters but in very low concentrations. The exact chemical composition of an ocean water sample is dependent on the physicochemical parameters of a given region.

Table 18-3 Concentration of the nine major ions in natural seawater

ION	PERCENT OF TOTAL
Chlorine	55.0
Sodium	30.6
Sulfate	7.7
Magnesium	3.7
Calcium	1.2
Potassium	1.1
Bicarbonate	0.4
Bromine	0.19
Strontium	0.02

Marine bacteria. The bacterial community of open ocean waters comprises numerous species. Although those species that grow on culture media have been well characterized, the degree to which they represent the entire community is unknown. The culturable portion of open ocean bacterial communities is usually dominated by gram-negative bacteria (80% to 90% of the total), especially members of the genera *Pseudomonas* and *Vibrio.* Other genera commonly isolated from marine waters include *Photobacterium, Flavobacterium,* and *Acinetobacter.* These latter genera make up a minor fraction of the total as compared with *Pseudomonas* and *Vibrio.* A larger proportion of the culturable bacteria isolated from the surface microlayer are gram-positive, since these bacteria tend to be more resistant to the effects of solar radiation. *Staphylococcus* and *Micrococcus* species are commonly isolated from the surface microlayer. (Species of these groups are also found on the skin surface of vertebrates.)

Members of the genus *Photobacterium* (and some *Vibrio* species) share a trait that is apparently restricted to a few species of marine bacteria—the ability to emit visible light (Plates 4 to 7) (see Chapter 4). These bioluminescent bacteria are primarily found in marine waters, usually in association with zooplankton or fish. The symbiotic relationships between bioluminescent bacteria and their host organisms are well documented. The "bioluminescent organs" of some fish consist of bioluminescent bacteria contained within an organ of the fish. The ecological advantage of bioluminescence is a subject of much debate. It is easy to recognize the advantage of a bacterium being associated with a fish, since the former is provided with a safe environment and a food source. However, many bioluminescent bacteria are found in the water column. Whether bioluminescence is an ecological advantage in this case is unknown.

Surface waters of the open ocean also contain significant populations of cyanobacteria (blue-green algae) (Table 18-4). In addition to contributing to the total quantity of the biomass, the cyanobacteria are responsible for the addition of significant quantities of nitrogen to the marine ecosystem as a result of their nitrogen-fixing activities (see Chapter 13).

In general, the number of bacteria in the marine environment is approximately 1×10^6/ml, but this number decreases with increasing distance from land. Increased bacterial loads of near-shore waters are due to increased nutrient concentrations that support growth of the ***autochthonous (native)*** marine bacteria. Near land masses nutrients enter the marine environment

Table 18-4 Cyanobacteria (blue-green algae) common to many aquatic environments

GENUS	CHARACTERISTIC
Anabaena	Common in blooms
Coelosphaerium	Common in blooms
Lyngbya	Common to most aquatic environments
Oscillatoria	Common to most aquatic environments

through surface runoff as well as through rivers and other freshwater systems emptying into the ocean. Not only do nutrients enter marine waters in this manner, but large numbers of ***allochthonous (foreign)*** bacteria are also contributed. Therefore, with increasing distance from land, nutrients are diluted, allochthonous bacteria die off, and the total number of bacteria decreases.

Many factors are known to influence the dying off of allochthonous bacteria in the marine environment. Among these are inhibition by increased solar radiation exposure, nutrient deprivation, predation by filter-feeding invertebrates, high salt concentrations, and heavy-metal toxicity; especially by nickel ions.

In addition to a horizontal distribution pattern, that is, decreasing numbers with increasing distance from land, the bacterial community also has a vertical distribution pattern. Relative to numbers of organisms in the open waters, the largest numbers of bacteria are found on the surface of the ocean. A thin microlayer covers the surface of ocean waters. This microlayer (microfilm) is approximately 30 μm thick and is composed of an assortment of hydrophobic materials, including hydrocarbons, lipids, and various organic and inorganic molecules. Many of the materials making up the surface microlayer can be used by bacteria as a growth substrate. Therefore relatively large numbers of bacteria are associated with this microlayer.

Below the surface microlayer the number of bacteria decreases significantly. Within the water column the number of bacteria decreases slightly with increasing depth. This is probably a result of a combination of factors, including decreased temperature (<4° C) and increased hydrostatic pressure. Within the upper levels of the water column the distribution of bacteria is relatively constant, since wave action and surface currents prevent stratification. This distribution pattern appears to be unchanged in the upper few hundred meters of many regions of the ocean. However, in some regions a second increase in the bacterial community has been observed to occur in conjunction with increases in the phytoplankton community. In such waters the phytoplankton are stratified at a certain depth (usually between 20 and 50 m). As the phytoplankton grow and release organics into the water column, the bacterial community increases.

Marine algae. The distribution of marine algae is similar to that of the bacteria—increased numbers of organisms in coastal waters. The composition of the phytoplankton community differs not only with increasing distance from land but also with latitude. For example, this community is dominated by diatoms in the cold Arctic and Antarctic waters. In the warm tropical waters the flagellated algae are dominant.

The phytoplankton community of open ocean waters is restricted to the surface waters that receive light—the ***photic zone.*** The photic zone varies in depth but is usually less than 100 m, depending on the characteristics of the water of a particular region. The depth of the photic zone is determined primarily by the concentration of suspended particulate materials, including planktonic organisms, in the water column. Planktonic organisms absorb and/or scatter (reflect) light in the water column. Those organisms that absorb light do so by virtue of intracellular pigments. Such pigments are employed by photosynthetic bacteria and algae to trap solar radiation for the photosynthetic process. Many marine (and freshwater) nonphotosynthetic bacteria contain pigments that serve to protect the cells from the adverse effects of solar radiation. The presence of pigments in organisms in the water column influences the color of the water column. For example, coastal waters are greenish colored as a result of the presence of yellow pigments in the phytoplankton.

On the average, the photosynthetic activities of the phytoplankton community annually contribute approximately 50 g of carbon per square meter of open ocean surface water. This amounts to a total annual contribution of more than 16 billion tons of carbon to the open waters of the marine environment. Much of this carbon is recycled within the marine system and returned to the atmosphere (see "Carbon Cycle").

Marine zooplankton. In waters containing large populations of bacteria and phytoplankton, the zooplankton are also present in increased numbers. These filter feeding organisms, which serve as the primary food source for many marine organisms of higher trophic levels, consume the smaller phytoplankton and bacteria. The composition of the zooplankton community, like that of the bacteria and phytoplankton, varies within the ocean environment, depending on the physicochemical characteristics of the water and the availability of food. The relationship between the occurrence of zooplankton and their food source is exemplified by the fact that many of the zooplankton carry out diurnal migration within the water column. Diurnal migration is beneficial to the zooplankton, since during the night they migrate to the photic zone and feed on the phytoplankton. During the daytime they migrate back to the dark, cool waters, where they hide from predators.

Marine zooplankton also demonstrate increased numbers near land masses—a result, no doubt, of the availability of food (bacteria and algae). The species composition of the zooplankton population is dependent on the region of the ocean in consideration.

Marine sediments. The sediments of the oceans contain a large microbial community that is dominated by the bacteria. There are significantly larger numbers of bacteria in the sediment than there are in the overlying waters. Also, the composition of the sediment community differs considerably from that of the water column. The presence of a large community is due to the high concentrations of nutrients in the sediment. The decomposition of chitin, the major structural component of the exoskeleton of invertebrates, and the tissues of other organisms occurs in the sediment (see ''Aquatic Nutrient Cycles,'' p. 411). Decomposition processes and their rates are depth dependent, since temperature decreases and hydrostatic pressure increases with increasing depth. Therefore recycling of materials occurs at a faster rate in coastal waters, especially in tropical or subtropical waters, than in the deep ocean. The sediment therefore supports a larger and more active community than the nutrient-poor water.

The number of bacteria as well as the composition of the community changes with depth into the sediment. With increasing depth in the sediment, the total number of bacteria decreases and the percentage of gram-positive bacteria increases.

The proportion of the bacterial community demonstrating a requirement for oxygen also changes with sediment depth. Only the uppermost levels of the sediment are aerobic. Therefore below the aerobic layer of the sediment this environment is anaerobic and is dominated by anaerobes, primarily of the genus *Clostridium*. Numerous *Bacillus* species have been isolated from marine sediments.

The deep ocean. Some of the trenches in the ocean floor have recorded depths of more than 10,000 m. At one time it was believed that the depths of the oceans were without life. The ocean depths appear to be an environment that cannot be inhabited. For example, this environment is characterized by low temperature (<4° C), complete absence of light, and extremely high hydrostatic pressures (as high as 1,100 atm; see Chapter 15). The microbial community of these waters is dominated by bacteria and zooplankton. Various crustaceans and some fishes also inhabit this seemingly hostile environment.

As early as 1884 the existence of bacteria in water samples collected from depths of 8400 m was demonstrated. Subsequent studies have demonstrated the existence of viable bacteria at depths of more than 10,000 m.

Bacteria inhabiting surface waters of the ocean have decreased metabolic activity when exposed to conditions characteristic of the deep ocean. Changes in activity are a result of the combined influence of low temperature and high pressure. Since bacteria have a slower metabolism in a low-temperture or low-temperature and high-pressure environment, the presence of viable bacteria in deep ocean water samples was believed to be a result of cells settling from the overlying surface waters. However, some investigators hypothesized that the deep ocean contained a resident bacterial community that had adapted to conditions of low temperature and high pressure. Such organisms would therefore be psychrophilic and barophilic. Studies of the activity of the bacterial community of deep ocean waters demonstrated very slow rates of metabolism—so slow, in fact, that an incubation period of many weeks was required to measure them. This led many to believe that there was not a true barophilic community, or if there was one, it was not present in the water column. Subsequent studies of the bacterial flora of the intestinal tract of invertebrates that also inhabit the deep ocean have demonstrated the existence of metabolically active bacteria. Furthermore, some of these bacteria are more metabolically active under high pressure. Recently the existence of bacteria that require high pressure for growth has been reported. Thus the deep ocean contains a barophilic bacterial community that is involved in the recycling of organic materials and that appears to reside primarily in the intestinal tracts of animals inhabiting these waters.

Estuarine environment

The estuarine environment is unique in the sense that it is a transition zone between marine and freshwater systems. In a typical estuary (Figure 18-6), fresh water enters in regions away from the ocean. Since there is a free exchange of waters with the ocean (tidal action), a salinity gradient exists that extends from the very low salt concentrations (<5 ppt) where fresh water enters to salt concentrations that are typical of ocean water (>30 ppt). At any point in the estuary, bottom waters will have a higher salinity than surface waters.

The microbial community of an estuary consists of organisms entering in fresh waters and in marine waters. The estuarine environment supports a very large and active biological community. The estuaries are highly productive and supply tremendous quantities of shellfish and other organisms to the food supply of humans.

The microbial community of an estuary is similar in composition to that of other aquatic systems. Species characteristic to the marine environment are common to the estuary, especially the regions that contain waters of high salinity. Some organisms are apparently restricted to the estuarine environment. For example, in the bacterial community, the dominant genera are *Pseudomonas* and *Vibrio.* Of the *Vibrio* species commonly isolated, *V. parahaemolyticus* appears to be a natural estuarine organism whose distribution is limited to this environment. Another closely related species, *V. cholerae,* the causative agent of epidemic cholera in humans, also appears to be a natural inhabitant of the estuarine environment.

Since estuaries receive inputs of nutrients from freshwater systems as well as surface runoff from the environs, estuarine waters are more densely populated with microorganisms than ocean waters. The bacterial community is one or more orders of magnitude larger than that of ocean waters. The phytoplankton community also contains more individuals than is the case with ocean waters. Likewise, estuaries support large communities of invertebrates and vertebrates. As a result of the presence of increased numbers of different organisms, the estuary is very productive and materials are rapidly recycled (see "Aquatic Nutrient Cycles").

Freshwater environments

The microbial community of freshwater environments differs considerably from that of marine waters. Organisms inhabiting freshwater environments do not have a requirement for high salt concentrations. Although many organisms inhabiting fresh waters and marine waters belong to the same genus, the species composition of communities of these systems differs markedly. For example, the bacterial community of freshwater environments is made up predominantly of members of the genera *Pseudomonas* and *Vibrio,* as is that of marine waters, but different species are characteristic of each type of system.

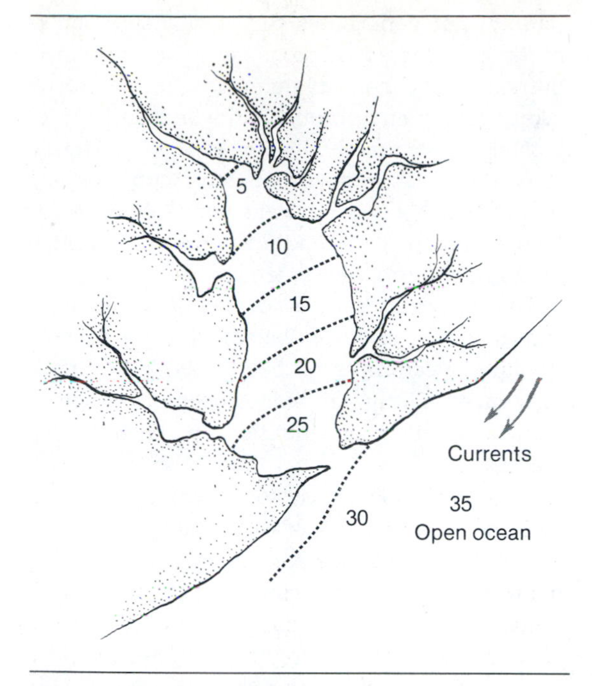

Figure 18-6 Pattern of salinity in hypothetical estuary. Numbers represent salinity (in parts per thousand).

The number of organisms in freshwater communities is dependent on the physicochemical characteristics of the system in question. For example, fewer organisms are present in systems containing nutrient-poor waters, such as pristine mountain streams, than those containing nutrient-rich waters, such as may be found in lowland streams or rivers.

Lakes. The occurrence and distribution of organisms within the water column is dependent on the physicochemical nature of the lake as well as the climate of the region in which it is located. As in other environments, the determining factor for the number of organisms in a lake is nutrient availability. Nutrients enter lakes in surface runoff as well as in any other waters that empty into the lake basin. In regions where the area is highly productive, that is, forests or farmlands, which also contain a productive photosynthetic community that contributes to the concentration of nutrients (organic carbon), the concentration of nutrients may be very high. In these types of lakes, termed ***eutrophic,*** often a large and active biological community is present. Conversely, lakes that receive small inputs of nutrients and that themselves are nutrient poor ***(oligotrophic lakes)*** support fewer and different types of organisms than eutrophic systems.

The surface of lakes is covered by a microlayer similar to that of ocean waters. A larger concentration of bacteria is present in the lake surface microlayer. Within the water column below the microlayer the concentration of bacteria decreases.

Within the lake ecosystem the phytoplankton carry out the majority of the photosynthetic addition of carbon to the environment. However, in many systems a significant fraction of the total primary production comes from photosynthetic bacteria. Zooplankton prey on the phytoplankton and bacteria and themselves serve as a food source for higher-trophic-level organisms (see "Aquatic Nutrient Cycles").

There is a seasonal pattern in the distribution of many organisms within lakes as a result of the influence of climate. Algae are usually restricted to the photic zone, and their abundance is controlled in large part by water temperature. Algae demonstrate increased metabolic activity during the warmer seasons.

Maximum populations of bacteria are present during the spring and fall. This is especially true for lakes in temperate regions. During winter the water temperature of a temperate region lake is somewhat uniform from the surface to the sediment, especially if the lake is covered with ice. Water has a maximum density of 4° C. At temperatures above or below 4° C, the density decreases. Therefore as ice melts or waters with a temperature of less than 4° C begin to warm, the density increases, resulting in a circulation pattern being established as the denser water moves to the bottom of the lake. During the fall, when the water temperature begins to decrease, the density increases and the denser water moves to the bottom of the lake, again establishing a circulation pattern. Such circulation patterns result in complete mixing of the water column, and materials on or in the sediments ***(benthic matter)*** are swept up into the water column. These events, occurring in the spring and fall, are referred to as spring and fall overturn, respectively. When an overturn occurs, the nutrient concentration in the water column increases dramatically. This results in bacterial and phytoplankton blooms, which are followed by blooms in the zooplankton community. Finally, as the zooplankton die as a result of the lack of food, a second bacterial bloom will occur as the zooplankton are decomposed.

Different physiological types of bacteria in the water column also occur on a seasonal basis. For example, during the winter the concentration of dissolved oxygen in a lake is very low. In some lakes, with the exception of the surface layer, the water column may be anaerobic. This is a result of little oxygen being added to the water by photosynthesis and what oxygen that is available being used by aerobic organisms. Therefore the water column may contain more anaerobic bacteria than aerobic bacteria during the winter. During the summer, when a lake becomes stratified, the colder waters of hypolimnion (lower region of a stratified lake) are anaerobic.

It is important to note that not all lakes (or ponds) undergo seasonal cycling and stratification as described above. Depending on the depth of the system as well as the climate, a system may never become stratified, as occurs in systems where the water temperature remains considerably more than 4° C. In some regions a system may be stratified during the summer, but only a fall overturn will occur. In any case when an overturn occurs, the result will be as described above—blooms in the bacterial, phytoplankton, and zooplankton communities. This is a result of the influx of nutrients to the water from the sediment. Most decomposition of organic materials occurs on or in the sediments. When particulate organic materials such as plant or animal tissues are broken down by bacterial activities, a portion of the material will be converted into a soluble form. Likewise, mineralization of plant and animal tissues releases inorganic nutrients such as nitrogen, phosphorus, and sulfur in soluble form (see "Aquatic Nutrient Cycles"). When an overturn occurs and the water circulation resuspends detritus (disintegrated debris) into the water column, the soluble nutrients are used by the planktonic communities.

Streams and rivers. The microbial community of flowing-water systems is similar in many ways to those of other aquatic systems. However, the majority of the members of communities of flowing-water systems exist in association with objects within these systems. The development of algal mats on the surface of rocks on the bottom of streams is well known. Within mats of attached algae epiphytic bacteria are common. It is easy to realize the necessity of organisms existing in association with solid surfaces in such systems. In fact, the existence of a true planktonic community in a flowing-water system can be questioned, since such organisms would be moved downstream. In order for a planktonic community to exist, organisms would have to be continually added to the system. Therefore the presence of organisms in the water column proper would be a result of attached species becoming dissociated from their substratum. Also, in flowing-water systems attached species would be favored, since they would only have to remain on a surface and have nutrients transported to them in the water currents.

In flowing-water systems the smallest numbers of microorganisms are in pristine mountain streams in which the nutrient concentration is very low. As the waters flow toward the low-lying areas, more nutrients enter the water column and the number of organisms increases.

Within the bacterial community the majority of species are gram-negative. The changes in species composition of the culturable proportion of the community depend on several factors, including the availability and types of nutrients and the composition of the land through which the system flows.

Freshwater sediments. The bacterial flora of freshwater sediments is similar to that of marine and estuarine sediments and is dominated by the bacteria. The proportion of gram-positive organisms increases with increasing depth into the sediment. This is due primarily to the increasing abundance of *Clostridium* species in the anaerobic zones of the sediment. *Bacillus* and other gram-positive bacteria, including coryneforms and filamentous forms, are also abundant in aquatic sediments.

AQUATIC NUTRIENT CYCLES

Carbon cycle

The cycling of carbon in the aquatic environment is a result of the combined metabolic activities of all the members of the biota (Figure 18-7). The majority of the carbon entering an aquatic system is converted into the biomass of photosynthetic organisms. Some of the photosynthetically fixed carbon is incorporated into the biomass of heterotrophs, which use dissolved organics released into the water column by the photosynthetic organisms. In a simplified model system such as that depicted in Figure 18-7, a large proportion of the biomass of these organisms is converted into the biomass of filter-feeding invertebrates. Likewise, higher-trophic-level organisms, that is, larger invertebrates and vertebrates, prey on the filter-feeders. At each step in the food chain as described above, the biomass of the prey organisms, which is converted into the predator biomass, returns to the atmospheric carbon pool as carbon dioxide or remains in the system in the form of waste products from the predator organisms. Waste products of aquatic organisms decomposed by portions of the microbial community are incorporated into the biomass of the decomposers or are lost from the system as carbon dioxide. Ultimately the biomass of all organisms is acted on by the microbial community involved in the decomposition of organic matter and is either converted into the biomass of the decomposers or lost from the system as carbon dioxide.

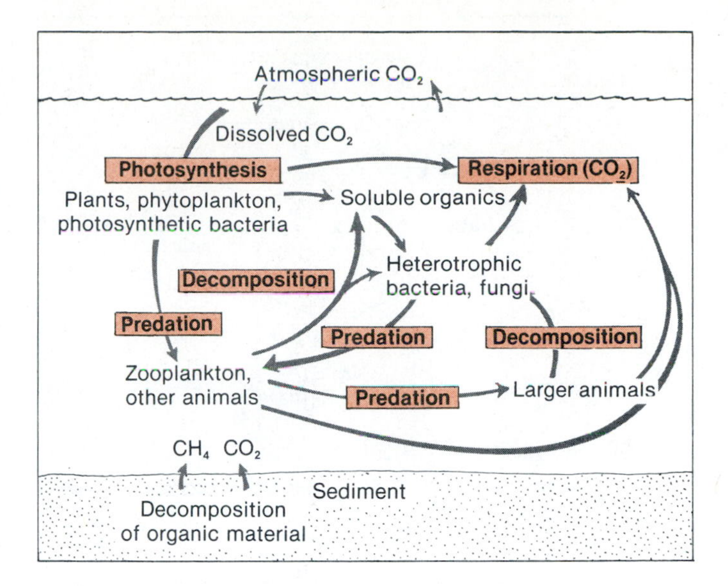

Figure 18-7 Carbon cycle in aquatic environment.

The majority of the plant and animal biomass becomes a part of the benthos, where it undergoes decomposition. The microbial communities responsible for the decomposition of the plant and animal biomass differ. The biomolecules of animal tissues (proteins, carbohydrates, lipids, etc.) are easily broken down by heterotrophic bacteria that produce the necessary extracellular enzymes. However, when the plant biomass contains a large proportion of cellulose and lignin, it is initially acted on by members of the fungal community. This is true in both marine and freshwater systems. The fungi colonize the lignocellulosic materials and, as a result of their activities, break these materials down into smaller particulates that are then colonized by bacteria, and the decomposition process continues. Therefore the decomposition of plant materials requires the presence of both fungi and bacteria.

The majority of the decomposition of organic materials, either of plant or animal origin, occurs

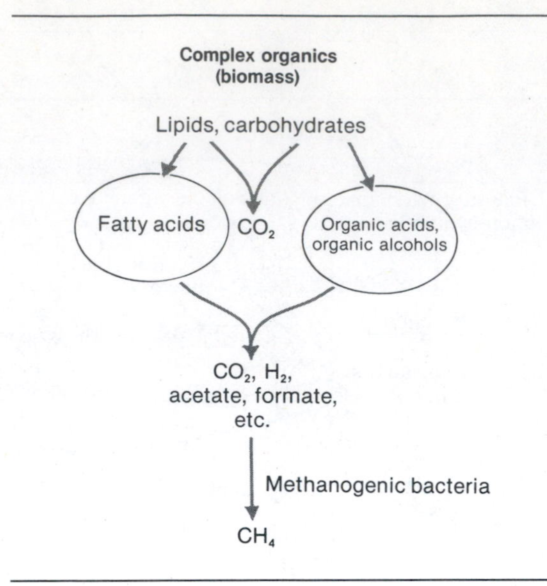

Figure 18-8 Biological production of methane from organic materials.

in the sediment or on the surface of the sediment. This environment is often partially or completely anaerobic. This results in populations of anaerobic and aerobic organisms being involved in decomposition processes. Indeed, both metabolic types are necessary for the cycling of organic materials in the aquatic environment.

One of the major products of these decomposition processes in many aquatic sediments is methane (CH_4). Methanogenesis is an important process, not only in carbon recycling in an ecosystem, but also from an economic standpoint, since this is an increasingly attractive alternative energy source. In order for methane to be produced in an aquatic sediment, at least three different metabolic groups of bacteria must be present (Figure 18-8). The first group is involved in the conversion of the substrate into simple compounds that are then fermented to yield a combination of fatty acids, organic acids and alcohols, and carbon dioxide. A second metabolic type oxidizes the fermentation products of the first group into carbon dioxide, hydrogen, and simple organic molecules such as acetate (CH_3COOH). These products are then metabolized by methanogenic bacteria (*Methanobacterium, Methanococcus,* and others) to yield methane. Approximately 65% of the methane produced in aquatic sediments is derived from acetate. Biologically produced methane either enters the atmosphere or is metabolized by another specialized bacterial group, methane oxidizers, which use single-carbon compounds as a carbon and energy source. Methanogenesis does not occur in all sediments, since different metabolic groups must be present and interact. Also, the presence of other bacterial groups can inhibit methane production. This occurs commonly in sediments containing an active sulfate-reducing community, since these organisms outrival the methanogens for hydrogen, which is the source of electrons in the reproduction of sulfate to hydrogen sulfide (see "Sulfur Cycle").

Nitrogen cycle

In some aquatic systems nitrogen has been demonstrated to be the limiting nutrient. Obviously, in such a system the availability of nitrogen controls productivity. Even in aquatic systems where nitrogen is not the limiting nutrient, it can become limiting unless organisms efficiently recycle what nitrogen is available within that system (Figure 18-9). The major source of nitrogen in some systems in surface runoff from the surrounding land, which contributes dissolved nitrogen compounds. In other waters the major source of nitrogen is the metabolic activities of free-living nitrogen-fixing bacteria and cyanobacteria. The majority of the nitrogen in an aquatic system is in the form of organic molecules, especially amino acids. A smaller fraction exists in dissolved form as nitrate, nitrite, and ammonia. Organic molecules such as amino acids may be incorporated directly into the biomass of decomposer organisms. Alternatively, nitrogen-containing organic molecules may be mineralized, with the release of ammonia, through various deamination reactions. Mineralization reactions make nitrogen available in a form (ammonia) that other organisms use. Some organisms incorporate ammonia into biomolecules, and some autotrophs use it as an energy source. Members of the genera *Nitrosomonas, Nitrosococcus,* and related genera oxidize ammonia to yield nitrite. Nitrites are then used as an energy source by members of the *Nitrobacter, Nitrococcus,* and *Nitrospira.* Nitrites are oxidized by these organisms to produce nitrates. Several species of bacteria in aquatic environments use nitrate as the terminal electron acceptor in anaerobic respiration reactions—denitrification reactions (Figure 18-9)—which primarily results in nitrate being reduced to nitrogen gas (N_2) and ammonia. The former reenters the atmospheric nitrogen pool, and the latter can be recycled with the food web.

Phosphorus cycle

As discussed in Chapter 14, phosphorus is used in all cells, primarily in high-energy molecules such as ATP and in nucleic acids. Since it is a required element, and in some aquatic environments may be the limiting nutrient, the cycling of phosphorus is important in maintaining the productivity of a system (Figure 8-10). The major reserve of phosphorus in the biosphere is in insoluble form in rocks and sediments. Phosphorus also exists in aquatic environments in organic molecules and to some degree in insoluble inorganic form as phosphate (PO_4). This inorganic form of phosphorus is incorporated into the biomass of microorganisms. Aquatic organisms are usually in direct competition for phosphates. Bacteria are more efficient at scavenging phosphates from the water column than most algae. Therefore the productivity of an aquatic system is often dependent directly on the bacterial community recycling phosphorus and making it available to other organisms. Many have developed mechanisms that allow them to carry out "luxurious uptake" of phosphates. When phosphates are present in concentrations above what are required for growth of the organisms, some will continue to take up and store phosphorus in the form of polyphosphate molecules (volutin). During periods of limited phosphorus supply, the intracellular reserve is used.

Inorganic phosphate enters an aquatic system in surface runoff from the surrounding terrestrial environment. Another source of phosphate in some systems is from the sediment. Phosphates that are contained in insoluble form in the sediment are solubilized by ***phosphoclastic*** bacteria (phosphate dissolving), which produce large quantities of organic acids as they metabolize an organic substrate. The insoluble phosphates are converted to soluble form as a result of the low pH. A similar result occurs in sediment (or soil, see Chapter 19) where active populations of ammonia and nitrite oxidizers exist. These organisms produce large quantities of nitric acid and other acidic products as they carry out oxidation of ammonia and nitrite.

Within an aquatic system phosphorus transfer between members of the food web is the usual mechanism for cycling. When an organism dies, phosphorus-containing organic molecules either are taken up in toto by the decomposers and incorporated into their biomass or are mineralized. When the latter process occurs, the soluble phosphates that are released during mineralization reenter the food web as other organisms incorporate them into their biomass. During mineralization, phosphorus can be lost from the system in the sediment as insoluble phosphates such as various iron phosphates.

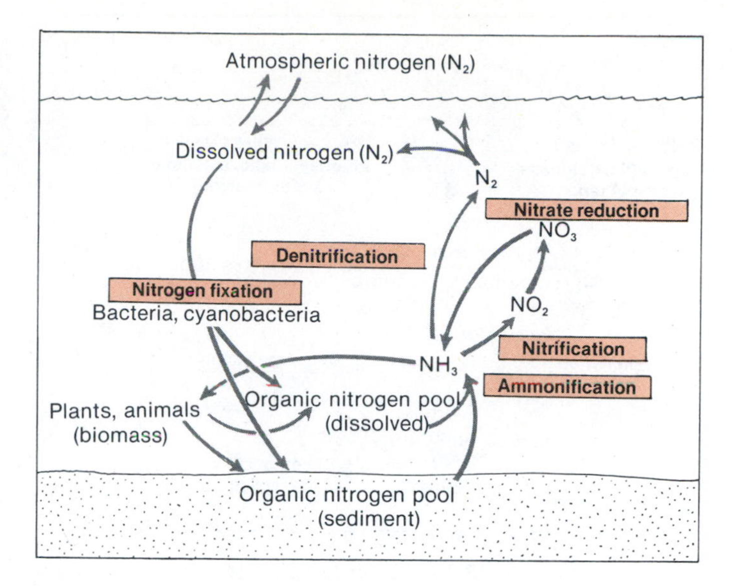

Figure 18-9 Nitrogen cycle in aquatic environment.

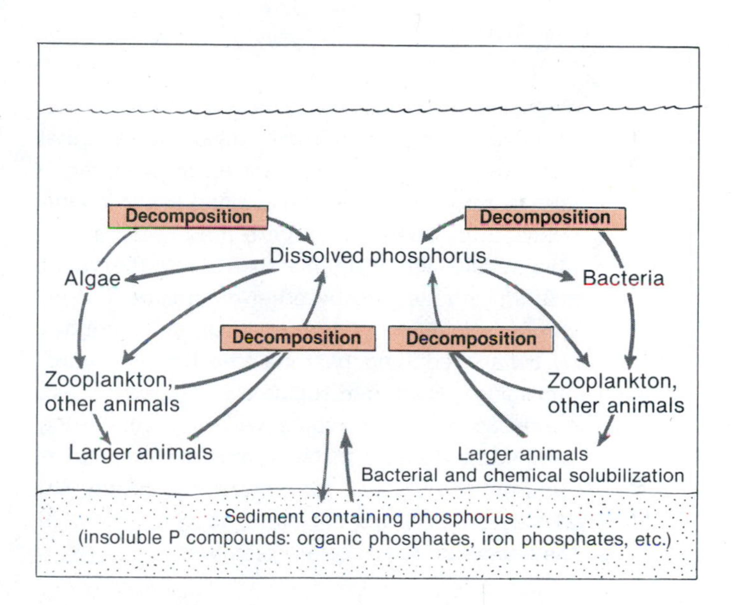

Figure 18-10 Phosphorus cycle in aquatic environment.

Sulfur cycle

Sulfur is found in various biomolecules such as cysteine and methionine, in addition to other sulfhydryl group–containing molecules. Although it is a required element, sulfur is rarely if ever the limiting nutrient in an aquatic environment.

Sulfur exists in biomolecules in reduced form, primarily as sulfhydryl groups (—SH). When an organism dies and is decomposed (Figure 18-11), sulfur is liberated in the form of hydrogen sulfide (H_2S) as the sulfur-containing molecules are mineralized. In the anaerobic zones of an aquatic environment, including the sediment, hydrogen sulfide is used as an energy source by sulfur-oxi-

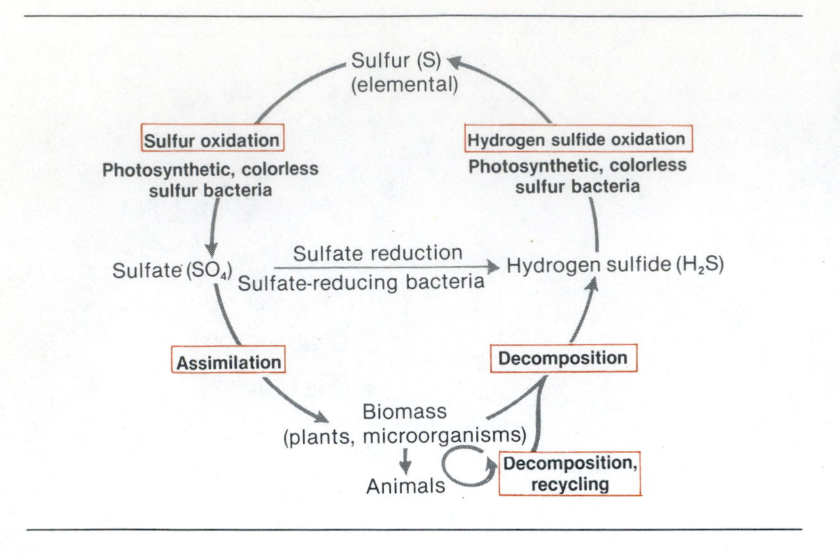

Figure 18-11 Sulfur cycle in aquatic environment.

EXAMPLES OF SPECIES OF BACTERIA INVOLVED IN TRANSFORMATIONS (OXIDATION AND REDUCTION) OF SULFUR

Sulfur oxidizers ($H_2S \longrightarrow SO_4^{2-}$)
- Nonphotosynthetic
  - *Beggiatoa*
  - *Thiobacillus*
  - *Thiomicrospira*
  - *Sulfolobus*
- Photosynthetic
  - *Chromatium*
  - *Thiocystis*
  - *Thiospirillum*
  - *Chlorobium*

Sulfate reducers ($SO_4^{2-} \longrightarrow H_2S$)
- *Desulfovibrio*
- *Desulfuromonas*
- *Desulfotomaculum*

dizing bacteria (see box). The majority of the sulfur bacteria oxidize hydrogen sulfide to elemental sulfur (S) and store it in intracellular or extracellular sulfur granules. The elemental sulfur is later oxidized to sulfate (SO_4), with the release of soluble sulfate, which then either enters the water column, is taken up by other organisms and incorporated into organic molecules, or is reduced by sulfate-reducing bacteria (see box) to hydrogen sulfide. Hydrogen sulfide either is recycled as described above or reacts with iron-containing compounds to form insoluble iron sulfates and is lost from the system. As portrayed in Figure 18-11, the major steps in sulfur recycling occur in the anaerobic zones of the sediment.

SEWAGE AND WASTEWATER MICROBIOLOGY

The beginning of aquatic microbiology as a discipline of microbiology is due in part to studies on sewage contamination of water sources. After bacteria were established as causative agents of many diseases, microbiologists noted that diseases such as typhoid fever, cholera, and bacillary dysentery were spread via contaminated water. The ability to detect these organisms in water sources and to treat wastewater to prevent the spread of disease became of primary public health importance. Procedures to monitor aquatic systems to determine to what degree, if any, a system is contaminated by sewage have resulted in dramatic reductions in the occurrence of waterborne diseases (Figure 18-12).

Sewage (fecal) contamination of water is determined on the basis of the presence of bacteria specific for the intestinal tract of mammals, that is, the ***coliforms.**** These organisms are detected with the use of various selective media designed for their isolation. All such media contain lactose as the primary substrate, since lactose fermentation is one of the defining characteristics of enteric bacteria. The coliforms are divided into two groups on the basis of the temperature at which they ferment lactose. Total coliforms ferment lactose at 37° C, whereas fecal coliforms ferment lactose at 44.5° C. Likewise, a related bacterial group, the fecal streptococci, members of the genus *Streptococcus* indigenous to the intestinal tract of mammals, are also useful in evaluating whether a water source is contaminated by sewage. The dramatic decline in waterborne diseases since the nineteenth century is a result of the development of methods to properly treat sewage and wastewater as well as the use of coliforms to monitor water sources for sewage contamination.

Sewage treatment

Microorganisms are used in the treatment of sewage and wastewater. It would be impossible to safely recycle water without the activities of

* The coliforms are a heterogeneous group of gram-negative rods found in the intestinal tract that includes *E. coli* and species of *Klebsiella, Enterobacter, Serratia,* and others.

these organisms. The three objectives of sewage treatment are (1) to mineralize organic material to reduce the concentration of organic materials; that is, to lower the ***biological (biochemical) oxygen demand*** (BOD)*; (2) to remove minerals from the water; and (3) to kill pathogenic organisms. These objectives are met by modern treatment processes (Figure 18-13).

Biological treatment of sewage and wastewater consists of three stages: primary, secondary, and tertiary (Figure 18-14). The objective of primary treatment is to remove materials such as sand and other particulates, whereas in secondary treatment the objective is to remove dissolved organic materials from the water by allowing microorganisms to metabolize (mineralize) them. The final stage of wastewater treatment, tertiary, which consists of the removal of minerals from the water, is expensive and not practiced in all regions. Mineralization of organic materials without the removal of the minerals is of limited value, since organisms in waters that receive sewage treatment effluents will use the minerals. In such cases algal and bacterial blooms often occur.

In a modern wastewater treatment facility the wastewater is initially collected in a settling tank, where particulate materials are removed by sedimentation. Materials such as oil or grease, which float on the surface, are removed with a skimmer. The materials collected at this stage (the primary sludge) can be further concentrated by the removal of excess water and at least partially degraded in a digester (see below). The wastewater liquor is then pumped to another system for secondary treatment, during which the particulate and dissolved organic materials are mineralized by microorganisms. In the mineralization process portions of the organic materials are converted to the microbial biomass and/or removed from the system as gaseous carbon compounds (carbon dioxide, methane, etc.).

Secondary treatment systems are either aerobic or anaerobic. During aerobic processing, the microorganisms mineralize the organic materials in an aerobic system. Different types of aerobic wastewater treatment systems are employed in different regions. However, of these the two most common methods of aerobic treatment, and the ones used by most cities, are activated-sludge and trickling-filter systems.

In ***activated-sludge systems*** (Figure 18-15) the wastewater liquor contains a large bacterial

* BOD is a measure of the amount of oxygen taken up by a water sample during a 5-day incubtion period at 20° C. It is a reflection of the quantity of oxidizable organic matter present in a sample.

Figure 18-12 Result of release of untreated sewage and wastewater.
Courtesy D.J. Grimes.

Figure 18-13 Aerial view of modern wastewater treatment plant. This treatment plant is located in Williamsburg, Va.
Courtesy Hampton Roads Sanitation District.

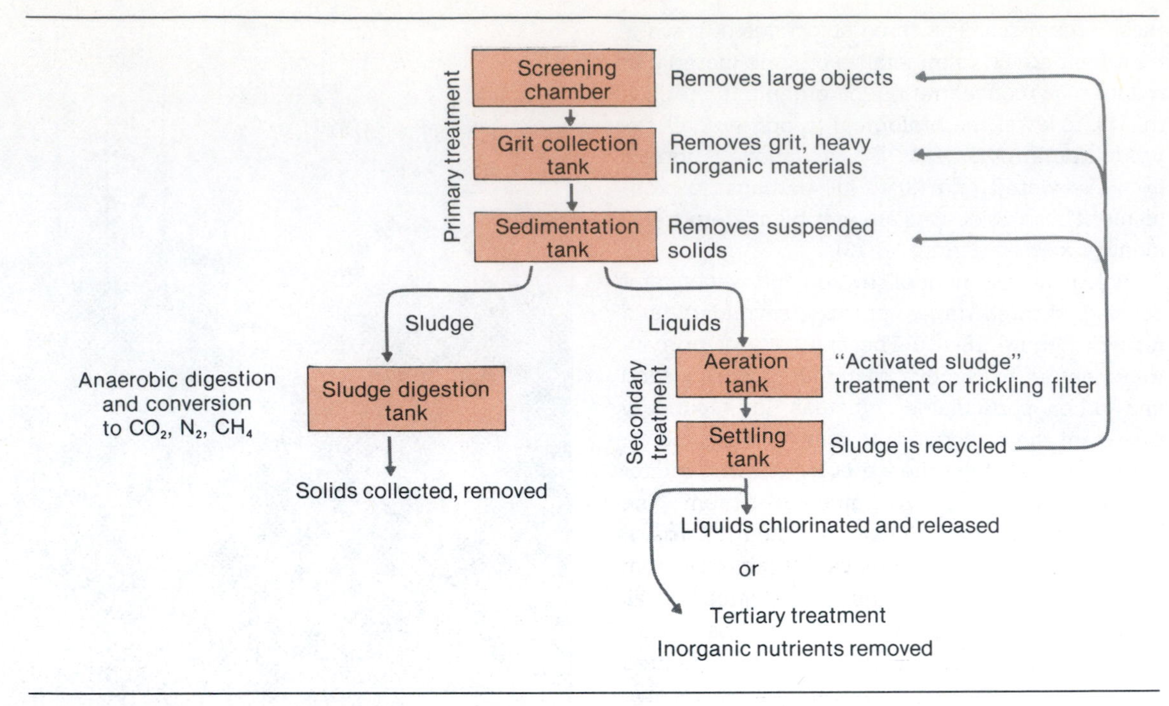

Figure 18-14 Flow diagram of typical sewage treatment plant.

community, most of which is within the "floc" (an organic matrix), which is believed to be produced by the bacterium *Zoogloea ramigera.* As *Z. ramigera* grows in the wastewater liquor, it is reported to produce an extracellular organic material in which other bacteria become embedded. Organisms within this zoogloeal mass metabolize dissolved organic materials in the wastewater liquor, and the mass increases in size. Within activated-sludge systems a large protozoan community also thrives. The protozoa, responsible for "grooming" the zoogloeal mass by grazing on it, are consumed by other organisms present in the system. Thus a food chain is created, and at each stage of the food chain a fraction of the original organic material is removed from the system as carbon dioxide.

After a period sufficient to allow the microbial community to incorporate the bulk of the organic materials into the biomass (or respire it off as carbon dioxide), aeration is ceased and the floc settles to the bottom of the system. The wastewater is then removed and either is chlorinated to kill any pathogenic microorganisms that survive the activated-sludge process and released into the environment or undergoes tertiary treatment (see below).

The bacterial community of the activated-sludge system, in addition to *Z. ramigera,* consists of bacteria typical of an aquatic system, that is, *Pseudomonas, Vibrio, Flavobacterium,* and others present in lower numbers. Bacteria common to the human intestinal tract (*Bacteriodes, Fusobacterium, Eubacterium,* and members of the Enterobacteriaceae) do not contribute significantly to the community of the zoogloeal mass.

A second aerobic system that is commonly employed for the treatment of wastewaters in many cities, the use of ***trickling filters*** (Figure 18-16), is also based on the concept of forming a short food chain within a closed system in which the bacteria use dissolved organic materials and protozoa consume the bacteria. In trickling-filter systems wastewater liquor is slowly sprayed over a bed of rocks covered by a film of microorganisms. In this system, however, fungi and algae make up the majority of the biomass. The biological film covering the rocks (or other solid support) is approximately 2 mm thick and consists of three layers, each with a characteristic microbial flora. The inner layer is composed of bacteria, algae, and fungi. The middle layer primarily contains fungi and algae. Fungi are the dominant organisms in the outermost layer. As the wastewater slowly trickles over the surface of the biofilm, dissolved organics are used by the microbial flora. Protozoa are present in large numbers and groom the biofilm.

This system is very effective in reducing the concentration of organic material in the wastewater. The role of algae in this system is not well understood. It is believed by many that algae are involved in the removal of some minerals from the water, especially phosphates. Water exiting the

Figure 18-15 Wastewater undergoing treatment in activated-sludge system.

trickling filter can be either chlorinated to kill pathogens or processed in tertiary treatment.

Anaerobic treatment processes can also be employed as a secondary treatment or, as often occurs, in conjunction with aerobic systems. More often than not, anaerobic systems are employed only to treat the primary sludge obtained during primary treatment, as well as the settled floc from activated-sludge systems. These systems tolerate larger quantities of organic materials than aerobic systems.

Once a system is filled with wastes and becomes anaerobic, which occurs in a matter of minutes if the system is closed from the atmosphere, the anaerobic bacteria begin to grow and metabolize organic materials. Two reactions that occur in anaerobic systems are particularly useful in wastewater treatment: conversion of some carbon to carbon dioxide and methane and conversion of organic-bound nitrogen and inorganic nitrogen compounds to nitrogen gas (N_2).

The major difficulties with anaerobic treatment systems are the long retention times required for mineralization and the fact that although these systems can tolerate high concentrations of organics, they are not as efficient as aerobic systems in lowering the concentration of organics, that is, lowering the BOD. Anaerobic systems do result in the production of methane, which can be vented from the system and used as a fuel source by the treatment plant or by other facilities.

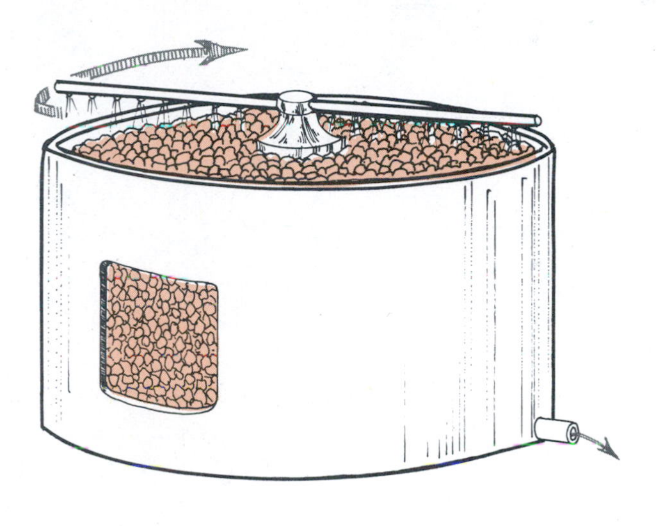

Figure 18-16 Diagram of trickling-filter wastewater treatment system. See text for details.

In general, after the secondary treatment process wastewaters are chlorinated to kill any remaining organisms and are released into the environment (usually into a river or other water source) (Figure 18-17). Any minerals that remain in the water supports the growth of organisms in the receiving waters. Therefore, to prevent harmful blooms of algae and bacteria, in limited situ-

Figure 18-17 Treatment plant outfall. Treated sewage and wastewater are released into environment after treatment.
Courtesy D.J. Grimes.

ations the treated wáters are processed through tertiary treatment to remove dissolved minerals. The addition of lime ($Ca[OH]_2$) will precipitate phosphates as insoluble calcium phosphate ($Ca_5OH[PO_4]_3$). Also, alum ($Al_2[SO_4]_3$) is used in this manner in some regions. Once a precipitating agent is added, the waters are allowed to stand, and the precipitate settles. The phosphate-free water is then removed from the system. It is important to remove phosphates from wastewater, since an algal bloom can be stimulated by excess phosphates. If a bloom is initiated, it can be perpetuated by nitrogen-fixing bacteria and cyanobacteria as they supply the system with nitrogen to support the growth of other organisms. Unfortunately, tertiary treatment is rarely practiced, since the expense is prohibitive.

SUMMARY

Homeostasis of aquatic ecosystems is maintained in large part by members of the microbial community. The combined metabolic activities of the various microbial populations function in the cycling and regeneration of materials within these systems. Within the microbial community aquatic bacteria were once thought to be important only in the decomposition of organic materials. However, they are now recognized as being an important component of the biotic community and are involved in the decomposition and recycling of organic materials, as well as serving as a food source for many invertebrates and furnishing organisms of other trophic levels with vitamins and other growth factors.

Aquatic bacteria are found either in the free-floating (planktonic) state, in association with other organisms, or in association with particulate materials. Such associations confer ecological advantages on members of the bacterial community. In associations between bacteria and other organisms, it appears that the bacteria are provided with a favorable microenvironment in which to live, and the host organism either is not affected by the resident bacterial flora or uses growth factors produced by the bacteria.

All natural aquatic systems have a resident microbial flora. Each type of system contains a characteristic community. Waters of the open ocean contain approximately 1×10^6 bacteria per milliliter. The bacterial community exhibits certain distributional patterns. Numbers of bacteria decrease with increasing distance from land masses, a result of decreasing nutrient concentrations. Also, the number of bacteria tends to decrease with increasing depth. Even so, the deep ocean contains a resident microbial population. It appears that much of the recycling of organic matter in the deep ocean (depths greater than 2000 m) occurs in the intestinal tracts of other organisms that also inhabit this environment.

The bacterial community of open ocean waters includes members of the genera *Pseudomonas, Vibrio, Photobacterium, Flavobacterium, Acinetobacter* primarily, in addition to lesser numbers of other gram-negative species and some gram-positive species (*Staphylococcus* and *Micrococcus*).

The bacterial community of freshwater systems is similar in composition to that of the marine environment. However, members of these bacterial communities do not have a requirement for high concentrations of salt. The composition of these systems is influenced by the type of system, as well as the physicochemical water parameters. Likewise, the size of freshwater bacterial communities depends on the concentration of available nutrients.

The microbial community of aquatic systems is essential to the various nutrient cycles within these systems. Each species is involved in some manner in one or more nutrient cycles. The ability of aquatic microorganisms to degrade and recycle organic matter is the basis of some stages of modern wastewater treatment processes.

Study questions

1. What are the major types of aquatic systems?
2. How does the microbial community of fresh waters differ from that of marine waters?
3. What are the predominant types of associations between microorganisms in the aquatic environment? How can such associations benefit the organisms?
4. What is a food chain? Describe one characteristic of an aquatic system and one of a sewage treatment plant.
5. What are the major objectives of sewage treatment?
6. How can the occurrence of microorganisms in a lake be influenced by season and climate?
7. What are the most common genera of bacteria in marine surface waters? In fresh waters?
8. What are allochthonous and autochthonous organisms?
9. Why is the "attached" microbial community in streams more representative of the true community than planktonic organisms?
10. Describe a typical aquatic carbon cycle. List organisms that would be involved in each stage of the cycle.
11. Describe the pattern of distribution of bacteria and phytoplankton in the marine environment. How do these differ from those of freshwater systems?
12. Describe the distribution of microorganisms in an aquatic sediment.
13. Why are more gram-positive bacteria than gram-negative bacteria found in the surface microlayer of an aquatic system?

Selected readings

Books

Cushings, D.H., and Walsh, J.J. (Eds.). *The ecology of the seas.* W.B. Saunders Co., Philadelphia, 1976.

Litchfield, C.D. (Ed.). *Marine microbiology.* Benchmark papers in microbiology/11. Dowden, Hutchinson and Ross, Inc., Stroudsburg, Pa., 1976. Distributed by Halsted Press, New York.

Reid, G.K., and Wood, R.D. (Eds.). *Ecology of inland waters and estuaries* (2nd ed.). Van Nostrand Reinhold Co., New York, 1976.

Rheinheimer, G. (Ed.). *Microbial ecology of a brackish water environment.* Springer-Verlag, New York, Inc., New York.

Tait, R.V. *Elements of marine ecology: an introductory course* (3rd ed.). Butterworths Publishers, Inc., Woburn, Mass., 1981.

Chapter 19 SOIL MICROBIOLOGY

Martin Alexander

The microbial changes that occur in the soil and the populations responsible for the transformations are the subjects of soil microbiology. The field includes many facets of microbiology, biochemistry, and soil science. The microbiological aspects focus on the types of organisms found below ground, and the biochemical facets are concerned with the metabolic activities and factors affecting the biochemical capabilities of the populations. From the viewpoint of soil science, interest is focused on the effects of the microbial community on its environment and on the plant, and consideration is also directed to the effect of this environment and its inhabitants on the changes brought about by the indigenous communities.

SOIL MICROBIOLOGY AND EXPERIMENTAL RESEARCH

Several areas of soil microbiology have been prominent for many years. Microbiologists have long been interested in an ecosystem that contains a vast number of dissimilar species and morphological types as well as a variety of beneficial and detrimental interactions among them. The indigenous communities have also achieved prominence because it has long been evident that they are essential for the nutrition of higher plants and the maintenance of soil structure. Plant pathologists and, to a lesser extent, medical and veterinary scientists have also examined the communities of this habitat because several of the native species cause disease. Furthermore, even the early soil microbiologists realized that the populations they were studying were critical to the maintenance of life and the functioning of the biosphere.

New areas of concern have arisen in recent years. One of these is the role of indigenous heterotrophs in detoxication. Soils receive herbicides, insecticides, fungicides, municipal and industrial wastes, and a variety of other toxic substances, and many of these are destroyed before the concentration of the toxicants rises to a point where they are of ecological concern. By thus acting as agents of biodegradation, microorganisms in the soil are a significant means of ridding natural environments of potentially hazardous chemicals and wastes. It has also been increasingly apparent that the indigenous communities frequently fail to degrade or detoxify many toxic substances. These failings are evident in the increasing pollution of ground and surface waters in many regions and in the persistence of certain types of pesticides, for example, DDT. Furthermore, recent research has established a role for microorganisms in creating pollutants. Thus, microorganisms in soil form nitrate, and the downward movement of nitrate from soil to the underlying groundwater leads to levels of this anion that may be toxic to infants consuming the water. If the microbiologically formed nitrate is assimilated by plants to levels that become extremely high, as occurs with certain vegetables, a precursor for human carcinogens may become abundant. The formation of nitrous oxide in soil, the effect of this nitrous oxide on the ozone shield in the atmosphere, and the relationship of these processes to skin cancer in humans are considered in Chapter 17.

BACTERIA

Bacteria are numerically the dominant group of microorganisms in soil. However, because their cells are small, the total biomass of bacteria frequently is less than that of fungi. However, in terms of metabolic activity, certain processes in well-aerated soils are dominated by bacteria, and under anaerobic conditions, bacteria are chiefly responsible for biochemical changes underground.

Because of the wide range of physiological and

Table 19-1. Typical morphology of bacteria obtained by plate counts of soil

MORPHOLOGICAL GROUP	BACTERIA IN CLASS (%)
Long rods, non-spore-forming	<1 to 7
Short rods, non-spore-forming	
Gram-positive	8 to 23
Gram-negative	13 to 27
Gram-variable	6 to 12
Rods that become coccoidal	2 to 13
Coccoidal rods	18 to 54
Cocci	<1 to 8
Rods, spore-forming	6 to 10

After A. G. Lochhead, Can. J. Res. **C18:**42, 1940.

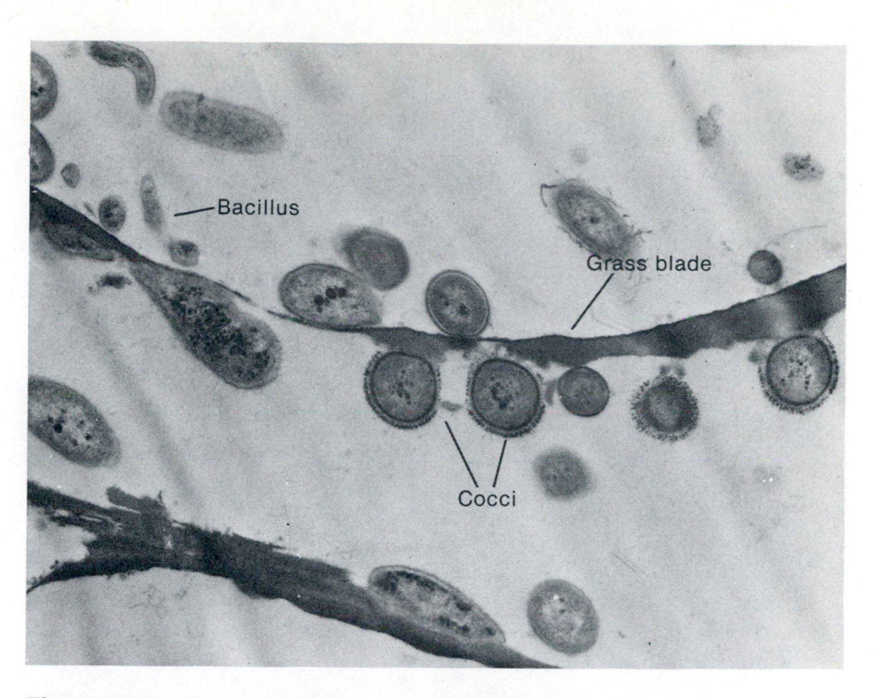

Figure 19-1 Transmission electron micrograph of encapsulated cocci and irregularly shaped bacilli attaching to and degrading grass. (×18,000.)
From Akin, D.E. Appl. Environ. Microbiol. ***39:**242, 1980.*

nutritional types of bacteria, no one culture medium and no single method are considered adequate for defining and estimating the size of the bacterial community of soil. Plate counts on agar media give large numbers of bacteria, frequently from 10^6 to 10^8 or more per gram of soil. The size of the community varies with the location and the environmental conditions that prevail. However, microscopic examination of soils indicates far greater population sizes, and often plate counts give values less than 1% or sometimes as much as 10% of the total count as indicated by microscopic means. Even if many of the cells seen under the microscope are nonviable, it is clear that plating procedures do not estimate the total community. The bacterial biomass even accounts for a significant part of the total soil mass, figures as high as 0.40% being common.

The bacterial community is dominated by 25 genera. Several other genera are present occasionally or in small numbers. The major species tend to be members of the genera *Arthrobacter, Pseudomonas, Agrobacterium, Alcaligenes, Flavobacterium,* and *Bacillus.* Although always present in small numbers, pathogens affecting humans, livestock and other animals, cultivated plants, and wild species of higher plants are present. Among the animal pathogens are species of *Clostridium, Bacillus, Listeria, Erysipelothrix, Coxiella,* and *Streptococcus.* Because roots of higher plants develop underground, the pathogens of higher plants are of particular significance. These include species of *Agrobacterium, Corynebacterium, Erwinia, Pseudomonas,* and *Xanthomonas.* The percentages of the major soil morphological types in a representative study are shown in Table 19-1.

It is often assumed that bacteria require no growth factors. This is untrue. A large percentage of the indigenous community is unable to grow in the absence of one or several B vitamins or amino acids. These growth factors are not present in soil in the absence of biological activity, so their presence indicates a role either for the plant in excreting growth factors or, more likely, for microorganisms in excreting the growth factors required by other inhabitats of the ecosystem.

Bacteria stand out because of the many different processes they carry out. Heterotrophic bacteria are able to metabolize and mineralize many low-molecular-weight organic compounds, and they carry out a varied set of biochemical processes and decomposition reactions under a wide range of conditions. They destroy many natural organic materials (Figure 19-1), with the possible notable exception of lignin, and they also decompose many synthetic organic compounds. The bacteria are particularly significant in oxidizing and reducing ions of nitrogen, sulfur, iron, and manganese. They are dominant members of the microflora in the absence of oxygen. Furthermore, many grow particularly quickly, and they thus often assume dominance. However, fast growth rates are only advantageous if the substrate being used is readily metabolized, and compounds that are quickly destroyed are not widely added to or generated in soil, except in the vicinity of plant roots. Microscopic observation of soil

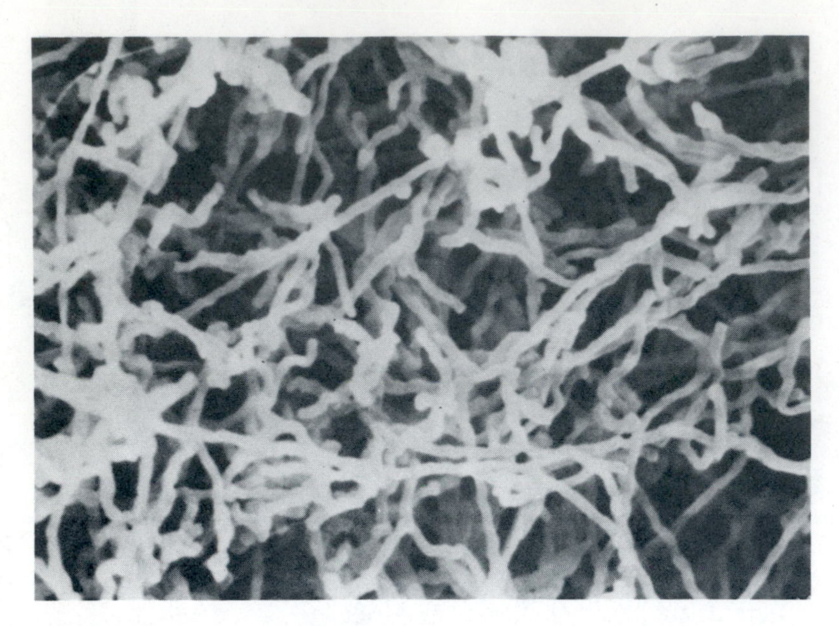

Figure 19-2 Scanning electron micrograph of filamentous actinomycete *(Nocardia)*. (×5000.)
Courtesy Yasuko Rikihisa.

shows the presence of small bacteria, often smaller in size than 0.3 μm. These organisms are largely unknown in culture, and hence their role in nature has yet to be established.

FILAMENTOUS BACTERIA AND FUNGI

Bacteria

Actinomycetes are especially common in soil. Taxonomically, the actinomycetes are bacteria, but they are often considered separately in soil microbiology because of their morphological characteristics, their slow growth rates, and the limited range of substrates for which they can effectively compete. Because of their slow growth, they often appear on agar media slowly, and the microbiologist who incubates plates for short periods of time frequently misses these organisms because their colonies only appear after 1 or more weeks of incubation. If the plates are incubated sufficiently long, it is evident that the actinomycetes are numerically second in abundance only to the true bacteria.

These organisms exist either in the form of the filamentous hypha (Figure 19-2) or as asexual spores, the ***conidia.*** Other specialized structures are formed by some actinomycetes, but these structures are reasonably uncommon. It is not clear to what extent the actinomycetes exist in soil as hyphae and to what extent the large numbers arise from the occurrence underground of conidia. However, it is generally believed that the filaments appear only occasionally and that the large numbers of colonies appearing on agar media inoculated with soil dilutions reflect the persistence of the asexual spores. If this view is correct, then the actinomycetes are not as significant contributors to many of the biochemical transformations in soil as their numbers would suggest. Their role would largely be linked to the availability of substrates or conditions particularly suited for their proliferation.

Fungi

Plate counts of fungi are of little value. The reason is that the fungi, as well as actinomycetes, on plates inoculated with soil dilutions may arise from long filaments, a fragment of a hypha, a single conidium, a cluster of conidia, or one of several types of structures produced in soil by fungi. Hence, although plate counts of fungi in soil are almost invariably low, this does not indicate their true biomass. Measurements of their biomass by a variety of techniques show that they are extremely abundant, and frequently the mass of fungi exceeds that of all other microbial groups in soil. A better appreciation of the fungal mass can be obtained by considering the lengths of their mycelium: as short as 10 meters per gram of surface soil, but sometimes as high as 100 meters per gram, and occasionally in excess of 1000 meters per gram.

Mycologists have devoted much time to identifying the dominant species and genera of fungi. Even now, new species and genera are still being described. The numbers of genera thus are still not known, but the species diversity of filamentous fungi is unquestionably impressive. The fungal flora includes species that lack sexual stages as well as those producing abundant sexual spores. Some species produce spores in ascospores, and others produce motile zoospores. Because of the many different fungal communities and the large number of highly dissimilar genera, no listing of fungal genera is here given.

Although the hyphae of some fungi develop quite slowly, and thus the organisms resemble the actinomycetes in not being effective competitors for readily available substrates, the hyphae of other species develop quickly. Organisms possessing these characteristics may dominate when substrates available to them are present. Fungi are particularly active by their use of the two major types of polysaccharides in plants—cellulose and hemicellulose—and they are the chief organisms in decomposing lignin aerobically. However, when the nutrient supply falls to low levels or the environment becomes hostile for other reasons, fungi endure because of the production of specialized resting structures. These structures include chlamydospores, sclerotia; oospores, ascospores, and rhizomorphs. Species producing these structures thus have a means for enduring adversity, and the hyphae reemerge from the resistant body once the conditions in soil become again suitable for fungal development.

Fungi are especially important in soil in the decomposition of resistant compounds, especially those that are not readily utilized by bacteria. These filamentous organisms decompose organic nitrogen compounds that are abundant in soil and convert them to ammonium, a process of significance to higher plants, which require inorganic forms of nitrogen. The fungi are important in breaking down the cellulose and many hemicelluloses that are prominent in the structure of higher plants. They are the dominant organisms in aerated conditions in the breakdown of lignin, a significant component of plants and a chief cause for the strength of woody tissues. They are also significant both in breaking down and in forming ***humus,*** the native organic fraction of soils. The filamentous organisms, by virtue of the filaments themselves, are able to bind soil particles together, and the binding of the small particles into larger aggregates improves soil aeration and water movement. These aggregates are helpful in allowing for the growth of roots and their extension through the soil matrix. A number of fungi are prominent because of their role in disease, often in diseases of higher plants but occasionally in those of humans and animals.

ALGAE AND CYANOBACTERIA

The presence of algae and cyanobacteria is easily shown. They are able to bring about photosynthesis, and thus their development is readily noted when soils are inoculated into an inorganic nutrient solution that is incubated for several weeks in the light. However, their significance in most soils and many areas of the world is limited because the algae and cyanobacteria seem to become metabolically active only where water is abundant and light is present. Below the soil surface, water is abundant, but the absence of sunlight restricts their capacity to grow in this dark habitat, especially when they must compete with fast-growng bacteria and fungi for the simple organic molecules that some algae and cyanobacteria can metabolize.

On the other hand, algae and cyanobacteria are often noted at the surface. These algae grow when moisture is abundant, a circumstance that tends to be confined to areas in which the soil is protected from high light intensities or to brief periods immediately following rainfall. Under these conditions, algae and cyanobacteria add organic matter to the soil as a result of photosynthesis. Once the organisms die, the dead cells serve as nutrients for bacteria and fungi, thus making available the organic compounds synthesized from carbon dioxide. Algae and cyanobacteria are important in starting the life cycle on rocks, volcanic areas, and in other regions largely free of life. The primary colonizers in these areas are free-living algae and cyanobacteria as well as lichens. In the lichen, one of a group of algae or cyanobacteria is accompanied by a fungal symbiont. In such areas, products excreted by the primary colonizers or substances released as they are being decomposed cause a slow but still appreciable destruction of the rock material. This destruction, over considerable periods of time, brings about weathering of the rocks and the initiation of soil formation.

Flooded paddy fields in southeast Asia often contain enormous masses of filamentous algae and cyanobacteria. In such ecosystems, light intensities are often high, the temperature is warm, inorganic nutrients are sometimes abundant, and moisture is not limited. Here, the algae and cyanobacteria have a prominent role in the ecosystem. As they carry out photosynthesis, they add appreciable organic matter to the water. The organic matter in the cells in turn supports the development of many herbivorous animals that grow in the water. The photosynthetic organisms also release oxygen, and this process may improve the aeration status of the otherwise poorly aerated soils and marshes. In addition, certain cyanobacteria are active in nitrogen fixation in many paddy fields. Rice has been grown almost every year for thousands of years in much of Asia, often with no signs of nitrogen deficiencies. The fields in which the rice is grown contain a rich flora of nitrogen-fixing cyanobacteria, and they add sufficient nitrogen to the soil and floodwater to permit the development of a crop of rice. However, the rice yield associated with reliance solely on cyanobacteria for nitrogen is low. Therefore, an international effort is now underway to find means to increase the rate of nitrogen fixation by the cyanobacteria.

PROTOZOA

Among the subterranean protozoa are representatives of many groups of flagellates, amebas, and

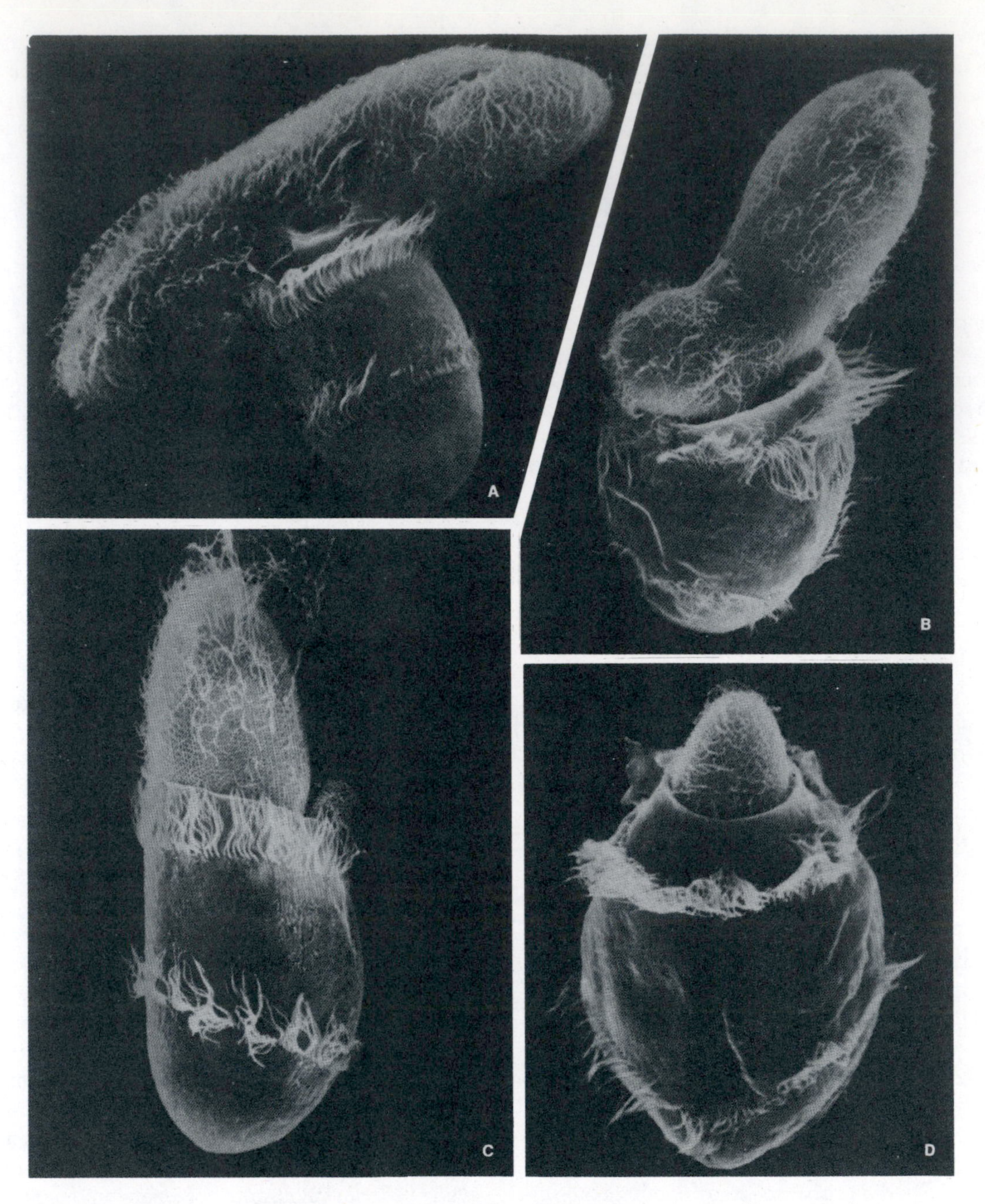

Figure 19-3 Scanning electron micrographs of *Didinium* species ingesting protozoan *(Paramecium)*. **A,** Cytosome of *Didinium* species has expanded. **B,** *Paramecium* is being folded as it is ingested. **C,** *Paramecium* has been partially engulfed. **D,** Most of *Paramecium* has been engulfed.
From Wessenberg, H., and Antipa. G.J. J. Protozool. ***17****:250, 1970.*

cilates. These animals have dissimilar methods of locomotion, namely by flagella, pseudopodia, and cilia, respectively. All three groups may be significant in soil, but their numbers vary appreciably from place to place and time to time. Because protozoa are unicellular organisms, determinations of their numbers have meaning. Such counts indicate that the numbers may be as low as 10,000, but sometimes the values may exceed 100,000 or more per gram of soil. Nevertheless, the counts may be misleading in attempting to assess protozoan contributions to the microbial biomass and activity, because their cells are frequently large and therefore one protozoan cell has a greater biomass than the cells of other unicellular organisms. On the other hand, many protozoa exist at any one time not in the metabolically more active forms but rather as cysts. Cysts have low metabolic activity. As a rule, the flagellates are abundant, and the ciliates are reasonably uncommon.

Certain protozoa grow in the absence of other

microorganisms. However, although able to decompose organic materials, these protozoa probably are not significant in this mode of nutrition because they do not compete effectively with bacteria and fungi in the same environment. Nevertheless, protozoa are avid predators (Figure 19-3), and they are able to consume thousands of bacteria per protozoan cell division. The predatory habit of protozoa appears to be more common than the nutritional pattern requiring feeding on nonviable organic materials.

Counts and microscopic examination of soil suggest that these small animals proliferate when the bacterial community becomes large, as occurs around plant roots or following the addition of significant quantities of readily available organic materials. Such proliferation is shown in Figure 19-4. At that time, the protozoan emerges from its cyst and begins to feed voraciously. Their abundance increases markedly, and at the same time, the number of bacteria falls. Nevertheless, the edible bacteria are not totally eliminated by predatory protozoa. Some mechanism exists to allow bacteria to coexist with species that feed on them. This coexistence is evident by the persistence in perpetuity of protozoa and bacteria in soil. A common explanation for the persistence of predator and prey is that the predation rate of protozoa declines with the fall in density of the bacteria they consume, and the two groups of organisms are able to match their growth and predation rates such that neither microbial type is eliminated. Predation may appear to reduce the biochemical activity of the community, but it does allow nutrients retained within the cell of the bacterial prey to be made available to succeeding populations. In the absence of any breakdown of the bacterial cells, elements of nutritional importance would not be recirculated, and the rate of biochemical transformations would then decline. In this way, protozoa may allow for a greater activity in soil than would be true in their absence.

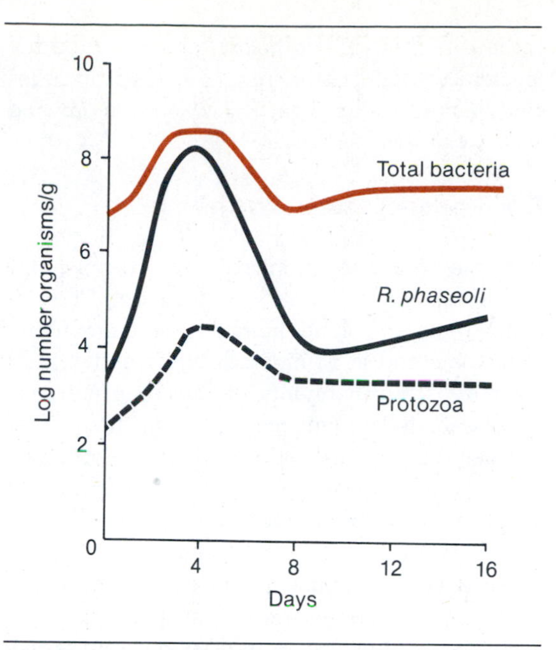

Figure 19-4 Growth of total bacteria and *Rhizobium phaseoli* and subsequent development of protozoa on bean roots. Seeds were inoculated with *R. phaesoli* and planted in soil on day 0.
Redrawn from Lennox, L.B., and Alexander, M. Appl. Environ. Microbiol. ***41:****404, 1981.*

ECOLOGICAL GROUPS

Indigenous heterotrophs

The terrestrial community contains fast-growing bacteria and fungi that use readily available nutrients in the environment as well as other heterotrophs that metabolize resistant materials in added organic substances. Because they use resistant compounds, the latter populations grow slowly in nature. Other heterotrophs use partially decayed organic matter, humus, microbial products, or microbial cells themselves. Parasites and predators are the inhabitants that live on the cells. The fast-growing heterotrophs respond immediately to nutrient additions, and these bacteria and fungi are active in the initial stages of many biochemical transformations, such as the decomposition of sugars, starch, organic acids, and proteins. The numbers of these fast-growing organisms fluctuate greatly, depending on the nutrient supply. Slow-growing populations using organic materials added to soil and those living on decayed organic matter and humus respond not too readily to recent nutrient additions, and thus the size of their populations or their biomass is not as variable as that of the fast-growing organisms.

Parasites

Many species are soil invaders. A variety of such transient species enter the environment in diseased plant or animal tissues, animal manure, or sewage sludge applied to the land. The transient behavior of these organisms is evident in the rapid decline of many of the species that are introduced onto or into soil together with sewage sludge. On the other hand, some soil invaders are not transient but still are not native. These are species that persist because they produce a resistant stage. Among this group are animal pathogens such as *Bacillus anthracis* or plant pathogens that survive below ground but do not proliferate. The latter group, which includes many fungi, infects their host when it is growing at that soil site.

Root-associated organisms

The true ***rhizosphere*** inhabitants are not really native to soil and are also not transient. The rhizosphere is the zone immediately surrounding plant roots and is dominated by the activities of the root. The inhabitants of this zone are truly root-associated organisms, although many exist in small numbers in soil apart from the root system. These species participate in a variety of chemical transformations, and they become prominent in the presence of a metabolically active root. Included among the root-associated or root-inhabiting populations are parasites that have a special ecological role within the tissues of some plant. In this zone are also found unique fungi that form symbiotic associations with roots of appropriate plant species. These are known as ***mycorrhizal*** fungi (see p. 430).

ABIOTIC FACTORS

Various nonbiological factors affect the composition of the soil community and its activities. The major abiotic determinants are nutrient supply, moisture level, pH, oxygen supply, oxidation-reduction potential, temperature, the presence of inhibitory substances, colloidal surfaces (such as those of clay and humus), and salts.

Nutrients

As a rule, the nutrients affecting the microbial community are almost only organic because few inorganic nutrients stimulate microbial populations and activity in soils. A specific type of organic nutrient will favor a specific group of populations that have the capacity to decompose that organic substrate. Thus, natural selection favors individual species that are able to metabolize sugars, cellulose, or proteins. In turn, the products of metabolism of these initial populations are used to satisfy the nutritional needs of others. In this way, the addition of one type of organic material leads to profound changes not only in the numbers of microorganisms but in the composition of the community. However, there is no direct correlation between the percent of organic matter found in soil by chemical analysis and the total number of microorganisms; too many other factors control population sizes. The populations that respond following addition of plant residues, animal wastes, or individual organic compounds may therefore live on the introduced nutrients, the excretions of the primary populations, cells or filaments that ***autolyze*** (that is, the cells or hyphae are digested by their own enzymes), or by utilizing the cells of the earlier populations as prey or parasites. The organisms that are active as predators on microbial cells in soils include protozoa, myxobacteria, and a variety of fungi.

Moisture

The moisture supply affects the community in four ways. First, it provides the water essential for replication. At low water levels, the increase in population sizes and activities with increasing moisture is a result of supplying the water that is needed for microbial development. Second, moisture affects the diffusion of oxygen into the soil. Thus, as soil approaches the water-logged condition, its pores become filled with water, and oxygen deficiency results as the spaces previously containing air become filled with water. Third, the soil solution carries many microbial nutrients, and moisture is therefore important in governing the supply of water-soluble nutrients. Finally, a soil with a high water content is not as subject to rapid temperature changes as a soil with little water.

Hydrogen ion concentration

The pH of soil is important because it alters the composition of the community and also directly influences the populations within the community. In acid soils, fungi are abundant, and bacteria and actinomycetes are less active than in neutral environments. The dominance of fungi under acid conditions is not a result of their being favored directly by the acidity but rather is a result of the elimination of competitors that do not tolerate the lower pH. Certain populations and the processes they carry out are completely inhibited under acid conditions. For example, the formation of nitrate is usually abolished as the pH of soil declines, although certain soils contain nitrate-forming populations that appear to be relatively less sensitive to acidity. The infection of legume roots by its symbionts, *Rhizobium,* is also affected deleteriously by acidity, and, although legumes continue to grow at moderate acidity, infection by the root-nodule bacteria is not initi-

Table 19-2 Microbial metabolism in anaerobic soils

	ORGANIC MATTER DECOMPOSITION	METABOLISM	EH VOLTS	ORGANIC ACIDS
Stage 1 (rapid)	Aerobic to semianaerobic	O_2 respiration, NO_3^- reduction, Mn^{4+}, Fe^{3+} reduction.	+0.6 to + 0.3	Only accumulted if fresh organic matter present
Stage 2 (slow)	Anaerobic	SO_4^{2-} reduction, CH_4 production	0 to −0.22	Rapid accumulation in early stages, rapid decrease in advanced stage

Takia, Y. *Folia Microbiol.* **11:**304, 1966.

ated, so that nitrogen-fixing nodules are not generated.

Aeration

Aeration affects the microbial community because most of its members are aerobes. Nearly all soil fungi and actinomycetes and most of the bacteria require oxygen for replication. The anaerobic members of the community are chiefly bacteria, which may be either facultatively or strictly anaerobic. The concentration of oxygen in the soil is in a state of flux—the gas coming in from the atmosphere, the microorganisms in the soil matrix using it up in line with their metabolic demand. If the supply of readily available organic nutrients is large, heterotrophs use up the oxygen present, and the demand for the gas exceeds the supply. Under these conditions, the size of the community and its activity are limited by the rate of entry of oxygen from the overlying atmosphere. Surprisingly, in what appears to be a well-aerated soil, anaerobic pockets exist. In these pockets are strictly anaerobic bacteria, so that a well-oxygenated soil still will support processes that are known to be carried out only by obligate anaerobes. Similarly, although denitrification proceeds only in the absence of oxygen, denitrification proceeds in well-aerated soils because the responsible bacteria exist in microenvironments where there is too little or no oxygen for their metabolism and proliferation.

Oxidation-reduction potential

The oxidation-reduction potential is a measure of the electropositivity or electronegativity of the environment. At a positive potential, conditions are oxidizing. At a negative potential, conditions are reduced. The chief oxidant is oxygen, and its presence raises the potential. Reducing agents lower it. Oxidation-reduction potential is ecologically important because aerobes need environments with high potential (that is, oxidized environments), although they lower the potential as a result of their metabolism. Anaerobes, in contrast, need low potentials or reduced environments (Table 19-2). It is difficult to differentiate, however, between the ecological effects of oxygen and those arising from oxidation-reduction potential. Aerobes and aerobic processes are dominant when the oxygen level and the potential are high, and anaerobes and anaerobic processes are prominent when the oxygen level is low or none is present and the potential is also low.

Temperature

Temperature affects community composition and its activities in soil, as in other natural environments. Each species has a minimum, an optimum, and a maximum temperature, but each is interacting with other members of the community. As a result, organisms are often not dominant at their optimum temperature but rather are prominent at those temperatures at which their activity is maximum relative to competing populations. Soils contain few strict psychrophiles, even in areas that have long, cold winters, and in these regions, the community is usually dominanted by cold-tolerant mesophiles. When soil is frozen, there is little if any microbial growth, but enzymes continue to function. Hence, low metabolic activity may be evident. Once frozen soil is thawed in the spring, the population size may increase rapidly. This increase is in part a result of the greater warmth, but it may in part be attributed to the breaking up of hyphae, cell clusters, or colonies by the thawing process. With increasing temperature, moreover, activity of the indigenous populations and their growth rates in-

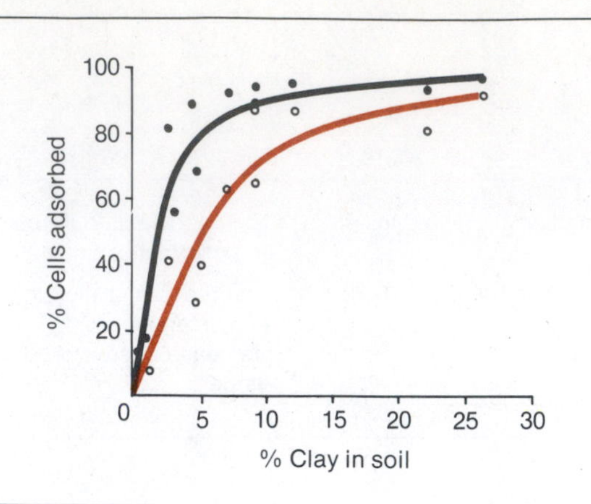

Figure 19-5 Increase in percent of cells of *Bacillus cereus* variety *mycoides* (closed circles) and *Serratia marcescens* adsorbed in soils with various amounts of clay.

Redrawn from Marshall, K.C. Interfaces in microbial ecology. *Harvard University Press, Cambridge, Mass. 1976. Reprinted by permission.*

crease, and the increase is proportional to the rise in temperature. In periods of the year when the habitat becomes hot, the previously dominant members of the community may be killed, so that areas with high temperatures are dominated by species able to cope with the heat stress. Nevertheless, few soils get sufficiently hot and still contain adequate water to permit the active replication of thermophiles; the high temperatures at the soil surface in the tropics are usually accompanied by drying so that opportunities for proliferation of thermophiles are not common.

Inhibitory substances

Inhibitory substances may be ecologically significant. The inhibitors include inorganic substances that may be effective only at high concentrations, such as the inorganic salts that are abundant in saline soils. Some inorganic inhibitors are ecologically significant at low concentrations; for example, ionic arsenic and copper as well as hydrogen sulfide are sometimes important. Similarly, toxic cations are present in some regions because of the return to soil of sludges that are rich in cationic forms of heavy metals.

In addition to inorganic inhibitors, a number of organic compounds that are produced biologically may be harmful to members of the community. Some of these compounds may be generated in large or small amounts by plant roots or microorganisms. Evidence exists that simple organic acids may be responsible for the resistance to decomposition of organic materials in flooded acid peats. Antibiotics may also have an ecological impact, although controversy still rages about their role in nature.

Colloidal substances

Soils are characterized by enormous surface areas. These large surface areas are the result of the presence of small particulate materials—clays and colloidal organic matter. The surface areas are highly reactive, but surface effects have only recently attracted widespread attention. They affect the community because the surfaces of clay and colloidal organic substances react with nutrients, microbial cells, and their extracellular enzymes. These colloidal materials have a charge, and they retain substances on their surfaces in part because of the charge on the colloid and in part by other mechanisms. The relation between the number of bacteria retained by soils and their clay content is shown in Figure 19-5. The activity of the adsorbed microorganism or its enzymes is different from that of the organism or enzyme in the free state. Similarly, adsorbed nutrients are often less available for use than nutrients that exist free in the soil solution. At low nutrient levels, surfaces may stimulate microbial activity because they concentrate the nutrients, which favors biological processes, and colloids may also be beneficial because they retain or adsorb toxins.

Salts

Inorganic salts are the chief agents of osmotic effects. Many land areas are extremely rich in salts, and some saline soils in the tropics are so rich in salts that higher plants are unable to grow. Nevertheless, microorganisms are still able to proliferate because of the presence of salt-tolerant or halophilic species. Among the halophiles in soil are both bacteria and actinomycetes.

INTERACTION AMONG POPULATIONS

The chief physical and chemical properties of these habitats that affect microorganisms are considered previously. These, together with the type of energy or carbon sources, are major factors governing the composition of the climax community. In addition, various interactions take place constantly among the indigenous populations. Because soils harbor vast numbers of microbial cells and filaments in small volumes, beneficial and harmful interactions naturally develop. These beneficial and antagonistic effects

Table 19-3 Examples of compounds toxic to one species but removed from the environment by a second species

COMPOUND	DETAILS OF INTERRELATIONSHIP
Concentrated sugar solutions	Osmophilic yeasts metabolize the sugar and thereby reduce the osmolarity, so allowing the growth of species that are sensitive to high osmotic pressure.
Oxygen	Aerobic organisms may reduce the oxygen tension, thus allowing anaerobes to grow.
Hydrogen sulfide	Toxic hydrogen sulfide is oxidized by photosynthetic sulfur bacteria, and the growth of other species is possible.
Food preservatives	The growth inhibitors benzoate and sulfur dioxide are destroyed biologically.
Lactic acid	The fungus *Geotrichum candidum* metabolizes the lactic acid produced by *Streptococcus lactis;* the acid would otherwise accumulate and inhibit the growth of the bacteria.
Mercury-containing germicides	*Desulfovibrio* species form hydrogen sulfide from sulfate, and the sulfide combines with mercury-containing germicides and permits bacterial growth.
Antibiotics	Enzymes are produced by some species of bacteria that break down antibiotics; thus the growth of antibiotic-sensitive species is allowed.
Phenols	Some bacteria can oxidize phenols and thereby allow other species to grow.
Tricholorphenol	A number of gram-negative bacteria can absorb trichlorophenol in their cell wall lipids and thereby protect *Staphylococcus aureus* from its action.

From Meers, J.L. Growth of bacteria in mixed culture. In Laskin, A., and Lechevalier, H. (Eds.). Reprinted with permission from *Microbial ecology*. Copyright CRC Press, Inc., Boca Raton, Florida, 1974.

are particularly pronounced in the microenvironment.

Beneficial interactions

Three types of beneficial interactions can be distinguished: ***commensalism, protocooperation,*** and ***mutualism.***

Commensalism. In commensalism, one of the two interacting species benefits. The second species is not helped or harmed by its associate. Many examples of commensalism have been described. For example, a facultative aerobe will remove oxygen to permit the proliferation of the strict anaerobe. This is not harmful to the first population, and it is beneficial to the second. Such an association may account for the finding of strict anaerobes in aerated soils (Table 19-3). Another example involves the decomposition of an organic compound by one population that converts that substrate to a product used by a second population. If the latter is unable to use the original substrate, the relationship is commensalism. Thus, when cellulose is added to soil, most of the many species appearing are unable to use the added polysaccharide but rather use the products of the initial populations.

Protocooperation. In protocooperation, both of the interacting species benefit, but the association is transitory and not obligate for the growth of the two species. In protocooperation, there are no fixed partners, but any population having the necessary physiological properties can interact with any other population for which the association is appropriate. A simple example involves a nitrogen-fixing bacterium that lives together with a green alga in environments that are poor in nitrogen and organic carbon. The former organism gets its nitrogen from dinitrogen, and the latter obtains its carbon from carbon dioxide. Should the first organism excrete a nitrogen compound that is useful to the second and the second releases an organic compound that is used for growth of the first population, the two grow well together. Another well-known example involves a species that converts a complex molecule to a product toxic to itself. The toxin producer cannot grow for long on the complex molecule because of the inhibitors it generates, but a second species may live in protocooperation if it cannot use the complex molecule but uses the toxin for growth. In this way, the second species grows and benefits the first, and the first is helpful to the second by giving it a waste product that is also a nutrient.

Mutualism. Mutualism, which is beneficial to both partners, is important in soil and is required either for carrying out some reaction or for the existence of the organisms in significant numbers in nature. In contrast with protocooperation, the partners in mutualism are fixed. Possibly the most important association in terrestrial environments is the one involving *Rhizobium* and leguminous plants. Although legumes may exist apart from these bacteria and *Rhizobium* persists in soil in the absence of the plant, dinitrogen is assimilated by neither of the organisms in nature when apart from their partners. It has recently been discovered that the bacteria may fix dinitrogen in culture, and they can use the gaseous form of the element as a nutrient under these conditions, but significant dinitrogen assimilation in nature appears to require that the two organisms live together. The seat of the symbiosis is the nodule, which is a swelling on the root of the host plant. The bacterium invades the root, causes root tissue to proliferate, and then begins to assimilate dinitrogen when protected in the nodule microenvironment.

A dinitrogen-fixing mutualism involving nodules on roots is also prominent among certain nonleguminous plants. In these instances, the microorganism is an actinomycete of the genus *Frankia*. The microorganism is not known to proliferate in soil, or to assimilate dinitrogen, in the absence of the higher plant, and the nonlegumes similarly are unable to make use of dinitrogen in the absence of the nodules engendered by the actinomycetes. The effects of both of these associations, that involving *Rhizobium* and that of *Frankia*, are evident in the development of plants bearing nodules in nitrogen-poor fields and the consequent increase in the nitrogen level of the ecosystem.

Mycorrhizae are associations of fungi with plant roots (Figure 19-6). The microorganism is known as the mycorrhizal fungus. This association is extremely common, and it has been estimated that 80% of the species of higher plants bear these root structures. As a rule, mycorrhizal fungi are poor competitors and are difficult to find in nature. For active and significant development, they must be in association with the root. Hence, the microorganism is benefited. Considerable evidence exists to show that the higher plant is benefited as well. For example, pine or spruce trees grow well in fertile but poorly in nonfertile areas. However, when the roots of these conifers are infected by appropriate fungi and mycorrhizae appear, the plants are able to grow at low nutrient levels. Particular attention has been given to the usefulness of mycorrhizal fungi in providing phosphorus to plants that are growing in soils that contain phosphorus in forms largely unusable by the fungus-free macrosymbiont.

Several other types of mutualistic associations may be prominent in specific areas. Lichens containing one of several fungal groups and either a green alga or a cyanobacterium are widespread in deserts, on mountainsides in cooled volcanic outflows, and in other areas that have little organic matter and low photosynthetic activity (Plates 22 and 23). Because of their nitrogen-fixing activity, an aquatic fern *(Azolla)* together with the cyanobacterium that the fern contains are prominent in paddy fields in Asia that are planted to rice. Another association that may be significant involves protozoa with their intracellular bacteria. The bacteria are located in the cytoplasm of the protozoan and do not exist apart from the animal. In this way, the bacterium finds its home. The protozoa appear to benefit because they will not grow in laboratory cultures if the bacteria are removed by means of antibacterial agents.

Antagonisms

Antagonisms are easily demonstrated. Thus, many bacteria multiply rapidly when introduced into samples of soil that have been sterilized. On the other hand, the same organisms fail to grow and may not even survive when added to nonsterile soils. Similarly, a pathogenic bacterium or fungus has a devastating effect on higher plants when the two are introduced into samples of sterilized soils, but the same pathogens have little or no effect when introduced together with their host plants in nonsterile samples of the same environment. The antagonism may result from ***competition, amensalism, parasitism,*** or ***predation.***

Competition. Competition is a rivalry between two species for a limiting factor in the environment. The less active species is suppressed, but the more active competitor is affected to a slight degree. Competition occurs when the demand for an environmental factor exceeds the supply. The major factors for which microorganisms may compete are carbon source, space, or oxygen. Most attention has been given to the competition for carbon because it is so often in limited supply. Although competition for space and oxygen may also be common, little information exists on their significance.

Many species introduced into soil probably are eliminated because they do not compete for the small supply of available carbon. An organism does not die from competition, but rather it is eliminated because of starvation. If other populations residing in the habitat are able to metabolize and use the supply of available carbon before an introduced species can, then the latter will not become a significant member of the community or it will soon perish. It is not certain what makes a microorganism a good competitor, but rapid growth rate, use of broad range of substrates, and the capacity to move from a depleted supply of organic matter to a site where there are still nutrients are likely to be important in competition (Figure 19-7).

Amensalism. In amensalism, one species is suppressed by the products of another. The product that is toxic may be a high hydrogen ion concentration, an organic acid that is toxic even apart from the acidity associated with it, carbon dioxide, or hydrogen sulfide. Surprisingly, carbon dioxide is ecologically important in some areas because many species are sensitive to the high levels of carbon dioxide that are common in soil. In addition, antibiotics represent another mechanism of amensalism. When tested under laboratory conditions, actinomycetes, fungi, and bacteria of many genera are able to produce antibiotic compounds. Yet, although most antibiotics of practical importance are derived from soil organisms, especially the actinomycetes, there still is not agreement on the significance of ***antibiosis*** in soil. The arguments against antibiotics being significant include the following: (1) antibiotic producers, especially the actinomycetes, often do not dominate the community and (2) many antibiotic-sensitive species are present and are often quite numerous in soil. Conversely, toxic substances have been obtained from soils, and these may be similar or identical to antibiotics of importance to chemotherapy. Nevertheless, no evidence exists that the toxins that have been obtained from soil that inhibit some fungi and bacteria are similar or identical to antibiotics that are used in medicine. Instead, those compounds that have been characterized as soil toxicants appear to be compounds like ammonia, ethylene, ethyl alcohol, certain simple fatty acids, and a number of high-molecular-weight organic compounds.

Parasitism or predation. Some of the predatory and parasitic organisms that are found in soil are given in Table 19-4. Viruses can be obtained from soil, but evidence for their significance in community activities is not rigorous. Protozoa, myxobacteria, and cellular slime molds are easily isolated by inoculating crumbs or dilutions of soil on an agar medium containing bacterial cells as the

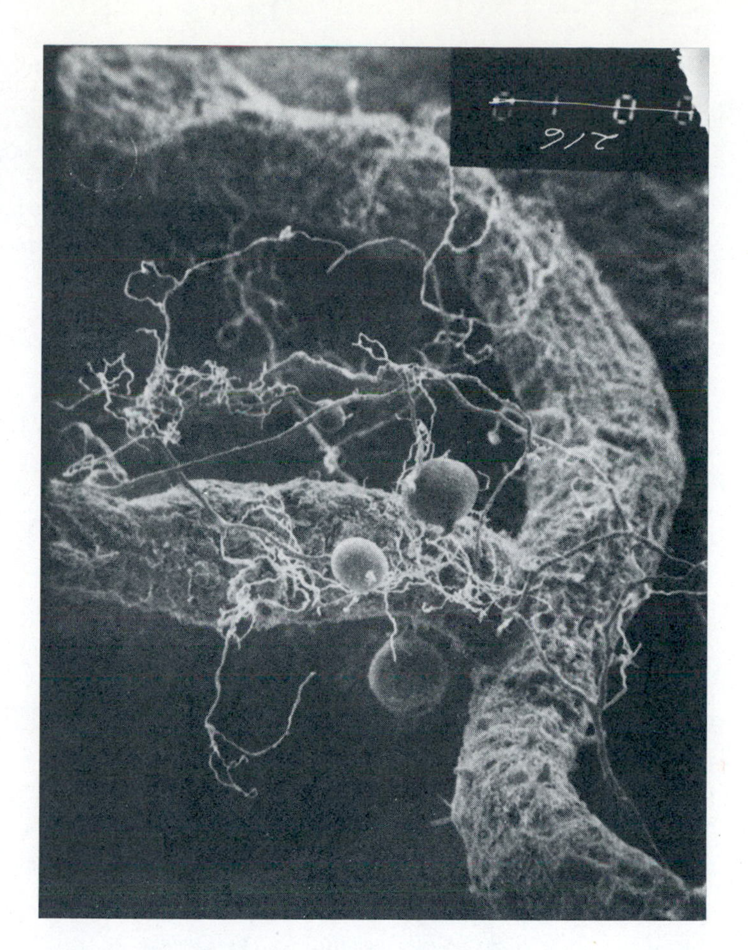

Figure 19-6 Light micrograph of portion of soybean root infected by fungus *Glomus mosseae*. Many filaments are seen to cover root surface. Spherical objects are chlamydospores of fungus.
Courtesy Merton F. Brown.

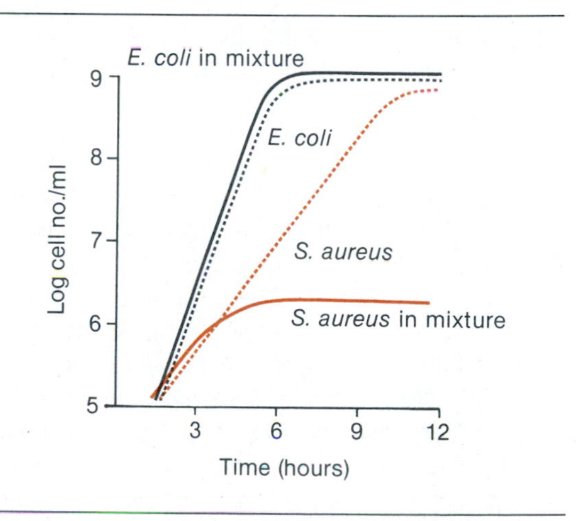

Figure 19-7 Competition between *Escherichia coli* and *Staphylococcus aureus* when species were grown either alone or as mixtures in batch cultures.
From Oberhofer, T.R., and Frazier, W.C. J. Milk Food Technol. ***24:****172, 1961.*

Table 19-4 Predatory and parasitic relationships in soil

FEEDER	SUSCEPTIBLE HOST	MECHANISM
Virus	Bacteria, actinomycetes, cyanobacteria, fungi	Intracellular parasitism
Protozoa	Bacteria, protozoa	Predation
Myxobacteria and cellular slime molds (Acrasiomycetes)	Bacteria	Lysis and predation
Fungi	Fungi	Hyphal penetration
Predaceous fungi	Nematodes and amebas	Hyphal appendages
Actinomycetes	Fungi	Antibiotic-induced lysis
Bacteria, *Actinomycetes*, fungi	Fungi	Enzymatic lysis
Actinomycetes, bacteria	Bacteria	Enzymatic lysis
Bacteria, fungi	Algae	Parasitism
Bdellovibrio	Bacteria	Cell penetration

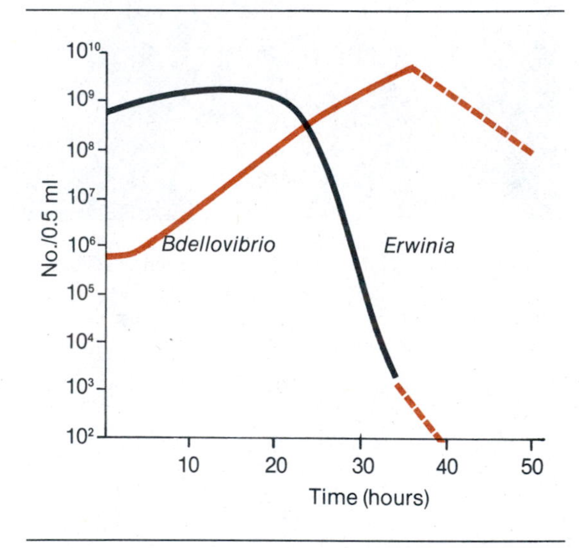

Figure 19-8 Growth of *Bdellovibrio bacteriovorus* on *Erwinia amylovora* in batch culture.
From Stolp, H., and Starr, M.P. Antonie van Leeuwenhoek **29**:217, 1963.

sole nutrient. It is believed that these predators obligately rely on predation in nature, and hence they must influence bacterial numbers or activities. Some fungi (for example, *Rhizoctonia*) parasitize other fungi, and their filaments penetrate and do injury to the hyphae of their hosts. This type of attack is rarely observed in nature. Nematode-trapping fungi attracted considerable interest at one time as a possible means for the biological control of plant-pathogenic nematodes, but although these fungi are evident in soils, they do not seem to be able to bring about the major decline in populations of nematodes that would be necessary for practical disease control. Fungi may occasionally penetrate the hyphae of amebas and bring about the digestion of the protozoan, but this phenomenon is only rarely observed in soils. The production by one species of an antibiotic affecting a second may lead to the lysis, or digestion, of cells or filaments of a second population, but the ecological significance of this type of interaction depends on whether antibiotics are produced in soils; as pointed out above, this issue is still not resolved.

On the other hand, enzymatic lysis is prominent. The organisms that are lysed include fungi and bacteria. Because of their large sizes, lysis of fungi, sometimes termed ***mycolysis,*** is readily detected by microscopic means. The resting structures of some fungi endure for long periods of time, and these are obviously, therefore, not readily prone to lysis. However, once the hyphae emerge from the resistant structures, enzymatic lysis may take place rapidly. This leads to a decline in the activity of the fungi. The hypha that is susceptible to lysis must be able to develop on some organic material and regenerate the resting structures or penetrate plant cells in order for the lysis-susceptible species to endure.

Algae are also prone to lysis by other microorganisms, and they also may be susceptible to parasitism by bacteria and fungi. Although such lysis and parasitism have been characterized in culture and, to some extent, in aquatic environments, their role in soil is still obscure. *Bdellovibrio* is a common inhabitat of many soils, but its population size is never great. Consequently, the significance of these vibrios is small at best (Figure 19-8).

Table 19-5 Compounds excreted by plant roots

	EXAMPLES
Sugars and oligosaccharides	
Pentoses (five-carbon sugars)	Xylose, arabinose, ribose, deoxyribose
Hexoses (six-carbon sugars)	Glucose, fructose, galactose, mannose
Disaccharides	Sucrose, maltose
Trisaccharides	Raffinose
Oligosaccharides	
Amino acids	About 20
Growth factors	Several B vitamins, choline, inositol
Organic acids	Tartaric, oxalic, citric, malic, acetic, propionic, unsaturated fatty acids, and *p*-hydroxybenzoic acid
Nucleic acid derivatives	Adenine, cytidine, guanine, uridine
Several enzymes	
"Factors" stimulating nematodes or fungi	Nematode-hatching factor, nematode-attracting factor, fungal spore–germination factors, fungal spore–germination inhibitors, fungal zoospore–attracting factors

RHIZOSPHERE

The rhizosphere is a plant-dominated and not really a soil-dominated habitat. In this underground environment that is influenced profoundly by roots, available nutrients are almost constantly available because of plant metabolism. The pH, oxygen and carbon dioxide level, osmotic pressure, and surface effects are also determined by the physiology and morphology of the root system and are little influenced by the surrounding soil. Some specialists distinguish between the ***rhizoplane,*** which is the root surface and its immediately adhering soil, and the rhizosphere. These zones can be distinguished in terms of the community size, the intensity of biochemical and microbial interactions, and chemical and physical properties. The rhizosphere and rhizoplane are important to the physiology and metabolism of the plant, to symbiotic associations between macroorganisms and microorganisms, and to plant pathogens.

The community in this restricted area is immense, even by plate counts, which are acknowledged to underestimate community size. The cell density often exceeds 10^9 per gram of rhizosphere soil. The populations, moreover, are especially active metabolically, and hence they probably have a major impact on the physiology and nutrition of the root system. Microscopic examination reveals vast numbers of bacterial cells on the surfaces of roots and root hairs. The bacteria are not distributed uniformly but rather exist as small colonies, presumably colonizing at the site of excretions. Fewer actinomycetes and fungi are evident by microscopic examination.

The chief effect of the plant in the rhizosphere zone probably is providing available carbon for microbial replication. Some of the compounds that are excreted by roots of higher plants are shown in Table 19-5. In young plants, these products are probably true excretions that are liberated by metabolically active roots. In contrast, older plants probably release considerable amounts of tissues and cells that are sloughed off as the root moves through the soil. If the composition of the community is largely governed initially by the types of substrates available, a comparison of the rhizosphere and nonrhizosphere soil is instructive. In the rhizosphere, simple organic molecules, several sugars, and amino acids are released, and their excretion is continuous. Conversely, the substances available to the soil community at a distance from roots are complex in structure; they are sugar polymers (polysaccharides) and amino compounds that are in organic complexes that are not readily available for microbial metabolism. These differences in the identities of the organic materials in soil at large and in the rhizosphere presumably are the chief reasons for the differences between the populations in the soil and the root habitat.

The stimulation in the rhizosphere is selective and affects only certain types of bacteria. The

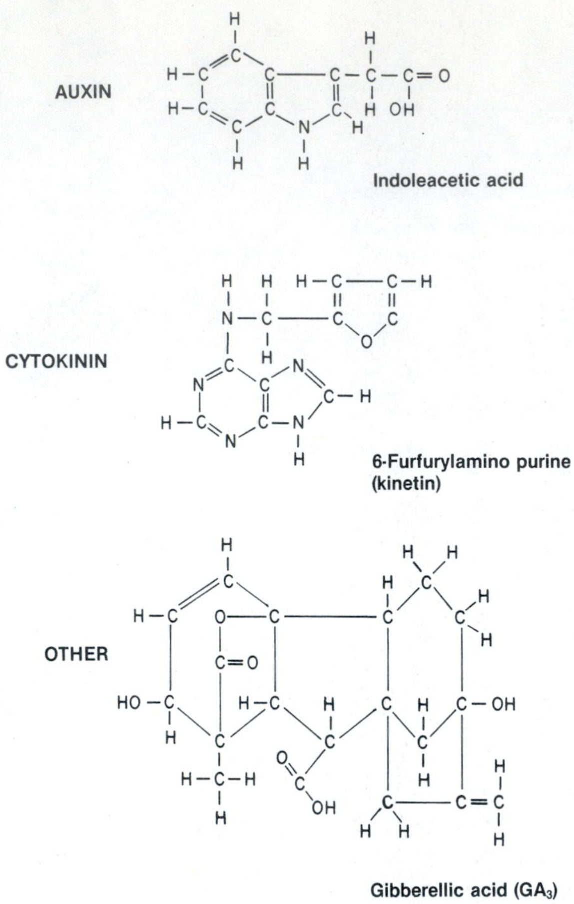

AUXIN

Indoleacetic acid

CYTOKININ

6-Furfurylamino purine (kinetin)

OTHER

Gibberellic acid (GA_3)

Figure 19-9 Structures of some important plant hormones and regulators.

groups that appear to be most benefited are species that require no growth factors and those that require just a few amino acids. These are characteristically fast-growing species that use simple organic molecules as carbon sources. The amino acids that allow for growth of the populations that need them are probably not derived from root excretions, which contain such compounds, but rather are excreted by the vast numbers of indigenous bacteria. As fast-growing organisms, these bacteria are much more biochemically active on an individual basis than the members of the surrounding soil community. The fungi, which are not especially numerous, may still be particularly important if they are agents of plant disease. The characteristic response in the rhizosphere among the fungi is to favor hyphal forms and to reduce the relative abundance of spores and other stages that are less metabolically active than the hyphae. In addition, mycorrhizal fungi are characteristic of this zone, although they are never abundant.

The rhizosphere community may be beneficial to the root and helpful to the development of the higher plant in several ways. The indigenous species mineralize the organic forms of various elements, and the conversion of organic nitrogen, phosphorus, and sulfur to inorganic products is crucial for plant development because plants cannot use, or use effectively, the organic complexes of the elements that are present in the soil. The bacteria in this zone may also aid the plant by solubilizing insoluble, inorganic forms of a number of elements. In soils, much of the phosphorus, iron, manganese, and calcium are insoluble, and the indigenous community has the capacity to convert the insoluble—and thus poorly available—forms of these elements to soluble products that plants assimilate readily. The solubilization may be the result of changes in pH, oxidation-reduction potential, or the excretion by bacteria of ***chelating agents*** that solubilize unavailable nutrients. These chelated nutrients then may be used by the plants. In culture, rhizosphere bacteria synthesize and excrete auxins, gibberellic acid, and other growth regulators (Figure 19-9), and these compounds bring about changes in cell development and root formation and also induce other responses in higher plants. It is believed that these products are also synthesized in the rhizosphere and presumably thus influence plants.

If antibiotics are indeed produced below ground, the likeliest environment for such activities is the rhizosphere because of the large supply of available carbon. These nutrients are needed to sustain the replication of antibiotic producers. If antibiotics are generated in the root zone, they may alter the composition of the community. In this regard, particular interest lies in the possibility that antibiotic producers alter the populations of pathogens.

DECOMPOSITION OF PLANT REMAINS

A major activity of the subterranean community is the decomposition of roots, leaves, stems, and other residues of higher plants. The process results in microbial replication because the heterotrophs are using the organic compounds they mineralize as nutrient sources. The conversions also are critical biogeochemically because they return to the atmosphere the limiting nutrient for photosynthesis, namely carbon dioxide. In the soil, the same processes of decomposition are also critical because the elements contained in the plant remains are transformed back to the inorganic state. Every element that exists in organic

Table 19-6 Major constituents of plants

CONSTITUTENT	COMPOSITION	OCCURRENCE
Cellulose	Polymer of glucose: straight chain $(C_6H_{10}O_5)_n$*	Cell wall
Hemicelluloses	Polymers of pentoses ($C_5H_{10}O_5$: xylose, arabinose), hexoses ($C_6H_{12}O_6$: mannose, glucose, galactose) or uronic acids ($C_6H_{10}O_7$: glucuronic, galacturonic)	Cell wall
Starch	Polymer of glucose: branched chain $(C_6H_{10}O_5)_n$	Storage carbohydrate
Lignin	Polymer of aromatic nuclei	Cell wall
Pectic substances	Polymer of galacturonic acid $(C_6H_8O_4)_n$	Cell wall
Inulin	Polymer of fructose	Storage carbohydrate
Hydrocarbons (aliphatic)	Straight chain compounds of carbon and hydrogen	Derived from fats, oils, and waxes

*In the formation of the polysaccharide, one water molecule is lost per repeating sugar unit.

complexes in plants is mineralized, and the nutrient elements are thus converted back to forms that are assimilated by root systems. In this way, the indigenous populations are required for the continued growth of higher plants. In the absence of chemical fertilizers, plants rely almost entirely on the microbial decomposition of organic materials to generate the nutrients that the roots must obtain. Only when fertilizers are used in large amounts is there another significant source of these nutrients.

The decomposition of carbonaceous materials is favored by an adequate supply of available nitrogen and phosphorus. Although other nutrient elements are needed by heterotrophic microorganisms for proliferation, their supply is usually greater than the microbial demand for them. As a rule, acidification tends to reduce the rate of decomposition. Good aeration and favorable moisture levels, by favoring aerobiosis, enhance carbon turnover. Furthermore, rapid decomposition is enhanced by a fine state of disintegration of the organic residue, and anything that increases the surface area for microbial attack—as does breaking down the physical structure of plant remains—tends to increase the rate of breakdown. In this connection, earthworms, insects, and other soil animals have a major impact on the degradation of organic substances.

The addition of carbonaceous materials to soil induces an immediate increase in microbial biomass and cell density. This is reflected in the rapid evolution of carbon dioxide and the disappearance of oxygen. Such additions of carbon also change the composition of the microflora, the community being dominated by fast-growing species if the plant remains are readily available and by slow-growing populations if the materials are largely cellulosic or are rich in lignin. After several days, the high initial rate of carbon dioxide evolution diminishes, and the large demand of the microflora for oxygen declines. The phase or rapid carbon dioxide liberation parallels the disappearance of readily available organic substrates. Then, when only the more resistant compounds remain, the rate of carbon dioxide evolution falls off. In the process, the simple, water-soluble compounds (such as sugars and amino acids), which are never abundant, disappear first. Subsequently, the polysaccharides, such as cellulose and several hemicelluloses, disappear more slowly. Metabolized at a slow rate under all circumstances is lignin. Every organic constituent of higher plants, so far as is known, can be decomposed by the microflora of soils, at least under aerobic conditions. Furthermore, regardless of the complexity or uniqueness of the organic constituent, it is metabolized by the same general metabolic pathways. The diversity of substances available to the microflora in this manner is shown in Table 19-6.

NUTRIENT CYCLES

Nitrogen

For plants and for agriculture, the most important of the elements coming from the soil is nitrogen. More often than any other element, it is nitrogen that limits plant growth and food production. Because the supply of this element is regulated almost entirely by microbial activity, soil microbiologists have focused considerable attention on the transformations of the nitrogen cycle. A diagram of that cycle is shown in Figure 19-10. Nearly every step in the cycle, except those involving

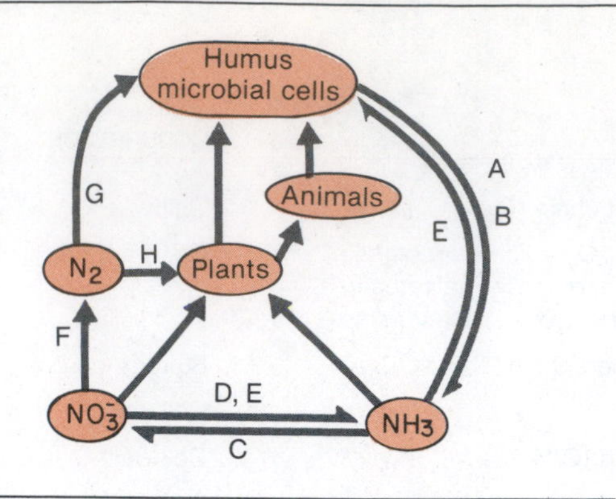

Figure 19-10 Nitrogen cycle. *A,* Ammonification, *B,* mineralization; *C,* nitrification; *D,* nitrate reduction; *E,* immobilization; *F,* denitrification; *G,* N_2 fixation, nonsymbiotic; *H,* N_2 fixation, symbiotic.
Redrawn from Alexander, M. Introduction to soil microbiology *(2nd ed.). John Wiley & Sons, Inc., New York, 1977.*

the uptake of inorganic nitrogen by plants and the consumption of plants by animals, is brought about by microorganisms.

The organic nitrogen in soil and that which is present in plants, animals, and microbial cells is converted by a large group of heterotrophic species to the inorganic form. In these mineralization reactions, the final product is ammonium. Some of the ammonium that is formed through mineralization is assimilated by microorganisms and is no longer available for plant growth. Some of the ammonium is used by higher plants. Varying amounts, depending on the conditions for microbial development, are oxidized to nitrate by nitrification. The nitrate thus produced may be assimilated by microorganisms and used for cell biosynthesis. Part of the nitrate that is generated is incorporated in plant tissues. A considerable part, when the oxygen supply is low, may be reduced to dinitrate or nitrous oxide by denitrifying bacteria. Variable amounts, sometimes large and sometimes small, are lost from the soil with drainage, and the nitrate that is thus lost reaches the groundwater that is used in many localities for drinking purposes. The dinitrate that is emitted from soil enters the atmosphere, and atmospheric dinitrate is a substrate for the nitrogen-fixing bacteria of soil. These bacteria may be free-living species of a variety of genera or they may be in symbiotic associations within root nodules. In a field growing nodule-bearing legumes, hundreds of kilograms of nitrogen may be fixed per hectare (roughly equivalent to pounds per acre). The free-living bacteria that fix dinitrate do not bring about as large nitrogen increases as do nodulated legumes, but their activity is present in lands even supporting nonleguminous crops or other types of vegetation.

Phosphorus

The community also decomposes organic compounds of phosphorus (Figure 19-11). Phosphorus mineralization leads to conversion of this element to the inorganic form, phosphate, that plants assimilate. The substrates for the mineralizing populations include the organic phosphorus of soil and the phytin, nucleic acids, and phospholipids of plant tissues. At the same time, microorganisms are assimilating phosphorus to satisfy their own nutritional demands, and this assimilation may be so great that the net effect of the decomposition of organic matter is not to make phosphorus more available but rather less so. If the plant remains contain small quantities of phosphorus, then microorganisms retain all of it for their own purposes, and none is available for plant growth. As a result, plants suffer even as microorganisms are bringing about decomposition of organic phosphorus compounds.

Sulfur

Sulfate is the major form of sulfur used by plants as well as by microorganisms (Figure 19-12). However, most of the sulfur in plant and animal tissues and in soil is not in the sulfate form, so that little of this element is available for the vegetation. Here, too, microorganisms play an important role in nature because they bring about the mineralization of the sulfur that is bound within plant remains and microbial tissues and that which is part of the soil organic fraction. Although microorganisms need some of the sulfate that they generate, the microbial demand for sulfur is usually small relative to the amount of sulfate that they produce, so that competition between macroorganisms and microorganisms for this element is rarely, if ever, of practical significance. *Desulfovibrio* may be prominent in soil when soil is so wet that it becomes anaerobic. At those times, sulfate-reducing bacteria may generate considerable hydrogen sulfide, and this metabolite is highly toxic to the root systems of growing plants. Sulfide accumulation is characteristic of waterlogged soils, especially when there is sufficient organic material available to sustain populations of the vibrios.

Iron

The community also contributes to the transformations of iron. However, this element is also subject to many abiotic transformations in nature. One of the more obvious transformations that microorganisms bring about is the reduction of ferric iron to the ferrous state. They perform this reaction when the soil is poorly drained so that it becomes anaerobic or when it contains sufficient available organic carbon to result in a diminished oxygen supply. Ferrous iron is preferred as a nutrient by crops and natural floras to ferric iron, but excess ferrous iron may result in phytotoxicity. Heterotrophic bacteria and fungi also decompose the organic compounds of iron that are present in plants and soil. These species use the organic portions of the compounds, and the remaining inorganic iron is only slightly soluble, and much precipitates from solution.

Other elements

Other elements of ecological or agricultural importance that are subject to microbial transformation are manganese, potassium, arsenic, selenium, and mercury.

BIOGEOGRAPHY

It is generally assumed that nearly all soil inhabitants are cosmopolitan and have a wide, and possibly nearly global, distribution. Present data indicate that this is true for many heterotrophs and autotrophs. However, a number of species or genera have a restricted occurrence and exhibit surprising geographical distributions.

An excellent example is *Beijerinckia*. This aerobic bacterium fixes dinitrate, and it does not appear to differ, in culture at least, from many other aerobic, dinitrate-assimilating bacteria other than by its acid tolerance. Nevertheless, this genus is found almost exclusively in soils of the tropics. Such a biogeography is evident from studies in Southeast Asia, Africa, South America, and Northern Australia (where it is found) and from other investigation in the temperate zones of Europe, Northern Asia, and the cooler areas of Southern Australia where it is relatively uncommon or is absent. Similarly, it is essentially absent from soils of the United States.

A few human pathogens also exhibit a striking biogeography. *Coccidioides immitis* is present in southern Texas, Arizona, and California and in northern Mexico. This fungus, which causes coccidioidomycosis (Plate 24), is thus restricted in location, although its host, humans, has a far wider distribution. The restriction of the organisms to arid areas with hot, dry summers and its absence from warm, humid regions

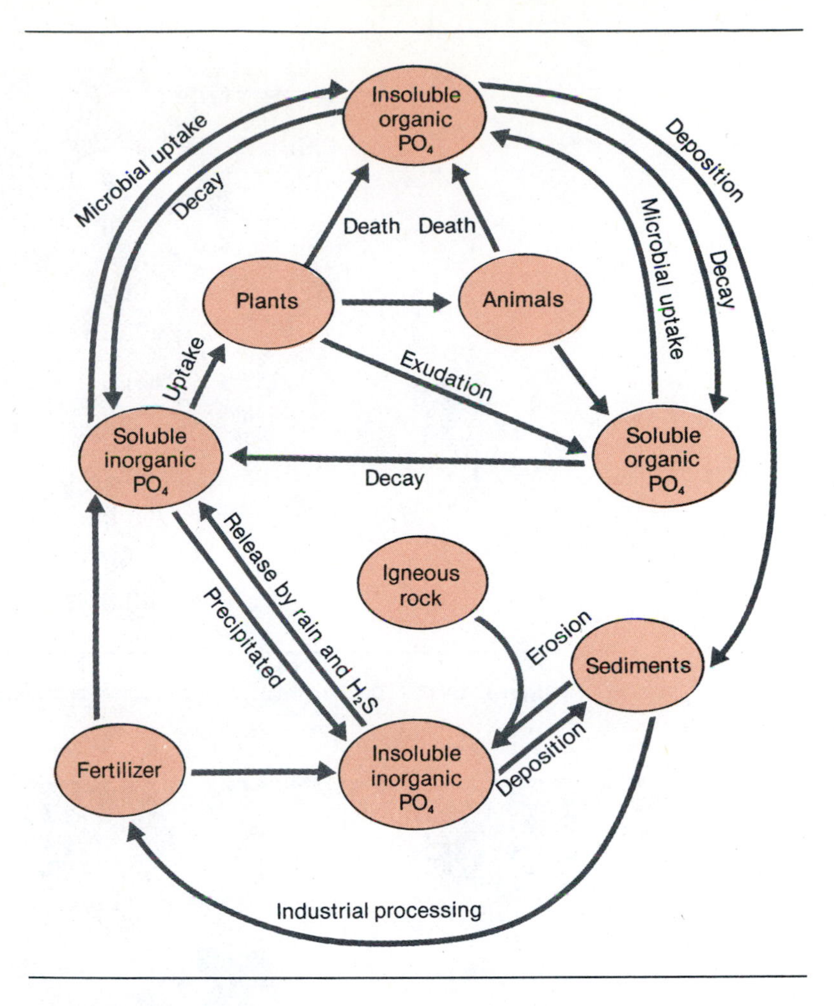

Figure 19-11 Phosphorus cycle showing various transformations with no change in original oxidation state of phosphorus.
Redrawn from Cambell, R. Microbial ecology. *Blackwell Scientific Publications, Oxford, England, 1977.*

of the United States and other countries is as yet unexplained. Similarly, *Histoplasma capsulatum,* the cause of histoplasmosis (Figure 19-13), has a noncosmopolitan distribution. Although this fungus has a worldwide distribution, it is locally common only in certain areas. In the United States, for example, it is abundant chiefly in the central Mississippi River area. In these regions, skin tests of 70% to 90% of adult humans are positive, suggesting their exposure to the fungus. The species has a predilection for areas in which bird droppings are abundant, but it is also found in other soils where such droppings are scarce. Moreover, bird droppings are present in many parts of the world, but the fungus is still abundant only in certain circumscribed geographical areas.

Biogeographical investigations have considerable ecological interest and, because some of the organisms are animal, human, or plant pathogens, they have practical importance as well.

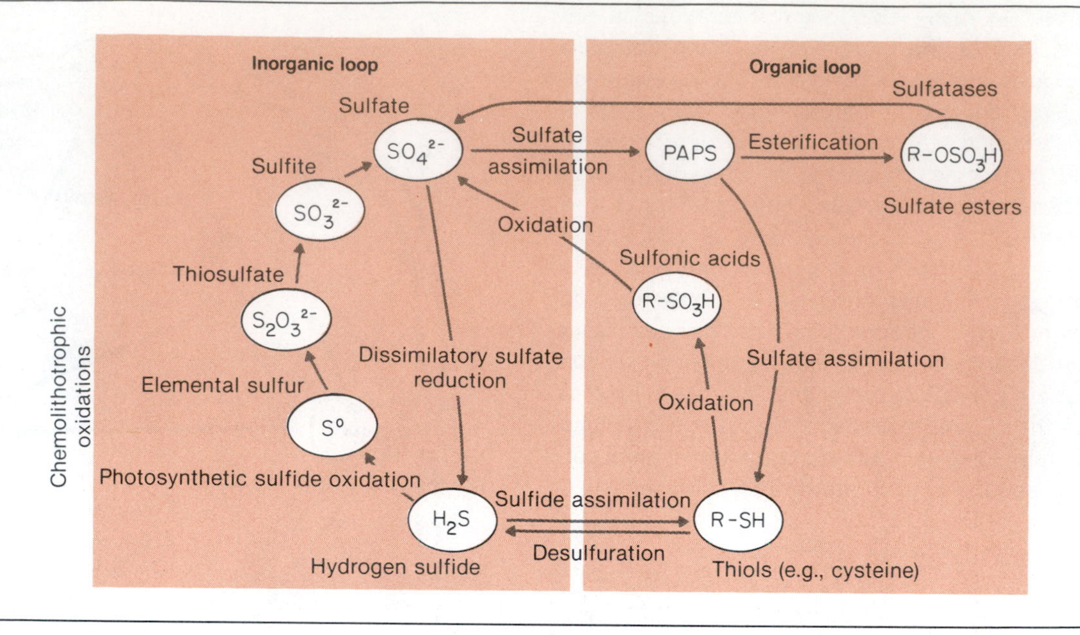

Figure 19-12 Detailed transformation process of sulfur. *PAPS,* phosphoadenosine phosphosulfate.

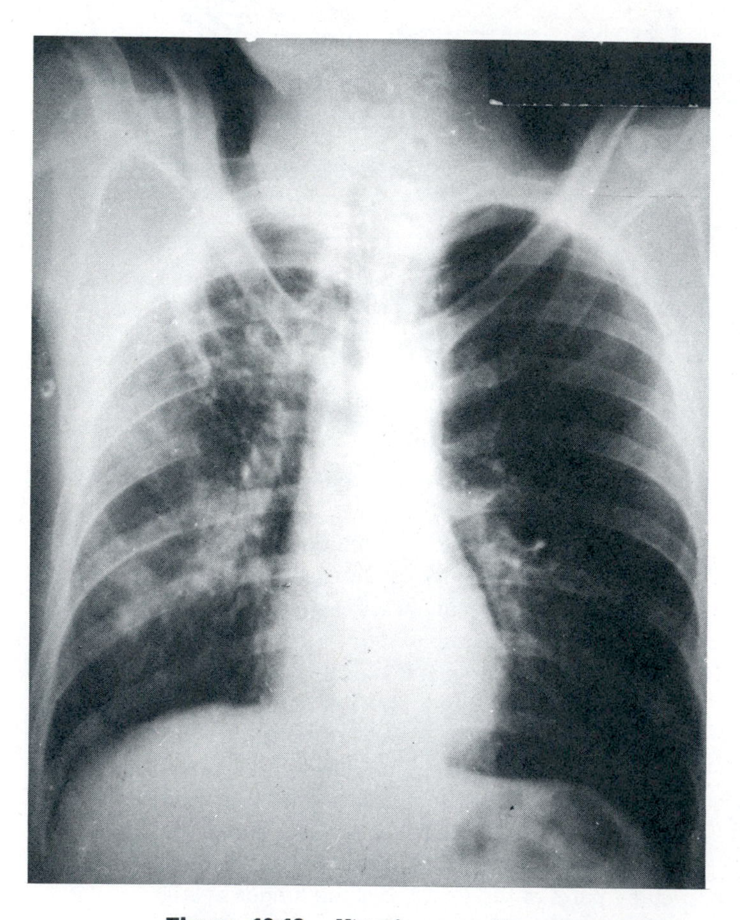

Figure 19-13 Histoplasmosis of lung on x-ray film. Extensive standlike and mottled infiltrate involving upper two thirds of right lung field represents infection with *Histoplasma capsulatum.*

Courtesy Jack Reynolds. The University of Texas Health Science Center at Dallas.

Nevertheless, the soil characteristics or physiological properties of these organisms that determine their distribution have yet to be established.

SUMMARY

The microbial community of soil contains many bacteria, actinomycetes, fungi, algae, and protozoa. No currently available method gives a true estimate of the numbers of viable bacteria. Many of these bacteria require growth factors. Actinomycetes are common in soil, but their role in this environment is uncertain. The biomass of fungi is large, and these organisms are important in the decomposition of plant polysaccharides. Algae and cyanobacteria often have little activity because of their need for both sunlight and water, but their activity may be pronounced where these two needs are satisfied. Protozoa assume prominence because of their voracious feeding on bacteria. The populations and activities of these various microorganisms are determined by the nutrient supply, moisture, pH, oxygen level, oxidation-reduction potential, temperature, inhibitors, and salts in the soil. Because of actions among these species, a microorganism is usually not most abundant or most active at the optimal intensity of these factors as determined in pure culture. The interactions that may be harmful to individual species in soil are competition, amensalism, parasitism, and predation. On the other hand, species may benefit from neighboring populations because of commensalism, protocooperation, or mutualism. In the rhizosphere, roots provide nutrients to the many bacteria present in that zone, and the microorganisms greatly affect the growing plant. A few soil microorganisms exhibit unique geographical patterns. Microorganisms are of special importance in soil because they decompose plant remains and, in the process, release the previously unavailable nutrient elements required by the vegetation. They are also important because they regenerate the limited supply of carbon dioxide in the atmosphere. Because the supply of available forms of phosphorus and nitrogen for plant growth is small, considerable attention has been given to the transformations of these elements. Microorganisms are also active in mineralizing, oxidizing, and reducing sulfur and iron compounds.

Study questions

1. What are the major groups of microorganisms in soil?
2. What are the problems in assessing quantitatively the abundance in soil of bacteria and actinomycetes?
3. In what processes are fungi and bacteria especially important?
4. What often limits the growth of algae and cyanobacteria?
5. Why may rice benefit from the growth of cyanobacteria?
6. Why do protozoa sometimes increase and sometimes decrease bacterial activity in soil?
7. What are the chief physical and chemical factors affecting microbial activity?
8. What is the effect on microorganisms of the large surfaces of clay and humus?
9. Describe how commensalism, protocooperation, and mutualism affect the interaction between two species?
10. Give arguments for and against the ecological significance of antibiotics.
11. Contrast the rhizosphere with soil at a distance from roots as habitats for microorganisms.
12. Which constituents of plant remains are decomposed rapidly and which slowly?
13. What microorganisms exhibit a geographic distribution in soils?

Selected readings

Books

Alexander, M. *Introduction to soil microbiology* (2nd ed.). John Wiley & Sons, New York, 1977.

Burges, A., and Raw, F. (Ed.). *Soil biology*. Academic Press, Inc., New York, 1967.

Dommergues, Y.R., and Krupa, S.V. (Eds.). *Interactions between non-pathogenic soil microorganisms and plants*. Elsevier Publishing Co., Amsterdam, 1978.

Gray, T.R.G., and Parkinson, D. (Eds.). *The ecology of soil bacteria*. University of Toronto Press, Toronto, 1968.

Hill, I.R., and Wright, S.J.L. (Eds.). *Pesticide microbiology*. Academic Press, Inc., London, 1978.

Marshall, K.C. *Interfaces in microbial ecology*. Harvard University Press, Cambridge, Mass., 1976.

Rovira, A.D., and Davey, C.B. Biology of the rhizosphre. In Carson, E.W. (Ed.). *The plant root and its environment*. University Press of Virginia, Charlottesville,

Vincent, J.M. (Ed.). *Nitrogen fixation in legumes*. Academic Press, Inc.; Sydney, 1982.

Wagner, G.H. Microbial growth and carbon turnover. In Paul, E.A. and McLaren, A.D. (Eds.). *Soil biochemistry* (Vol. 3). Marcel Dekker, Inc., New York, 1975.

Walker, N. (Ed.). *Soil microbiology*. John Wiley & Sons, New York, 1975.

Journal articles

Alexander, M. Why microbial predators and parasites do not eliminate their prey and hosts. *Annu. Rev. Microbiol.* **35:**113, 1981.

Lynch, J.M. Products of soil microorganisms in relation to growth. *Crit. Rev. Microbiol.* **5:**67, 1976.

Mosse, B., Stribley, D.P., and Letacon, F. Ecology of mycorrhizae and mycorrhizal fungi. *Adv. Microb. Ecol.* **5:**137,

Stout, J.D. The role of protozoa in nutrient cycling and energy flow. *Adv. Microb. Ecol.* **4:**1, 1980.

Chapter 20 FOOD MICROBIOLOGY

Robert Deibel and Robert F. Boyd

It has been variously estimated that one third of our food is lost to spoilage during growth, harvesting, processing, and distribution. The food microbiologist is primarily associated with the manufacturing segment of the food production industry, whose functions are to ensure a wholesome food supply free from disease-provoking potential and to endeavor to extend the shelf life of foods.

Every so often, previously unrecognized pathogens *(Yersinia, Campylobacter)* or toxins (mycotoxins) are associated with foods, and methods of detection and quantitation as well as procedures to eliminate an offending organism must be investigated. Improved methods of preservation are a never-ending quest. Relatively recent technological advances involving anaerobic and controlled-atmosphere packaging, intermediate-moisture foods, and the use of preservatives have extended the shelf life of some foods, but the quest continues.

This chapter provides a profile of the food industry and the role and problems of the food microbiologist. This is a dynamic aspect of microbiology, and the need for basically trained, competent food microbiologists is continuous.

MICROBIOLOGICAL EXAMINATION OF FOODS

Sampling

The procedures employed in the sampling of foods for bacteriological analyses are important, because bacteria are not randomly distributed in foods. Thus, measures must be taken to enhance the procurement of an analytical sample that approaches a true microbiological profile of the product. If the product is in discrete units (drums, bags, cartons), the square root of the number of units in a shipment or manufacturing code is sampled and composited. After blending, the analytical sample is withdrawn. Multiple sites are sampled when bulk products are examined, and the individual samples are blended before the analytical sample is withdrawn.

Analyses

For qualitative tests (sometimes referred to as presence/absence tests) a given amount of the analytical unit is introduced into a growth medium. Incubation in the growth medium is usually followed by transfer of an aliquot of the culture to selective-differential media to detect the organism in question. Qualitative tests are frequently used to detect enteric pathogens such as *Salmonella* and *Shigella.* The isolation of *Salmonella,* for example, first involves transferring a food sample to a nonselective enrichment medium such as lactose broth. This procedure enhances the isolation of salmonellas, but other microorganisms are also increased. A selective enrichment medium such as tetrathionate brilliant green broth is then used to select for only *Salmonella.* Following incubation in the selective medium an aliquot of the culture is transferred to brilliant green agar where the salmonellas produce a characteristic colonial appearance. The suspected salmonella colony is then subjected to biochemical testing by inoculation into triple sugar iron agar (TSIS) or lysine agar. These media indicate biochemical reactions typical of the salmonellas. If the biochemical tests are positive for salmonella, confirmation is made by subjecting isolates to serological testing. For quantitative tests such as those employed to enumerate coliform bacteria and *Staphylococcus aureus* and to ascertain the "total count" in a food product, the analytical sample is mechanically blended with a diluent such as dilute phosphate buffer or 0.1% peptone. Many foods are acidic, and after the initial blending the pH value is adjusted to approximately 7.0 to avoid acidification of the growth medium in the lower dilutions. Microbial counts can be determined by transferring aliquots of various dilutions to appropriate growth media employing such techniques as the spread plate and pour plate, or the most probable number can be determined by turbidity evaluations. The enumeration of *S. aureus* from a diluted food sample first involves growth on a medium that is high in salt or other inhibitor of nonstaphylococcal organisms, for example, mannitol salt agar. Pathogenic *S. aureus* is characterized by its ability to produce the enzyme ***coagulase.*** A positive coagulase test occurs when a suspension of the organisms is capable of clotting rabbit plasma that contains an anticoagulant. Further confirmation is provided by testing for thermostable nuclease (DNase), which is present in *S. aureus* and absent in other staphylococcal species.

Specifications and standards

Many manufacturers impose a specific limit on the number of total bacteria or specific organisms that occur in their finished product. These limits, or specifications as they are called in the industry, are also placed on raw incoming products, and often they are included in purchasing contracts. If a regulatory agency such as the Food and Drug Administration or the United States Department of Agriculture sets a limit, it is referred to as a standard (as opposed to a specification), and it is an enforcable law. Products that exceed the standard are subject to seizure and destruction.

Bacterial numbers in foods

The specific organism and its numerical occurrence are factors in determining the wholesomeness of the food. For some specific pathogens, such as *Salmonella,* the limit in a ready-to-eat food product is essentially zero or undetectable in a given analytical unit. On the other hand, the occurrence of 10^3 to 10^4 *S. aureus*/g in a cheese or sausage product may not be significant. *Salmonella* species are intestinal pathogens capable of invading the intestinal tract, and as few as seventeen organisms have been shown to cause disease. Staphylococci are not invasive and are unable to penetrate healthy tissue. Certain strains of *S. aureus* are capable of producing an enterotoxin in food, but they are usually only human strains (for example a food handler could contaminate food), and at least 10^6 or more organisms are required to cause disease symptoms.

The ***shelf life*** of a product is often a function of the holding temperature and the number of bacteria after manufacturing or harvesting. Significant effort is expended to keep the number to a minimum to enhance a long shelf life. With perishable products, especially those requiring refrigeration, the lower the total count (or load), the longer the shelf life. Marketed products such as pasteurized milk or sausage will often have total counts of 10^3 or 10^4 bacteria/ml or gram. The product is still quite acceptable, and no off flavors or odors can be detected. At a level of 10^7 bacteria/ml or gram the product may be in the incipient stage of spoilage, and counts of 10^8 to 10^{10} usually but not always give rise to souring or off odors. Thus, relatively large numbers of bacteria must be present in a food before detection by the senses (organoleptically).

In the manufacture of some foods such as cheese and fermented sausage, a starter culture (pure culture of a specific organism) is employed, and the growth of large number of bacteria is encouraged in the manufacturing process. The finished products often contain total counts of 10^8 to 10^{10} bacteria/g, and the absence of high counts may reflect manufacturing failure. Thus, the organism, its numbers, and the type of product must be considered when the significance of bacteria in a food product is evaluated.

CONTAMINATION AND MICROBIAL GROWTH IN FOODS

It is virtually impossible to avoid microbial contamination of foods during harvest and subsequent processing. Our only salvation is to keep the contamination to a minimum throughout processing as well as to control the subsequent growth of the contaminating microorganisms. Sanitation and good management help control the level of contamination, and rapid movement of the product through the food chain as well as a variety of processing procedures (pasteurization, canning, refrigeration, salting, drying, and acidification) are used to control the subsequent growth.

Sources of contamination

Many raw food items have a normal flora that is often associated with the immediate environment from which it was harvested (Table 20-1). Subsequent handling and processing allow growth of the normal flora or a portion of it. Quite often the unrestricted growth of the normal flora will develop into the spoilage flora. For example, when harvesting milk, the producer endeavors to reduce the load during harvest by cleansing the udder and by cleaning and sanitizing the milking equipment, pipelines, and holding tanks. The

Table 20-1 Primary sources of microorganisms that contaminate foods

FOOD ENVIRONMENT	MOST FREQUENTLY ENCOUNTERED GENERA IN FOOD
Soil and water	Bacteria *Alcaligenes, Bacillus, Citrobacter, Clostridium, Enterobacter, Micrococcus, Pseudomonas, Serratia* Fungi *Aspergillus, Rhizopus, Penicillium, Fusarium*
Plants and plant products	Bacteria *Acetobacter, Pseudomonas, Erwinia, Flavobacterium, Lactobacillus, Leuconostoc, Pediococcus, Streptococcus* Fungi *Saccharomyces, Torula*
Intestinal tract of humans and animals	Bacteria *Escherichia, Salmonella, Bacteroides, Lactobacillus, Proteus, Vibrio, Shigella, Staphylococcus, Streptococcus, Clostridium, Citrobacter, Pseudomonas*
Air and dust	Any of the genera mentioned above but particularly bacteria such as species of *Bacillus* and *Micrococcus* and fungal species such as *Cladosporium, Penicillium, Aspergillus*

milk in the udder is essentially sterile, and the initial contamination is derived from the teat canal and the milking machine, pipelines, and holding tanks. The milk parlor environment also contributes to the initial load. Producers refrigerate the holding tank and generally keep the milk 1 to 2 days. After pickup on the farm, the milk is transferred to large holding silos at the dairy plant. The sanitation of these silos, as well as the ambient temperature and the length of the holding time, may cause an increase in the load. The milk is pasteurized to kill the pathogens, and some nonpathogens are also killed. Many nonpathogens survive this milk thermal processing, and these organisms constitute the potential spoilage flora (Plate 25).

Processing equipment. The processing environment provides sources of contamination. Good management and proper sanitation help reduce this source. In the past the Food and Drug Administration has recalled or seized a product when it was observed that the product was produced in an unsanitary environment. Processing equipment such as machines, cooling tunnels, conveyor belts, and pipes are important sources of contamination. Food manufacturers expend considerable time and money cleaning and sanitizing their manufacturing facilities. Bacteria in the environment or on the machinery inevitably are transferred to the product.

Water and air. The microbiological quality of water and air that comes into direct contact with food must be monitored for quality assurance. Neither water nor air has a normal flora—both become contaminated from the environment before their introduction to the food or food processing plant. Species of *Pseudomonas* are a common water contaminant because of their ability to survive in limited nutrient environments. Water used in foods and for cleaning purposes must meet drinking water standards (500 organisms or less/ml). Quite often, water that is recirculated in a plant is filtered to remove food particles and rechlorinated to reduce the microbial load. Many food processing plants use large quantities of air for drying, in cooling tunnels and in the general food processing environment. Although the air is coarsely filtered (chiefly to remove particulate contamination), spores and vegetative cells readily pass through the filters and contaminate the processing areas.

Food ingredients. The chief sources of microbial contamination of a processed food are the food ingredients. The food processor must therefore maintain a substantial quality assurance program on the raw ingredients. Various microbial specifications are set for each incoming ingredient, and frequently the product is released for manufacture only after laboratory clearance. This is one of the several reasons for the continuing interest in rapid bacteriological methods, because it is costly to warehouse incoming ingredients before laboratory release.

Rework. Another potential source of microbial contamination is ***rework.*** Rework is partially or completely manufactured product that does not fully comply with various specifications set by the company. For example, candy bars that are not completely enrobed with chocolate, defectively packaged product, and the end portions of bologna and other cured meat loaf products that do not yield a uniform slice are returned to the beginning of the process and incorporated into new product. Just about all segments of the food industry use rework because it would be exorbitantly expensive to discard it. Reworking of a chemically or microbiologically defective product is not condoned because the load of the new product often results in process failure. Generally, product that has not left the manufacturing facility can be reworked; however, product that has been in the distribution chain is destroyed.

Control of contamination and growth in food processing plant

Food manufacturers must be aware of four cardinal considerations associated with the control of microbial contamination and growth in their plants: plant layout, compartmentalization, traffic control, and water control.

Plant layout. This first consideration centers on an orderly sequential processing scheme in which the "flow" of the product originates at one end of the plant and proceeds stepwise throughout, terminating with the finished product at the opposite end. Generally, the raw products and early processing mixtures will contain the highest load, and the finished product will have the lowest load. The decrease in load is usually associated with processing barriers such as heat, drying, and salting.

Compartmentalization. Physical barriers are placed at strategic points between processing areas to preclude cross-contamination between high- and low-load product during manufacture. For example, in the manufacture of chocolate, the raw cocoa bean is received in large jute sacks. The sacks are opened and inspected for discolored beans, twigs, stones and other undesirable matter. The beans are known to have a high microbial load and occasionally to be contaminated with *Salmonella.* The beans are conveyed to a large hopper that feeds the roasting machine. A pipe or chute through the wall that separates the raw and roasting areas radically decreases the potential for cross-contamination of roasted product with raw product. Roasting does not kill all bacteria, and the subsequent dehulling step generates dust and offers potential airborne contamination of the finished product. The dehulled bean or nib is then ground and heated, yielding the chocolate liquor (finished product), which is the basic ingredient for milk chocolate.

Traffic control. Traffic control in compartmented areas is difficult to maintain. Plant personnel may frequently have functions in two or more compartmented areas. Lift or fork trucks and hand carts also can carry contamination from one area to another. Some manufacturers place foot baths containing a bactericidal agent between critical areas to reduce cross-contamination.

Water control. The growth of microorganisms in a food processing plant depends on three conditions: the presence of organic matter (the food being processed), the presence of microorganisms, and water. We cannot remove the food being processed, and we cannot totally exclude microorganisms, but we can control the availability of water in the processing environment. Water is an absolute requirement for all life and by controlling it we can also control microbial growth. Roof leaks, pipe leaks, faulty water pump seals, and the unwarranted use of water (especially for cleaning purposes) are the chief violations of water control. When the above three conditions are present, large microbial populations can be the result. The organisms can be disseminated by foot and vehicular traffic, or the mixture can dry, aerosolize, and be spread by air currents. In many instances water must be used for cleaning; however, its judicious use must be practiced, and drying procedures after cleanup should be instituted. Many manufacturers of dried products (such as candies, flour, starch, and bakery items) use central or portable vacuum cleaners to avoid a wet cleanup. Water control is an important and often overlooked aspect of controlling microbial growth in a food processing plant.

Role of technology in microbial contamination and growth

Whenever technological alterations are made in the processing and packaging of a food product, the potential exists that an alteration of the microbiology of the product can take place. The advent of vacuum packaging, for example, has increased the industry's awareness of the potential for *Clostridium botulinum* growth in many of these anaerobically packaged products. The so-called ***controlled atmosphere packaging,*** wherein the product is packaged in a gaseous environment other than air, decreases the growth potential for certain groups of spoilage bacteria. Thus, the concept that technology dictates microbiology is pervasive in the food industry, and food microbiologists analyze technological changes before, during, and after their innovation to assess the microbiological alterations that may occur.

Investigation of spoilage problems in the manufacture of foods

Several approaches may be employed in investigating a microbial spoilage problem. Assume that the spoilage organism is known or the associated condition reflects the growth of a particular group of bacteria. For example, the souring of sliced, prepackaged ham is associated with the growth of lactic acid bacteria of the genera *Streptococcus, Lactobacillus,* and *Leuconostoc.* If the organism is not known, then it must be identified (usually the predominant organism in the spoiled product). The processing train of the product is charted, and the time and temperature throughout the processing are recorded. At significant (or the so-called critical) points in the processing train, samples are taken. These samples, as well as samples of raw ingredients and finished product, are then analyzed for the offending microorganism.

Assume that a sliced ham has been observed to have a short shelf life, and souring is evident after 1 week in the display case. Investigation at the plant indicates that the raw and pickled ham were within normal limits and the time and temperature of cooking were adequate. Examination of the slicing machine, however, indicates bacterial counts of 10^9/g. Thus, as the ham was being sliced it was also being inoculated with large numbers of microorganisms. A more frequent and careful cleaning of the machine during the manufacturing day would probably solve the problem. If the ham had been insufficiently cooked during processing, lactobacilli would be present in high numbers on the product. Remedial measures in this instance center on control of established thermal processing procedures.

FOOD PROCESSING AND METHODS OF PRESERVATION

Shelf life and coding data

The shelf life of a product is defined as the expected time that a product will remain organoleptically acceptable. Sometimes the shelf life of the product is dictated by chemical rather than microbiological considerations.

Many food manufacturers code date their product. The code date (or lot or code number) reflects a given manufacturing lot of product made in a given location during a specified period of time. For example, the code A241B2 could indicate Plant A, the 241st day of the year (Julian calendar), B shift, and the second line in the plant. In this manner manufacturers can identify specific manufacturing lots. If a lot is found to be defective, the incriminated lot can be removed easily from the market. In another type of code dating the manufacturer will indicate the expected shelf life of the product (for example, ''good until blank date'' or ''must be sold before blank date'').

Thermal processing of foods

Heat is one of the most common methods of food preservation. The thermal processing schedules of foods are generally calculated to kill a given segment of the contaminating microbial flora; in most instances, the food is cooked as well. In certain procedures, such as the blanching of vegetables, the thermal process is calculated only to inactivate the plant's enzymes. The heat and the cleansing of the product also help reduce the microbial load.

Effect of high temperature. Increases in temperature above the optimum range of growth for the organism result in cell injury and eventually death of the cell. One of the most important considerations in a processing plant is to determine the amount of heat required to produce a commercially sterile product. First, the most heat-resistant mesophilic spore-former that is encountered in the food is isolated. A given number of spores are mixed with a slurry of the food. The slurry is dispensed into glass vials that are sealed and then heated in an oil bath. The latter achieves temperatures above 100° C. The vials are removed from the bath at definitive time in-

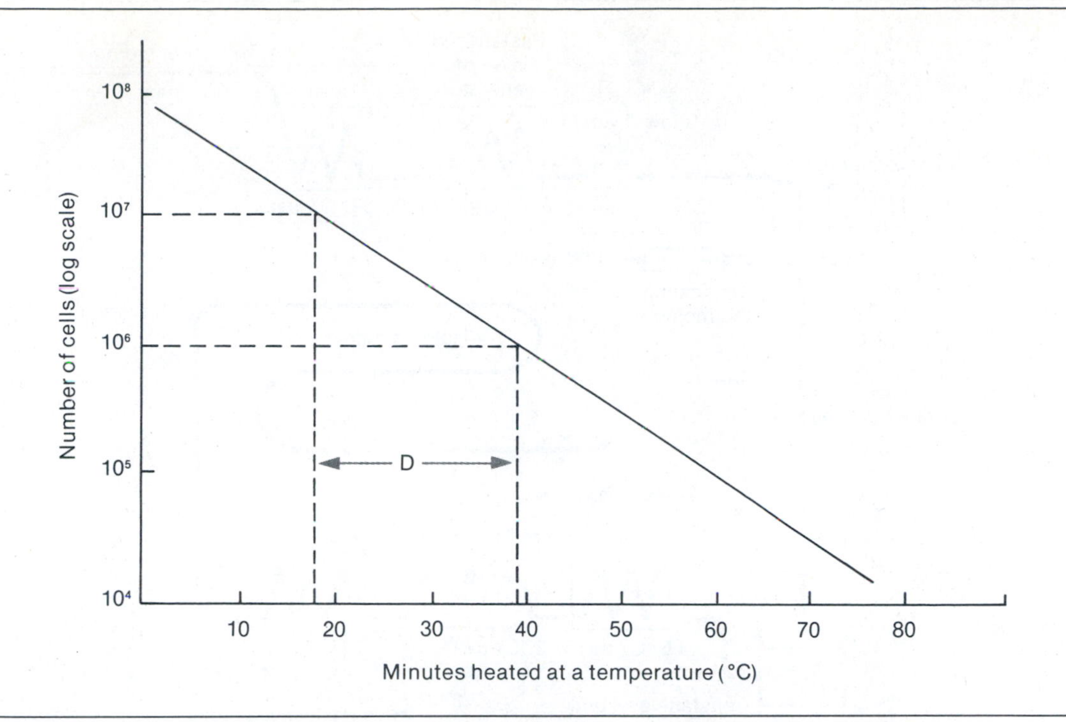

Figure 20-1 Decimal reduction time (D). Extrapolation on one log unit, for example between 6 and 7, gives the D value, which in this instance is 20 minutes, that is, 20 minutes to reduce the survivors by a factor of 10. The straight line in this figure theoretically extends below the base line into negative logarithms so that $\log_{10}{}^{-2}$ = one cell in 100 g of food. Thus the population of cells never reaches zero.

tervals and cooled, and each slurry is plated quantitatively to determine the number of survivors. A typical survival curve is illustrated in Figure 20-1. The plot of the logarithm of the number of microbial survivors against time of exposure to a particular temperature yields a "straight line." From this curve the ***decimal reduction time,*** D, may be determined. The D value is numerically equal to the minutes required to destroy 90% of the population, that is, the time required to reduce the population by one log cycle or by a factor of 10. Increases in temperature result in decreases in the D value. The D value is a measure of heat resistance, which makes it possible to determine beforehand a heating process that will result in a specific level of killing. This information is particularly useful for processing canned foods. The D value has use in many other aspects of microbiology, because it can be used to estimate or compare survival in the presence of germicides, acids, or any adverse chemical or physical environment where death is incurred.

Thermal death time (TDT). When D values are plotted against corresponding temperatures, a thermal death time curve is produced (Figure 20-2). Thermal death time is the time necessary to kill a given number of microorganisms at a

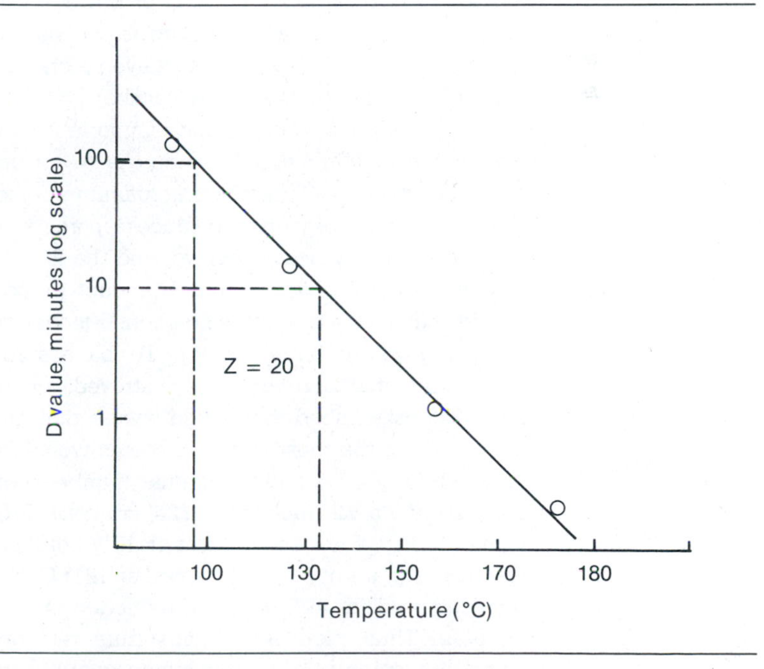

Figure 20-2 Thermal death time (TDT) curve. Four D values (○) have been graphed. *Z* is the degree centigrade required for the TDT to traverse one log cycle.

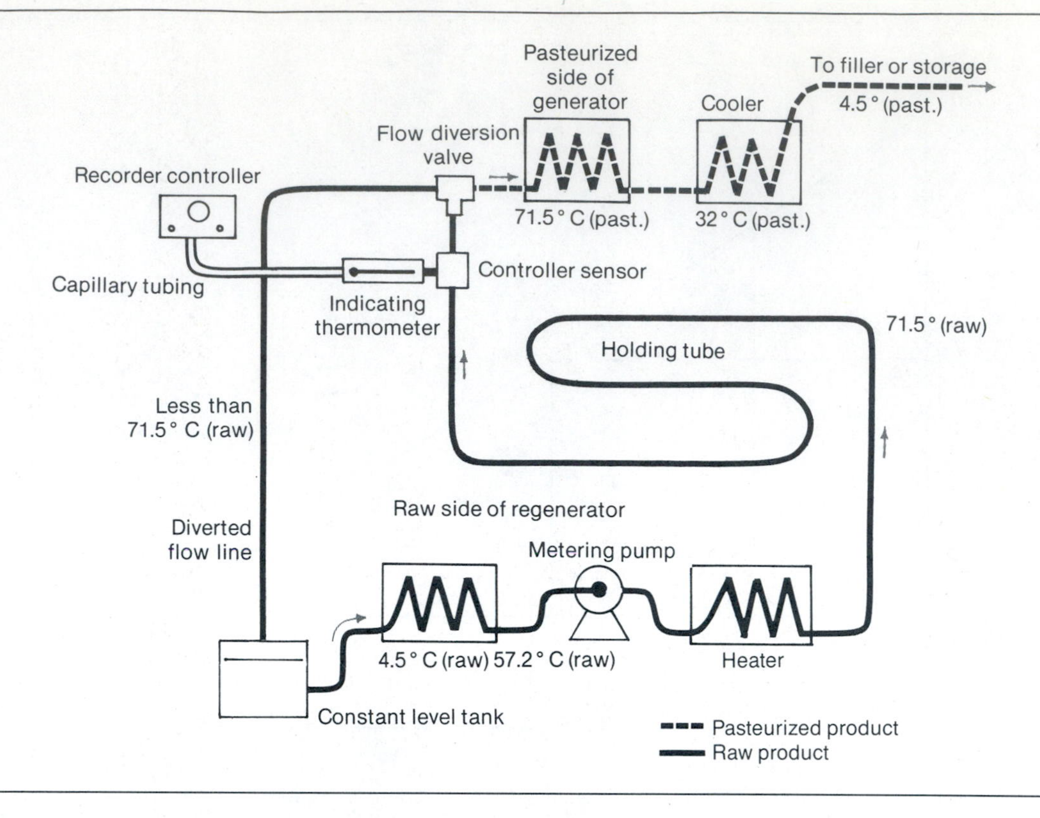

Figure 20-3 High-temperature short-time (HTST) pasteurization technique.

specified temperature. Examination of various points on the TDT curve (also a straight line) will determine the resistance of the microorganism at various temperatures. The slope of the curve is expressed by the term, Z, which is the degree Celsius required for the TDT to traverse one log cycle (or tenfold change in the D value).

12D concept. One of the most important considerations in commercial canning is to ensure the destruction of *Clostridium botulinum* spores. This mesophilic organism produces spores that are comparatively heat sensitive, and the finding of botulinal spores in a retorted product (retorts are devices for heat processing canned foods) indicates gross underprocessing. To be assured that spores of *C. botulinum* are destroyed, a concept was established that would ensure that any population of the most resistant spores would be reduced to 10^{-12} of their original numbers—in other words, a ***12 decimal*** or ***12D concept.*** Previous studies have shown that if 10^{12} botulinal spores per gram of food are heated at 121° C, the time required for a 12 decimal reduction is 2.52 minutes. Thus, time and temperature relationships are important in producing sterility in canned food products.

Pasteurization. Any thermal process in the food industry in which the highest temperature reached in the food is under 100° C is considered a pasteurization. Many foods such as dairy products, cured meats (bologna, weiners), fruit juices, bakery goods, vinegar, and wines and beers are pasteurized. Some pasteurization procedures are relatively mild, and many vegetative cells are not killed, but in others only bacterial spores survive. Thus, throughout the food industry there is a broad spectrum of pasteurization procedures (Figure 20-3).

The amount of heat that is used to pasteurize milk is designed to kill the most heat-resistant pathogen (vegetative cell) that occurs in the product. Previously, the processes were designed to kill *Mycobacterium tuberculosis,* but in more recent times it has been observed that the rickettsial agent of Q fever, *Coxiella burnetii,* is more heat resistant (see Chapter 4), thus necessitating a revision of the thermal processing procedure. There are two general procedures that are employed for milk pasteurization. ***Low-temperature holding (LTH)*** is used for vat-type or batch pasteurization, and milk is heated to 62.9° C for 30 minutes. The previously employed 61.7° C for 30

Figure 20-4 Steam-heated retort for thermal processing of canned foods.
Courtesy Malo, Inc., Tulsa, Okla.

Figure 20-5 Spoilage in canned goods. **A,** Can with leaky seam, subject to microbial spoilage. **B,** Normal can. **C,** "Swell" caused by growth of gas-producing bacteria in can or chemical production of hydrogen from acid in fruit reacting with metal container. **D,** Spoilage of canned fruit caused by gas-producing yeasts (pressure has forced fruit juice through seal of lid).
From Anderson, D.A., and Sobieski, R.J. Introduction to microbiology *(2nd ed. St. Louis, The C.V. Mosby Co., 1980.)*

minutes did not kill *Coxiella burnetii.* The ***high-temperature short-time*** (HTST) procedure is more amenable to continuous, modern processing, and the milk is heated to 71.5° C for 15 seconds. This thermal treatment kills *Coxiella burnetii,* and no revision is required. Both procedures ensure a palatable product that is free of all known pathogens. However, many other vegetative bacteria can survive, such as certain nonpathogenic streptococci, lactobacilli, and some micrococci. Eventually, these bacteria will grow and spoil (sour) the milk if it is held beyond its normal shelf life.

Steam-heated retorts. The majority of canned foods are thermally processed in large, steam-heated retorts (Figure 20-4), which operate on the same principles as a pressure cooker or autoclave. The temperature may vary depending on the composition of the product, size of the container, type of retort, and other factors. The amount of heat administered to the product is designed to kill the most heat-resistant mesophilic sporeformers such as *Clostridium botulinum,* the cause of botulism. The temperature for canning high-acid foods may be 100° C and 121° C for low-acid foods. The reason for the difference in temperature is that acids are inhibitory to the growth of *C. botulinum.* The recent development of less acid strains of tomatoes has contributed to more cases of botulism from home canning. Steam-heated retorts are not designed to kill thermophilic sporeformers, because in most instances the product would be overcooked and organoleptically unacceptable. In addition, thermophilic sporeformers are not agents of disease, and the growth of thermophiles does not take place under normal ambient temperatures of holding (less than 50° C). It has been estimated that 10% to 30% of the retorted canned goods on the market are not sterile and contain thermophilic spores. The term ***commercial sterility*** has been used to designate this situation. One of the most commonly encountered thermophilic sporeformers is *Bacillus stearothermophilus.* Studies have shown that 10^6 spores of this species may survive

Table 20-2 Some of the major microorganisms involved in the spoilage of canned foods

CANNED FOOD	SPOILAGE ORGANISM(S)	PRODUCT DEFECT
Red meats		
Commercially sterile	Sporeformers (species of *Bacillus* and *Clostridium*)	Gas, putrefaction
Semipreserved	*Streptococcus, Bacillus, Clostridium*	Souring, gas discoloration, putrefaction
Vegetables		
Corn, green beans, peas	*Bacillus stearothermophilus, Clostridium sporogenes*	Flat-sour
	C. thermosaccharolyticum	Gas
	Desulfotomaculum nigrificans	Sulfide (rotten egg) smell
Tomatoes	*Bacillus coagulans*	Flat-sour
	Clostridium pasteurianum, C. butyricum	Butyric acid smell
Fruit	*Clostridium butyricum*	Butyric acid smell

100° C for up to 20 hours. They produce flat-sour spoilage of low-acid canned foods.

Microbial spoilage of canned goods. Microbial spoilage of canned goods falls into three categories: preprocessing, underprocessing, and can leakage (Figure 20-5). Preprocessing spoilage occurs before retorting when the food has a high load and it is not thermally processed soon enough to prevent extensive bacterial growth. The condition is detected by observing large numbers of bacteria in stained preparations coupled with the absence of viable bacteria. Heating does not alter the staining propensities of bacteria. Underprocessed products usually contain large numbers of one or two species of viable, mesophilic sporeformers such as *C. botulinum* and *C. sporogenes.* Spoilage caused by can leakage merits special attention. Areas where the body seam intersects with either of the lead seams are the most susceptible to leakage. The leaks often reseal on cooling after heating, but occasional contamination occurs (estimated at about 1/5000 cans). Non-heat-resistant, viable vegetative microbes are encountered in these instances. Swollen cans (caused by microbial gas production) and sour or off tastes are the usual type of defects encountered (Table 20-2).

The vast majority of swollen cans that are seen in the marketplace are not associated with microbial spoilage but with ***hydrogen swells.*** This condition is associated with acid foods (such as fruits and fruit juices) wherein the acid reacts with the tin of the can producing hydrogen gas (Figure 20-5, *C*), as represented by the equation:

$$Sn^{\circ} + 2\,H^{+} \longrightarrow Sn^{++} + H_2$$

This is essentially an electromotive series replacement reaction, and the amount of gas produced depends on the acidity of the food and the composition and construction of the can. Foods containing large amounts of nitrate (spinach, beans) can also cause a hydrogen swell, wherein the nitrate reacts within the tin, producing hydrogen gas. In both types of hydrogen swells there is an absence of viable bacteria and usually a pitting and blackening of the inside of the can. The food microbiologist must be able to differentiate between hydrogen and microbial swelling, because each involves different approaches for rectification. Usually, canned foods have a shelf life of about 14 months, after which, the incidence of hydrogen swells increases significantly.

Preservation of foods by refrigeration and freezing

The generation time of microorganisms and the rates of chemical reactions have temperature optimums. Deviations from the optimums decrease the rates, and if the deviations are extreme, zero rates are approached. Decreases in temperature increase the generation time of the microorganisms and extend the shelf life of the food. Freezing generally incurs a complete cessation of growth and radically extends the shelf life from a microbiological standpoint.

Chilling temperatures. Chilling may be defined as reducing the temperature of a food to a range of 0° to 10° C. Psychrophilic bacteria, especially *Pseudomonas* and allied aerobic gram-negative rods, such as *Acinetobacter, Alcaligenes,* and

Flavobacterium, can grow in this temperature range and eventually spoil the food. There are also several genera of mesophilic microorganisms that are capable of growth between 0° and 10° C and are referred to as ***psychrotrophs***. As a generality, if a food spoils in 20 days at 0° C, it will spoil in 10 days at 5° C and in 5 days at 10° C. Thus, relatively small temperature variations markedly influence the growth rate of the psychrophilic bacteria (Figure 20-6). Most pathogenic bacteria do not grow below 10° C, and at 5° C almost all pathogens that occur in foods are inhibited. There is one type of *Clostridium botulinum* (type E) that can grow at 3.3° C, but it takes several months to produce toxin.

Low-temperature control is frequently a problem in industry. Large coolers lack temperature uniformity, and fans must be strategically located to assist in air distribution. Humidity control in these coolers may also be difficult. If the humidity is too high, water may accumulate on the product (such as sides of pork and beef) and accelerate bacterial growth. On the other hand if the humidity is too low, the product is dehydrated, thus incurring financial losses.

As a final word regarding low-temperature control in the food industry, it is axiomatic that more spoilage problems occur in the summer months as compared with the winter months.

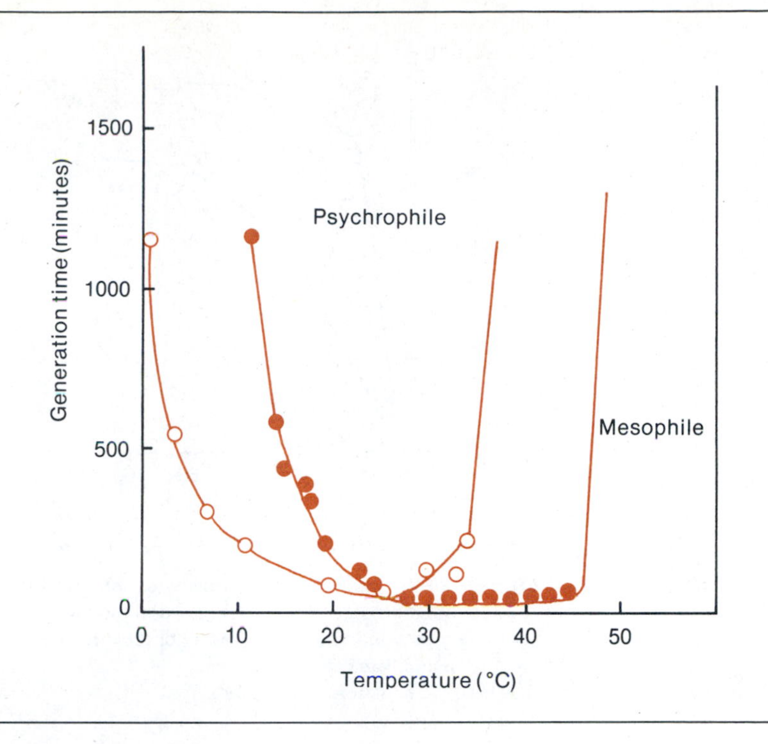

Figure 20-6 Effect of temperature on generation time of mesophilic (●) and psychrophilic (○) microorganism.
From Ingraham, J.L. Proceedings low temperature microbiology symposium. *Cambell Soup Company, Camden, N.J., 1961.*

Freezing temperatures. The freezing of foods essentially stops microbiological growth. The shelf life is then dictated by chemical (for instance rancidity) and physical (moisture migration) limits. Freezing does not kill all bacteria. Gram-positive bacteria, yeasts, molds, and microbial spores are relatively resistant to killing by freezing, whereas gram-negative bacteria are more susceptible to freezing. Initially, there is a large mortality from freezing, but a proportion of the cells survive. For instance, 30% to 70% of *E. coli* and 50% to 80% of *Pseudomonas* populations are killed by freezing. Those viable cells that remain die gradually when stored in the frozen state. Freezing inhibits growth but has no practical killing effect on microorganisms, nor does it affect the activity of any microbial toxin that may be present. Microbiologists use freezing to maintain bacterial cultures (for instance, frozen in milk) for years, especially gram-positive bacteria as well as yeasts and molds. Some of the effects of freezing on microorganisms follow.

1. Loss of cytoplasmic gases, such as oxygen and carbon dioxide
2. Change of pH in the cell
3. Loss of cellular electrolytes
4. Some denaturation of proteins
5. Metabolic injury

With regard to metabolic injury, on thawing, some microorganisms have increased nutritional requirements, at least for several cell divisions and then revert to their normal state of growth.

Microorganisms can grow during the freezing (and thawing) of a food. Consequently, many procedures and mechanical innovations have been devised to enhance a rapid freezing. Thawing, especially at the consumer level, should be as rapid as possible to preclude microbial multiplication. The refreezing of foods should be approached with caution because of the potential for increasing the load during repeated freezing and thawing.

Preservation by controlling the water content of food

All forms of life require water—some more, some less. Bacteria, yeasts, and some molds require a definitive aqueous environment, but some molds can grow in the absence of an aqueous environment if sufficient humidity is present. The water content of food can be a controlling factor for the inhibition of microbial growth. This is accomplished by water removal (drying) or by chemi-

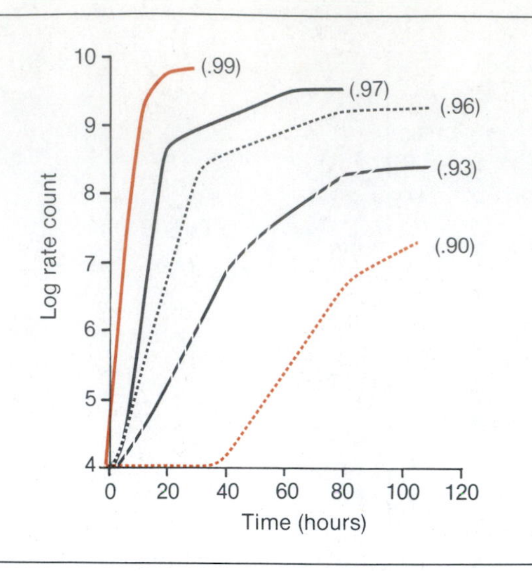

Figure 20-7 Effect of water activity (a_w) on growth of strain of *Staphylococcus aureus* grown in medium containing protein hydrolysate. The a_w of growth medium is indicated in parentheses.

From Troller, J.A. Appl. Microbiol. ***21**:435, 1971.*

Figure 20-8 Freeze dryer. One of many ways of drying foods or microorganisms. It differs from other equipment by reducing moisture content without use of heat. Temperature used is at least −20° C.

Photo by Karen Siebert. Courtesy Ralston Purina Co.

cally binding the water with water-soluble substances (solutes). The amount of water that is available for microbial growth can be measured accurately using the concept of ***water activity*** **(a_w)**.

When a volume of pure water is placed in a closed container, an equilibrium is established between the water molecules in the air and in the liquid phase. The water in the air creates a vapor pressure (P_w). When a solute, sodium chloride, is added to an identical container with an equal volume of water, again, an equilibrium is established, and the air above the solution will always be less than that of pure water, that is, $P_s < P_w$. The water activity of the solution is defined by the ratio of the vapor pressures, or $a_w = P_s/P_w$. Pure water has an a_w value of $P_w/P_w = 1.000$, and the water activity values of foods and solutions (or broth media) can never exceed unity.

All food has a water phase, and it is in this phase that microbial growth takes place. The solutes in the food are concentrated in the water phase, and the water activity of the food can be measured in a manner analogous to that described above, wherein the food is placed in a closed container, and after equilibrium the vapor pressure is measured, and the a_w is calculated.

Each microbe has an optimum a_w for growth, as well as a maximum and a minimum (Figure 20-7). In food microbiology the minimum a_w values are of paramount concern. Deviations from the optimum increase the generation time, and extreme deviations may cause death. An exception to this generality is the ***lyophilization,*** or freeze-drying, of microorganisms, which is an excellent procedure to maintain them for years (Figure 20-8). In this instance the drying is accomplished from the frozen state (sublimation). The limiting a_w value for certain microbes is presented in Table 20-3 and should be compared with the a_w values of some common foods (Table 20-4). In those foods with a high a_w most microorganisms will grow, whereas in those foods with a low a_w most microorganisms are inhibited.

Two foods can have identical water concentrations, but if one of the foods has a higher content of solute, its a_w will be lower (that is, the greater the solute concentration, the lower the a_w value), and its shelf stability will be greater. In more recent times pet feeds and a limited number of foods (such as toaster pastries) have appeared on the market wherein the a_w of the product is adjusted to about 0.84 or lower. These are referred to as intermediate-moisture foods. On occasion if the a_w is not carefully controlled, strict ***osmophilic*** (requiring high concentration of solute) molds and yeasts can grow and spoil the product.

Table 20-3 Minimum levels of water activity (a_w) of some microorganisms at temperatures near their optimum

MICROORGANISM	A_W
Molds	
Aspergillus candidus	0.75
Mucor plumbeus	0.93
Penicillium chrysogenum	0.79
Chryosporium fastidium	0.69
Yeasts	
Saccharomyces cerevisiae	0.90
S. rouxii	0.62
Bacteria	
Bacillus cereus	0.95
Bacillus subtilis	0.90
Clostridium botulinum type E	0.97
Escherichia coli	0.95
Staphylococcus aureus	0.86
Salmonella species	0.95
Lactobacillus viridescens	0.95

Adapted from Troller, J.A., and Christian, J.H.B. *Water activity and food.* Academic Press, Inc., New York, 1978.

Table 20-4 Water activity of some major groups of foods

FOOD	A_W
Fresh meat, fish, fruits and vegetables	0.98 and above
Canned fruit in light syrup	
Canned fruits in heavy syrup	0.93 to 0.97
Evaporated milk	
Fermented sausages	
Processed cheese	
Lightly salted fish, pork and beef	
Raw ham	0.85 to 0.92
Dried beef	
Dry sausage	
Intermediate-moisture foods	0.60 to 0.84
Dried fruit	
Flour	
Jams and jellies	
Nuts	
Cereals	
Chocolate	less than 0.60
Honey	
Dried milk and vegetables	

The growth of these organisms in the laboratory is relatively easy if 40% sucrose is added to the medium. The organisms will not grow in common laboratory media unless high concentrations of a solute are added. These osmophiles are ubiquitous, and quality control procedures must be instituted to assure their control.

Many food products are dried, both at the producing and manufacturing levels. Cereals, onions, and nuts are examples of products that are solar dried. Manufacturers employ a variety of methods, including spray, tunnel, and drum drying for the production of instant coffee and tea, dried milk, and a spectrum of other products. In all instances the goal is to reduce the free water content of the food, thus increasing the solute concentration. At some point, depending on the food, a sufficiently low a_w value is reached, and the shelf stability of the product is radically increased. It should be remembered that although microorganisms are killed by drying, many, including pathogens, survive. Thus the situation is equivalent to the freezing of foods. One of the most troublesome groups of microorganisms in dried foods is molds, particularly species of *Aspergillus.*

Preservation by acidification

Many of our food products are acidified to enhance their shelf stability. Some pickles, specialty meat products, certain vegetables, as well as mayonnaise and a variety of salad dressings, are chemically ***acidulated.*** (The biologically acidulated or fermented products such as cheese, sausage, and sauerkraut are presented in another section.) The most commonly used acidulant is acetic acid or vinegar, although citric acid and adipic and fumaric acids are sometimes employed. The principle involved is to lower the pH value of the food such that microbial growth is inhibited. Lowering the pH can also reduce microbial heat resistance in foods that are to be eventually processed by heat. In general the minimum pH for the growth of bacteria is 4.4, but in yeasts and molds the minimum pH may be as low as 1.5. On occasion spoilage of acidulated foods occurs if the pH is not controlled; the spoil-

Table 20-5 Common preservatives used in the food industry

PRESERVATIVE	MICROORGANISM AFFECTED	FOOD
Benzoic acid and benzoates	Yeasts and molds	Soft drinks, relishes, pickles, margarine, salad dressings
Sorbic acid and sorbates	Molds	Syrups, jellies, cakes, apple cider, semipreserved fish and meat products
Propionic acid and propionates	Molds	Bread, cakes
Sulfur dioxide, sulfites	Microorganisms	Wine, dried fruit, dehydrated mashed potatoes
Sodium diacetate	Molds	Bread and bakery products
Polymyxin B (antibiotic)	Bacteria	Yeast cultures for beer
Sodium nitrite	Bacteria	Meat curing, smoked salmon
Sodium chloride and sugars	Microorganisms	Sodium chloride for meats; sugars for jellies and preserves
Ethyl formate	Fungi	Dried fruits, nuts

age organisms involved are generally yeasts and gas-producing lactic acid bacteria.

Preservation by vacuum packaging and modified atmospheres

The advent of vacuum packaging radically enhanced the shelf life of a wide variety of food products. All molds, some yeasts, and many of the gram-negative, psychrophilic bacteria are strict aerobes, and their growth in vacuum-packaged products is precluded. In addition, the growth rate of facultative organisms is diminished under anaerobic conditions. This technological triumph is not without some shortcomings, because anaerobiosis affords the growth of *Clostridium botulinum*. This is a drawback, and unless other means of precluding growth and toxin formation are incorporated, vacuum packaging cannot be used.

In the immediate past substantial effort has been expended in investigating the packaging of foods in atmospheres other than air. The most effective gas used for this purpose is carbon dioxide, and commercial application of the technology is found in the packaging of cheese, fresh meats, and sea foods. Here again, oxygen is excluded, but the carbon dioxide itself has an inhibitory effect on many spoilage microorganisms.

Preservation with chemicals

The addition of chemicals to foods is closely regulated by the Food and Drug Administration, as well as individual state regulatory agencies. The FDA publishes and strictly enforces a list of accepted chemicals, called the ***GRAS list.*** GRAS is an acronym for Generally Recognized As Safe. The list not only specifies what chemicals can be used but also their maximum concentration and the specific foods to which the chemical can be added. In order to have a new chemical added to the GRAS list, extensive acute and chronic toxicity tests must be performed.

Some of the more common preservatives allowed in foods are listed in Table 20-5. In many instances, without these chemicals the production of the food involved would be impossible (dehydrated mashed potatoes containing sulfite is a good example), or the decreased shelf life would render it economically impossible to market the product. ***Nitrite*** is a preservative that deserves special attention. Nitite is converted to a variety of compounds as a result of either interaction with components of the meat or microbial metabolism. Nitrite at acid pH ionizes to yield nitrous acid (HNO_2), which is converted to nitric oxide (NO). Nitric oxide reacts with meat pigments such as myoglobin to produce water-insoluble pigments (red nitric oxide-myoglobin). Heating produces a pink nitrosyl hemochrome, which is stabilized by ascorbates (ascorbates are chelating agents). Nitrite also adds flavor to certain meats such as pork, but the chemical basis of this reaction is unknown. Nitrites can react with secondary amines, tertiary amines, and quaternary ammonium compounds to form ***nitrosamines*** (Figure 20-9). The legal limits of nitrite have been lowered in recent years when some nitrosamines were found to be carcinogenic. Nitrites have

$$(H_3C)_2N{-}H + NO_2 \longrightarrow (H_3C)_2N{-}N{=}O$$

Figure 20-9 Formation of nitrosamine from dimethylamine.

Figure 20-10 Milking machine.
Courtesy Ralston Purina Co.

never been shown to cause cancer in humans. One of the most important functions of nitrite, in conjunction with decreased pH, salt content, and temperature, is as an inhibitor of *Clostridium botulinum*. This function is particularly important in anaerobically packaged products such as hams and sausages.

Barrier concept. In many instances a single preservative or preservative condition (that is, a ***barrier***) will not afford the requisite inhibition to provide a reasonable shelf life. Therefore, combinations of these preservatives are frequently employed. As an example, consider a 10-pound canned ham (small 1- to 2-pound hams can be heated to commercial sterility because of heat penetration considerations). The ham is cured with sodium chloride and sodium nitrite. The can is evacuated (anaerobiosis), and the ham is heated to 74° C. After heating, the product is kept at refrigerator temperatures until consumed. None of the preservatives or preservative conditions alone would be sufficient to provide shelf stability, but together a relatively stable product with at least a 6-month shelf life is produced.

MICROBIOLOGY OF SPECIFIC FOODS

Dairy products

The microbiology of raw milk in the past 30 years has been radically altered by technological innovations such as the mechanical milking of the animal, the closed milking system to decrease barn contamination, and on-the-farm refrigeration during holding. Previously, lactic acid bacteria *(Lactobacillus, Leuconostoc, Streptococcus,* and *Pediococcus)* were the predominant flora, but in today's milk supply gram-negative rods are predominant. Milk residues left on equipment used to handle milk result in the growth of many gram-negative rods such as *Pseudomonas, Alcaligenes, Flavobacterium,* and *Chromobacterium.*

Milk. Milk is divided into grades on the basis of the milking facilities at the producer level (Figure 20-10). There is no other food that is regulated as closely as fluid milk. Fluid grade A milk for human consumption has the strictest standards (Table 20-6). Most communities throughout the United States permit the sale of only grade A milk

Table 20-6 Microbiological standards for Grade A dairy products, manufacturing milk, and certified milk

PRODUCT	MAXIMUM NUMBER PER GRAM	
	STANDARD PLATE COUNT	COLIFORM COUNT
Grade A Products		
Raw milk, at pick up	100,000	—
Raw milk, commingled	300,000	—
Pasteurized milk	20,000	10
Pasteurized, condensed milk	30,000	10
Pasteurized, condensed whey	30,000	10
Nonfat dry milk	30,000	10
Dry whey	30,000	10
Milk for manufactured dairy products		
Grade 1	500,000	—
Grade 2	3,000,000	—
Grade 3	over 3,000,000	—
Certified milk		
Raw certified	10,000	10

Table 20-7 Some of the major bacterial species used as starter cultures in cheese making

CULTURE	FOODS USED IN	FUNCTION
Streptococcus lactis subspecies *diacetyllactis*	Butter, sour cream, cultured buttermilk	Acid and flavor
Streptococcus lactis	Cottage cheese, cultured buttermilk	Acid and flavor
Streptococcus cremoris	Cottage cheese, cultured buttermilk	Acid and flavor
Lactobacillus acidophilus and *Streptococcus thermophilus*	Acidophilus milk, Swiss cheese, yogurt, Emmenthaler cheese	Acid
Leuconostoc cremoris	Cultured buttermilk, cottage cheese, butter	Flavor
Lactobacillus bulgaricus, *L. lactis*, and *L. helveticus*	Yogurt, kefir, kumiss, Swiss cheese, and related cheeses	Acid and flavor
Propionibacterium freundenreichii	Emmenthaler, Swiss, and related varieties of cheese	Flavor and eye formation

for human consumption. Some allow the sale of "certified milk." This type of milk is obtained from herds certified by testing to be free of *Mycobacterium tuberculosis* and *Brucella* species) but is not pasteurized. Grades 1, 2, and 3 are used for manufactured dairy products such as cheese.

The enzyme ***alkaline phosphatase*** normally occurs in cow's milk (but not in all milks—goat's milk is devoid of it), and it has the unique characteristic of just being thermally inactivated by the processes employed for pasteurization. A simple test has been devised such that a manufacturer can rapidly verify the adequacy of the pasteurization process. The pathogens *M. tuberculosis* and *Coxiella burnetii* are destroyed before the enzyme, thus providing a margin of safety. On rare occasions some bacteria may grow in the raw milk and produce a thermally stable phosphatase that will give a false-positive result.

Dried milk. A large amount of milk is dried for

manufacturing purposes as well as human consumption. Two products, nonfat dry milk and whole dry milk, account for most of the tonnage. They are used in the manufacture of candy (milk chocolate), ice cream, some sausages, and a variety of preblended bakery items such as cake mixes. The milk is preheated and concentrated to half of its original water content. It is then sprayed as a fine mist into a hot air chamber (115° to 125° C), and the dried particles collect at the bottom and are conveyed to the packaging room. Although the incoming air is sufficiently hot to kill vegetative forms, the effect of evaporative cooling on the individual particles precludes a complete killing. In addition, water problems in the driers and the equipment subsequently used sometimes affords bacterial growth or contamination of the product. The isolation of *Salmonella* from these products is not rare.

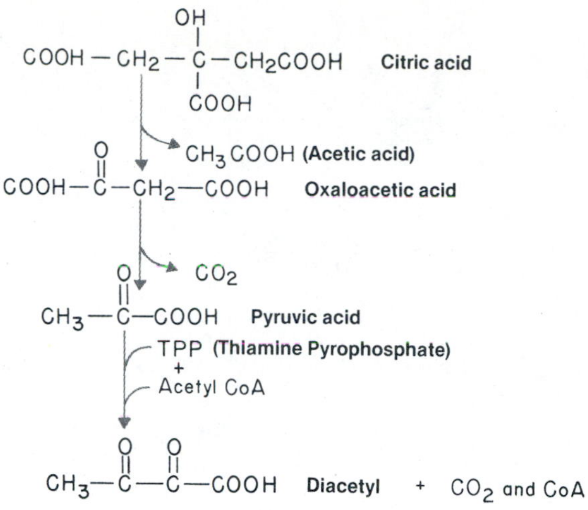

$COOH-CH_2-C(OH)(COOH)-CH_2COOH$ Citric acid

↓ → CH_3COOH (Acetic acid)

$COOH-C(=O)-CH_2-COOH$ Oxaloacetic acid

↓ → CO_2

$CH_3-C(=O)-COOH$ Pyruvic acid

↓ ← TPP (Thiamine Pyrophosphate) + Acetyl CoA

$CH_3-C(=O)-C(=O)-COOH$ Diacetyl + CO_2 and CoA

Figure 20-11 Pathway for formation of diacetyl from citric acid.

Milk starter cultures. Many years ago the diary industry pioneered the use of ***starter cultures,*** the purposeful addition of bacteria to milk, to effect unique organoleptic characteristics in the product as well as to aid in its preservation. In modern times this has become a science (Table 20-7). Not only do the starter cultures acidify the milk, but also they produce specific flavoring compounds during the fermentation, and the enzymes liberated by the billions of lysed cells during the subsequent ripening period also contribute to flavor. Starter cultures are used in the production of milk products (buttermilk, yogurt), some types of butter, and a variety of cheese, both unripened (cottage, cream) and ripened (cheddar, Swiss). The terms ripening, aging, and curing refer to the holding of a cheese for a period of time (variable with the type) at less than ambient temperature. The exact contribution of the enzymes liberated by lysis during ripening is not understood. Lipolytic and proteolytic activities are readily detected, but the exact subtle flavoring reactions have defied description.

The dominant flavoring compounds produced during the fermentation are ***diacetyl*** and ***acetoil*** (also termed acetylmethylcarbinol). The origin of these compounds is the citric acid in the milk, and although the concentration of citrate in milk is only 0.04% to 0.1%, its importance is not commensurate with its concentration. It is against the law to add chemically synthesized diacetyl or acetoin. The pathway from citrate to diacetyl is illustrated in Figure 20-11.

The starter cultures employed in fermented milk and cheese production fall into two categories based on the temperature of fermentation. The low-temperature fermentations are conducted at about 30° to 38° C and employ homofermentative lactic streptococci and heterofermentative *Leuconostoc* organisms (Figure 20-12). The high-temperature fermentations are conducted at 49° to 54° C and generally use lactobacilli and *Streptococcus thermophilus.* Some of the more common started cultures are listed in Table 20-7. In most instances two strains are used, and the balance, or proportion, of the strains contributes to the flavor of the product. *Lactobacillus bulgaricus* and *Streptococcus thermophilus* are used in the production of yogurt. *S. thermophilus* stimulates *L. bulgaricus* by producing formic acid, and the proteolytic activity of *L. bulgaricus* releases certain amino acids, particularly valine, which stimulates *S. thermophilus.* Thus more acid is produced by the combination of organisms (synergism) than by the two species grown independently of each other.

Bacteriophage and starter cultures. Cheese and fermented milk manufacturers are constantly on the alert for bacteriophage development in the starter cultures during scale up or in the fermentation vat. Bacteriophages are specific for the strain of the starter. For example, if there are 100 strains (each strain constitutes an isolate from a different source) of *S. lactis,* there is an excellent probability that a given strain of bacteriophage will infect and lyse only several strains of *S. lactis.* The host range of the bacteriophage is easily determined in the laboratory, and to circumvent starter failure, the starter strains are constantly rotated to avoid bacteriophage buildup in the plant.

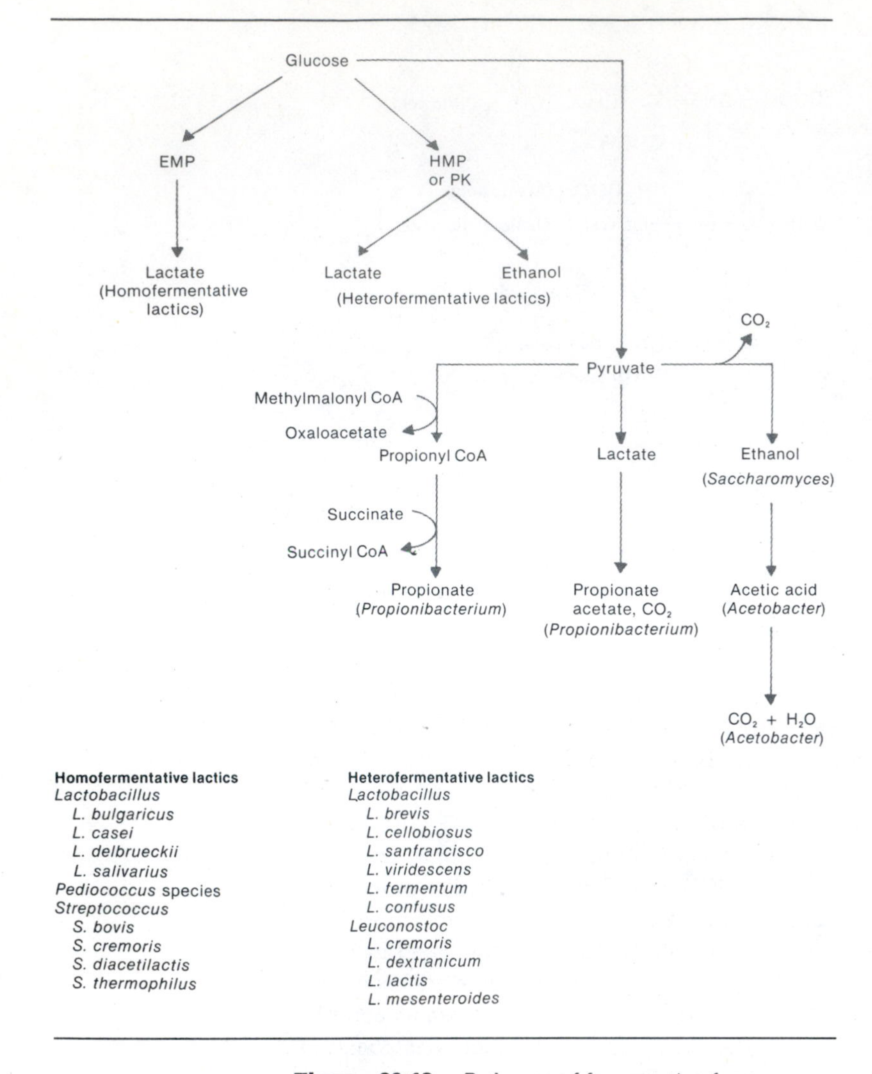

Figure 20-12 Pathways of fermentation for heterofermentative and homofermentative lactic acid bacteria plus some other species. List of some major homofermentative and heterofermentative microorganisms is also included. *EMP*, Embden-Meyerhoff-Parnas pathway; *HMP*, hexose monophosphate shunt pathway; *PK*, phosphoketolase pathway.

Cheese. All cheeses are produced by lactic fermentation of milk. In cheesemaking the protein-precipitating enzyme mixture, rennet extract, is added shortly after the starter culture decreases the pH value. These enzymes, in conjunction with the development of acid, aid in curd formation; the remaining fluid portion or whey is separated. The subsequent treatment of the curd varies with the type of cheese to be made. Regardless of type, most cheese is ripened for at least 60 days. Considerable microbiological activity takes place during the ripening of some cheeses, whereas in others it is radically diminished. For example, during ripening, gas-producing propionibacteria grow in Swiss cheese producing the "eyes" as well as the characteristic flavor, whereas in cheddar ripening microbiological activity is relatively minimal.

Another example of the use of microorganisms to produce a variety of products is seen in the manufacture of surface-ripened cheese. Previously, examples were given of internally ripened cheese. The surface-ripened varieties (Limburger, brick, Camembert, Brie, and others) usually depend on the growth of aerobic organisms and the diffusion of enzymes or end products of metabolism into the cheese during ripening. In the manufacture of Limburger, which starts with the normal lactic fermentation of the milk, the curd is cut, formed into small rectangles, and rubbed lightly with salt. At the beginning of the ripening period the mold *Geotrichum* and aerobic yeasts grow, oxidize the lactic acid, and raise the pH value to approximately 6.0. This affords the subsequent growth of the acid-sensitive, orange-pigmented, salt-tolerant, proteolytic organism *Brevibacterium linens*. As this organism grows, its proteolytic enzymes diffuse into the cheese, giving it the characteristic soft texture, flavor, and odor, and the surface assumes an orange coloration. Brick cheese is essentially a milk form of Limburger. In the surface-ripened varieties of cheese, the pH value must be returned to near neutral values by lactic acid–oxidizing organisms to facilitate the growth of the surface-ripening organisms. In the production of Camembert cheese the surface-ripening organism is *Penicillium camembertii*. The exact flavoring contributing principle is unknown.

Frequently the ripening organisms are sprayed on the curd or into the air of the ripening room. Sometimes the individual pieces of cheese are turned and rotated by hand to ensure surface inoculation and distribution of the ripening organism.

The ripening of blue or Roquefort cheese is peculiar (Plate 26). After the initial lactic fermentation, *Penicillium roquefortii* is mixed with the curd, and small circular "wheels" are formed by compression. Each wheel is multiply pierced with small, solid needles, and the cheese is allowed to ripen for 3 to sometimes 10 months. At first it was thought that the piercing was necessary to afford the entry of oxygen for the mold's metabolism, but subsequently it was observed that the escape of metabolic carbon dioxide was equally important. As the mold grows, it produces blue areas in the cheese and blue "veins" along the pierced channels. During growth the mold produces lipolytic and proteolytic enzymes that are associated with flavor; it also sporulates. Mold spores, unlike bacterial spores, are metabolically active (this is why they are heat sensitive). In

Table 20-8 Some of the major genera of microorganisms found on meats

GENERA	FRESH MEATS	PROCESSED MEATS	VACUUM PACKAGED MEATS	BACON
Bacteria				
Acinetobacter	++	+	+	+
Aeromonas	++			+
Lactobacillus	+	++	++	
Moraxella	++			+
Pseudomonas	++		+	
Streptococcus	+	++	+	
Pediococcus	+	+	+	
Corynebacterium	+	+	+	+
Molds				
Cladosporium	++	+		
Aspergillus	+	++		
Geotrichum	++	+		
Mucor	++	+		
Penicillium	+	++		
Rhizopus	++	+		
Sporotrichum	++			
Yeast				
Candida	+	+		
Rhodotorula	+			
Torulopsis	+	+		
Trichosporon		+		

Adapted from Jay, J.M. *Modern food microbiology* (2nd ed.). D. Van Nostrand, New York, 1978.
+, Known to occur; ++, most frequently found.

fact, it is the mold spore and only the spore that produces the enzyme for the flavoring reaction:

$$CH_3(CH_2)_6COOH + O_2 \longrightarrow CH_3(CH_2)_4COCH_3 + CO_2$$

Caprylic acid **Methyl-n-amyl ketone**

The caprylic acid is derived from the hydrolysis of fat, and the ketone gives the characteristic peppery, sharp flavor that is associated with blue cheese.

In many of the fermentations discussed above, there are definitive, stepwise appearances of specific microbial species or groups. These can be referred to as floral succession, which sometimes proceeds without human intervention.

Meat products

The meat industry can be conveniently divided into segments on the basis of their products. Using this scheme, fresh, cured, and fermented meats constitute the major products; and animal byproducts such as gelatin, animal feeds, and pharmaceuticals are the minor products. The meat industry is constantly concerned with the perishability and safety of their products because of the inherent susceptibility of the product to microbial contamination and growth. The physical nature of meats does not afford a streamlined processing approach such as that enjoyed in fluid milk processing and other segments of the food industry. Some of the major genera of microorganisms that are associated with meats are outlined in Table 20-8.

Each time meat is cut, the two surfaces are essentially inoculated. Further contamination from cutting boards, chutes, and conveyor belts adds additional bacteria to the product. Temperature is the only barrier that can be imposed on fresh meats. The short shelf life of fresh meats necessitates adequate sanitation, excellent temperature control, and rapid distribution.

Figure 20-13 Refrigerated meat carcasses.
Courtesy National Livestock and Meat Board, Chicago.

Fresh meats. The microbiology of fresh meats is similar to that of milk in that psychrophilic species, especially *Pseudomonas* (the others are *Acinetobacter* and *Moraxella*), predominate. After slaughter the sides of pork and beef are chilled to facilitate the subsequent cutting or breakdown of the carcass (Figure 20-13). Various fluids from the hanging carcasses and the carcasses themselves provide nutrient sources for psychrophiles in the large coolers. Airborne contamination is unavoidable, and the subsequent cutting and sawing of the carcass plus handling and contamination from knives and cutting blocks serve to spread the contamination. The low temperatures of holding enrich for psychrophiles.

The shelf life of fresh meats is directly proportional to the *Pseudomonas* or total count. In fact, many processes assess proper sanitation and holding and handling procedures in a plant with a *Pseudomonas* count. Most, but not all *Pseudomonas* species produce a pigment (fluorescein) that fluoresces under ultraviolet light. *Pseudomonas* colonies can be differentiated using this technique, and an approximate estimate of their numbers can be made. The count is easily determined with a peptone medium that is devoid of carbohydrate and enriched with magnesium and sulfate ions. The medium enhances the production of the pigment, but it offers no selectivity.

The meat sample is quantitatively plated in the same manner as that used for determining a total count. After incubation, only the fluorescing colonies are counted and the number of pseudomonads are estimated.

The shelf life of fresh meats can be as short as 3 and as long as 10 days. It is a function primarily of the initial load and the temperature of holding. Eventually, the meat will spoil regardless of the chilling temperature (Plate 27). An early indication of spoilage is the loss of the red color, or "bloom." The meat appears brown and may assume a surface slime that is chiefly bacteria. People often purchase meat on the basis of its color. This is the paramount reason mitigating against frozen, fresh meat because freezing destroys the bloom. If the bacterial population is sufficiently high, an off taste can be detected in the cooked meat.

Pathogenic bacteria can grow or survive in fresh meats. For example, various surveys have indicated that about 50% of marketed pork and poultry cuts contain *Salmonella*. The incidence of these organisms in beef is somewhat less. This is one reason why consumers and caterers must be careful of cross-contamination between fresh meats and prepared foods. *Salmonella* are ubiquitous, and their elimination from the meat supply is impossible.

Cured meats. Cured meats such as ham, corned beef, and bacon as well as comminuted or blended meats, including weiners, bologna, and sausage-loaf products, are manufactured using nitrite and usually salt. In many products sugar (primarily sucrose) is added for taste. The functions of nitrite have already been discussed.

The thermal process administered to bacon (maximum temperature is about 53° C) during smoking is not sufficient to kill the nematode, *Trichinella spiralis*, the etiological agent of trichinosis (Plate 28). Consequently, this is considered to be a fresh meat, and it must be cooked before consumption. Other cured meats are heated above 59° C, which kills this organism and cooks the meat. In most instances, the thermally processed meats are sliced or prepackaged, and in these operations postprocessing contamination is incurred. The microbial barriers in these products are salt, nitrite, anaerobic conditions, and low temperatures of holding. In many instances shelf lives of 60 days (weiners, sausage) or 6 months (perishable canned products) are readily achievable.

Cured meat spoilage problems usually result from inadequate sanitation or improper temperature control. The chief spoilage organisms are the lactic acid bacteria (streptococci, leuconostocs, and pediococci). These facultatively anaerobic organisms are salt tolerant, some produce gas (heterofermentative), and all produce lactic acid as an end product of carbohydrate fermentations. Spoilage results from souring, gas production in prepackaged items that bloat or puff the package, and slime production. Some species, notably those of *Leuconostoc* and *Lactobacillus*, are capable of synthesizing copious amounts of viscous polysaccharide (both levans and dextrans) when grown with sucrose in the medium. Although the slime is not a health hazard, it is organoleptically unacceptable. The lactic acid bacteria lack the enzyme catalase and acumulate hydrogen peroxide if they grow in the presence of oxygen. Some of the salt-tolerant lactobacilli and leuconostocs can grow on cured meats, and after the consumer opens the vacuum package or if there is a leaker, the organisms produce hydrogen peroxide, which oxidizes the cured red meat pigment to a derived green porphyrin. This defect is commonly encountered, and again, although it is not harmful to health, it is unsightly.

Fermented meats. Ancient humans are credited with the discovery that if ground and salted meat with added sugar is held at warm temperatures after a period of time, the meat develops a pleasing tang and its perishability is decreased. This would be called a "wild fermentation" because it reflects the growth of inherent contaminants. Relatively recently the fermented sausage industry used starter cultures, and currently several *Pediococcus* and *Lactobacillus* cultures are commercially available. Their chief function in the manufacturing process is the rapid and characteristic production of lactic acid to produce tang. No other end products of metabolism have been identified as enhancing flavor. Although the pH of the products (summer sausage, thuringer, salami) is in the range of 4.8 to 5.0, like cheese they are still perishable and must be kept under refrigeration to ensure a long shelf life.

Fish

Fish, along with red meats and poultry, constitutes a major source of animal protein in the worldwide human diet. The microbiology of fish is complex and not completely understood. The manner in which microbes penetrate the skin after death and initiate growth in the flesh remains to be elucidated. Data accumulated in the last few decades have firmly established that the growing environment of the fish is the major determinant of the organism's microbiology. Other determinants are the species of fish, the season of harvest, and the salinity of the water (fresh water versus marine water). It is axiomatic that if the growing and harvesting environment of fish is polluted chemically or microbiologically, the fish are polluted. Another dictum is that fish caught close to land masses have a significantly higher microbial load as compared with fish caught a distance from land or in the open waters.

The skin is covered by a mucilaginous slime that is secreted by cells in the skin, and the amount and composition of the slime varies with the fish species. The slime layer affords a luxurient growth of many different bacteria, and these bacteria, plus organisms from the gut and gill area, constitute significant sources of contamination.

After harvest, the larger fish are dressed aboard the ship and then iced or mechanically

Table 20-9 Stages in fish spoilage

STAGE	TISSUE CHANGES	ORGANOLEPTIC CHANGES	BACTERIAL COUNT
Stage 1 (0 to 5 days in ice)	Rigor mortis; ATP ⟶ inosine; slight increase in TMA; changes in bacterial types	Eyes bright, flesh firm, color good, gills bright, odor fresh	10^2 to 10^3/cm^2
Stage II (5 to 10 days in ice)	Bacterial growth becomes apparent; inosine ⟶ hypoxanthine; TMAO ⟶ TMA; NH_3 increases	Eyes begin to dull, gill color fades, skin color fades, odor neutral to slightly fishy, texture softening	10^3 to 10^6/cm^2
Stage III (10 to 14 days in ice)	Rapid bacterial growth and penetration of tissues; hypoxanthine ⟶ xanthine, uric acid, etc.; TMA increases rapidly; TVB and TVA increase	Eyes sunken, gills discolored and slimy, skin bleached, odor sour and fishy, texture soft	10^6 to 10^8/cm^2
Stage IV (over 14 days in ice)	Bacterial numbers stationary; some changes of species; general deterioration of flesh; TVB and TVA increase rapidly; TMA increases or levels off; proteolysis begins; H_2S and other products formed	Eyes opaque and sunken, gills bleached and slimy, skin very slimy, texture very soft, odor offensive	approx 10^8/cm^2

Adapted from Liston, J. In *Microbial ecology of foods* (Vol. 2). The International Commission on Microbiological Specifications for Foods, Academic Press, Inc., New York, 1980.
ATP, Adenosine triphosphate; *TMA,* trimethylamine; *TVB,* total volatile base; *TMAO,* trimethylamine oxide

chilled. Smaller varieties are similarly treated but not dressed. The fish may remain aboard ship for several days to several weeks, and during this time, as well as subsequently, the chilling temperature selects for psychrophiles such as *Pseudomonas, Acinetobacter, Moraxella.* The fish industry employs wooden boxes, bins and tables for processing and handling, and after several uses the wood surfaces become pitted and uneven. These surfaces are practically impossible to clean and sanitize, and huge bacterial populations develop and serve as important sources of contamination. Only recently (and slowly) has the adoption of stainless steel been instituted. During human handling, depending on the degree and the prevailing sanitary conditions, fish may be significantly contaminated with mesophiles.

Aside from chilling, fish are preserved by salting, drying, heating, and freezing. These methods are employed to facilitate and extend shelf life so that inland markets can be made available. The rapid rate of spoilage of fresh fish precludes their distant distribution without some form of further processing, including freezing.

The spoilage of fish occurs in stages and is associated with proteolysis and strong odors that may include sulfide, fecal, and ammoniacal descriptions (Table 20-9). Fish muscle, unlike mammalian, contains high concentrations of trimethylamine oxide. Some bacteria, such as *Proteus,* possess enzymes to convert the oxide to trimethylamine, which is associated with the characteristic fishy odor. Other spoilage organisms associated with seafoods are outlined in Table 20-10.

Food poisoning associated with fish. There are three types of human food poisoning that are peculiar to fish. Scombroid food poisoning is associated with fish of the families Scomberesocidae and Scombridae, which include tuna and mackerel. These fish contain high concentrations of the amino acid histidine in their tissues. The gut and surface of fish, as well as the processing environment, are contaminated with enteric bacteria (especially *Proteus morganii;* and if the organisms grow extensively in or on the fish, the histidine can be decarboxylated to histamine. The amine is stable to cooking temperatures and the acidity of the stomach. The ingestion of sufficient amounts of the amine produces character-

Table 20-10 Spoilage organisms associated with seafoods

SEAFOOD	DEFECT	MICROORGANISM(S)
Fresh fish	Off odors	*Pseudomonas, Acinetobacter, Proteus, Vibrio, Moraxella*
	Fruity	*Pseudomonas*
	Ammoniacal	*Pseudomonas*
	Rotten egg (H_2S)	*Pseudomonas*
Salted fish	Pink	*Halobacterium, Halococcus*
Oysters	Pink	Yeasts
Shrimp	Off odor	*Pseudomonas*
Squid	Yellow discoloration	*Pseudomonas putida*
	Red discoloration	*Serratia marcescens*
Crayfish	Sweet to foul odor	*Pseudomonas, Lactobacillus,* Coryneform bacteria

Adapted from Banwart, G.J. *Basic food microbiology*, AVI Publishing, Westport, Conn., 1979.

istic allergic symptoms (facial flushing, tingling of the extremities, and shortness of breath), commonly called scombroid food poisoning.

Vibrio parahaemolyticus, a halophile (see Chapter 28) requiring about 5% sodium chloride in laboratory media, was first isolated and associated with food poisoning in Japan. The symptoms of vomiting, diarrhea, headache, and fever are common. The vibrio gains entry with the consumption of raw fish. Generally, the afflicted recover within 1 week. The organism and outbreaks of its associated food poisoning have been described throughout the world. Several outbreaks in the United States have occurred as a result of cross-contamination between raw and cooked fish or fish dishes.

Clostridium botulinum (see Chapter 28) organisms that produce type E toxin appear to be restricted to fresh and marine coastal areas, especially the coastal muds. The organisms are free living and contaminate the slime layer and gut of fish. After heating or smoking of the fish, all vegetative forms of all bacteria are destroyed, but the spores remain viable. If temperature abuse occurs after heating, the normal spoilage flora does not grow and alert the consumer (that is, odor) of a potentially dangerous situation. If botulinal spores are present, they germinate and the organism grows with the concomitant formation of toxin. The product will appear quite acceptable, but it is lethal. The fish do not have to be vacuum packaged for growth to occur, because penetration of the flesh provides sufficient anaerobiosis for growth. The distribution of this organism is worldwide. Fortunately, the disease is not common.

Vegetables and fruits

The surfaces of fresh fruits and vegetables harbor a predominantly gram-negative flora *(Erwinia* and *Pseudomonas),* which may range from a few thousand to 100,000/cm^2, depending on the product and the conditions of growth. Also encountered with a high incidence but in low numbers are lactic acid bacteria, yeasts, and molds. Frequently, various members of the resident flora are encountered within the tissues of sound fruits and vegetables. In some manner these microorganisms are incorporated during plant growth; and although they do not multiply extensively, they remain viable. The chief procedures to maintain fruits and vegetables in the fresh state are chilling and the use of controlled atmospheres. During the storage period various bacteria and mold populations can develop and cause spoilage. Spoilage may consist of acrid and putrid odors, texture defects including various types of "rots", and discolorations (Plate 29). Chapter 35 discusses in detail some of the diseases associated with vegetables and fruits.

Some vegetables are fermented to produce characteristic flavors as well as to extend shelf life. Cabbage, cucumbers, and olives undergo a lactic acid fermentation, and the floral succession of these fermentations is similar. In the manufacture of sauerkraut the cabbage is shredded, salted, and added to large tierces, or vats. Oxygen is excluded from the tierce by covering with plastic sheeting and adding water or weights. The salt disrupts the osmotic balance in the shredded cabbage, and nutrient-containing juice is removed from the tissues. Sufficient carbohydrates, vitamins, and other nutrients are in the juice to afford luxurient growth of the nutritionally fastidious lactic acid bacteria. In the floral succession resident coliforms rapidly appear, followed by a burst of *Leuconostoc*. After a brief period the acid-tolerant, homofermentative species such as *Lactobacillus plantarum* predominate. The bursts in the floral succession are controlled by the pH of the juice.

Attempts have been made to employ lactic starter cultures to replace the "wild" vegetable fermentations. In most instances a product with at best equal acceptability was produced and at considerably more expense because of the cost of making the starter. Currently, all fermented vegetables use the wild fermentation procedure for commercial production.

Bread

Breadmaking has been practiced for centuries, and in many areas throughout the world a bread or breadlike product is produced that is peculiar to that area. This is made possible by the many raw ingredients and processing procedures that can be used. Most, but not all, breads are leavened. Leavening is the production of gas in the dough before baking, and this gives the finished product a light, porous texture. Many baked products such as pretzels, doughnuts, crackers, wafers, and bread sticks are leavened.

Two types of leavening are used in industry, and both produce carbon dioxide. Chemical leavening usually involves the addition of a weak acid and sodium bicarbonate to the dough. Fermentative yeasts, principally *Saccharomyces cerevisiae*, are used in biological leavening. The yeasts also produce aroma and flavoring compounds that are unknown and cannot be duplicated chemically. The yeasts are produced by a separate industry. They are grown on molasses from sugar processing, and most commercial yeast used today is dried and packaged as a powder. In breadmaking flour, water, and salt are mixed, and the yeast is added. After an initial fermentation, the dough is cut, placed in pans, and held for the final fermentation. During baking, all vegetative forms are killed, and only bacterial spores survive. The most important of these is *Bacillus subtilis*. This species is a cause of spoilage within the loaf called "rope" in which the bread appears stringy, soft, brown, and has an odor of ripe cantaloupe. The mucoid nature of the contamination is caused by the production of a capsular-like material by the offending microorganism. Most commercial bakeries add propionate to the product to inhibit molds. Essentially, this adds 1 to 2 days to the shelf-life. Molds are more evident on the surface of the bread, and the type of defects they cause are outlined in Table 20-11.

Table 20-11 Microorganisms causing spoilage in bread

MICROORGANISM	DEFECT
Bacteria	
Bacillus subtilis	Ropy, slime
Lactic acid species	Sour
Coliforms	Sour
Serratia marcescens	Red
Molds	
Rhizopus nigricans	Black mold
Penicillium	Blue mold
Neurospora	Pink mold

Sourdough breads are made using a lactic fermentation with an acid-tolerant yeast. Usually a portion of the first fermentation dough is saved, and the following day small amounts of it are used continuously to initiate the fermentation again (termed back inoculation). All sourdoughs are made using a wild fermentation.

The organisms involved in the manufacture of San Francisco sourdough breads have been isolated recently and identified. The lactic is a new species of *Lactobacillus (L. sanfrancisco)*, and this heterofermentative organism is peculiar in that it ferments maltose but not glucose or other common sugars. Two acid-tolerant yeasts were identified as *Saccharomyces exiguus* and *S. inusitatus*. Typical baker's yeast *(S. cerevisiae)* was not detected, probably because of its acid sensitivity.

FOOD POISONING

Food poisoning is a general term reflecting the passage of any deleterious chemical substance, microorganism, or microbial toxin by foods. The chemical poisons include those produced by poisonous plants and fish as well as the inadvertent addition of toxic chemicals such as insecticides and germicides. Microbial food poisonings may result from ingestion of preformed toxins (intoxi-

Table 20-12 Foodborne disease outbreaks confirmed in the United States, 1979

ETIOLOGY	NUMBER OF OUTBREAKS	%	NUMBER OF CASES	%	NUMBER OF DEATHS
Bacterial					
Brucella	2	(1.2)	18	(0.2)	none
C. botulinum	7	(4.0)	9	(0.1)	none
C. perfringens	20	(11.6)	1110	(15.0)	5
Enterobacter cloacae	1	(0.6)	37	(0.5)	none
Salmonella	44	(25.6)	2794	(37.9)	none
Shigella	7	(4.0)	356	(4.8)	1
S. aureus	34	(19.8)	2391	(32.4)	none
Streptococcus Group G	1	(0.6)	73	(1.0)	none
V. cholerae	1	(0.6)	5	(0.1)	none
V. parahaemolyticus	2	(1.2)	14	(0.2)	none
TOTAL	119	(69.2)	6806	(92.3)	6
Chemical					
Heavy metals	1	(0.6)	18	(0.2)	none
Ciguatoxin	18	(10.4)	85	(1.2)	none
Scombrotoxin	12	(6.9)	132	(1.8)	none
Mushroom poisoning	1	(0.6)	2	(0.03)	none
Other chemical	4	(2.3)	13	(0.2)	none
TOTAL	36	(20.9)	250	(3.4)	0
Parasitic					
Trichinella spiralis	11	(6.4)	93	(1.3)	none
Viral					
Hepatitis (non-B)	5	(2.9)	74	(1.0)	none
Other viral	1	(0.6)	155	(2.1)	none
TOTAL	6	(3.5)	229	(3.1)	none
CONFIRMED TOTAL	172	(100.0)	7378	(100.0)	6

From the Centers for Disease Control, Atlanta, 1981.

cations) or from toxin production by the microorganism in the intestinal tract. *Clostridium botulinum* and *Staphylococcus aureus* produce toxins in foods before ingestion, whereas *Clostridium perfringens* and species of *Salmonella* produce toxins by growth in the intestinal tract.

By far the majority of abuses that cause food poisoning outbreaks occur at the end of the food chain and involve temperature abuse. Most poisonings occur as a result of ignorance or indifference on the part of food handlers. In industry this is generally recognized, and considerable attention has been focused on the education of food handlers, especially restaurant personnel, caterers, and those in the home. The available data indicate that only occasionally does the insult occur at the food manufacturing level.

Bacterial food poisoning (Table 20-12)

Clostridium botulinum. The finding of botulinal toxin in a commercially canned or home-preserved food indicates gross underprocessing. In commercial practice the occurrence of underprocessing is much less frequent than in home-processed foods.

The botulinal toxins are quite heat labile, and heating a contaminated food to boiling destroys the toxin. Some foods, such as cooked (but not cured) meats and smoked fish, afford the growth of *C. botulinum* in the deep tissue areas. These products need not be vacuum packaged to support growth, and refrigeration must be employed to inhibit growth. Aside from thermal retorting, salt, nitrite, and chilling temperatures (or combination thereof) are used to inhibit the growth of this organism. More detailed explanations of botulism and the action of the toxin appear in Chapters 28 and 22, respectively.

Staphylococcus aureus. The cause of the majority of *S. aureus* food poisonings is contamination of cooked foods by human food handlers, followed by temperature abuse. The organism is consistently found on the skin and mucous membranes of humans, and avoidance of food contamination is impossible. Therefore, rigid temperature control must be practiced. The organism produces at least six serologically different exotoxins, which are referred to as enterotoxins A through F. Generally, type A is found in most food poisoning outbreaks. As little as 0.15 μg of the protein toxin will elicit the characteristic symptoms in humans. In extreme cases treatment of the afflicted consists of intravenous administration of fluids. Otherwise, the disease is allowed to run its normal 8- to 16-hour course with a 1- to 2-day recovery period. The mechanism of the toxin's action in the body is unknown.

Clostridium perfringens. Food poisoning associated with *C. perfringens* almost invariably reflects temperature abuse of a cooked product. The product is usually cooked insufficiently; vegetative cells are killed but not spores. Lysis of sporangia in the intestine liberates large amounts of an enterotoxin that produces symptoms (see Chapter 28 for a discussion of *C. perfringens*).

Salmonella. There are three clinical types of salmonellosis; however, the food microbiologist is concerned chiefly with one—gastroenteritis. Discussion of salmonella infection appears in Chapter 28.

Other bacterial agents (Table 20-12). Many pathogens that cause foodborne illness escape detection because of late or incomplete laboratory investigation. *Escherichia coli, Bacillus cereus, Yersinia enterocolitica, Vibrio parahaemolyticus,* and *Campylobacter fetus* subspecies *jejuni* cause foodborne illness, but the extent of their role in disease has not been determined. Except for *Bacillus cereus* these agents are discussed earlier in the chapter or will be discussed in Chapter 28. *Bacillus cereus* is an aerobic sporeformer found in the soil. The primary food vehicle for this organism is cereal dishes. The most severe symptoms have resulted from ingestion of contaminated fried or boiled rice. The incubation period of the disease is between 2 and 5 hours. There are two toxins elaborated by the organism that are important in disease—one associated with vomiting and the other with diarrhea.

Other bacteria suspected of being involved in foodborne disease are Group D *Streptococci, Citrobacter, Enterobacter, Klebsiella,* and *Pseudomonas.* These genera are infrequently detected because they are not considered during investigation and laboratory diagnosis of foodborne illness. The confirmed cases of foodborne disease caused by bacterial agents in 1979 are listed in Table 20-12. This table confirms the pattern alluded to earlier—the primary offenders are *Salmonella, S. aureus,* and *C. perfringens.*

Animal parasite food poisoning

Trichinella spiralis is a nematode that is associated with a foodborne illness called trichinosis. The organism is commonly found in carnivorous mammals where it encysts in muscle tissue (Plate 28). The organism is transmitted to humans when the latter ingests undercooked or underfrozen bear, wild pig, walrus, and other meats. The ingested parasite penetrates the intestinal wall and is carried via the hepatic-portal circulation to the liver, heart, lungs, and arterial system. Once in the arterial system the organism is circulated throughout the body, and any organ or tissue may be invaded. The symptoms of infection can occur within 2 days and resemble ordinary food poisoning (gastroenteritis). Once the organs and tissues are invaded, however, severe clinical symptoms may be manifested, such as pneumonia, meningitis, nephritis, muscular pain, difficulty in breathing, or heart damage. This disease left untreated can result in death.

Viral food poisoning (Table 20-12)

The principal viral agent associated with foodborne illness is the hepatitis A virus (HAV). HAV, which is not a common cause of foodborne illness, is discussed in Chapter 29.

Fungal food poisoning

Certain molds produce substances that are toxic to humans and animals. Although these toxins were known to veterinarians for some time, it was not until recently that the potential for involvement in human health was noted. Most of the mold toxins are called mycotoxins, and one that is commonly encountered is aflatoxin, which is produced by strains of *Aspergillus flavus* and species of *Penicillium*. These organisms grow on grains and peanuts and produce the toxin, which cannot be destroyed by normal thermal processes (see Chapter 31). Extremely sensitive methods for the detection and quantitation of aflatoxin in foods have been developed. Currently, the standard for aflatoxin in peanuts, peanut butter, and peanut products is less than 15 parts per billion.

Chemical food poisoning (Table 20-12)

Heavy metals. Various chemical, such as heavy metals (antimony, lead, cadmium, copper, tin, and zinc) can cause gastrointestinal illness when present in excessive amounts in foods.

Monosodium glutamate. Monosodium glutamate is a chemical that is used as a flavor enhancer in various manufactured foods and excessive amounts can cause intoxication. The clinical syndrome includes burning sensation in chest, neck, abdomen, and extremities plus a sensation of lightness and pressure over the face.

Fish poisoning

Ciguatera. Ciguatera fish poisoning may result from the ingestion of a variety of marine species such as barracuda, grouper, sea bass, and sharks. These fish become toxic by feeding on herbivorous fish that have ingested toxic algae or toxic matter found in coral reefs. The incubation period is usually 2 to 8 hours, followed by symptoms of gastrointestinal upset and numbing of facial areas and extremities. Death may occur from respiratory paralysis.

Paralytic and neurotoxic shellfish poisoning. This type of poisoning results from the ingestion of shellfish such as oysters, mussels, and clams that have fed upon certain dinoflagellates such as *Gonyaulax catenella*. Blooms of these organisms on the ocean produce a characteristic "red tide" (see Chapter 5). The incubation period is 30 minutes to 3 hours. Symptoms of disease are similar to those for ciguatera poisoning. Death results from cardiovascular collapse.

Scombroid poisoning. Scombroid fish poisoning results from ingestion of fish of the order Scombrodei, discussed earlier.

Mushroom poisoning. A variety of mushrooms are toxic when ingested by humans, and some are particularly lethal. Mushroom poisoning is discussed in Chapter 31.

SUMMARY

Foods that are processed and sold to the public undergo careful scrutiny before sale. Samples of food are first composited and then subjected to bacteriologic analysis. A variety of tests are used on the sample to detect pathogens as well as to determine the total number of nonpathogenic species that may be characteristically found in the food. The limit of bacteria is specified by the Federal Drug Administration and is called a standard, which is enforcable by law.

The major sources of contamination and microbial growth in foods are many. The first source is the environment from which the food was harvested such as soil or water. Many raw foods have a microbial flora that under proper conditions can act as a cause of spoilage. Further sources of contamination are provided in the processing plant and include the processing equipment, water, and air in the plant, the food ingredients being used to produce a processed food, and rework.

Control of microbial contamination requires a concerted effort on the part of all divisions in the plant. Plant layout is important to ensure that the product passes without interruption through its various processing phases. Physical barriers (compartmentalization) are placed at specific points in the processing train to prevent cross-contamination. Traffic control is maintained at a minimum to prevent contamination. Last, control of water leaks from the roof, pipes, and water pumps helps prevent one of the major ingredients for growth, water, from reaching the microbial environment.

Several techniques are used to preserve foods. Each food product is coded so the producer and consumer know how long the product can be stored before normal spoilage is encountered. The manufacturer preserves foods in a variety of individual ways but usually a combination of techniques is used. Food can be preserved by heating, which may include employing temperatures as low as 62.9° C (pasteurization) to kill *Mycobacterium tuberculosis* and *Coxiella burnetii* in milk. Canned foods may be heated to 121° C or higher to destroy *Clostridium botulinum,* the organism that produces one of the most lethal toxins known to humans. Low temperature (below 10° C) inhibits the growth of almost all pathogens but does not inhibit the growth of psychrophilic microorganisms such as *Pseudomonas*. Reducing the water content of foods can prevent microbial growth and can be accomplished by drying or by the addition of high concentrations of solutes such as sucrose and salt. The pH of the food can be reduced to prevent microbial growth. Foods that are acidified do not support the growth of most bacteria, but organisms such as molds and yeasts can grow at acid pH and cause spoilage. Vacuum packaging prevents growth of aerobic microorganisms but does not inhibit anaerobic microorganisms such as *C. botulinum*. Many foods are treated with a variety of chemicals which are inhibitors of growth. They include benzoates, propionates, nitrite, sulfites, and others.

Foods have a typical microbial flora that may be pathogenic for humans or may give rise to spoilage problems. The major pathogens are *M. tuberculosis* and *Coxiella burnetii* (milk and milk products), *Salmonella* (meats and meat-containing products), and *C. botulinum* (meats, fruits, and vegetables). Many foods are also produced from the action of specific microorganisms called starter cultures. The metabolic products of these organisms are responsible for the formation of processed foods. Milk products (yogurt, cheese, butter), meats (sausage), vegetables (sauerkraut), and bread are examples of foods produced from starter cultures.

The microorganisms responsible for most outbreaks of food poisoning in the United States are bacteria and include *S. aureus, C. perfringens,* and *Salmonella*. The only animal parasite of consequence is *Trichinella spiralis,* the agent of trichinosis. Some fungi, such as species of *Aspergillus* and *Penicillium,* produce a toxin, usually in grains, called aflatoxin that can cause serious illness. The only viral agent of consequence is the hepatitis A virus (HAV). Various chemicals can also be a cause of foodborne illness, and these include heavy metals, monosodium glutamate (used as a flavoring agent in many foods), mushroom poisoning, and fish poisoning, resulting from the ingestion of types of fish that themselves have ingested toxic organisms.

Study questions

1. Many spices, such as pepper, contain huge numbers of spore-forming bacteria. Outline a procedure that you could use to isolate a strictly thermophilic *Clostridium* species that is proteolytic. Be certain to include media as well as incubation temperatures and conditions in your outline.
2. Differentiate between (1) sporulation and germination; (2) enterotoxin and endotoxin; (3) standard and specification; and (4) carcinogenic and cariogenic.
3. Your technician has submitted the following report for your approval:

Quantitation of specific organisms in product XYZ

ORGANISM QUANTITATED	COUNT/GRAM OF PRODUCT
E. coli (MPN)	15
Coliforms (MPN)	1
Yeast	1500
Mold	10
Staphylococcus aureus	520
TOTAL COUNT	1432

Can you detect at least three problems with this report?

4. Assume that you have isolated several different organisms from defective yellow wieners. Also assume that the cause of the defect is bacterial. Outline the steps involved in definitively proving that one of the isolates is *the* cause of the discoloration.
5. A suspension of spores was prepared from a pure culture and mixed with a food product. After 10 minutes of heating at 230° F, the food contained 420 spores/g. After 15 minutes of heating at 230° F, the food contained 42 spores/g. What is the D value? How many spores will survive 20 minutes of heating? What was the initial spore count/g before heating?

Selected Readings

Books

Anonymous. *Prevention of microbial and parasitic hazards associated with processed foods.* Printing and Publishing Office, National Academy of Sciences, Washington, D.C., 1975.

Ayres, J.C., Mundt, J.O., and Sandine, W.E., *Microbiology of foods,* W.H. Freeman & Co., San Francisco, 1980.

Banwart, G.J. *Basic food microbiology.* Avi Publishing Co., Westport, Conn., 1975.

Defigueiredo, M.P., and Splittstoesser, D.F., *Food microbiology: public health and spoilage aspects.* Avi Publishing Co., Westport, Conn., 1976.

Fields, M.L. *Fundamentals of food microbiology.* Avi Publishing Co., Westport, Conn., 1979.

Jay, J.N. *Modern food microbiology* (2nd ed.). D. Van Nostrand Co., New York, 1978.

Journal articles

Speck, M.L., et al. Overview: use of microbial cultures to increase the safety, shelf life and nutritive value of food products. *Food Technol.* **35:**71, 1981.

Chapter 21
INDUSTRIAL MICROBIOLOGY

Glenn Chambliss

The use of microorganisms or microbial products to carry out processes of commercial importance has been going on for thousands of years. Of the earliest known such processes are the conversion of grape juice to wine, the production of beer from grain starch, the use of yeast to cause bread to rise, and production of yogurt from milk.

The early processes developed empirically without any knowledge that living microorganisms were the causative forces driving the processes. It was not until the mid 1800s when Pasteur showed that yeast is the causative agent in alcohol fermentation that this changed. Because of Pasteur's work it was discovered that many natural processes of industrial importance were caused by microorganisms. As the causative organisms for the various industrial processes were identified, it became common practice to use pure cultures of the organisms to carry out the processes. This resulted in a more uniform product with less batch-to-batch variability. The use of pure culture also opened the possibility that genetic variants of the causative organism could be isolated—which would be more efficient or would give a better product, or both—than did the original culture.

The knowledge that microorganisms can mediate chemical conversions such as glucose to ethanol prompted microbiologists to search for new and useful microbial processes. For example, at the beginning of the twentieth century a great demand for organic solvents such as acetone, butanol, and ethanol arose, especially with the coming of World War I. The chemical production of these solvents was not efficient and was expensive at that time. Consequently a program was undertaken for screening microorganisms for their ability to produce organic solvents as a result of starch fermentation. Chaim Weizmann isolated a bacterium, today called *Clostridium acetobutylicum,* which did efficiently produce acetone, butanol, and ethanol from starch fermentation. This organism served as the basis for industrial production of acetone and later butanol for several years until it became cheaper to produce them chemically from petrochemicals. At the height of the microbial production of acetone and butanol there were production plants in the U.S., Canada, and England, and as late as the 1940s fermentation provided 65% of the butanol and 10% of the acetone produced. Had it not been for Pasteur's pioneering work, in all likelihood, this industry would never have come into being.

During the same time that advances in the use of intact microorganisms for industrial purposes were being made, it was realized that microbial products, such as enzymes, had industrial potential. One of the earliest pioneers in this area was Jokichi Takamine who received a U.S. patent on the "Process of Making Diastatic Enzyme" in 1894. The enzyme preparation that was obtained by extracting moldy bran with water, precipitating the protein with alcohol and drying . . . "possesses the power of transforming starch into sugar for use in various industries. . ." This early work of Takamine and that of others has led to the development of today's $300,000,000 microbial enzyme industry.

The development of genetic engineering technology during the 1970s was the dawning of a new era in industrial microbiology. In the past to obtain a microorganism to carry out a specific function (for example, for use in cleaning up oil spilled in the ocean), one had to screen naturally occurring species for that ability, and often a species with the desired characteristics could not be found. With the advent of genetic engineering it has become possible to take desirable genetic characteristics from several organisms and combine them into a single microorganism tailor-made to carry out a specific function. Indeed a patent has been issued on a microorganism that was constructed to clean up oil spills. At this early stage of the game the future potential for tailoring microorganisms for specific functions seems to be limited only by the inventiveness of the industrial microbiologist.

During the past 100 years or so since Pasteur's time industrial microbiology has made tremendous advances. Entire industries, based on the ability of microbes to produce desired products, such as antibiotics, have developed. The rapid developments in the field of enzyme technology and the potential for development in the genetic engineering field promise that the field of industrial microbiology has a bright future.

Industrial microbiologists use the word *fermentation* differently than do microbiologists in general. To the classical microbiologist fermentation is an anaerobic process by which facultative or anaerobic microorganisms obtain energy from an organic compound, often a carbohydrate. Industrial microbiologists apply the term *fermentation* to any process by which a microorganism produces a desired product regardless of whether the process is aerobic or anaerobic. For example, industrial microbiologists refer to the production of penicillin by the aerobic mold *Penicillium chrysogenum* as the penicillin fermentation.

ANTIBIOTICS

The primary industrial importance of antibiotics is in their use as chemotherapeutic agents in the treatment of diseases caused by microorganisms. Since the discovery of penicillin in 1928 several thousand antibiotics have been isolated, but less than 100 have proved useful as chemotherapeutic agents for human use (Table 21-1). The specific effects of antibiotics on microorganisms in the treatment of human disease and their possible usefulness to the organisms producing them are discussed in Chapter 27.

Commercial production of antibiotics (Table 21-1)

Today penicillin (Figure 21-1) and other antibiotics are produced predominately in huge batch cultures. The producing organism is grown in ***fermenters,*** which are cylindrical and have capacities of 35,000 gallons or more (Figure 21-2). Most fermenters are made of stainless steel and contain large impellors for mixing the culture. Antibiotic producers are aerobic, so provision must be made for supplying large amounts of air to the culture. This is one of the biggest engineering problems to overcome. The fermentation vessel must also contain coils through which a coolant or steam can circulate for temperature control and sterilization. Generally the large fermentations produce so much heat that the culture medium must be cooled to maintain the ideal temperatures for growth and antibiotic production. Often the optimum temperature for antibiotic production is different from the optimum growth temperature. Through the use of computers the temperature of the fermentation can be programmed such that both growth and antibiotic production occur at their optimum temperatures.

Table 21-1 1979 U.S. sales of antibiotics produced by American companies

CATEGORY OF ANTIBIOTIC	MARKET VALUE (U.S. DOLLARS)
Penicillins	$220,943,000
Other antibiotics such as cephalosporins, tetracycline, erythromycin, and streptomycin	$638,297,000
Antibiotic-sulfonamide combinations	$ 16,921,000
Topical antibiotics such as bacitracin or neomycin	$ 17,064,000

Culture medium. The composition of the medium can radically affect antibiotic production, and a medium that might be good for the production of one antibiotic may not be at all good for the production of another antibiotic. An important goal of industrial microbiologists is to design the ideal culture medium for the production of each antibiotic; however, the ideal culture medium in economic terms may not be the medium that generates the greatest yield of antibiotic. This is because costs must be considered. The best medium is the one where the cost per unit of antibiotic produced is least. A culture medium for antibiotic production, like any other culture medium must provide a source of carbon, energy, nitrogen, phosphorus, and trace elements. Most industrial culture media are complex, containing organic components such as ***corn steep liquor,*** a byproduct of corn starch manufacture, as a source of growth factors and a crude form of dextrose, ***sucrose,*** or a sugar derivative such as molasses as the carbon source. Meals such as soybean meal are often the main nitrogen source. Inorganic nutrients include calcium, magnesium,

acyl—NH—CH—CH(—S—C(CH_3)$_2$—)
O=C—N—CHCOOH

GENERAL PENICILLIN STRUCTURE

NATURAL PENICILLINS

Acyl group	Name
C_6H_5—CH_2CO	**Penicillin G**
$CH_3(CH_2)_4CO$	**Penicillin K**
C_6H_5—OCH_2CO	**Penicillin V**
$CH_3CH_2CH{=}CHCH_2CO$	**Penicillin F**

SEMISYNTHETIC PENICILLINS

Acyl group	Name
C_6H_5—CHCO (with NH_2)	**Ampicillin**
$C_6H_3(OCH_3)_2$—CO	**Methicillin**
C_6H_5—CHCO (with COONa)	**Carbenicillin**
HO—C_6H_4—CHCO (with NH_2)	**Amoxicillin**

Figure 21-1 Structure of molecules of penicillin and some of its derivatives.

iron, manganese, phosphate, potassium, and others. Often the medium is mixed in a concentrated form and sterilized. It is then pumped, with dilution, into the presterilized fermenter. During the fermentation process the pH of the culture medium is monitored and controlled by the addition of acid or base.

Fermentation process. Maximum antibiotic production occurs after growth of the producing organism has ceased, which means that antibiotic synthesis occurs in a period when the growth medium is usually depleted of nitrogen sources and limited for carbon and energy sources. Indeed it has been observed that if glucose is fed slowly to penicillin-producing cultures of *Penicillium chrysogenum,* penicillin production greatly increases (Figure 21-3). If glucose is fed to the culture too rapidly, penicillin synthesis is inhibited; therefore, the rate at which glucose is fed to the culture must be closely monitored. One of the important recent advances in fermentation technology is the use of computers to control various parameters of a fermentation such as the rate of glucose feeding, temperature, rate of aeration, and stirring rate.

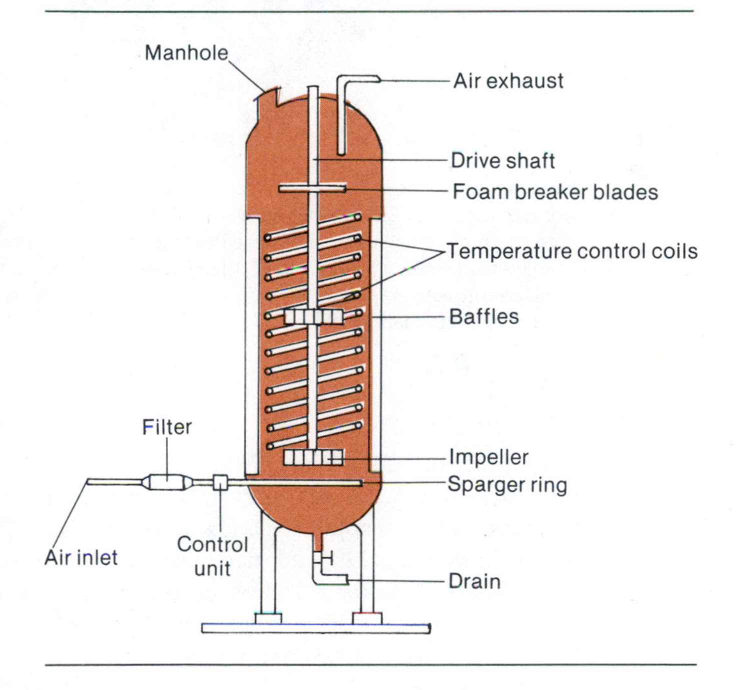

Figure 21-2 Fermenter used in production of antibiotics.

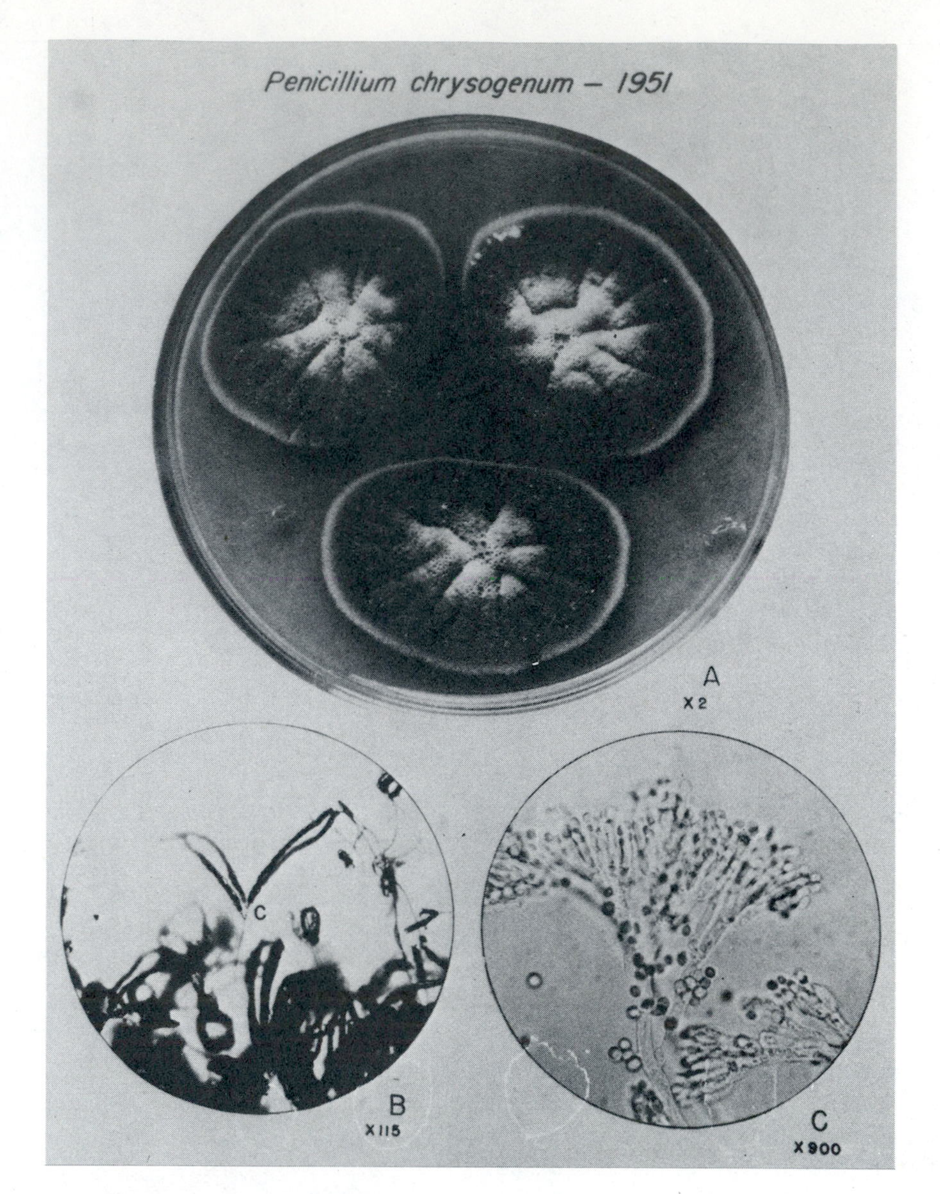

Figure 21-3 *Penicillium chrysogenum* Thom NRRL 1951, strain isolated from moldy cantaloupe at Northern Regional Laboratory and parent source of high-yielding strains now used for commercial production of penicillin by submerged culture process. **A,** Isolated colonies showing cultural characteristics. **B,** Small portion of colony margin in **A** magnified, showing character of single spore-bearing apparatus, or "penicillus." **C,** Single penicillus, highly magnified, showing details of structure and manner in which spores are borne.
Courtesy K. Raper.

An antibiotic fermentation typically lasts several days to about a week. Exactly when a fermentation is terminated depends on several variables; however, they are normally terminated when the rate of production of the particular antibiotic under synthesis falls below a certain level. This level is usually determined by a cost-benefit–type analysis.

Product recovery. Most antibiotics are excreted into the culture medium by the producing cells. To recover the antibiotics from the culture fluids, the cells and other particulate matter are first removed by filtration. The exact method for recovering an antibiotic from the clarified culture fluid varies from antibiotic to antibiotic. Some of the techniques used include solvent extraction, adsorption of the antibiotic to and elution from activated charcoal or diatomaceous earth, and selective precipitation of the antibiotic. Two or more of these and other techniques are usually required to obtain an antibiotic in pure form. The purification procedure is extremely important because antibiotics must contain a minimum of impurities if they are to be used on humans.

Inoculum production

Strains that have been selected for high antibiotic production tend to lose potency when carried for many generations on agar. For that reason stock cultures are made and stored away, either frozen or lyophilized.

In preparation for inoculating a large fermentation vessel, a frozen or lyophilized stock culture is retrieved and spread on solid growth medium. Growth from the solid medium is used to inoculate liquid medium in a shake flask. When the shake flask culture is near maturity, it is used to inoculate a larger culture (10 to 100 times the size of the first shake flask culture). The second culture is in turn incubated and used to inoculate larger and larger cultures until, finally, enough inoculum is available for a large production fermenter. The size and number of cultures needed is dictated by the size of the final fermenter vessel. Usually the volume of each culture is 10 to 100 times greater than that of the previous culture. The latter inoculum cultures are grown in stainless steel stirred-fermenters, similar to the ultimate fermentation vessel except smaller.

Strict sterile conditions and techniques are maintained throughout inoculum buildup to ensure that the culture does not become contaminated. The culture is also checked for contamination at each stage of scale-up. If contamination does occur, the process is usually aborted, since contamination adversely affects antibiotic production.

Strain improvement

When first isolated from nature, most antibiotic-producing microbes produce little antibiotic, usually only a few micrograms per milliliter of culture fluid. With so little antibiotic being produced, the production costs per unit of antibiotic produced are high. One way to reduce this cost per unit is to develop variants of the organism that produce greater amounts of the antibiotic. Programs for doing this are called strain improvement or strain development programs.

Strain improvement programs are based on improving antibiotic production by genetically altering the producing organism. This is done by subjecting a culture of the organism to a mutagen such as ultraviolet light, x-rays, or a chemical mutagen so that a certain percentage of the cells in the culture are killed. Among those that survive the mutagenesis, a small percentage will be so mutated that they will produce more of the antibiotic than did the wild type strain, so-called overproduction. The difficulty with this is finding the few mutants that overproduce the antibiotic among the millions of survivors that do not. In the first strain improvement programs each survivor of mutagenesis had to be tested. This is done by spraying a suspension of an indicator organism onto agar medium containing isolated colonies of survivors. After growth of the indicator organism, colonies of survivors that overproduce the antibiotic give greater zones of inhibition of the indicator organism than do the colonies of the other survivors. Also, the survivors can be tested by individually culturing each of survivors in liquid medium and testing it for amount of antibiotic produced. Often it is necessary to test thousands of survivors of a mutagenesis to find a single overproducing variant. As can be imagined this is laborious and time consuming.

Over the years correlations between certain types of mutations and increased antibiotic production have been found. For example, it has been observed that mutants of *Penicillium chrysogenum* that overproduce penicillin also have altered colony color or morphology. Conversely, not all mutants with altered colony color or morphology overproduce the antibiotic, but the percent of them that do is usually much higher than that among total survivors. Thus by testing survivors with altered colony morphology, the chance of finding an overproducer is greatly increased. Another mutant phenotype that has been correlated with antibiotic overproduction is auxotrophy. Certain mutations causing amino acid auxotrophy tend to cause antibiotic overproduction. Again not all such mutations affect antibiotic synthesis.

Many antibiotics inhibit the growth of the organism that produces it. It has been found that mutants resistant to the antibiotic they produce tend to overproduce the antibiotic. Today one of the first things that is done in strain improvement programs is to isolate mutants resistant to the antibiotic they produce.

In a strain improvement program when an overproducing mutant is isolated, it is subjected to mutagenesis. The survivors of this mutagenesis are tested for variants that produce more antibiotic than did the original mutant. If one is found, it will be subjected to mutagenesis, and the whole process repeated. A strain improvement using *Penicillium chrysogenum* has been carried through 20 rounds of mutagenesis. A variant was isolated from the last mutagenesis that produced more than 1000 times as much penicillin as did the original strain.

Search for new antibiotics

Pathogenic microorganisms continue to develop resistance to currently used antibiotics; thus pharmaceutical companies are constantly searching for new antibiotics. Their search takes at least two directions. There is the neverending search for organisms that produce novel antibiotics. This is done by examining samples, most commonly from soil, for microbes that produce antibiotics. In the initial screening a soil sample is diluted sufficiently to give isolated colonies and spread on solid growth medium. This growth medium is incubated to allow colony formation of the microbes present in the sample of soil. After colonies have formed, the surface of medium is carefully sprayed with a suspension of an indicator organism such as *Staphylococcus aureus* or *Bacillus subtilis* and reincubated. During this incubation period the indicator organism will form a "lawn" of microbial growth covering the entire surface of the solid medium except around those colonies that produced during the first incubation an antibiotic that inhibits growth of the indicator organism. The antibiotic producers are isolated in pure form and tested against a large battery of indicator organisms. By comparing the responses of these indicator strains to the unknown antibiotic with their responses to known antibiotics, one can deduce with accuracy whether the unknown antibiotic has been previously identified. As more and more antibiotics are identified, it becomes more and more difficult to find new ones. To increase the probability of finding new antibiotics, more exotic samples must be screened and more imaginative new screening methods must be developed.

The other major approach to developing new antibiotics is to chemically alter the existing ones. The development of the semisynthetic penicillins has shown this to be a fruitful approach. These are penicillins in which the penicillin "nucleus" produced by fermentation has side chains chemically added to it. Many such penicillins have been synthesized and tested, and several have proved worthwhile for one reason or another (see Chapter 27).

ENZYMES

Only a few of the thousands of enzymes produced by microorganisms have widespread commercial application (Table 21-2). The number of microbial species producing enzymes of commerical importance is limited and includes both fungi and bacteria. From an industrial point of view there are distinct advantages to having microbes produce enzymes. First, microbes are generally easy to grow, and with the huge fermenters already in existence, large quantities can be grown. Second, in contrast to plants and animals, microbes are easy to manipulate genetically for strain improvement. Third, some of the enzymes are excreted into the medium where they are easily separated from the cells.

Table 21-2 Worldwide sales of enzymes in 1980

ENZYME	AMOUNT (POUNDS)	SALES (U.S. DOLLARS)
Bacterial protease	1,060,000	$66,000,000
Glucamylase	700,000	$36,000,000
α-Amylase	640,000	$12,000,000
Glucose isomerase	140,000	$56,000,000
Rennin	52,000	$64,000,000

Industrially useful enzymes

α-Amylase. α-Amylase is an enzyme that attacks internal α-1,4-glucosidic links in starch molecules, producing dextrins, oligosaccharides, and maltose as the major products. The principal microorganisms used for the commerical production of α-amylase are the bacterium *Bacillus subtilis* and the molds *Aspergillus oryzae* and *Aspergillus niger*. The major uses of α-amylase follow.

Breadmaking. Yeast lack amylases for the hydrolysis of starch; therefore, α-amylase is often added to bread dough to promote the breakdown of sufficient starch to provide sugar, which yeast can convert to carbon dioxide and cause the dough to rise. The amylase must be used judiciously to prevent excessive starch breakdown and the consequent reduction in bread quality.

Brewing. Many American beer makers use adjuncts such as corn or rice to augment the starch content of malt. To aid the amylases already present in malt in the breakdown of the adjunct starches, α-amylase is often added. *Bacillus subtilis* α-amylase is frequently used for this purpose because of its great heat stability.

Alcohol production. α-Amylase is used in the distilling industry to reduce the viscosity of the gelatinized starch produced after the cooking of grains used in the production of whiskey and other distilled alcoholic beverages. Here as in the brewing industry heat stable amylases are desired. α-Amylase is also used in the production of gasohol from corn starch. Corn is ground into a meal that is cooked to gelatinize the starch, which is in turn treated with α-amylase to reduce

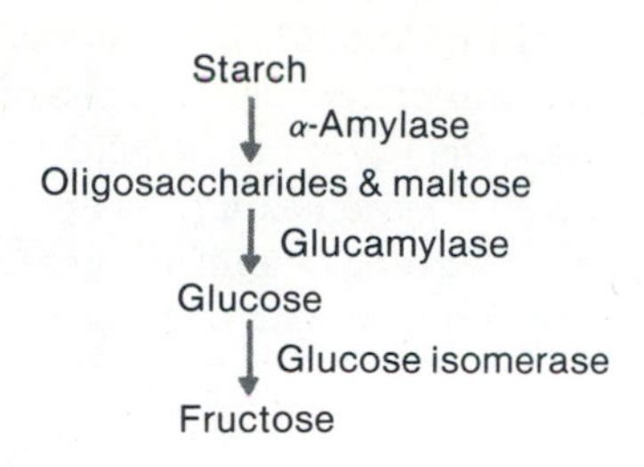

Figure 21-4 Enzymatic conversion of starch to fructose.

the viscosity of the starch and to provide yeast with fermentable sugars.

Sugar syrups. One of the major uses of α-amylase is in the hydrolysis of starch for the production of sugar syrups. In recent years the demand, most notably by the soft drink industry, for sugar syrups with a high fructose content has increased tremendously. Production of these syrups from corn starch has become a multibillion-dollar industry. The process involves several enzymes including α-amylase, which convert starch to a mixture of oligosaccharides and maltose (Figure 21-4). These are reduced to glucose by the enzyme glucamylase. The glucose is isomerized to fructose by glucose isomerase. The last enzyme does not convert all the glucose to fructose, so the final product contains a mixture of glucose and fructose. Since fructose is considerably sweeter than glucose, it is desirable to have as much glucose converted to fructose as possible.

Additional uses. α-Amylase has been used in spot removers. Instant hot breakfast cereals such as oatmeal are treated with α-amylase, which partially degrades the starch present in the oat flakes and consequently reduces cooking time. α-amylase has been used to aid in the removal of silver from used photographic film, and it is also used in the desizing of textiles.

Proteases. Proteases are enzymes that hydrolyze the peptide bonds in proteins. Those proteases that only hydrolyze internal peptide bonds are called endoproteases (or endopeptidases), and those that hydrolyse terminal peptide bonds are called exoproteases. Certain proteases are quite specific and only hydrolyze peptide bonds between specific amino acids; consequently, these proteases only partially degrade proteins to short peptides or oligopeptides. Industrially proteases have many uses, several of which follow.

Baking industry. The gluten proteins in wheat flour are important in determining dough mixing time and bread loaf quality; however, the quantity of these proteins varies considerably from one batch of flour to another. Bakers have found that careful addition of protease will reduce dough mixing time and improve loaf quality; however, overaddition of protease will reduce loaf quality. It has been estimated that over half the bread baked in the United States has protease added to the dough. The protease produced by *Aspergillus oryzae* is most commonly used in the baking industry.

Meat tenderizer. Proteases that only partially degrade protein are used to tenderize meat. The tenderizers can be applied to the surface of a piece of meat shortly before it is cooked. Proteases in the tenderizer preparation diffuse into the meat and cleave protein molecules in the muscle tissue. This partial degradation of the muscle protein tenderizes the meat without seriously affecting its quality. There are several tenderizing preparations available to the consumer, most of which contain about 2% commerical papain or 5% fungal protease as the active ingredient. The proteases in the meat are inactivated by heat during cooking and by stomach acids. A second method of tenderization that is reportedly used on a fairly large scale is the direct injection of a proteolytic enzyme solution into the vascular system of animals shortly before slaughter. In this case the proteases are distributed throughout the animal's body, thus giving a more uniform tenderization of large pieces of meat than can be achieved with surface application.

Laundry products. A major use of proteases is in laundry products. The bacterial proteases, which are active in alkaline conditions (so called alkaline proteases), are the most widely used, and the most commonly used alkaline protease is produced by *Bacillus subtilis*. The protease attacks proteinaceous material such as milk, egg, or blood that may soil clothes and facilitates its removal. Initially when proteases were incorporated into laundry powders, there were problems especially among workers handling the proteases in the soap factories. Apparently dust containing the proteases was inhaled by the workers. The inhaled proteases irritated the mucous membranes of the respiratory tract and also caused allergic reactions. These problems apparently have been controlled by new formulations of the enzyme preparation that eliminate dust formation.

Additional uses. Proteases are used extensively in the tanning of leather to remove proteinacious material from the fresh skin and in the softening of skins. Proteases are used as diges-

$$2\text{ Glucose} + 2O_2 \xrightarrow{\text{Glucose oxidase}} 2\text{ Gluconic acid} + 2H_2O_2$$

$$2H_2O_2 \xrightarrow{\text{Catalase}} 2H_2O + O_2$$

$$\text{Net:} \quad 2\text{ Glucose} + O_2 \longrightarrow 2\text{ Gluconic acid} + 2H_2O$$

Figure 21-5 Reactions by which glucose oxidase and catalase remove atmospheric oxygen.

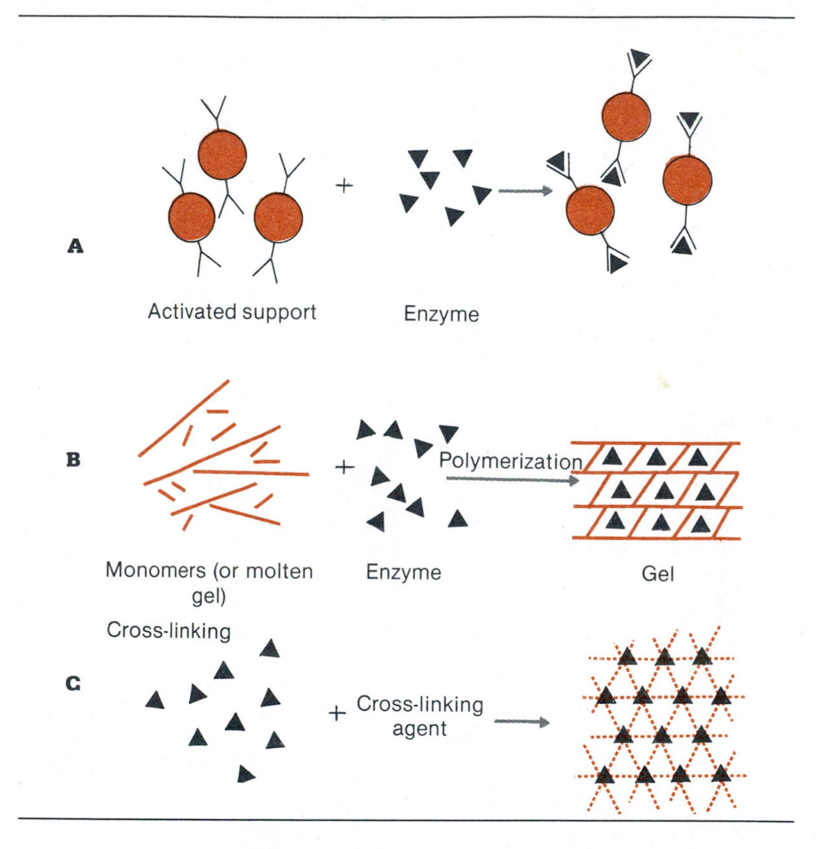

Figure 21-6 Methods for enzyme immobilization. **A,** Soluble enzyme molecules are covalently attached to insoluble support, such as tiny glass, polyacrylamide, or cellulose beads or iron filings. **B,** Soluble enzyme is mixed with acrylamide monomers or molten gel such as starch or agar. When acrylamide is polymerized or gel is set, enzyme becomes entrapped within matrix of resulting gel. For employment, gel is crushed to small particles. **C,** Soluble enzyme molecules are cross-linked to each other, forming insoluble aggregates that still retain enzymatic activity.

tive aids as well as in the chillproofing of beer. A haze can form in beer on chilling as a result of precipitation of protein. The formation of this haze is prevented by treating the beer before it is chilled with protease to remove the protein.

Rennin. Rennin is a proteolytic enzyme found in the stomach of suckling calves. It causes the casein in milk to form curds, which is one of the first steps in cheese manufacture. Recently the use of bovine rennin in cheesemaking has been partially replaced by the use of renninlike proteases produced by the fungi *Mucor pusillus* and *Mucor miehei.*

Glucose oxidase. Glucose oxidase is an enzyme that oxidizes glucose to gluconic acid with the concomitant generation of hydrogen peroxide. Its primary uses are in the food industry where it is used to remove from dried eggs and other dried products glucose, which forms undesirable crystals during the drying process. In combination with catalase, glucose oxidase is used to remove oxygen from mayonnaise, fruit juices, and other preserved products to prevent their deterioration and discoloration. Figure 21-5 shows reactions.

Industrially glucose oxidase is produced by *Aspergillus niger.* It is an intracellular enzyme, which means that the *A. niger* cells must be broken to release the enzyme before it can be purified. Because cell breakage releases all the cellular enzymes, purification of glucose oxidase is a complex process, involving many steps.

Miscellaneous enzymes. ***Pectinases*** are enzymes that breakdown pectins found in plant tissue. They are used commercially for clarifying fruit juices. The protease ***streptokinase*** and the nuclease ***streptodornase*** are both produced by ***Streptococcus hemolyticus*** and are used medically in the cleansing of wounds. ***Invertase*** from yeast is used in the confectionary industry to prevent the crystallization of sucrose and to make soft-centered candies. It hydrolyses sucrose to glucose and fructose, which is sweeter than sucrose.

Enzyme production

Today essentially all commercially important bacterial enzymes are produced by ***submerged culture*** in large fermenters by procedures essentially similar to those used for antibiotic production. Most fungal enzymes are also produced by submerged culture; however, some fungal amylase is produced by the ***Koji process*** in Japan. In the Koji process for amylase production moist sterile cereal bran (either wheat or rice) is inoculated with spores of ***Aspergillus oryzae.*** The inoculated bran is either spread in shallow trays or placed in windrows in sterile, constant-temperature rooms. The spores germinate and the mold beings to grow, producing mycelia, which per-

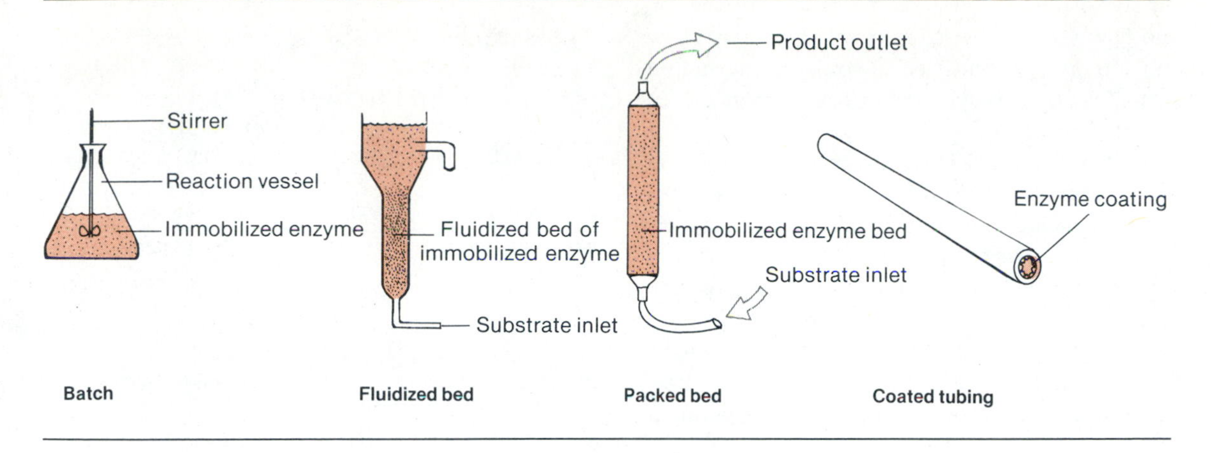

Figure 21-7 Methods by which immobilized enzymes are used.

meate the bran. The mold produces extracellular amylases that diffuse into the bran, breaking down the starch for use by the mycelia. Every so often the moldy bran is turned to maintain relatively constant conditions throughout the heap. After several days to a week, the enzyme is removed from the moldy bran by water extraction and is further purified or concentrated.

Immobilized enzymes

Until recently it was practical to use enzymes only once because of the difficulty involved in recovering the enzyme from the reaction mixture where it is employed. To reduce the costs connected with their use, ways were sought to make enzymes recyclable. In recent years this has resulted in the development of techniques that allow easy recovery of enzymes from reaction mixtures, thus permitting their reuse. This technology involves affixing enzymes to inert, insoluble matrices such as small polyacrylamide beads, tiny glass beads, or cellulose particles. Enzymes affixed to insoluble supports or encapsulated within a membrane are called ***immobilized enzymes*** (Figure 21-6). Another method of enzyme immobilization is to mix the purified enzyme with a cross-linking agent such as glutaraldehyde. The cross-linking agent binds enzyme molecules together, producing insoluble aggregate. These aggregates retain enzymatic activity. A third method is to entrap enzyme molecules within the matrix of a gel such as agarose or polyacrylamide.

Currently there are several methods of employing immobilized enzymes (Figure 21-7). The most widely used is the batch method, in which an immobilized enzyme is mixed with a batch of substrate. While this mixture is stirred, the enzyme reacts with the substrate, converting it to product. When the reaction is complete, the immobilized enzyme is removed from the mixture by filtration or centrifugation, leaving a clean solution of product that can be further purified if necessary. The immobilized enzyme after filtration or centrifugation can be washed with buffer and reused. The batch method is relatively inexpensive, efficient, easily scaled up to large batches, and technologically simple.

In the packed bed method, a suspension of an immobilized enzyme is placed in a column (usually of glass) that is plugged at each end. The immobilized enzyme is allowed to settle and form a packed bed. Substrate solution is pumped into the column through an inlet pore that is covered with a screen to prevent escape of the immobilized enzyme through the bottom plug. As the substrate percolates up through the immobilized enzyme bed, it encounters enzyme molecules and is converted into product that eventually emerges from the outlet pore, which is also covered with a screen in the upper plug of the column. An advantage of this method is that it can be used as a continuous process in which substrate is continuously pumped into the bottom of the column, and product is continuously exiting at the top of the column. A disadvantage of this method is that the packed bed tends to become clogged, preventing efficient flow of the substrate through the column.

The fluidized bed method was developed to overcome the clogging problem of the packed bed method. In this method immobilized enzyme is placed in a cylinder that has a larger diameter at the top than at the bottom. The substrate solution is pumped into the bottom of the cylinder at a sufficient rate to prevent the immobilized enzyme particles from becoming packed. As the diameter of the cylinder increases toward the top, the flow rate of the solution will decrease, allowing the enzyme particles to settle so that no enzyme particles reach the product outlet on the top of the cylinder.

A fourth method is being developed in which enzyme is affixed to the inside of a tube. As a substrate solution is passed through the tube, the enzyme converts the substrate molecules to product so that a solution of product emerges from the other end of the tube. Potentially, each of the latter three methods can be used as a continuous process. Such continuous processes have been maintained for several months on an experimental scale.

There are several important advantages to the use of immobilized enzymes. It has been found that the immobilization of certain enzymes also stabilizes them. Since immobilized enzymes are so easy to remove from the reaction mixture, purification of the reaction product is facilitated. The potential of making a continuous process is an important economic advantage in that it provides a continuous supply of product.

AMINO ACID PRODUCTION

The production of amino acids by microbial fermentation has become an important industrial process in recent years. Of the three most important commercial amino acids—glutamic acid, lysine, and methionine—glutamic acid and over 80% of the lysine are produced predominantly by fermentation.

Until the late 1950s most amino acids used commercially were obtained by hydrolysis of plant proteins such as wheat gluten and soybean protein. At that time the amino acid in greatest demand was ***glutamic acid,*** which was used as a flavoring agent in Oriental and to a lesser extent other cuisines. The high cost of glutamic acid production prompted Japanese researchers to search for alternative methods of producing amino acids. In 1957 Kinoshita and his colleagues reported the isolation of a bacterium that excreted glutamic acid in large amounts. The organism was originally called *Micrococcus glutamicus* but has subsequently been renamed *Corynebacterium glutamicum.* Shortly after its discovery it was successfully used to produce glutamic acid on an industrial scale. Today the production of amino acids by microbial fermentation is a billion-dollar industry (Table 21-3), especially in Japan where glutamic acid, lysine, proline, and valine are produced by microbial fermentation.

Table 21-3 Worldwide sales of amino acids in 1980*

AMINO ACID	AMOUNT (POUNDS)	SALES (U.S. DOLLARS)
Glutamic acid	600,000,000	$1,080,000,000
Methionine	210,000,000	$ 246,000,000
Lysine	100,000,000	$ 258,000,000

*Industrially glutamic acid is made entirely by fermentation. Approximately 80% of the commercial lysine is made by fermentation. Methionine and the remaining lysine are synthesized chemically.

Glutamic acid: case study

When Kinoshita began his search for microorganisms that excreted glutamic acid, he knew that most bacteria synthesize glutamic acid but that few excrete it in significant amounts. It was therefore important for him to develop a simple method of screening large numbers of bacterial isolates for their ability to excrete glutamic acid. In developing such a screening procedure, he took advantage of the fact that the bacterium *Leuconostoc mesenteroides* must be supplied with glutamic acid to grow. The screening procedure that was developed was beautifully simple. Many bacterial sources were screened in the search for glutamic acid excreters including soil, sewage, water, and strains from stock cultures. The general procedure that was developed was to take a sample such as soil, dilute it, and plate it for isolated colonies on solid medium lacking glutamic acid. The plates were incubated to allow well-defined colonies to form. These colonies were replicated onto a second plate by the replica-plating technique devised by Lederberg, which involves pressing a piece of sterile velveteen cloth onto the surface of the plate containing the colonies to be replicated (see Chapter 9). In the Kinoshita screening procedure, once the original plate has been replicated, the bacteria remaining on it were killed by ultraviolet irradiation. It was then overlaid with cooled, molten agar growth medium that lacked glutamic acid but had been seeded with *Leuconostoc mesenteroides.* After the seeded agar overlay had solidified, the plate was reincubated. If any of the bacteria in the original colonies had excreted glutamic acid, it would diffuse up into the overlay, providing *L. mesenteroides* with glutamic acid and allowing it to grow. After sufficient incubation of the overlay plates there would be growth of the *L. mesenteroides* cells above colonies that had excreted glutamic acid. There would be no growth of the *L. mesenteroides* above the colonies where no glutamic acid was excreted. The extent of *L. mesenteroides* growth was directly proportional to the amount of glutamic acid that had been excreted by the colony below it. Using this screening procedure, not only could Kinoshita identify glutamic acid excreters, but he could get an idea of how much glutamic acid they excreted.

This procedure allowed the screening of thousands of microorganisms from nature and from culture collections. Many of the organisms were found to excrete small amounts of glutamic acid but not enough for industrial purposes. The screening program did, however, turn up *Corynebacterium glutamicum,* which produces large amounts of glutamic acid.

Studies on glutamic acid production by *C. glutamicum* revealed that it required biotin for growth. When there were growth-limiting amounts of biotin in the growth medium, glutamic acid was excreted; however, with excess biotin in the growth medium, little or no glutamic acid would be excreted. This proved to be an important factor in developing the culture medium for glutamic acid production. Glucose is a precursor in the synthesis of glutamic acid (Figure 21-8), and growth substrates containing glucose were preferred for maximum glutamic acid production. Many cheap glucose-containing growth substrates, such as molasses, contained sufficient biotin to prevent glutamic acid excretion. This meant that the cheapest growth substrate could not be used. To control the biotin concentration, more expensive synthetic media had to be developed, consequently adding to the cost of glutamic acid production; yet it was still cheaper to produce it by microbial fermentation than by extraction from wheat or soy beans.

The basis for the biotin effect is not well understood but there is evidence that suggests that limiting biotin affects the chemical composition of the cytoplasmic membrane. Membrane synthesized during biotin limitation permits glutamic acid to be excreted from the cell.

Industrially, glutamic acid fermentation is carried out in submerged culture in large fermenters (Figure 21-9) similar to those used in antibiotic production. During fermentation 60% or more of the glucose or other saccharide growth substrate is converted to glutamic acid. In addition to the biotin concentration in the growth medium, several factors can dramatically affect glutamic acid production. In the course of a fermentation, if the cells do not receive adequate oxygen, little glutamic acid is produced; instead, lactic and succinic acids accumulate. With too much oxygen,

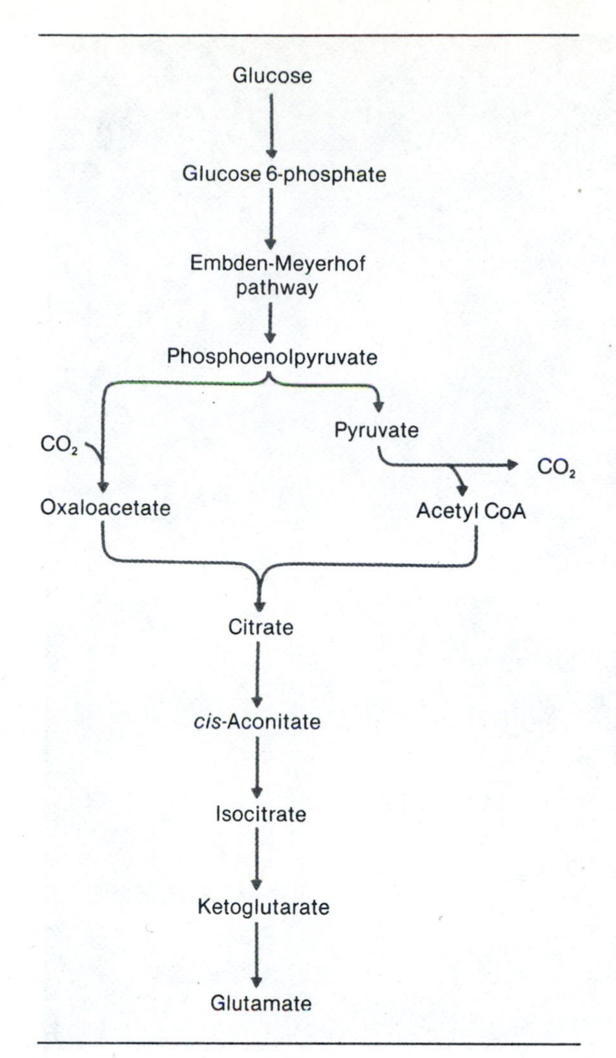

Figure 21-8 Glutamic acid pathway of *Corynebacterium glutamicum.*

α-ketoglutaric acid is produced instead of glutamic acid. The pH of the fermentation must be carefully controlled; if the pH drops too low, glutamic acid production will cease. Another important variable that must be controlled is temperature. The optimum temperature for glutamic acid production ranges from 30° to 35° C.

Lysine

Lysine is an essential amino acid and as such is an important food additive. Bread is fortified with lysine in some Third World countries, and in the United States certain breakfast cereals are fortified with this amino acid. Lysine is also used as a supplement in animal food. Because of these uses the demand for lysine is strong, and in excess of 50,000 tons is produced annually, predominantly by Japanese companies.

The microorganism that is used to produce lysine is a mutant of the glutamic acid producer

Figure 21-9 Outdoor fermentation tanks used by the Kyowa Hakko Kogyo Co., Ltd., of Japan for production of the amino acids, glutamic acid, and lysine. This picture shows seven of 20 such fermentation tanks at this site that are employed by the company in their annual production of at least 20,000 tons of glutamic acid and 10,000 tons of lysine. Most glutamic acid is used for production of monosodium glutamate (MSG), a flavor enhancer.
Courtesy H. Samejima, Kyowa Hakko Kogyo Co., Ltd.

Corynebacterium glutamicum. Lysine is a product of a branched metabolic pathway that, on the other branch, also produces methionine, threonine, and isoleucine (Figure 21-10). The concerted action of both lysine and threonine is required to regulate lysine formation in *C. glutamicum.* When both of these amino acids are present in excess, this pathway is shut down and no lysine or threonine is synthesized. When either lysine or threonine is absent or is present in a low amount, the pathway continues to function no matter how much of the other is present. This characteristic was taken advantage of for the development of a strain of *C. glutamicum* that would excrete lysine. To keep the concentration of threonine low, a mutant of *C. glutamicum* was isolated that could not make homoserine from aspartic semialdehyde (Figure 21-9). This mutant had the ability to make lysine but was unable to make methionine, threonine, and isoleucine; therefore it could not grow unless homoserine or methionine and threonine were added to the culture medium. The isoleucine requirement was met by the conversion of threonine to isoleucine (see Figure 21-9). When either homoserine or methionine and threonine were added in large amounts, no lysine was excreted by the mutant; but when the medium contained a growth-limiting amount of threonine and no homoserine even in the presence of excess methionine and isoleucine, the mutant would excrete large amounts of lysine. Under these conditions of limited threonine, the regulatory controls of the pathway perceived a need for threonine and continued to function in an effort to supply the needed threonine. The mutation prevented any threonine synthesis and caused all the aspartic semialdehyde to be shunted into lysine synthesis. In this way vast excesses of lysine are produced.

The strain of *C. glutamicum* used for lysine production retains the biotin requirement of the parent. Under conditions of biotin excess (30 μg/liter) lysine is excreted, but when biotin is limited in the culture medium, glutamic acid is excreted instead of lysine, even by the mutant.

Cane blackstrap molasses is the most commonly used growth substrate in the industrial fermentation of lysine. Yields of 3 g and more of lysine per 10 g of substrate used have been reported.

ORGANIC ACIDS AND SOLVENTS

In the past, many organic acids and solvents were produced commercially by microbial fermentation. In the early part of this century as much as 65% of the ***butanol*** and 10% of the ***acetone*** produced in the United States was by microbial fermentation. The organism used for the industrial production of acetone and butanol was isolated by Weizmann in 1912 and called *Bacillus granulobacter pectinovorum.* The name of this organism was later changed to *Clostridium acetobutylicum,* Weizmann. Because of the great demand during World War I for acetone for use in the production of explosives, distilleries were taken over in England and Canada and converted to acetone production by *C. acetobutylicum.* In the United States two distilleries in Terre Haute, Indiana were converted to acetone production. When the war as over, the demand for acetone dropped considerably, but a demand for butanol for use in the manufacture of automobile lacquers developed. Although butanol continued to be produced for several years by microbial fermentation, chemical synthesis eventually replaced fermentation.

The early production of butanol and acetone by microbial fermentation led to the development of techniques for production of several organic acids by microbial fermentation, some of which continue to be so produced today.

Citric acid

Citric acid (Figure 21-11) is used in large volumes as an industrial chemical. In 1980 in the United States alone 180,000 metric tons of citric acid were produced by two companies, Miles Laboratories and Pfizer. It is produced by cultures of *Aspergillus niger*. At 1980 prices the value of the citric acid produced in the United States was approximately $100,000,000.

Citric acid is a strong ***sequestering agent*** of cations in addition to being a strong organic acid. Based on these two properties it has many applications. In the food industry it is used as an ***acidulant*** in soft drinks, jams, jellies, candies, desserts, frozen fruits, and wines. In the pharmaceutical industry it is used in effervescent products such as Alka-Seltzer. Another reason it is so widely used in both the food and pharmaceutical industries is that it is rapidly and completely metabolized in the body.

There are several uses of citric acid based on its sequestering properties, including electroplating of metals, leather tanning, and reactivation of clogged oil wells. One reason oil wells become unproductive is that iron deposits clog the sands around the well mouth, preventing passage of the oil. The iron deposits can be removed by pumping into the well a solution of citric acid, which sequesters iron ions causing solubilization of the deposits.

Historically citric acid was produced from citrus fruits such as lemons. In the late nineteenth and early twentieth century, Italy, because of its capacity to produce lemons, had a virtual monopoly on citric acid production and consequently demanded an artificially high price for it. This monopoly was broken in 1923 when a microbial process for citric production was developed in the United States. Since that time the microbial process has become the dominant method for citric acid production. *Aspergillus niger*, which is the producing organism, is a highly aerobic filamentous fungus and must be continually supplied with air (oxygen). Consequently the early production methods relied on growing *A. niger* in shallow pans containing liquid growth medium.

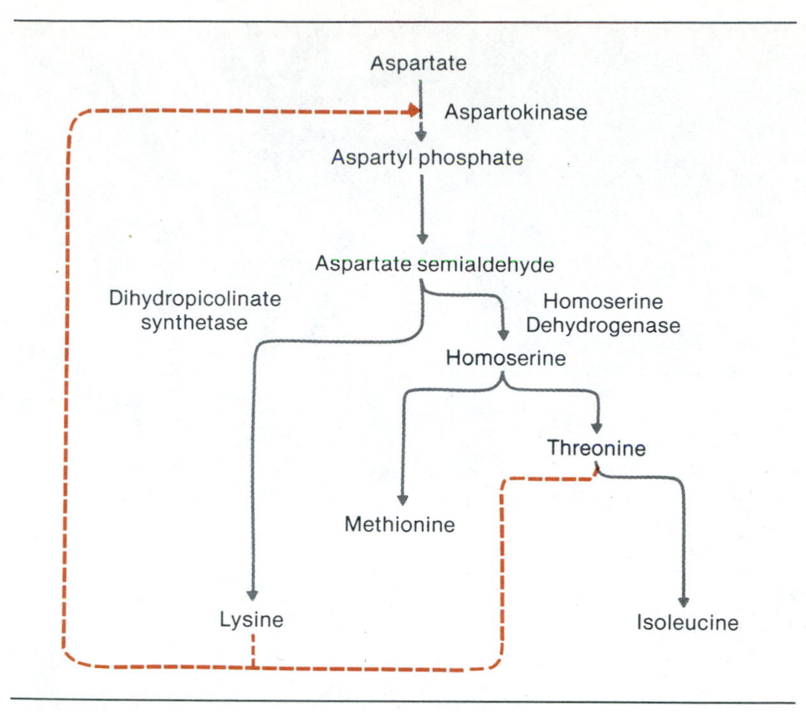

Figure 21-10 Branched pathway leading to lysine synthesis in *C. glutamicum*. Dashed lines leading from threonine and lysine to aspartokinase depict concerted feedback inhibition of this enzyme by these two amino acids. In mutant of *C. glutamicum* used in lysine production, enzyme homoserine dehydrogenase is missing; thus, required amino acids methionine, threonine, and isoleucine cannot be synthesized by this mutant.

Gluconic acid:
O
‖
C—OH
|
HCOH
|
HOCH
|
HCOH
|
HCOH
|
H_2COH

Citric acid:
CH_2COOH
|
HOC·COOH
|
CH_2COOH

Itaconic acid:
H
|
CH
‖
C·COOH
|
H_2C·COOH

Figure 21-11 Structures of citric acid, gluconic acid, and itaconic acid.

The *A. niger* would form floating mats on the surface of the liquid, and as the mold extracted nutrients from the growth medium, it would excrete citric acid into it. Glucose, the preferred substrate for citric acid production, was provided by adding molasses to the growth medium. This method is still used by Pfizer today.

Citric acid is also produced by submerged culture in large fermenters of up to 25,000 liters capacity. The carbohydrate sources for submerged fermentation are molasses and hydrolysed grain (especially corn) starches. These are treated with

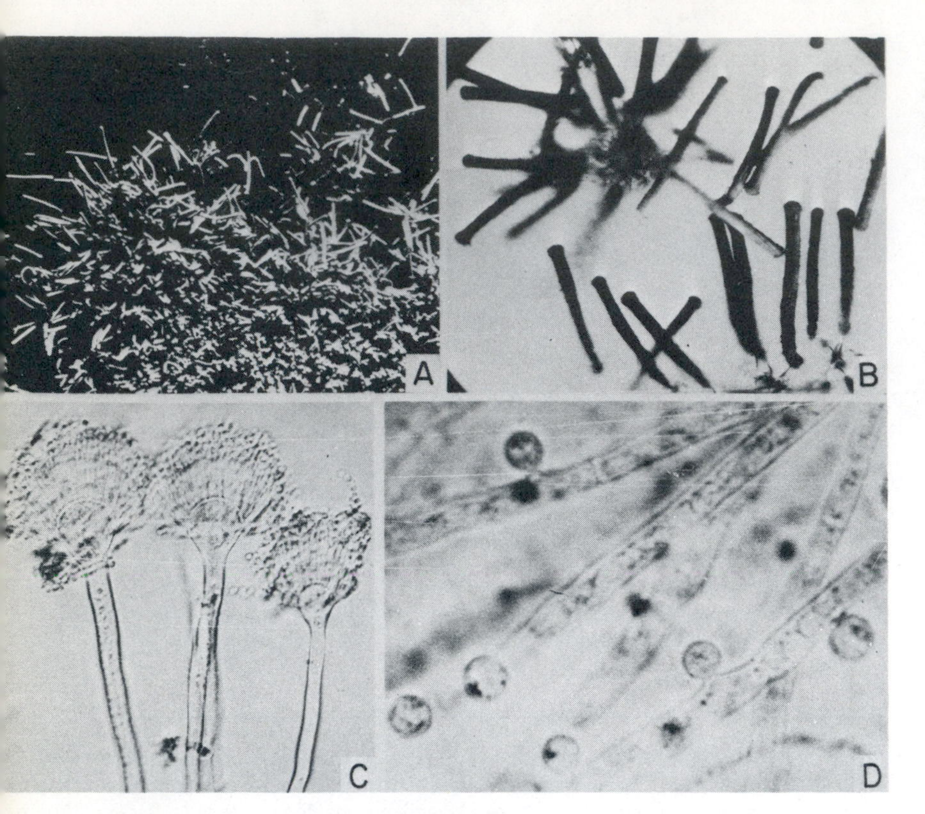

Figure 21-12 *Aspergillus terrus,* isolated from Texas soil by Dr. Kenneth Raper, is used for production of itaconic acid.
Courtesy K. Raper.

ion exchange resins to removed iron and manganese ions, which inhibit citric acid production. The culture is aerated by pumping air into the bottom of the fermenter and allowing it to bubble up through the stirred culture. The pH of the culture must be maintained at 3.5 or less to prevent the formation of oxalic and gluconic acids at the expense of citric acid. The length of the fermentation can vary from as few as 4 to as many as 15 days with as much as 95% of the glucose in the substrate being converted to citric acid.

The exact metabolic pathway by which *A. niger* produces citric acid has not been elucidated, but it is assumed that glucose is converted to pyruvic acid by glycolysis. The pyruvate is converted to acetyl CoA, which is condensed with oxaloacetate by the first enzyme of the Kreb's cycle to give citric acid, which is then excreted. Why citric acid should be excreted rather than being further metabolized via the Kreb's cycle is as yet unknown.

Gluconic acid

Another organic acid produced by a microbial process that is of commercial importance is gluconic acid (Figure 21-11). Although precise figures are not available, it has been estimated that annual worldwide production exceeds 30,000 metric tons. The price of gluconic acid is in the neighborhood of $1.00/kg, making its production a $30,000,000 a year industry.

Gluconic acid is also a sequestering agent, and because of that it is widely used in cleaning products. It is incorporated into detergents used in cleaning soft drink, beer, and milk bottles. It is also used in dishwashing soaps for domestic dishwashers to prevent mineral deposits on glassware. Incorporation of gluconic acid into soap helps prevent the formation of "bathtub ring" in hard water. Gluconic acid is an excellent metal cleaner because it aids in rust removal without adversely affecting the integrity of the unaffected metal. In electroplating, the addition of gluconates to the baths enhances the gloss of the plated metal.

In the food industry gluconic acid is used as an acidulant in baking powders, breads, sausages, cheeses, and other foods. In the pharmaceutical industry, gluconic acid has many uses. Calcium gluconate is used to treat or prevent calcium deficiency especially during pregnancy. Antibiotics such as tetracycline are formulated with gluconates to extend their stability and reduce toxicity.

A potentially large market for gluconic acid is as an additive for concrete. It has been reported that the addition of salts of gluconic acid to cement favorably affects several parameters including settling time, strength, volume, and water resistance.

Gluconic acid is produced in submerged culture under increased atmospheric pressure by a strain of *Aspergillus niger.* This organism contains the enzyme glucose oxidase, which converts glucose directly to gluconic acid. Molecular oxygen is required not only for growth of the fungus but also for gluconic acid synthesis. Inadequate provision of oxygen can limit gluconic acid production; thus copious quantities of air are pumped through the culture in the form of small bubbles that facilitate dissolution of oxygen into the culture medium. Glucose is the preferred substrate and is provided to the culture either as glucose crystals or as a syrup produced by treatment of corn starch with α-amylase and glucamylase, which together hydrolyse starch to glucose. It has been reported that as much as 95% of substrate glucose is converted to gluconic acid.

Itaconic acid

Itaconic acid (Figure 21-11) is another organic acid produced by microbial culture that is of commerical importance. It is a "specialty item" and consequently commands a premium price. Its primary use is as a co-monomer in emulsion latexes, which are used in paper adhesives, caulking compounds and putty and in the manufacture of carpet backing.

Itaconic acid is produced by submerged culture of a strain of *Aspergillus terreus* isolated from a sample of Texas soil by Dr. K.B. Raper (Figure 21-12). The preferred substrate for itaconic acid production is molasses, which yields 50% to 60% of the usable sugar being converted to itaconic acid. Control of the pH of the culture is important during production. When the pH is kept below neutral, itaconic acid accumulates in the culture medium; however, if it rises above neutral, itaconic acid becomes a choice growth substrate for *A. terreus* and is quickly consumed.

Lactic acid

Lactic acid is a large-volume, specialty chemical that is produced in part by microbial fermentation. It is also synthesized chemically from lactonitrile and acetaldehyde by the Monsanto Chemical Corporation. Lactic acid is used in the manufacture of plastics and in the food industry as an acidulant and a flavoring agent. It is also used in the pharmaceutical industry and in leather tanning.

Industrially *Lactobacillus delbrueckii* is used for the microbial production of lactic acid (Figure 21-13). During fermentation, glucose is converted to two molecules of lactic acid. Since this is a true microbial fermentation, oxygen is not required; thus there is no need to pump air into the culture. The optimum temperature of the lactic acid fermentation is 49° C, which inhibits the growth of most other organisms. For this reason the maintenance of strict sterile conditions is not as important as with most other fermentations. Substrates that can be fermented for the production of lactic acid include molasses, whey, and glucose. Yields of lactic acid equalling 85% of the carbohydrate in the substrate have been obtained.

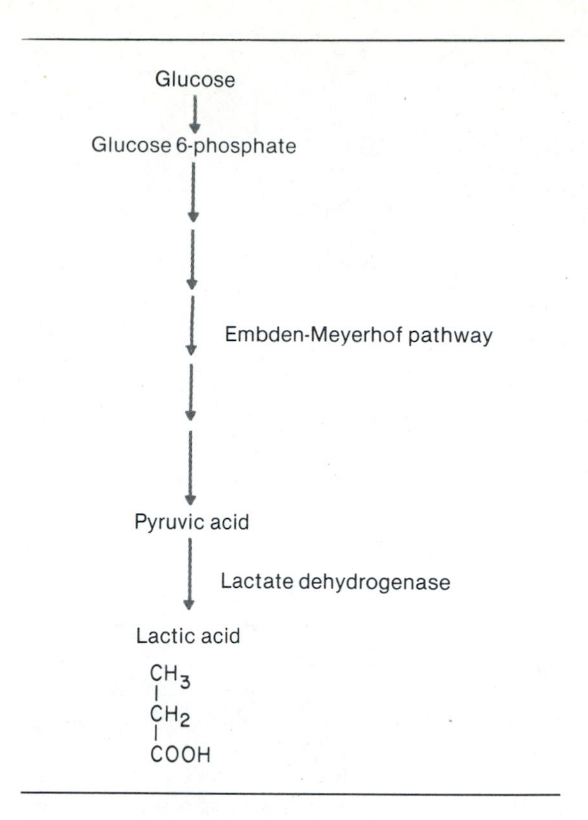

Figure 21-13 Lactic acid pathway of *Lactobacillus delbrueckii.*

Ethanol

Recently there was great interest, especially in the Midwest, in producing ethanol from grain starch by microbial fermentation. This ethanol would be blended with gasoline in a ratio of one part ethanol to nine parts gasoline and be used to fuel motor vehicles. At the height of interest in this process there was a bumper corn crop and fuel prices were increasing rapidly. The rationale for producing the ethanol-gasoline mixture called ***gasohol*** was to create a market for the excess corn and perhaps help reduce or stabilize fuel prices. The production of gasohol was not successful from an economic point of view because the cost of ethanol production was more than that of gasoline, so gasohol was more expensive than straight gasoline. Gasohol, however, continues to be sold in isolated areas of the Midwest.

Ethanol production for gasohol is based primarily on the fermentation of glucose from corn starch by the yeast *Saccharomyces cerevisiae.* Since this yeast lacks amylases, corn starch must first be treated with amylase to provide fermentable substrates. The substrates are fermented to two ethanol and two carbon dioxide molecules per glucose equivalent of substrate used. The fermentation stops when the concentration of ethanol in the culture liquid approaches about 15%; thus the culture fluid must be distilled to obtain the pure alcohol that must be used to blend with gasoline. The distillation step is expensive in terms of energy required and is the primary obstacle in preventing mass production of ethanol for fuel purposes by fermentation.

The production of ethanol from cellulose, a component of plant material, is being seriously researched today, but a major problem with this is the conversion of cellulose to a fermentable substrate. Many organisms have been described

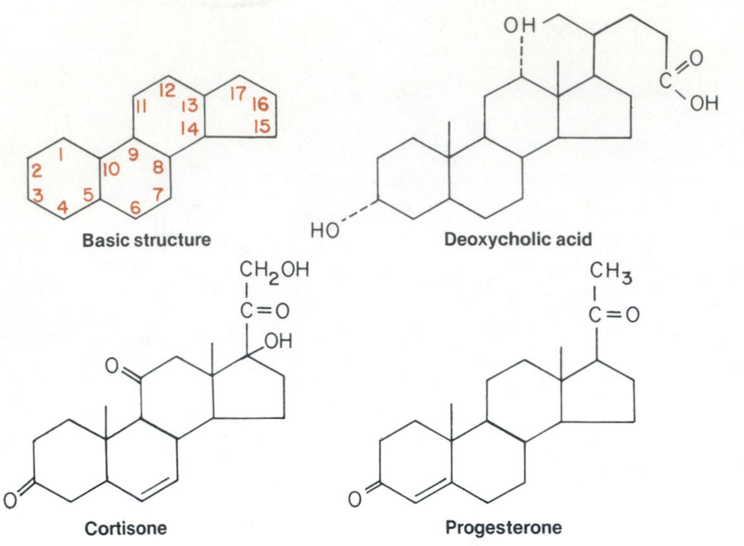

Basic structure
Deoxycholic acid
Cortisone
Progesterone

Figure 21-14 Steroid structures.

that produce enzymes that attack cellulose, but so far no enzymatic process for the conversion of cellulose to glucose has been developed on a commercial scale. The most promising at this point is based on the complex of cellulolytic enzymes produced by the mold *Trichoderma reesei* (formerly *T. viride*).

MICROBIAL TRANSFORMATION OF STEROIDS

The demonstration by Hench and co-workers in 1949 and 1950 that the cortical hormones, cortisone and cortisol, isolated from adrenal glands are effective in the treatment of rheumatoid arthritis created a tremendous demand for these steroids. Initially most of the steroid hormones were obtained from animal tissue, but a chemical synthesis was developed by Merck and Company in 1949. The starting material for this synthesis was deoxycholic acid, and it took 30 steps to synthesize cortisone (Figure 21-14). The most difficult portion of the synthesis was the attachment of an oxygen atom to carbon atom 11, which took 10 to 12 steps. The yield of cortisone was initially very low. In 1952 workers at the Upjohn Company reported that *Rhizopus nigricans* would specifically hydroxylate carbon 11 of progesterone to yield 11 α-hydroxyprogesterone. This introduced the use of microorganisms as catalysts to carry out specific transformations of steroid compounds. It also initiated a large-scale search for additional organisms to carry out other steroid transformations. That search has been fruitful, and today there are hundreds of organisms known to carry out specific modifications of steroids.

There are important advantages to using microbes to carry out chemical reactions in the synthesis of steroid hormones. The transformations are extremely specific, with yields approaching 100% of the substrate being converted to product. Additionally the transformations are relatively rapid and easy to carry out on an industrial scale. These advantages were immediately recognized by industry; thus microbial transformations became an integral part of steroid hormone synthesis. Examples of microbial transformations that are of industrial importance are the C-11 oxygenation by *Rhizopus nigricans* and other fungi, the C-16 α-hydroxylation by streptomycetes, the dehydrogenation of the A ring of glucocorticoids and analogs by *Arthrobacter simplex* and others, and the 3-hydroxydehydrogenation by many microorganisms.

In general the transformations are carried out by growing the transforming organism in cultures of up to 30,000/gallon in volume. When the culture becomes dense, near the end of exponential growth, the steroid compound to be transformed is added to the culture. After the transformation is complete, the steroid can be recovered from the culture medium by selective extraction with organic solvents. Since steroids are only slightly soluble in water, they are generally dissolved in water-miscible solvents before addition to the culture. It has been found that solid steroids in the form of fine powders can be added directly to the culture, and transformation will occur almost quantitatively.

Microbial transformation continues to play an important role not only in the manufacture of cortisone, cortisol, and their derivatives but also in the manufacture of steroid hormones used as antifertility agents.

MICROBIAL INSECTICIDES

Insects have been a serious threat to food production and storage since the earliest days of agriculture. They are also important vectors in the transmission of disease such as malaria, encephalitis, yellow fever, and bubonic plague. The development of chemical insecticides such as DDT provided the first major hope that insects could be eliminated as an important threat to the

Plate 20 Algal bloom in cyprus swamp.
Courtesy William R. Walker, Virginia Polytechnic Institute and State University.

Plate 21 Massive amounts of algae scooped from waterway that has received excessive nutrients.
Courtesy William R. Walker, Virginia Polytechnic Institute and State University.

Plate 22 *Herpothallon sanguineum* lichen growing on tree trunk in Amazon Valley.
Photo by Kjell Sandved.

Plate 23 *Cora pavonia* and *Parmelia enderythrea* lichen.
Photo by Kjell Sandved.

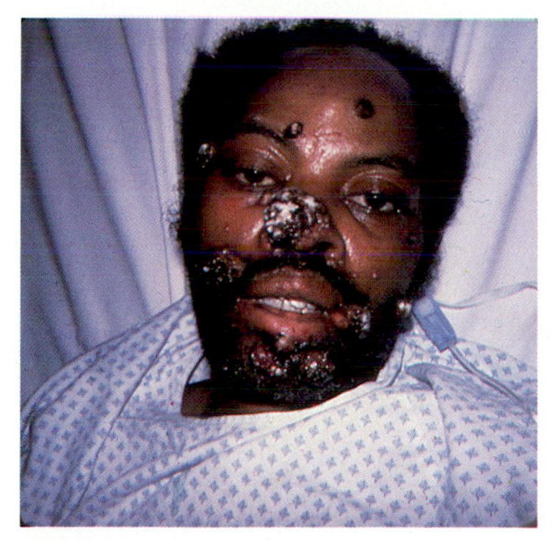

Plate 24 Disseminated coccidioidomycosis.
Courtesy William Martin.

Plate 25 Spoiled milk.
Photo by Karen Siebert.

Plate 26 Blue cheese.

Courtesy Nauvoo Cheese Co.

Plate 27 Spoiled meat.

Photo by Karen Siebert.

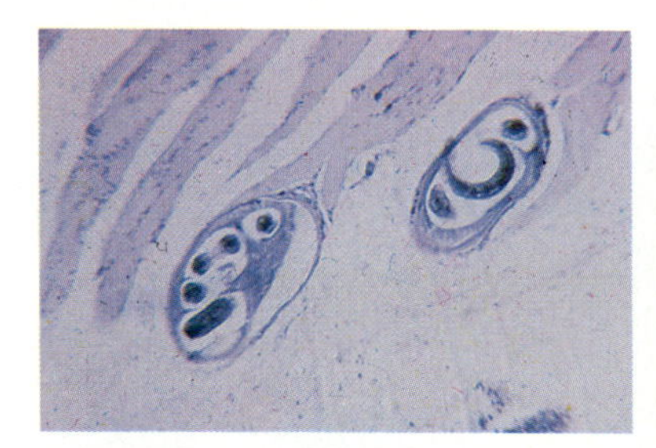

Plate 28 *Trichinella spiralis* larvae encysted in muscle.

From Finegold, S.M., and Martin, W.J. Diagnostic microbiology *(6th ed.).*
The C.V. Mosby Co., St. Louis, 1982.

Plate 29 Spoiled fruit and vegetables.

Photo by Karen Siebert.

Plate 30 Pinot Noir with *Botrytis*, also called "noble rot." This fungus causes increase in sugar levels of grapes, facilitating production of sweet dessert wines.

Photo by R.G. Peterson. Courtesy Monterey Vineyards, Gonzales, Calif.

wellbeing of humankind. This hope has not been entirely realized for several reasons. One important reason is that the target insects developed resistance to the insecticides. Second, the insecticides lacked specificity; that is, they affected a wide variety of insects, both harmful and helpful species, as well as other animals. Massive application of the long-lived chemical insecticides has caused their accumulation in the food chain. As a result of this, populations of several predator animals were seriously depleted to the verge of extinction. Consequently, the use of some of the most effective insecticides, such as DDT, was banned in the United States. These and other reasons have prompted a search for effective alternatives to chemical insecticides.

Several microbes are either insect pathogens or they produce toxins that affect insects and consequently are being considered or are used as insecticides. Microbial insecticides offer several potential advantages. In general they are specific, affecting only a limited group of insects. They are relatively short-lived in the environment, which is an advantage because they do not accumulate, but a disadvantage in that their insecticidal action is of short duration. Perhaps most important, many potent microbial insecticides appear to affect only insects and have no harmful effect on a variety of test animals, including both mammals and birds.

Probably the most widely used microbial insecticide is the bacterium *Bacillus thuringiensis,* although viruses, fungi, and other bacteria have potential for widespread use. *B. thuringiensis* is prepared by several companies and sold under trade names such as Thuricide. It appears that the insecticidal agent is not the bacterium itself but a protein toxin that it produces. In the susceptible insect the toxin disrupts the gut cells and induces paralysis. It is effective against more than 100 caterpillar pests including the cabbage worm, cabbage looper, alfalfa caterpillar, gypsy moth, and tobacco budworm.

Other *Bacillus* species are also effective insecticides. *B. popilliae* is effective in the control of the Japanese beetle and has been produced commercially for several years for this purpose. This bacterium apparently invades the Japanese beetle grub and multiplies profusely, causing what has been called "milky disease," which is generally fatal. Experimentally *Bacillus sphaericus* has been shown to be toxic to several species of mosquitoes.

Growth of microbes for the preparation of commercial insecticides varies from organism to organism. Certain microbes such as *Bacillus thuringiensis* will grow on artificial media and can be grown in large fermenters in submerged culture. When the culture reaches the proper stage, the cells or active agent are harvested and used to compound the insecticide. Other microbes, such as viruses, some fungi, and *Bacillus popilliae,* must be grown in living insects, which is technologically much more difficult than growing cells in submerged culture. A colony of the host insect must be maintained. Insects from the colony are infected with the microbial agent, which multiplies in the host insects. The microbial material used in compounding the insecticides is extracted from the infected insects. In the case of *B. popilliae* the material extracted from the infected insects is composed of endospores, whereas with viruses it is the infectious viral particles themselves.

With more and more attention being given to natural methods to control insect populations, more and more interest is being placed on microbial insecticides. As research continues and expands, it is likely that many important new microbial insecticides will be developed.

MICROBIAL POLYSACCHARIDES

Polysaccharides are produced by a wide variety of microorganisms. Those of primary industrial interest are found on the outside of the microbial cell wall and membrane and are called exopolysaccharides. They occur in two distinct forms: attached to the microbial cells that synthesize them as discrete physical structures termed capsules; or secreted from the cell into the surrounding environment in the form of soluble slime.

The chemical structures of the microbial polysaccharides fall into two classes: those composed of a single repeating structural unit, homopolysaccharides; and those constructed of two or more monomers, heteropolysaccharides. Homopolysaccharides of microbial origin are less abundant than heteropolysaccharides. Dextrans that are polymers of glucose are essentially the only group of microbial homopolysaccharides to have significant industrial applications at present. Heteropolysaccharides, which are composed of neutral sugars and, commonly, uronic acids, have numerous industrial uses.

The industrial uses of polysaccharides are based primarily on their ability to alter the rheological properties of aqueous solutions, either by gel formation or by alteration of flow characteristics. Polysaccharides are widely used by the food industry as suspending agents, as thickeners and as inducers of gel formation. The annual consumption of polysaccharides by this industry in the U.S. alone exceeds 36,000 tons. Generally, polysaccharides are used in instant foods, salad dressings, sauces, whips, toppings, processed cheeses, and dairy products.

An important potential use of microbial polysaccharides is in the recovery of tertiary oil by the petroleum industry. Pumping an aqueous polysaccharide solution into an oil well that has had most of the readily accessible oil removed significantly enhances the production of the well. Another use of polysaccharides by the oil industry is in drilling muds. The size of this latter market was recently quoted as 1830 tons annually.

The two microbial polysaccharides that are most widely used by industry today are dextran and xanthan. Other polysaccharides that are of potential use are curdlan, pullulan, and microbial alginate.

Dextrans are α-linked polymers of glucose produced by several bacterial species including *Klebsiella* spp., *Acetobacter* spp., streptococci, and *Leuconostoc* spp. Most dextrans for industrial purposes are produced by *Leuconostoc mesenteroides*. Dextrans differ in molecular weight, ranging from 1×10^7 to 3×10^8, and in the varying amounts of different α-linkage (e.g., α1-3-, α1-4-, or α1,6-). The two major uses of dextrans are as blood expanders and as the basis for a wide range of absorbents for use in the biochemical and pharmaceutical industries as well as in the research laboratories.

Xanthan is produced by *Xanthomonas campestris*. It is a complex heteropolysaccharide consisting of a substituted cellulose molecule with trisaccharide side-chains on alternate glucose residues. Xanthan is widely used for both food and nonfood uses. The major food use of xanthan is as a stabilizer in French dressing, fruit-flavored beverages, processed cheese, and other dairy products. It is also used in instant desserts, toppings and whips, and frozen and canned foods. Nonfood uses include as a stabilizer of paints and a suspending agent for laundry starch and in metal pickling baths and for paper finishing.

Alginates are currently produced commercially from brown algae. They consist of alternating D-mannuronic acid and L-glucuronic acid units. Recently it was discovered that *Azotobacter vinelandii* and *Pseudomonas aeruginosa* each produce alginates. There are several potential advantages to producing alginates by microbial fermentation. There would be no seasonal variations in production as occur with harvesting marine algae. Alginates produced by microbial fermentation would be free from the pollution now found in algal beds, and yields of known composition could be better assured.

Alginates have a wide range of applications with about 50% of that produced being used in the food industry where it serves as a stabilizer for ice creams, instant desserts, frozen custards, and cake mixes. It is also used in the brewing of beer for its ability to enhance "head" or foam formation. Alginates also have numerous applications in the pharmaceutical industry, primarily because of their value in forming stable emulsions.

Curdlan is a homopolysaccharide compound of β1-3-linked glucose units that is produced by *Alcaligenes faecalis*. It has been proposed for a number of applications including a gelling agent, a thickener, and a stabilizer in foods. Because it is not degraded in the human body, it has been suggested as an ingredient in low-calorie foods.

Pullulan is a homopolysaccharide composed of maltotriose units linked by β1-6-bonds. It is produced by the fungus *Aureobasidium pullulans*. Films made from pullulan are claimed to have antioxidant properties and have been proposed as a biodegradable material for food packaging and coating.

Although the present production of microbial polysaccharides is relatively small, the potential market is large, and new polysaccharides are continuously being developed.

PRODUCTION OF SINGLE-CELL PROTEIN

For many years scientists have toyed with the idea of producing edible protein from microbes. There are several outstanding advantages to doing so. In contrast to conventional sources of edible protein, such as poultry, pork, or beef, microbes double in weight in a matter of hours or less, whereas a baby pig or chicken may take weeks to do so. The facilities to produce large amounts of microbial or single-cell protein (SCP) require very little land, at most only a few acres.

Since the production of single-cell protein is capital intensive, it is normally uncompetitive pricewise with more conventional forms of protein such as soya and fish meal protein. The climatic changes in the 1960s that limited soybean and anchovy production caused the prices of these more conventional forms of protein to escalate dramatically. This opened the door for the production of SCP. Several projects were started primarily in Western Europe where the protein shortage was especially critical. These projects were based on growing microbes in large fermenters using a variety of growth substrates including starch, methanol, and hydrocarbons derived from petroleum. Technologically these projects were viable, but the climatic changes of the 1960s reversed themselves, allowing increased production of soya protein and fish meal, driving down their prices. At about the same time the price of oil rose sharply. These factors combined to make the cost of SCP production uncompetitive again. In addition to these factors there were problems with contamination of the final protein product in the processes that rely on hydrocarbons as growth substrate.

Today much of the SCP being produced commercially is done so on a small basis and for the most part uses waste materials as the growth substrate, although some is still produced using methanol as the growth substrate. In one process waste water from potato processing, which contains as high as 3% solids, much of which is starch, is used to grow a mixed culture of the yeasts *Candida utilis* and *Endomycopsis fibuliger.* The reason for the mixed culture is that *E. fibuliger* produces extracellular amylases, which break down the starch, allowing *C. utilis,* which produces no amylase, to grow. The *C. utilis* is the more desirable protein source, and it grows much faster than *E. fibuliger,* so that in the end it accounts for about 98% of the microbial mass in the mixed culture. The final product in which the biomass from the fermentation has been concentrated and dried is used primarily as an animal feed supplement.

The future of large-scale SCP production will depend on the price of competing protein sources, its acceptance as a human food, and the availability of inexpensive growth substrates.

PRODUCTION OF ALCOHOLIC BEVERAGES

Brewing

Humans have been making alcoholic beverages, including beer, for thousands of years, perhaps since the dawn of civilization. Today beer is consumed in vast quantities. In 1977 the annual per capita consumption of beer in the United States was 22.7 gallons for the population as a whole and 35.1 gallons when only persons over 21 years of age were considered. In terms of 12-ounce bottles, each adult in America consumed, on the average, about 370 bottles of beer in 1977, or slightly more than a bottle of beer every day.

Beer ingredients. Historically, beer was made from malt, hops, water, and yeast. Today many American breweries substitute adjuncts such as corn or rice starch for a portion of the malt.

Malt. Malt is made from germinated barley grains in facilities called malt houses. In Europe, most breweries have their own malt houses where they make malt by their own recipes. For the most part American breweries buy their malt from companies that specialize in malt production; however, some breweries do continue to make their own malt.

Malt is important in determining the character of beer. It contains enzymes, α and β amylases, that convert malt and adjunct starch into maltose and dextrins. The maltose is fermented to ethanol and carbon dioxide by the yeast. The ***dextrins,*** which are not fermented, contribute to the "body" of the beer; the more dextrins, the "heavier" the beer. Dextrins are also a major source of calories in beer. In the low-calorie beers, which have become popular in recent years, the dextrins have been removed by treatment with special enzymes, ***glucamylases,*** that convert the dextrins to sugars, which are fermented by the yeast. The low-calorie beers are often called "lite" or "light" beers, not only because they are low in calories but also because they are light bodied because of their lack of dextrins.

In addition to the amylase enzymes, malt contains proteins that are important in "head" formation and contribute to beer flavor. Beer color is also determined by the malt used. All beer, even the dark beers, is made primarily from light malt, which gives beer a yellow color. The darker beers are made by mixing darker malts with the light malt.

In the malting process barley grains are cleaned of extraneous material and placed in a steeptank where they are soaked with water for 2 days, during which time the grains swell. Afterward excess water is removed, and the swollen barley is transferred to the malting chamber for 5 to 7 days. The barley grains germinate in the malting chamber where high humidity and constant temperature are maintained. When the embryo reaches the proper size, the germinated barley is transferred to a kiln where the embryo is heat killed and the germinated grains, now called malt, are dried. The type of malt produced is determined by the amount of heat applied in the kiln. Light malt is heated to 80° C and dark malt to 105° C. Very dark malt is heated to over 200° C.

During the malting process the starch in the barley seeds is partially broken down into maltose and dextrins by the amylases, which are activated when the seeds germinate. On the average light malt contains 50% starch and 10% sugars. During light malt production the heating in the kiln does not totally inactivate the malt amylases; therefore when the malt is later mixed with water in the mash tub they continue to degrade the malt starch into dextrins and fermentable sugars. Dark malts provide mainly color and taste to beer and some starch.

Hops. Hops are added to beer primarily as a flavoring agent. They give beer its characteristic bitter taste and its aroma. The bitter taste is caused by acids and resins contained in the hops. The essential oils in hops contribute to the development of beer aroma and to some extent to the beer flavor. Hops also contain ***tannins,*** which contribute some bitterness to beer, but which serve mainly to cause coagulation of undesirable malt proteins, forming a sediment called ***break.*** Hops are the female flower of the hop plant, ***Humulus lupulus.***

Water. Beer is 85% to 92% water, which consequently is a very important factor in determining beer quality. Generally hard water makes the best beer. Water hardness is generally expressed as the concentration of calcium salts it contains. In Germany most beer is brewed with water with a hardness of 600 ppm, whereas the water used in English ales has a hardness of around 1800 ppm. In America most breweries adjust the water hardness to about 350 to 500 ppm. In addition to the proper hardness, the water used for beer should be unpolluted and pleasant to the taste. Because of this many breweries advertise that they use water from virgin springs or some other pristine source.

Yeast. Another critical element in determining the final quality of beer is the yeast used to ferment the sugar from the malt or adjunct to carbon dioxide and ethanol. Two general types of yeast are used: bottom-fermenting yeast and top-fermenting yeast. For the production of lager beers a bottom-fermenting strain of yeast is used. One such strain, *Saccharomyces carlsbergensis,* was discovered by workers at the Carlsberg Laboratories in Copenhagen, Denmark. Also strains of *Saccharomyces cerevisiae* are used. During fermentation the yeast cells settle to the bottom of the fermentation vessel, where they continue fermentation until the fermentable sugars are exhausted from the medium. Top-fermenting yeast strains are used in the production of ales. Generally speaking top-fermenting yeast work better at warmer temperatures (15° to 18° C) than do the bottom-fermenting strains (10° to 15° C) and are hardier. After a batch of beer has been fermented, the yeast are harvested and reused to inoculate fresh batches of beer. The harvested yeast are monitored carefully for vigor and contamination, and as long as they remain vigorous and uncontaminated, they may be reused many times.

Adjuncts. Adjuncts are added during the brewing process to increase the alcohol content of the beer. They are starchy materials that are enzymatically converted to fermentable sugars. Adjuncts add little or nothing to the taste, aroma, or color of beer. In Europe adjuncts are not used in brewing; hence malt is the only source of fermentable sugars. The all-malt beers have an intense yellow color and are heavy bodied. Americans seem to prefer beer that is pale yellow in color and lighter in body than European beer. These qualities can only be obtained if less malt is used. Because malt is the source of fermentable sugars a reduction in malt content would reduce the amount of sugar added and consequently would reduce the alcohol content of the beer if adjuncts were not added to compensate for the lost sugar. The most commonly used adjuncts are corn or rice starch; however, other products such as potatoes, cassava, soybeans, and sorghum have been used. Generally about 1.5 pounds of adjunct starch is added for each 3.5 pounds of light malt.

Brewing process. At the brewery malt is cleaned, milled to a powder, weighed, and then placed in the mash tub where it is mixed with water. Independently the adjunct, either ground corn or rice, is placed in a cooker with water and a small amount of milled malt. The temperature of the cooker is slowly brought to boiling. During the slow temperature rise amylases and proteases from the malt act, to some extent, on the adjunct starch and protein until these enzymes are inactivated when the temperature rises above 72° C. Boiling gelatinizes the adjunct starch, making it more amenable to breakdown by the malt amylases. After boiling, it is mixed with the wetted malt in the mash tub. The temperature is raised in the mash tube to stimulate starch breakdown by the malt amylases while proteases from the malt help breakdown undesirable malt proteins. When all the starch from

Figure 21-15 **A,** Brew kettles. Boiling of wort with hops in brew kettles is important in determining beer flavor. Additionally, it causes inactivation and precipitation of enzymes and protein in wort. **B,** Primary fermentation tanks. After being boiled in brew kettle, wort is cooled, filtered, and put in primary fermentation tanks where it is pitched, or seeded, with yeast. In these tanks, yeast consume fermentable sugars and produce carbon dioxide and ethanol.

Courtesy Anheuser-Busch, Inc.

the malt and the adjunct has been converted to dextrins and maltose, the temperature is raised to 72° C to inactivate the amylases. This temperature also causes some protein precipitation.

The solids in the mash are removed by transferring the mash to a ***Lauter Tub,*** where the solids are allowed to settle and the liquid is drawn off. The liquid, which is now called ***wort,*** is transferred to the brew kettle (Figure 21-15, *A*), usually made of copper, where hops are added and the mixture is boiled for about 1½ hours. The boiling serves several purposes. It promotes the extraction of the acids, resins, and tannins from the hops; it causes coagulation of unwanted proteins; it concentrates the wort; and finally it sterilizes the wort. The hot wort from the brew kettle is passed through a strainer to remove the hop reminants and into a settling tank to allow the coagulated protein and other insoluble material to settle out. The sediment, known as ***trub,*** is discarded. The clarified wort is passed through coolers to bring the temperature to about 20° C and is then passed to the fermenter (Figure 21-14, *B*), the yeast are added and fermentation takes place. Fermentation using bottom-fermenting yeast is carried out at 6 to 12° C and normally takes 8 to 10 days for completion. With top-fermenting yeast the temperature is higher (14 to 23° C), and the fermentation takes less time (5 to 7 days). When the fermentation is complete, the ***green beer,*** as it is now called, is carefully removed from the settled yeast in the fermenter to a storage tank, which is refrigerated to 0° C. The beer is stored here for a few weeks to several months, during which time unstable proteins, yeast, resins, and other undesirable substances precipitate from the beer. Esters are produced during this storage period, and the harsh taste of the green beer disappears.

After aging the mature beer is filtered, carbonated under pressure, pasteurized by heating for 5 minutes or more at 60° C, and canned or bottled. Some beer is not pasteurized but is filter-sterilized before being canned or bottled.

Types of beer. The most common type of beer made in the United States is ***lager beer.*** The word *lager* comes from the German verb *lagern,* which means to store. Lager beer quite literally is stored beer. It is produced by the process described above in which the green beer is stored in a refrigerated tank to allow it to mellow and mature. American lager beer is made from light malt and adjuncts so it has a pale yellow color and is light bodied. ***Bock beer*** is a lager beer that is made from a mixture of light and dark malts with little or no adjunct added. Generally bock beer is darker colored and heavier bodied than regular lager beer. European lager beers are generally made from light malt with little or no adjunct added. ***Pilsener beer*** is a type of lager beer that is made in Pilsen, Czechoslovakia. It is light in color, not sweet, with good hop aroma.

Ale, porter, and ***stout*** are all malt beverages made with top-fermenting yeast. Generally they are darker in color, heavier bodied, and hopped at a higher rate than is lager beer.

The low calorie beer being sold today is lager beer that has been treated to remove the dextrins by adding glucamylase to the malt-adjunct mixture while it is in the mash tub.

Malt liquor is brewed from malt without added adjunct and usually has a higher alcohol content than lager beer.

Wine making

Wine, like beer, has been produced for thousands of years and is produced by the fermentation of sugars to ethanol by yeast. The wine production process is much simpler than that for the production of beer because wine is made from fruit juices, which contain the fermentable sugars glucose and fructose and little or no starch, whereas beer requires that grain starch be converted to fermentable sugar.

The most important wine-producing regions of the world include Western Europe, the United States, and several South American countries. The major producing countries in Western Europe are Italy, France, West Germany, and Spain. In the United States, California produces by far the most wine, with New York state being the second most important wine-producing state. Both Chile and Argentina are developing into important wine producers in South America. They produce some very fine quality wines.

Production of grape wine. Grape wine is produced from selected varieties of grapes. White wines are generally produced from white grapes such as the ***Chardonnay*** variety, whereas red wine is made from red grapes such as ***Cabernet Sauvignon.*** Following is an outline of the production of red table wine.

Grapes are harvested when they reach the proper degree of ripeness. The sugar-to-acid ratio is an important determinant of the degree of ripeness. If the grapes are too ripe, there will be too much sugar and too little acid so that wine produced from such grapes will have a high alcohol content and will possess an insipid taste due to insufficient acid. Wines made from underripe grapes have reduced alcohol content and excess acid and consequently are not very desirable. Grapes at the optimum degree of ripeness usually give wine with an alcohol content of about 12.5% by volume (Plate 30).

The harvested grapes are crushed carefully so that only the skins are broken and not the stems and seeds which contain bitter tannins. The stems are removed from the broken grapes, now called must. ***Sulfite*** is added to the must to about 100 ppm to inhibit the growth of bacteria, undesirable yeast, and molds that may have been on the grapes when they were crushed. ***Saccharomyces cerevisiae*** var. ***ellipsoideus,*** which is the strain of yeast used to ferment the sugars in the grape juice to ethanol and carbon dioxides, is not inhibited by this concentration of sulfite.

Approximately 6 hours after sulfite treatment the crushed grapes in the fermentation tank are inoculated with a pure culture of the yeast. The contents of the fermentation tank are mixed, or stirred, twice a day during the early stages of fermentation to facilitate temperature equalization, the extraction of color and tannin from the grape skins, and the aeration of the must which is necessary for rapid yeast multiplication. Stirring is ceased when the right density of yeast has been reached, and the fermentation proceeds anaerobically.

The temperature of the fermentation is carefully controlled. The optimum temperature range for bouquet and aroma development is from 21° to 24° C. The fermentation proceeds for 3 to 5 days until the proper amount of color and tannin has been extracted from the grape skins. The winemaker decides when that point has been reached, and the wine is carefully drawn away from the skins and seeds and transferred to a second tank. The fermentation continues anaerobically in the second tank until the fermentable sugars are exhausted, usually within 11 days at a temperature of 21° C. As the alcohol content of the wine increases, various components such as yeast cells, tannins, proteins, and pectins form a sediment on the floor of the tank. After the fermentation is complete, the wine is carefully drained from the tank without disturbing the sediment and is transferred to ***aging casks*** where it is aged for several months to several years. The aging casks are generally made of white oak or sometimes of redwood. The bouquet of wine develops through the formation of esters from alcohols and acids during the aging process. The flavor, taste, and color of the wine are also affected by the aging process, which for high quality red wines may be as long as 5 years.

After the wine has been aged, it is filtered, bottled, and sold. Good wines continue to age for several years in the bottle.

OTHER INDUSTRIAL PRODUCTS

It would be impossible to discuss at length all of the industrial products that are produced by microorganisms. Some of these products have been discussed in Chapter 20, Food Microbiology. To make our list more complete, Table 21-4 outlines the various microbial species and products with which they are associated in industrial processes.

GENETIC ENGINEERING AND INDUSTRIAL MICROBIOLOGY

The industrial production of compounds may be accomplished in one of three ways: chemical synthesis, biological synthesis, or a combination of the two. The type of process to be used, if all three are available, usually is governed by the cost factor. The rising costs of raw materials such as petroleum for chemical synthesis and the advent of recombinant DNA technology (see Chapter 11) have made biological synthesis more attractive. It is expected in the near future that biological synthesis will replace many of the current synthetic chemical techniques.

The basic advantage of recombinant DNA technology is that the development of superior strains of an organism need not be based upon the randomness of spontaneous mutation or even induced mutations, which are time consuming and costly. Genetic engineering offers a technique by which genes can be exchanged among unrelated organisms, making possible the replacement of an existing metabolic pathway with one that will be more efficient in synthesizing a product for industrial use. The improvement of strains by genetic engineering may take different forms and includes the following.

Gene amplification

Enzyme production may be increased by increasing the number of copies of the gene (gene amplification) that codes for the enzyme. A microorganism containing one copy of a gene will produce 1X amount of the enzyme, but if the number of copies of the gene is increased to 20,

Table 21-4 Some important industrial products obtained with the aid of microbial activity

ORGANISM	PRODUCT
Pharmaceuticals	
Penicillium chrysogenum (mold)	Penicillins
Cephalosporium acremonium (mold)	Cephalosporins
Streptomyces species (bacterium)	Streptomycin, tetracyclines, kanamycin and others
Bacillus brevis (bacterium)	Gramicidin S
Bacillus subtilis (bacterium)	Bacitracin
Bacillus polymyxa (bacterium)	Polymixin B
Arthrobacter simplex (bacterium)	Steroid transformation
Rhizopus nigricans (mold)	Steroid transformation
Escherichia coli (bacterium, genetically engineered)	Insulin, interferon, human growth hormone, somatostatin
Polysaccharides	
Xanthomonas campestris (bacterium)	Xanthan gum
Leuconostoc mesenteroides (bacterium)	Dextran
Single-cell proteins	
Saccharomycopsis lipolytica (yeast)	Microbial protein from petroleum alkanes
Candida utilis (yeast)	Microbial protein from paper-pulp waste
Methylophilus methylotrophus (bacterium)	Microbial protein from growth on methane or methanol
Organic acids and solvents	
Saccharomyces cerevesiae (yeast)	Ethanol from glucose
Clostridium actobutylicum (bacterium)	Acetone and butanol
Aspergillus niger (mold)	Citric acid and gluconic acid
Aspergillus terreus (mold)	Itaconic acid
Lactobacillus delbrueckii (bacterium)	Lactic acid
Insecticides	
Bacillus thuringiensis (bacterium)	Bacterial protein toxin used against cabbage worm, cabbage looper, gypsy moth, and others
Bacillus popilliae (bacterium)	Endospores used against Japanese beetle
Vitamins	
Eremothecium ashbyi (yeast)	Riboflavin
Pseudomonas denitrificans (bacterium)	Vitamin B_{12}
Amino acids	
Corynebacterium glutamicum (bacterium)	L-Lysine and L-glutamic acid
Food and beverages	
Saccharomyces cerevesiae (yeast)	Wine, ale, sake, baker's yeast
Saccharomyces carlsbergensis (yeast)	Lager beer
Saccharomyces rouxii (yeast)	Soy sauce
Penicillium roquefortii (mold)	Blue cheeses
Penicillium camembertii (mold)	Camembert and brie cheeses
Streptococcus thermophilus (bacterium)	Yogurt
Lactobacillus bulgaricus (bacterium)	Yogurt
Propionibacterium shermanii (bacterium)	Swiss cheese
Gluconobacter oxydans subsp. *suboxydans* (bacterium)	Vinegar
Lactobacillus sanfrancisco (bacterium)	Sour french bread

the amount of enzyme produced will be 20X. A consequence of this would be a twentyfold reduction in the unit cost of the product of the enzyme. Recently the gene for α-amylase was cloned in *Bacillus subtilis*. The hybrid plasmid containing the cloned α-amylase gene was maintained at 50 copies per cell. The production of α-amylase by the strain carrying the cloned gene was 50 times that of the same strain lacking the cloned gene.

Vaccine production

Many disease-causing organisms and viruses are difficult to grow and often extremely dangerous to manipulate. The production of vaccines, on an industrial scale, from these organisms is often expensive and hazardous if possible at all. Removing the antigen-coding gene(s) from the disease-causing organism and placing them in nonpathogenic bacteria will permit less expensive and less hazardous vaccine production. The production of a vaccine for hoof-and-mouth disease in cattle has been accomplished through the use of genetic engineering and is commercially available, as is a vaccine against swine diarrhea.

Fermentation potential

There is a widespread demand for light beers that have reduced carbohydrate content. Malt wort is made up of approximately 22% dextrins, which are not fermented by naturally occurring strains of *S. cerevesiae*. Efforts are now underway to introduce genetic material for dextrin fermentation into *S. cerevesiae*.

Many of the fungi that carry out industrial fermentations are filamentous and form mats. Fungal mats become so intertwined in the large fermenter that efficient nutrient absorption is lost. Recombinant DNA technology may be applied to the development of single-cell fungi, which would make fermentation more efficient.

Product tolerance

Organisms that are selected because of their ability to produce copious amounts of a product are sometimes destroyed by the same product, for example, butanol. Recombinant DNA technology is being used to develop product-tolerant strains.

Product excretion

Some microorganisms do not excrete products, such as amino acids, in large quantities. It appears possible that through genetic engineering microbial strains with leaky membranes for product excretion can be developed.

The inability to excrete products such as enzymes is associated with gram-negative bacteria such as *Escherichia coli*, which is the host for several industrial products. Generally *E. coli* retains extracellular enzymes in the periplasm, and the cells must be treated to release them. This treatment also releases other soluble proteins present in *E. coli* and makes purification more difficult and expensive. *E. coli* also produces endotoxin, which contaminates the desired product and can be even more difficult to remove. Serious efforts are underway to develop host-vector systems using gram-positive bacteria that would excrete enzymes into the culture medium and would not produce endotoxins or other toxic substances.

Microbial production of nonmicrobial products

One of the most important potential uses of genetic engineering is to have microbes serve as factories for the production of specific products not normally produced by microbes. Many proteins produced in the human body have great potential in the treatment of human disease. For example, insulin is the only treatment for certain types of diabetes, and interferon has potential in the treatment of some forms of cancer. Fortunately insulin can be obtained from other than human sources, but interferon active in humans must be obtained from humans. To obtain interferon in sufficient amounts for evaluation, human interferon genes have been cloned into bacteria where interferon can now be produced in large amounts. This increased production makes it possible to investigate its potential as a therapeutic agent. Other nonmicrobial proteins of interest are human peptide hormones, such as growth hormone, which are difficult to isolate in sufficient quantity to study fully their effects.

SUMMARY

The products of industrial microbiology can be divided into the following groups (Table 21-4).

1. Pharmaceuticals, which include antibiotics, steroids, insulin, human growth hormone, somatostatin, and interferon
2. Bioinsecticides
3. Polysaccharides such as dextran
4. Vitamins such as riboflavin
5. Single-cell proteins
6. Industrial chemicals such as ethanol and critic acid
7. Food and beverages such as cheeses, alcoholic beverages, yogurt, and vinegar

Industrial products are synthesized by chemical and biological means or by a combination of the two. Petroleum products, which are becoming more expensive, are substrates for many chemical syntheses, and therefore biological processes are beginning slowly to replace them. Microorganisms can be used for industrial purposes if the cost of making the product is less than that for chemical synthesis. Factors that influence the cost are many and include (1) the cost of the raw materials, (2) the ease of isolation of the product from the organism, and (3) the amount of product. Genetic engineering promises to be the major force that will promote the transition from chemical synthesis to biological synthesis. Microbiologists now have the tools for gene amplification, replacement of regulatory mechanisms, and creating microorganisms with unlimited properties for efficient product synthesis.

Today microorganisms are used in the production of a wide variety of commercial products, the yearly value of which is measured in the billions of dollars worldwide. Many of these are end use products; i.e., the microbial product itself is sold directly to the consumer, for example, antibiotics. Some microbial products are used in processes for the production of consumer products. Many of the microbial enzymes fall into this category. α-Amylase, an example of one such enzyme, is used in several industrial processes such as the production of fructose from starch and the production of alcohol from starch. Microbial cells themselves are used in certain industrial processes for the production of consumer products of nonmicrobial origin. An example of this is the use of microbial cells to carry out specific transformation reactions in the synthesis of steroid hormones such as cortisone. In some instances microbial cells are used as consumer products. Microbial insecticides often are based on viable preparations of bacterial endospores.

Study questions

1. What are the three most important industrial amino acids? Which of these amino acids is produced by microbial fermentation, and what are the primary uses of industrial amino acids?
2. Why was it important to Kinoshita to develop a simple method for screening microbes for their ability to excrete glutamic acid? What is the basis for the screening procedure he developed?
3. Name four methods of employing immobilized enzymes. What are the advantages of each?
4. What are the two major schemes being followed to develop new antibiotics? Which holds more promise, and why?
5. What is the importance of microbial transformations of steroids?
6. Why are hops added to beer? What is an adjunct, and why are adjuncts added to beer? In the making of malt why are the barley seeds allowed to germinate?
7. Why is winemaking a simpler process than beermaking? What do the yeast cells contribute in winemaking? Why is it important to pick the grape for winemaking at the proper degree of ripeness?
8. Name the industrial organic acids that are produced by microbial fermentation today. What are the major uses of each? Why are they produced by microbes and not chemically?
9. What are the major industrial microbial polysaccharides and their uses?
10. Why is there little single-cell protein being produced on a large-scale basis today?
11. What are the advantages of microbial insecticides over chemical insecticides such as DDT? The disadvantages? Why is *Bacillus thuringiensis* insecticidal?

Selected readings

Books

Biomass Process Handbook. Technical Insights, Inc., Fort Lee, N.J., 1981.

Bull, A.T., Ellwood, D.C., and Ratledge, C. Microbial technology: current state, future prospects. Cambridge University Press, Cambridge, England, 1979.

Ghose, T.K., and Fiechter, A. Advances in biochemical engineering. I. Springer-Verlag New York, Inc., New York, 1971.

Hugo, W.B., and Russell, A.D. Pharmaceutical mirobiology. Blackwell Scientific Publications, Oxford, England, 1977.

Miller, B.M., and Litsky, W. Industrial microbiology. McGraw-Hill Book co., New York, 1976.

Norris, J.R., and Richmond, M.H. Essays in applied microbiology. John Wiley & Sons, Inc., Chichester, England, 1981.

Peppler, H.J. Microbial technology. Reinhold Publishing Corp, New York, 1967.

Peppler, H.J., and Perlman, D. *Microbial technology.* Vol. 1, *Microbial processes;* Vol. 2, *Fermentation technology.* New York, Academic Press, Inc., 1979.

Prescott, S.C., and Dunn, C.G. Industrial microbiology (3rd ed.). McGraw-Hill Book Co., New York, 1959.

Smith, J.E., and Berry D.R., The filamentous fungi (Vol. 1). Industrial mycology. Halsted Press, New York, 1975.

Wang, D.I-C., et al. Fermentation and enzyme technology. John Wiley & Sons, Inc., New York, 1979.

Yamada, K., Kinoshita, S., Tsunoda, T. and Aida, K. The microbial production of amino acids. Halsted Press, New York, 1972.

Journal article

Industrial microbiology. Sci. Am. **245:** Entire issue, 1981.

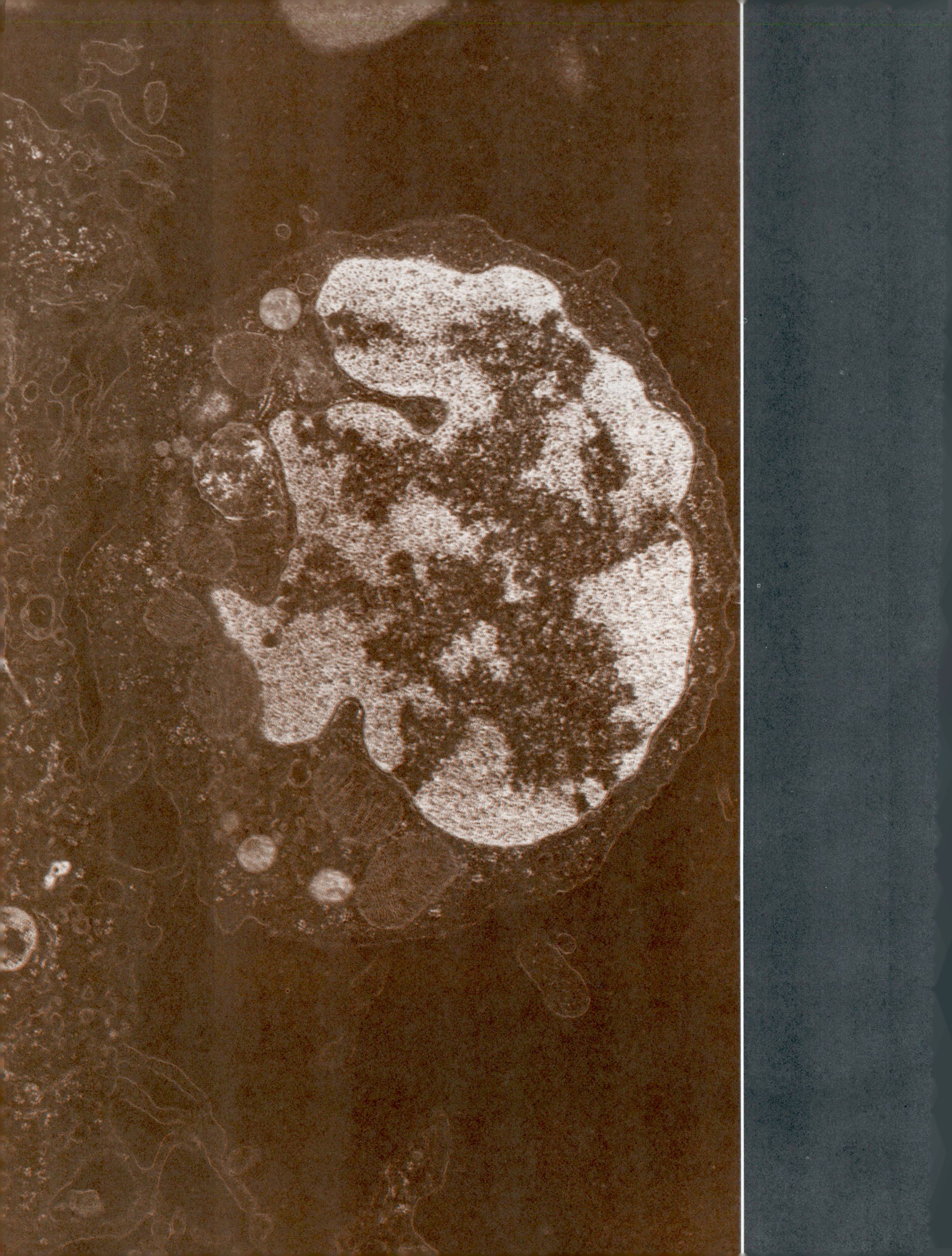

PART 6

INFECTION AND IMMUNITY

Lymphocyte adhering to macrophage. (×23,500.)
R. Rodewald, University of Virginia/BPS.

Chapter 22
HOST-PARASITE INTERACTIONS

A silent but constant "war" takes place on and within the human body. Our bodies are colonized and can be colonized by microorganisms whose activities, if not controlled by certain natural host defense measures, could lead to disease. This is a concept that is sometimes difficult to imagine for the healthy individual because the "silent war" produces no overt bodily responses or damage to the host. Every microorganism is equipped to cause disease, but some, to our consternation, are better equipped than others. Similarly, the host is equipped with an arsenal of physical and chemical factors to prevent any deleterious effects that might arise from microbial colonization. This host resistance can vary from individual to individual, depending on genetic (innate) as well as environmental factors. The purpose of this chapter is to elaborate on those methods used by microorganisms to cause disease and the nonspecific (do not involve specific antibodies) techniques employed by the host to resist infection and disease. The specific imune mechanisms used by the host to resist infection and disease are discussed in succeeding chapters.

CONCEPTS OF HOST-PARASITE RELATIONSHIPS

Commensals

Areas of the human body such as the skin, mouth, gastrointestinal tract, nose, throat, and genitals are the abode for a large number of microorganisms, especially the bacteria. Very few species of fungi or protozoa are indigenous to humans.* Organisms indigenous to the host are called ***commensals,*** and the relationship is one in which the commensal benefits and the host is neither helped nor harmed. Each ecological niche in the body that is inhabitable by microorganisms is made up of a large number of commensal species. The kind of microorganisms that can inhabit each niche is influenced by the chemical and physical characteristics of the microorganism, the chemical and physical characteristics of the host habitat, and other microorganisms already present in the habitat. The area between the gums and the tooth surface (gingival crevice), for example, is an anaerobic environment in which only anaerobic or facultatively anaerobic microorganisms are equipped to handle low levels of oxygen. If the organism residing in the area has some specific nutritional requirement, the nutrient must be supplied by the host or by other microorganisms that reside in the same habitat. The bacterium *Bacteroides melaninogenicus,* for example, is an inhabitant of the gingival crevice and requires the growth factor vitamin K, which is produced by other resident microorganisms (see Chapter 17).

The total number of commensal species, within certain limits, remains relatively constant at specific sites in the host. This constancy is maintained by the limitation of nutrients produced in the host or by metabolic products excreted by neighboring commensals. The products of some microorganisms, which are inhibitory to the growth of other microorganisms, help to maintain certain population levels. Many commensals ferment sugars, which results in the production of acids that reduce the pH of the environment to a level that is inhibitory to the growth of some microorganisms. Other commensals produce compounds, such as hydrogen peroxide or fatty acids, that are lethal to some microorganisms attempting to colonize a surface. Usually, as a rule of thumb, gram-positive bacteria control the number of gram-negative species and vice

* Those microorganisms that are part of the normal microflora of humans are described in Chapter 33, Tables 33-1 through 33-4.

versa, and the presence of both groups influences the concentration of yeasts. How this system works can be best explained by an example. Certain chemotherapeutic drugs, such as antibiotics, can severely reduce the population of gram-positive and gram-negative bacteria in many body sites, such as the oral cavity. Yeast cells, without the controlling influences of gram-positive and gram-negative bacteria, exhibit unrestricted growth and create a disease condition.

Parasites

A parasitic relationship is one in which the microorganism benefits but the host is harmed. Most parasitic relationships occur when nonindigenous microorganisms colonize the host, but occasionally commensals can also become parasites if the host has some underlying condition or disease that impairs normal defense mechanisms. Microorganisms that cause disease are called ***pathogens,*** but some are more pathogenic than others. The ability to become a pathogen is often related to many factors of either host or microbial origin. Certain strains of *Escherichia coli,* for example, that colonize the intestinal tract owe their pathogenicity not only to their ability to adhere to intestinal epithelium but also to the production of a toxic substance. The absence of one of these properties usually renders the microorganism harmless.

Two terms that are sometimes misinterpreted when considering parasitic interactions are infection and disease. ***Disease*** is defined as any disturbance of the structure or function of an organism. Diseases caused by microbial agents are called ***infectious.*** When you go to a doctor to find out the cause of your runny nose or sore throat, he invariably will say that you have a viral or bacterial infection. An infection, however, is not the same as a disease, even though the two terms are used interchangeably. For example, a considerable percentage of the population carry staphylococci in the anterior nares. The carriage may last a few days, weeks, or even months. If the organisms do not cause the formation of antibodies or elicit a hypersensitive response in the host, they are merely ***colonizers.*** If the colonizers cause the host to respond by producing antibodies or elicit a hypersensitive response, the organisms have caused an ***infection.*** Organisms that colonize the host and initiate overt manifestations such as fever, swelling, abscess formation and other responses produce a condition called ***disease.*** The distinction between disease and infection is important because treatment may be required for the former and not the latter. Unnecessary treatment using toxic drugs, expensive drugs, or drugs that may give rise to secondary infections is not in the best interest of the patient. The distinction between infection and disease is not always clearcut. The distinction may depend on the microbial species as well as on the condition of the host.

The degree to which a pathogen causes damage is sometimes referred to as ***virulence,*** but this quantitative aspect is difficult to measure. The term ***pathogenicity,*** or ***pathogenesis,*** is not quantitative, but it is often used interchangeably with virulence to indicate a microorganism's relative disease-producing qualities. These disease-producing traits can be maintained while the microorganism replicates in the host, but they may be lost if the organism is cultured outside the host. Bacteria can lose their virulence by repeated passage on laboratory media, and the virulence of viruses is affected by repeated passage in cell culture. Some bacteria owe their virulence to the production of a capsule or slime layer that prevents digestion by certain cellular elements in the blood called phagocytes. Repeated passage on laboratory media can result in the formation of mutants that lose the property of capsule production and are better suited to growth on artifical media than in the host. This reduction or loss of virulence is called ***attenuation.*** Attenuated microorganisms exhibit nearly all of the other properties of the virulent strain except that they are unable to cause disease. Attenuation is a technique used in the production of vaccines, which are microorganisms or material derived from microorganisms or other parasites such as viruses that induce the host's immune system to resist disease. Vaccines have reduced the incidence of infectious diseases and prevented early deaths from such afflictions as smallpox, polio, and diphtheria. Vaccines are discussed in Chapter 23.

MECHANISMS OF PATHOGENESIS AND HOST RESISTANCE

There are four conditions that must be satisfied before an infectious agent can cause disease: it must (1) gain entry into the host, (2) colonize host tissue, (3) resist the defenses of the host, and (4) damage host tissue.

Entry of microorganisms into the host

There are several avenues for the entry of microorganisms into the host: the skin, respiratory tract, genitourinary tract, and intestinal tract.

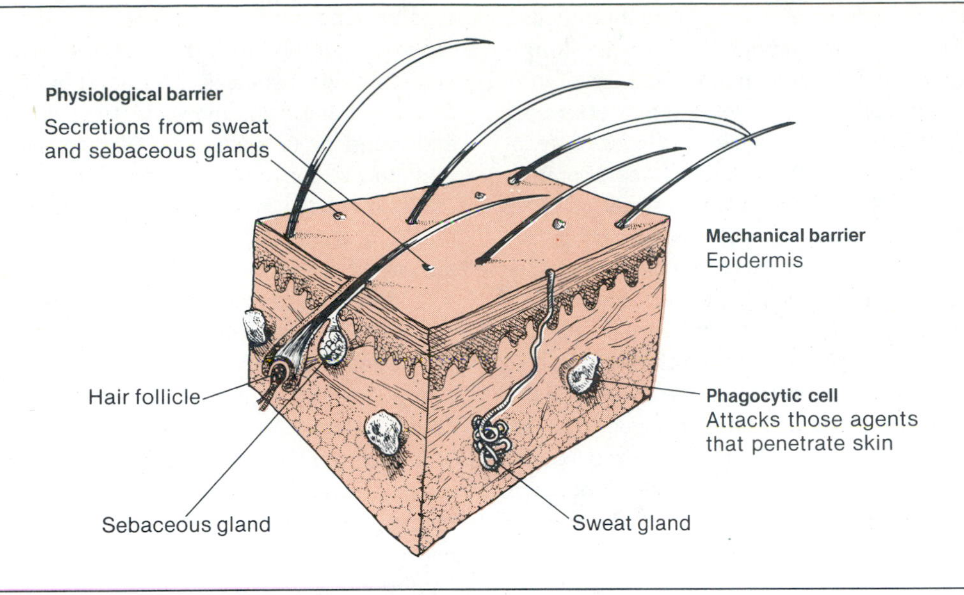

Figure 22-1 Cross section of skin demonstrating barriers to infectious agents.

Skin and mucous membranes. The host's first line of defense is the intact skin (Figure 22-1) and mucous membranes. The skin, because of its thick keratinized layer, is the most effective barrier to entry by foreign agents. Microorganisms that do adhere to the skin are easily removed by washing or by the natural shedding (desquamation) of epithelial cells. Secretions of the sebaceous glands and sweat glands contain antimicrobial factors, in the form of fatty acids or other organic acids, that reduce the pH. Acidic pH is an inhibitor to the growth of many microorganisms, particularly bacteria. The continuity of the skin is broken in such areas as the eye, genitourinary tract, and mouth, but these areas possess other barriers to infection. The conjunctiva is a transparent layer that protects the eye from invasion by microbial agents by the flushing action of tears. Tears, as well as other body fluids, also contain ***lysozyme,*** an enzyme capable of breaking down the peptidoglycan layer of bacterial cell walls. The mucous membranes of the respiratory tract, genitourinary tract, and intestinal tract are barriers to microbial invasion and are discussed later.

The hard epithelial layer of the skin can be penetrated when there are abrasions or wounds, or when a skin condition, such as eczema, alters the continuity of the skin and makes it susceptible to infection. Most infections of the eye are also due to abrasions of the conjunctiva followed by touching the eye with contaminated fingers or objects. The epithelial barrier can also be broken by biting insects or animals. Fleas, ticks, mosquitoes, and lice are arthropods that can penetrate the skin and inject the agents of such diseases as malaria, plague, and encephalitis into the bloodstream. Animal bites or scratches are responsible for such diseases as rabies and cat scratch fever. Indigenous species of microorganisms on the skin can cause infections such as acne, but only under abnormal physiological conditions in the host.

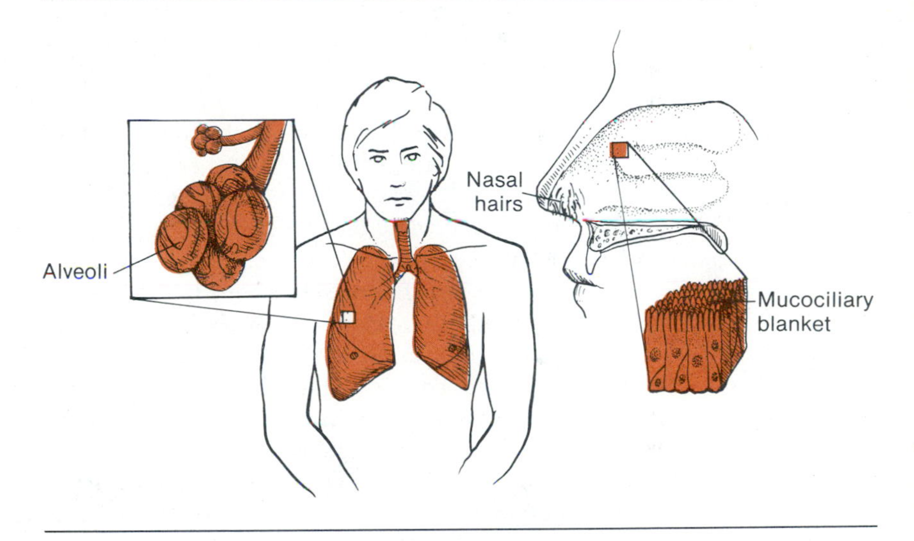

Figure 22-2 Barriers to infectious agents in respiratory tract.

Respiratory tract. A very large proportion of infectious diseases seen in the world today are respiratory in nature (Figure 22-2). The respiratory tract possesses several innate mechanisms for handling the presence of foreign agents, including microorganisms suspended in the air. A ***mucociliary blanket*** covers much of the respiratory tract and acts as a trap for foreign matter. It consists of mucus-secreting cells that contain ***lysozyme,*** and ***cilia,*** which are projections of epithelial cells. The mucociliary blanket traps particles up to 10 μm in diameter and expels them by using the sweeping motion of the cilia. Particles larger than 10 μm are usually trapped by hairs lining the nasal cavity. Particles in the lower respiratory tract are swept up to the throat; particles in the upper respiratory tract are swept to the back of the throat and then swallowed. The ***alveoli,*** which are the terminal portions of the lower respiratory tract, do not possess a mucociliary blanket, but they are lined with ***phagocytic cells (macrophages)*** that ingest foreign matter.

The successful colonization of the respiratory tract by microorganisms is often related to abnormal physiological conditions in the host. Disturbance or removal of the mucociliary blanket by prior conditions, such as bronchitis or viral influenza, can lead to microbial colonization of the unprotected epithelial cell layer.

Intestinal tract. Infections of the intestinal tract occur most often from the ingestion of contaminated food or water. Any exogenous (noncommensal) microorganisms that enter the intestinal tract in this manner will encounter many obstacles before they can establish a site for colonization. Epithelial surfaces, such as gingiva (gums) and the hard palate are ***keratinized*** and offer a mechanical barrier to infectious agents. The ***mucus*** coating the oral cavity, including the teeth, helps to trap microorganisms, and the saliva that is formed carries microorganisms to the back of the throat, where they are swallowed. Once microorganisms are swallowed, many are rapidly destroyed by ***stomach acids.*** Those microorganisms, however, that do survive this acidic cauldron are transported to the intestinal tract, where ***bile salts*** and ***intestinal enzymes*** can attack them. Even if the infectious agent survives these barriers, other host factors can contribute to its demise. Mucus, for example, lines the epithelium of the intestinal tract and acts as an obstruction to microbial adherence and penetration. Part of the intestinal secretion contains a special class of antibodies called ***secretory IgA (sIgA)*** that protects the immune individual against infection. (See Chapter 23 for a discussion of IgA.) These factors, combined with the ***peristaltic action*** of the intestine, which moves intestinal contents from the upper to the lower intestine, are major obstacles to colonization by infectious agents. Indigenous microorganisms can also be of importance in protecting the host from invasion by noncommensal species. The number of commensals that inhabit the intestinal tract is staggering, and some groups, such as the *Bacteroides,* can be found in concentrations as high as 1×10^{10}/g of feces. These species release products of metabolism, such as fatty acids, that prevent "outsiders" from colonizing the intestinal surface.

The elimination of commensal species or the reduction in their number by certain antimicrobial agents can reduce the intestinal population to a point where the commensal species exert no effect on ingested microorganisms.

Microorganisms can infect the intestinal tract if they possess special properties that enable them to resist host defense mechanisms, or if the natural barriers of the host have been altered by an underlying disease or condition. Many viruses, such as the enteroviruses (the poliovirus is an enterovirus), are resistant to acid, bile salts, and the action of intestinal enzymes. There are conditions in which the acid content of the stomach may be reduced, thereby permitting the entrance of acid-sensitive microorganisms. Microorganims sensitive to acid can also be protected by entrapment in large particles of food that escape the effects of acid in the stomach. Infection may also be influenced by the number of pathogens ingested, as in the case of salmonellal food poisoning, where the symptoms appear only after ingestion of microorganisms above a minimal number.

Genitourinary tract. The urinary tract and bladder are essentially sterile environments that, in the healthy individual, are not invaded by microorganims. The absence of microorganisms in these environments is due primarily to the flushing action of urine. The epithelial cells of the vagina, between puberty and menopause, contain large amounts of glycogen, which is metabolized to acids that prevent colonization by microorganisms other than commensals. Urinary tract infections are still more prevalent in females because of the shortness of the urethra and the proximity of the urethra to the anus, where gram-negative bacterial species are abundant. The vaginal secretions before puberty and after menopause are slightly alkaline, which also provides an environment for the growth of unwanted microbial species. Obstruction of the urinary tract, which prevents the free flushing action of urine, is an important reason for the appearance of urinary tract diseases. Abrasions resulting from the insertion of catheters and other instruments as well as from sexual intercourse can also lead to the introduction of potentially pathogenic microorganisms and their colonization of the genitourinary tract. Some microorganisms, such as the agent of gonorrhea, can overcome the natural barriers to colonization because of special attachment mechanisms on the microbial surface.

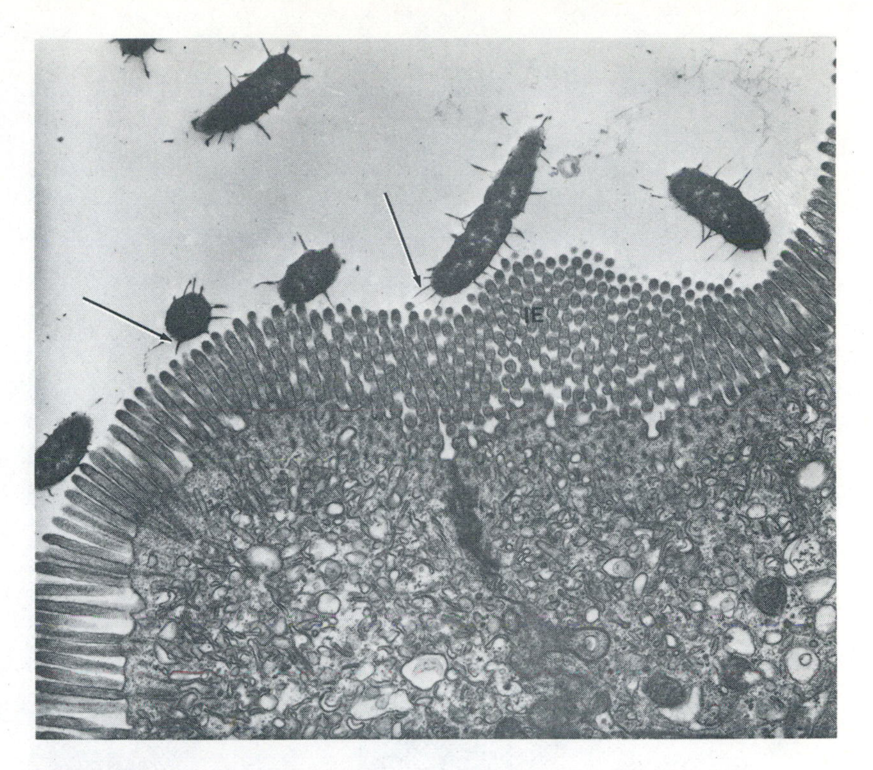

Figure 22-3 Electron micrograph demonstrating attachment of *E. coli* to intestinal epithelium *(IE)*. Arrows point to attachment-pili of microorganism.
From Moon, H. J. Infect. Dis. ***136:****S124, 1977.*

Colonization and multiplication of pathogens in the host

Growth on epithelial surfaces. The indigenous microbial population of the host as well as nonindigenous parasites colonize epithelial surfaces by highly specific interactions. These interactions involve microbial surface components referred to as ***adhesins.*** The adhesin may be a capsular sugar complex, fimbriae, or other components. Little is known of the molecular architecture at the site of microbial adherence and only now are microbial adhesins being isolated, purified, and characterized as to their role in the infectious disease process. Some of the parasites in which surface components have been implicated in the attachment process follow.

Escherichia coli. The bacterium *E. coli* is normally a commensal in the intestinal tract, but pathogenic strains also exist. Both pathogenic and nonpathogenic strains produce fimbriae. Chromosome-mediated fimbriae called type I enable the organism to adhere to specific epithelial cells (Figure 22-3). The host receptor is believed to be a mannose-containing carbohydrate that can be found on several eukaryotic cells. The fact that leukocytes also possess mannose-containing receptors implies that adhesion may also contribute to the organism's destruction by phagocytosis. In other words, *E. coli* may use fimbriae for attachment in the intestine, but in deeper tissue the fimbriae may contribute to its demise. Pathogenic *E. coli* require other factors to cause disease in humans, and these are discussed later.

Some *E. coli* strains also possess adhesins called the K88 and K99 antigens, which are plasmid-controlled. *E. coli* strains containing K88 antigen are infectious for pigs, whereas *E. coli* strains containing K99 antigen are infectious for calves (see Chapter 34). These two adhesins have been purified, and a vaccine (Vicogen) has been developed.

Influenza virus. The influenza virus, as well as those viruses that cause the common cold (rhinoviruses and coronaviruses), confines its activities to the epithelial surface. The influenza virus and other enveloped viruses possess glycoproteins projecting from the envelope surface (see Figure 6-5). These ***glycoproteins*** are important for absorption and appear to play a role in penetration and virulence. The hemagglutinin glycoprotein spike of the influenza virus adheres to the host cell by binding to the neuraminic acid component of the cytoplasmic membrane.

Streptococcus pyogenes. The bacterium *S. pyogenes* is a cause of streptococcal sore throat and its sequela, rheumatic fever. The ability of the microorganism to adhere to the respiratory epithelium is correlated with the presence of a protein called the ***M protein*** and ***lipoteichoic acid*** on the cell envelope. These two components form a fibrillar network on the surface of the streptococcus in which the lipid moiety of lipoteichoic acid serves as the binding factor to the host cell (Figure 22-4). The receptor on the host cell is believed to be an albumin-like protein. Antiserum to lipoteichoic acid but not other surface components inhibits adherence of the streptococci.

Streptococcus mutans. The bacterium *S. mutans* is the agent of dental decay (caries). The ability of some strains to adhere to the enamel of the tooth appears to be related to several surface components. *S. mutans* is known to interact with several types of glycoproteins found in whole saliva and secretions from the parotid gland. In addition, the microorganism produces an insoluble glucan polymer from sucrose called ***dextran.*** In vitro experiments have demonstrated the importance of dextran for attachment to a variety of

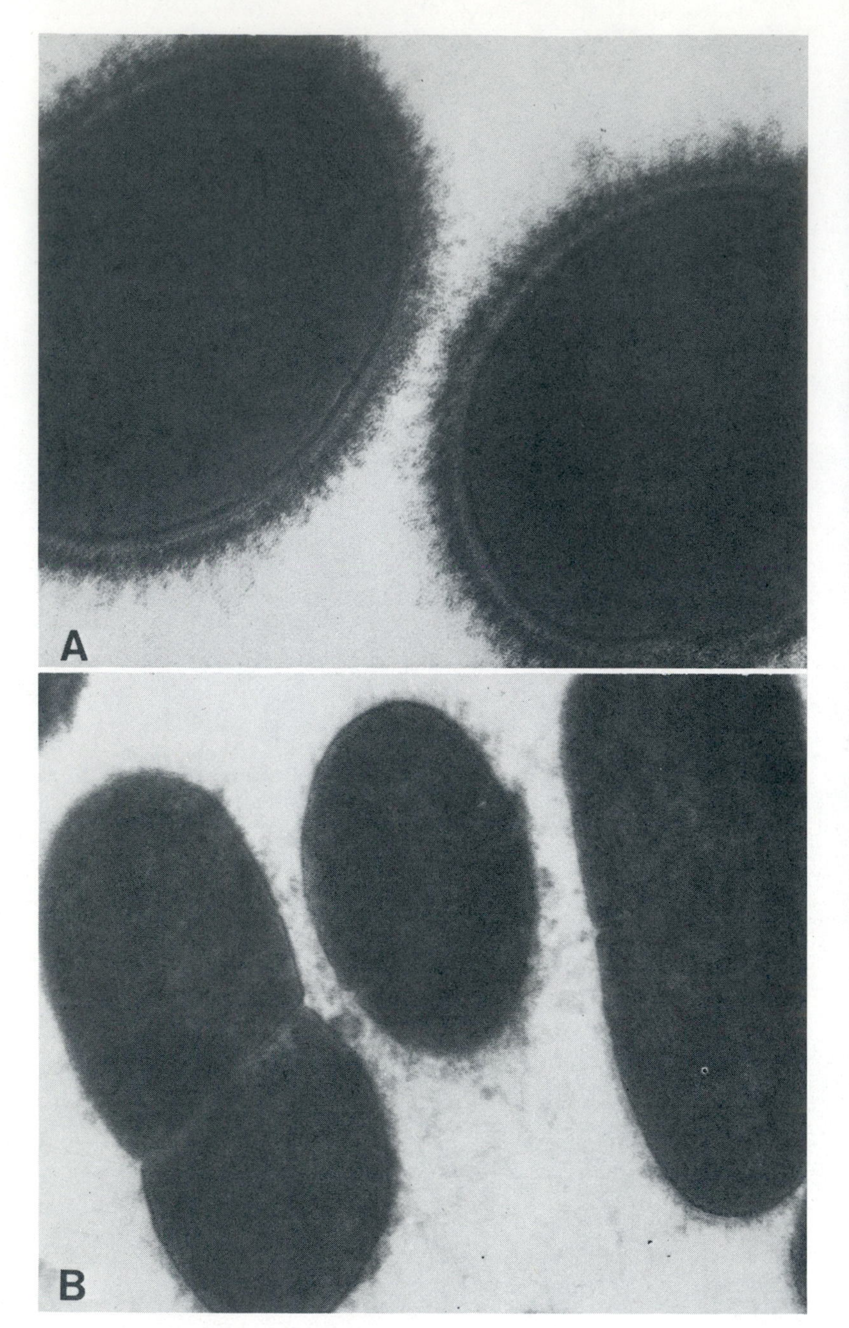

Figure 22-4 Electron micrographs of cross sections of *Streptococcus pyogenes* with M protein, **A,** and without M protein, **B.** (×84,000.)

From Beachey, E.H., and Stollerman, G.H., in Dumonde D.C. (Ed.). Infection and immunology in the rheumatic diseases. *Blackwell Scientific Publications, Ltd., Oxford, 1976.*

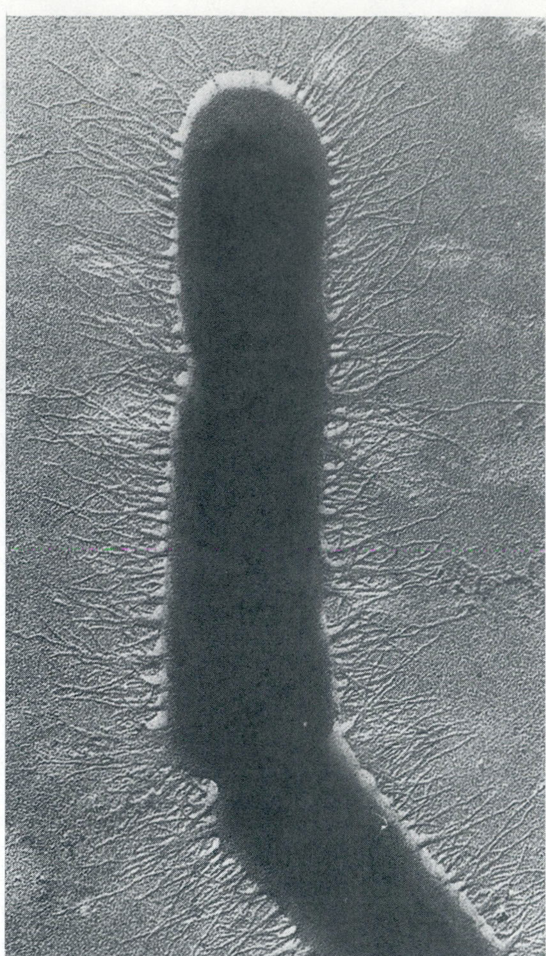

Figure 22-5 Electron micrograph of *Actinomyces viscosus.* Note long appendages emanating from surface.

From Masuda, N., et. al. J. Bacteriol. ***147:****1095, 1981.*

surfaces. Those strains of *S. mutans* that produce a fructan (fructose-containing polymer) adhere poorly to similar surfaces. It is believed that the enzyme ***glucosyl transferase,*** which catalyzes the formation of dextran, forms a complex with the dextran and the microbial surface and promotes attachment to the tooth surface.

Neisseria gonorrhoeae. Substantial evidence indicates that fimbriated *N. gonorrhoeae* adhere more strongly to mucosal surfaces of the urogenital tract than nonfimbriated strains. Adhesion is an important factor in the pathogenesis of the gonorrhea-producing organism. Adhesion permits the bacterium to more effectively release toxic secretions to host cells and also facilitates its invasion of epithelial cells. The receptors on the host cell are believed to be galactose β 1-3 *N*-acetylgalactosamine β 1-4 galactose.

Actinomyces viscosus. *A. viscosus* is an inhabitant of the oral cavity and is capable of adhering to the tooth surface and becoming a component of the aggregated material called ***dental plaque.*** The attachment function appears to be associated with long surface appendages called ***fibrils,*** which are made up of polysaccharide (Figure 22-5).

Campylobacter species. Campylobacter is an intestinal pathogen that is a cause of diarrhea in humans and other animals. It is capable of attachment to the epithelial surface and penetration into the mucosa (Figure 22-6). Its ability to

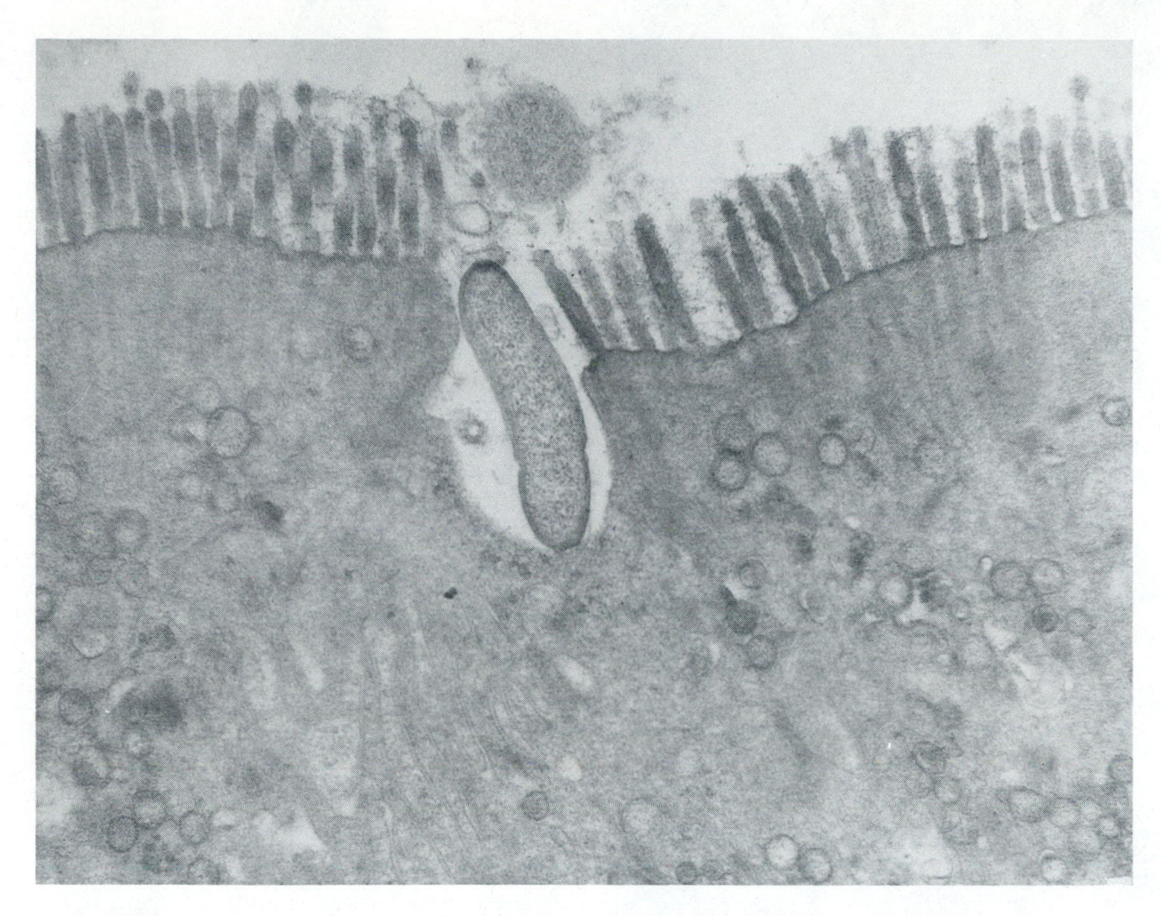

Figure 22-6 Electron micrograph of a *Campylobacter* species that has invaded intestinal mucosa and disrupted membrane.
From Ruiz-Palacios, G.M. Infect. Immun. ***34:**250, 1981.*

penetrate is apparently due to the release of cytotoxic enzymes.

Mycoplasma pneumoniae. *M. pneumoniae* is a respiratory pathogen that has an affinity for epithelial cells (Figure 22-7). The adhesin is believed to facilitate the transfer of toxic peroxides produced by the microorganism to host cells and prevent their destruction by catalase and peroxidase—enzymes that are present in mucosal secretions.

Spread of microorganisms in the host. Microorganisms that are confined to epithelial surfaces can be spread to other epithelial surfaces by a variety of means (Table 22-1). Surfaces covered with fluid, such as those in the respiratory and intestinal tracts, allow free diffusion of microorganisms, especially if there are excess secretions due to infection. Cilia, if they are not damaged, can sweep fluids to other parts of the respiratory tract, and coughing and sneezing also permit spread to uninfected areas. Microorganisms on the skin are usually spread by scratching and touching uninfected areas of the body with contaminated fingers. Microorganisms that remain on epithelial surfaces can produce symptoms in the host long before antimicrobial factors come into play. The immune system, which is discussed in Chapter 23, is slow to respond to any microbial agent that remains at the epithelial surface and is effective only during the later stages of infection. One of the most important nonspecific resistance mechanisms employed by the host against viral infection is interferon production. Interferon is a protein produced by viral-infected cells that acts to protect uninfected cells by inhibiting viral replication. A more detailed discussion of interferon is presented in Chapter 27.

Some microorganisms are virulent because of their ability to penetrate as well as adhere to epithelial tissue, but the microbial factors responsi-

Figure 22-7 Scanning electron micrograph of *Mycoplasma pneumoniae* demonstrating attachment to membrane of ciliated epithelial cell. (×8000.) *C*, Cilia; *M*, cluster of mycoplasma cells. Arrows point to individual filaments of mycoplasma.
Courtesy Michael G. Gabridge.

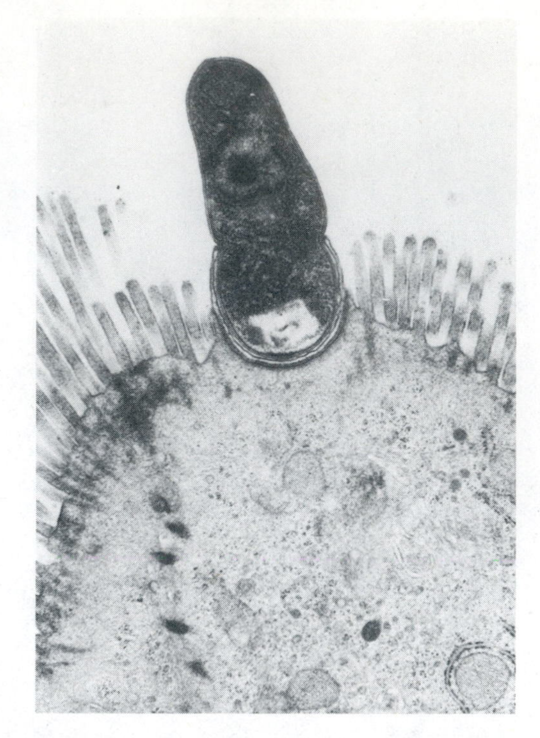

Figure 22-8 Electron micrograph of protozoan penetrating mucosal surface of guinea pig mucosal cell.
From Vetterling, J.M., et al. J. Protozool. **18**:255, 1971. Reprinted with permission of The Society of Protozoologists.

ble for penetration are not entirely understood. Experiments with avirulent and virulent strains of some intestinal parasites have demonstrated that a component of the cell envelope, such as a sugar moiety, may be one type of determinant. One mechanism for penetration by a bacterial intestinal parasite is believed to be due to the release of a microbial product that causes degeneration of the microvilli of the intestinal epithelial cells. The bacterial cells are then ingested by the epithelial cells of the host, a process similar to phagocytosis. Viruses, as previously indicated, demonstrate a specificity between the envelope and the host's cytoplasmic membrane. Surface specificity, however, does not ensure infection for some microorganisms. Certain protozoa, for example, can penetrate host cells only through the action of ***lysosomal*** enzymes located at the end of the parasite that is in contact with the host cell (Figure 22-8). Microorganisms in the subepithelial surfaces can be spread to uninfected areas, but the mechanisms have been difficult to evaluate in the whole animal. Laboratory studies have indicated that several enzymes, including hyaluronidase and collagenase, (Table 22-1), can promote the spread of microorganisms that produce them.

Those microorganisms that are able to penetrate the epithelial surface are subject to many host defense measures that could destroy them. The microbial agent must be equipped to resist these host factors and must find susceptible cells on which or in which it can replicate. Once the epithelial surface has been penetrated, the parasite can enter the ***lymphatics*** and conceivably infect any part of the host. The lymphatic system is a series of vessels and nodes that carry fluid into the circulatory system (Figure 22-9). The vessels of the lymph system are in close proximity to blood vessels, and at very strategic points along their path they become associated with large nodes called ***lymph nodes.*** Lymph nodes are particularly numerous at body joints and in areas where the arms and legs join the body. The lymph nodes are reticulated structures that serve to filter lymph passing through them. Specific areas within the nodes serve as repositories for special cellular elements of the blood called ***T*** and ***B lymphocytes.*** Lymph leaves the nodes

Table 22-1 Bacterial products believed to be important in the spread of microorganisms in tissue

MICROBIAL PRODUCT	MICROORGANISM(S) INVOLVED	ACTIVITY OF MICROBIAL PRODUCT
Hyaluronidase	Streptococci, staphylococci, and clostridia	Degrades hyaluronic acid, a component of connective tissue
Coagulase	Staphylococci	Acts on clotting blood, which results in formation of fibrin; is believed to prevent phagocytosis
Collagenase	Species of *Clostridium*	Digests collagen, a component of muscle
Hemolysins	Many bacterial species especially streptococci and staphylococci	Destroys red blood cells; importance is uncertain
Fibrinolysin (streptokinase)	Streptococci	Converts plasminogen in human sera to plasmin, an enzyme that digests fibrin; microorganisms trapped in fibrin clot are free to spread
Lecithinase	Species of *Clostridium*	Destroys lecithin, a component of cell membranes
DNase	Staphylococci	Causes lowering of viscosity of exudates (pus), giving microorganism greater mobility

through efferent vessels that eventually drain into a large lymphatic vessel called the ***thoracic duct.*** The latter empties directly into the venous system.

Microorganisms can be transported to a variety of body sites by entering the blood via direct or indirect means. Some microorganisms are transmitted directly into the blood stream following the bite of some arthropod, but most microorganisms find their way into the bloodstream via the lymphatics. During their sojourn through the blood stream, microorganisms may remain free or be carried within certain cellular elements of the blood, particularly macrophages (monocytes) and, infrequently, red blood cells. Once in the bloodstream, microorganisms may be disseminated to localize in various organs, the central nervous system, and the skin. Measles, for example, is caused by a virus that initially infects the respiratory tract but later penetrates into the bloodstream and is then disseminated to cells of the skin, where it multiplies and produces a rash.

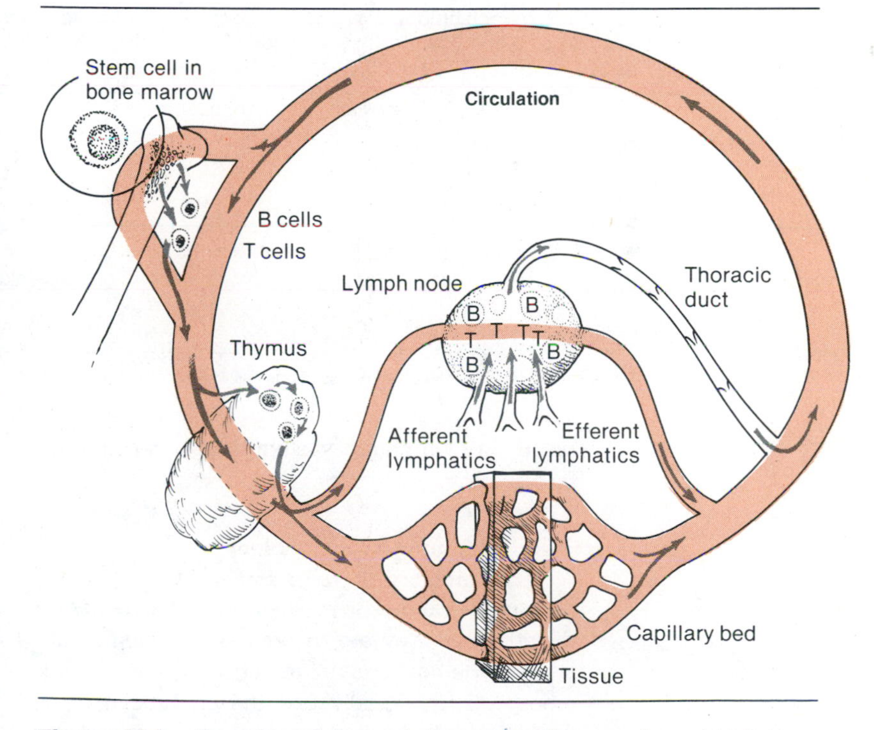

Figure 22-9 Flow of lymph in lymphatic system of humans. Interstitial fluid (lymph) collects in capillary beds and is transported into afferent lymphatics. Lymph is filtered in nodes and then passed into efferent vessels and finally thoracic duct, which empties into venous circulation. B and T lymphocytes produced in bone marrow enter circulation, and T lymphocytes are further processed in thymus. Lymphocytes enter lymph nodes through circulation and can be released again into circulation.

Host's second line of defense

Once parasites penetrate the subepithelial layer, they become prone to the host's second line of defense, which includes cellular elements of the

Table 22-2 Characteristics of the various leukocytes found in humans

CELL TYPE*	SIZE	MORPHOLOGY	FUNCTION
Lymphocytes (32%to 35%)	7 to 8 μm	Has very little cytoplasm and is composed primarily of nucleus	Has receptors for interaction with antigen; principal types are T and B cells, which increase in size after interaction with antigen; B cells give rise to plasma cells
Blast cell	18 μm	Large nucleus with abundant cytoplasm that contains large amounts of free and aggregated ribosmes	Intermediate cell type derived from lymphocyte; may give rise to plasma cell or special sensitized cell involved in hypersensitivity
Plasma cell	9 to 12 μm	Contains small nucleus but abundant cytoplasm with large amounts of endoplasmic reticulum	End cell derived from stimulated B lymphocyte via blast cell; is actual producer of antibody molecules
Mast cell 18 to 24 μm		Shape is often irregular compared with the similar basophil; characterized by large concentrations of granules	Similar to basophil but is found only in tissue, usually adjacent to arterioles; is involved in allergic reactions by release of active substances such as histamine
Macrophage (3% to 4%)	12 to 20 μm	Contains bilobed nucleus; cytoplasm contains large number of organelles; tissue macrophage or histiocyte is larger than blood macrophage	Primary function is phagocytosis; degradation of microorganisms releases antigens that can be processed for reactions involved in specific immune resistance
Polymorphonuclear leukocytes			
Neutrophil (60% to 65%)	9 to 12 μm	Contains multilobed nucleus; granules in cytoplasm do not stain strongly acidic or basic	Engages in phagocytosis; cells rapidly migrate to areas of acute inflammation
Basophil (0.1% to 0.3%)	18 to 24 μm	Lobulated nucleus; large number of granules stain basic (blue)	Granules contain pharmacologically active compounds, such as histamine, that may be involved in allergic reactions
Eosinophil (2% to 3%)	10 to 14 μm	Lobulated nucleus; cytoplasmic granules stain acidic (red)	Appears to aid in destruction of metozoan parasites, such as parasitic worms (helminths)

*The percentage of each in the blood is in parentheses.

blood, the lymphatic system, and other nonspecific resistance factors.

Cellular elements of the blood

Red blood cells (erythrocytes). Red blood cells carry oxygen to tissues and remove carbon dioxide, but they are in no way engaged in defending the host against infection. A very few microorganisms actually use the erythrocyte as a vehicle for dissemination to various parts of the body. The red blood cell is most important for the parasite causing malaria, because it uses the cell not only as a site for growth and replication but also as a ''hideaway'' from the phagocytic cells of the blood.

White blood cells (leukocytes). The term *white* refers to the sedimentation characteristics of unclotted blood, in which leukocytes appear as a white layer between the red blood cells and the fluid portion of the blood (plasma). Leukocytes (Table 22-2) are found in both blood and tissue and are divided into two major groups: granulocytes and agranulocytes.

Granulocytes are so called because of the presence of aggregates of hydrolytic enzymes, which are involved in the phagocytic process (see ''Phagocytic Process''). They also possess a multilobed nucleus and are frequently referred to as ***polymorphonuclear leukocytes*** (PMNs). Granulocytes are divided into three types based

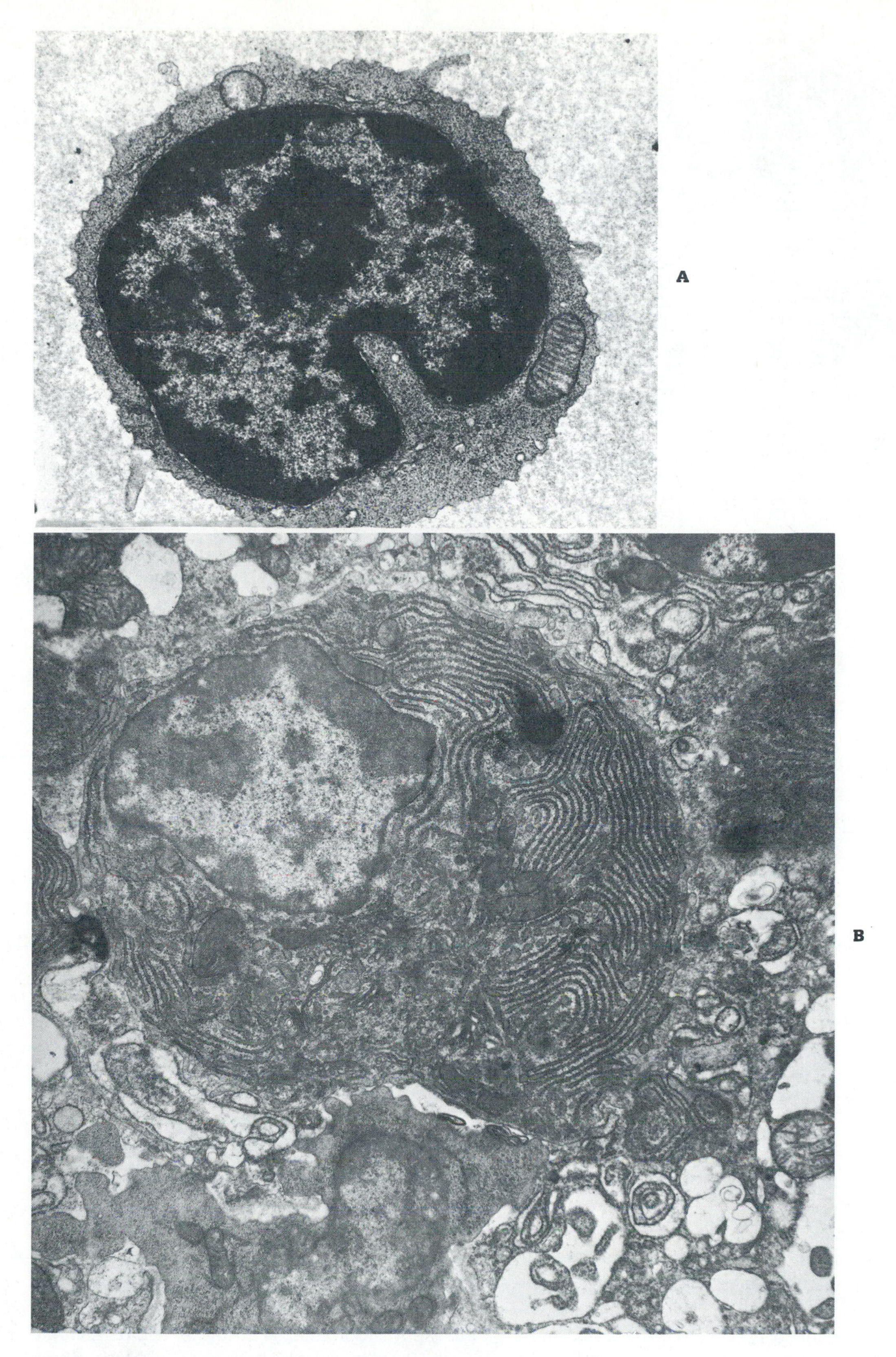

Figure 22-10 Electron micrographs of lymphocytes. **A,** T cell lymphocyte in which there is very little cytoplasm and indented nucleus. (×13,000.) **B,** Plasma cell derived from stimulated B lymphocyte. Cytoplasm is filled with endoplasmic reticulum *(curving lines),* which is site of immunoglobulin (antibody) synthesis.

A, *From Roitt, I.* Essential immunology *(4th ed.). Blackwell Scientific Publications, Ltd., Oxford, 1980.* ***B,*** *Courtesy E. Adelstein. From Barnett, J.T.* Textbook of immunology *(4th ed.). The C.V. Mosby Co., St. Louis, 1983.*

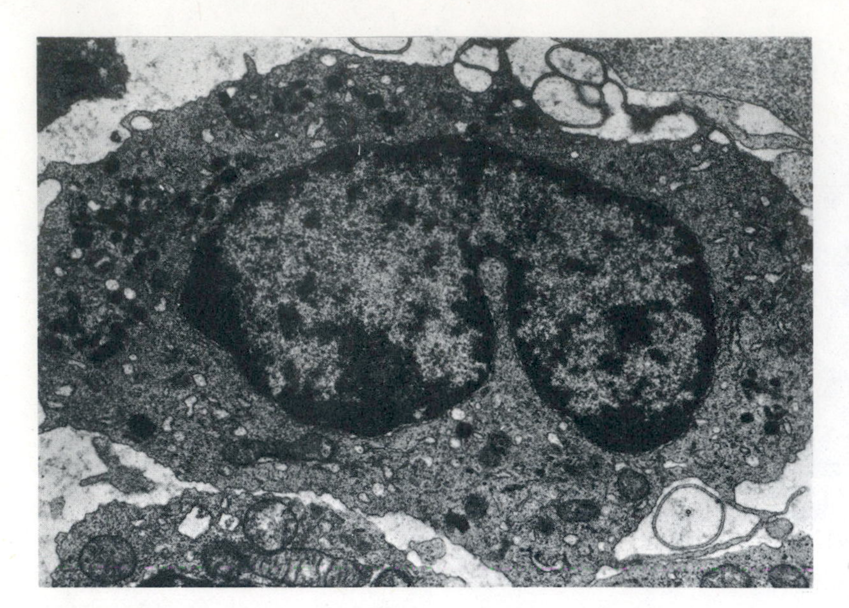

Figure 22-11 Electron micrograph of macrophage demonstrating horseshoe nucleus. (×10,000.)

From Roitt, I. Essential immunology (4th ed.). Blackwell Scientific, Publications, Ltd., Oxford, 1980.

Table 22-3 Names of the macrophages according to their tissue location

TISSUE	NAME OF MACROPHAGE
Liver	Kupffer cell
Connective tissue	Histiocyte
Lung	Alveolar macrophage or "dust cell"
Spleen	Sinusoidal lining cell
Bone	Osteoclast
Nervous system	Microglial cell

on their staining characteristics: neutrophils, eosinophils, and basophils (Plates 31 to 33). ***Neutrophils,*** like macrophages (discussed later), are the most active phagocytic cells and are therefore important in attacking the infectious agent as well as being responsible for the pathological features of inflammation.

Agranulocytes are divided into two groups: lymphocytes and monocytes. Lymphocytes represent a population of cells with different functions. They are composed of T and B cells, which have their origin in the bone marrow and fetal liver (Figure 22-10). They are the principal cells responsible for the immune response of the host (Table 22-2) and are discussed at length in Chapter 23. ***Monocytes*** and macrophages (Figure 22-11) are the largest of the lymphoid cells and are found in the blood as well as in tissue* (Table 22-3). Those macrophages that are part of the liver, spleen, bone marrow, and adrenal glands make up a system called the ***mononuclear phagocytic system.*** They process, by means of phagocytosis, damaged cellular material as well as parasitic microorganisms that appear in the circulation or at the site of infection. Their role in infection is discussed later.

Platelets. Platelets, which are nonnucleated cells, are derived from cells in the bone marrow and vascular sinuses. They have a life span of approximately 10 days and produce a component that is involved in the clotting of blood. They are also involved in reactions that are unfavorable to the host, such as hypersensitivity reactions. ***β-Lysin*** is a protein released from platelets that is believed to have antimicrobial properties, particularly against gram-positive bacteria.

*Macrophages are actually monocytes that have undergone differentiation and left the blood to reside in various body tissues.

Fluid elements of the blood. The fluid portion of the blood is called ***plasma*** and is over 90% water. Proteins, which are produced in the liver, make up 6% to 8% of the plasma and include such molecules as albumin, globulin, prothrombin, fibrinogen and complement. Globulins play the most important role in immunity, particularly ***immunoglobulins (antibodies),*** which are produced in response to antigenic stimulation and are associated with acquired immunity (see Chapter 23).

Complement is a series of glycoproteins that can be activated in a manner similar to that of the clotting factors in blood. Complement is involved in specific immune reactions involving antibody but can also take part in nonspecific immune reactions. Some complement proteins cause the release of vasoactive compounds, which dilate blood vessels and contract smooth muscle. Complement can also generate chemotactic factors that attract PMNs, and, finally, complement can bind to the bacterial surface and provide help in the phagocytic process. Complement is discussed at more length in Chapter 23.

Lymphatic system. Microorganisms that cannot be walled off and are not killed by leukocytes at the site of infection proliferate and pass into regional lymph nodes. The lymph nodes try to filter out the microorganisms with the help of resident macrophages and circulating neutrophils. If the number of microorganisms entering the node is very large, the interaction between parasite and leukocyte becomes so intense that the nodes become inflamed, swollen, and tender (lymphadenitis). Sometimes the nodes become greatly enlarged and necrotic and are of diagnostic significance. These enlarged nodes are called ***buboes*** and are characteristic of such diseases as bubonic plague. Localized infections may be terminated if the macrophages in the lymph nodes are capable of destroying the parasite, but some parasites survive and even multiply within the macrophage. The macrophage has a considerably longer life span than the neutrophil and provides the parasite with a vehicle for transport and protection in the circulatory system of the host. Some of the infectious agents that are known to multiply in macrophages are the poxviruses, herpesviruses, and bacterial agents that cause such diseases as tuberculosis, leprosy, and Rocky Mountain spotted fever. Once the macrophage reaches the bloodstream, it proceeds to localize in one or more organs and then engages in reseeding progeny microorganisms into the bloodstream to infect other organs of the body.

Phagocytosis

Phagocytosis and inflammation. Tissue in which an inflammatory response has taken place is characterized by being red, warm, swollen, and usually painful. Foreign agents including microorganisms, are irritants that cause ***mast cells*** (mast cells are similar to basophils) to release vasoactive compounds, such as histamine, which effect a change in the permeability of vessels in the infected area. The affected vessels become dilated, and plasma passes into the tissues, carrying with it certain proteins such as fibrinogen, a blood-clotting factor, and lymphocytes. ***Fibrinogen*** in the infected tissue is converted into a gel-like material called ***fibrin,*** which acts as a trap by walling off the foreign agent and preventing its spread. Circulating neutrophils are also attracted to the infected area, where they begin to destroy the foreign agent through the action of hydrolytic enzymes. Neutrophils, which have a short life span, appear in large numbers at the site of infection. Their lysis results in the release of compounds that reduce the pH of the surrounding area. The accumulation of dead and dying neutrophils produces the characteristic creamy pus associated with such localized infections. Macrophages clean up the debris left by the dead neutrophils and any microorganisms still present in the area. Macrophages working together to phagocytize a microorganism may fuse to form ***giant cells.*** Giant cells accumulate as they try to wall off the infection and in so doing produce nodules called ***granulomas.*** The organism causing tuberculosis, for example, survives the attack of neutrophils and macrophages and produces nodules called ***tubercles,*** which can be observed in many organs but especially by radiographs of the lungs.

Very little is known concerning the mechanisms used by parasites to interfere with the inflammatory response. It is believed, however, that the staphylococci possess a cell wall component that has antiinflammatory activity and thereby prevents the mobilization of host antimicrobial factors at the site of infection.

Phagocytosis and fever. A rise in body temperature above the normal 98.6° F is a frequent and normal response to infection. Fever is mediated by proteins, called ***endogenous pyrogens,*** that are produced in phagocytic cells such as PMNs and macrophages. Pyrogens act on the temperature-regulating centers of the brain in the anterior hypothalmus. The phagocytic cells releasing the pyrogens can be activated by a number of microbial products, as well as some nonmicrobial components. The suppression of fever is considered an important aspect in the treatment of infectious disease, but the height of the fever in most patients is not sufficient to be detrimental. Prolonged fever may be detrimental to the host by causing an increased heart rate or by causing convulsions, particularly in children. Fever of short duration may actually be beneficial to the host. The use of drugs (antipyretics) to reduce fever in some infections may be detrimental. Recent observations suggest that aspirin (acetylsalicylic acid) used to reduce fever following certain infections may depress the immune response and make the patient more susceptible to complications such as Reye's syndrome (see Chapter 29).

Many infections in the preantibiotic era were treated by inducing hyperthermia in humans. Treatment of gonococcal endocarditis, syphilis, and pneumococcal meningitis for example, included methods for inducing fever. Those earlier studies suggested that fever contributed greatly to the recovery from infection. Many of the investigators in those studies suggested that hyperthermia stimulated host defense mechanisms and did not directly affect the microorganism causing infection. Not enough controlled studies were developed, however, to provide specific answers.

Phagocytic process

Phagocyte chemotaxis. Chemotactic factors released by infectious agents or damaged tissue cells are believed to attract phagocytic cells. The chemotactic factors supposedly cause the contraction and relaxation of microfilaments in the leukocyte, which in turn orients the leukocyte to

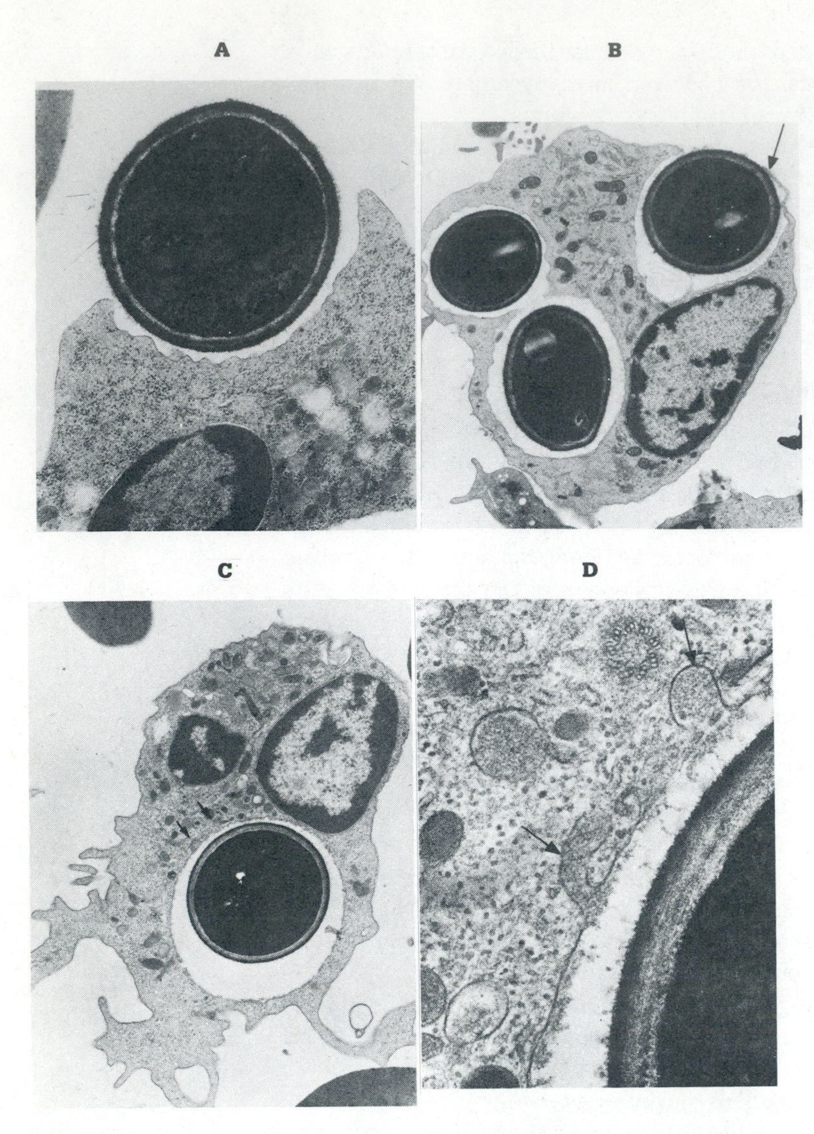

Figure 22-12 Electron micrograph demonstrating phagocytosis of yeast *Candida albicans*. **A,** Organism is being surrounded by cytoplasmic arms. (×15,000.) **B,** Three organisms are phagocytized; one *(arrow)* shows nearly complete phagosome formation, and other two are completely ingested. (×5000.) **C,** Thirty minutes after ingestion two lysosomal granules *(arrows)* are fusing with phagocytic vacuole. (×5000.) **D,** Higher magnification of **C** showing fusing granules discharging their contents into phagocytic vacuole *(arrow)*. (×33,000.)

Courtesy H. Valdimarsson, In Roitt, I. Essential immunology *(4th ed.). Blackwell Scientific Publications, Ltd., Oxford, 1980.*

move in a specific direction. The leukocyte must first determine the concentration difference between its tail and its head before it moves in the chemical gradient. This spatial gradient-detection mechanism is believed to be the result of cell-surface receptors covering the leukocyte. Once the concentration gradient has been determined, leukocytes move up the gradient. The importance of leukocyte locomotion in disease has been noted in a child whose leukocytes lacked this capacity. The child suffered from severe and repeated infections, and examination of the defective leukocytes revealed a microfilament dysfunction.

Mechanisms of phagocytosis. Once the phagocyte and the microorganism, or cellular debris, make contact, a series of events are set in motion whose purpose is to engulf and degrade the target material. This series of events can be divided into two basic processes: degranulation and respiratory burst activity.

The phagocytic membrane engulfs the target (Figure 22-12, *A*) and is pinched off inside the cell into a vesicle called the ***phagosome*** (Figure 22-12, *B*). The phagosome fuses with many ***lysosome*** granules, which contain hydrolytic enzymes, to produce a ***phagolysosome*** (Figure 22-12, *C* and *D*). Once fusion has taken place, the hydrolytic enzymes are discharged (a process called ***degranulation***) into the vesicle containing the material targeted for destruction. The hydrolytic enzymes in conjunction with certain microbicidal agents released during metabolism of the leukocyte are used to digest the cellular debris or kill the invading parasite.

Energy is required by the phagocyte to carry out the ingestion process, as well as the subsequent degradative processes. Following membrane contact of the phagocyte with the targeted material, there is a respiratory burst of activity in which oxygen consumption is increased significantly. During this respiratory burst certain metabolic reactions take place that result in the formation of microbicidal agents. Glucose oxidation in the phagocyte, for example, occurs via the hexose monophosphate shunt, resulting in the formation of reduced NAD and NADP. It is currently believed that an enzyme NADPH oxidase catalyzes the one electron reduction of oxygen to superoxide (see Chapter 13), using NADH as the electron donor. In a subsequent reaction 80% of the superoxide is converted to hydrogen peroxide according to the following reactions:

$$2\,O_2 + \text{NADPH} \longrightarrow \underset{\text{Superoxide}}{2\,O_2^-} + \text{NADP} + H^+$$

$$2\,O_2^- + 2\,H^+ \longrightarrow O_2 + \underset{\text{Hydrogen peroxide}}{H_2O_2}$$

Persons with chronic granulomatous disease* possess leukocytes with defects in these reactions, thus indicating their importance in resistance to infection. The superoxide ion is toxic, but it is not believed to be the major microbicidal agent. Microbicidal activity is believed to be as-

* Chronic granulomatous disease is a sex-linked recessive disease of males in which infections from pathogens of low virulence recur. The tissue in any organ of the body may reveal granulomas.

Table 22-4 Microorganisms in which surface components interfere with digestion by phagocytes

MICROORGANISM	SURFACE COMPONENT	DISEASE IN WHICH ORGANISM IS INVOLVED
Cryptococcus neoformans	Capsule (polysaccharide)	Cryptococcosis, a systemic fungal disease
Streptococcus pyogenes	M protein	Rheumatic fever
Streptococcus pneumoniae	Capsule (polysaccharide)	Bacterial pneumonia
Bacillus anthracis	Capsule (polypeptide)	Anthrax
Salmonella typhi	Vi antigen (amino sugar)	Typhoid fever
Escherichia coli	K antigen (polysaccharide)	Diarrhea
Pseudomonas aeruginosa	Slime layer (polysaccharide)	Infection as cause of death in burn patients
Yersinia pestis	Polysaccharide-protein complex	Bubonic plague
Treponema pallidum	? Hylauronic acid in cell envelope	Syphilis
Staphylococcus aureus	Protein A	Skin infections, pneumonia

sociated with hydrogen peroxide formation and the enzyme ***myeloperoxidase,*** since defects in their formation often lead to recurrent infections in the host. The myeloperoxidase system catalyzes the oxidation of halide ions (chlorine, iodine, and bromine) to ***hypohalite ions*** in the presence of hydrogen peroxide. For example, chlorine, which is the most predominant halide in the cell, takes part in the following reaction:

$$Cl^- + H_2O_2 \longrightarrow \underset{\text{Halite ion}}{ClO^-} + H_2O$$

Experimental evidence indicates that the myeloperoxidase system may contribute to microbicidal activity in the following ways:

1. Hypohalite ions interact with the bacterial cell wall, causing it to lose its integrity and eventually lyse.
2. The system causes the decarboxylation of amino acids to aldehydes, which are toxic and could kill the cell.
3. The hypohalite produced in the system can react with hydrogen peroxide to produce ***singlet oxygen*** ($^1O_2$), which is simply a higher energy form of oxygen. Singlet oxygen has the same molecular formula as atmospheric oxygen ($^3O_2$) but differs in the distribution of the electrons around the two oxygen atoms. Singlet oxygen is a highly reactive product and is capable of breaking double bonds that are normally unreactive to atmospheric oxygen. It may be produced in the following way:

$$ClO^- + H_2O_2 \longrightarrow {}^1O_2 + Cl + H_2O$$

4. The superoxide produced in the myeloperoxidase system reacts with hydrogen peroxide to produce the ***hydroxyl radical*** (OH·) a highly unstable but extremely toxic product:

$$O_2^- + H_2O_2 \longrightarrow (OH\cdot) + OH^- + O_2$$

The molecular events occurring within the phagocyte that are necessary for microbicidal activity have been identified because of investigations of cells from patients with congenital disorders of phagocytes. The ability of the phagocyte to kill the parasite is to a great extent dependent on microbial surface structures. Gram-positive bacteria, because of their simple cell wall structure, are more easily digested than gram-negative bacteria, and certain microbial species appear to be more sensitive to one phagocytic activity than to another. It is hoped that as a consequence of these investigations it will be possible in the near future to stimulate the phagocytes of patients who have increased susceptibility to infection.

Resistance of the parasite to phagocytosis

Interference with chemotaxis. Phagocytes are attracted to microbial agents by chemotactic substances produced by metabolizing microorganisms. Some microorganisms, such as the agent of tuberculosis, are capable of interfering with this process. The exact mechanism is not known, but a surface lipid called the "cord factor" is believed to be the interfering agent. A comparison of the chemotactic power of the causative agent of gonorrhoea *(Neisseria gonorrhoeae)* has been made. The infectious agent in most patients remains localized in the genitourinary tract, but in a very few instances a strain of this agent can break through the local barriers

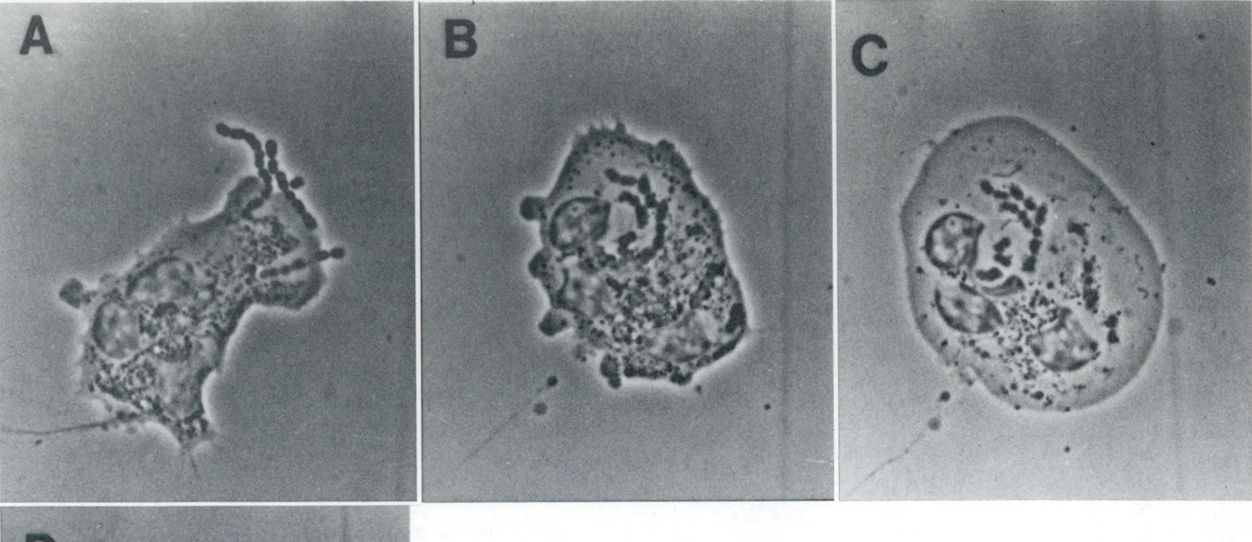

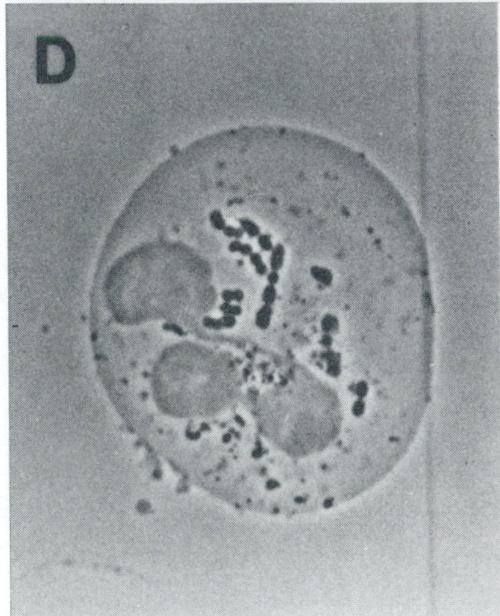

Figure 22-13 Phase-contrast micrograph demonstrating leukotoxic action of streptococci on human PMNs. **A,** PMN is seen to be engulfing three streptococcal chains. **B,** PMN has ceased moving and has begun to round up. Cytoplasmic blebs are seen to emanate from surface of PMN. **C,** PMN is swollen and immobile. **D,** Most internal structures of PMN have disappeared, and nucleus appears swollen.

From Sullivan, G.W., and Mandell, G.L. Infect. Immunol. ***30****:272, 1980.*

and enter the bloodstream to be disseminated throughout the body. Comparison of the strains from localized and disseminated infections has revealed that disseminated strains are poor stimulators of chemotaxis as compared with active stimulation by strains causing localized infections.

Interference with ingestion by phagocytes. The ingestion of microorganisms into the phagocytic cell can be prevented by some bacteria. These inhibitors of phagocytosis are surface components, usually capsules belonging to agents described in Table 22-4. The manner in which these components interfere with phagocytosis is not known, but they may prevent adsorption of immune components, called opsonins, that are known to aid in phagocytosis. (See Chapter 23 for a discussion of opsonins.) It is important to remember that not all capsule-bearing microorganisms are resistant to phagocytosis.

Some infectious agents acquire their capsules only after infection. Plague, for example, is a disease in which the infectious agent *(Yersinia pestis)* is harbored by fleas that infest rats and other rodents. The infectious agent does not produce a capsule while harbored in the flea, but when the flea bites a human, some of the transferred organisms are ingested by PMNs and killed, whereas those ingested by macrophages produce a capsule and multiply.

Interference with intracellular digestion by phagocytes. Microbial surface components are apparently involved in the prevention of phagocytic digestion. Microorganisms surviving in the short-lived PMN can be disseminated to other parts of the body. Microorganisms surviving in the long-lived macrophage multiply and are transported to organs where they are responsible for chronic infections such as leprosy, tuberculosis, and brucellosis. It appears that the principal technique for avoiding digestion in the macrophage is to prevent the release of lysosomal enzymes into the phagocytic vacuole (phagosome) that contains the intact microorganisms.

Destruction of the phagocyte. Some microorganisms have the ability to cause the premature death of phagocytes. The streptococci and staphylococci release soluble factors that bring about the destruction of phagocytes (Figure 22-13). The streptococci produce a hemolysin called ***streptolysin,*** and the staphylcocci produce a compound called ***leukocidin.*** These bacterial products cause the release of lysosomal enzymes into the cell cytoplasm of the phagocyte and bring about its destruction.

Defective phagocytic mechanisms. The ability of the microorganisms to withstand phagocytosis is greatly increased if there are inherited abnormalities in the killing mechanism of the phagocyte. Conditions such as myeloperoxidase deficiency and chronic granulomatous disease result in increased susceptibility to bacterial infections because of the inability of the phagocyte to generate superoxide as well as hydrogen peroxide.

Damage to host tissue

The symptoms of a disease or the damage to host tissue in most cases is in part related to the rate at which the microorganisms divide or multiply in a given tissue. Some members of the Enterobacteriaceae, which cause a form of bacterial gastroenteritis, produce symptoms within 18 to 24 hours following infection and have division times of 30 minutes or less. The agent of leprosy may take 2 to 3 weeks to divide and many months are required before disease symptoms are visible. Damage to host tissue may be due to one or more of the following conditions: (1) direct damage by microbial multiplication, (2) microbial toxin production, or (3) immunopathological reactions.

Direct damage by microbial multiplication. Many viruses damage host tissue as a result of replication and the inhibition of host protein, DNA, and RNA synthesis. Inclusion bodies, which represent an accumulation of virally associated components, can often be seen microscopically (Figure 22-14). The site of viral multiplication will determine the type of tissue damage. The poliovirus, for example, multiplies in the anterior horn cells of the spinal cord, where damage to the nerve tissue results in a loss of function to the muscles innervated by them. Damage to the host from protozoal infections may be due to the sheer number of parasites, which causes blockage of some physiological process. Bacteria may multiply within cells as well as in tissues and tissue fluids. Some bacteria have the ability to destroy the phagocytes in which they find themselves, but damage from most bacterial infections is due to toxic products such as microbial toxins.

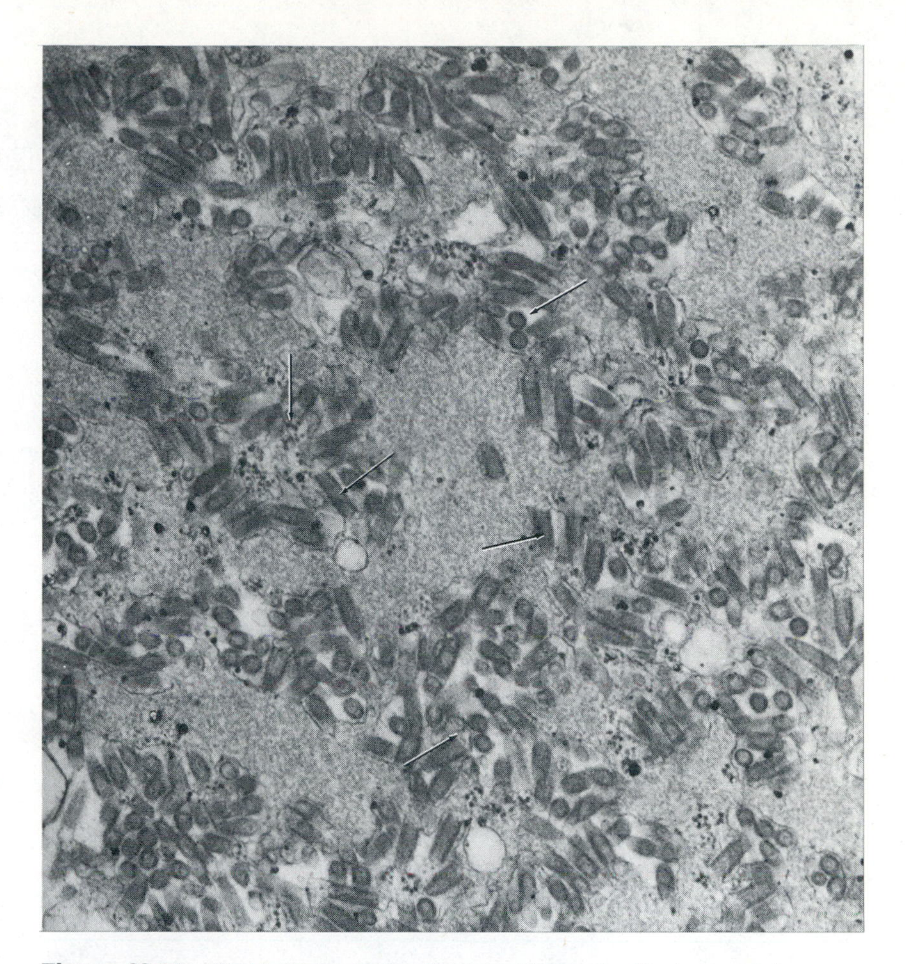

Figure 22-14 Viral inclusion bodies. Electron micrograph of brain neuron in suckling mouse demonstrating rabies virus inclusion body within cytoplasm. (×49,000.) Many viral particles *(arrows)* can be seen lengthwise and in cross section.

Courtesy Alyne K. Harrison.

Microbial toxins. Many microorganisms, especially bacteria, produce toxins. Bacterial toxins are of two types: ***exotoxins,*** which are produced by the cell and released into the surrounding tissues, and ***endotoxin,*** which is part of the cell wall and is released during growth or when the bacterial cell lyses (Table 22-5). Toxins are considered virulence determinants, but only in a few instances has it been unequivocally established that toxins are the sole or major determinant of the disease process. Often there are a number of other factors that together with the toxic factor play a role in the disease process. Viruses do not produce toxins per se, although some of their surface components may be cytotoxic (toxic to mammalian cells). The adenovirus, for example, possesses a surface projection, called a penton fiber (Figure 22-15), that may be considered cytotoxic, but other viral components with

Table 22-5 Characteristics that differentiate exotoxins from endotoxin

EXOTOXINS	ENDOTOXIN
Excreted by living cells	Part of cell wall of gram-negative bacteria and usually released on lysis of cell
Most are polypeptides with molecular weights between 1×10^4 to 9×10^5 daltons	Lipopolysaccharide (LPS) complex; lipid A is believed to be toxic factor; low molecular weight
Relatively unstable to temperatures above 60° C	Relatively stable to temperatures above 60° C for several hours with no loss of toxicity
Very antigenic and stimulate formation of antitoxic antibodies	Does not stimulate formation of antitoxin but does stimulate formation of antibodies to polysaccharide of LPS
Can be converted to toxoid* by treatment with formalin, acids, heat or other agents	Cannot be converted to toxoid
Do not produce fever in host	Does produce fever in host
Very toxic in small microgram quantities to laboratory animals	Weakly toxic; lethal for laboratory animals only in hundreds of micrograms

* A toxoid is a toxin that has been treated to remove its toxicity but not its antigenicity.

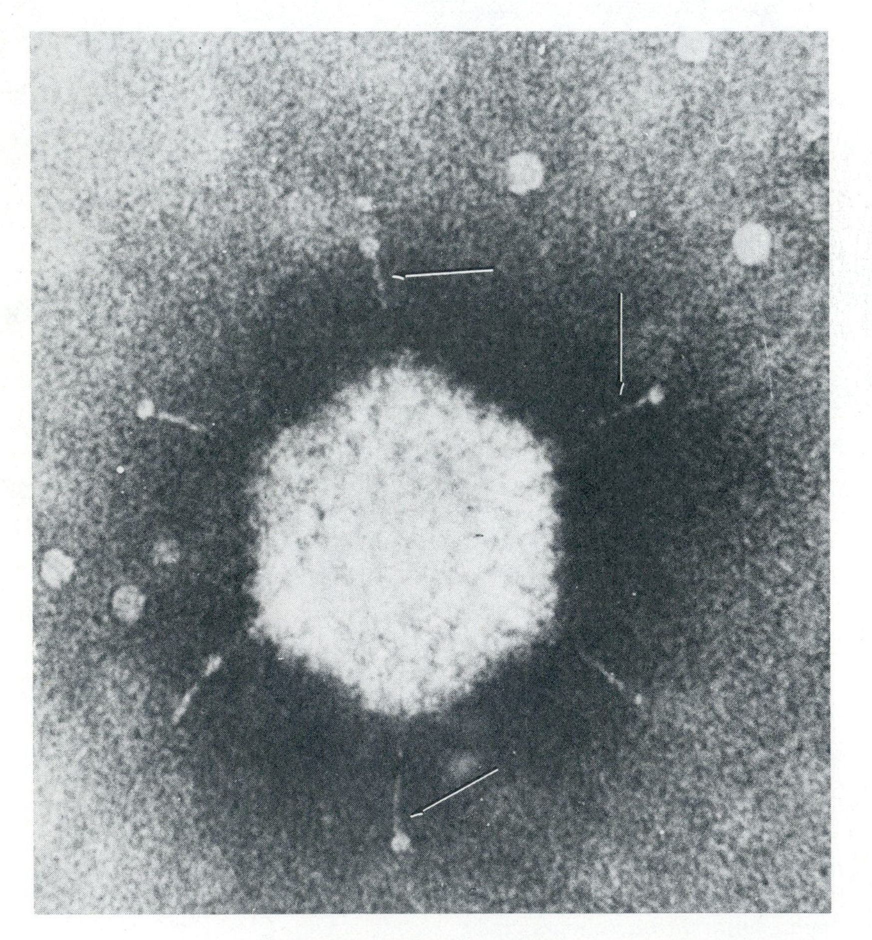

Figure 22-15 Electron micrograph of adenovirus demonstrating penton fibers *(arrows)* considered to be cytotoxic.

From Valentine, R.C., and Pereira, H.G. J. Mol. Biol. ***13:****13, 1965. Copyright Academic Press, Inc. (London), LTO.*

cytotoxic properties have not been identified.

Exotoxins. Exotoxins are proteins that are produced primarily by gram-positive bacteria (Table 22-6). Exotoxins are highly immunogenic (stimulate the formation of antibodies) and, when treated with agents that inactivate or temper their toxic properties, can be used as ***toxoids.*** Many exotoxins are enzymes that cause damage to a number of different components of the cell and during infection are spread into surrounding tissues and/or into the bloodstream. The relationship between toxin production and disease is not always a simple or direct one. The mechanism of exotoxin production in the agent of diphtheria *(Corynebacterium diphtheriae)* is multifaceted and is associated with a temperate virus. Toxin is synthesized only by strains of *C. diphtheriae* lysogenic for bacteriophage carrying the tox gene. The level of inorganic iron in the cell controls the production of exotoxin, and as the level of iron increases, the synthesis of exotoxin is decreased. The exotoxin released at the site of infection enters the bloodstream and can inhibit protein synthesis in many different organs but has its most devastating effects on the heart. The complete exotoxin molecule consists of two polypeptides: A and B. Polypeptide B is required for attachment of the toxin to the cytoplasmic membrane, and after the toxin is engulfed by the cell, fragment A, which is the actual inhibitor of protein synthesis, is released into the cytoplasm of the cell. Fragment A inhibits protein synthesis by preventing the transfer of amino acids to a growing polypeptide chain.

Table 22-6 Bacterial exotoxins

MICROORGANISM	NAME OF DISEASE OR TOXIN	ACTION	SYMPTOMS PRODUCED IN HOST
Clostridium botulinum	Botulism	A neurotoxin that blocks release of acetylcholine at nerve synapses	Double vision (diplopia), respiratory paralysis
Clostridium perfringens	Gas gangrene, alpha toxin	A lecithinase that splits lecithin in cell membranes	Hemolysis, toxemia as cause of death
	Food poisoning, enterotoxin	Stimulates adenyl cyclase and increases AMP, resulting in hypersecretion of water	Diarrhea with loss of water and monovalent ions
Clostridium tetani	Tetanus (lockjaw)	Inhibits antagonists of motor neurons in spinal cord and brain	Hyperreflex action of muscles and violent spasms of skeletal muscles; respiratory failure as cause of death
Corynebacterium diphtheriae	Diphtheria	Inhibits protein synthesis	Heart damage as most common cause of death
Staphylococcus aureus	Food poisoning	A neurotoxin that affects brain center that induces vomiting	Vomiting
	Scalded skin syndrome, exfoliation	Causes intradermal separation	Erythema and exfoliation of skin
Streptococcus pyogenes	Scarlet fever, erythrogenic toxin	Causes vasodilatation	Maculopapular rash
Bacillus anthracis	Anthrax; toxin is a three-component complex	Causes increase in vascular permeability	Pulmonary edema and hemorrhage
Vibrio cholerae	Cholera, enterotoxin (choleragen)	Stimulates adenyl cyclase and increases cAMP in gut	Diarrhea with massive losses of water and ions
Escherichia coli	Traveler's diarrhea, (enterotoxin)	Stimulates adenyl cyclase and increases cAMP in gut	Diarrhea
Pseudomonas aeruginosa	Exotoxin A	Similar to diphtheria toxin	?
Bacielus cereus	Enterotoxin	Similar to cholera toxin	Diarrhea

Exotoxin production by microorganisms is not always associated with infection in the host. Some toxins are produced in foods that have been contaminated by microorganisms. Ingestion of preformed toxins is responsible for the disease symptoms associated with food poisoning.

The exotoxin of *Clostridium botulinum* is a neurotoxin and one of the most potent toxins known to humankind. Less than $0.1 \times 10^{-3} \mu g$ can kill a mouse. The disease botulism is acquired by ingestion of foods contaminated with toxin excreted by the organisms. The toxin blocks the passage of nerve stimuli to nerve endings and causes paralysis, which frequently ends in death due to respiratory failure (see Chapter 28 for details of this disease.) There are seven types of botulinum toxins. Of these, types C_1 and D are controlled by temperate bacteriophages, and type A is the most frequently incriminated in fatal episodes. The toxin has been recovered in crystalline form and is a complex consisting of a ***toxic fraction*** and a ***hemagglutinin fraction.*** The toxic fraction consists of two dissimilar polypeptides that together have a molecular weight of 1.5×10^5 daltons. The chains are held together by a disulfide bond, but other forces also contribute to the integrity of the toxin. Toxin activity is lost if the sulfides in the disulfide linkage are reduced. The toxin complex (most types) is inactive until activated by proteases (such as trypsin) produced either by the bacterial cell or by a combination of bacterial and host enzymes. Activation takes place when the chains of the toxin assume a particular configuration. The toxin blocks cholinergic (mediated by acetylcholine) innervations but not adrenergic (mediated by norepinephrine) innervations. It acts at the junction formed by the nerve and the muscle or structure it innervates but not the affected structure per se. The inability of toxin-affected muscle to respond to stimuli is believed to be due to either the inhibition of acetylcholine synthesis (perhaps by preventing uptake of choline) or prevention of its release from motor nerves. Calcium ions appear to play an important role, because they are normally required for acetylcholine release. Studies with intact animals have demonstrated that bathing poisoned muscle in solutions containing high concentrations of calcium ions improves the release of acetylcholine.

Exotoxins that directly affect the tissue of the intestinal tract and cause intestinal disturbances are often referred to as ***enterotoxins.*** Two diarrheal diseases, one caused by *Escherichia coli* and the other by *Vibrio cholerae,* are the result of enterotoxin production (Table 22-6). The enterotoxin binds to a biochemical component of the intestinal tract and activates the enzyme adenyl cyclase found in mammalian cytoplasmic membranes. This enzyme causes the conversion of adenosine 5-triphosphate to cyclic adenosine monophosphate (cAMP), which in turn causes a change in the electrical potential of the cytoplasmic membrane, leading to a loss of fluids and electrolytes. Twenty liters of fluid may be lost per day in such diseases as cholera, caused by *V. cholerae.*

The fungi also produce exotoxins, the most active of which is called ***aflatoxin,*** produced by certain species that contaminate grains. The toxin, when ingested, can cause severe illness and death and has been implicated as a cause of liver cancer in populations of Southeast Asia. The fungal species that produces aflatoxin in grains can infect humans, but it is uncertain if toxin production can occur in vivo or if it is involved in the disease process. Aflatoxins are discussed in Chapter 31.

Endotoxins. Endotoxins are part of the cell wall of gram-negative bacteria. The most important constituent of the endotoxin is the lipopolysaccharide (LPS), which is composed of three components: a core of polysaccharide whose composition in species of the same genus may be the same but different in other genera; the O-specific polysaccharide region, which gives serological specificity to the species and whose alteration may change a virulent species to a nonvirulent one; and the lipid component, which is the actual toxic element of the LPS complex (see Figure 7-6).

Endotoxin has been studied extensively, but no direct role in the pathogenesis of any human diseases has been established for it. Studies comparing the symptoms of gram-negative infection in humans and laboratory animals indicate that endotoxin may play an important role in fever production and vascular change. Endotoxin injected directly into animals causes a release of fever-inducing agents from macrophages 60 to 90 minutes after injection. Injection of endotoxin can initiate a series of changes in the circulation: vascular permeability, decrease in blood flow, and shock. Disseminated intravascular coagulation, in which blood elements are clotted, leading to occlusion of the small peripheral blood vessels, can also be a complication of gram-negative infections. Shock and disseminated intravascular coagulation can lead to death, and both phenomena have been observed in gram-negative infections in humans.

Not all gram-negative bacteria possess endotoxin of equal potency, and their effect on the host is variable.

Damage via the immune response (immunopathology). One of the first events that normally occurs in the host following contact with an infectious agent is inflammation. Inflammatory materials are released by both host and microbe, and in most instances the response does not cause major damage to the host. A few infectious diseases are characterized by pathological changes due to immune responses. Rheumatic fever, a sequela of infection by *Streptococcus pyogenes,* is an example of tissue damage via the immune response and is discussed in Chapter 28.

Recovery from infection

Part of this chapter is devoted to an explanation of the nonspecific mechanisms used by the host to resist infection, but the final outcome in many instances is dependent on immunological factors, which are discussed in succeeding chapters. Some of these immunological factors are also associated with nonspecific mechanisms. Macrophages, for example, phagocytize parasites nonspecifically, but during the degradation process the digested microbial components are processed for use by those cells of the immune system (lymphocytes) that will produce antibodies (see Chapter 23). Complement is also involved in specific immune reactions and thus has a profound effect on limiting infections. Complement is also discussed at more length in Chapter 23.

Despite all the efforts of the host in defending against infection, there are some infections in which the microorganism is not eliminated but instead persists in the host. ***Persistent infections*** usually cause only minor illness, but their most important contribution to the host-parasite relationship is that they help the infectious agent to remain in the community and to be spread to others. Transmission of microorganisms that cause persistent infections is more appropriately discussed in Chapter 26.

SUMMARY

In the vertebrate host microorganisms may exist as commensals or parasites. Many parts of the body harbor commensal species, most of which are bacteria. Many factors influence an organism's ability to act as a commensal, including the presence or absence of oxygen at the site of colonization, nutritional requirements of the organism, pH of the environment, and the effect of other microorganisms inhabiting the colonization site. Commensals, by their sheer numbers or the products they elaborate, benefit the host by preventing the colonization of pathogens.

Most parasites are nonindigenous species; however, under certain conditions in which the host's defense mechanisms have been reduced, commensals may also become parasitic and harm the host. The ability to harm the host is related to the virulence of the microorganism, that is, its disease-producing traits.

Parasites can cause disease when they (1) gain entry into the host, (2) colonize the host tissue and multiply, (3) resist host defense mechanisms, and (4) damage host tissue. Entry into the skin is usually by abrasions or by bites from parasite-carrying insects such as ticks, mosquitoes, lice, and flies. Entry into the respiratory tract is through the inhalation of organisms suspended in air or directly from secretions expelled by infected individuals or animals. Entry into the intestinal tract is usually by the ingestion of contaminated food or water. Entry into the normally sterile genitourinary tract is often from prior conditions that produced obstruction of the tract.

The ability to colonize and multiply in the host is often related to the nature of the surface components on the microorganism, as well as the nature of the surface components on the tissue to be colonized. Several surface components have been discovered on bacteria and viruses that promote adherence to host tissue. The ability to penetrate the host cell may be due to the release of toxic products that destroy the epithelial surface, or the microorganism may be taken up by epithelial cells in a process similar to phagocytosis.

The parasite can resist host defense mechanisms in a number of different ways. It can interfere with nonspecific host responses such as leukocyte chemotaxis, the inflammatory response, ingestion by phagocytes, intracellular digestion by phagocytes, and phagocytic survival. The ability of the parasite to interfere with these defense mechanisms is related to surface components or products released by the parasite.

The parasite damages tissue through the act of multiplication and through the release of toxins—both exotoxins and endotoxins. Toxins are released primarily by bacteria. Vaccines have been prepared against some of the most potent toxins, such as diphtheria and tetanus. Damage to host tissue can also result from inflammatory events (immunopathology) that occur in response to disease. For example, rheumatic fever is believed to be due to antistreptococcal antibodies produced during streptococcal pharyngitis. The antibodies react with heart muscle and cause disease.

Parasites are also aided in the infection process by the release of various products that promote the spread of the parasite in tissue. These include enzymes such as hyaluronidase, collagenase, and coagulase, as well as the products hemolysin and fibrinolysin.

Study questions

1. Discuss the ways in which the indigenous population of microorganisms in humans is maintained at relatively constant numbers.
2. What is a pathogenic microorganism? Does a pathogenic microorganism always cause disease? Explain.
3. Discuss the ways in which microorganisms can infect humans through the skin, respiratory tract, intestinal tract, and genitourinary tract. Discuss the ways in which the host can resist the organism.
4. What pathogenic determinants provide the microorganism with the ability to colonize and invade the host?
5. Give some examples in which attachment to an epithelial surface provides a pathogenic potential. What is the chemical nature of these adherence factors?
6. How can the invading microorganism resist phagocytosis?
7. Give some examples in which attachment to an epithelial surface provides a pathogenic potential.
8. What is the difference between a toxoid and a vaccine?
9. Explain the mechanism of action of diphtheria toxin, cholera toxin, and botulinum toxin.
10. Of what value are commensal microorganisms to the host?
11. In what three ways do microorganisms damage host tissue?
12. Distinguish between the properties of exotoxins and endotoxins. What is their relationship to enterotoxins?
13. Discuss the mechanism of action of cholera enterotoxin
14. What is meant by an immunopathological disease. Give some examples.
15. Define the following: fibrinolysin, hemolysin, aflatoxin, desquamation, and pathogenicity.

Selected readings

Books

Bitton, G., and Marshall, K.C. (Eds.). *Adsorption of microorganisms to surface.* Wiley - Interscience, New York, 1980.

Braude, A.I. (Ed.). *Medical microbiology and infectious disease.* W.B. Saunders Co., Philadelphia, 1981.

Burnet, M., and White, D.O. *Natural history of infectious disease* (4th ed.). Cambridge University Press, New York, 1972.

Mims, C. *The pathogenesis of infectious disease.* Grune & Stratton, Inc., New York, 1976.

Randall, L.L., and Philipson, L. *Virus receptors.* Chapman and Hall, Ltd., London, 1981.

Journal articles

Beachey, E.H. Bacterial adherence: adhesion-receptor interactions mediating the attachment of bacteria to mucosal surfaces. *J. Infect. Dis.* **143:**325, 1981.

Bennett, J.V. Toxins and toxic shock syndrome. *J. Infect. Dis.* **143:**631, 1981.

Bizzini, B. Tetanus toxin. *Microbiol. Rev.* **43:**224, 1979.

Bullen, J.J. The significance of iron in infection. *Rev. Infect. Dis.* **3**(6):1127, 1981.

Costerton, J.W., Geesey, G.G., and Cheng, K.J. How bacteria stick. *Sci. Am.* **238:**86, 1978.

Densen, L.P., and Mandell, G. Phagocytic strategy vs. microbial tactics. *Rev. Infect. Dis.* **2**(5):817, 1980.

Elsbach, P. Degradation of microorganisms by phagocytic cells. *Rev. Infect. Dis.* **2**(1):106, 1980.

Field, M. Modes of action of enterotoxins from *Vibrio cholerae* and *Escherichia coli. Rev. Infect. Dis.* **1**(6):918, 1979.

Gabig, T.G., and Babior, B.M. The killing of pathogens by phagocytes. *Annu. Rev. Med.* **32:**313, 1981.

Gallin, J.I. Disorders of phagocyte chemotaxis. *Ann. Intern. Med.* **92:**520, 1980.

Gibbons, R.J., and Van Houte, J. Bacterial adherence in oral microbial ecology. *Annu. Rev. Microbiol.* **29:**19, 1975.

Johnston, R.B., et al. The role of superoxide anion generation in phagocytic bactericidal activity. *J. Clin. Invest.* **55:**1357, 1975.

Mills, E.L., and Quie, P.G. Congenital disorders of the functions of polymorphonuclear neutrophils. *Rev. Infect. Dis.* **2**(3):505, 1980.

Mogensen, S.C. Role of macrophages in natural resistance to virus infection. *Microbiol. Rev.* **43:**1, 1979.

Peterson, P.K., and Quie, P.G. Bacterial surface components and the pathogenesis of infectious disease. *Annu. Rev. Med.* **32:**29, 1981.

Roberts, N.J. Temperature and host defense. *Microbiol. Rev.* **43:**241, 1979.

Root, R.K., and Cohen, M.S. The microbicidal mechanisms of human neutrophils and eosinophils. *Rev. Infect. Dis.* **3**(3):565, 1981.

Schlievert, P.M. Identification and characterization of an exotoxin from *Staphylococcus aureus* associated with toxic shock syndrome. *J. Infect. Dis.* **143:**509, 1981.

Smith, H. Microbial surfaces in relation to pathogenicity. *Bacteriol. Rev.* **41:**475, 1977.

Wilkinson, P.C. Leukocyte locomotion and chemotaxis: effects of bacteria and viruses. *Rev. Infect. Dis.* **2**(2):293, 1980.

Chapter 23 IMMUNOLOGY Factors and Functions of the Immune System

Bryan Hoerl

Immunology focuses on the body's ability to specifically recognize and interact with certain substances that for the most part are foreign to the body. The immune system forms products that specifically interact with that foreign substance, called an ***antigen*** or an ***immunogen,*** during the initial encounter and in future encounters. The effects that an immune response initiates have a wide and varied range—from lifesaving protective actions such as occur in specific resistance to infectious agents and their products to life-threatening allergic reactions such as anaphylactic shock.*

The immune system is based in lymphoid tissue such as lymph nodes, bone marrow, thymus, and spleen. The ***lymphocyte*** is the prime cell type of the immune system because it forms immune products that are specific for foreign substances. Lymphoid tissue and the lymphatic system consist of a network of vessels and tissues whose functions include production and circulation of lymphocytes, capture of foreign substances by macrophages, and production of specific immune products by lymphocytes.

Nonspecific resistance is often inadequate to cope with foreign agents or substances. This is especially so when the foreign agent is a microorganism that can multiply without challenge in the host. Individuals with serious defects of the immune system are easy prey to microorganisms, even to microorganisms that ordinarily are nonpathogenic. The immune system possesses three characteristics that may be expressed when it encounters a foreign agent: ***specificity, memory,*** and ***recognition of self-antigens.*** These cardinal features are explained more fully under "Nature of the Immune System."

The specificity of responses by the immune system make possible swift and efficient actions that nonspecific resistance responses are incapable of, or that they perform inadequately. Immune sysem products specifically facilitate the phagocytosis of microorganisms, neutralization of toxins, lysis of microorganisms, destruction of host cells and the microorganisms they contain, and the production of other beneficial actions.

SCOPE OF IMMUNOLOGY

Immnology has ties to all of the disciplines and subdisciplines of life sciences. The ties with microbiology are especially strong because the major foreign agents that enter the body are microorganisms and the principal mechanism in protecting against them is the immune system. The immune system also has a central role in allergies. Among the substances that one can be

* Anaphylactic shock is a hypersensitive reaction resulting from a second exposure to a substance. The affected individual exhibits smooth muscle contraction, dilation of the vascular system, and edema. Hypersensitivity to penicillin, for example, results in over 200 deaths per year.

allergic to are microorganisms and the drugs that are used to combat them. Products of the immune system, furthermore, are used routinely to aid in the laboratory identification of microorganisms and their products.

Immunology transcends microbiology. A number of noninfectious diseases stem from a malfunctioning or an overresponding immune system. There is a strong alliance of immunology to hematology (the study of blood), especially in those aspects that pertain to transfusion and to pregnancy (for example, the Rh system). The response of the host to the transplantation of organs and tissue is a function of the immune system. Aside from the applied areas just described, immunology makes contributions to basic science areas such as genetics, biochemistry, and other disciplines that employ antigen-immune product interactions in identification procedures. The immune system is also used as a model system in the study of differentiation and cellular interactions.

STUDYING THE IMMUNE SYSTEM

It is helpful to conceptually divide the study of the immune system into four major topics: the nature of the immune system (Chapter 23), the components of the immune system (Chapter 23), immunoidentification (Chapter 24), and clinical immunobiology (Chapter 25).

Nature of the immune system

As we have already stated, there are three distinctive properties that make the immune system an improvement over nonspecific resistance mechanisms: specificity, self-recognition, and memory.

Specificity. The immune system specifically recognizes certain foreign substances ***(antigenic substances)*** and forms products that interact with those substances.

Self-recognition. Many of the foreign substances with which the immune system interacts are similar to the body's own molecules, cells, or tissues. The immune system, however, can discriminate between the foreign substances and self-substances. It does not normally form products against its own tissues, but there are exceptions (see Chapter 25).

Memory. During responses to foreign substances some of the proliferated lymphocytes become ***memory cells.*** Memory cells are primed to act with speed and vigor the next time the same foreign agent is encountered. The existence of this memory capacity makes active immunization (vaccination) feasible (see later under "Secondary Immune Response"). The first time a foreign agent is encountered, there is a short lag time before the immune system can generate enough immune products to overcome, for example, an infectious agent. There is such a multitude of antigenic determinants in nature that the body could not possibly contain enough cells of each specificity to respond speedily on the first encounter with each and every foreign agent. Memory cells make it possible for the immune system to remember and to be alerted to specific foreign agents that have been previously encountered—for example, through infection or immunization.

Components of the immune system

The molecular and cellular components of the immune system and the principal components that interact with the immune system include immunogens-antigens, macrophages and lymphocytes, immune products, complement, and other mediators. These are defined below and are more fully explained later in the chapter.

Immunoidentification

Immune system products are used extensively in the laboratory identification of microorganisms and their products. A great variety of sensitive identification tests are available because immune system products are highly specific for the substances that elicit their producion and because most antigen-immune product reactions are detectable in vitro (see Chapter 24).

Clinical immunobiology

The consequence of in vivo reactions between a foreign substance and an immune product are varied, but in most cases they are beneficial. The immune system specifically protects against microorganisms, foreign molecules and cells, and cancer. Disease or injury to the host's own tissue can occur if the immune system is defective or if the immune system produces an inappropriate or an overexuberant response, such as occurs in allergic responses (see Chapter 25).

DEFINITIONS AND EXPLANATIONS OF FREQUENTLY USED TERMS AND CONCEPTS

1. ***Antigen (Ag):*** A substance that incites the immune system to form immune products that are specific for that substance. Antigens have two properties: immunogenicity and the ability to react specifically with the immune products. Antigens are sometimes called immunogens when the emphasis is on the first property, immunogenicity, which is the ability to incite the formation of immune products.
2. ***Antigenic determinant:*** Small chemical groups on the surface of immunogens or antigens with which immune products interact.
3. ***Immune product:*** The products of immunocompetent lymphocytes that have been stimulated by appropriate antigens (immunogens). The immune products of B lymphocytes are called ***antibodies.*** T lymphocytes, on antigen stimulation, proliferate and differentiate into ***effector T cells.*** In a general sense, the effector T cells can be regarded as the immune products of a T cell response.
4. ***Immunoglobulin (Ig):*** A glycoprotein that is composed of H chains and L chains and functions as the antibody molecule. Functionally, there are two types of immunoglobulins: surface immunoglobulins, which are present on the surface of lymphocytes, where they act as specific receptors (recognition molecules) for the antigenic determinants of immunogens, and secreted immunoglobulins, which are the products of B lymphocytes and appear in the body fluids (humors) as antibodies.
5. ***Antibody (Ab):*** An immunoglobulin that has a known specificity for a given antigen; that is, an antibody has binding sites for an identified antigen.
6. ***B lymphocytes (B cells):*** A major class of lymphocytes that produce immunoglobulins and are primarily involved in humoral immunity.
7. ***T lymphocytes (T cells):*** A major class of lymphocytes that are thymus dependent and that form effector T lymphocytes on stimulation by antigens. Effector T lymphocytes destroy (as "killer" cells) cells that have specific surface antigens. They also produce lymphokines and have immunoregulatory functions.
8. ***Humoral immunity:*** Immunity provided by antibodies; B cell immunity; antibody-mediated immunity (AMI). Examples include toxin neutralization and the augmentation of phagocytosis.
9. ***Cell-mediated immunity (CMI):*** Immunity provided by effector T cells; T cell immunity. Examples include intracellular parasite elimination, graft rejection, and protection from cancer.
10. ***Macrophage:*** A phagocytic mononuclear cell that is involved in the input phase of an immune response through its role of capturing immunogens and presenting them to antigen-reactive lymphocytes. The macrophage is also prominently involved in cell-mediated immune responses through its role as the final effector. Immune products frequently are not engaged in the final outcome, such as removal or destruction of the antigen or of the host's tissues. Macrophages frequently are called on by immune products to produce the final effect.
11. ***Immune response:*** A series of events encompassing the time when an immunogen enters the body to the formation of immune products. An immune response involves (a) the capture of macrophages of an immunogen and the presentation of the immunogen to the lymphocytes; (b) the attachment of the immunogen to specific receptors on the surface of lymphocytes, and (c) the formation of immune products by clones of the contacted lymphocytes.
12. ***Complement system:*** A system of serum proteins that once activated proceeds in a serial manner (cascade). Complement fragments that are generated during the cascade have a variety of biological activities, including cell lysis, attraction of phagocytes (chemotaxis), aiding phagocytes to capture substances (opsonization), and abetting acute inflammation.
13. ***Serology:*** Study of the serum; the use of immune products in identification procedures.

IMMUNE SYSTEM COMPONENTS AND RELATED FACTORS

T lymphocytes and B lymphocytes

Two major populations. It is important to understand that there are two major populations of lymphocytes: T lymphocytes, or T cells, and B

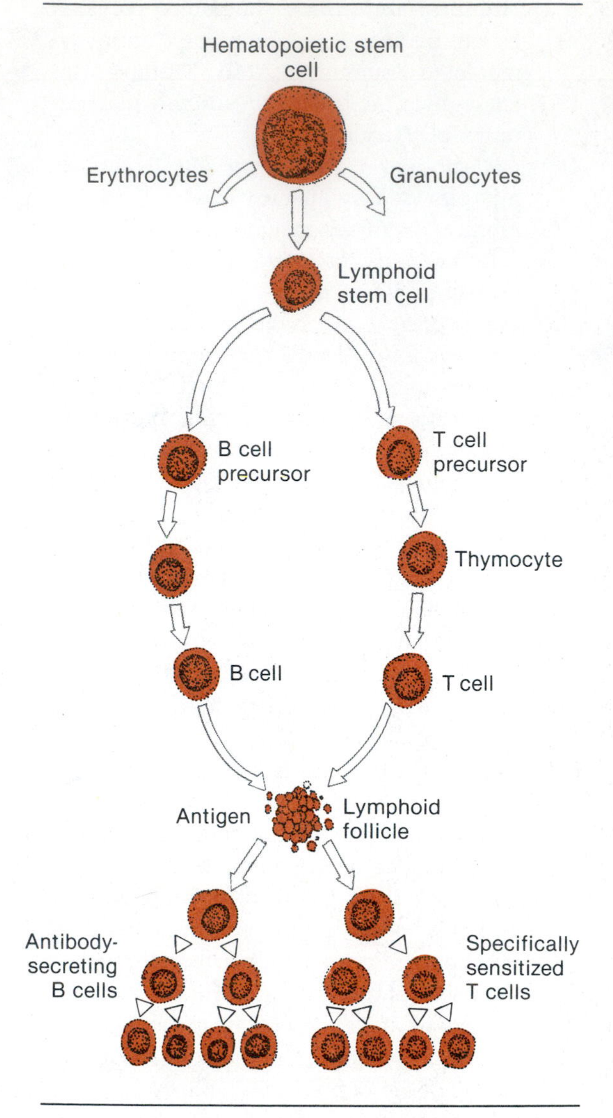

Figure 23-1 Development of B and T cells. Common precursor in bone marrow gives rise to pre-B and pre-T cells, which subsequently develop under as yet ill-defined influences to become mature B and T cells.

Redrawn from Boyd, R.F., and Marr, J.J. Medical microbiology, *Little, Brown & Co., Boston, 1980.*

lymphocytes, or B cells. They differ not only in ontogeny (development) and distribution, but also antigenically and functionally. The *T* of the T cells signifies ***thymus dependency,*** since the thymus gland at one stage is critical to the development of this population of lymphocytes. The B cells are symbolized with the letter *B* because in early experiments it was discovered that a cloaca-associated lymphoid organ in chickens, called the ***bursa of Fabricus,*** was critical to the development of immunocompetence of this population of lymphocytes. Evidence indicates that B cells in mammals differentiate in hematopoietic tissue (blood-forming tissue), which is primarily found in bone marrow; hence B cells are often referred to as being bone marrow derived.

Table 23-1 Percentage of B and T lymphocytes in various human lymphoid depots

SOURCE	B CELLS	T CELLS
Peripheral blood	1 to 15	70 to 75
Thymus	1	99
Spleen	40 to 45	40 to 45
Lymph node	15 to 20	65 to 70
Gut-associated lymphoid tissue	60	20 to 25

From Moticka, E.J., In *Medical microbiology*, Boyd, R.F., and Marr, J.J. (Eds.). Little, Brown & Co., New York, 1980

Stem cells in the bone marrow. Four functionally different cell types emerge from primitive stem cells of the bone marrow: cells involved in the transport of oxygen and carbon dioxide, namely red blood cells; cells for phagocytic defense, namely monocytes and granulocytes; cells involved in blood clotting, namely platelets (thrombocytes); and cells involved in immune responses, namely T cells and B cells (Figure 23-1).

T cell development. The stem cells of the lymphoid series that are destined to become T cells require either temporary residence in the thymus or exposure to humoral factors derived from the thymus. The thymus, in this role of T cell processing, is called a primary or central lymphoid organ. The majority of T cells while in the thymus are not fully competent to perform their immunological assignment. They travel primarily to lymph nodes and the spleen, which are referred to as secondary or peripheral lymphoid organs. The T cells preferentially locate in certain areas of the lymph nodes (paracortical areas) and of the spleen (periarteriolar sheaths), where they await activation by immunogens (antigens).

B cell development. The discrete organ for the differentiation of mammalian B cells, like the bursa in chickens (or the thymus for T cells), has not been identified, but it is most probably bone marrow. After differentiation in the central lymphoid organ, B cells migrate to peripheral lymphoid tissue, such as the lymph nodes and spleen, where they preferentially locate in the follicles.

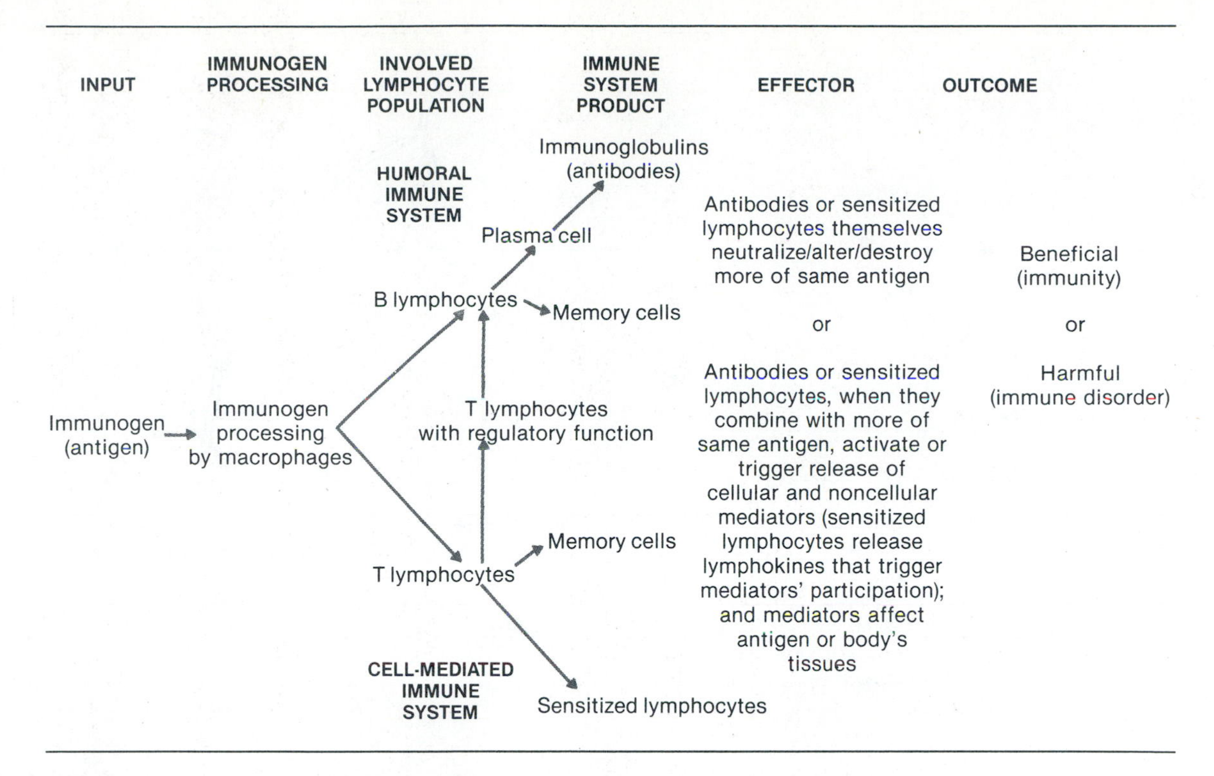

Figure 23-2 Overview of immune response.
Redrawn from Boyd, R.F., and Hoerl, B.G. Basic medical microbiology (2nd ed.). Little, Brown & Co., Boston, 1981.

Other differences between T cells and B cells. Resting cells of both T cells and B cells are morphologically indistinguishable, but there are differences in surface markers that can be detected by appropriate tests. T cells constitute the greater percentage of circulating small lymphocytes (Table 23-1). T cells usually have a life span of months, and this varies according to the subclass of T cell, whereas B cells have a life span of days to weeks. Functional differences will be apparent as the description of the immune system unfolds.

Overview of a typical immune response and its consequences

It is a good idea at this juncture to have an overview of a typical immune response and its consequences (Figure 23-2). When a substance of a type that can elicit a response by the immune system (immunogen or antigen) enters the tissues, it is usually engaged and "processed" by macrophages. The processed immunogen is then presented to either a B cell or a T cell, depending on the nature of the immunogen. The lymphocyte has genetically determined receptors on its surface that are specific for the immunogen. The lymphocyte, when stimulated by its corresponding immunogen, forms a clone of cells, most of which go on to produce an immune product or products specific for that immunogen. Some of the cells, however, become memory cells.

The clone of immunogen-stimulated B cells differentiates into another cell type, the ***plasma cell.*** Plasma cells secrete an immune product called an antibody, which is a biomolecule of a type called an immunoglobulin. It is the antibody that can specifically interact with molecular groups (antigenic determinants) that are located on the surface of the antigen. The antibody may have a direct or an indirect effect on the antigen. The neutralization of a toxin (the antigen) by an antibody, called an ***antitoxin,*** is an example of a direct effect. In many cases, however, the final effect on the antigen is not directly attributable to the antibody. Instead, other cells, systems, or pharmacologically active substances (referred to as intermediates and mediators) are triggered to go into action by the antigen-antibody union. When an antibody directed against a bacterial capsule, for example, combines with the capsule, the bacterium is not destroyed. Rather, the antibody serves to bind the bacterium to a phagocyte. The phagocyte, which was not very likely to capture the bacterium in the absence of helping substances such as antibodies, is now able to ingest and subsequently destroy the bacterium.

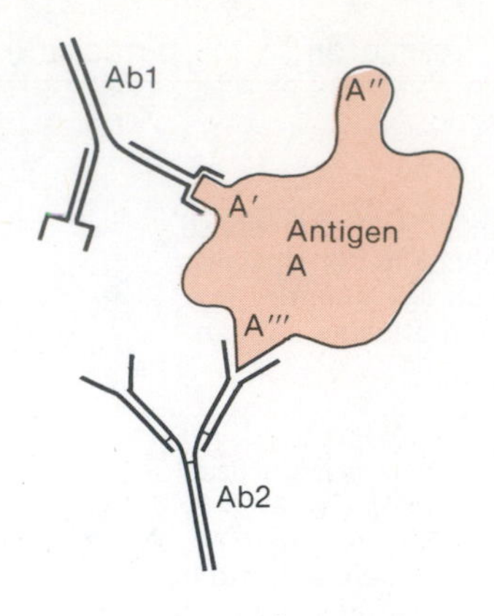

Figure 23-3 Antigenic determinants. Part of antigen bound by antibody is called antigenic determinant (A′, A″, A‴). Most antigens possess many determinants that are bound by different antibody molecules. Ab1 is specific for A′; whereas Ab2 is specific for determinant A‴.

Redrawn from Thaler, M.S., Klausner, R.D., and Cohen, H.J. Medical immunology, J.B. Lippincott Co., Philadelphia, 1977.

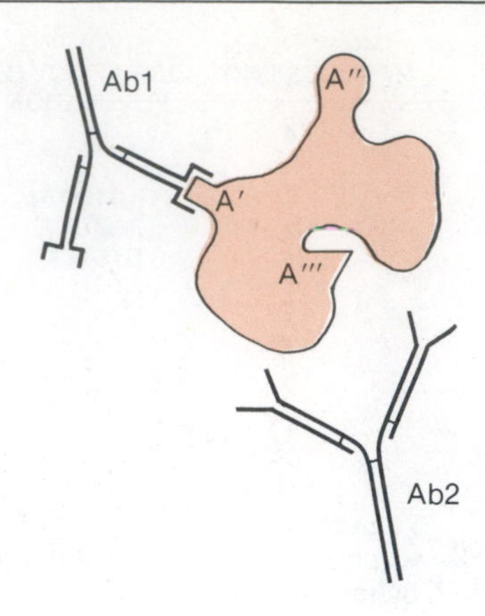

Figure 23-4 "Hidden" determinants. Only determinants expressed on surface of antigen can be bound by antibody. Thus antigen in figure has in effect only two determinants, and Ab2 does not have access to "hidden" determinant A‴.

In other words, the antibody did not directly eliminate the bacterium—the phagocyte was the final effector.

T cells, when contacted by their specific immunogen, multiply and increase in size, but they do not convert into another cell type as the B cells do. They remain as lymphocytes. Immunogen-stimulated lymphocytes are commonly referred to as ***sensitized lymphocytes.*** Sensitized lymphocytes sometimes directly destroy ("killer" cells) the antigen, or they produce substances called ***lymphokines.*** The lymphokines activate other cells (especially macrophages), systems, or pharmacologically active substances, which in turn exercise the final effect on the antigen.

The final outcome of an immune response for the body is either beneficial (immunity) or harmful (an immune disorder such as an allergy).

Immunogens/antigens

Attributes

Foreignness. The immune system's ability to discriminate between self- and nonself-antigens occurs during development. Most antigens are foreign and include such substances as bacteria, viruses, fungi, toxins, red blood cells, serum, and vaccines.

Autoantigens. There are exceptions to the rule that antigens are foreign. There are human and animal diseases, for instance, that are called autoallergies or autoimmune diseases (see Chapter 25). In autoimmune diseases products are directed against the person's own molecules, tissues, or cells that are damaging to the self-antigen.

Determinant groups. Specific immune products are formed only against limited portions of an immunogenic molecule, and it is with these portions that the immune products interact (Figure 23-3). An immunogenic molecule, or antigen, is composed of two parts: a carrier portion that makes up the bulk of the molecule and chemical groups called antigenic determinants. Functional antigenic determinants (those that interact with antibody) project from the surface of the molecule (Figure 23-4). Chemical groups that are internal (hidden) can come to serve as determinants if the immunogen or antigen is structurally altered, for example, through enzyme action. The carrier portion also has determinants to which lymphocytes respond. Immunogens contain multiple determinants (multivalent), but only those determinants that are on the surface are functional. Immunogens-antigens may have repeating copies of the same determinant, or there may be distinct determinants, each of which may be present in varying numbers. Determinants are no larger than 5 to 7 amino acids of a protein or 5 to 7 glucose residues of a polysaccharide. The antigen-binding region (Fab) of the antibody molecule binds to the antigen determinant group and because of its small size cannot accommodate structures larger than its own binding sites. The antibodies formed are complementary for a given antigenic determinant, and this accounts for specificity, but specificity is not absolute. It is possible for the Fab region to bind to a determinant group that is chemically similar to but not exactly like the specific determinant group. In other words, different antigens A and B may bind the same antibody, a property referred to as cross-reactivity. A large degree of cross-reactivity indicates a structural similarity between the two antigens.

Size. The size of the entire immunogenic molecule influences its ability to incite an immunogenic response. Usually, the larger the molecule, the greater the immunogenic potential. Apparently, the larger the molecule, the greater the number of available determinant groups. Generally, an immunogen must have a size of at least 1000 daltons to stimulate the immune-response mechanism.

Chemical composition and complexity. The greater the molecular complexity of the macromolecule, the greater its antigenicity. Most natural substances are constructed of a large number of low molecular weight constituents rather than one or two repetitive determinants. The majority of immunogens are natural proteins, such as the proteins found in serum, cells and tissues, enzymes, protein components of microorganisms, and microbial products. Polysaccharides such as those found in microbial capsules are immunogenic, but lipids are not immunogenic. Organic molecules composed of combinations of proteins, carbohydrates, and lipids such as lipoproteins (cell membrane), glycoproteins (the A and B blood cell group substance), polypeptides (insulin), lipopolysaccharides (endotoxins), and nucleic acids are immunogenic.

Haptens. Certain molecules, because of their small size, are not immunogenic. When coupled with a carrier molecule (the coupling can be done artificially through laboratory manipulations, or it can occur naturally), they can be immunogenic and are referred to as ***haptens.*** Antibodies or sensitized lymphocytes formed against the haptens will give a visible reaction when they interact with hapten-carrier conjugates, but not if they are allowed to interact with unconjugated hapten molecules. Haptens are sometimes referred to as partial or incomplete antigens. Some of the naturally occurring substances involved in allergic contact dermatitis, such as the catechols of poison ivy, are haptens (see Chapter 25).

Immunoglobulins

Immunoglobulins are glycoproteins that make up about 20% to 25% of serum proteins. They are found primarily in the gamma globulin fraction of the serum and on the surfaces of B lymphocytes. Immunoglobulins function in the immune response mechanism in two ways: as surface receptors and as immune products. ***Surface immunoglobulins*** located on B lymphocytes function as specific receptors for the determinant groups of immunogens. When the appropriate immunogen contacts the specific immunoglobulin of the lymphocyte surface, the latter proliferates and differentiates into a ***plasma cell.*** Plasma cells go on to produce immunoglobulins that can be found in body fluids and on mucous surfaces, but most are found in the serum. An immunoglobulin is called an antibody when the antigen for which the immunoglobulin is specific is known. Antibodies may be more specifically named, for example, ***agglutinins, precipitins, antitoxins,*** and ***opsonins,*** depending on the effect they have on the antigen.

Physicochemical properities

Lines of investigation. Much of the understanding of the nature of the immunoglobulin molecule stemmed from four lines of investigation:

1. The experiments of Edelman and co-workers culminated in 1959 with the revelation that the basic immunoglobulin molecule is composed of two heavy (H) and two light (L) polypeptide chains.
2. Porter in 1959, using controlled enzymatic degradation of the immunoglobulin molecule, obtained two major fragments: the ***Fab,*** or antigen-binding fragment, and the ***Fc,*** or crystalline fragment. Functional activities are relatable to those fragments.
3. Detailed information about immunglobulin composition was obtained through a third line of investigation, the examination of myeloma proteins. The cells that produce immunoglobulins sometimes become malignant and form clones of cells that yield large amounts of homogenous (monoclonal) nonspecific immunoglobulins called myeloma proteins. These diseases are called immunoproliferative disorders. Because the cancer cells are plasma cells, the cancers are also called plasmacytomas. Portions of the molecule, the ***L chain*** of the immunoglobulin molecule, may appear in the urine. These are known as ***Bence-Jones proteins.*** There are five major classes of immunoglobulins, all of which appear in serum in varying quantities. The immunoglobulins that appear in animals or humans with myeloma tumors usually belong to a single class of the five immunoglobulin classes. Investigations of myeloma proteins have also re-

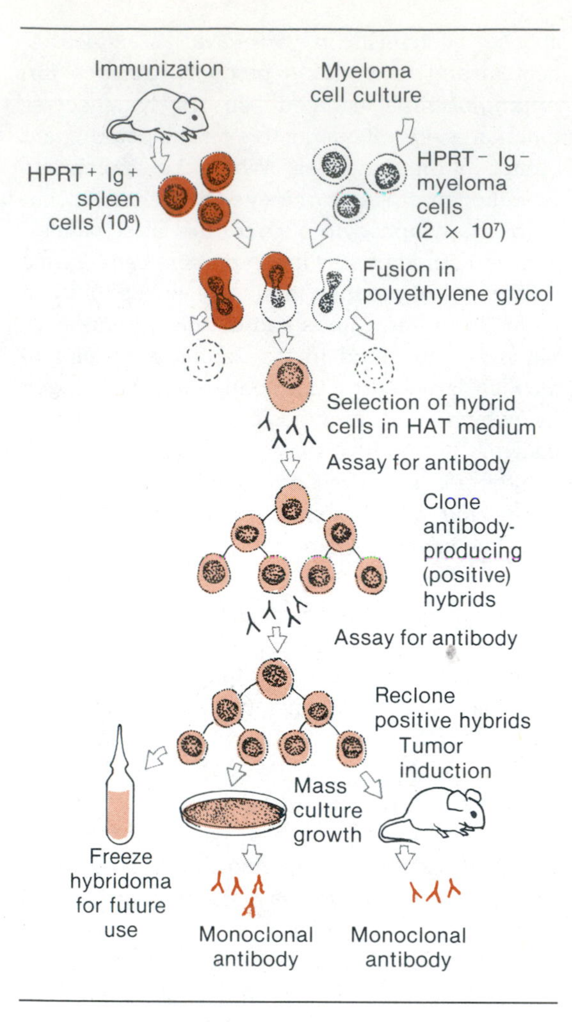

Figure 23-5 Hybridoma technique. Hybridoma production begins by immunizing mouse with selected antigen. Animal's spleen cells are harvested and mixed with mouse myeloma cells and briefly incubated with polyethylene glycol to promote fusion. Cells are transferred to growth medium containing hypoxanthine, aminopterin, and thymidine (HAT). Unfused myeloma cells and myeloma-myeloma hybrids, which lack enzyme HPRT (hypoxanthine phosphoribosyl transferase), cannot survive in HAT medium. Unfused spleen cells and spleen-spleen hybrids die naturally after a few replications. Remaining spleen cell–myeloma hybrids are assayed for antibody production against immunogen, and positive hybrids are cloned. Later they are again assayed for antibody production, and the positive hybrids recloned. At this point, hybridoma can be frozen and stored. To amplify antibody production, hybridomas are injected into mice, in which they produce ascites tumors that generate large amounts of monoclonal antibody. In human spleen cell–human myeloma hybridomas, amplification would be done by growing cells in mass culture or in immunosuppressed animals. Myeloma cells that are used in these experiments have a loss mutation for HPRT. This is important in selecting out myeloma cells that have hybridized with antibody-producing spleen cells. Only myeloma cells that have fused to spleen cells containing HPRT are able to use the hypoxanthine and thymidine in the HAT medium and, therefore, survive.

Redrawn from Scharff, M.D., Roberts, S., and Thammana, P. Hosp. Pract. ***16****:64, 1981.*

vealed how the five classes differ from each other, their amino acid sequences, and other features of the immunoglobulin molecule.

4. The most recent mode of investigation of immunoglobulins, called cell fusion or the ***hybridoma technique,*** is being actively explored. It is expected to further illuminate the composition and the genetics of immunoglobulins and to yield important clinical applications. In the hybridoma technique (Figure 23-5) the nucleus of a spleen cell that is producing antibodies specific for a particular antigen is fused in vitro with the nucleus of a ***myeloma tumor cell.*** The tumor cell multiplies and produces antibodies that are specific for a given antigen when the fused cell is transferred into a suitable experimental animal or is mass cultured. Two qualities are combined in the hybridoma: virtually unrestricted multiplication of immunoglobulin-producing cells and large yields of a monospecific monoclonal antibody. Stated another way, the hybridoma technique makes possible the mass production of selected antibodies of a single specificity. The importance of this technique cannot be overstated, both for the potential it offers for improving the understanding of underlying immunological and genetic phenomena and the potential for clinical applications (see box). As an example of a possible application of cell fusion, anticancer antibodies produced by this technique could have valuable diagnostic and immunotherapeutic applications.

Overall structure of the immunoglobulin molecule. The basic immunoglobulin molecule has four polypeptide chains, two of which are designated as heavy (H) chains and two as light (L) chains (Figure 23-6). The H chains of any particular immunoglobulin molecule are identical to each other, as are the L chains identical to each other. Each H chain of a representative immunoglobulin molecule has three regions or domains (C_H1, C_H2, and C_H3) and a variable region (V_H). The L chain, which is approximately one half the length of an H chain, has one constant region (C_L) and a variable region (V_L).

Immunoglobulin can be fragmented by enzyme treatment. One of the fragments is fab, the antigen-building fragment, which actually exists in duplicate on the basic immunoglobulin molecule. The variable region of one heavy chain (V_H) and the variable region of one light chain (V_L) constitute one antigen-binding site. The Fab end of the

APPLICATIONS OF MONOCLONAL ANTIBODIES

REPORTED TO DATE

- Routine diagnostic and investigative serology and tissue typing
- Identification and epidemiology of infectious agents
  - Viruses
    - Influenza: Monoclonal antibodies create genetic drift in vitro; map antigenic domains on hemagglutinin; compare in vitro mutants and in vivo variants; refine epidemiology
    - Rabies: Monoclonal antibodies reveal antigenic instability; reveal serological subtypes in street virus; refine epidemiology
  - Bacteria
  - Parasites
- Identification of tumor antigens; classification of leukemias and lymphomas
- Identification of functional subpopulations of lymphoid cells

ANTICIPATED IN HUMANS

- Passive immunization against
  - Infectious agents
  - Drug toxicity
- Provision of graft protection
- Potentiation of tumor rejection
- Manipulation of the immune response
- Targeting of diagnostic or therapeutic agents in vivo for
  - Detection of metastases
  - Delivery of cytotoxic agents in tumor cells

From Scharff, M.D., Roberts, S., and Pallaiah, T. *Hosp. Pract.* **16**(1):61, 1981.

Ig molecule is referred to as the N-terminus. The other end of the molecule is called Fc, the crystalline fragment, and is referred to as the C-terminus.

Porter found that the Fc fragment crystallized when placed in cold buffer and exhibited no antigen-binding capability. There are biological activities associated with the Fc fragment: antibodies attach to phagocytes and mast cells through the Fc portion of the molecule; the Fc portion is the attachment site for those immunoglobulins (IgG and IgM) that bind complement (see "Complement System"); and the capacity of crossing the placenta and entering the fetal circulation is associated with the Fc portion of immunoglobulin of class IgG.

The basic immunoglobulin molecule has a ***Y configuration*** (and is represented as such in illustrations) when the antibody molecule combines with appropriate antigenic determinants and has a ***T configuration*** when not binding to antigen. There is a flexible ***hinge region*** at about the middle of the H chains that demarcates the Fab segment from the Fc segment (Figure 23-7).

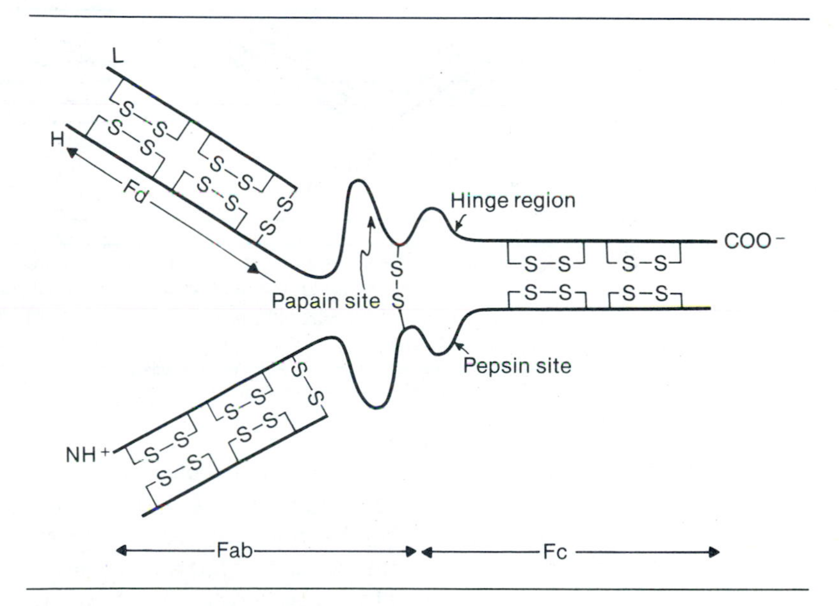

Figure 23-6 Structure of basic immunoglobulin molecule. Sites of enzyme (papain and pepsin) cleavage are indicated.

Redrawn from Golub, E.S. The cellular basis of the immune response. (2nd ed.) Sinauer Associates, Inc., Sunderland, Mass., 1981.

Immunoglobulin classes. Five major classes of immunoglobulins have been described in humans: IgG, IgM, IgA, IgE, and IgD. Some of the classes, furthermore, have subclasses that are de-

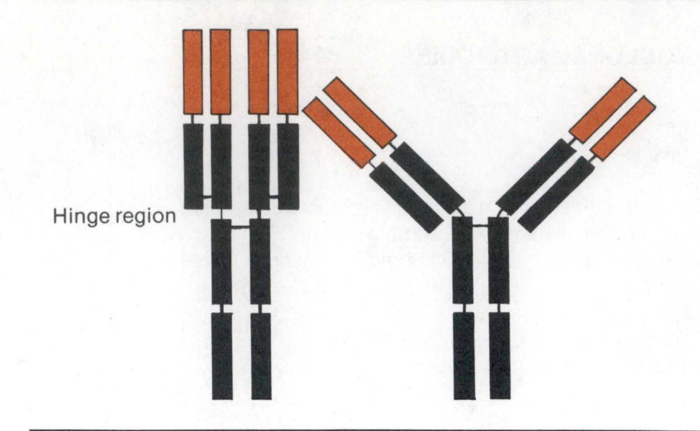

Figure 23-7 Hinge region of immunoglobulin molecule, where arms of antibody molecule may be free to rotate on binding antigen.

Redrawn from Thaler, M.S., Klausner, R.D., and Cohen, H.J. Medical immunology, J.B. Lippincott Co., Philadelphia, 1977.

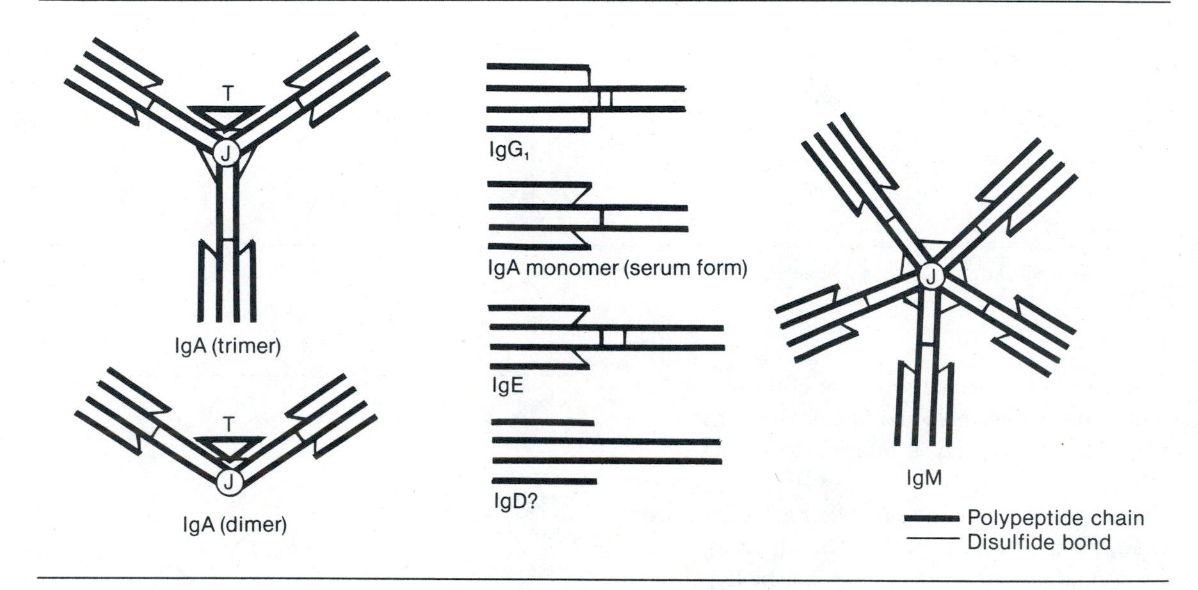

Figure 23-8 Representative structures for major human immunoglobulin classes. Each contains basic four-chain polypeptide unit, and some may exist as polymers (for example, IgM and IgA). Accessory polypeptides J (joining piece) and T (transport piece) may also be present in polymeric forms.

Redrawn from Joklik, W.K., Willett, H.P., and Amos. D.B. Microbiology *(17th ed.). Appleton-Century-Crofts, New York, 1980.*

termined on the basis of antigenic differences, such as IgG_1, IgG_2, etc. All but class IgD are known to exist in other mammals. All five classes are found in humans. IgE and IgD are present at very low levels. The classes differ chemically (in their amino acid content and sequences), structurally (Figure 23-8), and in biological functions (Table 23-2).

IgG. IgG is present in the serum in the greatest concentration of the five classes and accounts for over 70% of serum immunoglobulins. Its structure is that described previously for the basic immunoglobulin molecule.

Functionally IgG antibodies agglutinate and precipitate antigens, and form links between microorganisms and phagocytes, thereby facilitating endocytosis. The Fc portion of IgG binds and activates complement, thus inducing an inflammatory response, and it attaches to macrophages, thus facilitating antigen uptake. IgG is the only immunoglobulin that crosses the placenta, and antibodies transmitted to the fetus provide the newborn with a passive immunity to some of the diseases against which the mother has immunity.

IgM. IgM is the largest of the immunoglobulin molecules, consisting of five basic immunoglobulin molecules. There are ten heavy chains, ten light chains, ten identical antigen-binding sites, and five Fc regions. IgM has an added chain,

Table 23-2 Selected physicochemical and biological properties of immunoglobulins

	IgG	IgM	IgA	IgE	IgD
Physicochemical properties					
Sedimentation	7S	19S	7S to 11S	8S	7S
Molecular weight	148,000	900,000	170,000 400,000	200,000	200,000
Serum concentration (mg/100 ml)	1,000 to 1,500	60 to 180	100 to 400	0.01	3 to 5
Half-life in serum (in days)	18 to 23	5	6	2.5	3
Number of four-chain units	1	5	1 to 3	1	1
Effective combining sites (valences)	2	5	2 to 3	2	?
Biological properties					
Types of antibody	Agglutinin Precipitin Opsonin Neutralizing	Agglutinin Precipitin Opsonin Neutralizing	Secretory antibody on mucous surfaces	Reagin; skin sensitizing	?
Fixes complement	+	+	−*	−*	−
Crosses placenta	+	−	−	−	−
Other distinctive properties	Blocking antibody of atopic allergy hyposensitization	First Ig formed; ABO antibodies, receptor on B cell membrane	Found in serum and external secretions; surface protection	Reaginic antibody of anaphylaxis and atopic allergies; attaches to mast cells	Receptor on B cell membrane

From Boyd, R.F., and Hoerl, B.G., *Basic medical microbiology* (2nd ed.). Little, Brown & Co., Boston, 1981.
*Activates complement through the alternative pathway.

called the ***J chain,*** that is believed to be instrumental in joining together the five-unit structure, which is referred to as a pentamer. Disulfide bonds also serve in binding the basic units together. Even though there are ten antigen-binding sites, experiments indicate that all ten sites rarely bind to ten antigenic determinants at one time. Many antigens are too large or complex to fit into all ten antigen-binding sites at one time. IgM, which is found primarily in the blood, accounts for 5% to 10% of the immunoglobulin in serum.

Functionally, IgM is usually the first immunoglobulin produced in response to immunogen stimulation, but the synthesis is not prolonged. The human fetus is capable of forming IgM, so that its presence in fetal serum indicates infection or exposure to immunogens. IgM antibodies are very efficient in bringing about agglutination and in activating complement via the classic pathway (See "Complement System Pathways"). IgM antibodies figure prominently in blood transfusion reactions because they can destroy the improperly transfused red blood cells of the donor. These IgM antibodies (in this context they are more specifically designated ***isohemagglutinins***) are naturally occurring, since they develop spontaneously in humans without a known stimulus.

IgM has an important function in the early immune response sequence. It is present as a surface immunoglobulin on most B lymphocytes that are the precursors of antibody-secreting cells. IgM as a surface immunoglobulin differs from secreted (serum) IgM. It is a monomer instead of a pentamer, and the Fc portion is longer than it is in the secreted IgM molecule. As a surface immunoglobulin IgM provides an antigen-binding receptor that recognizes a specific antigen on contact.

IgA. IgA appears regularly in serum and in body secretions. It is structurally similar to IgG when it appears in the serum, consisting of the typical four-chain immunoglobulin unit. IgA in body secretions such as the gastrointestinal tract, saliva, colostrum, tears, and urine is called secretory IgA or ***sIgA.*** Secretory IgA consists of

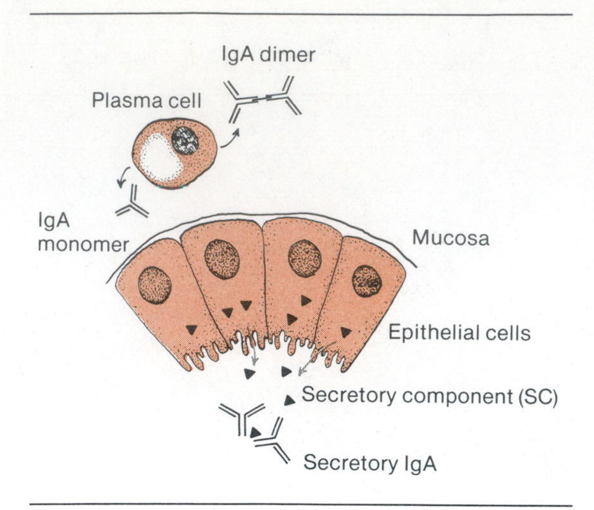

Figure 23-9 Steps involved in production of secretory IgA.

Redrawn from Hyde, R.M., and Patnode, R.A. Immunology, *Reston Publishing Co., Inc., Reston, Va., 1978.*

two and sometimes three basic immunoglobulin units joined together by a polypeptide J chain and one other polypeptide, called the ***secretory piece.*** IgA is formed in plasma cells located beneath the epithelial lining of mucosal surfaces, and the secretory piece is added to the IgA molecule as it is transported across the mucous membrane (Figure 23-9). Something about sIgA's structure prevents enzymes of the mucosal surface from degrading it.

IgA antibodies in the serum appear to have only a minor protective function. Individuals with serum IgA deficiencies appear not to suffer an impairment of their immunity. Secretory IgA on the mucosal surfaces, however, is immensely important as the first line of specific defense. A major defensive role it performs on the mucosal surface is to prevent the attachment of microorganisms. Attachment of an invading microorganism is a prelude to its penetrating the mucosa and subsequently entering the tissue. Some microorganisms, not to be outdone, produce proteases that destroy secretory IgA, Secretory IgA is regarded as the agent of topical, local immunity.

IgE. The overall structure of IgE is similar to that of IgA with the exception that the heavy chains are larger and they have an additional domain. IgE antibodies are best known for their role in acute allergic disease. They are variously named as ***reagin, atopic antibody,*** or ***skin-sensitizing antibody.*** In humans they are involved in certain allergies ***(hypersensitivities)*** of the immediate type that are referred to as ***atopic allergies, local and systemic anaphylaxis,*** or ***type I reactions*** (see Chapter 25). IgE antibodies have a distinctive quality: the predilection to attach to ***mast cells*** and ***basophils*** that are rich in pharmacologically active mediators such as histamine, serotonin, heparin, and SRS-A (slow-reacting substance of anaphylaxis). In an allergic individual IgE antibodies that have been induced as a result of previous antigen (the antigen may be more specifically called an allergen) exposure are attached to mast cells in the tissue and to basophils in the blood. The allergen combines with the IgE antibodies that are located on the mast cell surface when the allergic individual is exposed to more of the same allergen. This antigen-antibody union initiates the release of the pharmacologically active mediators. The events resulting from the release of these secretory granules include smooth muscle spasms, dilatation and increased permeability of blood vessels, secretion of mucus by exocrine glands, attraction (chemotaxis) of neutrophils and eosinophils, stimulation of cyclases, and anticoagulation. The seriousness of the allergic reaction caused through these effects is determined by the amount of the substance released, as well as the suddenness and location of their release.

The beneficial functions of IgE antibodies are not entirely clear. The very existence of antibodies of this type suggests that the antibodies contribute in some way to protection of the individual. Near the body surfaces IgE antibodies initiate a beneficial inflammatory response when irritants penetrate the integument. Many of the allergens that elicit an overexuberant response, such as in anaphylactic shock, enter the host via injection (antibiotics, antisera), an unnatural route that evolution could not adapt to and guard against. IgE-mediated reactions possibly provide protection against intestinal parasites, since persons with chronic roundworm infections of the intestines have comparatively high levels of IgE in their serum. The IgE, however, does not appear to be worm specific.

IgD. IgD is present in serum in trace amounts and does not have a known protective antibody function. It acts as a surface immunoglobulin on B lymphocytes, just as surface IgM does. IgD recognizes and interacts with specific antigen that initiates B cell proliferation and differentiation. This ultimately leads to the secretion of antibody that is specific for the contacted antigen and to the formation of memory cells.

T cells and their immune product, sensitized lymphocytes

The immunocompetent T cell has antigen-binding sites that are specific for a given antigen, but the nature of the antigen recognition molecules is not known. When the appropriate antigen is contacted, the T cell undergoes blast transformation. The ***blast cell,*** which is called an ***immunoblast*** or ***lymphoblast,*** divides rapidly and forms an expanded population of T cells that are specific for the same antigen that triggered blast transformation (Figure 23-10). Some of the blast cells develop into memory cells.

Cell-mediated immunity. The clones of T cells that have resulted from blast cell proliferation are called end cells and are the immune product of the T cell immune response mechanism. The T cell immune product is commonly referred to as a sensitized T cell. The immunity conferred by T cells is called cell-mediated immunity (CMI).* T cells are also involved in that type of hypersensitivity (allergy) referred to by the following interchangeable terms: ***delayed hypersensitivity, cell-mediated hypersensitivity,*** and ***type IV hypersensitivity.***

Elimination of antigens. How does the T cell accomplish immunity in the host? During a T cell–mediated response the T cell is responsible for the specific recognition event, but often other cells bring about the final effect on the antigen or on the host's tissue. T cells destroy, eliminate, or alter antigens in a less direct way than B cells. Sensitized T cells release lymphokines, which, unlike antibodies, are not specific for antigens. T cells that are specifically sensitized to a given antigen infiltrate the area where the antigen is located. Other cell types, especially macrophages, are recruited and activated through the intervention of released lymphokines. The recruited cells come to constitute as much as 95% of the cell infiltrate of the affected area. It is the recruited cell, not the T lymphocyte, that ultimately is the final effector on the antigen. It should be noted that sensitized lymphocytes do not always eliminate antigens via the lymphokine mechanism. Some sensitized T cells, called 'killer'' cells (T_k cells) directly contact and destroy the antigen. The protective assignments of T cells include (1) immunity against those infectious agents that have an intracellular habitat (viruses and certain bacteria such as the bacillus causing tuberculosis) and those that tend to cause chronic infections (fungi); (2) recognition and elimination of aberrant "self" cells, namely cancer cells (the process is called cancer surveillance; and (3) recognition and elimination of foreign cells and tissues such as occur, for example, in graft rejection.

* CMI is primarily associated with T cells, but B cells and "natural killer" lymphocytes (NK cells) can also evoke or directly participate in cell-mediated reactions. It has been proposed that these various CMIs be more specifically designated; that T cell–mediated immunity be symbolized as TCMI, B cell–mediated immunity as BCMI, and NK cell–mediated immunity as NKCMI.

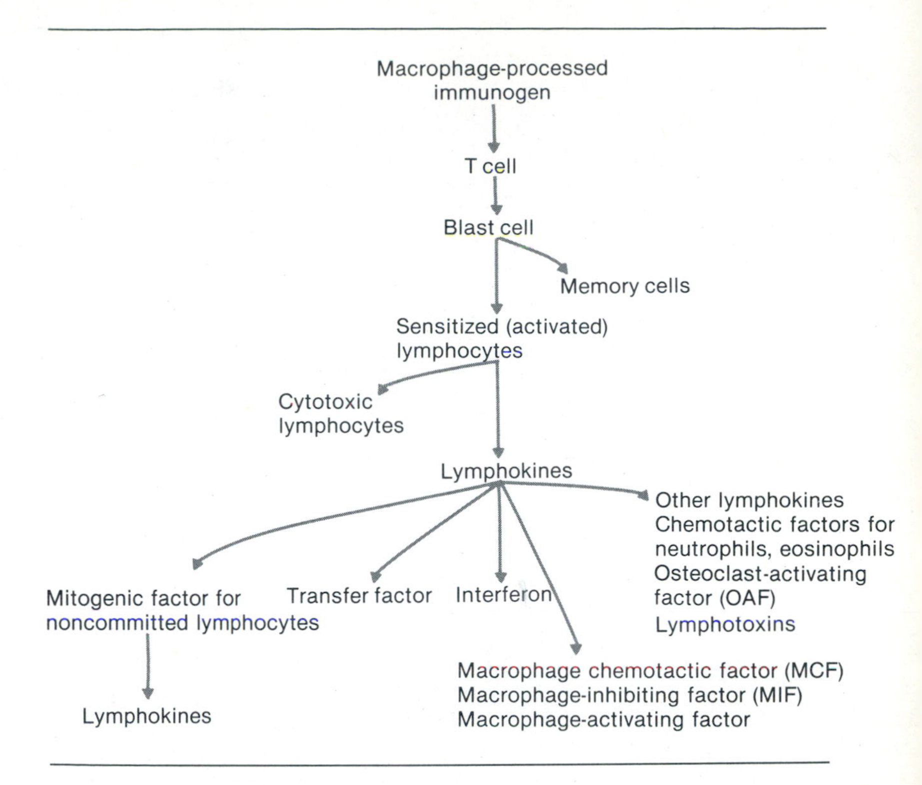

Figure 23-10 Responses of immunogen-exposed T cell. T cells with specific receptor (nature of receptor not known) for given antigenic determinant transform into blast cells on immunogen exposure. Lymphoblast cells transform into sensitized T lymphocytes and memory cells. Sensitized T lymphocytes may directly deal with antigen (cytotoxic lymphocytes), or they may release lymphokines. Lymphokines have not been chemically characterized. They are named according to their biological activity.

Redrawn from Boyd, R.F., and Hoerl, B.G. Basic medical microbiology *(2nd ed.). Little, Brown & Co., Boston, 1981.*

Lymphokines. Lymphokines, which are produced in minute quantities, act locally in the host. They have been difficult to purify and identify chemically and have therefore been named according to their biological activity. Anywhere from 50 to 100 have been named, and each has several biological activities. The action of most lymphokines is on cells: macrophages, other T cells, eosinophils, neutrophils, and basophils. These cells are attracted to the site of action by lymphokines to carry out their contribution to an inflammatory reaction. Some of the better known lymphokines are macrophage-activating factor (MAF), migration inhibitory factor (MIF), and chemotactic factors.

Macrophage-activating factor. Macrophages activated by this lymphokine are larger, have more lysosomes than nonactivated macrophages, secrete enzymes involved in inflammation, and have increased phagocytic activity. They have

LYMPHOCYTE MEDIATORS (LYMPHOKINES)

- Mediators affecting macrophages
  - Migration inhibitory factor (MIF)
  - Macrophage-activating factor (MAF)—indistinguishable from MIF
  - Chemotactic factors for macrophages
  - Antigen-dependent MIF
- Mediators affecting polymorphonuclear (PMN) leukocytes
  - Chemotactic factors
  - Leukocyte inhibitory factor (LIF)
  - Eosinophil stimulation promoter (ESP)
- Mediators affecting lymphocytes
  - Mitogenic factors
  - Factors enhancing antibody formation (antigen dependent and antigen independent)
  - Factors suppressing antibody formation (antigen dependent and antigen independent)
- Mediators affecting other cells
  - Cytoxic factors, lymphotoxin (LT)
  - Growth inhibitory factors (? same as LT)
  - Osteoclastic factor (OAF)
  - Collagen-producing factor
  - Colony-stimulating factor
  - Interferon
- Immunoglobulin-binding factor (IBF)
- Procoagulant (tissue factor)

From Rocklin, R.E., et al. *Adv. Immunol.* **29:**55, 1980.

increased killing capacities over resting cells because they release relatively large amounts of hydrogen peroxide and superoxide ion, both of which help to destroy microorganisms and possibly tumor cells.

Migration inhibitory factor. Migration of macrophages out of a capillary tube is inhibited when sensitized lymphocytes and their corresponding antigen are allowed to interact at the mouth of the capillary tube. Macrophage inhibition is due to a lymphokine produced by lymphocytes, but there is also a migration inhibitory factor for leukocytes (LIF).

Chemotactic factors. Lymphocytes produce lymphokines that attract macrophages, neutrophils, lymphocytes, eosinophils, basophils, and fibroblasts. Collectively, these are referred to as lymphocyte-derived chemotactic factors (LDCF). Some of these are more specifically named according to the cell type that is attracted: MCF for macrophage chemotactic factor and ECF for eosinophil chemotactic factor.

Other factors. Among other lymphokines that have been reported are specific macrophage-arming factor (SMAF), osteoclast-activating factor (OAF), lymphocyte-transforming factor (LTF), colony-stimulating factor (CSF), and interferon.

Table 23-3 T cell subsets and their functions

T CELL TYPE	FUNCTION
T_h,* T helpers	Aid antigen-stimulated development of B cells and of T effector cells (T_{DTH} and T_k)
T_s, T suppressors	Block induction and/or activity of T_h cells
T_r, T regulators	Develop into T_h or T_s cells and control balance between enhancement and suppression of responses to antigen
T_{DTH}	Mediate inflammation and nonspecific increased resistance to many infectious agents associated with activated macrophages in delayed-type hypersensitivity
T_k, T killer cells; cytotoxic lymphocytes, CTLs	Lyse target cells

From Eisen, H.N. *Immunology* (2nd ed.). Harper & Row, Publishers, Inc., New York, 1980.

*It is not clear whether the T_h cells that cooperate with B cells are the same as the T cells (sometimes called T amplifiers or T inducers) that aid in the development of various effector T cels (T_{DTH}, T_k).

Immunoregulatory functions of T cells

T cells have other functions aside from the roles described above. There are subpopulations, or subsets, of T cells that regulate the functions of B cells and other T cells (Table 23-3). There are T cells that serve as helper cells (T_h) for B cells and for effector T cells. Experimentally it has been shown that the inoculation of certain antigens (called thymus-dependent antigens) into thymus-deprived animals leads to the production of only small amounts of IgM but no other immunoglobulins. The inoculation of the same antigen into thymus-intact animals results in the production of the usual amounts of IgG and IgM. In other words, T_h cells are needed in order for B cells to respond to certain antigens. Another immunogenicity role of T cells is as ***T suppressor (T_s) cells*** to control B cells and T cells so that they do not continue to respond to just one antigen. Without such limits too much of the immune system might be committed to just a few predominant antigen specificities. There are also ***T regulatory (T_r) cells*** that control the maturation and activity of T_h and T_s cells.

Complement system

There are mediator-producing systems in the blood of vertebrates that operate in a sequential or cascade fashion when the precursors of the system are activated. Examples of such systems are the blood-clotting system, the kinin systems, and the complement system. Various complexes and factors that are generated in the activated complement sequence have important biological functions. In order to understand the biological activities, it is helful to know about the components and the pathways of the complement system (Figure 23-11).

Components. About 20 discrete proteins are recognized as components of the complement system. Thirteen of these belong to the cascade proper, and seven are inhibitors or inactivators. The latter have a regulatory role as fail-safe controls of this rapidly activated system.

Complement system pathways. There are two pathways of activation of the complement system: the ***classic pathway*** and the ***alternative pathway*** (the latter is also called the ***properdin pathway*** or the ***alternate pathway***). The early steps of the two pathways differ, but the proteins in both act in sequential order (cascade effect). In some of the steps one of the complement system proteins cleaves the protein that is next in the sequence. Some of the smaller fragments that result from the cleavage have various biological activities (see below).

Classic pathway. The numerical sequence of participation of the complement components is C1, C4, C2, C3, C5, C6, C7, C8, and C9. The union of antibodies of immunoglobulin classes IgG and IgM with the antigen is the best-known activator of the classic pathway. The other immunoglobulin classes apparently do not activate complement by this pathway. When the entire sequence of complement components is activated, the final biologically active fragment that is formed is called the ***membrane attack complex (MAC).*** Other biologically active fragments are formed along the pathway (see below).

Alternative pathway. The alternative pathway is entered at the C3 level, and C1, C4, and C2 of the classic pathway are bypassed. Plant, fungal, and bacterial polysaccharides and bacterial lipopolysaccharides that are in particulate form are examples of substances that can activate the alternative pathway. Substances that activate this pathway first activate certain serum proteins, and it is these activated proteins that cleave C3 into the same fragments as are produced in the classic pathway.

Biological activities of complement. There are important differences in what activates the classic pathway and what activates the alternative pathway (Table 23-4). The classic pathway is activated when certain immune system products, namely antibodies of immunoglobulin classes IgG or IgM interact with their antigens. The complement system is not directly involved in the immune system, but it is one effector mechanism triggered by antigen-immune product interaction. An immune system product is not required for activation of the alternative pathway, which provides a mechanism for amplifying nonspecific resistance. Thus activation of the alternative pathway provides immediate defense against infection before antibody-activated immunity has a chance to develop.

Complement components are found in the serum before they are activated. Once the complement components of the serum are activated, many of the biologically active fragments that are generated attach to the surface of cells. The biological activity that ensues is dependent on which fragment is attached and to what cell type it attaches. The capacity of complement fragments to attach to cells is important in several ways: for marking and preparing particles for phagocytosis, for causing the release of pharma-

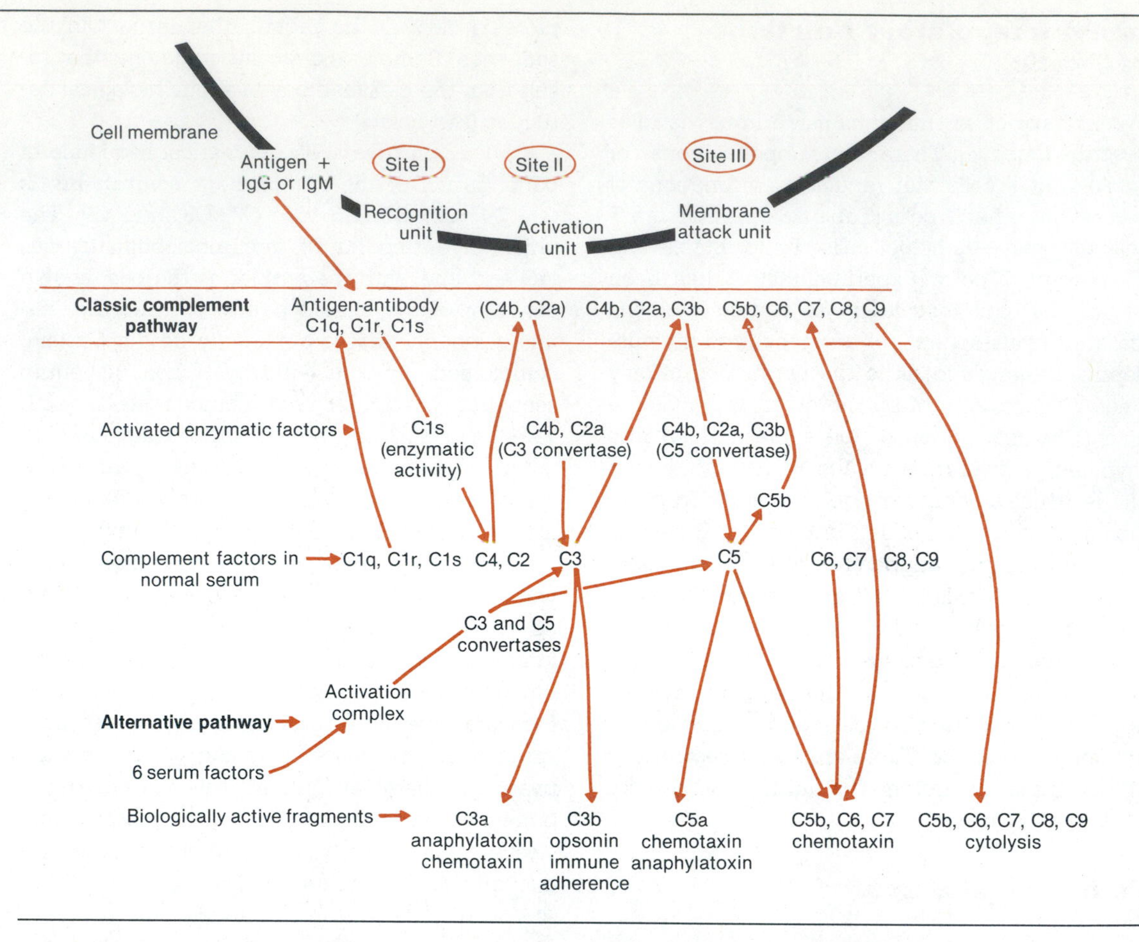

Figure 23-11 Complement system. In classic pathway nine complement factors act in sequence, with end result being immune cytolysis. Antigen and IgG or IgM bind C1q. Recognition by C1q leads to formation of activated recognition unit at site I on cell membrane. Enzyme C1s, formed by recognition unit, converts serum factors C4 and C2 to enzyme C4b, C2a, which converts pivotal C3 into full activation unit at site II. Activation unit C4b, C2a, C3b activates C5 of serum to form C5b and C5a. C5b, with C6, C7, C8, and C9, forms membrane attack unit at site III. In alternative pathway (or properdin pathway) six serum proteins form activation and stabilization complex that enters complement sequence at pivotal C3 level. Most biological activities of complement can be activated by this pathway. Antibodies are not required for activation, thus providing mechanism of engaging complement biological activities in nonimmunized state. Enzymes that activate C3 and C5 are different from convertases of classic pathway. Biological activities are generated by both pathways during cascade.

Redrawn from Boyd, R.F., and Hoerl, B.G. Basic medical microbiology *(2nd ed.). Little, Brown & Co., Boston, 1981.*

Table 23-6 Activation of the complement (C) system

	CLASSIC	ALTERNATIVE
Immunological	IgG, IgM	IgA, IgG, IgE
Nonimmunological	Trypsinlike enzymes DNA Staphylococcal protein A C-reactive protein	Trypsinlike enzymes Lipopolysaccharides Plant and bacterial polysaccharides Cobra venom factors

From Cooper, N.R. In Fudenberg, H.H., et. al. *Basic and clinical immunology.* Lange Medical Publications, Los Altos, Calif., 1976.

cologically active substances, and for causing lysis of cells.

Some biological activities of complement are membrane damage and cytolysis, chemotaxis, opsonization, and anaphylatoxin activity.

Membrane damage and cytolysis. Lesions form in the cell membrane where the membrane attack complex of complement is located.

The membrane attack complex is composed of C5b, C6, C7, C8, and C9 and causes osmotic lysis of the cell. When antibodies are involved in the process, the phenomenon is called ***immune cytolysis.***

Chemotaxis. Complement fragments that attach to bacterial cells and other antigens are the chemical attractants that "mark" the antigen and attract migrating phagocytic cells. The complement system is a source of the principal chemotactic factor C5a, which is a cleavage product of complement factor C5.

Opsonization. Some microorganisms are able to exercise their virulence because of their ability to escape ingestion by phagocytes. Antibodies, called ***opsonins,*** coat these evaders and provide a bridge to attach the evaders to phagocytes. The phagocytes can now more readily ingest the attached bacteria or other antigens. Complement fragment C3b also aids the phagocyte by binding to antigen-antibody complexes that are present, for example, on the surface of a bacterial cell. Leucocytes possess receptors on their membranes for C3b, which forms a bridge between the bacterium and the leucocyte and thus facilitates phagocytosis.

Anaphylatoxin activity. Anaphylatoxins, such as fragments C3a and C5a, are substances produced in the complement cascade that adhere to the membrane of mast cells and platelets, causing them to release histamine and other pharmacologically active substances. A similar response occurs during anaphylactic reactions. The pharmacologically active substances released from mast cells produce, among their more prominent effects; vascular permeability and smooth muscle contraction. The activities induced by these substances in modest amounts are beneficial. They allow phagocytes and serum proteins, including antibodies, to pour into the surrounding tissue.

Pathogenesis associated with complement. Complement-mediated reactions can lead to host cell destruction and tissue damage largely through intense inflammatory responses. Immune complex diseases and some of the blood abnormalities in which there is destruction of red blood cells and platelets are examples of complement-mediated diseases (see Chapter 25).

Complement in immunodiagnosis. A well-known test, the complement fixation test, is based on the involvement of complement in immune cytolysis. The test has an indicator system that is advantageous when an in vitro antigen-antibody interaction has occurred but a visible reaction is not observable (see Chapter 24).

IMMUNITY

States of immunity

The principal function of the immune system is to provide protection against foreign substances and, apparently, against nonnormal self-materials such as cancer cells. There are terms that are employed to describe an individual's state and type of protection. These especially pertain to protection against infectious agents. The two major categories of immune status are innate immunity and acquired immunity.

Innate immunity. Innate immunity is also called ***natural immunity, genetic immunity,*** and ***constitutional immunity.*** Immune system products, such as antibodies and sensitized T lymphocytes, are not involved. Permanent, genetically endowed physiological, anatomical, and biochemical differences in individuals and between species account for the resistance. These differences may be due to the lack of suitable receptor sites for the attachment of infectious agents, to differences in body temperature, to oxidation-reduction conditions of the tissues, to toxicity and pH of the body fluids, and to the availability of certain nutritional requirements. For example, resistance of chickens to anthrax infection is based on body temperature. If the temperature is lowered from the normal 39° C to 37° C, chickens become susceptible to anthrax. The bacterium that causes contagious abortion in cattle, *Brucella abortus,* requires erythritol, which is available in sufficient quantities in cattle placenta but not in human placenta. Contagious abortion does not occur in humans.

Acquired immunity. Acquired immunity involves specific immune products, antibodies, and sensitized T lymphocytes and includes the following types:

1. ***Artificially acquired.*** Artificially acquired immunity takes place when antigens or im-

mune products are introduced into the individual by man-made means. Ordinarily the attainment of the artificially acquired status involves the use of a vaccine or an antiserum.

2. ***Naturally acquired.*** Naturally acquired immunity is observed when the antigen or the immune product enters the host under natural conditions. Thus the infectious agent that incites the immune system to respond enters the host under natural circumstances during day-to-day contacts.
3. ***Active.*** Active immunity is obtained when the individual produces the immune products, and it most often results from infections or from immunizations with vaccines.
4. ***Passive.*** Immunity is considered passive when the individual did not form the immune products but received them from an outside source. The administration of antisera or gamma globulin obtained from other humans or from other animals and the transfer of maternal immune products via the placenta or colostrum lead to the passive immune state. The passive immune state is of short duration—weeks or a few months.

It is customary to combine the above terms when describing the acquired immune state. Examples of how the terms are combined include:

1. Artificially acquired active immunity results from immunization with vaccines and toxoids.
2. Naturally acquired active immunity is obtained after recovery from an infection.
3. Artificially acquired passive immunity is the transitory immunity that occurs following the administration of gamma globulin or an antiserum.
4. Naturally acquired passive immunity is the transitory immunity that occurs in a newborn who has received maternal immune products during placental transfer or from ingestion of mother's milk (colostrum).

The terms need not be given in the sequence cited in the examples. A naturally acquired passive immunity can as well be called a passive immunity, naturally acquired. Note that examples often cite the beforehand conditions, such as infections, immunization, and antiserum administration, that lead to a given immune status. The immune status is adjudged after the individual has been exposed to the antigen or has received the antibody.

Immunization practices

It is possible and practical to artificially induce a protective immune status in an individual or to transfer a protective immune status.

Active immunization. Materials variously named as immunogen, antigen, or immunizing material are used to induce active immunity. More specifically, such a material may be a ***toxoid*** if it is a modified toxin, or a ***vaccine*** if it is made up of living or inactivated microorganisms.

Commercially available immunogens have been processed in the laboratory in order to modify their disease-producing capability. Bacterial exotoxins such as the tetanus and diphtheria toxins are converted to toxoids by treatment with formalin. Whooping cough (pertussis) vaccine consists of inactivated whooping cough bacteria *(Bordetella pertussis).* The most effective immunogens are those that mimic the natural infection. These consist of living laboratory-developed strains that have been selected because of their lost or greatly lessened virulence. They are said to be ***attenuated,*** because they cause an infection but not the disease. Most of the virus vaccines consist of attenuated strains.

The level of antibody that is obtained and the persistence of antibody can be increased with some immunogens by the addition of ***adjuvants*** (helping materials) to the immunogen. Adjuvants supposedly help prolong the antigenic stimulus, or they foster the involvement of increased numbers of immune cells, thus creating conditions that lead to elevated antibody production. Toxoids that are alum precipitated, for example, produce a better immunity than toxoids alone.

Vaccines: whole vs. subunit. The use of whole pathogen preparations as vaccines has its limitations. Some bacterial vaccines, such as the pertussis vaccine for whooping cough, are believed to be responsible for brain damage (encephalitis). Some scientists also fear that whole-virus vaccines may be responsible for the release of viral DNA, which may be carcinogenic for humans. In addition, virus from attenuated preparations can be passed from person to person. A vaccine-related disease may not appear in the immunized person but can appear in a close contact, such as a family relative. The serious side effects of a vaccine are the result of abnormal immunological responses. Some people have different-functioning immune systems that result in disease following vaccination. Although the instances of vaccine-associated disease may be one in several thousand, a totally safe vaccine is the ultimate goal.

KINDS OF IMMUNITY

INNATE IMMUNITY (also termed natural or native immunity): Based on physiological, anatomical, and biochemical differences.

Species immunity: Many microorganisms can infect only a certain species; the meningococcus bacterium, the syphilis bacterium, and the rubeola (measles) virus naturally infect only humans; other species are resistant.

Racial immunity: Certain classes or ethnic stocks of a species are resistant or susceptible to given infectious agents; blacks with sickle cell anemia (genetically determined) are resistant to falciparum malaria.

Individual immunity: Individuals have differing levels of resistance or susceptibility. Differences are due to such factors as genetic defects, age, hormonal balance, nutritional status, and unexplained, unknown factors.

ACQUIRED IMMUNITY: Based on responses by the immune system.

Artificially acquired, active immunity: Follows exposure to prepared antigens such as toxoids, killed or inactivated microorganisms, and attenuated microorganisms.

Naturally acquired, active immunity: Follows infection that has been acquired in day-to-day contacts.

Artificially acquired, passive immunity: Follows injection of immune products that were obtained from another species or from the same species (for example, antitoxins, antisera, gamma globulin).

Naturally acquired, passive immunity: Follows from the transfer of immune products from one individual to another under natural circumstances (for example, the transplacental transfer of maternal antibodies of class IgG to the fetus and the ingestion of antibodies of class IgA in colostrum).

A more ideal vaccine is one in which the specific antibody-stimulating component of the pathogen can be used instead of the whole pathogen and its myriad of antigens. Specific microbial components that can be used as vaccines are called ***subunit vaccines.*** The organism causing gonorrhoea *(Neisseria gonorrhoeae),* for example, produces fimbriae that, when removed from the organism and purified, can be used as a vaccine. This subunit vaccine, however, works only if the individual is infected by the same strain of *N. gonorrhoeae* (there are over 100 strains). In order to produce subunit vaccines, however, the specific antigen must be identified and purified, and this is particularly difficult in some microorganisms such as bacteria, because the antigenic component is not always evident or it is part of a complex of antigens located in the cell wall or capsule. Even if the antigen is located, it must still be extracted and purified. The vaccine for the prevention of pneumonia caused by *Streptococcus pneumoniae* consists of purified capsular material extracted from 14 different types of the organism. The 14 extracted polysaccharides are combined in a final product and represent the types most responsible for pneumonia caused by *S. pneumoniae.*

Identification of subunits on viruses is being actively pursued and applied to the production of synthetic vaccines. One technique for which a patent has been applied involves determining the composition of the different protein molecules residing on the viral surface. This information is then fed into a computer that takes into account various molecular interactions that occur between protein molecules. The computer then selects which of the protein subunits is most likely to protrude from the viral surface. It is usually the protruding component on the microbial surface that is most likely to be recognized by the immune system. The number of subunit possibilities can be reduced by using mutant viruses that are not destroyed by specific antibodies. The subunit regions of normal and mutant virus are compared, and the subunit recognized by the immune system can be identified. These subunits are called ***neutralizing epitopes*** because they stimulate antibodies but cause no disease. The synthesis of the protein subunit in the laboratory is sometimes difficult, but hopefully the use of recombinant DNA technology will resolve this problem. In other words, the gene coding for the subunit could be isolated and inserted into a microbial cell for unrestricted subunit production. One licensed subunit vaccine that has been developed from recombinant DNA technology is the foot-and-mouth disease (FMD) vaccine for cattle. A subunit vaccine for hepatitis in humans is also available but is not produced by recombinant DNA technology.

One of the problems with subunit vaccines is their reduced immunogenicity as compared with whole-pathogen vaccines. The neutralization of viruses, for example, is related to several factors,

Table 23-5 Principal bacterial and viral vaccines in current use

DISEASE	VACCINE TYPE	COMMENT
Bacterial		
Pneumonia caused by *Streptococcus pneumoniae*	Mixture of 14 different capsular polysaccharide types	For older adults compromised by other illnesses
Whooping cough (pertussis)	Killed *Bordetella pertussis*	Begun at 3 months of age, followed by boosters
Tuberculosis	Attenuated *Mycobacterium tuberculosis*	Used primarily in Europe
Meningitis caused by *Neisseria meningitidis*	Purified polysaccharide from types A and C	Used primarily for servicemen
Cholera	Crude fraction of *Vibrio cholerae*	For those in areas endemic for the disease
Typhoid fever	Killed *Salmonella typhi*	For those in areas endemic for the disease
Plague	Killed *Yersinia pestis*	For those working in areas endemic for the disease
Viral		
Rabies	Inactivated virus	For those working in areas where rabies is endemic in wildlife
Smallpox	Attenuated virus	Disease has been eradicated, but vaccine is recommended for laboratory personnel who cultivate virus
Poliomyelitis	Attenuated or inactivated virus	Oral attenuated virus is most widely used; begun at 4 to 5 months of age plus boosters
Mumps	Attenuated virus	Given to infants 15 to 19 months of age
Measles	Attenuated virus	Given to infants 15 to 19 months of age
Influenza	Inactivated virus or live attenuated virus	Inactivated virus is used in United States; attenuated virus is used in Europe
Rubella	Attenuated virus	Given to infants 15 to 19 months of age and to women of childbearing age who are not pregnant
Yellow fever	Attenuated virus	For those traveling to areas endemic for the disease
Hepatitis	Subunit vaccine	Hopefully will reduce the large number of hepatitis carriers

including the molecular size and spatial conformation of the immunogen. Many subunit vaccines require some type of carrier molecule to be more immunogenic. One technique that has been used is to attach viral subunit proteins to synthetic bubbles called ***liposomes.*** The disguised liposome is more immunogenic than the solitary subunit and does not give rise to side effects in the host.

A list of some of the major bacterial and viral vaccines can be found in Table 23-5.

Passive immunization. It is sometimes necessary to artificially provide a passive immunity when a nonimmune individual has suffered an exposure to an infectious agent or to a microbial product. The individual may not have been actively immunized or may have been exposed to an agent for which it is not practical or possible to actively immunize. Temporary and immediate protection can be provided in some instances by the administration of appropriate antisera or of gamma globulin preparations. These products usually contain high levels of specific antibodies. Passive immunization is also referred to as serotherapy or immunotherapy. On a related note, it should be realized that there are foreign proteins present in an antiserum that are developed in other vertebrate species. The antiserum may temporarily protect the recipient against a given harmful agent, but at the same time it may sensitize the recipient to the serum proteins. Thus equine tetanus antitoxin (serum from a horse that has been actively immunized against tetanus toxin) administered to a human may temporarily protect the human against tetanus, but it also sensitizes him to horse serum proteins. A later injection of horse serum into the same individual could very well lead to an allergic reaction of so serious a consequence as death due to an anaphylactic reaction.

Dynamics of the immune response

Primary immune response. The primary immune response occurs the first time the immune system is incited to respond to a given immunogen (Figure 23-12). The response varies according to the nature of the immunogen, the competence of the host's immune system, the size of the dose, and the route of adminstration of the immunogen. Despite these variations, there is a fairly characteristic pattern of antibody production following the initial exposure to the immunogen. Readily detectable antibody does not appear for about 5 days. This is the lag or induction period. The first antibody to appear belongs to immunoglobulin class IgM. IgM level rises exponentially for several days and then begins to fall until very little residual IgM remains after several weeks. IgG antibody has a longer induction period, and detectable levels start about the time that the IgM level peaks. IgG antibody levels peak later and may remain elevated for a long period of time, or the IgG antibody may disappear.

Some of the proliferating lymphocytes evoked on immunogen stimulation are set aside as memory cells, an important consideration in the secondary immune response. Because of the lag period a protective level of antibody does not develop soon enough to prevent a natural infection from occurring. The significance of artificially inducing an active immunity whenever practical should be evident.

Secondary immune response. A secondary immune response occurs when there are exposures to the same immunogen subsequent to the first exposure. Natural exposure to an infectious agent or subsequent administration of the same vaccine material can serve to engender the secondary immune response. The ensuing doses that are used in artificial immunizations are referred to as ***booster doses.*** There are marked differences between the primary and the secondary immune responses. Reexposure to the same immunogen leads to (1) a shorter lag period for antibodies to appear, (2) a higher and more persistent level of antibodies, (3) a predominance of IgG antibodies, and (4) an increased affinity of the antibodies for the antigen.

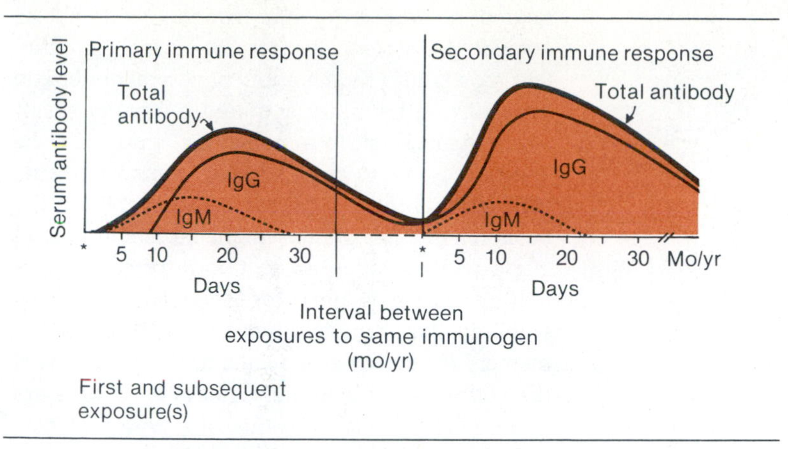

Figure 23-12 Dynamics of immune response. In primary immune response IgM is first to appear. It appears on fifth to seventh day and peaks at about 14 days. IgG appears on about tenth day and peaks several days later. In secondary immune response both IgM and IgG appear or increase within 2 to 3 days. Lag period is shorter, total antibody level far surpasses that of primary immune response, antibody is formed over longer period of time, and much more IgG is formed than IgM.

Redrawn from Boyd, R.F., and Hoerl, B.G. Basic medical microbiology (2nd ed.). Little, Brown & Co., Boston, 1981.

The prompt response to repeated immunogenic stimuli described above is due to memory cells. Memory cells that are preprimed with the first antigen exposure respond promptly and efficiently to subsequent exposures. This phenomenon is referred to as the ***anamnestic response*** or the recall phenomenon. Memory cells may persist for many years. It has been shown in humans that tetanus toxoid administered 20 years after primary immunization continues to elicit a prompt and vigorous antitoxin response.

COMPARATIVE IMMUNOLOGY

Most of the immunology in this textbook pertains to the human immune system. When examining immunity in other life forms it is natural to make comparisons with the human immune system. We use as the basis for comparison the major components of the human immune system and its cardinal features: specificity, self-recognition, and memory.

Plant immunity

Plants are well adapted to survive in their environments. This implies that there are anatomical, biochemical, and immunological bases for their resistance to certain diseases. They can to some degree recognize foreign invaders and defend against them (see Chapter 35). Their immunity falls mostly into the category of natural or genetic immunity rather than acquired immunity. Plant cells possess surface receptors that can be viewed as one mode of natural immunity. Plants can also be bred for disease resistance. What the basis for that resistance might be is not known, but it can be regarded as a form of natural immunity. There is no evidence that plants have anything that compares with the specificity and memory aspects that are such integral features of the vertebrate animal immune system. There are no circulating cells in plants of a type that play such key roles in both natural and acquired immunity in animals.

Plants really do not require an elegant immune system as do the multiorganed species of the higher vertebrates, with their highly specialized tissues and organs. Plants recover more readily from injury because of great regenerative powers and relatively unlimited growth, and because their somatic tissues can differentiate on a broad scale.

Some products and reactions of plants can be interpreted as manifestations of an immune response. Plants are known to produce ***phytoalexin*** (see Chapter 35) that is nonspecifically toxic to invading fungi. Grafting experiments reveal that ***autografts, allografts,*** and some ***xenografts*** are tolerated. Autografts are self-grafts; allografts are grafts between nonidentical members of the same species; and xenografts are grafts between members of different species. Graft rejection is regarded as a prime indicator of cellular immunity.

Invertebrate immunity

Invertebrates, because of their high reproductive capacity, short life span, and less-complex anatomical makeup, probably require a less-elegant immune system than more highly developed vertebrate forms. All animals, including invertebrates, exhibit some process that enables them to discriminate self from nonself. There is evidence for immunological memory in some invertebrates, especially in respect to transplantation immunity. Accelerated second-set graft reactions occur in sponges, annelids, coelenterates, and echinoderms. In regard to specificity, invertebrates do not have immunoglobulins, and there is no evidence of an inducible humoral immunity. It is possible that another type of molecule might be discovered that has specific immune function. Graft rejection and fusion incompatability are observed in most of the invertebrate phyla. The second time the recipient is exposed to a graft from the same source, there is often an infiltration of the graft rejection site by cells that have a morphological resemblance to lymphocytes.

Invertebrates protect themselves essentially by three mechanisms: phagocytosis, encapsulation, and nonspecific humoral substances. Phagocytosis is universal among invertebrates. ***Archeocytes, amebocytes, plasmatocytes,*** and ***granulocytes,*** are examples of phagocytic cells among the invertebrate phyla. Foreign objects that are too large to be endocytosed are encapsulated. Amebocytes of mollusks and earthworms and hemocytes of insects, for example, collect around the involved area, and a capsule, whose description varies according to the species involved, comes to surround the foreign object. A number of invertebrates have a "solid body," and there are no body fluids (humors) or circulatory system. Their defense depends on mobile leukocytes. Humoral components do appear in invertebrates that have a body cavity (coelomates) (Table 23-6). Nonspecific humoral substances that are not well characterized are found among the coelomates. The substances include lysozyme and other types of bacteriocidins, agglutinins for vertebrate red blood cells and some bacteria, and opsonin-like molecules in mollusks.

Vertebrate immunity

All vertebrates possess cell-mediated immunity that is highly developed and specific. They also produce at least IgM, and they usually possess immunological memory. All have a thymus and spleen, although of a primitive form in the cyclostomes (hagfish and lamprey), and all have lymphocytes. Vertebrates reject allografts and xenografts, but in the more primitive vertebrates the rejection of the first-set graft (the first time the recipient is exposed to a certain graft) is much slower than in the higher vertebrate forms. This difference is probably related to weaker transplantation ***(histocompatability)*** antigens among the lower forms. Graft rejection is also slower among the cold-blooded animals; therefore the speed of rejection may be related to body temperature. Second-set grafts are typically rejected in an accelerated manner.

Immunoglobulins with the specificity and inducibility characteristics of mammalian antibodies are found in all the classes of vertebrates. The general structure of the immunoglobulin is comparable to that of mammals. They consist of the typical heavy chains and light chains. IgM, which is the only class of immunoglobulins that appears in all vertebrates, is a pentamer, just as

Table 23-6 Immunopotentialities of major phylums of invertebrates*

PHYLUM OR SUBPHYLUM	EXTENSIVE ALLOGENEIC MACROMOLECULAR POLYMORPHISM EVIDENT	SPECIALIZED LEUKOCYTES OR DEFENSIVE CELLS PRESENT	SPECIFIC XENOGRAFT REJECTION	SPECIFIC ALLOGRAFT REJECTION	IMMUNOLOGICAL MEMORY DEMONSTRABLE
Protozoans	Yes	No	Yes	No	No
Sponges	Yes	Yes?	Yes	Yes	Yes
Coelenterates	Yes	Yes?	Yes	Yes	Yes
Annelids	Yes	Yes	Yes	Yes	Yes
Arthropods	Yes	Yes	Yes	Yes?	?
Mollusks	Yes	Yes	Yes	Yes	?
Echinoderms	Yes	Yes	Yes	Yes	Yes
Tunicates (Protochordates)	Yes	Yes	Yes	Yes	?

From Hildemann, W.H., and Reddy, A.L. *Fed. Proc.* **32:**2188, 1973.
*Vertebrate-type immunoglobulin antibodies have *not* been demonstrated in any invertebrate species. Inducible circulating "antibodies" found in annelids and arthropods are mainly bacteriocidins and are not strictly specific.

Table 23-7 Comparative phylogeny of immune responsiveness

	CYCLOSTOMES (LAMPREY, HAGFISH)	ELASMOBRANCHS (SHARKS, RAYS, GUITARFISH)	FISHES (CHONDROSTEAN, HOLOSTEAN, AND TELEOSTEAN)	AMPHIBIANS (FROGS)	BIRDS	MAMMALS
Serum gamma globulin	Lamprey—yes Hagfish—yes	Yes, IgM only	Yes	Yes	Yes	Yes
Circulating immunoglobulin	Lamprey—yes Hagfish—yes	Yes	Yes	Yes	Yes	Yes
Thymus	Not in adult forms	Yes	Yes	Yes	Yes	Yes
Spleen	Lamprey—yes Hagfish—no	Yes	Yes	Yes	Yes	Yes
Peripheral (lymphoid tissue)	Sparse, unlike mammalian	Yes	Yes	Yes	Bursa of fabricius; lymph nodes rare or absent	Several organs
Delayed hypersensitive response	Yes	Yes	Yes	Yes	Yes	Yes

From Barrett, J.T. *Textbook of immunology* (3rd ed.). The C.V. Mosby Co., St. Louis, 1978.

in mammals. In a few species, however, IgM has a tetrameric or hexameric configuration. Membrane (receptor) IgM is found in B cells (or their presumed equivalents) of all vertebrates. The number of immunoglobulin classes increases as one proceeds up the phylogenetic scale (Cyclostomes — Ichthyes — Amphibia — Reptilia — Aves—Mammalia). Fishes and urodelian amphibians have IgM only. Vertebrates from anuran amphibians to mammals have two to three classes of immunoglobulins. Birds produce a third class that is homologous to mammalian IgA.

From a phylogenetic viewpoint it would appear that higher forms built on the nonspecific resistance mechanisms of phagocytosis and inflammation that existed in the invertebrate forms. The higher forms added specific humoral immune mechanisms and the amplification systems of coagulation and complement (Table 23-7).

SUMMARY

The immune system specifically interacts with foreign substances (immunogens) that have certain characteristics. Memory cells of the immune system are prealerted to that same substance for future encounters. These features, specificity and memory, enable the body to cope with foreign substances much more effectively than can be done by nonspecific mechanisms. This system, with its ability to recognize specific chemical groupings, called antigenic determinants, on the surfaces of molecules must of necessity be able to recognize self-substances and refrain from attacking them.

Two major classes of lymphocytes, B cells and T cells, specifically recognize immunogens and form immune products called immunoglobulins and sensitized lymphocytes. The products specifically interact with antigens.

B cells in mammals develop in the bone marrow. Many of them emigrate to peripheral lymphoid organs, especially to the spleen and to the lymph nodes. B cells have receptor immunoglobulins on their membranes that are specific for given antigenic determinants. Contact with the antigenic determinants causes the B cell to multiply and differentiate into a cell type called a plasma cell. The plasma cell secretes immune products called immunoglobulins that specifically interact with the antigens that elicited their production.

The basic immunoglobulin molecule consists of four polypeptide chains: two heavy chains and two light chains. The chains and domains within the molecules are linked by sulfhydryl bonds. One end of the basic molecule has two antigen-binding sites (Fab), each of which is made up of the variable regions of a light chain and a heavy chain. The other end of the Ig molecule, depending on the Ig class, called the Fc fragment, has various biological functions: attachment to phagocytes and mast cells, binding site for complement, and passage through the placenta. There are five Ig classes based on physicochemical and biological activity differences. Antibodies are immunoglobulins with a known antigenic specificity; that is, the substance with which they interact is known. The immune activities of antibodies include neutralization of toxins and viruses, promotion of phagocytosis of antigens, and lysing of cells.

T cells derive from bone marrow stem cells, whence they emigrate to the thymus. The sojourn through the thymus conditions them to function as T cells. They emigrate to peripheral lymphoid organs where, through specific receptors, they are activated by immunogens. Immunogen contact causes a T lymphocyte to proliferate and form sensitized lymphocytes. Sensitized lymphocytes may directly destroy the antigen. They secrete nonspecific products called lymphokines. T cells also have immunoregulatory roles as helpers, suppressors, and regulator cells.

Immune products very often are not the final effectors on an antigen. The union of an antigen and the immune product frequently serves to initiate the participation of other body elements such as macrophages and the complement system. It is these elements that determine the ultimate fate of the antigen.

The complement system consists of a number of serum proteins that exhibit a major beneficial (and harmful) role, both in nonspecific resistance and in immunity. Complement's activities include cell lysis, phagocytosis, chemotaxis, and stimulation of inflammatory responses.

The prime function of the immune system is to provide specific protection (immunity). Terminology has been developed that signifies the various immune states: innate vs. acquired, natural vs. artificial, and active vs. passive. The development of the active immune state and the level of immunity attained are related to such considerations as the use of adjuvants; the type of immunogen, and the frequency and spacing of the exposure to the immunogen; types and order of appearance of immune products during the first immunogen exposure (primary immune response) and during subsequent exposures (secondary immune response); and the role of memory cells.

Questions

1. Why are specificity and memory important features of the immune system?
2. Why is the recognition of self-antigens a necessary corollary to the functioning of the immune system?
3. What is the basis for naming lymphocytes as either T cells or B cells?
4. What effect would the removal of the thymus gland of a newborn animal have on the immune system?
5. Differentiate between immunogen and antigen.
6. What is an antigenic determinant and how does it relate to an immunogen? To an antigen? To a hapten?
7. Differentiate between secreted immunoglobulins and receptor immunoglobulins.
8. With which immunoglobulin class are the following associated? Crossing of the placenta? First Ig to be formed after immunogen exposure? Topical immunity? Surface or receptor immunoglobulin? Predilection for mast cells?
9. What is the difference between an immunoglobulin and an antibody?
10. Describe the structure of a basic immunoglobulin molecule and cite the functions of the principal fragments.
11. What is meant by humoral immunity? By cell-mediated immunity? Relate these to B cells and T cells.
12. What are the secreted products of sensitized lymphocytes called? Are they antibodies? What is their function in cell-mediated immunity?
13. B cells produce antibodies against some antigens only after activated T cells have stimulated the proliferation of the antigen-bearing B cells. What is the subset of T cells called that has this kind of effect on B cells?
14. Strictly speaking, the complement system is not an integral part of the immune system, yet is always described in association with it. Why is this so?
15. Other substances besides antigen-antibody complexes can activate the complement system. They do so usually through which complement pathway?
16. How do complement system fragments help in eliminating substances via phagocytosis?
17. Differentiate between active immunity and passive immunity.
18. A patient was bitten by a dog suspected of being rabid. The patient received two kinds of "shots": a shot of antibody-containing gamma globulin and a shot (series of shots) that contained rabies virus antigen. What type of immunity does each kind of shot produce?
19. What is the purpose of booster shots?

Selected readings

Books

Barrett, J.T.: *Textbook of immunology* (3rd ed.). The C.V. Mosby Co., St. Louis, 1978.

Barrett, J.T.: *Basic immunology and its medical application* (2nd ed.). The C.V. Mosby Co., St. Louis, 1980.

Bellanti, J.A. *Immunology: basic processes.* W.B. Saunders Co., Philadelphia, 1978.

Bier, O.G., et al. *Fundamentals of Immunology.* Springer-Verlag, New York, Inc., New York, 1981.

Eisen, H.N. *Immunology* (2nd ed.). Harper & Row, Publishers, Inc., New York, 1980.

Fudenberg, H.H., et al. (Eds.). *Basic and clinical immunology* (3rd ed.). Lange Medical Publications, Los Altos, Calif., 1980.

Hyde, R.M., and Patnode, R.A. *Immunology.* Reston Publishing Co., Inc., Reston, Va., 1978.

King, L.S. (Ed.). *Immunology: readings from Scientific American.* W.H. Freeman & Co, Publisher, San Francisco, 1976.

Roitt, I. *Essential immunology* (4th ed.). Blackwell Scientific Publications, Ltd., Oxford, 1981.

Tizard, I.R. *Veterinary immunology.* W.B. Saunders, Co., Philadelphia, 1977.

Journal articles

Hanson, L.A., et al. Human milk: defense against infection. *Prog. Clin. Biol. Res.* **61:**147, 1981.

Hayward, A.R. Development of lymphocyte responses and interactions in the human fetus and newborn *Immunol. Rev* **57:**39, 1981.

Klinman, N.R., et al. Defining the immune mechanism with monoclonal antibodies. *In Vitro* **17:**1029, 1981.

Leder, P. The genetics of antibody diversity. *Sci. Am.* **246:**102, 1982.

Milstein, C. Monoclonal antibodies. *Cancer* **49:**1953, 1982.

Müller-Eberhard, H.J. Chemistry and function of the complement system. *Hosp. Pract.* **12:**33, 1977.

Nossaman, J. Immunoglobulins: structure, genetics, diversity. *Lab. Med.* **12:**284, 1981.

Reinherz, E.L., and Schlossman, S.F. The characterization and function of human immunoregulatory T lymphocyte subsets. *Immunol. Today* **2:**69, 1981.

Rocklin, R.E., Bendtzen, K., and Greineder, D. Mediators of immunity: lymphokines and monokines. *Adv. Immunol.* **29:**55, 1980.

Scharff, M.D., Roberts, S., and Thommana, P. Hybirdomas as a source of antibodies. *Hosp. Pract.* **16:**61, 1981.

Talmage, D.W. Recognition and memory in the cells of the immune system. *Am. Sci.* **67:**73, 1979.

Warr, G.W. Evolution of the lymphocyte. *Immunol. Today* **2:**63, 1981.

Wehrle, P.F., and Wilkins, J. Immunizing agents: potential for controlling or eradicating infectious diseases. *Annu. Rev. Pub. Health* **2:** 1981.

Chapter 24 ANTIGEN-ANTIBODY INTERACTIONS Their Detection and Quantitation

Bryan Hoerl

Products of the immune system, that is, antibodies and sensitized lymphocytes, are used extensively in identification and in monitoring of test procedures. The facts that immune products have high specificity for antigens and that their reactions with antigens are usually readily detectable make immune product–antigen interactions highly useful for identification and test monitoring. The techniques described in this chapter are mostly in vitro procedures that are encompassed under headings such as serology, immunodiagnosis, and immunoidentification. The term ***serology*** implies a study of serum, but serological tests also make use of material containing immune products other than serum.

APPLICATION OF IN VITRO TESTING

There is a broad range of procedures and applications in serological testing. Most infectious agents and other immunogenic substances induce the production of immune products in animals. If the induction period is long enough, suitable amounts of the immune products can be obtained from the patient and appropriate tests conducted. There are situations in which antigen and not the immune product is recovered from the patient or other source. Whenever a serological test is being performed, either the antigen or the immune product is known. Serological tests are used for the following purposes:

1. Confirmation of identifications made by other clinical or laboratory procedures. An infecting organism, for example, may be tentatively identified by biochemical tests, but a question may remain as to its full identity. Serological tests are sometimes used to substantiate identifications made by other means.
2. Immunodiagnosis. Sometimes it is not possible to isolate the antigen, such as the infectious agent, from the patient or it is impractical to do so. The immunogen, however, leaves its mark in the form of an immune product. Thus a child may have had an infection that clinically appears to be mumps. Many diagnostic laboratories are not equipped to isolate the mumps virus from the patient, but they can recover antibodies. After the child's immune system has had time to produce a detectable level of antibodies, blood is drawn and the serum is harvested. The serum is tested for the presence of antimumps antibodies.
3. Direct examination of clinical material. A

preliminary presumptive diagnosis is made by performing an immunodiagnostic test directly on a clinical specimen such as a body tissue. Sometimes this is the only option available. Immunofluoresence tests and other immunohistochemical tests, for example, are versatile tools for locating "unknown" antigens on cells and tissues and other clinical materials. Parts of the specimen that are used for immunodiagnosis may be submitted for culturing and the cultures used for subsequent routine identification procedures.

4. Following the course of a disease. It is good clinical management in some diseases to obtain serum specimens periodically and to examine the serum for changes in ***titer*** (amount of antibody). A stabilization of or decrease in titer may indicate that the disease has been overcome or is waning, whereas an increase in titer usually indicates that the disease is continuing. The determination of antistreptococcal antibodies during streptococcal disease is an instance of this application.
5. Determination of the serotype or serogroup. In some cases the identification of a pathogen has been made to the species level, and that information is sufficient to determine the treatment for a given patient. Strains of a species of microorganisms sometimes develop significantly different characteristics because of mutation and natural selection. The identification of such strains through serological testing may be useful information for epidemiological purposes and for determining the mode of control to be used against the strain.
6. Determining immune status. It is sometimes useful to know if a person has antibodies to a given microorganism or to a given microbial product. A woman of childbearing age, for example, may want to know if she has antibodies to the German measles virus. A blood sample is obtained and a suitable test conducted for serum antibodies. The serological tests may also be used on populations in order to determine their past or present exposures to infectious agents.

GENERAL PRINCIPLES OF SEROLOGICAL TESTING

Most serological tests involve serum antibodies; thus most of the following discussion pertains to tests for antibodies rather than tests for sensitized lymphocytes. The fact that antibodies specifically interact with a given antigen (specificity is not necessarily absolute) makes it possible to use in vitro antigen-antibody reactions for identification.

The components of a serological test consist basically of an antigen and an antiserum or serum sample. One of the components of a serological test, either the antigen or the antibody, usually must be known (identified). The antigen of the test might be a microorganism whose identity is unknown or whose identity, based on other laboratory tests, is to be confirmed. This "unknown" microorganism is tested against known antisera. The specificity and antibody content of the antiserum is known because it was obtained from animals that were deliberately immunized with a known antigen. (Most antisera and antigens remain serologically active for long periods of time when properly stored.) Thus a suspected typhoid bacterium isolated from a patient can be tested against various antibacterial antisera. Suspensions of the isolated typhoid *(Salmonella)* bacteria, are tested against antisera specific for suspected microorganisms, including an antisalmonellal serum (antiserum). A visible reaction occurs only if the specimen containing typhoid organsims is mixed with the antityphoid serum; thus the causative agent of infection is identified.

The antibody may be the unknown and unidentified component in a serological test. It is possible in many instances to determine what caused the patient's infection by testing his serum against known antigens. Thus the serum from a patient who had typhoid fever (assume that the causative agent was not isolated and therefore not identified) is checked against known suspensions of different bacteria, including *Salmonella* (typhoid) bacteria. Again, because antigen-antibody reactions are specific, the serum of the patient gives a visible reaction with the suspension of *Salmonella* (typhoid) bacteria but not with the other test antigens.

Quantitation in serological tests

Serological tests may be as simple as placing an antigen and some serum onto a slide or into a tube and observing whether or not a reaction occurs (Figure 24-1). The testing of just one concentration of both components, however, is not always satisfactory. Single-concentration reactions may represent, for example, cross-reactivity or former disease. It is important in some instances to show a change in the antibody level against a

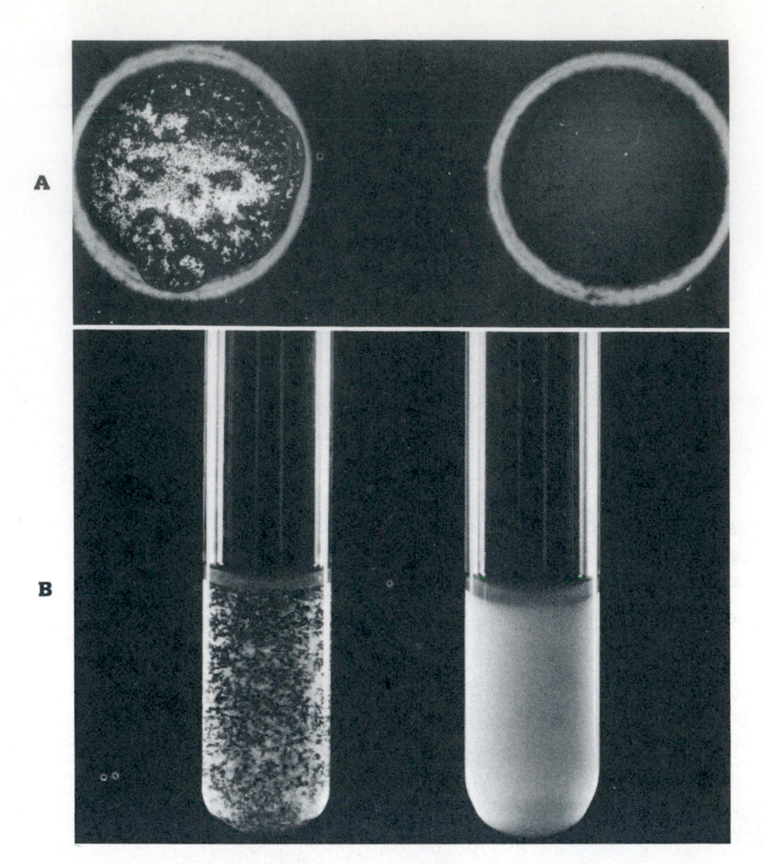

Figure 24-1 **A,** Slide agglutination of bacteria is rapid method of identifying unknown bacteria with known antisera or the reverse. **B,** Tube agglutination can be used for same purpose. In each case positive test is seen at left and negative control at right.

From Barrett, J.T. Textbook of immunology (4th ed.). The C.V. Mosby Co., St. Louis, 1983.

given antigen. Serological tests therefore, for these and other reasons, usually have a quantitative aspect. Serial dilutions of either the serum or the antigen are made with the purpose of determining a titer. Measured dilutions (usually of the serum) are made, and the highest dilution that gives a detectable reaction is recorded as the titer. Suppose that in a six-tube series of doubling dilutions of serum (1:20, 1:40, 1:80, 1:160, etc.), tubes 1, 2, 3, and 4 gave a detectable reaction (were positive). The titer of the serum would be reported as 160 (Figure 24-2). (The titer is usually expressed as the reciprocal of the highest dilution that gives a positive reaction.) The titer level is in many instances diagnostic, particularly when a significant change in titer can be shown between a serum sample that is collected early (an acute-phase specimen) in the course of an illness and a sample collected late in the illness (a convalescent-phase specimen). There is enough experience with most serological test types to interpret the significance of a titer for a given antigen-antibody system. Titers can also signify levels of protection, and they may be indicative of the severity and of the progress of a disease.

TYPES OF SEROLOGICAL TESTS

The union of antigen and antibody takes place under a variety of conditions, and the reaction may be measured in different ways. Longstanding serological tests are ***agglutination, precipitation, immune cytolysis, virus neutralization, toxin neutralization, complement fixation,*** and ***capsular swelling (quellung).*** These are what might be called "classical" tests and have been used for many years. The test names are mostly descriptive of what happens to the antigen. More recently designed tests are named according to the nature of the technique that is involved: ***fluorescent antibody, immunodiffusion, immunoelectrophoresis, immunoperoxidase,*** and ***radioimmunoassay,*** to name the more important ones.

The test antigen from the standpoint of its physical makeup is described as either ***particulate,*** such as a bacterium or red blood cell, or ***soluble,*** such as molecules in solution. Very often what we call the test antigen is a composite of many antigens; for example, bacterial cells that serve as the antigen in an agglutination test are actually composed of many antigens. These antigens include flagella, pili, and other components that could be separated from the bacterial cell to serve as individual antigens. A brief description of the more frequently used serological tests follows.

Agglutination

When suspensions of particulate antigens, such as bacteria or red blood cells, are mixed with sera that contain antibodies against antigens of those cells, aggregates or clumps ***(agglutinates)*** develop that are visible to the naked, unaided eye (Figure 24-2). The antibodies serve as links or bridges among the particulate antigens, and visible aggregates appear. Agglutination of red blood cells is more specifically termed ***hemagglutination.***

Agglutination tests of an indirect nature (passive agglutination) can be performed by adsorbing antibodies or soluble antigens to cells (usually red blood cells) or to appropriately sized particles such as latex particles. The cell surface of red blood cells, when treated with dilute tannic acid or chromic acid, is altered in some way so that antigen adsorbs to the surface. The red blood cell or

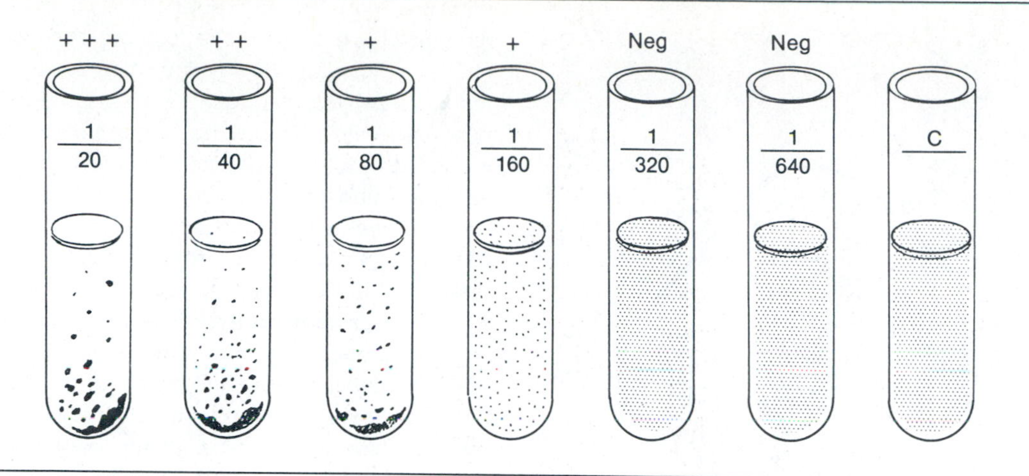

Figure 24-2 Bacterial agglutination, test tube method. Dilute 0.2 ml of patient's serum serially in isotonic saline through first six tubes *(left to right)*. Add 1 ml of suspension of bacteria to each tube through all seven. Final dilutions are indicated by numbers on each tube (such as 1 part serum in 20 parts, etc., suspension). Control tube on far right contains only bacterial suspension and saline. Note maximal agglutination of bacteria (large masses) in first three tubes on left, some in fourth tube, and none in fifth and sixth tubes. Agglutination test is positive; titer of serum is 1:160.

From Smith, A.L. Microbiology and pathology *(12th ed.). The C.V. Mosby, Co., St. Louis, 1980.*

latex particles with the adsorbed antigen given an agglutination reaction in the presence of antibodies that are specific for the attached soluble antigen.

Precipitation

The antigen in precipitation tests is soluble, and the antigen preparation is usually a clear solution. Examples of soluble antigens are egg albumin, bacterial extracts, and serum. (NOTE: Depending on the context, serum, with its many protein, can also be an antigen.) In precipitation reactions molecules of antigen ***(precipitinogen)*** are aggregated together by antibodies, and a visible precipitate is observed. Various techniques are used in antigen-antibody precipitation tests: the ***ring test,*** the ***capillary test,*** and ***diffusion tests.*** In the ring test a column of antiserum is placed into the bottom half of a small-diameter tube. A column of the antigen solution is carefully placed over the serum column. A positive test is characterized by a region of precipitate developing at the interface (where the two columns meet). The capillary test is performed by sequentially drawing, by capillary action, antiserum and antigen solutions into the tube. A precipitate usually develops in minutes in a positive test (Figure 24-3).

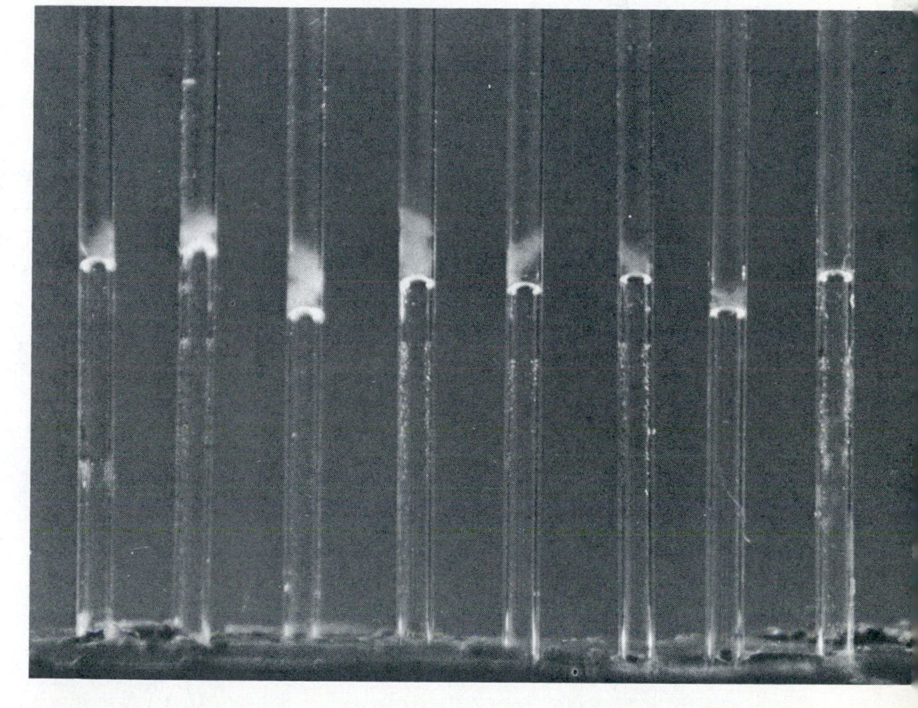

Figure 24-3 Capillary tube precipitin test. Dilutions of soluble antigen and constant amount of specific antiserum were drawn into capillary tubes. Tubes 3 and 4 show greatest amount of precipitate. White semicircle in each tube is due to light reflection from interface where antiserum and antigen dilution meet. Final tube is negative control.

Courtesy William Krass. From Barrett, J.T. Textbook of immunology *(4th ed.). The C.V. Mosby Co., St. Louis, 1983.*

Immunodiffusion. Many serological precipitation tests are now conducted in gels or other support media. Distinct lines of precipitate form in

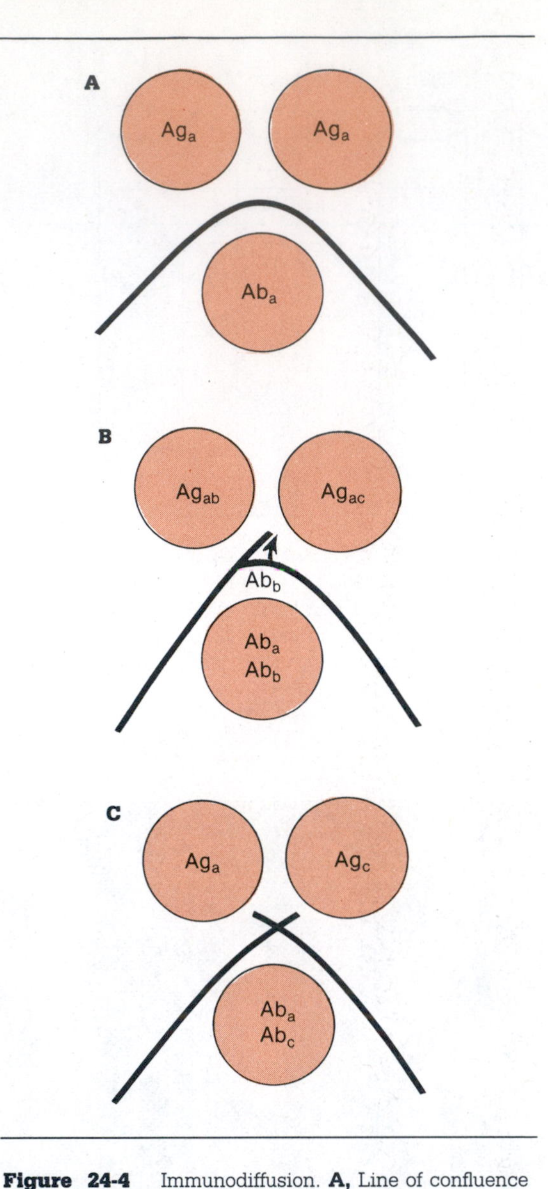

Figure 24-4 Immunodiffusion. **A,** Line of confluence obtained with two antigens that cannot be distinguished by antiserum used. **B,** Spur formation by partially related antigens having common determinant a but individual determinants b and c reacting with mixture of antibodies directed against a and b. Antigen with determinants a and c can only precipitate antibodies directed to a. Remaining antibodies (Ab$_b$) cross precipitin line to react with antigen from adjacent well, which has determinant b, giving rise to "spur" over precipitin line. **C,** Crossing lines formed with unrelated antigens.

Redrawn from Roitt, I.M. Essential immunology (2nd ed.). Blackwell Scientific Publications, Ltd., Oxford, 1974.

gel where optimal portions of diffusing antigen and antibody molecules meet.

The gels used today include ***agarose,*** a neutral polysaccharide that is extracted from agar; ***cellulose acetate film;*** and ***polyacrylamide gels.*** These are usually placed in Petri dishes or as a thin film on a slide or other support. Wells or troughs are usually made in the gel and filled with the antigen and antibody solutions. The number of wells used, the location and spacing of the wells, and the configuration of the wells (circular, rectangular, etc.) are dictated by the type of reagents and the information being sought and to some degree by personal preferences. Immunodiffusion tests can be varied as to whether one or both of the reagents diffuse ***(single diffusion, double diffusion)*** and as to the directions of the diffusing reagents ***(radial diffusion, one-dimensional diffusion).*** Single diffusion means that only one of the reagents diffuse from the deposit site. The other reagent is uniformly distributed throughout the gel because it was incorporated into the molten gel before the gel hardened. Double diffusion requires that both antigen and antiserum are placed in separate wells or troughs, resulting in both reagents diffusing through the gel. In radial diffusion the reagent diffuses in all directions (360 degrees) from the deposit site, whereas in one-dimensional diffusion the reagent diffuses in only one direction.

In a typical radial immunodiffusion test two wells are made in the gel. Antiserum is placed in one well and antigen in the other. The molecules from each of the wells diffuse in all directions (radial) in every-widening circles. Some of the molecules meet someplace between the two wells. A line (band) of precipitate forms if the serum contains antibodies specific for the antigen. If the antigen preparation contains three different antigens and if the antiserum has antibodies against the three antigens, theoretically three distinct lines of precipitate will develop between the two wells. Another basic immunodiffusion setup consists of three equidistantly spaced wells, one well containing antiserum and the other two containing antigen. Three principal types of reaction (precipitation patterns) may be detected. Assume that in this example each antigen-containing well contains only a single antigen. The reactions show whether the antigens in the two wells are serologically ***nonidentical, identical, indistinguishable,*** or ***partially identical*** (Figure 24-4 and 24-5). If the antigens are nonidentical, two separate lines of precipitate form and the lines merely cross in the common area where the lines of precipitate intersect. If the two antigens are serologically identical, a single line of precipitate develops that is curved in the common area where the two lines of precipitate meet. In the reaction of partial identity a single curved line develops, but a spur also occurs in the common area. The single curved line indicates that the two antigens share certain determinants. The spur indicates that the antigens are nonidentical.

Figure 24-5 Double-diffusion–double-dimension (Ouchterlony) test. Complete fusion of precipitates from antigens at lower and upper left wells indicates reaction of identity. (Center well contains antiserum; peripheral wells contain antigens.) Precipitates at upper left and right wells cross in reaction of nonidentity. Broad upper *(right)* band fuses in single line with one of several bands from right well in reaction of identity. Multiple bands of precipitate can be seen between lower right reservoir and antiserum well. Bottom well was filled with saline as negative control.

From Barrett, J.T. Textbook of immunology (3rd ed.). The C.V. Mosby Co., St. Louis, 1978.

Complement fixation

The complement fixation test was devised before the true physicochemical and biological nature of the complement system was discovered. *Complement,* when used in the context of complement fixation test, refers to a thermolabile component that is present in fresh serum. The feature of complement that is made use of in the test is its ability to attach IgG and IgM antibodies that have interacted with their specific antigens. If the antigen is a cell, the final result of the antigen-antibody-complement union is lysis of the cell, which is a visible reaction. The test consists of an ***indicator system*** and a ***test system*** (Figure 24-6). The complement fixation test is especially useful when an antigen-antibody union by itself does not give a visible reaction. A typical situation might be in a diagnostic laboratory where the question is, "Does the patient have antibodies against the mumps virus?" When the patient's serum is mixed with the mumps virus test antigen, a visible reaction does not occur, even though there are antimumps antibodies in the serum. It is necessary at this point to know the nature of the indicator system of the complement fixation test. The indicator system consists of sheep red blood cells (SRBCs) as the antigen of the indicator system and an antiserum that contains antibodies against the SRBCs. The antiserum is called ***hemolysin*** or ***amboceptor.*** When SRBCs and the hemolysin are allowed to interact in the presence of complement, the SRBCs lyse. The lysis is a visible reaction. The complete test is set up in this sequence: the patient's serum and the test antigen are mixed. A measured amount of complement (commercially available lyophilized guinea pig serum) is added to the test system (patient's serum and the known or suspected antigen). The indicator system (SRBCs and hemolysin) is added to the tube containing the test system and complement. Either the test system or the indicator system bind the complement. In a positive test there are antibodies in the patient's serum specific for the test antigen. The complement in the tube is fixed to the anti-

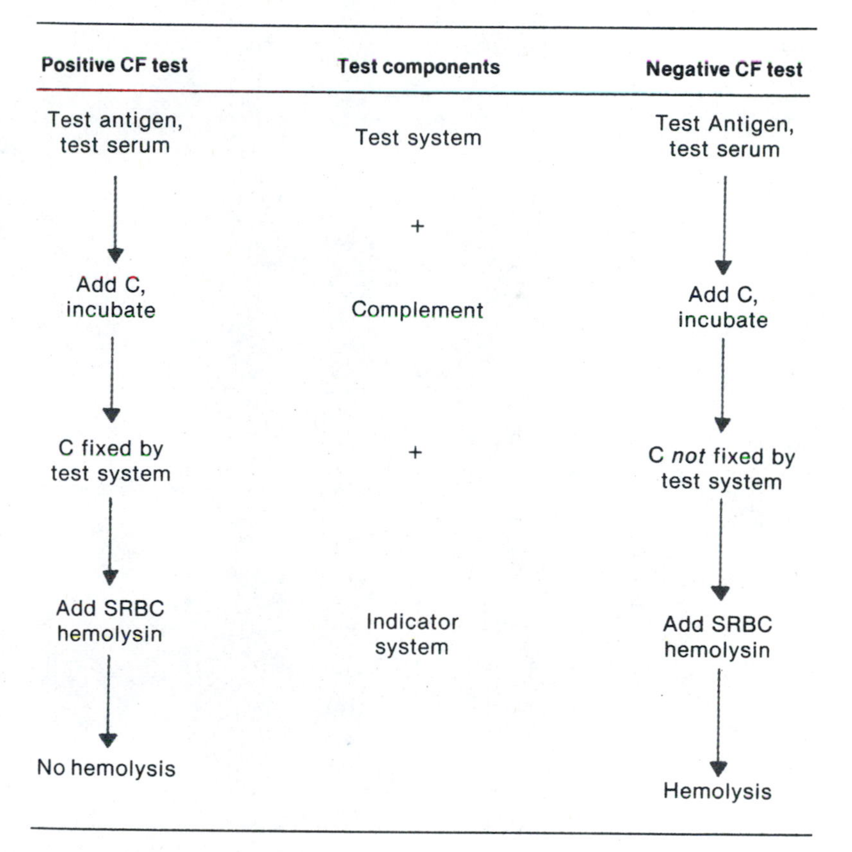

Figure 24-6 Complement fixation test. *C,* Complement; *SRBC,* sheep red blood cells.

Redrawn from Boyd, R.F., and Hoerl, B.G. Basic medical microbiology (2nd ed.). Little, Brown & Co., Boston, 1981.

A

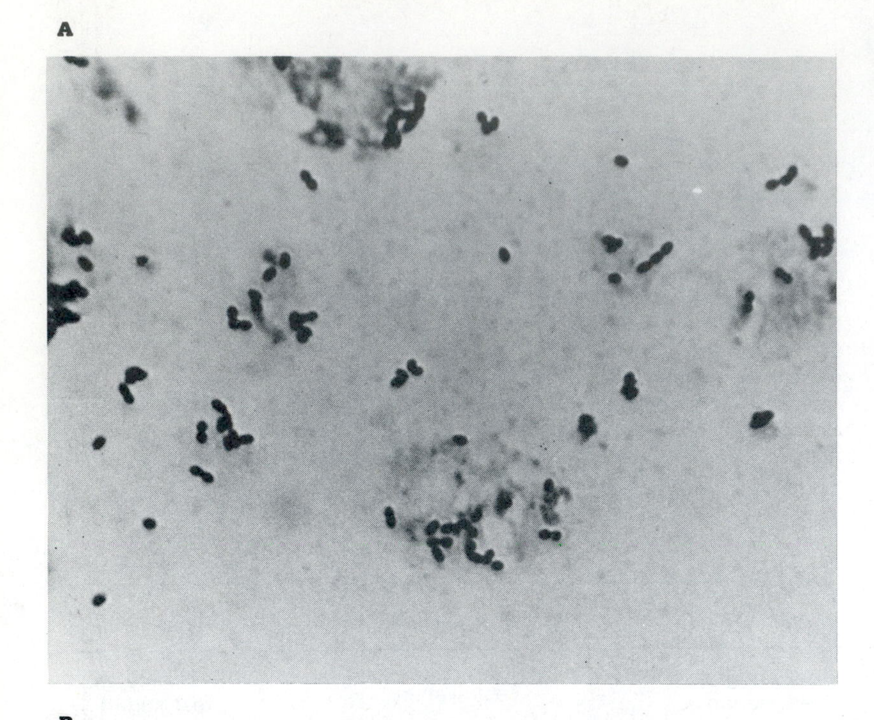

B

Figure 24-7 Capsular swelling (quellung) of bacteria. **A,** Before application of antiserum. Note very small halo around cells. **B,** After application of antiserum. Note enlarged capsule.

Courtesy R. Austrian.

gen-antibody complex and is not available to the indicator system when the latter is added to the tube. Hemolysis therefore does not occur in a positive test. In a negative test, because the patient's serum does not contain antibodies for the test antigen, the complement is not fixed. The complement can therefore interact with the indicator system and produce hemolysis.

Capsular swelling (quellung)

Antibodies against the capsular material of bacteria can bring about an apparent enlargement of the capsule (Figure 24-7), a reaction that is visible under the microscope. This phenomenon is useful in identification procedures and in serotyping. ***Serotyping*** refers to serological procedures that are used to differentiate strains of a species that have differences in the antigenic composition of a structure, product, etc. There are, for example, over 80 strains of the bacterial species *Streptococcus pneumoniae,* which is a leading cause of pneumonia. The strains or serotypes in this instance are based on differences in the antigenic composition of the capsular material of each strain. These differences can be detected by capsular swelling reactions using antisera that are specific for the capsular types. Serotyping is also possible with other serological techniques.

Neutralization tests

Neutralization tests are of a type that determine whether the activity of a toxin or of an infectious agent has been neutralized by an antibody. It is sometimes impossible or impractical to detect serum antibodies to a bacterial exotoxin or to a given virus in vitro. Animals or tissue culture cells are used as the "indicator" system in neutralization tests. The toxin or the virus (the antigens of the test) have a known effect on an animal or on tissue culture cells. The effect in the indicator-animal might be death, paralysis, or skin lesions. The exotoxin of *Clostridium botulinum,* for example, causes an often-fatal type of food intoxication. Diagnosis is made by collecting filtrates of the suspect food or of the serum, stool or vomitus of the patient and then injecting the filtrates into mice, some of which have received botulinum antitoxin. If there is toxin in the filtrate, all the mice except those that received antitoxin will be killed. The objective of a neutralization test, rather than testing for a toxin or a virus, might be to determine whether a patient's serum has antibodies to a given toxin or virus. The patient's serum and a measured amount of toxin or virus are mixed in the test tube and injected into a suitable laboratory animal that gives a known reaction to the toxin or virus. If the patient's serum contains antibodies specific for the toxin or virus, the toxin or virus is no longer free to bring about the effect it would typically produce if it had not been neutralized by antibodies.

Tests that incorporate chemical markers

Antigen-antibody interactions are also detectable in immunological tests in which chemical markers have been conjugated to antibodies or antigen. These tests are useful in cases where (1) an antigen-antibody union by itself is not readily detectable, (2) antibodies or antigens are more easily detected directly in clinical material, and (3) greater sensitivity and/or faster results are attainable. Three very important tests in this category are ***immunofluorescence (fluorescent antibody technique), enzyme-linked immunosorbent assay (ELISA),*** and ***radioimmunoassay (RIA).***

Immunofluorescence. Certain dyes, called ***fluorochromes,*** when irradiated with ultraviolet rays emit visible light. The dyes can be conjugated (coupled) to antibody molecules without changing the antibody's capacity to bind to the antigen, and they can also be attached to antigens. One application of this technique is in determining whether or not an animal had rabies. Laboratory diagnosis of rabies includes the examination of brain tissue of the rabies-suspected animal for ***Negri bodies*** and the injection of homogenized brain tissue into mice. Negri bodies, which are viral inclusion bodies elicited by the rabies virus, are not always visible in the brain tissue of the infected animal. In addition, with injections of homogenized brain tissue into mice, 3 weeks must elapse before it can be determined whether or not the mice have acquired rabies. A diagnosis, however, can be made within hours by using the fluorescent antibody technique. Sections or impression smears of the animal's brain tissue are exposed to an antiserum that contains conjugated antibodies specific for rabies virus. Antibodies will attach to the virus if rabies virus is present in the tissue. When the preparation is irradiated with ultraviolet light rays, the fluorochrome attached to the antibodies emits light that is readily visible when the preparation is examined with a fluorescent microscope.

ELISA and RIA tests. The ELISA and RIA tests are similar in principle to immunofluorescence, but they are easier to interpret. Enzymes, such as ***peroxidase*** and ***alkaline phosphatase,*** are conjugated to antibodies or to antigens in the ELISA tests. The enzymes are detected by adding a specific chromogenic enzyme substrate to the preparation. This produces a colored product that can be read visually or measured spectrophotometrically. ELISA tests are less expensive and not as laborious as immunofluorescence tests. The RIA technique uses antibodies or antigens that have been labled with radioactive isotopes. Radioactivity is detected by such means as radioisotope analyzers and by autoradiography (photographic emulsions that show up areas of radioactivity). RIA tests are easier to interpret than immunofluorescence, but they are expensive, and there is some risk because of radioactive material.

Immunoelectrophoresis

Charged molecules, as found, for instance, in antigen preparation and in serum, when subjected to direct current in a gel matrix, are separated because their ***electrophoretic mobilities*** differ. It is very unlikely that different molecules have the same electrophoretic mobility. The proteins that are present in a solution or in serum will be deposited in certain sites based on each protein's electrophoretic mobility. The electrophoresed protein can be stained and the location noted on the support medium. The eletrophoretic pattern of the proteins of normal serum is well known. Specific bands or peaks are associated with specific serum proteins. Electrophoretic analyses may reveal significant departures from the normal serum pattern, and certain diseases can be linked to the absence or change in specific peaks (Figure 24-8).

The specificity that exists between an antigen and an antibody can be applied to the electrophoretic characteristics to make the reaction immunoelectrophoretic. Immunoelectrophoretic techniques offer certain advantages over immunodiffusion methods because lines of precipitate can develop sooner, and diagnosis or identification can be speeded up. Better separations can be obtained by this method when there are multiple antigens in the antigen preparation. Two antigens may have mobilities that are not easily differentiated by diffusion. Electrophoresed molecules can in many instances be made to move in a specific direction. This latter feature is important when only small quantities of the reactants are available. The test is more sensitive because the available molecules are concentrated in one area, whereas in radial diffusion the molecules diffuse in all directions.

In a typical test multiple antigens are sepa-

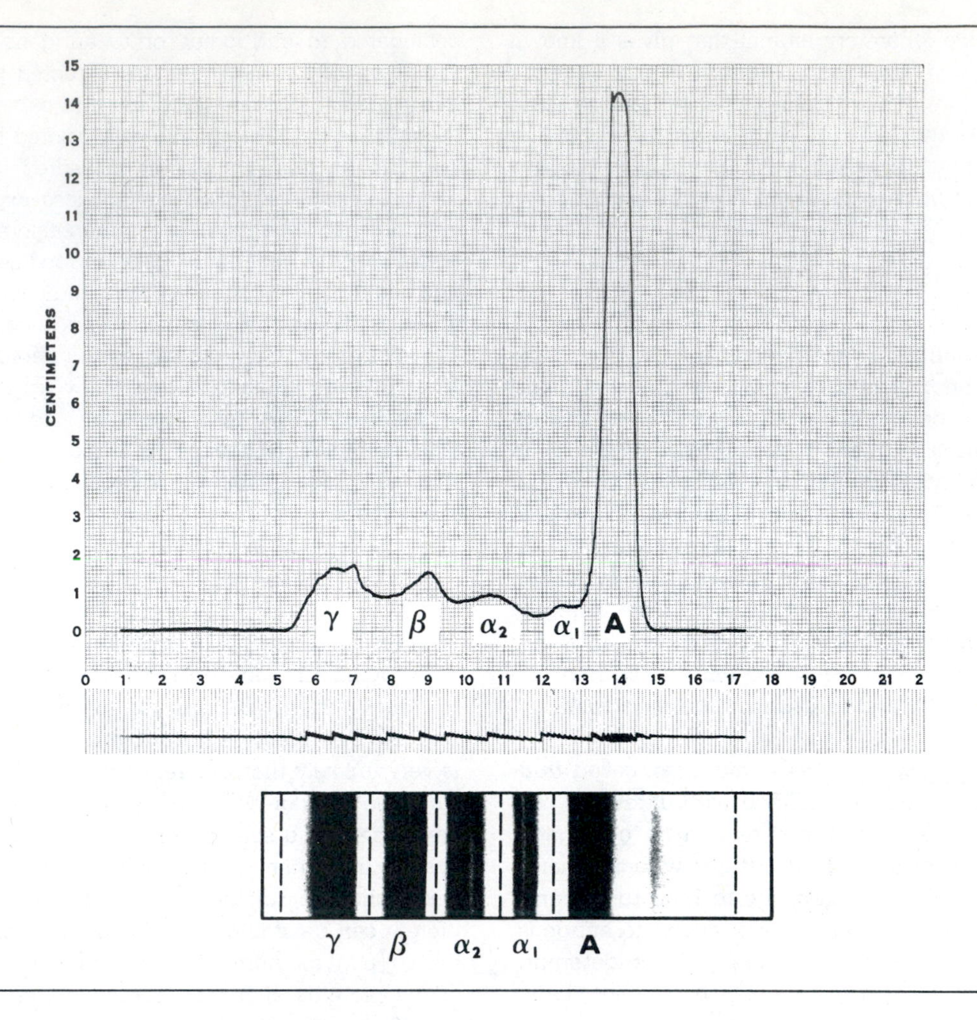

Figure 24-8 Electrophoresis of normal human serum. Human serum was electrophoresed on cellulose acetate strip *(lower part of illustration).* Recording densitometer was then used to scan strip, and curve was obtained, as shown on graph. *A,* Albumin; α_1 and α_2, α-globulins; *β,* β-globulins; *γ,* γ-globulins.

From Bauer, J.D. Clinical laboratory methods (9th ed.). The C.V. Mosby, Co., St. Louis, 1974.

rated by electrophoresis (Figure 24-9), but the separated antigens are not visible at this point. After the electrophoretic separation a trough, which parallels the line along which the antigen molecules were deposited, is filled with antiserum. The antigen molecules that diffuse through the gel in a direction that is perpendicular to the antigen deposit line make contact with antibody molecules. Lines of precipitate develop at the contact point where specific antibody and antigen meet. There are many modifications to the immunoelectrophoretic technique: ***counterimmunoelectrophoresis (CIE), one-dimensional rocket immunoelectrophoresis,*** and ***two-dimensional immunoelectrophoresis.*** Only counterimmunoelectrophoresis is described here.

The antigens and antibodies in counterimmunoelectrophoresis are made to flow in opposing directions toward each other and to form precipitate bands where their paths meet (Figure 24-10). The antigen and antibody molecules must have different ***isoelectric**** pHs, and the isoelectric pHs need to be known or empirically determined in order for this sensitive and rapid technique to be applicable. The gel is buffered to a pH that is intermediate to the isoelectric pHs of the two reactants. Under these conditions the antigen molecules have a net charge that is opposite that of the antibody molecules, and hence they migrate in opposing directions toward each other: one to the positive pole (cathode) and the other to the negative pole (anode).

* The isoelectric pH is the pH at which electrophoretic migration does not occur.

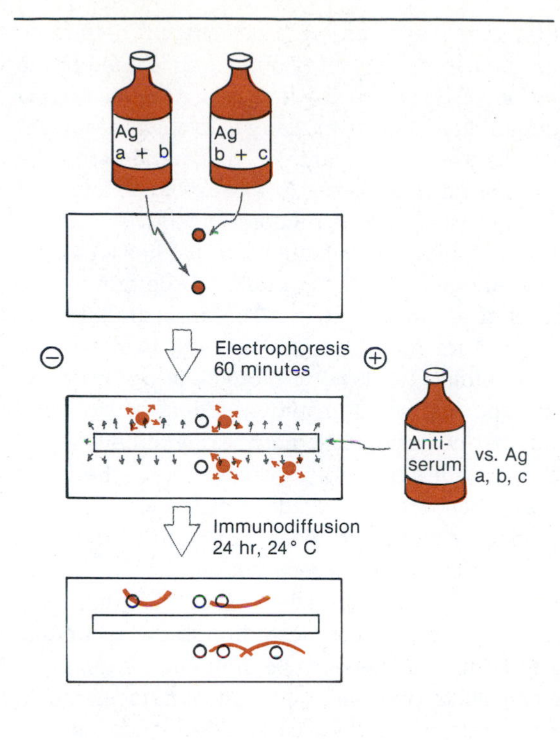

Figure 24-9 Immunoelectrophoresis. Antigens a and b *(lower black circle)* and antigens b and c *(upper black circle)*, on being electrophoresed for 60 minutes, are separated according to their differing electrophoretic mobilities. After antigens are separated, antiserum that contains antibodies to antigens a, b, and c is placed into central trough. Antigen and antibody molecules diffuse, and lines of precipitate *(arcs)* form where appropriate proportions of antigen and specific antibody meet.

Redrawn from Hyde, R.M., and Patnode, R.A. Immunology, *Reston Publishing Co., Inc., Reston, Va., 1978.*

Figure 24-10 Counterimmunoelectrophoresis. Antigen a, on electrophoresis, flows unindirectionally toward cathode. Anti-a antibody flows unindirectionally toward anode. Lines of precipitate form where appropriate proportions of antigen and specific antibody meet.

Redrawn from Hyde, R.M., and Patnode, R.A. Immunology, *Prentice-Hall, Inc., Englewood Cliffs, N.J., 1978.*

IN VIVO TESTS

Some immunological tests, usually in the form of a diagnostic skin test, are conducted in patients. Measured amounts of antigens are injected and the reaction noted. In some tests there is no detectable reaction, because the patient has antibodies specific for the antigen. For example, in the Shick test, which is used to determine if a person has antibodies to diphtheria toxin, a small amount of toxin is injected into the skin. There is no reaction in the patient who has sufficient antibodies to diphtheria toxin, but the patient who lacks antibodies to diphtheria toxin develops a skin reaction that is typically caused by the unneutralized toxin. In other tests the patient develops a reaction if he has formed immune products against the test antigen. This is especially

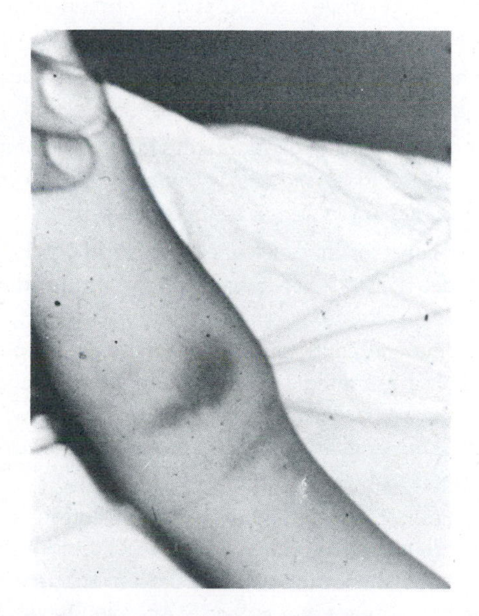

Figure 24-11 Positive tuberculin skin test.

From Nakamura, R.M. Immunopathology: clinical laboratory concepts and methods. *Little, Brown & Co., Boston, 1974.*

the case when cell-mediated immunity is involved, as in ***tuberculin testing*** (Figure 24-11) and in testing for fungus infections. In vivo tests are usually named after the individual who devised the test or after the test material that is extracted from the microorganism: ***Schick test, Mantoux test, Prausnitz-Kuestner test,*** and ***tuberculin test,*** to name a few. In vivo tests are usually adjudged positive or negative. They do not lend themselves to determining a titer.

SUMMARY

The union of antigen and immune products in vitro is generally a visible reaction, or it can be made visible by a variety of test techniques. Since there is a high degree of specificity of an immune product for its corresponding antigen, antigen-immune product reactions are a very sensitive means of identification and monitoring. In microbiology serological tests are used not only to identify microorganisms and their products or the antibodies they evoke, but also to (1) confirm a diagnosis made by other laboratory procedures, (2) follow the course of a disease, (3) determine the serotype and serogroup, and (4) determine the level of protection from disease. There is a quantitative aspect to many serological tests. Dilutions of either the antigen preparation or of the test serum make possible the establishment of a titer. Not only can it be determined that a given serum, for example, contains antibodies for a specific antigen, but it is possible to determine the level of antibody present. The antibody level (titer) in turn may be significant for diagnosis, for determining the level of protection, or for following the course of infection. There are a large number of basic serological test techniques, and there are a variety of modifications of the basic test procedures. The older "classical" tests are named according to what happened to the antigen: agglutination, precipitation, cytolysis, neutralization, and capsular swelling. The more recent tests are named according to the technique used: immunofluorescence, immunodiffusion, immunoelectrophoresis, and radioimmunoassay. In vivo tests, which are largely performed as skin tests, yield significant diagnostic information.

Study questions

1. What two features of an in vitro antigen-immune product interaction make those interactions so useful for identification and monitoring procedures?
2. What is the value of establishing a titer? Why not just use a single dilution and determine whether the reaction is positive or negative?
3. Why is it usually necessary to perform a neutralization test in animals or in tissue culture?
4. In a complement-fixation test why is the complement added to the test system before the indicator system is added to the same tube?
5. The capsular material of each of the 80 plus serotypes of ***Streptococcus pneumoniae*** differs only slightly from one serotype to another, yet serological reactions are sensitive enough to discriminate among the many serotypes. This discriminatory sensitivity is a reflection of which distinctive features of the immune system? (Specificity?, Self-recognition? Memory?)
6. Explain the statement, "The electrophoretic technique is sometimes used to obtain valuable diagnostic information without an antigen-antibody reaction taking place." After you have completed your study of immunodeficiency diseases in the next chapter, see if you can relate the statement to certain immunodeficiencies.
7. Marker chemicals are conjugated to antibodies or sometimes to antigens in certain immunological tests. What are the advantages of this technique over tests such as agglutination and cytolysis?

Selected readings

Books

Bauer, J.D. *Clinical laboratory methods* (9th ed.). The C.V. Mosby, Co., St. Louis, 1982.

Bryant, N.J. *Laboratory immunolgy and serology.* W.B. Saunders, Co., Philadelphia, 1979.

Hudson, L., and Hay, F.C. *Practical immunology* (2nd ed.). Blackwell Scientific Publications, Ltd., Oxford, 1980.

Rose, N.R., and Bigazzi, P.E. (Eds.) *Methods in immunodiagnosis* (2nd ed.). John Wiley and Sons, Inc., New York, 1980.

Rose, N.R., and Friedman, H. *Manual of clinical immunology* (2nd ed.). American Society for Microbiology, Washington, D.C., 1980.

Stansfield, W.D. *Serology and immunology.* Macmillan Publishing Co., Inc., New York, 1981.

Thaler, M.S., Klausner, R.D., and Cohen, H.J. *Medical immunology.* J.B. Lippincott Co., Philadelphia, 1977.

Journal articles

Gordon, D.S. Immunoglobulin quantitation and characterization as a diagnostic tool. *South. Med. J.* **70:**236, 1977.

Peterson, E.M. ELISA: a tool for the clinical microbiologist. *Am. J. Med. Tech.* **47:**905, 1981.

Rytel, M.W. Counterimmunoelectrophoresis: a diagnostic adjunct in clinical microbiology. *Lab. Med.* **11:**655, 1980.

Yolken, R.H. ELISA: enzyme-linked immunosorbent assay. *Hosp. Pract.* **13:**121, 1978.

Chapter 25 IMMUNE DISORDERS Immunodeficiencies and Hypersensitivities

Bryan Hoerl

The term *immune* implies protection, which is indeed the prime and expected function of the immune system. There are, however, genetic and/or developmental defects or acquired defects that cause the immune system to overreact or underreact and that can lead to immune disorders. Immune disorders can be placed into two broad categories: immunodeficiencies and hypersensitivities. Included among the immune disorders discussed in this chapter are immunodeficiency diseases, immunoproliferative disorders, hypersensitivities (allergies), and autoimmune disease. In addition, graft rejection and cancer are discussed. Graft rejection is customarily discussed with allergies because the immune system damages the transplanted tissue that is intended to become part of the graft recipient's tissues. Cancer is also discussed because the development of cancer is probably due to the proliferation of aberrant cells that have escaped the postulated cancer surveillance action of the immune system.

IMMUNODEFICIENCY DISEASES

Deficiency diseases involving the immune system proper can be subdivided into two categories: primary immunodeficiencies that are of genetic origin and secondary immunodeficiencies that are due to underlying nonspecific causes, including malnutrition, drugs, x-rays, and cancer.

Primary immunodeficiencies

Genetic and/or developmental disorders involving lymphoid tissue may affect the B cell component (humoral immunity, the T cell component (T cell–mediated immunity), or both. Combined B and T cell disorders affect the lymphoid stem cell population of the bone marrow. Primary immunodeficiencies often make the individual easy prey for infectious agents, and the type of infection can be a clue as to the type of immune dysfunction (Table 25-1). Primary immunodeficiencies are usually detected within the first 6 to 8 months of life because of recurrent infections or because of disorders of the skin, intestines, connective tissue, or vascular tissue in the newborn child. The naturally acquired passive immunity obtained through the transplacental transfer of maternal IgG antibodies has waned by 6 to 8 months following birth.

Defects in immune system accessory elements. Strictly speaking, some of the diseases that are categorized as immune disorders do not belong to the elements of the immune system proper, but to cells and systems that interact with the immune system.

Defects in phagocytes. Phagocytic cells may be defective at some point in the chain of chemical events that lead to intracellular killing. As discussed in Chapter 22, children who have chronic granulomatous disease (CGD) have poly-

Table 25-1 Infection as a clue to immune dysfunction

TYPE OF INFECTION	POSSIBLE DYSFUNCTION
Severe and/or recurrent infection with pyogenic (pus-forming) bacteria such as species of *Streptococcus* and *Staphylococcus*	B cells (except in IgA deficiency); phagocytic system (PMN-killing defect, splenic dysfunction); complement (deficiency of C1, C2, C3, C5, C6, C7, or C8)
Severe infection with herpesvirus, cytomegalovirus, chickenpox virus, live vaccine	T cells (SCID*, treated malignancy)
Severe infection with hepatitis virus, echovirus, vaccine-strain poliomyelitis virus	B cells (deficiency of IgG, IgM, or IgA)
Resistant superficial candidiasis (a fungal infection caused by species of *Candida*)	T cells (SCID, thymic hypoplasia, chronic mucocutaneous candidiasis, steroid therapy)
Systemic infection with opportunistic fungi such as species of *Nocardia, Candida,* and *Aspergillus*	T cells (Hodgkin's disease); phagocytic system (PMN-killing defect)
Pneumonia caused by *Pneumocystis carinii*	T cells (SCID, treated leukemia); B cells (very rare)
Giardiasis (an infection by the animal parasite *Giardia lamblia*)	B cells (deficiency of IgG, IgM, or IgA)
Sudden severe sepsis	B cells (deficiency of IgM); spleen; complement

Modified from Meuwissen, H. *Postgrad. Med.* **66**(5):116, 1979.
PMN, Polymorphonuclear neutrophil leukocyte; *SCID,* severe, combined immunodeficiency disease.
*SCID is an inherited disease that is sex linked and involves both the humoral and cellular arms of the immune system. Children often die before 1 year of age as a result of microbial infections.

MANIFESTATIONS OF PRIMARY IMMUNODEFICIENCIES

Infection (often by microorganisms that are weakly pathogenic)
Failure to thrive (those suffering from SCID)
Intestinal disorders: chronic diarrhea, malabsorption,* giardiasis
Skin disorders: eczema, seborrhea,† warts (IgM deficiency)
Collagen-vascular disorders: vasculitis‡—disease resembling systemic lupus erythematosus (see "Defects in the Complement System")

Modified from Meuwissen, H. *Postgrad. Med.* **66**(5):116, 1979.
*Malabsorption is the inability to absorb nutrients from the intestine.
†Seborrhea is a disease of the sebaceous glands in which secretions (sebum) collect on the skin and form an oily coating or scales.
‡Vasculitis is an inflammation of blood vessels

morphonuclear leukocytes (PMNs) that are morphologically normal, but they are unable to generate the superoxide ion that is necessary for the production of an intracellular bactericidal substance.

Defects in the complement system. Genetic defects are known for most of the complement factors. Such defects are associated with recurrent and disseminated bacterial infections, lupus erythematosus, and hereditary angioneurotic edema. ***Lupus erythematosus*** (Latin: *red wolf*) is an immune complex disease that may affect only the skin or the skin as well as internal organs. The disease gets its name from the characteristic red rash that appears across the nose and upper checks. One of the complications of systemic lupus erythematosus is the formation of antibodies against various cellular components, especially of the kidneys (glomerulonephritis). ***Hereditary angioneurotic edema*** is due to a defect in the regulatory protein C1 inhibitor. Edema fluid accumulates locally in various parts of the body and if localized in the larynx creates an obstruction of the airway.

B cell immunodeficiency. Sex-linked agammaglobulinemia, or ***Bruton's disease,*** is a congenital immunodeficiency involving B lymphocytes and is seen almost exclusively in males. Once a child with this immunodeficiency lives beyond 6 months of age (the age when maternal antibodies become depleted), he is subject to a succession of infections by pyogenic (pus-forming) cocci, such as streptococci and staphylococci, as well as *Haemophilus influenzae* and *Pseudomonas aeruginosa.* The classic case of Bruton's disease has these characteristics: there is a decrease (hypogammaglobulinemia) and occasionally an absence of all five serum immunoglobulin (agammaglobulinemia) classes; plasma cells and B cells are absent or decreased in number; the responses mediated by the T cell system are normal; areas of the secondary lymphoid tissue (lymph nodes, for example) where B lymphocytes localize are underdeveloped. Victims are able to reject grafts, however, because the T cell system is fully functional, and they can cope with tuberculosis as well as fungal and viral infections that the T cell system normally controls.

T cell immunodeficiency. The prototype of this deficiency is ***DiGeorge's syndrome,*** in which the thymus gland fails to develop properly. Spasms (tetany), due to low calcium levels, are an early indication of this disease because the parathyroid gland, which controls calcium metabolism, also develops from tissue in this same area (third and fourth pharyngeal pouches). The immune system deficiency is characterized by a poorly developed thymus gland and an absence or diminution in numbers of T cells in the circulation and in the thymus-dependent areas of the peripheral lymphoid organs. Serum Ig levels may be normal, but antibody-mediated reactions are affected because of the T cell–B cell interactions that are required in order for B cells to respond to T cell–dependent antigens. T cell–mediated immune functions, such as graft rejection and the development of delayed hypersensitivity reactions to skin test antigens, are markedly impaired. Infants with this defect are especially vulnerable to viral and fungal infections. Extensive infections of the skin and mucous membranes with the fungus *Candida* (mucocutaneous candidiasis) often occurs (see Figure 31-21). (NOTE: An extract called transfer factor, obtained from human T lymphocytes, is sometimes used therapeutically in ameliorating this condition.) Serious respiratory tract infections caused by *Pneumocystis carinii* are not uncommon and are generally an indication of an immunodeficiency, because this organism is ordinarily nonpathogenic in individuals with a normal immune system. Treatment of the immunodeficiency aspect of the disease involves the transplantation of thymus tissues obtained from fetuses.

B and T cell immunodeficiencies. Defects in the stem cell population of B and T lymphoid cells result in a condition referred to as a ***severe, combined immunodeficiency disease (SCID).*** Swiss-type agammaglobulinemia is one type of SCID. A child suffering from this type of immunodeficiency usually succumbs within the first 2 years to any one of a variety of infectious agents, including microorganisms of low pathogenicity. Some patients can be successfully reconstituted immunologically through the administration of compatible (from a sibling with well-matched histocompatibility antigens) bone marrow cells.

Secondary immunodeficiencies

Defects in what was originally a normal immune system may occur secondarily as a result of malignancies, infections, drugs, malnutrition, and therapies that affect the lymphoid system (Table 25-2). These deficiencies usually occur later in life and are not ordinarily as severe as the primary immunodeficiences. Graft recipients have a secondary immunodeficiency when they are immunosuppressed. Immunosuppressive measures are employed in graft recipients so that their immune system is less able to reject (immunoreject) the graft, but this also makes the immunosuppressed graft recipient vulnerable to infections.

A disease referred to as acquired immune deficiency syndrome (AIDS) was first recognized in 1981 among the homosexual community. Affected individuals would initially show weight loss, fatigue, swelling of lymph nodes, and diarrhea. Six to eight months later the immune system would fail, and the patient would fall prey to severe disease such as pneumonia caused by a rare infectious agent, *Pneumocystis carinii* (a protozoan), or Kaposi's sarcoma. Kaposi's sarcoma is a rare neoplastic malignancy seen primarily in elderly men and is rarely fatal. The AIDS victims, however, are generally young, and case fatality ratios are as high as 60% for cases first diagnosed a year previously. Epidemiological studies now indicate that AIDS victims fall into four general categories

1. Seventy-five percent are white homosexual men in their 30s or 40s. They have had prior infections with mononucleosis and venereal disease and have had many sexual partners. There have been two cases so far of AIDS among female sex partners of men who had the disease.

Table 25-2 Some factors predisposing to secondary immunodeficiency

FACTOR	IMPAIRED IMMUNE COMPONENT
Age	
Newborn (especially if premature)	B and T cells; phagocytic system; complement (severe bacterial and viral infections)
Elderly	B and T cells (anergy,* autoimmunity)
Lack of breast-feeding	IgA (decreased mucosal protection in intestines, increased frequency of respiratory and gastrointestinal tract infections)
Immunosuppressive therapy	
Corticosteroids	B and T cells; phagocytic system (decreased PMN chemotaxis)
Cyclophosphamide and 6-mercaptopurine	B and T cells; phagocytic system (decreased number of PMNs)
Irradiation	T cells; phagocytic system (decreased number of PMNs)
Diseases involving lymphoid system	
Hodgkin's disease	T cells
Lymphoma, chronic lymphatic leukemia	B cells
Multiple myeloma	B cells
Thymoma	B cell (hypogammaglobulinemia)
Splenectomy (anatomical or functional as in sickle cell disease)	Phagocytic system (decreased clearance of pneumococci and other bacteria by spleen and macrophages)
Other	
Severe infections	T cells (decreased delayed hypersensitivity); phagocytic system (PMN-killing defect following severe bacterial infections)
Burns	B and T cells
Malnutrition	T cells
Malignancy	T cells
Diabetes mellitus	Phagocytic system (decreased PMN chemotaxis)
Systemic lupus erythematosus	T cells; complement

From Meuwissen, H. *Postgrad. Med.* **66**(5):116, 1979.

*Anergy: lack of immunity to an antigen.

2. Twenty to twenty-five percent are intravenous drug abusers who are black or Hispanic and heterosexual.
3. Three to five percent are Haitian immigrants who are believed to have been infected before arriving in the United States.
4. Less than 1% are hemophiliacs who receive blood transfusions because of the need for blood-clotting factor.

The agent of AIDS is believed to be transmitted through lesions caused by anogenital sex, through dirty hypodermic needles, or possibly through the administration of blood or blood products. To date no person-to-person transmission has been identified other than through intimate contact or blood transfusion.

The type of immunosuppression caused by the agent of AIDS is unique. Ordinarily, healthy individuals have twice as many helper T cells as suppressor T cells, but in AIDS victims the ratio is reversed, and the helper T cells, which promote antibody formation, are depleted. Thus the patient is subject to infection by a multitude of opportunistic microorganisms that ordinarily do not harm the healthy individual.

The agent of AIDS has not as yet been discovered, but it is believed to be a virus. The hepatitis B virus, for example, is transmitted sexually and through the administration of blood and blood products. Recently (1983) human T cell leukemia virus infection was deteced in some patients with AIDS. This virus is a retrovirus—an RNA virus that contains the enzyme reverse transcriptase. This enzyme makes a DNA copy from an RNA genome. Retroviruses in animals are known to be a cause of immune suppression. There may be some etiological relationship between the T cell leukemia virus and AIDS, but further study is required.

HYPERSENSITIVITIES (ALLERGIES)

The in vivo reactions of immune products, with their homologous antigens, are beneficial to the host by virtue of their effect on the foreign agent (neutralization, cytolysis, phagocytosis). Sometimes, however, in vivo immune product-antigen reactions damage the host's tissues, cells, or systems. Hypersensitivities (or allergies—the terms are usually used interchangeably) are characteristic of an immune system that responds in a manner outside its normal response. Most allergic reactions are responses of a normal immune system, but an important exception should be noted. There is a hereditary predisposition in humans to form IgE, the immunoglobulin that is involved in anaphylactic (detrimental reactions resulting from a second exposure to antigen) and atopic allergic reactions. Whether or not an overresponse (allergic reaction) is going to occur and the nature of the response are usually determined by a combination of conditions, including (1) the nature of the antigen, (2) the amount of antigen that is introduced, (3) the route by which the antigen exposure occurs, (4) the chronological spacings and frequency of antigen exposure, and (5) the immune responsiveness of the host.

Different kinds of allergic reactions can be demonstrated in experimental animals by manipulating such factors as the amount of antigen and the routes of administration, and by the chronological spacing of antigen exposure. The injection of the same antigen ovalbumin (egg albumin), for example, under the varied conditions cited can lead to different allergic responses, among which are ***anaphylactic shock,*** the ***Arthus reaction,*** and ***serum sickness.*** A fatal anaphylactic shock reaction can be elicited by first sensitizing the animal with a small amount of ovalbumin administered by a parenteral route (other than orally—usually by injection), waiting 10 to 21 days for antibodies to be induced, and then challenging the sensitized (to ovalbumin) animal by a direct route (for example, intracardially) with a larger amount of the same antigen. An Arthus reaction, which is characterized by an intense inflammatory reaction of the skin that leads to localized tissue necrosis, can be elicited by giving a series of 3 to 5 ovalbumin shots about every 5 to 7 days into the skin at approximately the same site. A serum sickness type of reaction can be elicited by the injection of a single, but large, dose of ovalbumin. The symptoms of serum sickness include elevated temperature, lameness due to inflamed and enlarged joints, and enlarged lymph nodes.

Categorization

Immediate and delayed hypersensitivities. *Immediate* and *delayed* refer to the time of appearance of skin lesions when a sensitized individual is skin challenged with the antigen that induced the hypersensitive state. If the reaction takes place within minutes of the administration of the skin test dose, the allergy is categorized as being of the immediate type; if it is hours to days, it is of the delayed type. Immediate hypersensitivities are triggered by antibodies interacting with their appropriate antigens; thus immediate hypersensitivities are also termed humoral hypersensitivities. Delayed hypersensitivities involve sensitized T lymphocytes and are termed cell-mediated hypersensitivities. Immediate hypersensitivities are transferable via serum, and delayed hypersensitivities are transferable via intact T lymphocytes or via an extract of T lymphocytes called transfer factor (Table 25-3).

Gell and Coombs classification. Another commonly used classification is reflective of the underlying mechanism of the hypersensitivity reaction. Allergic disorders are divided into four types. The types are modifications of the original proposal by Gell and Coombs (Table 25-4).

- Type I: anaphylactic, reagin-dependent reactions
- Type II: cytotoxic and cytolytic reactions
- Type III: immune or toxic complex reactions
- Type IV: delayed hypersensitivity, T cell–mediated reactions

Descriptions in the literature quite commonly refer to allergic reactions as a type I reaction, a type II reaction, etc. The referral is to the Gell-Coomb types just cited and described here.

Table 25-3 Principal characteristics of immediate and delayed hypersensitivities

	TYPE OF ALLERGY	EXAMPLES OF ANTIGENS	COUPLER: IG CLASS OR T CELLS	PRINCIPAL EFFECTOR CELLS AND THEIR MEDIATORS	PRINCIPAL PATHOPHYSIOLOGICAL EFFECT
Immediate: humoral hypersensitivity; antibody mediated; passively transferable via serum	Systemic anaphylaxis	Penicillin, hymenoptera stings, foreign serum	IgE	Mast cells: Histamine SRS-A Kinins	Capillary permeability Smooth muscle contraction Exocrine secretions
	Atopic allergies (localized anaphylaxis)	Pollens, house dust, animal epidermoids	IgE	Mast cells: Histamine SRS-A Kinins	Exocrine secretions
	Arthus reaction	Ovalbumin, foreign serum	IgG and/or IgM plus complement	Cells of acute inflammation	Induration and dermonecrosis due to localized destruction of blood vessels
	Serum sickness	Foreign serum, depot penicillin	IgG and/or IgM plus complement	Cells of acute inflammation; mast cells	Vasoactive, edema, inflammation; hives, arthralgia, lymph node enlargement
	Immune (toxic) complex	Soluble antigens, viruses	IgG or IgM plus complement	Cells of acute inflammation	Kidney (basement membrane) damage; damage to other organs and tissues; vasculitis
Delayed: cell-mediated hypersensitivity; sensitized lymphocyte mediated; passively transferable via lymphocytes and transfer factor	Tuberculin (infectious allergy)	Tuberculin, fungi	T cells	Lymphocytes: Lymphokines Macrophages	Antigen lysis or isolation Destruction of host cells and tissues that surround antigen
	Allergic contact dermatitis	Poison ivy, industrial chemicals, cosmetics			
	Graft rejection	Foreign tissues, organs, cells			
	Cancer surveillance	Cancer cells			
	Autoimmune disease	Self-antigens	T cells and antibodies	Lymphokines and immune complexes	

From Boyd, R.F., and Hoerl, B.G. *Basic medical microbiology* (2nd ed.). Little Brown & Co., Boston, 1981.

Type I: anaphylactic, reagin-dependent reactions

Antibodies of class IgE (see Chapter 23) are involved in type I allergic reactions that are of the immediate type. Recall that IgE antibody is cytotropic for mast cells and basophils; that is, it preferentially binds to those cell types. IgE is also called ***reagin, reaginic antibody, skin-sensitizing antibody,*** and ***anaphylactic antibody.*** There is a hereditary predisposition to produce antibodies of this class. A positive history of some clinical allergy usually can be found among close relatives of an individual with reaginic types of allergy. It is estimated that 10% of the population have overt atopic allergies (common, clinical allergies) and that as high as 40% to 50% have minor, undetected allergies of this type.

Type I reactions are the result of the patient at some time becoming sensitized to an immunogen (allergen). An allergic reaction may occur on a subsequent exposure to the same allergen. The combination of the allergen and the antibody on the mast cell surface causes the mast cell to release (degranulate) the pharmacologically active substances that cause capillary permeability, vessel dilation, smooth muscle contraction, and mucous membrane responses (Figure 25-1). These effects may remain essentially localized, as in the common clinical (atopic) allergies, or they

Table 25-4 Classification of hypersensitive reactions

	ANAPHYLACTIC	CYTOTOXIC	IMMUNE COMPLEX	T CELL DEPENDENT
Immunoglobulin	IgE	IgG, possibly other	IgG, IgM, etc.	None
Antigens involved	Heterologous	Autologous or hapten modified	Autologous or heterologous	Autologous or heterologous
Complement involved	No	Yes	Yes	No
Cellular involvement	Mast cells and basophils	Red and white blood cells, platelets, etc., as targets	Host tissue cells	Host tissue cells
Chemical mechanism	Mast cell products and others	Complement-dependent cytolysis	Complement-dependent reactions	Lymphokines
Examples	Anaphylaxis, hay fever, food allergy	Transfusion reactions, Rh disease, thrombocytopenia	Arthus reaction, serum sickness, pneumonitis	Allergy of infection, contact dermatitis

From Barrett, J.T. *Textbook of immunology* (4th ed.). The C.V. Mosby Co., St. Louis, 1983.

may occur systemically on a grander scale, as in an anaphylactic shock reaction.

Anaphylaxis. The word ***anaphylaxis*** signifies "unprotected" or "unguarded." The sensitized individual is in an unguarded state in respect to a reaction to a given allergen. Systemic anaphylactic reactions usually develop with suddenness and often result from the injection of drugs and antisera* and from stings of *Hymenoptera* insects, such as wasps, hornets, and bees. The reactions include combinations of reddening (erytema) of areas of the skin, hives (urticaria), and itching; severe respiratory difficulties due to the accumulation of fluids and cells in lung tissue and constriction of the respiratory bronchioles; airway obstruction from laryngeal edema; hypotension (shock resulting from vascular permeability and collapse); and abdominal cramps, vomiting, and diarrhea (Table 25-5). A systemic anaphylactic reaction can be fatal in minutes. Countermeasures to be used in systemic anaphylactic reactions include (1) administration of epinephrine to counteract the effects of histamine and other vasoactive mediators, (2) maintenance of an airway, and (3) drugs to maintain adequate blood pressure whenever shock occurs.

Atopic allergies (local anaphylaxis). Atopic allergies are "out-of-place" reactions, as the word *atopy* signifies, and their underlying immune

* The term *iatrogenic* is sometimes applied to the procedure of drug injection; it means induced by a physician or through a physician-prescribed treatment.

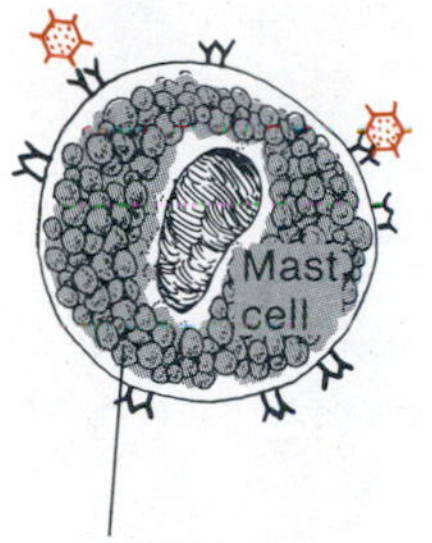

Figure 25-1 Type I (anaphylactic) hypersensitivity.

Table 25-5 Clinical data from six cases of human anaphylaxis

CASE NO.	SEX	AGE (YR)	AGENT	DOSE (ML)	ROUTE OF ADMINISTRATION	ESTIMATED TIME FROM CHALLENGE TO DEATH (MIN)	SYMPTOMS AND SIGNS	KNOWN PRIOR EXPOSURE	"ALLERGIC HISTORY"
1	F	39	Penicillin	1.5	Intramuscular	60	Generalized warmth; tightness of throat; respiratory distress; cyanosis; convulsion; respiratory failure	Yes	"Hives" 2 weeks before death; allergen unknown
2	F	21	Guinea pig hemoglobin	0.2	Subcutaneous	16	Headache; wheezing; cyanosis	No	"Asthma"; skin sensitivity to dog hair, kapok, shell fish, ragweed, timothy, orris root, and house dust
3	M	52	Bee venom	—	Subcutaneous	20	Unknown	Yes	Severe local reaction to bee sting 20 years previously
4	M	45	Penicillin	—	Intramuscular	60	Dyspnea	Unknown	Unknown
5	M	56	Hay fever desensitization vaccine	1/16	Subcutaneous	45	Difficulty breathing	Yes	"Hay fever" injection was eleventh in series of weekly desensitization injections
6	F	38	Penicillin and streptomycin	—	Intramuscular	120	Chest pain; cough; collapse; hypotension; cardiac arrest	Unknown	

From James, L.P., and Austen, K.F. Reprinted, by permission of the *New England Journal of Medicine* **270:**597, 1964.

COMMON CLINICAL ALLERGIES

Allergic respiratory disease (ARD)

Pollinosis—seasonal allergic rhinitis. Pollens are from trees and grasses in spring and early summer, weeds in late summer. Pollens are mostly anemophilous (wind borne) but also entomophilous (insect borne).

Perennial allergic rhinitis (PAR)—year-round. Common allergens are house dust and animal epidermoids (hair, feathers, dander). Among the many allergens of house dust are mold spores, animal epidermoids, and mites.

Bronchial asthma. Common allergens are those cited under ARD and PAR. The air sacs are overdistended, plugs of mucus fill bronchial passages, and smooth muscles enlarge (hypertrophy) to thicken and narrow the walls of the bronchi. Symptomatic relief is obtained by bronchodilators that relax the muscles of the bronchi, and by expectorants and liquefacients that dissolve and expel the mucus plugs and the other accumulations.

Gastrointestinal allergies—colic and possibly ulcerative colitis

Allergic skin disorders

Atopic dermatitis. This can be divided into stages: infant, childhood, and adolescent-adult.

Allergic contact dermatitis (ACD). A variety of allergens (contactants), many serving as haptens, elicit nonatopic skin allergies via the delayed hypersensitivity mechanism.

From Boyd, R.F., and Hoerl, B.G. *Basic medical microbiology* (2nd ed.). Little, Brown & Co., Boston, 1981.

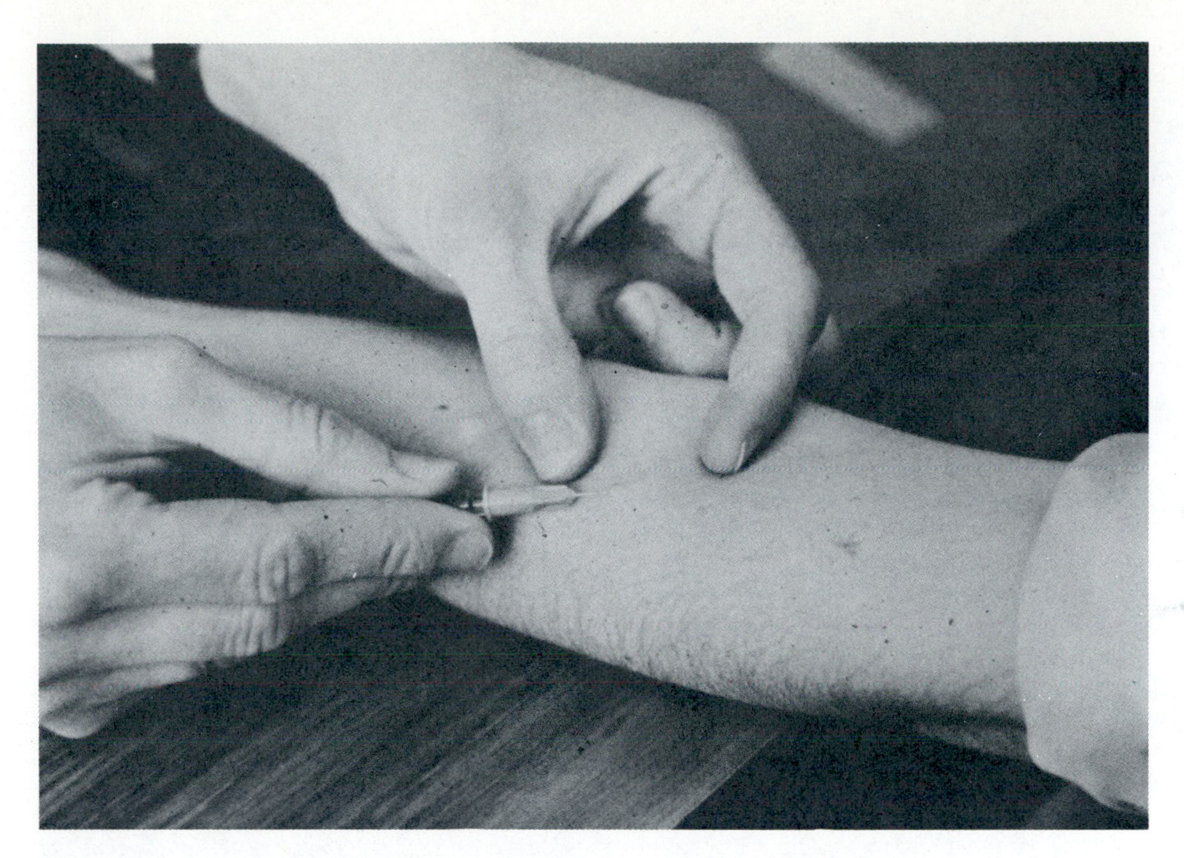

Figure 25-2 Skin test for atopic allergy.
From Middleton, E. Jr., Reed, C.E., and Ellis, E.F. (Eds.) Allergy: principles and practice *(Vol. 2). The C.V. Mosby Co., St. Louis, 1978.*

mechanism is the same as for systemic anaphylaxis. Atopic allergies in humans are IgE mediated; therefore there is a hereditary predilection toward forming these allergies. Atopic allergies are the common, spontaneously occurring, clinical allergies such as hay fever, hives, asthma, and food allergies.

The management of atopic allergies includes appropriate testing (Figure 25-2), avoidance of the identified allergen whenever possible, hyposensitization and/or desensitization, and the use of drugs appropriate to the allergic condition. Hyposensitization/desensitization procedures (''allergy shot'') attempt to decrease (hyposensitize) or eliminate (desensitize) the allergic state through the periodic administration of the same allergenic substance that caused the allergy. One might expect such a procedure to increase the patient's allergic reactivity, but in certain atopic allergies, notably in ragweed pollen hay fever, the allergic reactivity is greatly reduced. Administration of the allergen parenterally evokes the formation of antibodies of class IgG, which are referred to as ***blocking antibodies.*** There are several hypotheses as to how hyposensitization and/or desensitization works. IgG antibodies are believed to have a greater affinity for the allergen, thereby preventing the allergen from uniting with the mast cell–bound IgE antibodies. The IgG antibodies may prevent the induction of additional IgE antibodies specific for the same allergen. IgG antibodies may not be involved at all, and suppressor T cell activity may cause a decrease in IgE synthesis.

The drugs that are used to manage or give relief in atopic allergies include antihistamines, bronchodilators, decongestants, expectorants, and antiinflammatory agents. Antihistamines, incidentally, are of little value in systemic anaphylactic reactions because of the extent and the explosiveness of the reaction.

Type II: cytotoxic and cytolytic reactions

Cytotoxic and cytolytic reactions are cell-damaging and cell-destroying reactions, respectively, in which essentially an antigen-antibody union occurs on the cell surface. The cell can be destroyed by phagocytosis or lysis. The complement system is frequently involved in these reactions.

Cytolytic reactions take place under a variety of conditions: (1) the antigen may be a structural component of the cell membrane, or it may be an extrinsic antigen that has attached to the cell; (2)

the antigen that evoked the production of the involved antibody may be a cross-reacting antigen; that is, antigenic determinants on the host's own cells may be identical or sufficiently similar to antigenic determinants of foreign antigens to cross-react with the invoved antibody; (3) drug-induced changes may cause the appearance of new antigens on the cell surface; and (4) an antigen may combine with an antibody in the circulation, and this complex attaches to red blood cells, which in turn leads to red blood cell damage.

Most of the diseases resulting from cytolytic reactions fall into the category of hematological cytolytic diseases. The cell types primarily damaged or destroyed are red blood cells, white blood cells of the granulocyte type, platelets (thrombocytes), and vascular endothelial cells (cells that form the interior layer of vessels). The cells that are destroyed are usually the patient's cells, but they may be transfused or translocated cells, as in red blood cell transfusion reactions or Rh incompatability disease. The destruction of red blood cells of the patient leads to hemolytic anemia; that is, the patient has an anemia due to the destruction of red blood cells. Penicillin, for example, in massive doses can cause hemolytic anemia in some patients because it binds to the proteins on the red blood cell membrane. The benzylpenicilloyl portion of the penicillin molecule acts as a hapten that becomes the target for antibody formaton. The antibodies are believed to mediate the phagocytosis of the red blood cells. The destruction of granulocytes produces a condition known as ***agranulocytosis.*** Granulocytes are phagocytic cells; thus their destruction creates conditions that allow infections to occur. The destruction of granular platelets produces a decrease in circulating platelets (the decrease is called ***thrombocytopenia).*** Platelets are involved in the blood-clotting mechanism and in the release of some physiologically active substances. One clinical manifestation of thrombocytopenia is ***purpura,*** which is a discoloration of tissue such as the skin because of hemorrhage through the damaged blood vessel wall.

Type III: immune or toxic complex reactions

Type III reactions occur when soluble antigen and the corresponding antibody unite and activate complement. This combination can lead to intense inflammation and tissue destruction. Conditions that promote this type of reaction include the slight excess of soluble antigen to antibody and the presence of IgG and IgM antibody. If either antigen or antibody are present in considerable excess of each other, the complexes that form are eliminated uneventfully. Complement fragments C3a and C5a, as well as the trimolecular complement complex C567, cause PMNs to accumulate in the affected areas. The lysosomal enzymes of the PMNs cause various types of damage, especially of blood vessels, basement membranes, and joints.

Arthus reaction. The Arthus reaction and serum sickness have long been known as immune diseases, but the underlying immune complex cause was not fully appreciated for many years. The Arthus reaction is primarily a laboratory-induced phenomenon, but there are clinical counterparts. Several injections of a soluble protein antigen, such as horse serum, spaced at 3 to 5 day intervals, are made in the skin of a rabbit at approximately the same site. Beginning about the third to fifth injection, an inflammatory reaction of increasing intensity develops at the site of injection. The lesion increases from a spot that was originally erythematous to an ever-enlarging edematous (fluid leaks into the tissues because of vascular damage) inflammation that progresses to an extensive area of necrosis (Figure 25-3).

Serum sickness. Historically, serum sickness was observed in patients who received large single amounts of serum for the treatment or prophylaxis of such diseases as diphtheria, tetanus, and pneumococcal pneumonia. The serum, more specifically named antiserum or antitoxin, was administered with the intent of providing passive immunity. The serum was whole-animal serum, which naturally contains foreign antigenic proteins. Manifestations of serum sickness appear about 7 or more days after the injection. Sufficient antibody has been produced by this time to form complexes with antigen that remain in the tissues. Small amounts of protein would ordinarily have been eliminated by this time, and thus no antigen would have remained to interact with the antibody. Clinical manifestations include arthralgia (painful and enlarged joints), fever, urticaria (hives), and lymphadenopathy (enlargemnt of lymphoid tissue). A distinctive feature of serum sickness is that only a single large or persistent dose of antigen serves both to "sensitize" the patient and later to trip off the allergic reactions. Serotherapy is no longer widely used. Currently, serum sickness–like reactions are associated with chemotherapy designed to maintain the chemotherapeutic agent in the tissues over a

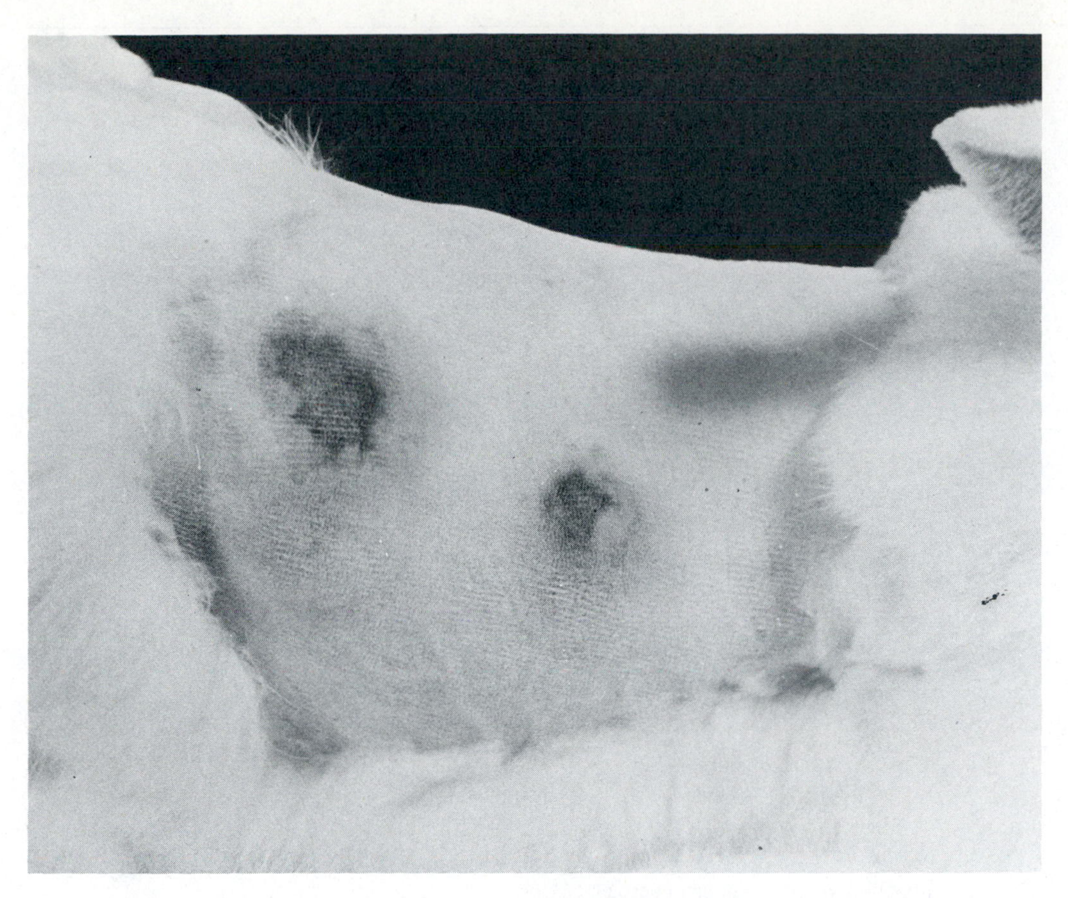

Figure 25-3 Two Arthus reactions in rabbit skin. Larger reaction has extensive zone of erythema and edema surrounding its necrotic center.

From Barrett, J.T. Textbook of immunology *(4th ed.). The C.V. Mosby Co., St. Louis, 1978.*

longer period of time for prolonged action (depot technique) and with the administration of antilymphocytic serum given as an immunosuppressant to graft recipients.

Immune complex disease. The cause of a number of human and animal diseases involves immune complex formation. The conditions are as previously described in the introduction to type III reactions. The antigens include soluble proteins, nucleic acids, viruses, and microbial proteins. Immune complex formation is at least partially involved in the pathogenesis of poststreptococcal glomerulonephritis, lupus erythematosus, periarteritis nodosa, pneumonitis (farmer's lung, pigeon breeder's lung), rheumatic fever, rheumatoid arthritis, scleroderma, a number of viral diseases (for example, serum hepatitis), and other diseases (Figure 25-4). The antigens in some of these diseases are derived from the patient, and they are additionally categorized as autoimmune disease.

Type IV: delayed hypersensitivity, T cell–mediated reactions

T lymphocytes are the recognition agent, coupler, or trigger for delayed hypersensitivity reactions such as the tuberculosis skin test. Delayed hypersensitivities can be passively transferred via intact sensitized T cells or in humans via an extract of sensitizied T cells. The extract is called ***transfer factor*** and has been used clinically with some success against conditions such as mucocutaneous candidiasis (see Chapter 31).

Exactly how immunological injury occurs in the host tissues because of a delayed hypersensitivity response is not known with certainty. As previously described, there is a heavy infiltrate of cells, notably macrophages, at the site of antigen deposit. T cells, through their secreted products, the lymphokines, attract and activate other cell types, and it is the other cell types that nonspecifically destroy the antigen and the host's cells and tissues. T cells can be di-

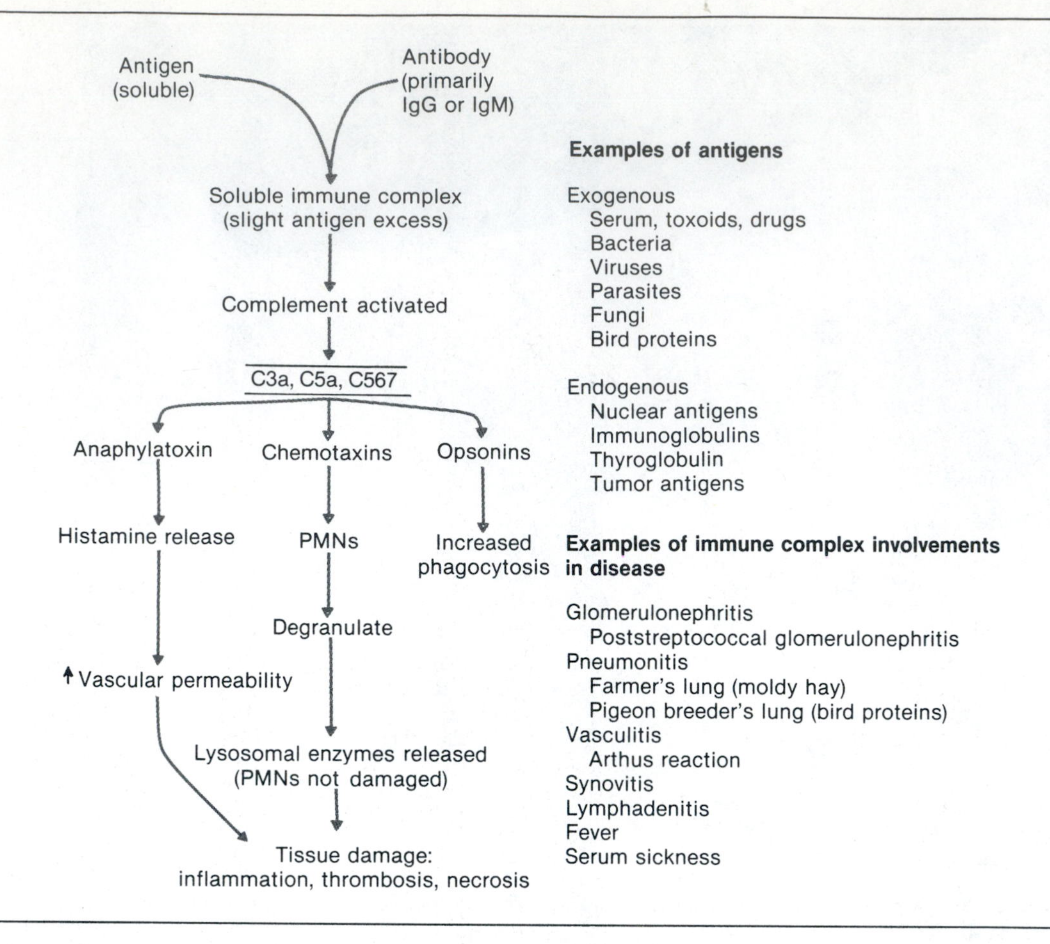

Figure 25-4 Immune complex disease. Antigen-antibody complexes (microprecipitates) are formed in blood and in tissue spaces and are deposited in walls of blood vessels, in basement membranes, and in joint synovia.

Redrawn from Boyd, R.F., and Hoerl, B.G. Basic medical microbiology *(2nd ed.). Little, Brown & Co., Boston, 1981.*

rectly toxic for the antigen. One can speculate that in T cell responses the nature of the antigen and the tissue locations of the antigen (intracellular, chronic, not easily lysed) are of such a type that the immune system must destroy some of the body's own cells and surrounding tissue in order to eliminate the antigen. The destruction of tissues may be sufficiently extensive to create a pathological clinical condition that falls into the category of delayed hypersensitivity.

Descriptions of delayed hypersensitivities classically include tuberculin hypersensitivity, allergic contact dermatitis (ACD), autoimmune diseases, graft rejection, and cancer surveillance. The student should take note that graft rejection and cancer surveillance are not adverse reactions, as their categorization, type IV reactions, would imply. The rejection of foreign tissue cells, despite our great desire to replace diseased organs and tissues with transplanted healthy foreign tissue, is basically a beneficial action on the part of the immune system. If it were not for immune rejection of foreign tissue cells, any viable foreign cell, no matter what the species, if it made appropriate contact, theoretically could survive and multiply. It should also be understood that the immunological events and the pathogenesis of a given allergic disease are not necessarily attributable to just a single underlying mechanism, as the categorization into the four types might imply. Combinations of types II, III, and IV reactions contribute to the pathogenesis of autoimmune diseases. Autoimmune diseases are described by some under delayed hyersensitivity diseases, even though antibodies are involved frequently. Stated in another way, in many cases a given immunogen evokes an immune response by both the B cell and T cell arms of the immune response mechanism. This applies to both beneficial (immunity) and harmful (hypersensitivity) outcomes. Indeed, beneficial and harmful response aspects to the same immunogen are not at all unusual.

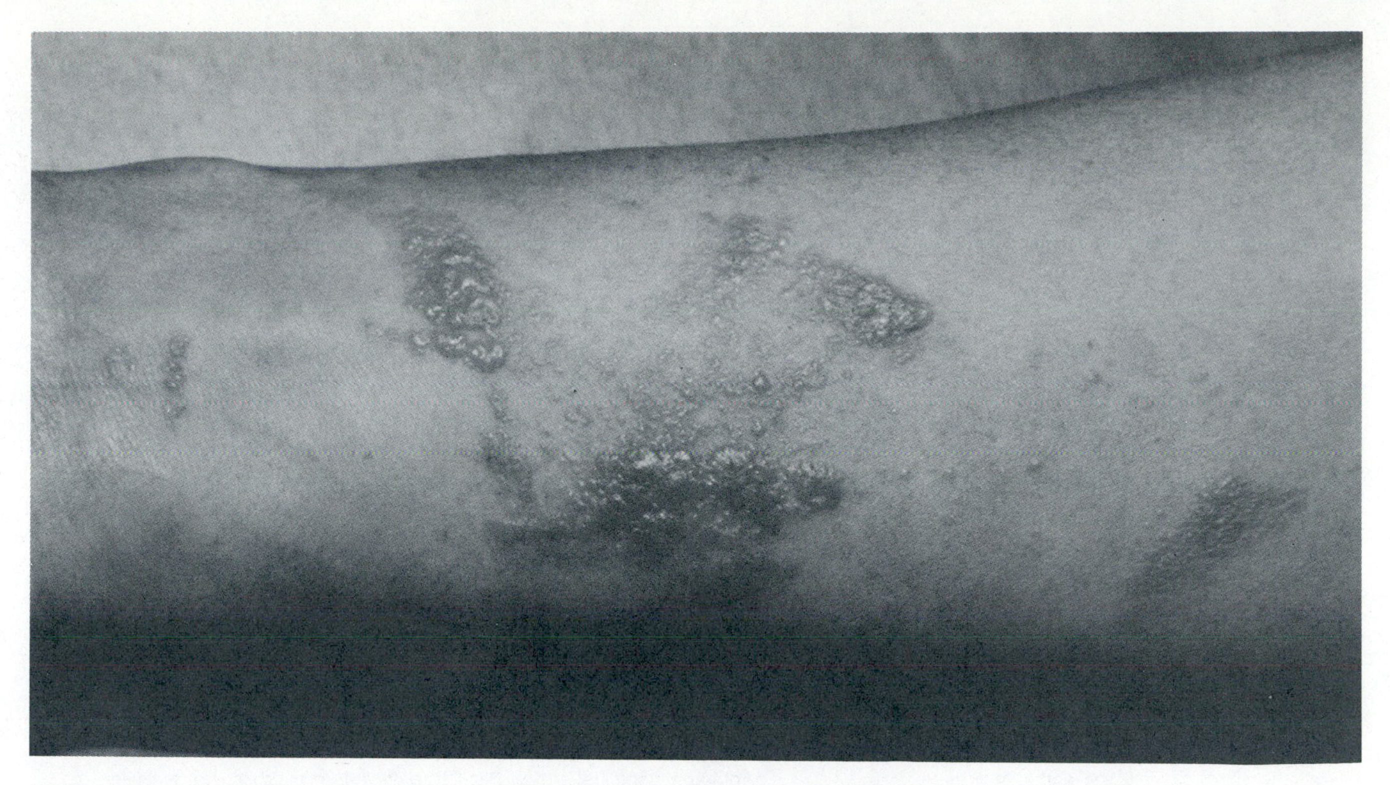

Figure 25-5 Poison oak contact dermatitis of forearm. Streaks and patches of acute dermatitis strongly suggest plant dermatitis.

From Middleton, E., Jr., Reed, C.E., and Ellis, E.F. Allergy: principles and practice *(Vol. 2), The C.V. Mosby Co., St. Louis, 1978.*

Tuberculin hypersensitivity. Many of the early studies of delayed hypersensitivity were conducted on responses to the bacilli that causes tuberculosis ***(Mycobacterium tuberculosis)*** and on extracts of tubercle bacilli. The extracts are called ***tuberculin*** or ***tuberculoprotein.*** The tissue damage that occurs in tuberculosis is primarily attributable to the reaction triggered by T cells that are sensitized to tuberculoprotein. Chronic infections and intracellular infections tend to evoke type IV reactions.

Diagnostic skin tests. There are useful diagnostic skin tests that are based on a person's delayed hypersensitivity reaction to the test microorganism or to an extract of the microorganism. The best known is the TB skin test (see Figure 24-11), in which tuberculin is introduced intradermally. The response in a tuberculin-positive individual begins in about 8 hours as an erythematous (reddened) area that becomes indurated (firm and hardened) in 12 to 24 hours. The reaction reaches its peak by 48 hours and then subsides. The size and the intensity of the reaction are related to the amount of antigen that was introduced and to the degree of hypersensitivity of the tested individual. Other microbial products that are used in delayed hypersensitivity skin testing are histoplasmin and coccidioidin for the fungal diseases histoplasmosis and coccidioidomycosis, respectively, and lepromin and brucellergen for the bacterial diseases leprosy and brucellosis, respectively.

Allergic contact dermatitis. Skin contact with a wide variety of simple chemicals can cause allergic contact dermatitis (ACD) (Figure 25-5). Allergens are found in cosmetics, industrial chemicals, dyes, ointments, plant materials, and topically applied chemotherapeutic agents, to name a few. Clinical allergists refer to ACD allergies as ***nonatopic allergies*** in order to distinguish them from the immediate-type, hereditarily predisposed atopic allergies.

The classic contact allergy is poison ivy hypersensitivity. The typical reaction develops 18 to 24 hours after a sensitized person contacts the allergen. Sharply defined erythematous areas of the skin are followed by induration and vesiculation (blistering). The blistering may be extensive, and the lesions are then called ***bullae.*** Oozing, crusting, intense itching, and occasionally infection are typical findings.

Immunogens are usually large molecules. How then, can the chemically simple allergens of ACD incite the immune system to develop a delayed hypersensitivity reaction? Many of the allergens

Table 25-6 Autoimmune phenomena in human disease

DISEASE	SPECIFICITY OF AUTOANTIBODY OR AUTOREACTIVE LYMPHOCYTES	PATHOLOGICAL EFFECT OF AUTOIMMUNE REACTION
Autoimmune hemolytic anemia	Anti-RBC	Red cell lysis; anemia
Thrombocytopenic purpura	Anti-platelets	Platelet destruction
Goodpasture's syndrome	Anti–basement membrane (kidney and lung)	Glomerulonephritis; pulmonary lesions
Pernicious anemia	Anti–intrinsic factor	Prevents absorption of vitamin B_{12} by ileum
Graves' disease	Anti–thyroid stimulating hormone receptor	Thyroid hyperfunction
Hashimoto's thyroiditis	Anti-thyroglobulin	Thyroid hypofunction
Addison's disease	Anti–adrenal cortex	Adrenalitis
Myasthenia gravis	Anti–acetylcholine receptor	Impairs neuromuscular transmission
Systemic lupus erythematosus	Anti-DNA	Multiple
Rheumatoid arthritis	Anti-immunoglobulin	Inflammation of synovial membranes
Glomerulonephritis	Various	Inflammation of glomerular basement membrane (kidney)

From Boyd, R.F., and Marr, J.J. *Medical microbiology*. Little, Brown & Co., Boston, 1980.

Table 25-7 Transplantation terminology

PREFIX	MEANING	COMBINING SUFFIXES	TRANSPLANTATION PARLANCE
Auto-	Self	-graft -geneic -antigen -antibody	An autograft is a self-graft (for example, skin from one site on the patient's body moved to another site on the patient's body)
Iso-	Equal; identical with another	-graft -geneic -antigen -antibody	An isograft is a graft between isogeneic individuals (that is, between genetically identical individuals such as identical (uniovular) twins)
Allo- (homo-)	Similar; like another	-graft -geneic -antigen -antibody	An allograft is a graft between allogeneic individuals (that is, between nonidentical members of the same species)
Xeno- (hetero-)	Dissimilar; unlike another; foreign	-graft -geneic -antigen -antibody	A xenograft is a graft between xenogeneics (that is, between members of different species, such as a graft from ape to human)

of ACD are believed to be haptens that combine with the patient's skin proteins to form the allergen. The haptens are the antigenic determinants, and the proteins of the skin are the carrier molecules for the haptens.

Autoallergies or autoimmune diseases. One of the essential features of the immune system is recognition of self-antigens. An immune system attack against the body's own antigens ordinarily does not occur. The immune system, however, does sometimes form immune products against self (auto)–antigens, and some grave diseases result. Autoantibodies have been demonstrated against basement membranes of kidneys (glomerulonephritis), against gamma globulin (the rheumatoid factor in rheumatoid arthritis), against nuclei (lupus erythematosus), against red blood cells (hemolytic anemia), and against other body molecules, cells, and tissues. What are some conditions that would lead to self-attack? Some possibilities are as follows: body proteins are altered so that they are similar to, or share antigen determinants with, foreign antigens; self-antigens were never "marked' or "recognized" as self during fetal development; antigens that are normally never exposed to the immune system (sequestered antigens) become exposed (through injury or surgery, for example); the immune system itself is in some way defective and responds inappropriately; and the immune system recognizes tissues that are diseased and eliminates them.

There is a lengthy list of human diseases, many of them of a less well known variety, whose cause at least includes autoimmune features: Hashimoto's thyroiditis, myasthenia gravis, scleroderma, Sjogren's syndrome, lupus erythematosus, and autoimmune hemolytic anemia, to name a few (Table 25-6).

Experimentally, the injection of homogenates of brain tissue, lens protein thyroglobulins, and testicular tissue into normal animals leads to diseases that resemble the human diseases of multiple sclerosis and myasthenia gravis, phakoanaphylactic endophthalmitis, thyroiditis, and aspermatogensis, respectively (Figure 25-6).

As noted earlier, autoimmune disorders involve a variety of immunopathological mechanisms: cytotoxic and cytolytic mechanisms, immune or toxic complex mechanisms, and delayed-typed hypersensitivity mechanisms. These are not solely delayed-type mechanisms, even though they are discussed here under that heading.

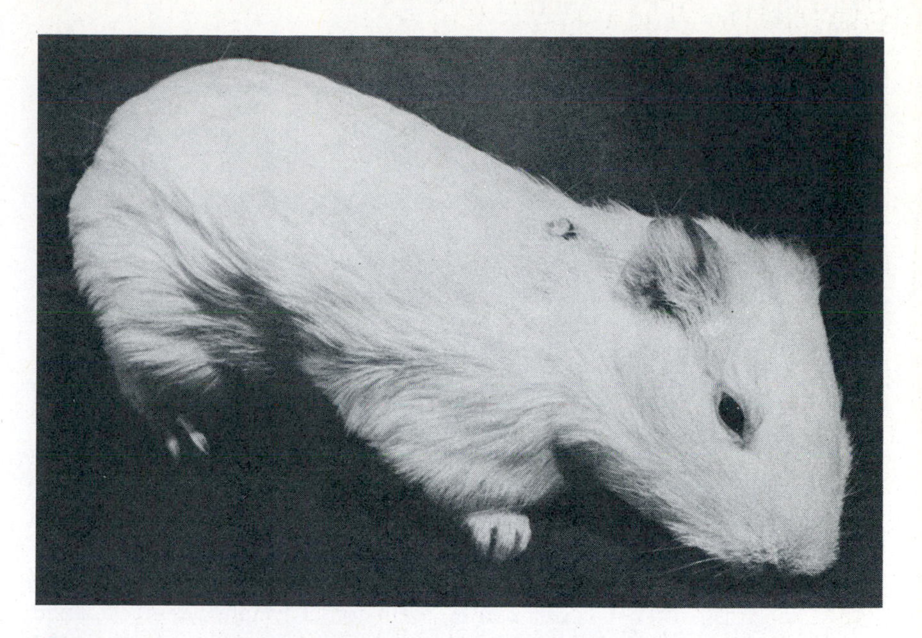

Figure 25-6 Autoimmune disease, animal model. Experimental autoimmune myasthenia gravis (EAMG) in guinea pig. Myasthenia gravis is skeletal muscle disorder characterized by fatigue and weakness after relatively mild exercise. Precise nature of disease is not clear. Immunological abnormalities are associated with disease. In this experiment guinea pig was injected with acetylcholine receptor (AChR) protein. It is hypothesized that autoantibodies were produced against acetylcholine receptor sites in neuromuscular junctions, resulting in neuromuscular block.

From Lennon, V.A., Lindstrom, J.M., and Seybold, M.E. J. Exp. Med. ***141**:1365, 1975.*

Transplantation rejection. The elimination of grafted tissue is due to immunorejection that is primarily T cell mediated. The terminology that is associated with grafting is outlined in Table 25-7. Autografts and isografts are not immunorejected, because the antigens of the transplant and of the recipient are identical. Most transplants in humans are allografts, which are grafts between individuals of the same species who are not genetically identical. Tissues contain multiple transplantation antigens whose composition is controlled by histocompatability genes. The graft recipient's immune system recognizes the foreign transplantation antigens of allografts and xenografts and immunorejects them. Some histocompatability antigens induce a stronger antigenic response than others and are referred to as major histocompatability antigens. A vigorous immunorejection response occurs when the donor's antigens possess major histocompatability antigens that are lacking in the recipient.

Maneuvers that are undertaken to prevent immunorejection include tissue typing and the use of immunosuppressive agents. Tissue typing is performed in the laboratory to identify the donor's transplantation antigens and to detect the recipient's immunoreactivity to the donor's antigens. Immunosuppressive measures are used to suppress the graft recipient's immune system's responsiveness, with the purpose of forestalling immunorejection. Immunosuppressive agents include drugs, irradiation, and antilymphocytic serum (ALS). Infections, including those by microorganisms that are ordinarily of low pathogenicity, are a constant threat in the immunosuppressed.

Fetal-maternal immunorelationship and graft-versus-host reaction. Two interesting phenomena are related to grafting: the fetal-maternal immunorelationship and the graft versus host reaction (GVHR). The fetus is not immunorejected by the mother's immune system despite the fact that half of the fetal tissue antigens are foreign to the mother. Why the fetus is not immunorejected is not well understood. Possible explanations are as follows: the placenta acts in part as a mechanical barrier to maternal lymphocytes; cytotoxic maternal lymphocytes bind to the placenta and therefore do not reach the fetus; the fetal immune system eliminates maternal lymphocytes that cross the placental barrier; and maternal antibodies attach to fetal membranes and block the access of maternal lymphocytes to fetal tissue. The last possibility, the inaccessibility of cells to lymphocytes because of the blocking effect of attached antibodies, is known to occur in at least one cancer, neuroblastoma. In neuroblastoma cancer patients the antibodies that are formed to the cancer cell antigens prevent the T cells from destroying the neuroblastoma cancer cells

GVHR sometimes occurs when immuncompetent tissue from the donor (for example, bone marrow) is transplanted into an immunologically depleted or defective recipient. The transplanted immunocompetent tissue forms immune products against the recipient's tissues. GVHR is sometimes severe enough to cause death.

Tumor immunology. Cancer cells are believed to develop regularly in everyone. The immune system apparently recognizes these cells as being different from normal body cells and eliminates them. The immune system, however, is unable to eliminate rapidly proliferating tumor cells. The importance of the immune system in tumor surveillance is supported by the observation that cancer incidence of the lymphoreticular system in individuals with primary immunodeficiency disease is considerably higher than that in the normal population. Various experimental animal cancers and some human cancers have been shown to have ***tumor-specific antigens (TSA),*** which make them different immunologically from normal cells. Viral oncogenic genes, for example, code for tumor-specific antigens that can be found in the cytoplasmic membrane of host cells (see Chapter 16).

The existence of tumor-associated antigens makes it possible to detect some cancers before they become clinically evident and to more precisely identify them as to cancer type. The presence of tumor-associated antigens also creates the possibility of controlling cancer by the administration of exogenous immune products and by manipulating the patient's immune system. These avenues continue to be actively explored with increasingly encouraging results.

SUMMARY

The immune system has as its primary function to provide specific and speedy protection against microorganisms, against foreign substances, and against the development of cancer. Problems associated with the immune system occur, however, when it becomes defective (immunodeficiencies) or when the system overreacts (hypersensitivities). The defects that lead to immunodeficiencies are hereditary (primary immunodeficiencies), or they are acquired (secondary immunodeficiencies) through malignancy, drug therapy, and infection. Individuals with defects in the immune system suffer a variety of distressing medical problems, many of which are due to infections.

Overreactions (hypersensitivities) occur under circumstances that include the nature of the immunogen, as well as the amount, route, and spacing of immunogen exposure. Hypersensitivities are of the immediate type or of the delayed type, depending on how soon a sensitized (allergic) individual develops a lesion after skin challenge with the allergen that is reponsible for the hypersensitivity. Immediate hypersensitivities are of the antibody-mediated or humoral type. Delayed hypersensitivities are of the cell-mediated type.

There are four basic immune mechanisms that account for the tissue damage that occurs in hypersensitivity reactions. In type I reactions (the anaphylactic type) antibodies (IgE in humans) located on the mast cells of sensitized individuals cause the mast cells to degranulate when antigen combines with the antibody. The physiologically active substances contained in the granules produce effects especially on blood vessels, smooth muscle, and exocrine glands. The seriousness of these effects ranges from the common clinical (atopic) allergies to death in a systemic anaphylactic reaction. There is a hereditary predisposition in humans to form immunoglobulins of class

IgE. In type II hypersensitivities cells are damged or destoyed when antigens and immune product react at the cell surface. Blood cells are the cell type frequently involved in these immune cytolytic reactions. In type III hypersensitivities (the immune complex type) the formation of complexes consisting of soluble antigen, antibodies of class IG or IgM, and complement initiate an intense, tissue-injuring inflammatory reaction. In type IV hypersensitivities (the delayed or cell-mediated type) tissue damage results from the activities of sensitized T cells or is mediated by lymphokines. Type IV hypersensitivities include the tuberculin type, allergic contact dermatitis, and autoimmune diseases. Graft rejections, fetal-maternal relationship, and cancer surveillance are frequently included in descriptions of type IV hypersensitivities. A distinctive feature of allergic contact dermatitis is that many of the allergens are haptens. In autoimmune diseases immune products are formed against the patient's own cells and tissues.

Study questions

1. How does a primary immunodeficiency differ from a secondary immunodeficiency?
2. How does an immunodeficiency differ from an allergy?
3. What is the role of IgE in local and systemic anaphylaxis in humans?
4. How does immunity differ from allergy from the standpoint of the final target of the reaction?
5. Did evolution produce IgE just to participate in adverse reactions?
6. What is the basis for categorizing hypersensitivities as immediate or delayed? What is the rationale for the Gell and Coombs classification scheme?
7. What is an atopic allergy?
8. Antihistamines are used in some atopic allergies but not in systemic anaphylaxis. Explain.
9. Define the following: anaphylaxis, iatrogenic, erythema, urticaria, hyposensitization, expectorant, cytolytic, edema, autoimmune disease, hypogammaglobulinemia, and allograft.
10. In hyposensitization the allergist injects more of the same allergen to which the person is allergic. It would seem that this would be the worst possible thing to do. Explain the rationale behind the allergist's actions.
11. Under what circumstances does immune complex disease develop?
12. Describe a positive TB skin test reaction. How does it differ from a positive atopic allergy skin test reaction?
13. How does allergic contact dermatitis differ from an atopic allergy?
14. What are some of the explanations that have been proposed for the development of autoimmune disease?
15. What is an allograft? Why are most grafts among humans of necessity allografts?
16. How does GVHR differ from graft rejection?
17. Why does the existence of tumor-specific antigens engender hope for controlling cancer by immunological methods? How might hybridomas fit into this picture?

Selected readings

Books

Feingold, B.F. *Introduction to clinical allergy.* Chares C. Thomas, Publisher, Springfield, Ill., 1973.

Lawlor, G.J., Jr., and Fischer, T.J. (eds). *Manual of allergy and immunology.* Little, Brown & Co., Boston, 1981.

Richter, M.A.: *Clinical immunology: a physician's guide* (2nd ed.). The Williams & Williams Co., Baltimore, 1981.

Journal articles

Billingham, R.E. Immunobiology of the maternal-fetal relationship. *Prog. Clin. Biol. Res.* **70:**339, 1981.

Georgitis, J.W.: Insect stings: responding to the gamut of allergic reactions. *Mod. Med.* **50:**106, 1982.

Glasser, R.J. How the body works against itself: autoimmune diseases. *Nursing* **7:**38, 1977.

Glassman, A.B. Current status of applied tumor immunology. *Lab. Med.* **10:**34, 1979.

Halpern, S.R., et al. Childhood allergy. *South. Med. J.* **73:**1601, 1980.

Hildreth. E.A.: Some common allergic emergencies. *Med. Clin. North Am.* **50:**1313, 1966.

Meuwissen, H.J.: Evaluating patients with suspected immunodeficiency. *Postgrad. Med.* **66:**116, 1979.

Solley, G.O.: Present status of immunotherapy (desensitization) for allergic disorders. *South. Med. J.* **72:**183, 1979.

Thomas, E.D.: The role of marrow transplantation in the eradication of malignant disease. *Cancer* **49:**1963, 1982.

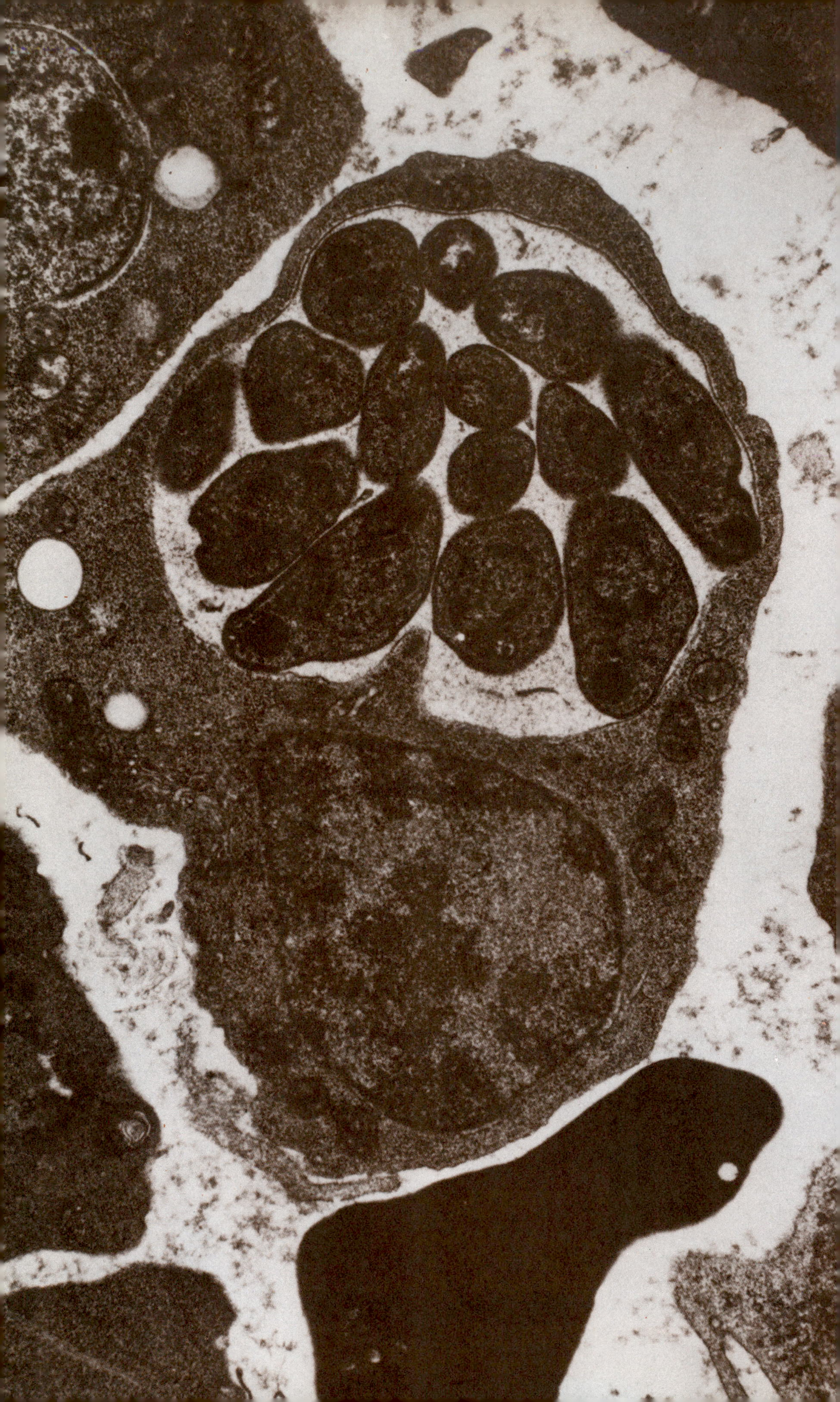

PART 7

MICROORGANISMS AND HUMAN INFECTIOUS DISEASE

Malaria organism infectng blood cells.

Omikron/Photo Researchers, Inc.

Chapter 26 EPIDEMIOLOGY OF INFECTIOUS DISEASE

Epidemiology is the science that deals with those factors associated with the transmission of disease as well as those factors that are responsible for its prevalence in a population. The epidemiology of infectious disease is concerned with determining what microbial agent has caused the disease (the etiology of the disease), the source of the disease, how it was transmitted, and what host and environmental factors could have caused the disease to develop. All of these factors are used to give us the natural history of a particular infectious disease. Epidemiological studies are important because from them we have been able to design measures that have been useful in the control, prevention, and eradication of disease. The best example of the value of epidemiological studies is in the case of smallpox. Epidemiological studies of small pox over the years have revealed that (1) there is no evidence of an animal reservoir for the infectious agent; (2) infected individuals who recover do not act as carriers after the infection has subsided; and (3) transmission is by close contact with the microbial agent from highly visible lesions. These characteristics permitted rapid identification of infected persons as well as their contacts. Immunization of contacts interrupted the cycle of transmission and prevented further spread of the infectious agent. In 1977, after intensive worldwide immunization programs, smallpox was declared eradicated.

The purpose of this chapter is to describe the influences of host and environment on the transmission of infectious disease and the methods or techniques being used to obtain epidemiological information. Identification of microbial strains is a practice common in the hospital environment (see Chapter 33).

EVOLUTION OF THE HOST-MICROBE RELATIONSHIP

It is difficult to determine the origin of infectious agents, since microorganisms were only first observed in the late eighteenth century and only in the nineteenth century was the relationship between microbes and infectious disease clearly established. The desriptions of some ancient diseases are similar to the clinical symptoms of certain modern infectious diseases, but we cannot be sure that they were caused by the same microbial agent; therefore a great deal of speculation exists as to their origin.

It is not unreasonable to assume that microorganisms first existed as free-living saprophytes in the soil and water. These primitive forms must have been highly specialized in terms of their biosynthetic potential and were capable of synthesizing all of the metabolites required for growth and multiplication. With the emergence of plants, animals, and eventually humans, these free-living microorganisms could have formed a close association with them. Such associations would be influenced by nutrition, environmental conditions, and various host factors. The relationship could be ***mutualistic,*** in which both host and microbe benefit; ***commensalistic,*** in which the microbe-benefits but the host is neither harmed nor benefitted; or ***parasitic,*** in which the microbe benefits but the host is harmed. A parasitic existence in which the host is killed, in the long run is not advantageous to the microorganism. Over the thousands of years of interaction between host and microorganism, genetic changes have occurred in both groups, but, generally speaking, they have evolved into a harmonious association. One consequence of this association is that organisms have lost much of their biosynthetic potential and are at the present time nutritionally dependent on their hosts. This nutritional dependency can sometimes result in the presence of microorganisms in only one ecological niche in the host, for example, the intestinal tract.

Early during the evolution of the host-microbe relationship there must have been instances in which the microorganism caused not only severe disease but even death. Some of the affected population undoubtedly developed, or were born with, a resistance to the microbial agent. In certain geographic areas isolated from the rest of civilization, tribes were ravaged by disease after their first contact with microorganisms from the outside world. We have only to recall the plight of the American Indian when first exposed to the agents of tuberculosis and syphilis, brought to them by the Europeans. Those infected individuals who survived their first encounter with the infectious agent often demonstrated traits that enabled them to coexist with the new invader. This coexistence, after many generations, seldom resulted in overt disease symptoms unless there had been a major genetic change in the microbe or the host.

Large epidemics of disease probably occurred infrequently during the early history of civilization. Small family units distributed over large geographic areas, and with little chance of communication with other tribes, kept disease localized. Over a long period of time they developed an immunity to the infectious agent, or a commensal relationship evolved. As human populations increased, urbanization also increased, and with it came the problems of sewage disposal, water purification, and overcrowding. In order to supply large amounts of food to the population, domestic animals were maintained in large numbers, and this also supported the existence of arthropods (ticks, mosquitoes, and mites) that parasitized them. Each represented a potential source of infection to humans. Any of the so-called local diseases could be spread by direct contact or by the transportation of local animals and their arthropods to other urban areas. And described earlier, the introduction of infectious agents into a virgin community can have devastating effects.

History has demonstrated, for some diseases, that the relationship between humans and microbes will eventually become stable and benign. Our environment is not a static one but is in a continual state of change, which means that there will be changes in the hosts as well as changes in the microorganisms that infect them. Today's benign commensal could become tomorrow's parasite, or vice versa. Although the host-microbe interaction may eventually develop into a nonparasitic one, waiting for evolution to resolve the problem could be devastating to humankind. This is evident when we consider the major epidemics that have decimated civilizations: cholera, plague, typhoid fever, yellow fever, diphtheria, etc. Modern technology and microbiological research have provided the means of interrupting the evolutionary process through the use of antibiotics and immunization procedures that treat and prevent diseases, respectively. Even with our advanced technology, however, it is not likely that all infectious disease will be eradicated. We must remember that the agent of smallpox exists in some laboratories and that in some animal populations smallpox virus or a related one may be smoldering and at some time could infect humans as they encroach on animal environments.

INFECTIOUS DISEASE CYCLE

The development of an infectious disease is associated with a cycle of events concerned with (1) the characteristics of the microbial agent, (2) the source or reservoir of the microbial agent, (3) the transmission of the microbial agent, (4) the susceptibility of the host, and (5) the mechanism of exit of the infectious agent from the body and its dissemination to a new host.

Microbial agents

The microbial agents capable of producing disease in humans are bacteria, viruses, fungi, protozoa, and helminths. Many microorganisms come in contact with the host during the hosts lifetime; some are transient visitors that are unable to become permanently established; others may live in or on the host but do no damage (by colonization or infection), whereas others invade the host and produce disease, that is, the development of symptoms of illness. In some infectious diseases, such as tuberculosis and meningococcal meningitis, the infected patient may be a source of infection to others; that is, the disease is ***communicable.*** Sometimes the infectious agent cannot be conveyed to the exterior of the host, and the infectious disease is noncommunicable. All microorganisms have the potential to produce disease, but this potential is controlled by such factors as (1) the number of microorganisms (dose), (2) the virulence or traits of the microorganism that permit it to invade tissue and cause damage, and (3) the susceptibility of the host.

Dose. The number or ***dose*** of microorganisms required to cause disease varies from one microbial species to another and is influenced by the susceptibility of the host as well as the route of infection. Some clear-cut examples of the effect of dosage have been demonstrated with enteric pathogens, such as *Salmonella* and *Shigella.* Human volunteers administered an inoculum of 10^3 *Salmonella typhi* developed no disease symptoms, but when the dose was increased to 10^7 or higher, at least 50% of the volunteers developed symptoms. Other studies have also demonstrated that the minimal infecting dose of *S. typhi* can be reduced considerably if bicarbonate is also administered with the bacteria. Many bacteria are destroyed by acidic conditions in the stomach, and any host condition (achlorhydria) or treatment (bicarbonate) that reduces stomach acidity favors microbial resistance. The dose is also affected by the route of infection. Studies of the virus causing the common cold (rhinoviruses) have shown that the dosage required to cause disease may vary by a factor of 200, depending on the site of inoculation in the respiratory tract. The nature of the epithelial surface, the flow of mucus, and the presence or absence of chemical and immune factors may be barriers to the adherence of microorganisms (see Chapter 22). Dosage plays no role in infectious disease if there is no specificity for the host. Many microorganisms can infect animals and plants and cause disease but cannot infect humans, and vice versa. The rabies virus, for example, which can cause disease in humans and some animals, can replicate in animals and cause no disease (skunks), but other animals do not support its colonization or replication.

Microbial traits. Some microorganisms are equipped with special determinants that permit the colonization and invasion of the host. These aspects of virulence are discussed in Chapter 22.

Sources and reservoirs of microbial agents

Living agents or inanimate objects that serve as the facilitator for contact between the host and the microorganism are referred to as ***sources of infection.*** Animate or inanimate objects that are necessary for the support of growth of microorganisms are termed ***reservoirs of infection.*** The reservoir and source may be the same for one microorganism, or they may be different. Defining the reservoirs and/or sources of microbial agents is an important aspect of epidemiological studies, because if they can be eliminated, the chain of events leading to the infectious state can be broken and disease transmission prevented.

Animate reservoirs. Humans represent the most important reservoirs of infectious agents. The microorganisms found on the host may produce recognizable symptoms that are characteristic of the disease, or the infected person may be a carrier of the infectious agent. Carriers may be categorized as ***healthy*** if they harbor the microorganism but do not develop characteristic disease symptoms at any time. A second type of carrier is the ***convalescent*** carrier, who has recovered from infection but continues to harbor large numbers of potentially infectious microorganisms. Some hepatitis patients may become convalescent carriers and remain carriers for life. The carrier is important in the cycle of transmission only if the infectious agents can be transmitted. The most potentially dangerous carriers are those who shed microorganisms via the respiratory or intestinal tract. Coughing or sneezing within 3 feet of a potential host is sufficient to transmit an infectious agent. Food, water, or ***fomites*** (inanimate objects that harbor microbes) may also be contaminated by the secretions or excretions of carriers and thus act as indirect sources of infection to others.

Typhoid fever is a disease frequently associated with carriers. Several members of a religious group in New York, for example, were hospitalized for typhoid fever. All had attended the same party, and a cake served at the party had been associated with the illness, but a carrier had not been identified among the bakery employees. Three months later other members of the religious group were hospitalized with typhoid fever following attendance at a second party. No food

Table 26-1 Microbial agents that cause disease primarily in animals and are secondarily transmitted to humans (zoonoses)

DISEASE	TRANSMISSION	ANIMAL RESERVOIR
Bacterial		
Salmonellosis	Ingestion of contaminated food or water	Dogs, cats, farm animals, poultry, reptiles, rodents
Brucellosis	Drinking raw milk, also occupational by direct contact with animal or carcass	Dogs, farm animals, rodents
Tularemia	Ingestion of or contact with contaminated meat	Rodents, particularly rabbits
Anthrax	Contact with contaminated animal products or hides	Farm animals
Leptospirosis	Contact with contaminated water	Dogs, cats, wild rodents, farm animals
Bubonic plague	Rat flea, but human-to-human transmission in pneumonic variety of disease	Rodents such as rats, ground squirrels, and chipmunks
Rickettsial disease (scrub typhus and murine typhus)	Ticks, fleas, mites	Primarily wild rodents
Q fever	Ingestion of raw milk, inhalation of contaminated dust, contact with contaminated parturition products	Farm animals (cattle, sheep, and goats)
Cat scratch fever	Scratch or bite of cat	Cats
Viral		
Rabies	Bite of animal or contact with infectious saliva or tissue	Dogs, cats, farm animals, wild animals (particularly fox, skunks, and racoons)
Equine encephalitis	Horses to mosquito to humans	Horses, pheasants, domestic pigeons
Parasitic (helminth infection)		
Echinococcosis	Ingestion of eggs deposited on raw fruit	Fox, dogs, and cats are principal hosts, but life cycle in nature is completed in mice and moles
Trichinosis	Ingestion of contaminated meat, usually pork	Pigs and bears are most important reservoirs

could be incriminated, and none of the guests had attended both parties. The second party, however, was held in the home of one of the food handlers from the first party. No positive bacterial cultures for the typhoid organism were isolated from the food handler, but one member of his family, a 72-year-old man, tested positive for typhoid organisms. The presumptive carrier agreed not to serve food to individuals outside his immediate family.

Some carriers harbor the infectious agents in the bloodstream, and bloodsucking insects, such as mosquitoes, may become infected by biting the host. The infected mosquito in turn can support the growth of the infectious agent and transmit it to susceptible hosts.

Animals play a very important role in human infections because of their ability to act as reservoirs of infectious agents. Infectious diseases of animals that are accidentally transmitted to humans are called ***zoonoses*** (Table 26-1). Humans acquire these diseases either by contact with infected animals or by eating contaminated meat. Infections may also be acquired indirectly through insects that prey on infected animals. The infectious agent may be acquired via the bite of the insect or by crushing of the insect in an abrasion or open wound of the host. Once humans are infected directly or indirectly by animals, they are usually unable to transmit the infectious agent to others, but there are exceptions. Flies, for example, may transmit the agent of the intestinal disease salmonellosis to humans. Infected humans, during the course of the disease, may transmit the infectious agent to others by way of contaminated hands.

Insects belong to the phylum Arthropoda, of which there are two important classes: Insecta and Arachnida. Members of the class Insecta include flies, mosquitoes, fleas, and lice, and those of the class Arachnida include ticks and mites. Insects that transfer infectious agents in a passive way from one host to another are called ***mechanical vectors.*** Flies, for example, can act as passive vectors because many of them do not parasitize humans or animals, but they do come in contact with them and their waste products, which may be contaminated. The infectious agents are usually carried on the insect's body parts. Some insects prey on animals or humans, and occasionally a part of the cell cycle of the microorganism develops within the insect. Insects that serve as a host and reservoir for infectious agents are called ***biological vectors,*** such as the *Anopheles* mosquito, which carries the agent of malaria.

Inanimate reservoirs. The primary ***inanimate reservoirs*** for infectious agents are soil, water, and food. Air is not considered a true inanimate reservoir, because it does not provide the organism with the nutrients required for growth or maintenance. Air can be contaminated by organisms that are derived from other sources such as soil, water, or animate hosts and does serve in the transmission of disease (discussed later).

Most saprophytic microorganisms whose primary habitat is the soil are not pathogenic for humans, but there are exceptions, such as some fungal agents that cause systemic infections. Most of the microorganisms found in the soil that cause disease in humans are inhabitants of plants and animals and are shed or excreted into the soil. Many of these organisms remain in the soil as spores and await transfer to a viable host before germinating. Anthrax, for example, is a bacterial disease of cattle and various herbivorous animals. The spores of the microorganism causing this disease are found in the soil and are ingested by grazing animals. The spores germinate in the animal host and release vegetative cells that multiply and produce disease. If the diseased cattle are destroyed and only buried, the vegetative cells sporulate and are returned to the soil as resistant spores. It is therefore necessary to either cremate the animals or bury them in lime to destroy the vegetative cells and prevent sporulation.

Foods are important reservoirs of microorganisms because they contain nutrients that support the growth and maintenance of microorganisms. Food may be contaminated if it is obtained from infected animals, or it may be contaminated during or after processing (see Chapter 20). Food-borne disease results from the ingestion of foods, including milk, contaminated by microorganisms that either directly or indirectly cause gastrointestinal symptoms. Indirectly, some contaminating microorganisms release potent toxins into the food, which when ingested cause intestinal symptoms. This is not a true infection but an ***intoxication.*** Some microorganisms that contaminate food are ingested replicate, and invade the intestine to cause disease—a direct course of events. Improper cooking or preservation of foods is the cause of infection or intoxication in nearly all food-borne disease. The vehicles for transmission, by order of importance, are meat and poultry; salads containing chicken, turkey, potatoes, or eggs (particularly mayonaise); shellfish; dairy products; and fruits and vegetables. Many animals harbor potentially infectious microbial species in their intestinal tract, such as *Salmonella* species, that can infect humans. Meat and particularly poultry contaminated by these species must be thoroughly cooked to prevent their multiplication. *Staphylococcus aureus,* which can be carried in the nasal cavity and can be a source of food contamination by food handlers and others, is a source of food intoxication. The microorganisms contaminating a salad, for example, will multiply and elaborate toxin, which may or may not be heat stable. Although not appropriate for salad, heating foods may be necessary to destroy the toxin and prevent gastrointestinal symptoms. The microbial agents involved in food-borne disease are outlined in Table 26-2.

Water, which is seldom pure distilled, usually contains inorganic ions and often is contaminated with organic material derived from the soil, air, animals, or plants. Saprophytes inhabiting the water are seldom pathogenic for humans, but water contaminated with organisms derived from fecal material may become important pathogens for humans. Waterborne disease results from consumption of contaminated water that has not been properly prepared for drinking. Most outbreaks involve semipublic and municipal water systems, including recreational areas. Several cases of severe gastrointestinal symptoms in a town in Colorado were traced to a sewer line that had become obstructed, resulting in leakage of sewage into a creek supplying water to the city. A protozoan called *Giardia lamblia* found contaminating the water was the cause of the outbreak of giardiasis. Other large outbreaks also result from leakage of sewage into private wells, with

Table 26-2 Microbial agents involved in food-borne disease

ETIOLOGICAL AGENT	CLINICAL SYNDROME
Bacterial	
Bacillus cereus	Gastrointestinal symptoms
Brucella species	Fever; enlargement of lymph nodes, spleen, and liver
Clostridium botulinum	Intoxication causing muscle paralysis, leading to double vision, difficulty in breathing and death
Clostridium perfringens	Primarily diarrhea
Escherichia coli	Primarily diarrhea
Salmonella species	Primarily diarrhea
Shigella species	Primarily diarrhea
Staphylococcus aureus	Intoxication causing gastrointestinal symptoms and vomiting
Group A streptococci	Fever, sore throat, other upper respiratory tract symptoms
Vibrio cholerae	Diarrhea
Vibrio parahaemolyticus	Diarrhea
Viral	
Hepatitis A	Jaundice and gastrointestinal symptoms
Parasitic (helminths)	
Trichinella spiralis	Fever, orbital (eye) edema, muscle soreness

Table 26-3 Microbial agents involved in waterborne disease

ETIOLOGICAL AGENT	CLINICAL SYNDROME
Bacterial	
Escherichia coli	Diarrhea and fever
Salmonella species	Diarrhea and fever
Shigella species	Diarrhea and fever
Vibrio cholerae	Diarrhea and vomiting
Campylobacter fetus	Diarrhea
Yersinia enterocolitica	Diarrhea
Leptospira species	Headache, conjunctivitis, meningitis
Parasitic (protozoa)	
Giardia lamblia	Chronic diarrhea, fatigue, weight loss, abdominal cramps
Entamoeba histolytica	Dysentery (severe diarrhea), fever, chills, or mild diarrhea
Viral	
Hepatitis A	Jaundice, gastrointestinal symptoms
Parvovirus-like agents	Vomiting, watery diarrhea, abdominal cramps
Rotavirus	Vomiting, watery diarrhea, abdominal cramps
Enterovirus (poliovirus is one example)	Variable depending on virus but may be paralytic polio to aseptic meningitis to common cold

the latter acting as a source of waterborne disease for those served by the well. Nearly all of the agents of waterborne disease cause gastrointesinal symptoms and are outlined in Table 26-3. Microbial contamination of water supplies is a very important aspect of hospital-associated infections (see Chapter 33).

Uncommon sources. Sources of infection sometimes go unrecognized because they are uncommon. Plague, for example, is normally transmitted to humans by the bite of infected fleas harbored by rats. Two cases of human plague associated with domestic cats were reported from the west coast of the United States in 1981. One of the patients became ill one morning, complaining of shortness of breath, vomiting, and diarrhea. The following day a dry, nonproductive cough developed, and the patient decided to go to a physician. The physician prescribed antibiotics for a presumed urinary tract infection. After 2 days the patient's condition worsened, the cough became productive, and the patient was admitted to a hospital. The patient died 4 hours after admission. Sputum samples revealed *Yersinia pestis,* the etiological agent of plague. Later investigations revealed that 8 days before the patient died, her cat had brought home a dead chipmunk. The cat had become ill with coughing and bloody nasal discharge and died 5 days later. The cat was exhumed and demonstrated to be infected with *Y. pestis.* Eating or mouthing of infected rodents (in this case, the chipmunk) by the cat was the most probable mode of infection. Contact with the nasal secretions from the cat was the source of infection for the patient.

Another unusual source of infection was discovered in 1978 among menstruating women. The disease, called toxic shock syndrome (TSS), was characterized by the sudden onset of fever, vomiting, and diarrhea, with rapid progression to shock. An investigation into practices and products associated with the menstrual cycle revealed that continual use of tampons favored acquisition of the disease. The tampons, particularly one brand, appeared to act as a cofactor in the disease and was believed to favor the growth of *Staphylococcus aureus.* How this association favors the disease process is not known, but a pyrogenic exotoxin, called type C, is produced more often by strains of *S. aureus* isolated from persons with toxic shock syndrome than from healthy individuals. The exotoxin is believed to be responsible for the symptoms of the disease. As many as 20% to 30% of the cases of TSS occur in individuals other than menstruating women.

Transmission

Microorganisms are transmitted to susceptible hosts by four routes: contact, air, common vehicle, and vector borne. Transmission may involve movement of microorganisms from animate or inanimate reservoirs to humans.

Contact

Direct contact. Direct contact implies actual physical interaction with the source of the microorganisms. Human-to-human transmission occurs most often through touching, kissing, or sexual intercourse. Contact with oral secretions or body lesions is one of the most common mechanisms for the spread of infectious agents. Many viral respiratory tract diseases, for example, are transmitted by touching the mucous membranes of the mouth, nose, or eyes and then touching another potentially susceptible host in these same areas. Enteric diseases, such as salmonellosis and hepatitis A can also be transferred by the hands, but transfer is primarily by the fecal-oral route. Venereal diseases are transmitted primarily by sexual contact, but infected nursing mothers may also transmit the microbial agent to their offspring by direct contact. The agent of syphilis, as well as some other microorganisms, can be transferred across the placental barrier to the developing fetus (in utero transfer). The agent of gonorrhea cannot be transferred across the placenta, but it can contaminate the conjunctiva of the newborn during passage through the birth canal. The agents that are transferred congenitally are primarily viruses (see Table 29-1).

Microorganisms may also be transmitted by direct contact with animals or animal products. The majority of infections in this category occur among those whose occupations place them in contact with animals. Farmers, dairymen, abattoir workers,* veterinarians, and hide and leather workers come in contact with infected animals or contaminated animal products. Ingestion of contaminated meat or animal products such as milk or cheese occurs more frequently in underdeveloped countries where domestic animals are not properly cared for or vaccinated. Scrofula caused by ingestion of milk contaminated with *Mycobacterium bovis* is almost unheard of in the United States because of pasteurization and dairy control, but in underdeveloped countries it is a frequent cause of infection because many persons drink raw milk. Raw milk, even when strictly controlled, may be contaminated with microorganisms indigenous to the cows. Two recent out-

* An abattoir is a slaughterhouse.

breaks in the United States occurred among families who drank raw milk. Species of *Salmonella* and *Campylobacter* were found in rectal swabs collected from the cows, and these organisms were implicated in the outbreak.

Indirect contact. Indirect contact implies that the infectious agent persists on an object (usually inanimate) from which there is transfer to a susceptible host. The intermediary object may be contaminated from an animate or an inanimate source. A variety of objects, such as clothes, eating utensils, hospital thermometers, drinking glasses, and bedding, may be the intermediary. Most infections acquired in this way usually involve a hand-to-mouth transfer of the infectious agent.

Air. Infectious microorganisms can reach the air in a number of different ways. They may be dispersed from the respiratory tract by sneezing or coughing or other related activities. These airborne mucous secretions are sometimes so small that they evaporate and remain suspended in the air for prolonged periods of time and are called ***droplet nuclei.*** Some microorganisms that are known to be transmitted by droplet nuclei are the viral agents causing smallpox and measles and the bacterial agent causing tuberculosis. The very large secretions expelled from the respiratory tract can contaminate clothing and various inanimate surfaces. These particles dry and become ***dust particles,*** and if the infectious agent is resistant to drying (smallpox), it may remain infectious for varying lengths of time. Dust particles may become airborne by dispersal from bedsheets or clothing, or by currents of air. Q fever, for example, is a respiratory tract infection in humans caused by contact with dried parturition (during birth) products of cattle. These particles remain on the floor of the barn, where they can be disseminated by air currents and inhaled. Infrequently, particles from the skin, urinary tract, or genitourinary tract may also be discharged into the environment to contaminate soil or inanimate objects. Once the particles dry, they may be carried by air currents to infect susceptible hosts.

Many soil microorganisms produce spores, some of which are easily disseminated in the air. In the United States fungal spores of *Coccidioides immitis* are spread over the desert floor in the lower Sonoran Life Zone after winter and spring rains. During the hot, dry summers violent dust storms spread the spores in the air, and persons traveling through the area during the storms are exposed to them. Inhalation of large numbers of spores can cause serious disease or allergy, even in the healthy host. Fortunately, most infections do not result in serious disease.

Common vehicle. Common-vehicle transmission refers to the acquisition of microorganisms by a large group of individuals from a common inanimate vehicle or source. This type of transmission is frequently associated with hospital-associated infections, which are discussed in Chapter 33.

Arthropod borne. Many animal parasites, such as helminths, go through a developmental and multiplication cycle in the arthropod before they are transferred to the vertebrate host. These cycles sometimes involve more than one intermediate host. The importance of arthropods in transmission is discussed briefly under "Animate Reservoirs" and is explained further in Chapter 32.

Host susceptibility

A multitude of factors can influence the outcome of an interaction between microorganism and host, and one of the most important is the genetic constitution of the host. The genetic traits that enable the host to resist infection cannot always be clearly delineated from other factors. Certain groups of individuals are resistant to tuberculosis, whereas others, such as blacks, are highly susceptible to the disease. Genetic factors probably operate at the level of the immune response. Aggressive macrophage function, for example, could conceivably be responsible for destruction of such intracellular parasites as the agents of tuberculosis and brucellosis. Individuals with a heterozygous form of sickle cell anemia, a genetic disease, are resistant to malaria. Malaria is caused by a protozoan that invades and multiplies in the normal red blood cell; in a sickle cell, however, the microbe is destroyed by the macrophage phagocytic system.

The very young and the aged are more susceptible to infection than other age groups. The immune response is greatly reduced in both groups, and some infections that would produce mild symptoms in the adult can cause death in these more susceptible groups. The mechanical as well as physiological barriers to infection are not developed in the very young, and in the very old these barriers have been compromised by the aging process as well as by any underlying conditions or diseases that may have occurred in the past or are currently present. Many infants, however, are less susceptible to certain infections than the adult because of maternal antibodies that are transferred to the fetus. This protection seldom exceeds 6 to 10 months following birth and is available only if the mother has come in contact with such infectious agents prior to the delivery of the infant. Transfer of both humoral and cellular factors can also occur via breast milk.

Other factors that may in part influence ones susceptibility to infection include hormones, nutrition, social customs, personal hygiene,and various environmental factors, such as temperature and humidity, and air quality. Underlying conditions or disease also predispose the host to infection, and these are discussed in Chapter 33.

Release of microorganisms from the host

The successful infectious agent finds a host that will support its growth and development, but to maintain itself in the environment, the microbe must be retained in the originally infected host or transferred to another susceptible host. Most microorganisms that are transferred from animals or arthropods to humans reach a dead end because they cannot be transmitted to another human host. Diseases such as anthrax or brucellosis, which are obtained from cattle, are maintained in nature because of animal reservoirs, but once a human is infected, there is no human-to-human transfer. Pathogenic microorganisms must be spread to other individuals if they are to be maintained in nature, and this spread may take one of two forms: horizontal or vertical. ***Horizontal spread,*** which is the most common mechanism, implies that one individual infects another individual by contact, through insect vectors, or by contaminated water, air, food, or fomites. ***Vertical spread*** implies that the infectious agent is transferred from parent to offspring via ovum, sperm, or by contact between placenta and offspring. The spread of microorganisms from one individual to another usually takes place within a few days during the infection, and before the immune system has a chance to destroy the microorganisms; however, there are exceptions. The immune system of some individuals is not sufficient to destroy the pathogen, and the microorganism persists in the host, sometimes for the lifetime of the host.

Persistence. Microbial persistence may give rise to mild symptoms and can involve shedding of the microorganisms—for example, from the gallbaldder (salmonella infection), the upper respiratory tract (diphtheria), or the intestinal tract (cholera). The individual harboring the microorganism is called a ***carrier*** (discussed briefly under "Animate Reservoirs"). The carrier state may exist for several months or several years, depending on the immune system response of the host. Microorganisms that persist in the host may give rise to disease symptoms over a long period of time, in which case the disease, such as tuberculosis or hepatitis B, is termed ***chronic,*** implying that the disease persists or is slowly progressive and that the host is generally unable to actively and efficiently respond to the infectious agent. Chronic infections can be the result of defective phagocytic activity or may be due to a lack of specific antibodies to aid in the phagocytosis of the pathogen. Many chronic diseases occur because the microbial agents elicit a poor humoral response from the host or the antibodies produced do not play a major role in preventing infection.

Sometimes the persistent microbial agent may remain in host tissue for years and then suddenly produce a progressive, fatal infection. The disease Kuru, for example, is caused by a virus that affects a tribe in New Guinea that practices cannibalism. The infected brains from dead tribal enemies are handled by women of the village and served uncooked as a ceremonial food. The virus is believed to be transmitted through abraded areas of the skin during the handling procedures. The first appearance of symptoms are from 3 to 18 years after contact with the infectious agent, and during this period the virus progressively destroys nervous tissue. Once symptoms have appeared, death occurs in 3 to 9 months.

The infectious agent in some persistent infections may enter a ***latent,*** state in which there is no shedding and the symptoms of infection are not present. This latency may be of the intermittent type, for example, the herpesvirus that is the causative agent of coldsores. The symptoms subside following the initial infection, but the virus travels to local neurons, where it remains concealed from the immune system. The virus can be intermittently activated by such factors as sunlight, hormonal changes (menstruation), and anxiety. The virus then migrates down the nerve fiber to the skin surface, where it multiplies and causes a recurrence of coldsores. Another type of latent infection is one in which the agent remains quiescent for a number of years and then becomes activated. The disease chickenpox, for example, is a childhood infection in which the viral agent remains latent for years until adulthood. At this time, under certain conditions, it may erupt into a disease called shingles.

CHARACTERISTICS OF INFECTION IN THE HOST

Clinical stages of infectious disease

Acute vs. chronic diseases. The clinical response of the host to an infectious disease agent may be characterized as acute or chronic. ***Acute*** diseases are those in which symptoms are manifested for a short period of time but then subside when the host's immune system has responded. ***Chronic*** diseases are characterized by the appearance of symptoms over an extended period of time due to an inefficient immune response.

Acute disease may be divided into four distinct periods: the incubation period, the prodromal period, the acute stage, and the convalescent period. The ***incubation period*** is that time between contact with the infectious agent and the appearance of the first symptoms. It is during this period of time that the infectious agent begins to multiply at the initial site of infection. Sometimes the initial site of infection acts as a source of dissemination of either microbial products or microbial cells to other parts of the body. Diseases in which organisms are spread throughout the body to infect organs or tissues are referred to as ***systemic.*** The length of the incubation period may be influenced by the initial dose of microorganisms, the rate of multiplication of the infectious agent, the immune response of the host, and the time for pathogenic factors, such as bacterial toxins, to reach a susceptible organ or tissue. The incubation period may be as short as a few hours, such as in food poisoning, or as long as several years, such as in leprosy.

The first sign of symptoms is called the ***prodromal period*** and usually lasts but 1 or 2 days. The symptoms might include a headache or upset stomach, but usually consist of just a general feeling of not being well.

The ***acute period*** represents the peak of microbial activity in the host and therefore the peak expression of host symptoms. ***Fever,*** which is characteristic of the acute stage in most infectious diseases, is often an indicator of the nature of the disease. Fluctuations in fever during the course of a disease are often associated with an increase of microorganisms in the bloodstream. The suffix *emia* denotes organisms in the bloodstream (for example, bacteremia, fungemia, or viremia) but if the bloodstream serves as a site for microbial multiplication, the term *septicemia* is applied.

The ***convalescent period*** represents a stage of recovery from disease and is characterized by a rapid decline in fever and/other symptoms because of an effective response to the infection.

Not all infections by a microorganism result in acute disease manifestations. Sometimes the host may demonstrate no symptoms or minor symptoms, such as a slight fever or cough. The immune system of the host, probably because of previous encounters with the microbe, responds rapidly, and the infection ceases. Such diseases are called ***subclinical*** or ***inapparent.***

CHARACTERISTICS OF INFECTIOUS DISEASE IN A POPULATION

Types of infections

Epidemiological investigations may be concerned with diseases that are divided into three types based on their occurrence in a population: epidemic, endemic, and pandemic. ***Epidemics*** are defined as either small or large numbers of cases of disease that are always in excess of what is expected or what is normal for that population. ***Endemic*** refers to disease that is continuously present in the population, and the frequency of disease remains at predictable levels (endemic levels) within the community. ***Pandemic*** implies overwhelming outbreaks of disease that involve countries or are world wide. Influenza, for example, has been implicated in several pandemics and is of current interest.

Patterns of infectious disease

The distribution pattern of infectious disease in a community may be influenced by many factors, including the number of susceptible hosts, the population density (the frequency and closeness of contacts), the communicability of the agent, and, if relevant, the density of the vector (mosquitoes, for example).

The introduction of a new infectious agent into a community can result in a very explosive epidemic, because few, if any, individuals will show resistance. Over a period of time, as the outbreak continues and people recover from the infection, a sufficient number of immune persons will be present to act as a barrier to the continued rapid spread of the disease among susceptible persons. Furthermore, if the same disease reappears, the immune individuals will prevent rapid spread of the disease, and only isolated cases among the susceptible hosts may appear. This type of immunity, which can protect susceptible persons, is called ***herd immunity.*** The protection is lost, however, if the infectious agent is highly communicable or if the community is in very close contact.

The frequency with which epidemics occur is a reflection of the immune status of the population. Influenza, for example, shows a cyclic nature and appears in epidemic proportions every 2 to 4 years. The reason for this is that infection by one strain of virus confers immunity against that strain for 1 to 2 years, but as new strains appear, the population as a whole shows little resistance.

How quickly an infectious disease spreads is usually related to the source of infection, the method of transmission, and the incubation period of the microbial agent. Diseases that have a common source, such as a contaminated water or food supply, and a short incubation period will appear quickly and involve large groups of individuals. Common-source outbreaks of food poisoning caused by bacteria, such as *Staphylococcus* and *Salmonella,* result in symptoms within hours of ingestion of the contaminated food or water and frequently involve large groups of people (picnics, hospitalized patients, etc.).

Epidemiological methods

There are three main types of epidemiological investigations: descriptive, analytical, and experimental.

Descriptive epidemiology. Descriptive epidemiology, which is the most frequent investigation used, examines the occurrence of disease in terms of persons affected, the place at which infection occurred, and the time period over which the disease appears.

Person. Any factor that can influence the development of disease in an individual must be considered and evaluated by the epidemiologist. One of the most important factors is age, because the very young and the very old are more susceptible to infection than other age groups. This susceptibility is due to either the immature nature or defectiveness of the immune system in the young and old, respectively. Other factors that are important are sex, race, socioeconomic status, marital status, and underlying disease or conditions.

Place. The number of specific diseases that occur within a population is influenced by the geographic area where contact was made between the host and the infectious agent. Many diseases involving parasitic worms, for example, are common in tropical areas, where vectors abound and where soil temperatures are high enough to support the existence of the parasite. Diseases acquired from animals would be more prevalent in individuals such as farmers, abattoir workers, and veterinarians than in those individuals whose occupations do not bring them in contact with animals. Determining the place of contact between host and infectious agent is not always sufficient in epidemiological investigations. A hospital-acquired disease, for example, may result from receiving contaminated medication, and the site of the contamination (factory) is more important than host contact.

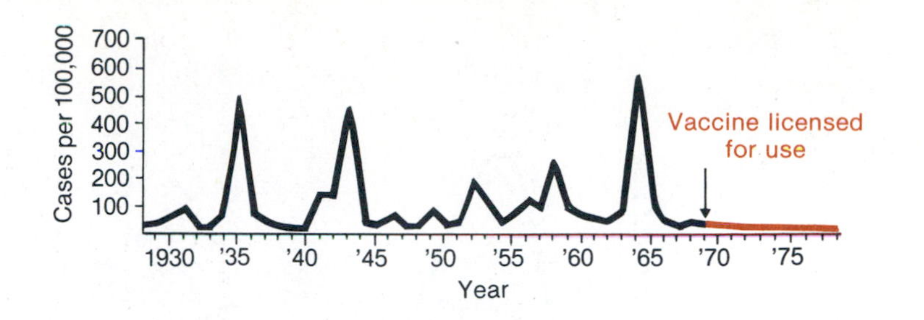

Figure 26-1 Rubella incidence in 10 selected areas of United States, 1928-1978. Peaks represent epidemic outbreaks, which appear in 6- to 9-year cycles. Last epidemic outbreak occurred in 1964, and no further outbreaks were seen, because rubella vaccine was licensed in 1969.

Time. Time may be considered by the epidemiologist in different ways. A seasonal trend may be examined in which variations in disease occur from season to season. Respiratory tract infections, for example, occur more frequently in the winter months, when conditions such as confinement and close proximity to other individuals favor droplet transmission. Variations in disease may also be related to changes that occur over a period of years. A disease such as polio may be examined over a period of several years to determine what effect hygienic practices or immunization practices have had on the course of the disease in a population. Time variations may also involve measurement of the acute or short-term epidemic and all the factors that influence it.

Epidemics are graphically represented as the number of cases of disease plotted against time. (Figure 26-1). The shape of the curve is influenced by many factors, including the characteristics of the infectious disease agent, the incubation period, the susceptibility of the host, environmental factors, and the method of transmission. The first case in the epidemic is called the ***index case.*** A sharp increase in the number of cases occurs with diseases that have a short incubation period, and there is a common source, such as food contaminated by *Salmonella.* The downslope of such a curve is relatively steep, but more gradual than the upswing (Figure 26-2). In diseases transmitted by person-to-person contact, such as many respiratory tract infections, the curve is characterized by a gradual upswing,

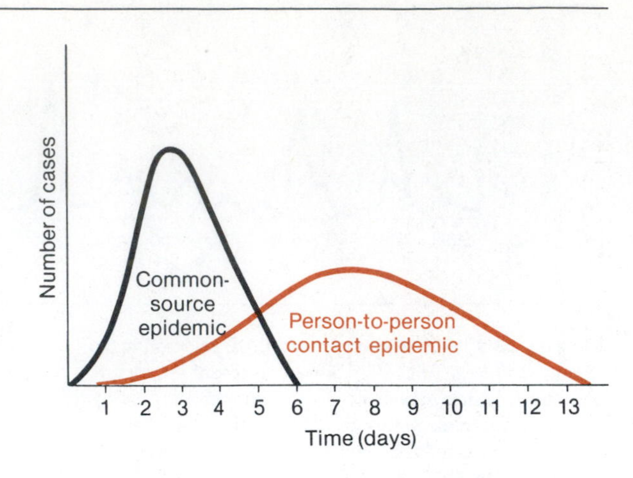

Figure 26-2 Comparison of epidemic curves in which there is common source and person-to-person transmission. Common source has rapid upswing in cases as compared with gradual upswing in epidemic caused by person-to-person transmission. Downswing of both curves is more gradual than upswing but is steeper in common-source epidemics.

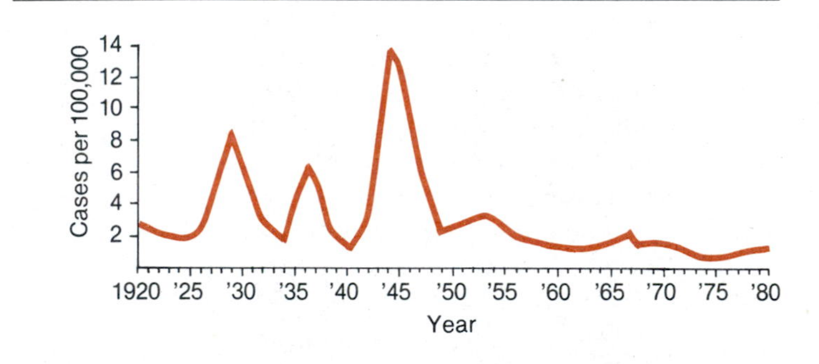

Figure 26-3 Reported meningococcal infection rates by year, United States, 1920-1980.
Redrawn from Morbidity and Mortality Weekly Report ***30:***114, 1981.

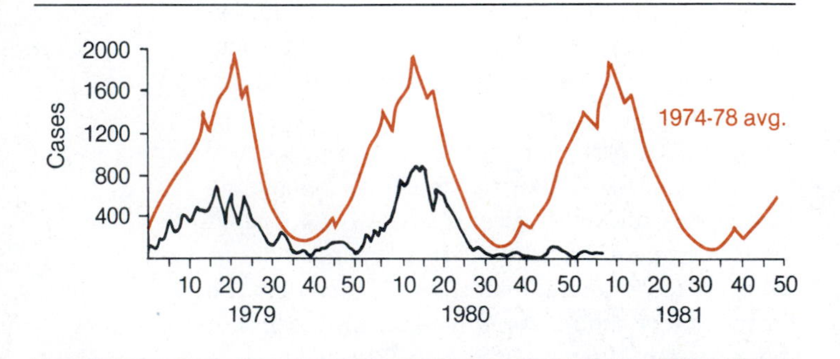

Figure 26-4 Measles incidence, United States, 1979-1981 and 1974-1978 average.
Redrawn from Morbidity and Mortality Weekly Report ***30:***183, 1981.

a flatter peak, and a very gradual downswing (Figure 26-2). The downslopes of both curves represent cases with longer incubation periods and a decreasing number of susceptible persons. In diseases such as hepatitis, in which the incubation period is several weeks, the number of diagnosed cases of disease may be so spread out that it gives the impression that no epidemic has taken place.

Analytical epidemiology. Analytical epidemiology seeks to evaluate the determinants of disease in terms of cause and effect. Two techniques are used. The first technique is retrospective and looks for factors that have preceded the epidemic. The epidemiologist usually performs his investigation by examining groups that have displayed the disease and compares them with a group of controls who do not have the disease. The epidemiologist is therefore looking for one or more determinants that may be responsible for infection. The second technique, which is prospective, involves the collection of data as the disease is in progress. In other words, at the outset of the investigation there are no cases of disease. In the prospective study two groups are compared: those that have an attribute or characteristic that may contribute to infection and those that do not have the characteristic.

Experimental epidemiology. In experimental epidemiology there is manipulation of a population in order to test a hypothesis. Such studies compare treated and untreated controls; that is, one group is exposed to a risk or benefit while the control group receives a standard treatment or ***placebo*** (a substance having no effect on the host). This approach can be used only when the risk factor is determined to cause no harm to the host.

An example of the experimental approach might involve the use of an immunoglobulin on a group of individuals with an induced viral respiratory tract infection. The group is divided so that some receive a placebo while others receive the immunoglobulin. The two groups are compared to see if the treatment has enhanced recovery from infection. If a statistically significant number of treated individuals recover much earlier from infection than the controls, then the difference can be attributed to the immunoglobulin, provided all other factors are equal.

Measurements in epidemiology

Once all of the data have been collected and tabulated, an ***infection rate*** can be determined.

Such a rate expresses the probability of occurrence of an event and is correlated with a specific time interval and population (Figure 26-3). The rate is expressed as k(x/y), where x is the number of times an event such as an infection has taken place during a specified time interval; y is the population from which the infectious disease is measured; and k is a multiplication factor based on the population measured (100, 1000, 10,000, etc.) and is used to describe the rate as a whole number rather than as a fraction. Sometimes the term ***ratio*** is used instead of rate to compare the frequency of one occurrence with another. For example, a ***case-fatality ratio*** compares the number of fatalities that has occurred in a specified number of cases. There are two types of rates: prevalence and incidence. The number of new cases that appear in a certain population over a specified period of time is referred to as ***incidence*** (Figure 26-4). The number of new cases in addition to the old cases that are recorded over a period of time is referred to as ***prevalence.*** Prevalence takes into account all of the new and old cases of disease and is influenced by the duration of infection. A patient with an infection that lasted 6 weeks would be counted only once in an incidence survey, but if a prevalence survey were taken twice during the 6-week period, the infection would be counted twice.

SUMMARY

Epidemiology is the science that deals with those factors that are associated with the frequency of disease and how disease is transmitted. These factors are rooted in the evolutionary relationship of humans and parasites. Microorganisms are believed to have first existed as free saprophytes that, after contact with humans, developed certain relationships that were either mutualistic, commensalistic, or parasitic. Many parasites, after their initial contact with higher organisms, evolved with their host into a relationship that would enable the host to survive and permit the parasite continued existence. Parasitic development in animals, as well as the arthropods harbored by them, resulted in a means of transfer of potentially infectious agents to humans, particularly following animal domestication. Disease transmission in humans increased greatly as overcrowding developed in the cities of the world. A lack of knowledge of the importance of sewage disposal and water purification compounded this problem.

The development of an infectious disease is dependent on:

1. The characteristics of the microbial agent, such as dose and virulence.
2. The source of the microorganisms, such as animate or inanimate sources. Among animate sources, human carriers are the most important means of transmission of infectious agents.
3. How the microorganisms are transmitted. Microorganisms can be transmitted by direct contact or indirect contact. Indirect contact includes air, common vehicle, or arthropod borne.
4. The susceptibility of the host. The age and the genetic constitution of the individual often determine susceptibility to infection. Other factors might include hormones, social customs, and personal hygiene.
5. How the microorganisms are disseminated from an infected to an uninfected host.

Disease in the host is characterized by various stages. Acute disease that terminates in recovery is divided into four periods: incubation, prodromal, acute, and convalescent or recovery period. Chronic diseases are characterized by the appearance of symptoms over extended periods of time, and subclinical infections are those in which the host is infected but disease symptoms do not appear because of resistance of the host.

Diseases in a population are divided into three groups based on their occurrence: epidemic, endemic, and pandemic. Epidemiological methods are used to determine the how, when, where, and why of an infectious disease so that techniques can be developed to prevent its spread. Descriptive epidemiology, which is the most frequently used epidemiological method, measures the occurrence of disease in terms of persons affected, the place at which infection occurred, and the time over which the disease appears. Epidemiological data are collected and translated into rates such as incidence rates, which are correlated with a specific time interval and population number.

Study questions

1. Give examples of host-microorganism associations that exhibit mutualism, commensalism, and parasitism.
2. What advantages did microorganisms acquire through their association with various vertebrate and invertebrate hosts?
3. What would be the value of having a disease run its course as compared with treating the infection with antibiotics?
4. What role did animals play in the evolution of the host-parasite relationship in humans?
5. What effect is there on infectious disease transmission if animals cannot serve as the reservoirs of the infectious agent?
6. Distinguish between chronic and acute infection.
7. List the various ways that humans can become infected by microbial agents.
8. What are the reasons for microbial persistence in the host?
9. What is taking place in the infected host during the incubation period?
10. What are latent diseases? Name some common diseases that are considered latent.
11. How does an infectious disease differ from an intoxication? Give examples.
12. What functions does the arthropod serve in the epidemiology of human infection?
13. Compare systemic diseases with focal diseases.
14. How does bacteremia differ from septicemia?
15. Explain how diseases that are epidemic in nature can become endemic, and vice versa. What are some reasons for these reversals?
16. Distinguish between the incidence of a disease and its prevalence in a population. Name some common diseases in which this distinction could be made.
17. What techniques can the epidemiologist use to trace the source of an infection?

Selected readings

Books

Austin, D.F., and Wener, S.B. *Epidemiology for the health sciences.* Charles C Thomas, Publisher, Springfield, Ill., 1974.

Burnet, F.M. *Natural history of infectious disease* (4th ed.), Cambridge University Press, Cambridge, 1972.

Centers for Disease Control. *Morbidity and mortality reports.* Atlanta, Ga. (A weekly report that discusses primarily infectious diseases, especially in the United States.)

Centers for Disease Control. *Surveillance.* (Annual summaries of specific infectious diseases such as food-borne, and water-borne diseases.)

Evans, A.S. *Viral infections of humans: epidemiology and control.* Plenum Medical Book Co., New York, 1976.

Friedman, G.D. *Primer of epidemiology.* McGraw-Hill Book Co., New York, 1974.

Fox, J.P., Hall, C.E., and Elveback, L.R. *Epidemiology—man and disease.* MacMillan Publishing Co., Inc., New York, 1971.

Lillienfeld, A.M. *Foundations of epidemiology.* Oxford University Press, New York, 1976.

MacMahon, B., and Pugh, T.F. *Epidemiologic principles and methods.* Little, Brown & Co., Boston, 1970.

Mausner, J.S., and Bahn, A.K. *Epidemiology.* W.B. Saunders Co., Philadelphia, 1974.

Mims, C. *The pathogenesis of infectious disease.* Grune & Stratton, Inc., New York, 1976.

Smith, C.E.G. *Epidemiology and infections.* Meadowfield Press, Ltd., Shildon, England, 1976.

Wehrle, P.F., and Top, F.H., Sr. *Communicable and infectious diseases* (9th ed.). The C.V. Mosby Co., St. Louis, 1981.

Journal articles

Black, F.L., et al. Evidence for persistence of infectious agents in isolated human populations. *Am. J. Epidemiol.* **100:**230, 1974.

Kaplan, M.M., and Webster, R.G. The epidemiology of influenza. *Sci. Am.* **237:**88, 1977.

Lilienfeld, A.M. Epidemiology of infectious and noninfectious disease: some comparisons. *Am. J. Epidemiol.* **97:**135, 1973.

Salk, D. Eradication of poliomyelitis in the United States. *Rev. Infect. Dis.* **2**(2):228, 1980.

Chapter 27 CONTROL OF INFECTIOUS MICROORGANISMS Chemotherapy, Sterilization, and Disinfection

The purpose of this chapter is to discuss the major antimicrobial agents that are used to inhibit or kill microorganisms both inside and outside the body. In addition, the mechanisms by which the antimicrobial destroys the infectious agent are discussed, as well as the mechanisms used by infectious agents to resist the action of antimicrobials.

CHEMOTHERAPY

The term ***chemotherapy*** was introduced into scientific literature by the German chemist Paul Ehrlich in the late 1890s. He defined it as the use of drugs to injure the invading organism without injury to the host. Before 1890 only three agents could be considered chemotherapeutic agents: chinchona, for the treatment of malaria; mercury, for the treatment of syphilis; and ipecacuanha, for the treatment of amebic dysentery. Ehrlich's initial research was concerned with the affinity of chemicals for tissues, and these studies led to significant discoveries in histological staining that are still used today. He was also intrigued by the selective binding occurring between antigens and antibodies, and this led him to propose that drugs bind to cellular components by bonds. He suggested that chemotherapeutic agents, like antigens, possess two reactive groups: one concerned with fixation or transport to the tissue and the other causing the fatal lesion in the parasite. Ehrlich predicted that the function of the drug would be to upset the parasite's metabolism, after which the host's natural defenses would take over.

The only chemotherapeutic agent of clinical significance discovered by Ehrlich was Salvarsan-606, an arsenical used in the treatment of syphilis. The number 606 refers to the trials employed by Ehrlich's research team to find the correct dosage for treatment of humans. Ehrlich's greatest contributions to chemotherapy were his original ideas, which provided the seed from which germinated new ideas and the inevitable discovery, by others, of a host of chemotherapeutic agents.

Another milestone in the history of chemotherapy was reached in 1935 by Domagk's discovery of prontosil, an azo-dye used to treat streptococcal infections. Prontosil in the mammalian body is converted to sulfanilamide, which inhibits microbial activity by binding to bacterial enzymes (Ehrlich's receptors). Sir Alexander Fleming in 1929 discovered a strain of mold called *Penicillium,* that excreted a substance capable of inhibiting the growth of the bacterium *Staphylococcus aureus.* The antibacterial agent was called penicillin. Penicillin was not recovered in a pure form until the 1940s when it was isolated by Florey, Chain, and others. Penicillin was used extensively during World War II to curb infection. The success of penicillin encouraged others to seek agents that could be used as chemotherapeutic agents, and that search is still continuing.

General concepts and terminology associated with chemotherapeutic agents

Selective toxicity. Ehrlich's definition of chemotherapy implies that the chemical agent is selectively toxic, but this is not always true, since most drugs show some degree of toxicity that is not severely detrimental to the host. Selective toxicity is best described by some examples in which the chemical agent inhibits microbial activity by interfering with metabolic processes. Sulfanilamide, for example, which interferes with folic acid synthesis in microorganisms, has no effect on humans, because folic acid is not synthesized in the body and must be supplied in the diet. Many chemotherapeutic agents affect protein synthesis because of their ability to bind certain ribosomal components. The binding of chemotherapeutic agents to microbial ribosomes is more pronounced than the binding to mammalian ribosomes because of variations in structure and composition between them. The cell membrane of mammalian cells represents to some degree a permeability barrier to chemotherapeutic agents. Microorganisms that invade mammalian cells (intracellular parasites) are much more difficult to combat with drugs than those microorganisms that do not invade cells (extracellular parasites). Finally, the rate of metabolism and reproduction of microbial cells is much faster than that of mammalian cells; therefore the inhibition of microbial function occurs more rapidly than that of host function. This usually leads to a faster rate of degradation of antimicrobials by microorganisms.

At this point in our discussion a distinction should be made between a chemotherapeutic agent and an antimicrobial agent. All chemotherapeutic agents are antimicrobials, but not all antimicrobials are chemotherapeutic. Antimicrobial agents can inhibit the growth of or kill microorganisms, but they are not necessarily selectively toxic. For example, many antimicrobial agents cannot be administered internally because of toxicity to the host. Sulfuric acid is certainly an antimicrobial, but taken internally it would destroy host tissue.

Sources. Many microorganisms produce compounds that in small quantities inhibit the growth of or kill unrelated species of microorganisms. These biologically produced compounds are called ***antibiotics.*** Species of *Streptomyces* and *Bacillus* are the principal bacterial producers of antibiotics, and the fungus *Penicillium* is the source of penicillin G. Compounds that have antimicrobial activity and chemotherapeutic value can also be synthesized in the laboratory and are called ***synthetic*** agents, for example, the sulfa drugs. Finally, antibiotics produced by some microorganisms can be modified in the laboratory by adding chemical groups to the existing nucleus, and these are called ***semisynthetic*** agents (Figure 27-1). Nearly all penicillins and cephalosporins used today are semisynthetic.

Activity spectrum. Chemotherapeutic agents do not affect all microbial species in the same way, and this variation can be described as an ***activity spectrum.*** A ***broad-spectrum*** chemotherapeutic agent is generally associated with antibacterial agents that are active against gram-positive as well as gram-negative bacterial species. A ***narrow-spectrum*** drug is active against one or only a few species within either the gram-positive or the gram-negative groups. The activity of a drug is also influenced by the concentration of microorganisms at the site of infection as well as by the type of tissue affected. The number of microorganisms at a body site may be as high as 1×10^9/g of tissue, and greater concentrations of drug may be required to destroy the infectious agents. Some tissues, such as nerve tissue, are relatively impermeable to many antimicrobial agents carried there by the bloodstream, and infection in those areas is very difficult to treat. The concentration of drug at the site of infection often will determine whether or not the organism can be destroyed.

An antimicrobial agent may be described as -cidal or -static. The suffix may be added to the type of microorganism affected by the drug, such as a bactericidal agent, fungicidal agent, etc. A *-cidal* agent kills the microorganism, whereas a *-static* agent inhibits growth. For some drugs it is the concentration that determines what the activity will be. If the concentration of a bactericidal agent is too low to kill microorganisms, it may act as a bacteriostatic agent on a population of microorganisms. ***Bactericidal*** agents are more useful in the treatment of those infections in which the normal immunological response of the host is depressed. One of the disadvantages of bactericidal agents is that gram-negative bacteria release a highly toxic endotoxin when lysed, which can cause serious complications in the host. Second, the effect of a bactericidal agent may be so rapid that it does not allow enough time for the host to respond immunologically and synthesize antibodies. Antibodies are beneficial not only in warding off the initial infection, but in helping to prevent future infections. The advantage of ***bacteriostatic*** agents is that they allow the immune system of the host to destroy the invading pathogen. One of the disadvantages of bacteriostatic agents is that the growth-inhibiting concentration of the drug must be maintained in the bloodstream for a relatively long period of time in order for it to be effective.

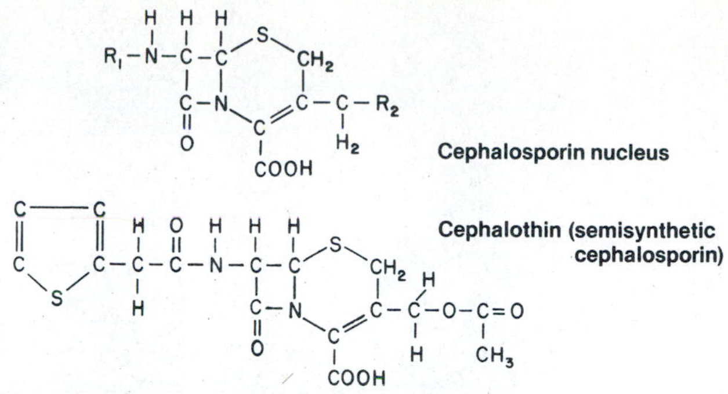

Toxicity. Many chemotherapeutic agents used currently are to some degree toxic to mammalian cells, but fortunately the drugs have a greater affinity for the microbial component than for mammalian cell structures and are thus less toxic to the latter. The degree of tissue toxicity is related to many factors: the type of drug being used, the dosage administered, the length of time of treatment, and the condition of the patient at the time of treatment. Under normal treatment procedures maximum therapeutic levels of chemotherapeutic agents are not injurious to mammalian tissue. Toxicity, when it does occur, may be direct or indirect. Indirectly, the drug may induce an allergic response in the host, resulting in the activation of components of the immune system, which can damage tissue. Direct toxicity may involve destruction or irritation of sensitive tissues such as those of the kidney, liver, blood, nervous tissue, etc. Some drugs, such as the aminoglycosides (gentamicin is an example) are toxic to kidneys (nephrotoxicity) and interfere with the function of the organ by damaging tubular cells. Once the cells have been damaged, the drug begins to accumulate to higher concentrations in the kidney, and this can cause permanent damage, necessitating dialysis for continued survival. There are some drugs that when used even at minimal therapeutic concentrations are very toxic to host tissue. In some instances the toxic drug may be the only one available, and its use may be necessary in life-threatening situations. Often drugs that are very toxic when administered for systemic infections can be used topically to treat infections with no harm to the sensitive internal organs. Bacitracin and neomycin, for example, are so toxic to internal organs that they cannot be administered for systemic infections, but they can be used in topical ointments for superficial skin infections.

Figure 27-1 Structure of cephalosporin nucleus as it is produced by microorganisms. R1 and R2 are sites on which chemical units can be added. Cephalothin is semisynthetic compound in which two chemical units have been added to R1 and R2 of cephalosporin nucleus.

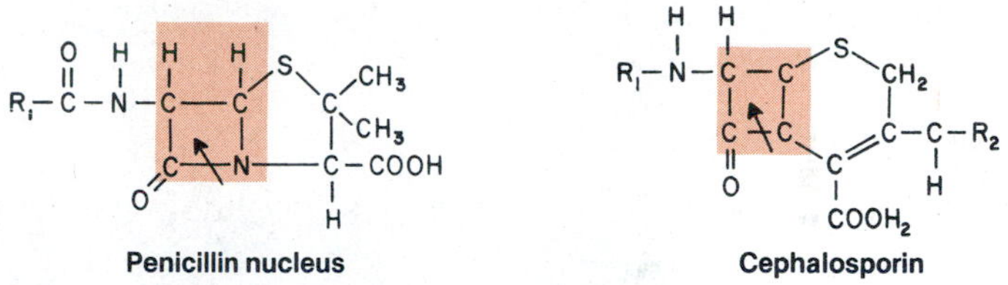

Figure 27-2 Structures of penicillin and cephalosporin nuclei demonstrating β-lactam rings *(colored squares)*. Arrows indicate sites of action of β-lactamase that cleave β-lactam rings.

Major chemotherapeutic agents based on mechanism of action

Inhibitors of cell wall synthesis. Agents that inhibit cell wall synthesis are essentially non-

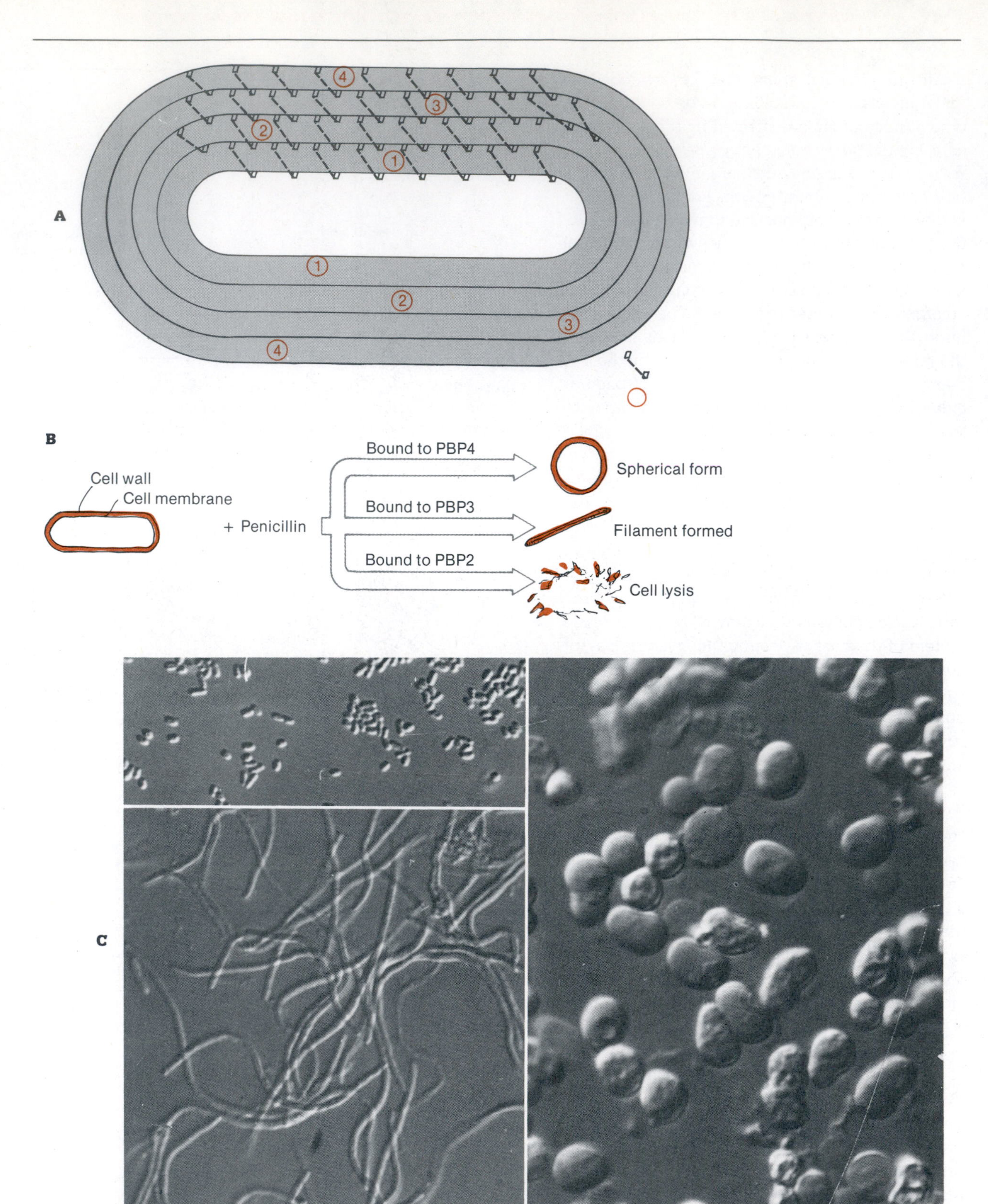

Figure 27-3 Hypothetical explanation of effects of penicillin on bacterial cell containing penicillin-binding proteins (PBPs). **A,** Bacterial cell showing transpeptidation linkage between peptidoglycan chains and penicillin-binding proteins. **B,** Possible effects of penicillin on different rod-shaped microorganisms. Penicillin produces spherical forms if it binds to PBP4, filamentous forms if it binds to PBP3, and cell lysis if it binds to PBP2. **C,** Interference phase-contrast microscopy showing effects of different penicillins on the bacillus *Proteus mirabilis*. (×2500.) Upper left cells are normal and untreated; lower left cells were treated with subinhibitory concentrations of ampicillin and produced filamentous forms; cells on right were treated with subinhibitory concentrations of mecillinam and produced round cells. Penicillin-binding proteins are located in cytoplasmic membranes although their activities are associated with cell wall synthesis.

C, *From Lorian, V., and Atkinson, B.A.* Rev. Infect. Dis. ***1****(5):797, 1979.*

Table 27-1 Characteristics of chemotherapeutic agents that inhibit cell wall synthesis

AGENT	SPECIFIC MECHANISM OF ACTION	ACTIVITY SPECTRUM	TOXICITY	COMMENTS
Penicillins*				
Penicillin G	Prevents cross-linking of peptidoglycan strands in cell wall	Gram-positive bacteria; bactericidal	Nontoxic	Sensitive to β-lactamase
Ampicillin	Same as above	Gram-positive and gram-negative bacteria; bactericidal	Nontoxic, but rashes are frequent	Same as above
Carbenicillin	Same as above	Gram-negative bacteria; bactericidal	Nontoxic	Same as above
Nafcillin	Same as above	Gram-positive bacteria; bactericidal	Nontoxic	Resistant to β-lactamase
Methicillin	Same as above	Gram-positive bacteria; bactericidal	Nontoxic	Resistant to β-lactamase
Cloxacillin	Same as above	Gram-positive bacteria; bactericidal	Nontoxic	Resistant to β-lactamase and acid resistant
Cephalosporins				
Cephalothin	Same as penicillin	Gram-positive and gram-negative bacteria; bactericidal	Nontoxic	Resistant to β-lactamase
Cephaloridine	Same as above	Gram-positive and gram-negative bacteria; bactericidal	Toxic to kidneys	More active than cephalothin against gram-positive bacteria
Cephalexin	Same as above	Gram-positive and gram-negative bacteria; bactericidal	Nontoxic	Can be taken orally
Cefotaxime	Same as above	Gram-positive and most gram-negative bacteria; bactericidal	Nontoxic	Greater spectrum of activity than all other cephalosporins
Bacitracin	Interferes with transfer of peptidoglycan components in cell wall	Gram-positive bacteria	Toxic to kidneys	Used in topical ointments
Vancomycin	Interferes with lipid carrier of cell wall components	Gram-positive bacteria	Toxic to kidneys and hearing mechanism	Used primarily for staphylococcal infections
Isoniazid	Inhibits synthesis of mycolic acids required in cell wall synthesis of mycobacteria	Mycobacteria	Toxic to liver and peripheral and central nervous system	Used only in treatment of tuberculosis

*The structure of various penicillins can be found in Chapter 21.

toxic* to the host, since cell walls are absent in mammalian tissue. The best examples of inhibitors of cell wall synthesis are the ***penicillins*** and ***cephalosporins,*** both of which possess a ***β-lactam*** ring (Figure 27-2). The mechanism of action of penicillin at one time was thought to be relatively simple and involve a single target site. Penicillin, because of its structural resemblance to the D-alanyl-D-alanine portion of the cell wall glycopeptide, specifically inhibits the linking of neighboring subunits via transpeptidation. The transpeptidation reaction lends structural support to the cell wall (Figure 27-3 *A*), but in the presence of penicillin the cell wall is mechanically weak and results in cell lysis. The mechanism of action of penicillin is apparently more involved and appears to vary between groups of bacteria and to be influenced by a variety of other factors. There now appears to be several possible target sites for penicillin, which are located in the cytoplasmic membrane and referred to as penicillin-binding proteins (PBPs). There are at least seven or eight PBPs, and some have been shown to exhibit different affinities for drugs possessing the

* Although not considered directly toxic to tissue, penicillin can elicit an allergic response in up to 5% of the human population.

β-lactam ring. In addition, these PBPs are associated with specific cell wall activities. The preferential binding of penicillin to one PBP may produce one kind of effect, whereas the binding of penicillin to other PBPs may produce other effects (Figure 27-3, *B* and *C*). These effects include a rod-shaped cell becoming spherical, a cell forming filaments of unusual length, an ovoid cell, and in some cases lysis of the cell with no obvious morphological changes preceding the lysis.

In *Escherichia coli* PBP4, PBP5, and PBP6 have been identified as forms of the enzyme carboxypeptidase. This enzyme removes an alanine residue from the two alanine residues in the murein peptide chain before its incorporation into the cell wall. PBP3 appears to be involved in septation; PBP2 is believed to be associated with bacterial shape; and PBP1 is believed to be a transpeptidase involved with cell elongation.

It now appears that in those organisms possessing murein hydrolases (autolytic enzymes that hydrolyze the bonds between the *N*-acetylmuramic acid residues and the L-alanine residues of the peptide chains of the cell wall), two steps may be involved in the irreversible destruction of the cell. First, penicillin inhibits growth by binding to PBPs and interfering with peptidoglycan synthesis. Second, a murein hydrolase is activated to cleave cell wall, resulting in the release of all the peptide units and leaving a structurally weak cell wall. This is followed by exposure of the plasma membrane to osmotic imbalance and eventual lysis of the cell. Penicillin, which lyses *Streptococcus pneumoniae,* only inhibits growth if this organism possesses a defective murein hydrolase. The relationship between lysis and loss of viability is not yet clear in other organisms, and other mechanisms for the action of penicillin may exist.

Commonly prescribed drugs that inhibit cell wall synthesis but by different mechanisms than the penicillins and cephalosporins are bacitracin, vancomycin, and isoniazid. Cycloserine also inhibits cell wall synthesis, but it is seldom used as a chemotherapeutic agent (Table 27-1). The stages in cell wall synthesis where the various antimicrobials exert their effects are outlined in Figure 27-4. The synthesis of cell wall in *S. aureus* can be broken down into four stages:

Stage 1. This stage is involved in cell wall–precursor formation and takes place in the cell cytoplasm. The first reaction involves the covalent binding of uridine triphosphate (UTP) to *N*-acetylglucosamine (NAG) to form UDP-*N*-acetylglucosamine. This latter compound goes through a series of reactions to form the precursor UDP-acetylmuramyl tripeptide, containing the amino acids L-alanine, D-glutamic acid, and L-lysine. An alanine dipeptide is then attached to the tripeptide to form muramyl pentapeptide. The two enzymes involved in these reactions are alanine racemase, which converts L-alanine to D-alanine, and D-alanyl-D-alanine synthetase, which attaches the alanyl-alanine dipeptide to the tripeptide. These enzymatic steps are inhibited by the antibiotic cycloserine, which is a structural analogue of D-alanine (Figure 27-5). The muramyl pentapeptide is now ready to take part in the second stage of cell wall synthesis.

Stage 2. This stage takes place on the cytoplasmic membrane. The UDP-acetylmuramyl pentapeptide formed in stage 1 is linked to UDP-*N*-acetylglucosamine (NAG) to produce a disaccharide. The linkage of these molecules is brought about by a lipid carrier molecule (a C_{55} isoprenoid alcohol called ***bactoprenol***). This results in the formation of a disaccharide-phospholipid complex. To this molecule is added five glycine residues, which are attached to the ε–amino group of lysine. Glycyl tRNA is the amino acid donor, but no ribosomes are involved in this peptide synthesis. The α–carboxyl group of the glutamic residue in the pentapeptide is amidated.

Stage 3. The phospholipid carrier transports the disaccharide across the cytoplasmic membrane to preexisting portions of the cell wall and in the process is cleaved. The transport of the disaccharide across the cytoplasmic membrane is believed to be greatly facilitated because of the hydrophobic nature of the lipid carrier. The phospholipid carrier with its two phosphates is dephosphorylated and can continue with another round of stage 2 cell wall synthesis. Vancomycin inhibits the reaction in which the disaccharide is coupled to preexisting cell wall, and this results in the premature release of the lipid carrier. The enzyme catalyzing the reaction is peptidoglycan synthetase. Vancomycin is believed to bind to the acyl D-alanyl-D-alanine portion of the cell wall precursors, and it prevents the latter's binding to peptidoglycan synthetase. Bacitracin is believed to form a complex with the lipid carrier and inhibit the dephosphorylation of the lipid, thus preventing its regeneration.

Stage 4. The last stage involved in cell wall synthesis is the cross-linking of peptidogly-

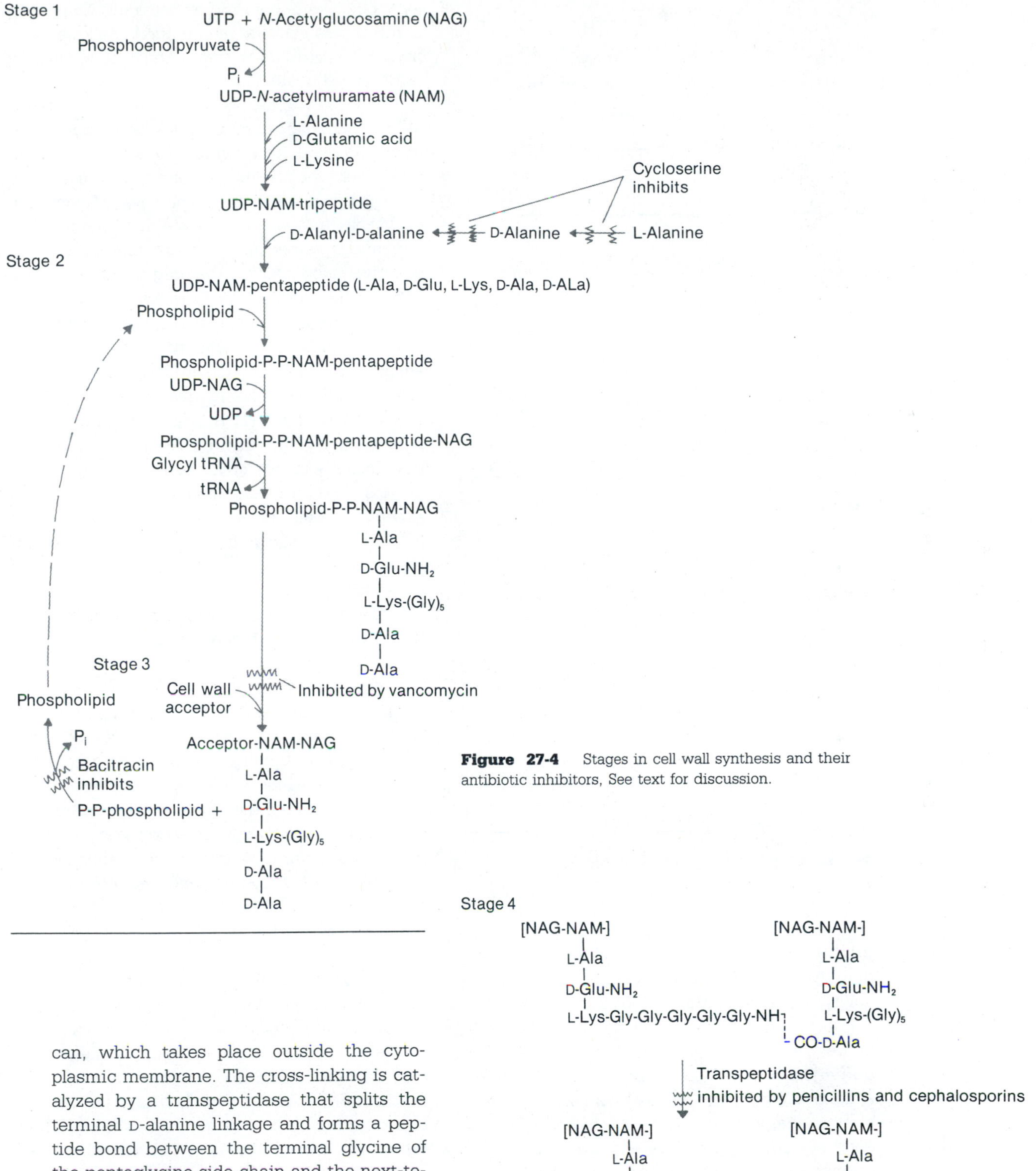

Figure 27-4 Stages in cell wall synthesis and their antibiotic inhibitors, See text for discussion.

can, which takes place outside the cytoplasmic membrane. The cross-linking is catalyzed by a transpeptidase that splits the terminal D-alanine linkage and forms a peptide bond between the terminal glycine of the pentaglycine side chain and the next-to-last (penultimate) alanine of an adjacent chain. The unused terminal D-alanines are removed by a carboxypeptidase. Penicillins and cephalosporins inhibit the transpeptidation reaction.

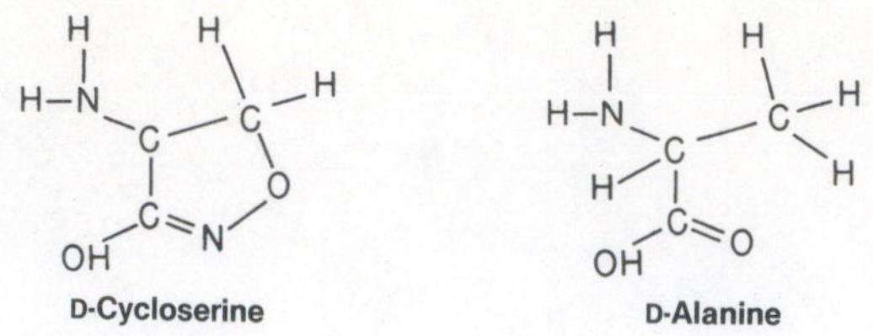

D-Cycloserine D-Alanine

Figure 27-5 Structural similarities of D-cycloserine and D-alanine.

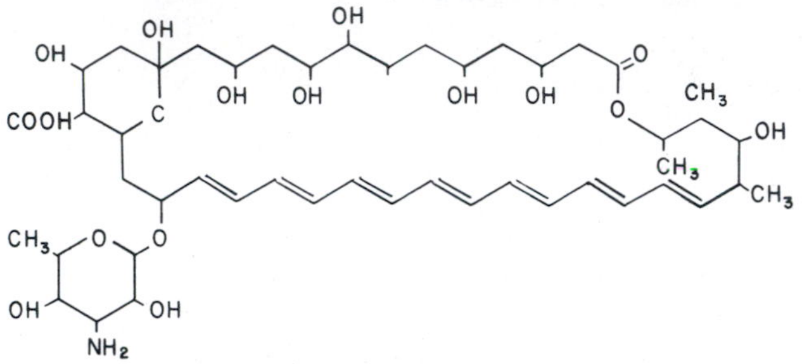

Figure 27-6 Structure of antifungal agent amphotericin B.

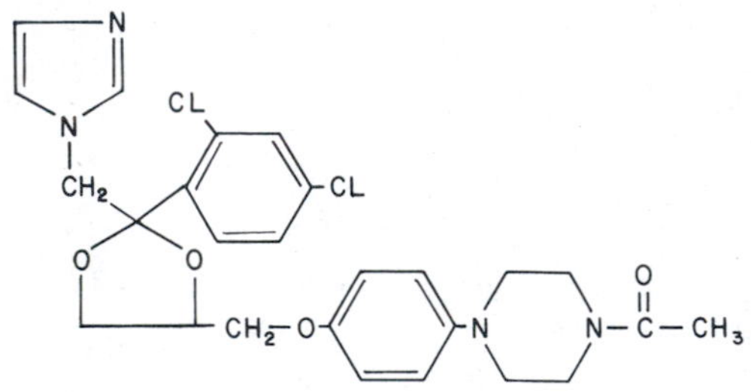

Figure 27-7 Structure of antifungal agent ketoconazole.

Inhibitors of cell membrane function. Drugs that affect the cell membrane bring about permeability changes in the microbial cell. The principal drugs currently used that affect the cell membrane are ***nystatin, amphotericin B,*** the ***polymyxins,*** and the ***imidazole derivatives.*** Amphotericin B and nystatin are polyene antibiotics characterized by a conjugated double-bond system (Figure 27-6) and are produced by species of *Streptomyces.* Amphotericin b binds irreversibly to the sterol component of the fungal membrane, ergosterol, but there is a lower degree of binding to the mammalian cell membrane because the principal sterol is cholesterol and not ergosterol. Amphotericin B and other membrane-binding antimicrobials alter the osmotic properties of the cell, which results in leakage of important cellular constitutents such as ammonium ions, potassium ions, and nucleotides. Amphotericin B brings about a deterioration of all membranous structures in the cell, and this eventually leads to cell death.

The imidazole derivatives include such drugs as ***miconazole, clotrimazole, econazole, isoconazole,*** and ***ketoconazole.*** They differ from other cell membrane inhibitors because they have a broad spectrum of activity. Miconazole, econazole, and isoconazole are used primarily as topical agents because they are toxic or because effective blood levels of the drug cannot be maintained when administered systemically. Ketoconazole (Figure 27-7) is one of the most recently developed imidazole derivatives and offers many advantages over other antifungal agents. The drug, in addition to being broad spectrum, is also important because (1) it is active after oral administration; (2) therapeutic levels of the drug in the bloodstream are obtained rapidly after a single dose and are maintained for several hours; (3) except for some unusual circumstances, such as in diabetes, it shows little toxicity when administered systemically; and (4) it has not given rise to fungal resistance.

Table 27-2 outlines the principal characteristics of the cell membrane inhibitors.

Inhibitors of protein synthesis. There are many drugs used today that are classified as inhibitors of protein synthesis, and in most instances they exhibit greater affinity for bacterial ribosomes than for mammalian ribosomes. The major inhibitors of protein synthesis are ***chloramphenicol, tetracycline, amikacin, streptomycin, neomycin, kanamycin, gentamicin, clindamycin, lincomycin, tobramycin,*** and ***spectinomycin.*** Many of these inhibitors belong to a group called the aminoglycosides, which are structurally similar (Figure 27-8), have similar spectrums of activity, and exhibit similar toxic reactions in the host.

Streptomycin is the prototype of the aminoglycosides. It is believed to produce two effects in the cell. It can bind to free 30S ribosomes before they engage the 50S ribosome to form a complex on the mRNA, and it can also bind to the 30S ribosome while it is engaged in chain elongation. When streptomycin binds to free 30S ribosomes, protein synthesis is completely blocked because chain elongation cannot occur, even though the 30S subunit complexes with the 50S subunit on the mRNA. When streptomycin binds to chain-elongating ribosomes, there is ribosome distortion, which slows down protein synthesis, pro-

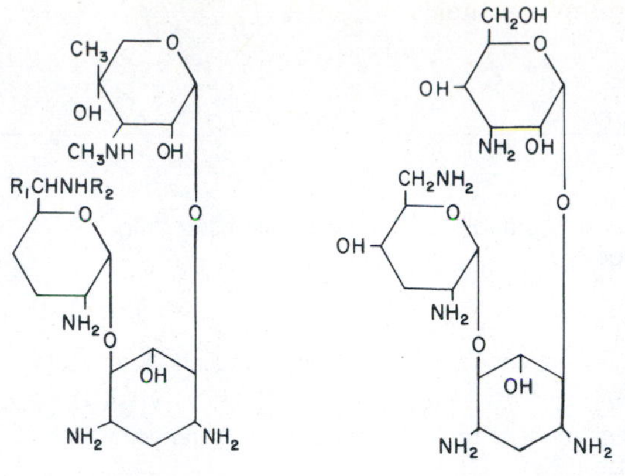

Gentamicin

Tobramycin

Figure 27-8 Structure of some important aminoglycosides.

Table 27-2 Characteristics of chemotherapeutic agents that inhibit cell membrane function

AGENT	SPECIFIC MECHANISM OF ACTION	ACTIVITY SPECTRUM	TOXICITY	COMMENTS
Nystatin	Binds to sterols in cell membrane	Narrow-spectrum antifungal agent; fungicidal	Cannot be administered IV or IM; relatively nontoxic taken orally or topically	Particularly useful in treatment of *Candida* infections
Amphotericin B	Binds to sterols in cell membrane	Broad-spectrum antifungal agent; fungicidal	Toxic to all tissues; renal toxicity common; thrombophlebitis with IV injections	Methyl ester of amphotericin B is less toxic
Polymyxin B and polymyxin E (colistin)	Competitively displaces magnesium and calcium from phosphate groups on membrane lipids	Gram-negative bacteria; bactericidal	Depresses muscle response to stimulation; impairs renal function	Important only as topical agent; rarely used systemically
Imidazole derivatives				
Miconazole, clotrimazole, econazole, isoconazole	Interferes with biosynthesis of lipids in fungal cell, especially sterols(?)	Primarily broad-spectrum antifungal agent, but also effective against some bacteria	Toxic if given systemically; drug interferes with enzyme activity of hepatic microsomes	Used primarily in topical ointments
Ketoconazole (under clinical investigation)	Same as other imidzole derivatives	Broad-spectrum antifungal agent	Very few side effects when administered orally	Potentially most important antifungal drug; has cured progressive fungal infections that were unresponsive to other drugs.

IV, intravenous; *IM*, intramuscular.

ducing a bacteriostatic effect. At high concentrations of streptomycin all the free 30S subunits can be bound with drug, thus preventing their initiation of protein chain elongation.

Table 27-3 describes the principal characteristics of the aminoglycosides.

Inhibitors of nucleic acid synthesis. The principal inhibitors of nucleic acid synthesis are ***rifampin*** and ***nalidixic acid,*** which are antibacterial agents, and ***griseofulvin,*** which is an antifungal agent (Table 27-4). Rifampin is an inhibitor of DNA-dependent RNA polymerase and

Table 27-3 Characteristics of chemotherapeutic agents that inhibit protein synthesis

AGENT	SPECIFIC MECHANISM OF ACTION	ACTIVITY SPECTRUM	TOXICITY	COMMENTS
Aminoglycosides				
Streptomycin	Binds to 30S ribosomes and causes incorrect codon-anticodon interaction	Bactericidal for gram-positive and gram-negative bacteria	Vestibular and auditory damage	Used with other drugs in tuberculosis
Neomycin	Similar to streptomycin	Same as streptomycin	Auditory and kidney damage	Cannot be used systemically
Kanamycin	Similar to streptomycin	Same as streptomycin	Auditory and kidney damage	Less toxic than neomycin
Gentamicin	Similar to streptomycin	Same as streptomycin, but also active against *Pseudomonas aeruginosa*	Auditory and kidney damage	Related drug called tobramycin is less toxic.
Amikacin	Similar to streptomycin	Same as streptomycin, but with wider range; also effective against *Pseudomonas aeruginosa*	Auditory and kidney damage	Is acetylated derivative of kanamycin
Tetracyclines	Binds to 30S ribosome and prevents binding of aminoacyl tRNA to ribosome	Bacteriostatic against gram-positive and gram-negative bacteria; has widest spectrum of activity of any antimicrobial	Causes discoloration of teeth if administered during first 6 years of life	Prolonged use can cause overgrowth of yeasts
Chloramphenicol	Inhibits peptide bond formation	Bacteriostatic against gram-positive and gram-negative bacteria	Prolonged used leads to effects on bone marrow; can cause aplastic anemia	Cannot be used to prevent infection in premature infants, because it is liver toxic and may cause death
Erythromycin	Prevents aminoacyl translocation reaction by inhibiting release of tRNA from P site	Bacteriostatic against gram-positive bacteria or bactericidal, depending on concentration	Relatively nontoxic	A substitute for those sensitive to penicillin
Lincomycin and clindamycin	Acts on 50S ribosome and inhibits initiation of protein synthesis	Similar to erythromycin	Relatively nontoxic	Can cause pseudomembranous colitis due to destruction of intestinal microflora and elaboration of toxin by clostridia
Spectinomycin	Similar to chloramphenicol	Bacteriostatic against gram-positive bacteria	Relatively nontoxic	Useful in treatment of gonorrhea caused by penicillin-resistant strains.

prevents RNA synthesis in the cell. It is believed that following the formation of the first phosphodiester bond during RNA synthesis, a rifampin-RNA polymerase complex is formed that prevents a further elongation of the RNA molecule.

Griseofulvin is believed to indirectly inhibit the synthesis of the hyphal cell wall of fungi by inhibiting nucleic acid synthesis and mitosis. The drug is effective only on growing cells, and it has been suggested that it interferes with microtubular activity. If this is true, then it might explain the drug's effect on cell wall synthesis, since it has been speculated that microtubules are involved in the transport of components that are used for cell wall synthesis (see Chapter 8).

Many nucleic acid inhibitors are antiviral agents, but they also adversely affect mammalian nucleic acid synthesis and cell replication. Most antiviral drugs are nucleoside analogues and include such agents as ***adenine arabinoside (ara-A), cytosine arabinoside (ara-C),*** and ***5-iodo-2-deoxyuridine (IDU).*** IDU, for example, is incor-

Table 27-4 Characteristics of chemotherapeutic agents that inhibit nucleic acid synthesis

AGENT	SPECIFIC MECHANISM OF ACTION	ACTIVITY SPECTRUM	TOXICITY	COMMENTS
Rifampin	Binds to DNA-dependent RNA polymerase and inhibits RNA synthesis	Bacteriostatic; used only in combination with other drugs in treatment of tuberculosis (bactericidal for *Mycobacterium tuberculosis*)	Relatively nontoxic	Drug-resistant forms emerge readily if used alone
Nalidixic acid	Inhibits DNA synthesis, presumably by affecting DNA gyrase	Gram-negative bacteria	Skin sensitivity to light; gastrointestinal upset	Used primarily for gram-negative urinary tract infections
Griseofulvin	Exact mechanism unknown	Dermatophytic fungi; fungistatic	Toxicity rare	Administered only orally
5-Iodo-2-deoxyuridine (IUDR)	Interferes with incorporation of thymidine into viral DNA	Antiviral agent used only topically in treatment of keratitis caused by herpesvirus	Prolonged use can damage corneal epithelium	
Adenine arabinoside (Ara-A)	Exact mechanism unknown, but does affect DNA polymerase	Antiviral agent is used systemically in treatment of herpes simplex, varicella-zoster, and cytomegalovirus infections	Relatively nontoxic but can cause gastrointestinal upset	
Methisazone	Exact mechanism unknown, but blocks formation of viral structural protein during formation of late proteins	Antiviral agent used only in treatment and prevention of smallpox	Gastrointestinal upset	

porated into DNA, resulting in a defective DNA that inhibits cellular as well as viral replication. Many of the antiviral drugs are therefore used as topical agents and are not used systemically. One of the most effective antiviral agents produced by the host during viral infection is called interferon.

Interferon. Isaacs and Lindemann in 1957 coined the term ***interferon*** as a designation for substances produced during exposure of chorioallantoic membrane to heat-inactivated influenza virus. They found that exposure of untreated chorioallantoic membranes to fluids containing interferon rendered them unable to support viral replication. Although possessing almost unbelievable potency, interferon is produced in incredibly small amounts in the mammalian cell, and purification at this time is difficult and expensive. What has made interferon research so exciting in recent times is that, in addition to its antiviral activity, it also possesses antitumor activity.

Interferon is a glycoprotein of low molecular weight produced in whole animals or in animal cell culture in response to viral infection. Interferon, with reduced activity, can also be produced in response to nonviral agents such as endotoxins, certain bacterial species, and other products. Interferon produced during infection by virus is excreted by the infected cell and interacts with other cells to alter their metabolic and immunological properties. These adjacent cells are not only protected from infection by the original invading virus, but are also protected from infection by unrelated DNA or RNA viruses. The mechanism of action of interferon is not completely understood, but the following sequence of events represents the most currently postulated model (Figure 27-9). The mammalian cell possesses genes that code for interferon production. Interferon, excreted by the originally infected cell, comes in contact with adjacent cells (interferon can also affect to some degree the cell that produced interferon originally) and shuts down viral protein synthesis. The manner in which interferon inhibits protein synthesis has been a subject of speculation for over 20 years, but it is now believed that it induces the synthesis of two enzymes that are initially inactive. The enzymes are activated by a double-stranded RNA molecule

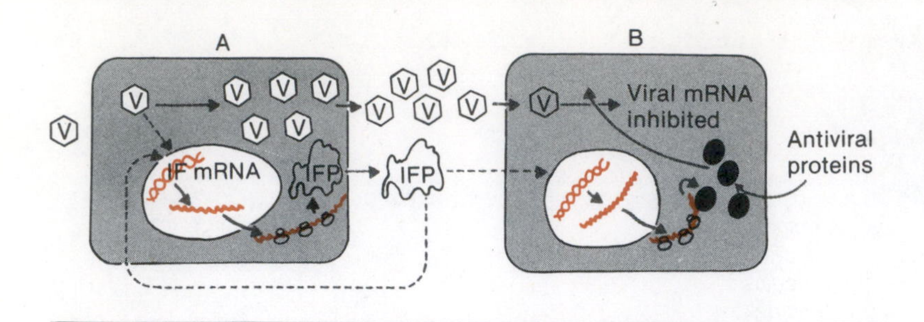

Figure 27-9 Postulated mechanism of interferon production and action following viral infection in mammalian cells. Virus *(V)* infects cell *A*, with subsequent formation of new viral particles, some of which infect cell *B*. Virus also stimulates *(broken arrow)* cellular DNA to produce interferon mRNA, which is produced in nucleus and released into cytoplasm, where it is translated into interferon protein *(IFP)*. IFP is released from cell *A* and induces cell *B* to produce mRNA, which is translated into antiviral proteins such as protein kinase and oligonucleotide synthetase, inhibiting translation of viral mRNA. See text for further details.

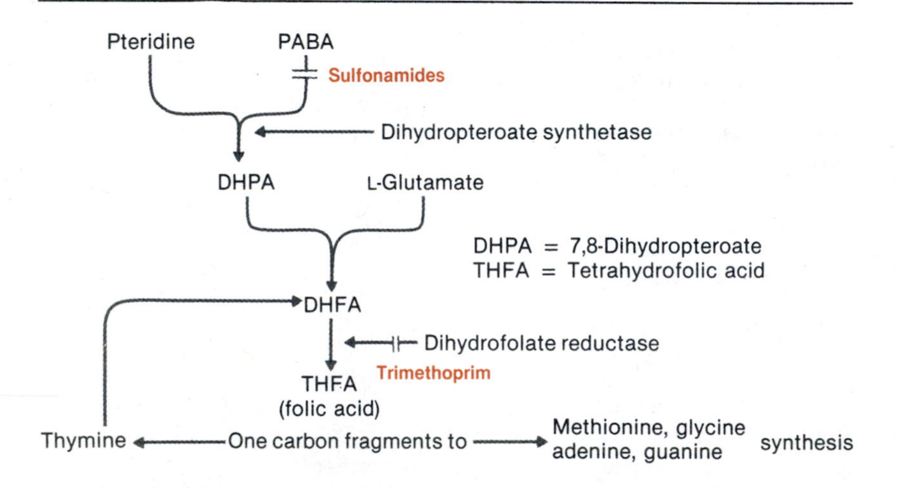

Figure 27-10 Mechanism of action of sulfonamides and trimethoprim. Figure shows reactions that are involved in synthesis of folic acid (tetrahydrofolic acid, THFA) and sites at which sulfonamides and trimethoprim act as inhibitors. Sulfonamides either prevent formation of dihydropteroate (DHPA) or are incorporated in place of paraaminobenzoic acid (PABA). Trimethoprim inhibits enzyme dihydrofolate reductase, which prevents conversion of dihydrofolic acid (DHFA) to folic acid.

produced during viral infection. The two enzymes—***protein kinase (PK)*** and ***oligoadenylate synthetase (OS)***—are the instruments for inhibition of protein synthesis. Protein kinase catalyzes the phosphorylation of a protein that functions as an initiator of protein synthesis. Phosphorylation inactivates this protein; thus protein synthesis is inhibited. Oligoadenylate synthetase catalyzes the synthesis of an unusual nucleotide called 2′5′-oligoadenylic acid, which appears to function as an activator of a ribonuclease that degrades viral mRNA.

The broad antiviral (RNA and DNA viruses) activity induced by interferon is species specific, and protection against virus occurs only in species in which the interferon is produced. Thus protection of humans from viral infection can occur only if the interferon is produced in human cells. One approach to interferon production is to induce it in humans by using a synthetically produced compound. One such inducer is a double-stranded RNA molecule called polyinosinic:polycyticylic acid (poly [I]:poly [C]), which is also very toxic to humans. Most of the interferon that is used clinically is produced in human cell culture, but it can also be produced in bacterial cells via genetic engineering techniques. Three different interferons can be produced from human cells, but only two of them have been purified: one is produced from leukocytes and the other is produced from skin fibroblasts.

Interferon, in addition to its antiviral activity, affects the mammalian cell by alteration of its cell surface, modification of the immune response, and inhibition of cell division. Clinical trials in 1980 with certain human cancers (osteogenic sarcomas and breast cancer) indicated that interferon might eventually play a role in antitumor therapy and prophylaxis. Preliminary results have shown that a high percentage of tumors regress following interferon treatment. Unless some means such as genetic engineering is found to increase the production of interferon and reduce its cost, however, its advantages will be available only to a limited number of patients.

Antimetabolites. Drugs that are structurally similar to cellular metabolites and compete with the natural substrate for incorporation into functionally important components of the cell are called ***antimetabolites.*** The drug ***sulfanilamide,*** for example, is structurally similar to the metabolite paraaminobenzoic acid (PABA). PABA is used by the cell in the synthesis of the vitamin folic acid. Sulfanilamide and other sulfa drugs can prevent the synthesis of folic acid, or they may be incorporated into the folic acid molecule in place of PABA, making the molecule defective and nonfunctional (Figure 27-10). The primary antimetabolites are the sulfa drugs, ***paraaminosalicylic acid (PAS), trimethoprim, ethambutol, nitrofurantoin,*** 5-***fluorocytosine,*** and ***pyrimethamine.*** The characteristics of these agents are outlined in Table 27-5.

Chemotherapeutic agents to prevent disease (prophylaxis)

Chemotherapeutic agents can be used not only to treat disease, but also to prevent disease; that

Table 27-5 Characteristics of chemotherapeutic agents that act as antimetabolites

AGENT	SPECIFIC MECHANISM OF ACTION	ACTIVITY SPECTRUM	TOXICITY	COMMENTS
Sulfonamides	Analogues of paraaminobenzoic acid (PABA); incorporated into folic acid precursors or inhibits incorporation of PABA	Gram-positive and gram-negative bacteria; bacteriostatic; antiprotozoal when used in combination with other drugs	Gastrointestinal; renal obstruction due to crystallization of drug	Many derivatives with variety of useful properties
Paraaminosalicylic acid (PAS)	Same as sulfonamides	Tuberculosis bacillus; bacteriostatic	Gastrointestinal disturbances	Always used in combination with other drugs
Trimethoprim	Analogue of portion of dihydrofolate molecule	Often used in combination with sulfonamides; bacteriostatic for gram-positive and gram-negative bacteria and similar to sulfonamides; also effective against protozoan that causes malaria	Same as sulfonamides	Important in treatment of urinary tract infections
Ethambutol	Unknown	Used only in combination with other agents in treatment of tuberculosis	Relatively nontoxic but can cause optic neuritis	
5-Fluorocytosine (flucytosine)	Analogue of cytosine; active only when converted to 5-fluorocytidine triphosphate	Antifungal agent used for systemic infections in combination with amphotericin B; fungistatic	Relatively nontoxic	When used alone, resistance to drug appears rapidly
Nitrofurantoin	Unknown—may affect carbohydrate metabolism	Gram-positive and gram-negative bacteria; bactericidal	Somewhat toxic, causing gastrointestinal disturbances	Often used to treat urinary tract infections
Pyrimethamine	Similar to trimethroprim	Active against protozoan that causes toxoplasmosis	Depression of bone marrow, causing thrombocytopenia	Can also be used with sulfonamides as alternate therapy for treatment of malaria and other protozoan infections

is, they can be used prophylactically. Prophylaxis, if not used with some discretion, can dead to severe problems because it can create an environment in the host in which sensitive microorganisms are removed and only resistant ones remain. Most chemotherapeutic agents are not so exactingly selective that they inhibit or kill only the infectious agents causing the disease. They also affect other microorganisms in the surrounding microflora that, although not considered primary pathogens, can cause serious and even fatal infections in the debilitated and weak patient. One of the most important aspects of prophylaxis is the time of drug administration. If the drug is used over too long a period of time, drug-resistant microorganisms invariably appear. Some of the instances in which prophylaxis may be useful are the following.

Tuberculosis. Isoniazid is a drug used to prevent tuberculosis and is administered to those individuals who will come in contact with others who have tuberculosis. Isoniazid is a bactericidal agent that inhibits the synthesis of mycolic acids, components of the cell wall of mycobacteria. Isoniazid competes with nicotinamide adenine dinucleotide (NAD) and may also act as an inhibitor of electron transport.

Surgery. Surgical procedures involving the intestinal tract invariably lead to some spillage of intestinal contents that can lead to infections.

These infections can be prevented by using chemotherapeutic agents just before surgery and 24 to 48 hours following surgery. Surgical procedures that are clean and do not involve spillage or exposure to microbial populations usually do not require the use of chemotherapeutic agents.

Heart abnormalities. Patients with heart abnormalities, such as those with rheumatic fever, should receive chemotherapeutic agents before intestinal, genitourinary, or oral surgery. Manipulations in those parts of the body can result in displacement of indigenous species, particularly streptococci from the oral cavity and enterococci from the intestinal tract, into the bloodstream. These microorganisms can colonize already-damaged heart valves and cause bacterial endocarditis (an inflammatory condition involving the endocardium (internal lining) of the heart).

Meningococcal meningitis. Meningitis caused by the bacterium *Neisseria meningitidis* is a highly communicable disease that can spread rapidly in military camps and other confining institutions. Prophylaxis can prevent the spread of the disease when a contact case is first recognized.

Chemotherapeutic agents in agriculture

A storm of controversy exists concerning the health hazard believed to exist from the use of antibiotics, such as penicillin and tetracycline, in animal feeds. Antibiotics have been used not only for therapeutic purposes, but also to prevent disease and to promote growth of the animal. The exact mechanism by which antibiotics promote growth has never been established. The reasons that antibiotics in animal feeds are considered potentially hazardous are twofold:

1. Low levels of antibiotics in animal feeds promote the emergence of resistant bacteria in the animal's intestinal tract, particularly the R factor (resistance factor)–containing group called the Enterobacteriaceae. When raw poultry products are consumed by humans, these resistant microorganisms, such as *Escherichia coli* or species of *Salmonella,* may colonize the human intestine and transfer their R factors to human strains. Any infections due to R plasmid–containing microorganisms are difficult to treat with our present arsenal of chemotherapeutic agents. Some evidence exists that animal diseases that can also appear in humans are much more difficult to treat in the latter when antibiotics have been used in animal feeds.
2. Low levels of antibiotics in feeds may also result in residual levels of the antibiotic in animal tissue. Ingestion of such meat could favor the emergence of resistant microorganisms in the intestinal tract of the meat consumer. In addition, these antibiotic residues could act as allergens, inducing an allergic response in the host. Penicillin, which is degraded to penicilloic acid during cooking, can conjugate with proteins, producing a sensitizing antigen that can induce an allergic response in the host. Some cases of allergic responses to milk contaminated by penicillin residues have been reported, particularly from foreign countries.

Proponents to the use of antibiotics in feed contend that most animal bacteria are poor colonizers of humans and that the exchange of R factor plasmids between animal and human strains of the Enterobacteriaceae is highly unlikely under normal conditions. Studies to determine the ability of microbial species found in animals to colonize humans have resulted in conflicting results, and the matter has yet to be settled.

Microbial resistance to drugs

The ability of microorganisms to develop resistance to drugs is one of the major reasons why new drugs are constantly being sought and evaluated. Resistance to chemotherapeutic agents may be the result of natural resistance, mutation, or plasmid-associated resistance.

Natural resistance. Natural resistance to a drug is characterized as a resistance that occurs among a population of microorganisms that are normally sensitive to the drug. Natural resistance can be exemplified in the following ways. Some microorganisms, such as bacteria, are inhibited by certain chemotherapeutic agents only during active growth and cell division. Penicillin, for example, which inhibits cell wall synthesis, is bactericidal only when the cell is actively synthesizing cell wall.

Drugs that inhibit metabolic processes within the cell must first penetrate the cell wall and cytoplasmic membrane. The envelope of some microbial cells is impermeable to certain drugs, and this impermeability may be the result of a structural component or a biochemical property. Some bacteria are resistant to the drug tetracycline, which is an inhibitor protein synthesis. The resistance to tetracycline is not always due to alterations in the protein-synthesizing machinery of the microbial cell, but to changes in the cell envelope. This can be proved by removing the cell envelope of the resistant cells and testing the effects of tetracycline on protein synthesis in cell extracts. Tetracycline inhibits protein synthesis of the cell extract, which shows that resistance of the whole cells is due to a permeability factor. It is assumed that other microbial species to which the resistant species is related are normally sensitive to the drug.

Mutation. Mutations in nature that result in microbial resistance to drugs are usually of the spontaneous type, but in the clinical setting they are of another type. Drug resistance in the clinical setting may be the result of a single mutational event or may result from a series of separate mutations. Single-step resistance implies that in a population of cells resistance to a specific concentration of drug has increased significantly with the mutation. For example, the concentration of drug required to inhibit a microbial population may be 100 times the normal concentration. Each mutation in multistep resistance results in a slight increase in resistance. Significant resistance to penicillin, for example, may occur only after four of five single-step mutations in the microorganism. Mutational resistance occurs more frequently in the clinical setting when there is prolonged therapy and a single drug is used. Treatment of tuberculosis in earlier times was confined to the use of one drug, and this invariably led to drug-resistant mutants. Today several drugs are used in combination to treat chronic diseases such as tuberculosis, because if resistance to one drug occurs, the second and third drugs will destroy any resistant mutants.

Plasmid-associated resistance. Clinically, plasmid-associated drug resistance is the most important mechanism or microbial resistance. This type of resistance is usually due to the formation of drug-inactivating or drug-modifying enzymes, but other mechanisms are also involved. As noted in Chapter 11, antibiotic-resistance genes, particularly those existing as transposons, are easily transferred to other cells during conjugation. They can also be transferred intracellularly between the different autonomous genetic units that are present in the cell (plasmid, chromosome, viral genome). In some instances a single plasmid may possess several drug-resistant genes, or the cell may contain several plasmids, with each possessing one or more antibiotic-resistant genes. Drug-resistant plasmids are usually maintained in the microbial population only as long as the drug is present in the patient's tissue. Once the patient is no longer using the drug, the resistant strains are replaced by drug-sensitive ones. Drug-resistant strains are often acquired by patients in the hospital, where they abound and where they can be transferred from nurse to patient, patient to patient, physician to patient, etc. Drug-resistant strains may also be acquired from domestic animals because their feed contains chemotherapeutic agents (see "Chemotherapeutic Agents in Agriculture").

Mechanisms of resistance (Table 27-6)

Altered transport. The ineffectiveness of certain antimicrobials (those that affect certain cytoplasmic properties of the cell) is usually due to alterations within the cell envelope. Resistance may result from changes in the lipid content of the cell wall, alteration in the porin size of the cell wall of gram-negative bacteria, changes in the cell wall configuration, alteration of a specific transport mechanism, or enzymatic modification of drugs before their entry into the cytoplasm. Many of the enterobacteria carry plasmids that apparently code for cell envelope proteins that prevent penetration of the tetracycline antibiotics. The enterobacteria and pseudomonads also carry plasmids that code for aminoglycoside (streptomycin, gentamicin, kanamycin, etc.)–modifying enzymes, which are located in the periplasmic space. They include acetylating enzymes, which catalyze the transfer of acetate from acetyl CoA to the amino group in the antibiotic; nucleotidylating enzymes, which use ATP or other nucleotides as substrates in the modification of hydroxyl and amino groups on the antibiotic; and phosphorylating enzymes, which add phosphate to a hydroxyl group on the antibiotic. Acetylating enzymes such as acetyltransferase also prevent the transport of chloramphenical across the cytoplasmic membrane.

Enzymatic inactivation of the drug. The most ubiquitous drug-inactivating enzymes are the β-lactamases (Figure 27-11). There are nearly as many different β-lactamases as there are β-lactam antibiotics, which explains the relentless search for new penicillins. β-lactamases are found in some gram-positive bacteria and in nearly all gram-negative bacteria. They may be encoded for by genes located in the chromosome, or they may be found in plasmids, but sometimes they are found in both types of replicons in the same cell. Gram-positive bacteria produce the enzyme extracellularly, whereas gram-negative bacteria retain the enzyme in the periplasmic

Table 27-6 Major mechanisms of microbial drug resistance

MECHANISM	DRUG(S)	ORGANISMS INVOLVED
Altered transport system		
Reduced uptake of drug	Tetracyclines	Gram-negative enterobacteria
Membrane not energized	Aminoglycosides	Anaerobes
Enzymatic modification	Aminoglycosides, chloramphenicol	Gram-negative enterobacteria, pseudomonads
Enzymatic inactivation		
β-lactamases	Penicillins and cephalosporins	Gram-positive and gram-negative bacteria
Chloramphenicol acetyltransferase	Chloramphenicol	Gram-positive and gram-negative bacteria
Alteration of antimicrobial targets		
DNA gyrase	Nalidixic acid	Gram-negative enterobacteria
RNA polymerase	Rifampin	Gram-negative enterobacteria
Penicillin-binding proteins	Penicillin	*Neisseria gonorrhoeae, Streptococcus pneumoniae, Staphylococcus aureus, Escherichia coli*
Methylated 23S RNA	Erythromycin and lincomycin	Staphylococci
30S ribosome	Streptomycin	Gram-negative enterobacteria
Synthesis of resistant metabolic pathways		
Dihydrofolate reductase	Trimethoprim	Gram-negative enterobacteria
Dihydropteroate synthetase	Sulfonamides	Gram-negative enterobacteria, staphylococci

space. The β-lactamases in gram-negative bacteria belong to one of two types: TEM-1 or TEM-2. The β-lactamases in these two groups are associated with transposons, and this may account for their widespread occurrence among the gram-negative bacteria.

A cytoplasmic-containing, drug-inactivating enzyme is chloramphenicol acetyltransferase, which catalyzes the following reaction:

Chloramphenicol + Acetyl-S-CoA ⟶ 3-Acetoxychloramphenicol + HS-CoA

The enzyme is encoded in a plasmid that is carried by both gram-positive and gram-negative bacteria.

Alteration of the target site. Several components involved in replication and protein synthesis are the targets of some antimicrobials, but microorganisms have also evolved mechanisms for resisting them. One type of alteration is plasmid associated and is found among the staphylococci. These organisms possess an enzyme that methylates the two adenine residues on the 23S RNA molecules of the 50S subunit of bacterial ribosomes. This type of alteration prevents the binding of erythromycin and lincomycin to the 50S subunit and prevents protein synthesis. Interestingly enough, the organism from which erythromycin is produced, *Streptomyces erythreus,* contains a dimethylated 23S RNA that protects it from the antibiotic.

Alteration of the 30S ribosome of the gram-negative enterobacteria makes these organisms resistant to streptomycin. A protein in the 30S subunit is the target of the antibiotic, and a single amino acid change in either of two positions on the protein prevents binding of the drug. Enzyme inactivation and not ribosomal alteration, however, is the most frequently encountered type of streptomycin resistance in clinical medicine.

An important mechanism of resistance is alteration of the penicillin-binding proteins in species such as *Neisseria gonorrhoeae. Streptococcus pneumoniae* (South Africa), *Streptococcus faecalis,* and *Staphylococcus aureus.* PBP1, PBP2, and PBP3 appear to be most important in resistance to penicillin in gram-negative bacteria, but this mechanism of alteration is not fully understood. Penicillin may induce a change in the PBPs that interfere with normal cell surface activities, such as cell division and septation. Thus the microbial cells become tolerant to the presence of the antibiotic, resulting in bacteriostasis rather than bactericidal events.

Synthesis of resistant metabolic pathways. Some plasmids found in the enterobacteria and staphylococci code for metabolic enzymes that substitute for the chromosome-mediated enzymes. The site of inhibition or sulfonamides, for example, is the enzyme dihydropteroate synthetase, which is required for dihydrofolate synthesis. The antimicrobial competes with PABA, the normal metabolite of the enzyme (see Figure 27-10), but sulfonamide-resistant bacteria possess plasmids that code for a dihydropteroate synthetase that is resistant to the drug.

The enzyme dihydrofolate reductase, which is also involved in folate synthesis, may also be coded for by a plasmid. The plasmid-associated enzyme is resistant to trimethoprim, which normally inhibits the chromosome-mediated enzyme.

Increased production of a drug-competing metabolite. An infrequent mechanism of resistance involves competition between sulfa drugs and PABA for the active site of the enzyme. Increased resistance to the sulfonamide may occur when a mutation causes an increase in the metabolite (in the case of sulfa drug resistance, PABA). This is a quantitative resistance, because the concentration of drug required to inhibit the microorganism may have to be elevated considerably. In fact, the concentration of drug may be so high that it becomes toxic to human tissue, and a substitute drug must be selected for therapy.

How new chemotherapeutic drugs are developed and evaluated

New chemotherapeutic drugs may be synthesized in the laboratory using the nucleus of the existing antimicrobial, or they may be obtained by screening the soil for antibiotic-producing species. Any new antimicrobial must be subjected to a battery of tests and clinical studies before it ever reaches the point of recommendation for human use. The selection of a drug as a chemotherapeutic agent is based initially on the following properties: (1) it should exert an inhibitory effect at low concentrations, that is, microgram quantities; (2) it should be nontoxic to the host (a relative quantity); and (3) it should be produced economically. The following discussion summarizes the steps that are used in evaluating new antimicrobials.

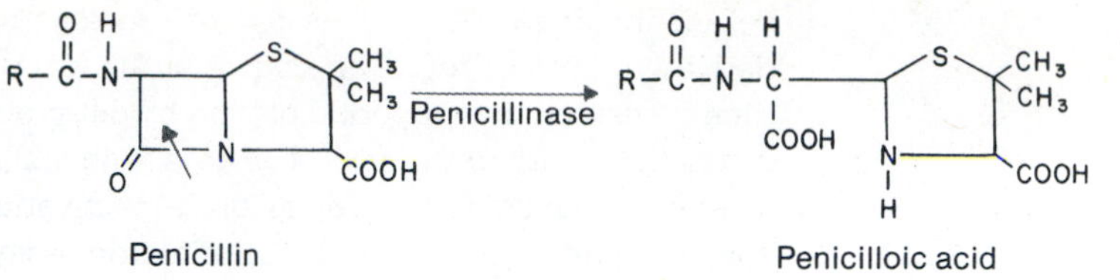

R–C(=O)–N(H)–[β-lactam ring]–S–C(CH_3)(CH_3)–COOH → Penicillinase → R–C(=O)–N(H)–C(H)(COOH)–[thiazolidine ring: S, CH_3, CH_3, COOH, N–H]

Penicillin — Penicilloic acid

Figure 27-11 Effect of penicillinase, a β-lactamase, on penicillin. Enzyme cleaves β-lactam ring *(arrow)* of penicillin molecule, converting it to penicilloic acid.

Screening of the antimicrobial in vitro. Whenever possible, the structure of the antimicrobial should be determined, because it will be helpful in evaluating the mode of action of the drug. The drug, for example, may be structurally similar to a cellular metabolite, and this would provide a clue to its inhibitory effects. For those antimicrobials in which the structure is not known, a primary screening is performed, using a specified group of aerobic and anaerobic bacteria, as well as fungi; that are sensitive to some of the major chemotherapeutic drugs. A secondary screening can be performed using pathogens that are known to be resistant to at least one of the commonly used antimicrobials. If these tests indicate that the antimicrobial may be useful, it is tested against up to 100 different pathogens commonly encountered in medical practice. The in vitro tests used to screen the antimicrobial are the agar dilution, broth dilution, and disk assays, which are discussed later in the chapter.

The antimicrobial, in addition to its activity against microorganisms, must also be evaluated for its biological and chemical stability. The drug is diluted and stored at temperatures of 4° C, 25° C, and 37° C in order to determine its biological activity. Ultraviolet (UV) or infrared spectrophotometry is also used on the drug before and after incubation in various solvents and at different temperatures to determine chemical stability.

Determination of bacterial resistance to the new drug. Many antimicrobials have been isolated over the years, but one of their major drawbacks has been their propensity for inducing the development of resistant strains during therapy. The two most important mechanisms of resistance are related to drug impermeability and the presence in the microbial cell of drug-inactivating enzymes, both of which can be tested. To determine drug permeability, a population of microbial cells is incubated in the presence of a radioactive derivative of the drug. The uptake of the drug is then measured over a period of time by determining radioactivity in the microbial cells. The ability of microorganisms to produce drug-inactivating enzymes can be measured by colorimetric assays that measure a product of the inactivation process. Some of the most important drug-inactivating enzymes are the β-lactamases that attack the β-lactam ring of certain penicillins and cephalosporins. One cannot always extrapolate in vitro drug inactivation with in vivo activity, because some microorganisms that produce β-lactamases are still sensitive to penicillin in vivo.

In vivo tests for determining drug toxicity and activity. The results of in vitro tests cannot always be extrapolated to the clinical situation, and for this reason certain animals are selected for testing the effectiveness of antimicrobials. In vivo tests measure the susceptibility of the infectious agent to the antimicrobial and also measure such variables as drug inactivation by host metabolic processes and the ability of the drug to reach the site of infection.

For in vivo screening techniques the Swiss albino mouse is used because of the excellent correlation that has been obtained over the years between clinical response of humans and activity in the mouse. Larger animals such as chimpanzees are used for studying the effectiveness of drugs against certain venereal diseases, and rabbits are used for studies involving acute infections of heart valves (endocarditis). Toxicity is the first drug characteristic measured in the mouse. The drug, at various concentrations, is injected via the oral, subcutaneous, or intraperitoneal route into healthy mice. The mice are observed for a period of days to determine the 100% lethal dose (LD_{100}), the 50% lethal dose (LD_{50}), and the maximum tolerated dose (LD_0). The LD_0 is the maximum dose of drug at which all animals survive. Next, a number of mice are injected with various dilutions of the virulent pathogens, and the minimum lethal dose (MLD) is determined. The MLD is the lowest dilution of the organisms at which all animals die. In the actual test, the MLD of microorgnisms is increased 100 to 1000 times to ensure that the measured drug activity is not borderline. Mucin is added to the inoculum of certain pathogens to enhance their virulence (mucin increases the virulence of some microorganisms). Once the mice have been infected, they are treated with the antimicrobials at the time of infection and at postinfection times of 5, 24, 29, 48, and 72 hours. The animals are observed for over 2 weeks, and the number of survivors is recorded. The 50% protective dose (PD_{50}), which is the dose at which 50% of the animals survive, is calculated and expressed in milligrams of antimicrobial per kilogram of body weight. These values are then extrapolated to the human clinical condition.

Clinical trials. Biochemists and microbiologists are called on to scale up the production of the drug after the preliminary studies of in vitro and in vivo activities have been performed. Antibiotics from soil microorgansims are normally produced in minimal amounts, but by using mutants and/or genetic recombinants, the antibiotic-producing microorganisms can be made to produce greater quantities of the antimicrobial. The microbial variants are cultivated on various test media containing different respirable or fermentable substrates. They are then subjected to environmental variables such as temperature, oxygen tension, and pH to find those conditions that will favor the production of maximum amounts of the antimicrobial.

The drug is submitted for use in clinical trials in human volunteers with and without infections. These studies determine the drug's effectiveness on the microorganisms as well as its safety in human subjects. Long-term toxicology studies are also carried out in animal models during these clinical trials, and all of the data from animal and human volunteer studies are finally compiled and submitted to federal drug agencies for scrutiny. The entire process, from discovery of a new drug to its application for use, takes anywhere from 5 to 10 years.

Measurement of antimicrobial activity. The activity of the drug can be measured quantitatively and qualitatively, and in each type of test certain standards must be maintained so that appropriate evaluations can be made. The tests are used not only to evaluate recently discovered drugs, but they are also used in the clinical laboratory to determine what chemotherapeutic agents will be useful in the treatment of specific infectious diseases.

Quantitative in vitro susceptibility tests. Quantitative in vitro susceptibility tests are used to determine more exactly the concentration of drug that will either inhibit the growth of microorganisms (the ***minimal inhibitory concentration,*** or ***MIC***) or will kill microorganisms (the ***minimal bactericidal concentration,*** or ***MBC***). Quantitative tests are not routinely used in the clinical laboratory but are more appropriate for the experimental laboratory. They have more

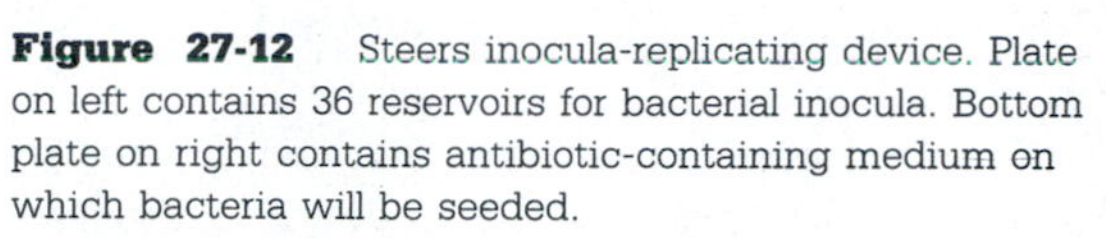

Figure 27-12 Steers inocula-replicating device. Plate on left contains 36 reservoirs for bacterial inocula. Bottom plate on right contains antibiotic-containing medium on which bacteria will be seeded.
From Washington, J.A. II. Mayo Clin. Proc. ***44****:811, 1969.*

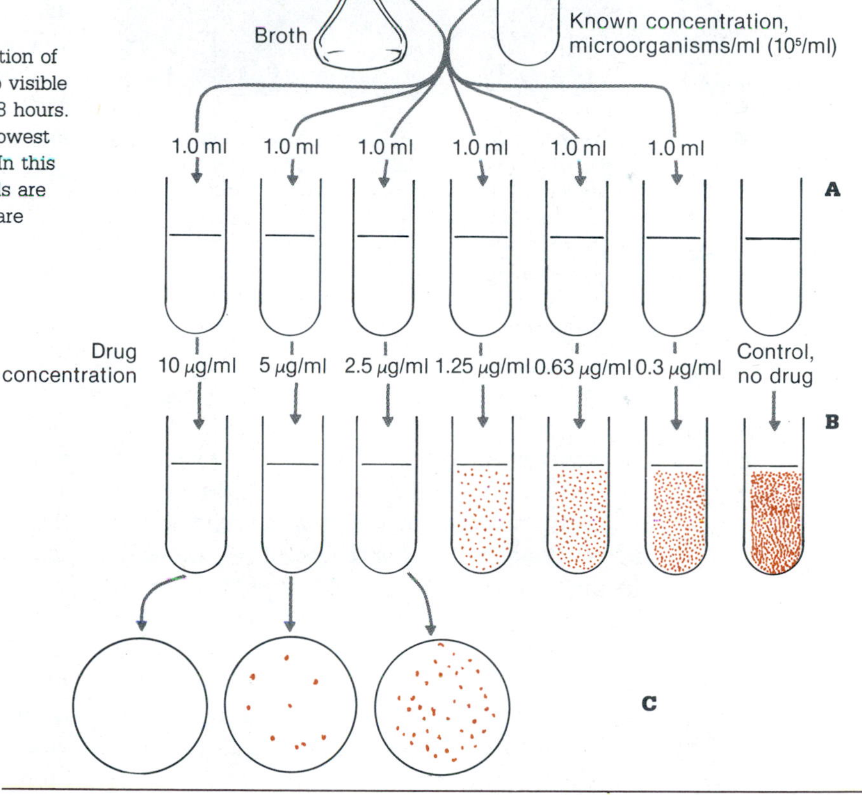

Figure 27-13 Broth dilution test used to determine minimal inhibitory concentration (MIC) and minimal bactericidal concentration (MBC) of antimicrobial compound. **A,** Each broth tube contains 9 ml and different concentration of antimicrobial. Concentration of organisms in each tube is so small that there is no visible turbidity. Tubes are incubated at 35° C for 12 to 18 hours. **B,** Tubes are examined for macroscopic growth. Lowest concentration of drug showing no growth is MIC. In this case it is 2.5 μg/ml. **C,** To determine MBC, loopfuls are removed from tubes in **B** showing no growth and are streaked on agar plates. Plate showing no growth indicates drug concentration that destroys microorganisms.

application in the pharmaceutical laboratory, where exact concentrations of drug must be known in order to establish levels of drug for future animal and human studies. Two in vitro tests are available for determining the MIC of a drug: the agar dilution test and the broth dilution test. In the ***agar dilution technique*** the antimicrobial is incorporated into a liquified agar medium (45° to 50° C), which is then mixed and poured into an agar plate. A series of plates are set up so that each will have a different concentration of drug. An inoculating device (Figure 27-12) is used that can seed up to 36 different microbial species on the surface of the agar. The microorganisms selected are commonly rapid-growing pathogens, such as members of the Enterobacteriaceae, *Staphylococcus aureus,* enterococci, *Pseudomonas* species, and other genera with similar growth rates. Slow growers give spurious results that cannot be extrapolated to the in

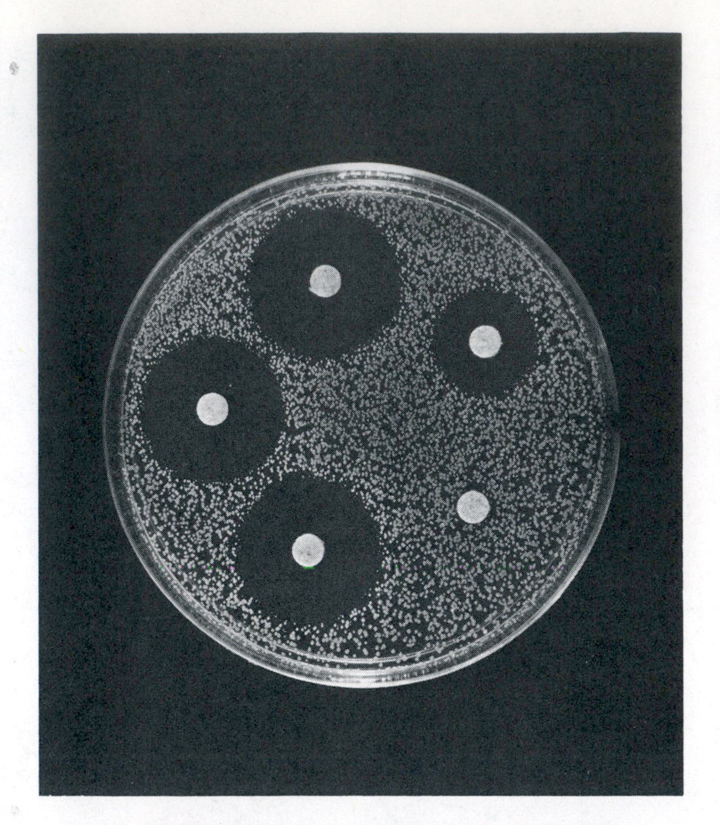

Figure 27-14 Disk diffusion test. Agar medium is inoculated with microbial species, followed by placing of antibiotic-containing disks on surface of medium. Plates are incubated for 18 to 24 hours and then examined for zones of inhibition. Plate demonstrates that zones of varying diameter surround disks except for disk in lower right. Seeded microorganisms are totally resistant to antimicrobial present in disk at lower right.

vivo situation. The seeded plates are incubated overnight at 35° to 37° C, and the MIC is read as the lowest concentration of drug that completely inhibits the growth of the seeded microorganism. The MIC and MBC can be calculated in the broth dilution tests, and the details of the design of this test are outlined in Figure 27-13. MICs obtained from these assays can be used in the clinical laboratory to establish the dosage that will be effective in treatment of human infections. (see below).

Qualitative antimicrobial sensitivity tests (disk diffusion). The disk diffusion test developed by Bauer, Kirby, Sherris, and Turck is an agar diffusion technique recommended by the Food and Drug Administration and is used widely in the hospital clinical laboratory for determining antimicrobial activity. The basic principle of the disk diffusion test is that a standard concentration of microorganisms is seeded on an agar plate, on which is placed a disk containing an antimicrobial of known potency. Those microorganisms inhibited by the antimicrobial, following suitable incubation, will show a zone of inhibition the diameter of which is proportional to the susceptibility of the microorganism. This proportionality is based on the standardization of certain variables: the agar, the type of disk, the antimicrobial, and the microorganisms tested (Figure 27-14).

The agar medium employed in the disk diffusion tests must possess the following characteristics: it must support the growth of a wide variety of microorganisms; it must not contain certain agents, such as trace metals, that would inactivate the antimicrobial; it must be buffered to prevent pH changes during growth, which would influence antimicrobial activity; and it must give reproducible results. Mueller-Hinton agar, which meets these requirements, is composed of beef infusion, casein, and cornstarch. The depth of the agar in the plastic culture plate must be 4 mm, and any variation in depth will influence the zone of inhibition because of changes in the rate of drug diffusion.

Filter paper disks of specified weight and size and absorbability contain a specific amount of antimicrobial. Once the disk is placed on the agar, the antimicrobial diffuses through the agar, producing a gradient of concentrations, with the highest concentration of drug nearest the disk. The rate of diffusion is related to the concentration and molecular size of the drug, the temperature of incubation, and the physical and chemical characteristics of the agar.

The antimicrobial selected for testing the susceptibility of a microorganism is usually a representative of a family of antimicrobials that have similar modes of action and toxicity. The selection of one antimicrobial is also based on the fact that there is cross-resistance among antimicrobials of the same family; for example, a microorganism resistant to one of the cephalosporins will usually be resistant to all the other cephalosporins. Of course, there are exceptions, since some semisynthetic antimicrobials are created in the laboratory expressly because they do not show cross-resistance with related drugs. Generally speaking, the type of antimicrobial selected for testing is dictated by the site of infection (urinary tract, spinal fluid, etc.), since certain species of microorganisms are frequently associated with such sites of infection in the human body. The potency of the antimicrobial in the disk is selected on the basis of previous determinations of MIC concentrations with over 100 strains of bacteria (recall that the agar dilution and broth dilution techniques were used to obtain these MICs) and by concentrations of the antimicrobial that

Table 27-7 Zone-size interpretive chart: a modified Kirby-Bauer table

CHEMOTHERAPEUTIC AGENT	DISK POTENCY (μg)	INHIBITION ZONE DIAMETER (MM)		
		RESISTANT	INTERMEDIATE*	SENSITIVE
Ampicillin (gram-negative bacteria)	10	11 or less	12 to 13	14 or more
Ampicillin (staphylococci and penicillin G–susceptible organisms)	10	20 or less	21 to 28	29 or more
Chloramphenicol	30	12 or less	13 to 17	18 or more
Clindamycin	2	14 or less	15 to 16	17 or more
Nitrofurantoin (urinary tract infections only)	300	14 or less	15 to 16	17 or more
Gentamicin	10	12 or less	13 to 14	15 or more
Tetracycline	30	14 or less	15 to 18	19 or more

*When the zone size falls in the intermediate range, the organism is said to be neither sensitive nor resistant to the drug. The test can be repeated to ensure that all standard conditions have been maintained. More often, another antimicrobial is tested—one that will be more effective against the microorganisms; that is, the zone diameter will be in the sensitive range.

are achievable in humans without causing severe toxicity.

If one were testing the activity of a new antimicrobial in the research laboratory, the microorganisms selected for determining susceptibility would be the rapid-growing pathogens mentioned previously (see "Quantitative in Vitro Susceptibility Tests"). Slow-growing microorganisms would always yield a large-zone diameter because of the disparity between growth rate and drug diffusion, and interpretation would be almost impossible. In the clinical setting the microbe is isolated from the patient and is tested against specific concentrations of the antimicrobial. The concentration of microorganisms used is approximately 3×10^8 cells/ml, which is obtained from an actively growing culture. A swab is used to spread an inoculum across the surface of the plate, and within 15 minutes after the inoculation the antimicrobial disks are placed on the surface of the plate so that they are at least 15 mm from the edge of the plate and at least 15 to 20 mm from each other. This procedure prevents overlapping of zones produced after incubation of the culture plates.

Interpretation of the results of disk diffusion. Inoculated plates used to demonstrate microbial susceptibility are incubated at 35° to 37° C for 18 hours, and any zones of inhibition around the disks are measured with a micrometer (Figure 27-14). The zone diameters are interpreted as ***resistant, intermediate,*** or ***susceptible.*** Such an interpretation is based on the assumption that a relationship between zone diameter and MIC has already been established. The relationship that they bear is an inverse one. The outer limit of the zone diameter indicating that the microorganism is susceptible takes into account results obtained with other strains as well as the level of antimicrobial achievable in vivo. The outer limit of the zone contains a concentration of antimicrobial that is similar to the MIC of that antimicrobial to a specific microorganism. Actually, there is no way to predict in vivo success based on in vitro results, and for this reason interpretive charts, which represent a compilation of clinical studies, have been established. These charts, which have been referred to as Kirby-Bauer tables, present a selection of zone diameters for resistant, intermediate, and susceptible responses of microorganisms to a specific concentration of antimicrobial. An example of such a chart is presented in Table 27-7. New drugs require publication of updated charts at frequent intervals. The selection of cutoff points for these three categories is finally based on years of clinical studies. As long as certain standards such as agar medium, inoculum size, and disk potency have been adhered to, the clinician can evaluate the potenital effectiveness of a drug in the treatment of infections by comparing the zone diameter of his patient's pathogen with the Kirby-Bauer tables.

Commercial testing systems. Commercial antimicrobial susceptibility testing systems are also available for clinical laboratories. There are basically two types of systems. One type uses a broth dilution technique for determining the MIC and MBC. The broth dilution technique, which gives a more definitive determination of MIC and MBC than the agar dilution procedure, is ordinarily very time consuming. A microdilution procedure has been developed commercially in which microtrays are filled with antibiotics added by the manufacturer. Some clinical laboratories use this technique instead of the disk diffusion technique. A second type of tests uses growth curve kinetics by measuring the turbidity of antibiotic-treated and untreated cell suspensions. These measurements are evaluated by a built-in computer.

STERILIZATION AND DISINFECTION

Under certain circumstances microbiologists endeavor to find ways to enhance microbial growth; for example, in industrial microbiology enhanced growth means an increase in a usable fermentation product, an enzyme, or an antibiotic. At other times it is important to inhibit microbial growth, such as during infectious disease, where the symptoms of infection are often directly related to the number of infectious microorganisms present in the host. Control of microbial growth is also important in the laboratory if one is performing experiments in which the microbiological aspects of growth, metabolism, genetics, or morphology are being examined and large populations of microorganisms are used. It is necessary that the microbial population being examined is composed of only one species type (an ***axenic*** or pure culture) and that unwanted microorganisms are prevented from contaminating it. Sterilization and disinfection, which are processes used to eliminate or control microorganisms, have applications, therefore, in industry and in the health sciences, as well as in research or teaching laboratories. The purpose of the remainder of the chapter is to describe the various chemical and physical techniques that can be applied to control microorganisms on the surface of the body or in the environment outside the body.

Terminology

Procedures used to control microbial growth do not always elicit the same results. Some agents destroy all forms of microbes, whereas other agents only inhibit their growth (and as soon as the chemical agent has been sufficiently diluted, the microorganisms resume growth). It is therefore important for the research microbiologist, nurse, physician, laboratory technician, and student laboratory worker to apply the appropriate procedures and to understand the terminology that characterizes these procedures.

Sterility is a term used to denote that the object or solution being treated is devoid of any viable form of life. A less precise term than sterility is ***disinfection,*** which implies that the object being treated is rendered devoid of infectious microorganisms. For example, an object may be contaminated by several different microbial species: some pathogenic and others nonpathogenic. A disinfectant would at the very least destroy any vegetative forms that could cause disease but not necessarily the nonvirulent types. The term ***antiseptic*** refers to the application of a nonchemotherapeutic agent to the surface of tissue that minimally would inhibit the growth of potentially pathogenic species. Antiseptics must be of such a nature that they do not harm tissue. ***Sepsis*** refers to the breakdown of living tissue due to the action of microorganisms, and any procedure that prevents this condition is called ***aseptic technique.*** Aseptic techniques imply that precautions have been taken to prevent the introduction of microorganisms into the body by eliminating or reducing the number of microorganisms from the environment as well as objects that come in contact with tissue. Germ-free animals (***gnotobiotic*** animals), for example, represent a sterile system that can be broken by introducing microorganisms.

Physical methods

Heat

Moist heat. Water is important in biochemical reactions because it can cause the hydrolysis of macromolecules such as proteins, nucleic acids, and lipids. Moisture at elevated temperatures causes the coagulation of macromolecules, but the lethal effects of heat are probably due first to the disintegration of the cytoplasmic membrane and inactivation of enzymes. Some microbial forms, such as the bacterial spore, are essentially devoid of free water, and this property, as well as the thick refractile outer coat of the spore, provides the cell with relative thermoresistance. Bacterial spores are the most heat resistant form of microbial life, and any procedure that destroys spores will also destroy other microbial life.

Moist heat at temperatures of 60° to 80° C and applied for 30 minutes is sufficient to kill vegetative bacteria, viruses, and fungi, but not bacterial spores. Boiling (100° C) can destroy bacterial spores, but the time required depends on the bacterial species and may range from 30 minutes to 5 to 6 hours. Boiling kills most vegetative forms of life in 3 to 5 minutes, but because of its corrosive effects it is not recommended for sterilization of metallic instruments. ***Pasteurization,*** which uses moist heat, is employed on palatable liquids such as milk, beer, and wine. Pasteurization destroys vegetative pathogens that may be derived from the soil, cattle, or humans but does not affect many of the microorganisms that spoil milk. There are two techniques for pasteurization of milk: in one a temperature of 62.9° C is applied for 30 minutes to large vats of material, whereas in the other thin films of milk pass over pipes heated to 71.5° C for 15 seconds (see Chapter 20).

The most preferred method of sterilization is steam under pressure, which is called ***autoclaving.*** The autoclave is a chambered device in which saturated steam is generated and placed under pressure. The temperature and pressure are adjustable, but the most satisfactory conditions for sterilization are 121° C and 15 psi applied for 15 minutes. These conditions are known to destroy any microbial life (provided that moisture touches every surface to be sterilized).

Dry heat. Dry heat does not inactivate biological molecules as quickly as moist heat. It is a process that dehydrates the cell, causing solute precipitation and oxidation of macromolecules rather than their coagulation. Dry heat is used in the range of 160° to 180° C for 1 to 2 hours and is appropriate wherever moist heat cannot be applied, such as the sterilization of certain glassware, powders, and oils.

Cold. Extremes of cold can destroy microorganisms, but it is not a practical technique for sterilization. The effect of cold is dictated by the extremes of temperature attained. The metabolism and growth of most microorganisms is reduced at refrigeration temperatures (2° to 6° C). This principal is effectively employed in the preservation of foods, biological solutions, and microorganisms. Mesophilic bacteria and fungi reproduce slowly at refrigeration temperatures, and some psychrophiles can divide every 1 to 2 hours. This type of response can be readily observed if food is left in the refrigerator for a few weeks. Microorganisms can be preserved at refrigeration temperatures by cultivating them on agar slants or slants covered with mineral oil. Microbial metabolism occurs rapidly enough on slants without mineral oil that the products of metabolism destroy many viable cells within a few days. The presence of mineral oil over the slant, however, not only reduces microbial metabolism, but also prevents the culture from drying out, and the culture can be preserved for several months.

Freezing at 0° C may result in destruction of vegetative microbial cells due to the formation of large ice crystals, which on warming enlarge and exert enough pressure on cell membranes to rupture them. There is an increase in solute concentration in the cell via dehydration, and it may be that solute precipitation (salting out of proteins) in some way also injures the cell during the freezing process. Many microbial cells can be "quick frozen" at temperatures of −50° to −90° C, even though many cells are killed. This technique is used commonly in the microbiological laboratory to preserve isolated species for long periods of time (months to years). Another technique, which employs liquid nitrogen (−76° C or lower) to freeze and maintain cells at temperatures near absolute zero, has given rise to the science called ***cryobiology.*** Cryobiology has its greatest application in higher animals, including humans, where the preservation of sperm, embryos, and various organs for future implantation is of genetic and economic significance. Microbial cultures can be maintained indefinitely, with no loss in the population, by using liquid nitrogen, but ***lyophilization*** (freeze-drying) is a more practical way of preserving microorganisms for storage as well as for shipment to other laboratories anywhere in the world. The basic principle of lyophilization is that the culture is dried in a glass vial while in the frozen state by removing the water through a process called ***sublimation;*** that is, water is removed from the frozen state as a vapor by using a high-vacuum system in which a desiccant absorbs water. The glass vial is sealed under a vacuum and stored. Bacteria can be removed from the vial by removing the tip and adding sterile distilled water to produce a suspension. An aliquot of the suspension can be removed and transferred to a suitable growth medium, and a percentage of the organisms that have survived will replicate.

Filtration. Filtration is a common technique for separating microorganisms of different sizes, but it is also useful for sterilization where other techniques are not appropriate, for example, solutions that would be harmed by heat (toxins, vaccines, and enzymes). Today most microbial filtration processes are performed with membrane filters that are composed of cellulose esters or other polymers. They can be obtained in different pore sizes, ranging from those that retain the smallest viruses to those that retain the largest mammalian cells. The solution is saved from the filtration process, and the filter retaining the contaminating microorganisms is discarded. Filters can also

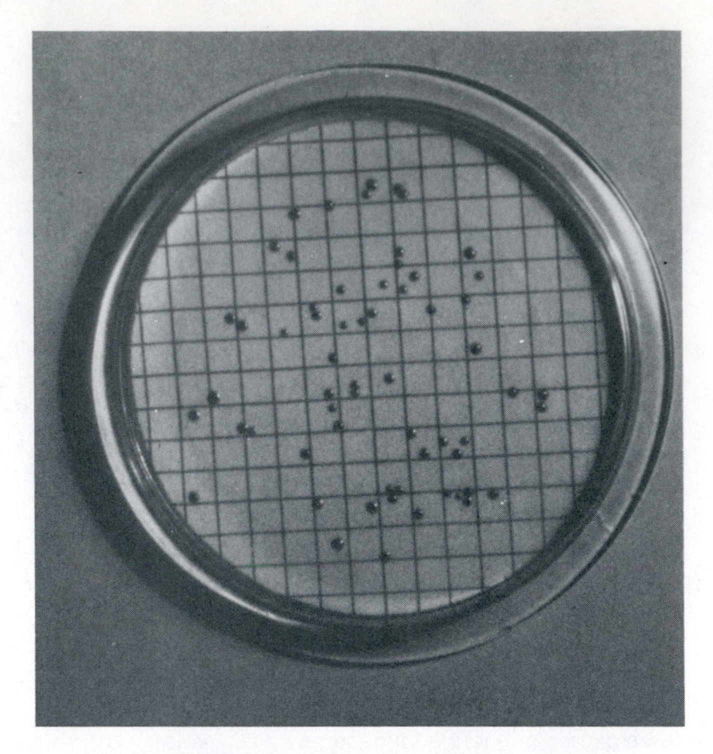

Figure 27-15 Bacterial colonies on gridded membrane filter.
Courtesy Gelman Sciences, Ann Arbor, Mich.

be used to measure the number of microorganisms in a solution, for example, when testing possible microbial contamination of a water supply. The water sample is filtered, and the microbial contaminants are retained on the filter. The membrane filter is then placed on an absorbant pad saturated with culture medium in a Petri dish. The nutrients penetrate the filter, and any deposited bacteria will reproduce and give rise to discrete colonies (Figure 27-15). The number of colonies can be correlated with the quantity of fluid filtered and is an indicator of the degree of contamination of the original water source.

Radiation. Radiation used for controlling microbial growth can be characterized as either nonionizing, such as UV light, or ionizing, such as x-rays and gamma rays. Each type under controlled conditions can be used to produce viable mutants (see Chapter 9), but each can also be microbicidal. UV light is absorbed by DNA at wavelengths between 240 and 280 nm. It has low penetrability and as a microbicidal agent is used primarily for irradiation of air and flat, nonporous surfaces. UV light can be used in hospitals, cafeterias, and meat-packing plants. The UV light fixtures can be so designed that any occupants in the room are protected from the harmful rays. Ultraviolet light generated by the sun is not a particularly effective microbicidal agent, because the earth's atmosphere absorbs most of the short penetrating rays and allows only the longer rays, such as those between 2850 and 2950 Å, to reach the earth's surface. Any microbicidal effects of the sun's rays are at their peak only on clear days that are dry and devoid of particles in the air. Some microorganisms can repair the effects of UV light, and the molecular mechanisms involved are discussed in Chapter 9.

Ionizing radiations are characterized by their ability to impart their energy to molecules and cause the dislodgment of electrons, resulting in the formation of ions. Their ability to act as microbicidal agents is related primarily to the ionization of water. Ionized water interacts with nonionized water to produce a ***hydroxyl radical*** (OH):

$$H_2O \xrightarrow{\text{Radiation}} H_2O^+ + e^-$$

$$H_2O^+ + H_2O \longrightarrow OH^- + H_3O$$

The hydroxyl radical is a strong oxidizing agent that reacts with DNA, causing breaks in the nucleotide chain. The effects of ionizing radiation appear to be also influenced by the presence or absence of oxygen. When equivalent amounts of radiation are applied to populations of anaerobes and aerobes, radiation damage is more evident in the aerobic culture than in the anaerobic one. It is believed that ionizing radiation converts molecular oxygen to the ***superoxide*** radical, which can react with DNA and cause single-stranded breaks and probably damage other biological molecules. (See Chapter 13 for a further discussion of the superoxide radical.) Ionizing radiations have been used for sterilization of pharmaceuticals and disposable medical supplies, such as gloves, plastic items, catheters, and sutures.

Chemical methods

Chemical agents used to control microorganisms on inanimate objects are called disinfectants, whereas those appropriate for mammalian tissue are called antiseptics. Often the two terms are used interchangeably. The response of microbial agents to chemical compounds is variable, and usually no single agent is equally effective against all microorganisms. The response of the microorganism to the antimicrobial is governed by the following:

1. Characteristics of the microorganism. The chemical makeup of the microorganism can influence the type of disinfectant used.

Some microorganisms possess capsules or slime, whereas others may be in spore state. The agent of tuberculosis *(Mycobacterium tuberculosis)* possesses high concentrations of lipid in its envelope and is consequently resistant to many water-soluble disinfectants. Certain viruses, such as the influenza virus and herpesvirus, possess lipid envelopes that protect them from some chemical agents.

2. Concentration of the disinfectant. Those disinfectants that are -cidal (kill) at high concentrations are often -static (inhibit) at lower concentrations. If the concentration is low enough, some disinfectants may actually stimulate growth. Even if the disinfectant is -cidal at several concentrations, the time required for cell death will sometimes increase by several factors at lower concentrations. For alcohols, however, too high a concentration may be ineffective. Ethyl alcohol at 95% concentration is less effective than at 75%, and this is related to the catalytic activity of water, whose concentration is less in the 95% alcohol solution.
3. pH of the disinfectant. Disinfectants appear to be most active at pHs slightly acidic to neutral. pH affects the degree of ionization, a property that is important for some disinfectants. If the pH of the disinfectant or the material being disinfected is at an extreme of pH, a reduction in activity can be expected. Chlorinated compounds, for example, release their active chlorine atoms at neutral or acidic pHs, but release is prevented at alkaline pHs.
4. Interfering matter. Most disinfectants interact with extraneous organic or inorganic material, as well as with the population of microorganisms. Disinfectants that inactivate enzymes can also be expected to interact with any protein. If, for example, a dissociable ion from a disinfectant is the active agent, it can be neutralized by interaction with organic matter, or it can be chelated by inorganic ions and removed from solution.
5. Time. No time limit can be established for the disinfection of objects, solutions, or body surfaces. Disinfection is controlled by the concentration of the disinfectant, the number of microorganisms present, the presence or absence of interfering organic matter, and other factors. The only way to determine the actual time required for complete disinfection is to perform a death curve in which the variables can be controlled. In such experiments the rate of death, like that of growth, is exponential.

Evaluation of disinfectants. Phenol and phenol derivatives for many years represented the major commercial disinfectants, and because of this, evaluations of other disinfectants based on time and concentration were related to comparisons with phenol. Phenol comparisons now have little meaning, since most of the new disinfectants are not phenols or their derivatives. Other factors play an important role in activity; therefore disinfectants must be tested under actual or simulated conditions for which they are being used.

Classes of disinfectants. There are many hundreds of disinfectants, and not all of them can be described in this book. Instead, they are divided into various classes in which only the most important ones of the group are discussed. The groups include phenols, halogens, alcohols, surfactants, alkylating agents, and heavy metals.

Phenols. Phenol (also called carboxylic acid) was first reported as a disinfectant by Joseph Lister in 1867. He applied phenol to surgical instruments and surgical dressings and reported a dramatic decrease in postoperative infections. Phenol is irritating to the skin and is not used today as an antiseptic, but it can be used as a disinfectant for organic matter, such as sputum and feces, that may harbor pathogenic species. Phenols have high lipid solubility and are thus active against the organism causing tuberculosis and related species. One of the most widely used phenolic derivatives is hexachlorophene, which consists of two phenol groups (Figure 27-16). It is an antiseptic that was used in many commercial soaps, surgical scrubs, deodorants, and for prophylactic bathing of infants. It is particularly active against gram-positive bacteria, especially *Staphylococcus aureus*, which is a common cause of serious infection in newborns. It was discovered, however, that hexachlorophene penetrates the skin and accumulates in brain tissue, causing neurological disorders after repeated use on infants. Hexachlorophene is now a prescription item and is used less widely. Cresol, a phenolic compound found in coal tar derivatives, has been used as a preservative for certain therapeutic preparations. Other uses have been in soaps and detergents.

Alcohols. Both ethyl and isopropyl alcohol are bactericidal agents whose -cidal activity is highest when diluted in water to concentrations of between 50% and 80%. Alcohols are lipid solvents and act on microbial membranes and lipids in the cell wall. For example, the differential mi-

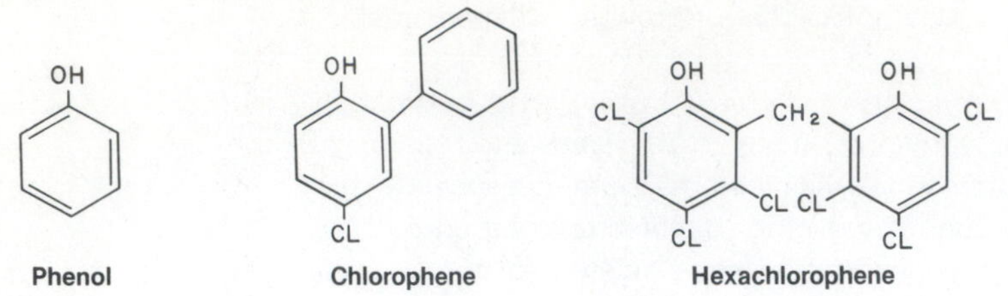

Phenol Chlorophene Hexachlorophene

Figure 27-16 Structure of phenol and two of its derivatives: hexachlorophene and chlorophene.

crobial stains (Gram stain and acid-fast stain) employ alcohol, which some investigators believe distinguishes microorganisms based on their concentrations of lipid in the cell envelope. Alcohols also cause dehydration of the cell and protein denaturation.

Halogens. The halogens—chlorine, bromine, fluorine, and iodine—are very strong oxidizing agents. Some commercial disinfectants contain either chlorine or iodine. They inactivate proteins by oxidizing sulfhydryl groups, resulting in the formation of sulfur-sulfur bonds, which cause changes in protein conformation and hence activity. Chlorine in liquid form is an effective bactericidal agent and is used in water treatment facilities as well as in swimming pools. Chlorine in water forms hypochlorous acid. Chlorinated lime (calcium hypochlorite) in 1% to 5% solutions is an effective disinfectant used in the sterilization of dairy barns and slaughterhouses, but it is also corrosive and has an offensive odor.

Iodine is poorly soluble in water and is found in preparations such as sodium iodide or potassium iodide. The most widely used antiseptic containing iodine is called tincture of iodine and consists of approximately 2% iodine and 4% sodium iodide in a 50% alcohol solution. The free iodine released in these preparations is toxic to the skin, and the alcohol solution causes a stinging sensation on open wounds. To combat these problems, iodine has been combined with an organic molecule and is called an ***iodophor.*** Iodophors (for example, povalone-iodine [Betadine]) are among the most frequently used antiseptics in the hospital.

Surface-active agents. ***Soaps*** and ***detergents*** belong to that class of agents called surface-active agents or ***surfactants.*** They have the property of lowering surface tension and increasing the miscibility of molecules. This is possible because they both have hydrophobic and hydrophilic groups, which help to separate molecules that have a tendency to be strongly attracted to each other. Soaps are sodium or potassium salts of fatty acids that function primarily as cleansing agents that increase the miscibility of oils and water and thus permit mechanical removal of microbes on the skin's surface. They are microbicidal for only a few species, such as species of *Streptococcus.* Many other disinfectants are inactivated by soaps, and the combination of the two in soap preparations must be carefully balanced. Detergents are synthetic agents that also possess hydrophobic and hydrophilic groups. They are divided into three types based on polarity, but only those that ionize are effective as disinfectants. ***Anionic detergents*** possess a hydrophobic hydrocarbon chain and a hydrophilic anion such as a carboxyl or sulfate group. They are more active in acidic solutions and affect the lipoprotein component of the cell membrane. ***Cationic detergents*** are the most microbicidal of the detergents. They are composed of cationic groups that dissociate to yield a positively charged ion, which is either ammonium, phosphonium, or sulfonium. The most popular cationic detergents are the ***quaternary ammonium*** ion compounds, also called ***quats*** (Figure 27-17). Their microbicidal activity is associated with their ability to interact with the bacterial cell membrane. The cationic portion is attracted to the negatively charged phosphate group of the membrane lipid, and the hydrophobic hydrocarbons of both the detergent and the membrane interact. Organic matter such as blood or tissue interferes with the activity of cationic detergents; thus surgical instruments must be thoroughly cleaned before disinfection with quats. Furthermore, quats are neutralized by anionic detergents and soaps. They are most active against gram-positive bacteria and some species of fungi. Today quats are used primarily in sanitizing glasses and flat, nonporous surfaces such as those in restaurants.

Alkylating agents. Some of the amino acids on proteins possess R groups, such as SH, COOH, and NH_3, that ionize and release hydrogen. When these hydrogens dissociate from the amino acid, their place can be occupied by alkyl groups. Nucleic acids can also be affected in the same way. The most active alkylating agents are formaldehyde, glutaraldehyde, β-propriolactone, and ethylene oxide (Figure 27-18). ***Formaldehyde*** is mainly used as a preservative and is a component of embalming fluid. Because of its irritation of tissue and odor, it is seldom used as a disinfectant. ***Formalin*** is a 37% solution of formaldehyde gas and is used for the preservation and fixation of tissue. Formalin is used in microbiology primarily in the preparation of vaccines, because it inactivates microorganisms but does not destroy their antigenic (antibody-stimulating) properties. ***Glutaraldehyde*** is a dialdehyde and is one of the preferred disinfectants because it is active against bacteria and viruses in a 2% solution but spores are destroyed only after 10 hours. It is too toxic for use as an antiseptic. The commercial product, which is called Cidex, is activated by the addition of sodium bicarbonate. ***Ethylene oxide*** is a gas that when placed under pressure is a most effective disinfectant for linens, sheets and bedding, and any articles or delicate instruments that cannot be sterilized by autoclaving. Its use is limited because of its toxicity if inhaled; explosiveness, if not diluted with inert gases such as carbon dioxide or freon; special equipment needs; and length of time to achieve sterility (4 to 12 hours) at room temperature. Furthermore, because of its toxicity, materials disinfected by this manner must be degassed before use.

Heavy metals. Some heavy metals are bacteriostatic because of their ability to bind to proteins and inactivate enzymes by tying up sulfhydryl groups. ***Mercury*** and ***silver*** are the only inorganic metal ions that are useful as disinfectants. Organic mercury compounds, such as thimerosal (Merthiolate) and merbromin (Mercurochrome) were once used in the treatment of superficial wounds, but they have been replaced by organic iodine compounds. ***Silver nitrate*** is used prophylactically for the prevention of gonococcal conjunctivitis, and many states require that newborns have a 1% solution of silver nitrate or a substitute instilled into the eyes at birth. Erythromycin is frequently used, because it is effective for gonococcal conjunctivitis plus *Chlamydia trachomatis,* which is a leading cause of neonatal conjunctivitis.

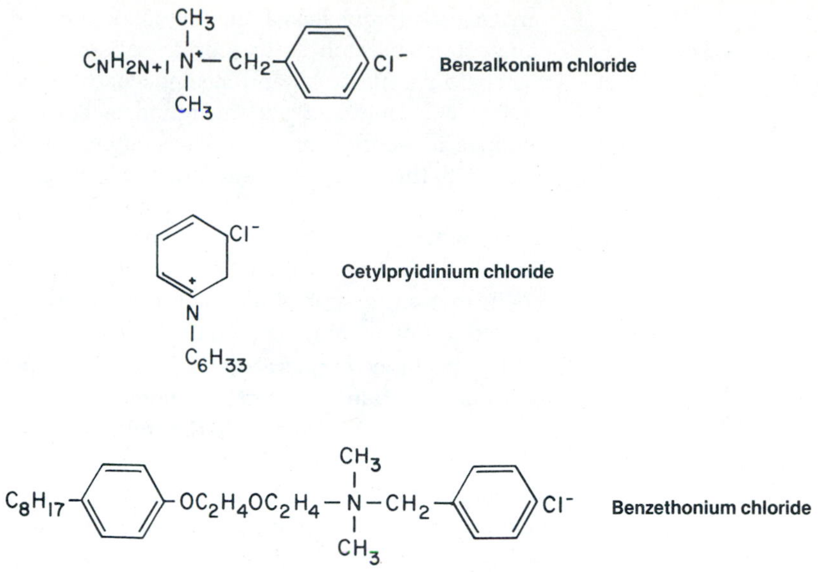

Benzalkonium chloride

Cetylpryidinium chloride

Benzethonium chloride

Figure 27-17 Three commonly used cationic detergents (quats).

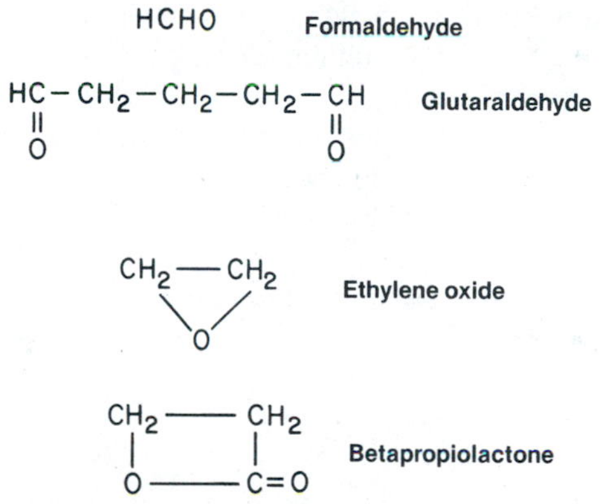

HCHO Formaldehyde

Glutaraldehyde

Ethylene oxide

Betapropiolactone

Figure 27-18 Structure of alkylating agents formaldehyde, glutaraldehyde, ethylene oxide, and betapropriolactone.

SUMMARY

The difference between antimicrobials and chemotherapeutic agents is that chemotherapeutic agents are selectively toxic for the parasite and relatively nontoxic to the host tissue, whereas antimicrobials may or may not possess these attributes. Most of the chemotherapeutic agents used in the treatment of infectious disease are produced by microorganisms; that is, they are antibiotics.

The spectrum of activity of chemotherapeutic agents may be narrow or broad. In addition, they may either kill the microorganism or simply inhibit their growth. Chemotherapeutic agents affect microorganisms in a number of ways; they may (1) inhibit cell wall synthesis (penicillins, cephalosporins, etc.), (2) inhibit protein synthesis (chloramphenicol, tetracycline, aminoglycosides, etc.), (3) inhibit cell membrane function (nystatin, amphotericin B, polymyxin, and imidazole derivatives), (4) inhibit nucleic acid synthesis (nalidixic acid, rifampin, and griseofulvin), and (5) act as antimetabolites (sulfa drugs). Interferon is an antiviral glycoprotein produced by the vertebrate host in response to viral infection. It is also produced in response to other agents, but in considerably smaller quantities. Interferon is an inhibitor of viral protein synthesis. It is currently produced from human cell lines in the laboratory and recently has been demonstrated to reduce malignant tumor growth.

Chemotherapeutic agents can be used to prevent disease (prophylaxis), but only under special circumstances, such as to prevent tuberculosis, postoperative infection, and meningococcal meningitis. The length of administration of chemotherapeutic agents for prophylaxis often determines whether drug-resistant forms will develop in the host.

Antibiotics have been used in animal feed to promote animal growth, but they may be hazardous in that antimicrobial-resistant species, infectious for humans, may be produced.

The resistance of microorganisms to antimicrobials may be due to natural phenomena, mutation, or transferable plasmids. Natural resistance is usually associated with the structural or biochemical properties of the organism. Mutation in microorganisms can produce resistance to drugs by (1) affecting the permeability of the cell wall, (2) altering the target site of the antimicrobial (such as the ribosome), and (3) increasing the synthesis of drug-inactivating enzymes. Plasmids containing genes that code for many drug-inactivating enzymes may be transferred from one microbial species to another via conjugation.

New drugs are evaluated and developed by using the following techniques: (1) determining the structure of the drug and then testing it in vitro against several pathogens, (2) testing the drug for its ability to induce resistant forms, (3) testing the toxicity of the drug using animals, and (4) developing procedures for mass production of the drug and then submitting it for clinical trials with humans.

Laboratory measurement of antimicrobial activity can be quantitative or qualitative. Most measurements are qualitative and determine the minimal inhibitory concentration (MIC) of the drug against a microorganism using agar or broth dilution tests. A qualitative test called the Kirby-Bauer test can be made quantitative by standardizing certain variables such as the agar type, agar pH, and concentration of the drug. The zone diameters produced on the agar surface can be correlated with interpretive charts and provide the physician with a method for determining what drug and what dosage to use in the treatment of the infection.

The process whereby an object or solution is rendered devoid of microorganisms is called sterilization. Agents that kill or control the growth of microorganisms are called antimicrobials and consist of two types: antiseptics and disinfectants. Disinfectants can kill microorganisms but are not necessarily applicable to the body surface. Antiseptics may inhibit the growth of microorganisms or kill them, but they can be applied to the surface of the body.

One of the most effective means of killing any type of microorganism is through the application of moist heat under pressure (autoclaving); however, this technique is not always applicable. Other physical methods include filtration and radiation. Chemical agents used as sterilizing agents are subject to certain criteria, including the type of microorganism involved, concentration of the chemical, temperature, pH, and presence or absence of interfering matter. Various classes of disinfectants exist, including phenols, alcohols, halogens, surface-acting agents, alkylating agents, and heavy metals.

Study questions

1. Describe some of the contributions of Paul Ehrlich to the field of chemotherapy.
2. From what sources are antibiotics obtained?
3. What is meant by broad spectrum and narrow spectrum? What are the advantages and disadvantages of each in chemotherapy?
4. What are the advantages of a microbicidal agent? What are the disadvantages?
5. Give some examples demonstrating selective toxicity.
6. Give at least two examples of chemotherapeutic agents that inhibit cell wall synthesis, cell membrane activity, protein synthesis, and nucleic acid synthesis.
7. What is the mechanism of action of interferon?
8. For what diseases is interferon currently being used?
9. What is the relationship of paraaminobenzoic acid (PABA) and the sulfa drugs?
10. Under what circumstances does one use chemotherapeutic agents prophylactically?
11. What is the advantage of chemotherapeutic agents in animal feed? What are the disadvantages?
12. What are the methods by which microorganisms develop resistance to antimicrobials?

13. Under what circumstances is more than one drug administered for therapy?
14. Describe the different mechanisms of mutational resistance.
15. What tests are used to screen potential antimicrobials?
16. After all the screening procedures have been performed, what attributes are most commonly sought?
17. How is drug toxicity for a potential antimicrobial determined?
18. How are the MIC and MBC of a drug determined?
19. Why can't slow-growing microorganisms be used as test species to determine the MIC of a drug?
20. What is the value of the Kirby-Bauer diffusion test in determining the sensitivity of microorganisms to a drug?
21. What must be standardized before the Kirby-Bauer technique can be interpreted?
22. Define the following: bacteriostasis, penicillin-binding proteins, single-step resistance, LD_{50}, and semisynthetic agents.
23. What is the difference between a bactericidal agent and a bacteriostatic one?
24. What conditions are required for sterilization by autoclaving? Are all microorganisms destroyed in this porcess? Under what circumstances do you not sterilize by autoclaving?
25. How does an antiseptic differ from a disinfectant? Are there situations where one could be substituted for the other?
26. What is the purpose of pasteurization? Does it affect viruses? Spores?
27. Under what conditions does one use filtration as a means of sterilization?
28. What molecules in the cell are directly affected by ionizing radiation? What is produced? In what way are they detrimental to the cell?
29. What are the most effective concentrations of alcohol for disinfection?
30. What types of disinfectants are effective against bacteria? Against viruses?
31. What factors should be considered before a disinfectant is used?
32. What are the advantages and disadvantages of the following antimicrobials: phenols, chlorine, tincture of iodine, quaternary ammonium compounds, soaps, formaldehyde, glutaraldehyde, and ethylene oxide?

Selected readings

Books

Barry, A.L. *The antimicrobic susceptibility test: principles and practices.* Lea & Febiger, Philadelphia, 1976.

Block, S.S. (Ed.). *Disinfection, sterilization and preservation* (2nd ed.). Lea & Febiger, Philadelphia, 1977.

Borick, P.M. (Ed.). *Chemical sterilization* (Vol. 2). Academic Press, Inc., New York, 1973.

Dineen, P. Local antiseptics. In Modell, W. (Ed.). *Drugs of choice 1980-1981.* The C.V. Mosby Co., St. Louis, 1980.

Dowding, J., and Davies, J. Mechanisms and origins of plasmid-determined antibiotic resistance. In Schlessinger, D. (Ed.). *Microbiology—1974.* American Society for Microbiology, Washington, D.C., 1974.

Hahn, F.E. (Ed.) *Antibiotics,* Vol. 1, *Mechanism of action of antibacterial agents.* Springer-Verlag, New York, Inc., New York, 1979.

Hammond, S.M., and Lambert, P.A. *Antibiotics and antimicrobial action.* Edward Arnold (Publishers), Ltd., London, 1978.

Kagan, B.M. *Antimicrobial therapy* (2nd ed.). W.B. Saunders Co., Philadelphia, 1974.

Kucers, A., and Bennett, N. *The uses of antibiotics* (3rd ed.). William Heinemann Medical Books, Ltd., London, 1979.

Lorian, V. (Ed.). *Antibiotics in laboratory medicine.* The Williams & Wilkins Co., Baltimore, 1980.

Perkins, J.J. *Principles and methods of sterilization in health sciences.* Charles C Thomas, Publisher, Springfield, Ill., 1973.

Perlman, D. (Ed.). *Structure-activity-relationship among the semisynthetic antibiotics.* Academic Press, Inc., New York, 1977.

Salton, M.R.J., and Shockman, G.D. (Eds.). Beta-lactam *antibiotics: mode of action, new Developments and future Prospects.* Academic Press, Inc., New York, 1981.

Schonfeld, H., and DeWeck, A. (Eds.). *Antibiotics: mode of Action.* S. Karger Publishers, Inc., New York, 1971.

Journal articles

Abraham, E.P. The beta-lactam antibiotics. *Sci. Am.* **244:**76, 1981.

Appel, G.B., and Neu, H.C. The nephrotoxicity of antimicrobial agents. *N. Engl. J. Med.* **296:**784, 1977.

Bennet, J.V. Antibiotic use in animals and human salmonellosis. *J. Infect. Dis.* **142:**631, 1980.

Davies, J., and Smith, D.I. Plasmid determined resistance to antimicrobial agents. *Annu. Rev. Microbio.* **32:**469, 1978.

Demain, A.L. How do antibiotic-producing microorganisms avoid suicide? *Ann. N.Y. Acad. Sci.* **235:**601, 1974.

Gordon, J., and Minks, M.A. The interferon renaissance: molecular aspects of induction and action. *Microbiol. Rev.* **45:**244, 1981.

Greenwood, D. In vitro veritas? Antimicrobial susceptibility tests and their clinical relevance. *J. Infect. Dis.* **144:**386, 1981.

Hirsch, M.S., and Swartz, M.N. Antiviral agents. *N. Engl. J. Med.* **302:**903, 1980.

Hopwood, D.A. Extrachromosomally determined antibiotic production. *Annu. Rev. Microbio.* **32:**373, 1978.

Koch, A.L. Evolution of antibiotic resistance gene function. *Microbiol. Rev.* **45:**267, 1981.

Murray, B.E., and Moellering, R.C. Cephalosporins. *Annu. Rev. Med.* **32:**59, 1981.

Rahal, J.J., and Simberkoff, M.S. Adverse reactions to anti-infective drugs, *DM* **25**(1): entire issue, 1979.

Sasinur, R., et al. Colitis associated with metronidazole therapy. *J. Infect. Dis.* **141:**772, 1980.

Smith, J.W. Proper use of antibiotics. *Tex. Med.* **73:**1, 1977.

Tomasz, A. The mechanism of the irreversible antimicrobial effects of penicillins. *Annu. Rev. Microbiol.* **33:**113, 1979.

Utz, J.P. Chemotherapy for the systemic mycoses: the prelude to ketoconazole. *Rev. Infect. Dis.* **2**(4):625, 1980.

Weinstein, L. Some principles of antimicrobial therapy. *Ration. Drug Ther.* **11**(3):1, 1977.

Chapter 28 BACTERIAL DISEASES

In developed countries of the world vaccines have led to dramatic reductions in many bacterial diseases such as diphtheria, tetanus, and whooping cough. Many of these diseases, however, remain major causes of death in underdeveloped countries, as well as in developed countries where lower socioeconomic conditions exist. Bacterial agents are still the most frequent cause of infectious disease because they are the most abundant commensals found on the vertebrate host. Any bacterial agent, under opportunistic conditions, can cause infection in the human host.

Bacterial disease can be acquired through different routes, including (1) the respiratory tract, (2) the alimentary tract, (3) the skin and mucosa (including the bite of insect vectors), (4) sexual transmission, and (5) congenital transfer. Four of these are discussed in this chapter; sexual transmission is discussed in Chapter 30.

GENERAL CHARCTERISTICS OF BACTERIAL DISEASES

Immunity to bacterial diseases involves both the humoral and cell-mediated responses of the host. The bacterial surface is composed of a number of antigens, which vary considerably from species to species and between various strains within a species. Many of these antigens induce the formation of specific protective antibodies in the host, which accounts for resistance to future infection as well as recovery from primary infections. In addition, many bacterial species produce toxins that induce specific antibodies called antitoxins. Those surface antigens that do not elicit a protective antibody response often belong to agents that cause chronic infections such as tuberculosis and brucellosis.

There is little discussion of specific laboratory diagnostic tests in this chapter or in succeeding chapters that deal with infectious diseases. Diagnostic tests are explained in Appendix B. There is a strategy used in the clinical laboratory for identification of bacteria that should be explained at this time. Effective treatment of a disease is best accomplished when the organism causing the disease, has been isolated, identified, and subjected to antimicrobial susceptibility testing. Material from the lesion is first inoculated onto specific media for isolation. The medium is usually one of two types: one that will support the growth of most bacteria or one that will support the growth of what is believed to be the infectious agent. The judgment as to what agent is causing infection is affected by several factors and includes the symptoms of disease, the site of infection, the age and state of health of the patient, what disease may be "going around" at the time, and the occupation of the patient. Following an incubation period of 18 to 48 hours or longer, the colonies that develop on the isolation medium are subjected to the following tests or observations that will aid in identification.

1. Gram stain. The ***Gram stain*** is one of the most important initial tests in identification of bacteria because it separates them into morphological as well as cell wall types: gram-positive or gram-negative cocci and gram-positive or gram-negative bacilli. The Gram stain provides enough information that treatment can be initiated during life-threatening infections, where time is important and other tests cannot be employed. Penicillin or related drugs can be used if the organism is a gram-positive coccus or bacillus; but if the organism is a gram-negative bacillus, gentamicin, kanamycin, or ampicillin are effective antimicrobials. The only gram-negative cocci of medical importance are species of *Neisseria,* which can cause meningitis or gonorrhea, for which there is a standard treatment.
2. Cultural characteristics. The characteristics

of the bacterial colony on the agar surface often aid in identification. The color, size, and consistency of the colony are valuable aids in helping differentiate one bacterial species from another. *Staphylococcus epidermidis,* for example, produces a white colony, whereas *S. aureus* produces a yellow colony. Cultural characteristics, however, are subject to considerable variations within a single species, and they cannot be used as the primary method of differentiation.

3. Biochemical characterization. The variation that exists in the metabolism of many bacterial species can also be used to differentiate them. Biochemical testing has more application to bacteria than to any other group of microorganisms because of their greater biochemical diversity. Many of the important biochemical tests are explained in Appendix B.
4. Serological characteristics. Some infectious bacterial agents, because they may be inaccessible, cannot be isolated from the patient for the purpose of biochemical testing. Serological testing can, therefore, be the most important and, sometimes, the only tool for the diagnosis of bacterial disease. Antibodies of known specificity can be used to detect the antigenic components of some bacteria that cannot be isolated intact from the host. Additionally, high titers of antibody produced by the patient during the course of infection are diagnostic for certain diseases. Antibody tests are used more routinely in the diagnosis of viral infections because the viral agent is seldom cultivated in the clinical laboratory (see Chapter 24).

Treatment of bacterial diseases, for the most part, involves the use of antibiotics; these are discussed in detail in Chapter 27.

DISEASES ACQUIRED THROUGH THE RESPIRATORY TRACT

Bacterial infections involving the respiratory tract are generally acquired by inhalation or ingestion of microorganisms from the air or by direct or indirect contact with respiratory secretions. The mechanism of spread of respiratory agents is discussed in Chapter 26. All the diseases discussed involve an infection of the respiratory tissue; they include pneumonia, whooping cough, diphtheria, tuberculosis, and pharyngitis. One bacterial agent, *Neisseria meningitidis,* which initiates infection in the upper respiratory tract but disseminates to cause infection of the meninges (meningitis), is also discussed.

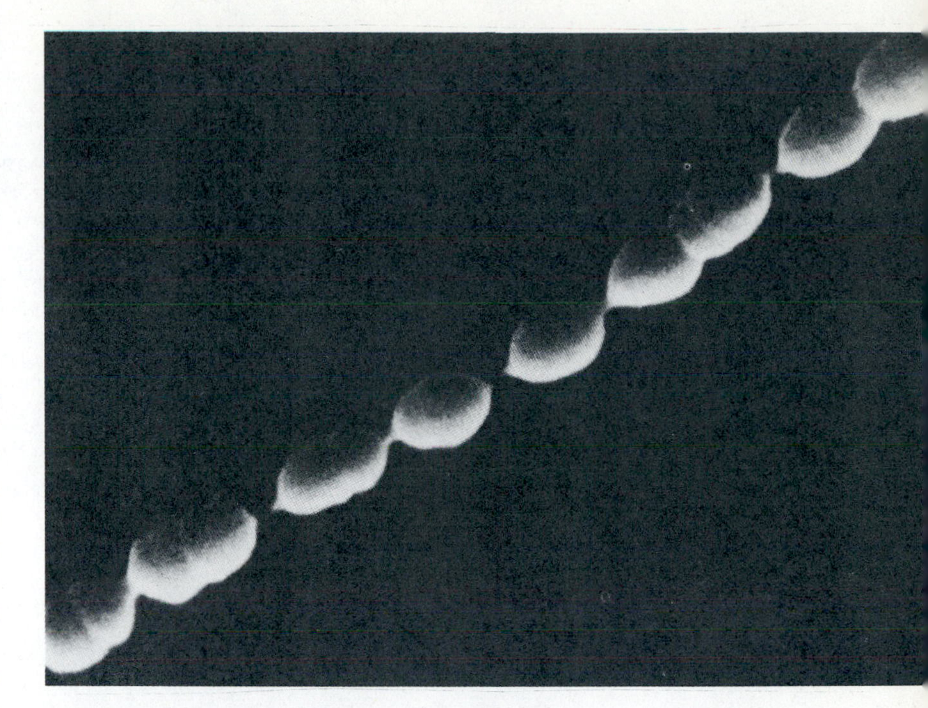

Figure 28-1 Scanning electron micrograph of *Streptococcus pneumoniae.* (×10,000.)
From Klainer, A.S. and Geis, I. Agents of bacterial disease, *Harper and Row, Publishers, Inc., New York, 1973.*

Pneumonia

Pneumonia is a condition in which the lungs are inflamed. Pneumonia, caused by microorganisms, may result from inhalation of contaminated dust or respiratory secretions, or it may be caused by aspiration of microorganisms that are normally present in the respiratory tract. Most pneumonias are secondary to conditions or diseases that exist in the patient, for example, bronchitis, viral infection, and alcoholism. An individual suffering from a viral disease of the upper respiratory tract will have a damaged respiratory epithelium, and his resistance to infection will be reduced. As a consequence, certain microorganisms, especially bacteria, are capable of unrestricted growth and initiation of disease. The principal bacterial agent causing pneumonia is *Streptococcus pneumoniae; Klebsiella pneumoniae* and a host of other agents can also cause pneumonia.

Streptococcus pneumoniae. *Streptococcus pneumoniae* is a natural inhabitant of the upper respiratory tract of humans and is the cause of over 50% of the bacterial pneumonias. It is a gram-positive coccus that occurs in chains (Figure 28-1), and virulent strains produce a capsule

Table 28-1 Characteristics of bacterial agents that cause pneumonia*

BACTERIAL AGENT	MORPHOLOGY	POPULATION MOST LIKELY AFFECTED	CHEMOTHERAPY	IMMUNIZATION	COMMENTS
Klebsiella pneumoniae	Gram-negative bacillus, capsule produced	Those with underlying respiratory conditions	Cephalosporins, but multiple drug resistance common	None	Causes necrosis of lung tissue
Staphylococcus aureus	Gram-positive coccus	Those infected with influenza virus	Multiple drug resistance; susceptibility tests required	None	Organism produces coagulase, important in identification
Haemophilus influenzae	Gram-negative coccobacillus	Those with underlying respiratory conditions	Ampicillin or chloramphenicol	None	Ampicillin-resistant strains common.
Legionella pneumophila (Legionnaires disease)	Gram-negative bacillus (Figure 28-2)	Those with underlying respiratory conditions	Erythromycin	None	Transmission perhaps by contaminated aerosols
Yersinia pestis (plague)	Gram-negative bacillus	Contact with respiratory secretions of those with pneumonic variety of plague	Streptomycin or tetracycline	Yes, but of limited value	See p. 635
Mycoplasma pneumoniae	Devoid of cell wall (Figure 28-3)	Those in crowded quarters (such as barracks)	Erythromycin or tetracycline	Experimental	Colonies look like fried egg on agar
Coxiella burnetii	Gram-negative bacillus	Those inhaling dried parturition products of barn animals	Tetracycline	None	Belongs to group Rickettsia (see p. 636)
Chlamydia psittaci	Gram-negative, exists in two morphological states: infectious elementary body and noninfectious reticulate body (Figure 4-41)	Primarily infects birds and secondarily infects humans (psittacosis); those inhaling particles contaminated with bird droppings	Tetracycline	None	Pneumonia an infrequent aspect of infection
Streptococcus pneumoniae	Gram-negative diplococcus	The elderly with underlying disease or poor health	Penicillin	Polyvalent vaccine available	Most common cause of bacterial pneumonia

*Many other microbial agents can produce pneumonia such as *Streptococcus pyogenes* and many gram-negative enteric species, but only the primary agents of pneumonia are listed in the table.

that enables them to resist phagocytosis. Infections usually involve the lower lobes of the lung where fluid and cells derived from blood vessels and surrounding tissue cause the lung to be engorged. Under these conditions proper gas exchange in the lungs is limited, but there is no necrosis of tissue as occurs with other microbial agents such as *K. pneumoniae.* The respiratory tissue, therefore, returns to normal following disease caused by *S. pneumoniae.*

Some of the morphological and biochemical characteristics of this organism are helpful in identification. *S. pneumoniae* ferments inulin, a carbohydrate that is not fermented by other streptococci, and is subject to autolysis when surface active agents such as bile salts are present. Its autolytic activity is the basis of the bile solubility test. The presence of a capsule permits the specific identification of *S. pneumoniae* when type-specific antiserum is applied to the cells. The capsule appears as a sharply delineated halo around the bacterial cells in a positive reaction, called a Quellung reaction (see Appendix B).

Penicillin is the most effective agent in chemotherapy. Recently a polyvalent vaccine containing 14 of the major pneumococcal types has been licensed for administration to those most susceptible to pneumonia—the elderly who are hospitalized.

Other causative agents of pneumonia are described in Table 28-1.

Whooping cough

Whooping cough, also called **pertussis** (pertussis means cough), was first described in 1578, but the causative agent was not isolated until 1906. Whooping cough is a highly contagious disease that affects predominately infants and children. The organism causing disease, *Bordetella pertussis,* is a gram-negative bacillus that inhabits the upper respiratory tract of humans. The only known source of infection is the infected human, who can easily transmit the disease to others. In the United States, between 1940 and 1948, whooping cough killed nearly three times as many infants under 1 year of age as measles, mumps, chickenpox, rubella, scarlet fever, diphtheria, poliomyelitis, and meningitis combined.

Whooping cough is transmitted by contact with respiratory secretions that contaminate the air or surrounding objects. During infection the organism erodes the mucosal lining of the respiratory epithelium, causing first a paroxysmal (repeated) cough. Later, mucus plugs the air passages, and inhalation of air, following a series of coughs, produces a characteristic "whoop." In several cases respiratory or cardiac arrest may occur. The disease is considered self-limiting, but the mortality rate may be high as a result of secondary bacterial or viral infections.

Whooping cough does not respond well to treatment, and antibiotics are used primarily to reduce the communicability of the disease by destroying existing organisms or preventing infections by secondary microbial invaders. The only approach to control is via immunization. Vaccination programs, initiated in the United States in the early 1950s on a large scale, resulted in dramatic decreases in disease. Before mass immunization there were over 100,000 cases of whooping cough per year in the United States, but today less than 1500 are recorded per year. The disease is still prevalent in most underdeveloped countries where few children have been vaccinated against the disease. In the 1980s Guatemala, for example, has recorded approximately 515 infant deaths due to whooping cough per 100,000 babies born. This figure is to be compared with 0.26 deaths per 100,000 babies born in the United States. In developed countries where immunizations have begun to decline because of parental apathy, there has been a dramatic increase in the cases of whooping cough. Even with the arsenal of antibiotics available to the physician, treatment is uncertain, and this makes immunizations even more important. The vaccine used today consists of *B. pertussis* cells and is administered as part of a mixture called

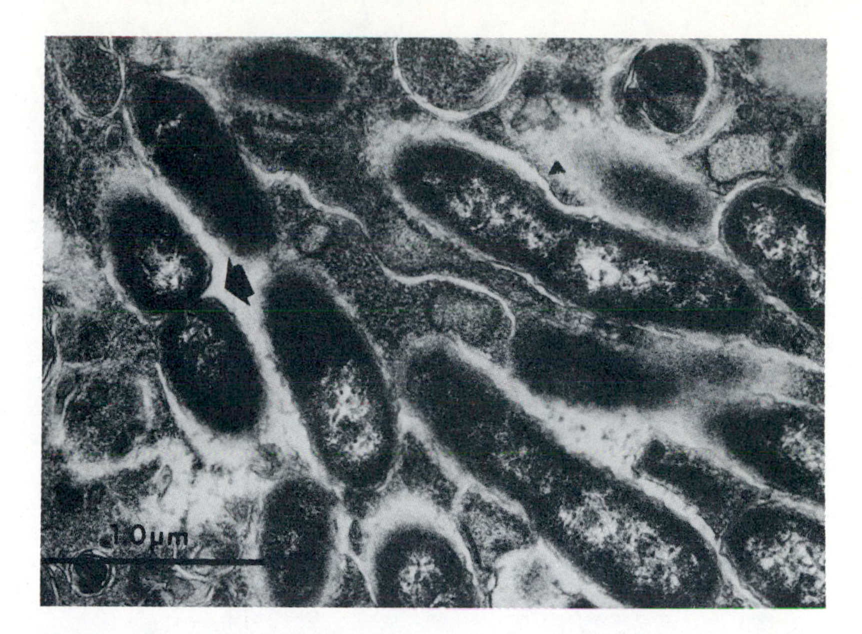

Figure 28-2 Electron micrograph of *Legionella pneumophila* in cytoplasm of alveolar macrophage. (×30,000.) Note evidence of binary fission *(solid arrow)* and cell wall of bacterium *(open arrow)* and vesicular membrane *(arrowhead).*
From Kishimoto, R.A. Infect. Immun. ***25:****761, 1978.*

Figure 28-3 Scanning electron micrograph of *Mycoplasma pneumoniae* attached to surface of human malignant cell. (×15,000.) Some cells are round, and others are ovoid or filamentous. Bar = 1 μm.
Courtesy E.S. Boatman.

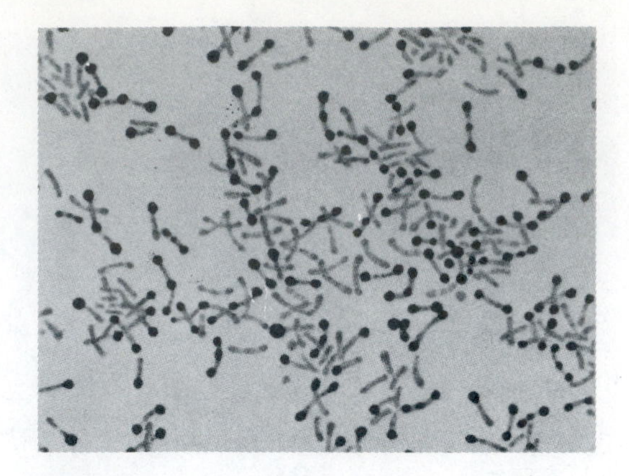

Figure 28-4 Photomicrograph of *Corynebacterium diphtheriae* showing stained metachromatic (polyphosphate) granules that give the cell a bulbous appearance.

From Gillies, R.R., and Dodds, T.C. Bacteriology illustrated (4th ed.). Churchill Livingstone, Edinburgh, 1976.

DTP (*D*, diphtheria toxoid; *T*, tetanus toxoid; *P*, pertussis vaccine). The vaccine-toxoid is usually administered as three doses at monthy or bimonthly intervals beginning 2 months after birth. Booster doses at age 1 to 2 and 5 to 6 years are also usual.

B. pertussis can be isolated from the upper respiratory tract and streaked on a special medium called Bordet-Gengou (potato-glycerol-blood agar)* for colony identification and other testing procedures. Isolation of the organism may be impossible if the disease has been in progress for 3 weeks or longer.

Diphtheria

The first clinical description of diphtheria was provided in 1826 by Pierre Bretonneau when he noted that a membrane formed in the pharynx during the course of the disease. He named the disease ***diphtheria,*** taken from the Greek word meaning membrane. The organism causing the disease was first described by Klebs, but it was Frederick Loeffler, in 1894, who isolated the organism and reproduced the disease in animals. Loeffler made a significant discovery during his experiments on diseased animals. He found that, although several different organs in the animals had hemorrhaged, the organisms causing infection could be isolated only from the initial site of inoculation. This suggested to him that the organisms may have excreted a soluble toxin that entered the bloodstream and traveled to distant organs. Roux and Yersin later grew the organism in culture and separated the organisms from the filtrate. They demonstrated that injection of the filtrate caused disease in experimental animals. In 1890 Von Behring and Kitasato injected the filtrate into horses and showed that specific neutralizing antibodies were produced in the serum of the infected animal. This discovery soon led the way to the formation of an antitoxin prepared from horses that, when injected into patients with diphtheria, reduced mortality by 50%. Later, in 1923 toxin was treated with formaldehyde, which destroyed its toxic activity but not its immunizing potential—a toxoid. The first active immunization program was started in New York in 1928, and this led to a decline in mortality in the United States from 780/100,000 population in 1894 to now less than 10/year.

Diphtheria is best described as an intoxication or poisoning caused by the aerobic bacterium *Corynebacterium diphtheriae* (corynebacterium, clubshaped bacterium; diphtheriae, membrane). The organism is rod shaped, and some appear to have swollen ends (Figure 28-4). *C. diphtheriae* causes superficial skin or respiratory infection, and transmission occurs by contact with the skin or respiratory secretions. The organism, during respiratory infections, excretes a toxin that initially causes an inflammatory response and the formation of a pseudomembrane that is made up of polymorphonuclear leukocytes, fibrin, bacterial cells, and dead epithelial cells. The membrane can become large enough to obstruct air passages and cause the patient to suffocate. The information for toxin production is found on the genome of a temperate bacteriophage harbored by the organism (see Chapter 22 for discussion of the mechanism of action of toxin). Toxin produced by bacteria passes into the bloodstream and can affect all major organs and is especially damaging to heart and nervous tissue. Death is most frequently caused by the effect of toxin on heart tissue.

Diagnosis of diphtheria requires isolation of the organism usually from the respiratory tract (by streaking on appropriate media such as tellurite and Loeffler's serum agar) and demonstration of toxin production (Figure 28-5).

There has been a dramatic decrease in diphtheria since immunization practices began in the 1920s. The average number of cases per year in the United States is now approximately 56, but diphtheria is still regarded as a serious disease because of the high case/fatality ratio, which has remained the same over the years (Figure 28-6). Diphtheria, at one time, was common in children but today is seen more often in older adults. This is the result of immunization practices and the fact that many older adults are unimmunized or

* Bordet and Gengou were the first to isolate *Bordetella pertussis*.

inadequately immunized. Treatment of diphtheria should include administration of antitoxin to neutralize any unadsorbed toxin in the patient's tissues and the use of antimicrobials, such as penicillin, to destroy existing microorganisms. Antibiotics, such as penicillin, may not affect the course of the disease but they do eliminate the carrier state. Immunization alone does not eliminate the carriage of *C. diphtheriae* in the pharynx or on the skin; therefore, those lacking immunization are always prone to infection. Immunization of older children and adults is usually preceded by injection of diluted diphtheria toxoid beneath the skin in order to determine toxoid sensitivity. A skin reaction indicates sensitivity, and the dosage for immunization is appropriately reduced.

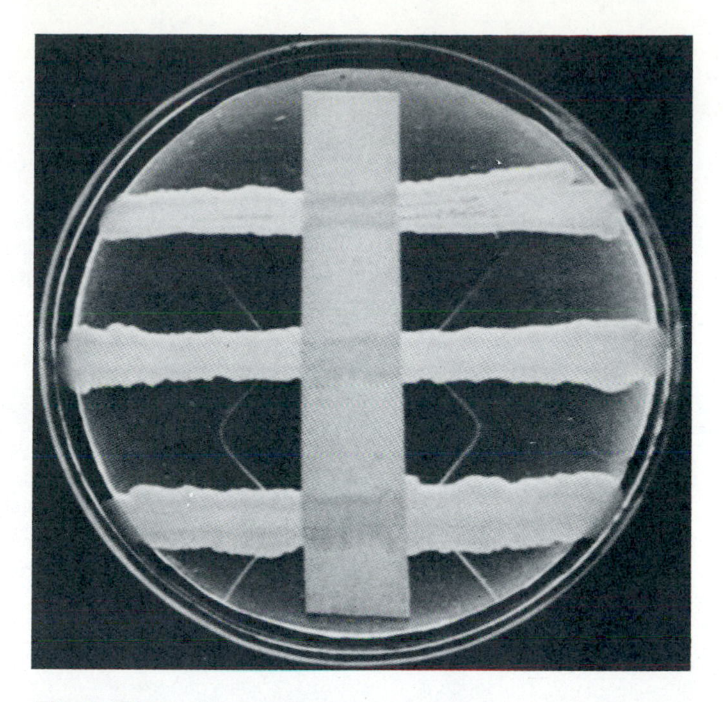

Figure 28-5 Gel diffusion plate for demonstrating toxigenicity of strains of *Corynebacterium diphtheriae*. Filter strip soaked in diphtheria antitoxin is placed in center and on surface of agar. Plate is inoculated on each side of strip and incubated for 24 to 48 hours. In this figure three test strains have been inoculated. A positive test (toxigenic strains) is evidenced by formation of one or more white lines of toxin-antitoxin precipitate. Two lower test cultures are positive.

From Olds, R.J. Color atlas of microbiology. Wolfe Medical Publishers Ltd., London, 1975.

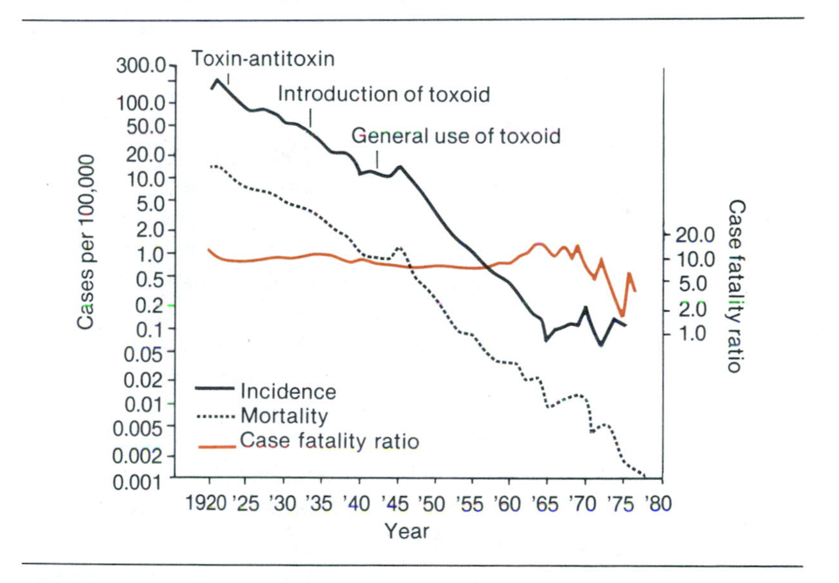

Figure 28-6 Diphtheria. Annual incidence, mortality rates, and case/fatality ratios in United States from 1920-1978.

Redrawn from the Centers for Disease Control, Atlanta, Georgia.

Meningitis

Meningitis is an inflammatory disease that involves the protective layers of tissue surrounding the spinal cord and brain, called the meninges. If the infection results in the formation of a purulent (pus-containing) exudate, it is called septic meningitis, and aseptic if nonpurulent. Spinal meningitis and cerebrospinal meningitis are terms used to describe the septic forms involving the central nervous system. One of the major problems in the treatment of meningitis is the relatively poor rate of diffusion of antibiotics from the bloodstream into the central nervous system. Meningitis is therefore regarded as a particularly severe disease that can result in a high percentage of fatalities. The two most important agents of bacterial meningitis are *Neisseria meningitidis* and *Haemophilus influenzae*. A wide variety of bacterial agents that cause primary diseases elsewhere in the body may also cause meningitis; they are outlined in Table 28-2.

Neisseria meningitidis. *Neisseria meningitidis* is a gram-negative diplococcus (also called meningococcus) (Figure 28-7) that resides in the upper respiratory tract. The meningococcus is the only bacterium causing epidemic meningitis and is divided into several immunological groups, the most invasive of which are groups B and C. Contact with respiratory secretions from carriers or active cases is the most important avenue of transmission. Infections usually begin with mild nasopharyngitis that may progress into septicemia (meningococcemia) that may be severe enough to cause death from hemorrhage (Figure 28-8). Septic meningitis is a consequence of the dissemination of the meningococcus from the bloodstream to the central nervous system. The

Table 28-2 Bacterial agents that are a primary cause of meningitis

BACTERIAL AGENT	MORPHOLOGY	POPULATION MOST LIKELY AFFECTED	CHEMOTHERAPY	IMMUNIZATION	COMMENTS
Streptococcus agalactiae (group B streptococci)	Gram-positive streptococci	Newborns during passage through birth canal	Penicillin or ampicillin plus gentamicin	None	Second only to *E. coli* as cause of neonatal meningitis
Haemophilus influenzae type B	Gram-negative coccobacillus, produces capsule	Children under 4 years of age	Ampicillin or chloramphenicol	None	Ampicillin-resistant varieties are arising
Escherichia coli	Gram-negative bacillus	Neonates and those debilitated as a result of surgery and catheter manipulation	Multiple antibiotic resistance; susceptibility tests required	None	Most frequent cause of neonatal meningitis
Streptococcus pneumoniae	Gram-positive streptococcus	Severely debilitated patients	Penicillin	Polyvalent vaccine containing 14 serotypes	Often follows lung infection
Neisseria meningitidis	Gram-negative diplococcus	Children less than 5 years of age	Penicillin	Polysaccharide vaccine against group A and C but not B serotypes	Vaccine recommended for contacts at day care centers

Many other microbial agents can cause meningitis, including *Listeria monocytogenes, Flavobacterium meningosepticum*, many gram-negative enteric species, as well as staphylococci and other streptococci. Only the primary agents of meningitis are listed in the table.

most common symptoms of disseminated disease are stiff neck or back, headache, and nausea.

One of the most important aspects of meningococcal disease is the carrier state. The organism colonizes the upper respiratory tract of carriers and stimulates the formation of antibodies that do not eliminate the organism but do prevent it from multiplying and disseminating to other tissues. It appears that epidemics occur only when the carrier rate exceeds 20%; no major epidemic has occurred in the United States in the past 36 years. Approximately 2800 cases of meningococcal disease were reported in 1980, and most of them were among those less than 4 years of age. Transmission is accelerated when there are crowded conditions such as in day care centers, military camps, and hospitals.

Identification is based on certain properties of the genus and species.

1. All *Neisseria* species are oxidase positive; that is, they produce an enzyme, indolphenol-oxidase, that can be detected by a color change when colonies on an agar surface are flooded with oxidase reagent (a solution of tetramethyl-p-phenylenediamine-hydrochloride). The colonies develop a pink color that becomes darker and finally black (Plate 34).
2. *Neisseria* species differ in ability to metabolize certain carbohydrates.
3. Spinal fluid can be examined for gram-negative diplococci.
4. Homologous group-specific antisera can be used to detect capsule-forming strains such as the more important ones belonging to groups A, B, and C.

Penicillin is the most effective drug in treatment of meningitis and related conditions caused by *N. meningitidis,* but chloramphenicol can be used for those sensitive to penicillin. Once a case of meningitis has been identified, it is important that intimate contacts of the index case be treated prophylactically. Sulfonamides (if the organism is shown to be sensitive) or rifampin is used prophylactically to reduce the carrier state. Vaccines prepared from strains of group A and C are also available as an adjunct to chemoprophylaxis. Group B *N. meningitidis* is the most prevalent type in the United States, but no vaccine is yet available.

Haemophilus influenzae. *H. influenzae* is a gram-negative organism that morphologically ranges from a coccobacillus to a filmaentous rod. It is part of the normal microbial flora of the respiratory tract of humans and many animals. Two

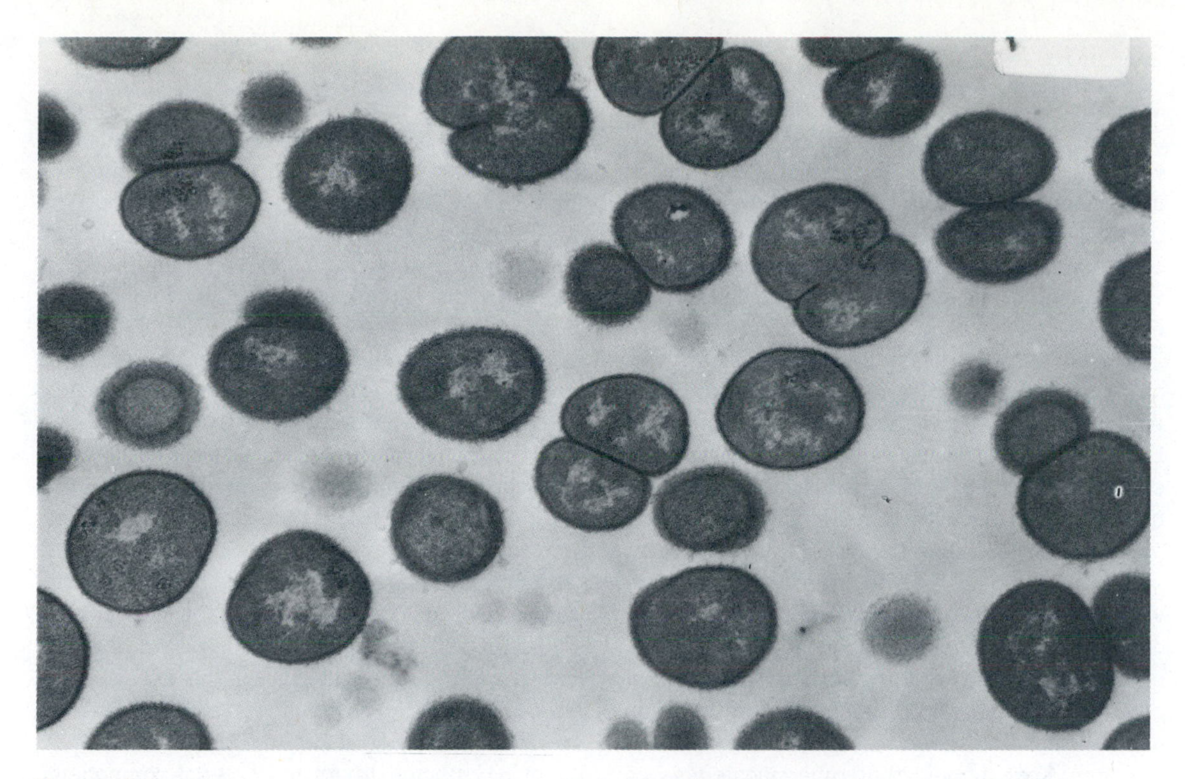

Figure 28-7 Electron micrograph of a *Neisseria* species demonstrating its diplococcal nature. (×27,000.)
From Dajani, A.S. Infect. Immun. ***14**:776, 1976.*

strains may be present in the respiratory tract: the encapsulated and nonencapsulated. The encapsulated strain, called *H. influenzae* type b, is the more virulent and also the more predominant type in children under 4 years of age. Older children and adults rarely carry the type b strain. Type b is one of the major causes of meningitis in children under four years of age but diseases in adults are rare.

Treatment of disease caused by *H. influenzae* type b should be initiated only after antibiotic sensitivity tests have been performed. Ampicillin at one time was universally used in treatment, but since 1974 many strains have acquired R-factor–associated β-lactamase and subsequent resistance to ampicillin. One of the most effective treatment regimens includes the combination of streptomycin and sulfisoxazole.

Laboratory diagnosis of meningitis can be obtained by performing a Gram stain on the cerebral spinal fluid. The bacteria will appear as gram-negative coccobacillary rods with rounded ends. The capsule of the organism can also be demonstrated by capsular swelling (Quellung reaction—see Appendix B), or the capsular antigens can be detected by radioimmunoassay or countercurrent immunoelectrophoresis.

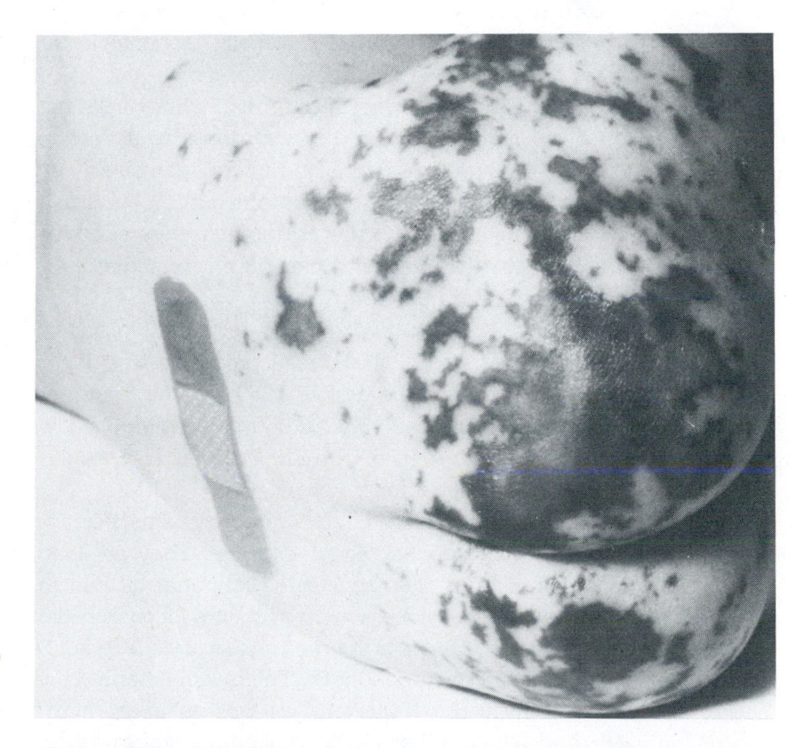

Figure 28-8 Fulminating meningococcemia. Cutaneous hemorrhages are evident before meningitis is recognized.
From Anderson, W.A.D., and Kissane, J.M. (Eds.). Pathology *(Vol. 1). The C.V. Mosby, Co., St. Louis, 1977.*

Tuberculosis

Tuberculosis has been the most important cause of death by infectious agents in the Western world for over 200 years. In the late 1800s villages in the eastern parts of the United States were decimated by tuberculosis, and as many as 30% of the deaths each year were caused by this disease. Even today nearly 4 million people throughout the world die each year from tuberculosis.

Originally called consumption, the disease was only referred to as tuberculosis in the early 1800s when two pathologists discovered swellings ***(tubercles)*** in lung tissue associated with the disease. Tuberculosis was originally thought to be hereditary because it usually afflicted several members of a single family. Many important members of society in the nineteenth century were affected by the disease, and the alabaster appearance of the skin associated with the disease was deemed a sign of beauty. It soon became evident that the disease affected both rich and poor alike.

Robert Koch proved that tuberculosis was an infectious disease in 1882 by applying his now famous Koch postulates. Later discoveries, such as the x-ray by Roentgen, provided diagnostic tools that enabled physicians to diagnose the disease before it could debilitate their patients. Several factors have led to a decline in tuberculosis.

1. Pasteurization of milk, which destroys the bovine organism, originally responsible for many cases of human disease
2. Development of sanitoriums where isolation, rest, and good diet contribute to a decline in transmission of the microorganisms
3. The discovery of streptomycin (by Waxsman in 1943) and other drugs that were effective in treatment and caused an immediate decline in the death rate

The success of all of these programs is evident from the statistics. Before 1850 the death rate for tuberculosis was 400/100,000 population. Today, the death rate in the United States is less than 2/100,000 population.

Tuberculosis is caused by organisms belonging to the genus *Mycobacterium.* They are long slender rods that have large amounts of lipid in their cell envelope, which accounts for their resistance to the usual staining procedures (see acid-fast stain in Appendix B). The mycobacteria grow slowly in the laboratory, as well as in the host, and it may take 2 to 3 weeks for colonies to develop on laboratory media.

Tuberculosis is most frequently acquired by contact with respiratory secretions of patients infected with *Mycobacterium tuberculosis.* Ninety percent of the cases in the United States are of the pulmonary type, but any organ or tissue may be involved. Ingestion of milk contaminated with *M. bovis* is still a major cause of disease in countries where pasteurization is not a routine practice. The first site of involvement, following infection, is the lung, where the lesion is called a tubercle (Figure 28-9). The course of the disease depends on many factors but ultimately relates to the cell-mediated immunity of the host. *M. bovis* or *M. tuberculosis* organisms are phagocytized but not killed by alveolar macrophages, where they multiply for several weeks. During this period the microorganisms may be disseminated to every major organ of the body. A hypersensitivity to the bacterial components in the macrophage develops primarily in the lung and results in necrosis around the tuberculous lesion. The necrotic lesion (Figure 28-10) is called ***caseation necrosis*** (caseation refers to a cheeselike consistency). Hypersensitivity to bacterial components is the basis of the tuberculin skin test (discussed later). The tubercles formed during infection will eventually heal to form calcified nodules, but may continue to harbor viable bacilli. These organisms can be reactivated—if at some later time the host's resistance is lowered—and produce disease a second time. The symptoms of reactivation depend on the location of the tubercle—lung, meninges, bone, or kidney.

The symptoms of tuberculosis are fever, fatigue, and loss of weight. The cough, which is characteristic of pulmonary involvement, may result in expectoration of bloody sputum. Clinical specimens (such as urine, bone, sputum) can be stained by the acid-fast technique or stained with fluorescent dyes and examined microscopically.

Disease caused by *M. tuberculosis* sensitizes the individual to the protein ***(tuberculin)*** of the tubercle bacillus, and this sensitivity can be detected by tuberculin skin tests. One tuberculin skin test uses a purified protein derivative (PPD) of the tubercle bacillus that is injected intradermally. Hypersensitivity is judged by measuring the amount of induration (hardening or infiltration of an area) at the site of injection after 48 to 72 hours. Generally, the larger the area of induration the more likely the person has tuberculosis. The tuberculin skin tests are an important technique for detecting tuberculosis in younger persons (see Figure 24-11).

Tuberculosis is a contagious disease, and con-

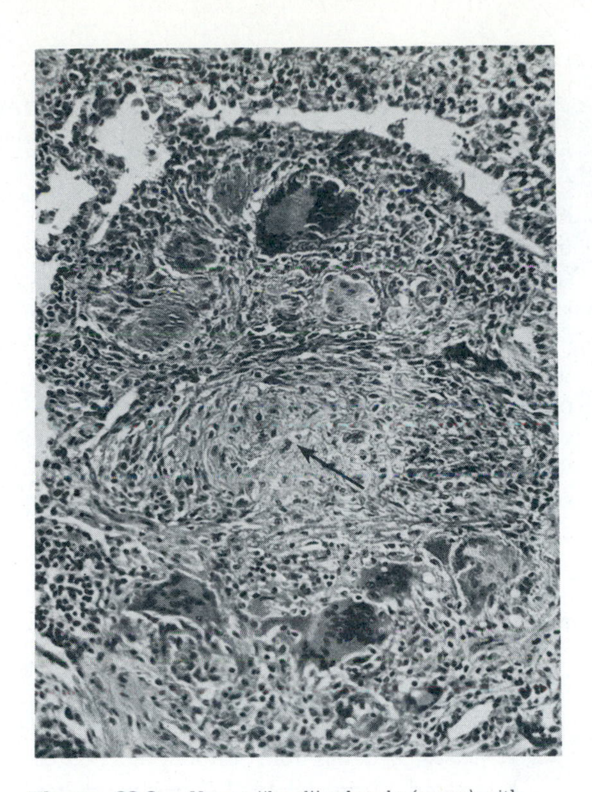

Figure 28-9 Young "hard" tubercle *(arrow)* with prominent epitheloid cells. (×120).

From Millard, M. In Anderson, W.A.D., and Kissane, J.M. (Eds.). Pathology *(Vol. 2). The C.V. Mosby Co., St. Louis, 1977.*

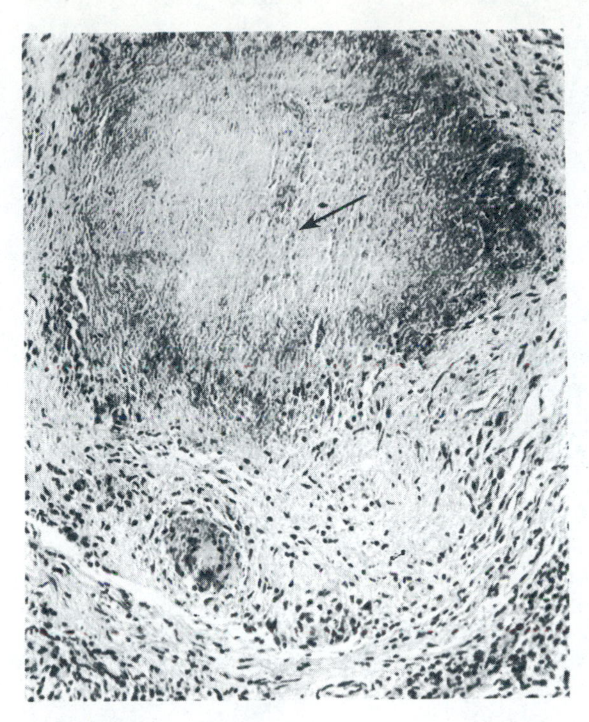

Figure 28-10 Soft tubercle demonstrating caseation necrosis *(arrow)*.

From Millard, M. In Anderson, W.A.D., and Kissane, J.M. (Eds.). Pathology *(Vol. 2). The C.V. Mosby Co., St. Louis, 1977.*

trol is maintained by isolation of the patient and use of chemotherapy. Both chemotherapy and chemoprophylaxis are carried out by the administration of ***isoniazid*** (INH). In addition to INH, other drugs are used in chemotherapy in some type of combination to prevent emergence of drug-resistant forms; they include ***streptomycin, ethambutol,*** and ***rifampin.*** Once the patient's test is bacteriologically negative for *M. tuberculosis,* isolation is not required. A vaccine called the ***BCG (Bacille-Calmette-Guerin)*** vaccine is available for those who do not have or have not had tuberculosis. An individual who has or has had tuberculosis is still sensitive to the bacillus or its products, and immunization with BCG vaccine can cause a hypersensitivity response and even death in those with immunological disorders.

Pulmonary diseases caused by mycobacteria other than *M. tuberculosis* and *M. bovis* are called nontuberculous. The organisms involved are saprophytes found in the soil and water and include *M. kansasii, M-avium-intracellulare, M. scrofulaceum,* and others. Infections by this group are not communicable and do not require isolation as in tuberculosis.

Pharyngitis and rheumatic fever

The oral cavity contains a large group of indigenous species belonging to the genus *Streptococcus.* One of these, *S. pneumoniae,* has already been discussed. Another species, *S. mutans,* is the etiological agent of tooth decay (caries), but the most important pathogens belonging to this group are called the ***group A β-hemolytic streptococci.*** There are several groups of streptococci that have been separated based on differences in carbohydrate cell wall antigens. The species name given to the β-hemolytic group A streptococci is *S. pyogenes,* which is responsible for pharyngitis and its sequelae, rheumatic fever and acute glomerulonephritis. If the strain of *S. pyogenes* causing pharyngitis produces an ***exfoliative toxin, scarlet fever*** may result, in which a red rash develops on the body and the tongue appears a bright red. The rash is caused by the toxin, whose genetic determinants are carried by a temperate bacteriophage. Pharyngitis occurs more frequently among school age children, and in about 3% of those who remain untreated, rheumatic fever may be a sequela to infection. ***Rheumatic fever*** is characterized by an inflammatory reaction in the heart in which

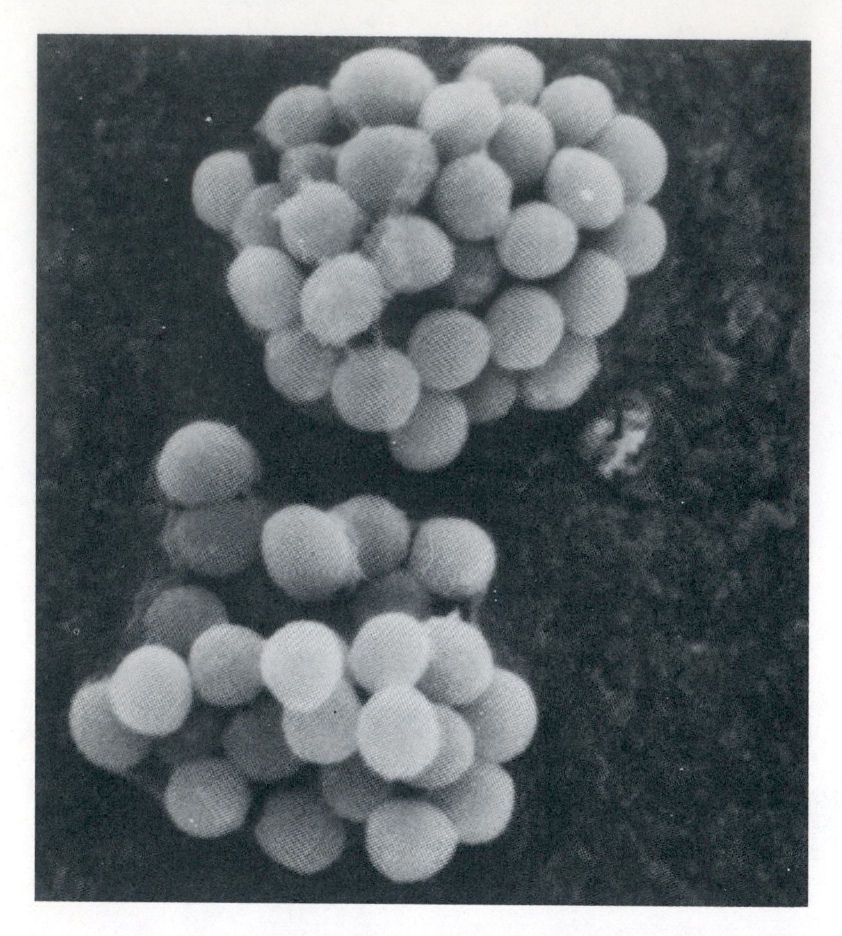

Figure 28-11 Scanning electron micrograph of *Staphylococcus aureus.* (×14,000.)
Courtesy Y. Zoshii.

the heart valves are permanently damaged.

Rheumatic fever is believed by many investigators to be primarily an immune disease in which antibodies to the streptococci react with heart muscle, presumably because of a similarity between heart antigen and streptococcal antigen. This is the most popular theory, although others have been suggested. Rheumatic fever is a particularly debilitating disease requiring continual antibiotic therapy because the damaged heart muscle can be easily colonized by various microorganisms from the circulation.

Group A streptococci are sensitive to penicillin. They can be identified by their sensitivity to bacitracin, by immunofluorescence, or by other serological techniques.

BACTERIAL DISEASES ACQUIRED THROUGH THE SKIN INCLUDING THE EYE

The skin and adjoining mucosal surfaces—confunctiva, nares, and genital tract—offer a natural barrier to invasion by pathogenic species (see Chapter 22). Some bacterial species such as staphylococci are indigenous to the skin and mucosal surfaces. Minor abrasions, insect bites, and traumatization by surgical procedures break the host's natural barriers and permit colonization by transient as well as indigenous microbial species. Once the barrier has been broken, microorganisms find an environment that is usually rich in nutrients and will support their growth. In addition, the organism may be endowed with certain properties, such as enzyme or toxin production, that permit degradation of tissue and promote the spread of infection to adjoining healthy tissue. Many infectious diseases are manifested by cutaneous lesions; however, the present discussion includes only those bacterial diseases that affect the skin or are acquired through the skin. They are divided into the following groups.

1. Sexually transmitted diseases. These are discussed separately in Chapter 30.
2. Diseases resulting from surgical procedures or related manipulations. These infections are discussed in Chapter 33.
3. Infectious diseases involving the skin or mucosa. Included in this group are the streptococci, the staphylococci, the agents of leprosy, yaws, and the agents causing diseases of the eye.
4. Infectious disease transmitted by animal contact or acquired through the bite of arthropod vectors harbored on the animals. Included in this group are the agents of anthrax, tularemia, leptospirosis, plague, brucellosis, relapsing fever, and rickettsial diseases.
5. Infections acquired by accidental injury. Included in this group are the agents of tetanus, gas gangrene, and those infecting burn wounds.

Diseases involving skin and mucosa

Staphylococcal skin diseases. If one had to select the "all-purpose" infectious agent, the award would probably be given to staphylococci (Figure 28-11). These organisms are physiologically suited to grow under most environmental conditions and can infect nearly all human tissues. They can be involved in diseases of the central nervous system, respiratory tract, cardiovascular system, and intestinal tract. They are best known for those diseases involving the skin, such as boils and carbuncles.

There are two principal species of *Staphylococ-*

cus: *S. epidermidis* and *S. aureus*. *S. epidermidis* is indigenous to the skin, whereas *S. aureus* may be present on the skin and in the nares of 30% of the population. *S. aureus* is considered the most pathogenic species, whereas *S. epidermidis* is considered an opportunistic species, causing disease only in the compromised host such as persons receiving heart transplants. A superficial staphylococcal disease that involves the hair follicles is called ***folliculitis,*** and those involving the subcutaneous tissue are called ***boils*** or ***furuncles*** (Figure 28-12). In young children a highly contagious staphylococcal disease affecting the superficial layers of the skin, particularly the face, is called ***impetigo*** (Figure 28-13). Impetigo is also caused by group A streptococci. A rare staphylococcal skin infection occurs in children under 6 years of age, especially infants, and is called ***scalded skin syndrome.*** This disease is caused only by certain strains of *S. aureus* that produce a toxin capable of causing the skin to peel (exfoliation) in large sheets, leaving the patient with a red glistening surface that appears like scalded skin (Figure 28-14).

Superficial skin diseases caused by the staphylococci require no chemotherapeutic treatment. Boils require drainage and chemotherapy to prevent dissemination of viable microorganisms to other parts of the body. *S. aureus* usually carries antibiotic-resistance plasmids; therefore, sensitivity testing is necessary before therapy is initiated.

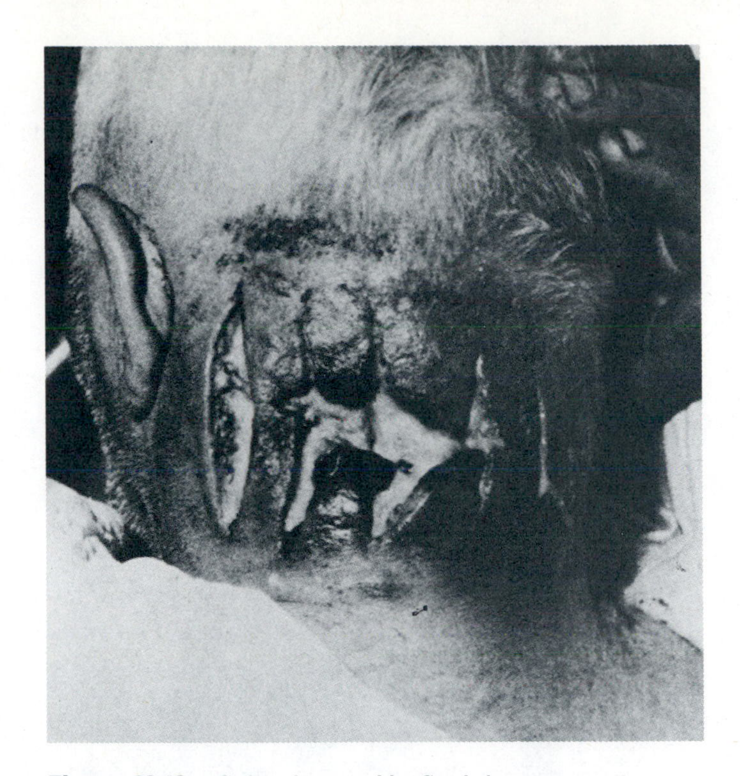

Figure 28-12 Carbuncle caused by *Staphylococcus aureus.*

From Norden, C.W., and Ruben, F.L. In Wehrle, P.F., and Top, F.H., Sr. (Eds.). Communicable and infectious diseases *(9th ed.). The C.V. Mosby Co., St. Louis, 1981.*

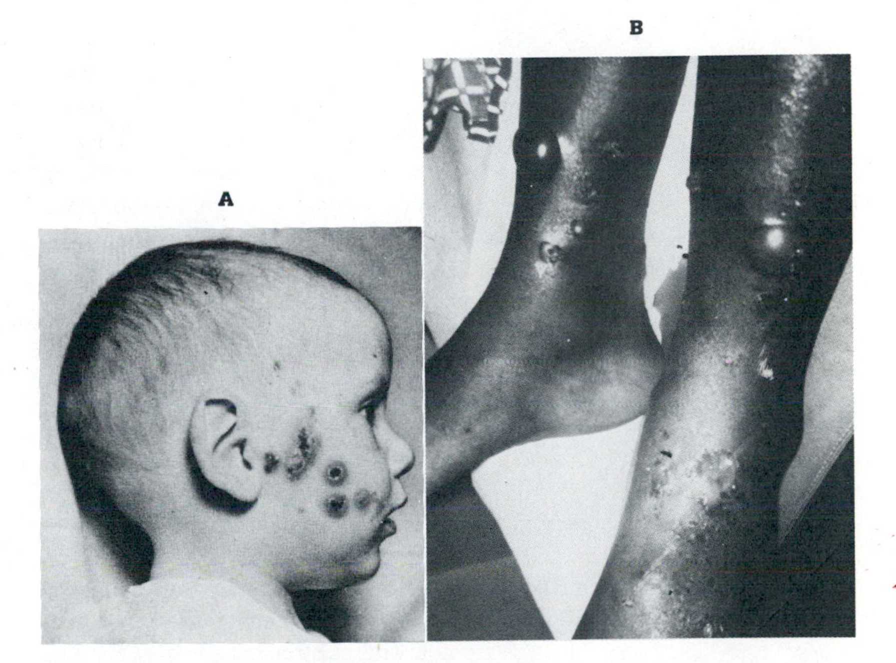

Figure 28-13 Impetigo. **A,** Facial lesions. **B,** Bullous lesions seen on lower extremities.

A, *From Dillon, H.C., Jr. In Wehrle, P.F., and Top, F.H., Sr. (Eds.).* Communicable and infectious diseases *(9th ed.). The C.V. Mosby Co., St. Louis, 1981.* **B,** *From Dillon, H.C., Jr. In Boyd, R.F., and Hoerl, B.G.* Basic medical microbiology *(2nd ed.). Little, Brown & Co., Boston, 1981.*

Leprosy. Leprosy is a disease that disfigures, cripples, and blinds. The origin of leprosy is unknown, but the disease was referred to in biblical writings. Leprosy was considered a disease of Divine punishment because of the revelations to Moses from God—revelations that the leper was unclean and would always dwell alone. Even today lepers are considered outcasts. For centuries leprosy was centered among Eastern civilizations, and during the earlier periods lepers were rounded up like cattle and then cast out of the cities. They were left to fend for themselves often without food or care. To ensure that the leper was easily recognized, local citizens clothed him in a uniform consisting of a black cloak and a tall hat. A mask was included in the attire if the face was disfigured. Leprosy spread rapidly to Europe following the return of Christians from the crusades. The number of lepers increased to such monumental proportions that hospitals were erected just to isolate them and prevent them from roaming the countryside. One of the first leprosariums (leper hospital) in Europe was erected in 1067 with the help of the famous Span-

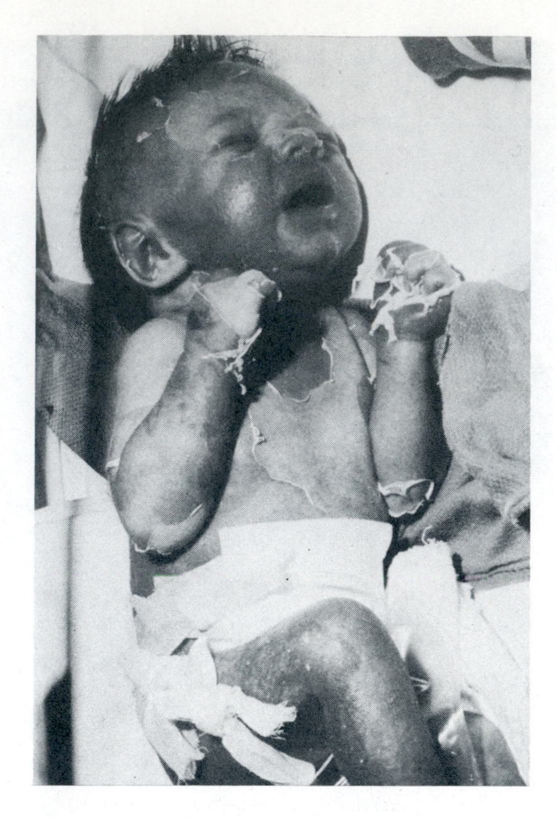

Figure 28-14 Scalded skin syndrome caused by *Staphylococcus aureus.*

Glasgow, L.A. Reprinted by permission of New England Journal of Medicine ***282:****114, 1970.*

ish hero El Cid. By the 13th century there were over 19,000 leper hospitals in Europe whose only purpose was isolation, not medical care. Leprosy became pandemic by the 16th century and then, just as suddenly, declined. Several reasons have been suggested for its decline: (1) all the lepers had died; (2) there was a loss of pathogenicity of the bacillus *Mycobacterium leprae;* and (3) the loss of pathogenicity of the bacillus coincided with the appearance of plague. The last theory appears to be the most reasonable, but has not been proved, and the decline of the disease is still a mystery. Plague is believed to have killed 25 million people between 1500 and 1700, and within the confines of the leprosarium plague could have been a devastating disease. Today, 11 million cases of leprosy persist in the world, but the disease is no longer pandemic and is confined to tropical and subtropical areas.

The agent of leprosy was discovered in biopsied tissue by Hansen in 1869, and for many years the disease was called Hansen's disease. The organism cannot be grown on laboratory media or in tissue culture, but it can be cultivated, albeit slowly, in the footpads of mice and in the armadillo. Another unusual characteristic of this microorganism is its incubation period in the human body, which may be 3 to 5 years or longer. Infection terminates in two types of response, depending on the cell-mediated immune response of the host. If the host has a relatively competent cell-mediated immune response, ***tuberculoid leprosy*** develops, in which the skin and associated nerves are affected. The lesions are relatively flat and not disfiguring. If the cell-mediated immune response of the host is incompetent, ***lepromatous leprosy*** occurs, in which nodules appear anywhere on the body. There is usually massive enlargement of the facial features, causing disfigurement (Figure 28-15).

Dapsone, a sulfa drug, is the principal chemotherapeutic agent in the treatment of leprosy and must be administered over long periods of time to eliminate all of the bacilli. The early cases of disease are subject to a greater cure rate with proper treatment. Even with modern treatment, leprosy, in many parts of the world, instills prejudice. In 1952, for example, 100 leper patients in an Asian country were removed from a hospital and burned alive.

Little is known of the epidemiology of leprosy. One member of a family may contract leprosy and never transmit it to other family members. Missionaries have worked for years with lepers and have not contracted the disease, whereas others have. A well-known example is Father Damian, who worked with lepers in Hawaii and eventually contracted the disease and died on the island.

Yaws. Yaws is a tropical communicable disease that is not seen in the United States but is of special interest because the organism causing disease, *Treponema pertenue,* cannot be morphologically distinguished from the agent causing syphilis, *Treponema pallidum.* In addition, the disease is usually contracted by those under 2 or 3 years of age by direct contact but not venereally. The disease is seldom fatal, but the lesions, which may appear anywhere on the body, can be disfiguring (Figure 28-16). Eighty percent of those with yaws have a positive serological test for syphilis. This cross-immunity implies a common ancestry for the agents of these two diseases. Syphilis is not common in tropical areas where yaws is endemic among the population; likewise, yaws is rare in temperate zones where syphilis predominates. One explanation for this unusual characteristic is that species of *Treponema* are sensitive to environmental conditions outside the host. Thus, in tropical areas, where few clothes are worn, the organism has a chance of infecting any area of the body and can be transmitted to others. In temperate zones, where clothes are worn to cover most of the body, only sexual con-

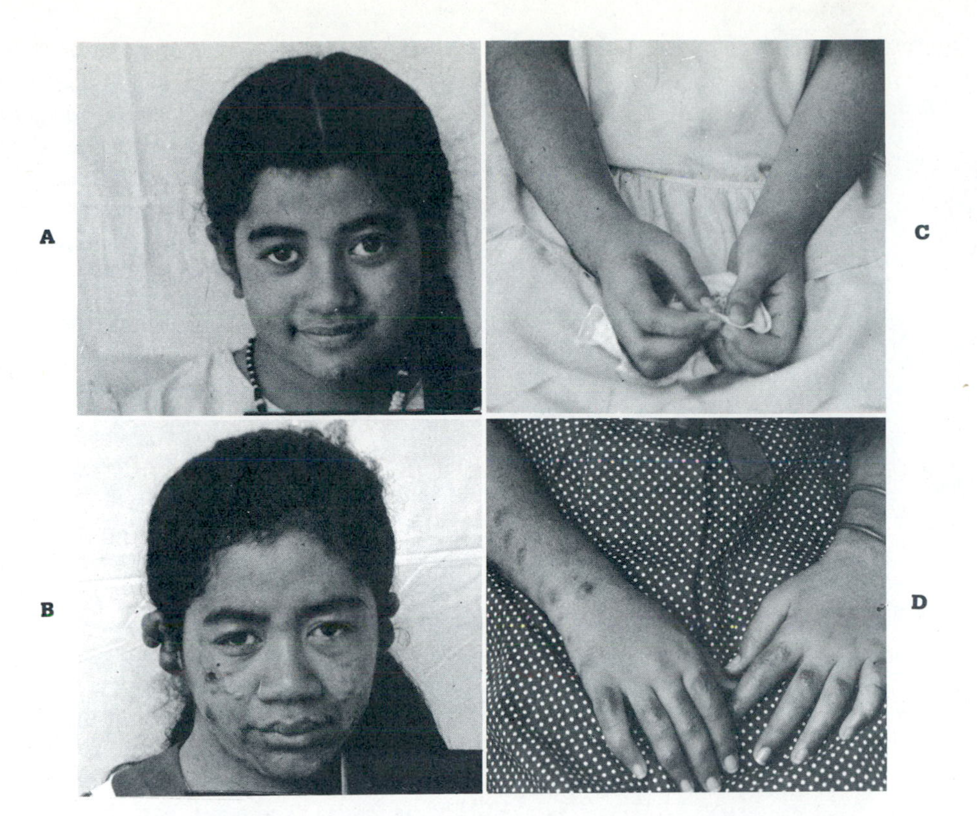

Figure 28-15 Lepromatous leprosy. **A,** Hawaiian girl, 11 years of age with early lepromatous leprosy. Inconspicuous lesions are seen on cheeks and chin. **B,** Same girl at 13 years of age. Observe nodular thickening of right ear, lesions of cheeks and chin, and dramatic change in facial appearance. **C,** Forearms and hands of patient at 11 years of age. No apparent lesions. **D,** Hands of patient at 13 years of age. Nodular lesions are apparent on forearm and fingers.

From Binford, C.H. In Anderson, W.A.D., and Kissane, J.M. (Eds.). Pathology *(Vol. 1). The C.V. Mosby Co., St. Louis, 1977.*

tact will permit transfer because treponemes cannot survive the colder outside temperatures. As in syphilis, penicillin is the drug of choice for treatment of yaws. The relationship of yaws to veneral disease is also discussed in Chapter 30.

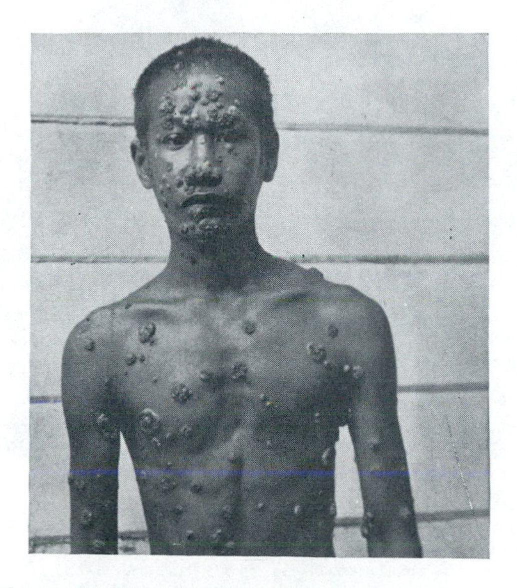

Figure 28-16 Yaws. Secondary lesions.

From Air Force Institute of Pathology, AFIP 39201.

Diseases of the eye

Haemophilus aegyptius and *Chlamydia trachomatis* are etiological agents of diseases involving the eye. *H. aegyptius* is an infrequent cause of ***conjunctivitis,*** or ***pinkeye,*** which is not a serious disease. *C. trachomatis* is the cause of two types of conjunctivitis: trachoma and inclusion conjunctivitis. ***Trachoma*** is the major cause of blindness in the world and is endemic in areas of India (Figure 28-17). The agent is transferred from person to person by fingers or fomites. The infection initially involves the conjunctiva but later progresses to also involve the cornea. Partial or permanent blindness can result if treatment is not instituted. *C. trachomatis* is also found in the female genital tract and can be transferred to the newborn to cause ***inclusion conjunctivitis,*** a less severe disease than trachoma. In addition,

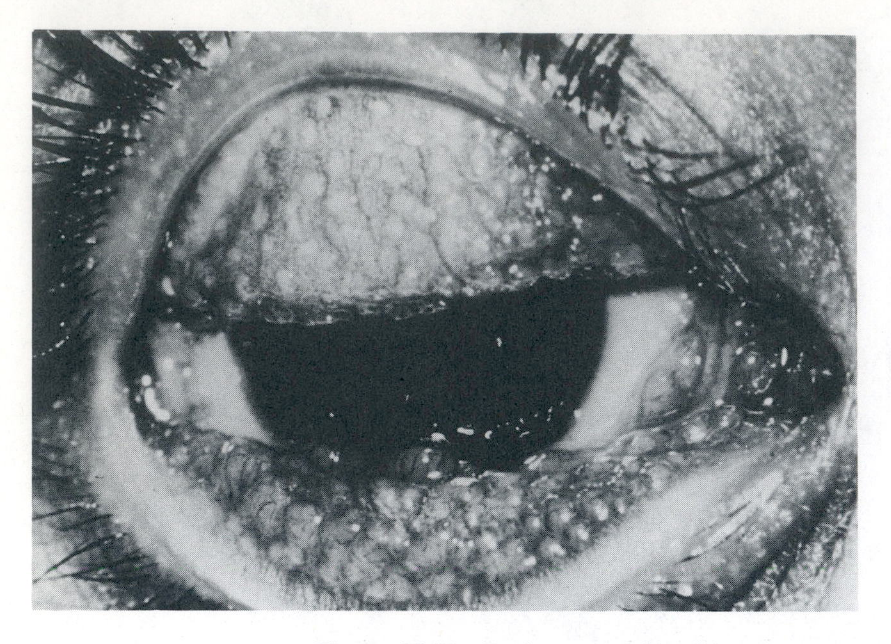

Figure 28-17 Trachoma caused by *Chlamydia trachomatis.*

Courtesy Phillips Thygeson, Los Altos, Calif. From Smith, A.L., Microbiology and pathology *(12th ed.). The C.V. Mosby Co., St. Louis, 1979.*

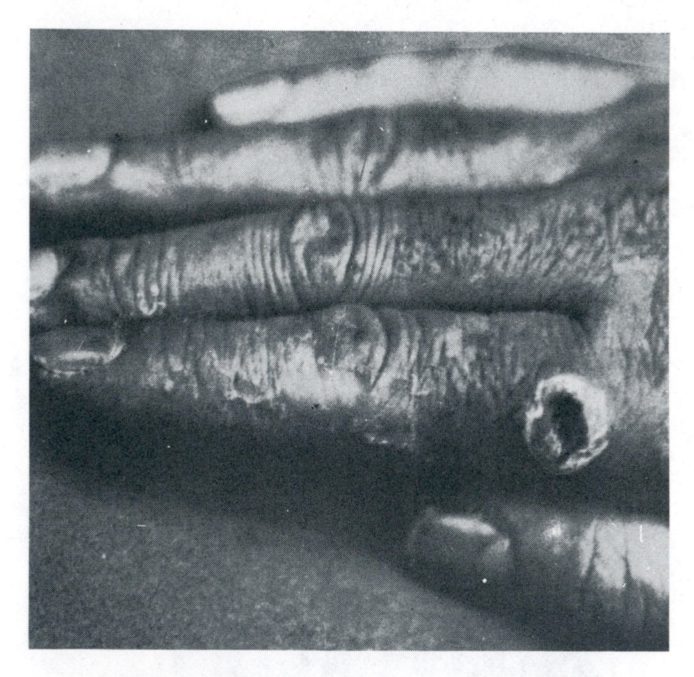

Figure 28-18 Tularemia.

Courtesy T. Cajigas, Washington, D.C., and Edward Francis, National Institutes of Health. From Top, F.H., Sr., and Werhle, P.F. (Eds.). Communicable and infectious disease *(9th ed.). The C.V. Mosby Co., St. Louis, 1981.*

genital secretions containing the chlamydial agent can contaminate swimming pools not properly chlorinated and cause conjunctivitis. Conjunctivitis caused by genital secretions is a mild and self-limiting disease. Tetracyclines are the drugs of choice in the treatment of chlamydial disease, and ampicillin or chloramphenicol are used against *H. aegyptius.*

Diseases resulting from animal contact or contact with arthropod vectors

Tularemia. Tularemia is a disease of rodents such as rabbits, muskrats, and moles. It is transmitted to humans through the bite of deer flies and ticks that have been in contact with infected animals, or by direct contact with infected meat, as during the skinning of rabbits ***(rabbit fever).*** The causative agent of disease is *Francisella tularensis,* a gram-negative motile bacillus. The most common manifestations of disease are the enlarged local lymph nodes and ulcerated lesions at the site of contact with the infectious agent (Figure 28-18). The agent can also be transferred to the eye by contact with contaminated fingers, resulting in the formation of conjunctival lesions. Streptomycin is the most effective agent in treatment.

An example of how individuals can be infected under unusual circumstances was illustrated in 1979. Nine members of a sheep-shearing crew became ill with fever and headache. Three of the patients had axillary lymph node enlargement, and all had lesions on the left hand. It was also noted that the sheep were covered with wood ticks. The lesions on the left hand were accounted for by the shearing process because shearers part fleece with their bare left hands while shearing with the right hand. This procedure apparently ruptures the ticks, spilling blood onto the left hand, thereby initiating infection.

Brucellosis. What today is called ***brucellosis*** was originally described as ***undulant fever, Malta fever,*** or ***Mediterranean fever.*** The disease was common among the peoples of the Mediterranean and also afflicted British servicemen stationed in the area. In 1866 Bruce isolated coccuslike organisms from infected humans and injected them into monkeys and reproduced the intermittent fever characteristic of human disease. The source of the human disease was not known at the time, but in 1915 was accidentally discovered. Goats were being prepared for inoculation with the organisms isolated from a human case of undulant fever when it was discovered that the goats already contained a high titer of antibody to the organisms. Further investigations revealed that the goats excreted the organisms in their milk and that the milk from such goats was the primary cause of the characteristic fever. Later it was discovered that not only goats but also other animals harbored the microorganisms. Earlier, in 1897, Bang isolated a microorganism responsible for contagious abortion in cattle and called the microorganism *Bacillus abortus.* These two diseases were studied independently and only in 1918 was the close morphological, cultural, and serological relationship between them noted. Four species have now been recognized that infect animals and can cause infections in humans, *Brucella melitensis* (goats), *B. abortus* (cattle), *B. suis* (swine), and *B. canis* (dogs). All brucellae are aerobic, gram-negative coccobacillary forms that can infect domestic and semidomestic animals (see Chapter 34).

A vaccination program for cattle has reduced the infection rate from 11.5% in 1934 to less than 1% in 1980. Consequently, the number of recorded cases of human infection in the United States has dropped from 6000 cases in 1947 to 200 cases in 1980. Contact with contaminated meat from cattle or swine is the most common cause of infection in the United States. Meat packers are at the highest risk of contracting the disease, but veterinarians and livestock producers are also subject to infection. Outside the United States brucellosis is frequently acquired by drinking raw contaminated milk or eating contaminated cheese.

Brucella, like the agent of tuberculosis, can survive and multiply in the macrophage and is a cause of chronic disease. Symptoms of disease include intermittent fever, chills, and body aches, which are often misdiagnosed as "stubborn influenza."

Serological tests, rather than isolation and identification of the organism, are used in diagnosis because the disease is often unrecognized for several weeks following infection. Tetracycline is the principal drug used in therapy but it is sometimes combined with streptomycin to prevent relapses.

Anthrax. Anthrax is a disease of herbivorous animals: sheep, horses, and cattle. It is caused by a spore-forming gram-positive bacillus called *Bacillus anthracis.* Anthrax in humans was originally called ***woolsorter's disease.*** Woolsorting in England was considered a healthy occupation until the mid 1800s when wool and hair from the Far East were introduced to the trade. John Henry Bell, a physician of the period, noted the relationship between the symptoms of disease and the occupation of woolsorting. He inoculated animals with blood from a victim of a fatal case of human anthrax and observed that all of the animals died. The anthrax bacillus was recovered from the dead animals, which confirmed the relationship between the organism and disease symptoms. This discovery, as well as the constant effort of others, prompted the development of measures for prevention of anthrax among woolsorters. One such measure was the soaking of wool and animal hair in formaldehyde solutions. Pasteur, in 1881, used an attenuated strain of the anthrax bacillus as an effective vaccine for cattle.

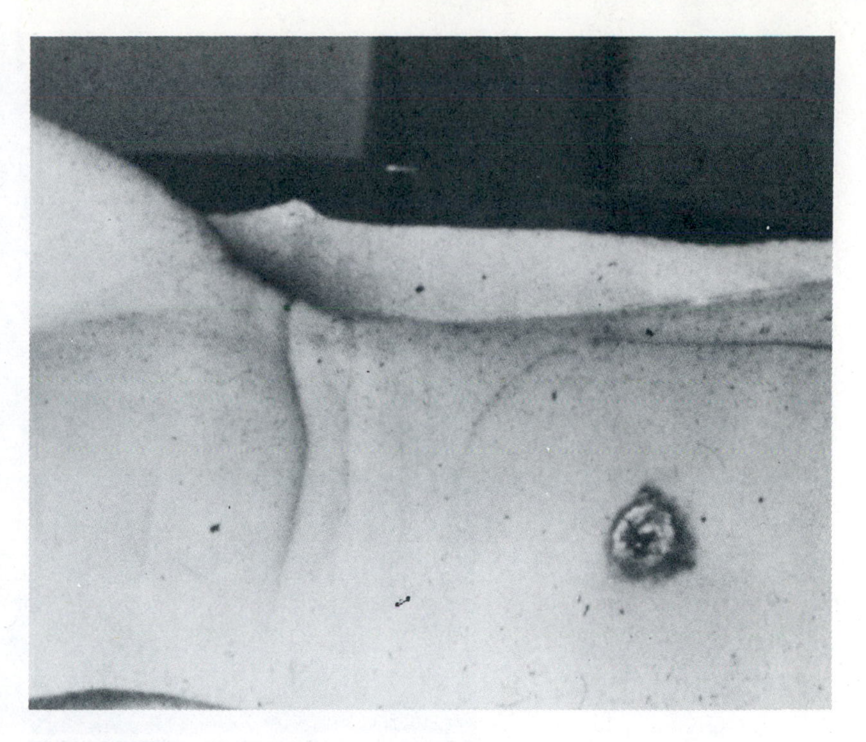

Figure 28-19 Anthrax. Cutaneous anthrax lesion. Note dark ulcerated area, called eschar.

From Brachman, P., In Boyd, R.F., and Hoerl, B.G. Basic medical microbiology *(2nd ed.). Little, Brown & Co., Boston, 1981.*

The anthrax spore is found in the soil and can be ingested by susceptible animals, which includes domesticated animals and wildlife, such as deer. The spores are often found on the hides of the animals, and most human infections result from contact with animal hides or products made from them. Recently a textile mill employee in the United States contracted anthrax while working with imported goat hair. Spores germinate in abraded areas of the skin or mucosa and cause the formation of depressed ulcerated lesions that have a dark necrotic center and are called ***eschars*** (Figure 28-19). If the bloodstream is invaded by the anthrax bacillus, the disease can be fatal.

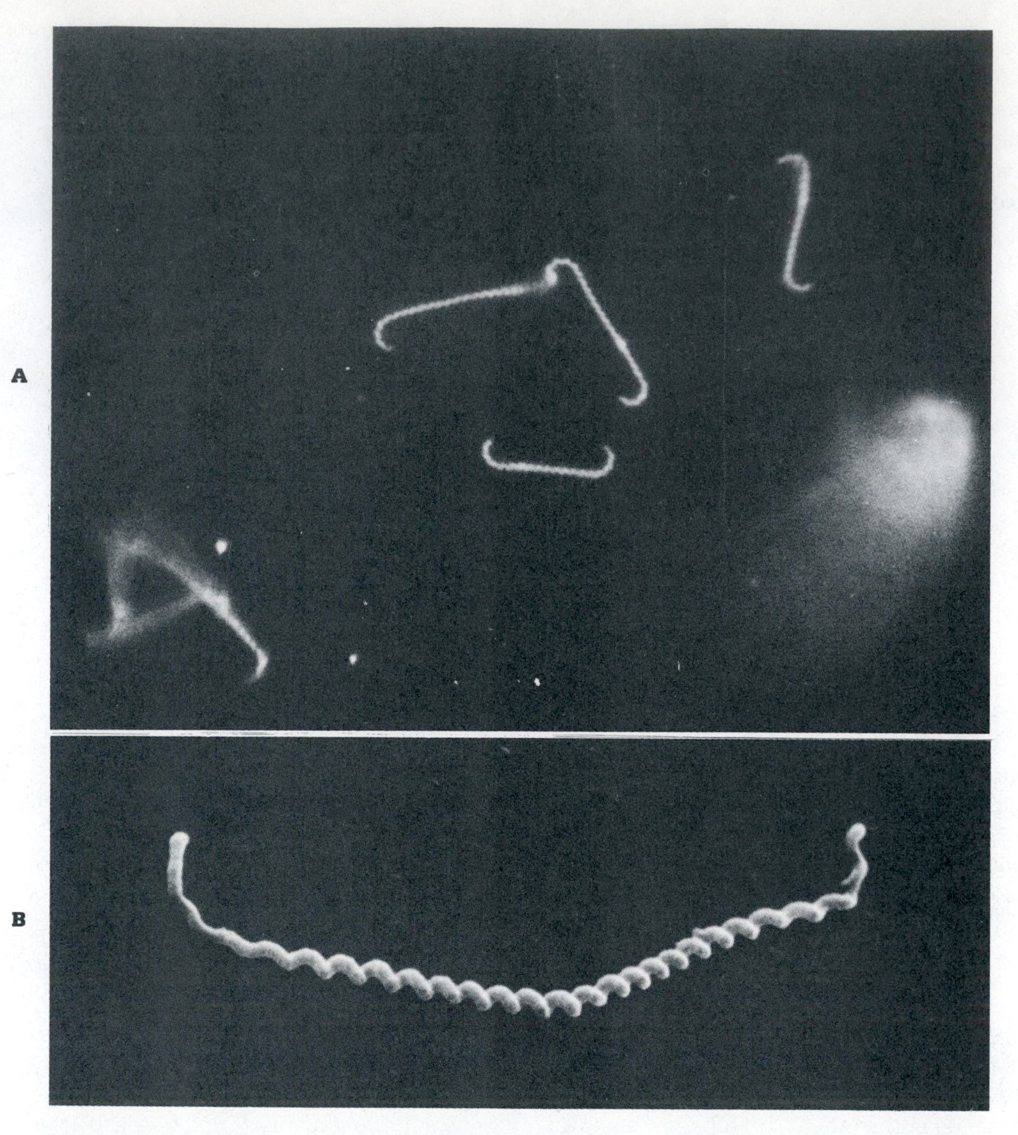

Figure 28-20 *Leptospira interrogans.* **A,** Darkfield microscopy. **B,** Scanning electron micrograph. (From Bromley, D., and Charon, N.W. J. Bacteriol. **137:**1406, 1979.)

Six or fewer cases of human anthrax are recorded each year in the United States, and the sources of infection are sometimes unusual but not unexpected. In one instance a woman on a Caribbean cruise purchased a souvenir drum made of animal hide. Anthrax lesions developed in the corner of her eye on the return trip to the United States. It was postulated that she touched the contaminated animal hide with her fingers and then transferred the spores to her eyes. Epizootics of anthrax in cattle are periodic in the United States, and once discovered the healthy cattle are quarantined while infected animals are destroyed and buried in quicklime or cremated. This process destroys spores and vegetative cells and prevents them from returning to the soil.

Vaccination of animals, particularly cattle, is the most important control measure. Penicillin is the primary chemotherapeutic agent used in the treatment of human anthrax.

Leptospirosis. Leptospirosis is an animal disease caused by microorganisms belonging to the genus *Leptospira,* such as *L. interrogans.* The organism is a spirochete with hooked ends (Figure 28-20), a characteristic that distinguishes it morphologcially from other spirochetes. Many wild rodents, particularly rats, are heavily infected with pathogenic leptospires, but many farm and domestic animals can also be sources of infection to humans.. Following animal infection the organism localizes in the kidneys and is excreted in large numbers during urination. Humans become infected by ingestion of contaminated water or by contact with articles or water contaminated by infected animals. Disease in humans is characterized by chills, fever, and muscular pain. The most severe form of the disease, called ***infectious jaundice,*** or ***Weil's disease,*** is characterized by jaundice and hemorrhage beneath the skin. Penicillin is effective in eradicating the microorganisms from the host.

Leptospirosis has been traditionally considered an occupational disease for sewer, rice fields, sugar cane, and abattoir workers, but today infections are frequently the result of recreational activities. Seven cases of leptospirosis were reported recently among children swimming in a stream in Tennessee. The stream was slow moving and recently visited by infected animals who probably urinated in the water.

Plague. Plague is a disease of ancient origin, appearing wherever rodents, such as rats, and their accompanying fleas dwell. Rats have been associated with humans for over 25,000 years, but the first recorded plague epidemic was in biblical times. The Philistines had conquered the Israelites and captured the Ark of the Covenant, but their victory proved to be short-lived since great epidemics of plague ravaged the population. The Philistine sorcerers told the population the disease was caused by their sins.

Plague was originally concentrated in Asia, but Europe became devastated by two pandemics in the sixteenth and fourteenth centuries. The reason for the 800-year hiatus has never been explained. The first pandemic, in 540 AD, lasted 50 to 60 years and was believed to have resulted in over 100 million deaths.* The second pandemic, called the Black Death (because of the subcutaneous hemorrhages that darkened the skin during infection) began in the 1300s and lasted for over 300 years. Nearly 43 million people in Europe are believed to have perished from the plague in the fourteenth century. After the great fire of London in 1662 epidemic plague disappeared from Europe but remained in Asia. The third pandemic, which began in China and included the United States, began in 1885 and was responsible for over 12 million deaths. The death rate dropped precipitously after 1917 because of rat control policies and the discovery of antibiotics.

There were many in the Middle Ages who thought plague was a contagious disease, and in the fourteenth century, for example, generals used to catapault plague victims into each other's cities, hoping, of course, to spread the disease. The plague bacillus was discovered in 1894 by Yersin and Kitasato, who also observed that the organism was transmitted by the rat flea, and

* Many historians believe that the disappearance of some ancient civilizations may have been the result of the plague.

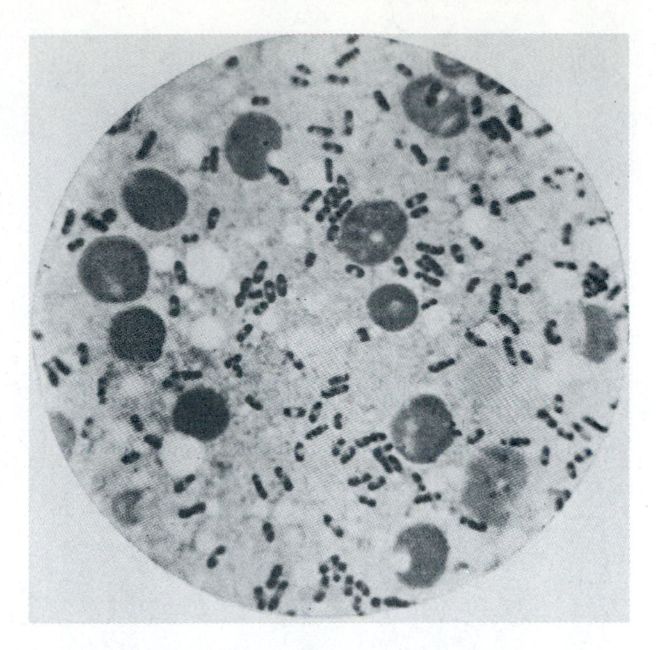

Figure 28-21 *Yersinia pestis* isolated from lymph gland and stained. Note bipolar staining.

From Gillies, R.R., and Dodds, T.C., Bacteriology illustrated. *Churchill Livingstone, Edinburgh, 1976.*

only when the rat died (rats are susceptible to plague) did the fleas seek other habitats, such as humans. The flea's stomach becomes blocked, and normal feeding is prevented after feeding on blood infected with plague bacilli. In an effort to feed normally the flea regurgitates the bacilli-laden stomach contents and passes them into the victim's punctured skin.

The etiological agent of plague is *Yersinia pestis,* a gram-negative bacillus that exhibits bipolar staining (Figure 28-21). In the United States wild rodents, such as ground squirrels, marmots, prairie dogs, wood rats, and rabbits carry the plague bacillus as well as their attendant fleas. Disease in these animals is called ***sylvatic plague.*** In addition to becoming infected from the bite of fleas, hunters have been shown to also acquire disease by contact with infected carcasses during the skining of rabbits or prairie dogs. Plague is not a contagious disease unless the lungs of the victims are infected (pneumonic plague) and bacilli, present in the saliva, can be transmitted by droplet spread. Plague is characterized by fever and the appearance of swollen lymph nodes (buboes) under the armpits, in the groin, and any lymph glands draining the areas in which the bite occurred. The bacilli migrate into the blood stream to infect all organs and in the skin cause hemorrhages. Death is usually caused by toxemia resulting from the heavy concentrations of bacilli in the blood. Streptomycin and tetracycline have proved effective in treatment of the disease.

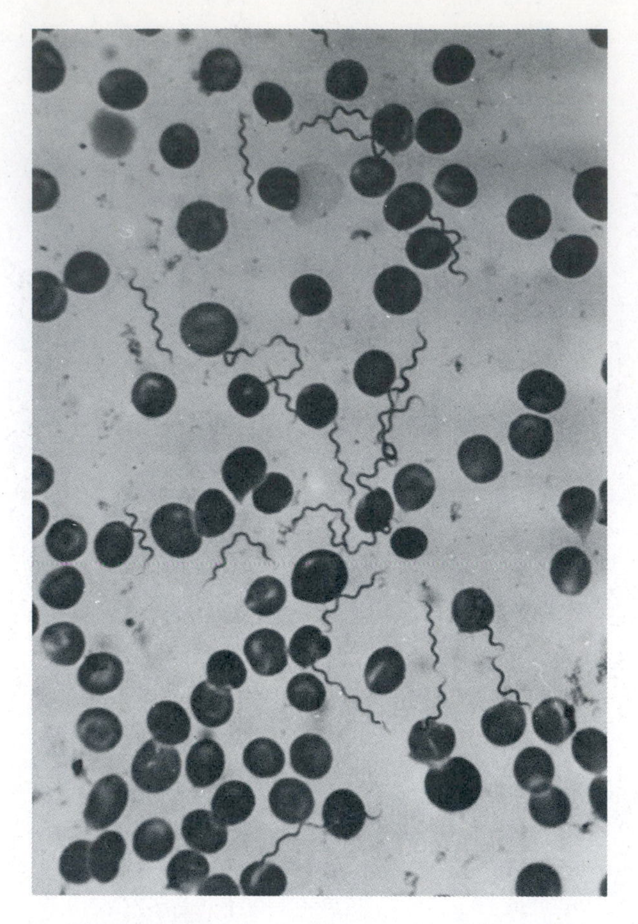

Figure 28-22 Photomicrograph of *Borrelia hermsii* in rodent blood.

From Burgdorfer, W. In Manual of clinical microbiology *(3rd ed.). American Society for Microbiology, Washington, D.C., 1980.*

Relapsing fever. Relapsing fever is a disease caused by the spirochete belonging to the genus *Borrelia* (Figure 28-22). The two most important human pathogens are *B. recurrentis* and *B. duttonii.* Humans are the only reservoir of the infecting borrellia, and human-to-human transmission occurs through the intervention of body lice or infected ticks. Lice become infected by feeding on infected humans. The salivary glands of the louse are not invaded by the pathogen, and the latter are not transmitted to humans by biting. Human infection occurs by scratching, which crushes the louse and hence contaminates the wound. Transmission of the etiological agent of tick-borne relapsing fever, however, does take place via biting and injection of contaminated tick saliva. Only tick-borne relapsing fever occurs in the United States and then infrequently (approximately 10 cases per year). The disease is characterized by the undulating nature of the fever, during which time large numbers of organisms are present in the bloodstream. The fever subsides because the immune system of the host is in the process of destroying most of the viable microorganisms. Some microorganisms, however, undergo mutation and are protected from the immune system. These mutants multiply and cause a recurrence of the fever until the immune system again responds to the mutants and destroys them. This type of response can occur repeatedly. Death, when it occurs, is usually caused by myocarditis. Tetracycline and chloramphenicol are the drugs of choice in the treatment of infection.

Rickettsial diseases. The genus *Rickettsia* contains a number of species that are pathogenic to humans. Except for one species, *Rochalimaea quintana,* the rickettsiae cannot be cultivated on ordinary laboratory media and require a living host. In addition, except for one species, *Coxiella burnetii,* all rickettsial diseases in humans are transmitted by arthropod vectors, such as lice, fleas, ticks, and mites. Nearly all rickettsial diseases are characterized by fever, rash, and enlargement of spleen and liver. Two rickettsial diseases are of particular significance: typhus fever and Rocky Mountain spotted fever.

Wherever human misery has resulted from war, famine, floods, an other disasters, typhus surfaces. The origins of typhus are unknown, probably because of its confusion with other fevers such as typhoid fever. Typhus is transmitted by lice, and lice prevail wherever humans congregate and filth predominates. The human louse can live only on humans and as long as clothes are changed infrequently or are not laundered, the louse has a chance to multiply and infect others. The louse is also affected by the typhus organisms and lives only long enough to contaminate neighboring lice and to bite its human host. Lice will leave the dying patient when the fever has become unbearable for both louse and host. Typhus prevailed in earlier civilizations where poverty, famine, and wars were a permanent staple of human existence. In the 16th century prisoners confined to jails for long periods of time, before coming to trial, often brought their jail-starved lice with them. The lice were transmitted to others during the trial and exacted a rather unusual type of justice. In 1577, one such prisoner, Robert Jencks was brought to trial and sentenced to have his ears removed. Shortly afterward the Chief Baron, the Sergeant at Law, two sheriffs, one knight, five justices of the peace, and most of the jury at his trial died of typhus. Robert Jencks survived 33 years after the trial.

The Spanish brought typhus to the New World

Table 28-3 Human rickettsial diseases

DISEASE	CAUSATIVE AGENT	MODE OF TRANSMISSION	GEOGRAPHIC DISTRIBUTION	RESERVOIR OF AGENT
Typhus group				
Epidemic typhus	*R. prowazekii*	Infected louse feces rubbed into abrasion	Worldwide	Humans
Murine typhus	*R. mooseri (typhi)*	Infected flea feces rubbed into abrasion	Worldwide	Rats and field mice
Spotted fever group				
Rocky Mountain spotted fever	*R. rickettsii*	Bite of infected tick	Western hemisphere	Rodents
Asian tick typhus	*R. siberica*	Bite of infected tick	Asia	Rodents and dogs
African tick typhus	*R. conori*	Bite of infected tick	Africa	Rodents and dogs
Ricekttsial pox	*R. australis*	Bite of infected mite	Worldwide	Marsupials or house mice
Scrub typhus	*R. tsutsugamushi*	Bite of infected mite	Japan, Southeast Asia, and Pacific islands	Rodents
Q fever	*Coxiella burnetii*	Inhalation of or contact with dust particles contaminated with parturition products of cattle	Worldwide	Cattle, sheep, or goats
Trench fever	*R. quintana*	Bite of human body louse	Worldwide	Humans

and as described by A.L. Baron* "In exchange for gold and silver the natives received malaria, dysentery, smallpox, tuberculosis, plague and typhus." It was recorded in 1576 that 2 million Aztecs died of typhus, and from the sixteenth to the nineteenth century typhus extracted a heavy toll in human lives, particularly during wars. Dead soldiers were relieved of their clothes to provide warmth for the living and unwittingly, the louse was provided a new viable host on which to feed. The scientific community did not realize the connection between the louse and the disease until the nineteenth century. The organism responsible for infection, *Rickettsia prowazekii,* was isolated and identified by Howard Ricketts in the early 1900s. At the age of 40, shortly after his discovery, Ricketts succumbed to the disease. During the same period a Serbian physician, Stanislas von Prowazek, forsook a career of medicine to study typhus fever, and he too died from the disease just before his fortieth birthday.

Epidemics of typhus still occur (seldom in the United States), but the death rate does not reach the proportions it once did. The greatest deterrent to typhus was the discovery of DDT which, just before the end of World War II, was used in delousing programs and saved countless lives.

Typhus is a particularly virulent disease. Two to three weeks after the bite the organism multiplies and reaches the vital organs. A fever ensues that is so high the patient is rendered feeble and barely able to move. There is an intense headache and finally a period of unconsciousness. Spots appear on the body that turn from pink to dark purple, and if the patient survives, several weeks of recuperation are required before the patient is able to assume a normal life.

The most common rickettsial disease in the United States is Rocky Mountain spotted fever, which is caused by *Rickettsia rickettsii.* Most cases are observed in the Southeast and Western United States. The primary vector in the Southeast is the dog tick, which is harbored by several mammals including the rabbit, fox, woodchuck, deer, and squirrel. The wood tick is the most common vector in the Western United States. Rocky Mountain spotted fever is often misdiagnosed because of its similarity to other diseases in which a rash and fever appear. The organism is disseminated to all organs and can cause intravascular coagulation, which is a frequent cause of death. Tetracycline is the recommended therapeutic agent for nearly all rickettsial infections.

* Baron, A.L. *Man against germs.* Scientific Book Club, London, 1958.

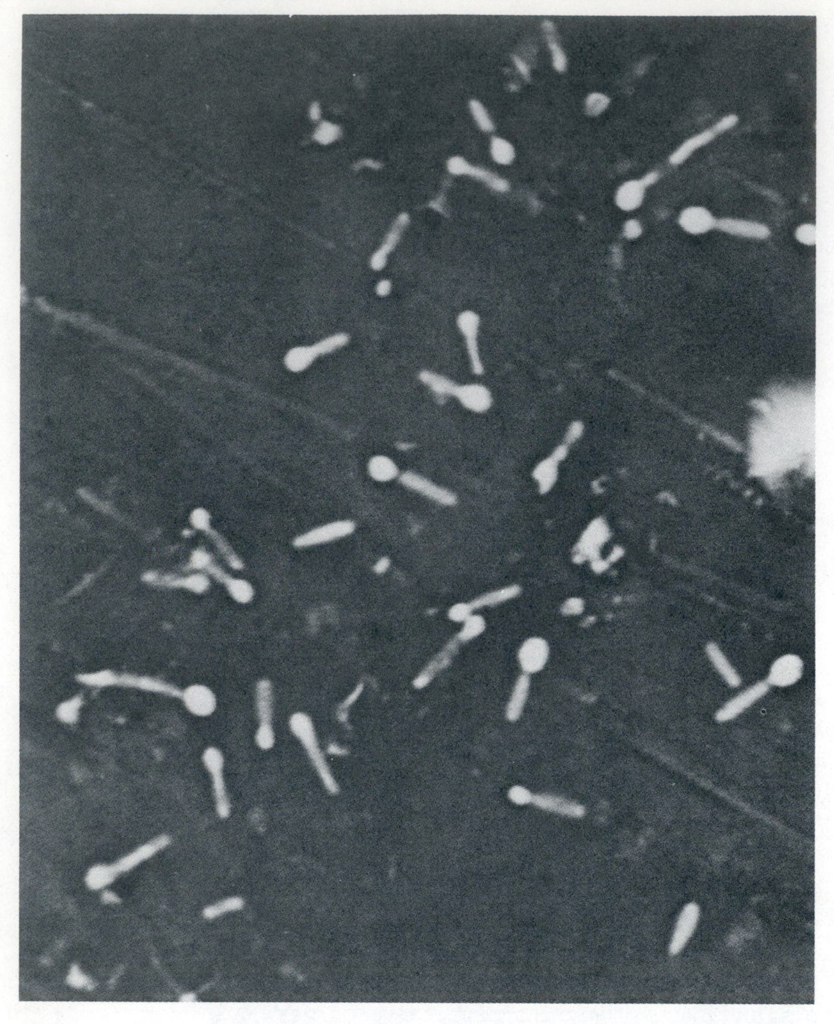

Figure 28-23 Negatively stained smear of *Clostridiuim tetani* revealing terminal spores.

From Olds, R.J. Color atlas of microbiology, Year Book Medical Publishers/Wolfe Medical Publications, Ltd., 1975.

The other rickettsial diseases are outlined in Table 28-3.

DISEASES RESULTING FROM MAJOR TRAUMA

It has already been noted that many microbial agents are capable of invading the body through minor abrasions. Some microbial agents are notorious for their ability to cause severe disease when they are introduced by penetrating wounds, burns, or accidental injuries into deeper tissues. The organisms that cause these diseases are found in the soil or water, sites from which they are accidentally introduced into the body. Three of the most important bacterial agents found in the soil are species of *Clostridium,* which are capable of causing death through the elaboration of potent toxins. They can contaminate wounds and are responsible for such diseases as tetanus, gas gangrene, and botulism. Botulism, which is ordinarily associated with food poisoning, can also contaminate wounds, but wound botulism is rare.

Tetanus

Tetanus is a disease whose clinical manifestations result from a toxin produced by the organism *Clostridium tetani.* The organism is an anaerobic, motile, gram-positive bacillus that produces spores which appear terminally in the vegetative cell, giving it the appearance of a tennis racket (Figure 28-23). It is found not only as a spore in the soil but also in the intestinal tract of some animals and humans. Puncture wounds produced in automobile accidents or in war, use of heroin by intravenous injection, and abortions are predisposing factors that may lead to infection. In developing countries, where most children are not born in hospitals but in earthen huts, tetanus ranks third among the major causes of infant mortality. Infection in this group is initiated by spore contamination of the umbilical cord. Tetanus spores germinate only in the absence of oxygen; therefore, deeper tissue, which is necrotic or contains a foreign object (such as a wood splinter or bullet), satisfies this requirement. The vegetative cells produced in the host release a toxin that affects motor nerves and causes their continued contraction, a condition called tetany. Convulsive contractions of the jaw, called trismus or lockjaw, are characteristic of disease and result in a sardonic grin, but other muscles of the body may also be affected, leading to uncontrolled spasms. Death is often caused by respiratory failure. Few cases of tetanus are observed today in the United States because of current immunization practices in which tetanus toxoid is administered shortly after birth in the form of a triple vaccine, DTP (see Chapter 23).

Gas gangrene

Gas gangrene is a condition that may be caused by a number of species belonging to the genus *Clostridium,* the most prominent of which is *C. perfringens.* This organism is a gram-positive bacillus that like *C. tetani* is an anaerobe but unlike *C. tetani* is nonmotile and produces a capsule. It is found in the soil as well as the intestinal tract of humans and animals. Infection usually results from deep wounds incurred accidentally or during surgery involving the gastrointestinal tract or genitourinary tract. *C. perfringens* requires anaerobic conditions for vegetative growth, which results in the production of a number of enzymes and toxins that have been implicated in its virulence. These enzymes include ***collagenase*** and ***hyaluronidase,*** which degrade components of tissue, collagen and hyaluronic acid, respectively. Tissue degradation leads to necrosis and a reduction in the blood supply which create anaerobic conditions and help promote the spread of infection. During tissue degradation carbon dioxide and hydrogen gases are some of the metabolic products released that cause swelling of the tissue. The primary virulence factor of *C. perfringens* is a ***lecithinase,*** called ***alpha toxin,*** which is capable of degrading the membrane of animal cells including blood cells. Death from gas gangrene is caused by severe toxemia. Treatment must include debridement of necrotic tissue from the wound (sometimes amputation is required), application of antiserum to neutralize any unadsorbed toxin, and penicillin therapy. Extreme cases are treated by using ***hyperbaric oxygen,*** a procedure in which the patient is placed in a chamber of pure oxygen. Oxygen is first introduced at 3 atmospheres and then over a period of time the pressure is reduced to 1 atmosphere. The basis of this treatment is that oxygen is lethal to the organism and the hyperbaric chamber forces oxygen into all tissues.

A condition similar to gas gangrene, called ***anaerobic cellulitis,*** may be caused by a variety of clostridial species. During anaerobic cellulitis infection the microorganisms invade uninjured connective tissue and damaged muscle but do not harm viable muscle. The prognosis for this condition is favorable, and amputation is not required.

Burn wounds

The principal barrier to microbial invasion is the intact skin, but in burn patients the skin is destroyed, and the exposed underlying tissue provides an excellent medium for the growth of bacteria. In addition, if the patient has 30% or more second- and third-degree burns, there is deficient neutrophil chemotaxis that may persist for weeks until new skin is grafted. Such conditions provide the opportunistic microbe with the chance to cause infection, and many deaths in burn patients result from bacterial diseases. The species most often incriminated is *Pseudomonas aeruginosa* an organism that is a common contaminant of water supplies, various medications, and hospital solutions (see Chapter 33). It is characterized by its ability to produce the pigments pyocyanin and fluorescein that give the medium in which they are growing a blue-green color and a particular earthy odor. *Pseudomonas* is resistant to many antimicrobials, and treatment is often difficult because the most effective antimicrobials are also toxic to the host. Most diseases today are treated with carbenicillin plus gentamicin. In hospitals where *P. aeruginosa* infection has been controlled, new invaders have emerged, such as certain species of *Providencia.*

DISEASES ACQUIRED THROUGH THE ALIMENTARY TRACT

Microorganisms are ingested daily, and most are destroyed either by enzymes or acids encountered in the stomach and intestinal tract or they are eliminated in the feces. These ingested bacterial pathogens that survive their sojourn through the intestinal tract may give rise to a variety of diseases. The ability to cause disease is related to a number of factors such as the number of microorganisms ingested, the pH of the intestinal tract, the ability of the microorganisms to produce toxins, and the immunological state of the host. Some ingested bacterial agents from contaminated food or water may remain localized in the intestinal tract giving rise to the typical symptoms of ***gastroenteritis,*** whereas others that infect the gastrointestinal tract may penetrate the epithelial barrier and invade the bloodstream to cause ***systemic infection.*** Gastrointestinal symptoms may also be brought about by ingestion of food containing preformed toxins released by multiplying microorganisms. Such conditions are usually referred to as ***intoxications.*** Most diseases that remain localized in the intestinal tract are characterized by nausea, vomiting, diarrhea, and abdominal cramps. Diarrhea at times can be severe, resulting in the loss of large volumes of water and electrolytes. Only replenishment of fluids and electrolytes is needed, and not antibiotics, because these diseases are usually self-limiting. Antibiotic therapy does not substantially alter the course of the disease and if administered for too long a period of time results in the selection of antibiotic-resistant species in the intestinal tract that may in turn cause disease. Organisms that do invade the bloodstream require antibiotic therapy.

Infections of the gastrointestinal tract

***Salmonella* infections.** *Salmonella* is a genus of gram-negative, motile, non-spore-forming bacilli that are found in the intestinal tract of animals

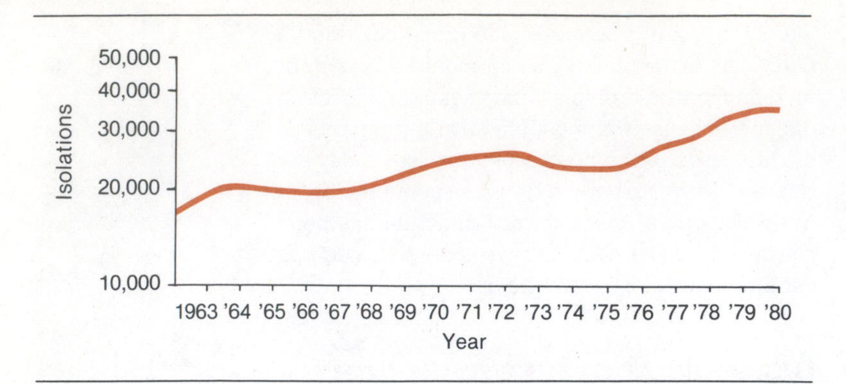

Figure 28-24 Reported human *Salmonella* isolating in the United States from 1963 to 1980.
Redrawn from the Centers for Disease Control, Atlanta, Georgia.

and humans. The infections they cause are referred to as ***salmonelloses.*** The majority of infections result in gastroenteritis, but a more severe infection called ***typhoid fever*** occurs when the organism penetrates the intestinal tract and invades the bloodstream (typhoid fever is discussed later in this chapter).

Animals such as turkeys, chickens, swine, and cattle are the primary reservoirs of the major pathogenic species. Infections in humans occur when products from these animals are cooked inadequately and are left at room temperature. These conditions support the rapid growth of *Salmonella* organisms, and when enough (1×10^6 or more)* are ingested, gastroenteritis results. More than 33,000 cases of salmonellosis were reported in the United States in 1980 (Figure 28-24), and most of these infections occurred as the result of contaminated food or food products. Asymptomatic carriers, who are food handlers, are responsible for most major outbreaks of salmonellosis. Outbreaks on cruise ships and airlines are a yearly occurrence, but in 1970 an unusually large number of cases were caused by infections from pet turtles bought by children. Early in 1981 two outbreaks of salmonellosis involving 62 teenagers and young adults occurred in two states. *S. muenchen* was isolated from all the patients. Investigations revealed that in over 75% of the patient households, there was a history of marijuana use. Saliva on shared marijuana was implicated in the spread of the infectious agent, but the precise mechanism of transmission was unknown. The patients may have been at additional risk because marijuana lowers gastric acidity, permitting lower doses of *Salmonella* to cause infection. The marijuana could have been contaminated by animal feces at the time of cultivation and storage. *Salmonella* organisms are resistant to drying and could have survived long storage.

There are three species of *Salmonella: S. cholerasuis, S. typhi,* and *S. enteritidis,* and of these, only *S. enteritidis* contains more than one serotype. *S. enteritidis* contains over 1600 serotypes, and it is the serotype that is referred to as the infectious agent and not the species. Some of the major serotypes causing disease in the United States are *S. typhimurium, S. heidelberg, S. enteritidis,* and *S. newport.* A large battery of biochemical tests is required to identify the *Salmonella* serotype because of the diversity in the cell envelope.

* Recent evidence indicates that ingestion of as few as one to five *Salmonella* organisms can cause food poisoning.

Shigella* infections.** The genus *Shigella,* like *Salmonella,* is a gram-negative, non-spore-forming bacillus, but unlike *Salmonella* is nonmotile. *Shigella* organisms cause disease referred to as ***shigellosis, a diarrheal disease.

An epidemic caused by a *Shigella* species *(S. dysenteriae)* was first recorded in 1898 by the Japanese bacteriologist, Shiga. *S. dysenteriae,* is responsible for a more severe illness (called bacillary dysentery) than that caused by other species of *Shigella.* ***Bacillary dysentery,*** which results in a bloody stool, is seldom seen in the United States and is more frequently associated with India, China, and Japan. In the United States *S. sonnei* and *S. flexneri* are the most common causes of shigellosis.

Since 1977 there has been an upswing in the number of cases of shigellosis in the United States, and in 1980 over 19,000 cases were recorded. *Shigella* organisms are harbored by humans, and diseases are caused by ingestion of fecally contaminated food or water. Most disease occurs in overcrowded areas such as hospitals, day care centers, and asylums. Most food-borne illness is caused by direct or indirect contact with food handlers. In a recent incident in the United States 32% of the employees in a children's hospital were afflicted with shigellosis. It was discovered that one of the employees who had come in contact with an infected child also happened to be responsible for the preparation of sandwiches and salads in the employees' cafeteria. Shigellosis remains a significant problem among American Indians who live in crowded living quarters and do not have sufficient running water.

The symptoms of shigellosis are abdominal pain, cramping, and watery diarrhea. Bacillary dysentery is characterized by invasion of the intestinal epithelium and the appearance of microabscesses, which later become ulcerated and bleed. Shigellosis is considered a self-limiting disease without the need for antimicrobials. Antimicrobials are administered when sanitation is minimal or nonexistant to diminish the duration of excretion of the *Shigella* organisms. Antimicrobials are, therefore, recommended in day care centers, Indian reservations, and among food handlers to prevent spread of the disease. The antibiotics used are ampicillin or the combination trimethoprim-sulfamethoxazole.

Presumptive identification of *Shigella* species can be accomplished by isolation on selective and differential media, but final identification requires serological testing.

Enteropathogenic *Escherichia coli*. *E. coli* is a gram-negative non-spore-forming bacillus that is an indigenous member of the intestinal tract of humans and animals (see Chapter 34). Enteropathogenic *E. coli* organisms differ from the normal *E. coli* residents of the intestinal tract because of their ability to invade the intestinal mucosa and to produce an enterotoxin. Most infections are associated with outbreaks of infant diarrhea in which the *E. coli* strains are acquired directly or indirectly from hospital attendants and others in the nursery. The symptoms of disease are fever, chills, abdominal pain, and dysentery. Antibiotics are not indicated for this type of disease since these strains are only mildly pathogenic.

Vibrio cholerae*.** There are many diarrheal diseases, but ***cholera has been the most devastating. The disease is usually referred to as Asiatic cholera because of its origin and predominance in that part of the world. The home of cholera is India, where the disease has been endemic for centuries. The origin of the disease is steeped in mystery but it appears that epidemic forms of cholera did not exist in biblical times. Cholera, until 1817, was thought of as a disease that was part of the Indian culture, but in 1817 six cholera pandemics appeared that made the rest of the world take notice. Even in India, where the first pandemic began, the disease appeared to be more virulent than ever before and was actually mistaken for a new disease. The disease killed countless millions primarily in Asia, Africa, the Middle East, and occasionally in Europe and the Western hemisphere. Cholera appeared periodically in the United States from 1826 to 1873 and accompanied the pandemics that appeared in Asia. The mortality rate for the disease ranged from 20% to 50%, and in 1832 nearly 20% of the population of New Orleans perished. In 1866 cholera had taken a heavy toll in the United States, and over 50,000 deaths were recorded. Today, cholera is seldom seen in the United States, but epidemics still appear in Asia.

American scientists in the early nineteenth century thought cholera to be a disease of gluttons, perverts, and boozers, but two important aspects of cholera came to pass in the nineteenth century. First, Felix Pouchet discovered the spiral-shaped bacterium in the stools of cholera patients. Second, in 1854 John Snow became interested in the epidemiology of disease. During an epidemic of cholera in England, Snow observed that those who used a certain water pump (called the Broad Street pump) developed disease, while those served by other pumps were unaffected unless they too drank water from the Broad Street pump. By removing the pump handle he noticed the epidemic was curtailed and therefore reasoned that sewage had leaked into the well water. Snow suggested that cholera was caused by a specific organism and that disease could be transmitted from person to person via contaminated food or water. Robert Koch in 1883 was encouraged to visit Egypt to study the disease. He and his colleagues isolated the causative agent from the stools of patients and demonstrated the importance of the carrier state in transmission. It became obvious that proper sanitation would be the most important measure to prevent epidemics, and yet in 1892 in Hamburg, Germany 8000 people died from cholera that had been traced to the use of water that had also been used to receive sewage.

The etiological agent of cholera is a gram-negative non-spore-forming curved rod called *Vibrio cholerae* (Figure 28-25), which is characterized by the presence of a polar flagellum. Humans are the only natural host for the organism, which is transmitted through fecally contaminated food or water. There have been few cases (10 in 1980) occurring in the United States, and most have been associated with ingestion of contaminated shellfish. The watery diarrhea (called rice water stool) associated with this disease is caused by an enterotoxin whose mechanism of action is discussed in Chapter 22. More fluids are lost in gastroenteritis caused by *V. cholerae* than in any other gastrointestinal infection. Up to 15 liters of fluid, containing important minerals such as sodium and potassium, can be lost per day. Treatment, therefore, requires rapid replenishment of water and electrolytes. Patients who do not receive treatment exhibit abnormal heart sounds, usually followed by shock, acidosis, and uremia, the major causes of death. Tetracycline is given

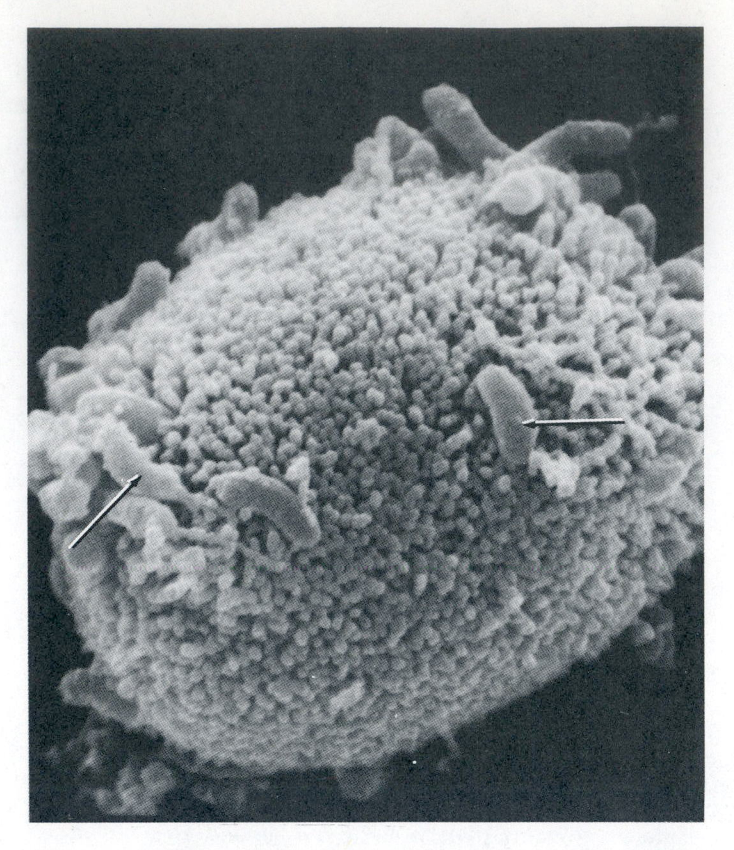

Figure 28-25 Electron micrograph of *Vibrio cholerae.* The rods *(arrows)* appear curved, but no flagella are observed in this photograph.
From Freter, R. Infect. Immun. ***14****:232, 1976.*

routinely to patients where cholera is endemic to decrease the duration of excretion of *Vibrio* organisms. Other drugs may have to be considered in treatment, however, because tetracycline-resistant strains have been recently isolated.

Confirmation of disease requires isolation of the microorganism, followed by biochemical and serological techniques. A vaccine is available for those traveling to areas endemic for the disease, but it is of limited value.

Vibrio parahemolyticus. *V. parahemolyticus* is morphologically similar to *V. cholerae.* It is a halophilic microorganism found in coastal waters and during the summer months contaminates shellfish. When shellfish are harvested and inadequately cooked or processed, the microorganisms multiply and at certain concentrations (1×10^6 to 1×10^8) can cause infection when ingested. Although originally discovered in Japan, several outbreaks have been reported in Florida, Maryland, Alabama, Louisiana, and Texas. The manifestations of disease are similar to other gastroenteritis infections and can be fatal.

Yersinia enterocolitica. *Yersinia enterocolitica* is a gram-negative non-spore-forming bacillus that is an infrequent cause of gastroenteritis in humans. The organism is transmitted to humans by contact with infected domestic animals or by ingestion of fecally contaminated food or water. In older children the symptoms of abdominal pain mimic appendicitis, and appendectomies have sometimes been needlessly performed.

Campylobacter fetus* subspecies *jejuni. *C. fetus* subspecies *jejuni* (formerly known as *Vibrio fetus*) is increasingly being recognized as a major cause of gastroenteritis. The organism is a gram-negative curved rod that contains a polar flagellum (Figure 28-26). The epidemiology of the organism is largely unknown, but it is found as a commensal or pathogen in many members of the animal kingdom, including cattle, sheep, fowl, swine, and rodents. Food and water appear to be important vehicles of transmission, but direct contact with infected animals may also be an important mode of spread. In 1981 and 1982 50% of all cases of campylobacteriosis were the result of ingestion of raw milk.

C. fetus subspecies *jejuni* is not part of the intestinal tract of normal adults, but in areas of Africa where standards of hygiene are low, the organism has been routinely found in 16% of young children.

The most common clinical features of the disease are diarrhea, abdominal pain, and fever. As many as 20 bowel movements may occur during the first 2 days of the illness with blood appearing in the stools the second or third day. The disease is self-limiting and seldom lasts more than 4 days, but erythromycin should be considered for treatment of patients with persistent symptoms.

Clostridium perfringens. *C. perfringens,* which is found in the soil and intestinal tract of humans and animals, is known not only as the etiological agent of gas gangrene but as a common cause of food poisoning. *C. perfringens* contaminates a variety of foods containing poultry and other meats that when left at room temperature for 12 to 24 hours lead to the germination of clostridial spores. Gravies, for example, provide a relatively anaerobic environment and an excellent growth medium. Food poisoning caused by *C. perfringens* differs from food poisonings caused by other microbial agents. Some vegetative cells of *C. perfringens* when ingested in food multiply in the intestinal tract, while others produce spores. It is during the sporulation process that an exotoxin (enterotoxin) is produced and enteritis ensues. Approximately 1×10^8 organisms or more must be ingested before gastrointestinal symptoms are manifested. The typical gastrointestinal symptoms begin 9 to 17 hours after ingestion, and recovery is complete within 12 to 24 hours.

Clostridial pseudomembranous colitis. *Clostridium difficile* has been found in the feces of every patient with pseudomembranous colitis (Plate 35). The organism produces a toxin that is cytotoxic when injected into hamsters. Symptoms of disease occur following the administration of antibiotics such as clindamycin, which destroys many intestinal flora but has no effect on some clostridia. The absence of competing commensals results in proliferation of clostridial species and toxin release. There is still some doubt as to whether the disease results from exogenous or endogenous infection with *C. difficile.*

Staphylococcal enterocolitis. Staphylococci are found more abundantly in the nares and on the skin surface, but a small population also resides in the intestinal tract. The intestinal population is kept in check by other commensals, particularly gram-negative species. The staphylococci can cause intestinal disease during antibiotic therapies in which the more abundant gram-negative indigenous members have been suppressed. Under these conditions the staphylococci increase their numbers rapidly and cause an enterocolitis that is characterized by fever and diarrhea.

Diseases caused by ingestion of preformed toxin

Staphylococcal food poisoning. Staphylococcal food poisoning has been recognized as a disease since an outbreak in the Philippines in 1914 was found to be caused by inadequate refrigeration of milk from a cow that had a chronic staphylococcal infection. Today, staphylococcal food poisoning remains the major type of food poisoning, and approximately 2400 cases were recorded in the United States in 1980. Intoxication from food poisoning is caused by strains of *Staphylococcus aureus* that produce a heat-stable enterotoxin. Toxin is usually produced by organisms that contaminate food, particularly those foods containing protein. In the United States ham is the most common vehicle and has been implicated in nearly 30% of the outbreaks. Contamination is usually due to food handlers who carry the organisms on their skin with or without skin conditions such as boils or pimples. Once the food is contaminated, improper handling temperatures allow multiplication of the staphylococci and subsequent elaboration of toxin. Outbreaks of food poisoning are often associated with local festivities such as weddings and large picnics, where food is prepared in advance and held at temperatures that permit bacterial growth. Outbreaks due to commercial procedures also occur, although infrequently. For example, in 1979 salami from a processing plant was implicated in several cases of food poisoning. In the production of fermented sausage, meat is intentionally temperature-controlled to permit the growth of lactobacilli, which inhibit the growth of other microorganisms. If the procedure is not moni-

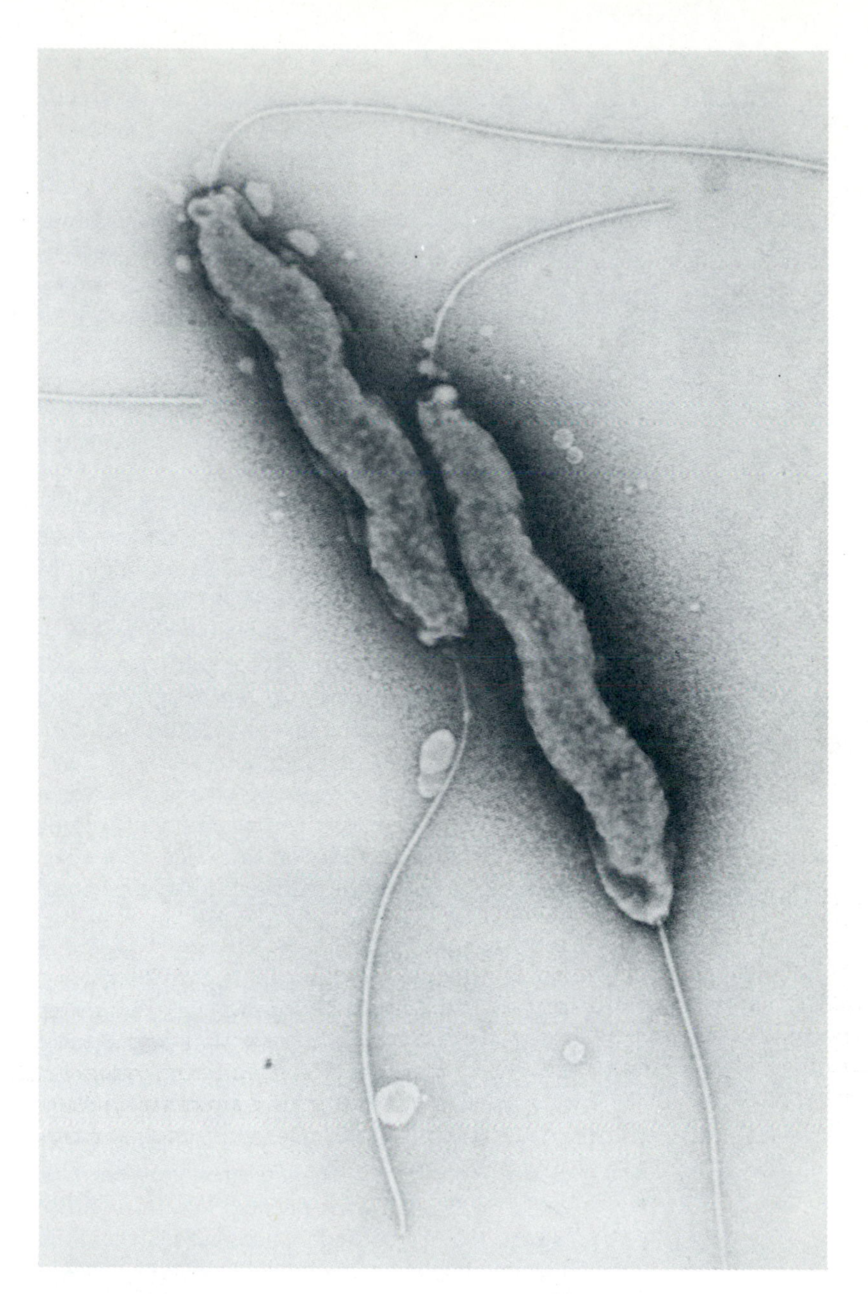

Figure 28-26 Electron micrograph of *Campylobacter fetus* subspecies *jejuni.* (×33,740.) Note curved shape of organism and polar flagellum.

From Pead, P.J., J. Med. Microbiol. ***12****:383, 1979.*

tored properly *S. aureus* organisms may grow on the surface of the sausage and produce enterotoxin. The curing procedure, which takes 1 to 2 months, will eventually kill the staphylococci, but the enterotoxin remains.

Symptoms may appear within 30 minutes following ingestion of enterotoxin, but may take as long as 8 hours. The primary symptom is vomiting, but diarrhea may also appear. The symptoms are short-lived and seldom fatal.

Botulism. There is little recorded information before the nineteenth century of outbreaks of the type of food poisoning we now call botulism. The first recorded outbreak of suspected botulism occurred in Germany in 1793 when 13 people were stricken with a food poisoning that produced a neuromuscular paralysis, and six died. All had eaten blood sausage that was prepared by washing out pig's stomachs, filling with blood and other ingredients, tying up the casings, boiling briefly, and then smoking the product. The sausages were stored at room temperature for several weeks. Uninvestigated outbreaks of sausage-related food poisonings continued until the early 1800s when a medical officer made a detailed study of the disease. He observed that if air pockets remained in the sausage, the latter did not give rise to toxic symptoms. He also noted that boiling and smoking in the manufacturing process led to poisoning of the sausages. These findings, although not obvious to the earlier investigator, in retrospect indicated that, if an organism was involved, it was capable of growing only under anaerobic conditions and was resistant to short periods of boiling and smoking. This type of food poisoning later became known as ***botulism,*** Latin for sausage. Outbreaks of botulism were later found to be associated with smoked fish and ham. In the late 1800s an incident of food poisoning involving ham resulted in neuromuscular paralysis and caught the attention of the bacteriologist, Van Ermengem. He procured the contaminated ham, prepared extracts from portions of it, and injected it into laboratory animals who developed a paralytic disease. He later isolated the causative agent by growing it in gelatin (agar had not yet been discovered) under hydrogen gas and called it *Bacillus botulinus.* From these preliminary studies several characteristics of the disease and the organism had been observed.

1. The disease is an intoxication and not an infection.
2. The organisms releasing the toxin grow only under anaerobic conditions.
3. Toxin is resistant to digestive enzymes.
4. The toxin can be inactivated if it is subjected to boiling for more than 5 minutes.
5. In the smoking process for meats if the salt concentration is high enough, the organism causing botulism cannot multiply.

The gram-positive spore-forming organism causing botulism is now classified as *Clostridium botulinum* (Figure 28-27). The heat-resistant spores of *C. botulinum* that contaminate food germinate under anaerobic conditions and during vegetative growth produce a toxin that is released when the bacterial cell lyses. The toxin from some *C. botulinum* strains is activated by enzymes found in the intestinal tract. The toxin, which is a neurotoxin, is one of the most potent toxins known to humans. It is absorbed into the bloodstream and carried to a number of organs and tissues where it affects motor nerves. The initial symptoms of disease include double vision and nausea, which are later followed by difficulty in breathing and finally cardiac arrest.

Food-borne botulism in the United States is most frequently associated with foods preserved at home and infrequently with commercial food products (Table 28-4). Home canning of vegetables, fish, or other meats has been most frequently implicated in botulism. The canning process creates anaerobic conditions, and if the sterilization process is inadequate any viable spores can germinate and produce toxin. Botulism can be prevented by careful sterilization of foods to be canned or by adequate cooking of foods before consumption. As a case in point, in 1976 a boy became ill after eating a frozen beef pie that had been heated to 218° C for just 20 minutes and then left overnight in the gas oven. The pie was later eaten without further heating. Apparently the cooking had heat-shocked clostridial spores and driven off any residual oxygen, thus permitting vegetative cells to multiply and produce toxin.

Treatment of botulism is accomplished by injection of an antitoxin made from three of the major toxins of *C. botulinum,* toxins of types A, B, and E. The antitoxin has no effect on toxin that is already adsorbed to nerve cells.

Botulinum toxin may also be produced within the host and is responsible for two types of syndromes; wound botulism and infant botulism. ***Wound botulism*** is characterized by spore germination in the wound and release of toxin to the systemic circulation. The symptoms of disease are the same as for food-borne botulism. Ingestion of spores, contaminating such foods as honey, vegetables, or fruits, is the apparent cause of ***infant botulism.*** The spores germinate in the intestinal tract and the toxin is released. There is reason to believe that infant botulism may be one of the causes of sudden infant death syndrome (also called crib death), in which infants unexpectedly demonstrate irritability, sluggishness, and inability to lift their heads. Infants can die within a few hours to a few days following the onset of symptoms.

Systemic disease following infection of the alimentary tract

Some intestinal pathogens following ingestion of contaminated food or water are not restricted to an intestinal phase and may either penetrate the bloodstream directly or indirectly via the lymph system. Those microorganisms most frequently associated with invasion of the bloodstream are *Brucella* and some species of *Salmonella*. *Brucella* are discussed earlier. Although the mechanism that permits mucosal penetration is not known, these species are equipped with surface components that inhibit phagocytosis. Localization in various organs following bacteremia may give rise to a variety of conditions such as meningitis, arthritis, endocarditis, and abscesses (Table 28-5).

***Salmonella* and typhoid fever.** Several *Salmonella* species and serotypes may penetrate the bloodstream to cause enteric fevers, but *S. typhi*, the causative agent of typhoid fever, is the most virulent.

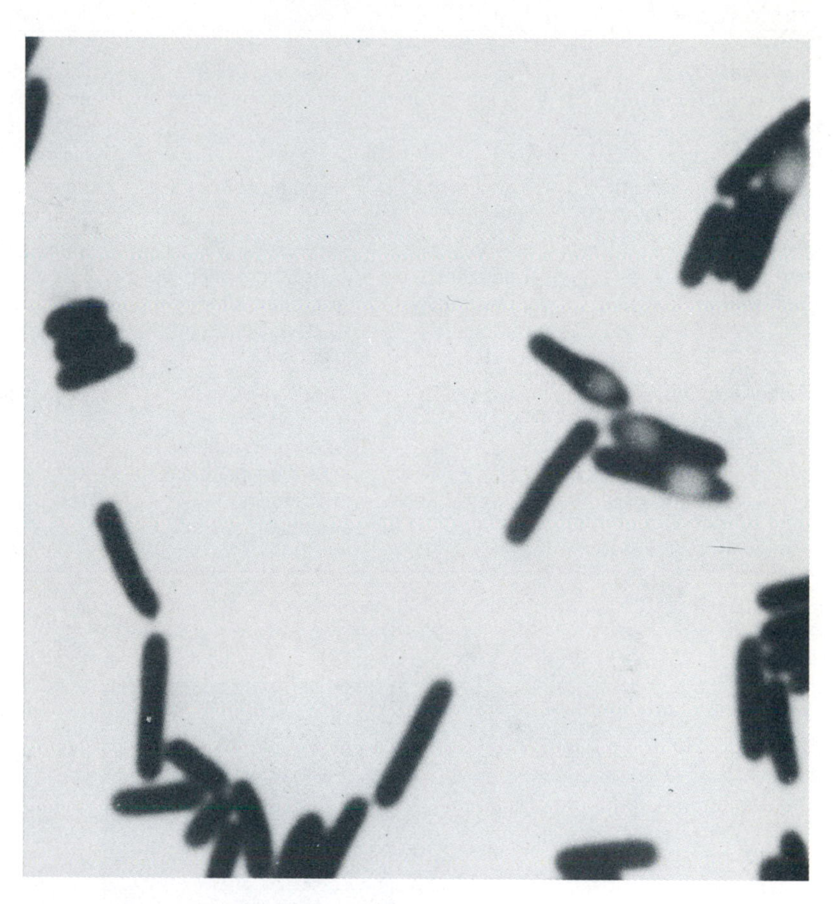

Figure 28-27 Sporulating cells of *Clostridium botulinum*.
Courtesy P.D. Walker, Wellcome Research Laboratories, Kent, England.

Table 28-4. Food products causing botulism outbreaks 1899-1977*

Botulinum toxin type	Vegetables	Fish and fish products	Fruits	Condiments†	Beef‡	Milk and milk products	Pork	Poultry	Other	Unknown[2]	Total
A	115	11	22	17	6	3	2	2	8	9	195
B	31	4	7	5	1	2	1	2	3	3	59
E	1	25							3	1	30
F					1						1
A & B	2										2
Unknown§	2	1		1						6	10
TOTAL	151	41	29	23	8	5	3	4	14	19	297

From the Centers for Disease Control, Atlanta, Georgia.
*For period 1899-1973 includes only outbreaks in which toxin type was determined and for period 1974-1977 includes all outbreaks.
†Includes outbreaks traced to tomato relish, chili peppers, chili cauce, and salad dressing.
‡Includes one outbreak of type F in venison, and one outbreak of type A in mutton.
§Categories added for period 1974-1977.

Table 28-5 Characteristics of agents that penetrate the bloodstream from the intestinal tract

ETIOLOGICAL AGENT	DISEASE	POPULATION AFFECTED	TREATMENT	IMMUNIZATION	COMMENTS
Salmonella typhi	Typhoid fever	Those ingesting contaminated food or water	Chloramphenicol, ampicillin, or trimethoprim-sulfamethoxazole	Killed vaccine	Carrier state important in transmission
Salmonella paratyphi A, *Salmonella paratyphi B*	Paratyphoid fever	Same as above	Same as above	Same as above	Less severe than typhoid
Salmonella typhimurium and other serotypes *Salmonella cholerasuis*	*Salmonella* bacteremia Enteric fever	Very young and those over 50 predisposed with blood disorders or diseases of lymph glands	Same as above	None	Infections may resemble typhoid fever or involve bone, meninges, and vascular system
Brucella abortus, *Brucella suis*, and *Brucella melitensis*	Brucellosis	Those in contact with infected animals (e.g., farm workers, veterinarians) those who ingest unpasteurized milk	Tetracycline	None	In the United States most infections among those handling pigs

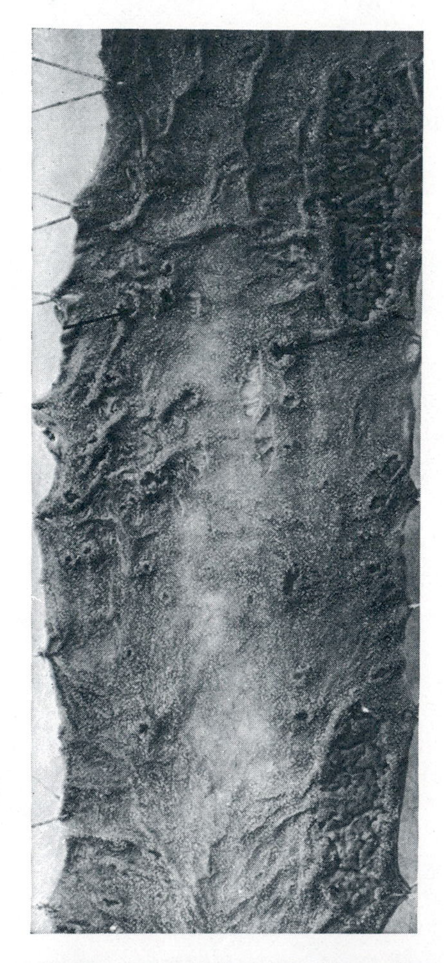

Figure 28-28 Typhoid ulceration of ileum. Note large number of ulcerations *(arrows)*
Air Force Institute of Pathology, AFIP 2803.

What was believed to have been typhoid fever was first described in ancient Greece. It was not until the nineteenth century that typhoid fever was accurately described and distinguished from other fevers such as typhus. The treatment for all fevers by medieval physicians was blood letting, and for persistent fevers, such as typhoid, as many as 25 to 30 blood lettings were recommended. Unfortunately, for those who could afford the physician's fee, death was the rule rather than the exception. The poor, who could not afford the treatment, had a better chance of survival. Some of the causes of typhoid that were alluded to during these times ranged from inhaling the fragrance of flowers, especially those from the mangrove swamps of Africa, to eating bacon.

Dr. William Budd in the 1860s discovered that contamination of water with human excreta was responsible for transmission of typhoid fever. Today, chlorination of water has practically eliminated typhoid from the developed countries of the world, and in the United States only 510 cases were recorded in 1980. Typhoid is still a major problem in Latin American countries.

Typhoid is associated with the carrier state. The classic example of the carrier state was Typhoid Mary, a cook, who in the early 1800s was incarcerated because she was the source of infection for several hundred people. One of the major concerns today is the potential for *S. typhi* to acquire resistance to the principal antibiotics used in treatment; chloramphenicol and ampicillin. Such resistant species could present problems to developed countries such as the United States because of the relative ease with which people travel from one geographic area to another. Many enteric species of bacteria, such as the strains of *E. coli* and species of *Shigella,* have been isolated in Latin America that possess plasmid-associated resistance to chloramphenicol and ampicillin. Thus, the conjugation between these species and *S. typhi* could result in antibiotic resistance transfer to the typhoid bacillus.

Typhoid fever is characterized by a persistent fever and abdominal pain. Later a rash (rose spots) appear on the upper abdomen (Plate 36). Two of the most important complications are intestinal hemorrhage and perforation (Figure 28-28). Recovery from disease may be associated with the carrier state and some of those who recover (about 3%) harbor *S. typhi* in the gall bladder. From this site the asymptomatic carrier sheds bacilli into the intestinal tract and hence can contaminate food and water. The other *Salmonella* serotypes that can also penetrate the intestinal tract are outlined in Table 28-5.

SUMMARY

Bacterial disease may involve any part of the body but most are acquired through the respiratory tract. Diseases involving the respiratory tract include pneumonia, diphtheria, whooping cough, meningitis, tuberculosis and rheumatic fever. Pneumonia is caused primarily by *Streptococcus pneumoniae* and secondarily by *Klebsiella pneumoniae.* A vaccine containing 14 pneumonococcal types has been developed for the elderly and infants. Diphtheria is an upper respiratory tract disease caused by *Corynebacterium diphtheriae.* During infection a toxin is released into the bloodstream that adversely affects many organs of the body, particularly the heart. Antitoxin to diphtheria is part of the DPT vaccine (*d*iphtheria, *p*ertussis, *t*etanus). Whooping cough is caused by *Bordetella pertussis.* The organisms are part of the DPT vaccine, which induces immunity to infection. Meningitis is a disease of the nervous tissue, that is, the meninges, which initially begins as an upper respiratory infection. *Neisseria meningitidis* is an important cause of meningitis. The organism normally resides as a commensal in the upper respiratory tract but can cause disease. Transmission is associated with the carrier state. Tuberculosis is the most important cause of death world wide and is caused by the organism *Mycobacterium tuberculosis.* The disease is characterized by the formation of swellings (tubercles) in lung tissue as well as other organs. Pharyngitis, caused by *Streptococcus pyogenes,* is important because of the potential for development of rheumatic fever, which is a sequela to upper respiratory tract infection. Rheumatic fever is believed to be an immune disease.

Those bacterial diseases that affect the skin and eye can be acquired by abrasions, sexual contact, bites of insects, or hospital procedures. Most skin diseases are caused by *Staphylococcus aureus,* which is harbored in the nares by carriers. Leprosy is a disfiguring disease of the skin caused by *Mycobacterium leprae.* The disease has a long incubation period, sometimes several years. Yaws is a tropical disease caused by the organism *Treponema pertenue,* which is morphologically indistinguishable from the organism causing syphilis, *Treponema pallidum.* Infections of the eye range from minor conjunctivitis to permanent blindness. *Chlamydia trachomatis* is the major cause of blindness in the world and appears more frequently in developing countries.

Diseases resulting from animal contact or contact with arthropod vectors include tularemia, anthrax, leptospirosis, relapsing fever, and rickettsial diseases. Tularemia is a disease of rodents that is transmissible to humans. The microbial agent, *Francisella tularensis,* is frequently acquired through the skinning of animals such as rabbits. Anthrax is a disease of herbivorous animals and is caused by *Bacillus anthracis.* The organism produces spores in the soil that contaminate animals, particularly their hides. Contact with products made from animal hides is the major source of human disease. Leptospirosis is caused by *Leptospira interrogans,* a spirochete harbored by rodents. Humans become infected primarily by contact with water contaminated by rat urine. Plague is caused by *Yersinia pestis.* The organism is carried by a variety of rodents in whom the disease may be fatal. Transmission to humans occurs via rat fleas. The disease is characterized by buboes (enlarged lymph nodes), hence the name bubonic plague. Relapsing fever is caused by a spirochete belonging to the genus *Borrelia.* Man is the reservoir of the spirochete and human-to-human transmission is carried out by infected ticks. The intermittent fever is caused by mutation involving the microorganisms cell envelope and the subsequent temporary evasion of the host's immune system. Rickettsial agents infectious for humans are transmitted by arthropod vectors such as fleas, ticks, mites, and lice. Disease is characterized by a rash and fever as well as intravascular coagulation. In the United States, Rocky Mountain spotted fever transmitted by the dog tick is the most common rickettsial disease.

Bacterial diseases resulting from major traumatization include tetanus, gas gangrene, and burn wound infections. The organisms causing tetanus and gas gangrene (*Clostridium tetani* and *Clostridium perfringens*, respectively) are found as spores in the soil. Conditions leading to traumatization of tissue are frequent causes of disease by these agents. Both organisms produce potent toxins that can cause death. *Pseudomonas aeruginosa* and species of *Providencia* are common contaminants of water supplies and are frequently invaders of burned tissue.

Bacterial diseases acquired through the alimentary tract are usually the result of ingestion of contaminated food or water. Most infections are characterized by gastroenteritis, especially diarrhea. Species of *Salmonella*, indigenous to the intestinal tract of many vertebrates including cattle and fowl, can contaminate foods; 1×10^6 organisms or more are required to cause disease symptoms. Species of *Shigella* are harbored by humans, and infections result from the ingestion of fecally contaminated food or water. Enteropathogenic *E. coli* are indigenous members of the intestinal tract and have been associated with outbreaks of infant diarrhea. Cholera is an intestinal disease usually associated with Asia. The cholera organism, *Vibrio cholerae* is harbored only by humans, and infections are the result of ingestion of fecally contaminated food and water. The diarrhea can be fatal because of the loss of many gallons of fluids per day. A species of *Vibrio*, *Vibrio parahemolyticus* is found in marine waters and contaminates shellfish. If inadequately cooked and eaten, the contaminated shellfish can cause gastroenteritis. *Yersinia enterocolitica* is found in domestic animals, and infection in humans occurs frequently by contact with infected animals or ingestion of fecally contaminated food or water. *Clostridium perfringens*, which is found in the intestinal tract of humans and animals, can contaminate meats. Ingested vegetative cells of *C. perfringens* form spores in the intestinal tract and then release a toxin that causes mild gastroenteritis.

Diseases caused by ingestion of preformed toxin is associated with such microbial species as *Staphylococcus aureus* and *Clostridium botulinum*. Staphylococcal food poisoning is the major type and is associated with foods containing high levels of protein. Food handlers carry toxin-producing organisms and are the major source of contamination of foods. Symptoms are seldom fatal. Botulism is often a fatal disease resulting from contamination of foods by spores of *Clostridium botulinum*. The spores germinate under anaerobic conditions in the food and release a potent toxin. Botulism is associated primarily with the canning industry, including home canning. Wound botulism results from contamination of open wounds with the spores of *C. botulinum*. Infant botulism is a newly discovered disease that is believed to be the result of ingestion of spores that germinate in the intestinal tract, releasing a potent enterotoxin.

Some diseases that originate in the intestinal tract invade the bloodstream and become systemic. Typhoid fever is a systemic disease caused by several species of *Salmonella*. Transmission is associated with the carrier state. Brucellosis is a disease caused by species of *Brucella* and is acquired from animals. Infections often terminate in chronic disease because of the ability of the organism to exist in a viable state within the phagocyte.

Study questions

1. List the ways in which diseases can be acquired through the skin.
2. What is the general sequence of events used in the isolation and identification of bacterial disease–producing agents?
3. What is pneumonia? What are the most important bacterial agents causing the disease?
4. What structural characteristics do *Bordetella pertussis, Klebsiella pneumoniae,* and *Streptococcus pneumoniae* have in common? What are their significance?
5. What is the DPT vaccine? What component is not a true vaccine?
6. Explain why diphtheria is called an intoxication. What is the significance of the temperate phage that inhabits *C. diphtheriae*?
7. Distinguish between aseptic and septic meningitis.
8. Why is tuberculosis such a problem in underdeveloped countries as opposed to developed countries?
9. What combinations of drugs are used in the treatment of tuberculosis? Can you give some reasons why this disease is difficult to treat?
10. Give the sequence of events that leads to infection by *Mycobacterium tuberculosis* and the pathological signs of disease.
11. What is the organism found in the throat or oral cavity that causes infections that may ultimately lead to rheumatic fever? How does rheumatic fever differ from pharyngitis?
12. What are the skin conditions that can be caused by staphylococcal infection?
13. Why is there such a social stigma associated with leprosy?
14. What is the distinction between trachoma and inclusion conjunctivitis?
15. How are the diseases tularemia and anthrax related?
16. In what organ of the body do the organisms causing leptospirosis lodge?
17. Why did plague become known as the Black Death?
19. How many varieties of plague are there? Are any contagious?
20. How is plague transmitted in the United States? In what geographic areas does it occur?
21. What is the reason for the repeated undulating nature of fever in the disease relapsing fever?
22. What is the most important rickettsial disease in the United States? How is it transmitted?
23. What are the characteristic symptoms and pathologic conditions of rickettsial diseases?
24. How are tetanus, gas gangrene, and botulism related?
25. Describe the manner in which *C. perfringens* produces the symptoms and pathologic conditions of gas gangrene.
26. What organism is most frequently associated with burn wound infections?
27. Why are most gastrointestinal diseases not treated with antibiotics?
28. What is the distinction between dysentery and diarrhea? What bacterial species are often associated with such conditions?
29. What important part did John Snow play in his studies of cholera?
30. Does cholera occur in the United States? Explain.
31. What is the standard treatment for most diarrheas?
32. List the various species of bacteria that cause food poisoning.
33. How does typhoid fever differ from gastrointestinal diseases caused by other species of *Salmonella?*
34. What do brucellosis and tuberculosis have in common?

Selected readings

Books

Acha, P.N. *Zoonoses and communicable diseases common to man and animals.* Pan American Health Organization, Washington, D.C. 1980.

Boyd, R.F. and Marr, J.J. *Medical microbiology,* Little, Brown & Co., Boston, 1980.

Davis, B., et al. *Microbiology* (3rd ed.). Harper and Row, Publishers, Inc., New York, 1980.

Freeman, B.A. *Burrows textbook of microbiology* (21st ed.). W.B. Saunders Co., Philadelphia, 1979.

Hoeprich, P.D. (Ed.). *Infectious diseases* (2nd ed.). Harper and Row, Publishers, Inc. New York, 1977.

Mandell, G.L. (Ed.). *Principles and practice of infectious disease,* John Wiley & Sons, Inc., New York, 1979.

Smith, L. de S. *The pathogenic anaerobic bacteria* (2nd ed.). Charles C Thomas, Publisher, Springfield, Ill., 1975.

The Pneumococcus, *Rev. Infect. Dis.* **3**(2): 1981.

Youmans, G., Paterson, P.Y., and Sommers, H.M. *The biologic and clinical basis of infectious disease* (2nd ed.). W.B. Saunders Co., Philadelphia, 1980.

Journal articles

Andrew, E.D., and Marrocco, G.R. Leptospirosis in New England. *J.A.M.A.* **239:**2027, 1977.

Aronon, S.S. Infant botulism. *Annu. Rev. Med.* **31:**541, 1980.

Baine, W.B., et al. Institutional salmonellosis. *J. Infect. Dis.* **128:**357, 1973.

Bartlett, J.G. Antibiotic-associated pseudomembranous colitis. *Rev. Infect. Dis.* **1**(3):530, 1979.

Beem, M.O., and Saxon, E.M. Respiratory tract colonization and a distinctive pneumonia syndrome in infants infected with *Chlamydia trachomatis. N. Engl. J. Med.* **296:**306, 1977.

Bizzini, B. Tetanus toxin, *Microbiol. Rev.* **43:**224, 1979.

Blake, P.A., et al. Cholera—a possible endemic focus in the United States. *N. Engl. J. Med.* **302:**305, 1980.

Blaser, M.J. Reservoirs for human campylobacteriosis. *J. Infect. Dis.* **141:**665, 1980.

Boyce, J.M. Recent trends in the epidemiology of tularemia in the United States. *J. Infect. Dis.* **131:**197, 1975.

Brezina, R., et al. Rickettsiae and rickettsial diseases. *Bull. W.H.O.* **49:**433, 1973.

Brucellosis in the United States, 1965-1974. *J. Infect. Dis.* **136:**312, 1977.

CDC. ACIP recommendation. Pneumococcal polysaccharide vaccine MMWR, *Morbid. Mortal. Week. Rep.* **30:**410, 1981.

CDC. Typhoid fever, San Antonio, Texas, 1981. *Morbid. Mortal. Week. Rep.* **30:**540, 1981.

CDC. Diphtheria, tetanus and pertussis: guidelines for vaccine prophylaxis and other preventive measures. *Morbid. Mortal. Week. Rep.* **30:**392, 1981.

Christie, A.B., The clinical aspects of anthrax. *Postgrad. Med. J.* **49:**565, 1973.

Echeverria, P., and Murphy, J.R. Enterotoxigenic *Escherichia coli* carrying plasmids coding for antibiotic resistance and enterotoxin production. *J. Infect. Dis.* **142:**273, 1980.

Field, M. Modes of action of enterotoxins from *Vibrio cholerae* and *Escherichia coli. Rev. Infect. Dis.* **1**(6):918, 1979.

Finkelstein, R.A. Cholera. *Crit. Rev. Microbiol.* **2:**553, 1973.

Fraser, D.W., and McDade, J.E. Legionellosis, *Sci. Am.* **241:**82, 1979.

Gray, B.M., Converse, G.M., and Dillon, H.C. Epidemiologic studies of *Streptococcus pneumoniae* in infants: acquisition, carriage and infection during the first 24 months of life. *J. Infect. Dis.* **142:**923, 1980.

Hattwick, M.A.W., et al. Rocky Mountain spotted fever: epidemiology of an increasing problem. *Ann. Intern. Med.* **84:**732, 1976.

Kaplan, M.H. Rheumatic fever, rheumatic heart disease and the streptococcal connection. *Rev. Infect. Dis.* **1**(6):988, 1979.

Kaufmann, A.F., et al. Trends in human plague in the United States: report from the Centers for Disease Control. *J. Infect. Dis.* **141:**522, 1980.

Keusch, G.T. and Jacewicz, M. The pathogenesis of shigella: toxins and antitoxins in *Shigella flexneri* or *Shegella sonnei* infection in humans. *J. Infect. Dis.* **135:**552, 1977.

Kryzhanovsky, G.N. Pathogenesis of tetanus. *Prog. Drug Res.* **19:**301, 1975.

Levine, M.M., et al. Pathogenesis of *Shigella dysenteriae. J. Infect. Dis.* **127:**261, 1973.

Linnemann, C.C., Jr., and Nasenbeny, J. Pertussis in the adult. *Annu. Rev. Med.,* **28:**179, 1977.

McCarty, M., An adventure in the pathogenetic maze of rheumatic fever, *J. Infect. Dis.,* **143:**375, 1981.

Meyer, R.D. and S.M. Finegold, Legionnaires' disease, *Annu. Rev. Med.* **31:**219, 1980.

Nelson, J.D. The changing epidemiology of pertussis in young infants: the role of adults as reservoirs of infections. *Am. J. Dis. Child.* **132:**371, 1978.

Pappenheimer, A.M., and Gill, D.M. Diphtheria. *Science* **182:**353, 1973.

Payne, D.J.H. Brucellosis. *Medicine* (Baltimore) **54:**123, 1975.

Powell, K.E., and Farer, L.S. The rising age of the tuberculosis patient: a sign of success and failure. *J. Infect. Dis.* **142:**946, 1980.

Razin, S. The mycoplasmas. *Microbiol. Rev.* **42:**414, 1978.

Sasinur, R., et al. Colitis associated with metronidazole therapy. *J. Infect. Dis.* **141:**772, 1980.

Schachter, J. Chlamydial infections. *N. Engl. J. Med.* **298:**428, 1978.

Schachter, J., and Caldwell, H.D. Chlamydiae. *Annu. Rev. Microbiol.* **34:**285, 1980.

Scheifele, D.W., et al. *Haemophilus influenzae* bacteremia and meningitis in infant primates. *J. Lab. Clin. Med.* **95:**450, 1980.

Silva, J., and Fekety, R. Clostridia and antimicrobial enterocolitis. *Annu. Rev. Med.* **32:**327, 1981.

Sippel, J.E. Meninigococci. *Crit. Rev. Microbiol.* **8**(3):267, 1981.

Smibert, R.M. The genus *Campylobacter. Annu. Rev. Microbiol.* **32:**673, 1978.

Sugiyama, H. *Clostridium botulinum* neurotoxin. *Microbiol. Rev.* **44:**419, 1980.

Underman, A.E., et al. Bacterial meningitis. *DM* **24:**5, 1978.

Wald, E.R., and Levine, M.M. *Haemophilus influenzae* type b pneumonia, *Arch. Dis. Child.* **53:**316, 1978.

Weinstein, L., and Barza, M.A. Gas gangrene. *N. Engl. J. Med.* **289:**1129, 1973.

Young, L.S. The role of exotoxins in the pathogenesis of *Pseudomonas aeruginosa* infections. *J. Infect. Dis.* **142:**626, 1980.

Chapter 29 VIRAL DISEASES

Improvements in hygiene, nutrition, and health services, including the development of some effective vaccines, has led to the virtual disappearance of some viral disease, such as smallpox and poliomyelitis, in developed countries of the world. Our inability to find effective antiviral drugs, however, still makes us prone to the potentially severe consequences of certain viral diseases. Viruses such as the hepatitis and influenza viruses are still an important cause of death, even in developed countries of the world.

The purpose of this chapter is to discuss some of the major viral diseases that afflict humans. These have been broken down into the following groups: diseases involving nervous tissue, diseases of the respiratory tract, diseases of the skin, diseases of the digestive glands, agents of diarrhea, and sexually transmitted diseases. Those agents transmitted by sexual contact are discussed in Chapter 30.

GENERAL CHARACTERISTICS

Viruses have been found that are capable of infecting practically every cell type. Their long-term survival within a population of humans depends on their transmissibility from cell to cell or from individual to individual. Unlike some bacteria and fungi that can survive in a resistant form, such as spores, most viruses cannot exist for long periods of time outside a living host and are susceptible to environmental factors. Slight variations in temperature, irradiation, and humidity can destroy most viral agents. Only a few viruses, such as the hepatitis viruses and the poliovirus are unusually resistant to environmental factors, including many chemical agents that would ordinarily destroy other microbial forms.

Viruses can be transmitted to humans by a number of routes: the respiratory tract, the intestinal tract, via insect vectors, via contaminated

Table 29-1 Viral agents that cause congenital infections

AGENT (COMMON SPECIES NAME)	CONSEQUENCE OF FETAL INFECTION
Cytomegalovirus	Hepatitis and jaundice; congenital heart disease, mental retardation
Rubella virus	Abortion, congenital malformation, deafness, cataracts, heart defects, encephalitis
Small pox virus (variola)	Abortion, stillbirth, congenital smallpox
Varicella-zoster virus (chickenpox-shingles)	Low birth weight, bilateral cataracts, atrophic limbs, mental retardation, congenital varicella.
Vaccinia virus	Abortion, generalized vaccinia
Poliovirus	Congenital poliomyelitis
Measles virus	Stillbirth, congenital measles
Mumps virus	Abortion, endocardial fibroelastosis
Herpes simplex virus (type II)	Abortion, excessive brain damage, hepatoadrenal necrosis with jaundice
Hepatitis B virus	Neonatal hepatitis, stillbirth, abortion
Echovirus	Hydrocephalus and neurological sequelae, jaundice; most infections are mild or asymptomatic
Coxsackievirus B	Myocarditis and central nervous system involvement

Figure 29-1 Plaque formation. Monolayer of cells was cultured in flat-surfaced bottles. Cell culture on right is uninfected, whereas culture on left was inoculated with a rotavirus. Clear areas are called plaques.
From Sattar, S.A., and Ramion, S. J. Clin. Microbiol. ***10:****609, 1980.*

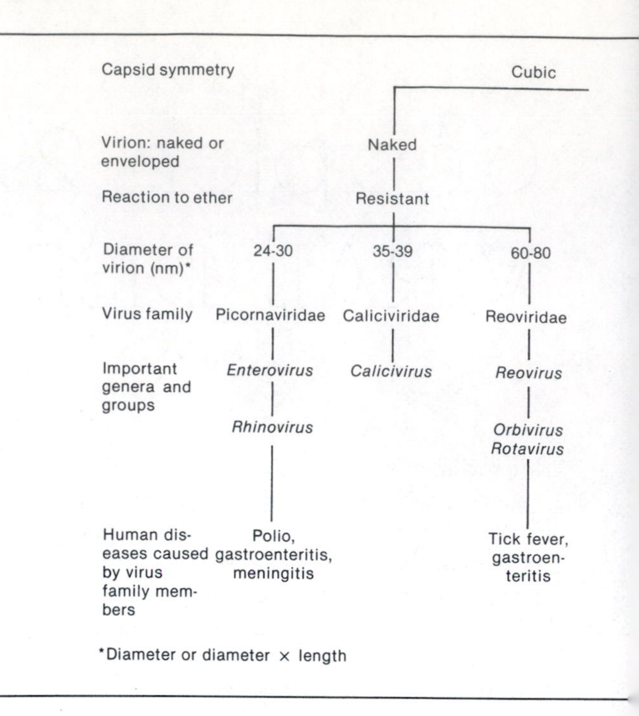

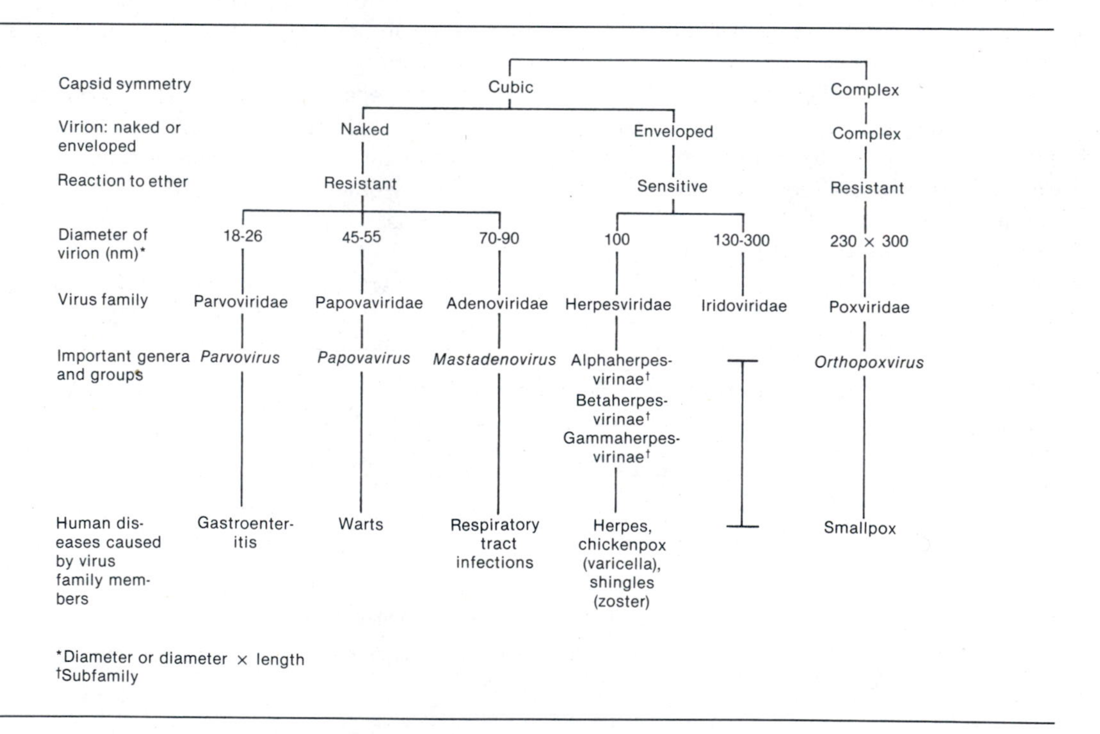

Figure 29-2 Classification of DNA-containing viruses of vertebrates

fomites, and venereally. In addition, congenital infections arise because of the ability of many viruses to cross the placenta or to be transferred in utero to the developing fetus (Table 29-1). The most common method of viral spread is the respiratory route.

The effect that viral infection has on human cells is variable. Most damage occurs when the rate of viral multiplication in the cell is high, and this is related to disruption of normal cellular metabolism. The changes that occur in the cell or tissue following viral disease (cytopathic effects) may be observed microscopically and may take a number of forms. There may be an accumulation of viral and/or cellular components that can be stained and that appear as ***inclusion bodies.*** Some of these inclusions are so distinctive that they are of diagnostic value, for example, during

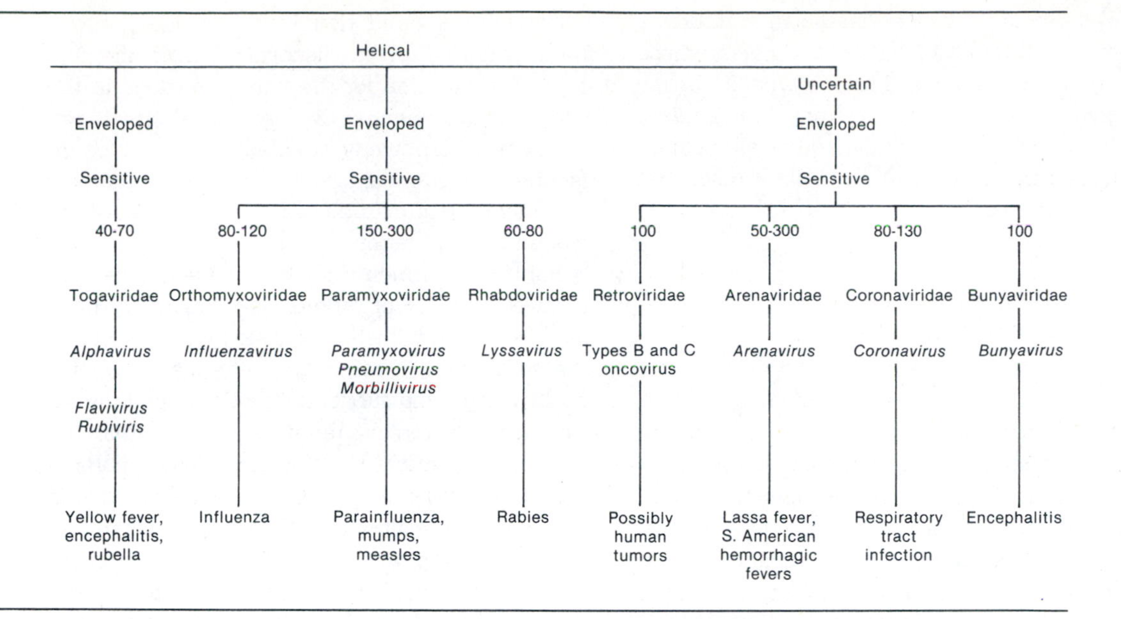

Figure 29-3 Classification of RNA-containing viruses of vertebrates.

rabies virus infection (see "Rabies"). Some cytopathic effects can be observed microscopically as morphological changes, such as displacement of the nucleus or nucleolus, or rearrangement of nuclear material. Virus-infected cells sometimes fuse to form ***giant cells,*** and in some cell cultures the cytopathic effects are distinctive enough to be observed with the naked eye as degenerative areas called ***plaques*** (Figure 29-1). Some viruses that possess envelopes incorporate host cell membrane components into their own structure (see Chapter 16). Thus antibodies against host-virus complexes may be generated during infection, resulting in damage to the host cell (immunopathology). Most viral diseases are acute, in which there is a short incubation period followed by a rapid recovery. Those viral agents that escape the immune system of the host may cause delayed manifestations, or they may be reactivated at a later time to cause disease. Some viruses escape detection by having their nucleic acid incorporated into the host DNA. Thus without a viral protein the virus nucleic acid, which is not antigenic inside the cell, is not recognized by the host's immune system. Animal cells harboring integrated viral DNA may also become ***transformed*** and exhibit properties not associated with normal cells. Transformed cells behave as neoplastic cells and on injection into susceptible hosts can cause tumors (see Chapter 16).

The host, following viral disease, may develop resistance to further infection by the same virus by using specific and nonspecific defense mechanisms. Specific resistance to viral infection can be obtained by antibodies at the site of infection as well as in the bloodstream. Local antibodies, such as secretory IgA, are particularly important in providing protection in the respiratory and gastrointestinal tracts. Cell-mediated immunity also plays a role in resistance to infection and appears to be more important than humoral antibodies. Individuals lacking globulins but possessing cell-mediated responses recover more rapidly from viral diseases than those possessing globulins but no cell-mediated response. The immune response may also be detrimental, because immune complexes develop that evoke an inflammatory reaction causing tissue destruction (see Chapter 25). The role of interferon production, probably the most important nonspecific factor in resistance to viral infections, is discussed in Chapter 27.

Treatment of viral diseases is difficult, because viral duplication is dependent on host metabolic pathways, and only a few viral processes—some enzymes involved in viral RNA and DNA synthesis—are independent of the host. The antiviral agents that are specific enough for viral processes and that are not toxic to the host are few in number. Most antiviral agents are seldom used in systemic diseases. Treatment of viral disease is concerned primarily with alleviation of symptoms and permitting the host's immune system to eliminate the virus, or an antiserum (hyperimmune globulin obtained from humans who have a high level of antibodies to the virus) may be administered. The most effective method for controlling viral disease is prevention through the use of vaccines. Today, many of the childhood diseases, such as measles, mumps, polio, and rubella are effectively prevented because of immunization practices.

The viruses that cause disease in vertebrates belong to a number of different families whose separation has been based on the type of nucleic acid, symmetry, and other factors. Figures 29-2 and 29-3 represent abbreviated classification schemes for the major RNA and DNA viruses that infect vertebrates.

VIRUSES THAT AFFECT NERVOUS TISSUE

Many viruses are capable of replicating in nervous tissue, but only a few find their way to the nervous system following naturally occurring infections. Viruses may reach the nervous system by three routes: ***olfactory*** (nasal mucosa), ***neural*** (nerve), or ***hematogenous*** (bloodstream). Spread via the nasal mucosa appears to be of minimal importance. A few viruses are spread by the neural route in which they travel along the axons of peripheral nerves and then multiply in the nucleus of the nerve cells. Animal studies have demonstrated that the poliovirus, rabies virus, and herpes simplex virus are capable of neural spread to the central nervous system. The most common method of spread to nervous tissue in naturally occurring infections is via the hematogenous route, where the virus multiplies outside nervous tissue, as in the epithelium of blood vessels or in lymphatics. This maintains a high concentration of virus in the blood (viremia) and is followed by the passage of virus into the brain or cerebral spinal fluid. Those viruses that use the hematogenous route to infect nervous tissue are the rabies virus, poliovirus, echoviruses, mumps virus, herpes simplex virus, and cytomegalovirus. Viruses are spread more readily by the hematogenous route, but the other avenues can also be used, depending on the initial site of inoculation.

Rabies

Rabies is a disease of certain wild animals, such as skunks, bats, racoons, and foxes, as well as domestic animals, particularly dogs, that can be transmitted to humans. The disease was described several centuries BC as an infection of dogs, and it was recorded that dogs bitten by infected animals became mad. Rabies was prevalent among wolves in Western Europe as early as 1271, and the first recognized epizootic of rabies among dogs was recorded in 1708 in Italy. Rabies is believed to have arrived in the New World in 1753 from Africa, where five viruses related to the rabies virus are known to exist. Rabies was proved to be an infectious disease in 1804 when saliva from a rabid dog was inoculated into a healthy animal. This discovery led to the first type of quarantine for the control of dogs in the countries of Denmark, Sweden, and Norway. Rabies was a particularly dreaded disease, since infection in dogs meant certain death and the only treatment for human disease was excision of the wound or amputation.

Rabies experiments in the nineteenth century involved the use of rabbits as a laboratory host, as well as dogs. Pasteur, for example, inoculated the saliva from a rabid dog into the brain of a healthy dog, who subsequently developed rabies. He then injected contaminated saliva from the diseased dog into a healthy rabbit. Later material from the rabid rabbit brain was inoculated into the brain of a healthy dog. The dog went through the usual symptoms of rabies but then miraculously recovered. This demonstrated to Pasteur that the virus could be "tamed." He then grew the virus in rabbit spinal cord and dried portions of it for varying periods of time to produce gradations of infectious material. Pasteur's vaccine was not tried on humans until a frightened mother asked the professor to help her son, Joseph Meister, a 9-year-old boy who had been gashed in 14 places, 2 days before, by a mad dog. Pasteur injected the boy with the "tamest" vaccine and followed this for the next 13 days with injections of increasingly stronger doses of vaccine. The boy survived. Pasteur died 10 years later in 1885, and Joseph Meister became the gatekeeper at the Pasteur Institute until his own death. In 1903 Negri examined rabies-infected neurons and described intracytoplasmic inclusions, which we now call ***Negri bodies*** (Plate 37). The presence of Negri bodies facilitates the early diagnosis of rabies in animals.

The rabies virus is maintained in wild animal populations, where the most common sources are skunks, foxes, raccoons, and bats. Infected dogs are an important domestic source of infection for humans, but vigorous immunization programs and leash laws have practically eliminated the number of cases of human rabies in the United States. The rabies virus is transmitted to humans via a bite or contact of an abraded area of the skin with the saliva from an infected animal. Occasionally inhalation of aerosolized virus in caves harboring large numbers of bats can be a means of transmission to humans.* The virus multiplies at the site of infection for several days and then travels along the peripheral nerves to the central nervous system, where it localizes and initiates

* In three instances rabies was transmitted during corneal transplantation in which the donor had died from a disease not recognized as rabies.

an inflammatory response ***(encephalitis)*** that is usually fatal. Degeneration of nervous tissue leads to the symptoms and manifestations of infection. The patient has difficulty swallowing, and the sight of liquids can induce painful contractions of the muscles involved in swallowing—a condition referred to as ***hydrophobia*** (fear of water). Death results from respiratory paralysis.

The incubation period for rabies is 2 to 16 weeks, and by the time the symptoms are recognized, it is too late to save the patient. Therefore prompt treatment following the bite of a potentially infected animal is important. The brain of the infected animal should be examined for the presence of virus. Infected nerve cells observed by light microscopy show the presence of Negri bodies (Plate 37), and virus can be detected by electron micrography (Figure 29-4). Smears of brain tissue can also be identified by fluorescent microscopy, which today is the principal method of identification

There have been only three recorded cases of survival from clinically apparent rabies, but post-exposure prophylaxis can be effective if instituted just following the bite of an infected animal. The wound should be flushed with soapy water, followed by the application of alchohol. Antirabies serum is applied to the wound, followed by active immunization with a recently developed rabies vaccine that is prepared from human fibroblast cultures. The vaccine, which requires only 5 or 6 doses (the old vaccine required 23 doses), has proved effective in Europe and in 1980 was made available for general use in the United States. Rabies can be prevented in most human populations by vaccination of domestic animals such as dogs. Those who are exposed to wild animals, such as veterinarians, can also be vaccinated.

Wild animals now constitute the most important source of infection for humans and domestic animals. Skunks, in particular, represent a threat to humans because of their contact with domestic animals. Several cases of rabies in cattle are recorded each year in the United States. Rabid dogs still present a great health hazard because of their close association with humans. In 1980, for example, over 103 children from an elementary school and a junior high school in Illinois came in contact with a dog with blood on its side, one half of its tail missing, and a limp in its left leg. Twenty students had significant exposure in which some were bitten, scratched, or had open wounds that had been licked by the dog. All 20 children received postexposure prophylaxis and none developed rabies, but the dog died of the disease shortly thereafter.

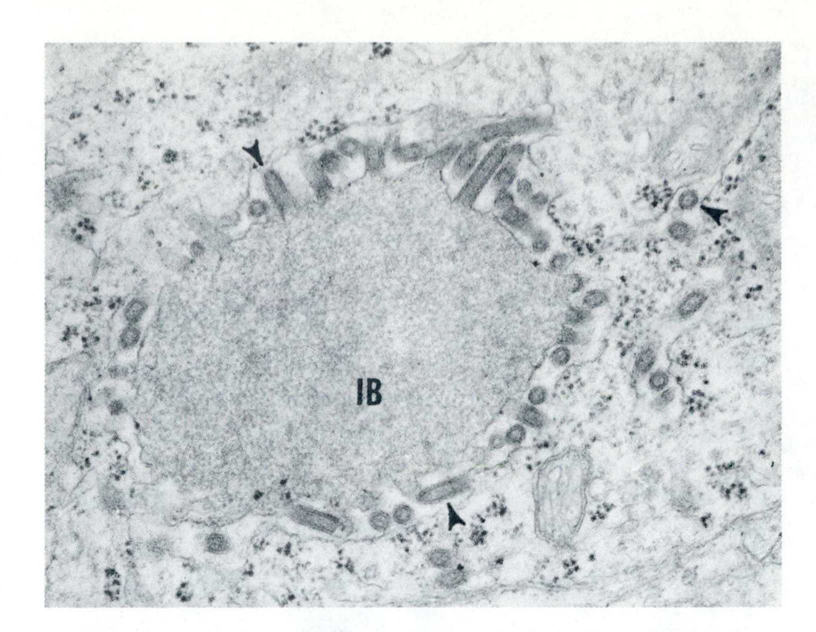

Figure 29-4 Rabies inclusion bodies (Negri bodies); electron micrograph of rabies virus infection in hamster brain neuron. (×50,000.) Virus *(arrows)* can be seen budding from intracytoplasmic membranes in association with inclusion body *(IB)*.

Courtesy Fred A. Murphy.

Polio

The disease that we now call poliomyelitis or polio is probably of ancient origin. Various Egyptian inscriptions depict individuals with withered limbs that many scientists believe to have been the result of poliomyelitis. The disease was not recognized as a distinct clinical entity until 1784. A German orthopedist, Jacob von Heine, in 1840 characterized the symptoms of the disease and associated them with an affliction of the spinal cord. Examinations of the gray matter of the spinal cord confirmed Heine's hypothesis. Later in 1905 investigators concluded that two forms of the disease existed: one, a short febrile illness of 3 to 4 days' duration that terminated in complete recovery, and another that lasted 8 to 10 days and terminated in paralysis. Landsteiner in 1908 showed that the disease was caused by a filterable agent that could be transferred to Old World monkeys in whom paralytic disease developed.

A great deal of public interest in polio resulted from the 1934 presidential election of Franklin D. Roosevelt, a victim of polio. A March of Dimes campaign was originated in 1938 whose sole purpose was to collect money to fight infantile paralysis (the common name used at that time, since the disease affected primarily the very young). John Enders, Thomas Weller, and Frederick Robbins in 1949 succeeded in culturing the poliovirus in human embryonic tissue—a discovery that

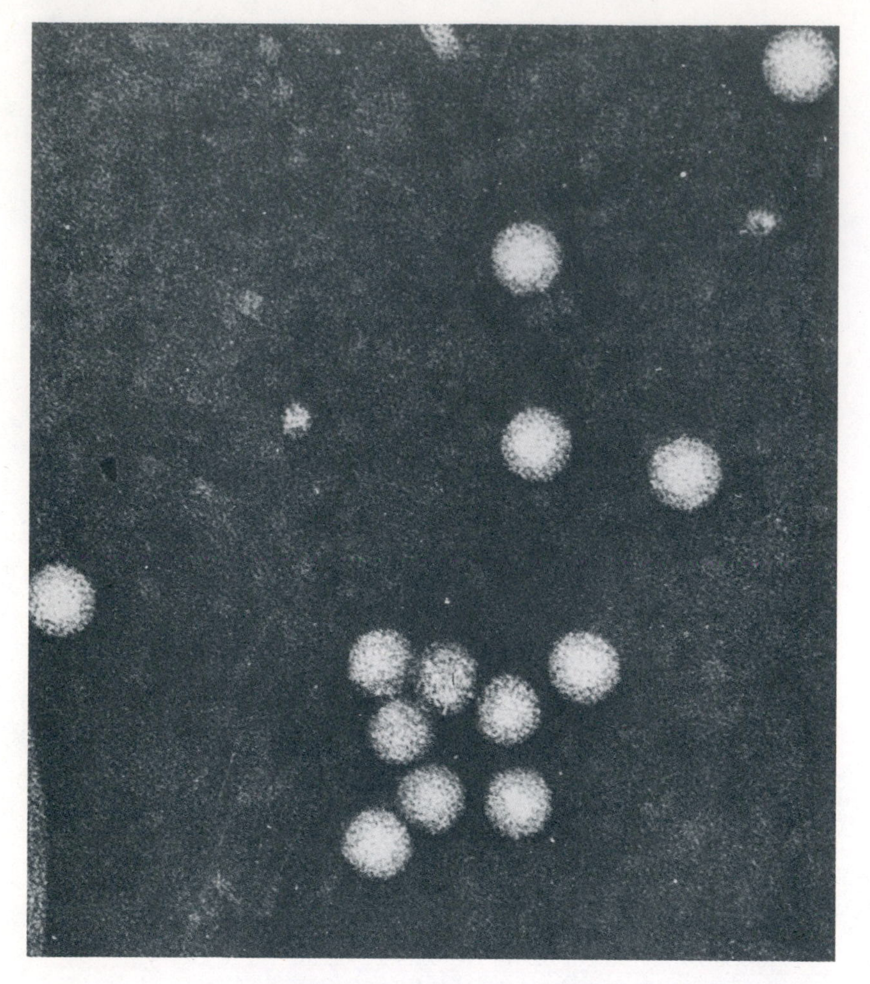

Figure 29-5 Electron micrograph of poliovirus. (×29,000.)
Courtesy R.C. Williams.

heralded the eventual development of viral vaccines. Jonas Salk in 1953 produced a formalin-inactivated vaccine that was licensed for use in 1958. Problems with quality control for the inactivated vaccine resulted in over 200 cases of vaccine-associated polio, but these problems were resolved, and the vaccine is still used today in combination with a live attenuated vaccine developed by Albert Sabin.

Polio is a disease caused by an RNA virus (Figure 29-5) belonging to the family Picornaviridae. A discussion of this disease is important because of certain aspects of epidemiology and prevention, even though vaccination has reduced the number of cases of polio in the United States from approximately 340,000 in 1955 to less than 30 in 1982.

The poliovirus is easily transmitted among infants and children under 4 years of age when hygienic conditions are lacking. The virus can be found in untreated sewage and water supplies, and ingestion of virus is followed by viral multiplication in the oropharynx or intestine. The disease is relatively mild with only minor symptoms and is rarely paralytic. Carriers represent a large pool of infectious virus, and large amounts of virus are shed from the intestinal tract of asymptomatic as well as symptomatic carriers. Children therefore become immune very early in life because of their contact with carriers. Improvement in sanitation (improved sewage disposal, rigid water treatment, as well as improved individual hygiene) brings about a change in the epidemiology of poliovirus infection, because few carriers are present and fewer children under 4 years of age become infected. Thus a much larger group of susceptibles over 5 years of age develops in the population.

Paralytic polio, which is very infrequent, results from invasion by the virus into the bloodstream and then into the central nervous system. Viral multiplication in the central nervous system, particularly in the spinal cord and brain, results in nerve cell damage that eventually affects muscle function. Paralysis may be temporary or permanent, but less than 2% of those infected with poliovirus show signs of paralysis.

The live oral vaccine for the past 12 years has been used exclusively in the United States. The oral vaccine offers the following advantages: (1) there is no need for subcutaneous injection; (2) there is an induction of intestinal immunity that prevents reinfection from wild* virus; and (3) there is no need for repeated booster doses. Probably the most important reason for use of the live vaccine is the fact that it protects the individual from intestinal infection from wild virus. An individual receiving the inactivated vaccine can later be infected by a carrier of active poliovirus and can remain a carrier and a source of infection to others. A case of vaccine-associated paralysis occurs with the live vaccine in rare instances (once for every 3 million doses of vaccine administered). There have been no recorded cases of vaccine-associated paralysis with the inactivated vaccine, which is one of the principal reasons why some public health officials still recommend its use.

Encephalitis caused by togaviruses

The most common cause of neural infections are the ***togaviruses*** (family: Togaviridae). Most of the viruses are characterized by their ability to multiply in both vertebrates and arthropods, particularly mosquitoes. They were originally classi-

**Wild* refers to virus that exists naturally in the environment and is not a laboratory strain that, for example, might be used in a vaccine.

Table 29-2 Viral infections that affect nervous tissue

VIRAL AGENT*	CLINICAL CONDITION	MODE OF TRANSMISSION	GEOGRAPHIC DISTRIBUTION	VACCINE	TREATMENT
Poliovirus *(Enterovirus)*	1% to 2% of those infected show paralysis of limbs	Fecal-oral route	Worldwide	Yes	None specific
Rabies virus *(Lyssavirus)*	Encephalitis	Contact of abraded skin with saliva from infected animal or animal bite	Worldwide	Yes (humans and animals)	Antirabies serum followed by vaccination
Western equine encephalitis virus *(Alphavirus)*	Encephalitis	Bite of mosquito	North and South America	None	None specific
Eastern equine encephalitis virus *(Alphavirus)*	Encephalitis	Bite of mosquito	Eastern and Northern United States	None	None specific
St. Louis encephalitis virus *(Flavivirus)*	Encephalitis	Bite of mosquito	Central United States	None	None specific
Venezuela equine encephalitis virus *(Alphavirus)*	Encephalitis	Bite of mosquito	South and Central America and United States	None	None specific
California encephalitis virus *(Bunyavirus)*	Encephalitis	Bite of mosquito	United States	None	None specific
Mumps virus† *(Paramyxovirus)*	Encephalitis	Respiratory secretions	Worldwide	Yes	None specific
Varicella (chickenpox) virus† (subfamily: Alphaherpesvirinae)	Encephalitis	Respiratory secretions	Worldwide	Yes	See Table 29-4
Measles (rubeola) virus† *(Morbillivirus)*	Encephalitis	Respiratory secretions	Worldwide	Yes	Gamma globulin can modify disease
Herpes simplex virus† (subfamily: Alphaherpesvirinae)	Encephalitis	Respiratory secretions	Worldwide	None	None specific
Echoviruses† ‡ *(Enterovirus)*	Meningitis	Fecal-oral route	Worldwide	None	None specific
Coxsackieviruses *(Enterovirus)*	Meningitis	Fecal-oral route and respiratory secretions	Worldwide	None	None specific
Mumps virus *(Paramyxovirus)*	Meningitis	Respiratory secretions	Worldwide	Yes	None specific

*The common species name of the viral agent is used with the genus in parentheses.
†The condition caused by the agent is a complication of infection and not the primary manifestation of infection.
‡Echoviruses: enteric cytopathogenic *h*uman orphan viruses.

fied as arboviruses (*ar*thropod *bo*rne). Two important genera of togaviruses are ***Alphavirus*** and ***Flavivirus*** (Figure 29-3), which are important causes of severe encephalitis (Eastern equine encephalitis, Western equine encephalitis, St. Louis encephalitis, Venezuelan equine encephalitis, and others). The term *equine* (horses) refers to the animals from which several of the togaviruses were originally isolated. Humans are usually incidental hosts for Eastern equine encephalitis (EEE) and Western equine encephalitis (WEE), which are the principal togavirus infections in the United States. The primary hosts for EEE and WEE are birds, both wild and domestic, but the mosquito, in which the virus multiplies, is the vector in transmission to humans and horses.

Togaviruses enter the human host following a bite of the mosquito or other appropriate vector and are injected into lymph or the bloodstream. Infection of endothelial cells in the lymph nodes or spleen is followed by dissemination of virus into the systemic circulation. The virus localizes primarily in the central nervous system, where multiplication leads to tissue hemorrhage and various stages of meningitis. EEE and WEE produce severe illness with a high mortality.

Control measures include avoiding exposure to mosquitoes or other arthropod vectors and spraying vector-breeding areas, such as stagnant waters.

Other viral agents that are known to affect nervous tissue are outlined in Table 29-2. They include agents that affect nervous tissue of the brain (encephalitis) or spinal cord (meningitis).

VIRUSES THAT AFFECT RESPIRATORY TISSUE

Inhalation of aerosolized respiratory secretions and direct contact with hands or fomites contaminated with respiratory secretions are the most common methods for infection of respiratory tissue. The site where the virus lodges in the respiratory tract is related to the size of the droplet. The larger droplets remain entrapped in the upper respiratory tract, whereas smaller droplets evade entrapment and are carried to the lower respiratory tract. The viscosity of the mucus and its rate of flow influence viral attachment to epithelial tissue. The lower the viscosity and the slower the rate of flow of mucus, the greater the chance for virus to attach to the epithelium. The influenza viruses can reduce the viscosity of mucus because of the presence of surface components, such as neuramidinase, which has enzymatic activity. Several viral agents have an affinity for respiratory tissue, in addition to the influenza viruses; these include the rhinoviruses, respiratory syncytial virus, parainfluenza viruses, and coronavirus. The most important of this group are the influenza viruses, because they cause considerable morbidity and mortality, and the rhinoviruses because of their association with the common cold.

Influenza

The word *influenza* is derived from the Italian, meaning influence and in the thirteenth century was used to describe a disease that appeared during the winter months. The winter-occurring disease was believed to be caused by the influence of the stars, and its characteristic recurrence during the winter months led to the permanent use of the word ***influenza.*** Influenza, unlike any other disease, is global and is not endemic to any one area. Most populations have been visited by the disease at one time or another. The first recorded pandemic in 1580 was followed by major outbreaks in 1645, 1743, 1782, 1830, 1837, and 1847. The entrance of influenza into the Americas was associated with the Spanish trade ships, and at one time the disease was called the Spanish influence. Other names included malignant catarrh, gentle corruption, and grip. Americans, during the pandemic of 1890, called the disease Chinese distemper because medical authorities believed that the disease was brought to the United States by dust blown from the dried banks of the Yellow River in China. Noah Webster (of dictionary fame) thought the causes of influenza to exist in air, fire, and water, since no other explanation could be made for its rapid spread within and among communities. The severest pandemic in which nearly 20 million lives were lost, mostly to one of the disease's major complications, bacterial pneumonia, appeared in 1918. Several explanations were given for the 1918 pandemic in the United States, including sugar deficiency caused by wartime rationing and eating fish infected with influenza germs spread underwater by German submarines. Large-scale epidemics still occur periodically, but none has ever matched the geographic scope and virulence of the 1918 strain of virus.

It was not until 1933 that scientists isolated an influenza virus from the throat washings of infected patients. The virus was called ***influenza virus type A.*** A second influenza virus, called ***type B,*** was isolated in 1940, but it does not cause as serious an illness as type A. A ***type C*** virus has also been isolated, but it causes very minor respiratory tract illness. A vaccine made from both A and B strains was prepared for use by men in the armed forces in 1943, and although it was not 100% effective, the vaccine gave a measure of protection.

The influenza viruses are RNA viruses that exhibit structural pleomorphism, but most particles are spherical (Figure 29-6). The capsid is surrounded by a lipid coat containing spikes that play an important role in the virulence of the virus.

Influenza is acquired by inhalation or ingestion of contaminated respiratory secretions. The virus adheres to the respiratory epithelium by the ***hemagglutinin*** spikes while the ***neuraminidase*** spikes make contact with their substrate, neuraminic acid, which is part of the mucosa. Viral multiplication leads to the hydrolysis of neuraminic acid and liquification of the mucosa, leaving an underlying epithelial layer unprotected from infection by other microbial agents and promoting the spread of virus to surrounding tissue. Unabated, the disease process may involve the bronchioles. The incubation period for influenza virus disease is 24 to 48 hours, with most infections showing no symptoms or mild symptoms such as fever, headache, sore throat, and conjunctivitis. Two percent of the cases are complicated by pneumonia, and most of these are associated with individuals having chronic lung or heart disease. Pneumonia caused by *Staphylococcus aureus* or *Streptococcus pneumoniae,* is the most frequent cause of death following infection by the virus.

One of the most interesting aspects of the influenza viruses are their genetic instability. It is this characteristic that is most responsible for the large-scale pandemics that occurred in 1918-1919, 1957-1958, and 1968-1969, as well as the more periodic epidemics. Changes in the influenza virus nucleic acid result in alterations in the chemical structure of the neuraminidase and hemagglutinin spikes that will affect the outcome of future infections. Antibodies to the hemagglutinin spikes interfere with viral infectivity. What this means is that the antibodies produced during infection by one strain of influenza A virus may be unable to ward off future infections by influenza A virus in which a genetic change has taken place. These genetic changes result in variations that may be classified as either minor or major. To understand this, one must first consider that following infection, antibodies to one virus protect the individual for approximately 2 to 3 years from infection by the same strain of virus. When a minor change, called ***antigenic drift,*** occurs in the viral spikes, the antibodies that were produced in response to the parent virus do not interact to the same degree with the mutant virus. When a major change, called ***antigenic shift,*** occurring in the viral spikes, results in antibodies being produced in response to the parent virus that not only show a reduced affinity, but also do not recognize the mutant's altered spikes. The result of this is that the individual or population originally infected by the parent influenza virus is unprotected from infection by the mutant virus in which the antigenic shift has occurred. Thus vaccines produced against influenza virus type A of one type might be, for example, only 60% effective against a virus that has undergone antigenic shift.

The reason for the genetic changes are not completely understood, but the changes may be due to the following: (1) a section of the viral nucleic acid is especially subject to mutation (that is, it is a hot spot for variation or mutation), or (2) genetic recombination may occur between viruses. For example, the infecting virus may retain certain segments* of its nucleic acid, and the other segments may be obtained from other viruses. It is known that various animals and birds harbor influenza viruses. The viruses circulating in the human population can also infect domestic animals such as birds, pigs, etc. It is suggested that an exchange of genetic segments may occur between these different viruses while still in the animal, resulting in the formation of a new influenza virus type that can later be transmitted to humans.

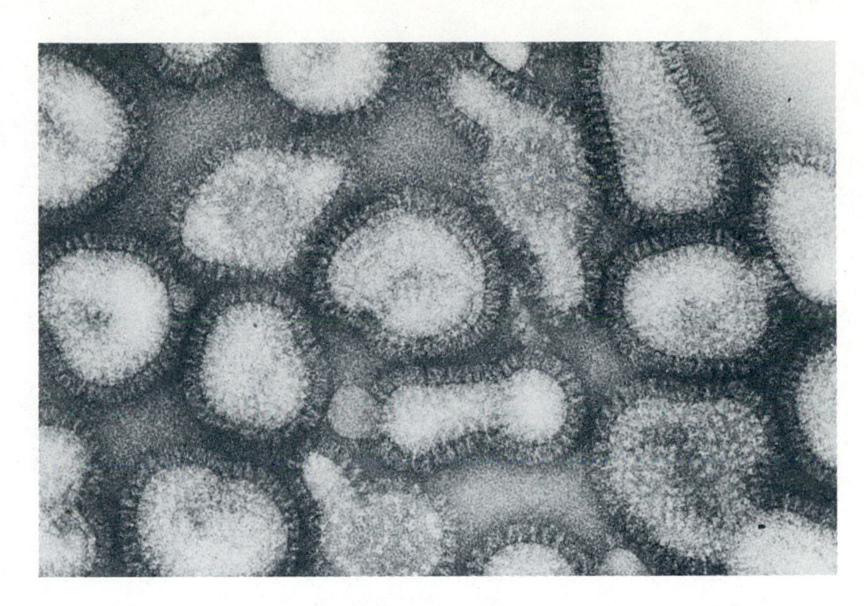

Figure 29-6 Electron micrograph of human influenza virus.(× 180,000.)
Courtesy R.C. Williams.

Epidemics of influenza A occur every 2 to 4 years, whereas pandemics occur every 10 years. Considerable genetic change can occur in the virus between epidemics and pandemics, and immunity to reinfection can be related to the genetic character of the virus, as well as to the dose of infecting virus. Influenza vaccines are recommended only for high-risk groups, such as the elderly over 65 years of age and those with chronic cardiovascular and pulmonary disorders. In recent years an antiviral drug called amantadine has been used with some success as a prophylactic agent in preventing influenza and as a therapeutic agent in decreasing the length of time for recovery from disease.

Influenza virus type B does not show the antigenic variation of its sister virus, influenza A. Influenza B seldom occurs in adults, but in children influenza B virus, as well as influenza A and some

*Influenza virus nucleic acid is made up of eight different segments.

Table 29-3 Viruses that affect respiratory tissue

VIRAL AGENT*	CLINICAL CONDITION	METHOD OF TRANSMISSION	VACCINE	TREATMENT
Rhinovirus types 1A to 114 *(Rhinovirus)*	Common cold observed in all age groups	Close contact with respiratory secretions	None	None specific
Parainfluenza virus *(Paramyxovirus)*	Common cold in older children and adults; in infants more serious illness, such as bronchitis and croup, requiring hospitalization	Close contact with respiratory secretions	Under investigation for infants	None specific
Respiratory syncytial virus *(Pneumovirus)*	Most frequent cause of infant pneumonia and bronchiolitis; infection in older children and adults is milder	Close contact with respiratory secretions	None	None specific
Human coronavirus *(Coronavirus)*	Colds; most infections are asymptomatic; no lower respiratory tract involvement	Close contact with respiratory secretions	None	None specific
Influenza virus types A, B, and C *(Influenzavirus)*	Mild symptoms—fever, headache, sore throat, and conjunctivitis; 2% develop pneumonia	Close contact with respiratory secretions	Yes	Amantadine has been used as prophylactic agent; therapeutically it reduces recovery time

*The common species name of the viral agent is used with the genus in parentheses.

other viruses, is believed to be responsible for the complication called ***Reye's syndrome.*** Reye's syndrome is characterized by encephalopathy and fatty degeneration of the viscera. The disease is fatal to approximately 25% of those affected. Nationwide outbreaks of Reye's syndrome (350 to 400 cases per year) have been associated with influenza B outbreaks during the 1973-1974 and 1976-1977 influenza seasons and with influenza A in the 1978-1979 season. It is suspected, but has not yet been proved, that salicylates such as aspirin administered for influenza infection contribute to the pathogenesis of Reye's syndrome.

Common cold (rhinoviruses)

The most important group of viruses associated with the common cold are the rhinoviruses, which belong to the family Picornaviridae. There are over 100 immunologically distinct types of rhinoviruses, with only chimpanzees and humans serving as hosts. Disease symptoms, however, appear only in humans.

Rhinovirus infection is believed to occur by close contact with respiratory secretions from an infected individual. Viable rhinovirus can be recovered from the hands of infected patients, and recent evidence suggests that hand contact may be a very important means of viral transmission, since the virus can remain viable on the skin for 3 to 4 hours. The most frequent symptoms associated with the common cold are sore throat, cough, and watery eyes (coryza), which seldom last more than 5 to 6 days. Disease caused by one type of virus confers immunity for up to 3 years against rechallenge by the same type of virus. Individuals, however, may have repeated colds during the year, not only because of the large number of immunologically distinct types, but also because of diseases by other viruses such as the coxsackieviruses.

The other viruses affecting respiratory tissue are outlined in Table 29-3.

VIRUSES THAT AFFECT THE SKIN

The intact skin is an effective barrier against penetration by viruses. Viruses can penetrate to the subepithelial layers when (1) the skin is abraded, (2) the skin is punctured by needles or by animal bites, (3) there is an underlying skin disease (eczema), or (4) the skin is bitten by arthropod vectors such as mosquitoes. The above represent avenues of entry for such agents as the rabies and yellow fever viruses, but these are not associated with skin lesions. The viruses discussed in this section of the chapter are those that cause skin lesions. They are usually acquired through the respiratory tract but are later disseminated in the blood and eventually find their way to the skin. The lesions or vesicles produced on the skin surface contain large numbers of virus, and they represent portals of exit from the body as well as sources of infection to others. Measles (rubeola), chickenpox, and smallpox are diseases that are transmitted by respiratory secretions but whose primary manifestations involve the skin.

Measles (rubeola)

Measles is an acute infectious disease associated primarily with children. It is highly communicable, and before the advent of a measles vaccine in 1967 over 500,000 cases were reported annually in the United States. Infection is acquired by inhalation or oral contact with respiratory secretions. The virus may also enter the body through the conjunctiva. Initially the virus multiplies in the respiratory mucosa and regional lymph nodes before it is disseminated through the bloodstream. A rash appears on the skin (Plate 38) about 2 weeks after contact with the virus, and lesions called ***Koplik's spots*** appear on the oral mucosa. A fever, which may reach as high as 105° F, precedes and accompanies the rash. Measles, although considered a relatively mild disease, can be fatal to those who are immunologically deficient. Most deaths are the result of complications following degeneration of tissue cells by the virus and include bacterial pneumonia and encephalitis. A very rare complication, following measles is called ***subacute sclerosing panencephalitis (SSPE).*** SSPE causes a progressive mental deterioration that does not appear until 5 to 10 years after apparent recovery from measles. It has not been determined whether SSPE is caused by the original measles virus or a variant of it.

Infants are protected from the measles virus for at least 6 months because of antibodies acquired from the mother. The measles vaccine, which is administered 15 months after birth, has reduced the incidence of measles from 500,000 in 1967 to 50,000 in 1975, but nearly 100,000 cases were reported in 1977. Some of the rise is due to a lack of parental concern in obtaining the vaccine for their children and to the fact that the original vaccine used between 1963 and 1967 was an inactivated one and was poorly immunogenic. Children vaccinated between 1963 and 1967, when later exposed to the measles virus, have manifested a disease that frequently terminates in severe and often fatal forms of pneumonia. A vigorous immunization campaign has resulted in the reduction of the number of cases of measles to less than 300 in 1982.

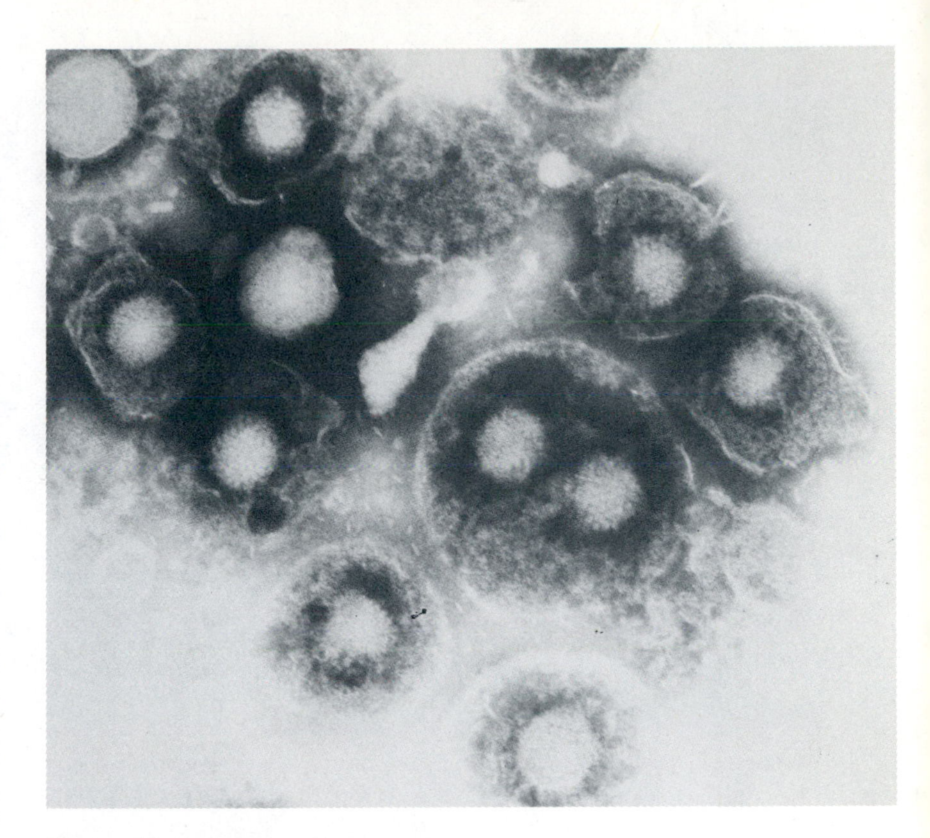

Figure 29-7 Herpes simplex virus; electron micrograph of herpesvirus particles, most containing envelopes.
Courtesy Erskine L. Palmer, Centers for Disease Control, Atlanta, Ga.

Herpes infection (herpes simplex virus)

Herpes simplex virus (Figure 29-7) belongs to a family of viruses that also includes the following agents:

1. Cytomegalovirus, which causes a number of clinical entities (see Table 29-6)
2. Varicella-zoster virus, which is the causative agent of two entities: ***varicella*** (***chickenpox,*** Figure 29-8) in children and ***zoster (shingles)*** in adults
3. Epstein-Barr virus, which is the suggested cause of ***infectious mononucleosis,*** as well as a carcinoma called ***Burkitt's lymphoma***

The herpes simplex virus can be divided into two antigenic types: I and II. Type I is associated primarily with lesions on the lips (coldsores) as well as with diseases of the eye and skin. Type II disease involves the genital area, and the virus is transmitted by sexual contact. The discussion of type II disease is found in chapter 30.

It is believed that most people are infected by type I virus at a very early age by close contact with oral secretions from family members. Facial

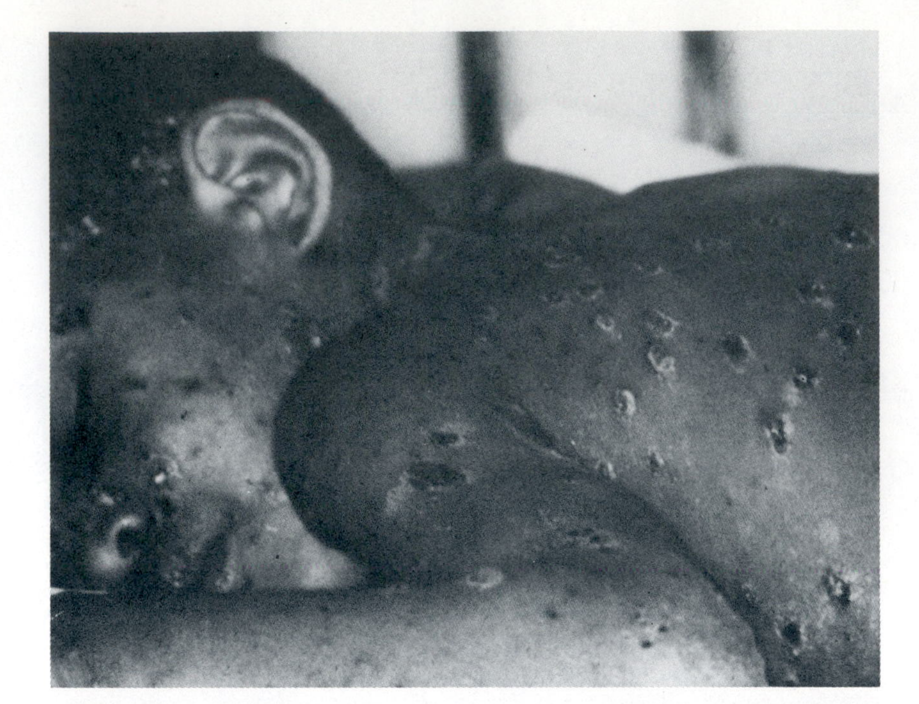

Figure 29-8 Chickenpox (varicella).
Courtesy Audiovisual Service Department of Pathology, University of Illinois Medical Center, Chicago, Ill.

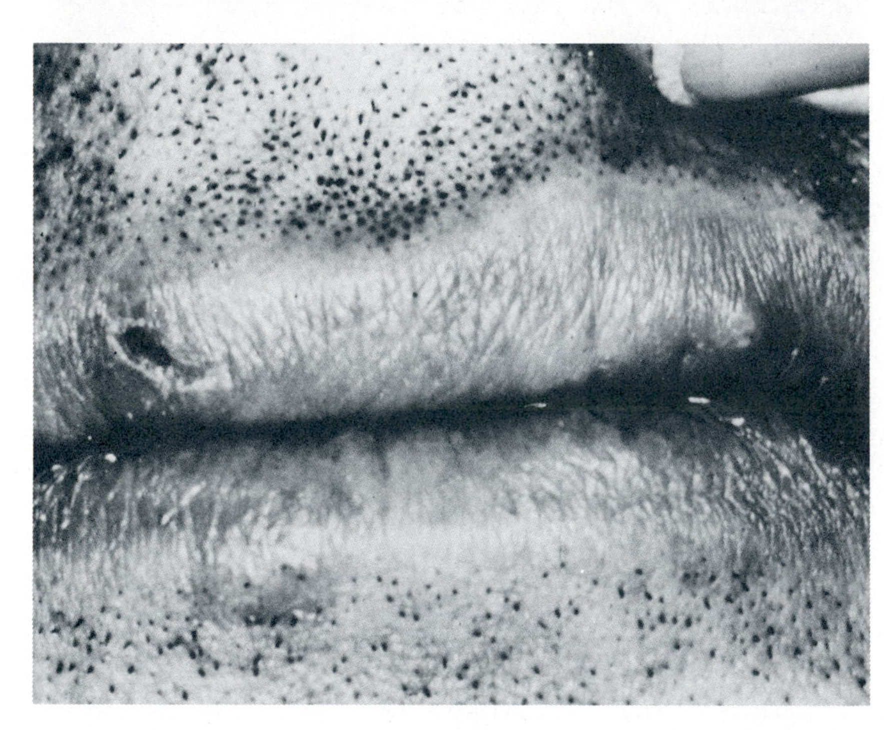

Figure 29-9 Cutaneous lesions on lips and skin around mouth of individual with recurrent herpes labialis.
Courtesy L.J. LeBeau, Department of Pathology, University of Illinois Medical Center, Chicago, Ill.

coldsores (Figure 29-9) are the result of reactivation of latent type I infection, and up to one third of the population have suffered from at least one episode of coldsores in their lifetime. A primary infection by type I virus may be followed by no symptoms, or the symptoms may be mild. The virus replicates at the site of infection and then travels to the trigeminal nerve (nerve that innervates the muscles of mastication), where it remains latent until reactivation, at some later time, by such factors as stress, hormonal change, exposure to sunlight, abrasions, etc. Type I virus is also involved in other clinical manifestations: lesions on the mucous membranes of the mouth resulting from primary disease (***gingivostomatitis,*** Figure 29-10); primary disease of the eye (keratoconjunctivitis, Figure 29-11), which can cause blindness if there are repeated episodes involving the cornea; primary disease of the skin, particularly eczema ***(eczema herpeticum);*** and ***meningoencephalitis,*** a frequently fatal infection that may result from dissemination of virus from

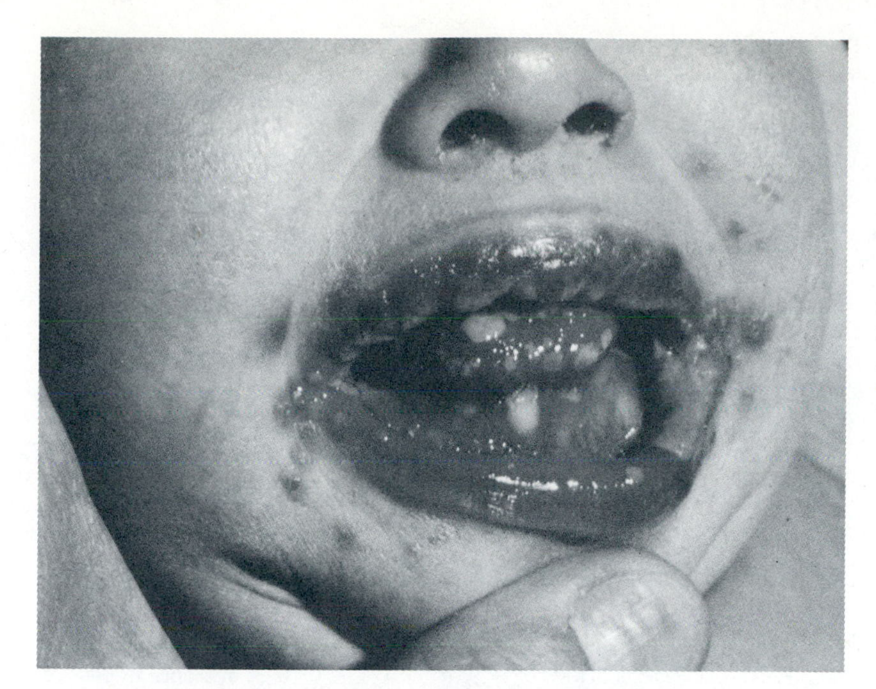

Figure 29-10 Gingiovostomatitis caused by herpesvirus infection.

From Overall, J.C., Jr. In Galasso, G.J., Merigan, T.C., and Buchanan, R.C. (Eds.). Artiviral agents and viral diseases of man. *Raven Press, New York, 1979.*

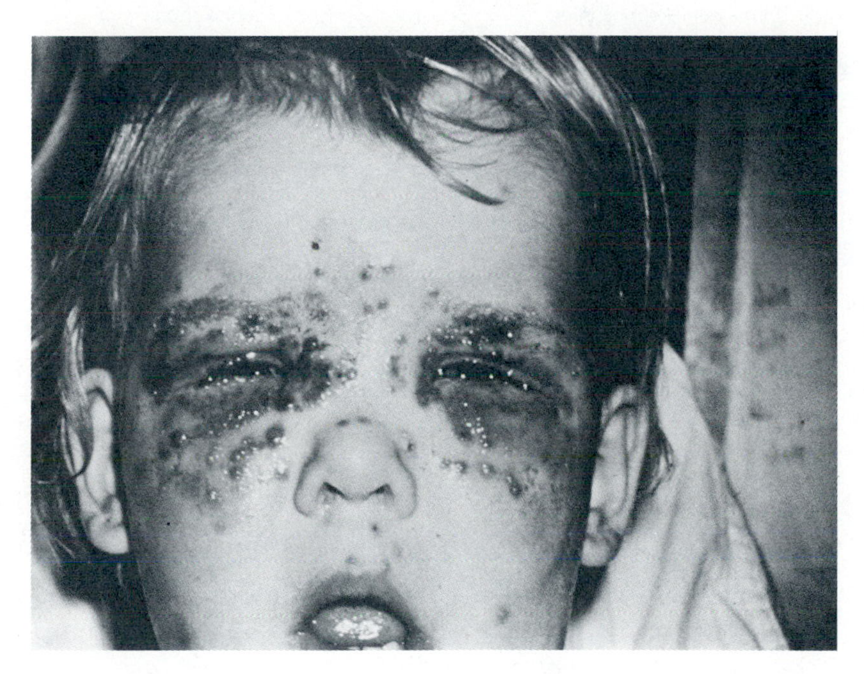

Figure 29-11 Primary infection of eye by herpesvirus. Lesions are seen on periorbital regions, as well as on conjunctiva and cornea.

From Pavan-Lanston, D. Int. Ophthalmol. Clin. ***15****(4):149, 1980.*

lesions that originally involved the mouth and skin.

One of the most promising antiviral drugs is ***acyclovir,**** which in preliminary trials has prevented recurrences of disease in animals. No vaccine is yet available, although one is being experimented with in Europe.

Viruses other than measles and herpes simplex that affect the skin are outlined in Table 29-4.

*See chapter 30 for further discussion of acyclovir.

VIRUSES THAT AFFECT DIGESTIVE GLANDS

Hepatitis viruses

The term ***hepatitis*** means inflammation of the liver. Several viruses, including the agents of cytomegalovirus infection, infectious mononucleosis, and yellow fever, infect the liver, but the most important are the hepatitis viruses. Today hepatitis is a major health problem, particularly among hospital personnel and hospitalized patients.

Table 29-4 Characteristics of viruses that affect the skin

VIRAL AGENT*	CLINICAL CONDITION	METHOD OF TRANSMISSION	VACCINE	TREATMENT
Rubella virus *(Rubivirus)*	German measles—mild infection with rash when acquired postnatally; when acquired by pregnant woman, virus can be transferred to fetus and cause fetal abnormalities or death	Contact with respiratory secretions or congenital transfer through placenta	Yes, for children over 12 months of age	None specific
Smallpox virus *(Orthopoxvirus,* Figure 29-12)	Smallpox—development of large pustules on skin (Figure 29-13); as of 1978 disease has been declared eradicated	Contact with pustules or various fomites contaminated by fluid from pustules	Yes, but is no longer recommended in United States; only those who may be exposed to virus (lab workers) receive vaccine	None specific
Varicella-zoster virus (subfamily: Alphaherpesvirinae)	Varicella (chickenpox)—highly communicable disease of children, producing rash; zoster (shingles) causes pain in skin area where nerves have been affected	Varicella by contact with fluid in skin lesions or contact with respiratory secretions; zoster is reactivation of latent virus from primary chickenpox infection	None	Zoster immune globulin is available for passive protection against varicella in high-risk adolescents exposed to chickenpox
Papilloma virus *(Papovavirus A)*	Warts	Direct contact or indirectly by contact with fomites; can be transferred from one part of body to another by autoinoculation	None	Not required, but surgery has been used
Dengue virus *(Flavivirus)*	Dengue fever; rash with joint pain; severe form of disease causes extreme vascular hemorrhage; prevalent in Caribbean	Bite of mosquito	Vaccine not yet available for commercial use	None specific
Measle virus *(Morbillivirus)*	Measles (rubeola); usually mild with skin lesions, as well as oral lesions called Koplik's spots	Inhalation of or oral contact with respiratory secretions	Yes, given 15 months after birth	No specific therapy
Herpes simplex (family: Alphaherpesvirinae)	Type I—facial lesions Type II—genital lesions	Type I—contact with oral secretions or facial lesions; Type II—sexual contact	None	Acyclovir appears to be a promising drug for treatment of primary disease

*The common species name of the viral agent is used with the genus in parentheses.

Diseases of the liver, particularly jaundice, have been described since ancient times, but it was not until 1791 that a description of an epidemic of jaundice was put forth. Epidemic jaundice occurred during the Middle Ages and, next to cholera and plague, was an important cause of pandemics. Epidemics of jaundice have been recorded in many wars, including World War II, in which epidemics of vast proportions swept through the armed forces in the Mediterranean sector. Despite the epidemic nature of jaundice, the concept that jaundice was hepatic in origin was not universally accepted. Most earlier medical practitioners and scientists believed that obstruction of the bile duct was the cause of jaundice. It was assumed that a microbial agent (not specific for the liver) caused disease in the intestines, which then spread to the bile duct and blocked its excretions. Major support for the view that epidemic jaundice was due to a specific microbial agent came in the late 1800s, when it was concluded that systemic infection ultimately involved cells of the liver. The disease was henceforth referred to as ***infectious*** or ***epidemic jaundice.***

In the early 1900s it was suspected that the agent of hepatitis was smaller than a bacterium. Since no animal could be infected by the hepatitis agent, the cause of the disease remained a mystery until 1939 when a clue as to its identity was revealed from a totally unrelated incident. A large number of individuals were administered a yellow fever vaccine, but shortly thereafter 27% of the vaccinees developed hepatitis. This result suggested that a filterable agent had contaminated the vaccine, since human serum was used in the viral culture medium and filtration would have removed a bacterial agent from the medium. This incident prompted a request that human serum not be used in the preparation of viral vaccines unless the history of all donors could be followed for at least 1 month (the incubation period for infectious hepatitis). The use of human serum for other viral cultivations or for transfusions also resulted in outbreaks of hepatitis and thus confirmed the earlier findings on the epidemiology of the disease. It became apparent that hepatitis could be transmitted by different routes. At least two clinically distinct diseases are associated with infection: ***infectious hepatitis (epidemic hepatitis)***, which has a long incubation period (hepatitis B virus), and ***serum hepatitis,*** which has a short incubation period (hepatitis A virus).

There are three hepatitis viruses: ***hepatitis A virus (HAV), hepatitis B virus (HBV),*** and ***non-A, non-B (NANB).*** Little success has been obtained in growing the viruses in culture, and most of our information concerning them comes from animal studies or from human infections. NANB has been recogized only by exclusion of HAV and HBV, and not by electron microscopy or serological results. Only in the past decade have HAV and HBV been characterized. HBV shows three distinct antigens (Figure 29-14): a spherical 22 nm particle, a 42 nm spherical particle called the ***Dane particle,*** and tubular forms. The Dane particle is believed to be the infective form of the virus. There exists, in addition to the three morphological entities, several antigenic types that become serologically detectable following infection. The most important of these is the surface antigen called HB_sAg, whose presence in the blood is an indicator of hepatitis B infection. HB_sAg is a 22 nm particle that in itself is not infectious. HAV was first detected in 1973 by using ***immune electron microscopy (IEM)*** on stool specimens. IEM is a technique in which immune serum (antibodies specific for the virus) is applied to a clinical specimen in order to aggregate the viral particles and make them more accessible to microscopic visualization (Figure 29-15). HAV is a single morphological entity—a 27 nm particle.

It may be very difficult to clinically distinguish which agent is the cause of hepatitis. Disease caused by each virus results in degeneration of liver tissue, with accompanying jaundice and the

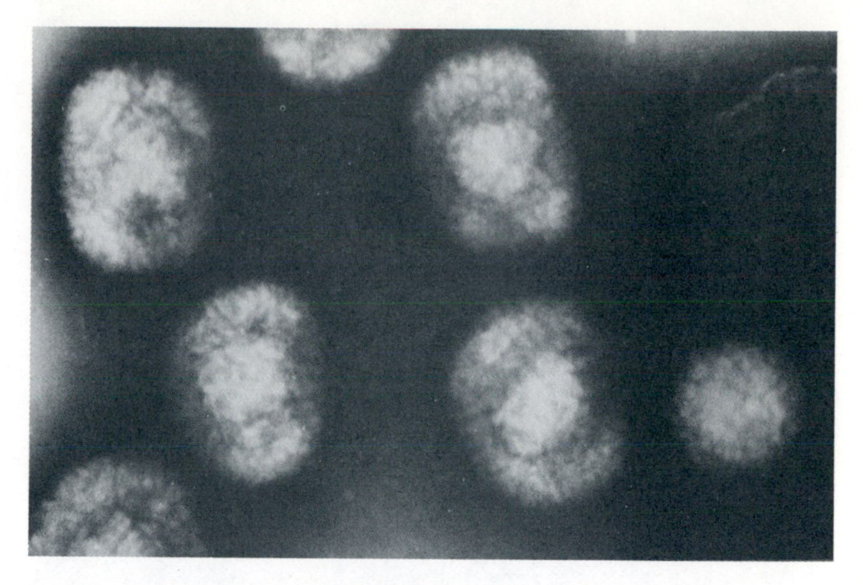

Figure 29-12 Electron micrograph of vaccinia (smallpox) virus. (×88,000.)
Courtesy R.C. Williams.

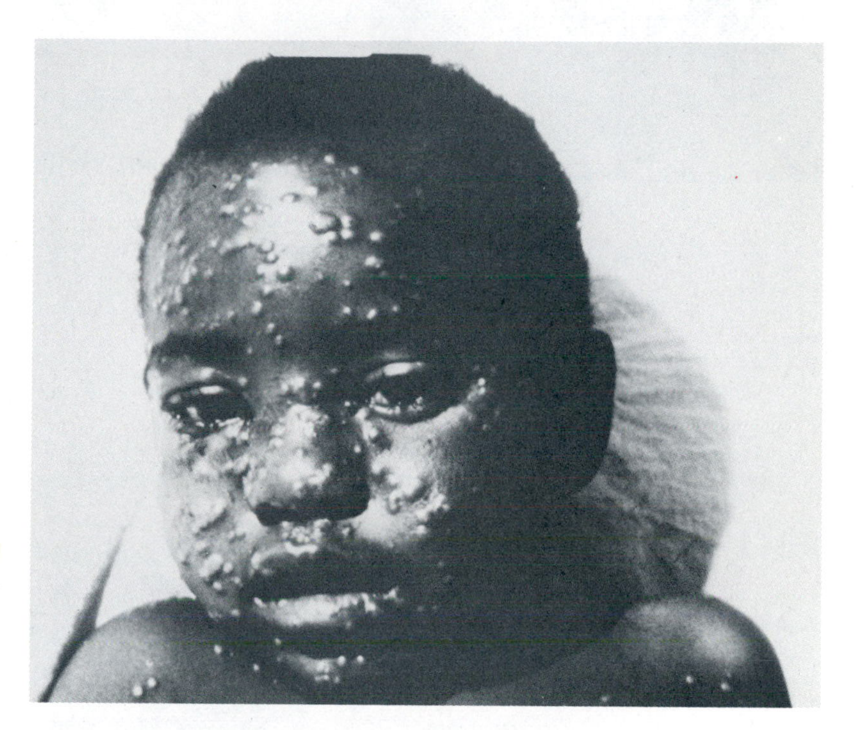

Figure 29-13 Smallpox pustules covering face of small boy.
Courtesy Stanley Foster, Centers for Disease Control, Atlanta, Ga.

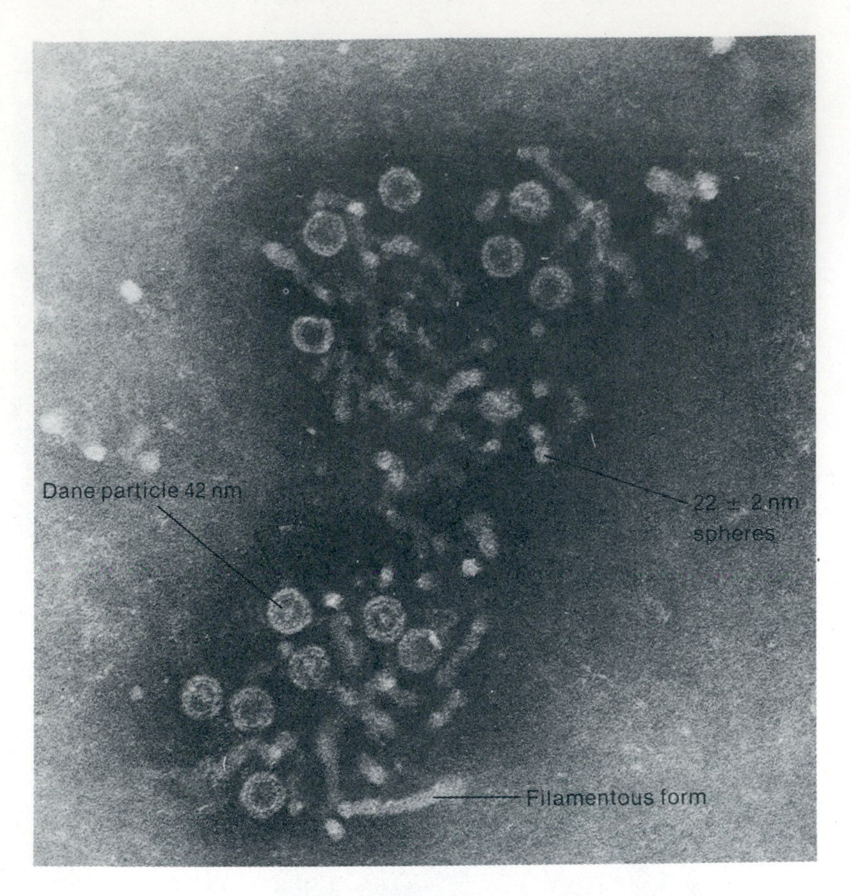

Figure 29-14 Electron micrograph demonstrating the three antigens of hepatitis B virus. (×210,000.)

From Cabral, G. In Boyd, R.F., & Hoerl, B.G. Basic Medical Microbiology. Little, Brown & Co., Boston, 1981.

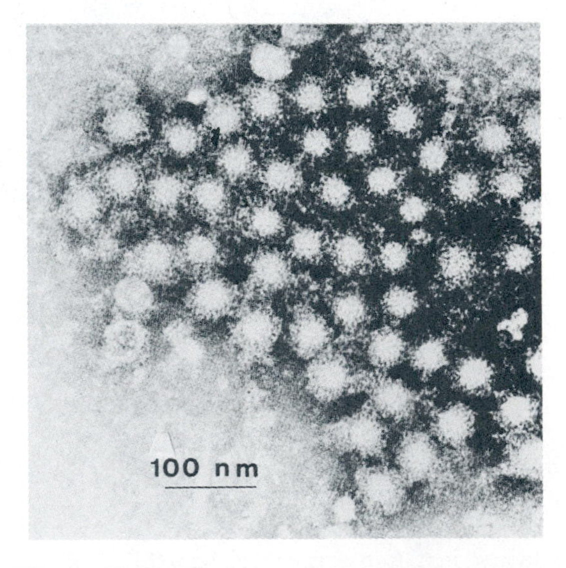

Figure 29-15 Electron micrograph of hepatitis A virus aggregated by human antibody.

Courtesy Centers for Disease Control, Bureau of Epidemiology, Hepatitis Laboratories Division, Phoenix, Ariz.

release of liver-associated enzymes and products into the bloodstream. Infection by HBV and possibly NANB results in up to 30% of infected individuals becoming chronic active viral carriers, and this represents an important health hazard. This situation is further complicated by the fact that all of the hepatitis viruses are unusually resistant to disinfectants, heat, and other chemical and physical agents that are normally detrimental to other viruses. Autoclaving, however, is effective in destroying the viruses. The characteristics of the hepatitis viruses are discussed below, and a comparison between HBV and HAV infection is outlined in Table 29-5.

Hepatitis A. Hepatitis A infection has been called infectious hepatitis or short-incubation hepatitis. HAV is transmitted by the fecal-oral route, with most infections occurring from the ingestion of contaminated food, water, and shellfish. Hepatitis A is not a major health hazard in the hospital. Asymptomatic HAV infection at one time was common among young children, a situation similar to that of early poliovirus infection. Today, because of improved hygienic conditions and sanitation, infections occur less frequently among children, and greater susceptibility now occurs among adolescents and adults, in whom the disease is more severe. HAV does not cause the chronic liver disease that is so often associated with hepatitis B infection; nor is the mortality for HAV disease very high. There is no specific treatment for hepatitis A, but immune serum globulin prepared from the plasma of adults who have recovered from the disease can temporarily prevent disease or, at the very least, reduce its severity.

Hepatitis B. Hepatitis B disease has also been called serum hepatitis or long-incubation hepatitis. HBV is transmitted primarily by contact with the infected blood of patients with active disease or with the blood of chronic carriers of the virus. HBV particles can also be found in many body fluids, including sweat, breast milk, semen, bile, tears, urine, feces, and saliva, but these are not the most efficient vehicles for transmission of the virus. Hepatitis B virus can be transmitted sexually and cause disease, especially in homosexuals (see Chapter 30). Many cases of hepatitis B occur as the result of administration of plasma or serum—or blood or blood products. Infection is especially common among patients and staff in hemodialysis units. Hepatitis B can be reduced by preventing persons known to have or have had hepatitis from being used as blood donors. Disinfection of materials to be used in procedures

Table 29-5 Characteristics of the hepatitis A and hepatitis B viruses*

CHARACTERISTIC	HEPATITIS A	HEPATITIS B
Principal mode of transmission	Fecal-oral route (contaminated food, water, shellfish, etc.)	Contact with blood or blood products (can also be transmitted venereally)
Incubation period	15 to 45 days	50 to 180 days
Fever	Common	Less common
Severity of disease	Mortality low (1%)	Mortality higher (1% to 10%)
Carrier state	Very low	High
Immunity	Lifetime	Lifetime
Prophylaxis	Gamma globulin prevents jaundice	High-titer gamma globulin prevents jaundice

*Non-A, non-B has not been included because of lack of information on the disease. Epidemiologically, it resembles hepatitis B.

Table 29-6 Characteristics of viral agents that affect digestive glands

VIRAL AGENT*	CLINICAL CONDITION	MODE OF TRANSMISSION	VACCINE	TREATMENT
Yellow fever virus *(Flavivirus)*	Yellow fever—virus multiplies in lymph nodes and later infects liver, spleen, and kidney; degeneration of liver or kidney is cause of death	Bite of mosquito *(Aedes aegyptii)*	Yes (attenuated vaccine); immunity lasts at least 10 years	None specific
Mumps virus *(Paramyxovirus)*	Mumps—virus multiplies in parotid glands, causing their enlargement (parotitis), and later disseminates to testes, ovaries, pancreas, and brain	Direct contact with contaminated saliva or indirectly by contact with contaminated fomites	Yes (attenuated vaccine); given with measles and rubella vaccines	None specific
Cytomegalovirus (subfamily: Betaherpesvirinae)	Cytomegalic inclusion disease—virus forms large intranuclear inclusions in cells of liver, salivary glands, kidneys, pancreas, and endocrine glands; fatalities are high in children under 2 years of age; fetal death or cytomegalic disease in infants	Unknown in general population; however, congenital transfer from infected mother to fetus during delivery; venereal transmission possible	None available in United States	None specific
Epstein-Barr virus (subfamily: Gammaherpesvirinae)	Infectious mononucleosis—occurs primarily in children and young adults; lymph node involvement and enlargement of spleen	Not entirely understood; close contact probably required.	None available	None specific
Hepatitis viruses	Hepatitis—degeneration of liver with accompanying jaundice	See Table 29-5	Vaccine available for hepatitis B	See Table 29-5

*The common species name of the viral agent is used with the genus in parentheses

involving blood or blood products can also prevent infection, as well as use of disposable items such as needles and syringes. A high level of hepatitis B immune globulin exerts a temporary protection against infection, and should infection occur, it will prevent the development of a chronic carrier state. In 1981 a hepatitis B vaccine was made available to the public and could not have been developed at a more opportune time. The world reservoir of hepatitis B carriers is estimated to be nearly 200 million. There is every reason to expect that large-scale immunization will reduce the pool of chronic carriers and also reduce the morbidity and mortality associated with chronic liver disease. The vaccine is prepared from human plasma obtained from healthy

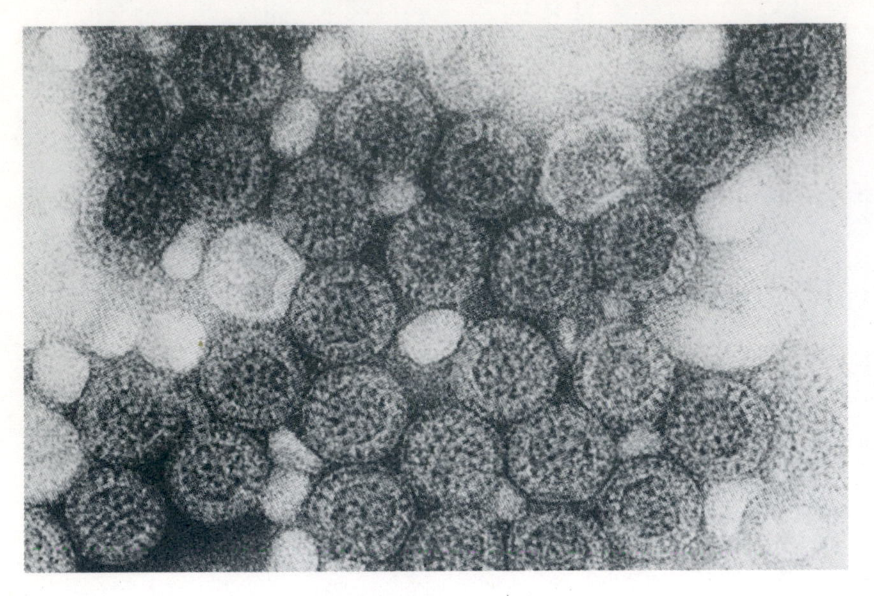

Figure 29-16 Electron micrograph of human rotavirus. Note spokelike appearance of capsid.
Courtesy R. Kogasaka.

persons who are chronic carriers of HB_sAg. The 22 nm particles are purified and treated with chemical agents to remove extraneous human liver cell and plasma host components. The antigen is then adsorbed to aluminum hydroxide, which acts as an adjuvant.

Non-A, non-B hepatitis. As mentioned previously, NANB hepatitis infection is diagnosed by exclusion of other viral agents. NANB virus disease is now believed to be the single most common cause of hepatitis following blood transfusion, since hepatitis B antigen can now be successfully detected in blood donors. Tests are currently being evaluated in humans and chimpanzees to fully determine the nature of infection by NANB.

Other viral agents that infect digestive glands are outlined in Table 29-6.

VIRUSES THAT CAUSE DIARRHEA

Rotaviruses

Diarrhea, especially in children, is a major cause of morbidity and mortality in the developing countries of the world. Of the many microbial agents causing diarrhea, the rotaviruses are among the most frequently isolated.

The rotaviruses are double-stranded RNA viruses that have icosahedral symmetry and in electron micrographs (Figure 29-16) give the appearance of a spokelike pattern (*rota* means wheel in Latin). The rotaviruses that cause human disease are morphologically indistinct from the agents of calf diarrhea, diarrhea in mice, and agents isolated from monkeys, sheep, and foals. Calf diarrhea is of economic importance, since it may be a major cause of mortality (see Chapter 34).

There are two antigenically distinct types of human rotaviruses, with type II being the major cause of disease during the winter months in the Northern Hemisphere. Disease is seen more frequently in infants aged 1 to 11 months and is rarely observed in those over 6 years of age. Infections are acquired by the fecal-oral route, with symptoms appearing within 48 hours. Vomiting and diarrhea, which are the principal symptoms, may lead to severe dehydration. The virus attacks the upper intestine and causes epithelial disturbances that lead to impairment of sodium transport and deficiency in enzymes such as maltase, sucrase, and lactase. These enzymes are important for breaking down the most common disaccharides found in the diet: maltose, sucrose, and lactose, respectively.

There are no antimicrobial agents that are effective in therapy. Parenteral rehydration using solutions containing sodium appears to be the most successful method for replenishing fluids and preventing mortality.

SUMMARY

Viral infection is usually associated with some cytopathic effect, such as the presence of inclusion bodies or other characteristics that can be observed with the unaided eye or by microscopic means. Most viral diseases are spread by the respiratory route.

Viruses that affect nervous tissue are usually carried to the site of infection via the blood stream. The viruses causing such diseases are the poliovirus, echoviruses, mumps virus, herpes simplex virus, and cytomegalovirus. The rabies virus is a bullet-shaped virus. Humans are usually infected by the bite of infected animals such as dogs. The disease is invariably fatal if left untreated. Wild animals, such as skunks and foxes, are the most important sources of infection for domestic animals. Postexposure prophylaxis now requires only 5 or 6 shots as a result of the development of a new vaccine, as opposed to the usual 21 shots required previously. Polio is a paralytic disease seldom seen today because of the development of vaccines by Sabin and Salk in the 1950s. Encephalitis caused by arthropod-borne viruses (arboviruses) have a high fatality rate. Horses are often accidentally infected. The mosquito is the vehicle for transmission of the virus to humans and horses.

Several viruses affect the respiratory tract, but the most important are the influenza viruses, types A and B. Influenza A pandemics, recorded throughout history, have caused the deaths of millions of humans. The most important characteristic of the virus is its ability to mutate, resulting in new strains for which vaccines are totally or partially ineffective. Influenza B disease occurs primarily in children and is not as virulent as the influenza A virus.

Viruses that affect the skin include the rubella virus, smallpox virus, varicella-zoster virus, papilloma virus, and flavivirus. These are acquired primarily through the respiratory tract but lodge in the skin, where the primary manifestations of disease are elicited. Because of vaccination procedures smallpox was declared eradicated worldwide in 1978 and is no longer the threat it originally was. Measles is a relatively mild childhood disease, but complications, such as encephalitis, can cause death. Measles vaccines have been available since 1963. Two herpes agents affect the skin: herpes simplex virus and varicella-zoster virus. Herpes simplex virus is divided into two types: I and II. Type I affects the oral cavity (coldsores), and type II affects the genital area and has been associated with carcinoma of the cervix in women. The virus causing chickenpox is varicella. It is believed to be the same agent that is later reactivated in adults to cause shingles (zoster). Rubella virus (German measles) is important because if it is acquired by pregnant women, the virus can be carried to the fetus, causing abnormalities at birth or fetal death.

The hepatitis viruses are the most important viral agents affecting the digestive glands. They include hepatitis A, hepatitis B, and non-A, non-B virus. All three viruses affect the liver. Hepatitis B is associated with the carrier state and is usually transmitted by contaminated blood or blood products. The mortality in hepatitis B disease is relatively high. Non-A, non-B virus is believed to be the most common cause of hepatitis following blood transfusions and is also associated with the carrier state. Hepatitis A is a milder disease with few fatalities. The cytomegalovirus can affect digestive glands and if transmitted to the fetus is frequently fatal to the developing infant. Yellow fever is caused by a flavivirus and was once a cause of many deaths; however, a vaccine is now available. Mumps is a childhood disease caused by a paramyxovirus. A vaccine is available that is usually administered with a measles and rubella vaccine.

Study questions

1. What are cytopathic effects? How can one determine if they are due to viruses?
2. What immune responses are used by the host to defend itself against viral infections?
3. Why is the treatment of viral diseases difficult? Why can't they be treated like bacterial diseases?
4. List those viruses that infect nervous tissue. What is the most common avenue that a virus takes to infect nervous tissue?
5. How did Pasteur "tame" or attenuate the rabies virus? What techniques are used today?
6. What animals are the most important sources of the rabies virus? Why is the disease so difficult to treat?
7. What is the relationship of hygienic and nonhygienic conditions to the epidemiology of poliomyelitis?
8. What are the advantages and disadvantages of live and inactivated poliovirus vaccine?
9. Discuss the probable mechanism of action of an influenza virus on respiratory tissue and the development of clinical symptoms. What is the relationship of influenza to bacterial pneumonia?
10. Explain why it is so difficult to develop influenza A virus vaccines. What is meant by antigenic drift? By antigenic shift?
11. Can you give some reasonable explanation as to why pandemics of influenza A occur every 10 years?
12. What is Reye's syndrome, and what is its relationship to influenza B infection?
13. What kinds of infections are caused by the various herpes agents?
14. What is the mechanism by which a herpes agent causes coldsores?
15. List the various ways in which the hepatitis A virus can be transmitted. What are the most common avenues of transmission?
16. Why has it been difficult to prepare hepatitis virus vaccines?
17. Which hepatitis virus is the most common cause of infection via blood transfusions?
18. How can one differentiate hepatitis caused by hepatitis A virus from that caused by hepatitis B virus?

Selected readings

Books

Acha, P.N. *Zoonoses and communicable diseases common to man and animals.* Pan American Health Organization, Washington, D.C., 1980.

Baer, G.M. (Ed.). *The natural history of rabies* (Vols. 1 and 2). Academic Press, Inc., New York, 1975.

Boyd, R.F., and Marr, J.J. *Medical microbiology.* Little, Brown & Co., Boston, 1980.

Davis, B., et al. *microbiology* (3rd ed.). Harper & Row, Pulishers, Inc., New York, 1980.

Evans, A.S. *Viral infections of humans* (2nd ed.) Plenum Publishing Corp., New York, 1982.

Fenner, F.J., and White, D.O. *Medical virology* (2nd ed.). Academic Press, Inc., New York, 1976.

Freeman, B.A. *Burrows textbook of microbiology* (21st ed.). W.B. Saunders Co., Philadelphia, 1979.

Hoeprich, P.D. (Ed.). *Infectious diseases* (2nd ed.). Harper & Row, Publishers, Inc., New York, 1977.

Joklik, W.K. (Ed.). *Principles of animal virology.* Appleton-Century-Crofts, New York, 1980.

Mandell, G.L. (Ed.). *Principles and practice of infectious diseases.* John Wiley & Sons, Inc., New York, 1979.

Stevens, J.G., Todaro, G.J., and Fox C.F. (Eds.). *Persistent viruses.* Academic Press, Inc., New York, 1978.

Youmans, G., Paterson, P.Y., and Sommers, H.M. *The biological and clinical basis of infectious disease* (2nd ed.). W.B. Saunders Co., Philadelphia, 1980.

Zuckerman, A.J., and Howard, C.R. *Hepatitis viruses of man.* Academic Press, Inc., New York, 1979

Journal articles

Alter, H.J., et. al. Transmissible agent in non-A, non-B hepatitis. *Lancet* **1:**459, 1978.

Arenaviruses in perspective. *Br. Med. J.* **1:**529, 1978.

Brunel, P.A. Protection against varicella. *Pediatrics* **59:**1, 1977.

Centers for Disease Control. Current status of measles in the United States. *J. Infect. Dis.* **137:**847, 1978.

Chanock, R.M. Impact of adenoviruses in human disease. *Prev. Med.* **3:**466, 1974.

Couch, R.B. The effect of influenza on host defenses. *J. Infect. Dis.* **144:**284, 1981.

Fenner, F. Portraits of viruses: the poxviruses. *Intervirol.* **11:**137, 1979.

Fenner, F. The eradication of smallpox. *Prog. Med. Virol.* **23:**1, 1977.

Henderson, B.E., and Coleman, P.H. The growing importance of California arboviruses in the etiology of human disease. *Prog. Med. Virol.* **13:**404, 1971.

Hendley, J.O., Wenzel R.P., and Gwaltney, J.M., Jr. Transmission of rhinovirus colds by self-inoculation. *N. Engl. J. Med.* **288:**1361, 1973.

Henle, W., Henle, G., and Lennette, E.T. The Epstein-Barr virus. *Sci. Am.* **241:**48, 1979.

Huang, A.S. Viral pathogenesis and molecular biology. *Microbiol. Rev.* **41:**811, 1977.

Kaplan, M.M., and Koprowski, J.H. Rabies. *Sci. Am.* **242:**120, 1980.

Morgan, E.M., and Rapp, F. Measles virus and its associated diseases. *Microbiol. Rev.* **41:**636, 1977.

Nahmias, A.J., and Norrild, B. Herpes simplex viruses 1 and 2—basic and clinical aspects. **DM, 25**(10):Entire issue, 1979.

Nahmias, A.J., and Roizman, B. Infection with herpes simplex virus 1 and 2. *N. Engl. J. Med.* **289:**667, 1973.

Osborn, J.E. Cytomegalovirus: pathogenicity, immunology, and vaccine initiatives. *J. Infect. Dis.* **143:**618, 1981.

Plotkin, S.A. Rabies vaccine prepared in human cell cultures: progress and perspectives. *Rev. Infect. Dis.* **2**(3):433, 1980.

Rawls, W.E. Congenital rubella: the significance of virus persistence. *Prog. Med. Virol.* **10:**238, 1968.

Reed, J.S., and Boyer, J.L. Viral hepatitis: epidemiologic, serologic and clinical manifestations. **DM 25**(4):Entire issue, 1979.

Sabin, A.B. Paralytic poliomyelitis: old dogmas and new perspectives. *Rev. Infect. Dis.* **3**(3):543, 1981.

Spector, D.M., and Baltimore, D. The molecular biology of poliovirus. *Sci. Am.* **232:**25, 1975.

Stuart-Harris, C.H. The influenza viruses and the human respiratory tract. *Rev. Infect. Dis.* **1**(4):592, 1979.

Sweet, C., and Smith, H. Pathogenicity of influenza virus. *Microbiol. Rev.* **44:**303, 1980.

Takemoto, K.K. Human papovaviruses. *Int. Rev. Exp. Pathol.* **18:**281, 1978.

Zuckerman, M.J. Hepatitis B: its prevention by vaccine. *J. Infect. Dis.* **143:**301, 1981.

Chapter 30 SEXUALLY TRANSMITTED DISEASES

One of the greatest concerns in the medical community is the increasing incidence of sexually transmitted diseases. Many in the general population have been lulled into a false sense of security because of our ability to rapidly treat and cure the venereal diseases syphilis and gonorrhea. These two diseases often produce overt symptoms or clinical manifestations that are easily recognized, rapidly confirmed by laboratory tests, and cured by treatment with drugs such as penicillin and related antibiotics. Today the sexual mores of many countries have changed, resulting in more promiscuous relationships without the fear of social stigma. Medical science has now recognized that agents other than those that cause syphilis and gonorrhea are becoming a more frequent cause of venereal disease. Many of these diseases are not of themselves a cause of discomfort to the sexual partners. The problem is that in the female partner many of these agents may be a cause of infertility or abortion and in some instances may cause damage to the fetus that results in physical disabilities or neurological disorders. Disease in the male partner is also of concern because he may transmit the disease to other female partners. In addition some of these diseases cause reduced sperm count, which also results in infertility. Table 30-1 describes the characteristics of the major and minor venereal diseases and the particular agents involved. Acquired immune deficiency syndrome (AIDS) is also transmitted sexually, but no microbial agent has as yet been implicated. See Chapter 25 for a discussion of this disease.

SYPHILIS

Origin

The origin of what we today call syphilis is still uncertain. The first epidemics, which were reported in the late fifteenth and sixteenth centuries, ravaged wide areas of Europe, frequently causing a slow and painful death. The reasons for these epidemics has never been established, but many, historians believe that Columbus's crew contracted the disease from women in the West Indies and on their return to Spain caused its spread among the Spanish population. The Spanish, to show their contempt for their European neighbors, called it the French disease, whereas the English called it the Spanish disease. An Italian pathologist, Fracastora, in 1530 wrote a poem that described a shepherd cursed by the Sun God, who had afflicted him with the French disease. Fracastora called the boy Syphilus, and the name has been universally adopted.

There are many who hold that the Columbian theory is not true and that syphilis had its origin long before the birth of Christ. The Chinese, for example, in 2637 BC were treating a disease whose symptoms resembled syphilis with mercury—a treatment that was used even in the late nineteenth century AD. The poet Horace in 148 BC speaks of a man worn out by debauchery who was covered with pimples and red eruptions over his entire body—a characteristic symptom of one of the stages of syphilis.

Syphilis is a disease caused by the organism belonging to the genus *Treponema (T. pallidum),* but there are nonvenereal diseases caused by organisms that cannot be morphologically distinguished from the one transmitted venereally. The nonvenereal diseases caused by species of *Treponema* are yaws, bejel, and pinta, which are restricted to specific areas of the world. All of the diseases caused by *Treponema* species are collectively called ***treponematoses.*** Their characteristics are described in Table 30-2. One of the currently held views on the origin of venereal syphilis involves an evolution from the nonvenereal types. The theory states that the primordial

Table 30-1 Characteristics of sexually transmitted diseases

DISEASE	ETIOLOGICAL AGENT	TYPE OF MICROORGANISM	CAUSE OF INFERTILITY, ABORTION, INFECTION TO FETUS	CASES REPORTED TO CDC, 1980	ESTIMATED NUMBER OF INFECTIONS PER YEAR
Gonorrhea	*Neisseria gonorrhoeae*	Bacterium	Yes	1,004,029	2.5 million
Nongonococcal urethritis	*Chlamydia trachomatis, Ureaplasma urealyticum*	Bacterium	No	NR	1.0 million
Trichomoniasis	*Trichomonas vaginalis*	Protozoa	No	NR	1.0 million
Venereal warts	Human papilloma virus	Virus	No	NR	400,000
Genital herpes	Herpes simplex virus type II	Virus	Yes	NR	340,00
Syphilis	*Treponema pallidum*	Bacterium	Yes	68,832	340,000
Lymphogranuloma venereum	*Chlamydia trachomatis*	Bacterium	No	199	?
Soft chancre (chancroid)	*Hemophilus ducreyi*	Bacterium	No	788	?
Granuloma inguinale	*Calymmatobacterium granulomatis*	Bacterium	No	51	?
Candidiasis	*Candida albicans*	Fungus	Yes	NR	?
Enteric disease					
	Salmonella species	Bacterium	No	NR	?
	Shigella species	Bacterium	No	NR	?
	Campylobacter fetus subspecies *jejuni*	Bacterium	Yes	NR	?
	Entamoeba histolytica	Protozoa	No	NR	?
	Giardia lamblia	Protozoa	No	NR	?

NR = Not reportable to the Centers for Disease Control
CDC, Centers for Disease Control, Atlanta, Ga.

Table 30-2 Characteristics of the treponematoses

	VENERAL SYPHILIS	BEJEL (ENDEMIC SYPHILIS)	YAWS	PINTA
Microbial agent	*Treponema pallidum*	*T. pallidum*	*T. pertenue*	*T. carateum*
Population initially infected	Adults	Children	Children	Children
Mechanism of transmission	Contact with or between genitalia	Contact with mouth or skin	Contact with skin	Contact with skin
Site of initial lesion	Genitalia	Mouth	Skin of legs	Exposed skin
Involvement of bone	Yes	Yes	Yes	No
Involvement of organs (heart, brain, liver)	Yes	Sometimes, but milder than venereal syphilis	No	No
Favoring climate				
Temperature	Cool (any)	Warm	Tropical	Tropical
Humidity	Any	Dry	Humid	Humid

Modified from Rosebury, T. *Microbes and morals.* The Viking Press, New York, 1971.

treponeme existed as a saprophyte in the soil and later established a commensal relationship with emerging animals. The human commensal is believed to have remained on the surface of the skin and thus resembled the organism causing ***pinta*** (Table 30-2). The commensal is believed to have become pathogenic for Europeans about 20,000 BC and to have been distributed worldwide about 15,000 BC. In 10,000 BC the warm and humid environment of Africa and Asia favored a mutation of the organism causing pinta to one that invaded the tissue and thus resembled the disease ***yaws.*** Humans, who migrated to more temperate and drier climates as the glaciers receded about 7000 BC brought clothing into use and probably created a dependence for transfer of the organism via mucosal surfaces rather than body skin. ***Bejel*** is believed to have emerged in the arid environment.

Venereal transmission probably began in large cities, where a cleaner and clothed society did not provide the kind of child-to-child contact found in the less urbane villages. Therefore the only intimate contact was by sexual intercourse, which provided the principal means of transfer of the organism. Venereal syphilis came to the coasts of western Mediterranean and Europe via shipping and then spread to the Americas, China, India, and Australia during and after the sixteenth century.

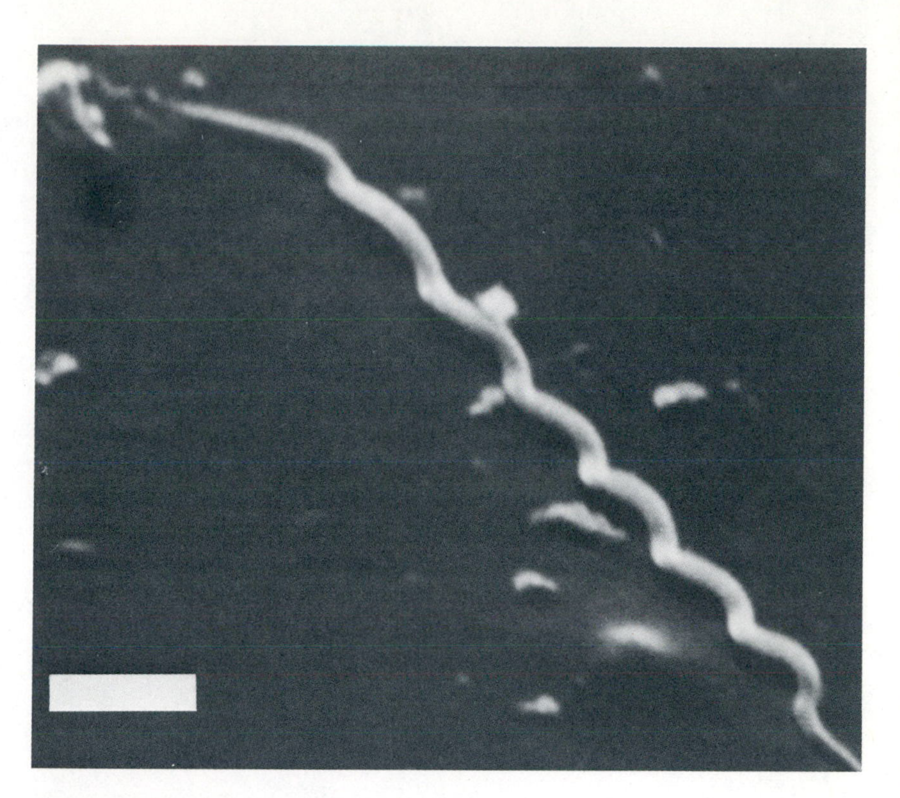

Figure 30-1 Electron micrograph of *Treponema pallidum.* (×33,300.)
From Fitzgerald, T.J. J. Bacteriol. ***130:****1333, 1977.*

Characteristics of the agent of syphilis

The agent of venereal syphilis *(T. pallidum)* belongs to a group of organisms called spirochetes (Figure 30-1). It is sensitive to heavy metals, such as antimony, bismuth, mercury, and arsenic, and before the advent of penicillin, these metals were used in treatment of syphilis. As one might suspect, the treatment was sometimes more deleterious than the disease.

T. pallidum cannot be cultivated on laboratory media, and little is known of its metabolic activities, chemical components, or biochemical reactions. If cultivation is required, the testicular tissue of rabbits is considered the best choice.

Epidemiology

Syphilis is a worldwide disease that is ordinarily transmitted by sexual contact. The number of cases of syphilis of all types in the United States in 1980 was 68,832. This figure was cut in half in 1981 and 1982 and is believed to be the result of a decline in promiscuity because of the fear of AIDS and herpes infections. Over the many years the disease has been reported, there have been major fluctuations in the number of cases reported (Figure 30-2). It is believed that only 20% to 30% of the actual number of cases are reported, since the very personal nature of transmission prevents disclosure to a physician. Infections are acquired primarily between 15 and 39 years of age, the sexually active years.

The organism causing syphilis can penetrate intact mucous membranes but not intact skin. Transmission occurs primarily by direct contact with mucous membranes in the genitalia, since the organism cannot live outside the body. Nearly 10% of the cases of syphilis may be transmitted nonsexually, because the organism causing disease may be present in lesions in the oral cavity and other parts of the body as well as the genitalia. Contact with such lesions, for example, by kissing, could result in transmission. Syphilitic lesions in the oral cavity or rectum are frequently found among members of the homosexual community. Syphilis can be considered an occupational hazard for oral surgeons, physicians, and other hospital personnel who come in contact with infected patients. The organism causing syphilis can also be transmitted congenitally, but this aspect is discussed later.

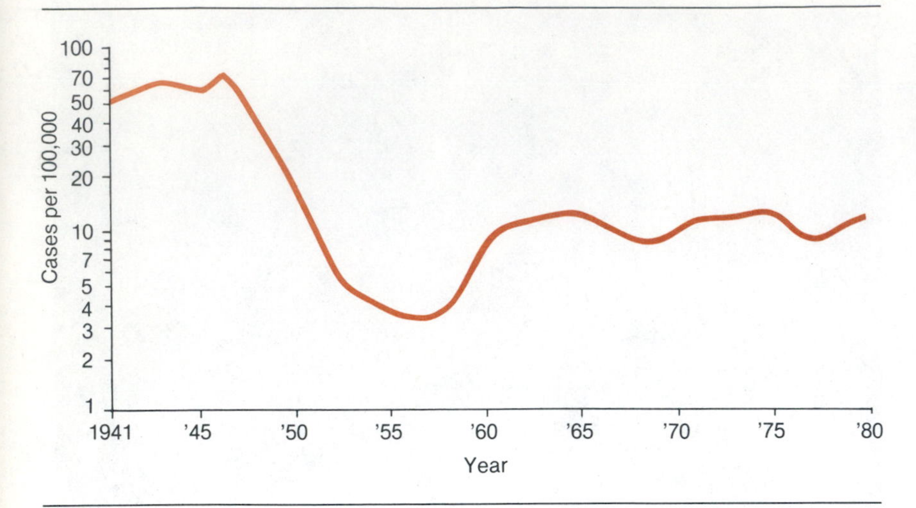

Figure 30-2 Reported civilian case rates by year of primary and secondary syphilis in United States, 1941-1980.

Redrawn from Centers for Disease Control, Atlanta, Ga.

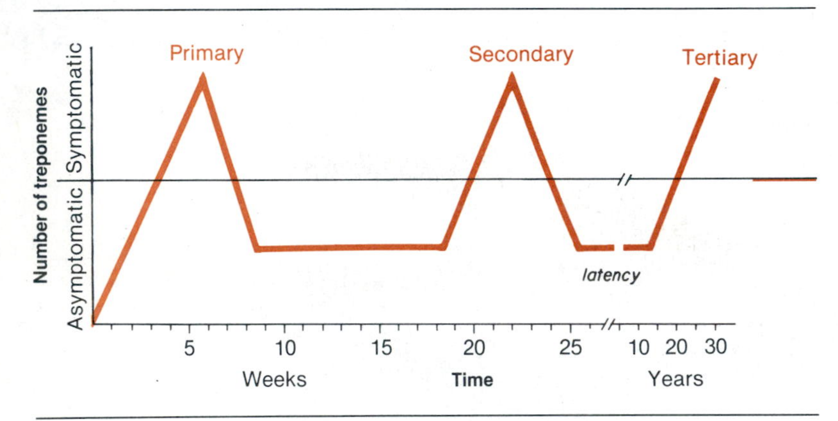

Figure 30-3 Average time intervals for stages of syphilis. Disease fluctuates between symptomatic and asymptomatic periods. Clinical manifestations of symptomatic syphilis result from multiplication of treponemes within various tissues. Effective host defenses then evolve, decreasing number of treponemes and resulting in asymptomatic periods.

Redrawn from Fitzgerald, T.J. Annu. Rev. Microbiol. ***35**:29, 1981.*
By annual reviews, Inc.

Clinical manifestations

T. pallidum is surrounded by a capsular-like mucopolysaccharide material that is believed to be important for attachment to the surface of mammalian cells. The agent of syphilis comes in contact with the genital mucosa during intercourse and multiplies at the site of entry before disseminating to the local lymph nodes. The disease is characterized by fluctuations between symptomatic and asymptomatic periods and is divided into stages: primary, secondary, latent, and tertiary (Figure 30-3).

Primary stage. Multiplication of treponemes at the portal of entry initiates an inflammatory response and the development of a small lesion called a papule (Figure 30-4). The lesion enlarges and erodes as the number of treponemes increases, and it eventually develops into a hollowed-out area called a ***chancre*** (Figure 30-5), which is very hard and feels like a button. The chancre appears anywhere from 2 to 4 weeks after infection and is filled with viable treponemes. The chancre, which represents a highly infectious stage, begins to heal 3 to 8 weeks after it appears. The healing process is associated with a decrease in the number of treponemes in the lesion. The chancre in men is easily recognized because it appears on the penis, but in women it usually appears on the cervix and may go unrecognized. Extragenital chancres may appear anywhere on the body (lip, toe, back, hand) where abraded skin has come in contact with viable treponemes (Figure 30-6). The healing of the chancre is followed by an asymptomatic period in which the remaining viable treponemes begin to invade the vascular system.

Secondary stage. The secondary stage is a symptomatic period that represents hematogenous dissemination of treponemes to body tissues. Any body tissue may exhibit lesions, including bone, skin, liver, kidney, heart, eye, and meninges. The secondary stage is recognized by a rash that may appear anywhere on the body but is seen more frequently on the trunk (Figure 30-7), face, palms, and soles. Lesions, called mucous patches, may also appear in the oral cavity, or lesions may appear as elevated papules, called condylomata lata (Plate 39), about the genitalia. Regardless of their site, all secondary lesions are teeming with treponemes and represent a highly infectious state. Syphilis may end with the secondary stage, but in one third of the cases a latent period occurs.

Latent stage. Latency represents a period of time following the secondary stage in which serological diagnosis indicates disease but there are no recognizable symptoms. The latent period may last for several weeks or several years and has been divided into early latency and late latency. Early latency indicates an asymptomatic period of 2 years or less, during which time relapses and secondary symptoms may appear. Late latency, an asymptomatic period of more than 2 years, may continue throughout the life of

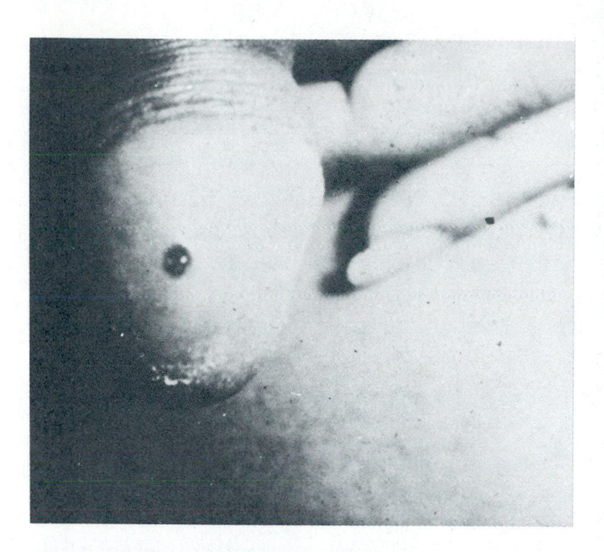

Figure 30-4 Papule formation on penis during primary stage of syphilis.
Courtesy Centers for Disease Control, Atlanta, Ga.

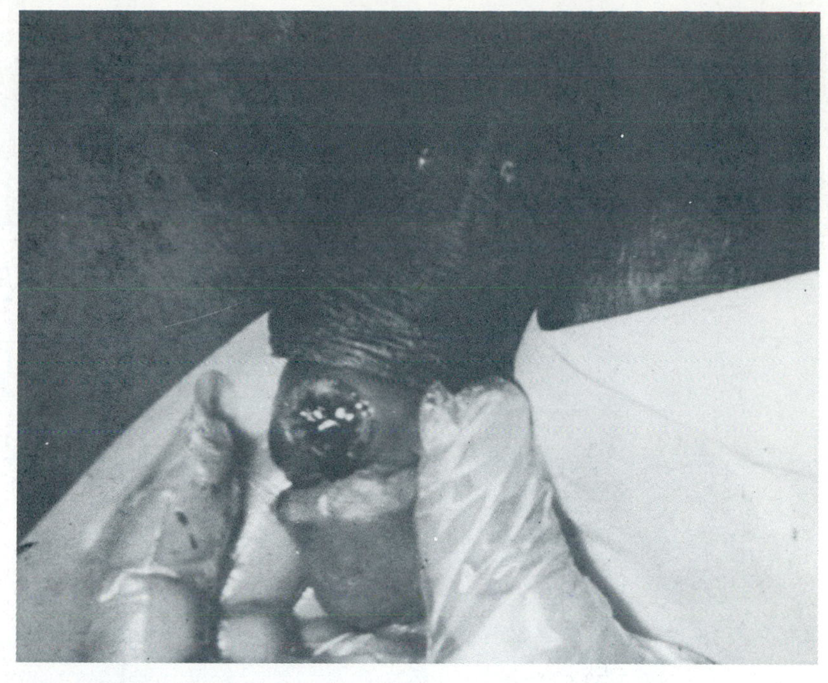

Figure 30-5 Chancres on penis during primary stage of syphilis.
Courtesy Centers for Disease Control, Atlanta, Ga.

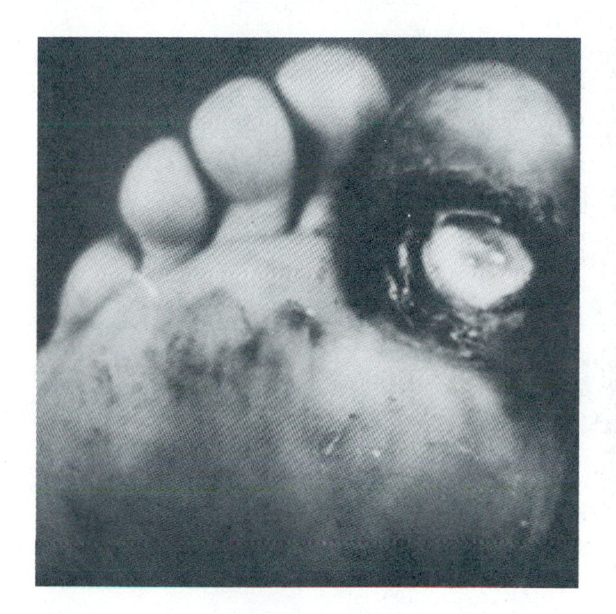

Figure 30-6 Chancres on toe during primary stage of syphilis.
Courtesy Centers for Disease Control, Atlanta, Ga.

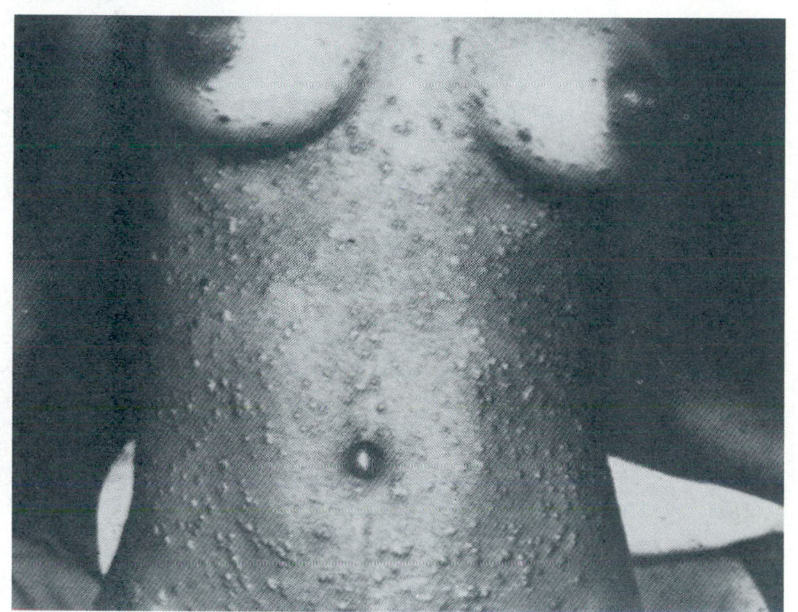

Figure 30-7 Papular rash observed during secondary stage of syphilis.
From Rudolph, A.H. In Wehrle, P.S., and Top., F.H., Jr. (Eds.). Communicable and infectious diseases (9th ed.). The C.V. Mosby Co., St. Louis, 1981.

the patient or may develop into a tertiary stage. Infectiousness is observed only when secondary symptoms appear.

Tertiary stage. The tertiary stage is not an infectious stage, but its symptoms may last for several years. Lesions called ***gummas*** (Figure 30-8), which appear on the skin, are the most recognizable symptoms, but the most damaging lesions are those involving the nervous system and cardiovascular system. Neurosyphilis can result in loss of motor control, blindness, or insanity. In cardiovascular syphilis the organism or its products erode the walls of the blood vessels, result-

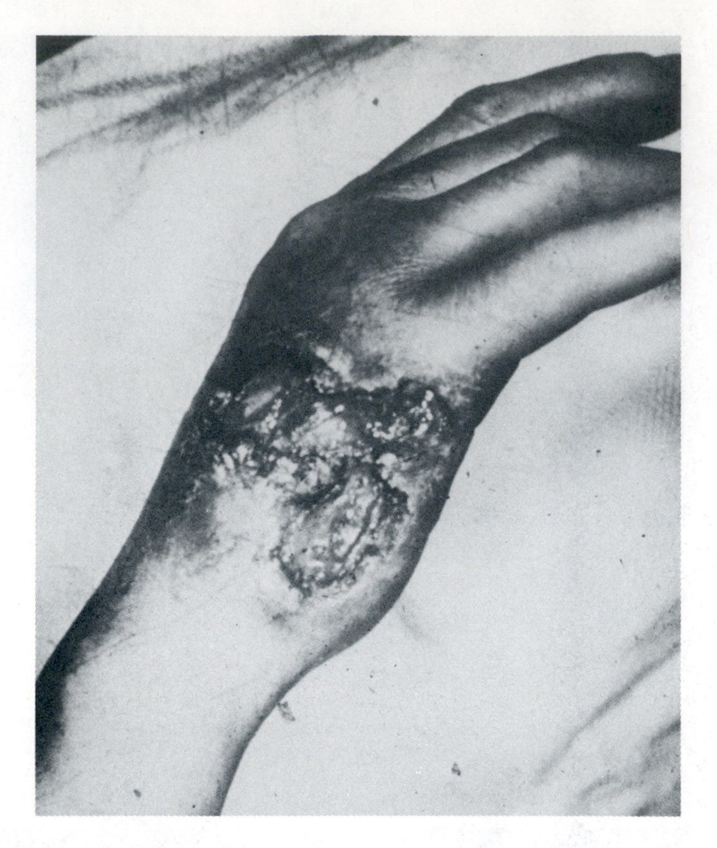

Figure 30-8 Gumma of hand observed during tertiary stage of syphilis.
Courtesy Centers for Disease Control, Atlanta, Ga.

ing in aneurysms, particularly of the aorta. The lesions of tertiary syphilis are considered to be the result of a hyperimmune response by the host to the presence of nonviable treponemes. Viable treponemes are seldom isolated from the patient during this stage of syphilis.

Congenital syphilis. Congenital syphilis is a nonvenereal disease resulting from the passage of *T. pallidum* from the mother's bloodstream across the placenta into the fetus during pregnancy. Transmission occurs only after the sixteenth week of gestation, and if the infection is left untreated, only 50% of the fetuses will be alive at birth. Those that survive often exhibit a rash or lesions characteristic of the secondary stage. Many syphilitic manifestations appear only after the second year of life. During the infant's first 2 years destruction of tissue, bone, and cartilage may result in later manifestations, such as perforated soft and hard palates (Figure 30-9), corneal scarring that impairs vision, and bowing of the legs (Figure 30-10). Congenital syphilis is therefore a more debilitating and sometimes life-threatening disease than venereal syphilis.

Laboratory diagnosis

The diagnosis of syphilis is based on microscopic as well as serological tests.

Microscopic tests. Treponemes can be stained, but their thinness makes them practically invisible by light microscopy. Darkfield microscopy is the recommended procedure for observation of treponemes, which appear white against a black background. The specimen for microscopic observation can be obtained from primary and secondary lesions only, but care must be taken in interpretation, since the oral cavity and genitalia contain large numbers of nonpathogenic treponemes. An alternative to darkfield microscopy is the direct fluorescent antibody test in which fluorescein-labeled antibody to *T. pallidum* is used to stain a serum sample.

Serological tests. Serological tests are of two types: those that measure ***nontreponemal antibodies*** and those that measure ***treponemal antibodies.*** Nontreponemal antibodies are produced and present in high concentrations during disease, but specific antibodies are slow to develop and may not appear until late in the primary stage.

Two of the most frequently used nontreponemal tests are the ***Venereal Disease Research Laboratory (VDRL)*** test and the ***rapid plasma reagin (RPR)*** test. These tests are used to detect a nonspecific serum component called reagin, which is produced during syphilis, as well as in other diseases and conditions, such as malaria and immune disorders. Reagin is detected in these tests by adding an antigen such as cardiolipin (a component of heart tissue and some plants) to the patient's serum. A positive test is indicated by a flocculation reaction. The VDRL and RPR are screening tests that have greater value for diagnosis during the early stages of disease. Confirmation of these tests can be best made by measuring specific treponemal antibodies.

The ***fluorescent treponemal antibody absorption (FTA-ABS)*** test and ***treponema palidum immobilization (TPI)*** tests are used to diagnose latency and to confirm equivocal reactions obtained with nontreponemal tests. The FTA-ABS test, which is the preferred method for confirming equivocal tests, uses fluorescent antibody to detect specific antibodies to *T. pallidum* in the patient's serum. (See Appendix B for details of this test.) The TPI test is based on the ability of specific antibodies in the patient's serum to inhibit the motility of live *T. pallidum* spirochetes. This test is used only for extreme cases, since it is difficult to perform and is very expensive.

Treatment and prevention

Treatments in the early twentieth century included the administration of heavy metals, such as mercury, bismuth, and arsenic. The treatments were prolonged and painful, with few patients willing to complete the full course of therapy. Fever therapy was also used as an alternative to heavy-metal administration. Some patients were injected with the parasites causing malaria, or were injected with typhoid vaccine, to induce a fever of 106° F for several hours, and others were placed in water boxes or an electric field. Penicillin, discovered in the 1940s, was found to be effective against the spirochete, and since that time has been used in the form of benzathine penicillin G or aqueous procaine penicillin for the treatment of primary, secondary, and tertiary syphilis. Tetracycline is an alternative drug if the patient is sensitive to penicillin. Tetracycline is toxic to the liver of pregnant women, and erythromycin should be substituted in those circumstances.

No syphilis vaccine is available, nor does it appear that one will be available in the near future. Prevention is the key to controlling syphilis, but that is not an easy task. The mildness of symptoms and the high degree of travel among the population make it difficult to locate and treat patients. One infected individual could indirectly be the source of infection for up to 100 others within a year. Once the diseased patient is located and interviewed, his or her contacts can also be sought out to determine if disease transmission has taken place. Testing programs such as those required before issuance of marriage licenses as well as preemployment examinations have detected many diseased individuals who were unaware of their condition. The introduction of prenatal laws in several states requiring syphilis serology of pregnant women has helped to decrease the number of cases of congenital syphilis. Four thousand cases of congenital syphilis were recorded in 1962, but in 1980 just over 270 cases were found in the United States. Perhaps the most important aspect of prevention is education of individuals, both young and old alike, concerning the signs and symptoms of infection as well as concerning the means of treatment and the professional care that are now available.

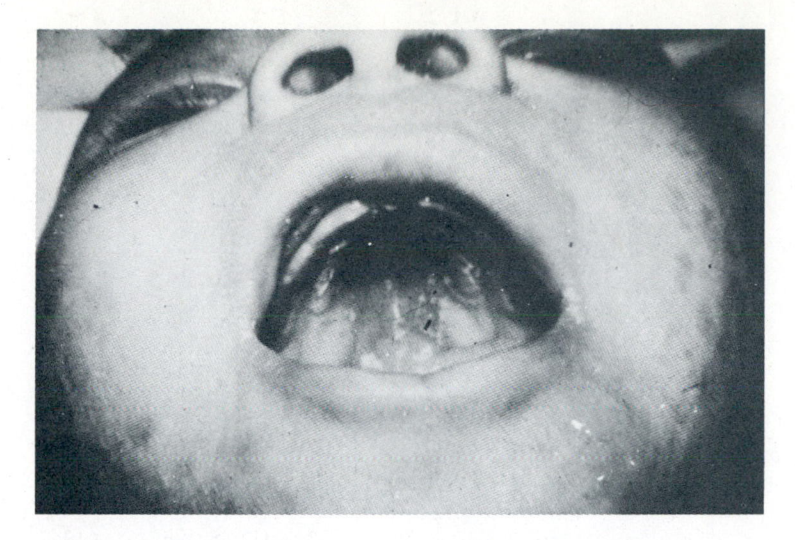

Figure 30-9 Perforated palate resulting from congenital syphilis.
Courtesy Centers for Disease Control, Atlanta, Ga.

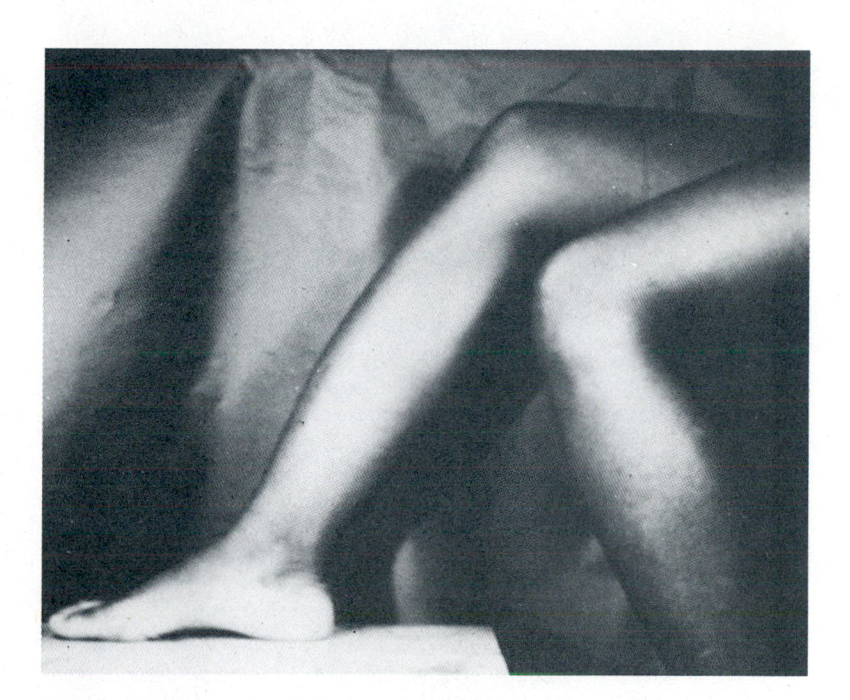

Figure 30-10 Saber shin resulting from congenital syphilis.
Courtesy Centers for Disease Control, Atlanta, Ga.

Immunity

Studies in Europe and with prisoners in the United States have confirmed many of the immunological aspects of disease obtained from infected rabbits. In studies of patients with untreated syphilis who were followed for more than 30 years, 75% did not develop secondary syphilis, and of those with secondary syphilis only 13% developed tertiary syphilis. Disease caused by the agent of syphilis does not confer permanent immunity. However, patients who remained un-

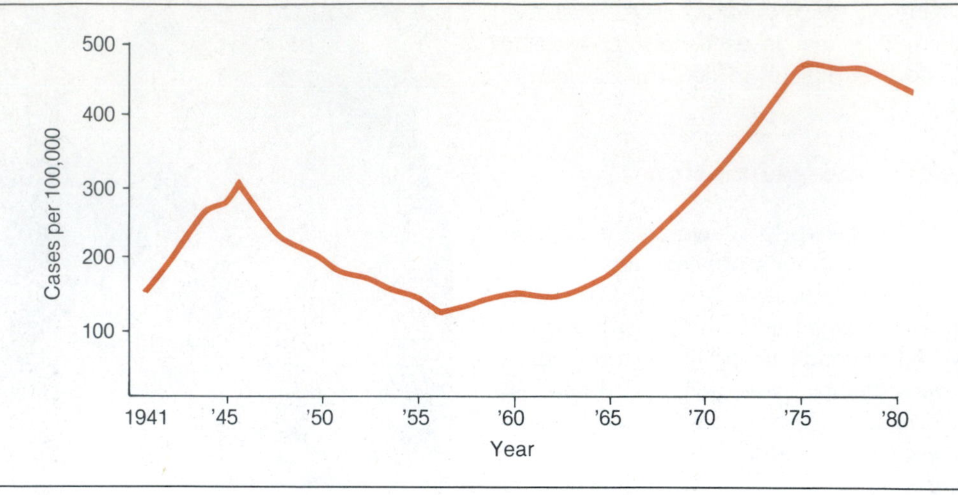

Figure 30-11 Reported civilian case rates of gonorrhea by year in United States for 1941-1980.
Redrawn from Centers for Disease Control, Atlanta, Ga.

treated during infection are difficult to reinfect when given a challenge dose of spirochetes. Immunity is slow to develop in the patient with syphilis and often takes 12 weeks or longer. A patient treated during the first 12 weeks can be easily infected by a challenge dose of *T. pallidum,* but after 12 weeks it becomes much more difficult.

It is interesting to note that in those areas of the world where bejel, yaws, and pinta predominate, the infected population is immune to the organism causing venereal syphilis. In turn, those individuals in temperate zones where venereal syphilis is endemic are resistant to the agents of yaws, bejel, and pinta. Recent investigations of villages where bejel, yaws, and pinta have been eradicated indicate that the population is now susceptible to infection with the agent of venereal syphilis. This has led some authorities to accept the theory that the treponeme was a common inhabitant of humans and that during evolution and separation into various ethnic and racial groups adapted to variations in climate and levels of sanitation.

GONORRHEA

Origin

The origin of gonorrhea is as obscure as that of syphilis. Suggestive references to gonorrhea are believed to have been made two to three centuries BC. Descriptions of a "cloudy urine" or drainage of "corrupt materials red or white" by the urethra imply that gonorrhea may be of ancient origin. In the fourth century AD a "burning disease" was associated with lying with a harlot, and in a chronicle published in 1546 the following statement was made: "If a man be burnt with a harlot and do meddle with another woman within a day, he shall burnt the woman he meddle with." The burning refers to the burning sensation associated with urination during gonorrhea, and such descriptions suggest that transmission of the disease was apparent to the writer. The more common synonyms used today to describe gonorrhea are clap, drip, and GC (gonococcus).

Gonorrhea was for many years thought to be a manifestation of syphilis, but in 1767 John Hunter, an English surgeon, sought to prove that gonorrhea was not related to syphilis. He inoculated himself with pus from a patient who had gonorrhea. The pus, unfortunately, also contained the organisms causing syphilis, and Hunter's career came to an unexpected halt. Later experiments showed the diseases to be separable and caused by different microorganisms.

Characteristics of the agent of gonorrhea

The agent of gonorrhea is a gram-negative diplococcus called *Neisseria gonorrhoeae* (Plate 40). It is a fragile microorganism that is sensitive to drying and temperature extremes and in culture is sensitive to lysis by its own enzymes (autolytic enzymes). The natural habitat of *N. gonorrhoeae* is the mucous membranes of warm-blooded animals. It is often referred to as the gonococcus.

Epidemiology

Most of the cases of gonorrhea are acquired by sexual contact, and clinical manifestations of dis-

Table 30-3 Gonorrhea: reported cases in civilians and cases per 100,000 by age and sex in the United States, 1980

	AGE GROUP	CASES	RATE
Male	0 to 14	2,877	11.0
	15 to 19	99,994	930.0
	20 to 24	224,091	2102.2
	25 to 29	140,673	1449.8
	30 to 39	95,345	613.7
	40 to 49	21,717	195.7
	50+	8,112	31.1
	TOTAL	592,809	538.8
Female	0 to 14	8,207	32.7
	15 to 19	147,245	1414.5
	20 to 24	155,365	1458.6
	25 to 29	62,448	636.3
	30 to 39	30,631	191.6
	40 to 49	5,277	45.2
	50+	2,047	6.2
	TOTAL	411,220	353.1
Total	0 to 14	11,084	17.7
	15 to 19	247,239	1168.3
	20 to 24	379,456	1780.5
	25 to 29	203,121	1040.7
	30 to 39	125,976	399.7
	40 to 49	26,994	115.6
	50+	10,159	17.2
	TOTAL	1,004,029	443.3

ease are usually associated with the genitalia. Gonorrhea is the most frequently reported communicable disease in the United States, and since 1966 it has increased sharply (Figure 30-11). Over 1 million cases of gonorrhea were reported in the United States in 1981 (Table 30-3), but estimates of unreported cases suggest that the actual number is closer to 3 million. Several reasons have been suggested for the increase in gonorrhea, including (1) increased promiscuity, which has not been proved; (2) use of the pill instead of the condom to prevent pregnancy; (3) increased resistance of the agent of gonorrhea to the usually prescribed drugs; and (4) increased mobility of the population, resulting in the inability to treat all contacts. A more-informed public has probably also contributed to an increase in the number of reported cases.

Gonorrheal disease is not limited to the genitalia and, depending on sexual practices and other conditions, can occur in the anal and oral cavities. Gonococcal disease may also occur by nonsexual means. The newborn may acquire ***gonococcal conjunctivitis (ophthalmia neonatorum)*** during passage through an infected birth canal. Gonococcal conjunctivitis can occur during the first year of life by accidental contamination from a parent or relative, although the same infection in older children is frequently acquired by sexual molestation or premature sexual activity. Gonococcal conjunctivitis in adults may also be acquired by touching the eyes with fingers that have been contaminated by gonococcal organisms.

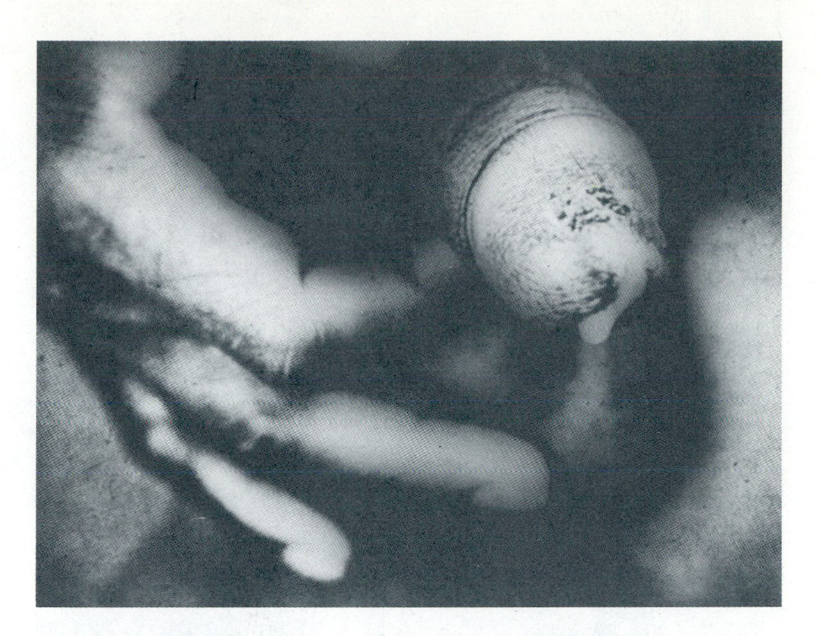

Figure 30-12 Purulent discharge observed during gonorrhea in man.
Courtesy Nicholas Fiumara, Department of Public Health, Boston. From Morton, B.N. VD: a guide for nurses and counselors. Little, Brown & Co., Boston, 1976.

The gonococcus lives only a few seconds on the skin, and transmission by doorknobs, towels, or toilets is unlikely.

Clinical manifestations

The incubation period for gonococcal disease is 2 to 5 days after sexual contact. The gonococci attach to mucosal tissue, where they multiply and spread to adjacent tissue. Penetration of the mucosa elicits an inflammatory response and further spread of the microorganisms. Occasionally the organisms penetrate into the bloodstream, causing disseminated disease.

Disease in the man is characterized by urethritis and the formation of a mucopurulent discharge (Figure 30-12), which usually resolves spontaneously in about 8 weeks. It has been estimated that from 20% to 40% of urethral infections are asymptomatic. In the untreated man infection may spread to the prostate, causing possible retention of urine, or to the epididymis, causing swelling and pain. Infection of the testes may result in sterility. The endocervix is the primary tissue affected in urogenital gonorrhea in

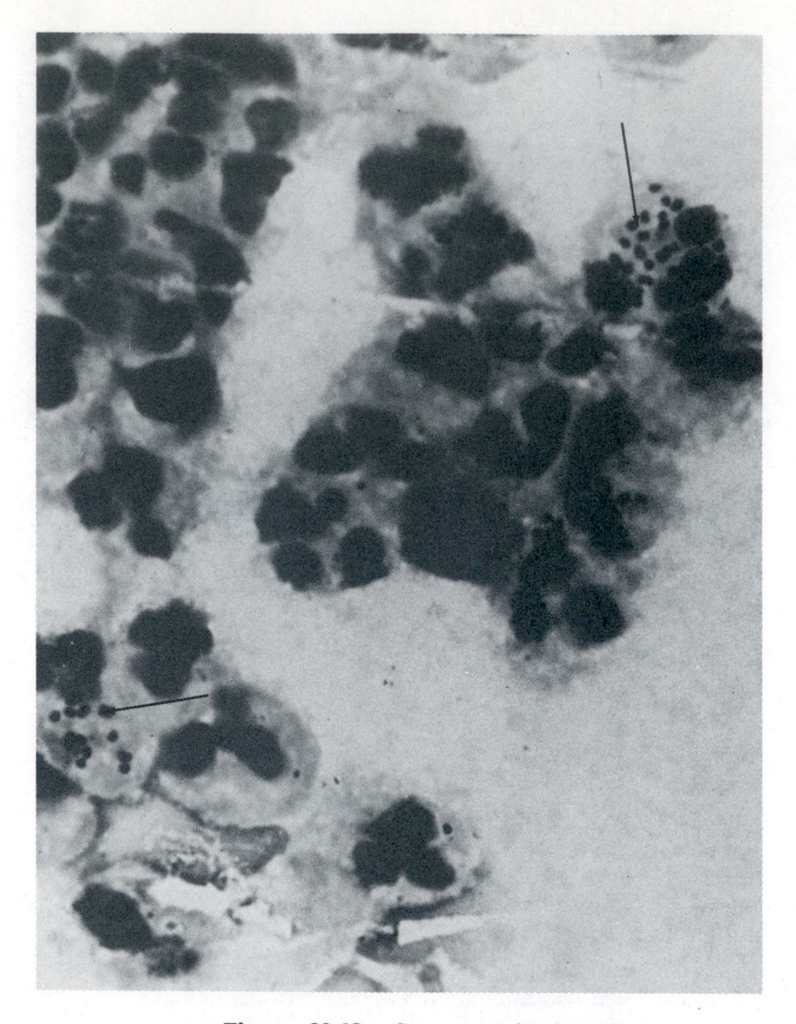

Figure 30-13 Gram stain of male urethral exudate. Some polymorphonuclear leukocytes contain several diplococci of *Neisseria gonorrhoeae (arrows)*.

From Morello, J.A., and Bohnhoff, M. Manual of clinical microbiology (3rd ed.). American Association for Microbiology, Washington, D.C., 1980.

women and is often associated with a vaginal discharge. Many infected women, however, remain asymptomatic and represent a reservoir of infection for their sexual partners. In about 15% of women infection may spread to the fallopian tubes, ovaries, and pelvic peritoneum, giving rise to an inflammatory response called ***pelvic inflammatory disease (PID).*** PID can be caused by a number of microbial agents: *Chlamydia trachomatis,* anaerobic bacteria, *Actinomyces israelii, Mycoplasma hominis,* and a host of gram-negative facultative rods such as *E. coli.* PID caused by gonococci appears to be more frequent in women wearing intrauterine devices to prevent pregnancy. The inflammatory response can cause necrosis and scarring of the fallopian tubes, which can lead to obstruction and infertility. PID recurs, and each episode increases the risk of infertility. Gonococcal infection in prepubertal girls presents the same clinical picture as in adults, except that the vagina and not the cervix is the primary site of infection.

Anorectal and oropharyngeal gonorrhea occurs in both men and women who practice rectal intercourse and fellatio but appears more frequently in homosexual men. The infections are usually asymptomatic, but occasionally pharyngitis or tonsillitis is reported.

Disseminated gonococcal infection appears in 1% to 3% of patients with gonorrhea. Initially organisms are found in high concentration in the bloodstream, and the disease is characterized by fever and hemorrhagic skin lesions. Later the infection settles in the joints, and organisms are found predominately in the synovial fluid and not the bloodstream. The gonococci are the major cause of septic arthritis in young adults. Very rarely, the gonococci infect the heart (endocarditis) and meninges (meningitis), and these diseases are often fatal. Many disseminated diseases have been reported to be caused by a particularly virulent strain that requires arginine, hypoxanthine, and uracil (AHU strain) for growth, but the reason for its virulence has not been determined.

Gonococcal conjunctivitis in the newborn (ophthalmia neonatorum) appears as a swelling of the conjunctiva and the formation of a purulent discharge. Untreated, the disease can involve the cornea and cause blindness. Disease in adults is less severe and involves primarily the conjunctiva.

Laboratory diagnosis

Clinical specimens from male patients with urethritis are of diagnostic value when Gram stained, because leukocytes can be observed to contain numerous diplococci (Figure 30-13). A Gram-stained smear from a female patient is not of diagnostic value because of the presence of other gram-negative diplococci that are commensals in the genital tract. In addition, 50% of women with negative smears are culturally positive. Clinical specimens should therefore be cultured by inoculation into a medium such as the Thayer-Martin medium (MTM) and incubated at 35° to 37° C in an atmosphere of 3% to 10% carbon dioxide. MTM is a chocolate agar medium (an enriched medium containing defibrinated blood that is heated to 80° C to impart the chocolate color) containing four antibiotics: vancomycin, colistin, nystatin, and trimethoprim lactate, which inhibit nonpathogenic *Neisseria* and other commensals but support the growth of pathogenic *Neisseria.* The culture plates are examined after 48 hours and subjected to colony observation, Gram staining, and oxidase testing (Plate 34). Confirmatory identification may include carbohydrate degradation and serological techniques such as fluorescent antibody. *N. gonorrhoeae* is oxidase positive and uses glucose but not sucrose, lactose, or maltose.

Treatment and prevention

The gonococcus has shown a relatively uniform sensitivity to penicillin since the drug's introduction in 1943. Penicillin resistance has been reported, but the resistance has been so slow to develop that higher dosages of pencillin are often adequate to curb infection. A penicillinase-producing gonococcus was isolated in 1976 that showed an unusually high resistance to penicillin. Most penicillinase-producing strains have been found in the Far East and Africa, but an increasing number have been recovered from diseases in the United States. The preferred treatment for uncomplicated gonococcal disease includes intramuscular administration of procaine penicillin G, or ampicillin and/or amoxicillin for both men and women. For those with uncomplicated diseases due to penicillinase-producing strains, spectinomycin is the recommended drug.

Ampicillin, amoxicillin, tetracycline, spectinomycin, erythromycin, and aqueous penicillin G can all effect a favorable outcome in disseminated gonococcal disease. Ophthalmia neonatorum can be prevented by applying ointment or drops containing tetracycline, erythromycin, or a 1% solution of silver nitrate (Credé's method) to the eyes of the infant following birth. Most states require that this procedure be performed.

Penicillin taken within a few hours before or after exposure does not provide safety from infection. It is true that the incidence of infection can be reduced by chemoprophylaxis, but when infection does occur, the gonococcal strain tends to be relatively drug resistant. In addition, the incubation period is prolonged, favoring transmission to others.

The most effective method for control is the education of the general public concerning the transmission of the disease. Control should include (1) diagnosing the asymptomatic patient who serves as a source of infection to others, particularly if he or she is promiscuous, (2) condom use, particularly for those who have several sexual partners, and (3) providing adequate treatment for the sexual partners of an individual who has already been diagnosed as having gonorrhea.

Immunity

Most diseases caused by *N. gonorrhoeae* involve the mucosal surface, and local antibodies, such as secretory IgA, are produced in high concentrations. Serum antibodies, which give longer protection, appear in relatively few cases of uncomplicated gonorrhea. There are several antigenically distinct gonococcal serotypes, and any serum antibodies produced are strain specific. Thus repeated infections from other serotypes would be expected to occur. The antibody response is even worse for those with disseminated gonococcal disease, because serum antibodies are not produced against the infecting strain. This lack of immunity may be due to the poor immunogenicity of gonococcal antigens by these strains. Studies with antibodies produced on mucosal surfaces of the genitourinary tract indicate that fimbriae are the principal antigens to which these antibodies are specific. Mixing specific IgA antibodies with virulent gonococci, for example, inhibits the latter's attachment to mucosal surfaces in the genitourinary tract. The results indicate that fimbriae could be used as a vaccine, however, the gonococcal attachment process is apparently due to other surface antigens as well, and these have not been clearly identified.

NONGONOCOCCAL URETHRITIS

A urethritis milder than that caused by *N. gonorrhoeae* is believed to be caused by two bacterial agents: *Chlamydia trachomatis* and *Ureaplasma urealyticum*. Nongonococcal urethritis (NGU) is a nonreportable disease, but the estimated number of cases in the United States is approximately 1 million.

Characteristics of the agents of nongonococcal urethritis

C. trachomatis is a spherical-shaped bacterium that ranges in size from 0.2 to 1.0 μm. This bacterium, as well as other members of the genus, are obligate intracellular parasites and require a living host in order to replicate.

U. urealyticum belongs to a group of bacteria called the Mycoplasmatales, which do not have cell walls but are capable of self-replication on laboratory media. *U. urealyticum* seldom exceeds 1 μm in length by 0.2 to 0.3 μm in diameter and is a common inhabitant of the vagina and cervix. The cell membrane of this organism contains lipids and proteins that make it resistant to condi-

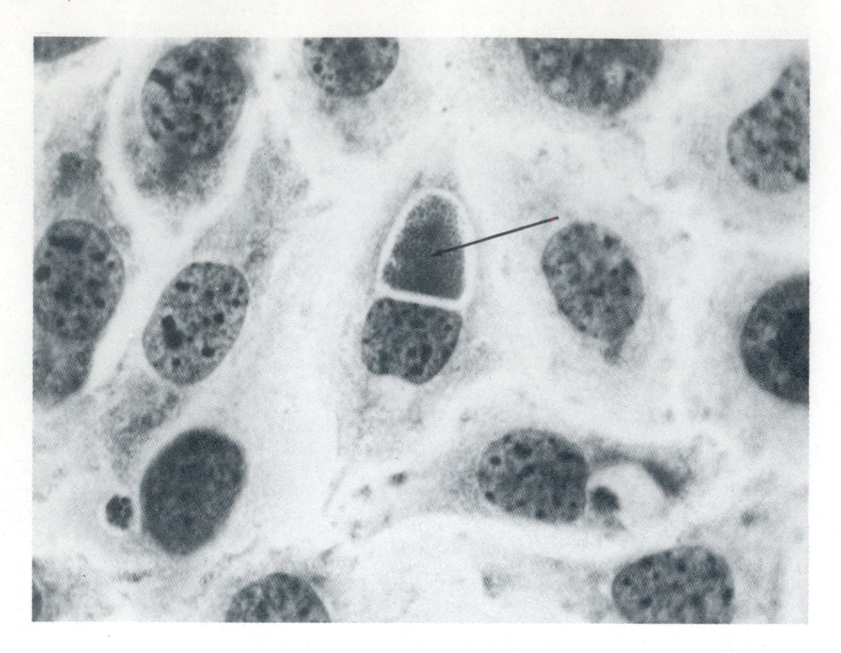

Figure 30-14 Light photomicrograph of stained *Chlamydia*-infected culture cells. Arrow points to inclusion body.
Courtesy P. Spears.

tions that might lyse other bacteria. *U. urealyticum,* as its name suggests, is capable of hydrolyzing urea, a characteristic that distinguishes it from other mycoplasmas.

Epidemiology

C. trachomatis has been recovered from 35% to 42% of men with nongonococcal urethritis as compared with 3% to 7% from those without nongonococcal urethritis. The higher recovery rate is associated with more promiscuous activity. The recovery rate in women approaches 60% to 70% if their sexual partner is positive for chlamydial infection.

The data implicating *U. urealyticum* at this time is only circumstantial. Reports have indicated that those with NGU, but with no chlamydiae present, possess high levels of *U. urealyticum* in clinical specimens. This group had also responded better to therapy specific for *Ureaplasma* than to therapy for chlamydial infection. Specific serotypes of *U. urealyticum* have also been demonstrated to cause urethritis when inoculated into human volunteers.

Clinical manifestations

Inflammation of the proximal end to the urethra, which is the most common manifestation of urethritis, is accompanied by pain and a burning sensation on urination (dysuria). Genital chylamydiae have also been incriminated in other syndromes including:

1. Cervicitis. As many as 50% of women with cervicitis have been shown to have positive chlaymydial infection.
2. Salpingitis or PID. Most of the evidence implicating *Chlamydia* in this syndrome has come from studies in England. Most cases in the United States appear to be associated with gonococci, although the wearing of intrauterine devices increases the risk of infection by microorganisms other than gonococci.
3. Neonatal infection. Approximately 11% of the general population of pregnant women have chlamydiae in the cervix, which offers an opportunity for infection of the neonate during passage through the birth canal. Conjunctivitis ***(inclusion blennorrhea)*** acquired during delivery is usually a benign infection. It has been documented that *C. trachomatis* is also the cause of ***neonatal pneumonia,*** an infection characterized by cough, congestion, and alveolar and interstitial infiltration. This organism is believed to account for 30% of all cases of pneumonia in hospitalized infants less than 6 months of age, but no deaths have as yet been attributed to this type of infection.

Laboratory diagnosis

C. trachomatis and *U. urealyticum* are best identified by isolation in culture. Clinical specimens suspected of containing chlamydiae are grown in tissue culture and later stained and examined for chlamydial inclusion bodies (Figure 30-14). Fluorescent antibody staining also provides an additional parameter for identification of *Chlamydia. U. urealyticum* (Figure 30-15), can be cultivated on laboratory media and, as mentioned earlier, is distinguished from other mycoplasmas by its ability to hydrolyze urea.

Treatment and prevention

Tetracycline and erythromycin are the most effective drugs for the treatment of any chlamydial or ureaplasmal infection. Erythromycin is the drug of choice for infants and pregnant women. The sexual partners of the infected individuals should also be treated.

VENEREAL HERPES

Characteristics of the agent of herpes

Several viral agents may be transmitted venereally (see Table 30-1), but the most commonly recognized is the herpes simplex virus (HSV). The disease caused by this virus is called ***genital herpes.*** The HSV is a DNA virus that is icosahedral and possesses an envelope composed of lipids derived from the nuclear membrane of the host and of glycoproteins that are virus specific. The herpes simplex virus can be divided into two types: I and II. Type II is associated with genital infection, but type I may occasionally be recovered from the genital tract and has been implicated in genital herpes.

Epidemiology

HSV type II infection can be acquired during birth, but venereal spread is the most common source of infection. First infections are usually associated with early sexual experience, whereas recurrent infections are more common in older adults, even though high levels of HSV antibody may be present. The high antibody level in this group indicates the ability of the virus to protect itself in the host in a latent state (see Chapter 16).

Clinical manifestations

Infection in the man. Lesions in the man appear as ulcers on the glans, prepuce, and shaft of the penis (Figure 30-16). Penile herpes is extremely painful; however, the condition resolves on its own. Lesions may also appear near the anus and/or oral cavity, depending on sexual practices. These genital diseases can be complicated by urethritis, neuralgia, or septic meningitis.

Infection in the woman. Most cases of genital herpes in the woman involve the cervix, but the condition is often asymptomatic and characterized only by inflammation. Women are more likely to engage in sexual intercourse when infected because of the asymptomatic nature of the disease and are therefore a more frequent cause of herpes transmission to others. Infections involving the vulva are more likely to result in more overt symptoms, such as fever, discomfort on urination, and genital pain.

Women who acquire primary genital herpes infection during pregnancy excrete large amounts of virus over a 3- to 4-week period, and cesarean section may be required to prevent disease and

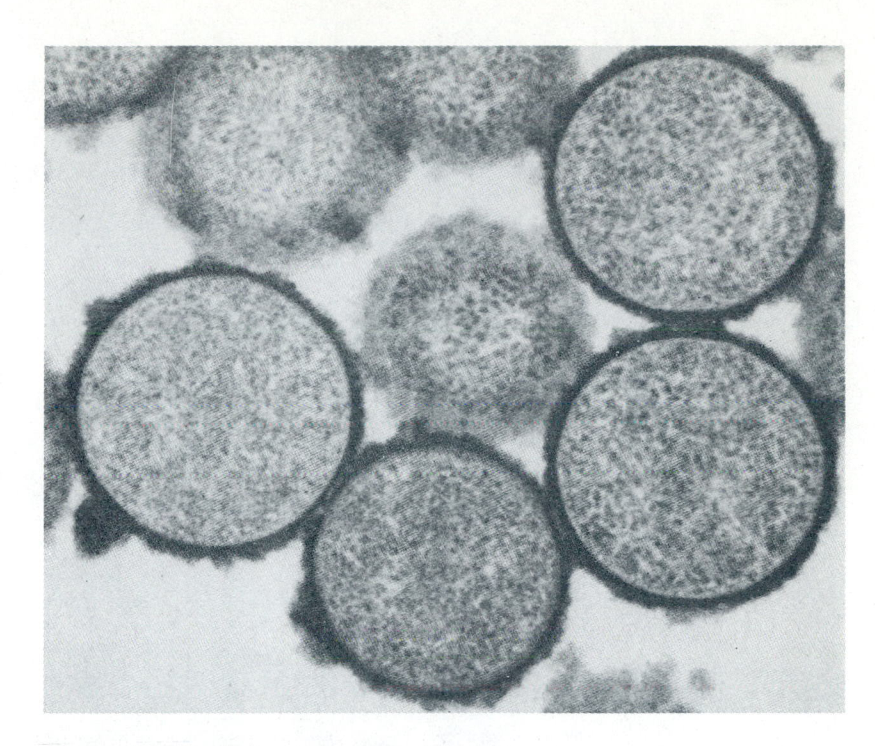

Figure 30-15 Electron micrograph of *Ureaplasma urealyticum.* (×73,920.)
From Robertson, J. J. Bacteriol. ***128:*** *658, 1976.*

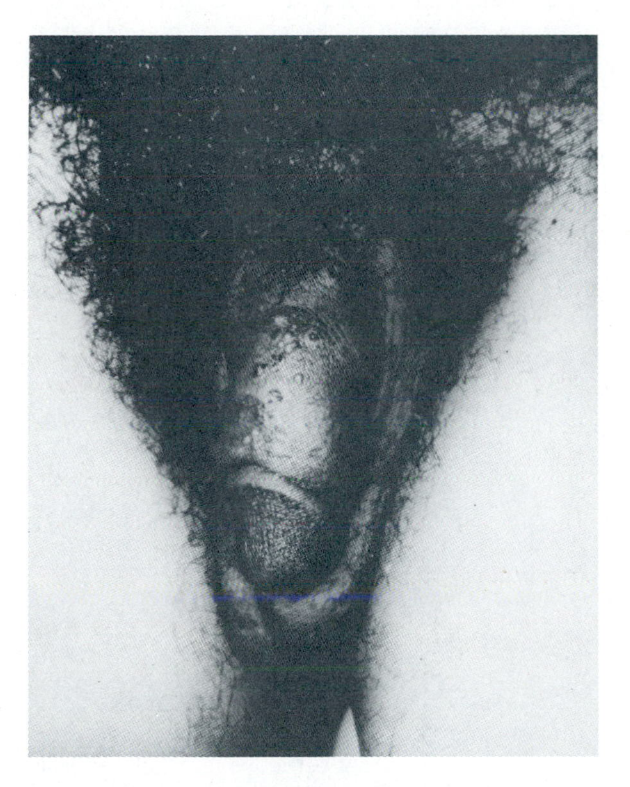

Figure 30-16 Genital lesions of penis due to herpes simplex virus.
Courtesy Nicholas Fiumara, Department of Public Health, Boston. From Morton, B.N. VD: a guide for nurses and counselors. Little, Brown & Co., Boston, 1976.

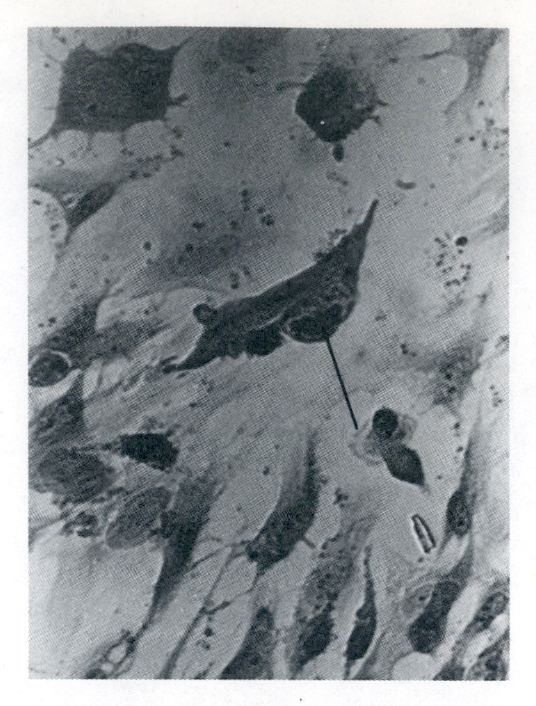

Figure 30-17 Herpes virus inclusion *(arrow)* observed in cultured epithelial cells. (×400.)
From Bia, F.J., et. al. J. Virol. ***36:****245, 1980.*

death of the fetus. Viral excretion during recurrent genital herpes infection is brief, and HSV antibody is present in the serum of the mother as well as in that of the fetus. The fetus is therefore protected and cesarean section is not required.

Neonatal infection. Herpes can result in abortion, and at least one half of those who survive will be afflicted with some neurological or ocular disorder. Systemic disease may also involve the viscera and the central nervous system.

Laboratory diagnosis

Scrapings of cells from the lesions can be examined for the appearance of intracellular inclusions (Figure 30-17); however, this does not distinguish between the two types of HSV. Fluorescent antibody specific for HSV type II can be applied to clinical specimens to detect the virus.

Treatment and prevention

There have been many studies employing various treatments for genital herpes, but few have proved effective. Anesthetics applied topically can reduce the pain associated with infection. Keeping the lesions dry helps to prevent secondary infection. One of the most promising drugs appears to be acyclovir, which is specific for viral enzymes and is not metabolically degraded following systemic administration. Acyclovir as a topical ointment has been marketed for public use and shows some effectiveness for first episodes of genital herpes but has no apparent usefulness for recurrent herpes. Treatment of ocular disease can be aided by the use of topically applied iododeoxyuridine (IDU) or adenine arabinoside (Ara-A). Cesarean section to prevent neonatal disease is discussed above.

Herpes simplex virus and cancer

Evidence has indicated that there is a positive association between cervical cancer and HSV type II infection. Some of the evidence includes the following:

1. Women with genital herpes stand a greater chance of contracting cervical cancer than those who are not infected.
2. Women having earlier and more frequent sexual experiences are more likely to have cervical cancer than virgins.
3. HSV type II can cause cervical cancer when inoculated into laboratory mice. The evidence to date is only circumstantial, but HSV appears to be a likely candidate as an etiological agent of cervical cancer.

TRICHOMONIASIS

Trichomoniasis of the genital tract is a disease caused by the protozoan *Trichomonas vaginalis.* The organism is flagellated and measures 10 to 15 μm in length (Figure 30-18). Disease in the man is asymptomatic, but in the woman there is frequently a yellow discharge. Diseased women also complain of a burning sensation. Little is known of the epidemiology, clinical picture, or laboratory diagnosis of trichomoniasis, but a recent study suggests that anywhere from 5% to 20% of the population are affected by this disease. The presence of *T. vaginalis* has been related to education, income, crowded living conditions, and sanitary facilities. The etiological agent can be diagnosed by microscopic observation of the vaginal specimen. Treatment should be given to both sexual partners and includes the use of metronidazole. Metronidazole is contraindicated during the first trimester of pregnancy.

LYMPHOGRANULOMA VENEREUM

Lymphogranuloma venereum is caused by the organism *Chlamydia trachomatis,* a strain similar to

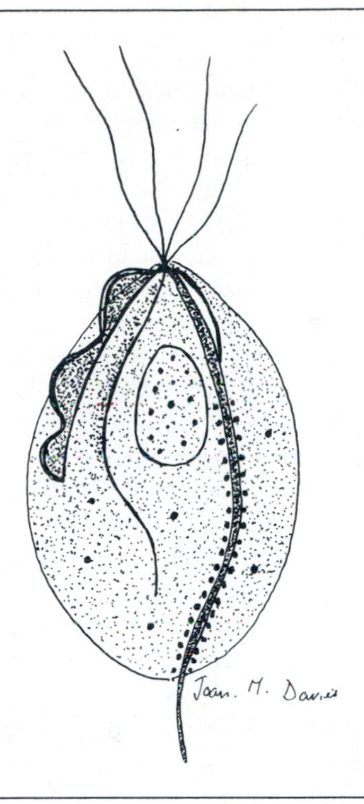

Figure 30-18 *Trichomonas vaginalis* trophozoite.
From Beck, J.W., and Davies, J.E. Medical parasitology (3rd ed.). The C.V. Mosby Co., St. Louis, 1981.

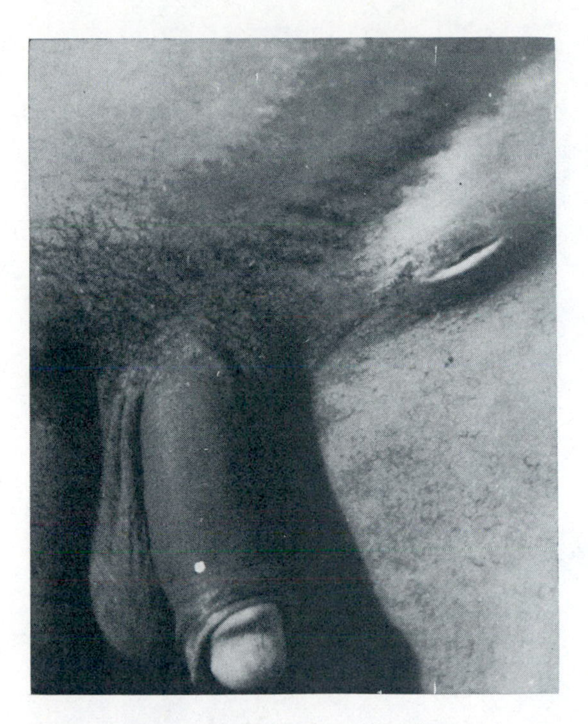

Figure 30-19 Bilateral buboes of lymphogranuloma venereum, one of which is shown to be surgically incised.
From Von Haam, E. In Anderson, W.A.D., and Kissane, J.M. (Eds.). Pathology (7th ed.). The C.V. Mosby, Co., St. Louis, 1977.

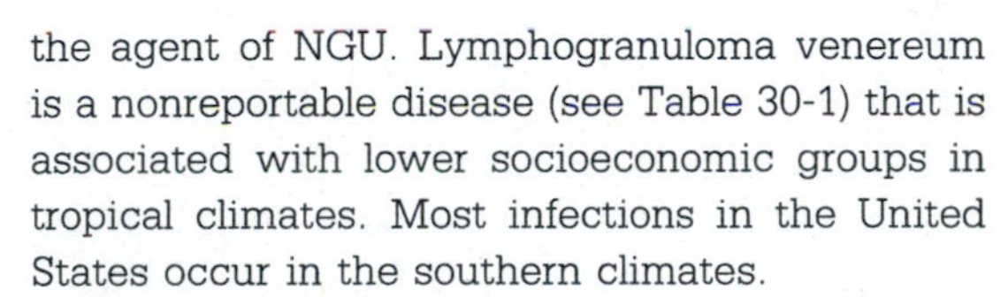

the agent of NGU. Lymphogranuloma venereum is a nonreportable disease (see Table 30-1) that is associated with lower socioeconomic groups in tropical climates. Most infections in the United States occur in the southern climates.

A small lesion or ulcer develops at the site of infection, which includes the glans penis, or labia or vagina, but infection can occur at extragenital sites such as the oral cavity. The microorganisms at the infection site migrate to regional lymph nodes, and within 7 to 30 days produce buboes (Figure 30-19). The superficial nodes, which enlarge and coalesce, exhibit a purplish color and sometimes become abscessed. Complications of disease may appear from a few weeks to several years after primary infection and include elephantiasis of the labia and perianal involvement. Diagnosis is based on clinical manifestations and serological tests such as the complement fixation test. Tetracycline is the drug of choice for treatment, but alternative drugs include doxycycline, erythromycin, and sulfamethoxasole.

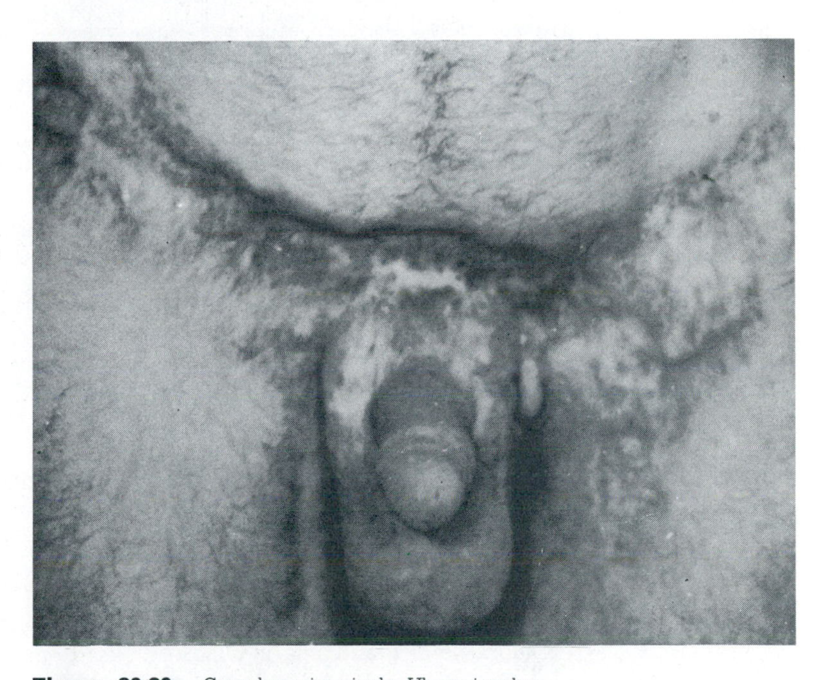

Figure 30-20 Granuloma inguinale. Ulcers involve inguinal regions and scrotum.
From Von Haam, E. In Anderson, W.A.D., and Kissane, J.M. (Eds.). Pathology (7th ed.). The C.V. Mosby Co., St. Louis, 1977.

GRANULOMA INGUINALE

Granuloma inguinale is a rarely contracted venereal disease caused by the gram-negative rod *Ca-*

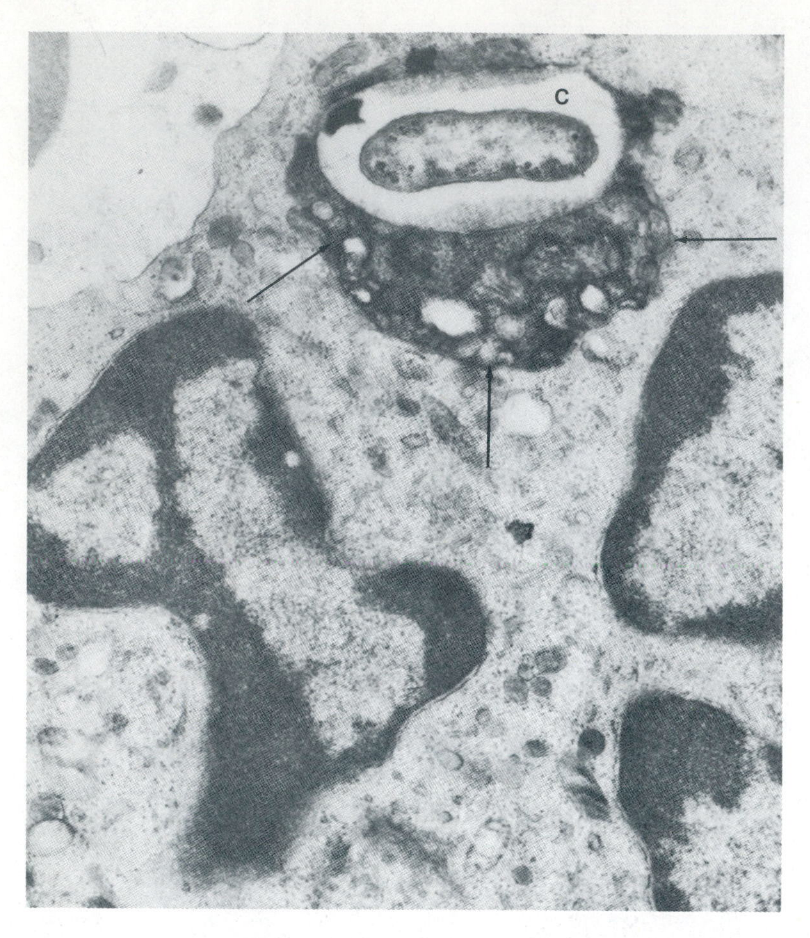

Figure 30-21 Polymorphonuclear leukocyte with phagosome *(arrows)* containing *Calymmatobacterium granulomatis* from patient with granuloma inguinale. (×18,200.). Clear area *(C)* surrounding organism is presumably a capsule.

From Kuberski, T. et. al. J. Infect. Dis. ***142:***744, 1980.

lymmatobacterium granulomatis (originally called *Donovania granulomatis*). The disease is seldom seen in temperate climates, and in the United States infections are observed more frequently in the South. Irregularly shaped ulcers, which are painless, are produced at the site of entry of the pathogen, which is usually the genitalia. The ulcers tend to coalesce (Figure 30-20), and as they heal they leave a depigmented area. The coalesced ulcerations can sometime be confused with those of venereal warts (see below).

Diagnosis of granuloma inguinale is based on the appearance of the lesions and by examination of scrapings from the lesions. These scrapings show mononuclear cells packed with the infectious agent (Donovan bodies; Figure 30-21). Streptomycin, chloramphenicol, tetracycline, erythromycin, and gentamicin can all effect a cure when administered in proper dosages.

VENEREAL WARTS

Venereal warts ***(condylomata acuminata)*** are soft pink cauliflower-like growths that appear singly or in clusters on the external genitalia and rectum. They are caused by a human papilloma virus (Figure 30-22), and as many as five distinct papilloma viruses may be involved in wart formation. This disease is one of the most common sexually transmitted diseases in the United States, and complications can be severe. The smaller growths usually make intercourse difficult or cause rectal pain. Larger growths can make urinating difficult and may result in transmission to neonates during childbirth. Neonatal transmission is thought to cause childhood laryngeal papillomatosis. There is some evidence that cervical neoplasia and perianal neoplasia occur in patients with a history of anogenital warts. Spontaneous regression can occur and is most often observed in women following childbirth.

Treatment involves the use of topically administered pharmacological agents such as podophyllin, dichloracetic acid, colchicine, idoxuridine, and other antiviral agents. If these are unsuccessful, surgery is recommended.

SOFT CHANCRE (CHANCROID)

Soft chancre is caused by the gram-negative bacterium *Haemophilus ducreyi.* This bacillus is associated with diseases that occur primarily in underdeveloped countries where personal cleanliness is at a minimum. Chancroid in the United States is more prevalent in the South, particularly among lower socioeconomic groups.

The incubation period for chancroid is about 4 days and is easily recognized in the man (Plate 41), in whom a papule develops at the site of infection (the penis), followed by ulceration. The ulcer is not hard (indurated), as in syphilis, but multiple ulcers often appear because of self-inoculation from the initial site of infection. The ulcer is very painful and bleeds easily. Inflammation of regional lymph nodes (buboes) is characteristic of infection and appears about 1 week after infection. Ulcers in the woman may appear on the clitoris and labia, but the only complaint is usually that of a burning sensation following urination. The disease often resolves on its own, but in the man deformities may appear on the penis.

Diagnosis of disease rests on identification of the microorganisms from scrapings from the ulcer or aspirates from the buboes. Darkfield microscopy is suggested, since syphilis is frequently associated with chancroid. Chancroid responds well to erythromycin or the combination trimetroprim-sulfamethoxazole. Healing is enhanced by cleansing with soap and water.

CANDIDIASIS

Candidiasis is caused by the yeastlike fungus *Candida albicans* (Plate 42) (see Chapter 31). The organism is indigenous to areas of the human body, including the oral cavity, intestinal tract, and genitourinary tract. During pregnancy or antibiotic therapy *C. albicans* can outgrow the indigenous vaginal flora and cause disease characterized by a thick white discharge. The organism can be transmitted to a sexual partner, but in men infection is usually asymptomatic. The oral cavity of the fetus may become infected during delivery, resulting in a condition called ***thrush*** (see Chapter 31). Treatment of candidiasis is by use of the drugs nystatin (Mycostatin), miconazole, or topical gentian violet.

SEXUALLY TRANSMITTED ENTERIC INFECTIONS

The use of the pill and a variety of intrauterine devices to prevent conception have resulted in an increase in transmission of agents normally associated with sexually transmitted diseases, syphilis and gonorrhea. Oral-genital and oral-anal sexual practices have resulted in the transfer of microbial agents that are not usually associated with venereal infections. This is not a surprising result if one considers the number and variety of microbial agents that can be found in the oral cavity, genitourinary tract, and intestinal tract. More and more cases of enteric infections, for example, caused by sexual transmission are appearing daily. The treatment of sexually transmitted enteric infections is usually the same as the treatment for this same disease when acquired by other means of transmission. Some of the infections are mild and self-limiting or asymptomatic; therefore, treatment must be carefully evaluated in order to prevent the emergence of antibiotic-resistant strains. Listed below are some of the characteristics of the major enteric infections transmitted venereally.

Campylobacter fetus had been known for many years by veterinarians as a cause of infertility and abortion in cattle. *Campylobacter fetus* subspecies *jejuni* is now recognized as a cause of abortion, meningitis, and gastrointestinal infections as well as other infections in humans. Transmission generally occurs by ingestion of contaminated water, food, or milk or close contact with infected animals or persons. Sexual transmission of the organism has been shown to result in cases of proctitis, particularly in homosexual men, and to cause abortion in infected women. *Campylobacter* organisms are sensitive to many antimicrobials including erythromycin, tetracycline, clindamycin, chloramphenicol, and aminoglycosides.

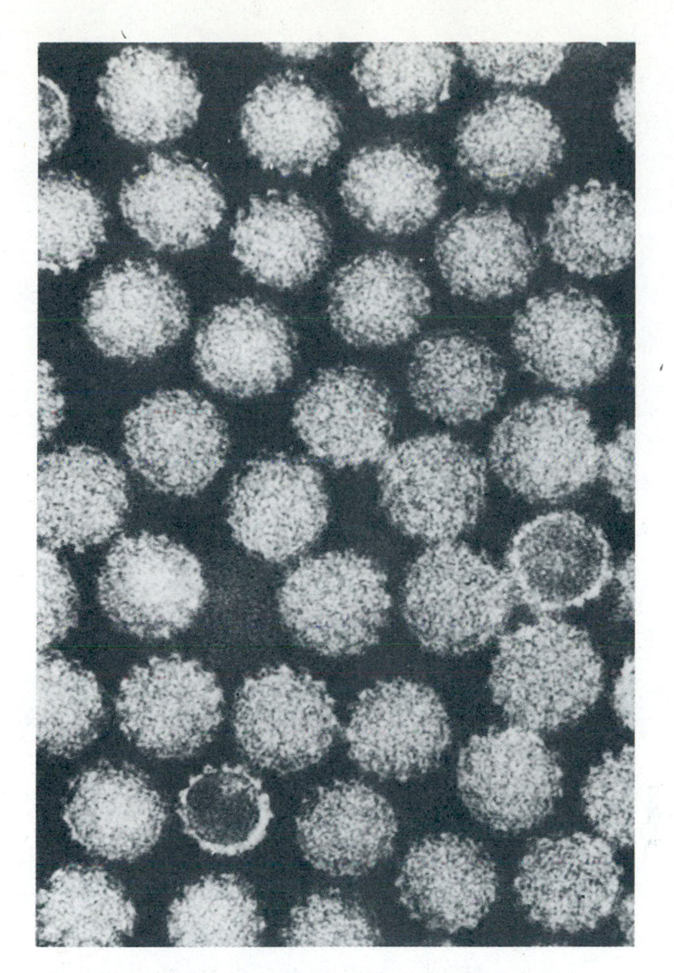

Figure 30-22 Electron micrograph of human papilloma virus.
From Viac, J. Br. J. Vener. Dis. ***54:****172, 1978.*

Shigella organisms, which are a cause of dysentery, are usually transmitted by ingestion of contaminated food or water. Since asymptomatic carriers may harbor up to 10^3 shigella per gram of stool, transmission can occur when oral-anal sex is practiced. Recent studies indicate that nearly 60% of the cases of shigellosis in adult men are found in homosexuals. Outbreaks of shigellosis have occurred frequently in gay communities located in large metropolitan areas. Treatment using antimicrobials is usually reserved for extremely ill patients because of the antibiotic resistance problem associated with *Shigella* strains. Some physicians, however, feel that antimicrobials should be administered to all cases because it reduces fecal excretion of the organism and prevents further spread of the disease.

Salmonella organisms are a cause of typhoid fever as well as intestinal illness. Sexually transmitted disease caused by this organism is rarely reported but can occur among those whose sexual practices involve direct fecal-oral contact. Antimicrobial therapy for uncomplicated enterocolitis caused by *Salmonella* organisms is usually not warranted. Replacement of fluids and electrolytes is the treatment of choice.

Entamoeba histolytica is a protozoa that was one time associated with infections of the poor who lived in rural populations. Today, infections (called amebiasis) are found in the middle-class as well as upper-class populations in urban areas. Sexual transmission is usually limited to homosexuals, but the disease can be spread by nonsexual activities such as handling foods. Untreated disease may result in severe diarrhea, weight loss, liver problems, and other symptoms. Treatment includes the use of many drugs including metronidazole, tetracycline, and paromomycin. Amebiasis is discussed in more detail in Chapter 32.

Giardia lamblia is also a protozoa that is usually contracted by ingestion of contaminated water, but sexual transmission of *G. lamblia* has been demonstrated to cause proctitis in homosexual men. The drugs used for treatment include atabrine, metronidazole, and furazolidone. Further discussion of infection by *G. lamblia* can be found in Chapter 32.

OTHER SEXUALLY TRANSMITTED DISEASES

The list of sexually transmitted diseases is not limited to those discussed above. Improved laboratory diagnostic techniques and instruments have facilitated epidemiological studies that implicate other microbial agents in sexually transmitted disease. These include group B streptococci, cytomegalovirus, and hepatitis B virus.

Group B streptococci are normally found in the intestinal tract, which represents a source of vaginal or urogenital colonization in women. Carriage of group B streptococci during pregnancy can lead to infection of neonates, which, if left untreated, can be fatal.

Cytomegalovirus is present in saliva and urine, and high titers have been found in semen and cervical secretions; thus venereal transmission can occur. Women who are infected with cytomegalovirus can transmit the virus to the fetus during pregnancy. Congenital infection is evidenced by enlargement of the spleen and liver and encephalitis as well as congenital abnormalities such as club foot, deafness, and high-arched palate.

The hepatitis B virus can be found in most body fluids and it is often sexually transmitted. Those at highest risk by sexual transmission are homosexuals. Anal-genital and oral-anal intercourse are the sexual practices most likely to increase the risk of transmission of the virus. Hepatitis B virus infection is discussed in Chapter 29.

Scabies and pediculosis are two diseases that occasionally are transmitted sexually. They are caused by animal parasites and are discussed in Chapter 32.

SUMMARY

The possibilities of deformed infants, infertility, and abortion resulting from sexually transmitted diseases have created concern among the medical community as well as the general public. Syphilis and gonorrhea are the most prevalent sexually transmitted diseases, but some other sexually transmitted diseases, although occurring less frequently, can also be devastating.

Many historians believe that syphilis was contracted by Columbus's crew during their visit to the West Indies and that, on their arrival in Spain, they transmitted the disease to the Spanish population. Syphilis may be of a venereal or nonvenereal type. Venereal transmission is believed to have begun in large cities, where a clothed society provided venereal contact as the primary means of transmission.

Venereal syphilis is caused by the bacterium *Treponema pallidum*. Disease is characterized by three stages: a primary stage in which a lesion called a chancre predominates, a secondary stage characterized by a rash, and a tertiary stage characterized by lesions called gummas. *Treponema pallidum* can penetrate the placenta and can infect the fetus, causing death or deformities in the infant.

The most widely used test for diagnostic purposes is the Venereal Disease Research Laboratory (VDRL) test. More specific tests include the fluorescent treponemal antibody absorption

(FTA-ABS) test and the treponema pallidum immobilization test.

Penicillin is the principal chemotherapeutic agent for treatment of syphilis. Other drugs that may be used under special circumstances include tetracycline and erythromycin. There is no permanent immunity to infection, nor is a vaccine available.

Gonorrhea is caused by the bacterium *Neisseria gonorrhoeae.* Over 1 million cases are recorded each year. During pregnancy the infectious agent can infect the conjunctiva of the fetus as it passes through the birth canal. Disease in women often goes unnoticed and can spread to the fallopian tubes or ovaries and cause sterility. Infrequently, *N. gonorrhoeae* penetrates the bloodstream and causes disseminated disease that results in polyarthritis.

Diagnosis of disease requires isolation of the organism from clinical specimens using special laboratory media. The isolated organisms are subjected to staining, oxidase reagent, and sugar fermentation tests.

Penicillin is an effective drug in the treatment of gonorrhea; however, resistance to penicillin has been noted throughout the country, and susceptibility testing is now recommended. It is also a state law that the eyes of neonates be flushed with 1% silver nitrate or a solution of some antibiotic such as erythromycin to prevent neonatal conjunctivitis due to *N. gonorrhoeae.*

Several other agents have been implicated in venereal disease (see Table 30-1).

Study questions

1. If venereal diseases are of little discomfort to the sexual partners, why are they so important?
2. What is the origin of syphilis? What is the relationship of syphilis to yaws and bejel?
3. Describe the clinical symptoms during the first, second, and third stages of syphilis. Which stages are contagious?
4. Can syphilis be acquired by contact with toilet seats, doorknobs, bedposts, etc.? Explain.
5. What are the consequences of transmission of *T. pallidum* to the fetus? What is the early clinical picture? The late clinical picture?
6. What is reagin? What is its importance in syphilis?
7. Would detection of organisms causing syphilis from lesions in the oral cavity be easy or difficult? Explain.
8. In which sexual partner is gonorrhea most evident? Does this present any epidemiological problems?
9. What is the most severe aspect of gonorrhea in the man? In the woman? In the neonate?
10. What is wrong with taking antibiotics before sexual contact in order to prevent infection?
11. What agents are believed to be involved in nongonococcal urethritis? What is the effect of infection on the man? On the woman? On the neonate?
12. What is the association of venereal herpes and cancer?
13. What is the affect of venereal herpes on the man? On the woman? On the neonate?

Selected readings

Books

Bender, S.J. *Venereal disease* (2nd ed.). William C. Brown Group, Dubuque, Iowa, 1975.

Boyd, R.F., and Marr, J.J. *Medical microbiology.* Little, Brown & Co., Boston, 1980.

Freeman, B.A. *Burrow's textbook of microbiology* (21st ed.). W.B. Saunders Co., Philadelphia, 1979.

Hoeprich, P.D. (Ed.). *Infectious diseases* (2nd ed.). Harper & Row, Publishers, Inc., New York, 1977.

Mandell, G.L. (Ed.). *Principles and practice of infectious diseases.* John Wiley & Sons, Inc., New York, 1979.

Nicholas, L. *Sexually transmitted diseases.* Charles C Thomas, Publisher, Springfield, Ill., 1973.

Youmans, G., Paterson, P.Y., and Sommers, H.M. *The biologic and clinical basis of infectious disease* (2nd ed.). W.B. Saunders Co., Philadelphia, 1980.

Journal articles

Centers for Disease Control. Syphilis: recommended treatment schedules. *Ann. Intern. Med.* **85:**94, 1976.

Chretien, J.H., McGinniss, C.G., and Muller, A. Venereal causes of cytomegalovirus mononucleosis. *J.A.M.A.* **238:**1644, 1977.

Corey, L., and Holmes, K.K. Sexual transmission of hepatitis A in homosexual men: incidence and mechanism. *N. Engl. J. Med.* **302:**435, 1980.

Edwards, L.E., et al. Gonorrhoea in pregnancy. *Am. J. Obstet. Gynecol.* **132**(6):637, 1978.

Fitzgerald, T.J. Pathogenesis and immunology of *Treponema pallidum. Annu. Rev. Microbiol.* **35:**29, 1981.

Fiumara, N.J. Syphilis in newborn children. *Clin. Obstet. Gynecol.* **18:**183, 1975.

Fiumara, N.J. The sexually transmissible diseases. *DM* **25**(3):(Entire issue), 1979.

Fluker, J.L. A 10-year study of homosexually transmitted infection. *Br. J. Vener. Dis.* **52:**155, 1976.

Jacobs, N.F., Jr., Arum, E.S., and Kraus, S.J. Nongonococcal urethritis: the role of *Chlamydia trachomatis. Ann. Intern. Med.* **86:**313, 1977.

Kuberski, T. Granuloma inguinale (donovanosis)—a review. *Sex. Transm. Dis.* **7:**29, 1980.

Mardh, P.A., and Westrom, L. Adherence of bacteria to vaginal epithelial cells. *Infect. Immun.* **13:**661, 1976.

McGee, Z.A., et al. Pathogenic mechanisms of *Neisseria gonorrhoeae. J. Infect. Dis.* **143:**413, 1981.

Nahmias, A.J., and Norrild, B. Herpes simplex viruses 1 and 2—basic and clinical aspects. *DM* **25**(10):(Entire issue), 1979.

Nahmias, A.J., and Roizman, B. Infection with herpes simplex virus 1 and 2. *N. Engl. J. Med.* **289:**667, 1973.

Novotny, P., et al. Studies on the mechanism and pathogenicity of *Neisseria gonorrhoeae. J. Med. Microbiol.* **10:**347, 1977.

Pearce, W.A., and Buchanan, T.A. Attachment role of gonococcal pili. *J. Clin. Invest.* **61:**931, 1978.

Richmond, S.J., and Sparling, P.F. Genital chlamydial infections. *Am. J. Epidemiol.* **103:**428, 1976.

Rothenberg, R. Ophthalmia neonatorum due to *Neisseria gonorrhoeae:* prevention and treatment. *Sex. Transm. Dis.* **6**(Suppl.):187, 1979.

Symposium: sexually transmitted disease. *Curr. Ther. Res.* **26**(6S):651, 1979.

Szmuness, E., et al. On the role of sexual behavior in the spread of hepatitis B infection. *Ann. Inter. Med.* **83:**489, 1975.

Urquhart, J. Effect of the venereal diseases epidemic on the incidence of ectopic pregnancy—implications for the evaluation of contraceptives. *Contraception* **19:**455, 1979.

Zelnick, M., et al. Probabilities of intercourse and contraception among U.S. teenage women 1971 and 1976. *Fam. Plann. Perspect.* **11:**55, 1979.

Chapter 31
FUNGAL DISEASES

The importance of the fungi in the cycle of nature, as well as their structure and genetics, is discussed in previous chapters. The purpose of this chapter is to describe those fungal agents that are pathogenic to humans. There are a 100,000 or more species of fungi known to us, but only 50 to 100 are pathogenic to humans, and only 20 of these cause fatal infections. Diseases caused by fungal agents are referred to as ***mycoses.*** They can be divided into four major subdivisions based on the site of infection: the superficial, the cutaneous, the subcutaneous, and the systemic mycoses. A fifth group is also included, called the opportunistic mycoses, which are seen in the compromised host. A general explanation of how mycotic agents are involved in human infection is presented before the individual agents of disease are discussed.

GENERAL CHARACTERISTICS

Epidemiology

Fungi are found primarily as saprobes on dead organic matter in the soil, but only a few fungal species have a developmental cycle requiring a mammalian host. The rest of the fungi develop normally in the soil, and humans are only an accidental host. Fungi can be found in nearly every ecological niche and at times are subjected to extremes of environmental forces such as temperature, pH, and salinity. One of the characteristics of the fungi, which enables them to survive environmental stress, is their ability to form spores. It is primarily the fungal spore that is transmitted to humans and is the cause of infection. We shall now examine the epidemiology of each of the mycoses.

Superficial and cutaneous mycoses. Superficial mycoses are diseases that affect only the cornified layers of the epidermis and the suprafollicular portions of the hair. The infectious agents are unable to parasitize nails or penetrate the keratinized layers.

Cutaneous mycoses are infections in which all of the keratinized layers may be parasitized and penetrated but not the deeper layers. The agents causing cutaneous mycoses are called ***dermatophytes,*** and the diseases they cause are referred to as ***dermatophytoses, ringworm,*** or ***tineas.*** Ringworm is a term that characterizes the physical appearance of the disease. Dermatophytes can be found in different ecological niches throughout the world; some parasitize animals, others infect only humans, and others are exclusive soil saprobes. One of the most common sources of infection for humans is by human-to-human contact, but infection may occur accidentally by contact with animals or by contact with fomites contaminated by hair or epithelial cells from animals. Dermatophytoses are the only contagious types of fungal infections known to occur in humans.

Subcutaneous mycoses. Some mycotic agents penetrate the epithelial layers but remain localized in the subcutaneous tissue to produce subcutaneous mycoses. They are not disseminated in the circulation and do not cause fatal infections. The agents of these diseases are found exclusively in the soil. Infection results from puncturing the skin with sharp objects contaminated by soil, such as wood splinters from rose bushes and contact with timbers in underground mines. Infections are confined primarily to the hands and feet and are observed in areas where inhabitants do not wear shoes, such as tropical and subtropical regions. Infections in the United States are of an occupational nature and are found among those who work in nurseries or are continually in contact with the soil. The most frequently encountered subcutaneous disease in the United States is ***sporotrichosis.***

Systemic mycoses. Systemic mycoses are caused by agents that can penetrate the epithelial tissue and can be disseminated throughout the body. They are most frequently the cause of pulmonary disease, which can become systemic and infect any organ of the body. The agents of systemic mycoses are found as saprobes in the soil and particularly in soil enriched by decaying organic matter, such as soil containing excrement from birds, bats, and other animals. The disease ***histoplasmosis,*** for example, has been associated with starling roosts. Several hundred thousand starlings in 1978 were observed to roost in one area of a small town in the southern United States. Bird excrement covered the ground after a few days and provided a selective medium for the germination of spores of the agent causing histoplasmosis *(Histoplasma capsulatum).* The vegetative cells, following germination, produced large numbers of spores that were disseminated in the air, and many individuals in contact with the starling roosts were infected by spore inhalation. The large number of diseases and the common environmental exposure made it appear that an epidemic had arisen; however, there was no human-to-human spread.

Opportunistic mycoses. The advances of medical science have enabled us to keep alive patients who, just a few years ago, might otherwise have died. New drugs, chemicals, and physical processes have not only kept the patient alive but have also created conditions in which he has become susceptible to disease by exogenous as well as opportunistic endogenous microbial agents. Opportunistic mycoses are caused by fungal agents that infect the host compromised by such conditions or agents that alter the immune response or upset the normal microbial flora. The major compromising conditions follow.

1. Diabetes. One of the side effects of diabetes, especially uncontrolled diabetes, is the formation of ketones, acids, and excess glucose in the bloodstream. These are conditions that favor tissue invasion by fungi, and in the extremities, where circulatory problems exist because of reduced oxygen tension, fungal (as well as bacterial) diseases often appear. Fungal species of the order *Mucorales* are responsible for disease in diabetics.
2. Immunodeficiencies. The principal immunological defense mechanism of humans to fungal disease is the cell-mediated response. Any disease, therapy, or condition that imperils cell-mediated immunity, such as steroid use or radiation treatment for cancer, predisposes the patient to fungal infections. Many fatal infections result from tissue invasion by indigenous fungal species, such as *Candida.*
3. Antibiotic therapy. Antibiotics such as tetracycline reduce the gram-negative or gram-positive flora of the host. The bacteria are natural competitors with fungi in the oral cavity, intestinal tract, and genitourinary tract, and without bacterial competition fungal species soon take over the habitat.
4. Other infections. Fungal spores are continually inhaled by humans, but disease is uncommon because of the host's normal defense mechanisms. A number of spores inhaled by an individual who already has a respiratory condition, such as tuberculosis, can give rise to secondary fungal disease. Species of *Aspergillus,* for example, are frequent causes of secondary infections of the respiratory tract.

Pathogenicity

Fungi, which cause human disease, must be able to reproduce at the temperature of the body, 37° C, and must be able to survive in tissue where the oxidation-reduction potential of the tissue may vary from 0 to + 300 millivolts. Most fungi have diverse enzyme systems and can metabolize many different substrates, but some pathogenic species have special requirements. For example, the dermatophytes can only infect keratinized tissue and prefer a temperature lower than 37° C. Consequently, they do not survive in other tissue. Such specialization for the majority of pathogenic fungi is not evident although there are certain environmental conditions that favor the growth of one species over another. The nutrition and physical requirements of the fungi are supplied by the host at the site of infection. Most fungi grow best when there are warm humid conditions but they are also able to adjust to different conditions.

What should not be overlooked, and must be emphasized, is that fungal diseases are the exception and not the rule. Unlike bacterial diseases, in which the severity of disease is determined by virulence factors such as toxins or enzymes, fungal diseases are usually not caused by such factors. Most of the time the fungal agent, which is either acquired by inhalation or by puncture wound, induces a cell-mediated immune response in which there is inflammation and a walling off of the fungal agent in a fibrous calcified deposit. Skin tests to determine hypersensitivity to a fungal agent have demonstrated that several million people have had contact with a fungal agent at sometime in their lives, but no disease symptoms have ever become apparent. What then is responsible for fungal disease?

The two most important factors that determine ones susceptibility to disease are the number of organisms to which the host is exposed and the state of health of the host at the time of exposure. Inhalation of a large number of spores can occur in a laboratory where fungal agents are processed and stored. Thousands of spores are produced by the fungi when cultivated in petri dishes, and air currents in the room passing over an exposed plate can disseminate them. Disease can also occur when the fungus is concentrated in nature and when there is continual human contact, such as diseases occurring as a result of one's occupation. Fungal diseases, like some bacterial diseases (such as tuberculosis where the cell-mediated immune response is critical for recovery), are often chronic and may persist in the host for years.

Several of the fungi, such as the dermatophytes and some species of *Candida,* act as allergens and cause allergic responses that are referred to as ***ids.*** The fungal allergies are manifested as skin lesions at sites distant from the initial infection site. Id is an abbreviation of ***dermatophytid,*** a term used to denote lesions appearing on the fingers from infection by a dermatophyte initially infecting the feet. Other mycotic agents that are inhaled, such as species of *Aspergillus* can cause asthmalike responses that are also believed to be related to the phenomenon of allergy. The allergic response in some individuals is not surprising because humans frequently inhale spores.

Immunity

Fungal diseases, as discussed earlier, occur under unusual circumstances, and normal host defense mechanisms are usually adequate to prevent them. The cell-mediated immune response appears to play the most important role in preventing disease. Conditions or agents—such as radiation, steroid therapy, antiinflammatory drugs, and immunosuppressive agents—that cause T lymphocyte defects predispose the patient to fungal infection. The fungi have surface antigens that elicit a poor antibody response, and antibodies that are produced do not appear to protect or aid in recovery from disease. Yet, in some systemic mycoses, the severity of disease appears to be greatest in those with defective antibody responses.

Diagnosis

The diagnosis of fungal disease, based on the presence of symptoms, may be particularly frustrating. This is especially true for systemic mycoses, because they initially begin as respiratory diseases and can be confused with bacterial and viral respiratory diseases. The diagnosis of cutaneous or subcutaneous disease, based on symptoms, is easier because the symptoms are associated with the body's surface. The diagnosis is enhanced if epidemiological data, such as the occupation of the patient, are known.

Definitive diagnosis depends on laboratory techniques and morphological features; sometimes biochemical characteristics are important laboratory tools. Determining the morphology of fungi in tissue and culture is extremely useful because of their ability to form spores and because of their dimorphic nature. Many agents of systemic mycoses are ***dimorphic.*** They exhibit mycelial growth and spore production in their natural state or on laboratory media cultivated at 30° C, but they appear in the yeast or spherical cell state, however, in human or animal tissue or on complex media incubated at 37° C.

The cultivation of fungi is not difficult, because they have minimal nutritional requirements. The most frequently used agar medium is called ***Sabouraud agar,*** which is characterized by a low pH (approximately 5.6) and is often supplemented with carbohydrates such as glucose or maltose. Some of the more enriched media contain antibiotics to inhibit the growth of bacteria that are also present in clinical specimens.

Pathologic materials can be examined directly by placing the specimen on a slide containing sodium hydroxide or potassium hydroxide followed by warming. This procedure separates the fungi from keratinaceous particles that are in the specimen and is especially useful when examining clinical specimens from cutaneous disease caused by dermatophytes. Diagnostic techniques are discussed further in Appendix B.

Treatment

The most important treatment for fungal disease involves the use of chemotherapeutic agents such as nystatin, amphotericin B, 5-fluorocytosine, and 2-hydroxystilbamidine, which are discussed in Chapter 27. A new drug, ketoconazole, appears to be an important addition to the antimicrobial arsenal because of its reduced toxicity compared with other antifungal agents and its wide spectrum of activity.

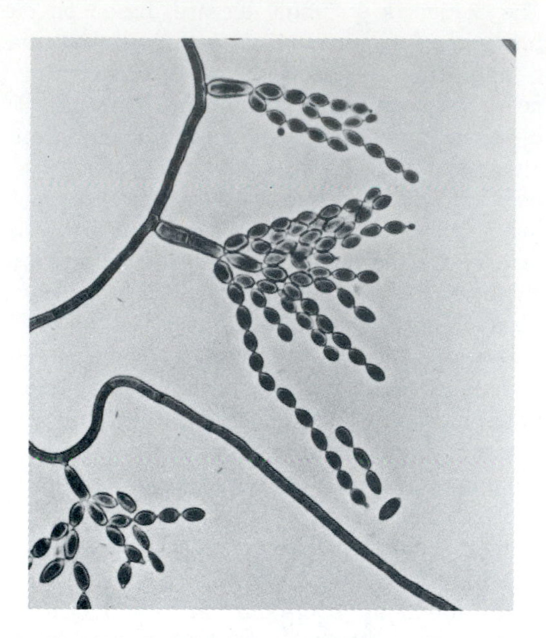

Figure 31-1 *Cladosporium* species showing conidia emanating from conidiophores.
From Roger, A.L. In Manual of clinical microbiology *(3rd. ed.). American Society for Microbiology, Washington, D.C., 1980.*

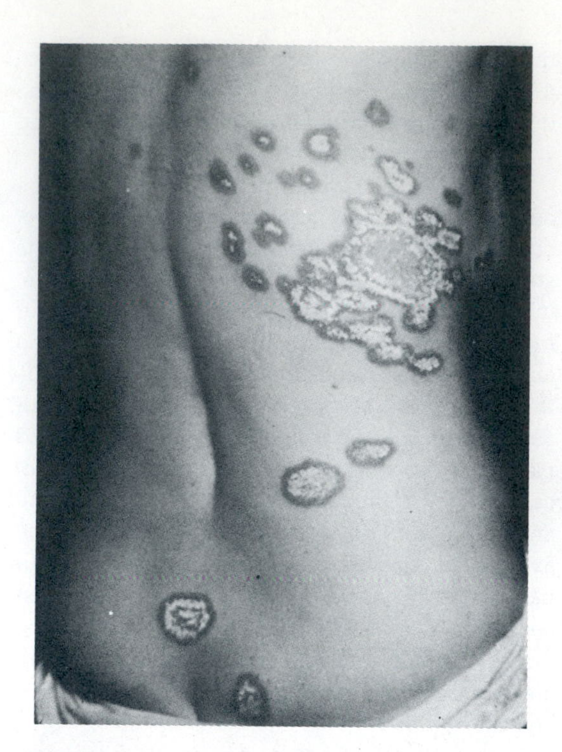

Figure 31-2 Tinea corporis (ringworm of body).
From Dolan, C.T., et al. Atlas of clinical mycology, *American Society of Clinical Pathologists, Chicago, 1975. Used by permission.*

Abscesses may be present in various organs and tissues of the body in systemic diseases, and successful treatment, especially of respiratory tissue, may require surgery and drainage. These techniques not only prevent blockage and hemorrhage of respiratory tissue, but encourage more rapid recovery from disease.

Vaccines have been used experimentally with some minor successes, but they are not part of the physician's arsenal.

SUPERFICIAL MYCOSES

Superficial mycoses are extremely rare in the United States; most occur in tropical or subtropical areas. There are four diseases that are classified in this group: white piedra, black piedra, tinea nigra, and tinea versicolor.

White and black piedra

The term *piedra* means stone in Spanish and is associated with the nodules formed by mycelia in the hair shaft during infection. The black or white color is due to the melanin-like pigment produced by the particular fungal species. Black piedra affects primarily the hair of the scalp and is caused by the agent *Piedraia hortae.* White piedra is caused by *Trichosporon begeli,* which infects the hairs of the axillary, facial, genital, and scalp regions. The hair shafts can become so weakened in both diseases that they are easily broken.

Tinea nigra and tinea versicolor

Tinea nigra and tinea versicolor both affect the smooth skin of the body. Tinea nigra is a disease most often involving the palmar surfaces and is caused by *Cladosporium werneckii* (Figure 31-1). The organism, which is found in the soil, humus, and decaying vegetation, occasionally causes disease among those living in the southern United States. Tinea versicolor, which is caused by *Malassezia furfur,* is characterized by scaly patches on the body that appear pigmented and are found primarily on the torso and upper limbs.

Laboratory diagnosis of superficial mycoses is accomplished by direct microscopic examination of scales, shed hair, or skin scrapings in wet mounts of 10% potassium hydroxide or by isolation on culture media.

CUTANEOUS MYCOSES

Cutaneous mycoses represent the most common fungal disease in humans and are important

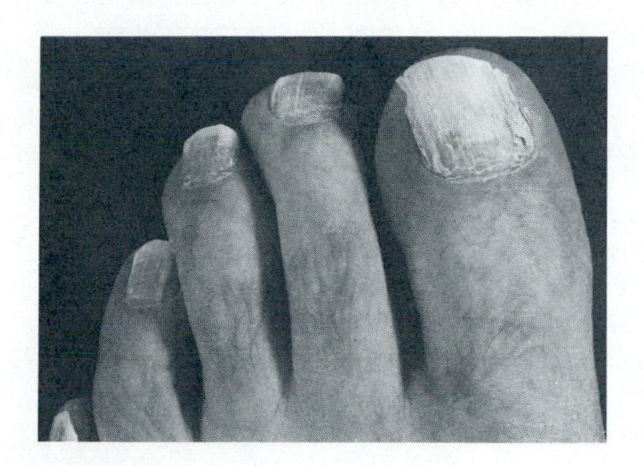

Figure 31-3 Tinea unguium (ringworm of the nails) produced by *Trichophyton rubrum*. Infection is manifested as thickened, white, and unattached nail.
Air Force Institute of Pathology, AFIP 74-19321.

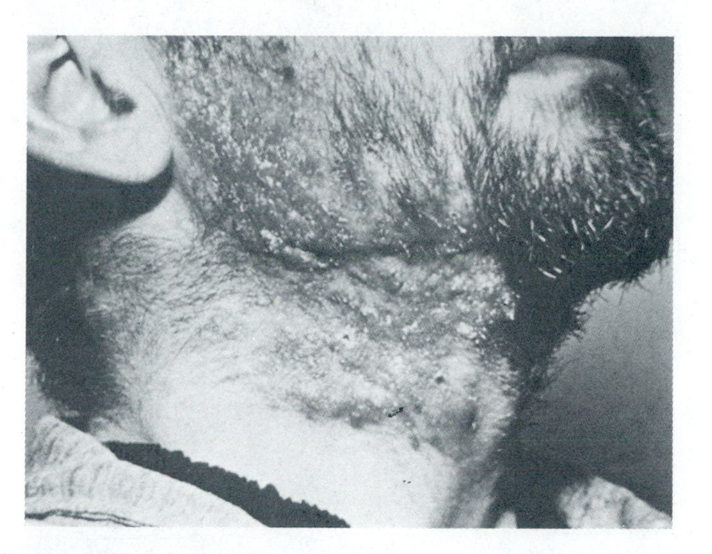

Figure 31-4 Tinea barbae (ringworm of beard).
From Dolan, C.T., et al. Atlas of clinical mycology, *American Society of Clinical Pathologists, Chicago, 1975. Used by permission.*

health problems in tropical countries where overcrowding and lack of simple hygiene exist. There are three genera of dermatophytes that cause infection: *Trichophyton, Microsporum,* and *Epidermophyton.* Cutaneous disease can be treated by oral administration of ***griseofulvin,*** which has affinity for keratinized tissue. Topical ointments, such as ***tolnaftate*** (Tinactin), ***haloprogrin,*** and ***miconazole*** are effective but require prolonged administration.

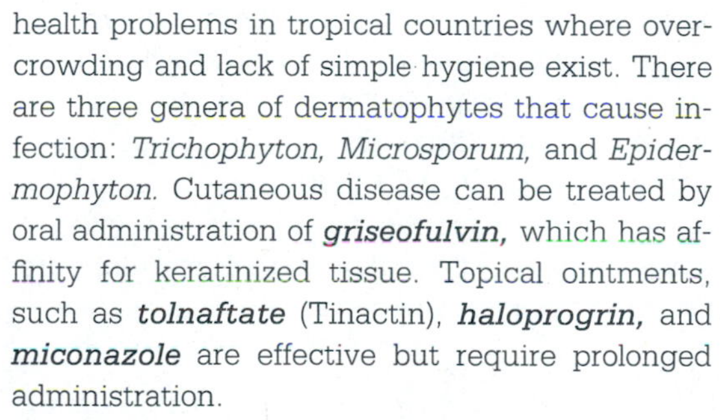

The characteristics of disease by dermatophytes depend on the body site, since some agents may be able to colonize more than one area or type of keratinized tissue. A variety of species may be associated with one type of cutaneous disease, but only the most frequently isolated ones are discussed here.

Tinea corporis (ringworm of the body)

Tinea corporis is a frequent disease in children, and the lesion, which is raised and circular (Figure 31-2), may appear on the face, arms, and other exposed parts of the body. Infection is usually by contact with dogs, horses, or other pet animals. *T. rubrum, M. canis,* and *T. mentagrophytes* are the agents most frequently associated with disease.

Tinea unguium (ringworm of the nails)

Ringworm of the nails may be secondary to disease by fungal agents causing tinea pedis (ringworm of the feet). Nails become thickened during disease as a result of hypertrophy of the nail bed, and they also appear discolored (Figure 31-3). *T. rubrum* is the most common cause of this type of disease (Plate 12).

Tinea barbae (ringworm of the beard)

Ringworm of the beard (Figure 31-4) is a disease of the hair follicles (folliculitis) that is usually caused by *T. rubrum* or *T. verrucosum* and is acquired by contact with animals such as horses, cattle, or dogs. The hair is shed during the course of the disease but returns shortly after recovery. Those dermatophytes that have a developmental cycle in humans can cause more severe and often chronic diseases of the beard.

Tinea cruris (ringworm of the groin)

Ringworm of the groin, also called ***jock itch,*** is caused by such species as *Epidermophyton floc-*

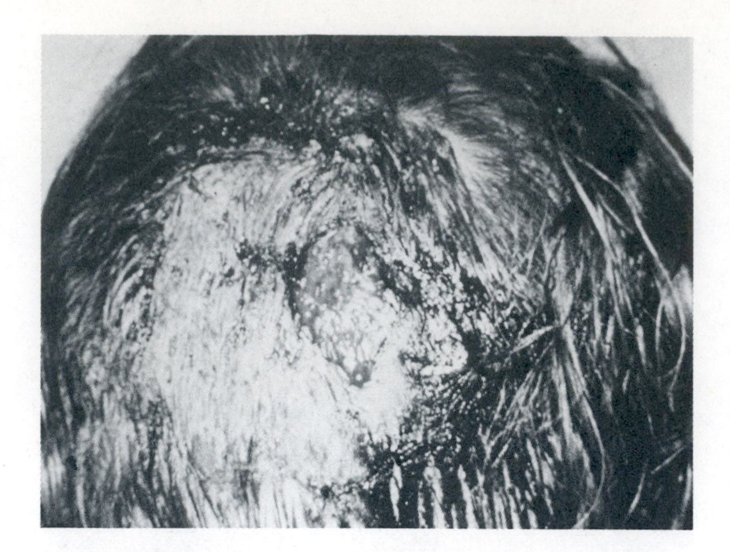

Figure 31-5 Tinea capitis (ringworm of scalp).
From Dolan, C.T., et al. Atlas of clinical mycology, *American Society of Clinical Pathologists, Chicago, 1975. Used by permission.*

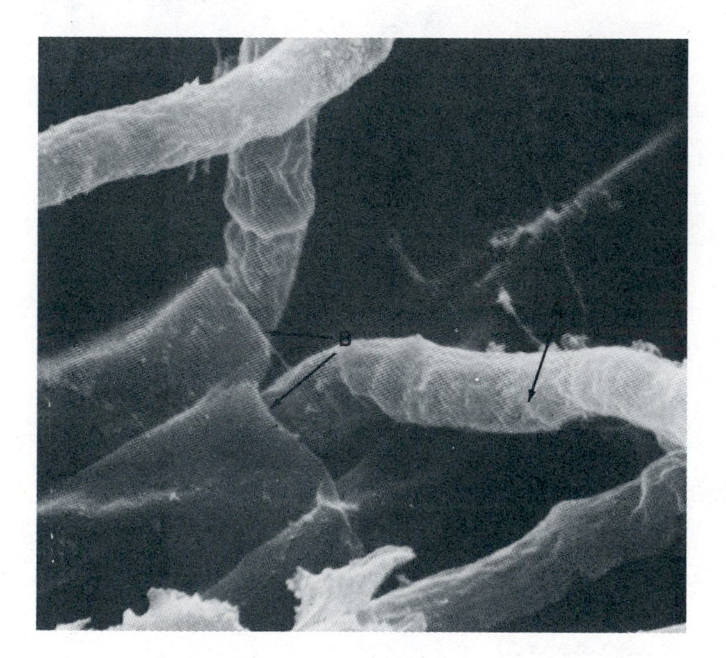

Figure 31-6 Scanning electron micrograph demonstrating hyphal *(A)* invasion of root hair *(B)* by *Trichophyton mentagrophytes*.
From Hutton, R.D. Infect. Immun. ***21****:247, 1978.*

cosum, T. rubrum, and *T. mentagrophytes.* Epidemics occur when there is careless exchange of locker room towels and other articles that may be contaminated.

Tinea capitis (ringworm of the scalp)

The lesions of tinea capitis (Figure 31-5) cause a breakage of the scalp hairs either just below or above the scalp surface. Invasion of the hair (Figure 31-6) may be associated with mycelial growth and spore formation in the interior of the hair shaft and is referred to as ***endothrix.*** Endothrix hairs are frequently infected by *T. soudanense, T. tonsurans, T. violaceum,* and *T. yaoundii.* Mycelial growth and spore formation on the surface of the hair shaft is referred to as ***ectothrix*** and is caused by such agents as *M. canis, M. audouinii, M. distortum,* and *M. ferrugineum.* The agents of endothrix produce a green fluorescence when exposed to ultraviolet light (Wood's lamp).

Tinea pedis (ringworm of the foot)

Tinea pedis, also called ***athlete's foot,*** is caused by species of *T. rubrum, T. mentagrophytes,* and *E. floccosum.* Disease is associated with wearing shoes, which provides a dark, humid environment, ideal for growth of dermatophytes. Lesions appear primarily between the toes and on the soles (Figure 31-7).

Tinea favosa (ringworm of the scalp and torso)

Tinea favosa is a severe form of ringworm (Figure 31-8) that when acquired during infancy and left untreated may lead to chronic disease characterized by heavy cup-shaped crusts ***(scutulae).*** Lesions on the scalp may heal with scarring and permanent loss of hair ***(alopecia).*** Flavus is seldom seen in the United States, and in countries where it is endemic the disease is less severe. *T. schoenleinii, T. violaceum,* and *M. gypseum* are the principal agents of disease.

SUBCUTANEOUS MYCOSES

The agents of subcutaneous mycoses are not invasive and require other factors to colonize subcutaneous tissue. Infections are caused by penetration of sharp objects contaminated with soil and are observed more often in countries where shoes are not worn and in persons whose occupation places them in contact with soil or objects contaminated with soil. Subcutaneous diseases are slow to develop, often taking years before manifestations are recognized. Treatment may sometimes require surgery plus antifungal agents such as amphotericin B. Sporotrichosis, the most common subcutaneous disease in the United States, is briefly discussed, and the remainder of the subcutaneous infections are outlined in Table 31-1.

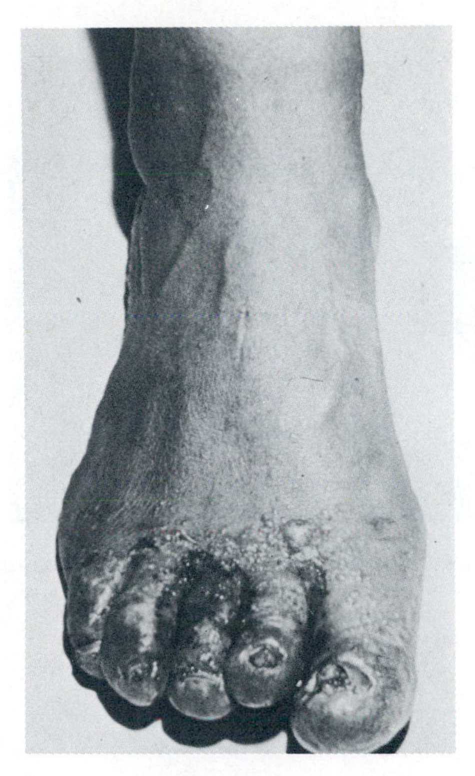

Figure 31-7 Tinea pedis (ringworm of the foot).
From Dolan, C.T., et al. Atlas of clinical mycology, *American Society of Clinical Pathologists, Chicago, 1975. Used by permission.*

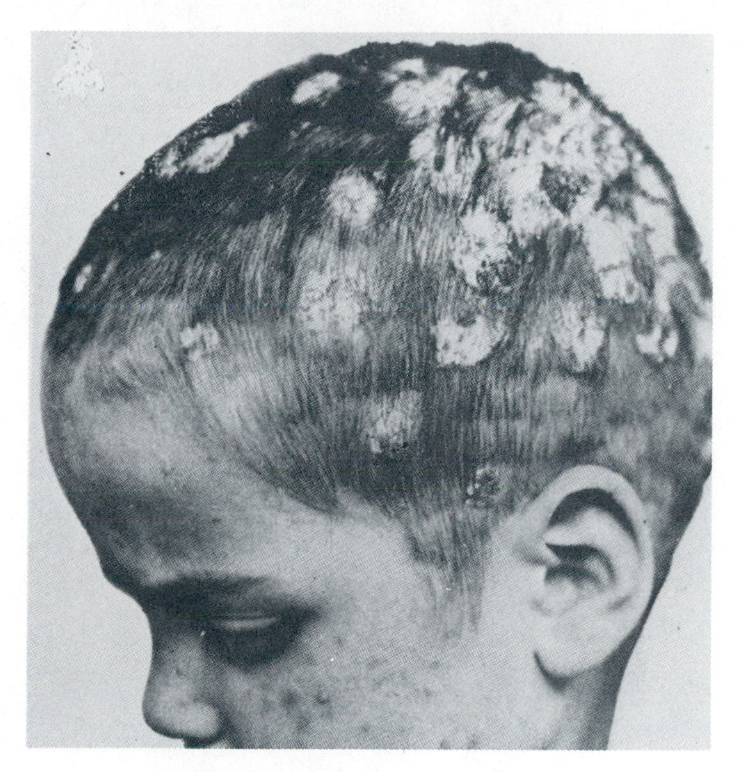

Figure 31-8 Tinea favosa (ringworm of scalp).
From Lewis, G.M., et al. An introduction to medical mycology *(4th ed.). Year Book Medical Publishers, Inc., Chicago, 1958.*

Table 31-1 Characteristics of the subcutaneous mycoses

DISEASE	MOST COMMON ETIOLOGICAL AGENT	GEOGRAPHIC LOCATION	TRANSMISSION	CHARACTERISTICS OF DISEASE
Sporotrichosis	*Sporotrichium schenckii*	Worldwide, but highest in United States, Mexico, South America and Africa	Most infections arise from contact with plants	Subcutaneous abscesses in local lymphatics
Lobomycosis	*Loboa loboi*	Northern South America	Contact with plants or soil	Chronic infection, fibrous tumors appear on feet, legs, and face (Figure 31-11)
Mycetoma (maduromycosis)	*Petrillidium boydii* and other species	Worldwide, especially the tropics	Wound contaminated with soil	Abscess formation usually of foot, chronic (Figure 31-12)
Chromomycosis	*Phialophora verrucosa* and other dematiaceous fungi	Worldwide, especially the tropics	Wound contaminated by soil	Most infections involve legs and feet (Figure 31-13); disease develops slowly; nodules appear that later enlarge
Rhinosporidosis	*Rhinosporidium seeberi*	India, South America, and United States	Associated with stagnant water	Polyps appear on mucosal surfaces, particularly nose and soft palate

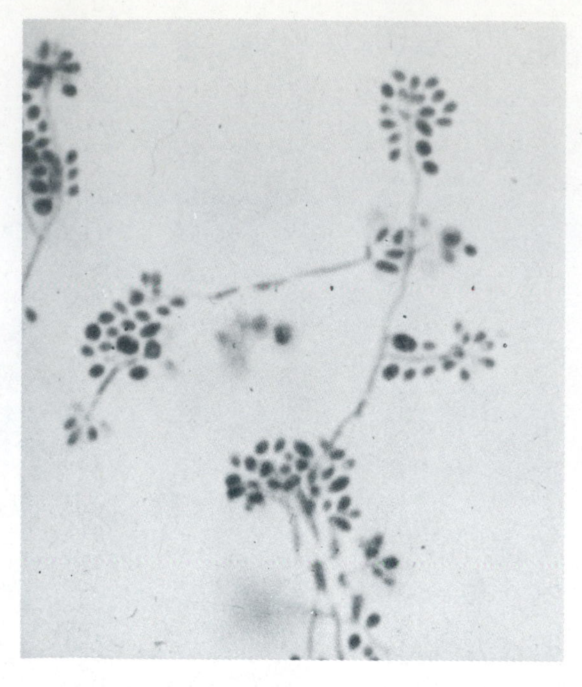

Figure 31-9 *Sporothrix schenckii.* Culture demonstrating clusters of microconidia along the sides and at the tips of conidiophores.
From Dolan, C.T., et al. Atlas of clinical mycology, *American Society of Clinical Pathologists, 1975. Used by permission.*

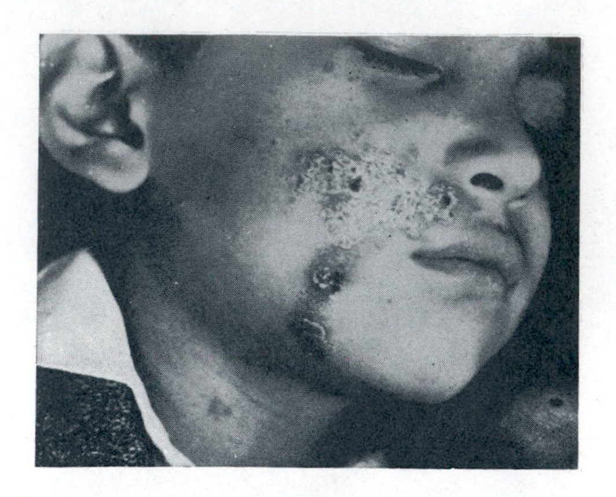

Figure 31-10 Sporotrichosis. Lesions on cheek with extension along lymphatics.
Courtesy Antonio Gonzales-Ochoa. In Anderson, W.A.D., and Kissane, J.M. (Eds.). Pathology *(7th ed.). The C.V. Mosby Co., St. Louis, 1977.*

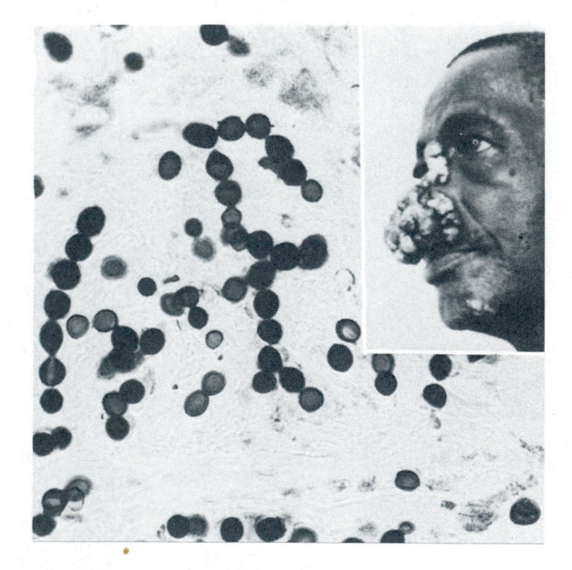

Figure 31-11 Lobomycosis. Strings of fungal cells. *Inset,* 60-year-old man with disseminated lesions of skin.
Air Force Institute of Pathology, AFIP 66-9099.

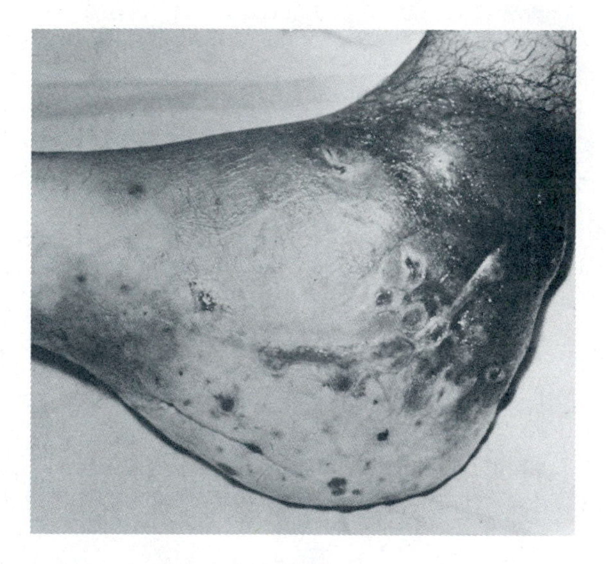

Figure 31-12 Mycetoma (maduromycosis).
From Dolan, C.T., et al. Atlas of clinical mycology, *American Society of Clinical Pathologists, Chicago, 1975. Used by permission.*

Sporotrichosis

Sporotrichosis is caused by the agent *Sporothrix schenckii* (Figure 31-9). Infection by a puncture wound from a thorn or wood splinter results in the formation of ulcerated areas within a period of 7 to 12 days (Figure 31-10). Nodules, which develop beneath the skin indicating lymph node involvement, appear purplish and may extend along the entire length of the arm. The lesions can remain localized for several years, but on rare occasions the fungus spreads into the circulatory system, causing widespread lesions over the body. Treatment of cutaneous sporotrichosis includes the use of potassium iodide, whereas amphotericin B is used for disseminated disease.

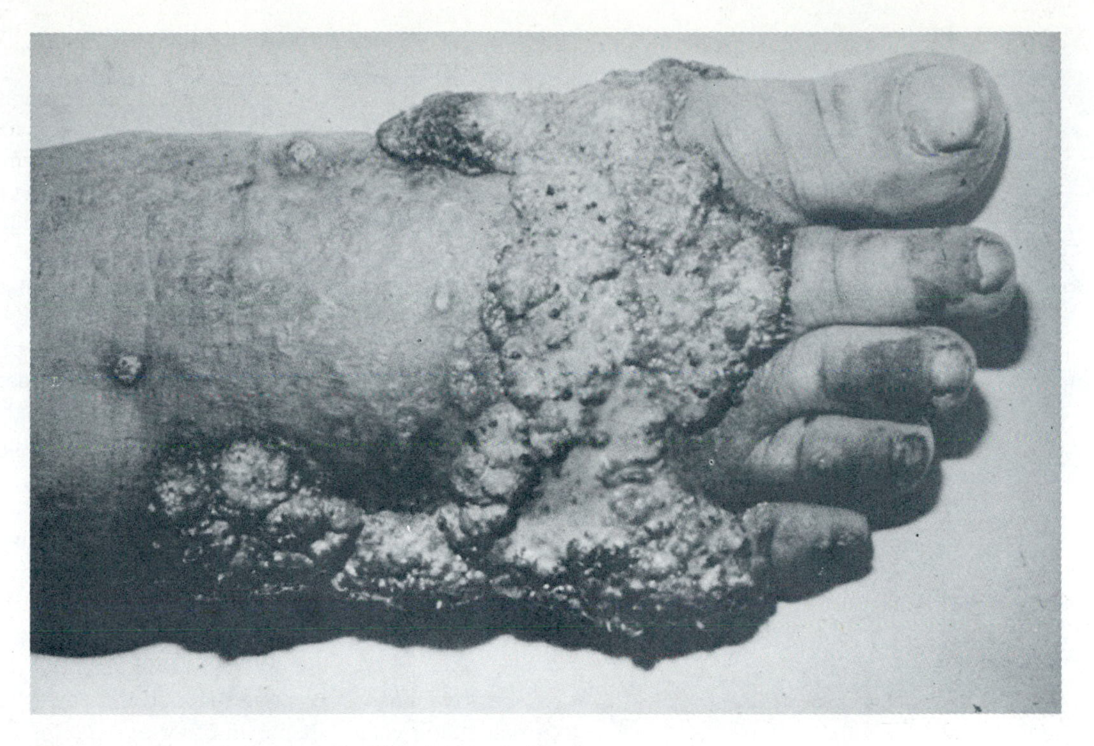

Figure 31-13 Chromomycosis. Chronic lesions of foot.
From Dolan, C.T., et al. Atlas of clinical mycology. *American Society of Clinical Pathologists, Chicago, 1975. Used by permission.*

SYSTEMIC MYCOSES

The agents of systemic mycoses are dimorphic except for *Cryptococcus neoformans,* which has yeast forms only. Infections result from inhalation of spores found in the soil, and if enough spores are inhaled, disease can ensue, even in the healthy host. Infection commences by development of lung lesions that usually resolve on their own with no further damage, but some diseases become chronic, and it is not unusual to confuse the disease with tuberculosis. The fungal agents in a few individuals may become disseminated into the bloodstream to infect other organs—the organs affected depending on the fungal agent involved. Direct microscopic examination of clinical specimens such as sputum, pus, exudates, cerebrospinal fluid, urine, or excised tissue for the fungal pathogen can save considerable time in the diagnosis of systemic disease. Selective and enriched media with antibiotics are also used to isolate the suspected agent and to examine for dimorphic properties at the required incubation temperatures.

The treatment of systemic disease includes amphotericin B alone or in combination with 5-fluorocytosine. One of the most common systemic mycoses in the United States—histoplasmosis—is discussed here; other systemic diseases are outlined in Table 31-2.

Histoplasmosis

The agent of histoplasmosis is *Histoplasma capsulatum* (Figure 31-14). It is found worldwide in the soil but is localized in areas that have been enriched with bird excreta, particularly from birds such as starlings, chickens, crows (but not pigeons), as well as bats. Barnyards are ideal areas for isolation of the fungus from the soil.

Histoplasmosis is endemic in the Mississippi River Valley of the United States. Infection results from the inhalation of spores, and the severity of disease is related to the number of spores inhaled. It is estimated that each year over 500,000 individuals in the United States become infected with *H. capsulatum;* 50,000 to 200,000 become ill and demonstrate some symptoms, and 3000 require hospitalization. Outbreaks in tropical countries are often associated with chicken coops or with dwellings or caves harboring bats. Both bird and bat guano are important for the growth of the fungus, but in nature only the bat demonstrates disease. The internal temperature of birds (approximately 42° C) appears to be above the maximum temperature for growth of the fungus.

Histoplasmosis is a disease of the macrophage phagocytic system and thus may involve many organs of the body. Most diseases (95%) cause only mild symptoms such as nonproductive cough, fever, and joint pain, and others afflicted

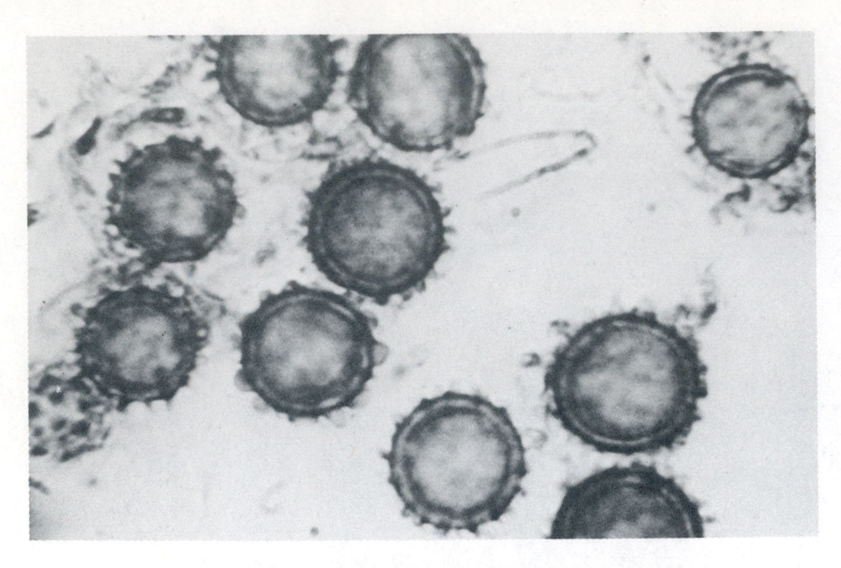

Figure 31-14 *Histoplasma capsulatum.* Microscopic appearance of mycelial phase of *H. capsulatum* demonstrating large tuberculate macroconidia, which are round, heavy-walled spores.

From Dolan, C.T., et al. Atlas of clinical mycology. *American Society of Clinical Pathologists, Chicago, 1975. Used by permission.*

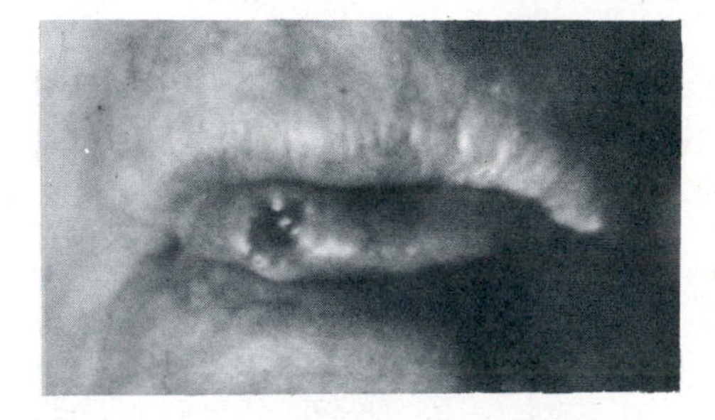

Figure 31-15 Histoplasmosis. Ulcerated lesion of histoplasmosis of tongue in chronic disseminated form in adult.

From Baker, R.D. In Anderson, W.A.D. and Kissane, J.M. (Eds.). Pathology *(7th ed.). The C.V. Mosby Co., St. Louis, 1977.*

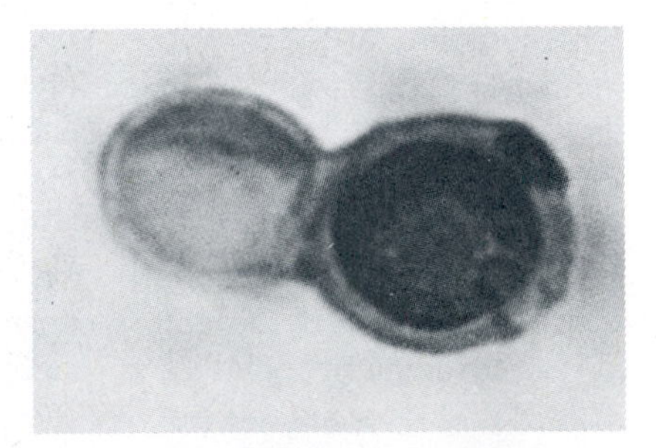

Figure 31-16 Blastomyces dermatitidis. Stained pus specimen showing charcteristic thick wall of mother cell and broad base bud.

From L. Laskowski. In Boyd, R.F., and Marr, J.J. Medical microbiology. *Little, Brown & Co., Boston, 1980.*

are asymptomatic. Lesions may appear in the lungs and show calcification and scarring, but most resolve on their own. On rare occasions there is dissemination of the disease involving the organs of the phagocytic system. Disseminated disease is recognized by nodular lesions of the mouth (Figure 31-15) and genitalia. The highest incidence of fatalities from disseminated disease appears among infants. A chronic form of the disease causes lesions that resemble tuberculosis but is actually an accentuation of the primary lesion seen in the acute form of the disease. The fungus is also disseminated during the chronic stage, and lesions may appear in other organs.

Most individuals with histoplasmosis exhibit a hypersensitive state, which can be demonstrated by a ***histoplasmin skin test.*** Antigens from *H. capsulatum* are injected intradermally into the forearm, where an area of induration and erythema 48 to 72 hours after injection indicates the sensitive state. A positive reaction in infants has diagnostic value but not in other age groups because the test cannot distinguish between past or present disease (unless the patient's test was negative before the onset of symptoms). The test has more value as an indicator of endemic disease in geographic areas.

OPPORTUNISTIC MYCOSES

Indigenous microbial species and certain soil saprobes cause disease only when the host's resistance is imperiled by underlying disease or when chemical or physical therapy is being used to treat certain pathological conditions. These microbial agents are called ***opportunistic pathogens. Candidiasis*** is the most frequently reported fungal disease that is a cause of death, but other diseases such as aspergillosis, zygomycosis, torulopsosis, and cryptococcosis are increasing in frequency.

Candidiasis

The agents of candidiasis are species of *Candida,* (Figure 31-19) which exist in nature on fruits and vegetables and can also be found as commensals on the skin, in the oral cavity, in the intestinal tract, and in the genitourinary tract of humans. Several species of *Candida* are indigenous to humans, but the most frequently isolated pathogen is *Candida albicans,* which may be found in the intestinal tract of up to 40% of the healthy population. Candidiasis takes several forms, depend-

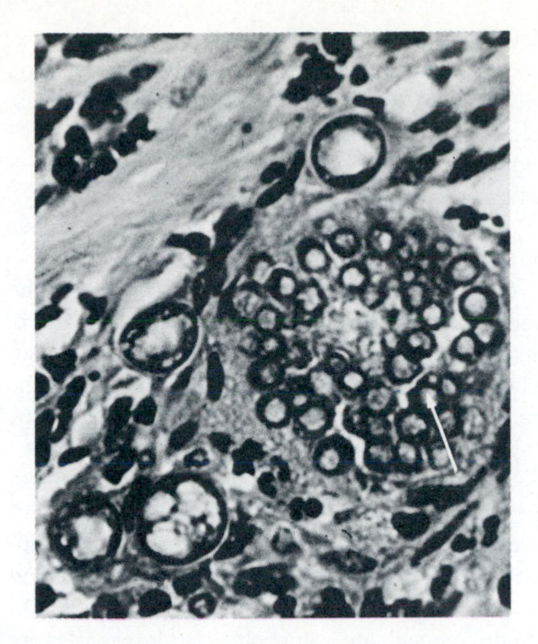

Figure 31-17 *Coccidioides immitis.* Microscopic appearance of *C. immitis* in tissue. (×900.) Fungus appears as spherical thick-walled structure filled with numerous endospores *(arrow).*

From Larsh, H.W., and Goodman, N.L. In Manual of clinical microbiology (3rd ed.). American Society for Microbiology, Washington, D.C., 1980.

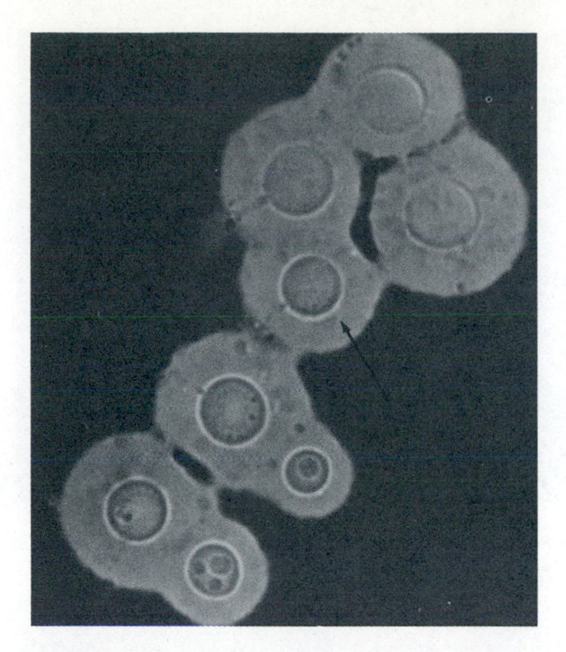

Figure 31-18 Cryptococcus neoformans. Nigrosin stain of *C. neoformans* demonstrating large casule *(arrow)* surrounding fungus.

From Dykstra, M. infect. Immun. ***16:****129, 1977.*

Table 31-2 Characteristics of the systemic mycoses

DISEASE	ETIOLOGICAL AGENT	TRANSMISSION	PRIMARY SITE OF INFECTION	CHARACTERISTICS OF THE AGENT
Blastomycosis	*Blastomyces dermatitidis*	Inhalation of spores spores from soil	Lung but disseminated with manifestations in skin and subcutaneous tissue (see Plate 43)	In tissue yeasts have thick wall (Figure 31-16)
Paracoccidioidomycosis	*Paracoccidioides brasilensis*	Inhalation of spores	Lung, but lesions more prominent in oral cavity and nasal mucosa; systemic spread involves spleen, intestines, and adrenal glands	Yeasts show multiple budding in tissue and may be 30 to 60 μm in diameter
Coccidioidomycosis	*Coccidiodes immitis*	Inhalation of only a few spores will produce disease; dark-skinned people more susceptible	Lung; dissemination likely if many spores are inhaled with lesions in many organs; meningitis fatal in dark-skinned people	Most virulent of systemic agents. Extremely hazardous in laboratory since arthrospores are easily spread; yeast phase is a spherule containing endospores (Figure 31-17)
Cryptococcosis	*Cryptococcus neoformans*	Inhalation of spores from soil or dust contaminated with pigeon droppings	Lung; dissemination to other tissues such as skin, bones, viscera, and central nervous system	Is not dimorphic, only yeast phase present; produces a capsule (Figure 31-18)
Histoplasmosis	*Histoplasma capsulatum*	Inhalation of spores found associated with bird dung, especially chickens and turkeys	Lung; dissemination to meninges, adrenal glands, heart, and bone	Mycelial stage shows tuberculate spores

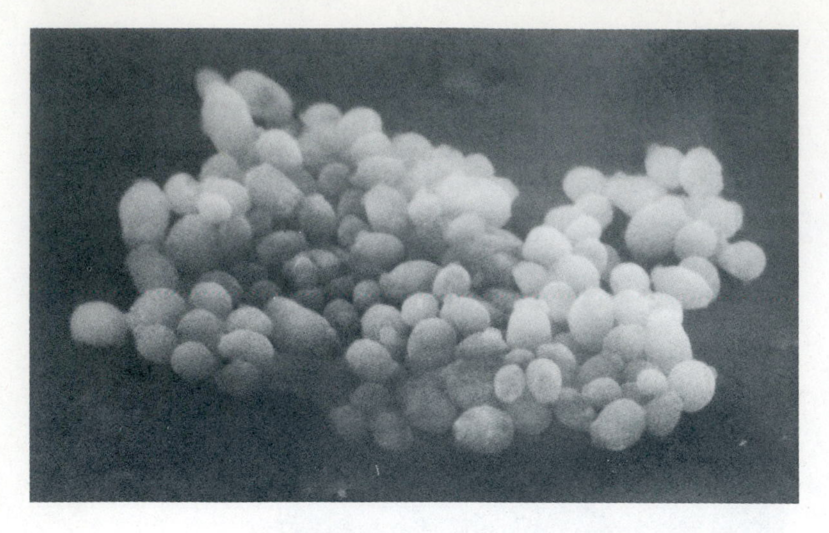

Figure 31-19 Electron micrograph of *Candida albicans*.
*From King, R. Infect. Immun. **27**:667, 1980.*

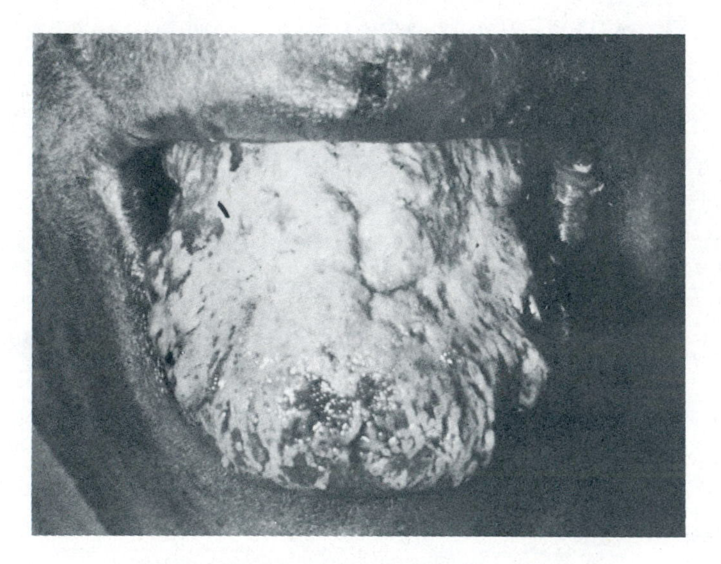

Figure 31-20 Oral thrush demonstrating white patches on surface of tongue.
From Dolan, C.T., et al. Atlas of clinical mycology, *American Society of Clinical Pathologists, Chicago, 1975. Used by permission.*

ing on the circumstances that have compromised the patient and include the following.

Cutaneous candidiasis. Cutaneous candidal diseases include oral thrush, onychia and paronychia, candida intertrigo, and vulvovaginitis. ***Oral thrush*** is characterized by the formation of gray white patches or pseudomembrane that may cover the tongue (Figure 31-20), soft palate, or other oral surfaces. The pseudomembrane is composed of spherical yeast cells, hyphal filaments (pseudohyphae) of *C. albicans* and cellular debris and may become so large that it obstructs breathing. Oral thrush is most often observed in newborns and in the debilitated and aged patient. Predisposing factors that lead to infection include low pH in the oral cavity in newborns, neoplasias, diabetes, and broad-spectrum antibiotic and steroid administration in older children and adults. *C. albicans* can also be transmitted to infants from mothers with genital candidiasis during passage through the birth canal.

Onychia and ***paronychia*** are infections of the subcutaneous tissue at the base of the nails of fingers or toes (Figure 31-21). Infections are usually the result of continued immersion of the hands or feet in water and are characterized by swelling, redness, and pain in the affected area.

Intertriginous candidiasis involves the moist areas of the body, such as axillae, groin, intermammary folds, interdigital spaces, and intergluteal folds. Lesions are red, scaling, or moist and may be painful, depending on their location. Diabetic, obese, and alcoholic patients are more prone to infection, but those with occupations such as barbers, bartenders, fruit canners, and dishwashers are also susceptible to infection.

Vulvovaginal candidiasis occurs primarily in pregnant women, diabetics, and patients on antimicrobial therapy or hormonal contraceptives. Lactobacilli, which are capable of controlling the population of candida in the vagina, are reduced in number or eliminated by the previously described predisposing factors. *Candida* organisms without competition proliferate and cause disease characterized by a yellow-white discharge and patches of pseudomembrane on the vaginal mucosa. The disease can also be transmitted to the sexual partner during intercourse.

Mucocutaneous candidiasis. Mucocutaneous candidiasis is a superficial disease that includes several clinical varieties of candidiasis such as vaginal, onychia and paronychia, oral, respiratory, and gastrointestinal (Figure 31-21). Disease is observed primarily in children who have defects in cell-mediated immunity, or occasionally in those with endocrine defects or leukemia. Treatment of superficial disease includes the use of such drugs as clotrimazole, miconizole, or nystatin.

Systemic candidiasis. Systemic candidiasis is the most frequent cause of death by any fungal agent. Predisposing factors that have increased the incidence of this disease are the use of antimicrobial and immunosuppressive agents, hyperalimentation fluids, organ transplantation, heroin abuse, and abdominal and heart surgeries. Once species of *Candida* have penetrated the bloodstream, any number of organs may be colo-

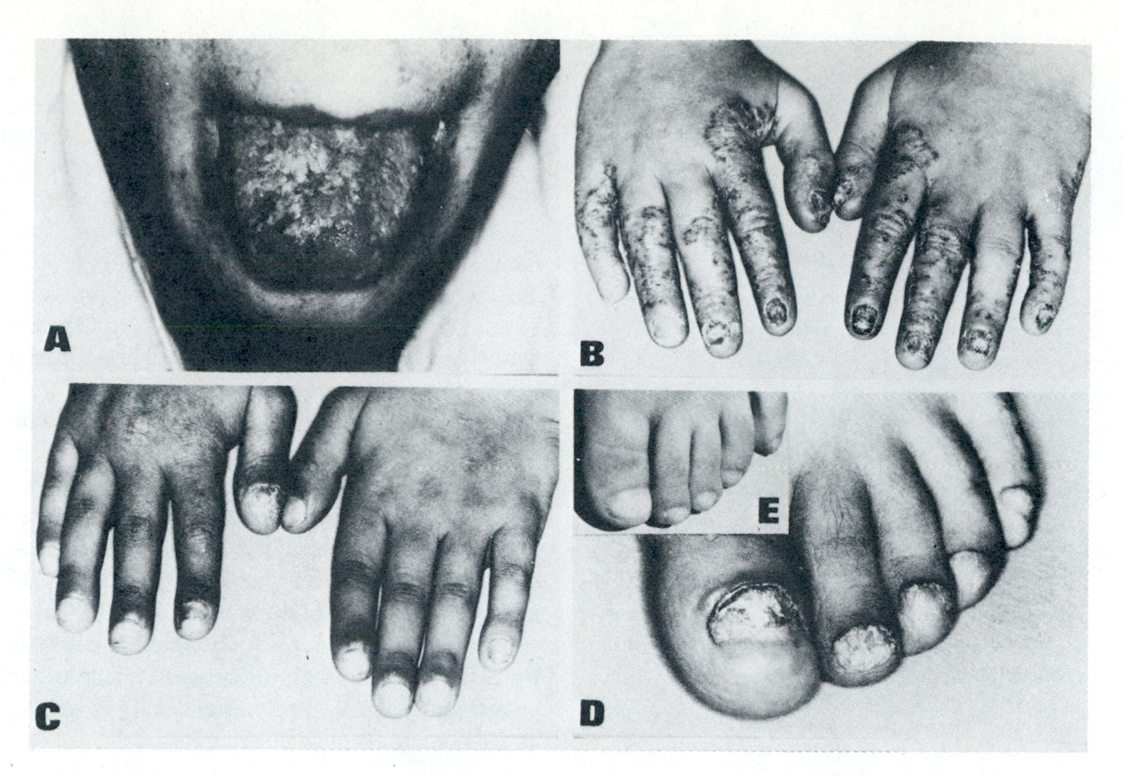

Figure 31-21 Chronic mucocutaneous candidiasis in children. **A.** Oral thrush **B.** Skin and nail lesions **C.** Paronychia **D.** Onychia. **E.** Toes of patients in **D** after treatment.
From Drouhet, E., and Dupont, B. Rev. Infect. Dis.***2****(4):606, 1980.*

nized and become diseased. The prognosis for systemic candidiasis is poor, and the death rate may be as high as 85% for those who are immunologically compromised. The diagnosis of systemic candidiasis is difficult, and examination of blood cultures and antibody determinations have been unreliable. Sometimes blood cultures are negative in disseminated disease, and positive cultures do not necessarily reflect invasive disease. Some patients are also incapable of generating an antibody response, making serodiagnosis useless. Constant monitoring, employing a variety of diagnostic techniques, should be employed for patients who are prone to systemic infections by *Candida* organisms. Treatment for systemic disease includes such drugs as amphotericin B and 5-fluorocytosine.

Aspergillosis

Aspergillosis is a fungal disease caused by species of *Aspergillus*. The aspergilli are filamentous fungi whose distinguishing characteristic is the swollen conidiophore (Figure 31-22). They are found primarily in decaying vegetation and frequently in homes or storage areas where grain is improperly stored. Three types of disease are caused by the aspergilli: (1) allergy such as bronchopulmonary aspergillosis by sensitization from

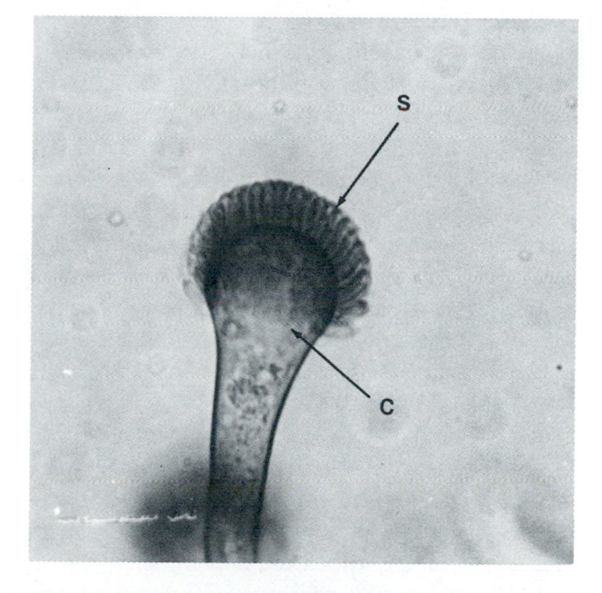

Figure 31-22 Microscopic appearance of swollen conidiophore *(C)* of *Aspergillus fumigatus*. *(×1170.) S,* single row of sterigmata.
From Austwick, P.K.C., and Longbottom, J.L. In Manual of clinical microbiology (3rd ed.). American Society for Microbiology, Washington, D.C., 1980.

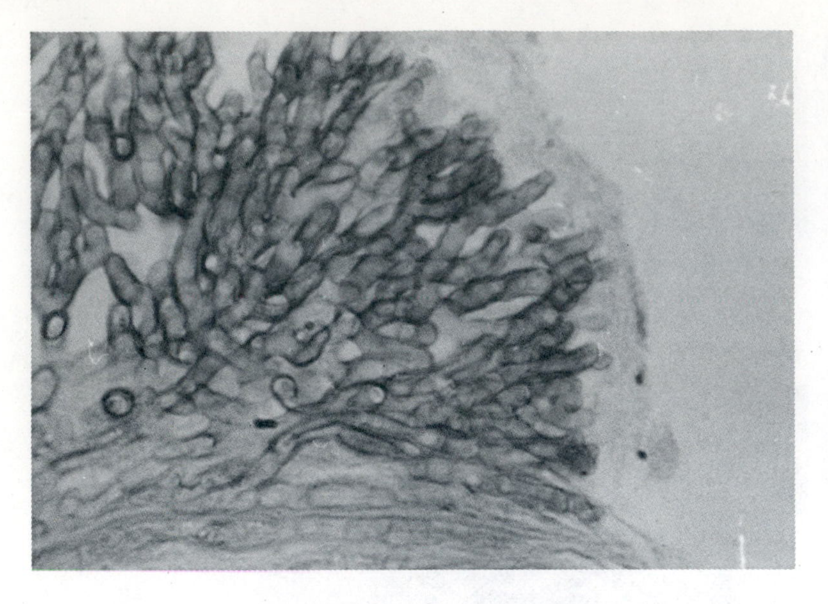

Figure 31-23 *Aspergillus fumigatus.* Microscopic appearance of hyphae at margin of human pulmonary aspergilloma (fungus ball). (×365.)
From Austwick, P.K.C., and Longbottom, J.L. In Manual of clinical microbiology *(3rd ed.). American Society for Microbiology, Washington, D.C., 1980.*

exposure to fungal antigens; (2) infection by species of *Aspergillus* that invade superficial as well as deep tissue such as the lungs (pulmonary aspergillosis) or ears; and (3) toxicosis resulting from the ingestion of mycotoxins produced by species of *Aspergillus.*

Pulmonary aspergillosis may frequently occur as an allergic response in those with occupations that bring them in contact with grains, such as farmers (farmer's lung) and malt workers (malt worker's lung). *Aspergillus fumigatus, A. clavatus* and *A. flavus* have been found to be the primary agents of disease whose symptoms include a dry cough and low fever. ***Aspergilloma*** is a chronic disease that often arises as a secondary disease following tuberculosis. The hyphae appear in the old tubercular cavity as a dense mass called a fungus ball (Figure 31-23).

Aspergillus species such as *A. terreus* occasionally cause superficial or subcutaneous diseases. Disease of the ear, for example, is characterized by fungal growth on the earwax leading to inflammation and hearing impairment.

Systemic aspergillosis is rare and occurs primarily in those with neoplasias and those undergoing heart surgery and renal transplantation. Systemic invasive aspergillosis is difficult to diagnose and if left untreated can cause death. Serologic immunodiagnosis represents the best technique for detecting disease but has not been perfected.

Zygomycosis (phycomycosis)

Genera of the class Zygomycetes are agents of disease for humans and lower animals called ***zygomycosis*** (originally called phycomycosis). Two orders of the class Zygomycetes are Mucorales and Entomophthorales. Three genera of the order Mucorales—*Rhizopus, Mucor,* and *Absidia*—are the most frequently isolated agents of zygomycosis in the United States. Predisposing factors that lead to disease are leukemia, diabetes acidosis, and corticosteroid therapy.

Zygomycosis caused by species in the order Mucorales ***(mucormycosis)*** occurs most commonly in those suffering from uncontrolled diabetes mellitus. The appearance of acidosis (whether from diabetes or other conditions) is an important condition for growth of the fungus. The most common form of the disease seen in diabetics is ***rhinocerebral,*** in which infection begins in the paranasal sinuses and spreads to the orbital area (Figure 31-24). The mortality rate is as high as 50%, even in those under treatment.

Cryptococcosis and torulopsosis

Cryptococcosis and torulopsosis are opportunistic yeast diseases that occur in patients compromised by neoplasias, leukemia, or corticosteroid and antibiotic therapy. Torulopsosis is caused by *Torulopsis (Candida) glabrata,* which is now considered synonymous with *Candida.* It is a cause of disease of the kidneys, lung, and meninges, particularly in the immunosuppressed patient.

Cryptococcosis is a disease caused by *Cryptococcus neoformans,* which is described under systemic mycoses (Table 31-2); however, the organism is also opportunistic in the compromised patient. The organism is rapidly disseminated in pigeon droppings in which it rapidly metabolizes urea and creatine. Windowsills and other pigeon roosts can be contaminated by large numbers of the organism, which is unusually resistant to drying. Diseases of the central nervous system are the most frequently diagnosed form of the disease, often with accompanying meningitis. The organism is easily discernible because of its large capsule (Figure 31-18).

MYCOTOXICOSIS AND MYCETISMUS

Toxins of fungal origin cause conditions referred to as mycotoxicosis and mycetismus. Mycotoxicosis results from the ingestion of toxins formed in foods, such as grains, that have been damaged

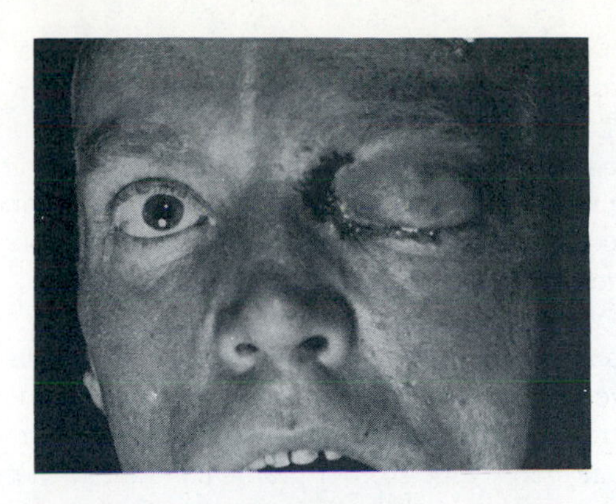

Figure 31-24 Mucormycosis of left orbital caused by a species of *Rhizopus* in diabetic patient.

From Baker, R.D. In Anderson, W.A.D., and Kissane, J.M. (Eds.). Pathology *(7th ed.). The C.V. Mosby Co., St. Louis, 1977.*

Figure 31-25 Structure of toxin, aflatoxin, produced by *Aspergillus flavus.*

because of the growth of fungi. Grain crops can become moldy when improperly stored and when conditions such as temperature and humidity are suitable for fungal growth. Mycotoxicosis is seldom reported in the United States, but many cases are reported from countries where grains are improperly stored. Many cases were reported in Russia following the Second World War, where grains left in the field during the winter were contaminated by toxin-producing fungi. The organisms most responsible for mycotoxicosis are species belonging to the genera *Penicillium, Aspergillus, Fusarium,* and *Streptomyces.* The first identification of a toxin from contaminated grains was not accomplished until 1960. In 1959 100,000 turkeys and other birds had perished under mysterious circumstances in England. It was discovered that the birds had ingested Brazilian peanut meal that had been contaminated with a toxin called ***aflatoxin*** (Figure 31-25), which is produced by *Aspergillus flavus.* Experiments with aflatoxin later showed that the toxin acts as a carcinogen and affects the liver and kidneys. There appears to be a correlation between lifetime ingestion of grains contaminated with aflatoxin and cancer of the liver in areas of Southeast Asia.

Mycetismus is known as mushroom poisoning. Most of the toxic species belong to the genus *Amanita,* one species of which is called the "angel of death" (Plate 13). The symptoms of toxicosis include nausea, vomiting, and other gastrointestinal disturbances. Some species affect the nervous system, causing confusion, delirium, and visual disturbances and can cause death. Some mushrooms produce hallucinogenic effects and were used in religious ceremonies for hundreds of years by certain Indian tribes of Mexico and South America. Treatment should be directed toward removing the fungus from the gastrointestinal tract by emetics or cathartics.

SUMMARY

Fungal diseases are called *mycoses* and are divided into the following types: systemic, superficial, subcutaneous, cutaneous, and opportunistic. Nearly all fungal diseases are acquired by contact with spores in the soil or by inhalation of spores dispersed in the air. The ability of the fungi to cause disease is related to the number of spores to which the host is exposed and the state of health of the individual. Fungal diseases in the healthy person are rare. Contact with a fungal agent usually results in the agent being walled-off in tissue as a result of cell-mediated immune response.

Systemic mycoses are caused by organisms that can penetrate epithelial tissue and can be disseminated throughout the body. Most infections are acquired by inhalation of spores resulting initially in respiratory symptoms that mimic tuberculosis. Histoplasmosis is the most common systemic disease in the United States and is acquired by contact with soil enriched with excreta from birds such as chickens. Most diseases are acute and seldom fatal.

Superficial mycoses, such as ringworm (tineas), are caused by agents called dermatophytes. The microbial agents colonize keratinized tissue on any body surface (scalp, beard, groin, feet). Many of the microorganisms parasitize animals with humans becoming accidentally infected. The most common source of infections for humans is via human-to-human spread. Some of the dermatophytes can cause allergic responses (ids).

Subcutaneous mycoses are often the result of puncture wounds. Diseases may take up to several years to develop. Sporotrichosis is the most common disease and is characterized by lymphatic nodule enlargement near the site of infection.

Opportunistic mycoses occur when the host has been compromised by diabetes, immunodeficiencies, antibiotic therapy, or underlying disease. Diseases result from the overgrowth of indigenous species, particularly species of *Candida,* which are commensals in the oral cavity, intestinal tract, vagina, and skin. Diseases involve moist areas of the body including the mucous membranes and nails.

Treatment of fungal diseases is divided between two types of agents: topical and systemic. Topical agents include nystatin, tolnaftate, and miconazole, and systemic agents include amphotericin B and flucytosine.

Fungal agents can be toxic when ingested, for example, mushrooms, and the condition is called mycetismus. Other fungal agents can produce toxic products when they grow in various stored grain crops. Ingestion of grains containing these toxic products, such as aflatoxin, can cause serious disease (mycotoxicosis).

Study questions

1. What are dermatophytic fungi?
2. Can you give a reasonable explanation as to why most fungal diseases are not contagious?
3. How are most fungal infections acquired? Does occupation play an important role?
4. What groups of fungi cause opportunistic diseases?
5. What is a common reason for indigenous fungal species such as *Candida* causing diseases in the oral cavity, genitourinary tract, and intestinal tract?
6. Are the virulence factors of fungi similar to those of bacteria? Explain?
7. What factors determine ones susceptibility to fungal diseases?
8. What kind of immune response does the host employ to ward off fungal diseases?
9. What morphological property is the most important in the laboratory identification of fungi?
10. What are tineas? List some common sites of infection by tineas.
11. What is the most common method for acquiring systemic mycoses?
12. What is the relationship of birds to histoplasmosis?
13. Describe the various conditions caused by the commensal *Candida albicans.*
14. What is the distinction between mycotoxicosis and mycetismus?
15. What are ids?

Selected readings

Books

Al-Doory, Y. (Ed.), *The epidemiology of human mycotic disease.* Charles C Thomas, Springfield, Ill., 1975.

Boyd, R.F., and Marr, J.J. *Medical microbiology,* Little, Brown & Co., Boston, 1980.

Davis, B., et al. *Microbiology* (3rd ed.). Harper and Row, Publishers, Inc., 1980.

Emmons, C.W., et al. *Medical mycology.* Lea & Febiger, Philadelphia, 1977.

Freeman, B.A. *Burrow's textbook of microbioloby* (21st ed.), W.B. Saunders Co., Philadelphia, 1979.

Hoeprich, P.D. (Ed.), *Infectious diseases* (2nd ed.), Harper and Row Publishers, Inc., New York, 1977.

Mandell, G.L. (Ed.). *Principles and practice of infectious disease.* John Wiley & Sons, Inc., New York, 1979.

Odds, F.A. *Candida and candidiasis.* University Park Press, Baltimore, 1979.

Rippon, J.W. *Medical mycology* (2nd ed.). W.B. Saunders Co., Philadelphia, 1982.

Sinski, J.T. *Dermatophytes in human skin, hair and nails.* Charles C Thomas, Springfield, Ill., 1974.

Uraguchi, K., and Yamazaki, M. (Eds.). *Toxicology: biochemistry and pathology of mycotoxins.* John Wiley & Sons, Inc., New York, 1980.

Youmans, G., Paterson, P.Y. and Sommers, H.M. *The biologic and clinical basis of infectious disease* (2nd ed.) W.B. Saunders Co., Philadelphia, 1980.

Journal articles

Ahearn, D.G. Medically important yeasts. *Annu. Rev. Microbiol.* **32:**59, 1978.

Ajello, L. Systemic mycoses in modern medicine. *Contrib. Microbiol. Immunol.* **3:**2, 1977.

Chu, F.S., Mode of action of mycotoxins and related compounds. *Adv. Appld. Microbiol.* **22:**83, 1977.

Dwyer, J.M. Chronic mucocutaneous candidiasis. *Annu. Rev. Med.* **32:**491, 1981.

Goodwin, R.A., Jr,. and DesPrez, R.M. Histoplasmosis. *Am. Rev. Respir. Dis.* **117:**929, 1978.

Hoeprich, P.D., Chemotherapy of systemic fungal diseases. *Annu. Rev. Pharmacol. Toxicol.* **18:**205, 1978.

Mariat, F., Destombes, P. and Segretain, G. The mycetomas: clinical features, pathology, etiology and epidemiology. *Contrib. Microbiol. Immunol.* **4:**1, 1977.

Meyer, R.D., et al. Aspergillosis complicating neoplastic disease. *Am. J. Med.* **54:**6, 1973.

Sobel, J.D., et al. Adherence of *Candida albicans* to human vaginal and buccal epithelial cells. *J. Infect. Dis.* **143:**76, 1981.

Wogan, G.N. Mycotoxins. *Annu. Rev. Pharm.* **15:**437, 1975.

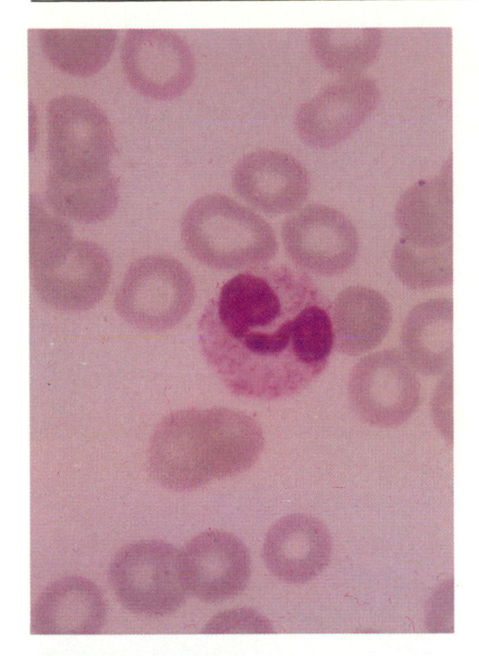

Plate 31 Band neutrophil.

From Miale, J.B. Laboratory medicine: hematology *(6th ed.). The C.V. Mosby Co., St. Louis, 1982.*

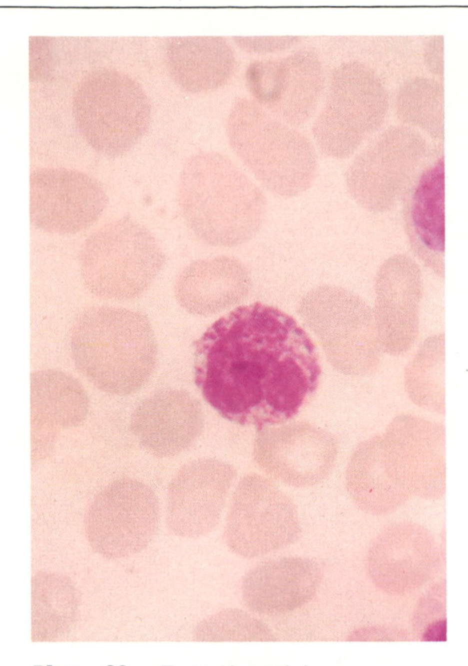

Plate 32 Typical basophil.

From Miale, J.B., Laboratory medicine: hematology *(6th ed.). The C.V. Mosby Co., St. Louis, 1982.*

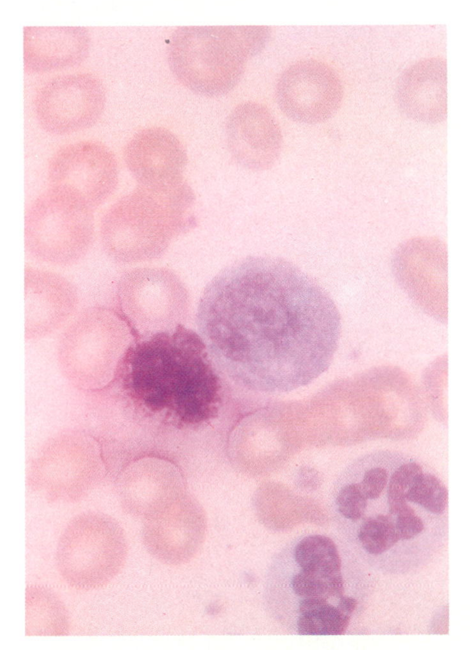

Plate 33 Basophil (left center), eosinophil (right center), segmented neutrophil (4 o'clock), and band neutrophil (5 o'clock). Note that eosinophil granules are uniform in size and have pale centers.

From Miale, J.B. Laboratory medicine: hematology *(6th ed.). The C.V. Mosby Co., St. Louis, 1982.*

Plate 34 *Neisseria* colonies, positive oxidase test.

From Finegold, S.M., and Martin, W.J. Diagnostic microbiology *(6th ed.). The C.V. Mosby Co., St. Louis, 1982.*

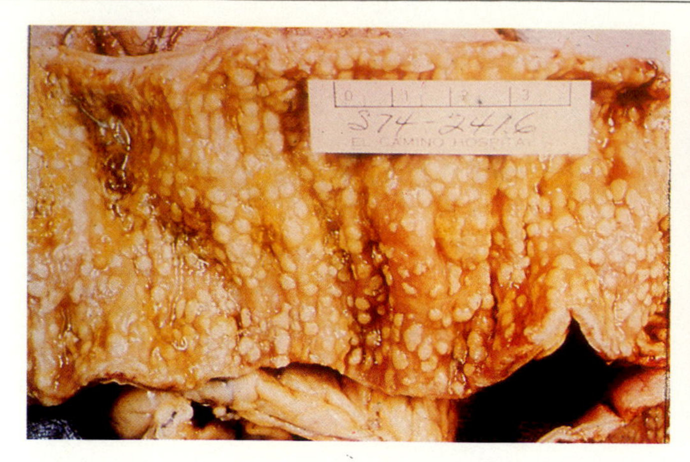

Plate 35 Autopsy specimen of colon demonstrating pseudomembranous colitis. Note numerous yellow elevated plaques.
From Finegold, S.M., and Martin, W.J. Diagnostic microbiology *(6th ed.). The C.V. Mosby Co., St. Louis, 1982.*

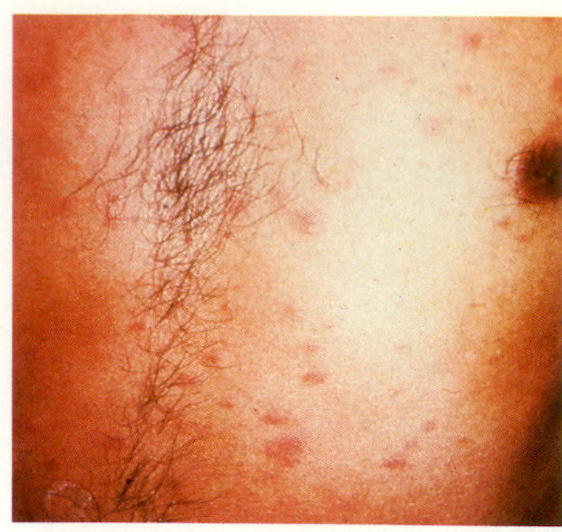

Plate 36 Rose spots on abdomen characteristic of typhoid fever.
Courtesy The Centers for Disease Control, Atlanta.

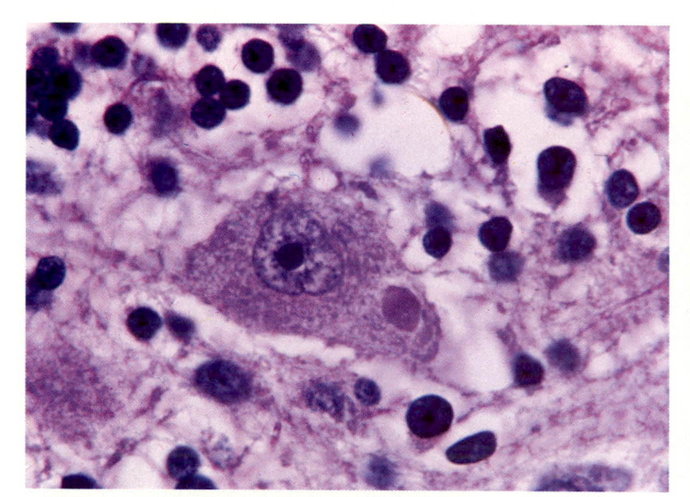

Plate 37 Rabies virus inclusion body. Purkinje cells from cerebellum of human rabies case using Tendrum's stain. (×2000.)
Courtesy F.A. Murphy.

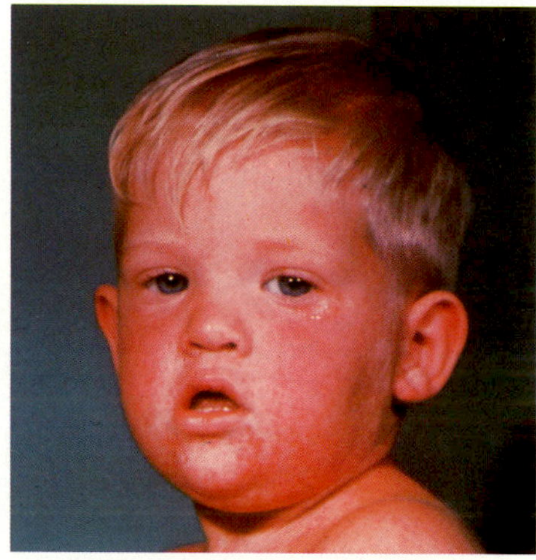

Plate 38 Measles rash.
Courtesy The Centers for Disease Control, Atlanta.

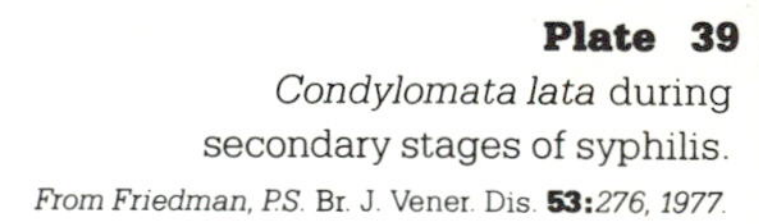

Plate 39 *Condylomata lata* during secondary stages of syphilis.
From Friedman, P.S. Br. J. Vener. Dis. **53:***276, 1977.*

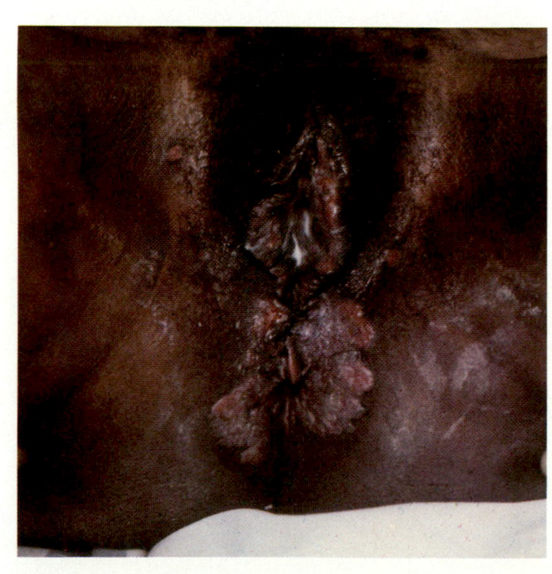

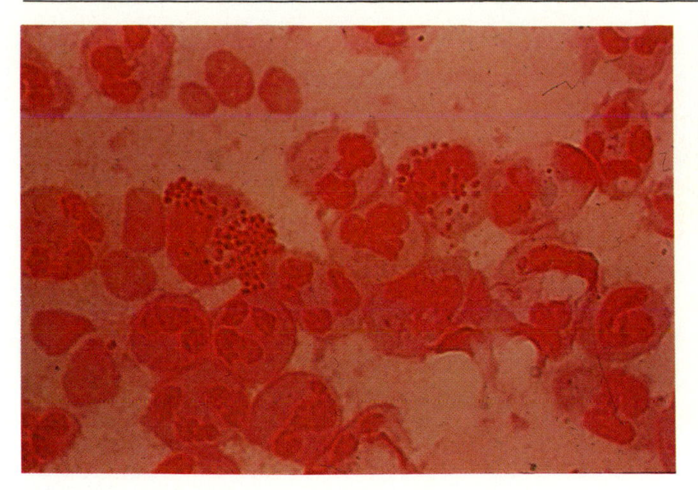

Plate 40 Urethral smear, gonorrhea; gram-negative intracellular diplococci.

From Finegold, S.M., and Martin, W.J. Diagnostic microbiology (6th ed.). The C.V. Mosby Co., St. Louis, 1982.

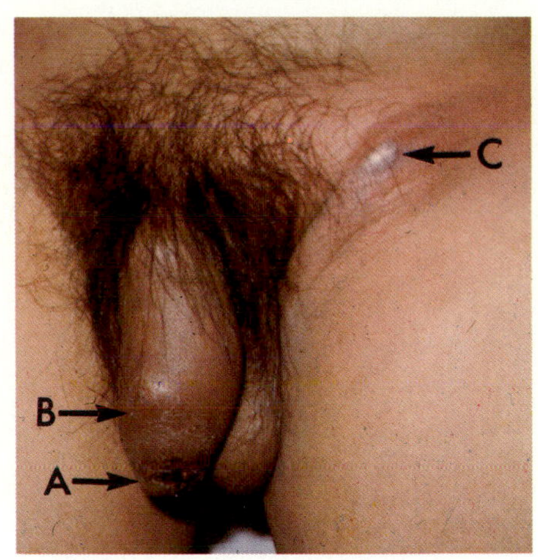

Plate 41 Chancroid in male patient. *A,* Ulceration on prepuce of penis; *B,* lymphadenopathy of dorsal penis; *C,* left inguinal bubo about to rupture.

From Hammond, G.W. Rev. Infect. Dis. **2***(6):867, 1980.*

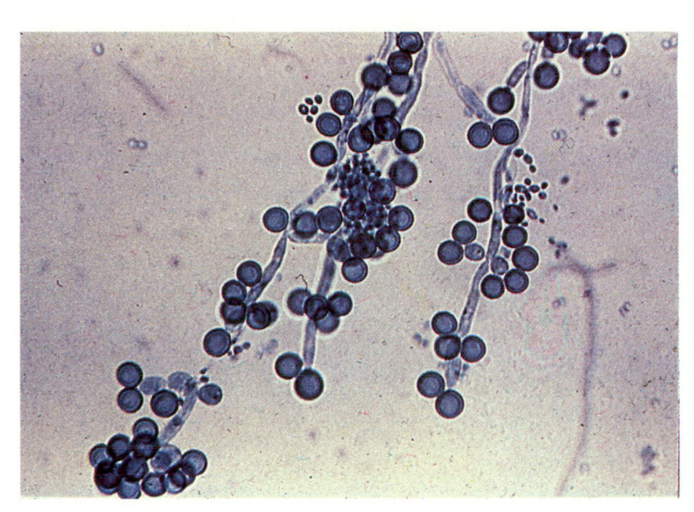

Plate 42 *Candida albicans* showing yeastlike blastospores and chlamydospores.

From Finegold, S.M., and Martin, W.J. Diagnostic microbiology (6th ed.). The C.V. Mosby Co., St. Louis, 1982.

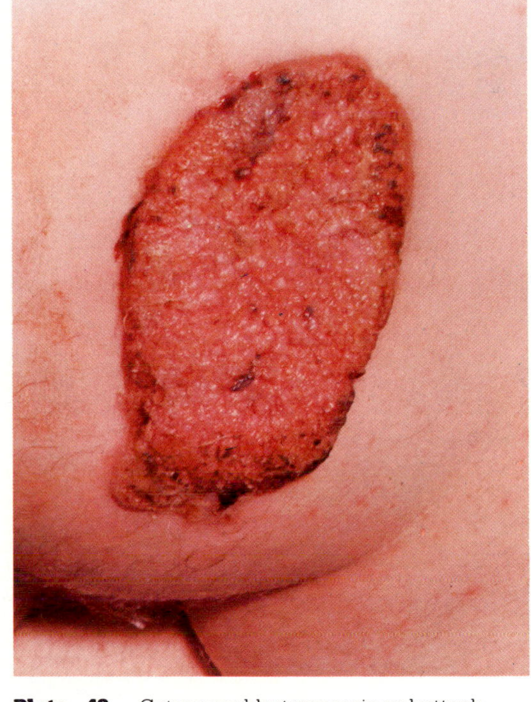

Plate 43 Cutaneous blastomycosis on buttock.

Courtesy John Utz, M.D.

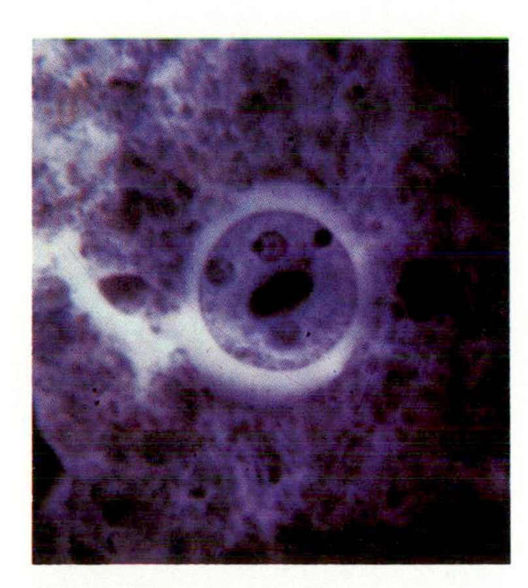

Plate 44 *Entamoeba histolytica* cyst. (× 4342.) Iron-hematoxylin stain.

From Stratford, B.C. An atlas of medical microbiology, Blackwell Scientific Publications, Oxford, 1977.

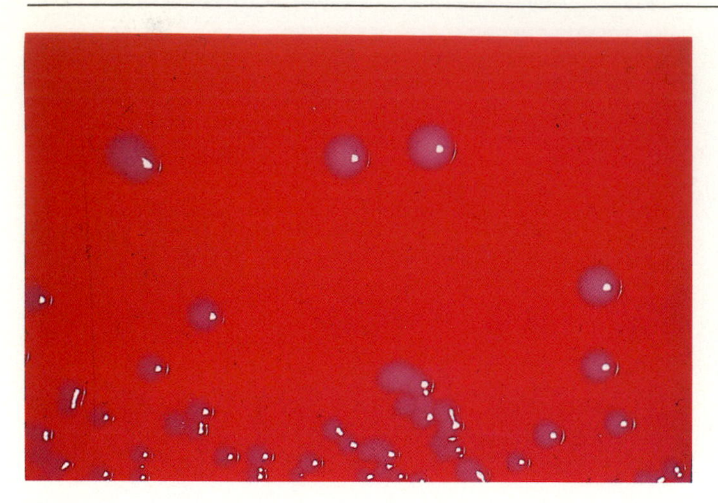

Plate 45 Bacterial colonies on agar showing no hemolysis.
From Finegold, S.M., and Martin, W.J. Diagnostic microbiology *(6th ed.).*
The C.V. Mosby Co., St. Louis, 1982.

Plate 46 Partial (alpha) hemolysis by bacterial colony embedded in agar. Compare with Plate 47.
From Finegold, S.M., and Martin, W.J. Diagnostic microbiology *(6th ed.).*
The C.V. Mosby Co., St. Louis, 1982.

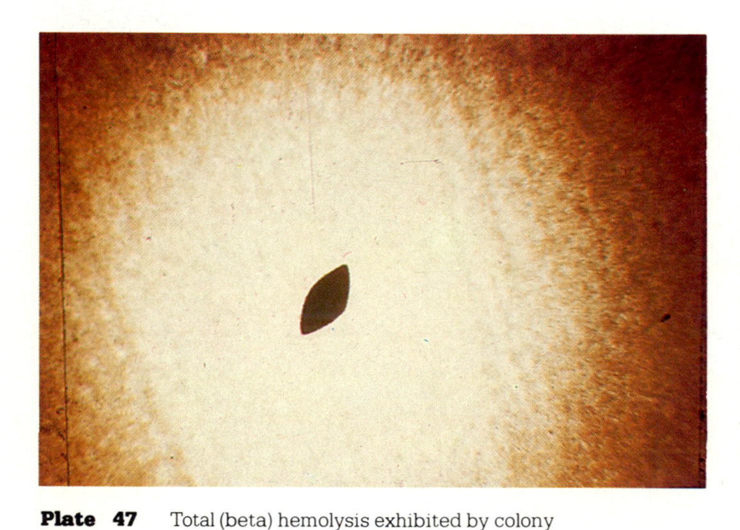

Plate 47 Total (beta) hemolysis exhibited by colony embedded in agar.
From Finegold, S.M. and Martin, W.J. Diagnostic microbiology *(6th ed.).*
The C.V. Mosby Co., St. Louis, 1982.

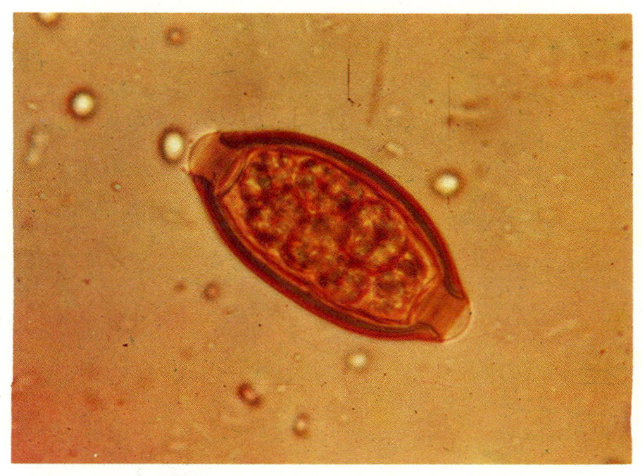

Plate 48 *Trichuris trichura* egg stained with iodine. Egg can be seen to possess thick wall and two bipolar plugs.
From Stratford, B.C. An atlas of medical microbiology.
Blackwell Scientific Publications, Oxford, 1977.

Chapter 32
ANIMAL PARASITE DISEASES

The parasites discussed up to this point are members of three of the four major groups of microorganisms, that is, the bacteria, viruses, and fungi. There are, however, parasites called animal parasites that are eukaryotic and belong to the animal kingdom. They exhibit biochemical as well as morphological differences from the other three microbial groups. The animal parasites include ***arthropods,*** such as ticks and mites, that normally live on the body surface and are referred to as ***ectoparasites.*** There are also the ***protozoa,*** which are single-celled, as well as the multicellular ***helminths,*** or worms. Protozoa and helminths both live within the host and are referred to as ***endoparasites.***

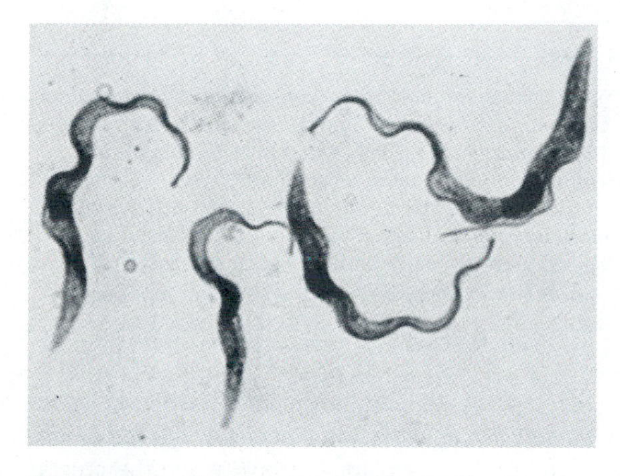

Figure 32-1 Trophozoite stage of animal parasite *Trypanosoma brucei.*
From Cunningham, I. J. Protozool. ***26:****429, 1979.*

CHARACTERISTICS OF ANIMAL PARASITE INFECTIONS

Life cycles and reproduction

During an animal parasite infection the parasite becomes adapted to its host and often goes through a complicated course of development. The stages in parasitic development are referred to as the ***life cycle.*** Many parasites colonize a number of different hosts and during evolution have developed varied mechanisms of reproduction to suit the environment. The host is termed the ***definitive host*** if the parasite undergoes sexual reproduction within it. If the parasite does not undergo sexual reproduction, the definitive host is the one that is believed to be the most important. An ***intermediate host*** is one in which the parasite does not reach a mature state. Some parasites do not undergo any developmental stage in the host, and the latter is referred to as a ***paretenic host.***

The life cycle may be relatively simple, such as in some protozoa, where there are only two stages; the motile parasitic stage, or ***trophozoite*** (Figure 32-1), and the inactive resistant form called the ***cyst*** (Plate 44). It is the cyst that is transmitted to susceptible hosts, whereas in nearly all cases, it is the trophozoite that causes the symptoms of disease. A simple life cycle also exists for some helminths; for example, some adult male and female worms live in the intestine of the host. The female deposits fertilized ***eggs,*** which pass into the feces. The eggs are ingested by humans, and ***larvae*** develop into adults to repeat the cycle in the intestine. In some parasites there may be stages involving both invertebrate and vertebrate hosts. The agent of malaria, for example, develops in the mosquito as well as in humans. The broadfish tapeworm also has a complex life cycle involving three hosts. Eggs in the intestine of humans are passed in the feces and

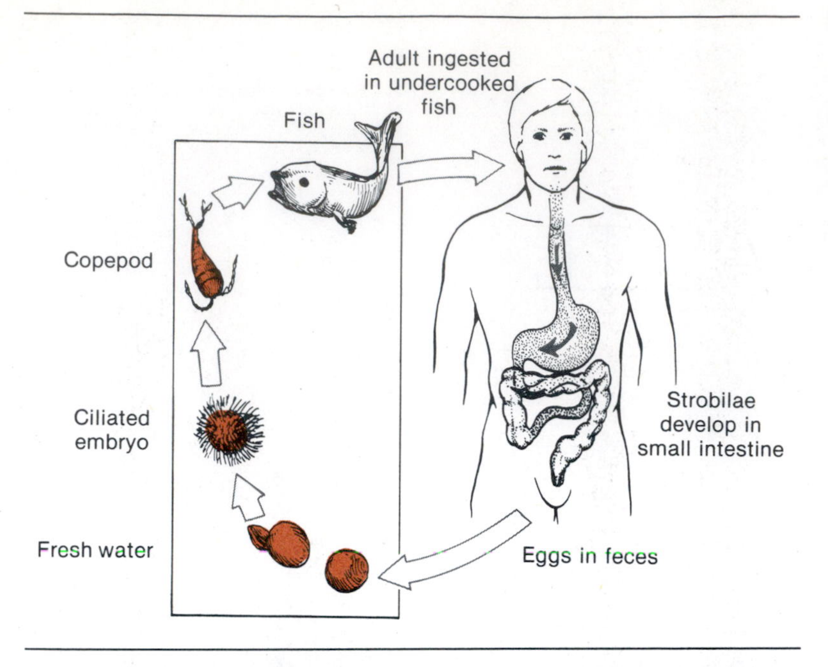

Figure 32-2 Life cycle of fish tapeworm *Dibothriocephalus latus*. Larvae in raw or insufficiently cooked freshwater fish when ingested develop into another larval form in small intestine of humans. Immature eggs passed in feces of humans into fresh water develop ciliated embryo within called coracidium. Hatching occurs, and appropriate freshwater crustaceans ingest larvae, which develop in body cavity. Small game fish ingest infected crustacean and in turn are eaten by larger game fish. Larvae develop in fish and serve as source of infections in humans.
Modified from Beck, J.W., and Davies, J.E. Medical parasitology, *(3rd ed.). The C.V. Mosby Co., St. Louis, 1981.*

hatch in water into a juvenile form that is ingested by copepods (crustaceans). The larvae undergo further development only when the crustaceans are eaten by large fish. Humans contract disease by ingesting the larval forms in uncooked fish. The adult tapeworm develops in the intestine of humans (Figure 32-2).

Parasites encounter many barriers to infection, and consequently many progeny must be produced to ensure that some will survive to cause disease. Asexual and sexual reproduction occur in animal parasites. ***Multiple fission*** (schizogony) is a characteristic asexual process among the protozoa that is discussed in Chapter 5. ***Budding*** is a type of asexual reproduction in some tapeworms in which a single larva may bud off up to 100 new larvae, each of which can develop into an adult. Asexual reproduction among the group called the flukes begins with an egg that goes through several larval stages before reaching the adult stage. During these intermediate stages many individuals are produced. Sexual reproduction is also common among the more advanced types of animal parasites. Some parasites are hermaphroditic and possess both male and female sex organs. This type of sexual reproduction is fairly common among the tapeworms and flukes. Tapeworms consist of many segments called ***proglottids,*** each of which contains both sex organs. Considering that the average number of proglottids per specimen is 3500, the tapeworm is a veritable egg-laying machine.

Epidemiology

Most animal parasite diseases are a consequence of poor hygiene. In addition, factors such as humidity and temperature also exert an influence on the probability of maintaining the parasite in the environment and acquiring disease. One, therefore, finds more diseases in tropical and subtropical regions. In developed countries of the world many of the severe parasitic diseases have been virtually eliminated. This has also resulted in a loss of acquired immunity because of lack of exposure to the parasite. With increased travel there has been a reintroduction of parasites into developed countries by refugees as well as illegal aliens, resulting in an increase in animal parasitic infections among inhabitants of the warmer climates of North America and Europe.

There are four avenues for the transmission of animal parasites to humans.

1. Active transmission. The parasite actively invades the host, such as hookworms. Hookworms make contact with the skin and burrow into it.
2. Passive. The parasite does not actively penetrate the host but is accidentally ingested in food or water or is injected into the host through the bite of an arthropod. Diseases such as amebic dysentery are acquired by ingestion of contaminated meat or contact with soil where cysts have been excreted by infected animals. Diseases such as malaria are transmitted by the bite of mosquitoes. Parasites may also be accidentally injected into the body by hypodermic injection of contaminated fluids. Malaria has been contracted by this method.
3. Veneral. Vaginitis in women and urethritis in men is caused by *Trichomonas vaginalis.*
4. Congenital. The agent causing toxoplasmosis is transmitted via the placenta.

In the United States there is a high prevalence of colonization by certain animal parasites, such as *Trichomonas vaginalis,* which has been estimated to be found in more than 30% of the population. Disease (vaginitis) is associated with decreased vaginal acidity. A protozoan called *Toxoplasma gondii,* which is the agent ***toxoplasmosis,*** has worldwide distribution and is believed to be prevalent in over 50% of the population of the United States. Yet, nearly all infections are not accompanied by symptoms and only occasionally, in congenital infections and in those with depressed immune systems, does serious illness develop. ***Pinworm*** *(Enterobius vermicularis)* is believed to infect nearly 20% of the population of the United States and is especially prevalent in children, but many cases are symptomless or only produce minor symptoms, such as perianal itching.

Many parasites may be beneficial because they provide cross protection; that is, infection by one species of parasite can provide protection from other more virulent species of the same or different genera. Experiments with mice and hamsters have demonstrated that antigens isolated from the tegument of *Fasciola hepatica* worms provide protection from disease caused by *Schistosoma mansoni.* This is not a new concept since BCG vaccine protects the individual against not only tuberculosis, but other diseases as well, by stimulating either cell-mediated or antibody responses in the host. The presence of an infectious agent such as *Toxoplasma gondii* can also protect the host against further infections by the same agent. Cysts of the parasite can persist for years in the host and prevent reinfection. This type of protection is called ***premunition,*** or concomitant immunity.

Several parasite diseases found in animals can be accidentally transmitted to humans. Many of the animals are pets (cats, dogs, and horses) and provide a ready avenue for transmission. The animals harbor the parasite in various stages of development with humans acting as an alternative host because of their close association with the animals involved. The pets pick up the parasite by eating infected rodents or scraps of meat taken from infected animals. Humans acquire infection by handling pets and ingesting the parasites. In some instances the disease is more detrimental to humans than to other hosts. For example, *Echinococcus granulosis* is a small tapeworm that infects dogs and is only a few milliliters long. The parasitic egg when ingested by humans develops into a form called the ***hydatid cyst*** and may increase to the size of a cantaloupe depending on its location in the body. Other tapeworms and hookworms are also harbored by domestic animals and can be transmitted to humans.

Pathogenicity

Protozoa, like viruses and bacteria, can multiply in the host and a single exposure can cause serious illness. Multiplication of the adult in the host is not characteristic of most helminths (except for *Strongyloides,* which multiply in the intestine), and the severity of disease is related to the number of worms ingested or the number of repeated exposures. Colonization by animal parasites seldom results in overt symptoms. Many intestinal amebas are harmless commensals, but occasionally a virulent form can cause tissue necrosis in the bowel. *Giardia lamblia* is an intestinal protozoa found among many human populations of the world. Diarrhea frequently occurs after returning from trips to foreign countries but may also occur among residents in established communities. In 1978 outbreaks of giardiasis occurring among residents of some Colorado ski villages were linked to contaminated water supplies.

Many pathological conditions caused by helminth disease are the result of inflammation or an immunological reaction with eosinophils. The large production of eosinophils is one of the characteristics of all helminth diseases. The degree of pathogenicity in helminth diseases is usually directly related to the number of worms harbored by the host. Many helminth diseases also have a less direct effect on the host than protozoal diseases. The broadfish tapeworm, for example, absorbs vitamin B^{12} from the host and may be a cause of anemia. Other worms may be in such great numbers that when attached to the intestinal epithelium they mask potential sites of nutrient absorption by the host. The host is literally robbed of nutrients, which pass unabsorbed through the feces.

Toxins do not appear to play as significant a role in the disease process of animal parasites as they do in the bacteria. A protein toxin has been isolated from a pathogenic strain of *Entamoeba histolytica* that has been shown to be cytotoxic for hamster kidney cells. Toxinlike substances have also been isolated from the agents of African sleeping sickness *(Trypanosomma brucei gambiense)* and Chaga's disease *(Trypanosoma cruzi),* but they have not been purified.

Some parasites have no apparent adaptations for penetration of tissue barriers in the host but those that do fall into two groups: those parasites that possess specialized structures such as hooks that are used to abrade tissue and those parasites that have enzymatic secretions that are capable of dissolving host membrane components. A parasite may possess one or both of these adaptive mechanisms. Invasion of the epithelial surface of the skin of vertebrates by certain parasites is believed to be caused by more than one factor, and few are capable of invading tissue directly. An example of a special mechanism for penetration can be observed in the helminth that causes schistosomiasis. A larval form of the parasite burrows from the snail and is capable of penetrating the skin of humans. The temperature difference between the environment and tissue attracts the parasite to the skin surface where lipids are believed to stimulate penetration by the larvae. Larvae move across the skin by alternatively attaching an oral and ventral sucker. Adhesion to the skin is assisted by mucus secreted from a gland and is deposited on the skin at each attachment site of the oral sucker. When a site is located for invasion, the larvae orient themselves perpendicular to the skin. They lash their tails and initiate muscular movement that actively forces the anterior end of the parasite into the entry site. After the stratum corneum of the skin is penetrated, the tail is shed and the parasite orients itself parallel to the skin. It moves through the epidermis until a hair follicle is reached and then penetrates the dermis, where it can enter veins or the lymphatic system. Different mechanisms of infection are involved for the protozoa that causes malaria. The infective stage of the parasite is transmitted to humans via the bite of the female *Anopheles* mosquito and is deposited directly into the bloodstream. The parasite migrates to the liver and undergoes multiple fission before reentering the bloodstream where it enters the red blood cell as a form called the ***merozoite.*** The outer coat of the merozoite possesses bristles that are approximately 20 nm long and are T- or Y-shaped at their distal ends. They can be stretched to 35 to 40 nm by attachment to another cell. The extension of these bristles is believed to overcome the repulsion between the negatively charged surfaces of the merozoite and red blood cell. The receptors on the red blood cell are important for this interaction to take place since blacks are more resistant to certain species of *Plasmodium* than are whites. Blacks apparently have red blood cells that are deficient in surface receptors. An interesting corollary can be made. About 70 years ago one of the treatments for syphilis was to increase the body temperature (pyrotherapy) because the organism causing syphilis was extremely sensitive to temperatures above 37° C. One technique was to infect the individual with an organism that caused sustained and elevated temperatures. The organism selected was the one causing malaria. The technique did not work on blacks. Fortunately for society penicillin was discovered in 1940 and eliminated the need for such drastic treatment. Resistance to malaria by many individuals in Africa and the Mediterranean has been known for some time. Genetic resistance among black persons is the result of certain heritable diseases such as sickle cell anemia, thalassemia, and glucose 6-phosphate dehydrogenase deficiency (favism). Each of these genetic defects is in some manner associated with the red blood cell, the target of the malaria agent. The reason for resistance by individuals with favism is clear because the malaria agent depends totally on host glucose 6-phosphate dehydrogenase for the conversion of hexoses to pentoses in the hexose monophosphate shunt. Why those with the sickle cell trait or thalassemia are resistant is not totally clear, but it appears that a red blood cell component may play an important role. A sugar-containing component is absent from the red blood cells of resistant individuals but has been found to be present in those susceptible to malaria.

Immunity

Little is known concerning the immune response of the host to animal parasites. Some of the parasites (for example, the helminths) possess a multitude of surface antigens, and many exist in a variety of different stages. Animal parasites do stimulate the humoral and cell-mediated immune responses of the host, but permanent immunity, with one or two exceptions, does not occur. In those infections that are asymptomatic or low level, immunity to reinfection is possible; but once the parasite has been expelled from the system, this immunity is lost. The infecting organisms of toxoplasmosis are intracellular parasites that travel through the circulation in monocytes until they encyst in various tissues of the body. Humoral antibodies do not appear to play any role in immunity by the host. In most cases toxoplasmosis infections are asymptomatic, and immunity is believed to be related to the cell-mediated response. Experiments have demonstrated that in individuals with overactive antibody production (hypergammaglobulinemia) but depressed cell-mediated immunity, *Toxoplasma* infection results in severe progressive disease. When T cells are activated by *Toxoplasma* antigens, they activate macrophages, and the organism is destroyed.

Many helminths that inhabit the gastrointestinal tract stimulate the production of secretory IgA. This antibody is believed to act nonspecifically by preventing attachment of the parasite to mucosal cells.

In many parasitic diseases there is the production of specific IgE. It has not been determined what part this antibody plays in parasite elimination. One suggestion is that the combination of IgE and parasite antigens form a complex with mast cells that leads to the production and release of vasoactive amines. The amines may provide a means for increased intestinal permeability and release of immune factors into the lumen of the gut.

The immune response to certain parasites is believed to be responsible for certain pathological conditions associated with the disease. In patients with chronic infections caused by the protozoa *Trypanosoma cruzi,* antibodies are produced that cross-react with antigens shared by the parasite as well as by heart and vascular tissue of the host. The heart damage in these patients is believed to be caused by this immunological response.

Laboratory diagnosis

Laboratory diagnosis is complicated by the fact that many animal parasitic diseases do not cause clinical manifestations, or the symptoms may vary substantially from one case to another. In most human diseases the parasite is shed in the feces in some morphologic form such as an adult- or intermediate-stage parasite. Direct microscopic examination of stained specimens is sufficient for identification of many clinical specimens such as the examination of morphological features of eggs as well as larvae or cysts. In stool specimens both trophozoite and cyst stages of protozoans can be identified, whereas in helminth examination the eggs, tapeworm proglottids, and larvae may be isolated from the feces. Identification of eggs is related to size as well as shape. Specimens are preserved by fixing in polyvinyl alcohol (PVA) and 10% formalin. This mixture also destroys any contaminating species of microorganisms. Less frequently urine and blood can also be examined for parasites such as *Trichomonas vaginalis* and the agent of malaria (species of *Plasmodium*), respectively.

Serological diagnosis can sometimes be employed when the parasite cannot be isolated. Several tests are available including indirect fluorescent antibody, complement fixation, and indirect hemagglutination. Serological data are also important in the field of epidemiology. Seroepidemiology is becoming an important tool in investigating infections. Serologic data can provide the clinician with information that will indicate through antibody titers an individual's past or present encounters with the parasitic agent. Such data prove invaluable when direct identification of the parasite is not possible.

Chemotherapy, prevention, and control

Diseases caused by animal parasites can be treated with a variety of chemotherapeutic agents (Table 32-1), but most are toxic to the host as well as the parasite, and drug levels must be carefully controlled. Many of the drugs are effective upon certain larval stages but have no effect upon cysts. Cysts are unusually resistant to chemicals even at high concentrations. In fact, some cysts that are to be identified in the laboratory can be stored in a solution of sulfuric acid. The most widely used drugs for helminth infections are ***diethylcarbamazine, niclosamide, mebenidazole,*** and ***thiabendazole,*** whereas ***atabrine, antimony*** compounds, and ***metronidazole*** are used specifically for protozoal diseases. One of the problems in trying to find suitable antiparasitic drugs is our inability to cultivate parasites and many of their stages in vitro. Parasitic enzyme systems are similar to those in humans, and differences in metabolism that could be exploited can be found only through in vitro studies. The importance of understanding parasitic life cycles and multiplication sites is exemplified in the treatment of leishmaniasis. Trivalent antimony compounds were thought to be too toxic for use in the treatment of mammalian infections. Research on the organism indicated that it multiplied in the macrophage. It had already been known that macrophages take up specific liposomes, so researchers encased the antimony compounds in liposomes and injected them into diseased animals. Once the liposomes were absorbed by macrophages in the animal, the antimony compounds were released and a severalfold increase in protozoal destruction was observed with a significant reduction in host toxicity. Future studies of parasite metabolism may reveal important enzyme differences from the host that can be utilized for the absorption of chemotherapeutic agents with more selective inhibitory properties.

There are no vaccines available for preventing parasitic diseases in humans, primarily because of the inability to define the antigens involved in antibody formation. Many of the parasites go through various stages in which their surface antigens change, thus making identification difficult. Research is now engaged in using nonvirulent variants of virulent parasites to stimulate antibodies and protect against infection by virulent strains. This is the same principle used by Edward Jenner in the eighteenth century, in which the cowpox virus was used to prevent smallpox.

Table 32-1 Characteristics of drugs used in the treatment of animal parasite infections

DRUG	PARASITE AFFECTED	TOXICITY	MECHANISM OF ACTION
Thiabendazole	Anthelmintic drug	May cause vomiting	Blocks production of ATP by inhibiting fumaric reductase, an enzyme involved in electron transport
Mebendazole	Anthelmintic drug (nematodes)	May cause nausea but less toxic than thiabenazole	Prevents uptake of glucose by parasite, which leads to glycogen depletion and eventually inability to produce energy
Metronidazole	Antiprotozoal drug (amebic dysentery, trichomoniasis, and giardiasis)	Gastrointestinal symptoms; possible carcinogen, not used during pregnancy or in asymptomatic infections	Mechanism of action unknown but it is believed that the drug is reduced in the parasite to a compound that interferes with DNA synthesis
Atabrine	Antiprotozoal drug (giardiasis)	Vomiting and dizziness	Mechanism of action unknown
Chloroquine	Antiprotozoal drug (malaria and amebiasis)	Ocular toxicity with corneal and retinal changes that may be permanent; also crosses placenta and can cause fetal damage	Drug forms complex with DNA and prevents DNA replication as well as transcription; exact mechanism is not known
Diethylcarbamazine	Anthelmintic drug (filiariasis)	Relatively nontoxic but can cause allergic reactions	Mode of action is not known

Parasitic disease is most common in individuals deprived of proper nutrition, sanitation, and suitable living conditions. For this reason improvement in hygiene is the answer to control and eradication of most parasitic diseases. To understand what control measures must be instituted, one needs to know the means of parasite transmission and parasite life cycles. Many infections result from ingestion or contact with parasites or cysts found in soil or water. Proper sewage disposal, wearing of shoes, avoidance of water known to contain certain parasites, and refraining from the use of raw feces as a fertilizer are measures that would reduce infection. Some helminth diseases, such as trichinosis, are prevented by avoiding the use of garbage as food for slaughter animals and by thorough cooking of meat that may be potentially contaminated. For those diseases that are transmitted by arthropods one can avoid arthropod-infested areas and use chemical or biological control methods.

PARASITIC HELMINTHS

The Helminths are divided into two phyla: (1) the Platyhelminthes, which include several groups, among which are the important parsites of humans, the tapeworms and the flukes, and (2) the Nematoda, or roundworms.

Tapeworms

Tapeworms (also called ***cestodes***) live in the intestine of vertebrates (Figure 32-3). The larval forms are found in the flesh of animals as well as in insects, crustaceans, and other intermediate hosts on which the vertebrate feeds. The adult parasite attaches to the intestinal wall and may be considered a family of individuals and not a single individual. This is a characteristic unique to the tapeworms and is not found in other multicellular organisms. The adult tapeworm is referred to as a ***strobilia,*** in which the head end is called a ***scolex*** (Figure 32-4), upon which are located suckers or hooks. The suckers or hooks are used to attach to the intestinal wall of the infected vertebrate. Located beneath the scolex are a series of segments called ***proglottids*** (Figure 32-5). Immature proglottids are produced immediately adjacent to the scolex, and as new ones are formed the older ones are pushed posteriorly. The more mature proglottids are hermaphroditic and develop male and female sex organs. The most mature proglottids (***gravid*** proglottids) are at the hind end and are filled with thousands of

Figure 32-3 Sheep tapeworm.
From Hickman, C.P., et al. Integrated principles of zoology *(6th ed.). The C.V. Mosby Co., St. Louis, 1979.*

eggs. Humans may be infected by ingesting the eggs but more commonly by ingesting larvae encysted in the flesh of animals they eat. The most common tapeworm affecting humans is the beef tapeworm *(Taeniarhynchus saginata).*

The tapeworm has no digestive tract, and all nutrients must be absorbed through its tegument from the host. Monosaccharides, such as glucose and galactose, are the most important energy sources that are absorbed by the parasite, but nutrients, such as vitamins and amino acids, can also be absorbed. Nutritional deficiency by the host is seldom observed in those with tapeworms even though the old notion that someone who ate constantly and gained no weight had such a parasite.

Taeniarhynchus saginata. *T. sagniata* is found throughout the world and especially in areas where raw beef is eaten. Eggs passed in the feces can survive on the ground for long periods of time. When ingested by herbivorous animals,

A

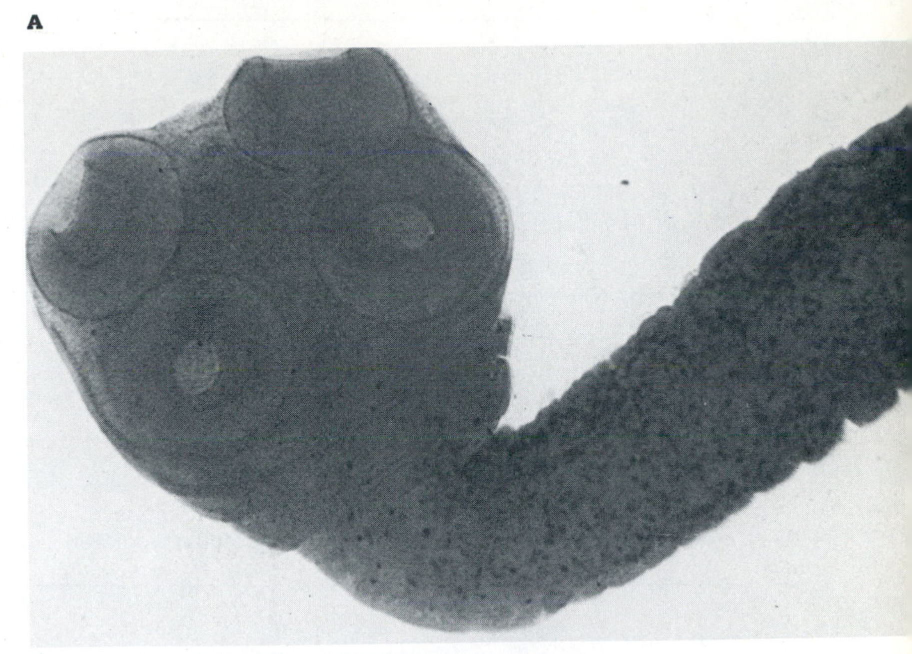

B

Figure 32-4 Head (scolex) of two tapeworms. **A,** Scolex has suckers but no hooks. **B,** Scolex has suckers plus hooks
A nonprofit cooperative endeavor by numerous colleagues under the editorship of Dr. Herman Zaiman, Valley City, N.D. From Beck, J.W., and Davies, J.E. Medical parasitology *(3rd ed.). The C.V. Mosby Co., St. Louis, 1981.*

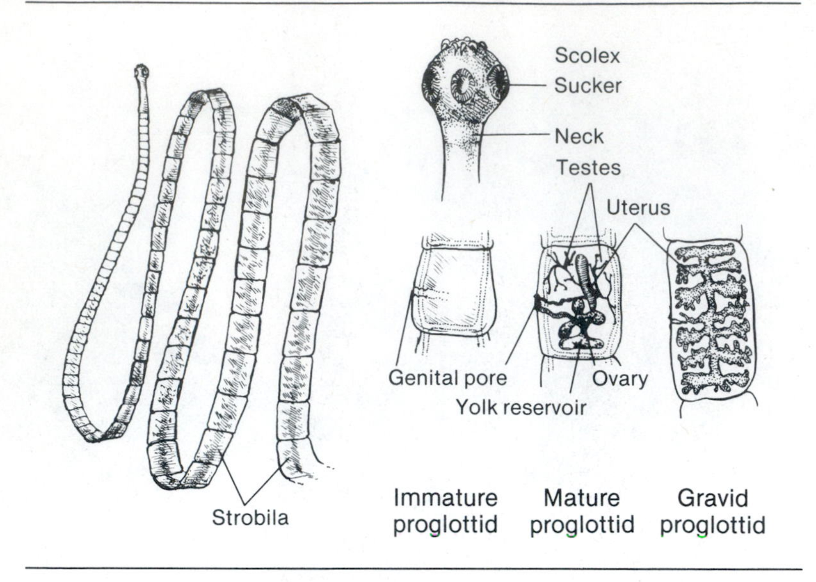

Figure 32-5 Morphologic features of tapeworm.
Redrawn from Medical protozoology and helminthology. *United States Naval Medical School, National Naval Medical Center, Bethesda, Md., 1955.*

such as the cow, from contaminated soil and vegetation, the larvae penetrate the intestinal wall and migrate through the circulatory system until they reach muscle fibers and develop into an infective form called the ***cysticercus.*** The cysticerci remain in the muscle tissue of the intermediate host as juveniles until ingested by beef eaters such as humans. The tough outer coat of the beef cysticercus is digested by intestinal enzymes releasing a juvenile tapeworm, which attaches to the small intestine. In 12 weeks it will mature and reach a length of 4 to 5 meters. Infection in humans results from eating raw or inadequately cooked beef containing cysticerci. The tapeworm seldom causes harm to the host, although mild symptoms of diarrhea and abdominal pain may occur. More severe forms of tapeworm infection result from ingestion of the cysticerci of the pork tapeworm *(T. solium).* The cysticerci of the pork tapeworm develop in the intestine of humans much as they do during infection by the beef tapeworm, but the human may also be an intermediate host in pork tapeworm infection. Humans infected by adult worms may pass proglottids in their feces with the subsequent release of eggs that can be ingested by other humans or by themselves, as well as swine. The ingested eggs hatch in the intestine and may migrate to many different body sites but usually lodge in the musculature.

Tapeworm infections can be prevented by thorough cooking of meat because the cysticerci are destroyed by heat above 56° C. Niclosamide is the drug of choice for treatment.

Diseases caused by other cestodes are outlined in Table 32-2.

Flukes

Unlike tapeworms, flukes (also called trematodes) are found in other areas besides the intestine, such as the bile ducts, the liver, blood—and rarely, the central nervous system. Except for the group called the schistosomes, all are hermaphroditic and possess a simple intestinal tract. Attachment to the host is facilitated by suckers—one surrounding the mouth and the other located on the ventral surface. Their life cycles are much more complex than tapeworms. Eggs contain a ciliated larval form called a ***miracidium.*** The miracidium hatches in water and seeks out an intermediate host, which is nearly always a snail but occasionally a clam or annelid. Asexual multiplication occurs in the snail, resulting in the formation of ***cercariae.*** Cercariae are shed by the snail and infect a second intermediate host such as a plant or crustacean. The cercariae encyst on their new hosts and develop into ***metacercariae.*** Humans are infected by ingesting metacercariae found on vegetation or in crustaceans. Once ingested, the larvae penetrate the intestine and enter the circulation, eventually reaching the lungs or liver where they mature into adults. Some species, following maturity and mating, pass their eggs into the intestine or urinary bladder to be carried out of the body. Some flukes mature in the lungs, where eggs are released to be carried to the oral cavity and then swallowed and finally passed in the feces. Most fluke infections are not found in the Western world. Two important flukes of medical interest are *Schistosoma japonicum* and *Fasciola hepatica.*

Schistosoma japonicum. *S. japonicum* is one of three schistosomes of medical significance; the others are *S. hematobium* and *S. mansoni.* The disease they cause is referred to as ***schistosomiasis*** and is second to malaria as the primary cause of illness and death in the tropics.

Eggs passed in the feces hatch in fresh water as ***miracidia*** and seek a snail host. Cercariae develop from miracidia in the snail and eventually leave to seek a human host. The cercariae penetrate wet skin and travel through the circulation and reach the liver where development into mature adults takes place. Male and female adults copulate and then migrate to the mesenteric venule of the small or large intestine where the female deposits eggs. The eggs pass into the lumen of the intestine and then into the feces (Figure 32-9).

Table 32-2 Characteristics of the parasitic worms

PARASITIC WORM	CLINICAL CONDITION	METHOD OF TRANSMISSION	CHEMOTHERAPY	CHARACTERISTICS OF LIFE CYCLE
Platyhelminthes				
Fasciolopsis buski	Fascioliasis; gastrointestinal condition endemic in Orient.	Ingestion of larval stage (metacercaria) on aquatic plants such as watercress	Hexylresorcinol	Miracidium penetrates snail; cercariae released by snail onto aquatic plants; adult development in human followed by egg laying
Fasciola hepatica (liver fluke)	Fascioliasis; pathologic changes in bile ducts, liver, or gall bladder; rare disease in humans	Same as above	Dichlorophenol	Same as above except adult worms inhabit bile duct
Clonorchis sinensis (Oriental liver fluke)	Clonorchiasis; bile ducts affected; chronic condition associated with secondary bacterial infection as well as carcinoma; endemic in Asia	Ingestion of metacercaria in fish	Chloroquine-phosphate	Ova released in human feces infect snails; larvae in snails pass in water and infect fish where larvae remain dormant until ingested by humans
Paragonimus westermani (lung fluke)	Paragonimiasis; respiratory condition; secondary bacterial pneumonia common complication	Ingestion of raw crustaceans	Dichlorophenol	Same as for *C. sinensis,* except crustaceans instead of fish are intermediate hosts
Schistosoma species (blood fluke) *S. mansoni* *S. haematobium* *S. japonicum*	Schistosomiasis; fever, lymph gland enlargement and liver enlargement; parasites can cause obstruction of hepatic veins; endemic in Asia, Africa, South America, and Caribbean	Contact of skin with contaminated water, as in swimming	Niridazole	Eggs in human feces reach water and infect snail; larvae released by snail into water
Diphyllobothriun latum (fish tapeworm)	Diphyllobothriasis; benign disease with abdominal pain, weight loss, nausea and anemia	Ingestion of larvae encysted in fish muscle	Niclosamide	Eggs from feces of humans eaten by crustaceans; if latter eaten by fish, larvae develop and become encysted and remain dormant until eaten by humans
Taenia species *T. solium* (pork tapeworm) *T. saginata* (beef tapeworm)	Taeniasis; mild diarrhea, abdominal pain, and weight loss; larvae may localize in brain, heart, or spinal cord, causing death	Ingestion of contaminated beef or pork'	Niclosamide	Eggs passed in feces; ingestion of eggs by pigs or cattle; larvae encyst in muscle of these intermediate hosts
Vampirolepsis nana (dwarf tapeworm)	Diarrhea, abdominal pain; endemic worldwide	Fecal contamination or eating infected insects, such as grain beetles	Niclosamide	Insects are intermediate hosts; rodents, fleas, and beetles are reservoirs of parasite
Trichinella spiralis	Trichinosis; asymptomatic to fatal infections depending on worm burden; fever, periorbital edema, severe muscular pain	Ingestion of viable larvae in improperly cooked pork, bear, or walrus	Thiabendazole	Larvae in striated muscle of pigs, bear, or walrus is ingested by humans (Plate 28)
Nematoda				
Enterobius vermicularis (pinworm)	Perianal itching in children; asymptomatic in adults; most common parasitic infection	Person-to-person and ingestion of eggs	Mebendazole	Life cycle only in humans

Continued.

Table 32-2 Characteristics of the parasitic worms—cont'd

PARASITIC WORM	CLINICAL CONDITION	METHOD OF TRANSMISSION	CHEMOTHERAPY	CHARACTERISTICS OF LIFE CYCLE
Trichuris trichiuria (whip worm) (Figure 32-6)	Mostly asymptomatic, but can be serious with rectal prolapse and death resulting	Ingestion of eggs found in the soil	Mebendazole	Life cycle in humans
Ascaris lumbricoides	Ascariasis; more common in children under 14; abdominal pain, vomiting, insomnia; intestinal obstruction most common complication	Ingestion of eggs from contaminated soil or water	Mebendazole	Eggs released in human feces develop into infectious state on soil
Toxocara species *T. canis* (dog) *T. cati* (cat)	Toxicariasis or visceral larva migrans; usually benign with fever, weight loss, irritability; in severe infection heart and central nervous system may be involved, but eyes most drastically affected	Ingestion of soil contaminated with eggs	No drug available	A zoonotic disease with humans, especially children, as accidental hosts
Necator americanus and *Ancylostoma duodenale* (hookworms)	Skin itch at site of invasion; abdominal symptoms of pain and diarrhea	Invasion of skin by larvae in the soil	Mebendazole	Larvae penetrate skin of humans and finally settle in intestine and mature; eggs released in feces
Strongyloides stercoralis	Strongylodiasis; common in United States; abdominal symptoms that mimic peptic ulcer; most severe in immunosuppressed patient	Invasion of skin by larvae in the soil	Thiabendazole	Larvae penetrate skin; eggs produced in intestine may develop into larvae that either pass in feces or may reinfect patient and disseminate in bloodstream
Wucheria bancrofti and *Brugia malayi* (filarial roundworms)	Filariasis; nausea and headache initially with lymph gland involvement; obstruction of lymph glands and enlargement (elephantiasis) more common in chronic cases (Figure 32-7)	Bite of infected mosquito	Diethylcarbamazine	Eggs develop into adults in human lymph glands; fertilized females release juveniles (microfilariae), and latter are picked up by mosquito from circulation
Loa loa (eye worm)	Loiasis; skin swellings on body especially on hands, legs, and around eyes; Worms can migrate to eye and cause blurred vision (Figure 32-8)	Bite of tabanid fly	Diethylcarbamazine	Larvae penetrate skin of humans and mature in connective tissue; eggs enter circulation
Onchocerca volvulus	Onchocerciasis (river blindness); nodules, dermatitis, or corneal opacities; Blindness most severe complication	Black flies	Diethylcarbamazine	Eggs develop in human connective tissue; flies pick up larvae from dermal fluid; larvae picked up from dermal fluid by flies

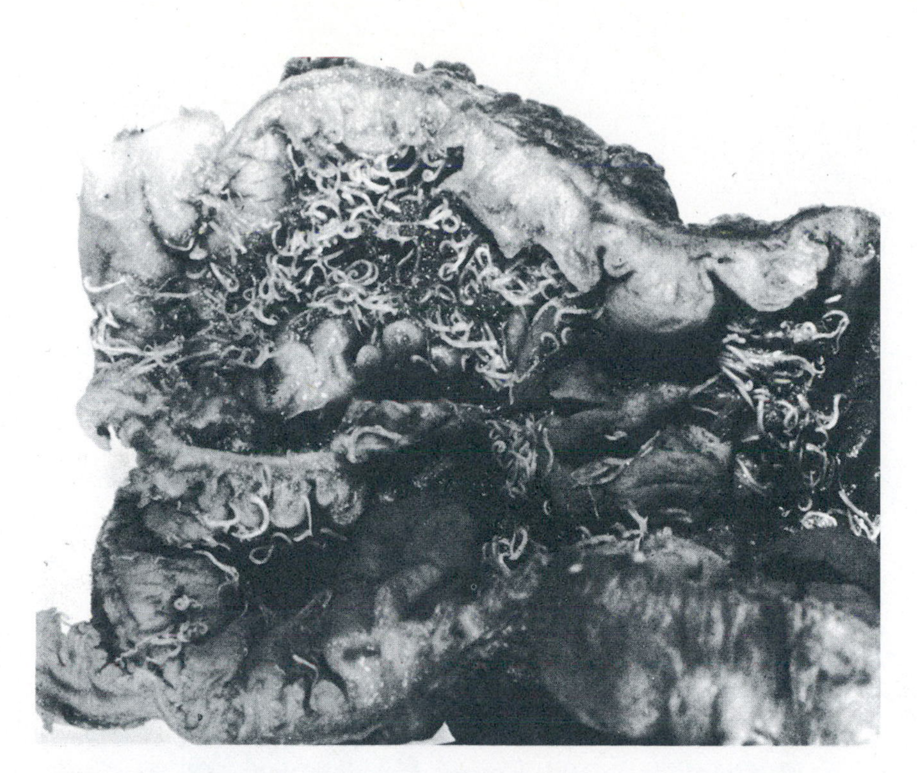

Figure 32-6 Whipworm *(Trichuris trichuria)*. Massive infection that produced severe hemorrhage, diarrhea, and death in child.
From Anderson, W.A.D., and Kissane, J.M. (Eds.). Pathology *(7th ed.). The C.V. Mosby Co., 1977.*

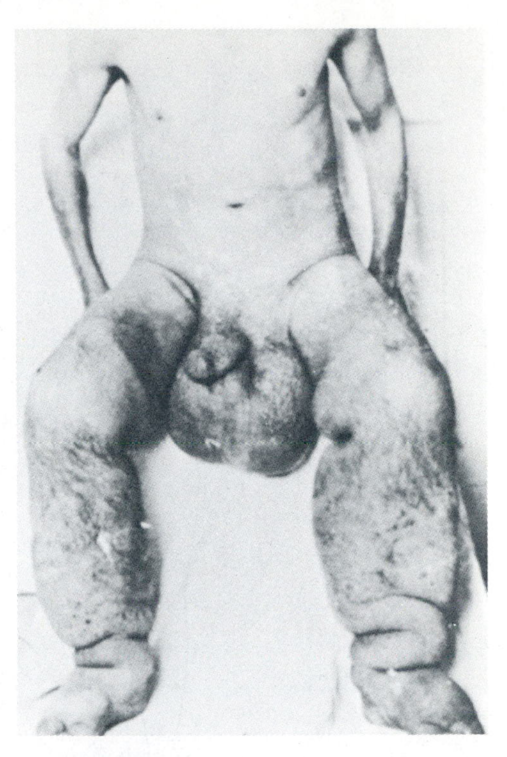

Figure 32-7 Elephantiasis involving both legs, scrotum, and penis as seen in chronic filariasis.
From Price, D.L., and Hoops, H.C. In Marcial-Rojas, R.A. (Ed.). Pathology of protozoal and helminthic diseases with clinical correlation. *Williams & Wilkins Co., Baltimore, 1971.*

The symptoms of infection can be divided into three phases and are associated with the sites of infection in the host. The first stage, in which the parasite penetrates the skin, may be asymptomatic, but occasionally there is minor dermatitis. The second set of symptoms is correlated with egg deposition in the intestine, which occurs about 6 weeks after the initial penetration of the skin. Fever, chills, and gastrointestinal symptoms such as cramps and dysentery may appear. The third stage is a tissue response to the parasite and is characterized by granulomas resembling those of tuberculosis. The adult worms may survive as long as 20 years and produce eggs that can initiate chronic responses in the host. Granulomas in the liver lead to cirrhosis. Eggs may also reach the brain to cause coma and paralysis. The later stages of infection are also characterized by accumulation of fluid in the abdominal cavity ***(ascites)***.

Treatment is not always warranted and must be evaluated according to the degree of infection, viability of eggs, and concentration of eggs in the feces. When the liver is severely affected, most antihelminthic drugs are contraindicated because of their toxicity. Some surgical techniques

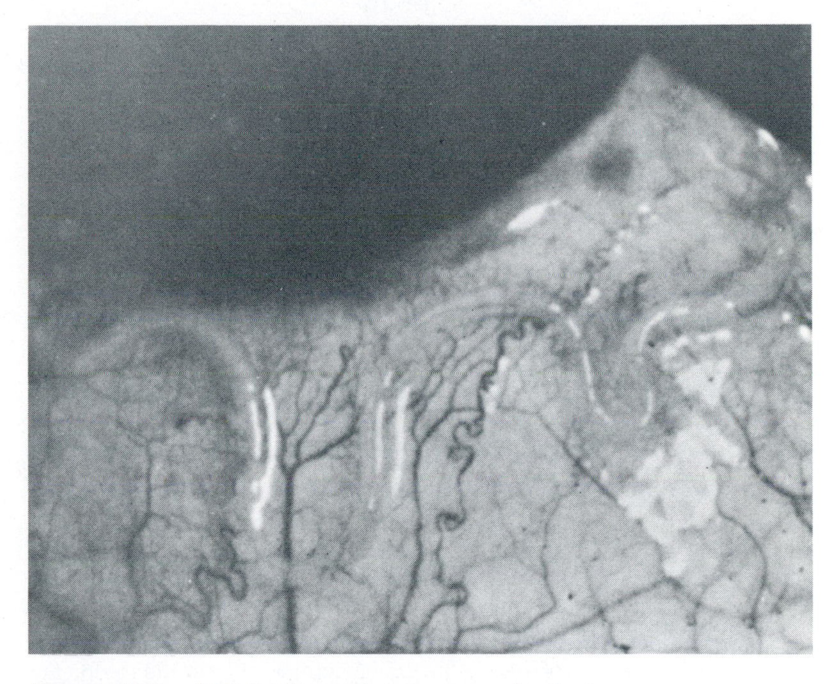

Figure 32-8 Adult *Loa loa* moving across eye of patient.
Air Force Institute of Pathology, AFIP 73-6554.

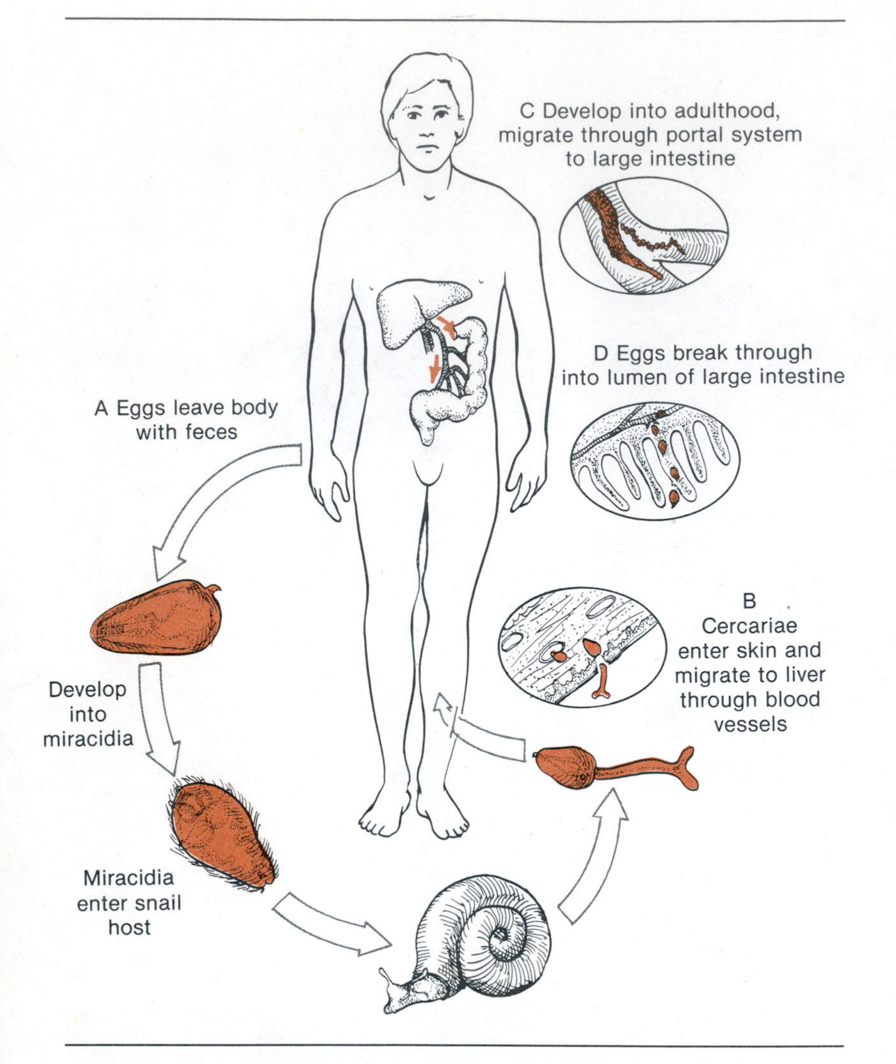

Figure 32-9 Life cycle of *Schistosoma mansoni. A,* Eggs passed in feces hatch in fresh water and produce miracidia, which penetrate snail. Cercariae produced in snail leave and penetrate skin of human, *B.* Larvae enter bloodstream and go to liver where they develop into adults. Copulating male and female adults migrate to venules in rectosigmoid area of large intestine, *C.* Eggs are deposited in venules, *D,* and break into lumen of intestine and pass into feces.
Modified from Beck, J.W., and Davies, J.E. Medical parasitology (3rd ed.). The C.V. Mosby Co., St. Louis, 1981.

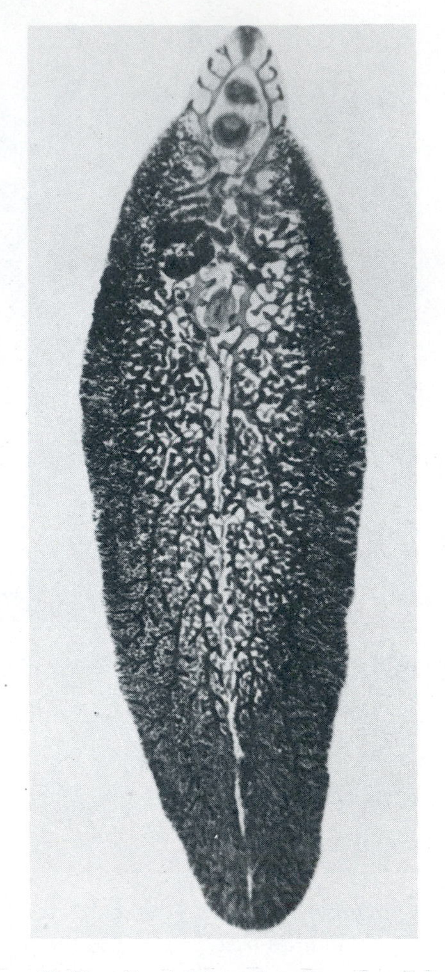

Figure 32-10 *Fasciola hepatica,* sheep liver fluke.
Courtesy Turtox/Cambosco. From Schmidt, G.D., and Roberts. L.S. Foundations of parasitology (2nd ed.). The C.V. Mosby Co., St. Louis, 1981.

have proved effective in flushing adult worms from the nervous system. Niridazole and antimony sodium dimercaptosuccinate are two drugs available from the Centers for Disease Control in Atlanta, Georgia.

Prevention is best obtained through control or elimination of snails by engineering methods or use of molluscicides. If these are not available, the elevation of hygienic standards through proper waste disposal facilities and avoidance of polluted waters will reduce the incidence of infection.

Fasciola hepatica. Adult *F. hepatica* (Figure 32-10) are found in the bile ducts, where approximately 3000 eggs are laid daily. The eggs are passed in the feces and develop into miracidiae only in the water. Further development can proceed only if the intermediate host, the snail, is located. Development in the snail takes 6 to 7 weeks, in which a single miracidium through sexual reproduction gives rise to about 1000 cercariae. The cercariae leave the snail to encyst on water vegetation and await ingestion by the definitive host. When ingested by humans, the cercariae lose their protective cyst coat, pass through the intestinal wall, and reach the liver by passage through the peritoneal cavity. The larvae feed in the liver and reach final maturity in the bile ducts, where eggs are laid about 12 weeks after infection.

The intensity of symptoms is related to the

number of parasites causing disease. There is pain over the hepatic region as well as liver enlargement. Diarrhea and intermittent fever may also be present. Infections can be prevented by cooking aquatic vegetation such as watercress or removing the snail by chemical or physical means. Dichlorophenol is one of the drugs of choice in treatment of human disease.

Fascioliasis is not a major disease in humans, but it is in domestic animals. Infected animals such as cows may die from the disease, or extensive liver damage may result in poor milk production.

Other diseases caused by flukes are outlined in Table 32-2.

Roundworms (Nematoda)

Nematoda constitute perhaps the largest number of animals that inhabit the earth. All, except one, share a common life cycle: egg stage, four larval stages, and adult. They are unsegmented, cylindrical, and taper at both ends. The worm is covered with a thick impermeable cuticle and possesses a large body cavity (pseudocoel) that separates the body musculature from the intestine. They can infect the intestine and other visceral organs as well as the eyes, bloodstream, and subcutaneous tissue. Most diseases occur in tropical or subtropical areas. In the United States pinworm *(Enterobius vermicularis)* is the most prevalent roundworm infection, but ascariasis *(Ascaris lumbricoides)* is of greater medical importance.

Figure 32-11 Intestinal obstruction secondary to masses of *Ascaris* adult worms in young child.

Courtesy Rafael Ramirez-Weiser, San Juan, Puerto Rico. From Anderson, W.A.D., and Kissane, J.M. (Eds.). Pathology *(7th ed.). The C.V. Mosby Co., St. Louis, 1977.*

Ascaris lumbricoides. *A. lumbricoides* is one of the largest of intestinal worms infecting humans, with the female reaching a length of nearly 30 cm (Figure 32-11). It has worldwide distribution but is more prevalent in tropical and subtropical areas. The infective eggs are passed in the feces (Figure 32-12). Following their deposition on the soil, infective larvae develop within the egg over a span of 3 weeks. Once the eggs are ingested by humans, they hatch in the small intestine, where larvae penetrate the mucosal wall and enter the circulation. One week after infection the larvae pass into the lungs, where further development takes place. At this stage the juveniles are too large to enter the circulation from the lungs and instead migrate up the bronchial tree to the pharynx, where they are swallowed and enter the intestine. Final development to the mature state takes place in the intestine. Egg production by adults in the intestine takes place from 5 to 8 weeks after infection.

The eggs of *A. lumbricoides* are extremely resistant to environmental factors and when passed in the soil have been known to be infective even after 10 years. They are resistant to 50% solutions of hydrochloric, nitric, and sulfuric acids but are easily destroyed by the sun. Most infections occur among young children who play in areas contaminated by human fecal material. In those areas of the world where human wastes are used as fertilizer (night soil), infection frequently occurs from ingestion of contaminated vegetables.

Penetrations of the intestinal wall by larvae causes few symptoms, but heavy infections of the intestinal tract by adult worms produces the most recognizable symptoms. Worms may ball up and cause intestinal obstruction. The adult worm has an affinity for small orifices and can migrate to areas such as the bile duct, pancreatic duct, urinary tract, and fallopian tubes. Blockage of such areas, as well as air passages can result in grave consequences—even death—to the patient. Heavy infection in the lung can lead to congestion and a condition called *Ascaris* pneumonitis, which is believed to be an allergic re-

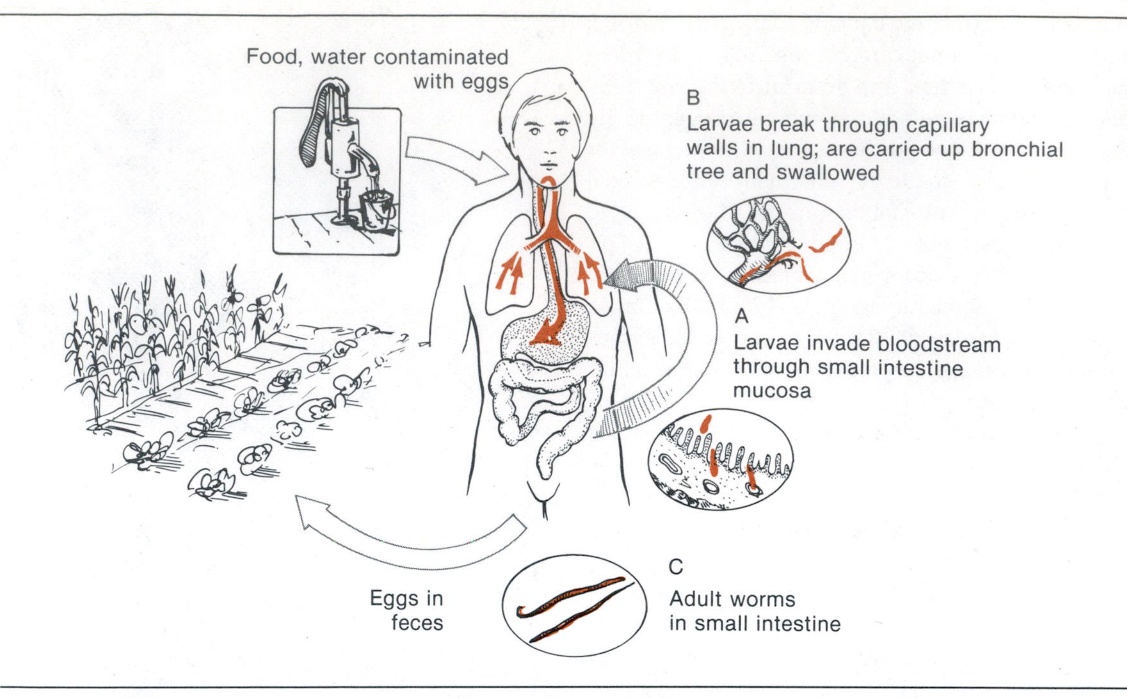

Figure 32-12 Life cycle of ascaris lumbricoides, an intestinal roundworm. Infective eggs, ingested in contaminated food and water hatch in small intestine, where penetration of mucosa, *A,* results in invasion of bloodstream by larvae, which are carried to lungs. Larvae, too large to cross capillary bed, break out into alveolar spaces, *B,* are carried up bronchial tree, are swallowed, and reach intestine, where they become adult worms, *C.*

Modified from Beck, J.W., and Davies, J.E. Medical parasitology *(3rd ed.). The C.V. Mosby Co., St. Louis, 1981.*

sponse to the parasite. Worms can migrate into the stomach to be vomited and provide the patient with his first clue to worm infection.

Intestinal obstruction may require intubation and drainage or even surgery. Only supportive care can be used during lung infection. The most effective drugs against *Ascaris* are piperazine and mebendazole.

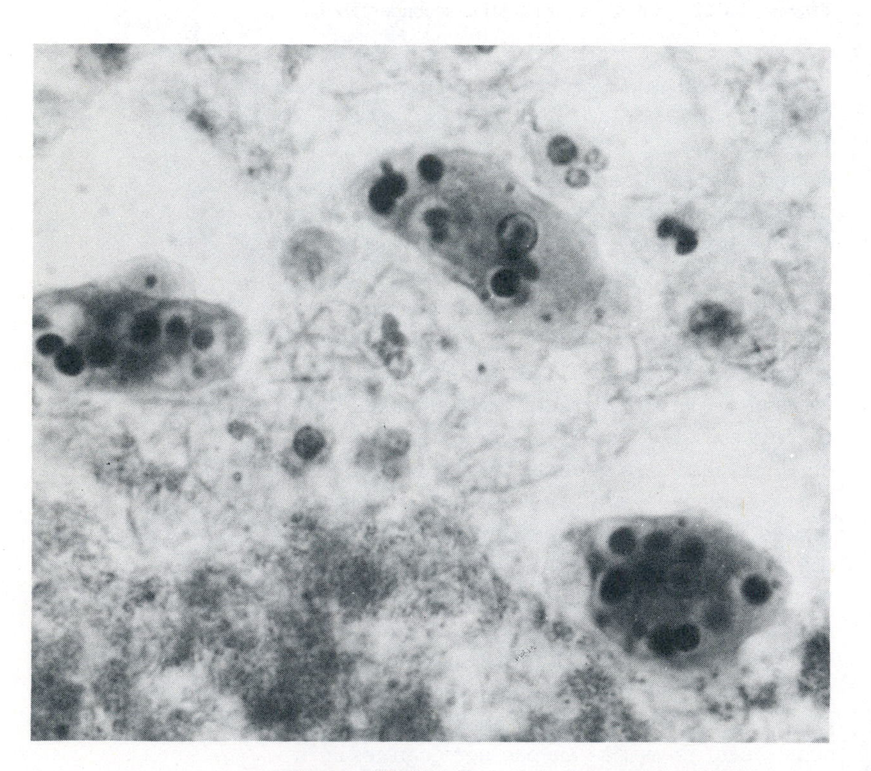

Figure 32-13 *Entamoeba histolytica* trophozoites showing ingested red blood cells and directional motility.

From Beck, J.W., and Davies, J.E. Medical parasitology *(3rd ed.). The C.V. Mosby Co., St. Louis, 1981.*

PARASITIC PROTOZOA

Protozoa are unicellular animals found whereever there are moist environments. Most are free living but a few are parasitic to humans. Depending on the species they can reproduce by asexual and sexual methods. The biology and classification of protozoa are discussed in Chapter 5. Only one protozoal disease, amebic dysentery, is discussed; the others are summarized in Table 32-3.

Amebic dysentery *(Entamoeba histolytica)*

Several amebas are commensals in the intestinal tract of humans, but only *Entamoeba histolytica* (Figure 32-13) is a pathogen of major significance. Humans are the most important reservoirs of the

Table 32-3 Characteristics of protozoal diseases

PROTOZOAL AGENT	CLINICAL CONDITION	METHOD OF TRANSMISSION	CHEMOTHERAPY	CHARACTERISTICS OF LIFE CYCLE
Naegleria fowleri	Primary amebic meningoencephalitis; symptoms resemble meningitis; not common but almost invariably fatal	Through nares following swimming	Amphotericin B	Infects humans and is free living outside host
Dientamoeba fragilis	Diarrhea in children; no fatalities	May be transmitted in pinworm eggs	Metronidazole	Commensal in humans
Giardia lamblia	Diarrhea	Transmission through contaminated water supplies	Atabrine and metronidazole	Humans are carriers; beavers, dogs, and other mammals are reservoirs
Trichomonas vaginalis	Symptomatic infection only in females	Venereal	Metronidazole	Found in urogenital tract
Leishmania species	Leishmaniasis	Bite of sand flies	Antimony gluconate	Dogs are the only important reservoir
L. tropica	Cutaneous ulcers	Same as above	Same as above	Dogs are reservoirs
L. braziliensis	Cutaneous lesions with parasites attacking cartilage of nasal septum	Same as above	Same as above	Sloths, anteaters, arboreal rodents, and marsupials are important reservoirs
L. donovani	Visceral involvement (kala-zar) enlargement of spleen and liver; high mortality	Same as above	Same as above	Dogs and foxes are important reservoirs
Balantidium coli	Common intestinal infection occurring in debilitated patients; severe dysentery	Ingestion of cysts from feces of humans and swine	Oxytetracycline	Infects humans and many lower primates; pigs most important reservoir
Entamoeba histolytica	Amebiasis; intestinal infection most common with severe dysentery; hepatic infection causes abscesses and pain over liver and may be fatal	Ingestion of food or water contaminated with cysts	Metronidazole; furamide for cyst passers	Trophozoite and cyst stage occur only in single host; humans are important reservoirs, but dogs, pigs, and monkeys also implicated
African *Trypanosoma* species *T. brucei gambiense* *T. brucei rhodesiense*	African sleeping sickness; central nervous system involvement; heart failure, pneumonia, or encephalitis frequent cause of death	Bite of tsetse fly	Suramin and pentamidine isethionate	Humans only reservoir of *T. gambiense;* cattle, hartebeast, bushbucks, and other mammals are reservoirs
American *Trypanosoma cruzi*	Chaga's disease; visceral organs affected as well as heart; acute disease in all age groups; chronic disease in adults with cardiac dysfunction	Feces of reduviid bugs	No satisfactory treatment available	Humans as well as domestic animals serve as hosts; cats and wild animals such as armadillo and oppossum are important reservoirs
Plasmodium species *P. vivax* *P. falciparum* *P. malariae* *P. ovale*	Malaria; chills and fever, Enlargement of spleen and liver in severe cases and anemia; fever related to release of parasites from infected red blood cells	Bite of female *Anopheles* mosquito	Chlorquine phosphate, primaquine	Life cycle takes place in two hosts: a vertebrate such as humans and an invertebrate, mosquito
Babesia microti	Nantucket fever (babesiosis); symptoms similar to malaria	Transmitted by ticks	Chlorquine phosphate	Life cycle similar to *Plasmodium;* ticks are vectors and intermediate hosts
Toxoplasma gondii	Toxoplasmosis; congenital infection most severe involving eye and central nervous system, sometimes resulting in mental retardation and other clinical conditions; adult infection usually asymptomatic	Ingestion of cysts from cat feces; ingestion of uncooked contaminated meat or transfer across placenta to fetus from infected mother	Sulfadiazone and pyrimethamine	Cats are final host, and they are infected by eating cysts or infected rodents; parasite infects a wide variety of birds and animals; humans are intermediate hosts, as well as wild and domestic animals

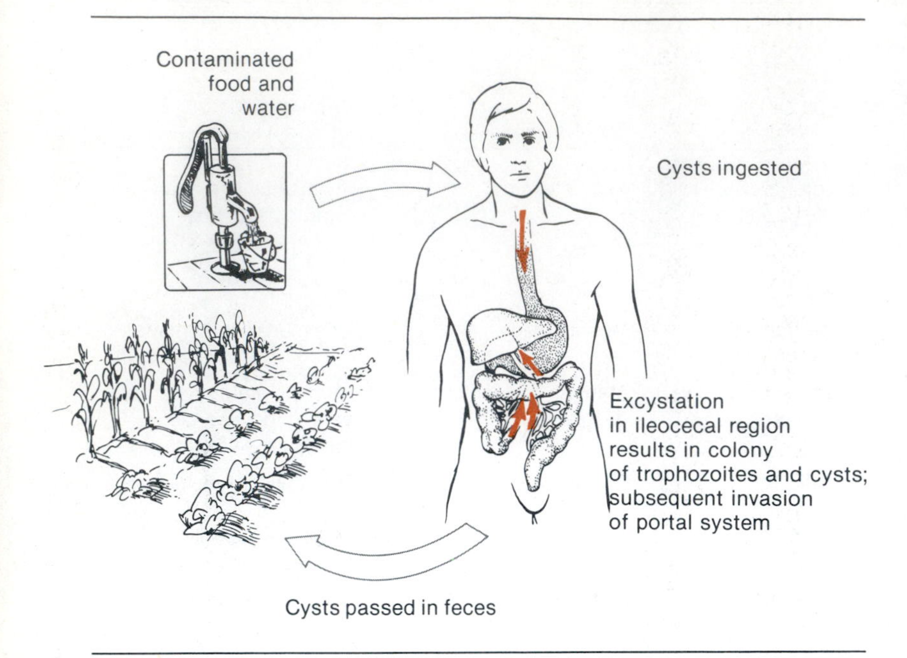

Figure 32-14 Life cycle of *Entamoeba histolytica.* Food and water contaminated with infective cysts are ingested. Excystation in ileocecal regions results in a colony of trophozoites and cysts becoming established in the cecum and colon. Penetration of mucosa by trophozoites with subsequent invasion of portal circulation may result in extraintestinal amebiasis. Cysts passed in feces contaminate food and water, which in turn are ingested by humans.

Modified from Beck, J.W., and Davies, J.E. Medical parasitology *(3rd ed.). The C.V. Mosby Co., St. Louis, 1981.*

infectious agent, but dogs, pigs, and monkeys can also be implicated. Most infections are the result of ingestion of food or water contaminated by cysts (Figure 32-14). Approximately 10% of the world population is affected by this disease; most cases occur in tropical or subtropical areas. In the United States between 3000 to 4000 cases are reported each year. The majority of infections result from poor sanitation and improper disposal of human wastes. Flies and cockroaches can carry cysts from contaminated material and are important vectors in transmission to humans. In areas where human feces is used as a fertilizer, infections among the human population are common. Asymptomatic human cyst shedders are prevalent and provide a ready avenue for transmission to family members.

The virulent strains of *E. histolytica* are cytolytic and invasive. Amebic enzymes have been implicated as the agents that allow the parasite to cause cytolytic effects and to penetrate tissue. The enzymes implicated, however, have also been found in nonvirulent strains of the parasite. More recent findings suggest that microfilaments on the amebic surface make contact with host cells and cause cytolysis. Biochemical characterization of the microfilaments is still lacking. The ingestion of food or water contaminated by cysts (Plate 44) initiates the life cycle. Excystation occurs in the ileocecal region of the intestine, and cysts as well as trophozoites can be found there. Trophozoites that penetrate the intestinal wall produce an ulcer that is usually located in the appendix, cecum, or colon. Death can result from a perforated colon and subsequent peritonitis. Trophozoites that invade the bloodstream can cause secondary disease especially in the liver. Abscesses that appear in the liver sometimes rupture and pass parasites into the body cavity to infect other organs.

Most disease develops slowly with slight nausea and mild gastrointestinal disturbances. Tissue invasion results in sharp abdominal pain sometimes accompanied by vomiting and constipation, with or without diarrhea. The disease may progress to the point that dysentery appears with as many as 10 to 20 or more bloody stools per day. It is during the latter period that extensive ulceration of the intestine takes place sometimes followed by secondary bacterial diseases.

Metronidazole is the drug of choice in treatment. Furamide is used almost exclusively for asymptomatic cyst shedders.

ARTHROPODS AS PARASITES AND VECTORS

The phylum Arthropoda is the largest of the animal phyla, consisting of several million species, half of which are insects. They are all characterized by having jointed appendages, chitinized exoskeletons, and a hemocele (main body cavity). Many of the arthropods implicated as vectors in the transmission of various disease-producing agents, such as bacteria, viruses, and protozoa, are discussed in previous chapters. Table 32-4 lists the various infectons in which arthropods act as vectors of disease-producing agents. Arthropods can also act as ectoparasites or endoparasites on mammals, including humans, in which the latter serve as definitive hosts. The term ***infestation*** is used to describe arthropods that are ectoparasites; however, some can also act as endoparasites and thereby cause disease.

Sucking lice

Human lice belong to the order Anoplura and spend their entire life cycles on their host. Depending upon the species they have preferences for certain parts of the body. Lice utilize piercing and sucking mouth parts to feed several times a day on their human hosts, injecting their saliva into the bloodstream. Repeated biting causes

Table 32-4 Major vector-borne diseases of humans

VECTOR	DISEASE	MICROBE CLASSIFICATION
Mosquito	Animal parasite diseases	
Culex, Anopheles, and *Ades* species	Filariasis	*Wucheria* and *Brugia* species
Anopheles species	Malaria	*Plasmodium* species
	Viral diseases	
Aedes species	Dengue fever	Flavivirus
	Yellow fever	Flavivirus
	St. Louis encephalitis	Flavivirus
Culex species	Eastern equine encephalitis	Alphavirus
	Western equine encephalitis	Alphavirus
	St. Louis encephalitis	Flavivirus
Tick	Bacterial diseases	
	Relapsing fever	*Borrelia recurrentis*
	Rocky Mountain spotted fever	*Rickettsia rickettsi*
	Q fever	*Coxiella burnetii*
	Tularemia	*Francisella tularensis*
	Viral diseases	
	Colorado tick fever	Orbivirus
	Crimean hemorrhagic fever	Ungrouped
Mite	Bacterial diseases	
	Q fever	*Coxiella burneti*
	Rickettsial pox	*Rickettsia akari*
Flies	Animal parasite diseases	
Deer fly	Eye worm	*Loa loa*
Tsetse fly	Sleeping sickness (African)	*Trypanosoma* species
Black fly	Leishmaniasis	*Leishmania* species
	Onchocerciasis	*Onchocerca volvulus*
	Bacterial diseases	
Deer fly	Tularemia	*Francisella tularensis*
Muscoid fly	Yaws	*Treponema pertenue*
	Viral diseases	
Sand fly	Sand fly fever	?
	Vesicular stomatitis	Vesiculovirus
Lice	Bacterial diseases	
	Epidemic typhus	*Rickettsia prowazeki*
	Trench fever	*Rochalimaea quintana*
Fleas	Bacterial diseases	
	Endemic typhus	*Rickettsia typhi*
	Bubonic plague	*Pasteurella pestis*
True bugs (reduviid)	Animal parasite diseases	
	Chaga's disease (American trypanosoiasis)	*Trypanosoma cruzi*

sensitization resulting in severe itching. The effects of infestation include weariness, aching of legs and feet, and irritability. Outside of war, in which typhus is a serious and often fatal disease (see Chapter 28), one of the most serious problems is the stigma that is associated with lice infestation, particularly among young children.

Originally lice were believed to be parasites of birds that fed on skin and feather debris. With the appearance of mammals and humans, lice developed mouth parts for piercing skin and sucking blood. As humans began to wear clothes and

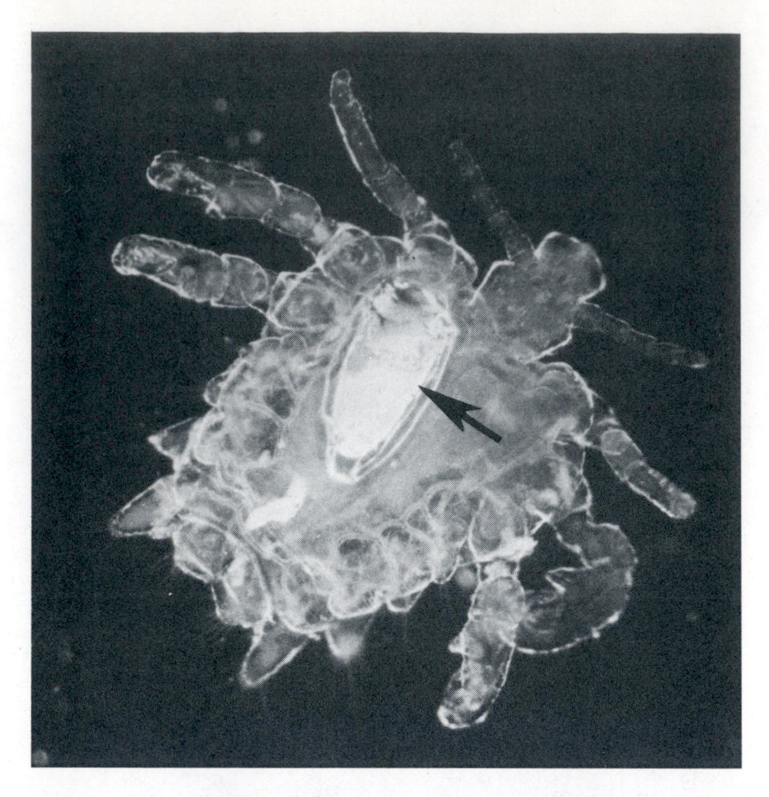

Figure 32-15 Crab louse *(Phthirus pubis)*. Arrow points to developing egg.

From Schmidt, G.D, and Roberts, L.S. Foundations of parasitology *(2nd ed.)*. The C.V. Mosby Co., St. Loius, 1981. Photograph by Warren Buss.

their hair became restricted to certain body parts, such as head, armpits, and pubic areas, the louse evolved and adapted to those areas.

Pubic lice (crab louse). The crab louse, *Phthirus pubis* (Fig. 32-15), prefers coarse hair and is found most frequently in pubic hair. The female, following copulation with the male, lays eggs (nits) on the hair, which hatch in about 8 days. Spread of the infestation is usually by sexual contact. Most methods of treatment include lotions that contain malathion, carbaryl, or gamma-benzene hexachloride.

Body lice. *Pediculus humanus humanus* is the cause of louse infestation involving the body. The female lays eggs on the clothing next to the skin and visits the skin only for blood meals. Eggs hatch in 20 days, and adulthood is reached in about 7 days. Their existence depends on the host wearing the same clothing at least part of each day. Transmission occurs only by close contact with the clothing of an infested person. To dispose of body lice, clothing can be placed in a dryer for several minutes, fumigated with methyl bromide, or washed in hot water. Body lice are vectors in the transmission of agents that cause epidemic typhus and trench fever, as well as relapsing fever (see Chapter 28).

Head lice. *Pediculus humanus capitis,* or head louse, confines its activities to the head. Most infestations occur in school children in whom transmission occurs by close contact such as when heads are close together. In the United States as many as 6 million infestations are believed to occur each year. Unlike the body louse, the head louse does not live on clothing. Its life span is about 30 days, and during that period the female lays eggs (6 to 8) every 24 hours. She secretes a sticky substance on the hair shaft where the eggs become attached. The eggs, which are large enough to be seen with the naked eye, develop into larvae that molt three times before reaching adulthood. Skin irritations do not appear until several hours after a blood meal by the parasites. Malathion, carbaryl, and lindane are the insectides of choice for treatment. Although DDT has been banned because of ecological reasons, it is still one of the most formidable agents for controlling head lice. In many areas of the world resistance to DDT may be encountered because of its frequent use.

True bugs

The true bugs belong to one of the largest insect orders, *Hemiptera.* Few species are ectoparasites of humans, and only two are of medical significance, ***bedbugs*** and ***triatomid bugs*** (assassin bugs). Bedbugs (Fig. 32-16) feed on the blood of their hosts and are well known for their painful bites, but they seldom serve as vectors of infectious agents for humans. They are nocturnal insects and feed on their hosts during sleep. In those households where infestation occurs, the bedbug prefers to hide out in mattresses, bedsprings, wall cracks, and floor crevices before venturing out for nocturnal meals.

The triatomid bugs are large and may be over an inch in length. They are often brightly colored and have well developed wings for flight. They are found in tropical and subtropical areas but can also be found in temperate climates. Many are found in cracks and crevices of wood piles, on the ground, or even in trees, but few are found in human dwellings. The bite of some triatomids is painful, and they may harbor the agent *Trypanosoma cruzi,* the causative agent of Chaga's disease (Table 32-3). *T. cruzi,* which is transferred to the host when the bug defecates, may penetrate intact tissue or enter through the puncture wound made by the bug.

A variety of insecticides such as dieldrin can

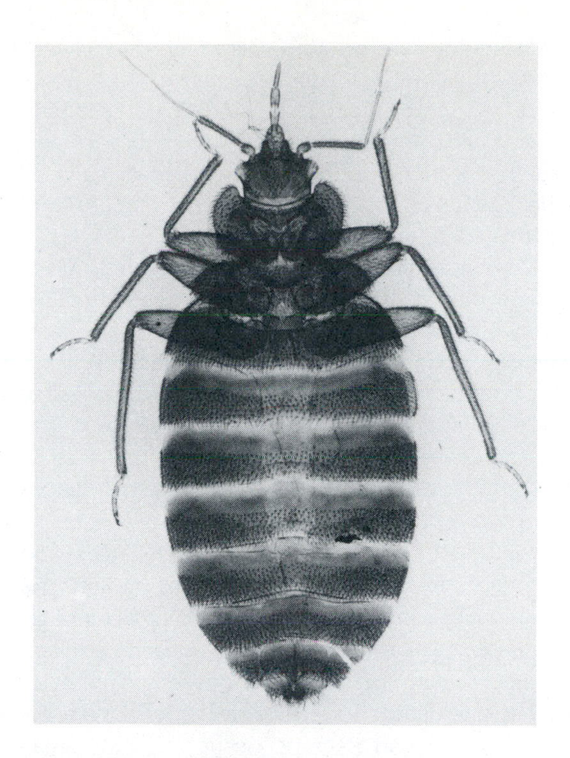

Figure 32-16 Bedbug.
Courtesy W. Henry Leigh, Department of Biology, University of Miami. From Beck, J.W., and Davies, J.E. Medical parasitology *(3rd ed.). The C.V. Mosby Co., St. Louis, 1981.*

Figure 32-17 Rat flea *(Xenopsylla)*. Most common vector of plague and murine typhus
From Schmidt, G.D., and Roberts, L.S. Foundations of parasitology *(2nd ed.). The C.V. Mosby Co., St. Louis, 1981. Photograph by Jay George.*

be applied around the hiding places of the true bugs, but household cleanliness and reducing the possible sites of insect habitation are equally important.

Fleas

Fleas belong to the order Siphonaptera, but unlike many other insects do not have wings. They have well developed legs for jumping and can jump 100 times their body length, which would be equivalent to a human jumping over most city buildings. Fleas are found in households because of the presence of three groups of hosts: humans, dogs and cats, and rats and mice. The bites of fleas are particularly annoying and usually painful, but their importance lies in the fact they can transmit agents of disease (Table 32-4).

Dog fleas can feed on human hosts when their preferred hosts are absent. These fleas are intermediate hosts of the dog tapeworm *Dipylidium caninum*, which can be transmitted to children.

The fleas of rats (Figure 32-17) and mice as well as other rodents such as ground squirrels, and prairie dogs are carriers of the causative agent of ***bubonic plague*** (Chapter 28) and ***murine typhus***. The human flea is also a carrier of the agent of plague and can be found on other hosts such as dogs, ground squirrels, and coyotes.

Flies

Flies are two-winged insects that are the most important arthropods in terms of medical significance. Only those groups that directly or indirectly cause disease in humans are discussed.

Black flies. Blackflies are worldwide and in North America are especially prevalent in the lakes of Canada and northern United States. Larval development requires well-oxygenated water. The adults feed both day and night, and fisherman who venture into northern lakes and streams will testify to their painful sting. When they appear in large numbers they are capable of killing animals by sucking large quantities of blood. Herds of livestock have perished from black fly infestations. In tropical areas of Central and South America and Africa species of black fly are intermediate hosts for the filarial worm *Onchocerca volvulus* (Table 32-2).

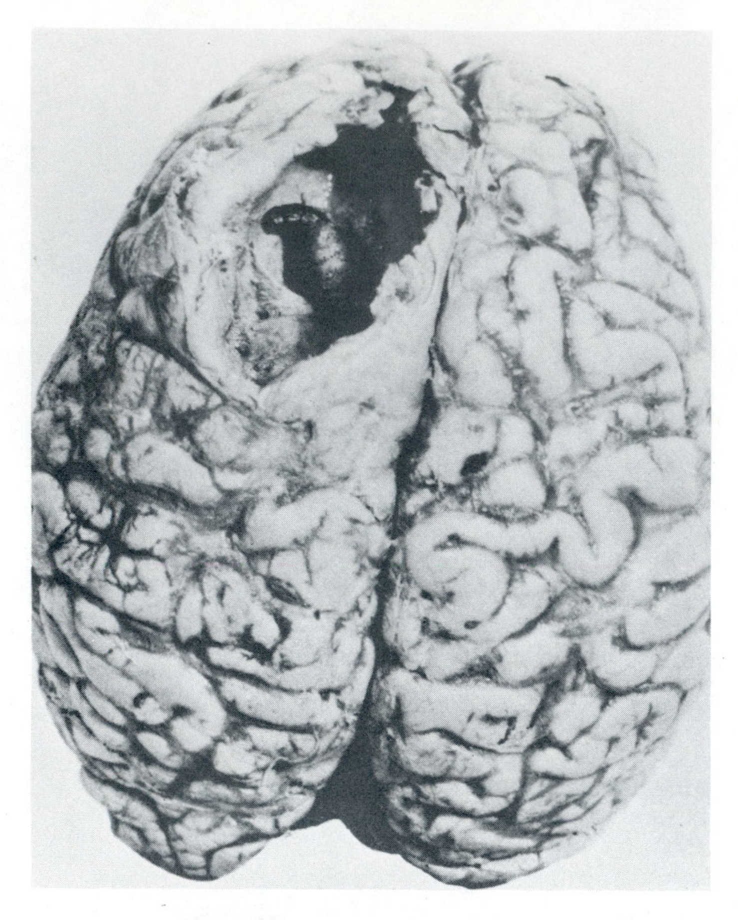

Figure 32-18 Myiasis of brain. Several larvae were found in scalp of patient. Larvae have bored through brain.
Air Force Institute of Pathology, AFIP N-50807.

Horse flies and deer flies. Horse flies and deer flies belong to the same family of arthropods, Tabanidae. Most feed during the day and attack animals but they can also bite humans. They become particularly annoying near water, which is often their breeding ground. Their mouth parts are large, and the bite they inflict is painful. During the feeding process they travel from one host to another and in so doing can easily transmit infectious agents. The deer fly is a vector for the agent of tularemia *Francisella tularensis* (Table 32-4) and is also an intermediate host for the filarial worm *Loa loa* (Table 32-2).

Muscoid flies. Muscoid flies include those flies most frequently associated with human disease, such as the housefly. The housefly and related species are important carriers of infectious agents. They feed on garbage and human excrement. Their bodies are covered with hairs and they have sticky pads upon which microbial agents are easily transported to a potential host. The microbial agents survive in the intestine of the fly and are freely dispersed when it vomits or defecates in the wound. Most of the disease agents that are transmitted cause enteric infections such as typhoid fever, shigellosis, salmonellosis, polio, and hepatitis, but other disease-producing agents may be carried as well, such as the agents of yaws *(Treponema pertenue),* and worms such as *Ascaris.*

The tsetse flies are restricted to areas in Africa where they feed on animals as well as humans. They are vectors for the agents of African sleeping sickness (trypanosomiasis) (Table 32-3).

Certain species of muscoid flies are responsible for the disease called myiasis. ***Myiasis*** is caused by the larvae (maggots) that live as parasites in human tissue (Figure 32-18). Myiasis can be associated with many areas of the body such as urogenital tract, intestinal tract, orbital areas, and cutaneous sites. In humans most infestations are cutaneous, which results in either invasion of lesions on the skin ***(traumatic myiasis)*** or invasion of intact skin ***(furuncular myiasis).*** Traumatic myiasis is characterized by deposited larvae living on dead and necrotic tissue. In some areas where hygienic conditions are substandard, the umbilicus of newborn babies is a site for infestation, as well as the nasal passages or ears. Most cases of traumatic myiasis are considered benign, but infestation by the screwworm can be deforming. In furuncular myiasis larvae may develop in subdermal tissue just below the entry site, or the larvae may migrate through the body exiting to the surface. At the site of exit a boil (furuncle) is formed, and the mature larvae fall to the ground. The boil is considerably painful and can become infected with bacteria. The developing larvae are large enough to be removed from the lesion, and dressings saturated with 5% to 10% chloroform in a light vegetable oil will cause the maggots to migrate to the skin surfaces where they can be physically removed. Sanitation and hygiene can prevent infestations, and for this reason they are not often seen in the United States.

Sand flies. Sand flies are important vectors of disease in tropical and subtropical areas. The adult female feeds on animal blood (including humans'), but the male like most other Diptera is restricted to feeding on plant juices because of underdeveloped mouth parts. Sand flies have soft exoskeletons that are easily damaged by dry, hot weather. They avoid such environments and feed when the humidity is very high. Sand flies are vectors for such diseases as leishmaniasis (Table 32-3) and bartonellosis, and the nonfatal viral disease, sand fly fever. Bartonellosis is found in Ecuador, Colombia, and Peru, and one clinical form

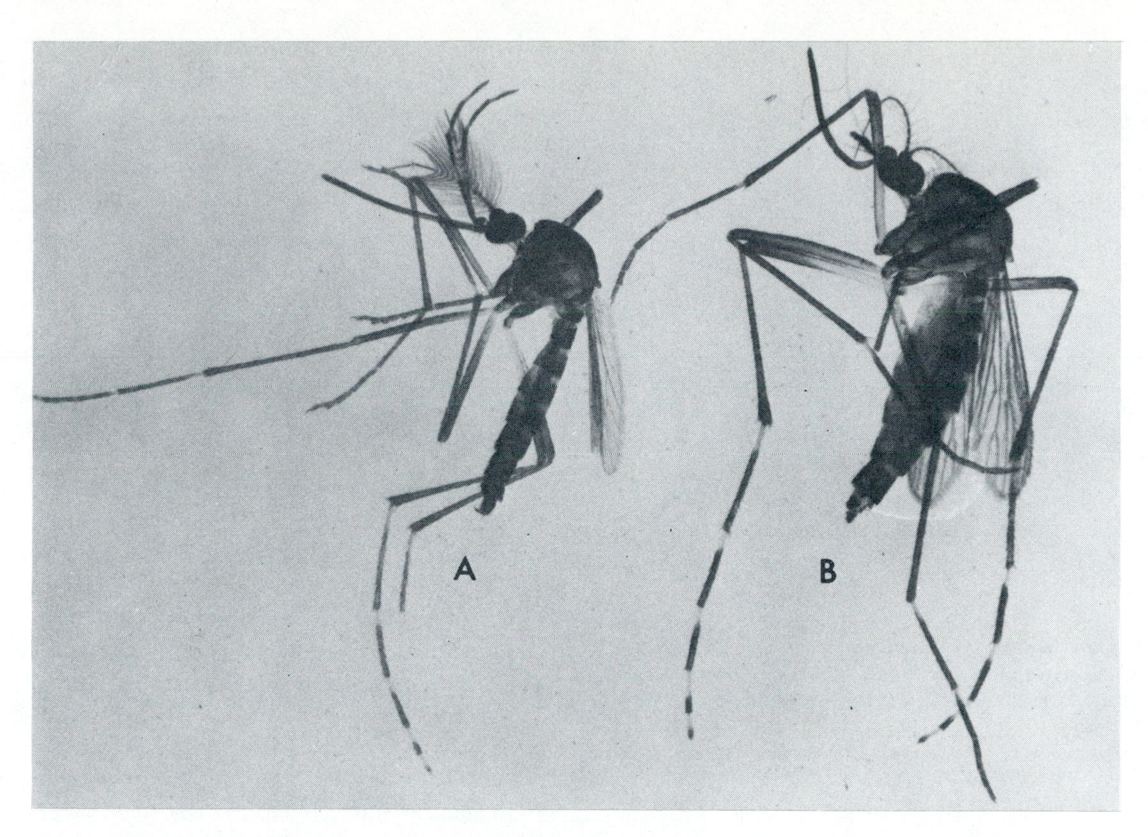

Figure 32-19 *Aedes aegypti.* Agent of yellow fever. **A,** Male. **B,** Female.
From Schmidt, G.D., and Roberts, L.S. Foundations of parasitology *(2nd ed.). The C.V. Mosby Co., St. Louis, 1981. Photograph by Warren Buss.*

of the disease, called Oroya fever, can cause a fatal disease.

Midges. Midges (gnats) are extremely small members of the order Diptera, seldom reaching 1 mm in length. Most species are merely pests for domestic livestock and humans, but some can transmit microbial parasites. The etiological agent of ***bluetongue,*** a virus disease of sheep and cattle, is transmitted by midges. Bluetongue is characterized as a hemorrhagic disease that can sometime be fatal. Some species transmit filarial worms to humans, but humans in the United States are not affected.

Mosquitoes. Mosquitoes are members of the family Culcidae, which are the most important vectors of human disease. Their life cycle consists of adult, egg, larva (wiggler), and pupa (tumbler). Part of their development takes place in water. The medically important genera are *Anopheles, Culex,* and *Aedes.*

The genus *Anopheles* is the sole vector for the agent of malaria in humans. Malaria has been practically eliminated from some developed countries of the world, but the incidence of infections in many tropical and subtropical countries is alarmingly high. Progress was made in the early 1970s in controlling malaria transmission, but laxity by government officials in maintaining eradication programs, as well as resistance of the mosquito to insecticides, has resulted in a dramatic increase in disease in the 1980s.

The genus *Culex* contains species that transmit the agents of such diseases as Western equine encephalitis, St. Louis encephalitis, and Eastern equine encephalitis (see Chapter 29). Birds may serve as reservoirs for the agents causing these three diseases, but humans and other mammals are fed upon by the mosquito. Horses, for example, are particularly susceptible to the agent of Western equine encephalitis, and fatality rates are high. The common mosquito found in the home is also a species of *Culex,* and in addition to its annoying bite is also a vector for the filarial worm *Wucheria bancrofti* (Table 32-2). *Culex* species are also vectors for species of *Plasmodium,* which are the cause of bird malaria.

Species in the genus *Aedes* are vectors for the agents of viral encephalitis and dengue fever (Table 32-4). *Aedes aegypti* (Figure 32-19) is the principal vector of the virus causing yellow fever.

Control of mosquitoes is best achieved by focusing on the breeding grounds, such as any area in the environment that holds water and becomes stagnant. Insecticides such as DDT, mal-

Figure 32-20 Wood tick *(Dermacentor)*. One species of this genus transmits agents of Rocky Mountain spotted fever and tularemia in humans as well as tick paralysis in pets.

From Hickman, C.P., et al., Integrated principles of zoology (6th ed.). The C.V. Mosby Co., St. Louis, 1981.

athion, and dieldrin can be used to fog areas harboring large mosquito populations.

Ticks and mites

Ticks and mites belong to the order Acarina. Unlike the insects that have segmented bodies, the Acarina possess a head, thorax, and abdomen that is fused into a single structure. Ticks are strictly terrestrial, but mites may be found in terrestrial and aquatic environments. Both groups can be found throughout the world and can act as parasites as well as vectors of microbial agents.

Ticks. Ticks (Figure 32-20) can be divided into two families: the hard ticks (Ixodidae) and the soft ticks (Argasidae). Hard ticks have a hard scutum that covers their back, whereas soft ticks possess a leathery cuticle. All ticks undergo four stages of development: egg, larva, nymph, and adult. Larvae are found on vegetation where they wait for a suitable host to pass by. Once the host is located, the larva feeds and then molts to an eight-legged nymph. Molting may take place on the same animal or on many hosts, depending on the species of tick. Eventually the nymph falls to the ground and molts for the last time to become an adult. The adult awaits a host upon which it feeds and engorges itself with blood. Soft ticks feed rapidly and fall off the host but hard ticks may remain on the host for several days.

Ticks can cause several conditions depending on the host, site of attachment, and tick species. These conditions include the following.

1. Some ticks attach to animals and can extract large volumes of blood over a period of time, causing anemia. Infestations may be so heavy that the animal dies.
2. The larvae of some ticks may attach to the ears of mammals, particularly domestic animals. The larvae molt on the animals and can cause serious complications such as secondary bacterial infections.
3. Tick paralysis is a condition resulting from ticks attaching to the base of the skull of mammals, including humans. The condition is believed to be the result of a poison injected by the tick into the host.
4. Inflammatory reactions at the site of a tick bite are relatively common. The host response may be caused by the presence of tick mouth parts in the wound following removal of the tick, as well as by microbial contamination. Once the tick has been removed, the wound should be treated with an antiseptic.
5. Ticks are vectors for many disease agents, including viruses, bacteria, and animal parasites. Human diseases that are transmitted by ticks include rickettsial infections such as Rocky Mountain spotted fever and Q fever, relapsing fever caused by *Borrelia recurrentis,* tularemia caused by *Francisella tularensis,* and viral diseases such as Colorado tick fever and Crimean-Congo hemorrhagic fever. Animals are also susceptible to infectious agents transmitted by ticks. Some of the infections are fatal to the host. East Coast fever, for example, is a protozoan disease of the red blood cells of cattle in which the mortality rate may be as high as 80%.

Mites. Mites are considerably smaller than ticks but like ticks possess both parasitic as well as vector properties. They affect wild and domestic animals as well as humans. Mites are vectors for such diseases as endemic typhus, rickettsial pox, and Q fever. Mites as parasites are important for two conditions: scabies and chiggers dermatitis.

Scabies. Scabies or "the itch" is believed to have been observed in biblical times, but it was not until 1687 that microscopical observations and identification of the mite were made. The agent of scabies *(Sarcoptes scabei)* is approxi-

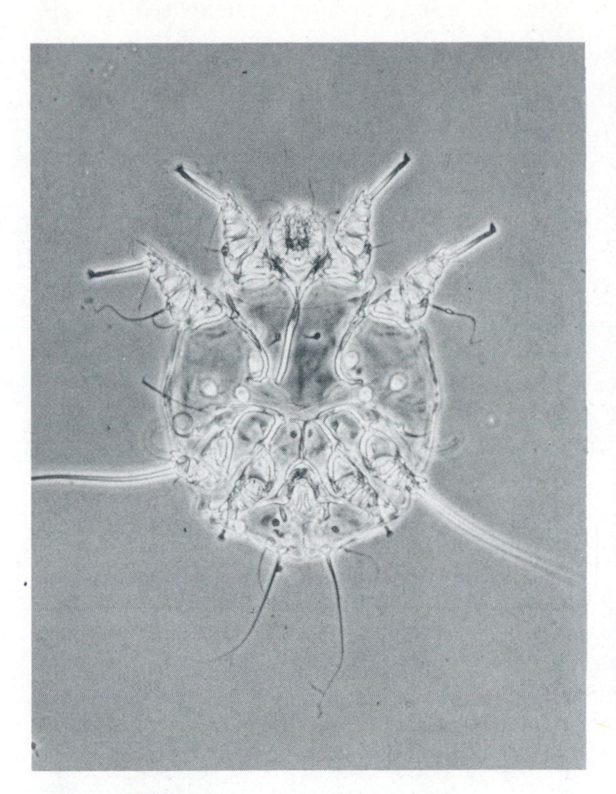

Figure 32-21 Itch mite *(Sarcoptes scabiei).*
From Beck, J.W., and Davies, J.E. Medical parasitology (3rd ed.). The C.V. Mosby Co., St. Louis, 1981.

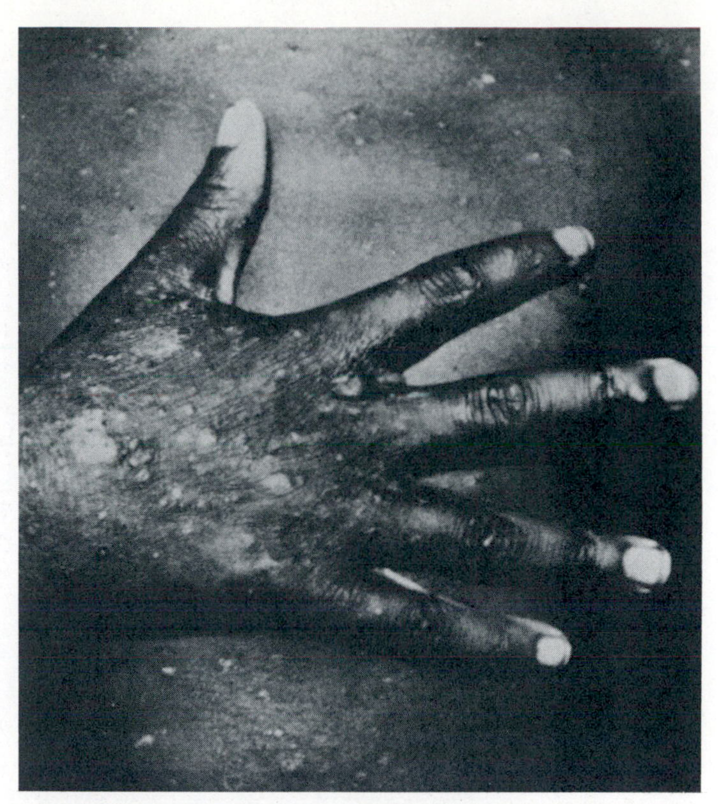

Figure 32-22 Scabies of hand. Lesions complicated by scratching and secondary bacterial infections.
Air Force Institute of Pathology, AFIP 68-7834-20.

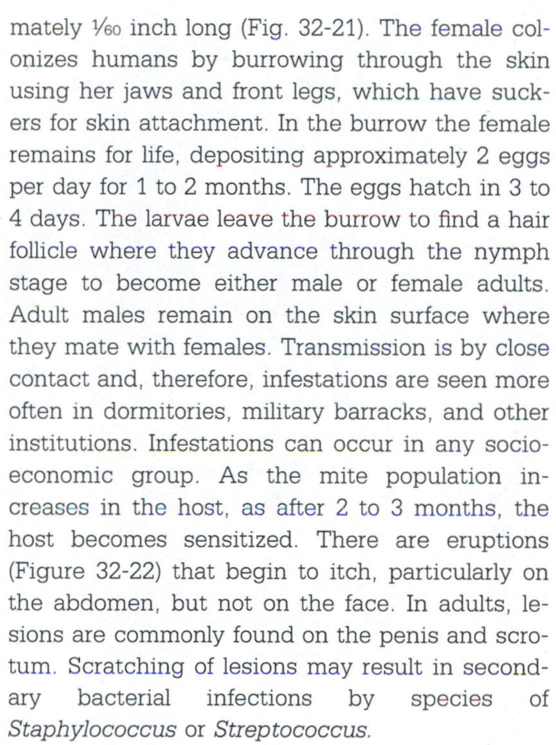

mately 1/60 inch long (Fig. 32-21). The female colonizes humans by burrowing through the skin using her jaws and front legs, which have suckers for skin attachment. In the burrow the female remains for life, depositing approximately 2 eggs per day for 1 to 2 months. The eggs hatch in 3 to 4 days. The larvae leave the burrow to find a hair follicle where they advance through the nymph stage to become either male or female adults. Adult males remain on the skin surface where they mate with females. Transmission is by close contact and, therefore, infestations are seen more often in dormitories, military barracks, and other institutions. Infestations can occur in any socioeconomic group. As the mite population increases in the host, as after 2 to 3 months, the host becomes sensitized. There are eruptions (Figure 32-22) that begin to itch, particularly on the abdomen, but not on the face. In adults, lesions are commonly found on the penis and scrotum. Scratching of lesions may result in secondary bacterial infections by species of *Staphylococcus* or *Streptococcus*.

Diagnosis is made by removing the mite from the burrow with a pin and observing it microscopically or with a hand lens. Scabies is a common infestation in the United States, and most cases occur in those under 14 years of age. Several drugs are available for treatment: gammabenzene hexachloride, crotamiton, and benzyl benzoate.

Chiggers dermatitis. The mite (species of *Trombicula*) is responsible for an infestation referred to as ***chiggers*** or ***trombiculiasis.*** Eggs are laid in the soil and eventually hatch into larvae, whose further development requires animal protein. It is at this stage that they become parasitic. The larvae initially attach to the tips of grass or weeds but attach to animals or humans when contact is made. Mites invade areas of constricted clothing such as the ankles and waist but do not burrow into the skin. They inject salivary secretions into the epidermis that digest host cells. The mite then sucks up the host juices until engorged and then drops off. Itching begins about 4 to 6 hours after contact with the mite. A day later small reddened areas (papules) develop that are associated with intense itching. The papules disappear within 1 to 2 weeks after exposure. Infestations can be prevented by avoiding tall grassy areas. Treatment consists of relieving the itching with various solutions such as 5% benzocaine.

SUMMARY

Animal parasites differ in many biological and morphological aspects from other microbial agents. For example, some are unicellular, and others are multicellular. The unicellular agents are called the Protozoa, and the multicellular types are called the Helminths. Their life cycles may be simple or complex, depending on the number of hosts involved. In many parasitic diseases it is the environmentally resistant cyst stage that is transmissible to humans, whereas in other diseases it is an active larval stage that is involved.

The acquisition of most animal parasite infections is brought about by poor hygiene. Most occur in tropical and subtropical regions of the world. The parasite may be transmitted to humans by: (1) active transmission, (2) passive transmission, (3) venereal means, and (4) congenital means. Whether one becomes infected is related to the number of parasites ingested or number of exposures and the ability of the parasite to penetrate epithelial barriers by special enzymes or specialized structures. Certain parasites have been found in up to 50% of the human population, such as *Toxoplasma gondii* and *Trichomonas vaginalis.* Some animal parasites are harmless commensals in humans, but they can be virulent if they invade an organ that is not their normal site of development. Many diseases in the United States are acquired by contact with pets, in whom the parasite is harbored.

Unlike other microbial agents animal parasites do not confer a permanent immunity following disease. During the disease process, however, the host is protected from reinfection. Animal parasite diseases do stimulate the production of specific IgE antibodies, but their part in immunity is not known.

Many, but not all, animal parasite diseases can be diagnosed by examination of the parasite in the feces. When the parasite is present in the body or is expelled from the body, it exists in some morphological form, such as a cyst, larva, or egg, that exhibits a unique morphological characteristic suitable for identification. Serological tests can also be used if the parasite cannot be identified by other means.

Most chemotherapeutic agents used in the treatment of disease are toxic, and drug levels must be controlled. The important drugs are diethylcarbamazine, niclosamide, mebenidazole, thiabendazole, atabrine, and metronidazole. No vaccines are available for humans, and the best method for prevention is to maintain proper hygiene and nutrition.

The parasitic helminths are divided into two groups: the tapeworms and flukes (Platyhelminthes) and roundworms (Nematoda). Tapeworms (species of *Taenia*) are found in the intestine of vertebrates. Humans are infected by ingestion of eggs or larvae encysted in the flesh of animals they eat. The parasites may infect any organ of the body. Flukes have more complex life cycles and unlike tapeworms are found in areas other than the intestine, such as liver, lung, blood, and central nervous system. Ingested by humans the parasites mature in the lung, liver, and heart with some later passing their eggs in the feces. Roundworms (nematodes) are unsegmented and free-living parasites that can infect the intestine, blood, and visceral organs. One of the most important roundworms is *Ascaris lumbricoides,* which is transmitted to humans primarily by eating eggs from contaminated soil or vegetables. The parasite is found in the intestinal lumen of humans, where it may cause obstruction or it may migrate to other areas such as the bile duct and pancreatic duct.

The parasitic protozoa are primarily free living, but a few are parasitic. *Entamoeba histolytica* is the most important human parasitic protozoa and is responsible for amebic dysentery. Transmission to humans is by fecally contaminated water or food. The parasite can be disseminated to many visceral organs.

Arthropods not only are vectors in the transmission of infectious microbial agents but by themselves may be parasites for vertebrates including humans. Infestation is a term used to describe their habitation of humans. The arthropods of medical significance as parasites and vectors in transmission of infectious agents are lice, mosquitoes, flies, midges, true bugs, and ticks and mites.

Study questions

1. Distinguish between trophozoite and cyst stages.
2. How does the transmission of animal parasites compare with the transmission of bacterial agents?
3. What animal parasite diseases are prevalent in the United States?
4. Discuss what is meant by parasite life cycles.
5. What structural or biochemical characteristics enable many animal parasites to invade tissue and cells of the human host?
6. What type of host immunity predominates during disease by animal parasites? How does this compare with bacterial diseases?
7. How do the identification procedures for animal parasites differ from those for bacteria?
8. What considerations must be taken into account before drugs are administered for animal parasite diseases? Why are there so few drugs available?
9. Why are vaccines unavailable for animal parasite infections, although they are for bacterial diseases?
10. What human endeavors or habits make animal parasite diseases so common in underdeveloped countries?
11. Why can tapeworms grow to such lengths?
12. How are tapeworms acquired by humans?
13. In terms of their site of infection in the human body, how do tapeworms differ from flukes? How do their life cycles differ?
14. What are the four major groups of protozoa that infect humans, and how are they distinguished?
15. What is the major cause of amebiasis? How is the disease transmitted?
16. What arthropods cause infestations of humans? Describe the various types of lice and how they are transmitted. What infectious agents can lice carry?
17. What is scabies? What are the symptoms of infestation?
18. What is the cause of myiasis?
19. Define: helminth, definitive host, giardiasis, cysticercus larvae, proglottid, miracidium, hermaphroditic.

Selected readings

Books

Arthur, D.R. *Ticks and disease.* Pergamon Press, Oxford, England, 1962.

Beck, J.W., and Davies, J.E. *Medical parasitology* (3rd ed.). The C.V. Mosby Co., St. Louis, 1981.

Blecka, L.J. *Concise medical parasitology.* Addison-Wesley, Reading, Mass., 1980.

Boyd, R.F., and Marr, J.J. *Medical Microbiology,* Little, Brown & Co., Boston, 1980.

Brown, H.W. *Basic clinical parasitology* (4th ed.). Appleton-Century-Crofts, New York, 1975.

Faust, E.C., Beaver, P.C. and Jung, R.C. *Animal agents and vectors of human disease* (4th ed.). Lea & Febiger, Philadelphia, 1975.

Freeman, B.A. *Burrows Textbook of Microbiology* (21st ed.). W.B. Saunders Co., Philadelphia, 1979.

Harrison, G. *Mosquitoes, malaria and man: a history of the hostiles since 1880,* E.P. Dutton & Co., New York, 1978.

Hoeprich, P.D. (Ed.). *Infectious diseases* (2nd ed.). Harper & Row Publishing Co. New York, 1977.

Kreier, J.P. *Parasitic protozoa* (Vol. 1). Academic Press Inc., New York, 1977.

Mandell, G.L. (Ed.). *Principles and practice of infectious disease.* John Wiley & Sons, New York, 1979.

Markell, E.K., and Voge, M. *Medical parasitology* (4th ed.). W.B. Saunders Co., Philadelphia, 1977.

Schell, S.C. *How to know the trematodes.* William C. Brown Co., Dubuque, Iowa, 1970.

Schmidt, G.D., and Roberts, L.S. *Foundations of parasitology* (2nd ed.). The C.V. Mosby Co., St. Louis, 1981.

Youmans, G., Paterson, P.Y. and Sommers, H.M. *The biologic and clinical basis of infectious disease* (2nd ed.). W.B. Saunders Co., Philadelphia, 1980.

Journal articles

Benach, J.L., and Habicht, G.S. Clinical characteristic of human babesiosis, *J. Infect. Dis.* **144:**481, 1981.

Blumenthal, D.S. Intestinal nematodes in the United States. *N. Engl. J. Med.* **297:**1437, 1977.

Cursons, R.T.M., Brown, T.J. and Keyes, E.A. Virulence of pathogenic free-living amebae. *J. Parasitol.* **64:**744, 1978.

Fouts, A.C., and Kraus, S.J. *Trichomonas vaginalis:* Reevaluation of its clinical presentation and laboratory diagnosis. *J. Infect. Dis.* **141:**137, 1980.

Jellison, W.L. Fleas and disease. *Annu. Rev. Entomol.* **4:**389, 1959.

Knight, R., and Wright, S.G. Progress report: intestinal protozoa, *Gut* **19:**940, 1978.

Krick, J.A., and Remington, J.S. Toxoplasmosis in the adult—an overview *N. Engl. J. Med.* **298:**550, 1978.

Lavoipierre, M.M.J. Feeding mechanisms of bloodsucking arthropods. *Nature* **208:**302, 1965.

Mahmoud, A.A.F., and Warren, K.S. Algorithms in the diagnosis and management of exotic disease. IV. American trypanosomiasis. *J. Infect. Dis.* **132:**121, 1975.

Maunder, J.W. Human lice—biology and control. *R. Soc. Health J.* **97:**29, 1977.

Most, H. Trichinosis—preventable yet still with us. *N. Engl. J. Med.* **298:**1178, 1978.

Rothschild, M. Fleas. *Sci. Am.* **213:**44, 1965.

Smith, J.W., and Wolfe, M.S. Giardiasis. *Annu. Rev. Med.* **31:**373, 1980.

Tan, J.S. Common and uncommon parasitic infections in the United States. *Med. Clin. North Am.* **62:**1059, 1978.

Chapter 33 HOSPITAL-ASSOCIATED DISEASES

Hospital-associated diseases are those that develop in the patient during his stay in the hospital and that were not present at the time of admission. Hospital-associated diseases are also called ***nosocomial,*** which is derived from the Greek: *nosos,* disease; *komeion,* to take care of. A recent study indicates that from 5% to 10% of the patients in acute-care hospitals will develop a nosocomial disease. Nosocomial diseases are an important health problem, not only because of the increased risk on the patient's life, but for economic reasons as well.* One of the ironies is that the improvements and advances in hospital medical care are also responsible for many nosocomial diseases. Increased antibiotic therapy, the use of therapeutic and diagnostic equipment (such as intravenous and urinary catheters), surgical procedures, and transplantation have increased the risk of nosocomial disease. It is important to understand, however, that the percentage of patients who might die without these therapies and procedures is much greater than the percentage of patients who acquire diseases from their use. The purpose of this chapter is to describe those factors that contribute to hospital-associated diseases and to briefly outline the techniques that are being used to prevent them.

EPIDEMIOLOGY

The epidemiological factors that are associated with diseases outside the hospital (community diseases) are also important in nosocomial diseases. A more detailed discussion of those factors that are common to nosocomial as well as community-associated diseases is found in Chapter 26.

Sources

The organisms causing disease in the hospital may be from endogenous or exogenous sources. ***Endogenous diseases*** are those caused by the patient's own flora, and ***exogenous diseases*** are those that arise from sources other than the patient. The microbial flora (see boxes) of the patient, which is made up of microorganisms that are considered "nonvirulent," can cause endogenous diseases under circumstances in which (1) the organism is displaced from its normal habitat (under such circumstances the new habitat often promotes the rapid multiplication of the organism); (2) conditions alter the environment of the organism, such as an irritant (foreign object) or alteration of the blood supply to a body site; (3) multiple or broad-spectrum antibiotics reduce the number of competing microorganisms; and (4) the immunological defenses of the host have been suppressed by drugs, such as steroids, or by radiation therapy,

Exogenous sources of infection may be classified into animate and inanimate. ***Animate*** sources of infection include (1) other patients in the hospital with infectious disease who may represent a risk to other patients as well as hospital staff, (2) hospital staff who serve as a reservoir of potentially infectious microorganisms, and (3) visitors who may not only serve as a source of infection but may themselves be at risk of acquiring an infection from the patient.

There are many ***inanimate*** objects within the hospital environment that are sources of infection to the patient (Table 33-1). These sources may also represent reservoirs where microorganisms,

*Approximately 34 million people are admitted to acute-care hospitals each year. If 5% acquire nosocomial diseases and require the average 7 days of additional care, the cost of the excess hospitalization is approximately 2.4 billion dollars per year.

Table 33-1 Types of microorganisms associated with certain exogenous sources

SOURCE	COMMON TYPES OF MICROORGANISMS
Outside air	Fungi, bacterial spores, soil-associated bacteria and fungi
Urinary catheters	Bacteria associated with contaminated antiseptics, instruments, drainage tubes, and/or collecting bags; *E. coli*, enterococci, and species of *Proteus, enterobacter, Klebsiella, Serratia, Pseudomonas*, and *Candida*
Intravenous therapy equipment	Gram-negative bacteria, species of *Enterobacter, Flavobacterium, Pseudomonas, E. coli, S. aureus*, species of *Candida*, and the hepatitis B virus
Respiratory therapy equipment	*S. aureus* and pneumococci
Water systems (such as water softeners, deionization units, hemodialysis machines, and hydrotherapy equipment)	Gram-negative bacteria and associated bacterial endotoxins, species of *Acinetobacter, Serratia, Aeromonas, Klebsiella, Escherichia*, and the hepatitis B virus
Antiseptics	Gram-negative bacteria, species of *Pseudomonas, Flavobacterium, Acinetobacter, Klebsiella, Serratia*, and *Candida*

Reprinted from Favero M.S. In Fahlberg, W.J., and Gröschel, D. (Eds.). *Occurrence, diagnosis, and sources of hospital-associated infections.* Marcel Dekker, Inc., New York, By courtesy of Marcel Dekker, Inc.

particularly bacteria, replicate and maintain themselves. Bacteria can grow in a variety of moist environments, including water softeners, humidifiers, whirlpools, sinks, soap dispensers, hemodialysis machines, and humidifiers. Moist environments support the growth of many gram-negative bacteria, such as species of *Pseudomonas, Klebsiella, Serratia,* and *Flavobacterium,* which can survive in water for up to 1 year. These organisms are especially troublesome because they are able to resist many disinfectants, such as chlorine and quaternary ammonium compounds, which are used as antimicrobials in the hospital. One of the organisms, *Pseudomonas aeruginosa,* deserves special mention. This organism can be regarded as a water organism that has become one of the most important causes of nosocomial infections. It is an extremely versatile organism capable of surviving for many months at ambient temperatures. Much of its versatility is related to its ability to metabolize a variety of organic substrates, much more than any other microbial species. Many strains of *P. aeruginosa* produce an exopolysaccharide that promotes the formation of large microcolonies in host tissue. The exopolysaccharide also serves to prevent antimicrobial penetration and thus hinders therapeutic as well as disinfecting measures in the hospital.

In addition to the usual patient-care supplies and equipment, food may be a direct or indirect source of infection. Food may be contaminated before coming to the hospital and when improperly cooked can be a source of enteric infections from such bacterial agents as *Salmonella* and *Shigella.* Food may also be contaminated in the hospital by food handlers who may be carriers of infectious agents. Nosocomial disease from contaminated food occurs infrequently.

Transmission of infectious agents

Microorganisms can be transmitted within the hospital environment by any of four routes: contact, common vehicle, air, or vectors. Some microorganisms may be transmitted by more than one route; therefore the epidemiology of nosocomial infections may vary from one epidemic to another.

Contact transmission. The hospitalized patient can be infected by direct or indirect contact with microorganisms. The hands of the physician or nurse may be inadequately washed following contact with a contaminated environment and microorganisms may be transferred directly to the patient. Indirect contact implies that the transfer of microorganisms to the patient takes

PRINCIPAL MICROORGANISMS INDIGENOUS TO THE ORAL CAVITY AND RESPIRATORY TRACT

BACTERIA
Actinomyces species
Acinetobacter species
Corynebacterium species
Branhamella species
Fusobacterium species
Lactobacillus species
Moraxella species
Mycoplasma species
Peptostreptoooccus species
Spirochetes
Streptococci such as enterococci and *S. pneumoniae*
Veillonella species
Spirochetes
Streptococci such as enterococci and *S. pneumoniae*
Veillonella species
Viridans streptococci

FUNGI
Candida albicans and other yeasts

PROTOZOA
Entamoeba gingivalis
Trichomonas tenax

PRINCIPAL MICROORGANISMS INDIGENOUS TO THE INTESTINAL TRACT

BACTERIA
Achromobacter species
Aeromonas species
Alcaligenes faecalis
Bacteroides species
Clostridium species
Corynebacterium species
Gram-negative enteric bacilli such as species of *Proteus* and *Enterobacter*
Enterococci
Flavobacterium species
Fusobacterium species
Lactobacillus species
Mycoplasma species
Peptrostrepococcus species
Pseudomonas aeruginosa
Staphylococcus aureus
Viridans streptococci

FUNGI
Candida albicans and other yeasts

PRINCIPAL MICROORGANISMS INDIGENOUS TO THE SKIN

BACTERIA
Corynebacterium species
Gram-negative enteric bacilli
Mycobacterium species
Propionibacterium acnes
Staphylcoccus aureus
Staphylococcus epidermidis
Viridans streptococci

FUNGI
Candida albicans and other yeasts
Trichophyton species

place by means of an intermediary object. Any type of patient-care supply or therapeutic and diagnostic device may be contaminated by microorganisms. These sources include such objects as needles, syringes, bedding, catheters, cystoscopes, rectal thermometers, bedpans, antiseptic solutions, medicines, and shaving brushes, to name a few.

Common-vehicle transmission. Common-vehicle transmission refers to the transmission of multiple cases of a disease in which the source of infection is usually contaminated food or water, or medicines or blood. One of the severest nosocomial epidemics in the United States occurred in 1970 and 1971 when 378 cases of septicemia resulted in 40 deaths. The common vehicle was contaminated infusion fluid in which a screw-cap closure was improperly sterilized. Blood and blood products can also be contaminated with microorganisms directly or indirectly and when used for transfusions are a frequent cause of fatal infections. Bacterial species such as species of *Staphylococcus* and *Pseudomonas* and viral agents such as the hepatitis B virus are the most common cause of diseases resulting from contaminated blood or blood products.

Air transmission. The air may serve to transmit any infectious agent that can survive outside the host for at least a short interval of time. Many fungal and bacterial spores are found naturally in outside air and can be spread in the hospital by improperly designed ventilating systems. Other microorganisms can be suspended in droplet nuclei or dust particles and, depending on air currents, may be transmitted for long distances.

PRINCIPAL MICROORGANISMS INDIGENOUS TO THE GENITOURINARY TRACT*

BACTERIA
- *Acinetobacter* species (V)
- *Bacteroides* species (EG)
- *Clostridium* species (V)
- *Corynebacterium* species (V, EG, AU)
- Gram-negative bacilli such as species of *Proteus, Escherichia, Serratia* (V, EG, AU)
- Enterococcus (V, EG, AU)
- *Haemophilus vaginalis* (V, AU)
- *Mycobacterium* species (V, AU, EG)
- *Neisseria* species (V, EG, AU)
- *Peptostreptococcus* species (V, EG)
- *Staphylococcus aureus* (V, EG)
- *Staphylococcus epidermidis* (V, EG)
- *Streptococcus agalactiae* (V)
- Viridans streptococci (V, EG)

FUNGI
- *Candida albicans* and other yeasts (V, EG, AU)

PROTOZOA
- *Trichomonas vaginalis* (V, AU)

*Specific sites are indicated as follows: *V*, vagina; *EG*, external genitalia; *AU*, anterior urethra.

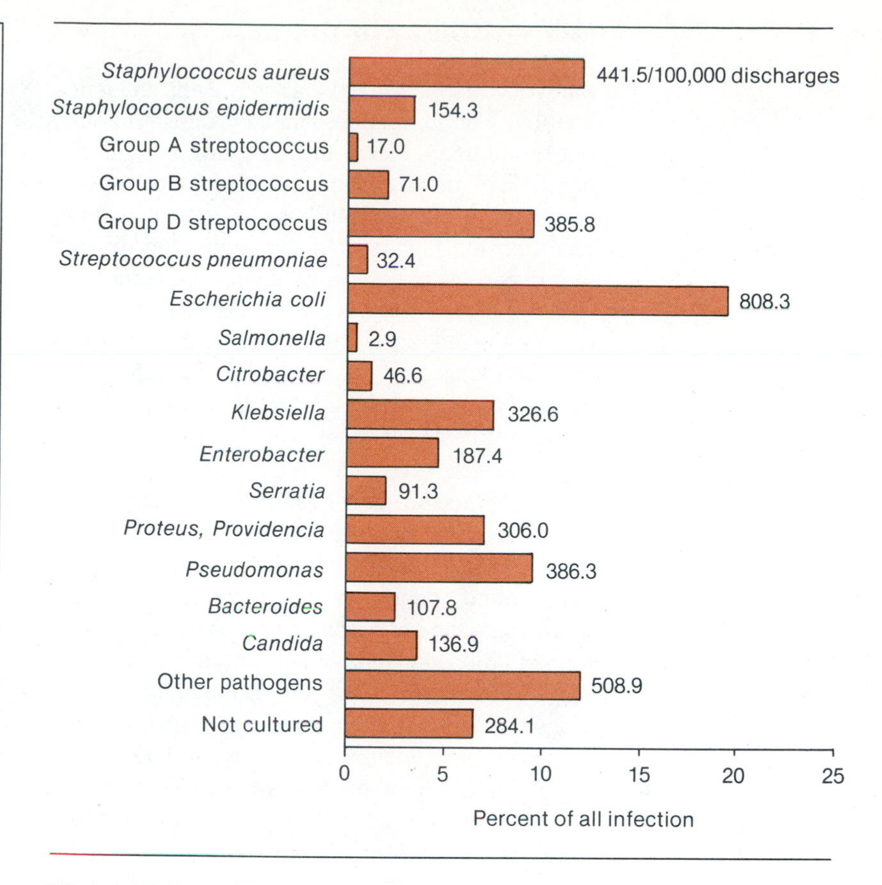

Figure 33-1 Pathogens causing nosocomial infections.
Redrawn from National Nosocomial Infection Study, 1978.

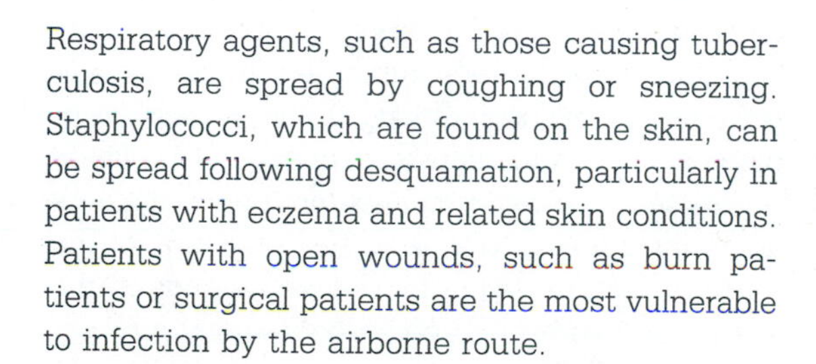

Respiratory agents, such as those causing tuberculosis, are spread by coughing or sneezing. Staphylococci, which are found on the skin, can be spread following desquamation, particularly in patients with eczema and related skin conditions. Patients with open wounds, such as burn patients or surgical patients are the most vulnerable to infection by the airborne route.

Vector-borne transmission. Vector-borne transmission is unreported in the United States but is important in many underdeveloped countries of the world. *Salmonella* and *Shigella,* for example, can be carried on the body parts of flies, and some vectors can transmit disease agents, such as those causing malaria or plague, via the bite of the arthropod.

Microorganisms involved

Bacteria are the leading cause of nosocomial diseases, and viruses are a distant second (Figure 33-1). Occasionally fungi cause disease, but rarely are protozoa involved. One of the factors that make the epidemiology of nosocomial diseases so much different from community-associated diseases is that the majority of microorganisms involved are considered noninvasive. Diseases can be caused by microorganisms that do not elaborate potent toxins or enzymes that permit invasion of epithelial tissue or spread within tissue. Most nosocomial diseases in the United States are caused by ***gram-negative bacilli,*** such as *Escherichia coli,* which under ordinary circumstances are harmless commensals in the intestinal tract. Disease is also related to the dose or number of microorganisms that are initially involved at the site of infection. The dose necessary to cause nosocomial disease is often significantly smaller than the dose required to cause community-associated diseases, but this is also influenced by other factors, such as the mode of transmission, host susceptibility, and site of infection. Those organisms that are naturally invasive and virulent will cause diseases whose symptoms appear rapidly and that often terminate in a fatal response. The specific microorganisms associated with various sites of infection are discussed later.

Susceptibility of the host

Many factors, as we have already seen, influence the ability of the microorganism to cause disease, but the ultimate factor is the susceptibility of the host. Each hospitalized patient demonstrates a variation in susceptibility to infectious agents, and a dose of microorganisms producing disease in one patient may have no effect on another patient. The role that the host plays in nosocomial diseases is influenced by the site of infection and the specific and nonspecific defense mechanisms that are present in the host at the time of contact with the infectious agent.

Site of infection. The host sites that can be infected are the skin, respiratory tract, intestinal tract, and genitourinary tract. Many microorganisms cannot penetrate the integument and require some type of abraison or break in the integrity of the skin in order to cause disease. *Staphylococcus aureus,* for example, is a commensal on the skin that does not cause disease unless it contaminates an open wound. Once beneath the integument, *S. aureus* can metabolize and replicate in practically any tissue of the body. Many microorganisms require specific sites before they cause disease. Ingestion of certain viruses such as types of adenoviruses causes no harm to the host, but displacement of the same virus from contaminated fingers to the eye can cause severe keratitis. Many gram-negative bacilli cause no diseases in the intestinal tract but if displaced into the bloodstream can cause serious diseases. The displacement of microorganisms from one body site to another occurs primarily through the use of diagnostic or therapeutic devices such as syringes or catheters.

Microorganisms may be inhaled and lodge in the respiratory tract, the site depending on the size of the particle. Organisms may also be transferred to the patient via contaminated equipment or fluids. Inhalation-therapy equipment, for example, can aerosolize contaminated fluids to produce droplets that carry potentially infectious microorganisms.

Diseases of the intestinal tract are often caused by the ingestion of contaminated food or water. *Salmonella, Shigella,* enterotoxigenic *Escherichia coli,* and the hepatitis A virus (HAV) are microbial agents frequently incriminated in intestinal diseases. The intestinal tract and especially the urinary tract can also be infected by contaminated diagnostic or therapeutic devices, such as the endoscope, which is inserted into the intestinal tract, and cystoscopes or catheters, which are inserted into the urethra.

Transplantation offers an avenue for the transfer of infectious agents. Organs, such as the kidney, frequently harbor endogenous viruses, such as the cytomegalovirus, Epstein-Barr virus, and herpes simplex virus, that exhibit latency and may be reactivated following transplantation.

Defense mechanisms of the host

Nonspecific immunity. Mechanical, physiological, and biochemical mechanisms provide the host with nonspecific measures for preventing infection. The mechanical barriers, such as the skin; biochemical factors, such as lysozyme in body fluids; and physiological processes, such as contraction of the intestinal tract, are discussed in Chapter 22. Nonspecific immunity that involves the inflammatory response and phagocytic processes are also discussed in Chapter 22.

Other nonspecific mechanisms that influence susceptibility to disease are age, as well as genetic, hormonal, and nutritional factors, but specific data have been obtained only when age is considered. The neonate, for example, acquires antibodies from the mother that provide limited protection against cetain infectious agents for up to 12 months after birth. Neonates, however, particularly those born prematurely or those who are sick, have polymorphonuclear leukocytes that are less functional than those of older individuals. Most diseases in the nursery are caused by *Staphylococcus aureus* and, less frequently, *E. coli*—organisms that colonize the skin of neonates almost immediately following birth. There is a natural decline in bodily functions in the very aged, including a decline in the efficiency of the immune system. Many tissues and organs in the aged may exhibit reduced blood supply and impaired immunological responses because of aging processes in the tissue, previous or current disease, and poor nutrition.

Specific immunity. Specific immunity of the host may be naturally or artificially acquired, and both types are extensively discussed in Chapter 23. Most of the diseases acquired in the hospital cannot be prevented by vaccination or immunoglobulin administration, and for this reason any condition or therapy that affects the normal immunological response puts an additional burden on the hospitalized patient. Surgery, burn wounds, and chronic diseases, such as diabetes and certain lymphomas, alter the host's resistance and make him susceptible to infection. In addition, the use of immunosuppressive drugs for certain neoplasias and transplantations, as well as the overuse of antibiotics, reduces the normal immunological response of the host and makes him prone to infection.

Hospital procedures contributing to disease

There has been an increase in recent years in the number of mechanical devices that are inserted into or come in contact with the hospitalized patient. A plethora of drugs and solutions have also been developed that keep people alive who otherwise would have died from illness, injury, or old age. Some of the more important hospital procedures and devices that contribute to nosocomial diseases are described below.

Catheters. Some of the most commonly used devices in the hospital are urinary catheters and catheters used for intravascular infusions. Catheters are flexible tubular devices used to withdraw fluid, such as urine, from the urinary bladder. They are also used to introduce fluids into a body cavity, such as in the intravenous feeding ***(parenteral nutrition)*** of a patient or the administration of medications. Urinary catheterization, which is the cause of 75% of nosocomial urinary tract diseases, may be used for short periods of time (a few hours to 1 or 2 days) to relieve temporary obstruction or inability to void urine or to obtain a clean specimen from patients who cannot give a clean-voided urine specimen. Urinary catheterization may be of the ***indwelling type,*** in which the catheter is inserted for 2 to 3 weeks or longer. This type of catheterization is frequently used on comatose patients or incontinent patients to provide a dry environment or to facilitate surgical repair of the urethra and associated structures. The indwelling catheter is responsible for the majority of induced nosocomial urinary tract diseases. Bacteria, which are the major cause of disease, can be derived from the patient's own flora or from an exogenous source. Some of the principal mechanisms that may permit infection are (1) inadequate preparation of the area around the point of catheter insertion (periurethral area), which can result in introduction of the patient's own flora into the urinary tract; (2) insertion of an already-contaminated catheter; (3) retrograde movement of microorganisms from a contaminated collection bag up the drainage tube; and (4) contamination when the junction between the catheter and the collection tube is broken (Figure 33-2). Bacteremia occurs in approximately 1% to 3% of those who have urinary catheterization, particularly those with indwelling catheters.

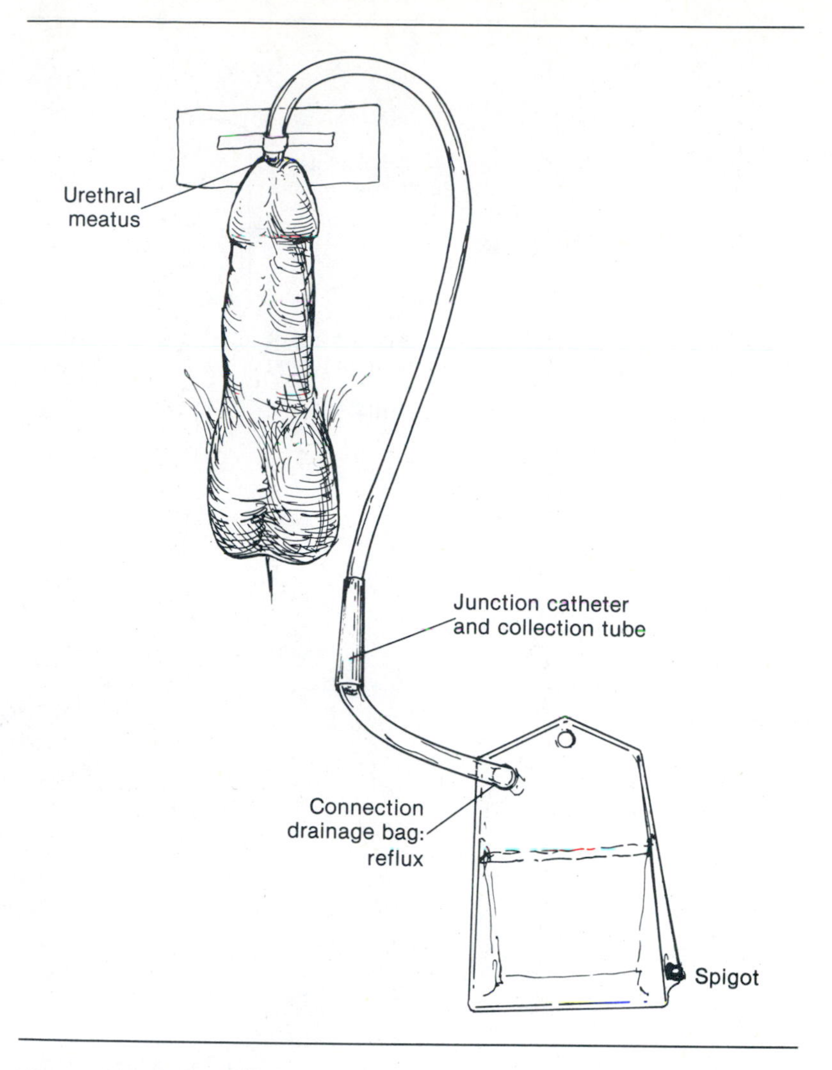

Figure 33-2 Potential sites of contamination in urinary drainage systems.

Intravenous infusion devices and the fluids they introduce into the patient may be sources of infection (Figure 33-3). The insertion of a ***cannula**** or catheter into the vein traumatizes the blood vessel and can result in inflammation at the site of injection—a condition called ***phlebitis,*** which may or may not be accompanied by microbial infection. The introduction of microorganisms into the bloodstream may be the result of contaminated infusion fluids or contaminated cannulas. Fluids may be contaminated (1) by miniscule breaks in the fluid container that permit the entrance of microorganisms, (2) during the injection of fluids into the tubing, and (3) during the administration of additional fluids into the IV container or during related procedures. The infusion fluid on very rare occasions may be con-

*The term *cannula* can be used interchangeably with *catheter*. A cannula is usually a steel needle, whereas a catheter is usually a plastic insertion device.

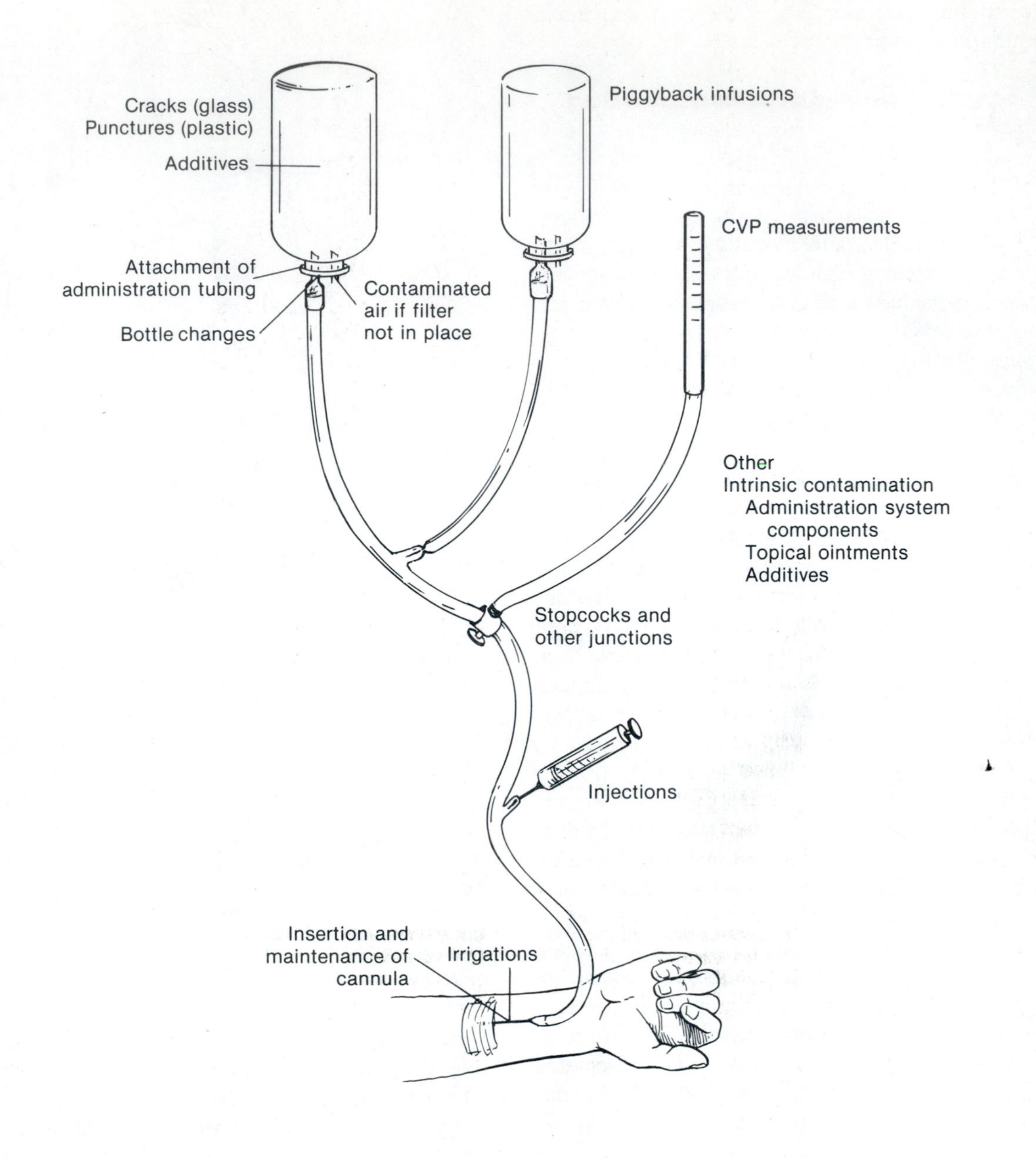

Figure 33-3 Potential sites for contamination of intravenous infusion systems. *CVP*, central venous pressure.

Redrawn from Goldmann, D.A.G., et. al. In Bennett, J.V., and Brachman, P.S. (Eds.). Hospital infections. *Little, Brown & Co., Boston, 1979.*

taminated during the injection process if the injection site is not treated with an antimicrobial or antiseptic, or the cannulas may be contaminated by hospital personnel before injection. Disease, particularly septicemia resulting from contaminated infusion fluids, is caused primarily by species of *Klebsiella, Enterobacter,* and *Serratia.* These organisms grow rapdily in a 5% glucose infusion solution because they can fix nitrogen from the air and do not require nitrogen compounds in solution. Diseases resulting from contaminated cannulas are caused primarily by organisms that contaminate the skin, such as *Staphylococcus aureus, Staphylococcus epidermidis,* and *Pseudomonas* species—organisms that slowly die in a 5% glucose solution. Patients who receive total parenteral nutrition (TNP) are infused with fluids that contain a synthetic amino acid solution, such as casein hydrolysate, and these fluids support the rapid growth of fungi, such as *Candida albicans* and *Torulopsis glabrata*. Fungal septicemia is therefore a more frequent complication of TPN than bacterial septicemia.

Respiratory devices. Many patients require oxygen and/or medications that are inhaled, and one of the devices used to dispense them is a ***nebulizer.*** A nebulizer saturates gases, such as oxygen or compressed air, with water vapor and dispenses the mixture as droplets (oxygen dispensed without water vapor could dry out the mucous membranes and make the patient more susceptible to infection). Large-volume nebulizers, which hold up to 500 ml, are used to deliver medication or moisture to the tracheobronchial tree and are the most important mechanism for disseminating microorganisms. The fluid-containing reservoir of the nebulizer may be contaminated in a variety of ways, including (1) contamination of the gas used to drive the nebulizer, (2) contamination of the reservoir by hospital personnel during cleaning or the replenishing of fluid, or (3) failure to sterilize the nebulizer jet.

Transducers. Transducers are pressure gauges connected to a closed space by a length of fluid-filled tubing that are used to monitor cardiovascular and cerebrospinal fluid pressures of the severely ill patient. These devices are known to be a source of infection by a variety of microorganisms, including bacterial species, such as *Pseudomonas* and *Klebsiella;* fungi, such as species of *Candida;* and viruses, such as the hepatitis and Epstein-Barr viruses. The transducer, which may be contaminated by organisms from the patient's bloodstream or by extrinsic sources, is a delicate instrument that cannot be autoclaved and must be sterilized chemically or by gas. If not sterilized after patient use, the transducer is a source of infection when used to monitor other patients.

Prostheses. The implantation of prostheses, such as heart valves; hip, knee, and finger joints; and intraocular lenses, can compromise the patient and lead to infections. The incidence of deep-wound diseases following the implantation of a skeletal prosthesis, such as hip replacement, varies from 1% to 3%. The staphylococci are responsible for nearly 80% of these diseases. The implanted device compromises the surrounding tissue and provides a site for microbial growth. Most diseases appear to be caused by contact-spread from the surgeon to the patient. Surgical gloves, which are often punctured by tools such as saws, hammers, and power tools used during surgery, provide an avenue for transmission.

Devices that are implanted in the cardiovascular system include total–heart valve prostheses, arterial grafts, and cardiac pacemakers. Unlike disease from skeletal prostheses, most disease from cardiovascular prostheses do not appear at the site of the implant but occur more frequently in the respiratory tract, urinary tract, and surgical wound. Direct contamination of the prosthesis in the operating room is believed to the major factor resulting in nosocomial disease. Staphylococci and *Pseudomonas aeruginosa,* which are the organisms most frequently associated with disease, often contaminate blood in the heart-lung machine—the most important source of contamination.

Hemodialysis units. Hemodialysis is used extensively for those with chronic renal failure. Many of these patients have been compromised by renal transplantation and the use of immunosuppressive drugs to prevent tissue rejection. Bacterial diseases, which are often acquired from the patient's own microbial flora, are usually localized, but bacteremia and septicemia can also occur and cause death. The staphylococci and species of *Pseudomonas* are the bacterial agents most commonly incriminated in bacterial disease. Viral hepatitis caused by the hepatitis B virus (HBV) is of particular concern in hemodialysis units. Once HBV is discovered in a patient, it is not unusual to find other patients, as well as hospital personnel, to also be affected. Infected patients and staff usually show no clinical illness, and contact with other individuals in the hemodialysis unit usually results in rapid spread of the disease.

Endoscopes. An endoscope is any type of instrument used to examine mucosal areas of various orifices, such as the stomach, oropharynx, colon, and bladder. They are constructed of lenses and a fiberoptic lighting system, which makes them difficult to sterilize. Without proper disinfection following patient use, these devices may become sources of contamination for other patients.

Antimicrobial and other drug therapy. The value of antibiotics in the treatment of disease cannot be questioned, but their overuse or inappropriate administration can compromise the pa-

MAJOR GROUPS AND SPECIES OF MICROORGANISMS CAUSING INFECTION IN TRANSPLANT PATIENTS

BACTERIA
Klebsiella species
Escherichia coli
Serratia species
Pseudomonas species
Staphylococcus species
Nocardia species

FUNGI
Aspergillus species
Candida species

VIRUSES
Epstein-Barr virus
Cytomegalovirus
Herpes simplex virus
Varicella-zoster virus
Hepatitis B virus

PROTOZOA
Toxoplasma gondii
Pneumocystis carinii

tient and place him in a life-threatening situation. Widespread use of antimicrobials has resulted in the selection of organisms with specific drug resistance and in some instances has assured their own obsolescence in the treatment of certain diseases. Penicillin G, one of the first antimicrobials, was so overused in the 1940s and 1950s that it caused the selection of penicillin-resistant strains of staphylococci, and now nearly all diseases caused by the staphylococci can no longer be treated with penicillin G. Staphylococcal infection caused by a nosocomial drug-resistant strain is therefore potentially the most difficult to treat. The staphylococci and aerobic gram-negative bacilli are the most drug-resistant organisms and are the most responsible for bacterial diseases in the hospital.

Antibiotics also affect the host's susceptibility to infection by direct or indirect means. A few drugs, such as chloramphenicol, can cause bone marrow depression, and drugs such as penicillin and the cephalosporins cause dermatitis, which increases the chances of infection. The major effect of antibiotics on host susceptibility is an indirect one and involves the alteration of the host's microbial flora. High doses of certain antibiotics, such as penicillin, the aminoglycosides, or broad-spectrum antibiotics, reduce the flora and permit colonization and subsequent infection by multiple drug-resistant organisms. Antibiotics are not the only determinant that influences colonization by drug-resistant microorganisms; other hospital procedures, discussed earlier, also play an important role. Various aspects of antimicrobial therapy are discussed in Chapter 27.

Heart, liver, bone marrow, and kidney transplantations require the use of drugs that suppress the cell-mediated immunological response, which is necessary to prevent graft rejection. There are two major classes of immunosuppressive drugs: ***corticosteroids*** and ***cytotoxic drugs.*** Corticosteroids appear to suppress the inflammatory response and are known to reduce the number of leukocytes migrating to a site of inflammation. Corticosteroids, such as prednisone, decrease the level of circulating T lymphocytes and also decrease the primary and secondary immunological responses of the host. Bacteria are the most common cause of infections in transplant patients, but other microorganisms may also be involved (see box).

Epidemiology and the identification of microorganisms

Precise identification of microorganisms during epidemiologic studies is an important aspect in the control of hospital-associated infections. If, for example, a cluster of patients in one or more wards of a hospital becomes severely ill from infection by a bacterial species, it is important to determine if there is a common source for the transmission of the infectious agent. Specimens are taken from the patients as well as attending physicians, nurses, and others coming in contact with the affected patients, and samples are also taken from various devices, solutions, and other inanimate objects with which the patient may come into direct or indirect contact. The organisms causing disease in the patient are isolated and identified. The same organism is sought from the other sources and is then subjected to a number of different tests to determine if there is a common source for this hospital epidemic. A variety of different tests can be used to "type" the organism, that is, produce a strain profile. Not all the tests are applicable for identification of all species. One can identify strains by (1) ***biotyping,*** or determining the pattern of biochemial activity of the organism; (2) ***antibiograms,*** or determining antibiotic susceptibility patterns; (3) ***serotyping,*** or identification of strain-specific an-

Table 33-2 Lytic groups of *Staphylococcus* typing phages

LYTIC GROUP	STAPHYLOPHAGES WITHIN GROUP
I	29, 42B, 52, 52A, 79, 80, 81, 83A
II	3A, 3B, 3C, 55, 71
III	6, 7, 42B, 42E, 47, 53, 54, 75, 77, 83A, 86
Miscellaneous	44A, 86, 94, 96, 292

tigens such as those on the cell surface; (4) ***bacteriocin typing,*** or determining the ability of strains to produce specific bacteriocins; and (5) ***bacteriophage typing,*** or determining a bacterial strain's sensitivity or resistance to bacteriophage. Bacteriocin typing has been used extensively for differentiating strains of *Pseudomonas aeruginosa* and *Serratia marcescens.* Antibiotic susceptibility patterns and phage typing have been used with much success in the identification of strains of *Staphylococcus aureus.*

Bacteriophage typing is carried out by taking an inoculum from the culture of the staphylococcal strain to be tested and spreading on an agar surface so that at the end of an 18-hour incubation period there will be confluent growth on the agar surface. Before the incubation period, however, a drop of different staphylophage is placed at properly identified and demarcated areas on the agar surface. The plate is examined after the incubation period. Areas of confluent growth at the site of bacteriophage inoculation indicate the bacterial strains resistance to infection by one type of staphylophage, whereas areas of clearing represent susceptibility to the staphylophage strain (see Chapter 6). There are three major lytic groups of staphylococci (Table 33-2). In the 1960s and 1970s the 80/81 complex was the principal epidemic strain; but currently the major epidemic strains belong to lytic group III and the 94/96 complexes.

PRINCIPAL SITES OF INFECTION AND THE ORGANISMS INVOLVED

Urinary tract

The urinary tract is the most common site of infection during hospitalization. Most disease results from instrumentation (catheterization) of the urinary tract and not from blood-borne infections from other sites. The urinary tract is considered infected when bacterial counts of 1×10^5/ml or more are obtained from properly collected urine specimens. Urinary tract infections may not of themselves be life threatening, but they can serve as a source of invasion of the bloodstream. Bacteremia may lead to damage of renal tissue or infection of other tissues, as well as endotoxic shock, which results from gram-negative sepsis.* The microorganisms most frequently associated with urinary tract disease are *Escherichia coli* and other gram-negative species (Table 33-3).

Table 33-3 Distribution of microorganisms found in urinary tract infections associated with a catheter or other instrument as the predisposing factor

ORGANISM	NUMBER OF ISOLATES	PERCENT
Escherichia coli	3369	33.4
Proteus species	1491	14.8
Enterococci	1356	13.5
Klebsiella species	1002	9.9
Pseudomonas species	894	8.8
Candida and related fungi	479	4.8
Enterobacter	418	4.2
Staphylococcus	429	4.3
Serratia species	142	1.4
Others*	498	<1.0 each

*From National Nosocomial Infections Study (NNIS), Centers for Disease Control, Atlanta; 1972.
Includes streptococci (groups A through D), lactobacilli, diphtheroids, *Clostridium perfringens, Bacillus subtilis, Acinetobacter, Providencia, Citrobacter, Bacteroides, Salmonella, Shigella, Edwardsiella,* and amebas.

Respiratory tract

The lower respiratory tract is the second most common site of nosocomial infection; and more patients die of pneumonia than of any other nosocomial disease. The lower respiratory tract is equipped with many mechanims to prevent microbial infection (see Chapter 22), but these natural defense barriers can be compromised by underlying disease or conditions, as well as by

* Sepsis is a condition resulting from the presence of microbes or microbial products in the body.

Table 33-4 Etiology of nosocomial pneumonia and lower respiratory tract infection

CAUSATIVE AGENT	NUMBER OF ISOLATES	PERCENT
Gram-positive bacteria		
Staphylococci (coagulase +)	1867	12.6
Streptococcus pneumoniae	1187	8.0
Gram-negative bacteria		
Klebsiella species	2273	15.4
Pseudomonas aeruginosa	1991	13.5
Escherichia coli	1627	11.0
Enterobacter species	1339	9.1
Serratia marcescens	294	2.0
Mycobacterium tuberculosis	1	0.1
Viruses		
Influenza virus	10	0.1
Other pathogens*	4183	28.3

From National Nosocomial Infections Study (NNIS), Centers for Disease Control, January 1970 through August 1973.
*Includes bacteria, such as *Staphylococcus epidermidis, Proteus* species, and *Haemophilus influenzae;* fungi, such as *Candida* species, *Cryptococcus* species, *Aspergillus* species, *Mucor,* and *Pneumocystis carinii;* and viruses, such as the adenoviruses and respiratory syncytial virus.

Table 33-5 Classification of surgical wounds

OPERATIVE PROCEDURE	DEFINITION
Clean wound (1)*	Nontraumatic wound in which neither respiratory, alimentary, nor genitourinary tracts have been entered; there is no inflammation
Clean-contaminated wound (2)*	Respiratory, alimentary, and genitourinary tracts are entered but without significant contamination
Contaminated wound (3)*	Open, fresh wound with breaks in aseptic technique or gross spillage from the gastrointestinal tract; incisions are inflamed but nonpurulent
Dirty wound (4)*	Old traumatic wound is encountered, including perforated viscera; pus is encountered, and contamination occurred before surgery

*The number represents the relative rate of infection for the procedure; 4 = highest rate of infection; 1 = lowest rate of infection.

hospital procedures. Pneumonia can be produced by (1) aspiration of microorganisms from the oropharynx, (2) inhalation of aerosols generated by coughing or sneezing or by nebulizing equipment, and (3) invasion of the lung by the hematogenous route.

The oropharynx is not ordinarily colonized by aerobic gram-negative bacteria, but in the compromised host these organisms can be found there in relatively high numbers. Patients in a coma, those with acidosis, or those with underlying respiratory conditions or disease in whom lung clearance mechanisms are altered are most susceptible to this type of colonization. The source of these gram-negative aerobic bacilli is usually the patient's own intestinal flora. The staphylococci can also colonize the oropharynx, but these organisms are acquired primarily through contact with other patients or hospital personnel. Both groups of microorganisms infect the lungs by aspiration and are the major cause of nosocomial pneumonia (Table 33-4). Microorganisms may also be spread from patient to patient by coughing or sneezing, but the most important mechanism for dissemination is the nebulizer, especially the large-volume type. The lung is rarely invaded by the hematogenous route unless there is a primary infection at a distant site, such as the urinary tract.

Surgical wounds

The incidence of disease in surgical incisions has not decreased appreciably in the past 30 years despite the extensive use of antimicrobials as a prophylaxis in surgical procedures. The misuse of antimicrobials has led not only to an increasing number of multiple drug-resistant microorganisms in the hospital, but to an overdependence on antimicrobials to prevent disease at the expense of proper surgical procedures. This problem has been compounded by the fact that modern techniques, such as implantation, often keep the patient in the operating room for extended periods of time. The rate of disease is associated with the type of surgical wound and whether it is clean, clean-contaminated, contaminated, or dirty (Table 33-5). Microbial contamination occurs more frequently following operations involving the alimentary, respiratory, and genitourinary

Table 33-6 Pathogens most frequently isolated from surgical wounds

PATHOGEN	PERCENT RECOVERED FROM INFECTIONS
Staphylococcus aureus	15.4
Escherichia coli	15.2
Enterococci	10.3
Proteus species	7.2
Klebsiella species	5.4
Staphylococcus epidermidis	4.8
Pseudomonas aeruginosa	4.6
Enterobacter species	4.1
Nongroup D streptococci	5.4
Group B streptococci	1.6
Group A streptococci	0.7
Bacteroides fragilis (an anaerobe)	2.9

From National Nosocomial Infections Study (NNIS), Centers for Disease Control Atlanta, 1975-1978.

tracts. The reason for this is that these sites harbor large numbers of microorganisms that, if displaced or spilled onto incised tissue, could cause serious disease. The incidence of disease at these sites also increases with the duration of the operation.

The major source of most incisional wound diseases is the patient. Endogenous microorganisms may be those brought with the patient at the time of admission to the hospital, or they may have been acquired in the hospital before surgery. Studies have indicated, for example, that patients carrying *Staphylococcus aureus* in the nares are more prone to surgical wound disease than those who do not carry the organisms. Airborne organisms play no significant role in surgical wound disease except when a member of the surgical team is a carrier and sheds microorganisms continually. The punctured surgical glove is the major mechanism of contact-spread, because microorganisms from sebaceous glands of the surgeon can gain access to the wound through a break in the glove. The wound can also be infected during surgery from inanimate sources such as contaminated instruments or in the post-operative ward by contact with contaminated dressings, irrigation solutions, or other hospital-associated devices or instruments. The microbial agents most frequently recovered from surgical wounds are listed in Table 33-6.

Nomicrobial factors that can predispose the patient to disease are many, but the more important ones are the following:

1. Tissue that is devitalized or necrotic. Such tissue should be removed because it supports the growth of microorganisms.
2. The presence of foreign bodies such as sutures or prosthetic devices. Such devices often carry microbial contaminants and, when left in the patient, serve as a site of infection.
3. The physical condition of the patient. Conditions such as shock, diabetes, and leukemia lower the host's resistance to disease.

Skin and eyes

Skin and eye diseases are among the least-encountered nosocomial diseases, but they have a greater tendency to cause secondary bacteremias than urinary tract or surgical wound infections. Newborns are particularly susceptible to superficial infections, and the nursery has an infection rate three times greater than any other hospital service.

Staphylococcus aureus causes approximately 34% of all nosocomial superficial diseases, with aerobic gram-negative bacilli causing the majority of the remaining diseases. Aerobic gram-negative bacilli are the most common cause of superficial diseases in adults. Superficial diseases can be classified into the following types: (1) ***pyoderma,*** the presence of cutaneous or subcutaneous pus in a lesion; (2) ***omphalitis,*** a disease of the umbilical cord that may or may not be accompanied by purulence; (3) ***cellulitis,*** an inflammatory condition of the subcutaneous tissue without purulence; (4) ***dermatitis,*** a superinfection* of skin that has been compromised by such conditions as eczema or psoriasis; (5) ***mastitis,*** an inflammatory condition of the breast that may lead to abscess formation; and (6) ***conjunctivitis,*** an inflammatory condition of the conjunctiva that is associated with mucopurulent discharge.

The sources of infection may be the patient's own flora, since staphylococci rapidly colonize the neonate, but hospital personnel with minor

* Superinfection refers to an infection superimposed on an already existing infection.

Table 33-7 Distribution of pathogens isolated from infected burns

PATHOGEN	TOTAL ISOLATES	PERCENT
Pseudomonas aeruginosa	336	23.1
Staphylococcus aureus	321	22.0
Escherichia coli	135	9.3
Klebsiella species	131	9.0
Proteus species	112	7.7
Enterobacter species	109	7.5
Candida species	67	4.6
Providenica species	28	1.9
Acinetobacter species	22	1.5
β-hemolytic streptococci and group A streptococci	16	1.1
Unspecified fungi	12	0.8
Aspergillus species	4	0.3
Mucor species	1	0.1
Other or unspecified	163	11.2
TOTAL	1457	100.1

National Nosocomial Infections Study (NNIS), Centers for Disease Control, Atlanta, January 1970 through August 1973.

skin lesions or carriers may also be involved. The most important method of spread is via the hands of hospital personnel. Patient crowding and inadequate handwashing are two important factors that encourage the spread of disease, and once a patient becomes infected, he serves as a reservoir from which hospital personnel become contaminated and other patients can subsequently become infected.

Burn wounds

Microbial disease is the leading cause of death in burn wounds when more than 40% of the body surface has been traumatized. It is practically impossible to prevent some microbial contamination, by one or more microorganisms, of burn wounds. The type of microbial infection that results is dependent on many factors, including the site of the burn; what other hospital procedures, such as intravenous feeding, are being performed; and what underlying disease or condition is present. One of the most important factors that contribute to the development of nosocomial disease in the burn patient is defects in the immunological response of the host caused by the burns. Burn wounds result in depressed circulating IgG levels, decreased chemotactic behavior of leukocytes, abnormal complement function, and lymphoid depletion.

Pseudomonas aeruginosa and *Staphylococcus aureus* are the two most frequently isolated microbial species from burn wounds, but other organisms, particularly many aerobic gram-negative bacilli, are also incriminated (Table 33-7). Many gram-negative infections are superinfections following therapy for diseases caused by gram-positive bacteria. The gram-negative organism is usually a multiple drug-resistant strain. *P. aeruginosa,* because of its resistance to many antimicrobials, was at one time an important cause of death in burn patients. The administration of a vaccine called ***Pseudogen,*** combined with hyperimmune globulin, has caused a 60% decrease in *Pseudomonas* septicemia, and mortality from this type of infection has almost been eliminated. The source of infection is usually the gastrointestinal tract of the patient, with direct contact between patient and hospital personnel being the most important mode of spread. *S. aureus* colonizes approximately 50% of all patients and is the primary cause of septicemia in the burn wound group. *S. aureus* is much more susceptible to antimicrobials than *P. aeruginosa,* and very few patients die as a result of its infection of the host. Most *S. aureus* diseases result from patient-to-patient contact or contact with hospital personnel who are carriers of the organism.

Bacteremia

Infections of the bloodstream have become more common with the increase in hospital-associated techniques. Procedures associated with intravenous infusion, total parenteral nutrition, indwelling catheters, hemodialysis, and vascular and gastrointestinal surgery have contributed significantly to increases in nosocomial bacteremia. Neonates and the elderly are the groups most often affected because of their natural susceptibility to disease and their need for hospital procedures.

Bacteremia can be divided into two types: primary and secondary. ***Primary bacteremias*** are those in which there is no underlying disease that is responsible for the introduction of organisms into the bloodstream. A bacteremia following the administration of contaminated infusion fluid would be considered a primary bacteremia. ***Secondary bacteremias*** are those that follow a disease at another site; for example, a patient having an underlying urinary or respiratory tract

Table 33-8 Organisms causing nosocomial gram-negative primary and secondary bacteremias

AGENT	PERCENT
Primary bacteremia	
Escherichia coli	29
Klebsiella species	22
Pseudomonas species	18
Enterobacter species	11
Proteus species	8
Serratia species	6
Acinetobacter species	3
Citrobacter species	1
Other	2
Total isolates	2177
Secondary bacteremia	
E. coli	35
Pseudomonas species	20
Klebsiella species	17
Proteus species	11
Enterobacter species	9
Serratia species	6
Other	2
Total isolates	1157

From National Nosocomial Infections Study (NNIS), Centers for Disease Control, Atlanta, Januay 1970 through August 1973.

disease could develop a bacteremia secondary to the underlying disease. The patient's own gastrointestinal tract is believed to be one of the primary sources of infectious agents causing primary bacteremia. Other sources of primary bacteremia include cannulation, intravenous infusion, transfusion of contaminated blood, and contaminated transducers and other devices. Secondary bacteremias result from underlying diseases that occur primarily in the urinary tract, respiratory tract, and surgical wounds.

Most bacteremias are caused by aerobic gram-negative bacilli. The symptoms of gram-negative bacteremia include chills and fever, and sometimes nausea, vomiting, and diarrhea. If shock develops from endotoxin release, the symptoms also include abnormal heart rhythms and confusion. Severe cases of endotoxic shock result in death due to disseminated intravascular coagulation, respiratory insufficiency, or heart failure. The organisms most frequently associated with nosocomial bacteremia are *Escherichia coli, Klebsiella* species, and *Pseudomonas* species. *Pseudomonas* bacteremias are responsible for more fatalities than bacteremias caused by other organisms, but underlying host factors also influence the outcome of disease. Table 33-8 lists the gram-negative organisms causing primary and secondary bacteremias.

Bacteremias occur primarily in neonates and the elderly, in whom natural defense mechanisms are compromised not only by age but also by lifesaving hospital procedures. More bacteremias occur in premature infants; *E. coli* is the principal cause of disease. Neonatal sepsis is fatal in up to 40% of the cases, and this figure is probably higher in premature infants. The highest case fatality ratio is 60%, which occurs in those over 60 years of age and is associated with chronic renal disease, diabetes mellitus, cirrhosis, and various malignancies.

PREVENTION

Accredited hospitals are now required to have an infection-control program that includes (1) surveillance of nosocomial diseases, (2) employee health program surveillance (3) an isolation policy for certain diseased patients, (4) education of hospital personnel concerning nosocomial diseases, (5) a microbiology laboratory, (6) environmental sanitation procedures, and (7) accepted techniques for certain hospital procedures, such as catheterization. A discussion of each of these aspects is beyond the scope of this book, but some general procedures that are being used to identify and control nosocomial diseases are highlighted below.

Surveillance

Hospitals employ infection-control practitioners, such as nurses, physicians, or public health officials, to perform surveillance tasks. Surveillance may use different sources to collect data that are processed and disseminated to an infection-control committee. The sources of nosocomial information should include (1) reports from the microbiology laboratory; (2) antimicrobial data from the pharmacy; (3) autopsy reports; (4) ward rounds to determine the presence of fevers, types of hospital procedures being performed, antibiotic regimens, and underlying disease of the patients; (5) the employee health clinic; and (6) the outpatient department. Special surveillance activities may also include determining the spread of antibiotic resistance in specific groups of patients or monitoring the occurrence of a viral infection, such as varicella, in a specific group. Once the data have been tabulated and assessed, appropriate measures can be taken to educate hospital personnel and to design methods of prevention.

Techniques to prevent nosocomial infections

Handwashing. The single most important procedure that can aid in the prevention of hospital-associated diseases is the practice of handwashing. Nurses, physicians, and other hospital personnel are frequently in contact with patients and thus serve as vectors in the transmission of infectious agents from one patient to another. The handwashing procedures vary depending on the task to be performed. Handwashing with soap and water is sufficient when the nurse is going to bathe the patient or merely change the bed linen. Handwashing with antiseptic soaps that contain either 3% hexachlorophene, an iodophor, or 4% chlorhexidene should be used before such procedures as surgery or the manipulation of an invasive device such as a catheter. Contact with a lesion or a draining skin infection requires the use of gloves. For special cases, such as patients with hepatitis, gloves are also recommended for persons performing such duties as drawing blood, dismantling dialysis machinery, or changing bedpans.

Disinfection and antiseptic procedures. Equipment and devices used in the diagnosis of patients require decontamination before use, but many of these cannot be autoclaved and require special disinfection techniques. Respiratory therapy equipment can be sterilized by soaking in glutaraldehyde or can be gas sterilized with ethylene oxide. Nebulizers, following an initial sterilization, can be disinfected daily by separating the unit from the patient and nebulizing with 0.25% acetic acid. Depending on the item used in hemodialysis, sterilization by heat, ethylene oxide, or chemical disinfection can be used, but because of the presence of HBV special disinfection techniques may be employed, such as immersion in (1) sodium hypochlorite (0.5% to 1.0%) for 30 minutes, (2) 20% formalin in 70% alcohol for 18 hours, or (3) 2% aqueous alkalinized glutaraldehyde.

The insertion of devices such as cannulas or catheters can result in infection at the skin-cannula or skin-catheter junction. Antibiotic-containing ointments are used to protect against cannula-associated sepsis and are particularly effective against infections caused by staphylococci. The antimicrobials used in the ointments include neomycin, bacitracin, polymyxin, and nitrofurazone, used singly or in combination. Iodine-containing ointments are more effective against diseases caused by *Candida* and are used especially for TPN therapy. Urinary catheters are often impregnated with antimicrobials, and sometimes antibacterial agents are instilled into the bladder to prevent disease.

Antimicrobial surveillance. There is no question that antimicrobials have had a profound influence on the character of nosocomial disease. Antimicrobials are administered to between 25% and 38% of hospitalized patients and account for up to 35% of the total pharmaceutical cost. The kind and amount of drugs used in different hospitals may vary considerably, but nearly one half of the antimicrobials used are either unnecessary or are used inappropriately. The surveillance of antimicrobials in the hospital is a very important task, because overuse of any drug results in the selection of organisms that are resistant to it. Not only will future diseases be difficult to treat with the drug, but the patient acts as a reservoir for the drug-resistant pathogen, which can be transmitted by hospital personnel to other patients. Fortunately, the resistant microorganisms in the host are at a disadvantage in terms of their ability to compete with other microorganisms unless the antimicrobial is present. Therefore removal of the antimicrobial for a period of time from the therapeutic regimen usually results in the selection of drug-sensitive strains in the patient.

The administration of antibiotics to prevent disease, under certain circumstances, is a worthwhile and often necessary procedure. Many operations and techniques involve areas of the body where the resident flora are in very high numbers, and even minor spillage following surgery is difficult to avoid. Antimicrobial prophylaxis reduces the resident flora, and any organisms that do gain entry to normally sterile sites are destroyed by the drug present in the tissue. The situations in which antimicrobial prophylaxis is justified are indicated in Table 33-9.

Even with advanced techniques and preventive measures, nearly 50% of all hospital-associated diseases are unavoidable. Each individual entering the hospital is equipped with physicial and chemical barriers to infection and responds differently to the presence of microorganisms. Until we are able to define all the factors that contribute to the virulence of the microorganism as well as the resistance of the host, hospital-associated diseases will be inevitable. All we can do at this time is be aware of the factors responsible for such diseases and use all the measures available to us for their prevention.

Table 33-9 Situations in which antimicrobial prophylaxis is justified and the antimicrobials used

PROCEDURE OR CONDITION	ANTIMICROBIAL USED
Vaginal hysterectomies	Cephalosporin or penicillin
Total abdominal hysterectomies	Cephalosporin
High-risk cesarean sections	Cephalosporin
Colorectal surgery	Oral erythromycin-neomycin, kanamycin-metronidazole, or doxycycline
Vascular grafts of the abdominal aorta or lower extremity vasculature	Cephalosporin
Total hip replacement	Cephalosporin or penicillinase-resistant penicillin
Head and neck cancer surgery	Cephalosporin
Traveler's diarrhea	Doxycycline
Prevention of pneumonia due to *Pneumocystis carinii* in susceptible cancer patients	Trimethoprim-sulfamethoxazole
Recurrent urinary tract infections in female patients	Trimethoprim-sulfamethoxazole

SUMMARY

Diseases acquired in the hospital are referred to as nosocomial. Bacteria are the most frequent etiological agents of hospital-associated diseases, but other microorganisms may also be involved. The gram-negative aerobic bacilli, such as members of the Enterobacteriaceae, are the most common cause of bacterial diseases.

The organisms causing disease in the hospital may be derived from the host (endogenous) or from sources, both animate and inanimate, outside the host (exogenous). The infectious agents can be transmitted in the hospital in the same way as in the community (see Chapter 26).

Many factors contribute to the acquisition of hospital-associated diseases and may be associated with the patient or with the hospital. The host's susceptibility to infection is influenced by the type of microorganism involved, the dose of the microorganism, the site of infection, and the efficiency of the specific and nonspecific immune mechanisms. Most diseases develop in patients who have some underlying disease or condition that impairs the normal immunological response. Hospital procedures, both diagnostic and therapeutic, have increased in recent years, and with their increase has been an increase in hospital-associated diseases. Catheterization and the use of intravenous infusion devices have contributed to the majority of diseases, but other sources such as respiratory therapy and anesthetic devices also play an important role. The improper use of drugs, primarily antimicrobials, has led to the development of resistant species of microorganisms, toxic reactions, and allergic responses in the hospitalized patient. Some antimicrobials destroy members of the indigenous microflora and create sites for infection by resistant species. Other drugs, particularly those used during leukemia, cancer therapy, or transplantation, suppress the immunological response and also make the patient susceptible to infection by his own microorganisms.

The hospitalized patient is also subject to cross-infection from other patients and from hospital staff who may be carriers of infectious microorganisms. Nurses and physicians are the most frequent carriers of infectious agents.

The principal sites of infection and the microorganisms involved are the urinary tract—*Escherichia coli* and other gram-negative aerobic bacilli; the respiratory tract—gram-negative aerobic bacteria and staphylococci; surgical wounds—staphylococci; skin and eyes—staphylococci and gram-negative aerobic bacilli; burn wounds—*Pseudomonas aeruginosa* and *Staphylococcus aureus;* and bacteremia—*E. coli* and other gram-negative aerobic bacilli.

The most important method for preventing hospital-associated diseases is handwashing by hospital staff. Since many hospitalized patients are subjected to hypodermic injection, disinfection of the site of injection is also an important means of preventing infection. In most hospitals an infection-control committee is in charge of surveying the number, types, and sites of infection in the hospitalized patients and promulgating this information to hospital personnel. It is also the purpose of the committee to suggest methods for reducing hospital-associated disease. Such proposals may include reducing the use of certain antimicrobials for treatment or suggesting the use of new disinfectants.

Study questions

1. Neonates in the hospital are most susceptible to infection by what bacterial agent?
2. Why are diabetics so susceptible to disease?
3. List the ways in which the host's immunological responses may be depressed in the hospital, making the host susceptible to infection.
4. Following your admission to a hospital for a condition involving an organ such as the kidney, heart, or lung, what hospital procedures would make you prone to infection?
5. What bacterial agents are the most common causes of hospital-associated diseases? What fungal agents? What viral agents?
6. Why does the use of chemotherapeutic agents give rise to hospital-associated diseases? Which of the gram-positive bacteria?
7. Why are most hospital personnel carriers of penicillin-resistant staphylococci?
8. Which of the gram-negative bacteria is most commonly involved in hospital-associated diseases? Which of the gram-positive bacteria?
9. Make a list of possible procedures that could result in infection, such as catheterization and transplantation, and give a reasonable explanation as to which microorganisms would cause infection during each procedure.
10. What are the recommended ways of preventing hospital-associated diseases?

Selected readings

Books

American Hospital Association. *Infection control in the hospital* (4th ed.). American Hospital Association, Chicago, 1979.

Bennett, J.V., and Brachman, P.S. *Hospital Infections.* Little, Brown & Co., Boston, 1979.

Boyd, R.F., and Marr, J.J. *Medical microbiology.* Little, Brown & Co. Boston, 1980.

Burke, J.F., and Heldrich-Smith, G.Y. *The infection-prone hospital patient.* Little, Brown & Co., Boston, 1978.

Castle, M. *Hospital infection control.* John Wiley & Sons, Inc., New York, 1980

Centers for Disease Control National Nosocomial Infections Study Report: annual summary 1978. Centers for Disease Control, Atlanta, 1981.

Journal articles

Betts, R.F. Cytomegalovirus (CMV) in the compromised host. *Annu. Rev. Med.* **28:**103, 1977.

Crossley, K., et al. An outbreak of infections caused by strains of *Staphylococcus aureus* resistant to methicillin and aminoglycosides. II. Epidemiologic studies. *J. Infect. Dis.* **139:**280, 1979.

Edelman, R., and Levine, M.M. Acute diarrhea infections in infants. II. Bacterial and viral causes. *Hosp. Proc.* **15:**97, 1980.

Finland, M., and McGowan, J.E. Nosocomial infections in surgical patients. *Arch. Surg.* **111:**143, 1976.

Gardner, I.D. The effect of aging on susceptibility to infection. *Rev. Infect. Dis.* **2**(5):801, 1980.

Garner, P., et al. Nosocomial management of resistant gram-negative bacilli. *J. Infect. Dis.* **141:**415, 1980.

Goldmann, D.A. Nosocomial infections in a neonatal intensive care unit. *J. Infect. Dis.* **144:** 449, 1981.

International symposium on control of nosocomial infection. *Rev. Infect. Dis.* **3**(4):(Entire issue), 1981.

Maki, D.G., et al. Prevention of catheter-associated urinary tract infection. *J.A.M.A.* **221:**1270, 1972.

Maki, D.G., et al. Nationwide epidemic of septicemia caused by contaminated intravenous products. I. Epidemiologic and clinical features. *Am. J. Med.* **60:**471, 1976.

Pierce, A.K., and Sanford, J.P. Bacterial contamination of aerosols. *Arch. Intern. Med.* **131:**155, 1973.

Schaberg, D.R., et al. Evolution of antimicrobial resistance and nosocomial infection: lessons from the Vanderbilt experience. *Am. J. Med.* **70:**445, 1981.

Steere, A.C. and Mallison, G.F. Handwashing practices for the prevention of nosocomial infections. *Ann. Intern. Med.* **83:**683, 1975.

Weinstein, R.A., et al. Pressure monitoring devices, overlooked source of nosocomial infection. *J.A.M.A.* **236:**936, 1976.

PART 8

INFECTIOUS DISEASES OF ANIMALS AND PLANTS

Tobacco mosaic virus (TMV). (×220,000.)

Omikron/Photo Researchers, Inc.

Chapter 34 INFECTIOUS DISEASES OF ANIMALS

Most of the general public becomes concerned about animal disease only if it results in the loss of the companionship of a family pet. Diseases of domesticated pets as well as farm animals should also be of concern for two other reasons. First, pets and farm animals are susceptible to diseases that are also transmissible to humans. These diseases are called ***zoonoses.*** Second, disease in farm animals used for human food production can result in tremendous economic loss and can affect the eating habits of our global population.

The purpose of this chapter is to describe some of the more important diseases of animals, particularly farm animals, with special attention given to those that often result in severe economic loss.

ZOONOSES

The diseases of lower animals may be transmitted to humans by direct or indirect contact or by insect vectors. The principal diseases transmitted by insect vectors that affect lower animals as well as humans include malaria, leishmaniasis, Rocky Mountain spotted fever, typhus fever, Q fever, tularemia, and plague. These are discussed in earlier chapters and are outlined in Table 32–4. Those animal diseases transmitted to humans by direct or indirect contact are outlined in Table 34-1.

SOME EPIDEMIOLOGICAL ASPECTS

The principles of host-parasite interaction that apply to humans, for the most part, also apply to animals. The sources of infection and the manner in which diseases are transmitted are also very similar for both groups, but they are often more complicated in animals, particularly farm animals. The major epidemiological difference between disease in humans and in farm animals is of a quantitative nature; this is due to the particular environment of the farm animal, the manner in which farm animals carry out biological processes, and the type of hygienic conditions to which farm animals are exposed.

Infected animals shed or excrete potential pathogens that contaminate the air, milk, water, food, litter, and dust and that are often infectious for humans. Infections from food and water occur less frequently in adult farm animals, but diarrhea is common in young animals. If the stable is not properly cleaned, or if railroad cars or motor trucks are not disinfected, infections may spread rapidly by direct or indirect contact. Respiratory tract infections occur frequently among penned animals, whose respiratory secretions or fecal excretions contaminate their surroundings. Diseases such as psittacosis and tuberculosis can be acquired by animals via inhalation of contaminated dust. Tuberculosis among humans is generally acquired by inhalation of contaminated respiratory aerosols or by ingestion of contaminated milk.

Table 34-1 Animal diseases transmitted to humans by means other than insect vectors*

DISEASE	TYPE OF ETIOLOGICAL AGENT AND SPECIES	ANIMAL RESERVOIR	METHOD OF SPREAD TO HUMANS
Tuberculosis	Bacterial: *Mycobacterium bovis*	Cattle	Ingestion of contaminated milk
Brucellosis	Bacterial: *Brucella* species	Goats, cattle, sheep, swine	Ingestion of contaminated milk, meat, or related products
Anthrax	Bacterial: *Bacillus anthracis*	Cattle, horses, sheep, goats, swine, wild animals	Contact with animal hides, wool, etc.; ingestion of contaminated meat
Q fever	Bacterial: *Coxiella burnetii*	Cattle	Contact with infectious contaminated parturient products; inhalation of contaminated dust
Psittacosis and ornithosis	Bacterial: *Chlamydia psittaci*	Domestic birds	Inhalation of contaminated dust
Leptospirosis	Bacterial: *Leptospira* species	Cattle, dogs, swine, rodents	Contact with infectious urine and contaminated water
Salmonellosis	Bacterial: *Salmonella* species	Cattle, swine, dogs, rodents	Ingestion of contaminated foods
Ringworm	Fungal: species of *Microsporum* and *Trichophyton*	Cats, dogs, cattle, horses	Contact with animal
Trichinosis	Animal parasite: *Trichinella spiralis*	Swine	Ingestion of infected pork
Echinococcosis	Animal parasite: *Echinococcus granulosis*	Wild animals and domestic dogs in endemic areas	Ingestion of cysts after handling dogs
Taeniasis	Animal parasite: *Taenia solium*	Swine	Ingestion of contaminated pork
Rabies	Virus: lyssavirus	Dogs, skunks, bats, foxes	Bite of infected animal; inhalation of aerosols in bat caves (rare)

*Diseases transmitted by insect vectors are described in Table 32-4.

Infections resulting from wounds are more common in farm animals than in humans. Animals acquire wounds in a variety of ways, including castration, calving, goring, punctures via barbed wire, and biting. Continual contact with the soil and contaminated fomites also makes the domestic animal subject to infection, particularly by species of *Clostridium* and the fungal agents of ringworm.

Vector-borne disease in the United States occurs more frequently among farm animals than in humans. Diseases such as equine viral encephalitis, tularemia, and swinepox are important in veterinary medicine. In underdeveloped countries, such as Africa, certain zoonoses are as common among humans as they are among domestic animals. Trypanosomiasis in Africa, for example, is responsible for killing much livestock and is also responsible for considerable morbidity and mortality in the human population.

Diseases associated with pregnancy and birth appear more commonly in farm animals than in humans. Transplacental infections caused by species of *Brucella, Vibrio, Listeria,* and *Chlamydia* and certain fungal agents may resut in abortion or cause extensive fetal damage. Umbilical infection in the newborn is also frequent, especially among lambs.

BACTERIAL DISEASES

Brucellosis

Brucellosis in humans is called undulant fever and is discussed in Chapter 28. The incidence of human brucellosis in the United States is very low, but its incidence in domestic animals is often quite high and is of great economic significance. The primary agents of infection, *Brucella abortus* (cattle), *B. suis* (swine), *B. melitensis* (goats and sheep), and *B. canis* (dogs) can cause a generalized bacteremia and fever in humans, but in domestic animals the most frequent manifestation of infection involves the reproductive tract, which results in abortion.

Healthy cattle herds first become infected by the introduction of already-diseased animals. The agents of disease are usually ingested in milk, food, or water that has been contaminated by aborted fetuses, placenta, or genital discharges. Infections may also be acquired during artifical insemination with infected semen or during coitus with diseased bulls. A generalized bacteremia follows ingestion of the organisms, with localization in the placenta of the gravid uterus, but in the nonpregnant cow the organisms colonize internal organs as well as the udders (mastitis). The presence of erythritol in the placenta of pregnant animals and the genitals of males but not in the human placenta suggests the importance of this sugar for colonization by the bacteria. Pregnant cows either abort their fetuses, usually during the last 3 months of gestation, or they become sterile. Diseased males show abscesses in the testes and associated structures that may result in infertility.

Many states in the United States have eradicated brucellosis from cattle by adhering to the following procedures. A serological test (agglutination) is performed on all cattle to determine brucella antibody levels. Those cattle with titers of 1:100 or greater are called reactors and are removed from the herd and slaughtered. To reduce the later possibility of reinfection, all female calves are vaccinated (attenuated live vaccine strain 19) between 3 and 8 months after birth. The vaccine usually affords protection for at least five pregnancies. The vaccine is not used in males, since it may cause infertility. An agglutination titer is determined on the calves following vaccination, and if a measurable agglutination titer appears but then disappears when measured later, the calves can be retained in the herd. If an agglutination titer of 1:200 or greater appears and then persists up to the time the calf is 30 months of age, the calf is considered a reactor and is slaughtered. Vaccination will not totally protect the herd, because the vaccine is effective on only 75% of those vaccinated. In some states where herds were once considered free of disease, a herd-rebuilding program has resulted in new cases of brucellosis.

Brucellosis in swine resembles the disease in humans because of its generalized nature. Brucella organisms can be found in most tissues of the swine; therefore infected animals represent a source of infection to humans working in slaughterhouses. The incidence of abortion is lower in swine than in cattle, but infertility in the sow and weak piglets are common manifestations of disease. There is no vaccine for swines, and control is dependent on the removal and slaughter of reactors and replacement with brucellosis-free swine.

Brucellosis in goats is similar to that in cattle, and control measures are approximately the same. Accurate data are not available on the incidence of the disease in these animals.

Brucellosis is also present in the general dog population and is similar to that in cattle. *B. canis* is the most common cause of disease in dogs, but *B. abortus* and *B. suis* can also cause infection. Treatment is not practiced, and control measures in breeding kennels consist of elimination of reactors. Infection is common among beagles, who are frequently raised in kennels and who are used in large numbers for research. No vaccine is yet available.

Brucellosis in sheep may involve both ewes and rams. Ewes occasionally abort, but after recovery there is permanent resistance. Rams, however, are more susceptible to infection, with the epididymes and testicles becoming involved. Ram infection leads to reduced fertility and is a source of infection to others in the herd. Diseased rams are slaughtered. Vaccination is the principal method of control.

Pasteurellosis

A respiratory tract disease complex called ***pasteurellosis*** is a major cause of economic loss to cattle feedlot operators. The disease is a very contagious bronchopneumonia that is usually associated with some type of stress placed on the animals during their transport from one place to another. The disease is thought to be caused by a combination of several microbial agents: *Pasteurella multocida* (plate 49), *P. haemolytica* (Plates 18 and 19), several viruses, and mycoplasmas. The disease is commonly referred to as ***shipping fever.*** The *Pasteurella* species are carried by a number of apparently normal animals and are not pathogenic until predisposing conditions such as stress and probably other environmental factors cause the organisms to multiply rapidly and initiate primary or, more often, secondary disease. Pasteurellosis is not limited to feeder cattle but can also affect other domestic animals placed under stressful conditions.

Pasteurellosis is initially manifested as a fever, which may reach 106° F, with coughing and nasal discharge. Nasal discharges enable the organisms to be expelled; thus the infected animal becomes a source of infection for others. As the disease progresses, there is obstruction of airflow, which can lead to cyanosis from lack of oxygen. There is considerable mortality from this disease, and even those that survive are so weak that they fail to gain weight at the same rate as normal cattle.

Treatment can be very effective when applied early during the course of the disease, before there is significant damage to the lungs. The *Pasteurella* species are sensitive to several antimicrobials, including penicillin, streptomycin, tetracycline, and sulfa drugs. More-advanced cases may recover from infection following treatment with antimicrobials, but lung tissue will be permanently impaired. Whole-broth killed cultures (bacterins) of *P. multocida* with or without *P. haemolytica* and the parainfluenza virus are available and provide some immunity for up to 1 year.

Disease may be prevented by buying healthy cattle that have not been subjected to stressful conditions. These stressed animals should be kept apart from the main herd for a few weeks so that if infection does occur, it can be contained.

P. multocida is also the primary agent of fowl cholera and hemorrhagic septicemia of cattle and water buffaloes.

Colibacillosis

Colibacillosis (scours) is a disease found primarily in calves, swine, and lambs. It is caused by serotypes of *Escherichia coli* that are referred to as ***enteropathogenic.*** Enteropathogenic *E. coli* as a cause of human diarrhea is discussed in Chapter 28. The enterotoxigenic *E. coli* produce one or two types of enterotoxin, depending on the serotype: one designated heat labile (LT), which is inactivated by heating at 60° C, and one designated heat stable (ST), which remains active after heating to 60° C. The effect of enterotoxin on intestinal epithelium and its mechanism of action are discussed in Chapter 22. The pathogenicity of the enterotoxigenic strains is related not only to their ability to procedure enterotoxin, but also to their capacity to colonize the intestinal epithelium. The attachment process is associated with certain fimbriae-like surface antigens, which have been designated K88 for swine and K99 for calves. The intestinal receptor that binds the K88 antigen has been isolated and appears to be a membrane glycolipid. Studies with the K88 strain of *E. coli* indicate that both pathogenicity factors are carried on plasmids. Pathogenic K88 strains are able to convert normal *E. coli* strains found in healthy swine to enteropathogenic strains, but only if the genes for both enterotoxin production and adherence antigen are transferred. The presence of either pathogenicity factor alone is not sufficient to make the strain virulent.

The animals ingest *E. coli* from their environment, and specific serotypes become part of the normal flora. Colostrum from the mother contains antibodies that confer a passive immunity to the young for that particular serotype. Moving an in-calf cow from one farm to another may result in the calf ingesting serotypes that were not present on the original farm. Consequently, the colostrum contains antibodies that are not specific for the calf's *E. coli* serotypes, and ingestion of pathogenic serotypes could result in severe diarrheal disease.

Two forms of disease are associated with enteropathogenic *E. coli:* enteric and septicemic. The enteric form is common in calves, swine, and lambs. The first sign of disease is a whitish yellow diarrhea. The animals lose considerable water and refuse to feed and become increasingly weak. Untreated animals usually die in 3 to 5 days. Piglets demonstrate a more acute disease and can die within 1 day following the appearance of symptoms. The septicemic form does not provide any clinical signs, and death occurs without warning. Septicemia is more common in lambs and foals and is usually preceded by infection of the navel.

The most effective measure for preventing colibacillosis is to ensure that all newborn animals receive colostrum as soon as possible. The young should also be kept under the best hygienic conditions available. Treatment must include replacement of fluid loss by the administration of an electrolyte solution. The solution can be administered orally, but if fluid losses have been severe, intravenous adminstration may be necessary. Antibiotics may be given as a prophylactic measure to prevent septicemia, but they are of limited value in diarrheal disease because of multiple antibiotic resistance.

Clostridial diseases

The clostridia are found primarily in the soil and in the intestinal tract of animals. Several species are causative agents of diseases that appear in animals, some of which (tetanus, botulism, and gas gangrene) are also associated with humans. They may be divided into three groups based on the site of action of their toxins: ***histotoxic, enterotoxic, and neurotoxic*** **(Table 34-2).**

Table 34-2 The principal species of *Clostridia* causing disease in domestic animals based on the site of action of their toxins

GROUP	CLOSTRIDIAL SPECIES AND TYPE	DISEASE
Histotoxic	*C. chauvoei*	Blackleg in ruminants
	C. septicum	Malignant edema (gas gangrene) in horses, cattle, sheep, and pigs; Braxy in sheep
	C. novyi	
	Type A	Big head in rams
	Type B	Black disease in sheep
	C. haemolyticum	Bacillary hemoglobinuria in cattle; necrotic liver disease in sheep
Enterotoxic	*C. perfringens*	
	Type B	Dysentery in lambs; enterotoxemia in goats, calves, sheep, and foals
	Type C	Enterotoxemia in sheep (struck), lambs, calves, and piglets
	Type D	Enterotoxemia: overeating disease or pulpy kidney disease in sheep
Neurotoxic	*C. botulinum* types A, B, C, D, and E	Botulism in cattle, sheep, and horses
	C. tetani	Tetanus in all domestic animals

Histotoxic clostridial diseases. ***Blackleg*** is a disease primarily of cattle, but it also occurs in sheep, goats, and pigs. Infections occur primarily during spring and summer months, but the reasons for this are not understood. The organisms causing disease, *Clostridium chauvoei* and *C. septicum,* may be found in the intestine of normal farm animals as well as in the liver of healthy dogs and cattle. The organisms are ingested as spores, and it has been suggested that in cattle the spores germinate only under predisposing conditions that would provide the proper anaerobic environment. Blackleg in sheep is believed to be due to contaminated wounds incurred by castration, shearing, or docking. Once the vegetative cells have been produced in the intestine or wound, the bloodstream is invaded, followed by the invasion of vascular organs and muscles, particularly of the hip and loin area. These areas swell as a result of gas production and exhibit twitching or trembling. At necropsy the affected muscles appear black; hence the name of the disease—blackleg (Plate 50). By the time the disease is recognized, it is too late to save the animal, but prompt treatment with penicillin used in conjunction with hyperimmune serum can save some animals. Active vaccination with a formalinized whole-broth culture of *C. chauvoei* can produce effective immunity.

Malignant edema is a commonly fatal disease of cattle, horses, sheep, swine, and goats. It is caused primarily by *C. septicum,* but other organisms may be involved, such as *C. chauvoei* and *C. novyi.* Infection occurs by contamination of a wound, particularly deep puncture wounds. *C. septicum* produces a disease in farm animals similar to blackleg, except there is little gas production. It also produces a disease in sheep called ***braxy,*** which is associated with eating frozen succulent feed. Necrotic lesions and edema appear in the abomasal (stomach) wall, flank, and abdomen. Braxy is more frequently observed in the United Kingdom, Scandinavia, and Iceland. Prevention and treatment are the same as for infection caused by *C. chauvoei.*

Bacillary hemoglobinuria is a disease of cattle and sheep sometimes referred to as ***redwater.*** The disease is caused by *C. haemolyticum* and is preceded by liver damage from fluke infection. The disease occurs only in areas where snails exist: snails are part of the life cycle of the animal parasite. The clostridia multiplying in the liver release toxins that destroy red blood cells, causing hemoglobinuria. The resulting anemia reduces oxygen transport, and the animal eventually dies of anoxia. Early treatment with penicillin or broad-spectrum antibiotics can save some animals. A vaccine (formalinized whole–broth cell culture) provides a short immunity of up to 6 months; boosters are required to maintain active immunity. The disease can be controlled by eliminating the intermediate host of the fluke, the snail.

Big head is a gangrene that appears in sheep and cattle and is caused by *C. novyi* type A. The infection develops from wounds on the face and head of rams, usually incurred from butting. The localized swellings are the reason for the name of the disease.

Black disease of sheep is caused by *C. novyi* type B. The disease is a chronic hepatitis that arises in fluke-infected areas such as Australia, New Zealand, the United Kingdom, and some areas of the western part of the United States. Damage by the liver fluke provides an ideal environment for germination of clostridial spores and toxin formation. Recovery from infection is rare. Elimination of the snail, which is a host for the liver fluke, is the best method of prevention. A whole-cell formalinized vaccine is available that produces immunity for 1 year.

Enterotoxic clostridial diseases. Enterotoxemia is an acute, highly fatal infection of animals, particularly sheep, caused by the same species of organisms causing gas gangrene in humans—*C. perfringens* (see Chapter 28). This organism is classified according to types, based on toxin production: A, B, C, D, and E. Type A, which is the agent of human disease is usually nontoxigenic for farm animals. Lamb dysentery, an enterotoxemia caused by *C. perfringens* type B, usually occurs in lambs less than 2 weeks old. Death may occur without any clinical signs, whereas the less acute cases are characterized by a profuse and bloody diarrhea. Type B also affects calves 7 to 10 days old, with symptoms similar to those in sheep. Type C causes an enterotoxemia of older sheep and is called ***struck***. Struck is restricted primarily to the United Kingdom. Type D causes an enterotoxemia in sheep of all ages and is referred to as the ***overeating disease*** or ***pulpy kidney disease.*** The disease is often associated with feedlots, where the animals may be overfed. The development of the disease is thought to result from concentrated feeding which results in a change in the environment of the animal's stomach and intestine that favors clostridial multiplication and toxin formation. Delayed postmortem examination of animals reveals a degenerated kidney, hence the name *pulpy kidney disease.*

Immunization, especially of sheep, with formalinized whole-broth culture of types C and D produces an active immunity for only 6 to 12 months; therefore boosters are required. Antitoxin from types B, C, and D are also available, but they are effective for only 2 to 3 weeks. Toxoids are administered to ewes 6 weeks before lambing to prevent infection of lambs by types B, C, and D.

Neurotoxic clostridial diseases. Neurotoxins are produced by *C. botulinum* and *C. tetani.* The symptoms of infection are the same as those occurring in humans (see Chapter 28), but there are different predisposing factors that lead to infection. Botulism in animals is an intoxication resulting from eating food containing toxins produced by *C. botulinum.* Intoxication occurs more frequently in wild waterfowl and is referred to as ***limber neck*** because of paralysis of the neck muscles. Botulism is infrequent in horses, cattle, sheep, and goats but is more likely to occur if the animals feed on carrion or improperly stored and cured silage. Immunization using a toxoid is employed only in special-risk groups, such as mink and cattle in South Africa.

Tetanus occurs among domestic animals such as horses, pigs, and sheep. The conditions that predispose animals to infection are castration, shearing, and docking. Horses acquire infection usually because of a wound on the hoof. One of the early diagnostic signs of disease in animals is a spasm or twitch of muscle in response to loud noises. Toxoids are used for active immunization of horses, which are the most susceptible of the domestic animals. Mortality among horses may be as high as 50%.

VIRAL DISEASES

Infectious bovine rhinotracheitis

Infectious bovine rhinotracheitis (IBR) is caused by a herpesvirus that is morphologically indistinguishable from other herpesviruses. The disease is found almost exclusively in cattle, although wild ruminants may also be susceptible to infection. The virus is transmitted primarily by inhalation or direct contact with contaminated animal nasal or ocular secretions. The virus is also present in placental tissue of cattle that abort, and this may represent a source of infection to other animals.

IBR can spread rapidly in a herd, but the mortality is very low. The most common manifestation of disease is an acute upper respiratory tract condition involving the sinuses, pharynx, larynx, and trachea (Plate 51). More severe respiratory tract disease can lead to bronchopneumonia, which is a cause of greater mortality in calves. The onset of respiratory tract disease is sudden, with a rise in body temperature from 105° to 106° F and inflammation of the mucous membranes. Most animals recover from infection in 4 to 7 days but may remain as shedders of the virus for a variable length of time.

IBR can also be manifested in a variety of other ways: conjunctivitis and meningoencephalitis in 2- to 3-month-old calves, abortion in pregnant cows, and a disease of the genital tract called infectious pustular vulvovaginitis. The virus can be transmitted in the semen, and bulls in artificial insemination units are routinely tested for IBR.

Two modified live vaccines that can be applied intramuscularly and intranasally are available for prevention of IBR. These vaccines are also available with inactivated *Pasteurella* organisms, which reduce the probability of pneumonic pasteurellosis (shipping fever). The intranasal vaccine may also be combined with the parainfluenza virus, but it must be repeated annually. Intranasal vaccines induce a rapid (within 40 hours) local immunity that is not inactivated by maternal antibody when administered to young calves and is less likely to produce abortion in pregnant cows.

Bovine virus diarrhea

Bovine virus diarrhea (BVD) is caused by a pestivirus that is a member of the family Togaviridae. The disease is thought to be one of the most prevalent viral infections of cattle. Transmission may be by carriers or contaminated fomites, with most infections resulting from the ingestion of virus.

The manifestations of infection may vary from (1) mild, with few symptoms, to (2) acute, with rapid recovery, to (3) chronic, with periods of disease lasting for several months. The first sign of disease is a high fever ranging from 104° to 106° F. The virus has an affinity for epithelial and lymphoid tissue, with lesions appearing in the upper respiratory tract, the digestive tract (such as in the esophagus), and Peyer's patches. Fever is often followed by nasal discharge and a profuse watery diarrhea resulting in dehydration and rapid weight loss. The mortality is usually between 4% and 8%, with most fatalities occurring in calves who demonstrate the most severe clinical symptoms.

A live modified vaccine is available for administration to calves between 6 and 9 months of age (maternal antibodies from colostrum would reduce vaccine effectiveness if it were administered earlier). Vaccine should not be administered to pregnant animals but can be administered after calving.

Transmissible gastroenteritis

Transmissible gastroenteritis is an infection of swine that is caused by a coronavirus of the family *Coronaviridae*. The disease has a particularly high mortality among infected piglets. The virus is present in the feces and nasal secretions of infected animals and is usually spread by ingestion of contaminated material.

The disease is characterized by a profuse watery diarrhea preceded by rapid dehydration and vomiting in piglets. Piglets may lose up to 20% of their body weight after a few days of the illness. In older swine vomiting is uncommon, and recovery from disease is complete in a few days. Older swine, however, remain as shedders of the virus for 2 to 3 months after recovery and become a constant source of infection to others in the herd. The intestinal tract demonstrates the most significant pathological changes, in which the eroded epithelium appears almost transparent. Atrophy of the intestinal villi in the jejunum and ileum are characteristic of disease. The mortality in piglets may range up to 100%.

Piglets born to sows that have been exposed to the virus 40 or more days before farrowing are immune to natural infections because of specific Ig_A in the colostrum. One of the methods for preventing the disease is to feed infected-swine intestines to pregnant sows 30 to 40 days before farrowing. A vaccine is available for the immunization of sows before farrowing, and in some European countries purified virus in capsular form is orally administered to cattle.

Foot-and-mouth disease

Foot-and-mouth disease (FMD) is not observed in Australia, the United States, or Canada but is among the most important diseases elsewhere in the world. The viral agent is an aphtovirus, a member of the Picornaviridae. The disease, which is one of the most contagious, affects cattle and all cloven animals, including sheep, swine, goats, deer, and water buffalo. FMD may be acquired by humans through contact with infected animals, but there are only a few recorded cases. Human disease is invariably mild and is characterized by localized lesions, usually on the hands. Human infection is usually by inhalation or ingestion of virus following contact with infected animals or fomites. The virus can remain viable in dried secretions for 2 months or longer, and this contributes to its contagiousness among cattle. Importation of meat and meat products used to feed farm animals has resulted in several outbreaks. FMD is not a fatal disease but one that results in severe economic loss because of rapid weight loss and lost milk production. The first signs of disease are fever and blisters on the mouth, on the tongue, at the bulb of the heel, between the toes, and on the teats. There is considerable lacrimation and foamy salivation by infected animals. The vesicles (blisters) usually rupture within 20 hours of their appearance, leaving raw and eroded areas. Lameness may be a sequela of infection involving the feet. Infection in adult animals results in a few fatalities, but calves may be more severely affected. In-calf cows may abort without demonstrating any clinical signs.

There are several distinct serotypes of FMD virus, and the use of vaccines confers only short-term immunity. Vaccination of potentially susceptible animals is a costly means of control, particularly if the disease occurs sporadically in the area; therefore other preventive measures are employed, which include (1) quarantine of infected animals, (2) slaughter and burial of infected animals, (3) prompt disinfection of obvious sites of contamination, and (4) notification of outbreaks. Vaccination is practiced in those areas in which the disease is enzootic. The most common vaccine consists of virus grown in cell cultures. The vaccine,* which is killed with formalin, provides immunity for only 5 to 6 months.

Pseudorabies (Aujeszky's disease)

Pseudorabies is primarily a disease of swine, which are the natural hosts, but other animals such as cattle, dogs, cats, horses, and sheep may also be infected. The agent of pseudorabies is a herpesvirus that morphologically resembles the herpes simplex virus infecting humans but is antigenically distinct from it. There have been no recorded accounts of pseudorabies in humans. The disease is prevalent in the midwestern part of the United States and in some European countries where pig rearing is a major occupation.

Swine are thought to acquire the disease by contact with nasal or oral secretions from other infected swine. Cattle are believed to acquire the infection by contact with swine carriers who have inapparent disease. Dogs and cats may acquire the disease by eating infected meat. The severity of the disease in swine is inversely related to the age of the affected animal. Older pigs, for example, often undergo subclinical or mild illness, whereas young piglets up to 1 month in age have severe disease symptoms that may result in 100% mortality. The disease in young piglets is characterized by high fever and central nervous system involvement that manifests itself in the form of a staggering gait, convulsions, paralysis leading to prostration, coma, and death. Older diseased pigs up to 6 months in age have few symptoms other than nervousness. Adult pigs may be symptomless, but infected sows are subject to abortion and stillbirths. Cattle and animals other than swine, in which the disease is almost always fatal, represent a dead end in terms of viral transmission. Affected animals suffer from intense itching at the site of infection and rub and lick this part of the body until the area is raw. This behavior was noted as early as the 1800s by farmers in the United States and was described as the ***mad itch*** in dairy cattle.The animals eventually lapse into a coma, with death occurring in 18 to 48 hours after the appearance of clinical signs.

Swine that recover from pseudorabies develop neutralizing antibodies that appear in the colostrum and protect newly born piglets for the first 2 months of life. A live modified vaccine is available in Europe and the United States, but there is evidence that vaccinated animals shed the virus or become latent carriers. A delayed hypersensitivity test is being used in some areas to screen and detect infected animals, which are later destroyed. Cattle can be protected from infection by being separated, and pigs and breeding sows can be potected by being separated from fattening pigs.

Newcastle disease

In 1926 a new disease of poultry was reported in Indonesia. An outbreak of a similar disease was reported near Newcastle-on-Tyne in England, where the agent of the new disease was identified. The Newcastle disease virus is a paramyxovirus belonging to the family Paramyxviridae. The mortality from disease by the earlier reported virus was nearly 100%. In the late 1930s an apparently new disease appeared in California and spread to other states, but this outbreak was later found to be caused by a less virulent type of the Newcastle disease virus. Newcastle disease today is worldwide and is caused by three virus strains that have been classified as ***velogenic, mesogenic,*** and ***lentogenic.*** The velogenic strain is the most virulent and is a more frequent cause of disease outside the United States. The mesogenic and lentogenic strains are associated with a milder disease and are found in the United States and some other countries.

Newcastle disease is a highly contagious and fatal disease affecting primarily poultry and

*A vaccine produced through recombinant DNA technology has recently been approved for administration to animals.

guinea fowl, but it may also affect a wide number of domestic and wild birds. Healthy birds acquire infection by inhaling dust from contaminated litter, foodstuffs, utensils, and feces. Infection may also be acquired by contamination of the incubator with broken eggs that had been infected in utero. Although the embryos invariably die, they represent a small but potential means of viral transmission.

The trafficking of birds and movements of dealers, workers, and slaughterers represent the major methods of viral spread from one area to another. Viral spread is rapid in poultry houses, where birds with subclinical forms of the disease shed the virus for long periods of time.

The virus enters the host via the respiratory tract or conjunctiva, and very mild forms of conjunctivitis have been reported in humans working in contact with infected animals. The incubation period in the avian disease is 5 days, during which time the virus multiplies in epithelial cells and penetrates the bloodstream. The virus then accumulates in specific tissues, particularly the respiratory tract, intestinal tract, and nervous system. The symptoms of disease differ depending on the strain involved in infection. Infections by the velogenic strain often result in death without clinical signs. Infections by the mesogenic and lentogenic strains result in congestion of the respiratory tract followed by nervous system signs such as paralysis of the legs and wings, trembling, and twisting of the neck. The mortality from the milder forms of disease range from 5% to 50%.

Slaughter of infected animals for eradication is effective in areas where the disease is virtually unknown, for example, Scotland, but vaccination is the most practical method of control for endemic areas. Live modified virus vaccine grown in the chicken embryo can be administered by dust, spray, or drinking water. In some countries the vaccine is often combined with either infectious bronchitis virus vaccine or fowlpox vaccine. An inactivated vaccine is also available but gives protection for only up to 6 months.

FUNGAL DISEASES

Brooder pneumonia

Some of the most widespread of all fungi are members of the genus *Aspergillus*. They are present in soil and decaying plant and animal matter, and their spores are present in the air. The most important pathogenic species is *Aspergillus fumigatus*, which can infect humans as well as domestic animals. The organism is also responsible for the production of ***aflatoxin*** in moldy grains (see Chapter 31). One important manifestation of *Aspergillus* infection in animals is brooder pneumonia.

Brooder pneumonia is a respiratory tract disease that has profound effects on all birds. Animals are infected by inhalation of spores from fungi growing in feed or litter. In the acute form of the disease many affected chickens may die within 24 to 48 hours of infection, with relatively few recognizable symptoms. These symptoms, when present, include listlessness, high temperature, increased respiration, and diarrhea. At autopsy the lungs contain many discrete yellowish white granulomas (plate 52). Other animals may also be infected by species of *Aspergillus*. Cattle may have placentitis, resulting in abortion during the second half of pregnancy. Most abortion in cattle occurs while the animal is housed during the winter, and infections apparently occur from the inhalation of spores that have contaminated straw and hay.

No effective treatment is available for pulmonary aspergillosis, and recovery must occur by natural means.

Ringworm

Ringworm (dermatomycoses) in humans is caused by three genera of fungi—*Trichophyton, Microsporum,* and *Epidermophyton*—but only the first two are encountered among lower animals. Human ringworm is discussed in Chapter 31.

Ringworm in animals may be caused by contact with organisms that are found primarily as saprophytes in the soil (geophilic) or by contact with organisms that prefer the habitat of the vertebrate host (zoophilic). Most infections in dogs and cats, for example, are caused by indigenous species of *Microsporum canis*. The lesions in dogs appear as slightly raised, circular, scaly areas in which stubs of hair may appear or the hair has completely fallen out. In cats the lesions are sometimes imperceptible to the naked eye, and very few hairs are affected. Ringworm in cattle is most frequently observed in calves, especially those that have been housed in the winter in overcrowded pens. The lesions in cattle are similar to those that appear in dogs. The lesions in horses are circular and similar to those of dogs and cattle (plate 53). They are found more often in friction areas, such as the saddle and girth regions.

Ringworm has a variable course in animals that may last for several weeks to months. The disease is transmissible to humans; therefore every effort should be made to control it in lower animals. Griseofulvin given orally is effective in the treatment of ringworm, but it is only fungistatic and is too expensive for administration to all but valuable large animals. The disease can be controlled if hair around the lesion is clipped to remove any contaminated hair or debris and if the lesion is treated with topical antimicrobial agents such as natamycin, miconazole, clotrimazole, and others. Cattle ringworm often clears up when the animal leaves the stable in the spring for pasture.

SUMMARY

Animal diseases are important for two reasons: some are transmissible to humans and some can result in severe economic loss from lost food production. Animals are infected by all major groups of microorganisms, but those of major economic significance in the developed countries of the world are the bacteria and the viruses. Animal diseases transmitted to humans by insect vectors are discussed in Chapter 32 (see Table 32-4), and those diseases transmitted by direct or indirect contact are outlined in Table 34-1.

The host-parasite interactions associated with human disease are similar to those in animal disease. The epidemiology of animal disease is similar to that in humans but is often more complex. The acquisition of infection by animals is directly related to their close association with the soil, insect vectors, and the hygienic conditions in which they are maintained. The size of the animal often precludes the use of some drugs for treatment of disease because of the cost factor. The treatment of domestic animals is based on the concept of saving the herd and not the individual animal. Large numbers of animals may therefore be sacrificed to save the herd or prevent the spread of disease to other herds.

The major bacterial diseases of economic consequence are brucellosis, pasteurellosis, colibacillosis, and clostridial disease. Brucellosis, which may involve cattle, swine, goats, sheep, and dogs, is particularly important because it is a cause of abortion and/or sterility. Pasteurellosis is referred to as shipping fever and is apparently due to infection by a complex of microbial agents, as well as environmental stress during the shipping process. A high mortality is associated with the disease. Colibacillosis is found mainly in calves, piglets, and lambs and is caused by enterotoxigenic *E. coli.* Enteric infection causes the animals to become weak and lose weight. Occasionally the infection becomes systemic, and death occurs rapidly. Clostridial disease is a result of an animal's close association with the soil. Animals ingest clostridial spores that germinate in the host and cause disease under certain predisposing conditions. Animals subjected to castration, shearing, docking, or many other procedures that can result in wound contamination can be infected by clostridial species. Clostridial disease may be histotoxic, enterotoxic, or neurotoxic.

Viral diseases are common in animals, and those of major economic importance are infectious rhinotracheitis (IBR), bovine virus diarrhea (BVD), transmissible gastroenteritis, foot-and-mouth disease, pseudorabies, and Newcastle disease. IBR is an upper respiratory tract infection that causes considerable morbidity but few fatalities. BVP is one of the most prevalent viral infections of cattle, and fatalities are very high in calves. Transmissible gastroenteritis is an infection of swine that has a high mortality among piglets. Foot-and-mouth disease, which is seen more frequently in underdeveloped countries of the world, is one of the most contagious of all diseases and affects all cloven animals. It is not a fatal disease, but infection causes rapid weight loss and lost milk production. Pseudorabies is primarily a disease of swine that may be contracted by other animals. The mortality is especially high if the disease occurs in cattle. Newcastle disease is an important disease of poultry that has a high mortality. It is a respiratory tract infection that can spread rapidly in poultry houses. Few fungal diseases are of major econommic importance. Two diseases discussed are brooder pneumonia and ringworm. Brooder pneumonia is caused by *Aspergillus fumigatus,* which is inhaled by animals in the form of spores. Chickens are severely affected by infection and may die within 24 to 48 hours of infection. Disease caused by *Aspergillus* organisms can also occur in cattle, particularly when they are penned during the winter months.

Study questions

1. What major diseases are transmitted from animals to humans by insect vectors?
2. What are some of the significant differences in the treatment of humans as compared with animals?
3. What predisposing conditions make cattle and other domestic animals susceptible to infection?
4. Both animals and humans are infected by *Brucella* species; how does the site of infection differ between them?
5. What microbial agents are believed to be involved in shipping fever. What predisposing factors contribute to the appearance of the disease?
6. Infection by enterotoxigenic *Escherichia coli* causes what two forms of disease in animals? What is the best measure for preventing this disease?
7. What is meant by histotoxic, enterotoxic, and neurotoxic clostridial disease? What microbial agents are involved in each type?
8. What are the predisposing factors and conditions that make animals more susceptible to disease by species of *Clostridia*?
9. Name two animal viral diseases caused by a herpesvirus.
10. What are the characteristic symptoms of foot-and-mouth disease?
11. What viral disease was originally called the mad itch?
12. What animal is primarily affected by the agent of pseudorabies? How is the pathogen transmitted to other domestic animals?
13. What group of animals is primarily affected by the Newcastle disease virus? How did the disease get its name?
14. Brooder pneumonia is a fungal disease caused by what species? Does this species have any importance in terms of human disease? Explain.

Selected readings

Books

Andrews, C., and Periera, H.G. *Viruses of vertebrates* (4th ed.). The Williams & Wilkins Co., Baltimore, 1978.

Buxton, A., and Fraser, G. *Animal microbiology* (Vols. 1, 2). Blackwell Scientific Publications, Ltd., Oxford, 1977.

Carter, G.R. *Essentials of veterinary bacteriology and mycology.* Michigan State University Press, East Lansing, 1982.

Fenner, F. et al. *The biology of animal viruses* (2nd ed.). Academic Press, Inc., New York, 1974.

Gillespie, J.H., and Timoney, J.F. *Hagan and Bruner's infectious diseases of domestic animals* (7th ed.). Cornell University Press, Ithaca, N.Y., 1981.

Johnson, R.C. (Ed.). *Biology of parasitic spirochetes.* Academic Press, Inc., New York, 1976.

Jungerman, P.F., and Schwartzman, R.M. *Veterinary medical mycology.* Lea & Febiger, Philadelphia, 1977.

Roberts, A.W., and Carter, G.R. *Essentials of veterinary virology.* Michigan State University Press, East Lansing, 1981.

Schwabe, C.W., Preman, H.P., and Franti, C.E. *Epidemiology in veterinary practice.* Lea & Febiger, Philadelphia, 1977.

Tully, J.G., and Whitcomb, R.F. (Eds). *The mycoplasmas,* Vol. 2, *Human and animal mycoplasmas.* Academic Press, Inc., New York, 1979.

Journal articles

Carter, G.R. Pasteurellosis: *Pasteurella multocida* and *Pasteurella haemolytica. Adv. Vet. Sci.* **11:**321, 1967.

Elwell, L.P., and Shipley, P.L. Plasmid-mediated factors associated with virulence of bacteria to animals. *Annu. Rev. Microbiol.* **34:**465, 1980.

Foggie, A. Chlamydial infections in mammals. *Vet. Rec.* **100:**315, 1977.

Luchsinger, D.W., and Anderson, R.K. Longitudinal studies of naturally acquired *Brucella abortus* infection in sheep. *Am. J. Vet. Res.* **40:**1307, 1979.

Moon, H.W. Mechanisms in the pathogenesis of diarrhea: a review. *J. Am. Vet. Med. Assoc.* **172:**443, 1978.

Oldenkamp, E.P. Treatment of ringworm of the horse with natamycin. *Vet. Rec.* **11:**36, 1979.

Prescott, J.F., and Brein-Mosch, C.W. Carriage of *Campylobacter jejuni* in healthy and diarrheic animals. *Am. J. Vet. Res.* **42:**164, 1981.

Chapter 35 INFECTIOUS DISEASES OF PLANTS

Robert F. Boyd and Gary Hooper

Plants are susceptible to diseases in ways very similar to situations in humans or other animals. Discussions of cause, symptoms, and control or cures for this subject are as appropriate as the preceding discussions on animal diseases. Microorganisms play a large part in plant pathology, or ***phytopathology*** as it is called by agriculturists. However, the host plants and the disease organisms involved in plant problems are sufficiently distinct from the animal situation as to require a separate discussion. In order to set the stage for a discussion of plant diseases, this chapter is organized with an introductory section on the art and science of plant pathology, followed by a description of the major causes of plant disease. The introductory material is concluded with a discussion of disease economic impact and control strategies. The remainder of the chapter is devoted to the host-parasite relationship and a discussion of some individual diseases of economic importance.

ART AND SCIENCE OF PLANT PATHOLOGY

Historical perspective

Plant pathology today is a discipline with broad application. The world of plant pathology ranges from field studies of resistance or control to molecular biology at the very forefront of sophisticated science. The field includes representives of agribusiness, mycologists, virologists, nematologists, bacteriologists, computer scientists, plant breeders, plant physiologists, meteorologists, and countless similar disciplines. Each group represents an important cog in a machine that has tremendous implications for the survival of humankind.

While diseases of plants have been known about as long as plants have been cultivated, most early plant problems, like human conditions, were ascribed to supernatural causes, such as God's displeasure. Theophrastus (371-287 BC) wrote of diseases of field and orchard and used terminology such as scabs, cankers, rusts, and blights. These terms are still in as common use today as when Theophrastus adapted them from the agriculturists of his day. The first actual studies linking fungi and other biological entities to plant disease came in the eighteenth and nineteenth centuries. Much of this early work was done by people trained in human medicine. These physicians were often familiar with fungi and extended their animal mycology expertise to cases of plant disease. The following are some notable examples of early physician-plant pathologists:

1. Giovanni Targioni-Tozzetti and Felice Fontana of Italy in the mid 1700s, who published comments associating small-plant parasites as the cause of rust diseases in cereal grains.
2. The Swiss-born naturalist-philosopher Jonas Benedict Prevost, who studied smut dis-

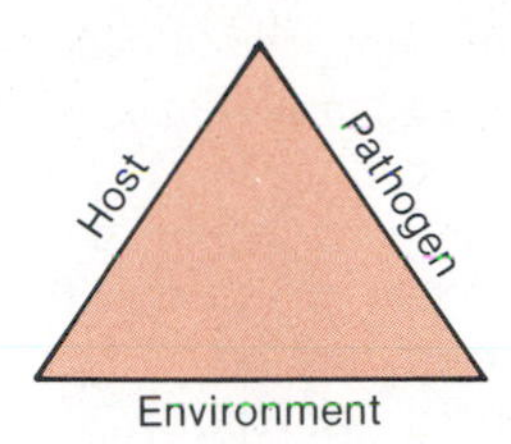

Figure 35-1 Disease triangle representing interrelationships between plant, environment, and plant pathogen.

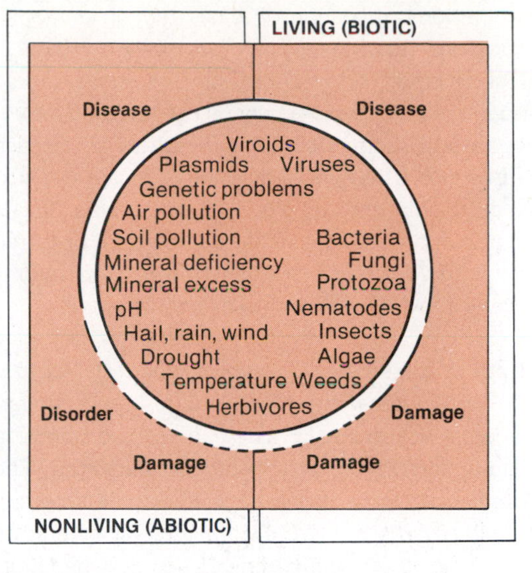

Figure 35-2 Factors that affect plants.

eases of grain in the late 1700s. His published conclusions demonstrate a clear understanding of the cause of this disease. His work, as well as that of most other early investigators, was not universally accepted or cited during his lifetime.

3. Anton de Bary (1831-1888) and Julius Kuhn (1825-1910) of Germany, who are considered the fathers of plant pathology. They provided the final impetus for establishing and continuing plant disease studies when serious food shortages, such as those caused by the potato blight in Europe and Ireland, occurred in the 1840s. Plant pathology, or phytopathology (Greek: *phyto,* plant; *pathos,* disease; *logos,* discourse) is defined as the science of understanding the nature of disease and its causes, as well as the arts of diagnosis and control of these diseases. Exactly what constitutes a plant disease and how to control it are often a cause of disagreement. Few can agree on the exact terminology and boundaries of disease, but all can recognize plants that differ in productivity, size, color, and form from their fellows or that have obvious dead or dying parts. Disease can be considered a "malfunctioning of some or all normal plant processes which continue to the detriment of plant growth, appearance, or production."*

*Definition of G.R. Hooper—similar to that accepted by the American Phytopathological Society.

Etiology

Plants are remarkably resistant to disease and are normally free of large populations of commensal organisms. They exist, however, in an environment that contains parasitic organisms and that is capable of causing problems to plants. The interrelationships between plant, environment, and plant pathogen are many and complex. They are normally depicted as a disease triangle (Figure 35-1), with time as an additional factor implied in the relationship.

The previous definition of disease fits most general situations. However, a plant's interactions with the biotic (living) and abiotic (nonliving) parts of its environment range from destructive situations such as breakage, ingestion by herbivores, or short-term injuries to very complex biological associations ranging from commensal to pathogenic relationships. The living and nonliving entities that affect plants create conditions described as damage, disorder, or disease (Figure 35-2). The line between damage and disease is always difficult to draw, whether we are discussing living organisms or environmental conditions. Many prefer to call problems associated with certain abiotic factors disorders rather than diseases to emphasize the noninfectious aspects of these relationships. Many living organisms establish very intimate associations with plants. Chief among these are other plants, fungi, bacteria, nematodes, insects, and viruses. Less commonly encountered or more imperfectly understood are associations involving viroids, mycoplasma-like organisms (MLOs), algae, and various protozoa. Most of these associations involve some degree of parasitism; however, not all are totally destructive associations. While a number of disease causes are listed in Figure 35-2, those of pertinence to this text are the microorganisms. These include most classes of fungi; certain bacteria, including *Rickettsia* and *Spiroplasma* organisms as well as MLOs; and viruses and viroids.

Phytopathogenic prokaryotes. Bacteria, including *Spiroplasma* organisms, MLOs, and actinomycetes, are normally listed under the common heading of ***phytopathogenic prokaryotes.*** The *Spiroplasma* organisms and MLOs are internal-vascular parasites, which are difficult to culture free from their plant host and are generally grouped as fastidious prokaryotes.

Historically, bacteria are of considerable interest to plant pathologists. The earliest disease discovered to be caused by a bacterium was fire blight of apples and pears. This disease, caused by *Erwinia amylovora,* was discovered by T.J. Burrill in the late 1800s. Bacteria at one time were not accepted as etiological agents of plant disease. In the 4 years between 1897 and 1901 the concept of bacteria as plant pathogens was hotly debated by an American, Erwin Smith (for whom the genus *Erwinia* was named), and a German, Alfred Fischer. Professor Fischer represented the established, prestigious European school and held that bacteria were simply contaminates in diseased plant tissue and not pathogens. Dr. Smith of the US Department of Agriculture had done extensive research with fire blight and other diseases. He viewed bacteria as causes of plant diseases. Eventually Dr. Smith's views were accepted by the scientific community.

Nonfastidious bacteria. Only a small percentage (about 10%) of the genera of fastidious bacteria are known to cause problems in plants. Most well-established pathogens are usually classified in six genera: *Pseudomonas, Xanthomonas, Agrobacterium, Erwinia, Corynebacterium,* and *Streptomyces.* However, pathogens of plants have tentatively been placed in several other genera, including *Acetobacter, Acetomonas, Aplanobacter, Bacillus, Clostridium, Methanobacterium,* and *Nocardia.* They can be differentiated by the usual microbial criteria, such as Gram stain, morphology, oxygen requirements, and growth on selected media. Species are often distinguished by their ability to produce lesions or other reactions on selected plant species. The general characteristics of the six important groups are described below:

1. *Pseudomonas.* About 80 species of these large, rod-shaped motile bacteria are associated with plant diseases. Many are distinguished in culture by their fluorescent behavior on certain media. Their most common effect on plants is the formation of leaf and fruit spotting.
2. *Xanthomonas.* All species of *Xanthomonas* (about 75) are plant pathogens. They are motile, medium-sized rods and usually exhibit a yellow to orange pigment in culture.
3. *Agrobacterium.* There are about six species of *Agrobacterium* (formerly *Rhizobium*), which cause galls or hairy-root symptoms on plants. They are nonpigmented, motile rods occurring frequently in agricultural soils. One organism, *A. tumefaciens,* causes crown gall of plants through a plasmid carried by virulent strains. Crown gall is discussed at more length under "Bacterial Diseases." Another plant-associated nonpathogenic soil-borne species, *A. radiobacter,* is useful in the control of some diseases caused by other agrobacteria.
4. *Erwinia.* This group contains facultative anaerobes that cause several important diseases of wilt, blight, or soft-rot types. About 12 to 16 species have been described as plant pathogens.
5. *Corynebacterium.* Twelve to fifteen species of *Corynebacterium* are known as plant pathogens. They are irregular rods, many nonmotile, that invade vascular tissues to cause wilts in some plants. They also cause fruit spotting, wilts, and a disease called ring-rot of potato.
6. *Streptomyces.* This is the only true filamentous bacterial form found associated with plant diseases. Two species occur as causes of wart and scab diseases of underground plant parts, such as potatoes and yams (sweet potatoes).

Fastidious bacteria. Fastidious vascular-limited bacteria (FVB), formerly called rickettsia-like organisms and MLOs, occur in plants and cause diseases. Only a few of these organisms, classified in the genus *Spiroplasma,* have been isolated and successfully cultured. In years past, diseases caused by these organisms were called scorch or yellows. They were thought to be caused by viruses because they were graft and insect transmissible. However, as time went by, no viral particles were found associated with diseased tissue, and ultimately either bacteria or bacteria-like organisms were found by electron microscopy in the vascular cells of the plants. Their classification was based on morphology and antibiotic sensitivity.

The FVBs are rod-shaped, obligate parasites with characteristically wavy walls in electron microscope preparations. In Pierce's disease they inhabit the xylem of grapevines, causing vascular occlusion, wilting, and plant death. The same organisms occur in alfalfa, causing alfalfa dwarf disease. Leafhoppers are the major vector transmitting the disease agent from one host to another. FVBs are inhibited by tetracycline and to some extent by penicillin and streptomycin, but they are unaffected by sulfonamides. Some FVBs are also phloem limited, like the organisms causing clover club.

MLOs and *Spiroplasma* organisms are restricted to the phloem cells of affected plants, where they cause dwarfing, yellowing, and various growth abnormalities. These organisms lack cell walls and may be pleomorphic during parts of their life cycle. Only a few have been cultured on artifical media. All appear to be transmitted by insect vectors, primarily leafhoppers. In some cases the microbial agents are known to multiply in the insect and to cause pathological changes. MLOs and *Spiroplasma* organisms are susceptible to various tetracyclines but are resistant to penicillins.

Phytopathogenic fungi. There are over 100,000 species of fungi; of these 500 to 600 are estimated to cause disease in plants, and many more are important as wood-rotting organisms or as part of mycorrhizal associations. Many fungi are obligate parasites, and in some cases the association between parasite and host is so intimate and of such long standing that a relationship between the genome of the fungus and that of the host plant has evolved. This association is described by the gene-for-gene hypothesis. In this view it appears that each gene for host resistance has a countering gene for pathogenic virulence in the parasitic fungus. Other fungal associations range from very distinctive rot- or canker-causing organisms, which use a variety of extracellular enzymes and toxins to invade tissues, to facultative species, which ordinarily exist on organic debris but given suitable host and environmental conditions are capable of causing plant disease.

Only the class *Myxomycetes* is a significant plant pathogen among the slime molds. Two important genera of slime molds are *Plasmodiophora* and *Polymyxa*. Many genera of the true fungi (Eumycota) are plant pathogens. The subclass Oomycetes is one of the most important of the lower Eumycota, with the genera *Pythium, Phytophthora, Albugo,* and *Peronospora* being of historic and economic importance. Representatives of the higher Eumycota include members of the classes Ascomycetes, Basidiomycetes, and Deuteromycetes. Genera in the class Ascomycetes, include *Taphrina, Erisiphe* and other powdery mildews, and *Ceratocystis, Endothia,* and *Claviceps*. Genera in the class *Basidiomycetes* include a large group of wood-rotting fungi, such as *Fomes* and *Armillaria,* as well as rusts, such as *Puccinia,* and smuts, such as *Ustilago.* Some of the most common plant pathogen genera occur in the class Deuteromycetes, for example, *Verticillium, Rhizoctonia, Botrytis* (Plate 30), and *Phomopsis.*

Viruses and viroids. Of some 500 transmissible agents that cause disease in plants, over 200 have been identified as viruses. It is likely that most of the other, unknown diseases will prove to have a viral origin as well. A few are suspected to be viroids, whereas others may prove to be plasmids or other infectious entities not yet characterized or known.

Virus-caused diseases rank second only to those induced by fungi in frequency of occurrence and economic importance. Since they are obligate submicroscopic entities, viruses are difficult to detect and characterize. They are frequently vectored by insects, nematodes, mites, or fungi and may persist in an infectious state on tools, debris, or even human hands. Many are passed from plant to plant by vegetative propagation and/or seeds and pollen.

Viruses are resistant to most classes of pesticides and therefore cannot be controlled by usual chemical means. Control measures are centered on prevention and early identification, rather than cure of diseased plants.

Plant virus characteristics such as size, morphology, and nucleic acid composition are discussed in Chapter 6.

Less than a dozen diseases are known to be caused by viroids, although several more diseases are suspect. The biology of the viroid is understood primarily in terms of chemical properties, which are discussed in Chapter 6. Most biochemical processes involved in infection, multiplication, and movement within plants are a matter of conjecture. Transmission is via plant propagation or mechanically on tools, but vectors may also be involved.

Economics and control

Losses from diseases of plants are a major production problem. Losses may be direct, as in the case of actual destruction of plants or parts of plants, or they may be indirect, as in the case of costs related to control practices. Worldwide crop losses due to fungi, viruses, bacteria, nematodes, and other disease problems are estimated at one third of the potential production. Losses occur from seed to storage. In the United States, disease losses (including cost of control but excluding postharvest losses) were set at 12% of the total farm value of 65.2 billion dollars in 1978. This loss is proportionally the same in most years. If it were possible to estimate losses such as land abandoned to agriculture because of diseases or changes in produce value due to minor disfiguration from pathogens, the cost would be even higher.

Control of diseases is motivated by economic considerations and is designed to fit pathogen biology, application technology, social factors, and political restraints. These practices are costly and complicated, but we eat today because of them.

Plants may be generally considered resistant to disease because of chemical and physical factors. They have physical barriers such as waxy layers on the surface of leaves or stems, cell walls, and closable stomas. There are preexisting chemical barriers to disease agents, as well as a variety of protective reactions of chemical or mechanical types to invasion. However, once the plant is invaded to any extent by pathogens so that disease occurs, it is difficult to effect a cure. Control measures are therefore designed to prevent disease by any means possible.

Plants may be protected simply by keeping known pathogens out of their environment by quarantines or just common sense. Plant pests of known destructive potential are regulated by statute. Farmers do not knowingly plant susceptible plants in areas known to be infested by certain pathogens. Crop rotation is practiced to keep pathogens at low levels, hence off balance. Chemicals are dusted or sprayed on surfaces, fumigated into soil or storage bins, or soaked into seeds as barriers to potential pathogenic contact. Where possible, environmental factors are manipulated, such as changes in planting dates, choice of irrigation methods, and field designs emphasizing or manipulating air or water drainage. A major tool in disease control is the incorporation of existing genes for resistance into crop plants. Modern plant breeding emphasizes the search for resistant genes in crop ancestors, as well as exploitation of genetic engineering techniques as a future method of incorporation or manufacture of disease-resistant genetic control.

HOST-PARASITE INTERACTIONS

Host-parasite interactions involving plants are similar to those found in vertebrates. All potential pathogens must obtain nutrients from their hosts in order to survive and replicate. Nonobligate plant parasites, such as most bacteria and fungi, accomplish this by secreting enzymes that degrade cells and release cytoplasmic nutrients. Obligate parasites, such as viruses and mycoplasmas, require a living host and therefore only penetrate the host cell. They divert metabolites, intended for the plant, for their own growth and replication, usually without destroying the cell. The basic problem for all plant pathogens is to find a way to penetrate the plant so that the underlying cells and the nutrients they hold can be used by the parasite.

Transmission of plant pathogens

The plant is part of an environment in which microorganisms, as well as vertebrate and invertebrate animals, are often in contact. Considering the number of environmental factors and surrounding animate species that can adversely affect them, plants are unusually resistant to disease. We have already discussed how human pathogens show specificities in regard to age or organs affected. These characteristics are also attributable to many plant pathogens—some infect only young plants, whereas others infect mature plants; some infect only roots, whereas others attack leaves; and there are some that colonize the plant but do not cause disease. The microscopic microbial agents that cause disease in plants include viruses, viroids, bacteria, fungi, and some protozoa. In addition, some macroscopic agents, such as nematodes and plants, act as parasites. The number of microbial agents or inoculum required to cause infection may vary from a single fungal spore to several million viruses, depending on the site of infection, the virulence of the pathogen, and environmental factors such as temperature and humidity. The microbial agents capable of infecting the plant can be found in the soil, in water, in infected plants, and in various vectors such as insects. Fungal spores, for example, are found in the soil and may be expelled from a sporocarp and disseminated into the air for many miles. Bacteria and nematodes may be disseminated by air, but this is not how they are usually transmitted. Water is an important method of dissemination of microorganisms, particularly of bacteria and fungi. Water droplets may splash on the ground and disseminate spores, or the pathogen may be carried long distances in ditches or irrigation canals. Many of the obligate parasites, such as viruses, viroids, and mycoplasmas, are disseminated in vectors such as aphids or leafhoppers. Obligate parasites have special requirements and will usually cause disease only in very specific hosts, whereas nonobligate pathogens are not as selective as to the type of host they infect.

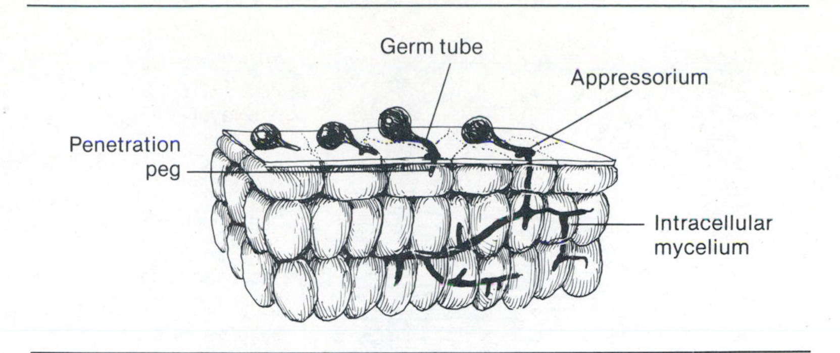

Figure 35-3 Formation of appressorium as means for direct penetration by fungus. Penetration peg forms at tip of appressorium and forces its way through epidermal layer.

Penetration of the plant

Once microorganisms have made contact with the plant surface, a means must be found to penetrate the surface and seek nutrients for growth. The penetration process may be accomplished in one of three ways: direct penetration, penetration through abrasions, or penetration through natural openings on the plant.

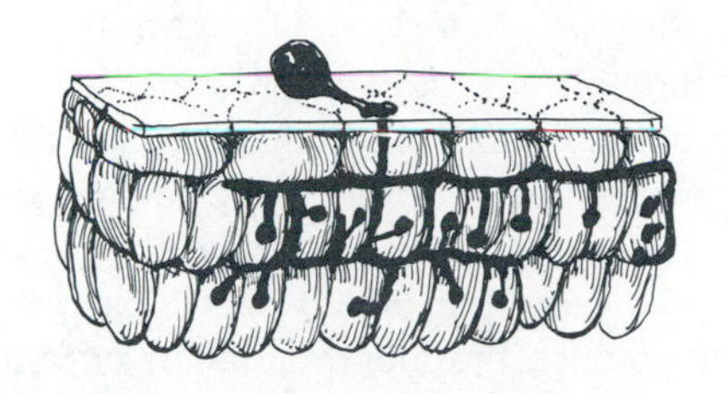

Figure 35-4 Direct penetration by fungi using haustoria as penetrating and absorbing organ. Mycelia are intercellular.

Direct penetration. Only fungi, among microscopic forms of life, are capable of direct penetration of the plant surface. The fungal spore germinates and forms a germ tube on the plant surface, provided that temperature and humidity are appropriate. The developing hypha may penetrate the surface in one of several ways, but most produce an ***appressorium.*** The appressorium (Figure 35-3) is a flat bulblike structure at the end of the hypha that secures the fungus to the plant surface. A very thin and tapered filament, called the penetration peg, develops from the hypha. The penetration peg, with the assistance of secreted enzymes, forces through the cuticle and cell wall of the underlying epidermal cell layer. Once the penetration peg has entered the cell wall and passes into the cytoplasm, it expands to a diameter that is comparable to the size of the normal hypha. In order to enter other cells from its position inside the cell, the hypha again assumes a tapered point and pierces through the opposite cell wall until another cell is reached. Some fungi produce haustoria that may penetrate underlying cells ***(intracellular mycelium)*** or penetrate between the epidermal cells ***(intercellular mycelium).*** The haustoria act as a nutrient-absorbing organ (Figure 35-4) (see also Figure 5-6).

Penetration through abrasions. Abrasions on the plant surface are caused by environmental factors, such as wind, hail, and sun; by human manipulations, such as grafting and pruning; or by animal or insect feeding. Most potential pathogens can enter the plant through these abrasions. Some obligate parasites require a vector for transmission: either natural vectors, such as insects, or unnatural vectors, such as contaminated fomites or contaminated human hands. Other obligate parsites, such as rusts, are wind borne, and no vector is involved. The abraded surface is usually filled with plant exudate, which can be used by the pathogen for growth and the production of enzymes or toxins. These products of metabolism affect the underlying tissue and help the pathogen invade healthy cells. Some plant exudates, for example, have been shown to induce microbial enzymes, which dissolve pectin, a component of plant cell walls.

Penetration through natural openings. There are four types of openings on the plant surface that may provide an avenue for penetration by plant pathogens, but the most important are the ***stomas.*** Stomas are small openings on the surface of leaves and stems through which gases pass. The greatest number of stomas are present on the underlying surface of leaves. Bacteria can

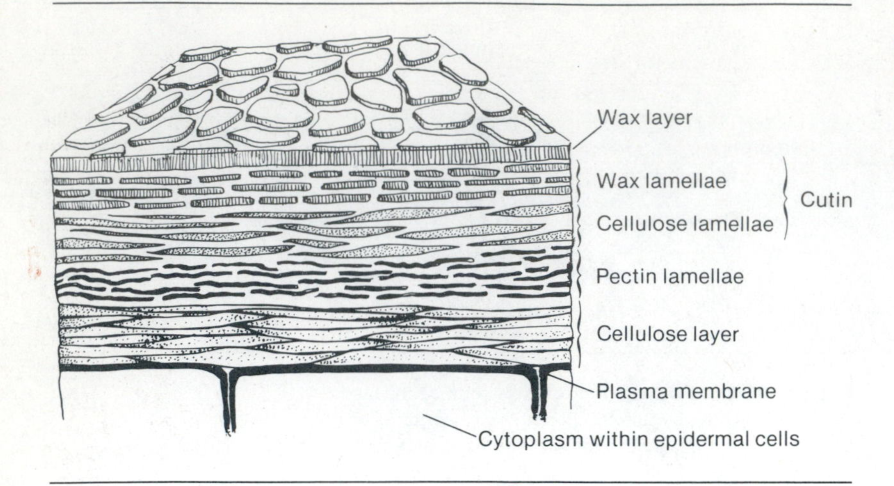

Figure 35-5 Cross section of leaf surface.

swim through the stomas on water films, but fungi produce germ tubes that migrate through them. Other openings on the plant surface include ***hydathodes,*** which are openings at the base of flowers from which nectar exudes; and ***lenticels,*** which are found on stems or fruit and allow the passage of air. These last three surface openings are very minor channels for entry into the plant.

Sites of infection and disease symptoms

Most pathogens penetrate the epidermal cells and spread into the major plant tissue, such as roots, stems, and leaves. Some fungal agents, such as those causing powdery mildew, remain on the surface of the plant, but even they must send haustoria into or between the epidermal cells to obtain nutrients. Obligate parasites require an intracellular environment, whereas nonobligate parasites can invade tissue by moving between cells and surviving extracellularly. Intercellular fungi without absorptive organs (haustoria) must intercept solutes diffusing into the cells. The ectomycorrhizal fungi develop extensive thalli without the use of haustoria to obtain nutrients. Many diseases caused by microbial agents remain localized and involve only a few cells or small patches of tissue on the plant. Certain pathogens, however, may move passively throughout the plant by passing through the minute plasmodesmata, which connect adjacent protoplasts (only viruses and viroids are small enough to do this), or through the transport vessels such as xylem and phloem. Active movement involving the use of flagella or related structures is possible for some fungi, bacteria,

Table 35-1 Some of the major symptoms of disease in infected plants

DISEASE SYMPTOM	DEFINITION
Blight	A disease in which there is rapid killing of the plant
Blotch	A disease characterized by irregular spots on fruit, leaves, stems, or shoots
Canker	A necrotic sunken lesion on a stem or branch
Chlorosis	yellowing of normally green chlorophyll-containing tissue
Damping off	Destruction of seedlings near the soil line
Gall	A swelling produced as a result of infection
Mildew	A fungal disease in which the fungus appears as a whitish growth on the plant surface
Ring spot	A circular area of chlorosis with a green center
Rot	Softening and discoloration of plant roots, stems, or storage organs as a result of bacterial or fungal infection
Rust	A disease caused by several fungi that gives a rusty color to the plant
Smut	A disease caused by certain fungi and characterized by loose, dark, powdery spores
Wilt	Dropping of plants due to insufficient water or toxic metabolites

protozoa, and nematodes. Once the disease process has been initiated, certain symptoms become visible and are often characteristic of specific types of infection. Some of the major symptoms of plant diseases are outlined in Table 35-1.

Microbial factors and pathogenicity

The plant pathogen is often endowed with a number of chemical or physical characteristics that enable it to invade the host and cause disease. Tough, protective layers are characteristic of the surface as well as the underlying epidermal cells of plants. Epidermal cells may be covered with a cellulose layer, such as on the roots, and are usually covered by a cuticle, which often possesses an outer waxy layer (Figure 35-5). In the majority of cases these tough outer layers can be

Plate 49 Hemorrhagic septicemia caused by *Pasteurella multocida* in water buffalo; submandibular edema is characteristic.
Courtesy Gordon Carter, Virginia Polytechnic Institute and State University.

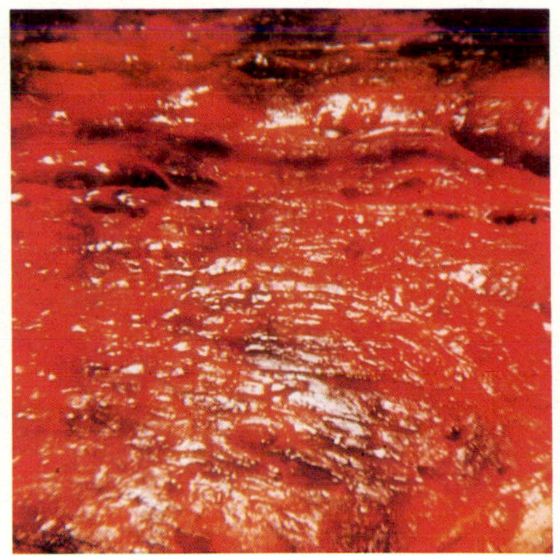

Plate 50 Muscular lesions in blackleg of cattle caused by *Clostridium chauveoi* are gangrenous, gaseous, dark, and dry.
Courtesy Gordon Carter, Virginia Polytechnic Institute and State University.

Plate 51 Infectious bovine rhinotracheitis (IBR).
Courtesy Robert Kahrs.

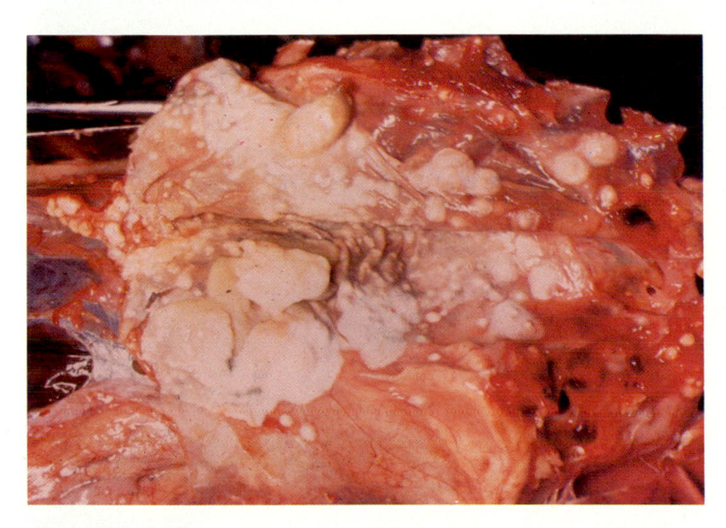

Plate 52 Lesions of aspergillosis involving lungs and thorax of penguin.
Courtesy Gordon Carter, Virginia Polytechnic Institute and State University.

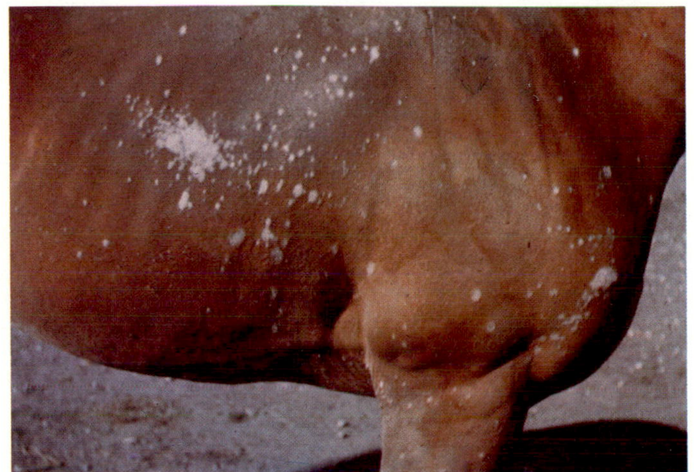

Plate 53 Ringworm affecting horse.
Courtesy Gordon Carter, Virginia Polytechnic Institute and State University.

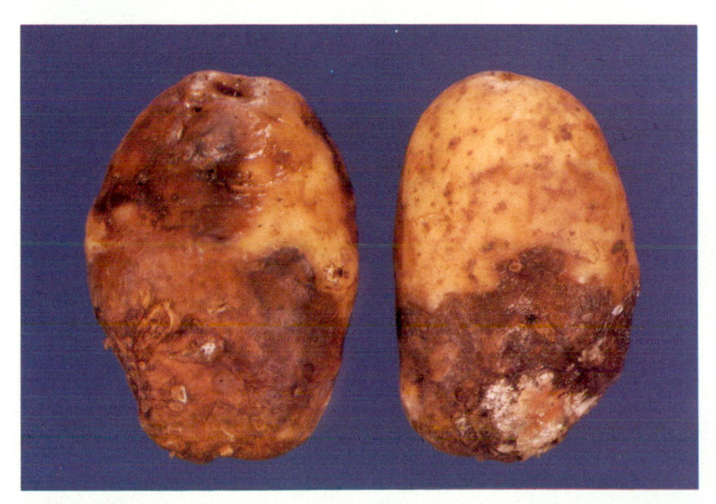

Plate 54 Potato blight caused by *Phytophthora infestans* showing storage rot.
Courtesy Randall C. Rowe.

Plate 55 Wheat rust. Wheat rust in field. Plants on right are resistant, whereas those on left (brown) are diseased.
Courtesy Cereal Rust Laboratory, USDA-ARS, University of Minnesota.

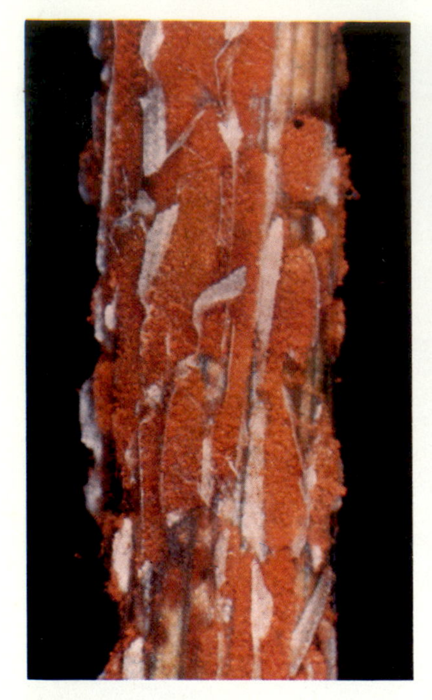

Plate 56 Wheat rust. Close-up of stalk.

Courtesy Cereal Rust Laboratory, USDA-ARS, University of Minnesota.

Plate 57 Tobacco plant wilt. Unilateral effect is characteristic of tobacco wilt caused by *Fusarium* species.

Courtesy Wirt H. Wills, Virginia Polytechnic Institute and State University.

Plate 58 Citrus root rot caused by species of *Phytophthora*. There is obvious deterioration of foliage. There is also increased fruit yield, characteristic of disease.

Courtesy Donald Erwin, University of California, Riverside.

Plate 59 Fire blight of pyracantha. Plant on right is uninfected. Plant on left has characteristic droop associated with infection.

Courtesy George H. Lacy and Robert C. Lamb, Virginia Polytechnic Institute and State University.

Plate 60 Crown gall caused by *Agrobacterium tumefaciens*. Galls appear at soil line of plant.

Courtesy George H. Lacy, Virginia Polytechnic Institute and State University.

Plate 61 Mottled appearance of field tobacco infected by tobacco mosaic virus.

Courtesy Sue A. Tolin, Virginia Polytechnic Institute and State University.

Table 35-2 Structural barriers to plant pathogens and the enzymes that degrade them

CHEMICAL COMPONENT	LOCATION IN CELL	DEGRADATIVE ENZYME AND PRODUCT OF DEGRADATION
Waxes (lipids)	Granular projections on cuticle or continuous covering of cuticle	No known degradative enzymes have been found in pathogens
Pectin (polysaccharide)	Cutin, epidermal cells, middle lamella, primary cell wall	Pectinase, galacturonic acid
Cellulose (polysaccharide)	Primary and secondary cell wall	Cellulase, glucose
Lignin (phenylpropanoid)	Primarily xylem vessels	Ligninase, phenylpropanoids; lignin is the most resistant plant polymer, but it is degraded by wood-rotting fungi

penetrated by microorganisms only if they are first softened up by enzymes. Plant pathogens, unlike human pathogens, are equipped with a greater arsenal of enzymes for penetrating the host. But even after outer layers are penetrated, the cell wall of underlying cells must also be pentrated and the cytoplasmic contents diverted for parasite metabolism. Plant parasites can penetrate the plant surface by using mechanical forces, but these mechanisms are usually coordinated with enzymatic processes. One group of organisms (species of *Plasmodiophora*), however, is capable of literally blasting a hole in the plant cell wall without the benefit of enzymes (Figure 35-6). Mechanical penetration without apparent enzymatic degradation is more characteristic of nematodes and higher plants that act as parasites.

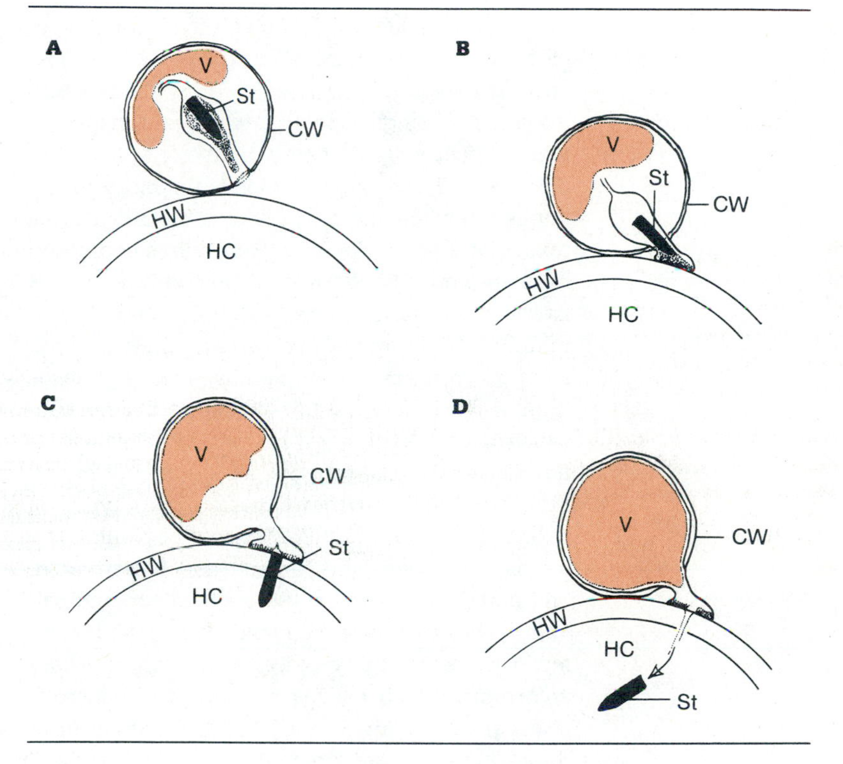

Figure 35-6 Mechanism of direct penetration of plant cell wall by species of *Plasmodiophora*. **A,** Organism encysts on host cell wall. Vacuole begins to develop in cyst. Most unusual feature of cyst is tubular cavity, within which develops sharp-pointed rod called stachel. **B,** Vacuole enlarges, and small structure called adhesorium develops and adheres to host cell wall. **C,** Stachel punctures host cell wall, and in 1 second, **D,** parasite enters host cytoplasm. *HW,* Host wall; *CW,* cyst wall; *HC,* host cytoplasm; *V,* vacuole; *St,* stachel.

Redrawn from Aist, J., and Williams, P.H. Can. J. Bot. ***49:****2023, 1971.*

Mechanical penetration. Fungi are the only microscopic forms that can penetrate the plant surface by mechanical means. They apparently bind to plant receptors and produce a hypha with a flat bulblike structure called the appressorium. This technique for penetration is discussed earlier.

Chemical factors. A variety of chemical compounds produced by the invading pathogen, or produced by the plant in response to infection, cause disease in plants. In many instances no individual chemical can be singled out as the initiating factor in disease, and combinations of factors are usually involved. The type of chemical involved in the disease process may be classified as an enzyme, toxin, growth regulator, or polysaccharide (some toxins may also be a polysaccharide).

Enzymes. The plant surface and underlying epidermal cell wall is made up of large molecular weight polymers that require some type of digestion before the pathogen can penetrate and cause disease. The plant substances that are degraded by enzymes produced by invading pathogens are outlined in Table 35-2.

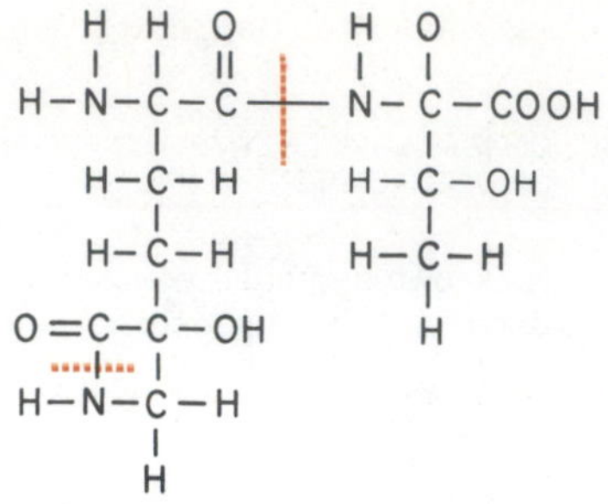

```
  H  H  O     H  O
  |  |  ||    |  |
H-N--C--C-----N--C--COOH
     |           |
   H-C-H       H-C-OH
     |           |
   H-C-H       H-C-H
     |           |
 O=C-C-OH        H
   |  |
 H-N--C-H
      |
      H
```

Figure 35-7 Structure of tabtoxin.

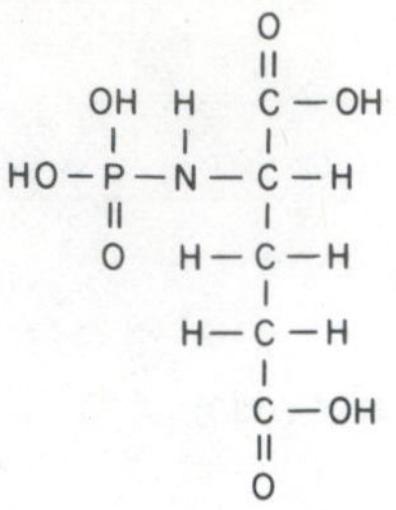

```
                 O
                 ||
  OH  H          C-OH
  |   |          |
HO-P--N----------C-H
  ||             |
  O            H-C-H
                 |
               H-C-H
                 |
                 C-OH
                 ||
                 O
```

Figure 35-8 Structure of phaesolotoxin A produced by *Pseudomonas phaseolicola.*

Microbial toxins. Compounds produced by parasites that are toxic to plants and produce disease symptoms are called ***phytotoxins.*** Phytotoxins are produced by bacterial and fungal species and may be classified as those that are host specific and those that are not. Host-specific toxins affect a single species, whereas nonspecific toxins affect a wide variety of different plant species, as well as animals. Three of the most widely studied nonspecific toxins are tabtoxin, fusarial toxins, and phaseolotoxin. ***Tabtoxin*** is composed of an amino acid derivative tabtoxinine. The β latam ring of tabtoxinine, which apparently is an inhibitor of glutamine synthetase, is released following hydrolysis of the toxin by plant peptidases (Figure 35-7). Tabtoxin is produced by the bacterium *Pseudomonas tabaci,* which causes wildfire disease of tobacco. Several species of *Fusarium,* a fungal pathogen, produce two types of toxins that cause wilt in a variety of plants. The toxins are called lycomarasim and fusaric acid, and both bind heavy metals such as iron and calcium. The toxin-metal complex is believed to affect the permeability of cell membranes. ***Phaesolotoxin,*** which causes halo blight of beans, is a tripeptide composed of phosphosulfamylornithine, alanine, and homoarginine (Figure 35-8). Apparently, phosphosulfamylornithine is the active component of the toxin.

Approximately 20 host-specific toxins have been identified, but little is known of their specific mechanisms of action. It is believed that they act directly on the cytoplasmic membrane or mitochondria of sensitive cells by binding to specific receptor sites. ***Victorin*** is a toxin produced by *Helminthosporium victoriae,* a fungal pathogen that affects some plants and causes necrosis of the roots and stem. The toxin is believed to affect the cytoplasmic membrane of cells. Treatment of cells with toxin has demonstrated that there is solute leakage, inhibition of uptake of molecules, and depolarization of electrical potential across the membrane. The site of toxin binding, however, has not been determined, and there is evidence suggesting that the cytoplasm may be affected before the effects of the cytoplasmic membrane are detected. In vivo and in vitro experiments with a toxin from another species of *Helminthosporium* indicate that mitochondria-binding sites may be involved. Toxin appears to make the inner membrane of mitochondria highly permeable to small molecules and ions, resulting in proton leakage and reduction in ATP levels in the cell. The bacterium *Erwinia amylovora* produces a polysaccharide toxin that affects the cytoplasmic membrane of the xylem parenchyma tissue and seems to be responsible for some of the symptoms of the disease called fire blight.

Growth regulators. The regulation of growth in plants is controlled by a variety of small molecular weight compounds that stimulate as well as inhibit growth processes. These growth regulators include ***indole acetic acid (IAA), gibberellins,*** and ***cytokinins*** (see Figure 19-19) Invading pathogens may cause an imbalance in these regulators by interfering with their production movement, or use by the plant, or they may produce the same or similar growth regulators. One of the more widely studied examples of growth regulation is crown gall tumor, which is introduced in Chapter 9. Briefly, crown gall is a hyperplasia of dicotyledonous plants induced by the bacterium *Agrobacterium tumefaciens.* The induced proliferation of tissue is due to a plasmid, ***pTi*** (tumor inducing), carried by the invading bacterium, that is integrated into plant nuclear DNA. The pTi plasmids are classified on the basis of their ability to determine what class of compounds, called ***opines,*** are synthesized in tumor tissue. Opines are amino acid derivatives (Fig. 35-9) synthesized in tumor tissue, but they are not produced by uninfected plant tissue or by the infecting bacterium. Many times, but not always, the opine that is synthesized is degradable by the infecting bacterium. It has not been determined whether the pTi plasmid codes for the structural genes for opine synthesis, or whether it codes for a regulatory gene that turns on a plant gene that is normally turned off in uninfected cells. The wounding of the plant appears to be necessary to tumor induction. The bacterium binds to the plant apparently via lipopolysaccharide interactions of their respective cell walls. The infecting bacterium remains outside the plant cell, but the pTi plasmid is passed through the cell wall into the cytoplasm. A portion of pTi DNA is believed to be integrated into plant nuclear DNA, but there does not appear to be a unique site for integration. Three types of proteins have been suggested to be coded for by tumor DNA: an opine synthase plus enzymes involved in auxin and cytokinin synthesis. The latter hormones are at abnormally high levels in tumor cells.

pTi plasmid DNA remains integrated into the plant genome for many years after subculturing. For this reason, there has been considerable interest in using *A. tumefaciens* for the purpose of genetic engineering in plants. Foreign genes could be introduced into the pTi plasmid DNA in place of the tumor genes and thus could easily be introduced into the plant genome. The introduction of genes conferring such properties as resistance to disease or rapid nitrogen fixation are exciting possibilities.

Polysaccharides. A limited number of fungal pathogens that cause vascular wilt produce extracellular polysaccharides during infection. The secreted polysaccharides cause the obstruction of vascular tissue such as xylem and phloem. Such obstructions prevent the transfer of water and solutes to various plant tissues.

Octopine

Histopine

Argopine

Figure 35-9 Structure of some opines synthesized in tumor tissue following infection by *Agrobacterium tumefaciens*.

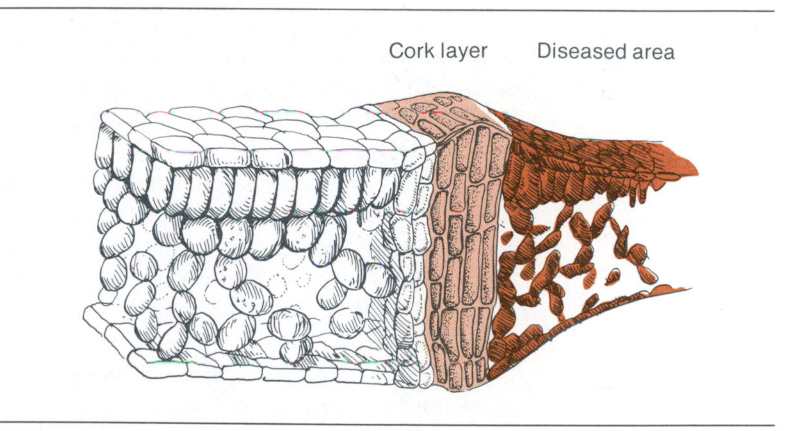

Figure 35-10 Formation of cork layer between infected and healthy tissue.

Plant defense mechanisms

Structural mechanisms

Natural barriers. The thickness and chemical composition of the plant surface as well as of the cell walls of underlying epidermal cells provide a hindrance to pathogen invasion. The waxy plant surface also acts as a water repellent and prevents the accumulation of water films that could provide the pathogen with a site for multiplication and potential spread. These structural barriers, however, are not the plant's best mechanism of defense, since many pathogens enter via abrasions on the plant surface or are injected by insect vectors. The most successful defense mechanisms are those that are a direct response to the invading pathogen.

Induced barriers. Although plants lack the immunological responses associated with vertebrates, they are still able to respond in a sensitive manner to the invading pathogen and/or its products. Some plant cells are stimulated to produce layers of ***cork*** (Figure 35-10) at the infected site and beyond it. The cork prevents spread of the infectious agent or its products beyond the lesion, and also reduces the flow of water and nutrients from healthy tissue into the infected site. The infected tissue is therefore isolated from healthy tissue and may eventually be sloughed off by the plant.

Some plants, such as peaches and plums, respond to certain infectious agents by producing ***abscission*** layers (Figure 35-11). Abscission layers are gaps produced between two circular layers of cells at the site of infection. A break between the two cell layers occurs as a result of

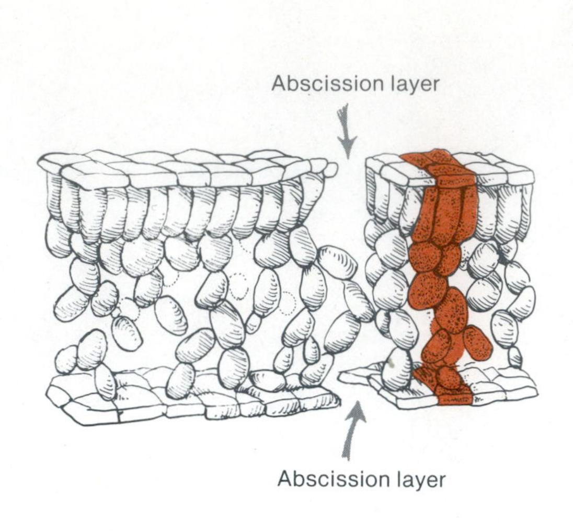

Figure 35-11 Formation of abscission layer around diseased area of plant. Diseased area is in color.

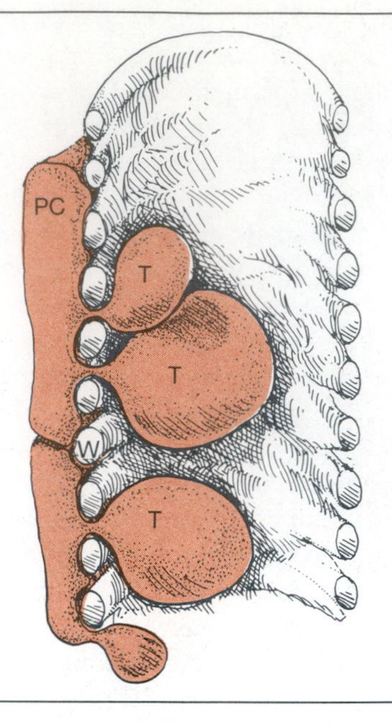

Figure 35-12 Tylose formation in lumen of vessel. *T*, Tylose; *W*, wall; *PC*, parenchyma cell.

cellular breakdown as the infection begins to progress. The continued dissolution of cells cuts off the diseased area from healthy tissue, and the diseased section is later sloughed off.

A very common response by most plants to invasion by pathogens is the formation of ***tyloses*** (Figure 35-12). Tyloses are protoplasmic outgrowths of parenchyma (thin-walled cells) that lie adjacent to the thicker-walled xylem vessels and protrude into them. The blockage of xylem vessels ahead of the site of pathogen invasion acts as a barrier to further spread of disease.

Many plants, in response to injured cells or pathogenic invasion, produce large amounts of polysaccharides called ***gums.*** The gums fill the intercellular spaces around the locus of infection and prevent further spread. The gums are particularly impenetrable by water and nutrients, and consequently the pathogen is starved to death.

Many parasites, such as obligate fungi and viruses, are subject to a hypersensitive response when the host cell is destroyed. Once the cytoplasmic membrane of the plant cell has been breached by a pathogen, the nucleus of the plant cell moves toward the site of invasion and disintegrates. This necrotic process results in the formation of toxic by-products, such as phenols, which, with certain physical barriers (gums, closed plasmodesmata), limit the invasiveness of fungi and viruses.

Biochemical mechanisms. A variety of useful metabolic compounds are produced by plants. Some of these have been isolated and used as antimicrobials in the treatment of animals and humans, as growth stimulators, in a variety of other biological activities. Plants also produce a number of primary and secondary metabolites that appear to be capable of preventing tissue invasion by pathogens, as well as pathogen spread. In some cases a potential pathogen is prevented from causing disease because the plant lacks specific nutrients (growth factors) that the pathogen requires.

Natural biochemical defense. Plants produce a number of biochemicals that have as a primary purpose plant growth and development but secondarily prevent plant invasion by microorganisms. ***Stilbenes,*** which have a resorcinol ring (Figure 35-13), are present in the dead tissue of heartwood and bark. They are secondary metabolites produced in situ in dying xylem parenchyma cells. Stilbenes exhibit considerable antifungal activity but are also toxic to many species of bacteria. They apparently exert their activity through the inactivation of fungal enzymes containing —SH groups at their active sites. Enzymes such as cellulase, pectinase, and xylanase are inhibited by stilbenes. Stilbenes are therefore important in the resistance of heartwood to fungal decay.

Chlorogenic acid is a phenolic compound that is produced by many plants. Those plants resistant to specific disease agents produce considerably more chlorogenic acid than their nonresistant counterparts. Chlorogenic acid, as well as

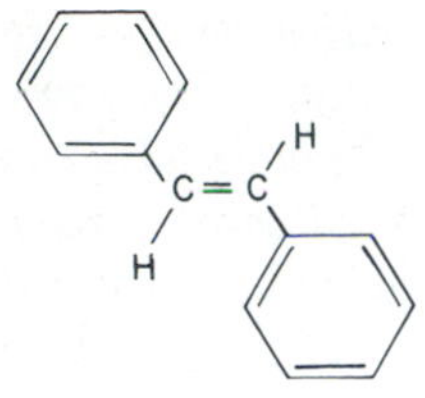

Figure 35-13 Structure of stilbenes.

other phenolic compounds, is derived from intermediates produced in the pentose pathway. Infections by pathogens stimulate a shift from glycolysis to the pentose pathway.

Induced biochemical defense. Any insult to plant tissue, whether by microbial attack or mechanical or chemical injury, will induce plant cells to respond biochemically. Those products that appear in response to the inducing agent but are not present in healthy tissue are called ***phytoalexins.*** Some of the principal phytoalexins are ipomeamarone, orchinol, pisatin, phaesolin, and rishitin. It has been observed in some diseases that the degree of resistance by the plant is directly related to the concentration of phytoalexins produced.

Some phenolic complexes or phenol compounds become more toxic when specific enzymes are synthesized by the plant in response to infection. These enzymes effect the release of more toxic derivatives. Occasionally the phenolic compounds affect the parasitic enzymes used to invade the cell, for example, pectinases.

Plant pathogens rely on specific plant substrates for their nutrition. Some plants can alter these substrates in such a way that they resist catabolism by the pathogen. Pectins, for example, can be complexed with certain metal ions to produce pectic salts that resist degradation by the invading pathogen.

SELECTED PLANT DISEASES

Fungal diseases

Potato blight. Potato blight, which is caused by the fungus *Phytophthora infestans,* was responsible for the great potato famine that struck Ireland and England in the 1840s. Potato famers in Ireland during that period were extremely poor because of the high rent paid to absentee landlords who lived in England. Their principal source of food was the potato, and each person consumed from 10 to 14 pounds per day. Long periods of cool and wet weather in the mid 1840s led to potato tuber rot, which prevented farmers from earning enough money to pay the rent. As a consequence of this, evictions were common. These conditions led to the starvation of over 1 million Irishmen between 1845 and 1860 and also led to the emigration of over 1.5 million to the United States.

Potato blight also had an impact on Germany during World War I. A comlex mixture called Bordeaux mixture, which was a combination of copper sulfate and lime, had been developed in France in 1882 in order to protect the potato from the agent of blight. Potato blight struck Germany in 1916, but the military would not release the copper needed for the Bordeaux mixture, because it was needed for shell casings and electric wire. Any grain and potatoes that were produced went to the army while the farmers went hungry, and in 1917 many farmers starved. The morale of the soldiers of these families declined, and it has been suggested that this may have been one of the reasons for the weakening of the German military establishment during World War I.

The environment plays an important role in potato blight. The pathogen remains viable over the winter as a mycelium in infected potato tubers. Potato farmers usually cull out or throw away tubers that are discolored or unsuitable for marketing. These potatoes are often infected with *P. infestans.* If culled potatoes are left undisturbed until spring, the tubers will sprout and the fungus can sporulate and infect nearby fields. Sporangia are produced when the air temperature is 18° C or lower and when a film of water exists on the plant surface. Sporangia germinate and produce two to eight infective zoospores and, in theory, eight new infections are therefore possible. When the temperature is above 18° C, the sporangium acts as a conidium and germinates by means of a single germ tube. Appresoria form at the tip of the germ tubes, and penetration pegs enter the cuticle of the leaf. A mycelium develops intercellularly, and within 2 to 3 days the infected plant cells die. The initial leaf symptoms are characterized by the appearance of brownish to purplish black lesions 5 to 7 days following infection (Plate 54). Sporangia and sporangiophores protrude from the stomas and can initiate a second cycle of infection. Spores are disseminated by the wind and can infect the same plant or new plants. The tubers, which are formed at the soil surface or just below the surface, become infected later during the season by spores that have dropped from infected leaves.

Protective spraying and dusting of plants is the

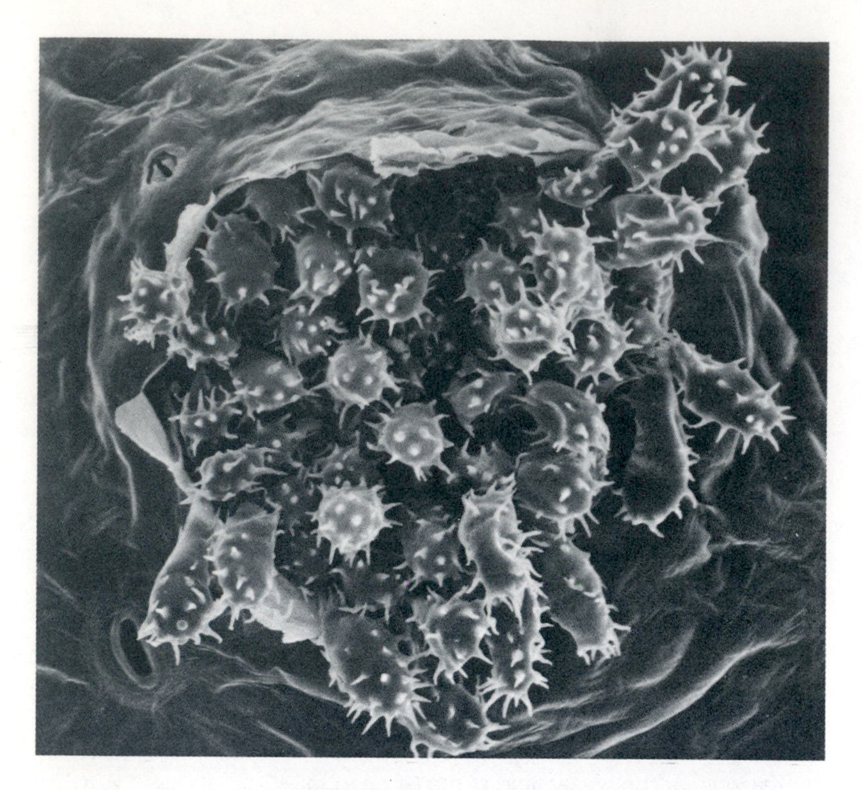

Figure 35-14 Scanning electron micrograph of teliospores of *Puccinia podophylia,* agent of Mayapple rust. (×750.0)
Courtesy Gary Hooper.

most reliable method of controlling disease. Potato blight is so affected by specific weather conditions that the time for spraying can be forecasted fairly accurately. Potato farmers should also destroy cull piles to prevent overwintering of the fungus.

Cereal stem and leaf rust. Plant diseases known as rusts have been recognized since ancient times. The fungal species responsible for such diseases belong to the subdivision Basidiomycotina, whose characteristics are briefly described in Chapter 5. Stem rust of wheat is caused by *Puccinia graminias,* which requires two hosts to carry out the disease cycle: barberry and wheat (Plates 55 and 56). During the disease cycle four types of spores are produced (see Figure 5-28). The fungus overwinters in a spore state on infected wheat debris. These spores, called ***teliospores*** (Figure 35-14), germinate and, following development, give rise to ***basidiospores*** that require the common barberry leaf as a host for germination and development. Development in the barberry leaf results in the formation of spores that are infective for wheat. These spores, called ***aeciospores,*** produce germ tubes that penetrate the wheat stem or leaf via stomas. A mycelial mat is produced that spreads intercellularly beneath the epidermis. The mycelium produces spores ***(urediospores)*** and exerts enough pressure to break through the epidermis, revealing a powdery film of rust-colored spores. The urediospores are easily disseminated by air currents and can infect healthy wheat plants or reinfect already-contaminated wheat plants. When the wheat plant reaches maturity, or when it dies, the mycelium produces thick-walled teliospores instead of uredospores, and this permits the fungus to overwinter.

Water and nutrients are lost to the plant during infection because of their use by the developing mycelium. Water and nutrients are also lost because of evaporation at the site of epidermal rupture. Photosynthesis is severely reduced because of tissue damage, and thus carbohydrate supply to the plant is drastically reduced. These manifestations of disease affect the number and size of seeds, and entire crops may be lost or severely damaged if the plants are infected before or at the time of flowering.

Many wheat-growing areas have been cleared of surrounding barberry to eliminate one of the hosts necessary for development of the fungus, but the most effective method of control has been the development of rust-resistant varieties of wheat.

Dutch elm disease. Up until the 1950s many streets of the United States were lined with the graceful and shady American elm tree. Today, except for some areas of the Northeast, the American elm has virtually disappeared. The trees were killed by a fungus, an ascomycete called *Ceratocystis ulmi,* which was transmitted by bark beetles. The disease is presumed to have originated in Asia, where the Chinese elm is resistant to the fungus. During World War I Chinese laborers were brought to Flanders to dig ditches, and they are believed to have brought with them wooden wicker baskets made from the Chinese elm. Apparently, the vector beetles, which were carried in the bark, escaped and proceeded to ravage millions of elms in Europe. The disease was eventually called Dutch elm disease because of its initial appearance in the Netherlands. Furniture manufacturers in the United States, despite importation quarantines, obtained diseased logs from Europe and helped unleash the disease that practically destroyed most American elms (Figure 35-15).

The female bark beetles, which carry spores of the fungus on or in their bodies, deposit eggs in the bark of the stems of weakened or recently killed elm branches. The beetle larvae eat the outside layer of wood and the inner layer of bark, which results in the formation of tunnels. The fungal spores carried by the adult beetle germinate and produce compact or fused conidiophores with clustered sticky spores. When the new adult beetles emerge from the tunnels to fly away, the sticky spores become attached to their bodies. The spore-laden beetles fly away to feed on live trees and in the process of boring through the bark infect the sapwood. Dutch elm disease is a vascular wilt, or upset in water balance, in which infected branches lose turgor, become limp, and eventually die. Spores can be produced that will be disseminated to other areas of the plant to initiate new sites of infection if the large xylem vessels are reached by the invading mycelium. Rapid death of infected trees is believed to be due to the formation of tyloses, gums, and fungal growth, obstructing xylem vessels and upsetting the water balance. The vascular tissue appears brown, and the outer layer of wood appears streaked, mottled, or brown to green. The tree may die within a few weeks if there has been a generalized vascular infection.

Dutch elm disease is best prevented by the removal of recently killed elm wood, since beetles lay their eggs in killed or diseased trees. Recent plantings can be protected by applying insecticide such as methoxychlor early in the spring to destroy newly emerging adult beetles. Control in individual trees has also been obtained by injecting systemic fungicides into the trunks or roots. This practice arrests the disease and prevents further spread.

Vascular wilts. At least three genera of fungi cause vascular wilts. One of them, *Ceratocystis ulmi,* the etiological agent of Dutch elm disease, is discussed above. The other two genera, *Fusarium* and *Verticillium,* affect a wide variety of plants, including annual vegetables, flowers, perennial ornamentals, fruit trees *(Verticillium),* and field crops such as cotton.

Fusarium and *Verticillium,* which are found as saprophytes in the soil, can infect the roots of susceptible plants. They enter through lateral root openings by direct penetration or by entrance through mechanically injured areas. The fungus migrates to the xylem and breaks down the walls of xylem vessels, tracheids, and parenchyma cells by the production of pectolytic enzymes. Affected parenchyma cells turn brown as a result of oxidation of polyphenols, which gives rise to dark melanin pigments that stain tracheids and xylem vessels. The symptoms of infection are similar for all vascular wilts regardless of the pathogen causing them. Initially the leaves lose their turgidity, become limp, turn light green to yellow, and then finally turn brown and die (Plate 57). The effects of infection are due to the occlusion of xylem vessels, which may be due to (1) the accumulation of gums and gels produced by the oxidation of plant degradation products via fungal enzymes; (2) the clogging of vessels by mycelia, spores, or fungal polysaccharides; and (3) the stimulation of parenchyma cells surrounding xylem vessels to pro-

A

B

Figure 35-15 Dutch elm disease. **A,** Uninfected tree. **B,** Tree infected by *Ceratocystis ulmi,* the agent of Dutch elm disease.

Courtesy R. Jay Stipes, Virginia Polytechnic Institute and State University.

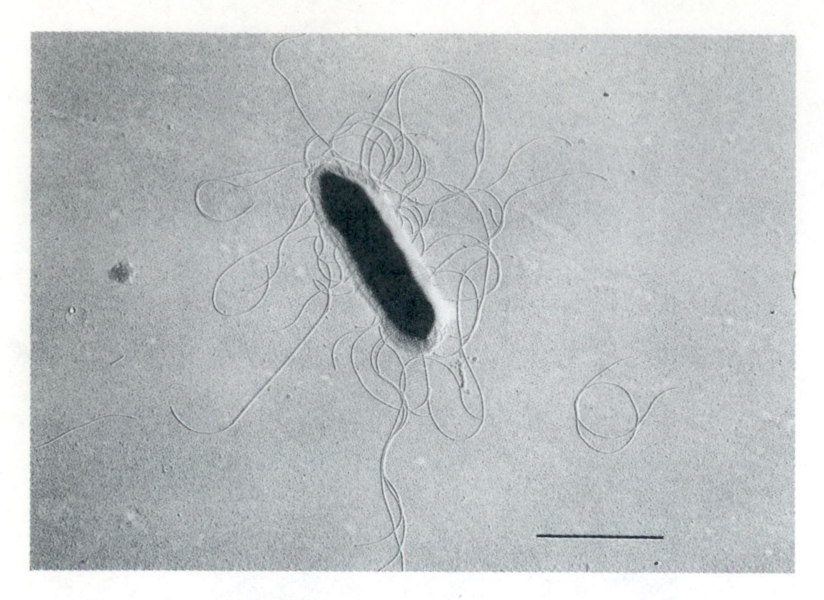

Figure 35-16 Electron micrograph of *Erwinia amylovora,* peritrichously flagellated gram-negative bacterium.
Courtesy George Lacy, Virginia Polytechnic Institute and State University.

duce tyloses. The flow rate of water in the xylem of infected plants may be reduced by as much as 95% of that occurring in healthy plants. There is also evidence that some wilting may be due to toxic substances released as the fungi grow in affected tissues. These toxins may affect membrane permeability in leaf cells.

Several techniques can be used to prevent disease, none of which is totally successful. The most successful technique is the development of wilt-resistant varieties of plants. The pathogens are relatively slow to develop new virulence traits, and resistant plant varieties may remain resistant for long periods of time.

Root rots of citrus. Root rots are caused by several groups of fungi that are characterized by their specialized requirements for cortical tissue. Several species in the genus *Phytophthora* are among the most important causes of root rot affecting such plants as annual and perennial flowers, fruit trees, and vegetables. Citrus plantings worldwide are affected by *Phytophthora* root rots. The most important species are *P. cinnamomi, P. cryptogea,* and *P. fragariae.*

Phytophthora root rot occurs anywhere in the world where the soil may be excessively moist and temperatures range between 15° and 23° C. The pathogen survives in the soil primarily in a mycelial state or as chlamydospores. As the soil warms, chlamydospores germinate and produce sporangia that release motile zoospores. The zoospores produce germ tubes that penetrate the feeder roots of plants. A mycelium develops in the roots, where it absorbs nutrients and destroys cortical tissue.

The roots of young citrus plants affected by *Phytophthora* become necrotic and may die within a few weeks or months. In older plants the destruction of the root system may take longer, and there is progressive deterioration characterized by reduced foliage, yellow leaves, and dieback in twigs and branches (Plate 58).

Root rot can be controlled by planting in well-drained soil or soil known to be free of the pathogen. Disease may also be prevented by using, whenever possible, disease-resistant varieties of plants. In citrus plants desirable ***cultivars*** (cultivated varieties) are customarily budded onto disease-resistant root stocks.

Bacterial diseases

Fire blight. Fire blight is a disease caused by the bacterium *Erwinia amylovora* (Figure 35-16), which is a peritrichously flagellated gram-negative bacterium belonging to the Enterobacteriaceae. The microorganism attacks the blossoms, leaves, stems, fruits, and limbs of many plants. It is most devastating to fruits such as pears and apples, but it may also attack other rosaceous plants, such as mountain ash, strawberry, blackberry, raspberry, rose, pyracantha, and quince (Plate 59).

Erwinia amylovora was the first bacterium proved to be pathogenic to plants. It overwinters in limb cankers and becomes active in wet weather during the spring. The bacteria ooze out of diseased limbs and are transferred to blossoms and leaves by splashing rain or insects. The bacteria penetrate the stomalike openings (nectaries) at the base of flowers, from which nectar is secreted. Inside the nectary the bacteria multiply and produce pectolytic enzymes that break down adjacent tissue. This is followed by necrosis and blackening of diseased tissues. The bacteria migrate intercellularly until twigs are reached, where they destroy cortical tissue. When larger branches are invaded, the movement of the bacteria is slowed and blight cankers occur. Cortical tissue is eventually destroyed, which results in death of the plant.

Some trees are naturally resistant to attack by *E. amylovora* because of the presence of a phenolic glucoside called arbutin. This compound is hydrolyzed by β-glucosidase to produce glucose and an antimicrobiologically active semiquinone. There is a positive correlation between the β-glucosidase content of tissue and resistance to infection. Factors such as wet weather and heavy nitrogen fertilization, which favor succulent growth, also increase the incidence of infection. A good insect control program in the postblossom period to reduce the spread of bacteria by insects to succulent twigs and avoidance of heavy nitrogen fertilization help prevent infection. Sprays for the dormant period do not appear to be effective, but there are satisfactory blossom sprays, such as Bordeaux or streptomycin.

Aster yellows. Aster yellows is a disease caused by an MLO that infects over 175 different species of plants but is particularly destructive to asters, lettuce, and carrots.

The etiological agent of disease overwinters in vegetables and weed plants, from which it is transmitted to healthy plants by several species of leafhoppers. Leafhoppers acquire the pathogen by feeding from the phloem of infected plants. The pathogen multiplies in the insect for several days before it can be effectively transmitted to a susceptible host. The insect inserts its stylet directly into the phloem and injects the MLO into the healthy plant.

Aster yellows is characterized by a general chlorosis of young leaves, dwarfing of the plant, and the formation of spindly shoots. The latter development may give the plant the appearance of a witch's broom. The greatest economic losses are incurred from infections of carrots, which are stunted and have an unpleasant taste and odor when processed.

The disease can be prevented to varying degrees by employing certain measures. Elimination of the overwintering host, such as perennial or biennial weeds, or the planting of crops at least 200 feet from weeds can reduce a large source of the pathogen. The application of insecticides to reduce leafhoppers also helps lessen the incidence of disease.

Crown gall. Crown gall, which is characterized by the formation of tumors, is discussed briefly earlier in this chapter, with particular attention given to the genetics and biochemistry of the disease process. The following discussion centers on the epidemiology and pathology of this disease.

The vector of the crown gall disease, *Agrobacterium tumefaciens*, is a gram-negative bacterium variously described as peritrichously to polarly flagellated. This bacterium harbors a plasmid (pTi) that carries a sequence of genes, called T-DNA, that are responsible for pathogenesis. The bacterium survives in the soil and attacks many plants. It is especially severe on stone fruits, grapes, roses, and cane fruits. The organism binds to the cells or roots or stems of tissue that have been abraded by cultural practices, insects, or grafting. Some or all of pTi appears, by an unknown process, in the plant cell while the pathogen remains outside the cell. Only part of the plasmid DNA, the T-DNA, becomes inserted into the chromosomal material of the plant. The T-DNA induces an imbalance of plant hormones: auxins, such as indolacetic acid, and cytokinins, such as zeatin, in the infected cell. T-DNA replicates with the chromosome during mitosis and, following the regeneration of plantlets, meiosis, and fertilization of the plant, can be inherited in a mendelian fashion. The hormones cause an imbalance leading to the proliferation of cells (hyerplasia) and the formation of tumors (galls).

Galls may be produced anywhere on the plant, but they occur most often on roots or in the stems near the soil line (plate 60). The young tumors have peripheral cells that are unprotected by tough epidermal layers and are attacked by insects and microorganisms. The affected tumor cells decay and discolor to impart a brown or black color to the tumor. Tumors may rot partially or completely, with new tumors appearing at the same or different sites in the next growing season. The size of the tumors ranges from less than 1 cm to 30 cm or more in diameter. Tumors may enlarge to the point where they exert enough pressure on surrounding xylem tissue to reduce the flow of water and photosynthetic products to healthy tissue. This condition ultimately leads to a reduction in plant size and fruit or flower yield.

Crown gall can be prevented by growing uninfected plant materials in fields that are known to be free of the pathogen. Those plants known to be susceptible to crown gall should be handled carefully so as to avoid damage to roots and crowns. A biological control measure has been developed that involves soaking seedlings or rootstock in a suspension of the bacterium *Agrobacterium radiobacter* strain 84. This organism can inhibit or kill *A. tumefaciens* by liberating a nucleotide antibiotic called agrocin 84. The genes for agrocin production are contained on a plasmid in *A. radiobacter*; however, the gene for agrocin sensitivity is located on the pTi plasmid of *A. tumefaciens*. The gene apparently codes for a protein that permits the transport of agrocin across the cytoplasmic membrane.

Viral disease: tobacco mosaic virus

The tobacco mosaic virus (TMV) is the cause of disease in over 15 families of plants, the most important of which are tomato and tobacco. The TMV is an RNA virus (see Chapter 6) that is among the most heat stable of all plant viruses. It is resistant to 90° C for 10 minutes and remains viable in dried plant tissue when heated to 120° C for 30 minute. The virus has been known to remain vaiable in dried plant tissue for several years.

TMV is usually transmitted mechanically to susceptible plants by workers whose hands have been contaminated by the virus. Tomato and tobacco leaves have leaf hairs that are easily broken off, and virus brought into contact with such abrasions can cause infection. Transplants can also be infected by virus that has overwintered in crop residues. Transmission by insect vectors, which is the most common method of transmission of all other viruses, is of minor consequence in the transmission of TMV.

TMV moves from cell to cell via plasmodesmata or connecting cytoplasmic strands. It may also move longer distances in vascular tissues. It is capable of infecting all plant tissues. Viral nucleic acid replication occurs in the nucleus of the cell, but viral assembly occurs in the cytoplasm, the site of protein synthesis. Mature viral particles, however, have been found in vacuoles, in the nucleus, and in chloroplasts. Two types of cytoplasmic inclusions can be observed following infection by TMV: first, an amorphous body consisting of viral particles, mitochondria, and other cellular debris and, second, crystal-containing layers of TMV particles. Symptoms of infection include chlorosis, mottling, dwarfism, discoloration, and sometimes necrosis. Mottled dark green and light green areas are the most frequent symptoms of infection (Plate 61). The light green areas represent sites where chloroplasts have been damaged. Reduction in photosynthesis leads to loss of carbohydrates and dwarfism. The tobacco plant is not killed, but the quality and quantity of the crop is severely reduced. Infection of tomato plants produces mottling and some reduction in crop yield. If the plant has been infected at a very early stage, there can also be internal discoloration of the fruit.

Sanitation and the use of disease-resistant varieties of plants are the best methods of control. Some disease-resistant varieties, however, do not yield a high-quality crop. Contaminated seedbeds or fields should not be used for planting a tobacco crop for at least 2 years. Workers handling and removing diseased plants should wash their hands with soap and water before handling healthy plants. Since milk inactivates TMV, it is often sprayed over plants during transplanting. Chewing and smoking of tobacco during cultivation practices should be avoided. Workers who tend tomatoes in glass houses are advised to keep a separate set of coveralls for each house to prevent the spread of TMV by their clothes.

SUMMARY

The study of plant diseases is called phytopathology. Infectious diseases of plants may be caused by all the major groups of microorganisms, but most are caused by fungi, bacteria, and viruses. Plants, unlike animals, have few indigenous microbial species, and diseases caused by opportunistic commensals are of little consequence. Many bacteria and fungi that infect plants are saprophytes of the soil, whereas others exist primarily as parasites. Bacteria are transmitted by water, insects, animals, and humans. Fungi are transmitted primarily by water and wind. Mycoplasmas are transmitted from one host to another by insect vectors such as leafhoppers. Viruses are transmitted by insects, nematodes, mites, or fungi and may be passed from plant to plant via contaminated hands, clothing, tools, and other fomites.

The plant surface, except for some limited openings such as stomas, is a barrier that is difficult for most microorganisms to breach. Some microorganisms, such as fungi, can directly penetrate the surface by producing specific structures and enzymes, whereas other pathogens require an abraded surface in order to penetrate. Sometimes the surface is abraded by insect vectors who inject the virus or mycoplasma directly beneath the epithelial surface.

Plant pathogens possess chemical factors that, combined with certain physical attributes, allow them to penetrate the plant's defenses. These chemical factors include enzymes, toxins, growth regulators, and polysaccharides. The plant is equipped to resist pathogens by structural and biochemical mechanisms. The waxy surface, for example, is a natural barrier, but the plant's best structural defenses are those that are induced. Plants can respond to infection by producing cork, abscission layers, or tyloses, all of which wall off the infectious agents and prevent them from spreading throughout the plant. The spread of infection can also be deterred by the production of polysaccharides called gums.

The biochemical mechanisms used by the plant to prevent disease may also be natural or induced. Natural biochemical mechanisms include the production of secondary metabolites, such as stilbenes, which inhibit tissue-degrading enzymes (pectinase, cellulase) produced by invading pathogens. Chlorogenic acid is also a compound produced naturally by the cell that inhibits the growth of plant pathogens. Biochemicals that are induced when the plant pathogen invades its host are called phytoalexins. The degree of resistance to infection is related to the concentration of phytoalexins in the plant.

- The bacterial agents causing infection can be broken down into two groups: ***fastidous*** and ***nonfastidious.*** Fastidious bacteria have considerable nutritional requirements and include fastidious vascular-limited bacteria and mycoplasma-like organisms. The majority of nonfastidious bacteria belong to the following genera: *Pseudomonas, Xanthomonas, Agrobacterium, Erwinia, Corynebacterium,* and *Streptomyces.* The phytopathogenic fungi are represented by all the major subdivisions. Viruses and viroids are second only to the fungi as the most frequent causes of infection and in economic importance.

Study questions

1. Explain some basic differences between plants and animals in terms of host-parasite interactions.
2. What are the major genera of bacteria that cause disease in plants? Why are genera such as *Staphylococcus* and *Streptococcus* not plant pathogens?
3. What theoretically is the best method for preventing disease in plants?
4. How are bacteria usually transmitted to plants? To fungi? To viruses?
5. What mechanisms do fungi use to directly penetrate the plant surface?
6. List some of the major symptoms of plant disease.
7. List the chemical factors that are used by the plant pathogen to cause disease.
8. What is so unusual about the plant disease caused by *Agrobacterium tumefaciens?*
9. How do some invading pathogens cause obstruction of plant vascular tissue?
10. What kinds of induced structural barriers are produced by plants following infection? How do they work?
11. Name two chemicals that are produced by plants in response to microbial invasion.
12. What is meant by overwintering by infectious disease agents?
13. What environmental factors are important in the fungal disease called potato blight?
14. What are the two hosts that are part of the infectious disease cycle in cereal stem and leaf rust?
15. Damage to chloroplasts will produce what kind of disease symptoms in the plant?
16. What is so unusual about the principal mechanism of transmission of the tobacco mosaic virus?
17. What biological control mechanism has been used to prevent infection by *Agrobacterium tumefaciens?*
18. Define the following: witch's broom, tylose, vascular wilt, appressorium, phytoalexin, abscission layer, opine, pT1 plasmid, fire blight, Bordeaux mixture, teliospores, and canker.

Selected readings

Books

Agrios, G.N. *Plant Pathology* (2nd ed.). Academic Press, Inc., New York, 1978.

Dodman, R.L. How the defenses are breached. In Horsfall, J.G., and Cowling, E.B. (Eds.). *Plant diseases,* Academic Press, Inc., New York, 1979.

Gibbs, A., and Harrison, B. *Plant virology: the principles.* John Wiley & Sons, Inc., New York, 1976.

Kelman, A. How bacteria induce disease. In Horsfall, J.G., and Cowling, E.B. (Eds.). *Plant diseases.* Academic Press Inc., New York, 1979.

Maramorosch, K., and Raychaudhuri, S.P. (Eds.). *Mycoplasma diseases of trees and shrubs.* Academic Press, Inc., New York, 1981.

Matthews, R.E.F. *Plant virology* (2nd ed.). Academic Press, Inc., New York, 1981.

Roberts, D.A., and Boothroyd, C.W. *Fundamentals of plant pathology.* W.H. Freeman & Co., Publishers, San Francisco, 1972.

Strobel, G.A., and Mathre, D.E. *Outlines of plant pathology.* Van Nostrand Reinhold Co., New York, 1970.

Vanderplank, J.E. *Genetics and molecular basis of plant pathogenesis.* Springer-Verlag, New York, Inc., New York, 1978.

Journal Articles

Davis, M.J., Purcell, A.H., and Thompson, S.V. Pierce's disease of grapevines: isolation of the causal bacterium. *Science* **199:**75, 1978.

Gibbs, J.N. Intercontinental epidemiology of Dutch elm disease. *Annu Rev. Phytopathol.* **16:**287, 1978.

Griffith, E. Iatrogenic plant disease. *Annu. Rev. Phytopathol.* **19:**69, 1981.

Hart, J.H. Role of phytostilbenes in decay and disease resistance. *Annu. Rev. Phytopathol.* **19:**437, 1981.

Kennedy, B.W., and Alcorn, S.M. Estimates of U.S. crop losses to procaryote plant pathogens. *Plant Dis.* **64**(7):674, 1980.

Kerr, A. Biological control of crown gall through production of Agrocin 84. *Plant Dis.* **65**(3):230, 1981.

Leben, C. How plant pathogenic bacteria survive. *Plant Dis.* **65**(8):633, 1981

Nester, E.W., and Kosuge, T. Plasmids specifying plant hyperplasias. *Annu. Rev. Microbiol.* **35:**531, 1981.

Schenck, N.C. Can mycorrhizae control root disease? *Plant Dis.* **65**(3):230, 1981.

Schmidt, E.L. Initiation of plant-root-microbe interactions. *Annu. Rev. Microbiol.* **33:**355, 1979.

Strobek, G.A. Plant phytotoxins. *Annu. Rev. Microbiol.* **31:**205, 1977.

Vidaver, A.K. The plant pathogenic corynebacteria. *Annu. Rev. Microbiol.* **36:**495, 1982.

Whitcomb, R.F. The genus *Spiroplasma*. *Annu. Rev. Microbiol.* **34:**677, 1980.

APPENDIXES

Appendix A SCIENTIFIC MEASUREMENTS AND MATHEMATICS

METRIC SYSTEM

Length

1 meter (m) = 10 decimeters (dm) = 100 centimeters (cm) = 1000 millimeters (mm) = 1,000,000 micrometers (μm) = 1,000,000,000 nanometers (nm)

1 angstrom (Å) = 0.1 nm = 0.0001 μm

1 kilometer (km) = 1000 m

Examples of interconversions:

1 cm = 10 mm; 1 mm = 0.1 cm; 1 mm = 1000 μm; 1 μm = 0.001 mm; 1 μm = 1000 nm; 1 nm = 0.001 μm; 1 nm = 10 Å

Volume

1 liter (L) = 1000 milliliters (ml)
= 1000 cubic centimeters (cc or cm^3)

Mass

1 kilogram (kg) = 1000 grams (g)
= 1,000,000 milligrams (mg)

1 mg = 1000 micrograms (μg)

ENGLISH EQUIVALENTS TO THE METRIC SYSTEM

Length

1 inch = 2.540 centimeters

Volume

1 liter = 1.056 quarts

1 cubic inch = 16.38 cubic centimeters

Mass

2.2 pounds = 1 kilogram

1 pound = 453 grams

1 ounce = 28.3 grams

Temperature Equivalents

$$\text{Degrees centigrade (°C)} = \frac{5}{9}(\text{°F} - 32)$$

$$\text{Degrees Fahrenheit (°F)} = \frac{9}{5}\text{°C} + 32$$

0° C = 32° F

100° C = 212° F

MATHEMATICAL OPERATIONS: EXPONENTS

An exponent is a number that indicates how many times the base appears as a factor. In most mathematical operations 10 is the most convenient base. The exponent appears as a superscript on the base. For example:

$$10^3 = 1 \times 10^3 = 1000$$

Therefore the exponent of 10^3 is 3, and 10^3 means $10 \times 10 \times 10$. The exponent of the base 10 determines where the decimal point must be shifted. If the shift is to the right, the exponent is positive:

$$1 \times 10^3 = 1000$$

If the shift is to the left, the exponent is negative:

$$1 \times 10^{-3} = 0.001 = \frac{1}{1000}$$

Multiplication of exponents

During multiplication exponents are added:

$$10^m \times 10^n = 10^{m+n}$$

Some examples are:

$$10^4 \times 10^2 = 10^6 = 1{,}000{,}000$$

$$5.3 \times 10^2 \times 10^3 = 5.3 \times 10^5 = 530{,}000$$

Division of exponents

During division exponents are subtracted:

$$\frac{10^m}{10^n} = 10^{m-n}$$

Some examples are:

$$\frac{2 \times 10^5}{1 \times 10^2} = 2 \times 10^{5-2} = 2 \times 10^3 = 2000$$

$$\frac{5 \times 10^4}{2 \times 10^6} = 2.5 \times 10^{4-6} = 2.5 \times 10^{-2} = 0.025$$

$$\frac{5 \times 10^{-2}}{2 \times 10^3} = 2.5 \times 10^{-2 + -3}$$

$$= 2.5 \times 10^{-5} = 0.000025$$

Addition and subtraction of exponents

During addition and subtraction of exponents the terms must have the same exponent; the terms are then either added or subtracted. Some examples are:

$$5 \times 10^{-2} + 3 \times 10^{-2} = 8 \times 10^{-2}$$

$$4.2 \times 10^2 + 2 \times 10^2 = 6.2 \times 10^2$$

$$5 \times 10^{-2} + 2 \times 10^{-3} = 5 \times 10^{-2} + 0.2 \times 10^{-2} = 5.2 \times 10^{-2}$$

$$4 \times 10^3 = 5 \times 10^4 = 0.4 \times 10^4 + 5 \times 10^4 = 5.4 \times 10^4$$

$$5 \times 10^{-2} - 2 \times 10^{-3} = 5 \times 10^{-2} - 0.2 \times 10^{-2} = 4.8 \times 10^{-2}$$

Exponents of exponents

The exponent of an exponent is multiplied by the exponent. For example:

$$(10^3)^2 = 10^6$$

LOGARITHMS (TO BASE 10)

The logarithm of a number is the power to which the base must be raised to equal the number. For our purposes, the base number is 10. For example:

$$\log 10 = 1 \text{ or } 10^1$$

$$\log 100 = 2 \text{ or } 10^2$$

$$\log 1000 = 3 \text{ or } 10^3$$

$$\log 0.1 = -1 \text{ or } 10^{-1}$$

$$\log 0.01 = -2 \text{ or } 10^{-2}$$

$$\log 0.001 = -3 \text{ or } 10^{-3}$$

Multiplication of logarithms

During the multiplication of logarithms the number is added. For example:

$$\log (1000 \times 1000) = \log (10^3 \times 10^3) = \log (10^3) + \log (10^3) = 3 + 3 = 6$$

Division of logarithms

During division logarithms are subtracted. For example:

$$\log (1000 \div 100) = \log (1000) - \log (100) = 3 - 2 = 1$$

Appendix B TECHNIQUES AND PROCEDURES USED IN IDENTIFICATION OF INFECTIOUS MICROORGANISMS

Microorganisms that infect humans can be found in many ecological habitats, such as the soil or water, or they may be commensals or parasites in vertebrates and invertebrates. Infectious microorganisms, particularly bacteria, coexist with noninfectious species in the host and must be distinguished from them if they are to be properly identified. Tests devised to differentiate microorganisms include morphological, biochemical, and serological procedures. In some instances all three types may be used in the differentiation process, whereas for some microorganisms only one type of test can be used. Time is a significant factor in the identification process of infectious agents, particularly in life-threatening situations. The selection of the proper chemotherapeutic agent for treatment is often dependent on the isolation and identification of the infectious agent. The isolated pathogen can then be subjected to antimicrobial sensitivity tests to determine what drugs it is resistant or sensitive to. Thus to ensure that the patient receives rapid treatment, the tests used in isolation and identification should be non–time consuming, as well as relatively specific and economical. Entire books are devoted to the type of tests used in the clinical laboratory. Our discussion here centers only on those tests that are more commonly employed in the laboratory. Very specific tests can be found in books that deal with diagnostic procedures, and the student is encouraged to examine them. Some of these books are listed in the references at the end of this appendix.

BACTERIAL IDENTIFICATION PROCEDURES

As discussed in Chapter 28, a specific strategy is used in identifying bacteria in the clinical laboratory. Rapid identification requires the use of staining procedures plus biochemical and/or serological tests.

Staining

Gram stain. The Gram stain, which is discussed in the text, is the single most important procedure in the identification of most bacteria in the clinical laboratory. Gram staining divides the bacterial world into groups that are either gram-positive or gram-negative. A smear of the organisms is prepared on a glass slide in the Gram-staining procedure. The smear is stained with a primary stain called crystal violet. The excess stain is washed off with water, and a solution of Gram's iodine is supplied to the smear. The iodine acts as a mordant and fixes the crystal violet to the cell wall. The smear is then destained by adding a few drops of an alcohol-acetone solution. Gram-positive bacteria can resist decolorization, but gram-negative bacteria are destained, and crystal violet is removed from the cell wall. A secondary stain, safranin (red), is applied to the smear. Gram-positive cells that have already been stained with crystal violet appear blue or purple (Plate 2). If the organisms are gram-negative, they take up the secondary stain and appear red (Plate 3).

Acid-fast stain. The acid-fast stain, like the Gram stain, is a differential stain, but its use is more limited. The acid-fast stain is used to distinguish species of *Mycobacterium* (agents of leprosy and tuberculosis), which are acid fast. The primary stain in the acid-fast–staining procedure is carbol-fuchsin (red). After the cells have been stained, a decolorized solution made up of acid and alcohol is added to the smear. Mycobacterial species contain as much as three times the amount of lipid as gram-negative species, and, unlike bacteria, they can resist decolorization and retain the primary stain. A secondary stain of methylene blue is applied to the smear after decolorization. Acid-fast organisms are not stained by the secondary stain but remain red or pink whereas non-acid-fast cells take up the secondary stain and appear blue.

Biochemical tests

Biochemical tests represent the most common method for identification of bacteria. There are a multitude of biochemical tests that may be applied individually or in combination. Special kits are now available that can measure several different properties at one time. The most commonly measured biochemical property is sugar fermentation. This test, coupled with others, can provide a tentative identification of the microorganism.

Fermentation tests. Fermentation tests are customarily carried out in special fermentation tubes. Each tube contains (1) a nutrient medium that will support the growth of the majority of bacteria; (2) a single fermentable carbohydrate such as glucose, lactose, or sucrose; (3) a pH indicator, such as phenol red; and (4) an inverted vial called a Durham tube that serves to trap gases. If the organism grows only on the nutrient medium and cannot ferment the sugar, the pH of the medium remains alkaline; for example, it is colored red if phenol red is the indicator. Acids and sometimes gases are produced if the organism can also catabolize the sugar. The acids act on the pH indicator, and the medium appears yellow (if phenol red is the indicator). Any gases that are produced displace fluid in the inverted vial, and this can be detected with the naked eye.

Hydrogen sulfide production. The sulfur found in such compounds as the amino acid cysteine and inorganic thiosulfate can be reduced to hydrogen sulfide gas by some microorganisms. The gas can be detected because of its ability to intereact with ferrous salts to produce ferrous sulfide, a black precipitate, according to the following reaction:

$$FeSO_4 + H_2S \longrightarrow FeS + H_2SO_4$$

Indole formation. Indole is a product released by the oxidation of the amino acid tryptophan according to the following reaction:

$$\text{Tryptophan } (C_8H_6N\text{-}CH_2CH(NH_2)COOH) \xrightarrow{H_2O} \text{Indole} + CH_3COCOOH \text{ (Pyruvic acid)} + NH_3 \text{ (Ammonia)}$$

Indole can be detected in tryptophan broth when Kovac's reagent (paradimethylaminobenzaldehyde) is added to the culture. It reacts with indole to produce a red-violet ring in the broth tube.

Tripe-sugar iron agar. Tripe-sugar iron agar (TSIA) contains three carbohydrates: glucose, sucrose, and lactose, with glucose at a concentration of one tenth that of the other sugars. The medium also contains a nutrient medium, which supports the growth of most bacteria, plus thiosulfate, which can be reduced to hydrogen sulfide. In addition, the indicator phenol red is present to reveal either acidic or alkaline conditions in the tube. The agar is slanted in the TSIA tube, and the clinical specimen is streaked on the slant and then stabbed into the tube. Stabbing is important because the butt of the tube maintains anaerobic conditions, a requirement for fermentation as well as hydrogen sulfide production. The following reactions can take place in the TSIA tube:

1. If glucose is the only sugar fermented, the slant appears red (alkaline)—the concentration of acid produced from glucose fermentation is small and is rapidly oxidized on the slant in the presence of oxygen. The butt of the tube is yellow, because the acids produced there cannot be completely oxidized in the absence of air.
2. If sucrose or lactose is metabolized, the slant and butt appear yellow (acidic), because the concentration of acid is very high and complete oxidation is not possible during the required period of incubation.
3. If no sugars are fermented, only the proteins in the nutrient medium are metabolized, with the release of alkaline products such as ammonia, and the slant and butt remain red (alkaline).
4. If hydrogen sulfide is produced, the gas reacts with ferrous salts in the medium to produce a black precipitate of ferrous sulfide.
5. If gases such as carbon dioxide and hydrogen are produced, there is displacement of the agar from the butt of the tube, and areas within the agar appear to contain bubbles.

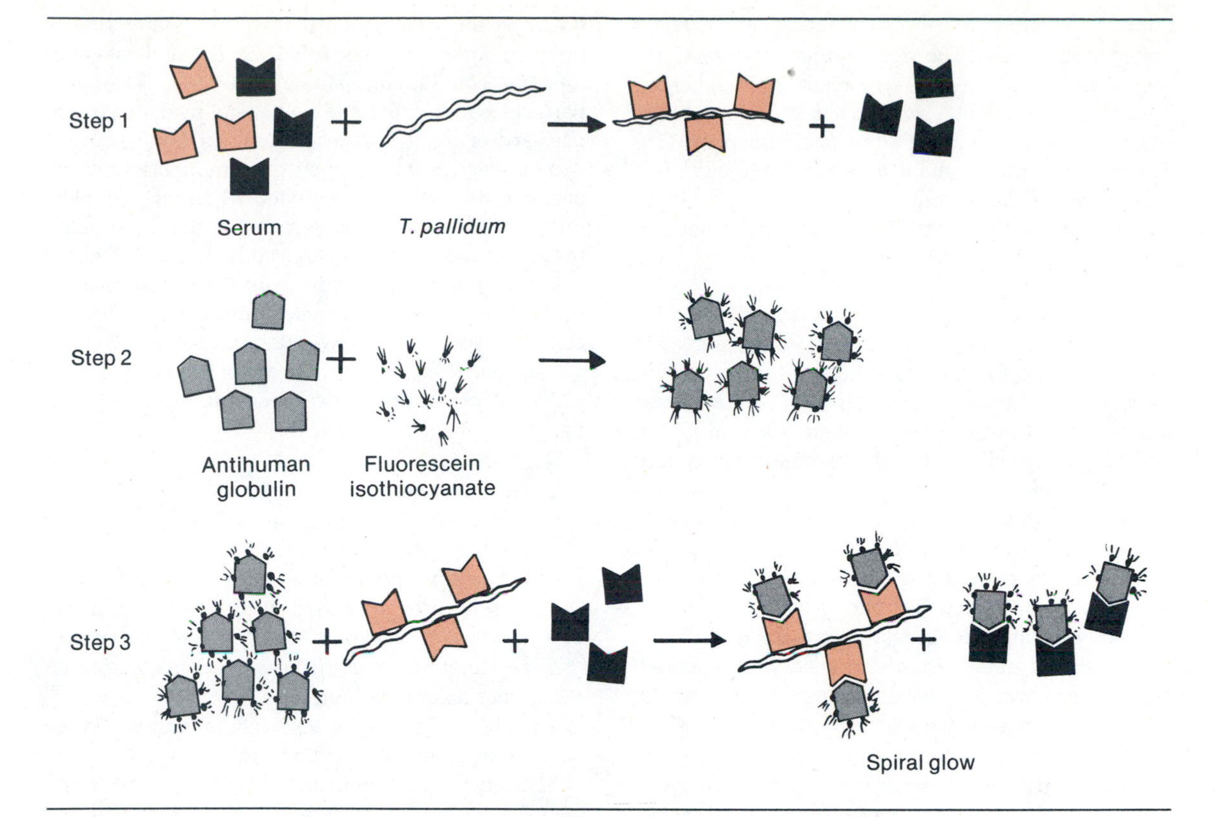

Figure B-1 Fluorescent treponemal absorption (FTA-ABS) test. *Step 1:* Patient's serum may contain specific antitreponemal antibodies ( ) or nonantitreponemal antibodies ( ). Serum is reacted with *Treponema pallidum. Step 2:* Antihuman globulin is conjugated with fluorescein isothiocyanate. *Step 3:* Conjugated antihuman globulin will coat any human antibodies, but treponemes will fluoresce only if patient's serum contains antitreponemal antibodies.

Decarboxylases. Certain amino acids can be decarboxylated by microorganisms. The decarboxylases are induced in broth tubes by their respective substrate–amino acids. The decarboxylases most frequently detected are lysine, ornithine, and arginine. Decarboxylation releases neutral products, and this process can be detected directly by using a pH indicator such as bromocresol purple, which is yellow under acidic conditions and purple in the alkaline state.

Hemolysis. The ability to lyse red blood cells (hemolysis) is a characteristic of certain bacteria. The compounds causing hemolysis are called hemolysins and may be organic products such as various toxins, or they may be inorganic, for example, hydrogen peroxide. To detect hemolysis, an agar medium containing intact red blood cells is streaked with the clinical specimen to obtain isolated colonies. Following incubation the red blood cells around the isolated colonies may exhibit no hemolysis (Plate 45), partial hemolysis (α-hemolysis), or complete hemolysis (β-hemolysis). When there is partial hemolysis, a greening around the colony occurs because of alteration of the hemoglobin in the red blood cell (Plate 46). In complete hemolysis the area around the colony is clear because of hemolysis of the red blood cells. The clear area extends outward from the colony to produce a clear, circular zone (Plate 47).

Serological procedures

Serological techniques in bacterial identification may be direct (that is, the organism is first isolated from the patient, and immune serum is applied to the specimen to determine specificity) or indirect (that is, serum (hopefully containing specific antibodies) is isolated from the patient, and antigen is applied). Indirect serological methods are referred to as immunoserological and include such tests as radioimmunoassay (RIA), counterimmunoelectrophoresis (CIE), and enzyme-linked immunosorbent assay (ELISA). These tests are discussed in Chapter 24. Two direct serological techniques are discussed here: the quelling reaction and one of the serological tests for syphilis.

Quellung reaction. The quellung reaction is used to identify *Streptococcus pneumoniae,* the primary etiological agent of pneumonia, as well as *Haemophilus influenzae* type b, the cause of meningitis in children under 4 years of age. To perform the test, a small aliquot of the clinical material, such as sputum, blood, spinal fluid, or organisms from an agar colony, is air dried on a microscopic slide. Antiserum containing antibodies specific for the microorganism, as well as a drop of methylene blue, is added to the slide. The slide is then examined under oil immersion. A positive reaction is indicated by the appearance of a swelled refractile capsule appearing around a blue-stained microorganism (see Figure 24-7). Since many different serotypes of *S. pneumoniae* exist, a polyvalent antiserum is used in which over 84 different types of the organism can be detected by capsular swelling.

Serological test for syphilis. Many serological tests for syphilis have been devised, but only a few are currently being used. One of these is the fluorescent treponemal antibody absorption (FTA-ABS) test (Figure B-1). In the test the patient's serum is first treated (absorbed) with a treponemal antigen that removes nonspecific antibodies. The absorbed serum is reacted with *Treponema pallidum,* the specific antigen. Any antitreponemal antibody that is present in the patient's serum will bind to the antigen. Next, antihuman globulin that has been conjugated with fluorescein isothiocyanate is added to the slide. The slide is then examined with a fluorescent microscope. The following results are possible when this test is performed:

1. If the patient does not have syphilis, antibodies to *T. pallidum* are not present in the serum, and no specific fluorescing *T. pallidum* can be observed. The test is negative.
2. If the patient has or recently had syphilis, antibodies to *T. pallidum* are present in the serum, and the treponemes are coated with specific antitreponemal antibodies plus the conjugated antihuman gamma globulin. The test is positive, and fluorescing treponemes can be observed.
3. Sometimes, because of some underlying illness or condition or immunological dysfunction, nonspecific antibodies may bind to the treponemes, resulting in what appears to be a positive test. These tests are called false-positive, and other, more specific tests are required for proper identification.

VIRAL DIAGNOSTIC PROCEDURES

Viral infections may be diagnosed in the clinical laboratory by (1) microscopic examination of tissue for cytopathic effects or for the presence of virus, (2) isolation of the virus from the patient and cultivation in cell or tissue culture, and (3) serological techniques. Viral diagnosis using laboratory techniques was for many years a very slow and tedious process because of the time required for cultivation and isolation of the virus, the long period of time required for virus-specific antibodies to be produced in the patient, and the absence of advanced microscopic techniques. Today many viral diseases can be rapidly diagnosed because of the emergence of the electron microscope and the development of biochemical and serological procedures. Many viruses produce surface lesions, which in the early stage of infection contain large amounts of virus. These viral particles can be detected by several available methods. For example, diagnosis can be obtained in 1 hour if electron microscopy or immunoelectron microscopy is used. Viruses such as herpes simplex, rotavirus, and hepatitis A can be detected using these methods. Viruses can also be rapidly detected using fluorescent antibody techniques. Either a direct or an indirect fluorescent antibody technique can be used (see below). Rapid identification requires that the source of the virus be easily accessible and not contain any excess, interfering material.

1. Direct fluorescent antibody technique. The direct method requires that a fluorescein-conjugated virus-specific antiserum be used. Thus a different antiserum is required for each virus that is to be identified. In addition, the direct method is one tenth as sensitive as the indirect method, thus making it unsatisfactory for most diagnostic purposes.
2. Indirect fluorescent antibody technique. In the indirect method virus-specific antibodies are used to react with the isolated viral antigen. The virus-specific antibodies are obtained from some animal such as a goat. The antigen-antibody complex is then reacted with fluorescein-labeled antiserum that is specific for the immunoglobulins of the animal in which the viral antiserum was prepared (the goat). Thus fluorescein-labeled antiserum can be used in the identification of any virus as long as the same animal is used for preparing the antiserum.

Other techniques, such as the ELISA method, immunoperoxidase, and immunoelectrophoresis, are discussed in Chapter 24.

FUNGAL DIAGNOSTIC PROCEDURES

Except for a few species, biochemical techniques cannot be used for the identification of the fungi. Conclusive evidence of fungal infection is best obtained by direct microscopic examination of clinical specimens or by culture techniques.

Direct microscopic examination

Among the important characteristics of the fungi are their distinctive spores (thallospores, conidia, and sporangiospores) and spore-bearing structures. Such structures can be identified microscopically from clinical material such as skin scrapings, pus, gastric washings, sputum, urine, and cerebrospinal fluid. Loopfuls of the specimen are placed on a microscopic slide to which is added potassium hydroxide. By gently heating this suspension, any opacity in the specimen (particularly when skin scrapings are used) can be reduced. The suspension is examined microscopically for characteristic mycelia, spores, or budding yeast cells. The specimen can also be stained using Wright's or Giemsa stain in order to highlight morphological characteristics, or with india ink to indicate capsule formation. Direct microscopic examination yields results that are only presumptive, and culture procedures must be applied for more conclusive evidence.

Culture techniques

Two types of culture media are used to isolate suspected fungal pathogens from clinical material. The principal agar media are Sabouraud, blood, birdseed, or Mycosel agar. For example, a combination of Sabouraud and blood agar are frequently used. These media also contain various antibiotics, including chloramphenicol, penicillin, tetracycline, streptomycin, and cycloheximide, to inhibit growth of bacterial species that might contaminate the clinical specimen. The clinical specimens are usually cultivated on agar media at two different temperatures: 25° to 30° C and 37° C. The purpose of using two different temperatures is to determine if the fungus is diphasic; that is, does it exhibit mycelial growth at room temperature and yeastlike growth at 37° C? After suitable incubation (which may be from 48 hours to several weeks) colonies develop on the agar that can be differentiated by their characteristic color and morphology. A portion of the fungal colony may be examined by placing it on a slide containing lactophenol blue. When observed under low or high power, the spores and spore-bearing structures may be identified. The best method for examining intact spores is to prepare a slide culture. In this technique the fungi are cultured on specially designed microscopic slides that contain an agar well. After the molten agar has been added to the well and allowed to harden, it is inoculated with the fungal specimen, and a coverslip is placed over the agar. After suitable incubation the slide can be observed microscopically for the appearance of undisturbed spores and various mycelial appendages. If the isolated fungus is a species of *Candida,* sugar assimilation and fermentation tests may be applied for differentiation.

Serodiagnosis

Serological techniques for identification of fungal infections present major problems because of the lack of commercially available antigens and the cross-reactivity between the antigens of different fungal species. Serological tests, such as complement fixation, immunodiffusion, and counterimmunoelectrophoresis, have proved useful in the diagnosis of some fungal infections.

IDENTIFICATION OF PROTOZOA AND HELMINTHS

As discussed in Chapter 32, parasites such as the protozoa and helminths have life cycles in which different stages are present in different fluids of the body. A great majority of protozoa and helminths can be recovered from the feces, blood, urine, and cerebrospinal fluids. Most specimens sent to the clinical laboratory for diagnosis are of fecal material. The proper collection and preservation of such body fluids is therefore a very important aspect in the diagnosis of parasitic infections. These body fluids may contain protozoan cysts or trophozoites, as well as the eggs or larvae of helminths. In addition to the parasites, other microorganisms such as fungi and bacteria may also contaminate the specimen, particularly fecal specimens. Specimens can be preserved by fixing them in polyvinyl alcohol and 10% formalin, although other preservatives are available. This mixture also destroys any contaminating species of microorganisms. Since most diagnostic procedures involve fecal specimens, only that aspect of identification is discussed here.

Fecal material may be examined directly by preparing wet mounts in which a drop of saline is added to the fecal specimen. The suspension is observed microscopically first under low power to scan the field and then under high power for individual detail. Wet mounts are used primarily to detect motile trophozoites of the protozoa. If a special iodine solution is added to the wet mount, protozoan eggs can be stained. The cytoplasm of the egg appears yellow, the glycogen brown, and the nucleus refractile. Permanent staining of specimens can also be obtained by using the trichrome- or iron-hemotoxylin–staining procedures. If fecal specimens contain too few of the parasites for microscopic identification, they can be concentrated by flotation or sedimentation. Both procedures separate the organisms from fecal debris by differences in specific gravity.

Differences in nuclear morphology are among the most important criteria for identification of ameba, either as trophozoites or cysts. A permanently stained specimen is particularly important to detect these differences. Flagellates are more easily recognized because of their motility. Each species has distinctive morphological characteristics, such as flagella, an undulating membrane, or sucking disks. Only one ciliate, *Balantidium coli,* infects humans, and it is easily recognized. The key to identification of the helminths is characterization of eggs found in the feces or recovery of adult worms (Plate 48). Helminth eggs exhibit specific lengths and widths; therefore microscopic examination requires that the microscope be equipped with an ocular micrometer and that each objective be calibrated.

Selected readings

Books

Bergan, T., and Norris, J.R. (Eds.) *Methods in microbiology* (Vol. 12). Academic Press, Inc., London, 1979.

Blazevic, D.N., and Ederer, G.M. *Biochemical tests in diagnostic microbiology.* John Wiley & Sons, Inc., New York, 1975.

Chernesky, M.A., Ray, C.G., and Smith, T.F. *Laboratory diagnosis of viral infections. Cumitech-15,* American Society for Microbiology, Washington, D.C., 1982.

Finegold, S.M., and Martin, W.J. *Diagnostic microbiology* (6th ed.). The C.V. Mosby Co., St. Louis, 1982.

Garcia, L.S., and Ash, L.R. *Diagnostic parasitology.* The C.V. Mosby Co., St. Louis, 1979.

Haley, L.D., Trandel, J., and Coyle, M.B. In Sherris, J.C. (Ed.). *Practical method for culture and identification of fungi in the clinical microbiology laboratory. Cumitech 11.* American Society for Microbiology, Washington, D.C., 1980.

Lennette, E.H., and Schmidt, N.J. *Diagnostic procedures for viral, rickettsial, and chlamydial infections* (5th ed.). American Public Health Association, Inc., Washington, D.C., 1979.

Lennette, E.H., et al. (Eds.). *Manual of clinical microbiology* (3rd ed.). American Society for Microbiology, Washington, D.C., 1980.

McFaddin, J.F. *Biochemical tests for identification of medical bacteria* (2nd ed.). The Williams & Wilkins Co., Baltimore, 1980.

Rose, N.R., and Friedman, H. *Manual of clinical immunology* (2nd ed.). American Society for Microbiology, Washington, D.C., 1980.

Suter, V.L., Citron, D.M., and Finegold, S.M. *Wadsworth anaerobic bacteriology manual* (3rd ed.). The C.V. Mosby Co., St. Louis, 1980.

Tilton, R.C. (Ed.) *Rapid methods and automation in microbiology.* American Society for Microbiology, Washington, D.C., 1981.

Appendix C BERGEY'S CLASSIFICATION SCHEME

Kingdom *PROCARYOTE*

Division I. The cyanobacteria

Division II. The bacteria

PART 1

Phototrophic bacteria

Order I. *Rhodospirillales*
- Family I. *Rhodospirillaceae*
  - Genus I. *Rhodospirillium*
  - Genus II. *Rhodopseudomonas*
  - Genus III. *Rhodomicrobium*
- Family II. *Chromatiaceae*
  - Genus I. *Chromatium*
  - Genus II. *Thiocystis*
  - Genus III. *Thiosarcina*
  - Genus IV. *Thiospirillum*
  - Genus V. *Thiocapsa*
  - Genus VI. *Lamprocystis*
  - Genus VII. *Thiodictyon*
  - Genus VIII. *Thiopedia*
  - Genus IX. *Amoebobacter*
  - Genus X. *Ectothiorhodospira*
- Family III. *Chlorobiaceae*
  - Genus I. *Chlorobium*
  - Genus II. *Prosthecochloris*
  - Genus III. *Chloropseudomonas*
  - Genus IV. *Pelodictyon*
  - Genus V. *Clathrochloris*

PART 2

The gliding bacteria

Order I. *Myxobacterales*
- Family I. *Myxococcaceae*
  - Genus I. *Myxococcus*
- Family II. *Archangiaceae*
  - Genus I. *Archangium*
- Family III. *Cystobacteraceae*
  - Genus I. *Cystobacter*
  - Genus II. *Melittangium*
  - Genus III. *Stigmatella*
- Family IV. *Polyangiaceae*
  - Genus I. *Polyangium*
  - Genus II. *Nannocystis*
  - Genus III. *Chondromyces*

Order II. *Cytophagales*
- Family I. *Cytophagaceae*
  - Genus I. *Cytophaga*
  - Genus II. *Flexibacter*
  - Genus III. *Herpetosiphon*
  - Genus IV. *Flexithrix*
  - Genus V. *Saprospira*
  - Genus VI. *Sporocytophaga*
- Family II. *Beggiatoaceae*
  - Genus I. *Beggiatoa*
  - Genus II. *Vitreoscilla*
  - Genus III. *Thioploca*
- Family III. *Simonsiellaceae*
  - Genus I. *Simonsiella*
  - Genus II. *Alysiella*
- Family IV. *Leucotrichaceae*
  - Genus I. *Leucothrix*
  - Genus II. *Thiothrix*
- Families and genera of uncertain affiliation
  - Genus *Toxothrix*
- Family *Achromatiaceae*
  - Genus *Achromatium*
- Family *Pelonemataceae*
  - Genus *Pelonema*
  - Genus *Achroonema*
  - Genus *Peloploca*
  - Genus *Desmanthos*

PART 3

The sheathed bacteria

- Genus *Sphaerotilus*
- Genus *Leptothrix*
- Genus *Streptothrix*
- Genus *Lieskeella*
- Genus *Phragmidiothrix*
- Genus *Crenothrix*
- Genus *Clonothrix*

PART 4
Budding and/or appendaged bacteria

Genus *Hyphomicrobium*
Genus *Hyphomonas*
Genus *Pedomicrobium*
Genus *Caulobacter*
Genus *Asticcacaulis*
Genus *Ancalomicrobium*
Genus *Prosthecomicrobium*
Genus *Thiodendron*
Genus *Pasteuria*
Genus *Blastobacter*
Genus *Seliberia*
Genus *Gallionella*
Genus *Nevskia*
Genus *Planctomyces*
Genus *Metallogenium*
Genus *Caulococcus*
Genus *Kusnezovia*

PART 5
The spirochetes

Order I. *Spirochaetales*
Family I. *Spirochaetaceae*
Genus I. *Spirochaeta*
Genus II. *Cristispira*
Genus III. *Treponema*
Genus IV. *Borrelia*
Genus V. *Leptospira*

PART 6
Spiral and curved bacteria

Family I. *Spirillaceae*
Genus I. *Spirillum*
Genus II. *Campylobacter*

Genera of uncertain affiliation

Genus *Bdellovibrio*
Genus *Microcyclus*
Genus *Pelosigma*
Genus *Brachyarcus*

PART 7
Gram-negative aerobic rods and cocci

Family I. *Pseudomonadaceae*
Genus I. *Pseudomonas*
Genus II. *Xanthomonas*
Genus III. Zoogloea
Genus IV. *Gluconobacter*
Family II. *Azotobacteraceae*
Genus I. *Azotobacter*
Genus II. *Azomonas*
Genus III. *Beijerinckia*
Genus IV. *Derxia*
Family III. *Rhizobiaceae*
Genus I. *Rhizobium*
Genus II. *Agrobacterium*
Family IV. *Methylomonadaceae*
Genus I. *Methylomonas*
Genus II. *Methylococcus*
Family V. *Halobacteriaceae*
Genus I. *Halobacterium*
Genus II. *Halococcus*

Genera of uncertain affiliation

Genus *Alcaligenes*
Genus *Acetobacter*
Genus *Brucella*
Genus *Bordetella*
Genus *Francisella*
Genus *Thermus*

PART 8
Gram-negative facultatively anerobic rods

Family I. *Enterobacteriaceae*
Genus I. *Escherichia*
Genus II. *Edwardsiella*
Genus III. *Citrobacter*
Genus IV. *Salmonella*
Genus V. *Shigella*
Genus VI. *Klebsiella*
Genus VII. *Enterobacter*
Genus VIII. *Hafnia*
Genus IX. *Serratia*
Genus X. *Proteus*
Genus XI. *Yersinia*
Genus XII. *Erwinia*
Family II. *Vibrionaceae*
Genus I. *Vibrio*
Genus II. *Aeromonas*
Genus III. *Plesiomonas*
Genus IV. *Photobacterium*
Genus V. *Lucibacterium*

Genera of uncertain affiliation

Genus *Zymomonas*
Genus *Chromobacterium*
Genus *Flavobacterium*
Genus *Haemophilus*
(H. vaginalis)
Genus *Pasteurella*
Genus *Actinobacillus*
Genus *Cardiobacterium*
Genus *Streptobacillus*
Genus *Calymmatobacterium*
Parasites of *Paramecium*

PART 9
Gram-negative anaerobic bacteria

Family I. *Bacteroidaceae*
Genus I. *Bacteroides*
Genus II. *Fusobacterium*
Genus III. *Leptotrichia*

Genera of uncertain affiliation

Genus *Desulfovibrio*
Genus *Butyrivibrio*
Genus *Succinivibrio*
Genus *Succinimonas*
Genus *Lachnospira*
Genus *Selenomonas*

PART 10
Gram-negative cocci and coccobacilli

Family I. *Neisseriaceae*
Genus I. *Neisseria*
Genus II. *Branhamella*
Genus III. *Moraxella*
Genus IV. *Acinetobacter*

Genera of uncertain affiliation

Genus *Paracoccus*
Genus *Lampropedia*

PART 11
Gram-negative anaerobic cocci

Family I. *Veillonellaceae*
Genus I. *Veillonella*
Genus II. *Acidaminococcus*
Genus III. *Megasphaera*

PART 12

Gram-negative, chemolithotrophic bacteria

a. Organisms oxidizing ammonia or nitrite
  Family I. *Nitrobacteraceae*
    Genus I. *Nitrobacter*
    Genus II. *Nitrospina*
    Genus III. *Nitrococcus*
    Genus IV. *Nitrosomonas*
    Genus V. *Nitrosospira*
    Genus VI. *Nitrosococcus*
    Genus VII. *Nitrosolobus*
b. Organisms metabolizing sulfur
    Genus *Thiobacillus*
    Genus *Sulfolobus*
    Genus *Thiobacterium*
    Genus *Macromonas*
    Genus *Thiovulum*
    Genus *Thiospira*
c. Organisms metabolizing sulfur
    Genus *Thiobacillus*
    Genus *Sulfolobus*
    Genus *Thiobacterium*
    Genus *Macromonas*
    Genus *Thiovulum*
    Genus *Thiospira*
c. Organisms depositing iron ormanganese oxides
  Family I. *Siderocapsaceae*
    Genus I. *Siderocapsa*
    Genus II. *Naumanniella*
    Genus III. *Ochrobium*
    Genus IV. *Siderococcus*

PART 13

Methane-producing bacteria

  Family I. *Methanobacteriaceae*
    Genus I. *Methanobacterium*
    Genus II. *Methanosarcina*
    Genus III. *Methanococcus*

PART 14

Gram-positive cocci

a. Aerobic and/or facultatively anaerobic
  Family I. *Micrococcaceae*
    Genus I. *Micrococcus*
    Genus II. *Staphylococcus*
    Genus III. *Planococcus*
  Family II. *Streptococcaceae*
    Genus I. *Streptococcus*
    Genus II. *Leuconostoc*
    Genus III. *Pediococcus*
    Genus IV. *Aerococcus*
    Genus V. *Gemella*
b. Anaerobic
  Family III. *Peptococcaceae*
    Genus I. *Peptococcus*
    Genus II. *Peptostreptococcus*
    Genus III. *Ruminococcus*
    Genus IV. *Sarcina*

PART 15

Endospore-forming rods and cocci

  Family I. *Bacillaceae*
    Genus I. *Bacillus*
    Genus II. *Sporolactobacillus*
    Genus III. *Clostridium*
    Genus IV. *Desulfotomaculum*
    Genus V. *Sporosarcina*

Genus of uncertain affiliation
    Genus *Oscillospira*

PART 16

Gram-positive, asporogenous rod-shaped bacteria

  Family I. *Lactobacillaceae*
    Genus I. *Lactobacillus*

Genera of uncertain affiliation
    Genus *Listeria*
    Genus *Erysipelothrix*
    Genus *Caryophanon*

PART 17

Actinomycetes and related organisms

  Coryneform group of bacteria
    Genus I. *Corynebacterium*
      a. Human and animal parasites and pathogens
      b. Plant pathogenic corynebacteria
      c. Nonpathogenic corynebacteria
    Genus II. *Arthrobacter*
    Genera incertae sedis
      Brevibacterium
      Microbacterium
    Genus III. *Cellulomonas*
    Genus IV. *Kurthia*
  Family I. *Propionibacteriaceae*
    Genus I. *Propionibacterium*
    Genus II. *Eubacterium*
Order I. *Actinomycetales*
  Family I. *Actinomycetaceae*
    Genus I. *Actinomyces*
    Genus II. *Arachnia*
    Genus III. *Bifidobacterium*
    Genus IV. *Bacterionema*
    Genus V. *Rothia*
  Family II. *Mycobacteriaceae*
    Genus I. *Mycobacterium*
  Family III. *Frankiaceae*
    Genus I. *Frankia*
  Family IV. *Actinoplanaceae*
    Genus I. *Actinoplanes*
    Genus II. *Spirillospora*
    Genus III. *Streptosporangium*
    Genus IV. *Amorphosporangium*
    Genus V. *Ampullariella*
    Genus VI. *Pilimelia*
    Genus VII. *Planomonospora*
    Genus VIII. *Planobispora*
    Genus IX. *Dactylosporangium*
    Genus X. *Kitasatoa*
  Family V. *Dermatophilaceae*
    Genus I. *Dermatophilus*
    Genus II. *Geodermatophilus*
  Family VI. *Nocardiaceae*
    Genus I. *Nocardia*
    Genus II. *Pseudonocardia*
  Family VII. *Streptomycetaceae*
    Genus I. *Streptomyces*
    Genus II. *Streptoverticillium*
    Genus III. *Sporichthya*
    Genus IV. *Microellobosporia*
  Family VIII. *Micromonosporaceae*
    Genus I. *Micromonospora*
    Genus II. *Thermoactinomyces*
    Genus III. *Actinobifida*
    Genus IV. *Thermomonospora*
    Genus V. *Microbispora*
    Genus VI. *Micropolyspora*

PART 18
The rickettsias

Order I. *Rickettsiales*
Family I. *Rickettsiaceae*
Tribe I. *Rickettsieae*
Genus I. *Rickettsia*
Genus II. *Rochalimaea*
Genus III. *Coxiella*
Tribe II. *Ehrlichieae*
Genus IV. *Ehrilichia*
Genus V. *Cowdria*
Genus VI. *Neorickettsia*
Tribe III. *Wolbachieae*
Genus VII. *Wolbachia*
Genus VIII. *Symbiotes*
Genus IX. *Blattabacterium*
Genus X. *Rickettsiella*
Family II. *Bartonellaceae*
Genus I. *Bartonella*
Genus II. *Grahamella*
Family III. *Anaplasmataceae*
Genus I. *Anaplasma*
Genus II. *Paranaplasma*
Genus III. *Aegyptionella*
Genus IV. *Haemobartonella*
Genus V. *Eperythrozoon*
Order II. *Chlamydiales*
Family I. *Chlamydiaceae*
Genus I. *Chlamydia*

PART 19
The mycoplasmas

Class I. *Mollicutes*
Order I. *Mycoplasmatales*
Family I. *Mycoplasmataceae*
Genus I. *Mycoplasma*
Family II. *Acholeplasmataceae*
Genus I. *Acholeplasma*

Genera of uncertain affiliation

Genus *Thermoplasma*
Genus *Spiroplasma*

Mycoplasma-like bodies in plants

GLOSSARY

abberation disturbance of the rays of light so they are unable to be brought to a sharp focus.

abscission separation of dead or dying plant tissue from living tissue.

acidophile acid-loving microorganism whose most active growth response is at low pH.

acidulent a substance used to increase acidity.

acquired immunity immunity that is based on prior exposure and the subsequent interaction of specific immune system products with antigens.

active immunity acquired immunity resulting from the presence of antibody or immune lymphoid cells produced in response to an antigenic stimulus.

active site a site on the enzyme for binding of substrate and promoting catalytic activity.

active transport transfer of molecules across a membrane by an expenditure of energy.

activated sludge method of sewage treatment where flocs absorb organic material.

acute having a short and relatively severe time course.

aerobe organism requiring oxygen for growth.

aerosol a suspension of particles in air.

aging cask a barrel-shaped vessel of staves, headings, and hoops usually of oak, used to store wine for a period during which taste and aroma develop before it is bottled.

akinete dormant type of structure found in cyanobacteria.

ale beer brewed with top-fermenting yeast.

allele one of the alternative forms of a gene that may occupy a particular site on the chromosome.

allergen a more specific term for the antigens that are involved in allergic or hypersensitivity reactions.

allergy an immediate or delayed type of hypersensitivity that follows exposure to an allergen.

allochthonous referring to organisms not indigenous to a given environment.

allograft (homograft) a tissue or organ graft between members of the same species who are not genetically identical.

allosteric an enzyme whose activity with another molecule, such as its substrate, is altered because of interaction of the enzyme with other molecules.

alternative (alternate) complement pathway pathway in which the complement cascade is activated at the C3 level; immune system products are not required for the activation of this pathway.

ameboid (movement) a type of movement in which the protoplasm of the cell changes shape to produce feet (pseudopodia) that help the cell move.

amylase an enzyme that hydrolyses the $\alpha-1,4$ glucosidic bonds of starch, yielding dextrins and maltose.

anabolism metabolic process involved in the synthesis of cell material.

anaerobe organism that lives in the absence of air (oxygen).

anamnestic response the recall or the "remembering" by the immune system of a prior response to a given antigen (see memory cell).

anaphylatoxin the activated C3 and C5 components of complement that cause mast cells to release histamine and other physiologically active amines that contribute to intense inflammmatory reactions.

anaphylaxis a form of immediate hypersensitivity that can occur in anyone.

anaplerotic a reaction or series of reactions that replenishes intermediates used in biosynthetic reactions.

anoxygenic refers primarily to photosynthetic processes in which oxygen is not present or produced.

antheridium structure responsible for production of male gametes.

antibiosis the production by one species of an organic compound that is toxic at low concentrations to another species.

antibiotic a chemical compound produced by microorganisms that can inhibit or kill other microorganisms.

antibody a glycoprotein of the Ig type that is produced by a plasma cell and specifically reacts with the known antigen that elicited its production.

anticodon the triplet of nucleotides in a transfer RNA molecule that complements a specific codon on the messenger RNA.

antigen a foreign substance capable of inducing an immunological response.

antigen-binding site the Fab portion of the antibody molecule that binds to an antigenic determinant.

antigenic capable of stimulating the formation of antibodies.

antigenic determinants small chemical groups on antigens that elicit the formation of specific antibodies and sensitized lymphocytes.

antimicrobial an agent (chemical or physical) that inhibits growth of or kills microorganisms.

antiseptic a chemical compound that can be used on the surface of living tissue and inhibits bacterial growth.

antitoxins antibodies that are formed in response to bacterial toxins or toxoids and neutralize or inactivate those toxins.

appressorium swollen or flattened portion of a fungal germ tube.

arthropod invertebrate with jointed limbs.

ascocarp sexual fruiting structure of higher ascomycetes.

ascogonium female sex organ.

ascospores a sexual spore of the Ascomycetes.

ascus a saclike structure that holds spores (ascospores) produced by species of fungi.

aseptic being free of microbial life.

assimilation the absorption of food and its conversion into cells or tissue.

atopic hypersensitivity immediate-type, IgE-mediated common clinical allergies for which there is a genetically determined predisposition for formation.

atopy a form of immediate hypersensitivity that tends to occur in those with a hereditary predisposition.

attenuation lessening; reduction of the virulence of microorganisms as in a vaccine.

autoallergy immune response against an individual's own antigens.

autochthonous referring to the indigenous members of an ecological site.

autoclave a chamber in which steam under pressure is used to sterilize objects and solutions.

autogamy union of gametes produced by one individual.

autoimmune disease a disease in which the host system destroys its own tissue.

autolysis digestion of a cell by enzymes produced by that cell at time of death.

autotrophic the ability to use carbon dioxiode as the sole source of carbon.

auxospore the reproductive cell of diatoms.

auxotroph a mutant microorganism that will grow only on minimal media supplemented with growth factors not required by the normal parent.

axenic pure or uncontaminated culture.

bacillus an organism of the genus *Bacillus;* also used to designate rod-shaped bacteria.

bacteriocin antimicrobial substance produced by bacteria that kills sensitive members of related strains.

bacteriophage a bacterial virus, sometimes referred to as phage.

bacteroid irregular shape of bacteria associated with symbiotic nitrogen fixation.

baeocyte reproductive cells produced by species of cyanobacteria.

basal body structure at the base of flagella and cilia.

basidiospore a sexual spore formed in the fungal group Basidiomycotina.

basidium a specialized cell of the Basidiomycetes in which the products of meiosis are formed.

basophil a blood leukocyte that is rich in vasoactive substances; the blood counterpart of mast cells. Stains with basic dyes.

B cell (B lymphocytes) a major population of lymphocytes; the B cell stem cells are found in bone marrow (in the bursa of Fabricius in avian species); mature B lymphocytes have cell-surface Ig receptors for specific antigenic determinants; antigen-stimulated B cells proliferate and differentiate into antibody-producing plasma cells.

benthos sediment, or pertaining to life within the sediment.

binary fission an asexual process in which a cell splits into two cells of equivalent size.

binomial two names.

biodegradation the biological destruction of organic compounds.

biogeochemistry the science that deals with the role of living organisms in affecting the chemistry of the earth and its atmosphere.

biomass all of the living organisms within a particular region.

biosynthesis the building up of chemical compounds.

blast cell a leukocyte that is an intermediate cell type of B and T lymphocytes.

bock beer a dark-colored lager beer brewed with more malt and hops than light-colored lager beers.

BOD biochemical oxygen demand; amount of oxygen required to convert organic matter to carbon dioxide and water.

boil a localized abscess resulting from infection of a hair follicle; also called a furuncle.

buboe a lymph node that has become enlarged.

budding an asexual reproductive process in which one pole of the cell gives rise to a body (daughter cell).

bulk starter method starting with a purchased mother culture, the progressive inoculation and growth of starter cultures in increasing volumes of milk culminating with the inoculating of the cheese fermentation vessel.

bursa of fabricius a cloaca-associated central lymphoid organ in avian species where B stem cells originate.

Cabernet Sauvignon a variety of purple grape used in the production of high-quality red wines.

canker a disease that kills definite and relatively localized areas of bark on branches or trunks of trees.

cannula an artifical tube used for insertion into the body, for example, an artery.

capsid protein coat of a virus.

capsomere an aggregate of polypeptides forming a unit of the viral capsid.

carcinogen an agent (chemical, physical, or biological) that can cause malignant tumor formation.

caries a disease of the calcified tissue of teeth; tooth decay.

carrier a host that harbors infectious microorganisms and can transmit them to others but shows no disease symptoms.

caseation necrosis a type of necrosis in which the tissue resembles an amorphous mass of cheese.

cassava any of several plants of the spurge family grown in the tropics for their fleshy edible rootstocks, which yield a nutritious starch.

catabolism a metabolic process in which foodstuffs are broken down to release energy.

catheter a tubular device used to withdraw fluids from, or introduce them into, body cavities.

cell-mediated immunity (CMI) T cell–dependent immunity.

cellulitis inflammatory condition of the subcutaneous tissue.

central lymphoid organs lymphatic organs where lymphoid cell subpopulations become committed to specific functions; the thymus and bone marrow for T and B cells, respectively; also called primary lymphatic organ.

centriole structure associated with nuclear division and involved in spindle organization.

centromere the point on a chromosome that segregates during the first division of meiosis.

chancre the primary lesion of syphilis.

chardonnay a variety of green grape used in the production of white wine.

chemiosmotic hypothesis the theory that states that the usage and synthesis of energy in a cell is the result of a proton electrical gradient across the cell membrane.

chemoautotroph an organism that derives its energy and carbon from the oxidation of inorganic chemicals.

chemoheterotroph an organism that derives its energy and carbon from the oxidation of preformed chemicals.

chemolithotroph an organism that obtains its energy and carbon from inorganic sources.

chemostat an apparatus used to maintain bacterial cultures in a state of continuous division.

chemotaxis phenomenon of movement toward or away from a chemical stimulus.

chemotherapeutic chemical agents used in treatment of disease.

chiasma crosslike configuration in a tetrad resulting from a genetic exchange between nonsister chromatides.

chimeric in genetics refers to a DNA molecule derived from different sources.

chitin a polysaccharide found in the shells of invertebrates such as lobsters and makes up the exoskeleton of insects.

chloroplast an organelle found only in photosynthetic organisms and carrying chlorophyll.

chlorosis yellowing of green leaf tissue through lack of green pigment.

chromatid one of two chromosome strands visibly present during mitosis or meiosis.

chromatin (body) a network of deoxyribonucleic acid fibers giving the appearance of a distinct body in the cell.

chronic over a long period of time.

cidal killing.

cilia small hairlike projections found on the surface of some microorganisms and used for movement; found on respiratory epithelium, cilia are used to direct flow of various secretions and particles.

cirri tuft of cilia used as organelle for motility and feeding in certain protozoa.

cistron a segment of the chromosome used as a pseudonym for gene; it codes for a single polypeptide.

climax community the stable community characteristic of a site.

coccus a spherical bacterial cell.

code date a mixture of letters or numbers that signify the manufacturing place and date of a product; also called a manufacturing lot number.

codon a nucleotide triplet in the messenger RNA that specifies for a particular amino acid.

coenocytic a mass of protoplasm in which several nuclei are present.

coenzyme an organic molecule that is loosely bound to the enzyme and necessary for the activity of the enzyme.

cofactor an inorganic molecule, such as a metal ion, that increases enzyme activity.

colicin a protein secreted by certain strains of *Escherichia coli* and lethal to other strains of the same species.

colitis inflammation of the colon.

colony a uniform mass of cells derived from a single cell growing on a solid surface.

cometabolism metabolism of substrates incapable of being used as a carbon or energy source in the presence of alternative, usable substrates.

commensal a relationship between organisms in which one benefits but the other neither benefits nor is harmed.

commercial sterility the thermal processing of canned foods such that only thermophilic spores survive.

communicable easily transmitted.

compensation depth point where photosynthesis and respiration balance in an aquatic environment.

competition a rivalry between two species for a limiting factor in the environment.

competitive inhibitor a molecule that competes with the substrate for attachment to the active site of an enzyme.

complement a complex system of serum proteins that are activated in a sequential fashion (in the complement cascade) by immune complexes or by other substances such as bacterial endotoxins; biological consequences of complement system activation include cell lysis, opsonization, chemotaxis, and intense inflammation (through anaphylatoxin).

congenital acquired by the fetus during pregnancy.

conidia asexual fungal spores produced on specialized mycelial branches called conidiophores.

conidiophore a hyphal filament on which spores, called conidia, are borne.

conjugation the act of joining together; in bacteria a process in which genetic information is transferred from one cell to another.

conjunctivitis inflammation of the conjunctiva of the eye.

constitutive enzyme an enzyme produced by a cell under any environmental conditions.

consumer organisms feeding on preformed organic material.

convalescent state of recovery from disease.

corn steep liquor a waste product of the corn starch industry; it contains both carbohydrates and nitrogenus compounds and can serve as an inexpensive source of each in the cultivation of microorganisms.

cortex a layer found in the spore coat.

counterimmunoelectrophoresis (CIE) electrophoretically directed movement of antigen and antibody toward each other through a support medium.

crown upper dome of tree, bearing leaves, flowers, and fruits.

cultivar cultivated variety (often not a valid taxon).

culture a particular strain of microorganism growing in a medium.

cyst any closed cavity or sac, or a type of resting cell in prokaryotes and eukaryotes.

cystoscope an instrument for examining the interior of the urinary bladder.

cytokinesis cytoplasmic division into daughter cells following nuclear division in eukaryotes.

cytolytic causing lysis and death of the cell.

cytopathic characterized by pathological change in a cell.

cytophilic antibodies antibodies that bind to cell surface receptors through the Fc portion of the Ig molecule.

cytoplasmic streaming the flow of the cytoplasm in stationary cells that is used to distribute nutrients to all parts of the cell.

cytotoxic capable of killing cells.

D value the time in minutes necessary for a lethal agent or environment to reduce the population.

damping-off collapse of seedling near soil line from necrotic infection induced by water and fungus.

debilitation loss of normal function.

decomposer organism converts organic material into its inorganic constituents.

degeneracy in genetics a term used to denote that an amino acid has more than one triplet code word.

dehydrogenation the removal of hydrogens from a compound.

delayed hypersensitivity T cell–mediated hypersensitivity in which the reaction reaches the maximum in about 24 to 48 hours after exposure to the antigen that elicited the original hypersensitive state.

denature to lose the natural configuration, for example, heating an enzyme in which there is change in conformation and usually biologic activity.

denitrification microbial reduction of nitrate to N_2, N_2O, and NO.
dermatophyte a fungus that affects the skin.
desiccate to dry.
desquamation shedding of the superficial layers of skin.
detoxication biological destruction of a toxic chemical.
dextrin a soluble polymer of glucose usually derived from starch.
dextrorotatory capable of rotating plane-polarized light to the right.
diarrhea abnormal fluidity of fecal discharge.
dikaryon a mycelium consisting of cells, each containing two nuclei of unlike mating type.
dimorphism the ability to exist in one of two morphological states; For example, some fungi may exist in filamentous or yeastlike state.
diploid containing twice the haploid chromosome number.
disease abnormal condition of the body, usually having characteristic symptoms.
disinfectant a substance that destroys vegetative pathogens.
dissimilation the act of breaking down a component into smaller parts.
domain folded regions or segments of H and L immunoglobulin chains that are stabilized with disulfide bonds. L chains have one domain in the variable region and one in the constant region. H chains have one domain in the variable region and three or four domains in the constant region.
dynein protein linkages found in the flagella of eukaryotes.
ecosystem the ecological site where an interaction between living and nonliving components exists.
ectosymbiosis a symbiotic state in which one organism attaches to the surface of another.
eczema inflammatory skin disease characterized by scales and crusts.
effector T cells T cells that have an effect on specific antigens (such as cytotoxic cells) as opposed to regulatory T cells that have regulatory effects on other lymphocytes.
elephantiasis a condition in which fluids accumulate in various tissues, causing grotesque features
endemic continually present in a community.
endergonic a reaction in which energy is used or absorbed in the system.
endocytosis the process in which soluble or insoluble materials are engulfed by ameboid cells.
endogenous produced within the cell.
endoplasmic reticulum a protoplasmic network in cells of higher animals and plants consisting of a continuous double membrane system that courses through the cytoplasm.
endospore a spore occurring within the cell.
endosymbiosis the state in which one cell lives within another, for example, a prokaryotic bacterium living in a eukaryotic protozoan.
endotoxin a toxin derived from the cell wall of gram-negative bacteria.
enterotoxin a toxin that gives rise to gastrointestinal symptoms when ingested.
enzyme a protein that can engage in specific reactions in which a product or products are formed.
eosinophil a type of leukocyte containing granules that have an affinity for acid dyes.
epidemic an outbreak of a disease affecting a large number of individuals in a community.
epidemiology the science that deals with the incidence, transmission, and prevention of disease.
epilimnion surface waters of oceans or lakes above the thermocline.
epilithic association of organism with inanimate materials.
epiphyte microorganism living attached to phytoplankton.
epiphytic living in association with plants or algae.
episome a piece of DNA that may be an autonomous unit in the cell or may be integrated into the chromosome.
epizootic outbreak of disease affecting a large number of animals.
erythema redness of skin.
eschar a dry mass of necrotic (dead) tissue.
etiology study of the cause of disease.
eucarpic type of fungal cell in which part of the thallus contains rhizoids.
euchromatin diffuse chromatin that appears during interphase.
eukaryotic of a cell type in which the nuclear material is bounded by a membrane.
eutroph a species that grows at high nutrient concentrations.
eutrophication biological enrichment of a lake.
exergonic a reaction in which energy is released.
exfoliation dropping off of scales or layers, for example, skin.
exocytosis extrusion of materials from the cell into the environment via vacuoles.
exon functional sequences on eukaryotic DNA.
exotoxin toxin produced in the cell and released into the environment.
exudate material (fluids or cells) that has escaped from blood vessels.
fab fragment the fragment of the enzyme-digested antibody molecule that binds antigen, consisting of the combined amino terminals of the variable regions of an H and an L chain.
Fc fragment the crystalizable fragment of the enzyme-digested antibody molecule. Consists of the carboxy terminal half of the two H chains of an antibody molecule. Biological functions include attachment to appropriate cell-surface receptors, complement binding, placental passage.
facultative anerobes organisms that grow best under aerobic conditions but can grow anaerobically.
feeder root system fine root hairs that absorb nutrients for a tree.
fermentation an anaerobic metabolic process that uses an organic compound as the final electron acceptor. In industrial microbiology refers to any aerobic or anaerobic process in which a product is produced.
fermenter a large cylindrical vessel, usually made of stainless steel, used to grow microbes on an industrial scale.
fertility (F) factor a genetic factor carried by some microorganisms that permits conjugation to occur with the same or related species.
fibrinogen a plasma protein involved in clotting.
fimbria fringe; in microbial genetics, a small nonconjugative projection on the surface of bacteria.
flagellin the protein making up the flagella.
flagellum organ of motility.
flavoprotein a protein component of the electron transport system.
flora resident organisms in an area.

fluorescein a fluorescing pigment produced by many pseudomonads and observed by illumination with ultraviolet light.

fluorescent capable of emitting light of one wavelength when exposed to light of another wavelength.

fluorescent antibody (FA) an antibody conjugated with a fluorescent dye that fluoresces on ultraviolet irradiation; used in an immunoidentification procedure; may be visualized by UV or fluorescent microscopy.

fluorochrome a compound that releases its energy following excitation in the form of light.

food chain a group of organisms and their interrelationships based on exchange of nutrients.

folliculitis inflammation of hair follicle.

fomite an inanimate object that may be involved in disease transmission.

fruiting body a specialized structure in some bacteria (Myxobacteria) and some fungi that produce spores.

frustule another term for the cell wall of diatoms.

furuncle a boil.

gall gross enlargement and deformity, usually globose, of plant tissue.

gamete a sex cell (either male or female) that unites with gamete of opposite sex to produce a zygote.

gamma globulin a fraction of globulin (a plasma protein); identified by electrophoresis; contains the majority of immunoglobulins (antibodies); also signifies the antibody-containing fraction of plasma that is used for prophylaxis or therapy.

gene a particular segment of a nucleic acid (DNA usually) that codes for a single polypeptide.

genome complete set of hereditary characteristics.

genotype the genetic constitution of an organism.

germination the process of vegetative cell formation following a period of dormancy (spore formation).

germ tube a filamentous protuberance produced by yeast cells.

gingiva (gums) soft tissue surrounding teeth.

globulins class of proteins found in blood.

glucan a general term given to polysaccharides in which glucose is a component.

glucoamylase an enzyme that removes successive glucose units from the nonreducing ends of dextrin chains.

gluconeogenesis formation of glucose from noncarbohydrate precursors.

glycolysis the anaerobic process in which carbohydrates such as glucose are oxidized to pyruvic or lactic acid.

golgi complex membranous complex found in eukaryotes that packages proteins and lipids, which are transferred to selected areas of cell.

graft a tissue to be transplanted.

grana condensed bodies of chlorophyll within plant cells.

granulocyte a type of white blood cell such as neutrophils, eosinophils and blasophils containing granules of hydrolytic enzymes involved in phagocytic digestion.

granuloma a nodule formed in tissue as a result of the accumulation of fused macrophages, which try to destroy infectious agents.

GRAS list an acronym for Generally Recognized as Safe. A list of chemicals that are allowed by law to be added to foods.

green beer freshly fermented beer before it has been aged.

gumma granuloma found in tertiary stage of syphilis.

H chain (heavy chain) the larger of the two types of polypeptide chains of the basic Ig molecule; there are two identical H chains per basic Ig molecule; H chain composition determines the class of the Ig molecule.

habitat physical environment of an organism.

halophile an osmophile that has a specific requirement for increased amounts of an inorganic salt (usually NaCl) for growth.

haploid containing but one set of chromosomes.

hapten a substance that by itself does not elicit an immune response unless conjugated with an immunogenic carrier molecule; may react specifically with immune products in the absence of a carrier molecule.

haustoria hyphae that penetrate host cells for the purpose of obtaining nutrients.

heartwood central cylinder of nonfunctional xylem in woody stem.

helminth parasitic worm.

helper T cells (T_h) a subset of regulatory T cells that facilitate the response of B and other T cells to antigens.

hemagglutination clumping of red blood cells.

hemolysins substances, including antibodies (immune hemolysins), that lyse red blood cells.

hepatitis inflammation of the liver.

heterocyst specialized structure found in some cyanobacteria that carries out nitrogen fixation.

heterokaryon a cell, usually a mycelium, containing more than one genetically distinct type of nucleus in a common cytoplasm.

heterotrophic requiring preformed organic compounds as a source of carbon.

histamine a vasoactive amine released (especially in type I IgE-mediated hypersensitivity reactions) from mast cells and basophils; causes smooth muscle contraction, increased capillary permeability, and increased glandular secretion.

histocompatability antigens tissue antigens whose similarities or dissimilarities determine graft acceptance or graft rejection. Also called transplantation antigens.

histone a basic protein associated with eukaryotic nuclear DNA.

holdfast an adhesive-containing structure that enables an organism to attach to a substrate.

holocarpic a type of fungal species in which the thallus has no reproductive structures.

hops female flower of the moraceous vine *Humulus lupulus.*

horizons the layers making up the vertical soil profile.

horizontal spread the spread of infectious agents from one person to another through contact, insect vectors, or by contaminated food or fomites.

humoral immunity antibody-mediated (B cell) immunity; antibodies are primarily located and transferable in body fluids (humors) such as serum.

humus organic fraction of the soil that is relatively stable.

hydrogen swell the nonmicrobial formation of hydrogen gas in a canned food product, usually resulting from a reduction of an acid food with the tin lining of the can.

hydrophilic readily attracted to water and dissolving it.

hydrophobia fear of water.

hydrophobic not attracted to water and unable to mix with it.

hydrosphere earth's aquatic environment.

hymenium spore-bearing (sporogenous) layer of a fungal fruiting body.
hypersensitivity a state of overresponsiveness by the immune system to an antigen.
hypertonic a solution in which there is a higher concentration of salt relative to that found in the interior of a cell (isotonic).
hypha (pl. **hyphae**) one of the filaments that make up a fungal mycelium.
hypolimnion deep waters of oceans or lakes below the thermocline.
hypotonic a solution having a lower salt concentration than an isotonic solution.
iatrogenic a disorder or disease caused by the diagnosis or treatment by a physician.
icosahedron twenty-sided object.
icterus jaundice; yellow discoloration of skin due to bile pigments.
id an allergic reaction to fungi or fungal products.
immediate hypersensitivity antibody-mediated (humoral) hypersensitivity reactions that occur within minutes after antigen is introduced into an individual who has been previously sensitized (i.e., has antibodies) to the antigen.
immobilized enzyme an enzyme that has been attached to or entrapped within an inert matrix.
immunity the state of being protected or resistant, especially to microorganisms and their products.
immunofluorescence fluorescence resulting from the interaction of an immunoglobulin, to which a fluorescent dye has been conjugated, and a specific antigen.
immunogen a substance capable of inducing an immune response.
immunoglobulins (Igs) plasma proteins of the gamma globulin class that are produced by or associated with B lymphocytes; the basic Ig molecule consists of two H chains and two L chains; antibodies are Igs.
immunosuppression the reduction in the capability of the immune system to respond to an antigen; obtained through immunosuppressive drugs, irradiation, and antilymphocytic serum (ALS).
impetigo infection of the skin usually caused by streptococci.
inclusion bodies in virus infections, the highly stainable components, usually virus, found in the cytoplasm or nucleus of the infected cell.
index case the first case observed or detected in an epidemic.
indigenous native to a particular place.
inducer a molecule capable of stimulating the formation of compounds such as enzymes involved in cellular metabolism.
induration a hardened area or lesion.
infectious able to cause disease.
infestation attachment of parasites to superficial layers of the skin.
inflammation nonspecific response to irritants characterized by pain, heat, redness, and swelling.
inoculum a substance (such as microorganisms or serum) introduced into the tissues or culture media.
insertion sequence in genetics, the addition of nucleotides or genes to the chromosome.
intercalary inserted between components.
interferon a class of proteins produced by vertebrate cells in response to viruses, endotoxins, and certain chemicals; associated primarily with antiviral activity.
intoxication state of being poisoned.
intron the nonfunctional sequences on eukaryotic DNA.
in utero within the uterus.
invasion able to invade or penetrate the body.
invertase an enzyme that converts sucrose to a mixture of glucose and fructose.
in vitro outside of the body or performed in artificial environments.
in vivo within the body or within a living organism.
iodophors disinfectants consisting of iodine combined with a carrier molecule.
isomer a molecule having the same atoms and groups as another molecule but having a different arrangement.
J chain (joining chain) glycoprotein whose function is believed to be to join together basic IgM molecules to each other, and basic IgA molecules to each other, and thus form polymers (IgM pentamer and IgA dimer and trimer).
jaundice see icterus.
karyogamy fusion of nuclei of cells.
killer T cell sensitized effector T cells that are cytolytic for corresponding target cells.
kinetochore see centromere
kinetosome see basal body
kinins a group of inactive peptides in the blood and tissue that, on activation, are involved in blood clotting and inflammatory processes.
Koji process a process by which molds are cultured on the surface of moist bran.
Koplik's spots small bluish red lesions appearing on the oral mucosa during measles infection
kuru chronic progressive degenerative disorder of the central nervous system caused by a virus and found among New Guinea natives.
L-form a wall-deficient form of bacteria.
lager beer beer that has been aged to develop its characteristic taste.
lambda (virus) a bacterial virus noted for its ability to integrate into bacterial chromosome.
latency a state of inactivity.
lauter tub in the brewing process it is the tank where insoluble spent brewers grains are removed after mashing is completed.
leafhopper delicate, leaf-sucking, homopterous insect; often a disease vector.
leavening the production of gas in the manufacture of a baked food product.
lectins glycoprotein found on the surface of plant cells.
legume a group of plants such as peas, beans, and clover, in which nodules form on the roots and contain bacteria that fix nitrogen.
leukocidin a substance produced by some pathogenic bacteria and toxic to white blood cells.
leukopenia a smaller than normal number of circulating leukocytes.
lichen an association of a fungus with an algae, resulting in the formaton of a biological unit morphologically distinct from its partners.
light chain (L chain) the smaller of the two types of polypeptide changes (the other is the H chain) that compose the basic Ig molecule; each basic Ig molecule contains two identical light chains.
lignification hardening of tissue from wall deposition of lignins.
lignin complex organic material that in part constitutes wood.
lipid organic molecule that is insoluble in water but soluble in organic solvents such as chloroform and ether.

lithotroph a microorganism that obtains its energy from the oxidation of inorganic compounds.

local anaphylaxis a type I hypersensitivity reaction that primarily affects a given organ or tissue (such as skin or nasal mucosa).

lophotrichous having a tuft of flagella at a pole.

lymph a fluid derived from tissues that contains white blood cells and is organized into vessels that drain into the blood stream.

lymphadenitis inflammation of a lymph node.

lymphocyte a white blood cell devoid of cytoplasmic granules, associated with immune response and chronic inflammation.

lymphokines soluble factors released by sensitized lymphoid cells (classically T cells) in response to antigenic contact; such factors modulate the biological activity of other cell populations, especially macrophage cells.

lyophilization process of freeze-drying.

lysis rupture or dissolution of a cell.

lysogenic conversion the change in properties of a bacterium resulting from the carriage of a prophage.

lysogeny the state of a bacterial cell in which a bacteriophage genome is in association with the bacterial genome.

lysosome a cell organelle containing hydrolytic enzymes collectively termed acid hydrolases.

macrophage a large mononuclear cell derived from the mononuclear phagocytic system and having phagocytic activity.

macrophage activation factor (MAF) a lymphokine that increases the capacity of macrophage cells to destroy antigens.

mast cell cell found in connective tissue that stores heparin and histamine.

matrix the ground substance of a component.

meiosis a eukaryotic reproductive process in which two mitotic divisions result in the formation of daughter cells (usually sex cells) receiving the haploid chromosome number.

memory cell a lymphocyte of either B or T cell lineage that was formed during a primary immune response and that is capable of synthesizing specific immune products at an increased rate during an anamnestic response.

meningitis inflammation of the meninges (membranes that surround the brain and spinal cord).

merozoite a trophozoite resulting from multiple fission.

mesophile a microorganism that grows best at temperatures between 20° and 45° C.

mesosome involuted membrane of the bacterial cell.

metabolism the sum total of all the physical and chemical changes that take place in a cell and maintain the cell's integrity.

metachromatic granules granules of certain bacteria that often contain phosphate chains, also called volutin.

methylotroph organisms that gain their energy from the oxidation of one-carbon compounds other than carbon dioxide.

Michaelis-Menten constant a constant that specifies the quantitative relationship between substrate concentration and maximum velocity for different enzymes.

microaerophile an organism that grows in air but at a concentration of oxygen less than our atmosphere.

microhabitat the microscopic environment that is important to a population or community in nature.

microtubules tubes that are the structural basis of eukaryotic flagella and are also part of the spindle organization during mitosis of eukaryotes.

mineralization conversion of the organic form of an element to the inorganic state.

mitosis the process in which genetic material is duplicated and distributed equally between daugher cells.

mixotroph organism that can obtain energy from inorganic sources but requires organic carbon.

mold another name for fungi that exhibit branching.

monocyte see macrophage.

monomer the simplest molecule of a compound.

morbidity sickness; the ratio of sick to well individuals in a community.

morphogenesis the change in structure as an organism develops.

mortality fatality; the ratio of the number of deaths to a given population in a defined situation.

mosaic disease symptom of leaves with mixed green, light green, and yellow patches.

mutagen a substance that increases the mutation rate of an organism.

mutation a process in which a gene undergoes structural changes.

mutualism a relationship between organisms in which both benefit.

mycelium a matlike organization of fungal hyphae.

mycetismus mushroom poisoning.

mycology the study of fungi.

mycolysis the enzymatic lysis of fungi.

Mycoplasma a taxonomic group of bacteria without cell walls.

mycorrhiza an association of a fungus with the roots of a plant.

mycoses diseases caused by fungi.

mycotoxicosis poisoning of humans and animals following the ingestion of foods containing toxins produced by fungi.

myeloma proteins Igs or portions of Igs produced in large quantities by neoplastic plasma cells (plasmacytoma).

myiasis infestation caused by maggots (fly larvae).

nebulizer a device used to produce fine sprays from liquids.

necrosis death of tissue or cells.

negri bodies inclusion bodies found in brain cells of those animals infected with rabies virus.

nematode roundworm.

neonate a newborn infant (up to 4 weeks old).

nephrotoxic toxic to the kidney.

neutrophil a leukocyte in which the granules do not have an apparent affinity for acid or basic dyes.

niche an organism's function within a habitat.

nitrification microbial formation of nitrate.

nitrogen fixation biological use of nitrogen gas (N_2).

nosocomial pertaining to the hospital.

nucleocapsid the capsid protein of a virus with its enclosed nucleic acid.

nucleoid a region in prokaryotic cells in which there is a high concentration of DNA.

nucleolus a small body residing in the nucleus that contains a high concentration of ribonucleic acid.

nucleosomes a structural arrangement on the eukaryotic chromosome in which DNA of 130 to 140 base pairs is associated with eight histone molecules.

nutrient substances extracted from the environment by cells and used in metabolic processes.

obligate necessary or required.

Okazaki fragments short pieces of DNA that are synthesized during replication and later polymerized.

oligotroph a species that grows at low nutrient concentrations.

oncogenic having the ability to cause tumors.

operon a unit of the chromosome consisting of adjacent genes controlled by an operator.

ophthalmic pertaining to the eye.

opsonin a substance (usually an antibody or complement fragment C3b) that adheres to bacteria, viruses, and antigen-antibody aggregated molecules and makes those particles more susceptible to ingestion by phagocytes.

organolepically detected by the senses.

osmophile a microorganism that requires a high solute concentration in the medium for growth.

osmosis passage of fluids through a semipermeable membrane.

oxidative phosphorylation the process of energy (ATP) formation during the transport of electrons, in the electron transport system, to the acceptor molecule oxygen.

oxygenic the ability to use a photosynthetic process in which oxygen is a byproduct.

PMN or PMNL polymorphonuclear leukocyte of a type called a neutrophil; it is a white blood cell with a multilobed nucleus that is highly phagocytic and the predominant cell type in acute inflammation.

palindromic sequences the same nucleotide sequences that appears on each strand but in opposite directions.

pandemic a worldwide epidemic.

parasite an organism that survives on or at the expense of another living host.

parenteral injection by a route other than oral, that is, intravenous, intramuscular, subcutaneous.

passive immunity immunity in which preformed immune products are acquired by an individual from an external source either naturally (via colostrum or transplacental passage) or artificially (via injection).

passive transport transfer of molecules without the need for energy.

pasteurization a process in which fluids are heated at temperatures below boiling to kill pathogenic microorganisms in the vegetative state.

pathogen an organism capable of producing disease.

pellicle thick covering over the cytoplasmic membrane.

peplomer a projection (spike) extending from the outer surface of a virus envelope.

peptidoglycan the rigid structural component of the bacterial cell wall consisting of *N*-acetylmuramic acid and *N*-acetylglucosamine.

permease an enzyme found in cell membranes that transports compounds into the cell cytoplasm.

pertussis whooping cough caused by the bacterium *Bordetella pertussis.*

phage an abbreviated term for bacterial virus (bacteriophage).

phagocytes cells (especially PMNs and macrophages) that are capable of ingesting (endocytosing) substances and subsequently digesting them.

phagocytosis a process in which a cell (phagocyte) engulfs other cells or particles.

phagosome a cell vacuole resulting from phagocytosis of particulate materials.

phagotroph organism that ingests living material as a source of food.

phenotype the genetic makeup of an organism; its observable properties.

phlebitis inflammation of blood vessels.

phloem food-conducting tissues in trees.

photoheterotroph an organism that uses light for energy and organic molecules as a source of carbon.

photon a particle of radiant energy.

photosynthesis a process in plants and some microorganisms in which light as an energy source is used to reduce carbon dioxide to organic material.

phototaxis movement toward light.

phragmoplast structure in some eukaryotes composed of microtubules that extend between recently divided nuclei.

phylogeny the ordering of species into taxa.

phytoplankton plantlike organisms that are floating or free swimming.

pili small protein projections on the bacterial cell involved in conjugation process.

pilsener beer lager beer made according to the beer prevailing in Pilsen, Czechoslovakia.

pinocytosis the engulfment of liquid droplets by a cell.

plaque (dental) the structureless debris found on the surface of teeth.

plasma fluid portion of the blood.

plasma cell cell type regarded as the principal producer of antibodies.

plasmid an extrachromosomal piece of DNA.

plasmodium a cellular mass of protoplasm associated with the slime molds.

plasmogamy fusion of protoplasts (such as gametes), often followed by nuclear fusion.

plasmolysis shrinkage of the cell caused by osmotic removal of water.

plasmoptysis swelling or bursting of a cell due to osmotic inflow of water.

platelet small blood components important in blood coagulation.

pleomorphism the state of having more than one form or shape.

pock a pit, spot, or pustule.

polycistronic a messenger RNA in which several coded genes are part of the molecule.

polyhedral inclusion body protein-encased insect virus complex.

polyhedral many-sided.

polymerase an enzyme that engages in the polymerization of monomeric units.

polyploid having a chromosome number more than twice the haploid number.

porter a weak stout that is rich in saccharine matter and contains about 4% alcohol.

Prausnitz-Küstner Test (P-K test) a passive transfer test that is used in atopic allergy diagnosis; Serum from the allergic individual is deposited in the skin of a volunteer and the suspected allergen is injected later into the same site.

precipitin antibody that forms precipitating lattice upon interaction with its homologous soluble antigen.

predation the relationship in which one animal feeds or preys on another.

proctitis inflammation of the rectum.

producer organism that synthesizes organic material.

profile vertical cross section of soil.

prokaryote a microorganism characterized by the absence of a nuclear membrane; the absence of cytoplasmic organelles.

promoter region the area on the operon that controls the synthesis of structural gene products.
properdin a bactericidal protein component of the blood.
prophage the state of a virus in which the viral genome is integrated into the host's genome and is duplicated each cell generation.
prophylaxis protection, for example, against disease.
prostheca a cytoplasmic protrusion from a cell.
prosthesis artificial component substituting for a defective body part.
protease an enzyme that hydrolyzes peptide bonds, thus converting proteins into polypeptides and eventually amino acids.
protista a taxonomic group used for grouping microrganisms.
protocooperation a transitory association between two species in which both benefit.
proton-motive force the force tending to pull protons across the cell membrane because of the electrochemical potential generated across the membrane.
protoplast a cell in which the cell wall has been removed.
prototroph an organism that has no organic growth requirements as compared with an auxotroph, which does have organic requirements.
protozoa eukaryotic unicellular microorganisms with animal characteristics.
provirus virus integrated into host chromosome and transmitted from one generation to another.
pseudopodium a "foot" or temporary extension of the cell membrane of ameboid cells and used for movement.
pseudowild the partial restoration of a function to an organism.
psychrophile microorganisms that grow best at temperatures between 0° and 30° C.
psittacosis a bird disease transmissible to humans.
purulent associated with the formation of pus.
pus a yellow white substance found in abscesses containing primarily living, dying, and dead leukocytes.
pyrogen an agent that induces fever.
quarantine to detain or isolate individuals because of suspicion of infection.
quellung reaction a swelling of encapsulated cells resulting from the interaction of specific immune serum with capsular antigens.
R-factor a transferrable plasmid that carries the information for resistance to antimicrobials or other agents.
race group of individuals within a cultivar or species distinguished by pathogenic behavior but not by morphology.
radioimmunoassay (RIA) an immunoassay technique in which one of the test components (antibody or antigen) is labeled with a radioactive isotope; radioactivity is used to indicate that an antigen-antibody reaction has occurred.
raphe a linear slit on the valves of diatoms.
reagin the skin-sensitizing antibody IgE; in syphilis the nonprotective antibody that has specificity for cardiolipin antigens.
recalcitrant molecule a compound that is slowly mineralized or that is not metabolized by microorganisms in nature.
recombination a process in which genetic information is exchanged between two organisms.
repression a genetic process in which the synthesis of an inducible enzyme is inhibited by a repressor molecule.
reservoir of infection animate or inanimate objects that support the growth of microorganisms.
resistance transfer factor (RTF) a group of genes that permits the transfer of genetic material during the conjugation process.
resolving power the ability to distinguish two objects separately when observed with the unaided eye or with a microscope.
respiration an oxidative process in which energy is released from foodstuffs.
retrovirus a group of RNA viruses that possess reverse transcriptase for transcribing the RNA genome into a DNA intermediate.
rhinitis inflammation of the nose.
rhizoid filamentous appendages by which organisms such as algae and plants attach to the substratum.
rhizoplane the root surface and its adhering soil.
rhizosphere microenvironment in soil immediately around plant roots.
ribosome a ribonucleoprotein particle found in the cell cytoplasm and involved in protein synthesis.
ringworm fungal infection caused by a group called dermatophytes.
ripening the aging of a food product to enhance the chemical or microbiological development of specific flavors and textures.
rubella German measles.
rubeola red measles.
saprobe an organism that derives its nourishment from dead or decaying material.
sapwood physiologically active zone of wood contiguous to cambium.
secretory IgA (sIgA) composed of two (dimer) or three (trimer) basic IgA molecules, a secretory piece, and a J chain; provides topical or local immunity via sIgA-containing secretions (mucus secretions, colostrum, saliva).
sepsis a toxic condition resulting from the presence of microbes or microbial products in the body.
septum a wall dividing biological units, such as cells in a fungal filament.
sequestering agent an agent that will hold metallic ions in solution usually by inclusion of the ions in an appropriate coordination complex.
serology the study of antigen-antibody reactions in vitro.
serotype subtype of a species that is based on antigenic differences.
serum the clear portion of the blood minus the factors necessary for clotting.
serum sickness a systemic, type III hypersensitivity that develops when soluble antigen that was administered in large quantities remains in the tissues and forms immune complexes with newly formed antibody.
sex factor (F factor) a genetic unit that may be part of the chromosome or may be extrachromosomal; it functions in the transfer of genes during conjugation.
siderophore metal-binding molecule.
sieve tube specialized phloem structure that conducts food in plants.
sludge digestion anaerobic treatment of sewage sludge.
sorghum an Old World tropical grass similar to Indian corn in habit.

source of infection animate or inanimate objects that serve as a contact between host and microorganism.

species a taxonomic group composed of individuals having closely related characteristics.

specification accepted value in the manufacture of a product.

spheroplast wall-deficient bacterium that retains part of the wall.

spirochete a corkscrew-shaped bacterium.

sporangium a structure that holds asexual spores.

spore the resistant form of a bacterium derived from the vegetative cell; the reproductive cell of certain organisms.

sporozoite spore produced after fertilization.

sporulation the process in which a spore is formed.

springwood early part of the yearly growth ring.

standard a law stating accepted value(s) for a manufactured product.

starter culture the inoculation of a food or beverage with an organism to ensure the development of a specific product.

static growth-inhibiting.

steady-state open system in which raw materials are converted into biomass.

sterilization the process that destroys all living microorganisms.

stoma regulated pore opening in the leaf epidermis for passage of gases and water vapor.

stout a heavy-bodied brew that is darker and sweeter than porter and is made with roasted malt and a relatively high percentage of hops.

stroma the supporting substance of an organelle or structure in which other components may be suspended.

stylet stiff, slender, hollow feeding organ of plant-parasitic nematodes.

subclinical without clinical manifestations of the disease.

submerged culture a technique for culturing molds and other microorganisms in which the microbes are distributed through the culture fluid by vigorous mixing. It is in contrast to surface culture where the organism grows on the upper surface of the culture fluid.

succession replacement of types of populations and communities with time.

suppressor T cells (Ts) a subset of regulatory T cells that suppresses the activity of B cells and other T cells.

surfactant substance that reduces surface tension; a wetting agent such as detergents and quaternary ammonium compounds.

sylvatic occurring in or affecting wild rodent populations.

symbiosis an association or relationship between two organisms.

synapse the junction between two processes such as two nerve cells.

syndrome a group of symptoms that characterize a disease.

syngamy union of gametes.

systemic not localized in any one part of the body but spread throughout the body.

T cell thymus-dependent lymphocytes involved in cell-mediated immunity.

tannin any of several soluble astringent complex phenolic substances of plant origin.

taxis movement of an organism in a particular direction in response to an external stimulus.

teichoic acid polysaccharide containing ribitol or glycerol phosphates found in gram-positive bacteria.

test shell or capsule secreted by a protozoan.

tetany spasms.

thallus the vegetative unit of a fungus.

thermal death time (TDT) the point on the abscissa resulting from an extrapolation of the death time curve.

thermocline temperature boundary in aquatic environments.

thermophiles microorganisms growing best at temperatures between 45° and 70° C.

thrush a condition affecting the oral mucous membranes caused by the fungus *Candida*.

thylakoids photosynthetic membrane found in eukaryotes and some prokaryotes, such as cyanobacteria, that contains pigments.

tinea a name applied to fungal infections of the skin.

titer the quantity of a substance required to produce a particular reaction.

toxoid an inactivated toxin that has lost its toxigenicity but retains its immunogenicity and thus can be used for active immunization.

tracheid thick-walled cell of xylem tissue that is capable of water transport.

transamination transfer of an amino group from an amino acid to a keto group and the formation of a new amino acid.

transcription the process by which the nucleic acid, acting as a template, is reproduced into a complementary molecule.

transduction the transfer of bacterial genetic information from one cell to another by a virus.

transfection infection of host cells with naked viral DNA.

transformation in bacterial systems the uptake of free nucleic acid (DNA) by a cell and its acquisition of new traits; in virology a process in which virus infection alters the eukaryotic cell, making it malignant.

transition a mutation caused by substitution of one purine for another or one pyrimidine for another.

translation the conversion of the mRNA into a protein product in the process of protein synthesis.

transplantation antigens cell surface antigens that induce an immunorejection response when the transplant recipient is not genetically identical to the transplant donor.

transpiration loss of water vapor through the leaves.

transposition the shift of a gene or chromosome segment to a new position on the chromosome.

transposon a genetic element that can be transposed to other sites on the same or different genome.

transversion a mutation in which there is a substitution of a purine for a pyrimidine or vice versa.

treponematoses all of the diseases caused by species of *Treponema*.

trichocyst an organelle found in certain eukaryotic microorganisms that ejects a filament through the cytoplasmic membrane.

trichome filamentous structure.

triplet code three nucleotide code specifying for one amino acid.

trophic level organisms in a food chain that have similar relationships to the source of their nutrients.

trophozoite the active vegetative stage of a protozoan.

tropism a growth response of an organism following an external stimulus.

trub the coagulate protein that settles out of freshly boiled wort.

tubercle a nodule.
tuberculin a bacteria-free protein fraction extracted from the bacilli causing tuberculosis *(Mycobacterium tuberculosis).*
tubulin the protein subunit of microtubles.
tumor mass of tissue resulting from uncontrolled growth.
turbidometry measurement of the turbidity of a solution.
tyloses balloonlike extrusions of parenchyma cells into lumina of contiguous vessels.
ulcer circumscribed area of inflammation found in epithelial lining of a body surface.
unsaturated presence of double or triple covalent bonds in a molecule.
urethritis inflammation of the urethra.
urticaria inflammatory reaction characterized by eruptions that itch.
vaccine a suspension of organisms usually killed or attenuated and used for immunization.
vacuole space or cavity in the cell usually filled with somme substance.
varicella chickenpox.
vascular containing blood vessels.
vasoactive affecting vessels. (especially blood vessels).
vector a carrier of pathogenic microorganisms.
vegetative concerned with the growing stage of a microorganism as opposed to the spore state.
venereal transmitted by sexual contact.
vertical spread transfer of infectious agent from parent to offspring via ovum or sperm or by contact between placenta and offspring.
vesicle blister.
virion a complete virus particle consisting of a core of nucleic acid and a protein capsid.
viroid an infectious subviral particle consisting of nucleic acid without a protein capsid.
viropexis a type of pinocytosis in which a virus enters the cell within a closed membrane.
virulence the relative ability of an organism to cause disease.
walling-off separation of diseased from healthy tissue by barrier tissues produced by diseased plant.
water activity (a_w) the ratio of existing vapor pressure to pure water vapor pressure.
wilt collapse of foliage from water deficiency; may be temporary or permanent.
witches'-broom broomlike proliferation of woody stems originating from closely spread nodes.
wort a dilute solution of sugars obtained from malt by infusion and fermented to form beer.
xenograft a graft between individuals of different species.
xylem water-conducting, woody tissue produced by the cambium.
xylem parenchyma living parenchyma cells associated with the xylem.
yeast a unicellular fungus.
Z value the degrees Fahreheit required to reduce the thermal death time by 90% or one log value.
zoa pertaining to animal.
zoonoses diseases of animals.
zooplankton a collective term for nonphotosynthetic microorganisms found in flora and fauna of a body of water.
zoospore a motile reproductive spore common to aquatic fungi.
zygote the diploid cell in eukaryotes resulting from union of male and female gametes.
zymogenous soil microorganisms, primarily rods and spore formers that readily decompose (ferment) organic matter.

INDEX

A

B

C

D

G

H

I

J

Q

R